CLINICAL VIROLOGY MANUAL

FIFTH EDITION

WITHDRAWN

CLINICAL VIROLOGY MANUAL

FIFTH EDITION

Editor in Chief

Michael J. Loeffelholz
Professor in the Department of Pathology, Director of the Clinical Microbiology Laboratory at the University of Texas Medical Branch (UTMB) at Galveston.

Editors

Richard L. Hodinka
Professor in the Microbiology Department and Chair of the Department of Biomedical Sciences at the University of South Carolina School of Medicine

Stephen A. Young
Director of Research and Clinical Trials at TriCore Reference Laboratories

Benjamin A. Pinsky
Assistant Professor in the Departments of Pathology and Medicine, Division of Infectious Diseases and Geographic Medicine, at the Stanford University School of Medicine

ASM PRESS

Washington, DC

Library of Congress Cataloging-in-Publication Data

Names: Loeffelholz, Michael J., editor. | Hodinka, Richard L., editor. |
Young, Stephen A., editor. | Pinsky, Benjamin A., editor.
Title: Clinical Virology Manual Fifth Edition / editor in chief, Michael J. Loeffelholz;
editors, Richard L. Hodinka, Stephen A. Young, Benjamin A. Pinsky.
Description: Fifth edition. | Washington, DC: ASM Press, [2016]
Identifiers: LCCN 2016020815 | ISBN 9781555819149 (hard cover) | ISBN 9781555819156 (e-ISBN)
Subjects: LCSH: Diagnostic virology—Handbooks, manuals, etc.
Classification: LCC QR387 .C48 2016 | DDC 616.9/101—dc23 LC record available at https://lccn.loc.gov/2016020815

ISBN 978-1-55581-914-9
e-ISBN 978-1-55581-915-6
doi:10.1128/9781555819156

Printed in Canada

10 9 8 7 6 5 4 3 2 1

Address editorial correspondence to: ASM Press, 1752 N St., N.W., Washington, DC 20036-2904, USA.
Send orders to: ASM Press, P.O. Box 605, Herndon, VA 20172, USA.
Phone: 800-546-2416; 703-661-1593. Fax: 703-661-1501.
E-mail: books@asmusa.org
Online: http://estore.asm.org

DEDICATION

We dedicate this edition of the Clinical Virology Manual to our families for their patience and support during this and our other professional endeavors. We are truly blessed to be part of their lives and to receive their unconditional love.

We would also like to thank, and gratefully acknowledge the support and leadership of, our close colleague, mentor, and friend, Dr. Steven Specter, who has worked tirelessly over the years in delivering the first four editions of the Manual, to advance the field of viral diagnostics, and to provide a forum for clinical virologists, academicians, and clinicians to present and discuss the latest scientific discoveries. We will be forever appreciative of his unwavering efforts.

Contents

Contributors

NADIM AJAMI
Baylor College of Medicine, Department of Molecular Virology and Microbiology, Alkek Center for Metagenomics and Microbiome Research Houston, TX 77030

KEVIN ALBY
University of Pennsylvania, Department of Pathology and Laboratory Medicine, Philadelphia, PA 19104

MATTHEW J. BANKOWSKI
Diagnostic Laboratory Services (The Queen's Medical Center) Aiea, HI 96701

JOHN A. BURNS
School of Medicine, Department of Pathology, University of Hawaii at Manoa Honolulu, HI 96813

JOANNE BARTKUS
Minnesota Department of Health, Public Health Laboratory St Paul, MN 55164

WILLIAM BELLINI
Centers for Disease Control and Prevention, MMRHLB/DVD/NCIRD Atlanta, GA 30329

GREGORY BERRY
University of Texas Medical Branch, Clinical Microbiology Laboratory Galveston, TX 77555

SEWERYN BIALASIEWICZ
Centre for Children's Health Research, Children's Health Queensland, and the Child Health Research Centre, The University of Queensland South Brisbane, QLD 4101 Australia

MATTHEW J. BINNICKER
Mayo Clinic, Division of Clinical Microbiology, Department of Laboratory Medicine and Pathology Rochester, MN 55905

MICHAEL D. BOWEN
Centers for Disease Control and Prevention, National Center for Immunization and Respiratory Diseases Atlanta, GA 30333

RICHARD BULLER
Washington University School of Medicine, Department of Pediatrics Saint Louis, MO 63110

CAREY-ANN BURNHAM
Washington University School of Medicine, Department of Pathology & Immunology Saint Louis, MO 63110

ANA MARIA CARDENAS
University of Pennsylvania, Department of Pathology and Laboratory Medicine Philadelphia, PA 19103

ROBERTA CAREY
Centers for Disease Control and Prevention, Laboratory Quality Management Atlanta, GA 30333

LAKSHMI CHANDRAMOHAN
Texas Children's Hospital, Department of Pathology Houston, TX 77030

DERRICK CHEN
Mayo Clinic, Laboratory Medicine and Pathology Rochester, MN 55905

ERIC C.J. CLAAS
Leiden University Medical Centre, Medical Microbiology Leiden 2333 ZA

APRIL DAVIS
New York State Health Department, Wadsworth Center Slingerlands, NY 12159

SHEILA C. DOLLARD
Centers for Disease Control and Prevention, Division of Viral Diseases Atlanta, GA 30329

STEVEN J. DREWS
ProvLab Alberta, Diagnostic Virology University of Alberta, Department of Laboratory Medicine and Pathology Edmonton, AB Canada

JAMES DUNN
Texas Children's Hospital, Department of Pathology Houston, TX 77030

CHARLOTTE GAYDOS
Johns Hopkins University School of Medicine, Division of Infectious Diseases Baltimore, MD 21205

JANE GETCHELL
Public Health Laboratory Consultant Bethany Beach, DE 19930

THOMAS GRYS
Mayo Clinic in Arizona, Department of Laboratory Medicine and Pathology Phoenix, AZ 85054

CAROLE HICKMAN
Centers for Disease Control and Prevention, MMRHLB/DVD/NCIRD Atlanta, GA 30329

REBECCA HORVAT
University of Kansas Medical Center, Department of Pathology Kansas City, Kansas 66160

JOSEPH ICENOGLE
Centers for Disease Control and Prevention, MMRHLB/DVD/NCIRD Atlanta, GA 30329

JORMA ILONEN
University of Turku, The Immunogenetics Laboratory Turku, Finland

SCOTT JAMES
University of Alabama at Birmingham, Department of Pediatrics Birmingham, AL 35233

TIMOTHY KARNAUCHOW
Children's Hospital of Eastern Ontario, Division of Virology Ottawa, Ontario K1H 8L1 Canada

ELIZABETH KAUFFMAN
New York State Department of Health, Wadsworth Center Slingerlands, NY 12159

SUE KEHL
Medical College of Wisconsin, Department of Pathology Milwaukee, WI 53226

HARALD KESSLER
Medical University of Graz Graz, Austria

REETI KHARE
Northwell Health Laboratories, Department of Pathology and Laboratory Medicine Lake Success, NY 11042

ASHLEY KONDAS
Centers for Disease Control and Prevention, Poxvirus and Rabies Branch Atlanta, GA 30333

LAURA KRAMER
New York State Department of Health, Wadsworth Center, and Department of Biomedical Sciences, SUNY Albany Slingerlands, NY 12159

MARIE LOUISE LANDRY
Yale University, Departments of Laboratory Medicine and Internal Medicine (Infectious Diseases) New Haven, CT 06520

DIANE LELAND
Indiana University School of Medicine, Department of Pathology and Laboratory Medicine Indianapolis, IN 46202

MICHAEL LOEFFELHOLZ
University of Texas Medical Branch, Clinical Microbiology Laboratory Galveston, TX 77555

STANLEY MARTIN
The Ohio State University Wexner Medical Center, Division of Infectious Diseases; Transplant Infectious Diseases Service Columbus, OH 43210

MICHAEL NISSEN
Children's Health Queensland, Queensland Children's Medical Research Institute Brisbane, Queensland 4029 Australia

SUSAN NOVAK-WEEKLEY
Southern California Permanente Medical Group, Regional Reference Laboratories, Microbiology North Hollywood, CA 91605

M. STEVEN OBERSTE
Centers for Disease Control and Prevention, Poxvirus and Rabies Branch Atlanta, GA 30333

VICTORIA OLSON
Centers for Disease Control and Prevention, Poxvirus and Rabies Branch Atlanta, GA 30333

GUSTAVO PALACIOS
USAMRIID, Center for Genome Sciences Frederick, MD 21702

MARK PALLANSCH
Centers for Disease Control and Prevention, Division of Viral Diseases Atlanta, GA 30329

PREETI PANCHOLI
The Ohio State University Wexner Medical Center, Department of Pathology Columbus, OH 43205

VILLE PELTOLA
Turku University Hospital, Department of Pediatrics and Adolescent Medicine Turku 20521 Finland

JOSEPH PETROSINO
Baylor College of Medicine, Department of Molecular Virology and Microbiology, Alkek Center for Metagenomics and Microbiome Research Houston, TX 77030

CHRISTOPHER PREAS
California Department of Public Health, Viral and Rickettsial Disease Laboratory Richmond, CA 94804

ROBERT PRETORIUS
Southern California Permanente Medical Group— Fontana, Obstetrics and Gynecology Fontana, CA 92445

MARK PRICHARD
University of Alabama at Birmingham, Department of Pediatrics Birmingham, AL 35233

MICHAEL PURDY
Centers for Disease Control and Prevention, Division of Viral Hepatitis Atlanta, GA 30329

RYAN RELICH
Indiana University School of Medicine, Department of Pathology and Laboratory Medicine Indianapolis, IN 46202

REBECCA ROCKETT
Centre for Infectious Diseases & Microbiology - Public Health (CIDM-PH) Institute of Clinical Pathology & Medical Research (ICPMR) Westmead Hospital, Westmead NSW, 2145, Sydney, Australia

MATTHEW ROSS
Baylor College of Medicine, Department of Molecular Virology and Microbiology, Alkek Center for Metagenomics and Microbiome Research Houston, TX 77030

ROBERT RUDD
New York State Health Department, Wadsworth Center Slingerlands, NY 12159

LINOJ SAMUEL
Henry Ford Health System, Department of Pathology Detroit, MI 48202

DAVID SCHNURR
California Department of Public Health, Viral and Rickettsial Disease Laboratory Richmond, CA 94804

JÖRG SCHÜPBACH
University of Zurich, Institute of Medical Virology, Swiss National Center for Retroviruses Zurich CH-8057 Switzerland

SUZANE SILBERT
Tampa General Hospital, Department of Pathology, Esoteric Testing/R&D Tampa, FL 33606

THEO SLOOTS
Centre for Children's Health Research, Children's Health Queensland, and the Child Health Research Centre, The University of Queensland South Brisbane, QLD 4101 Australia

EVELYN STELZL
Medical University of Graz Graz, Austria

ARIEL SUAREZ
IACA Laboratorios, Molecular Biología Bahia Blanca, Buenos Aires B8000FIB Argentina

NORMA TAVAKOLI
New York State Department of Health, Wadsworth Center, And Department of Biomedical Sciences, State University of New York Albany, NY 12208

ERIN MCELVANIA TEKIPPE
University of Texas Southwestern Medical Center, Department of Pathology and Pediatrics Dallas, TX

MARTHA T. VAN DER BEEK
Leiden University Medical Centre, Medical Microbiology Leiden 2333 ZA Netherlands

BARBARA VAN DER POL
University of Alabama at Birmingham School of Medicine, Division of Infectious Diseases Birmingham, AL 35294

GILBERTO VAUGHAN
Centers for Disease Control and Prevention, Division of Viral Hepatitis Atlanta, GA 30329

CRISTINA VIDELA
CEMIC Virology Buenos Aires C1431FWO Argentina

DEBRA WADFORD
California Department of Public Health, Viral and Rickettsial Disease Laboratory Richmond, CA 94804

YUN F. WANG
Emory University School of Medicine, Department of Pathology & Laboratory Medicine Atlanta, GA 30303

NATALIE WHITFIELD
OpGen Clinical Services Laboratory Gaithersburg, MD 20878

RAYMOND WIDEN
Tampa General Hospital, Pathology Department, Esoteric Testing/R&D Tampa, FL 33606

DONNA WOLK
Geisinger Health Systems, Dept. of Laboratory Medicine Danville, PA 17822

BELINDA YEN-LIEBERMAN
Cleveland Clinic, Laboratory Medicine Cleveland, OH 44195

DONGXIANG ZIA
California Department of Public Health, Viral and Rickettsial Disease Laboratory Richmond, CA 94804

Author Biographies

Michael J. Loeffelholz is a Professor in the Department of Pathology, Director of the Clinical Microbiology Laboratory at the University of Texas Medical Branch (UTMB) at Galveston, and Director of the American Society for Microbiology (ASM) CPEP-accredited Medical Microbiology Fellowship program at UTMB. He is an editor of the *Journal of Clinical Microbiology* and a diplomate of the American Board of Medical Microbiology (ABMM).

Richard L. Hodinka is a Professor in the Microbiology Department and Chair of the Department of Biomedical Sciences at the University of South Carolina School of Medicine, Greenville. He has served as President and a Council Member for the Pan American Society for Clinical Virology and, as a member of the International Scientific Advisory Committee for the Asia Pacific Congress of Medical Virology.

Stephen A. Young is Director of Research and Clinical Trials at TriCore Reference Laboratories and Professor (*emeritus*) in the Department of Pathology at the University of New Mexico. He is a diplomate of the ABMM.

Benjamin A. Pinsky is an Assistant Professor in the Departments of Pathology and Medicine, Division of Infectious Diseases and Geographic Medicine, at the Stanford University School of Medicine and is the Medical Director of the Clinical Virology Laboratory for Stanford Health Care and Stanford Children's Health. He is board certified in Clinical Pathology by the American Board of Pathology.

Preface to the Fifth Edition

The aims of the fifth edition of the *Clinical Virology Manual* remain the same as prior editions and include serving as a reference source to healthcare professionals and laboratorians in providing clinical and technical information regarding viral diseases and the diagnosis of viral infections.

This new edition includes 40 chapters and 3 appendices and, similar to the organization of prior additions, consists of the four sections: general topics, laboratory procedures, viral pathogens, and the appendices. We have modified the content of the appendices to provide basic but practical information on reference virology laboratories at both the national and international levels. The viral pathogen chapters have a consistent organization, with proportionally more content dedicated to diagnostics and testing. Additionally, a new section, with the heading of "Diagnostic Best Practices", has been included in each viral pathogens chapter. The section summarizes recommendations for diagnostic testing and cites evidence-based guidelines when available.

The past several years have been very challenging, as well as exciting, for diagnostic virologists, with outbreaks of enterovirus D68, measles virus, mumps virus, norovirus, Ebola virus, and, most recently, Zika virus. In addition, there is continued emergence of chikungunya, dengue, and influenza viruses, highlighted by the influenza pandemic of 2009. The landscape of hepatitis C virus has changed, and will continue to change dramatically, with the availability of new classes of direct-acting antiviral drugs that provide an excellent probability of cure.

This edition has incorporated these significant events to the extent allowed by the production schedule. We thank the authors for their contributions, particularly during this very busy time for virologists. We also thank the staff of the American Society for Microbiology Press for their support and hard work in bringing this edition to fruition.

The fifth edition of the Manual also brings a major change in editors, as a new editor has been added and a previous editor has cycled off. Also, after successfully leading this series through four editions, Dr. Steven Specter has passed on the reins of Editor-in-Chief. We hope that this edition is a credit to Dr. Specter, as well as to other prior editors, Drs. Lancz and Wiedbrauk.

MICHAEL J. LOEFFELHOLZ
RICHARD L. HODINKA
STEPHEN A. YOUNG
BENJAMIN A. PINSKY

SECTION I

General Topics in Clinical Virology

The Taxonomy, Classification, and Characterization of Medically Important Viruses

STEVEN J. DREWS

1

Viruses are a complex and diverse group of organisms that may have incredibly diverse and ancient origins. Their interaction with humans not only involves disease processes, but also evolutionary pressures that shape viral characteristics. Viral taxonomy, classification, and characterization is not a simple academic exercise but practically improves our ability to diagnose, track, and compare viruses of medical importance and develop a better understanding of pathophysiologic processes. Over the last 5 years, there have been significant changes in the proper names of some commonly identified viruses of medical importance, relationships between these medically relevant viruses, technologic tools, as well as websites and bioinformatics tools. Changes, including what constitutes the definition of a viral species, have already had an impact on how viruses are characterized and classified. The expanded utilization of whole genome sequence analysis and metagenomic approaches has increased the amount of biological information available to the scientific community for virus characterization and categorization. With these newer molecular approaches for virus identification and characterization, as well as enhanced bioinformatics approaches, viral classification is as dynamic and challenging as ever, requiring continuous monitoring, reassessment, and updating to achieve a rational taxonomic framework.

WHAT ARE VIRUSES?

Historically, viruses have been a difficult group of pathogens to describe, and there is continuous and vibrant discussion on whether they should be included in the tree of life, and if so where their places are within that tree (1). The dominant theory, the "escape theory", postulates that viruses evolved recently and arose from genetic elements that escaped from cellular hosts and evolved independent replication processes. In contrast, the "reduction hypothesis" suggests that viruses are the remnants of cellular organisms (2). Finally, the virus "first hypothesis" suggests that viruses have ancient origins and arose before the last universal cellular ancestor (3). Regardless of the theory, it is apparent that mammals evolved in a world with viral threats and that viruses have co-evolved with humans and our cellular ancestors (4). However, the differences in how viruses encode genetic information (DNA versus RNA), or how that information is stored (double stranded versus single stranded) suggests that viruses are polyphyletic; that is, they lack a common origin and are developing along multiple evolutionary pathways. These tensions between polyphyletic and monophyletic characters, although evolutionary focused, also have an impact on viral taxonomy. The key question that arises is, how is it that a group of pathogens that are relatively simply designed so difficult to characterize and categorize?

As living organisms, viruses are also extremely divergent and have great diversity in a variety of other characteristics. In contrast to all other forms of life, viruses can be described as the only organisms that replicate in the form of information (5). The most noticeable difference from other organisms and one of the more variable characters of this group of pathogens is their diversity in how they encode this genetic information (Fig. 1, Tables 1 to 7). In contrast to other forms of life that encode genetic information within double-stranded (ds)DNA, virus genomes may be composed of dsDNA, single-stranded (ss)DNA, dsRNA, and ssRNA. The form of the genome has a direct correlation to factors such as substitution and mutation rate that are associated with viral evolution. In general, mutation rates (mutations/site/replication) are highest in ssRNA viruses, followed by retrotranscribing viruses, ssDNA viruses, and finally dsDNA viruses. In contrast, although substitution rates (substitution/site/year) for ssDNA viruses are greater than dsDNA viruses, the substitution rates of ssDNA, ssRNA, and retrotranscribing viruses may overlap. Retrotranscribing viruses may have wide ranges of substitution rates. Mutation and substitution rate trends in dsRNA viruses are well established (6). Variables impacting substitution rates can include generation time, transmission, and selection, while variables impacting mutation rate can include genomic architecture, replication speed, viral enzymes, host enzymes, and environmental effects (6). However, these trends do not always occur as expected, and mutation rate variation also exists among RNA viruses, which may be due to a variety of factors, particularly of the host (7). Virus genomes are also arranged in a variety of different topologies including linear, circular, single segment, or multiple segments, and this

doi:10.1128/9781555819156.ch1

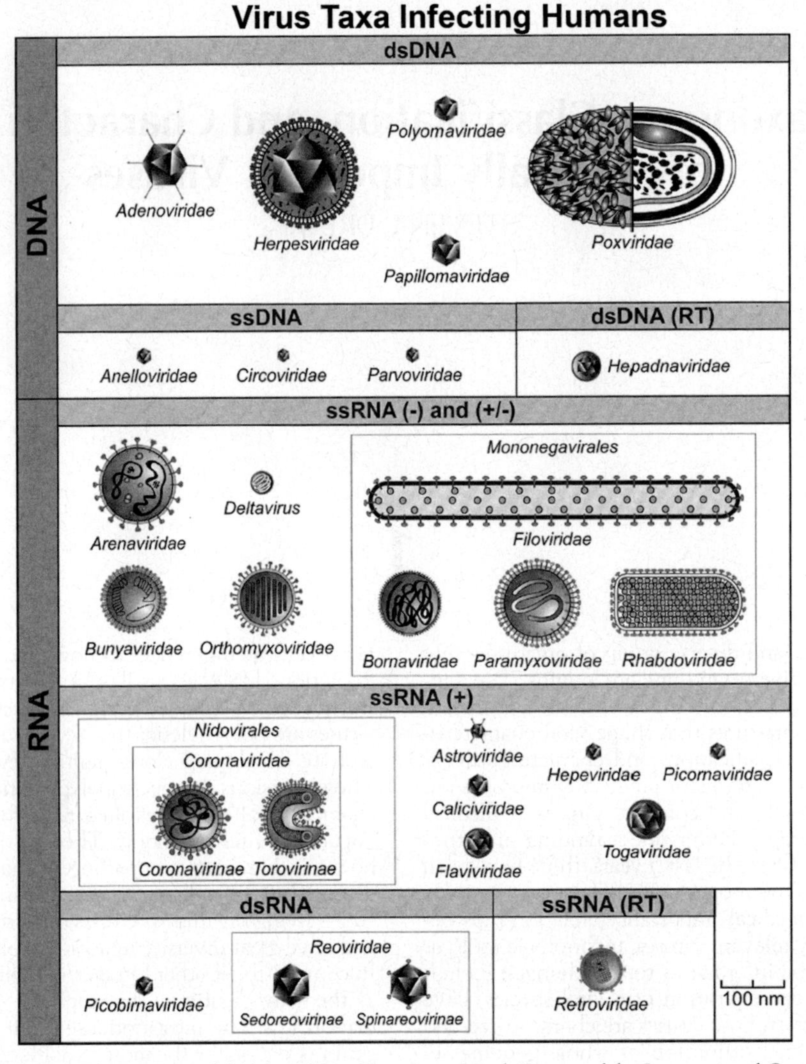

FIGURE 1 Virus taxa infecting humans. Modified from Virus Taxonomy, Ninth Report of the International Committee on Taxonomy of Viruses (Reprinted from Elsevier Books, *Virus Taxonomy*, 2002, with permission from Elsevier.)

organization can have an impact on horizontal transfer of genetic information between individuals of different lineages. Each form of maintenance of the viral genome has its own evolutionary benefits and drawbacks (8, 9). Viruses can also be divided into pathogens that only infect humans, those that infect other mammalian species, and those that infect nonmammalian vectors.

Several factors separate viruses from other forms of life, and these factors are often characterized by vertical but not horizontal gene transfer. Although viruses contain information, their evolution requires host cells (1). They are parasitic agents that infect cells to reproduce virions and disseminate genes (10), and they cannot maintain or replicate themselves without hosts (1). The virally encoded genes that are required for carbon metabolism, energy metabolism, and protein synthesis are postulated to have a cellular origin (1). Multiple differences from other life forms have been presented and include their polyphyletic origins, the lack of a common gene shared by all viruses, the lack of membrane heredity, the cellular origin of translation genes, and a biased one-way direction of horizontal gene transfer (1). However, four factors have been described that viruses share with other living organisms: (i) the ability of genomes and gene products to produce progeny genomes, (ii) the possession of self-regulation, (iii) the ability to adapt and respond to changing environments, and (iv) maintenance of structural organization (11).

Multiple biological pressures drive virus evolution and shape key viral characteristics. Selective processes include positive selection (increases prevalence of adapting traits), negative selection (decreases the prevalence of adapting traits), or neutral selection (random neutral occurrences with no evolutionary advantage). Temporally, evolutionary pressures may not be consistent, and organisms may emerge from long periods of evolutionary stasis and enter periods of heavy selective pressure from factors such as the host immune system (12). Biologic pressure may not be applied equally on all regions of a gene, or genome, with some epitopes under more pressure than others, and the selective pressures that impact one gene may depend on the genetic background of the virus at other gene locations (13). There may also be differences in evolutionary pressure on viruses of the same species, and genotype may be influenced by the impact of climate, vector, and host on the organism, as seen,

TABLE 1 Taxonomy and characterization of double-stranded DNA viruses of human medical importance (Baltimore classification I)

Genome composition	Order	Family	Subfamily	Genus	Species (ICTV or other common names)	Related primary ICD-10 codes
Linear	*Herpesvirales*	*Herpesviridae*	*Alphaherpesvirinae*	*Simplexvirus*	*Human herpesvirus 1* (herpes simplex virus 1)	B00 herpesviral (simplex) infection A60 anogenital herpes virus infection
					Human herpesvirus 2 (herpes simplex virus 2)	P35.2 congenital herpes virus infection
					Macacine herpesvirus 1 (B virus)	B00.4+ Herpesviral encephalitis
				Varicellovirus	*Human herpesvirus 3* (varicella zoster virus)	B01 Varicella (chickenpox) B02 Zoster (herpes zoster)
			Betaherpesvirinae	*Cytomegalovirus*	*Human herpesvirus 5* (HHV-5; cytomegalovirus)	B25 Cytomegalovirus disease B27.1 Cytomegaloviral mononucleosis P35.1 Congenital cytomegalovirus infection
				Roseolovirus	*Human herpes virus* 6A (HHV-6A)	B08.2 Exanthema subitum (sixth disease)
					Human herpes virus 6B (HHV-6B)	T86.0 Bone marrow transplant rejection
					Human herpes virus 7 (HHV-7)	
			Gammaherpesvirinae	*Lymphocryptovirus*	*Human herpes virus 4* (Epstein-Barr virus; HHV-4)	B27.0 Gammaherpesviral mononucleosis C11 Nasopharyngeal carcinoma C83.7 Burkitt lymphoma D82.3 X-linked lymphoproliferative disease
				Rhadinovirus	*Human herpes virus* 8 (HHV-8, Kaposi's sarcoma associated herpes virus [KHSV])	C46 Kaposi sarcoma
	Unassigned	*Adenoviridae*	NA	*Mastadenovirus*	*Human mastadenovirus A–G*	B34.0 Adenovirus infection, unspecified site B30.0+ Keratoconjunctivitis due to adenovirus B30.1+ Conjunctivitis due to adenovirus B97 Adenovirus as the cause of diseases classified to other chapters A08.2 Adenovirus enteritis A85.1+ Adenovirus encephalitis A87.1+ Adenovirus meningitis J12.0 Adenoviral pneumonia
	Unassigned	*Poxviridae*	*Chordopoxvirinae*	*Molluscipoxvirus*	*Molluscum contagiosum virus*	B08.1 Molluscum contagiosum
				Orthopoxvirus	*Cowpox virus*	B08.0 Other orthopox infections
					Monkeypox virus	B04 Monkeypox
					Vaccinia virus	B08.0 Other orthopox infections
					Variola virus	B03 Smallpox (for surveillance purposes only)

(*Continued on next page*)

TABLE 1 Taxonomy and characterization of double-stranded DNA viruses of human medical importance (Baltimore classification I) (*Continued*)

Genome composition	Order	Family	Subfamily	Genus	Species (ICTV or other common names)	Related primary ICD-10 codes
				Parapoxvirus	*Orf virus*	B08.0 Other orthopox infections
					Pseudocowpox virus	B08.0 Other orthopox infections
				Yatapoxvirus	*Tanapox virus*	B08.8 Unspecified viral infections characterized by skin and mucous lesions
					Yaba monkey tumor virus	B08.8 Unspecified viral infections characterized by skin and mucous lesions
Circular	Unassigned	*Papillomaviridae*	NA	*Alphapapillomavirus*	*Alphapapillomavirus 3* (human papillomavirus 6)	B07 Viral warts A63 Anogenital (venereal) warts
					Alphapapillomavirus 9 (human papillomavirus 16)	D26.0 Papilloma of cervix N87 Dysplasia of cervix uteri D00-09 *In situ* neoplasms, Bowen's disease
					Alphapapillomavirus 7 (human papillomavirus 18)	D26.0 Papilloma of cervix N87 Dysplasia of cervix uteri D00-09 *In situ* neoplasms, Bowen's disease
					Alphapapillomavirus 1 (human papillomavirus 32)	D00-09 *In situ* neoplasms, Bowen's disease
				Betapapillomavirus	*Betapapillomavirus 1* (human papillomavirus 5)	D04 Carcinoma *in situ* of skin; possible association
				Gammapapillomavirus	*Gammapapillomavirus 1* (human papillomavirus 4)	B07 Viral warts D04 Carcinoma *in situ* of skin; possible association
				Mupapillomavirus	*Mupapillomavirus 1* (human papillomavirus 1)	B07 Viral warts
				Nupapipillomavirus	*Nupapillomavirus 1* (human papillomavirus 41)	B07 Viral warts
	Unassigned	*Polyomaviridae*	NA	*Polyomavirus*	*BK polyomavirus* *JC polyomavirus*	B34.4 Papovavirus infection, unspecified site

ICD, International Statistical Classification of Diseases and Related Health Problems; ICTV, International Committee on Taxonomy of Viruses; NA, not applicable.

TABLE 2 Taxonomy and characterization of single-stranded DNA viruses of human medical importance (Baltimore classification II)

Genome composition	Order	Family	Subfamily	Genus	Species (ICTV or other common names)	Related primary ICD-10 codes
Linear	Unassigned	*Parvoviridae*	*Parvovirinae*	*Bocaparvovirus*	*Primate bocaparvovirus 1–2*	J06 Acute upper respiratory infections of multiple and unspecified sites
				Dependoparvovirus	*Adeno-associated dependoparvovirus virus A*	—
					Adeno-associated dependoparvovirus B	—
				Erythroparvovirus	*Primate erythroparvovirus 1*	B34.3 Parvovirus unspecified site
				Tetraparvovirus	*Primate tetraparvovirus 1*	—
Circular	Unassigned	*Anelloviridae*	NA	*Alphatorquevirus*	*Torque teno virus 1*	—

ICD, International Statistical Classification of Diseases and Related Health Problems; ICTV, International Committee on Taxonomy of Viruses; NA, not applicable.

for example, in the pressures encountered by temperate and tropical genotypes of *Japanese encephalitis virus* (14). Another key driving pressure behind viral evolution causing human disease includes the immunologic niche or immune-mediated interactions of the human host (15). Differences in pressures on subgroups of viruses may be ameliorated by the differences in numbers of strains or subgroups of a virus below the species level and how often strains are replaced within a specific population or time period (16).

Several definitions, including taxonomy, classification, and characterization, will be used extensively in this chapter. Viral taxonomy has been defined as an approach to arranging viruses into related clusters, defining relatedness within and between clusters, and naming clusters or taxa (17). In contrast, classification can be thought of as an exercise in which one decides to use characters, features, or variables to place a particular virus within a taxonomic system. Characterization can be described as a process in which specific

TABLE 3 Taxonomy and characterization of double-stranded RNA viruses of human medical importance (Baltimore classification III)

Genome composition	Order	Family	Subfamily	Genus	Species (ICTV or other common names)	Related primary ICD-10 codes
Linear, segmented	Unassigned	*Reoviridae*	*Sedoreovirinae*	*Orbivirus*	*Changuinola virus*	A93.8 Other specified arthropod-borne viral fevers
					Lembobo virus	A93.8 Other specified arthropod-borne viral fevers
					Orungo virus	A93.8 Other specified arthropod-borne viral fevers
				Rotavirus	*Rotavirus A, B, and C*	A08.0 Rotaviral enteritis
				Seadornavirus	*Banna virus*	A85.2 Arthropod-borne viral encephalitis, unspecified; possible association
			Spinareovirinae	*Coltivirus*	*Colorado tick fever virus*	A93.2 Colorado tick fever
				Orthreovirus	*Mammalian orthoreovirus*	J06 Acute upper respiratory infections of multiple and unspecified sites; possible association
						A08.3 Other viral enteritis; possible association
						A08.4 Viral intestinal infection, unspecified; possible association
	Unassigned	*Picobirnaviridae*	NA	*Picobirnavirus*	*Human picorbirnavirus*	A08.3 Other viral enteritis; possible association
						A08.4 Viral intestinal infection, unspecified; possible association

ICD, International Statistical Classification of Diseases and Related Health Problems; ICTV, International Committee on Taxonomy of Viruses; NA, not applicable.

TABLE 4 Taxonomy and characterization of positive sense single-stranded RNA viruses of human medical importance (Baltimore classification IV)

Genome composition	Order	Family	Subfamily	Genus	Species (ICTV or other common names)	Related primary ICD-10 codes
Linear	*Nidovirales*	*Coronaviridae*	*Coronavirinae*	*Alphacoronavirus*	*Human coronavirus 229E*	B34.2 Coronavirus infection, unpsecified site
					Human coronavirus NL63	B97.2 Coronavirus as the cause of diseases classified to other chapters
				Betacoronavirus	*Human coronavirus HKU1*	B34.2 Coronavirus infection, unspecified site
						B97.2 Coronavirus as the cause of diseases classified to other chapter
					Severe acute respiratory syndrome-related coronavirus	U04 Severe acute respiratory syndrome (SARS)
					Middle Eastern respiratory syndrome coronavirus	B34.2 Coronavirus infection, unpsecified site
						B97.2 Coronavirus as the cause of diseases classified to other chapter
			Torovirinae	*Torovirus*	*Human torovirus*	A08.3 Other viral enteritis
	Picornavirales	*Picornaviridae*	NA			
				Cardiovirus	*Theilovirus*	A88 Other viral infections of central nervous system, not classified elsewhere; possible role
				Cosavirus	*Cosavirus A*	A08.3 Other viral enteritis; possible association
						A08.4 Viral intestinal infection, unspecified; possible association
				Enterovirus	*Enterovirus A*	B08.4 Enteroviral vesicular stomatitis with exanthem
						B08.5 Enteroviral vesicular pharyngitis
						B97.1 Enterovirus as the cause of disease classified to other chapters
						G02.0* Enteroviral meningitis
						G05.1* Enteroviral encephalomyelitis
					Enterovirus B	B08.4 Enteroviral vesicular stomatitis with exanthem
						B08.5 Enteroviral vesicular pharyngitis
						B97.1 Enterovirus as the cause of disease classified to other chapters
						G02.0* Enteroviral meningitis
						G05.1* Enteroviral encephalomyelitis
						H13.1* Acute epidemic hemorrhagic conjunctivitis (enteroviral)
					Enterovirus C (e.g., CV-A24)	B08.5 Enteroviral vesicular pharyngitis
						B97.1 Enterovirus as the cause of disease classified to other chapters
						G02.0* Enteroviral meningitis
						A80 Acute poliomyelitis
						G05.1* Enteroviral encephalomyelitis
						H13.1* Acute epidemic hemorrhagic conjunctivitis (enteroviral)

(*Continued on next page*)

				Enterovirus D	B97.1 Enterovirus as the cause of disease classified to other chapters
					G02.0* Enteroviral meningitis
					G05.1* Enteroviral encephalomyelitis
					H13.1* Acute epidemic hemorrhagic conjunctivitis (enteroviral)
				Rhinovirus A,B,C	J00 Acute nasopharyngitis (common cold)
					J20.6 Acute bronchitis due to rhinovirus
			Hepatovirus	*Hepatitis A virus*	B15 Acute hepatitis A
			Kobuvirus	*Aichivirus A*	-
			Parechovirus	*Human parechovirus*	A88 Other viral infections of the central nervous system, not elsewhere classified
					A41.8 Other specified sepsis
			Salivirus	*Salivirus A*	08.3 Other viral enteritis; possible association
					A08.4 Viral intestinal infection, unspecified; possible association
Unassigned	*Astroviridae*	NA	*Mamastrovirus*	*Mamoastovirus* 1 (human astrovirus)	A08.3 Other viral enteritis
Unassigned	*Caliciviridae*	NA	*Norovirus*	*Norwalk virus*	A08.1 Acute gastroenteropathy due to Norwalk virus
			Sapovirus	*Sapporo virus*	A08.3 Other viral enteritis
Unassigned	*Flaviviridae*	NA	*Flavivirus*	*Dengue virus*	A90 Dengue fever
					A91 Dengue hemorrhagic fever
				Japanese encephalitis virus	A83.0 Japanese encephalitis
				Kyasanur Forest disease virus	A98.2 Kyasanur Forest disease
				Langat virus	-
				Louping ill virus	A84.8 Other tick-borne encephalitis
				Murray Valley encephalitis virus	A83.4 Australian encephalitis
				Omsk hemorrhagic fever virus	A98.1 Omsk hemorrhagic fever
				Powassan virus	A84.8 Other tick-borne encephalitis
				St. Louis encephalitis virus	A83.3 St. Louis encephalitis
				Tick-borne encephalitis virus	A88 Other viral infections of the central nervous system, not elsewhere classified
				Wesselsbron	A92 Other mosquito-borne viral fevers
				West Nile virus	A92.3 West Nile infection
				Yellow fever virus	A95 Yellow fever
				Zika virus	A94 Unspecified arthropod-borne viral fever
			Hepacivirus	*Hepatitis C virus*	B17.1 Acute hepatitis C
					B18.2 Chronic hepatitis C
Unassigned	*Hepeviridae*	NA	*Hepevirus*	*Hepatitis E virus*	B17.2 Acute hepatitis E

(*Continued on next page*)

TABLE 4 Taxonomy and characterization of positive sense single-stranded RNA viruses of human medical importance (Baltimore classification IV) (*Continued*)

Genome composition	Order	Family	Subfamily	Genus	Species (ICTV or other common names)	Related primary ICD-10 codes
	Unassigned	*Togaviridae*	NA	*Alphavirus*	*Barmah Forest virus*	A92.8 Other specified mosquito-borne viral fevers
						B33.8 Other specified viral diseases
					Chikungunya virus	A92.0 Chikungunya virus disease
					Eastern equine encephalitis virus	A83.2 Eastern equine encephalitis
					Madariaga virus	A83.2 Eastern equine encephalitis, attenuated
					Mayaro virus	A92.8 Other specified mosquito-borne viral fevers
					O'nyong-nyong virus	A92.1 O'nyong-nyong fever
					Ross River virus	B33.1 Ross River disease
					Semliki Forest virus	B33.8 Other specified viral diseases
					Sinbis virus	A92.8 Other specified mosquito-borne viral fevers
						B33.8 Other specified viral disease
					Venezuelan equine encephalitis virus	A92.2 Venezuelan equine fever
						• Encephalitis
						• Encephalomyelitis virus disease
					Western equine encephalitis virus	A83.1 Western equine encephalitis
				Rubivirus	*Rubella virus*	B06 Rubella (German measles)

ICD, International Statistical Classification of Diseases and Related Health Problems; ICTV, International Committee on Taxonomy of Viruses; NA, not applicable.

TABLE 5 Taxonomy and characterization of negative sense single-stranded RNA viruses of human medical importance (Baltimore classification V)

Genome composition	Order	Family	Subfamily	Genus	Species (ICTV or other common names)	Related primary ICD-10 codes
Linear segmented	Unassigned	*Arenaviridae*[a]	NA	*Arenavirus*	*Guanarito virus*	A96.8 Other arenaviral hemorrhagic fevers
					Junin virus	A96.0 Junin hemorrhagic fever
					Lujo virus	A96.8 Other arenaviral hemorrhagic fevers
					Lassa virus	A96.2 Lassa fever
					Lymphocytic choriomeningitis virus	A87.2 Lymphocytic choriomeningitis
					Machupo virus	A96.1 Machupo hemorrhagic fever
					Sabiá virus	A96.8 Other arenaviral hemorrhagic fevers
	Unassigned	*Bunyaviridae*	NA	*Hantavirus*	*Andes virus*	B33.4+ Hanta(cardio)-pulmonary syndrome
						J17.1 Pneumonia in viral diseases classified elsewhere
						N17.9 Acute renal failure, unspecified
					Bayou virus	B33.4+ Hanta(cardio)-pulmonary syndrome
						J17.1 Pneumonia in viral diseases classified elsewhere
						N17.9 Acute renal failure, unspecified
					Black Creek Canal Virus	B33.4+ Hanta (cardio)-pulmonary syndrome
						J17.1 Pneumonia in viral diseases classified elsewhere
						N17.9 Acute renal failure, unspecified
					Hantaan virus	A98.5 Hemorrhagic fever with renal syndrome
					New York virus	B33.4+ Hantavirus (cardio)-pulmonary syndrome
					Puumala virus	A98.5 Hemorrhagic fever with renal syndrome
					Sin Nombre virus	B33.4+ Hantavirus (cardio)-pulmonary syndrome
					Seoul virus	A98.5 Hemorrhagic fever with renal syndrome
					Thottapalayam virus	—
				Nairovirus	*Crimean-Congo hemorrhagic fever*	A98.0 Other viral hemorrhagic fever not classified elsewhere
					Dugbe virus	A93.8 Other specified arthropod-borne viral fevers
				Orthobunyavirus	*Bwamba virus*	A92.8 Other specified mosquito-borne viral fevers
					California encephalitis virus	A83.5 California encephalitis
					Guama virus	A92.8 Other specified mosquito-borne viral fevers
					Madrid virus	A92.8 Other specified mosquito-borne viral fevers
					Oropouche virus	A93.0 Oropouche virus disease
					Tacaiuma virus	A92.8 Other specified mosquito-borne viral fevers
				Phlebovirus	*Rift Valley fever virus*	A92.4 Rift Valley fever

(*Continued on next page*)

TABLE 5 Taxonomy and characterization of negative sense single-stranded RNA viruses of human medical importance (Baltimore classification V) (*Continued*)

Genome composition	Order	Family	Subfamily	Genus	Species (ICTV or other common names)	Related primary ICD-10 codes
					Sandfly fever Naples virus	A93.1 Sandfly fever
						A93.8 Other specified arthropod-borne viral fevers
						A87.8 Other viral meningitis
Linear, segmented		*Orthomyxoviridae*	NA	*Influenzavirus A*	*Influenza A virus*	J09 Influenza due to certain identified influenza virus
						J10 Influenza virus not identified
					Influenza B virus	J09 Influenza due to certain identified influenza virus
						J10 Influenza virus not identified
					Influenza C virus	J09 Influenza due to certain identified influenza virus
						J10 Influenza virus not identified
Linear nonsegmented	*Mononegavirales*	*Bornaviridae*	NA	*Bornavirus*	*Borna disease virus*	—
		Filoviridae	NA	*Ebolavirus*	*Bundibugyo ebolavirus*	A98.4 Ebola virus disease
					Reston ebolavirus	—
					Sudan ebolavirus	A98.4 Ebola virus disease
					Tai forest ebolavirus	A98.4 Ebola virus disease
					Zaire ebolavirus	A98.4 Ebola virus disease
				Marburgvirus	*Marburg marburgvirus*	A98.3 Marburg virus disease
	Orthomyxoviridae	*Paramyxoviridae*	*Paramyxovirinae*	*Henipavirus*	*Hendra virus*	B33.8 Other specified viral diseases
					Nipah virus	B33.8 Other specified viral diseases
				Morbillivirus	*Measles virus*	B05 Measles
				Respirovirus	*Human parainfluenza virus 1*	J00 Acute nasopharyngitis
						J05.0 Acute obstructive laryngitis (croup)
						J06 Acute respiratory infections of multiple and unspecified sites
						J12.2 Parainfluenza virus pneumonia
						J20.4 Acute bronchitis due to parainfluenza virus
						J21.8 Acute bronchiolitis due to other specified organism
					Human parainfluenza virus 3	J00 Acute nasopharyngitis
						J05.0 Acute obstructive laryngitis (croup)
						J06 Acute respiratory infections of multiple and unspecified sites
						J12.2 Parainfluenza virus pneumonia
						J20.4 Acute bronchitis due to parainfluenza virus
						J21.8 Acute bronchiolitis due to other specified organism

(*Continued on next page*)

				Rubulavirus	*Human parainfluenza virus 2*	J00 Acute nasopharyngitis
						J05.0 Acute obstructive laryngitis (croup)
						J06 Acute respiratory infections of multiple and unspecified sites
						J12.2 Parainfluenza virus pneumonia
						J20.4 Acute bronchitis due to parainfluenza virus
						J21.8 Acute bronchiolitis due to other specified organism
					Human parainfluenza virus 4	J00 Acute nasopharyngitis
						J05.0 Acute obstructive laryngitis (croup)
						J06 Acute respiratory infections of multiple and unspecified sites
						J12.2 Parainfluenza virus pneumonia
						J20.4 Acute bronchitis due to parainfluenza virus
						J21.8 Acute bronchiolitis due to other specified organism
					Mumps virus	B26 Mumps including parotitis: epidemic, infectious
			Pneumovirinae	*Metapneumovirus*	*Human metapneumovirus*	J00 Acute nasopharyngitis
						J06 Acute respiratory infection of multiple and unspecified sites
						J12.3 Human metapneumovirus pneumonia
						J21.1 Acute bronchiolitis due to human metapneumovirus
				Pneumovirus	*Human respiratory syncytial virus*	J00 Acute nasopharyngitis
						J05.0 Acute obstructive laryngitis (croup)
						J12.1 Respiratory syncytial virus pneumonia
						J20.5 Acute bronchitis due to respiratory syncytial virus
						J21.0 Acute bronchiolitis due to respiratory syncytial virus
						B97.4 Respiratory syncytial virus as the cause of disease classified to other chapters
		Rhabdoviridae	NA	*Lyssavirus*	*Australian bat lyssavirus*	A82 Rabies
					Rabies virus	A82 Rabies
				Vesiculovirus	*Chandipura virus*	A85.8 Other specified viral encephalitis, possible association
					Isfahan virus	—
					Vesicular stomatitis Indiana virus	A93.8 Other specified arthropod-borne viral fevers
					Vesicular stomatitis New Jersey virus	A93.8 Other specified arthropod-borne viral fevers
Circular	Unassigned	Unassigned	NA	*Deltavirus*	*Hepatitis delta virus*	B17.0 Acute delta-(super) infection of hepatitis B carrier

ICD, International Statistical Classification of Diseases and Related Health Problems; ICTV, International Committee on Taxonomy of Viruses; NA, not applicable.
[a]Have been described as ambisense.

TABLE 6 Taxonomy and characterization of positive sense single-stranded RNA with a DNA replication intermediate viruses of human medical importance (Baltimore classification VI)

Genome composition	Order	Family	Subfamily	Genus	Species (ICTV or other common names)	Related primary ICD-10 codes
Linear	Unassigned	*Retroviridae*	*Orthoretrovirinae*	*Deltaretrovirus*	*Primate T-lymphotropic virus 1* (human T-lymphotropic virus 1 [HTLV-1])	C84.1 Cezary disease C84.5 Other mature T-cell/NK cell lymphomas C91.4 Hairy-cell leukemia C91.5 Adult T-cell lymphoma/leukemia (HTLV-1 associated) G04.9 Encephalitis, myelitis, encephalomyelitis, unspecified
					Primate T-lymphotropic virus 2 (human T-lymphotropic virus 2 [HTLV-2])	—
				Lentivirus	*Human immunodeficiency virus 1* (HIV-1)	B20-B24 Human immunodeficiency virus (HIV) disease Z21 Asymptomatic human immunodeficiency virus (HIV) infection status O98.7 Complicating pregnancy, childbirth, and puerperium
					Human immunodeficiency virus 2 (HIV-2)	B20-B24 Human immunodeficiency virus (HIV) disease Z21 Asymptomatic human immunodeficiency virus (HIV) infection status O98.7 Complicating pregnancy, childbirth, and puerperium

ICD, International Statistical Classification of Diseases and Related Health Problems; ICTV, International Committee on Taxonomy of Viruses.

characters (e.g., factors, features, or variables, as described later in this chapter) are attributed to a virus in order to classify it into a structured taxonomy. The practice of a taxonomic approach is not just an academic exercise whereby we develop a better understanding of how viruses are related or just place names to living things. Instead, taxonomy and the exercises described above improve our knowledge of molecular biology, pathogenesis, and epidemiology, as well as our ability to respond to newly emergent viruses with new diagnostics and therapies or preventive approaches (18). Taxonomy creates a common language that aids in how we communicate with colleagues and discuss viral pathogens. We can all quickly understand that we are discussing a specific and definable organism. For example, the Ebolavirus species affecting West Africa in 2014 can be further discussed and characterized as a member of the species *Zaire ebolavirus*, or the enterovirus infecting patients in North America in the summer and fall of 2014 is in fact a member of the species *Enterovirus D*.

The taxonomic grouping of viruses often relies on utilization of a variety of defined characters, and early systems of classification would have utilized characters as seen in Fig. 1. One of the most widely utilized methods for viral classification is the Baltimore classification, a nonhierarchical approach, named after the Nobel Prize winner David Baltimore. This system of categorizing viruses was originally divided into six groups, but with the inclusion of hepatitis B virus, it is now divided into seven groups and is based on the genome present in virions and type of replication (http://viralzone.expasy.org/all_by_species/254.html) (19). As seen in Tables 1 to 7, group I comprises dsDNA viruses, while group II comprises ssDNA viruses. Group III is composed of dsRNA viruses. Group IV is composed of positive sense ssRNA while group V is negative sense ssRNA. Group VI is composed of positive sense ssRNA viruses that replicate by means of a DNA intermediate. Group VII is composed of dsDNA viruses that replicate by means of a ssRNA intermediate (20). However, this method alone does not permit for stratified classification of viruses, and thus does not give a sense of hierarchies of relationships down to species or the subspecies level. An approach like the Baltimore system is also arbitrary in its division of viral characteristics and may miss key attributes such as the ambisense nature of the genomes of arenaviruses or *Rift valley fever virus* within the family *Bunyaviridae* in which an S segment uses an ambisense strategy (Table 5) (21).

There are other historic but less widely used systems of viral classification, and the hierarchical principles seen in some of these earlier systems can be seen as laying the ground work for current hierarchical approaches. The principles identified in these approaches have been utilized for decades but are still used today to help us characterize viruses of medical importance. Early approaches still seen today include elements of the Holmes classification, an early hierarchical classification approach from the 1940s for insect viruses that attempted to classify viruses largely on the basis of their morphology, the physical characteristics of their inclusions (or lack of inclusions), their host insect population, and disease processes (22). Two early approaches that took into account the physical characteristics of viruses were the L.H.T. (Lwoff Horne Tournier) system from the 1960s, a hierarchical classification system focusing on shared physical characteristics (nucleic acid, symmetry, presence/absence of an envelope, diameter of capsid, and number of capsomers) (23) and the Casjens and Kings classification from the 1970s, a nonhierarchical system that classified viruses on the

TABLE 7 Taxonomy and characterization of DNA reverse-transcribing viruses of human medical importance (Baltimore classification VII)

Genome composition	Order	Family	Subfamily	Genus	Species (ICTV or other common names)	Related primary ICD-10 codes
Partially double-stranded, circular genome	Unassigned	*Hepadnaviridae*		*Orthohepadnavirus*	*Hepatitis B virus*	B16 Acute hepatitis B B18 Chronic viral hepatitis

ICD, International Statistical Classification of Diseases and Related Health Problems; ICTV, International Committee on Taxonomy of Viruses.

basis of nucleic acid, symmetry, presence or absence of an envelope, and site of assembly (24).

Modern taxonomy came into being with the formation of the International Committee on Taxonomy of Viruses (ICTV). The ICTV is a committee of the Virology Division of the International Union of Microbiological Societies with activities governed by statutes. These statutes are intended to (i) develop internationally agreed taxonomy for viruses, (ii) develop internationally agreed names for virus taxa, (iii) communicate taxonomic decisions to the international virology community, and (iv) maintain an index of agreed names for virus taxa (25). The principles of nomenclature identified by the ICTV include (i) essential principles to aim for stability, avoid or reject names that might cause error or confusion, and avoid the unnecessary creation of names; (ii) viral nomenclature that is independent of other biological nomenclature and is a recognized exception; (iii) the primary purpose of a taxon being to supply a means of referring to the taxon rather than to indicate the characters or history of a taxon; and (iv) the name of a taxon having no official status until approved by the ICTV (http://www.ictvonline.org/codeOFVirusClassification.asp). Since 1971, nine reports have been released by the ICTV. Historically, this group decided to use species to classify viruses along with genus and family and set about to create working groups to develop plans to demark these species within a hierarchical structure when possible (e.g., http://www.ictvonline.org/proposals/2005.020-72.04.Herpes.pdf) (26).

The ninth report of the ICTV identified six orders, 87 families, 19 subfamilies, 349 genera, and 2,284 virus and viroid species. Representative viruses of medical importance are outlined in Tables 1 to 7 of this chapter. Within the report, each genus contains a type species and often other species, and some ICTV study groups worked to define "type isolates." Species may or may not be included within a genus, but all species are assigned to a subfamily or family. Genera and families are defined on a phylogenetic basis, and thus most genera are assigned to families, although some are unassigned until they can be further defined in terms of status and relationship. By the ninth report, it became apparent that classification of viruses would need to account for the increasing amount of genetics information available and the strategies used for making decisions about classification (27). In some less common cases, ICTV study groups have also worked on developing standards for naming strains and genetic variants that are becoming more evident with partial and whole genome analysis (28). An extensive and relatively up-to-date species master list is available at the ICTV website (http://talk.ictvonline.org/files/ictv_documents/m/msl/default.aspx).

Viral taxonomy is a dynamic field, and this is evident by recent updates that have occurred in the ninth report or since that time. In particular, multiple recent changes were ratified by the ICTV in March 2014, some key ones of which are identified in this chapter and described on the ICTV website (http://talk.ictvonline.org/files/ictv_official_taxonomy_updates_since_the_8th_report/m/vertebrate-official/default.aspx) but may not be yet identified in the master species list (http://talk.ictvonline.org/files/ictv_documents/m/msl/default.aspx). There are some striking and very important changes within the family *Parvoviridae*, with five new genera, five names expanded, a decrease in the identity required for species determination, new species introduced, and binomial species names used. Most notably, the species *Human parvovirus B19* was removed from the genus *Erythrovirus* in the subfamily *Parvovirinae*, family *Parvoviridae*. The species *Human parvovirus B19* was renamed *Primate erythroparvovirus 1* and placed in the genus Erythroparvovirus. (http://talk.ictvonline.org/files/ictv_official_taxonomy_updates_since_the_8th_report/m/vertebrate-official/default.aspx) (29).

Other important changes were also included in the 2014 ratification. Within the family *Adenoviridae* multiple changes occurred, including renaming the genus *Adenovirus* to *Mastadenovirus* and renaming the species *Human adenovirus* A-G to *Human mastadenovirus* A-G. These changes were intended to be on the species level and were not intended to impact colloquial virus, strain, or isolate names. To prevent confusion, uppercase letters were proposed to be retained, but in the future, there would be an understanding that the uppercase letters would not be considered sequential, nor would they imply a sense of completeness within a series (30). In the family *Papillomaviridae*, genus *Gammapapillomaviridae*, 10 new species *Gammapapillomavirus 11* to *Gammapapillomavirus 20* were created, and multiple changes were made in this family (http://talk.ictvonline.org/files/ictv_official_taxonomy_updates_since_the_8th_report/m/vertebrate-official/default.aspx). A new species, *Bokeloh bat lyssavirus*, in the genus *Lyssavirus*, family *Rhabdoviridae* was created (31). This virus has been identified as a potential emerging human pathogen, and a fatal cause of rabies in a Natterer's bat was reported, but a link to human disease has not been identified; this virus is not included in Table 5 at this time (32).

Several recent changes should be noted in the ninth report, or following in the species master list. Within the *Picornaviridae*, the species *Human enterovirus* A to *D* were renamed as *Enterovirus* A to *D*, and the species *Human rhinovirus* A to C were renamed *Rhinovirus* A to C. A new genus *Salivirus* (Stool Aichi-like Virus) was created, with a new type species, *Salivirus A*, created to encompass the previous Salivirus NG-J1. The previous possible species Human cosavirus A was re-assigned with the new species *Cosavirus A* and the Human cosaviruses B to D were left unassigned. Also, the species "Aichi virus" was named *Aichivirus A* within the genus *Kobuvirus*, family *Picornaviridae* (30).

Key taxonomic changes (http://talk.ictvonline.org/files/ictv_official_taxonomy_updates_since_the_8th_report/m/vertebrate-official/default.aspx) also occurred in a variety of families and are seen in the species master list. Following a proposal in 2010, within the family *Astroviridae*, genus *Mamastrovirus*, the species *Human astrovirus* was changed to *Mamastrovirus 1*. *Lujo virus* was designated as a new species in the genus *Arenavirus*, and it has been described to be associated with viral hemorrhagic fevers in South Africa and Zambia (33). In 2012, a proposal was initiated to create a new species *Madariaga virus* within the genus *Alphavirus*, which comprised strains of the species *Eastern equine encephalitis virus* from Central and South America and the Eastern Caribbean lineages II to IV. Multiple reasons justify this discrimination, including an attenuated illness in *Madariaga virus* disease compared to illness caused by *Eastern equine encephalitis virus* (34). A new species, *Sangassou virus*, was created within the genus *Hantavirus* to describe a murine virus with amino acid sequence similarity to hantaviruses that are possibly associated with fever of unknown origin in patients in Africa (http://talk.ictvonline.org/files/ictv_official_taxonomy_updates_since_the_8th_report/m/vertebrate-official/default.aspx) (35).

There have been multiple discussions and disagreement about how virologists should be define a species. The sixth report of the ICTV in 1995 defined species as a "polythetic class of viruses that constitutes a replicating lineage and occupies a particular ecological niche." This focus on a polythetic origin was a controversial topic even by the time of the ninth ICTV report (27), and in 2011 a proposed species definition that "A virus species should be defined on the basis of a range of criteria to ensure that the viruses assigned to it form a phylogenically distinct lineage" was introduced. Another proposed definition of species was introduced in 2012, which suggested that "A species is a monophyletic group of viruses whose properties can be distinguished by multiple criteria" (36). These multiple criteria could include properties such as natural or experimental host cell range, cell and tissue tropism, pathogenicity, vector specificity, antigenicity, and degree of relatedness in genes and genomes. Further to this, Gibbs commented that a species should consist of viruses that are linked with "a single 'type genomic sequence'" and "should be predominately monophyletic," which would lead to a definition of species that is more informative and acts as a quality assurance measure (37).

Below the level of species, there is no widespread, consistent, generalized, or systemized approach to naming and identifying viruses. However, some well-established approaches do exist, including those that account for variation due to laboratory-originated recombination. For example, for filoviruses, the genetic variant naming takes the approach, <virus name> ("strain>/)<isolation host-suffix>/<country of sampling>/<year of sampling>/<genetic variant designation>-<isolate designation> with the proposal to add a "rec" suffix for laboratory-derived recombinants (38). This is a similar approach to the nomenclature for influenza A strains, but use of geographic and temporal variables can be difficult to maintain due to a lack of standardization. In 2011, the World Health Organization (WHO) suggested revising how highly pathogenic influenza H5N1 is named to create a unified system that would allow for interpretation of data from different laboratories, replace geographic labeling with a more representative system, and create a system that accounts for antigenic variation and reassortment in multiple genotypes (http://www.who.int/influenza/gisrs_laboratory/h5n1_nomenclature/en/). Segmented viruses such as influenza A or rotavirus also have an additional level of characterization based on individual gene segments. The rotavirus working groups have taken a nucleotide-sequencing approach and utilized percentage cutoff values to identify strains. They have also given descriptors to each of the 11 gene segments (Gx-P[x]-Ix-Rx-Cx-Mx-Ax-Nx-Tx-Ex-Hx) and have proposed that strains are named as "RV group/species of origin/country of identification/common name/year of identification/G- and P-type" (39).

The ICTVdb was a curated virus database initiated following a decision by the ICTV executive in 1991, and it is still accessible (http://ictvdb.bio-mirror.cn/Ictv/ICTVindex.htm). The database used a decimalized numbering system to allow for the easy and unique identification of a virus at the level of species, genus, subfamily, and family. The ICTVdb was integrated with other databases containing genome sequence such as NCBI GenBank and EBI EMBL. Unfortunately, following the retirement of its curator, the ICTVdb became out of date, and by 2011 the ICTV suspended the ICTVdb project. With the suspension of the ICTVdb, other forums have arisen to provide continuity in taxonomic activities (Table 8). Some of these, such as the ExPASY Bioinformatics Resource Portal, are general in nature and provide a quick overview of viral characterization. Others such as the NCBI viral genomes database or the Viral Bioinformatics Resource Centre (University of Victoria), the VIDA 3.0 database, the Icosahedral virus capsid structure database, the RNAs and proteins of dsRNA viruses website and are broadly focused and can be used to study, characterize, and classify a broad variety of viral pathogens. Other websites may focus on one specific virus or smaller clusters of viruses as listed in Table 8. A disease-focused taxonomy involving viruses can also be created using the WHO ICD-10 database for identifying direct and indirect characters associated with human viral pathogens.

The International Statistical Classification of Diseases and Related Health Problems (ICD) is a standardized tool developed by the WHO to organize and code mortality and morbidity data that are then used for statistics, epidemiology, health care management, health care resource planning and allocation, monitoring, evaluation, research, primary care, and treatment. This tool can also be used to characterize the general health of a country or population as well as the impact that viruses have on the morbidity and mortality of individuals and populations (http://www.who.int/classifications/icd/revision/icd11faq/en/). The 10th revision was endorsed by the World Health Assembly in 1990 and is expected to be utilized until work on the current 11th revision is complete around 2017 (http://www.who.int/classifications/icd/en/). The 2010 English version is available online (http://apps.who.int/classifications/icd10/browse/2010/en) and allows for easy searching of viral diseases, syndromes, and viruses themselves and is supported by a user guide (http://www.who.int/classifications/icd/ICD-10_2nd_ed_volume2.pdf).

With the ICD-10, diseases are classified using an alphanumeric system that allows for assigning primary and secondary disease codes. These codes are provided as examples of diseases caused by or associated with specific viruses in Tables 1 to 7. Table 9 outlines how ICD-10 codes focused on a character, in this case viral hemorrhagic fever in humans, could be used to categorize arthropod-borne viral hemorrhagic fevers and create a disease-focused taxonomy (40) separate from one focused on viral order, family, genus, and species. Some codes such as A91 (dengue hemorrhagic

TABLE 8 Websites for online taxonomy databases

Focus	Working or other group	Title/topic	Website
Specific viruses			
Astroviruses	Pirbright Institute	The Astrovirus Pages	http://www.iah-virus.org/astroviridae/
Bat-associated viruses	Institute of Pathogen Biology, Beijing, China	dBatVir/Viral genome database	http://www.mgc.ac.cn/DBatVir/
Coronaviridae	VIPR: Virus Pathogen Resource	CoVDB/Viral genome database	www.viprbrc.org/brc/home.spg?decorator=corona
Dengue virus	Broad Institute	Dengue virus portal	http://www.broadinstitute.org/annotation/viral/Dengue/Home.html
Group A rotaviruses	Multiple authors	RotaC2.0 automated genotyping tool	http://rotac.regatools.be
Hepatitis B	Multiple groups	The Hepatitis B Virus Database (HBVdb)	https://hbvdb.ibcp.fr/HBVdb/
HIV	Los Alamos	HIV resistance mutation database	http://www.hiv.lanl.gov/content/sequence/RESDB/
HIV	Los Alamos	HIV sequence database	http://www.hiv.lanl.gov/content/sequence/HIV/mainpage.html
HIV	Stanford University	HIV drug resistance database	http://hivdb.stanford.edu/
Human adenovirus	Comparative Virology Team	Adenovirus Genetics and Taxonomy	www.vmri.hu/~harrach/
Influenza	Chinese Academy of Sciences	IVDB/Viral genome database	http://influenza.psych.ac.cn/
Influenza	Swiss Institute of Bioinformatics	Open Flu Database	http://openflu.vital-it.ch/browse.php
Picornaviruses	European study group on the molecular biology of Picornaviruses	Europic	http://www.europic.org.uk/
Picornaviruses	ICTV Picornaviridae study group	*Picornaviridae* Study Group Pages	http://www.picornastudygroup.com/
Picornaviruses	Pirbright institute	The Picornavirus Pages	http://www.picornaviridae.com/
General			
Bioinformatics	ExPASy Bioinformatics Resource Portal	Viral zone	http://viralzone.expasy.org/all_by_species/677.html

(*Continued on next page*)

TABLE 8 Websites for online taxonomy databases (*Continued*)

Focus	Working or other group	Title/topic	Website
Bioinformatics	University of Victoria	Viral bioinformatics resource centre	http://athena.bioc.uvic.ca/
Poxviruses			
Disease-focused taxonomy	World Health Organization	ICD-10 Version:2010	http://apps.who.int/classifications/icd10/browse/2010/en
Genomics	NCBI	Viral genomes	http://www.ncbi.nlm.nih.gov/genomes/GenomesHome.cgi?taxid=10239
Icosahedral virus capsid structure database	The Scripps Research Institute	Viperdb: Virus Particle ExploreR^2	http://viperdb.scripps.edu/
Taxonomy	ICTV	Virus taxonomy:2013 release	http://ictvonline.org/virusTaxonomy.asp
Universal protein database	UniProt consortium	UniProt	http://www.uniprot.org/
Multiple			
Herpesviridae	University College London	VIDA 3.0	http://www.biochem.ucl.ac.uk/bsm/virus_database/VIDA3/VIDA.html
Poxviridae			
Papillomaviridae			
Coronaviridae			
Arteriviridae			
Influenza virus	NCBI	Virus variation database	http://www.ncbi.nlm.nih.gov/genomes/VirusVariation/index.html
Dengue virus			
West nile virus			

ICTV, International Committee on Taxonomy of Viruses; NCBI, National Center for Biotechnology Information.

TABLE 9 Arthropod-borne viral fevers and viral hemorrhagic fevers (A90–A99)

A91 Dengue hemorrhagic fever

A92 Other mosquito-borne viral fevers
- **Excluding:** Ross River disease (B33.1)
- **A92.0 Chikungunya virus disease**
  - Chikungunya (hemorrhagic) fever
- **A92.4 Rift Valley fever**

A95 Yellow fever
- **A95.0 Sylvatic yellow fever**
  - Jungle yellow fever
- **A95.1 Urban yellow fever**
- **A95.9 Yellow fever, unspecified**

A96 Arenaviral hemorrhagic fever
- **A96.0 Junin hemorrhagic fever**
- **Including:** Argentinian hemorrhagic fever
- **A96.1 Machupo hemorrhagic fever**
  - Bolivian hemorrhagic fever
- **A96.2 Lassa fever**
- **A96.8 Other arenaviral hemorrhagic fevers**
- **A96.9 Arenaviral hemorrhagic fever, unspecified**

A98 Other viral hemorrhagic fevers, not elsewhere classified
- **Excluding:** Chikungunya hemorrhagic fever (A92.0), dengue hemorrhagic fever (A91)
- **A98.0 Crimean-Congo hemorrhagic fever**
  - Central Asian hemorrhagic fever
- **A98.1 Omsk hemorrhagic fever**
- **A98.2 Kyasanur Forest disease**
- **A98.3 Marburg virus disease**
- **A98.4 Ebola virus disease**
- **A98.5 Hemorrhagic fever with renal syndrome**
  - Hemorrhagic fever:
    - Epidemic
    - Korean
    - Russian
    - Hantaan virus disease
    - Hantaan virus disease with renal manifestations
    - Nephropathia epidemica
  - **Excluding:** Hantavirus (cardio-)pulmonary syndrome (B33.4+, J17.1*)
- **A98.8 Other specified viral hemorrhagic fevers**

fever) and A95 (yellow fever) are tightly linked to an easily identifiable viral species. Other codes, such as A96 arenaviral hemorrhagic fever identify a genus associated with disease but may not identify all species such as *Sabia virus* (Brazilian hemorrhagic fever) or *Guanarito virus* (Venezuelan hemorrhagic fever). Yet codes, such as A92 (other mosquito-borne viral fevers, excluding Ross River disease), may be vector associated and include different genera such as alphaviruses and phleboviruses. Other genera and species not characterized elsewhere would be lumped into A98 other viral hemorrhagic fevers, not classified elsewhere.

ICD-10 codes are considered administrative health data, and there are concerns about how well these data can characterize illness as well as their accuracy. Administrative health data have value in helping us understand clinical outcomes associated with viral diseases at a population level as well as risk factors for disease. The current version is thought to provide both a better description of clinical situations as well as more specificity in describing health care problems than ICD-9 (41). However, using chart reviews, it was found that ICD-9 and ICD-10 had roughly equal sensitivity for coding conditions in general (42). ICD codes, in this case ICD-9 codes, have been shown to be highly predictive of determining pneumonia, herpes simplex virus infections, cirrhosis with hepatitis C virus, and HIV or hepatitis B co-infections with hepatitis C virus when administrative databases were analyzed (43). However, validations need to be undertaken to ensure each code is accurately describing a viral disease process.

Character-based description allows for the use of descriptors, variables, or characters to classify and compare viruses. The ICTV uses an extensive and comprehensive listing of different characters, and these generally include isolation details, historic ICTVdb virus codes, classification at taxonomic level, virion properties, morphology, physiochemical and physical properties, nucleic acid, proteins, genome organization and replication, antigenicity, and biological properties including natural host range and pathology. The ICD-10 codes described earlier could also be considered pathology-focused characters. As an example, the following species demarcation criteria would be used within the genus *Flavivirus*: nucleotide and deduced amino acid sequence data, antigenic characteristics, geographic association, vector association, host association, disease association, and ecological characteristics (http://ictvdb.biomirror.cn/Ictv/fs_flavi.htm). Use of these multiple and diverse characters allows for the systematic understanding of how viruses compare to each other, and it could be argued that they are a natural progression of other historical methods while still ensuring that a hierarchical classification based on a modern multidisciplinary approach can be undertaken. One of the issues with using a character-based system and character-based descriptors is that their demarcation criteria can vary greatly within and between families and as such they lack a single unifying property. This variability is required to ensure that each virus is classified (44). However, as described earlier, there now seems to be a greater role for a genetics-based approach in defining virus taxonomy.

Molecular phylogenetics is an approach that allows for the comparison of nucleic acid and/or protein sequences to investigate evolutionary relationships. The multiple issues with non-sequence-based viral taxonomy, including the subjective nature of other characters, poor clinical characterization, or more practical factors, such as the lack of adequate tissue culture propagation systems or animal infection models for certain viruses, suggests that nucleic acid or protein sequence should be the primary driver of taxonomic decisions (45). The most common method used is a pairwise analysis of a particular gene, amino acid sequence, or subgenomic marker and the creation of a "tree" that allows for an estimation of genetic relatedness; this has traditionally been a method for comparing sequences to determine phylogeny at the subgenomic level (46). Much of this work will be described in chapter 15 and several previously reviewed approaches to genome tree formation include (i) alignment-free trees, (ii) gene content trees, (iii) chromosomal gene order trees, (iv) average sequence similarity trees, and (v) meta-analysis trees (47, 48). As described later in this chapter there are some examples in which classification systems are based largely, or even purely, on sequence homology including human papillomaviruses.

Different approaches in terms of target, such as amino acids versus nucleotides, as well as genes sequenced and whether to include hypervariable regions in the analysis, can impact taxonomic classifications. One important choice is whether to use nucleotide or amino acid sequences within the analysis. It has been argued that phylogenetic relationships based on nucleotide sequences alone may be misleading since they analyze sites with saturated substitutions, and it has been suggested that these biases should be compensated for by using Bayesian methods or maximum likelihood methods or by analyzing aligned amino acid sequences (49). However, amino acid analysis alone may not be sufficient because some taxonomic or phylogenetic approaches may take into account noncoding regions. Another key choice is whether to include partial or full genome sequences. For obvious reasons, including earlier technologic issues with sequencing long regions of nucleic acid and the management of sequencing information, earlier classification approaches were often based on partial genome sequences of viruses. For example, the RNA-dependent RNA-polymerase (RdRp) protein sequence was used as one tool to understand relatedness of families within the order *Picornavirales* and could be used to distinguish members of different genera within the family *Reoviridae* (27). Subgenomic analysis of one or multiple genes will not reveal the nature of all genetic changes within a virus and may not confidently classify a virus that is being studied within an appropriate taxonomic framework. The increased use of whole genome sequencing rather than sequencing only subgenomic regions has led to instances in which greater diversity or variants are identified from previously studied viral populations (49). Whole genome approaches have also uncovered previously undescribed evolutionary relationships, including evidence of interspecies transmission and related recombination events (50), that can then assist in how viruses are classified. When these approaches are applied, they can be used to generate more consistent nomenclature (39). This new information identified by analysis of a complete genome is important because it increases our awareness of relatedness between individual viruses being studied and improves our knowledge of viral epidemiology and pathophysiology.

The impact of the viral metagenome on understanding the virome and characterizing virus components within primary specimens or natural samples should also be noted. High-throughput deep-sequencing approaches have played important roles in the discovery of viruses and viral communities, or the virome, within primary specimens and biological samples (51). However, one of the issues with this approach is the incredibly large amount of information produced and how to manage this information as it significantly increases on a yearly basis (52). Other key problems include concerns in the bioinformatics community about how to account for factors such as their small genomes, fast mutation rates, and low conservation (53), and how to assign taxonomy to very short reads of nucleic acid sequence (54).

Once phylogenetic approaches are used, questions then arise as to how comparisons between viruses will be made, and whether these approaches will be consistent or inconsistent with the previously defined taxonomy (55). These questions have not only been faced by virologists but are universal when phylogenetic approaches are taken to classify organisms. Multiple factors will impact phylogenetic analysis, including how trees are established and how they change as new sequences are added (56). In some cases a tree model may not be used, and phylogenetic networks may instead be used for investigating evolutionary relationships to establish relationships; however, these often require extensive full genome sequences (57). Other methods such as the calculation of genetic distances between nucleotide sequences of full genome sequences can be used without construction of trees and can correlate well to subgenomic regions, without the requirement of extensive full genome sequences being available (57). Regardless of the approaches to determine phylogenetic relationships, the conclusions may still be biased if they do not account for recombination and convergence (58).

Descriptions of viral taxonomy and categorization can easily diverge from clinically relevant viruses unless a strong effort is made to link the virological information to information describing disease processes. Furthermore, viral infections may not actually be linked to any disease processes, or infections may be associated with disease processes but may not be confirmed with Koch's postulates. Part of this problem may be that until recently we had very limited tools for diagnosing viral diseases and the age of viral discovery is now outstripping our ability to show causality with exercises such as the use of Koch's postulates. Tables 1 to 7 show a summary of viruses of medical importance and use ICD-10 codes to indicate the associated disease processes attributed to these viruses. These codes act as the disease- or pathology-focused character associated with viral infection. A framework of these relationships can also be seen in Table 9, which uses a viral hemorrhagic disease as an example. However, it should be noted that the disease-focused taxonomy provided in Tables 1 to 7 is not intended to be an exhaustive description of the diseases caused by each pathogen but is shown to indicate medical relevance and to identify specific disease-focused characters.

As seen in the Tables 1 to 7, if the virus is not directly listed with an ICD-10 code then the correlation becomes more complicated. For example, the pathophysiology linked to *Human torovirus* could be linked to A08.3 other viral enteritis. Other disease processes may not be related to all species of a genus, and the diseased-focused taxonomy may not be entirely specific. In the case of code B30.0 + keratoconjunctivitis due to adenoviruses, it would be simplistic to link this disease to all types of adenovirus because types 8, 19, and 37 are usually involved, while type 5 can be involved with severe disease. In contrast, B30.1 + conjunctivitis is mostly caused by types 3, 4, and 7, but most types can cause this disease. Similarly, enterovirus categorization is complex and examples given use a previous review on enterovirus infection (59). ICD-10 coding to describe a viral infection may primarily link a virus to a specific disease process, while other secondary disease processes may be described later, sometimes as footnotes. For examples, *Venezuelan equine encephalitis virus* disease is described in ICD-10 as a viral fever, but in a minority of cases they lead to viral encephalitis as described in a footnote in ICD-10 coding.

Other infectious processes may be hard to define in terms of an ICD-10 code or another disease- or pathology-focused character and may not currently fulfill Koch's postulates. *Betapapillomavirus 1* may play a role in carcinoma *in situ* of the skin and in actinic keratitis, *Mupapillomavirus 1* is sometimes found in warts and other times on normal skin (60), while the role for gamma papilloma viruses in human disease is even less obvious (61). *Banna virus* has been identified in patients with viral encephalitis, and there may be a possible association with illness (62). Mammalian orthoreoviruses have been identified in humans with multiple illnesses (63); however, evidence on causation is not strong, and these are listed in the table as associations. The role of *Borna disease virus* in human disease, including viral encephalitis and neurologic or psychiatric disorders, is still controversial (64). There is also a possible association of *Cosavirus A* with diarrhea in immune-compromised and pediatric patients (65). *Aichivirus A*, *Salivirus A*, and *Theilovirus* or "Saffold virus" have been shown to circulate in children with diarrhea, but their roles are not well understood. *Theilovirus* virus may also be an emerging viral cause of central nervous system disease (66). *Human picobirnavirus* also has an associated role in diarrheal illness (67). Although *Torquetenovirus 1* has been identified in human specimens, its role in human disease is unclear (68), as is the role for *Thottapalayam virus* (69).

The following scenarios describe the issues faced by the scientific community in determining taxonomy. Some are relatively straightforward, while others have required significant discussion or are still points of discussion. Examples are described for the papillomaviruses, picornaviruses, adenoviruses, and noroviruses. A common theme that appears in all examples, and one that has been described previously, is the impact of whole genome sequence analysis on categorization of viruses within a taxonomic framework. Primarily, much of these discussions focuses on what criteria should be used to classify these viruses, with the understanding that these criteria are key because typing needs to be consistent across methods to ensure continuity in understanding the epidemiology and clinical presentation of these viruses and to allow for the effective identification of new strains or types that may cause severe illness.

Papillomaviruses

Multiple characteristics can be used to develop a taxonomic approach; however, in some cases the taxonomic approach is restricted to genetic approaches, and the question still arises about which genetic approach to use. With human papillomaviruses, genetic approaches were required because of a lack of reliable cell culture systems and animal models of infection for these particular viruses (27). As a result of these pressures, taxonomy developed on two basic themes: host specificity and the use of phylogenetic analysis. Also, some categorization focused on whether the HPV type could be grouped as cutaneous or mucosal, but this approach was not maintained following more extensive phylogenetic analysis (45). Coordination within the scientific community studying human papillomaviruses emerged early, and in the 1980s, the community established a reference center in Germany. Basic rules established that identifying a new type required storing the full-length cloned genome at this reference center. Even with this strong coordination, there was no consensus on which gene targets to utilize for taxonomic classification, and for a considerable period of time, there was significant discussion on the gene targets or sequences (e.g., L1, and E6 and E7), whether open reading frames (ORFs) and partial gene sequences or full gene sequences should be used, and what level of similarity should be used for each target to classify a new species (70). As new technologies increased the output of sequence available to be analyzed, there was an increased need to standardize approaches to classification (71). Currently, human papillomaviruses are classified by phylogenetic analysis of the L1 gene ORF with variations in the percentage of difference used to determine if a newly identified sequence belongs to a new species, a new subtype, or a new variant (71). Following this approach, new discussions have now moved onto whether to accept new types that are sequenced and identified by metagenomic approaches (45).

Picornaviruses

The following illustrative example describes the issues the scientific community may need to deal with when transitioning from a taxonomic approach involving multiple potentially variable characteristics to one using potentially more objective characteristics. Current picornavirus taxonomic classification is carried out by the Picornavirus Study Group on behalf of the ICTV. Classification of picornaviruses involves a number of rules that take into account several different characteristics, including polyprotein sequence homology, genome organization, genome base composition, host range, host cell receptor variety, and replicative processes. Multiple molecular markers may also be used to create a picornavirus taxonomy (27, 72). At the species level the use of VP1 pairwise sequencing can often be used to determine relationships between viruses (73). However, for the purposes of developing hierarchal categorizations, it is argued that this approach lacks a gold standard, and a growing number of picornaviruses are not assigned to any taxonomic grouping or are in provisional groupings (72). Also, the identification of clades and relationships between strains at the subspecies level, such as those within human enterovirus 68 (EV-D68), requires the analysis of several other non-VP1 targets including the 5′-untranscribed region and VP4 (74). The inability to assign specific viruses to a particular taxonomic grouping is problematic because there is a need to link clinical disease with specific types, as well as a need to develop and define the characteristics of new tests that may need to account for the absence of current assay targets. As described previously, the increased utilization of whole genome sequencing has allowed for the characterization of viruses to identify new relationships within the picornaviruses (75). New bioinformatics approaches for comparing whole genome sequences, including quantitative procedures to hierarchically classify picornaviruses based on intervirus genetic divergence, are now being attempted by some scientists (76). A side-by-side comparison with ICTV classification has already been undertaken using this approach, with the authors proposing that the genome contains enough information to act as the sole demarcation criterion for the picornaviruses (72).

Adenoviruses

As stated earlier, the lowest level of taxonomic classification that the ICTV undertakes is the species, and multiple criteria are used to determine a species within the genus *Mastadenovirus* (http://www.vmri.hu/~harrach/AdVtaxlong.htm). Below the level of species, serotype has been used to understand the clinical epidemiology and pathophysiology of these viruses in humans; however, in the case of adenoviruses, serotype/type has played a key role in linking a species to a disease process. Traditional adenovirus typing involved the isolation and propagation of the virus followed by serotyping, which in the case of a suspect novel type would require an extremely large number of virus neutralization assays. However, for almost a decade, the amplification of the hexon gene provided a reasonable surrogate to the traditional approaches (77). Recently, major points of discussion include how to define type, how to deal with recombination events (including intertypic recombination), the extent of sequences required for comparison, and how to manage and identify new strains as well as storage of sequence information, and it is clear that a typing method focusing on one gene target will not be operationally viable going into the future (78). There are already significant criteria being introduced at the Human Adenovirus Working Group to address the use of sequencing information, link species to type in a new nomenclature system, require the use of complete genome sequencing and phylogenetics in the creation of a new type identifier, provide a rule for naming priorities, and deal with the issue of recombination (79). Some researchers have already proposed that whole genome analysis should be used to identify new lineages of adenoviruses and provide the evidence for either a new species or a new type number (80), and this approach has also been used to speculate on viral evolution and search for potentially emerging types and subtypes (81). Regardless, these issues will definitely create changes in how adenoviruses are characterized over the next 5 to 10 years and will push consensus groups further into the realm of subspecies classification.

Norovirus

Norovirus genogroups I, II, and IV are clinically important for humans, with recent novel strains emerging and data suggesting that strain variation can be driven by positive selection during chronic infection within immunocompromised hosts (82). Currently, real-time PCRs to determine genogroups I and II are in broad use and the ability to genotype has also been widely established, but multiple approaches exist and there is a need for consistency for genotyping as well as identification of new strains (83). These genogroups are further divided into genotypes (84). Since the mid-1990s, norovirus genotypes have been based on the complete VP1 gene sequence (ORF2; open reading frame 2), encoding the 60 kDa capsid protein, with new genotypes being designated when more than 20% of VP-1 amino acids differed using pairwise analysis. In 2011, researchers who were part of the Food-Borne Viruses in Europe Network proposed a molecular epidemiologic approach focused on the analysis of ORF2 (85). This focus on ORF2 and its epitopes B, C, and D is still used to characterize new strain variants (86). However, the primary focus on ORF2 as a sole target for genotyping has begun to shift within the last 5 years. During the 4th International Conference on Caliciviruses in 2010, a need for common classification of noroviruses was identified and a norovirus working group was established. This group was influenced by the *Picornaviridae* and *Flaviviridae* working groups described earlier in this chapter that had created practical standards for universal nomenclature and typing systems. By 2013, members of this working group proposed a phylogenetic analysis of the full VP1 sequence as well as the partial 3′ ORF1 sequence being utilized to generate new genotypes (84). The ORF1 encodes for a nonstructural polyprotein that undergoes proteolytic cleavage to release six nonstructural proteins (87). An expanded approach has been shown to allow for identification of recombination events at the ORF1/ORF2 overlap (88, 89), recombinations within VP1 (ORF2) (90), and possibly within the ORF2/ORF3 boundary (90) in emerging variants, which would not be identified if only ORF1 or ORF2 sequences were analyzed (91). As seen previously, other groups have gone to full genome analysis to characterize the emergence of new strains within their jurisdictions (92, 93). These whole genome approaches have already been used in outbreak settings to identify minor genetic variations that could suggest transmission events and might be utilized to suggest a direction of transmission (94).

In conclusion, in spite of their simplicity, viruses are a complex and diverse group of organisms that may have equally diverse origins and evolutionary pressures. Their interaction with their human hosts may cause disease but also impacts viral evolution and shapes key viral characteristics. Viral taxonomy, classification, and characterization can be thought of as important tools that improve our ability to diagnose and compare viruses of medical importance. This framework also allows us to place newly identified viruses within the tree of life and may provide clues to pathophysiology when they may not yet be completely evident. Linkage to well-understood disease processes also allows for the characterization and classification of viruses into disease-focused frameworks that may not be completely driven by the biology of the organisms. Over the last 5 years, there have been significant changes in the field of viral taxonomy, and this includes changes in the proper name of some commonly identified viruses as well as realignment of relationships between these medically relevant viruses. Some resources, such as databases, have ceased to exist as up-to-date tools, while new databases have emerged or been strengthened to support viral taxonomy. Some of these changes, such as with the nature of what constitutes a viral species, have led to vibrant discussion and are critical for the development of taxonomy in the future. Related to this discussion with the nature of species, the changes in taxonomic approaches and even our understanding of viral evolution have also changed and are now being driven by significant increases in genetic information created by whole genome analysis and metagenomic approaches as well as the bioinformatics tools to support this information. These molecular approaches now allow for the classification of viruses in new ways, which will in turn also impact how currently known and yet to be discovered viral pathogens are characterized and classified. No doubt, we will continue to see a greater role for phylogenetics in the placement of viruses within a structured framework, while other more subjective or historic characters of these viruses will have a lesser impact on viral taxonomy.

REFERENCES

1. **Moreira D, Lopez-Garcia P.** 2009. Ten reasons to exclude viruses from the tree of life. *Nat Rev Microbiol* **7:**306–311.
2. **Forterre P.** 2006. The origin of viruses and their possible roles in major evolutionary transitions. *Virus Res* **117:**5–16.
3. **Holmes EC.** 2011. What does virus evolution tell us about virus origins? *J Virol* **85:**5247–5251.
4. **Nasir A, Kim KM, Caetano-Anolles G.** 2012. Viral evolution: primordial cellular origins and late adaptation to parasitism. *Mob Genet Elements* **2:**247–252.
5. **Rohwer F, Barott K.** 2013. Viral information. *Biol Philos* **28:**283–297.
6. **Duffy S, Shackelton LA, Holmes EC.** 2008. Rates of evolutionary change in viruses: patterns and determinants. *Nat Rev Genet* **9:**267–276.
7. **Jenkins GM, Rambaut A, Pybus OG, Holmes EC.** 2002. Rates of molecular evolution in RNA viruses: a quantitative phylogenetic analysis. *J Mol Evol* **54:**156–165.
8. **Chan JM, Carlsson G, Rabadan R.** 2013. Topology of viral evolution. *Proc Natl Acad Sci U S A* **110:**18566–18571.
9. **Iranzo J, Manrubia SC.** 2012. Evolutionary dynamics of genome segmentation in multipartite viruses. *Proc Biol Sci* **279:** 3812–3819.

10. **Forterre P, Prangishvili D.** 2009. The origin of viruses. *Res Microbiol* **160:**466–472.
11. **Ludmir EB, Enquist LW.** 2009. Viral genomes are part of the phylogenetic tree of life. *Nat Rev Microbiol* **7:**615.
12. **Henquell C, Mirand A, Richter J, Schuffenecker I, Bottiger B, Diedrich S, Terletskaia-Ladwig E, Christodoulou C, Peigue-Lafeuille H, Bailly JL.** 2013. Phylogenetic patterns of human coxsackievirus B5 arise from population dynamics between two genogroups and reveal evolutionary factors of molecular adaptation and transmission. *J Virol* **87:**12249–12259.
13. **Ward MJ, Lycett SJ, Avila D, Bollback JP, Leigh Brown AJ.** 2013. Evolutionary interactions between haemagglutinin and neuraminidase in avian influenza. *BMC Evol Biol* **13:** 222.
14. **Schuh AJ, Ward MJ, Brown AJ, Barrett AD.** 2013. Phylogeography of Japanese encephalitis virus: genotype is associated with climate. *PLoS Negl Trop Dis* **7:**e2411.
15. **Cobey S.** 2014. Pathogen evolution and the immunological niche. *Ann N Y Acad Sci* **1320:**1–15.
16. **Chi H, Liu HF, Weng LC, Wang NY, Chiu NC, Lai MJ, Lin YC, Chiu YY, Hsieh WS, Huang LM.** 2013. Molecular epidemiology and phylodynamics of the human respiratory syncytial virus fusion protein in northern Taiwan. *PLoS One* **8:** e64012.
17. **Haenni A-L.** 2008. Virus evolution and taxonomy, p. 205–217. *In* Roossinck M (ed), *Plant virus evolution*. Springer-Verlag, Berlin.
18. **Lefkowitz E.** 2014. Taxonomy and classification of viruses, p. 1265–1275. *In*: Versalovic J, Carroll KC, Funke G, Jorgensen JH, Landry ML, Warnock DW (ed), *Manual of clinical microbiology*, 10th ed. ASM Press, Washington, DC.
19. **Baltimore D.** 1971. Expression of animal virus genomes. *Bacteriol Rev* **35:**235–241.
20. **Pringle CR.** 1999. Virus taxonomy–1999. The universal system of virus taxonomy, updated to include the new proposals ratified by the International Committee on Taxonomy of Viruses during 1998. *Arch Virol* **144:**421–429.
21. **Brennan B, Welch SR, Elliott RM.** 2014. The consequences of reconfiguring the ambisense S genome segment of Rift Valley fever virus on viral replication in mammalian and mosquito cells and for genome packaging. *PLoS Pathog* **10:** e1003922.
22. **Steinhaus EA.** 1949. Nomenclature and classification of insect viruses. *Bacteriol Rev* **13:**203–223, illust.
23. **Lwoff A, Horne R, Tournier P.** 1962. A system of viruses. *Cold Spring Harb Symp Quant Biol* **27:**51–55.
24. **Casjens S, King J.** 1975. Virus assembly. *Annu Rev Biochem* **44:**555–611.
25. **The International Code of Virus Classification and Nomenclature.** 2013. 25-3-2014. http://ictvonline.org/codeofvirusclassification_2012.asp
26. **van Regenmortel MH, Mahy BW.** 2004. Emerging issues in virus taxonomy. *Emerg Infect Dis* **10:**8–13.
27. **King AMQ, Lefkowitz E, Adams MJ, Carstens EB.** 2011. *Virus Taxonomy: Ninth Report of the International Committee on Taxonomy of Viruses*. Elsevier, London, United Kingdom.
28. **Kuhn JH, Bao Y, Bavari S, Becker S, Bradfute S, Brister JR, Bukreyev AA, Chandran K, Davey RA, Dolnik O, Dye JM, Enterlein S, Hensley LE, Honko AN, Jahrling PB, Johnson KM, Kobinger G, Leroy EM, Lever MS, Mühlberger E, Netesov SV, Olinger GG, Palacios G, Patterson JL, Paweska JT, Pitt L, Radoshitzky SR, Saphire EO, Smither SJ, Swanepoel R, Towner JS, van der Groen G, Volchkov VE, Wahl-Jensen V, Warren TK, Weidmann M, Nichol ST.** 2013. Virus nomenclature below the species level: a standardized nomenclature for natural variants of viruses assigned to the family Filoviridae. *Arch Virol* **158:**301–311.
29. **Drexler JF, Reber U, Muth D, Herzog P, Annan A, Ebach F, Sarpong N, Acquah S, Adlkofer J, Adu-Sarkodie Y, Panning M, Tannich E, May J, Drosten C, Eis-Hübinger AM.** 2012. Human parvovirus 4 in nasal and fecal specimens from children, Ghana. *Emerg Infect Dis* **18:**1650–1653.
30. **ICTV.** ICTV Official Taxonomy: Updates since the 8th report. 26-6-2014. http://talk.ictvonline.org/files/ictv_official_taxonomy_updates_since_the_8th_report/default.aspx
31. **Adams MJ, Lefkowitz EJ, King AM, Carstens EB.** 2014. Ratification vote on taxonomic proposals to the International Committee on Taxonomy of Viruses (2014). *Arch Virol* **159:** 2831–2841.
32. **Freuling CM, Beer M, Conraths FJ, Finke S, Hoffmann B, Keller B, Kliemt J, Mettenleiter TC, Mühlbach E, Teifke JP, Wohlsein P, Müller T.** 2011. Novel lyssavirus in Natterer's bat, Germany. *Emerg Infect Dis* **17:**1519–1522.
33. **Briese T, Paweska JT, McMullan LK, Hutchison SK, Street C, Palacios G, Khristova ML, Weyer J, Swanepoel R, Egholm M, Nichol ST, Lipkin WI.** 2009. Genetic detection and characterization of Lujo virus, a new hemorrhagic fever-associated arenavirus from southern Africa. *PLoS Pathog* **5:** e1000455.
34. **Arrigo NC, Adams AP, Weaver SC.** 2010. Evolutionary patterns of eastern equine encephalitis virus in North versus South America suggest ecological differences and taxonomic revision. *J Virol* **84:**1014–1025.
35. **Klempa B, Witkowski PT, Popugaeva E, Auste B, Koivogui L, Fichet-Calvet E, Strecker T, Ter MJ, Kruger DH.** 2012. Sangassou virus, the first hantavirus isolate from Africa, displays genetic and functional properties distinct from those of other murinae-associated hantaviruses. *J Virol* **86:**3819–3827.
36. **Van Regenmortel MH, Ackermann HW, Calisher CH, Dietzgen RG, Horzinek MC, Keil GM, Mahy BW, Martelli GP, Murphy FA, Pringle C, Rima BK, Skern T, Vetten HJ, Weaver SC.** 2013. Virus species polemics: 14 senior virologists oppose a proposed change to the ICTV definition of virus species. *Arch Virol* **158:**1115–1119.
37. **Gibbs AJ.** 2013. Viral taxonomy needs a spring clean; its exploration era is over. *Virol J* **10:**254.
38. **Kuhn JH, Bào Y, Bavari S, Becker S, Bradfute S, Brauburger K, Rodney Brister J, Bukreyev AA, Caì Y, Chandran K, Davey RA, Dolnik O, Dye JM, Enterlein S, Gonzalez JP, Formenty P, Freiberg AN, Hensley LE, Hoenen T, Honko AN, Ignatyev GM, Jahrling PB, Johnson KM, Klenk HD, Kobinger G, Lackemeyer MG, Leroy EM, Lever MS, Mühlberger E, Netesov SV, Olinger GG, Palacios G, Patterson JL, Paweska JT, Pitt L, Radoshitzky SR, Ryabchikova EI, Saphire EO, Shestopalov AM, Smither SJ, Sullivan NJ, Swanepoel R, Takada A, Towner JS, van der Groen G, Volchkov VE, Volchkova VA, Wahl-Jensen V, Warren TK, Warfield KL, Weidmann M, Nichol ST.** 2014. Virus nomenclature below the species level: a standardized nomenclature for filovirus strains and variants rescued from cDNA. *Arch Virol* **159:**1229–1237.
39. **Matthijnssens J, Ciarlet M, McDonald SM, Attoui H, Bányai K, Brister JR, Buesa J, Esona MD, Estes MK, Gentsch JR, Iturriza-Gómara M, Johne R, Kirkwood CD, Martella V, Mertens PP, Nakagomi O, Parreño V, Rahman M, Ruggeri FM, Saif LJ, Santos N, Steyer A, Taniguchi K, Patton JT, Desselberger U, Van Ranst M.** 2011. Uniformity of rotavirus strain nomenclature proposed by the Rotavirus Classification Working Group (RCWG). *Arch Virol* **156:** 1397–1413.
40. **Cookson WO.** 1999. Disease taxonomy–polygenic. *Br Med Bull* **55:**358–365.
41. **Topaz M, Shafran-Topaz L, Bowles KH.** 2013. ICD-9 to ICD-10: evolution, revolution, and current debates in the United States. *Perspect Health Inf Manag* **10:**1d.
42. **Quan H, Li B, Saunders LD, Parsons GA, Nilsson CI, Alibhai A, Ghali WA, IMECCHI Investigators.** 2008. Assessing validity of ICD-9-CM and ICD-10 administrative data in recording clinical conditions in a unique dually coded database. *Health Serv Res* **43:**1424–1441.
43. **Kramer JR, Davila JA, Miller ED, Richardson P, Giordano TP, El-Serag HB.** 2008. The validity of viral hepatitis and chronic liver disease diagnoses in Veterans Affairs administrative databases. *Aliment Pharmacol Ther* **27:**274–282.

44. **Lauber C, Gorbalenya AE.** 2012. Genetics-based classification of filoviruses calls for expanded sampling of genomic sequences. *Viruses* **4:**1425–1437.
45. **de Villiers EM.** 2013. Cross-roads in the classification of papillomaviruses. *Virology* **445:**2–10.
46. **Trask SA, Derdeyn CA, Fideli U, Chen Y, Meleth S, Kasolo F, Musonda R, Hunter E, Gao F, Allen S, Hahn BH.** 2002. Molecular epidemiology of human immunodeficiency virus type 1 transmission in a heterosexual cohort of discordant couples in Zambia. *J Virol* **76:**397–405.
47. **Snel B, Huynen MA, Dutilh BE.** 2005. Genome trees and the nature of genome evolution. *Annu Rev Microbiol* **59:**191–209.
48. **Yu ZG, Chu KH, Li CP, Anh V, Zhou LQ, Wang RW.** 2010. Whole-proteome phylogeny of large dsDNA viruses and parvoviruses through a composition vector method related to dynamical language model. *BMC Evol Biol* **10:**192.
49. **Smith DB, Purdy MA, Simmonds P.** 2013. Genetic variability and the classification of hepatitis E virus. *J Virol* **87:**4161–4169.
50. **Wang YH, Pang BB, Zhou X, Ghosh S, Tang WF, Peng JS, Hu Q, Zhou DJ, Kobayashi N.** 2013. Complex evolutionary patterns of two rare human G3P[9] rotavirus strains possessing a feline/canine-like H6 genotype on an AU-1-like genotype constellation. *Infect Genet Evol* **16:**103–112.
51. **Wylie KM, Weinstock GM, Storch GA.** 2013. Virome genomics: a tool for defining the human virome. *Curr Opin Microbiol* **16:**479–484.
52. **Radford AD, Chapman D, Dixon L, Chantrey J, Darby AC, Hall N.** 2012. Application of next-generation sequencing technologies in virology. *J Gen Virol* **93:**1853–1868.
53. **Marz M, Beerenwinkel N, Drosten C, Fricke M, Frishman D, Hofacker IL, Hoffmann D, Middendorf M, Rattei T, Stadler PF, Topfer A.** 2014. Challenges in RNA virus bioinformatics. *Bioinformatics* **30:**1793–1799.
54. **Liu B, Gibbons T, Ghodsi M, Treangen T, Pop M.** 2011. Accurate and fast estimation of taxonomic profiles from metagenomic shotgun sequences. *BMC Genomics* **12**(Suppl 2)**:** S4.
55. **Ulitsky I, Burstein D, Tuller T, Chor B.** 2006. The average common substring approach to phylogenomic reconstruction. *J Comput Biol* **13:**336–350.
56. **Van der Auwera S, Bulla I, Ziller M, Pohlmann A, Harder T, Stanke M.** 2014. ClassyFlu: classification of influenza A viruses with discriminatively trained profile-HMMs. *PLoS One* **9:**e84558.
57. **Wang S, Luo X, Wei W, Zheng Y, Dou Y, Cai X.** 2013. Calculation of evolutionary correlation between individual genes and full-length genome: a method useful for choosing phylogenetic markers for molecular epidemiology. *PLoS One* **8:**e81106.
58. **Doyle VP, Andersen JJ, Nelson BJ, Metzker ML, Brown JM.** 2014. Untangling the influences of unmodeled evolutionary processes on phylogenetic signal in a forensically important HIV-1 transmission cluster. *Mol Phylogenet Evol* **75:**126–137.
59. **Tapparel C, Siegrist F, Petty TJ, Kaiser L.** 2013. Picornavirus and enterovirus diversity with associated human diseases. *Infect Genet Evol* **14:**282–293.
60. **De Koning MN, Quint KD, Bruggink SC, Gussekloo J, Bouwes Bavinck JN, Feltkamp MC, Quint WG, Eekhof JA.** 2014. High prevalence of cutaneous warts in elementary school children and ubiquitous presence of wart-associated HPV on clinically normal skin. *Br J Dermatol* **172:**196–201.
61. **Gottschling M, Goker M, Kohler A, Lehmann MD, Stockfleth E, Nindl I.** 2009. Cutaneotropic human beta-/gamma-papillomaviruses are rarely shared between family members. *J Invest Dermatol* **129:**2427–2434.
62. **Liu H, Li MH, Zhai YG, Meng WS, Sun XH, Cao YX, Fu SH, Wang HY, Xu LH, Tang Q, Liang GD.** 2010. Banna virus, China, 1987–2007. *Emerg Infect Dis* **16:**514–517.
63. **Steyer A, Gutierrez-Aguire I, Kolenc M, Koren S, Kutnjak D, Pokorn M, Poljsak-Prijatelj M, Racki N, Ravnikar M, Sagadin M, Fratnik SA, Toplak N.** 2013. High similarity of novel orthoreovirus detected in a child hospitalized with acute gastroenteritis to mammalian orthoreoviruses found in bats in Europe. *J Clin Microbiol* **51:**3818–3825.
64. **Zhang L, Xu MM, Zeng L, Liu S, Liu X, Wang X, Li D, Huang RZ, Zhao LB, Zhan QL, Zhu D, Zhang YY, Xu P, Xie P.** 2014. Evidence for Borna disease virus infection in neuropsychiatric patients in three western China provinces. *Eur J Clin Microbiol Infect Dis* **33:**621–627.
65. **Campanini G, Rovida F, Meloni F, Cascina A, Ciccocioppo R, Piralla A, Baldanti F.** 2013. Persistent human cosavirus infection in lung transplant recipient, Italy. *Emerg Infect Dis* **19:**1667–1669.
66. **Nielsen AC, Böttiger B, Banner J, Hoffmann T, Nielsen LP.** 2012. Serious invasive Saffold virus infections in children, 2009. *Emerg Infect Dis* **18:**7–12.
67. **Ganesh B, Nataraju SM, Rajendran K, Ramamurthy T, Kanungo S, Manna B, Nagashima S, Sur D, Kobayashi N, Krishnan T.** 2010. Detection of closely related Picobirnaviruses among diarrhoeic children in Kolkata: evidence of zoonoses? *Infect Genet Evol* **10:**511–516.
68. **Jartti T, Jartti L, Ruuskanen O, Söderlund-Venermo M.** 2012. New respiratory viral infections. *Curr Opin Pulm Med* **18:**271–278.
69. **Song JW, Baek LJ, Schmaljohn CS, Yanagihara R.** 2007. Thottapalayam virus, a prototype shrewborne hantavirus. *Emerg Infect Dis* **13:**980–985.
70. **Chan SY, Delius H, Halpern AL, Bernard HU.** 1995. Analysis of genomic sequences of 95 papillomavirus types: uniting typing, phylogeny, and taxonomy. *J Virol* **69:**3074–3083.
71. **Chouhy D, Bolatti EM, Piccirilli G, Sanchez A, Fernandez BR, Giri AA.** 2013. Identification of human papillomavirus type 156, the prototype of a new human gammapapillomavirus species, by a generic and highly sensitive PCR strategy for long DNA fragments. *J Gen Virol* **94:**524–533.
72. **Lauber C, Gorbalenya AE.** 2012. Toward genetics-based virus taxonomy: comparative analysis of a genetics-based classification and the taxonomy of picornaviruses. *J Virol* **86:**3905–3915.
73. **Naeem A, Hosomi T, Nishimura Y, Alam MM, Oka T, Zaidi SS, Shimizu H.** 2014. Genetic diversity of circulating Saffold viruses in Pakistan and Afghanistan. *J Gen Virol* **95**(Pt 9)**:** 1945–1957.
74. **Tokarz R, Firth C, Madhi SA, Howie SR, Wu W, Sall AA, Haq S, Briese T, Lipkin WI.** 2012. Worldwide emergence of multiple clades of enterovirus 68. *J Gen Virol* **93:**1952–1958.
75. **Yakovenko ML, Gmyl AP, Ivanova OE, Eremeeva TP, Ivanov AP, Prostova MA, Baykova OY, Isaeva OV, Lipskaya GY, Shakaryan AK, Kew OM, Deshpande JM, Agol VI.** 2014. The 2010 outbreak of poliomyelitis in Tajikistan: epidemiology and lessons learnt. *Euro Surveill* **19:**20706.
76. **Lauber C, Gorbalenya AE.** 2012. Partitioning the genetic diversity of a virus family: approach and evaluation through a case study of picornaviruses. *J Virol* **86:**3890–3904.
77. **Ebner K, Pinsker W, Lion T.** 2005. Comparative sequence analysis of the hexon gene in the entire spectrum of human adenovirus serotypes: phylogenetic, taxonomic, and clinical implications. *J Virol* **79:**12635–12642.
78. **Aoki K, Benko M, Davison AJ, Echavarria M, Erdman DD, Harrach B, Kajon AE, Schnurr D, Wadell G.** 2011. Toward an integrated human adenovirus designation system that utilizes molecular and serological data and serves both clinical and fundamental virology. *J Virol* 85:5703–5704.
79. **Seto D, Chodosh J, Brister JR, Jones MS.** 2011. Using the whole-genome sequence to characterize and name human adenoviruses. *J Virol* **85:**5701–5702.
80. **Seto D, Jones MS, Dyer DW, Chodosh J.** 2013. Characterizing, typing, and naming human adenovirus type 55 in the era of whole genome data. *J Clin Virol* **58:**741–742.
81. **Hage E, Huzly D, Ganzenmueller T, Beck R, Schulz TF, Heim A.** 2014. A human adenovirus species B subtype 21a associated with severe pneumonia. *J Infect* **69:**490–499.

82. **Hoffmann D, Hutzenthaler M, Seebach J, Panning M, Umgelter A, Menzel H, Protzer U, Metzler D.** 2012. Norovirus GII.4 and GII.7 capsid sequences undergo positive selection in chronically infected patients. *Infect Genet Evol* **12:**461–466.
83. **Huynen P, Mauroy A, Martin C, Savadogo LG, Boreux R, Thiry E, Melin P, De MP.** 2013. Molecular epidemiology of norovirus infections in symptomatic and asymptomatic children from Bobo Dioulasso, Burkina Faso. *J Clin Virol* **58:**515–521.
84. **Kroneman A, Vega E, Vennema H, Vinjé J, White PA, Hansman G, Green K, Martella V, Katayama K, Koopmans M.** 2013. Proposal for a unified norovirus nomenclature and genotyping. *Arch Virol* **158:**2059–2068.
85. **Verhoef L, Kouyos RD, Vennema H, Kroneman A, Siebenga J, van Pelt W, Koopmans M, Foodborne Viruses in Europe Network.** 2011. An integrated approach to identifying international foodborne norovirus outbreaks. *Emerg Infect Dis* **17:**412–418.
86. **Giammanco GM, De GS, Terio V, Lanave G, Catella C, Bonura F, Saporito L, Medici MC, Tummolo F, Calderaro A, Banyai K, Hansman G, Martella V.** 2014. Analysis of early strains of the norovirus pandemic variant GII.4 Sydney 2012 identifies mutations in adaptive sites of the capsid protein. *Virology* **450–451:**355–358.
87. **Belliot G, Sosnovtsev SV, Mitra T, Hammer C, Garfield M, Green KY.** 2003. In vitro proteolytic processing of the MD145 norovirus ORF1 nonstructural polyprotein yields stable precursors and products similar to those detected in calicivirus-infected cells. *J Virol* **77:**10957–10974.
88. **Bull RA, Tanaka MM, White PA.** 2007. Norovirus recombination. *J Gen Virol* **88:**3347–3359.
89. **Martella V, Medici MC, De GS, Tummolo F, Calderaro A, Bonura F, Saporito L, Terio V, Catella C, Lanave G, Buonavoglia C, Giammanco GM.** 2013. Evidence for recombination between pandemic GII.4 norovirus strains New Orleans 2009 and Sydney 2012. *J Clin Microbiol* **51:**3855–3857.
90. **Eden JS, Tanaka MM, Boni MF, Rawlinson WD, White PA.** 2013. Recombination within the pandemic norovirus GII.4 lineage. *J Virol* **87:**6270–6282.
91. **Hoffmann D, Mauroy A, Seebach J, Simon V, Wantia N, Protzer U.** 2013. New norovirus classified as a recombinant GII.g/GII.1 causes an extended foodborne outbreak at a university hospital in Munich. *J Clin Virol* **58:**24–30.
92. **Lee GC, Jung GS, Lee CH.** 2012. Complete genomic sequence analysis of norovirus isolated from South Korea. *Virus Genes* **45:**225–236.
93. **Wong TH, Dearlove BL, Hedge J, Giess AP, Piazza P, Trebes A, Paul J, Smit E, Smith EG, Sutton JK, Wilcox MH, Dingle KE, Peto TE, Crook DW, Wilson DJ, Wyllie DH.** 2013. Whole genome sequencing and de novo assembly identifies Sydney-like variant noroviruses and recombinants during the winter 2012/2013 outbreak in England. *Virol J* **10:**335.
94. **Kundu S, Lockwood J, Depledge DP, Chaudhry Y, Aston A, Rao K, Hartley JC, Goodfellow I, Breuer J.** 2013. Next-generation whole genome sequencing identifies the direction of norovirus transmission in linked patients. *Clin Infect Dis* **57:**407–414.

Quality Assurance and Quality Control in Clinical and Molecular Virology

MATTHEW J. BANKOWSKI

2

The clinical virology laboratory provides important and often critical information to the health care provider in order to support the diagnosis or monitoring of viral disease for the patient. Testing results will often serve as a guide for optimal treatment of the disease, contribute to infection control and prevention of a hospitalized patient or offer insight into the prognosis for the disease. Therefore, the quality of the virology laboratory testing has to be highly accurate and offered in a timely fashion in order to achieve optimal patient management. A well-structured and ongoing quality assurance (QA) program will provide the framework for maintaining accuracy in all phases of the testing process. These phases include the preanalytical, analytical, and postanalytical stages of the testing. However, no process is perfect, and every QA program should include a surveillance component that continuously identifies and corrects any weakness in the system. This corrective action should also be followed by preventative action in order to eliminate weaknesses and improve the entire QA program.

REGULATORY REQUIREMENTS

In the United States all clinical laboratories have to be certified under the Clinical Laboratory Improvement Amendments (CLIA) (1, 2). This amended USA federal law governing clinical laboratory testing is listed in Section 353 of the Public Health Service Act (42 U.S.C. 263a) as published in the Federal Register on 28 February 1992 as a final rule. The CLIA regulations established three levels of complexity corresponding to minimal quality standards for the type of laboratory. These categories consist of waived, moderate-complexity, and high-complexity. CLIA was established to ensure the quality of laboratory services based on these complexity levels. The CLIA regulations incorporate provisions for clinical laboratory personnel, facilities, quality assurance, quality control, proficiency testing, record keeping, and record retention.

Subsequently, the Department of Health and Human Services (HHS) published a revised final rule in the Federal Register on 24 January 2003. This revised final rule contained clarifications and reorganization to make the document more concise. Importantly, this revised final rule incorporated the quality system concept into clinical laboratory testing. All clinical laboratories must be certified under CLIA. However, depending upon the state in which the clinical laboratory is located, other agencies, such as the Centers for Medicare and Medicaid Services (CMS), may approve the laboratory licensure. CLIA-certified laboratories are also subject to biennial inspections, which are intended to be educational and aid in improving testing and optimizing patient care. Clinical laboratories may also meet the CLIA requirements through being inspected by CMS-approved nonprofit organizations (e.g., College of American Pathologists [CAP] or The Joint Commission).

VIROLOGY QUALITY ASSURANCE

Quality assurance in the clinical laboratory is a multifaceted process. QA includes quality control, proficiency testing, technical staff training and competency, instrument calibration, and clinical correlation. It is an ongoing process that maintains optimum test performance that is controlled at every stage of the testing process. This includes testing personnel from preanalytical to analytical and postanalytical test procedures. Quality control reagents are included in the day-to-day testing process, and frequent challenging of the process is also carried out by proficiency testing. A troubleshooting process is instituted when tests fail, which is followed up by investigation, corrective action, and preventive action. Useful documents that serve as guidelines for maintaining quality assurance in the clinical virology laboratory can be obtained from the Clinical and Laboratory Standards Institute (CLSI) (Table 1) and the American Society for Microbiology (ASM) (Table 2). General documents from these two reference sources include CLSI QMS02A6 and Cumitech 3B.

CLINICAL LABORATORY PERSONNEL

All laboratory staff involved in any part of the testing process need to be qualified according to CLIA and applicable state licensure requirements. The laboratory director is responsible for defining the qualifications and responsibilities in written form for all of the staff involved in this process.

Virology testing is considered as moderate or high complexity according to CLIA-88. Therefore, any staff involved

doi:10.1128/9781555819156.ch2

TABLE 1 Guideline documents from the Clinical and Laboratory Standards Institute (CLSI)[a]

Document no.	Date	Document title and description
General laboratory		
GP17A3	06/29/12	Clinical Laboratory Safety; Approved Guideline—Third Edition
GP27A2	02/22/07	Using Proficiency Testing to Improve the Clinical Laboratory; Approved Guideline—Second Edition
GP29A2	08/29/08	Assessment of Laboratory Tests When Proficiency Testing Is Not Available; Approved Guideline—Second Edition
GP31A	08/22/12	Laboratory Instrument Implementation, Verification, and Maintenance; Approved Guideline
QMS02A6	02/28/13	Quality Management System: Development and Management of Laboratory Documents; Approved Guideline. Sixth Edition.
QMS03A3	05/02/09	Training and Competence Assessment; Approved Guideline Third Edition.
QMS04A2	02/22/07	Laboratory Design; Approved Guideline—Second Edition
QMS05A2	09/28/12	Quality Management System: Qualifying, Selecting, and Evaluating a Referral Laboratory; Approved Guideline—Second Edition
QMS12A	12/29/10	Development and Use of Quality Indicators for Process Improvement and Monitoring of Laboratory Quality; Approved Guideline
Method evaluation		
EP12A2	01/25/08	User Protocol for Evaluation of Qualitative Test Performance; Approved Guideline—Second Edition.
EP15A3	09/11/14	User Verification of Precision and Estimation of Bias; Approved Guideline—Third Edition
EP23A	10/25/11	Laboratory Quality Control Based on Risk Management; Approved Guideline
EP25A	09/23/09	Evaluation of Stability of *In Vitro* Diagnostic Reagents; Approved Guideline
EP26A	09/30/13	User Evaluation of Between-Reagent Lot Variation; Approved Guideline
Microbiology		
M41A	11/30/06	Viral Culture; Approved Guideline
M53A	06/30/11	Criteria for Laboratory Testing and Diagnosis of Human Immunodeficiency Virus Infection; Approved Guideline
Molecular methods		
MM03A2	02/17/06	Molecular Diagnostic Methods for Infectious Diseases; Approved Guideline—Second Edition
MM06A2	11/30/10	Quantitative Molecular Methods for Infectious Diseases; Approved Guideline—Second Edition
MM09A2	02/28/14	Nucleic Acid Sequencing Methods in Diagnostic Laboratory Medicine; Approved Guideline—Second Edition
MM13A	01/06/06	Collection, Transport, Preparation, and Storage of Specimens for Molecular Methods; Approved Guideline
MM14A2	05/23/13	Design of Molecular Proficiency Testing/External Quality Assessment; Approved Guideline—Second Edition
MM17A	03/21/08	Verification and Validation of Multiplex Nucleic Acid Assays; Approved Guideline

[a]Clinical and Laboratory Standards Institute (CLSI), Wayne, PA, http://clsi.org

in the actual testing of specimens are required to be qualified under these categories. Staff involved in the analytical phase need to be adequately trained on a test in order to ensure that there is a complete understanding of the test procedure. In order to ensure fulfillment of this step, an evaluation by actual observation of the technologist performing the test on a recurrent basis (i.e., operator competency assessment) is instituted. This approach to testing is to be unaltered, and strict adherence to the procedure manual, biosafety training and awareness, patient confidentiality, result interpretation, reporting, and quality control are to be maintained at all times. Competency assessment is instituted to identify employee performance issues. Documentation of problems, especially a pattern of performance issues, is to be addressed using remediation. Testing personnel also need to be knowledgeable enough to recognize unusual results and to be proficient in troubleshooting of failed runs. In addition, laboratory personnel are to have documented evidence of continuing education and active licensure. Refer to Table 1 (CLSI) and Table 2 (Cumitech) for further information and guidance from documents QMS03A3, Cumitech 39 and 41.

PROCEDURE MANUAL

The procedure manual is one of the most important documents in the laboratory. It is customized to the individual laboratory but is standardized to contain procedures with sections that are required as described in the CLIA document QMS02A6 (Table 1). It is required that the procedure manual contain directions and guidance for all three stages of the testing process: preanalytical, analytical, and postanalytical. The procedure is not just a rewritten form of the package insert but a highly organized, concise, step-by-step document customized to the individual laboratory.

TABLE 2 Cumitech—Cumulative Techniques and Procedures in Clinical Microbiology documents from the American Society for Microbiology (ASM)[a]

Document no.	Date	Document title and description
3B	2005	Quality Systems in the Clinical Microbiology Laboratory
29	1996	Laboratory Safety in Clinical Microbiology
31A	2009	Verification and Validation of Procedures in the Clinical Microbiology Laboratory
39	2003	Competency Assessment in the Clinical Microbiology Laboratory
41	2004	Detection and Prevention of Clinical Microbiology Laboratory-Associated Errors
44	2006	Nucleic Acid Amplification Tests for Detection of *Chlamydia trachomatis* and *Neisseria gonorrhoeae*

[a]American Society for Microbiology (ASM), Washington, DC, http://www.asm.org

An outline of the clinical virology procedure manual sections with examples is listed in Table 3. Note that it covers all three parts of the testing process. It is specific and very precise in describing each of these sections. For example, the analytical section discusses specific specimen types, and media used for the collection, and transport conditions. All tests have to be validated/verified for test performance and signed off by the clinical laboratory director in order to assure the claims issued by the test kit or reagent manufacturer. Any change or deviation in these materials or conditions will require revalidation/reverification by the laboratory. Refer to reference 3 (see Table 5 in Chapter 51) for a suggested guide to the verification/validation process.

The discussion of the analytical process is also very specific in the test procedure. It includes specimen preparation, quality control, step-by-step test procedure, calculation,

TABLE 3 Outline of the clinical virology procedure manual sections including examples and comments

Clinical procedure manual section	Section example(s) and/or comment(s)
General Information	Date of acceptance for the test procedure signed by the laboratory director or designee
Effective dates and signatures	
Purpose	Brief description of the purpose of the test and fundamentals of the test principle.
Principle	
Department/Section	Microbiology, Virology Section
Test Frequency (e.g. performed daily)	Performed daily
Preferred Container Type	Viral transport media (VTM)
Acceptable Sources	Cervical swab
Other Acceptable Specimen(s)	Pleural fluid and cerebrospinal fluid
Ambient Stability	One hour at room temperature in VTM
Refrigerated Stability	Two days at 2–8°C in VTM
Frozen Stability	One month at −70°C to −80°C
Local Transport Temperature	Refrigerated (2–8°C) or frozen (<−70°C)
Long Distance Transport Temperature	Refrigerated (2–8°C) or frozen (<−70°C)
Specimen Rejection	Specimen not received in VTM
Equipment, Supplies, and Reagents (ESR's)	Briefly and completely list all ESR's
Quality Control	List the specific QC used in the testing.
	Positive and negative QC on each run (or less frequently under Equivalent Quality Control or Individualized Quality Control Plan)
Frequency of Use	Repeat the test run if QC is unacceptable.
Corrective Actions	Quality control documented daily, reviewed weekly, and stored for two years
QC Data Storage	
Procedure	Completely and clearly list all steps used in the test procedure
Specimen Preparation	
Procedure	Cell culture inoculation and incubation
Troubleshooting	Repeat run failed, retest with new lot
Test Limitations	Performance is unknown for <14 yrs age
Calculation and Interpretation	Software calculation is automated
Results and Interpretation	HSV type 1 virus detected
References	Isenberg, et al. 2004. *Clinical Microbiology Procedures Handbook*. ASM Press, Washington, DC
Attachment(s)	Include all pertinent forms, algorithms, tables, and figures as attachments

TABLE 4 Molecular virology infectious disease testing quality control sources[a]

Infectious disease agent	AM	BR	EG	MM	SC	QU	WHO	ZM
Bacteria								
Chlamydia trachomatis		X			X			X
Viruses								
BK Virus (BKV)	X		X					
Cytomegalovirus (CMV), HSV (H), VZV (V)			CMV			H,V	CMV	X
Epstein Barr Virus (EBV)	X		X					X
Hepatitis B Virus (HBV)	X				X		X	X
Hepatitis C Virus (HCV)					X		X	X
Human Immunodeficiency Virus type 1 (HIV-1)	X				X		X	X
Human Papillomavirus (HPV)		X			X		X	
Influenza Virus types 1 and 2						X		X
Parvovirus B-19							X	X
West Nile Virus (WNV)					X			
Other (Adenovirus [Ad], enterovirus [E], HSV)			Ad, E			Ad		
Multiple ID agents and resistance genes								
Gastrointestinal (GI panel)				X				X
Ebola Virus (Control panel)				X				
Respiratory pathogens (RP panel)				X				X
Negative molecular control								
Blood and/or specific matrix control	X	X			X			

[a](AM) AcroMetrix, Inc. (Life Technologies), Benicia, CA, http://www.lifetechnologies.com
(BR) Bio-Rad Laboratories, Irvine, CA, http://www.bio-rad.com
(EG) ELITech Group, Princeton, NJ, http://www.elitechgroup.com
(MM) Maine Molecular Quality Controls, Inc., Scarborough, ME, http://mmqci.com
(SC) SeraCare Life Sciences, Inc., West Bridgewater, MA, http://www.seracare.com
(QU) Quidel Corporation, San Diego, CA, http://www.quidel.com
(WHO) World Health Organization, Geneva, Switzerland, http://www.who.int/bloodproducts/ref_materials/en/
(ZM) ZeptoMetrix Corporation, Buffalo, NY, http://www.zeptometrix.com

results, and troubleshooting a failed test run. It is always advisable to include an algorithm, table, and/or figures of the testing procedure for clarity. Once the results are obtained, the postanalytical process of the results' confirmation, interpretation, and reporting to the ordering healthcare provider through the laboratory's information technologies is instituted by qualified laboratory personnel.

TABLE 5 Molecular virology infectious disease testing controls and calibrators

Molecular test category	Controls and calibrators
Qualitative	Internal control (IC)
	Positive (low, LP)
	Positive (high, HP)
	Negative control (NC)
	No nucleic acid control (NNA)
Quantitative	Internal control (IC)
	Positive (low, LP)
	Positive (high, HP)
	Negative control (NC)
	Calibrators (CAL)
Multiplex	Internal control (IC)
	Multiple positive controls (PC)
	Negative control (NC)

SPECIMEN COLLECTION AND TRANSPORT

Specimens, such as blood or CSF, need to arrive in the laboratory as soon as possible. However, others that are collected on a swab and placed in transport media can sustain a delay in transport and processing. Virus viability can usually be maintained from collection to testing in viral transport media (VTM). VTM in the most basic formulation contains a balanced salt, a pH indicator, and a virus stabilizing protein. VTM also contain antimicrobials, such as gentamicin, streptomycin, or nystatin. Strict adherence to the intended use, storage conditions, and shelf life of the VTM should ensure virus viability until the specimen is processed. Lastly, the use of rayon or synthetic polyester fiber-tipped swabs is recommended. Calcium alginate or cotton swabs should never be used because they will usually result in loss of virus viability and recovery. Refer to CLSI M41A for more detailed information on recommended specimens for specific virus isolation.

Transport and storage conditions are equally important in the preanalytical process. Ideally, transport time should not exceed two hours. However, this is often not possible, and temperature then becomes critically important in the transport and storage process. The preferred transport temperature is usually in the range of 2 to 8°C for up to 72 hours. If extended transport conditions are encountered, freezing is sometimes recommended at -70°C or below, depending upon the specimen type. Strict adherence to the freezing temperature is required because lower temperatures are detrimental to virus viability and result in an unacceptable specimen. Likewise, freeze-thaw cycles often are detrimental to the viability of the virus (i.e., especially in the case of CMV recovery).

Inadequate adherence to collection, storage, and transport of clinical specimens for virus culture should always be addressed with the ordering healthcare provider (HCP). Ideally, the specimen should be re-collected or if the specimen cannot be re-collected, the HCP should be made fully aware of the possible adverse effect on virus viability and the effect on viral isolation in culture.

REAGENTS, MATERIALS, AND CELL CULTURES

Reagents should be examined for breakage, acceptable dating on shelf life, acceptable shipping temperature, and time to arrival in the laboratory and the overall expected physical integrity of the material. Each lot or shipment has to be challenged for expected performance using control materials and known specimens prior to use in patient testing. Lastly, all reagents should be stored according to the label or package insert requirements.

Acceptance of materials (e.g., disposable plasticware tubes or plates) should also be examined for breakage and overall physical integrity. Sterile supplies should be thoroughly inspected for any package perforations upon receipt and also prior to use. The manufacturer or distributor should be immediately notified of any damage noticed upon receipt or prior to use in clinical testing.

Virus isolation requires the use of susceptible target cells that are viable and optimized for virus isolation. Cell culture microscopic assessment of cell monolayers (e.g., acceptable cell density), media contamination, and pH assessment should be performed upon receipt of the cells from the distributor or manufacturer and prior to use. The manufacturer should supply a record indicating the product was checked for mycoplasma and endogenous viral contamination. If an extraneous agent is present, a possible false negative or false positive result may occur.

It is often the case that, before use of cell culture media, various additives need to be incorporated into the media. These cell culture additives are to be free of contamination, and any animal sera used in the media should be checked for toxicity. This is usually supported by an attestation certificate supplied by the product manufacturer.

Optimal viral isolation depends upon the appropriate use of cell lines for viral culture. There are many cell types available for the isolation of most viruses in culture. The choice is up to the laboratory, but it should involve the use of an adequate number and variety of cell types to at least cover the common types of viruses encountered in the clinical setting. An extensive list of both cell types and viruses is outlined in more detail in CLSI M41A.

It is also necessary that tube cultures are incubated for a sufficient time period in order to recover viruses suspected in a particular specimen type. For example, herpes simplex virus should be held for at least five days (CAP) and respiratory viruses for at least 10 to 14 days in order to allow sufficient time for viral recovery.

Absence of contamination, optimal cell density and cell line virus sustainability are all quality indicators for the successful isolation of virus in culture. Viral culture conditions should be monitored on a daily basis for incubation time and temperature within the acceptable range stated in the procedure manual. Appropriate monitoring of cells for cytopathic effect (CPE) and hemadsorption by a trained and competent technologist should be conducted frequently according to the times stated in the procedure manual and recorded. Any deviation from this schedule may result in a reporting delay for viral detection and less than optimal patient care. Appropriate recording of cell types, passage number, source, culture media and control viruses used to challenge the cell types should all be contained within a written and retained record.

MOLECULAR TESTING

Quality management of molecular virology testing begins with the preanalytical phase. The use of molecular virology testing is often more costly than culture but has many advantages. These include an increase in test performance, a decrease in turnaround time, the detection of uncultivable viruses, and a much higher level of patient care. Molecular testing is often more complex and requires the ordering of an appropriate test from a HCP. Consultation between the laboratory director and the HCP is highly recommended if the intended use is not clear.

The collection, handling, and transport of an appropriate specimen should always strictly follow the procedure manual instructions. Any deviation from the stated protocol has the potential to cause patient harm or even result in unnecessary and costly expense.

Molecular amplified nucleic acid testing may be ordered as qualitative, quantitative, multiplex, or even nucleic acid sequencing, if available: The type of test required will depend upon the particular virus(es) sought in the diagnosis.

Nucleic acid amplification testing (NAAT) is categorized as target- or signal-based. The polymerase chain reaction (PCR), transcription mediated amplification (TMA), and strand displacement amplification (SDA) are all target-based compared to the signal-based methods of Hybrid Capture or Invader (Clevase) technology. Depending upon the technology and the test platform, careful consideration in containing and monitoring the amplified product should be clearly written into the procedure manual. This may include a physical separation of the specimen processing with the extraction and purification of nucleic acid (e.g., DNA or RNA) from the amplification and detection stage. This is accomplished by the use of a positive and negative pressure room, respectively. Likewise, each of these areas should have dedicated lab coats, reagents, instruments, and disposables. Ideally, a unidirectional workflow should exist from positive to negative pressure rooms in order to avoid contamination carryover. The reader is referred to the chapter on "Laboratory Design" in this book, which provides more specific details.

It is beyond the scope of this chapter to include any extensive detail on quality assurance using molecular testing. Recommended guidelines and references can be found in CLSI documents MM03A2, MM06A2, MM09A2, and MM17, Cumitech 44, *Molecular Microbiology: Diagnostic Principles and Practice* (3), and the *Manual of Clinical Microbiology* (4).

VIRAL SEROLOGY AND DIRECT ANTIGEN DETECTION

In the preanalytical phase, collection of the blood specimen in the correct type of tube should always be confirmed (e.g., EDTA, ACD or no preservative) by both the phlebotomist and the accessioning personnel. It is also important to organize the collection tubes and draw specimens from only a single patient at a time in order to avoid mislabeling specimens. If the blood specimen is collected in an incorrect tube, documentation of the collection error and describing the specific details on the employee's record should always be a part of the QA process. In some cases, this may even be followed up by retraining of the employee, if warranted.

In the analytical phase, viral serology involves both antigen and antibody test methods. Most of these tests involve screening for exposure or infection from a particular virus. Screening tests are usually optimized for test sensitivity and confirmatory tests for test specificity. Therefore, screening tests will show more false positives than confirmatory tests. The percentage of false positives will also vary among the different test kits. An appropriate quality assurance measure for the laboratory is to monitor the percentage of false positives for a particular antigen or antibody test over time. One course of action might be to consider the replacement of a particular test kit that is exhibiting a high level of false positives. In some cases, consultation by the laboratory director with the ordering HCP may even be warranted.

Direct antigen detection by either direct immunofluorescence antibody (IFA) or enzyme immunoassay (EIA) requires adequate cell numbers from the patient for testing. Grading the cells present for IFA testing as a quality assurance measure is suggested. EIA is more difficult to access, but the incorporation of a human cell antigen control by the test kit manufacturer may serve the purpose.

INSTRUMENTATION

Assurance that all instrumentation in the clinical virology laboratory is performing according to the manufacturers' specifications requires operational validation checks. Preventive maintenance at specified time intervals will also serve to extend the life of the instruments and prevent breakdowns. Instrument analytical measurement range (AMR) determination is a CAP requirement and is also required on a biannual basis or more frequently depending upon the manufacturers' recommendations or major instrument changes (e.g., major software version changes). Other recommendations may be found in the CLIA document GP31A.

QUALITY CONTROL AND STANDARDS

Each lot or shipment of media for viral culture from a manufacturer should have a specification sheet. This sheet will list the particular components of the media in addition to any quality control performed by the manufacturer on the particular lot in the shipment. A written record should exist to verify that the media was received without breakage and the appearance of the media is as expected. Any rejected media and the reason for the rejection should be noted in the written record. The laboratory is also responsible for the development of a procedure for verification of the media performance for the intended use in the laboratory. In most cases, other components are also added to the media before use and the verification of these materials should also be documented. A written procedure to ensure that sterility has been maintained following the introduction of all required components to the media and a record showing that the procedure has been followed need to be available in the laboratory. In addition to sterility, animal sera used for cell growth media have to be checked for the absence of cell toxicity. This information is to be contained in a written procedure, which also shows the records of animal sera checks.

Cell lines (i.e., primary, diploid, continuous, or genetically altered) have to be checked for mycoplasma contamination. If cell lines are infected by *Mycoplasma spp*., they are subsequently rendered noninfectable to a virus. For this reason, virus and mycoplasma cultures should never exist in the same physical area of the laboratory. Proof of mycoplasma-free media can be accomplished by documentation received from a commercial vendor or alternatively by the monitoring of a negative, uninoculated control cell line.

Continuous cell lines should also be checked for endogenous viral contamination (e.g., foamy virus). If contamination is noted, it has to be recorded in the laboratory written record and the appropriate follow-up steps listed (e.g., rejection of the cell line lot). Another check by the commercial vendor is written documentation of testing for monkey virus. Commercial vendors will always supply evidence that a particular cell line has also been checked for monkey B virus. This is a lethal virus for humans (70% mortality if not treated with antivirals), and any handling of monkey-derived cell lines is to be handled under strict biosafety conditions.

Every type of cell line lot used for viral culture in patient testing has to be checked for infectivity by a representative virus or viruses expected to grow in the particular cell line. Demonstration of growth includes, but is not limited to, showing the correct cytopathic effect (CPE) or hemadsorption under the correct incubation conditions in an appropriate time period.

Each new lot and shipment of reagents and test kits have to be verified for the expected performance. If the reagents or test kits are intended to detect multiple viruses, each individual virus detection needs to be assessed prior to patient testing. However, pooled controls can be used for the daily quality control check following the initial individual virus detection by the reagent or test kit.

In the case of molecular virology testing, quality control material may be difficult to obtain. A listing of quality control sources is provided in Table 4, which includes multiplex testing materials. Recommendations for qualitative, quantitative, and multiplex testing using various controls and calibrators are outlined in Table 5. Daily QC testing for both clinical and molecular virology testing is to be reviewed by qualified personnel for acceptability before reporting patient results.

PROFICIENCY TESTING

The establishment of an external, graded proficiency testing (PT) program was implemented with CLIA-88. All laboratories must participate in PT programs for each analyte or test used for patient testing. If proficiency testing is not available for a particular analyte or test, internal or external PT testing can be established. Reference materials should be carefully considered, and interlaboratory variation should be carefully assessed before implementing such a program (5, 6). CLIA documents GP27A2 and GP29A2 offer guidance on using PT in the clinical laboratory and also include recommendations when a PT program is not available for a particular analyte or test (Table 2).

A written procedure describing the PT process must be available and understood by all testing personnel. Every PT challenge is to be treated exactly as you would treat testing for a patient specimen. In addition, an internal or external PT program should include a minimum of five samples tested three times per year for regulated analytes. However, if the analyte is unregulated (i.e., not listed in CLIA), one or more

TABLE 6 Corrective action and preventive action (CAPA) checklist for test failures

Investigation and Corrective Action (CA)[a]	Preventive Action (PA)[a]
Investigation	**Course of action plan**
List the type of test	Metrics supporting success or failure
List the reason for the test failure	If unsuccessful—propose next plan
Methodology, technical, clerical, quality control	Integrate successful PA plan into workflow
Summary of findings	Document the PA change in the procedure
Conclusion from the findings	
Corrective Action (CA)	
Step(s) or action(s) to correct the error(s)	
Summary of CA success or lack of success	
If unsuccessful—discuss the next step(s)	

[a]Document all personnel involved and the date of action(s) by the involved staff.

TABLE 7 Section outline of the clinical virology test verification/validation protocol

Section	Description
Purpose	Briefly describe the purpose of the test
Principle	Briefly describe the test principle (e.g., real-time PCR)
Responsibility	Who can perform and report the test as examples
Materials, Reagents, and Equipment	Briefly and completely list all supplies used in testing.
Controls and Specimens	What specimens/numbers are included with controls?
Precautions	Safety and MSDS[a] considerations
Procedure	List the precise steps used in the procedure.
Acceptance and Rejection Criteria	What results are acceptable compared to the standard?
Results	List all results in the test report, including failed runs.
Definitions of Terms	Define all terms (e.g. analytical sensitivity)
Discrepant Analysis	How will discordant results be resolved?
Data Retention	How long and in what location will the data be stored?
Technical References	Include all pertinent references with a complete citation.
Addendum	For example, include 2 x 2 test performance tables.

[a]Materials Safety Data Sheet (MSDS).

samples (i.e., no minimum) are required per event twice a year. In either case, both accuracy and reliability of the system still has to be verified at least twice a year. A laboratory or referee response showing at least 80% correct is considered acceptable. Any test failures should be listed on a written form, investigated, and the corrective action described. If appropriate, preventive action should also be instituted and written into the procedure.

CORRECTIVE AND PREVENTIVE ACTION (CAPA)

Quality management in the laboratory includes quality indicators for process improvement and overall monitoring of the laboratory's quality. Quality control of equipment and testing are considered examples of quality indicators. They are a part of quality improvement described in the ISO-9000 requirements. Corrective action is defined as the process that identifies and eliminates the cause(s) of a problem in an effort to prevent a recurrence through preventive action. Corrective action is a way of supporting a quality management system and preventive action is a way of ensuring that a particular error or breach in quality may be prevented in the future. A sample checklist for corrective action and preventive action is outlined in Table 6. The interested reader is referred to the CLIA QMS12A document, which describes this in more detail.

TEST VERIFICATION AND VALIDATION

A suggested outline for the clinical virology verification/validation protocol is shown in Table 7. A protocol document should be brief and concise. It should contain all of the sections listed, and it is highly advisable to develop a template for the protocol. This will standardize the process and save time for the technologists involved in the protocol. The final protocol should be signed off by the author, appropriate management, and the director of the department. Testing should not begin until all have agreed to the content of the protocol by signatures. The test performance characteristics to be determined will depend upon the test category (e.g., FDA-cleared or -approved, FDA modified and laboratory-developed test or LDT). For instance, an LDT should include analytical and clinical sensitivity and specificity, accuracy, precision, interfering substances, and reportable and normal reference ranges.

Upon completion of the verification or validation, the sections should be filled in under the appropriate subsections according to all the data accumulated from the testing. Any deviations from the original signed protocol should be listed in the Appendix section and discussed under the results section of the report. Finally, and most importantly, patient testing should *not* begin until the final report is again signed off by the same individuals or designees involved or familiar with the details in the protocol. The final approval signature is always from the laboratory director or designee. Upon final approval, a statement in the report is included that indicates the method can be used to test patient specimens: "This verification/validation has been reviewed and the test performance is considered acceptable for patient testing."

A guide to verification/validation of newly introduced or modified molecular tests under various FDA categories is shown in reference 3 (see Table 5 in Chapter 51). It is beyond the scope of this chapter to discuss the details of how to approach verification and validation in the clinical and molecular virology laboratory. However, it should be noted that the laboratory developed test (LDT) is the most complex and time-consuming in the verification or validation process. Unlike the FDA-cleared or approved tests, the LDT test performance characteristics must be established rather than simply validated. For more specific details and information on the verification and validation process, the reader is referred to CLSI documents EP15A3 and EP26A (Table 2), Cumitech 31A (Table 3), and references 3,7, and 8.

REFERENCES

1. **Bachner P, Hamlin W.** 1993. Federal regulation of clinical laboratories and the Clinical Laboratory Improvement Amendments of 1988—Part II. *Clin Lab Med* **13:**987–994.
2. **Bachner P, Hamlin W.** 1993. Federal regulation of clinical laboratories and the Clinical Laboratory Improvement Amendments of 1988—Part I. *Clin Lab Med* **13:**739–752, discussion 737–738.
3. **Bankowski MJ.** Molecular microbiology test quality assurance and monitoring. *In* Persing DH, Tenover FC, van Belkum A, Hayden RT, Ieven M. Miller MB, Nolte FS, Tang Y-W (ed), *Molecular Microbiology: Diagnostic Principles and Practice*, 3rd ed. ASM Press, Washington, DC, in press.
4. **Versalovic J, Carroll KC, Funke G, Jorgensen JH, Landry ML, Warnock DW (ed). American Society for Microbiology.** 2011. *Manual of Clinical Microbiology*, 10th ed. ASM Press, Washington, DC.
5. **Fryer JF, Baylis SA, Gottlieb AL, Ferguson M, Vincini GA, Bevan VM, Carman WF, Minor PD.** 2008. Development of working reference materials for clinical virology. *J Clin Virol* **43:**367–371.
6. **Fryer JF, Minor PD.** 2009. Standardisation of nucleic acid amplification assays used in clinical diagnostics: a report of the first meeting of the SoGAT Clinical Diagnostics Working Group. *J Clin Virol* **44:**103–105.
7. **Halling KC, Schrijver I, Persons DL.** 2012. Test verification and validation for molecular diagnostic assays. *Arch Pathol Lab Med* **136:**11–13.
8. **Puppe W, Weigl J, Gröndahl B, Knuf M, Rockahr S, von Bismarck P, Aron G, Niesters HG, Osterhaus AD, Schmitt HJ.** 2013. Validation of a multiplex reverse transcriptase PCR ELISA for the detection of 19 respiratory tract pathogens. *Infection* **41:**77–91.

Regulatory Compliance

LINOJ SAMUEL

3

Clinical laboratories have come a long way since the 1900s when concerns were raised that they were too expensive and testing was too time consuming to be of practical use (1). In spite of those objections, laboratories have become the cornerstone of medical decision making. The beginning of laboratory regulation can be traced to the 1940s when Sunderman and Belk published findings from a voluntary survey of proficiency testing of regional laboratories that showed significant variation in laboratory performance (2). At around the same time, the College of American Pathology (CAP) was established, and one of its first functions was the initiation of national proficiency surveys in 1947 and 1948. The results further confirmed the need for standardization and regulation of clinical laboratories (3). In subsequent years, participation in proficiency surveys became standard practice among large hospital and reference laboratories (4). In 1967, Congress passed the Clinical Laboratory Improvement Amendment of 1967 (CLIA 67), which mandated certain minimum performance standards for reference laboratories involved in interstate commerce (4, 5). Similar regulation for hospital laboratories that were funded by Medicare followed in 1968 (6). The CLIA 67 regulations mandated that laboratories participate in "state approved or state operated proficiency testing programs" (5).

In 1987, a series of articles in the media brought attention to erroneous results generated by cytopathology labs reading Pap smears (7). These media reports eventually resulted in the passage of the CLIA Act of 1988 (CLIA 88). The final regulations of CLIA 88 were published on February 28, 1992, with minor modifications on January 24, 2003 (8). The Centers for Medicare and Medicaid Services (CMS) was charged by Congress with implementing this legislation. CLIA 88 "sets forth the conditions that all laboratories must meet to be certified to perform testing on human specimens" (8). This legislation brought all laboratories that perform clinical testing under the umbrella of CLIA regulatory requirements (9).

The regulatory landscape in the United States involves several agencies. The Clinical and Laboratory Standards Institute (CLSI), which was formerly known as the National Committee for Clinical Laboratory Standards, is a nonprofit voluntary organization that develops consensus standards and guidelines for clinical laboratories. CLSI draws on volunteers from industry, government, and health care services to accomplish this goal. CLSI standards are widely accepted both in the United States and around the world and span a wide range of topics from validation and verification of tests to veterinary medicine and laboratory informatics. In addition, CLSI plays a significant role in the development of International Organization for Standardization (ISO) standards for clinical laboratories (http://clsi.org/standards/iso-standards/).

The Centers for Disease Control and Prevention (CDC) is a federal agency under the authority of the Department of Health and Human Services (HHS) that is responsible for tracking, monitoring, and responding to disease trends and outbreaks in the United States. The CDC also participates in the development of laboratory standards and guidelines and advises the HHS on technical and scientific issues via the Clinical Laboratory Improvement Advisory Committee (http://wwwn.cdc.gov/cliac/).

The Food and Drug Administration (FDA) is another federal agency of the HHS. It oversees the regulation of human and veterinary drugs, vaccines, and medical devices, and it is responsible for food and cosmetic safety and regulation of tobacco products.

The Centers for Medicare and Medicaid Services (CMS) started out as the Health Care Financing Administration, and it is another federal agency within the HHS. CMS is responsible for administration of the Medicare program, and it works with the states for administration of the Medicaid program. Among its other roles, CMS is responsible for the regulation of clinical laboratories that perform testing on humans in the United States (http://www.cms.gov).

The level of regulation as per CLIA 88 depends on the complexity of testing performed by the laboratory. The higher the complexity of testing, the more stringent the regulatory requirements. Three categories of test complexity have been established:

1. Waived
2. Moderate complexity, including provider-performed microscopy (PPM)
3. High complexity

While CLIA describes the categories of test complexity, the FDA is responsible for determining whether a test falls into a particular category. For a test to receive a certificate of waiver (waived test status), the FDA must approve it for

doi:10.1128/9781555819156.ch3

home use. In addition, it must also determine that it is relatively simple to perform and poses limited risk to patients if performed incorrectly (8). Determination of whether a test falls into moderate or high complexity status is determined by the FDA using scoring criteria in seven different categories (Table 1) (8).

Tests are scored in each category on a scale of 1 to 3 based on complexity, and a total score of greater than 12 will result in a test being categorized as high complexity (8). For a test to qualify as PPM, it must be performed by either a physician, mid-level practitioner, or dentist and be limited to moderately complex microscopy procedures utilizing either a bright field or phase contrast microscope in situations in which delay could potentially result in inaccurate test results. Examples of PPM procedures include direct wet mounts and potassium hydroxide preparations. The FDA database showing test complexity can be accessed at http://www.accessdata.fda.gov.

A laboratory may limit the testing performed to just one category or multiple categories of complexity. Per CLIA 88, the laboratory must be either CLIA-exempt or possess one of the following CLIA certificates:

1. Certificate of registration, which enables the laboratory to conduct moderate or high complexity laboratory testing or both until the laboratory is determined to be in compliance through a survey by CMS or its agent.
2. Certificate of waiver, which enables a laboratory to perform only the waived tests.
3. Certificate for PPM procedures, which is issued to a laboratory in which a physician, mid-level practitioner, or dentist performs no tests other than PPM procedures and/or waived tests.
4. Certificate of compliance, which is issued to a laboratory after an inspection that finds the laboratory to be in compliance with appropriate CLIA requirements.
5. Certificate of accreditation, which is issued by an accreditation organization approved by CMS indicating that the laboratory meets all applicable CLIA requirements.

For laboratories performing moderate and/or high complexity testing, CLIA also outlines the minimum requirements for quality control, proficiency testing, quality assurance, and personnel. CMS is responsible for the implementation of the various facets of CLIA, while the FDA is responsible for the classification of test complexity (8).

CLIA is funded by fees charged to the laboratories for registration and inspection. Laboratories may register for a CLIA certificate by filing an application and undergoing an inspection if necessary. Laboratories that only perform waived and PPM testing are not subject to routine inspection, although inspections may occur randomly as part of representative sampling or in the setting of allegations of misconduct. The purpose of these inspections is to ensure that the laboratories are performing testing appropriate to the certificate of waiver and are following manufacturers' test instructions. Laboratories that perform moderate or high complexity testing may choose to be inspected by either CMS or a private nonprofit accreditation program that is granted deemed status by CLIA. In addition, licensure by state regulatory agencies in states where the requirements are determined to be at least as stringent as CLIA may serve in lieu of CLIA requirements. Currently, two states meet these requirements: New York and Washington (http://www.cms.gov/clia).

TABLE 1 Categories for determination of test complexity

Categories
Knowledge required to perform test
Training and experience required
Reagent and materials preparation
Nature of operational steps
Calibration, quality control, and proficiency testing materials
Troubleshooting and maintenance
Interpretation and judgment for test performance

TABLE 2 List of accrediting agencies with deemed status under CLIA

AABB (formerly known as American Association of Blood Banks)
American Association for Laboratory Accreditation
American Osteopathic Association
American Society for Histocompatibility and Immunogenetics
COLA (formerly known as Commission of Office Laboratory Accreditation)
College of American Pathologists
Joint Commission

Source: http://www.cms.gov/clia; accessed October 2014.

Laboratories that have passed inspection by CMS receive a certificate of compliance, while those inspected by organizations having deemed status under CMS receive a certificate of accreditation. Accrediting agencies with deemed status under CMS include the CAP and The Joint Commission (TJC) among others (Table 2). Rather than focus on a methodical evaluation of the compliance of the laboratory with each regulatory standard, the CMS surveys take a quality assurance approach that assesses the ability of the laboratory to monitor its own processes and provide accurate test results in a reliable and timely manner. As part of the educational approach, the laboratory is given time to adequately address any deficiencies that have been identified, unless the deficiencies pose the risk of immediate harm to patients. Failure to resolve these deficiencies may result in sanctions commensurate with the nature and scope of the situation (http://www.cms.gov/clia).

To remain compliant with CLIA 88, a laboratory performing moderate or high complexity testing must address the following CLIA requirements:

1. Personnel: CLIA sets the minimum qualifications for personnel performing or supervising PPM, moderate, or high complexity testing. The requirements vary based on the level of testing complexity and can be found at http://www.cms.gov/clia/. There are no requirements for personnel performing waived testing.
2. Proficiency testing: CLIA law Section 353(f)(3) states that laboratories performing nonwaived testing should participate in graded proficiency testing for the analytes listed under 42 CFR part 493, Subpart I. Under 42 CFR part 493, Subpart H, proficiency testing for nonwaived processes or analytes requires an 80% passing score and should include at least three annual events composed of five challenges per event. Proficiency testing materials may be obtained from CMS-approved vendors (http://www.cms.gov/clia). For analytes or test processes not listed under Subpart I, CLIA requires that the laboratory check the accuracy of the

test at least twice annually. Repeated failure to obtain minimum satisfactory score in successive or two out of three consecutive proficiency test events for a particular analyte, test, specialty, or subspecialty may result in limitations being placed on the laboratory's ability to perform testing for the analyte or subspecialty in question unless appropriate remedial action is taken. Proficiency testing is not required for waived tests (8).

3. Quality system: CLIA Subpart K mandates that laboratories have a quality system that controls and monitors all phases of testing: pre-analytic, analytic, and postanalytic as well as general laboratory systems. This system should be elaborated in written policies and procedures that outline the process of continuous improvement that serves to identify, evaluate, and resolve problems with the testing process (8).

As of 2013, there were 244,564 laboratories registered under CLIA, with the vast majority of these (201,842) performing either waived or PPM procedures (http://www.cms.gov/clia). Among accredited laboratories, most are certified by the CAP, TJC, or COLA.

The CAP is the largest organization in the world composed solely of board-certified pathologists. It currently accredits 7,600 laboratories and provides proficiency testing for over 20,000 laboratories worldwide (http://www.cap.org). CAP's proficiency testing program is the largest in the world and allows laboratories to compare themselves against their peers in a wide range of analytes and test systems.

TJC (formerly known as Joint Commission on Accreditation of Healthcare Organizations) is an independent nonprofit group that certifies and accredits 20,500 health care organizations in the United States. TJC inspects health care facilities on a 3-year accreditation cycle and laboratories on a 2-year cycle (http://www.jointcommission.org).

COLA, formerly known as the Commission on Office Laboratory Accreditation, was originally established in 1988 to accredit physician office laboratories. COLA is a physician-directed organization and accredits over 8,000 laboratories (http://www.cola.org).

For laboratories that are looking to go beyond the regulatory requirements to establish a quality management system, ISO provides laboratory-specific standards. ISO is an independent nongovernmental organization, headquartered in Geneva, Switzerland, whose members are the national standards bodies of 165 countries. ISO develops standards to ensure the quality of products, services, or systems. The organization is the world's largest developer of standards, with 19,500 at last count (http://www.iso.org). ISO standards are the product of collaboration between experts from every facet of a particular industry or field who come together under the aegis of a technical committee. The U.S. representative in ISO governance is the American National Standards Institute (ANSI). In the United States, compliance with ISO standards is voluntary, but many industries seek ISO accreditation due to the implications for quality, reliability, and efficiency. While ISO provides the standards, it does not certify that any product, organization, or service meets these standards; that function is performed by third-party organizations that serve as certifying or accrediting agencies. ISO certification implies that a "product, system, or service meets specific requirements." ISO accreditation on the other hand, indicates that an organization operates according to specific standards (http://www.iso.org).

The most commonly referenced ISO standard is ISO 9001, which sets out the general criteria for a quality management system and can be adapted to most industries and services. ISO 9001 contains general requirements that are open to significant interpretation, however, and they may be difficult to adapt to the laboratory setting. In contrast, ISO/IEC 17025 pertains to the competence of general testing and calibration laboratories. In 1994, at the request of the CLSI via the ANSI, ISO created Technical Committee 212 (ISO/TC 212) with the mandate to develop laboratory-specific standards for "Clinical diagnostic testing and in vitro diagnostic systems" (http://clsi.org/standards/iso-standards/). The scope of the guidance for ISO/TC 212 is "Standardization and guidance in the field of laboratory medicine and in vitro diagnostic test systems." This includes quality management, pre-analytical and postanalytical procedures, analytical performance, laboratory safety, reference systems, and quality assurance (http://www.iso.org). CLSI serves as the secretariat for ISO/TC 212, a responsibility that was delegated to CLSI by ANSI. CLSI works with the 33 participating countries to develop standards using the ISO consensus process (http://www.clsi.org). Under the auspices of ANSI, CLSI also administers the U.S. technical advisory group (TAG), which allows any organization or individual within the United States to participate in development of ISO laboratory standards via ISO/TC 212. The criteria for membership of U.S. TAG can be found at http://clsi.org/standards/iso-standards/.

The first ISO standard that specifies particular requirements for quality and competence for medical laboratories is ISO 15189, which was based on ISO/IEC 17025 and ISO 9001 and released in 2003, with revisions in 2007 and 2012 (10). In some countries, ISO accreditation is the standard for reimbursement, but in the United States, accreditation is considered voluntary (9). The process of ISO 9001 accreditation involves assessment of quality management systems, whereas with ISO 15189, the focus of the audit is on technical competence of the laboratory in addition to its quality management system. By this measure, the successful completion of the ISO 9001 audit confers certification, whereas establishment of compliance with ISO 15189 results in accreditation under that standard. In addition, ISO 15189 contains requirements for routine assessment of laboratory performance in specific tests (technical competence) via proficiency testing, whereas ISO 9001 contains no requirement for determining technical competence (10).

The structure of ISO 15189:2012 is similar to that of ISO 15189:2007 and ISO/IEC 17025/2005 on which it is based. The standards are broken down into the following components:

1. Scope
2. Normative references
3. Terms and definitions
4. Management requirements
5. Technical requirements

The bulk of the ISO 15189 standards are covered under sections 4 (management standards) and 5 (technical standards). The ISO 15189 management standards cover a wide range of aspects, but the primary focus revolves around establishment of a quality management system (QMS) under the oversight of a quality manager. The quality manager is responsible for effective implementation and monitoring of the performance of the QMS. Under the QMS, ISO 15189 fosters a system of continuous improvement by requiring the laboratory to maintain a comprehensive system for identification and tracking of defects within the testing process. In addition, the laboratory should have a process for determining root cause, implementing corrective action, and monitoring the effectiveness of changes. A documented mechanism should also be in place to capture and resolve customer complaints. Regularly scheduled internal audits assess the effectiveness of the QMS via management review as part of the process of continuous improvement required by the ISO standards (Table 3). However, the audits also serve the purpose of ensuring that the laboratory processes effectively serve the needs of the users and conform to the requirements of the QMS (10).

Because of its focus on continuous improvement, compliance with ISO 15189 serves as an excellent foundation for implementation of other quality initiatives such as LEAN and Six Sigma (9). The management standards also outline specific criteria for service agreements, evaluation of reference laboratories, and advisory services. Results of management review should be communicated to laboratory staff, and the administration is responsible for effective follow-up.

Document control is also a major component of the ISO 15189 management standards, with requirements for a document management system that encompasses all documents under the QMS, including policies, flowcharts, posters, calibration charts, and requisitions. The standards also prescribe a naming convention for documents and requires that steps be taken to ensure that obsolete documents are removed from service. A structured communication system should ensure that changes in documentation are communicated to relevant staff at all levels and that actual practice matches protocol (10).

The technical component of ISO 15189 standards spans a range of requirements covering topics from personnel qualifications and competency assessment to the laboratory testing process itself (Table 4). The term "examination process" refers to any laboratory test method. ISO 15189 standards outline the need for clear documentation of laboratory protocol for every step of the testing process from specimen collection and transport to electronic results reporting. It also describes the structure, format, and scope of laboratory procedures. While the standards describe the need for validation and verification of laboratory tests, it does not have specific requirements for performing either. The ISO 15189 technical standards also cover the quality control requirements (10).

TABLE 3 Scope of ISO 15189 management review

(a) The periodic review of requests, and suitability of procedures and sample requirements
(b) Assessment of user feedback
(c) Staff suggestions
(d) Internal audits
(e) Risk management
(f) Use of quality indicator
(g) Reviews by external organizations
(h) Results of participation in interlaboratory comparison programs
(i) Monitoring and resolution of complaints
(j) Performance of suppliers
(k) Identification and control of nonconformities
(l) Results of continual improvement including current status of corrective actions and preventive actions
(m) Follow-up actions from previous management reviews
(n) Changes in the volume and scope of work, personnel, and premises that could affect the quality management system
(o) Recommendations for improvement, including technical requirements

TABLE 4 Technical components of ISO 15189:2012

5.1 Personnel: Qualifications, training, competency, performance review and continuing education
5.2 Accommodation and environmental conditions: Physical space and environment (for patients and staff)
5.3 Laboratory equipment, reagents, and consumables: Equipment acquisition, calibration, maintenance and repair. Reagent storage, new lot validation and inventory management
5.4 Pre-examination processes: Patient preparation, requisition format, sample acceptability criteria and laboratory user guide, sample collection and transport
5.5 Examination processes: Verification and validation of laboratory testing, measurement uncertainty, reference ranges, and protocols
5.6 Ensuring quality of examination results: Quality control and proficiency testing requirements
5.7 Postexamination processes: Specimen storage, results review and acceptability criteria
5.8 Reporting of results: Reporting format and contents
5.9 Release of results: Critical values and revised reports
5.10 Laboratory information management: Criteria for implementation, validation, and operation of a laboratory information system

While the technical committees under ISO are responsible for developing the ISO standards, the best practices to assess compliance with the standards are outlined by the ISO Committee on Conformity Assessment (CASCO). The committee does not perform assessments but sets out policy and standards on conformity assessment (http://www.iso.org/iso/Casco). Third-party agencies that perform ISO accreditation can use these standards for assessing compliance with ISO standards. Organizations such as the International Laboratory Accreditation Cooperation (ILAC), which is a collaboration between a number of national laboratory-accrediting agencies, play a role in assessing the competence of ISO accreditors (http://www.ilac.org). In addition to promoting laboratory standards and accreditation, ILAC has the stated goal of promoting acceptance of inspection results. ILAC recognition of an ISO accrediting organization may be obtained by inspection by a peer inspector under the ILAC umbrella. ISO recommends the use of certification bodies that use the appropriate CASCO standard and are accredited under ILAC (http://www.iso.org/iso/home/standards/certification.htm). However, this accreditation is not required at this time, and some major organizations that offer ISO 15189 accreditation, such as the CAP, have chosen not to be accredited under ILAC (http://www.cap.org).

The process of adopting the ISO 15189 standards begins with selection of an accrediting agency. Laboratories in the United States can select ILAC-recognized accrediting agencies via http://www.ilac.org. Alternatively, laboratories may select accrediting agencies that are not recognized by ILAC, such as the CAP, since ILAC recognition is not a requirement for an ISO accreditor. The steps in the process of ISO accreditation are outlined in Table 5.

TABLE 5 Steps in the ISO 15189 accreditation pathway

Step	Summary
Application submission	Laboratory submits application along with supporting documentation
Assessment and gap analysis	Accrediting agency reviews documentation and determines whether additional information is required. This stage may require on-site visit depending on the accrediting agency
Onsite assessment	Inspection by accrediting agency, which determines compliance with both the management and technical component of ISO 15189 standards
Corrective action	Laboratory submits documentation of corrective action to address nonconformances identified during accreditation inspection
Assessment of compliance	Accrediting agency reviews response to nonconformances and determines if the laboratory is in compliance with standards
Accreditation	Accreditation committee votes on whether to grant accreditation

The key differences between ISO accreditation and laboratory accreditation programs (LAPs), such as those administered by the CAP, are as follows:

- ISO 15189 may serve as the regulatory standard in some countries, but it does not address U.S. regulatory requirements and as such does not replace laboratory accreditation programs such as those offered by the CAP and other deemed organizations under CLIA.
- LAP accreditation programs such as the CAP may involve either volunteer peer or professional assessors. The ISO 15189 accreditation program typically uses professional assessors to determine compliance with the ISO standards (http://www.cap.org/apps/docs/laboratory_accreditation/15189/15189_accreditation_faq.pdf, http://www.a2la.org).
- ISO inspections are on a 3-year cycle, whereas the LAP inspections under CLIA are on a 2-year cycle. Under the CAP 15189 program, once a laboratory is accredited, it has to undergo annual surveillance assessments during the first and second year with a full reaccreditation inspection in the third year (http://www.cap.org/apps/docs/laboratory_accreditation/15189/15189_accreditation_faq.pdf).
- LAP inspectors focus on the technical competence of the clinical laboratory, whereas ISO accreditors such as the CAP typically provide separate assessors for the quality management system and the technical component. However, this may vary based on the accreditor involved and the scope of accreditation requested (http://www.cap.org, http://www.a2la.org).
- ISO accreditors focus on the quality management system under ISO, which differs from LAP requirements in that it emphasizes a continuous improvement process that captures defects, performs root cause analysis, and implements and monitors corrective action.

As mentioned previously, ISO 15189 may serve as the regulatory standard in some countries, but it does not address U.S. regulatory requirements and does not replace laboratory accreditation programs such as those offered by the CAP and other deemed organizations under CLIA. In the current climate of limited resources, this may place a significant burden on clinical laboratories that have to satisfy two separate and distinct regulatory standards. However, while compliance with LAP requirements establishes that a laboratory meets the basic technical standards, compliance with the ISO 15189 standards can serve to improve the efficiency, quality, and productivity of clinical laboratories. For example, the requirements for reagent management ensure that the laboratory can streamline inventory management in a manner that avoids waste due to expired reagents and reduces inventory costs. The document control requirements are often best implemented using electronic systems that can help avoid the use of large volumes of paper documents that can be difficult to track and maintain. The overall focus of the ISO 15189 standards on continuous improvement processes ensures that the laboratory is continually working towards improvement in all aspects of performance, rather than focusing on meeting the minimum required standards. Adoption of ISO 15189 standards will go a long way towards meeting the goals of improving laboratory standards, optimally utilizing limited resources, delivering a higher quality of patient care, and meeting customer expectations.

REFERENCES

1. **Camac C.** 1900. Hospital and clinical ward laboratories. *J Am Med Assoc* **XXXV:**219–227.
2. **Belk WP, Sunderman FW.** 1947. A survey of the accuracy of chemical analyses in clinical laboratories. *Am J Clin Pathol* **17:**853–861.
3. **Sunderman FW Sr.** 1992. The history of proficiency testing/quality control. *Clin Chem* **38:**1205–1209, discussion 1218–1225.
4. **Laessig RH, Ehrmeyer SS, Lanphear BJ, Burmeister BJ, Hassemer DJ.** 1992. Limitations of proficiency testing under CLIA '67. *Clin Chem* **38:**1237–1244, discussion 1245–1250.
5. Department of Health, Education, and Welfare, Public Health Service, Centers for Disease Control. 1967. Clinical Laboratory Improvement Act of 1967. Part F, Title ifi, Public Health Service Act, sect. 353.
6. 1968. *Federal health insurance for the aged: regulations for coverage of service of independent laboratories, Public Health Rep-13.* Department of Health, Education, and Welfare, Washington, DC.
7. **Bogdanich W.** 1987. Lax laboratories: the Pap test misses much cervical cancer through labs' errors. *Wall Street Journal* **210(88):**1,20.
8. Public Law No. 100-578—100th Congress, Oct. 31, 1988. 2903, Clinical Laboratory Improvement Amendments of 1988.
9. **Garcia LS, Bachner P.** 2014. *Clinical laboratory management,* 2nd ed. ASM Press, Washington, DC.
10. International Organization for Standardization (ISO) (ed.). 2012. Medical laboratories—requirements for quality and competence. *In* ISO 15189:2012, 3rd ed.

Laboratory Safety

K. SUE KEHL

4

This chapter outlines the requirements for a safe environment in the clinical virology laboratory. This begins with development of a culture of safety which identifies the risks, develops a system to mitigate these risks, and encourages ongoing evaluation of the environment and continuous risk reduction. Chemical and fire safety, and decontamination and waste disposal, in addition to biosafety, are important components of an overall safety program. Classification of organisms by risk group and the corresponding biosafety containment levels are described. Routine work practices in clinical virology are identified along with recommended safe practices.

SAFETY IN THE CLINICAL VIROLOGY LABORATORY

The risk of laboratory-acquired viral infections is unknown as there is no systematic reporting system or surveillance mechanism. Infections may be subclinical, have atypical incubation periods, unusual routes of acquisition, or unusual disease presentations. Most reports identify bacterial agents as the most common cause of laboratory-acquired infections and cannot identify a specific accident or incident associated with the infection (1). However, an extensive survey of research and clinical laboratories published by Pike in 1976 identified viral infections as the cause of 26.7% of laboratory-acquired infections, a large percentage of these due to hepatitis B virus (2). The majority of reported viral infections occur in research laboratories despite the larger number of laboratory personnel at risk in diagnostic laboratories. A recent systematic review (3) focused on reports of research laboratory-acquired viral infections from 1935 to 2006. During the most recent time period analyzed, 1983–2006, aerosol exposure/inhalation was the leading mode of transmission (92%) with Hantavirus accounting for 70% of all laboratory-acquired viral infections. Data from research laboratories may not apply directly to clinical laboratories since facilities and methods often differ significantly. However, it does highlight aerosol exposure/inhalation as a primary mode of transmission. Other common modes of exposure include hand to mouth (e.g., eating, drinking, mouth pipetting, transfer via contaminated fingers, or articles), skin (e.g., needles or other sharps, cuts), or mucous membranes (splashes or transfer via contaminated fingers). National guidelines have progressed since 1974, when graded levels of biosafety were first introduced by the Centers for Disease Control and Prevention (CDC).

Principles of Biosafety

The fifth edition of Biosafety in Microbiological and Biomedical Laboratories (BMBL5) (1) outlines four biosafety levels with increasing requirements for containment, safety equipment, microbiological practices, and facility safeguards. Biosafety level 1 is suitable for Risk Group 1 organisms, which are agents unlikely to cause disease in immunocompetent humans or animals and present little hazard to personnel or the environment. Biosafety level 1 relies on standard microbiologic techniques with no special requirements for containment, safety equipment, or facility safeguards. Secondary education or undergraduate teaching laboratories can often perform at a biosafety level 1. Biosafety level 2 is appropriate for work with Risk Group 2 organisms, which are agents that can cause human or animal disease. The organisms handled at this level are unlikely to be a serious hazard; however, exposure may result in infection. Effective treatment is available and the risk of spread to others or the environment is minimal. Many clinical virology laboratories operate at biosafety level 2. Biosafety level 3 is appropriate for work with Risk Group 3 organisms, which are agents that usually cause serious human or animal disease. Exposure may result in serious infection, effective treatment is available, and the risk of spread to others or the environment is minimal. There are numerous clinical laboratories with the capability to operate at biosafety level 3. For example, detection, identification and susceptibility testing of *Mycobacterium tuberculosis* is performed at biosafety level 3. Biosafety level 4 is appropriate for work with Risk Group 4 organisms, which are agents that usually cause serious human or animal disease. Exposure may result in serious infection; effective treatment is usually not available, with the potential for spread to others or the environment. There are a limited number of these facilities in the United States and worldwide.

A Culture of Safety

It is the responsibility of the laboratory director and the entire laboratory management team to develop a "Culture of Safety" within the laboratory. This requires ongoing assessment of

doi:10.1128/9781555819156.ch4

work processes and procedures to identify risks and to develop plans to mitigate those risks. In the clinical virology laboratory, the biohazard risks are often unknown so policies and procedures must be in place to mitigate the most serious of these potential biohazards.

The first step in this process is performance of a risk assessment by the laboratory director or other members of the management team. While there is no standardized procedure for performing a risk assessment, that outlined by the CDC (4) is an excellent process. The first step is to identify the biohazards associated with the infectious agents or specimens. One must consider the agents likely to be recovered, their infectious dose, and the route of infection. Agent summary statements in BMBL5 are an excellent source for this information. Pathogen safety data sheets are also available from the Public Health Agency of Canada (http://www.phac-aspc.gc.ca/lab-bio/res/psds-ftss/index-eng.php). These include information on the pathogenicity, mode of infection, potential laboratory handling risks, and recommended containment. Second, identify the activities likely to expose workers to the biohazards. Table 1 lists activities that might expose workers to hazards and highlights work practices and engineering controls that can be employed to mitigate these risks. Consider the likelihood of generating aerosols and the environment in which these activities are performed. The concentration of organism and the equipment utilized affect the likelihood of aerosol generation. Scraping cells from tissue cultures demonstrating viral induced cytopathic effects (CPE) and subsequent spotting of potentially infected cells onto the surface of slides can be aerosol-generating procedures. Aliquoting and preparation of specimens prior to nucleic acid amplification testing can also be aerosol-generating procedures. Consider the experience and training of the personnel performing these activities. This is of ever-increasing importance as molecular testing is implemented outside the clinical laboratory. Inexperienced personnel may be more likely to generate aerosols. Then, evaluate the likelihood of occurrence of the risk and the severity of the consequences. This information is used to determine the appropriate biosafety level to be employed. It is important to note that there may be instances where, based on the risk assessment, necessary precautions exceed those for the suspected risk group organism and associated biosafety level.

Based on this assessment, the laboratory director and the management team should develop policies and procedures to mitigate these risks. Whenever possible, engineering controls should be used to mitigate risks. This decreases reliance on personnel to properly perform tasks. Implementation of safe work practices and utilization of personnel protective equipment further decrease risk. It is a requirement for certification by the College of American Pathologists (5) that laboratories comply with national, state, and local guidelines on occupational exposure to bloodborne pathogens, as well as have written procedures for the safe handling of microbiological samples (6). These procedures should be included in staff training and be readily available for their use.

A key component of this culture of safety is a comprehensive Biosafety Manual. This manual should address not only biosafety, but also disinfection and sterilization, chemical hygiene, fire safety, waste disposal, and biosecurity.

TABLE 1 Activities associated with exposure to hazards and work practices and engineering controls that can be employed to mitigate these risks

Activity	Risk mitigation
Mouth pipetting	Handheld pipetting device
Splash	Work in biosafety cabinet, behind Plexiglas shield, or use face mask
	Use of disinfectant-soaked, absorbent, plastic-backed material on bench surface
Needlestick	Do not recap or bend needles; if necessary, use one-hand method for recapping
	Use nonsharp device or safety needles
	Safe sharp disposal
Broken tube in centrifuge	Sealed centrifuge cups opened in biosafety cabinet
Aerosol from cell spot preparation	Work in Biosafety cabinet
	Fix cells prior to removal from biosafety cabinet
Aerosol from nucleic acid extraction	Inoculate lysis buffer with pipet tip below surface of buffer
Aerosol from direct specimen testing	Work in biosafety cabinet, behind Plexiglas shield or use face mask

Disinfection and Sterilization

An essential component to safe laboratory practices is disinfection and sterilization. Disinfection is the process where most, but not necessarily all, microorganisms are destroyed. This can be accomplished by physical or chemical means and is usually preceded by a decontamination or cleaning step. Physical methods of disinfection are boiling at 100°C for 15 minutes, pasteurizing, or UV light irradiation. It is not recommended that UV light be used for disinfection of biosafety cabinets. UV light does not penetrate effectively and organisms not in the direct path are not affected. Also, UV light deteriorates plastic in the biosafety cabinet and exposes employees to risk. UV lights must be monitored as intensity decreases over time with life expectancy of 9,000 hours (4).

Chemical methods of disinfection are more commonly used. The choice of disinfectant used should be based upon its effectiveness against expected pathogens. In the virology laboratory, a high-level disinfectant which has activity against nonlipid or small viruses should be used. In the United States, the disinfectant should be one approved by the U.S. Environmental Protection Agency (EPA) for environmental surfaces. The CDC has issued guidelines (7) which include recommendations for disinfection and sterilization. Ethyl alcohol is nonsporicidal and evaporates quickly such that adequate exposure time is difficult to achieve. At concentrations of 60% to 80%, ethanol is active against enveloped viruses (e.g., herpesviruses, vaccinia virus, and influenza virus) and many nonenveloped viruses including adenovirus, enterovirus, rhinovirus, and rotaviruses, but not hepatitis A virus or poliovirus. It is effective against bacteria, fungi, and mycobacteria as well. A 70% ethanol concentration has been used to disinfect external surfaces of equipment and small surfaces. Alcohols do not penetrate protein rich environments and thus should not be used to decontaminate spills. Phenolic compounds are labeled as bactericidal, fungicidal, tuberculocidal, and virucidal. However, they have been reported not to have activity against enterovirus. Quaternary ammonium compounds are also effective surface disinfectants. They are fungicidal, bactericidal, and virucidal against enveloped viruses but are not sporicidal, tuberculocidal, or virucidal against nonenveloped viruses. Their use should be avoided in a clinical virology laboratory. Chlorine bleach is an effective disinfectant and is recommended for surface disinfection. It has a broad spectrum of activity being bactericidal, fungicidal, sporicidal, tuberculocidal, and virucidal. A 1:10 dilution of 5.15% to 6.25% sodium hypochlorite (household bleach) can be used for decontamination of blood spills (7). For disinfection of equipment, the disinfectant recommended by the manufacturer must be employed (8). If no instructions are provided, one should consider the potentially contaminating organisms, the composition of the equipment, and the potentially caustic nature of the disinfectant. Bleach disinfection followed by a rinse with 70% alcohol or water is often recommended to decontaminate stainless steel surfaces (1, 4, 9). The length of contact time required depends on the organism load and the disinfectant employed. Cleaning surfaces or equipment to remove organic debris decreases microbial load and thus can result in more effective contact time. It is imperative that the disinfectant be prepared according to the manufacturer's recommendations and that the disinfectant be used for the recommended contact time.

The laboratory environment should be clean prior to performing any testing. All laboratory benches, supplies, and equipment should be disinfected whenever there is a spill and laboratory surfaces should be disinfected after performing work and at the end of the day (4).

Each laboratory must have a procedure to disinfect and clean up spills (10). This procedure may vary depending upon the location of the spill, the volume of the spill, and the infectious agent. If a spill occurs within the biosafety cabinet, the biosafety cabinet should be left on. If the spill occurs in an occupied area, others should be warned to leave. If the area is under negative pressure, the bioaerosol should be allowed to settle for 30 minutes prior to cleanup. If the area is not under negative pressure, spill cleanup should begin immediately. In general, spills can be disinfected with tuberculocidal agents, agents active against nonenveloped viruses, or bleach. The spill should be covered by towels to prevent aerosolization when the disinfectant is added or the disinfectant should be added from the edge of the spill toward the center. The disinfectant must be allowed to act for its required contact time prior to clean-up of the area. All spill materials are then collected using a squeegee or dust pan and decontaminated prior to disposal (11).

Sterilization is the process where all forms of microorganisms are killed. This can also be accomplished by physical or chemical means. The physical methods of sterilization are incineration, moist heat or autoclaving, dry heat, filtration, and ionizing radiation. The most common chemical method is ethylene oxide; however, glutaraldehyde and peracetic acid are also used. Incineration is a method commonly used by medical waste handlers. It is a common practice to package laboratory medical waste for incineration offsite by a medical waste management company. Moist heat is also used for sterilization of media and supplies by autoclaving at 121°C for 15 minutes and for sterilization of medical waste by autoclaving at 132°C for 30 to 60 minutes. Filtration using a 0.2 μm filter can be used to remove bacteria from solutions; however, this will not remove viral agents. Regardless of the method employed it is important that the sterilization process be monitored. This can be accomplished by either chemical or biological indicators (11).

Chemical Safety

Each laboratory in the United States is required to have a Chemical Hygiene Plan (CHP) as outlined by the Occupational Safety and Health Administration (OSHA)'s Occupational Exposure to Hazardous Chemicals in Laboratories standard (29 CFR 1910.1450) (12). This standard requires that all hazardous chemicals are inventoried yearly and labeled according to the United Nations' Globally Harmonized System of Classification and Labelling of Chemicals stating the health risks. The standard requires that the CHP include criteria that the laboratory will use to determine and implement engineering controls, and personal protective equipment to reduce exposure to hazardous materials. The CHP must include when to use a fume hood and the proper procedures for its use. Laboratories are also required to maintain Safety Data Sheets for every chemical utilized. Safety Data Sheets are available from manufacturers or suppliers. Instructions for accessing Safety Data Sheets are available at http://www.ors.od.nih.gov/sr/dohs/LabServices/MSDS/Pages/material_safety_data_main.aspx. Safety Data Sheets provide information on the chemical and precautions to be taken in case of exposure or spill. The standard also requires that laboratories inform employees of the hazardous chemicals they work with, the signs and symptoms due to exposure, and provide training in the appropriate measures to take to protect themselves. The Safety Data Sheets must be available to employees at all times in case of exposure or spill.

The hazardous chemicals handled in the clinical virology laboratory will vary depending on the methods employed (4). Alcohols are commonly employed to fix cells, extract nucleic acids, and disinfect surfaces. Acetone is also used to fix cells. Both should be stored in a flammable cabinet. If cold acetone is required, it must be stored in an explosion-proof refrigerator. Several chemicals used in the clinical virology laboratory are toxic when handled as powders. These include antibiotic powders, solutions of which are routinely used in culture and viral transport media, and Cycloheximide, used in culture of *Chlamydia trachomatis*. Evans Blue, used as a counterstain in fluorescence methods, is a potential carcinogen. Exposure to these powders can be reduced by purchasing prepared solutions or handling powders with the use of gloves, masks, or within fume hoods. Sodium azide is used as a preservative to prevent bacterial growth in many reagents. It is a poison, and even minute amounts can cause symptoms. It is also explosive when in contact with metal. When disposed of in drains, the drains must be flushed with copious amounts of water (13). Thimerosal is an organomercury compound also used as an antibacterial and antifungal agent. Thimerosal is very toxic by inhalation, ingestion, and in contact with skin (14).

Guanidinium compounds used in nucleic acid extraction produce toxic fumes when mixed with bleach (15). Care should be taken to not mix. Ethidium bromide is an intercalating agent commonly used as a fluorescent nucleic acid stain in molecular biology laboratories for techniques such as agarose gel electrophoresis. Ethidium bromide is a potential mutagen due to its intercalation with DNA. It is not regulated as hazardous waste at low concentrations, but is often treated as hazardous waste. Ethidium bromide can be degraded chemically, or collected and incinerated (16). Low concentrations are often disposed of by pouring it down a drain or treating with bleach before disposal. Ethidium bromide can also be removed from solutions with activated charcoal or ion exchange resin. Various commercial products are available for this use. Safer alternatives, such as SYBR Safe DNA Gel Stain are available and should be considered for use.

Dimethyl sulfoxide (DMSO) is a cryoprotectant used to freeze cell cultures. It is a mutagenic agent and can readily penetrate cells. Double gloving should be considered when handling DMSO as it will eventually penetrate latex gloves and dissolves nitrile gloves (4).

Sodium hypochlorite, or bleach, at a dilution of 1:10 is commonly used as a laboratory disinfectant. However, it is important to note that it is also a toxic chemical. Undiluted, it is a strong oxidizer and capable of causing burns. It is also not compatible with many chemicals used in the laboratory. For example, mixing with acid can cause the release of toxic chlorine bleach.

Electrical Safety

Faulty electrical equipment is both a fire hazard and a personal health hazard. The appropriate number and type of electrical outlets should be present and distributed in such a manner that extension cords are unnecessary. Prior to installation, equipment should be checked for proper ground. It is imperative that electrical equipment undergo periodic electrical safety checks as part of preventive maintenance. Faulty equipment must be taken out of service (8, 10).

Fire Safety

Personnel should know how to respond to a fire. The acronym Race (Rescue, Alarm, Contain, and Extinguish) provides simple instruction on proper response. Evacuation plans should be posted. Fires are classified according to the material involved. The fire extinguisher appropriate for the class of fire should be readily available (17). A fire extinguisher that can extinguish class A (ordinary solid combustibles), class B (flammable liquids and gases), and class C (energized electrical equipment) fires is most frequently used in laboratories. Personnel should be trained in its use. The acronym PASS (which stands for Pull, Aim, Squeeze, and Sweep) provides clear instruction on proper usage.

Biosafety

Laboratory directors and the management team are responsible for identifying potential hazards, identifying activities that might expose workers to those hazards, assessing the risks associated with those hazards, and implementing procedures to mitigate those risks. However, all laboratory personnel must assume responsibility for identifying potential hazards in their work environment and bringing these to the attention of the laboratory director and management team. To classify potential hazards, the World Health Organization in the third edition of the WHO Laboratory Biosafety Manual (18) has defined risk groups for infectious organisms that are based on their risk to laboratory workers, animals, and to the environment. Murine leukemia virus, murine sarcoma virus, and adeno-associated virus are examples of Risk Group 1 viruses. Many viruses, including adenovirus, herpesviruses, orthomyxoviruses and paramyxoviruses, picornaviruses, hepatitis B virus, and hepatitis C virus are Risk Group 2 viruses (19). These viruses can cause human or animal disease but are unlikely to be a hazard to laboratory workers. HIV, Hantaviruses, Japanese B encephalitis virus, Rift Valley fever virus, Yellow Fever virus, and rabies virus are examples of Risk Group 3 viruses. Risk Group 3 viruses usually cause serious or fatal disease and are a hazard to laboratory workers. Risk Group 4 organisms, such as the hemorrhagic fever viruses, usually cause serious or fatal disease but therapy is not available.

Exposure to biohazards can occur through a variety of routes. Ingestion, inoculation, contamination of skin or mucous membranes, and inhalation are the most common routes. Ingestion occurs through mouth pipetting and eating, practices prohibited in most laboratories, or transfer of organisms by placing contaminated items into the mouth. Staff should be trained not to place pens in their mouth and avoid hand contact with eyes and mucous membranes. Contamination of skin or mucous membranes also occurs through splashes or spills. Working in a biosafety cabinet, wearing a mask, or working behind a splash shield can mitigate risks associated with activities likely to generate aerosols or splashes. In addition to wearing gloves when contact with blood or other potentially infectious material is expected, gloves should also be worn when there are cuts in the skin to eliminate risk of exposure.

Inoculation occurs through needle stick or sharps injuries. Adherence to OSHA's Bloodborne Pathogen Standard (29 CFR 1910.1030) (20) greatly reduces risk associated with needlestick or sharps injuries. This standard requires that an employer identify employees at risk and the activities that put them at risk. Engineering controls must be identified and used. These are devices that isolate or remove the bloodborne pathogen hazards from the workplace. They include sharps disposal containers, self-sheathing needles, and safer medical devices, such as sharps with engineered sharps-injury protection and needleless systems. All employees with occupational exposure must be offered the hepatitis B vaccine within 10 days of assignment. An exposure incident is a specific eye, mouth, other mucous membrane, nonintact skin, or parenteral contact with blood or other potentially infectious material. Postexposure evaluation and follow-up must be provided immediately and at no cost to the worker. The exposure control plan must be updated annually. In up to 20% of laboratory-associated infections the route is unknown, but believed to be infectious aerosols. There are a variety of laboratory activities that are commonly associated with aerosol generation (2). Handling tissue culture cells by pouring or decanting fluids, opening culture containers, mixing infected cell suspensions with pipettes, and aspirating cultures using vacuum are common practices in the clinical virology laboratory which are capable of generating aerosols. Aliquoting or preparing specimens for molecular testing or processing samples for antigen detection are also activities that are capable of generating aerosols. Most Risk Group 3 and 4 organisms can be acquired through respiratory transmission of aerosol droplets (2). Aerosol droplets >0.1 mm in size settle out quickly while smaller droplets (<0.05 mm) evaporate and remain suspended in air. Aerosol droplets can be removed from the air within 30 to 60 minutes with an air exchange rate of 6 to 12 changes per hour (2).

Standard Precautions, first described by the CDC in 1987 (21), are guidelines identifying work practices to reduce the risk of transmission of bloodborne pathogens. Now referred to as Universal Precautions, these precautions include the routine use of appropriate barriers to prevent skin and mucous membrane exposure, including the use of gloves when contact with blood or other body fluids is expected, the use of masks and face shields when procedures that are likely to generate droplets are anticipated, and the use of gowns when splashes are anticipated. Table 2 lists safe work practices specifically identified for laboratories.

TABLE 2 Safe laboratory work practices

Universal Precautions—Safe work practices for laboratories

- Specimens should be placed in secure, leak-proof containers
- Gloves should be worn when handling specimens
- Masks and protective eyewear should be worn if mucous membrane contact is anticipated
- Biological safety cabinets should be used whenever droplet generating procedures are conducted
- Mechanical pipetting devices should be used
- Needles and sharp use should be limited. If required, safe sharp practices should be employed
- Laboratory work surfaces should be decontaminated
- Contaminated materials should be decontaminated prior to disposal or disposed of as infective waste
- Equipment should be decontaminated before being repaired
- Staff should wash their hands and remove protective clothing when leaving the laboratory
- Staff should not eat, drink, smoke, or apply cosmetics in the laboratory
- Staff should assume all patients and patient's samples are infectious

Biosafety containment levels (BSL) range from BSL1 to BSL4. They are a combination of safe microbiological work practices, primary barriers (safety equipment and personal protective equipment), and secondary barriers (facilities design) required to safely handle infectious agents. Safety equipment includes biological safety cabinets and safety centrifuge containers. Laboratory coats, gowns, gloves, respirators, face shields, booties, etc., are examples of personal protective equipment. Directional airflow and controlled access are examples of facility design.

Table 3 is a summary of recommended BSL practices, safety equipment, and facilities. Detailed description of the BSL recommendations can be found in BMBL5 (1). In addition to safe microbiological practices performed at BSL-1, BSL-2 safe practices require limited access to the laboratory, biohazard warning signs at entrances to the laboratory, utilization of safe "sharps" precautions, and a biosafety manual outlining waste disposal and medical surveillance, if necessary. Additionally, BSL-3 safe practices require controlled access to the laboratory, decontamination of all waste prior to disposal, and decontamination of all laboratory clothing prior to laundering. Collection and storage of serum samples, or serological testing, prior to performing work with Risk Group 3 agents should be considered to facilitate exposure investigation.

Primary containment barriers physically separate the worker from the biohazard and are required for all BSL levels except BSL-1. A biosafety cabinet must be utilized as primary containment in the clinical virology laboratory. Class I biological safety cabinets provide containment of the hazard with no protection of the samples. Air is pulled into the cabinet through a HEPA filter and vented to the outdoors. They may be used to enclose aerosol-generating equipment such as centrifuges or perform procedures such as necropsy where sample protection is not essential. Class II biological safety cabinets provide personal protection, sample protection, and containment of the hazard. HEPA filtered air flows downward and splits over the work area. The HEPA filtered air directed toward the front of the cabinet prevents contaminants from entering the work area. Air pulled into the cabinet prevents aerosols from exiting the cabinet. All of the air leaving the work area is filtered through a HEPA filter and either exhausted to the outdoors or into the room. Because balance between intake and downflow air is critical, certification and proper operation are required to ensure safe performance. Class II biosafety cabinets are most appropriate for routine use in the clinical virology laboratory (1). Class III biological safety cabinets provide maximum containment of high-risk agents, protecting both the worker and the environment. They are specialized glove boxes with HEPA filtered intake air and double HEPA filtered exhaust air operated under negative pressure with a complete barrier between the worker and the agents. Class III biological safety cabinets are utilized in BSL-4 laboratories.

In addition to the biosafety cabinet, BSL-2 and BSL-3 primary barriers should also include an autoclave for sterilization and waste disposal. The personal protective equipment (PPE) to be used should be based on the risk. PPE should be worn only in the work area. PPE in the form of a laboratory coat is recommended for BSL-1. Laboratory coats should be knee-length and fluid repellant. Attire should minimize the amount of skin exposed. The use of laboratory coats and gloves is recommended at BSL-2. Laboratory coats must not be altered in any way and sleeves should not be rolled or pushed up. Gloves should be pulled over the cuff of the laboratory coat to reduce skin exposure. Gloves should be single use and changed when visibly soiled, torn, or punctured. Additionally, aerosol-generating activities must be performed in a biosafety cabinet or eye and face protection, such as goggles and face masks or splash shields, must be used. When necessary, such as working at BSL-3, additional respiratory protection may be required. The use of N-95 mask or powered air purifying respirator, as determined by risk assessment, may be necessary. BSL-4 PPE requires a positive pressure personnel suit with a segregated air supply.

Secondary containment barriers are those barriers incorporated into the facility design to protect both the workers in the laboratory as well as the community. They provide additional containment to prevent spread outside the primary barrier. Refer to Table 1 in the chapter on Laboratory Design for secondary barriers required for the different biosafety levels. Impervious laboratory benches and hand washing sinks are a requirement at all biosafety levels. BSL-1 secondary barriers include separation of the laboratory from public areas, and easily decontaminated laboratory space and furniture. BSL-2 secondary barriers additionally include a biosafety cabinet and a method for waste decontamination. Consideration of airflow such that there is an inward flow of air without recirculation outside the laboratory should be included in facility design. BSL-3 secondary barriers include physical separation from corridors, self-closing double door access, negative airflow into the laboratory, and nonrecirculating exhaust air. BSL-4 secondary barriers include multiple showers (both to enter and exit) and other safety precautions designed to destroy all traces of the biohazard. The laboratory must have double door access with airlocks and electronic security to prevent both doors from opening at the same time. All air and water going to and coming from the laboratory must be decontaminated.

The biosafety level utilized usually corresponds to the risk group of the organisms; however, it is imperative that a risk assessment be performed to consider other factors. BSL-1 is usually recommended when working with agents not known to cause disease in adults, such as occurs in teaching laboratories. BSL-2 practices are used in diagnostic laboratories

TABLE 3 Biosafety level recommendations

Biosafety level (BSL)	Work practices	Primary barriers	Facility design
BSL 1	Standard microbiologic methods • Hands must be washed after handling potentially infectious material and upon leaving the laboratory • Eating, drinking, applying cosmetics, handling contact lenses, and storing food are prohibited • Mouth pipetting is prohibited • Splashing or aerosol generation must be minimized • All cultures and other potentially infectious material must be decontaminated prior to disposal • A universal biohazard sign must be posted at entry to laboratory • Pest management program must be in place • All laboratory staff must be trained	None	Open laboratory bench Hand-washing sink
BSL 2	BSL 1 plus: • Limited access, no unauthorized personnel • Biohazard warning sign identifying pathogens • "Safe sharps" practices	• Class II Biosafety cabinet for potential aerosol-generating procedures including specimen processing for antigen, antibody, and molecular testing as well as work with tissue cultures • Laboratory coats, gloves, face shield as necessary	BSL 1
BSL 3	BSL 2 plus: • Controlled access • Decontamination of waste prior to disposal • Decontamination of PPE prior to disposal or laundering • Baseline serum, if applicable	BSL 2 plus: • Class II Biosafety cabinet for all procedures • Respiratory protection	BSL 2 plus: • Physical separation from outside corridor • Self-closing double door access • Air exhausted directly outside • Negative airflow
BSL4	BSL 3 plus: • Change of clothing on entry • Shower out • Decontamination of all material prior to disposal	BSL 3 plus: All procedures performed in Class III biosafety cabinet with full body, air-supplied, positive pressure suit	BSL 3 plus: • Separate building or isolated area • Dedicated vacuum, supply and exhaust, and decontamination system

that handle organisms not transmitted via aerosols. BSL-3 practices are recommended when handling organisms that are highly infectious and transmitted via aerosols. BSL-4 practices are required when handling life-threatening agents for which no therapy is available such as Ebola virus and Crimean-Congo hemorrhagic fever virus.

Recommended Work Practices

For the clinical virology laboratory, it is recommended that minimally BSL-2 work practices and facilities be employed. However, BSL work practices or facilities may be increased based upon the risk assessment. All human specimens, regardless of test requested, must be handled employing Universal Precautions and minimally at BSL-2. Laboratory coats and gloves should be worn by all laboratory personnel. Processing of specimens for viral testing has the same requirements as other laboratory test requests. Specimens must be transported to the laboratory in a secondary container; most commonly a zip-lock bag is employed. Leaky containers may lead to contamination of the specimen as well as be a hazard to laboratory personnel. Specimens with gross external contamination that cannot be safely disinfected should not be handled (11, 22). Occasionally it may be necessary to perform testing on an irretrievable or precious specimen. If the leak is contained within the secondary container, the external surface of the container can be disinfected, taking care to not disinfect the specimen. This can be accomplished by placing the specimen in a petri dish and applying a disinfectant such as 10% bleach to the external surface for the appropriate period of time. Work should be performed within a biosafety cabinet with appropriate PPE. If the specimen has leaked out of the secondary container, immediate disinfection with appropriate handling, as for any spill of infectious material, is required. Test requests and specimens must include proper patient identification. It is helpful if the request for viral testing indicates the suspect agent to ensure selection of the appropriate method, whether it be molecular detection, culture with inoculation of proper cell lines, antigen detection, or serologic testing. The laboratory should have a list of viral agents indicating the preferred sample and the preferred method of testing. This list should include those agents for which culture should not be performed. This list includes hemorrhagic fever viruses such as Lassa virus, Marburg and Ebola viruses, smallpox virus, and other CDC Select Agents. If requests are received for these agents, contact the local public health department and/or the CDC.

Processing of specimens for viral testing should be performed in a Class II biosafety cabinet. Cabinets should be cleaned prior to and after use, should be documented to be performing properly, and certified prior to use. Prior to work in the biosafety cabinet, any items within the cabinet should be disinfected with a 1:10 dilution of bleach and removed. The interior surface should be disinfected with bleach and rinsed with water or 70% alcohol. Bleach alone will corrode the surface of the biosafety cabinet. Alcohol alone has little effect on nonenveloped viruses. All items to be used in the biosafety cabinet should be disinfected prior to being placed back in the cabinet (4). The biosafety cabinet should be allowed to run for at least 4 minutes (or that recommended by the manufacturer) prior to use (1, 4, 9). After completion of work, all items should be disinfected and removed. The interior surface should be disinfected with bleach and rinsed with water or 70% alcohol. The blower should be allowed to run for 15 to 20 minutes; however, many laboratories leave biosafety cabinet blowers running continuously.

Cell cultures, whether inoculated or not, are potentially infectious (4). Cell cultures may contain unintended or adventitious agents. Primate cell lines are commonly contaminated with SV-5 and SV-40. Herpesvirus simiae (B virus) may infect some Old World monkey cell lines. Human cell lines may also contain unintended infectious agents. These may go undetected as they do not produce cytopathic effect. When handling cell cultures, adherence to BSL-2 precautions is required. PPE should include laboratory coats and gloves. A biosafety cabinet should be used for handling all specimens received for viral testing as well as for handling tissue culture cells, whether inoculated or not. Only fixed tissue culture cells can safely be handled outside the biosafety cabinet. This includes, but is not limited to, inoculation, feeding, passing cells, hemadsorption or hemagglutination, preparation of cell spots, and fixation of cells. It is a good practice to inoculate specimens to cell cultures in a separate cabinet from that used to handle infected cell lines and controls. If this is not possible, then these activities should be performed at separate times within the cabinet, with the cabinet cleaned between sessions.

Shell vials and microtiter plates are centrifuged during inoculation. Centrifuge safety cups should be employed and the safety cups opened in the biosafety cabinet.

Cell cultures may be contaminated on the outside during inoculation, feeding, or passing. The exterior surface of tubes, shell vials, and microtiter plates should be disinfected prior to removal from the biosafety cabinet. Gloves and a laboratory coat should be worn whenever handling the tubes, including reading cultures (4). When removing the lids of culture tubes, the use of a single sterile gauze for each lid can reduce contamination. Only one tube should be opened at a time. Handheld pipetting devices should be used for feeding cells. When inoculating and feeding cells, working with shell vials or microtiter cultures, or working with cultures demonstrating CPE, tissue culture media is often aspirated from the cells. If using a vacuum apparatus to do this, the vacuum system should be protected with a HEPA filter. A liquid disinfectant trap should also be utilized such that the final concentration is the recommended dilution of disinfectant. A two-bottle system should be used to avoid aspiration of contaminated materials into the system.

A molecular diagnostic laboratory or the molecular section of the clinical virology laboratory for the detection of infectious agents from human specimens must operate at BSL-2. Work practices employed for molecular testing to prevent amplicon contamination include many of the same safe work practices, safety equipment, and laboratory facilities as described for BSL-2 laboratories (23). Specimens should be aliquoted for extraction or added to lysis buffer within a biosafety cabinet. Dead air boxes used in the performance of molecular tests function as a shield or splashguard and do not provide protection to the operator. They should not be used for handling unknown or hazardous agents. Mechanical pipettes with barrier tips must be used. Specimen should be added below the level of the lysis buffer to prevent aerosol generation. Due to the sensitivity of molecular methods, it is good practice to change gloves between each specimen to prevent contamination. Automated sample to answer molecular instruments often require decontamination of the instrument between samples. This should be performed according to the manufacturer's recommendation.

Tests performed directly on patient specimens, such as rapid antigen tests, should also be performed in a biosafety cabinet. If a biosafety cabinet is not available for direct specimen testing, the testing can be performed behind a protective shield or splashguard. Gloves must be worn when handling patient specimens (4).

Specimens submitted for viral testing may involve unknown viruses, select agents, or prions. Specimens containing prions can be handled safely at BSL-2 following special work practices. Work should be performed in a biosafety cabinet with the exterior surface of containers disinfected with 2N NaOH. Solid waste should be autoclaved prior to disposal. Liquid waste should be treated with 2N NaOH prior to disposal (24). Unusual cytopathic effect, CPE in an unusual cell type, or failure of a confirmatory test, using either a fluorescent or a molecular method, can be indicative of an unusual agent, an agent of bioterrorism, or an agent of public health importance. These cultures should be handled at BSL-3, and a reference laboratory or local Laboratory Response Network should be consulted for recommendations.

Safe Transport of Specimens/Biological Agents

Specimens or biological agents transported within the facility should be contained within a leak-proof secondary container (11, 20). Zip-lock bags are commonly employed, some with an outer sleeve for separate requisition storage. Specimens transported outside the facility are required to be placed within a secondary leak-proof container labeled with a biohazard warning label in accordance with OSHA Bloodborne Pathogens Standard 1910.1030. Absorbent material capable of absorbing the specimen must be placed between the primary and secondary container. The secondary container must be placed in a durable outer package. Transport of specimens outside the facility is regulated by governmental and regulatory agency regulations. The International Air Transport Association (IATA) and the Department of Transportation (DOT) mandate minimum requirements for shipping within the United States (25, 26). Shipping regulations vary depending upon the dangerous good classification of the specimens. Biological materials fall into the categories of Category A infectious substances, Category B infectious substances, diagnostic specimens, biological products, genetically modified organisms and microorganisms, and unregulated biological materials. Category A infectious substances are capable of causing permanent disability, life-threatening or fatal disease. Many select agents are classified as Category A substances. Category B infectious substances are not generally capable of causing life-threatening or fatal disease. Infectious substances that are not considered to be in Category A are generally classified within Category B. Patient specimens may be Category A, Category B, or exempt human diagnostic specimens (26). Any human excreta, secreta, blood and its components, and tissue and tissue fluids being transported for diagnostic purposes are normally classified as diagnostic and clinical specimens. Packaging and labeling requirements vary depending on the category. Refer to your shipper's requirements for proper shipping and labeling instructions.

Specimens shipped on dry ice must be shipped as Class 9 dangerous goods. Dry ice must be placed outside the secondary container. It should never be handled with bare hands as it will freeze cells and cause injury similar to a burn. Dry ice can be disposed of by allowing it to sublimate in a well-ventilated area. It is an explosion hazard, so the outer package must allow for the release of CO_2. Packages containing dry ice must be labeled as such. Refer to your shipper's requirements for proper shipping and labeling instructions.

Waste Disposal

A waste management plan should identify potentially infectious liquids and solids and outline the proper disposal of each. It should also identify chemical hazards and outline the proper disposal. A "Hazardous laboratory chemicals disposal guide" (16) and chemical safety data sheets are good references for chemical disposal. Specimens must be disposed of as infectious waste and disinfected by autoclaving or disposed as infectious waste to be treated offsite by a medical waste management company. Solid waste such as specimen containers, transport bags, swabs, pipettes, and tissue cultures can also be disinfected by autoclaving or disposed of as infectious waste to be treated offsite by a medical waste management company. Infectious liquid waste, such as waste wash solutions, can be mixed with a disinfectant to its final concentration and disposed of in the sanitary sewer. Most infectious liquid waste from performing viral culture, and antigen and antibody testing can be mixed with household bleach to a final 1:10 dilution prior to disposal. Liquid waste from performing molecular methods may not be compatible with bleach. Guanidine salts found in many lysis buffers and utilized in nucleic acid extraction can generate the toxic fume hydrogen cyanide and hydrochloric acid (15). It is imperative that compatibility between the liquid waste and the disinfectant be determined prior to use. Care must be taken to treat spills and dispose according to manufacturer guidelines.

Biosecurity

A biosafety manual that addresses good laboratory work practices, safety equipment, personal protective equipment, and facility design is no longer adequate without a discussion of biosecurity. Biosecurity addresses the institutional security measures to prevent the loss, theft, or misuse of pathogens. It begins with a thorough risk assessment that includes the identification of pathogens, their quantity, their location, and identification of personnel responsible for them. The CDC and the Animal and Plant Health Inspection Services have identified agents which have the potential to pose a severe threat to the public, animal, or plant health. These agents are designated as Select Agents. Laboratories must be registered to possess, use, or transfer select agents. Additional information on select agents is available at www.selectagents.gov. Biosecurity measures should apply not only to select agents, but to all pathogens. A protocol should be developed for the identification, reporting, investigation, and remediation for breaches in biosecurity, including discrepancies in inventory.

Summary

Clinical virology laboratories often perform a variety of methods for the detection and identification of viral agents, including antigen and antibody detection, traditional viral cell culture, and molecular methods. These different methods encompass a wide variety of hazards, both biological and chemical. It is imperative that provision of a safe work environment include an ongoing assessment of risks as new technologies and methods are implemented in the laboratory. Development of a "Culture of Safety" within the laboratory requires that all workers, not only laboratory directors and management staff, become aware of the hazards and work collectively to mitigate the risks.

REFERENCES

1. **CDC/National Institues of Health.** 2007. *Biosafety in microbiological and biomedical laboratories*, 5th ed. CDC and NIH, Washington, D.C.
2. **Sewell DL.** 1995. Laboratory-associated infections and biosafety. *Clin Microbiol Rev* **8:**389–405.
3. **Pedrosa PB, Cardoso TA.** 2011. Viral infections in workers in hospital and research laboratory settings: a comparative review of infection modes and respective biosafety aspects. *Int J Infect Dis* **15:**e366–e376.
4. **Miller JM, Astles R, Baszler T, Chapin K, Carey R, Garcia L, Gray L, Larone D, Pentella M, Pollock A, Shapiro DS, Weirich E, Wiedbrauk D, Biosafety Blue Ribbon Panel, Centers for Disease Control and Prevention (CDC).** 2012. Guidelines for safe work practices in human and animal medical diagnostic laboratories. Recommendations of a CDC-convened, Biosafety Blue Ribbon Panel. *MMWR Surveill Summ* **61**(Suppl)**:**1–102.
5. **College of American Pathologists.** 2015. Laboratory General Checklist. College of American Pathologists, Northfield, IL.
6. **College of American Pathologists.** 2015. All Common Checklist. **College of American Pathologists,** Northfield, IL.
7. **Rutala WA, Weber DJ, Healthcare Infection Control Practices Advisory Committee.** 2008. Guideline for Disinfection and Sterilization in Healthcare Facilities, 2008. website http://www.cdc.gov/hicpac/Disinfection_Sterilization/3_4surface-Disinfection.html.
8. **Forbes BA, Sahm DF, Weissfeld AS.** 2007. Laboratory Safety, p 45–60. *In Bailey & Scott's Diagnostic Microbiology*, 12th ed. Elsevier, Philadelphia.
9. **Esco Technologies.** 2009. *Inc. A guide to Biosafety & Biological Safety Cabinets. V3 02/09.* Esco Micro Pte. Ltd.
10. **CLSI.** 2012. *Clinical Laboratory Safety; Approved Guideline*—Third edition. CLSI Document GP 17-A3. Clinical and Laboratory Standards Institute, Wayne, PA.
11. **CLSI.** 2011. *Protection of laboratory workers from occupationally acquired infections; Approved guideline*—fourth edition. CLSI

document M29-A4. Clinical and Laboratory Standards Institute, Wayne, PA.

12. **Occupational Safety and Health Administration.** 2016. *Occupational safety and health standards. Exposure to Hazardous Chemicals in Laboratories*. Occupational Safety and Health Administration.
13. **Sigma-Aldrich**. Safety Data Sheet. Sodium azide. V 4.10. 8-21-2015. Sigma-Aldrich Corporation.
14. **Sigma-Aldrich.** Safety Data Sheet. Thimerosal. V. 3.7. 12-29-2015. Sigma-Aldrich Corporation.
15. **Wu X, Paik S.** 2005. Measuring toxic gases generated from reaction of guanidine isothiocyanate-containing reagents with bleach. *Chem Health Saf* **12:**33–38.
16. **Armour M-A.** 2003. *Hazardous Laboratory Chemicals Disposal Guide*. CRC Press, Boca Raton, Florida.
17. **Occupational Safety and Health Administration.** 2011. *Laboratory Safety Guidance*. Occupational Safety and Health Administration.
18. **World Health Organization.** 2004. *Laboratory Biosafety Manual*, 3rd ed. World Health Organization, Geneva, Switzerland.
19. **Public Health Agency of Canada.** 2015. Pathogen Data Safety Sheets and Risk Assessment. http://www.phac-aspc.gc.ca/lab-bio/res/psds-ftss/ http://www.phac-aspc.gc.ca/lab-bio/res/psds-ftss/.
20. **Occupational Safety and Health Administration**. 2014. Occupational safety and health standards. Toxic and hazardous substances. Bloodborne pathogens. Standard no. 1910.1030.
21. **Centers for Disease Control.** 1987. Recommendation for prevention of HIV transmission in health-care settings. *Morbidity and Mortality Weekly Report* **21:**1S–18S.
22. **Baron EJ.** 2015. Specimen Collection, Transport, and Processing: Bacteriology, p 270–315. *In* Jorgensen JH, Pfaller MA, Carroll KC, Funke G, Landry ML, Richter SS, Warnock DW (ed), *Manual of Clinical Microbiology*, 11th ed. ASM Press, Washington, D.C.
23. **Mifflin TE.** 2003. Setting up a PCR laboratory, p 5–14. *In* Dieffenbach CW, Dveksler GS (ed), *PCR Primer: a Laboratory Manual*. Cold Spring Harbor Laboratory Press, Cold Spring Harbor, NY.
24. **World Health Organization.** WHO Infection Control Guidelines for Transmissible spongiform Encephalopathies. WHO/CDS/RAPH/**2000**.3. 2000. Geneva, Switzerland.
25. **Denys GA, Gray LD, Snyder JW.** 2004. *Cumitech 40, Packing and Shipping of Diagnostic Specimens and Infectious Substances*. ASM Press, Washington, D.C.
26. **Gray L. Snyder JW.** Packing and Shipping Infectious Substances. 2012. American Society for Microbiology. *Sentinel Level Clinical Laboratory Guidelines for Suspected Agents of Bioterrorism and Emerging Infectious Diseases*.

5

Laboratory Design

MATTHEW J. BINNICKER

Planning and executing the design of a clinical laboratory is a unique and challenging task. This process not only involves coordinating how clinical samples will be received and tested, but also how efficiency can be maximized, safety ensured, and flexibility maintained. The design of a modern-day clinical virology laboratory can be especially challenging, given the continued application of traditional viral cell culture in some laboratories and the increasing use of molecular techniques (e.g., real-time PCR) for the diagnosis of viral infections. If there is one truth regarding laboratory design, it is that there is no "one size fits all" approach. Before a laboratory can begin to discuss the ideal approach for virologic testing at its institution, a number of important issues must be considered. These issues include the space that is (or will be) available, the number of laboratory staff that will occupy the space, the testing that will be performed, and the patient population from whom samples will be collected and submitted for testing. Furthermore, important decisions need to be made by laboratory and institutional leadership regarding whether testing should be performed in a centralized (i.e., consolidated) laboratory, a decentralized (i.e., specialized) laboratory, or a combination of the two. Another important consideration, driven by the increasing development of rapid, sample-to-answer molecular devices, centers around whether testing should be performed "near the patient" or at the "point of care." Addressing each of these important issues is outside the scope of this chapter. However, key components of laboratory design, especially as they pertain to clinical virology, will be discussed to provide laboratory professionals with a foundation and guide to help ensure that test results are accurate, laboratory staff are safe, and future growth can be accommodated.

THE PHASES OF LABORATORY DESIGN

Planning

The first, and potentially most important, phase of designing a clinical laboratory is planning and programming (Fig. 1). In the beginning of this phase, it is essential to identify a project team that will coordinate the planning efforts and oversee each step of the process. At a minimum, the project team should consist of representation from laboratory leadership (e.g., laboratory director and supervisor), laboratory administration, the facilities department, and the architectural/engineering firm. Ideally, the architectural firm will have experience in the design of clinical laboratory space and be knowledgeable about the requirements from state, federal, and accrediting agencies (e.g., College of American Pathologists, Occupational Safety and Health Administration) regarding accepted guidelines for laboratory design (1). Once the project team is formed, it is important to assess the advantages and limitations of existing space, if applicable. This will allow the future design to incorporate features that are known to work well, while improving on areas where the current operation falls short. In addition to reviewing existing space, it may also be beneficial to perform a benchmarking analysis of similar laboratories in the region. This analysis may involve the project team participating in "site visits" to observe other laboratory operations and discuss features of laboratory designs at other institutions. Another critical component of the planning phase is to gather feedback from the laboratory staff that will be occupying the future space (2). Laboratory technologists are often best suited to provide guidance on the features of a clinical laboratory that ensure proper workflow and employee safety. The planning phase will allow for the completion of a space program, which will summarize the estimated square footage required for each area in the laboratory. From the details outlined in the space program, representatives from the facilities department will be able to develop an initial cost estimate for the project. It is imperative that the project team enter the planning phase with an open mind, willing to "think outside the box" and always keeping in mind the impact that future changes (e.g., test volume changes, new technology) may have on the laboratory operation.

Design

Following the planning phase, the project enters a more detailed stage in which ideas, concepts, and data are translated into a working schematic design (Fig. 1). During the design phase, the project team should maintain flexibility and continue to critically assess new ideas; however, the ultimate goal of the design phase is to develop a floor plan that will illustrate the relative position of laboratory equipment, benches, doors, and windows. Furthermore, the design phase will begin to define the layout of the laboratory, as well as the proposed workflow of clinical samples and laboratory

doi:10.1128/9781555819156.ch5

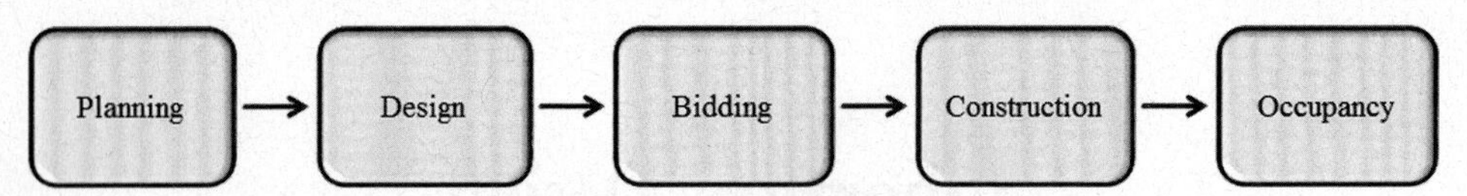

FIGURE 1 The phases of laboratory design.

personnel. An important consideration during the design phase is whether fixed or moveable casework will be used. Moveable casework is advantageous because it allows for future redesigns and the incorporation of new laboratory equipment at a fraction of the cost compared to fixed casework. The schematic design should be reviewed by the project team and laboratory staff to further refine the plan. Subsequently, the project enters the design development phase, in which more specific details are discussed, including the types of finishes, casework, utilities, and laboratory equipment that will be incorporated into the space.

Bidding

Once the schematic and design development stages are complete, the information should be distributed to at least three construction firms to initiate the bidding process. During this phase, cost estimates for completing the project are submitted from potential construction firms. Once the bids are received, modifications to the design and scope of the project may possibly be required, especially if the cost estimate exceeds the budgeted allocation. When the construction firm is selected, the project team may also complete a process known as value engineering, in which the budget for the project can be reduced by deferring certain unnecessary items (e.g., intercom systems) or identifying less expensive options that serve a similar function to more costly alternatives (e.g., stainless steel versus Corian countertops). The construction bid should always include a contingency fund (typically 5% to 10% of the total estimate) to cover any unforeseen problems or expenses that arise during construction.

Construction

The construction phase of a new laboratory can represent one of the most exciting, yet stressful, stages of the project (Fig. 1). It is important for the project team to develop a detailed timeline, with approved milestones and deadlines for the construction firm. Regular updates should be provided to the laboratory staff and institutional leadership to assure these groups that the project is on budget and within scope and is meeting established milestones. Regular communication with the laboratory staff is extremely important so that they are aware of the progress being made. In cases in which an existing laboratory is being remodeled or expanded, the construction will likely need to be completed in a phased approach so that laboratory testing can continue while the construction is in progress. This can be especially challenging, but developing a detailed phasing plan will help ensure that testing is not compromised, employee safety is maintained, and the project is completed on time.

Occupancy

The project team should begin planning for the occupancy phase well in advance of the initiation of construction. This process should include a detailed plan of how and when equipment, instrumentation, and reagents will be relocated

TABLE 1 Biosafety level requirements in the clinical virology laboratory

	Biosafety level[a]			
	1	2	3	4
Airlock present	No	No	Yes	Yes
Anteroom	No	No	Yes	Yes
Autoclave				
• On site	No	Optional	Yes	Yes
• In-room	No	No	Optional	Yes
• Double-ended	No	No	Optional	Yes
Biological safety cabinet	No	Yes	Yes	Yes
Physical separation of laboratory	No	No	Yes	Yes
Room sealed for decontamination	No	No	Yes	Yes
Ventilation				
• Negative pressure	No	Yes	Yes	Yes
• HEPA-filtered	No	No	Yes	Yes
Viruses that can be manipulated[b]		Adenovirus, herpes viruses (CMV, EBV, HSV, VZV), enteroviruses, hepatitis viruses (A–E)	Arboviruses (WNV), MERS, Nipah, rabies, HIV, avian influenza	Ebola virus, Marburg, Lassa, smallpox

[a]CMV, cytomegalovirus; EBV, Epstein-Barr virus; HSV, herpes simplex viruses 1 and 2; MERS, Middle Eastern Respiratory Syndrome coronavirus; VZV, varicella zoster virus; WNV, West Nile virus.
[b]Propagated in viral cell culture and/or worked with as a live virus in clinical samples.

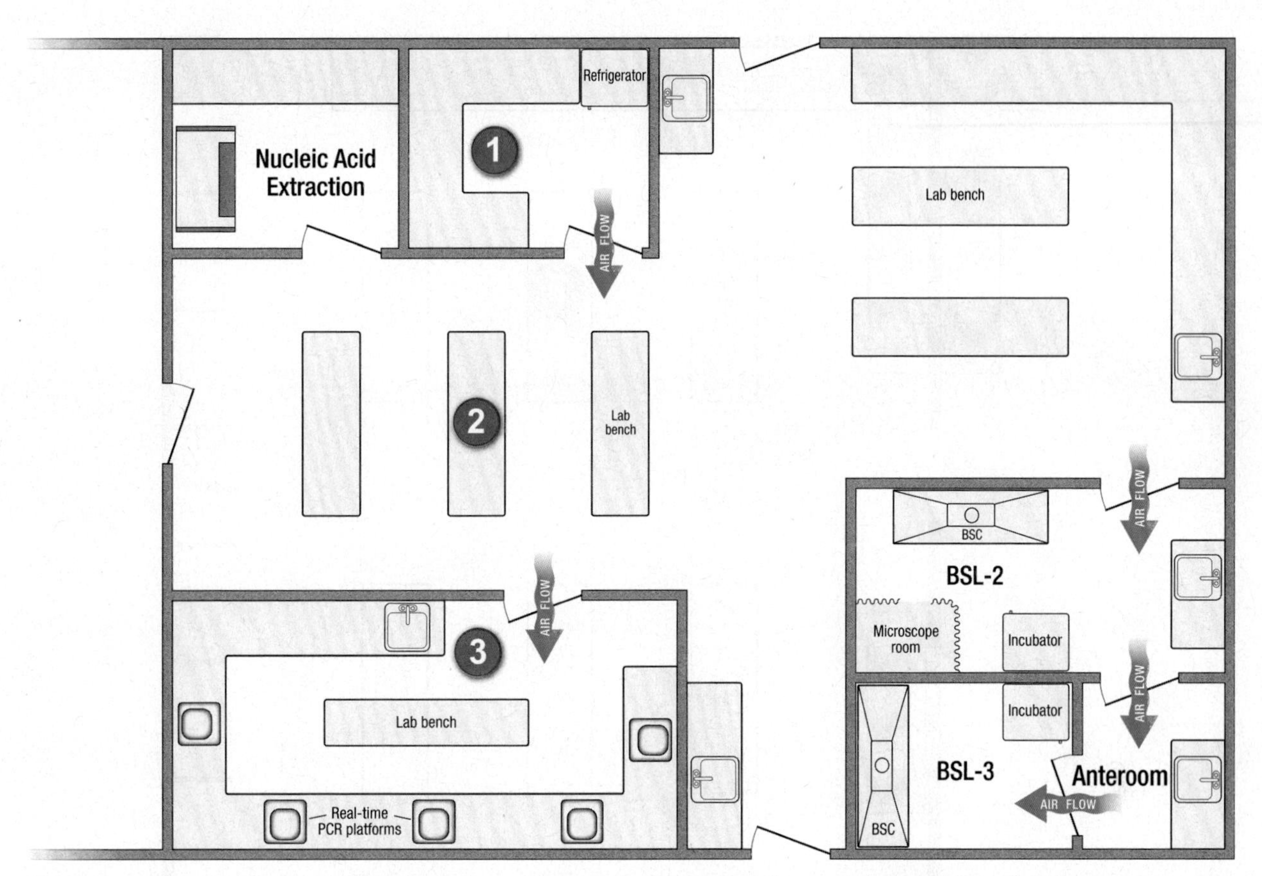

FIGURE 2 Laboratory design schematic for performing both molecular testing and routine viral cell culture. Sample processing is often performed inside a biosafety cabinet (BSC). For molecular testing, reagent and master mix preparation is ideally performed in a separate walled-in room that has positive pressure in relation to the remainder of the laboratory (1). Nucleic acid extraction may be performed in a unique or shared space as long as strict unidirectional workflow is followed. Test setup, which includes loading the master mix and nucleic acid extract into the test cartridge, is then completed in a separate area (2). Finally, PCR is performed in a third walled-in space that has negative air pressure (3). No material (e.g., laboratory coats, gloves, work books, equipment) in the PCR analysis room should be removed from this space without proper decontamination. For viral cell culture testing, work should be performed in a biosafety level 2 (BSL2) or BSL3 facility using an appropriate BSC. The airflow should maintain negative pressure in relation to the remainder of the laboratory.

to the new laboratory space. Furthermore, it is critical that the project team develop a phasing plan, so that laboratory services are not disrupted during the move. This approach will allow the team to identify the areas that should be operational first and plan accordingly. Note that instrumentation and general laboratory equipment must be validated following installation (or relocation) in the new space. The validation process must be completed prior to the commencement of patient testing (3). A final inspection of the new laboratory space should be performed before occupancy to commission the laboratory and ensure that the facility, including ventilation, biosafety cabinets, and airflow, is working properly.

IMPORTANT ELEMENTS OF LABORATORY DESIGN FOR CLINICAL VIROLOGY

A number of important issues must be carefully considered when designing a clinical virology laboratory. These include the test menu, the patient population, and whether the laboratory will be part of a centralized (consolidated) or decentralized (specialized) operation (4). In regards to the test menu, the clinical virology director and/or supervisor should determine whether testing will consist of conventional methods (e.g., viral cell culture, microscopy, antigen testing), contemporary techniques (e.g., molecular testing), or a combination of the two. Defining the test menu will help determine the type of laboratory infrastructure that is required, as well as the ideal workflow. A second important consideration is the patient population from which the laboratory will receive samples for testing. Important questions include, "Will the primary patient population be from the local community, or will the laboratory serve as a reference laboratory for regional, national, or international patients?" Answering these questions will help define the types of infectious diseases the laboratory may encounter and therefore assist the project team in planning for an appropriate laboratory space.

A third element of laboratory design that must be discussed in detail is ensuring the safety of laboratory personnel. Viral pathogens such as influenza and novel coronaviruses (e.g., severe acute respiratory syndrome, Middle Eastern respiratory syndrome [MERS]) pose a substantial risk to laboratory staff, and the appropriate precautions must be taken to prevent laboratory-acquired infections (5). If a virology laboratory will be manipulating raw clinical samples and propagating viruses in cell culture, then the work must be performed in the appropriate biosafety level (BSL) using

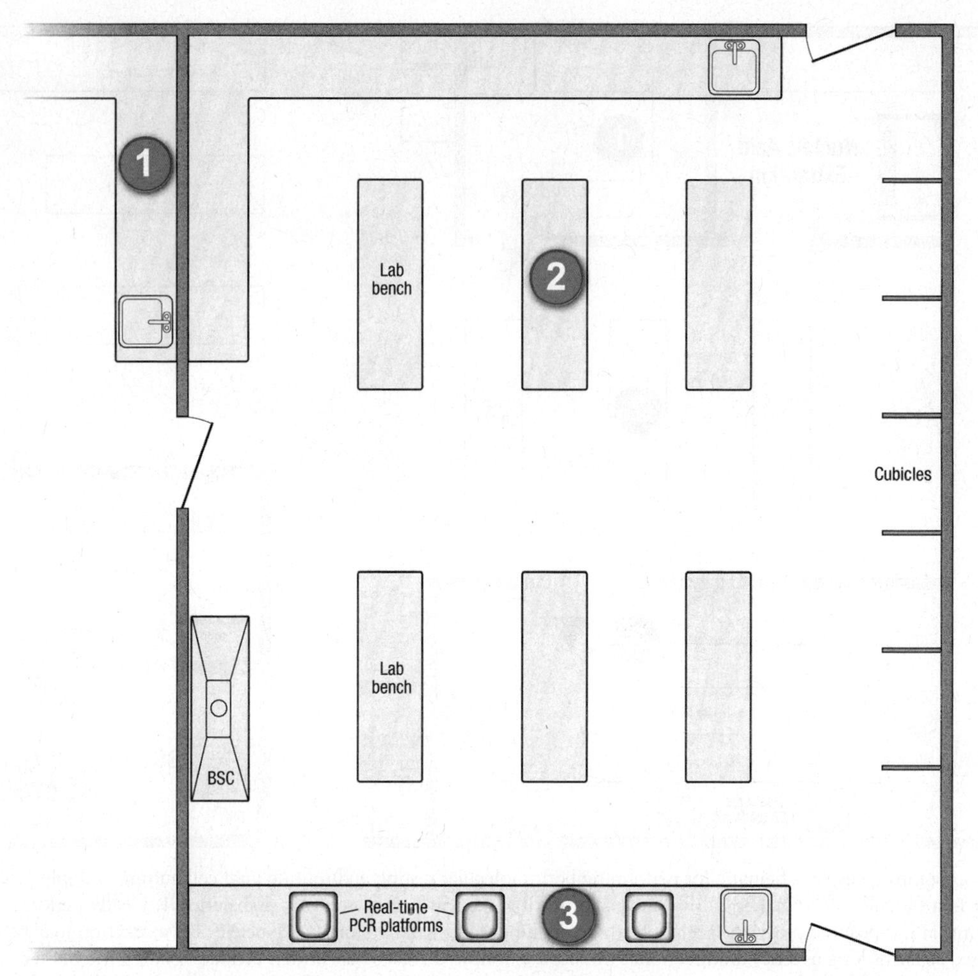

FIGURE 3 Laboratory design schematic for performing both molecular testing and routine viral cell culture in a common area. For molecular testing, reagent and master mix preparation as well as nucleic acid extraction should be performed in a separate area, preferably in a different room (1). Test setup (2) and PCR amplification/analysis (3) may be performed in a common space, but testing should follow a strict unidirectional workflow. BSC, biosafety cabinet.

the correct tier biosafety cabinet (BSC) (Table 1). At a minimum, a diagnostic virology laboratory must meet the requirements to operate at a BSL 2 or above (5). This involves laboratory staff wearing personal protective equipment and performing testing inside a BSC if there is the potential that aerosols will be generated. For laboratories that perform viral cell culture, the manipulation of clinical samples and inoculated cell culture tubes must be performed inside a BSC in a room with negative air pressure in relation to surrounding work units and hallways. Furthermore, the laboratory design should include eye washes and safety showers to protect laboratory personnel in the event of a spill or splash with chemicals or biohazardous material.

Specimens collected from patients under investigation for emerging viral pathogens (e.g., MERS-coronavirus, avian influenza) should be performed in a BSL3 facility (Table 1). Testing of samples collected from patients under investigation for a BSL4 pathogen (e.g., Ebola virus) may be performed in a BSL2 facility; however, laboratories should follow Centers for Disease Control and Prevention (CDC)-specific guidance based on patient risk factors. If preliminary testing suggests the infection may be due to a BSL4 pathogen, confirmatory testing should be completed at an approved state or local health department laboratory or at the CDC where BSL4 precautions can be applied.

WORKFLOW

Designing a laboratory with proper workflow can help maximize efficiency, reduce contamination, and enhance safety. Taking the time to carefully consider workflow can also have a significant impact on reducing expenses and improving staff satisfaction. Issues such as the proximity of the clinical virology laboratory to other areas in microbiology, especially the specimen receiving and initial processing areas, should be discussed. Integrating the virology laboratory with other areas that share specimens and instrumentation can reduce turnaround time and increase communication among laboratory personnel.

The issue of proper workflow has become of paramount importance as molecular diagnostic techniques (e.g., real-time PCR) have become commonplace in today's clinical virology laboratory. This is due to the exquisite sensitivity of molecular methods and the potential that trace amounts of raw clinical samples containing viral particles (e.g., BK virus) or amplicon contamination from prior rounds of PCR may impact the results of patient testing. Because of these concerns, a number of guidance documents have been published on the subject of proper workflow in the area of molecular diagnostics (6–9). It is now accepted that a laboratory performing molecular testing should have separate areas for reagent preparation, nucleic acid extraction, test setup, and molecular amplification. Ideally, these areas would be physically separated, with individual rooms for each activity (Fig. 2). Preparing reagents and master mix in a separate room with positive pressure in relation to the remainder of the laboratory can help prevent the contamination of reagents, which may render significant downstream consequences on testing and patient results. Similarly, having a specified BSC for nucleic acid extraction and a walled-in area for PCR amplification and post-PCR analysis is ideal. This room should have negative pressure in relation to surrounding areas, and all material inside this room, including laboratory coats, gloves, and equipment should not leave the space unless thoroughly decontaminated. If space and resources are limited, these activities can be completed in a common space, but staff should make every effort to physically separate the areas as much as possible (Fig. 3). Regardless, when molecular testing is performed in the clinical virology laboratory, it is essential that laboratory staff follow a unidirectional workflow (reagent preparation → nucleic acid extraction → test setup → post-PCR) to minimize the potential for specimen and/or amplicon contamination (Fig. 2 and 3). Unidirectional workflow should also be followed for viral culture techniques to prevent the contamination of uninoculated reagents with clinical specimens or amplified virus (10).

In recent years, clinical laboratories have implemented design concepts that reduce cost and waste by optimizing efficiency (11). One philosophy to accomplish these goals is the concept of LEAN design, which was first introduced in the manufacturing industry but has gained significant interest in the clinical arena (2). Laboratories introducing LEAN concepts attempt to design a facility that reduces or eliminates waste, standardizes work practices, and maximizes productivity (12). This process uses data to drive decisions. For example, while designing a new clinical laboratory, the project team decides to complete an exercise known as "value stream mapping." In this exercise, the team gathers data on how different designs impact efficiency. In Design A, it is determined that a laboratory technologist will be required to visit three different stations, and therefore, take approximately 150 steps to complete a certain task. In comparison, Design B allows the technologist to finish the task while visiting only two stations and taking less than 50 steps. These data assist the project team in determining which design would yield maximal efficiency. Other examples in which LEAN principles can impact design include the storage of reagents/supplies and planning for redundancy (back-ups) in the laboratory. Rather than storing months of reagents in the laboratory, a LEAN approach would be to store minimal supplies on hand and to have an automated process for ordering new supplies when they reach a predefined level. This approach would minimize the amount of casework that is required and the amount of space that is taken up by laboratory supplies. Similarly, reducing the amount of redundancy in the laboratory (e.g., back-up instrumentation) to only that which is required can maximize the amount of usable space.

SUMMARY

Designing a clinical virology laboratory is a challenging task that requires a significant amount of planning and coordination. Virology labs that perform traditional cell culture should have the appropriate facilities and infrastructure to keep personnel safe and prevent laboratory-acquired infections. The adoption of molecular techniques for the diagnosis of viral infections has revolutionized the field; however, it also creates substantial challenges in terms of laboratory design and workflow so that contamination risks are mitigated. Laboratories using molecular methods should have separate areas for reagent preparation, test setup, and post-PCR analysis. Furthermore, the laboratory should follow a strict unidirectional workflow to minimize the risk of specimen and/or amplicon contamination. In the future, it will be interesting to observe how new technologies impact the design and operation of the clinical virology laboratory. It is likely that testing will migrate closer to the patient as new technologies become even more rapid, less expensive, and provide direct "sample-to-answer" results.

REFERENCES

1. **College of American Pathologists.** Commission on Laboratory Accreditation. 2013. Laboratory general checklist. College of American Pathologists, Northfield, IL.
2. **Clinical and Laboratory Standards Institute (CLSI).** 2007. *Laboratory design; approved guideline*, 2nd ed. CLSI document QMS04-A2. Clinical and Laboratory Standards Institute, Wayne, PA.
3. **College of American Pathologists.** Commission on Laboratory Accreditation. 2013. Microbiology checklist. College of American Pathologists, Northfield, IL.
4. **Kurec AS, Lifshitz MS.** 2011. General concepts and administrative issues, p. 7–11. *In* McPherson RA, Pincus MR(ed), *Henry's clinical diagnosis and management by laboratory methods*, 22nd ed. Elsevier Saunders, Philadelphia, PA.
5. **World Health Organization (WHO).** 2008. *Guidelines on establishment of virology laboratory in developing countries*. World Health Organization Regional Office for Southeast Asia, New Delhi, India.
6. **Dieffenbach CW, Dveksler GS.** 1993. Setting up a PCR laboratory. *PCR Methods Appl* **3:**S2–S7.
7. **Mifflin TE.** 2003. Setting up a PCR laboratory, p. 5–14. *In* Dieffenbach CW, Dveksler GS (ed), *PCR primer*, 2nd ed. Cold Spring Harbor Laboratory Press, Cold Spring Harbor, NY.
8. **Viana RV, Wallis CL.** 2011. Good clinical laboratory practice (GLCP) for molecular based tests used in diagnostic laboratories, p. 29–52. *In* Akyar I (ed), *Wide spectra of quality control*. Intech, Rijeka, Croatia.
9. **Public Health England.** 2013. UK standards for microbiology investigations; Good laboratory practice when performing molecular amplification assays. Standards Institute, Microbiology Services, Public Health England, London, United Kingdom.
10. **Warford A.** 2000. Quality assurance in clinical virology, p. 9. *In* Specter S, Hodinka RL, Young SA (ed), *Clinical virology manual*, 3rd ed. ASM Press, Washington, DC.
11. **Gomah ME, Turley JP, Lu H, Jones D.** 2010. Modeling complex workflow in molecular diagnostics: design specifications of laboratory software for support of personalized medicine. *J Mol Diagn* **12:**51–57.
12. **Mitchell PS, Mandrekar JN, Yao JD.** 2014. Adoption of lean principles in a high-volume molecular diagnostic microbiology laboratory. *J Clin Microbiol* **52:**2689–2693.

SECTION II

Laboratory Procedures for Detecting Viruses

Specimen Requirements Selection, Collection, Transport, and Processing

REETI KHARE AND THOMAS E. GRYS

6

Proper specimen collection is essential for interpreting test results because of the wide range of viruses, the complexity of virus–host interactions, and changes in testing methodology. Some viruses are part of our normal flora and cause no symptoms. Others, such as human herpes virus 6 (HHV-6), can integrate into germ line cells and be transmitted vertically to children (1, 2). Chronic, suppressed, or latent viruses like human immunodeficiency virus (HIV) or BK virus (BKV) can be reactivated to cause transient, low-level viremia that may or may not need clinical attention (3, 4). Asymptomatic shedding can also result in detection of virions for long periods of time (5–7). In order to recognize and interpret infectious episodes, it is crucial for diagnostic laboratories to receive appropriate specimens for viral testing.

Suitable specimens are dependent on the methods used for testing. Molecular, antigen, culture, serology, or histopathology assays each have different performance characteristics and volume requirements. Importantly, each method detects viral infections in different ways and therefore might require different specimen sources. For instance, using a readily obtained specimen source (serum), serology is a valuable tool to measure a host response (antibodies) against some pathogens like chikungunya virus that can cause a systemic viremia (8). On the other hand, for localized disease, such as a herpetic lesion, culture is useful for detecting the presence of actively replicating virus particles. Molecular assays can be particularly useful when serum is unavailable and there is no chance of a viable virus, such as when only formalin-fixed, paraffin-embedded tissue is available (9). Currently, the landscape of viral diagnostics is changing rapidly and an increasing number of laboratories are transitioning to molecular methods. Early application of methods like polymerase chain reaction (PCR) to the field of clinical virology tended to be in the form of laboratory-developed tests with varying performance characteristics and high interassay variability (10, 11). However, the utilization of international standards and commercial assays cleared or approved by the Food and Drug Administration (FDA) for targets such as HIV, hepatitis C virus (HCV), hepatitis B virus (HBV), and cytomegalovirus (CMV) has enabled consistency and standardization across facilities (12).

Recent advances in molecular technologies have also encouraged development of “syndromic” testing. Viral culture using mammalian cells was once the main method for a broad investigation of viral pathogens, but today, a single multiplex PCR assay can detect 10 to 20 infectious organisms from a small amount of sample with high sensitivity and same-day results (13). This approach to diagnostic testing significantly affects the types and amounts of specimens accepted by the laboratory.

IMPORTANCE OF TEST SELECTION, SPECIMEN SELECTION, AND SPECIMEN COLLECTION

The primary purpose of the diagnostic laboratory is to provide actionable results to providers of patient care. For this to happen, the entire process of test ordering, specimen collection and transport, analytical testing, and reporting needs to happen with efficiency and quality. Laboratories can only generate useful and reliable results when they receive patient specimens from an appropriate source, in the right amount, and at the correct time of illness. These parameters should be collated into an electronic ordering process, a test catalog, or another similar resource that is readily available to providers. Information regarding clinical symptoms, associated viral etiologies, optimal specimen types, methods of collection, methods of detection, volumes required, and containers for transport should be included. An overview of specimens collected from different viral syndromes is represented in Tables 1–12. Together, this information gives clinicians guidelines for interpretation of assay results, and should therefore be updated periodically. Ideally, electronic ordering of laboratory tests will require that a specimen can be submitted only according to the appropriate conditions and correctly labeled so that the specimen is processed for optimal testing. With the advent of multiplex syndromic panels, additional efforts are required by the laboratory to guide providers to use them appropriately. These tests are typically costly to the patient, but when used appropriately, the multiplex panels may save time, lead to a rapid treatment, and create opportunities to reduce

doi:10.1128/9781555819156.ch6

TABLE 1 Specimen Information for respiratory disease [e.g., pharyngitis, laryngotracheobronchitis (croup), bronchitis, bronchiolitis, pneumonitis]

Viral pathogen	Nucleic acid testing	Serology	Culture	Other comments
Influenza virus Parainfluenza virus Respiratory syncytial virus Human metapneumovirus Coronavirus Rhinovirus Enterovirus Adenovirus	Nasal swab NP swab, wash, or aspirate Throat swab BAL Lung tissue Sputum		Nasal swab NP swab, wash, aspirate Throat swab BAL Lung tissue Sputum	Coronaviruses are not cultured on routine laboratory cell lines.
Severe acute respiratory syndrome virus	Tracheal aspirate NP aspirate Throat, nasal swab Throat wash Stool		Culture is not recommended	Tracheal aspirate or stool is the preferred specimen. Consult with public health department.
Middle Eastern respiratory syndrome virus	NP and OP swabs NP wash, aspirate Sputum BAL Tracheal aspirate Pleural fluid Serum, blood	Serum	Culture is not recommended	Serology should be used for epidemiological rather than diagnostic purposes. Consult with public health department.
Herpes simplex virus	Respiratory washes BAL Throat swab Tissue		Respiratory washes BAL Throat swab Tissue	
Cytomegalovirus	Throat washes Blood Plasma Urine Ocular Amniotic fluid BAL		Respiratory washes Throat swab BAL	
Mumps virus	Buccal swab Stensen's duct exudates Throat swab Urine	Serum	Buccal swab Stensen's duct exudates Throat swab Urine	

NP, nasopharyngeal; BAL, bronchoalveolar lavage; OP, oropharyngeal.

inappropriate use of antimicrobials. Laboratory staff plays a critical role in consulting on the use of these tests.

HOST FACTORS THAT AFFECT SPECIMEN COLLECTION

Selecting the appropriate specimens depends heavily on the host, the clinical presentation, and the suspected viral pathogen. Host factors, such as age and immune status, can significantly affect specimen collection conditions. For instance, several genotypes of human papillomavirus (HPV), like HPV16 and HPV18, are implicated in the development of cervical cancer. The incidence of HPV in adolescent women is extremely common, but 91% of HPV-positive patients with abnormal cytology results will regress to normal cytology within 3 years (14). Therefore, the American Society of Colposcopy and Cervical Pathology (ASCCP) does not recommend HPV cytological screening for women <21 years old or routine HPV DNA testing for women <30 years old (15). The laboratory should have specimen rejection criteria to reflect these recommendations because unnecessary testing of young women may result in unneeded colposcopies, biopsies, and undue harm.

Herpes simplex virus (HSV) is another pathogen for which the age and immune status of the patient may influence the diagnostic approach. Testing for HSV in the peripheral blood of immunocompetent individuals is another example of unnecessary evaluation as the yield is generally low and results do not always correlate with clinical findings. In pediatric patients with primary HSV gingivostomatitis,

TABLE 2 Specimen information for exanthems (e.g., maculopapular, diffuse, dermatomal, vesicular)

Viral pathogen	Nucleic acid testing	Serology	Culture	Other comments
Herpes simplex virus	Vesicular fluid Dermal and genital swabs, scrapings, tissue from base or unroofed lesion	Serum	Vesicular fluid Dermal and genital swabs, scrapings, tissue from base or unroofed lesion	Vesicle scrapings for Tzanck smears have low sensitivity and specificity. Compared to PCR, culture has poor sensitivity and turnaround time.
Varicella-zoster virus	Vesicular fluid Dermal and genital swabs, scrapings, or tissue from base or unroofed lesion	Serum	Vesicular fluid Dermal and genital swabs, scrapings, or tissue from base or unroofed lesion	Vesicle scrapings for Tzanck smears have low sensitivity and specificity. Compared to PCR, culture has poor sensitivity and turnaround time.
Enterovirus	Throat swab Vesicular fluid Blood, plasma, serum Rectal swab/stool CSF		Throat swab Rectal swab/Stool Vesicular fluid	
Parvovirus B19	Bone marrow Amniotic fluid Blood, plasma, serum	Serum		Serology is the preferred method of diagnosis. Appearance of characteristic "slapped cheek" rash can be diagnostic.
Rubella virus	Throat or nasal swab Amniotic fluid Urine	Serum. The presence of IgM may indicate acute infection, but can be negative during early infection.	Viral culture may be performed on throat, nasal, urine, and amniotic fluid specimens for epidemiologic purposes.	Serology is the most common method of diagnosis and should be performed with paired and convalescent sera. PCR is available at the CDC.
Measles virus (rubeola)	Throat, nasal, or nasopharyngeal swabs Urine	Serum. The presence of IgM may indicate acute infection, but can be negative during early infection and can persist for up to 2 months.	Viral culture may be performed on oropharyngeal or urine specimens for epidemiologic purposes.	Serology should be performed with paired and convalescent sera. Collect clinical specimens within 7 days of onset of the rash. PCR is available at the CDC.
Human T lymphotropic virus type I and II	Blood	Serum. Serology is generally used as the primary screen.		Consider performing PCR in cases where serology is negative but clinical suspicion of infection is high.
Human herpesvirus 6	PBMCs Blood, plasma, serum CSF	Serum. Serology is only useful for primary diagnosis of young infants.		PCR of PBMCs can be used to detect primary infection, reactivation, or chromosomally integrated viral DNA.
Human herpesvirus 7	PBMCs Blood, plasma, serum	Serum		Serology may cross-react with HHV-6.
Human herpesvirus 8	Blood, plasma, serum	Serum		PCR is typically positive during active infection.
Molluscum contagiosum virus				Usually diagnosed clinically (e.g., classic appearance of waxy, convex papules with central umbilication) Histologic/cytologic analysis may be performed from biopsy or fine needle aspiration of the lesions.
Human papillomavirus	Viral typing can be performed from lesion biopsies, but is not typically done.			Lesion biopsy can be assessed by histologic methods.
Adenovirus	Throat swab		Throat swab	

CSF, cerebrospinal fluid; IgM, immunoglobulin M; PCR, polymerase chain reaction; CDC, Centers for Disease Control and Prevention; PBMCs, peripheral blood mononuclear cells; HHV-6, human herpesvirus 6.

TABLE 3 Specimen information for neurological infections (e.g., meningitis, meningoencephalitis, encephalitis, encephalopathy)

Viral pathogen	Nucleic acid testing	Serology	Culture	Other comments
Herpes simplex virus	CSF Brain biopsy	CSF Serum	Brain biopsy	Culture of CSF has poor sensitivity. PCR is the preferred method of detection.
Enterovirus	CSF Brain biopsy Blood, plasma, serum Urine Stool or rectal swab Throat swab		Brain biopsy Serum Urine Stool or rectal swab Throat swab Rectal swab	Culture of CSF has poor sensitivity. PCR is the preferred method of detection.
West Nile virus	Blood, plasma, serum CSF Tissue from brainstem and spinal cord	Serum CSF		Viremia may be brief, so molecular assays from blood and CSF should only be performed within 4 days of symptom onset. Serology may cross-react with other flaviviruses or may be negative if tested during early disease; retesting may be indicated.
California (La Crosse) encephalitis St. Louis encephalitis Western equine encephalitis Eastern equine encephalitis Powassan virus Bourbon virus Heartland virus Venezuelan encephalitis	Serum CSF	Serum CSF		PCR assays are available at some local and state public health laboratories and CDC.
Dengue virus	Serum CSF	Serum		Viremia may be brief, so molecular assays should only be performed within 5 days of symptom onset. Serology may cross-react with other flaviviruses or may be negative if tested during early disease; retesting may be indicated.
Chikungunya virus	Serum CSF	Serum		PCR and serological assays are available at some state public health laboratories, CDC, and commercial laboratories. Viremia may be brief, so molecular assays should only be performed within 8 days of symptom onset. Serology may cross-react with other alphaviruses.

(*Continued on next page*)

Epstein-Barr virus	Blood, plasma, serum CSF PBMCs	Serum		
Varicella-Zoster virus	CSF Dermal			Viral culture of CSF has poor sensitivity.
Cytomegalovirus	CSF Amniotic fluid			Viral culture of CSF has poor sensitivity.
Measles virus (Rubeola)	CSF Brain biopsy	Serum CSF		
Rabies virus	Skin biopsy Saliva	Serum CSF	Saliva	Immunohistochemistry and immunofluorescence can be performed on brain biopsies. Antigen testing can be performed on skin biopsies taken from the base of hair follicles. Contact the CDC for further testing information.
Mumps virus	Buccal swab Stensen's duct exudates Throat swab Urine	Serum	Buccal swab Stensen's duct exudates Throat swab Urine	
JC virus	CSF			PCR from CSF may not detect all cases of progressive multifocal leukoencephalopathy. Follow up a negative PCR result with histopathology of brain biopsy if clinical suspicion is high.
Human T-lymphotropic virus type I and II	Blood	Serum		Serology is generally used as the primary screen. Consider performing PCR in cases where serology is negative but clinical suspicion of infection is high.
Lymphocytic choriomeningitis virus		Serum CSF		

CSF, cerebrospinal fluid; PCR, polymerase chain reaction; CDC, Centers for Disease Control and Prevention; PBMCs, peripheral blood mononuclear cells.

TABLE 4 Specimen information for infections with joint pain (e.g., fever, polyarthralgia, polyarthritis, tenosynovitis)

Viral pathogen	Nucleic acid testing	Serology	Culture	Other comments
Dengue virus	Serum CSF	Serum. Serology may cross-react with other flaviviruses or may be negative if tested during early disease; retesting may be indicated.		Viremia may be brief, so molecular assays should only be performed within 5 days of symptom onset.
Chikungunya virus	Serum	Serum. Serology may cross-react with other alphaviruses		PCR and serological assays are available at some state public health laboratories, CDC, and commercial laboratories. Viremia may be brief, so molecular assays should only be performed within 8 days of symptom onset.
Parvovirus B19	Synovial fluid Blood, plasma, serum	Serum		Serology is the preferred method of diagnosis.
Epstein-Barr virus	Blood Plasma	Serum		Serologic testing for EA IgG, EBNA IgG, VCA IgG, VCA IgM is indicated for those patients with heterophile-negative (rapid test method) determinations.
Human immunodeficiency virus type 1 and 2	Blood, plasma, serum	Blood, plasma, serum. Serology with an antigen/antibody immunoassay is recommended as the initial screen.		Serology results should be confirmed with an HIV1/2 differentiation immunoassay. Nucleic acid assays can be used for confirmation of acute HIV infection as well as for monitoring.
Rubella virus	Throat or nasal swab Urine	Serum. The presence of IgM may indicate acute infection, but can be negative during early infection.	Viral culture may be performed on throat, nasal, urine and amniotic fluid specimens for epidemiologic purposes.	Serology is the most common method of diagnosis and should be performed with paired and convalescent sera. PCR is available at the CDC.

CSF, cerebrospinal fluid; PCR, polymerase chain reaction; CDC, Centers for Disease Control and Prevention; EA, early antigen; IgG, immunoglobulin G; EBNA, Epstein-Barr virus nuclear antigen; VCA, viral capsid antigen; HIV-1/2, human immunodeficiency virus-1/2.

TABLE 5 Specimen information for hepatic infections (e.g., jaundice, elevated liver enzymes, dark urine, abdominal pain)*

Viral pathogen	Nucleic acid testing	Serology	Culture	Other comments
Hepatitis A virus	Serum Stool Saliva	Serum. The presence of IgM indicates acute infection. IgG only likely represents vaccination or past infection.		Serology is the current gold standard method of diagnosis.
Hepatitis B virus	Serum	Serum		Quantitative molecular assays are available for monitoring disease progression and efficacy of therapy. Both antigen and antibody markers may be detected from serological studies.
Hepatitis C virus	Serum	Serum		Reactive serology is most commonly confirmed with a quantitative viral load assay and can be used for monitoring disease progression and efficacy of therapy.
Hepatitis D virus	Serum	Serum		Serology can detect antigen or antibody. Hepatitis D virus is dependent on hepatitis B surface antigen, so patient must also be positive for HBV.
Hepatitis E virus	Stool Serum	Serum		

*EBV and CMV occasionally cause hepatitis, especially in immunocompromised patients.
IgM, immunoglobulin M; HBV, hepatitis B virus; EBV, Epstein-Barr virus; CMV, cytomegalovirus.

no cases of viremia were detected by culture (16). In a separate study, only 34% of patients with PCR-positive oral lesions also had detectable HSV in the blood (17, 18). Therefore, HSV PCR or culture from the blood of immunocompetent patients >30 days old may yield false-negative results (19, 20). On the other hand, asymptomatic reactivation of HSV-1 and HSV-2 in oral and anogenital specimens leads to results of limited clinical meaning. Asymptomatic reactivation of HSV in these sources occurs with high frequency, and shedding can result in transmission during subclinical phases of infection (6, 21). If HSV-2 is detected by PCR from lesions of immunocompetent patients, treatment with oral valacyclovir has been shown to reduce shedding and decrease transmission to seronegative partners (22).

Host immune status can play a significant role in interpretation of results. Neonates and immunosuppressed hosts (e.g., transplant recipients, persons with malignancy or HIV infection, or those taking immunosuppressive medication) are at increased risk of significant disseminated HSV disease. In these patients, detection of HSV in blood components is correlated with increased mortality (19, 20, 23).

For some pathogens, host immune status can also have implications for specimen collection by infection control. Norovirus causes acute but self-limited gastroenteritis in most patients, for which the Centers for Disease Control and Prevention (CDC) recommends contact precautions for approximately 48 hours after cessation of symptoms to prevent further transmission (24). However, symptomatic or asymptomatic immunocompromised patients may shed viral RNA in their stool for extended periods of time (25). These patients may require repeat specimen collection for the purpose of prolonged isolation precautions, although asymptomatic shedders contribute minimally to transmission (24, 26). The significance of shedding is not always clear because the testing modalities are often non-culture based, as in the example of

TABLE 6 Specimen information for gastroenteritis (e.g., diarrhea, abdominal pain, vomiting)

Viral pathogen	Nucleic acid testing*	Serology	Culture	Other comments
Rotavirus	Stool			Antigen detection assays can be used to detect rotavirus in stool specimens, but these tests have lower sensitivity and specificity compared to PCR.
Norovirus I and II Astrovirus Sapovirus	Stool			Antigen testing on stool may be useful in outbreaks (lower cost with high throughput and adequate sensitivity). PCR is more sensitive and may continue being positive well beyond the symptomatic period.
Adenovirus	Stool Rectal swab		Stool Rectal swab	PCR assays may specifically target adenovirus 40/41, which are the serotypes most frequently associated with gastrointestinal disease.
Hepatitis A virus	Serum Stool Saliva	Serum. The presence of IgM indicates acute infection. IgG only likely represents vaccination or past infection.		Serology is the current gold standard method of diagnosis.

*PCR assays are more sensitive and specific and are becoming more widely available for gastrointestinal pathogens, particularly as multiplex assays.
PCR, polymerase chain reaction; IgM, immunoglobulin M; IgG, immunoglobulin G.

TABLE 7 Specimen information for genital infections (e.g., lesions, atypical squamous cells)

Viral pathogen	Nucleic acid testing	Serology	Culture	Other comments
Herpes simplex virus	Genital swabs Scrapings of vesicles, pustules, and ulcerative lesions	Serum	Genital swabs Scrapings of vesicles, pustules, and ulcerative lesions	Serology is rarely informative, unless used to detect primary genital infection. Vesicle scrapings for Tzanck smears have low sensitivity and specificity.
Human papillomavirus	Endocervical/vaginal brush, broom, or swab in SurePath™ or ThinPrep® solution Biopsy tissue			Molecular methods are typically used for secondary screening and confirmation and should be performed on women ≥21 years of age. Some commercial platforms are endorsed by the FDA for primary screening and are currently recommended for testing women ≥30 years old.

FDA, Food and Drug Administration

norovirus, which is typically diagnosed by enzyme immunoassay (EIA) or PCR. Paradoxically, there is often limited information gained regarding infectivity of the patient from the testing methods that are optimal for diagnosis.

VIRAL FACTORS THAT AFFECT SPECIMEN COLLECTION

The characteristics of viral pathogens, like incubation period, duration of infection, and cellular tropism, affect the type and time of specimen collection. For example, BKV is a latent virus that establishes asymptomatic, chronic infection in renal epithelial cells (27). It is highly prevalent in humans, with greater than 80% of the population showing serological evidence of exposure (28). In the event of immunosuppression, BKV may reactivate in the kidneys to cause polyomavirus-induced nephropathy (PVAN) and can produce high titers (>2.0×10^7 viral copies/ml) in the urine as detected by PCR (29). It is important for clinicians to note, however, that detection of BKV viruria is not definitive evidence of disease. Studies have shown that urine samples can be positive for BKV DNA in up to 14% of immunocompetent controls (28, 30–32). Therefore, high viral load in the urine and in the blood, correlation with the clinical picture, as well as biopsy-proven histological evidence, must be taken together for a definitive diagnosis of BKV reactivation (33).

If infection with arboviruses such as dengue virus and West Nile Virus (WNV) is suspected, serology often has higher clinical utility than PCR (34). This is due to the short period of viremia after symptom onset, which may last a week or less. The levels of viremia are often reduced significantly by the time a patient presents to a physician and submits a specimen. One study showed that within 8 days of symptom onset, PCR testing identified 56% of cases of WNV infection whereas only 4.3% of cases were detected after this time. On the other hand, immunoglobulin M (IgM) assays identified 54% of cases within 8 days and 98% of cases after a week of infection (35, 36). Therefore, PCR is useful in detection of WNV in asymptomatic blood donor populations, but a rise in titers of WNV-specific IgM and immunoglobulin G (IgG) from blood and cerebrospinal fluid (CSF) is generally more useful in identifying exposure to WNV after presentation of symptoms. Antibody may also remain elevated for months, such that exposure to a virus in the recent past (e.g., 6–9 months previous) could confound the interpretation of current clinical symptoms.

The kinetics of infection is necessary for understanding the most appropriate time of specimen collection. In addition to viremic and symptomatic phases of infection, the length of virus incubation is helpful for determining the route of exposure. For instance, a CMV-positive urine, saliva, blood, or tissue specimen collected during the first 2–3 weeks of life from an infant with congenital disease is strong evidence supporting the viral etiology of the congenital anomalies, presumably because the agent was acquired *in utero* and active viral infection and replication of the virus is present (37). Conversely, isolation of CMV from the same source after 2–3 weeks would not discriminate between congenitally or postnatally acquired CMV.

IMPORTANCE OF TEST PLATFORM FOR SPECIMEN COLLECTION

The decision of which specimen to collect depends on the most appropriate test method for the viral pathogen under

TABLE 8 Specimen information for retroviruses

Viral pathogen	Nucleic acid testing	Serology	Culture	Other comments
Human immunodeficiency virus types 1 and 2	Blood, plasma, serum	Blood, plasma, serum. Serology with an antigen/antibody immunoassay is recommended as the initial screen.		Serology results should be confirmed with an HIV-1/2 differentiation immunoassay. Nucleic acid assays may be used for diagnosis of acute HIV infection as well as for monitoring.
Human T lymphotropic virus types I and II	Blood	Serum		Serology is generally used as the primary screen. Consider performing PCR in cases where serology is negative but clinical suspicion of infection is high.

TABLE 9 Specimen information for congenital infections (e.g., intrauterine, perinatal, postnatal)

Viral pathogen	Nucleic acid testing	Serology	Culture	Other comments
Cytomegalovirus	Plasma Blood Urine Ocular Respiratory washings Amniotic fluid for *in utero* detection	Serum	Urine Throat swab	Detection of CMV in urine from neonates within 2–3 weeks of birth indicates congenital infection. Presence of IgM in cord or neonatal blood can also indicate congenital infection.
Herpes simplex virus	CSF Eye swab Anal/rectal swab Nasal swab Amniotic fluid for *in utero* detection Multisite swab Genital swabs and scrapings of active lesions from the mother	Serum from pregnant mother is only useful in identifying primary exposure to HSV. Cord blood. The presence of IgM indicates congenital infection.	Vesicle, pustule or lesion swab Throat swab Amniotic fluid for *in utero* detection Multi-site swab	Blood is a poor source for HSV detection but can be used to in babies ≤30 days old to assess disseminated disease. A single swab of multiple sites (e.g., eye, oral mucosa, nasal cavity, axilla, and rectum) collected 12–24 hours after birth can be used to identify active neonatal infection.
Rubella virus	Amniotic fluid for *in utero* detection NP or throat swab Urine Conjunctival swab	Serum CSF Cord blood	Viral culture may be performed on throat, nasal, urine, conjunctiva, CSF, and amniotic fluid specimens for epidemiologic purposes.	A 4-fold rise in maternal IgG may indicate infection in asymptomatic mothers. In babies up to 6 months old, only IgM antibody should be assayed because IgG reflects passive transfer from the mother.

(*Continued on next page*)

TABLE 9 Specimen information for congenital infections (e.g., intrauterine, perinatal, postnatal) (*Continued*)

Viral pathogen	Nucleic acid testing	Serology	Culture	Other comments
	CSF			In countries of very low prevalence, IgM testing has such low positive predictive value that has little to no clinical value.
Enterovirus	Throat swab Vesicular fluid Blood, plasma, serum Rectal swab, stool CSF Amniotic fluid for *in utero* detection		Throat swab Vesicular fluid Blood, plasma, serum Rectal swab, stool CSF Amniotic fluid for *in utero* detection	Cord blood is a useful and productive specimen for recovering enteroviruses.
Parvovirus B19	Amniotic fluid for *in utero* detection Blood, serum, plasma	Serum		Acute infection in the mother is most commonly evaluated by serological methods.
Human immunodeficiency virus type 1 and 2	Blood, plasma, serum (RNA) Whole blood, PBMC (DNA)	Serology should not be used for diagnosis of infants less than 18 months old because of transfer of maternal antibodies.		PCR of integrated HIV DNA can be detected in PBMCs.
Varicella-zoster virus	Placental tissue and amniotic fluid for *in utero* detection. Neonatal tissue, blood, serum, plasma	Serum. The presence of IgM can indicate congenital infection.	Placental tissue and amniotic fluid for *in utero* detection Neonatal tissue	Diagnosis can be made with clinical findings in the neonate and confirmation of maternal infection during early pregnancy.

CMV, cytomegalovirus; IgM, immunoglobulin M; CSF, cerebrospinal fluid; HSV, herpes simplex virus; NP, nasopharyngeal; IgG, immunoglobulin G; PBMCs, peripheral blood mononuclear cells; HIV, human immunodeficiency virus.

TABLE 10 Specimen information for infectious mononucleosis

Viral pathogen	Nucleic acid testing	Serology	Culture	Other comments
Epstein-Barr virus	Blood, plasma, serum	Serum. Serologic testing is recommended for diagnosis.		Serologic testing can be done using heterophile antibody testing. In the context of classical mononucleosis symptoms and negative heterophile results, additional EBV markers such as EA IgG, VCA IgG, and VCA IgM should be evaluated.
Cytomegalovirus	Blood, plasma, serum Throat swab	Serum	Throat swab	Detection of IgM can indicate primary infection.
Human immunodeficiency virus types 1 and 2	Blood, plasma, serum	Blood, plasma, serum. Serology with an antigen/antibody immunoassay is recommended as the initial screen.		Serology results should be confirmed with an HIV-1/2 differentiation immunoassay. Nucleic acid assays may be used for diagnosis of acute HIV infection as well as for monitoring.

EBV, Epstein-Barr virus; EA, early antigen; IgG, immunoglobulin G; VCA, viral capsid antigen.

question. Despite the shift toward increased molecular testing, traditional test methods like viral culture, serology, histopathology, and antigen assays still have essential roles in viral diagnostics. For instance, biopsy specimens containing histopathological evidence of adenoviral inclusions within epithelial cells of the gastrointestinal tract suggest invasive gastrointestinal disease (38). On the other hand, detection of adenoviral DNA in stool samples cannot differentiate between viral shedding and active disease (39). A well-defined differential diagnosis of potential viral etiologies will guide appropriate test utilization and is fundamental for interpretation of laboratory results.

Compared to culture-based assays, molecular methods have significantly improved sensitivity and specificity of detection for HSV and varicella-zoster virus (VZV) and can provide same-day detection and differentiation of these viruses from clinical specimens (40, 41). VZV replicates slowly and poorly in cell cultures, and real-time PCR is 91% more sensitive than detection of the virus in shell vials and other diagnostic methods (42, 43). At our institution, culture requests for HSV and VZV from dermal, oral mucosa, and genital lesions are automatically converted to molecular assays to enhance overall clinical sensitivity of viral detection. From dermal virus culture, HSV (70%), VZV (29%), coxsackievirus type A (1%), and some echoviruses are the most common agents that are identified. Similarly, recovery of respiratory viruses from patient specimens was significantly enhanced from a positivity rate of just 9.2% with viral culture to 16.5% with real-time PCR (44).

Cell culture is still popular with some providers because specimens from a wide range of sources are generally acceptable, such as tissue, swabs, and sterile and nonsterile body fluids. Viral culture has also traditionally been used for relatively broad coverage of viral pathogens. At our institution, viral culture is performed on rhesus monkey kidney (RhMK) and human fetal lung fibroblast cells (MRC-5). Respiratory specimens can be tested for influenza A and B, parainfluenza virus (types 1, 2, and 3), respiratory syncytial virus, and rhinovirus, but culture in these cell lines does not identify human metapneumovirus or coronaviruses. Similarly, CSF is a poor source for viral culture. Many neurotropic viruses like JC virus (JCV), WNV, St. Louis encephalitis virus, and other neurotropic viruses are not detected by standard cell lines used in diagnostic laboratories. Some viruses like HSV and enteroviruses may be cultured, but PCR is the most sensitive approach for detection of viruses in CSF. It is important for laboratories to specify in their lab test catalogs and in the report which pathogens are screened in an assay, because providers are likely not aware of the viruses that grow in commonly used primary and diploid cell culture lines.

As molecular methods have been developed and integrated into clinical virology laboratories over the years, increasing emphasis has been placed on detecting multiple pathogens from a single specimen. These multiplex assays combine the breadth of viral culture with the specificity and sensitivity of PCR and are useful when different viruses can cause indistinguishable clinical symptoms. However, commercial multiplex tests may be limited by specimen source, such as testing only on nasopharyngeal swabs for respiratory pathogens. In some cases, this validated source is not ideal for testing of all pathogens on the panel, which can result in different levels of clinical sensitivity for each target. Laboratories may consider validating additional sources such as nasopharyngeal aspirates, nasal washes, throat swabs, tracheal aspirates, and bronchoalveolar lavage (BAL) fluid to make these assays more useful and appropriate.

As technology continues to develop, sensitivity of virus detection from patient specimens will increase. However, laboratories will need to consider more carefully the value of a result in the context of test platform and types of specimens selected. For example, like other members of the herpesvirus family, HHV-6 is a mild pathogen in immunocompetent hosts and can cause fever and exanthema subitum rash in children under 24 months old (45). After primary infection, the virus causes life-long latency in many cell types, such as mononuclear cells, salivary glands, bone marrow, lung, and liver, although some periods of benign reactivation may occur (46, 47). When reactivation occurs in immunocompromised patients, like transplant or acquired immunodeficiency syndrome (AIDS) patients, HHV-6 can be a serious pathogen; it can cause encephalitis, pneumonitis, and may be involved in a host of other complications (48). Diagnostic tests must be able to differentiate between active and quiescent infection with HHV-6. Serological assays are generally unhelpful because 85% of the human population has antibodies against the virus by the time they are 3 years old (49). Furthermore, viral DNA is chromosomally

TABLE 11 Specimen information for hemorrhagic fevers

Viral pathogen	Nucleic acid testing	Serology	Culture	Other comments
Dengue virus	Serum	Serum CSF		Viremia may be brief, so molecular assays should only be performed within 5 days of symptom onset. Serology may cross-react with other flaviviruses. Additional personal protective equipment may be required.
Hantavirus	Lung or bone marrow aspirate Tissue (lung, kidney, spleen, heart, brain, lymph nodes, pancreas, pituitary) biopsy	Serum Blood		Organ tissue can be tested by immunohistochemistry. PCR assays are available at the CDC.
Marburg virus	Blood, serum Tissue	Blood, serum		Contact the CDC or state public health labs for further testing information.
Ebola virus	Blood, plasma, serum Semen, vaginal fluid Vomitus, stool Aqueous humor Other body fluids and tissue	Blood, serum	Culture is not recommended in diagnostic laboratories due to risk of exposure.	Contact the CDC or state public health labs for further testing information. Serology for IgM and IgG may be used for monitoring. Additional personal protective equipment required. Tissue may also be analyzed with histologic techniques. In acute clinical disease compatible with Ebola, at least two negative molecular results are necessary to consider the patient negative. Ebola survivors may shed virus in semen and other fluids for extended periods.
Lassa virus	Blood/serum Tissue (frozen)	Blood Serum		Contact the CDC for further testing information. Additional personal protective equipment may be required.

CSF, cerebrospinal fluid; PCR, polymerase chain reaction; CDC, Centers for Disease Control and Prevention; IgG, immunoglobulin G; IgM, immunoglobulin M.

TABLE 12 Specimen information for other infections, like ocular (e.g., conjunctivitis, retinitis, endophthalmitis), heart (e.g., myocarditis, pericarditis), and renal infections (e.g., hemorrhagic cystitis, nephropathy)

Viral pathogen	Nucleic acid testing	Serology	Culture	Other comments
Cytomegalovirus	Aqueous or vitreous humor	Detection of CMV IgG from serum in HIV+ patients may indicate the possibility of reactivation.		CMV retinitis is typically diagnosed by identifying characteristic retinal features by fundoscopy and histologic features from retinal biopsy. PCR can be performed from ocular fluid but it may have low sensitivity.
Herpes simplex virus Adenovirus Enterovirus Varicella-zoster virus	Corneal scraping or biopsy Conjunctival swab Aqueous or vitreous humor Tear film		Corneal scraping or biopsy Conjunctival swab	Keratitis may be assessed by histology of corneal specimens.
Adenovirus Epstein-Barr virus Parvovirus B19 Enteroviruses Human herpesvirus 6	Myocardial, endocardial, or pericardial tissue biopsy Pericardial fluid Blood	Serum	Biopsy tissue	Histologic examination of biopsy tissue can confirm the diagnosis of carditis.
BK virus JC virus Adenovirus	Plasma, blood Urine Urine		Urine	Detection of viruses may represent shedding without active renal disease, so results should be correlated with the clinical presentation.

CMV, cytomegalovirus; IgG, immunoglobulin G; PCR, polymerase chain reaction; HIV, human immunodeficiency virus.

integrated into tissues and peripheral blood mononuclear cells (PBMCs) of normal controls. These integrated genomes can be detected by molecular assays (48) and, as a result, the interpretation of qualitative PCR assays for diagnosis of latent versus active HHV-6 disease can be difficult. In situations of latent infection, quantitative assays can provide additional context in terms of viral dynamics in relation to clinical symptoms and treatment.

SPECIMEN TYPE

Detection of viruses from blood specimens may indicate systemic invasive infections in the right host with the right clinical picture. Serology testing on serum or plasma is the primary screening method for antibodies against bloodborne pathogens like HCV. However, serology alone cannot distinguish between acute, persistent, and resolved HCV infection because the antibodies that are formed are lifelong. For patients with reactive screening for HCV, confirmatory reverse-transcription PCR should be performed to detect the viremia present in active infections (50–52). The PCR result both confirms an actual infection and also provides a baseline viral load.

Many viruses can invade the central nervous system (CNS) to cause inflammation of the brain, brainstem, and spinal cord. In these cases, CSF is an important specimen type for testing and will typically demonstrate abnormal characteristics, such as lymphocytic pleocytosis and elevated protein levels, but normal glucose levels (53). Molecular assays have dramatically increased the detection of viral pathogens from CSF compared with viral culture, as PCR can rapidly identify common causes of viral meningoencephalitis like enteroviruses and herpesviruses. However, for some viruses like dengue virus and WNV, serological analysis from CSF can play a significant role in diagnosis of acute CNS infections. For instance, detection of IgM antibodies against WNV or dengue virus in the CSF and serum, or detection of a 4-fold rise in IgG antibodies titers from paired acute and convalescent specimens, supports the diagnosis of acute infection (54).

Specimen types should be selected carefully because detection of viruses in some patient specimens may not have real clinical significance. For instance, JCV is a cause of fatal progressive multifocal leukoencephalopathy (PML) in immunosuppressed hosts (55). However, in both normal and symptomatic hosts, JCV DNA can be detected in stool or urine specimens. Therefore, testing should be limited to symptomatic patients. Even then, results from urine and stool specimens should be interpreted with caution. For patients presenting with neurologic symptoms, infection with JCV should be assessed by appearance of characteristic histopathological features on brain biopsy or PCR of the virus from CSF (56, 57). Similarly, nongastrointestinal serotypes of adenovirus, such as adenovirus types 2 and 5, typically cause upper respiratory tract infections, but they can be shed asymptomatically in the stool for extended periods of time (58). Enteroviruses are also transmitted by the fecal–oral route and can be isolated from the feces of individuals with asymptomatic, acute, or past gastroenteritis (59, 60). If an enterovirus is suspected to be the cause of neurological or upper respiratory tract symptoms, CSF or throat samples should be submitted for analysis by PCR, although supplementary testing from stool may be indicated (61). In general, specimens should be collected from the foci of infection, and waste specimens like urine and stool should be evaluated together with other clinical evidence.

In the immunocompetent host, HSV (corneal infections) and adenoviruses (conjunctivitis) are the most common ocular viral pathogens. Detection of either virus is best determined from a conjunctival swab or corneal scrapings by molecular testing or culture. Molecular testing can provide the most sensitive and specific diagnosis of CMV retinitis in immunosuppressed individuals, especially HIV-infected persons who are most at risk. For example, Yamamoto et al. determined that in a group of AIDS patients with retinitis, 88.1% of aqueous humor from eyes with active retinitis were positive for CMV DNA, and 78.4% of those specimens from ocular sites became DNA-negative after treatment (62).

SPECIMEN COLLECTION

The utility of diagnostic virology testing depends not just on the sensitivity of the test but also on the quality of the patient specimen. Specimens can be collected in a variety of ways that may affect specimen quality. For instance, anticoagulants used in collection of blood specimens have been shown to affect detection of viral nucleic acid. In early-generation HIV assays, heparinized plasma resulted in lower HIV RNA values in comparison with EDTA plasma (63, 64). In studies of newer HIV platforms that use different extraction methods, no difference in viral load was observed between EDTA and heparin plasma (65).

Swabs are popular collection devices because they are easy to use, inexpensive, and minimally invasive. Swabs can be composed of different materials, such as synthetic rayon, nylon, or Dacron tips with plastic or aluminum shafts. Wood-shaft swabs containing preservatives and swabs with calcium alginate tips can inhibit viral recovery and should not be accepted for testing (66, 67). Traditional swabs are made by wrapping long strands of fiber onto the end of the shaft, whereas flocked swabs are made with short nylon fibers that are electrostatically charged and bonded perpendicularly to each other. This latter technique substantially increases the swab's surface area for more efficient collection and release of particulate matter (68, 69).

Swabs are convenient to use and, despite the small volume of material collected, a single swab specimen may be enough to test for multiple viral pathogens if the sample is collected appropriately and is tested on a suitable assay platform. In evaluations of three commercially available multiplex assays, only a single nasopharyngeal swab is necessary to test simultaneously for 10–20 viruses (70, 71).

Despite the convenience of swabs, the overall sensitivity of detection for most test modalities is enhanced by increasing specimen volume. A study comparing paired nasopharyngeal aspirates and nasopharyngeal swabs showed that the combined sensitivity of detection for adenovirus, influenza, parainfluenza, and respiratory syncytial virus (RSV) was statistically lower for the swabs regardless of whether immunofluorescence assays, culture, or PCR were used (72). In some cases where pathogens cause lower respiratory tract infections, more invasive bronchoalveolar lavage (BAL), bronchial wash, or biopsy procedures may further improve sensitivity of detection.

REPEAT SPECIMEN COLLECTION

In some cases, submission of a single specimen is sufficient for clinical and laboratory assessment. For example, a dermal specimen positive for HSV should be considered a meaningful result that requires immediate clinical attention. In other cases, such as HIV diagnosis, where a positive or negative result can have significant implications for the patient, repeat or additional specimen collection may be warranted for confirmatory testing. False-negative HIV results from testing very early in infection may occur as there is a window period of ~10 days before HIV RNA titers begin to rise and ~15 days before p24 increases (73). Therefore, repeat specimen collection and testing may be necessary to ensure that a diagnosis of HIV is not missed in these patients or in patients with repeated exposures. If possible, a dedicated tube should be collected if molecular assays are used for confirmation and follow-up testing to avoid any potential contamination events.

For therapeutic, rather than diagnostic, purposes, laboratories should be prepared to receive multiple specimens from a single patient over time. The Infectious Disease Society of America (IDSA) recommends monitoring HIV viral load in untreated patients or therapeutically stable patients every 3–6 months and more frequently (2–4 weeks) if therapy is recently initiated or changed (74). Similarly, transplant patients such as hematopoietic stem cell recipients, will have specimens collected and sent to the laboratory for routine monitoring of CMV and Epstein-Barr virus (EBV) viral load (75). Pre-emptive monitoring can help to determine whether transplantation needs to be delayed due to active infection. After transplant, it is recommended that high-risk (CMV-seropositive or patients with CMV-seropositive donors) patients be screened for CMV in the blood approximately once a week from day 10 up to day 100 to determine appropriate therapy (75). As another example, detailed guidelines exist for the treatment of HCV genotypes 1 through 6, whereby the viral load informs the response to treatment and probability of sustained virologic response (52).

Laboratories may consider restricting testing of repeat specimens due to high cost. For example, some commercial multiplex tests are associated with a high charge to the patient ($500–$1000/test or more). These tests are valuable for diagnostic purposes due to their broad coverage of pathogens from a single specimen, but may be unwarranted for monitoring purposes. Rapidly conveying this information to clinicians, or creating an automatic system for preventing these orders, can help minimize unnecessary laboratory charges and costs and optimize overall test utilization. A helpful reminder to care providers is the question of whether a test will provide actionable information. If the result will not affect patient care, the test may not be needed.

SPECIMEN PROCESSING

Body fluid specimens like CSF, urine, synovial fluid, secretions, and other aspirates can be applied directly to antigen assays and culture cell lines. For PCR-based assays, nucleic acid is generally extracted prior to testing. Although nucleic acid extraction from specimens requires an additional processing step, time, and expense, it can enhance detection of viral pathogens by removing inhibitors from the patient sample. Inhibitors are molecules like bile salts, bilirubin, hemoglobin, lactoferrin, IgG, and other complex carbohydrates that decrease the efficiency of PCR, such as inhibiting the action of DNA polymerase (76–78). Purification of nucleic acids prior to amplification may reduce the amount of inhibitors in the final extract, but in some cases they may be co-extracted along with nucleic acids (79). As a result, PCR on some patient specimens (especially whole blood, stool, and bile) may be predisposed to false-negative results. To ensure that PCR is able to proceed correctly, inclusion of an internal amplification control is one way to assess each sample. If a control gene does not amplify, this indicates the potential presence of inhibitory substances and therefore the potential for false negatives. A study performed at Mayo Clinic laboratories showed that inhibition rates are generally ≤1.05% for most specimen types, except for formalin-fixed, paraffin-embedded tissue (80). Also, fresh tissue and biopsy specimens may require additional processing before diagnostic PCR assays can be performed. In our laboratory, the tissue is digested with proteinase K at 55°C and shaken (e.g., 500 rpm) for up to 24 hours to disrupt the cellular material and enhance recovery of viral pathogens. The supernatant is then extracted prior to nucleic acid testing. Therefore, microbiology laboratories should only accept complex specimen sources for testing by PCR after thorough validation for potential inhibitors (81).

Cell culture may also be susceptible to inhibitory substances. For instance, whole blood and urine specimens are difficult to culture because they can cause lysis of the cellular monolayer. Bacterial overgrowth may also disrupt mammalian cell growth. Contamination events can be caused by inadequate quality control of commercial culture reagents, nonadherence to aseptic technique during processing, or from the specimen itself (82). Respiratory and fecal specimens are examples of specimens that are likely to contain significant bacterial flora. If microbial overgrowth is seen, the original samples are filtered through a 0.45-µm filter and then reinoculated onto fresh cell monolayers. At the Mayo Clinic, these filter and repeat procedures need to be performed on approximately 30% of respiratory and 10% of fecal samples.

Specimens containing liquid-based formalin, ethanol, formaldehyde, acetone, or other fixatives are generally rejected because these compounds inactivate viable organisms and cross-link nucleic acid, thus rendering PCR ineffective. However, there are two exceptions for viral testing. One is PCR from formalin-fixed, paraffin-embedded tissue, which is an accepted and validated source for some pathogens in some laboratories (83, 84). If fixatives are used, specimens should spend minimal time in formalin (85).

The second exception to performing molecular assays on samples containing fixatives is HPV nucleic acid detection. Both high- and low-risk genotypes can be detected from patient specimens collected in ThinPrep® fluid, which contains an alcohol fixative, as well as SurePath™ preservative fluid, which contains alcohol and formaldehyde (86–88).

SPECIMEN TRANSPORT AND STORAGE

The type of transport medium and transportation time to the laboratory are important considerations for viability of the specimen matrix, the pathogen, viral antigens, and nucleic acid. A laboratory-based study showed that inoculation of various swabs with enveloped and nonenveloped RNA and DNA viruses could be detected by PCR after 7 days at room temperature (4°C and 37°C). However, dry swabs, swabs with Amies gel containing charcoal, and swabs maintained at 37°C for >3 days yielded consistently lower amounts of viral nucleic acids (89).

To maintain stability, most swabs, brushes, and tissue specimens should be transported in viral transport media. Viral transport medium (VTM) is used to stabilize viral particles and nucleic acids, prevent specimen drying, maintain pH, and inhibit bacterial overgrowth in culture. Several formulations are available that contain a combination of buffered solution, protein stabilizers, antibiotics, and minimal nutrients (66). As an exception, nonsterile (e.g., stool, urine, and respiratory secretions) and sterile (e.g., blood/plasma, CSF, and pleural, ocular, and joint fluids) body fluids should be tested without dilution in viral transport medium to optimize sensitivity of detection.

In general, viruses that are enveloped, like blood-borne (e.g., HIV, HBV, and HCV) or respiratory (e.g., the orthomyxo- and paramyxoviruses, like influenza, RSV, human metapneumovirus, and parainfluenza viruses) pathogens are relatively labile compared with those without envelopes (66). Most viruses persist longer at cooler temperatures, and even highly labile pathogens such as CMV, RSV, and VZV can survive transit for 1 to 3 days if maintained at 4°C (90, 91). To increase viral recovery, specimens for viral culture should be transported to the laboratory under refrigerated conditions, and frozen samples should be discouraged.

Unlike viral culture, where each hour of delay before inoculation translates to progressively lower probability of successful isolation, molecular and antigen testing abrogates the need for viral viability. This is a distinct advantage, as an unexpected delay in sample processing due to an error in communication, handling, or due to the weather will minimally affect the quality of the test and result for the patient. In a study performed by Jerome et al., swabs from a variety of anatomical sites were placed in viral transport medium (92). Part of this material was extracted immediately for quantitative real-time PCR testing and the remainder of the nucleic acid extract was stored at 4°C. The rest of the original sample was frozen at −20°C. After 16 months in these conditions, the stored extracts were reevaluated by real-time PCR; the viral titers were essentially the same, with 93% of the extracts agreeing within 1 log of the initial quantitative result. Similarly, raw samples stored for 16 months and then re-extracted for PCR analysis showed 94% agreement within 1 log of the initial results. These results demonstrate that freezing patient specimens and refrigerating nucleic acid extracts of specimens do not significantly affect viral quantification for at least 16 months. Another group showed that quantification of concentrations of nucleic acids from HBV (DNA) and HCV (RNA) remained stable in patient samples after eight freeze–thaw cycles, as determined by a commercial detection platform (93).

An alternative approach was taken for storage of respiratory swab specimens in a study by Krafft et al. (94). Swabs were stored in ethanol at ambient temperatures for 1 month or 6 months, and the results of molecular testing (for influenza viruses A and B and adenovirus) were compared to culture results at the time of specimen collection. The results of real-time PCR tests on the stored specimens demonstrated a high correlation with initial culture results, and PCR could even detect positive results in specimens that were culture-negative. In another study, real-time PCR was 82% sensitive (and more sensitive than repeat culture) when used to detect influenza in nasal aspirates stored frozen for 1–3 years at −70°C compared to viral culture at the time of the aspirate (95). At our institution, ambient, refrigerated and frozen samples are acceptable for molecular testing for up to 7 days. After the result is reported, patient specimens are generally stored for 7 days in the laboratory at 2–8°C.

Specimens from patients suspected of being infected with highly virulent and transmissible viruses like avian influenza, filovirus (e.g., Ebola and Marburg), arenavirus (e.g., Lassa), bunyaviruses (e.g., Hantaan), rabies, severe acute respiratory syndrome (SARS), or Middle East respiratory syndrome (MERS) coronaviruses should be handled with particular caution. To minimize exposures in the laboratory, procedures should be in place to triage specimens suspected of having these pathogens to be tested at a state laboratory or the Centers for Disease Control and Prevention (CDC). Some limited testing may be performed on site to rule out more common pathogens presenting with similar symptoms. If rule-out testing is necessary, viral culture should not be performed to avoid viral amplification and specimens should be handled according to the appropriate biosafety level guidelines (66).

During the Ebola virus disease outbreak in 2014, the CDC described safety measures to protect hospital workers. Specimens from potentially infected patients should be transported to the laboratory in plastic tubes within a secondary, leakproof container. To minimize the risk of breakage, specimens must be transported manually instead of using the pneumatic tube system. Upon receipt of these specimens, safe handling measures include the use of a certified class II biosafety cabinet and personal protective equipment like double gloves, impermeable gowns, eye protection, and an N95 mask or respirator (96). Testing should be limited to what is absolutely necessary, and the testing that is needed should take place in the patient room or in a nearby contained testing area. It is safest to use point-of-care systems to minimize transport, handling, and number of staff exposed to the specimens. Laboratories should have plans and procedures in place to address the collection, processing, transport, and possible testing of these suspect specimens. State laboratories and the CDC are also vital resources to be consulted when these diagnoses are being considered. Check with local and federal authorities for the latest during ongoing outbreaks as information regarding recommendations can change daily.

FUTURE CONSIDERATIONS

Technology has improved dramatically over the past three decades. The increasing use of rapid, highly sensitive, and specific nucleic acid amplification assays has affected which specimens are the most appropriate for testing, how much is needed, and at what point of the infection the specimen should be collected for maximum yield. In turn, this has improved the ability of a clinician to recognize and interpret acute, quiescent, chronic, and reactivated viral infections.

Currently, nucleic acid amplification methods have largely displaced more traditional methods of virus detection, like virus culture, rapid antigen assays, and immunofluorescence-based antigen assays. Replacement of viral culture with molecular testing may affect regulations and requirements for specimen collection. As laboratories transition away from viral culture, they will no longer be generating relatively pure, high-titer viral isolates for submission to regional and public health laboratories. Instead, public health laboratories may expect to see a significant rise in the number of clinically collected specimens sent directly to them for further characterization and analysis.

Increasing attention is being given to the ability of molecular testing to detect multiple pathogens from a single specimen based on clinical syndromes, like respiratory or gastrointestinal diseases. In an evaluation of multiplex testing of gastrointestinal pathogens at our institution, we determined that patients with possible infectious diarrhea had an average of three bacterial, viral, or parasitic tests performed, with only an 8.3% positivity rate per patient. Using a single multiplex molecular assay for 15 or 23 targets, depending on the assay used, the positivity rate increased to >30%. Notably, the two multiplex panels evaluated in this study detected a significant number of viruses for which we currently do not have routine test methods, such as astrovirus and sapovirus. This type of "syndromic" testing from a single specimen collection allows clinicians to screen for a panel of pathogens that present with overlapping clinical symptoms (97).

The convenience and benefit of testing multiple pathogens from a single specimen is becoming popular with clinicians, and, as a result, some laboratories are striving to provide broad testing panels from single-specimen collection devices, such as a single swab for diagnosis of gynecologic pathogens. Similarly, increasing use of technology like microfluidics, as well as the miniaturization of diagnostic testing, is being investigated for identification of multiple analytes from just a few drops of blood. These new methods and devices can potentially minimize incorrect specimen collection, reduce the number of invasively collected specimens, and dramatically reduce the volume of material that needs to be collected from a patient. However, despite the value in such testing, it is necessary for laboratories to ensure that changing specimen type and volume does not compromise the sensitivity of detection of any analytes. Also, it is crucial to provide panels only containing analytes that provide relevant, meaningful clinical information. The risk of overtesting from single samples is misinterpretation of the clinical importance of a result, which may lead to overtreatment and patient harm.

Further developments and breakthroughs in technology will increase our ability to characterize viral infections. For example, Sanger sequencing of HIV is used routinely to monitor genotype resistance mutations, but this platform may be replaced by deep sequencing techniques for greater coverage of minor resistance mutations (98) and ability to detect low-prevalence mutations within the patient's viral population. Influenza virus nucleic acid is also highly dynamic, and next-generation sequencing of the neuraminidase gene is being used to monitor oseltamivir resistance (99, 100).

The type of testing platforms used in clinical laboratories for viral detection has evolved rapidly and will continue to do so as newer technology emerges. The technological changes that increase our ability to detect multiple analytes from a single specimen have, and will, affect the specimen type, collection devices, and amount requirements that are needed for virus testing.

REFERENCES

1. **Popgeorgiev N, Temmam S, Raoult D, Desnues C.** 2013. Describing the silent human virome with an emphasis on giant viruses. *Intervirology* **56:**395–412.
2. **Arbuckle JH, Medveczky MM, Luka J, Hadley SH, Luegmayr A, Ablashi D, Lund TC, Tolar J, De Meirleir K, Montoya JG, Komaroff AL, Ambros PF, Medveczky PG.** 2010. The latent human herpesvirus-6A genome specifically integrates in telomeres of human chromosomes in vivo and in vitro. *Proc Natl Acad Sci USA* **107:**5563–5568.
3. **Razonable RR, Brown RA, Humar A, Covington E, Alecock E, Paya CV, PV16000 Study Group.** 2005. A longitudinal molecular surveillance study of human polyomavirus viremia in heart, kidney, liver, and pancreas transplant patients. *J Infect Dis* **192:**1349–1354.
4. **Jones LE, Perelson AS.** 2005. Opportunistic infection as a cause of transient viremia in chronically infected HIV patients under treatment with HAART. *Bull Math Biol* **67:**1227–1251.
5. **Kalu SU, Loeffelholz M, Beck E, Patel JA, Revai K, Fan J, Henrickson KJ, Chonmaitree T.** 2010. Persistence of adenovirus nucleic acids in nasopharyngeal secretions: a diagnostic conundrum. *Pediatr Infect Dis J* **29:**746–750.
6. **Mark KE, Wald A, Magaret AS, Selke S, Olin L, Huang ML, Corey L.** 2008. Rapidly cleared episodes of herpes simplex virus reactivation in immunocompetent adults. *J Infect Dis* **198:**1141–1149.
7. **Tronstein E, Johnston C, Huang ML, Selke S, Magaret A, Warren T, Corey L, Wald A.** 2011. Genital shedding of herpes simplex virus among symptomatic and asymptomatic persons with HSV-2 infection. *JAMA* **305:**1441–1449.
8. **Yap G, Pok KY, Lai YL, Hapuarachchi HC, Chow A, Leo YS, Tan LK, Ng LC.** 2010. Evaluation of Chikungunya diagnostic assays: differences in sensitivity of serology assays in two independent outbreaks. *PLoS Negl Trop Dis* **4:**e753.
9. **Dries V, von Both I, Müller M, Gerken G, Schirmacher P, Odenthal M, Bartenschlager R, Drebber U, Meyer zum Büschenfeld KH, Dienes HP.** 1999. Detection of hepatitis C virus in paraffin-embedded liver biopsies of patients negative for viral RNA in serum. *Hepatology* **29:**223–229.
10. **Von Müller L, Hampl W, Hinz J, Meisel H, Reip A, Engelmann E, Heilbronn R, Gärtner B, Krämer O, Einsele H, Hebart H, Ljubicic T, Löffler J, Mertens T.** 2002. High variability between results of different in-house tests for cytomegalovirus (CMV) monitoring and a standardized quantitative plasma CMV PCR assay. *J Clin Microbiol* **40:**2285–2287.
11. **Hayden RT, Yan X, Wick MT, Rodriguez AB, Xiong X, Ginocchio CC, Mitchell MJ, Caliendo AM, College of American Pathologists Microbiology Resource Committee.** 2012. Factors contributing to variability of quantitative viral PCR results in proficiency testing samples: a multivariate analysis. *J Clin Microbiol* **50:**337–345.
12. **Madej RM, Davis J, Holden MJ, Kwang S, Labourier E, Schneider GJ.** 2010. International standards and reference materials for quantitative molecular infectious disease testing. *J Mol Diagn* **12:**133–143.
13. **Popowitch EB, O'Neill SS, Miller MB.** 2013. Comparison of the Biofire FilmArray RP, Genmark eSensor RVP, Luminex xTAG RVPv1, and Luminex xTAG RVP fast multiplex assays for detection of respiratory viruses. *J Clin Microbiol* **51:**1528–1533.
14. **Moscicki AB, Shiboski S, Hills NK, Powell KJ, Jay N, Hanson EN, Miller S, Canjura-Clayton KL, Farhat S, Broering JM, Darragh TM.** 2004. Regression of low-grade squamous intra-epithelial lesions in young women. *Lancet* **364:**1678–1683.
15. **Saslow D, Solomon D, Lawson HW, Killackey M, Kulasingam SL, Cain JM, Garcia FA, Moriarty AT, Waxman AG, Wilbur DC, Wentzensen N, Downs LS Jr, Spitzer M, Moscicki AB, Franco EL, Stoler MH, Schiffman M, Castle PE, Myers ER, Chelmow D, Herzig A, Kim JJ, Kinney W, Herschel WL, Waldman J.** 2012. American Cancer Society, American Society for Colposcopy and Cervical Pathology, and American Society for Clinical Pathology screening guidelines for the prevention and early detection of cervical cancer. *J Low Genit Tract Dis* **16:**175–204.
16. **Halperin SA, Shehab Z, Thacker D, Hendley JO.** 1983. Absence of viremia in primary herpetic gingivostomatitis. *Pediatr Infect Dis* **2:**452–453.
17. **Harel L, Smetana Z, Prais D, Book M, Alkin M, Supaev E, Mendelson E, Amir J.** 2004. Presence of viremia in patients with primary herpetic gingivostomatitis. *Clin Infect Dis* **39:** 636–640.
18. **Zuckerman RA.** 2009. The clinical spectrum of herpes simplex viremia. *Clin Infect Dis* **49:**1302–1304.
19. **Malm G, Forsgren M.** 1999. Neonatal herpes simplex virus infections: HSV DNA in cerebrospinal fluid and serum. *Arch Dis Child Fetal Neonatal Ed* **81:**F24–F29.
20. **Kimura H, Futamura M, Kito H, Ando T, Goto M, Kuzushima K, Shibata M, Morishima T.** 1991. Detection of viral DNA in neonatal herpes simplex virus infections: frequent and prolonged presence in serum and cerebrospinal fluid. *J Infect Dis* **164:**289–293.

21. **Benedetti J, Corey L, Ashley R.** 1994. Recurrence rates in genital herpes after symptomatic first-episode infection. *Ann Intern Med* **121:**847–854.
22. **Corey L, Wald A, Patel R, Sacks SL, Tyring SK, Warren T, Douglas JM Jr, Paavonen J, Morrow RA, Beutner KR, Stratchounsky LS, Mertz G, Keene ON, Watson HA, Tait D, Vargas-Cortes M, Valacyclovir HSV Transmission Study Group.** 2004. Once-daily valacyclovir to reduce the risk of transmission of genital herpes. *N Engl J Med* **350:** 11–20.
23. **Berrington WR, Jerome KR, Cook L, Wald A, Corey L, Casper C.** 2009. Clinical correlates of herpes simplex virus viremia among hospitalized adults. *Clin Infect Dis* **49:**1295–1301.
24. **MacCannell T, Umscheid CA, Agarwal RK, Lee I, Kuntz G, Stevenson KB.** 2011. Guideline for the prevention and control of norovirus gastroenteritis outbreaks in healthcare settings. *Infect Control Hospital Epidemiol* **32:**939-969.
25. **Frange P, Touzot F, Debré M, Héritier S, Leruez-Ville M, Cros G, Rouzioux C, Blanche S, Fischer A, Avettand-Fenoël V.** 2012. Prevalence and clinical impact of norovirus fecal shedding in children with inherited immune deficiencies. *J Infect Dis* **206:**1269–1274.
26. **Sukhrie FH, Teunis P, Vennema H, Copra C, Thijs Beersma MF, Bogerman J, Koopmans M.** 2012. Nosocomial transmission of norovirus is mainly caused by symptomatic cases. *Clin Infect Dis* **54:**931–937.
27. **Sawinski D, Goral S.** 2014. BK virus infection: an update on diagnosis and treatment. *Nephrol Dial Transplant.*
28. **Egli A, Infanti L, Dumoulin A, Buser A, Samaridis J, Stebler C, Gosert R, Hirsch HH.** 2009. Prevalence of polyomavirus BK and JC infection and replication in 400 healthy blood donors. *J Infect Dis* **199:**837–846.
29. **Costa C, Cavallo R.** 2012. Polyomavirus-associated nephropathy. *World J Transplant* **2:**84–94.
30. **Drachenberg CB, Papadimitriou JC, Hirsch HH, Wali R, Crowder C, Nogueira J, Cangro CB, Mendley S, Mian A, Ramos E.** 2004. Histological patterns of polyomavirus nephropathy: correlation with graft outcome and viral load. *Am J Transplant* **4:**2082-2092.
31. **Behzad-Behbahani A, Klapper PE, Vallely PJ, Cleator GM, Khoo SH.** 2004. Detection of BK virus and JC virus DNA in urine samples from immunocompromised (HIV-infected) and immunocompetent (HIV-non-infected) patients using polymerase chain reaction and microplate hybridisation. *J Clin Virol* **29:**224-229.
32. **Randhawa P, Uhrmacher J, Pasculle W, Vats A, Shapiro R, Eghtsead B, Weck K.** 2005. A comparative study of BK and JC virus infections in organ transplant recipients. *J Med Virol* **77:**238–243.
33. **Garces JC.** 2010. BK virus-associated nephropathy in kidney transplant recipients. *Ochsner J* **10:**245–249.
34. **Campbell GL, Marfin AA, Lanciotti RS, Gubler DJ.** 2002. West Nile virus. *Lancet Infect Dis* **2:**519–529.
35. **Tilley PA, Fox JD, Jayaraman GC, Preiksaitis JK.** 2006. Nucleic acid testing for west nile virus RNA in plasma enhances rapid diagnosis of acute infection in symptomatic patients. *J Infect Dis* **193:**1361–1364.
36. **Tilley PA, Walle R, Chow A, Jayaraman GC, Fonseca K, Drebot MA, Preiksaitis J, Fox J.** 2005. Clinical utility of commercial enzyme immunoassays during the inaugural season of West Nile virus activity, Alberta, Canada. *J Clin Microbiol* **43:**4691–4695.
37. **Leruez-Ville M, Vauloup-Fellous C, Couderc S, Parat S, Castel C, Avettand-Fenoel V, Guilleminot T, Grangeot-Keros L, Ville Y, Grabar S, Magny JF.** 2011. Prospective identification of congenital cytomegalovirus infection in newborns using real-time polymerase chain reaction assays in dried blood spots. *Clin Infect Dis* **52:**575–581.
38. **Legrand F, Berrebi D, Houhou N, Freymuth F, Faye A, Duval M, Mougenot JF, Peuchmaur M, Vilmer E.** 2001. Early diagnosis of adenovirus infection and treatment with cidofovir after bone marrow transplantation in children. *Bone Marrow Transplant* **27:**621–626.
39. **Allard A, Albinsson B, Wadell G.** 1992. Detection of adenoviruses in stools from healthy persons and patients with diarrhea by two-step polymerase chain reaction. *J Med Virol* **37:**149–157.
40. **Stránská R, Schuurman R, de Vos M, van Loon AM.** 2004. Routine use of a highly automated and internally controlled real-time PCR assay for the diagnosis of herpes simplex and varicella-zoster virus infections. *J Clin Virol* **30:**39–44.
41. **Rübben A, Baron JM, Grussendorf-Conen EI.** 1997. Routine detection of herpes simplex virus and varicella zoster virus by polymerase chain reaction reveals that initial herpes zoster is frequently misdiagnosed as herpes simplex. *Br J Dermatol* **137:**259–261.
42. **Johnson G, Nelson S, Petric M, Tellier R.** 2000. Comprehensive PCR-based assay for detection and species identification of human herpesviruses. *J Clin Microbiol* **38:**3274–3279.
43. **Lilie HM, Wassilew SW, Wolff MH.** 2002. Early diagnosis of herpes zoster by polymerase chain reaction. *J Eur Acad Dermatol Venereol* **16:**53–57.
44. **Espy MJ, Uhl JR, Sloan LM, Buckwalter SP, Jones MF, Vetter EA, Yao JD, Wengenack NL, Rosenblatt JE, Cockerill FR III, Smith TF.** 2006. Real-time PCR in clinical microbiology: applications for routine laboratory testing. *Clin Microbiol Rev* **19:**165–256.
45. **Hall CB, Long CE, Schnabel KC, Caserta MT, McIntyre KM, Costanzo MA, Knott A, Dewhurst S, Insel RA, Epstein LG.** 1994. Human herpesvirus-6 infection in children. A prospective study of complications and reactivation. *N Engl J Med* **331:**432–438.
46. **De Bolle L, Naesens L, De Clercq E.** 2005. Update on human herpesvirus 6 biology, clinical features, and therapy. *Clin Microbiol Rev* **18:**217–245.
47. **Campadelli-Fiume G, Mirandola P, Menotti L.** 1999. Human herpesvirus 6: an emerging pathogen. *Emerg Infect Dis* **5:**353–366.
48. **Corti M, Villafañe MF, Trione N, Mamanna L, Bouzas B.** 2011. Human herpesvirus 6: report of emerging pathogen in five patients with HIV/AIDS and review of the literature. *Rev Soc Bras Med Trop* **44:**522–525.
49. **Levy JA, Ferro F, Greenspan D, Lennette ET.** 1990. Frequent isolation of HHV-6 from saliva and high seroprevalence of the virus in the population. *Lancet* **335:**1047–1050.
50. **Ghany MG, Strader DB, Thomas DL, Seeff LB, American Association for the Study of Liver Diseases.** 2009. Diagnosis, management, and treatment of hepatitis C: an update. *Hepatology* **49:**1335–1374.
51. **CDC.** 2013. Testing for HCV infection: an update of guidance for clinicians and laboratorians. *MMWR Morb Mortal Wkly Rep* **62:**362–365.
52. **Ghany MG, Nelson DR, Strader DB, Thomas DL, Seeff LB, American Association for Study of Liver Diseases.** 2011. An update on treatment of genotype 1 chronic hepatitis C virus infection: 2011 practice guideline by the American Association for the Study of Liver Diseases. *Hepatology* **54:**1433–1444.
53. **DeBiasi RL, Tyler KL.** 2004. Molecular methods for diagnosis of viral encephalitis. *Clin Microbiol Rev* **17:**903–925.
54. **Peeling RW, Artsob H, Pelegrino JL, Buchy P, Cardosa MJ, Devi S, Enria DA, Farrar J, Gubler DJ, Guzman MG, Halstead SB, Hunsperger E, Kliks S, Margolis HS, Nathanson CM, Nguyen VC, Rizzo N, Vázquez S, Yoksan S.** 2010. Evaluation of diagnostic tests: dengue. *Nat Rev Microbiol* **8** (Suppl):S30–S37.
55. **Bag AK, Curé JK, Chapman PR, Roberson GH, Shah R.** 2010. JC virus infection of the brain. *AJNR Am J Neuroradiol* **31:**1564–1576.
56. **Berger JR, Khalili K.** 2011. The pathogenesis of progressive multifocal leukoencephalopathy. *Discov Med* **12:**495–503.
57. **Berger JR, Miller CS, Mootoor Y, Avdiushko SA, Kryscio RJ, Zhu H.** 2006. JC virus detection in bodily fluids: clues to transmission. *Clin Infect Dis* **43:**e9–e12.

58. **Garnett CT, Erdman D, Xu W, Gooding LR.** 2002. Prevalence and quantitation of species C adenovirus DNA in human mucosal lymphocytes. *J Virol* **76:**10608–10616.
59. **Chung PW, Huang YC, Chang LY, Lin TY, Ning HC.** 2001. Duration of enterovirus shedding in stool. *J Microbiol Immunol Infect* **34:**167–170.
60. **Cinek O, Witso E, Jeansson S, Rasmussen T, Drevinek P, Wetlesen T, Vavrinec J, Grinde B, Ronningen KS.** 2006. Longitudinal observation of enterovirus and adenovirus in stool samples from Norwegian infants with the highest genetic risk of type 1 diabetes. *J Clin Virol* **35:**33-40.
61. **Kupila L, Vuorinen T, Vainionpää R, Marttila RJ, Kotilainen P.** 2005. Diagnosis of enteroviral meningitis by use of polymerase chain reaction of cerebrospinal fluid, stool, and serum specimens. *Clin Infect Dis* **40:**982–987.
62. **Yamamoto N, Wakabayashi T, Murakami K, Hommura S.** 2003. Detection of CMV DNA in the aqueous humor of AIDS patients with CMV retinitis by AMPLICOR CMV test. *Ophthalmologica* **217:**45–48.
63. **Ginocchio CC, Wang XP, Kaplan MH, Mulligan G, Witt D, Romano JW, Cronin M, Carroll R.** 1997. Effects of specimen collection, processing, and storage conditions on stability of human immunodeficiency virus type 1 RNA levels in plasma. *J Clin Microbiol* **35:**2886–2893.
64. **Holodniy M, Kim S, Katzenstein D, Konrad M, Groves E, Merigan TC.** 1991. Inhibition of human immunodeficiency virus gene amplification by heparin. *J Clin Microbiol* **29:**676–679.
65. **Jagodzinski LL, Weston HR, Liu Y, O'Connell RJ, Peel SA.** 2012. Efficient quantification of HIV-1 in heparin plasma spiked with cultured HIV-1 by the Roche Cobas TaqMan and Abbott RealTime HIV-1 tests. *J Clin Microbiol* **50:**2804–2806.
66. **Forman MS, Valsamakis A.** 2011. Specimen collection, transport, and processing: virology, p 1276–1296. *In* Landry ML (ed), *Manual of Clinical Microbiology*, 10th ed, **vol 2**. ASM Press, Washington, D.C.
67. **Levin MJ, Leventhal S, Masters HA.** 1984. Factors influencing quantitative isolation of varicella-zoster virus. *J Clin Microbiol* **19:**880–883.
68. **Hernes SS, Quarsten H, Hagen E, Lyngroth AL, Pripp AH, Bjorvatn B, Bakke PS.** 2011. Swabbing for respiratory viral infections in older patients: a comparison of rayon and nylon flocked swabs. *Eur J Clin Microbiol Infect Dis* **30:**159-165.
69. **Daley P, Castriciano S, Chernesky M, Smieja M.** 2006. Comparison of flocked and rayon swabs for collection of respiratory epithelial cells from uninfected volunteers and symptomatic patients. *J Clin Microbiol* **44:**2265–2267.
70. **Loeffelholz MJ, Pong DL, Pyles RB, Xiong Y, Miller AL, Bufton KK, Chonmaitree T.** 2011. Comparison of the FilmArray Respiratory Panel and Prodesse real-time PCR assays for detection of respiratory pathogens. *J Clin Microbiol* **49:**4083–4088.
71. **Rand KH, Rampersaud H, Houck HJ.** 2011. Comparison of two multiplex methods for detection of respiratory viruses: FilmArray RP and xTAG RVP. *J Clin Microbiol* **49:**2449–2453.
72. **Sung RY, Chan PK, Choi KC, Yeung AC, Li AM, Tang JW, Ip M, Tsen T, Nelson EA.** 2008. Comparative study of nasopharyngeal aspirate and nasal swab specimens for diagnosis of acute viral respiratory infection. *J Clin Microbiol* **46:**3073–3076.
73. **Lee K, Park HD, Kang ES.** 2013. Reduction of the HIV seroconversion window period and false positive rate by using ADVIA Centaur HIV antigen/antibody combo assay. *Ann Lab Med* **33:**420–425.
74. **Aberg JA, Gallant JE, Ghanem KG, Emmanuel P, Zingman BS, Horberg MA.** 2014. Primary care guidelines for the management of persons infected with HIV: 2013 update by the HIV medicine association of the Infectious Diseases Society of America. *Clin Infect Dis* **58:**e1–e34.
75. **Tomblyn M, Chiller T, Einsele H, Gress R, Sepkowitz K, Storek J, Wingard JR, Young JA, Boeckh MJ,** 2009. Guidelines for preventing infectious complications among hematopoietic cell transplantation recipients: a global perspective. *Biol Blood Marrow Transplant* **15:**1143–1238.
76. **Holland JL, Louie L, Simor AE, Louie M.** 2000. PCR detection of Escherichia coli O157:H7 directly from stools: evaluation of commercial extraction methods for purifying fecal DNA. *J Clin Microbiol* **38:**4108–4113.
77. **Kermekchiev MB, Kirilova LI, Vail EE, Barnes WM.** 2009. Mutants of Taq DNA polymerase resistant to PCR inhibitors allow DNA amplification from whole blood and crude soil samples. *Nucleic Acids Res* **37:**e40.
78. **Schrader C, Schielke A, Ellerbroek L, Johne R.** 2012. PCR inhibitors - occurrence, properties and removal. *J Appl Microbiol* **113:**1014–1026.
79. **Shulman LM, Hindiyeh M, Muhsen K, Cohen D, Mendelson E, Sofer D.** 2012. Evaluation of four different systems for extraction of RNA from stool suspensions using MS-2 coliphage as an exogenous control for RT-PCR inhibition. *PLoS One* **7:**e39455.
80. **Buckwalter SP, Sloan LM, Cunningham SA, Espy MJ, Uhl JR, Jones MF, Vetter EA, Mandrekar J, Cockerill FR III, Pritt BS, Patel R, Wengenack NL.** 2014. Inhibition controls for qualitative real-time PCR assays: are they necessary for all specimen matrices? *J Clin Microbiol* **52:**2139–2143.
81. **Hoorfar J, Malorny B, Abdulmawjood A, Cook N, Wagner M, Fach P.** 2004. Practical considerations in design of internal amplification controls for diagnostic PCR assays. *J Clin Microbiol* **42:**1863–1868.
82. **Stacey GN.** 2011. Cell culture contamination. *Methods Mol Biol* **731:**79–91.
83. **Steinau M, Patel SS, Unger ER.** 2011. Efficient DNA extraction for HPV genotyping in formalin-fixed, paraffin-embedded tissues. *J Mol Diagn* **13:**377–381.
84. **McKinney MD, Moon SJ, Kulesh DA, Larsen T, Schoepp RJ.** 2009. Detection of viral RNA from paraffin-embedded tissues after prolonged formalin fixation. *J Clin Virol* **44:**39-42.
85. **Inoue T, Nabeshima K, Kataoka H, Koono M.** 1996. Feasibility of archival non-buffered formalin-fixed and paraffin-embedded tissues for PCR amplification: an analysis of resected gastric carcinoma. *Pathol Int* **46:**997–1004.
86. **Zhao FH, Hu SY, Bian JJ, Liu B, Peck RB, Bao YP, Pan QJ, Frappart L, Sellors J, Qiao YL.** 2011. Comparison of ThinPrep and SurePath liquid-based cytology and subsequent human papillomavirus DNA testing in China. *Cancer Cytopathol* **119:**387–394.
87. **Tirumala R, Clary KM.** 2013. *Dual Sample Collection and Comparison of COBAS HPV Testing Using SurePath and ThinPrep Media.* Rochester General Hospital, Rochester, New York, USA, Daytona Beach, Florida.
88. **Dickinson B.** 2011. *BD SurePath™ Preservative Fluid. Material Safety Data Sheet:Becton Dickinson.* Franklin Lakes, NJ.
89. **Druce J, Garcia K, Tran T, Papadakis G, Birch C.** 2012. Evaluation of swabs, transport media, and specimen transport conditions for optimal detection of viruses by PCR. *J Clin Microbiol* **50:**1064–1065.
90. **Johnson FB.** 1990. Transport of viral specimens. *Clin Microbiol Rev* **3:**120–131.
91. **Abad FX, Pintó RM, Bosch A.** 1994. Survival of enteric viruses on environmental fomites. *Appl Environ Microbiol* **60:**3704–3710.
92. **Jerome KR, Huang ML, Wald A, Selke S, Corey L.** 2002. Quantitative stability of DNA after extended storage of clinical specimens as determined by real-time PCR. *J Clin Microbiol* **40:**2609–2611.
93. **Krajden M, Minor JM, Rifkin O, Comanor L.** 1999. Effect of multiple freeze-thaw cycles on hepatitis B virus DNA and hepatitis C virus RNA quantification as measured with branched-DNA technology. *J Clin Microbiol* **37:**1683–1686.
94. **Krafft AE, Russell KL, Hawksworth AW, McCall S, Irvine M, Daum LT, Connoly JL, Reid AH, Gaydos JC, Taubenberger JK.** 2005. Evaluation of PCR testing of ethanol-fixed nasal swab specimens as an augmented surveillance strategy for influenza virus and adenovirus identification. *J Clin Microbiol* **43:**1768–1775.

95. **Frisbie B, Tang YW, Griffin M, Poehling K, Wright PF, Holland K, Edwards KM.** 2004. Surveillance of childhood influenza virus infection: what is the best diagnostic method to use for archival samples? *J Clin Microbiol* **42:**1181–1184.
96. **CDC.**2014. How U.S. Clinical Laboratories Can Safely Manage Specimens from Persons Under Investigation for Ebola Virus Disease, on Department of Health and Human Services. http://www.cdc.gov/vhf/ebola/hcp/safe-specimen-management.html. Accessed 9/11/14.
97. **Khare R, Espy MJ, Cebelinski E, Boxrud D, Sloan LM, Cunningham SA, Pritt BS, Patel R, Binnicker MJ.** 2014. Multiplex detection of gastrointestinal pathogens: a comparative evaluation of two commercial panels using clinical stool specimens. *J Clin Microbiol* **Electronically published ahead of print**.
98. **Mohamed S, Penaranda G, Gonzalez D, Camus C, Khiri H, Boulmé R, Sayada C, Philibert P, Olive D, Halfon P.** 2014. Comparison of ultra-deep versus Sanger sequencing detection of minority mutations on the HIV-1 drug resistance interpretations after virological failure. *AIDS* **28:**1315–1324.
99. **Téllez-Sosa J, Rodríguez MH, Gómez-Barreto RE, Valdovinos-Torres H, Hidalgo AC, Cruz-Hervert P, Luna RS, Carrillo-Valenzo E, Ramos C, García-García L, Martínez-Barnetche J.** 2013. Using high-throughput sequencing to leverage surveillance of genetic diversity and oseltamivir resistance: a pilot study during the 2009 influenza A(H1N1) pandemic. *PLoS One* **8:**e67010.
100. **Orozovic G, Orozovic K, Lennerstrand J, Olsen B.** 2011. Detection of resistance mutations to antivirals oseltamivir and zanamivir in avian influenza A viruses isolated from wild birds. *PLoS One* **6:**e16028.

Primary Isolation of Viruses

MARIE L. LANDRY AND DIANE LELAND

7

Viruses are obligate intracellular parasites and thus are propagated using living cells in the form of cultured cells, embryonated hen's eggs, or laboratory animals. Culture has long been considered the "gold standard" for viral diagnosis because it secures an isolate for further analysis, is more "open-minded" than methods that target single agents, and allows the unexpected or even novel agent to be recovered. In practice, use of specialized cell culture systems, embryonated eggs, and laboratory animals is confined to research or major public health reference laboratories, with cell cultures in monolayers the sole isolation system utilized in routine diagnostic laboratories. The past two decades have seen conventional cell culture methods supplemented or even replaced by more rapid and targeted cell culture methods. Rapid culture methods can be performed by less experienced personnel, with less labor, and with results reported within 1 to 5 days of inoculation.

Isolating viruses in cell cultures in various configurations has been the main diagnostic approach in most clinical virology laboratories, but this approach is being used less frequently now that viral antigen detection assays and molecular techniques have been improved in analytical sensitivity and specificity and simplified to allow their performance in routine laboratories. At this writing isolation in cell culture is most often incorporated into viral diagnostic algorithms to confirm optimal performance of other methods, to validate newer methods, or to aid the discovery of new viruses, rather than as the primary diagnostic method. This change in the role of virus isolation in cell cultures is reflected in the lack of recent publications in innovative cell culture methods for clinical diagnosis. Regardless of the breadth of application across smaller clinical laboratories, virus isolation in culture will continue to be used in larger specialized virology and public health laboratories and by those interested in virus discovery. Thus this chapter will provide details of current cell culture isolation methods used for viral diagnosis and describe various clinical applications of virus isolation in cell culture.

VIRUS ISOLATION IN CELL CULTURES

Types of Cell Culture

The discovery by Enders, Weller, and Robbins in the late 1940s that poliovirus replicates in cultivated mammalian cells revolutionized and simplified procedures for the isolation of viruses (1). After that landmark discovery, cell cultures were prepared for virus studies from a wide variety of animal and human tissues, and many of the common viruses we are familiar with today were discovered. Cell cultures are generally separated into three types (Table 1): primary cells, which are prepared directly from animal or human tissues and can usually be subcultured for only one or two passages; diploid cell cultures, which are usually derived from human tissues, either fetal or newborn, and can be subcultured 20 to 50 times before senescence; and continuous cell lines, which can be established from human or animal tissues, from tumors, or following the spontaneous transformation of normal tissues. These have a heteroploid karyotype and can be subcultured an indefinite number of times. However, sensitivity to virus infection may change after serial passage and after passage in different laboratories.

Variation in Sensitivity to Different Viruses

Cell cultures vary greatly in their sensitivity to different viruses (Table 2). If a virus is inoculated into an insensitive cell culture, the virus will not be able to replicate and a negative result will be obtained. When small amounts of virus are present in a clinical sample, a positive result may be obtained only when the most sensitive systems are used. Therefore, it is important that providers caring for the patients inform the laboratory of the clinical syndrome and/or virus(es) suspected, so that the most sensitive cell cultures can be used and appropriate detection methods employed. Laboratories should periodically monitor the sensitivity characteristics of cell cultures, since significant changes can occur over time or even from season to season for rapidly changing viruses such as influenza (2–4).

Supplies and Equipment Needed

The materials needed for the isolation of viruses in cell culture are given in Table 3. Maintaining different cell cultures in healthy condition is absolutely necessary to ensure good results. A wide variety of cell cultures are available commercially and can be purchased and delivered once or twice a week according to the needs of the laboratory. Vendors of prepared cell cultures in the United States include Diagnostic Hybrids (DHI, Quidel, Athens, OH) and CellPro Labs (Golden Valley, MN). Both the quantity and

doi:10.1128/9781555819156.ch7

TABLE 1 Types of cell cultures commonly used in a clinical virology laboratory

Cell culture[a]	Examples	No. of subpassages
Primary	Kidney tissues from monkeys (RhMK), rabbits (RK), etc. Embryos from chickens (CE), guinea pigs (GPE), etc.	1 or 2
Diploid	Human embryonic lung (MCR-5) or human newborn foreskin (HFF)	20–50
Continuous	Human epidermoid carcinoma of lung (A549), mink lung (ML or Mv1Lu)	Indefinite

[a]Mixtures of different cell cultures within a single tube or well are now commonly used. Some continuous cell lines have been genetically engineered to provide a reporter system for rapid and simplified detection, or for greater sensitivity.

types of cell cultures used will vary with the seasonal variations in virus activity. The use of cryopreserved cell cultures, that can be stored at −70°C and thawed for use as needed, can provide additional flexibility (5, 6).

Obtaining and Processing Specimens

It is important to reiterate that without appropriate specimens that are properly collected early in illness and promptly transported to the laboratory, the subsequent time and effort spent in isolation attempts will be wasted. Accomplishing this is an important task of the clinical virology laboratory and requires continuing communication with and education of the providers.

Conventional Cell Culture

Conventional cell cultures in clinical laboratories have traditionally been grown as monolayers in screw-capped roller tubes and been inoculated and incubated as described in the following section. However, cultures in 24-well plates and shell vials (1-dram vials) can also be used and either fixed and stained at a predetermined time after inoculation or observed for longer periods for cytopathic effects (CPE). To isolate a spectrum of viruses in conventional culture, cells of several different types are inoculated, such as human diploid fibroblasts (HDF), a human heteroploid cell line (e.g., A549), and a primary monkey kidney cell culture. Alternatively, for specific indications (e.g., herpes simplex virus [HSV] infection), limited cultures intended to detect only one or two virus types can be performed. The cell type(s) most sensitive to the suspected viruses in the clinical specimen should be included. Ideally, only actively dividing cultures should be used because aged cells are less sensitive to virus infection. All cell cultures should be examined under the microscope before inoculation to ensure that the cells are in good condition.

Inoculation and Incubation

Although techniques may vary somewhat for different viruses, in general, the following procedures apply for noncentrifuged conventional cultures:

1. Pour off or aspirate culture media and inoculate specimens, 0.1 to 0.3 ml, into each culture tube. Uninoculated cultures should be kept in parallel for comparison.
2. Allow specimen to adsorb (in a horizontal position in a stationary rack) in the incubator at 35°C to 37°C for 30 to 60 minutes. Then, 1.0 to 1.5 ml of maintenance medium should be added and the inoculated cultures returned to the incubator. Inoculated cultures in roller tubes can be placed in a rotating drum if available, which is optimal for the isolation of respiratory viruses, especially rhinoviruses, and results in the earlier appearance of CPE for many viruses. If stationary racks are used, it is critical that culture tubes be positioned so that the cell monolayer is bathed in nutrient medium; otherwise, the cells will degenerate, especially at the edge of the monolayer.
3. Microscopically (light microscope, ×10 objective, reduced light) evaluate inoculated culture tubes daily for the first week, then every other day for virus-induced CPE. Compare the appearance of the inoculated tubes with uninoculated control tubes from the same lot of cell cultures. Also note the color of the medium in the inoculated tubes. Culture medium containing a phenol red indicator should appear light orange or peach colored. Bright magenta color signals pH that is too basic, and bright yellow signals acid

TABLE 2 Conventional cell cultures for viruses commonly isolated in a clinical virology laboratory[a,b]

Virus	Sensitivity in cell culture[c]			Average time to CPE (range)	Characteristic CPE
RNA viruses	PMK	HDF	A549		
Influenza	+++	+	−	2 (1–7)	Cellular granulation, HAd-positive
Parainfluenza	+++	+	+/−	6 (1–14)	Rounded cells, some syncytia, HAd-positive
RSV	++	+	+/−	6 (2–14)	Syncytia in Hep-2 and RhMK
Rhinovirus	+/−	++/−	−	5 (2–14)	Degenerating rounded cells
Enterovirus	+++	++	+/−	2 (1–8)	Small round refractile cells
DNA viruses					
Adenovirus	+	++	+++	6 (1–14)	Grape-like clusters
CMV	−	+++	−	8 (1–28)	Foci of rounded cells; slow progression
HSV	+/−	++	+++	2 (1–7)	Foci large rounded cells; HSV-2 produces syncytia; rapid progression
VZV	+	+++	+++	6 (3–14)	Foci of pyknotic, degenerating cells

[a]PMK, primary monkey kidney; HDF, human diploid fibroblasts; A549, human heteroploid cell line; CPE, cytopathic effects; HAd, hemadsorption.
[b]Viruses that are not routinely isolated in conventional cultures include HMPV, coronaviruses, rhinovirus group C, EBV, HHV-6.
[c]Degree of sensitivity: +++, highly sensitive; ++, moderately sensitive; +, low sensitivity; +/−, variable; −, not sensitive.

TABLE 3 Supplies and equipment needed for isolation of viruses in cell culture

Process	Supplies and equipment needed
Inoculation of cell cultures	Laminar flow hood, centrifuge, pipettes, automatic pipetting device, pipette jar and discard can, disposable gloves, disinfectant, and sterile glass and plastic ware
Maintenance of cell cultures	Culture media, serum, antibiotics, 4°C refrigerator, test tube racks, and/or rotating drum, shell vial racks, room air incubators, CO_2 incubator, waterbath, and upright and inverted microscopes
Staining of shell vials and identification of virus isolates	Centrifuge, centrifuge tubes, PBS, Teflon-coated microscope slides, forceps, incubator, monoclonal antibodies and reagents, mounting medium, fluorescence microscope
Preservation and storage of viruses	Freezer vials, ultra-low temperature freezer (−70°C), and dimethyl sulfoxide as stabilizer

pH, which is often a marker of bacterial or fungal contamination. Samples showing altered pH should be evaluated and the pH corrected. Steps 4 and 5 describe corrective action.

4. Certain specimens, such as urine and stool, will frequently be toxic to the cell cultures, and this toxicity can be confused with virus-induced CPE. With such specimens it is a good practice to check inoculated tubes, either after adsorption or within 24 hours of inoculation, and refeed with fresh medium if necessary. If toxic effects are extensive, it may be necessary to subpassage the inoculated cells in order to dilute toxic factors and provide viable cells for virus growth.
5. Bacterial or fungal contamination will require filtration, using a 450-μm filter, of either the inoculated culture supernatant fluid or of the original specimen, followed by inoculation of fresh cultures.
6. Inoculated cultures and the uninoculated cell culture control tubes are generally kept for observation for virus-induced effects for 10 to 14 days. Exceptions include cultures for HSV only, which may be terminated at 7 days, and for cytomegalovirus (CMV), which are commonly kept for 3 to 4 weeks. During this time, cell cultures may need to be refed to maintain the cells in good condition. Some continuous cell lines may require refeeding or subculturing every few days. Great care must be taken when refeeding cultures to ensure that cross contamination from one specimen to another does not occur. Separate pipettes should be used for separate specimens. Aerosols, spatter, and contamination of test tube caps and gloves should be avoided.
7. When virus-induced CPE occurs and progresses to include 25% to 50% of the monolayer, specific identification can usually be accomplished by immunofluorescence (IF) staining of infected cells. Passage of infected cultures into a fresh culture of the same cell type may be necessary to ensure recovery of sufficient virus for further identification of the isolate or ensure that a virus is present. For certain cell-associated viruses, such as CMV or varicella zoster virus (VZV), it is necessary to trypsinize and passage intact infected cells. Adenovirus can be subcultured after freezing and thawing infected cells, which disrupts the cells and releases intracellular virus.
8. For certain fastidious viruses, when the amount of infectious virus in the specimen is low, or when the patient has received antiviral therapy, blind passage (i.e., subculture of the inoculated culture in the absence of viral CPE) into a set of fresh culture tubes may be necessary before virus proliferation can be detected.

Detection of Virus-Induced Effects

Cytopathic Effects

Many viruses can be identified by the characteristic cellular changes they induce in susceptible cell cultures. These can be visualized under the light microscope. Examples of CPE characteristic for a number of common viruses are shown in Figure 1 and described in greater detail in the sections on individual viruses. The degree of CPE is usually graded from + to ++++ based on the percentage of the cell monolayer infected: 25% of the cell monolayer (+), 50% (++), 75% (+++), and 100% (++++). There are two important points that should be emphasized regarding CPE induced by virus:

1. The *rate* at which CPE progresses may help to distinguish similar viruses; for example, HSV progresses rapidly to involve the entire monolayer of several cell systems. In contrast, two other herpesviruses, CMV and VZV, grow primarily in HDF cells and progress slowly over a number of days or weeks.
2. The *type* of cell culture(s) in which the virus replicates is important factor in identification (Table 2).

Caution must be exercised to be certain that a virus-induced CPE is distinguished from "nonspecific" CPE caused by cell age or toxicity of specimens, bacteria, fungi, or parasites. A subculture onto fresh cells should amplify virus effects and dilute toxic effects. On occasion, foci of cells inoculated onto the culture monolayer from the original specimen or from another cell culture can be mistaken for viral CPE. With experience, the appearance of the cellular changes, taken together with the susceptible cell systems, the specimen source, and clinical disease, usually allows a presumptive diagnosis to be made as soon as the virus-induced cellular changes occur.

Hemadsorption (HAd)

Parainfluenza and sometimes influenza virus replication may not induce distinctive cellular changes; however, these viruses express hemagglutinins that are expressed on the infected cell membrane and have an affinity for red blood cells (RBCs). The addition of a guinea pig RBC suspension to the infected cultures allows the RBCs to adsorb onto the infected cells, resulting in the observations shown in Figure 2B. However, if older guinea pig RBCs are used, nonspecific HAd may occur in an uninoculated culture (Fig. 2C) and should be distinguished from that resulting from a specific viral infection. Furthermore, the HAd test is usually performed at 4°C to 22°C because the RBCs will elute when incubated at 37°C. When HAd is observed on the monolayer, the infected cells are transferred to a slide and stained by IF to identify the causative virus. Alternatively, the culture fluid can be subcultured into a fresh culture either to confirm the virus isolation or to permit further identification.

It should be noted that not all viruses that agglutinate RBCs can adsorb them onto infected cell monolayers. HAd

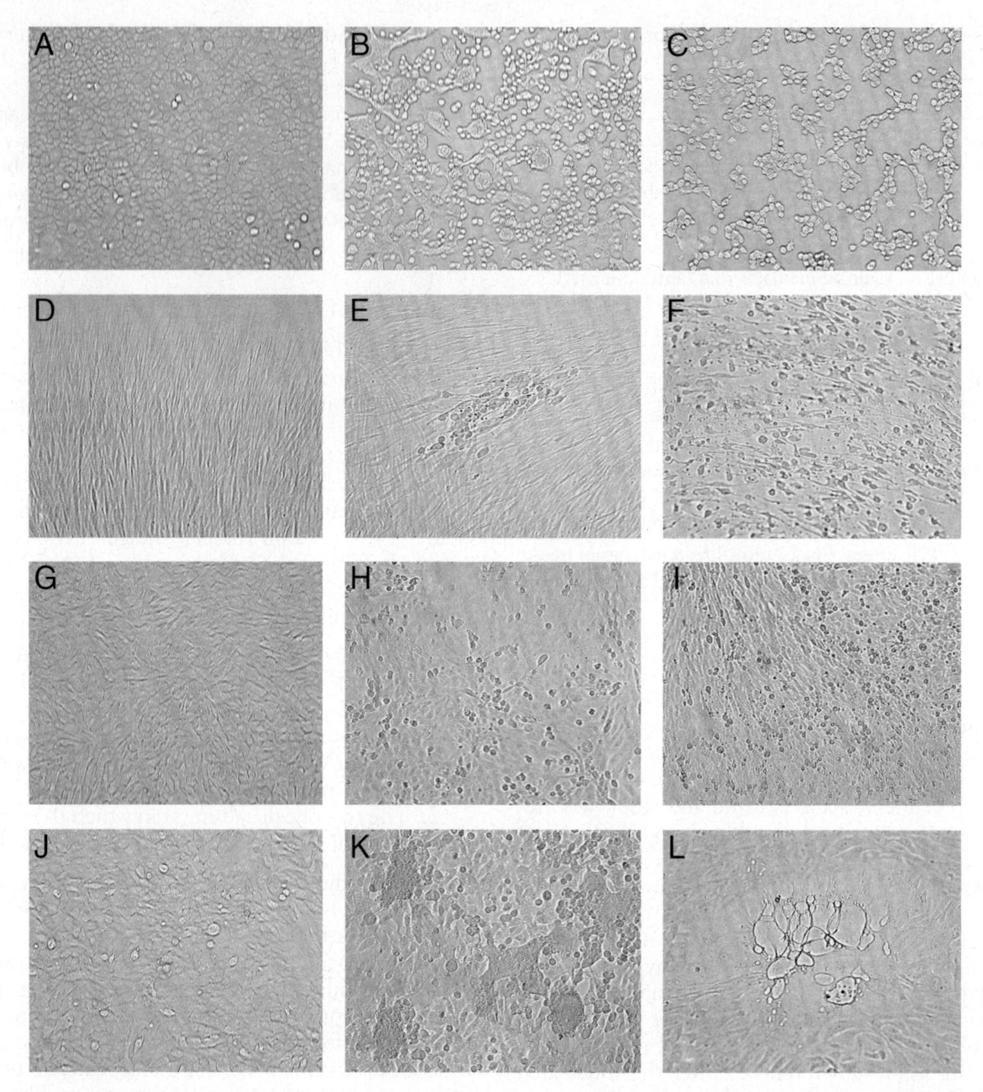

FIGURE 1 Uninfected cell cultures and cell cultures showing CPE of viruses commonly isolated. (A) Uninfected A549 cells, (B) HSV-2 in A549, (C) adenovirus in A549, (D) uninfected MRC-5 fibroblasts, (E) CMV in MRC-5, (F) rhinovirus in MRC-5, (G) uninfected RhMK, (H) enterovirus in RhMk, (I) influenza A in RhMk, (J) uninfected HEp-2, (K) RSV in HEp-2, and (L) monkey virus contaminant in RhMk. ×85. (Photos by permission, Leland and Ginocchio [6].)

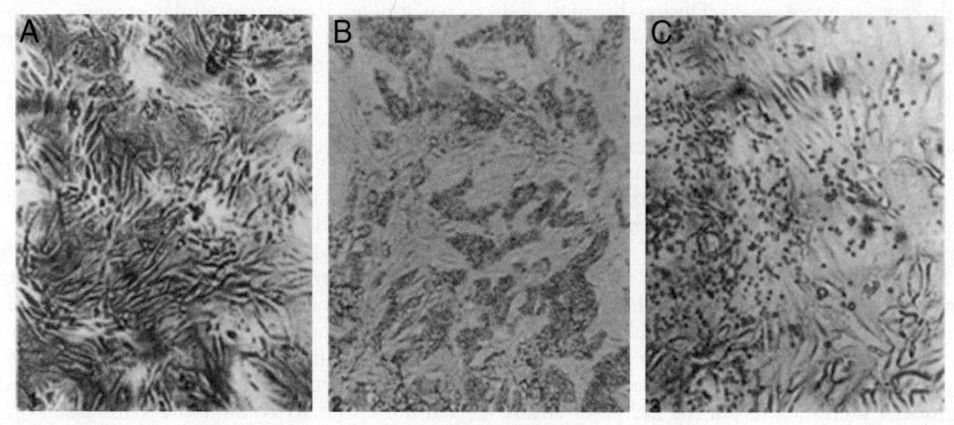

FIGURE 2 Hemadsorption of guinea pig red blood cells by parainfluenza virus in monkey kidney (MK) cells. (A) Uninfected MK cells; (B) specific hemadsorption in parainfluenza infected MK cells; (C) nonspecific hemadsorption seen with aged red blood cells in uninfected cell cultures. Modified from Hsiung et al. (72). (Reprinted from last edition.)

is a property of those viruses that bud from the host cell membrane during maturation and thus express viral hemagglutinin on the surface of the infected cell.

Blind Immunostaining

To more rapidly detect virus growth, shorten the observation period, and avoid repeated examinations for CPE, intact cell culture monolayers can be fixed and stained with fluorescein or horseradish peroxidase-labeled antibodies to viral antigens usually within 1 to 6 days after inoculation. Alternatively, at the end of the observation period and in the absence of positive CPE or HAd, cells can be scraped or trypsinized from roller tubes, affixed to glass slides, and blindly stained for detection of viral antigens prior to discard.

Identification of Virus Isolates

Presumptive identification of virus isolates usually can be made on the basis of characteristic virus-induced effects (e.g., type of CPE or HAd) and selective cultured cell sensitivity. Definitive identification most frequently uses virus-specific fluorescein-labeled antibodies. Monolayers showing CPE or HAd can be trypsinized or dislodged via scraping. The cells are transferred and fixed to a glass slide with individual wells, and the fixed preparations are stained with antibodies specific to viral antigens. In more difficult cases, PCR also can be used for virus identification. When determination of specific type is requested for enteroviruses or adenoviruses, neutralization of virus-induced cytopathology in cell culture may be used; this testing is rarely performed outside of the research setting. Typing is now performed by PCR, followed by nucleic acid sequencing (7, 8).

Occasionally, a virus isolate cannot be identified by the standard tests. The morphologic properties of the infecting virus can be determined by electron microscopy, if available, with subsequent identification by molecular techniques (9). Some viruses may be known but rarely diagnosed in human disease, for example, neurologic disease due to Cache Valley Virus, which has been isolated from cerebrospinal fluid in Vero E6 (10), A549 and RD (11), or Buffalo green monkey kidney (BGMK) cells (12). "New" viruses recovered by isolation from clinical samples include human metapneumovirus in tertiary monkey kidney cells (13), severe acute respiratory syndrome (SARS) coronavirus in Vero E6 cells (14), NL-63 coronavirus in tertiary monkey kidney cells (15, 16), Middle East respiratory syndrome (MERS) coronavirus isolated in Vero and LLC-MK2 (17), and severe fever and thrombocytopenia syndrome (SFTS) and Heartland phleboviruses isolated in DH82 cells, a canine macrophage-monocyte cell line (18, 19).

RAPID CULTURE METHODS

Centrifugation Culture (Shell Vial Technique)

The rapid diagnosis of viral infections is important in patient management. However, conventional virus isolation requires observation of monolayer cultures for CPE, which can take days to weeks to appear. The application of centrifugation cultures to viral diagnosis can shorten time to diagnosis to 1 to 2 days.

It has long been recognized that low-speed centrifugation of cell cultures enhances infectivity of viruses (20) as well as chlamydia. The mechanism for this effect is unclear and may involve centrifugation of virus aggregates or virus attached to cell debris or an effect on cell membranes to enhance virus entry.

FIGURE 3 Centrifugation culture: detection of CMV immediate early antigens in infected nuclei at 16 to 24 hours postinoculation (immunofluorescence stain). (Reprinted from last edition.)

In 1984 the use of centrifugation cultures followed by staining with a monoclonal antibody at 24 hours postinoculation was first reported for CMV (21) (Fig. 3). Subsequent reports documented its usefulness in rapid diagnosis of other viruses, including HSV (22), VZV (23), adenovirus (24), respiratory viruses (25), and polyomavirus BK (26). In addition, when the inoculum is standardized, semiquantitative results can be obtained by counting the number of virus-positive cells (27).

The shell vial technique usually combines (i) cell culture to amplify virus in the specimen, (ii) centrifugation to enhance viral infectivity, and (iii) early detection of virus-induced antigen (before CPE) by the use of high-specificity antibodies. It can be used for any virus that replicates in cell culture and for which a specific antibody is available. For viruses with a long replication cycle, such as CMV, viral antigens produced early in the replication cycle can be detected many days before CPE are apparent using light microscopy. For viruses that replicate faster (e.g., HSV) or if the available antibodies are directed toward late rather than early replication products, less time is gained for detecting positives using the shell vial technique. However, shell vial cultures can be terminated and negatives reported at 1 to 5 (VZV) days rather than at 7 to 14 days.

It should be noted that centrifugation cultures can also be employed without early IF staining, but rather can be monitored for CPE and tested by HAd, similar to conventional cultures in roller tubes. Centrifugation cultures can also be performed using 24- or 48-well tissue culture plates, instead of in shell vials.

The overall sensitivity of the shell vial technique varies with the type of specimen (28), the length and temperature of centrifugation (29), the virus, the cell cultures, the antibody employed, and the time of fixation and staining. The use of young cell monolayers (30) and inoculation of multiple shell vials enhance the recovery rate (31). Toxicity, particularly problematic with blood and urine specimens, can lead to cell death and the loss of the monolayer, necessitating blind passage of the specimen or specimen reinoculation.

Furthermore, with all rapid techniques that target specific viruses, only the viruses sought will be detected. Although conventional isolation using a spectrum of cell cultures can detect a variety of virus types, including unanticipated agents (32), maximal sensitivity and virus recovery is obtained by performing both conventional culture and centrifugation cultures in parallel (25,33–35). To conserve resources, this combined approach might be limited to selected patient groups and sample types.

Inoculation of Shell Vials (Traditional Single Cell Culture Type, Stained for One Virus)

Note: Manufacturer's instructions should be followed. Fixatives, amount of antibody, and staining times may vary.

Reagents and Equipment: Antibodies to specific viral types, usually fluorescein labeled
Cold acetone
Cell cultures grown on coverslips in shell vials, sensitive to the suspected viruses
Low-speed centrifuge with adapters for shell vials
Humidified chamber
Rotator or rocker
Suction flask and vacuum source

Test Procedure

1. Prepare two shell vials.
2. Remove cap and aspirate medium from shell vial.
3. Inoculate prepared specimen onto monolayer, 0.2 to 0.3 ml per vial.
4. Replace cap and centrifuge (30 to 60 minutes at 700 × *g*).
5. Aspirate inoculum for blood, urine, and stool samples, then rinse with 1 ml of medium to reduce toxicity.
6. Add 1.0 ml of maintenance medium to each shell vial and incubate at 35°C for 1–2 days.

Fixation of coverslips in shell vials.

1. Before fixation, inspect the coverslips for toxicity, contamination, and so forth. If necessary, passage the cell suspension to a new vial and repeat incubation before staining.
2. If monolayer is intact, aspirate medium from shell vials and rinse once with 1.0 ml of phosphate-buffered saline (PBS) (pH 7). If monolayer appears fragile, do not rinse with PBS.
3. Aspirate medium completely, add 1.0 ml of 100% cold acetone or 50/50 acetone/methanol to each shell vial and allow cells to fix for 10 minutes.
4. Aspirate the acetone and allow the coverslips in the shell vial to dry completely.

Staining of coverslips.

1. Add 1.0 ml of PBS to each coverslip, then aspirate the PBS.
2. Pipet 150 μl (five drops) of antibody reagent (appropriately titrated and diluted) into the shell vial. Replace the cap.
3. Rock the tray holding shell vials to distribute the reagent; then check to see that coverslips are not floating above the reagent.
4. Place rack holding the shell vials in a humidified chamber in the 35°C incubator.
5. Incubate for 30 minutes.
6. Add 1.0 ml of PBS to the shell vial, then aspirate. Repeat wash step two additional times.

For *direct assays* (primary antibody is labeled), go directly to step 9. For *indirect assays* (primary antibody is not labeled):

7. Pipet 150 μl (five drops) of labeled conjugate onto the monolayer.
8. Repeat steps 3 to 6, except do not aspirate the last 1.0 ml of PBS.
9. Using forceps and a wire probe, remove coverslip, and blot on tissue or absorbent paper (e.g., Kimwipe).
10. Add one drop of mounting fluid to a properly labeled slide and place coverslip on mounting fluid with cell side down, being careful not to trap air bubbles.

Reading procedure. Coverslips are examined using a ×20 objective with a fluorescence microscope equipped with the appropriate filters to maximize detection of the fluorescein isothiocyanate (FITC) label (or a light microscope if a peroxidase label is used). A known positive control is run for each viral antigen with each assay. Noninfected monolayers are fixed and stained as negative antigen controls. For indirect IF, normal goat serum, or PBS plus FITC conjugate, is used as a negative serum control.

The pattern of fluorescence varies depending upon the suspected virus, the antibody used, the cell culture, and the stage of virus replication. Even a single cell, characteristically stained, is considered a positive result. The test should be repeated in the following circumstances:

1. The staining pattern is not typical for the suspected virus,
2. Nonspecific staining is observed on the negative control, or
3. The staining color is more yellow than green.

Mixed (Co-Cultivated) Cell Cultures and Monoclonal Antibody Pools

The great success of rapid shell vial cultures for detection of individual viruses led to an impetus to simplify the process. In order to detect more viruses with fewer cell cultures, antibodies to more than one virus, often with two or three different fluorescent labels, were pooled (25, 36, 37), and two to three different cell cultures were combined in one vial, thus described as co-cultivated cells (38–40). This concept was embraced by a commercial supplier and the cell cultures further enhanced through genetic engineering (Table 4). Numerous investigators have reported the value of these mixed cell cultures. Some laboratories have eliminated conventional cell culture tubes and converted to shell vials with mixed cells. When combined with IF staining using antibody pools, detection of common respiratory viruses is simplified, labor is reduced, results are reported more rapidly on both positives and negatives, and the need for primary monkey kidney cells is eliminated. Learning to read and interpret IF staining of shell vial cultures is much easier than reading direct IF on clinical samples or CPE in conventional cell cultures. Mixed cell cultures also can be maintained longer than the 1 to 5 days typically employed for IF staining and observed for CPE for 1 to 2 weeks, if desired.

There are at present a variety of combinations of cultures to choose from, depending upon the suspected viruses (Table 4). R-mix (Mv1Lu and A549) and R-mix Too (Madin Darby Canine Kidney [MDCK] and A549) are commonly used with monoclonal antibody pools to rapidly detect selected respiratory viruses—that is, adenovirus; influenza A and B; parainfluenza 1, 2, and 3; and respiratory syncytial virus (RSV) (40–46). Human metapneumovirus also can be detected in this system. Other viruses can be detected by observing the cultures for CPE. R-mix Too was developed to avoid inadvertently growing SARS coronavirus. In contrast to Mv1Lu cells, MDCK cells will not support the growth of SARS coronavirus.

TABLE 4 Mixed cell culture methods used in clinical laboratories

Culture source/cell culture composition	Targeted viruses	Principle	Reference no.
R-mix[a]/Mink lung (Mv1Lu) and A549	RSV; influenza A and B; parainfluenza 1, 2, and 3; adenovirus Can detect HMPV by IF CPE for other viruses can be observed (HSV, CMV, enteroviruses); can monitor third vial for CPE or stain by IF	Inoculate three shell vials for each specimen. Stain with monoclonal antibody (MAb) pool at 24 hours; if positive, scrape second vial, make multiwell slide, and identify pathogen by individual MAbs. If negative, scrape second shell vial at 48 hours, stain, and make single-well and multiwell slides. Stain single well with pooled reagent; if positive, identify using multiwell slide and individual MAbs. Third vial can be observed for other CPE for 5 days if desired or used for identification of positives on day 2.	40–46
R-mix Too[a]/MDCK and A549	Same as R-mix, except not susceptible to SARS coronavirus More sensitive than R-mix for some respiratory viruses especially adenovirus and influenza B	Same as R-mix	
H&V mix[a] African green monkey kidney (strain CV-1) and MCR-5 cells	HSV-1 and -2, VZV Can also isolate CMV; mumps; measles; rotavirus; polio type 1; rhino-, adeno-, and enteroviruses; RSV	Combination of cells highly sensitive to HSV and VZV Can observe for CPE then confirm by IF, or stain by IF before CPE develops Can stain shell vials for HSV at days 1 and 2, and for VZV at days 2 and 5	48
Super E-mix[a] BGMK-hDAF and A549	Enteroviruses	Same as above	
ELVIS[a] BHK cell line with UL39 promoter and *E. coli* *LacZ* gene	HSV-1 and HSV-2	Positive cells stain blue with X-Gal Can type positive cultures by adding two type-specific MAbs	51,53–57

[a]Available from Diagnostic Hybrids, Inc.

E-mix targets enteric viruses. The original E-mix A (RD and H292) and E-mix B (BGMK and A549) cells provided an advantage over the three to five conventional cell cultures traditionally inoculated to detect enteroviruses. However, E-mix A and B cells were discontinued and replaced by the more sensitive, genetically engineered Super E-mix described in the section Genetically Modified Cell Lines (47).

H&V mix (CV-1 and MRC-5) was developed for isolation of HSV 1 and 2 and VZV, and it can also detect CMV. Although all of these viruses can be detected after 1 or 2 days of incubation via IF staining, optimal detection of VZV requires staining at 2 and 5 days. Many other viruses can also replicate and be detected by CPE (48).

The protocols for inoculation, incubation, and staining for commercially obtained mixed cell cultures are generally those recommended by the supplier and modified as needed by the user. Steps in inoculation, fixing, staining, and reading of mixed cell cultures in shell vials are similar to those described for shell vial cell cultures for a single virus (described previously). In general for mixed cultures, two to three shell vials are inoculated. For respiratory viruses, one shell vial is stained with the antibody pool one day postinoculation (Fig. 4A). If this is positive, a second shell vial is scraped, and cells are spotted onto an eight-well slide for identification by staining with individual antibodies (Fig. 4B). If the day 1 screen is negative, a second shell vial can be scraped on day 2 of incubation and spotted onto both a single well and an eight-well slide. If the screening reagent applied to the single-well slide is positive, the multiwell slide is then stained with individual antibodies. If the screening reagent is negative, the multiwell slide is discarded. If a third vial was inoculated, this vial can be observed for CPE for a longer period to allow some slower growing and low titered viruses to be detected. In another approach, some laboratories may stain the second shell vial *in situ* and, if positive, use the third shell vial to prepare a multiwell slide.

For enteroviruses, two shell vials are needed and staining at 2 and 5 days of incubation is recommended. Samples that contain high titers of virus, such as stools, are generally positive by day 2, but for spinal fluids up to 5 days may be required (47).

Genetically Modified Cell Lines

In the Super E-mix for enteroviruses (previously described), human decay-accelerating factor (hDAF) or CD55, a cellular receptor for several enteroviruses, was transfected into BGMK cells to enhance cell susceptibility to enterovirus isolation (49, 50). BGMK-hDAF cells were then combined with the human colon adenocarcinoma cell line (CaCo-2) in a mixed cell culture. The resulting Super E-mix cells in one culture vessel were reported to be more sensitive for enterovirus recovery than inoculation of three separate

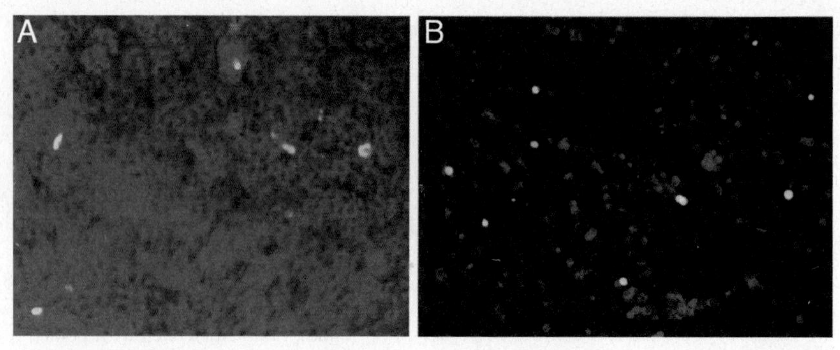

FIGURE 4 R-mix cells in shell vial format. (A) Stained with respiratory virus screen reagent at day 1 postinoculation; (B) identified as influenza A by spotting shell vial cells onto a multiwell slide and staining with individual antibodies (immunofluorescence stain). Photos courtesy of Diagnostic Hybrids, Inc. (Reprinted from last edition.)

conventional tube cultures using primary rhesus monkey kidney, A549, and fetal foreskin (SF) cells (47). In the current Super E-mix, CaCo-2 cells have been replaced by A549 cells.

In another approach, genetic elements derived from viral, bacterial, or cellular sources can be stably introduced into a cell, and when the target virus enters the cell, an event in the viral replication cycle triggers the production of a measurable enzyme. In a simple histochemical assay, infected cells stain a characteristic color (Fig. 5). This approach has been shown to be feasible for both DNA and RNA viruses (51, 52), although different strategies are necessary for enzyme induction. In contrast to CPE, infected cells stained in the inducible system can be read by an untrained observer and the earliest stages of infection reliably detected.

The acronym ELVIS (enzyme-linked inducible system) has been given to a commercially available cell system for isolation of HSV (53). Transgenic baby hamster kidney (BHK) cells have been altered to include an HSV-specific promoter and an *Escherichia coli* LacZ reporter gene. HSV-positive cells form a blue precipitate when reacted with a chromogenic substrate (X-Gal). Both positive and negative results can be reported in as little as 16 hours. It is simple, sensitive, and rapid and can be used for the simultaneous detection, identification, and typing of HSV isolates from clinical specimens (54–57). However, ELVIS is somewhat less sensitive than the most sensitive of conventional cell culture systems when specimens contain only a few infectious HSV particles.

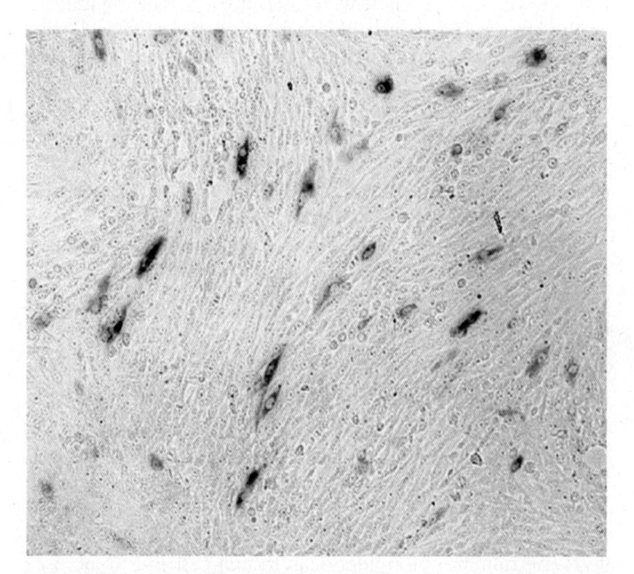

FIGURE 5 Detection of HSV in ELVIS cells. Blue cells positive for HSV infection (X-Gal stain). Photo courtesy of Diagnostic Hybrids, Inc. (Reprinted from last edition.)

VIRUSES COMMONLY ISOLATED IN CONVENTIONAL CELL CULTURE

Herpes Simplex Viruses (HSV-1 and HSV-2)

Vesicular fluids, throat swabs, and genital lesions are the most common sources for virus isolation. Both HSV-1 and HSV-2 infect a wide variety of cell cultures. Early studies demonstrated rabbit kidney (RK) and human embryonic kidney (HEK) to be very sensitive to HSV infection (2). Subsequent evaluations found MRC-5 to be more sensitive than RK cells from commercial suppliers (4). Four continuous cell lines, mink lung (ML), RD, A549, and H292 cells, are also highly sensitive (58–60). Differences in sensitivity are more evident when specimens contain low titers of virus (61) or are collected late in illness or if transport is delayed by error or distance. Variations in susceptibility over time or between suppliers can be problematic. The recently introduced mixtures of two sensitive cell lines (H & V Mix—described previously) in one culture may help alleviate this problem.

HSV produces a rapid degeneration of cells, often appearing within 24 hours of inoculation of the cell culture (Fig. 1B). CPE begins as clusters of enlarged, rounded, refractile cells and spread to involve the entire monolayer, usually within 48 hours. The formation of multinucleated giant cells also can be seen with HSV-2 and is more apparent in epithelial than in fibroblast cells. Subcultures are performed by passaging 0.2 ml of supernatant fluid to a fresh culture tube. Over 90% of positives will be identified within 3 to 5 days. Occasionally, CPE develops later when very low titers are present or if the patient is on antiviral therapy.

Centrifugation cultures in shell vials or 24-well plates can be stained from 1 to 3 days after inoculation; however, staining at 1 day may miss some low-titered samples (34).

Identification of virus as HSV and differentiation as type 1 or 2 is most readily done by IF using monoclonal antibodies (62, 63). As already discussed, genetically engineered cell lines (e.g., ELVIS) also can be used to isolate and identify HSV.

Varicella Zoster Virus (VZV)

Vesicle fluid and lesion swabs are the usual sources for VZV isolation. VZV is difficult to grow and less sensitive than direct IF even when optimally performed (64). If attempted, prompt inoculation into cell culture is desirable.

HDF and A549 cells are the most sensitive cells for the isolation of VZV, although the virus also has been isolated in other human epithelial cells, primary MK cells and CV-1 cells.

Cytopathology starts as foci of rounded enlarged cells, as seen with HSV; however, the onset and progression are much slower, and the foci of CPE tend to progress linearly along the axis of the cells similar to CMV. However, VZV-infected foci degenerate more rapidly than those of CMV. CPE first appears 3 to 7 days after inoculation but may take 2 to 3 weeks. The virus is cell associated, and subpassages are performed by trypsinization and passage of infected intact cells to fresh monolayers of cells. Final identification is by IF using monoclonal antibodies.

Centrifugation cultures are significantly more sensitive than conventional roller tubes for VZV isolation (3, 23). Staining of monolayers at 2 days and again at 5 days after inoculation is recommended for optimal sensitivity. Use of mixed cell cultures (48) and antibody pools with different fluorescent labels may optimize detection of both HSV and VZV in a single culture (36).

Cytomegalovirus (CMV)

CMV can be isolated from a variety of specimens including urine, saliva, tears, milk, semen, stools, vaginal or cervical secretions, peripheral blood leukocytes, and bronchoalveolar lavage fluid, as well as biopsy tissue from lung, liver, and gastrointestinal sites. HDF are the single most successful conventional culture system for the isolation of CMV. The source of the fibroblasts can be either human embryonic tissues or newborn foreskin. The latter, however, lose their sensitivity after the 10th to 15th passage.

CPE may develop within a few days to many weeks, depending on the amount of virus in the specimen. Characteristic CPE consists of foci of enlarged, refractile cells that slowly enlarge over weeks and often do not involve the entire monolayer (Fig. 1E). Thus, it is important that the monolayers be maintained in good condition for at least 3 weeks. On the other hand, when a high titer of CMV is inoculated, as contained in urine samples from congenitally infected babies, generalized rounding at 24 hours—which can be confused with an HSV CPE—may be seen. For subculture, early passage of intact infected cells is essential. Monolayers should be trypsinized and then dispensed onto fresh uninfected cells. Identification of isolates can be accomplished with IF. Since CPE is slow to advance, passage of trypsinized monolayers into centrifugation cultures, followed by staining at 24 to 48 hours can provide a more rapid confirmation.

Centrifugation cultures have had a major impact on the rapid diagnosis of CMV infections. However, for optimal CMV recovery from samples other than urine, conventional cultures should be performed in parallel (33). ML cells, though not permissive for CPE in conventional cultures, have proved very useful for CMV centrifugation cultures, especially because ML cells are less susceptible to toxicity from blood samples (65).

Adenovirus

For adenovirus detection, throat swabs, nasopharyngeal swabs, eye swabs, and stool are good sources of virus, the choice depending on the clinical syndrome. In general, human adenoviruses produce CPE in continuous human cell lines, such as A549, HEK, and HDF cell cultures. Each of these cell systems has its disadvantages: the continuous cell lines may be difficult to maintain; HEK often are not readily available and are expensive; and HDF are less sensitive and the changes produced are not characteristic (66). Nonhuman cells, such as rhesus monkey kidney (RhMK) cells infected with SV40, are of variable sensitivity, and virus growth is slower.

Characteristic CPE consists of grape-like clusters of rounded cells (Fig. 1C), which appear in 2 to 7 days with types 1, 2, 3, 5, 6, and 7. Other adenovirus types may require 3 to 4 weeks or blind passage. Adenovirus is cell-associated, similar to VZV and CMV; however, adenovirus is non-enveloped and stable to freezing and thawing. Therefore, two to three cycles of freezing at −70°C and thawing disrupts the cells and releases infectious virus. Enteric adenovirus types 40 and 41, associated with gastroenteritis, do not grow readily in A549 cells or HDF, but can be isolated in H292 cells (60).

Identification of isolates as adenoviruses can be done by IF using anti-hexon antibody. Neutralization tests with type-specific antiserum or molecular analysis will identify virus types.

Centrifugation cultures can provide a more rapid diagnosis, but staining at 2 days and again at 5 days may be needed for optimum sensitivity (24, 66, 67).

Enterovirus

Enteroviruses were originally classified by their growth in different types of cell culture and suckling mice. The use of nucleotide sequencing has resulting in the identification of new strains, the reclassification of enteroviruses into four groups (A to D) and the reclassification of echoviruses types 22 and 23 into a new genus, *Parechovirus*, which has now expanded to include 16 virus types (68).

Enteroviruses can be recovered from feces, throat swabs, cerebrospinal fluid, blood, vesicle fluid, conjunctival swabs, and urine. In general, enteroviruses grow best in epithelial cells of primate origin. Polio and coxsackie B viruses grow well in primary monkey kidney (MK), HEp-2, and BGMK cells, and echovirus grows well in primary MK and RD cells (a rhabdomyosarcoma cell line), but not in HEp-2 cells. The universal host for coxsackie group A is the newborn mouse; however, some strains grow in HDF, HEK, MK, or RD cells. Since inoculation of multiple cell types optimizes enterovirus detection (69, 70), the use of mixed cell cultures may result in greater yield while conserving time and resources.

Characteristically, infected monolayer cells round up, become refractile, shrink, degenerate, and then detach from the surface of the culture vessel (Fig. 1H). Virus in the supernatant fluid can be subpassaged. Preliminary identification can be determined by characteristic CPE and cell susceptibility (71, 72). Shell vial centrifugation cultures using multiple single cell types in separate vials (67, 73) or mixed cells in a single vial (47) have shortened time to detection. Fluorescein-labeled monoclonal antibodies available for identification of enteroviruses, either in shell vial cultures or isolates from conventional cultures, have been plagued with nonspecific staining (74). Newer monoclonal antibodies are reported to be more sensitive and specific (75). Final identification and serotyping by microneutralization tests in cell culture using antiserum pools is expensive and time-consuming and has largely been replaced by PCR and nucleotide sequencing (7, 8, 76).

Rhinovirus

Rhinoviruses are classified as picornaviruses and are now considered within the enterovirus genus. Rhinoviruses have traditionally been separated from the latter by their sensitivity to low pH and preference for a lower growth temperature. Sources of virus include nasal swabs or washes and throat swabs.

Many rhinovirus types were originally isolated in organ cultures of human embryonic trachea. Rhinoviruses can be isolated in cells of human origin (usually HDF), certain strains of HeLa cells, and human fetal tonsil cells; however, varying sensitivity of different lots of cells can be a problem. WI-38 and HeLa-I cells have been identified as the most sensitive cells (77) and cultivation at 33°C in a roller drum apparatus is optimal.

CPE may occur from the first to the third week of incubation. CPE is similar to those of the enteroviruses, start as foci of rounded cells, and spread gradually (Fig. 1F). However, CPE may not progress and may even disappear; if they are not progressing, subpassage of supernatant fluids from infected cells should be performed. Identification of isolates is by characteristic CPE and inactivation at pH 3. Reverse transcriptase (RT)-PCR can be used to identify an isolate as a picornavirus; however, identification as a rhinovirus is less reliable due to similarity to enterovirus. Typing is reserved for public health or research laboratories, and neutralization tests have largely been supplanted by sequencing (78).

Nevertheless, only a minority of rhinovirus infections are identified by cell culture. With wider use of molecular methods, the true prevalence of rhinovirus infections has been shown to be much greater than shown by isolation in cell culture (79). Indeed, group C rhinoviruses have not yet been grown in cell culture (80).

Influenza

Nasopharyngeal aspirates and swabs, nasal washings, and throat swabs are good sources for virus and should be collected early in illness, preferably in the first 24 to 48 hours. Primary MK is the most widely used cell culture for isolation of influenza, although the host range may be increased by addition of trypsin to the medium (81). MDCK, MRC-5, and ML cells (82) have all been used successfully, especially in centrifugation cultures (83). Influenza is also reliably isolated in embryonated eggs (84).

Serum components may inhibit influenza virus from replicating. Therefore, serum should be removed from cell cultures by rinsing with Hanks' balanced salt solution (HBSS) before inoculation, and cultures should be maintained in serum-free medium after inoculation. Incubation at 33°C in a roller drum is optimal for isolation. The presence of virus is generally detected by HAd of guinea pig RBCs onto infected monolayers (Fig. 2B). CPE is seen with influenza (Fig. 1I), but they usually occur later than the detection of virus by HAd. Subcultures can be performed by passaging the supernatant fluids. Isolates can be identified as influenza A or B by IF using monoclonal antibodies. Subtyping is feasible by hemagglutination inhibition (HI) or monoclonal antibody staining (85) but is now most commonly determined by RT-PCR and other molecular methods (86). Strain identification is still determined by HI.

Parainfluenza (PIV)

Nasopharyngeal aspirates and swabs, nasal washings, and throat swabs are good sources for virus. Primary MK culture is the most sensitive system for isolation of these viruses. HEK, HDF, and HEp-2 cells are less sensitive. Some success has recently been reported with H292 cells (60).

Cell cultures should be washed with HBSS before inoculation and refed with medium without serum. Incubation at 33°C to 36°C in a roller drum is optimal. The presence of virus is detected by HAd (Fig. 2B), which occurs before CPE. PIV 2 may produce syncytia, especially in HEp-2 cells. On subculture PIV 3 may also induce syncytia formation. In those instances when high levels of virus are present, HAd may be detected in the infected cultures within a few days; with specimens containing less virus, 10 days or more of incubation may be necessary. PIV 4 requires HAd at room temperature instead of 4°C, which is commonly used for types 1 to 3. Identification of parainfluenza viruses is done by either IF or HAd inhibition.

Respiratory Syncytial Virus (RSV)

RSV is found in respiratory secretions from the nose and oropharynx. Sample collection is important, and RSV is more reliably detected from nasopharyngeal aspirates than from swabs in children (87). RSV grows best in continuous cell lines, such as HEp-2, in which it produces characteristic syncytia (Fig. 1K). However, because syncytium formation is variable, using it as the sole criterion for RSV replication could cause viral replication to be missed. If HEp-2 cells are confluent and 5 to 7 days old when inoculated, syncytia may not form. Rather, nonspecific rounding may occur. Syncytia formation is also dependent upon the presence of adequate levels of glutamine and calcium in the medium (88, 89). Primary MK cells show CPE and HDF cells support RSV growth, but the cytopathology is not as characteristic (90). HEp-2 cells have become so difficult to work with that many laboratories rely on other cell cultures such as A549 or MK, with suboptimal results. Use of centrifugation cultures can improve detection. Identification of RSV in cell culture is done by IF.

QUALITY ASSURANCE AND TROUBLESHOOTING

Based on the assumption that most clinical virology laboratories purchase rather than produce their own cell cultures, the following components are important for assuring quality of cultured cells (91). Careful documentation of all quality assurance activities is required. Shipments of commercial cell cultures must be examined for breakage, and the cell monolayers must be examined microscopically to ensure that the monolayer is well-established, with cells attached to the substratum, confluence appropriate for the method and cell line, and cell appearance that is typical. If additives, such as antibiotics, are added to commercial cell culture medium, the culture medium must be tested for sterility after they are introduced. Animal sera used for cell growth media must be checked to ensure absence of toxicity to cells. The laboratory must maintain records of the cell types, passage number, source, and media that are used for specimen testing and virus culture.

The laboratory must ensure that appropriate cell lines are available for all types of specimens tested and for all viruses reported. Appropriate incubation times and temperatures for target viruses must be used, and inoculated cultures must be checked for CPE to optimize the time to detection of viral pathogens. Uninoculated cell monolayers or monolayers inoculated with sterile material must be available for comparison with cultures of clinical material. There must be a procedure for handling cell culture showing unusual CPE.

APPLICATIONS IN DIAGNOSTIC VIROLOGY

As clinical virologists attempt to provide the most effective viral diagnostic services, the advantages and limitation of detection methods must be considered before choosing a combination that is right for a particular setting. The advantages of cell culture for virus diagnosis include broad spectrum, biologic amplification of the input virus, ability to differentiate infectious virus, greater sensitivity than antigen detection methods, and the recovery of unknown or unexpected infectious virus(es) that may be present in the specimen (92).

One particularly appealing aspect of virus isolation in cell culture is the "open" nature of specimen types that can be tested. With effective sample processing prior to culture inoculation (e.g., addition of antibiotics and antimycotics to samples from sites contaminated with bacterial and fungal flora and attention to pH and other factors that are potentially toxic to cell culture monolayers), cell culture systems can be effective for isolating virus from most body sites. Individual extensive validation and revalidation are not usually required for each sample type. Such validation are essential for each specimen type for antigen detection and molecular methods, especially those that received FDA clearance based on testing of a single sample type (e.g., nasopharyngeal swab only); this is a factor that has dramatically complicated utilization of these methods compared to cell culture isolation.

An additional consideration with cell culture methods is that viral proliferation confirms the viability of the virus and differentiates viable from nonviable virus. Detection of infectious virus in culture may have a better predictive value for clinical significance than detecting viral DNA or RNA with a highly sensitive molecular methods. For CMV, molecular methods are highly sensitive and have a strong negative predictive value, but there is as yet no clear consensus on what level of CMV DNA correlates with disease in lung or gastrointestinal tissue or bronchoalveolar lavage (93). In a study of PCR of urine versus culture to diagnosis congenital CMV, PCR detected more positives, but discrepant samples were collected 10 to 17 days after birth, raising the issues of whether these were postnatal infections, and with PCR, whether samples must be collected closer to birth (94).

Cell culture continues to facilitate the discovery and rapid molecular characterization of new viruses (9). Recent examples include hPMV (13) and NL63 coronavirus (15, 16) in tertiary MK cells, SARS coronavirus (14) in Vero E6 cells, MERS coronavirus in Vero and LLC-MK2 (17), Cache Valley virus in Vero E6, A549 and BGMK (12), and Heartland and SFTS in DH82 (18, 19).

Virus isolation is generally more sensitive than rapid antigen detection assays, detects a broader spectrum of viruses, and is less costly than molecular assays. When performed locally, HSV culture of lesion swab specimens tested was recently shown to be 92.8% sensitive compared to direct HSV PCR using a commercial kit (95). Culture can also detect positives missed by molecular methods due to sequence variation (95). Although conventional virus isolation with observation for CPE has more limited application in the diagnostic laboratory than in the past, rapid culture innovations have increased viral diagnostic capabilities, shortened turnaround times to 1 to 2 days in most cases, and significantly reduced the technical expertise, labor, and quality control required.

There are definitely disadvantages of virus isolation in cell culture. It is limited by the inherent time delay required for virus growth, which can extent to weeks for some viruses, by the difficulty in maintaining cell cultures, by the sometimes variable quality of cultures, and by the decreased sensitivity of cell lines at higher passage levels. To get the best results from primary isolation in cell culture, use of healthy cell cultures susceptible to a spectrum of viruses is essential. Traditionally this has required the inoculation of at least three separate cell cultures such as a primary MK cell culture, an HDF cell strain, and a human heteroploid cell line (e.g., A549), with observation for CPE for 1 to 3 or 4 weeks. This time delay is one of the major disadvantages of virus isolation in cell cultures. If patient management is to be affected, results must be available quickly.

Some common viruses neither replicate nor produce identifiable effects in readily available cell cultures. Examples include hepatitis viruses, gastroenteritis viruses, group C rhinoviruses, and some group A coxackieviruses.

Contamination with adventitious agents occurs, including bacteria, fungi, and mycoplasma, which can inhibit the growth of viruses in clinical specimens (96–101). Viruses that are latent in the tissue or contaminating the calf serum can begin to replicate during cultivation and can cause CPE or HAd (Fig. 1L), and thus, they can be confused with virus isolated from the patient's specimen (102). Viruses such as herpes B virus potentially present in primary MK cell cultures can pose a serious health risk to laboratory personnel.

The presence of inhibitory substances and/or antibodies in calf serum used in the cell culture media can reduce the isolation of certain viruses, especially of the orthomyxo- and paramyxovirus groups (103). Ideally, maintenance media for inoculated cultures should be serum-free; however, serum is required for long-term maintenance of cells. Using fetal or agammaglobulin calf serum reduces this problem, but it adds to expense. To date, no completely satisfactory, chemically defined medium is available (104).

In view of the disadvantages of virus isolation in cell culture and the advancements in molecular methods, the following question has been asked (105): "Is the era of viral culture over in the clinical microbiology laboratory?" This question must be answered by each clinical laboratory after a careful analysis of each unique situation. Laboratory size, patient population, technical expertise of personnel, level and range of viral diagnostic services desired, and access to molecular testing must all be considered. Reducing the time to result has become increasingly important in order to limit unnecessary tests and antibiotics, initiate antiviral therapy, implement infection control measures, and shorten hospital stays. The need to lower costs and do more with fewer personnel has created additional pressures. Innovations in viral culture methods that have allowed both positive and negative cultures to be reported in 48 hours include the following: rapid shell vial centrifugation cultures, use of pooled antibodies for detection of multiple viruses, mixtures of two cell systems in one culture vessel, and genetic engineering to enhance cell culture susceptibility to particular viruses. The greater use of continuous cell lines and more stringent quality control has reduced the incidence of adventitious agents in the cultures. Furthermore, cryopreserved cell cultures can now be purchased and stored at −70°C, to be thawed as needed for inoculation of clinical samples, thus making culture more economical and user-friendly.

In a setting where a limited range of viral diagnostic services is desired (e.g., a moderate-sized microbiology laboratory serving primarily sexually transmitted disease clinics), HSV may be the only target virus. This laboratory may use the ELVIS cell culture system to provide reliable HSV detection without requiring personnel experienced in detection of CPE. A laboratory desiring to provide detection of the common respiratory viruses as well as HSV might incorporate mixed cell lines in shell vials for respiratory virus detection, along with ELVIS for HSV detection. Such scenarios are practical and "do-able" for most microbiology laboratories. Even large laboratories with the capacity to offer a broad range of molecular viral detection methods as well as cell culture may employ appropriate molecular methods as the first line in diagnostics, using cell culture for certain specialized purposes: for detection of less commonly encountered agents for which molecular methods are unavailable on site (e.g., measles, mumps), for testing in parallel with molecular methods to ensure detection of viruses that have the capacity for genetic changes (e.g., influenza A), for detection of viruses known to be less effectively detected by molecular method (e.g., adenovirus), to provide a more open detection—especially with clinical specimens such as biopsies and bronchoalveolar lavage—which are collected via a surgical procedure.

Ultimately, the combination of culture and nonculture methods will provide the most effective viral diagnostic service, empowering laboratories with multiple approaches. Initial viral detection may be by rapid antigen detection testing, with reflexing of negative samples for culture or molecular testing. Rapid cell cultures provide a more broad spectrum and sensitive diagnosis than antigen tests, with results in 1 to 2 days. In addition, culture is valuable for lower respiratory tract and tissue biopsy samples due to better correlation with disease and to detect additional or unsuspected viruses. A discussion of algorithms incorporating virus isolation, antigen detection, molecular methods, electron microscopy, and antibody detection for detection and identification of viruses is available (106).

As molecular methods become increasingly user-friendly and economical and syndromic molecular panels (e.g., respiratory and gastrointestinal pathogens panels) are further developed, use of cell culture is expected to decline. In the interim, viral diagnostic laboratories continue to find culture helpful. Since cultures can be inoculated every day, the time to result can be faster than molecular methods that are not performed as frequently or are sent to a distant reference laboratory. Lastly, for public health and new or unusual virus detection, cell culture will continue to play an essential role.

REFERENCES

1. **Enders JF, Weller TH, Robbins FC.** 1949. Cultivation of the Lansing strain of poliomyelitis virus in cultures of various human embryonic tissues. *Science* **109:**85–87.
2. **Landry ML, Mayo DR, Hsiung GD.** 1982. Comparison of guinea pig embryo cells, rabbit kidney cells, and human embryonic lung fibroblast cell strains for isolation of herpes simplex virus. *J Clin Microbiol* **15:**842–847.
3. **Coffin SE, Hodinka RL.** 1995. Utility of direct immunofluorescence and virus culture for detection of varicella-zoster virus in skin lesions. *J Clin Microbiol* **33:**2792–2795.
4. **McCarter YS, Robinson A.** 1997. Comparison of MRC-5 and primary rabbit kidney cells for the detection of herpes simplex virus. *Arch Pathol Lab Med* **121:**122–124.
5. **Huang YT, Yan H, Sun Y, Jollick JA Jr, Baird H.** 2002. Cryopreserved cell monolayers for rapid detection of herpes simplex virus and influenza virus. *J Clin Microbiol* **40:**4301–4303.
6. **Leland DS, Ginocchio CC.** 2007. Role of cell culture for virus detection in the age of technology. *Clin Microbiol Rev* **20:** 49–78.
7. **Oberste MS, Maher K, Kilpatrick DR, Flemister MR, Brown BA, Pallansch MA.** 1999. Typing of human enteroviruses by partial sequencing of VP1. *J Clin Microbiol* **37:**1288–1293.
8. **Oberste MS, Maher K, Michele SM, Belliot G, Uddin M, Pallansch MA.** 2005. Enteroviruses 76, 89, 90 and 91 represent a novel group within the species Human enterovirus A. *J Gen Virol* **86:**445–451.
9. **Goldsmith CS, Ksiazek TG, Rollin PE, Comer JA, Nicholson WL, Peret TC, Erdman DD, Bellini WJ, Harcourt BH, Rota PA, Bhatnagar J, Bowen MD, Erickson BR, McMullan LK, Nichol ST, Shieh WJ, Paddock CD, Zaki SR.** 2013. Cell culture and electron microscopy for identifying viruses in diseases of unknown cause. *Emerg Infect Dis* **19:**886–891.
10. **Sexton DJ, Rollin PE, Breitschwerdt EB, Corey GR, Myers SA, Dumais MR, Bowen MD, Goldsmith CS, Zaki SR, Nichol ST, Peters CJ, Ksiazek TG.** 1997. Life-threatening Cache Valley virus infection. *N Engl J Med* **336:**547–549.
11. **Campbell GL, Mataczynski JD, Reisdorf ES, Powell JW, Martin DA, Lambert AJ, Haupt TE, Davis JP, Lanciotti RS.** 2006. Second human case of Cache Valley virus disease. *Emerg Infect Dis* **12:**854–856.
12. **Nguyen NL, Zhao G, Hull R, Shelly MA, Wong SJ, Wu G, St George K, Wang D, Menegus MA.** 2013. Cache valley virus in a patient diagnosed with aseptic meningitis. *J Clin Microbiol* **51:**1966–1969.
13. **van den Hoogen BG, de Jong JC, Groen J, Kuiken T, de Groot R, Fouchier RA, Osterhaus AD.** 2001. A newly discovered human pneumovirus isolated from young children with respiratory tract disease. *Nat Med* **7:**719–724.
14. **Ksiazek TG, Erdman D, Goldsmith CS, Zaki SR, Peret T, Emery S, Tong S, Urbani C, Comer JA, Lim W, Rollin PE, Dowell SF, Ling AE, Humphrey CD, Shieh WJ, Guarner J, Paddock CD, Rota P, Fields B, DeRisi J, Yang JY, Cox N, Hughes JM, LeDuc JW, Bellini WJ, Anderson LJ, SARS Working Group.** 2003. A novel coronavirus associated with severe acute respiratory syndrome. *N Engl J Med* **348:**1953–1966.
15. **van der Hoek L, Pyrc K, Jebbink MF, Vermeulen-Oost W, Berkhout RJ, Wolthers KC, Wertheim-van Dillen PM, Kaandorp J, Spaargaren J, Berkhout B.** 2004. Identification of a new human coronavirus. *Nat Med* **10:**368–373.
16. **Fouchier RA, Hartwig NG, Bestebroer TM, Niemeyer B, de Jong JC, Simon JH, Osterhaus AD.** 2004. A previously undescribed coronavirus associated with respiratory disease in humans. *Proc Natl Acad Sci USA* **101:**6212–6216.
17. **Zaki AM, van Boheemen S, Bestebroer TM, Osterhaus AD, Fouchier RA.** 2012. Isolation of a novel coronavirus from a man with pneumonia in Saudi Arabia. *N Engl J Med* **367:** 1814–1820.
18. **Yu XJ, Liang MF, Zhang SY, Liu Y, Li JD, Sun YL, Zhang L, Zhang QF, Popov VL, Li C, Qu J, Li Q, Zhang YP, Hai R, Wu W, Wang Q, Zhan FX, Wang XJ, Kan B, Wang SW, Wan KL, Jing HQ, Lu JX, Yin WW, Zhou H, Guan XH, Liu JF, Bi ZQ, Liu GH, Ren J, Wang H, Zhao Z, Song JD, He JR, Wan T, Zhang JS, Fu XP, Sun LN, Dong XP, Feng ZJ, Yang WZ, Hong T, Zhang Y, Walker DH, Wang Y, Li DX.** 2011. Fever with thrombocytopenia associated with a novel bunyavirus in China. *N Engl J Med* **364:**1523–1532.
19. **McMullan LK, Folk SM, Kelly AJ, MacNeil A, Goldsmith CS, Metcalfe MG, Batten BC, Albariño CG, Zaki SR, Rollin PE, Nicholson WL, Nichol ST.** 2012. A new phlebovirus associated with severe febrile illness in Missouri. *N Engl J Med* **367:**834–841.
20. **Hudson JB, Misra V, Mosmann TR.** 1976. Cytomegalovirus infectivity: analysis of the phenomenon of centrifugal enhancement of infectivity. *Virology* **72:**235–243.
21. **Gleaves CA, Smith TF, Shuster EA, Pearson GR.** 1984. Rapid detection of cytomegalovirus in MRC-5 cells inoculated

with urine specimens by using low-speed centrifugation and monoclonal antibody to an early antigen. *J Clin Microbiol* **19:** 917–919.
22. **Gleaves CA, Wilson DJ, Wold AD, Smith TF.** 1985. Detection and serotyping of herpes simplex virus in MRC-5 cells by use of centrifugation and monoclonal antibodies 16 h postinoculation. *J Clin Microbiol* **21:**29–32.
23. **Gleaves CA, Lee CF, Bustamante CI, Meyers JD.** 1988. Use of murine monoclonal antibodies for laboratory diagnosis of varicella-zoster virus infection. *J Clin Microbiol* **26:**1623–1625.
24. **Espy MJ, Hierholzer JC, Smith TF.** 1987. The effect of centrifugation on the rapid detection of adenovirus in shell vials. *Am J Clin Pathol* **88:**358–360.
25. **Olsen MA, Shuck KM, Sambol AR, Flor SM, O'Brien J, Cabrera BJ.** 1993. Isolation of seven respiratory viruses in shell vials: a practical and highly sensitive method. *J Clin Microbiol* **31:**422–425.
26. **Marshall WF, Telenti A, Proper J, Aksamit AJ, Smith TF.** 1990. Rapid detection of polyomavirus BK by a shell vial cell culture assay. *J Clin Microbiol* **28:**1613–1615.
27. **Slavin MA, Gleaves CA, Schoch HG, Bowden RA.** 1992. Quantification of cytomegalovirus in bronchoalveolar lavage fluid after allogeneic marrow transplantation by centrifugation culture. *J Clin Microbiol* **30:**2776–2779.
28. **Paya CV, Wold AD, Smith TF.** 1987. Detection of cytomegalovirus infections in specimens other than urine by the shell vial assay and conventional tube cell cultures. *J Clin Microbiol* **25:**755–757.
29. **Shuster EA, Beneke JS, Tegtmeier GE, Pearson GR, Gleaves CA, Wold AD, Smith TF.** 1985. Monoclonal antibody for rapid laboratory detection of cytomegalovirus infections: characterization and diagnostic application. *Mayo Clin Proc* **60:**577–585.
30. **Fedorko DP, Ilstrup DM, Smith TF.** 1989. Effect of age of shell vial monolayers on detection of cytomegalovirus from urine specimens. *J Clin Microbiol* **27:**2107–2109.
31. **Paya CV, Wold AD, Ilstrup DM, Smith TF.** 1988. Evaluation of number of shell vial cell cultures per clinical specimen for rapid diagnosis of cytomegalovirus infection. *J Clin Microbiol* **26:**198–200.
32. **Blanding JG, Hoshiko MG, Stutman HR.** 1989. Routine viral culture for pediatric respiratory specimens submitted for direct immunofluorescence testing. *J Clin Microbiol* **27:**1438–1440.
33. **Rabella N, Drew WL.** 1990. Comparison of conventional and shell vial cultures for detecting cytomegalovirus infection. *J Clin Microbiol* **28:**806–807.
34. **Espy MJ, Wold AD, Jespersen DJ, Jones MF, Smith TF.** 1991. Comparison of shell vials and conventional tubes seeded with rhabdomyosarcoma and MRC-5 cells for the rapid detection of herpes simplex virus. *J Clin Microbiol* **29:**2701–2703.
35. **Rabalais GP, Stout GG, Ladd KL, Cost KM.** 1992. Rapid diagnosis of respiratory viral infections by using a shell vial assay and monoclonal antibody pool. *J Clin Microbiol* **30:** 1505–1508.
36. **Brumback BG, Farthing PG, Castellino SN.** 1993. Simultaneous detection of and differentiation between herpes simplex and varicella-zoster viruses with two fluorescent probes in the same test system. *J Clin Microbiol* **31:**3260–3263.
37. **Engler HD, Preuss J.** 1997. Laboratory diagnosis of respiratory virus infections in 24 hours by utilizing shell vial cultures. *J Clin Microbiol* **35:**2165–2167.
38. **Brumback BG, Wade CD.** 1994. Simultaneous culture for adenovirus, cytomegalovirus, and herpes simplex virus in same shell vial by using three-color fluorescence. *J Clin Microbiol* **32:**2289–2290.
39. **Navarro-Marí JM, Sanbonmatsu-Gámez S, Pérez-Ruiz M, De La Rosa-Fraile M.** 1999. Rapid detection of respiratory viruses by shell vial assay using simultaneous culture of HEp-2, LLC-MK2, and MDCK cells in a single vial. *J Clin Microbiol* **37:**2346–2347.
40. **Huang YT, Turchek BM.** 2000. Mink lung cells and mixed mink lung and A549 cells for rapid detection of influenza virus and other respiratory viruses. *J Clin Microbiol* **38:**422–423.
41. **Fong CKY, Lee MK, Griffith BP.** 2000. Evaluation of R-Mix FreshCells in shell vials for detection of respiratory viruses. *J Clin Microbiol* **38:**4660–4662.
42. **Barenfanger J, Drake C, Mueller T, Troutt T, O'Brien J, Guttman K.** 2001. R-Mix cells are faster, at least as sensitive and marginally more costly than conventional cell lines for the detection of respiratory viruses. *J Clin Virol* **22:**101–110.
43. **Dunn JJ, Woolstenhulme RD, Langer J, Carroll KC.** 2004. Sensitivity of respiratory virus culture when screening with R-mix fresh cells. *J Clin Microbiol* **42:**79–82.
44. **St George K, Patel NM, Hartwig RA, Scholl DR, Jollick JA Jr, Kauffmann LM, Evans MR, Rinaldo CR Jr.** 2002. Rapid and sensitive detection of respiratory virus infections for directed antiviral treatment using R-Mix cultures. *J Clin Virol* **24:**107–115.
45. **Fader RC.** 2005. Comparison of the Binax NOW Flu A enzyme immunochromatographic assay and R-Mix shell vial culture for the 2003–2004 influenza season. *J Clin Microbiol* **43:**6133–6135.
46. **Weinberg A, Brewster L, Clark J, Simoes E, ARIVAC Consortium.** 2004. Evaluation of R-Mix shell vials for the diagnosis of viral respiratory tract infections. *J Clin Virol* **30:** 100–105.
47. **Buck GE, Wiesemann M, Stewart L.** 2002. Comparison of mixed cell culture containing genetically engineered BGMK and CaCo-2 cells (Super E-Mix) with RT-PCR and conventional cell culture for the diagnosis of enterovirus meningitis. *J Clin Virol* **25**(Suppl 1)**:**S13–S18.
48. **Huang YT, Hite S, Duane V, Yan H.** 2002. CV-1 and MRC-5 mixed cells for simultaneous detection of herpes simplex viruses and varicella zoster virus in skin lesions. *J Clin Virol* **24:** 37–43.
49. **Lublin DM, Atkinson JP.** 1989. Decay-accelerating factor: biochemistry, molecular biology, and function. *Annu Rev Immunol* **7:**35–58.
50. **Huang YT, Yam P, Yan H, Sun Y.** 2002. Engineered BGMK cells for sensitive and rapid detection of enteroviruses. *J Clin Microbiol* **40:**366–371.
51. **Stabell EC, O'Rourke SR, Storch GA, Olivo PD.** 1993. Evaluation of a genetically engineered cell line and a histochemical beta-galactosidase assay to detect herpes simplex virus in clinical specimens. *J Clin Microbiol* **31:**2796–2798.
52. **Lutz A, Dyall J, Olivo PD, Pekosz A.** 2005. Virus-inducible reporter genes as a tool for detecting and quantifying influenza A virus replication. *J Virol Methods* **126:**13–20.
53. **Proffitt MR, Schindler SA.** 1995. Rapid detection of HSV with an enzyme-linked virus inducible system (ELVIS) employing a genetically modified cell line. *Clin Diagn Virol* **4:** 175–182.
54. **Patel N, Kauffmann L, Baniewicz G, Forman M, Evans M, Scholl D.** 1999. Confirmation of low-titer, herpes simplex virus-positive specimen results by the enzyme-linked virus-inducible system (ELVIS) using PCR and repeat testing. *J Clin Microbiol* **37:**3986–3989.
55. **Turchek BM, Huang YT.** 1999. Evaluation of ELVIS HSV ID/Typing System for the detection and typing of herpes simplex virus from clinical specimens. *J Clin Virol* **12:**65–69.
56. **Kowalski RP, Karenchak LM, Shah C, Gordon JS.** 2002. ELVIS: a new 24-hour culture test for detecting herpes simplex virus from ocular samples. *Arch Ophthalmol* **120:**960–962.
57. **Crist GA, Langer JM, Woods GL, Procter M, Hillyard DR.** 2004. Evaluation of the ELVIS plate method for the detection and typing of herpes simplex virus in clinical specimens. *Diagn Microbiol Infect Dis* **49:**173–177.
58. **Woods GL, Young A.** 1988. Use of A-549 cells in a clinical virology laboratory. *J Clin Microbiol* **26:**1026–1028.
59. **Johnston SLG, Wellens K, Siegel CS.** 1990. Rapid isolation of herpes simplex virus by using mink lung and rhabdomyosarcoma cell cultures. *J Clin Microbiol* **28:**2806–2807.

60. **Hierholzer JC, Castells E, Banks GG, Bryan JA, McEwen CT.** 1993. Sensitivity of NCI-H292 human lung mucoepidermoid cells for respiratory and other human viruses. *J Clin Microbiol* **31:**1504–1510.
61. **Zhao LS, Landry ML, Balkovic ES, Hsiung GD.** 1987. Impact of cell culture sensitivity and virus concentration on rapid detection of herpes simplex virus by cytopathic effects and immunoperoxidase staining. *J Clin Microbiol* **25:**1401–1405.
62. **Miller MJ, Howell CL.** 1983. Rapid detection and identification of herpes simplex virus in cell culture by a direct immunoperoxidase staining procedure. *J Clin Microbiol* **18:** 550–553.
63. **Balkovic ES, Hsiung GD.** 1985. Comparison of immunofluorescence with commercial monoclonal antibodies to biochemical and biological techniques for typing clinical herpes simplex virus isolates. *J Clin Microbiol* **22:**870–872.
64. **Wilson DA, Yen-Lieberman B, Schindler S, Asamoto K, Schold JD, Procop GW.** 2012. Should varicella-zoster virus culture be eliminated? A comparison of direct immunofluorescence antigen detection, culture, and PCR, with a historical review. *J Clin Microbiol* **50:**4120–4122.
65. **Gleaves CA, Hursh DA, Meyers JD.** 1992. Detection of human cytomegalovirus in clinical specimens by centrifugation culture with a nonhuman cell line. *J Clin Microbiol* **30:** 1045–1048.
66. **Mahafzah AM, Landry ML.** 1989. Evaluation of immunofluorescent reagents, centrifugation, and conventional cultures for the diagnosis of adenovirus infection. *Diagn Microbiol Infect Dis* **12:**407–411.
67. **Van Doornum GJJ, De Jong JC.** 1998. Rapid shell vial culture technique for detection of enteroviruses and adenoviruses in fecal specimens: comparison with conventional virus isolation method. *J Clin Microbiol* **36:**2865–2868.
68. **Esposito S, Rahamat-Langendoen J, Ascolese B, Senatore L, Castellazzi L, Niesters HG.** 2014. Pediatric parechovirus infections. *J Clin Virol* **60:**84–89.
69. **Dagan R, Menegus MA.** 1986. A combination of four cell types for rapid detection of enteroviruses in clinical specimens. *J Med Virol* **19:**219–228.
70. **Kok TW, Pryor T, Payne L.** 1998. Comparison of rhabdomyosarcoma, buffalo green monkey kidney epithelial, A549 (human lung epithelial) cells and human embryonic lung fibroblasts for isolation of enteroviruses from clinical samples. *J Clin Virol* **11:**61–65.
71. **Johnston SLG, Siegel CS.** 1990. Presumptive identification of enteroviruses with RD, HEp-2, and RMK cell lines. *J Clin Microbiol* **28:**1049–1050.
72. **Hsiung GD, Fong CKY, Landry ML.** 1994. *Hsiung's Diagnostic Virology*, 4th ed. Yale University Press, New Haven.
73. **She RC, Crist G, Billetdeaux E, Langer J, Petti CA.** 2006. Comparison of multiple shell vial cell lines for isolation of enteroviruses: a national perspective. *J Clin Virol* **37:**151–155.
74. **Rigonan AS, Mann L, Chonmaitree T.** 1998. Use of monoclonal antibodies to identify serotypes of enterovirus isolates. *J Clin Microbiol* **36:**1877–1881.
75. **Miao LY, Pierce C, Gray-Johnson J, DeLotell J, Shaw C, Chapman N, Yeh E, Schnurr D, Huang YT.** 2009. Monoclonal antibodies to VP1 recognize a broad range of enteroviruses. *J Clin Microbiol* **47:**3108–3113.
76. **Muir P, Kämmerer U, Korn K, Mulders MN, Pöyry T, Weissbrich B, Kandolf R, Cleator GM, van Loon AM, The European Union Concerted Action on Virus Meningitis and Encephalitis.** 1998. Molecular typing of enteroviruses: current status and future requirements. *Clin Microbiol Rev* **11:**202–227.
77. **Arruda E, Crump CE, Rollins BS, Ohlin A, Hayden FG.** 1996. Comparative susceptibilities of human embryonic fibroblasts and HeLa cells for isolation of human rhinoviruses. *J Clin Microbiol* **34:**1277–1279.
78. **Bochkov YA, Grindle K, Vang F, Evans MD, Gern JE.** 2014. Improved molecular typing assay for rhinovirus species A, B, and C. *J Clin Microbiol* **52:**2461–2471.
79. **Miller EK, Lu X, Erdman DD, Poehling KA, Zhu Y, Griffin MR, Hartert TV, Anderson LJ, Weinberg GA, Hall CB, Iwane MK, Edwards KM, New Vaccine Surveillance Network.** 2007. Rhinovirus-associated hospitalizations in young children. *J Infect Dis* **195:**773–781.
80. **Ashraf S, Brockman-Schneider R, Bochkov YA, Pasic TR, Gern JE.** 2013. Biological characteristics and propagation of human rhinovirus-C in differentiated sinus epithelial cells. *Virology* **436:**143–149.
81. **Frank AL, Couch RB, Griffis CA, Baxter BD.** 1979. Comparison of different tissue cultures for isolation and quantitation of influenza and parainfluenza viruses. *J Clin Microbiol* **10:** 32–36.
82. **Schultz-Cherry S, Dybdahl-Sissoko N, McGregor M, Hinshaw VS.** 1998. Mink lung epithelial cells: unique cell line that supports influenza A and B virus replication. *J Clin Microbiol* **36:**3718–3720.
83. **Reina J, Fernandez-Baca V, Blanco I, Munar M.** 1997. Comparison of Madin-Darby canine kidney cells (MDCK) with a green monkey continuous cell line (Vero) and human lung embryonated cells (MRC-5) in the isolation of influenza A virus from nasopharyngeal aspirates by shell vial culture. *J Clin Microbiol* **35:**1900–1901.
84. **Smith TF, Reichrath L.** 1974. Comparative recovery of 1972–1973 influenza virus isolates in embryonated eggs and primary rhesus monkey kidney cell cultures after one freeze-thaw cycle. *Am J Clin Pathol* **61:**579–584.
85. **Tkácová M, Varecková E, Baker IC, Love JM, Ziegler T.** 1997. Evaluation of monoclonal antibodies for subtyping of currently circulating human type A influenza viruses. *J Clin Microbiol* **35:**1196–1198.
86. **Spackman E.** 2014. Influenza subtype identification with molecular methods. *Methods Mol Biol* **1161:**119–123.
87. **Ahluwalia G, Embree J, McNicol P, Law B, Hammond GW.** 1987. Comparison of nasopharyngeal aspirate and nasopharyngeal swab specimens for respiratory syncytial virus diagnosis by cell culture, indirect immunofluorescence assay, and enzyme-linked immunosorbent assay. *J Clin Microbiol* **25:**763–767.
88. **Marquez A, Hsiung GD.** 1967. Influence of glutamine on multiplication and cytopathic effect of respiratory syncytial virus. *Proc Soc Exp Biol Med* **124:**95–99.
89. **Shahrabadi MS, Lee PWK.** 1988. Calcium requirement for syncytium formation in HEp-2 cells by respiratory syncytial virus. *J Clin Microbiol* **26:**139–141.
90. **Arens MQ, Swierkosz EM, Schmidt RR, Armstrong T, Rivetna KA.** 1986. Enhanced isolation of respiratory syncytial virus in cell culture. *J Clin Microbiol* **23:**800–802.
91. **College of American Pathologists:** 2014. *Microbiology Checklist 04.21.2014*. Virology Section. CAP, Northfield, IL.
92. **Mackenzie JS.** 1999. Emerging viral diseases: an Australian perspective. *Emerg Infect Dis* **5:**1–8.
93. **Boeckh M.** 2011. Complications, diagnosis, management, and prevention of CMV infections: current and future. *Hematology (Am Soc Hematol Educ Program)* **2011:**305–309.
94. **de Vries JJ, van der Eijk AA, Wolthers KC, Rusman LG, Pas SD, Molenkamp R, Claas EC, Kroes AC, Vossen AC.** 2012. Real-time PCR versus viral culture on urine as a gold standard in the diagnosis of congenital cytomegalovirus infection. *J Clin Virol* **53:**167–170.
95. **Gitman MR, Ferguson D, Landry ML.** 2013. Comparison of Simplexa HSV 1 & 2 PCR with culture, immunofluorescence, and laboratory-developed TaqMan PCR for detection of herpes simplex virus in swab specimens. *J Clin Microbiol* **51:**3765–3769.
96. **Hsiung GD.** 1968. Latent virus infections in primate tissues with special reference to simian viruses. *Bacteriol Rev* **32:**185–205.
97. **Smith KO.** 1970. Adventitious viruses in cell cultures. *Prog Med Virol* **12:**302–336.
98. **Stanbridge E.** 1971. Mycoplasmas and cell cultures. *Bacteriol Rev* **35:**206–227.

99. **Chu FC, Johnson JB, Orr HC, Probst PG, Petricciani JC.** 1973. Bacterial virus contamination of fetal bovine sera. *In Vitro* **9:**31–34.
100. **Whiteman MD.** 2006. Scope and practicality of in vivo testing for adventitious agents. *Dev Biol (Basel)* **123:**147–152, discussion 183–197.
101. **Purfield A, Ahmad N, Park BJ, Kuhles D, St George K, Ginocchio C, Harris JR.** 2013. Epidemiology of commercial rhesus monkey kidney cells contaminated with *Coccidioides posadasii*. *J Clin Microbiol* **51:**2005.
102. **Fong CKY, Landry ML.** 1992. An adventitious viral contaminant in commercially supplied A549 cells: identification of infectious bovine rhinotracheitis virus and its impact on diagnosis of infection in clinical specimens. *J Clin Microbiol* **30:**1611–1613.
103. **Križanová O, Rathová V.** 1969. Serum inhibitors of myxoviruses. *Curr Top Microbiol Immunol* **47:**125–151.
104. **Merten OW.** 2002. Development of serum-free media for cell growth and production of viruses/viral vaccines—safety issues of animal products used in serum-free media. *Dev Biol (Basel)* **111:**233–257.
105. **Hodinka RL.** 2013. Point: is the era of viral culture over in the clinical microbiology laboratory? *J Clin Microbiol* **51:** 2–4.
106. **Landry ML, Caliendo AM, Ginocchio CC, Tang Y, Valsamakis A.** 2015. Algorithms for detection and identification of viruses, p 1432–1435. *In* Jorgensen JH, Pfaller MA, Carroll KC, Funke G, Landry ML, Richter SS, Warnock DW (ed), *Manual of Clinical Microbiology*, 11th ed. ASM Press, Washington, DC.

Viral Antigen Detection

DIANE S. LELAND AND RYAN F. RELICH

8

Laboratorians continually seek methodologies that yield accurate results in a timely fashion, are cost effective, and require less technical expertise. Diagnosis of viral infections via viral antigen detection methods such as immunofluorescence (FA), immunochromatography (lateral flow) (IC), and enzyme immunoassays (EIA) offer many of these attractive features and are useful for direct detection of viral antigens in an array of clinical specimens and for identification of cultivated viruses. Whether the detection method is FA, rapid IC, or EIA, detection of antigens of the common respiratory viruses (i.e., adenovirus; influenza virus [Flu] A and B; parainfluenza virus [PIV] −1, −2, and −3 and respiratory syncytial virus [RSV]), has been shown to be more useful in patient management than either traditional virus isolation (1, 2, 3) or viral detection in rapid culture using centrifugation-enhanced inoculation (4). There is considerable variability in the sensitivity, specificity, technical considerations, and turnaround time among the various methods, and each method may perform differently depending on the viral target. This chapter deals with principles of FA, IC, and EIA and their contemporary applications in viral antigen detection.

BASIC CONCEPTS OF VIRAL ANTIGEN DETECTION METHODS

Immunofluorescence (FA)

Assays utilizing antibodies tagged with fluorescent dyes have been used for viral antigen detection for many years. Most employ U.S. Food and Drug Administration (FDA)-cleared monoclonal antibodies (MAbs) commercially marketed in the United States for use in detecting viral antigens directly in clinical specimens and in virus-infected cells from cell cultures. This approach is popular for detecting and differentiating herpes simplex (HSV) type 1 and type 2, varicella zoster virus (VZV), and eight common respiratory pathogens: adenovirus; human metapneumovirus (hMPV); Flu A and B; PIV −1, −2, and −3 and RSV. Distributors of these reagents in the U.S. include, but are not limited to, the following: DakoCytomation USA, Carpinteria, CA; Diagnostic Hybrids/Quidel (DHI), Athens, OH; Millipore Corporation Light Diagnostics, Temecula, CA; and Trinity Biotech, Carlsbad, CA.

For FA testing for viral antigens, cells from skin or genital lesions and various respiratory samples (nasopharyngeal washes, aspirates, and swabs) are fixed on a microscope slide, usually in several cell spots or dots or by cytocentrifugation (5). Intact cells must be present if the assay is to be valid. Cells scraped from an infected cell culture monolayer are prepared on microscope slides; cells growing on a shell vial monolayers may also be stained. The instructions provided by the manufacturer of the specific reagents must be followed.

Most procedures are direct immunofluorescence (DFA) methods. In a typical DFA, specimens are fixed in acetone and dried prior to staining. Fixed smears may be stored for several days at 2 to 8°C or frozen at −20°C for up to a year when stored in an air-tight container. For smears on microscope slides, 1 or 2 drops of fluorescein-labeled MAbs are added. For shell vial monolayers, 3 or 4 drops of MAbs are added. The preparations are incubated for 15 to 30 minutes at 37°C in a humidified environment. The smears are washed with a phosphate-buffered saline (PBS) solution and then air dried. Mounting fluid and coverslips are added, and the smears are examined microscopically at a magnification of ×200 or higher, using a fluorescein isothiocyanate (FITC) filter system. Virus-infected cells demonstrate bright apple-green fluorescence against a background of red fluorescing material stained by the Evans Blue counterstain contained in most MAb preparations. The technologist must evaluate the intensity of fluorescence as well as the distribution of antigens within the stained cells.

Characteristic distribution of fluorescence varies from virus to virus. Fluorescence may be cytoplasmic, nuclear, or both, with either uniform, even, punctate, or speckled staining. Fluorescence may be found in association with syncytia. For hMPV, MAbs for DFA staining were not commercially available until 2008. The hMPV fluorescence is speckled and predominantly cytoplasmic (6). Examples of fluorescence staining of several respiratory viruses are shown in Fig. 1.

Some MAbs for viral antigen detection are marketed in an indirect immunofluorescence (IFA) format. IFA is similar to DFA and is appropriate for the same types of samples with the same smear preparation and fixation guidelines. In IFA, two staining steps are required. First, unlabeled antiviral MAbs are added to fixed cells. Following incubation and

doi:10.1128/9781555819156.ch8

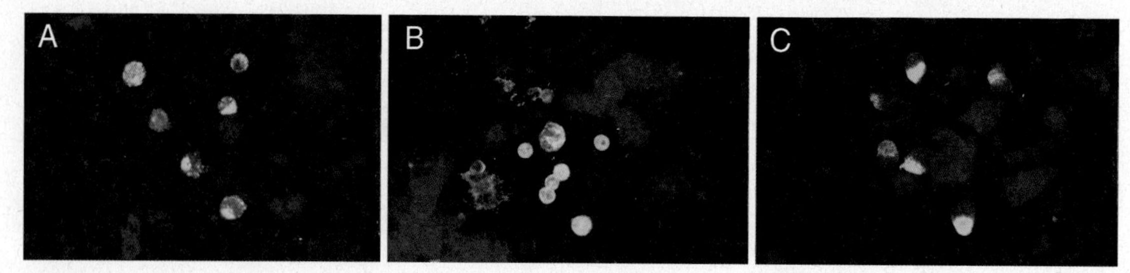

FIGURE 1 DFA staining with FITC-labeled monoclonal antibodies of virus-infected cells, 400x. (A) influenza B–infected cells, (B) herpes simplex–infected cells, and (C) respiratory syncytial virus–infected cells. Green fluorescence indicates antigen detected. Red fluorescence is background material stained with Evans Blue counterstain included in the stain preparation. Photos courtesy of Indiana Pathology Images.

washing, FITC-labeled antimouse antibodies are added to cover the cells. Following a second incubation and washing, slides are dried and a coverslip is added. Stained smears are observed as described above for the DFA procedure. IFA is described in detail in this volume in chapter 9, Serologic Methods. MAbs marketed in the IFA format are used most frequently for confirming the identity of viral cell culture isolates. Measles virus, mumps virus, PIV-4, and enteroviruses are examples of viruses often confirmed by IFA staining methods.

MAb pools that screen for common respiratory viruses are marketed in a DFA format that allows definitive identification of more than one virus simultaneously through the use of two different fluorescent dyes with overlapping spectra (Light Diagnostics SimulFluor reagents; Millipore Corp.). These reagents are FDA-cleared for direct specimen testing and for culture confirmation. When stained preparations are examined with a fluorescence microscope with an FITC filter set, one antibody will produce apple-green fluorescence and the second will appear gold or golden orange. SimulFluor reagents include the following combinations: RSV/six other respiratory viruses, RSV/PIV-3, PIV-1, -2, -3/adenovirus, PIV-1, -2/PIV-3, Flu A/Flu B, RSV/Flu A, HSV/VZV, and HSV-1/HSV-2. SimulFluor reagents have shown excellent sensitivities and specificities comparable to those of individual stains, for the respiratory viruses (5).

DHI also markets two MAb preparations with two different fluorescent labels in the DFA format (D^3 Duet DFA Kits). Staining with one of these preparations identifies Flu A with golden fluorescence while showing green fluorescence for the other six common respiratory viruses (D^3 Duet Influenza A/Respiratory Virus Screening Kit). The second Duet preparation identifies RSV with golden fluorescence, whereas the other six respiratory viruses showed green fluorescence (D^3 Duet RSV/Respiratory Virus Screening Kit).

A rapid DFA format for staining nasal and nasopharyngeal (NP) cells in solution (D^3 FastPoint L-DFA; DHI) features three dual-labeled (R-phycoerythrin versus FITC) MAb preparations also containing propidium iodide and an Evans Blue counterstain: Flu A/B, RSV/hMPV, and PIV 1-3/adenovirus. Cells infected with the first virus named for each preparation will fluoresce yellow, and cells infected with the second virus named for each preparation will fluoresce green. Uninfected cells and background material will fluoresce red. PIV−1, −2, and −3 are not differentiated in this system. Examples of FastPoint staining are shown in Fig. 2. The staining procedure is expedited because the specimen material is not fixed to a microscope slide. After a short incubation of portions of liquid specimen material with each of the three MAb preparations, the three samples are placed on a microscope slide and examined in the wet state with a fluorescence microscope with an FITC filter set. Results are available in 25 minutes or less, with accuracy comparable to traditional DFA (Product information, D^3 FastPoint L-DFA, DHI).

Immunochromatography (Lateral Flow)

The test principle of a typical IC assay is shown in Fig. 3. In this testing, antiviral MAbs labeled with "visualizing particles," such as colloidal gold nanoparticles, carbon black or blue polystyrene—sometimes called the "signal" or "detection" antibodies—are adsorbed nonspecifically onto one end of the strip. This end of the strip is where the patient's sample is applied (Fig. 3A). At the opposite end of the test strip, there is an area, the test area, in which unlabeled antiviral antibodies, usually polyclonal, are immobilized in a thin line. Further toward the opposite end of the strip, there is an area, the control area, in which unlabeled polyclonal antispecies (usually antimouse) IgG is immobilized.

In IC assays the patient's sample may be one that has been collected on a swab and diluted in a small amount of buffer or extracting compound or may be a drop of liquid sample from various aspirates or washes, etc. If the sample contains the antigen in question, the labeled MAbs on the

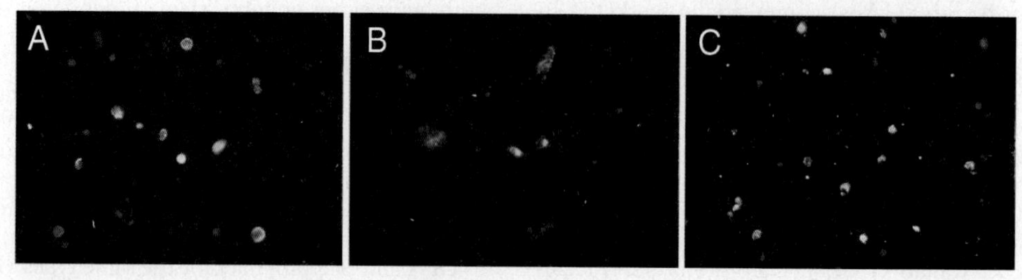

FIGURE 2 Diagnostic Hybrids (DHI) proficiency panel samples stained with the DHI D^3 FastPoint L-DFA method, 200x. (A) parainfluenza (yellow) and adenovirus (green), (B) human metapneumovirus (green), and (C) influenza A (yellow) and B (green). Red fluorescence is uninfected cells and debris stained with propidium iodide and Evans Blue counterstain. Photos courtesy of Indiana Pathology Images.

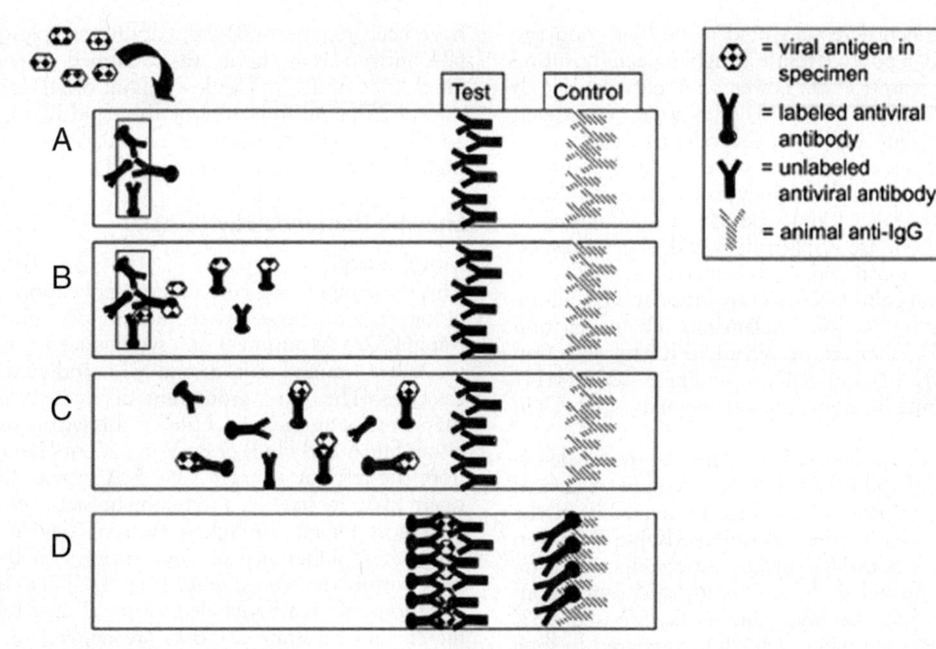

FIGURE 3 IC (lateral flow) testing mechanism. (A) Viral antigen in the specimen is added to the sample pad of a nitrocellulose strip; labeled antiviral antibodies are bound nonspecifically on the sample pad. The nitrocellulose strip also includes a test area of unlabeled antiviral antibodies and a control area of unlabeled animal antihuman IgG; (B) Labeled antiviral antibodies on the sample pad bind to viral antigen in the sample, and the complexes migrate along the strip; (C) Migration continues; (D) Viral antigens, in complex with labeled antiviral antibodies, are recognized and captured by the unlabeled antiviral antibodies in the test area of the strip, forming a visible line; excess labeled antiviral antibodies continue to migrate and are captured at the control line by anti-IgG. This control ensures that the specimen migrated the entire length of the strip and that the strip is functioning properly.

sample pad bind to the viral antigen, and the labeled antiviral MAb-viral antigen complexes travel laterally along the test strip membrane (Fig. 3B, C). When they reach the test area in which the unlabeled antiviral antibodies are bound, these antibodies bind to the viral antigens—the ones that were initially bound to the labeled MAbs, resulting in a visible colored line (usually pink or red). Unbound or excess labeled antiviral antibodies pass through the test line and are bound at a control line by the antimouse IgG, also resulting in a visible colored line (Fig. 3D). Gross visual assessment of presence/absence of colored lines determines the test result. The presence of a line at the test area and at the control area indicates a positive result. The appearance of a line at the control area in the absence of a line in the test area indicates a negative result. The assay result cannot be interpreted unless a line is seen in the control area. Many IC kits for Flu A and B and RSV detection are available commercially. This includes, but is not limited to, the following: BinaxNOW (Alere Scarborough, Scarborough, ME), Directigen (Becton-Dickinson [BD], Sparks, MD), SAS FluAlert (SA Scientific, Inc., San Antonio, TX), QuickVue (Quidel, San Diego, CA), and Xpect (Remel, Inc., Lexena, KS).

In contrast to the visual examination of results in traditional IC assays, the interpretation of results of some newer IC assays is enhanced by use of an automated readout device or analyzer. In the BD Veritor IC (Becton-Dickinson) a readout device displays the assay results 10 seconds after the test cartridge—which has undergone 10 minutes of color development—is inserted. The reader detects a proprietary enhanced colloidal gold particle, which is deposited at the test line. A fluorescence-based IC system, the Sofia (Quidel), features an analyzer that detects unique polystyrene microparticles impregnated with a europium compound. Elimination of the subjectivity of visual evaluation of results enhances accuracy and reproducibility of these methods, and these assays routinely show higher sensitivity and specificity than traditional ICs (7, 8, 9). The Sofia analyzer also tracks quality control functions and offers additional operator oversight.

Most IC assays are one-step procedures. Following addition of the patient's sample, the test requires no further manipulation other than observation of the result at the end of the time period. Many of the traditional IC assays have been granted waived status according to the Clinical Laboratory Improvement Act (CLIA) guidelines. This facilitates performance of these assays in physicians' offices and clinics. They are user-friendly, require only 10 to 30 minutes to complete, are stable in the long term, and cost $20 to $30 per test.

Of the common respiratory viruses, only RSV and Flu A and B virus antigens are frequently detected by ICs. Although all three of these viruses proliferate in standard cell cultures, and antigens of all three can be identified by DFA, the rapid ICs are popular because they can be performed quickly and require little time or technical expertise. Studies comparing these rapid IC antigen assays to virus isolation in cell culture show that the rapid methods have very good specificity, averaging 98%, but moderate to poor sensitivity, averaging 62% (10). IC systems may require as many as 1,000,000 viral particles to yield a positive result; this is in contrast to viral culture, which may require as few as 10 infectious virus particles for successful virus isolation (11). The sensitivity of RSV antigen detection by IC compared to cell culture is higher than that of similar tests that detect Flu antigens. In general, the specificities (compared to virus isolation in cell culture) reported for rapid assays for RSV and Flu virus antigen detection tend to be high, and the predictive value of a positive result is high, especially during respiratory virus season (12). Lower sensitivity for Flu B antigen has been reported (10, 13) with some IC antigen detection systems, but sensitivity equivalent to that of Flu A detection has been seen (9). However, there is considerable variation in findings from study to study—with differences relating to the level and type of virus circulating in the particular season, age of patients tested, skill of testing personnel, IC method tested, and format of the reference method (rapid viral culture versus conventional tube culture vs. reverse transcriptase [RT] polymerase chain reaction [PCR]) (9, 13, 14). Comparisons of various ICs for detection of Flu A (H3N2)v (8) and of DFA for Flu A 2009 H1N1 (15) showed definite differences among popular rapid ICs in detecting different strains.

Although IC assays are used most frequently for detection of Flu A and B and RSV antigens, several FDA-cleared ICs for detecting adenovirus antigens are available in the U.S. A CLIA-waived IC, the AdenoPlus (Rapid Pathogen Screening, Inc., Sarasota, FL), used to test conjunctival samples primarily as an aid in differentiating adenoviral conjunctivitis from HSV keratoconjunctivitis, showed 93% sensitivity and 90% specificity detection compared to shell vial culture and PCR (16). Other FDA-cleared ICs are marketed for detection of adenovirus in eye swabs, nasal and pharyngeal secretions, fecal samples, and cell culture supernatants. At this writing, the literature is sparse regarding performance of these IC tests. However, a pseudo-outbreak of adenovirus infection in a neonatal intensive care unit due to false-positive results from an adenovirus antigen detection assay was reported (17).

FDA-cleared ICs are available for detection of rotavirus antigens. These compare favorably in sensitivity and specificity with microwell-based rotavirus EIAs and with electron microscopy for detecting group A rotavirus (18). These ICs would be suitable for use in a point-of-care setting.

One new IC, a fourth generation human immunodeficiency virus (HIV)-1/-2 antibody/HIV-1 p24 antigen assay (Alere Determine HIV-1/2 Ag/AB Combo), simultaneously detects both antigens and antibodies in the same sample. Specimen (serum, plasma, fingerstick, or venous whole blood) is added to the sample pad on the nitrocellulose strip. The specimen mixes with a biotinylated anti-p24 antibody, selenium colloid-antigen conjugate and selenium colloid anti-p24 antibody. This mixture continues to migrate through the solid phase to the immobilized avidin recombinant antigens and synthetic peptides at the patient window sites. If antibodies to HIV-1 and/or HIV-2 are present in the specimen, the antibodies bind to the antigen-selenium colloid and to the immobilized recombinant antigens and synthetic peptides, forming one red bar. If HIV antibodies are absent, no red bar forms. If free nonimmunocomplexed HIV-1 p24 antigen is present, the antigen binds to the biotinylated anti-p24 from the sample pad and the selenium colloid anti-p24 antibody, and it binds to an immobilized avidin, forming a red bar at the patient HIV antigen window site. If p24 antigen is not present, both the biotinylated anti-p24 and selenium colloid anti-p24 antibody flow past the patient window, and no red bar is formed at the patient HIV antigen site (19). Detection of HIV-1 and HIV-2 antibodies with this IC has been reported to be equal to that of other HIV-1/-2 antibody detection ICs (20). Because of the antigen detection component of this assay, it is expected to offer earlier detection of new HIV infections. However, at this writing, very poor sensitivity and specificity for antigen detection have been reported (20–22). Of five samples positive by a p24 antigen assay, none was identified as positive in the rapid IC assay (20). The lower limit of HIV antigen detection of 25 pg/ml indicated by the rapid IC manufacturer—established with control serum dilutions—appears to be higher in actual clinical samples (21).

Enzyme Immunoassay (EIA)

Membrane EIAs

The most rapid and convenient application of EIA technology in viral antigen detection involves membrane EIAs. Rapid (20 to 30 minutes) EIA systems are available as single-use, self-contained units assembled in individual modules or cassettes. The most prominent application of membrane EIAs in viral antigen detection at this writing is detection of either Flu A and Flu B or RSV in patients' samples collected from the respiratory tract. One EIA system, the BD Directigen EIA, includes a pretreatment step, after which the sample is forced through a focusing device, resulting in nonspecific adherence of viral antigen on the membrane held within the test cassette (Fig. 4). Then sequential applications of enzyme-labeled antiviral antibodies, washing buffer, and substrate solution are applied to the cassette. Color development occurs on the pad in the packet if viral antigen was present in the sample. This pad is often prepared in a unique shape that facilitates interpreting the color change reaction. The EIA test area also contains an internal control to monitor the performance of both the assay and the user. The internal control is actually a dot of viral antigen. There should be color development of this control if the reagents are working properly and the test is performed correctly, regardless of whether the patient's sample is positive or negative for the analyte. These assays have several timed steps as part of the procedure. Although necessary technical expertise is minimal, most of these assays are classified as moderately complex according to CLIA. Costs for membrane EIAs may range from $25 to $35 per test.

In general, membrane EIAs for detection of Flu A and B and RSV yield sensitivities that are lower than those of DFA. However, sensitivities are usually slightly higher than those of most ICs (23, 24).

Immunohistochemical Staining

Immunohistochemical (IHC) staining is a type of EIA that permits visual detection and localization of antigens in tissue-thin sections by conventional light microscopy. IHC employs antibodies, either monoclonal or polyclonal, coupled to reporter enzymes, such as horseradish peroxidase and alkaline phosphatase. These enzymes catalyze chromogenic chemical reactions, when combined with a substrate, resulting in the formation of an insoluble, visually detectable colored reaction product that marks the locations where the antibodies have bound within the specimen. IHC staining in modern histology laboratories is, for the most part, performed using automated systems on formalin-fixed, paraffin-embedded specimens mounted on microscope slides. Briefly, thin sections of tissue are transferred onto microscope slides and deparaffinized prior to antigen retrieval pretreatment, which exposes epitopes that may have been masked by formalin fixation and specimen processing. Specimens are next blocked, usually with normal serum, to minimize nonspecific antibody binding that could hamper detection of specific staining.

Antibodies of known specificity, such as those raised against viral antigens, are next added to the specimen. In

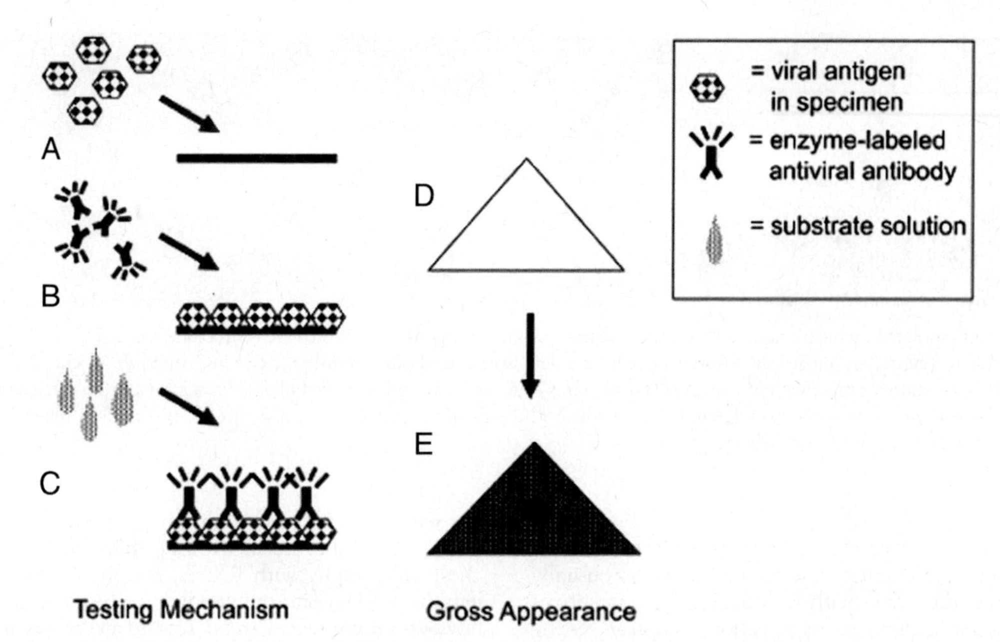

FIGURE 4 Membrane EIA testing mechanism and gross appearance. Testing mechanism: (A) viral antigen in specimen is nonspecifically adhered to the membrane by filtration through a focusing device; (B) enzyme-labeled antiviral antibody is added and binds to viral antigen present on the membrane; (C) a substrate solution is added and changes color when acted upon by the enzyme. Gross appearance: (D) a test area is defined within the cassette—this is colorless at the beginning of the assay; (E) the test area changes color due to the action of the enzyme on the substrate solution. The colored dot in the center of the test area is a built-in control to ensure proper function of the device.

direct IHC staining, the antibodies are labeled with the reporter enzyme. After an incubation period and rinsing, a substrate, often 3, 3′-diaminobenzidine (DAB) or 5-bromo-4-chloro-3-indolyl phosphate/nitro blue tetrazolium (BCIP/NBT), is added. In areas where enzyme-labeled antibodies have bound to their cognate viral antigen(s), the enzymes act on the substrate to produce the insoluble colored reaction product. Direct IHC staining usually requires only 20 to 30 minutes. For indirect IHC staining, the antibodies applied initially to the specimen are unlabeled. After incubation and rinsing, a preparation of enzyme-labeled antispecies antibodies (directed against the species in which the primary antibody was raised) is added. These "detection" antibodies bind to unlabeled antibodies that were bound in the first step. After incubation and rinsing, a substrate solution is added and color development occurs. For amplification of the analytical signal, an alternate indirect staining method that employs a biotin-labeled primary antibody that binds with very high affinity to enzyme-labeled streptavidin can be used. Indirect staining requires approximately 90 minutes. Finally, specimens are counterstained, and the slides are examined with a standard light microscope. The intensity, distribution, and pattern of the staining are evaluated.

Antibodies for IHC-based detection of several viral antigens, including those of adenovirus, BK polyomavirus (using anti-SV40 large T antigen), Epstein-Barr virus (EBV), HSV, cytomegalovirus (CMV), human papillomaviruses, parvovirus B19, and West Nile virus, among others, are commercially available from a variety of manufacturers such as Thermo Fisher Scientific, Dako and Chemicon International, Inc. (25). Typically, IHC for virus detection is performed secondary to the observation of intracellular inclusions or other morphological clues seen during examination of hematoxylin and eosin-stained specimens. In our institution, IHC is most frequently used to detect antigens of CMV and EBV, but, overall, IHC staining is not used as frequently as other viral antigen detection methods, such as immunofluorescence.

The obvious advantage of IHC over immunofluorescence is that a fluorescence microscope is not required for examination of IHC stains; a standard light microscope is all that is required. Also, many histology laboratories are already equipped to perform viral antigen IHC stains, as most perform IHC stains for detection of other, nonviral, antigens. Disadvantages of IHC methods include the time required for color development during the staining process and nonspecific staining that may be due to endogenous peroxidases in some types of clinical specimens. Another disadvantage is a rather limited pool of peer-reviewed data regarding the analytical parameters of viral antigen IHC. However, Lu et al. (26) demonstrated that automated IHC for CMV yielded a sensitivity of 75.7% and a specificity of 100%. In contrast, a study conducted by Fanaian and colleagues (27) determined that the sensitivity of automated EBV IHC was 44% and the specificity was 93%. A review of the analytical parameters of specific viral antigen IHCs is provided by Elston et al. (28). Examples of IHC staining for viral antigens in tissues are shown in Fig. 5.

Tube or Microwell-Based EIA

EIAs requiring more analytical steps and often performed by automated systems are used for detection of some viral antigens. In most of these systems, viral antibody is mounted on the inner surface of a microwell or test tube, and sequential addition of patient's serum and various detection reagents, one of which is labeled with an enzyme, produces a color change when viral antigen is detected. These systems are described in detail in chapter 9 of this volume. Often the antigens of bloodborne pathogens such as HIV or hepatitis B (HBV) are detected with such systems. Neither HIV nor HBV proliferates in standard cell cultures, and their antigens are not detected through FA methods. Because both HIV

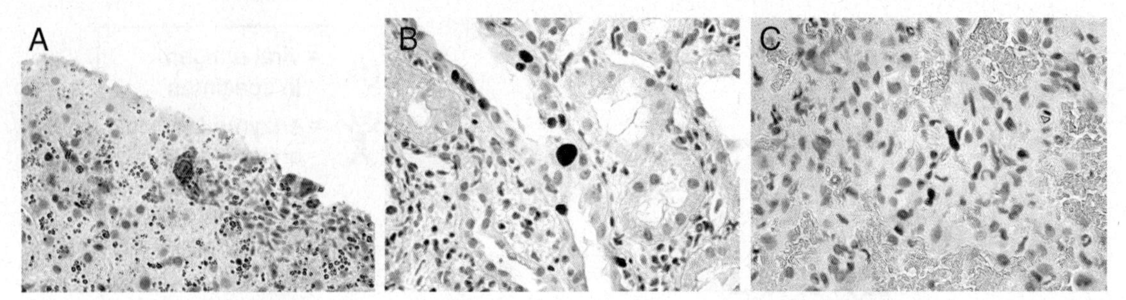

FIGURE 5 Immunohistochemical staining reveals the presence of viral antigens (dark brown areas) in tissue-thin sections. (A) Adenovirus-infected hepatocyte (center of image) demonstrating both nuclear and cytoplasmic staining. Scattered throughout the tissue are coarse, refractile, and brown-staining bile pigment granules; (B) BK virus–infected renal tubular epithelial cells demonstrating nuclear staining; (C) Cytomegalovirus-infected lung tissue reveals viral antigens in both nuclear and cytoplasmic compartments. Original magnification 400x. Photos courtesy of Indiana Pathology Images.

and HBV antigens can be detected in peripheral blood, this is the specimen of choice for detection of these antigens. Both HIV and HBV antigen detection EIAs are usually performed in conjunction with EIA testing for a variety of additional antibody markers of infection, and this testing represents some of the highest volume viral marker testing offered in clinical laboratories. Although HIV-1 p24 antigen testing has largely been replaced by more sensitive nucleic acid tests, newer fourth generation HIV screening methods include detection of this analyte, along with HIV-1 and HIV-2 antibodies. Fourth generation combination screening tests are the first step in the 2014 CDC recommended HIV screening algorithm (29). Most of these screening tests are highly automated and have been shown to provide sensitive and specific detection of both antibodies and p24 antigen of HIV (30).

Microwell-based EIAs for detection of antigens of several nonculturable gastrointestinal viral pathogens are available. These may be manual or automated, but the specimen of choice is stool, which is a specimen type that is not easily processed with automated systems, and, also, is not a satisfactory specimen for FA testing. Fortunately, EIA works well for detecting viral antigens in stool samples. Antigens of rotavirus, adenovirus types 40/41, and norovirus are commonly detected in these systems. Rotavirus antigen EIAs are widely used and have high sensitivity and specificity. A review of EIAs for testing for the antigens of rotavirus, adenovirus 40/41, and norovirus is available (31). Molecular methods for detecting these gastrointestinal viral pathogens provide even more sensitive detection (31, 32).

APPLICATION OF VIRAL ANTIGEN DETECTION IN DIAGNOSTIC VIROLOGY

Numerous investigators have attempted to evaluate the performance of rapid viral antigen detection assays. The focus of many of these investigations has been Flu A and B antigen detection by rapid influenza diagnostic tests (RIDT). A recent meta-analysis (10) of results obtained with 26 different RIDTs showed high specificity with modest sensitivity, better performance with samples from children than specimens from adults, higher sensitivity for Flu A than Flu B, and no single commercial brand performing markedly better or worse than others. A second meta-analysis of rapid viral diagnosis of acute febrile respiratory illness in children in the emergency department (ED) focused primarily on RIDTs and their impact on patient management and outcomes (33). Four studies (3, 34, 35, 36) were included in the analysis with the following findings: rapid viral testing did not reduce antibiotic use in the ED, there were lower rates of chest radiography with RIDTs, but there was no effect on length of ED visits or blood or urine testing in the ED; however, there was a trend toward decreases in the latter three criteria. Viewed individually, one study (3) showed decreases in antibiotic use, shorter ED visits, and lower rates of chest radiography while another (36) showed no significant differences in any of these parameters.

Such conflicting reports in the literature regarding performance characteristics and possible clinical impact of viral antigen detection methods make it difficult for laboratorians to determine which, if any, of these assays to choose. Nonetheless, during the last three decades viral antigen detection assays have become a part of diagnostic algorithms in most settings where viral diagnoses are made. In view of the many outpatient visits and hospitalizations of young children with acute respiratory infections, most of which are caused by viral agents, the availability of assays that detect the common respiratory viruses is essential. The ICs and membrane EIAs for RSV and FLU A and B virus antigens are especially useful when a rapid result is needed and highly skilled virologists are not available.

Specimen collection/transport, technical expertise, equipment, turnaround time, target viruses, sensitivity/specificity, and cost are just a few of the parameters that must be considered in selecting these assays. Some of these factors are compared for the various viral antigen detection methods in Table 1.

With more and more diagnostic testing being performed in point-of-care settings and physician offices, the IC assays are very popular. In general, these require no equipment and virtually no technical expertise to perform. These simple, one-step assays require only a few minutes to set up and are ready for a final reading after a 10- to 30-minute incubation. Because testing can be completed while the patient remains available in the facility, results can be used in patient management. Viral diagnostic testing in the point-of-care setting is described in detail in this volume in chapter 17.

FDA-cleared IC assays are available commercially in the U.S. for detecting antigens of very few viruses: adenovirus, Flu A/B, rotavirus, and RSV, with Flu A/B and RSV IC assays the most widely used of these. Sensitivities and specificities vary from virus to virus and from season to season. The CDC website (12) provides guidance for clinical laboratory directors for selection of rapid diagnostic tests for influenza, indicating that for this highly seasonal virus,

TABLE 1 Characteristics of Viral Antigen Detection Methods

Method	Viral Antigens Detected**	Typical Turnaround Time	Equipment Required/ Steps in Assay	CLIA Status
Immuno-Fluorescence				
DFA	adeno, HSV-1/-2, hMPV, Flu A & B, PIV-1, -2, -3, VZV	25–30 min	Fluorescence Microscope, incubator, multiple steps in assay	Mod or Highly Complex
IFA	Used to confirm isolate ID: entero, measles, mumps, PIV-4	2–3 hours	(same as DFA)	(same as DFA)
Immuno-chromatography	adeno, Flu A & B, rotavirus, RSV	15–30 min	One step assay, no equipment (some have readout device)	Waived or Mod Complex
EIA Membrane Cassette	Flu A & B, RSV	30 min	No equipment, multiple steps in assay	Mod Complex
EIA Immuno-histochemistry	BK, CMV, EBV, HSV-1/-2, others	24–48 h to process tissue, staining performed in batches, read by pathologist	Facilities for tissue processing, imbedding, cutting, staining, light microscope, multiple steps in assay	Highly complex
EIA Microwell	hep B surface Ag, hep Be Ag, HIV-1 p 24, rotavirus	~24 hours, usually performed in batches with related Ab EIAs	Usually automated instrumentation or spectrophotometric plate readers, multiple steps in assay	Mod Complex

Abbreviations: Ab=antibody, adeno=adenovirus, CLIA=Clinical Laboratory Improvement Act, EBV=Epstein-Barr Virus, entero=enteroviruses, Flu=influenza, hep=hepatitis, HIV=human immunodeficiency virus, hMPV=human metapneumovirus, PIV=parainfluenza virus, RSV=respiratory syncytial virus.
**Common utilization, may not be all inclusive.

sensitivities vary from 50 to 70%, and specificities typically range between 90 and 95% when compared to virus isolation or RT-PCR. However, when disease prevalence is low (e.g., beginning and end of the influenza season), false positive results are seen more often, and, when disease prevalence is high, false negative results are more likely to occur. Disease prevalence is an important factor with all viral diagnostic testing.

Because most viral antigen detection methods detect only Flu A and B or RSV antigens, viruses other than the target viruses that may be present in the sample will remain undetected. In addition, a positive IC or membrane EIA test result does not eliminate the possibility that patients may be coinfected with another virus that may be contributing to their symptoms (37). This is of particular significance when testing persons with impaired immune function and children with severe respiratory illness (38, 39). Therefore, if dual infections are suspected, additional testing must be done.

Sample type must be considered. The approved specimens for each IC or membrane EIA may be few in number and different from those of other comparable assays. For example, some IC Flu A/B assays are approved for testing of nasal swabs only, whereas others accurately test NP washes/aspirates/swabs and throat swabs, thus providing more options for testing. Approved sample types also vary for adenovirus and RSV antigen ICs. However, both ICs and EIAs for viral antigen detection, in comparison to the FA methods, have less stringent specimen requirements. For FA, samples must contain intact infected cells. ICs and EIAs can detect free virus, so intact infected cells are not essential for accurate results. For virus isolation, viable infectious virus must be present in the sample. FAs, ICs, and EIAs successfully detect nonviable viruses and viral antigen fragments, which would fail to proliferate in cell cultures. However, falsely negative viral antigen IC results have been reported with bloody samples and those with high viscosity (40). In addition to sample type, the age of the patient contributes to the expected sensitivity of the test. Young children, under the age of 5 years, are known to shed virus in higher titer than adults.

Although the need for technical expertise is advertised as being minimal for performance of most rapid viral antigen detection methods, testing that is carried out by personnel who are less experienced with test kits, especially in reading results that are weakly positive, yields lower sensitivity and specificity than that advertised by the manufacturer (41). There are a number of concerns for FDA-cleared CLIA-waived testing performed outside the laboratory at point-of-care or near-patient waived sites in physicians' offices and clinics. With simple test procedures, less emphasis is placed on training and fewer restrictions are placed on the number of health care workers allowed to carry out testing. It is likely that test sensitivity and specificity will suffer if those who perform waived testing are not properly trained and provided with adequate oversight. These issues and others involved with waived testing have been reviewed (42) and should be considered if results of waived status viral antigen tests performed outside the laboratory are to be used in patient management.

Both IC and membrane EIA methods are sometimes performed within the clinical laboratory because they offer a short turnaround time and may be used during off-shifts when the virology staff is unavailable. In some clinical laboratories, the diagnostic algorithm for respiratory virus detection begins with either IC or EIA testing for viral antigens of Flu A/B and RSV. Then, based on the result of this testing, the sample may be reflexed for testing by more sensitive methods (43). Product information for most viral antigen detection ICs and EIAs includes the suggestion that samples with negative results should be tested with another more sensitive method, such as viral culture or a molecular method. The sensitivity and specificity obtained with the rapid IC and membrane EIA tests when put to use in the routine diagnostic virology laboratory are often lower than that stated by the manufacturer and lower than that reported in studies that were conducted under tightly controlled circumstances targeting a particular group of patients (44).

FA methods, both DFA and IFA, have been standards in the clinical laboratory for many years for viral antigen detection. Specimen preparation requirements and staining procedures, in addition to the requirement for a fluorescence microscope, keep FA from being an attractive method for the point-of-care setting. FA methods require considerable expertise for performing testing and reading results; however, these methods typically yield substantially higher sensitivity and specificity than IC and EIA methods. The added benefit of this testing is that the quality of the sample can be determined to ensure that adequate numbers and types of cells are present. Likewise, the staining intensity and distribution ensure specificity. FA testing may be the next step in a viral respiratory testing algorithm to follow either IC or EIA screening (43).

An added benefit of FA testing over IC and EIA is the larger test menu. All of the common respiratory viruses (adeno, Flu A/B, hMPV, PIV1-3, and RSV) as well as HSV-1/-2 and VZV can be detected with DFA methods. The menu is further enlarged with IFA methods for confirming the identity of viral culture isolates of enterovirus, measles, mumps, and PIV-4. The relatively short turnaround time of 30 minutes to 1 hour for DFA and 2 to 2.5 hours for IFA also contribute to the usefulness of this method. Modified FA methods that detect more than one virus at a time (Millipore Simulfluor and DHI Duet) and those that expedite staining with centrifugation and reading in suspension (DHI FastPoint) represent alternatives to simplify and speed up FA technology while retaining the advantages of FA assay sensitivity and specificity.

Immunohistochemical stains that are performed on fixed tissue sections represent another viral antigen detection approach. These methods, which are actually simply EIAs performed on tissue sections mounted on microscope slides, are a great aid to anatomic pathologists who are challenged with the task of making preliminary viral diagnosis based largely on the morphology of virus-infected cells in tissues. Through immunohistochemical staining, the identity of the viral infection can be confirmed in a much shorter time than would be the case if isolation of the virus in culture was required.

Viral antigen detection performed in more traditional tube or microwell-based EIA procedures are needed for detecting and quantitating viral antigens that circulate in peripheral blood. These are definitely not the types of assays involved in rapid viral detection, with testing confined largely to antigens of bloodborne pathogens and usually highly automated. Panels of antibody and related antigen detection EIA are often performed together for these pathogens.

When compared with molecular methods that include target amplification, the viral antigen detection assays described here invariably have lower sensitivity. As molecular assays are improved to be more user-friendly, require less specialized equipment and technologists' time, are developed as syndromic panels that include both bacterial and viral pathogens, and receive FDA-clearance, they will replace some of the viral antigen detection methods. At this writing, viral antigen methods continue to play a major role in assisting with rapid viral diagnoses.

QUALITY ASSURANCE AND REPORTING

Membrane-based EIAs, ICs, and FAs for viral antigen detection are qualitative in nature, with results reported as positive or negative. Microwell-based EIAs with spectrophotometric readouts may convert readings to numerical values (continuous scale), so it is possible to report results in either a qualitative (positive vs. negative) or quantitative format. Samples containing standardized levels of antigens are not widely available, making nation- or world-wide standardization of quantitative assays difficult. However, most commercially available microwell or tube-based EIA products are marketed with calibrator samples and positive and negative control samples to ensure each run of testing is valid.

One additional type of control that must be incorporated into testing is an external lot-to-lot control that is tested when new lots of reagents are put into service. The purpose of this type of control is to ensure that each new lot of reagents produces results comparable to those of the previous lot. This type of control is especially important in tests that provide a quantitative value. Lot-to-lot control materials are not routinely provided by manufacturers and must be purchased or otherwise obtained by the laboratory. Patients' samples (or pooled patient materials) previously tested may be used for this purpose. Prior to being put into service, all lot-to-lot control material must be tested in duplicate or replicated in several runs of testing, including more than one lot number of reagents, in order to define an acceptable range. Most laboratories prefer to use a range that includes ±2 standard deviations, although some laboratories use ±3 standard deviations as their cutoff.

An additional consideration for performance of viral antigen tests is a relatively new addition to the laboratory inspection checklist of the College of American Pathologists (CAP). The item is part of the CAP Immunology Checklist and is identified as IMM.41850 Direct Antigen Test QC (45). This item applies to "nonwaived direct antigen tests on patient specimens that DO include internal controls" and indicates that "a positive and negative external control are tested and documented with each new kit lot number or shipment, and as frequently as recommended by the manufacturer, or every 30 days (whichever is more frequent)." This checklist item likely should be applied to EIA and IC assays (performed in non-CLIA-waived formats) because they meet the criteria of "nonwaived tests with internal controls."

Further recommendations by CAP for this checklist item indicate that acceptability studies must be done to include daily comparison of external controls to built-in controls for at least 20 consecutive days when patients' samples are tested. This requirement is effective for studies performed after January 31, 2012. External control samples must include each antigen sought if the assay includes more than one analyte.

As with all quality assurance and quality control procedures and processes, careful documentation must be created and maintained. Such documentation must also be reviewed at acceptable intervals by the laboratory director.

CONCLUSIONS

As the need for more rapid "real-time" diagnoses continues to increase, careful and creative selecting and combining of viral antigen detection methods with other viral diagnostic techniques is needed. Rapid viral antigen detection methods continue to provide both timely and accurate results.

REFERENCES

1. **Woo PC, Chiu SS, Seto WH, Peiris M.** 1997. Cost-effectiveness of rapid diagnosis of viral respiratory tract infections in pediatric patients. *J Clin Microbiol* **35:**1579–1581.
2. **Barenfanger J, Drake C, Leon N, Mueller T, Troutt T.** 2000. Clinical and financial benefits of rapid detection of respiratory viruses: an outcomes study. *J Clin Microbiol* **38:**2824–2828.
3. **Bonner AB, Monroe KW, Talley LI, Klasner AE, Kimberlin DW.** 2003. Impact of the rapid diagnosis of influenza on physician decision-making and patient management in the pediatric emergency department: results of a randomized, prospective, controlled trial. *Pediatrics* **112:**363–367.
4. **Adcock PM, Stout GG, Hauck MA, Marshall GS.** 1997. Effect of rapid viral diagnosis on the management of children hospitalized with lower respiratory tract infection. *Pediatr Infect Dis J* **16:**842–846.
5. **Landry ML, Ferguson D.** 2000. SimulFluor respiratory screen for rapid detection of multiple respiratory viruses in clinical specimens by immunofluorescence staining. *J Clin Microbiol* **38:**708–711.
6. **Landry ML, Cohen S, Ferguson D.** 2008. Prospective study of human metapneumovirus detection in clinical samples by use of light diagnostics direct immunofluorescence reagent and real-time PCR. *J Clin Microbiol* **46:**1098–1100.
7. **Bell JJ, Anderson EJ, Greene WH, Romero JR, Merchant M, Selvarangan R.** 2014. Multicenter clinical performance evaluation of BD Veritor™ system for rapid detection of respiratory syncytial virus. *J Clin Virol* **61:**113–117.
8. **Centers for Disease Control and Prevention (CDC).** 2012. Evaluation of rapid influenza diagnostic tests for influenza A (H3N2)v virus and updated case count–United States, 2012. *MMWR Morb Mortal Wkly Rep* **61:**619–621.
9. **Peters TR, Blakeney E, Vannoy L, Poehling KA.** 2013. Evaluation of the limit of detection of the BD Veritor™ system flu A+B test and two rapid influenza detection tests for influenza virus. *Diagn Microbiol Infect Dis* **75:**200–202.
10. **Chartrand C, Leeflang MM, Minion J, Brewer T, Pai M.** 2012. Accuracy of rapid influenza diagnostic tests: a meta-analysis. *Ann Intern Med* **156:**500–511.
11. **St George K, Patel NM, Hartwig RA, Scholl DR, Jollick JA Jr, Kauffmann LM, Evans MR, Rinaldo CR Jr.** 2002. Rapid and sensitive detection of respiratory virus infections for directed antiviral treatment using R-Mix cultures. *J Clin Virol* **24:**107–115.
12. **Centers for Disease Control and Prevention.** Rapid diagnostic testing for influenza: information for clinical laboratory directors. www.cdc.gov/flu/professionals/diagnosis/rapid lab.htm (accessed November 7, 2014).
13. **Hassan F, Nguyen A, Formanek A, Bell JJ, Selvarangan R.** 2014. Comparison of the BD Veritor System for Flu A+B with the Alere BinaxNOW influenza A&B card for detection of influenza A and B viruses in respiratory specimens from pediatric patients. *J Clin Microbiol* **52:**906–910.
14. **Lewandrowski K, Tamerius J, Menegus M, Olivo PD, Lollar R, Lee-Lewandrowski E.** 2013. Detection of influenza A and B viruses with the Sofia analyzer: a novel, rapid immunofluorescence-based in vitro diagnostic device. *Am J Clin Pathol* **139:**684–689.
15. **Pollock NR, Duong S, Cheng A, Han LL, Smole S, Kirby JE.** 2009. Ruling out novel H1N1 influenza virus infection with direct fluorescent antigen testing. *Clin Infect Dis* **49:**e66–e68.
16. **Sambursky R, Trattler W, Tauber S, Starr C, Friedberg M, Boland T, McDonald M, DellaVecchia M, Luchs J.** 2013. Sensitivity and specificity of the AdenoPlus test for diagnosing adenoviral conjunctivitis. *JAMA Ophthalmol* **131:**17–22.
17. **Faden H, Ramani R, Lamson D, St George K.** 2010. Pseudo-outbreak of adenovirus infection in a neonatal intensive care unit due to a false-positive antigen detection test. *J Clin Microbiol* **48:**4251–4252.
18. **Dennehy PH, Hartin M, Nelson SM, Reising SF.** 1999. Evaluation of the ImmunoCardSTAT! rotavirus assay for detection of group A rotavirus in fecal specimens. *J Clin Microbiol* **37:**1977–1979.
19. **World Health Organization.** 2012 WHO prequalification of diagnostics programme public report PQDx 0034-013-00 Alere Determine HIV-1/2 Ab/Ab Combo, version 2.0. www.who.int/diagnostics_laboratory/evaluations/120320_0034_013_00_final_public_report_version2.pdf (accessed November 17, 2014).
20. **Rosenberg NE, Kamanga G, Phiri S, Nsona D, Pettifor A, Rutstein SE, Kamwendo D, Hoffman IF, Keating M, Brown LB, Ndalama B, Fiscus SA, Congdon S, Cohen MS, Miller WC.** 2012. Detection of acute HIV infection: a field evaluation of the determine® HIV-1/2 Ag/Ab combo test. *J Infect Dis* **205:**528–534.
21. **Conway DP, Holt M, McNulty A, Couldwell DL, Smith DE, Davies SC, Cunningham P, Keen P, Guy R, Sydney Rapid HIV Test Study.** 2014. Multi-centre evaluation of the Determine HIV Combo assay when used for point of care testing in a high risk clinic-based population. *PLoS One* **9:** e94062. Correction published 7-16-2014. DOI: 10.137/journal.pone.0103399
22. **Duong YT, Mavengere Y, Patel H, Moore C, Manjengwa J, Sibandze D, Rasberry C, Mlambo C, Li Z, Emel L, Bock N, Moore J, Nkambule R, Justman J, Reed J, Bicego G, Ellenberger DL, Nkengasong JN, Parekh BS.** 2014. Poor performance of the determine HIV-1/2 Ag/Ab combo fourth-generation rapid test for detection of acute infections in a National Household Survey in Swaziland. *J Clin Microbiol* **52:**3743–3748.
23. **Rahman M, Kieke BA, Vandermause MF, Mitchell PD, Greenlee RT, Belongia EA.** 2007. Performance of Directigen flu A+B enzyme immunoassay and direct fluorescent assay for detection of influenza infection during the 2004–2005 season. *Diagn Microbiol Infect Dis* **58:**413–418.
24. **Landry ML, Ferguson D.** 2003. Suboptimal detection of influenza virus in adults by the Directigen Flu A+B enzyme immunoassay and correlation of results with the number of antigen-positive cells detected by cytospin immunofluorescence. *J Clin Microbiol* **41:**3407–3409.
25. **Eyzaguirre EJ, Walker DH, Zaki SR.** 2014. Immunohistology of infectious diseases, p. 56–72. *In* Dabbs DJ (ed.), *Diagnostic immunohistochemistry*, 4th ed. Elsevier/Saunders, Philadelphia, PA.
26. **Lu DY, Qian J, Easley KA, Waldrop SM, Cohen C.** 2009. Automated in situ hybridization and immunohistochemistry for cytomegalovirus detection in paraffin-embedded tissue sections. *Appl Immunohistochem Mol Morphol* **17:**158–164.
27. **Fanaian NK, Cohen C, Waldrop S, Wang J, Shehata BM.** 2009. Epstein-Barr virus (EBV)-encoded RNA: automated in-situ hybridization (ISH) compared with manual ISH and immunohistochemistry for detection of EBV in pediatric lymphoproliferative disorders. *Pediatr Dev Pathol* **12:**195–199.
28. **Elston DM, Gibson LE, Kutzner H.** 2011. Infectious diseases, p. 501–520. *In* Lin F, Prichard J (ed.), *Handbook of practical immunohistochemistry*. Springer Science+Business Media, LLC, New York, NY.
29. **Centers for Disease Control and Prevention.** 2014. Laboratory testing for the diagnosis of HIV infection updated

recommendations. http://www.cdc.gov/hiv/pdf/HIVtestingAlgorithmRecommendation-Final.pdf (accessed February 6, 2015).

30. **Centers for Disease Control and Prevention.** 2014. Advantages and disadvantages of FDA-approved HIV immunoassays used for screening by generation and platform. http://www.cdc.gov/hiv/pdf/testing_AdvDisadvHIVtesting.pdf (accessed February 6, 2015).
31. **Pang X, Jiang X.** 2011. Gastroenteritis viruses, p. 1456–1469. *In* Versalovic J, Carroll KC, Funke G, Jurgensen JH, Landry ML, Warnock DW (ed.), *Manual of clinical microbiology,* 10th ed. ASM Press, Washington, DC.
32. **Centers for Disease Control and Prevention.** 2014. Diagnostic methods. http://www.cdc.gov/norovirus/lab-testing/diagnostic.html (accessed February 6, 2015).
33. **Doan Q, Enarson P, Kissoon N, Klassen TP, Johnson DW.** 2012. Rapid viral diagnosis for acute febrile respiratory illness in children in the Emergency Department. *Cochrane Database Syst Rev* **5:**CD006452.
34. **Doan QH, Kissoon N, Dobson S, Whitehouse S, Cochrane D, Schmidt B, Thomas E.** 2009. A randomized, controlled trial of the impact of early and rapid diagnosis of viral infections in children brought to an emergency department with febrile respiratory tract illnesses. *J Pediatr* **154:**91–95.
35. **Poehling KA, Zhu Y, Tang YW, Edwards K.** 2006. Accuracy and impact of a point-of-care rapid influenza test in young children with respiratory illnesses. *Arch Pediatr Adolesc Med* **160:**713–718.
36. **Iyer SB, Gerber MA, Pomerantz WJ, Mortensen JE, Ruddy RM.** 2006. Effect of point-of-care influenza testing on management of febrile children. *Acad Emerg Med* **13:**1259–1268.
37. **Cazacu AC, Demmler GJ, Neuman MA, Forbes BA, Chung S, Greer J, Alvarez AE, Williams R, Bartholoma NY.** 2004. Comparison of a new lateral-flow chromatographic membrane immunoassay to viral culture for rapid detection and differentiation of influenza A and B viruses in respiratory specimens. *J Clin Microbiol* **42:**3661–3664.
38. **Boivin G, Baz M, Côté S, Gilca R, Deffrasnes C, Leblanc E, Bergeron MG, Déry P, De Serres G.** 2005. Infections by human coronavirus-NL in hospitalized children. *Pediatr Infect Dis J* **24:**1045–1048.
39. **Foulongne V, Guyon G, Rodière M, Segondy M.** 2006. Human metapneumovirus infection in young children hospitalized with respiratory tract disease. *Pediatr Infect Dis J* **25:** 354–359.
40. **Kuroiwa Y, Nagai K, Okita L, Ukae S, Mori T, Hotsubo T, Tsutsumi H.** 2004. Comparison of an immunochromatography test with multiplex reverse transcription-PCR for rapid diagnosis of respiratory syncytial virus infections. *J Clin Microbiol* **42:**4812–4814.
41. **Noyola DE, Clark B, O'Donnell FT, Atmar RL, Greer J, Demmler GJ.** 2000. Comparison of a new neuraminidase detection assay with an enzyme immunoassay, immunofluorescence, and culture for rapid detection of influenza A and B viruses in nasal wash specimens. *J Clin Microbiol* **38:**1161–1165.
42. **Howerton D, Anderson N, Bosse D, Granade S, Westbrook G, Centers for Disease Control and Prevention.** 2005. Good laboratory practices for waived testing sites: survey findings from testing sites holding a certificate of waiver under the clinical laboratory improvement amendments of 1988 and recommendations for promoting quality testing. *Morbidity and Mortality Weekly Report.* **54:**1–25.
43. **Leland DS, Ginocchio CC.** 2007. Role of cell culture in the age of technology. 2007. *Clin Microbiol Rev* **20:**49–78.
44. **Newton DW, Mellen CF, Baxter BD, Atmar RL, Menegus MA.** 2002. Practical and sensitive screening strategy for detection of influenza virus. *J Clin Microbiol* **40:**4353–4356.
45. **College of American Pathologists**, Northfield, IL. Immunology Checklist 4.21.2014. Item IMM.41850.

Serologic (Antibody Detection) Methods

DONGXIANG XIA, DEBRA A. WADFORD,
CHRISTOPHER P. PREAS, AND DAVID P. SCHNURR

9

For communicable diseases, clinical management, and public health response, it is often important to know the body's immune response following exposure and infection with pathogens. Although humoral and cell-mediated immunity both play roles in the body's specific immunity against viral pathogens, testing antibody response for humoral immunity is much more common and is also easier in clinical virology laboratories than testing for cell-mediated immunity because of the convenience of antibody serological testing methods. There is a long history of using various serologic methods in clinical virology laboratories for antibody detection. Some methods, such as complement fixation test and immunodiffusion test, have been gradually phased out and replaced by faster and less laborious methods (1, 2). In this chapter, we will focus on antibody detection methods used in clinical virology laboratories: neutralization, hemagglutination inhibition, indirect immunofluorescence, enzyme immunoassay, and Western blot.

Antibody detection tests can be used to diagnose current or past infection, acute or chronic disease, and evaluate immune status; however, diagnosis is dependent on the timing of specimen collection, the immunoglobulin (Ig) class tested, and the patient's clinical history. Some viral diseases rely on serologic detection not only to assess infection and the body's immunity, but also to predict prognosis, such as with hepatitis B virus (HBV) infections in combination with antigen detection, and for human immunodeficiency virus (HIV) infections in combination with CD4 cell counts and virus load testing. Serologic antibody testing results are widely used as a laboratory criterion for clinical diagnosis of diseases, and are used to define cases for epidemiological investigation and surveillance. In the United States, the Council of State and Territorial Epidemiologists (CSTE) and the Centers for Disease Control and Prevention (CDC) have developed and annually update case definitions for notifiable infectious conditions, which are a set of uniform criteria used to define an infectious condition or disease for public health surveillance (3). For some viral diseases, the CSTE case definitions list antibody detection test results with specification of Ig classes and titers as laboratory criteria for diagnosis. Case classification and definitive diagnosis of infection must be based on the combination of clinical findings, patient history, laboratory test results, and epidemiological data. Vaccination history should also be considered for some conditions, particularly for vaccine preventable diseases.

There are 5 Ig classes or isotypes of antibodies in humans: IgG, IgM, IgA, IgE, and IgD. Specific IgA detection has been reported in infections with several viruses (4–6), including hepatitis E virus, dengue virus, and Epstein-Barr virus (EBV). There are some reports showing the detection of human IgE antibodies to HIV, HBV, and influenza virus (7–9). However, the role, onset, level, and duration of IgA, IgD, and IgE are less predictable than those of either IgG and IgM, and serologic tests for the former isotypes are not performed routinely in diagnostic laboratories (2). The majority of antibody detection assays performed in the clinical virology laboratories are directed at IgG and IgM, which will be the focus of this chapter.

In general, during a primary viral infection, IgM appears 1 to 2 days after the onset of illness, peaks at 7 to 10 days, and declines to an undetectable level within 1 to 3 months, or even longer for some conditions. Following natural infection or vaccination, IgG usually appears several days after IgM and increases to higher levels than IgM. IgG then gradually declines but remains detectable for years or even lifelong. IgG will anamnestically respond and rise to high titer upon reinfection, revaccination, or reactivation, while IgM may or may not respond. Therefore, for diagnosis of viral infection, detection of IgM in the acute phase of infection from a serum specimen collected several to 14 days after disease onset indicates current or recent infection; a 4-fold rise in IgG titer between acute and convalescent sera, collected 2 to 3 weeks apart, or demonstration of seroconversion from acute to convalescent sera, suggests current or recent infection. However, antibody responses following reactivation of latent viruses (e.g., herpesviruses) are often less predictable than responses following primary acute infection or reinfection (1). For immune status, only one specimen is required and detection of IgG indicates past infection or immunization. The interpretation of IgG and IgM as described is generally applicable to all serologic assays covered in this chapter with the exception of the neutralization test.

For clinical manifestations that suggest infection by a group of possible viruses, serology panels of candidate

doi:10.1128/9781555819156.ch9

antigens may be employed to test for antibodies to these viruses. These viruses may cause clinical syndromes such as central nervous system (CNS) infection, myocarditis-pericarditis, respiratory conditions, and rashes, which require timely laboratory differential diagnosis. Clinical consultation with physicians may help to decide if a panel is necessary or whether to choose targeted pathogens included within the panel (10).

NEUTRALIZATION

Overview

Virus neutralization is defined as the blocking or prevention of viral infection of a susceptible host or cell system by a specific antibody or other reagent. This assay can be used to identify antibody response or the specific virus. Early on, it was recognized that the neutralization assay has high specificity as it could distinguish between closely related viruses or antibody responses, such as those against polioviruses 1, 2, and 3. Among the measures of antibody response, specific neutralization is considered as the best indicator of protection from infection with certain agents. This assay has been a valuable test for diagnostic purposes and for surveillance of exposure or immunity for many years.

Historically, the neutralization test was one of the first tests used for virus identification and measuring host response to viral infection. Although no longer used for routine diagnostic purposes, protocols for identifying a wide variety of viruses and host responses to viral infection can be found in older editions of viral texts such as the 5th edition of *Diagnostic Procedures for Viral, Rickettsial and Chlamydial Infections* (11). A more up-to-date discussion of the topic is available (12).

Methods

The neutralization assay has 3 main components: the test specimen, usually serum, the target antigen or agent, and the host system. To measure antibody response to a virus, a well-characterized, pre-titrated virus is required. This is determined in a host system that supports replication of the virus, usually cell culture, although embryonated eggs or an animal host such as mice may also be used. The virus is prepared in the same host system in which the test will be run. The type of cells used in cell culture is critical as the use of different cell types may result in different measures of neutralizing antibody, for example, antibody titers to dengue virus differ when determined in Vero cells as compared to Raji cells (13).

The test for antibody detection can be run on a single serum, paired acute-convalescent sera, or CSF obtained from the subject being tested. The test serum specimen is heated at 56°C for 30 minutes to inactivate complement or other nonspecific inhibitors, serially diluted, and mixed with standardized dose of virus. The virus titer used in the test is checked simultaneously. After incubation of the serum-virus mixtures at a defined temperature and time, the cells of choice are added to each dilution of the serum–virus mixture, or susceptible host animals are inoculated with each serum–virus mixture. The cells or host animals are examined daily for evidence of viral growth: cytopathic effect (CPE), plaque formation, fluorescent foci, or cell death for cell culture systems; also a lethal dose or other effects on the host when using an animal system. The highest dilution of the test serum protecting the host against the virus is the serum neutralization titer. Evidence of previous exposure or infection to a virus can be inferred from a single serum sample demonstrating a neutralization titer, typically of 1:8 or greater. With paired acute and convalescent sera, a 4-fold or greater rise in the antibody titer between the acute and convalescent sera is defined as evidence for recent exposure or infection, and is usually considered diagnostically significant. Measureable titers in acute and convalescent sera without a 4-fold increase between them are interpreted as previous exposure or infection at an undetermined time. No measurable titer is interpreted as no previous infection or exposure.

For viruses that do not replicate in cell culture, such as human papillomavirus (14) and hepatitis C virus (15), or that are of high risk or present safety concerns such as Middle East respiratory syndrome coronavirus (MERS-CoV) (16), certain strains of influenza (17), or dengue virus (18), pseudoviruses have been developed and employed in the neutralization assay for determining specific antibody titer. Pseudoviruses are constructs that contain the surface proteins and neutralizing epitopes of the target virus, and the core of a second replication competent virus capable of at least one round of replication in the host system but is unable to replicate outside that system. Applications such as use of microtiter plates and use of vital stains and spectrophotometric measuring instruments have increased throughput and reduced the requirement for reagents, and have in some cases increased the objectivity of test results (19).

Applications

The application of the virus neutralization assay is to use single or paired specimens to determine evidence of current or recent infection, infection or exposure at some time, or no evidence of exposure for the given specimen(s). The availability of other more rapid diagnostic and serologic assays has all but eliminated the use of the neutralization assay for immediate diagnostic purposes. However this assay does have important applications. The high specificity of neutralization tests is useful for diagnostic differentiation of infection of closely related viruses, where a 4-fold or greater difference in neutralizing antibody titers of a patient specimen between 2 targeted viruses is considered significant for differential diagnosis. For example, plaque reduction neutralization test (PRNT) and colorimetric microneutralization assay are able to distinguish infection between closely related flaviviruses such as West Nile virus and St. Louis encephalitis virus (20). For viruses with many serotypes, neutralization may be the only method for discriminating infection with a particular serotype, such as serologic surveys for particular enteroviruses (21) or adenoviruses (22).

Virus neutralization data can be used for determining recent infection in a particular subject or for surveys of population exposure or immunity to the agent (23). For instance, serosurveys for exposure or successful vaccination to viruses such as rubella and mumps have been based on plaque reduction neutralization tests (24, 25). In some cases where it is important to measure immunity or protection, as for rabies virus, the presence of neutralizing antibodies at some minimum titer is considered the best measure of protection or immunity to that virus. Although the Advisory Committee on Immunization Practices (ACIP) does state that there is no one measure for protection against rabies virus, a neutralizing antibody titer of 0.1 IU or 1:5 dilution is accepted as a safe titer for a person at risk for potential exposure to the virus (26, 27).

Quality Assessment and Troubleshooting

Before setting up a neutralization test, the reagents must be carefully standardized. The virus is propagated in the same living host system as will be used for the test and titrated and must be verified as to type by neutralization or another test of high specificity. Aliquots containing the virus at the target titer, 100 $TCID_{50}$ for example, are prepared, stored frozen, and the titer verified in each test. Positive control serum obtained from the CDC, commercial sources, or produced in-house is titered against the standard dose of the virus. This serum should continue to neutralize the virus to the established titer. Long-term storage of serum at −20°C preserves the antibody titer.

Each time the test is performed, internal controls that give expected results must be included for the test to be valid. These controls include (i) the virus control demonstrating that the virus replicates in the host cells of the test as expected; (ii) a test serum control demonstrating that it is not toxic to the host cells by itself; and (iii) a cell control demonstrating that uninoculated host cells do not exhibit any effects that could interfere with reading or interpreting the test results. Additionally, these standards should be periodically checked for run to run reliability.

HEMAGGLUTINATION INHIBITION

Overview

Hemagglutination is the aggregation of erythrocytes or red blood cells (RBCs) into a lattice-like formation that causes the RBCs to form a diffuse reddish solution in PBS. Viruses that express hemagglutinin (HA) proteins on their surface can cause hemagglutination when viral HA molecules bind receptors on RBCs. Hemagglutination is a rapid physical measure of virus and historically this effect has had a wide impact on the study of viruses (28). The hemagglutination assay is commonly used to determine relative concentration or titer of influenza virus. Building upon the phenomenon of hemagglutination, hemagglutination-inhibition (HI) assays were developed to detect serum antibodies to viruses that express HA such as influenza virus, arboviruses, adenovirus, measles virus (29), and hantavirus (30), as well as to identify these viruses. Virus-specific serum antibodies bind antigenic sites of viral HA molecules, which prevents binding of the virus to the RBCs. This effect inhibits hemagglutination and is the basis for the HI assay. Although this assay is reliable, relatively simple, and inexpensive, it is time-consuming and labor-intensive and thus is no longer routinely used in most clinical virology laboratories. The most common use of the HI assay is for subtyping and/or strain-typing of influenza virus isolates by some public health laboratories. However, the HI assay for antibody detection is discussed in this chapter.

Methods

The HI assay requires three basic components: (i) virus or HA, (ii) test serum, and (iii) RBCs. The virus is prepared from cell culture, embryonated chicken eggs, or animal tissue depending on the target viruses. The patient serum is pretreated to remove non-specific viral inhibitors and RBC agglutinins. The RBCs are prepared from appropriate animal or human blood, depending on the target virus. Before the HI test is performed, the viral HA titer is first determined by the hemagglutination assay, where serial dilutions of virus or HA are mixed with RBCs to yield the HA titer of the virus corresponding to the highest dilution showing hemagglutination of RBCs. For the HI assay, a predetermined amount of virus or standardized HA is added to serial dilutions of pretreated test serum in wells and incubated for 30 minutes. Then red blood cells are added to the virus-serum mixture and incubated for 30 minutes. The presence of specific anti-HA antibodies will inhibit hemagglutination, which would otherwise occur between the virus and the RBCs. If the serum contains no antibodies to the virus, then hemagglutination will be observed in all wells. If antibodies to the virus are present, then hemagglutination will be inhibited. The highest dilution of serum that prevents hemagglutination is called the HI titer of the serum.

Applications

The HI assay indicates presence of antibody in human serum to the virus being tested, but does not indicate infection status. However, paired acute-convalescent sera yielding a 4-fold rise by the HI assay is indicative of recent or current infection. The influenza-specific HI assay is primarily employed for vaccine efficacy studies, where sera from individuals who have received influenza vaccine are tested for the presence of strain-specific antibodies to the particular influenza virus in question (31, 32). The HI assay has also been used to conduct serosurveillance studies, to determine the prevalence of antibodies to particular strains of influenza A virus, such as avian H5N1 and 2009 H1N1 (33–35). In addition to influenza, the HI assay has been commonly used for paramyxoviruses and arboviruses, as well as for serodiagnosis and serosurveillance of hemorrhagic fever with renal syndrome caused by Hantaan virus (36, 37). Although the HI assay can be time-consuming and labor-intensive, it is best used in high throughput situations and for such settings it may be the best option to determine serum antibody levels, as opposed to other serology assays such as enzyme immunoassay (EIA) or immunofluorescence assay (IFA), because specific virus strains are tested to directly indicate the specificity of the antibody response. The HI assay is not a universal serology assay because it only works for viruses that express hemagglutinin.

Quality Assessment, Troubleshooting and Limitations

It is imperative to run in parallel with each test the proper controls for the virus used, the RBCs used, and the serum samples being tested. The virus control ensures the hemagglutination capacity of the virus with the test RBCs. The RBC control determines whether the RBCs in use agglutinate in PBS on their own without virus, which would invalidate the test. Finally, the serum control consists of the test serum diluted at 1:10 and added to RBCs. If the serum alone agglutinates the RBCs, this may indicate interfering agglutinins in the test serum that may yield incorrect results.

The HI assay can be affected by nonspecific inhibitors of agglutination, which naturally occur in sera, and may give rise to false-positive results and nonspecific agglutinins in some serum samples that may cause false-negative results. These nonspecific factors must be removed in order for the HI assay to give accurate results. Other limitations of the HI assay include the need to standardize the virus concentration each time a test is performed, obtaining a reliable source of fresh RBCs, appropriate preparation and storage of RBC suspension, and the need for experienced analysts for interpretation of test results.

Species choice of RBCs is critical for accurate HI results. For different influenza A virus subtypes, guinea pig, turkey, or horse RBCs work best. Rat RBCs are best for adenoviruses, goose RBCs for arboviruses and hantaviruses, adult chicken RBCs for mumps, and monkey RBCs for measles virus HI assays. For reoviruses and some enteroviruses, human RBCs are required (38). The challenge of having fresh RBCs from different animal species in stock in the laboratory, as well as the laborious procedures required to prepare the RBCs, limits its common use in clinical laboratories.

IMMUNOFLUORESCENCE ASSAYS

Overview

As the previous chapter discussed, both direct and indirect immunofluorescence (IIF) assays can be used for antigen detection. In addition to antigen detection, IIF assays are employed to detect serum antibody levels of IgM and/or IgG to determine a current, recent, or past infection. This inexpensive and rapid method relies on known viral antigens, specificity of patient serum antibodies, and labeled reagent antibodies to diagnose viral infections. Antihuman immunoglobulins, specific for human IgG or IgM, are labeled with a fluorophore, most commonly fluorescein isothiocyanate (FITC), to yield what is termed a fluorescent antibody "conjugate." A mixture of virus-infected and uninfected cells are affixed to a glass slide, test serum is applied to the slide, and if specific antibodies are present they will bind to the viral antigen. Bound antibodies can then be detected by application of the conjugate followed by visual examination using fluorescence microscopy. Conjugates are available from many commercial sources and can also be adapted for laboratory-developed IIF assays. Commercial IIF antigen slides are available for the detection of antibodies to several viruses including measles, HIV, mumps, and West Nile virus.

When certain viruses, such as some herpesviruses, are tested by IIF, nonspecific staining reactions may occur due to expression of Fc receptors on virus-infected cells (39–42), which cause nonspecific binding of serum antibody (mainly IgG). A modified version of the IIF, the anticomplement immunofluorescence (ACIF) assay, mitigates these nonspecific reactions when virus-infected cell preparations and a fluorescent-labeled anticomplement antibody are used to test for antibodies to these viruses. ACIF measures total antibody binding to the C3 component of complement, which binds to antigen-antibody complexes within infected cells. For EBV, a member of *Herpesviridae, in vitro* diagnostic kits are available that contain EBV-antigen slides and the necessary ACIF test reagents including complement and control sera.

A fluorescence microscope must be used to read and interpret results of IIF tests. There are many different commercial sources for this type of microscope and innovations in lighting technology have vastly improved the ease of use of these microscopes in the diagnostic field, including systems using liquid light guides and LED light sources, which minimize lamp replacement and eliminate lamp alignment. In addition to light source, critical components for high quality, reproducible, and reliable reading of immunofluorescence include high quality optical lenses and objectives, the proper filters for the specific fluorophores in use, and routine preventive maintenance. A depiction of the functional parts of a fluorescence microscope is shown in Figure 1. In this case, the filters and mirror shown are specific for FITC, but other filters and mirrors are available depending on which fluorophore is used. The excitation or absorption wavelength for FITC is 495 nm, while its emission wavelength is 519 nm, which is visualized as brilliant apple green.

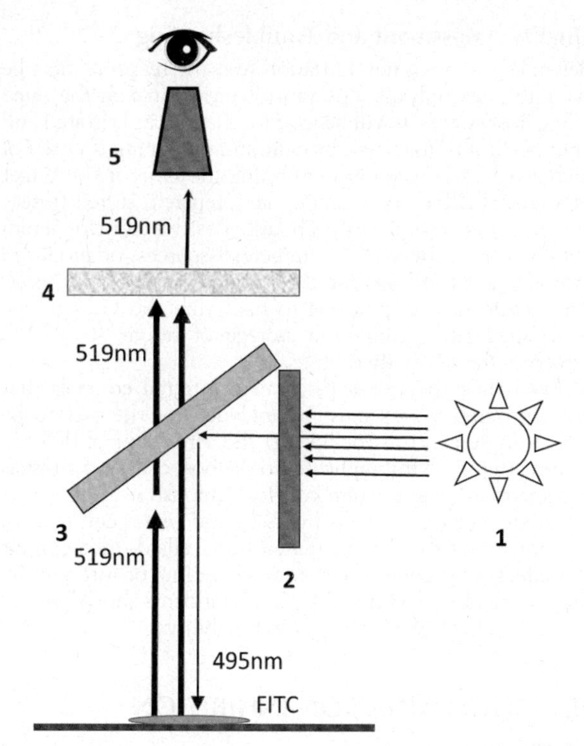

FIGURE 1 Diagram of the functional aspects of a fluorescent microscope used to detect FITC-labeled antibodies. 1) Broad spectrum light from mercury arc lamp or fiber optic light source emits light of all wavelengths, 2) bandpass excitation filter allows ONLY 495 nm to pass through for FITC excitation, 3) dichroic mirror: reflects light at 495 nm; transmits light at all other wavelengths, 4) longpass emission filter (Barrier filter) allows >500 nm to pass through, and 5) eyepiece.

Methods

IIF Assay. This is a 2-step method and can be performed for detection of either IgM or IgG antiviral antibodies from clinical specimens (43, 44). IgM determinations require pretreatment of serum to remove IgG, which will bind specific epitopes and compete with specific IgM. Removal of IgG will also eliminate IgM false-positive reactions due to the presence of rheumatoid factor (discussed later in this chapter) in the serum. For IgG or IgM antibody titer determinations, the first step is to add serial dilutions of patient serum to separate cell spots on the antigen slide. Positive and negative control sera are also added to respective spots on the slide. Test and control sera are incubated in a humid chamber at 35 to 37°C for 30 to 45 minutes, then washed in PBS to remove sera and unreacted antibody. Antihuman IgG or IgM conjugate is then applied to each well and incubated as in step 1. Following a final PBS wash, mounting medium is applied to each well and a cover slip put on the slide. The slide is read using a fluorescent microscope. Positive reactions reveal specific brilliant green staining, which matches the pattern observed in the positive control. Interpretation of antibody titer correlates with the highest dilution of serum showing specific fluorescence, provided controls yield acceptable results.

ACIF Assay. ACIF is a 3-step method compared to the IIF and consists of viral antigen on a slide, inactivated test and control sera, guinea pig complement, and FITC-labeled anti-guinea pig complement antibodies. Inactivation of sera is required to remove endogenous complement from human serum that would interfere with the outcome of the test. All incubations occur in a humid chamber. The first step is to add inactivated patient and control sera to the antigen slide and incubate at 36°C for 30 minutes, followed by PBS wash. The second step is to add guinea pig complement to the wells and incubate at 36°C for 45 minutes, followed by a gentle PBS wash. The final step is to add anti-guinea pig complement conjugate and incubate 36°C for 30 minutes. Similar to the IIF assay, following the final PBS wash, mounting medium is applied to each well and a cover slip put on the slide. The slide is read using a fluorescent microscope. Positive reactions reveal specific brilliant green staining, which matches the pattern observed in the positive control. Interpretation of antibody titer correlates with the highest dilution of serum showing specific fluorescence, provided controls yield acceptable results.

Applications

The IIF assay is the test of choice for a rapid, low throughput method to determine serum antibody titer for both IgG and IgM. The IIF assay may be especially useful for rapid determination of infection or immunity to highly contagious agents such as measles virus and some other notifiable infectious conditions, such as arboviral diseases. The IIF assay is recommended by the CDC as a confirmatory serologic testing for several rickettsial diseases (3), which is helpful for differential diagnosis of viral rash syndromes. It has also been useful as an alternative confirmatory testing method for HIV and human T-cell lymphotropic virus (HTLV) infections. Factors limiting the choice of this test include availability of test reagents, such as antigen slides and secondary conjugates. ACIF is more sensitive than IIF because total antibody is measured, which amplifies the number of complement binding sites and is therefore capable of detecting lower amounts of antibodies or antibodies of low avidity. The ACIF assay method has been used in virology laboratories for the detection of antibodies to the nuclear antigen of EBV (45) and varicella-zoster virus (46); however, enzyme immunoassays (EIAs) are now more commonly used by clinical laboratories for EBV diagnosis than the ACIF.

Quality Assessment, Troubleshooting, and Limitations

Each virus/antigen IIF test must include a known positive control serum and a negative control. The positive control should exhibit a 3 to 4+ staining intensity with characteristic staining pattern. The negative control should exhibit no fluorescence. One caveat to working with FITC conjugates is that they are pH sensitive, so it is imperative to use an aqueous mounting medium in the range of pH 8.5 to 9.0. Fading of FITC fluorescence will become evident when pH falls below 8.5, while pH levels greater than 9.0 may cause nonspecific fluorescence, which can hamper interpretation of results.

For IgM testing by IIF, sera must be pre-treated with an IgG absorbent to avoid false positive and false negative IgM reactions. Refer to the IgM Determinations section of this chapter for further discussion.

Staining patterns observed in the positive control can be used as a reference for specific staining. If control sera do not yield expected staining results, the test should be repeated. Should unacceptable results occur for either control upon repeat testing, then the control may have deteriorated or be contaminated. In either case, a new control should be used and the old control discarded. Nonspecific staining can be an issue when performing the IIF assay. This assay is subjective and more experienced analysts are able to distinguish specific from nonspecific staining. Nonspecific staining may be inherent to the specimen in which case the test should be reported as unsatisfactory. There may be instances when the cells on the slide have sloughed off. This situation may be mitigated by using slides that are at room temperature and not removing slides from their desiccant pouch until just prior to testing. If there are fewer than 20 cells per well, the antigen slide is unsatisfactory and the test must be repeated. The manufacturer of the slides should be contacted with the appropriate lot number for follow-up.

ENZYME IMMUNOASSAY

Overview

The introduction of labeled components into immunoassays for antigen or antibody detection was a revolution in diagnostic immunology and virology. In the 1950s, Yalow and Berson developed the radioimmunoassay (RIA) (47). Although the RIA technique is extremely sensitive and specific, it has been replaced by immunoassays that use enzymes or fluorophores as label markers rather than the hazardous radioisotope labels for most of clinical laboratory tests. EIAs are widely used in clinical laboratories because they are suitable for high throughput testing, require little technical expertise, and in general are cost effective. EIAs include various methods based on specific antigen-antibody reactivity, which is detected through using enzyme conjugates; the enzyme subsequently acts on its substrate to produce color change. One advantage of EIA tests is that reactions can be measured by spectrophotometer and are less subjective than an assay that needs to be visually interpreted such as IFA. EIAs used for antigen or antibody detection can be either solid-phase based (using a solid surface such as tubes, microwell plates, or beads) or membrane-based. Readers are referred to the 3rd edition of *The Immunoassay Handbook* (48) for a more complete description of a wide variety of EIA designs. Because antigen detection assays are discussed in another chapter, EIAs discussed in this chapter only focus on antibody detection methods currently used in clinical virology laboratories.

Methods

Indirect EIA. The indirect EIA is a rapid method that can be applied to a high volume of patient samples for the accurate detection of virus-specific immunoglobulin in human body fluid (e.g., serum, plasma, or CSF). The indirect EIA system is predicated on the immobilization of viral proteins by hydrophobic forces, to a solid phase such as plastic microwells. Patient specimen is added to the solid phase and homologous antibodies present in the sample will bind viral epitopes. Unbound antibodies in the patient sample are washed away. The addition of anti-human Ig conjugated with an enzyme is then added to the solid phase that will specifically bind to the patient antibody that is attached to viral antigen. A wash step removes unbound conjugate and is followed by the addition of substrate specific for the conjugated enzyme. Color change in the substrate occurs in the presence of specific target host antibody, either IgG or IgM, determined by the anti-class specific conjugate (IgM detection will be discussed in detail later). Color development can be measured spectrophotometrically and reactivity determined according to an established cutoff, or read

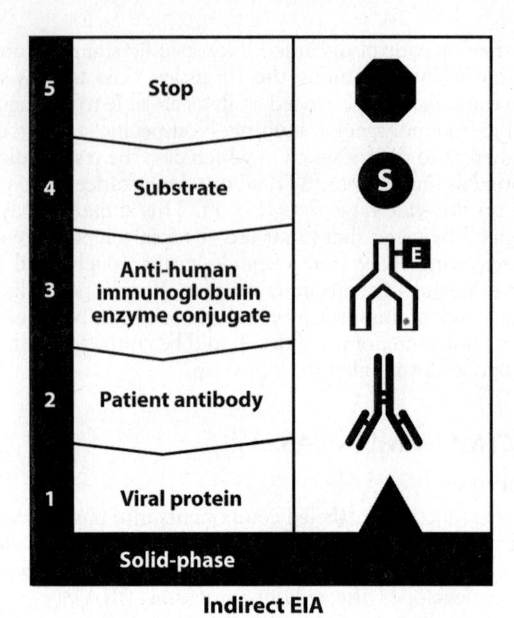

FIGURE 2 Sequential addition of reagents in the indirect EIA.

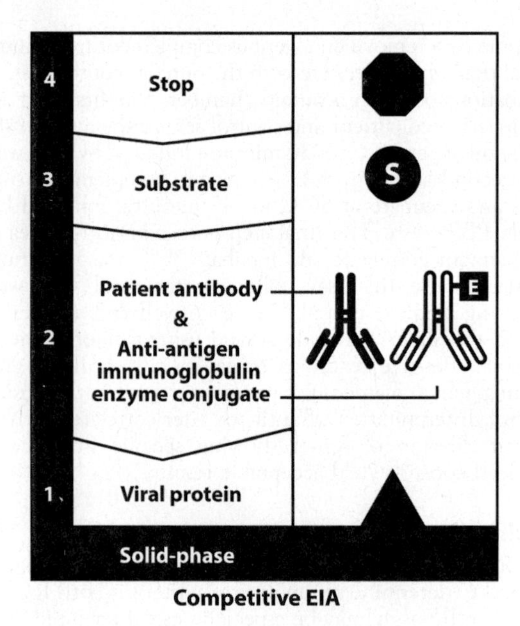

FIGURE 3 Sequential addition of reagents in the competitive EIA.

visually to qualitatively differentiate reactivity from nonreactivity. The sequential addition of reagents in the indirect EIA is depicted in Figure 2.

Competitive EIA. The competitive EIA differs from the indirect EIA in that the patient sample and a measured amount of known conjugated antibody specific for the immobilized antigen are added simultaneously to the solid phase. In this format, patient homologous antibody, if present, competes with the conjugated antibody for binding to the immobilized antigen. Color development in this format is an indication that patient sample does not contain specific antibody against the target antigen. The sequential addition of reagents in the Competitive EIA is depicted in Figure 3.

Multiplexing EIA. There is an ever-increasing demand in clinical virology to boost the number of samples analyzed and to decrease the turnaround time. Innovative testing platforms have been developed that attain this goal without sacrificing accuracy or sensitivity. One such platform is known as multiplexing, a technology that allows for multiple analyte determinations tested simultaneously in a single well as opposed to a singleplex EIA with one analyte reaction per well. Multiplexing platforms vary; the description here will focus on the method using a suspension of polystyrene microspheres (beads) as the target analyte solid phase. The microspheres are color-coded and each color represents a specific analyte. The analyte can be directly coupled to the microsphere, or by first attaching monoclonal antibodies, and then incubating with respective antigen target. Diluted serum or CSF is added followed by anti-human antibody with a fluorescence-based reporter dye such as R-phycoerythrin. The microspheres are identified and read by an instrument with dual lasers: one laser identifies a color set corresponding to a unique analyte, and the other laser excites the reporter dye captured during specimen testing. Some of the advantages of this system include a reduction in specimen and reagent volume needed, and the ability to test multiple, up to 100 analytes, all in the same well at the same time. To perform such testing by a singleplex EIA would theoretically require 100 different commercial kits, and the volume of patient specimen needed would be impractically high. Although such a high throughput system has desirable attributes, the system requires additional controls and the quality control for the vast array of analytes can be complex. Additionally, there are a limited number of FDA cleared multiplex assays currently available for clinical use.

Applications

The EIA is an ideal assay for accurate high throughput determination of antibody to target antigen. It can be used for IgG and IgM testing for various conditions and the acceptable specimen could be serum, CSF, or other types depending on conditions. As in most clinical virology laboratories, commercial and laboratory-developed EIAs are used in the chapter authors' laboratory to detect antibodies against viruses from a number of families including *Bunyaviridae*, *Adenoviridae*, *Retroviridae*, *Flaviviridae*, *Paramyxoviridae*, *Togaviridae*, and *Herpesviridae*. The interpretation of test results for single or paired acute-convalescent serum samples is discussed in the introduction of this chapter, and is the same as with other serologic antibody detection methods. IgM detection by EIA is discussed in the IgM determinations section of this chapter.

In the absence of paired sera, or when the suspected disease is one in which IgM is known to persist, e.g., West Nile (49), hepatitis A (50), and hepatitis B (51), a single IgG response can be diagnostically significant when assaying the avidity strength of the Ig to the target antigen (52). An IgG avidity assay measures IgG maturation and corresponding antibody-antigen reaction stability or binding strength. Over time, as the IgG response matures, its binding strength and avidity to the specific epitope increases. To measure binding strength, the avidity EIA uses a denaturant, e.g., urea, to disrupt the hydrogen bonds that form antibody-antigen complexes. The denaturant is added in the wash step following the patient sera incubation, low avidity IgG readily dissociates from the antigen in the presence of urea, whereas high avidity IgG does not. A low avidity index is indicative of recent exposure to the virus while a high avidity index can demonstrate past exposure (52).

Membrane based EIAs for antibody testing, especially immunochromatography, are commercially developed platforms in which the system is a single-use, self-contained unit assembled in individual cassettes. They are mostly performed at clinics as rapid or screening point-of-care tests, e.g., HIV testing. Positive results usually need to be verified by confirmatory analysis at a laboratory.

Quality Assessment and Troubleshooting

Reagent optimization for each EIA platform is critical to ensure consistent, accurate, and reproducible results. In the indirect method, the level of viral antigen adsorbed to the microtiter well solid-phase must be determined by previous titration using a highly characterized group of patient serum samples. The samples in the group should range from nonreactive to highly reactive, with an emphasis on inclusion of low-positive samples with signals near the established cutoff level of the assay. Microwell surfaces should not be overcoated with viral antigen as this may cause steric hindrance of available binding sites. By not over-coating solid-phase surfaces, there will be remaining uncoated surface area that should be blocked with a buffered blocking reagent such as casein. A blocking reagent will help mitigate the potential of nonspecific proteins in a patient sample binding uncoated surfaces of the solid-phase; in turn this will enhance assay sensitivity. The use of an uninfected cell control antigen (nonspecific control) coated well can help decrease the potential that a patient sample may contain components that nonspecifically react with proteins from cells used to propagate the virus. A nonspecific control well is not typically used in commercially available assays as the manufacturer may be able to produce purified antigens that would not contain potential interfering proteins from cells used to amplify virus. Inclusion of a nonspecific control well reduces the number of samples that can be assayed in a commercial kit and would in turn be less cost effective for the manufacturer. However, a nonspecific control can assist in improving assay specificity with in-house assays where crude antigen lysates are developed from cell culture.

Enzyme conjugates must also be optimized with antigen to determine the optimal working dilution of both reagents. This can be done by checkerboard titration (53). The success of the EIA is dependent on the performance of thorough efficient washes between each reagent addition and incubation step. The wash reagent should be made in a physiologic buffer solution such as PBS and a detergent such as Tween 20 should be part of the solution. The detergent will optimize the wash and enhance the removal of non-bound or nonspecifically attached material. When washing the microtiter wells, it is helpful to allow a convex wash buffer meniscus to form in the well to remove any remaining reagent throughout the steps of the assay. Multiple rounds of wash cycles help ensure the removal of unbound reagents. This process can be enhanced with the use of a properly maintained automated washer that will provide consistent, reproducible, thorough washes. Additional steps, such as programming soaks and shaking of the plate, can enhance the thoroughness of the wash. Each assay run must include a known positive and negative control serum to ensure the reagents in the test worked as expected. When using a commercial EIA kit, a laboratory may wish to include an in-house known reactive serum (continuity control) to monitor kit lot-to-lot variation. Since the spectrophotometer is a critical mechanical component of a high throughput EIA, it must be routinely maintained as outlined by the manufacturer of the product.

The laboratory may adopt a membrane-based EIA platform in which color development is visually determined by the technician. One such assay utilizes a self-contained test card with reaction ports, and a membrane solid-phase that the reagents migrate across (54). The disadvantage of such a system is that the reading becomes subjective and the throughput of samples is decreased. Another disadvantage of the test card system is that in some of the platforms the patient sample is added undiluted, and therefore uses a larger volume of sample. In many cases an undiluted sample can cause problems in the sample migration across reaction ports.

IMMUNOGLOBULIN M DETERMINATIONS

Overview

In general, detection of virus specific IgM in a single acute serum sample has been accepted as an indication for or evidence of current or recent viral infection. A variety of methods have been developed for IgM determination in diagnostic virology. These methods can generally be separated into 3 groups: (i) those based on comparing IgM titers before and after chemical inactivation of serum IgM, (ii) those based on the physicochemical separation of IgM from other serum Ig classes, and (iii) those based on solid-phase immunologic detection of IgM antibodies (55). The first 2 groups are rarely used in clinical virology laboratories, and the IgM determination by IFA has been discussed previously, thus we will focus our discussion on solid-phase immunologic detection of IgM by EIA.

Methods

The 2 most common EIA methods for IgM determinations are the indirect IgM EIA and the IgM Capture Immunoassay. The Indirect EIA method for IgM is similar to the platform used for IgG determinations, with the exception that anti-human IgM conjugate is used instead of an anti-human IgG conjugate and serum incubation times may be longer for the IgM determination.

The IgM Capture EIA starts with immobilization of anti-human IgM antibody, adsorbed by hydrophobic forces, to the solid-phase, e.g., plastic microwells. Patient serum or CSF is added and the affixed anti-human IgM antibody captures patient IgM antibodies, if present. Unbound non-IgM antibodies are removed by washing. The addition of specific target antigen follows. Anti-viral immunoglobulin enzyme conjugate (detection antibody) is next added and reacts with target host antigen. Unbound conjugate is removed by washing, followed by the addition of substrate specific for the conjugated enzyme. Color change in the substrate occurs in the presence of specific target host antibody, the reaction is then terminated by the addition of stop reagent. The level of IgM detection is quantitated by reading the microtiter well with a spectrophotometer set at the proper absorbance for the specific substrate. The sequential addition of reagents in the IgM Capture EIA is depicted in Figure 4.

Applications

While properly timed acute and convalescent phase sera are necessary to demonstrate evidence of current or recent infection for the IgG assay, a single acute phase serum sample can be all that is needed for the IgM EIA. However, the diagnostic value of the specific IgM is dependent on the respective virus and the infection targeted. Detection of IgM in a single serum must be interpreted with caution, and should be evaluated with consideration of clinical

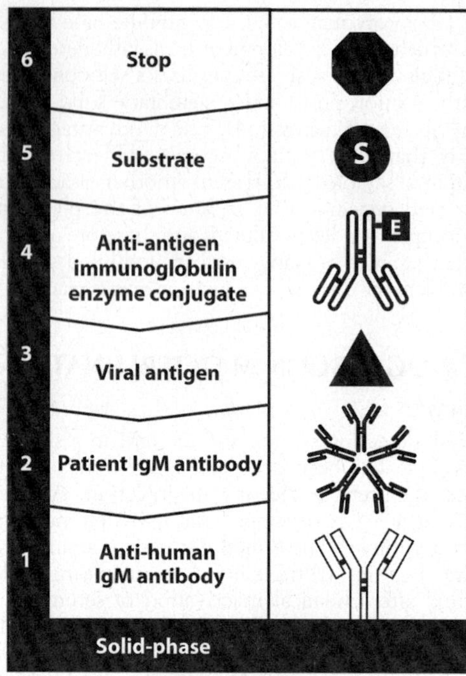

FIGURE 4 Sequential addition of reagents in the IgM capture EIA.

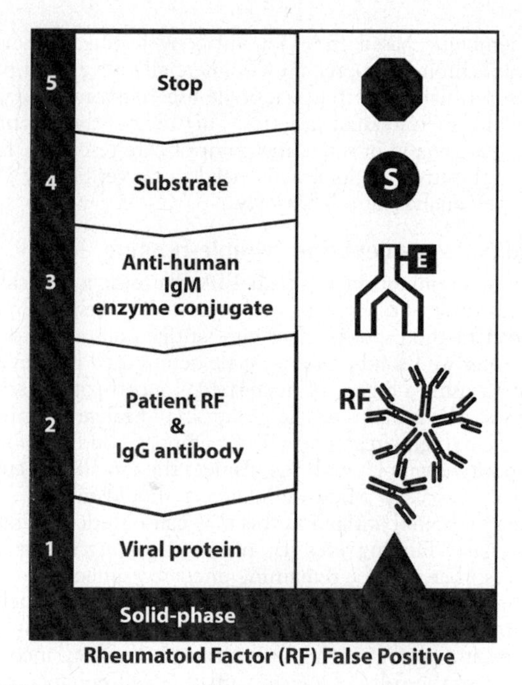

FIGURE 5 Depiction of rheumatoid factor (RF) false positive in indirect IgM EIA.

presentation in the patient. Generally, pronounced transient IgM responses are characteristic of acute virus infections caused by viruses that elicit long-lasting immunity, such as with rubella, measles, mumps, parvovirus B19, and hepatitis A viruses. In these infections, a diagnosis can usually be made by specific antibody testing of a single serum specimen taken early in the illness. In infections with viruses belonging to groups of closely related strains or serotypes (e.g., herpesviruses, adenoviruses, enteroviruses, parainfluenza viruses, alphaviruses, and flaviviruses), IgM serodiagnosis may be complicated by the possible absence of a specific IgM response, as well as by possible false positive reactions to related viruses (55). False positive reactions can also occur due to the presence of rheumatoid factor or heterophilic antibodies in the patient serum or CSF. IgM detection may also be a false indication of current or recent infection due to its persistence following some viral infections (discussed later).

Quality Assessment and Troubleshooting

The presence of rheumatoid factor (RF), primarily an IgM class immunoglobulin that complexes endogenous IgG, can be problematic with the indirect IgM EIA, and to a lesser degree with an IgM capture format. Removal of RF is paramount to the success of the assay. In the indirect EIA, RF present in the patient serum can cause a false positive reaction by attaching to IgG bound to antigen on the solid phase. When this occurs, the anti-class IgM enzyme conjugate can bind RF and produce color mimicking specific IgM class detection in the patient sample (as shown in Figure 5). In order to reduce this type of nonspecific reaction, the patient specimen should be pretreated with an IgG absorbent. The pretreatment of patient specimens can also improve assay sensitivity. High levels of antigen specific IgG in patient specimens can outcompete specific IgM for antigen epitopes on the solid phase. This interference may prevent IgM from binding and cause a false negative IgM reaction.

For IgM capture EIA, one possible interfering factor is the presence of heterophilic antibodies in a patient sample, serum, or CSF, which can lead to false positive reactions (56). Heterophilic antibodies in an individual are produced against animal species' immunoglobulins, and often are found in animal handlers or in individuals who have been treated with animal immunoglobulin (57). The antibody is a bridging antibody that can cross-link the reagent antibodies, i.e., the capture antibody and the detection antibody. The problem occurs when the patient has produced IgM antibody to the animal species from which the capture and detection antibody have been derived. This nonspecific cross-linking can produce a false positive reaction (as shown in Figure 6). One way to mitigate this problem is to use reagent antibodies from different species, e.g., mouse capture antibody and goat detection antibody. This, however, is not foolproof, as individuals may have developed antibody to multiple animal species. To determine whether heterophilic antibody may have produced a false-positive reaction, any reactive result must be repeated and tested in 2 microwells. One well is tested with all reagents as in the original test; however, viral antigen is omitted from the second well. The retest wells are then compared. If heterophilic antibodies are present and cross-link reagent antibodies, color development will occur in the second well containing no antigen, indicating a nonspecific heterophilic antibody reaction.

Another confounding problem with IgM assays, across all platforms, can be that of IgM persistence in the host. IgM has been noted to persist in chronic infections such as hepatitis B (43) and in congenital rubella infections (58). It has also been noted with West Nile virus, where host IgM antibody has been shown to persist for 90 days and in some cases longer (41). In these examples, the overall interpretation of the patient's results can be better defined by testing paired acute-convalescent sera and looking for a significant

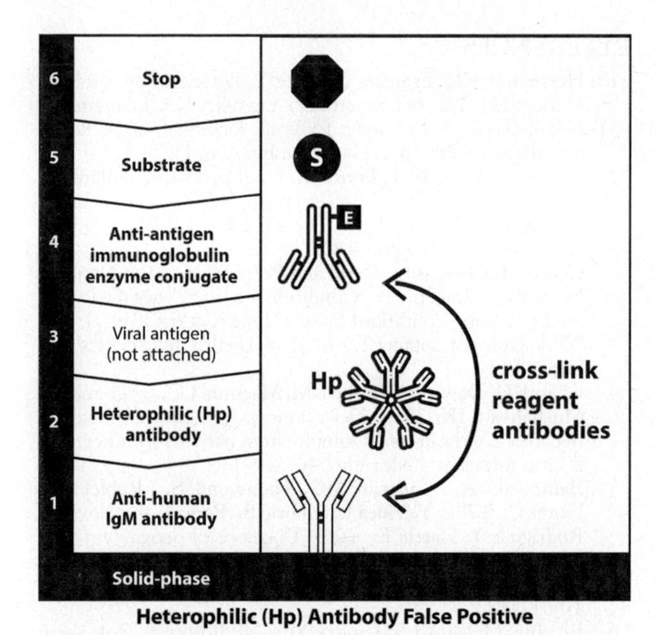

FIGURE 6 Depiction of heterophilic (Hp) antibody false positive in IgM capture EIA.

rise in specific IgG, by seroconversion of IgG, or by using a more specific assay such as PRNT.

WESTERN BLOT

Overview

Although some serologic assays, such as IgG and IgM EIA for certain viral infections, are very sensitive and highly specific, false positive reactions may occur, particularly in low prevalence populations. Given the medical and social significance of particular viral infections, such as HIV and hepatitis, it is important that the diagnostic tests or algorithms be specific, accurate, and as sensitive as possible (59). The immunoblot or Western blot (WB) assay is as sensitive as and more specific than standard colorimetric EIA (60). The WB was developed on the basis of dot immunobinding assay and protein electrophoresis, a combination of molecular and immunologic laboratory technologies. WB can discriminate antibodies to individual viral proteins and has the potential to distinguish between antibody responses to closely related viruses. It has been used as the confirmatory test for diagnosis of viral infections, such as those caused by HIV and HTLV (61–63). Once prepared, WB strips are noninfectious and may be stored and transported, and their use in WB is rapid and inexpensive and requires no specialized equipment.

Methods

WB for antiviral antibody testing consists of 3 steps. The first step is to grow and purify viruses and then to separate viral proteins by electrophoresis. Viral lysates are made by mechanical and chemical disruption of virus and loaded on 2 layer gels. Then the virion proteins are separated by gel electrophoresis according to their size. The second step is the transblot, which transfers the migrated viral proteins from the separation gel to a nitrocellulose membrane. The third step is a modified EIA, where strips of the membrane are first incubated with patient specimens (serum, plasma, dried blood spot elute, oral fluid, etc.), then incubated with anti-IgG or IgM conjugate, and lastly with substrate (59, 60). After all steps are completed, stained bands, which indicate a specific antibody reaction with the viral protein on the membrane, are compared with controls, and results are reported provided controls are acceptable.

Most clinical laboratories purchase WB kits with viral lysates pre-loaded on nitrocellulose strips and only perform the final step, the modified EIA of the strips, to test patient specimens (59, 60). Some laboratories purchase disrupted virus proteins from commercial sources, and prepare their own strips.

Quality Assessment and Limitations

Quality assessment of WB should be site and procedure specific and appropriate for the sample type. If electrophoresis is performed, run parameters should be within normal range and protein molecular weight standards should be included with every gel run, to determine relative molecular weights of viral proteins. Negative and low- and high-titer controls should be included for each test. If commercial materials or kits are used for any portion of the test, procedures must be conducted exactly as described in package insert. If laboratory-developed WB materials and procedures are used, pertinent CLIA requirements for quality assessment must be followed. The WB technique requires well-trained and experienced personnel to perform the test and interpret results, with the understanding of the limitation of the procedure used and performance characteristics.

While WB is much more specific than EIA, it can produce "indeterminate" reactions in which often only a single band is seen, alternatively multiple bands but of a banding pattern that does not meet criteria for a positive interpretation, or bands of reaction with nonspecific cellular proteins may appear. These results may be attributed to an underlying autoimmune disease, among other causes, or precede a truly positive antibody response. Any indeterminate reaction requires testing a second specimen collected at a later date. Similarly, false negative reactions may occur but these are relatively rare (53, 59).

Applications

In the U.S., for clinical virology laboratories, WB is commonly used for diagnosis of retrovirus infections. WB has been used as a supplemental and confirmatory test for HIV diagnostics according to CDC guidelines since 1988. Recently, the "fourth generation" testing algorithm replaces the WB with an HIV-1/HIV-2 antibody differentiation immunochromatographic assay as the supplemental test (64). But the WB for HIV diagnostics will still be used by laboratories that do not have fourth generation testing capabilities. In addition, FDA approved rapid HIV tests are widely used as a more accessible point-of-care test for oral fluid, urine, and finger-stick blood. All HIV reactive rapid test results must be confirmed by either WB or IFA (65). For HTLV, WB is the confirmatory serology diagnostic testing method. In some specialized laboratories, laboratory-developed procedures are used for detection of HIV-1/-2 and HTLV I/II antibodies from plasma and serum specimens (63), and commercial test kits are used for detection of HIV antibodies from oral fluid specimens.

Glycoprotein G (gG)-based WB is also used to detect both type specific and type common antibodies to herpes simplex virus (HSV)-1 and HSV-2 (66). This assay is useful to diagnose subclinical infections and to supplement virus detection methods in cases of atypical presentation or when sampling conditions are not optimal for virus recovery, as diagnosis can indicate treatment with appropriate antiviral medication (e.g., acyclovir). The WB may be used to distinguish maternal from infant antibody profiles and to distinguish serum and CSF profiles of HSV antibodies. WB is the current gold standard HSV serologic test and provides a highly accurate confirmatory test for positive results by EIA (67). WB is used to measure antibody to human herpesvirus (HHV)-6 mainly to identify and analyze the role of specific proteins in the immune response to HHV-6 (68). WB is also designed to detect antibody response to HHV-8-specific antigens expressed during latency or lytic infection, but are of limited use in the diagnosis and management of acute clinical disease (68).

In recent years, WB has been reported to differentiate infections caused by serologically cross-reactive flaviviruses, such as West Nile virus, St. Louis encephalitis virus, and dengue virus, as well as different alphaviruses, such as Eastern equine encephalitis virus and Chikungunya virus. WB analytical results are comparable to the PRNT and the WB may supplant the neutralization assay as the confirmatory test for virus-specific antibodies from specimens, especially for laboratories lacking cell culture capabilities (69, 70). In fact, WB correlates more closely with neutralizing antibodies than the EIA, and has been employed for parvovirus B19 virus IgM and IgG antibody determinations (71).

For RSV, WB provides serologic evidence of a recent infection even if a single serum is used. It also helps to discriminate immune response associated with subunit RSV vaccine from those associated with natural RSV infection (72). WB is employed in serologic antibody testing for hantavirus diseases. It has been reported that recombinant antigen-based WB is more sensitive than EIA, especially for IgG testing (59). A WB assay using recombinant antigens and isotype-specific conjugates for IgM-IgG differentiation has also been developed, and its results are generally in agreement with those of the IgM-capture format (73).

SUMMARY

Serologic assays are very important in diagnosing viral infections, especially when specimens of infected tissue are not available (e.g., brain tissue), when the viruses under consideration cannot be cultured or isolated, and when causative viruses are no longer detectable in the host. As reviewed in this chapter, serologic methods are diverse, yet similar in their ability to determine current, recent, or past infections. In clinical virology laboratories, classic and labor-intensive assays of antibody detection such as the neutralization assay and HI have largely given way to more rapid, sensitive, accurate, and economical tests, such as IFA, EIA, and WB. Technology is ever-advancing and in the future, multiplex platforms or other advanced methods may overtake conventional formats to provide even more improved serologic diagnostics.

David P. Schnurr, PhD, passed away in November 2014 shortly after completing the Neutralization portion of this Chapter. He authored the Neutralization Chapter of the 4th edition of this book. Dr. Schnurr was an impactful scientist who contributed greatly to the field of clinical and public health virology and he served as a mentor to many laboratorians at public health laboratories in California. He is dearly missed.

REFERENCES

1. **Herrmann KL, Erdman DD.** 1995. Diagnosis by serologic assay, p. 121–138. *In* Lennette EH, Lennette DA, Lennette TL (eds.), *Diagnostic Procedures for Viral, Rickettsial, and Chlamydial Infections*, 7th ed. APHA, Washington, DC.
2. **Boivin G, Mazzulli T, Petric M, Couillard M, Couillard M.** 2009. Diagnosis of viral infections, p 265–294. *In* Richman DD, Whitley RJ, Hayden FG (eds.), *Clinical Virology*, 3rd ed. ASM Press, Washington, DC.
3. **Centers for Disease Control and Prevention.** 2014. National Notifiable Infectious Conditions. http://wwwn.cdc.gov/NNDSS/script/ConditionList.aspx?Type=0&Yr=2014. 1/ 17/ 2014. Content source: Division of Health Informatics and Surveillance
4. **Chau KH, Dawson GJ, Bile KM, Magnius LO, Sjogren MH, Mushahwar IK.** 1993. Detection of IgA class antibody to hepatitis E virus in serum samples from patients with hepatitis E virus infection. *J Med Virol* **40:**334–338.
5. **Balmaseda A, Guzmán MG, Hammond S, Robleto G, Flores C, Téllez Y, Videa E, Saborio S, Pérez L, Sandoval E, Rodriguez Y, Harris E.** 2003. Diagnosis of dengue virus infection by detection of specific immunoglobulin M (IgM) and IgA antibodies in serum and saliva. *Clin Diagn Lab Immunol* **10:**317–322.
6. **Bhaduri-McIntosh S, Landry ML, Nikiforow S, Rotenberg M, El-Guindy A, Miller G.** 2007. Serum IgA antibodies to Epstein-Barr virus (EBV) early lytic antigens are present in primary EBV infection. *J Infect Dis* **195:**483–492.
7. **Fletcher M, Miguez-Burbano MJ, Shor-Posner G, Lopez V, Lai H, Baum MK.** 2000. Diagnosis of human immunodeficiency virus infection using an immunoglobulin E-based assay. *Clin Diagn Lab Immunol* **7:**55–57.
8. **Smith-Norowitz TA, Tam E, Norowitz KB, Chotikanatis K, Weaver D, Durkin HG, Bluth MH, Kohlhoff S.** 2014. IgE anti Hepatitis B virus surface antigen antibodies detected in serum from inner city asthmatic and non asthmatic children. *Hum Immunol* **75:**378–382.
9. **Smith-Norowitza TA, Kusonruksa M, Wong D, Norowitz MM, Joks R, Durkin HG, Bluth MH.** 2012. Long-term persistence of IgE anti-influenza A HIN1 virus antibodies in serum of children and adults following influenza A vaccination with subsequent H1N1 infection: a case study. *J Inflamm Res* **5:**111–116.
10. **Tille PM (ed).** 2014. *Overview of the Methods and Strategies in Virology*, p. 819. *In Bailey & Scott's Diagnostic Microbiology*, 13th ed. Elsevier/Mosby, St. Louis, MO.
11. **Hawkes RA.** 1979. General principles underlying laboratory diagnosis of viral infections, p. 26–27. *In* Lennette EH, Schmidt NJ (eds.), *Diagnostic Procedures for Viral, Rickettsial and Chlamydial Infections*, 5th ed. American Public Health Assn, Washington, DC.
12. **Schnurr D.** 2009. Neutralization, p. 110–118. *In* Specter S, Hodinka RL, Young SA, Wiedbrauk DL (eds.), *Clinical Virolgoy Manual*, 4th ed. ASM Press, Washignton, DC.
13. **Mukherjee S, Dowd KA, Manhart CJ, Ledgerwood JE, Durbin AP, Whitehead SS, Pierson TC.** 2014. Mechanism and significance of cell type-dependent neutralization of flaviviruses. *J Virol* **88:**7210–7220.
14. **Roden RB, Greenstone HL, Kirnbauer R, Booy FP, Jessie J, Lowy DR, Schiller JT.** 1996. In vitro generation and type-specific neutralization of a human papillomavirus type 16 virion pseudotype. *J Virol* **70:**5875–5883.
15. **Meyer K, Basu A, Przysiecki CT, Lagging LM, Di Bisceglie AM, Conley AJ, Ray R.** 2002. Complement-mediated enhancement of antibody function for neutralization of pseudotype virus containing hepatitis C virus E2 chimeric glycoprotein. *J Virol* **76:**2150–2158.
16. **Zhao G, Du L, Ma C, Li Y, Li L, Poon VK, Wang L, Yu F, Zheng BJ, Jiang S, Zhou Y.** 2013. A safe and convenient pseudovirus-based inhibition assay to detect neutralizing antibodies and screen for viral entry inhibitors against the novel human coronavirus MERS-CoV. *Virol J* **10:**266.

17. **Qiu C, Huang Y, Zhang A, Tian D, Wan Y, Zhang X, Zhang W, Zhang Z, Yuan Z, Hu Y, Zhang X, Xu J.** 2013. Safe pseudovirus-based assay for neutralization antibodies against influenza A(H7N9) virus. *Emerg Infect Dis* **19:**1685–1687.
18. **Yamanaka A, Suzuki R, Konishi E.** 2014. Evaluation of single-round infectious, chimeric dengue type 1 virus as an antigen for dengue functional antibody assays. *Vaccine* **32:** 4289–4295.
19. **Crawford-Miksza LK, Schnurr DP.** 1994. Quantitative colorimetric microneutralization assay for characterization of adenoviruses. *J Clin Microbiol* **32:**2331–2334.
20. **Taketa-Graham M, Powell Pereira JL, Baylis E, Cossen C, Oceguera L, Patiris P, Chiles R, Hanson CV, Forghani B.** 2010. High throughput quantitative colorimetric microneutralization assay for the confirmation and differentiation of West Nile Virus and St. Louis encephalitis virus. *Am J Trop Med Hyg* **82:**501–504.
21. **Lu QB, Wo Y, Wang HY, Wei MT, Zhang L, Yang H, Liu EM, Li TY, Zhao ZT, Liu W, Cao WC.** 2014. Detection of enterovirus 68 as one of the commonest types of enterovirus found in patients with acute respiratory tract infection in China. *J Med Microbiol* **63:**408–414.
22. **Schmitz H, Wigand R, Heinrich W.** 1983. Worldwide epidemiology of human adenovirus infections. *Am J Epidemiol* **117:**455–466.
23. **Drexler JF, Grard G, Lukashev AN, Kozlovskaya LI, Böttcher S, Uslu G, Reimerink J, Gmyl AP, Taty-Taty R, Lekana-Douki SE, Nkoghe D, Eis-Hübinger AM, Diedrich S, Koopmans M, Leroy EM, Drosten C.** 2014. Robustness against serum neutralization of a poliovirus type 1 from a lethal epidemic of poliomyelitis in the Republic of Congo in 2010. *Proc Natl Acad Sci USA* **111:**12889–12894.
24. **LeBaron CW, Forghani B, Beck C, Brown C, Bi D, Cossen C, Sullivan BJ.** 2009. Persistence of mumps antibodies after 2 doses of measles-mumps-rubella vaccine. *J Infect Dis* **199:**552–560.
25. **LeBaron CW, Forghani B, Matter L, Reef SE, Beck C, Bi D, Cossen C, Sullivan BJ.** 2009. Persistence of rubella antibodies after 2 doses of measles-mumps-rubella vaccine. *J Infect Dis* **200:**888–899.
26. **Manning SE, Rupprecht CE, Fishbein D, Hanlon CA, Lumlertdacha B, Guerra M, Meltzer MI, Dhankhar P, Vaidya SA, Jenkins SR, Sun B, Hull HF, Advisory Committee on Immunization Practices Centers for Disease Control and Prevention (CDC).** 2008. Human rabies prevention–United States, 2008: recommendations of the Advisory Committee on Immunization Practices. *MMWR Recomm Rep* **57**(RR-3): 1–28.
27. **Rupprecht CE, Briggs D, Brown CM, Franka R, Katz SL, Kerr HD, Lett S, Levis R, Meltzer MI, Schaffner W, Cieslak PR.** 2009. Evidence for a 4-dose vaccine schedule for human rabies post-exposure prophylaxis in previously non-vaccinated individuals. *Vaccine* **27:**7141–7148.
28. **Fields BN, Knipe DM.** 1990. Introduction, p. 6. *In* Fields BN, Knipe DM, Chanock RM, Melnick J, Roizman B, Shope R (eds.), *Fields Virology*, 2nd ed. Raven Press, New York, NY.
29. **Young SA.** 2009. Hemadsorption and Hemagglutination Inhibition, p. 120–122. *In* Specter S, Hodinka RL, Young SA, Wiedbrauk DL (eds.), *Clinical Virology Manual*, 4th ed. ASM Press, Washington, DC.
30. **Tsai TF, Tang YW, Hu SL, Ye KL, Chen GL, Xu ZY.** 1984. Hemagglutination-inhibiting antibody in hemorrhagic fever with renal syndrome. *J Infect Dis* **150:**895–898.
31. **Wood JM, Gaines-Das RE, Taylor J, Chakraverty P.** 1994. Comparison of influenza serological techniques by international collaborative study. *Vaccine* **12:**167–174.
32. **Glathe H, Bigl S, Grosche A.** 1993. Comparison of humoral immune responses to trivalent influenza split vaccine in young, middle-aged and elderly people. *Vaccine* **11:**702–705.
33. **Rowe T, Abernathy RA, Hu-Primmer J, Thompson WW, Lu X, Lim W, Fukuda K, Cox NJ, Katz JM.** 1999. Detection of antibody to avian influenza A (H5N1) virus in human serum by using a combination of serologic assays. *J Clin Microbiol* **37:**937–943.
34. **Reed C, Katz JM, Hancock K, Balish A, Fry AM, H1N1 Serosurvey Working Group.** 2012. Prevalence of seropositivity to pandemic influenza A/H1N1 virus in the United States following the 2009 pandemic. *PLoS One* **7:**e48187.
35. **Nasreen S, Uddin Khan S, Azziz-Baumgartner E, Hancock K, Veguilla V, Wang D, Rahman M, Alamgir AS, Sturm-Ramirez K, Gurley ES, Luby SP, Katz JM, Uyeki TM.** 2013. Seroprevalence of antibodies against highly pathogenic avian influenza A (H5N1) virus among poultry workers in Bangladesh, 2009. *PLoS One* **8:**e73200.
36. **Xia D, Wang M, Ni D, Jiang S, Ma W.** 1988. Application of hemagglutination inhibition test to the sero-diagnosis of hemorrhagic fever with renal syndrome. *J Med Colleges PLA* **3:**144–148.
37. **Tang YW, Li YL, Ye KL, Xu ZY, Ruo SL, Fisher-Hoch SP, McCormick JB.** 1991. Distribution of hantavirus serotypes Hantaan and Seoul causing hemorrhagic fever with renal syndrome and identification by hemagglutination inhibition assay. *J Clin Microbiol* **29:**1924–1927.
38. **Lennette EH, Schmidt NJ (ed).** 1979. *Diagnostic Procedures for Viral, Rickettsial and Chlamydial Infections*, 5th ed. American Public Health Assn, Washington, DC.
39. **Furukawa T, Hornberger E, Sakuma S, Plotkin SA.** 1975. Demonstration of immunoglobulin G receptors induced by human cytomegalovirus. *J Clin Microbiol* **2:**332–336.
40. **Schmitz H, Kampa D, Heidenreich W.** 1975. Sensitive method to detect non-complement-fixing antibodies to Epstein-Barr virus nuclear antigne. *Int J Cancer* **16:**1030–1034.
41. **Gallo D, Schmidt NJ.** 1981. Comparison of anticomplement immunofluorescence and fluorescent antibody-to-membrane antigen tests for determination of immunity status to varicella-zoster virus and for serodifferentiation of varicella-zoster and herpes simplex virus infections. *J Clin Microbiol* **14:**539–543.
42. **Feorino PM, Shore SL, Reimer CB.** 1977. Detection by indirect immunofluorescence of Fc receptors in cells acutely infected with Herpes simplex virus. *Int Arch Allergy Appl Immunol* **53:**222–233.
43. **Schutzbank TE, McGuire R, Scholl DR.** 2009. Immunofluorescence, p. 80–81. *In* Specter S, Hodinka RL, Young SA, Wiedbrauk DL (eds.), *Clinical Virology Manual*, 4th ed. ASM Press, Washington, DC.
44. **Janda WM.** 2010. Immunologic assays used in the serologic diagnosis of infectious diseases, p 11.1.2.7-11.1.2.8 Section 11, Immunology, Plaeger SF, Denny TN (ed.). *In* Garcia LS, Isenberg HD (eds.), *Clinical Microbiology Procedures Handbook*, 3rd ed. ASM Press, Washington, DC.
45. **Gartner BC.** 2011. Epstein-Barr Virus, p. 1578–1581. *In* Versalovic J, Carroll KC, Funke G, Jorgensen JH, Landry ML, Warnock DW (eds.), *Manual of Clinical Microbiology*, 10th ed. ASM Press, Washington, DC.
46. **Puchhammer-Stockl E, Aberle SW.** 2011. Varicella-Zoster Virus, p. 1552. *In* Versalovic J, Carroll KC, Funke G, Jorgensen JH, Landry ML, Warnock DW (eds.), *Manual of Clinical Microbiology*, 10th ed. ASM Press, Washington, DC.
47. **Yalow RS, Berson SA.** 1960. Immunoassay of endogenous plasma insulin in man. *J Clin Invest* **39:**1157–1175.
48. **Davies C.** 2005. Introduction to Immunoassay Principles, p. 3-40. In Wild D (ed.), *The Immunoassay Handbook*, 3rd ed. ELSEVIER Ltd, Kidington, Oxford.
49. **Roehrig JT, Nash D, Maldin B, Labowitz A, Martin DA, Lanciotti RS, Campbell GL.** 2003. Persistence of virus-reactive serum immunoglobulin m antibody in confirmed west nile virus encephalitis cases. *Emerg Infect Dis* **9:**376–379.
50. **Kao HW, Ashcavai M, Redeker AG.** 1984. The persistence of hepatitis A IgM antibody after acute clinical hepatitis A. *Hepatology* **4:**933–936.
51. **Chau KH, Hargie MP, Decker RH, Mushahwar IK, Overby LR.** 1983. Serodiagnosis of recent hepatitis B infection by IgM class anti-HBc. *Hepatology* **3:**142–149.

52. **Fox JL, Hazell SL, Tobler LH, Busch MP.** 2006. Immunoglobulin G avidity in differentiation between early and late antibody responses to West Nile virus. *Clin Vaccine Immunol* **13:**33–36.
53. **Herrmann KL, Erdman DD.** 1995. Diagnosis by serologic assays, p. 134. Lennette EH, Lennette DA, Lennette ET (eds.), *Diagnostic Procedures for Viral, Rickettsial and Chlamydial Infections*, 7th ed. American Public Health Association, Washington, DC.
54. **Talkington DF, Shott S, Fallon MT, Schwartz SB, Thacker WL.** 2004. Analysis of eight commercial enzyme immunoassay tests for detection of antibodies to Mycoplasma pneumoniae in human serum. *Clin Diagn Lab Immunol* **11:**862–867.
55. **Erdman DD, Haynes LM.** 2009. Immunoglobulin M Determinations, p. 124–133. *In* Specter S, Hodinka RL, Young SA, Wiedbrauk DL (eds.), *Clinical Virolgoy Manual*, 4th ed. ASM Press, Washington, DC.
56. **Bolstad N, Warren DJ, Nustad K.** 2013. Heterophilic antibody interference in immunometric assays. *Best Pract Res Clin Endocrinol Metab* **27:**647–661.
57. **Tate J, Ward G.** 2004. Interferences in immunoassay. *Clin Biochem Rev* **25:**105–120.
58. **Cradock-Watson JE, Ridehalgh MK, Chantler S.** 1976. Specific immunoglobulins in infants with the congenital rubella syndrome. *J Hyg (Lond)* **76:**109–123.
59. **Meads MB, Medveczky PG.** Application of Western blotting to diagnosis of viral infections, p. 150–155. *In* Specter S, Hodinka RL, Young SA, Wiedbrauk DL (eds.), *Clinical Virology Manual*, 4th ed. ASM Press, Washington, DC.
60. **Janda WM.** 2010. Immunologic assays used in the serologic diagnosis of infectious diseases, p11.1.2.9-11.1.2.10. Section 11, Immunology, Plaeger SF, Denny TN (ed). In Garcia LS, Isenberg HD (ed.), *Clinical Microbiology Procedures Handbook*, 3rd ed. ASM Press, Washington, DC.
61. **Centers for Disease Control (CDC).** 1988. Update: serologic testing for antibody to human immunodeficiency virus. *MMWR Morb Mortal Wkly Rep* **36:**833–840, 845.
62. **Centers for Disease Control.** 1989. Interpretation and use of the Western blot assay for serodiagnosis of human immunodeficiency virus type 1 infection. *MMWR Morb Mortal Wkly Rep* **38**(Suppl. 7)**:**1–7.
63. **Gallo D, Diggs JL, Hanson CV.** 1990. Comparison of Western immunoblot antigens and interpretive criteria for detection of antibody to human T-lymphotropic virus types I and II. *J Clin Microbiol* **28:**2045–2050.
64. **Centers for Disease Control and Prevention (CDC).** 2013. Detection of acute HIV infection in two evaluations of a new HIV diagnostic testing algorithm-United States, 2011–2013. *MMWR Morb Mortal Wkly Rep* **62:**489–494.
65. **Centers for Disease Control and Prevention.** 2004. Notice to readers: Protocols for confirmation of reactive rapid HIV tests. *MMWR Morb Mortal Wkly Rep* **53:**221–222.
66. **Schmid DS.** 2006. Herpes Simplex Virus, p. 626–630. *In* Detrick B, Hamilton RG, Folds JD (eds.), *Manual of Molecular and Clinical Laboratory Immunology*, 7th ed. ASM Press, Washington, DC.
67. **University of Washington. Department of Laboratory Medicine.** HSV Western Blot Serology. http://depts.washington.edu/rspvirus/documents/hsv_western_blot.pdf. 7/16/2013.
68. **Hodinka RL.** 2006. Human Herpesviruses 6,7, and 8, p. 658–668. *In* Detrick B, Hamilton RG, Folds JD (eds.), *Manual of Molecular and Clinical Laboratory Immunology*, 7th ed. ASM Press, Washington, DC.
69. **Oceguera LF III, Patiris PJ, Chiles RE, Busch MP, Tobler LH, Hanson CV.** 2007. Flavivirus serology by Western blot analysis. *Am J Trop Med Hyg* **77:**159–163.
70. **Patiris PJ, Oceguera LF III, Peck GW, Chiles RE, Reisen WK, Hanson CV.** 2008. Serologic diagnosis of West Nile and St. Louis encephalitis virus infections in domestic chickens. *Am J Trop Med Hyg* **78:**434–441.
71. **Erdman DD, Durigon EL.** Human Parvovirus B19, p. 500–501. *In* Lennette EH, Lennette DA, Lennette TL (eds.), *Diagnostic Procedures for Viral, Rickettsial, and Chlamydial Infections*, 7th ed. APHA, Washington, DC.
72. **Piedra PA, Cron SG, Jewell A, Hamblett N, McBride R, Palacio MA, Ginsberg R, Oermann CM, Hiatt PW, Purified Fusion Protein Vaccine Study Group.** 2003. Immunogenicity of a new purified fusion protein vaccine to respiratory syncytial virus: a multi-center trial in children with cystic fibrosis. *Vaccine* **21:**2448–2460.
73. **Centers for Disease Control and Prevention.** Hantavirus Diagnostics, http://www.cdc.gov/hantavirus/technical/hps/diagnostics.html, 8/29/2012.

Nucleic Acid Extraction in Diagnostic Virology

RAYMOND H. WIDEN AND SUZANNE SILBERT

10

Nucleic acid (NA) extraction is a critical step used in molecular biology and molecular diagnostics (1–5). Successful extraction of NA depends on the quantitative recovery of pure molecules of RNA and DNA in an undegraded form. Salts, for example, are common impurities in NA samples, and it is important that they are removed from NA before any downstream processes and analyses can be performed (1). Therefore, single or multiple separation and/or purification steps are needed to desalt the sample containing the NA. The process of extraction and purification of NA also removes a variety of inhibitors of downstream NA amplification procedures. The first step of NA extraction and purification involves cell lysis to liberate NA from cell nuclei or pathogens. Effective NA extraction methods include reagents that inactivate nucleases (DNase and RNase) to preserve the NA in an intact state. The final steps involve separation and recovery of the NA free of cellular debris, proteins, and various potential inhibitors of downstream assays (1–5).

In the past, the process of extraction and purification of NA used to be complicated, time consuming, and labor intensive (6–11). Currently, there are innumerable specialized methods that can be used to extract pure NA, including manual techniques and automated systems (12–17). The objective of this chapter is to present a review of the methods and systems designed to be used in small, medium, and large molecular microbiology laboratories.

MANUAL AND SEMI-AUTOMATED METHODS FOR NA EXTRACTION IN DIAGNOSTIC VIROLOGY

Phenol-Chloroform

The phenol-chloroform method represents an organic solvent liquid-liquid extraction that is widely used in isolating NA. A liquid-liquid extraction separates mixtures of molecules based on the differential solubilities of the individual molecules in two different immiscible liquids (18). This method requires that a cell lysate be prepared with the addition of proteinase K. The extraction of NA involves adding an equal volume of phenol-chloroform to an aqueous solution of lysed cells, mixing the two phases, and separating the phases by centrifugation. At the proper pH, the hydrophobic phase will settle on the bottom, with the hydrophilic phase on the top. Lipids and other hydrophobic components will be dissolved in the lower hydrophobic phase, while DNA will be dissolved in the upper aqueous phase. Amphiphilic components, which have both hydrophobic and hydrophilic properties, and cell debris will collect as a white precipitate at the interface between the two layers (2, 3, 7).

To minimize protein contamination of the final DNA preparation, the aqueous phase can be repeatedly reextracted with phenol-chloroform until the interface is translucent rather than thick and opaque in appearance. The crude DNA solution is then mixed with an equal volume of chloroform-isoamyl alcohol solution with shaking to remove any residual phenol from the aqueous part. Finally, the upper phase is precipitated by mixing with ethanol or isopropanol and salt. The ethyl or isopropyl alcohol is added to the upper phase solution at a 2:1 or 1:1 ratio, respectively. The DNA forms a solid precipitate, which is collected by centrifugation. Excess salt is removed by rinsing the pelleted NA in 70% ethanol, centrifuging, discarding the supernatant, and then dissolving the DNA pellet in rehydration buffer, usually 10 mM Tris, 1 mM EDTA, or water (2, 3, 7).

The phenol-chloroform extraction method can be adapted to process a wide variety of patient specimens from paraffin-embedded tissue to body fluids. It produces highly concentrated DNA with high purity, making it a reliable method for any molecular diagnostic assay. However, the manual phenol-chloroform extraction method is labor intensive and requires some user expertise. In addition, phenol-chloroform solutions are chemical hazards, and the waste from this method needs to be handled appropriately. Finally, contamination by residual phenol and/or chloroform will inhibit downstream enzymatic reactions and consequently must be carefully and completely removed (2, 3, 7).

Guanidinium Thiocyanate-Phenol-Chloroform Extraction

Although phenol, a flammable, corrosive, and toxic carbolic acid, can denature proteins rapidly, it does not completely inhibit RNase activity, which can interfere with the RNA extraction (1). The use of guanidinium isothiocyanate in RNA extraction was first mentioned by Ullrich et al. (6). The method was laborious, and it has therefore been replaced by guanidinium thiocyanate-phenol-chloroform extraction (8), which is a single-step technique whereby the

doi:10.1128/9781555819156.ch10

homogenate is extracted with phenol-chloroform at reduced pH. Guanidinium thiocyanate is a chaotropic agent used in protein degradation. The principle of this single-step technique is that RNA is separated from DNA after extraction with an acidic solution consisting of guanidinium thiocyanate, sodium acetate, phenol, and chloroform (11). Under these conditions, total RNA remains in the upper aqueous phase of the whole mixture, while DNA and proteins remain in the interphase or lower organic phase. Recovery of total RNA is then done by precipitation with isopropanol (5).

Acid phenol-chloroform-isoamyl alcohol (25:24:1) solution efficiently extracts RNA. Chloroform enhances the extraction of the NA by denaturing proteins and promoting phase separation, and isoamyl alcohol prevents foaming. For RNA, the organic phase must be acidic (pH 4 to 5). The acidity of the organic phase is adjusted by overlaying it with buffer of the appropriate pH. In some isolation procedures, DNase is added at the lysis step to eliminate contaminating DNA. Alternatively, RNase-free DNase may also be added directly to the isolated RNA at the end of the procedure. After phase separation, the upper aqueous phase containing the RNA is removed to a clean tube, and the RNA is precipitated by addition of two volumes of ethanol or one volume of isopropanol. Glycogen or yeast transfer RNA may be added at this step as a carrier to aid RNA pellet formation. The RNA precipitate is then washed in 70% ethanol and resuspended in RNase-free buffer or water (1, 5, 7–9, 11).

TRIzol (Life Technologies, Carlsbad, CA) and QIAzol (Qiagen, Valencia, CA) are complete ready-to-use reagents for the isolation of high-quality total RNA or the simultaneous isolation of RNA, DNA, and protein from a variety of biological samples. These reagents are an improvement from the single-step RNA isolation technique developed by Chomczynski and Sacchi (8). TRIzol and QIAzol perform well with small quantities of tissue and cells as well as with large quantities of tissue and cells. They come with purification protocols and maintain the integrity of the RNA due to highly effective inhibition of RNase activity while disrupting cells and dissolving cell components during sample homogenization. The simplicity of these methods allows simultaneous processing of a large number of samples. The entire procedure using these reagents can be completed in approximately 1 hour. Total RNA isolated by TRIzol and QIAzol is free of protein and DNA contamination (TRIzol and QIAzol package inserts).

QIAamp Viral RNA Mini Kit

The QIAamp Viral Mini Kit (Qiagen, Hilden, Germany) provides reliable isolation of high-quality viral RNA (16, 19–22). The kit combines the selective binding properties of a silica-gel-based membrane with the speed of microspin or vacuum technology and is ideally suited for simultaneous processing of multiple samples. The entire extraction procedure includes four different steps.

First, the sample is lysed under highly denaturing conditions to inactivate RNases and to ensure isolation of intact viral RNA. Next, viral RNA is adsorbed onto the QIAamp silica-gel membrane during centrifugation or by vacuum. The third step includes a couple of washes to ensure complete removal of any residual contaminants without affecting RNA binding. In the end, the sample is eluted in RNase-free water buffer that contains 0.04% sodium azide to prevent microbial growth and subsequent contamination with RNases (19).

The QIAamp viral mini kit provides a fast and easy way to purify viral RNA for reliable use in amplification technologies and can be fully automated on the QIAcube (Qiagen). Viral RNA can be purified from plasma (untreated or treated with anticoagulants other than heparin), serum, and other cell-free body fluids. Samples may be fresh or frozen; if frozen, samples should not be thawed more than once. Repeated freeze-thawing of plasma samples will lead to reduced viral titers and should be avoided for optimal sensitivity. This kit can be used for isolation of viral RNA from a wide variety of viruses including human immunodeficiency virus (HIV), hepatitis A virus, hepatitis C virus, hepatitis D virus, enterovirus, herpes simplex virus type 2, and more (16, 19, 21, 22). However, performance cannot be guaranteed for every virus, so in-house verification of performance is necessary. Also, QIAamp kits are not designed to separate viral RNA from cellular DNA, and both will be purified in parallel if present in the samples. Therefore, to avoid copurification of cellular DNA, using cell-free body fluids for preparation of viral RNA is recommended. Samples containing cells, such as cerebrospinal fluid, bone marrow, urine, and most swabs, should first be filtered or centrifuged for 10 minutes at 1500 × *g* and the supernatant used. However, in most cases it is not absolutely necessary to remove cellular material from any sample since the downstream nucleic acid amplification test (NAAT) protocols will generally be highly specific for the target viral NA sequences and will not cross react with cellular nucleic acids (RNA or DNA). If the protocol used requires pure RNA, and RNA and DNA have been isolated in parallel, the eluate can be DNase digested using RNase-free DNase, followed by heat treatment to inactivate the DNase (19).

QIAamp DNA Mini and Blood Mini Kit

QIAamp DNA Mini and QIAamp DNA Blood Mini Kits (Qiagen) provide fast and easy methods for purification of total DNA. Total DNA (i.e., genomic, viral, mitochondrial) can be purified from whole blood, plasma, serum, buffy coat, bone marrow, other body fluids, lymphocytes, cultured cells, tissue, and forensic specimens. The procedure is similar to the QIAamp Viral RNA Mini Kit, following the four main steps, which are ideal for simultaneous processing of multiple samples and yield pure DNA ready for direct amplification in just 20 minutes. Purification of viral DNA is possible with QIAamp DNA Mini or QIAamp DNA Blood Mini Kits. However, since cellular DNA co-purifies with viral DNA, cell-free samples (e.g., plasma, serum) are necessary to obtain pure viral DNA (20, 23, 24). However, depending on the application and targeted virus, it may be advantageous to retain the cells if they are likely infected with the target virus, in which case one can detect either free or cell-associated virus.

In addition to Qiagen, several other vendors offer column (spin or vacuum) based NA extraction systems including the GeneJET RNA and DNA kits (ThermoFisher, Waltham, MA), EconoSpin (Epoch Life Science, Sugar Land, TX), Genelute (Sigma Aldrich, St. Louis, MO), Pe+NAD (Genereach USA, Lexington, MA), PureLink (LifeTechnologies, Grand Island, NY), and PureYield (Promega, Madison, WI). There are systems available for manual NA sample preparation that includes a bead beating step to break up cells. Bead beating is not generally required for isolation of viral NA; however, the systems could be useful if one is attempting to recover NA from samples that could include difficult-to-lyse organisms (fungi, mycobacteria, other bacteria) and there is also a need to test for these pathogens. Available systems include those from MoBio (Carlsbad, CA), MP Biochemicals (Santa Ana, CA), and Bertin (Rockville, MD).

NucliSens miniMAG System

The NucliSens miniMAG system (bioMérieux, Inc., Durham, NC) is a semi-automated extraction platform that uses magnetic silica particles and a magnetic device (13, 15, 25–28). The magnetic device has a rack for 12 tubes, allowing 12 extractions to be processed at one time. The system is based on the silica-based technology described by Boom et al. (12), using magnetic silica particles to avoid the need for microcentrifugation during multiple washes and the final elution of NA containing both DNA and RNA molecules.

The NucliSens miniMAG platform is a simple, rapid, and clean extraction procedure, with no pre-extraction ultracentrifugation step, that yields NA of sufficient quantity and quality for downstream NAAT (13–15, 25–28). The extraction workflow includes two main steps. The first is to add the lysis buffer to the sample and second, to mix the magnetic silica with the lysis buffer-sample mixture. The lysis buffer-silica-sample mixture is pelleted, and the supernatant is aspirated. The pellet is resuspended in wash buffer 1 and transferred to a 1.5-ml centrifuge tube. Several wash steps are performed using the miniMAG semi-automated instrument. After the last wash buffer is aspirated, the elution buffer is added, followed by a short incubation at 60°C. Tubes are moved against a magnetic rack while the eluted DNA is pipetted. A maximum of 12 specimens can be processed during each run, and the entire procedure takes approximately 1 hour (13). The extraction method is universal and can be applied for a broad range of different specimens and volumes (26, 27). Furthermore, miniMAG is flexible with its generic extraction chemistry and allows parallel processing of different sample types in the same run. One eluate is compatible with multiple downstream applications and is ready for immediate use after the extraction procedure is concluded. No use of ethanol or other organic solvents is required, and the method presents a highly effective removal of inhibitors (13–15, 25–28).

Several studies have shown that the performance of this semi-automated system is satisfactory (13–15, 25–28) and sometimes even superior to other more automated systems (13). The NucliSens miniMAG instrument is compact enough that it can fit on a benchtop or inside a hood, and it can simultaneously extract 12 to 24 samples in about 60 to 90 minutes, by using two instruments in parallel. However, there is no walk-away time during the extraction procedure. This is inconvenient for handling large numbers of specimens; nevertheless, for small or mid-sized laboratories, this semi-automated but convenient and efficient method of NA extraction can be a good choice (13–15).

TruTip Technology

The TruTip (Akonni Biosystems, Frederick, MD) is a new manual NA extraction technology (29–31) that also uses the BOOM silica binding chemistry (12) for NA binding and elution. TruTip consists of a porous rigid silica monolith inserted into a pipette tip that connects to standard laboratory pipettors (single or multi-channel) or automated liquid handling systems. The geometry of the monolith can be adapted for specific pipette tips ranging in volume from 1.0 to 5.0 ml. The porosity of the TruTip monolith is available in configurations to allow low viscosity or highly viscous or complex samples to readily pass through it with minimal fluidic backpressure. Bidirectional flow maximizes residence time between the monolith and sample and enables large sample volumes to be processed within a single TruTip. Figure 1 describes the TruTip process for purifying NA.

The fundamental steps, irrespective of sample volume or TruTip geometry, include cell lysis, NA binding to the inner pores of the TruTip monolith, washing away unbound sample components and lysis buffers, and eluting purified and concentrated NAs into an appropriate buffer. TruTip kits accommodate a wide range of sample sources (i.e., blood, saliva, urine, sputum, and nasopharyngeal aspirate), sample viscosities, and sample types from human genomic DNA to microorganisms (bacterial or viral pathogens).

Few studies so far reported results using this new technology in different formats (29–31). The first one (29) showed the development and application of a manual TruTip procedure for purifying influenza RNA from nasopharyngeal samples using a single or multi-channel pipettor. This simple, pipette-operated manual sample preparation technology was as effective as the clinically validated QIAcube or easyMAG (bioMérieux) automated sample preparation systems for extracting and purifying influenza RNA from nasopharyngeal samples. A second study (30) validated the TruTip extraction technology for an automated extraction of influenza RNA from clinical specimens, using the epMotion 5070 liquid handling system (Eppendorf, Hamburg, Germany). The authors described a TruTip/epMotion protocol that successfully extracted influenza RNA in 99% of previously positive nasopharyngeal samples and presented comparative clinical efficacy between the automated TruTip/epMotion procedure and the easyMAG and QIAcube (Qiagen) instruments. Last, automated protocols using the TruTip technology described in another study (31) emphasized the utility of the TruTip for processing diverse clinical samples and how it can be adapted for large volumes and specific liquid handling extraction robots.

The simplicity of the TruTip technology also affords some cost advantages for those interested in purchasing a new automated NA purification system because the primary hardware required for automating TruTip procedures is the pipette channel arm itself rather than magnetic rods, vacuum systems, or on-board centrifuges. Minimizing deck space with TruTip protocols also enables advanced users to integrate upstream or downstream automated processes with the TruTip. Postextraction processes such as NA quantitation, normalization, polymerase chain reaction (PCR) setup, or DNA sequencing are also readily integrated with TruTip on the larger liquid handling platforms (29–31).

AUTOMATED METHODS FOR NA EXTRACTION IN DIAGNOSTIC VIROLOGY

Automation of NA extraction for detection of viral agents using molecular diagnostic methods has resulted in numerous improvements in the field. One important advantage is the reduction in labor time dedicated to the extraction process. By being able to load samples on an instrument and walk away, the technologist is able to better multitask, performing other procedures while the samples are processed. Other advantages to the automated extraction methods include

- Potentially greater reproducibility due to the precision robotics versus manual pipetting.
- Potentially lower risk of contamination compared to performing extraction with multiple manual steps where errors could occur and lead to contamination of samples.
- Lower probability for sample mix-ups during processing.

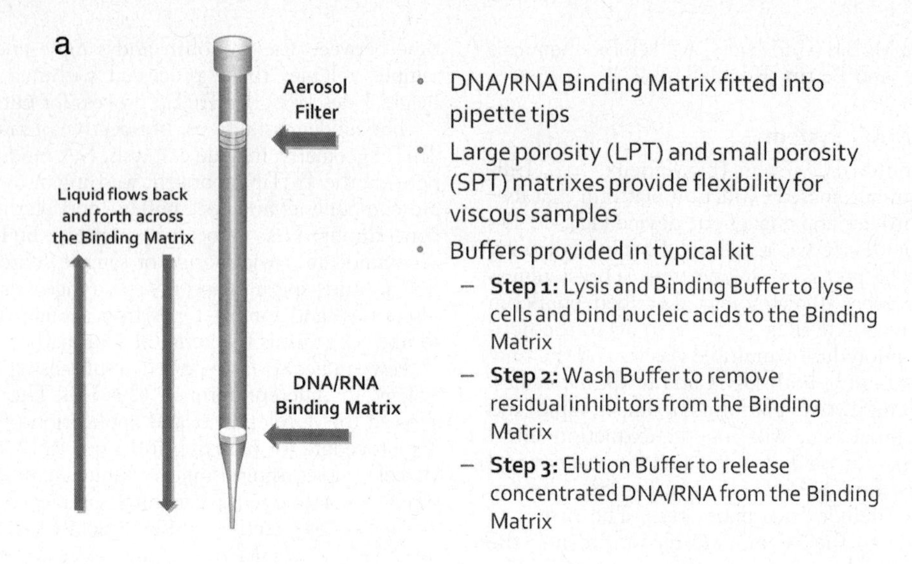

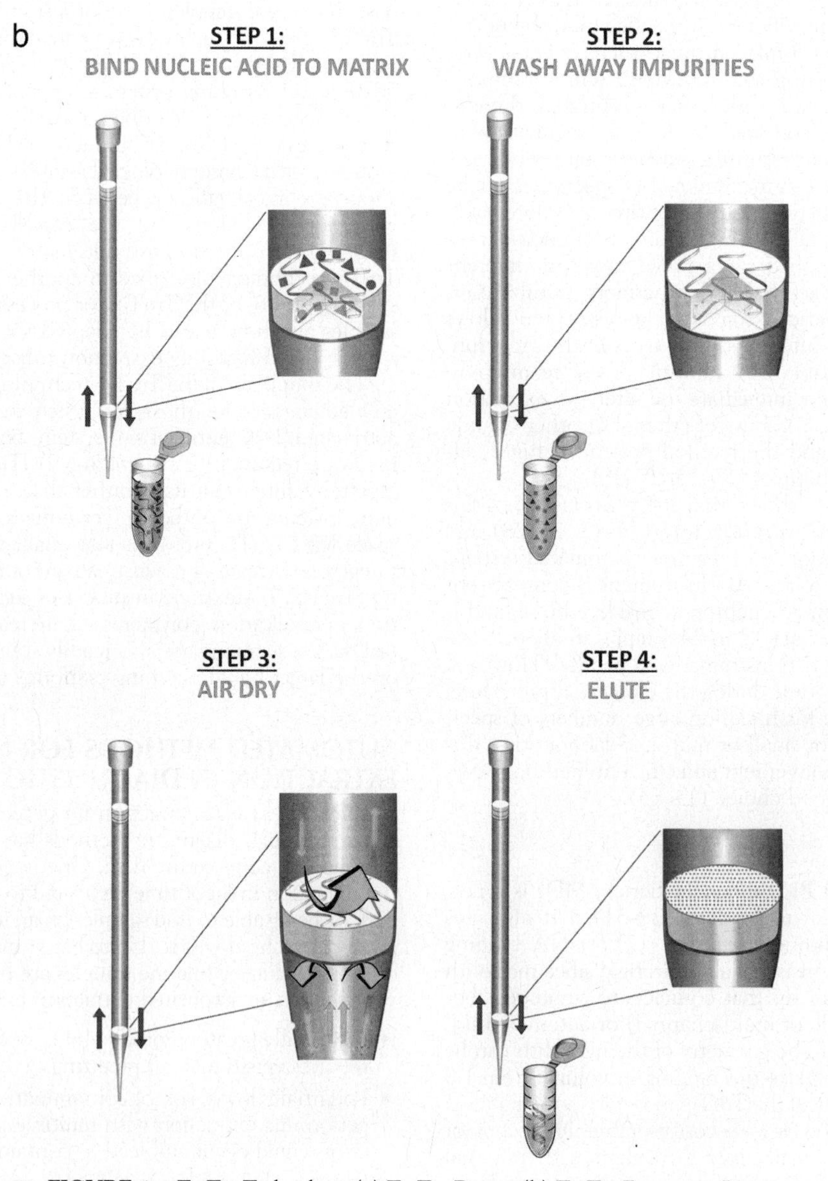

FIGURE 1 TruTip Technology. (a) TruTip Design (b) TruTip Extraction Process.

There are some disadvantages with automated extraction systems. These include

- More expensive because they require either a capital expenditure up front or addition of costs to disposables to cover reagent rental.
- Maintenance and repair costs for the systems (service contracts).
- Potentially less flexibility in sample type, sample size, and elution volume compared with a manual extraction process.

However, in many laboratories the positive impact on productivity and reproducibility make the automated approaches a positive fit for the lab. In addition, most of the Food and Drug Administration (FDA)-cleared viral molecular diagnostics assays are specifically cleared for use on an automated extraction instrument as part of the system. Using a different method would require validation, which is not particularly difficult but nevertheless does add effort for assay implementation. Some molecular assays are FDA cleared on an extractor sold by the assay vendor, while others are cleared on third party extraction systems.

Automation of the NA extraction process can take several forms ranging from automating a manual process all the way to building systems that provide completely automated sample in–answer out instrumentation platforms.

Automation of Manual Processes

Qiagen has been successful at automating their filter spin-vacuum column procedures using three different instruments. Characteristics of the instruments are shown in Table 1. The QIAcube automates a centrifugation-based protocol, eliminating the multiple hands-on steps associated with the manual spin column method. The QIAcube HT and the BioRobot instruments achieve automation of the column-based manual process using a vacuum filtration method and robotic liquid handling. Each system significantly reduces hands-on time for sample NA extraction processing. All of them can use many of the column-based extraction protocols designed for different sample types, providing a high level of flexibility for the laboratory. The Qiagen automated column-based extraction protocols have been evaluated for viral molecular diagnostic applications (21, 32).

Automation of NA Extraction Using Magnetic Particle Capture and Robotics

Numerous systems are available for fully automated NA extraction that provide an eluate of pure NA that is ready for use in viral molecular diagnostics applications. The instruments all utilize a lysis buffer that releases viral NAs from the viral capsid, inactivates any nucleases present in the sample, and captures the NAs on magnetic silica particles or another magnetic particle matrix to bind the released NAs. The QIACube, QIAcube HT, and Biorobot systems discussed earlier use a solid-phase silica-based matrix to capture NAs using spin or vacuum filtration methods. Some of the instruments (Qiagen EZ1 series [Qiagen], MAXwell [Promega, Madison, WI], Abnova [BioGX, Birmingham, AL], MagNA Pure Compact [Roche Diagnostics, Indianapolis, IN], and BD MAX [BD Diagnostics, Sparks, MD]) use a reaction strip format that is essentially self-contained, with all reagents in wells within the strip. Figure 2 describes the configuration of the BD MAX reaction strip. Most others will have a similar configuration with tips and reagents for the extraction process; however, unlike the BD MAX strip they do not have a position for the PCR master mix. Others (NucliSENS easyMAG system [bioMérieux], KingFisher [Thermo Fisher Scientific Inc., Waltham, MA], MagNA Pure and Ampliprep [Roche Diagnostics], Abbott *m*2000 sp [Abbott Molecular, Des Plaines, IL], Sentosa SX101 [Vela Diagnostics, Singapore], epMotion [Eppendorf], and QIAsymphony [Qiagen]) utilize robotics to process reagent trays with disposable pipetting tips to transfer and wash magnetic particles. The Sentosa SX101 is based on the Eppendorf epMotion platform. The systems and some of their operating characteristics are summarized in Table 2. All have been evaluated and shown to have the ability to effectively recover viral NAs from a variety of pathogens in numerous sample types (13, 16, 17, 21, 27, 32–49). The instruments provide throughput options from single samples per run up to 96 samples per run, with numerous options for sample input volumes and elution volumes.

Automation of Extraction Plus Setup of Master and Reaction Mix

The next step in the development of automated NA extraction instrumentation involves the addition of setting up the PCR or other reactions on-board and dispensing the complete reaction mix into 96-well trays or reaction disks that are ready to be placed in amplification instruments. The instruments that provide this functionality are listed in Table 3. Some of these instruments were included in Table 2 because it is optional for the reaction mix setup function to be utilized. Some instruments require an additional module to allow for automated master mix and reaction setup (Roche Ampliprep and Qiagen QIAsymphony). Automating preparation of master mix and setting up the PCR in the amplification vessel (vial, plate, or disk depending on the amplification platform) serve to further reduce labor time. These steps also have the potential to reduce the likelihood of human error in setting up reactions due to pipetting an incorrect amount of one of the reaction components or

TABLE 1 Instruments that automate the column extraction process

Instrument	Samples per run	Processing time (h)	Sample input volume (μl)	Elution volume (μl)	Sample types[a]	Vendor contact
Qiagen Qiacube	Up to 12	1–1.5	200–600	20–200	B, S, P, CSF, U, T, UTM, SW	www.qiagen.com
Qiagen Qiacube HT	8–96 in sets of 8	1–1.25	200–350	30–150	B, S, P, CSF, U, T, UTM, SW	www.qiagen.com
Qiagen Biorobot Universal	Up to 96	2.20–3	Up to 200	20–150	B, S, P, CSF, U, T, UTM, SW	www.qiagen.com

[a]B, whole blood; CSF, cerebral spinal fluid; P, plasma; S, serum; SW, swab; T, tissue; U, urine; UTM, universal or viral transport medium.

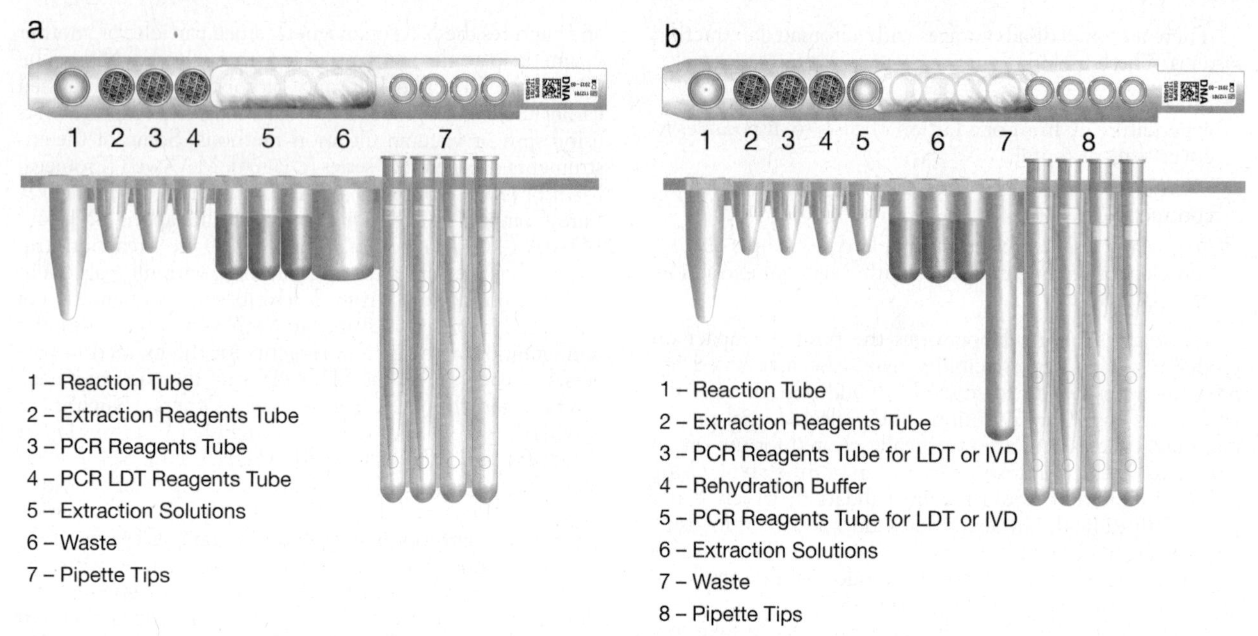

FIGURE 2 Reaction strip from the BD MAX Instrument. (a) Current configuration: one extraction and one polymerase chain reaction (PCR) amplification. (b) Future configuration: one extraction and two PCR amplifications.

neglecting to add extracted NAs to reaction wells. Some of the systems offer connectivity with the real-time PCR instrument to transfer patient identifiers without the need for manually entering the information on the cycler. These include the *m*2000 sp (Abbott), Sentosa SX 101 (Vela DX), Ampliprep (Roche), and the QIAsymphony (Qiagen). The Sentosa SX101 and the QIAsymphony possess the flexibility to perform this with either vendor-supplied reagents or with laboratory-developed tests. The ability to utilize manufacturer-supplied kits and laboratory-developed procedures on the same platform increases efficiency by reducing the need for staff to learn how to use and maintain multiple instruments.

Systems with Full Sample In–Answer Out Automation

Several systems, listed in Table 4, offer sample in–answer out automation that combines NA extraction, reaction mix setup, and on-board amplification with minimal hands-on time. The instruments operate in several formats, including real-time PCR, PCR followed by array detection, and transcription-mediated amplification (TMA). The Panther and Tigris systems (Hologic Gen-Probe, San Diego, CA) provide for sample in–answer out automation of the TMA process. The instruments utilize capture oligos rather than true extraction for recovery of NAs. The systems are currently available for only vendor-supplied assays. Hologic is developing an add-on module for the Panther instrument that will provide automated setup of their PCR-based kits (ProFlu+, etc.) and will have the ability to accommodate user-developed PCR assays. There are several PCR-based systems that use varying detection technologies and provide sample in–answer out automation. The GeneXpert (Cepheid, Sunnyvale, CA) system uses self-contained cartridges that require only the addition of sample and in some cases reagents to the cartridge. The instruments range from only one cartridge slot for very-low-volume applications to 16 position systems for medium-throughput applications and up to 48 positions for high-throughput applications. The GeneXpert system is designed to use only tests supplied by Cepheid and cannot be used for laboratory-developed tests. The FilmArray (bioMérieux) system performs highly multiplexed assays in a sample-to-answer format using two-stage nested PCR and array detection in a specially designed pouch that contains all reagents on-board. It requires only addition of sample and addition of a buffer to hydrate the reagents. Each FilmArray instrument contains a single processing station, and the time to result is 60 to 70 minutes. The FilmArray system cannot be used for laboratory-developed procedures. Focus Diagnostics (Cypress, CA) has the 3M Integrated Cycler (Focus Diagnostics), a sample-to-answer system that utilizes a specially formulated reagent mix to allow for PCR amplification on samples without full NA purification. A proprietary mix of reagents in the master mix decreases the effect of inhibitors in the samples. Focus has several *in vitro* diagnostics (IVD) assays using this technology. The sample-to-answer format is not currently available for laboratory-developed procedures. Nanosphere (Northbrook, IL) also has a sample-to-answer automation format, the Verigene system, that uses fluidics to perform extraction, PCR amplification, and then array-based detection of amplified products, incorporating nano-sized gold particles and optics for enhancement of detection.

The BD MAX (BD Diagnostics) instrument is the only system currently available that can accommodate either IVD-cleared assays or laboratory-developed procedures in the fully automated sample-to-answer format. It provides sample-to-answer automation of real-time PCR assays using flexible batch size with extraction/reagent strips and PCR cards with two sets of 12-well positions. The time to result for a run of 24 samples is about 2.5 hours.

Table 5 lists sample-to-answer instruments that are currently in development and likely to be available in 2015 to 2016. Luminex (Austin, TX) is developing the Aries MDx

TABLE 2 Instruments that use variations of magnetic particle capture for nucleic acid extraction

Instrument	Samples per run	Processing time	Sample input volume	Elution volume	Sample types[a]	Vendor contact
EZ1 Advanced	1–6	20–50 min	200–400 μl	50–100 μl	B, S, P, CSF, U, T, UTM, SW	www.qiagen.com
EZ1 Advanced XL	1–14	20–50 min	200–400 μl	50–100 μl	B, S, P, CSF, U, T, UTM, SW	www.qiagen.com
QIAsymphony SP	1–96 in batches of 24	3.5–6 h	200–1,000 μl	50–400 μl	B, P, U, UTM, SW	www.qiagen.com
EasyMag	1–24, best in sets of 8	45 min	Up to 2 ml depending on sample type[b]	Variable from 25 to 110 μl	B, P, S, UTM, T, CSF, U, BAL/BW, ST, FL, SP	www.biomerieux.com
Maxwell	1–16	28–52 min	100–300 μl	50 μl	P, S	www.promega.com
KingFisher	Up to 24	35–60 min	20–200 μl	100 μl	P, S, SW, CSF, UTM, ST	www.thermofisher.com
KingFisher ML	Up to 15	35–60 min	50–1,000 μl	100 μl	P, S, SW, CSF, UTM, ST	www.thermofisher.com
KingFisher Flex	Up to 24 or up to 96	35–60 min	20–1,000 μl or 200–5,000 μl	100 μl	P, S, SW, CSF, UTM, ST	www.thermofisher.com
Abnova	16 or 32	55–90 min	50–200 μl	100 μl	B, P, S, UTM, T, CSF, ST, BAL/BW	www.biogx.com
Sentosa SX101	1–96	90–120 min	1–1,000 μl	80 μl	B, P, S, UTM, SW, CSF, SP	www.veladx.com
*m*2000 sp	24–96 in batches of 24	90–250 min	0.4–4 ml	15–190 μl	B, P, S, SW, T, CSF, ST, BAL/BW	www.abbottmolecular.com
Magnapure 96	1–96	50–90 min	50–1,000 μl	50–100 μl	B, P, S, ST, CSF, SW, BAL/BW	www.molecular.roche.com
Ampliprep	1–48, optimal batches of 12	1.5–2 h	50–850 μl	75 μl	S, P	www.molecular.roche.com
Magnapure Compact	1–8	20–45 min	100–1,000 μl	50–200 μl	B, S, P, T, FL, CSF, U, UTM	www.molecular.roche.com
QuickGene 810	1–8	6–20 min	100–200 μl	50–200 μl	B, P, S, UTM, SW, ST, T	www.autogen.com
Arrow	1–12	30–40 min	200–500 μl	50–200 μl	B, P, S, UTM, SW, ST, T	www.autogen.com

[a]B, whole blood; S, serum; P, plasma; U, urine; CSF, cerebral spinal fluid; T, tissue; ST, stool; SW, swab; UTM, universal or viral transport medium; BAL/BW, bronchial alveolar lavage/bronchial washing; FL, other fluid; SP, sputum.

System that will use a strip format to combine extraction, PCR amplification, and real-time detection of the PCR product. It is designed to perform six sets of reactions at a time with dual six-position reaction bays. The vendor's plans include development of IVD assays on the system, but it will also support the vendor's analyte-specific reagents and will allow for adaption of laboratory-developed procedures on the system. GenMark Diagnostics, Inc. (Carlsbad, CA) is developing a sample-to-answer platform that will combine extraction, PCR amplification, and detection, using their eSensor technology on arrays. It will be similar to the GeneXpert in that the system will consist of blocks of six independent processing positions with the ability to link up to six of the blocks to a single computer providing throughput of up to 36 reactions. It will not accommodate laboratory-developed procedures. ELITech (Puteaux, France) is also developing a sample-to-answer system, the inGenius system, that will combine extraction with real-time PCR in a reagent strip format similar to the approach in the BD MAX. ELITech is planning to use the platform for IVD assays along with their analyte-specific reagents, and it is reported to be amenable to adapting laboratory-developed procedures to the platform.

Performance of the Automated Systems

Our laboratory has validated the bioMérieux easyMAG system and the Sentosa SX101 with a large variety of sample types. These include

- Whole blood, bone marrow, plasma, and serum
- Urine
- Cerebrospinal fluid and other fluids, including bronchoalveolar lavage, bronchial washing, vitreous, amniotic, pleural
- Swabs in universal transport media (UTM) or Copan ESwabs

TABLE 3 Automated extraction systems with PCR mix setup function

Instrument	Samples per run	Processing time	Sample input volume	Elution volume	Sample types[a]	IVD kits or LDT[b]	Vendor contact
m2000sp	1–96	50–90 min	50–1,000 μl	50–100 μl	B, P, S, ST, CSF, SW, BAL/BW	IVD kits	www.abbottmolecular.com
Qiagen Biorobot	Up to 96	2.20–3 h	Up to 200 μl	20–150 μl	B, S, P, CSF, U, T, UTM	Both	www.qiagen.com
QIAsymphony	1–96 in batches of 24	3.5–6 h	200–1,000 μl	50–400 μl	B, P, U, UTM	Both	www.qiagen.com
Ampliprep	1–48 in batches of 12	1.5–2 h	50–850 μl	75 μl	P, S	IVD kits	www.molecular.roche.com
Magnapure 96	1–96	50–90 min	50–1,000 μl	50–100 μl	B, P, S, ST, CSF, SW, BAL	Both	www.molecular.roche.com
Sentosa SX101	1–96	90–120 min	1–1,000 μl	80 μl	B, P, S, UTM, SW, CSF, SP	Both	www.veladx.com
epMotion®	Variable depending on exact model	60–120 min	1–1,000 μl	Variable	B, P, S, U	LDT	www.eppendorfna.com

[a]B, whole blood; BAL/BW, bronchial alveolar lavage/bronchial washing; CSF, cerebral spinal fluid; P, plasma; S, serum; SP, sputum; ST, stool; SW, swab; T, tissue; U, urine; UTM, universal or viral transport medium.
[b]IVD, *in vitro* diagnostics; LDT, laboratory-developed test.

We also have validated a number of sample types that require pretreatment:

- Sputum: pretreat with Copan SL to liquefy
- Tissue: pretreat with protease K
- Stool: suspend in UTM and perform a low-speed spin to remove larger matter

Both systems provide for automation of NA extraction, and as noted earlier, the Sentosa SX101 provides the ability to set up master mixes and PCR amplifications with the purified NA. We have compared the two systems for extracting a variety of pathogens including BK virus from urine and plasma; cytomegalovirus (CMV) from plasma, bronchoalveolar lavage, cerebrospinal fluid, and other samples; and *Mycobacterium tuberculosis* from liquefied sputum. There was excellent correlation of results between the two extraction systems. For example, the performance of a Laboratory Developed Test (LDT) for *M. tuberculosis* complex using easyMAG (bioMérieux) extraction and manual PCR setup followed by PCR on a Rotor-Gene 3000 (Qiagen) was identical to the performance of the LDT on the Sentosa SX101 (Vela DX) extraction and automated master mix/PCR setup with PCR on a Rotor-Gene Q (R. H. Widen and S. Silbert, unpublished). Similar results were obtained for CMV. Our routine CMV PCR is performed using the easy-

TABLE 4 Systems that provide total automation with sample-in, result-out functionality

Instrument	Samples per run	Processing time	Random access	Sample types[a]	IVD kits only or LDT[c]	Vendor contact
Tigris	180 (450 in 8 h)	3.5 h to first results	No	SW, LC, U, P	IVD kits only	www.hologic.com
Panther	Up to 275	3.5 h to first results	Yes	SW, LC, U	IVD kits only currently	www.hologic.com
GeneXpert	Variable dependent on format	< 1–2 h depending on test	Yes	CSF, UTM, SW, LB, ST	IVD kits only	www.cepheid.com
FilmArray	1	About 1 h	No; 1 module per instrument	BC, UTM, ST, BAL/BW[b]	IVD kits only	www.biomerieux.com
Verigene	1 (in modules)	2.5 h	Yes	UTM, SW, ST	IVD kits only	www.nanosphere.us
3M Integrated Cycler	1–8 or up to 96 depending on test	1–1.25 h	No	CSF, SW, ST	IVD kits only currently	www.focusdx.com
BD MAX	1–24	2–3 h	No	SW, UTM, LB, CSF, BAL/BW, SP, ST	IVD kits and LDTs	www.bd.com

[a]BC, positive blood culture; BAL/BW, bronchial alveolar lavage/bronchial washing; CSF, cerebral spinal fluid; LB, Lim broth; LC, liquid cytology; P, plasma; SP, sputum; ST, stool; SW, swab; U, urine; UTM, universal or viral transport medium.
[b]BAL/BW validated in house on the FilmArray for their respiratory pathogen panel.
[c]IVD, *in vitro* diagnostics; LDT, laboratory-developed test.

TABLE 5 Fully automated sample-to-answer instruments in development

Instrument	Samples per run	Processing time	Random access	Sample types[a]	IVD kits only or LDT[b]	Vendor contact
GenMark	1–6 in a module	To be determined	Yes	Projected SW, CSF, UTM, ST	IVD kits only	www.genmarkdx.com
Aries	1–12 with 2 independent 6-sample bays	Approximately 2 h	No	Projected SW, ST, UTM, LB	IVD kits and LDTs	www.luminex.com
InGenius system	1–12	<90 min	No	Projected SW, CSF, UTM, ST	IVD kits and LDTs	www.elitechgroup.com

[a]CSF, cerebral spinal fluid; LB, Lim broth; ST, stool; SW, swab; U, urine; UTM, universal or viral transport medium.
[b]IVD, *in vitro* diagnostics; LDT, laboratory-developed test.

MAG for extraction and Focus DX CMV quantitative PCR. Similar results were obtained with an LDT for BK virus with excellent correlation of both qualitative and quantitative (no values were >0.5 log different) results for BK from urine or plasma extracted with the easyMAG or Sentosa SX101 instruments with PCR performed on the RG3000 or Rotor-Gene Q instruments, respectively (our unpublished data). The only situations in which discrepancies occurred were with samples that had very low viral load (500 IU/ml or less), and these were most likely due to sampling error since there was no trend for one system being consistently better.

Literature References Comparing Performance of Extraction Systems

Numerous published studies have compared many of the automated extraction systems, often including comparison to a manual extraction system (13, 16, 17, 21, 27, 32–49). General observations include

- The automated systems generally demonstrated at least equal performance in recovery of NA compared to a manual processing method, especially the liquid phase manual extraction methods. Many demonstrated greater recovery and purification than obtained with manual methods (16, 17, 27, 33–35, 42, 49).
- For most of the studies, there were minimal differences in the ability to detect pathogens in samples. Some studies demonstrated greater analytic sensitivity of one system versus another, with variations dependent on specific targets and sample types (16, 21, 32, 37).
- When differences in recovery were noted, it was often with samples that were close to the limit of detection for the PCR assay (13, 16, 37, 39, 48).
- For most studies the differences in analytic sensitivity or quantitative recovery with the automated platforms were less than 0.5 log, indicating overall good correlation of the various systems (21, 34, 35, 38, 45).
- Studies also revealed that protocols optimized for one sample type may not be optimal for different samples (36).
- Many of the studies emphasized the need for individual laboratories to include an analysis of performance of the NA extraction method as part of the validation of the overall assay performance.

Emerging Extraction Technology

Rogacs et al. recently published an article reviewing the potential application of a technology called isotachophoresis (ITP) for isolation and recovery of NAs (50). ITP does not require special chemistries or matrices for DNA binding, and the process does not require instruments with multiple moving parts. As the authors noted, the process is compatible with numerous lysing chemistries and sample types. ITP takes advantage of the fact that NA has ionic electrophoretic mobility compared to other materials present in crude lysates. It uses this electrophoretic mobility to preconcentrate, purify, and then recover NA. By setting the parameters of the buffers and the electrical field correctly, it is possible to selectively separate the NA. ITP has been demonstrated to be effective in recovering human DNA and pathogen-specific DNA and RNA from biological samples. The authors noted that challenges remain in adapting ITP to routine NA extraction protocols. We are not aware of any systems that incorporate ITP in NA purification that are commercially available at this time.

Another variation being examined for NA extraction involves transport of the NA bound to magnetic particles through an immiscible liquid to remove inhibitors and recover NA as described by Sur et al. (51). The concept essentially involves two wells overlaid with liquid wax. The first well contains the sample lysate with the NA and paramagnetic NA binding particles. The bound particles are moved through the liquid wax and carried to the elution well (second well) by magnets. Impurities and inhibitors are captured in the liquid wax matrix. The authors evaluated the method for extracting and recovering HIV-1 RNA from plasma, HIV-1 proviral DNA in whole blood, and bacterial (*Chlamydia trachomatis* and *Neisseria gonorrhoeae*) DNA in urine samples. There was no statistically significant difference in recovery of any of the targets in comparison to an alternative standard extraction protocol. The challenge the authors discussed is determining methods to minimize carry-over of potential inhibitors and determining tolerance of the amplification system to the materials. They were able to determine specific characteristics of the paramagnetic particles that led to minimal carryover. The amount of time to extract and recover NA is far less with their approach as compared to methods that utilize multiple wash steps of the particles with the bound NA. Finally, the authors note that this approach may provide a method for developing self-contained point of care molecular diagnostics assays due to the simplicity of the extraction and recovery process. It is not clear if this technology has been or is being incorporated into commercial products at this time.

The final area with ongoing research involves the lab-on-a-chip process in which the goal is to continually miniaturize the components and use microfluidics to accomplish extraction and molecular amplification (52–54). In reality the lab-on-a-chip approach has already been implemented to some extent with commercial systems, specifically the Cepheid GeneXpert and the Nanosphere Verigene systems. These systems use microfluidics to combine extraction, recovery, amplification, and detection. The efforts to further miniaturize the systems are ongoing. Kemp et al. (53) presented data in 2012 that demonstrated recovery of plasmid herpes simplex virus 2 DNA could be achieved from urine by using a chip that successfully extracted 400 µl of urine. They utilized an adaptation of the immiscible barrier concept to process the bound paramagnetic particles. The chip did not include the PCR amplification step; however, it did demonstrate the ability to rapidly extract and recover NA on a microfluidic chip. Shin et al. (54) demonstrated the utility of the microfluidic lab-on-a-chip approach to extract and recover human genomic DNA from blood and urine. The concept is reviewed by Price and colleagues (52).

Validation of NA Extraction Systems as Part of a Clinical Laboratory Viral Molecular Diagnostics Assay

As noted earlier, an important component of new assay validation, especially for LDTs or when validating an IVD-cleared assay with a noncleared sample type, is to demonstrate acceptable performance characteristics of the modified assay, including recovery of the NA from the sample in a state amenable to use in the assay. One approach is to utilize patient samples run in parallel with two different assays with the same or different NA extraction protocols and document equivalent performance. However, there may be too few positive samples for some targets, and the other limitation is the fact that the absolute quantity of the viral target may not be known. An acceptable approach would be to use spiked samples in a series of dilutions to document the ability of the NA extraction system to recover the NA. If there is quantitated material present, a true limit of detection can be defined by using the extraction protocol with any particular sample matrix. However, even if there is not a quantitated standard, one can still design a validation study by using serial dilutions of the virus and comparing recovery in the "standard" or FDA-cleared sample type with the sample type to be validated. This will allow the laboratory to verify equal recovery in a variety of matrices and equivalent performance in the downstream PCR or other NAAT assay. If the recovery is not equivalent (not within one to two cycle numbers or 0.5 log), one may attempt to modify the extraction parameters (pretreat, add more sample, reduce the elution volume) or simply report a different limit of detection with different sample types. Interestingly the *in vitro* analytic performance does not always correlate with detection of infection. For example, in studies comparing the Focus DX first generation influenza A/B and RSV assay, which incorporated extraction on the easyMAG, with performance of the Focus Simplexa Direct FluA/B/RSV, we found that the two assays had equal ability to detect positive versus negative samples. The ability was equal even though the direct assay had reduced analytical sensitivity, with three to four cycles more, indicating either reduced recovery or possibly partial inhibition. But nonetheless all positive samples were detected with both approaches. As a final point, it is important to state that regulations require laboratories to validate several performance characteristics of modified tests, including both analytical and diagnostics performance.

REFERENCES

1. **Tan SC, Yiap BC.** 2009. DNA, RNA, and protein extraction: the past and the present. *J Biomed Biotechnol* **2009:**574398.
2. **Sun W.** 2010. Nucleic extraction and amplification, p 35–47. *In* Grody WW, Nakamura RM, Kiechle F, Strom C (eds.), *Molecular Diagnostics Techniques and Applications for the Clinical Laboratory.* Elsevier, Oxford, UK.
3. **Buckingham L.** 2012. Nucleic acid extraction methods, p. 69–86. *In* Buckingham L (ed), *Molecular Diagnostics: Fundamentals, Methods and Clinical Applications,* 2nd ed. F.A. Davis Company, Philadelphia, PA.
4. **Lisby G.** 1999. Application of nucleic acid amplification in clinical microbiology. *Mol Biotechnol* **12:**75–99.
5. **Sambrook J, Fritsch EF, Maniatis T.** 1989. *Molecular Cloning: A Laboratory Manual,* 2nd ed. Cold Spring Harbor Laboratory Press, New York.
6. **Ullrich A, Shine J, Chirgwin J, Pictet R, Tischer E, Rutter WJ, Goodman HM.** 1977. Rat insulin genes: construction of plasmids containing the coding sequences. *Science* **196:**1313–1319.
7. **Slater RJ.** 1985. The extraction of total RNA by the detergent and phenol method. *Methods Mol Biol* **2:**101–108.
8. **Chomczynski P, Sacchi N.** 1987. Single-step method of RNA isolation by acid guanidinium thiocyanate-phenol-chloroform extraction. *Anal Biochem* **162:**156–159.
9. **Chomczynski P.** 1993. A reagent for the single-step simultaneous isolation of RNA, DNA and proteins from cell and tissue samples. *Biotechniques* **15:**532–534, 536–537.
10. **Casas I, Powell L, Klapper PE, Cleator GM.** 1995. New method for the extraction of viral RNA and DNA from cerebrospinal fluid for use in the polymerase chain reaction assay. *J Virol Methods* **53:**25–36.
11. **Chomczynski P, Sacchi N.** 2006. The single-step method of RNA isolation by acid guanidinium thiocyanate-phenol-chloroform extraction: twenty-something years on. *Nat Protoc* **1:**581–585.
12. **Boom R, Sol CJ, Salimans MM, Jansen CL, Wertheim-van Dillen PM, van der Noordaa J.** 1990. Rapid and simple method for purification of nucleic acids. *J Clin Microbiol* **28:**495–503.
13. **Tang YW, Sefers SE, Li H, Kohn DJ, Procop GW.** 2005. Comparative evaluation of three commercial systems for nucleic acid extraction from urine specimens. *J Clin Microbiol* **43:**4830–4833.
14. **Petrich A, Mahony J, Chong S, Broukhanski G, Gharabaghi F, Johnson G, Louie L, Luinstra K, Willey B, Akhaven P, Chui L, Jamieson F, Louie M, Mazzulli T, Tellier R, Smieja M, Cai W, Chernesky M, Richardson SE, Ontario Laboratory Working Group for the Rapid Diagnosis of Emerging Infections.** 2006. Multicenter comparison of nucleic acid extraction methods for detection of severe acute respiratory syndrome coronavirus RNA in stool specimens. *J Clin Microbiol* **44:**2681–2688.
15. **McClernon DR, Ramsey E, StClair MS.** 2007. Magnetic silica extraction for low-viremia human immunodeficiency virus type 1 genotyping. *J Clin Microbiol* **45:**572–574.
16. **Dundas N, Leos NK, Mitui M, Revell P, Rogers BB.** 2008. Comparison of automated nucleic acid extraction methods with manual extraction. *J Mol Diagn* **10:**311–316.
17. **Esona MD, McDonald S, Kamili S, Kerin T, Gautam R, Bowen MD.** 2013. Comparative evaluation of commercially available manual and automated nucleic acid extraction methods for rotavirus RNA detection in stools. *J Virol Methods* **194:**242–249.
18. **Stenesh J.** 1989. *Dictionary of Biochemistry and Molecular Biology,* 2nd ed. John Wiley & Sons, New York, NY.
19. **QIAGEN, Inc.** 2012. *QIAamp® Viral RNA Mini Handbook,* 3rd ed. Qiagen, Valencia, CA.
20. **QIAGEN, Inc.** 2012. *QIAamp® DNA Mini and Blood Mini Handbook,* 3rd ed. Qiagen, Valencia, CA.

21. **Knepp JH, Geahr MA, Forman MS, Valsamakis A.** 2003. Comparison of automated and manual nucleic acid extraction methods for detection of enterovirus RNA. *J Clin Microbiol* **41:**3532–3536.
22. **Shafer RW, Levee DJ, Winters MA, Richmond KL, Huang D, Merigan TC.** 1997. Comparison of QIAamp HCV kit spin columns, silica beads, and phenol-chloroform for recovering human immunodeficiency virus type 1 RNA from plasma. *J Clin Microbiol* **35:**520–522.
23. **Klein A, Barsuk R, Dagan S, Nusbaum O, Shouval D, Galun E.** 1997. Comparison of methods for extraction of nucleic acid from hemolytic serum for PCR amplification of hepatitis B virus DNA sequences. *J Clin Microbiol* **35:**1897–1899.
24. **Kleines M, Schellenberg K, Ritter K.** 2003. Efficient extraction of viral DNA and viral RNA by the Chemagic viral DNA/RNA kit allows sensitive detection of cytomegalovirus, hepatitis B virus, and hepatitis G virus by PCR. *J Clin Microbiol* **41:**5273–5276.
25. **Rutjes SA, Italiaander R, van den Berg HH, Lodder WJ, de Roda Husman AM.** 2005. Isolation and detection of enterovirus RNA from large-volume water samples by using the NucliSens miniMAG system and real-time nucleic acid sequence-based amplification. *Appl Environ Microbiol* **71:** 3734–3740.
26. **Loens K, Ursi D, Goossens H, Ieven M.** 2008. Evaluation of the NucliSens miniMAG RNA extraction and real-time NASBA applications for the detection of Mycoplasma pneumoniae and Chlamydophila pneumoniae in throat swabs. *J Microbiol Methods* **72:**217–219.
27. **Loens K, Bergs K, Ursi D, Goossens H, Ieven M.** 2007. Evaluation of NucliSens easyMAG for automated nucleic acid extraction from various clinical specimens. *J Clin Microbiol* **45:**421–425.
28. **Schuurman T, de Boer R, Patty R, Kooistra-Smid M, van Zwet A.** 2007. Comparative evaluation of in-house manual, and commercial semi-automated and automated DNA extraction platforms in the sample preparation of human stool specimens for a Salmonella enterica 5′-nuclease assay. *J Microbiol Methods* **71:**238–245.
29. **Griesemer SB, Holmberg R, Cooney CG, Thakore N, Gindlesperger A, Knickerbocker C, Chandler DP, St George K.** 2013. Automated, simple, and efficient influenza RNA extraction from clinical respiratory swabs using TruTip and epMotion. *J Clin Virol* **58:**138–143.
30. **Holmberg RC, Gindlesperger A, Stokes T, Brady D, Thakore N, Belgrader P, Cooney CG, Chandler DP.** 2013. High-throughput, automated extraction of DNA and RNA from clinical samples using TruTip technology on common liquid handling robots. *J Vis Exp* **11:**e50356.
31. **Chandler DP, Griesemer SB, Cooney CG, Holmberg R, Thakore N, Mokhiber B, Belgrader P, Knickerbocker C, Schied J, St George K.** 2012. Rapid, simple influenza RNA extraction from nasopharyngeal samples. *J Virol Methods* **183:**8–13.
32. **Yang G, Erdman DE, Kodani M, Kools J, Bowen MD, Fields BS.** 2011. Comparison of commercial systems for extraction of nucleic acids from DNA/RNA respiratory pathogens. *J Virol Methods* **171:**195–199.
33. **Kim Y, Han M-S, Kim J, Kwon A, Lee K-A.** 2014. Evaluation of three automated nucleic acid extraction systems for identification of respiratory viruses in clinical specimens by multiplex real-time PCR. *BioMed Res Int* **2014:**430650.
34. **Mengelle C, Mansuy J-M, Da Silva I, Davrinche C, Izopet J.** 2011. Comparison of 2 highly automated nucleic acid extraction systems for quantitation of human cytomegalovirus in whole blood. *Diagn Microbiol Infect Dis* **69:**161–166.
35. **Laus S, Kingsley LA, Green M, Wadowsky RM.** 2011. Comparison of QIAsymphony automated and QIAamp manual DNA extraction systems for measuring Epstein-Barr virus DNA load in whole blood using real-time PCR. *J Mol Diagn* **13:**695–700.
36. **Waggoner JJ, Pinsky BA.** 2014. Comparison of automated nucleic acid extraction methods for the detection of cytomegalovirus DNA in fluids and tissues. *PeerJ* **2:**e334.
37. **Verheyen J, Kaiser R, Bozic M, Timmen-Wego M, Maier BK, Kessler HH.** 2012. Extraction of viral nucleic acids: comparison of five automated nucleic acid extraction platforms. *J Clin Virol* **54:**255–259.
38. **Miller S, Seet H, Khan Y, Wright C, Nadarajah R.** 2010. Comparison of QIAGEN automated nucleic acid extraction methods for CMV quantitative PCR testing. *Am J Clin Pathol* **133:**558–563.
39. **Bravo D, Clari MA, Costa E, Muñoz-Cobo B, Solano C, José Remigia M, Navarro D.** 2011. Comparative evaluation of three automated systems for DNA extraction in conjunction with three commercially available real-time PCR assays for quantitation of plasma Cytomegalovirus DNAemia in allogeneic stem cell transplant recipients. *J Clin Microbiol* **49:** 2899–2904.
40. **Chan KH, Yam WC, Pang CM, Chan KM, Lam SY, Lo KF, Poon LLM, Peiris JS.** 2008. Comparison of the NucliSens easyMAG and Qiagen BioRobot 9604 nucleic acid extraction systems for detection of RNA and DNA respiratory viruses in nasopharyngeal aspirate samples. *J Clin Microbiol* **46:**2195–2199.
41. **Fieblekorn KR, Lee BG, Hill CE, Caliendo AM, Nolte FS.** 2002. Clinical evaluation of an automated nucleic acid isolation system. *Clin Chem* **48:1613–1615.**
42. **Espy MJ, Rys PN, Wold AD, Uhl JR, Sloan LM, Jenkins GD, Ilstrup DM, Cockerill FR III, Patel R, Rosenblatt JE, Smith TF.** 2001. Detection of herpes simplex virus DNA in genital and dermal specimens by LightCycler PCR after extraction using the IsoQuick, MagNA Pure, and BioRobot 9604 methods. *J Clin Microbiol* **39:**2233–2236.
43. **Petrich A, Mahony J, Chong S, Broukhanski G, Gharabaghi F, Johnson G, Louie L, Luinstra K, Willey B, Akhaven P, Chui L, Jamieson F, Louie M, Mazzulli T, Tellier R, Smieja M, Cai W, Chernesky M, Richardson SE; Ontario Laboratory Working Group for the Rapid Diagnosis of Emerging Infections.** 2006. Multicenter comparison of nucleic acid extraction methods for detection of severe acute respiratory syndrome coronavirus RNA in stool specimens. *J Clin Microbiol* **44:**2681–2688.
44. **Beuselinck K, van Ranst M, van Eldere J.** 2005. Automated extraction of viral-pathogen RNA and DNA for high-throughput quantitative real-time PCR. *J Clin Microbiol* **43:** 5541–5546.
45. **Forman M, Wilson A, Valsamakis A.** 2011. Cytomegalovirus DNA quantification using an automated platform for nucleic acid extraction and real-time PCR assay setup. *J Clin Microbiol* **49:**2703–2705.
46. **Raggam RB, Bozic M, Salzer HJ, Hammerschmidt S, Homberg C, Ruzicka K, Kessler HH.** 2010. Rapid quantitation of cytomegalovirus DNA in whole blood by a new molecular assay based on automated sample preparation and real-time PCR. *Med Microbiol Immunol (Berl)* **199:**311–316.
47. **Schuurman T, van Breda A, de Boer R, Kooistra-Smid M, Beld M, Savelkoul P, Boom R.** 2005. Reduced PCR sensitivity due to impaired DNA recovery with the MagNA Pure LC total nucleic acid isolation kit. *J Clin Microbiol* **43:**4616–4622.
48. **Fahle GA, Fischer SH.** 2000. Comparison of six commercial DNA extraction kits for recovery of cytomegalovirus DNA from spiked human specimens. *J Clin Microbiol* **38:**3860–3863.
49. **Kessler HH, Mühlbauer G, Stelzl E, Daghofer E, Santner BI, Marth E.** 2001. Fully automated nucleic acid extraction: MagNA Pure LC. *Clin Chem* **47:**1124–1126.
50. **Rogacs A, Marshall LA, Santiago JG.** 2014. Purification of nucleic acids using isotachophoresis. *J Chromatogr A* **1335:** 105–120.
51. **Sur K, McFall SM, Yeh ET, Jangam SR, Hayden MA, Stroupe SD, Kelso DM.** 2010. Immiscible phase nucleic acid purification eliminates PCR inhibitors with a single pass of

paramagnetic particles through a hydrophobic liquid. *J Mol Diagn* **12:**620–628.

52. **Price CW, Leslie DC, Landers JP.** 2009. Nucleic acid extraction techniques and application to the microchip. *Lab Chip* **9:**2484–2494.
53. **Kemp C, Wojciechowska JM, Esfahani MN, Benazzi G, Shaw KJ, Haswell SJ, Pamme N.** 2012. On-chip processing and DNA extraction from large volume urine samples for the detection of Herpes simplex virus type 2. Presented at the 16th International Conference on Miniaturized Systems for Chemistry and Life Science, Ginowan City Okinawa, Japan.
54. **Shin Y, Perera AP, Wong CC, Park MK.** 2014. Solid phase nucleic acid extraction technique in a microfluidic chip using a novel non-chaotropic agent: dimethyl adipimidate. *Lab Chip* **14:**359–368.

Nucleic Acid Amplification by Polymerase Chain Reaction

ANA MARÍA CÁRDENAS AND KEVIN ALBY

11

Nucleic acid detection methods have been rapidly evolving and play an important role in viral detection and quantification. One technology which emerged in the 1990s, polymerase chain reaction, better known as PCR, has established itself as a primary tool for molecular biology. In fact, this technology has been so widely adopted in the clinical virology laboratory it has, in many cases, completely replaced culture. At its core, PCR is a straightforward chemical reaction whereby one strand of template DNA is exponentially amplified. This chapter will give an overview of different PCR technologies available, as well as the strength and weaknesses of each. For a more in-depth review of their applications in the clinical laboratory, the reader is directed to the appropriate chapter(s) elsewhere in this text.

CONVENTIONAL PCR

Although the detection of PCR products varies widely, the mechanics of the reaction stay the same. First, DNA, which is in a double-stranded state, needs to be denatured. This is typically accomplished by heating the reaction to high temperature such as 95°C. After denaturation, the reaction is cooled to allow for primers to anneal. Primers are short, typically 15 to 30 nucleotides, pieces of DNA that are complementary and specific to the target of interest. There are traditionally 2 primers in any PCR reaction, one designed for the 5′ end of the sequence, and the other designed for the 3′ end. The reaction is cooled to allow the primers to anneal. Depending on the reaction, this temperature may be appropriate for the subsequent extension step. For other reactions, the temperature is changed again to allow for the DNA polymerase to generate a copy of the DNA by extending the primer after it has bound. After each round of extension, the reaction is again heated to allow for the newly synthesized double-stranded DNA fragments to separate from one another, making both old and new strands available for additional rounds of PCR. In this manner, one strand of DNA can turn into millions of copies of the fragment of interest. The more cycles performed, the greater the amplification. Too many cycles, however, increases the risk of a non-specific product being amplified (Fig. 1).

Arguably the most important breakthrough in the overall development of PCR was the discovery of a heat stable DNA polymerase. The temperature required for the DNA polymerase to work depends on the polymerase itself. First used by Kary Mullis, DNA polymerase derived from *Thermus aquaticus*, or Taq for short, has become the standard for PCR amplification (1). The use of Taq allowed for the enzyme to be added to a reaction before the denaturation step, and eliminated the need to add additional enzyme throughout the process. Since the original discovery of the utilization, dozens of other polymerases from other organisms including *Pyrococcus* spp. and *Thermococcus* spp. as well as *Thermus* spp. have been studied and commercialized for use in amplification reactions. Researchers as well as commercial manufacturers have also modified the native Taq polymerase in a number of ways to change the temperature at which the reaction is most efficient, alter the fidelity of the enzyme, and even influence the speed at which the polymerase is capable of copying the DNA. The fidelity of the polymerase determines how many errors are introduced during each cycle of the reaction. Basic low fidelity enzymes introduce one incorrect nucleotide at a rate of about 2 per 10^4 base pairs, whereas high fidelity enzymes only introduce one incorrect nucleotide per 10^6 base pairs (2). For applications where the end point is the nonspecific intercalation of a dye, a loss in fidelity may not be critically important. When the endpoint involves secondary sequence specificity, such as in probe-based detection, a low fidelity enzyme may generate more falsely negative results. Different polymerases also have different rates in which they can copy 1 kb of sequence. For example, Taq polymerase has an extension rate of 1.4 to 4.8 kb per minute, depending on reaction conditions and manufacturer modifications, whereas KOD1 polymerase from *Thermacoccus kodacaraensis* has an extension rate of 6.0 to 7.8 kb per minute (2). With the advent of "rapid" molecular diagnostics, the need for polymerases that have faster rates of extension has grown. Because each of these factors (temperature, fidelity, and speed) can greatly influence the performance and usability of a clinical assay, many commercial assays utilize specific proprietary enzymes for their applications.

While many of the polymerases mentioned are very useful for amplifying DNA, many clinically relevant viruses are RNA viruses. Therefore, in order to detect these viruses or any other RNA template using PCR, a DNA copy must be made first. This is achieved by utilizing a reverse transcriptase

doi:10.1128/9781555819156.ch11

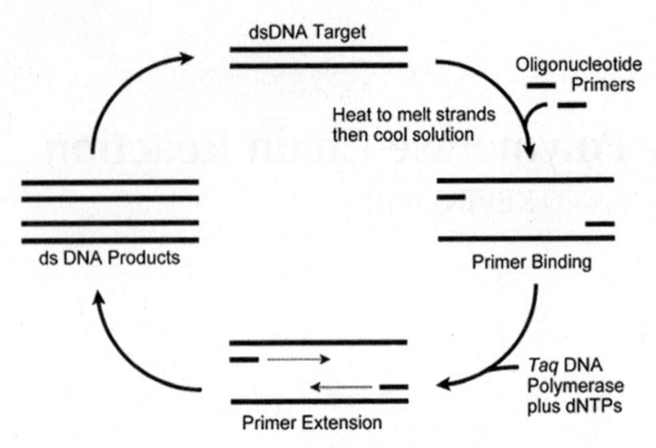

FIGURE 1 Conventional PCR. Start at top, the dsDNA strands are heated causing them to denature. As the solution cools, the two primers anneal to opposite strands of the target DNA (right). The DNA polymerase then extends the primers creating new DNA strands complementary to the target sequence (bottom). The old and new strands serve as templates for further DNA synthesis during the next cycle. dNTPs, deoxynucleoside triphosphates.

enzyme. Commonly used reverse transcriptases come from the Moloney murine leukemia virus (M-MuLV or MMLV), avian myeloblastosis virus (AMV), or the commercially manufactured Superscript family of enzymes from Life Technologies (Carlsbad, CA) (3, 4). Reverse-transcriptase PCR can be done in either a one- or two-step approach. The two-step approach is not commonly used in the clinical laboratory and requires manipulation of the sample and addition of new primers and reagents after the reverse transcriptase step. The one-step approach, however, is commonly used in the diagnostic virology laboratory for everything from arboviruses to zoonotic influenzas (5, 6), and combines both reverse transcription and DNA amplification/detection in the same reaction. Newer two-step approaches combine a "traditional" one-step with a second nested assay or sequencing application. One example of this is sequencing protease and polymerase regions of hepatitis C to determine genotype 1 subtypes (7).

One of the key steps in developing any assay that utilizes PCR is to design appropriate primer pairs. An important consideration in primer design is for the two primers to contain a similar percentage of G/C nucleotides so that they will anneal at a similar rate at a given temperature. This annealing temperature will vary from primer set to primer set, but typically is in the range of 55 to 65°C. Too high of an annealing temperature will decrease the sensitivity of the reaction, whereas too low of an annealing temperature will decrease the specificity of the reaction. This latter feature is used when designing reactions to look for novel or nonidentical targets. Additionally, nonspecific nucleotide analogs can be incorporated into primers to allow for lower stringency reactions, so-called degenerate PCR. This application is especially suitable when querying for something that may have a conserved amino acid sequence but differ at the nucleotide level.

Optimization of PCR conditions is important for the development of a robust assay. Indications of poor optimization are a lack of reproducibility between replicates as well as inefficient and insensitive assays. There are three main approaches for optimization, either changing primer/probe concentrations or reaction mixtures, and/or trying different annealing temperatures. The most common aspect of a reaction mixture that is changed to optimize a PCR reaction is the amount of magnesium added to the reaction. Magnesium serves as an important cofactor for Taq and other polymerases and the amount of Mg^{2+} in the reaction influences the fidelity of the reaction (8). Gradient PCR, as its name indicates, enables testing a range of temperatures simultaneously in a single experiment. It is often used to determine the optimal annealing temperature for a given reaction. To find the optimal annealing temperature of the reaction, a range of temperatures both above and below the calculated Tm of the primers/probes are tested. The PCR master mix stays the same, with each sample having the same concentration of all the ingredients (enzyme, template, primers, etc.). Therefore, the only value that changes is the annealing temperature for each sample. The optimal annealing temperature is the one that results in the largest signal amplitude difference, measured via the intensity of the product bands upon gel electrophoresis, between the positives and the negatives and that also avoids nonspecific amplification.

Though it was the technology that started the move from culture to molecular methods for the detection of viruses, conventional single-plex endpoint PCR has been widely replaced with multiplex PCR, often using real-time detection. This endpoint method is particularly useful if one is simply looking for the presence or absence of a particular gene/mutation in a sample. Conventional PCR also provides the template for gene sequencing, a method used to detect resistance in cytomegalovirus or human immunodeficiency virus for example (9–11). In fact, a conventional PCR still serves as the first step for many emerging technologies (discussed below).

In terms of assay design, instrumentation, and detection, conventional single-plex PCR has several advantages over other amplification methods. Because only one target is being queried in any single reaction, sacrifices to optimize large sets of primers and probes do not need to be made. Thermocyclers used for basic single-plex reactions cost significantly less than cyclers used for other applications, such as real-time PCR (discussed later). Finally, detection methods provide straightforward positive or negative results, using electrophoresis equipment common to any molecular biology laboratory.

Although PCR is quite versatile, it is not necessarily the best method of nucleic acid amplification for all situations. Alternative methods of amplification such as transcription mediated amplification (TMA) or isothermal amplification are discussed elsewhere. Isothermal amplification has particular advantages over PCR in that it can potentially typically be done using less sophisticated equipment because no temperature cycling is needed, making the technology more amenable for products for resource limited settings. Additionally, detection of conventional PCR products may require an additional step, sometimes using carcinogens such as ethidium bromide to visualize the results of the reaction in an agarose gel.

NESTED PCR

Nested PCRs are secondary PCRs that utilize the diluted product of an initial reaction as the template for a second PCR (Fig. 2). By nesting the second PCR within the product of the first PCR, the specificity of the reaction is increased because the second PCR should only work if the specific product is available from the first reaction.

One commonly used commercial product that heavily uses nested PCR is the FilmArray (BioFire Diagnostics, Salt

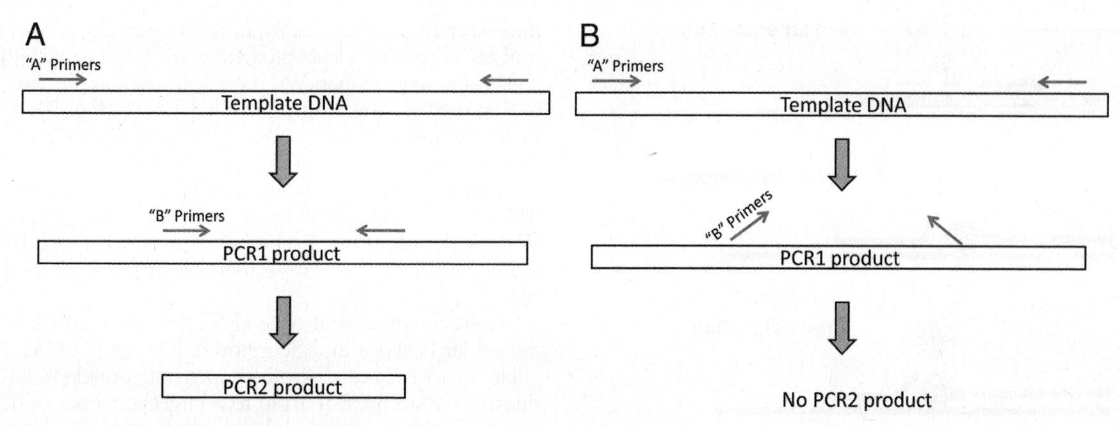

FIGURE 2 Nested PCR. The first primer set is used to amplify the target DNA. The product of this first PCR is diluted and serves as the template for a second PCR reaction. The primers of the second reaction are "nested" more centrally than those of the first reaction. If the first reaction amplified the specific target of interest, the second set of primers anneal and generate a new product (A). If the first reaction generated a nonspecific amplicon, the second set of primers will not anneal and no product will be formed (B).

Lake City, UT). These multiplex panels contain between 20 and 27 targets and provide sample-to-answer analysis. The nested reactions are the key in providing adequate specificity while performing such a large number of reactions simultaneously. Nested PCRs can also be used to distinguish between closely related viruses after amplification of a common region. This approach has been utilized to distinguish BK from JC polyomavirus in the urine samples of transplant recipients (12). This approach is also utilized for the diagnosis of dengue virus infections (13). Nested PCRs are also useful in diagnosing rare/difficult manifestations of infectious diseases. For example, the diagnosis of occult hepatitis B infection is quite difficult and oftentimes controversial using traditional markers of hepatitis B infection (serum viral load, HBsAg, antibody profile). The gold standard molecular method for diagnosis of occult hepatitis B is nested PCR on both a liver biopsy and blood sample (14).

One of the biggest advantages of nested PCR is that it greatly improves the specificity of the reaction. Similar to probe based techniques, a successful second round of amplification requires that the initial product contained the appropriate target sequence. Another benefit of nested PCRs is the variety of detection methods available. The detection of nested PCR products can be achieved via a number of ways including gel electrophoresis, incorporation of intercalating dyes, probe hybridization, and melting curve analysis.

The design of a nested PCR reaction is only as good as the primers it utilizes and the database of strains that generate those primers. Nested PCRs are especially prone to sequence variation, as there are two different sets of primers that may be affected. Recent examples include genetic variation of dengue virus leading to decreased sensitivity (13). Another potential downside of nested PCRs is that they often require a dilution step between the first and second PCRs. This may require the handling of amplified material in the laboratory, which can lead to contamination in the laboratory. Some applications, such as the FilmArray, mitigate this risk by performing the dilution within the assay itself, meaning no amplified material is handled.

REAL-TIME DETECTION OF NUCLEIC ACIDS BY PCR

Real-time PCR is a variation of the conventional PCR technique where nucleic acid amplification is measured as it occurs, rather than collecting results after the reaction is complete. In real-time PCR, the amount of DNA is measured after each cycle, commonly via fluorescent dyes. The increase in fluorescent signal over the course of the reaction is directly proportional to the number of PCR amplicons generated. Data collected in the exponential phase of the reaction gives quantitative information on the starting amount of amplification target. The fluorescent signal is plotted against cycle number to generate an amplification plot that represents the accumulation of product over time. If a particular sequence is abundant in the sample, amplification is observed in earlier cycles; if the sequence is scarce, amplification is observed in later cycles.

Many real-time fluorescent PCR chemistries exist, but the most widely used are 5' nuclease based assays such as TaqMan reporter oligonucleotides, and melting curve analysis utilizing SYBR green dye. In 1995, SYBR green was introduced as being safer and more sensitive than ethidium bromide for staining agarose gels (15). SYBR green is a fluorescent dye that binds to the minor groove of dsDNA. Unbound SYBR green barely fluoresces in solution, but the conformation changes brought about by DNA binding lead to a 100-fold increase in fluorescence intensity. When the reaction is heated after each cycle, the change in fluorescence of the intercalating dye is monitored. It is a simple and inexpensive approach to real-time amplification and detection. Although few commercial technologies utilize SYBR green, it remains a fixture of many laboratory developed tests, including those for commonly encountered viruses such as human coronavirus and West Nile virus (16, 17).

One of the main disadvantages of using an intercalating dye is that SYBR green will bind to any dsDNA in the reaction, including primer-dimers and other nonspecific reaction products. Therefore, when utilizing SYBR green for PCR, the procedure must be carefully optimized to prevent mispriming and primer dimer formation. Additionally, the use of intercalating dyes in a real-time application limits the ability to multiplex as the product which is being amplified may not be discernable. Multiplexing can only be achieved by dye-based methods if the products have different characteristics such as size or a melting point that can be independently resolved, since the dye itself intercalates nonspecifically.

Another widely used technology is based on TaqMan reporter oligonucleotides. Detection by this method employs

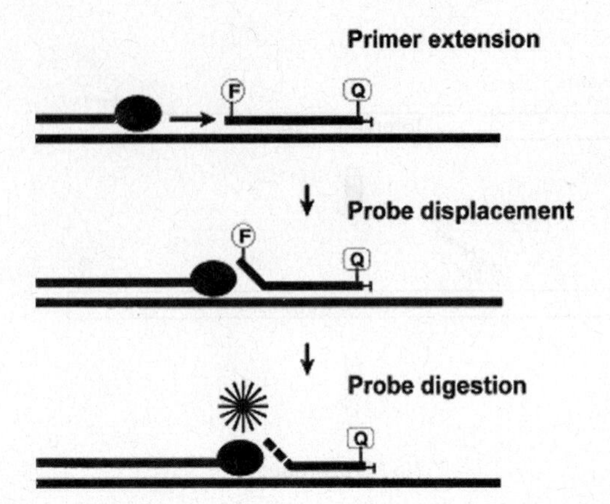

FIGURE 3 TaqMan probes. As the upstream primer is extended, the DNA polymerase (black oval) reaches the probe, which is displaced and digested by the 5′ to 3′ nuclease activity of the polymerase. Once the reporter and quencher molecules are separated an increase in fluorescence is observed from the reporter molecule.

two primers and an internal reporter oligonucleotide known as a probe. The probe has a fluorescent reporter molecule coupled to the 5′ end and a quencher molecule at the 3′ end (18). When the probe is intact, the proximity of the quencher molecule suppresses the fluorescence of the reporter molecule (19). This phenomenon is referred to as fluorescence resonance energy transfer (FRET), where the fluorescence of the fluorophore is quenched by energy transfer when in close proximity to a quencher (20). If complementary sequence is present in the sample, the probe will anneal to the single stranded DNA (Fig. 3). During PCR, the reporter probe is displaced and digested using the exonuclease activity of the Taq polymerase, separating the reporter and quencher molecules. This causes an increase in fluorescence (21) and allows extension to continue normally (Figure 3). Additional probes are cleaved with each thermal cycle, causing a cumulative increase in fluorescence intensity.

There are other real-time FRET systems available such as molecular beacons and Scorpion molecules (22–24). Molecular beacons are hairpin-shaped oligonucleotides that fluoresce upon hybridization to a target sequence. The loop sequence is the probe, and the stem is formed by annealing of complementary arm sequences located on either side of the probe sequence. A fluorophore is linked to the end of one arm and a quencher to the end of the other arm. In the absence of target, the stem keeps the fluorophore and quencher in close proximity causing quenching by energy transfer. However, when they hybridize to a target strand they undergo a conformational change that enables fluorescence to be detected (Fig. 4A). Molecular beacons have been used to detect both RNA and DNA viruses such as respiratory pathogens, West Nile virus, and hepatitis B (23).

Scorpion molecules are similar to molecular beacons in that they are also stem loop structures with a fluorophore on one end and a quencher on the other end of the stem loop. The difference is that they are bi-functional molecules that carry both probe and priming functions on the same

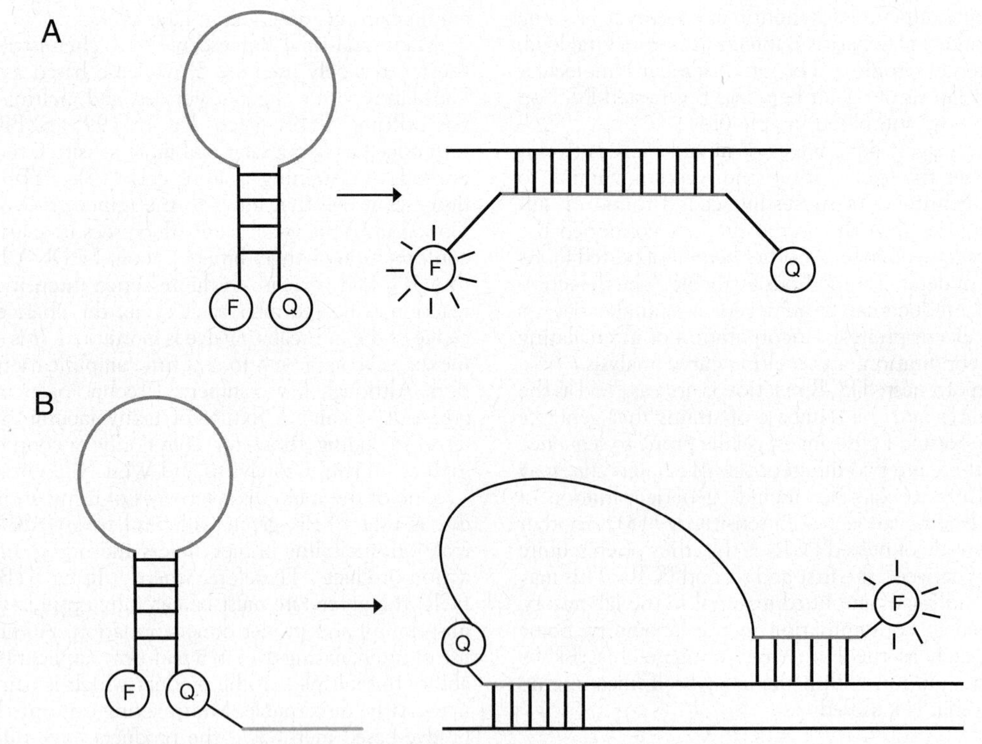

FIGURE 4 Molecular beacon and Scorpion molecules. Molecular beacons are stem cell loop structures where the ends are self-complementary and the center portion is complementary to the target sequence. In the presence of a target sequence, the reporter and quencher molecules are separated and fluorescence is detected (A). Scorpion molecules are also stem-cell loop structures but have a primer added to their 5′ end. After initial extension of the target sequence, the loop unfolds and the loop-region of the probe hybridizes to the newly synthesized target sequence. Since the reporter is no longer in close proximity to the quencher, fluorescence emission takes place (B).

oligonucleotide construct. The primer element is attached to the 5′ end (Fig. 4B). For more information see (25).

The most common applications of real-time PCR are pathogen detection and viral quantitation in clinical specimens. One of the main advantages of real-time PCR is the ease in which a quantitative answer can be obtained. By generating a standard curve and using cycle threshold values, the quantity of an unknown target can be determined. Another major advantage of real-time PCR over conventional PCR is an increased dynamic (linear) detection range. It is not unusual for viral quantitative assays to detect from tens to hundreds of copies/ml up to 10^9 copies/ml. Quantitative real-time PCR (qPCR) is used in a myriad of viral infections to not only aid in determining patient prognosis, but for monitoring therapeutic response in HIV positive patients receiving antiretroviral therapy, for example. They are also widely used in transplant recipients for determining CMV, EBV, BKV, and HHV-6 viral loads over time. For an in-depth discussion of quantitative PCR, please see chapter 13.

Real-time PCR is not only used quantitatively. Many real-time-PCR assays are performed for qualitative results since they require less sample manipulation in comparison to conventional PCR and allow for faster turnaround times having no post-amplification processing. Less sample and amplicon manipulation also leads to less contamination and laboratory errors. Probe based real-time PCR also offers increased specificity over conventional end-point PCR, as a product must not only be generated but the probe must be able to anneal to the amplicon in order for a signal to be detected. These are some of the reasons why real-time based PCR is the staple of many commercial and laboratory developed methods.

There are many different types and models of real-time PCR instruments available but each must have an excitation source, which excites the fluorescent dye(s), a detector to detect the fluorescent emission(s), and a thermal cycler to control the temperature at each step of the reaction. More advanced models allow for additional manipulations such as melting point determination or gradient cycling. The type of instrument needed for any one particular assay can vary by laboratory based on other needs. While there exist a multitude of commercial kits available for use with traditional real-time instruments, a number of manufacturers have started to develop assays for use with specialty instruments that frequently do all of the necessary processing steps, in addition to the real-time amplification and detection. Examples of these platforms include the Cepheid GeneXpert (Cepheid, Sunnyvale, CA) and Focus Simplexa Direct (Focus Diagnostics, Cypress, CA). These types of platforms allow for the decentralization of molecular testing. Offering speed and ease of use, these assays still perform comparably to traditional molecular assays (26, 27). This approach will likely continue to expand the footprint and accessibility of molecular testing in laboratory medicine. In addition to commercially available assays, numerous commercial companies have developed real-time PCR analyte specific reagents (ASRs) for many commonly encountered viral pathogens. This reduces the development time necessary for new assays while allowing laboratories to continue to use the equipment for molecular diagnostics that they already have. The performance characteristics of ASRs tend to vary compared to tests using laboratory developed protocols (28, 29). It is therefore important that each ASR is carefully evaluated by the clinical virology laboratory before put into use.

Multiplex PCR

Probe based real-time PCR also allows for assay multiplexing in which there is amplification and specific detection of two or more nucleic acid target sequences in the same reaction. The most common type of multiplex is a duplex, but higher-order multiplexes are also run. A common application of multiplex pathogen detection is PCR assays for herpes simplex viruses that can differentiate between HSV-1 and HSV-2 in the same reaction. Typically, in multiplex reactions separate primers and probes are used for each reaction. The use of differentially labelled probes is what allows for detection of multiple targets in the same reaction.

Benefits to multiplexing include increased throughput, since more samples and targets can be assayed per reaction, reduced sample usage, and reduced reagent usage. It also allows for the use of an internal control within the reaction to check for inhibition or other problems with the reaction. This is a key feature of many commercial assays and provides added assurance that negative results are not the result of technical error or artifact. The choice of quencher for each probe in a multiplexed assay becomes more important as the number of probes being multiplexed increases. There are both fluorescent quenchers such as TAMRA, and dark quenchers such as BHQ or QSY. Fluorescent quenchers work by releasing the energy of the fluorophore at a different wavelength while dark quenchers release energy in the form of heat rather than fluorescence, keeping the overall fluorescent background lower. They are also useful to increase the number of targets in a reaction, since a quencher is not occupying a usable channel.

Although multiplexing has a number of benefits, drawbacks exist as well. They are more difficult assays to design as only a single set of cycling parameters is used. This may require selecting a different primer target to make sure all primer/probe combinations have similar melting and annealing profiles. If this is not done properly, false-positive results occur due to less specific binding or false-negative results occur due to inappropriate annealing temperatures. The greater the number of targets multiplexed, the more difficult this process becomes. As mentioned previously, the ability to multiplex is also limited by the detection method. Conventional multiplex detection requires amplicons of different sizes or G/C content. Real-time detection is limited by the ability to separate fluorescent signals. Target probe or microarray based detection methods require additional steps and cannot be performed in real-time.

There are numerous examples of commercial multiplex assays in both plate based and so called "sample-to-answer" formats. Plate based multiplex assays are more useful for batching large numbers of specimens while sample-to-answer formats allow for testing outside of the molecular laboratory. Additionally, the development of alternatives to real-time detection has allowed for the development of large, massively multiplexed panels that test ten or more different targets simultaneously. This approach, sometimes termed "syndromic testing" has been particularly impactful in the virology laboratory as it pertains to respiratory virus testing. Examples of commonly used assays are the Luminex RVP (Luminex, Austin, TX), the BioFire FilmArray Respiratory Panel (bioMerieux, Durham, NC), the Genmark eSensor RVP (Genmark Diagnostics, Carlsbad, CA), and the Nanosphere RP Flex (Nanosphere,Northbrook, IL). These panels allow not only for the detection of influenza and respiratory syncytial virus, but also parainfluenza, rhinovirus, enterovirus, coronavirus, adenovirus, and human metapneumovirus using the

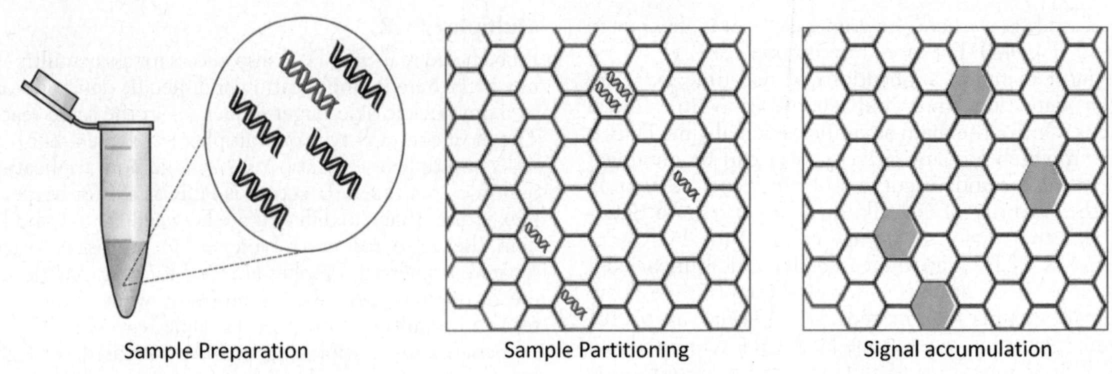

FIGURE 5 Digital PCR. A sample of nucleic acid containing the target sequence (red) is partitioned into many individual cells. The partitions containing the target sequence are amplified causing signal to accumulate. The fraction of positive partitions is determined and the target concentration estimated.

same assay. These assays show high, but not equivalent, sensitivity/specificity compared to single-plex molecular assays (30–32). One concern for the use of these panels is whether or not a full panel is needed for every patient, especially as these tests can be significantly more expensive for both the patient and the laboratory than targeted assays (33).

EMERGING TECHNIQUES

PCR-ESI/MALDI

A number of technologies are being developed to couple PCR with sophisticated detection methods to increase the utility of the technology. One such application is that of PCR followed by mass spectrometry. Two methods of mass spectrometry have been used to analyze PCR products, electrospray ionization (ESI) and matrix assisted laser/desorption ionization time of flight (MALDI-TOF). The major advantage of these methods is their ability to provide additional information about the PCR product. In the case of PCR-ESI, the mass of the product is specific such that the base composition can be determined. For PCR-MALDI, weighted nucleotides are used to determine the sequence of a product. Although these detection methods are quite sophisticated, both utilize conventional PCR as a starting point. In PCR-ESI, the PCR product is nebulized via solvent evaporation in the presence of an electric field. This generates ions of single stranded DNA molecules. The mass of these molecules is directly proportional to the nucleotide composition. By analyzing multiple primer sets, the identity of the pathogen can be deduced. This approach has been used for respiratory infections, demonstrating the ability to detect commonly encountered pathogens such influenza or RSV, but also emerging pathogens such as SARS coronavirus (30, 31). In PCR-MALDI, products are mixed with an organic matrix and then excited with laser energy. This excitation results in the formation of ions, which then migrate through a drift field according to their size and charge. The incorporation of nonextendable weighted nucleotides and resulting mass shifts allows for the identification of the

TABLE 1 Comparison of different PCR techniques

	Conventional PCR	Real-time PCR	Digital PCR
Overview	Detects PCR product at the end of the reaction	Detects PCR amplification in real-time	Measures fraction of negative droplets to calculate absolute copies
Quantitative	Not routinely	Can be if standard curve is used	Yes, utilizes a Poisson statistical algorithm
Applications	Qualitative pathogen detection Amplification of DNA for sequencing Amplification of DNA for mass spectrometry analysis	Qualitative pathogen detection Quantification of viral load	Absolute quantification of viral load without standards
Strengths	Simple design Lower cost compared to other molecular methods Ability to be used in multiple different downstream applications	Increased dynamic range No post-amplification processing Fluorescence based discrimination of products Ability to be used in multiple different downstream applications	Does not rely on references or standards Increased precision Higher tolerance to inhibitors Detects small fold-change differences
Weaknesses	Post-amplification processing needed Poor precision Low resolution Narrow dynamic range	Depends on standard curve for analysis Limited number of targets in any one reaction	Limited sensitivity Low throughput High cost

nucleotide sequence. Most clinically available solutions harness the power of these technologies by using them as a way to offer a limited bias approach to diagnostics using large numbers of broad range primers either to detect large groups of pathogens (e.g., bacteria) or heavily query a specific one (e.g., influenza). Studies have demonstrated the ability of PCR-ESI to rapidly identify and differentiate influenza strains into clades and subtypes. This analysis can be performed directly from patient sample with highly accurate results, further reducing the need to maintain culture for epidemiologic purposes (32). PCR-MALDI has been successfully used to differentiate closely related viruses such as enteroviruses using a limited number of PCR reactions (33). Similarly, PCR-MALDI has also been used to detect and characterize hepatitis B virus mutations that arise during antiviral therapy (34). With the development of new direct acting antiviral agents for common viral infections such as hepatitis C, the need for these types of assays may increase in the future. The biggest challenge to widespread adoption of these techniques, however, will remain their ability to provide a cost-benefit ratio that exceeds that of other emerging technologies such as next generation sequencing.

Digital PCR (dPCR)

Digital PCR was originally described in 1992 (35) as an alternative method to nucleic acid detection and quantification that uses molecular counting. In a typical digital PCR experiment, a sample of nucleic acid is randomly partitioned into many individual parallel reactions. Some of these partitions will contain one or more template copies while others will have none. During amplification, TaqMan chemistry is utilized to detect sequence specific targets as in real-time PCR. If no target is present then no signal will accumulate (Fig. 5). The partitions are PCR amplified simultaneously and read using a droplet reader to determine the fraction of positive partitions. The target concentration is estimated by modeling as a Poisson distribution:

$$\text{Copies per droplet} = -\ln(1-p); \quad p = \text{fraction of positive droplets}$$

where the numbers of positive and negative droplets are used to calculate the concentration of the target sequence and their Poisson-based 95% confidence intervals.

An advantage of digital PCR over real-time PCR is that the fraction of negative reactions is used to generate an absolute count of the number of target molecules in the sample without the need for references or standards. For more information about dPCR see chapter 13.

CONCLUSIONS

The impact of PCR on the clinical virology field has been immense. It has become a key player not only for the diagnosis of viral diseases, but also on detecting drug resistance, aiding in management of patients on antiviral therapies, and monitoring of immunocompromised individuals such as transplant recipients. Rapid time-to-result of PCR has also improved patient management, ultimately reducing length of hospital stay. In addition, PCR provides rapid diagnosis with improved sensitivity when dealing with viruses that otherwise would be dangerous to amplify in culture such as Ebola, and has been used to help contain the deadly epidemics that have occurred.

PCR has a great amount of flexibility in the different types of assays available (Table 1). Each application comes with its own strengths and weaknesses, but together allow for a robust method of detecting and investigating pathogens. Looking forward, advances in PCR chemistry will allow for the continued improvement of diagnostic assays not only in accuracy, but also in speed and efficiency. This will allow for the continued development of sample-to-answer PCR tests for the clinical laboratory, which will help foster broader use of these tests. Furthermore, the coupling of PCR techniques to new detection methods will allow reduction in bias and contribute to the discovery of novel viruses that affect humans.

REFERENCES

1. **Saiki RK, Gelfand DH, Stoffel S, Scharf SJ, Higuchi R, Horn GT, Mullis KB, Erlich HA.** 1988. Primer-directed enzymatic amplification of DNA with a thermostable DNA polymerase. *Science* **239:**487–491.
2. **Terpe K.** 2013. Overview of thermostable DNA polymerases for classical PCR applications: from molecular and biochemical fundamentals to commercial systems. *Appl Microbiol Biotechnol* **97:**10243–10254.
3. **Fuchs B, Zhang K, Rock MG, Bolander ME, Sarkar G.** 1999. High temperature cDNA synthesis by AMV reverse transcriptase improves the specificity of PCR. *Mol Biotechnol* **12:**237–240.
4. **Roth MJ, Tanese N, Goff SP.** 1985. Purification and characterization of murine retroviral reverse transcriptase expressed in Escherichia coli. *J Biol Chem* **260:**9326–9335.
5. **Chen Y, Liu T, Cai L, Du H, Li M.** 2013. A one-step RT-PCR array for detection and differentiation of zoonotic influenza viruses H5N1, H9N2, and H1N1. *J Clin Lab Anal* **27:**450–460.
6. **Zink SD, Jones SA, Maffei JG, Kramer LD.** 2013. Quadraplex qRT-PCR assay for the simultaneous detection of Eastern equine encephalitis virus and West Nile virus. *Diagn Microbiol Infect Dis* **77:**129–132.
7. **Koletzki D, Pattery T, Fevery B, Vanhooren L, Stuyver LJ.** 2013. Amplification and sequencing of the hepatitis C virus NS3/4A protease and the NS5B polymerase regions for genotypic resistance detection of clinical isolates of subtypes 1a and 1b. *Methods Mol Biol* **1030:**137–149.
8. **Eckert KA, Kunkel TA.** 1990. High fidelity DNA synthesis by the Thermus aquaticus DNA polymerase. *Nucleic Acids Res* **18:**3739–3744.
9. **Hall Sedlak R, Castor J, Butler-Wu SM, Chan E, Cook L, Limaye AP, Jerome KR.** 2013. Rapid detection of human cytomegalovirus UL97 and UL54 mutations directly from patient samples. *J Clin Microbiol* **51:**2354–2359.
10. **Kim YJ, Boeckh M, Cook L, Stempel H, Jerome KR, Boucek R Jr, Burroughs L, Englund JA.** 2012. Cytomegalovirus infection and ganciclovir resistance caused by UL97 mutations in pediatric transplant recipients. *Transpl Infect Dis* **14:**611–617.
11. **Stürmer M, Berger A, Preiser W.** 2004. HIV-1 genotyping: comparison of two commercially available assays. *Expert Rev Mol Diagn* **4:**281–291.
12. **Comerlato J, Campos FS, Oliveira MT, Cibulski SP, Corrêa L, Kulmann MI, Arantes TS, Hentges LP, Spilki FR, Roehe PM, Franco AC.** 2015. Molecular detection and characterization of BK and JC polyomaviruses in urine samples of renal transplant patients in Southern Brazil. *J Med Virol* **87:**522–528.
13. **Klungthong C, Manasatienkij W, Phonpakobsin T, Chinnawirotpisan P, Rodpradit P, Hussem K, Thaisomboonsuk B, Ong-ajchaowlerd P, Nisalak A, Kalayanarooj S, Buddhari D, Gibbons RV, Jarman RG, Yoon IK, Fernandez S.** 2015. Monitoring and improving the sensitivity of dengue nested RT-PCR used in longitudinal surveillance in Thailand. *J Clin Virol* **63:**25–31.
14. **Raimondo G, Navarra G, Mondello S, Costantino L, Colloredo G, Cucinotta E, Di Vita G, Scisca C, Squadrito G,**

Pollicino T. 2008. Occult hepatitis B virus in liver tissue of individuals without hepatic disease. *J Hepatol* **48:**743–746.
15. **Singer VL, Lawlor TE, Yue S.** 1999. Comparison of SYBR Green I nucleic acid gel stain mutagenicity and ethidium bromide mutagenicity in the Salmonella/mammalian microsome reverse mutation assay (Ames test). *Mutat Res* **439:** 37–47.
16. **Goka EA, Vallely PJ, Mutton KJ, Klapper PE.** 2015. Pan-human coronavirus and human bocavirus SYBR Green and TaqMan PCR assays; use in studying influenza A viruses co-infection and risk of hospitalization. *Infection* **43:**185–192.
17. **Kumar JS, Saxena D, Parida M.** 2014. Development and comparative evaluation of SYBR Green I-based one-step real-time RT-PCR assay for detection and quantification of West Nile virus in human patients. *Mol Cell Probes* **28:**221–227.
18. **Livak KJ, Flood SJ, Marmaro J, Giusti W, Deetz K.** 1995. Oligonucleotides with fluorescent dyes at opposite ends provide a quenched probe system useful for detecting PCR product and nucleic acid hybridization. *PCR Methods Appl* **4:**357–362.
19. **Holland PM, Abramson RD, Watson R, Gelfand DH.** 1991. Detection of specific polymerase chain reaction product by utilizing the 5′——3′ exonuclease activity of Thermus aquaticus DNA polymerase. *Proc Natl Acad Sci USA* **88:**7276–7280.
20. **Cardullo RA, Agrawal S, Flores C, Zamecnik PC, Wolf DE.** 1988. Detection of nucleic acid hybridization by nonradiative fluorescence resonance energy transfer. *Proc Natl Acad Sci USA* **85:**8790–8794.
21. **Lee LG, Connell CR, Bloch W.** 1993. Allelic discrimination by nick-translation PCR with fluorogenic probes. *Nucleic Acids Res* **21:**3761–3766.
22. **Tyagi S, Kramer FR.** 1996. Molecular beacons: probes that fluoresce upon hybridization. *Nat Biotechnol* **14:**303–308.
23. **Tyagi S, Kramer FR.** 2012. Molecular beacons in diagnostics. *F1000 Med Rep* **4:**10.
24. **Whitcombe D, Theaker J, Guy SP, Brown T, Little S.** 1999. Detection of PCR products using self-probing amplicons and fluorescence. *Nat Biotechnol* **17:**804–807.
25. **Carters R, Ferguson J, Gaut R, Ravetto P, Thelwell N, Whitcombe D.** 2008. Design and use of scorpions fluorescent signaling molecules. *Methods Mol Biol* **429:**99–115.
26. **Binnicker MJ, Espy MJ, Irish CL.** 2014. Rapid and direct detection of herpes simplex virus in cerebrospinal fluid by use of a commercial real-time PCR assay. *J Clin Microbiol* **52:**4361–4362.
27. **Popowitch EB, Rogers E, Miller MB.** 2011. Retrospective and prospective verification of the Cepheid Xpert influenza virus assay. *J Clin Microbiol* **49:**3368–3369.
28. **Sails AD, Saunders D, Airs S, Roberts D, Eltringham G, Magee JG.** 2009. Evaluation of the Cepheid respiratory syncytial virus and influenza virus A/B real-time PCR analyte specific reagent. *J Virol Methods* **162:**88–90.
29. **Stellrecht KA, Espino AA, Nattanmai SM, Jackson WF, Conti DJ.** 2013. Comparison of three real-time PCR for the quantification of polyomavirus BK. *J Clin Virol* **56:**354–359.
30. **Pierce VM, Elkan M, Leet M, McGowan KL, Hodinka RL.** 2012. Comparison of the Idaho Technology FilmArray system to real-time PCR for detection of respiratory pathogens in children. *J Clin Microbiol* **50:**364–371.
31. **Pierce VM, Hodinka RL.** 2012. Comparison of the GenMark Diagnostics eSensor respiratory viral panel to real-time PCR for detection of respiratory viruses in children. *J Clin Microbiol* **50:**3458–3465.
32. **Popowitch EB, O'Neill SS, Miller MB.** 2013. Comparison of the Biofire FilmArray RP, Genmark eSensor RVP, Luminex xTAG RVPv1, and Luminex xTAG RVP fast multiplex assays for detection of respiratory viruses. *J Clin Microbiol* **51:**1528–1533.
33. **Schreckenberger PC, McAdam AJ.** 2015. Point-Counterpoint: Large Multiplex PCR Panels Should Be First-Line Tests for Detection of Respiratory and Intestinal Pathogens. *J Clin Microbiol* **53:**3110–3115.
34. **Chen KF, Rothman RE, Ramachandran P, Blyn L, Sampath R, Ecker DJ, Valsamakis A, Gaydos CA.** 2011. Rapid identification viruses from nasal pharyngeal aspirates in acute viral respiratory infections by RT-PCR and electrospray ionization mass spectrometry. *J Virol Methods* **173:**60–66.
35. **Sampath R, Hofstadler SA, Blyn LB, Eshoo MW, Hall TA, Massire C, Levene HM, Hannis JC, Harrell PM, Neuman B, Buchmeier MJ, Jiang Y, Ranken R, Drader JJ, Samant V, Griffey RH, McNeil JA, Crooke ST, Ecker DJ.** 2005. Rapid identification of emerging pathogens: coronavirus. *Emerg Infect Dis* **11:**373–379.
36. **Sampath R, Russell KL, Massire C, Eshoo MW, Harpin V, Blyn LB, Melton R, Ivy C, Pennella T, Li F, Levene H, Hall TA, Libby B, Fan N, Walcott DJ, Ranken R, Pear M, Schink A, Gutierrez J, Drader J, Moore D, Metzgar D, Addington L, Rothman R, Gaydos CA, Yang S, St George K, Fuschino ME, Dean AB, Stallknecht DE, Goekjian G, Yingst S, Monteville M, Saad MD, Whitehouse CA, Baldwin C, Rudnick KH, Hofstadler SA, Lemon SM, Ecker DJ.** 2007. Global surveillance of emerging Influenza virus genotypes by mass spectrometry. *PLoS One* **2:**e489.
37. **Piao J, Jiang J, Xu B, Wang X, Guan Y, Wu W, Liu L, Zhang Y, Huang X, Wang P, Zhao J, Kang X, Jiang H, Cao Y, Zheng Y, Jiang Y, Li Y, Yang Y, Chen W.** 2012. Simultaneous detection and identification of enteric viruses by PCR-mass assay. *PLoS One* **7:**e42251.
38. **Luan J, Yuan J, Li X, Jin S, Yu L, Liao M, Zhang H, Xu C, He Q, Wen B, Zhong X, Chen X, Chan HL, Sung JJ, Zhou B, Ding C.** 2009. Multiplex detection of 60 hepatitis B virus variants by maldi-tof mass spectrometry. *Clin Chem* **55:**1503–1509.
39. **Sykes PJ, Neoh SH, Brisco MJ, Hughes E, Condon J, Morley AA.** 1992. Quantitation of targets for PCR by use of limiting dilution. *Biotechniques* **13:**444–449.

Isothermal Nucleic Acid Amplification Methods

HARALD H. KESSLER AND EVELYN STELZL

12

In clinical virology, molecular diagnostics based on the direct detection of specific genetic material in a specimen through nucleic acid testing (NAT) has largely replaced antigen testing by immunoassays and has become the leading technology. Molecular test systems are more specific and sensitive. They are able to detect the presence of a pathogen earlier than an antigen immunoassay and are thus mainly used for diagnosing and screening patients for numerous viral diseases today.

Among NAT techniques, the polymerase chain reaction (PCR) is the most widely used tool for qualitative and quantitative detection of viral nucleic acids present in clinical specimens. Today, in the clinical virology laboratory, molecular test systems based on real-time polymerase chain reaction (qPCR) have largely replaced those based on end-point PCR. Numerous qPCR and reverse transcription (RT)-qPCR assays have been developed to allow the identification of DNA and RNA viruses. For all PCR-based assays, nucleic acid amplification through a thermal cycling program with specific temperatures is common. In contrast, isothermal nucleic acid amplification assays require only a single optimized reaction temperature. A large number of techniques based on isothermal nucleic acid amplification have been developed; however, just a few are used in the clinical virology laboratory. This chapter highlights those techniques that have been shown to be useful for molecular diagnostics in the clinical virology laboratory.

TRANSCRIPTION-BASED AMPLIFICATION METHODS: NUCLEIC ACID SEQUENCE-BASED AMPLIFICATION (NASBA) AND TRANSCRIPTION MEDIATED AMPLIFICATION (TMA)

Technique:

Nucleic acid sequence-based amplification (NASBA) and transcription mediated amplification (TMA) are isothermal RNA amplification methods that mimic retroviral replication (1–3). Both methods use a combination of multiple enzymes: NASBA employs three enzymes (avian myeloblastosis virus reverse transcriptase/DNA polymerase, T7 RNA polymerase, and a separate RNase H), whereas TMA makes use of the intrinsic RNaseH activity of the reverse transcriptase. The polymerization process generates RNA amplification products; these have the theoretical advantage of reducing the possibility of carry-over contamination since RNA is less stable than DNA. The methods use a primer that contains target-specific as well as T7 RNA polymerase promoter sequences. Following hybridization of this primer to the target RNA, reverse transcriptase generates a cDNA copy of the target RNA, the resulting RNA:DNA duplex is degraded by RNaseH activity and a second primer binds to the cDNA. A new strand of DNA is synthesized from the end of this primer by reverse transcriptase, generating a double-stranded DNA molecule with a T7 RNA polymerase promoter at the beginning. The T7 RNA polymerase recognizes the promoter sequence and initiates transcription. Each of the newly synthesized RNA amplification products re-enters the amplification process and serves as a template for a new round of replication leading to an exponential expansion of the RNA amplification products. A target-specific probe using a real-time format detects the amplification products generated in these reactions. In contrast to RT-PCR, amplification by NASBA and TMA works at isothermal conditions, usually at a constant temperature of 41°C.

Application of NASBA in the Clinical Virology Laboratory

Assays for the clinical virology laboratory based on the NASBA technology have been developed by bioMerieux (Marcy l'Etoile, France). Currently, the NucliSENS EasyQ HIV-1 v2.0 and HPV kits are commercially available (Table 1). Furthermore, bioMerieux provides the NucliSENS EasyQ basic kit, allowing for establishment of laboratory developed assays based on the NASBA technology. Following either manual nucleic acid extraction with the NucliSENS miniMAG kit or automated nucleic acid extraction on the NucliSENS easyMAG platform, NASBA is performed on the NucliSENS EasyQ platform.

The NucliSENS easyQ HIV-1 v2.0 kit has been designed for detection and quantitation of human immunodeficiency virus type 1 (HIV-1) RNA in plasma, or dried blood spot specimens. This assay has been *in vitro* diagnostics (IVD)/Conformite Europeene (CE)-marked. bioMerieux was the first manufacturer to offer CE/IVD marking for extraction and quantitation of HIV-1 RNA in dried blood spot specimens with the NucliSENS easyMAG and NucliSENS EasyQ platforms. The NucliSENS EasyQ HIV-1 v2.0 kit targets the HIV-1 p24 *gag* gene. Quantitation is achieved

doi:10.1128/9781555819156.ch12

TABLE 1 Isothermal nucleic acid amplification methods: Manufacturers and commercially available kits for detection and/or quantitation of viral DNA or RNA (updated August 2015)

Method	Manufacturer	Kits commercially available
NASBA	bioMerieux (Marcy l'Etoile, France)	NucliSENS easyQ HIV-1 v2.0; NucliSENS easyQ HPV
TMA	Hologic (Bedford, MA)	Procleix Ultrio; Procleix Ultrio Plus; Procleix Ultrio Elite; Procleix HEV; Procleix Parvo/HAV; Procleix WNV; Aptima HIV-1 Quant Dx; Aptima HPV
SDA	BD (Franklin Lakes, NJ)	BD ProbeTec Herpes Simplex Viruses (HSV 1 & 2) Q^x Amplified DNA Assays
LAMP / Q-LAMP	DiaSorin (Saluggia, Italy)	Iam BKV; Iam HSV1&2; Iam VZV; Iam CMV; Iam Toxo; Iam Parvo
	Meridian Bioscience (Cincinnati, OH)	*illumi*gene® HSV 1&2
NEAR	Alere (Waltham, MA)	Alere I Influenza A & B
RPA	None	None

NASBA, nucleic acid sequence-based amplification; TMA, transcription mediated amplification; SDA, strand displacement amplification; LAMP, loop-mediated isothermal amplification; Q-LAMP, quantitative loop-mediated isothermal amplification; NEAR, nicking endonuclease amplification reaction; RPA, recombinase polymerase amplification.

through an individual, internal calibrator that is already added at the sample lysis step (and serves as internal control in parallel). The calibrator is extracted simultaneously and co-amplified on the NucliSENS EasyQ platform. According to the manufacturer, the linear range of the NucliSENS easyQ® HIV-1 v2.0 kit is 2.5×10^1 to 1.0×10^7 copies/ml when a plasma input volume of 1 ml is used. When a total volume of 100 µL of a re-suspended dried blood spot (whole blood) is used, the lower limit of quantitation is 800 copies/ml. The turnaround time from sample to result is less than 3 hours per run, with less than 1 hour hands-on time. The clinical performance of the NucliSENS easyQ HIV-1 v2.0 kit has been investigated in several studies. While plasma HIV-1 subtype B concentrations were found to be comparable to alternative assays including the m2000rt RealTime HIV-1 (Abbott Laboratories, Abbott Park, IL) and the COBAS AmpliPrep/COBAS TaqMan HIV-1 Test v2.0 (Roche Molecular Diagnostics, Pleasanton, CA), discrepant results were obtained when samples containing HIV-1 non B subtypes and those containing circulating recombinant forms were tested (4–8). For quantitation of HIV-1 RNA in dried blood spot specimens, the NucliSENS easyQ HIV-1 v2.0 kit has been shown to be useful providing reliable results (9, 10). Recently, a superior specificity of this assay was shown with this kind of specimens when compared to the Roche COBAS AmpliPrep/COBAS TaqMan HIV-1 Test v2.0 (11).

The IVD/CE-labeled NucliSENS easyQ HPV kit has been designed to detect the expression of human papillomavirus (HPV) oncogenic factors E6 and E7 from the five most prevalent HPV genotypes in cervical cancer. For minor cervical cytology, this mRNA-based assay was found to provide a higher specificity but lower sensitivity when compared to DNA-based assays and was proposed to serve as a triage test (12–14). For anal dysplasia in HIV-positive men who have sex with men, the NucliSENS easyQ HPV kit was found to have an increased specificity and positive predicted value but similar or lower sensitivity when compared to DNA-based assays (15, 16).

Furthermore, the NASBA technique has been used for laboratory developed test systems for detection of viruses including human bocavirus, SARS coronavirus, enteroviruses, and rhinoviruses (17–20).

Application of TMA in the Clinical Virology Laboratory

Assays for the clinical virology laboratory based on the TMA technology have originally been developed by Gen-Probe. In 2012, Hologic (Bedford, MA) announced its successful acquisition of Gen-Probe, the latter being a wholly-owned subsidiary of Hologic since then (Table 1).

In order to increase safety on donated blood, the TMA-based Procleix assays have been developed. The Procleix assay procedure consists of three main steps—target capture, amplification, and detection—and is performed either on the Panther or Tigris systems. Both CE-marked platforms are fully-automated, no pretreatment or manual sample handling is required. Capture probes hybridize simultaneously with the internal control and viral nucleic acids and bind them to magnetic particles. After amplification, the dual kinetic assay technology allows for the simultaneous detection of both IC-encoded and viral-encoded RNA. The classic Procleix Ultrio assay has been designed to simultaneously detect hepatitis B virus (HBV) DNA, hepatitis C virus (HCV) RNA, and HIV-1 RNA in donated blood and has been available worldwide since 2006. Following IVD/CE-labeling, the assay was FDA-approved for use on the Tigris system in 2007. The Procleix Ultrio Plus assay has been designed to enhance the Ultrio assay by improving the sensitivity for HBV and can be used for plasma and serum specimens. It was CE-marked in 2009 and approved by the FDA in 2012. When the Procleix Ultrio assay was compared with the Procleix Ultrio Plus assay using dilution panels of WHO international standards, both assays showed an equal analytical sensitivity for the detection of the WHO HCV subtype 1 and the WHO HIV-1 subtype B standards; however, the Procleix Ultrio Plus assay was found to have a 3-fold higher analytical sensitivity to detect HBV DNA than the Procleix Ultrio assay (21). Those results were confirmed in other studies showing that the Procleix Ultrio Plus assay was found to have a significantly higher clinical sensitivity to detect HBV DNA when compared to the Procleix Ultrio assay while HCV and HIV-1 RNA detection were not compromised (22, 23). More recently, further assays have been brought to the market: The Procleix Ultrio Elite assay (for detection of HBV DNA, HCV RNA, HIV-1 RNA, and

HIV-2 RNA), the Procleix HEV assay (for detection of hepatitis E virus RNA), the Procleix Parvo/HAV assay (for detection of parvovirus B19 DNA and hepatitis A virus RNA), and the Procleix WNV assay (for detection of West Nile virus RNA), all of these assays being CE-marked with the Procleix WNV assay additionally being FDA-approved. In a recent study, it was shown that the change in the oligonucleotide design of the Procleix Ultrio Elite assay in order to enable HIV-2 RNA detection has not affected the analytical sensitivity for the other viruses regardless of the genotype (24). The Procleix HEV assay has recently been shown to provide evidence of HEV RNA presence in Catalan blood donors (25). Finally, the Procleix WNV assay was reported to perform well when used for detection of both West Nile virus lineages in an Italian external quality assessment program (26). Very recently, the Aptima HIV-1 Quant Dx assay, based on real-time TMA, providing quantitative results, has been introduced. This assay has been IVD/CE-labeled while currently not approved for use in the USA. It is designed for performance on the Panther system. The Aptima HIV-1 Quant Dx assay is useful for both diagnosis of HIV-1 infection and antiretroviral treatment monitoring. Corresponding assays for detection and quantitation of HBV and HCV will be available shortly.

Similar to the NASBA-based HPV kit (see above), the TMA-based Aptima HPV assay targeting high-risk HPV mRNA from the E6/E7 oncogenes has been brought on the market. Additionally, the Aptima HPV 16 18/45 Genotype assay which detects HPV 16 and a pool of HPV 18 and 45 is commercially available. Both assays have been IVD/CE-labeled and FDA-approved. In recent studies, the Aptima HPV assay performed similar to (DNA-based) competitor assays and no cross-contamination was detected with this assay in a challenge experiment (27–29). Furthermore, it was reported that the Aptima HPV assay has a high negative predictive value for future cervical intraepithelial neoplasia grade 3 (CIN3), indicating that HPV-mRNA-negative women are at low risk of progression to high grade CIN (30, 31). The Aptima HPV assay meets the criteria required for the UK National Health Service (NHS) cervical screening program (32). Both the Aptima HPV and the Aptima HPV 16 18/45 Genotype assays were found to be useful for cervical cancer risk stratification in women with an "atypical squamous cells of undetermined significance" (ASC-US) cytology diagnosis and may provide a more cost-effective patient care (33–35). In contrast to the NASBA-based technology, no laboratory developed test systems for detection of viruses based on the TMA technology have been published recently.

STRAND DISPLACEMENT AMPLIFICATION (SDA)

Technique:

Strand displacement amplification (SDA) is a somewhat complex isothermal amplification technique (36, 37). Amplification consists of two steps: a target generation step followed by an exponential amplification phase that replicates the target sequence through a series of complex extension, nicking, and strand displacement steps. The products contain a detector probe-annealing region and amplification product detection occurs simultaneously with amplification. The detector probe consists of a target specific hybridization region at the 3′ end and a hairpin structure at the 5′ end. The loop of the hairpin contains a restriction enzyme recognition sequence. The 5′ base is conjugated to a fluorophore donor molecule, while the 3′ base of the hairpin stem is conjugated to an acceptor molecule in close proximity. When the donor is excited, the fluorescent energy is transferred to the acceptor molecule, and little fluorescence is observed. As the hairpin anneals to the target, another complex set of displacement, extension, and restriction steps frees the donor from the quenching effects of the acceptor, and allows fluorescence to be observed.

Application of SDA in the Clinical Virology Laboratory

Assays for routine use in the clinical laboratory based on the SDA technology have been developed by Becton Dickinson (Franklin Lakes, NJ). The majority of these assays are designed for detection of bacteria and parasites. For the clinical virology laboratory, there are assays for the detection of herpes simplex virus type 1 (HSV-1) and type 2 (HSV-2) currently available, the BD ProbeTec herpes simplex viruses (HSV 1 & 2) Q^x amplified DNA assays (Table 1). These qualitative IVD/CE-marked and FDA-cleared assays are indicated for use with symptomatic individuals to aid in the diagnosis of anogenital HSV-1 and HSV-2 infections. When these assays are used with the BD Viper System in extracted mode, specimens must be collected with specially designed collection and transport devices, the BD Universal Viral Transport Standard Kit including a 3 ml vial containing BD Universal Viral Transport medium and 2 pieces of BD Universal Viral Transport polyester-tipped swabs (vials and swabs available separately if required) or the identical Copan (Brescia, Italy) kit including a 3 ml vial containing Copan Universal Transport Medium and 2 pieces of polyester-tipped swabs (vials and swabs available separately if required). After extraction including a pre-warm step in the BD Viper Lysing Heater, followed by cooling, all further steps are performed automatically by the BD Viper System. To date, only one major evaluation study regarding the BD ProbeTec herpes simplex viruses (HSV 1 & 2) Q^x amplified DNA assays has been published (38). In this study, the clinical performance of these assays was compared to that of a laboratory developed PCR-based assay. For anogenital HSV-1 and HSV-2 infections, sensitivities and specificities were found to be comparable.

LOOP-MEDIATED ISOTHERMAL AMPLIFICATION (LAMP) AND Q-LAMP

Technique:

The loop-mediated isothermal amplification (LAMP) technique has been developed by members of Eiken Chemical (Tokyo, Japan). This technology employs a set of four primers that bind to six distinct sequences on the target nucleic acid, resulting in high specificity (39). An inner primer initiates LAMP and subsequent strand-displacement DNA synthesis, primed by an outer primer, releases a single-stranded DNA. This acts as a template for DNA synthesis primed by the second inner and outer primers. These hybridize to the opposite end of the target sequence, producing a stem-loop DNA structure. Subsequently, as LAMP progresses, an inner primer hybridizes to the loop on the product and initiates displacement DNA synthesis, yielding the original stem-loop DNA and a new stem-loop DNA that is twice as long. This is a rapid and stable isothermal process, producing up to 10^9 copies of the target in less than one hour. The final stem-loop DNA products contain multiple loops formed by the annealing of alternately inverted repeats of the target in the same strand.

The LAMP technology has been further developed to a quantitative (fluorescence-quenching) loop-mediated isothermal amplification method (Q-LAMP). Q-LAMP is a rapid real-time fluorescent technique based on the recognition of multiple primer binding regions on the target nucleic acid and amplification of the target sequence, facilitated by a polymerase with strand displacement activity. Quantitation is achieved through the use of fluorophore-labeled primers and known calibrators within the reaction solution, which fluoresce in their free form but are quenched when bound. Because the fluorophore-labeled primers are consumed during amplification of the target sequence, a decrease in fluorescence is observed. This produces a characteristic real-time quenching profile when measured over time. Notably, unlike PCR, this method achieves single tube RNA amplification without the need for an additional RT step.

Application of LAMP/Q-LAMP in the Clinical Virology Laboratory

Eiken Chemical has signed license agreements with DiaSorin (Saluggia, Italy) and Meridian Bioscience (Cincinnati, OH) for the LAMP/Q-LAMP technologies. Assays for the clinical virology laboratory have been developed by these manufacturers (Table1).

DiaSorin has brought several assays for detection of viral DNAs in different specimens on the market including Iam BKV, Iam HSV1&2, Iam VZV, Iam CMV, and Iam Parvo. After automated extraction of nucleic acids on the Liason IXT system (DiaSorin), amplification and detection is performed on the Liason IAM analyzer (DiaSorin). Both instruments and the kits have been CE-marked; however, they are not available in the USA and Canada currently. No major evaluation studies regarding these new assays have been published yet.

Meridian Bioscience has brought the *illumi*gene HSV 1&2 DNA amplification assay on the market, besides several assays for detection of bacteria based on the LAMP technology. This CE/IVD-marked and FDA-cleared assay detects and differentiates HSV-1 and HSV-2 DNA in cutaneous and mucocutaneous lesion specimens that can be collected by using several collection devices (see package insert). After manual extraction of nucleic acids including the *illumi*gene SMP PREP III (Meridian) solution, amplification and detection is performed on the *illumi*pro-10 Incubator Reader (Meridian). To date, no published studies on performance are available.

Several studies describing laboratory developed assays based on the LAMP technology have been published, mainly by Japanese scientists. Articles focus on detection of dengue viruses, enterovirus 71, herpes simplex viruses, influenza viruses, measles virus, Middle East Respiratory Syndrome Coronavirus, mumps virus, respiratory syncytial virus, rubella virus, and varicella-zoster virus (40–51). Recently, the establishment and evaluation of a multiplex (M-) LAMP assay for the detection of influenza A/H1, A/H3, and influenza B has been reported (52).

NICKING ENDONUCLEASE AMPLIFICATION REACTION (NEAR)

Technique:

The nicking endonuclease amplification reaction (NEAR) technique, originally patented by members of Envirologix (Portland, ME), provides exponential amplification of DNA or RNA with a somewhat complex procedure (53). In case of an RNA target, two templates and three enzymes (a thermostable DNA polymerase, a reverse transcriptase, and a thermostable nicking endonuclease) are employed. The NEAR technique can be used on crude specimens, including different kinds of respiratory specimens, without any requirement for nucleic acid extraction. The whole reaction including detection in real time, using fluorescence-labeled molecular beacons, is finished usually within 15 minutes.

Application of NEAR in the Clinical Virology Laboratory

A commercially available IVD/CE-marked and FDA cleared assay based on the NEAR technology has been brought on the market by Alere (Waltham, MA). The Alere I influenza A & B assay is performed on the specially designed Alere I instrument and provides fully automated detection and discrimination between influenza A and B viruses (Table 1). When this assay was compared to PCR-based detection of influenza viruses, sensitivity was found to be between 73.2 and 89.4% for influenza A virus and between 75.0 and 100% for influenza B virus, while specificity was found to be between 98.6 and 100% for influenza A virus and between 99.0 and 100% for influenza B virus (53–55). No major reports describing laboratory developed assays based on the NEAR technology useful in the clinical virology laboratory have been published so far.

RECOMBINASE POLYMERASE AMPLIFICATION (RPA)

Technique:

The recombinase polymerase amplification (RPA) was developed and launched by TwistDx Ltd. (formerly known as ASM Scientific Ltd.), based in Cambridge, UK. RPA is a multienzyme isothermal amplification technique that makes use of the DNA binding and unwinding properties of prokaryotic recombinases. Target-specific primers and a recombinase are added to a sample and the recombinase/oligonucleotides complex scans for complementary sequences within target DNA samples (56). If the target sample is present, the recombinase displaces one DNA strand with the oligonucleotides that serve as primers for the initiation of DNA synthesis by a DNA polymerase. The assay also contains single-stranded DNA-binding proteins that attach to and stabilize the displaced strand. These interactions result in the duplication of the original template from primer-defined points in the DNA sample. Repetition of this process results in exponential DNA amplification. Assays can be performed in a real-time fluorimeter with up to eight RPA reactions to be monitored simultaneously. Alternative detection techniques include standard fluorescence plate readers and qPCR instruments (with a DNA binding dye or an additional component, exonuclease III, which will degrade the hybridization probe and generate a real-time readout). The RPA technique can be used on crude specimens, including blood and different kinds of swabs, without any requirement for nucleic acid extraction. An RPA-based assay operates from typical ambient temperatures (albeit more slowly) to temperatures as high as 45°C and tolerates fluctuations within these limits. The whole reaction including detection in real-time is finished usually within 15 minutes, if combined with nucleic acid extraction the turnaround time will usually not exceed 30 minutes. The entire reaction system is stable as a dried formulation and can be transported safely without refrigeration.

Application of RPA in the Clinical Virology Laboratory

Currently, no commercial assay based on the RPA technology useful in the clinical virology laboratory has been available. However, several studies describing laboratory developed assays based on the RPA technology have been published. Articles include detection of dengue virus, HIV-1 DNA, Middle East Respiratory Syndrome Coronavirus, and yellow fever virus (57–61).

PRESENT STATE AND FUTURE ASPECTS OF DIFFERENT ISOTHERMAL NUCLEIC ACID AMPLIFICATION METHODS USEFUL FOR THE CLINICAL VIROLOGY LABORATORY

Today, detection and quantitation of viral nucleic acids has become state-of-the-art in the clinical virology laboratory. In contrast to PCR- and qPCR-based assays, isothermal nucleic acid amplification assays require only a single optimized reaction temperature. They provide simpler and more effective reaction conditions, some of them with significantly less expensive equipment. Isothermal nucleic acid amplification assays should thus have become a reasonable part of molecular diagnosis in the clinical virology laboratory. However, latest data obtained from commercially available international proficiency programs including QCMD, NEQAS, and INSTAND do not indicate a frequent use of isothermal nucleic amplification assays in the clinical virology laboratory. When results obtained from laboratories participating are broken down by technology types employed for detection and quantitation of different viruses, it currently turns out that for each viral panel less than 5% of all results obtained were generated using an isothermal nucleic acid amplification method. Reasons for this low number include the limited number of commercial assays available currently and the often complicated establishment of a laboratory developed assay due to the complexity of reagents (including primers) required. In the event of introduction of a new commercially available assay in a clinical virology laboratory, it may be desirable to run it together with pre-existing assays on the identical platform in order to facilitate instrument maintenance and calibration, training the laboratory staff, and connection of the instrument to the LIS. For instance, in immunosuppressed patients, usually a panel of viral DNAs/RNAs to be screened and/or monitored is requested. This panel may include the detection and/or quantitation of DNAs/RNAs of adenoviruses, CMV, EBV, enteroviruses, human herpes virus type 6 (HHV-6), HSV-1, HSV-2, parvovirus B19, and varicella zoster virus (VZV). With this situation, it appears to be most convenient to run the assays required together with pre-existing assays on the identical platform. While manufacturers providing kits based on qPCR usually offer a panel of kits for detection and/or quantitation of various viruses, manufacturers focusing on isothermal nucleic acid amplification methods have brought a very limited number of kits on the market until now. In case of establishment of laboratory developed assays, it may be necessary to clarify whether a license must be purchased to use the technology. Furthermore, the demand for many different enzymes may cause inconveniences such as enzyme storage and the need to find assay conditions that are permissive for all of the enzymes being used as they might require different additives or temperatures for optimal performance. Isothermal nucleic amplification methods may thus be still considered as niche techniques in the clinical viral laboratory; however, data mainly refer to laboratories located in Europe and the USA. Indeed, the worldwide situation may be diverse for different techniques and must be analyzed individually for each technique. Table 2 gives an overview about advantages and disadvantages of each technique.

Advantages and Disadvantages of Transcription-Based Amplification Methods (NASBA and TMA)

Although NASBA and TMA are the worldwide most commonly used isothermal nucleic acid amplification methods in the clinical virology laboratory, these technologies play only a minor role in the clinical virology laboratory, especially in Europe and in the USA. Two major NASBA-based kits, the NucliSENS EasyQ HIV-1 v2.0 and the NucliSENS easyQ HPV, are commercially available. Both of them can be

TABLE 2 Strengths and weaknesses of isothermal nucleic acid amplification methods

Method	Advantage(s)	Disadvantage(s)
NASBA	Largely automated; manufacturer provides a basic kit that allows generation of laboratory developed assays; well-documented in literature	Very few commercial assays available; not ideal for DNA analytes
TMA	Commercially available kits fully automated; number of kits available increases continuously	The majority of virus kits aimed at laboratories that support transfusion medicine
SDA	Commercially available kits fully automated	Only kits for detection of the two subtypes of a single virus available; establishment of laboratory developed assays rather difficult
LAMP / Q-LAMP	Largely automated; manufacturer provides an increasing number of kits; laboratory developed assays can be established relatively easily	Regarding commercially available kits, no independent literature about assay performance available; complex primer design
NEAR	Robust technique tolerating crude samples; very quick providing results within 15 minutes; commercially available test well-documented in literature; suitable for point-of-care diagnostics	Only a single test commercially available currently; no medium- to high-throughput instrument available
RPA	Robust technique tolerating crude samples; very quick providing results within 15 minutes; suitable for point-of-care diagnostics in the future	No test commercially available currently

NASBA, nucleic acid sequence-based amplification; TMA, transcription mediated amplification; SDA, strand displacement amplification; LAMP, loop-mediated isothermal amplification; Q-LAMP, quantitative loop-mediated isothermal amplification; NEAR, nicking endonuclease amplification reaction; RPA, recombinase polymerase amplification.

performed easily using automated platforms for nucleic acid extraction (NucliSENS easyMAG) and amplification/detection (NucliSENS EasyQ), which are benchtop instruments, within less than 3 hours. Although NucliSENS EasyQ HIV-1 v2.0 is widely used for detection and quantitation of HIV-1 RNA in Africa, very few laboratories utilize this assay in other parts of the world with the qPCR-based Abbott or Roche kits being the major assays used for detection and quantitation of HIV-1 RNA in the clinical virology laboratory. In view of the fact that the distribution of HIV-1 subtypes is significantly different in Africa, with predominantly HIV-1 non-B subtypes and a growing number of circulating recombinant forms, identical amplification efficiency for all subtypes and recombinant forms is of major importance to warrant reliable and comparable results. When compared to results obtained from competitor assays, the NucliSENS EasyQ HIV-1 v2.0 has been shown to provide reliable results. Detection of HPV mRNA can also be performed easily employing automated platforms. Currently, this assay is utilized in only a few specialized laboratories but may gain further importance in the future as soon as detection of HPV mRNA will be implemented into international HPV guidelines.

The majority of currently available virus kits based on the TMA technique focus on detection of transfusion-transmitted viruses. All Procleix kits can easily be performed on fully automated platforms. Both instruments, the larger Tigris, and the smaller Panther, are floor models with a rather large footprint. The Tigris is designed for high-throughput laboratories being especially useful for transfusion medicine facilities while the Panther may also be useful for a lower throughput clinical virology laboratory. In the near future, Hologic plans to introduce kits for quantitation of HBV DNA, HCV RNA, and HIV-1 RNA to be performed on the same instruments as the Procleix kits. This will allow to screen donated blood as well as to detect disease and monitor therapy in infected patients on the identical instrument.

Advantages and Disadvantages of SDA

The BD ProbeTec Herpes Simplex Viruses (HSV 1 & 2) Q^x amplified DNA assays can easily be performed on a fully automated platform. However, no further kit for detection or quantitation of any additional viral DNA or RNA is commercially available currently. No independent data on assay performance is available. The establishment of laboratory developed assays based on the SDA technique may be complicated because SDA is performed at low non-stringent temperatures that may lead to generation of high background signals with clinical samples (due to the abundance of human genetic material).

Advantages and Disadvantages of LAMP and Q-LAMP

Several kits based on the LAMP/Q-LAMP technologies have recently been brought on the market. These assays can easily be performed by extraction of nucleic acids followed by amplification/detection on two rather small benchtop instruments. Unfortunately, no major evaluation study has been published yet making comparison of this new technique to alternative techniques impossible. Furthermore, independent data on assay performance is not yet available. However, several studies have shown that laboratory developed assays based on the LAMP technology may be established rather easily in the clinical virology laboratory. In these studies, it has been reported that the LAMP technology has a good sensitivity. This technique was shown to be less sensitive to both the presence of nontarget DNA in samples and well-known PCR inhibitors. Moreover, the specificity of the LAMP technique must be considered extremely high because primers must bind six distinct regions on the target DNA. Otherwise, the demand for perfectly designed primers may be a reason for less application of LAMP in practice. To overcome this issue, a web-based software is available for designing LAMP primers (https://primerexplorer.jp/e).

Advantages and Disadvantages of NEAR

The NEAR technique tolerates crude samples, i.e., samples may be amplified without preceding extraction procedure. Only a single test based on the NEAR technique has been brought on the market so far. This test has been evaluated in several studies and shown to reliably detect influenza A and B RNAs. It can be run on a benchtop instrument with small footprint. Results are provided within 15 minutes making this molecular test suitable for point-of-care diagnostics facilitating effective patient management and enabling prompt initiation of infection control measures. It may thus be useful for testing on influenza viruses during epidemic outbreaks; however, it must be considered that only one test at a time can be run on a single instrument. Furthermore, outside epidemic outbreaks, the benefit of this assay may be questionable because usually a panel of viral DNAs/RNAs for diagnosis of respiratory disease is requested.

Advantages and Disadvantages of RPA

Since there is no requirement to melt double-stranded DNA or to unfold RNA, RPA is a true one-step amplification procedure. Furthermore, similar to the NEAR technique, RPA tolerates crude samples, i.e., samples may be amplified without preceding extraction procedure. No commercial test based on the RPA technique has been brought on the market so far. However, studies describing laboratory developed assays based on the RPA technique have shown that results can be provided within 15 minutes making this technology suitable for point-of-care diagnostics. In the future, an assay designed for detection of viral DNAs or RNAs based on the RPA technique may even be run on a lab-on-a-chip platform that may facilitate utilizing such an assay for field use.

CONCLUSION AND FUTURE PERSPECTIVES

Molecular diagnostics of viral diseases has evolved dramatically over the last few decades. Compared to alternative detection methods, NAT has superior sensitivity and specificity. Many different technologies have been developed but none of them have become as popular as PCR-based techniques. Isothermal amplification techniques have remained less-utilized for development of viral detection assays.

Because of the excellent amplification efficiencies, the rapid performance, and the ongoing simplification of procedures, some of the isothermal nucleic acid amplification techniques may be regarded as ideal candidate techniques for molecular point-of-care diagnostics and field use. These techniques are more tolerant to crude samples making amplification without the need for a preceding extraction procedure possible. Furthermore, the integration of isothermal nucleic acid amplification techniques with microfluidic platforms will facilitate easier automation on smaller platforms. The ongoing incorporation of these technologies in commercial products will simplify procedures through automation making them commonplace in the clinical virology laboratory. In the more distant future, miniaturized reactors, advanced microfluidics, and sophisticated detection technologies will provide the development of a molecular lab-on-a-chip device.

REFERENCES

1. **Compton J.** 1991. Nucleic acid sequence-based amplification. *Nature* **350:**91–92.
2. **Weusten JJ, Carpay WM, Oosterlaken TA, van Zuijlen MC, van de Wiel PA.** 2002. Principles of quantitation of viral loads using nucleic acid sequence-based amplification in combination with homogeneous detection using molecular beacons. *Nucleic Acids Res* **30:**e26.
3. **Hill CS.** 2001. Molecular diagnostic testing for infectious diseases using TMA technology. *Expert Rev Mol Diagn* **1:**445–455.
4. **Xu S, Song A, Nie J, Li X, Wang Y.** 2010. Performance of NucliSens HIV-1 EasyQ Version 2.0 compared with six commercially available quantitative nucleic acid assays for detection of HIV-1 in China. *Mol Diagn Ther* **14:**305–316.
5. **Xu S, Song A, Nie J, Li X, Meng S, Zhang C, Wang Y.** 2012. Comparison between the automated Roche Cobas AmpliPrep/Cobas TaqMan HIV-1 test version 2.0 assay and its version 1 and Nuclisens HIV-1 EasyQ version 2.0 assays when measuring diverse HIV-1 genotypes in China. *J Clin Virol* **53:**33–37.
6. **Gomes P, Carvalho AP, Diogo I, Gonçalves F, Costa I, Cabanas J, Camacho RJ.** 2013. Comparison of the NucliSENS EasyQ HIV-1 v2.0 with Abbott m2000rt RealTime HIV-1 assay for plasma RNA quantitation in different HIV-1 subtypes. *J Virol Methods* **193:**18–22.
7. **Muenchhoff M, Madurai S, Hempenstall AJ, Adland E, Carlqvist A, Moonsamy A, Jaggernath M, Mlotshwa B, Siboto E, Ndung'u T, Goulder PJ.** 2014. Evaluation of the NucliSens EasyQ v2.0 assay in comparison with the Roche Amplicor v1.5 and the Roche CAP/CTM HIV-1 Test v2.0 in quantification of C-clade HIV-1 in plasma. *PLoS One* **9:** e103983.
8. **Ndiaye O, Diop-Ndiaye H, Ouedraogo AS, Fall-Malick FZ, Sow-Sall A, Thiam M, Diouara AA, Ndour CT, Gaye-Diallo A, Mboup S, Toure-Kane C, Study Group.** 2015. Comparison of four commercial viral load techniques in an area of non-B HIV-1 subtypes circulation. *J Virol Methods* **222:**122–131.
9. **van Deursen P, Oosterlaken T, Andre P, Verhoeven A, Bertens L, Trabaud MA, Ligeon V, de Jong J.** 2010. Measuring human immunodeficiency virus type 1 RNA loads in dried blood spot specimens using NucliSENS EasyQ HIV-1 v2.0. *J Clin Virol* **47:**120–125.
10. **Napierala Mavedzenge S, Davey C, Chirenje T, Mushati P, Mtetwa S, Dirawo J, Mudenge B, Phillips A, Cowan FM.** 2015. Finger prick dried blood spots for HIV viral load measurement in field conditions in Zimbabwe. *PLoS One* **10:** e0126878.
11. **Mercier-Delarue S, Vray M, Plantier JC, Maillard T, Adjout Z, de Olivera F, Schnepf N, Maylin S, Simon F, Delaugerre C.** 2014. Higher specificity of nucleic acid sequence-based amplification isothermal technology than of real-time PCR for quantification of HIV-1 RNA on dried blood spots. *J Clin Microbiol* **52:**52–563.
12. **Oliveira A, Verdasca N, Pista Â.** 2013. Use of the NucliSENS EasyQ HPV assay in the management of cervical intraepithelial neoplasia. *J Med Virol* **85:**1235–1241.
13. **Perez Castro S, Iñarrea Fernández A, Lamas González MJ, Sarán Diez MT, Cid Lama A, Alvarez Martín MJ, Pato Mosquera M, López-Miragaya I, Estévez N, Torres Piñón J, Oña Navarro M.** 2013. Human papillomavirus (HPV) E6/E7 mRNA as a triage test after detection of HPV 16 and HPV 18 DNA. *J Med Virol* **85:**1063–1068.
14. **Verdoodt F, Szarewski A, Halfon P, Cuschieri K, Arbyn M.** 2013. Triage of women with minor abnormal cervical cytology: meta-analysis of the accuracy of an assay targeting messenger ribonucleic acid of 5 high-risk human papillomavirus types. *Cancer Cytopathol* **121:**675–687.
15. **Silling S, Kreuter A, Hellmich M, Swoboda J, Pfister H, Wieland U.** 2012. Human papillomavirus oncogene mRNA testing for the detection of anal dysplasia in HIV-positive men who have sex with men. *J Clin Virol* **53:**325–331.
16. **Sendagorta E, Romero MP, Bernardino JI, Beato MJ, Alvarez-Gallego M, Herranz P.** 2015. Human papillomavirus mRNA testing for the detection of anal high-grade squamous intraepithelial lesions in men who have sex with men infected with HIV. *J Med Virol* **87:**1397–1403.
17. **Capaul SE, Gorgievski-Hrisoho M.** 2005. Detection of enterovirus RNA in cerebrospinal fluid (CSF) using NucliSens EasyQ Enterovirus assay. *J Clin Virol* **32:**236–240.
18. **Keightley MC, Sillekens P, Schippers W, Rinaldo C, George KS.** 2005. Real-time NASBA detection of SARS-associated coronavirus and comparison with real-time reverse transcription-PCR. *J Med Virol* **77:**602–608.
19. **Loens K, Goossens H, de Laat C, Foolen H, Oudshoorn P, Pattyn S, Sillekens P, Ieven M.** 2006. Detection of rhinoviruses by tissue culture and two independent amplification techniques, nucleic acid sequence-based amplification and reverse transcription-PCR, in children with acute respiratory infections during a winter season. *J Clin Microbiol* **44:**166–171.
20. **Böhmer A, Schildgen V, Lüsebrink J, Ziegler S, Tillmann RL, Kleines M, Schildgen O.** 2009. Novel application for isothermal nucleic acid sequence-based amplification (NASBA). *J Virol Methods* **158:**199–201.
21. **Grabarczyk P, van Drimmelen H, Kopacz A, Gdowska J, Liszewski G, Piotrowski D, Górska J, Kuśmierczyk J, Candotti D, Lętowska M, Lelie N, Brojer E.** 2013. Head-to-head comparison of two transcription-mediated amplification assay versions for detection of hepatitis B virus, hepatitis C virus, and human immunodeficiency virus Type 1 in blood donors. *Transfusion* **53:**2512–2524.
22. **Vermeulen M, Coleman C, Mitchel J, Reddy R, van Drimmelen H, Ficket T, Lelie N.** 2013. Sensitivity of individual-donation and minipool nucleic acid amplification test options in detecting window period and occult hepatitis B virus infections. *Transfusion* **53:**2459–2466.
23. **Stramer SL, Krysztof DE, Brodsky JP, Fickett TA, Reynolds B, Dodd RY, Kleinman SH.** 2013. Comparative analysis of triplex nucleic acid test assays in United States blood donors. *Transfusion* **53:**2525–2537.
24. **Grabarczyk P, Koppelman M, Boland F, Sauleda S, Fabra C, Cambie G, Kopacz A, O'Riordan K, van Drimmelen H, O'Riordan J, Lelie N.** 2015. Inclusion of human immunodeficiency virus Type 2 (HIV-2) in a multiplex transcription-mediated amplification assay does not affect detection of HIV-1 and hepatitis B and C virus genotypes: a multicenter performance evaluation study. *Transfusion* **55:**2246–2255.
25. **Sauleda S, Ong E, Bes M, Janssen A, Cory R, Babizki M, Shin T, Lindquist A, Hoang A, Vang L, Piron M, Casamitjana N, Koppelman M, Danzig L, Linnen JM.** 2015. Seroprevalence of hepatitis E virus (HEV) and detection of HEV RNA with a transcription-mediated amplification assay in blood donors from Catalonia (Spain). *Transfusion* **55:**972–979.
26. **Pisani G, Pupella S, Cristiano K, Marino F, Simeoni M, Luciani F, Scuderi G, Sambri V, Rossini G, Gaibani P, Pierro A, Wirz M, Grazzini G.** 2012. Detection of West Nile virus RNA (lineages 1 and 2) in an external quality assessment programme for laboratories screening blood and blood components for West Nile virus by nucleic acid amplification testing. *Blood Transfus* **10:**515–520.
27. **Rebolj M, Lynge E, Ejegod D, Preisler S, Rygaard C, Bonde J.** 2014. Comparison of three human papillomavirus DNA assays and one mRNA assay in women with abnormal cytology. *Gynecol Oncol* **135:**474–480.
28. **Iftner T, Becker S, Neis KJ, Castanon A, Iftner A, Holz B, Staebler A, Henes M, Rall K, Haedicke J, Hann von Weyhern C, Clad A, Brucker S, Sasieni P.** 2015. Head-to-head comparison of the RNA-based Aptima(R) HPV assay and the DNA-based HC2 HPV test in a routine screening population of women aged 30 to 60 years in Germany. *J Clin Microbiol* JCM.01013-15.

29. **Pyne MT, Hamula CL, Tardif K, Law C, Schlaberg R.** 2015. High-risk HPV detection and genotyping by APTIMA HPV using cervical samples. *J Virol Methods* **221:**95–99.
30. **Arbyn M, Snijders PJ, Meijer CJ, Berkhof J, Cuschieri K, Kocjan BJ, Poljak M.** 2015. Which high-risk HPV assays fulfil criteria for use in primary cervical cancer screening? *Clin Microbiol Infect* **21:**817–826.
31. **Johansson H, Bjelkenkrantz K, Darlin L, Dillner J, Forslund O.** 2015. Presence of High-Risk HPV mRNA in relation to future high-grade lesions among high-risk HPV DNA positive women with minor cytological abnormalities. *PLoS One* **10:**e0124460.
32. **Moss SM, Bailey A, Cubie H, Denton K, Sargent A, Muir P, Vipond IB, Winder R, Kitchener H.** 2014. Comparison of the performance of HPV tests in women with abnormal cytology: results of a study within the NHS cervical screening programme. *Cytopathology.*
33. **Persson M, Elfström KM, Brismar Wendel S, Weiderpass E, Andersson S.** 2014. Triage of HR-HPV positive women with minor cytological abnormalities: a comparison of mRNA testing, HPV DNA testing, and repeat cytology using a 4-year follow-up of a population-based study. *PLoS One* **9:**e90023.
34. **Sauter JL, Mount SL, St John TL, Wojewoda CM, Bryant RJ, Leiman G.** 2014. Testing of integrated human papillomavirus mRNA decreases colposcopy referrals: could a change in human papillomavirus detection methodology lead to more cost-effective patient care? *Acta Cytol* **58:**162–166.
35. **Castle PE, Cuzick J, Stoler MH, Wright TC Jr, Reid JL, Dockter J, Giachetti C, Getman D.** 2015. Detection of human papillomavirus 16, 18, and 45 in women with ASC-US cytology and the risk of cervical precancer: results from the CLEAR HPV study. *Am J Clin Pathol* **143:**160–167.
36. **Walker GT, Fraiser MS, Schram JL, Little MC, Nadeau JG, Malinowski DP.** 1992. Strand displacement amplification–an isothermal, in vitro DNA amplification technique. *Nucleic Acids Res* **20:**1691–1696.
37. **Little MC, et al.** 1999. Strand displacement amplification and homogeneous real-time detection incorporated in a second-generation DNA probe system, BDProbeTecET. *Clin Chem* **45:**777–784.
38. **Van Der Pol B, Warren T, Taylor SN, Martens M, Jerome KR, Mena L, Lebed J, Ginde S, Fine P, Hook EW III.** 2012. Type-specific identification of anogenital herpes simplex virus infections by use of a commercially available nucleic acid amplification test. *J Clin Microbiol* **50:**3466–3471.
39. **Notomi T, Okayama H, Masubuchi H, Yonekawa T, Watanabe K, Amino N, Hase T.** 2000. Loop-mediated isothermal amplification of DNA. *Nucleic Acids Res* **28:**E63.
40. **Enomoto Y, Yoshikawa T, Ihira M, Akimoto S, Miyake F, Usui C, Suga S, Suzuki K, Kawana T, Nishiyama Y, Asano Y.** 2005. Rapid diagnosis of herpes simplex virus infection by a loop-mediated isothermal amplification method. *J Clin Microbiol* **43:**951–955.
41. **Fujino M, Yoshida N, Yamaguchi S, Hosaka N, Ota Y, Notomi T, Nakayama T.** 2005. A simple method for the detection of measles virus genome by loop-mediated isothermal amplification (LAMP). *J Med Virol* **76:**406–413.
42. **Okafuji T, Yoshida N, Fujino M, Motegi Y, Ihara T, Ota Y, Notomi T, Nakayama T.** 2005. Rapid diagnostic method for detection of mumps virus genome by loop-mediated isothermal amplification. *J Clin Microbiol* **43:**1625–1631.
43. **Ushio M, Yui I, Yoshida N, Fujino M, Yonekawa T, Ota Y, Notomi T, Nakayama T.** 2005. Detection of respiratory syncytial virus genome by subgroups-A, B specific reverse transcription loop-mediated isothermal amplification (RT-LAMP). *J Med Virol* **77:**121–127.
44. **Mori N, Motegi Y, Shimamura Y, Ezaki T, Natsumeda T, Yonekawa T, Ota Y, Notomi T, Nakayama T.** 2006. Development of a new method for diagnosis of rubella virus infection by reverse transcription-loop-mediated isothermal amplification. *J Clin Microbiol* **44:**3268–3273.
45. **Imai M, Ninomiya A, Minekawa H, Notomi T, Ishizaki T, Van Tu P, Tien NT, Tashiro M, Odagiri T.** 2007. Rapid diagnosis of H5N1 avian influenza virus infection by newly developed influenza H5 hemagglutinin gene-specific loop-mediated isothermal amplification method. *J Virol Methods* **141:**173–180.
46. **Chen HT, Zhang J, Sun DH, Ma LN, Liu XT, Cai XP, Liu YS.** 2008. Development of reverse transcription loop-mediated isothermal amplification for rapid detection of H9 avian influenza virus. *J Virol Methods* **151:**200–203.
47. **Higashimoto Y, Ihira M, Ohta A, Inoue S, Usui C, Asano Y, Yoshikawa T.** 2008. Discriminating between varicella-zoster virus vaccine and wild-type strains by loop-mediated isothermal amplification. *J Clin Microbiol* **46:**2665–2670.
48. **Nakauchi M, Yoshikawa T, Nakai H, Sugata K, Yoshikawa A, Asano Y, Ihira M, Tashiro M, Kageyama T.** 2011. Evaluation of reverse transcription loop-mediated isothermal amplification assays for rapid diagnosis of pandemic influenza A/H1N1 2009 virus. *J Med Virol* **83:**10–15.
49. **Shi W, Li K, Ji Y, Jiang Q, Shi M, Mi Z.** 2011. Development and evaluation of reverse transcription-loop-mediated isothermal amplification assay for rapid detection of enterovirus 71. *BMC Infect. Dis.* **11:**197-2334-11-197.
50. **Teoh BT, Sam SS, Tan KK, Johari J, Danlami MB, Hooi PS, Md-Esa R, AbuBakar S.** 2013. Detection of dengue viruses using reverse transcription-loop-mediated isothermal amplification. *BMC Infect. Dis.* **13:**387-2334-13-387.
51. **Shirato K, Yano T, Senba S, Akachi S, Kobayashi T, Nishinaka T, Notomi T, Matsuyama S.** 2014. Detection of Middle East respiratory syndrome coronavirus using reverse transcription loop-mediated isothermal amplification (RT-LAMP). *Virol. J.* **11:**139-422X-11-139.
52. **Mahony J, Chong S, Bulir D, Ruyter A, Mwawasi K, Waltho D.** 2013. Multiplex loop-mediated isothermal amplification (M-LAMP) assay for the detection of influenza A/H1, A/H3 and influenza B can provide a specimen-to-result diagnosis in 40 min with single genome copy sensitivity. *J Clin Virol* **58:**127–131.
53. **Nie S, Roth RB, Stiles J, Mikhlina A, Lu X, Tang YW, Babady NE.** 2014. Evaluation of Alere i Influenza A&B for rapid detection of influenza viruses A and B. *J Clin Microbiol* **52:**3339–3344.
54. **Bell JJ, Selvarangan R.** 2014. Evaluation of the Alere I influenza A&B nucleic acid amplification test by use of respiratory specimens collected in viral transport medium. *J Clin Microbiol* **52:**3992–3995.
55. **Hazelton B, Gray T, Ho J, Ratnamohan VM, Dwyer DE, Kok J.** 2015. Detection of influenza A and B with the Alere™ i Influenza A & B: a novel isothermal nucleic acid amplification assay. *Influenza Other Respi Viruses* **9:**151–154.
56. **Piepenburg O, Williams CH, Stemple DL, Armes NA.** 2006. DNA detection using recombination proteins. *PLoS Biol* **4:** e204.
57. **Boyle DS, Lehman DA, Lillis L, Peterson D, Singhal M, Armes N, Parker M, Piepenburg O, Overbaugh J.** 2013. Rapid detection of HIV-1 proviral DNA for early infant diagnosis using recombinase polymerase amplification. *MBio* **4:**e00135-13.
58. **Abd El Wahed A, Patel P, Heidenreich D, Hufert FT, Weidmann M.** 2013. Reverse transcription recombinase polymerase amplification assay for the detection of Middle East respiratory syndrome coronavirus. *PLoS Curr* **5:**ecurrents.outbreaks.62df1c7c75ffc96cd59034531e2e8364 10.1371/currents.outbreaks.62df1c7c75ffc96cd59034531e2e8364.
59. **Escadafal C, Faye O, Sall AA, Faye O, Weidmann M, Strohmeier O, von Stetten F, Drexler J, Eberhard M, Niedrig M, Patel P.** 2014. Rapid molecular assays for the detection of yellow fever virus in low-resource settings. *PLoS Negl Trop Dis* **8:**e2730.
60. **Crannell ZA, Rohrman B, Richards-Kortum R.** 2014. Quantification of HIV-1 DNA using real-time recombinase polymerase amplification. *Anal Chem* **86:**5615–5619.
61. **Teoh BT, Sam SS, Tan KK, Danlami MB, Shu MH, Johari J, Hooi PS, Brooks D, Piepenburg O, Nentwich O, Wilder-Smith A, Franco L, Tenorio A, AbuBakar S.** 2015. Early detection of dengue virus by use of reverse transcription-recombinase polymerase amplification. *J Clin Microbiol* **53:**830–837.

Quantitative Molecular Methods

NATALIE N. WHITFIELD AND DONNA M. WOLK

13

Diagnostic information describing the actual or relative density of specific viral nucleic acids in a human blood sample is most easily ascertained using quantitative molecular methods (aka viral loads), whose historical use spans more than two decades. A common function of quantitative viral load testing is to support clinical strategies and practices that monitor patients for human immunodeficiency virus (HIV) and most of the hepatitis viruses. For transplant patients, viral loads are monitored to detect numerous viruses classified in the herpes virus group, including cytomegalovirus (CMV), Epstein-Barr virus (EBV), and human herpes virus 6 (HHV-6), among others. It seems that each year, there is a new application or interpretation for results of viral load assays. Bacterial load assays are uncommon; however, there is some discussion of its use for certain conditions.

In the early 1990s, the feasibility of quantitative detection of nucleic acids was demonstrated by establishing polymerase chain reaction (PCR) using a traditional block thermal cycler and gel electrophoresis as suitable detection methods for viral loads. This detection method easily demonstrated a linear relationship between input copy number of the targeted virus and the extent of amplification over a quantitative range of approximately 3 to 4 orders of magnitude. The historical gel-based methods were often cumbersome and required stringent control of sample processing, amplification, and endpoint gel electrophoresis detection. Moreover, their dynamic range was limited.

Over the last 30 years, techniques for performing quantitative molecular testing have improved dramatically with the evolution of various quantitative methods including quantitative PCR (qPCR), quantitative reverse transcriptase PCR (qRT-PCR) and commercially available alternative nucleic acid amplification tests (NAAT). Alternative quantitative methods that rely on target amplification include transcription-mediated amplification (TMA) and similar nucleic acid sequenced-based amplification (NASBA) (1). Commercially available quantitative signal and probe amplification methods include the branched DNA (bDNA) method and hybrid capture assays, respectively. Several publications have reviewed historical versions and test characteristics of quantitative methods (2, 3); therefore, they will not be completely revisited. Rather, the clinical utility of viral quantitation using molecular methods (4–6) is provided briefly in this chapter.

In the last decade, quantitative real-time PCR has become the mainstay for quantitative microbial testing. PCR-based quantitative methods routinely used in clinical virology can be divided into the following categories: (i) Q-PCR, typically used to determine the microbial density or "load" of DNA viruses or bacteria in clinical specimens; (ii) QRT-PCR, used to determine the viral load of RNA viruses; and (iii) "gene expression" assays, QRT-PCR targeted at human host genetic targets, which are used to determine the relative mRNA expression levels present in different disease states.

A number of FDA-approved and laboratory-developed procedures (LDPs) for detection of viruses have become available in the years since quantitative molecular amplification was first described. Viruses remain the disease agents for which quantitative molecular testing is most commonly used, with several viral load assays representing the best characterized of the quantitative assays, including: methods for HIV, hepatitis C virus (HCV), hepatitis B virus (HBV), and recently CMV (7–12).

Given the limited availability of FDA-approved assays (Table 1), there has been a rapid expansion of clinical applications for LDP used in clinical laboratories without the prior rigor of the FDA approval process (12–36). LDPs used in clinical virology laboratories incur additional responsibilities for many aspects of method verification. Laboratory personnel (37), who verify and validate these methods for patient care, do so under the guidelines set forth in the Clinical Laboratory Improvement Act, 42 CFR Part 493 (CLIA '88), as discussed later in this chapter. In order to produce quality analytical data, laboratories that use LDP methods in their clinical practice must weigh the benefits of implementation against the effort and expense required for verification and validation of an LDP within a health care setting. Because precision of quantitative assays is critical for those who rely on the measurement of quantitative differences in microbial load over the course of the disease, careful planning, critical assessment of statistical data from method verification and validation, statistical comparisons of parallel methods, and availability of assay controls and standards are essential (38, 39).

For FDA-approved quantitative methods, typical assay variability ranges from 0.2 to 0.5 $\log_{10}$, so relevant clinical differences are often >2 to 3 $\log_{10}$. For quantitative LDPs, interassay variability can be large, and it is often difficult to

doi:10.1128/9781555819156.ch13

TABLE 1 FDA-approved quantitative assays by type, a list of FDA-approved viral load assays (Adapted from http://www.fda.gov/MedicalDevices/ProductsandMedicalProcedures/InVitroDiagnostics/ucm330711.htm.)

Target	Test or reagent (manufacturer)	Method	Measurable range (IU/ml or copies/ml)	IU/Ml to copies/ml conversion*
HIV-1	Cobas Amplicor HIV-1 Monitor v. 1.5 (Roche)	Real-time RT-PCR	Standard, 400 × 10^5 – 7.5 × 10^5 (copies/ml) Ultrasensitive, 50 × 10^5 – 1 × 10^5 (copies/ml)	
	Cobas AmpliPrep/ Amplicor HIV-1 Monitor v. 1.5 (Roche)	Real-time RT-PCR	Standard, 500 × 10^6 – 1 × 10^6 (copies/ml) Ultrasensitive, 50 × 10^6 – 1 × 10^6 (copies/ml)	
	NucliSens HIV-1 QT (bioMeriéux, Inc)	NASBA	176 × 10^6 – 3.47 × 10^6 (copies/ml)	
	Cobas Amplipre/Cobas TaqMan HIV-1 (Roche)	Real-time RT-PCR	48 × 10^7 – 1 × 10^7 (copies/ ml)	
	RealTime TaqMan HIV-1 (Abbott)	Real-time RT- PCR	40 × 10^7 – 1 × 10^7 (copies/ ml)	
HCV	Cobas AmpliPrep/Cobas TaqMan (Roche)	Real-time RT-PCR	43 × 10^7 – 6.9 × 10^7# (IU/ml)	
	Realtime HCV Assay (Abbott)	Real-time RT-PCR	12 × 10^8 (IU/ml)	
HBV	RealTime HBV PCR (Abbott)	Real-time PCR	9 × 10^9 – 4 × 10^9 (IU/ml)	
	Cobas TaqMan HBV (Roche)	Real-time PCR	20× 10^8 – 2.3 × 10^8 (IU/ ml)	
CMV	artus CMV RGQ (Qiagen)	Real-time PCR	159 × 10^7 – 7.94 × 10^7 (IU/ml)	1.259
	Cobas TaqMan CMV (Roche)	Real-time PCR	1.37 × 10^2 – 9.10 × 10^6 (IU/ml)	1.1

*1 IU/ml= × copies/ml.
Expanded clinical reportable range with maximum dilution of 1:100 (43 × 10^9–6.9 × 10^9 IU/ml).

compare quantitative results between different laboratories (40–42). There is much work yet to be done to harmonize viral load testing by implementation of additional quantitative virus standards, by which to benchmark and harmonize both FDA-approved methods and LDPs.

This chapter will provide a review of the technological features unique to the quantitative molecular methods used in clinical virology laboratories today, including key performance issues associated with use of these assays and current applications. The advantages and limitations of these methods are also reviewed. For further information regarding the underlying principles and mechanisms of quantitative assays, the reader is referred to several published reviews (43–50).

COMMON QUANTITATIVE METHODS

PCR

In a typical PCR reaction, a DNA sequence, which serves as the target or template, is amplified using a thermal cycler instrument that heats and cools the reaction allowing for interaction between the nucleic acids and the assay reaction components. For DNA amplification to occur, each temperature cycle includes the following stages: (i) denaturation (heating to high temperatures to separate DNA into single strands, typically at 95°C); (ii) annealing, i.e., lowering the temperature to allow the primers (synthetic oligonucleotide strands that are complementary to the ends of the original target gene sequence) to anneal to a specific region(s) of the single-stranded DNA and create a partial double strand; and (iii) extension (historically at 72°C), in which deoxynucleotide triphosphates are added to the 3′ ends of the bound primers by a DNA polymerase, thereby creating a new strand of synthetic DNA. The synthesized oligonucleotide sequence is complementary to that of its original template strand. Double-stranded synthetic DNA is called the amplicon, or PCR product. Once the amplicon is created, it can then serve as a template and be amplified further. This allows doubling of the number of amplicons with each cycle and an exponential increase, a main feature of PCR. Of note, some current methods enable primer annealing and amplicon extension to be performed at one common temperature (45).

In the PCR, a DNA sequence (template) is amplified in a buffered reaction solution containing oligonucleotide primers, thermostable *Taq* DNA polymerase, deoxynucleotide triphosphates, and magnesium or manganese ions. The reaction solution is placed into a thermal cycler, which heats and cools the reaction components, exposing them to consecutive cycles of varying temperatures. In each temperature cycle three steps occur: denaturation, primer annealing, and primer extension A mathematical description for the product accumulation within each cycle is: $Y_n = Y_{n-1} * (1+E_v)$ with $0 \leq E_v \leq 1$.

In this equation, E_v represents the efficiency of the amplification, Y_n the number of molecules of the PCR product after cycle n, and Y_{n-1} the number of molecules of the PCR product after cycle $n-1$ (85). Historically, amplicon or product would be visualized using gel electrophoresis, which has been replaced with real-time PCR methods that combine target amplification and fluorescent detection of probe hybridization within a single reaction vessel.

Real-time PCR

Real-time PCR methods utilize fluorescence tags whereby the number of fluorescent signals increase until the reaction components are depleted, reaching a plateau phase of amplification. The amplification process is displayed as a plotted curve of the generated fluorescent signals and is commonly called a response curve (45). The number of PCR cycles required to exceed the background fluorescence is called the cycle threshold (CT), crossover value, crossing threshold, or crossing point, depending on the manufacturer of the system. A direct and inverse relationship between the concentration of nucleic acid and the CT values in the original specimen extract make it possible to quantify viral loads.

Since the fluorescent signal generated by the reaction is proportional to the concentration of DNA in the reaction, real-time PCR methods can incorporate standards that enable results to be reported in quantitative measures of nucleic acid concentration (e.g., viral copies/mL or International units IU/ml). In general, measurements and readings that are to be used for quantitation should be gathered during the exponential phase of PCR amplification, where the amplification plot crosses the threshold (Figure 1A). Measurements taken during the lag phase or the plateau phase of amplicon production will yield misleading results. Measurement of results during the exponential phases is referred to as threshold analysis and contrasts to that of traditional gel electrophoreses, referred to as endpoint analysis. At low concentrations of target, Gaussian distribution effects will impact the sampling accuracy and increase variability while diminishing the ability to accurately quantitate the target at low density (51). Such variability affects the assay's lower limit of quantitation (LLOQ). Assays with high precision, even at the low end of quantitation, can prove to be extremely variable for monitoring antiretroviral and anti-HCV therapies (52).

Several fluorescent chemistries exist for use in real-time PCR assays to detect PCR amplicon and fall into two main types: (i) intercalating dyes, and (ii) sequence-specific DNA probes. Intercalating dyes are nonspecific fluorescent dyes that intercalate with any double-stranded DNA, i.e., SYBR Green, while sequence-specific DNA probes consist of oligonucleotides labeled with a fluorescent reporter which permits detection only after hybridization of the probe with its complementary nucleic acid sequence has occurred. In general, real-time PCR assays have the following commonalities. A light source (e.g., LED, halogen lamp, or lasers with differing wavelengths), housed within a specialized thermal cycler, stimulates the reporter dye(s) to emit fluorescence, or in some cases, diminish fluorescence, which can then be monitored by various detection systems. Target-specific hybridization can be assessed by incorporating additional sequence-specific fluorescent probes into the assay; these bind to target sites internal to the primer sites in the amplicon, and in some cases, to the primers themselves, and serve as the detection method. Dual specificity of both primers and probes allows for a highly specific fluorescent signal to be generated. The fluorescent signal generated as amplicon is produced, measured, and converted by software to generate an amplification plot. Several historical internal probe formats are still used today, including hybridization probes (53–55), hydrolysis probes (56) like those used in TaqMan chemistry (57–59), and molecular beacon probes (60, 61). Newly developed probes or primers include scorpion probes (62), major groove binding (MGB) probes (63), fluorescence-labeled locked nucleic acid (LNA) hydrolysis probes (64, 65), and scalable target analysis routine (STAR) technology, measuring amplicon accumulation through incorporation of labeled primers (66).

TaqMan chemistry, fluorescence resonance energy transfer, and molecular beacon technology can be used to monitor real-time endpoint (qualitative) PCRs as well as real-time quantitative PCRs. The amount of microbe-specific nucleic acid in patient specimens is determined by comparing target amplification signal to internal or external quantitative standards or calibrators. Through mathematical algorithms, linear regression analysis compares the CT at which amplicon is detected with the known concentration of the standard. The resulting standard curve can be used to predict the quantity of nucleic acid present in the test samples (Figure 1B). Typically, real-time quantitative PCR assays have wider linear ranges than conventional quantitative PCR assays. Several automated real-time PCR instruments that use real-time PCR chemistries are available, and many can use more than one fluorescent chemistry platform. Refer to the section titled "Quantification" for more information.

The efficiency of the PCR is calculated based on the slope of the regression analysis of these data. Typically, each real-time PCR cycler has its own equation to calculate amplification efficiency, which is based on the slope of the line plotting CT values versus known viral densities, where at least four to five different densities are tested (Figure 1B). Assays utilizing intercalating dyes as the reporter are limited in efficiency since the dye binds to both specific and nonspecific amplification products produced in the reaction. Hence, dye incorporation alone does not ensure detection of a specific PCR product and nonspecific amplification may affect quantitative results, a limitation certainly overcome by the use of sequence-specific DNA probes. s addition, PCR inhibitors and purity of the nucleic acid extract can lower the efficiency of Q-PCR.

Monitoring trends in the CT of a positive control of known concentration is also recommended, since it will allow for comparison of different reagent lots and extraction methods. Real-time PCR data are nothing more than fluorescent signal/noise ratios, perhaps more precise, but similar in concept to colorimetric signal/noise ratios generated by other laboratory assays.

Nucleic Acid Sequence-based Amplification (NASBA)

NASBA is an isothermal (41°C), transcription-based amplification method that sensitively detects RNA targets. Some NASBA techniques less efficiently amplify DNA, requiring that the target DNA be in excess (>1,000-fold) over the RNA target or only in the absence of the corresponding RNA target. As NASBA is primer dependent and amplicon detection is based on probe binding, following

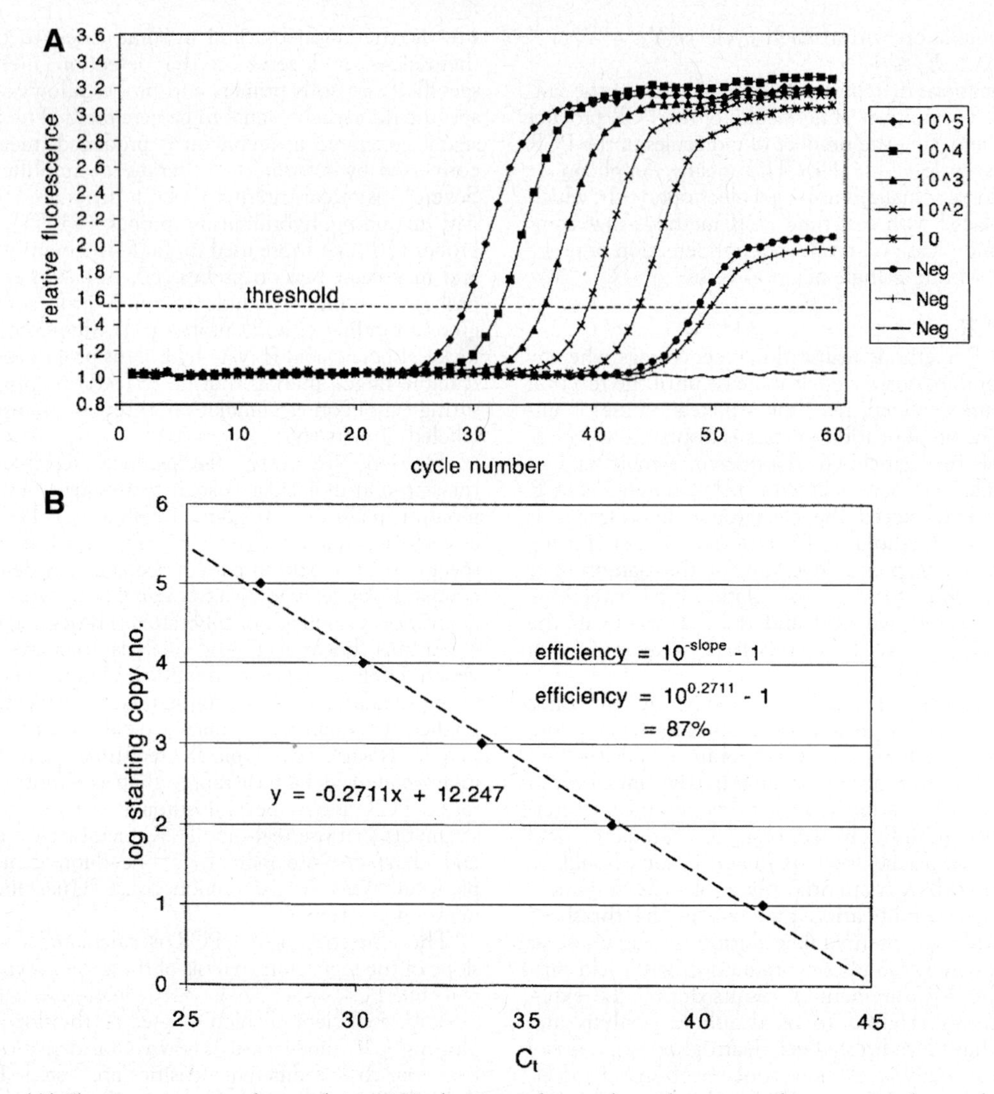

FIGURE 1 (A) Fluorescence output from real-time PCR showing eight samples with corresponding copy numbers and corresponding standard curves derived from the C_T values. Reactions occurring after cycle 45 could depict the formation and amplification of primer-dimers, so sequence analysis of these types of reaction may be necessary. (B) Example of a standard curve derived from real-time Q-PCR with calculations for PCR efficiency. The log of the starting copy number of nucleic acid is plotted against C_T. Samples of unknown concentration can be assessed in comparison to the standard curve. Efficiency can be calculated from the slope of the regression line.

primer and probe design rules are imperative to successful amplification. A major advantage of NASBA is attributed to the production of single-stranded RNA amplicons. This ssRNA can then be directly used as a template for another round of amplification, much like PCR, or it can be probed for detection without the need for preliminary denaturation or strand separation. Therefore, the amount of RNA product obtained in NASBA exceeds the concentration of primers by at least one order of magnitude, whereas for amplification processes like PCR, the initial primer concentrations limit the maximum yield of the product. The fidelity of NASBA is comparable to that of other amplification processes using DNA polymerases that lack the 3′ exonuclease activity.

Disadvantages of NASBA include: (i) NASBA enzymes are not thermostable and thus can only be added after the melting step, (ii) primers are not incorporated in the amplicon and thus labeled primers cannot be used for detection, and (iii) the length of the target sequence to be amplified efficiently is limited to approximately 100 to 250 nucleotides. In addition, any nonspecific interactions of the primers can be increased if the reaction temperature is lowered.

Quantification by NASBA is achieved by including a coextracted calibrator in the assay, which is added prior to isolation of the nucleic acids and is coamplified with the target nucleic acid. The calibrator, in turn, acts as an extraction control. The calibrator sequence to be amplified is identical to that of the target's sequence; however, the detection probe binding sequence of the calibrator is a randomized version of that of the target sequence, with comparable hybridization efficiency for the interaction between the probe and the amplicon of the target, and for the calibrator as well. Likewise, during amplification, the same primers are used for the calibrator as for the target, resulting in competition between the calibrator and the target. The competition that takes place determines the quantitative range of the assay, which is approximately 5 logs. Calibrator concentration determines which target concentrations are included in the quantitative range. The lowest amount of target that still results in an efficient isolation and amplification of both target and calibrator nucleic acids is designated as the assay's lower detection limit (67).

NASBA target detection is accomplished by two main methods, electro-chemiluminescence (ECL), an endpoint analysis whereby biotin is attached to the 5′ end of the capture probe to immobilize it on streptavidin-coated paramagnetic beads, and real-time analysis using molecular beacons. The ECL-based quantitative NASBA assays include three calibrators (68) added in fixed, but different, concentrations, with each having its own detection probe binding sequence and accompanying detection probe with amplification taking place in the same test tube as that of the target nucleic acid. After amplification has completed, four independent hybridization reactions are performed with the specific detection probes, and the results of the calibrator reactions are used to quantify the target input. The molecular beacon-based assays include only one calibrator with a fixed concentration to quantify the target input, with the amplification and detection both taking place in a single test tube. The relationship between the fluorescence kinetics results obtained from the calibrator beacon and the target nucleic acid beacon are then used to calculate results for quantification.

kPCR

Branched DNA (bDNA), previously used for viral load testing, was removed from the market by Siemens Healthcare USA. To take its place, Siemens introduced the VERSANT kPCR Molecular System (not currently available in the United States, CE marked) with assays available for HIV-1, HBV, and HCV viral loads offering a wider quantitation range than the former bDNA VERSANT 440 assays. The performance of the VERSANT kPCR HCV RNA 1.0 assay substantially under quantitates HCV RNA concentrations for genotypes 2 and 3, a limitation which will undoubtedly be improved in future versions of the assay (69, 70).

Digital PCR (dPCR)

Successfully used for quantitation of HIV and CMV, digital PCR is noted to have potential for highly accurate and reproducible quantitation of viruses and other genetic targets without the need for a standard curve (10, 71–76). Massive sample partitioning is a key aspect of the dPCR technique. In contrast to real-time PCR, dPCR uses an alternate method that is not dependent on CTs. Instead, the samples are partitioned into thousands of independent endpoint PCR reactions prior to amplification. Each reaction well is scored as either positive or negative for amplification of the viral target sequence of interest. Counting of the positive wells and conversion to a concentration of target in the original sample provides the quantification. Assigning each well as positive or negative prior to quantification greatly minimizes the measurement's dependency on parameters such as assay efficiency and instrument calibration. Therefore different laboratories can compare viral load measurement results in a standardized manner without interference from external factors, such as reaction chemistry, instrument variability, or extraction method.

Droplet Digital PCR (ddPCR) from BioRad (Hercules, CA) is a technique that is based on partitioning of many PCR reactions into water-oil emulsion droplets housed in multiwell plates. With individual nucleic acid molecules amplified in droplets, this technology can overcome the dependency on calibrants to determine copy number concentrations (77). Bio-Rad's QX100 system fractionates each sample into 20,000 individual nanoliter-sized droplets, and PCR amplification of the template molecules occurs in each individual droplet. PCR-positive and PCR-negative droplets from every sample are then counted to provide absolute target quantification in digital form. Highly precise quantitation is achieved by Poisson algorithms (78) that determine absolute copy numbers independently of a standard curve as the DNA fragments are independently segregated into droplets (79); more detailed information is provided in chapter 11 describing digital PCR. Random partitioning of the amplification fluorescence data fits a Poisson distribution (78) that can be used to determine the actual number of molecules present in the droplet. The equation used for calculation is the value of $\ln(1-p)$, where p is the fraction of positive droplets. ddPCR technology uses reagents and workflows similar to those used for most standard TaqMan probe-based real-time PCR assays. Though there are currently no FDA-approved assays available for clinical use today, the potential of ddPCR as an LDP holds great promise (10, 76, 80).

ASSESSMENT OF QUANTITATIVE ASSAY PERFORMANCE AND LIMITATIONS

General Concepts

Several important aspects of a quantitative nucleic acid assay can affect the use, results, and interpretation of a test; therefore, all assay conditions and characteristics, and their impact, should be individually taken into consideration for every assay and every technology (81). In order to understand the technical difficulties, limitations, and advantages of quantitation strategies for specific organisms, one must first understand some generalities that apply to all quantitative methods. Topics for consideration include: (i) inherent variability of testing methods, (ii) specimen integrity and matrix issues, (iii) technical issues related to the method verification and validation and comparative performance to other platforms, and (iv) statistical analysis of quantitative data. A complete discussion of all these aspects is outside the breadth of this chapter; however, some key aspects are discussed herein. The reader is referred to this chapter's additional references for more detail.

Quantification

In the standard curve approach, simple serial dilutions of a known organism density (e.g., CFU or PFU/ml), a known concentration of a synthetic oligonucleotide (e.g., copies/ml), or a known concentration of a cloned plasmid (e.g., copies/ml) are amplified in parallel with the samples of interest. Calibration curves, commonly called standard curves, are created based on amplification data obtained from these standards (typically through linear regression analysis) and used to calculate unknown concentrations of the target sequence using the CT value of the unknown and the equation that defines the slope of the standard curve batches. Historically, standard curves were tested with each quantitative real-time PCR batch tested; however, depending on the manufacturer, some new software options have the capability to "store" standard curves to be used for multiple batches over various time periods.

For the standard curve approach, repeated testing of external positive controls (EPC) of documented quantity with each batch tested is often called "trend analysis." There are several reasons to perform quantified controls when using external calibration curves. One limitation of the standard curve strategy is that the method cannot control for sample-to-sample variation. Even small differences in sample substrates may affect amplification efficiency and therefore skew the quantitative results; therefore, analysis of EPC results can often identify assay problems more easily, before they become major failures. In addition, cycling conditions of each run may vary slightly, also influencing results. Finally, since target quantities are calculated based on the CT of the densities within the standard curve, imprecision, which commonly occurs with low copy numbers of target, will alter the standard curve and therefore alter quantitative results between different batches. In summary, minor changes in a number of factors may result in assay imprecision; therefore, quantitative PCR results should be viewed as indicative of relative rather than absolute quantitation.

For microbial quantitation, whole-organism comparisons are best but are not always feasible; therefore, plasmids or oligonucleotides are often used as substitutes. Coamplified homologue templates and external standards are often used to control for amplification efficiency in an attempt to normalize results from one reaction to another. The most reliable quantitation can be achieved when coamplification of a synthetic oligonucleotide and an internal reference standard is performed. Ideally, the reference standard is designed with the same primer binding sites as, and sequence composition similar to, those of the target such that amplification efficiency is very similar for both the target and the standard. In this scenario, the internal reference control is added to the sample, and if target is present, both are coamplified in proportion to the relative amounts of control and the target template in the original sample (82–85). During method verification, this standard can be compared to the results of serial dilutions of organisms or target with known concentrations such that the performance of the synthetic control can be compared and, by mathematical conversion, adjusted (i.e., harmonized) to better reflect that of a live organism control (82–85).

External quality control and calibration material is not always sufficient to assess quantitative method performance due to differences between the composition of the control material and that of the clinical specimens. Plasmids typically do not behave in the same way as whole organisms in fresh specimens from patient samples. As such, plasmid-based calibrators may therefore introduce bias, which should be assessed prior to interpreting patient results (44, 86).

Despite the fact that manufacturers typically consider some matrix effects in their method evaluations, analytical differences in the results can consistently occur, even for FDA-cleared methods. The differences can be even wider for LDPs. For these reasons it is recommended that controls, standards, and calibrators that are used for method verification and assessment be as analogous to the actual specimen as possible. If this is not possible, an assessment of matrix effects and interfering substances is prudent (43, 86).

Assay Variability

For HIV, HCV, and other viral load testing, there are test utilization scenarios with immediate and profound clinical and financial consequences. Therapy may be terminated or extended based on viral load results, and a medically important error in a test result for viral load could lead to a patient remaining on failing therapy, which could lead to resistance and could increase the possibility of transmission.

The clinicians and patients assume that the quality of the assay allows for the correct answer to be delivered with just one test, but given the new medical decision points expected for some viral load assays, like HCV and HIV, that assumption may be best confirmed through objective and rigorous analysis such as Sigma metrics (refer to the section entitled "Sigma Metrics for Assessing Accuracy").

Many variables can interfere with quantitative molecular methods (Table 2). For *in vitro* amplification processes such as PCR or RT-PCR, variability can be introduced via preanalytical variables, which include the presence of amplification inhibitors in the samples, as well as improper collection, transport, and storage, among others. The effects of certain anticoagulants or alternate transport devices may also adversely affect the results of quantitative testing, as will delay in the specimen's transport to the testing laboratory. For these reasons a laboratory's quality program should include assessment and controls for preanalytical variables in addition to analytical variables.

Host factors may also contribute to assay variability (Table 2). For HIV type 1 (HIV-1) viral loads, vaccinations or other infections may cause transient increases in the viral loads, while female ovulation may decrease viral load. Nontarget amplification-based quantitative methods such as hybrid capture are also affected by many of these variables, but are not affected by the presence of enzyme amplification inhibitors since they are hybridization-based technologies.

Analytical variables may also pose problems and add to assay variability (Table 2). Variables include: (i) the quality

TABLE 2 Factors contributing to quantitative assay variability

Host factors	Preanalytical factors	Analytical factors	Postanalytical factors
Vaccination	Improper transport device	Quality of water, reagents, or consumables	Report interpretation
Other infections	Improper collection	Extraction or detection method	Diagnostic algorithms
Ovulation	Improper storage (i.e., temperature)	Pipette calibration error or pipetting variability	Noncompliance to clinical standards
Anticoagulants or other interfering substances or inhibitors in samples		Biologic or genetic variability of organisms	
		Human technical error	
		Temperature	

of laboratory water, (ii) the quality of plastic resins used in PCR tubes, (iii) calibration of pipettes and other equipment, (iv) temperature deviations of the heating devices or any room temperature incubation, (v) differences in extraction methods, and (vi) the types of detection methods used (i.e., plate washing for PCR–enzyme-linked immunosorbent assay [ELISA] or reader calibration of visible and UV light sources).

Minor technical or human error can compound the variability of these relatively complex quantitative molecular assays. Particularly problematic is pipetting variability of individuals and the variability inherent to the practice of pipetting small volumes (87). In addition, there is inherent error related to the statistical improbability of pipetting the target of interest when target concentration is very low. This variability can be characterized by the statistical analysis of the Poisson distribution (78). The lower the concentration of microorganisms or target, the greater the heterogeneity of the sample with regard to organism concentration, and the greater the likelihood of low-volume sampling error. Assay precision may be improved at higher concentrations and worsened at lower concentrations. At lower copy numbers, the statistical variation is larger, primarily because of sampling errors described by the Poisson distribution (78). In short, at the low end of quantitation, large variations may be due to the variability in the assay itself, rather than to true variations in viral load.

The biological variability of microorganisms may also play an important role in the inherent variability exhibited in quantitative molecular assays. For example, the genetic composition of a virus may play a role in the accuracy of its quantitation. While it is common and practical for synthetic quantitative HCV standards to be used in many commercial HCV assays, these standards are composed of only one HCV genotype; therefore, viral quantitation may vary when different genotypes are isolated (21). Synthetic viral particles offer some utility for standardization of diagnostic assays for HCV (88), but use of intact viral particles represents a more realistic condition with which to assess extraction and amplification processes (89).

Important components of quantitative molecular assays include precision, accuracy, and tolerance limit. The tolerance limit is: (i) the difference between two sequential samples that can be considered to be significantly different, and (ii) the sum of the biological variation in quantitation combined with intra-assay variability. For example, due to analytical and biological variables inherent with HIV-1 quantitative assays, changes in viral load are not considered to be significant until the change reflects an HIV concentration that is at least 3-fold (90) greater than the previous results (91). This tolerance limit of an assay should take assay variability into account and reflect only biologically relevant changes in the level of viral replication (91). Similar situations occur with HCV; due to variability in quantitative HCV viral loads, only a 3-fold (90) increase or decrease in viral load is considered to be significant. In the laboratory, precision can be monitored by using trend analysis of quantitative test controls. This practice is essential to provide a measure of reproducibility over time. Use of a low external positive control can provide assurance of overall function and employee competency and warn of upcoming issues with controls or instruments. Finally, assay accuracy can be best determined by the use of well-characterized standards and controls.

Considering all factors, it is more valid to consider that each amplification reaction generates quantitative results which are relative to the individual run and to other runs of the same method rather than to absolute measures. Highly reproducible correlation coefficients may be generated and suggest a precise result but not necessarily an accurate one. True accuracy depends on accuracy of the quantitative standards (21, 92, 93).

A thorough understanding of molecular microbiology assay parameters is important for interpretation of laboratory results and for comparison of results generated with different assays (94) or in different locations (91). Significant differences occur among laboratories, even when commercially available assays are used (21, 92, 95–97). Quality controls to assess kit-to-kit and lot-to-lot variations are helpful. Trend analysis can be performed on control parameters such as optical density readings for PCR-ELISA, relative light units for bDNA, and CT values for real-time PCR, to provide early warning for problems that may occur with reagent kits or equipment.

Postanalytical variables are mostly related to interpretation of reports (98), diagnostic algorithms for subsequent testing and therapy (99, 100), and the limited ability to compare methods to each other for assays that are not harmonized by international units. The copy numbers of target may not be equivalent if different assay platforms are used. While conversion factors for HIV assays currently exist and allow limited comparison between different assays and different reporting units (i.e., copies/ml versus international units (IU)/ml), the most accurate reflection of the viral load still comes from those results that were performed by the same assay platform and with the same version number of the assay. While all three of the current commercial assays for HIV (RT-PCR, bDNA, and NASBA) are significantly correlated, it is still not prudent to interchange assays for quantitation of viral load in the same patient (101–103); it should be avoided if possible. HIV-1 quantitative standards are available with density recorded as IU/ml and help to improve assay standardization and future comparability of these methods.

Impact of Specimen Integrity

Specimen preparation techniques contribute to the overall utility of molecular technologies. Optimal specimen preparation efficiently releases nucleic acid from the organism and places the target into an aqueous environment suitable for use in amplification and other molecular assays. The choice of nucleic acid extraction methods may enhance or detract from test performance, as will the choices for input extraction and output volumes. Test characteristics and results may vary considerably if any conditions are altered. Specimen volumes of 200 μl or less are common; however, in some cases larger volumes may be required, especially when the density of infectious organisms is low. These variables and their lack of standardization in specimen processing are important reasons why results are difficult to compare from laboratory to laboratory. Choice of extraction methods can cause variability of assay results (46, 104–108).

For quantitation, it is important that the physiological level of the virus in the sample be preserved. High levels of RNase enzymes in blood specimens make RNA targets very susceptible to degradation. In general, serum or plasma should be separated within 4 to 6 hours of collection and stored, ideally at −70°C for long-term storage, or at refrigerated temperature for short-term storage. Repeated freeze-thaws of specimens should be avoided as they can degrade nucleic acid and make the viral load appear falsely low. Each assay will have its own specific requirements for specimen collection, transport, and storage, and adherence to these requirements is crucial to quantitative assay performance.

In addition to the extraction method, the choice of specimen or dilution matrix for quantitative standards may also alter the standard curve results (21, 92, 109). Cloned DNA may behave differently than viral particles (23, 110, 111), and whole organisms may produce a better indicator of extraction efficiency than plasmid DNA (21, 92).

Whether the method is commercial or noncommercial, certain specimen collection and storage parameters apply. For example, EDTA and sodium citrated plasma are the preferred specimens for HCV PCR (112). Serum is also an acceptable specimen for some assays if it is centrifuged immediately after clot formation and frozen. Refrigerated (4°C) short-term storage of serum or plasma is also acceptable (113–116). For HIV-1 assays (Roche RT-PCR, NASBA, and bDNA), EDTA plasma is preferred, and if frozen within 8 hours of collection, plasma can be frozen and thawed up to three times without substantial loss. In one study, Amplicor HIV-1 Monitor results were maintained within 0.5 $\log_{10}$ (3-fold) for plasma at 4°C for up to 3 days and for long term at −70°C (90).

Another important aspect of assay limitations is the presence of inhibitors and interfering substances present in patient specimens or introduced during specimen collection or processing. Certain substances, such as bilirubin, hemoglobin, lipids, heparin, and food by-products, can be inhibitory or interfere with nucleic acid extraction or the assay method itself. Interaction with inhibitors and nucleic acid or critical enzymes, especially DNA polymerases, can prevent amplification of target. Likewise, inhibitors may remove reaction components (117), which affect enzymatic substrates.

The presence of such inhibitors can be determined by several methods, as described previously (118). Controls should be designed to detect the presence of inhibitors and to evaluate the quality and quantity of nucleic acid. One approach is to incorporate amplification controls into the design of the molecular assay. This can be accomplished by spiking the patient's specimen with characterized concentrations of intact organisms, or cloned nucleic acid target, that have/has a sequence composition similar but not identical to that of the microbial target. This approach assumes that the lower limit of detection and other assay performance characteristics are documented in matrices without inhibitors prior to patient testing. Then, when both target and amplification control are tested under the same conditions, inhibition may be present if the spiked patient control does not amplify or produces a weaker result than is typical in noninhibitory matrices.

When spiked specimens are assayed, the negative predictive value of the assay is generally enhanced and results are more reliable; however, the effectiveness of spiked controls varies depending on how closely control conditions mimic test conditions. For example, since cloned DNA is often easier to amplify than its genomic counterparts, the concentration chosen for cloned nucleic acid controls should be selected based on their lower limit of detection characteristics and documented performance, not on the performance of the corresponding intact pathogen. Furthermore, the control target sequence is optimally similar in size and guanine-cytosine (GC) ratio to the microbial target but should not be identical, so as not to compete directly with the microbial target.

Other approaches to identify inhibitors may be incorporated into assays. In certain circumstances the use of analyte-specific capture probes, which can be coupled to a solid matrix and washed to separate target DNA from inhibitors in specimens, may be particularly useful. Nucleic acid capture to a variety of solid matrices, such as magnetic beads or silica membranes, is also useful to bind nucleic acids while inhibitors are washed away.

Another important specimen issue is that of specimen adequacy. To assess the adequacy of specimen collection, primer sets for human housekeeping genes like human β-globin, β-actin, or GAPDH (glyceraldehyde-3-phosphate dehydrogenase), are designed and included in the assay. Measurements like these are useful only if cellularity is an indicator of adequacy. DNA concentration within a patient sample may also be measured spectrophotometrically, to ensure samples contain sufficient human DNA, which should be indicative of infected human cells. The choice of targets depends on the type of specimen collected and the type of disease state being tested. While not all are appropriate for all specimens, amplification of these targets allows the test to assess the presence of human cellular DNA, which should be present if specimen collection was appropriate.

Standardization

In an additional effort to standardize quantitative testing, the World Health Organization (WHO) and collaborators have established the WHO International Standards, which are described in their website, available at http://www.who.int/biologicals/en/. The first international standard for HCV RNA was established in 1997, based on the results of an international collaborative study (119, 120). Since then, calibration of working reagents has become possible and harmonization of data from individual laboratories is feasible. Standard reference materials with concentration expressed in IU/ml (as opposed to viral copies/ml) can be used to calibrate, validate, and compare commercially available quantitative molecular assays. However, the commutability of such reference materials for quantitative viral loads may not be as simple as described for other analytes (121). Commutability, described as the equivalence of the mathematical relationships between the results of different measurement procedures for a reference material and native clinical samples, is a critical property that ensures reference materials are fit for use across laboratories (121). Commutability is necessary to reduce laboratory-to-laboratory variability, yet recent data evaluating the commutability of the WHO standard for CMV using 10 different real-time PCR assays and run by 8 different laboratories indicated that the reference material showed poor or absent commutability for up to 50% of the assays (11). True consensus for commutable reference materials will require further assessment and analysis.

To date, quantitative international standards exist for seven viruses: CMV, EBV, Hepatitis A, HBV, HCV, HIV-1, and Parvovirus B19 (39, 119, 120, 122–124) with a collaborative study planned to evaluate the proposed 1st WHO International Standard for human cytomegalovirus (HCMV) with the HCMV standard created by the National Institute for Science and Technology (NIST). Further information can be found by contacting the National Institute for Biological Standards and Controls at http://www.nibsc.ac.uk, and at http://www.nist.gov/clinical-diagnostics.cfm. Proficiency panels for quantitative HCV and HBV assays have also been developed and are being used in clinical testing (125–127). Other widely used assays, including BK virus and the herpesviruses, do not have the advantage of established standards, and assay variability is a common limitation (41, 42, 86).

Laboratory Developed Procedures (LDP)

Akin to any laboratory assay, all assays do not have identical performance characteristics, whether FDA approved or not. In general, LDPs may have very different performance characteristics, but certain clinical parameters apply to all laboratory assays. Their definition is important to the reader's understanding of the advantages and limitations of molecular test methods. A recent publication (118) describes parameters such as analytical and clinical specificity and sensitivity and their relevance to LDP molecular testing. For quantitative molecular microbiological diagnostic testing, several other characteristics, such as the linear range and the upper and lower limits of quantification, are essential to physicians' understanding of test results.

Variance in the quality and characterization of non-FDA-cleared commercial methods and user-developed assays may be even greater. It is commonly known that there may be wide variability in the performance of some defined LDPs. Statistical assessment of variability is essential for the practical application of these assays (128). Since some degree of variability is accepted and inherent to commercial quantitative assays, it is difficult to decide how much variability is acceptable. Unfortunately, there is no firm answer to that question. To the extent possible, LDPs should maintain variability standards, which are similar to the current and accepted standard of care for similar assays in the marketplace. Assay design, verification, and validation are critical to the performance and use of quantitative molecular LDPs (128). Experimental design elements should include some consideration of the following: demographics of the patient population to be tested (since not all geographical locations will have the same genotypes), the purpose and proposed clinical utility of the new assay, and the scientific and biological background of the assay or the disease (128). In addition, response variables and control variables for statistical analysis of assay results should be defined prior to the actual method development and performance testing. Possible interferences, potential PCR inhibitors, and other matrix effects of the samples, controls, or standards used must also be considered and evaluated during the assay development process (39, 129, 130).

Sigma Metrics for Assessing Precision and Accuracy

Disease status and therapy decisions are often defined by a single positive or negative result; therefore, precision and diagnostic accuracy are critical to optimize patient care. Statistical assessment tools derived from testing quality control material, collectively called Sigma metrics, can help molecular diagnostic laboratories improve the assessment of their viral load assay's accuracy. Assays that can achieve a Six Sigma level of performance will by their nature have fewer false positives and reduce unnecessary repeat testing (131). To accomplish this, one must critically assess the results and know how to analyze quality control data to prove the assay is functioning properly.

The core requirement of any method used for quality assessing is the ability to numerically define what we consider good performance and acceptable quality; therefore, we must also define unacceptable quality and poor performance, i.e., we must define an unacceptable error that is a defect in the method's performance. For viral loads, we can define a defect or an error as any instance in which a patient result is misclassified as it related to medical decisions. In that context, there can be two kinds of defects: one occurs either when a positive result is misclassified as negative based on viral load; the other occurs when a normal, disease-free patient result is misclassified as positive or diseased (131).

The quality management techniques collectively known as "Six Sigma" have been practiced in health care for several decades, and for even longer in manufacturing, business, and industry. The core concept of Six Sigma is to identify defects (false negative or false positive reactions) and then reduce or eliminate as many of those defects as possible until a nearly defect-free operation is achieved.

Within the constructs of Sigma metrics, a defect occurs whenever a process outcome (for example, a viral load result) deviates beyond a predefined tolerance limit. That is to say, there is a numerical value that the test process should produce (sometimes called the "true value") and there is a defined amount of variation that is considered by experts to be acceptable. The acceptable variation bounds the true value on either side of the true value (e.g., a mean or an International Standard, etc.).

For diagnostic testing, laboratorians, clinicians, or statisticians place a reasonable limit on the amount of variation that can be accepted or tolerated, depending on the pathogen detected and therapeutic options available. The boundary of that acceptable variation is called the tolerance limit, which is generally specific to the assay, the disease, or the therapeutic decision points. Therefore, whenever the process variation exceeds the tolerance limit, a process defect is said to occur. In Sigma metrics, the number of defects that occur over time is reported on a scale of defects per million (DPM) or defects per million opportunities (DPMO).

When adapted for clinical laboratories, the "Six" in Six Sigma comes from the idea that six standard deviations (SD) of process variation must fit between the true value of a test result and the defined tolerance limit. When variation is limited to this degree of precision (Six Sigma), the process generates only approximately 3.4 DPMO, on what is called the short-term Sigma scale. When a process achieves Five Sigma on the short-term scale, approximately 233 DPM occur. With Four Sigma, the number rises to 6,210 DPM, and with Three Sigma, it rises again to 66,807. The rapid rise in defects observed as the Sigma metric declines helps to explain why Six Sigma is considered the ideal performance standard and Three Sigma is typically considered the minimum acceptable performance level for a process or assay. Processes operating at or below Three Sigma are typically the most error prone, resource consuming, difficult-to-maintain, and expensive processes. Within health care laboratories, test methods that operate at or below Three Sigma typically generate significant rework in the form of repeated controls, repeated calibrations, troubleshooting, and even a need for repeated testing of the patient in order to determine a diagnosis. In some cases, diagnostic methods performing significantly below Three Sigma have been recalled from the market. But the concepts of Six Sigma are relatively new to molecular diagnostic laboratories and are not yet commonly calculated. Awareness of the Sigma metrics should help molecular laboratories better define assay performance and improve overall quality (131).

Most molecular laboratories are not yet fully accustomed to discussing the "tolerance limits" of our test methods. Laboratorians are more familiar with a similar concept, known as the allowable total error (TEa). With the TEa, the net analytical error is defined as a combination of method imprecision (random error) and method inaccuracy (systematic error, or bias) (131, 132). When a test method exceeds the TEa, the method or process begins to generate defects (result errors), which could be classified as either false positive or false negative results. When we want to prevent our analytical method from exceeding the TEa we can rely on Sigma metrics. If we can achieve Six Sigma performance, in other words, maintain our imprecision at approximately less than one-sixth of the TEa, we keep our defect rate below 3.4 DPM.

For analytical testing processes, it is more difficult to determine whether a test result is a defect without doing parallel testing against a reference method or performing retrospective clinical chart review with correlation to the longitudinal disease state of the patient. Simple observation is not sufficient to identify all defects in viral load assays. For analytical methods, like viral load assays, an alternative approach can be deployed in order to calculate the Sigma metric. Rather than count defects, we instead calculate the expected Sigma metric through an equation that combines the impacts of imprecision and inaccuracy. The Sigma metric equation (131, 132) is as follows: Sigma metric = (TEa − bias)/CV. In this equation, the TEa, bias, and coefficient of variation (%CV) are all expressed as percentages and the absolute value of the bias is used. For any molecular method, as long as all the variables are expressed in either logarithmic or nonlogarithmic numerical scales, it is possible to calculate a Sigma metric.

For viral loads, the TEa value is established based on clinical guidelines, so the percent TEa reflects a medically important change (i.e., 0.5 log copies/ml in an HIV viral load). The %CV and bias values are obtained from repeated measurement of laboratory control material used to assess precision during the method verification or by analyzing longitudinal quality control data. The %CV and bias values can also be obtained from comparisons with peer group surveys or proficiency testing surveys. Examples of the relationships between Tea, bias, CV and defects are depicted in Figure 2 and are more fully described in references 131 and 132.

The Sigma metric can be calculated with the variables expressed in unit-based measurements, as long as all the terms of the equation are kept consistent. The graph in Figure 3 is a visual explanation of the Sigma metric equation, showing how both bias and imprecision affect the distribution of test results. Whenever the curve of the distribution exceeds the TEa (the lines drawn on either side of the true value), the area remaining under the curve represents the number of expected defects.

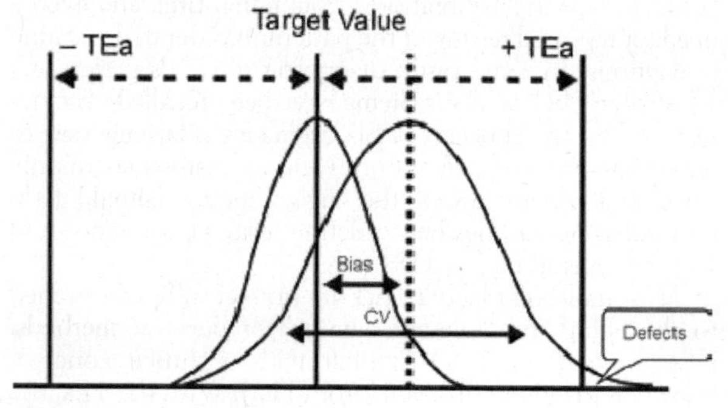

FIGURE 2

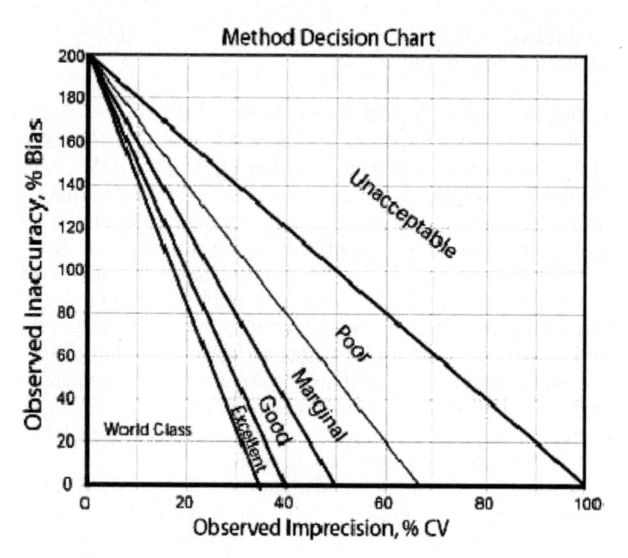

FIGURE 3

While the Sigma metric can be calculated, it can also be visually displayed using a method decision chart (Figure 3). This chart plots the performance of a test method as a combination of the observed imprecision and bias, specifically, using imprecision as the x coordinate and bias as the y coordinate. The diagonal lines on the chart represent different Sigma zones, from world class quality performance (better than Six Sigma) near the graph's origin to excellent performance (Five Sigma), to good performance (Four Sigma), and so on, until the last zone, the upper right quadrant of the graph, represents less than Two Sigma performance, which is considered unacceptable, unstable, and unreliable. The visual simplicity of the graph shows the relationship between the various Sigma conditions: the closer to the origin, the fewer defects and the more confidence the laboratory and clinicians can place in the test result. Conversely, as imprecision and bias grow, they progressively impair a method's performance, generating more defects and confusing the clinician instead of confirming a diagnosis. In one recommended approach, laboratories can choose to adopt the more demanding quality requirements at either end of the quantitative spectrum, i.e., that the highest negative patient never exceeds the cutoff, and likewise, the lowest positive patient never falls below the cutoff (131, 132).

Most importantly, for a growing number of medical therapies, the historical range of acceptable variability (0.5 $\log_{10}$ copies/ml) is no longer the most important decision point used by clinicians for disease management. For many treatment regimens, a far lower cutoff is used, one that defines the success of a drug therapy as the achievement of eradication or near eradication of the viral load (133–135). In these cases, the numerical value reported is often the actual unit measured, not a number converted to the logarithmic scale. To determine the quality of method performance in those scenarios, it will be more appropriate to use the actual units to define the TEa. Thus, we may express the TEa for a test on either the logarithmic or nonlogarithmic scale, depending on the clinical decision level we are interested in assessing.

Six Sigma statistical concepts have much broader implications than those of a single patient management decision. From a therapy management perspective, a virus with resistance-associated mutations is more challenging to treat, with narrower windows for therapy options. In addition, false positive rates (blip frequency) can add unnecessary costs due to extensive repeat testing or more expensive resistance testing, and increase anxiety for the patient. Tracking an assay and instrument performance with Sigma metrics allows laboratories to identify defects in assays prior to downstream patient management decisions and impact.

Sigma metrics can be implemented only with clinically defined TEa limits, which must relate directly to the treatment guidelines for the disease. When these TEa limits are defined, a Sigma metric can be calculated and an objective judgment can be made about method precision and acceptability (52). Once the Sigma metric is calculated, that assessment can be leveraged into additional actionable steps for the laboratory, for example, design of the appropriate quality control procedures (i.e., the number of controls, the control limits, and, to some extent, the frequency of controls), as well as estimations of the number of tests that will be required to detect an error within the viral density range where medical decisions are to be made. In summary, the Sigma metric is not just a benchmark; it is the first step in a process of optimizing an assay's performance in the laboratory and determining its most effective use in diagnostic and treatment pathways. Six-Sigma analyses provide a useful tool to assay precision and overall quality for viral load testing. The Sigma metric translates abstract analytical performance characteristics into tangible measures that can impact laboratory operations and patient outcomes. In previous decades, extreme precision may not have been as necessary, but today's medical treatments demand higher precision. Without setting a higher bar for analytical performance, more blips and errors will occur in viral loads, and fewer desirable outcomes will be achieved (52).

APPLICATIONS OF VIRAL LOADS

Human Immunodeficiency Virus 1 (HIV-1)

Viral load testing is recommended as the preferred monitoring method to diagnose and confirm antiretroviral (ART) treatment failure in the latest World Health Organization guidelines, updated in 2013. However, if viral load is not routinely available, CD4 count and clinical monitoring should be used to diagnose treatment failure. The WHO guidelines do not recommend HIV load testing initially, at the time of HIV diagnosis, but recommend testing 6 months into ART treatment and every 12 months thereafter to detect any treatment failure.

A plasma viral load >1,000 copies/ml on two consecutive measurements at least 3 months apart, in the setting of adherence counseling, is indicative of a treatment failure and the need to change to second-line ART. The rationale for the threshold of 1,000 copies/ml was based on two main sources of evidence. First, viral blips or intermittent low-level viremia (50 to 1,000 copies/ml) can occur during effective treatment but have not been associated with an increased risk of treatment failure unless low-level viremia is sustained (136, 137). Second, clinical and epidemiological studies show that the risk of HIV transmission and disease progression is very low when the viral load is lower than 1,000 copies/ml (138–140).

The guidelines advocate that treatment should not be withheld if laboratory capabilities are not available and both CD4 and viral load testing should be performed only if resources permit. The development of new technologies, like point-of-care tests, for these resource-limited areas has led to the use of certain technologies that use whole blood as a sample type in the form of dried blood spots, which is unreliable at lower thresholds. Therefore, where these are used a higher threshold should be adopted since the preferred specimen type for HIV-1 quantification is plasma. Data provided in the WHO Interim Technical Update on Implementing HIV Viral Load Testing indicates the threshold of 1,000 copies/ml can be effectively utilized for dried blood spots used on multiple laboratory platforms (141). Most standard blood and plasma viral load platforms available and being developed have good diagnostic accuracy at this lower threshold. However, the sensitivity of dried blood spots for viral load determination at this threshold may be reduced (141). Diagnostic testing that relies on dried blood spot technology for viral load assessment should consider retaining the higher threshold (3,000 to 5,000 copies/ml) until sensitivity at lower thresholds is established (141).

For HIV viral loads, examples of the ways in which many of the described parameters can affect the performance and result interpretation of the HIV-1 RT-PCR assay are well known. It is the responsibility of the laboratory to provide clinicians with an awareness of an assay's inherent variability, as QRT-PCR HIV assays exhibit differences among laboratories. At best, a 0.5 $\log_{10}$ variance (a 3-fold difference) is documented to exist between repeats of the same concentration, even with an FDA-cleared method. Inherent assay variability makes more accurate quantitation possible only within the same run of the same method; therefore, changes of small magnitude should not take on assumed relevance. Current therapy guidelines (Department of Health and Human Services [DHHS]) for HIV-1 state that virologic failure occurs when the viral load exceeds 200 copies/ml; therefore, a physician would need to confirm this change in the viral load with a subsequent viral load measurement (142–145).

As previously described, precision is reduced at lower concentrations and increased variability is typically exhibited when target concentration is less than 200 copies/ml, so an HIV patient with consecutive undetectable viral load results (146) may, at some point, have a viral load result of 250 IU/ml and still be well within the normal variability of the quantitative RT-PCR. A rise in the viral load may simply indicate a "viral blip," that is, a temporary rise that will resolve itself or it may continue to rise and indicate an actual change in the patient's status (virologic failure). For these reasons, a >3-fold difference in viral load is often required before decisions that would affect therapy are made (147–151). Although assay improvements have removed some variability, HIV-1 quantitation may also vary with genotype or subtypes (152–155). Most of the Group M genotypes are now accurately quantitated, but not all assays will accurately quantitate or even detect Group O genotypes (152, 153, 155–159).

There are two recent reports that serve as examples of the way in which HIV-1 viral load assay performance, specifically imprecision, can impact patient management and therapy. Naeth et al. (160) evaluated patient samples at densities of 40, 80, and 90 copies/ml. Using the study data and applying the Sigma metric, the authors were able to predict how likely one could identify a real change in the viral load with a value between 40 to 200 cps/ml. This viral load change from 40 to 200 cps/ml (0.7 log copies/ml) represents a clinically significant change as it is greater than the 0.5 $\log_{10}$ unit currently required by the DHHS guidelines. In this example, the %CVs, which were used to assess viral load changes, were 26% and 51%, respectively. Sigma analysis, as described previously in this chapter, demonstrated that within this clinical decision interval (40 to 200 cps/ml) one assay's performance achieves a Six Sigma level (less than 3.4 defects or errors per million) while the comparator performance achieved only the Three Sigma level (66,800 defects or errors per million).

Ruelle et al. (161) provides another example of viral load performance at the clinical cutoff. Similar to the previous example, after applying the Sigma metric, authors were able to ascertain the ability of an assay to identify a significant change in the viral load from 25 to 400 cps/ml (a 1.2-log-copies/ml change). This change in the viral load represents a larger shift than DHHS considers virologic failure. In this study, the %CVs were 41% and 83%, respectively. Sigma analysis demonstrated that one assay performance achieved a Six Sigma level (less than 3.4 defects or errors per million) while the comparator performance achieved Four Sigma performance (6,210 defects or errors per million).

Hepatitis C Virus (HCV)

Several commercially available assays and LDPs are routinely in use for quantification of HCV RNA, with results standardly reported in international units per milliliter (IU/ml). In contrast to HIV-1 infection, HCV viral loads remain relatively stable in untreated patients with chronic HCV; consequently, the viral load and HCV genotype are not good predictors of disease severity or progression (162, 163). Instead, quantitation of HCV viral load provides important information about the treatment response of patients undergoing antiviral therapy (135, 164).

The treatment goal for chronic HCV infection is to prevent complications as a result of normalizing alanine aminotransferase levels and achieving a sustained virologic response (SVR), which is defined as the absence of detectable viremia at the end of treatment and 6 months later. Achievement of an SVR is heavily dependent on the genotype of the hepatitis C virus (165). Quantitative HCV assays provide a baseline value for assessment of the kinetics of therapy in order to counsel patients about the likelihood of an SVR and likely duration of the treatment. Patients with genotype 1 viruses tend to be treated with longer therapeutic courses, whereas those infected with genotypes 2 or 3 infections have better SVR rates. Established standard of care for chronic HCV infection is a combination of weekly injections of PEGylated interferon alpha (Peg-IFN) and oral ribavirin, with a rate of treatment success of 40% to 50% for genotype 1 (165, 166). The recent introduction of telaprevir and boceprevir increased the SVR rate from approximately 50% to >70%, but this therapy is only approved for treating HCV genotype 1 infection.

Several other novel therapies, including simeprevir, faldaprevir, and asunaprevir, have already been, or will soon be, approved for treating genotype 1. Interferon-free treatment regimens for HCV genotype 1 are gaining traction, with newer regimens having shorter durations, fewer side effects, low pill burden, and efficacy approaching 90% to 100% (see Table 3) (167). The three new drug regimens currently available in the United States are ledipasvir/sofosbuvir (LDV/SOF), simeprevir/sofosbuvir (SIM+SOF), and paritaprevir/ritonavir/ombitasvir/dasabuvir (OBV/PTV/r +DSV). Potential for cure of each of the new regimens is patient, subtype, and duration specific, with previous treatment experience or failure, HIV-1 confection, and cirrhosis all impacting SVRs (167).

Novel treatments or combination therapies for genotypes 2 and 3, daclatasvir, and sofosbuvir, show great promise with high SVR rates and reductions in treatment duration to 12 weeks (168). Sofosbuvir with ribavirin treatment for 12 weeks for genotype 2 HCV and 24 weeks for genotype 3 resulted in overall SVRs of 93% and 84%, respectively. More detailed information on the indications for use of sofosbuvir in patients with HIV or those awaiting liver transplant is addressed in Lam et al. (167). In addition, a combination of daclatasvir and sofosbuvir can be used for treatment of genotypes 2 and 3 (169).

Improvements in HCV therapy require that the precision of HCV assays, particularly at the low end of the dynamic range, be exceedingly accurate. Current HCV treatment guidelines state that sustained virologic response at the end of treatment is achieved if the viral load is <25 IU/ml (170). Therefore precision near the LLOQ must be as precise as possible because new clinical decision points are in place.

In a study by Wiesmann et al. (171), HCV patients were monitored for a 2-log drop from baseline to ensure therapeutic response. They evaluated HCV assay reproducibility at 25 IU/ml using two commercial HCV assays and demonstrated that precision ranged between 23% CV and 51% CV. Using the analytical performance at 25 IU/ml, the authors applied Sigma analysis, where total error was the change in viral load from 25 IU/ml (the assay LLOQ) to 50 IU/ml. Assay performance ranged from Six Sigma to Three Sigma, respectively, with the lower sigma rightly associated with the higher %CV assay. The use of a highly precise Six Sigma method could benefit patients, as it can document a more accurate assessment of sustained virologic response and clearance of HCV near the LLOQ.

Similar HCV assay precision was documented by Kessler et al. (172) when the World Health Organization International Standard was used to evaluate precision at 100, 50, and 25 IU/ml. This study demonstrated that at 25 IU/ml, one assay's %CV was 36% while the comparator's CV was 95%. The former assay performed at the Five Sigma level (230 errors per million) while the comparator performed at the Two Sigma level (308,000 errors per million, or 31% of all assay's with errors at the LLOQ), an error rate not generally considered acceptable in the clinical laboratory. Five Sigma assays can more accurately detect changes in viral

TABLE 3 Newer regimens available for genotype 1 with FDA-approved indications. Adapted from Lam, 2015 (167)

Regimen name	Brand name (manufacturer)	Duration (weeks)*	Potential for cure*	FDA approval date
LDV/SOF	Harvoni (Gilead Sciences)	8–24	95–100%	10 October 2014
SMI+SOF	Olysio (Janssen Therapeutics)	12–24	95–100%	6 November 2014
OBV/PTV/r +DSV**	Viekira Pak (AbbVie Inc.)	12–24	93–100%	19 December 2014

*Dependent on patient population; see reference for details.
**Other indications include HCV/HIV-1 coinfection or liver transplant recipients with mild fibrosis.

load, which confirm sustained virologic response, assess viral clearance of HCV, or indicate virologic failure.

Therapeutic monitoring has been simplified by real-time quantitative PCR applications due to greater sensitivity compared to qualitative TMA-based assays or bDNA-based formats. In general, the sensitive qualitative assays are typically used to evaluate the end-of-treatment response (EOT), since low levels of residual RNA can be found in a proportion of conventional PCR-negative EOT samples (173–175). The highly sensitive real-time quantitative PCR assays can be used to evaluate the early virologic response of infections to discontinue treatment in nonresponders and modify the treatment strategy. However, when less-sensative quantitative test methods, like conventional PCR, are used, use of the qualitative test methods in combination is recommended to assess the rapid virologic response, EOT, and SVR. Quantification and genotyping of HCV RNA will continue to evolve as a key component of the therapeutic strategy for chronic HCV treatment, especially as new therapeutics receive FDA approval for routine clinical use.

Using the current medical decision points at 25 IU/ml, a large amount of inherent imprecision can only be mitigated by testing duplicate or triplicate samples and using the average value as the viral load results. While this approach will improve the accuracy of methods with high inherent error, if also adds cost to the process. Selection of the most accurate and precise method to detect changes in viral load will more readily confirm sustained virologic response, assess viral clearance, identify virologic failure, and limit overall testing costs.

Hepatitis B Virus (HBV)

The risk of development of cirrhosis and hepatocellular carcinoma in chronic hepatitis B infected patients is a priority in the management strategy for treatment, with HBV viral load testing playing a vital role. Traditionally, the HBV e antigen (HBeAg) has been the marker for active viral replication; however, with advances in molecular testing methods, chronic HBV disease outcome has been well correlated with HBV viral loads. Though monitoring HBV DNA levels provides the best evaluation of disease, it is the achievement of either HBsAg seroconversion or HBsAg loss that are the ideal outcomes of antiviral treatment, since low levels of HBV DNA may persist in some patients that have serologically recovered (176). Serial monitoring of HBV DNA levels is the best approach to determine the need for treatment as opposed to an arbitrary cutoff value; however, it is now realized that low levels of HBV DNA (3 $\log_{10}$ IU/ml to 5 $\log_{10}$ IU/ml) may warrant treatment and be linked to liver disease in those who are HBeAg negative or have developed cirrhosis (176). To predict likelihood of a sustained response to HBV treatment, one proposed treatment plan has recommended viral load testing at 24 weeks of treatment to characterize the virologic response as complete, partial, or inadequate (177). The 2009 update of the American Association for the Study of Liver Diseases (AASLD) Practice Guidelines for Management of Chronic Hepatitis B defines a virologic response to therapy as a decrease in serum HBV DNA to undetectable levels by PCR assays, and a loss of HBeAg in patients who were initially HBeAg positive. A complete response is defined as fulfilling the criteria for the virologic response, loss of HBsAg, as well as a biochemical response or decrease in serum ALT to within normal ranges (176). Thus, measurement of HBV viral loads is critical to determining the efficacy of therapy in chronic HBV patients, whether it is assessing for a total reduction or the kinetics of a decrease in HBV DNA levels.

Cytomegalovirus (CMV)

Clinical utility of quantitative viral loads for CMV has most extensively been demonstrated in solid organ transplant (SOT) recipients for prognostication of CMV disease, to guide preventative treatment, assess efficacy of treatment, guide treatment duration, and indicate risk of CMV clinical relapse or antiviral drug resistance (9). CMV infection is widespread; seroprevalence rates vary depending on several factors, including age, geography, and socioeconomic status. In the United States, the estimated overall CMV seroprevalence is 50.4%, yet some studies have described rates approaching 100% in some populations (178, 179). Exposure to CMV via infected saliva and body secretions is the primary route of infection in immunocompetent persons and presents as an asymptomatic or self-limited illness. CMV may reactivate throughout life in these individuals, yet a functional immune system fights off the infection, and rarely does an immunocompetent individual exhibit clinical illness. These latently infected individuals can, however, transmit CMV to susceptible individuals. Hence, hematopoietic cell (HCT) and solid-organ transplant (SOT) recipients that receive infected donor organs, cells, or cellular blood products are quite commonly at risk for CMV infection.

Though the incidence of CMV infection is dependent on the type of organ, up to 75% of SOT recipients who do not receive antiviral prophylaxis will experience a CMV infection (180). Those at highest risk for CMV infection are lung, intestinal, and pancreas recipients; moderately at risk are heart and liver recipients; and the lowest risk of CMV infection and disease is in kidney recipients (180, 181). After SOT, CMV infection may occur as a primary infection, reactivation infection (reactivation of latent virus in the recipient), or superinfection (reactivation of latent virus in donor cells) (181). Primary infection is defined as receipt of an allograft from a CMV-seropositive donor (CMV D+/R-) by a CMV-seronegative recipient, acquiring CMV by natural transmission or from a CMV-seropositive blood product (181). Risk of CMV disease is higher for CMV Donor seropositive/Recipient seropositive (D+/R+) patients than for CMV D-/R+ SOT patients (9).

Most HCT patients will reactivate latent CMV rather than acquire a primary infection (182). In HCT patients, the paradigm exhibited in SOT recipients is reversed. Although, a greater risk of CMV reactivation and progression to CMV disease occurs in CMV R+ HCT patients, irrespective of CMV donor status (183), CMV D-/R+ HCT patients experience further major complications compared to CMV D+/R+ patients (184–186).

The onset of CMV infection is delayed when a prophylactic prevention strategy is employed within the first 3 months after SOT, deferring CMV infection occurrence until sometime during the first 3 months, after completion of CMV prophylaxis. Generally the risk of CMV disease beyond this period is low unless SOT recipients are under a greater degree of immunosuppression.

Three major strategies exist for prevention of CMV disease: antiviral prophylaxis, monitoring viral reactivation, and a hybrid approach. Antiviral prophylaxis as a prevention strategy is managed most commonly with valganciclovir administered for a defined duration, generally 3 to 6 months to all patients at risk for CMV disease. In contrast to SOT recipients, use of ganciclovir as a prophylactic treatment strategy in HCT recipients is limited by the myelosuppression induced by the drug. Use of aciclovir or valganciclovir for prophylaxis has the following potential benefits when using these drugs in selected patients: reducing the need for hospital admission and/or IV preemptive therapy, reducing indirect effects of CMV reactivation on immune status post-transplant, and delaying CMV reactivation until the patient no longer requires immunosuppression and has recovered from transplant-associated toxicity (187).

Monitoring SOT and HCT patients for CMV reactivation using highly sensitive nucleic acid amplification tests is an alternative approach. Upon detection of a positive viral threshold, treatment is initiated until the viral level falls below a clinically relevant threshold. Monitoring of CMV DNA load in HCT recipients should be performed at least weekly for the first 3 months posthematopoietic stem cell transplant and should continue to 6 to 12 months if the patient has chronic graft versus host disease (GvHD) or prolonged immunodeficiency (187). A hybrid approach to reduce the incidence of late-onset CMV disease uses antiviral prophylaxis that is followed by the preemptive approach of monitoring viral loads and initiating treatment in those who develop CMV reactivation above a predefined viral threshold. CMV treatment should be continued until the quantitative viral load declines below the predefined threshold or to an undetectable level (9). Viral load thresholds to initiate preemptive therapy that can be applied universally in all patients post-transplantation have not been established. Thresholds up to 1,000 copies/ml have been suggested, yet other thresholds dependent on the recipient's risk have also been recommended (188). Recently published data for HCT patients recommends a threshold of 135 IU/ml as the trigger for early preemptive therapy to reduce the time to resolution of viremia (median time of 15 days for 135 to 440 IU/ml, 18 days for 441 to 1,000 IU/ml, 21 days at >1,000 IU/ml) and duration of therapy (median time of 28 days for 135 to 440 IU/ml, 34 days for 441 to 1,000 IU/ml, 37 days at >1,000 IU/ml) (189).

CMV disease and infection severity correlates with the viral replication (i.e., viral load); high absolute viral load values represent active CMV replication, and low viral load values are representative of latent viral DNA (9, 181). The higher CMV viral loads correlate with symptomatic infections and tissue-invasive disease compared to asymptomatic CMV syndrome. Sensitivity and specificity of CMV assays are greatly impacted by the target nucleic acid, RNA versus DNA, and the specimen type, whole blood versus serum or plasma, and should be considered when deciding which assay to utilize. RNA targeted assays, though less sensitive than DNA, have been developed to serve as a better indicator of clinical disease; however, RNA is readily degraded and proper transport and processing is critical. CMV DNA is rather stable in specimens and is rarely affected by delayed sample processing, yet is less specific for active CMV infection partly due to the ability of such assays to detect latent viral DNA. Specificity issues have been overcome with the development and routine use of quantitative CMV assays. With the availability of two FDA-cleared assays, as well as the recently released WHO international standard for CMV quantitation, harmonization of CMV quantitative viral loads is promising. In this regard, the recent evaluation of the CMV WHO standard across multiple quantitative assays, which clearly indicates a lack of commutability of the WHO reference material across assay and laboratories, suggests that though these advances are promising for standardizing viral load thresholds and treatment strategies, interlaboratory variability will still be an issue and requires further study (11).

Quantitative CMV viral loads may be used to assess treatment response utilizing viral load decline and viral load suppression as indicators. Higher viral loads at diagnosis have been associated with longer treatment duration. A recent study showed SOT recipients with a CMV load at diagnosis of <18,200 (4.3 $\log_{10}$) IU/ml had a shorter time to disease resolution compared to those with a higher CMV viral load, whose time to clinical disease resolution was 5 days longer (7 days versus 12 days, respectively) (190). Current treatment guidelines recommend viral eradication from blood prior to discontinuing antiviral therapy. Viral eradication occurs later than clinical resolution, therefore viral load suppression should be assessed beyond the resolution of clinical disease, and until the viral load is suppressed to levels below a safe threshold, before discontinuance of treatment (9). Use of assays calibrated to the WHO international reference standard is supported by studies indicating that the presence of detectable virus in the blood is highly associated with disease relapse (191). Therefore, viral load suppression to levels <137 (2.14 $\log_{10}$) IU/ml as the end of treatment is significant of clinical disease resolution (190).

Epstein-Barr Virus

Quantitative detection of EBV is commonly performed in T-cell-depleted allogeneic stem cell transplant patients (7, 192) and many other disease conditions (7, 193–198). Use of EBV viral loads enabled development of models for preemptive anti-B-cell immunotherapy for EBV reactivation, and for reducing not only the incidence of EBV lymphoproliferative disease (EBV-LPD) (199–201), but also the virus-related mortality (202, 203). Quantitative analyses of circulating EBV DNA in nasopharyngeal carcinomas have demonstrated a positive correlation with disease stage and a strong relationship with clinical events, as well as being of prognostic importance (204). For EBV-associated lymphomas, quantitative EBV DNA analysis has also been found to correlate closely with clinical progress (205). High variability existed (206, 207) until recently, when EBV quantitative standards became available (124, 208).

Other Herpesviruses

Quantitative viral load assays are useful for establishing active virus replication of other ubiquitous herpesviruses, like human HHV-6, HHV-7, and HHV-8. The high rate of reactivation and asymptomatic excretion of these viruses is difficult to correlate with clinical disease without the use of quantitative assays, as qualitative assays cannot differentiate latent from active viral infection. HCT and SOT recipients are at risk for developing symptomatic HHV-6 infections, commonly through reactivation of latent virus (209). HHV-6 DNA detected in plasma has been a proposed marker of active viral infection; however, differentiating the source of the DNA, whether from cell-free viral particles produced by active infection of lymphoid tissues or from the lysis of circulating infected cells in the blood, has not been successful. Later, it was demonstrated that HHV-6 in plasma likely originates from the lysis of infected circulating blood cells as opposed to circulating cell-free viral particles, thus quantification of HHV-6 DNA in whole blood has a higher sensitivity for diagnosis of active HHV-6 infection originating in lymphoid tissues rather than measuring DNA load in purified peripheral blood mononuclear cells (PBMCs) or plasma (210–212). Like many of the quantitative viral load assays, HHV-6 viral load assays are not standardized or interchangeable; furthermore, thresholds to guide treatment have not yet been established. The literature supports a viral load of $>10^3$ targets per 10^6 PBMCs in stem cell transplant patients associated with clinical symptoms (209). It should be noted that the use of quantitative viral load assays to assess HHV-6 infection can be confounded by the possibility of cross-reactivity with chromosomally integrated HHV-6.

BK virus (BKV)

Originally described in 1971, BK virus (BKV) was isolated from urine of a renal transplant patient with ureteral stenosis and bearing the initials "B. K."(213). Primary infection of BKV occurs in early childhood, 65% to 90% of children age 5 to 9 are seropositive, with chronic latent carriage in several organs and tissues (kidneys, urothelium, leukocytes, etc.) (214). Reactivation of latent virus occurs during episodes of immunosuppression, allowing the virus to replicate to pathogenic levels. As such, immunocompromised patients can experience a host of complications attributed to BKV including most commonly, BK virus associated nephropathy (BKVAN) and hemorrhagic cystitis. Less common complications of BKV reactivation include pneumonitis, retinitis, liver disease, and meningoencephalitis (215). Because the most significant clinical impact of BKV is the complications associated with BKVAN, the reader is referred to several other publications that discuss the further clinical complications of BKV (215–217).

Reactivation of BKV starts within three months after transplantation and occurs in 30% to 50% of renal transplant recipients (218). Among renal transplant recipients, 80% will exhibit BK viruria, with 5% to 10% developing BKVAN. Widespread use of PCR for quantitation has confirmed that BK viruria and viremia are reliably identified before the development of BKVAN (219). Screening for persistent BK virus shedding, presence of decoy cells in urine, or urine viral loads of $>10^7$copies/ml, has been shown to identify patients at risk for developing BKVAN (216, 220). BK viral load is usually 1,000-fold higher in the urine than in the plasma (220); however, patients may present with asymptomatic viruria without disease. Therefore, positive urine viral loads should be confirmed with plasma viral loads and renal biopsy (215). Though highly specific and definitive, the "gold standard" renal biopsy can result in false negatives due to focal involvement, especially during early stages of BKVAN. "Presumptive" BKVAN, defined as a plasma viral load of $>10^4$ copies/ml, is 93% sensitive and specific for associated positive tissue histology (221). Evaluation of a renal biopsy to exclude other coexisting pathologic processes, or exclude BKVAN altogether as the cause of allograft dysfunction, as well as staging for prognosis, remains an important component of therapeutic management (219).

No specific antiviral to BKV exists; therefore, decreasing immunosuppression is the most common form of therapy for BKVAN. Other pharmaceutical options that have shown activity, primarily *in vitro*, though have not been formally approved for the treatment of BKV associated nephropathy, include cidofovir, leflunomide, quinolones, and intravenous IgG (215). In two studies, Kadambi et al. and Vats et al., BKV clearance in patients was observed when treated with cidofovir; however, since cidofovir is transported into tubular epithelial cells and excreted renally, the risk of nephrotoxicity should be considered (222, 223). Use of leflunomide and quinolone as therapeutic options have successfully been shown to result in viral clearance or decrease of viremia, yet intravenous Ig treatment has shown mixed results with some effect on BK viremia but no significant improvement in graft survival (224–226).

Monitoring peripheral blood for evidence of BK viremia through quantitative real-time PCR testing is an important management tool that allows for interventions that prevent nephropathy in renal allograft patients. Significant genetic heterogeneity of BKV exists and that variability dramatically affects the assays performance. The implications of BKV DNA sequence variation for the performance of molecular diagnostic assays are only beginning to be studied (220, 227, 228).

SUMMARY AND FUTURE IMPLICATIONS

In little more than a decade, quantitative molecular methods have become an integral part of standard medical practice for prognosis and treatment of many viral infections. Clinical laboratories now have an abundance of technologies including PCR-ELISA, RT-PCR, real-time PCR, bDNA, NASBA, TMA, and other non-PCR amplification methods for quantitation of infectious agents. Other applications, including gene expression assays and assays to model pathogenesis, also rely on quantitative methods. Applications of these methods will increase dramatically in the near future as host profiling and genetic factors related to the process of infection and treatments are defined.

Many variables exist and influence the utility of current quantitative molecular assays. Variables such as platform choice, condition and concentration of target, specific microorganism characteristics, and specimen processing must be taken into consideration in order to validate and interpret quantitative methods. Other important aspects of method design include primer and probe selection, selection of the type of controls and standards used to enable accurate quantitation, and determination of assay threshold. As with all laboratory assays, limitations related to aspects as basic as sampling error still exist. All these variables combined make statistical analysis of method validation and routine performance crucial to the accuracy of quantitative methods.

Recent advances in technology have resulted in rapid, user-friendly, automated, contamination-resistant testing platforms that will allow some quantitative molecular testing methods to replace certain conventional microbiology laboratory assays. As new equipment and technologies continue to evolve, traditional issues related to method verification and validation for nonmolecular quantitative methods will be considered for their potential application to quantitative diagnostic techniques. The ultimate success of these technologies depends on their successful application to patient care and their related overall cost. A quality systems approach to verification and validation of molecular microbiologic diagnostic assays is in the early stages and will require further development. Some standardized methods and controls exist, but more will be needed to provide a basis for method comparisons. Further analysis of the clinical utility, test utilization, and pathogen-or-disease-targeted algorithms will also enable better use of these technologies.

Quantitative methods and viral loads are already an integral part of disease management, and the future of quantitative testing depends on our ability to apply quality practices and statistical analyses and quality metrics to commercial or user-defined methods and to incorporate standardization into our testing methods. The effectiveness of technology depends on careful attention to quality-based practices and evidence-based medicine to ensure that results from quantitative methods may be more easily compared. Comparison of quantitative methods is difficult and will be improved with the development of more internationally accepted standards and controls. Automation will add to reproducibility and may reduce overall costs. When user-developed quantitative methods are applied to clinical testing, assay performance characteristics must be carefully documented and scrutinized. Quantitative data must be carefully correlated to disease in order to assess the clinical utility of the methods by determining clinical sensitivity and specificity. Adherence to standard quality practices is essential, as practitioners of molecular diagnostic methods exercise great care to maintain the highest standards when incorporating molecular methods into existing diagnostic algorithms. Speed, accuracy, and utilization of results will be paramount to the future of quantitative technology as new methods extend our understanding of pathogenesis and advance our ability to improve diagnosis and disease management.

REFERENCES

1. **Wolk D, Mitchell S, Patel R.** 2001. Principles of molecular microbiology testing methods. *Infect Dis Clin North Am* **15:**1157–1204.
2. **Hayden R, Persing DH.** 2001. Diagnostic molecular microbiology. *Curr Clin Top Infect Dis* **21:**323–348.
3. **Hodinka RL.** 1998. The clinical utility of viral quantitation using molecular methods. *Clin Diagn Virol* **10:**25–47 .
4. **Jungkind D, Kessler HH.** 2002. Molecular Methods for Diagnosis of Infectious Disease, p 306–323. *In* Truant AL (ed), *Manual of Commercial Methods in Clinical Microbiology.* ASM Press, Washington, D.C.
5. **Wolk DM, Persing DH.** 2002. Clinical Microbiology: Looking Ahead, p 429–450. *In* Truant AL (ed), *Commercial Methods in Clinical Microbiology.* ASM Press, Washington, D.C.
6. **Cumitech A.** 1998. *Laboratory Diagnosis of Hepatitis Virus.* ASM Press, Washington, D.C.
7. **Gulley ML, Tang W.** 2008. Laboratory assays for Epstein-Barr virus-related disease. *J Mol Diagn* **10:**279–292.
8. **Halfon P, Bourlière M, Pénaranda G, Khiri H, Ouzan D.** 2006. Real-time PCR assays for hepatitis C virus (HCV) RNA quantitation are adequate for clinical management of patients with chronic HCV infection. *J Clin Microbiol* **44:**2507–2511.
9. **Razonable RR, Hayden RT.** 2013. Clinical utility of viral load in management of cytomegalovirus infection after solid organ transplantation. *Clin Microbiol Rev* **26:**703–727.
10. **Hayden RT, Gu Z, Ingersoll J, Abdul-Ali D, Shi L, Pounds S, Caliendo AM.** 2013. Comparison of droplet digital PCR to real-time PCR for quantitative detection of cytomegalovirus. *J Clin Microbiol* **51:**540–546.
11. **Hayden RT, Preiksaitis J, Tong Y, Pang X, Sun Y, Tang L, Cook L, Pounds S, Fryer J, Caliendo AM.** 2015. Commutability of the first WHO international Standard for Human Cytomegalovirus. *J Clin Microbiol.*
12. **Vincent E, Gu Z, Morgenstern M, Gibson C, Pan J, Hayden RT.** 2009. Detection of cytomegalovirus in whole blood using three different real-time PCR chemistries. *J Mol Diagn* **11:**54–59.
13. **Abe A, Inoue K, Tanaka T, Kato J, Kajiyama N, Kawaguchi R, Tanaka S, Yoshiba M, Kohara M.** 1999. Quantitation of hepatitis B virus genomic DNA by real-time detection PCR. *J Clin Microbiol* **37:**2899–2903.
14. **Broccolo F, Locatelli G, Sarmati L, Piergiovanni S, Veglia F, Andreoni M, Buttò S, Ensoli B, Lusso P, Malnati MS.** 2002. Calibrated real-time PCR assay for quantitation of human herpesvirus 8 DNA in biological fluids. *J Clin Microbiol* **40:**4652–4658.
15. **de Baar MP, Timmermans EC, Bakker M, de Rooij E, van Gemen B, Goudsmit J.** 2001. One-tube real-time isothermal amplification assay to identify and distinguish human immunodeficiency virus type 1 subtypes A, B, and C and circulating recombinant forms AE and AG. *J Clin Microbiol* **39:**1895–1902.
16. **Kawazu M, Kanda Y, Goyama S, Takeshita M, Nannya Y, Niino M, Komeno Y, Nakamoto T, Kurokawa M, Tsujino S, Ogawa S, Aoki K, Chiba S, Motokura T, Ohishi N, Hirai H.** 2003. Rapid diagnosis of invasive pulmonary aspergillosis by quantitative polymerase chain reaction using bronchial lavage fluid. *Am J Hematol* **72:**27–30.
17. **Kleiber J, Walter T, Haberhausen G, Tsang S, Babiel R, Rosenstraus M.** 2000. Performance characteristics of a quantitative, homogeneous TaqMan RT-PCR test for HCV RNA. *J Mol Diagn* **2:**158–166.
18. **Loeb KR, Jerome KR, Goddard J, Huang M, Cent A, Corey L.** 2000. High-throughput quantitative analysis of hepatitis B virus DNA in serum using the TaqMan fluorogenic detection system. *Hepatology* **32:**626–629.
19. **Machida U, Kami M, Fukui T, Kazuyama Y, Kinoshita M, Tanaka Y, Kanda Y, Ogawa S, Honda H, Chiba S, Mitani K, Muto Y, Osumi K, Kimura S, Hirai H.** 2000. Real-time automated PCR for early diagnosis and monitoring of cytomegalovirus infection after bone marrow transplantation. *J Clin Microbiol* **38:**2536–2542.
20. **Martell M, Gómez J, Esteban JI, Sauleda S, Quer J, Cabot B, Esteban R, Guardia J.** 1999. High-throughput real-time reverse transcription-PCR quantitation of hepatitis C virus RNA. *J Clin Microbiol* **37:**327–332.
21. **Niesters HG.** 2001. Quantitation of viral load using real-time amplification techniques. *Methods* **25:**419–429.
22. **Niesters HG, Krajden M, Cork L, de Medina M, Hill M, Fries E, Osterhaus AD.** 2000. A multicenter study evaluation of the digene hybrid capture II signal amplification technique for detection of hepatitis B virus DNA in serum samples and testing of EUROHEP standards. *J Clin Microbiol* **38:**2150–2155.
23. **Pas SD, Fries E, De Man RA, Osterhaus AD, Niesters HG.** 2000. Development of a quantitative real-time detection assay for hepatitis B virus DNA and comparison with two commercial assays. *J Clin Microbiol* **38:**2897–2901.
24. **Pham AS, Tarrand JJ, May GS, Lee MS, Kontoyiannis DP, Han XY.** 2003. Diagnosis of invasive mold infection by real-time quantitative PCR. *Am J Clin Pathol* **119:**38–44.
25. **Sashihara J, Tanaka-Taya K, Tanaka S, Amo K, Miyagawa H, Hosoi G, Taniguchi T, Fukui T, Kasuga N, Aono T, Sako M, Hara J, Yamanishi K, Okada S.** 2002. High incidence of human herpesvirus 6 infection with a high viral load in cord blood stem cell transplant recipients. *Blood* **100:**2005–2011.
26. **Si HX, Tsao SW, Poon CS, Wang LD, Wong YC, Cheung AL.** 2003. Viral load of HPV in esophageal squamous cell carcinoma. *Int J Cancer* **103:**496–500.
27. **Varma M, Hester JD, Schaefer FW III, Ware MW, Lindquist HD.** 2003. Detection of Cyclospora cayetanensis using a quantitative real-time PCR assay. *J Microbiol Methods* **53:** 27–36.
28. **Weinberger KM, Wiedenmann E, Böhm S, Jilg W.** 2000. Sensitive and accurate quantitation of hepatitis B virus DNA using a kinetic fluorescence detection system (TaqMan PCR). *J Virol Methods* **85:**75–82.
29. **Yates S, Penning M, Goudsmit J, Frantzen I, van de Weijer B, van Strijp D, van Gemen B.** 2001. Quantitative detection of hepatitis B virus DNA by real-time nucleic acid sequence-based amplification with molecular beacon detection. *J Clin Microbiol* **39:**3656–3665.
30. **Baldanti F, Grossi P, Furione M, Simoncini L, Sarasini A, Comoli P, Maccario R, Fiocchi R, Gerna G.** 2000. High levels of Epstein-Barr virus DNA in blood of solid-organ transplant recipients and their value in predicting posttransplant lymphoproliferative disorders. *J Clin Microbiol* **38:**613–619.
31. **Drew WL.** 2007. Laboratory diagnosis of cytomegalovirus infection and disease in immunocompromised patients. *Curr Opin Infect Dis* **20:**408–411.

32. **Engelmann I, Petzold DR, Kosinska A, Hepkema BG, Schulz TF, Heim A.** 2008. Rapid quantitative PCR assays for the simultaneous detection of herpes simplex virus, varicella zoster virus, cytomegalovirus, Epstein-Barr virus, and human herpesvirus 6 DNA in blood and other clinical specimens. *J Med Virol* **80:**467–477.
33. **Gunson RN, Maclean AR, Shepherd SJ, Carman WF.** 2009. Simultaneous detection and quantitation of cytomegalovirus, Epstein-Barr virus, and adenovirus by use of real-time PCR and pooled standards. *J Clin Microbiol* **47:**765–770.
34. **Jebbink J, Bai X, Rogers BB, Dawson DB, Scheuermann RH, Domiati-Saad R.** 2003. Development of real-time PCR assays for the quantitative detection of Epstein-Barr virus and cytomegalovirus, comparison of TaqMan probes, and molecular beacons. *J Mol Diagn* **5:**15–20.
35. **Lin MH, Chen TC, Kuo TT, Tseng CC, Tseng CP.** 2000. Real-time PCR for quantitative detection of Toxoplasma gondii. *J Clin Microbiol* **38:**4121–4125. In Process Citation.
36. **Razonable RR, Brown RA, Wilson J, Groettum C, Kremers W, Espy M, Smith TF, Paya CV.** 2002. The clinical use of various blood compartments for cytomegalovirus (CMV) DNA quantitation in transplant recipients with CMV disease. *Transplantation* **73:**968–973.
37. **Mansfield E, O'Leary TJ, Gutman SI.** 2005. Food and Drug Administration regulation of in vitro diagnostic devices. *J Mol Diagn* **7:**2–7.
38. **Foy CA, Parkes HC.** 2001. Emerging homogeneous DNA-based technologies in the clinical laboratory. *Clin Chem* **47 (6):**990–1000.
39. **NCCLS.** Quantitative Molecular Methods for Infectious Diseases, NCCLS, 940 West Valley Road, Suite 1400, Wayne, PA 19087.
40. **Hayden RT, Hokanson KM, Pounds SB, Bankowski MJ, Belzer SW, Carr J, Diorio D, Forman MS, Joshi Y, Hillyard D, Hodinka RL, Nikiforova MN, Romain CA, Stevenson J, Valsamakis A, Balfour HH Jr, U.S. EBV Working Group.** 2008. Multicenter comparison of different real-time PCR assays for quantitative detection of Epstein-Barr virus. *J Clin Microbiol* **46:**157–163.
41. **Hoffman NG, Cook L, Atienza EE, Limaye AP, Jerome KR.** 2008. Marked variability of BK virus load measurement using quantitative real-time PCR among commonly used assays. *J Clin Microbiol* **46:**2671–2680.
42. **Wolff DJ, Heaney DL, Neuwald PD, Stellrecht KA, Press RD.** 2009. Multi-Site PCR-based CMV viral load assessment-assays demonstrate linearity and precision, but lack numeric standardization: a report of the association for molecular pathology. *J Mol Diagn* **11:**87–92.
43. **Bustin SA, Mueller R.** 2005. Real-time reverse transcription PCR (qRT-PCR) and its potential use in clinical diagnosis. *Clin Sci (Lond)* **109:**365–379.
44. **Hoorfar J, Wolffs P, Rådström P.** 2004. Diagnostic PCR: validation and sample preparation are two sides of the same coin. *APMIS* **112:**808–814.
45. **Kubista M, Andrade JM, Bengtsson M, Forootan A, Jonák J, Lind K, Sindelka R, Sjöback R, Sjögreen B, Strömbom L, Ståhlberg A, Zoric N.** 2006. The real-time polymerase chain reaction. *Mol Aspects Med* **27:**95–125.
46. **Mengelle C, Legrand-Abravanel F, Mansuy JM, Barthe C, Da Silva I, Izopet J.** 2008. Comparison of two highly automated DNA extraction systems for quantifying Epstein-Barr virus in whole blood. *J Clin Virol* **43:**272–276.
47. **Nolte FS.** 1998. Branched DNA signal amplification for direct quantitation of nucleic acid sequences in clinical specimens. *Adv Clin Chem* **33:**201–235.
48. **Pang XL, Martin K, Preiksaitis JK.** 2008. The use of unprocessed urine samples for detecting and monitoring BK viruses in renal transplant recipients by a quantitative real-time PCR assay. *J Virol Methods* **149:**118–122.
49. **Paraskevis D, Haida C, Tassopoulos N, Raptopoulou M, Tsantoulas D, Papachristou H, Sypsa V, Hatzakis A.** 2002. Development and assessment of a novel real-time PCR assay for quantitation of HBV DNA. *J Virol Methods* **103:**201–212.
50. **Paredes R, Marconi VC, Campbell TB, Kuritzkes DR.** 2007. Systematic evaluation of allele-specific real-time PCR for the detection of minor HIV-1 variants with pol and env resistance mutations. *J Virol Methods* **146:**136–146.
51. **Reischl U, Haber B, Vandezande W, De Baere T, Vaneechoutte M.** 1997. Long-term storage of unamplified complete PCR mixtures. *Biotechniques* **23:**580–582, 584.
52. **Westgard SLD, Lucic D.** 2015. Sigma metrics for assessing accuracy of molecular testing. *Clin Microbiol Newsl* **37 (13):**103–110.
53. **Nazarenko IA, Bhatnagar SK, Hohman RJ.** 1997. A closed tube format for amplification and detection of DNA based on energy transfer. *Nucleic Acids Res* **25:**2516–2521.
54. **Wittwer CT, Herrmann MG, Moss AA, Rasmussen RP.** 1997. Continuous fluorescence monitoring of rapid cycle DNA amplification. *Biotechniques* **22:**130–131, 134–138.
55. **Chen X, Zehnbauer B, Gnirke A, Kwok PY.** 1997. Fluorescence energy transfer detection as a homogeneous DNA diagnostic method. *Proc Natl Acad Sci USA* **94:**10756–10761.
56. **Livak KJ, Flood SJA, Marmaro J, Giusti W, Deetz K.** 1995. Oligonucleotides with fluorescent dyes at opposite ends provide a quenched probe system useful for detecting PCR product and nucleic acid hybridization. *PCR Methods Appl* **4:**357–362.
57. **Gibson UE, Heid CA, Williams PM.** 1996. A novel method for real time quantitative RT-PCR. *Genome Res* **6:**995–1001.
58. **Heid CA, Stevens J, Livak KJ, Williams PM.** 1996. Real time quantitative PCR. *Genome Res* **6:**986–994.
59. **Lee LG, Livak KJ, Mullah B, Graham RJ, Vinayak RS, Woudenberg TM.** 1999. Seven-color, homogeneous detection of six PCR products. *Biotechniques* **27:**342–349.
60. **Tyagi S, Bratu DP, Kramer FR.** 1998. Multicolor molecular beacons for allele discrimination. *Nat Biotechnol* **16:**49–53.
61. **Tyagi S.** 1996. Taking DNA probes into a protein world. *Nat Biotechnol* **14:**947–948.
62. **Saha BK, Tian B, Bucy RP.** 2001. Quantitation of HIV-1 by real-time PCR with a unique fluorogenic probe. *J Virol Methods* **93:**33–42.
63. **Baker JL, Ward BM.** 2014. Development and comparison of a quantitative TaqMan-MGB real-time PCR assay to three other methods of quantifying vaccinia virions. *J Virol Methods* **196:**126–132.
64. **Wenzel JJ, Walch H, Bollwein M, Niller HH, Ankenbauer W, Mauritz R, Höltke HJ, Zepeda HM, Wolf H, Jilg W, Reischl U.** 2009. Library of prefabricated locked nucleic acid hydrolysis probes facilitates rapid development of reverse-transcription quantitative real-time PCR assays for detection of novel influenza A/H1N1/09 virus. *Clin Chem* **55:**2218–2222.
65. **Leo E, Venturoli S, Cricca M, Musiani M, Zerbini M.** 2009. High-throughput two-step LNA real time PCR assay for the quantitative detection and genotyping of HPV prognostic-risk groups. *J Clin Virol* **45:**304–310.
66. **Hlousek L, Voronov S, Diankov V, Leblang AB, Wells PJ, Ford DM, Nolling J, Hart KW, Espinoza PA, Bristol MR, Tsongalis GJ, Yen-Lieberman B, Slepnev VI, Kong LI, Lee MC.** 2012. Automated high multiplex qPCR platform for simultaneous detection and quantification of multiple nucleic acid targets. *Biotechniques* **52:**316–324.
67. **Cai T, Lou G, Yang J, Xu D, Meng Z.** 2008. Development and evaluation of real-time loop-mediated isothermal amplification for hepatitis B virus DNA quantification: a new tool for HBV management. *J Clin Virol* **41:**270–276.
68. **Deiman B, van Aarle P, Sillekens P.** 2002. Characteristics and applications of nucleic acid sequence-based amplification (NASBA). *Mol Biotechnol* **20:**163–179.
69. **Grüner N, Viazov S, Korn K, Knöll A, Trippler M, Schlaak JF, Gerken G, Roggendorf M, Ross RS.** 2015. Performance characteristics of the VERSANT hepatitis C virus RNA 1.0 (kPCR) assay. *Int J Med Microbiol* **305:**627–635.

70. **Kessler HH, Hübner M, Konrad PM, Stelzl E, Stübler MM, Baser MH, Santner BI.** 2013. Genotype impact on HCV RNA levels determined with the VERSANT HCV RNA 1.0 assay (kPCR). *J Clin Virol* **58:**522–527.
71. **Bizouarn F.** 2014. Clinical applications using digital PCR. *Methods Mol Biol* **1160:**189–214.
72. **Dingle TC, Sedlak RH, Cook L, Jerome KR.** 2013. Tolerance of droplet-digital PCR vs real-time quantitative PCR to inhibitory substances. *Clin Chem* **59:**1670–1672.
73. **Sedlak RH, Jerome KR.** 2013. Viral diagnostics in the era of digital polymerase chain reaction. *Diagn Microbiol Infect Dis* **75:**1–4.
74. **Petriv OI, Heyries KA, VanInsberghe M, Walker D, Hansen CL.** 2014. Methods for multiplex template sampling in digital PCR assays. *PLoS One* **9:**e98341.
75. **Strain MC, Lada SM, Luong T, Rought SE, Gianella S, Terry VH, Spina CA, Woelk CH, Richman DD.** 2013. Highly precise measurement of HIV DNA by droplet digital PCR. *PLoS One* **8:**e55943.
76. **Henrich TJ, Gallien S, Li JZ, Pereyra F, Kuritzkes DR.** 2012. Low-level detection and quantitation of cellular HIV-1 DNA and 2-LTR circles using droplet digital PCR. *J Virol Methods* **186:**68–72.
77. **Corbisier P, Pinheiro L, Mazoua S, Kortekaas AM, Chung PY, Gerganova T, Roebben G, Emons H, Emslie K.** 2015. DNA copy number concentration measured by digital and droplet digital quantitative PCR using certified reference materials. *Anal Bioanal Chem* **407:**1831–1840.
78. **Paulson DS.** 2008. *Biostatistics and Microbiology: A Survivial Manual.* Springer, New York, NY.
79. **Pinheiro LB, Coleman VA, Hindson CM, Herrmann J, Hindson BJ, Bhat S, Emslie KR.** 2012. Evaluation of a droplet digital polymerase chain reaction format for DNA copy number quantification. *Anal Chem* **84:**1003–1011.
80. **Hall Sedlak R, Jerome KR.** 2014. The potential advantages of digital PCR for clinical virology diagnostics. *Expert Rev Mol Diagn* **14:**501–507.
81. **Dailey PJ, Hayden D.** 1999. Viral load assays: methodologies, variables, and interpretation. In: Cohen PT, Sande MA, Volberding PA, eds. The AIDS Knowledge Base 3rd ed. Philadelphia, PA: Lippincott-Raven Publishers. p 119–129.
82. **Ferre F.** 1992. Quantitative or semi-quantitative PCR: reality versus myth. *PCR Meth Appl* **2:**1–9.
83. **Persing DH.** 1993. In Vitro Nucleic Acid Amplification Techniques, p 51–87. *In* Persing DH, Smith TF, Tenover FC, White TJ (ed), *Diagnostic Molecular Microbiology, Principles and Applications.* ASM Press, Washington, D.C.
84. **Martin GP, Timmers E.** 1997. PCR and Its Modifications for the Detection of Infectious Diseases, p 79–99. *In* Lee H, Morse S, Olsvik O (ed), *Nucleic Acid Amplification Technologies: Application to Disease Diagnosis.* Eaton Publishing, Natick, MA.
85. **Udo Reischl U, Kochanowski B.** 1995. Molecular Biotechnology. *Vol 3, Issue 1 pp 55–71* Humana Press Inc. New York, New York.
86. **Murphy J, Bustin SA.** 2009. Reliability of real-time reverse-transcription PCR in clinical diagnostics: gold standard or substandard? *Expert Rev Mol Diagn* **9:**187–197.
87. **Nebbia G, Mattes FM, Cholongitas E, Garcia-Diaz A, Samonakis DN, Burroughs AK, Emery VC.** 2007. Exploring the bidirectional interactions between human cytomegalovirus and hepatitis C virus replication after liver transplantation. *Liver Transpl* **13:**130–135.
88. **Cartwright CP.** 1999. Synthetic viral particles promise to be valuable in the standardization of molecular diagnostic assays for hepatitis C virus. *Clin Chem* **45:**2057–2059. editorial; comment.
89. **Poljak M, Seme K, Koren S.** 1997. Evaluation of the automated COBAS AMPLICOR hepatitis C virus PCR system. *J Clin Microbiol* **35:**2983–2984.
90. **Sebire K, McGavin K, Land S, Middleton T, Birch C.** 1998. Stability of human immunodeficiency virus RNA in blood specimens as measured by a commercial PCR-based assay. *J Clin Microbiol* **36:**493–498.
91. **Kellogg JA, Atria PV, Sanders JC, Eyster ME.** 2001. Intra- and interlaboratory variabilities of results obtained with the Quantiplex human immunodeficiency virus type 1 RNA bDNA assay, version 3.0. *Clin Diagn Lab Immunol* **8:**560–563.
92. **Niesters HG.** 2001. Standardization and quality control in molecular diagnostics. *Expert Rev Mol Diagn* **1:**129–131.
93. **Sambrook J, Russell DW.** 2001. *Molecular Cloning: A Laboratory Manual,* **vol 3**. Cold Spring Harbor Press, Cold Spring Harbor, New York.
94. **Detmer J, Lagier R, Flynn J, Zayati C, Kolberg J, Collins M, Urdea M, Sánchez-Pescador R.** 1996. Accurate quantification of hepatitis C virus (HCV) RNA from all HCV genotypes by using branched-DNA technology. *J Clin Microbiol* **34:**901–907.
95. **Quint WG, Heijtink RA, Schirm J, Gerlich WH, Niesters HG.** 1995. Reliability of methods for hepatitis B virus DNA detection. *J Clin Microbiol* **33:**225–228.
96. **Schmitt Y.** 2001. Performance characteristics of quantification assays for human immunodeficiency virus type 1 RNA. *J Clin Virol* **20:**31–33.
97. **Zaaijer HL, Cuypers HT, Reesink HW, Winkel IN, Gerken G, Lelie PN.** 1993. Reliability of polymerase chain reaction for detection of hepatitis C virus. *Lancet* **341:**722–724.
98. **Centers for Disease Control and Prevention (CDC).** 2014. National HIV Testing Day and new testing recommendations. *MMWR Morb Mortal Wkly Rep* **63:**537.
99. **Sheorey H, Waters MJ.** 2001. Viral hepatitis? Which test should I order? *Aust Fam Physician* **30:**433–437.
100. **Wolk DM, Jones MF, Rosenblatt JE.** 2001. Laboratory diagnosis of viral hepatitis. *Infect Dis Clin North Am* **15:**1109–1126.
101. **Centers for Disease Control and Prevention.** 2001. Guidelines for laboratory test result reporting of human immunodeficiency virus type 1 ribonucleic acid determination. Recommendations from a CDC working group. *MMWR Recomm Rep* **50**(RR-20)**:**1–12.
102. **Centers for Disease Control and Prevention.** 2001. Revised guidelines for HIV counseling, testing, and referral. *MMWR Recomm Rep* **50**(RR-19)**:**1–57, quiz CE1, a1–CE6, a1.
103. **Centers for Disease Control and Prevention.** 2001. Revised recommendations for HIV screening of pregnant women. *MMWR Recomm Rep* **50**(RR-19)**:**63–85, quiz CE1, a2–CE6, a2.
104. **Dundas N, Leos NK, Mitui M, Revell P, Rogers BB.** 2008. Comparison of automated nucleic acid extraction methods with manual extraction. *J Mol Diagn* **10:**311–316.
105. **Fafi-Kremer S, Brengel-Pesce K, Barguès G, Bourgeat MJ, Genoulaz O, Seigneurin JM, Morand P.** 2004. Assessment of automated DNA extraction coupled with real-time PCR for measuring Epstein-Barr virus load in whole blood, peripheral mononuclear cells and plasma. *J Clin Virol* **30:**157–164.
106. **Fafi-Kremer S, Morand P, Barranger C, Barguès G, Magro S, Bés J, Bourgeois P, Joannes M, Seigneurin JM.** 2008. Evaluation of the Epstein-Barr virus R-gene quantification kit in whole blood with different extraction methods and PCR platforms. *J Mol Diagn* **10:**78–84.
107. **Fredricks DN, Smith C, Meier A.** 2005. Comparison of six DNA extraction methods for recovery of fungal DNA as assessed by quantitative PCR. *J Clin Microbiol* **43:**5122–5128.
108. **Miller S, Seet H, Khan Y, Wright C, Nadarajah R.** 2010. Comparison of QIAGEN automated nucleic acid extraction methods for CMV quantitative PCR testing. *Am J Clin Pathol* **133:**558–563.
109. **Niesters HG, van Esser J, Fries E, Wolthers KC, Cornelissen J, Osterhaus AD.** 2000. Development of a real-time quantitative assay for detection of Epstein-Barr virus. *J Clin Microbiol* **38:**712–715.
110. **Kato N, Yokosuka O, Hosoda K, Ito Y, Ohto M, Omata M.** 1993. Quantification of hepatitis C virus by competitive reverse transcription-polymerase chain reaction: increase of the virus in advanced liver disease. *Hepatology* **18:**16–20.

111. **Nitsche A, Steuer N, Schmidt CA, Landt O, Ellerbrok H, Pauli G, Siegert W.** 2000. Detection of human cytomegalovirus DNA by real-time quantitative PCR. *J Clin Microbiol* **38:**2734–2737.
112. **Dufour DR, Lott JA, Nolte FS, Gretch DR, Koff RS, Seeff LB.** 2000. Diagnosis and monitoring of hepatic injury. I. Performance characteristics of laboratory tests. *Clin Chem* **46:**2027–2049.
113. **Bukh J, Purcell RH, Miller RH.** 1992. Importance of primer selection for the detection of hepatitis C virus RNA with the polymerase chain reaction assay. *Proc Natl Acad Sci USA* **89:**187–191.
114. **Busch MP, Wilber JC, Johnson P, Tobler L, Evans CS.** 1992. Impact of specimen handling and storage on detection of hepatitis C virus RNA. *Transfusion* **32:**420–425.
115. **Davis GL, Lau JY, Urdea MS, Neuwald PD, Wilber JC, Lindsay K, Perrillo RP, Albrecht J.** 1994. Quantitative detection of hepatitis C virus RNA with a solid-phase signal amplification method: definition of optimal conditions for specimen collection and clinical application in interferon-treated patients. *Hepatology* **19:**1337–1341.
116. **Dohner DE, Dehner MS, Gelb LD.** 1995. Inhibition of PCR by mineral oil exposed to UV irradiation for prolonged periods. *Biotechniques* **18:**964–967.
117. **Liu J, Geng Y, Pound E, Gyawali S, Ashton JR, Hickey J, Woolley AT, Harb JN.** 2011. Metallization of branched DNA origami for nanoelectronic circuit fabrication. *ACS Nano* **5:**2240–2247.
118. **Kawai S, Yokosuka O, Imazeki F, Saisho H, Mizuno C.** 2002. Evaluation of the clinical usefulness of COBAS AMPLICOR HCV MONITOR assay (ver2.0): comparison with AMPLICOR HCV MONITOR assay (ver1.0) and HCV core protein level. *J Med Virol* **68:**343–351.
119. **Saldanha J, Heath A, Lelie N, Pisani G, Nübling M, Yu M, The Collaborative Study Group.** 2000. Calibration of HCV working reagents for NAT assays against the HCV international standard. *Vox Sang* **78:**217–224.
120. **Saldanha J, Lelie N, Heath A, WHO Collaborative Study Group.** 1999. Establishment of the first international standard for nucleic acid amplification technology (NAT) assays for HCV RNA. *Vox Sang* **76:**149–158.
121. **Vesper HW, Miller WG, Myers GL.** 2007. Reference materials and commutability. *Clin Biochem Rev* **28:**139–147.
122. **Saldanha J.** 1999. Standardization: a progress report. *Biologicals* **27:**285–289.
123. **Schnepf N, Scieux C, Resche-Riggon M, Feghoul L, Xhaard A, Gallien S, Molina JM, Socié G, Viglietti D, Simon F, Mazeron MC, Legoff J.** 2013. Fully automated quantification of cytomegalovirus (CMV) in whole blood with the new sensitive Abbott RealTime CMV assay in the era of the CMV international standard. *J Clin Microbiol* **51:**2096–2102.
124. **Abeynayake J, Johnson R, Libiran P, Sahoo MK, Cao H, Bowen R, Chan KC, Le QT, Pinsky BA.** 2014. Commutability of the Epstein-Barr virus WHO international standard across two quantitative PCR methods. *J Clin Microbiol* **52:**3802–3804.
125. **Gilbert N, Corden S, Ijaz S, Grant PR, Tedder RS, Boxall EH.** 2002. Comparison of commercial assays for the quantification of HBV DNA load in health care workers: calibration differences. *J Virol Methods* **100:**37–47.
126. **Jorgensen PA, Neuwald PD.** 2001. Standardized hepatitis C virus RNA panels for nucleic acid testing assays. *J Clin Virol* **20:**35–40.
127. **Valentine-Thon E, van Loon AM, Schirm J, Reid J, Klapper PE, Cleator GM.** 2001. European proficiency testing program for molecular detection and quantitation of hepatitis B virus DNA. *J Clin Microbiol* **39:**4407–4412.
128. **Wolk DM, Marlowe EM.** 2015. Molecular Method Verification. *In* Persing DH (ed), *Molecular Microbiology: Diagnostic Principles and Practice*, **vol 3**. ASM Press, Washington, DC.
129. **Krouwer JS.** 2002. *Assay Development and Evaluation—a manufacturer's perspective*. American Association for Clinical Chemistry.
130. **Saunders GC.** 1999. *Analytical Moecular Biology—Quality and Validation*. Thomas Graham House, Science Park, Milton Road, Cambridge CB4 0WF, UK.
131. **Westgard JO, Westgard SA.** 2015. Assessing quality on the Sigma scale from proficiency testing and external quality assessment surveys. *Clin Chem Lab Med* **53:**1531–1535.
132. **Westgard S.** 2013. Prioritizing risk analysis quality control plans based on Sigma-metrics. *Clin Lab Med* **33:**41–53.
133. **Ambrosioni J, Nicolas D, Agüero F, Manzardo C, Miro JM.** 2014. HIV treatment outcomes in Europe and North America: what can we learn from the differences? *Expert Rev Anti Infect Ther* **12:**523–526.
134. **Coppola N, Martini S, Pisaturo M, Sagnelli C, Filippini P, Sagnelli E.** 2015. Treatment of chronic hepatitis C in patients with HIV/HCV coinfection. *World J Virol* **4:**1–12.
135. **Shaheen MA, Idrees M.** 2015. Evidence-based consensus on the diagnosis, prevention and management of hepatitis C virus disease. *World J Hepatol* **7:**616–627.
136. **Havlir DV, Bassett R, Levitan D, Gilbert P, Tebas P, Collier AC, Hirsch MS, Ignacio C, Condra J, Günthard HF, Richman DD, Wong JK.** 2001. Prevalence and predictive value of intermittent viremia with combination hiv therapy. *JAMA* **286:**171–179.
137. **Gale HB, Gitterman SR, Hoffman HJ, Gordin FM, Benator DA, Labriola AM, Kan VL.** 2013. Is frequent CD4+ T-lymphocyte count monitoring necessary for persons with counts >=300 cells/μL and HIV-1 suppression? *Clin Infect Dis* **56:**1340–1343.
138. **Ioannidis JP, Abrams EJ, Ammann A, Bulterys M, Goedert JJ, Gray L, Korber BT, Mayaux MJ, Mofenson LM, Newell ML, Shapiro DE, Teglas JP, Wilfert CM.** 2001. Perinatal transmission of human immunodeficiency virus type 1 by pregnant women with RNA virus loads <1000 copies/ml. *J Infect Dis* **183:**539–545.
139. **Mayaux MJ, Dussaix E, Isopet J, Rekacewicz C, Mandelbrot L, Ciraru-Vigneron N, Allemon MC, Chambrin V, Katlama C, Delfraissy JF, Puel J, SEROGEST Cohort Group.** 1997. Maternal virus load during pregnancy and mother-to-child transmission of human immunodeficiency virus type 1: the French perinatal cohort studies. *J Infect Dis* **175:**172–175.
140. **Townsend CL, Cortina-Borja M, Peckham CS, de Ruiter A, Lyall H, Tookey PA.** 2008. Low rates of mother-to-child transmission of HIV following effective pregnancy interventions in the United Kingdom and Ireland, 2000–2006. *AIDS* **22:**973–981.
141. **Organization WH.** Technical and operational considerations for implementing HIV viral load testing.
142. **Althoff KN, et al, North American AIDS Cohort Collaboration on Research and Design (NA-ACCORD).** 2014. Disparities in the quality of HIV care when using US Department of Health and Human Services indicators. *Clin Infect Dis* **58:**1185–1189.
143. **Matthews T, DeLorenzo L, Matosky M, Young S, Huang A, Feit B, Malitz F, Cheever L.** 2012. National performance measures within a changing environment: how a federal agency developed and improved the measurement for HIV care and treatment. *J Health Care Poor Underserved* **23** (Suppl)**:**225–235.
144. **Panneer N, Lontok E, Branson BM, Teo CG, Dan C, Parker M, Stekler JD, DeMaria A Jr, Miller V.** 2014. HIV and hepatitis C virus infection in the United States: whom and how to test. *Clin Infect Dis* **59:**875–882.
145. **Valdiserri RO, Forsyth AD, Yakovchenko V, Koh HK.** 2013. Measuring what matters: development of standard HIV core indicators across the U.S. Department of Health and Human Services. *Public Health Rep* **128:**354–359.
146. **Nübling CM, Unger G, Chudy M, Raia S, Löwer J.** 2002. Sensitivity of HCV core antigen and HCV RNA detection in the early infection phase. *Transfusion* **42:**1037–1045.
147. **Duclos-Vallée JC, Roche B, Samuel D.** 2009. [Liver transplantation in patients with hepatitis B virus, hepatitis C virus, or human immunodeficiency virus]. *Presse Med* **38:**

1281–1289. Liver transplantation in patients with hepatitis B virus, hepatitis C virus, or human immunodeficiency virus.

148. **Lisker-Melman M, Sayuk GS.** 2007. Defining optimal therapeutic outcomes in chronic hepatitis. *Arch Med Res* **38:**652–660.
149. **Martin P, Jensen DM.** 2008. Ribavirin in the treatment of chronic hepatitis C. *J Gastroenterol Hepatol* **23:**844–855.
150. **Pol S, Mallet VO.** 2006. Improving anti-hepatitis C virus therapy. *Expert Opin Biol Ther* **6:**923–933.
151. **Santantonio T, Wiegand J, Gerlach JT.** 2008. Acute hepatitis C: current status and remaining challenges. *J Hepatol* **49:**625–633.
152. **Bürgisser P, Vernazza P, Flepp M, Böni J, Tomasik Z, Hummel U, Pantaleo G, Schüpbach J.** 2000. Performance of five different assays for the quantification of viral load in persons infected with various subtypes of HIV-1. Swiss HIV Cohort Study. *J Acquir Immune Defic Syndr* **23:**138–144.
153. **Lyamuya E, Olausson-Hansson E, Albert J, Mhalu F, Biberfeld G.** 2000. Evaluation of a prototype Amplicor PCR assay for detection of human immunodeficiency virus type 1 DNA in blood samples from Tanzanian adults infected with HIV-1 subtypes A, C and D. *J Clin Virol* **17:**57–63.
154. **Swanson P, Harris BJ, Holzmayer V, Devare SG, Schochetman G, Hackett J Jr.** 2000. Quantification of HIV-1 group M (subtypes A-G) and group O by the LCx HIV RNA quantitative assay. *J Virol Methods* **89:**97–108.
155. **Swanson P, Soriano V, Devare SG, Hackett J Jr.** 2001. Comparative performance of three viral load assays on human immunodeficiency virus type 1 (HIV-1) isolates representing group M (subtypes A to G) and group O: LCx HIV RNA quantitative, AMPLICOR HIV-1 MONITOR version 1.5, and Quantiplex HIV-1 RNA version 3.0. *J Clin Microbiol* **39:**862–870.
156. **Berger A, Rabenau HF, Stief A, Troonen H, Doerr HW.** 2001. Evaluation of the new LCx HIV RNA quantitative assay: comparison with the Cobas Amplicor HIV Monitor assay. *Med Microbiol Immunol (Berl)* **190:**129–134.
157. **Clarke JR, Galpin S, Braganza R, Ashraf A, Russell R, Churchill DR, Weber JN, McClure MO.** 2000. Comparative quantification of diverse serotypes of HIV-1 in plasma from a diverse population of patients. *J Med Virol* **62:**445–449. In Process Citation.
158. **de Mendoza C, Lu W, Machuca A, Sainz M, Castilla J, Soriano V.** 2001. Monitoring the response to antiretroviral therapy in HIV-1 group O infected patients using two new RT-PCR assays. *J Med Virol* **64:**217–222.
159. **Glencross DK, Stevens G, Scott LE, Mendelow BV, Stevens W.** 2002. The challenge of laboratory monitoring of HIV. *S Afr Med J* **92:**248.
160. **Naeth G, Ehret R, Wiesmann F, Braun P, Knechten H, Berger A.** 2013. Comparison of HIV-1 viral load assay performance in immunological stable patients with low or undetectable viremia. *Med Microbiol Immunol (Berl)* **202:**67–75.
161. **Ruelle J, Debaisieux L, Vancutsem E, De Bel A, Delforge ML, Piérard D, Goubau P.** 2012. HIV-1 low-level viraemia assessed with 3 commercial real-time PCR assays show high variability. *BMC Infect Dis* **12:**100.
162. **Adinolfi LE, Utili R, Andreana A, Tripodi MF, Rosario P, Mormone G, Ragone E, Pasquale G, Ruggiero G.** 2000. Relationship between genotypes of hepatitis C virus and histopathological manifestations in chronic hepatitis C patients. *Eur J Gastroenterol Hepatol* **12:**299–304.
163. **Yeo AE, Ghany M, Conry-Cantilena C, Melpolder JC, Kleiner DE, Shih JW, Hoofnagle JH, Alter HJ.** 2001. Stability of HCV-RNA level and its lack of correlation with disease severity in asymptomatic chronic hepatitis C virus carriers. *J Viral Hepat* **8:**256–263.
164. **Anonymous.** 2003. National Institutes of Health Consensus Development Conference statement Management of Hepatitis C: 2002 June 10–12, 2002. *HIV Clin Trials* **4:** 55–75.
165. **Strader DB, Wright T, Thomas DL, Seeff LB, American Association for the Study of Liver Diseases.** 2004. Diagnosis, management, and treatment of hepatitis C. *Hepatology* **39:**1147–1171.
166. **McHutchison JG, Lawitz EJ, Shiffman ML, Muir AJ, Galler GW, McCone J, Nyberg LM, Lee WM, Ghalib RH, Schiff ER, Galati JS, Bacon BR, Davis MN, Mukhopadhyay P, Koury K, Noviello S, Pedicone LD, Brass CA, Albrecht JK, Sulkowski MS, Team IS, IDEAL Study Team.** 2009. Peginterferon alfa-2b or alfa-2a with ribavirin for treatment of hepatitis C infection. *N Engl J Med* **361:**580–593.
167. **Lam BP, Jeffers T, Younoszai Z, Fazel Y, Younossi ZM.** 2015. The changing landscape of hepatitis C virus therapy: focus on interferon-free treatment. *Therap Adv Gastroenterol* **8:**298–312.
168. **Hayes CN, Chayama K.** 2015. Emerging treatments for chronic hepatitis C. *J Formos Med Assoc* **114:**204–215.
169. **Sulkowski MS, Gardiner DF, Rodriguez-Torres M et al.** 2014. Daclatasvir plus sofosbuvir for HCV infection. *N Engl J Med* **370:**1560–1561.
170. **AASLD-IDSA.** HCV Guidance: Recommendations for Testing, Managing, and Treating Hepatitis C. http://www.hcvguidelines.org/full-report/hcv-testing-and-linkage-care. Accessed Jan. 19.
171. **Wiesmann F, Naeth G, Sarrazin C, Berger A, Kaiser R, Ehret R, Knechten H, Braun P.** 2014. Variation analysis of six HCV viral load assays using low viremic HCV samples in the range of the clinical decision points for HCV protease inhibitors. *Med Microbiol Immunol (Berl).*
172. **Kessler HH, Cobb BR, Wedemeyer H, Maasoumy B, Michel-Treil V, Ceccherini-Nelli L, Bremer B, Hübner M, Helander A, Khiri H, Heilek G, Simon CO, Luk K, Aslam S, Halfon P.** 2015. Evaluation of the COBAS(®) AmpliPrep/COBAS(®) TaqMan(®) HCV Test, v2.0 and comparison to assays used in routine clinical practice in an international multicenter clinical trial: the ExPECT study. *J Clin Virol* **67:**67–72.
173. **Ferraro D, Giglio M, Bonura C, Di Marco V, Mondelli MU, Craxì A, Di Stefano R.** 2008. Assessment of hepatitis C virus-RNA clearance under combination therapy for hepatitis C virus genotype 1: performance of the transcription-mediated amplification assay. *J Viral Hepat* **15:**66–70.
174. **Gerotto M, Dal Pero F, Bortoletto G, Ferrari A, Pistis R, Sebastiani G, Fagiuoli S, Realdon S, Alberti A.** 2006. Hepatitis C minimal residual viremia (MRV) detected by TMA at the end of Peg-IFN plus ribavirin therapy predicts post-treatment relapse. *J Hepatol* **44:**83–87.
175. **Morishima C, Morgan TR, Everhart JE, Wright EC, Apodaca MC, Gretch DR, Shiffman ML, Everson GT, Lindsay KL, Lee WM, Lok AS, Dienstag JL, Ghany MG, Curto TM, HALT-C Trial Group.** 2008. Interpretation of positive transcription-mediated amplification test results from polymerase chain reaction-negative samples obtained after treatment of chronic hepatitis C. *Hepatology* **48:**1412–1419.
176. **Lok AS, McMahon BJ.** 2009. Chronic hepatitis B: update 2009. *Hepatology* **50:**661–662.
177. **Keefe.** 2007. Report of an international workshop: Roadmap for management of patients receiving oral therapy for chronic hepatitis B.
178. **Bate SL, Dollard SC, Cannon MJ.** 2010. Cytomegalovirus seroprevalence in the United States: the national health and nutrition examination surveys, 1988–2004. *Clin Infect Dis* **50:**1439–1447.
179. **Cannon MJ, Schmid DS, Hyde TB.** 2010. Review of cytomegalovirus seroprevalence and demographic characteristics associated with infection. *Rev Med Virol* **20:**202–213.
180. **Razonable R.** 2010. Direct and indirect effects of cytomegalovirus: can we prevent them? *Enferm Infecc Microbiol Clin* **28:**1–5.
181. **Razonable RR.** 2013. Management strategies for cytomegalovirus infection and disease in solid organ transplant recipients. *Infect Dis Clin North Am* **27:**317–342.
182. **Emery VC.** 2013. CMV infected or not CMV infected: that is the question. *Eur J Immunol* **43:**886–888.
183. **Ljungman P, Hakki M, Boeckh M.** 2011. Cytomegalovirus in hematopoietic stem cell transplant recipients. *Hematol Oncol Clin North Am* **25:**151–169.

184. **Ozdemir E, Saliba RM, Champlin RE, Couriel DR, Giralt SA, de Lima M, Khouri IF, Hosing C, Kornblau SM, Anderlini P, Shpall EJ, Qazilbash MH, Molldrem JJ, Chemaly RF, Komanduri KV.** 2007. Risk factors associated with late cytomegalovirus reactivation after allogeneic stem cell transplantation for hematological malignancies. *Bone Marrow Transplant* **40:**125–136.
185. **Zhou W, Longmate J, Lacey SF, Palmer JM, Gallez-Hawkins G, Thao L, Spielberger R, Nakamura R, Forman SJ, Zaia JA, Diamond DJ.** 2009. Impact of donor CMV status on viral infection and reconstitution of multifunction CMV-specific T cells in CMV-positive transplant recipients. *Blood* **113:**6465–6476.
186. **Ugarte-Torres A, Hoegh-Petersen M, Liu Y, Zhou F, Williamson TS, Quinlan D, Sy S, Roa L, Khan F, Fonseca K, Russell JA, Storek J.** 2011. Donor serostatus has an impact on cytomegalovirus-specific immunity, cytomegaloviral disease incidence, and survival in seropositive hematopoietic cell transplant recipients. *Biol Blood Marrow Transplant* **17:**574–585.
187. **Chawla JS, Ghobadi A, Mosley J III, Verkruyse L, Trinkaus K, Abboud CN, Cashen AF, Stockerl-Goldstein KE, Uy GL, Westervelt P, DiPersio JF, Vij R.** 2012. Oral valganciclovir versus ganciclovir as delayed pre-emptive therapy for patients after allogeneic hematopoietic stem cell transplant: a pilot trial (04-0274) and review of the literature. *Transpl Infect Dis* **14:**259–267.
188. **Peres RM, Costa CR, Andrade PD, Bonon SH, Albuquerque DM, de Oliveira C, Vigorito AC, Aranha FJ, de Souza CA, Costa SC.** 2010. Surveillance of active human cytomegalovirus infection in hematopoietic stem cell transplantation (HLA sibling identical donor): search for optimal cutoff value by real-time PCR. *BMC Infect Dis* **10:**147.
189. **Tan SK, Waggoner JJ, Pinsky BA.** 2015. Cytomegalovirus load at treatment initiation is predictive of time to resolution of viremia and duration of therapy in hematopoietic cell transplant recipients. *J Clin Virol* **69:**179–183.
190. **Razonable RR, Åsberg A, Rollag H, Duncan J, Boisvert D, Yao JD, Caliendo AM, Humar A, Do TD.** 2013. Virologic suppression measured by a cytomegalovirus (CMV) DNA test calibrated to the World Health Organization international standard is predictive of CMV disease resolution in transplant recipients. *Clin Infect Dis* **56:**1546–1553.
191. **Asberg A, Humar A, Jardine AG, Rollag H, Pescovitz MD, Mouas H, Bignamini A, Töz H, Dittmer I, Montejo M, Hartmann A, Group VS, VICTOR Study Group.** 2009. Long-term outcomes of CMV disease treatment with valganciclovir versus IV ganciclovir in solid organ transplant recipients. *Am J Transplant* **9:**1205–1213.
192. **García-Cadenas I, Castillo N, Martino R, Barba P, Esquirol A, Novelli S, Orti G, Garrido A, Saavedra S, Moreno C, Granell M, Briones J, Brunet S, Navarro F, Ruiz I, Rabella N, Valcárcel D, Sierra J.** 2015. Impact of Epstein Barr virus-related complications after high-risk allo-SCT in the era of pre-emptive rituximab. *Bone Marrow Transplant* **50:**579–584.
193. **Visco C, Falisi E, Young KH, Pascarella M, Perbellini O, Carli G, Novella E, Rossi D, Giaretta I, Cavallini C, Scupoli MT, De Rossi A, D'Amore ES, Rassu M, Gaidano G, Pizzolo G, Ambrosetti A, Rodeghiero F.** 2015. Epstein-Barr virus DNA load in chronic lymphocytic leukemia is an independent predictor of clinical course and survival. *Oncotarget* **6:**18653–18663.
194. **Ciccocioppo R, Racca F, Paolucci S, Campanini G, Pozzi L, Betti E, Riboni R, Vanoli A, Baldanti F, Corazza GR.** 2015. Human cytomegalovirus and Epstein-Barr virus infection in inflammatory bowel disease: need for mucosal viral load measurement. *World J Gastroenterol* **21:**1915–1926.
195. **Yip TT, Ngan RK, Fong AH, Law SC.** 2014. Application of circulating plasma/serum EBV DNA in the clinical management of nasopharyngeal carcinoma. *Oral Oncol* **50:**527–538.
196. **Gill H, Hwang YY, Chan TS, Pang AW, Leung AY, Tse E, Kwong YL.** 2014. Valganciclovir suppressed Epstein Barr virus reactivation during immunosuppression with alemtuzumab. *J Clin Virol* **59:**255–258.
197. **Lo YM.** 2001. Quantitative analysis of Epstein-Barr virus DNA in plasma and serum: applications to tumor detection and monitoring. *Ann N Y Acad Sci* **945:**68–72.
198. **Jakovljevic A, Andric M.** 2014. Human cytomegalovirus and Epstein-Barr virus in etiopathogenesis of apical periodontitis: a systematic review. *J Endod* **40:**6–15.
199. **Fox CP, Burns D, Parker AN, Peggs KS, Harvey CM, Natarajan S, Marks DI, Jackson B, Chakupurakal G, Dennis M, Lim Z, Cook G, Carpenter B, Pettitt AR, Mathew S, Connelly-Smith L, Yin JA, Viskaduraki M, Chakraverty R, Orchard K, Shaw BE, Byrne JL, Brookes C, Craddock CF, Chaganti S.** 2014. EBV-associated post-transplant lymphoproliferative disorder following in vivo T-cell-depleted allogeneic transplantation: clinical features, viral load correlates and prognostic factors in the rituximab era. *Bone Marrow Transplant* **49:**280–286.
200. **Wang LH, Ren HY, Sun YH, Qiu ZX, Cen XN, Ou JP, Xu WL, Wang MJ, Wang WS, Li Y, Dong YJ, Yin Y, Liang ZY.** 2010. Quantitative monitoring of mononucleated cell Epstein-Barr virus (EBV)-DNA for predicting EBV associated lymphoproliferative disorders after stem cell transplantation. *Zhonghua Xue Ye Xue Za Zhi* **31:**73–76.
201. **Bocian J, Januszkiewicz-Lewandowska D.** 2014. Utility of quantitative EBV DNA measurements in cerebrospinal fluid for diagnosis and monitoring of treatment of central nervous system EBV-associated post-transplant lymphoproliferative disorder after allogenic hematopoietic stem cell transplantation. *Ann Transplant* **19:**253–256.
202. **Stevens SJ, Verkuijlen SA, Brule AJ, Middeldorp JM.** 2002. Comparison of quantitative competitive PCR with LightCycler-based PCR for measuring Epstein-Barr virus DNA load in clinical specimens. *J Clin Microbiol* **40:**3986–3992.
203. **Stevens SJ, Verschuuren EA, Verkuujlen SA, Van Den Brule AJ, Meijer CJ, Middeldorp JM.** 2002. Role of Epstein-Barr virus DNA load monitoring in prevention and early detection of post-transplant lymphoproliferative disease. *Leuk Lymphoma* **43:**831–840.
204. **Leung SF, Chan KC, Ma BB, Hui EP, Mo F, Chow KC, Leung L, Chu KW, Zee B, Lo YM, Chan AT.** 2014. Plasma Epstein-Barr viral DNA load at midpoint of radiotherapy course predicts outcome in advanced-stage nasopharyngeal carcinoma. *Ann Oncol* **25:**1204–1208.
205. **Hasselblom S, Linde A, Ridell B.** 2004. Hodgkin's lymphoma, Epstein-Barr virus reactivation and fatal haemophagocytic syndrome. *J Intern Med* **255:**289–295.
206. **Loginov R, Aalto S, Piiparinen H, Halme L, Arola J, Hedman K, Höckerstedt K, Lautenschlager I.** 2006. Monitoring of EBV-DNAemia by quantitative real-time PCR after adult liver transplantation. *J Clin Virol* **37:**104–108.
207. **Niesters HG.** 2004. Molecular and diagnostic clinical virology in real time. *Clin Microbiol Infect* **10:**5–11.
208. **Rychert J, Danziger-Isakov L, Yen-Lieberman B, Storch G, Buller R, Sweet SC, Mehta AK, Cheeseman JA, Heeger P, Rosenberg ES, Fishman JA.** 2014. Multicenter comparison of laboratory performance in cytomegalovirus and Epstein-Barr virus viral load testing using international standards. *Clin Transplant* **28:**1416–1423.
209. **Boutolleau D, Fernandez C, André E, Imbert-Marcille BM, Milpied N, Agut H, Gautheret-Dejean A.** 2003. Human herpesvirus (HHV)-6 and HHV-7: two closely related viruses with different infection profiles in stem cell transplantation recipients. *J Infect Dis* **187:**179–186.
210. **Lautenschlager I, Razonable RR.** 2012. Human herpesvirus-6 infections in kidney, liver, lung, and heart transplantation: review. *Transpl Int* **25:**493–502. review.
211. **Lautenschlager I, Loginov R.** 2011. HHV-6 infection and its clinical significance. *Duodecim* **127:**1204–1211.
212. **Achour A, Boutolleau D, Slim A, Agut H, Gautheret-Dejean A.** 2007. Human herpesvirus-6 (HHV-6) DNA in plasma reflects the presence of infected blood cells rather than circulating viral particles. *J Clin Virol* **38:**280–285.

213. **Gardner SD, Knowles WA, Hand JF, Porter AA.** 1989. Characterization of a new polyomavirus (Polyomavirus papionis-2) isolated from baboon kidney cell cultures. *Arch Virol* **105:**223–233.
214. **Siguier M, Sellier P, Bergmann JF.** 2012. BK-virus infections: a literature review. *Med Mal Infect* **42:**181–187.
215. **Pinto M, Dobson S.** 2014. BK and JC virus: a review. *J Infect* **68**(Suppl 1)**:**S2–S8.
216. **Hirsch HH, Brennan DC, Drachenberg CB, Ginevri F, Gordon J, Limaye AP, Mihatsch MJ, Nickeleit V, Ramos E, Randhawa P, Shapiro R, Steiger J, Suthanthiran M, Trofe J.** 2005. Polyomavirus-associated nephropathy in renal transplantation: interdisciplinary analyses and recommendations. *Transplantation* **79:**1277–1286.
217. **Hirsch HH, Drachenberg CB, Steiger J, Ramos E.** 2006. Polyomavirus-associated nephropathy in renal transplantation: critical issues of screening and management. *Adv Exp Med Biol* **577:**160–173.
218. **Bressollette-Bodin C, Coste-Burel M, Hourmant M, Sebille V, Andre-Garnier E, Imbert-Marcille BM.** 2005. A prospective longitudinal study of BK virus infection in 104 renal transplant recipients. *Am J Transplant* **5:**1926–1933.
219. **Drachenberg CB, Papadimitriou JC, Ramos E.** 2006. Histologic versus molecular diagnosis of BK polyomavirus-associated nephropathy: a shifting paradigm? *Clin J Am Soc Nephrol* **1:**374–379.
220. **Randhawa P, Ho A, Shapiro R, Vats A, Swalsky P, Finkelstein S, Uhrmacher J, Weck K.** 2004. Correlates of quantitative measurement of BK polyomavirus (BKV) DNA with clinical course of BKV infection in renal transplant patients. *J Clin Microbiol* **42:**1176–1180.
221. **Drachenberg RC, Drachenberg CB, Papadimitriou JC, Ramos E, Fink JC, Wali R, Weir MR, Cangro CB, Klassen DK, Khaled A, Cunningham R, Bartlett ST.** 2001. Morphological spectrum of polyoma virus disease in renal allografts: diagnostic accuracy of urine cytology. *Am J Transplant* **1:**373–381.
222. **Vats A, Shapiro R, Singh Randhawa P, Scantlebury V, Tuzuner A, Saxena M, Moritz ML, Beattie TJ, Gonwa T, Green MD, Ellis D.** 2003. Quantitative viral load monitoring and cidofovir therapy for the management of BK virus-associated nephropathy in children and adults. *Transplantation* **75:**105–112.
223. **Kadambi PV, Josephson MA, Williams J, Corey L, Jerome KR, Meehan SM, Limaye AP.** 2003. Treatment of refractory BK virus-associated nephropathy with cidofovir. *Am J Transplant* **3:**186–191.
224. **Williams JW, Javaid B, Kadambi PV, Gillen D, Harland R, Thistlewaite JR, Garfinkel M, Foster P, Atwood W, Millis JM, Meehan SM, Josephson MA.** 2005. Leflunomide for polyomavirus type BK nephropathy. *N Engl J Med* **352:**1157–1158.
225. **Sener A, House AA, Jevnikar AM, Boudville N, McAlister VC, Muirhead N, Rehman F, Luke PP.** 2006. Intravenous immunoglobulin as a treatment for BK virus associated nephropathy: one-year follow-up of renal allograft recipients. *Transplantation* **81:**117–120.
226. **Wadei HM, Rule AD, Lewin M, Mahale AS, Khamash HA, Schwab TR, Gloor JM, Textor SC, Fidler ME, Lager DJ, Larson TS, Stegall MD, Cosio FG, Griffin MD.** 2006. Kidney transplant function and histological clearance of virus following diagnosis of polyomavirus-associated nephropathy (PVAN). *Am J Transplant* **6**(5p1)**:**1025–1032.
227. **Greer AE, Forman MS, Valsamakis A.** 2015. Comparison of BKV quantification using a single automated nucleic acid extraction platform and 3 real-time PCR assays. *Diagn Microbiol Infect Dis* **82:**297–302.
228. **Randhawa P, Kant J, Shapiro R, Tan H, Basu A, Luo C.** 2011. Impact of genomic sequence variability on quantitative PCR assays for diagnosis of polyomavirus BK infection. *J Clin Microbiol* **49:**4072–4076.

Signal Amplification Methods

YUN (WAYNE) WANG

14

Signal amplification methods were initially designed as an alternative to the target amplification technologies such as polymerase chain reaction (PCR) so as to minimize the possibility of contamination by target amplification products. Unlike target amplification, signal amplification methods (as defined herein) do not rely on enzymes for the amplification. Probe-based amplification techniques such as cleavage-based amplification (36) and rolling-circle amplification (1) rely on enzymes, and will not be covered in this chapter. Signal amplification increases or amplifies the signal generated from the probe molecule hybridized to the target nucleic acid sequence. The advantages of signal amplification methods include specific detection, dynamic range, ease-of-use, and reproducibility. To date these methods have met the challenge from advanced or automated target amplification methods such as real time PCR. Signal amplification technologies include hybrid capture (HC) and branched DNA (bDNA) assays (32, 33). The HC method was developed and marketed initially by Digene Corporation (Gaithersburg, MD) which was acquired by Qiagen (Valencia, CA) in 2007. The bDNA method was initially developed by Chiron (Emeryville, CA), marketed by Bayer Diagnostics (Emeryville, CA) which diagnostic division was acquired by Siemens (Tarrytown, NY) in 2006. Due to the growing demand for quick time to detection, automation, and multiplexing, the popularity of commercial signal amplification methods has declined in clinical virology laboratories.

HYBRID CAPTURE TECHNOLOGY

Principles and Characteristics of HC

Utilizing antibody capture and chemiluminescent signal detection, HC combines the nucleic acid technology such as RNA probes for RNA:DNA hybridization with the simplicity of an immunoassay using monoclonal antibodies RNA/DNA hybrid for rapid gene detection. HC technology detects nucleic acid targets directly and uses signal amplification to provide sensitivity that is comparable to target amplification methods. HC has been successfully applied to the detection of human papillomavirus (HPV), *Chlamydia trachomatis*, *Neisseria gonorrhoeae*, and human cytomegalovirus (CMV).

The current commercially available HC test is second generation, the so-called HC2 test. For improved convenience and workflow, HC2 uses plate wells, whereas the first generation HC test used tubes as the reaction vessel. Assays using the HC2 technology take approximately 3.5 hours to complete. Same-day results can be achieved in a chemiluminescent microplate format. Automation of the HC methods may extend the use for HC testing.

The HC2 technology uses RNA probes to detect DNA targets (Figure 1). Following are the steps of HC technology:

(1) Release of DNA from cells and denaturation of nucleic acids. Alkali such as sodium hydroxide is added to the specimen to disrupt the virus, release target DNA, and make the target DNA molecules single stranded and accessible for hybridization.
(2) Hybridization of target DNA with RNA probe. The specimen is transferred to a container and a single-stranded RNA probe that is complementary to the target DNA sequence is added to the solution and heated. The RNA probe finds its complementary DNA target sequence and hybridizes to it, forming a double-stranded RNA–DNA hybrid complex.
(3) Capture of RNA–DNA hybrids onto a solid phase. The sample is then transferred to a plate well that has been coated with antibodies (i.e., goat anti-RNA–DNA hybrid antibody) that specifically recognize and bind RNA–DNA hybrids. Multiple RNA–DNA hybrids are captured or bound onto the microplate surface by the coated antibodies specific for RNA–DNA hybrids.
(4) Reaction of captured hybrids with multiple antibody conjugates and label for detection. A second antibody is added to the solution, which recognizes and binds to the RNA–DNA hybrids that are captured onto the surface of the plate well. This anti-RNA–DNA antibody is conjugated with alkaline phosphatase (AP), an enzyme that in the presence of chemiluminescent substrate produces light and acts as a signal amplification. Several AP molecules are conjugated to each antibody and multiple conjugated antibodies bind to each captured hybrid, which in turn results in substantial (about 3,000-fold) signal amplification.
(5) Detection of amplified chemiluminescent signal. The plate well is washed to remove the unbound or free components while the RNA–DNA hybrids and the labeled antibody remain bound. Chemiluminescent

doi:10.1128/9781555819156.ch14

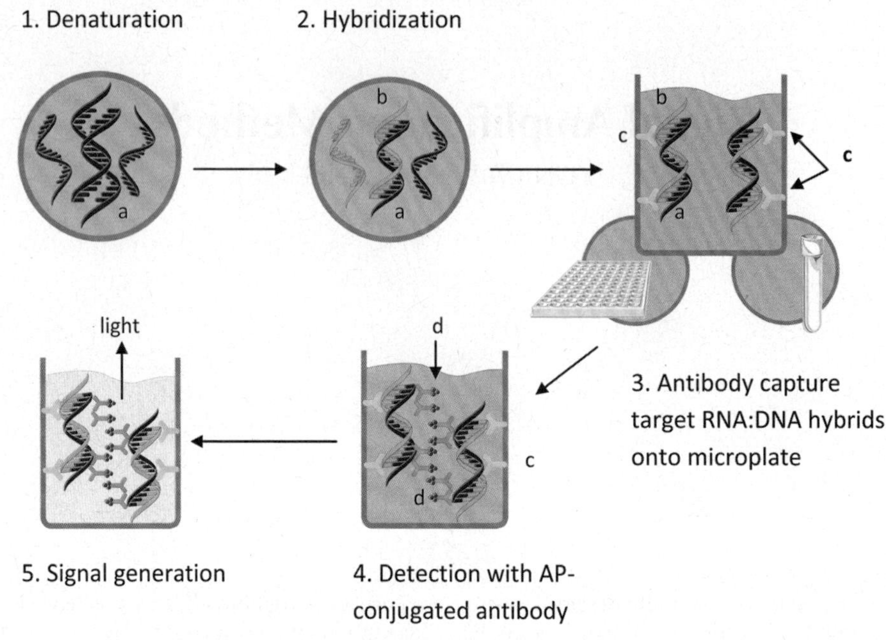

FIGURE 1 Scheme for Hybrid Capture 2 (HC2) technology.

dioxetane substrate is added, which is cleaved by the bound AP to produce light. The emitted light is detected and measured as relative light units (RLUs) on a luminometer (Microplate Luminometer DML 2000 Instrument). The intensity of the light is evidence of target DNA in the specimen (4, 32).

Application of HC in the Clinical Virology Laboratory

HC2 is the current technology for hybrid capture-based detection of HPV (21, 32). The HC method was first introduced by Digene in 1995; the second generation HC2 was approved by the U.S. Food and Drug Administration (FDA) for the detection of high-risk HPV types in thin preparation, liquid-based cervical specimens (9). HPV nucleic acid tests, such as the HC2 HPV DNA test (Qiagen) have now become established as part of the standard of care for cervical cancer screening. An HC2 assay has also been used to detect CMV. The HC2 CMV DNA test was the first molecular diagnostic test to be cleared by the FDA for the qualitative detection of CMV DNA in peripheral white blood cells isolated from whole blood (23). This test is no longer commercially available, and has been replaced largely by quantitative PCR for CMV viral load monitoring. Thus, this chapter only focuses on HPV testing.

HPV is a small DNA tumor virus in the family *Papovaviridae* that is a causative agent of cervical cancer (24). There are more than 100 types of HPV. Low-risk (LR) types of HPV may cause genital warts. High-risk (HR) types have been shown to cause most cases of cervical cancer. Thirteen types are implicated in the pathogenesis of high-grade squamous intraepithelial lesions and invasive cancer: 16, 18, 31, 33, 35, 39, 45, 51, 52, 56, 58, 59, and 68. The HC2 HPV test uses RNA probe cocktails to detect carcinogenic high-risk HPV types as well as low-risk HPV types. Five probes detect low-risk viral types associated with low-grade squamous intraepithelial lesions: 6, 11, 42, 43, and 44. The LR-HPV DNA test has little clinical usefulness and is rarely used. The HC2 HR HPV DNA test was initially approved by the FDA in 2000 for follow-up evaluation in women with inconclusive Pap smear results, so-called atypical squamous cells of undermined significance (ASCUS), to determine the need for referral to colposcopy. With 99% negative predictive value, the HC2 HR HPV DNA test can reliably exclude HPV-associated dysplasias in postmenopausal women diagnosed with ASCUS (22, 29). The HC2 HR HPV DNA test can be performed on liquid-based cytology samples using a specific sample collection kit. Specimens collected and rinsed in the Cytyc's ThinPrep Pap Test vial (Hologic, Marlborough, MA) can be used for both the Pap test and the HC2HPV test. Additionally, cervical biopsies collected in a specific collection kit can be tested (2, 20).

The HC2 HPV test has been shown to have similar sensitivity and negative predictive value to PCR or transcription-mediated amplification methods for HPV DNA detection (6, 10, 19, 26, 27, 34, 35).

BRANCHED DNA (BDNA) TECHNOLOGY

Principles and Characteristics of bDNA

bDNA is a signal amplification technology that detects the presence of specific nucleic acids by measuring the signal generated by specific hybridization of many branched, labeled DNA probes on an immobilized target nucleic acid. Signal amplification is achieved by sequential (or simultaneous) hybridization of synthetic oligonucleotides, assembling a branched complex structure on the immobilized target nucleic acid (32). One end of bDNA binds to a specific target and the other end has many branches of DNA. The branches amplify detection signals with linear amplification. The end result is a target molecule with several hundred labels; this provides the analytical sensitivity. The final detection step uses alkaline phosphatase (AP) to generate chemiluminescence. The amplified signal on the target molecules is related to the number of target molecules. Creation of a standard curve in each assay allows calculation of the number of targets in the samples and thus bDNA is a quantitative technology and is used in the determination of viral load (7, 11, 18, 32, 33). The bDNA assay is based on a few key steps (4, 17) which are listed below and shown in Figure 2.

(1) Release of nucleic acid from the target such as a virus, the so-called target nucleic acid release. A lysis buffer containing proteinase K disrupts virus, degrades nucleases (RNases), and releases viral target RNA or DNA (DNA targets require additional denaturation to yield single-stranded target).
(2) Capture of target nucleic acid. Oligonucleotide capture probes in solution hybridize to multiple sequence-specific sites on the target viral nucleic acid. These hybrids in turn are captured by specific probes that are immobilized on microwells.
(3) Preamplification probe hybridization (to target probes and thus to the microwell) can be performed on the first or second day. Following the incubation, the microwells are washed to remove unbound capture probes, target probes, lysis reagent, and cellular debris. Target probes mediate the binding of preamplifier probes to the hybrid complex. Preamplifier probes are added to the microwells. Each preamplifier probe hybridizes to two adjacent target probes in a cruciform configuration or cruciform design.
(4) Amplifier probe hybridization. Amplifier probes are added to the microwells, and hybridize to preamplifiers. There are multiple amplifier-binding sites present on each preamplifier for the amplifier probe to hybridize to the preamplifier and form a branched DNA complex (bDNA) or so-called signal amplification multimer for amplification. Thus, the amplifier molecule is the key to bDNA technology.
(5) Alkaline phosphatase (AP) labeled probe hybridization. AP-conjugated probes called label probes are added to the microwells and hybridize to the immobilized amplifier complex. There are multiple label probe-binding sites present on each amplifier. Dioxetane substrate is added to the microwell for signal generation. The dioxetane substrate chemically reacts with the AP from the label probes, which excites an electron, resulting in emission of a photon of light producing chemiluminescence.
(6) Recording of chemiluminescence by a photomultiplier tube in an analyzer. The amount of light produced by dioxetane substrate is proportional to the initial target RNA concentration. Results are recorded as RLUs by the analyzer. Data management software creates a standard curve from standards of known concentrations. The concentration of viral material in specimens is determined by comparing the RLU of each sample with this standard curve.

The first generation of bDNA was first introduced in 1990. Subsequent bDNA generations were modified to increase sensitivity. Two of the probe design features for the bDNA assay are cruciform target probes or binding design and base isomers (Figure 2). Two target probes are required to stabilize binding of the preamplifier probe (32). This reduces background by minimizing hybridization of amplification molecules to nonspecifically bound target probes. Interaction between oligonucleotides is minimized by incorporating non-natural bases in the sequences of the target probes, preamplifiers, amplifiers, and AP-conjugated label probes (Figure 2). Isocytosine (Iso5MeC) and isoguanosine (IsoG) are non-natural isomers of cytosine (C) and guanosine (G). Iso5MeC and IsoG participate in Watson–Crick base pairing with each other but have unstable interactions with DNA sequences containing natural bases. Approximately every 4th nucleotide in selected probes is Iso5MeC or IsoG. Use of a 6-base code allows the design of amplification sequences that do not interact with target sequences or other bDNA components (11).

The configuration of the amplification complex (preamplifier, amplifier, and AP-conjugated label probes), reduces potential hybridization to nontarget nucleic acids, and increases the signal to noise ratio 30-fold, thus improving signal amplification with equivalent sensitivity to some target amplification technologies like PCR (32).

Changes incorporated into the third generation (3.0 version) bDNA assays have increased the sample volume and the signal-to-noise ratio to such a high level that the

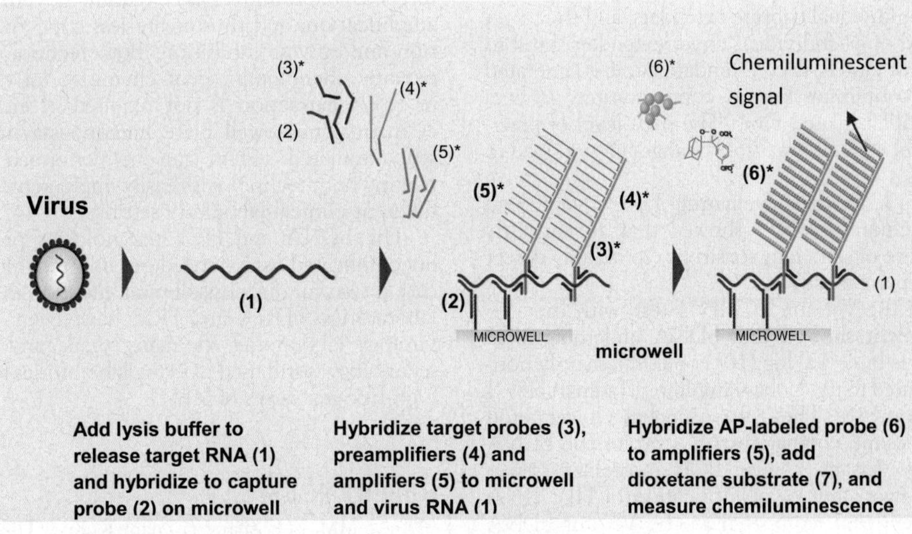

FIGURE 2 Scheme for branched DNA (bDNA) technology.

TABLE 1 Comparison of bDNA and HC2 technologies

Test feature	Hybrid Capture II (HC2)	Branched DNA (bDNA)
Signal amplification	Anti-RNA:DNA hybrid antibody	Many probes including capture target, preamplifier, amplifier probes
Detection	AP-conjugated anti-RNA:DNA hybrid antibody	AP-conjugated label probe
Chemiluminescent	Dioxetane (for HPV)	Dioxetane
Substrate	Lumi Phos 530	Lumin-Phos Plus, Lumi Phos 530
Common features	No enzymes involved for target amplification, less contamination concern	No DNA or RNA extraction is needed
Semiautomation application	Rapid Capture System: up to 352 samples (4 microwell plates) in 8 hr.	System 340/440: up to 168 samples in 1–2 days

analytical sensitivity of bDNA approaches that of PCR. Nonspecific hybridization can be reduced by using effective blockers for the solid phase or by redesigning the amplifier molecule or the solid phase itself (18, 32).

Assays were initially performed in a 96-microwell format using the System 340 Analyzer (Siemens), a semiautomated instrument that performs incubations, washes, and detection. Automation of specimen preparation (Versant 440) improves the consistency of results.

Application of bDNA in the Clinical Virology Laboratory

bDNA is a quantitative signal amplification method for the measurement of viral load. Virus assays currently commercially available in the United States include Versant HIV-1 RNA 3.0 and HCV RNA 3.0. These assays target the polymerase (*pol*) gene of the HIV-1 viral RNA, and the 5′-untranslated (UTR) and core regions of HCV, respectively.

The Versant HIV-1 RNA 3.0 assay is able to quantify HIV-1 RNA in plasma over the range of 75 to 500,000 copies/ml. In addition, HIV Group M subtypes A to G have been validated for quantitation by the assay (7, 11, 12, 13, 25). The test does not require viral RNA extraction steps. HIV is denser than HCV and therefore can be concentrated by centrifugation. Beads are added to the sample before centrifugation to make the HIV pellet more visible. In addition, a set of target probes hybridizes to both the viral RNA and the preamplifier probes. The capture probes, comprised of 17 individual capture extenders, and the target probes comprised of 81 individual target extenders, bind to different regions of viral RNA. A standard curve is generated from standards containing known concentrations of beta propiolactone (BPL)-treated virus. The high level of precision afforded by bDNA allows 3-fold changes in viral load to be distinguished.

A review of 37 studies evaluating HIV-1 viral load technologies, including bDNA, showed that all currently available assays are of sufficient sensitivity to reliably detect 1,000 copies/ml plasma (28).

Evaluation of the Versant 3.0 HIV-1 test, with the Versant 440 instrument showed that bDNA underquantified some HIV-1strains by ≥ 1.0 log (10) copies/ml, mainly non-B subtypes compared to the Cobas Ampliprep/Taqman HIV-1 viral load (VL) assay (8). The Cobas assay has a lower cutoff of 20 RNA copies/ml, compared with < 50 for the bDNA assay. The new Versant kinetic PCR molecular system (kPCR) has a lower limit of quantification of HIV RNA viral load at 1 copy/mL compared to the Versant bDNA system (16). Clinicians need to be aware that changes in assay could result in difficulties in interpretation of patient results (30).

An overnight incubation is a significant drawback of highly sensitive bDNA assays. The Versant HIV-1 RNA 3.0 assay was modified to allow shorter target incubation, enabling the viral load assay to be run in a single day. The vendor modified the composition of the Lysis Diluent reagent to allow reduced target incubation time from 16 to 18 h to 2.5 h, which was comparable to PCR (3).

The HCV RNA 3.0 assay is approved by the FDA for the quantitation of HCV RNA in the serum or plasma of HCV-infected individuals. The assay measures HCV RNA levels at baseline and during therapy and is useful in predicting nonsustained response to HCV therapy. The HCV RNA 3.0 Assay accurately quantitates all HCV RNA genotypes (genotypes 1 to 6) (5, 14, 15, 31).

Commercial bDNA technology has been gradually replaced by real time quantitative PCR, or kPCR (28), due to improved automation, less hands-on time, and capability of multiplexing.

SUMMARY

Signal amplification technology has unique features and even some advantages over target amplification systems for direct detection or quantification of target nucleic acid sequences. As summarized in Table 1, signal amplification methods HC2 and bDNA require no enzymes for target amplification, and thus create less concern for contamination and enzyme inhibition. Both technologies use the dioxetane chemiluminescent chemistry for detection. DNA or RNA extraction is not required of either technology. A manual microwell plate immunoassay-like format or a semiautomated system can be performed in one room, making both technologies easily implemented in a variety of different clinical laboratory settings.

The bDNA and HC2 technologies provide relatively uncomplicated assay procedures and reliable signal amplification tests for diagnosis of viral infection in routine clinical laboratories. bDNA and HC2 have been used in clinical virology laboratories for many years, and hybrid capture technology is still used in many laboratories for the detection high-risk genotypes of HPV.

REFERENCES

1. **Ali MM, Li F, Zhang Z, Zhang K, Kang DK, Ankrum JA, Le XC, Zhao W.** 2014. Rolling circle amplification: a versatile

tool for chemical biology, materials science and medicine. *Chem Soc Rev* **43:**3324–3341.

2. **Arbyn M, Snijders PJ, Meijer CJ, Berkhof J, Cuschieri K, Kocjan BJ, Poljak M.** 2015. Which high-risk HPV assays fulfil criteria for use in primary cervical cancer screening? *Clin Microbiol Infect* **21:**817–826.
3. **Baumeister MA, Zhang N, Beas H, Brooks JR, Canchola JA, Cosenza C, Kleshik F, Rampersad V, Surtihadi J, Battersby TR.** 2012. A sensitive branched DNA HIV-1 signal amplification viral load assay with single day turnaround. *PLoS One* **7:**e33295.
4. **Beck S, Köster H.** 1990. Applications of dioxetane chemiluminescent probes to molecular biology. *Anal Chem* **62:**2258–2270.
5. **Beld M, Sentjens R, Rebers S, Weegink C, Weel J, Sol C, Boom R.** 2002. Performance of the New Bayer VERSANT HCV RNA 3.0 assay for quantitation of hepatitis C virus RNA in plasma and serum: conversion to international units and comparison with the Roche COBAS Amplicor HCV Monitor, Version 2.0, assay. *J Clin Microbiol* **40:**788–793.
6. **Binnicker MJ, Pritt BS, Duresko BJ, Espy MJ, Grys TE, Zarka MA, Kerr SE, Henry MR.** 2014. Comparative evaluation of three commercial systems for detection of high-risk human papillomavirus in cervical and vaginal ThinPrep PreservCyt samples and correlation with biopsy results. *J Clin Microbiol* **52:**3763–3768.
7. **Cao Y, Ho DD, Todd J, Kokka R, Urdea M, Lifson JD, Piatak M Jr, Chen S, Hahn BH, Saag MS, Shaw GM.** 1995. Clinical evaluation of branched DNA signal amplification for quantifying HIV type 1 in human plasma. *AIDS Res Hum Retroviruses* **11:**353–361.
8. **Church D, Gregson D, Lloyd T, Klein M, Beckthold B, Laupland K, Gill MJ.** 2011. Comparison of the RealTime HIV-1, COBAS TaqMan 48 v1.0, Easy Q v1.2, and Versant v3.0 assays for determination of HIV-1 viral loads in a cohort of Canadian patients with diverse HIV subtype infections. *J Clin Microbiol* **49:**118–124.
9. **Clavel C, Masure M, Bory JP, Putaud I, Mangeonjean C, Lorenzato M, Gabriel R, Quereux C, Birembaut P.** 1999. Hybrid Capture II-based human papillomavirus detection, a sensitive test to detect in routine high-grade cervical lesions: a preliminary study on 1518 women. *Br J Cancer* **80:**1306–1311.
10. **Clavel C, Masure M, Putaud I, Thomas K, Bory JP, Gabriel R, Quereux C, Birembaut P.** 1998. Hybrid Capture II, a new sensitive test for human papillomavirus detection. Comparison with Hybrid Capture I and PCR results in cervical lesions. *J Clin Pathol* **51:**737–740.
11. **Collins ML, Irvine B, Tyner D, Fine E, Zayati C, Chang C, Horn T, Ahle D, Detmer J, Shen LP, Kolberg J, Bushnell S, Urdea MS, Ho DD.** 1997. A branched DNA signal amplification assay for quantification of nucleic acid targets below 100 molecules/ml. *Nucleic Acids Res* **25:**2979–2984.
12. **Elbeik T, Charlebois E, Nassos P, Kahn J, Hecht FM, Yajko D, Ng V, Hadley K.** 2000. Quantitative and cost comparison of ultrasensitive human immunodeficiency virus type 1 RNA viral load assays: Bayer bDNA quantiplex versions 3.0 and 2.0 and Roche PCR Amplicor monitor version 1.5. *J Clin Microbiol* **38:**1113–1120.
13. **Elbeik T, Alvord WG, Trichavaroj R, de Souza M, Dewar R, Brown A, Chernoff D, Michael NL, Nassos P, Hadley K, Ng VL.** 2002. Comparative analysis of HIV-1 viral load assays on subtype quantification: Bayer Versant HIV-1 RNA 3.0 versus Roche Amplicor HIV-1 Monitor version 1.5. *J Acquir Immune Defic Syndr* **29:**330–339.
14. **Elbeik T, Surtihadi J, Destree M, Gorlin J, Holodniy M, Jortani SA, Kuramoto K, Ng V, Valdes R Jr, Valsamakis A, Terrault NA.** 2004. Multicenter evaluation of the performance characteristics of the bayer VERSANT HCV RNA 3.0 assay (bDNA). *J Clin Microbiol* **42:**563–569.
15. **Germer JJ, Heimgartner PJ, Ilstrup DM, Harmsen WS, Jenkins GD, Patel R.** 2002. Comparative evaluation of the VERSANT HCV RNA 3.0, QUANTIPLEX HCV RNA 2.0, and COBAS AMPLICOR HCV MONITOR version 2.0 assays for quantification of hepatitis C virus RNA in serum. *J Clin Microbiol* **40:**495–500.
16. **Gianotti N, Galli L, Racca S, Salpietro S, Cossarini F, Spagnuolo V, Barda B, Canducci F, Clementi M, Lazzarin A, Castagna A.** 2012. Residual viraemia does not influence 1 year virological rebound in HIV-infected patients with HIV RNA persistently below 50 copies/mL. *J Antimicrob Chemother* **67:**213–217.
17. **Horn T, Urdea MS.** 1989. Forks and combs and DNA: the synthesis of branched oligodeoxyribonucleotides. *Nucleic Acids Res* **17:**6959–6967.
18. **Kern D, Collins M, Fultz T, Detmer J, Hamren S, Peterkin JJ, Sheridan P, Urdea M, White R, Yeghiazarian T, Todd J.** 1996. An enhanced-sensitivity branched-DNA assay for quantification of human immunodeficiency virus type 1 RNA in plasma. *J Clin Microbiol* **34:**3196–3202.
19. **Kulmala SM, Syrjänen S, Shabalova I, Petrovichev N, Kozachenko V, Podistov J, Ivanchenko O, Zakharenko S, Nerovjna R, Kljukina L, Branovskaja M, Grunberga V, Juschenko A, Tosi P, Santopietro R, Syrjänen K.** 2004. Human papillomavirus testing with the hybrid capture 2 assay and PCR as screening tools. *J Clin Microbiol* **42:**2470–2475.
20. **Laudadio J.** 2013. Human papillomavirus detection: testing methodologies and their clinical utility in cervical cancer screening. *Adv Anat Pathol* **20:**158–167.
21. **Lorincz A, Anthony J.** 2001. Hybrid capture method of detection of human papillomavirus DNA in clinical specimens. *Papillomavirus Rep.* **12:**145–154.
22. **Manos MM, Kinney WK, Hurley LB, Sherman ME, Shieh-Ngai J, Kurman RJ, Ransley JE, Fetterman BJ, Hartinger JS, McIntosh KM, Pawlick GF, Hiatt RA.** 1999. Identifying women with cervical neoplasia: using human papillomavirus DNA testing for equivocal Papanicolaou results. *JAMA* **281:**1605–1610.
23. **Mazzulli T, Drew LW, Yen-Lieberman B, Jekic-McMullen D, Kohn DJ, Isada C, Moussa G, Chua R, Walmsley S.** 1999. Multicenter comparison of the digene hybrid capture CMV DNA assay (version 2.0), the pp65 antigenemia assay, and cell culture for detection of cytomegalovirus viremia. *J Clin Microbiol* **37:**958–963.
24. **Munoz N, Bosch FX, de Sanjose S, Herrero R, Castellsague X, Shah KV, Snijders PJ, Meijer CJ, International Agency for Research on Cancer Multicenter Cervical Cancer Study Group.** 2003. Human papillomavirus and cancer: the epidemiological evidence. *N Engl J Med* **348:**518–527.
25. **Pachl C, Todd JA, Kern DG, Sheridan PJ, Fong S-J, Stempien M, Hoo B, Besemer D, Yeghiazarian T, Irvine B, Kolberg J, Kokka R, Neuwald P, Urdea MS.** 1995. Rapid and precise quantification of HIV-1 RNA in plasma using a branched DNA signal amplification assay. *J Acquir Immune Defic Syndr Hum Retrovirol* **8:**446–454.
26. **Peyton CL, Schiffman M, Lörincz AT, Hunt WC, Mielzynska I, Bratti C, Eaton S, Hildesheim A, Morera LA, Rodriguez AC, Herrero R, Sherman ME, Wheeler CM.** 1998. Comparison of PCR- and hybrid capture-based human papillomavirus detection systems using multiple cervical specimen collection strategies. *J Clin Microbiol* **36:**3248–3254.
27. **Phillips S, Garland SM, Tan JH, Quinn MA, Tabrizi SN.** 2015. Comparison of the Roche Cobas 4800 HPV assay to Digene Hybrid Capture 2, Roche Linear Array and Roche Amplicor for detection of high-risk human papillomavirus genotypes in women undergoing treatment for cervical dysplasia. *J Clin Virol* **62:**63 – 65.
28. **Sollis KA, Smit PW, Fiscus S, Ford N, Vitoria M, Essajee S, Barnett D, Cheng B, Crowe SM, Denny T, Landay A, Stevens W, Habiyambere V, Perrins J, Peeling RW.** 2014. Systematic review of the performance of HIV viral load technologies on plasma samples. *PLoS One* **9:**e85869.
29. **Solomon D, Schiffman M, Tarone R, ALTS Study Group.** 2001. Comparison of three management strategies for patients with atypical squamous cells of undetermined significance: baseline results from a randomized trial. *J Natl Cancer Inst* **93:**293–299.

30. **Tipple C, Oomeer S, Dosekun O, Mackie N.** 2014. Service impact of a change in HIV-1 viral load quantification assay. *J Int AIDS Soc* **17**(Suppl 3)**:**19677.
31. **Veillon P, Payan C, Picchio G, Maniez-Montreuil M, Guntz P, Lunel F.** 2003. Comparative evaluation of the total hepatitis C virus core antigen, branched-DNA, and amplicor monitor assays in determining viremia for patients with chronic hepatitis C during interferon plus ribavirin combination therapy. *J Clin Microbiol* **41:**3212–3220.
32. **Wang YF.** 2006. Signal amplification techniques, p 228–242. In Y.-W. Tang and C.W. Stratton (ed), *Advanced Techniques in Diagnostic Microbiology.* Springer, New York, NY.
33. **Wang YF, Eaton ME, Schuetz AN, Nesheim SR.** 2008. Human Immunodeficiency Virus, p 47–67. *In* Hayden RT, Carroll K, Tang YW, Wolk DM (ed), *Diagnostic Microbiology of the Immunocompromised Host.* ASM Press, Washington, DC.
34. **Wong AA, Fuller J, Pabbaraju K, Wong S, Zahariadis G.** 2012. Comparison of the hybrid capture 2 and cobas 4800 tests for detection of high-risk human papillomavirus in specimens collected in PreservCyt medium. *J Clin Microbiol* **50:**25–29.
35. **Yu S, Kwon MJ, Lee EH, Park H, Woo HY.** 2015. Comparison of clinical performances among Roche Cobas HPV, RFMP HPV PapilloTyper and Hybrid Capture 2 assays for detection of high-risk types of human papillomavirus. *J Med Virol* **87:**1587–1593.
36. **Zhao Y, Zhou L, Tang Z.** 2013. Cleavage-based signal amplification of RNA. *Nat Commun* **4:**1493.

DNA Sequencing for Clinical and Public Health Virology: Some Assembly Required

JOANNE BARTKUS

15

Recent advances in sequencing technology, coupled with the relatively small genomes of viruses, make routine sequencing of entire genomes in clinical and public health settings increasingly feasible. The first two widely adopted DNA sequencing methodologies described, the chemical cleavage method of Maxam and Gilbert (1) and the chain termination method of Sanger et al., were both published in the 1970s (2, 3). Coincidental to the subject of this chapter, the first full genome to be sequenced by Sanger was that of a virus, albeit a bacteriophage, PhiX 174 (4). The Sanger method proved to be the more durable sequencing technology and, especially after the process was automated in 1996, was the most widely used method for DNA sequencing for more than a decade. Beginning in 2005, however, advances in sequencing technology, the so-called next generation sequencing (NGS) methodologies, resulted in a dramatic increase in the amount of sequence that can be generated and a concomitant dramatic decrease in the cost of sequencing. These factors have led to the widespread implementation of NGS in place of the Sanger method for typical sequencing applications and also for some novel purposes, for example, replacing microarrays to study gene expression. The increased use of NGS technologies has, not surprisingly, resulted in a rapid increase in the number of sequences submitted to the National Center for Biotechnology Information's (NCBI) Genbank database. In particular, the number of viral sequences submitted since 2012 has increased 22.9% as measured in nucleotide base pairs (5), and the number of publications based on NGS is increasing at an impressively rapid pace (6).

While the improvements in DNA sequencing technologies have been dramatic, a number of significant challenges remain that have prevented the widespread adoption of DNA sequencing in clinical and public health virology laboratories. Some of these challenges are related to the technologies, but some of the most daunting are related to the biology of viruses themselves. As obligate parasites, viruses cannot be propagated outside of a host system. This necessitates that, prior to sequencing, either the virus be purified or target regions be specifically or nonspecifically amplified and/or that host genomic sequences be removed during the postsequencing data analysis. An additional complication is that viruses with RNA genomes require a reverse transcription step in the sequencing workflow in order to generate cDNA that is used in the sequencing reaction. Utilization of DNA sequencing in a clinical setting requires that assays either be cleared by the U.S. Food and Drug Administration (FDA) or be validated in the laboratory as a laboratory developed test (LDT) to meet the requirements laid out in the Clinical Laboratory Improvement Amendments and the College of American Pathology standards. At this writing, there are very few FDA-cleared assays and sequencing platforms. However, it is the amount and complexity of the data generated by next generation sequencing (NGS) methodologies, in particular, that pose the most formidable barriers to implementation for most laboratories. These challenges are not insurmountable, and it is likely that continued technological advancements and the creation and adoption of regulatory and quality assurance standards will increase the feasibility and likelihood that DNA sequencing will become a routine addition to the clinical virology repertoire. The remainder of this chapter will be devoted to discussing the available technologies, applications to clinical and public health virology, data analysis and storage strategies, and quality assurance and other considerations for laboratories that are considering or have implemented DNA sequencing technology. There is quite a bit of terminology that is unique to DNA sequencing in general and to NGS in particular. Table 1 provides definitions for some of the more commonly encountered terms used in the field of DNA sequencing.

DNA SEQUENCING TECHNOLOGIES

Sanger Sequencing

Since the advent of NGS, the Sanger sequencing method has also been referred to as first-generation sequencing technology. The chemistry used in Sanger sequencing is termed dye- or chain-termination methodology because incorporation of a labeled, modified nucleotide substrate (dideoxynucleotide triphosphate or ddNTP) blocks subsequent elongation of the DNA molecule. The ddNTPs are typically fluorescently labeled, and, in most automated systems, DNA

doi:10.1128/9781555819156.ch15

TABLE 1 Next generation sequencing terminology

Term	Definition
Bar coding	Addition of unique sequence tags to template DNA so that the multiplexed sequence data can be sorted during analysis of the data and assigned back to the correct sample
Bioinformatics pipeline	A bioinformatics pipeline is a series of data processing scripts that are linked together to perform a desired analysis, essentially a computational workflow
Contig	Short for contiguous sequence, i.e., a long stretch of DNA sequence composed of overlapping sequence fragments
De novo assembly	Assembly of a contig or whole genome in the absence of a reference sequence. Easiest to do with long reads. Short reads require very high levels of sequence coverage and a large amount of computing power to enable *de novo* assembly
Deep sequencing	Sequencing with high depth of coverage in order to detect variants present at low concentrations
Depth of coverage	Number of reads spanning a defined region of DNA sequence, e.g., 5x, 10x, etc.
DNA library	A collection of DNA fragments used for DNA sequencing. A common method of generating a DNA library is to fragment DNA or cDNA into specific size pieces and ligate on adapter sequences. Amplification of the fragmented DNA is accomplished in different ways for different sequencing platforms.
Emulsion PCR	Amplification of DNA in water droplets immersed in oil, used by some NGS platforms for amplification of template DNA. Each droplet contains a single immobilized DNA molecule that is amplified to produce a clonal colony that becomes the template for DNA sequencing.
FASTA/BFA	A sequence read file format that does not include quality score information. BFA is binary FASTA.
FASTQ	A sequence read file format that includes phred quality score information for each base-call (225)
Heat map	A heat map is a graphical way of displaying a table of numbers by using colors to represent the numerical values
Indel	Insertions and deletions
Insert	The DNA fragment to be sequenced is sometimes referred to as an insert because it is flanked by adapter sequences
kmer	Short DNA subsequences of k-length. Kmers are frequently used to facilitate data analysis by processing fragments as opposed to single nucleotides, for example for sequence assembly and generation of phylogenetic trees.
Mate pair read	Read obtained from both ends of a DNA strand where originates from a single strand, accomplished by circularizing DNA using biotinylated adapters, shearing, and then capture of the biotinylated adapters. Useful for genome finishing and *de novo* sequencing, especially when elucidating the sequence of complex regions of DNA
Metagenomics	Refers to study of microbial communities, but often applied to sequencing of a biological specimen for pathogen detection
Multiplexed run	Sequencing multiple samples simultaneously in the same reaction; requires barcoding of sample for data analysis
Paired-end read	Read obtained from both ends of a defined region of DNA, but not necessarily from the same strand
Phred quality score	A quality or Q score is a quality metric generated during sequencing and reported in the sequence read FASTQ file. A score of Q20 indicates base call accuracy of 99% while a score of Q30 indicates a base call accuracy of 99.9% (226)
Quality trimming	The process of trimming low-quality bases from the ends of sequence reads
Read	Reads are referred to as either raw reads, which are the clusters identified by the instrument, or as filtered reads, which are the reads accepted after base-calling. Low numbers of filtered reads indicates a problem either with preparation of the sample or with the sequencing run itself. Filtered reads are used for subsequent analysis
Read length	The length of DNA sequence produced by a sequencing platform. The Illumina platform typically provide sequence with short reads of <100 bp while Roche 454 produces read lengths of up to 1 kb.
Reference guided assembly	Assembly of unknown sequence reads based on comparison, or mapping, to a reference sequence
Resequencing	Sequencing of a genomic region for which a reference sequence or comparator is available. Resequencing may be targeted, i.e., a specific region is amplified and sequenced, or it may include the whole genome.
SAM/BAM	Sequence alignment file formats: SAM format stores text data in readable tab delimited ASCII columns while BAM stores data in binary form
SNP analysis	Computational steps to identify single nucleotide polymorphisms, i.e., differences in DNA sequence between the experimental
Systematic error	Error that tends to repeat and is reproduced in replicate studies
Virome	The viral component of the microbiome (149)

fragments are separated by capillary electrophoresis. Detection of the fragments is accomplished by laser-induced fluorescence of the tagged nucleotides. Instrument software analyzes the raw data, which can be exported in file formats that are compatible with commonly used analysis applications. Template DNA for Sanger sequencing typically consists of purified PCR amplicons, which are sequenced using primers specific to the region of interest, or of cloned fragments that may be sequenced using universal primers that hybridize to the flanking sequences present in the cloning vector. Due to the high degree of accuracy and long read lengths (typically 800 and up to 1,000 base pairs) produced by Sanger sequencing, this technology remains the gold standard for sequencing and remains useful for targeted resequencing to confirm variants identified by NGS and for low-throughput applications.

Pyrosequencing

Pyrosequencing, first described in 1996 (7), relies on detection of pyrophosphate (PPi) released upon incorporation of a nucleotide into a growing DNA chain. Template DNA is prepared by amplification of the target sequence by PCR using one biotinylated primer. The biotinylated primer is used to immobilize the template DNA by binding to streptavidin-coated sepharose beads. The ideal amplicon ranges between 80 and 200 base pairs in length; however it may also be possible to perform pyrosequencing on longer templates, up to 500 base pairs. Individual nucleotides are added sequentially to the immobilized DNA template. Detection of incorporation of a nucleotide is accomplished by monitoring the conversion of PPi to ATP, which is catalyzed by the enzyme ATP sulfurylase. The ATP produced by this reaction is used as the energy source by the enzyme luciferase, which in turn catalyzes generation of a chemiluminescent signal, from the substrate luciferin, that is captured by the instrument. Finally, the enzyme apyrase is added to degrade unincorporated nucleotides and ATP prior to the addition of each subsequent nucleotide and the cycle is repeated. Pyrosequencing generally produces shorter reads than Sanger sequencing and has been used primarily for detection of mutations in defined DNA targets, such as identification of antiviral resistance mutations in influenza strains (8–11). Pyrosequencers are available either as low-throughput instruments (e.g., Pyromark Q24, Qiagen) or in a NGS high-throughput format (e.g., Roche 454), which will be discussed in the section on Next Generation Sequencing.

Next Generation Sequencing

NGS technologies increase the amount of sequence data that can be generated in a single run by allowing parallel sequencing of many strands simultaneously, often referred to as massively parallel sequencing by synthesis. These technologies are categorized as either second generation, which indicates that amplification of the template molecules prior to sequencing is required, or third generation, which are NGS technologies that allow for sequencing of a single DNA molecule without requiring amplification (12, 13). Although the various NGS systems rely on different template preparation and sequencing chemistries, they share the common feature of the immobilization of template DNA to a solid substrate, which allows for the simultaneous sequencing of a large number of molecules (14). NGS technologies have been reviewed extensively in the literature, so descriptions here will be brief and will be provided in the context of the sequencing workflow (12–17).

NGS Workflow

The workflow for NGS involves a number of steps, each of which is critical to ensure the generation of high-quality, unbiased DNA sequence. Following nucleic acid extraction, the first step for second generation sequencing technologies (as well as for some third generation technologies) is preparation of a DNA library. Library preparation typically involves fragmentation of DNA into pieces short enough to be sequenced and ligation of adapter sequences that enable immobilization of the DNA onto a solid surface. DNA fragmentation may be accomplished by sonication or enzymatically. Sizing of fragments prior to sequencing may be required depending on the technology used to generate the library. Preparation of the DNA library is a critical step in the sequencing process and can introduce substantial bias and errors into the sequence. Strategies for addressing the potential for bias during template preparation will be addressed below in the section on Quality Assurance. Library preparation can be time-consuming and laborious, employing many manual steps. However, techniques for template preparation are evolving along with the sequencing technologies, and commercial products that simplify the process are available. For example, Illumina's Nextera technology allows for DNA to be fragmented and tagged with sequencing adapters simultaneously (18). Library preparation in this system is very efficient, requiring only 50 nanograms of DNA input for the standard DNA kit and as low as 1 nanogram for the Nextera XT kit.

Sequencing of RNA templates, including RNA viruses, typically requires conversion of RNA to cDNA prior to library preparation. This may be accomplished in multiple steps by generation of cDNA by reverse transcription of the RNA template, followed by library preparation. Kits used for transcriptome analysis, such as TruSeq RNA (Illumina) (19, 20), simplify the sample preparation for RNA viruses. The Ovation RNA-Seq (Nugen) kit (21, 22), which is used for generating amplified cDNA, can be used with samples containing low-copy number RNA viruses. Epicentre's TotalScript utilizes an optimized reverse transcriptase, which limits rRNA contamination as compared to standard buffers used for reverse transcription. This is very helpful for viral RNA templates. Similar to the Nextera technology, the cDNA generated by reverse transcription is tagged using an *in vitro* transposase (23).

For second generation NGS technologies, the next step in the workflow is amplification of the DNA template. Different platforms use different amplification strategies. Ion Torrent and Roche 454 utilize emulsion PCR to amplify the template, while Illumina uses a technique referred to as bridge amplification (Fig. 1). After template amplification, if required, the next step in the workflow is the actual sequencing reaction. The sequencing technology used by Illumina is similar to Sanger sequencing in that it relies on use of a fluorescently labeled reversible chain terminator, which is cleaved after imaging to allow for addition of the next nucleotide. Sequencing in the Roche 454 system is accomplished by pyrosequencing, analogous to that of the Pyromark platforms, but on a massively parallel scale. The technology used by Ion Torrent is similar to that of the 454 pyrosequencing method; however, instead of using luminescence to detect release of inorganic phosphate, the Ion Torrent uses semiconductor technology to detect hydrogen ions released during the incorporation of each nucleotide base (24) (Fig. 2). The Life Technologies SOLiD system, which employs ligation of fluorescently labeled di-base

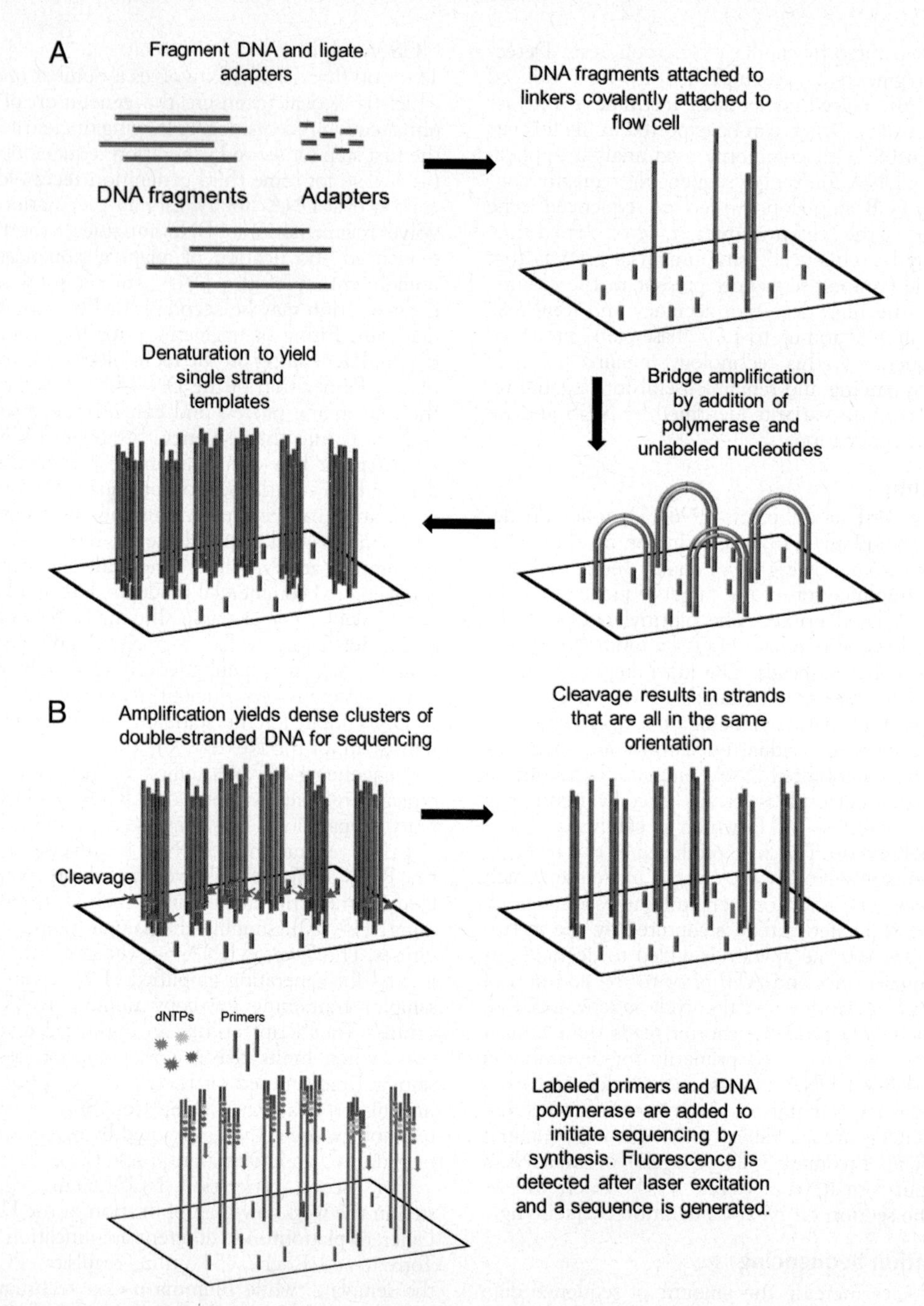

FIGURE 1 Illumina Sequencing Technology. A) Libraries for Illumina sequencing are prepared by ligation of adaptors to both ends of random fragments of DNA. DNA fragments are attached to the Illumina flow cell by hybridization of the adaptors to complimentary linker oligonucleotides on the surface of the flow cell. The hybridized DNA fragments are next amplified by a process referred to as solid phase bridge amplification. Each DNA strand has an attached and a free terminus. The strands are then denatured, resulting in the generation of millions of single-stranded DNA clusters. B) The reverse strands are released from the flow cell by cleavage and washed away, leaving only the forward strands for sequencing. Sequencing of these DNA clusters is done simultaneously by adding a sequencing primer and fluorescently labeled nucleotides. Laser excitation results in fluorescence of the last base incorporated, the label and a blocking group are removed and washed away, and the cycle is repeated. Fluorescence is detected by way of CCD camera image capture.

probes and provides data in "color space" as opposed to base space (base calls), is being discontinued, and the instrument is no longer being sold (25). The review article by Metzker provides a color diagram of the process used by the SOLid system, as well as the Illumina and 454 technologies (14).

Third-generation sequencing methods do not require a template amplification step. The PacBio RS system utilizes what is referred to as single-molecule, real-time (SMRT) technology. In this technology the DNA polymerase is immobilized in what is referred to as a zero-mode waveguide (ZMW). A ZMW guides light into a very small observation

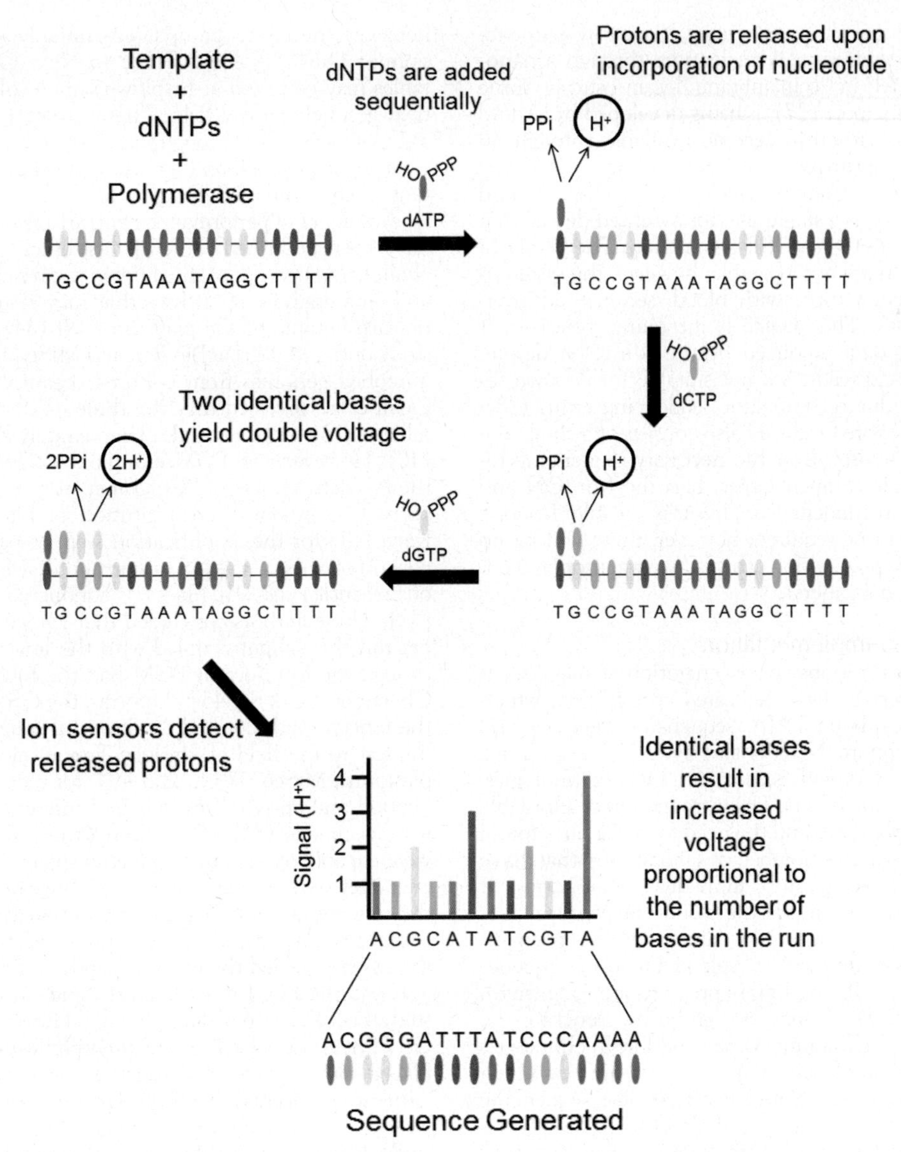

FIGURE 2 **Ion Torrent Sequencing Technology.** Libraries for Ion Torrent sequencing are prepared by ligating adaptors onto DNA fragments. Fragments are clonally amplified by emulsion PCR. The particles are applied to the Ion Torrent chip for sequencing. Sequencing is accomplished by addition of a sequencing primer, DNA polymerase, and the sequential addition of dNTPs. The Ion Torrent chip is a semiconductor, and each microwell contains millions of copies of the amplified DNA template. Incorporation of a nucleotide results in the release of a proton, which results in a detectable change in pH in the microwell, which is converted to digital information. If no base is incorporated, then there is no voltage change. When multiple sequential nucleotides with the same base are present in the sequence, the result is an increase in the voltage that is proportional to the number of identical bases in the homopolymer run. So for two bases the voltage is doubled, for three it is tripled, and so on. Signals are processed and bases are called by the Ion Torrent software.

window that is just large enough to visualize a single fluorescently tagged nucleotide being incorporated by the polymerase. Once the nucleotide has been incorporated, the tag is cleaved off and the fluorescence diffuses outside of the visualization field. In this way, the addition of nucleotide can be measured as it occurs. Library preparation consists of DNA fragmentation and ligation of hairpin adapters. At this writing, the PacBio RS technology is the only third generation NGS platform currently on the market. The Helicos NGS technology, developed by HeliScope, was the first instrument capable of single-molecule sequencing, but the instrument is no longer sold, and the technology is now available only as a service. Helicos technology uses what are referred to as virtual terminator nucleotides, which have a cleavable indicator tag. Samples are immobilized via a 3′ poly (A) tail added during the library preparation step, and the labeled nucleotides are added, incorporated, washed and imaged, and then the cycle is repeated following cleavage of the fluorescent tag. The poly(A) tail also serves as the primer for sequencing. Unique among NGS technologies, the DNA polymerase used in this method can be replaced by reverse transcriptase, which enables direct sequencing of RNA molecules without prior conversion to cDNA (26). The requisite poly(A) tails may be attached to RNA molecules enzymatically using a poly-A polymerase. There are, however, no publications demonstrating the utility of this technique for direct sequencing of RNA viruses. Nanopore sequencing, which analyzes transit of nucleotides through a nanopore-sized channel in a membrane by measuring ionic current through the pore (27), is being developed by Oxford Nanopore and has recently become available through an early access program (https://nanoporetech.com/community/the-minion-access-programme-philosophy). The Oxford Nanopore MinION, is a single-use, miniaturized device that plugs into a computer via a USB connection. The details of sample preparation are not available; however, the system is designed to be compatible with blood, serum, and environmental samples. This device is intriguing; however, it appears that the data produced by the MinION, despite providing very long reads, are not suitable for *de novo* sequence assembly due to systematic sequencing errors (28). The Oxford Nanopore GridION also consists of a single-use cartridge that contains all of the necessary sequencing reagents; however, it is much larger than the MinION and requires separate instrumentation, but it is scalable. Double-stranded DNA will be required; however, there will be no need for library preparation (https://www.nanoporetech.com/about-us/for-customers).

Options for NGS Implementation

In many large institutions, next-generation sequencing is performed as a service by a dedicated core facility, which typically provides both DNA sequencing services and bioinformatics support. NGS is also available as a service from a number of providers, such as Life Technologies, SeqLL, ChunLab, and others. The introduction of affordable bench top NGS platforms and the need for rapid turnaround times has increased the number of laboratories that have brought the technology into individual laboratories. It should be noted, however, that the cost of the platform does not reflect the total cost, as the laboratory also needs to consider costs associated with sample preparation, sequencing reagents (flow cells and library prep kits are expensive), and data analysis, which may be significant. Benchtop sequencers, such as the Illumina MiSeq and Life Technologies Ion Torrent PGM and Roche 454 GS Junior, are best suited to amplicon, resequencing, and small genome sequencing applications due to throughput limitations. The various platforms and models differ in throughput, accuracy, read lengths, purchase cost, and suitability for particular applications. Numerous options are available, and the list in Table 2 highlights the most commonly used, currently available, or under development (e.g., Oxford Nanopore) NGS platforms. The field is evolving rapidly, and, in fact, the first NGS technology to be introduced to the market, the 454 pyrosequencing platform, is about to be discontinued (http://www.genomeweb.com/sequencing/roche-shutting-down-454-sequencing-business). The ABI SOLid instrument has also been discontinued and will be replaced by the Ion Torrent Proton semiconductor sequencing system. At this writing, the Illumina platforms dominate the NGS market, with the Illumina MiSeq and Ion Torrent PGM sharing much of the benchtop sequencer segment. Manufacturers offer both affordable benchtop and more expensive ultra-high throughput instruments, providing a great deal of flexibility for the laboratory to choose an instrument best suited to their applications and throughput needs. Because of the rapid pace of technological innovation, any summary is almost immediately out of date; however, an excellent source of information relevant to choosing a NGS platform is Travis Glenn's Field Guide to Next Generation DNA Sequencers (12). This article was published initially in 2011; however, the tables are updated annually and are available online. The 2014 Field Guide to Next Generation DNA tables may be found at http://www.molecularecologist.com/next-gen-fieldguide-2014/. These tables provide comparisons of accuracy, throughput, instrument cost, cost per gigabase sequenced, and a variety of other variables important to choosing a NGS platform.

A number of performance comparisons of NGS platforms have been published (29–33). Instrument comparisons are challenging because of the differences in sample preparation and data analysis algorithms that may result in differences not attributable to the platform itself (34). A comparative study of the PGM, PacBio RS, and MiSeq for sequencing of microbial genomes from cultured organisms conducted by Quail et al. revealed that the three platforms were comparable for GC-rich, neutral, or moderately AT-rich genomes (30). However, the PGM exhibited a significant bias with the AT-rich genome of *Plasmodium falciparum*. This bias was partially overcome by substitution of Platinum Taq with Kapa HiFi for the amplification step. A performance comparison of MiSeq, 454 GS Junior, and PGM for resequencing of *Escherichia coli* O104:H4 was conducted by Loman et al. (32). These authors concluded that MiSeq had the highest per run throughput coupled with the lowest error rates, although the Ion Torrent PGM had the highest throughput. Characteristic of the 454 platforms, the GS Junior generated the longest reads but had the lowest throughput. Of greater interest to the field of virology, Frey et al. compared three platforms, MiSeq, PGM, and 454, for metagenomic identification and antiviral resistance of influenza H1N1 in clinical specimens (33). They found that all three platforms were capable of detecting influenza spiked into blood; however, they were unable to detect resistance mutations at any of the concentrations tested. The performance characteristics of each platform differed: the 454 produced the longest reads, the MiSeq yielded the greatest depth and breadth of coverage, and the PGM was the most rapid. In another virology study, Li et al. compared the Illumina HiSeq 2000, Roche 454 GS-FLX, and Roche 454 GS Junior platforms for detection of HIV-1 minority variants associated with resistance to the antiretroviral agents raltegravir in pretreatment specimens from patients who ultimately failed therapy (34). The results of this study showed that the Illumina platform provided 1,000 times greater coverage compared to the 454 platform, along with higher sensitivity for variant detection with fewer false positives. The cost of sequencing the multiplexed samples was similar for the two platforms; however, the authors concluded that the higher coverage produced by the Illumina system would allow for multiplexing of specimens to reduce cost. Yet another study compared four NGS platforms, Illumina, Ion Torrent, PacBio, and 454, for determination of HIV coreceptor tropism (35). The performance of all four systems was comparable with regard to error rates and ability to detect virus variants. There were minor differences in the ability of the platforms to identify non-R5 HIV viruses (i.e., those that did not utilize chemokine receptor CCR5 for entry into target cells) however, the authors concluded that all four platforms could be used to reliably predict HIV receptor tropism.

Ultimately, the choice of NGS platform will be based on considerations of throughput, multiplexing capacity, read length, accuracy, instrument cost, reagent costs, turnaround time, and laboratory space limitations. In general, higher throughput is required for deep sequencing for viral population analysis and metagenomic applications, while longer read lengths are helpful for applications requiring *de novo* sequence assembly.

TABLE 2 Overview of next generation sequencing technologies and platforms

Platform	Sequencing technology	Advantages	Disadvantages
GenomeLab GeXP (Beckman) ABI 310, 3130, 3730, 3500 (Life Technologies)	Dye terminator	High accuracy rate Long reads Data analysis easily performed	Cost prohibitive for high-throughput applications
PyroMark Q24, Q96 (Qiagen)	Pyrosequencing	Rapid	Short reads Errors in homopolymeric regions
454, 454 GS Junior (Roche)	Massively parallel Pyrosequencing	Long reads Fast run time	High reagent cost Errors in homopolymeric repeats Roche to discontinue 454 sequencers and will no longer support product starting in 2016
MiSeq, NextSeq 500, HiSeq (Illumina)	Reversible Terminator	Low cost per base Moderate read lengths Very high throughput, for HiSeq models	Requires heterogeneity of sequence template, which is problematic for amplicon sequencing
Ion Torrent PGM, Ion Proton (Life Technologies)	Semiconductor	Low-cost instrumentation Simple machine	Errors in homopolymeric and AT regions High reagent cost, depending on application
ABI SOLiD (Life Technologies)	Ligation	Very high throughput High accuracy	Very short reads Long run times Instrument no longer commercially available
PacBio RS (Pacific Biosciences)	Single Molecule Real-time	Extremely long reads Fast run times	High error rate Expensive Relatively low throughput
MinION, GridION (Oxford Nanopore)	Nanopore (27)	Small footprint Simple design	Available through an access program Performance characteristics for virology application not yet evaluated
SeqLL http://seqll.com/ Helicose sequencing performed as a service ChunLab Numerous other	Sequencing as a service	Does not require purchase of instrumentation Ideal for sporadic sequencing needs	May be cost prohibitive Turnaround time may not be sufficient for diagnostic or public health investigation purposes

NGS Strategies

Because of the small size of viral genomes, Sanger and small-scale pyrosequencing work well for many applications, especially if the virus is known and the region of interest is small. However, NGS may be preferable for large-scale sequencing of amplicons (36) or when the goal is detection and identification of an unidentified virus, especially if it is present in a complex matrix (i.e., metagenomic sequencing). There are a number of strategies for optimizing NGS sequencing to make sequencing cost effective and to facilitate data analysis, and the strategy selected will depend on the application.

Given the small genome size of viruses relative to the amount of data generated in a NGS run, multiplexing of samples would make NGS more cost effective for most laboratories. Multiplexing refers to combining, or pooling, multiple samples in a single sequencing lane. In multiplexing, template DNA from individual samples is "barcoded" by adding a unique tag to the DNA. This is typically accomplished by adding the unique sequence to the PCR primer used during generation of amplicons for targeted sequencing, or to the adaptor ligated to fragments during library preparation. The bar codes are used to sort the sequence data during the analysis step so that the reads can be assigned to the correct sample. Multiplexing may not be suitable on benchtop sequencers for applications requiring high coverage levels, such as virus population analysis or metagenomics.

For applications requiring *de novo* assembly (i.e., assembly in the absence of a reference sequence) or for sequencing regions of DNA with repetitive elements, mate-paired and paired-end sequencing, as opposed to single-end sequencing strategies, increase the ability to accurately assemble the sequencing reads. In paired-end sequencing, different adapters are attached to each end of the template DNA such that the reads originate from both ends of the same molecule being sequenced; however, they are not complimentary unless the fragment being sequenced is very short. The advantage to having paired ends is that it enables better alignment of the reads because the distance between the two ends of the insert being sequenced is known, making it easier to identify gaps, insertions and deletions. While similar in concept (i.e., a sequence is obtained from both ends of a DNA template) mate-end sequencing is slightly different in execution. For mate-paired sequencing, the DNA fragment created during library preparation is circularized using biotinylated adapters, and then sheared. The biotinylated adapters are used to capture the fragments, and the adapters are used to initiate sequencing. Mate-paired sequencing typically allows for use of larger inserts than does paired-end sequencing. A good overview of these two sequencing strategies can be found on the Illumina website, http://technology.illumina.com/technology/next-generation-sequencing.ilmn.

Host Contamination

Regardless of the sequencing technology employed, it is necessary to address the issue of contamination of virus samples with host DNA and RNA that complicate the detection and identification of viral sequence reads. Decreasing host contamination (host depletion) by centrifugation, filtration, or nuclease treatment or increasing viral nucleic acid (target enrichment) by amplification are two approaches used to enable enrichment of viruses that may be present at low levels in a sample. These methodologies are frequently used in combination to provide the largest increase in the relative amount of viral nucleic acid relative to host contamination. The problem of host contamination is lower for cultured viruses as centrifugation may be used to remove most of the host cells (37).

The simplest method for dealing with host contamination is to amplify specific genomic regions of the viral target by PCR, followed by amplicon sequencing (i.e., targeted sequencing). Examples of this approach are numerous. Amplification of a single region of the genome is commonly used for highly targeted applications, such as assessment of antiviral resistance mutations or virus genotyping (34, 38–41). Whole viral genomes may be amplified by PCR using degenerate primers or overlapping primers that generate amplicons covering the entire genome (42–46). A commercial product, the PathAmpFluA kit (Life Technologies), is available for whole genome analysis of influenza A (https://tools.lifetechnologies.com/content/sfs/brochures/Influenza_A_Typing_App_Note.pdf). The kit includes universal influenza primers for multiplexed amplification of all eight influenza A genome segments. The cDNA amplicons produced are then sequenced on the Ion Torrent PGM. The advantage of targeted amplification is that the methodology is fairly standard, and sequencing may be accomplished by either Sanger or NGS methods, depending on the throughput required. The disadvantage of targeted amplification is that it is sequence-dependent, requiring a priori knowledge of the identity of the virus as well as genomic sequence information.

Sequence independent amplification methods can be used to increase the amount of viral nucleic present in a sample prior to sequencing to identify novel viruses without *a priori* sequence information. Random amplification of RNA viruses has been accomplished using random hexamer primers (47–49), or more recently using whole transcriptome or RNA-seq commercial kits as described previously in the section above on Amplification and Sequencing (21, 22, 50, 51). Treatment of specimens with DNase to remove contaminating host DNA, followed by sequence-independent single-primer amplification (SISPA), has been used for amplification of both DNA and RNA viruses prior to cloning and sequencing (52–54). Rolling circle amplification may be used for isothermal amplification of viruses that contain a circular DNA genome, for example human papilloma virus (HPV) (55, 56). VIDISCA-454 is a technique that couples VIDISCA with Roche 454 sequencing. VIDISCA (virus discovery based on cDNA-AFLP [amplified fragment length polymorphism]) is a method whereby adaptors ligated to restriction fragments are used as primers for PCR amplification. This method has been successfully applied to detect Norovirus in feces and HIV-1 in serum (57). It should be noted, however, that random amplification methods have the potential to introduce bias and errors, which need to be addressed in the experimental design and/or during data analysis (58–60).

Viral targets may also be separated from host contamination by a technique referred to as target capture. Target capture (sometimes referred to as "baiting") is a hybridization-based method that uses long (120 nucleotide) RNA sequences as capture probes to enrich specific sequences, which are then subjected to NGS. This methodology can be used to capture either DNA or RNA sequences and has been used for enrichment of herpeseviruses from clinical specimens (61) and for identification of virus integration sites from formalin-fixed, embedded human tissue (62) using the Agilent SureSelectXT Target Enrichment System.

Physical methods including ultracentrifugation and filtration are commonly used to concentrate viral particles present in a sample (63–66). These methods are frequently used in conjunction with other methods, such as nuclease digestion, to remove any unencapsidated nucleic acid present in the sample, or random amplification to enrich low abundance sequences as described above. Marston et al. (37) showed that RNA viruses could be enriched relative to host contamination by polyethylene glycol precipitation, followed by RNA extraction with Trizol, DNase digestion to remove host genomic DNA, and rRNA depletion by exonuclease digestion, although there was a 3- to 100-fold decrease in the amount of extracted RNA following DNase treatment. This study also demonstrated that the extraction method used had a significant impact on the amount of RNA extracted, with Trizol providing greater RNA yields than a spin column extraction method (RNeasy). A recent study by Hall et al. (67) compared five combinations of centrifugation, filtration and nuclease digestion for virus enrichment. Their results indicated that large amounts of contaminating nucleic acids remained and that the relative abundance of viral sequences within a metagenomics data set was not substantially increased except when a three-step enrichment method, consisting of centrifugation, filtration, and nuclease digestion, was used. While capable of significantly reducing host background sequences, physical as well as enzymatic methods may result in loss of viral sequences and are not capable of enriching viruses that are integrated into the host genome.

With the exception of target-specific PCR, none of the methodologies described above are capable of complete removal of contaminating sequence reads. Because of host contamination of specimens for metagenomic NGS sequencing, virus detection and identification typically require over-sampling during sequencing to provide enough sequence coverage to detect viruses present at low concentrations, which poses a significant challenge to analysis of the sequencing data. Removal of host sequences during the post sequencing data analytical step is typically accomplished by computational subtraction, which involves mapping the sequencing reads to a host reference genome and then filtering the mapped reads so that they are no longer included in subsequent data analysis. This process will be described in more detail below in the section on Bioinformatics.

Bioinformatics

The analytical processes for data generated using first generation Sanger sequencing technologies are well developed, and numerous tools are available, either commercially or in the public domain, for various steps in the analytical pipeline. The first step in DNA sequence analysis, regardless of whether it is from a first, second, or third generation platform, is the processing of instrument data into base calls. This step is typically accomplished by software integral to, and provided with, the instrument platform, and the output is the DNA sequence read. The output file from the instrument typically also includes PHRED, or quality scores, for each base called. Sanger sequencing instruments typically provide the option to export data in a number of sequence file formats, including those that are specific to the instrument, for example, ABI or CEQ (Beckman), as well as formats that are used for subsequent data analysis, such as FASTA or FASTQ. The next step in the analysis of sequencing data typically involves editing or filtering the sequence to remove low-quality reads. The quality score for each base can be used to facilitate the trimming process, as well as to identify positions in the sequence containing mixed bases that may indicate a polymorphism at an individual position in the sequence or a mixed virus population. For Sanger sequences, visual inspection of the sequencing trace may be performed to identify regions requiring editing, which can be done manually. This is not possible, however, for sequences generated on NGS platforms. Trimmed sequences are then used as input in other analytical steps. For some applications, such as identification of sequence variants, the sequence may be plugged directly into the appropriate analytical program. However, if longer sequences are required, it may be necessary to compile overlapping sequences into longer pieces, also referred to as contigs (short for contiguous sequence). The type of computational tool used for the analysis will depend on the application. Some common DNA sequence analyses include performing a BLAST search (http://www.ncbi.nlm.nih.gov/BLAST/) against a database of virus sequences to search for homology to a known virus, comparison to a reference sequence to identify population variants and quasispecies or phylogenetic analysis using ClustalW (http://www.ebi.ac.uk/clustalw/), or other alignment programs to determine evolutionary relationships (68), as well as numerous other downstream applications.

The magnitude of the data generated by next generation sequencers and the shorter read lengths necessitate different computer algorithms than those used for assembly and analysis of sequences from first generation platforms. The review by Gogol-Döring and Chen provides a very nice overview of the general steps involved in the analysis of NGS data (69). As for Sanger sequencing, primary analysis of NGS data involves converting the instrument data into DNA sequences (base-calling). The sequence data are first demultiplexed, if multiplexing is used as part of the sequencing strategy (as described previously). Sequence-read statistics, such as the number of raw reads and the number of filtered reads (e.g., reads accepted after base-calling) should be reviewed to assess the overall success of the sequencing reaction. Sequences are next trimmed to remove low-quality portions of reads and adaptor sequences prior to further analysis. Low-complexity reads (e.g., homopolymeric regions) may also be filtered out and removed from subsequent analysis.

Secondary analysis includes sequence assembly, which may be accomplished either by mapping or alignment of reads to a reference sequence (reference-guided assembly), or *de novo* assembly in the absence of a reference. Read mapping involves finding the region of the reference sequence that best matches the newly sequenced fragment. Because reads may not match the reference sequence exactly, the user can define the number of mismatches that will be allowed between the read and the reference sequence. Setting this number too high may result in improperly mapped reads, while setting it too low may preclude mapping of reads when there is a mismatch due to sequencing error or sequence variants. While mapping tools may provide output in a variety of formats, SAM/BAM is rapidly becoming the standard mapping file format. Mapping statistics can be used as a quality indicator, with a low number of mapped reads indicating poor sequence quality or contamination. Additionally, mapping data provide information on sequencing coverage, or the number of reads that map to a specific portion of the reference sequence. For example, if 10 reads map to the same region, the coverage is said to be 10-fold for that region. Nonuniform coverage of the reference genome may indicate the potential for bias in the data or the presence of additional repeated regions in the newly sequenced genome.

De novo assembly is compilation of an assembled sequence in the absence of a reference sequence. This is accomplished by alignment of overlapping reads to form contigs and is extremely challenging for platforms that yield short sequence reads. Therefore, an increased amount of coverage is required to ensure that there are enough overlapping reads to assemble into a full-length genome (64). For this reason, platforms that yield longer sequence reads (e.g., 454) enable *de novo* assembly at lower coverage levels. The need to generate a full-genome sequence will depend on the application. Because of the expense and time needed to complete a full-length genome, many sequences are left as a collection of contigs.

Similar to first generation sequencing, post-alignment or downstream data analysis, for example, variant detection, depends on the clinical or research application and will require specialized analytical tools. Many of the algorithms used for analysis of NGS data are customized by the end user, require specialized expertise to execute, and are not as well standardized as those used for Sanger sequencing. In addition to the need for specialized bioinformatics tools and expertise, the amount of sequence data produced by NGS and the number of processes involved in data analysis result in the need for much greater computer processing and storage capacity and may be problematic for transfer of data electronically due to bandwidth limitations. Bioinformatics and computing needs represent the biggest barrier to broader implementation of NGS technology in individual laboratories. Recently, there has been an increase in the availability of commercially available software packages, including some that are included by the NGS system manufacturers when the system is purchased or are available from the vendor at an additional cost. In addition, bioinformatics is now increasingly being offered as a service, further decreasing the need for laboratories to have specialized expertise in bioinformatics. However, the data analysis remains complex, and laboratories will likely still need to have in-house expertise to review and interpret sequencing results. These changes in bioinformatics offerings, along with provision of server and cloud-based storage and analytical services, have enabled laboratories to begin to at least explore the potential of NGS for clinical applications and are described further below.

Bioinformatics Tools and Options

There are a wide array of bioinformatics tools and options currently available either as free, downloadable, public-domain tools, as well as commercially available software packages, either stand-alone or provided by the instrument manufacturer (70, 71). Table 3 provides a partial list of the variety of bioinformatics and software solutions that a laboratory can select from, a few examples of offerings in each category, and the advantages and disadvantages of each. The field is rapidly evolving; therefore laboratories interested in pursuing any of these options should conduct their own research to find a package that best suits their needs, computing and informatics structure, bioinformatics expertise, and budget. NGS platform vendors offer computing storage and analytical packages. Many basic data analyses can be performed using software provided by the vendor.

Primary sequence analysis, such as base-calling and assessment of sequence quality, is typically accomplished by software integral to the sequencing platform. Secondary analyses may be accomplished in a variety of different ways and can be tailored to the abilities and resources of the individual laboratory. Illumina's MiSeq Reporter is the bioinformatics software built into the MiSeq that can be used to perform secondary computational functions, such as alignments, structural variants, and contig assemblies. The analysis is launched automatically based on the data analysis workflow, or pipeline, specified by the user during instrument setup. In addition to software, Illumina also offers storage in the form of either a standard computational system that provides storage for up to 400 genomes, or as an advanced system for high-capacity storage (up to seven HiSeq instruments). Illumina's BaseSpace is a cloud-based system for analysis, archiving, and sharing of NGS data and is provided via Amazon's Web Services. BaseSpace is also available as a local system. Base Space users can select from a variety of applications tools, or apps, some of which are standard and others that are customized by end users and made available to other users. An example of a user provided app is DeepChek-HIV, which is an app for antiviral resistance typing of HIV (http://blog.basespace.illumina.com/2014/07/23/deepchek-hiv-app-for-genotyping-by-ngs-and-inferred-drug-resistance-testing-for-research-use-only/). PathSEQ Virome is a BaseSpace app that is capable of detecting more than 50,000 virus genomes in approximately one hour (http://www.pathgendx.com/_asset/PathSEQ_Virome-Getting_Started.pdf). Likewise, Ion Torrent provides both computational software and storage options. As with Illumina, these computational solutions are scalable and can be adapted to the needs of the individual laboratory. The Ion Reporter Software consists of a set of preconfigured bioinformatics tools for automated data analysis workflows and provides for both customization and data and workflow sharing via a collaboration space. The PacBio Sequencing System also includes a software package that automates data analysis and integrates with LIMS and third-party analysis tools. Roche also provides a suite of software for the GS Junior and GS FLX Systems at no extra cost for applications including *de novo* assembly, reference mapping, and amplicon variant analysis. Information on all of these computational options may be found on the vendor websites. Data output from all of these systems includes standard formats, such as FASTA (sequence file), FASTQ (sequence file with PHRED quality scores), and SAM/BAM (standard alignment format).

Commercial software for analysis of sequencing data, such as Sequencher, Bionumerics, Lasergene, CLC Bio and Vector NTI (Table 3), have been widely used for analysis of first generation sequencing, and NGS options are also available or are being developed for many of these software packages. These software packages are affordable and run on most commonly used operating systems, such as Windows, Mac OSX, and Linux. CLC Bio offers several different software options for analyzing both Sanger and NGS data, including desktop and enterprise (server system) solutions. CLC Genomics Workbench is a desktop application that provides a user-friendly graphical user interface and supports all major NGS platforms. CLC Genomics Workbench includes tools for many of the most commonly performed NGS data analysis operations, including resequencing, read mapping, *de novo* assembly, and variant detection and has been used for analysis of hepatitis C drug-resistant mutations (72), characterization of Epstein Barr virus in human B lymphocytes (73), and typing of human papillomavirus (HPV) (56), among others. DNASTAR Lasergene also provides an integrated software package for assembly and analysis of NGS data. In addition to desktop software packages, many vendors offer cloud-based options as well.

For those laboratories with sophisticated bioinformatics expertise and advanced informatics infrastructure, there are a number of open source NGS software tools available to choose from that are freely available in the public domain. These include single-purpose tools that may be used to develop an analysis pipeline that can be tailored to a specific application, as well as pipelines that have already been developed. For example, Bowtie2 (74) is a tool commonly used in bioinformatics pipelines that performs short-read alignments to reference genomes, and SOAPsnp (75) is a tool for variant detection that identifies single nucleotide polymorphisms (SNPs) by comparing a consensus sequence to a reference sequence. These single-purpose tools, when performed in series, form the basis of Crossbow, an automated analysis pipeline (or workflow) used to detect SNPs in high-coverage, short-read genome sequence data (76, 77). There are many open-source bioinformatics pipelines that have been described in the literature. These programs typically run on a Linux operating system and require some level of expertise to create the pipeline and/or download, install, and run the scripts, as well as bioinformatics expertise to interpret the data. The programs are often run as shell scripts that combine a sequence of commands into a single script. This is useful for processes that are performed repeatedly; however, informatics expertise is needed to write the scripts and to troubleshoot when the script fails. A tool for simplifying this process for bioinformatics pipelines, Bpipe (78), has been developed to address some of the problems related to running a shell script. In addition to pipelines developed for a specific purpose, there are software packages that provide access to a wide variety of NGS tools. One example is Galaxy (79, 80), which is available to install locally or can be used as a web-based interface that provides a single interface to some of the commonly used NGS tools. Likewise, the Broad Institute has developed a Genome Analysis Toolkit (GATK) that offers a wide variety of tools for analysis of resequencing data with an emphasis on quality assurance (81). Cloud-based resources are becoming increasingly available, which allow access to single-purpose

TABLE 3 Bioinformatics software options

Application	Platform/ Availability	Examples[a]	Advantages	Disadvantages
Single-purpose tools	Downloadable Public Domain	Numerous, see reviews (227, 228)	Free of charge Configurable, create your own pipeline	Does only one specific job Requires bioinformatics expertise for scripting and maintenance on Linux Time consuming
Bioinformatics workflows	Downloadable Public Domain	GATK (81) CG-Pipeline See Table 4 for pipelines developed for virology applications	Free of charge Available for a variety of general or specific purposes	Require extensive validation Pipelines for the same purpose using different component tools may yield different answers Requires Linux and bioinformatics expertise
Bioinformatics workflows	Cloud-based Public Domain	Galaxy (74, 79) http://galaxyproject.org/ Rainbow (97) http://s3.amazonaws.com/jnj_rainbow/index.html Crossbow (76) CloVR (83)	Does not require Linux expertise	Require extensive validation Transfer of large data packages can be problematic Privacy and security concerns
Bioinformatics infrastructure	Downloadable or cloud-based Public Domain	CloVR(83)	No need for computational core	Requires informatics/ bioinformatics expertise
Sequence viewing and data analysis	Desktop software Public Domain	GENtle http://gentle.magnusmanske.de/ MEGA6 http://www.megasoftware.net/ BioEdit http://www.mbio.ncsu.edu/bioedit/bioedit.html NCBI Genome Workbench http://www.ncbi.nlm.nih.gov/tools/gbench/	Free of charge Runs on local machine	May not be suitable for analysis of NGS data Limited analytical capability
Sequence viewing and data analysis	Desktop software Commercial	CLC Genomics Workbench (CLCbio) Lasergene (DNASTAR) Sequencher (Gene Codes) Bionumerics 7.0[b] (Applied Maths) Vector NTI (Life Technologies) Geneious 8.0 (Biomatters) NextGENe (SoftGenetics) CodonCode (CodonCode Corporation) Geospiza (Perkin Elmer)	User friendly Less bioinformatics expertise needed User support provided Automated Integrated tools	Analysis options may be limited May be cost prohibitive May not support NGS applications
Instrument-based data analysis	Instrument software Commercial	Varies, see text	User friendly Some software provided with instrument platform Enables basic analysis Ability for end users to develop and share applications	May not include tools for all applications Cloud storage may be cost prohibitive
Bioinformatics as a service	Commercial	Integrated Analysis Inc. http://www.i-a-inc.com/proteomics-saas ChunLab http://www.chunlab.com/ Accura Science http://www.accurascience.com/	Does not require informatics equipment of expertise	May be cost prohibitive (prices range from $75 to $260 per hour)

[a]Examples are not all inclusive, nor do they represent an endorsement of any particular, program, product, or service.
[b]Bionumerics 7.5 will include NGS capability.

tools, bioinformatics pipelines or bioinformatics infrastructure through the Internet via cloud-hosted services (76, 79, 82–86). Although software with a web interface simplifies usability of the programs, bandwidth limitations and concerns about data security may preclude widespread use of these bioinformatics options in a clinical setting.

Pipelines for Viral Bioinformatics

Viruses pose a number of bioinformatics challenges due primarily to the fact that viruses are obligate intracellular parasites and cannot be cultivated in the absence of host cells. Additionally, viruses display a high degree of sequence heterogeneity due to the tendency to undergo recombination at relatively high rates and because of RNA viruses' lack of proof-reading during nucleic acid synthesis (87). From a taxonomic perspective, viruses are challenging because, unlike the bacterial 16S ribosomal RNA gene, there are no genes that can be used as a phylogenetic marker across all virus families (88). The applications of NGS to virology vary widely, and the bioinformatics tools and databases needed are dependent on the specific application and task, which may include structure/function correlation, virus detection and taxonomy, phylogenetic analysis, and variant analysis, among others (see Applications of Sequencing to Virology section). There are a number of standard computational tools that have been used for analysis of viral DNA sequences for many years, for example BLAST for homology searching or ClustalW for phylogenetic analysis (see review of Yan for additional examples [89]). Recently, an increasing number of bioinformatics pipelines have been developed for specific virology applications, and most of them have been made freely available in the public domain (Table 4). Many of these pipelines are downloadable Linux-based programs, which typically require expertise in informatics for downloading, installation, configuration, and maintenance of the software, as well as bioinformatics expertise to interpret the data output. Some of these pipelines are developed for a relatively narrow purpose, for example, variant calling or detection of virus integration sites, while others provide general purpose tools and pipelines. In addition, some of the pipelines are specific to a particular sequencing platform, for example, VirusHunter (90), which was developed for the longer reads generated by the 454 sequencers, and snp-assess (34), which was developed to characterize viral populations using the Illumina pipeline. However, many of the tools that have been developed recently accept standard FASTA or FASTQ formats as input, making them usable for output from any of the first or next generation sequencing platforms.

As for analysis of human genomic sequences, an increasing number of bioinformatics pipelines and tools for viral sequence analysis are web- or cloud-based, requiring little or no informatics expertise to install and provide output that requires less bioinformatics expertise than many of the Linux-based programs. The J. Craig Venter Institute has developed Cloud BioLinux to provide viral genomic data analysis pipelines. Cloud BioLinux is a public virtual machine on the pay-by-the-hour Amazon Elastic Compute Cloud (EC2) that provides preinstalled analysis pipelines for viral "end-to-end, sequencing-to-annotation" (84). Users without computational infrastructure or expertise can access more than 100 bioinformatics tools via a web browser and run them without needing to purchase software or to modify the pipelines. Galaxy CloudMan (79) is included in the Cloud BioLinux package to provide the suite of Galaxy tools. The Virus Pathogen Database and Analysis Resource (ViPR) is a web-based database that includes metadata on multiple virus families categorized as priority pathogens or that impact public health (91, 92). Data present in ViPR include information from publicly available sources, direct submission, and novel data generated by ViPR. In addition to the database, ViPR includes a number of tools for comparative genome analysis that enables multiple sequence analysis, phylogenetic analysis, sequence variation, and visualization of 3D structures.

Metagenomic analysis of viruses is a particularly challenging problem requiring a series of time-consuming computational steps, specialized analytical tools, and a large amount of computer memory and processing capacity. An overview of the computational issues and tools is provided in a relatively recent review by Fancello et al. (93); however, there are a number of newly described tools and approaches to the analysis of viral metagenomic data that have appeared in the literature since that publication. Typical metagenomic pipelines utilize mapping of reads to a host reference sequence followed by removal of those sequences from subsequent analyses. Many of these pipelines utilize a BLAST or MEGABLAST algorithm for the final pathogen identification step, which is very slow and may be unable to detect pathogens that are highly divergent at the nucleotide level. The Metagenomics Pathogen Identification for Clinical Specimens (MePIC) pipeline (94) is a cloud-based analysis pipeline that accepts the standard FASTQ file generated by most sequencing platforms as input. The MePIC pipeline first processes the sequence to remove adapter sequences and low-quality bases, maps reads to the human genome and removes them from the analysis, and uses the remaining sequences as input to query the NCBI database using MEGABLAST. The database search result is downloadable by the user and includes annotated sequence reads that can be visualized in freely available metagenome browsers. Run times for the MEGABLAST search vary from 10 hours using a single core computer to 6 minutes for 100 cores running in parallel to analyze 1 million reads that are 200 base pairs in length. The run time depends on the specimen type, with longer run times for highly complex specimens, such as stool or sputum samples, which contain microbial sequences that are not removed by filtering of the host sequences. Petty et al. (95) have recently published a bioinformatics pipeline, ezVIR, for virus detection and identification by NGS that is sequencing-platform agnostic and that provides a user-friendly report. The first step in the ezVIR pipeline is to map reads to the human genome to enable computational subtraction of host sequences. Following filtering of the mapped host sequences, the remaining, presumably nonhuman, reads are mapped to a curated database of virus whole-genome sequences. Mapped reads are analyzed, and a report is generated that provides information on positive hits to the virus database, as well as sequencing statistics, such as percent genome coverage, maximum coverage depth, and total genome length covered. While the output of ezVIR represents an improvement in the utility of this particular pipeline, there are still hurdles to the routine use of this approach to virus detection in a clinical setting. Although the report is user-friendly relative to other programs, ezVIR is Linux-based and requires expertise to install and run the program. In addition, the data analysis took approximately 4 days, and the cost of sequencing per paired-end run was $1,500, posing a challenge to implementation in a clinical setting. Naccache et al. (85) have described the development and testing of a Linux-based, cloud-compatible, bioinformatics pipeline, SURPI (sequence-based ultrarapid

TABLE 4 Bionformatics pipelines for virology applications

Application	Program	Computer platform	Data input	Additional information[a]
Assembly of heterogeneous virus populations	VGA (Viral Genome Assembler) (165)	Linux	Sequence reads aligned to a consensus sequence generated using VICUNA	http://genetics.cs.ucla.edu/vga/
Metagenomic detection of pathogens	SURPI (85)	Linux, cloud-compatible	FASTQ	http://chiulab.ucsf.edu/surpi
Pathogen detection	PathSeq (229)	Linux or Cloud-based	FASTA/BFA	http://www.broadinstitute.org/software/pathseq/
Pathogen detection	RINS (230)	Linux	Mate-paired or unpaired sequencing reads	http://khavarilab.stanford.edu/resources.html
Pathogen detection	READSCAN (231)	Linux	FASTA/FASTQ	http://cbrc.kaust.edu.sa/readscan
Variant-calling in deeply sequenced viral populations, illumina pipeline	snp-assess (34)	Linux	Illumina sequencing reads	Center for Health Informatics, Harvard School of Public Health, https://github.com/hbc/projects/tree/master/snp-assess
Viral integration site discovery	ViralFusionSeq(232)	Linux	FASTQ	http://sourceforge.net/projects/viralfusionseq/
Virome analysis	VIROME (233)	Web interface	FASTA + Qual, FASTQ, or 454 sequencing .sff format	http://virome.dbi.udel.edu/
Virome analysis and metagenome comparison	Metavir (234), Metavir2 (235)	Web interface	FASTQ	Only accepts assembled sequence, http://metavir-meb.univ-bpclermont.fr.
Virus bioinformatics, general purpose	Cloud BioLinux (84)	Cloud-based virtual machine or run on local machine (Linux, Windows, Mac OSX)	Sequencing reads	http://www.cloudbiolinux.org
Virus detection (known viruses) and identification of integration sites	VirusSeq (236)	Linux	FASTQ	http://odin.mdacc.tmc.edu/~xsu1/VirusSeq.html
Virus detection, virus integration sites, and sequence variants	VirusFinder 2 (237)	Linux	FASTQ or BAM	http://bioinfo.mc.vanderbilt.edu/VirusFinder/
Virus genotyping, influenza A	FluGenome	Web tool	FASTA	http://www.flugenome.org/
Virus identification	ViPR (91)	Web interface	FASTA	http://www.viprbrc.org/brc/home.spg?decorator=vipr
Virus identification in clinical specimens	MePIC (94)	Cloud-based	FASTQ	https://mepic.nih.go.jp/
Virus identification in clinical specimens	ezVIR (95)	Linux	FASTQ	http://cegg.unige.ch/ezvir/
Virus identification in clinical specimens	VirusHunter(90)	Linux	Roche 454 or longer sequences (Sanger or assembled contigs)	http://pathology.wustl.edu/VirusHunter/
Virus population analysis	VICUNA (167)	Linux	Paired-end read FASTQ	http://www.broadinstitute.org/scientific-community/science/projects/viral-genomics-analysis-software

[a]Websites last accessed 10/13/2014.

pathogen identification), for detection of and identification of pathogens by NGS directly from clinical samples that addresses the need for rapid turnaround of NGS data analysis. SURPI speeds up the analytical process by comparison of nucleotide sequences against viral and bacterial databases in a rapid mode and against the full NCBI nucleotide database in a more comprehensive mode. To enhance the detection of novel or highly divergent viruses, reads are also translated and aligned to viral and NCBI protein databases. These processes are completed in minutes to hours, as compared to days or even weeks for more traditional analysis algorithms using rapid alignment tools. The output of the SURPI pipeline is a report that lists all reads that mapped to a known pathogen along with taxonomic assignment and coverage statistics and maps. A graphical user interface is currently under development, and SURPI is being incorporated into a clinical workflow to develop a validated NGS workflow that may be used in a CLIA-certified laboratory.

Viruses lack a single common ancestor, and there is no gene that can be used for comparison across all virus families, which makes virus taxonomy challenging. The taxonomic principles and challenges in classification of viruses are reviewed in Chapter 1 of this volume, which also provides an updated list of taxonomic databases. Accurate, annotated databases are necessary for virus identification and variant analysis, predicting open reading frames, and finding information on gene function, as well as other downstream data analyses. In order to be useful to clinical virology applications, there needs to be a curated and validated database, a standardized set of bioinformatics tools, and common nomenclature. Underwood and Green (96) have called for a quality standard for sequence-based assays in clinical microbiology that includes the need for curated and quality-controlled databases. There are numerous virus-specific databases available that differ in the type of information that is provided and the degree to which the database is curated and controlled for quality. The databases and tools for virus identification and classification are constantly changing, and any comprehensive listing is soon outdated; however, the review by Yan (89) provides an overview of many of the types of bioinformatics tools and databases used by researchers for comparative genomics of viral genomic and protein sequences.

Data Storage and Management

Data generated by first generation sequencers is easily managed by commercially available software such as Vector NTI and Bionumerics, to name but two. However, NGS has the potential to generate massive amounts of data that rapidly exceed the storage capacity of even the largest institutions. As mentioned previously, instrument vendors provide both cloud-based and local server solutions for data storage, and a number of third-party vendors also offer data storage solutions. However, it is not clear that these solutions will be cost effective or, in the case of cloud-based storage, that it will be feasible to move large amounts of data via the Internet. In fact, some users of cloud-based bioinformatics pipelines actually write the data to a disk and send it by courier to the cloud vender (e.g., Amazon EC2) (97). There are a number of evolving strategies to reduce the amount of data that needs to be stored. The image or signal files produced by the instrument, depending on the platform used, may be quite large. For example, a single sequencing run on the Illumina Hi-Seq platform generates 2 terabytes of raw data (98). Once base-calling has been accomplished, however, it may be feasible for the image or signal data files to be deleted and for the laboratory to retain the FASTA or FASTQ sequence files, which are much smaller (99). However, no data standards have been established, and laboratories must ensure that whatever files are retained contain enough information to enable reanalysis of the data as new software algorithms become available. In addition, data compression algorithms may be used to reduce the amount of data to be stored. Challenges to compression include scalability, compression rates, and the need to account for quality scores when compressing read files (100). Association of metadata with the sequence and the need for protection of patient privacy will also be taken into consideration, as well as integration with a laboratory information management system (101).

Public access to data for research and comparison purposes is important, and a Sequence Read Archive (SRA) has been created to act as a publicly available archive of next generation sequence data (102). The SRA is operated by the International Nucleotide Sequence Database Collaboration, whose partners include NCBI, the European Bioinformatics Institute, and the DNA Data Bank of Japan. The SRA is a "raw data" archive, and data submitted to the SRA must include both base calls and quality scores. For example, BAM and FASTQ are supported file formats, although binary data are preferred over text data. Data are organized into BioProjects, which are studies or research initiatives that may contain data generated from multiple biological sources or BioSamples. The SRA Handbook provides instructions on how to submit sequencing data to the SRA (http://www.ncbi.nlm.nih.gov/books/NBK47528/). Data stored in the SRA may be viewed using the downloadable NCBI Genome Workbench (link to download provided in Table 3).

Quality Assurance

The rapid advancement of sequencing technologies has far outpaced our ability to develop guidelines and standards for validation, quality assurances, and interpretive standards. These standards are necessary if these technologies are to be used routinely in the clinical setting. There are many potential sources of bias and error in generation and interpretation of both Sanger and next generation DNA sequencing data that need to be addressed and controlled for by implementation of appropriate quality assurance parameters.

Errors and bias may be introduced at all stages of the sequencing process, including nucleic acid extraction, reverse transcription, PCR amplification of sequence targets or during library preparation and sequencing and at many steps in the bioinformatics pipeline. This is true for all generations of DNA sequencing technologies. Different sequencing platforms have different sources of error that are specific to the platform. For example, Ion Torrent and Roche 454 errors tend to be systematic, with indel errors occurring frequently in homopolymeric regions, while the errors generated in the Illumina platforms tend to be random substitution errors (103). In addition, sequencing errors tend to be greatest at the each end of the sequencing read, although the effect is more pronounced at the end of the sequencing run, especially in NGS platforms where the sequencing process can become unsynchronized (referred to as dephasing) during the later cycles of sequencing, resulting in incorrect base calling. The first quality assurance parameter in the pipeline is the quality score provided for each base in the sequence read, although the Q-score has been shown to vary with regard to reliability and may not adequately indicate systematic errors or errors introduced during template preparation or amplification. Still, it is an important quality indicator that can be used to monitor and correct for problems with the sequencing reaction and to filter out obviously low-quality sequences. Errors may also be introduced in the bioinformatics pipeline. Alignment programs also provide quality metrics, including the fraction of uniquely mapped reads, the distance of mate pairs and duplicate reads that may indicate PCR or other artifacts, although the reliability of read-mapping quality tools has been questioned (104, 105). Errors are also possible during *de novo* assembly (106) and variant analysis (107) as well as other downstream applications.

As noted previously, library preparation represents the potential to introduce bias into NGS sequencing, and a variety of strategies may be used to reduce bias (58–60). As is the case with sequencing error, it is imperative to identify and address the potential causes of bias in NGS sequencing data and to take any unavoidable bias into account when the data are analyzed. Template or library preparation that involves amplification tends to result in bias due to preferential amplification of certain sequences, especially for GC- or AT-rich sequences. Reverse transcription for sequencing of RNA viruses using random hexamers is another common source of bias because the binding of the hexamers is not totally random. Other causes of bias include differences in the efficiency of adapter ligation, pipetting accuracy and reagent batches, pooling of barcoded samples prior to purification, and biases introduced during nucleic acid extraction. Strategies to reduce these sources of bias will depend on the cause. Bias related to PCR amplification may be addressed by use of PCR enzymes that have increased fidelity with GC-rich sequences, such as Kapa HiFi (Kapa Biosystems), and by reducing the number of amplification cycles (58). Oyola et al. (108) have described a method for reducing bias during library preparation of AT-rich genomes on the Illumina platform by optimizing the library preparation conditions by addition of a DNA-binding agent, TMAC, to the PCR reaction.

The different sources and types of errors necessitate different strategies to account for sequencing errors; therefore it is imperative that the laboratory understand the sources of error unique to their processes and platform and must determine what level of error is acceptable. For example, for resequencing applications, a consensus sequence may be generated from the reads covering each region of the genome, which corrects for occasional random errors, assuming that the sequencing coverage is deep enough. However, for deep sequencing for viral population analysis, variants may be present at levels approaching the error rate, making it difficult to distinguish true variants from sequencing errors. Sequence resampling, which refers to the repeated sequencing of the same template, has been addressed by tagging PCR primers with a random sequence tag known as a Primer ID (109). Sequences with the same tag are used to generate a consensus sequence for that template, enabling correction of a variety errors introduced during the sequencing process. While this is a promising approach to improving the assessment of complex viral populations, Primer ID is not without its own challenges, including tagging of different templates with the same Primer ID and overestimation of sequence templates due to PCR errors in the Primer IDs (110, 111). The choice of reference genome and alignment parameters is also critical, and different selections may yield different interpretations (99). The review of McElroy et al. provides an overview of sources of error, their impact on deep sequencing of pathogen populations, and bioinformatics solutions to correct for some common sequencing errors (112).

Sanger sequencing is also prone to errors and bias, but in spite of the long history of the use of this technology for sequencing in clinical settings, there are few studies that have evaluated the ability of laboratories to generate high-quality sequences and to correctly analyze the data. The European Union has launched an initiative to develop a methodologic External Quality Assurance (EQA) program for DNA sequencing, EQUALseq (113). The program conducted a study to assess DNA sequencing quality whereby participating laboratories were provided with a four-sample set to assess the laboratory's ability to generate accurate sequences and to interpret the data correctly. The results of this study revealed variation in both the ability of laboratories to generate the correct sequence and to analyze the information. In some cases the sequences generated contained multiple errors, or the laboratory failed to resolve ambiguous base calls. Only 33% of participating laboratories generated sequences without any ambiguous base calls. Laboratories also frequently failed to identify mixed samples and heterozygous nucleotide positions. Many of the laboratories utilized the same sequencing platform and chemistry; however, the amount of template used in the sequencing reactions varied from 1 to 1,000 nanograms per microliter, which may explain, in part, the variable results observed. The European Molecular Genetics Quality Network conducted a study in which four DNA samples of known genotype (450-bp PCR-amplified fragments of exon 10 of the cystic fibrosis transmembrane conductance regulator [CFTR] gene involved in cystic fibrosis) were sent to laboratories for sequencing (114). The authors observed a 5% error rate in identification of variants and an 8% error rate in naming mutations among the data from participating laboratories. The data from laboratories that provided acceptable results were used to generate a consensus benchmark to enable laboratories to assess their performance. Taken together, these studies indicate the need for an external quality assessment program to enable laboratories to assess their performance and to provide a benchmark for improvement. Yang et al. (115) conducted a failure mode analysis of Sanger sequencing on the ABI 3700 and 3730XL platforms. Some runs failed completely due to process issues such as blocked capillaries, problems with automated liquid handling systems, or loss of the DNA template during precipitation. Other failures were related to cross contamination and challenging templates, such as those containing homopolymeric regions or repetitive regions. Tracking failure and error rates and performing a root-cause analysis is critical to developing corrective and preventive actions to address errors and prevent recurrence of the problem. The publication of Holm-Hansen and Vainio provides a detailed protocol for Sanger sequencing of viral PCR products, as well as tips for troubleshooting problems that occur during sequencing, the most common of which are the quality and amount of template material (116).

There is clearly a need to develop quality standards and method validation parameters that can be applied to DNA sequencing in clinical and public health laboratories. The Centers for Disease Control convened a workgroup to develop approaches for establishing the elements of a quality management system to ensure the analytical validity of NGS (117). The workgroup focused on detection of variants associated with human genetic disorders; however, their recommendations are applicable to clinical virology, as well as strictly following CLIA guidelines for validation of laboratory developed tests (LDTs). CLIA requires that laboratories establish the performance characteristics of LDTs, including accuracy, precision, analytical sensitivity, analytical specificity, reportable range, and reference range. In addition, CLIA requires adequate quality controls be run with each test and that proficiency testing (PT) and competency testing be conducted to ensure that laboratory staff is capable of generating accurate results. The workgroup provided a translation table that adapted the CLIA requirements for test-method validation to fit with both NGS and Sanger methodologies. Quality control metrics recommended for monitoring DNA sequencing include depth of coverage, uniformity of coverage, and quality scores for base calling and alignment, among other parameters. Because there are no formal PT programs for NGS and because of the logistic challenges and expense associated with NGS, the workgroup developed guidelines for combining a formal PT challenge, once one is available, with an alternative process such as an interlaboratory program in which laboratories could exchange well-characterized specimens for sequencing and/or electronic sequence files to assess or validate data analysis pipelines. Underwood and Green (96) have proposed that a common language for DNA sequence quality be developed, as well as development of an EQA, similar to the EQUALseq program developed by the European Union (113). Ladner et al. have proposed standard nomenclature for conveying the completeness and quality of viral genomic sequences, as well as an indication of what quality of sequence would be appropriate for specific applications (118). For example, description of a novel virus may require a more complete genome sequence than identification of a known viral pathogen in a clinical sample. The U.S. Food and Drug Administration (FDA) convened a workshop in 2011 to address approaches to determining the analytical validity of NGS (119). Workshop participants stressed the need for a flexible approach that "accommodates a rapidly evolving field both at the bench and in bioinformatics analyses." This, however, is easier said than done as evidenced by the fact that, 3 years later, we are still in need of a regulatory framework and standards for quality assurance and validation of DNA sequencing in the clinical virology laboratory. The National Institute of Standards and Technology held a Workshop to Identify Standards Needed to Support Pathogen Identification via Next-Generation Sequencing in October, 2014; however, the recommendations from that workshop are not yet available. Development and adoption of standards will be critical to enable more widespread adoption of DNA sequencing in clinical virology and to allow for interlaboratory comparison of sequence data.

There are currently very few DNA sequencing tests that are cleared by the FDA on first generation platforms, and only two of those are for virology applications, the ViroSeq HIV-1 genotyping system (Celera Diagnostics) and the TruGene HIV-1 genotyping kit (Siemens Healthcare Diagnostics). The first FDA clearance of a next generation sequencing platform, the MiSeqDx, as well as a universal sequencing kit, was granted in late 2013 (120) and should lead to the development of new tests; however, many of the manufacturer-developed tests that will go through the FDA clearance process are likely to be for detection of human genetic disorders. Therefore virology laboratories will need to rely on LDTs for many DNA sequencing applications. The FDA has recently provided draft guidance for regulation of LDTs, as well as a notice for requiring that laboratories notify the FDA of all LDTs and report an adverse events associated with a LDT. Because of these regulatory changes, the validation requirements in the CLIA regulations may no longer apply, and any future validation and quality assurance parameters for the test will likely be mandated by the FDA.

Applications of Sequencing to Virology

Application of DNA sequencing, particularly NGS, in virology has been reviewed in a number of recent publications. (6,121–126) The intent of this section is not to reproduce the information in these reviews but to provide examples of the use of the various sequencing methodologies, strategies, and bioinformatics pipelines described in previous sections of this chapter. DNA sequencing of viruses has utility in several areas, including clinical care, public health, and research settings. DNA sequencing applications fall into two broad categories: those for virus detection and identification and those for virus characterization. Sequencing for virus detection and identification includes both targeted sequencing, as well as metagenomics for detection of known and novel viruses. Virus characterization may include virus genotyping for epidemiologic investigation, population analysis, assessment of tumorigenic potential and identification of antiviral resistance mutations, as well as for many research applications, such as evaluation of host-pathogen interactions, identification of targets for development of antiviral therapies, and studies of virus evolution. Any of the DNA sequencing platforms will work for the applications described below, although there is a clear advantage to using next generation sequencing technologies for metagenomic applications and in applications where there is no *a priori* knowledge of the identity of the virus.

Virus Detection and Identification

The most obvious application of DNA sequencing in clinical virology is for infectious disease diagnostics and detection of emerging viral infections (121, 127). Molecular methods have rapidly replaced culture-based methods for detection and identification of viruses, and, although rapid, sensitive, and specific, are typically limited to detection of a small number of specific viral pathogens. More highly multiplexed assays are becoming available, but these are still limited to a defined set of viral pathogens, and the viruses included are typically grouped in panels according to the syndrome with which they are most commonly associated. So a respiratory panel would include targets for influenza A and B, coronavirus, human metapneumovirus, etc. These panels, therefore, do not detect novel agents associated with these syndromes and, because of the sequence diversity within virus families, the test may not be able to detect all members of the virus, depending on the degree of conservation of the molecular target used in the assay. DNA sequencing provides a comprehensive and unbiased method for detection and identification of viruses in clinical specimens.

Detection and identification of viral pathogens from complex matrices (metagenomics or metagenomic diagnostics) are becoming more common and have been the subjects of a number of recent reviews and commentaries (128–134). Both Sanger and NGS technologies have been used for detection of novel viruses (135). In a public health setting, our virology laboratory at the Minnesota Department of Health, we cultured a virus from a case of severe respiratory illness that produced a cytopathogenic effect when grown in cell culture but which we were unable to identify using either classical or PCR-based methods. Using the method of Victoria et al. (49) we were able to use random amplification to enrich for virus RNA, followed by Sanger sequencing to detect human enterovirus 71 in the specimen (Gongping Liu, unpublished). The presence of this virus was subsequently confirmed by virus-specific PCR. We have also used Sanger DNA sequencing to identify Saffold virus as the putative etiology of an outbreak of unknown etiology (136). In this case, standard PCR, using primers performed in our laboratory designed to detect calicivirus, yielded the same non-specific band for all of the outbreak specimens. Cloning and sequencing of this fragment yielded a nucleotide sequence that did not match any sequences in GenBank; however, a protein query of the translated sequence showed homology to mouse theilovirus (revealing the need to conduct a peptide search to identify highly divergent viruses). The DNA sequence of this virus was subsequently shown to be homologous to Saffold virus, a newly-identified cardiovirus (137). Svraka et al. also used random amplification and cloning to identify Saffold virus, as well as BK polyomavirus and herpes simplex virus, in cell cultures that exhibited cytopathic effect but for which the etiology was unknown (138). Sanger sequencing has also been used to identify viruses from patients with acute respiratory syndrome (52), novel picornaviruses associated with gastroenteritis and febrile illness (54, 137), and a variety of viruses associated with non-polio flaccid paralysis in children (139).

While Sanger sequencing has been used quite successfully for metagenomic diagnostics, there are some clear advantages of NGS, especially in terms of the reduced cost to generate large amounts of sequence data, but NGS may also increase the sensitivity of detection. de Vries et al. compared VIDISCA followed by either Sanger sequencing of cloned fragments or by NGS sequencing without cloning (140). They found that reduction of rRNA amplification, coupled with NGS sequencing (Roche 454), increased the sensitivity of the assay markedly compared to Sanger sequencing and enabled detection of viruses in 11 of 18 clinical specimens known to contain respiratory viruses. A recent study by Prachayangprecha et al. (66) utilized random PCR followed by NGS for detection of respiratory viruses directly from 81 clinical specimens and compared the results to RT-PCR. Viruses were identified by NGS in the majority of the specimens, with NGS revealing the presence of multiple viruses in the majority (60%) of the samples. In many cases at least one of the viruses identified was not known to cause respiratory illness (e.g., anellovirus); however, mixed infections with two or more viruses known to cause respiratory infections were found in 25% of the specimens. The sensitivity of NGS approached that of RT-PCR, and the number of reads mapping to viral reference sequences correlated with the Ct value of the RT-PCR. Both Sanger and NGS sequencing enabled detection of a novel astrovirus (VA1) associated with an outbreak of acute gastroenteritis in Virginia (141). Both methods yielded similar results, revealing the presence of the novel adenovirus in two of six specimens and yielding nearly identical sequences. However, a RT-PCR assay using primers designed specifically to detect astrovirus VA1 was more sensitive than the sequencing methods and yielded positive results for all six samples. These three studies utilized Roche 454 technology. Advantages to the longer sequencing reads produced by this platform include simplifying *de novo* assembly and enabling translation of nucleotide to peptide sequences, which allows for better identification of highly divergent viruses. However, the higher throughput of the short-read platforms, such as Illumina, compensates, at least in part, for the short read length (142). Examples of metagenomic diagnostics utilizing Sanger, NGS (various platforms), or some combination of these methodologies include identification of yellow fever virus in Uganda (143), polyomavirus associated with Merkel cell carcinoma (144), *Trichodysplasia spinulosa* (145), acute respiratory tract infections (146), arenavirus as the cause of febrile illness associated with organ transplant (147), and a novel rhabdovirus associated with acute hemorrhagic fever (148), to name but a few.

Viromics

The human virome consists of all viruses found in association with the human microbiome, including acute and chronic disease-causing viruses, endogenous viruses that are integrated into the human genome, orphan viruses (those not known to cause disease in humans) and viruses that infect bacteria (bacteriophages) (149). Sequences derived from viruses comprise 8% of the human genome, most of which are retroviral in origin; however, sequences with homology to Borna-like viruses have recently been identified as endogenous within mammalian genomes, including humans (150, 151). The composition of the virome may be an important factor in human health. For example, bacteriophages are capable of transmitting virulence factors between bacteria (152) and have the potential to impact bacterial populations and human health (153), and alterations in bacteriophage population have been shown to be associated with periodontal disease (154) Viruses, including beta- and gamma-papillomaviruses, polyomavirus and circovirus have been found to be a component of the flora of apparently healthy human skin, although some have the potential to cause human disease (e.g., Merkel cell and squamous cell carcinomas) (155–157). A number of viruses have been found in blood of healthy people, including Torque teno Virus (TTV), Epstein-Barr virus, and parvovirus B19 (65). TTV is an anellovirus that is found at high prevalence, up to 100% depending on the population screened, in apparently healthy individuals (158). The significance of these viruses in apparently healthy individuals is not well understood; however, there is speculation that they may be pathogenic under certain circumstances (159). It is quite likely that the composition of the virome will differ depending on the human population studied, the anatomical location from which samples are taken, and the health of the individual tested. Understanding the role of the virome in human health will be critical to enable interpretation of metagenomic data derived from individuals, especially when the goal is to identify the etiology of an infectious (or chronic) disease.

Viral Population Analysis

DNA sequencing has been an indispensable tool for analyzing viral intra-species heterogeneity. Virus populations, RNA viruses in particular, consist of a mix of variants that are referred to as quasispecies. Characterization of these quasispecies has utility in guiding antiviral therapy, development of vaccines, understanding disease transmission, virus evolution, and epidemiologic investigation. It is of greater importance for viruses of particular concern such as HIV, HBV, and HCV as mutations present at low levels in the virus population may lead to treatment failure (72, 160–162). As with metagenomic analysis of viruses, a number of different sequencing strategies have been utilized to assess viral populations; however, unlike metagenomic analysis, the goal is to characterize variation within a specific virus population as opposed to detection and identification of novel or unknown viruses. Because the virus species typically is known, population analysis is often done on DNA that has been amplified using primers specific to the species of interest, and sequencing is performed using either first or next generation technologies. While this simplifies the procedure and the analysis, there may be a bias introduced due to mismatches between the PCR primer and certain variants in the virus population. Performing a limiting dilution, followed by cloning and Sanger sequencing, has been a traditional approach to assessing population sequence heterogeneity; however, the procedure is laborious and may lack sensitivity to detect variants present at low concentrations. For these reasons, NGS is an attractive option for population analysis; however, there are numerous challenges with that approach as well. Errors may be introduced during sample preparation due to the need for amplification used in second generation technologies and during the sequencing reaction due to misincorporation of nucleotides. In addition, sequencing coverage of the genome may not be uniform, resulting in an incomplete picture of the true diversity of the population. Paradoxically, this actually reveals an advantage of amplicon sequencing, which enables more uniform coverage of the entire virus genome or targeted regions. Lastly, to state that the bioinformatics of virus population analysis is challenging, is an understatement. Because the goal is to assess variants, errors introduced during sequencing and data analysis may lead to inaccurate and misleading conclusions. Therefore avoiding the introduction of errors and correcting errors during data analysis are critical. Beerenwinkel et al. provide a detailed review of the sources of error and challenges associated with virus population analysis (163). A number of approaches have been developed to minimize and identify sequencing artifacts and to better enable assembly of virus genomes to allow for accurate population analysis (164–166). To cite the most recent example, Mangul et al. (165) utilize a library preparation technique in which template DNA fragments are barcoded, which enables comparison of sequences derived from a common fragment to detect sequencing errors. Following elimination of inaccurate reads, VICUNA (167) is used to generate a *de novo* consensus sequence, which allows for comparison in the absence of a suitable reference genome. Assembly of individual genomes is accomplished with Viral Genome Assembler, which first maps reads to the consensus sequence and then performs the viral population assembly. The authors demonstrated that this approach reduced the number of sequencing errors enabling accurate assembly of HIV populations and identification of low-level variants from millions of sequencing reads.

Antiviral Drug Resistance

There are numerous applications of DNA sequencing for the detection of antiviral resistance in viruses, including detection of resistance to adamantanes and neuraminidase inhibitors in influenza A (8, 11, 38), nucleotide/nucleoside analogs in Hepatitis B (162, 168), protease inhibitors and other direct-acting antivirals in Hepatitis C (72, 169, 170), among others. DNA sequencing has also been used to demonstrate that late relapse of Hepatitis C infection following a sustained virological response resulted from a relapse of the initial infection as opposed to reinfection with a different Hepatitis C virus (171). However, the bulk of the literature on sequencing for detection of antivirals pertains to the many classes of antivirals used to treat HIV, including protease inhibitors, nucleoside and nonnucleoside reverse transcriptase inhibitors, CCR5 agonists and integrase strand-transfer inhibitors, to name but a few (34, 35, 172, 173). Sequencing for detection of antiviral drug resistance testing may be performed on a viral population directly from a clinical specimen as described above or on viruses grown in culture and, as for population analysis, may be accomplished using first or next generation technologies. The challenges surrounding population-based analysis of antiviral resistance were described above.

The simplest and most common method of assessing antiviral resistance is to amplify the target gene and sequence using either first or next generation platforms, depending on the throughput needed and whether population analysis is being performed. The only two FDA-cleared sequencing assays for antiviral resistance are the ViroSeq HIV-1 Genotyping System (Celera Diagnostics) and the TruGene HIV-1 Genotyping Kit (Siemens Healthcare Diagnostics). Both of these systems utilize first generation sequencing technology to detect resistance mutations in the protease and reverse transcriptase genes. Both of these systems have been found to work well with subtype B strains of HIV-1, which is the predominant subtype in the United States and Europe. However, there are numerous other subtypes of HIV-1 throughout the developing world, and many of these subtypes have been introduced into the United States. We have characterized the diversity of HIV-1 subtypes in African-born residents of Minnesota (40), and it has been demonstrated that the TruGene assay is capable of providing both accurate antiviral resistance information on the non-B subtypes and subtype identity (174). It has been reported that the ViroSeq system works well for detection of resistance mutations in nonsubtype B strains of HIV-1 (175, 176); however, a more recent study reveals a high failure rate of the system for genotyping of non-B HIV-1 subtypes in Cameroon (177). The reason for these failures is not clear, although variation in the primer binding sites is one possibility. However, this could not be confirmed because the sequence of the ViroSeq primers is not public.

Use of NGS to perform unbiased sequencing may overcome the limitations associated with amplification-based sequencing; however, at this time there are no systems for HIV-1 typing that are commercially available and laboratories must develop their own tests, which is not a trivial task. Roche was developing an HIV resistance typing assay for the 454 GS Junior, and a prototype of the assay was provided to a small number of laboratories for clinical evaluation (178, 179). The accuracy and reproducibility of NGS and Sanger were comparable, while the advantage of the NGS assay over Sanger sequencing (TruGene) was the ability to detect resistance mutations present in low abundance (less than 20%). However, the turnaround time was much slower (4-fold) for the NGS method than for Sanger, and there were many manual steps (179). As was the case with Sanger sequencing, widespread adoption of the NGS technology for HIV-1 sequencing will likely not occur in the clinical laboratory until the assay is automated. It is not clear what the status of development of this assay is since Roche has announced discontinuation of the 454 platform.

Molecular detection of antiviral resistance requires knowledge of the specific mutations associated with a resistance phenotype. This necessitates that phenotypic tests be conducted and the results correlated with genetic mutations. Given the rapid mutation rate of some viruses, this presents a significant challenge to ensuring that the data analysis provides an accurate assessment of resistance mutations. In some cases, influenza, for example, there are a relatively small number of drugs and drug targets, and it is relatively easy to keep track of the mutations associated with resistance. HIV, however, poses more of a bioinformatics challenge due to the number of drugs available, the diversity of the targets and the potential for discordance between phenotypic and genotypic resistance determinations (180, 181). The International Antiviral Society-USA publishes a list of mutations reported in the scientific literature that lead to antiviral resistance in HIV, and this list is updated annually (182). TruGene and ViroSeq both provide software that allows for assessment of sequence quality and produces a resistance genotype report. Laboratory developed tests must rely on publicly available databases and software. One widely used tool is the HIV Drug Resistance Database, HIVdb. HIVdb was developed by Stanford University for assessment of protease, reverse transcriptase, and integrase mutations (183) and can be accessed by individuals and institutions via Sierra, the Stanford HIV web service (http://sierra2.stanford.edu/sierra/servlet/JSierra). The British Columbia Centre for Excellence recently developed an automated drug resistance genotyping pipeline, RECall, which has been made freely available (http://pssm.cfenet.ubc.ca/) (184). RECall compared favorably with the Stanford algorithm and has the advantage of automating the process of trimming low-quality data, alignment to a reference standard, and exporting a FASTA file. While these software solutions were designed for use with Sanger sequencing data, they may be used with NGS platforms as well since the input into the genotyping tool is a standard FASTA format.

Virus Genotyping

In addition to antiviral resistance genotyping, there are numerous other applications in clinical virology. Although genotyping in the clinical setting is currently limited to a relatively small number of applications, there are numerous public health applications, which will be detailed below. One important clinical application of genotyping in the clinical setting is for characterization of human papillomavirus. HPV is the etiologic agent of cervical cancer and genital warts, with different genotypes associated with each of these conditions. Out of more than 200 genotypes, approximately a dozen have been associated with cervical cancer, with HPV-16 and HPV-18 responsible for the largest proportion of cases (185). NGS methods have recently been described for typing of HPV (39, 56). NGS was found to be sensitive and specific as compared to the INNO-LiPA HPV Genotyping Extra assay. Depending on the HPV type, there were some differences in sensitivity, with NGS being better at detecting variants present at low concentration (39). NGS is capable of detecting variants not included in commercial assays (56); therefore it will be increasingly important to develop interpretive guidelines and continue to evaluate the oncogenic potential of those types not known to be associated with cervical cancer. Genotyping is also important for assessment of HCV, as certain genotypes may be less susceptible to interferon or other therapies (186), and for HBV, to assess risk factors for development of liver cirrhosis (187). Detection of enterovirus in clinical laboratories is routine; however, genotyping of enterovirus is less common. Recently, genotyping by DNA sequencing has been used to identify enterovirus D68 (EV-D68) as the cause of clusters of cases of acute respiratory illness in children (188). While the recent increase and severity of illness was striking, increased incidence of EV-D68 has previously been reported to have occurred in North America (including the United States), Europe, and Asia (189, 190) and in New Zealand (191).

In the public health arena, virus genotyping has become critical for identification and investigation of outbreaks, conducting surveillance for rare and emerging diseases, understanding the vectors, reservoirs and transmission routes of emerging diseases and, occasionally, for conducting forensic investigations. Outbreaks may be due to relatively common viruses, such as norovirus, or rare or emerging viruses. For rarely encountered viruses, even a single case can trigger an investigation. The examples provided below include a variety of epidemiologic investigation scenarios in which genotyping has been demonstrated to have utility.

Norovirus is the most common etiology of foodborne disease outbreaks in the United States (192). Although norovirus outbreaks have been linked to shellfish that have been contaminated in nature, most transmission occurs due to contamination of food by a food handler or by person-to-person contact and contaminated surfaces. Because of the relatively short incubation period for norovirus (24 to 48 hours), distinctive symptoms, and the fact that outbreaks tend to involve clusters of illness, sequencing has limited utility for outbreak detection. However, genotyping has been helpful in detecting single-transmission sources linked to multiple outbreaks, including staff working in multiple long-term care facilities (LTCF) (193), a post-symptomatic food handler (194), and symptomatic oyster harvesters (195). While most norovirus sequencing is done by Sanger sequencing, NGS has been used to characterize nosocomial transmission of norovirus within a health care facility (196). The CDC has established CaliciNet, which, similar to the PulseNet system for bacterial foodborne pathogens, allows for submission of genotyping data (in this case, DNA sequences of norovirus and sapovirus) to a central database to enable calicivirus surveillance (192). Recent genotype analysis of norovirus outbreaks in the United States from 2009 to 2013 revealed cyclic emergence of new norovirus strains and indicated that some genotypes were more likely than others to be associated with foodborne outbreaks (197). Sequencing has also been used to characterize sapovirus, a calicivirus related to norovirus, outbreaks in long-term care facilities in Oregon and Minnesota (198). Similar to norovirus, multiple genotypes of sapovirus were found to be associated with outbreaks in LTCFs, with the majority being attributed to genogroup IV.

Molecular genotyping, usually by RT-PCR, has become common for public health surveillance of influenza; however, novel viruses are only detected as viruses that do not amplify using primers targeted to a limited number of currently circulating strains. DNA sequencing may be used to identify and characterize novel influenza viruses, as well as reassortment events, and is increasingly used to assess antiviral resistance and to assess transmission pathways. NGS is an ideal tool to enable surveillance for oseltamivir resistance in large numbers of viruses by sequencing pooled amplicons of the neuraminidase gene (38). NGS also enables analysis of full-length influenza genomes by sequencing of RT-PCR amplicons spanning the entire genome, which has been used on both cultured virus and directly on human specimens (10, 43) or on cultured virus using a random amplification method (20). As mentioned previously, a commercial kit, PathAmp FluA, is available for amplification and sequencing of influenza A on the Ion Torrent platform.

Because of under-vaccination in some populations, there has been an increase in outbreaks due to vaccine-preventable diseases, most notably measles. Measles is an airborne disease, and transmission occurs readily upon face-to-face contact and spreads rapidly in an unvaccinated population. Genotyping by DNA sequencing was used to aid in the investigation of a large measles outbreak that occurred in Minnesota in 2011(199). This outbreak began with importation of the virus in a child returning to Minnesota from Kenya. Sequencing confirmed the genotype as B3, a genotype that is endemic in sub-Saharan Africa (200). Genotyping at CDC and at the Minnesota Department of Health (MDH) was also used to rule out a case that had a nonoutbreak genotype, to rule in a case with no known exposure but who had the outbreak genotype, and to rule out several individuals who were found to be carrying the vaccine strain. This demonstrated the utility of genotyping to help focus the epidemiologic investigation to enable better use of scarce resources for contact tracing and follow-up.

DNA sequencing has been used for characterization of outbreaks of disease involving emerging viral pathogens such as SARS, MERS, and other coronaviruses (42, 128, 201, 202) and, most recently, Ebola. NGS was used to determine that the 2014 epidemic of hemorrhagic fever in Guinea was caused by a strain of Ebola related to a lineage that has caused previous outbreaks (203). More recent sequencing of 99 Ebola virus genomes from Guinea, Liberia, and Sierra Leone revealed that transmission likely crossed from Guinea to Sierra Leone and that human-to-human contact has been the primary mode of transmission and that there have been no reintroductions from animals (204). Concurrent to the outbreak in Guinea, Sierra Leone, and Liberia, there were several cases of Ebola in the Democratic Republic of Congo, which were demonstrated by sequencing not to be linked to the cases in the other countries (http://www.who.int/mediacentre/news/ebola/2-september-2014/en/). This distinction was important to dispel the fear that the outbreak had spread to central Africa.

Phylogenetic analysis utilizing calculation of evolutionary rates can act as a "molecular clock," providing an estimate of viral divergence (205). Phylogenetic analysis has been instrumental in the investigation of two cases of illness in Minnesota caused by live attenuated oral poliovirus vaccine (OPV). The first case involved introduction and transmission of vaccine-derived poliovirus (VDPV) in an Amish community, whose members were unvaccinated (206). Sequencing of the VP1 region of the enterovirus genome (41) at MDH was used to identify an enterovirus isolated from an infant with an underlying immunodeficiency, as a Sabin strain of poliovirus. Subsequent whole genome sequencing at CDC provided phylogenetic evidence that the VDPV had been circulating in the community for about 2 months prior to detection of the infant's infection, and that the ancestral virus (from the initiating OPV dose) had been circulating for about 8 months prior to introduction into the community. The second case involved a death caused by a type 2 VDPV in which two attenuating substitutions had reverted to wild-type sequence leading to neurovirulence (207). Whole genome sequencing indicated that the patient who had an underlying immunodeficiency had likely been infected with the VDPV at the time her child was immunized with OPV approximately 12 years earlier.

Phylogenetic analysis has also been critical in understanding the transmission of HIV (208–210). HIV transmission can be challenging to investigate due to the reticence or inability of individuals to provide accurate information to enable contact tracing, as well as to the relatively long period of time during which individuals are infectious. Using "molecular phylodynamics," Lewis et al. determined that in the population that they studied, HIV transmission among men who have sex with men (MSM) was episodic and that in 25% of cases, transmission occurred within 6 months of initial infection (209). These data indicate that early detection of infection and rigorous follow-up to identify contacts may be effective strategies to reduce the rate of HIV transmission (210). Phylogenetic relationships have been used both to identify an individual as the source of HIV infection (211) as well as to rule out an individual (212). Molecular phylogeny has also been used as a tool for the investigation of criminal cases involving HIV transmission with the "intent to inflict bodily harm." Scaduto et al. describe two criminal cases in which phylogenetic data were used to reveal an index case to be the defendant on trial (213). The use of phylogenetic analysis to evaluate HIV transmission raises ethical concerns and reveals the need for standards for data interpretation and maintaining data privacy (214).

Sequencing is also important to understand host-pathogen relationships and to identify potential reservoirs of infection. Hepatitis E virus (HEV) infections have been linked to pork consumption and are a particular problem for transplant patients, especially in southern France where HEV is hyperendemic. Sequencing of a genotype 3e HEV strain from a chronically infected kidney transplant patient revealed that the sequence most closely resembled HEV sequences from swine and, in fact, the patient had reported consumption of wild boar meat (215). Bats have been demonstrated to be important reservoirs for many emerging viruses (216). In the United States, bats are a major reservoir for the rabies virus. While rabies virus infects many different mammals, rabies variants replicate most efficiently in their reservoir host and sometimes within a defined geographic distribution. Interspecies transmission is typically a single event (e.g., bat variant to skunk), and secondary transmission (skunk to skunk) is rare (217, 218), although secondary transmission has been documented (219). Control measures for rabies are dependent on the host reservoir, and understanding bat rabies is important, especially in areas where rabies has been eliminated from carnivores (217). Raccoon rabies, which has emerged as a particular concern due to the close proximity of raccoons to human populations (220), may be controlled by using bait containing a live vaccine (221, 222), while skunk rabies has been controlled by trapping and parenteral vaccination (219). Genotyping of rabies' variants aids in determining the appropriate control measures given the likely reservoir host of the virus. Rabies virus may have a very long incubation period, and it may take several years after exposure for a person to develop symptoms. Boland et al. used phylogenetic analysis to show that a Massachusetts resident developed rabies 8 years after having been exposed in Brazil (223).

CONCLUSIONS

DNA sequencing has become an important tool in virology; however, widespread implementation in clinical settings will depend on the development of guidelines and standards for performing the sequencing reactions and interpretation of the data. Additionally, most laboratories will likely be unable to implement sequencing unless an automated, FDA-cleared platform becomes available. While nothing is currently available on the U.S. market for virology applications, Illumina has an FDA-cleared version of the MiSeq platform (MiSeqDx), and Ion Torrent has listed the Ion PGM Dx System with the FDA. It important to note, however, that although the platforms are FDA-cleared, absent an FDA-cleared informatics pipeline, the laboratory will need to validate the test as an LDT (224). Still, given the pace of developments in the field of next generation sequencing, it seems likely that there will be assays developed for virology applications in the not-too-distant future and that these assays will be integrated into the workflow of the clinical virology laboratory.

Thanks to Bill Wolfgang and Pascal Lapierre from New York's Wadsworth Center for lending their expertise to conduct a critical review of this chapter to ensure accuracy and readability. Their contribution is greatly appreciated.

REFERENCES

1. **Maxam AM, Gilbert W.** 1977. A new method for sequencing DNA. *Proc Natl Acad Sci USA* **74:**560–564.
2. **Sanger F, Coulson AR.** 1975. A rapid method for determining sequences in DNA by primed synthesis with DNA polymerase. *J Mol Biol* **94:**441–448.
3. **Sanger F, Nicklen S, Coulson AR.** 1977. DNA sequencing with chain-terminating inhibitors. *Proc Natl Acad Sci USA* **74:**5463–5467.
4. **Sanger F, Air GM, Barrell BG, Brown NL, Coulson AR, Fiddes CA, Hutchison CA, Slocombe PM, Smith M.** 1977. Nucleotide sequence of bacteriophage phi X174 DNA. *Nature* **265:**687–695.
5. **Benson DA, Clark K, Karsch-Mizrachi I, Lipman DJ, Ostell J, Sayers EW.** 2014. GenBank. *Nucleic Acids Res* **42**(D1): D32–D37.
6. **Quiñones-Mateu ME, Avila S, Reyes-Teran G, Martinez MA.** 2014. Deep sequencing: becoming a critical tool in clinical virology. *J Clin Virol* **61:**9–19.
7. **Ronaghi M, Karamohamed S, Pettersson B, Uhlén M, Nyrén P.** 1996. Real-time DNA sequencing using detection of pyrophosphate release. *Anal Biochem* **242:**84–89.
8. **Deyde VM, Nguyen T, Bright RA, Balish A, Shu B, Lindstrom S, Klimov AI, Gubareva LV.** 2009. Detection of molecular markers of antiviral resistance in influenza A (H5N1) viruses using a pyrosequencing method. *Antimicrob Agents Chemother* **53:**1039–1047.
9. **Chantratita W, Sukasem C, Sirinavin S, Sankuntaw N, Srichantaratsamee C, Pasomsub E, Malathum K.** 2011. Simultaneous detection and subtyping of H274Y-positive influenza A (H1N1) using pyrosequencing. *J Infect Dev Ctries* **5:**348–352.
10. **Deng Y-M, Caldwell N, Barr IG.** 2011. Rapid detection and subtyping of human influenza A viruses and reassortants by pyrosequencing. *PLoS One* **6:**e23400.
11. **Bright RA, Shay DK, Shu B, Cox NJ, Klimov AI.** 2006. Adamantane resistance among influenza A viruses isolated early during the 2005–2006 influenza season in the United States. *JAMA* **295:**891–894.
12. **Glenn TC.** 2011. Field guide to next-generation DNA sequencers. *Mol Ecol Resour* **11:**759–769.
13. **Schadt EE, Turner S, Kasarskis A.** 2010. A window into third-generation sequencing. *Hum Mol Genet* **19**(R2):R227–R240.
14. **Metzker ML.** 2010. Sequencing technologies–the next generation. *Nat Rev Genet* **11:**31–46.
15. **van Dijk EL, Auger H, Jaszczyszyn Y, Thermes C.** 2014. Ten years of next-generation sequencing technology. *Trends Genet* **30:**418–426.
16. **Pareek CS, Smoczynski R, Tretyn A.** 2011. Sequencing technologies and genome sequencing. *J Appl Genet* **52:**413–435.
17. **Mardis ER.** 2011. A decade's perspective on DNA sequencing technology. *Nature* **470:**198–203.
18. **Caruccio N.** 2011. Preparation of next-generation sequencing libraries using Nextera™ technology: simultaneous DNA fragmentation and adaptor tagging by *in vitro* transposition. *Methods Mol Biol* **733:**241–255.
19. **Illumina.** Application Note. Culture-free detection and identification of unknown RNA viruses. https://www.illumina.com/content/dam/illumina-marketing/documents/products/appnotes/appnote_afrims_rna_viruses.pdf
20. **Rutvisuttinunt W, Chinnawirotpisan P, Simasathien S, Shrestha SK, Yoon I-K, Klungthong C, Fernandez S.** 2013. Simultaneous and complete genome sequencing of influenza A and B with high coverage by Illumina MiSeq Platform. *J Virol Methods* **193:**394–404.
21. **Hang J, Forshey BM, Kochel TJ, Li T, Solórzano VF, Halsey ES, Kuschner RA.** 2012. Random amplification and pyrosequencing for identification of novel viral genome sequences. *J Biomol Tech* **23:**4–10.
22. **Malboeuf CM, Yang X, Charlebois P, Qu J, Berlin AM, Casali M, Pesko KN, Boutwell CL, DeVincenzo JP, Ebel GD, Allen TM, Zody MC, Henn MR, Levin JZ.** 2013. Complete viral RNA genome sequencing of ultra-low copy samples by sequence-independent amplification. *Nucleic Acids Res* **41(1):**e13.
23. **Kuersten S.** 2012. A transposable approach to RNA-seq from total RNA. *Nat Methods* 9:i-ii.
24. **Merriman B, Ion Torrent R&D Team, Rothberg JM.** 2012. Progress in ion torrent semiconductor chip based sequencing. *Electrophoresis* **33:**3397–3417.
25. **McKernan KJ, Peckham HE, Costa GL, McLaughlin SF, Fu Y, Tsung EF, Clouser CR, Duncan C, Ichikawa JK, Lee CC, Zhang Z, Ranade SS, Dimalanta ET, Hyland FC, Sokolsky TD, Zhang L, Sheridan A, Fu H, Hendrickson CL, Li B, Kotler L, Stuart JR, Malek JA, Manning JM, Antipova AA, Perez DS, Moore MP, Hayashibara KC, Lyons MR, Beaudoin RE, Coleman BE, Laptewicz MW, Sannicandro AE, Rhodes MD, Gottimukkala RK, Yang S, Bafna V, Bashir A, MacBride A, Alkan C, Kidd JM, Eichler EE, Reese MG, De La Vega FM, Blanchard AP.** 2009. Sequence and structural variation in a human genome uncovered by short-read, massively parallel ligation sequencing using two-base encoding. *Genome Res* **19:**1527–1541.
26. **Ozsolak F, Milos PM.** 2011. Single-molecule direct RNA sequencing without cDNA synthesis. *Wiley Interdiscip Rev RNA* **2:**565–570.
27. **Branton D, Deamer DW, Marziali A, Bayley H, Benner SA, Butler T, Di Ventra M, Garaj S, Hibbs A, Huang X, Jovanovich SB, Krstic PS, Lindsay S, Ling XS, Mastrangelo CH, Meller A, Oliver JS, Pershin YV, Ramsey JM, Riehn R, Soni GV, Tabard-Cossa V, Wanunu M, Wiggin M, Schloss JA.** 2008. The potential and challenges of nanopore sequencing. *Nat Biotechnol* **26:**1146–1153.
28. **Check Hayden E.** 2014. Data from pocket-sized genome sequencer unveiled. *Nature* **521.**
29. **Suzuki S, Ono N, Furusawa C, Ying B-W, Yomo T.** 2011. Comparison of sequence reads obtained from three next-generation sequencing platforms. *PLoS One* **6:**e19534.
30. **Quail MA, Smith M, Coupland P, Otto TD, Harris SR, Connor TR, Bertoni A, Swerdlow HP, Gu Y.** 2012. A tale of three next generation sequencing platforms: comparison of Ion Torrent, Pacific Biosciences and Illumina MiSeq sequencers. *BMC Genomics* **13:**341.
31. **Liu L, Li Y, Li S, Hu N, He Y, Pong R, Lin D, Lu L, Law M.** 2012. Comparison of next-generation sequencing systems. *J Biomed Biotechnol* **2012:**251364.
32. **Loman NJ, Misra RV, Dallman TJ, Constantinidou C, Gharbia SE, Wain J, Pallen MJ.** 2012. Performance comparison of benchtop high-throughput sequencing platforms. *Nat Biotechnol* **30:**434–439.
33. **Frey KG, Herrera-Galeano JE, Redden CL, Luu TV, Servetas SL, Mateczun AJ, Mokashi VP, Bishop-Lilly KA.** 2014. Comparison of three next-generation sequencing platforms for metagenomic sequencing and identification of pathogens in blood. *BMC Genomics* **15:**96.
34. **Li JZ, Chapman B, Charlebois P, Hofmann O, Weiner B, Porter AJ, Samuel R, Vardhanabhuti S, Zheng L, Eron J, Taiwo B, Zody MC, Henn MR, Kuritzkes DR, Hide W, and the ACTG A5262 Study Team.** 2014. Comparison of illumina and 454 deep sequencing in participants failing raltegravir-based antiretroviral therapy. *PLoS One* **9:**e90485.

35. **Archer J, Weber J, Henry K, Winner D, Gibson R, Lee L, Paxinos E, Arts EJ, Robertson DL, Mimms L, Quiñones-Mateu ME.** 2012. Use of four next-generation sequencing platforms to determine HIV-1 coreceptor tropism. *PLoS One* **7:**e49602.
36. **Ninomiya M, Ueno Y, Funayama R, Nagashima T, Nishida Y, Kondo Y, Inoue J, Kakazu E, Kimura O, Nakayama K, Shimosegawa T.** 2012. Use of illumina deep sequencing technology to differentiate hepatitis C virus variants. *J Clin Microbiol* **50:**857–866.
37. **Marston DA, McElhinney LM, Ellis RJ, Horton DL, Wise EL, Leech SL, David D, de Lamballerie X, Fooks AR.** 2013. Next generation sequencing of viral RNA genomes. *BMC Genomics* **14:**444.
38. **Téllez-Sosa J, Rodríguez MH, Gómez-Barreto RE, Valdovinos-Torres H, Hidalgo AC, Cruz-Hervert P, Luna RS, Carrillo-Valenzo E, Ramos C, García-García L, Martínez-Barnetche J.** 2013. Using high-throughput sequencing to leverage surveillance of genetic diversity and oseltamivir resistance: a pilot study during the 2009 influenza A(H1N1) pandemic. *PLoS One* **8:**e67010.
39. **Barzon L, Militello V, Lavezzo E, Franchin E, Peta E, Squarzon L, Trevisan M, Pagni S, Dal Bello F, Toppo S, Palù G.** 2011. Human papillomavirus genotyping by 454 next generation sequencing technology. *J Clin Virol* **52:**93–97.
40. **Sides TL, Akinsete O, Henry K, Wotton JT, Carr PW, Bartkus J.** 2005. HIV-1 subtype diversity in Minnesota. *J Infect Dis* **192:**37–45.
41. **Oberste MS, Maher K, Kilpatrick DR, Flemister MR, Brown BA, Pallansch MA.** 1999. Typing of human enteroviruses by partial sequencing of VP1. *J Clin Microbiol* **37:**1288–1293.
42. **Cotten M, Lam TT, Watson SJ, Palser AL, Petrova V, Grant P, Pybus OG, Rambaut A, Guan Y, Pillay D, Kellam P, Nastouli E.** 2013. Full-genome deep sequencing and phylogenetic analysis of novel human betacoronavirus. *Emerg Infect Dis* **19:**736–42B.
43. **Höper D, Hoffmann B, Beer M.** 2009. Simple, sensitive, and swift sequencing of complete H5N1 avian influenza virus genomes. *J Clin Microbiol* **47:**674–679.
44. **Gardner SN, Jaing CJ, Elsheikh MM, Peña J, Hysom DA, Borucki MK.** 2014. Multiplex degenerate primer design for targeted whole genome amplification of many viral genomes. *Adv Bioinforma* **2014:**101894.
45. **Kapoor A, Simmonds P, Slikas E, Li L, Bodhidatta L, Sethabutr O, Triki H, Bahri O, Oderinde BS, Baba MM, Bukbuk DN, Besser J, Bartkus J, Delwart E.** 2010. Human bocaviruses are highly diverse, dispersed, recombination prone, and prevalent in enteric infections. *J Infect Dis* **201:** 1633–1643.
46. **Fan X, Xu Y, Di Bisceglie AM.** 2006. Efficient amplification and cloning of near full-length hepatitis C virus genome from clinical samples. *Biochem Biophys Res Commun* **346:** 1163–1172.
47. **Froussard P.** 1992. A random-PCR method (rPCR) to construct whole cDNA library from low amounts of RNA. *Nucleic Acids Res* **20:**2900.
48. **Tan V, Van Doorn HR, Van der Hoek L, Minh Hien V, Jebbink MF, Quang Ha D, Farrar J, Van Vinh Chau N, de Jong MD.** 2011. Random PCR and ultracentrifugation increases sensitivity and throughput of VIDISCA for screening of pathogens in clinical specimens. *J Infect Dev Ctries* **5:**142–148.
49. **Victoria JG, Kapoor A, Dupuis K, Schnurr DP, Delwart EL.** 2008. Rapid identification of known and new RNA viruses from animal tissues. *PLoS Pathog* **4:**e1000163.
50. **Nakamura S, Yang C-S, Sakon N, Ueda M, Tougan T, Yamashita A, Goto N, Takahashi K, Yasunaga T, Ikuta K, Mizutani T, Okamoto Y, Tagami M, Morita R, Maeda N, Kawai J, Hayashizaki Y, Nagai Y, Horii T, Iida T, Nakaya T.** 2009. Direct metagenomic detection of viral pathogens in nasal and fecal specimens using an unbiased high-throughput sequencing approach. *PLoS One* **4:**e4219.
51. **Batty EM, Wong THN, Trebes A, Argoud K, Attar M, Buck D, Ip CLC, Golubchik T, Cule M, Bowden R, Manganis C, Klenerman P, Barnes E, Walker AS, Wyllie DH, Wilson DJ, Dingle KE, Peto TEA, Crook DW, Piazza P.** 2013. A modified RNA-Seq approach for whole genome sequencing of RNA viruses from faecal and blood samples. *PLoS One* **8:** e66129.
52. **Jones MS, Kapoor A, Lukashov VV, Simmonds P, Hecht F, Delwart E.** 2005. New DNA viruses identified in patients with acute viral infection syndrome. *J Virol* **79:**8230–8236.
53. **Karlsson OE, Belák S, Granberg F.** 2013. The effect of preprocessing by sequence-independent, single-primer amplification (SISPA) on metagenomic detection of viruses. *Biosecur Bioterror* **11**(Suppl 1):S227–S234.
54. **Li L, Victoria J, Kapoor A, Blinkova O, Wang C, Babrzadeh F, Mason CJ, Pandey P, Triki H, Bahri O, Oderinde BS, Baba MM, Bukbuk DN, Besser JM, Bartkus JM, Delwart EL.** 2009. A novel picornavirus associated with gastroenteritis. *J Virol* **83:**12002–12006.
55. **Johne R, Müller H, Rector A, van Ranst M, Stevens H.** 2009. Rolling-circle amplification of viral DNA genomes using phi29 polymerase. *Trends Microbiol* **17:**205–211.
56. **Meiring TL, Salimo AT, Coetzee B, Maree HJ, Moodley J, Hitzeroth II, Freeborough M-J, Rybicki EP, Williamson A-L.** 2012. Next-generation sequencing of cervical DNA detects human papillomavirus types not detected by commercial kits. *Virol J* **9:**164.
57. **de Vries M, Oude Munnink BB, Deijs M, Canuti M, Koekkoek SM, Molenkamp R, Bakker M, Jurriaans S, van Schaik BDC, Luyf AC, Olabarriaga SD, van Kampen AHC, van der Hoek L.** 2012. Performance of VIDISCA-454 in feces-suspensions and serum. *Viruses* **4:**1328–1334.
58. **Head SR, Komori HK, LaMere SA, Whisenant T, Van Nieuwerburgh F, Salomon DR, Ordoukhanian P.** 2014. Library construction for next-generation sequencing: overviews and challenges. *Biotechniques* **56:**61–64, 66, 68 passim.
59. **van Dijk EL, Jaszczyszyn Y, Thermes C.** 2014. Library preparation methods for next-generation sequencing: tone down the bias. *Exp Cell Res* **322:**12–20.
60. **Ross MG, Russ C, Costello M, Hollinger A, Lennon NJ, Hegarty R, Nusbaum C, Jaffe DB.** 2013. Characterizing and measuring bias in sequence data. *Genome Biol* **14:**R51.
61. **Depledge DP, Palser AL, Watson SJ, Lai IY-C, Gray ER, Grant P, Kanda RK, Leproust E, Kellam P, Breuer J.** 2011. Specific capture and whole-genome sequencing of viruses from clinical samples. *PLoS One* **6:**e27805.
62. **Duncavage EJ, Magrini V, Becker N, Armstrong JR, Demeter RT, Wylie T, Abel HJ, Pfeifer JD.** 2011. Hybrid capture and next-generation sequencing identify viral integration sites from formalin-fixed, paraffin-embedded tissue. *J Mol Diagn* **13:**325–333.
63. **Szpara ML, Parsons L, Enquist LW.** 2010. Sequence variability in clinical and laboratory isolates of herpes simplex virus 1 reveals new mutations. *J Virol* **84:**5303–5313.
64. **Kapranov P, Chen L, Dederich D, Dong B, He J, Steinmann KE, Moore AR, Thompson JF, Milos PM, Xiao W.** 2012. Native molecular state of adeno-associated viral vectors revealed by single-molecule sequencing. *Hum Gene Ther* **23:** 46–55.
65. **Breitbart M, Rohwer F.** 2005. Method for discovering novel DNA viruses in blood using viral particle selection and shotgun sequencing. *Biotechniques* **39:**729–736.
66. **Prachayangprecha S, Schapendonk CME, Koopmans MP, Osterhaus ADME, Schürch AC, Pas SD, van der Eijk AA, Poovorawan Y, Haagmans BL, Smits SL.** 2014. Exploring the potential of next-generation sequencing in detection of respiratory viruses. *J Clin Microbiol* **52:**3722–3730.
67. **Hall RJ, Wang J, Todd AK, Bissielo AB, Yen S, Strydom H, Moore NE, Ren X, Huang QS, Carter PE, Peacey M.** 2014. Evaluation of rapid and simple techniques for the enrichment of viruses prior to metagenomic virus discovery. *J Virol Methods* **195:**194–204.

68. **Lam TT-Y, Hon C-C, Tang JW.** 2010. Use of phylogenetics in the molecular epidemiology and evolutionary studies of viral infections. *Crit Rev Clin Lab Sci* **47:**5–49.
69. **Gogol-Döring A, Chen W.** 2012. An overview of the analysis of next generation sequencing data. *Methods Mol Biol* **802:** 249–257.
70. **Gilbert D.** 2004. Bioinformatics software resources. *Brief Bioinform* **5:**300–304.
71. **Kerr A.** 2011. *Desktop Sequence Analysis: software review.* Bioinforma Knowledgeblog.
72. **Lauck M, Alvarado-Mora MV, Becker EA, Bhattacharya D, Striker R, Hughes AL, Carrilho FJ, O'Connor DH, Pinho JRR.** 2012. Analysis of hepatitis C virus intrahost diversity across the coding region by ultradeep pyrosequencing. *J Virol* **86:**3952–3960.
73. **Lei H, Li T, Hung G-C, Li B, Tsai S, Lo S-C.** 2013. Identification and characterization of EBV genomes in spontaneously immortalized human peripheral blood B lymphocytes by NGS technology. *BMC Genomics* **14:**804.
74. **Langmead B, Trapnell C, Pop M, Salzberg SL.** 2009. Ultrafast and memory-efficient alignment of short DNA sequences to the human genome. *Genome Biol* **10:**R25.
75. **Li R, Li Y, Fang X, Yang H, Wang J, Kristiansen K, Wang J.** 2009. SNP detection for massively parallel whole-genome resequencing. *Genome Res* **19:**1124–1132.
76. **Gurtowski J, Schatz MC, Langmead B.** 2012. Genotyping in the Cloud with Crossbow. *Curr Protoc Bioinforma* Ed Board Baxevanis A, **CHAPTER:** 15. Unit: 15.3.
77. **Langmead B, Schatz MC, Lin J, Pop M, Salzberg SL.** 2009. Searching for SNPs with cloud computing. *Genome Biol* **10:** R134.
78. **Sadedin SP, Pope B, Oshlack A.** 2012. Bpipe: a tool for running and managing bioinformatics pipelines. *Bioinformatics* **28:**1525–1526.
79. **Afgan E, Baker D, Coraor N, Chapman B, Nekrutenko A, Taylor J.** 2010. Galaxy CloudMan: delivering cloud compute clusters. *BMC Bioinformatics* **11**(Suppl 12)**:**S4.
80. **Cock PJA, Grüning BA, Paszkiewicz K, Pritchard L.** 2013. Galaxy tools and workflows for sequence analysis with applications in molecular plant pathology. *PeerJ* **1:**e167.
81. **McKenna A, Hanna M, Banks E, Sivachenko A, Cibulskis K, Kernytsky A, Garimella K, Altshuler D, Gabriel S, Daly M, DePristo MA.** 2010. The Genome Analysis Toolkit: a MapReduce framework for analyzing next-generation DNA sequencing data. *Genome Res* **20:**1297–1303.
82. **Dai L, Gao X, Guo Y, Xiao J, Zhang Z.** 2012. Bioinformatics clouds for big data manipulation. *Biol Direct* **7:**43, discussion 43.
83. **Angiuoli SV, Matalka M, Gussman A, Galens K, Vangala M, Riley DR, Arze C, White JR, White O, Fricke WF.** 2011. CloVR: a virtual machine for automated and portable sequence analysis from the desktop using cloud computing. *BMC Bioinformatics* **12:**356.
84. **Krampis K, Booth T, Chapman B, Tiwari B, Bicak M, Field D, Nelson KE.** 2012. Cloud BioLinux: pre-configured and on-demand bioinformatics computing for the genomics community. *BMC Bioinformatics* **13:**42.
85. **Naccache SN, Federman S, Veeraraghavan N, Zaharia M, Lee D, Samayoa E, Bouquet J, Greninger AL, Luk K-C, Enge B, Wadford DA, Messenger SL, Genrich GL, Pellegrino K, Grard G, Leroy E, Schneider BS, Fair JN, Martínez MA, Isa P, Crump JA, DeRisi JL, Sittler T, Hackett J Jr, Miller S, Chiu CY.** 2014. A cloud-compatible bioinformatics pipeline for ultrarapid pathogen identification from next-generation sequencing of clinical samples. *Genome Res* **24:** 1180–1192.
86. **Liu B, Madduri RK, Sotomayor B, Chard K, Lacinski L, Dave UJ, Li J, Liu C, Foster IT.** 2014. Cloud-based bioinformatics workflow platform for large-scale next-generation sequencing analyses. *J Biomed Inform* **49:**119–133.
87. **Marz M, Beerenwinkel N, Drosten C, Fricke M, Frishman D, Hofacker IL, Hoffmann D, Middendorf M, Rattei T, Stadler PF, Töpfer A.** 2014. Challenges in RNA virus bioinformatics. *Bioinformatics* **30:**1793–1799.
88. **Mokili JL, Rohwer F, Dutilh BE.** 2012. Metagenomics and future perspectives in virus discovery. *Curr Opin Virol* **2:**63–77.
89. **Yan Q.** 2008. Bioinformatics databases and tools in virology research: an overview. *In Silico Biol* **8:**71–85.
90. **Zhao G, Krishnamurthy S, Cai Z, Popov VL, Travassos da Rosa AP, Guzman H, Cao S, Virgin HW, Tesh RB, Wang D.** 2013. Identification of novel viruses using VirusHunter—an automated data analysis pipeline. *PLoS One* **8:**e78470.
91. **Pickett BE, Sadat EL, Zhang Y, Noronha JM, Squires RB, Hunt V, Liu M, Kumar S, Zaremba S, Gu Z, Zhou L, Larson CN, Dietrich J, Klem EB, Scheuermann RH.** 2012. ViPR: an open bioinformatics database and analysis resource for virology research. *Nucleic Acids Res* **40**(D1)**:**D593–D598.
92. **Pickett BE, Greer DS, Zhang Y, Stewart L, Zhou L, Sun G, Gu Z, Kumar S, Zaremba S, Larsen CN, Jen W, Klem EB, Scheuermann RH.** 2012. Virus pathogen database and analysis resource (ViPR): a comprehensive bioinformatics database and analysis resource for the coronavirus research community. *Viruses* **4:**3209–3226.
93. **Fancello L, Raoult D, Desnues C.** 2012. Computational tools for viral metagenomics and their application in clinical research. *Virology* **434:**162–174.
94. **Takeuchi F, Sekizuka T, Yamashita A, Ogasawara Y, Mizuta K, Kuroda M.** 2014. MePIC, metagenomic pathogen identification for clinical specimens. *Jpn J Infect Dis* **67:** 62–65.
95. **Petty TJ, Cordey S, Padioleau I, Docquier M, Turin L, Preynat-Seauve O, Zdobnov EM, Kaiser L.** 2014. Comprehensive human virus screening using high-throughput sequencing with a user-friendly representation of bioinformatics analysis: a pilot study. *J Clin Microbiol* **52:**3351–3361.
96. **Underwood A, Green J.** 2011. Call for a quality standard for sequence-based assays in clinical microbiology: necessity for quality assessment of sequences used in microbial identification and typing. *J Clin Microbiol* **49:**23–26.
97. **Zhao S, Prenger K, Smith L, Messina T, Fan H, Jaeger E, Stephens S.** 2013. Rainbow: a tool for large-scale whole-genome sequencing data analysis using cloud computing. *BMC Genomics* **14:**425.
98. **Willner D, Hugenholtz P.** 2013. From deep sequencing to viral tagging: recent advances in viral metagenomics. *BioEssays* **35:**436–442.
99. **Budowle B, Connell ND, Bielecka-Oder A, Colwell RR, Corbett CR, Fletcher J, Forsman M, Kadavy DR, Markotic A, Morse SA, Murch RS, Sajantila A, Schmedes SE, Ternus KL, Turner SD, Minot S.** 2014. Validation of high throughput sequencing and microbial forensics applications. *Investig Genet* **5:**9.
100. **Wandelt S, Rheinländer A, Bux M, Thalheim L, Haldemann B, Leser U.** 2012. Data Management challenges in next generation sequencing. *Datenbank-Spektrum* **12:**161–171.
101. **Mariette J, Escudié F, Allias N, Salin G, Noirot C, Thomas S, Klopp C.** 2012. NG6: integrated next generation sequencing storage and processing environment. *BMC Genomics* **13:**462.
102. **Leinonen R, Sugawara H, Shumway M, International Nucleotide Sequence Database Collaboration.** 2011. The sequence read archive. *Nucleic Acids Res* **39**(Database)**:**D19–D21.
103. **Yang X, Chockalingam SP, Aluru S.** 2013. A survey of error-correction methods for next-generation sequencing. *Brief Bioinform* **14:**56–66.
104. **Li H, Ruan J, Durbin R.** 2008. Mapping short DNA sequencing reads and calling variants using mapping quality scores. *Genome Res* **18:**1851–1858.
105. **Ruffalo M, Koyutürk M, Ray S, LaFramboise T.** 2012. Accurate estimation of short read mapping quality for next-generation genome sequencing. *Bioinformatics* **28:**i349–i355.
106. **Haiminen N, Kuhn DN, Parida L, Rigoutsos I.** 2011. Evaluation of methods for de novo genome assembly from high-throughput sequencing reads reveals dependencies that affect the quality of the results. *PLoS One* **6:**e24182.

107. **Li H.** 2014. Toward better understanding of artifacts in variant calling from high-coverage samples. *Bioinformatics* **30:**2843–2851.
108. **Oyola SO, Otto TD, Gu Y, Maslen G, Manske M, Campino S, Turner DJ, Macinnis B, Kwiatkowski DP, Swerdlow HP, Quail MA.** 2012. Optimizing Illumina next-generation sequencing library preparation for extremely AT-biased genomes. *BMC Genomics* **13:**1.
109. **Jabara CB, Jones CD, Roach J, Anderson JA, Swanstrom R.** 2011. Accurate sampling and deep sequencing of the HIV-1 protease gene using a Primer ID. *Proc Natl Acad Sci USA* **108:**20166–20171.
110. **Sheward DJ, Murrell B, Williamson C.** 2012. Degenerate Primer IDs and the birthday problem. *Proc Natl Acad Sci USA* **109:**21.E1330; author reply E1331.
111. **Brodin J, Hedskog C, Heddini A, Benard E, Neher RA, Mild M, Albert J.** 2015. Challenges with using primer IDs to improve accuracy of next generation sequencing. *PLoS One* **10:** e0119123.
112. **McElroy K, Thomas T, Luciani F.** 2014. Deep sequencing of evolving pathogen populations: applications, errors, and bioinformatic solutions. *Microb Inform Exp* **4:**1.
113. **Ahmad-Nejad P, Dorn-Beineke A, Pfeiffer U, Brade J, Geilenkeuser W-J, Ramsden S, Pazzagli M, Neumaier M.** 2006. Methodologic European external quality assurance for DNA sequencing: the EQUALseq program. *Clin Chem* **52:** 716–727.
114. **Patton SJ, Wallace AJ, Elles R.** 2006. Benchmark for evaluating the quality of DNA sequencing: proposal from an international external quality assessment scheme. *Clin Chem* **52:**728–736.
115. **Yang GS, Stott JM, Smailus D, Barber SA, Balasundaram M, Marra MA, Holt RA.** 2005. High-throughput sequencing: a failure mode analysis. *BMC Genomics* **6:**2.
116. **Holm-Hansen C, Vainio K.** 2009. Sequencing of viral genes. *Methods Mol Biol* **551:**203–215.
117. **Gargis AS, Kalman L, Berry MW, Bick DP, Dimmock DP, Hambuch T, Lu F, Lyon E, Voelkerding KV, Zehnbauer BA, Agarwala R, Bennett SF, Chen B, Chin ELH, Compton JG, Das S, Farkas DH, Ferber MJ, Funke BH, Furtado MR, Ganova-Raeva LM, Geigenmüller U, Gunselman SJ, Hegde MR, Johnson PLF, Kasarskis A, Kulkarni S, Lenk T, Liu CSJ, Manion M, Manolio TA, Mardis ER, Merker JD, Rajeevan MS, Reese MG, Rehm HL, Simen BB, Yeakley JM, Zook JM, Lubin IM.** 2012. Assuring the quality of next-generation sequencing in clinical laboratory practice. *Nat Biotechnol* **30:**1033–1036.
118. **Ladner JT, Beitzel B, Chain PSG, Davenport MG, Donaldson EF, Frieman M, Kugelman JR, Kuhn JH, O'Rear J, Sabeti PC, Wentworth DE, Wiley MR, Yu G-Y, Sozhamannan S, Bradburne C, Palacios G, Threat Characterization Consortium.** 2014. Standards for sequencing viral genomes in the era of high-throughput sequencing. *MBio* **5:** e01360–14.
119. **US Food and Drug Administration (FDA).** Ultra High Throughput Sequencing for Clinical Diagnostic Applications–Approaches to Assess Analytical Validity, June 23, 2011 http://www.fda.gov/MedicalDevices/NewsEvents/WorkshopsConferences/ucm255327.htm
120. **Collins FS, Hamburg MA.** 2013. First FDA authorization for next-generation sequencer. *N Engl J Med* **369:**2369–2371.
121. **Barzon L, Lavezzo E, Militello V, Toppo S, Palù G.** 2011. Applications of next-generation sequencing technologies to diagnostic virology. *Int J Mol Sci* **12:**7861–7884.
122. **Barzon L, Lavezzo E, Costanzi G, Franchin E, Toppo S, Palù G.** 2013. Next-generation sequencing technologies in diagnostic virology. *J Clin Virol* **58:**346–350.
123. **Capobianchi MR, Giombini E, Rozera G.** 2013. Next-generation sequencing technology in clinical virology. *Clin Microbiol Infect* **19:**15–22.
124. **Li L, Delwart E.** 2011. From orphan virus to pathogen: the path to the clinical lab. *Curr Opin Virol* **1:**282–288.
125. **Radford AD, Chapman D, Dixon L, Chantrey J, Darby AC, Hall N.** 2012. Application of next-generation sequencing technologies in virology. *J Gen Virol* **93:**1853–1868.
126. **Köser CU, Ellington MJ, Cartwright EJP, Gillespie SH, Brown NM, Farrington M, Holden MTG, Dougan G, Bentley SD, Parkhill J, Peacock SJ.** 2012. Routine use of microbial whole genome sequencing in diagnostic and public health microbiology. *PLoS Pathog* **8:**e1002824.
127. **Marston HD, Folkers GK, Morens DM, Fauci AS.** 2014. Emerging viral diseases: confronting threats with new technologies. *Sci Transl Med* **6**(253)**:253ps10.**
128. **Holmes KV, Dominguez SR.** 2013. The new age of virus discovery: genomic analysis of a novel human betacoronavirus isolated from a fatal case of pneumonia. *MBio* **4:**e00548–12.
129. **Pallen MJ.** 2014. Diagnostic metagenomics: potential applications to bacterial, viral and parasitic infections. *Parasitology* **141:**1856–1862.
130. **Quan P-L, Briese T, Palacios G, Ian Lipkin W.** 2008. Rapid sequence-based diagnosis of viral infection. *Antiviral Res* **79:**1–5.
131. **Bexfield N, Kellam P.** 2011. Metagenomics and the molecular identification of novel viruses. *Vet J* **190:**191–198.
132. **Chiu CY.** 2013. Viral pathogen discovery. *Curr Opin Microbiol* **16:**468–478.
133. **Ambrose HE, Clewley JP.** 2006. Virus discovery by sequence-independent genome amplification. *Rev Med Virol* **16:**365–383.
134. **Delwart EL.** 2007. Viral metagenomics. *Rev Med Virol* **17:** 115–131.
135. **Bibby K.** 2013. Metagenomic identification of viral pathogens. *Trends Biotechnol* **31:**275–279.
136. **Fuller C, Cebelinski E, Bartkus J, Juni B, Smith K, Besser J.** 2008. *Enhanced laboratory testing of enteric disease outbreaks of unknown etiology in Minnesota.* Abstr Int Conf Emerg Infect Dis.
137. **Jones MS, Lukashov VV, Ganac RD, Schnurr DP.** 2007. Discovery of a novel human picornavirus in a stool sample from a pediatric patient presenting with fever of unknown origin. *J Clin Microbiol* **45:**2144–2150.
138. **Svraka S, Rosario K, Duizer E, van der Avoort H, Breitbart M, Koopmans M.** 2010. Metagenomic sequencing for virus identification in a public-health setting. *J Gen Virol* **91:** 2846–2856.
139. **Shaukat S, Angez M, Alam MM, Jebbink MF, Deijs M, Canuti M, Sharif S, de Vries M, Khurshid A, Mahmood T, van der Hoek L, Zaidi SSZ.** 2014. Identification and characterization of unrecognized viruses in stool samples of non-polio acute flaccid paralysis children by simplified VIDISCA. *Virol J* **11:**146.
140. **de Vries M, Deijs M, Canuti M, van Schaik BDC, Faria NR, van de Garde MDB, Jachimowski LCM, Jebbink MF, Jakobs M, Luyf ACM, Coenjaerts FEJ, Claas ECJ, Molenkamp R, Koekkoek SM, Lammens C, Leus F, Goossens H, Ieven M, Baas F, van der Hoek L.** 2011. A sensitive assay for virus discovery in respiratory clinical samples. *PLoS One* **6:**e16118.
141. **Finkbeiner SR, Li Y, Ruone S, Conrardy C, Gregoricus N, Toney D, Virgin HW, Anderson LJ, Vinjé J, Wang D, Tong S.** 2009. Identification of a novel astrovirus (astrovirus VA1) associated with an outbreak of acute gastroenteritis. *J Virol* **83:**10836–10839.
142. **Yang J, Yang F, Ren L, Xiong Z, Wu Z, Dong J, Sun L, Zhang T, Hu Y, Du J, Wang J, Jin Q.** 2011. Unbiased parallel detection of viral pathogens in clinical samples by use of a metagenomic approach. *J Clin Microbiol* **49:**3463–3469.
143. **McMullan LK, Frace M, Sammons SA, Shoemaker T, Balinandi S, Wamala JF, Lutwama JJ, Downing RG, Stroeher U, MacNeil A, Nichol ST.** 2012. Using next generation sequencing to identify yellow fever virus in Uganda. *Virology* **422:**1–5.
144. **Feng H, Taylor JL, Benos PV, Newton R, Waddell K, Lucas SB, Chang Y, Moore PS.** 2007. Human transcriptome subtraction by using short sequence tags to search for tumor viruses in conjunctival carcinoma. *J Virol* **81:**11332–11340.

145. **van der Meijden E, Janssens RWA, Lauber C, Bouwes Bavinck JN, Gorbalenya AE, Feltkamp MCW.** 2010. Discovery of a new human polyomavirus associated with trichodysplasia spinulosa in an immunocompromized patient. *PLoS Pathog* **6:**e1001024.
146. **Gaynor AM, Nissen MD, Whiley DM, Mackay IM, Lambert SB, Wu G, Brennan DC, Storch GA, Sloots TP, Wang D.** 2007. Identification of a novel polyomavirus from patients with acute respiratory tract infections. *PLoS Pathog* **3:**e64.
147. **Palacios G, Druce J, Du L, Tran T, Birch C, Briese T, Conlan S, Quan P-L, Hui J, Marshall J, Simons JF, Egholm M, Paddock CD, Shieh W-J, Goldsmith CS, Zaki SR, Catton M, Lipkin WI.** 2008. A new arenavirus in a cluster of fatal transplant-associated diseases. *N Engl J Med* **358:**991–998.
148. **Grard G, Fair JN, Lee D, Slikas E, Steffen I, Muyembe J-J, Sittler T, Veeraraghavan N, Ruby JG, Wang C, Makuwa M, Mulembakani P, Tesh RB, Mazet J, Rimoin AW, Taylor T, Schneider BS, Simmons G, Delwart E, Wolfe ND, Chiu CY, Leroy EM.** 2012. A novel rhabdovirus associated with acute hemorrhagic fever in central Africa. *PLoS Pathog* **8:**e1002924.
149. **Wylie KM, Weinstock GM, Storch GA.** 2012. Emerging view of the human virome. *Transl Res* **160:**283–290.
150. **Feschotte C.** 2010. Virology: bornavirus enters the genome. *Nature* **463:**39–40.
151. **Horie M, Honda T, Suzuki Y, Kobayashi Y, Daito T, Oshida T, Ikuta K, Jern P, Gojobori T, Coffin JM, Tomonaga K.** 2010. Endogenous non-retroviral RNA virus elements in mammalian genomes. *Nature* **463:**84–87.
152. **Boyd EF.** 2012. Bacteriophage-encoded bacterial virulence factors and phage-pathogenicity island interactions. *Adv Virus Res* **82:**91–118.
153. **De Paepe M, Leclerc M, Tinsley CR, Petit M-A.** 2014. Bacteriophages: an underestimated role in human and animal health? *Front Cell Infect Microbiol* **4:**39.
154. **Ly M, Abeles SR, Boehm TK, Robles-Sikisaka R, Naidu M, Santiago-Rodriguez T, Pride DT.** 2014. Altered oral viral ecology in association with periodontal disease. *MBio* **5:** e01133–14.
155. **Foulongne V, Sauvage V, Hebert C, Dereure O, Cheval J, Gouilh MA, Pariente K, Segondy M, Burguière A, Manuguerra J-C, Caro V, Eloit M.** 2012. Human skin microbiota: high diversity of DNA viruses identified on the human skin by high throughput sequencing. *PLoS One* **7:**e38499.
156. **Bzhalava D, Johansson H, Ekström J, Faust H, Möller B, Eklund C, Nordin P, Stenquist B, Paoli J, Persson B, Forslund O, Dillner J.** 2013. Unbiased approach for virus detection in skin lesions. *PLoS One* **8:**e65953.
157. **Spurgeon ME, Lambert PF.** 2013. Merkel cell polyomavirus: a newly discovered human virus with oncogenic potential. *Virology* **435:**118–130.
158. **Simmonds P, Prescott LE, Logue C, Davidson F, Thomas AE, Ludlam CA.** 1999. TT virus—part of the normal human flora? *J Infect Dis* **180:**1748–1750.
159. **Delwart E.** 2013. A roadmap to the human virome. *PLoS Pathog* **9:**e1003146.
160. **Metzner K.** 2006. The significance of minority drug-resistant quasispecies, p. *In* Geretti AM (ed), *Antiretroviral Resistance in Clinical Practice*. Mediscript, Chapter 11. London.
161. **Hedskog C, Mild M, Jernberg J, Sherwood E, Bratt G, Leitner T, Lundeberg J, Andersson B, Albert J.** 2010. Dynamics of HIV-1 quasispecies during antiviral treatment dissected using ultra-deep pyrosequencing. *PLoS One* **5:** e11345.
162. **Lindström A, Odeberg J, Albert J.** 2004. Pyrosequencing for detection of lamivudine-resistant hepatitis B virus. *J Clin Microbiol* **42:**4788–4795.
163. **Beerenwinkel N, Günthard HF, Roth V, Metzner KJ.** 2012. Challenges and opportunities in estimating viral genetic diversity from next-generation sequencing data. *Front Microbiol* **3:**329.
164. **Eriksson N, Pachter L, Mitsuya Y, Rhee S-Y, Wang C, Gharizadeh B, Ronaghi M, Shafer RW, Beerenwinkel N.** 2008. Viral population estimation using pyrosequencing. *PLOS Comput Biol* **4:**e1000074.
165. **Mangul S, Wu NC, Mancuso N, Zelikovsky A, Sun R, Eskin E.** 2014. Accurate viral population assembly from ultra-deep sequencing data. *Bioinformatics* **30:**i329–i337.
166. **Skums P, Mancuso N, Artyomenko A, Tork B, Mandoiu I, Khudyakov Y, Zelikovsky A.** 2013. Reconstruction of viral population structure from next-generation sequencing data using multicommodity flows. *BMC Bioinformatics* **14**(Suppl 9)**:**S2.
167. **Yang X, Charlebois P, Gnerre S, Coole MG, Lennon NJ, Levin JZ, Qu J, Ryan EM, Zody MC, Henn MR.** 2012. De novo assembly of highly diverse viral populations. *BMC Genomics* **13:**475.
168. **Han Y, Zhang Y, Mei Y, Wang Y, Liu T, Guan Y, Tan D, Liang Y, Yang L, Yi X.** 2013. Analysis of hepatitis B virus genotyping and drug resistance gene mutations based on massively parallel sequencing. *J Virol Methods* **193:**341–347.
169. **Svarovskaia ES, Martin R, McHutchison JG, Miller MD, Mo H.** 2012. Abundant drug-resistant NS3 mutants detected by deep sequencing in hepatitis C virus-infected patients undergoing NS3 protease inhibitor monotherapy. *J Clin Microbiol* **50:**3267–3274.
170. **Jabara CB, Hu F, Mollan KR, Williford SE, Menezes P, Yang Y, Eron JJ, Fried MW, Hudgens MG, Jones CD, Swanstrom R, Lemon SM.** 2014. Hepatitis C Virus (HCV) NS3 sequence diversity and antiviral resistance-associated variant frequency in HCV/HIV coinfection. *Antimicrob Agents Chemother* **58:**6079–6092.
171. **Hara K, Rivera MM, Koh C, Demino M, Page S, Nagabhyru PR, Rehermann B, Liang TJ, Hoofnagle JH, Heller T.** 2014. Sequence analysis of hepatitis C virus from patients with relapse after a sustained virological response: relapse or reinfection? *J Infect Dis* **209:**38–45.
172. **Ji H, Li Y, Graham M, Liang BB, Pilon R, Tyson S, Peters G, Tyler S, Merks H, Bertagnolio S, Soto-Ramirez L, Sandstrom P, Brooks J.** 2011. Next-generation sequencing of dried blood spot specimens: a novel approach to HIV drug-resistance surveillance. *Antivir Ther* **16:**871–878.
173. **Swenson LC, Däumer M, Paredes R.** 2012. Next-generation sequencing to assess HIV tropism. *Curr Opin HIV AIDS* **7:**478–485.
174. **Hirigoyen DL, Cartwright CP.** 2005. Use of sequence data generated in the Bayer Tru Gene genotyping assay to recognize and characterize non-subtype-b human immunodeficiency virus type 1 strains. *J Clin Microbiol* **43:**5263–5271.
175. **Church JD, Jones D, Flys T, Hoover D, Marlowe N, Chen S, Shi C, Eshleman JR, Guay LA, Jackson JB, Kumwenda N, Taha TE, Eshleman SH.** 2006. Sensitivity of the viroseq hiv-1 genotyping system for detection of the k103n resistance mutation in hiv-1 subtypes a, c, and d. *J Mol Diagn* **8:**430–432.
176. **Eshleman SH, Hackett J Jr, Swanson P, Cunningham SP, Drews B, Brennan C, Devare SG, Zekeng L, Kaptué L, Marlowe N.** 2004. Performance of the celera diagnostics viroseq hiv-1 genotyping system for sequence-based analysis of diverse human immunodeficiency virus type 1 strains. *J Clin Microbiol* **42:**2711–2717.
177. **Aghokeng AF, Mpoudi-Ngole E, Chia JE, Edoul EM, Delaporte E, Peeters M.** 2011. High failure rate of the ViroSeq HIV-1 genotyping system for drug resistance testing in Cameroon, a country with broad HIV-1 genetic diversity. *J Clin Microbiol* **49:**1635–1641.
178. **Avidor B, Girshengorn S, Matus N, Talio H, Achsanov S, Zeldis I, Fratty IS, Katchman E, Brosh-Nissimov T, Hassin D, Alon D, Bentwich Z, Yust I, Amit S, Forer R, Vulih Shultsman I, Turner D.** 2013. Evaluation of a benchtop HIV ultradeep pyrosequencing drug resistance assay in the clinical laboratory. *J Clin Microbiol* **51:**880–886.
179. **Stelzl E, Pröll J, Bizon B, Niklas N, Danzer M, Hackl C, Stabentheiner S, Gabriel C, Kessler HH.** 2011. Human immunodeficiency virus type 1 drug resistance testing: evaluation of a new ultra-deep sequencing-based protocol and comparison

with the TRUGENE HIV-1 Genotyping Kit. *J Virol Methods* **178:**94–97.

180. **Van Laethem K, Vandamme A-M.** 2006. Interpreting resistance data for HIV-1 therapy management—know the limitations. *AIDS Rev* **8:**37–43.
181. **Sen S, Tripathy SP, Paranjape RS.** 2006. Antiretroviral drug resistance testing. *J Postgrad Med* **52:**187–193.
182. **Wensing AM, Calvez V, Günthard HF, Johnson VA, Paredes R, Pillay D, Shafer RW, Richman DD.** 2014. 2014 Update of the drug resistance mutations in HIV-1. *Top Antivir Med* **22:** 642–650.
183. **Tang MW, Liu TF, Shafer RW.** 2012. The HIVdb system for HIV-1 genotypic resistance interpretation. *Intervirology* **55:** 98–101.
184. **Woods CK, Brumme CJ, Liu TF, Chui CKS, Chu AL, Wynhoven B, Hall TA, Trevino C, Shafer RW, Harrigan PR.** 2012. Automating HIV drug resistance genotyping with RECall, a freely accessible sequence analysis tool. *J Clin Microbiol* **50:**1936–1942.
185. **de Sanjose S, Quint WG, Alemany L, Geraets DT, Klaustermeier JE, Lloveras B, Tous S, Felix A, Bravo LE, Shin HR, Vallejos CS, de Ruiz PA, Lima MA, Guimera N, Clavero O, Alejo M, Llombart-Bosch A, Cheng-Yang C, Tatti SA, Kasamatsu E, Iljazovic E, Odida M, Prado R, Seoud M, Grce M, Usubutun A, Jain A, Suarez GA, Lombardi LE, Banjo A, Menéndez C, Domingo EJ, Velasco J, Nessa A, Chichareon SC, Qiao YL, Lerma E, Garland SM, Sasagawa T, Ferrera A, Hammouda D, Mariani L, Pelayo A, Steiner I, Oliva E, Meijer CJ, Al-Jassar WF, Cruz E, Wright TC, Puras A, Llave CL, Tzardi M, Agorastos T, Garcia-Barriola V, Clavel C, Ordi J, Andújar M, Castellsagué X, Sánchez GI, Nowakowski AM, Bornstein J, Muñoz N, Bosch FX, Retrospective International Survey and HPV Time Trends Study Group.** 2010. Human papillomavirus genotype attribution in invasive cervical cancer: a retrospective cross-sectional worldwide study. *Lancet Oncol* **11:**1048–1056.
186. **Zein NN.** 2000. Clinical significance of hepatitis C virus genotypes. *Clin Microbiol Rev* **13:**223–235.
187. **Chen C-H, Lee C-M, Lu S-N, Changchien C-S, Eng H-L, Huang C-M, Wang J-H, Hung C-H, Hu T-H.** 2005. Clinical significance of hepatitis B virus (HBV) genotypes and precore and core promoter mutations affecting HBV e antigen expression in Taiwan. *J Clin Microbiol* **43:**6000–6006.
188. **Midgley CM, Jackson MA, Selvarangan R, Turabelidze G, Obringer E, Johnson D, Giles BL, Patel A, Echols F, Oberste MS, Nix WA, Watson JT, Gerber SI.** 2014. Severe respiratory illness associated with enterovirus D68 - Missouri and Illinois, 2014. *MMWR Morb Mortal Wkly Rep* **63:**798–799.
189. **Tokarz R, Firth C, Madhi SA, Howie SRC, Wu W, Sall AA, Haq S, Briese T, Lipkin WI.** 2012. Worldwide emergence of multiple clades of enterovirus 68. *J Gen Virol* **93:**1952–1958.
190. **Lu Q-B, Wo Y, Wang H-Y, Wei M-T, Zhang L, Yang H, Liu E-M, Li T-Y, Zhao Z-T, Liu W, Cao W-C.** 2014. Detection of enterovirus 68 as one of the commonest types of enterovirus found in patients with acute respiratory tract infection in China. *J Med Microbiol* **63:**408–414.
191. **Todd AK, Hall RJ, Wang J, Peacey M, McTavish S, Rand CJ, Stanton J-A, Taylor S, Huang QS.** 2013. Detection and whole genome sequence analysis of an enterovirus 68 cluster. *Virol J* **10:**103.
192. **Vega E.** 2011. Novel Surveillance Network for Norovirus Gastroenteritis Outbreaks, United States. *Emerg Infect Dis* **17:**8
193. **Nguyen LM, Middaugh JP.** 2012. Suspected transmission of norovirus in eight long-term care facilities attributed to staff working at multiple institutions. *Epidemiol Infect* **140:**1702–1709.
194. **Thornley CN, Hewitt J, Perumal L, Van Gessel SM, Wong J, David SA, Rapana JP, Li S, Marshall JC, Greening GE.** 2013. Multiple outbreaks of a novel norovirus GII.4 linked to an infected post-symptomatic food handler. *Epidemiol Infect* **141:**1585–1597.
195. **McIntyre L, Galanis E, Mattison K, Mykytczuk O, Buenaventura E, Wong J, Prystajecky N, Ritson M, Stone J, Moreau D, Youssef A, Outbreak Investigation Team.** 2012. Multiple clusters of norovirus among shellfish consumers linked to symptomatic oyster harvesters. *J Food Prot* **75:**1715–1720.
196. **Kundu S, Lockwood J, Depledge DP, Chaudhry Y, Aston A, Rao K, Hartley JC, Goodfellow I, Breuer J.** 2013. Next-generation whole genome sequencing identifies the direction of norovirus transmission in linked patients. *Clin Infect Dis* **57:**407–414.
197. **Vega E, Barclay L, Gregoricus N, Shirley SH, Lee D, Vinjé J.** 2014. Genotypic and epidemiologic trends of norovirus outbreaks in the United States, 2009 to 2013. *J Clin Microbiol* **52:**147–155.
198. **Lee LE, Cebelinski EA, Fuller C, Keene WE, Smith K, Vinjé J, Besser JM.** 2012. Sapovirus outbreaks in long-term care facilities, Oregon and Minnesota, USA, 2002–2009. *Emerg Infect Dis* **18:**873–876.
199. **Gahr P, DeVries AS, Wallace G, Miller C, Kenyon C, Sweet K, Martin K, White K, Bagstad E, Hooker C, Krawczynski G, Boxrud D, Liu G, Stinchfield P, LeBlanc J, Hickman C, Bahta L, Barskey A, Lynfield R.** 2014. An outbreak of measles in an undervaccinated community. *Pediatrics* **134:** e220–e228.
200. **Centers for Disease Control and Prevention (CDC).** 2011. Notes from the field: Measles outbreak—Hennepin County, Minnesota, February-March 2011. *MMWR Morb Mortal Wkly Rep* **60:**421.
201. **Holmes EC.** 2007. Viral evolution in the genomic age. *PLoS Biol* **5:**e278.
202. **Holmes EC, Rambaut A.** 2004. Viral evolution and the emergence of SARS coronavirus. *Philos Trans R Soc Lond B Biol Sci* **359:**1059–1065.
203. **Dudas G, Rambaut A.** 2014. Phylogenetic analysis of guinea 2014 ebov ebolavirus outbreak. *PLoS Curr* 1st ed.
204. **Gire SK, et al.** 2014. Genomic surveillance elucidates Ebola virus origin and transmission during the 2014 outbreak. *Science* **345:**1369–1372.
205. **Drummond A, Pybus OG, Rambaut A.** 2003. Inference of viral evolutionary rates from molecular sequences. *Adv Parasitol* **54:**331–358.
206. **Alexander JP, Ehresmann K, Seward J, Wax G, Harriman K, Fuller S, Cebelinski EA, Chen Q, Jorba J, Kew OM, Pallansch MA, Oberste MS, Schleiss M, Davis JP, Warshawsky B, Squires S, Hull HF, Vaccine-Derived Poliovirus Investigations Group.** 2009. Transmission of imported vaccine-derived poliovirus in an undervaccinated community in Minnesota. *J Infect Dis* **199:**391–397.
207. **DeVries AS, Harper J, Murray A, Lexau C, Bahta L, Christensen J, Cebelinski E, Fuller S, Kline S, Wallace GS, Shaw JH, Burns CC, Lynfield R.** 2011. Vaccine-derived poliomyelitis 12 years after infection in Minnesota. *N Engl J Med* **364:**2316–2323.
208. **Castro-Nallar E, Crandall KA, Pérez-Losada M.** 2012. Genetic diversity and molecular epidemiology of HIV transmission. *Future Virol* **7:**239–252.
209. **Lewis F, Hughes GJ, Rambaut A, Pozniak A, Leigh Brown AJ.** 2008. Episodic sexual transmission of HIV revealed by molecular phylodynamics. *PLoS Med* **5:**e50.
210. **Pilcher CD, Wong JK, Pillai SK.** 2008. Inferring HIV transmission dynamics from phylogenetic sequence relationships. *PLoS Med* **5:**e69.
211. **Ou CY, Ciesielski CA, Myers G, Bandea CI, Luo C-C, Korber BTM, Mullins JI, Schochetman G, Berkelman RL, Economou AN, Witte JJ, Furman LJ, Satten GA, MacInnes KA, Curran JW, Jaffe HW.** 1992. Molecular epidemiology of HIV transmission in a dental practice. *Science* **256:**1165–1171.
212. **Jaffe HW, McCurdy JM, Kalish ML, Liberti T, Metellus G, Bowman BH, Richards SB, Neasman AR, Witte JJ.** 1994. Lack of HIV transmission in the practice of a dentist with AIDS. *Ann Intern Med* **121:**855–859.

213. **Scaduto DI, Brown JM, Haaland WC, Zwickl DJ, Hillis DM, Metzker ML.** 2010. Source identification in two criminal cases using phylogenetic analysis of HIV-1 DNA sequences. *Proc Natl Acad Sci USA* **107:**21242–21247.
214. **Brooks JI, Sandstrom PA.** 2013. The power and pitfalls of HIV phylogenetics in public health. *Can J Public Health* **104:** e348–e350.
215. **Moal V, Ferretti A, Devichi P, Colson P.** 2014. Genome sequence of a hepatitis e virus of genotype 3e from a chronically infected kidney transplant recipient. *Genome Announc* **2:** e01156–e0115613.
216. **Calisher CH, Childs JE, Field HE, Holmes KV, Schountz T.** 2006. Bats: important reservoir hosts of emerging viruses. *Clin Microbiol Rev* **19:**531–545.
217. **Kuzmin IV, Shi M, Orciari LA, Yager PA, Velasco-Villa A, Kuzmina NA, Streicker DG, Bergman DL, Rupprecht CE.** 2012. Molecular inferences suggest multiple host shifts of rabies viruses from bats to mesocarnivores in Arizona during 2001–2009. *PLoS Pathog* **8:**e1002786.
218. **Borucki MK, Chen-Harris H, Lao V, Vanier G, Wadford DA, Messenger S, Allen JE.** 2013. Ultra-deep sequencing of intra-host rabies virus populations during cross-species transmission. *PLoS Negl Trop Dis* **7:**e2555.
219. **Leslie MJ, Messenger S, Rohde RE, Smith J, Cheshier R, Hanlon C, Rupprecht CE.** 2006. Bat-associated rabies virus in Skunks. *Emerg Infect Dis* **12:**1274–1277.
220. **Chang H-GH, Eidson M, Noonan-Toly C, Trimarchi CV, Rudd R, Wallace BJ, Smith PF, Morse DL.** 2002. Public health impact of reemergence of rabies, New York. *Emerg Infect Dis* **8:**909–913.
221. **Blanton JD, Self J, Niezgoda M, Faber M-L, Dietzschold B, Rupprecht C.** 2007. Oral vaccination of raccoons (Procyon lotor) with genetically modified rabies virus vaccines. *Vaccine* **25:**7296–7300
222. **Rupprecht CE, Hanlon CA, Slate D.** 2004. Oral vaccination of wildlife against rabies: opportunities and challenges in prevention and control. *Dev Biol (Basel)* **119:**173–184.
223. **Boland TA, McGuone D, Jindal J, Rocha M, Cumming M, Rupprecht CE, Barbosa TFS, de Novaes Oliveira R, Chu CJ, Cole AJ, Kotait I, Kuzmina NA, Yager PA, Kuzmin IV, Hedley-Whyte ET, Brown CM, Rosenthal ES.** 2014. Phylogenetic and epidemiologic evidence of multiyear incubation in human rabies. *Ann Neurol* **75:**155–160.
224. **Gargis AS, Kalman L, Bick DP, da Silva C, Dimmock DP, Funke BH, Gowrisankar S, Hegde MR, Kulkarni S, Mason CE, Nagarajan R, Voelkerding KV, Worthey EA, Aziz N, Barnes J, Bennett SF, Bisht H, Church DM, Dimitrova Z, Gargis SR, Hafez N, Hambuch T, Hyland FCL, Luna RA, MacCannell D, Mann T, McCluskey MR, McDaniel TK, Ganova-Raeva LM, Rehm HL, Reid J, Campo DS, Resnick RB, Ridge PG, Salit ML, Skums P, Wong L-JC, Zehnbauer BA, Zook JM, Lubin IM.** 2015. Good laboratory practice for clinical next-generation sequencing informatics pipelines. *Nat Biotechnol* **33:**689–693.
225. **Cock PJA, Fields CJ, Goto N, Heuer ML, Rice PM.** 2010. The Sanger FASTQ file format for sequences with quality scores, and the Solexa/Illumina FASTQ variants. *Nucleic Acids Res* **38:**1767–1771.
226. **Ewing B, Green P.** 1998. Base-calling of automated sequencer traces using phred. II. Error probabilities. *Genome Res* **8:**186–194.
227. **Bao S, Jiang R, Kwan W, Wang B, Ma X, Song Y-Q.** 2011. Evaluation of next-generation sequencing software in mapping and assembly. *J Hum Genet* **56:**406–414.
228. **Magi A, Benelli M, Gozzini A, Girolami F, Torricelli F, Brandi ML.** 2010. Bioinformatics for next generation sequencing data. *Genes (Basel)* **1:**294–307.
229. **Kostic AD, Ojesina AI, Pedamallu CS, Jung J, Verhaak RGW, Getz G, Meyerson M.** 2011. PathSeq: software to identify or discover microbes by deep sequencing of human tissue. *Nat Biotechnol* **29:**393–396.
230. **Bhaduri A, Qu K, Lee CS, Ungewickell A, Khavari PA.** 2012. Rapid identification of non-human sequences in high-throughput sequencing datasets. *Bioinformatics* **28:**1174–1175.
231. **Naeem R, Rashid M, Pain A.** 2013. READSCAN: A fast and scalable pathogen discovery program with accurate genome relative abundance estimation. *Bioinformatics* **29:**391–391.
232. **Li J-W, Wan R, Yu C-S, Co NN, Wong N, Chan T-F.** 2013. ViralFusionSeq: accurately discover viral integration events and reconstruct fusion transcripts at single-base resolution. *Bioinformatics* **29:**649–651.
233. **Wommack KE, Bhavsar J, Polson SW, Chen J, Dumas M, Srinivasiah S, Furman M, Jamindar S, Nasko DJ.** 2012. VIROME: a standard operating procedure for analysis of viral metagenome sequences. *Stand Genomic Sci* **6:**427–439.
234. **Roux S, Faubladier M, Mahul A, Paulhe N, Bernard A, Debroas D, Enault F.** 2011. Metavir: a web server dedicated to virome analysis. *Bioinformatics* **27:**3074–3075.
235. **Roux S, Tournayre J, Mahul A, Debroas D, Enault F.** 2014. Metavir 2: new tools for viral metagenome comparison and assembled virome analysis. *BMC Bioinformatics* **15:**76.
236. **Chen Y, Yao H, Thompson EJ, Tannir NM, Weinstein JN, Su X.** 2013. VirusSeq: software to identify viruses and their integration sites using next-generation sequencing of human cancer tissue. *Bioinformatics* **29:**266–267.
237. **Wang Q, Jia P, Zhao Z.** 2013. VirusFinder: software for efficient and accurate detection of viruses and their integration sites in host genomes through next generation sequencing data. *PLoS One* **8:**e64465.

Phenotypic and Genotypic Antiviral Susceptibility Testing

MARTHA T. VAN DER BEEK AND ERIC C. J. CLAAS

16

Where the use of antibiotics goes back to the discovery of penicillin by Alexander Fleming in 1928, application of antiviral treatment was not achieved until the early 1960s with the use of the nucleoside analogue idoxuridine for treatment of herpetic keratitis (1) and methisazone for treatment of smallpox, variola, and cowpox. The first major advances in antiviral treatment were obtained for herpes viruses with the discovery of another nucleoside analogue, acyclovir, by the Burroughs Wellcome Company in the early 1980s. Despite these advances, the real wave of antiviral drug discovery was the result of the HIV epidemic, and started with the development of azidothymidine (AZT) as the first antiretroviral for AIDS patients (2). Nowadays a wide spectrum of antiviral agents are used for a variety of infections. However, similarly to bacteria, development of resistance is an important complication when using antiviral agents. In addition, many viral pathogens have an RNA genome and use RNA polymerases that lack proofread activity for their replication. Therefore, mutations will be introduced into the viral genome in every replication cycle and as a consequence an altered susceptibility or even resistance to an antiviral agent may develop. These resistant variants are easily selected under pressure of an antiviral treatment as shown by the rapid development of resistance when using AZT (3, 4). The viral reverse transcriptase incorporates this thymidine analogue into the viral genome, inhibiting proper replication. Resistant viruses rapidly evolve by acquiring resistance-associated mutations (RAM) in the gene encoding for the enzyme. With HIV being a retrovirus, these mutations are also incorporated in the host DNA and therefore the resistant viral genome is stored in the DNA of the patient and will emerge upon reintroducing the drug.

Over the last decades, considerable progress has been made in the development of antiviral agents and consequently in the treatment of viral infections. Increased use of antiviral agents has led to an increase in the development of resistant viruses as well. Due to the relatively small genome sizes, the limited number of genes, and the presence of *in vitro* cultivation systems, the mechanisms of resistance have been elucidated for many drugs. Once the mechanism is known, diagnostic tests for antiviral susceptibility testing can be introduced to optimize patient treatment.

In this chapter, laboratory methods are described for phenotypic and genotypic susceptibility testing for the major viruses where antiviral treatment is applied and thus there may be a need for analyzing antiviral susceptibility or resistance. Assays for resistance testing largely overlap and will be discussed in more detail for antiviral resistance testing of herpes viruses. A more general overview will be provided for HIV, influenza viruses, and hepatitis viruses.

ANTIVIRAL SUSCEPTIBILITY TESTING

The classical approach of testing for antiviral resistance is based on replication of the virus in cell culture and, therefore, application is limited to those viruses that can be propagated *in vitro*. Reduction of growth in the presence of an antiviral agent is indicative of resistance and can be measured in different ways, including the inhibition of virus-induced plaque formation or cytopathic effect or a decrease in the production of viral antigens, enzyme activities, or nucleic acid synthesis. In the molecular era, genotypic assays that provide rapid analyses of genetic markers that are associated with antiviral resistance have been developed and now, in many cases, have replaced the more traditional phenotypic approaches. The choice of which molecular method to use depends on the complexity of the resistance mechanism. If only a few mutations result in resistance, real time PCR assays or hybridization tests such as the LineProbe assay (LiPA, Fujirebio Inc., Belgium) can be used. If a few mutations result in resistance, real time PCR assays can identify single nucleotide polymorphisms. However, if the mechanism is more complex or the number of mutations that potentially can result in antiviral resistance is higher, nucleotide sequence analysis is the preferred method. Sanger sequencing has been applied in many diagnostic assays to identify RAMs. For a more in-depth analysis of viruses consisting of quasispecies or for detection of minor variants, next generation sequencing (NGS) applications can be used.

Culture-Based Systems

With culture-based systems, the mode of action of antiviral agents is based on interference with viral replication. One well-known mode of action of antiviral agents is that the molecule, sometimes after processing, acts as a defective nucleotide that is incorporated in the viral genome and aborts further replication or expression. The large group of

doi:10.1128/9781555819156.ch16

nucleoside/nucleotide polymerase inhibitors (NPI) belong to this group. Another mechanism is direct interaction of the antiviral agent with viral proteins or receptors being used by the virus. In all cases the antiviral agents interfere with replication or propagation of viruses. As a result, for those viruses that can be grown in cell or tissue culture, *in vitro* models examining antiviral activity on viral strains are readily available.

An *in vitro* model can be used to monitor antiviral activity of new compounds or old compounds on new viruses (5–7). However, they can also be used to establish antiviral resistance in patients not responding to antiviral therapy. Inhibition of a particular strain can be quantitated by determining the reduction of growth. The 50% tissue culture infectious dose ($TCID_{50}$) is a measure that indicates what concentration of virus is required to infect 50% of the cells in tissue culture. By inoculating multiple culture plate wells with dilutions of a virus, the $TCID_{50}$ can be determined by an algorithm that includes those dilutions that do not infect all replicates of the cells in the number of wells (8). As a result, when a virus with a known $TCID_{50}$ is grown in the presence of a dilution series of an antiviral agent, the reduction of the $TCID_{50}$ is a measure for the antiviral activity of the compound. In the same way, the antiviral resistance can be measured by an increase of $TCID_{50}$ in the presence of an inhibitory concentration of antiviral agent.

If diluted viruses are grown in cell culture with an agarose overlay, a single infectious virus particle will infect a single cell and spread only to its neighboring cells. In this way, a small defined part of the cell layer is affected and can be made visible as a plaque. The number of plaques is related to the growth capacity of the viral strain and is indicated as plaque forming units (pfu). If a viral strain is incubated with an antiviral agent, the number of pfu will be reduced as a result of the antiviral activity. On the other hand, no reduction in growth as measured by no decrease in the number of pfu in the presence of antiviral agent is indicative for antiviral resistance.

Currently the diagnostic use in determining antiviral resistance by these laborious cell culture methods is limited. By using quantitative PCR to determine an increase in viral DNA or RNA rather than increase in $TCID_{50}$ of pfu in resistant viruses, some improvement in turnaround time is achieved. Still, these DNA/RNA reduction assays remain laborious (9). With an increase in technological possibilities, a shift towards molecular methods as PCR and nucleotide sequencing has been observed.

PCR-Based Methods

Hybridization assays were the first molecular resistance tests that became available. PCR amplification of the viral genomic regions carrying RAMs, as for example the hepatitis B YMDD motif, with subsequent hybridization to immobilized probes that represented all possible wild type and resistant sequences provided rapid information on the antiviral susceptibility (10). Initially these assays were also used for HIV (11) but the rapid evolution of the number of RAMs in the HIV genome together with the rapid increase in antiviral agents made it impossible to maintain an appropriate test. As a result, nucleotide sequence analysis became the standard for diagnosing these more complicated resistance patterns.

PCR or real time PCR methods can be combined with culture-based methods. By assessing the amount of virus grown that is grown in the presence or absence antiviral agent by PCR amplification, laborious $TCID_{50}$ or pfu assays can be avoided. DNA or RNA can be extracted from the cell culture supernatant and subsequently amplified and quantified using real time PCR. Although still laborious, resistance could be determined independently on the mechanism of resistance, comparable to phenotypic assays.

If only a limited number of RAMs are responsible for antiviral resistance, as for example amantadine resistance of influenza viruses, a real time PCR can be designed that can differentiate the susceptible virus (with amino acid S31) from the resistant virus (with aminoacid mutation S31N) by using probes with different fluorophores or a difference in melting temperatures (12). PCR-based methods have revolutionized clinical microbiology, not only for diagnosing the infection, but also in antiviral susceptibility testing. However, for detailed analysis of antiviral resistance, the information provided by PCR amplification alone is usually insufficient. Identification of a set of mutations associated with resistance requires detailed information that can be obtained by nucleotide sequence analysis.

Sequencing Methods

The exact nucleic acid sequence of a gene can be determined by nucleotide sequence analysis. Most widely used is the Sanger chain termination method. By using the correct mixture of regular deoxy nucleoside triphosphates (dNTPs), that result in chain elongation, and dideoxy nucleoside triphosphates (ddNTPs), which halt elongation, the exact sequence can be measured. By analyzing the viral genes that harbor mutations that are associated with antiviral resistance, detailed information on potential resistance is obtained. The length of the sequence required is dependent on the available antiviral agents and knowledge about the mechanism of resistance with respect to the mode of action and, more specifically, what mutations result in resistance. In HIV, patients are treated with combination antiretroviral therapy to control the viruses. If the dosage is inadequate, for example because of incompliance of the patient in taking the medication, mutations may occur that will result in resistance to the antivirals used. The majority of the mutations that will lead to resistance are detected in the first 1,500 nucleotides of the polymerase gene of HIV. This gene codes for the viral protease, reverse transcriptase, and integrase, enzymes that are the main targets to be inhibited by antiviral treatment. In treatment of herpes viruses, the main targets are the thymidine kinase (TK) and polymerase (pol) which are two crucial proteins in viral replication. In development of resistance, the range of RAMs observed in the affected TK and *pol* gene genes may require up to 3,000 nucleotides of sequence for proper analysis. A limitation of Sanger sequencing is that only the major variant of the virus is being analyzed. Presence of minor variants, with minor being less than 25% of the total population, will remain unnoticed.

Next generation sequencing is able to provide complete information of viral genomes and enables in depth analysis of minor variants of resistance mutations, i.e., deep sequencing.

For HIV resistance, NGS deep sequencing applications have been used to identify minor variants of RAMs (13, 14). These NGS application studies for HIV resistance suggest an increased risk for viral failure by the presence of minor variant RAMs. However, these findings are the result of population based analysis. For individual patients the added value of this approach remains to be established. As only short fragments of sequence are generated, it is hard to determine if they are part of viable viruses or part of viral genomes with reduced fitness. NGS is still in development and with improvement of turnaround time and read length, NGS application in resistance testing will become an important tool. However, further evolution of the methodology is required before it can be implemented.

ANTIVIRAL TREATMENT AND SUSCEPTIBILITY TESTING FOR SELECT VIRUSES

Herpes Viruses

Antiviral Treatment

Antiviral treatment is indicated in various, but not all, herpes virus infections. Infections caused by herpes simplex virus (HSV), varicella-zoster virus (VZV), and cytomegalovirus (CMV) are most commonly treated with antivirals and therefore will be discussed here. Antiviral preparations for herpes virus ophthalmological infections will not be discussed.

Severe primary infections with HSV and oral or genital ulcerations due to reactivation of HSV patients can be treated with systemic, oral or topical acyclovir (ACV), or its oral prodrug valacyclovir (15, 16). Likewise, chickenpox and herpes zoster can be treated with the same agents. Other antivirals such as pencyclovir (topical administration) and famcyclovir (oral administration) can be used for the same indications. Primary infections with CMV are usually only treated in the immunocompromised hosts. In this group, reactivation of CMV can also be treated with ganciclovir or its oral prodrug valgancyclovir (vGCV) (17–21). Second line agents for treating all of the above mentioned herpes virus infections, in case of antiviral resistance or drug intolerance, are foscarnet (FOS) and cidofovir (CDV), which are administered intravenously. The efficacy of maribavir in the prevention and treatment of CMV is still under investigation and, hence, antiviral susceptibility testing regarding this agent will not be discussed (22–25).

Clinical Indications for Antiviral Susceptibility Testing

Antiviral susceptibility testing is mainly indicated in immunocompromised patients, since antiviral resistance is rare in herpes virus infections in immunocompetent individuals (26–30). In immunocompromised patients, persistent herpes virus infection despite antiviral treatment may be caused by antiviral resistance. However, such treatment failure, either clinical or virological, can have various causes since antivirals merely suppress viral replication awaiting restoration of antiviral immunity. Hence, treatment failure may be due to a profound state of immunodeficiency in which the patient's immune system is unable to control viral replication to any extent. Also, inadequate dosing or impaired drug absorption of antivirals can play a role. Lastly, resistance of the virus to antivirals can cause failure of treatment. Therefore, susceptibility testing is required to guide further treatment in patients with clinical or virological treatment failure. Figure 1 provides a flow chart on the suggested diagnostics and treatment of persistent herpes virus infections (31).

Definitions of Antiviral Resistance and Variables of Antiviral Susceptibility Testing

Optimally, breakpoints to identify resistant viruses take into account the *in vitro* established susceptibility, pharmacokinetics of the antiviral agents, and clinical response to the given treatment. Only for HSV, such clinically validated susceptibility breakpoints have been established for plaque reduction assay by Clinical and Laboratory Standards Institute (CLSI) (32). Both technical and clinical issues have

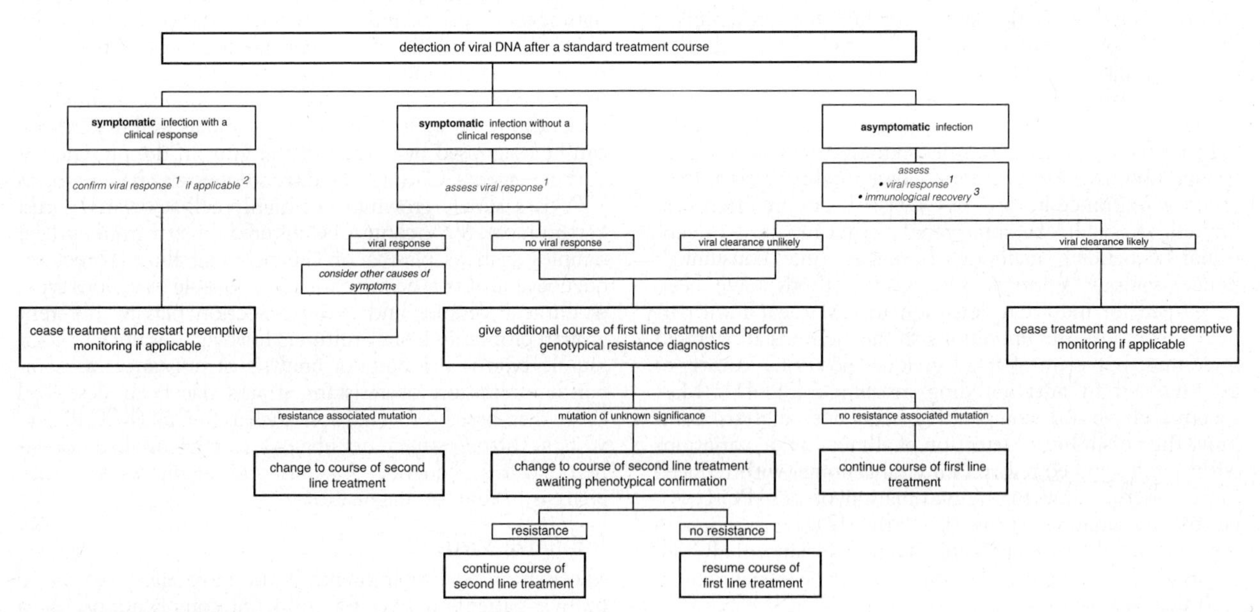

FIGURE 1 Flow chart for diagnosis and treatment in case of failure of antiviral therapy of DNA virus infections in immunocompromised patients. Global approach to diagnosis and treatment in case of persistent DNA virus infections in immunosuppressed patients.[1]Assessment of viral response may include the viral load after treatment and the decrease in viral load in response to the given treatment.[2]In case of severe symptomatic disease.[3]Assessment of immunological recovery may include total and virus-specific T-cell counts and the possibility to decrease the use of immunosuppressive medication. Details may differ per virus (31).

hampered the development of standardized assays and breakpoints for CMV and VZV. Technical issues are the slow and cell-associated growth of CMV and VZV and the difficulties isolating virus from relevant body sites such as blood and cerebrospinal fluid. In addition, it is difficult to reliably assess clinical outcome in severely immunocompromised patients in whom failure may be only partly related to antiviral treatment. Hence, in many publications, phenotypic antiviral resistance is merely defined as a reduced *in vitro* susceptibility compared to a reference isolate or a pretreatment or wild-type isolate from the same patient, with rather arbitrary cut-offs to define significant reduction in susceptibility in the various assays. At least intra- and interassay variation of a specific assay should be taken into account when defining relevant changes in susceptibility. The subsequent clinical relevance of a diminished susceptibility remains hard to determine from these results. Furthermore, resistance is often defined on the basis of sequence analysis of a viral isolate showing mutations in relevant parts of the associated viral genes. The subsequent annotation of such *in silico* changes is optimally performed by transferring specific mutations into a susceptible virus to establish the effect on *in vitro* susceptibility, but should at least be confirmed by *in vitro* susceptibility testing of the viral isolate containing the mutation. Often, this marker transfer is not technically feasible and/or viral isolates cannot be cultured, hampering this analysis. Also, in isolates containing various mutations it can be difficult to separate the mutation(s), contributing to resistance from polymorphisms. All relevant genes in HSV, VZV, and CMV contain numerous polymorphisms. Hence, antiviral resistance studies based solely on genotypic analysis should be interpreted with caution.

Assays

GCV resistance mutations in CMV mainly map to the viral kinase gene UL97 (33–37). After prolonged treatment and after treatment with foscarnet or cidofovir, mutations in the viral polymerase gene UL54 can also emerge (36, 38). Sequencing analysis is the fastest method for susceptibility testing of CMV, which cannot be easily cultured. Furthermore, many mutations in the CMV genome have been accurately characterized phenotypically by marker transfer (Tables 1 and 2) (49–58). The relevance and extent of the reduced susceptibility varies per mutation and depends on the method used for phenotypic susceptibility testing. Data on cross-resistance in cases of UL54-mutants are often very limited, so should be interpreted with caution. Conventional sequencing methods are rather time consuming; hence, various genotypic screening methods have been developed for mutation detection in CMV, each with its own advantages and limitations. Some methods allow rapid screening, but only of fixed genome positions known to be involved in antiviral drug resistance (39–41). Mass spectrometry-based comparative sequence analysis combines the possibility of detection of all nucleotide variations within a designated region of a viral genome, with reduced hands on time due to the automation of post-PCR processing and analysis, but is still costly (42). Deep-sequencing methods allow for detection of minor variants of HCMV in patients (43), but are laborious and may be less suitable in a clinical setting.

Resistance to antiviral drugs for HSV 1 is primarily caused by mutations in the UL23 gene of the viral TK (resistance to ACV) or, rarely, in the UL30 gene of the viral DNA polymerase (resistance to ACV, CDV, or FOS) (30, 44, 45). Also, for HSV-2, mutations in TK and DNA polymerase genes can cause resistance. Sequencing of these genes may reveal a resistance conferring mutation (Tables 3 and 4), but since nucleotide variations are common, mutations of unknown significance are frequently detected (30, 44, 45). Included in the tables are mutations whose phenotypical significance has either been characterized by transferring specific mutations into susceptible virus or has been confirmed by a phenotypic susceptibility test of the mutated isolate (44–87). In the latter situation, isolates containing more than one mutation of unknown significance were excluded. The relevance and extent of the reduced susceptibility varies per mutation and depends on the method used for phenotypic susceptibility testing. Data on cross-resistance in cases of UL30 mutants are limited, so they need to be interpreted with caution. In such cases, phenotypic susceptibility testing of HSV is required and traditionally performed by a standard plaque reduction assay or some variation of this method (32). This methodology is labor intensive and time consuming and, hence, results are usually not available in a clinically relevant time frame. Faster versions of the plaque reduction assays have been developed (88–90). Real time PCR based phenotypical susceptibility assays may also overcome these limitations and facilitate timely diagnosis of antiviral resistance (9, 91, 92). However, clinical breakpoints for these assays need standardization.

Similar to HSV-1 and HSV-2, VZV resistance to antivirals is mainly due to mutations in the viral TK gene of VZV (resistance to ACV), or, in rare cases, in the viral DNA polymerase (Pol, resistance to ACV, CDV, or FOS) (93–98). However, mutations of unknown significance can occur, due to infrequent antiviral resistance and the scarcity of phenotypic data (Tables 5 and 6) (94, 98, 99). Included in the tables are mutations whose phenotypic significance has either been characterized by marker transfer or has been confirmed by a phenotypic susceptibility test of the mutated isolate (93, 94, 96–98, 100–114). In the latter situation, isolates containing more than one mutation of unknown significance were excluded. The relevance and extent of the reduced susceptibility varies per mutation and depends on the method used for phenotypic susceptibility testing. Data on cross-resistance in cases of Pol mutants are limited, so they need to be interpreted with caution as well. Resistance can be diagnosed by culture of the virus in the presence of antiviral agents. Culture-based techniques are challenging as VZV is a slowly growing and highly cell-associated virus. Furthermore, VZV cannot be cultured directly from clinical samples such as plasma or cerebrospinal fluid. Direct sequence analysis of the target genes is possible in various types of clinical samples and avoids selection bias by different growth properties in cell culture. However, it is unclear what sample type is the best for analysis of resistance, as compartmentalization of resistant strains has been described (93). Web-based software tools, such as ReCall (RECall beta v2.6 at http://pssm.cfenet.ubc.ca) can be applied for all herpes virus sequence analyses and facilitates sequence analysis in routine diagnostics.

Influenza Viruses

Vaccination against influenza is the most effective way to protect patients at risk for influenza complications. As a second line of defense, antiviral agents can be used. Antiviral treatment of influenza infections dates back to 1966 when Symmetrel (Amantadine) received clearance from the U.S. Food and Drug Administration (FDA). Amantadine blocks the influenza M2 ion channel, a product of the matrix

TABLE 1 Mutations in the UL97-gene of CMV

AA position	Mutation	Interpretation	Remark
19	Q19E	susceptible to (val)ganciclovir	
68	N68D	susceptible to (val)ganciclovir	
95	T95S	susceptible to (val)ganciclovir	
108	S108N	susceptible to (val)ganciclovir	
112	R112C	susceptible to (val)ganciclovir	
126	L126Q	susceptible to (val)ganciclovir	
137	R137C	susceptible to (val)ganciclovir	
227	E227D	susceptible to (val)ganciclovir	
244	I244V	susceptible to (val)ganciclovir	
329	D329H	susceptible to (val)ganciclovir	
337	L337M	susceptible to (val)ganciclovir	
342	F342S	***reduced susceptibility to (val)ganciclovir***	
353	V353A	susceptible to (val)ganciclovir	
355	DEL355	***reduced susceptibility to (val)ganciclovir***	
356	V356G	***reduced susceptibility to (val)ganciclovir***	
397	L397R	susceptible to (val)ganciclovir	
405	L405P	***reduced susceptibility to (val)ganciclovir***	
409	T409M	susceptible to (val)ganciclovir	
411	H411L	susceptible to (val)ganciclovir	
411	H411N	susceptible to (val)ganciclovir	
411	H411Y	susceptible to (val)ganciclovir	
427	A427V	susceptible to (val)ganciclovir	
449	Q449K	susceptible to (val)ganciclovir	
460	M460I	***reduced susceptibility to (val)ganciclovir***	
460	M460T	***reduced susceptibility to (val)ganciclovir***	
460	M460V	***reduced susceptibility to (val)ganciclovir***	
466	V466G	***reduced susceptibility to (val)ganciclovir***	
466	V466M	susceptible to (val)ganciclovir	
469	H469Y	susceptible to (val)ganciclovir	
478	A478V	susceptible to (val)ganciclovir	
510	N510S	susceptible to (val)ganciclovir	
518	C518Y	***reduced susceptibility to (val)ganciclovir***	
520	H520Q	***reduced susceptibility to (val)ganciclovir***	
521	P521L	***reduced susceptibility to (val)ganciclovir***	
550	M550I	susceptible to (val)ganciclovir	
582	A582T	susceptible to (val)ganciclovir	
582	A582V	susceptible to (val)ganciclovir	
587	H587Y	susceptible to (val)ganciclovir	
588	A588V	susceptible to (val)ganciclovir	
590	DEL590-593	***reduced susceptibility to (val)ganciclovir***	
591	A591V	susceptible to (val)ganciclovir	possible low-level resistance
591	DEL591-594	***reduced susceptibility to (val)ganciclovir***	
591	DEL591-607	***reduced susceptibility to (val)ganciclovir***	
592	C592G	***reduced susceptibility to (val)ganciclovir***	
594	A594E	***reduced susceptibility to (val)ganciclovir***	
594	A594G	***reduced susceptibility to (val)ganciclovir***	
594	A594P	***reduced susceptibility to (val)ganciclovir***	
594	A594T	***reduced susceptibility to (val)ganciclovir***	
594	A594V	***reduced susceptibility to (val)ganciclovir***	
595	DEL595	***reduced susceptibility to (val)ganciclovir***	
595	DEL595-603	***reduced susceptibility to (val)ganciclovir***	

(*Continued*)

TABLE 1 Mutations in the UL97-gene of CMV (*Continued*)

AA position	Mutation	Interpretation	Remark
595	L595F	***reduced susceptibility to (val)ganciclovir***	
595	L595S	***reduced susceptibility to (val)ganciclovir***	
595	L595T	susceptible to (val)ganciclovir	
595	L595W	***reduced susceptibility to (val)ganciclovir***	
596	E596G	***reduced susceptibility to (val)ganciclovir***	
597	N597D	susceptible to (val)ganciclovir	
598	G598S	***reduced susceptibility to (val)ganciclovir***	
599	K599R	susceptible to (val)ganciclovir	
599	K599T	***reduced susceptibility to (val)ganciclovir***	
600	DEL600	***reduced susceptibility to (val)ganciclovir***	
600	L600I	susceptible to (val)ganciclovir	
601	DEL601	***reduced susceptibility to (val)ganciclovir***	
601	DEL601-603	***reduced susceptibility to (val)ganciclovir***	
603	C603R	***reduced susceptibility to (val)ganciclovir***	
603	C603S	***reduced susceptibility to (val)ganciclovir***	possible low-level resistance
603	C603W	***reduced susceptibility to (val)ganciclovir***	
605	D605E	susceptible to (val)ganciclovir	
607	C607F	***reduced susceptibility to (val)ganciclovir***	possible low-level resistance
607	C607Y	***reduced susceptibility to (val)ganciclovir***	
615	M615V	susceptible to (val)ganciclovir	
617	Y617H	susceptible to (val)ganciclovir	
623	G623S	susceptible to (val)ganciclovir	
634	L634Q	susceptible to (val)ganciclovir	
659	T659I	susceptible to (val)ganciclovir	
665	V665I	susceptible to (val)ganciclovir	
674	A674T	susceptible to (val)ganciclovir	

gene segment of influenza A (but not influenza B) viruses. Resistance is induced by mutations in the M2 gene coding for the ion channel pore, most frequently an S31N mutation. Resistance testing is possible by testing for this mutation by allele specific PCR analysis (12, 115). However, in the present time, over 99% of all circulating human influenza virus A strains, including subtypes influenza A(H1N1) as well as A(H3N2), are now resistant to adamantanes, and therefore these antivirals are no longer of use in clinical practice (116, 117).

Nowadays, neuraminidase inhibitors (NAI) are the most effective antivirals for treatment of influenza infections, although antivirals targeting the polymerase are under evaluation. Oseltamivir and zanamivir have been available for over a decade and the newer NAI as peramivir and lanimivir are now licensed for use in some countries (118). In 2009, nearly all seasonal influenza A(H1N1) viruses were resistant to oseltamivir as they contained the H275Y mutation in the neuraminidase (NA) gene. Emergence of influenza virus A (H1N1)pdm09 almost completely reduced the proportion of resistant A viruses to less than 1%. Occasionally, isolates or clusters of resistant viruses emerge and worldwide surveillance is implemented to monitor their prevalence (119). Resistance can also be induced by prolonged treatment with NAI (120) requiring susceptibility testing for analysis of treatment efficacy.

For measuring resistance to NAI, actual phenotypic analysis of the NA activity provides the most accurate information. The function of the NA enzyme is to release newly synthesized viruses from the cell by cleaving the viral hemagglutinin from the sialic acid receptor of the cell. An inactive NA results in accumulation of viruses at the cell surface. Commercial chemiluminescent (121) or fluorescent (122) neuraminidase inhibition (NI) assays are available and measure the concentration of the NAI which results in 50% reduction of enzymatic NA activity. For daily diagnostic use, these assays are quite laborious as viruses need to be propagated and clinically relevant thresholds are difficult to define. Therefore, in most laboratories, genetic analysis is implemented as the primary diagnostic tool for resistance testing.

In the influenza NA gene, mutations have been identified that are associated with the majority of resistant viruses. For influenza A virus (H1N1)pdm09, the H275Y mutation has been most frequently found. For influenza A virus (H3N2), the 3 most common substitutions resulting in NAI resistance are E119V, R292K, and N294S. So by molecular analysis of 600 nucleotides of the NA gene, the susceptibility or resistance to NAI can be established. In influenza B viruses, changes at E117, D197, H273, and R374 are associated with resistance, but other RAMs and combinations of RAMs have been identified as well (116). For surveillance purposes, large scale analysis of resistance in influenza viruses is required and pyrosequencing would be an option as short sequences can be rapidly generated. As a general rule, using molecular methods, information is collected for only those

TABLE 2 Mutations in the UL54-gene of CMV

AA position	Mutation	Interpretation	Remark
19	Q19E	susceptible to (val)ganciclovir	
68	N68D	susceptible to (val)ganciclovir	
95	T95S	susceptible to (val)ganciclovir	
108	S108N	susceptible to (val)ganciclovir	
112	R112C	susceptible to (val)ganciclovir	
126	L126Q	susceptible to (val)ganciclovir	
137	R137C	susceptible to (val)ganciclovir	
227	E227D	susceptible to (val)ganciclovir	
244	I244V	susceptible to (val)ganciclovir	
329	D329H	susceptible to (val)ganciclovir	
337	L337M	susceptible to (val)ganciclovir	
342	F342S	***reduced susceptibility to (val)ganciclovir***	
353	V353A	susceptible to (val)ganciclovir	
355	DEL355	***reduced susceptibility to (val)ganciclovir***	
356	V356G	***reduced susceptibility to (val)ganciclovir***	
397	L397R	susceptible to (val)ganciclovir	
405	L405P	***reduced susceptibility to (val)ganciclovir***	
409	T409M	susceptible to (val)ganciclovir	
411	H411L	susceptible to (val)ganciclovir	
411	H411N	susceptible to (val)ganciclovir	
411	H411Y	susceptible to (val)ganciclovir	
427	A427V	susceptible to (val)ganciclovir	
449	Q449K	susceptible to (val)ganciclovir	
460	M460I	***reduced susceptibility to (val)ganciclovir***	
460	M460T	***reduced susceptibility to (val)ganciclovir***	
460	M460V	***reduced susceptibility to (val)ganciclovir***	
466	V466G	***reduced susceptibility to (val)ganciclovir***	
466	V466M	susceptible to (val)ganciclovir	
469	H469Y	susceptible to (val)ganciclovir	
478	A478V	susceptible to (val)ganciclovir	
510	N510S	susceptible to (val)ganciclovir	
518	C518Y	***reduced susceptibility to (val)ganciclovir***	
520	H520Q	***reduced susceptibility to (val)ganciclovir***	
521	P521L	***reduced susceptibility to (val)ganciclovir***	
550	M550I	susceptible to (val)ganciclovir	
582	A582T	susceptible to (val)ganciclovir	
582	A582V	susceptible to (val)ganciclovir	
587	H587Y	susceptible to (val)ganciclovir	
588	A588V	susceptible to (val)ganciclovir	
590	DEL590-593	***reduced susceptibility to (val)ganciclovir***	
591	A591V	susceptible to (val)ganciclovir	possible low-level resistance
591	DEL591-594	***reduced susceptibility to (val)ganciclovir***	
591	DEL591-607	***reduced susceptibility to (val)ganciclovir***	
592	C592G	***reduced susceptibility to (val)ganciclovir***	
594	A594E	***reduced susceptibility to (val)ganciclovir***	
594	A594G	***reduced susceptibility to (val)ganciclovir***	
594	A594P	***reduced susceptibility to (val)ganciclovir***	
594	A594T	***reduced susceptibility to (val)ganciclovir***	
594	A594V	***reduced susceptibility to (val)ganciclovir***	
595	DEL595	***reduced susceptibility to (val)ganciclovir***	
595	DEL595-603	***reduced susceptibility to (val)ganciclovir***	

(Continued)

TABLE 2 Mutations in the UL54-gene of CMV (*Continued*)

AA position	Mutation	Interpretation	Remark
595	L595F	***reduced susceptibility to (val)ganciclovir***	
595	L595S	***reduced susceptibility to (val)ganciclovir***	
595	L595T	susceptible to (val)ganciclovir	
595	L595W	***reduced susceptibility to (val)ganciclovir***	
596	E596G	***reduced susceptibility to (val)ganciclovir***	
597	N597D	susceptible to (val)ganciclovir	
598	G598S	***reduced susceptibility to (val)ganciclovir***	
599	K599R	susceptible to (val)ganciclovir	
599	K599T	***reduced susceptibility to (val)ganciclovir***	
600	DEL600	***reduced susceptibility to (val)ganciclovir***	
600	L600I	susceptible to (val)ganciclovir	
601	DEL601	***reduced susceptibility to (val)ganciclovir***	
601	DEL601-603	***reduced susceptibility to (val)ganciclovir***	
603	C603R	***reduced susceptibility to (val)ganciclovir***	
603	C603S	***reduced susceptibility to (val)ganciclovir***	possible low-level resistance
603	C603W	***reduced susceptibility to (val)ganciclovir***	
605	D605E	susceptible to (val)ganciclovir	
607	C607F	***reduced susceptibility to (val)ganciclovir***	possible low-level resistance
607	C607Y	***reduced susceptibility to (val)ganciclovir***	
615	M615V	susceptible to (val)ganciclovir	
617	Y617H	susceptible to (val)ganciclovir	
623	G623S	susceptible to (val)ganciclovir	
634	L634Q	susceptible to (val)ganciclovir	
659	T659I	susceptible to (val)ganciclovir	
665	V665I	susceptible to (val)ganciclovir	
674	A674T	susceptible to (val)ganciclovir	

markers known to confer resistance and therefore phenotypic testing remains the gold standard (118).

Real time PCR-based methods are more suitable for rapid, immediate screening of influenza patients, as they can provide result in only hours. Due to the limited number of mutations to be analyzed, SNP analysis can be performed using real time PCR by using two Minor Groove Binder (MGB) probes specific for each SNP or by analyzing differences in melting temperatures of the PCR products by high resolution melting (12, 115).

Hepatitis Viruses

Hepatitis B Virus

Approximately 240 million people are chronically infected with hepatitis B virus (123). First-line monotherapy treatment regimens of nucleoside/nucleotide analogues lamivudine (LAM), telbivudine, and adefovir resulted in significant development of resistance. After 5 years of treatment, almost 80% of patients treated with LAM carried resistant virus. Mutations in the reverse transcriptase part of the polymerase gene cause resistance and testing is performed exclusively by molecular methods. All relevant mutations are grouped around the catalytic site of the reverse transcriptase between amino acids 180 and 250. The line probe assay INNO-LiPA HBV DR (Innogenetics, Belgium) specifically targets a set of mutations conferring resistance (124). Amplification and sequence analysis of a region of just over 200 bp of the polymerase (pol) gene enables analysis of all relevant mutations present. Analysis of sequence data can be performed by the Geno2pheno drug resistance interpretation algorithm (http://hbv.geno2pheno.org).

Presently, first-line treatment with entecavir (125, 126) or tenofovir (127, 128), greatly reduces the risk of developing resistance and testing for resistance is not commonly indicated.

Hepatitis C Virus

An estimated 150 million people have a chronic hepatitis C virus (HCV) infection resulting in a significant number of patients developing liver cirrhosis or hepatocellular carcinoma (129). Until a few years ago, the main treatment option for those patients was pegylated interferon-α (PEG-IFN) and ribavirin (RBV). Success of treatment, as indicated by a sustained viral response (SVR) after 12, 24, or 48 weeks, varied from 50% for HCV genotype 1 to over 80% for HCV genotype 2 (130). However, from 2011 until the present time, new classes of antiviral agents have been approved, increasing treatment options and success rates (131). These antiviral drugs act directly on the nonstructural HCV proteins and therefore are referred to as direct-acting antiviral agents (DAAs). The three classes of DAAs comprise the protease inhibitors, polymerase inhibitors, and the NS5A replication complex inhibitors. The first approved DAAs were the protease inhibitors telaprevir and bocepre-

TABLE 3 Mutations in the UL23-gene of HSV

HSV type	AA position	Mutation	Interpretation	Remark
1	6	C6G	susceptible to (val)aciclovir	
1	12	A12P	susceptible to (val)aciclovir	
1	14	D14Y	susceptible to (val)aciclovir	
1	17	A17V	susceptible to (val)aciclovir	
1	19	S19P	susceptible to (val)aciclovir	
1	20	R20S	susceptible to (val)aciclovir	
1	23	S23N	susceptible to (val)aciclovir	
2	26	R26H	susceptible to (val)aciclovir	
2	27	A27T	susceptible to (val)aciclovir	
1	30	R30C	susceptible to (val)aciclovir	
1	32	R32C	susceptible to (val)aciclovir	
2	34	R34C	***reduced susceptibility to (val)aciclovir***	
1	36	delE36	***reduced susceptibility to (val)aciclovir***	
1	36	E36K or D	susceptible to (val)aciclovir	
2	39	E39G	susceptible to (val)aciclovir	
1	41	R41H	susceptibility to (val)aciclovir varyingly reported	
1	42	L42P	susceptible to (val)aciclovir	
1	44	Q44H	susceptible to (val)aciclovir	
1	51	R51W	***reduced susceptibility to (val)aciclovir***	
1	53	Y53 stop or D or C or H	***reduced susceptibility to (val)aciclovir***	
1	55	D55N	***reduced susceptibility to (val)aciclovir***	
1	56	G56S	***reduced susceptibility to (val)aciclovir***	
1	57	P57H	***reduced susceptibility to (val)aciclovir***	
1	58	H58R	***reduced susceptibility to (val)aciclovir***	
2	59	G59P	***reduced susceptibility to (val)aciclovir***	
1	59	G59W	***reduced susceptibility to (val)aciclovir***	
1	60–61	frameshift (deletion of G in 180–183)	***reduced susceptibility to (val)aciclovir***	
1	61	G61A	susceptible to (val)aciclovir	
1	62	K62N	***reduced susceptibility to (val)aciclovir***	
1	62–63	frameshift (deletion of A in 184–187)	***reduced susceptibility to (val)aciclovir***	
1	63	T63I or A	***reduced susceptibility to (val)aciclovir***	
1	65	T65N	***reduced susceptibility to (val)aciclovir***	
2	66	S66P	***reduced susceptibility to (val)aciclovir***	
1	68	L68P	***reduced susceptibility to (val)aciclovir***	
2	72	A72S	***reduced susceptibility to (val)aciclovir***	
2	73–74	frameshift (deletion of G in 219–222)	***reduced susceptibility to (val)aciclovir***	
1	74	S74stop	***reduced susceptibility to (val)aciclovir***	
1	77	D77N	susceptible to (val)aciclovir	
2	78	D78N	susceptible to (val)aciclovir	
1	78	I78F	susceptible to (val)aciclovir	
1	83	E83K	***reduced susceptibility to (val)aciclovir***	
1	84	P84L or S	***reduced susceptibility to (val)aciclovir***	
1	85	M85I	susceptible to (val)aciclovir	
1	89	Q89R	susceptible to (val)aciclovir	
2	96	E96D	susceptible to (val)aciclovir	
2	102	S102N/ D/ R?	***reduced susceptibility to (val)aciclovir***	exact amino acid change unknown
1	103	T103P or stop	***reduced susceptibility to (val)aciclovir***	
1	104	Q104stop	***reduced susceptibility to (val)aciclovir***	
1	105	H105P	***reduced susceptibility to (val)aciclovir***	
2	105	Q105P	***reduced susceptibility to (val)aciclovir***	

(*Continued*)

TABLE 3 Mutations in the UL23-gene of HSV (*Continued*)

HSV type	AA position	Mutation	Interpretation	Remark
1	108	108stop	***reduced susceptibility to (val)aciclovir***	
2	110	I101S	***reduced susceptibility to (val)aciclovir***	
1	111	E111K	susceptible to (val)aciclovir	
1	111	frameshift (deletion of G at 333)	***reduced susceptibility to (val)aciclovir***	
2	119	A119T	susceptible to (val)aciclovir	
1	121	M121R	***reduced susceptibility to (val)aciclovir***	
1	123	R123S	***reduced susceptibility to (val)aciclovir***	
1	125	Q125H or E or L	***reduced susceptibility to (val)aciclovir***	
1	125	Q125N	susceptible to (val)aciclovir	
1	128	M128L	***reduced susceptibility to (val)aciclovir***	
1	129	G129D	***reduced susceptibility to (val)aciclovir***	
2	131	T131P	***reduced susceptibility to (val)aciclovir***	
2	133	Y133F	***reduced susceptibility to (val)aciclovir***	
2	135	A135P	susceptible to (val)aciclovir	
2	137	D137E	susceptible to (val)aciclovir	
2	137	D137stop	***reduced susceptibility to (val)aciclovir***	
1	138	V138I	susceptible to (val)aciclovir	
1	139	L139V	***reduced susceptibility to (val)aciclovir***	
2	140	F140L	susceptible to (val)aciclovir	
1	144–146	frameshift (deletion of G in 430–436)	***reduced susceptibility to (val)aciclovir***	
1	144–146	frameshift (insertion of G in 430–436)	***reduced susceptibility to (val)aciclovir***	
2	145–147	frameshift (deletion of G in 433–439)	***reduced susceptibility to (val)aciclovir***	
2	145–147	frameshift (insertion of G in 433–439)	***reduced susceptibility to (val)aciclovir***	
2	145–147	frameshift (insertion of GG in 433–439)	***reduced susceptibility to (val)aciclovir***	
1	146	E146G	susceptible to (val)aciclovir	
2	151	frameshift (deletion of C in 452)	***reduced susceptibility to (val)aciclovir***	
1	151	H151Y	susceptible to (val)aciclovir	
1	152–153	frameshift (deletion of C in 455–458)	***reduced susceptibility to (val)aciclovir***	
1	154–155	frameshift (deletion of C in 460–464)	***reduced susceptibility to (val)aciclovir***	
1	154–155	frameshift (insertion of C in 460–464)	***reduced susceptibility to (val)aciclovir***	
2	155–156	frameshift (deletion of C in 463–467)	***reduced susceptibility to (val)aciclovir***	
1	156	A156V	***reduced susceptibility to (val)aciclovir***	
1	156	frameshift (deletion of C in 467)	***reduced susceptibility to (val)aciclovir***	
1	161	F161L	susceptible to (val)aciclovir	
1	162	D162A	***reduced susceptibility to (val)aciclovir***	
1	163	R163H	***reduced susceptibility to (val)aciclovir***	
1	167	A167V	***reduced susceptibility to (val)aciclovir***	
1	168	A168T	susceptible to (val)aciclovir	
1	170	L170P	***reduced susceptibility to (val)aciclovir***	
1	172	Y172C	***reduced susceptibility to (val)aciclovir***	
1	173	P173L or R	***reduced susceptibility to (val)aciclovir***	
1	174	A174P	***reduced susceptibility to (val)aciclovir***	
1	175	A175V	***reduced susceptibility to (val)aciclovir***	
1	176	R176Q or W	***reduced susceptibility to (val)aciclovir***	
2	177	R177W	***reduced susceptibility to (val)aciclovir***	
1	178	L178R	***reduced susceptibility to (val)aciclovir***	
1	181	S181N	***reduced susceptibility to (val)aciclovir***	
2	183	M183I	***reduced susceptibility to (val)aciclovir***	
1	183–185	frameshift (deletion of C in 548–553)	***reduced susceptibility to (val)aciclovir***	
1	183–185	frameshift (insertion of C in 548–553)	***reduced susceptibility to (val)aciclovir***	
2	184–186	frameshift (deletion of C in 551–556)	***reduced susceptibility to (val)aciclovir***	

(*Continued*)

TABLE 3 Mutations in the UL23-gene of HSV (*Continued*)

HSV type	AA position	Mutation	Interpretation	Remark
2	184–186	frameshift (insertion of C in 551–556)	***reduced susceptibility to (val)aciclovir***	
1	187	V187M	***reduced susceptibility to (val)aciclovir***	
1	189	A189V	***reduced susceptibility to (val)aciclovir***	
1	191	V191A	susceptible to (val)aciclovir	
1	192	A192V	susceptible to (val)aciclovir	
1	194	delI194	***reduced susceptibility to (val)aciclovir***	
2	196–197	frameshift (deletion of C in 586–590)	***reduced susceptibility to (val)aciclovir***	
2	196–197	frameshift (insertion of C in 586–590)	***reduced susceptibility to (val)aciclovir***	
1	200	G200D or S or C	***reduced susceptibility to (val)aciclovir***	
2	201	G201D	***reduced susceptibility to (val)aciclovir***	
1	201	T201P	***reduced susceptibility to (val)aciclovir***	
1	204	V204G	***reduced susceptibility to (val)aciclovir***	
1	206	G206R	***reduced susceptibility to (val)aciclovir***	
1	207	A207P	***reduced susceptibility to (val)aciclovir***	
1	208	L208H	***reduced susceptibility to (val)aciclovir***	
1	210	E210D	susceptible to (val)aciclovir	
2	210	frameshift (insertion of T at 628)	***reduced susceptibility to (val)aciclovir***	
1	214	I214T	susceptible to (val)aciclovir	
2	215	A215T	susceptible to (val)aciclovir	
1	216	R216H or C	***reduced susceptibility to (val)aciclovir***	
1	219	K219T	susceptible to (val)aciclovir	
1	220	R220C or H	***reduced susceptibility to (val)aciclovir***	
2	220	R220K	susceptible to (val)aciclovir	
2	221	R221H	***reduced susceptibility to (val)aciclovir***	
2	222	Q222stop	***reduced susceptibility to (val)aciclovir***	
1	222	R222C	***reduced susceptibility to (val)aciclovir***	
1	222–224	frameshift (deletion of C in 665–670)	***reduced susceptibility to (val)aciclovir***	
2	223	R223H	***reduced susceptibility to (val)aciclovir***	
1	227	L227F	***reduced susceptibility to (val)aciclovir***	
1	239	Y239S	***reduced susceptibility to (val)aciclovir***	
2	239	Y239stop	***reduced susceptibility to (val)aciclovir***	
1	240	G240E	susceptible to (val)aciclovir	
1	243	A243S	susceptible to (val)aciclovir	
1	245	T245M	***reduced susceptibility to (val)aciclovir***	
1	250	Q250stop	***reduced susceptibility to (val)aciclovir***	
1	251	G251C	susceptible to (val)aciclovir	
2	251	Q251R	***reduced susceptibility to (val)aciclovir***	
1	256	R256W	***reduced susceptibility to (val)aciclovir***	
1	259	W259R	susceptible to (val)aciclovir	
1	260–261	frameshift (deletion of G in 779–782)	***reduced susceptibility to (val)aciclovir***	
1	261	Q261stop	***reduced susceptibility to (val)aciclovir***	
1	265	A265T	susceptible to (val)aciclovir	
1	267	V267L	susceptible to (val)aciclovir	
1	268	P268T or S	susceptible to (val)aciclovir	
2	270–271	frameshift (deletion of C in 808–811)	***reduced susceptibility to (val)aciclovir***	
1	272	A272T	susceptible to (val)aciclovir	
1	273	E273K	susceptible to (val)aciclovir	
1	273	E273Q	susceptible to (val)aciclovir	
2	274	D274G	***reduced susceptibility to (val)aciclovir***	
1	274	P274T	susceptible to (val)aciclovir	
1	275	Q275P	susceptible to (val)aciclovir	

(*Continued*)

TABLE 3 Mutations in the UL23-gene of HSV (*Continued*)

HSV type	AA position	Mutation	Interpretation	Remark
1	276	S276R	susceptible to (val)aciclovir	
1	280	280stop	***reduced susceptibility to (val)aciclovir***	
1	281	R281stop	***reduced susceptibility to (val)aciclovir***	
1	281	unknown	susceptible to (val)aciclovir	
1	285–286	frameshift (deletion of G in 853–856)	***reduced susceptibility to (val)aciclovir***	
1	286	D286E	susceptible to (val)aciclovir	
1	287	T287M	***reduced susceptibility to (val)aciclovir***	
2	288	T288M	***reduced susceptibility to (val)aciclovir***	exact position unclear, possibly T287M
1	289	L298A	***reduced susceptibility to (val)aciclovir***	
1	291	L291P or R	***reduced susceptibility to (val)aciclovir***	
1	293	R293W	susceptible to (val)aciclovir	
1	293–294	frameshift (deletion of G in 878–880)	***reduced susceptibility to (val)aciclovir***	
1	294	A294V	susceptible to (val)aciclovir	
1	295	P295T	susceptible to (val)aciclovir	
1	299–300	frameshift (deletion of C in 896–900)	***reduced susceptibility to (val)aciclovir***	
1	302	G302S	susceptible to (val)aciclovir	
1	315	L315S	***reduced susceptibility to (val)aciclovir***	
1	322	M322L	susceptible to (val)aciclovir	
1	323	H323Y	susceptible to (val)aciclovir	
1	327	L327M	susceptible to (val)aciclovir	
1	328	D328E	susceptible to (val)aciclovir	
1	332	S332P	susceptible to (val)aciclovir	
1	334	A334T	susceptible to (val)aciclovir	
1	336	C336Y	***reduced susceptibility to (val)aciclovir***	
2	336	C337Y	***reduced susceptibility to (val)aciclovir***	exact position unclear, possibly C336Y
1	341	Q341stop	***reduced susceptibility to (val)aciclovir***	
1	345	S345P	susceptible to (val)aciclovir	
1	348	V348I	susceptible to (val)aciclovir	
1	350	T350S	susceptible to (val)aciclovir	
1	355	frameshift (deletion of A at 1065)	***reduced susceptibility to (val)aciclovir***	
1	355	P355Q	susceptible to (val)aciclovir	
1	357	S357C	susceptible to (val)aciclovir	
1	364	L364P	***reduced susceptibility to (val)aciclovir***	
1	365	A365T	***reduced susceptibility to (val)aciclovir***	
1	367	M367T	susceptible to (val)aciclovir	
1	373–374	frameshift (deletion of G in 1117–1121)	***reduced susceptibility to (val)aciclovir***	
1	374	E374A	susceptible to (val)aciclovir	
1	376	N376H or P or T	susceptible to (val)aciclovir	

vir, inhibitors of the NS3/NS4A encoded protease of HCV genotype 1. Later, the second generation protease inhibitors simeprevir (SPV) and asunaprevir became available and showed better tolerability profiles and a higher genetic barrier, i.e., an increased number of RAMs required to result in resistance. Unfortunately, these DAAs still only have limited effect on genotypes other than HCV genotype 1.

A second class of DAAs are available and inhibit the NS5B encoded polymerase. As in HIV, there are nucleoside/NPI and non-nucleoside polymerase inhibitors (NNPI). NNPIs, as for example asabuvir (DAS), have a low genetic barrier to resistance, activity to only HCV genotype 1, and a variable tolerability. NPIs have a broader range of activity to other genotypes, with sofosbuvir (SOF) being the most widely used NPI as it is well tolerated and has a pangenotypic activity (132).

Finally, inhibitors of the NS5A encoded phosphoprotein are now available. The protein is involved in replication and assembly, but has no enzymatic activity. The NS5A inhibitors daclatasvir (DCV) and ledipasvir (LDV) are used in combination with other DAAs and carry a broader genotypic coverage of at least genotypes 1, 3, and 4 (131).

TABLE 4 Mutations in the UL30-gene of HSV

HSV type	AA Position	Mutation	Interpretation	Remark
1	3	S3A	susceptible to (val)aciclovir and foscarnet	cidofovir susceptibility unknown
2	9	A9T	susceptible to (val)aciclovir, cidofovir, and foscarnet	
1	12	G12E	susceptible to (val)aciclovir and foscarnet	cidofovir susceptibility unknown
2	15	P15R	susceptible to (val)aciclovir, cidofovir, and foscarnet	
2	15	S15P	susceptible to (val)aciclovir, cidofovir, and foscarnet	
1	20	A20V	susceptible to cidofovir and foscarnet	(val)aciclovir susceptibility unknown
1	23	F23I	susceptible to (val)aciclovir and foscarnet	cidofovir susceptibility unknown
1	25	A25V	susceptible to (val)aciclovir and foscarnet	cidofovir susceptibility unknown
1	27	A27T	susceptible to cidofovir and foscarnet	(val)aciclovir susceptibility unknown
1	29	P29H	susceptible to (val)aciclovir and foscarnet	cidofovir susceptibility unknown
1	33	G33S	susceptible to (val)aciclovir, cidofovir, and foscarnet	
1	33	S33R	susceptible to (val)aciclovir and foscarnet	cidofovir susceptibility unknown
2	34	Q34H	susceptible to cidofovir and foscarnet	(val)aciclovir susceptibility unknown
2	37	P37L	susceptible to (val)aciclovir and foscarnet	cidofovir susceptibility unknown
2	40	C40W	susceptible to (val)aciclovir and foscarnet	cidofovir susceptibility unknown
2	41	R41H	susceptible to (val)aciclovir, cidofovir, and foscarnet	
1	42	Q42K	susceptible to (val)aciclovir and foscarnet	cidofovir susceptibility unknown
2	43	Q43R	susceptible to (val)aciclovir and foscarnet	cidofovir susceptibility unknown
2	49	H49Y	susceptible to (val)aciclovir and foscarnet	cidofovir susceptibility unknown
2	60	P60L	susceptible to (val)aciclovir and foscarnet	cidofovir susceptibility unknown
1	70	E70D	susceptible to (val)aciclovir, cidofovir, and foscarnet	
1	72	D72N	susceptible to cidofovir and foscarnet	(val)aciclovir susceptibility unknown
1	78	A78D	susceptible to cidofovir and foscarnet	(val)aciclovir susceptibility unknown
1	78	A78V	susceptible to (val)aciclovir, cidofovir, and foscarnet	
1	98	H98Y	***possibly reduced susceptibility to foscarnet***	(val)aciclovir and cidofovir susceptibility unknown
1	102	A102T	susceptible to (val)aciclovir, cidofovir, and foscarnet	
1	104	K104Q	susceptible to cidofovir and foscarnet	(val)aciclovir susceptibility unknown
1	112	R112H	susceptible to (val)aciclovir and foscarnet	cidofovir susceptibility unknown
1	116	R116H	susceptible to (val)aciclovir and foscarnet	cidofovir susceptibility unknown
1	132	G132C	susceptible to (val)aciclovir and foscarnet	cidofovir susceptibility unknown
1	137	P137Q	susceptible to (val)aciclovir and foscarnet	cidofovir susceptibility unknown
1	138	A138V	susceptible to (val)aciclovir and foscarnet	cidofovir susceptibility unknown
2	139	E139K	susceptible to (val)aciclovir, cidofovir, and foscarnet	
1	145	T145A	susceptible to (val)aciclovir and foscarnet	cidofovir susceptibility unknown
1	146	V146I	susceptible to (val)aciclovir and foscarnet	cidofovir susceptibility unknown

(*Continued on next page*)

TABLE 4 Mutations in the UL30-gene of HSV (*Continued*)

HSV type	AA Position	Mutation	Interpretation	Remark
1	164	F164L	susceptible to (val)aciclovir and foscarnet	cidofovir susceptibility unknown
1	171	F171S	susceptible to (val)aciclovir	susceptibility foscarnet and cidofovir unknown
2	203	R203Q	susceptible to (val)aciclovir and foscarnet	cidofovir susceptibility unknown
1	219	R219S	susceptible to cidofovir and foscarnet	(val)aciclovir susceptibility unknown
1	229	A229T	susceptible to cidofovir and foscarnet	(val)aciclovir susceptibility unknown
2	232	A232T	susceptible to (val)aciclovir, cidofovir, and foscarnet	
1	248	F248L	susceptible to (val)aciclovir, cidofovir, and foscarnet	
2	250	E250Q	***reduced susceptibility to (val)aciclovir, susceptible to foscarnet***	cidofovir susceptibility unknown
1	280	S280A	susceptible to (val)aciclovir and foscarnet	cidofovir susceptibility unknown
2	286	N286Y	susceptible to (val)aciclovir, cidofovir, and foscarnet	
1	289	A289D	susceptible to (val)aciclovir and foscarnet	cidofovir susceptibility unknown
2	291	I291V	***reduced susceptibility to foscarnet***	(val)aciclovir and cidofovir susceptibility unknown
2	324	G324E	susceptible to (val)aciclovir and foscarnet	cidofovir susceptibility unknown
1	330	R330A	susceptible to (val)aciclovir, cidofovir, and foscarnet	
1	352	I352V	susceptible to (val)aciclovir and foscarnet	cidofovir susceptibility unknown
2	370	D370N	***reduced susceptibility to (val)aciclovir, susceptible to foscarnet***	cidofovir susceptibility unknown
1	421	E421D	susceptible to (val)aciclovir and foscarnet	cidofovir susceptibility unknown
1	425	N425T	susceptible to (val)aciclovir, cidofovir, and foscarnet	
1	428	A428V	susceptible to (val)aciclovir, cidofovir, and foscarnet	
1	429	A429T	susceptible to cidofovir and foscarnet	(val)aciclovir susceptibility unknown
1	434	T434M	susceptible to (val)aciclovir and foscarnet	cidofovir susceptibility unknown
1	479	L479M	susceptible to (val)aciclovir and foscarnet	cidofovir susceptibility unknown
1	494	N494S	***reduced susceptibility to foscarnet, susceptible to (val)aciclovir and cidofovir***	
1	498	V498G	susceptible to cidofovir and foscarnet	(val)aciclovir susceptibility unknown
1	498	V498M	susceptible to (val)aciclovir and foscarnet	cidofovir susceptibility unknown
2	507	Q507H	susceptible to (val)aciclovir and foscarnet	cidofovir susceptibility unknown
1	507	S507R	susceptible to (val)aciclovir and foscarnet	cidofovir susceptibility unknown
1	522	N522S	susceptible to (val)aciclovir	susceptibility foscarnet and cidofovir unknown
1	561	P561S	susceptible to (val)aciclovir, cidofovir, and foscarnet	
1	562	A562T	susceptible to (val)aciclovir, cidofovir, and foscarnet	
1	566	A566T	susceptible to (val)aciclovir, cidofovir, and foscarnet	
1	573	V573M	susceptible to (val)aciclovir and foscarnet	cidofovir susceptibility varyingly reported
1	577	Y577H	***reduced susceptibility to foscarnet, susceptible to (val)aciclovir***	cidofovir susceptibility unknown
1	597	E597D	***reduced susceptible to (val)aciclovir and foscarnet***	cidofovir susceptibility unknown

1	597	E597K	***reduced susceptible to (val)aciclovir***	foscarnet resistandtie borderline, cidofovir susceptibility unknown
1	599	S599L	***reduced susceptibility to foscarnet***	(val)aciclovir and cidofovir susceptibility unknown
1	605	A605V	***reduced susceptibility to (val)aciclovir and foscarnet, susceptible to cidofovir***	
1	618	Q618H	***reduced susceptibility to (val)aciclovir, susceptible to cidofovir and foscarnet***	
1	639	T639I	susceptible to (val)aciclovir, cidofovir, and foscarnet	
1	645	A645T	susceptible to foscarnet	(val)aciclovir and cidofovir susceptibility unknown
1	646	A646T	susceptible to (val)aciclovir and foscarnet	cidofovir susceptibility unknown
1	657	A657T	***reduced susceptibility to (val)aciclovir, susceptible to cidofovir and foscarnet***	
1	660	D660E	susceptible to (val)aciclovir and foscarnet	cidofovir susceptibility unknown
1	671	D671N	susceptible to foscarnet	(val)aciclovir and cidofovir susceptibility unknown
2	671	D671N	susceptible to (val)aciclovir and foscarnet	cidofovir susceptibility unknown
1	672	D672N	susceptible to (val)aciclovir and cidofovir	foscarnet susceptibility varyingly reported
1	675	E675A	susceptible to (val)aciclovir and foscarnet	cidofovir susceptibility unknown
2	676	D676G	susceptible to (val)aciclovir and foscarnet	cidofovir susceptibility unknown
2	676	D676N	susceptible to (val)aciclovir and foscarnet	cidofovir susceptibility unknown
2	676–677	del DD676-677	susceptible to (val)aciclovir and foscarnet	cidofovir susceptibility unknown
2	677–679	del DED677-679	susceptible to (val)aciclovir and foscarnet	cidofovir susceptibility unknown
2	678	E678G	susceptible to foscarnet	cidofovir susceptibility unknown, (val)aciclovir susceptibility varyingly reported
2	680	G680R	susceptible to (val)aciclovir and foscarnet	cidofovir susceptibility unknown
2	681–682	del GD681-682	susceptible to (val)aciclovir and foscarnet	cidofovir susceptibility unknown
2	682	E682K	susceptible to (val)aciclovir and foscarnet	cidofovir susceptibility unknown
2	683–686	del DGDE683-686	susceptible to (val)aciclovir and foscarnet	cidofovir susceptibility unknown
2	683–686	ins DGDE683-686	susceptible to (val)aciclovir and foscarnet	cidofovir susceptibility unknown
1	684	E684A	susceptible to (val)aciclovir and foscarnet	cidofovir susceptibility unknown
1	684	E684D	susceptible to (val)aciclovir	susceptibility foscarnet and cidofovir unknown
1	688	E688K	susceptible to (val)aciclovir and foscarnet	cidofovir susceptibility unknown
1	700	R700G	***reduced susceptible to (val)aciclovir and foscarnet***	cidofovir susceptibility unknown
1	700	R700M	***reduced susceptibility to cidofovir, susceptible to (val)aciclovir and foscarnet***	
1	702	L702H	***reduced susceptibility to (val)aciclovir and foscarnet, susceptible to cidofovir***	
1	702	L702I	***reduced susceptibility to foscarnet, susceptible to (val)aciclovir and cidofovir***	
1	703	D703N	susceptible to (val)aciclovir and foscarnet	cidofovir susceptibility unknown

(*Continued on next page*)

TABLE 4 Mutations in the UL30-gene of HSV (*Continued*)

HSV type	AA Position	Mutation	Interpretation	Remark
1	711	N711K	susceptible to (val)aciclovir and foscarnet	cidofovir susceptibility unknown
1	711	N711T	susceptible to cidofovir and foscarnet	(val)aciclovir susceptibility unknown
1	714	V714M	***reduced susceptibility to (val)aciclovir and foscarnet, susceptible to cidofovir***	
1	715	V715M	***reduced susceptible to (val)aciclovir, susceptible to foscarnet and cidofovir***	
2	716	D716N	susceptible to (val)aciclovir	susceptibility foscarnet and cidofovir unknown
1	716	F716L	***reduced susceptibility to (val)aciclovir and foscarnet, susceptible to cidofovir***	
1	719	A719T	***reduced susceptible to (val)aciclovir and foscarnet, susceptible to cidofovir***	
1	719	A719V	***reduced susceptibility to (val)aciclovir***	foscarnet and cidofovir susceptibility varyingly reported
2	721	F721L	susceptible to cidofovir and foscarnet	(val)aciclovir susceptibility unknown
2	724	A724T	***reduced susceptibility to foscarnet, susceptible to (val)aciclovir and cidofovir***	foscarnet resistandtie borderline
2	724	A724V	***reduced susceptibility to foscarnet***	(val)aciclovir and cidofovir susceptibility unknown
1	724	S724N	***reduced susceptible to (val)aciclovir and foscarnet***	cidofovir susceptibility varyingly reported
2	729	S729N	***reduced susceptibility to foscarnet, susceptible to (val)aciclovir and cidofovir***	
2	732	Q732R	***reduced susceptibility to foscarnet***	(val)aciclovir and cidofovir susceptibility unknown
1	740	A740S	susceptible to (val)aciclovir and foscarnet	cidofovir susceptibility unknown
1	743	V743M	susceptible to cidofovir and foscarnet	(val)aciclovir susceptibility unknown
2	746	E746K	susceptible to cidofovir and foscarnet	(val)aciclovir susceptibility unknown
1	749	G749D	susceptible to (val)aciclovir and foscarnet	cidofovir susceptibility unknown
1	753	L753M	susceptible to (val)aciclovir and foscarnet	cidofovir susceptibility unknown
1	771	E771Q	***reduced susceptibility to (val)aciclovir, susceptible to foscarnet***	cidofovir susceptibility unknown
1	775	S775N	***reduced susceptibility to (val)ciclovir, susceptible to foscarnet***	cidofovir susceptibility unknown
1	778	L778M	***reduced susceptible to (val)aciclovir and foscarnet***	cidofovir susceptibility varyingly reported
1	780	D780N	***reduced susceptible to (val)aciclovir and foscarnet, susceptible to cidofovir***	
1	782	L782I	***reduced susceptible to (val)aciclovir and foscarnet, susceptible to cidofovir***	
2	783	L783M	***reduced susceptibility to foscarnet, susceptible to (val)aciclovir***	susceptibility to cidofovir unclear
2	785	D785N	susceptible to (val)aciclovir and cidofovir	foscarnet susceptibility varyingly reported
1	797	P797T	***reduced susceptible to (val)aciclovir and foscarnet***	cidofovir susceptibility unknown
1	798	E798K		

			reduced susceptibility to (val)aciclovir and foscarnet, susceptible to cidofovir	
2	799	Q799R	susceptible to (val)aciclovir and foscarnet	cidofovir susceptibility unknown
2	801	P801T	susceptible to (val)aciclovir, cidofovir, and foscarnet	
1	802	L802F	***reduced susceptibility to (val)aciclovir, cidofovir, and foscarnet***	
1	805	K805Q	susceptible to (val)aciclovir, cidofovir, and foscarnet	
1	813	V813M	***reduced susceptible to (val)aciclovir and foscarnet***	cidofovir susceptibility unknown
1	815	N815S	***reduced susceptible to (val)aciclovir, susceptible to foscarnet and cidofovir***	
1	817	V817M	***reduced susceptibility to (val)aciclovir and foscarnet, susceptible to cidofovir***	
1	818	Y818C	***reduced susceptibility to (val)gancivloir, cidofovir, and foscarnet***	
1	821	T821M	***reduced susceptible to (val)aciclovir and cidofovir***	foscarnet susceptibility varyingly reported
2	829	Q829R	***reduced susceptibility to (val)aciclovir, susceptible to foscarnet***	cidofovir susceptibility unknown
1	834	A834S	***reduced susceptibility to foscarnet, susceptible to (val)aciclovir and cidofovir***	
2	837	H837R	***reduced susceptibility to foscarnet***	(val)aciclovir and cidofovir susceptibility unknown
1	839	T839I	***reduced susceptibility to (val)aciclovir and foscarnet, susceptible to cidofovir***	
1	841	G841C	***reduced susceptibility to (val)aciclovir and foscarnet, susceptible to cidofovir***	
1	841	G841C	***reduced susceptibility to (val)aciclovir and foscarnet, susceptible to cidofovir***	
1	841	G841S	***reduced susceptibility to (val)aciclovir, susceptible to cidofovir and foscarnet***	foscarnet susceptibility varyingly reported
1	842	R842S	***reduced susceptibility to (val)aciclovir and foscarnet***	cidofovir susceptibility unknown
2	844	T844I	***reduced susceptibility to foscarnet***	(val)aciclovir and cidofovir susceptibility unknown
1	846	L846I	susceptible to (val)aciclovir, cidofovir, and foscarnet	
2	850	L850I	***reduced susceptibility to foscarnet, susceptible to (val)aciclovir and cidofovir***	
1	870	A870G	***reduced susceptibility to foscarnet***	(val)aciclovir and cidofovir susceptibility unknown
1	871	D871N	susceptible to cidofovir and foscarnet	(val)aciclovir susceptibility unknown
1	875	P875S	susceptible to (val)aciclovir and foscarnet	cidofovir susceptibility unknown
2	880	P880H	susceptible to (val)aciclovir, cidofovir, and foscarnet	
1	891	F891C	***reduced susceptible to (val)aciclovir and foscarnet, susceptible to cidofovir***	
2	903	T903M	susceptible to (val)aciclovir and foscarnet	cidofovir susceptibility unknown

(Continued on next page)

TABLE 4 Mutations in the UL30-gene of HSV (*Continued*)

HSV type	AA Position	Mutation	Interpretation	Remark
2	904	A904G	susceptible to (val)aciclovir	susceptibility foscarnet and cidofovir unknown
1	904	V904M	susceptible to (val)aciclovir and foscarnet	cidofovir susceptibility unknown
2	905	A905E	susceptible to (val)aciclovir	susceptibility foscarnet and cidofovir unknown
1	905	V905M	susceptible to (val)aciclovir, cidofovir, and foscarnet	
2	906	G906A	susceptible to (val)aciclovir	susceptibility foscarnet and cidofovir unknown
1	906	M906V	susceptible to (val)aciclovir	susceptibility foscarnet and cidofovir unknown
1	907	D907V	***reduced susceptible to (val)aciclovir and foscarnet, susceptible to cidofovir***	
1	908	K908R	susceptible to (val)aciclovir	susceptibility foscarnet and cidofovir unknown
1	910	A910V	***reduced susceptibility to foscarnet, susceptible to (val)aciclovir and cidofovir***	
2	912	D912V	***reduced susceptibility to foscarnet, susceptible to cidofovir***	(val)aciclovir susceptibility unknown
1	914–916	SRA914-916LCV	***reduced susceptibility to (val)aciclovir and foscarnet, susceptible to cidofovir***	
1	916	A916T	susceptible to (val)aciclovir, cidofovir, and foscarnet	
1	920	P920S	susceptible to (val)aciclovir, cidofovir, and foscarnet	
1	922	I922N	***reduced susceptibility to foscarnet, susceptible to cidofovir***	(val)aciclovir susceptibility unknown
1	941	Y941H	***reduced susceptible to (val)aciclovir, susceptible to foscarnet and cidofovir***	
1	948	G948C	susceptible to cidofovir and foscarnet	(val)aciclovir susceptibility unknown
1	958	V958L	***reduced susceptibility to (val)aciclovir and foscarnet, susceptible to cidofovir***	
1	959	R959H	***reduced susceptibility to (val)aciclovir and foscarnet, susceptible to cidofovir***	
1	959	R959H	***reduced susceptibility to (val)aciclovir and foscarnet, susceptible to cidofovir***	
1	961	N961K	***reduced susceptible to (val)aciclovir, susceptible to cidofovir***	foscarnet susceptibility varyingly reported
2	987	T987S	susceptible to (val)aciclovir and foscarnet	cidofovir susceptibility unknown
1	995	A995T	susceptible to (val)aciclovir and foscarnet	cidofovir susceptibility unknown
1	995	L995F	susceptible to cidofovir and foscarnet	(val)aciclovir susceptibility unknown
2	1000	A1000T	susceptible to (val)aciclovir and foscarnet	cidofovir susceptibility unknown
1	1005	E1005K	susceptible to (val)aciclovir and cidofovir	foscarnet susceptibility varyingly reported
1	1007	L1007M	susceptible to (val)aciclovir and foscarnet	cidofovir susceptibility varyingly reported
1	1012	A1012V	susceptible to (val)aciclovir and foscarnet	cidofovir susceptibility unknown
1	1019	R1019G	susceptible to (val)aciclovir and foscarnet	cidofovir susceptibility unknown
1	1020	R1020C	susceptible to cidofovir and foscarnet	(val)aciclovir susceptibility unknown

1	1026	I1026S	susceptible to cidofovir and foscarnet	(val)aciclovir susceptibility unknown
1	1028	I1028T	***reduced susceptibility to cidofovir, susceptible to (val)aciclovir and foscarnet***	
1	1077	V1077L	susceptible to (val)aciclovir and foscarnet	cidofovir susceptibility unknown
1	1082	E1082K	susceptible to (val)aciclovir and foscarnet	cidofovir susceptibility unknown
2	1083	A1083T	susceptible to (val)aciclovir and foscarnet	cidofovir susceptibility unknown
1	1086	T1086M	susceptible to (val)aciclovir, cidofovir, and foscarnet	
1	1099	A1099T	susceptible to (val)aciclovir and foscarnet	cidofovir susceptibility unknown
1	1100	A1100T	susceptible to (val)aciclovir, cidofovir, and foscarnet	
1	1103	D1103H	susceptible to cidofovir and foscarnet	(val)aciclovir susceptibility unknown
2	1103–1111	del PGDEPA1106-1111	susceptible to (val)aciclovir and foscarnet	cidofovir susceptibility unknown
1	1104	E1104D	susceptible to (val)aciclovir, cidofovir, and foscarnet	
1	1113	S1113C	susceptible to (val)aciclovir and foscarnet	cidofovir susceptibility unknown
2	1114	A1114V	susceptible to (val)aciclovir and foscarnet	cidofovir susceptibility unknown
1	1123	S1123A	susceptible to (val)aciclovir and foscarnet	cidofovir susceptibility unknown
1	1123	S1123L	susceptible to (val)aciclovir, cidofovir, and foscarnet	
1	1124	H1124P	susceptible to (val)aciclovir, cidofovir, and foscarnet	
1	1128	P1128L	susceptible to (val)aciclovir and foscarnet	cidofovir susceptibility unknown
1	1129	G1129V	susceptible to (val)aciclovir and foscarnet	cidofovir susceptibility unknown
1	1147	A1147T	susceptible to cidofovir and foscarnet	(val)aciclovir susceptibility unknown
2	1149	E1149G	susceptible to (val)aciclovir and foscarnet	cidofovir susceptibility unknown
1	1166	L1166W	susceptible to (val)aciclovir, cidofovir, and foscarnet	
1	1168	A1168V	susceptible to cidofovir and foscarnet	(val)aciclovir susceptibility unknown
1	1175	A1175D	susceptible to (val)aciclovir and foscarnet	cidofovir susceptibility unknown
1	1181	A1181T	susceptible to (val)aciclovir and foscarnet	cidofovir susceptibility unknown
1	1194	E1194D	susceptible to (val)aciclovir and foscarnet	cidofovir susceptibility unknown
1	1199	P1199Q	susceptible to (val)aciclovir, cidofovir, and foscarnet	
1	1203	A1203T	susceptible to (val)aciclovir, cidofovir, and foscarnet	
1	1204	A1204T	susceptible to (val)aciclovir and foscarnet	cidofovir susceptibility unknown
1	1208	A1208T	susceptible to (val)aciclovir, cidofovir, and foscarnet	
1	1209	A1209T	susceptible to (val)aciclovir and foscarnet	cidofovir susceptibility unknown
1	1219	T1219M	susceptible to (val)aciclovir and foscarnet	cidofovir susceptibility unknown
2	1223	A1223V	susceptible to (val)aciclovir and foscarnet	cidofovir susceptibility unknown
1	1229	R1229I	susceptible to (val)aciclovir	cidofovir susceptibility unknown, foscarnet susceptibility varyingly reported
1	1234	L1234I	susceptible to (val)aciclovir, cidofovir, and foscarnet	

TABLE 5 Mutations in the TK gene of VZV

AA position	Mutation	Interpretation
5–101	frameshift (deletion of nt 14–303)	***reduced susceptibility to (val)aciclovir***
16–17	frameshift (deletion of ATTT at 47–50)	***reduced susceptibility to (val)aciclovir***
24	G24E	***reduced susceptibility to (val)aciclovir***
24	G24R	***reduced susceptibility to (val)aciclovir***
24–26	frameshift (deletion of A in 72–76)	***reduced susceptibility to (val)aciclovir***
25	K25R	***reduced susceptibility to (val)aciclovir***
41	N41S	susceptible to (val)aciclovir
48	E48G	***reduced susceptibility to (val)aciclovir***
59	E59G	***reduced susceptibility to (val)aciclovir***
86	T86A	***reduced susceptibility to (val)aciclovir***
90	Q90stop	***reduced susceptibility to (val)aciclovir***
125–126	frameshift (deletion of TA at 375–376)	***reduced susceptibility to (val)aciclovir***
129	D129N	***reduced susceptibility to (val)aciclovir***
130	R130Q	***reduced susceptibility to (val)aciclovir***
130	R130Q	***reduced susceptibility to (val)aciclovir***
138	C138R	***reduced susceptibility to (val)aciclovir***
138	frameshift (insertion of TA at 412–413)	***reduced susceptibility to (val)aciclovir***
143	R143G	***reduced susceptibility to (val)aciclovir***
143	R143G	***reduced susceptibility to (val)aciclovir***
143	R143K	***reduced susceptibility to (val)aciclovir***
154	L154P	***reduced susceptibility to (val)aciclovir***
165–166	frameshift (deletion of C in 493–498)	***reduced susceptibility to (val)aciclovir***
165–166	frameshift (insertion of C in 493–498)	***reduced susceptibility to (val)aciclovir***
179	S179N	susceptible to (val)aciclovir
179–180	frameshift (deletion of nt 535–539)	***reduced susceptibility to (val)aciclovir***
192	E192stop	***reduced susceptibility to (val)aciclovir***
214	frameshift (deletion of T at 641)	***reduced susceptibility to (val)aciclovir***
225	225stop	***reduced susceptibility to (val)aciclovir***
225	W225R	***reduced susceptibility to (val)aciclovir***
226	frameshift (deletion of AC at 677–678)	***reduced susceptibility to (val)aciclovir***
227–228	frameshift (deletion of AC at 681–682)	***reduced susceptibility to (val)aciclovir***
228	frameshift (deletion of CT at 682–683)	***reduced susceptibility to (val)aciclovir***
256	T256A	***reduced susceptibility to (val)aciclovir***
266	C266I	susceptible to (val)aciclovir
269	frameshift (deletion of GA at 805–806)	***reduced susceptibility to (val)aciclovir***
288	S288L	susceptible to (val)aciclovir
297	frameshift (insertion of TC at 889–890)	***reduced susceptibility to (val)aciclovir***
303	Q303stop	***reduced susceptibility to (val)aciclovir***
308	E308Q	susceptible to (val)aciclovir
317	frameshift (deletion of C at 950)	***reduced susceptibility to (val)aciclovir***
319	A319V	susceptible to (val)aciclovir
329–330	frameshift (insertion of GAAA at 987–990)	***reduced susceptibility to (val)aciclovir***
332	L332P	***reduced susceptibility to (val)aciclovir***

Following the implementation of DAAs, treatment guidelines for HCV are changing. Clinical trials show the success of combination therapy of different classes DAA where a high percentages of SVR is achieved. However, combination with RBV and or PEG-IFN treatment is still required for SVR with some HCV genotypes (130).

HCV infection results in an extremely high replication rate of a trillion genomes a day and with the high mutation rate of its polymerase, the HCV genome pool is likely to accumulate mutations that cause resistance upon treatment. Protease inhibitors have been shown to have a low genetic barrier towards selection of resistance mutations. Most of these mutations are detected between amino acids 36 and 170 of the protein and, as a result, sequence analysis of approximately 400 bp of the gene will provide information on the presence of RAMs (133, 134). For SPV, a well-tolerated protease inhibitor (PI) for HCV genotype 1, a very general Q80K mutation resulting in reduced susceptibility is present

TABLE 6 Mutations in the POL gene of VZV

AA position	Mutation	Interpretation	Remark
186	G186C	susceptible to foscarnet	(val)aciclovir susceptibility unknown
286	M286I	susceptible to foscarnet	(val)aciclovir susceptibility unknown
298	Q298K	susceptible to foscarnet	(val)aciclovir and cidofovir susceptibility unknown
512	E512K	***reduced susceptibility to foscarnet***	(val)aciclovir and cidofovir susceptibility unknown
662	K662E	***reduced susceptibility to (val)aciclovir, cidofovir, and foscarnet***	effect on cidofovir and aciclovir seems limited
665	R665G	***reduced susceptibility to foscarnet***	(val)aciclovir and cidofovir susceptibility unknown
666	V666L	***reduced susceptibility to (val)aciclovir, cidofovir, and foscarnet***	effect on cidofovir seems limited
668	D668Y	***reduced susceptibility to (val)ciclovir and cidofovir, susceptible to foscarnet***	effect on cidofovir and aciclovir seems limited
692	Q692R	***reduced susceptibility to foscarnet***	(val)aciclovir and cidofovir susceptibility unknown
762	E762D	susceptible to (val)aciclovir, foscarnet, and cidofovir	
767	L767S	***reduced susceptibility to (val)aciclovir, cidofovir, and foscarnet***	effect on cidofovir seems limited
779	N779S	susceptible to foscarnet	(val)aciclovir susceptibility unknown
805	G805C	***reduced susceptibility to foscarnet***	(val)aciclovir susceptibility unknown
806	R806S	***reduced susceptibility to foscarnet***	(val)aciclovir and cidofovir susceptibility unknown
808	M808V	***reduced susceptibility to foscarnet***	(val)aciclovir and cidofovir susceptibility unknown
809	L809S	***reduced susceptibility to foscarnet***	(val)aciclovir and cidofovir susceptibility unknown
824	E824Q	susceptible to foscarnet	(val)aciclovir susceptibility unknown
855	V855M	***reduced susceptibility to foscarnet***	(val)aciclovir susceptibility unknown
863	S863G	susceptible to (val)aciclovir and foscarnet	susceptibility cidofovir unknown
984	R984H	susceptible to foscarnet	(val)aciclovir susceptibility unknown
1089	H1089Y	susceptible to foscarnet	(val)aciclovir susceptibility unknown
1095	L1095M	susceptible to foscarnet	(val)aciclovir and cidofovir susceptibility unknown
1159	C1159R	susceptible to foscarnet	(val)aciclovir susceptibility unknown

in about 30% of the HCV genotype 1 strains. Therefore, testing for the presence of this mutation is indicated before treating with SPV (134, 135). Also, the initial NS5B polymerase inhibitors had a low genetic barrier for selection of RAMs, with the S282T being the most general mutation observed. Fortunately, most RAMs also result in a reduction of the replication fitness of the virus. The recently licensed SOF, however, shows an excellent activity against most genotypes and appears to have a higher genetic barrier to resistance (135, 136).

NS5A inhibitors like DCV and LDV accumulate RAMs that are located between amino acids 28 and 93 of the protein (135, 137), which can be easily analyzed by nucleotide sequence analysis of less than 250 nucleotides of the genome.

In summary, antiviral resistance testing for DAAs is based on nucleotide sequence analysis of a relatively small part of the genome. Because of on-going improvements in reaching a SVR by combination therapy in an increasing number of HCV patients, resistance testing may not be indicated for most patients, with the exception of Q80K testing in HCV genotype 1 infected patients if use of SPV is considered (136, 138). However, SPV is not present in the recently released DAA combinations. Oral only, interferon-free DAA combinations as Harvoni (SOF/LDP) (139) and the three dose Viekira Pak that contains newer DAAs such as paritaprevir (PTV-R), ombitasvir (OMB) and DAS further improve treatment for HCV infected patients. With more DAAs in development, and the treatment success of current combinations already up to 99% in combination with ribavirin (140), resistance testing for HCV will only be indicated for a limited number of patients.

HIV

Monitoring drug resistance is part of standard care for HIV-infected patients in the Western world. Phenotypic testing by measuring the inhibition of replication of HIV by antiviral drugs provides the most detailed information. In practice, genes or mutations of interest are cloned into recombinant or pseudoviruses and their effect on replication is measured (141). In 1998, the Antivirogram Phenotype assay (Virco, Mechelen, Belgium) was launched, but discontinued some years ago as it was replaced by the VircoTYPE HIV-1 assay to determine the virtual phenotype based on genomic sequences. The PhenosenseGT plus Integrase (Monogram Biosciences, San Francisco, CA) is still being offered to determine the phenotype of HIV strains. The same company also offers an assay (Phenosense Entry) to determine resistance to the entry inhibitor Fuzeon. In addition, assays (Trofile and Trofile DNA) are available to determine the tropism of the virus, required if Maraviroc is considered as the drug is only active for viruses using the CCR5 coreceptor and not the CXCR4 coreceptor. Actual phenotypic assays are laborious and largely replaced by genotypic assays.

Sequence analysis of the *pol* gene (coding for protease, reverse transcriptase, and integrase proteins) and parts of the *env* gene (coding for structural envelope proteins) provide information on resistance against all classes of antiretroviral therapy. Most widely used commercial kits are the Trugene HIV-1 genotyping kit (142, 143) (Siemens Healthcare), the Viroseq HIV-1 Genotyping system (144, 145) (Abbott Molecular), and the GenoSure PRIme and MG assays (Monogram Biosciences). In addition, in-house assays have been used as well (146, 147). Once a nucleotide sequence has been determined, many web-based tools are available for interpretation of the results (148–150). Currently, a large number of antivirals are present for treatment of HIV patients and information on induction of RAMs and their role in the development of resistance is accumulating. Therefore, web-based algorithms need frequent updates with the latest insights as provided by the International Antiviral Society on a yearly basis (151). Using conventional sequencing, minor variants (MVs) up to 25% can be reliably detected. Recently, NGS has been applied for deep sequencing of HIV and provides information on MVs below 1% (152, 153). However, with current technology, only very short fragments are being sequenced which makes it difficult to interpret these MV mutations. The MV RAMs that are detected may not be present on viable viral genomes and thus have no effect on the therapy of the patient. However, recently it was shown that pre-existing MVs increase the risk of virological failure (154).

Discussion

With an increase of treatment with antiviral agents and the development of new antivirals, testing for antiviral susceptibility will rapidly evolve as a routine diagnostic tool, most likely for specialized laboratories only. The recent revolution in HCV treatment may prove beneficial for treatment of other flaviviruses. Dengue virus, for example, is an important flavivirus with increased spread over the world. An estimate of 50 to 100 million infections annually with half a million severe cases (155, 156) shows the necessity of exploring possibilities to treat dengue virus infections. Currently, no antiviral treatment is available for this infection, nor for other, sometimes fatal, infections of flaviviruses such as West Nile virus, Yellow Fever virus, and Japanese encephalitis virus. The new DAAs for hepatitis C virus have been developed to some targets that are also present in other emerging flaviviruses. There is no guarantee that successful treatment protocols will be developed for these viruses and, therefore, it is too early to speculate on potential requirements for antiviral resistance testing. If testing is indicated, it will follow the general strategy as outlined in this chapter, with focus on nucleotide sequence analysis of the gene products most suitable as targets for treatment (7). A similar phenomenon was observed for favipiravir, developed for treatment of influenza but potentially effective for treatment and prophylaxis of Ebola infection (157).

For enteroviruses, antivirals may be needed to enable the final step of polio eradication. Apart from polio, other enteroviruses can cause severe infections for which treatment would be indicated. Human rhinoviruses, another species of the enterovirus genus, are the most frequently detected pathogen in acute respiratory infections for which broad reactive antivirals would be both clinically and economically beneficial. So far, compounds have been developed that are directed against the capsid of these viruses with Pleconaril as a known example. Despite promising efficacy, Pleconaril has some safety issues preventing it from being FDA cleared. Unfortunately, the capsid proteins are the most variable proteins in this virus group and mutations are quite common. Therefore, resistance may readily develop following monotherapy with this kind of antiviral drug.

Altogether, the next decade will greatly enhance the possibilities for treatment of viral infections and, therefore, potentially increase the need for antiviral resistance testing. Sequence analysis will be the most important method and, with increasing technical developments, next-generation sequencing will most likely play an important role in the near future.

REFERENCES

1. **Davidson SI.** 1962. 5-Iodo-2-deoxy-uridine in herpetic keratitis. *Lancet* **2**:1326–1327.
2. **Richman DD, Fischl MA, Grieco MH, Gottlieb MS, Volberding PA, Laskin OL, Leedom JM, Groopman JE, Mildvan D, Hirsch MS, Jackson GG, Durack DT, Nusinoff-Lehrman S.** 1987. The toxicity of azidothymidine (AZT) in the treatment of patients with AIDS and AIDS-related complex. A double-blind, placebo-controlled trial. *N Engl J Med* **317**:192–197.
3. **Larder BA, Darby G, Richman DD.** 1989. HIV with reduced sensitivity to zidovudine (AZT) isolated during prolonged therapy. *Science* **243**:1731–1734.
4. **Bach MC.** 1989. Failure of zidovudine to maintain remission in patients with AIDS. *N Engl J Med* **320**:594–595.
5. **Zumla A, Memish ZA, Maeurer M, Bates M, Mwaba P, J Al-Tawfiq A, Denning DW, Hayden FG, Hui DS.** 2014. Emerging novel and antimicrobial-resistant respiratory tract infections: new drug development and therapeutic options. *Lancet Infect Dis* **14**:1136–1149.
6. **Hilgenfeld R.** 2014. From SARS to MERS: crystallographic studies on coronaviral proteases enable antiviral drug design. *FEBS J* **281**:4085–4096
7. **Debing Y, Jochmans D, Neyts J.** 2013. Intervention strategies for emerging viruses: use of antivirals. *Curr OpinVirol.* **3**:217–224.
8. **Reed LJ, Muench H.** 1938. A simple method for estimating fifty percent end-points. *Am J Hyg* **27**:493–497.
9. **van der Beek MT, Claas EC, van der Blij-de Brouwer CS, Morfin F, Rusman LG, Kroes AC, Vossen AC.** 2013. Rapid susceptibility testing for herpes simplex virus type 1 using real-time PCR. *J Clin Virol* **56**:19–24.
10. **Wolters LM, Niesters HG, Hansen BE, van der Ende ME, Kroon FP, Richter C, Brinkman K, Meenhorst PL, de Man RA.** 2002. Development of hepatitis B virus resistance for lamivudine in chronic hepatitis B patients co-infected with the human immunodeficiency virus in a Dutch cohort. *J ClinVirol* **24**:173–181.
11. **Stuyver L, Wyseur A, Rombout A, Louwagie J, Scarcez T, Verhofstede C, Rimland D, Schinazi RF, Rossau R.** 1997. Line probe assay for rapid detection of drug-selected mutations in the human immunodeficiency virus type 1 reverse transcriptase gene. *Antimicrob Agents Chemother* **41**:284–291.
12. **Liu CM, Driebe EM, Schupp J, Kelley E, Nguyen JT, McSharry JJ, Weng Q, Engelthaler DM, Keim PS.** 2010. Rapid quantification of single-nucleotide mutations in mixed influenza A viral populations using allele-specific mixture analysis. *J Virol Methods* **163**:109–115.
13. **Lataillade M, Chiarella J, Yang R, Schnittman S, Wirtz V, Uy J, Seekins D, Krystal M, Mancini M, McGrath D, Simen B, Egholm M, Kozal M.** 2010. Prevalence and clinical significance of HIV drug resistance mutations by ultra-deep sequencing in antiretroviral-naïve subjects in the CASTLE study. *PLoS One* **5**:e10952.
14. **Dudley DM, Chin EN, Bimber BN, Sanabani SS, Tarosso LF, Costa PR, Sauer MM, Kallas EG, O'Connor DH.** 2012. Low-cost ultra-wide genotyping using Roche/454 pyrosequencing for surveillance of HIV drug resistance. *PLoS One* **7**:e36494.

15. **Weller S, Blum MR, Doucette M, Burnette T, Cederberg DM, de Miranda P, Smiley ML.** 1993. Pharmacokinetics of the acyclovir pro-drug valaciclovir after escalating single- and multiple-dose administration to normal volunteers. *Clin Pharmacol Ther* **54:**595–605.
16. **Glenny AM, Fernandez Mauleffinch LM, Pavitt S, Walsh T.** 2009. Interventions for the prevention and treatment of herpes simplex virus in patients being treated for cancer. *Cochrane Database Syst Rev* **(1):**CD006706.
17. **Brown F, Banken L, Saywell K, Arum I.** 1999. Pharmacokinetics of valganciclovir and ganciclovir following multiple oral dosages of valganciclovir in HIV- and CMV-seropositive volunteers. *Clin Pharmacokinet* **37:**167–176.
18. **Hodson EM, Barclay PG, Craig JC, Jones C, Kable K, Strippoli GF, Vimalachandra D, Webster AC.** 2005. Antiviral medications for preventing cytomegalovirus disease in solid organ transplant recipients. *Cochrane Database Syst Rev* **(4):**CD003774.
19. **Matthews T, Boehme R.** 1988. Antiviral activity and mechanism of action of ganciclovir. *Rev Infect Dis* **10**(Suppl 3)**:** S490–S494.
20. **Strippoli GF, Hodson EM, Jones CJ, Craig JC.** 2006. Preemptive treatment for cytomegalovirus viraemia to prevent cytomegalovirus disease in solid organ transplant recipients. *Cochrane Database Syst Rev* **(1):**CD005133.
21. **Zhang LF, Wang YT, Tian JH, Yang KH, Wang JQ.** 2011. Preemptive versus prophylactic protocol to prevent cytomegalovirus infection after renal transplantation: a meta-analysis and systematic review of randomized controlled trials. *Transpl Infect Dis* **13:**622–632.
22. **Winston DJ, Saliba F, Blumberg E, Abouljoud M, Garcia-Diaz JB, Goss JA, Clough L, Avery R, Limaye AP, Ericzon BG, Navasa M, Troisi RI, Chen H, Villano SA, Uknis ME.1263-301 Clinical Study Group.** 2012. Efficacy and safety of maribavir dosed at 100 mg orally twice daily for the prevention of cytomegalovirus disease in liver transplant recipients: a randomized, double-blind, multicenter controlled trial. *Am J Transplant* **12:**3021–3030.
23. **Marty, FM, Ljungman P, Papanicolaou GA, Winston DJ, Chemaly RF, Strasfeld L, Young JA, Rodriguez T, Maertens J, Schmitt M, Einsele H, Ferrant A, Lipton JH, Villano SA, Chen H, Boeckh M.** 2011. Maribavir prophylaxis for prevention of cytomegalovirus disease in recipients of allogeneic stem-cell transplants: a phase 3, double-blind, placebo-controlled, randomised trial. *Lancet Infect Dis* **11:**284–292.
24. **Winston DJ, Young JA, Pullarkat V, Papanicolaou GA, Vij R, Vance E, Alangaden GJ, Chemaly RF, Petersen F, Chao N, Klein J, Sprague K, Villano SA, Boeckh M.** 2008. Maribavir prophylaxis for prevention of cytomegalovirus infection in allogeneic stem cell transplant recipients: a multicenter, randomized, double-blind, placebo-controlled, dose-ranging study. *Blood* **111:**5403–5410.
25. **Marty FM, Boeckh M.** 2011. Maribavir and human cytomegalovirus-what happened in the clinical trials and why might the drug have failed? *Curr OpinVirol* **1:**555–562.
26. **Cole NL, Balfour HH Jr.** 1986. Varicella-Zoster virus does not become more resistant to acyclovir during therapy. *J Infect Dis* **153:**605–608.
27. **Danve-Szatanek C, Aymard M, Thouvenot D, Morfin F, Agius G, Bertin I, Billaudel S, Chanzy B, Coste-Burel M, Finkielsztejn L, Fleury H, Hadou T, Henquell C, Lafeuille H, Lafon ME, Le Faou A, Legrand MC, Maille L, Mengelle C, Morand P, Morinet F, Nicand E, Omar S, Picard B, Pozzetto B, Puel J, Raoult D, Scieux C, Segondy M, Seigneurin JM, Teyssou R, Zandotti C.** 2004. Surveillance network for herpes simplex virus resistance to antiviral drugs: 3-year follow-up. *J Clin Microbiol* **42:**242–249.
28. **Ozaki T, Nishimura N, Kajita Y, Ida M, Shiraki K.** 1998. Susceptibilities to aciclovir in viral isolates from children with varicella. *Arch Dis Child* **78:**95.
29. **Stránská R, Schuurman R, Nienhuis E, Goedegebuure IW, Polman M, Weel JF, Wertheim-Van Dillen PM, Berkhout RJ, van Loon AM.** 2005. Survey of acyclovir-resistant herpes simplex virus in the Netherlands: prevalence and characterization. *J Clin Virol* **32:**7–18.
30. **Piret J, Boivin G.** 2011. Resistance of herpes simplex viruses to nucleoside analogues: mechanisms, prevalence, and management. *Antimicrob Agents Chemother* **55:**459–472.
31. **van der Beek MT.** 2012. Doctoral thesis. *Herpesvirus infections in immunocompromised patients: treatment, treatment failure and antiviral resistance*, Leiden University. ISBN/EAN: 978-90-9026850-7
32. **Clinical and Laboratory Standards Institute.** *Antiviral Susceptibility Testing: Herpes Simplex Virus by Plaque Reduction Assay; Approved Standard* M33-A. 2004. CLSI, Wayne, PA.
33. **Chou S, Waldemer RH, Senters AE, Michels KS, Kemble GW, Miner RC, Drew WL.** 2002. Cytomegalovirus UL97 phosphotransferase mutations that affect susceptibility to ganciclovir. *J Infect Dis* **185:**162–169.
34. **Erice A, Gil-Roda C, Pérez JL, Balfour HH Jr, Sannerud KJ, Hanson MN, Boivin G, Chou S.** 1997. Antiviral susceptibilities and analysis of UL97 and DNA polymerase sequences of clinical cytomegalovirus isolates from immunocompromised patients. *J Infect Dis* **175:**1087–1092.
35. **Jabs, DA, Martin BK, Forman MS, Dunn JP, Davis JL, Weinberg DV, Biron KK, Baldanti F.** 2001. Mutations conferring ganciclovir resistance in a cohort of patients with acquired immunodeficiency syndrome and cytomegalovirus retinitis. *J Infect Dis* **183:**333–337.
36. **Lurain NS, Bhorade SM, Pursell KJ, Avery RK, Yeldandi VV, Isada CM, Robert ES, Kohn DJ, Arens MQ, Garrity ER, Taege AJ, Mullen MG, Todd KM, Bremer JW, Yen-Lieberman B.** 2002. Analysis and characterization of antiviral drug-resistant cytomegalovirus isolates from solid organ transplant recipients. *J Infect Dis* **186:**760–768.
37. **Myhre HA, Haug Dorenberg D, Kristiansen KI, Rollag H, Leivestad T, Asberg A, Hartmann A.** 2011. Incidence and outcomes of ganciclovir-resistant cytomegalovirus infections in 1244 kidney transplant recipients. *Transplantation* **92:**217–223.
38. **Smith IL, Cherrington JM, Jiles RE, Fuller MD, Freeman WR, Spector SA.** 1997. High-level resistance of cytomegalovirus to ganciclovir is associated with alterations in both the UL97 and DNA polymerase genes. *J Infect Dis* **176:**69–77.
39. **Göhring K, Mikeler E, Jahn G, Rohde F, Hamprecht K.** 2008. Rapid semiquantitative real-time PCR for the detection of human cytomegalovirus UL97 mutations conferring ganciclovir resistance. *Antivir Ther* **13:**461–466.
40. **Liu JB, Zhang Z.** 2008. Development of SYBR Green I-based real-time PCR assay for detection of drug resistance mutations in cytomegalovirus. *J Virol Methods* **149:**129–135.
41. **Yeo AC, Chan KP, Kumarasinghe G, Yap HK.** 2005. Rapid detection of codon 460 mutations in the UL97 gene of ganciclovir-resistant cytomegalovirus clinical isolates by real-time PCR using molecular beacons. *Mol Cell Probes* **19:**389–393.
42. **Posthuma CC, van der Beek MT, van der Blij-de Brouwer CS, van der Heiden PL, Marijt EW, Spaan WJ, Claas EC, Nederstigt C, Vossen AC, Snijder EJ, Kroes AC.** 2011. Mass spectrometry-based comparative sequencing to detect ganciclovir resistance in the UL97 gene of human cytomegalovirus. *J Clin Virol* **51:**25–30.
43. **Görzer I, Guelly C, Trajanoski S, Puchhammer-Stöckl E.** 2010. Deep sequencing reveals highly complex dynamics of human cytomegalovirus genotypes in transplant patients over time. *J Virol* **84:**7195–7203.
44. **Burrel S, Deback C, Agut H, Boutolleau D.** 2010. Genotypic characterization of UL23 thymidine kinase and UL30 DNA polymerase of clinical isolates of herpes simplex virus: natural polymorphism and mutations associated with resistance to antivirals. *Antimicrob Agents Chemother* **54:**4833–4842.
45. **Chibo D, Druce J, Sasadeusz J, Birch C.** 2004. Molecular analysis of clinical isolates of acyclovir resistant herpes simplex virus. *Antiviral Res* **61:**83–91.
46. **Andrei G, Snoeck R, De Clercq E, Esnouf R, Fiten P, Opdenakker G.** 2000. Resistance of herpes simplex virus type 1 against different phosphonylmethoxyalkyl derivatives of pu-

rines and pyrimidines due to specific mutations in the viral DNA polymerase gene. *J Gen Virol* **81:**639–648.

47. **Andrei G, Balzarini J, Fiten P, De Clercq E, Opdenakker G, Snoeck R.** 2005. Characterization of herpes simplex virus type 1 thymidine kinase mutants selected under a single round of high-dose brivudin. *J Virol* **79:**5863–5869.
48. **Andrei G, Fiten P, Goubau P, van Landuyt H, Gordts B, Selleslag D, De Clercq E, Opdenakker G, Snoeck R.** 2007. Dual infection with polyomavirus BK and acyclovir-resistant herpes simplex virus successfully treated with cidofovir in a bone marrow transplant recipient. *Transpl Infect Dis* **9:**126–131.
49. **Andrei G, Snoeck R.** 2013. Herpes simplex virus drug-resistance: new mutations and insights. *Curr Opin Infect Dis* **26:**551–560.
50. **Bestman-Smith J, Schmit I, Papadopoulou B, Boivin G.** 2001. Highly reliable heterologous system for evaluating resistance of clinical herpes simplex virus isolates to nucleoside analogues. *J Virol* **75:**3105–3110.
51. **Bestman-Smith J, Boivin G.** 2003. Drug resistance patterns of recombinant herpes simplex virus DNA polymerase mutants generated with a set of overlapping cosmids and plasmids. *J Virol* **77:**7820–7829.
52. **Bohn K, Zell R, Schacke M, Wutzler P, Sauerbrei A.** 2011. Gene polymorphism of thymidine kinase and DNA polymerase in clinical strains of herpes simplex virus. *Antivir Ther* **16:**989–997.
53. **Burrel S, Bonnafous P, Hubacek P, Agut H, Boutolleau D.** 2012. Impact of novel mutations of herpes simplex virus 1 and 2 thymidine kinases on acyclovir phosphorylation activity. *Antiviral Res* **96:**386–390.
54. **Burrel S, Boutolleau D, Azar G, Doan S, Deback C, Cochereau I, Agut H, Gabison EE.** 2013. Phenotypic and genotypic characterization of acyclovir-resistant corneal HSV-1 isolates from immunocompetent patients with recurrent herpetic keratitis. *J Clin Virol* **58:**321–324.
55. **Burrel S, Aime C, Hermet L, Ait-Arkoub Z, Agut H, Boutolleau D.** 2013. Surveillance of herpes simplex virus resistance to antivirals: a 4-year survey. *Antiviral Res* **100:**365–372.
56. **Duan R, de Vries RD, Osterhaus AD, Remeijer L, Verjans GM.** 2008. Acyclovir-resistant corneal HSV-1 isolates from patients with herpetic keratitis. *J Infect Dis* **198:**659–663.
57. **Duan R, de Vries RD, van Dun JM, van Loenen FB, Osterhaus AD, Remeijer L, Verjans GM.** 2009. Acyclovir susceptibility and genetic characteristics of sequential herpes simplex virus type 1 corneal isolates from patients with recurrent herpetic keratitis. *J Infect Dis* **200:**1402–1414.
58. **Frobert E, Ooka T, Cortay JC, Lina B, Thouvenot D, Morfin F.** 2005. Herpes simplex virus thymidine kinase mutations associated with resistance to acyclovir: a site-directed mutagenesis study. *Antimicrob Agents Chemother* **49:**1055–1059.
59. **Frobert E, Ooka T, Cortay JC, Lina B, Thouvenot D, Morfin F.** 2007. Resistance of herpes simplex virus type 1 to acyclovir: thymidine kinase gene mutagenesis study. *Antiviral Res* **73:**147–150.
60. **Frobert E, Cortay JC, Ooka T, Najioullah F, Thouvenot D, Lina B, Morfin F.** 2008. Genotypic detection of acyclovir-resistant HSV-1: characterization of 67 ACV-sensitive and 14 ACV-resistant viruses. *Antiviral Res* **79:**28–36.
61. **Gaudreau A, Hill E, Balfour HH Jr, Erice A, Boivin G.** 1998. Phenotypic and genotypic characterization of acyclovir-resistant herpes simplex viruses from immunocompromised patients. *J Infect Dis* **178:**297–303.
62. **Gibbs JS, Chiou HC, Bastow KF, Cheng YC, Coen DM.** 1988. Identification of amino acids in herpes simplex virus DNA polymerase involved in substrate and drug recognition. *Proc Natl Acad Sci USA* **85:**6672–6676.
63. **Harris W, Collins P, Fenton RJ, Snowden W, Sowa M, Darby G.** 2003. Phenotypic and genotypic characterization of clinical isolates of herpes simplex virus resistant to aciclovir. *J Gen Virol* **84:**1393–1401.
64. **Horsburgh BC, Chen SH, Hu A, Mulamba GB, Burns WH, Coen DM.** 1998. Recurrent acyclovir-resistant herpes simplex in an immunocompromised patient: can strain differences compensate for loss of thymidine kinase in pathogenesis? *J Infect Dis* **178:**618–625.
65. **Hwang YT, Smith JF, Gao L, and Hwang CB.** 1998. Mutations in the Exo III motif of the herpes simplex virus DNA polymerase gene can confer altered drug sensitivities. *Virology* **246:**298–305.
66. **Kakiuchi S, Nonoyama S, Wakamatsu H, Kogawa K, Wang L, Kinoshita-Yamaguchi H, Takayama-Ito M, Lim CK, Inoue N, Mizuguchi M, Igarashi T, and Saijo M.** 2013. Neonatal herpes encephalitis caused by a virologically confirmed acyclovir-resistant herpes simplex virus 1 strain. *J Clin Microbiol* **51:**356–359.
67. **Knopf CW.** 1987. The herpes simplex virus type 1 DNA polymerase gene: site of phosphonoacetic acid resistance mutation in strain Angelotti is highly conserved. *J Gen Virol* **68:** 1429–1433.
68. **Kudo E, Shiota H, Naito T, Satake K, Itakura M.** 1998. Polymorphisms of thymidine kinase gene in herpes simplex virus type 1: analysis of clinical isolates from herpetic keratitis patients and laboratory strains. *J Med Virol* **56:**151–158.
69. **Kussmann-Gerber S, Kuonen O, Folkers G, Pilger BD, Scapozza L.** 1998. Drug resistance of herpes simplex virus type 1—structural considerations at the molecular level of the thymidine kinase. *Eur J Biochem* **255:**472–481.
70. **Larder BA, Kemp SD, Darby G.** 1987. Related functional domains in virus DNA polymerases. *EMBO J* **6:**169–175.
71. **Malartre N, Boulieu R, Falah N, Cortay JC, Lina B, Morfin F, Frobert E.** 2012. Effects of mutations on herpes simplex virus 1 thymidine kinase functionality: an in vitro assay based on detection of monophosphate forms of acyclovir and thymidine using HPLC/DAD. *Antiviral Res* **95:**224–228.
72. **Morfin F, Souillet G, Bilger K, Ooka T, Aymard M, Thouvenot D.** 2000. Genetic characterization of thymidine kinase from acyclovir-resistant and -susceptible herpes simplex virus type 1 isolated from bone marrow transplant recipients. *J Infect Dis* **182:**290–293.
73. **Pan D, Kaye SB, Hopkins M, Kirwan R, Hart IJ, Coen DM.** 2014. Common and new acyclovir resistant herpes simplex virus-1 mutants causing bilateral recurrent herpetic keratitis in an immunocompetent patient. *J Infect Dis* **209:**345–349.
74. **Saijo M, Suzutani T, De Clercq E, Niikura M, Maeda A, Morikawa S, Kurane I.** 2002. Genotypic and phenotypic characterization of the thymidine kinase of ACV-resistant HSV-1 derived from an acyclovir-sensitive herpes simplex virus type 1 strain. *Antiviral Res* **56:**253–262.
75. **Saijo M, Suzutani T, Morikawa S, Kurane I.** 2005. Genotypic characterization of the DNA polymerase and sensitivity to antiviral compounds of foscarnet-resistant herpes simplex virus type 1 (HSV-1) derived from a foscarnet-sensitive HSV-1 strain. *Antimicrob Agents Chemother* **49:**606–611.
76. **Sauerbrei A, Deinhardt S, Zell R, Wutzler P.** 2010. Phenotypic and genotypic characterization of acyclovir-resistant clinical isolates of herpes simplex virus. *Antiviral Res* **86:**246–252.
77. **Sauerbrei A, Bohn K, Heim A, Hofmann J, Weissbrich B, Schnitzler P, Hoffmann D, Zell R, Jahn G, Wutzler P, Hamprecht K.** 2011. Novel resistance-associated mutations of thymidine kinase and DNA polymerase genes of herpes simplex virus type 1 and type 2. *Antivir Ther* **16:**1297–1308.
78. **Sauerbrei A, Liermann K, Bohn K, Henke A, Zell R, Gronowitz S, Wutzler P.** 2012. Significance of amino acid substitutions in the thymidine kinase gene of herpes simplex virus type 1 for resistance. *Antiviral Res* **96:**105–107.
79. **Schmit I, Boivin G.** 1999. Characterization of the DNA polymerase and thymidine kinase genesof herpes simplex virus isolates from AIDS patients in whom acyclovirand foscarnet therapy sequentially failed. *J Infect Dis* **180:**487–490.
80. **Sergerie Y, Boivin G.** 2008. Hydroxyurea enhances the activity of acyclovir and cidofovir against herpes simplex virus type 1 resistant strains harboring mutations in the thymidine kinase and/or the DNA polymerase genes. *Antiviral Res* **77:** 77–80.

81. **Stránská R, van Loon AM, Polman M, Beersma MF, Bredius RG, Lankester AC, Meijer E, Schuurman R.** 2004. Genotypic and phenotypic characterization of acyclovir-resistant herpes simplex viruses isolated from haematopoietic stem cell transplant recipients. *Antivir Ther* **9:**565–575.
82. **Suzutani T, Saijo M, Nagamine M, Ogasawara M, Azuma M.** 2000. Rapid phenotypic characterization method for herpes simplex virus and Varicella-Zoster virus thymidine kinases to screen for acyclovir-resistant viral infection. *J Clin Microbiol* **38:**1839–1844.
83. **Suzutani T, Ishioka K, De Clercq E, Ishibashi K, Kaneko H, Kira T, Hashimoto K, Ogasawara M, Ohtani K, Wakamiya N, Saijo M.** 2003. Differential mutation patterns in thymidine kinase and DNA polymerase genes of herpes simplex virus type 1 clones passaged in the presence of acyclovir or penciclovir. *Antimicrob Agents Chemother* **47:**1707–1713.
84. **van Velzen M, van Loenen FB, Meesters RJ, de Graaf M, Remeijer L, Luider TM, Osterhaus AD, Verjans GM.** 2012. Latent acyclovir-resistant herpes simplex virus type 1 in trigeminal ganglia of immunocompetent individuals. *J Infect Dis* **205:**1539–1543.
85. **van Velzen M, van Loenen FB, Meesters RJ, de Graaf M, Remeijer L, Luider TM, Osterhaus AD, Verjans GM.** 2013. Acyclovir-resistant herpes simplex virus type 1 in intra-ocular fluid samples of herpetic uveitis patients. *J Clin Virol* **57:**215–221.
86. **Wang LX, Takayama-Ito M, Kinoshita-Yamaguchi H, Kakiuchi S, Suzutani T, Nakamichi K, Lim CK, Kurane I, Saijo M.** 2013. Characterization of DNA polymerase-associated acyclovir-resistant herpes simplex virus type 1: mutations, sensitivity to antiviral compounds, neurovirulence, and invivo sensitivity to treatment. *Jpn J Infect Dis* **66:**404–410.
87. **Wang Y, Wang Q, Zhu Q, Zhou R, Liu J, Peng T.** 2011. Identification and characterization of acyclovir-resistant clinical HSV-1 isolates from children. *J Clin Virol* **52:**107–112.
88. **Danve C, Morfin F, Thouvenot D, Aymard M.** 2002. A screening dye-uptake assay to evaluate in vitro susceptibility of herpes simplex virus isolates to acyclovir. *J Virol Methods* **105:**207–217.
89. **Stránská R, Schuurman R, Scholl DR, Jollick JA, Shaw CJ, Loef C, Polman M, van Loon AM.** 2004. ELVIRA HSV, a yield reduction assay for rapid herpes simplex virus susceptibility testing. *Antimicrob Agents Chemother* **48:**2331–2333.
90. **Tardif KD, Jorgensen S, Langer J, Prichard M, Schlaber RG.** 2014. Simultaneous titration and phenotypic antiviral drug susceptibility testing for herpes simplex virus 1 and 2. *J Clin Virol* **61:**382–386.
91. **Stránská R, van Loon AM, Polman M, Schuurman R.** 2002. Application of real-time PCR for determination of antiviral drug susceptibility of herpes simplex virus. *Antimicrob Agents Chemother* **46:**2943–2947.
92. **Thi TN, Deback C, Malet I, Bonnafous P, Ait-Arkoub Z, Agut H.** 2006. Rapid determination of antiviral drug susceptibility of herpes simplex virus types 1 and 2 by real-time PCR. *Antiviral Res* **69:**152–157.
93. **Brink AA, van Gelder M, Wolffs PF, Bruggeman CA, van Loo IH.** 2011. Compartmentalization of acyclovir-resistant varicella zoster virus: implications for sampling in molecular diagnostics. *Clin Infect Dis* **52:**982–987.
94. **Hatchette T, Tipples GA, Peters G, Alsuwaidi A, Zhou J, Mailman TL.** 2008. Foscarnet salvage therapy for acyclovir-resistant varicella zoster: report of a novel thymidine kinase mutation and review of the literature. *Pediatr Infect Dis J* **27:** 75–77.
95. **Morfin F, Frobert E, Thouvenot D.** 2003. [Contribution of the laboratory in case of resistance to acyclovir of herpes simplex and varicella zoster virus]. *Ann Biol Clin (Paris)* **61:**33–40.
96. **Sahli R, Andrei G, Estrade C, Snoeck R, Meylan PR.** 2000. A rapid phenotypic assay for detection of acyclovir-resistant varicella-zoster virus with mutations in the thymidine kinase open reading frame. *Antimicrob Agents Chemother* **44:**873–878.
97. **Saint-Léger E, Caumes E, Breton G, Douard D, Saiag P, Huraux JM, Bricaire F, Agut H, Fillet AM.** 2001. Clinical and virologic characterization of acyclovir-resistant varicella-zoster viruses isolated from 11 patients with acquired immunodeficiency syndrome. *Clin Infect Dis* **33:**2061–2067.
98. **Sauerbrei A, Taut J, Zell R, Wutzler P.** 2011. Resistance testing of clinical varicella-zoster virus strains. *Antiviral Res.* **90:**242–247.
99. **van der Beek MT, Vermont CL, Bredius RG, Marijt EW, van der Blij-de Brouwer CS, Kroes AC, Claas EC, Vossen AC.** 2013. Persistence and antiviral resistance of varicella zoster virus in hematological patients. *Clin Infect Dis* **56:**335–343.
100. **Andrei G, Topalis D, Fiten P, McGuigan C, Balzarini J, Opdenakker G, Snoeck R.** 2012. In vitro-selected drug-resistant varicella-zoster virus mutants in the thymidine kinase and DNA polymerase genes yield novel phenotype-genotype associations and highlight differences between antiherpesvirus drugs. *J Virol* **86:**2641–2652.
101. **Boivin G, Edelman CK, Pedneault L, Talarico CL, Biron KK, Balfour HH Jr.** 1994. Phenotypic and genotypic characterization of acyclovir-resistant varicella-zoster viruses isolated from persons with AIDS. *J Infect Dis* **170:**68–75.
102. **Bryan CJ, Prichard MN, Daily S, Jefferson G, Hartline C, Cassady KA, Hilliard L, Shimamura M.** 2008. Acyclovir-resistant chronic verrucous vaccine strain varicella in a patient with neuroblastoma. *Pediatr Infect Dis J* **27:**946–948.
103. **Fillet AM, Dumont B, Caumes E, Visse B, Agut H, Bricaire F, Huraux JM.** 1998. Acyclovir-resistant varicella-zoster virus: phenotypic and genetic characterization. *J Med Virol* **55:**250–254.
104. **Ida M, Kageyama S, Sato H, Kamiyama T, Yamamura J, Kurokawa M, Morohashi M, Shiraki K.** 1999. Emergence of resistance to acyclovir and penciclovir in varicella-zoster virus and genetic analysis of acyclovir-resistant variants. *Antiviral Res* **40:**155–166.
105. **Kamiyama T, Kurokawa M, Shiraki K.** 2001. Characterization of the DNA polymerase gene of varicella-zoster viruses resistant to acyclovir. *J Gen Virol* **82:**2761–2765.
106. **Kodama E, Mori S, Shigeta S.** 1995. Analysis of mutations in the thymidine kinase gene of varicella zoster virus associated with resistance to 5-iodo-2′-deoxyuridine and 5-bromo-2′-deoxyuridine. *Antiviral Res* **27:**165–170.
107. **Levin MJ, Dahl KM, Weinberg A, Giller R, Patel A, Krause PR.** 2003. Development of resistance to acyclovir during chronic infection with the Oka vaccine strain of varicella-zoster virus, in an immunosuppressed child. *J Infect Dis* **188:** 954–959.
108. **Morfin F, Thouvenot D, De Turenne-Tessier M, Lina B, Aymard M, Ooka T.** 1999. Phenotypic and genetic characterization of thymidine kinase from clinical strains of varicella-zoster virus resistant to acyclovir. *Antimicrob Agents Chemother* **43:**2412–2416.
109. **Ng TI, Shi Y, Huffaker HJ, Kati W, Liu Y, Chen CM, Lin Z, Maring C, Kohlbrenner WE, Molla A.** 2001. Selection and characterization of varicella-zoster virus variants resistant to (R)-9-[4-hydroxy-2-(hydroxymethy)butyl]guanine. *Antimicrob Agents Chemother* **45:**1629–1636.
110. **Roberts GB, Fyfe JA, Gaillard RK, Short SA.** 1991. Mutant varicella-zoster virus thymidine kinase: correlation of clinical resistance and enzyme impairment. *J Virol* **65:**6407–6413.
111. **Sawyer MH, Inchauspe G, Biron KK, Waters DJ, Straus SE, Ostrove JM.** 1988. Molecular analysis of the pyrimidine deoxyribonucleoside kinase gene of wild-type and acyclovir-resistant strains of varicella-zoster virus. *J Gen Virol* **69:**2585–2593.
112. **Suzutani T, Saijo M, Nagamine M, Ogasawara M, Azuma M.** 2000. Rapid phenotypic characterization method for herpes simplex virus and Varicella-Zoster virus thymidine kinases to screen for acyclovir-resistant viral infection. *J Clin Microbiol* **38:**1839–1844.
113. **Talarico CL, Phelps WC, Biron KK.** 1993. Analysis of the thymidine kinase genes from acyclovir-resistant mutants of

varicella-zoster virus isolated from patients with AIDS. *J Virol* **67:**1024–1033.

114. **Visse B, Dumont B, Huraux JM, Fillet AM.** 1998. Single amino acid change in DNA polymerase is associated with foscarnet resistance in a varicella-zoster virus strain recovered from a patient with AIDS. *J Infect Dis* **178**(Suppl 1)**:**S55–S57.
115. **Arvia R, Corcioli F, Azzi A.** 2013. High resolution melting analysis as a tool to detect molecular markers of antiviral resistance in influenza A viruses. *J Virol Methods* **189:**265–270.
116. **Boivin G.** 2013. Detection and management of antiviral resistance for influenza viruses. *Influenza Other Respi Viruses* **7** (Suppl 3)**:**18–23.
117. **Gubareva LV, Trujillo AA, Okomo-Adhiambo M, Mishin VP, Deyde VM, Sleeman K, Nguyen HT, Sheu TG, Garten RJ, Shaw MW, Fry AM, Klimov AI.** 2010. Comprehensive assessment of 2009 pandemic influenza A (H1N1) virus drug susceptibility in vitro. *Antivir Ther* **15:**1151–1159.
118. **Okomo-Adhiambo M, Sheu TG, Gubareva LV.** 2013. Assays for monitoring susceptibility of influenza viruses to neuraminidase inhibitors. *Influenza Other Respi Viruses* **7**(Suppl 1)**:** 44–49.
119. **Hurt AC.** 2014. The epidemiology and spread of drug resistant human influenza viruses. *Curr Opin Virol* **8:**22–29.
120. **Gooskens J, Jonges M, Claas EC, Meijer A, Kroes AC.** 2009. Prolonged influenza virus infection during lymphocytopenia and frequent detection of drug-resistant viruses. *J Infect Dis* **199:**1435–1441.
121. **Okomo-Adhiambo M, Hurt AC, Gubareva LV.** 2012. The chemiluminescent neuraminidase inhibition assay: a functional method for detection of influenza virus resistance to the neuraminidase inhibitors. *Methods Mol Biol* **865:**95–113.
122. **Hurt AC, Okomo-Adhiambo M, Gubareva LV.** 2012. The fluorescence neuraminidase inhibition assay: a functional method for detection of influenza virus resistance to the neuraminidase inhibitors. *Methods Mol Biol* **865:**115–125.
123. **World Health Organization.** Hepatitis B. *Factsheet 204*, update July 2015. http://www.who.int/mediacentre/factsheets/fs204/en/.
124. **Libbrecht E, Doutreloigne J, Van De Velde H, Yuen MF, Lai CL, Shapiro F, Sablon E.** 2007. Evolution of primary and compensatory lamivudine resistance mutations in chronic hepatitis B virus-infected patients during long-term lamivudine treatment, assessed by a line probe assay. *J Clin Microbiol* **45:**3935–3941.
125. **Colonno RJ, Rose R, Baldick CJ, Levine S, Pokornowski K, Yu CF, Walsh A, Fang J, Hsu M, Mazzucco C, Eggers B, Zhang S, Plym M, Klesczewski K, Tenney DJ.** 2006. Entecavir resistance is rare in nucleoside naïve patients with hepatitis B. *Hepatology* **44:**1656–1665.
126. **Tenney DJ, Rose RE, Baldick CJ, Pokornowski KA, Eggers BJ, Fang J, Wichroski MJ, Xu D, Yang J, Wilber RB, Colonno RJ.** 2009. Long-term monitoring shows hepatitis B virus resistance to entecavir in nucleoside-naïve patients is rare through 5 years of therapy. *Hepatology* **49:**1503–1514.
127. **Buti M, Tsai N, Petersen J, Flisiak R, Gurel S, Krastev Z, Schall RA, Flaherty JF, Martins EB, Charuworn P, Kitrinos KM, Subramanian GM, Gane E, Marcellin P.** 2015. Seven-year efficacy and safety of treatment with tenofovir disoproxil fumarate for chronic hepatitis B virus infection. *Dig Dis Sci* **60:**1457–1464.
128. **Snow-Lampart A, Chappell B, Curtis M, Zhu Y, Myrick F, Schawalder J, Kitrinos K, Svarovskaia ES, Miller MD, Sorbel J, Heathcote J, Marcellin P, Borroto-Esoda K.** 2011. No resistance to tenofovir disoproxil fumarate detected after up to 144 weeks of therapy in patients monoinfected with chronic hepatitis B virus. *Hepatology* **53:**763–773.
129. **World Health Organization.** Hepatitis C. *Factsheet 164*, update July 2015. http://www.who.int/mediacentre/factsheets/fs164/en/.
130. **deLemos AS, Chung RT.** 2014. Hepatitis C treatment: an incipient therapeutic revolution. *Trends Mol Med* **20:**315–321.
131. **Feeney ER, Chung RT.** 2014. Antiviral treatment of hepatitis C. *BMJ* **349:**g3308.
132. **Rodríguez-Torres M.** 2013. Sofosbuvir (GS-7977), a pan-genotype, direct-acting antiviral for hepatitis C virus infection. *Expert Rev Anti Infect Ther* **11:**1269–1279.
133. **Vermehren J, Sarrazin C.** 2012. The role of resistance in HCV treatment. *Best Pract Res Clin Gastroenterol* **26:**487–503.
134. **Kieffer TL, George S.** 2014. Resistance to hepatitis C virus protease inhibitors. *Curr Opin Virol* **8:**16–21.
135. **Poveda E, Wyles DL, Mena A, Pedreira JD, Castro-Iglesias A, Cachay E.** 2014. Update on hepatitis C virus resistance to direct-acting antiviral agents. *Antiviral Res* **108:**181–191.
136. **Schneider MD, Sarrazin C.** 2014. Antiviral therapy of hepatitis C in 2014: do we need resistance testing? *Antiviral Res* **105:**64–71.
137. **Nakamoto S, Kanda T, Wu S, Shirasawa H, Yokosuka O.** 2014. Hepatitis C virus NS5A inhibitors and drug resistance mutations. *World J Gastroenterol* **20:**2902–2912.
138. **Reddy KR, Zeuzem S, Zoulim F, Weiland O, Horban A, Stanciu C, Villamil FG, Andreone P, George J, Dammers E, Fu M, Kurland D, Lenz O, Ouwerkerk-Mahadevan S, Verbinnen T, Scott J, Jessner W.** 2015. Simeprevir versus telaprevir with peginterferon and ribavirin in previous null or partial responders with chronic hepatitis C virus genotype 1 infection (ATTAIN): a randomised, double-blind, non-inferiority phase 3 trial. *Lancet Infect Dis* **15:**27–35.
139. **Bourlière M, Bronowicki JP, de Ledinghen V, Hézode C, Zoulim F, Mathurin P, Tran A, Larrey DG, Ratziu V, Alric L, Hyland RH, Jiang D, Doehle B, Pang PS, Symonds WT, Subramanian GM, McHutchison JG, Marcellin P, Habersetzer F, Guyader D, Grangé JD, Loustaud-Ratti V, Serfaty L, Metivier S, Leroy V, Abergel A, Pol S.** 2015. Ledipasvir-sofosbuvir with or without ribavirin to treat patients with HCV genotype 1 infection and cirrhosis non-responsive to previous protease-inhibitor therapy: a randomised, double-blind, phase 2 trial (SIRIUS). *Lancet Infect Dis* **15:**397–404.
140. **Krishnan P, Tripathi R, Schnell G, Reisch T, Beyer J, Irvin M, Xie W, Larsen L, Cohen D. Podsadecki T, Pilot-Matias T, Collins C.** 2015. Resistance analysis of baseline and treatment-emergent variants in hepatitis C virus genotype 1 in the AVIATOR study with paritaprevir-ritonavir, ombitasvir, and dasabuvir. *Antimicrob.Agents Chemother* **59:**5445–5454.
141. **Ross L, Boulmé R, Fisher R, Hernandez J, Florance A, Schmit JC, Williams V.** 2005. A direct comparison of drug susceptibility to HIV type 1 from antiretroviral experienced subjects as assessed by the antivirogram and PhenoSense assays and by seven resistance algorithms. *AIDS Res Hum Retroviruses* **21:**933–939.
142. **Grant RM, Kuritzkes DR, Johnson VA, Mellors JW, Sullivan JL, Swanstrom R, D'Aquila RT, Van Gorder M, Holodniy M, Lloyd RM Jr, Reid C, Morgan GF, Winslow DL.** 2003. Accuracy of the TRUGENE HIV-1 genotyping kit. *J Clin Microbiol* **41:**1586–1593.
143. **Jagodzinski LL, Cooley JD, Weber M, Michael NL.** 2003. Performance characteristics of human immunodeficiency virus type 1 (HIV-1) genotyping systems in sequence-based analysis of subtypes other than HIV-1 subtype B. *J Clin Microbiol* **41:** 998–1003.
144. **Eshleman SH, Hackett J Jr, Swanson P, Cunningham SP, Drews B, Brennan C, Devare SG, Zekeng L, Kaptué L, Marlowe N.** 2004. Performance of the Celera Diagnostics ViroSeq HIV-1 Genotyping System for sequence-based analysis of diverse human immunodeficiency virus type 1 strains. *J Clin Microbiol* **42:**2711–2717.
145. **Cunningham S, Ank B, Lewis D, Lu W, Wantman M, Dileanis JA, Jackson JB, Palumbo P, Krogstad P, Eshleman SH.** 2001. Performance of the applied biosystems ViroSeq human immunodeficiency virus type 1 (HIV-1) genotyping system for sequence-based analysis of HIV-1 in pediatric plasma samples. *J Clin Microbiol* **39:**1254–1257.
146. **Parkin N, de Mendoza C, Schuurman R, Jennings C, Bremer J, Jordan MR, Bertagnolio S.** 2012. Evaluation of in-house genotyping assay performance using dried blood spot specimens in the Global World Health Organization laboratory network. *Clin Infect Dis* **54**(Suppl 4)**:**S273–S279.

147. **Aitken SC, Kliphuis A, Wallis CL, Chu ML, Fillekes Q, Barth R, Stevens W, Rinke de Wit TF, Schuurman R.** 2012. Development and evaluation of an assay for HIV-1 protease and reverse transcriptase drug resistance genotyping of all major group-M subtypes. *J Clin Virol* **54:**21–25.
148. **Woods CK, Brumme CJ, Liu TF, Chui CK, Chu AL, Wynhoven B, Hall TA, Trevino C, Shafer RW, Harrigan PR.** 2012. Automating HIV drug resistance genotyping with RECall, a freely accessible sequence analysis tool. *J Clin Microbiol* **50:**1936–1942.
149. **Liu TF, Shafer RW.** 2006. Web resources for HIV type 1 genotypic-resistance test interpretation. *Clin.Infect.Dis.* **42:** 1608–1618.
150. **Tang MW, Liu TF, Shafer RW.** 2012. The HIVdb system for HIV-1 genotypic resistance interpretation. *Intervirology* **55:**98–101.
151. **Wensing AM, Calvez V, Günthard HF, Johnson VA, Paredes R, Pillay D, Shafer RW, Richman DD.** 2014. 2014 Update of the drug resistance mutations in HIV-1. *Top Antivir Med* **22:**642–650.
152. **Stelzl E, Pröll J, Bizon B, Niklas N, Danzer M, Hackl C, Stabentheiner S, Gabriel C, Kessler HH.** 2011. Human immunodeficiency virus type 1 drug resistance testing: evaluation of a new ultra-deep sequencing-based protocol and comparison with the TRUGENE HIV-1 Genotyping Kit. *J Virol Methods* **178:**94–97.
153. **Avidor B, Girshengorn S, Matus N, Talio H, Achsanov S, Zeldis I, Fratty IS, Katchman E, Brosh-Nissimov T, Hassin D, Alon D, Bentwich Z, Yust I, Amit S, Forer R, Vulih Shultsman I, Turner D.** 2013. Evaluation of a benchtop HIV ultradeep pyrosequencing drug resistance assay in the clinical laboratory. *J Clin Microbiol* **51:**880–886.
154. **Cozzi-Lepri A, Noguera-Julian M, Di Giallonardo F, Schuurman R, Däumer M, Aitken S, Ceccherini-Silberstein F, D'Arminio Monforte A, Geretti AM, Booth CL, Kaiser R, Michalik C, Jansen K, Masquelier B, Bellecave P, Kouyos RD, Castro E, Furrer H, Schultze A, Günthard HF, Brun-Vezinet F, Paredes R, Metzner KJ.** 2015. Low-frequency drug-resistant HIV-1 and risk of virological failure to first-line NNRTI-based ART: a multicohort European case-control study using centralized ultrasensitive 454 pyrosequencing. *J Antimicrob Chemother* **70:**930–940.
155. **Ferreira GL.** 2012. Global dengue epidemiology trends. *Rev Inst Med Trop Sao Paulo* **54**(Suppl 18)**:**S5–S6.
156. **World Health Organization.** *Dengue: Guidelines for Diagnosis, Prevention and Control—New Edition.* 2009. WHO Press.
157. **Van Herp M, Declerck H, Decroo T.** 2015. Favipiravir—a prophylactic treatment for Ebola contacts? *Lancet* **385:**2350.

Point-of-Care Diagnostic Virology

JAMES J. DUNN AND LAKSHMI CHANDRAMOHAN

17

A point-of-care test (POCT) may be defined as an analytical or diagnostic test that is performed at the bedside or in a near-patient setting, a location distinct from a typical hospital laboratory (1). POCTs are technically less complex than traditional laboratory tests and therefore can be performed by health care professionals or nonmedical personnel. Some laboratories may also use POCTs as a rapid alternative to conventional methods or in facilities where complex testing is limited. Key features of POCTs are listed in Table 1. To be of value, POCTs should afford rapid results and have a high degree of sensitivity and specificity compared to more complex traditional testing performed in a laboratory. With reliable POCT results in hand, providers can make patient management decisions that improve outcomes for the patient or hospital. In some instances the impact may also be a more cost-effective solution compared to laboratory-based testing. POCTs should be simple to perform and interpret by nonlaboratory personnel using uncomplicated instrumentation, contain internal controls to ensure validity of results, have temperature-stable components that allow easy and prolonged storage, and be relatively inexpensive.

POCTs have been developed for a number of disciplines in clinical medicine. Commonly used assays include those for biochemical analytes (e.g., glucose, electrolytes, cardiac markers), hematologic assays (e.g., prothrombin time/INR, D-dimer, complete blood count), hormonal assays (e.g., parathyroid hormone, urine pregnancy testing), and drug testing (e.g., drugs of abuse). POCTs for diagnosis of viral infections rely on detection of specific antigens from the site of infection (e.g., a rapid influenza diagnostic test), specific antibodies from blood or other body fluids (e.g., a rapid HIV test), or specific detection of viral nucleic acids (e.g., enterovirus in cerebrospinal fluid). This chapter examines the performance characteristics of POCT and near-POCT systems when performed exclusively by nonlaboratory personnel as well as their impact on patient care and outcomes. Refer to chapters 8 through 14 (Antigen Detection Methods, Serologic Methods, and Nucleic Acid Amplification Methods) for detailed descriptions of the design and configuration for these types of assays.

In the United States, with the implementation of the Clinical Laboratory Improvement Amendments of 1988 (CLIA), performance standards for laboratory testing are regulated by the Centers for Medicare and Medicaid Services. The classification of the laboratory tests being performed is determined by complexity (waived, moderate, and high). Those tests that are considered CLIA-waived must use direct, unprocessed specimens and be easy to perform with negligible chance of error. A full list of CLIA-waived tests can be found on the U.S. Food and Drug Administration (FDA) website (http://www.accessdata.fda.gov/scripts/cdrh/cfdocs/cfClia/analyteswaived.cfm). CLIA-waived tests can be performed by individuals without formal laboratory training and outside traditional laboratories at the point-of-care. Results from POCTs are needed for making many medical decisions and are particularly important in settings where either the urgency of the clinical problems or the difficulties of patient follow-up make them clinically valuable and economically feasible. Although the number of publications describing POCTs of this kind for viral diagnostics is increasing, there remains a paucity of data describing their utility and impact at bedside or in a near-patient clinical setting with clear outcomes-based and cost-effectiveness information.

ANTIGEN DETECTION ASSAYS: RESPIRATORY VIRUSES

Respiratory viruses are responsible for a wide range of acute respiratory tract infections, including the common cold, influenza, and croup, and represent the most common cause of acute illness in the United States. Studies using viral culture as the diagnostic method estimated that infants and preschool children experienced a mean of 6 to 10 viral infections annually, and school-age children and adolescents experienced three to five illnesses annually (2). Accurate and timely diagnosis of the cause of lower respiratory tract infections has many benefits. Diagnosis of viral infection can improve patient management by ensuring appropriate antiviral treatment and decrease the overall cost of care (3, 4). The ability to differentiate between viral and bacterial infections and treat them appropriately may reduce selection for antibiotic-resistant bacteria (4). In addition, accurate diagnosis of the cause of respiratory infections provides information that can be used by infection control personnel and public health authorities to decrease the spread of disease and help form public health policies (3).

doi:10.1128/9781555819156.ch17

TABLE 1 Essential features of point-of-care and near point-of-care testing

- Provides accurate results with high sensitivity and specificity
- Results can be obtained in a timely manner
- Simple and easy for nonlaboratory personnel to perform and interpret the test
- Provides a result that impacts patient or hospital outcomes or provides overall health care cost savings
- Contains internal controls to assure the validity of results
- Includes test components that are stable at room temperature and have a long shelf life
- Does not require highly complex or elaborate instrumentation
- Cost of testing is relatively inexpensive
- Reduces the risk of pre-analytical errors such as handling, transporting, and labeling of specimens

Rapid antigen detection tests for influenza viruses and respiratory syncytial virus (RSV) are immunoassays that utilize a sample of secretions from the nasopharyngeal cavity to qualitatively detect specific viral antigens. Many commercially available test formats are available and include immunochromatographic tests, enzyme immunoassays, and optical immunoassays. For many, no specific virology or laboratory expertise is required (CLIA-waived), and results can be obtained within 10 to 30 minutes. Many factors can influence the performance characteristics of rapid antigen detection tests. Technically, false negatives may occur due to poor analytical sensitivity. The assay's limit of detection may vary depending on the type or strain of virus. Factors inherent to the host such as age, disease severity, and time to presentation and testing also affect test performance. The type and quality of respiratory specimen influences test accuracy, and infections that are primarily of the lower respiratory tract are less likely to be identified by sampling through the nasal cavity. Furthermore, administration of immunoprophylaxis (e.g., palivizumab) may compete with immunoassay antibodies for binding the viral target protein and thereby reduce sensitivity.

Influenza Virus

In both pediatric and adult patients, the diagnosis of influenza infection based solely on clinical signs and symptoms can be challenging (5–7). Patients may present with influenza-like illness due to myriad causes other than infection due to influenza. The availability of rapid and accurate diagnostic testing specifically for influenza makes it possible to establish the etiology of infection and impact medical decision making. Rapid influenza diagnostic tests (RIDTs) have been used extensively for many years in a variety of settings including physicians' offices, urgent care centers, and small laboratories where more complex viral diagnostic capabilities may not be available (8). During periods of higher prevalence, positive results by these rapid methods generally correlate well with actual influenza virus infection. However, many RIDTs have suffered from poor performance in terms of analytical and clinical sensitivity and low negative predictive values compared to culture and/or molecular detection methods; pooled sensitivity has been calculated to be 62.3%, while pooled specificity is 98.2% (9, 10). These findings are particularly apparent for novel or pandemic influenza strains (11–13). Because of these limitations, several organizations and professional societies have cautioned clinicians about the utility of RIDTs for certain patient populations and how results should be interpreted (14–16). Their opinion is that since a negative RIDT result may not reliably exclude influenza infection and misdiagnosis can have potentially serious consequences, follow-up testing with a more sensitive and specific method such as RT-PCR or viral culture should be considered to confirm the result. Unfortunately, this practice will have the consequence of delaying decisions about patient management. In addition, the correct interpretation and appropriate use of RIDTs should be considered in the context of the prevalence of circulating influenza strains in the community since this affects the positive and negative predictive values of the tests. If the prevalence is unknown, RIDT results become difficult to interpret and are of limited use in making patient management decisions. Also, RIDTs do not give information on the influenza subtype, a situation that becomes important when deciding among choices of antiviral treatment.

A protocol for rapid influenza testing at the point-of-care in the emergency department by nonlaboratory personnel for febrile infants and young children significantly decreased additional testing (e.g., complete blood count, blood culture, chest x-ray), the total length of time spent in the emergency department, and total medical charges (17–20). Prompt diagnosis of influenza has also been shown to facilitate the appropriate use of antiviral therapy. Health care providers in outpatient facilities such as community health centers and physician offices rely heavily on results of RIDT performed at the point of care for determining whether a patient receives antiviral medications, particularly for those at higher risk for influenza complications (8, 21). Studies have noted the effect of RIDTs in reducing the unnecessary use of antibacterials among children and adults. Doctors who collected and tested nasal swabs during house-call visits were less likely to prescribe antibiotics than those who did not use an RIDT (22). Fewer antibiotics were prescribed when the diagnosis of influenza in emergency departments and physician offices was made in association with results of an RIDT (18, 21, 23, 24). Early identification of cases of influenza also allows for the institution of infection control measures within hospitals or long-term care facilities to reduce secondary cases (25, 26).

The FDA has recently proposed up-classifying RIDTs from class I to class II devices subject to special controls and performance standards (27). The intent of the reclassification is to assure the reliability and accuracy of RIDTs, reduce the likelihood of false negative results, aid clinicians in making appropriate treatment decisions, and enable effective infection control and public health response during influenza outbreaks. This new rule would "1) identify the minimum acceptable performance criteria, 2) identify the appropriate comparator for establishing performance of new assays, and 3) call for mandatory annual analytical reactivity testing of contemporary influenza strains, including testing of newly emerging strains that pose a danger of public health emergency." The proposed minimum performance criteria are listed in Table 2. Currently, many RIDTs do not meet these standards.

Respiratory Syncytial Virus

Respiratory syncytial virus (RSV) is responsible for annual winter outbreaks of respiratory tract infection among children in temperate climates. RSV infection accounts for 50% to 80% of all bronchiolitis in the United States and has the highest incidence of identification as the etiology of serious

TABLE 2 The FDA-proposed minimal performance criteria for RIDTs

Comparator	RIDT characteristic	Influenza A	Influenza B
Viral culture	Sensitivity	≥90% point est.	≥80% point est.
		≥80 (lower 95% CI)	≥70% (lower 95% CI)
	Specificity	≥95% point est.	≥95% point est.
		≥90% (lower 95% CI)	≥90% (lower 95% CI)
Molecular	Sensitivity	≥80% point est.	≥80% point est.
		≥70% (lower 95% CI)	≥70% (lower 95% CI)
	Specificity	≥95% point est.	≥95% point est.
		≥90% (lower 95% CI)	≥90% (lower 95% CI)

CI, confidence interval; FDA, U.S. Food and Drug Administration; RIDT, rapid influenza diagnostic tests.

lower respiratory tract disease in infants and young children, frequently requiring hospitalization (28). Hospital bed capacity may become limited in these situations. Clinically, RSV is virtually indistinguishable from other viral respiratory infections. Traditional laboratory methods for diagnosis of RSV have relied on moderate to complex techniques such as cell culture, nucleic acid amplification, and immunofluorescence assays, procedures that require trained laboratory personnel and specialized equipment. Given the burden of disease in pediatric hospitals during the RSV season, rapid testing may assist with infection control, patient isolation, and cohorting, thus reducing nosocomial transmission. Rapid diagnosis of RSV in pediatric patients has been associated with reductions in antibiotic use, hospital costs, and length of stay. A decrease in nosocomial RSV transmission has been noted following the introduction of POCT (29). Rapid diagnosis of RSV may also enable early institution of antiviral therapy (e.g., ribavirin) in high-risk patients. In most adults, RSV causes a mild illness, but in the elderly or those with comorbid lung or cardiac conditions, RSV infection can be disabling and even fatal (30). Due to partial prior immunity, the viral load present in nasal secretions during RSV infection decreases with age, as does the duration of viral shedding. These factors contribute to the lower sensitivity of RSV-POCT when used in adults (30, 31).

When performed by trained nurses in a pediatric hospital unit, RSV-POCT was highly specific and allowed prompt infection control and clinical care measures to be instituted (32). However, due to the low sensitivity of the RSV-POCT, follow-up testing by RT-PCR was required, a process that delayed some results by several hours. RSV-POCT performed by trained emergency department staff allowed for RSV-positive children <2 years of age to be admitted directly to a designated cohort area, a process that improved bed management such that an average of five inpatient treatment areas could be left free each day (33). Thus significant extra capacity was generated through rapid RSV testing, and the additional cost of testing was relatively small compared to costs such as the savings achieved through less cleaning of more areas in the emergency department. RSV-POCT has been used to identify outbreaks rapidly so that appropriate infection control measures could be implemented in a timely manner (34).

Unlike other respiratory viruses such as influenza, concomitant bacterial superinfection in RSV-infected children is very rare (≤1.6%) (35, 36). Therefore confirmation of RSV infection should prompt treating physicians to discontinue antibiotics. In some studies rapid diagnosis of RSV has led to a reduction in antibiotic use in children (37), while others failed to show a positive impact on antimicrobial stewardship among children hospitalized for respiratory infection (38).

ANTIGEN DETECTION ASSAYS: GASTROINTESTINAL VIRUSES

Rotavirus

Prior to the availability of vaccines, rotavirus was the most important etiological agent of diarrhea in children under 5 years of age, accounting for 111 million episodes of gastroenteritis in the world with approximately 2 million hospitalizations and 611,000 deaths annually, mainly in developing countries (39). Many cases of rotavirus infection require hospitalization, which can lead to a high prevalence of health care–acquired diarrhea in other pediatric patients. Although the rotavirus vaccine has been shown to reduce both childhood infections and hospitalizations, outbreaks may still occur in unvaccinated children or elderly adults due to the emergence of nonvaccine strains or waning immunity.

Immunochromatographic antigen tests for rotavirus have been used to identify infected inpatients in a timely manner to allow for prompt institution of infection control measures to prevent nosocomial transmission (40, 41). Limiting the spread of rotavirus within health care institutions is important to improve patient outcomes and to avoid additional health care costs. These types of assays are also suitable for testing in developing countries where the burden of pediatric diarrheal disease is associated with high morbidity and mortality and laboratory resources are limited.

Norovirus

Noroviruses have emerged as one of the most important pathogens causing acute gastroenteritis in children and adults. Although rarely fatal, norovirus cases add a huge burden to the health care system in terms of doctor visits, hospitalizations, and loss of productivity. After an incubation period of 1 to 2 days, there is a rapid onset of vomiting and nonbloody diarrhea, which typically last for 48 to 72 hours. Noroviruses are highly contagious since very low doses of viral particles can cause infection. Norovirus outbreaks occur in health care settings (including nursing homes and hospitals), cruise ships, restaurants, and schools (42–44). The control of norovirus outbreaks in hospitals requires rapid and strict infection control procedures that efficiently limit the spread of the infection among patients as well as staff. Prompt initiation of infection control measures requires a rapid diagnosis of the infection.

Although no norovirus rapid antigen assays are currently FDA-cleared in the United States, several studies have described their potential in identifying an outbreak situation (45–47). These are immunochromatographic tests that detect antigens specific for genotype I, genotype II, or both and can be completed in 15 to 20 minutes. Assay sensitivity is highly genotype-dependent with higher sensitivities generally noted for genotype II noroviruses (47, 48). Based on worldwide surveillance, most norovirus outbreaks are caused by the GII.4 strain. Specificities have been reported to be >97%. In general, because the sensitivities of norovirus rapid antigen tests are much lower than those of molecular methods, RT-PCR is recommended to confirm negative immunochromatographic test results.

ANTIBODY DETECTION ASSAYS

Hepatitis C Virus

The current testing algorithm recommended by the Centers for Disease Control and prevention for hepatitis C virus (HCV) relies on testing whole blood specimens for anti-HCV screening, using enzyme immunoassays followed by supplemental confirmatory assays if the screening assay is reactive but below a defined signal-to-cutoff ratio threshold (31, 49, 50). Although this algorithm effectively detects active infection, its limitations include higher likelihood of false positives in low prevalence populations, high test costs, and long turnaround time (TAT). Several HCV antibody detection-based POCTs that are convenient and quality-assured are in use, offering rapid TAT fewer than 30 minutes and facilitating preliminary screening; however, these tests cannot differentiate between acute and chronic infections (51). These tests provide opportunities to expand screening, and their rapid turnaround time limits the loss of patients to follow-up and facilitates earlier establishment of patient care programs (52).

Of the several POCTs available worldwide, only the OraQuick HCV Rapid Antibody Test (OraSure Technologies Inc., Bethlehem, PA, USA) is currently FDA-approved and was granted a CLIA waiver for testing in nontraditional settings. OraSure is a qualitative immunoassay that detects antibodies to HCV associated with HCV genotypes 1 to 4 and may be performed using fingerstick whole blood and venipuncture whole blood specimens from symptomatic or at-risk individuals older than 15 years; it is not approved for use in pregnant women (20, 53, 54). Although OraSure is not FDA-approved for use with oral fluid, it has been shown to detect anti-HCV antibodies from this type of specimen (55, 56). This is useful for epidemiological purposes, especially when blood collection is difficult or for field collection in nonclinical settings by personnel with minimal training. Antibody detection with the OraSure assay employs core NS3 and NS4 antigens immobilized on a nitrocellulose membrane with the resultant antigen-antibody complex directly visualized using colloidal gold labeled with protein A, which generates a reddish-purple line within 20 minutes in the presence of HCV-specific antibodies.

In a number of studies published to date, OraSure testing of blood specimens has displayed clinical performance characteristics comparable to those of laboratory-based methods (enzyme immunoassay [EIA], recombinant immunoblot assay, and PCR) with better specificity and sensitivity compared to four other HCV POCTs (51). OraSure has a manufacturer-reported accuracy of >99% compared to standard laboratory-based EIA tests and has been shown to detect HCV specific antibodies up to 4.9 days earlier than some conventional EIA tests (53). The overall sensitivity and specificity of OraSure for whole blood specimens (finger stick, venous) was reported as 99.8% and 99.9%, respectively (51, 53, 54). A lower sensitivity ranging from 83.3% to 98.1% and a specificity of 99.6% have been observed for oral fluids (56, 57).

Currently, limited data are available describing the clinical utility or the cost effectiveness and economic impact of HCV point-of-care testing in field settings where they are likely to be implemented. Dried blood spot HCV testing in drug users was shown to be associated with an average increased acceptance and allowance of HCV testing by 14.5% compared to the 8% prior to the randomized trial. Although this study was laboratory-based, the increased uptake in HCV testing with dried blood spot sampling is indicative of the potential impact of HCV point-of-care testing. Integration of HCV point-of-care testing using the OraSure assay in a sexually transmitted disease (STD) clinic resulted in an increased awareness of HCV infection and the rapid determination of test results to the clients provided the opportunity to educate them about their potential risks for HCV acquisition and the consequences of infection (58). Additionally, it allowed the linkage-to-care counselors an immediate opportunity to inform the HCV-positive clients on the process of appropriate medical evaluation, treatment, and care. Similarly, field screening of a high-risk urban population, using the OraSure oral swab, and in people who inject drugs, using a fingerstick blood sample, demonstrated that the test was well-accepted and appropriate in community settings (59, 60). The reasons for acceptability were attributed to participants wanting quicker results regarding their anti-HCV status and with potentially less pain associated with testing from a venous blood draw. Of note, the test preference was unaffected by the anti-HCV results. This is important since those testing HCV positive will require a blood draw for follow-up HCV RNA testing.

A critical characteristic of point-of-care assays in field settings is readability (i.e., ease of interpretation of immunochromatographic test lines). In a study assessing the readability of OraSure and two other rapid anti-HCV tests, the agreement score for OraSure and a second rapid test were very high based on the comparison of interoperator agreement, implying that the same test interpretation could be obtained regardless of who reads the assay (54). Further evaluations of preference and acceptability for anti-HCV test type (i.e., fingerstick at point-of-care versus venous collection and standard lab-based testing) in real-world settings and at other community settings such as drop-in services, military camps, and needle-exchange programs are needed to assess the true impact of point-of-care HCV testing. Recently the FDA cleared the Home Access Hepatitis C test system (Home Access Health Corp.), which can be purchased online, by fax, or by mail. This kit is based on collecting a blood sample on a blood specimen card using a fingerstick at home and shipping the card to an accredited laboratory. Test results are usually available within 14 days. There are provisions for pre- and post-test counseling and referrals. The performance of this kit remains to be determined.

Currently available rapid, point-of-care anti-HCV testing is limited by its inability to distinguish between acute/past and active/resolved infections, distinctions that currently require laboratory-based RNA testing. It also remains hampered by the need for counseling and further medical evaluation, including confirmatory testing of positive results. Despite these limitations, point-of-care anti-HCV testing, by being accessible in nonmedical settings with lay workers, by not initially requiring phlebotomy, and by providing immediate results, provides an important opportunity to enhance prevention and treatment programs as well as streamline access to care.

Epstein-Barr Virus

In primary health care, the classic rapid diagnostic test for infectious mononucleosis (IM) caused by Epstein-Barr virus (EBV) is the Paul-Bunnell or "monospot" heterophile antibody (HA) test (61). The conventional Paul-Bunnell test has now been replaced by rapid qualitative latex agglutination tests or solid phase chromatographic immunoassays that use antigens of purified bovine red blood cell extracts. These rapid heterophile antibody tests are frequently used in a physician's office and have facilitated a more rapid diagnosis of IM with reduced cost, no extensive blood analyses, and minimal training of nonlaboratory personnel who can perform the test with results available in about 5 minutes. There are numerous FDA-approved POCTs for EBV based on HAs that have been in use for a number of years and include (among many others) Clearview Mono-plus II test (Inverness Medical Innovations), OSOM Mono Test (Genzyme Diagnostics), QuickVue + Infectious Mononucleosis Test (Quidel Corporation), BioStar Acceava Mono II test kit (Alere), and Meridian ImmunoCard STAT Mono (Applied Biotech) (43). These commercially available kits (e.g., "Mono spot" tests) are CLIA-waived for use with whole blood specimens and CLIA-nonwaived for serum or plasma specimens.

Several studies have shown varying sensitivities and specificities of the different point-of-care HA tests compared to EBV-specific antibody tests. In individuals 13 years or older, the overall sensitivity and specificity of various "Mono spot" tests ranged from 71% to 91% and from 82% to 100%, respectively (62–66). It has been repeatedly demonstrated that HA tests have low sensitivities ranging from 25% to 50% when used in children younger than 13 years of age (62). Lower specificities have been reported in EBV uninfected patient populations and those with autoimmune conditions. Furthermore, tests with purified antigens have increased sensitivity and specificity compared with tests that used whole red blood cell agglutination (62, 67). Although relatively specific, HA tests suffer from limitations of occasional false positives coupled with a false-negative rate as high as 25% during the first week of illness, particularly in patients less than 4 years of age who do not produce appreciable HAs. Therefore, a negative HA test does not rule out the diagnosis of IM, and further testing using EBV-specific antibody detection by conventional EIA is needed.

Newer POCTs based on EBV-specific antibodies have been developed and are available for use in Europe. These tests may be a more sensitive and specific alternative to the HA-based POCTs. EBV-specific antibody POCTs include Monolert Rapid Elisa test, TruEBV-M, and TruEBV-G tests (Meridian Bioscience). Monolert is a qualitative assay that detects anti-EBV nuclear antigen-1 (anti-EBNA-1) IgM and IgG antibodies in human serum or plasma. The TAT is 10 minutes, and the manufacturer-reported sensitivity is 98.8% with a specificity of 94%. TruEBV-M and TruEBV-G are rapid immunochromatographic assays that detect anti-viral capsid antigen (VCA) IgM or anti-VCA IgG and anti-EBNA-1 antibodies in human serum, respectively. The manufacturer-reported sensitivities for TruEBV-M and EBV-G are 91.6% to 96.5% with specificities of 95.8% to 96.4%. The TATs for TruEBV-M and EBV-G tests are 20 minutes. These tests are not currently FDA-approved for use in the United States.

Further studies evaluating the outcomes of using rapid HA tests at point-of-care sites are needed to assess the impact on the number of clinic visits, length of stay, antibiotic usage, and number of recovery days or days of work or school lost. Additionally, although these tests could currently replace the Paul-Bunnell test in many instances, further studies are required to understand the downstream effects of a correct or incorrect diagnosis of IM by point-of-care HA tests. It is important that clinicians be aware of these limitations, and before any rapid test is instituted as a diagnostic tool, a thorough validation of its clinical performance and outcomes must be performed to ensure that the most accurate test is used.

Human Immunodeficiency Virus

Rapid testing for antibodies and/or antigens to HIV provides a highly beneficial solution for the need to provide at-risk populations and those unaware of their positive serostatus with accessible testing and treatment options (68). To lower transmission rates and to link infected patients with appropriate care, public and private health organizations rely on rapid point-of-care HIV testing in clinical and nonclinical settings. Although point-of-care HIV testing is not designed for screening the general population, it can be used in high-risk populations and may be most useful in resource-limited settings or outreach settings with poor infrastructure, lack of trained personnel, extremes of climate, and lack of uninterrupted power supply, all of which impact conventional laboratory testing. In general, these rapid POCTs are less expensive (US$0.5 to US$1.2 per test), portable and easy to use, require few or no reagents or instrumentation, and provide results within 30 minutes.

Rapid POCTs for HIV diagnosis include those used for screening, initial diagnosis, disease staging, treatment/viral load monitoring, and early infant diagnosis. The majority of rapid HIV tests used at a point-of-care are antibody-based tests for screening, and the description and performance of these in point-of-care settings are discussed below. Numerous FDA-approved, highly sensitive, rapid HIV antibody tests are now available with TATs of 10 to 20 minutes. Currently, there are six CLIA-waived rapid antibody tests for use in nonlaboratory settings. These include OraQuick Advance (OraSure Technologies), INST HIV-1 Antibody test (bioLytical Laboratories), Uni-Gold Recombigen (Trinity Biotech PLC), and Clearview HIV-1/2 Stat Pak and Complete tests (Chembio Diagnostic Systems) (69). These rapid test kits are lateral flow or flow-through cassette tests for use on a small volume of whole blood (fingerstick or venipuncture) or oral fluid specimens (70, 71). These tests are categorized as moderately complex when used with serum or plasma. The common targets detected by these antibody tests include envelope proteins glycoprotein (gp) 41, gp120, and gp60 in HIV-1; gp36 in HIV-2; and core antigen p24 (72). Some of the previously mentioned rapid POCTs detect antibodies to HIV-1 and 2 (e.g., OraQuick, Clearview HIV Stat Pak and Complete, Uni-Gold Recombigen), while others are specific only for HIV-1 (e.g., INSTI HIV-1).

Recent evaluation of all FDA-approved rapid antibody tests on fingerstick specimens documented high accuracy with similar sensitivity and specificity to current laboratory-based EIA methods. Branson (72) evaluated the performance of the OraQuick Rapid HIV-1 Antibody test on whole blood and plasma specimens, and the device was shown to be 98.6% sensitive and 99.9% specific on whole blood and 100% sensitive and 99.8% specific on plasma. Rapid antibody detection tests using oral fluids or saliva display variable performance characteristics but are generally less sensitive than tests using whole blood; test sensitivity in pregnant women is as low as 75% (73–76). However, the noninvasiveness of specimen collection makes this type of test an attractive alternative for screening as well as surveillance purposes. Oral fluid sampling could be highly beneficial in young children and injection drug users with collapsed blood vessels, as a backup for patients who decline fingerstick testing, or for settings where high-volume fingerstick testing might be less feasible. The ease of use of the oral fluid-based HIV tests also makes them amenable for home-based HIV self-testing, as shown in a recent study in Africa (77).

Since rapid HIV tests have become available, they have proven to be extremely useful in many settings beyond the traditional hospital laboratory. A number of potential benefits of rapid HIV antibody testing are listed in Table 3. These tests have been used in many types of outpatient clinics, in obstetric wards, during military field operations, and in population screening in developing countries. Rapid HIV testing in nonclinical settings has been shown to detect a substantial number of newly infected persons with high accuracy in community-based organizations, transgender populations, college and university screening programs, and STD clinics and in pregnant women with unknown HIV status during delivery. Several studies have demonstrated improved acceptance of HIV screening with point-of-care testing in outpatient and perinatal settings and in resource-limited countries (72, 78, 79). In one outpatient setting, rapid testing was associated with a 73% rate of return to the emergency department for confirmation of results among those who were HIV-positive compared to 62% for those tested by conventional methods (80). Similar results were also observed in STD clinics. The use of a simple oral-fluid test in a labor ward was successful in reducing mother-to-child HIV transmission (81, 82). Additionally, the median TAT for rapid tests performed in the maternal unit was 45 minutes versus 3 hours for tests performed in the laboratory, leading to faster initiation of prophylaxis to prevent vertical transmission (83). Furthermore, the uptake of rapid HIV testing (reported as acceptability) ranged from 67% to 97% in pregnant women and from 83% to 93% in youth seen in emergency departments and adolescent primary care clinics (82, 84). In pregnant women the increased uptake was mostly associated with age greater than 21 years, higher education status, gestational age less than 32 weeks, and lack of prenatal care during pregnancy. These studies highlight the benefits of point-of-care HIV testing in unique populations. Interestingly, the acceptance rates were unaffected by the type of rapid HIV test offered in the emergency department; the acceptance rate was 69% for oral-fluid testing compared to 67% for fingerstick blood testing (85). However, oral-fluid testing was preferred for home-based testing and tablet-based kiosk testing in the emergency department as shown in the systematic review by Pant Pai et al. (86). In 2012, the FDA approved the first rapid self-administered over-the-counter HIV test, OraQuick In-Home HIV Test. The OraQuick In-Home HIV Test is a manually performed, visually read 20-minute immunoassay for the qualitative detection of antibodies to HIV-1 and HIV-2 in human oral fluid. Although self-testing at home does not fall under CLIA regulations, it is important to note that all tests used outside of professional health care settings have to be CLIA-waived and approved by the FDA for home use.

TABLE 3 Benefits of rapid HIV antibody testing

- Decentralization of testing with community or home-based services
- Increased uptake in HIV testing
- Improved recognition and diagnosis of individuals with HIV infection
- Faster initiation of prophylaxis and reduced transmission
- Decreased number of patients lost to follow-up
- Increased linkage to and retention in care
- Reduced mother-to-child HIV transmission
- Decreased length of hospital stay

Evaluation of quality-of-care measures for inpatients diagnosed with HIV via point-of-care testing in the emergency department versus patients diagnosed via conventional testing after admission revealed that the mean length of stay was 6 days for the rapid test group compared to 13 days for the conventional test group and 16% of the conventional test group, but none of the rapid test group patients were discharged without receiving their HIV diagnosis (87). These results suggest that rapid point-of-care HIV testing can play an important role in inpatient settings. For the newly identified HIV-infected patient, inpatient rapid HIV testing may contribute to better care and decreased chances of being discharged without knowledge of infection status. Economic analyses demonstrated lower health care costs with the use of point-of-care HIV tests since only a single visit was required and the cost effectiveness of HIV screening using rapid HIV tests was similar to that of commonly accepted interventions, even in relatively low-prevalence populations (88). In outreach settings, the mean overall cost of new HIV diagnoses varied extensively depending on the recruitment strategy for testing (outreach, partner notification, or social networks), type of testing, and whether variable and/or fixed costs were included in the testing (89).

As with all tests, rapid point-of-care HIV testing has several limitations. Antibody-based tests are insensitive for detecting acute infections, and whenever possible, patients who are screened should be evaluated for high-risk exposures and signs and symptoms of acute retroviral infection that may warrant testing using a more highly sensitive test (e.g., HIV RNA detection). Additionally, all reactive antibody test results must be confirmed using a conventional laboratory-based test at an approved HIV testing laboratory to rule out the likelihood of a false-positive result. Health care workers at the point-of-care, in contrast to a clinical laboratory, are responsible for specimen collection, testing, reporting, and counseling; hence, adequate resources, appropriate training, and implementation of quality assurance practices are critical for proper administration of a rapid point-of-care HIV test. Additionally, HIV testing in a point-of-care setting requires pre- and post-test counseling that is different from the usual HIV counseling with conventional testing. Counseling ensures that the individual understands the implications of the test and risk of transmission and returns for confirmatory testing and referral for anti-retroviral therapy.

Newer advancements in the field of rapid HIV antibody testing include the Alere Determine HIV-1/2 Ag/Ab Combo test, which is an FDA-approved CLIA moderately complex medical device for simultaneous detection of free HIV-1 p24 antigen as well as antibodies to both HIV-1 and HIV-2 in human serum, plasma, and venous or fingerstick whole blood specimens. The detection of both antibodies and the p24 antigen not only enables detecting nearly 90% of infected individuals who were missed by an initial antibody-only screening test, but also reduces the diagnostic window period. While the Alere Determine test is already in use in hospitals, laboratories, and some physician's offices, it has not yet been CLIA-waived for use in facilities and point-of-care settings that lack certification for higher complexity testing.

NUCLEIC ACID DETECTION ASSAYS

The current market for POCT for viral diseases has been dominated by immunological-based diagnostic tests using lateral flow technology that can be performed outside clinical laboratories. Unfortunately, assay performance for some analytes has been suboptimal in these settings (90, 91). Until recently, molecular or nucleic acid–based testing for the diagnoses of viral infections has not been incorporated into POCT settings. The current nucleic acid–based testing for clinical diagnostic applications is comprised of highly complex PCR-based tests that can be exclusively performed in centralized laboratories, requiring high-end instrumentation with highly skilled personnel. Therefore, at present there are currently no true point-of-care or CLIA-waived molecular tests, although at least one company has applied for CLIA-waived status for a PCR-based platform.

A successful molecular POCT would be one that provides access to much-needed diagnostic methods in low-resource areas with a high disease burden, includes sample-to-result analysis with no off-instrument processing or reagent preparation, features maintenance-free instrumentation, is easy to use by minimally trained personnel, and yields clear actionable results in addition to being suitably sensitive and specific. There have been significant technological advancements focusing on platforms that are fully automated and integrated to generate "sample in to result out," offering significant advantages in a nonlaboratory setting. These tests are often moderately complex, require less hands-on time, are frequently more user friendly to perform, and reduce the requirement for highly trained personnel. With automated recording and reporting of results, they are typically less subjective to interpret than immunochromatographic immunoassays. In recent years, many commercial entities have endeavored to provide such platforms that can be used near-patient or near point-of-care in a rapid manner to provide clinicians with actionable results.

Enteroviruses

Rapid diagnosis of enteroviral (EV) meningitis by real-time molecular assays is now recognized as the "gold standard" and is a critical tool in the assessment of patients with central nervous system disease. The GeneXpert EV assay (Cepheid) is a moderately complex FDA-approved qualitative test for detection of enterovirus RNA in cerebrospinal fluid specimens. Since the test can be completed within 2.5 hours, it can often be performed and provide results during the time the patient remains in the clinic or emergency department. The GeneXpert system is a random access, on-demand test platform that utilizes a single-use microfluidic cartridge that integrates the sample preparation, amplification, and detection steps. Each cartridge contains a syringe and rotary drives that move the sample into reaction chambers for the addition and mixing of reagents.

The Xpert EV assay has a reported sensitivity of 97.1% to 98.8% and a specificity of 100% (3). A study by Nicosia et al. (92) demonstrated that implementation of Xpert EV assay performed in adult and pediatric emergency rooms was associated with reductions in the average response time from 5 days to 3 hours, time to perform the test from 5 to 2.45 hours, the technical hands-on time from 2 hours to 10 minutes, and the test cost per cerebrospinal fluid sample compared to a laboratory-based RT-PCR assay. These differences also reduced the average hospital stay from 6.6 to 1.2 days (infants) and from 7.3 to 1.9 days (adults), findings corroborated by other investigators (93). Use of Xpert EV method afforded overall health care savings of more than 33% compared to conventional batch testing. These costs included a reduction in antibiotic therapy (3.5 to 2 days) and a reduction in the need for additional diagnostic tests (from 72% to 26% of patients). In a second study the empiric antibiotic treatment for bacterial meningitis was stopped in 40% of young infants, and 25% of these patients were discharged following a positive EV result (94).

Respiratory Viruses

The FilmArray rapid respiratory panel (RP) (BioFire Diagnostics, Salt Lake City, UT) uses a self-contained, multiplex PCR, microarray detection system to detect 20 viral and bacterial agents, including adenovirus, coronavirus (229E, HKU1, NL63, OC43), human metapneumovirus, rhinovirus/EV, FluA, FluA subtype (H1, 2009 H1, or H3), FluB, parainfluenza (1, 2, 3, 4), RSV, *Bordetella pertussis*, *Chlamydophila pneumoniae*, and *Mycoplasma pneumoniae*. The test is run on the FilmArray instrument, a small benchtop, closed single-piece flow real-time PCR system. The platform analyzes a single-use cartridge with the assay and integrates automation of nucleic acid extraction, an initial reverse transcription and multiplex PCR, followed by single-plex second-stage PCR amplifications with microarray detection of specific agents in about an hour, with a hands-on time of approximately 2 minutes.

The published sensitivity and specificity of the FilmArray RP are 84% to 100% and 98% to 100%, respectively. A study assessing the impact of the FilmArray RP on patient outcomes showed that the mean time to test result was shorter (383 vs. 1,119 minutes for laboratory PCR), and 51.6% of patients in the emergency department received their results during their visit compared to 13.4% with laboratory-based PCR (95). Furthermore, the identification of a viral pathogen by the FilmArray RP within 4 hours of receipt of sample resulted in a decrease in length of stay of about a quarter of a day and decreased duration of antibiotic use, and those who received a positive result were discharged about 6 hours earlier when the FilmArray test was used. The implementation of the FilmArray also resulted in savings of $231 in hospital costs and $17 in antibiotic use per patient.

The Verigene Respiratory Virus Plus (RV+) (Nanosphere Inc.) is an FDA-cleared multiplex microarray-based technology for the detection of FluA (H1, 2009 H1), FluB, and RSV within 2.5 hours. The Verigene system carries out automated nucleic acid extraction, RT-PCR, and hybridization through nanoparticle-conjugated target-specific probes. The clinical utility and performance of the RV+ testing by pediatricians at near-patient level was assessed in a recent study (96). Compared to rapid antigen tests, RV+ testing had overall sensitivity and specificity for influenza viruses of 100% and 96.2%, respectively; sensitivity and specificity of RSV were 100% and 100%, respectively. Based on the time of fever onset, the sensitivity of RV+ for influenza viruses was 100% compared to 37.5% for rapid antigen tests. This study highlights that the administration of RV+ at near-patient level represents a major development in point-of-care molecular testing. With only 5 minutes of hands-on time for test setup and loading, the RV+ test has the potential to change respiratory virus diagnostics in the primary care setting, especially where fast and accurate results are needed, such as during the index visit.

The FDA-cleared Simplexa Flu A/B & RSV Direct assay (Focus Diagnostics) is a real-time RT-PCR system that detects FluA, FluB, and RSV from nasopharyngeal swabs in about 1 hour and does not require nucleic acid extraction. The assay consists of the Simplexa Flu A/B & RSV Direct reagents, a 3M Integrated Cycler, and direct amplification discs. The published Simplexa sensitivities for FluA, FluB, and RSV are 82.8%, 76.2%, 94.6%, while specificities are 100%, 94.6%, and 100%, respectively (97). The short run time in combination with high specificity and good sensitivity make it an option for near-patient rapid detection of these three important viral respiratory pathogens.

The Alere i Influenza A&B is an isothermal amplification-based, simple-to-use automated test for FluA and B viruses with the potential to generate results within 15 minutes. The Alere instrument is relatively small and portable, requires minimal training, and can be ideally placed in near-patient settings for most minimal TAT for results. When evaluated using pediatric frozen respiratory specimens, the Alere i Influenza A&B assay displayed overall sensitivity and specificity of 93.3% and 94.5% for FluA and 100% and 100% for FluB compared to viral culture (98). In comparison to real-time RT-PCR, the overall sensitivity and specificity were 88.8% and 98.3% for FluA and 100% and 100% for FluB virus, respectively. Additional studies evaluating the clinical performance in direct comparison with other molecular tests and rapid antigen tests in the point-of-care setting are needed to fully understand its potential. At the time of writing, CLIA-waived status of the Alere i Influenza A&B assay was pending.

The IQuum Lab-In-A-Test (LIAT) Influenza A/B assay (Roche Molecular Diagnostics) performed on the LIAT analyzer is a moderately complex isothermal automated multiplex real-time RT-PCR developed for the rapid qualitative detection of FluA/B viruses in nasopharyngeal swab specimens. The IQuum LIAT system fully integrates multiple features, such as sample volume metering, on-board internal control, advanced error diagnostics, and automated data interpretation to ensure the quality of results when operated by minimally trained users. The LIAT analyzer is a sample-to-result, relatively small bench-top system that can be battery operated and does not require an external computer. However, the LIAT analyzer can only process one sample at a time. A slightly larger LIAT workstation that can process eight samples in random access mode has recently been developed. Having an operator hands-on time of less than 1 minute and a total time-to-result of approximately 20 minutes, the LIAT assay can be performed on-demand in near-patient settings, providing physicians with accurate and timely results; however, this assay does not subtype influenza A virus–positive specimens and scant information is currently available regarding the clinical performance of this assay for detecting influenza viruses.

The SAMBA (simple amplification-based assay) Flu duplex test is a dipstick-based molecular assay developed for diagnosis of influenza A and B viruses (targets matrix gene of influenza A and a nonstructural gene of influenza B) intended for near-patient testing. This test couples isothermal amplification with visual detection of nucleic acid on a test strip. The entire test procedure, including extraction, amplification, and detection, is integrated into an enclosed semi-automated system. This test is currently available only in Europe. The single study that has evaluated the test's performance demonstrated that the clinical sensitivity and specificity using nasal/throat or nasopharyngeal swab specimens were 100% and 97.9% for FluA and 100% and 100% for FluB, respectively, compared to RT-PCR (99). However, the current TAT for the SAMBA Flu duplex test is 2 hours and 15 minutes, longer than a number of other commercially available influenza molecular tests. The capacity to process multiple (up to four) samples in one test run, high sensitivity of molecular detection, low technical requirement, along with the result interpretation by eye on the dipstick make it particularly suitable in the near-patient setting.

Gastrointestinal Viruses

The FilmArray Gastro Intestinal (GI) panel (BioFire Diagnostics), which was recently FDA approved, allows for the detection of 23 viral, bacterial, and parasitic pathogens, including adenovirus F40/41, astrovirus, norovirus GI/GII, rotavirus A, and sapovirus (I, II, IV, and V) among viral agents. The TAT is approximately 1 hour with approximately 2 minutes hands-on time. A high sensitivity of >90% has been reported for most targets (100). The Verigene Enteric Pathogens (EP) Test (Nanosphere Inc.) recently received FDA clearance for the detection of two viral (noroviruses and rotaviruses) and seven bacterial agents (and associated toxins) that cause gastroenteritis infections, and it is reported to provide a result in approximately 2 hours. Advantages include the little hands-on time required to perform the test, coupled with the integrated sample-to-result format. Studies evaluating its performance in clinical and nonclinical settings are needed to assess its potential as a near-patient test for diagnosing GI infections.

QUALITY MANAGEMENT

The expansion of POCT outside the clinical laboratory necessitates implementation of new technologies and training of new operators who may not be familiar with regulatory requirements. In the United States, compliance with regulatory guidelines for POCT is mandatory for institutions to maintain their accreditation status. Federal and state requirements along with accreditation standards developed by the College of American Pathologists, Commission on Office Laboratory Accreditation, and The Joint Commission have resulted in major improvements in the point-of-care testing environment. A complete POCT program includes supervision and management, written procedures, operator training and competency, instrument evaluation, quality control, and proficiency testing, all with the necessary documentation. A POCT management program should ensure quality of testing and regulatory compliance and promote efficient management of resources such as devices, consumables, and staff time. The management team is responsible for oversight of the program, evaluation and implementation of new technologies, and training and education of staff that perform testing.

In the United States, CLIA specifies minimum testing requirements for every site examining "materials derived from the human body for the purpose of providing information for the diagnosis, prevention, or treatment of any disease . . ." (101). POCT sites in hospitals usually meet the CLIA requirements and carry an appropriate and current CLIA certificate by being accredited and inspected by the College of American Pathologists or The Joint Commission. Free-standing POCT sites, such as physician office laboratories, must obtain their own CLIA certificates by applying to the Centers for Medicare and Medicaid Services. These certificates are based in part on the complexity of the methods as determined by the technical difficulty and the knowledge, skills, and experience needed to perform the testing. Because of personnel and quality requirements, usually only waived or moderately complex methods are used at the point-of-care. For a laboratory performing only waived tests, CLIA has no specific requirements other than to follow the manufacturers' instructions.

Quality assurance is the process that monitors the entire test process to ensure that the test results are reliable. A quality assurance program identifies indicators that can be tracked and monitored over time to determine improvements for the test process. This involves ensuring that test procedures and processes are effectively implemented, test results are acceptable, the test system provided by the manufacturer meets quality requirements, and the risk of erroneous results is minimized. POCT has many of the same overall quality assurance requirements as testing in a central laboratory. Training of staff and assurance of competency are key to making POCT work. Control processes of POCT should assess equipment, specimens, reagents, and testing personnel (102). Some devices that are interpreted by direct visual inspection have internal procedural controls that indicate whether the individual device is operational and whether there are inadequate specimens or interferences with the test. External quality control and proficiency testing are important checks for equipment, reagents, and operator performance and are required for maintaining accreditation.

Because those performing POCT are not specialists in laboratory testing, and because these tests are frequently performed in settings where other medical and nonmedical activities compete for attention, ensuring careful quality oversight can be challenging in maintaining a POCT operation. The aim of POCT is to provide accurate, reliable, and clinically useful services to improve the speed, accessibility, and quality of patient care. The goal of quality control is to ensure a reliable and correct test result is reported to the health care provider. The manufacturers' instructions describe the minimum conditions or recommended frequencies for testing internal and external controls. At a minimum, test controls should be run with each new lot or shipment of test reagents or kits and quality control testing should be performed by the same personnel who perform the patient test.

POCT operator proficiency is critical to obtaining quality test results. Operators need adequate training on the specific test and must demonstrate a minimum level of competency to ensure their ability to achieve an appropriate test result. POCT operator training must emphasize not only proper device operation but should also include common sources of error in sample collection, patient identification, reagent storage, quality control, test analysis, reporting results, and instrument maintenance. Individuals performing tests for patient care need to be checked periodically to make sure they are still performing the test correctly. This competency assessment must be documented and retained for a period of 2 years. Every year at a minimum, the testing personnel should be assessed on the following: properly collecting and handling samples, doing quality control, performing the test according to the procedure, reading the test properly and obtaining correct answers, and recording and reporting the results properly (103).

SUMMARY

POCT is one of the most rapidly growing and changing segments of clinical diagnostic testing. Turnaround time usually is perceived as the key advantage of POCT over testing in a centralized laboratory. POCT can expedite treatment decisions and provide convenience for the patient. The drive to shorten patient encounters and provide care during a single visit makes POCT an important and attractive patient assessment tool. POCT can also provide access to testing for populations that otherwise may not be tested.

In order to justify the cost of the implementation of a POCT, the target pathogen should give rise to significant cost burden and pressure on the health care system. To ensure optimum POCT performance characteristics (sensitivity, specificity, positive predictive value, negative predictive value), the target pathogen should be present in the sample at a high titer and be highly prevalent in the target population (or have periods when it has high prevalence). POCT has proven most beneficial in situations in which more rapid institution of appropriate patient management can limit ongoing damage and produce better medical outcomes. Improved operational efficiency is a potential outcome and can also lead to lower costs, especially if it results in decreased length of stay in high-expense facilities such as emergency departments, intensive care units, and operating rooms.

Rapid antigen tests for influenza viruses and RSV have been shown to improve patient outcomes but still suffer from deficiencies in analytical and clinical sensitivity. Numerous studies demonstrate the benefits of rapid diagnosis of influenza, both in directing the appropriate use of antivirals and in reducing unnecessary diagnostic tests. The greatest cost benefit for rapid influenza virus testing is achieved when unnecessary antibiotics are not prescribed for patients with positive test results. It is generally accepted that these tests are most useful at the point-of-care when the virus is prevalent in the community. During peak prevalence of influenza in the community, negative results cannot be used to rule out infections.

Rapid antibody tests for HCV and HIV have documented exceptional sensitivities and specificities in both traditional and point-of-care settings. Importantly, the ease of use and affordability of these rapid tests have made them critical diagnostic tools in resource-limited and developing countries, where facilities may not be optimal and formal education programs for laboratorians are absent. The clinical impact of rapid HCV and HIV screening and diagnosis has been demonstrated by the reduction in transmission of infection, reduced loss of patients to follow-up, and more effective triage to treatment. While most of the current rapid HIV antibody tests are insensitive for detecting acute infections and require additional confirmatory testing, newer versions of these POCTs seek to shorten the window between infection and antibody production by using a combination of p24 antigen and antibody detection.

Most molecular platforms are best suited for placement in the laboratory for testing to be performed by medical technologists, particularly those of high complexity. However, several on-demand sample-to-result molecular testing platforms that have been simplified in complexity such that point-of-care or "near point-of-care" diagnostics are feasible and in fact anticipated. Fully automated on-demand or single-test formats are often significantly more expensive on a per test basis than batched testing formats. However, the rapid results provided by these systems often enable patient management decisions that can reduce the total cost of care. To reap the greatest benefit from these technologies, the assays must be able to be conducted and results reported 24 hours a day, 7 days a week or the benefit of rapid TAT to patient care will be lost.

It is necessary to ensure the clinical utility for POCT and the training and competence of all health care personnel performing the assay in all quality aspects, including appropriate documentation of tests performed, sample collection, performance of the tests, and interpretation of their results (104). Previous studies have shown that failure to train staff in each of these aspects can have adverse outcomes in terms of assay performance (105). Sometimes, the most difficult aspect of many POCTs is the interpretation of the results, particularly for immunochromatographic tests in which any color change at the sample line should be interpreted as positive. This can be quite subjective, with very faint reactions often missed by the user or misinterpreted as being false positive and thus reported as negative.

The expanding menu of POCT makes it desirable to improve access to testing, quality of patient care, and convenience for clinical care providers. Before we can reliably determine whether viral POCT warrants widespread utilization, we need stronger evidence that the provision of prompt results for a particular virus improves patient or hospital outcomes and is cost effective. Ultimately, the state-of-the-art bedside diagnostic technique will be able to rapidly and accurately identify nearly all clinically significant pathogens on a single panel.

REFERENCES

1. **Plebani M.** 2009. Does POCT reduce the risk of error in laboratory testing? *Clin Chim Acta* **404:**59–64.
2. **Glezen P, Denny FW.** 1973. Epidemiology of acute lower respiratory disease in children. *N Engl J Med* **288:**498–505.
3. **Mahony JB.** 2008. Detection of respiratory viruses by molecular methods. *Clin Microbiol Rev* **21:**716–747.
4. **Pavia AT.** 2011. Viral infections of the lower respiratory tract: old viruses, new viruses, and the role of diagnosis. *Clin Infect Dis* **52**(Suppl 4)**:**S284–S289.
5. **Babcock HM, Merz LR, Dubberke ER, Fraser VJ.** 2008. Case-control study of clinical features of influenza in hospitalized patients. *Infect Control Hosp Epidemiol* **29:**921–926.
6. **Heinonen S, Peltola V, Silvennoinen H, Vahlberg T, Heikkinen T.** 2012. Signs and symptoms predicting influenza in children: a matched case-control analysis of prospectively collected clinical data. *Eur J Clin Microbiol Infect Dis* **31:**1569–1574.
7. **Poehling KA, Edwards KM, Weinberg GA, Szilagyi P, Staat MA, Iwane MK, Bridges CB, Grijalva CG, Zhu Y, Bernstein DI, Herrera G, Erdman D, Hall CB, Seither R, Griffin MR, New Vaccine Surveillance Network.** 2006. The underrecognized burden of influenza in young children. *N Engl J Med* **355:**31–40.
8. **Williams LO, Kupka NJ, Schmaltz SP, Barrett S, Uyeki TM, Jernigan DB.** 2014. Rapid influenza diagnostic test use and antiviral prescriptions in outpatient settings pre- and post-2009 H1N1 pandemic. *J Clin Virol* **60:**27–33.
9. **Chartrand C, Leeflang MM, Minion J, Brewer T, Pai M.** 2012. Accuracy of rapid influenza diagnostic tests: a meta-analysis. *Ann Intern Med* **156:**500–511.
10. **Petrozzino JJ, Smith C, Atkinson MJ.** 2010. Rapid diagnostic testing for seasonal influenza: an evidence-based review and comparison with unaided clinical diagnosis. *J Emerg Med* **39:**476–490.e1.
11. **Babin SM, Hsieh YH, Rothman RE, Gaydos CA.** 2011. A meta-analysis of point-of-care laboratory tests in the diagnosis of novel 2009 swine-lineage pandemic influenza A (H1N1). *Diagn Microbiol Infect Dis* **69:**410–418.
12. **Ginocchio CC, Zhang F, Manji R, Arora S, Bornfreund M, Falk L, Lotlikar M, Kowerska M, Becker G, Korologos D, de Geronimo M, Crawford JM.** 2009. Evaluation of multiple test methods for the detection of the novel 2009 influenza A (H1N1) during the New York City outbreak. *J Clin Virol* **45:**191–195.
13. **Chu H, Lofgren ET, Halloran ME, Kuan PF, Hudgens M, Cole SR.** 2012. Performance of rapid influenza H1N1 diagnostic tests: a meta-analysis. *Influenza Other Respi Viruses* **6:**80–86.
14. **Harper SA, Bradley JS, Englund JA, File TM, Gravenstein S, Hayden FG, McGeer AJ, Neuzil KM, Pavia AT, Tapper ML, Uyeki TM, Zimmerman RK, Expert Panel of the Infectious Diseases Society of America.** 2009. Seasonal influenza in adults and children—diagnosis, treatment, chemoprophylaxis, and institutional outbreak management: clinical practice guidelines of the Infectious Diseases Society of America. *Clin Infect Dis* **48:**1003–1032.
15. **World Health Organization.** 2010. *Use of influenza rapid diagnostic tests*. Available at: http://whqlibdoc.who.int/publications/2010/9789241599283_eng.pdf.
16. **Centers for Disease Control and Prevention.** *Guidance for clinicians on the use of rapid influenza diagnostic tests*. Available at: http://www.cdc.gov/flu/professionals/diagnosis/clinician_guidance_ridt.htm.
17. **Abanses JC, Dowd MD, Simon SD, Sharma V.** 2006. Impact of rapid influenza testing at triage on management of febrile infants and young children. *Pediatr Emerg Care* **22:**145–149.
18. **Benito-Fernández J, Vázquez-Ronco MA, Morteruel-Aizkuren E, Mintegui-Raso S, Sánchez-Etxaniz J, Fernández-Landaluce A.** 2006. Impact of rapid viral testing for influenza A and B viruses on management of febrile infants without signs of focal infection. *Pediatr Infect Dis J* **25:**1153–1157.
19. **Hojat K, Duppenthaler A, Aebi C.** 2013. Impact of the availability of an influenza virus rapid antigen test on diagnostic decision making in a pediatric emergency department. *Pediatr Emerg Care* **29:**696–698.
20. **Poehling KA, Zhu Y, Tang Y-W, Edwards K.** 2006. Accuracy and impact of a point-of-care rapid influenza test in young children with respiratory illnesses. *Arch Pediatr Adolesc Med* **160:**713–718.
21. **Jennings LC, Skopnik H, Burckhardt I, Hribar I, Del Piero L, Deichmann KA.** 2009. Effect of rapid influenza testing on the clinical management of paediatric influenza. *Influenza Other Respi Viruses* **3:**91–98.
22. **Theocharis G, Vouloumanou EK, Rafailidis PI, Spiropoulos T, Barbas SG, Falagas ME.** 2010. Evaluation of a direct test for seasonal influenza in outpatients. *Eur J Intern Med* **21:**434–438.
23. **Blaschke AJ, Shapiro DJ, Pavia AT, Byington CL, Ampofo K, Stockmann C, Hersh AL.** 2014. A national study of the impact of rapid influenza testing on clinical care in the emergency department. *J Pediatric Infect Dis Soc* **3:**112–118.
24. **Özkaya E, Cambaz N, Coşkun Y, Mete F, Geyik M, Samanci N.** 2009. The effect of rapid diagnostic testing for influenza on the reduction of antibiotic use in paediatric emergency department. *Acta Paediatr* **98:**1589–1592.
25. **Bonner AB, Monroe KW, Talley LI, Klasner AE, Kimberlin DW.** 2003. Impact of the rapid diagnosis of influenza on physician decision-making and patient management in the pediatric emergency department: results of a randomized, prospective, controlled trial. *Pediatrics* **112:**363–367.
26. **Church DL, Davies HD, Mitton C, Semeniuk H, Logue M, Maxwell C, Donaldson C.** 2002. Clinical and economic evaluation of rapid influenza a virus testing in nursing homes in Calgary, Canada. *Clin Infect Dis* **34:**790–795.
27. **U.S. Department of Health and Human Services.** 2014. Microbiology devices; reclassification of influenza virus antigen detection test systems intended for use directly with clinical specimens. 21 CFR Part 866 Available at: https://www.federalregister.gov/articles/2014/05/22/2014-11635/microbiology-

devices-reclassification-of-influenza-virus-antigen-detection-test-systems-intended-for.
28. **Shay DK, Holman RC, Newman RD, Liu LL, Stout JW, Anderson LJ.** 1999. Bronchiolitis-associated hospitalizations among US children, 1980–1996. *JAMA* **282:**1440–1446.
29. **Mackie PL, Joannidis PA, Beattie J.** 2001. Evaluation of an acute point-of-care system screening for respiratory syncytial virus infection. *J Hosp Infect* **48:**66–71.
30. **Walsh EE.** 2011. Respiratory syncytial virus infection in adults. *Semin Respir Crit Care Med* **32:**423–432.
31. **Casiano-Colón AE, Hulbert BB, Mayer TK, Walsh EE, Falsey AR.** 2003. Lack of sensitivity of rapid antigen tests for the diagnosis of respiratory syncytial virus infection in adults. *J Clin Virol* **28:**169–174.
32. **Khanom AB, Velvin C, Hawrami K, Schutten M, Patel M, Holmes MV, Atkinson C, Breuer J, Fitzsimons J, Geretti AM.** 2011. Performance of a nurse-led paediatric point of care service for respiratory syncytial virus testing in secondary care. *J Infect* **62:**52–58.
33. **Mills JM, Harper J, Broomfield D, Templeton KE.** 2011. Rapid testing for respiratory syncytial virus in a paediatric emergency department: benefits for infection control and bed management. *J Hosp Infect* **77:**248–251.
34. **Dizdar EA, Aydemir C, Erdeve O, Sari FN, Oguz S, Uras N, Dilmen U.** 2010. Respiratory syncytial virus outbreak defined by rapid screening in a neonatal intensive care unit. *J Hosp Infect* **75:**292–294.
35. **Purcell K, Fergie J.** 2002. Concurrent serious bacterial infections in 2396 infants and children hospitalized with respiratory syncytial virus lower respiratory tract infections. *Arch Pediatr Adolesc Med* **156:**322–324.
36. **Titus MO, Wright SW.** 2003. Prevalence of serious bacterial infections in febrile infants with respiratory syncytial virus infection. *Pediatrics* **112:**282–284.
37. **Flaherman V, Li S, Ragins A, Masaquel A, Kipnis P, Escobar GJ.** 2010. Respiratory syncytial virus testing during bronchiolitis episodes of care in an integrated health care delivery system: a retrospective cohort study. *Clin Ther* **32:**2220–2229.
38. **Thibeault R, Gilca R, Côté S, De Serres G, Boivin G, Déry P.** 2007. Antibiotic use in children is not influenced by the result of rapid antigen detection test for the respiratory syncytial virus. *J Clin Virol* **39:**169–174.
39. **Parashar UD, Gibson CJ, Bresee JS, Glass RI.** 2006. Rotavirus and severe childhood diarrhea. *Emerg Infect Dis* **12:** 304–306.
40. **Borrows CL, Turner PC.** 2014. Seasonal screening for viral gastroenteritis in young children and elderly hospitalized patients: is it worthwhile? *J Hosp Infect* **87:**98–102.
41. **Festini F, Cocchi P, Mambretti D, Tagliabue B, Carotti M, Ciofi D, Biermann KP, Schiatti R, Ruggeri FM, De Benedictis FM, Plebani A, Guarino A, de Martino M.** 2010. Nosocomial Rotavirus Gastroenteritis in pediatric patients: a multi-center prospective cohort study. *BMC Infect Dis* **10:**235.
42. **Trivedi TK, DeSalvo T, Lee L, Palumbo A, Moll M, Curns A, Hall AJ, Patel M, Parashar UD, Lopman BA.** 2012. Hospitalizations and mortality associated with norovirus outbreaks in nursing homes, 2009–2010. *JAMA* **308:**1668–1675.
43. **Kak V.** 2007. Infections in confined spaces: cruise ships, military barracks, and college dormitories. *Infect Dis Clin North Am* **21:**773–784, ix–x.
44. **Beersma MF, Sukhrie FH, Bogerman J, Verhoef L, Mde Melo M, Vonk AG, Koopmans M.** 2012. Unrecognized norovirus infections in health care institutions and their clinical impact. *J Clin Microbiol* **50:**3040–3045.
45. **Battaglioli G, Nazarian EJ, Lamson D, Musser KA, St George K.** 2012. Evaluation of the RIDAQuick norovirus immunochromatographic test kit. *J Clin Virol* **53:**262–264.
46. **Geginat G, Kaiser D, Schrempf S.** 2012. Evaluation of third-generation ELISA and a rapid immunochromatographic assay for the detection of norovirus infection in fecal samples from inpatients of a German tertiary care hospital. *Eur J Clin Microbiol Infect Dis* **31:**733–737.
47. **Kirby A, Gurgel RQ, Dove W, Vieira SC, Cunliffe NA, Cuevas LE.** 2010. An evaluation of the RIDASCREEN and IDEIA enzyme immunoassays and the RIDAQUICK immunochromatographic test for the detection of norovirus in faecal specimens. *J Clin Virol* **49:**254–257.
48. **Ambert-Balay K, Pothier P.** 2013. Evaluation of 4 immunochromatographic tests for rapid detection of norovirus in faecal samples. *J Clin Virol* **56:**194–198.
49. **Centers for Disease Control and Prevention.** 1998. Recommendations for prevention and control of hepatitis C virus (HCV) infection and HCV-related chronic disease. *MMWR Recomm Rep* **47**(RR-19)**:**1–39.
50. **Smith BD, Morgan RL, Beckett GA, Falck-Ytter Y, Holtzman D, Teo CG, Jewett A, Baack B, Rein DB, Patel N, Alter M, Yartel A, Ward JW, Centers for Disease Control and Prevention.** 2012. Recommendations for the identification of chronic hepatitis C virus infection among persons born during 1945–1965. *MMWR Recomm Rep* **61**(RR-4)**:**1–32.
51. **Shivkumar S, Peeling R, Jafari Y, Joseph L, Pant Pai N.** 2012. Accuracy of rapid and point-of-care screening tests for hepatitis C: a systematic review and meta-analysis. *Ann Intern Med* **157:**558–566.
52. **Mabey D, Peeling RW, Perkins MD.** 2001. Rapid and simple point of care diagnostics for STIs. *Sex Transm Infect* **77:**397–398.
53. **Lee SR, Kardos KW, Schiff E, Berne CA, Mounzer K, Banks AT, Tatum HA, Friel TJ, Demicco MP, Lee WM, Eder SE, Monto A, Yearwood GD, Guillon GB, Kurtz LA, Fischl M, Unangst JL, Kriebel L, Feiss G, Roehler M.** 2011. Evaluation of a new, rapid test for detecting HCV infection, suitable for use with blood or oral fluid. *J Virol Methods* **172:**27–31.
54. **Smith BD, Drobeniuc J, Jewett A, Branson BM, Garfein RS, Teshale E, Kamili S, Weinbaum CM.** 2011. Evaluation of three rapid screening assays for detection of antibodies to hepatitis C virus. *J Infect Dis* **204:**825–831.
55. **Lee SR, Yearwood GD, Guillon GB, Kurtz LA, Fischl M, Friel T, Berne CA, Kardos KW.** 2010. Evaluation of a rapid, point-of-care test device for the diagnosis of hepatitis C infection. *J Clin Virol* **48:**15–17.
56. **Cha YJ, Park Q, Kang ES, Yoo BC, Park KU, Kim JW, Hwang YS, Kim MH.** 2013. Performance evaluation of the OraQuick hepatitis C virus rapid antibody test. *Ann Lab Med* **33:**184–189.
57. **Smith BD, Teshale E, Jewett A, Weinbaum CM, Neaigus A, Hagan H, Jenness SM, Melville SK, Burt R, Thiede H, Al-Tayyib A, Pannala PR, Miles IW, Oster AM, Smith A, Finlayson T, Bowles KE, Dinenno EA.** 2011. Performance of premarket rapid hepatitis C virus antibody assays in 4 national human immunodeficiency virus behavioral surveillance system sites. *Clin Infect Dis* **53:**780–786.
58. **Hickman M, McDonald T, Judd A, Nichols T, Hope V, Skidmore S, Parry JV.** 2008. Increasing the uptake of hepatitis C virus testing among injecting drug users in specialist drug treatment and prison settings by using dried blood spots for diagnostic testing: a cluster randomized controlled trial. *J Viral Hepat* **15:**250–254.
59. **Drobnik A, Judd C, Banach D, Egger J, Konty K, Rude E.** 2011. Public health implications of rapid hepatitis C screening with an oral swab for community-based organizations serving high-risk populations. *Am J Public Health* **101:**2151–2155.
60. **Hayes B, Briceno A, Asher A, Yu M, Evans JL, Hahn JA, Page K.** 2014. Preference, acceptability and implications of the rapid hepatitis C screening test among high-risk young people who inject drugs. *BMC Public Health* **14:**645.
61. **Hurt C, Tammaro D.** 2007. Diagnostic evaluation of mononucleosis-like illnesses. *Am J Med* **120:**911.e1–911.e8.
62. **Linderholm M, Boman J, Juto P, Linde A.** 1994. Comparative evaluation of nine kits for rapid diagnosis of infectious mononucleosis and Epstein-Barr virus-specific serology. *J Clin Microbiol* **32:**259–261.
63. **Elgh F, Linderholm M.** 1996. Evaluation of six commercially available kits using purified heterophile antigen for the rapid

diagnosis of infectious mononucleosis compared with Epstein-Barr virus-specific serology. *Clin Diagn Virol* **7:**17–21.

64. **Gerber MA, Shapiro ED, Ryan RW, Bell GL.** 1996. Evaluations of enzyme-linked immunosorbent assay procedure for determining specific Epstein-Barr virus serology and of rapid test kits for diagnosis for infectious mononucleosis. *J Clin Microbiol* **34:**3240–3241.
65. **Svahn A, Magnusson M, Jägdahl L, Schloss L, Kahlmeter G, Linde A.** 1997. Evaluation of three commercial enzyme-linked immunosorbent assays and two latex agglutination assays for diagnosis of primary Epstein-Barr virus infection. *J Clin Microbiol* **35:**2728–2732.
66. **Llor C, Hernández M, Hernández S, Martínez T, Gómez FF.** 2012. Validity of a point-of-care based on heterophile antibody detection for the diagnosis of infectious mononucleosis in primary care. *Eur J Gen Pract* **18:**15–21.
67. **Bruu AL, Hjetland R, Holter E, Mortensen L, Natås O, Petterson W, Skar AG, Skarpaas T, Tjade T, Asjø B.** 2000. Evaluation of 12 commercial tests for detection of Epstein-Barr virus-specific and heterophile antibodies. *Clin Diagn Lab Immunol* **7:**451–456.
68. **Keenan PA, Keenan JM, Branson BM.** 2005. Rapid HIV testing. Wait time reduced from days to minutes. *Postgrad Med* **117:**47–52.
69. **Centers for Disease Control and Prevention.** 2014. List of CLIA-waived rapid HIV tests. Available at: http://www.accessdata.fda.gov/scripts/cdrh/cfdocs/cfclia/analyteswaived.cfm?start_search=H.
70. **Centers for Disease Control and Prevention.** 2014. FDA-approved Rapid HIV Antibody Screening Tests. Available at: https://www.cdc.gov/topics/testing/rapid/rt-comparison.htm.
71. **World Health Organization.** 2013. HIV Assays: Operational Characteristics (Phase I). Report 17: HIV rapid diagnostic tests. WHO, Geneva. Available at: http://www.who.int/diagnostics_laboratory/evaluations/hiv/131107_hiv_assays17_final.pdf.
72. **Branson B.** 2003. Point-of-care rapid tests for HIV antibodies. *Laboratoriums Medizin* **27:**288–295.
73. **Pant Pai N, Balram B, Shivkumar S, Martinez-Cajas JL, Claessens C, Lambert G, Peeling RW, Joseph L.** 2012. Head-to-head comparison of accuracy of a rapid point-of-care HIV test with oral versus whole-blood specimens: a systematic review and meta-analysis. *Lancet Infect Dis* **12:**373–380.
74. **Garg N, Gautam V, Gill PS, Arora B, Arora DR.** 2006. Comparison of salivary and serum antibody detection in HIV-1 infection by ELISA and rapid methods in India. *Trop Doct* **36:**108–109.
75. **Delaney KP, Branson BM, Uniyal A, Kerndt PR, Keenan PA, Jafa K, Gardner AD, Jamieson DJ, Bulterys M.** 2006. Performance of an oral fluid rapid HIV-1/2 test: experience from four CDC studies. *AIDS* **20:**1655–1660.
76. **Holguín A, Gutiérrez M, Portocarrero N, Rivas P, Baquero M.** 2009. Performance of OraQuick Advance Rapid HIV-1/2 Antibody Test for detection of antibodies in oral fluid and serum/plasma in HIV-1+ subjects carrying different HIV-1 subtypes and recombinant variants. *J Clin Virol* **45:**150–152.
77. **De Cock L, Hutse V, Verhaegen E, Quoilin S, Vandenberghe H, Vranckx R.** 2004. Detection of HCV antibodies in oral fluid. *J Virol Methods* **122:**179–183.
78. **Pai NP, Tulsky JP, Cohan D, Colford JM Jr, Reingold AL.** 2007. Rapid point-of-care HIV testing in pregnant women: a systematic review and meta-analysis. *Trop Med Int Health* **12:** 162–173.
79. **Bulterys M, Jamieson DJ, O'Sullivan MJ, Cohen MH, Maupin R, Nesheim S, Webber MP, Van Dyke R, Wiener J, Branson BM; Mother-Infant Rapid Intervention At Delivery (MIRIAD) Study Group.** 2004. Rapid HIV-1 testing during labor: a multicenter study. *JAMA* **292:**219–223.
80. **Kelen GD, Shahan JB, Quinn TC.** 1999. Emergency department-based HIV screening and counseling: experience with rapid and standard serologic testing. *Ann Emerg Med* **33:** 147–155.
81. **Pai NP, Barick R, Tulsky JP, Shivkumar PV, Cohan D, Kalantri S, Pai M, Klein MB, Chhabra S.** 2008. Impact of round-the-clock, rapid oral fluid HIV testing of women in labor in rural India. *PLoS Med* **5:**e92.
82. **Pai NP, Klein MB.** 2009. Rapid testing at labor and delivery to prevent mother-to-child HIV transmission in developing settings: issues and challenges. *Womens Health (Lond Engl)* **5:**55–62.
83. **Puro V, Francisci D, Sighinolfi L, Civljak R, Belfiori B, Deparis P, Roda R, Modestino R, Ghinelli F, Ippolito G.** 2004. Benefits of a rapid HIV test for evaluation of the source patient after occupational exposure of healthcare workers. *J Hosp Infect* **57:**179–182.
84. **Turner SD, Anderson K, Slater M, Quigley L, Dyck M, Guiang CB.** 2013. Rapid point-of-care HIV testing in youth: a systematic review. *J Adolesc Health* **53:**683–691.
85. **Donnell-Fink LA, Arbelaez C, Collins JE, Novais A, Case A, Pisculli ML, Reichmann WM, Katz JN, Losina E, Walensky RP.** 2012. Acceptability of fingerstick versus oral fluid rapid HIV testing: results from the universal screening for HIV infection in the emergency room (USHER Phase II) randomized controlled trial. *J Acquir Immune Defic Syndr* **61:** 588–592.
86. **Pant Pai N, Sharma J, Shivkumar S, Pillay S, Vadnais C, Joseph L, Dheda K, Peeling RW.** 2013. Supervised and unsupervised self-testing for HIV in high- and low-risk populations: a systematic review. *PLoS Med* **10:**e1001414.
87. **Lubelchek R, Kroc K, Hota B, Sharief R, Muppudi U, Pulvirenti J, Weinstein RA.** 2005. The role of rapid vs conventional human immunodeficiency virus testing for inpatients: effects on quality of care. *Arch Intern Med* **165:**1956–1960.
88. **Sanders GD, Bayoumi AM, Sundaram V, Bilir SP, Neukermans CP, Rydzak CE, Douglass LR, Lazzeroni LC, Holodniy M, Owens DK.** 2005. Cost-effectiveness of screening for HIV in the era of highly active antiretroviral therapy. *N Engl J Med* **352:**570–585.
89. **Shrestha RK, Clark HA, Sansom SL, Song B, Buckendahl H, Calhoun CB, Hutchinson AB, Heffelfinger JD.** 2008. Cost-effectiveness of finding new HIV diagnoses using rapid HIV testing in community-based organizations. *Public Health Rep* **123**(Suppl 3):94–100.
90. **Holland CA, Kiechle FL.** 2005. Point-of-care molecular diagnostic systems–past, present and future. *Curr Opin Microbiol* **8:**504–509.
91. **Niemz A, Ferguson TM, Boyle DS.** 2011. Point-of-care nucleic acid testing for infectious diseases. *Trends Biotechnol* **29:**240–250.
92. **Nicosia A, Bandi M, DiFatta T, Ravanini P.** 2013. Usefulness of a real-time polymerase chain reaction (RT-PCR) point-of-care testing (POCT) instrument for viral emergencies. Abstr R2694. 23rd Eur Soc Clin Microbiol Infect Dis Congress.
93. **Ninove L, Nougairede A, Gazin C, Zandotti C, Drancourt M, de Lamballerie X, Charrel RN.** 2011. Comparative detection of enterovirus RNA in cerebrospinal fluid: GeneXpert system vs. real-time RT-PCR assay. *Clin Microbiol Infect* **17:** 1890–1894.
94. **Menasalvas-Ruiz AI, Salvador-García C, Moreno-Docón A, Alfayate-Miguélez S, Pérez Cánovas C, Sánchez-Solís M.** 2013. Enterovirus reverse transcriptase polymerase chain reaction assay in cerebrospinal fluid: an essential tool in meningitis management in childhood. *Enferm Infecc Microbiol Clin* **31:**71–75.
95. **Rogers B, Shankar P, Jerris R, Kotzbauer D, Anderson E, Watson J, O'Brien L, Uwindatwa F, McNamara K, Bost J.** 2015. Impact of a rapid respiratory panel test on patient outcomes. *Arch Pathol Lab Med.* **139:**636–641.
96. **Nakao A, Hisata K, Matsunaga N, Fujimori M, Yoshikawa N, Komatsu M, Kikuchi K, Takahashi H, Shimizu T.** 2014. The clinical utility of a near patient care rapid microarray-based diagnostic test for influenza and respiratory syncytial virus infections in the pediatric setting. *Diagn Microbiol Infect Dis* **78:**363–367.

97. **Alby K, Popowitch EB, Miller MB.** 2013. Comparative evaluation of the Nanosphere Verigene RV+ assay and the Simplexa Flu A/B & RSV kit for detection of influenza and respiratory syncytial viruses. *J Clin Microbiol* **51:**352–353.
98. **Bell JJ, Selvarangan R.** 2014. Evaluation of the Alere I influenza A&B nucleic acid amplification test by use of respiratory specimens collected in viral transport medium. *J Clin Microbiol* **52:**3992–3995.
99. **Wu LT, Thomas I, Curran MD, Ellis JS, Parmar S, Goel N, Sharma PI, Allain JP, Lee HH.** 2013. Duplex molecular assay intended for point-of-care diagnosis of influenza A/B virus infection. *J Clin Microbiol* **51:**3031–3038.
100. **Khare R, Espy MJ, Cebelinski E, Boxrud D, Sloan LM, Cunningham SA, Pritt BS, Patel R, Binnicker MJ.** 2014. Comparative evaluation of two commercial multiplex panels for detection of gastrointestinal pathogens by use of clinical stool specimens. *J Clin Microbiol* **52:**3667–3673.
101. **Centers for Medicare and Medicaid Services, Department of Health and Human Services.** 1992. Clinical laboratory improvement amendments of 1988; final rule. *Fed Regist* **57:**7139–7186. Codified at 7142 CFR 49342 CFR part 49493.
102. **CLSI.** 2006. *Point-of-care in vitro diagnostic (IVD) testing; Approved Guideline. CLSI document POCT4-A2.* CLSI, Wayne, PA.
103. **CLSI.** 2010. *Quality practices in noninstrumental point-of-care testing: and instructional manual and resources for health care workers: approved guideline. CLSI document POCT08-A.* CLSI, Wayne, PA.
104. **Lewandrowski K, Gregory K, Macmillan D.** 2011. Assuring quality in point-of-care testing: evolution of technologies, informatics, and program management. *Arch Pathol Lab Med* **135:**1405–1414.
105. **Fox JW, Cohen DM, Marcon MJ, Cotton WH, Bonsu BK.** 2006. Performance of rapid streptococcal antigen testing varies by personnel. *J Clin Microbiol* **44:**3918–3922.

Future Technology

ERIN MCELVANIA TEKIPPE AND CAREY-ANN D. BURNHAM

18

This is a very exciting time for the field of clinical diagnostics. In recent years, there has been an absolute explosion in new methodologies and instrumentation to diagnose infection, with a focus on optimization of therapy and antimicrobial stewardship (1, 2). The result is new technology that improves the diagnosis, characterization, and monitoring of viral infections. These technologies are changing almost every aspect of the laboratory workflow. For example, viral culture, once an important diagnostic tool for viral infections, is now commonly absent from diagnostic virology laboratories and reserved for public health and research laboratories. The transition away from viral culture was the consequence of stepwise innovations in viral diagnostics, including serologic testing, traditional end-point PCR, real-time PCR, multiplex PCR syndromic panels, and ultimately automated molecular methods (3). Technologies such as next-generation sequencing are making their way into the clinical space, offering potential advantages such as an unbiased approach to pathogen detection.

Most of these methods achieve marked improvements in turnaround time compared to conventional methods, but this often comes at an increased cost, might be technically demanding, or may require specialized equipment. It is unclear how many of these emerging technologies will rise to widespread, routine clinical use. There are a number of challenges that preceded widespread adoption, including regulatory approval and demonstration of adequate analytical performance characteristics.

This chapter will highlight and summarize some of the emerging technologies for the diagnosis of viral infections, including digital PCR, next-generation sequencing methods, mass spectrometry, surface plasmon resonance assays, and novel approaches to point-of-care diagnostics. The strengths and limitations of each methodology, as well as potential clinical diagnostic applications, will be described.

APPLICATIONS IN DIAGNOSTIC VIROLOGY

Next-Generation Sequencing

Approaches to DNA sequencing have evolved dramatically over the past several decades. In 1977, Frederick Sanger developed a chain-termination sequencing method (also known as Sanger sequencing); this technology was widely adopted as the primary "first-generation" DNA sequencing methodology. Although the initial iterations of this method were slow and laborious, the technology evolved and automated sequencers incorporating capillary electrophoresis were ultimately adopted, improving the speed and accuracy of this method (4).

Next-generation sequencing technologies represent major leaps forward, especially with regard to throughput and cost. That said, significant variability exists between methods, especially with regards to accuracy, quality, speed, throughput, sequence read length, and cost per base sequenced. We refer the reader to Chapter 15, Table 2 for an extensive overview of next-generation sequencing methods. Next-generation sequencing methods are typically further classified into "second-generation" and "third-generation" approaches. Different approaches to second-generation sequencing technologies emerged, such as pyrosequencing, based on detection of pyrophosphate release during nucleotide incorporation, or "sequencing by synthesis" whereby each type of nucleotide is labeled with a different cleavable fluorescent dye and a removable blocking group which complements the template one base at a time. Imaging systems are used to photograph fluorescently labeled nucleotides. An example of this type of technology would be the Illumina system (Illumina Inc., San Diego, CA). Another platform is the Ion Torrent system (ThermoFisher Scientific, Asheville, NC) based on semi-conductor sequencing, using the production of hydrogen (i.e., change in pH as a result of proton release) as a marker of the sequence. Although this approach can generate up to 25 Mb of sequence in a single run with a 2-hour runtime, accuracy of long repeats are very challenging for this method, and the read length is relatively short at approximately 50 to 100 nucleotides (4, 5).

At the time of writing, third-generation sequencing methods are based on detection of a fluorescent signal (e.g., the Pacbio system, Pacific Bioscience, Menlo, CA) or electric current (e.g., the Nanopore system, Oxford Nanopore Technologies, Oxford, UK). The Pacbio system has the advantage of producing an average read length of 1300 bp, which is much longer than second-generation methodologies; however, this comes with the tradeoff of a higher error

doi:10.1128/9781555819156.ch18

rate, a large instrumentation footprint, and increased cost (both capital cost and reagent cost) (4, 5).

There is a great deal of excitement around creating sequencing technologies that are even smaller and faster, with the ability to be used in almost any setting. For example, the Nanopore MinION is poised to be a "disruptive technology," based on the ability of this platform to produce sequences using a hand-held device that can connect to nearly any computer using a USB connection and is approximately the size of a cellular phone (6–9).

Next-generation sequencing approaches have played an important role in viral discovery as well as in providing an unbiased approach to viral detection in clinical specimens (10). For example, some of the work in this area has been focused on describing the so-called human virome—that is, the collection of viruses found in or on humans. The virome may consist of viruses causing active/acute infection, or latent viruses (11, 12). Other applications include unbiased surveys to identify a viral etiology for diseases that have been postulated to be attributed to a virus, but, to date, no virus has been identified (e.g., multiple sclerosis or Kawasaki disease) (13–17).

A number of bioinformatics pipelines are emerging to facilitate the use of next-generation sequencing for diagnosis of viral infections directly from clinical specimens. Viral applications of next-generation sequencing have some unique challenges. In contrast to bacteria and fungi, which have conserved genomic regions that can be used to query the overall community of taxa present, viruses present a special challenge as they lack such a conserved genomic segment. As an additional complication to analysis, viral sequences may be in low abundance compared to human or other microbial sequences. In addition, it may be difficult to classify viral sequences as a consequence of a lack of reference genomic sequences in databases to with which viral sequences can be aligned.

At this time, it is unclear if next-generation sequencing of clinical specimens will itself become a widely adopted and utilized methodology, or rather if it will be used to discover and define new viruses and provide detailed genomic information to inform targets for alternative diagnostic approaches, such as multiplex PCR, antigen detection, or array technologies.

Digital Polymerase Chain Reaction

Digital PCR (dPCR) is similar to traditional real-time PCR, in that primers, probes, and taq polymerase are used to amplify target DNA present within a sample. Unlike traditional PCR, in which the PCR reaction takes place in a single tube, dPCR is a collection of hundreds to thousands of individual, independent PCR reactions that take place in a tiny droplets (also referred to as droplet digital PCR) or wells of a specially designed chamber or plate. Each reaction is then analyzed for the presence or absence of DNA amplification using a fluorescent reporter and the total viral load is calculated based on the number of individual positive and negative reactions. Although not FDA approved, dPCR is commercially available for research applications from four manufacturers, including Bio-Rad (Hercules, CA), RainDance Technologies (Billerica, MA), Life Technologies (Carlsbad, CA), and Fluidigm (South San Francisco, CA). This is a rapidly evolving field, and many of these platforms are at an early stage of development; a summary comparing the systems can be found in a recent review (18).

Digital PCR is useful for detection and quantification of very low levels of virus in patient specimens. Traditional qPCR contains a great deal of interassay variability, making it a challenge to compare different commercial and laboratory-developed tests, or the same assay between different laboratories. In contrast, dPCR produces an absolute quantification of viral load without the requirement for a standard calibration curve, making interassay comparisons feasible (19). To date, this technology has been used to detect human immunodeficiency virus (HIV) (20, 21), cytomegalovirus (CMV) (22, 23), *Chlamydia trachomatis* (24), human herpesvirus type 6 (HHV-6) (25), and Hepatitis B virus (26) from clinical specimens. There is debate as to the relative sensitivity of viral detection of dPCR compared to quantitative PCR (qPCR), with studies showing that dPCR is more sensitive (20), the methods are equivalent (22), or even that real-time PCR is more sensitive for viral detection (23, 26). These discrepancies are likely influenced by the virus which is being assayed, the specimen type, and the study design (19).

The superior precision of dPCR has been widely reported, but the benefit of increased precision is unknown or unproven in most clinical scenarios. One area where a highly precise assay may be clinically useful is long-term monitoring of chronic viral diseases such as HIV, Hepatitis C virus, CMV, Epstein-Barr virus (EBV), adenovirus, BK virus, and HHV-6, especially in specialized patient populations. Often the decision to treat patients with antiviral medications and/or modulate immunosuppressive regimens are made based on changes in viral load at the low end of the quantifiable range of the PCR assay, where there is known to be a great deal of interassay variation. Precise measurement of the viral load in these patient specimens would be useful for monitoring therapy and making treatment decisions (22). Another niche where dPCR could have clinical benefit is in detection of proviral DNA as a measure of the cellular reservoir of HIV in HIV-infected patients. These reservoirs are frequently small and can have sequence variation, thus accurate quantification may be difficult using traditional viral load assays. Strain et al. found that dPCR had greater sensitivity and precision for detection of low levels of HIV proviral DNA compared to qPCR due to reduced background noise and increased detection of sequence variants (20).

Given the described technology, it is not likely that dPCR will supplant more traditional qPCR assays for routine virus detection. Some of the major reasons for this are that dPCR has a lower throughput and higher reagent costs than qPCR. However, dPCR could be a very useful tool for the specialized scenarios described above. Comparison studies of dPCR and qPCR as well as comparisons between different dPCR platforms are an area of active research which are likely to continue as dPCR technology is more often utilized for research and specialized clinical testing (23, 27).

Polymerase Chain Reaction/Electrospray Ionization Mass Spectrometry

Polymerase Chain Reaction/Electrospray Ionization Mass Spectrometry (PCR/ESI-MS) is a relatively new approach for detecting established and novel pathogens directly from clinical specimens. In brief, PCR amplification is performed on a sample using either target-specific or broad-range PCR primers, and ESI-MS is then performed on the resulting amplified nucleic acids. The mass to charge ratio detected by MS is analyzed by companion software and is translated into the base pair composition of the PCR product. PCR/ESI-MS is highly sensitive for detection of minute quantities of nucleic acids in patient specimens, and the technology is able to detect multiple pathogens from a single patient specimen.

Although first described for identification of bacterial pathogens, the PCR/ESI-MS and reverse transcriptase (RT)-PCR/ESI-MS have been used to detect both DNA and RNA viruses, respectively. Broad-range PCR primers that target highly conserved sequences allow for detection of both known and novel viruses. PCR-ESI/MS has been used to detect viruses that undergo rapid mutation, allowing for detection and subtyping of novel influenza viruses (28–31). It has also been used in humans to detect and serotype adenoviruses (32), identify Severe Acute Respiratory Syndrome-associated coronavirus (SARS-CoV) and other pathogenic coronaviruses (33), and identify other respiratory pathogens (34, 35). Additionally, PCR-ESI/MS has been used to identify monkeypox virus from clinical specimens obtained from experimentally infected macaques (36) as well as alphaviruses and flaviviruses from tick and mosquito vectors (37, 38).

PCR/ESI-MS testing requires specialized, expensive equipment with a large footprint, specific containment requirements to control for potential contamination, and highly skilled technicians to perform the assay. The technology is primarily used for infectious disease research, epidemiology, surveillance, and in the food industry. At the time of writing, no FDA-approved PCR/ESI-MS platforms are available for clinical use. Efforts have been made to automate and streamline this type of testing, and the Ibis 5000 BioSensor System was the first PCR-ESI/MS instrument introduced for identification of pathogens from clinical specimens (39). The technology was acquired in 2008 by Abbott Molecular and the platform was renamed PLEX-ID (40). PLEX-ID has a turnaround time of 6 to 8 hours for detection and identification of bacterial, viral, or fungal pathogens from patient specimens (41). Although PCR/ESI-MS is a powerful technology with enormous potential, it is not broadly applicable for routine clinical use in its current form.

Viral Microarray-Based Assays

The detection of novel viral pathogens in the clinical virology laboratory using molecular testing methods is a major challenge because assays only detect specific sequences for established viruses. Three microarray-based assays exist for detection of known and novel viral pathogens. These include the ViroChip (University of California San Francisco, CA) (42, 43), GreeneChip system (Columbia University, NY) (44), and Lawrence Livermore Microbial Detection Array (Lawrence Livermore National Laboratory, CA) (45).

Viral microarrays target conserved sequences established for known viruses, permitting simultaneous screening for all known viruses, as well as detection of novel viruses related to known viral families based on sequence homology. As of 2009, the ViroChip array contained >36,000 probes from >1,500 known viruses (43). For known viruses, the ViroChip pan-viral microarray has been shown to have a sensitivity that is comparable, if not superior in some cases, to conventional laboratory testing by real-time PCR for viral pathogens (46–48). The real value of viral microarray assays is in the ability to detect viruses from patient samples that are negative following routine PCR testing, either because the patient is infected with a novel virus, or because testing for the virus present is not widely available. Some examples of the latter include a 2005 case of parainfluenza virus type 4 detected from an immunocompromised adult with respiratory failure (49) and a 2006 case of human metapneumovirus detected from an adult with cancer and an acute respiratory illness (50). In addition, the ViroChip assay has been used to identify emerging viral pathogens including SARS coronavirus (51) and the 2009 pandemic influenza A H1N1 virus (52).

The ViroChip microarray assay is technically demanding, requires specialized instrumentation, and contains several analytical steps. Specimens undergo nucleic acid extraction, PCR amplification using random primers to amplify all DNA present in a specimen, fluorescent dye incorporation, microarray hybridization and scanning, and extensive data analysis. Although turnaround time is estimated to be 24 hours, the analysis requires several complicated steps, and, as a result, repeat analysis and/or troubleshooting may be required for a large percentage of samples. Therefore, viral microarray-based assays are not widely used for routine diagnosis of viral pathogens in the clinical microbiology laboratory at this time. These assays may be used as a tool for viral discovery from specimens from critically ill patients from which conventional testing fails to yield a diagnosis.

VirScan is a novel array approach to comprehensively analyze the collection of antiviral antibodies in human sera. The assay combines DNA microarray synthesis and bacteriophage display to create a uniform, synthetic representation of peptide epitopes comprising all know viruses that have human tropism. Immunoprecipitation and high-throughput DNA sequencing reveal the antigenic peptides recognized by antibodies in the human samples (53). This combination of a serologic- and sequencing-based approaches is a novel and powerful method for interrogating interactions between the human immune system and the virome. Additional studies evaluating how this type of approach may be incorporated into routine diagnostic applications in the future are eagerly awaited.

Surface Plasmon Resonance

Surface Plasmon Resonance (SPR) technology has been widely used for many years to investigate molecules and receptor interactions, such as biomolecular interactions between viruses and host cells, with the goal of developing new antiviral agents. Only recently has it been used for the direct detection of viruses from clinical specimens. SPR uses highly localized electromagnetic fields and a biosensor to detect very small changes in the refractive index of molecules in a signal chamber. To detect viruses or other pathogens, one signal chamber contains a glass slide coated with a virus-specific antibody while a second reference chamber contains a glass slide coated with nonspecific viral antibodies. Specimen is placed within the flow cell system, and as viral antigens pass through the flow cell, they are bound by virus-specific antibodies that are immobilized on the glass slide. As the sample runs through both flow cells simultaneously, any bound antigen will alter the surface plasmon waves, resulting in a change in the refractive index as measured by a detector. The viral load of a specimen can then be calculated based on the ratio of refractive index between the chambers. SPR technology is ideal for measuring viral load because it can be used to detect unlabeled virus and it is sensitive enough to detect minute amounts of antigen-antibody binding (54). To date, the literature is scarce regarding SPR and diagnosis of viral infections (55–60). A study by Su et al. found that SPR was not as sensitive as traditional quantitative PCR for detection of BK virus in urine, but that sensitivity of the assay was sufficient for detection of BK at subclinical levels in renal transplant patients (59).

Currently there are no commercially available SPR assays for detection of viruses in the clinical setting. Testing requires expensive and specialized equipment, no commercial reagents are available, and performing the assay requires a very high level of technical expertise. Technically, it may be difficult to maintain a high level of sensitivity with SPR given that specimen matrices such as plasma, serum, and urine, may nonspecifically absorb onto the sensing surface, which appears to greatly affect the lower limit of detection of this system. If these obstacles could be overcome, detection of viruses from clinical specimens is hypothetically a simple and rapid process, with a turnaround time of approximately 20 minutes. SPR is a flexible application, which can be adapted to detect any virus to which a capture antibody can be designed. On the other hand, SPR testing was not as sensitive for detection of BK virus detection compared to real-time PCR and the decrease in turnaround time (e.g., 20 minutes as opposed to 60 minutes for standard real-time PCR) is not clinically actionable. Therefore, at this time SPR remains a research tool rather than a test that is ready to be moved to the clinical laboratory for routine diagnosis of viral infection.

Digital Immunoassays for Rapid Detection and Differentiation of Influenza Viruses

Commercially available rapid influenza diagnostic tests (RIDTs) are a mainstay of influenza detection in outpatient physician offices and emergency rooms due to their rapid turnaround time, ease of use, and low cost (61). Unfortunately, these tests have been shown to be less sensitive than culture and real-time PCR-based detection methods, have reduced sensitivity in specimens with low concentrations of virus, and have variable sensitivity across influenza subtypes which can lead to the inability to detect novel or emerging influenza strains (62–66). The technological advancement and advantage of emerging digital immunoassays is not necessarily in methodology of the assay, but rather in the machinery used to interpret the assays. Traditionally, RIDTs are read by eye, resulting in reduced sensitivity and variability in interpretation among operators. The use of digital readers for assay interpretation permits a lower limit of detection than can be achieved by a manual interpretation of the assay. In addition, the objective digital readout of positive or negative results reduces interoperator variability, and results in increased specificity due to the presence of a negative control which measures nonspecific binding. Although nearly as user friendly as traditional RIDTs, digital immunoassays require test cartridges to be placed into the digital reader, which increases turnaround time and hands-on time of performing the assay, obligates an equipment purchase, and requires readers to be available wherever assays are resulted. The following section highlights new testing that has been designed to increase the sensitivity for detection of seasonal, pandemic, and novel influenza A and B detection provided by RIDTs, while retaining the desirable qualities of rapid turnaround time and ease of use.

Sofia Influenza A+B Fluorescent Immunoassay

The Sofia Influenza A+B Fluorescent Immunoassay (FIA) (Quidel, San Diego, CA) is a fluorescence-based lateral flow immunoassay which detects viral nucleoproteins (67). The assay uses polystyrene microbeads impregnated with europium, a compound which efficiently converts energy from 365 to 618 nm in wavelength, increasing the separation of excitation from emission wavelengths. This method reduces problems associated with traditional fluorescent compounds, such as background and interfering substances found in patient samples, and results in higher analytical sensitivity. In a study of over 2,000 patients at 16 testing sites, the Sofia influenza A+B FIA was 78% sensitive for detection of influenza A virus and 86% sensitive for influenza B virus, and specificity was >95% for both influenza A and B viruses compared to PCR-based testing (68). Similar to other types of influenza immunoassays, the Sofia assay performed well with samples containing high viral loads, but lacked sensitivity when testing samples with low viral titers. These results match those of two other published studies which found the Sofia influenza A+B FIA to be 80% and 74.78% sensitive for influenza A and B viruses, respectively (69), and 82.2 and 77.8% sensitive for influenza A and B viruses, respectively (70), compared to PCR testing of clinical isolates.

The Sofia influenza A+B FIA is an FDA cleared, Clinical Laboratory Improvement Amendments (CLIA) waived test that occurs in a self-contained cassette, requires minimal hands-on time, and 15 minutes of incubation. For analysis, the test cartridge is placed into the Sofia FIA for one minute to generate a result, eliminating the need for visual interpretation by a technologist. The Sofia FIA can also be used to test for respiratory syncytial virus (RSV), Group A *Streptococcus*, and *Legionella* (not currently available in the United States), making the analysis instrumentation versatile for clinical laboratory and point-of-care testing.

BD Veritor System Flu A+B Immunoassay

The BD Veritor System Flu A+B (Becton, Dickinson, and Company, Franklin Lakes, NJ) is a digital immunoassay system that uses a reader and test device for direct detection and differentiation of influenza A and B viruses from nasopharyngeal specimens, nasal washes, nasopharyngeal aspirates, and swabs from symptomatic patients (71). The BD Veritor System is a reading instrument which analyzes the line signal intensity in comparison to background signal to increase the sensitivity of testing. A study comparing BD Veritor System Flu A+B, BinaxNOW Influenza A+B, and QuickVue influenza testing found the BD Veritor System Flu A+B to be the most sensitive of the assays for detection of the three influenza strains tested (two influenza A viruses, one influenza B virus) (72). BD Veritor System Flu A+B achieved approximately a one log greater sensitivity of detection compared to the QuickVue Influenza Test and 1.5 logs greater sensitivity of detection compared to BinaxNOW Influenza A+B. Studies using clinical specimens have shown that the BD Veritor System Flu A+B has improved sensitivity compared to rapid influenza diagnostic tests. In one study, the assay was 72% and 69.3% sensitive for detection of influenza A and B viruses, respectively, compared to real-time PCR (73). The sensitivity of influenza detection using digital immunoassays appears to be higher in pediatric populations compared to their adult counterparts. A study of 200 pediatric clinical specimens found an overall sensitivity and specificity of 89.6% and 98.8%, respectively, compared to PCR (74).

The BD Flu A+B Immunoassay is an FDA-cleared and CLIA-waived rapid assay that requires a 10-minute incubation period once the patient specimen has been added. Analysis using the BD Veritor System takes approximately 10 seconds per assay and reports a positive or negative result, completely eliminating the need for a technologist visualization and interpretation. In addition to the Flu A+B Immunoassay, the BD Veritor System is able to analyze other BD immunoassays for RSV and Group A *Streptococcus*, making it versatile in both clinical laboratories and outpatient settings. The reduced incubation, analysis, and hands-on times required by the BD Veritor System Flu A+B compared to the Sofia influenza A+B FIA have been shown to reduce the turnaround time to results when testing single samples or batching samples in groups of ten (75).

In summary, implementation of digital analyzers may increase the sensitivity of influenza immunoassays while still providing a rapid turnaround time compared to more sensitive real-time PCR assays.

Emerging Point-of-Care Diagnostics

Recent advances in clinical virology testing have focused on developing testing devices and assays that are smaller, portable, technically simple to use, and can be performed one test at a time with high sensitivity and specificity. Ideally, these tests would be performed near to the patient. The following point-of-care assays expand viral testing from within the confines of the clinical microbiology laboratory to outpatient physician offices and remote locations around in the world.

Influenza Viruses

Alere I Influenza A&B Assay

The Alere i Influenza A&B assay (Alere, Waltham, MA) is a new, rapid molecular assay for identification and differentiation of influenza virus subtypes A and B from clinical specimens. This assay is the first CLIA-waived molecular rapid influenza test to be cleared by the FDA. The assay uses isothermal nucleic acid amplification to amplify short regions of target RNA from clinical respiratory specimens, followed by a thermostable nicking endonuclease to differentiate between influenza virus subtypes A and B. Fluorescently labeled molecular beacons are used to detect the presence of sequences specific for influenza A virus and influenza B virus in real time.

In studies by Bell et al. and Nie et al., a combined total of >900 adult and pediatric nasopharyngeal and nasal swab specimens were tested using the Alere i assay (76, 77). It was found to be 73.2 to 99.3% sensitive and 98.1 to 100% specific for detection of all influenza A viruses and 97% sensitive and 100% specific for detection of influenza B virus compared to real-time PCR. The assay performed very well overall, but struggled to identify influenza A viruses outside of the subtypes H1N1 and H3N2 (76).

The major advantage of the Alere i Influenza A&B assay is an increase in sensitivity over RDITs. Testing requires approximately 2 minutes of hands on time for specimen handling and setup. Results are available in less than 15 minutes, providing molecular detection of influenza with a turnaround time comparable to RIDTs. The major limitation of this assay is that only one specimen can be run on the instrument at a time. Also, although molecular testing occurs in a single-use cartridge, there are a few steps necessary to dilute and load the correct amount of patient specimen into the test cartridge. Overall, these data suggest that the Alere i Influenza A&B assay is a viable alternative to point-of-care RIDTs.

Roche Cobas Liat System

The Cobas Liat or "lab in a tube" system (Roche Diagnostics, Indianapolis, IN) is a point-of-care molecular platform that provides highly sensitive detection of influenza A and B viruses in less than 20 minutes, which is similar to the time it takes to run a rapid antigen test. Briefly, patient sample is loaded into a test strip and inserted into the small Cobas Liat instrument. A real-time PCR assay is performed using a series of plungers and clamps which move sample to different areas of the test strip, allowing for cell lysis, nucleic acid amplification, and detection within a single test strip. The assay is a closed system and an instrument can run one specimen at a time. The assay is FDA-cleared and CLIA-waived. Studies have found the Cobas Liat System to be >97% sensitive and >99% sensitive for detection of influenza A and B viruses compared to the Prodesse ProFlu +real-time PCR assay (78), and >99% sensitive and 100% specific compared to the Focus Simplexa Flu A/B and RSV Direct test (79).

Cepheid GeneXpert Omni

The Cepheid GeneXpert Omni (Cepheid Inc., Sunnyvale, CA) is a small, portable, molecular point-of-care testing device. The instrument weighs only 1 kg, is wireless, runs off battery power, and uses Cepheid's preexisting GeneXpert testing cartridges. The novelty of this instrument is not in the decreased turnaround time of results, but in its portability. The instrument size and the ability to perform testing without the need for electricity allow highly sensitive real-time PCR testing to take place nearly anywhere in the world. Like the original GeneXpert system, nucleic acid extraction, amplification, and detection occur in a close system and the Omni runs one test cartridge at a time. The Omni is expected to debut in the United States and emerging markets in 2016 with an expected test menu of Xpert MTB Ultra, Xpert HIV-1 Qualitative, Xpert HIV-1 Viral Load, Xpert HCV Viral Load, and Xpert Ebola (Cepheid website. Accessed January 7, 2016) (80).

Focus Simplexa FluA/B & RSV Direct

The Focus Simplexa Flu A/B & RSV Direct assay (Focus Diagnostics, Inc., Cypress, CA) is an FDA-approved, multiplex, real-time PCR assay for the detection of influenza A and B viruses, and RSV from nasopharyngeal specimens. Compared to traditional real-time PCR assays requiring nucleotide extraction prior to PCR amplification, the Focus Simplexa FluA/B & RSV assay is an extraction-independent assay. Specimen and master mix are added directly to the assay well with no further processing steps, which reduces the time it takes to perform the assay to just 60 minutes and reduces the hands-on time to just minutes. In a study of 210 clinical specimens from patients of all ages, the Focus Simplexa FluA/B & RSV Direct assay was found to have specificities of 92.2% for influenza A virus, 89.1% for influenza B virus, and 80.7% for RSV compared to a laboratory developed real-time PCR assay (81). The specificity was 100% for all targets. Many other comparison studies have since been published on the topic showing high positive and negative percent agreement compared to real-time PCR (82–85).

Enigma MiniLab Influenza A/B & RSV

Enigma MiniLab influenza A/B & RSV (Enigma Diagnostics Inc, San Diego, CA) is a point-of-care molecular assay for detection of influenza A and B viruses, and RSV from nasopharyngeal swabs. The RT-PCR assay requires minimal hands-on time and provides a result in 90 minutes. The instrument is random access and scalable, with the ability to run a maximum of six samples at a time. In a prospective study, 698 specimens from children <16 years with respiratory symptoms were tested using the Enigma MiniLab Influenza A/B & RSV system and the Luminex xTag Respiratory Virus Panel. Very few of the children had influenza infection during the testing period, with only 22 being positive for influenza A virus and only 6 being positive for influenza B virus by the Enigma assay. In comparison to the Luminex assay, there were 18 specimens positive by both methods, but 10 discrepant results which were positive by only one of the assays. Thus, there is a positive percent agreement of 81.8% between the methods for influenza A virus with 98.8% negative agreement. For influenza B virus the positive and negative percent agreement were 100% and 99.8%, respectively (86, 87). The Enigma MiniLab influenza A/B & RSV has obtained CE-IVD designation in Europe, but is not currently available in the United States.

Human Immunodeficiency Virus

The development of HIV diagnostics for point-of-care use has been a focus with two groups of users in mind. In developed countries, highly sensitive point-of-care tests are used to rapidly diagnose persons with acute HIV in the comfort of their physician's office and immediately place them into counseling and establish a treatment plan and follow-up care. In rural settings in underdeveloped international countries, there is a great need to rapidly diagnose persons with HIV because patients who test positive for HIV can be lost to follow-up during the weeks to months it takes for results to be returned from a centralized testing laboratory.

Alere Determine HIV-1/2 Ag/Ab Combo Rapid Test

Previously, screening for HIV-1 & HIV-2 antibodies has been the mainstay of point-of-care HIV testing, but this method does not detect HIV in the "window period" of 3 to 4 weeks in which infected persons will test negative for HIV-1 & HIV-2 antibodies, have a high viral load, and are highly infectious to others. The Alere Determine HIV-1/2 Ag/Ab Combo Rapid Test (Alere, Waltham, MA) is a 4th generation point-of-care assay developed to detect both HIV-1 & HIV-2 antibodies as well as p24 antigen. Studies have shown that the assay was able to detect 24 of 39 (62%) banked blood specimens from patients diagnosed with early HIV infection based on p24 antigen and/or HIV RNA presence, but not HIV-1 & HIV-2 antibodies (88). In a separate study, the same assay detected 55% of early HIV cases, classified as patients with a negative or indeterminate 3rd generation HIV screening assay or a negative antibody confirmation test following a positive 4th generation screening assay (89). These studies show that, although not as sensitive as laboratory-based 4th generation HIV Ag/Ab screening assays, the Alere Determine HIV-1/2 Ag/Ab Combo Rapid Test is a viable option for rapid detection of early HIV infection in outpatient international settings.

Alere Q NAT

The Alere Q NAT system (Alere, Scarborough, ME) is a point-of-care test that detects HIV-1 and 2 RNA from 25 µl of whole blood in 60 minutes. In a study of 827 HIV-exposed infants aged 4 weeks to 18 months being seen at a primary health center in Mozambique, the Alere Q NAT was 98.5% sensitive and 99.9% specific for detection of HIV when compared to the Roche Cobas Ampliprep/Cobas TaqMan HIV-1 Qualitative Test, which detects HIV-1 total nucleic acids (RNA and DNA) (90). Having point-of-care testing on site allows for immediate follow up and treatment initiation for HIV positive patients.

Biobarcode Amplification (BCA) Assay

A new research method under development for point-of-care HIV testing, the biobarcode amplification (BCA) assay, is a europium nanoparticle-based microtiter-plate immunoassay. Tang et al. have previously developed a traditional europium chelate-dyed nanoparticle assay placed on a lateral flow immunoassay platform that is able to detect HIV p24 antigen in amounts less than 1 pg/ml (91). To adapt europium nanoparticle-based technology for detection of HIV in low resource settings, the assay was miniaturized to a microchip format. The miniaturized assay was slightly less sensitive than the original assay, with a limit of detection of 5 pg/ml of HIV-1 p24 antigen, but the sample and reagent volumes required for testing have been reduced by 4.5-fold and the assay run time has been decreased from around 2.5 hours to 60 to 75 minutes (92). Although currently at the concept stage of development, with further optimization this assay has the potential to be used as a highly sensitive point-of-care method of HIV detection in resource-limited settings.

Coinfections with HIV and *Treponema pallidum*, the causative agent of syphilis, are on the rise. An emerging group of point-of-care diagnostic tests for HIV and syphilis are currently being developed for rapid detection of both sexually transmitted diseases at once. A study comparing three research-use only assays, MedMira Multiplo TP/HIV test (MedMira Inc., Halifax, Nova Scotia, Canada), Standard Diagnostics (SD) Bioline HIV/Syphilis Duo test (Standard Diagnostics Inc., Gyeonggi-do, Republic of Korea), and Chembio DPP HIV-syphilis assay (ChemBio Diagnostics Inc., Medford, NY), was performed using 150 serum specimens previously tested for the presence of HIV and *T. pallidum* using routine laboratory methods (81). The three dual detection assays were 98 to 99% sensitive and 94 to 100% specific for detection of HIV compared to the Siemens Advia HIV enzyme immunoassay. For *T. pallidum* detection, the dual assays were 93 to 95% sensitive and 97 to 100% specific compared to the TP-PA assay. Having a dual assay available at the point-of-care for both HIV and syphilis may prevent further spread of these sexually transmitted diseases due to earlier diagnosis and initiation of treatment.

Dried Blood Spot Testing

In developed countries, patients are monitored by HIV viral load determination and HIV drug resistance genotyping prior to initiation of antiretroviral therapy and again in the setting of treatment failure. Unfortunately, this type of monitoring is difficult to provide to the majority of HIV-infected persons, as they reside in health care limited settings where the majority of HIV testing is performed in distant centralized laboratories. A successful method to store and ship blood specimens has been the dried blood spot, in which patient blood samples are blotted onto absorbent filter paper and dried. The dried blood spot is noninfectious, relatively stable, light weight, and can be shipped without refrigeration, making it an ideal method to transport patient specimens to distant laboratories for testing. A study by Parry et al. was initiated comparing use of dried blood spot specimens to the gold standard of plasma for HIV drug resistance genotyping. The study demonstrated that genotyping was feasible from dried blood spots as long as they had not been stored at ambient temperatures for more than two weeks (93). The testing performance was best if the HIV RNA viral load in the blood specimen was >50,000 copies/ml. Based on this data, dried blood spot testing is a reasonable alternative as a specimen type for HIV resistance genotyping and other HIV testing in resource limited settings. It is hoped that the ease and cost effectiveness of dried blood spot testing will promote more HIV resistance genotyping as well as measurements of viral load and early infant diagnosis, thereby improving care of patients in health care–limited settings.

Ebola Virus

A growing area of rapid diagnostic testing (Table 1) has been prompted by the major Ebola virus outbreak that started in

TABLE 1 Summary of diagnostic assays for Ebola virus

	Test	Specimens	Method	FDA status	WHO status
Antigen Assays	ReEBOV™ Antigen Rapid Test (Corgenix, Inc.)	Fingerstick (capillary) whole blood, venous whole blood, or plasma	Rapid chromatographic immunoassay for the qualitative detection of VP40 antigen from Ebola viruses	Emergency use authorization, March 23, 2015	Emergency Use Assessment and Listing
Antigen Assays	Ebola virus rapid antigen diagnostic test (Defense Science and Technology Laboratory)	Whole blood (capillary)	Lateral flow immunoassay	None	None
Antigen Assays	Ebola Rapid Test (Senova)	Whole blood	Lateral flow immunoassay	None	None
Molecular Assays	FilmArray NGDS BT-E Assay	Whole-blood, plasma and serum	Multiplexed PCR and DNA Melting Analysis	Emergency use authorization, October 25, 2014	None
Molecular Assays	FilmArray Biothreat-E test	Whole blood or undiluted urine	Multiplexed PCR and DNA Melting Analysis	Emergency use authorization, October 25, 2014	None
Molecular Assays	Xpert® Ebola Assay (Cepheid)	Venous whole blood	Real-time reverse transcription polymerase chain reaction (RT-PCR)	Emergency use authorization, February 24, 2015	Emergency Use Assessment and Listing
Molecular Assays	LightMix® Ebola Zaire rRT-PCR Test (Roche Molecular Systems, Inc.)	Whole blood	Real-time reverse transcription polymerase chain reaction (rRT-PCR)	Emergency use authorization, December 23, 2014	None
Molecular Assays	Liferiver™ - Ebola Virus (EBOV) Real Time RT-PCR (Shanghai ZJ BioTech Co., Ltd.)	Blood, serum, plasma	Real-time RT-PCR	None	Emergency Use Assessment and Listing
Molecular Assays	RealStar® Ebolavirus RT-PCR Kit 1.0 (altona Diagnostics, GmbH)	Plasma	Real-time reverse transcription polymerase chain reaction (RT-PCR)	Emergency use authorization, November 10, 2014	None
Molecular Assays	RealStar® Filovirus Screen RT-PCR Kit 1.0 (altona Diagnostics, GmbH)	Plasma	Real-time reverse transcription polymerase chain reaction (RT-PCR)	None	Emergency Use Assessment and Listing
Molecular Assays	EZ1 Real-time RT-PCR Assay (DoD)	Whole blood, plasma	Real-time reverse transcription (rRT) polymerase chain reaction (PCR) (TaqMan®) assay	Emergency use authorization, August 5, 2014	None
Molecular Assays	Ebola Virus NP Real-Time RT-PCR Assay (CDC)	Whole blood, serum, plasma, and urine	Real-time (TaqMan®) RT-PCR (rRT-PCR) assay	Emergency use authorization, October 10, 2014	None
Molecular Assays	Ebola Virus VP40 Real-Time RT-PCR Assay (CDC)	Whole blood, serum, plasma, and urine	Real-time (TaqMan®) RT-PCR (rRT-PCR) assay	Emergency use authorization, October 10, 2014	None

West Africa in 2014. Traditionally, regional laboratories proximal to outbreak areas have employed reverse transcriptase PCR-based testing for Ebola virus detection. PCR-based detection is highly sensitive and specific, and is rapid compared to viral culture-based methods of detection. Unfortunately, outbreaks tend to occur in low-resource settings which lack infrastructure such as mechanisms for specimen transport, biological safety cabinets, thermocyclers and other laboratory instrumentation, skilled personnel, as well as refrigeration and electricity. These factors are all barriers to Ebola testing in this setting.

To combat these challenges, several companies have developed lateral flow immunochromatographic assays for diagnosis of Ebola virus antigen. These tests are rapid and can be performed as point-of-care tests with blood from a simple finger stick. They do not require specialized equipment or highly trained laboratory personnel. The assays are easy to transport and fairly inexpensive, ideal characteristics for use in remote locations with limited access to clinical laboratories. Although antigen tests are generally not as sensitive as amplified PCR testing, there is hope that these rapid immunoassays could be used in outbreak settings to triage patients (94).

ReEBOV Antigen Rapid Test

The ReEBOV Antigen Rapid Test is an immunochromogenic dipstick immunoassay manufactured by Corgenix (Broomfield, CO). It is the first rapid antigen test for detection of Ebola virus to be granted Emergency Use Assessment and Listing (EUAL) through the World Health Organization (WHO) and Emergency Use Authorization (EUA) by the U.S. FDA. Briefly, blood from a finger stick is placed on the pad of the diagnostic strip and a chemical solution is added to inactivate the virus. Lateral flow action wicks the specimen up the test strip, which contains antibodies specific for Ebola virus VP40 antigen. Results of the testing are available within 15 to 25 minutes. This assay was developed in cooperation with the Viral Hemorrhagic Fever Consortium, a group of researchers with ties to industry and academics. ReEBOV technology is based on the group's previous development of a rapid diagnostic assay for Lassa fever. Testing of nearly 300 fresh whole blood and archived plasma specimens showed that the ReEBOV was 91.8% sensitive and 84.6% specific for detection of Ebola Zaire virus compared to reverse transcriptase-PCR (WHO emergency use assessment public report-ReEBOV).

Defense Science and Technology Laboratory Ebola Virus Rapid Antigen Diagnostic Test

Similar to the ReEBOV rapid test, the British-designed Defense Science and Technology Laboratory Ebola Virus Rapid Antigen Diagnostic Test (DSTL EVD RDT) is a rapid, lateral flow assay that can be performed at point-of-care. The assay uses capillary blood to test for the presence of Ebola virus antigen and results are available in 20 minutes. The DSTL EVD RDT is unique in that it produces semi-quantitative results based on color intensity of the developed line. Intensity is scored between 2 and 10, and any number 4 or greater is considered positive. In a study of 131 patients from Sierra Leone on which Ebola testing was performed, 15 patients were positive for Ebola virus by RT-PCR. The DSTL EVD RDT analysis results in a sensitivity of 96.6% and a specificity of 100%. The lower sensitivity was due to specimens with lower intensity ratings. For those assays scoring an 8 or greater led to a rise in sensitivity to 99% (95).

There are currently no FDA-approved tests for detecting Ebola. In October 2014, the FDA granted Emergency Use Authorization (EUA) for Ebola virus testing (Table 1) in the face of a public health crisis. EUA status was first granted for the use of RT-PCR assays used by the Department of Defense and the Centers for Disease Control and Prevention. Shortly after, private companies were granted EUA status for their Ebola diagnostic testing products starting with the FilmArray NGDS BT-E and Biothreat-E tests (Biofire, Salt Lake City, UT).

FilmArray BioThreat Panel

Biofire is best known for multiplex PCR assays that detect pathogens associated with a specific disease syndrome. They offer a respiratory panel for detection of 20 respiratory pathogens, a gastrointestinal panel for detection of common viral, bacterial, and parasitic causes of infectious diarrhea, a meningitis/encephalitis panel for detecting viruses, bacteria, and yeast responsible for central nervous system diseases, and a blood culture identification panel for the differentiation of 24 bacteria and yeast and three antibiotic resistance genes. The company has recently developed a panel to be used for surveillance of bioterrorism agents and pandemic diseases called the FilmArray BioThreat Panel. It detects 17 bacterial and viral pathogens as well as toxin genes. The assay is easy to set up, requiring less than five minutes of hands-on time for technologists, and it has an analysis time of approximately an hour. The BioThreat panel is performed on the same instrumentation as other FilmArray assays, making it easily adaptable to the clinical laboratory for institutions running other FilmArray assays. The instrument itself is small enough to fit into a biosafety cabinet, providing an extra level of safety for laboratories that wish to pursue testing.

Summary

Diagnostic virology is a rapidly evolving discipline, both because the patient population is changing and because of the emergence of new technology. These new technologies have variable complexity, instrumentation requirements, and analytical performance characteristics and, as a result, variable potential for translation into routine use in the clinical laboratory. That said, molecular testing in the clinical virology laboratory is rapidly evolving and is being made smaller and simpler due to advances in microelectronics, microfluidics, and microfabrication. The number of approved "sample-in-answer-out" assays is growing and more are under development. As a result of these advances, we may see a change in not only how testing is done, but where testing is performed. It is expected that some testing will move out of the laboratory and into physician offices and community pharmacies. The dichotomy between highly technical, highly complex testing requiring expert interpretation (e.g., next-generation sequencing) and rapid testing requiring minimal expertise to perform, is expanding. This is an exciting time of tremendous growth and rapid development, which will enhance our understanding of viral pathogens and enhance patient care.

REFERENCES

1. **Caliendo AM, Gilbert DN, Ginocchio CC, Hanson KE, May L, Quinn TC, Tenover FC, Alland D, Blaschke AJ, Bonomo RA, Carroll KC, Ferraro MJ, Hirschhorn LR, Joseph WP, Karchmer T, MacIntyre AT, Reller LB, Jackson AF, Infectious Diseases Society of America (IDSA).** 2013. Better tests, better care: improved diagnostics for infectious diseases. *Clin Infect Dis* **57**(Suppl 3)**:**S139–S170.
2. **Zumla A, Al-Tawfiq JA, Enne VI, Kidd M, Drosten C, Breuer J, Muller MA, Hui D, Maeurer M, Bates M, Mwaba P, Al-Hakeem R, Gray G, Gautret P, Al-Rabeeah AA, Memish ZA, Gant V.** 2014. Rapid point of care diagnostic tests for viral and bacterial respiratory tract infections—needs, advances, and future prospects. *Lancet Infect Dis* **14:**1123–1135.
3. **Boonham N, Kreuze J, Winter S, van der Vlugt R, Bergervoet J, Tomlinson J, Mumford R.** 2014. Methods in virus diagnostics: from ELISA to next generation sequencing. *Virus Res* **186:**20–31.
4. **Liu L, Li Y, Li S, Hu N, He Y, Pong R, Lin D, Lu L, Law M.** 2012. Comparison of next-generation sequencing systems. *J Biomed Biotechnol* **2012:**251364.
5. **Buchan BW, Ledeboer NA.** 2014. Emerging technologies for the clinical microbiology laboratory. *Clin Microbiol Rev* **27:** 783–822.
6. **Rusk N.** 2015. MinION takes center stage. *Nat Methods* **12:** 12–13.
7. **Loman NJ, Quinlan AR.** 2014. Poretools: a toolkit for analyzing nanopore sequence data. *Bioinformatics* **30:**3399–3401.
8. **Mikheyev AS, Tin MM.** 2014. A first look at the Oxford Nanopore MinION sequencer. *Mol Ecol Resour* **14:**1097–1102.
9. **Watson M, Thomson M, Risse J, Talbot R, Santoyo-Lopez J, Gharbi K, Blaxter M.** 2015. poRe: an R package for the visualization and analysis of nanopore sequencing data. *Bioinformatics* **31:**114–115.
10. **Stephenson J.** 2015. Genomic technologies speed pathogen detection. *JAMA* **314:**212–214.
11. **Wylie KM, Weinstock GM, Storch GA.** 2012. Emerging view of the human virome. *Transl Res* **160:**283–290.
12. **Wylie KM, Weinstock GM, Storch GA.** 2013. Virome genomics: a tool for defining the human virome. *Curr Opin Microbiol* **16:**479–484.
13. **Yim D, Curtis N, Cheung M, Burgner D.** 2013. Update on Kawasaki disease: epidemiology, aetiology and pathogenesis. *J Paediatr Child Health* **49:**704–708.

14. **Owens GP, Gilden D, Burgoon MP, Yu X, Bennett JL.** 2011. Viruses and multiple sclerosis. *The Neuroscientist: a review journal bringing neurobiology, neurology and psychiatry* **17:**659–676.
15. **Tselis A.** 2011. Evidence for viral etiology of multiple sclerosis. *Semin Neurol* **31:**307–316.
16. **Voumvourakis KI, Kitsos DK, Tsiodras S, Petrikkos G, Stamboulis E.** 2010. Human herpesvirus 6 infection as a trigger of multiple sclerosis. *Mayo Clin Proc* **85:**1023–1030.
17. **Saglam O, Samayoa E, Somasekar S, Naccache S, Iwasaki A, Chiu CY.** 2015. No viral association found in a set of differentiated vulvar intraepithelial neoplasia cases by human papillomavirus and pan-viral microarray testing. *PLoS One* **10:** e0125292.
18. **Baker M.** 2012. Digital PCR hits its stride. *Nat Methods* **9:** 541–544.
19. **Hall Sedlak R, Jerome KR.** 2014. The potential advantages of digital PCR for clinical virology diagnostics. *Expert Rev Mol Diagn* **14:**501–507.
20. **Strain MC, Lada SM, Luong T, Rought SE, Gianella S, Terry VH, Spina CA, Woelk CH, Richman DD.** 2013. Highly precise measurement of HIV DNA by droplet digital PCR. *PLoS One* **8:**e55943.
21. **Eriksson S, Graf EH, Dahl V, Strain MC, Yukl SA, Lysenko ES, Bosch RJ, Lai J, Chioma S, Emad F, Abdel-Mohsen M, Hoh R, Hecht F, Hunt P, Somsouk M, Wong J, Johnston R, Siliciano RF, Richman DD, O'Doherty U, Palmer S, Deeks SG, Siliciano JD.** 2013. Comparative analysis of measures of viral reservoirs in HIV-1 eradication studies. *PLoS Pathog* **9:** e1003174.
22. **Sedlak RH, Cook L, Cheng A, Magaret A, Jerome KR.** 2014. Clinical utility of droplet digital PCR for human cytomegalovirus. *J Clin Microbiol* **52:**2844–2848.
23. **Hayden RT, Gu Z, Ingersoll J, Abdul-Ali D, Shi L, Pounds S, Caliendo AM.** 2013. Comparison of droplet digital PCR to real-time PCR for quantitative detection of cytomegalovirus. *J Clin Microbiol* **51:**540–546.
24. **Roberts CH, Last A, Molina-Gonzalez S, Cassama E, Butcher R, Nabicassa M, McCarthy E, Burr SE, Mabey DC, Bailey RL, Holland MJ.** 2013. Development and evaluation of a next-generation digital PCR diagnostic assay for ocular Chlamydia trachomatis infections. *J Clin Microbiol* **51:**2195–2203.
25. **Sedlak RH, Cook L, Huang ML, Magaret A, Zerr DM, Boeckh M, Jerome KR.** 2014. Identification of chromosomally integrated human herpesvirus 6 by droplet digital PCR. *Clin Chem* **60:**765–772.
26. **Boizeau L, Laperche S, Désiré N, Jourdain C, Thibault V, Servant-Delmas A.** 2014. Could droplet digital PCR be used instead of real-time PCR for quantitative detection of the hepatitis B virus genome in plasma? *J Clin Microbiol* **52:**3497–3498.
27. **Dong L, Meng Y, Sui Z, Wang J, Wu L, Fu B.** 2015. Comparison of four digital PCR platforms for accurate quantification of DNA copy number of a certified plasmid DNA reference material. *Sci Rep* **5:**13174.
28. **Deyde VM, Sampath R, Gubareva LV.** 2011. RT-PCR/electrospray ionization mass spectrometry approach in detection and characterization of influenza viruses. *Expert Rev Mol Diagn* **11:**41–52.
29. **Metzgar D, Baynes D, Myers CA, Kammerer P, Unabia M, Faix DJ, Blair PJ.** 2010. Initial identification and characterization of an emerging zoonotic influenza virus prior to pandemic spread. *J Clin Microbiol* **48:**4228–4234.
30. **Deyde VM, Sampath R, Garten RJ, Blair PJ, Myers CA, Massire C, Matthews H, Svoboda P, Reed MS, Pohl J, Klimov AI, Gubareva LV.** 2010. Genomic signature-based identification of influenza A viruses using RT-PCR/electrospray ionization mass spectrometry (ESI-MS) technology. *PLoS One* **5:**e13293.
31. **Zhang C, Xiao Y, Du J, Ren L, Wang J, Peng J, Jin Q.** 2015. Application of multiplex PCR couple with MALDI-TOF analysis for simultaneous detection of 21 common respiratory viruses. *J Clin Microbiol.*
32. **Blyn LB, Hall TA, Libby B, Ranken R, Sampath R, Rudnick K, Moradi E, Desai A, Metzgar D, Russell KL, Freed NE, Balansay M, Broderick MP, Osuna MA, Hofstadler SA, Ecker DJ.** 2008. Rapid detection and molecular serotyping of adenovirus by use of PCR followed by electrospray ionization mass spectrometry. *J Clin Microbiol* **46:**644–651.
33. **Sampath R, Hofstadler SA, Blyn LB, Eshoo MW, Hall TA, Massire C, Levene HM, Hannis JC, Harrell PM, Neuman B, Buchmeier MJ, Jiang Y, Ranken R, Drader JJ, Samant V, Griffey RH, McNeil JA, Crooke ST, Ecker DJ.** 2005. Rapid identification of emerging pathogens: coronavirus. *Emerg Infect Dis* **11:**373–379.
34. **Chen KF, Blyn L, Rothman RE, Ramachandran P, Valsamakis A, Ecker D, Sampath R, Gaydos CA.** 2011. Reverse transcription polymerase chain reaction and electrospray ionization mass spectrometry for identifying acute viral upper respiratory tract infections. *Diagn Microbiol Infect Dis* **69:**179–186.
35. **Ecker DJ, Sampath R, Blyn LB, Eshoo MW, Ivy C, Ecker JA, Libby B, Samant V, Sannes-Lowery KA, Melton RE, Russell K, Freed N, Barrozo C, Wu J, Rudnick K, Desai A, Moradi E, Knize DJ, Robbins DW, Hannis JC, Harrell PM, Massire C, Hall TA, Jiang Y, Ranken R, Drader JJ, White N, McNeil JA, Crooke ST, Hofstadler SA.** 2005. Rapid identification and strain-typing of respiratory pathogens for epidemic surveillance. *Proc Natl Acad Sci USA* **102:**8012–8017.
36. **Grant RJ, Baldwin CD, Nalca A, Zoll S, Blyn LB, Eshoo MW, Matthews H, Sampath R, Whitehouse CA.** 2010. Application of the Ibis-T5000 pan-Orthopoxvirus assay to quantitatively detect monkeypox viral loads in clinical specimens from macaques experimentally infected with aerosolized monkeypox virus. *Am J Trop Med Hyg* **82:**318–323.
37. **Eshoo MW, Whitehouse CA, Zoll ST, Massire C, Pennella TT, Blyn LB, Sampath R, Hall TA, Ecker JA, Desai A, Wasieloski LP, Li F, Turell MJ, Schink A, Rudnick K, Otero G, Weaver SC, Ludwig GV, Hofstadler SA, Ecker DJ.** 2007. Direct broad-range detection of alphaviruses in mosquito extracts. *Virology* **368:**286–295.
38. **Grant-Klein RJ, Baldwin CD, Turell MJ, Rossi CA, Li F, Lovari R, Crowder CD, Matthews HE, Rounds MA, Eshoo MW, Blyn LB, Ecker DJ, Sampath R, Whitehouse CA.** 2010. Rapid identification of vector-borne flaviviruses by mass spectrometry. *Mol Cell Probes* **24:**219–228.
39. **Ecker DJ, Sampath R, Massire C, Blyn LB, Hall TA, Eshoo MW, Hofstadler SA.** 2008. Ibis T5000: a universal biosensor approach for microbiology. *Nat Rev Microbiol* **6:**553–558.
40. **Wolk DM, Kaleta EJ, Wysocki VH.** 2012. PCR-electrospray ionization mass spectrometry: the potential to change infectious disease diagnostics in clinical and public health laboratories. *J Mol Diagn* **14:**295–304.
41. **Emonet S, Shah HN, Cherkaoui A, Schrenzel J.** 2010. Application and use of various mass spectrometry methods in clinical microbiology. *Clin Microbiol Infec* **16:**1604–1613.
42. **Wang D, Coscoy L, Zylberberg M, Avila PC, Boushey HA, Ganem D, DeRisi JL.** 2002. Microarray-based detection and genotyping of viral pathogens. *Proc Natl Acad Sci USA* **99:** 15687–15692.
43. **Chen EC, Miller SA, DeRisi JL, Chiu CY.** 2011. Using a pan-viral microarray assay (Virochip) to screen clinical samples for viral pathogens. *J Vis Exp* **(50):**2536.
44. **Palacios G, Quan PL, Jabado OJ, Conlan S, Hirschberg DL, Liu Y, Zhai J, Renwick N, Hui J, Hegyi H, Grolla A, Strong JE, Towner JS, Geisbert TW, Jahrling PB, Büchen-Osmond C, Ellerbrok H, Sanchez-Seco MP, Lussier Y, Formenty P, Nichol MS, Feldmann H, Briese T, Lipkin WI.** 2007. Panmicrobial oligonucleotide array for diagnosis of infectious diseases. *Emerg Infect Dis* **13:**73–81.
45. **Gardner SN, Jaing CJ, McLoughlin KS, Slezak TR.** 2010. A microbial detection array (MDA) for viral and bacterial detection. *BMC Genomics* **11:**668.
46. **Nicholson TL, Kukielka D, Vincent AL, Brockmeier SL, Miller LC, Faaberg KS.** 2011. Utility of a panviral microarray

for detection of swine respiratory viruses in clinical samples. *J Clin Microbiol* **49:**1542–1548.

47. **Kistler A, Avila PC, Rouskin S, Wang D, Ward T, Yagi S, Schnurr D, Ganem D, DeRisi JL, Boushey HA.** 2007. Pan-viral screening of respiratory tract infections in adults with and without asthma reveals unexpected human coronavirus and human rhinovirus diversity. *J Infect Dis* **196:**817–825.
48. **Chiu CY, Urisman A, Greenhow TL, Rouskin S, Yagi S, Schnurr D, Wright C, Drew WL, Wang D, Weintrub PS, Derisi JL, Ganem D.** 2008. Utility of DNA microarrays for detection of viruses in acute respiratory tract infections in children. *J Pediatr* **153:**76–83.
49. **Chiu CY, Rouskin S, Koshy A, Urisman A, Fischer K, Yagi S, Schnurr D, Eckburg PB, Tompkins LS, Blackburn BG, Merker JD, Patterson BK, Ganem D, DeRisi JL.** 2006. Microarray detection of human parainfluenzavirus 4 infection associated with respiratory failure in an immunocompetent adult. *Clin Infect Dis* **43:**e71–e76.
50. **Chiu CY, Alizadeh AA, Rouskin S, Merker JD, Yeh E, Yagi S, Schnurr D, Patterson BK, Ganem D, DeRisi JL.** 2007. Diagnosis of a critical respiratory illness caused by human metapneumovirus by use of a pan-virus microarray. *J Clin Microbiol* **45:**2340–2343.
51. **Rota PA, Oberste MS, Monroe SS, Nix WA, Campagnoli R, Icenogle JP, Peñaranda S, Bankamp B, Maher K, Chen MH, Tong S, Tamin A, Lowe L, Frace M, DeRisi JL, Chen Q, Wang D, Erdman DD, Peret TC, Burns C, Ksiazek TG, Rollin PE, Sanchez A, Liffick S, Holloway B, Limor J, McCaustland K, Olsen-Rasmussen M, Fouchier R, Günther S, Osterhaus AD, Drosten C, Pallansch MA, Anderson LJ, Bellini WJ.** 2003. Characterization of a novel coronavirus associated with severe acute respiratory syndrome. *Science* **300:** 1394–1399.
52. **Greninger AL, Chen EC, Sittler T, Scheinerman A, Roubinian N, Yu G, Kim E, Pillai DR, Guyard C, Mazzulli T, Isa P, Arias CF, Hackett J, Schochetman G, Miller S, Tang P, Chiu CY.** 2010. A metagenomic analysis of pandemic influenza A (2009 H1N1) infection in patients from North America. *PLoS One* **5:**e13381.
53. **Xu GJ, Kula T, Xu Q, Li MZ, Vernon SD, Ndung'u T, Ruxrungtham K, Sanchez J, Brander C, Chung RT, O'Connor KC, Walker B, Larman HB, Elledge SJ.** 2015. Viral immunology. Comprehensive serological profiling of human populations using a synthetic human virome. *Science* **348:** aaa0698.
54. **Safina G.** 2012. Application of surface plasmon resonance for the detection of carbohydrates, glycoconjugates, and measurement of the carbohydrate-specific interactions: a comparison with conventional analytical techniques. A critical review. *Anal Chim Acta* **712:**9–29.
55. **Suenaga E, Mizuno H, Kumar PK.** 2012. Influenza virus surveillance using surface plasmon resonance. *Virulence* **3:** 464–470.
56. **Suenaga E, Mizuno H, Penmetcha KK.** 2012. Monitoring influenza hemagglutinin and glycan interactions using surface plasmon resonance. *Biosens Bioelectron* **32:**195–201.
57. **Wang HN, Fales AM, Zaas AK, Woods CW, Burke T, Ginsburg GS, Vo-Dinh T.** 2013. Surface-enhanced Raman scattering molecular sentinel nanoprobes for viral infection diagnostics. *Anal Chim Acta* **786:**153–158.
58. **Hoang V, Tripp RA, Rota P, Dluhy RA.** 2010. Identification of individual genotypes of measles virus using surface enhanced Raman spectroscopy. *Analyst (Lond)* **135:**3103–3109.
59. **Su LC, Tian YC, Chang YF, Chou C, Lai CS.** 2014. Rapid detection of urinary polyomavirus BK by heterodyne-based surface plasmon resonance biosensor. *J Biomed Opt* **19:**011013.
60. **Rusnati M, Chiodelli P, Bugatti A, Urbinati C.** 2015. Bridging the past and the future of virology: surface plasmon resonance as a powerful tool to investigate virus/host interactions. *Crit Rev Microbiol* **41:**238–260.
61. **Petrozzino JJ, Smith C, Atkinson MJ.** 2010. Rapid diagnostic testing for seasonal influenza: an evidence-based review and comparison with unaided clinical diagnosis. *J Emerg Med* **39:** 476–490.
62. **Centers for Disease Control and Prevention.** 2012. Evaluation of 11 commercially available rapid influenza diagnostic tests—United States, 2011–2012. *MMWR Morb Mortal Wkly Rep* **61:**873–876.
63. **Gao F, Loring C, Laviolette M, Bolton D, Daly ER, Bean C.** 2012. Detection of 2009 pandemic influenza A(H1N1) virus Infection in different age groups by using rapid influenza diagnostic tests. *Influenza Other Respi Viruses* **6:** e30–e34.
64. **Kumar S, Henrickson KJ.** 2012. Update on influenza diagnostics: lessons from the novel H1N1 influenza A pandemic. *Clin Microbiol Rev* **25:**344–361.
65. **Cheng VC, To KK, Tse H, Hung IF, Yuen KY.** 2012. Two years after pandemic influenza A/2009/H1N1: what have we learned? *Clin Microbiol Rev* **25:**223–263.
66. **Cheng PK, Wong KK, Mak GC, Wong AH, Ng AY, Chow SY, Lam RK, Lau CS, Ng KC, Lim W.** 2010. Performance of laboratory diagnostics for the detection of influenza A(H1N1) v virus as correlated with the time after symptom onset and viral load. *J Clin Virol* **47:**182–185.
67. August 2014. *Sophia Influenza A+B FIA,* **Quidel Corp.** Pagckage Insert *In* 1219103EN01 (ed.).
68. **Lewandrowski K, Tamerius J, Menegus M, Olivo PD, Lollar R, Lee-Lewandrowski E.** 2013. Detection of influenza A and B viruses with the Sofia analyzer: a novel, rapid immunofluorescence-based in vitro diagnostic device. *Am J Clin Pathol* **139:**684–689.
69. **Leonardi GP, Wilson AM, Zuretti AR.** 2013. Comparison of conventional lateral-flow assays and a new fluorescent immunoassay to detect influenza viruses. *J Virol Methods* **189:** 379–382.
70. **Lee CK, Cho CH, Woo MK, Nyeck AE, Lim CS, Kim WJ.** 2012. Evaluation of Sofia fluorescent immunoassay analyzer for influenza A/B virus. *J Clin Virol* **55:**239–243.
71. April 2012. *BD Veritor System for Rapid Detection of Flu A+B. In* Company BDa (ed.), Package Insert 8087667(03).
72. **Peters TR, Blakeney E, Vannoy L, Poehling KA.** 2013. Evaluation of the limit of detection of the BD Veritor™ system flu A+B test and two rapid influenza detection tests for influenza virus. *Diagn Microbiol Infect Dis* **75:**200–202.
73. **Nam MH, Jang JW, Lee JH, Cho CH, Lim CS, Kim WJ.** 2014. Clinical performance evaluation of the BD Veritor System Flu A+B assay. *J Virol Methods* **204:**86–90.
74. **Hassan F, Nguyen A, Formanek A, Bell JJ, Selvarangan R.** 2014. Comparison of the BD Veritor System for Flu A+B with the Alere BinaxNOW influenza A&B card for detection of influenza A and B viruses in respiratory specimens from pediatric patients. *J Clin Microbiol* **52:**906–910.
75. **Dunn J, Obuekwe J, Baun T, Rogers J, Patel T, Snow L.** 2014. Prompt detection of influenza A and B viruses using the BD Verito System Flu A+B, Quidel Sofia Influenza A+B FIA, and Alere BinaxNOW Influenza A&B compared to real-time reverse transcription-polymerase chain reaction (RT-PCR). *Diagn Microbiol Infect Dis* **79:**10–13.
76. **Nie S, Roth RB, Stiles J, Mikhlina A, Lu X, Tang YW, Babady NE.** 2014. Evaluation of Alere i Influenza A&B for rapid detection of influenza viruses A and B. *J Clin Microbiol* **52:** 3339–3344.
77. **Bell J, Bonner A, Cohen DM, Birkhahn R, Yogev R, Triner W, Cohen J, Palavecino E, Selvarangan R.** 2014. Multicenter clinical evaluation of the novel Alere i Influenza A&B isothermal nucleic acid amplification test. *J Clin Virol* **61:** 81–86.
78. **Chen L, Tian Y, Chen S, Liesenfeld O.** 2015. Performance of the Cobas influenza A/B assay for rapid PCR-based detection of influenza compared to Prodesse ProFlu+ and viral culture. *Eur J Microbiol Immunol (Bp)* **5:**236–245.
79. **Binnicker MJ, Espy MJ, Irish CL, Vetter EA.** 2015. Direct detection of influenza A and B viruses in less than 20 minutes using a commercially available rapid PCR assay. *J Clin Microbiol* **53:**2353–2354.

80. **Drain PK, Garrett NJ.** 2015. The arrival of a true point-of-care molecular assay-ready for global implementation? *Lancet Glob Health* **3:**e663–e664.
81. **Svensson MJ, Lind I, Wirgart BZ, Östlund MR, Albert J.** 2014. Performance of the Simplexa Flu A/B & RSV Direct Kit on respiratory samples collected in saline solution. *Scand J Infect Dis* **46:**825–831.
82. **Hindiyeh M, Kolet L, Meningher T, Weil M, Mendelson E, Mandelboim M.** 2013. Evaluation of Simplexa Flu A/B & RSV for direct detection of influenza viruses (A and B) and respiratory syncytial virus in patient clinical samples. *J Clin Microbiol* **51:**2421–2424.
83. **Selvaraju SB, Bambach AV, Leber AL, Patru MM, Patel A, Menegus MA.** 2014. Comparison of the Simplexa Flu A/B & RSV kit (nucleic acid extraction-dependent assay) and the Prodessa ProFlu+ assay for detecting influenza and respiratory syncytial viruses. *Diagn Microbiol Infect Dis* **80:**50–52.
84. **Woodberry MW, Shankar R, Cent A, Jerome KR, Kuypers J.** 2013. Comparison of the Simplexa FluA/B & RSV direct assay and laboratory-developed real-time PCR assays for detection of respiratory virus. *J Clin Microbiol* **51:**3883–3885.
85. **Selvaraju SB, Tierney D, Leber AL, Patel A, Earley AK, Jaiswal D, Menegus MA.** 2014. Influenza and respiratory syncytial virus detection in clinical specimens without nucleic acid extraction using FOCUS direct disc assay is substantially equivalent to the traditional methods and the FOCUS nucleic acid extraction-dependent RT-PCR assay. *Diagn Microbiol Infect Dis* **78:**232–236.
86. **Hardie A, Mackenzie L, Guerendiain D, Templeton K.** 2015. Evaluation of Enigma MiniLabTM Influenza A/B RSV assay designed for rapid testing and point of care diagnosis. *J Clin Virol* **70:**S68.
87. **Douthwaite ST, Walker C, Adams EJ, Mak C, Vecino Ortiz A, Martinez-Alier N, Goldenberg SD.** 2016. Performance of a novel point-of-care molecular assay for detection of influenza A and B viruses and respiratory syncytial virus (Enigma MiniLab) in children with acute respiratory infection. *J Clin Microbiol* **54:**212–215.
88. **Kania D, Truong TN, Montoya A, Nagot N, Van de Perre P, Tuaillon E.** 2015. Performances of fourth generation HIV antigen/antibody assays on filter paper for detection of early HIV infections. *J Clin Virol* **62:**92–97.
89. **Pilcher CD, Louie B, Facente S, Keating S, Hackett J Jr, Vallari A, Hall C, Dowling T, Busch MP, Klausner JD, Hecht FM, Liska S, Pandori MW.** 2013. Performance of rapid point-of-care and laboratory tests for acute and established HIV infection in San Francisco. *PLoS One* **8:** e80629.
90. **Jani IV, Meggi B, Mabunda N, Vubil A, Sitoe NE, Tobaiwa O, Quevedo JI, Lehe JD, Loquiha O, Vojnov L, Peter TF.** 2014. Accurate early infant HIV diagnosis in primary health clinics using a point-of-care nucleic acid test. *J Acquir Immune Defic Syndr* **67:**e1–e4.
91. **Tang S, Zhao J, Wang A, Viswanath R, Harma H, Little RF, Yarchoan R, Stramer SL, Nyambi PN, Lee S, Wood O, Wong EY, Wang X, Hewlett IK.** 2010. Characterization of immune responses to capsid protein p24 of human immunodeficiency virus type 1 and implications for detection. *Clin Vaccine Immunol* **17:**1244–1251.
92. **Liu J, Du B, Zhang P, Haleyurgirisetty M, Zhao J, Ragupathy V, Lee S, DeVoe DL, Hewlett IK.** 2014. Development of a microchip Europium nanoparticle immunoassay for sensitive point-of-care HIV detection. *Biosens Bioelectron* **61:**177–183.
93. **Parry CM, Parkin N, Diallo K, Mwebaza S, Batamwita R, DeVos J, Bbosa N, Lyagoba F, Magambo B, Jordan MR, Downing R, Zhang G, Kaleebu P, Yang C, Bertagnolio S.** 2014. Field study of dried blood spot specimens for HIV-1 drug resistance genotyping. *J Clin Microbiol* **52:**2868–2875.
94. **Vogel G.** 2014. Infectious Diseases. Testing new Ebola tests. *Science* **345:**1549–1550.
95. **Walker NF, Brown CS, Youkee D, Baker P, Williams N, Kalawa A, Russell K, Samba AF, Bentley N, Koroma F, King MB, Parker BE, Thompson M, Boyles T, Healey B, Kargbo B, Bash-Taqi D, Simpson AJ, Kamara A, Kamara TB, Lado M, Johnson O, Brooks T.** 2015. Evaluation of a point-of-care blood test for identification of Ebola virus disease at Ebola holding units, Western Area, Sierra Leone, January to February 2015. *Euro surveillance: bulletin Europeen sur les maladies transmissibles = European communicable disease bulletin* **20.**

SECTION III

Viral Pathogens

Respiratory Viruses

CHRISTINE ROBINSON, MICHAEL J. LOEFFELHOLZ, AND BENJAMIN A. PINSKY

19

Respiratory tract illnesses are one of the most common health conditions affecting humans. Most of these illnesses are caused or triggered by viruses. Clinical presentations and severity range from mild and self-limited upper respiratory tract illnesses (URTI) to serious or fatal lower respiratory tract disease. Respiratory viruses exert considerable pressure on health care systems, are significant drivers of antibiotic overuse, and contribute significantly to loss of productivity. Accurate identification of these infections and appropriate care of affected patients are therefore priorities for health care systems.

Prior to 2000, six "classic" human respiratory viruses were known: respiratory syncytial virus (RSV), influenza virus, parainfluenza virus (PIV) 1–3, rhinovirus (RHV), the human coronaviruses (HCoV) OC43 and 229E, and adenovirus. Molecular methods have now permitted identification of human metapneumovirus (HMPV), new coronaviruses (CoV), including two with pandemic potential, a fourth PIV, and noncultivatable RHVs. This chapter describes the collective list. Viruses that affect the respiratory tract, but primarily involve other organ systems, are described in their respective chapters.

CLASSIFICATION AND BIOLOGY

Influenza Virus

Of all respiratory pathogens, influenza viruses cause the largest number of serious acute infections in humans. The most notable is "influenza," which affects 10% or more of the population during epidemics and significantly more during a pandemic. The name "influenza" stems from the Latin *influentia* because influenza outbreaks were once considered under astrological influence. Influenza viruses belong to the *Orthomyxoviridae* family of complex, moderately sized, enveloped RNA viruses. Within this family are 3 genera with eponymous species (or "types") that infect humans: influenza A, influenza B, and influenza C. Species are distinguished by antigenic differences in their internal matrix (M1) proteins and nucleoproteins (NP). Influenza A was initially described in the 1920s as a transmissible agent in human mucus capable of infecting swine; influenza B and C viruses were identified decades later. Besides humans, influenza A naturally infects many species of birds and nonhuman mammals, most notably swine. Infections with influenza B and C mostly involve humans.

Virions are approximately 100 nm in size and spherical or pleomorphic in shape. Projecting from the host-derived envelope are glycoprotein spikes of two types known as hemagglutinin (HA) and neuraminidase (NA). HA permits attachment to specific sialic acid-containing receptors on the surface of cells. Similar binding to guinea pig red blood cells causes hemagglutination. HA also contains the major (and most variable) antigenic determinants and is the most important component in influenza vaccines. NA enzymatically removes viral receptors, enhancing release of progeny from infected cells. Influenza C lacks NA but has glycoprotein spikes with similar functions.

The RNA genome is negative-sense, single-stranded and segmented. Influenza A and B contain 8 RNA segments; influenza C contains 7. Each segment of influenza B and C encodes a single protein; some influenza A segments have additional coding capacity. Total genome length is 10–14.6 kb. Internal proteins include the nucleoprotein and the RNA polymerase complex proteins PB1, PB2, and PA. The M1 protein is interposed between the ribonucleoprotein complexes and the viral envelope; a second matrix (M2) protein in influenza A creates an ion channel that aids viral uncoating. Replication takes place primary in the nucleus, an unusual feature for an RNA virus.

Influenza A viruses are further classified into subtypes based on the properties of their HA or NA surface glycoproteins. Eighteen antigenic forms of HA (H1-18) and 11 of NA (N1-11) are now recognized. Only viruses with H1, H2, or H3 and N1or N2 commonly infect humans; the others circulate among other mammals or birds, with the greatest variety in wild waterfowl. Influenza B and C lack distinct subtypes, but influenza B has evolved into two lineages (Yamagata and Victoria); lineages of influenza C are also described. Within each grouping are antigenically distinct strains. Strains are denoted by virus type, species of origin (except for human), geographic location of first isolation, strain number, and year of first isolation, with the influenza A subtype in parentheses, as in A/California/7/2009 (H1N1). Influenza A variants that naturally infect swine and occasionally humans are identified by a "v" before the subtype, as in v(H3N2).

RSV and HMPV

RSV was first isolated in 1956 from a chimpanzee with respiratory symptoms and later named for its characteristic

doi:10.1128/9781555819156.ch19

cytopathic effect (CPE) of syncytia formation in cell culture. HMPV was identified in 2001, among a collection of slow-growing paramyxovirus-like isolates from young children with RSV-like illnesses, and later characterized as the first human species within the *Metapneumovirus* genus. Both viruses belong to the *Pneumoviridae* family of nonsegmented, negative-sense, enveloped RNA viruses. Within the *Pneumoviridae* are two genera. The *Orthopneumovirus* genus consists of three species, including RSV, and its bovine and murine relatives. The *Metapneumovirus* genus contains HMPV and two avian *Metapneumovirus*.

RSV particles are irregularly spherical or filamentous and 150–300 nm in diameter. The 15 kb RNA genome contains 10 genes that encode 11 proteins. The nucleoprotein (N) surrounds the viral RNA and forms the nucleocapsid. The RNA-dependent RNA polymerase complex consists of the large (L) protein, the phospho (P) protein, and transcription/replication factors M2-1 and M2-2. The matrix protein (M) surrounds the nucleocapsid and links it to glycoprotein (G) and the fusion glycoprotein (F). The short hydrophobic (SH) protein spans the viral envelope and forms short spikes on the surface. There are also two nonstructural proteins, NS1 and NS2. The G protein, and perhaps the F protein as well, mediate attachment to host cells. The F protein induces membrane fusion, which is important for virus entry into host cells and spread of infection by syncytia formation. The F protein may also aid viral attachment to cellular receptors. The F and G proteins are the primary targets of the immune response. HMPV virions resemble RSV but are slightly larger and have longer spikes. The genome is also 2 kb shorter, has a different gene order, and lacks the NS1 and NS2 genes. After correcting for this difference, the two viruses share 50% nucleotide-sequence homology. Since the NS proteins of RSV may have anti-interferon activity, their absence may contribute to HMPV's somewhat lesser pathogenicity. Both viruses replicate in the cytoplasm and are highly cell-associated.

Parainfluenza Virus

PIV was first isolated in the 1950s from cultures demonstrating CPE or hemadsorption when inoculated with specimens from individuals with acute respiratory illnesses. The name reflects the early observations that some symptoms induced by PIV are influenza-like and that PIV and influenza particles have hemagglutination and neuraminidase activities. Subsequently, the two viruses were found to differ significantly and were classified into separate families.

The PIVs belong to the *Paramyxovirinae* subfamily of *Paramyxovirdae*. Viruses in this subfamily differ from members of the *Pneumovirinae* subfamily like RSV and HMPV in several ways. The most notable difference is that the PIVs have an attachment molecule with combined hemagglutination and neuraminidase activities that *Pneumovirinae* lack. PIV virions are of medium size, 150–200 nm in diameter, enveloped, and mostly spherical in shape. They contain a nonsegmented, single-stranded, negative-sense RNA genome of approximately 16 kb that encodes a basal complement of six structural proteins. A nucleoprotein (N) encases the RNA, forming the helical nucleocapsid. The large (L) protein and phospho (P) protein form the polymerase complex. The matrix protein (M) connects the nucleocapsid to two surface glycoproteins, hemagglutinin-neuraminidase (HN) and fusion (F), which project through the viral envelope and form short spikes. The HN glycoprotein confers hemagglutinating (sialic acid-binding) and neuraminidase (sialic acid receptor-cleaving) activities. Functions of HN include mediating virus attachment to sialic acid–containing cellular receptors, triggering of the F glycoprotein to initiate fusion between virus and host cells, and cleaving sialic acid residues from receptors on the viral envelope so that particles do not self-aggregate during release from infected cells. The HN and F proteins are the major antigenic determinants.

There are four PIV types that are further categorized into two genera based primarily on differences in genetic organization and antigenicity. PIV-1 and -3 are grouped within the *Respirovirus* genus, whereas PIV-2 and -4 are in the *Rublavirus* genus. PIV-4 is further classified into subtypes 4A and 4B based on differences in antigenicity and organization of P and HN genes.

Rhinovirus

Rhinoviruses are members of the *Picornaviridae* family, *Enterovirus* genus. Their name is derived from the Greek word for nose, an important site of virus replication. Rhinoviruses cause more common colds (minor URTIs) than any other virus infecting the upper respiratory tract. The first isolate was obtained in the mid 1950s from nasal washings of individuals with colds. Additional recoveries were made by mimicking nasal conditions, including reducing the culture temperature to 33°C, lowering the pH to neutral, and providing aeration.

There are three rhinovirus species, A, B, and C, and more than 100 distinct serotypes. Historically, neutralization with type-specific antibody was used to identify serotypes. Current classification is genotypic (based on capsid-protein sequences) but retains the serotype designation. This approach is necessary to classify serotypes of rhinovirus C, which do not replicate in standard cell culture. Individual serotypes are designated by the letter of their species and genotype number, as in A23. Strains of each type are also recognized. The 2014 International Committee on Taxonomy of Viruses lists 80 rhinovirus A types, 32 rhinovirus B types, and 55 rhinovirus C types (http://www.picornaviridae.com/ [Accessed 3 May 2016]).

The rhinovirus virion is a small (15–30 nm), nonenveloped, icosahedral particle encasing a single-stranded, positive-sense RNA of approximately 7.2 kb. The genome contains a noncoding region at the 5′ end and a single open reading frame. The capsid consists of four viral proteins, VP1, VP2, VP3, and VP4, which are created by posttranslational cleavage of a precursor polyprotein. Nonstructural proteins consist of 2 proteases and an RNA-dependent RNA polymerase. The VP1, VP2, and VP3 proteins are on the capsid surface. VP4 is inside and helps anchor viral RNA to the capsid. The capsid surface contains hydrophobic "pockets" that function in binding to cellular receptors and are targets for some antiviral medications. Variation in the surface proteins leads to antigenic diversity and stimulates durable, type-specific host immune responses. There is no group antigen, so sequential infections with different genotypes are frequent. Unlike enteroviruses, most rhinovirus serotypes lose infectivity upon exposure to mild acid, which explains their general inability to infect the gut.

Coronavirus

Human coronaviruses (HCoV) were discovered in the 1960s from nasal secretions of persons with mild URTI and were initially considered relatively harmless causes of colds. By contrast, animal CoV have long been known to cause many significant and diverse illnesses, including respiratory disease and diarrhea. Interest in all CoV increased substantially once animal CoV were found to cross species barriers and cause serious human respiratory diseases, such as severe acute respiratory syndrome (SARS) and Middle Eastern Respiratory Syndrome (MERS).

CoV are medium-sized, enveloped, single-stranded, positive-sense RNA viruses of the *Coronaviridae* family. Their name derives from their appearance, which resembles a crown (or *corona* in Latin) due to numerous club-shaped spike (S) proteins radiating from the viral envelope. The *Coronavirinae* subfamily contains two genera, the Alphacoronaviruses and Betacoronaviruses, with species that infect humans. The human Alphacoronavirus species are 229E and NL63. Betacoronavirus species that infect humans cluster into 3 lineages. OC43 and HKU1 are in lineage a, SARS-CoV is in lineage b, and MERS-CoV is in lineage c.

The CoV are pleomorphic, mostly spherical particles, 80–160 nm in diameter. They contain a helical nucleocapsid surrounding a nonsegmented genome of approximately 27–32 kb, the largest among RNA viruses. Genes at the 5′ end specify proteins for RNA replication; genes at the 3′ end encode the structural S, envelope, membrane, and nucleocapsid proteins, and accessory proteins that may interfere with the immune response. A hemagglutinin esterase (HE) gene found only in some betacoronaviruses, including HKU1, encodes a second set of short spike proteins. Following attachment to specific cell surface receptors via their S (and HE) proteins, CoV enter the cell and replicate in the cytoplasm. Replication is error-prone, leading to significant rates of mutation and recombination, which aid virus evolution.

Adenovirus

Adenoviruses are medium-sized (70–90 nm), nonenveloped, icosohedral DNA viruses within the *Adenoviridae* family. The name originates from their initial isolation from cultures of human tonsils and adenoids in 1953. With rare exception, only members of the Mastadenovirus genus infect humans. Within the genus are 7 species (previously known as groups), designated A-G, and termed *Human adenovirus A*, *Human adenovirus B*, and so forth. To date, 51 serotypes are accepted. Additional types are described, some of which are probably intertypic recombinants whose clinical relevance remains to be established. Only a third of recognized serotypes cause appreciable human illness.

Adenoviruses contain a single, linear, double-stranded DNA molecule of about 36 kb that encodes approximately 35 genes. Mutations and recombination within types aid virus evolution. The icosahedral capsids consist of proteins called hexons and pentons. Hexons form the particle's triangular faces, and a penton occupies each vertex. From each penton projects a fiber (two in some serotypes) with a terminal knob. Hexons contain epitopes, which are common to all adenoviruses, but these epitopes reside within the capsid, so neutralizing antibodies are not generated. Species-specific determinants are on the fiber. Type-specific determinants on external regions of the hexon and the fiber stimulate neutralizing antibodies and serotype-specific immunity. Attachment of viral knobs to specific receptors on cell surfaces initiates infection, followed by viral DNA replication and assembly in the nucleus.

EPIDEMIOLOGY

Influenza Virus

Influenza viruses cause annual epidemics in temperate climates globally during cool weather. Seasonality is less pronounced in the tropics. Epidemics last 6–8 weeks in a locale and typically involve one or two subtypes of influenza A, with or without influenza B. Strains of each virus may persist temporarily but are invariably replaced due to "antigenic drift." This gradual process, which is most pronounced with influenza A, is due to slow accumulation of spontaneous point mutations in HA and NA genes generated during viral replication, which results in strains with new antigenic features. Strains emerge if incompletely neutralized by existing antibodies in the population. Antigenic drift necessitates frequent reformulation of influenza vaccines.

Attack rates vary by age, with the highest rates in young children and the elderly. In unimmunized populations, attack rates range from 10–70%. The morbidity and mortality associated with epidemics is considerable, with 80,000 or more influenza-associated hospitalizations likely to occur annually in U.S. adults (1). Populations at highest risk for complications include children 6–59 months, adults older than 50 years, persons with underlying chronic conditions, pregnant women, the morbidly obese, and American Indian or Alaska natives. The highest fatality rates are found in the elderly and when influenza A(H3N2) predominates. Influenza B causes a lesser burden of disease by causing fewer illnesses.

Pandemics occur unpredictably and less frequently, spread faster and globally, involve more individuals, and have higher fatality rates than epidemics. Pandemics result from the sudden appearance of an antigenically distinct and transmissible new influenza A subtype or strain, which is poorly recognized, if at all, by preexisting immunity. Six pandemics have occurred since 1900. The most devastating was the 1918 "Spanish Flu" pandemic that killed 25–50 million adults worldwide. Its cause was most likely a novel A(H1N1) originating from birds (2). By contrast, the recent 2009 "Swine Flu" pandemic, also due to an A(H1N1) virus, was relatively mild, perhaps because many elderly retained immunologic memory to the viral 1918-like antigens (3). Panolemic viruses are generated by "antigenic shift," an abrupt and dramatic change in viral HA (and to a lesser extent in other proteins) due to genetic reassortment occurring during co-infections with human influenza A and avian influenza A (AI) viruses. Pigs often serve as intermediaries in this process because AI viruses rarely infect humans directly, but swine can host human influenza and AI virus infections. Swine may also contribute genetic information from their own endemic influenza viruses. Indeed, the 2009 A(H1N1) "swine flu" pandemic virus was a triple reassortant containing sequences of human influenza A (H3N2), AI viruses, and swine influenza A viruses. Swine occasionally generate influenza A variants that can infect humans and have pandemic potential as well (4).

Although human infections with AI viruses were once rare, a 2003 epizootic of a highly pathogenic AI among wild waterfowl in Asia spilled over to domestic flocks, causing millions of poultry to die and deaths in humans in contact with infected birds (5). Birds have now spread this A(H5N1) AI virus to at least 15 countries in Asia, Africa, and Europe, resulting in more than 700 mostly sporadic human fatalities. The A(H5N1) virus is now endemic among birds in these regions and reports of human cases have slowed. Another novel AI virus, A(H7N9), emerged in China in 2013 and so far has caused nearly 300 human deaths. Other novel AI viruses continue to emerge in Asia and cause sporadic human fatalities, a situation that remains closely watched. In late 2014, AI "H5-like" viruses appeared in North American wild birds and infected domestic flocks.

These viruses have now been detected in more than 25 U.S. states, causing mortality and significant financial losses to poultry production facilities. No human cases are yet recorded. Updates on avian influenza can be found at http://www.cdc.gov/flu/avianflu/index.htm (last accessed 16 May 2016).

RSV and HMPV

RSV is the major cause of bronchiolitis and pneumonia in young children and other age groups. Infections are exceedingly common, with virtually all children infected by age 2. Worldwide, RSV causes an estimated 33.8 million LRTI annually in children younger than age 5. In the U.S., more than 20,000 children younger than age 5 are hospitalized annually for these conditions (6). Estimated infection rates range from 4–10% in high-risk adults with underlying conditions, to nearly 35% among children with preterm birth. Hospitalizations in children are most frequent in the first 6 weeks–6 months of life, with most children infected by age 3. Although most children who acquire RSV were previously healthy, predispositions to severe illness include premature birth, underlying neuromuscular or cardiopulmonary disorders, or being an Alaska or Native American Indian. HMPV is likewise an important cause of respiratory illnesses at all ages. Initial infections occur at a slightly older age (a median of 22 months for HMPV compared to 1–3 months for RSV), with most children infected by age 5. The overall burden of HMPV disease is somewhat less than that of RSV, with 4–10% of severe acute LRTI in young children associated with HMPV compared to 23% with RSV. Together, RSV and HMPV cause more LRTI in young children than other common respiratory viruses, including influenza (7).

RSV re-infections during childhood are common due to the impermanence of the initial immune response. First re-infections are usually symptomatic and can again involve lower airways. Subsequent re-infections are milder as protective immunity begins to develop but can continue throughout life due to waning immunity. In one prospective study, repeat infections occurred in 7% of adults, with the highest rates observed in individuals who have frequent contact with children (8). Morbidity and mortality can be substantial in the frail elderly. Indeed, RSV has been linked to over 17,000 U.S. deaths annually in persons over age 65, resulting in more deaths in this age group than in children (9). Repeat infections with HMPV are likewise frequent, occurring in an estimated 1–9% of adults annually. Unlike RSV, most recurrent HMPV infections after infancy and in otherwise healthy nonelderly adults are asymptomatic or mild. As with RSV, however, HMPV re-infections in the frail elderly can be severe (10).

RSV (and to a lesser extent HMPV) is an important cause of severe lower airway disease in persons with underlying immune system defects, particularly those involving T-cells so recipients of hematopoietic stem cell transplants (HSCT) are at highest risk. Mortality higher than 50%, due mostly to pneumonia, was reported in early studies but is now lower due mostly to improved management (11). Also at risk are lung-transplant and cystic fibrosis patients and children from nonaffluent countries with HIV infection.

In temperate climates, RSV circulates in regular midwinter epidemics lasting 4–5 months; in equatorial areas virus circulation is continuous. Onset weeks and duration of epidemics vary by year and location. Transmission is highly efficient, with as many as 50% of babies born the previous spring and summer infected during an epidemic. Spread is facilitated by shedding of virus from infants for as long as 4 weeks and from immunocompromised individuals for months after infection; older children and adults shed only a few days. Circulating viruses belong to one of two subgroups, designed A and B, based primarily on antigenic differences of the G protein. Both subgroups can co-circulate or one may predominate for several seasons before being replaced. Local strains also appear in patterns that change with successive seasons. Subtypes and strains can differ between adjacent locales, suggestive of constant virus evolution and local immunologic pressure. Viruses of subgroup A replicate to the highest titers in culture and the human respiratory tract and are more common than subgroup B, perhaps due to a more transient immunity.

HMPV also circulates in annual epidemics in cool weather, with peak periods of RSV and HMPV illnesses often coinciding. HMPV, however, is more likely than RSV to appear year-round. The amount of HMPV circulating season-season also varies more than with RSV. HMPV has two antigenic subgroups, A and B, with the amount of diversity similar to RSV. The F protein is highly conserved between the two subgroups and is the primary target of protective antibodies. The G protein is highly variable and is minimally conserved between subgroups. Four HMPV lineages (A1, A2, B1, and B2) are described. Like RSV, circulation is community-based with several lineages often co-circulating and varying seasonally. Re-infections can involve homologous and heterologous lineages.

Parainfluenza Virus

PIVs are the most significant cause of croup (laryngotracheitis) worldwide and second only to RSV (or perhaps HMPV) as a cause of viral bronchiolitis and pneumonia in infants and young children. In the U.S. alone, from 6,000 to almost 30,000 children are hospitalized annually due to PIV-associated respiratory tract disease (12). Seasonality of the four PIV types is type-specific and interactive. In temperate climates, PIV cause large biennial epidemics of croup in children between 6 months and 6 years of age in the autumn of odd-numbered years. Up to 250,000 U.S. children may be affected in such outbreaks, with PIV-1 responsible for half of the cases and PIV-2 and -3 for a smaller but appreciable number. Circulation of PIV-3 is more endemic, with springtime peaks. A second smaller autumn peak of PIV-3 sometimes appears when PIV-1 is not prominent. As many as 20,000 U.S. neonates and young infants acquire symptomatic PIV-3 infections annually, and many are hospitalized. Overall, nearly 14% of all hospitalizations for LRTIs in U.S. children younger than 5 years are attributable to PIV-1, -2, or -3. Less is known about PIV- 4. but inital studies suggest that the virus circulates year round, with biennial peaks in the autumn and spring of odd numbered years, like PIV-1 (13).

Primary PIV infections occur during infancy or early childhood, with most individuals infected by age 5. Re-infections are frequent, especially during the first two years of life, which is the age range for most PIV-associated hospitalizations. In otherwise healthy adults, PIV re-infections account for 1–15% of all acute respiratory illnesses. PIVs are among the four most common viral pathogens in hospitalized adults but are less frequently associated with exacerbations of asthma or chronic obstructive pulmonary disease (COPD) than are rhinoviruses or HCoV. Prospective studies have documented PIV in 2–14% of elderly residents of nursing homes, but infections are usually of lesser significance than with influenza viruses, rhinovirus, or HCoV, although fatalities have been reported.

PIV infections are reported in 2–7% of patients with cancer or HSCT, with PIV-3 most often involved. Infections can be especially severe in pediatric HSCT patients with mortality rates of 4–75% (14). Asymptomatic shedding is frequent and may contribute to nosocomial infections (15, 16). Lung transplant recipients may acquire significant PIV infections as well, resulting in acute graft rejection or bronchiolitis obliterans. Lower airway involvement is also reported in subjects with HIV infection or cancer.

Rhinovirus

Rhinoviruses are the most frequent cause of the common cold in all age groups. Species A and C are responsible for most infections (17). Rhinovirus infections occur year-round with increased circulation in spring and fall. During peaks, more than 20 rhinovirus types can be identified, with complete changeover in subsequent seasons. Throughout the year, about half of all colds are due to rhinoviruses, but as many as 80% of all URTI may be rhinovirus-associated during the fall-spring peaks. Infants, young children, and the elderly are especially susceptible. Although most rhinovirus-associated colds are trivial, their sheer number produces more episodes of illness, greater restriction of activities, and more physician visits than any other human pathogen. Complications of URTI include acute otitis media, which develops in as many as a third of symptomatic or asymptomatic infections in young children, as well as acute and chronic sinusitis.

Although rhinoviruses are traditionally associated with URTIs, they can also cause lower airway disease with considerable morbidity, an association that recently became apparent by molecular methods. Rhinovirus C in particular is associated with lower respiratory tract disease (18). Rhinovirus LRTIs are most frequent in young children and in persons with chronic respiratory diseases, such as COPD and cystic fibrosis, and the elderly. High hospitalization rates occur in preterm infants and children with a history of wheezing or asthma. In the elderly and young children, the burden of rhinovirus LRTI may exceed that of influenza virus. Other rhinovirus-associated manifestations are croup and bronchiolitis. An influenza-like syndrome is a frequent reason for hospitalization in children (19). Fatalities due to rhinoviruses in immune-competent individuals are rare.

Rhinoviruses may be a significant viral trigger of preexisting pulmonary conditions, such as asthma in genetically predisposed individuals, COPD, or cystic fibrosis, and may worsen symptoms due to other pathogens. Analysis of these studies, however, is complicated by the high rate of rhinovirus detection in asymptomatic subjects and frequent coinfections. Indeed, asymptomatic rhinovirus infections are identified in 20% or more of healthy individuals, and coinfections are detected in 8% to more than 50% of symptomatic persons (20). Highly immunocompromised patients, particularly HSCT recipients, can also acquire serious rhinovirus infections, particularly if acquired pretransplantation (21). Coinfections and prolonged shedding are frequent in these complex populations, so the contribution of rhinoviruses to direct pulmonary damage, in predisposing to secondary bacterial infections, or causing death remains unclear. Rhinovirus infections in lung transplant patients can predispose them to graft rejection. Persistent shedding of rhinovirus RNA is common in patients with primary hypogammaglobulinemia and other immunodeficiencies (22). Shedding of rhinovirus in otherwise healthy individuals, however, rarely persists beyond one month (23). Fatalities in some immunocompromised patients are documented.

Coronavirus

Collectively, HCoV 229E, OC43, HKU1, and NL63 are the most frequent cause of URTI, behind rhinovirus. The highest rates of infection are in young children, although repeat infections occur throughout life, with as many as half of all school age children and 80% of adults having serologic evidence of previous infection. About 30% of all URTIs are due to NL-63 and OC43, with 229E and HKU1 responsible for about 10% (24). HCoV circulation is markedly restricted to winter in temperate climates, with the prevalence of individual species varying annually. HCoV-associated LRTIs occur, and can be severe in children with underlying medical conditions, the elderly, immunocompromised hosts, and when OC43 is the cause. Mortality is uncommon. Establishing the degree of HCoV involvement in LRTI, however, is difficult due to frequent coinfections with other respiratory viruses and asymptomatic shedding.

In 2002–2003, an unprecedented pandemic of severe LRTI known as SARS affected a total of 8,000 individuals in 26 countries, including Canada and the United States (25). Most cases were in China, where the virus arose, and in Hong Kong. All infections were in adults but were especially severe in persons older than 60 years or with underlying conditions. The fatality rate was 9.6%. The causative agent, SARS-CoV, probably originated from a bat CoV. Humans were probably not infected by this virus. Rather, exotic mammals purchased for consumption in live animal markets may have been infected first, with subsequent mutations permitting human infection. Human-human transmission then spread the virus worldwide, facilitated by lack of experience with CoV-associated pandemics and high rates of nosocomial infections. Infection required close contact with contaminated fomites, droplets, or aerosols from the respiratory or gastrointestinal tract, so most cases were in families or in hospitals. Laboratory-related infections also occurred. Intense quarantining and inefficient person-person transmission aided cessation of the pandemic. It is now apparent that some Chinese were infected with similar viruses on multiple occasions before SARS was recognized, suggestive of frequent species-crossing events. Therefore concerns for re-emergence continue.

MERS was first identified in 2012 in a patient with severe LRTI in Saudi Arabia. As of March 2016, more than 1,700 confirmed cases of MERS and more than 600 fatalities have been reported worldwide (http://www.who.int/emergencies/mers-cov/en/). Most affected individuals are from the Arabian Peninsula, with some cases appearing in Europe, Africa, Korea, the Philippines, and the United States. So far, all infections have been found in recent travelers or are linked to travelers from the Middle East. Persons with the most severe involvement have had significant comorbidities whereas only 10–30% of patients with severe SARS had underlying conditions (26). The source of the causative agent, MERS-CoV, is probably an animal reservoir, most likely dromedary camels, although bats may have been the original source. Many camels have antibodies to MERS-CoV in serum and milk and virus in nasal secretions. The exact mode of animal-human transmission remains uncertain, and new cases continue to appear. Outbreaks and small clusters of human-human transmission, mostly in health care settings, have also occurred. Asymptomatic shedding and unrecognized URTIs may have contributed to the magnitude of these outbreaks. Spread likely occurs via large droplets and contact. Transmission has occurred in communities but so far has not been sustained. Unlike SARS, all age groups are affected by MERS. Like SARS, most MERS fatalities are in the elderly.

Adenovirus

Adenoviral respiratory infections are common. From 2–5% of all acute respiratory viral infections and as many as 18% of such infections in children are caused by adenoviruses. Infections begin in childhood once maternal antibodies wane; by late childhood, almost all individuals have been infected with one or more types. Many infections are subclinical. Adenoviruses of species B, C, and E cause respiratory disease, most types 1, 2, 5, and 6 are primarily endemic and types 4, 7, 14, and 21 are responsible for small epidemics, mostly in winter-spring. Spread is facilitated by the stability of the particles and persistent shedding from the gastrointestinal tract and tonsils. Outbreaks are frequent within families and closed settings.

Adenovirus types 4 and 7, and to a lesser extent type 21, have caused epidemics of acute respiratory distress syndrome (ARDS) among new military recruits. As many as 80% of trainees can be affected, with many hospitalizations and some fatalities. Stress, fatigue, crowing, and lack of immunity are contributing factors to these adenovirus types in young adults. In 2006–2007 a new strain of adenovirus type 14 (14a) emerged and caused severe respiratory disease and some fatalities in several U.S. military camps. The strain then spread to the general population, causing serious infections and deaths among civilians, with and without comorbidities, throughout the United States (27).

CLINICAL SIGNIFICANCE

Respiratory viruses are commonly associated with the clinical syndromes of first description, such as rhinoviruses with the common cold, parainfluenza viruses with croup, and RSV with bronchiolitis. Each virus, however, can cause disease along the pulmonary tract, depending on viral and host characteristics, as well as environmental contributors such as air pollution. The relative involvement of each virus in clinical entities is summarized in Table 1.

Influenza Virus

The classic "influenza" syndrome consists of a triad of abrupt fever that peaks early, respiratory symptoms, such as a nonproductive cough, sore throat, conjunctivitis, rhinitis, and systemic symptoms, including malaise, myalgia, and headache. Influenza in the elderly can present solely as confusion or listlessness. Infections are acquired by inhaling large droplets or perhaps small particle aerosols, or by autoinoculation. The incubation period is 1–4 days. Shedding is detectable before symptoms develop; peaks in 2–3 days when symptoms are maximal, and ends 6–8 days later. In uncomplicated cases, symptoms resolve about a week after onset, but cough and malaise can linger for weeks. The clinical picture, morbidity, and mortality are similar in illnesses caused by influenza A and B (28).

Influenza in children can be a more significant illness than in adults and difficult to distinguish from other respiratory virus infections. Presentation can include a sepsis-like illness with high fevers and febrile seizures, fever with few respiratory symptoms, URTI, croup, otitis media, an asthma exacerbation, or an influenza-like syndrome. Gastroenteritis is found more frequently in children and with influenza B (29). Children also shed influenza viruses longer and at higher titer than adults. Influenza C affects mostly children in whom it causes mostly sporadic mild URTI, although some LRTI are reported.

Serious complications are most frequent in the elderly, children younger than 2 years, and other high-risk populations. Complications include viral pneumonia, which can rapidly progress to death, and secondary bacterial pneumonia, which can also be fatal. Extra-pulmonary manifestations are rare but can be significant. Central nervous system involvement ranges from influenza encephalopathy, postinfluenza encephalitis, acute disseminated encephalomyelitis, Guillain-Barré syndrome, transverse myelitis, to Reye's syndrome (associated with aspirin use). Encephalopathy is more frequent in children of Japanese descent and with influenza A. Myositis, and myocarditis can complicate influenza B infections in children (30). Influenza during the 2009 pandemic affected slightly younger age groups. Gastrointestinal symptoms were common in all ages. Complications included viral pneumonitis accompanied by shock and renal failure (31). New groups at high risk of complications were the morbidly obese and pregnant women.

Human infections with AI virus usually present as LRTI due to preferential replication of these viruses in lower airways. Viral loads are usually low in the nasopharynx (NP), which may be why person-person transmission is rare. Gastrointestinal symptoms are frequent, and acute encephalitis can occur. Infections with A(H7N9) tend to exacerbate underlying medical conditions and are frequently complicated by secondary bacterial pneumonia (32). In most AL virus infections, virus shedding is prolonged and high levels of cytokines are released, with the highest levels and viral loads in fatalities. Death occurs in more than half of cases in

TABLE 1 Relative importance of major respiratory viruses in upper and lower respiratory tract disease

Virus	Importance in[a]:				
	Common cold	Influenza or Flu-like Illness	Croup	Bronchiolitis	Pneumonia
Influenza	+++	++++	+	+	++++
RSV	+++	+	++	++++	++++
HMPV	++	+	+	++++	++++
Parainfluenza	+++	+	++++	+++	++++
Rhinovirus	++++	++	+	–	+++
HCoV	+++	+	+	+	+
Adenovirus	+++	++	+	–	++++

[a]+ to ++++, minimal to major importance;—, no, or negligible, importance.

2 weeks or less due to progressive viral pneumonia or ARDS related to unrestrained virus replication, host inflammatory responses, and slow development of specific immunity.

RSV and HMPV

Most RSV (and probably HMPV) infections are acquired by contact with contaminated secretions in droplets or on fomites. Transmission is highly efficient, facilitated by survival of virus on surfaces for as long as 6 hours and by shedding from infants for 4 weeks or longer from immunocompromised hosts. The median incubation period is 4–8 days. Acute infections in most healthy term infants manifest as a classic URTI, with symptoms of cough, sneezing, and rhinorrhea. Other indicators are lethargy, irritability, and poor feeding. Fever may be absent. Few initial infections are asymptomatic, except in very young infants who are protected by maternal antibodies. RSV is significant risk factor for otitis media without bacterial pathogens (33).

Spread to lower airways occurs in 25–40% of RSV and 5–15% of HMPV infections in children. Many cases require hospitalization. Major manifestations are bronchiolitis, pneumonia, and sometimes croup. Significant wheezing, cough, coryza, intermittent fever, and apneic spells can develop, requiring intensive care or mechanical ventilation in some cases. Atelectasis is reported with some infections. Most patients improve with supportive care and recover in 8–15 days. Some prematurely born infants with RSV LRTI can have residual lung dysfunction. Another 10–50% can have recurrent wheezing or asthma episodes, which can persist for years and are triggered by re-infection. Whether these viruses cause reactive airway disease, or whether infection reveals preexisting host abnormalities remains unclear. The pathogenesis of LRTI in infants is also unclear. Concurrent invasive bacterial co-infections are infrequent unless patients are immunocompromised or intubated. Viral factors may be important, with RSV nonstructural proteins NS1 and NS2 and glycoprotein G impeding both innate and adaptive immune responses. It was once believed that viral replication triggers an inappropriately excessive immune response, but during natural RSV infection in infants, cytotoxic T cells in the airways correlate with recovery rather than with disease (34). Additional evidence now favors an inadequate adaptive immune response and immunosuppressive effect of maternal antibodies, followed by rapid and profound viral replication leading to massive apoptotic sloughing of respiratory cells and airway obstruction (35). Extrapulmonary manifestations may develop with RSV infection, but whether they represent direct or indirect effects of virus remains unclear. Gastrointestinal symptoms and encephalitis have been reported, with viral RNA (and, less frequently, infectious virus) detectable in blood, CSF, or tissue other than lung in severe or fatal cases.

In response to infection, only half of infected infants produce specific IgM, IgG, and IgA in respiratory secretions. Titers and avidity are also low, perhaps due to protection by maternal antibodies and immunologic immaturity. Antibodies also wane rapidly and may disappear completely by age 1, especially if infection occurred in early infancy. Durable protective responses begin to develop after re-infection. Cell-mediated immunity is also important for recovery, as demonstrated by the severe consequences of RSV and HMPV infection in immunocompromised patients.

Repeat infections with RSV in normal adults tend to be symptomatic and associated with rhinorrhea, pharyngitis, cough, and bronchitis. Recovery is prolonged in many cases and prompts many adults to seek medical attention, but fatalities are rare. Most HMPV infections in working-age adults are asymptomatic, although clinically apparent illness resembles RSV. In frail elderly and immunocompromised individuals as many as 50% of RSV or HMPV URTI infections can progress to the lower airways, requiring hospitalization due to pneumonia or exacerbation of underlying medical conditions. Viral pneumonia in HSCT recipients can also be severe or fatal. Persistent viral shedding by immunocompromised patients can prolong symptoms and lead to progressive pulmonary damage.

Parainfluenza Virus

PIV infections are acquired, like other respiratory virus infections, by inoculation of mucous membranes of the respiratory tract with virus-laden secretions from fomites or large droplet aerosols. After an incubation period of 2–6 days, replication is detected in the NP, and symptoms appear. Shedding continues for 3–10 days but can be longer, especially in immunocompromised patients, or persons with chronic respiratory disease. Survival of virus on fingers is brief but can last as long as 10 hours on surfaces. This hardiness, a low infectious dose, and prolonged shedding contribute to efficient transmission to susceptible individuals and frequent nosocomial spread.

URTIs are the most common clinical manifestation. In young subjects, PIV ranks as the third or fourth most common cause of URTIs when molecular tests are used. Most of these illnesses are mild and self-limited, although otitis media can complicate as many as a third of cases in infants and young children. Common clinical features include rhinitis, bronchitis, pharyngitis, cough, and fever. Re-infections in children and adults usually manifest as mild URTIs that cannot be distinguished from other respiratory virus infections.

Approximately 15% of these illnesses spread to the lower airways. Persistent fever and productive cough usually signal this involvement. The signature manifestation is croup in children, with a barking cough that identifies airway obstruction. Most cases are due to PIV-1, with lesser involvement of PIV-2 or -3. Progressive obstruction, requiring supportive care, can develop, but most cases can be managed without hospitalization. PIV-3 infections are the most severe, appearing during the first year of life or soon thereafter, and are responsible for most fatalities (36). Bronchiolitis and pneumonia, with wheezing, tachypnea, retractions, and cyanosis, are frequent clinical presentations. Less is known about the clinical presentation of PIV-4, but one recent U.S. study identified more PIV-4 than PIV-2 or PIV-3 in hospitalized children. Most PIV-4 associated illnesses resembled those induced by PIV-3, and no child had croup (13). PIV infections are usually self-limited, but poor outcomes can occur in infants and children, individuals with underlying medical problems, or HSCT patients (14). Spread beyond the respiratory tract is rare, but PIV-associated parotitis, aseptic meningitis, and encephalitis have been reported. Fatal infections involving gastrointestinal and urinary tract tissues have occurred in children with congenital immunodeficiencies.

Rhinovirus

Rhinovirus URTIs usually begin by autoinoculation of virus into the nose or eyes. Aerosol transmission may occur with prolonged contact. Psychological stress may increase the occurrence of symptomatic infections. The incubation period is approximately 2–3 days. Symptoms typically peak 2–4 days after onset and resolve in 10–14 days. Illness begins with a profuse watery nasal discharge, which may become mucopurulent and viscous. Other symptoms can include malaise, nasal congestion, sneezing, mild headache, pharyngitis, and cough. Fever is infrequent. Peak viral shedding from the nose occurs at 2–3 days after symptoms develop. Shedding usually ceases by 7–10 days, but infectious virus can be recovered for as long as 3 weeks in some individuals. Viral RNA is shed for about a month, except in immunocompromised patients, who can shed for prolonged periods (21, 23, 37). Re-infections are more common than prolonged infections, however, and contribute to the appearance of persistent shedding (38). Replication occurs in the cytoplasm of mucosal cells of the nose, sinuses, and eustachian tubes with surprisingly little tissue damage. Symptoms largely derive from induction of inflammatory mediators.

Symptoms of rhinovirus LRTI can range from an influenza-like syndrome to bronchiolitis and pneumonia (39). Clinical characteristics do not distinguish rhinovirus-associated LRTI from those caused by other common respiratory viruses, but are often preceeded by asthma or wheezing episodes. Rhinovirus LRTI, like those due to other respiratory viruses, may predispose immunocompromised patients to bacterial super-infections (40). In patients with chronic asthma, COPD, or cystic fibrosis, illnesses that begin as a URTI can progress to exacerbations accompanied by severe cough, wheezing, and sometimes hypoxia. Fatalities are rare but reported. Viremia has been detected in severe cases (41). Virus can also be identified in the feces of children with respiratory illness (42).

Type-specific immunity consists of IgG neutralizing antibodies in serum and a local IgA response, which is probably most important for long-lasting immunity. Interferon and other innate responses may also contribute to recovery because individuals who have one rhinovirus infection rarely get another within the following month. Type-specific serologic responses are mounted even by infants and persist for years, with the number of individual responses increasing throughout life. This pattern suggests that repeat infections are related in large part to the number of viral types encountered.

Coronavirus

The most frequent clinical manifestation in otherwise healthy individuals is a self-limited cold. Infections are acquired through inhalation of large droplets or by autoinoculation of mucous membranes. The incubation period is 2–5 days. Symptoms last about a week (range 3–18 days). Typical clinical features are coryza, rhinorrhea and nasal congestion, sore throat, and pharyngeal edema. Fever can be absent. Croup is a frequent manifestation of NL63 in children younger than 3 years. Otitis media complicates as many as half of the infections in young children. Underlying respiratory disorders can be exacerbated, although exacerbation seems to be more of an issue with rhinoviruses. Virus shedding can persist 3 weeks or longer after onset, and asymptomatic infections are frequent. Species-specific antibody appears but lasts for a mean of 4 months, and there is no heterotypic antibody response. These factors probably contribute to a high re-infection rate.

In some studies, as many as a third of young children hospitalized with LRTI have evidence of HCoV infection, but causal roles are difficult to establish due to frequent asymptomatic shedding in asymptomatic controls (43, 44). Symptoms are usually not as severe as RSV. Clinical manifestations are bronchitis, bronchiolitis, and pneumonia and occur most often in young children, the elderly, and persons with immunodeficiencies. LRTI involving OC43 may be more severe than with other HCoV. Fatalities are rare but have been reported in the very young, the very old, and in high-risk groups. Co-infections with other respiratory viruses are frequent. Disease may extend beyond the respiratory tract, with enteritis, hepatitis, and neurologic manifestations noted in some studies. Involvement in multiple sclerosis and Kawasaki disease were once proposed but are now discounted.

SARS usually presents as a LRTI without preceding upper respiratory tract symptoms (25). Onset is 4–7 days after exposure and begins with signs of systemic involvement, such as fever, myalgia, malaise, and headache. A nonproductive cough or dyspnea follows. Pneumonia appears in 7–10 days and progresses to severe acute respiratory distress in 20–30% of patients. Diarrhea develops in as many as 70% of individuals. Fatal cases have virus in multiple organs, including lung, bowel, liver, kidney, and brain. The highest upper airway viral loads appear late in the course of infection when patients are in intensive care, so many nosocomial infections occur. Most survivors fully recover. Seroconversion is demonstrable 2–3 weeks after onset of illness, with IgM and IgG appearing together. The IgG response, however, is not always sufficient for recovery and does not appear to be durable.

The clinical spectrum of MERS is more variable than that of SARS, ranging from asymptomatic infection or a URTI to full-blown ARDS with septic shock, and multi-organ failure (26). Case fatality rates are currently 35%. Mean incubation period for cases transmitted among humans is 5 days. Progression is more rapid than with SARS. Extrapulmonary manifestations involve the gastrointestinal tract, kidney, and liver; hypotension and septic shock can also occur. Co-infection with other respiratory viruses and bacteria are reported. Viral loads in the upper airway are higher than with SARS, but highest in lower airway specimens. Virus can also be detected in blood, urine, and stool. In some cases, diarrhea precedes respiratory symptoms. Shedding after recovery can be prolonged. The immune response resembles that in SARS. Secondary cases tend to be milder.

Adenovirus

Acute infections are acquired by contact with contaminated respiratory secretions, stool, or fomites or inhalation of aerosols. Initial sites of infection are usually the conjunctivae, oropharynx, or gut. The incubation period is 5–6 days. Following a brief viremia, virus spreads to regional lymphoid tissue where it can persist for weeks to months after recovery, especially in children. URTI manifest as colds, exudative tonsillitis, pharyngitis, pharyngoconjunctival fever, and occasionally croup. Discrete flecks of exudate on tonsils are often observed. Systemic symptoms include fever, headache, and malaise. Bacterial otitis media often complicates upper airway illnesses in young children. Follicular conjunctivitis (or "pink eye") accompanies many adenoviral URTI but can be the sole presentation. Severe adenoviral ocular infection can also be acquired by swimming in contaminated water or by contact with contaminated ophthalmologic instruments.

Adenoviral LRTI can be severe, prolonged, and associated with significant sequelae, such as bronchiectasis or bronchiolitis obliterans or bacterial super-infection. Fatalities in newborns and young children are not uncommon. Manifestations include tracheobronchitis, bronchiolitis, ARDS, and an influenza-like illness. Previous associations with pertussis-like symptoms and Kawasaki disease were most likely due to coincidental detection of persistent virus. Radiographic findings sometimes mimic a bacterial process. Adenoviral LRTI can be especially severe in solid-organ transplant

(SOT) or HSCT recipients (45). Sources are latent or reactivated virus or new exposures. Pneumonia can be an isolated finding or can accompany disseminated adenoviral disease. Mortalities of 18–52% in HSCT patients and as high as 35% in SOT recipients are described, with the highest rates in children. Acute infections in lung transplant patients can be severe and can induce bronchiolitis obliterans. Co-infections with other respiratory viruses are frequent in immunocompromised patients but can also occur in other populations.

After recovery, virus can be shed for 3–6 weeks in the throat and as long as 18 months in the stool of otherwise healthy children; shedding can occur for more than a year in some immunocompromised individuals. Virus can also be detected in about 5% of respiratory tract specimens from asymptomatic subjects by molecular methods (20). Genus- and type-specific IgG antibodies appear 7–10 days after onset in older children and adults; neutralizing antibodies may last a decade or more. The IgG response in young children can be delayed and is often directed only at the infecting virus type. A rise in titer to heterologous viruses within a species occurs in about 25% of adults. Virus-specific IgM appears in only 20–50% of cases and may reappear upon re-infection.

TREATMENT AND PREVENTION

Influenza Virus

Influenza is the only respiratory viral illness for which there are effective treatments and vaccines. Three neuraminidase inhibitors are licensed and approved for treatment and chemoprophylaxis of influenza A and B: orally administered oseltamivir (Tamiflu); zanamivir (Relenza) which is inhaled; and peramivir (Rapivab), which is given intravenously. Treatment, if initiated within 48 hours of symptom onset, may shorten duration of fever and other symptoms by about a day. It can be considered for otherwise healthy persons with recent onset of illness and is recommended for high-risk patients with confirmed or suspected influenza or in severe, complex, or progressive illness. Treatment should be initiated as soon as possible and not await laboratory confirmation of virus; treatment of hospitalized patients 4–5 days (or longer) after illness onset may improve outcomes. Side effects include nausea, vomiting, renal syndrome, headache, and occasional psychiatric syndromes. Infections with AI viruses may respond, even when treatment is implemented late in disease. Resistance is currently negligible, but a few 2009 A(H1N1) viruses have acquired resistance due mostly to a H275Y mutation in the NA gene. Resistance develops most often in immunocompromised patients, who can be a reservoir of resistant virus for others. Amantadine and rimantadine, antiviral medications that block viral uncoating, are still approved but are no longer recommended for treatment of typical influenza A virus infections because currently circulating influenza A viruses are resistant. However, the adamantanes may still have benefit to treat novel or neuraminidase inhibitor-resistant strains.

Vaccination is the most effective management strategy. Current influenza vaccines are given annually due to viral antigenic drift and the transient nature of vaccine-induced immunity. Vaccination is recommended for persons older than 6 months, lacking specific contraindications, individuals with risk factors for complications, and pregnant women for their protection and to "cocoon" their infants with maternal antibodies until the infants are eligible for vaccination. Close contacts of high-risk individuals should also be immunized to protect themselves and their patients. Indeed, annual influenza vaccination of U.S. health care workers is mandated by many facilities as a condition of employment.

Influenza vaccines are prepared using strains deemed most likely to circulate during the subsequent season. Vaccine preparation and distribution are lengthy processes, so mismatch can occur if antigenic drift happens late in the season. Effectiveness is usually 40–60% in healthy adults but can be lower when A(H3N2) predominates and in the elderly, who have difficulty mounting protective antibody responses. Four types of vaccine are currently approved: inactivated influenza vaccines (IIV); live inactivated influenza vaccines (LAIV); cell-culture inactivated influenza vaccines (ccIIV); and recombinant influenza vaccines (RIV). IIV is administered by injection and induces systemic immunity. LAIV is attenuated and is administered by nasal spray, so the live viruses can replicate in the lower temperature of the nasal passages and induce local immunity. Shedding of LAIV is detectable by culture, immunoassay (IA), or nucleic acid amplification techniques (NAAT) for about a week after administration, so recent immunization should be considered when interpreting positive influenza test results. For persons with severe allergy to eggs, ccIIV and RIV are made in canine and insect cell cultures, respectively. Both are delivered by injection and induce systemic immunity. They can be made more quickly than egg-derived vaccines, which could prove useful in a pandemic. Candidate vaccine viruses (CVVs) have also been prepared against A(H5N1) and A (H7N9) in case the vaccines are needed.

Current influenza vaccines are either trivalent or quadrivalent. Trivalent vaccines contain the selected strains of influenza A(H3N2) and A(H1N1) plus a single clade of influenza B. Quadrivalent vaccines contain both influenza A subtypes plus both clades of influenza B virus. Vaccine nomenclature denotes the vaccine type with a subscript numeral denoting valency, as in $LAIV_4$, a quadrivalent live virus vaccine. A high-dose IIV for persons 65 years or older is available and effective in this age group. Two doses of vaccine are needed for children 6 months to 8 years to achieve adequate immune responses. Protective responses usually appear about 2 weeks after vaccination of healthy individuals. For more information, see http://www.cdc.gov/flu/professionals/index.htm.

RSV and HMPV

Treatment of most RSV and HMPV infections is currently supportive. Corticosteroids or bronchodilators are generally ineffective. The synthetic nucleoside ribavirin has antiviral activity against both RSV and HMPV *in vitro*. Aerosolized ribavirin was once used to treat severe RSV-associated LRTI in pediatric patients, but routine use is no longer recommended due to concerns about efficacy, high cost, teratogenicity, and difficulties with drug delivery. Ribavirin does have modest therapeutic benefit, so it is sometimes used in aerosol or intravenous form in severe cases. Ribavirin also inhibits HMPV *in vitro*, and there are anecdotal reports of clinical benefit. GS-5806, a new antiviral that interferes with virus-cell fusion, reduces RSV viral loads and severity of disease in experimentally infected adult subjects. Phase 2 clinical trials are under way to assess safety and efficacy (46). Other compounds under investigation to treat or prevent RSV or HMPV LRTI include small interfering (si) RNA molecules to reduce viral RNA replication and novel nucleoside analogues.

Efforts to prevent infection involve active and passive immunization. To date, no approved RSV vaccine is available because of significant hurdles that include a requirement for efficacy in young infants in the presence of maternal antibody and a poor immune response of infants to initial infection. One experimental, formalin-inactivated vaccine actually enhanced disease in young children after natural infection, resulting in hospitalization and deaths of some recipients. Progress toward a safe and effective RSV vaccine has been aided by improved understanding of viral pathogenesis, the immune response to RSV, and the crystal structure of the F-protein, which is the major component of many proposed vaccines. Vaccines currently in development or advancing to early clinical trials now include live attenuated, live chimeric, replication defective, and nucleic acid-based preparations (47). Similar approaches are under investigation for HMPV.

By contrast, passive immunization with a humanized mouse monoclonal antibody known as palivizumab or Synagis (Medimmune), to prevent RSV LRTI in high-risk infants and young children, is approved and effective at reducing RSV severity in infants and young children with chronic lung disease of prematurity, preterm birth, or congenital heart disease. Palivizumab can prevent 40–80% of RSV-associated hospitalizations in these high-risk children, although its economic benefits are not firmly established (48). It has no efficacy as a therapeutic agent against severe RSV or HMPV disease. A second generation palivizumab antibody (motavizumab), with enhanced RSV F protein binding, significantly reduced hospital admissions and serious adverse events, compared with placebo, when given as prophylaxis, but like palivizumab, it has little benefit for treating severe disease. Resistance to palivizumab develops rarely, with only a single report to date of a palivizumab-resistant RSV mutant isolated from an infant. Similar monoclonal antibodies specific for the HMPV F-protein and MEP8, a monoclonal antibody that can neutralize RSV and HMPV viruses, are under investigation, but clinical trials in humans are not yet reported.

Parainfluenza Virus

No proven antiviral treatment for PIV infections is available. Ribavirin has activity against the virus *in vitro* and in animal models but is of uncertain benefit in humans. Reports of use are mostly limited to treatment of HSCT or lung transplant recipients in anecdotal reports or small case series. Other promising treatments include neuraminidase inhibitors, activators of interferon, polyoxometalates, small interfering RNAs, and the promising attachment inhibitor, DAS 181 (49). Dexamethasone and corticosteroids are established treatment for symptomatic relief of severe croup, but corticosteriod use in PIV-infected HSCT patients is associated with the development of viral pneumonia.

Currently no PIV vaccines are licensed. Most efforts have been directed toward developing a vaccine for PIV-3, although vaccines against PIV-1 and formulations effective against all types are under consideration. Development has been slowed by tactical questions of whether a stand-alone vaccine is indicated, whether the effort should be directed toward a combined PIV-3/RSV vaccine, or if vaccination against PIV should await proof of principle that an RSV vaccine will work. Current efforts are focused on intranasally administered vaccines using the classical technique of cold-passage, host-range attenuation, or reverse genetics to introduce specific attenuating mutations. The PIV vaccines with the most accumulated data include cp45, a live-attenuated, cold-passaged PIV-3 vaccine, and rHPIV3cp45, and a cDNA-derived recombinant version of cp45. Trials are also under way or planned with a chimeric vaccine of PIV-3 HN and F genes inserted into a bovine PIV-3, PIV-3 HN genes inserted into a murine PIV-1 (Sendai virus), a combined RSV/PIV vaccine, and a PIV-1 vaccine.

Rhinovirus

Antiviral therapy or vaccination for rhinovirus infections would have widespread application, but no therapies or vaccines are currently approved. Many promising antiviral compounds have been evaluated, including substances blocking virus attachment, uncoating, RNA replication, and protein synthesis. Medications interfering with cellular susceptibility and immunomodulators have also been assessed. Many putative antivirals show activity in the laboratory, but issues of drug delivery often reduce their clinical benefit. Among the compounds that have been studied are soluble ICAM-1, inhibitors of ICAM-1 upregulation, interferon alpha-2, intranasal imiquimod, intranasal ipratropium bromide, and pyrrolidine dithiocarbamate. One especially promising substance was pleconaril, which binds to a pocket in the VP1 protein and blocks uncoating of most picornavirus serotypes. It was not approved for use in humans due to interference of the compound with the efficacy of oral contraceptives. In volunteer trials, a viral protease inhibitor, rupintrivir, was effective but caused nasal irritation. Vapendavir inhibits the release of viral RNA into target cells. A phase 2 clinical trial to assess safety and efficacy of vapendavir in asthmatic adults with rhinovirus infection is currently under way. Alternative substances, such as Echinacea, have been evaluated, but most trials have been inconclusive or contradictory. Oral zinc sulfate may have some benefit for prevention and treatment of rhinovirus infections in children. Limited progress has been made in the development of an effective cross-serotype HRV vaccine due at least in part to the substantial antigenic heterogeneity observed in HRVs (50). Prevention of infection and symptomatic relief are the most effective interventions.

Coronavirus

Currently there are no specific antiviral medications or vaccines for the common HCoV, so treatment is supportive. Many attempts were made to treat SARS during the pandemic, but few large studies were conducted, and none was controlled. Compounds tested included ribavirin, corticosteroids, lopinavir/ritonavir, interferon, intravenous gamma globulin, convalescent plasma, and traditional Chinese medications. Some were active against SARS-CoV *in vitro*, but none was beneficial *in vivo*. More promising candidates developed recently include monoclonal antibodies and other compounds to prevent virus attachment, SARS-CoV specific protease inhibitors, and small interfering RNAs. Several vaccines have now been prepared and tested for efficacy in animal models. Neutralizing antibodies against SARS-CoV have been stockpiled to provide passive immunity should the virus resurface.

Treatment for MERS remains intensively supportive, although various interferon regimes, with or without ribavirin or mycophenolate, have been tried *ad hoc*. Antiviral compounds coming under more systematic investigation include GS5734, which was developed against Ebola and is now in clinical trials, as well as BCX4430, a potential treatment for filioviruses (51). Convalescent plasma and several new monoclonal antibodies are available for prophylaxis in health care settings and may be useful for treatment as well. Vaccines that induce neutralizing antibody responses to the S protein in humans, as well as a novel vaccine to immunize camels, have been developed (52). Initial assessment of treatments and prevention has been hampered by lack of small animal models that mimic MERS, but novel technology to generate "partially humanized mice" may aid this effort and interventions for other emerging infections (53). The most effective means to prevent SARS and MERS, however, may be surveillance for the early identification of cases and the rapid implementation of infection control.

Adenovirus

No treatment is currently approved for respiratory adenovirus infections. Investigational approaches include pharmacotherapy, biologic agents, and adoptive immunotherapy. Promising antiviral medications for severe respiratory illness in immunocompromised or immune normal individuals are the nucleoside analogues cidofovir and its lipid-linked and less-toxic derivative, brincidofovir (CMX001) (54, 55). Ganciclovir and ribavirin have been less effective but may be useful as adjuncts. A promising approach is silencing of adenoviral receptor and protein-gene expression by small interfering RNA molecules given alone or in combination with soluble receptor traps to inhibit virus attachment. Adoptive immunotherapy, using related or unrelated donor or recipient T cells sensitized to adenoviral structural proteins, is in clinical trials (56). Currently, no medications are approved to treat adenoviral ocular disease.

A live virus, enteric-coated, oral vaccine against adenovirus types 4 and 7 was given to U.S. military recruits in 1971–1996 and was effective in preventing epidemics of ARDS. When the vaccine program lapsed, outbreaks predictably resumed. A similar vaccine was redeveloped and introduced to troops in 2011; results are encouraging so far (57). No vaccine for civilians is planned.

DIAGNOSIS

Diagnosis of common respiratory virus infections can be made by culture, detection of viral antigens by immunoassays (IA), such as direct immunofluorescence (DIF) also referred to as direct fluorescence antibody tests (DFA), enzyme linked immunosorbent assay (ELISA), or immunochromatography, or identification of viral RNA or DNA by NAAT. Serology plays a minor role because other methods are more rapid and accurate. It can be useful if other specimens are not available and to detect immune response to novel viruses.

Influenza Virus

Swabs, aspirates, or washes of the NP are suitable for influenza virus detection by most methods; nasal and oropharyngeal swabs will yield virus in some methods but especially by NAAT (58). Lower respiratory tract specimens, such as tracheal aspirates, bronchoalveolar lavage (BAL), or lung tissue, are usually necessary for diagnosis in adults or immunocompromised patients. CSF rarely yields virus, but stool can be positive, particularly in 2009 A(H1N) infections (29). Influenza virus can be detected in blood by NAAT when disease is severe (59). Blood and feces are often positive in AI virus infections (5).

Influenza viruses can be identified by culture, IA, or molecular methods. Culture is typically performed on primary monkey kidney cells or the Madin Darby kidney (MDCK) cell line, but Mv1LU, LLC MK2, BGMK, or human A549 cell lines or lung fibroblasts are suitable. CPE usually develops in 4–5 days but can take 2 weeks. Occult replication can be revealed by hemadsorption of guinea pig (influenza A, B) or chicken (influenza C) red blood cells to infected monolayers. Isolates are usually confirmed and typed by immunofluorescence testing (IF) of infected cells. Sensitivity varies season-season but is less sensitive than NAAT. Hemagglutination inhibition with specific antisera can identify virus strains, although molecular analysis of HA or NA genes is increasingly used. Centrifugation-enhanced shell vial culture followed by IF (SVCC) on A549, monkey, or mixed cells (R-Mix, Quidel) can detect influenza viruses in 1–3 days with 50–94% sensitivity compared to culture in tubes (60, 61).

Two IA formats are available, (DIF), and rapid influenza diagnostic tests known as RIDT. DIF is performed on exfoliated NP cells using individual or pooled fluorophore-labeled monoclonal antibodies. Kits approved in the United States by the Food and Drug Administration (FDA) to detect influenza and other respiratory viruses by DIF include the D3 Ultra Respiratory Virus Screening and ID Kit (Diagnostic Hybrids), Bartel's Viral Respiratory Screening and Identification Kit (Intracel), and PathoDx Respiratory Virus Panel (ThermoFisher). Sensitivity of DIF for influenza virus was considered adequate relative to culture until molecular methods became widely available for comparison. Subsequently antigen detection methods were found to be of highly variable sensitivity, particularly for the 2009 A (H1N1) virus (62). Detection of AI viruses was poor by IA and culture (63).

RIDT are widely used by many laboratories and physician offices to obtain results in fewer than 30 minutes. The 10 or more RIDT currently approved in the U.S. are listed at http://www.cdc.gov/flu/professionals/diagnosis/rapidlab.htm. Most detect viral antigens by immunochromatography on paper strips and are read visually. Sensitivities are only 50–70% for the 2009 virus (H1N1), slightly better for seasonal strains of influenza A, and vary in their ability to detect influenza B and AI viruses relative to NAAT. A high false positive rate was also observed with one kit (64). Two new lateral flow RIDT, the Sofia Influenza A and B Fluorescent Immunoassay (Quidel) and the Veritor System for Flu A+B (BD Diagnostics), have been developed for better accuracy and to provide traceable results by instrument-based digital scans of their test strips. Accuracy is somewhat improved, but assessments have been performed mostly in children and adults and not the elderly, who have the highest rates of influenza morbidity and mortality and low viral loads (65).

Detection of viral RNA can be accomplished with sensitivity exceeding 95% and high specificity by NAAT. To date, numerous such assays are FDA-approved and are listed at http://www.cdc.gov/flu/professionals/diagnosis/molecular-assays.htm. Most require an NP specimen, and all distinguish type A from type B. Some also identify influenza A subtypes, which could be helpful if antiviral resistance develops in a subtype. At least one assay specifically detects A(H5N1). Co-detection of influenza A and B and RSV is provided by Xpert Flu/RSV (Cepheid), Simplexa Flu A/B & RSV Direct Kit (Focus Diagnostics), and Prodesse ProFlu+ (Hologic/GenProbe). Highly multiplexed panels approved in the United States to detect influenza A and B and other respiratory viruses include Film Array Respiratory Pathogen Panel (Biomerieux), eSensor RVP (GenMark Diagnostics), xTag and NexTag Respiratory Virus Panels (Luminex), and Verigene Repiratory Pathogens+ (Nanosphere). Time to results ranges from 1–8 hours. In head-to-head comparisons of multiplexed NAAT assays for respiratory viruses, sensitivities for influenza A and B ranged from 73–100% (66, 67). Methods are described or available outside the United States to differentiate the two clades of influenza B or influenza A variants from typical strains, to quantify virus loads, assess antigenic drift and shift, and to identify all known subtypes, clades, and vaccine viruses.

Three NAATs, the Alere i Flu A/B (Alere), COBAS Influenza A/B (Roche Diagnostics), and Xpert Flu+RSV Xpress (Cepheid), have recently received waivers to enable testing at the point of care. All tests are performed on small instruments and provide results in fewer than 30 minutes. Independent studies report that Alere i is 78–98% and 75–100% sensitive for influenza A and B, respectively, and COBAS is >99% sensitive for both viruses compared to other NAATs. Specificities exceed 99%. Accuracy is therefore significantly higher than for culture or RIDT (68, 69). Influenza serology is infrequently used for diagnosis for reasons previously discussed. Interpretation is also complicated by the "doctrine of original antigenic sin" characterized by a higher anamnestic response to the first infecting virus after an infection with a related one.

RSV and HMPV

Appropriate specimens from infants and children are aspirates or swabs of the NP. Lower airway specimens, such as BAL, can also be tested and are the best samples to diagnose LRTI in adults or immunocompromised patients. RSV grows in human A549, Hep-2, or HeLa cell lines; yield is lower using monkey kidney and human fibroblast cells. CPE of rounded cells is evident in 3–8 days and may not include syncytia formation. Hemadsorption of guinea pig red blood cells to cell monolayers should be absent. Confirmation by IF or a similar method should be performed because other respiratory viruses can induce syncytia. Sensitivity of culture ranges from 50–90%, with variations due to virus subtype, laboratory expertise, quality of the cells, stage of infection, and comparison assay. Alternatively, SVCC on individual or mixed cells can be performed. It is more sensitive than tube culture and can provide results in 1–3 days (60, 61). Tube culture for HMPV is insensitive and takes weeks and sometimes blind passages before a subtle RSV-like CPE is evident. Vero and LLC-MK2 cells are most often used. A low concentration of trypsin can enhance growth in some cell types. HMPV can also be identified by SVCC, with one study claiming more than 90% recovery relative to molecular methods (70).

Several IA formats are available in the United States for RSV antigen detection, including DIF, ELISA, and lateral flow immunochromatography. When correctly performed, DIF for RSV in exfoliated NP cells can be more sensitive than ELISA or immunochromatographic assays and 60–94% sensitive relative to NAAT. When the aforementioned multiplex DIF kits are used, RSV and other respiratory viruses can be detected. HMPV can also be detected by DIF. Sensitivities of 85% and 95% are reported for the D3 DFA (Diagnostic Hybrids) and Light Diagnostics Immunofluorescence (Millipore) kits, respectively, relative to NAAT. Both kits are >99% specific (71).

Lateral flow immunochromatographic assays are more rapid and easier to perform but less accurate than DIF. Sensitivities for RSV in most studies are 50–80% and specificities are 93–97% relative to NAAT. Two newer IA, Sofia RSV (Quidel) and Veritor RSV (Becton Dickinson) kits, have been developed, as with influenza viruses, for higher sensitivity. A recent study comparing these tests to NAAT in fresh NP specimens from infants and toddlers reports modest improvement; sensitivities were 85% for Sofia RSV and 73% for Veritor RSV compared with 57–70% for two conventional IA. (72). Similar assays for HMPV are described but are no longer available in the United States.

Currently, NAAT is the most accurate method to detect RSV and HMPV in all age groups and patient populations. Many good RSV and HMPV assays using different chemistries and platforms are FDA-approved for U.S. use. Most provide results within hours. No commercial assay detects only RSV, but Pro-HMPV+ (Hologic) detects only HMPV; RSV and HMPV can be identified by the Lyra RSV and HMPV Assay (Quidel). The previously described Xpert Flu/RSV, ProFlu+, and Simplexa Flu/RSV assays detect RSV and influenza virus well, with greater than 97% agreement among them (73, 74). The highly multiplexed approved NAAT assays from Biofire, Genmark, Luminex, and Nanosphere also detect RSV and HMPV exceedingly well (66, 67). RSV A and B subgroups can be separately identified by the Genmark, Luminex, and Nanosphere assays. No HMPV subgroup data are provided by commercial NAAT, although such assays are reported. At this time, no RSV or HMPV NAAT is approved in the United States for point of care. Reported sensitivities for commercial RSV NAAT, variously compared to tube culture, SVCC, IF, or other NAAT, range from 88–100%; sensitivity of commercial HMPV NAAT relative to other molecular assays is 98–100%. There is no cross-reaction between RSV and HMPV. Typing of RSV or HMPV is relevant mostly for epidemiology or research. Serology is insensitive and difficult to interpret due to frequent re-infection and therefore is rarely used for diagnostic purposes.

Parainfluenza Virus

PIV can be isolated from many upper- or lower-airway specimen types. Whereas specimens from the NP are best, swabs of the nose or oropharynx can also be used for culture and particularly for NAAT. Growth in culture is slow, with 3–5 days or longer being required. Replication is detected by CPE, which differs subtly by serotype, or by hemadsorption of guinea pig red blood cells to infected monolayers. PIV-4 can require 10 days or more to express CPE, and hemadsorption may be weak. Confirmation can be performed by IF with the aforementioned commercial group- or type-specific antibodies. SVCC, using human A549, monkey kidney, or mixed cells, can detect 60–90% or more of PIV-1, -2, or -3 positive specimens in 1–3 days (60, 61).

Rapid detection of PIV 1–3 antigen by DIF in exfoliated cells is only slightly less sensitive than culture. Testing by molecular methods increases the detection rate considerably. For example, a 41% increase in PIV yield using NAAT, compared to culture, was reported with nasal or oropharyngeal swabs from ill children in one study (75). NAAT also detected 30% more PIV-3, 64% more PIV-1, and 88% more PIV-2 than DIF in specimens from pediatric patients in another assessment (76). An FDA-approved multiplex molecular assay, Prodesse ProParaFlu+ (Hologic/GenProbe), detects and separately reports PIV-1, -2, and -3 (67). All four PIV types are typically included on the previously mentioned highly multiplexed respiratory virus panels (66, 67). Serology is of little diagnostic importance because of the requirement for acute and convalescent specimens. Serologic cross-reactions between PIV and mumps virus can also hamper interpretation of results.

Rhinovirus

Rhinoviruses can be identified in most respiratory specimen types, including swabs or aspirates from the nose, pharynx, or NP, as well sputum, tracheal aspirates, BAL, and lung tissue

(77). Culture was the initial method of virus detection, but it is now known to be insensitive and will not recover species C. Culture can be performed on human cells, such as MRC-5 or fetal tonsil, under conditions simulating those in the nose, i.e., pH of 7.0, temperature of 33–35°C, and gentle rolling of tubes. Typical CPE of rounded small and large cells appears in 1–4 days but can appear later, evolve slowly, or regress. Most isolates can be presumptively identified by CPE, although some resemble enteroviruses. Sensitivity to brief acid treatment was once used to distinguish rhinoviruses from the acid-stable enteroviruses. The assay is cumbersome, and interpretation can be confounded by respiratory enteroviruses, such as D68, that are partially acid labile. Commercial "rhinovirus-specific" monoclonal antibodies for confirmation by IF can also detect enteroviruses, so confirmation (and typing) of an isolate requires neutralization using type-specific antibodies or molecular analysis (78). The lack of a common antigen and a plethora of rhinovirus types also limits use of serology for diagnostic work.

Molecular methods are the most sensitive and rapid means to detect all three rhinovirus species. Most NAAT identify a region in the 5′ untranslated region shared by rhinoviruses and enteroviruses without differentiating between them. Assays purported to detect only rhinoviruses may co-detect enteroviruses, may not detect all serotypes, or may require some form of postamplification processing to detect all serotypes. Improved assays are reported but not commercially available (79). Detection of rhinoviruses (co-detected with enteroviruses) is also included in the previously mentioned commercial multiplexed NAAT panels. Given the frequent detection of rhinovirus RNA in asymptomatic individuals, quantitative PCR has been used to evaluate clinically relevant cutoff values (80). Analysis of host transcriptome patterns has the potential to differentiate between symptomatic rhinovirus disease and asymptomatic infection (81).

Coronavirus

Respiratory coronaviruses were initially detected by electron microscopy, which is cumbersome and insensitive for diagnostic work. Culture is likewise challenging because, although 229E grows on human diploid cells and NL63 on monkey cells, OC43 requires organ culture and HKU1, primary human airway epithelial cells. Antigen detection methods are not commercially available. Currently, NAAT offers the most practical mean of detection. The common HCoV can now be detected in upper and lower airway specimens by several previously mentioned FDA-approved multiplexed NAAT, including the Luminex NexTag and Biofire kits.

SARS-CoV will exhibit CPE on primary monkey or Vero cells, but culture should not be attempted by routine clinical laboratories for safety reasons. Commercially available multiplex NAAT, that detect common HCoV, do not identify SARS-CoV. Testing should be requested through public health laboratories, although some commercial laboratories offered testing when SARS-CoV circulated. NAAT was the most useful test during the pandemic (25). Sensitivity was about 80% on day 3 of illness but will likely be higher with contemporary methods. Early in the course of illness, NP and oropharyngeal swabs plus serum should be tested. Specimens from lower airway are most important, as replication tends to be highest there. Stool can also yield virus. The slow development of antibodies limits the immediate usefulness of serology, so serum should be collected initially and 28 days after onset to confirm or rule out a case.

The full spectrum of MERS infection remains to be defined, so collection of lower and upper respiratory tract specimens, serum, and stool for MERS-CoV-specific NAAT is recommended early and throughout the illness (26). As with SARS-CoV, the virus can be isolated on monkey cells, but culture should not be attempted in clinical diagnostic laboratories. Commercial multiplex NAAT that detect common HCoV also do not detect MERS-CoV, so specific testing in the United States must be accessed through public health departments. Serodiagnosis requires paired specimens collected early in the course of infection and 14–21 days later. If onset is greater than 14 days, serology may be essential because NAAT could be negative. An initial positive screening serology should be confirmed by neutralization since cross-reactivity with antibodies to other coronaviruses can occur.

Adenovirus

Suitable respiratory specimens are swabs, aspirates, or washes of the NP, as well as tracheal aspirates, BAL, lung tissue, and swabs or scrapings from the eye. Adenovirus DNA is also detectable in blood by molecular methods during some respiratory infections. Plasma or serum is usually preferred for analysis because whole blood white cells, especially from children, can contain low levels of virus (82).

Adenoviruses can be cultivated in tubes of human epithelial cells, such as A549, Hep-2, HeLa, or KB. Recovery is lower on human fibroblasts and monkey kidney cells. A characteristic CPE, usually of grapelike clusters, appears in 2–14 days and can be confirmed by IF. Some types, particularly those in species D, may not grow in routine culture. SVCC on individual cell types considerably reduces time to detection. The yield is only 50% compared to tube culture at 24 hours, which is lower than for other respiratory viruses, but can increase to almost 100% if incubation time is extended. Mixed cells do not recover adenoviruses well (61).

Several immunoassay kits are approved in the United States to detect viral antigens in exfoliated respiratory cells or fluids or ocular swabs. All are less sensitive than culture and molecular methods in particular. The SAS Rapid Adeno Test (SAS Scientific), a rapid immunochromatographic assay, can detect common types in NP specimens or oropharyngeal swabs. Initial reports suggested high sensitivity that could not be confirmed in a subsequent study (83). The AdenoPlus (Rapid Pathogen Screening Inc.) is another rapid immunochromatographic assay for ocular swabs that can be performed at the point of care. Sensitivity was only 40% in a recent prospective study compared to NAAT (84).

Molecular tests are now widely used to identify adenoviruses in respiratory specimens and blood during acute infection. An FDA-approved commercial NAAT kit, Adenovirus R-gene (Biomerieux), is available for qualitative and quantitative viral detection in respiratory specimens and blood. A recent multicenter assessment, using mixed cells and another NAAT, reported greater than 98% sensitivity and high specificity for qualitative detection in NP specimens compared with SVCC culture (85). Adenoviruses are detected by most of the aforementioned highly multiplexed commercial NAAT kits, but high sensitivity for all relevant virus types has been difficult to achieve. Indeed, the adenovirus component of the BioFire kit has already been reformulated but still may not detect specimens with low viral loads (86). Qualitative or quantitative detection of adenovirus DNA in blood may also have utility in identifying patients with adenoviral respiratory disease (45, 87).

Adenovirus typing can be performed by serology or NAAT. Its major use is for epidemiology or to study virus evolution, not for diagnosis. Detection of an IgG response to adenovirus in paired acute and convalescent specimens can help identify infections in otherwise normal adults, but it is not useful in children or in immunocompromised hosts.

BEST PRACTICES

Rapidly establishing the viral etiology of a respiratory illness can be beneficial in many ways. Improved antimicrobial use is the positive outcome most often demonstrated in observational studies (88, 89). Other advantages may include reduced hospital costs, fewer ancillary tests, less time spent in isolation, and shorter stays (90). Rapid testing also provides important data for hospital preparedness efforts and may reduce excess morbidity and mortality in the severely immunocompromised (91). Indeed, many transplant centers now provide nonapproved treatment for RSV and adenovirus LRTI in HSCT patients and delay conditioning treatments if any respiratory virus, including rhinovirus, is detected prior to transplant (92, 93). Clinical care guidelines also recommend rapid detection of respiratory viruses, particularly for hospitalized children with community-acquired pneumonia and patients with hematologic disorders (94, 95). Studies unable to verify significant benefits have cited difficulties with providers responding quickly or appropriately to test results and the expense of testing (96, 97). It should be noted, however, that, to our knowledge, only a single randomized controlled trial, that of rapid influenza testing in an ED that included children with underlying diseases, has demonstrated the clinical utility of rapid respiratory virus testing (98). Other randomized controlled trials of rapid testing have shown no statistical difference in antibiotic use, ancillary testing, hospital admissions, or length of stay (99–106). Thus, the data supporting the cost-effectiveness and clinical utility of rapid respiratory virus testing are suggestive but far from conclusive. Additional randomized controlled studies are critically important to develop respiratory virus testing algorithms that ensure quality, cost-effective clinical care of patients with suspected respiratory virus infections.

Considerations that affect the quality of rapid results include patient age, timing of specimen collection, collection site, and sampling method, as well as assay design. Yield is always best with specimens from children due to their higher viral loads and longer duration of shedding. Specimens should be collected within the first few days after onset of symptoms, although molecular assays can extend the detection window. Specimens from the posterior NP are most often sent because collection is straightforward, and the viruses detected there usually predict those replicating in lower airways (107, 108). Swabs of the NP can be easier to collect than aspirates or washes, but the latter result in a higher yield, even from children, but especially from adults and immunocompromised patients (109, 110). Oropharyngeal swabs can recover some respiratory viruses, but NP or nasal swabs are usually more sensitive (111). Oropharyngeal and nasal swabs contain mostly squamous cells, so are less useful for DIF but may increase yield of culture or NAAT when combined with NP specimens (112). Saliva, gargles, and nasal mucus can yield respiratory viruses by NAAT but are less informative than other specimens.

Lower tract specimens, such as BAL, are optimal to detect viral LRTI in adults, the elderly, or immunocompromised patients due to low viral loads. Sputum can yield common respiratory viruses by NAAT and may be more sensitive than nasal and oropharyngeal swabs, but special processing may be needed prior to testing (113). Sputum from young children is usually saliva and of less diagnostic use. Nucleic acids of many respiratory viruses, including rhinoviruses, are detectable by NAAT in the blood of otherwise healthy or immunocompromised symptomatic patients. Though virus load in blood appears to correlate with the amount of virus in BAL fluid, at least for some respiratory viruses, and may reflect disease severity (114–116), quantitative respiratory virus NAAT in blood specimens is not widely available; qualitative detection in BAL provides sufficient information for most clinical decision making.

Test methods for rapid detection of respiratory viruses are in transition. Use of culture (tube and centrifugation) and antigen detection continues to diminish. Molecular methods are increasingly favored due to high sensitivity and specificity, rapidity, ease of use with newer commercial kits and instruments, operator safety, and the ability to detect cultivatable and noncultivatable viruses. Concerns for contamination are also decreasing with the advent of molecular closed systems. Remaining challenges include the expense of testing and interpretation of results. Cost-benefit ratios can be optimized by limiting testing to populations most likely to benefit. Interpretation of results is more problematic because molecular methods can detect viral nucleic acids from previous, incipient, asymptomatic infections. Firmly establishing the etiology may ultimately require identification of virus-specific host-response biomarkers or use of quantitative assays with thresholds to discriminate active from incidental virus infections. An interim solution may be to consider virus type and patient age. Indeed, several studies suggest that influenza virus, RSV, and HMPV are more likely to be etiologic agents than PIV, HCoV, rhinovirus, and adenovirus, especially in children (80, 117). Yet severe and even fatal illnesses with these latter viruses are well documented in many populations.

Influenza

Once seasonal influenza is widespread, a clinical diagnosis can reasonably guide treatment in most low-risk outpatients. Testing is important, however, for sound decision making in other settings. Unlike specimen requirements for many respiratory viruses, oropharyngeal and nasal swabs can be informative, especially by NAAT. Collection of oropharyngeal swabs or BAL is recommended for AI viruses, which replicate to highest titer in lower airways. NP specimens should also be obtained in case the virus is a seasonal strain.

The most appropriate diagnostic method is controversial. Culture, even SVCC, is too slow to affect patient care. Sensitivity and timeliness are concerns for DIF. RIDT continue to be used by many providers due to low cost and a body of older literature supporting their benefits. Many RIDT, however, now have unacceptably low sensitivity and specificity, prompting the FDA to propose that manufacturers regularly demonstrate adequate detection of currently circulating and AI viruses relative to culture and NAAT. Participation in this program, however, is currently voluntary. Meanwhile the CDC has issued guidelines, available at (http://www.cdc.gov/flu/professionals/diagnosis/rapidlab.htm [last accessed 3 May 2016]) recommending NAAT for confirmation of RIDT-negative specimens during periods of high influenza prevalence, and for RIDT-positive specimens when virus circulation is low. How widely these guidelines are followed is uncertain. Studies on the performance and acceptance of rapid influenza NAAT as an alternative to RIDT are pending.

Another consideration is the need for reliable detection of human infections with AI or swine variant influenza viruses. With the recent appearance of highly pathogenic AI viruses in U.S. poultry, awareness should be high that persons involved in poultry outbreak response or with exposure to infected birds are at risk of infection. Most commercial influenza assays cannot detect AI viruses well or differentiate avian from human subtypes, so suspect cases should promptly be referred to public health laboratories, even if initial influenza test results are negative, where such tests can be obtained. Guidelines to determine who should be evaluated are at http://www.cdc.gov/flu/avianflu/clinicians-evaluating-patients.htm (last accessed 16 May 2016). Similar considerations apply for persons frequently exposed to swine.

RSV and HMPV

A presumptive clinical diagnosis of RSV or HMPV can be made in an infant or young child with classic bronchiolitis during the wintertime. Due to lack of antiviral treatment for RSV or HMPV, establishing a specific diagnosis is less pressing than for influenza. Indeed, current U.S. clinical care guidelines discourage viral testing for infants and young children with suspected RSV bronchiolitis (118). This recommendation, however, is controversial, cohort pediatric patients by specific viral etiology, as many institutions with open wards (119). Clinical care guidelines also recommend testing of infants who are hospitalized and receiving monthly palivizumab because detection of RSV permits discontinuation of prophylaxis due to low probability of a second RSV infection in the same year. Diagnosis is also recommended for immunocompromised patients to inform decisions about clinical care, including delay of conditioning treatment or chemotherapy or to consider ribavirin therapy (92). Considerable data also support testing for hospitalized children with pneumonia, as well as adults and the elderly with LRTI and no evidence of bacterial infection, and perhaps in emergency room settings (88).

RSV and HMPV can be detected in the same NP or lower airway specimens that identify other common respiratory viruses. Nonculture methods are best because both viruses are labile and grow slowly. Antigen detection methods, particularly DIF and, to a lesser extent, immunochromatography, for RSV can be used, but NAAT is preferred for reasons previously mentioned. Highly multiplexed NAAT have the advantage of detecting both RSV and HMPV and differentiating these viruses from others that cause similar symptoms. Typing of RSV or HMPV is unnecessary for clinical work. Detection of virus in blood may be a marker of severe RSV disease, but its usefulness in management of individual patients remains to be determined.

Parainfluenza Virus

A reasonably accurate clinical diagnosis of PIV can be made in a child with croup when PIV-1 is circulating. A specific viral diagnosis is usually unnecessary because management of croup is symptom-based. In pneumonia or other unusual presentations, testing may be indicated if a viral etiology could improve outcomes. As with RSV, PIV infections in HSCT and other immunocompromised patients can be especially severe, so a viral diagnosis can contribute to appropriate management, provide access to novel antiviral treatments, and help limit nosocomial outbreaks (16, 49). Virus can be detected in the same NP or lower airway specimens used to identify other common respiratory viruses, although nasal or oropharyngeal swabs may be adequate if NAAT is used. Detection of PIV by molecular methods is currently the most sensitive and practical approach to detect all four virus types. Due to the nonspecific nature of symptoms in PIV-associated LRTI, other than croup, multiplex NAAT rather than PIV-specific tests may be preferable.

Rhinovirus

Diagnosis of rhinovirus-associated URTI or exacerbations of asthma or COPD is currently unnecessary due to lack of specific antiviral therapy or virus-specific interventions. Since rhinoviruses are now known to cause serious, although rarely fatal, LRTI, rapid identification is likely to improve the same outcomes associated with diagnosis of other respiratory viruses. Rhinovirus yield was once thought to be highest from the nose, so nasal swabs were the sample most often recommended, but upper and lower respiratory tract specimens are now considered suitable. Diagnosis of rhinovirus disease, however, is more challenging than for other respiratory viruses due to frequent co-detection with other respiratory viruses, difficulty differentiating rhinoviruses from enteroviruses, and frequent asymptomatic infections. Methods to distinguish clinically relevant rhinovirus-associated illnesses from asymptomatic or persistent infections are under investigation, including determinations of relevant viral loads or molecular signatures of host response to rhinovirus disease (80, 81). Meanwhile, if a specific diagnosis is sought, NAAT should be used. Culture is insensitive and cannot recover species C, and antigen detection methods are not available. Currently, the only commercial NAATs approved for rhinovirus detection in the U.S. are the large multiplex NAAT panels. These tests can also determine if a pathogen other than rhinovirus is present.

Coronavirus

Detection of the common HCoV in appropriate clinical scenarios is best made by NAAT due to the insensitivity of culture and lack of antigen detection tests. Virus can be detected in the same NP or lower-airway specimens useful for the other common respiratory viruses. As with rhinoviruses, the significance identifying HCoV is especially difficult to interpret due to the high rate of asymptomatic carriage and frequent co-detection of other viruses. Rapid diagnosis of human infections with MERS-CoV (and SARS-CoV if it re-emerges) is essential to prevent human-human spread. Awareness of MERS should be high because cases continue to emerge globally; SARS is less likely to be considered since the virus has not circulated since 2004. Early symptoms of SARS and MERS can also be confused with illnesses due to common respiratory viruses, which has been a significant problem in the Middle East where MERS is prominent. To prevent over-testing, recommendations to identify probable cases based on epidemiologic risk factors and clinical presentations are available at http://www.cdc.gov/coronavirus/mers/hcp.html (last accessed 16 May 2016) for MERS and http://www.cdc.gov/sars/surveillance/absence.html (last accessed 16 May 2016) for SARS. Recommendations include asking about contact with travelers who live or have visited regions where MERS is currently circulating or SARS once spread, or about contact with dead birds. Indeed, questioning visitors for possible exposure to MERS has been added to assessment of risk for Ebola in many U.S. health care facilities. Public health officials should also be notified promptly about persons of interest, even before testing them for common respiratory viruses. Suitable specimens to diagnose MERS and SARS differ from those recommended to recover common HCoV in that oropharyngeal swabs and lower airway specimens, as well as serum and stool, should be submitted for MERS- and SARS-CoV.

Adenovirus

Laboratory testing is necessary when a specific diagnosis is sought because adenovirus-induced symptoms closely resemble those due to other respiratory viruses; ocular symptoms may be more specific. Typical NP, ocular, and lower respiratory tract specimens are suitable for diagnosis of respiratory disease. Throat swabs should be avoided because of shedding from adenoids and tonsils. Blood may have diagnostic utility in immunocompetent children and immunocompromised hosts with respiratory illnesses. Histopathologic evidence in lung tissue is more suggestive of adenoviral disease than of other respiratory virus illnesses.

Culture and antigen detection are less helpful for adenovirus testing than for many other respiratory viruses and have largely been displaced by molecular methods. Test methods vary in sensitivity and may miss important virus types, so caution with assay selection is warranted. Interpreting positive results from respiratory specimens is also challenging due to the tendency of adenovirus to cause asymptomatic infections, to be shed for prolonged periods after acute infection, or to reactivate, particularly in children and immunocompromised hosts. In this regard, assessment of viral load in respiratory specimens by NAAT has been proposed to distinguish active from incidental respiratory tract infections (120), although this approach has not been widely implemented. Qualitative detection of adenovirus in serum may also be useful (87). Quantitative assessment of adenovirus DNA in blood is an established method to diagnose disseminated disease and other clinical manifestations in immunocompromised patients. It is therefore possible that relevant values can be derived that distinguish active respiratory disease from asymptomatic shedding (45, 80, 87). Institution-specific values, however, should be established because consensus cutoff values are unreliable due to lack of international quantitative standards and wide variations in assay performance.

CONCLUSIONS

Respiratory virus diagnostics have been energized by significant changes, particularly the development of highly sensitive molecular assays that have altered our understanding of the clinical spectrum, epidemiology, and pathogenesis of respiratory disease. The value of diagnostic methods to help control virus spread and provide early warning of impending pandemics is undisputed. It is also becoming clear that a specific viral diagnosis in the right clinical setting can improve many clinical and economic outcomes. Yet considerable challenges remain, most notably a need for more effective antiviral therapies and vaccines for prevention. The medical burden of respiratory viral disease is significant and will continue to increase with population growth, increasing numbers of susceptible individuals in the population, and globalization. Continued efforts to improve diagnostic strategies and prevent or treat respiratory virus infections should therefore be priorities.

REFERENCES

1. **Ortiz JR, Neuzil KM, Cooke CR, Neradilek MB, Goss CH, Shay DK.** 2014. Influenza pneumonia surveillance among hospitalized adults may underestimate the burden of severe influenza disease. *PLoS One* **9:**e113903.
2. **Morens DM, Taubenberger JK, Harvey HA, Memoli MJ.** 2010. The 1918 influenza pandemic: lessons for 2009 and the future. *Crit Care Med* **38**(Suppl)**:**e10–e20.
3. **Jhung MA, Swerdlow D, Olsen SJ, Jernigan D, Biggerstaff M, Kamimoto L, Kniss K, Reed C, Fry A, Brammer L, Gindler J, Gregg WJ, Bresee J, Finelli L.** 2011. Epidemiology of 2009 pandemic influenza A (H1N1) in the United States. *Clin Infect Dis* **52**(Suppl 1)**:**S13–S26.
4. **Jhung MA, Epperson S, Biggerstaff M, Allen D, Balish A, Barnes N, Beaudoin A, Berman L, Bidol S, Blanton L, Blythe D, Brammer L, D'Mello T, Danila R, Davis W, de Fijter S, Diorio M, Durand LO, Emery S, Fowler B, Garten R, Grant Y, Greenbaum A, Gubareva L, Havers F, Haupt T, House J, Ibrahim S, Jiang V, Jain S, Jernigan D, Kazmierczak J, Klimov A, Lindstrom S, Longenberger A, Lucas P, Lynfield R, McMorrow M, Moll M, Morin C, Ostroff S, Page SL, Park SY, Peters S, Quinn C, Reed C, Richards S, Scheftel J, Simwale O, Shu B, Soyemi K, Stauffer J, Steffens C, Su S, Torso L, Uyeki TM, Vetter S, Villanueva J, Wong KK, Shaw M, Bresee JS, Cox N, Finelli L.** 2013. Outbreak of variant influenza A(H3N2) virus in the United States. *Clin Infect Dis* **57:**1703–1712.
5. **Su S, Bi Y, Wong G, Gray GC, Gao GF, Li S.** 2015. Epidemiology, evolution, and recent outbreaks of avian influenza viruses in China. *J Virol* **89:**8671–8676.
6. **Díez-Domingo J, Pérez-Yarza EG, Melero JA, Sánchez-Luna M, Aguilar MD, Blasco AJ, Alfaro N, Lázaro P.** 2014. Social, economic, and health impact of the respiratory syncytial virus: a systematic search. *BMC Infect Dis* **14:**544–574.
7. **Jain S, Williams DJ, Arnold SR, Ampofo K, Bramley AM, Reed C, Stockmann C, Anderson EJ, Grijalva CG, Self WH, Zhu Y, Patel A, Hymas W, Chappell JD, Kaufman RA, Kan JH, Dansie D, Lenny N, Hillyard DR, Haynes LM, Levine M, Lindstrom S, Winchell JM, Katz JM, Erdman D, Schneider E, Hicks LA, Wunderink RG, Edwards KM, Pavia AT, McCullers JA, Finelli L, CDC EPIC Study Team.** 2015. Community-acquired pneumonia requiring hospitalization among U.S. children. *N Engl J Med* **372:**835–845.
8. **Falsey AR, Hennessey PA, Formica MA, Cox C, Walsh EE.** 2005. Respiratory syncytial virus infection in elderly and high-risk adults. *N Engl J Med* **352:**1749–1759.
9. **Falsey AR.** 2007. Respiratory syncytial virus infection in adults. *Semin Respir Crit Care Med* **28:**171–181.
10. **Walsh EE, Peterson DR, Falsey AR.** 2008. Human metapneumovirus infections in adults: another piece of the puzzle. *Arch Intern Med* **168:**2489–2496.
11. **Avetisyan G, Mattsson J, Sparrelid E, Ljungman P.** 2009. Respiratory syncytial virus infection in recipients of allogeneic stem-cell transplantation: a retrospective study of the incidence, clinical features, and outcome. *Transplantation* **88:** 1222–1226.
12. **Abedi GR, Prill MM, Langley GE, Wikswo ME, Weinberg GA, Curns AT, Schneider E.** 2016. Estimates of parainfluenza virus-associated hospitalizations and cost among children aged less than 5 years in the United States, 1998–2010. *J Pediatric Infect Dis Soc* **5:**7–13.
13. **Frost HM, Robinson CC, Dominguez SR.** 2014. Epidemiology and clinical presentation of parainfluenza type 4 in children: a 3-year comparative study to parainfluenza types 1–3. *J Infect Dis* **209:**695–702.
14. **Shah DP, Shah PK, Azzi JM, Chemaly RF.** 2016. Parainfluenza virus infections in hematopoietic cell transplant recipients and hematologic malignancy patients: A systematic review. *Cancer Lett* **370:**358–364.
15. **Peck AJ, Englund JA, Kuypers J, Guthrie KA, Corey L, Morrow R, Hackman RC, Cent A, Boeckh M.** 2007. Respiratory virus infection among hematopoietic cell transplant recipients: evidence for asymptomatic parainfluenza virus infection. *Blood* **110:**1681–1688.
16. **Sydnor ER, Greer A, Budd AP, Pehar M, Munshaw S, Neofytos D, Perl TM, Valsamakis A.** 2012. An outbreak of human parainfluenza virus 3 infection in an outpatient hematopoietic stem cell transplantation clinic. *Am J Infect Control* **40:**601–605.
17. **Rahamat-Langendoen JC, Riezebos-Brilman A, Hak E, Schölvinck EH, Niesters HG.** 2013. The significance of

rhinovirus detection in hospitalized children: clinical, epidemiological and virological features. *Clin Microbiol Infect* **19:** E435–E442.

18. **Linder JE, Kraft DC, Mohamed Y, Lu Z, Heil L, Tollefson S, Saville BR, Wright PF, Williams JV, Miller EK.** 2013. Human rhinovirus C: Age, season, and lower respiratory illness over the past 3 decades. *J Allergy Clin Immunol* **131:**69–77.e1-6.
19. **Messacar K, Robinson CC, Bagdure D, Curtis DJ, Glodé MP, Dominguez SR.** 2013. Rhino/enteroviruses in hospitalized children: a comparison to influenza viruses. *J Clin Virol* **56:**41–45.
20. **Jartti T, Jartti L, Peltola V, Waris M, Ruuskanen O.** 2008. Identification of respiratory viruses in asymptomatic subjects: asymptomatic respiratory viral infections. *Pediatr Infect Dis J* **27:**1103–1107.
21. **Piralla A, Zecca M, Comoli P, Girello A, Maccario R, Baldanti F.** 2015. Persistent rhinovirus infection in pediatric hematopoietic stem cell transplant recipients with impaired cellular immunity. *J Clin Virol* **67:**38–42.
22. **Peltola V, Waris M, Kainulainen L, Kero J, Ruuskanen O.** 2013. Virus shedding after human rhinovirus infection in children, adults and patients with hypogammaglobulinaemia. *Clin Microbiol Infect* **19:**E322–E327.
23. **Loeffelholz MJ, Trujillo R, Pyles RB, Miller AL, Alvarez-Fernandez P, Pong DL, Chonmaitree T.** 2014. Duration of rhinovirus shedding in the upper respiratory tract in the first year of life. *Pediatrics* **134:**1144–1150.
24. **Gaunt ER, Hardie A, Claas EC, Simmonds P, Templeton KE.** 2010. Epidemiology and clinical presentations of the four human coronaviruses 229E, HKU1, NL63, and OC43 detected over 3 years using a novel multiplex real-time PCR method. *J Clin Microbiol* **48:**2940–2947.
25. **Muller MP, McGeer A.** 2007. Severe acute respiratory syndrome (SARS) coronavirus. *Semin Respir Crit Care Med* **28:** 201–212.
26. **Zumla A, Hui DS, Perlman S.** 2015. Middle East respiratory syndrome. *Lancet* **386:**995–1007.
27. **Kajon AE, Lu X, Erdman DD, Louie J, Schnurr D, George KS, Koopmans MP, Allibhai T, Metzgar D.** 2010. Molecular epidemiology and brief history of emerging adenovirus 14-associated respiratory disease in the United States. *J Infect Dis* **202:**93–103.
28. **Su S, Chaves SS, Perez A, D'Mello T, Kirley PD, Yousey-Hindes K, Farley MM, Harris M, Sharangpani R, Lynfield R, Morin C, Hancock EB, Zansky S, Hollick GE, Fowler B, McDonald-Hamm C, Thomas A, Horan V, Lindegren ML, Schaffner W, Price A, Bandyopadhyay A, Fry AM.** 2014. Comparing clinical characteristics between hospitalized adults with laboratory-confirmed influenza A and B virus infection. *Clin Infect Dis* **59:**252–255.
29. **Minodier L, Charrel RN, Ceccaldi PE, van der Werf S, Blanchon T, Hanslik T, Falchi A.** 2015. Prevalence of gastrointestinal symptoms in patients with influenza, clinical significance, and pathophysiology of human influenza viruses in faecal samples: what do we know? *Virol J* **12:**215.
30. **Paddock CD, Liu L, Denison AM, Bartlett JH, Holman RC, Deleon-Carnes M, Emery SL, Drew CP, Shieh WJ, Uyeki TM, Zaki SR.** 2012. Myocardial injury and bacterial pneumonia contribute to the pathogenesis of fatal influenza B virus infection. *J Infect Dis* **205:**895–905.
31. **Reed C, Chaves SS, Perez A, D'Mello T, Daily Kirley P, Aragon D, Meek JI, Farley MM, Ryan P, Lynfield R, Morin CA, Hancock EB, Bennett NM, Zansky SM, Thomas A, Lindegren ML, Schaffner W, Finelli L.** 2014. Complications among adults hospitalized with influenza: a comparison of seasonal influenza and the 2009 H1N1 pandemic. *Clin Infect Dis* **59:**166–174.
32. **Yu L, Wang Z, Chen Y, Ding W, Jia H, Chan JF, To KK, Chen H, Yang Y, Liang W, Zheng S, Yao H, Yang S, Cao H, Dai X, Zhao H, Li J, Bao Q, Chen P, Hou X, Li L, Yuen KY.** 2013. Clinical, virological, and histopathological manifestations of fatal human infections by avian influenza A(H7N9) virus. *Clin Infect Dis* **57:**1449–1457.
33. **Ruohola A, Pettigrew MM, Lindholm L, Jalava J, Räisänen KS, Vainionpää R, Waris M, Tähtinen PA, Laine MK, Lahti E, Ruuskanen O, Huovinen P.** 2013. Bacterial and viral interactions within the nasopharynx contribute to the risk of acute otitis media. *J Infect* **66:**247–254.
34. **Lukens MV, van de Pol AC, Coenjaerts FE, Jansen NJ, Kamp VM, Kimpen JL, Rossen JW, Ulfman LH, Tacke CE, Viveen MC, Koenderman L, Wolfs TF, van Bleek GM.** 2010. A systemic neutrophil response precedes robust CD8(+) T-cell activation during natural respiratory syncytial virus infection in infants. *J Virol* **84:**2374–2383.
35. **Meng J, Stobart CC, Hotard AL, Moore ML.** 2014. An overview of respiratory syncytial virus. *PLoS Pathog* **10:** e1004016.
36. **Henrickson KJ, Hoover S, Kehl KS, Hua W.** 2004. National disease burden of respiratory viruses detected in children by polymerase chain reaction. *Pediatr Infect Dis J* **23**(Suppl):S11–S18.
37. **Peltola V, Waris M, Kainulainen L, Kero J, Ruuskanen O.** 2013. Virus shedding after human rhinovirus infection in children, adults and patients with hypogammaglobulinaemia. *Clin Microbiol Infect* **19:**E322–E327.
38. **Zlateva KT, de Vries JJ, Coenjaerts FE, van Loon AM, Verheij T, Little P, Butler CC, Goossens H, Ieven M, Claas EC; GRACE Study Group.** 2014. Prolonged shedding of rhinovirus and re-infection in adults with respiratory tract illness. *Eur Respir J* **44:**169–177.
39. **Iwane MK, Prill MM, Lu X, Miller EK, Edwards KM, Hall CB, Griffin MR, Staat MA, Anderson LJ, Williams JV, Weinberg GA, Ali A, Szilagyi PG, Zhu Y, Erdman DD.** 2011. Human rhinovirus species associated with hospitalizations for acute respiratory illness in young US children. *J Infect Dis* **204:**1702–1710.
40. **Jacobs SE, Lamson DM, Soave R, Guzman BH, Shore TB, Ritchie EK, Zappetti D, Satlin MJ, Leonard JP, van Besien K, Schuetz AN, Jenkins SG, George KS, Walsh TJ.** 2015. Clinical and molecular epidemiology of human rhinovirus infections in patients with hematologic malignancy. *J Clin Virol* **71:**51–58.
41. **Fuji N, Suzuki A, Lupisan S, Sombrero L, Galang H, Kamigaki T, Tamaki R, Saito M, Aniceto R, Olveda R, Oshitani H.** 2011. Detection of human rhinovirus C viral genome in blood among children with severe respiratory infections in the Philippines. *PLoS One* **6:**e27247.
42. **Harvala H, McIntyre CL, McLeish NJ, Kondracka J, Palmer J, Molyneaux P, Gunson R, Bennett S, Templeton K, Simmonds P.** 2012. High detection frequency and viral loads of human rhinovirus species A to C in fecal samples; diagnostic and clinical implications. *J Med Virol* **84:**536–542.
43. **Prill MM, Iwane MK, Edwards KM, Williams JV, Weinberg GA, Staat MA, Willby MJ, Talbot HK, Hall CB, Szilagyi PG, Griffin MR, Curns AT, Erdman DD, New Vaccine Surveillance Network.** 2012. Human coronavirus in young children hospitalized for acute respiratory illness and asymptomatic controls. *Pediatr Infect Dis J* **31:**235–240.
44. **Talbot HK, Shepherd BE, Crowe JE Jr, Griffin MR, Edwards KM, Podsiad AB, Tollefson SJ, Wright PF, Williams JV.** 2009. The pediatric burden of human coronaviruses evaluated for twenty years. *Pediatr Infect Dis J* **28:**682–687.
45. **Echavarría M.** 2008. Adenoviruses in immunocompromised hosts. *Clin Microbiol Rev* **21:**704–715.
46. **DeVincenzo JP, Whitley RJ, Mackman RL, Scaglioni-Weinlich C, Harrison L, Farrell E, McBride S, Lambkin-Williams R, Jordan R, Xin Y, Ramanathan S, O'Riordan T, Lewis SA, Li X, Toback SL, Lin SL, Chien JW.** 2014. Oral GS-5806 activity in a respiratory syncytial virus challenge study. *N Engl J Med* **371:**711–722.
47. **Graham BS.** 2016. Vaccines against respiratory syncytial virus: the time has finally come. *Vaccine* **34:**3535–3541.
48. **Andabaka T, Nickerson JW, Rojas-Reyes MX, Rueda JD, Bacic Vrca V, Barsic B.** 2013. Monoclonal antibody for reducing the risk of respiratory syncytial virus infection in children. *Cochrane Database Syst Rev* (4):CD006602.

49. **Salvatore M, Satlin MJ, Jacobs SE, Jenkins SG, Schuetz AN, Moss RB, Van Besien K, Shore T, Soave R.** 2016. DAS181 for Treatment of Parainfluenza Virus Infections in Hematopoietic Stem Cell Transplant Recipients at a Single Center. *Biol Blood Marrow Transplant* **22:**965–970.
50. **Glanville N, Johnston SL.** 2015. Challenges in developing a cross-serotype rhinovirus vaccine. *Curr Opin Virol* **11:**83–88.
51. **Modjarrad K.** 2016. Treatment strategies for Middle East respiratory syndrome coronavirus. *J Virus Erad* **2:**1–4.
52. **Perlman S, Vijay R.** 2016. Middle East respiratory syndrome vaccines. *Int J Infect Dis* **Apr 7. pii:** S1201-9712(16)31021-9.. [Epub ahead of print]
53. **Pascal KE, Coleman CM, Mujica AO, Kamat V, Badithe A, Fairhurst J, Hunt C, Strein J, Berrebi A, Sisk JM, Matthews KL, Babb R, Chen G, Lai KM, Huang TT, Olson W, Yancopoulos GD, Stahl N, Frieman MB, Kyratsous CA.** 2015. Pre- and postexposure efficacy of fully human antibodies against Spike protein in a novel humanized mouse model of MERS-CoV infection. *Proc Natl Acad Sci USA* **112:**8738–8743.
54. **Kim SJ, Kim K, Park SB, Hong DJ, Jhun BW.** 2015. Outcomes of early administration of cidofovir in non-immunocompromised patients with severe adenovirus pneumonia. *PLoS One* **10:**e0122642.
55. **Florescu DF, Keck MA.** 2014. Development of CMX001 (Brincidofovir) for the treatment of serious diseases or conditions caused by dsDNA viruses. *Expert Rev Anti Infect Ther* **12:**1171–1178.
56. **O'Reilly RJ, Prockop S, Hasan AN, Koehne G, Doubrovina E.** 2016. Virus-specific T-cell banks for 'off the shelf' adoptive therapy of refractory infections. *Bone Marrow Transplant* **2016:**1–10.
57. **O'Donnell FL, Taubman SB.** 2015. Follow-up analysis of the incidence of acute respiratory infections among enlisted service members during their first year of military service before and after the 2011 resumption of adenovirus vaccination of basic trainees. *MSMR* **22:**2–7.
58. **Spencer S, Gaglani M, Naleway A, Reynolds S, Ball S, Bozeman S, Henkle E, Meece J, Vandermause M, Clipper L, Thompson M.** 2013. Consistency of influenza A virus detection test results across respiratory specimen collection methods using real-time reverse transcription-PCR. *J Clin Microbiol* **51:**3880–3882.
59. **Lee N, Chan PK, Hui DS, Rainer TH, Wong E, Choi KW, Lui GC, Wong BC, Wong RY, Lam WY, Chu IM, Lai RW, Cockram CS, Sung JJ.** 2009. Viral loads and duration of viral shedding in adult patients hospitalized with influenza. *J Infect Dis* **200:**492–500.
60. **Rabalais GP, Stout GG, Ladd KL, Cost KM.** 1992. Rapid diagnosis of respiratory viral infections by using a shell vial assay and monoclonal antibody pool. *J Clin Microbiol* **30:**1505–1508.
61. **LaSala PR, Bufton KK, Ismail N, Smith MB.** 2007. Prospective comparison of R-mix shell vial system with direct antigen tests and conventional cell culture for respiratory virus detection. *J Clin Virol* **38:**210–216.
62. **Ginocchio CC, Zhang F, Manji R, Arora S, Bornfreund M, Falk L, Lotlikar M, Kowerska M, Becker G, Korologos D, de Geronimo M, Crawford JM.** 2009. Evaluation of multiple test methods for the detection of the novel 2009 influenza A (H1N1) during the New York City outbreak. *J Clin Virol* **45:**191–195.
63. **Chan KH, To KK, Chan JF, Li CP, Chan KM, Chen H, Ho PL, Yuen KY.** 2014. Assessment of antigen and molecular tests with serial specimens from a patient with influenza A (H7N9) infection. *J Clin Microbiol* **52:**2272–2274.
64. **Stevenson HL, Loeffelholz MJ.** 2010. Poor positive accuracy of QuickVue rapid antigen tests during the influenza A (H1N1) 2009 pandemic. *J Clin Microbiol* **48:**3729–3731.
65. **Leonardi GP, Wilson AM, Mitrache I, Zuretti AR.** 2015. Comparison of the Sofia and Veritor direct antigen detection assay systems for identification of influenza viruses from patient nasopharyngeal specimens. *J Clin Microbiol* **53:**1345–1347.
66. **Popowitch EB, O'Neill SS, Miller MB.** 2013. Comparison of the Biofire FilmArray RP, Genmark eSensor RVP, Luminex xTAG RVPv1, and Luminex xTAG RVP fast multiplex assays for detection of respiratory viruses. *J Clin Microbiol* **51:**1528–1533.
67. **Loeffelholz MJ, Pong DL, Pyles RB, Xiong Y, Miller AL, Bufton KK, Chonmaitree T.** 2011. Comparison of the FilmArray Respiratory Panel and Prodesse real-time PCR assays for detection of respiratory pathogens. *J Clin Microbiol* **49:**4083–4088.
68. **Nguyen Van JC, Caméléna F, Dahoun M, Pilmis B, Mizrahi A, Lourtet J, Behillil S, Enouf V, Le Monnier A.** 2016. Prospective evaluation of the Alere i Influenza A&B nucleic acid amplification versus Xpert Flu/RSV. *Diagn Microbiol Infect Dis* **85:**19–22.
69. **Binnicker MJ, Espy MJ, Irish CL, Vetter EA.** 2015. Direct detection of influenza A and B viruses in less than 20 minutes using a commercially available rapid PCR assay. *J Clin Microbiol* **53:**2353–2354.
70. **Jun KR, Woo YD, Sung H, Kim MN.** 2008. Detection of human metapneumovirus by direct antigen test and shell vial cultures using immunofluorescent antibody staining. *J Virol Methods* **152:**109–111.
71. **Aslanzadeh J, Zheng X, Li H, Tetreault J, Ratkiewicz I, Meng S, Hamilton P, Tang YW.** 2008. Prospective evaluation of rapid antigen tests for diagnosis of respiratory syncytial virus and human metapneumovirus infections. *J Clin Microbiol* **46:**1682–1685.
72. **Leonardi GP, Wilson AM, Dauz M, Zuretti AR.** 2015. Evaluation of respiratory syncytial virus (RSV) direct antigen detection assays for use in point-of-care testing. *J Virol Methods* **213:**131–134.
73. **Popowitch EB, Miller MB.** 2015. Performance characteristics of Xpert Flu/RSV XC assay. *J Clin Microbiol* **53:**2720–2721.
74. **Selvaraju SB, Bambach AV, Leber AL, Patru MM, Patel A, Menegus MA.** 2014. Comparison of the Simplexa™ Flu A/B & RSV kit (nucleic acid extraction-dependent assay) and the Prodessa ProFlu+™ assay for detecting influenza and respiratory syncytial viruses. *Diagn Microbiol Infect Dis* **80:**50–52.
75. **Weinberg GA, Erdman DD, Edwards KM, Hall CB, Walker FJ, Griffin MR, Schwartz B, New Vaccine Surveillance Network Study Group.** 2004. Superiority of reverse-transcription polymerase chain reaction to conventional viral culture in the diagnosis of acute respiratory tract infections in children. *J Infect Dis* **189:**706–710.
76. **Kuypers J, Wright N, Ferrenberg J, Huang ML, Cent A, Corey L, Morrow R.** 2006. Comparison of real-time PCR assays with fluorescent-antibody assays for diagnosis of respiratory virus infections in children. *J Clin Microbiol* **44:**2382–2388.
77. **Waris M, Österback R, Lahti E, Vuorinen T, Ruuskanen O, Peltola V.** 2013. Comparison of sampling methods for the detection of human rhinovirus RNA. *J Clin Virol* **58:**200–204.
78. **Lee WM, Kiesner C, Pappas T, Lee I, Grindle K, Jartti T, Jakiela B, Lemanske RF Jr, Shult PA, Gern JE.** 2007. A diverse group of previously unrecognized human rhinoviruses are common causes of respiratory illnesses in infants. *PLoS One* **2:**e966.
79. **Osterback R, Tevaluoto T, Ylinen T, Peltola V, Susi P, Hyypiä T, Waris M.** 2013. Simultaneous detection and differentiation of human rhino- and enteroviruses in clinical specimens by real-time PCR with locked nucleic Acid probes. *J Clin Microbiol* **51:**3960–3967.
80. **Jansen RR, Wieringa J, Koekkoek SM, Visser CE, Pajkrt D, Molenkamp R, de Jong MD, Schinkel J.** 2011. Frequent detection of respiratory viruses without symptoms: toward defining clinically relevant cutoff values. *J Clin Microbiol* **49:**2631–2636.
81. **Heinonen S, Jartti T, Garcia C, Oliva S, Smitherman C, Anguiano E, de Steenhuijsen Piters WA, Vuorinen T, Ruuskanen O, Dimo B, Suarez NM, Pascual V, Ramilo O, Mejias A.** 2016. Rhinovirus Detection in Symptomatic and

Asymptomatic Children: Value of Host Transcriptome Analysis. *Am J Respir Crit Care Med* **193:**772–782.

82. **Flomenberg P, Gutierrez E, Piaskowski V, Casper JT.** 1997. Detection of adenovirus DNA in peripheral blood mononuclear cells by polymerase chain reaction assay. *J Med Virol* **51:**182–188.
83. **Levent F, Greer JM, Snider M, Demmler-Harrison GJ.** 2009. Performance of a new immunochromatographic assay for detection of adenoviruses in children. *J Clin Virol* **44:**173–175.
84. **Kam KY, Ong HS, Bunce C, Ogunbowale L, Verma S.** 2015. Sensitivity and specificity of the AdenoPlus point-of-care system in detecting adenovirus in conjunctivitis patients at an ophthalmic emergency department: a diagnostic accuracy study. *Br J Ophthalmol* **99:**1186–1189.
85. **Manji R, Zheng X, Patel A, Kowerska M, Vossinas M, Drain A, Todd KM, Lenny N, DeVincenzo JP, Ginocchio CC.** 2014. Multi-center evaluation of the adenovirus R-gene US assay for the detection of adenovirus in respiratory samples. *J Clin Virol* **60:**90–95.
86. **Song E, Wang H, Salamon D, Jaggi P, Leber A.** 2016. Performance characteristics of FilmArray respiratory panel v1.7 for detection of adenovirus in a large cohort of pediatric nasopharyngeal samples: one test may not fit all. *J Clin Microbiol* **54:**1479–1486.
87. **Aberle SW, Aberle JH, Steininger C, Matthes-Martin S, Pracher E, Popow-Kraupp T.** 2003. Adenovirus DNA in serum of children hospitalized due to an acute respiratory adenovirus infection. *J Infect Dis* **187:**311–314.
88. **Rogers BB, Shankar P, Jerris RC, Kotzbauer D, Anderson EJ, Watson JR, O'Brien LA, Uwindatwa F, McNamara K, Bost JE.** 2015. Impact of a rapid respiratory panel test on patient outcomes. *Arch Pathol Lab Med* **139:**636–641.
89. **Byington CL, Castillo H, Gerber K, Daly JA, Brimley LA, Adams S, Christenson JC, Pavia AT.** 2002. The effect of rapid respiratory viral diagnostic testing on antibiotic use in a children's hospital. *Arch Pediatr Adolesc Med* **156:**1230–1234.
90. **Nelson RE, Stockmann C, Hersh AL, Pavia AT, Korgenksi K, Daly JA, Couturier MR, Ampofo K, Thorell EA, Doby EH, Robison JA, Blaschke AJ.** 2015. Economic analysis of rapid and sensitive polymerase chain reaction testing in the emergency department for influenza infections in children. *Pediatr Infect Dis J* **34:**577–582.
91. **Reich NG, Cummings DA, Lauer SA, Zorn M, Robinson C, Nyquist AC, Price CS, Simberkoff M, Radonovich LJ, Perl TM.** 2015. Triggering interventions for influenza: the ALERT algorithm. *Clin Infect Dis* **60:**499–504.
92. **Campbell AP, Guthrie KA, Englund JA, Farney RM, Minerich EL, Kuypers J, Corey L, Boeckh M.** 2015. Clinical outcomes associated with respiratory virus detection before allogeneic hematopoietic stem cell transplant. *Clin Infect Dis* **61:**192–202.
93. **Beaird OE, Freifeld A, Ison MG, Lawrence SJ, Theodoropoulos N, Clark NM, Razonable RR, Alangaden G, Miller R, Smith J, Young JA, Hawkinson D, Pursell K, Kaul DR.** 2016. Current practices for treatment of respiratory syncytial virus and other non-influenza respiratory viruses in high-risk patient populations: a survey of institutions in the Midwestern Respiratory Virus Collaborative. *Transpl Infect Dis* **18:**210–215.
94. **Bradley JS, Byington CL, Shah SS, Alverson B, Carter ER, Harrison C, Kaplan SL, Mace SE, McCracken GH Jr, Moore MR, St Peter SD, Stockwell JA, Swanson JT, Pediatric Infectious Diseases Society and the Infectious Diseases Society of America.** 2011. Executive summary: the management of community-acquired pneumonia in infants and children older than 3 months of age: clinical practice guidelines by the Pediatric Infectious Diseases Society and the Infectious Diseases Society of America. *Clin Infect Dis* **53:**617–630.
95. **Hirsch HH, Martino R, Ward KN, Boeckh M, Einsele H, Ljungman P.** 2013. Fourth European Conference on Infections in Leukaemia (ECIL-4): guidelines for diagnosis and treatment of human respiratory syncytial virus, parainfluenza virus, metapneumovirus, rhinovirus, and coronavirus. *Clin Infect Dis* **56:**258–266.
96. **Brittain-Long R, Westin J, Olofsson S, Lindh M, Andersson LM.** 2011. Access to a polymerase chain reaction assay method targeting 13 respiratory viruses can reduce antibiotics: a randomised, controlled trial. *BMC Med* **9:**44.
97. **van de Pol AC, Wolfs TF, Tacke CE, Uiterwaal CS, Forster J, van Loon AM, Kimpen JL, Rossen JW, Jansen NJ.** 2011. Impact of PCR for respiratory viruses on antibiotic use: theory and practice. *Pediatr Pulmonol* **46:**428–434.
98. **Esposito S, Marchisio P, Morelli P, Crovari P, Principi N.** 2003. Effect of a rapid influenza diagnosis. *Arch Dis Child* **88:**525–526.
99. **Doan Q, Enarson P, Kissoon N, Klassen TP, Johnson DW.** 2012. Rapid viral diagnosis for acute febrile respiratory illness in children in the Emergency Department. *Cochrane Database Syst Rev* **5:**CD006452.
100. **Bonner AB, Monroe KW, Talley LI, Klasner AE, Kimberlin DW.** 2003. Impact of the rapid diagnosis of influenza on physician decision-making and patient management in the pediatric emergency department: results of a randomized, prospective, controlled trial. *Pediatrics* **112:**363–367.
101. **Doan QH, Kissoon N, Dobson S, Whitehouse S, Cochrane D, Schmidt B, Thomas E.** 2009. A randomized, controlled trial of the impact of early and rapid diagnosis of viral infections in children brought to an emergency department with febrile respiratory tract illnesses. *J Pediatr* **154:**91–95.
102. **Poehling KA, Zhu Y, Tang YW, Edwards K.** 2006. Accuracy and impact of a point-of-care rapid influenza test in young children with respiratory illnesses. *Arch Pediatr Adolesc Med* **160:**713–718.
103. **Iyer SB, Gerber MA, Pomerantz WJ, Mortensen JE, Ruddy RM.** 2006. Effect of point-of-care influenza testing on management of febrile children. *Acad Emerg Med* **13:**1259–1268
104. **Wishaupt JO, Russcher A, Smeets LC, Versteegh FG, Hartwig NG.** 2011. Clinical impact of RT-PCR for pediatric acute respiratory infections: a controlled clinical trial. *Pediatrics* **128:**e1113–e1120.
105. **Cohen R, Thollot F, Lécuyer A, Koskas M, Touitou R, Boucherat M, d'Athis P, Corrard F, Pecking M, de La Rocque F.** 2007. Impact of the rapid diagnosis downtown in the assumption of responsibility of the children in period of influenza. (In French). *Arch Pediatr* **14:**926–931.
106. **Oosterheert JJ, van Loon AM, Schuurman R, Hoepelman AI, Hak E, Thijsen S, Nossent G, Schneider MM, Hustinx WM, Bonten MJ.** 2005. Impact of rapid detection of viral and atypical bacterial pathogens by real-time polymerase chain reaction for patients with lower respiratory tract infection. *Clin Infect Dis* **41:**1438–1444.
107. **Hakki M, Strasfeld LM, Townes JM.** 2014. Predictive value of testing nasopharyngeal samples for respiratory viruses in the setting of lower respiratory tract disease. *J Clin Microbiol* **52:**4020–4022.
108. **Azadeh N, Sakata KK, Brighton AM, Vikram HR, Grys TE.** 2015. FilmArray Respiratory Panel Assay: comparison of nasopharyngeal swabs and bronchoalveolar lavage samples. *J Clin Microbiol* **53:**3784–3787.
109. **Chan KH, Peiris JS, Lim W, Nicholls JM, Chiu SS.** 2008. Comparison of nasopharyngeal flocked swabs and aspirates for rapid diagnosis of respiratory viruses in children. *J Clin Virol* **42:**65–69.
110. **Öhrmalm L, Wong M, Rotzén-Östlund M, Norbeck O, Broliden K, Tolfvenstam T.** 2010. Flocked nasal swab versus nasopharyngeal aspirate for detection of respiratory tract viruses in immunocompromised adults: a matched comparative study. *BMC Infect Dis* **10:**340–343.
111. **Hernes SS, Quarsten H, Hamre R, Hagen E, Bjorvatn B, Bakke PS.** 2013. A comparison of nasopharyngeal and oropharyngeal swabbing for the detection of influenza virus by real-time PCR. *Eur J Clin Microbiol Infect Dis* **32:**381–385.
112. **Dawood FS, Jara J, Estripeaut D, Vergara O, Luciani K, Corro M, de León T, Saldaña R, Castillo Baires JM, Rauda-Flores R, Cazares RA, Brizuela de Fuentes YS, Franco D, Gaitan M, Schneider E, Berman L, Azziz-Baumgartner E, Widdowson MA.** 2015. What is the added benefit of

oropharyngeal swabs compared to nasal swabs alone for respiratory virus detection in hospitalized children aged <10 years? *J Infect Dis* **212**:1600–1603.

113. **Branche AR, Walsh EE, Formica MA, Falsey AR.** 2014. Detection of respiratory viruses in sputum from adults by use of automated multiplex PCR. *J Clin Microbiol* **52**:3590–3596.

114. **Campbell AP, Chien JW, Kuypers J, Englund JA, Wald A, Guthrie KA, Corey L, Boeckh M.** 2010. Respiratory virus pneumonia after hematopoietic cell transplantation (HCT): associations between viral load in bronchoalveolar lavage samples, viral RNA detection in serum samples, and clinical outcomes of HCT. *J Infect Dis* **201**:1404–1413.

115. **Fuji N, Suzuki A, Lupisan S, Sombrero L, Galang H, Kamigaki T, Tamaki R, Saito M, Aniceto R, Olveda R, Oshitani H.** 2011. Detection of human rhinovirus C viral genome in blood among children with severe respiratory infections in the Philippines. *PLoS One* **6**:e27247.

116. **Tse H, To KK, Wen X, Chen H, Chan KH, Tsoi HW, Li IW, Yuen KY.** 2011. Clinical and virological factors associated with viremia in pandemic influenza A/H1N1/2009 virus infection. *PLoS One* **6**:e22534.

117. **Self WH, Williams DJ, Zhu Y, Ampofo K, Pavia AT, Chappell JD, Hymas WC, Stockmann C, Bramley AM, Schneider E, Erdman D, Finelli L, Jain S, Edwards KM, Grijalva CG.** 2016. Respiratory viral detection in children and adults: comparing asymptomatic controls and patients with community-acquired pneumonia. *J Infect Dis* **213**:584–591.

118. **Ralston SL, Lieberthal AS, Meissner HC, Alverson BK, Baley JE, Gadomski AM, Johnson DW, Light MJ, Maraqa NF, Mendonca EA, Phelan KJ, Zorc JJ, Stanko-Lopp D, Brown MA, Nathanson I, Rosenblum E, Sayles III S, Hernandez-Cancio S.** 2014. Clinical practice guideline: the diagnosis, management, and prevention of bronchiolitis. *Pediatrics* **134**:e1474–502.

119. **Mills JM, Harper J, Broomfield D, Templeton KE.** 2011. Rapid testing for respiratory syncytial virus in a paediatric emergency department: benefits for infection control and bed management. *J Hosp Infect* **77**:248–251.

120. **Song E, Wang H, Kajon AE, Salamon D, Dong S, Ramilo O, Leber A, Jaggi P.** 2016. Diagnosis of Pediatric Acute Adenovirus Infections: Is a Positive PCR Sufficient? *Pediatr Infect Dis J* **Mar 11**; Epub ahead of print.

Enteroviruses and Parechoviruses

M. STEVEN OBERSTE AND MARK A. PALLANSCH

20

Human enteroviruses (EV) are members of the *Enterovirus* genus of the family *Picornaviridae* and are among the most common human viral infections. Their discovery had implications for all of virology because they indicated, first, that poliovirus (PV) grew in various tissue culture cells that did not correspond to the tissues infected during the human disease and, second, that PV destroyed cells with a specific cytopathic effect (CPE). The infectious virus is relatively resistant to many common laboratory disinfectants, including 70% ethanol, isopropanol, dilute Lysol, and quaternary ammonium compounds. Poliomyelitis should be considered in all cases of pure motor paralysis and is usually associated with a normal or slightly elevated value for protein, normal sugar value, and moderate mononuclear pleocytosis in cerebrospinal fluid (CSF). Real-time PCR is the primary test used to detect, even with very small amounts of clinical specimens, such as CSF. All important information about a virus could potentially be obtained directly by PCR in conjunction with nucleic acid sequencing if all the molecular correlates of viral phenotypic determinants were understood. The most common molecular typing system is based on reverse transcription (RT)-PCR and nucleotide sequencing of a portion of the genomic region encoding VP1. Another group of pathogenic picornaviruses, the human parechoviruses (genus *Parechovirus*), cause a similar array of illnesses as the EV, and they are detected clinically by distinct molecular assays that are analogous to those used for the EV.

VIRUS CLASSIFICATION AND BIOLOGY

Human enteroviruses are members of the *Enterovirus* genus of the family *Picornaviridae* and are among the most common human viral infections. Although most EV infections are asymptomatic, millions of symptomatic EV infections are estimated to occur each year in the United States (1–3). Most EV infections cause only mild nonspecific disease, but infections can infrequently lead to serious illness and hospitalization, especially in infants and in those who are immune-compromised. In addition, this group of viruses is the most common cause of aseptic meningitis, the most frequent central nervous system (CNS) infection (4). Enteroviruses have also been implicated in other acute and chronic diseases. The human parechoviruses (family *Picornaviridae*, genus *Parechovirus*) also have a somewhat similar clinical spectrum and physical properties, but they are genetically distinct and present particular diagnostic challenges.

History of Virus Discovery

The history of EV begins with the history of poliovirus (PV). Many of the early PV studies are landmarks in the study of EV and all of virology. The first clinical descriptions of poliomyelitis were made in the 1800s, with reports of cases of paralysis with fever. Understanding of the infectious nature of this disease began with studies in the early 20th century. These studies recognized the communicable nature of poliomyelitis, the importance of asymptomatic infection in the transmission of PV, and the role of enteric infection in disease pathogenesis. In a classic study, Viennese investigators Landsteiner and Popper proved the infectious nature of poliomyelitis by inoculation of CNS tissue homogenates from human cases into monkeys and successfully transmitting the clinical disease and its pathology to a nonhuman host (5).

Building on studies of others, Enders et al. showed that PV could be propagated in nonneural tissue culture (6). These investigations had implications for all of virology because they indicated, first, that PV grew in various tissue culture cells that did not correspond to the tissues infected during the human disease and, second, that PV destroyed cells with a specific cytopathic effect (CPE).

The control of polio began with the production of two different vaccines, the Salk inactivated polio vaccine (IPV) delivered via intramuscular injection (licensed in 1955 in the United States [U.S.]) and the orally-delivered Sabin live, attenuated vaccine (oral polio vaccine [OPV], licensed in 1961–1962).

Coxsackieviruses (CV) were first isolated from the feces of paralyzed children following a poliomyelitis outbreak in 1947 in Coxsackie, New York (7), and from cases of aseptic meningitis in 1948 (8). These isolates were detected by observing paralysis following intracerebral inoculation of suckling mice. The original CV group A (CVA) isolates produced myositis with flaccid hind-limb paralysis in newborn mice, whereas the CVB produced a spastic paralysis and generalized infection in newborn mice, with myositis as well as involvement of the brain, pancreas, heart, and brown fat (9, 10).

doi:10.1128/9781555819156.ch20

In 1951, echoviruses (E) were first isolated in tissue culture from the stools of asymptomatic individuals (11). Echoviruses received their name because they were enteric isolates, cytopathogenic in tissue culture, isolated from humans, and orphans (i.e., not associated with a known clinical disease). Subsequent studies have shown that echoviruses, in fact, do cause a variety of human diseases (12–18).

Virus Classification

Historically, the classification of EV into the subgroups of polioviruses, CVA, CVB and echoviruses was based on the empirical observations of their association with certain clinical syndromes or disease, tissue tropism, the nature of disease in suckling mice, growth in certain specific cell cultures and, in some cases, antigenic similarities (19–22). Using these criteria, 67 different serotypes were recognized and classified, despite the fact that several antigenically-related viruses had different pathogenic properties in mice (22). This last discrepancy eventually led to the dropping of the old designations; thereafter, new serotypes were simply termed "enterovirus" followed by a number, beginning with EV68 (23, 24). A more satisfactory, and completely new, classification system followed from the advent of comprehensive genetic studies. From these new data, and to avoid the inconsistencies of the previous classification scheme, the human enteroviruses have been reclassified into four species: *Enterovirus* (EV) A through *D* (25). The current classification for picornaviruses affecting humans is given in Table 1.

Continued molecular characterization of enteroviruses has also identified many new members of the genus, with the naming continuing sequentially from enterovirus 73, extending up to EV121, including some viruses from nonhuman primates that are related to the human EV (26). These have been assigned to one of the four species and are included in Table 1 for reference. Based upon similar genetic comparisons, the human rhinoviruses have been classified as three species in the *Enterovirus* genus (Table 1).

Molecular studies also demonstrated that echoviruses 22 and 23 are genetically distinct from the EV (27–29). On the basis of a very low genetic relationship, differences in viral proteins and processing, and a novel protease, E22 and E23 were reclassified as members of the new picornavirus genus, *Parechovirus*, and renamed human parechoviruses 1 and 2, respectively (25, 30). Additional members of this genus have been recently identified, including both additional serotypes of human parechovirus (31–34), as well as a separate species first isolated in Swedish bank voles, Ljungan virus (35). Other recently described picornavirus genera associated with human infection are *Cosavirus*, *Kobuvirus*, and *Salivirus* (36, 37). Although little information is currently available about these viruses, it appears that kobuviruses and saliviruses are associated with gastroenteritis in young children and infection is common. It is unknown whether cosaviruses cause any disease in humans. What is notable about the newer genera is that the molecular reagents for the detection of EV do not detect these viruses (see Laboratory Diagnosis).

Biology

Structure

Enteroviruses are small RNA viruses consisting of a small (~30 nm) spherical, nonenveloped capsid that contains a single-stranded positive-sense RNA genome. The genomes

TABLE 1 Picornavirus genera, species,[a] and serotypes affecting humans

Genus and species	No. of types	Types
Genus *Enterovirus*		
Enterovirus A	21	CVA2-8, 10, 12, 14, 16; EV-A71, A76, A89-92, A114, A119, A120, A121
Enterovirus B	59	CVA9; CVB1-6; E1-7, 9, 11–21, 24–27, 29–33; EV-B69, B73-75, B77-88, B93, B97-98, B100-101, B106-107, B111
Enterovirus C	23	CVA1, 11, 13, 17, 19–22, 24; PV1-3; EV-C95-96, C99, C102, C104-105, C109, C116-118
Enterovirus D	4	EV-D68, D70, D94, D111
Rhinovirus A	80	HRV-A1-2, A7-13, A15-16, A18-25, A28-34, A36, A38-41, A43, A45-47, A49-51, A53-68, A71, A73-78, A80-82, A85, A88-90, A94, A96, A98, A100-109
Rhinovirus B	32	HRV-B3-6, B14, B17, B26-27, B35, B37, B42, B48, B52, B69-70, B72, B79, B83-84, B86, B91-93, B97, B99-106
Rhinovirus C	55	HRV-C1-55
Genus *Cardiovirus*		
Encephalomyocarditis virus	1	Encephalomyocarditis virus
Theilovirus	11	Saffold viruses 1–11
Genus *Cosavirus*	24	Human cosavirus A1-24
Genus *Hepatovirus*	1	Human hepatitis A virus
Genus *Kobuvirus*		
Aichivirus A	1	Aichivirus 1
Genus *Parechovirus*		
Parechovirus A	16	HPeV1-16
Genus *Salivirus*	2	Salivirus viruses A1-2

[a]The classification scheme shown is adapted from the Picornavirus Study Group of the International Committee for the Taxonomy of Viruses (http://www.picornastudygroup.com/) (25) and distributes serotypes among different species, based on genome organization, sequence similarity and other physical properties. Many of the listed species and genera also contain members not listed that affect mammalian hosts other than humans, and that to date have no documented role in human disease.

of the EV and HPeV are approximately 7,500 nucleotides long and encode a single open reading frame for all viral capsid and functional proteins. The genomes of almost all EV types have been sequenced completely (for updated information, see "The Picornavirus Pages," http://www.picornaviridae.com/). A small virus protein (VPg, for "virus protein—genome-linked") is covalently bound to the 5′ end of the RNA molecule, and the 3′ end of the virion RNA is polyadenylated. The 5′ and 3′ termini of the viral RNA include nontranslated regions (NTRs) of different lengths (5′ NTR, 711 to 755 nucleotides; 3′ NTR, 69 to 109 nucleotides), each showing a high degree of secondary structure.

The icosahedral virus capsid, consisting of 60 protomers, is assembled from 12 identical pentamers. Each protomer is composed of the four virus proteins, VP1, VP2, VP3, and VP4. VP1, VP2, and VP3 constitute most of the capsid surface, whereas VP4 is located in the interior of the capsid in close contact with the viral RNA (38). In the case of parechoviruses, the protomer consists of only three proteins, VP0, VP1, and VP3, since VP0 does not undergo the maturation cleavage to VP4 and VP2 that occurs in the EV (30, 39). Additional differences in genome organization and protein function differentiate the parechoviruses from the enteroviruses (3).

Antigenicity and Neutralization

The picornaviruses are among the simplest RNA viruses, having a highly-structured capsid with limited surface elaboration. Yet, despite the limited genetic material and structural constraints, evolution within the picornaviruses has resulted in a large number of readily distinguishable members. This variability has been categorized antigenically as serotype. Each of the serotypes correlates with the immunologic response of the human host, protection from disease, receptor usage and, to a lesser extent, the spectrum of clinical disease. Molecular studies have provided a framework in which the EV antigenic relationships can be better understood. These studies suggest that the nucleotide sequence of VP1 is an excellent surrogate for antigenic typing by means of neutralization tests in order to distinguish EV serotypes (40–44). When virions are disrupted by heating, particularly in the presence of detergent, nonsurface antigens are exposed that are shared broadly among many EV (45). Despite these limitations, the serotype remains the single most important physical and immunologic property that distinguishes the different EV.

Although there are measurable *in vitro* antigenic differences among strains within a serotype, the significance of these differences during natural infection has not been determined. Several PV isolated during outbreaks have demonstrated different antigenic properties when compared with the reference vaccine strains (46) but, in all cases, immunity derived from vaccination has been sufficient to provide protection and to control the circulation of these strains. Even in the face of massive PV immunization campaigns, no antigenic escape mutants resistant to neutralization have ever been observed, and successive genotypes of PV have been eliminated.

Replication in Cell Culture

One of the prominent characteristics of EV is the cytolytic nature of growth in cell culture. For many years, PV was the prototype of a lytic viral infection. At the microscopic level, infection is usually manifested within 1 to 7 days by the appearance of a characteristic CPE, which features visible cell rounding and shrinking, nuclear pyknosis, refractility, and cell degeneration. The earliest effects can be seen in less than 24 hours if the inoculum contains many infectious particles. With fewer virions, however, visible changes are not recognizable for several days, although a sufficient number of cells are infected. In addition, some EV either do not cause CPE at all, or do so only after several passages. In general, once focal CPE is detected, infection spreads rapidly throughout the cell sheet with total destruction of the monolayer, sometimes in a matter of hours.

All known EV (to the extent the newer members have been tested) can be propagated in either cell culture or in suckling mice (1, 47). Most of the serotypes can be grown in at least one human or primate continuous cell culture (48–56). No cell line, however, can support the growth of all cultivable EV. Even after many years of experimentation, a few serotypes (e.g., CVA19) can be propagated only in suckling mice. The typical host range of human EV in cell cultures or animals is not clearly associated with a given virus species. At least half of the known HPeV grow poorly, if at all, in culture, so their characterization has mainly relied on molecular methods (57).

Infection of target cells depends on viruses binding to specific receptors on the cell surface. Collectively, the EV use at least seven different receptors, including two different integrins, decay-accelerating factor (DAF; CD55), the "coxsackievirus-adenovirus receptor" (CAR), intracellular adhesion molecule 1 (ICAM-1), sialic acid, and the poliovirus receptor (PVR; CD155) (58, 59). Some EV are able to use more than one receptor, and other, unidentified, receptors may also exist. HPeV-1 uses α_v integrins for cell entry, but specific receptors are not known for the other HPeVs.

EPIDEMIOLOGY

Mode of Transmission

EV can be isolated from both the lower and upper alimentary tract and can be transmitted by both fecal-oral and respiratory routes. Fecal-oral transmission may predominate in areas with poor sanitary conditions, whereas respiratory transmission may be more important in more developed areas (60). The relative importance of the different modes of transmission probably varies with the particular EV type and the environmental setting. It is believed that almost all EV can be transmitted by the fecal-oral route; however, it is not known whether most can also be transmitted by the respiratory route. EV70 and CVA24v, the agents that cause acute hemorrhagic conjunctivitis, are seldom isolated from the respiratory tract or stool specimens and are probably primarily spread by direct or indirect contact with eye secretions (61). EV that cause a vesicular exanthema presumably can be spread by direct or indirect contact with vesicular fluid, which contains infectious virus. It is likely that EV are transmitted in the same manner as are other viruses causing the common cold—that is, by hand contact with secretions (e.g., on the hand of another person) and autoinoculation to the mouth, nose, or eyes. Direct bloodstream inoculation, usually by laboratory accidents (e.g., needle sticks), can result in EV infection. EV are efficiently amplified and transmitted among humans without intermediaries, such as arthropods or other animals. Although several species of nonhuman primates may become experimentally or even naturally infected with some human EV, there is no evidence that they play any significant role in virus circulation or constitute an effective animal reservoir. EV have been detected in pooled donor blood (62, 63), but neither blood transfusion nor mosquito or other insect bite appears to be a significant route of transmission.

Transmission within households has been well studied for both PV and nonpolio EV. Small children generally introduce EV into the family, although young adults make up the majority of index cases in some outbreaks of AHC (64). Intrafamily transmission can be rapid and relatively complete, depending on duration of virus excretion, household size, number of siblings, socioeconomic status, immune status of household members, and other risk factors (65). Transmission generally has been found to be greatest in large families of lower socioeconomic status, with a greater number of children 5 to 9 years of age and with no evidence of immunity to the virus type. Not surprisingly, infections in different family members can result in different clinical manifestations.

Household secondary attack rates in susceptible members may be greatest for the agents of AHC and for the PV and of lesser magnitude for the CV and echoviruses. In some studies, secondary attack rates may be greater than 90%, although they are typically lower.

Transmission occurs within the neighborhood and community, particularly where people congregate. EV can be rapidly transmitted within institutions when circumstances permit (e.g., crowding, poor hygiene, or contaminated water), similar to many other viruses. School teams or activity groups and institutionalized ambulatory retarded children or adults may be at special risk (66). Despite crowding, EV transmission is not usually accelerated to a noticeable degree in institutions where good sanitation is found.

Nosocomial transmission of various EV and parechoviruses also has been well documented, typically in newborn nurseries. Hospital staff may have been involved in mediating transmission in some of these outbreaks. EV70 and CVA24v are highly transmissible and can cause outbreaks in ophthalmology clinics when instruments are inadequately cleaned between patients. An apparent outbreak of CVA1, which included some fatal cases, has been reported in bone marrow transplant recipients (67).

Although little evidence suggests that EV found in the environment are of public health importance, concern has been expressed about the possible dangers of contaminated water sources. Recreational swimming water has been investigated in several studies, and EV have been isolated from swimming and wading pools in the absence of fecal coliforms and in the presence of recommended levels of free residual chlorine (68). Reports have suggested that swallowing of contaminated pool or lake water theoretically may account for transmission, but no proof exists that this type of transmission is significant in recreational settings. In industrialized countries, EV transmission from potable water is apparently uncommon but is a constant source of concern for public health investigators because the usual conditions under which municipal drinking water is chlorinated may be insufficient to completely inactivate EV.

Enteroviruses have been isolated from raw or partly cooked mollusks and crustaceans and their overlying waters (69). Shellfish rapidly concentrate many viruses, including EV. These viruses can survive in oysters for 3 weeks at temperatures of 1 to 21°C but, to date, no outbreak of EV disease has been attributed to the consumption of contaminated shellfish. Other foodborne transmission has been documented but is thought to be uncommon.

Enteroviruses are more prevalent in sewage from areas with low socioeconomic conditions or with large proportions of young children. In addition, sewage workers have been shown to have a higher prevalence of serum antibodies to EV, consistent with an occupational risk (70). Soil and crops also provide conditions favorable to EV. Enteroviruses survive well in sludge and remain on the surface of sludge-treated soil and even on crops. Air samples from aerosolized spray irrigants using contaminated effluents also have been found to contain EV (71). Survival of EV on vegetable food crops exposed to contaminated water or fertilizer has not been proved to be associated with virus transmission.

In tropical regions, especially where sanitation is poor, the efficiency of transmission is high. Consequently, not only is the overall prevalence of EV infections higher but also the average age of infection is younger. During infancy and preschool, children get frequent infections with many EV (72, 73). It is not uncommon in these areas to detect two or three simultaneous infections of different EV serotypes, often causing no disease. Since children likely become infected with nearly all EV in the tropics, most adults are immune and nearly all babies are born with maternal antibodies to most EV (74). Normally, neonates not protected by maternal antibodies are at risk for serious illness when infected by EV. Because of the protective maternal antibodies, neonatal EV disease is extremely rare in the tropics, despite the high prevalence of EV infections in older children. In contrast, in temperate regions where the overall prevalence of EV infections and maternal antibodies are much lower, neonatal EV diseases are relatively more common (75).

Shedding may be intermittent and is affected by the immune status of the individual. Past natural infection with the same EV serotype (or immunization, in the case of poliovirus) serves to significantly reduce the extent and duration of virus shedding (76). Immunity will protect against disease but does not form an absolute block that will prevent future infection. Therefore, immune individuals can also contribute to virus transmission, while not being at risk for significant disease.

Geography, Season, Socioeconomic Factors, Sex, Age, and Risk Groups

Enterovirus excretion does not necessarily imply association with disease because most such excretion is asymptomatic. Enterovirus activity in populations can be either sporadic or epidemic, and certain EV types are associated with both sporadic and epidemic disease occurrences (2). The reported incidence or prevalence of a given EV disease may artifactually be increased in an outbreak situation when a sudden focus of attention improves diagnosis and reporting of cases, but this may also increase reporting of noncases. In addition, there may be a tendency for other strains to be excluded when a particular strain is predominant in a community; however, large communities with summer enteroviral disease typically support cocirculation of several different types simultaneously and in no particular pattern (77).

An important concept in understanding the epidemiology of the EV is variation: by serotype, by time, by geographic location, and by disease. This concept is illustrated in surveillance studies of nonpolio EV infections. For example, during 1970 to 2006, several major E30 epidemics occurred in the U.S., from 1981 to 1982, 1990 to 1994, and 1997 to 1998 (78–80). By contrast, endemic viruses (e.g., CVB3) are isolated nearly every year and in similar numbers each year. Even with endemic viruses, larger outbreaks do occasionally occur, as with CVB3 in 1980. Similar endemic and epidemic patterns are seen for the other echoviruses and CVA (2).

Variation by location is also a major characteristic of EV. Outbreaks can be restricted to small groups (e.g., schools and day care centers) or to select communities, or they may become widespread at the regional, national, or even international level. Outbreaks in small groups can sometimes be linked epidemiologically to a breakdown in hygiene practices. Even during national outbreaks of a specific serotype, the location of virus activity may not be uniform. During the period of 1990 to 2005, E30 was the most commonly isolated EV in the U.S., however, not all parts of the country had E30 isolates during the entire period. Some areas, such as the New England states, had extensive circulation in only one year, whereas other areas, such as the entire western U.S., had extensive virus circulation at least three of the four years. It is important to note, therefore, that aggregate national data can obscure significant regional and local variation in viral prevalence.

In temperate climates, EV are characteristically found during the summer and early autumn, although outbreaks can continue into the winter. In fact, naturally occurring EV have a distinct seasonal pattern of circulation that varies by geographic area; in contrast, live attenuated PV vaccine strains are isolated year round in OPV-using countries, reflecting the routine administration of polio vaccine to children. In tropical and semitropical areas, circulation tends to be year-round or associated with the rainy season. In the U.S., 36 years of surveillance indicated that 82% of EV isolations were made during the five summer or fall months of June to October (2).

Molecular Epidemiology

Study of the molecular variation of viral proteins or nucleic acid may contribute significant epidemiologic information on viral diseases. Molecular epidemiologic studies have helped in our understanding of EV by providing the opportunity for unequivocal strain identification, providing insights into EV classification and taxonomy, clarifying the origins of outbreaks and allowing identification of strains transmitted between outbreaks. For the EV, and in particular PV, the primary method used to generate epidemiologic information is direct analysis of genomic variation using nucleic acid sequencing (81, 82). Nucleic acid sequencing technology has been most comprehensively applied to studies of PV where the information has proved valuable for supporting the global PV eradication program. The introduction of the technique of genomic nucleic acid sequencing and its application to the study of wild PV isolates from different parts of the world has significantly extended the epidemiologic power of molecular studies. By analyzing the random mutations that occur in the genome of different PV, closely-related viruses were easily differentiated and, in addition, more distantly-related viruses were clustered into distinct geographic groupings of endemic circulation (83). This approach allowed epidemiologic links to be extended beyond those identified with other techniques. Building on a nucleic acid sequence database of PV strains worldwide, it has been possible to develop rapid approaches to tracking wild PV strains (84).

Studies on the molecular epidemiology of nonpolio EV have focused on the evolutionary inference derived from the comparison of virus isolates within a serotype over time, as well as the comparison of isolates from different serotypes and even between different genera within the *Picornaviridae*. One of the studies of CVB5 isolates examined the pattern of genetic changes over three separate outbreaks in the U.S. (85). The nucleotide sequence from multiple isolates from the epidemics showed that each of the epidemics was caused by a single genotype. The genotype of CVB5 observed in the 1967 epidemic showed more similarity to the virus observed in the 1983 epidemic than to viruses isolated during the intervening years, suggesting discontinuous transmission of epidemic CVB5 in the U.S. during this time. In an analogous manner, E30 genotypes have demonstrated an overlapping succession among the isolates characterized in the U.S. (86).

More than one genotype may be found in certain periods, and the displaced genotype can be found in other parts of the world after isolations have ceased in the U.S. for many years. In studies of EV71 isolates, at least three distinct genotypes, and possibly a fourth, have been characterized (87–89). Unlike the situation with E30 and more similar to the CVB5 example, the transition from one genotype to another occurred during a single year, 1987, and the older genotype has not been isolated in the U.S. since then, despite isolation in other parts of the world.

PATHOGENESIS AND CLINICAL SYNDROMES

Typically, the primary site of infection is the epithelial cells of the respiratory or gastrointestinal tract and the lymphoid follicles of the small intestine (90). Replication at the primary site of infection may be followed by viremia, leading to a secondary site of tissue infection. Secondary infection of the CNS results in aseptic meningitis or, rarely, encephalitis or paralysis. Other tissue-specific infections can result in pleurodynia or myocarditis. Disseminated infection can lead to exanthems, nonspecific myalgias, or severe multiple-organ disease in neonates.

Incubation Times

All polioviruses, group A and B coxsackieviruses, and echoviruses have incubation times ranging from 2 to 35 days, with an average of 7 to 14 days (47). The shortest incubation period, 12 to 72 h, has been reported for local infections of the eye by EV70 (91).

Asymptomatic Infections

The link between an EV infection and a disease syndrome should be made with caution. Inapparent infections and prolonged excretion of virus, especially in stools, are common. A definitive link cannot be made between infection and disease based solely on isolating virus from the stool of an individual patient. A link can be inferred if the virus is isolated from a site that corresponds to the clinical symptoms and if that site is normally sterile. Most associations between EV infection and disease have been made from studies of outbreaks in which a large number of persons with the same clinical signs and symptoms have evidence of infection with the same serotype. Such studies have clearly demonstrated that EV infection can cause aseptic meningitis, pericarditis, pleurodynia, myocarditis, acute hemorrhagic conjunctivitis (AHC), acute respiratory illness, and encephalitis. When an individual patient has a disease syndrome shown clearly to be associated with EV infection and there is no evidence of involvement by another agent, infection implies probable causation.

Clinical Syndromes

It is neither necessary nor practical to enumerate all diseases caused by each of the EV serotypes. A limited number of viruses cause a few clinically-distinct diseases (e.g., poliomyelitis, AHC, and herpangina). They are relatively easily recognized, and etiologic confirmation by laboratory tests, if required, can be directed at a few specific EV. Certain syndromes (e.g., meningitis, encephalitis, and myocarditis) have varied causes, including EV, other viruses, bacteria, non-infectious causes, etc. In such cases etiologic diagnosis is important for selecting appropriate treatment or avoiding inappropriate treatment. With some notable exceptions, most EV generally are capable of causing a variety of clinical diseases, and for any specific disease it is difficult to predict the serotype from signs and symptoms alone (Table 2).

Poliomyelitis

The term poliomyelitis refers to the inflammatory damage due to infection of the anterior horn cells of the spinal cord, recognized clinically as acute-onset, lower-motor-neuron paralysis (or paresis) of one or more muscles. When its viral cause was recognized, the agent was called poliovirus, thereby redefining poliomyelitis as spinal cord disease caused specifically by one or another poliovirus serotype. Polioviruses may cause other diseases, but muscle paralysis due to myelitis is the most important. Until poliovirus infections were controlled by immunization, they were the most common cause of acute flaccid paralysis (AFP), but this is no longer the case in almost all countries.

Poliomyelitis may vary widely in severity from paresis of one or a few muscles or paralysis of one or more limbs to quadriplegia and paralysis of the muscles of respiration (diaphragm, intercostal muscles). Most children recover from the acute illness, but some 70% continue to have some residual motor weakness, which may vary from mild impairment to complete flaccid paralysis. The permanent loss of motor neurons results in denervation atrophy of the affected muscles.

While the vast majority of poliovirus infections are either asymptomatic or associated with nonspecific febrile illnesses, the case-fatality rate of those who develop poliomyelitis is 2% to 5% and in epidemics it can be as high as 10%. Death is most often due to respiratory paralysis or arrest in children with bulbar poliomyelitis.

Poliomyelitis should be considered in all cases of pure motor paralysis and is usually associated with a normal or slightly-elevated value for protein, normal sugar value, and moderate mononuclear pleocytosis in CSF. Early in the illness polymorphonuclear cells may predominate in the CSF, followed by a shift to mononuclear cells. Defects in the ventral horns of the spinal cord can be observed by magnetic resonance imaging (MRI). The MRI lesion corresponds to the innervation pattern of the affected extremity. Electromyography and nerve conduction velocities (NCVs) generally fail to show evidence of a conduction block. The differential diagnosis includes spinal cord compression, stroke, neuropathy, and Guillain-Barré syndrome (GBS).

Delayed progression of neuromuscular symptoms (post-polio syndrome) may occur 20 years or longer after the initial paralysis due to poliovirus (92). Post-polio syndrome is characterized by new muscle weakness associated with dysfunction of surviving motor neurons. The illness is usually associated with deterioration of those nerves involved in reinnervation during recovery from the original poliovirus infection. It does not appear that reactivation or replication of poliovirus is involved, but current data are inconclusive, since no infectious virus has ever been isolated (93, 94).

Paralytic Myelitis Caused by Other Enteroviruses

A clinical syndrome of AFP may be caused infrequently by certain EV other than polioviruses. In children, EV71 may cause AFP either sporadically or in outbreaks (95–98). Several other EV and parechoviruses have been found to be associated with AFP on rare occasions (31, 99–102). The clinical picture is usually one of a mild disease, occasionally with paralysis of a single muscle, such as the deltoid, and often with complete recovery. Although most patients recover completely, some may continue to have residual paralysis, as in poliomyelitis.

Viral Meningitis

Fever, headache, and nuchal rigidity, often with Brudzinski's sign, are characteristic of enteroviral meningitis in children and adults. The CSF is usually clear and under normal or mildly to moderately increased pressure and with mild-to-moderate pleocytosis (usual range, 100 to 1000 cells per microliter) (103–105). Although on the first or second day of illness, CSF cells may be predominantly neutrophils, they are predominantly lymphocytes when evaluated 1 or 2 days later (106). Enteroviral meningitis usually occurs sporadically, while some children are infected with the same virus without neurologic disease or even asymptomatically. Occasionally EV meningitis may occur as small outbreaks. Enteroviruses are by far the most frequent cause of viral meningitis in most locations (4, 107, 108).

Many other viruses, such as mumps virus, herpes simplex virus (HSV), Epstein-Barr virus, arenavirus, and several arboviruses, also may cause viral meningitis (109). The clinical picture and the laboratory findings on CSF examination lead to a diagnosis of aseptic meningitis. Viral isolation in cell culture or suckling mice from CSF is usually successful

TABLE 2 Clinical syndromes associated with enterovirus infection

CNS
Aseptic meningitis
Encephalitis
Flaccid paralysis
Respiratory
Mild upper respiratory tract illness (common cold)
Lymphonodular pharyngitis
Bronchiolitis
Bronchitis
Pneumonia
Exanthems
Hand, foot, and mouth disease (HFMD)
Herpangina
Cardiac
Myocarditis
Pericarditis
Other
Pleurodynia
Acute hemorrhagic conjunctivitis (AHC)
Neonatal disseminated disease
Chronic infection of agammaglobulinemic patients

only early in the course of illness. Virus isolation and detection by molecular methods are the only definitive agent-specific diagnostic results.

Encephalitis

Infection of the brain parenchyma is a relatively rare manifestation of EV/parechovirus infection (1, 110, 111). The encephalitis may be global or focal. It is probably more common in children than in adults in both tropical and temperate regions. The illness starts with fever and constitutional symptoms. After a few days, confusion, irritability, lethargy, or drowsiness develops and usually progresses rapidly to generalized convulsions and coma. In some children, focal encephalitis is characterized by focal seizures, very much as in herpes simplex virus encephalitis. Other clinical manifestations are usually related to elevated intracranial pressure and to cranial nerve or cerebellar involvement. Occasionally myelitis may also occur, with lower motor neuron paralysis of muscles (112).

More recently, a syndrome of fatal brain stem encephalitis that is associated with EV71 infection has been described in several countries of Southeast Asia (113–115). Although sporadic cases of infection with this virus have been described since the virus was first recognized in 1971, the deaths in Southeast Asia occurred in the context of widespread hand-foot-and-mouth disease (HFMD) outbreaks. The fatal outcome had few clinically predictive symptoms but was specifically associated with young children, most of whom were less than 2 years of age. The onset of neurologic symptoms was particularly rapid, and death often occurred within 24 hours as a result of cardiopulmonary failure, presumably of neurogenic origin (113).

As a rule, CSF culture does not yield an EV. Brain tissue is seldom obtained by biopsy for virus isolation. It is generally believed that herpes simplex virus, several arthropod-transmitted viruses, and EV are the most common causes of viral encephalitis (116, 117).

Acute Myocarditis and Pericarditis and Their Chronic Sequelae

Acute myocarditis, with or without pericarditis, caused by several EV can occur in infants, children, adolescents and young adults (118). The most common serotypes implicated in acute myocarditis are the CVB (119–121).

Acute Hemorrhagic Conjunctivitis

Mild conjunctival hyperemia is noted with many EV diseases. However, a severe form of conjunctivitis, usually occurring in rapidly-spreading epidemics and characterized by subconjunctival hemorrhage in nearly half the subjects, is caused by two EV serotypes. This disease is different from other enteroviral illnesses, having occurred in global pandemics since its introduction around 1969, when both EV70 and "CVA24 variant" (CVA24v) emerged as causes of acute hemorrhagic conjunctivitis (AHC) (61, 122–128). To date, AHC epidemics have occurred largely in the tropical and subtropical countries of Asia, Africa, and Latin America. Only sporadic cases or small outbreaks have occurred in temperate climates.

The illness has a sudden onset. The incubation period for these agents is shorter than for other EV (24 to 72 hours), systemic illness is much less common and conjunctival replication of virus is the rule. Spread is mainly through direct contact, via fingers or fomites. Eye pain, photophobia, excessive lacrimation, and congestion of the conjunctiva are almost always present. The characteristic subconjunctival hemorrhage in a proportion of cases in an epidemic is an important diagnostic feature. The disease is usually bilateral. Adults and school-age children are more affected than infants and preschool children, although household spread is efficient regardless of age. After a few days, symptoms abate, but the hemorrhage resolves slowly and recovery is complete in 5 to 10 days.

Pleurodynia

Pleurodynia, also known as Bornholm disease, is a distinct illness with mild or high fever of short duration and chest pain located on either side of the sternum or retrosternally and may occur sporadically or in outbreaks. The pain is usually intermittent or spasmodic and sometimes excruciating, often exacerbated by deep breathing. Intercostal muscle tenderness and a pleural rub, when present, are important signs distinguishing the illness from myocardial infarction, which is often suspected in adults. The chest radiograph and electrocardiogram are normal. Symptoms usually last for a few days to more than 2 weeks, with occasional relapses. In children severe abdominal pain, apparently arising from the diaphragm, may occur.

This syndrome has primarily been associated with CVB (129, 130), particularly CVB3 and CVB5, although sporadic cases may be caused by other EV. Rarely, pleurodynia may be accompanied by another clinical manifestation of EV infection, such as aseptic meningitis, or even myocarditis.

Hand-Foot-and-Mouth Disease

The distinguishing feature of this illness is the characteristic vesicular eruption on the hands and feet and in the mouth (131). The oral lesions, mostly on the buccal mucosa, become shallow ulcers. CVA16 is the most frequent etiologic agent, occasionally causing large outbreaks. Other EV, especially CVA10 and EV71, may also cause HFMD outbreaks (88, 132). The etiologic agent can be isolated or detected from the vesicles and from the throat and feces. Occasionally, HFMD may occur with other EV involvement such as meningitis. CVA6 recently emerged as a frequent cause of HFMD worldwide (133–136). In these cases, the skin lesions were unusually severe, causing significant concern among parents and health care providers (137–140), and a significant fraction of cases occurred among adults (141, 142).

Herpangina

This illness is characterized by a typical crop of vesicles on the soft palate, uvula, other parts of the oropharynx, or the tongue (143). Each vesicle is about 1 to 2 mm in diameter, with a surrounding red areola. It usually occurs in children younger than 10 years old. Fever and sore throat are the common symptoms. Careful examination of the oropharynx reveals 1 to 12 discrete lesions, which usually subside without ulcerating.

Short Fever with Maculopapular Rash

Many EV may cause a short febrile illness with a maculopapular rash resembling rubella or mild measles, particularly in infants and very young children (143). The distribution of the rash on the face, neck, and chest, and occasionally on the arms and thighs, may mimic other well-recognized exanthems. Sometimes there may be mild upper respiratory symptoms, adding to the difficulty of accurate clinical diagnosis. Outbreaks of EV exanthem also have been called Boston exanthem (144).

Diarrhea

Mild diarrhea is a common accompanying symptom in many enteroviral diseases, and EV occasionally may cause a short acute diarrheal illness. Although occasional outbreaks of diarrhea without other typical EV symptoms have been attributed to EV or parechovirus infections, these are uncommon.

Neonatal Enterovirus Diseases

Neonates are more vulnerable to invasive EV diseases than are older children and adults. Infection may occur *in utero* or, more commonly, perinatally. Many of the clinical features described previously, as well as the more sinister lesions in the CNS or the heart, may cluster together in the infected neonate (145, 146). Such illnesses very much resemble other severe systemic infectious diseases, such as bacterial septicemia. Thus, the infant may present with lethargy, feeding difficulty, vomiting, tachycardia, dyspnea, cyanosis, jaundice, and diarrhea, with or without fever. Clinical evidence for aseptic meningitis, encephalitis, myocarditis, hepatitis, or pneumonia may be present in any combination. The case fatality rate is high. Sometimes death may occur rapidly.

Respiratory Disease

Like the related human rhinoviruses, EV are often associated with mild respiratory illness, most frequently restricted to the upper respiratory tract (common cold) (147). However, EV and parechoviruses may also cause more serious, lower respiratory tract illness, including bronchitis, bronchiolitis, and pneumonia (148, 149). In particular, EV-D68 appears to be almost exclusively restricted to the respiratory tract and associated with respiratory disease (150). Since the mid-2000s, EV-D68 has been increasingly recognized as a cause of acute respiratory illness clusters and outbreaks, highlighted by a nationwide outbreak of more than 1000 confirmed cases in the United States in 2014 (151, 152).

TREATMENT AND PREVENTION

Antivirals

A number of specific antiviral compounds have been developed to target enteroviral proteins and steps in the virus' life cycle; none of these has received regulatory approval, so there are currently no drugs available to treat enterovirus or parechovirus infection.

Polio Eradication

The epidemiology of poliovirus infection has been radically altered by the widespread use of both IPV and OPV. The recent activities of the Polio Eradication Initiative have eliminated endemic poliovirus from most of the world (84). Since 1988, poliomyelitis from wild poliovirus has declined dramatically and in the year 2015 remains endemic in only three countries in Africa and South Asia. Four regions of the world—the Americas, the Western Pacific, Europe and Southeast Asia—have been certified to be free of endemic poliovirus transmission (153). One of the three serotypes of wild poliovirus (type 2) has already been eliminated from the entire world (154). The only remaining wild type 2 viruses are now found in laboratories and vaccine manufacturing facilities. The eradication goal is attainable because humans are the only known reservoir for poliovirus. The current global eradication program launched by WHO relies exclusively on OPV for mass vaccination, but IPV is being introduced globally in 2015-2016.

To achieve high vaccine coverage in all regions of the world, the Polio Eradication Initiative also relies on supplemental immunization campaigns and aggressive investigation of all suspected cases of AFP to identify wild PV circulation (155). To ensure that no case of poliomyelitis is missed, every case of AFP in every location in every country must be detected through clinical surveillance and investigated virologically. Two stool samples should be collected on consecutive days, within 2 weeks after the onset of paralysis, and processed for virus isolation.

Any area with endemic wild PV can serve as a reservoir for reintroduction of PV to areas that have no endemic PV circulation (156). In addition to several documented long-range importations over the past 25 years, wild PV spread from an endemic reservoir in Nigeria to cause cases in 18 additional countries during 2002 to 2005 and re-establish virus circulation in six of these countries. The frequency and ease of international travel probably results in frequent introduction of wild PV into all regions of the world. High rates of polio-vaccine coverage are necessary to prevent poliomyelitis epidemics.

At least two dozen poliomyelitis outbreaks in 18 countries have been associated with circulating vaccine-derived polioviruses (cVDPV) (157). The outbreak strains were unusual because their capsid sequences were derived from OPV. These viruses had recovered the capacity to cause paralytic poliomyelitis in humans and to be transmitted efficiently among human populations (158). Intense investigations suggest that circulation of vaccine-derived virus is a rare event, occurring only in populations with low immunization rates and high population densities. The discovery of cVDPVs has created urgency in planning a comprehensive post-eradication immunization strategy and emphasizes the fact that the risk of polio will not be eliminated until OPV vaccination stops (159). As a consequence, the eradication effort will need to address issues of future vaccination policy, the containment of laboratory and vaccine production strains and a coordinated strategy to achieve the cessation of OPV immunization as vital parts of the eradication effort following the successful elimination of wild virus circulation (159).

LABORATORY DIAGNOSIS

A general diagnostic caveat is shared among EV and other ubiquitous pathogens. Since EV infections are quite common, especially in childhood, and since most infections are noninvasive and prolonged, the detected EV infection need not be the cause of the illness under investigation. Rather, it is an issue of probabilities. If the clinical syndrome is already known to be associated with the detected agent, then infection is taken as reasonable evidence of causation. If the agent is found in diseased tissue or a relevant body fluid (such as CSF), that constitutes concrete evidence of invasion, hence causation. To rule out poliovirus infection in a paralytic case, two stool specimens should be collected at least 24 hours apart during the first 14 days following onset of paralysis. Isolation in culture remains the gold standard for poliovirus detection (160). The preferred methods for laboratory diagnosis of picornavirus infections are outlined in Table 3.

TABLE 3 Preferred methods for Picornavirus detection and characterization

	Sensitivity	Specificity	Ease of use	Speed
Detection				
Cell culture	++	+	++	++
Suckling mouse inoculation	++	+	+	+
Real-time RT-PCR	+++	++	+++	+++
Identification/ characterization				
Indirect immunofluorescence (of culture isolates)	++	++	+++	++
PCR + sequencing	+++	+++	+	++

Virus Isolation and Identification

Many of the detailed procedures for the laboratory diagnosis of EV and parechovirus infections using virus isolation have been described (47). The traditional techniques for detecting and characterizing EV rely on the time-consuming and labor-intensive procedures of viral isolation in cell culture and neutralization by reference antisera. Isolation of EV from specimens using appropriate cultured cell lines is often possible within 2 or 3 days and remains a very sensitive method for detecting these viruses. The best specimens for isolation of virus are, in order of preference, stool specimens or rectal swabs, throat swabs or washings, and CSF. Throat swabs or washings and CSF are most likely to yield virus isolates if they are obtained early in the acute phase of the illness. For cases of AHC, the best specimens are conjunctival swabs (47), although occasionally virus can be isolated from tears (161).

The procedure for virus isolation involves inoculation of appropriate specimens onto susceptible cultured cells. No single cell line, however, exists that is capable of growing all human EV. It is common practice to use several types of human and primate cells to increase the spectrum of viruses that can be detected (47, 162). Even with a variety of cells, however, several CVA serotypes fail to propagate in culture (55). The CV, including those that do not grow in cell culture, can be isolated and propagated by intracerebral inoculation of suckling mice (55). The nature of the CPE in various cell cultures is so characteristic that a cautious presumptive diagnosis of EV infection can often be made at initial detection of cytopathic changes. Some components of a sample inoculated onto cell cultures (from fecal samples, for example) can produce cell toxicity in the first 24 hours, which can be confused with viral CPE. To distinguish toxicity from virus CPE, an additional passage can be performed so that toxic components of a specimen can be diluted; the passage will allow cells to maintain viability or, alternatively, for the virus to amplify and produce CPE.

As a consequence of current PV eradication activities and the importance of PV as a public health problem, specific diagnostic procedures have been developed to detect this virus (163). In general, PV grows well on a variety of primate and human cell culture lines, but it cannot be distinguished from other EV solely on the basis of CPE. Polioviruses are unique in their use of CD155, which is distinct from receptors used by all other EV to infect cells. This receptor has been transfected and expressed in a murine cell line that normally cannot be infected by most EV but is permissive to viral replication when the viral genome is present within the cell. One of these transfected murine cells, L20B, can grow PV and has been exploited selectively to isolate PV even in the presence of other EV (164, 165). When a specimen is inoculated onto these cells and a characteristic EV CPE is seen, the virus can be presumptively identified as a PV. A few strains of certain nonpolio EV serotypes are able to grow on the parent murine L cells, however, and therefore growth on L20B cells is not a definitive identification of PV, and confirmatory testing is required.

Type-specific monoclonal antibodies may also be used for EV typing, typically in indirect immunofluorescence assays to identify viruses isolated in cultured cells (166, 167). Commercially-available monoclonal antibodies can be used to detect relatively common serotypes, including PV1-3, CVA9, CVA24, CVB1-6, E4, E6, E9, E11, E30, EV70, and EV71. Additional monoclonal antibodies have recently been developed for CVA2, CVA4-5, and CVA10 (167). Indirect immunofluorescence is faster and easier to perform than neutralization, and the reagents can be produced in large quantity as needed, but the method still suffers from the same limitations as other antigenic typing methods, namely the requirement for a virus isolate in culture prior to typing and the need for a large number of reagents to identify all serotypes. Despite these limitations, the method has been adopted as the standard EV typing method in a large number of clinical and reference laboratories.

Molecular Detection and Characterization

Molecular procedures are now the methods of choice for EV detection in CSF and are widely used for a patient with a clinical presentation of meningitis.

As with virus isolation and serotyping, the molecular methods attempt to detect the presence of EV in a specimen and, in some procedures, to further characterize the detected virus. The techniques can be grouped on the basis of their infrastructure and technical requirements and the types of specimens to which they are applied. Additionally, the tests can be grouped on the basis of the type of answer the test provides and the predictive value of a positive and a negative finding. The first broad group of tests is based on PCR, which is used primarily to detect EV genomes in cell cultures, clinical specimens, and biopsy or autopsy tissues. The second and newest procedure utilizes genomic sequencing for the characterization of EV at the highest levels of specificity. While it has not yet been integrated into routine clinical virology workflows, so-called "next-generation" sequencing is an exciting technology that promises to revolutionize pathogen molecular identification without prior knowledge of viral targets (168–171). However, considerable development work is required for this vision to become a reality.

By far the most common use of PCR for EV diagnosis is the direct detection of virus in clinical specimens (172, 173). Numerous variations on the details of the procedures are found, but all methods that can generically detect EV have several common features. The most important property of these tests is that the primers are targeted to amplify the 5'-NTR of the virus genome. Many of these assays, however, have not been completely evaluated on a large number of clinical isolates to confirm reactivity with all EV serotypes and strains within serotypes, and therefore they have not been validated sufficiently for diagnostic use (174). The major advantage of the pan-EV PCR is that rapid detection of an EV is possible, even with very small amounts of clinical specimens such as CSF. It is also possible to detect EV that do not readily grow in cell culture. As with all PCR, the sensitivity of amplification of RNA from biologic specimens is extremely variable, depending on the nature of the specimen. Although the PCR procedure can be shown to give a positive result even from only one or a few copies of viral RNA, it is not unusual for the sensitivity to be many orders of magnitude lower in certain specimens (e.g., stool). The introduction of "real-time" PCR methods has largely solved this issue as the fluorescence detection systems generally increase the sensitivity significantly. A small number of "panEV" real-time PCR assays have received regulatory approval for clinical use in the U.S. or Europe, but some of these are labeled for narrow indications or specimen types, so many clinical laboratories still rely on laboratory-developed tests. A more recent innovation is the development of molecular respiratory virus panels that allow simultaneous detection of a dozen or more common respiratory viruses (175). Some of these incorporate a "rhinovirus-enterovirus" assay targeting conserved, shared sites in the 5′-NTR. Sensitivity varies among the available systems and among targets within a given platform, but the ability to rapidly test for a wide range of common respiratory pathogens offers a distinct advantage over single-assay systems.

A goal in virus identification is knowledge of the sequence of the viral genome. Encoded within this sequence are determinants for all the biologic properties that are attributable to a given virus. Therefore, the nucleic acid sequence information of a virus represents its ultimate characterization. All important information about a virus could potentially be obtained directly by PCR in conjunction with nucleic acid sequencing if all the molecular correlates of viral phenotypic determinants were understood. At present, however, the genetic location for many properties of the virus remains uncertain. Nevertheless, it is possible to use sequence information to assign an EV isolate to a particular serotype (40–44, 174). The most common molecular typing system is based on RT-PCR and nucleotide sequencing of a portion of the genomic region encoding VP1. The serotype of an unknown isolate is inferred by comparison of the partial VP1 sequence with a database containing VP1 sequences for the prototype and variant strains of all human EV serotypes (174). Using this approach, strains of homologous serotypes can be easily discriminated from heterologous serotypes and new serotypes can be identified. The technique is also useful to rapidly determine whether viruses isolated during an outbreak are epidemiologically related. The most sensitive version of this approach uses semi-nested PCR to allow molecular typing directly from clinical material and is approximately equal in sensitivity to real-time PCR using hydrolysis probes (176–179). In some cases, it may be advantageous to use VP1 sequence of a few viruses in an outbreak to develop a simple real-time PCR assay to facilitate rapid identification of cases that are part of the outbreak and differentiate them from sporadic cases infected with other co-circulating EV types. For example, several such assays were developed during the 2014 North American EV-D68 outbreak (180–182).

Serologic Diagnosis

Serologic diagnosis of EV infection can be made by comparing antibody titers in acute and convalescent phase (*paired*) serum specimens. In general, however, EV serodiagnosis is more relevant to epidemiologic studies than to clinical diagnosis. The most basic serologic test is that of neutralization in cell culture. The mechanics of the assay are similar to the test used for identification of isolates, except that a known virus is mixed with serial dilutions of antisera from a patient (183). A 4-fold or greater rise in type-specific neutralizing antibody titer is considered diagnostic of recent infection. Antibody, however, may already be present at the time the original specimen is obtained because of the extended incubation period and prodromal period of many enteroviral illnesses, which complicates interpretation of results. The neutralization assay also requires knowledge of a suspected serotype as the specific virus must be used as antigen. With more than 100 known EV types, this approach is impractical if the serotype is unknown or cannot be reduced to a very small number.

Several groups have developed an enzyme-linked immunosorbent assay (ELISA) for EV-specific IgM (184–187). These tests have been found positive for nearly 90% of culture-confirmed CVB infections and can be performed rapidly. The ELISA has been successfully applied for epidemiologic investigations of outbreaks (188), as well as for specific diagnostic use (189–191). In most cases, the IgM ELISA test is not completely serotype-specific. Depending on the configuration and sensitivity of the test, from 10% to nearly 70% of serum samples show a heterotypic response caused by other EV infections. This heterotypic response has been exploited to measure broadly reactive antibody, and the assay has been used to detect EV infection generically (192, 193). In attempting to characterize the exact nature of the response using different antigens, it is clear that the human immune response to EV infection includes antibodies that react with both serotype-specific epitopes and shared epitopes (194). In summary, the IgM assays that are generally used in epidemiologic studies have very good sensitivity and appear to be very specific for EV infection; however, these assays detect heterotypic antibodies resulting from other EV infections and therefore cannot be considered strictly serotype-specific. A positive result with either the neutralization test or IgM ELISA indicates a recent viral infection; however, the infecting serotype found with the IgM assay may not be the same one determined by the neutralization test.

Parechovirus Diagnostics

Despite their original classification, sequencing and PCR studies demonstrated that parechoviruses 22 and 23 were distinct from the enteroviruses, resulting in their reclassification as members of a new picornavirus genus, *Parechovirus* (25, 27–29). Since then, 14 additional human parechovirus types have been identified (26, 31–33, 195–199). Like the enteroviruses, human parechoviruses were traditionally detected and identified by virus isolation and antigenic typing (200). By these methods, HPeV1 (formerly E22) consistently accounted for 2% to 4% of "enteroviruses" reported to CDC from 1975 to 1995 (2). RT-PCR began to supplant virus culture as the method of choice for EV detection in clinical diagnostic laboratories in the mid-1990s. Since that time, the number of HPeV1 reports has declined to under 1%, probably because HPeV-containing specimens are usually reported as EV PCR-negative and not further characterized.

A number of investigators have developed real-time PCR assays to detect HPeVs (196, 201–211). These methods target conserved sites in the 5'-NTR that are analogous to those targeted by EV-specific PCR assays. A number of recent studies have applied molecular methods to the detection of HPeVs in patients with enteritis, respiratory illness, and neonatal sepsis-like syndrome (31, 33, 35, 146, 195, 196, 201, 212, 213). As these methods are increasingly integrated into the diagnostic routine of clinical and reference laboratories, a better estimate will emerge of the burden of disease attributable to this group of picornaviruses. Molecular typing assays, using RT-PCR and sequencing targeted to VP1 (and analogous to approaches used for the enteroviruses), have been developed to identify parechoviruses in clinical specimens (214, 215).

DIAGNOSTIC BEST PRACTICES

Virus isolation in cell culture is still sometimes used to detect enteroviruses and parechoviruses, but intracranial inoculation of suckling mice is rarely attempted these days (Table 3). In current practice, the most sensitive assays to detect enteroviruses and parechoviruses use molecular amplification methods such as RT-PCR. Enteroviruses and rhinoviruses cannot always be distinguished using molecular assays that target the conserved 5′-NTR, so additional PCRs targeting the capsid region and coupled with sequencing are generally used to differentiate these two closely related virus groups. Conversely, detection of parechoviruses requires a separate set of assays as the parechovirus genome sequences are distinct from those of the enteroviruses. Identification of enterovirus types by indirect immunofluorescence with commercially-available monoclonal antibodies is relatively simple (though reagents are limited to a small number of type specificities), but as virus isolation has become less common, most enterovirus and parechovirus typing is accomplished through amplification and sequencing of a portion of the VP1 capsid region.

REFERENCES

1. **Pallansch MA, Roos R.** 2006. Enteroviruses: polioviruses, coxsackieviruses, echoviruses, and newer enteroviruses, p 839–893. *In* Knipe DM, Howley PM, Griffin DE, Lamb RA, Martin MA, Roizman B, Straus SE (ed), *Fields Virology*, 5th ed. Lippincott Williams & Wilkins, Philadelphia, PA.
2. **Khetsuriani N, Lamonte-Fowlkes A, Oberst S, Pallansch MA, Centers for Disease Control and Prevention.** 2006. Enterovirus surveillance—United States, 1970–2005. *MMWR Surveill Summ* **55:**1–20.
3. **Pallansch MA, Oberste MS, Whitton JL.** 2013. Enteroviruses: polioviruses, coxsackieviruses, echoviruses, and newer enteroviruses, p 490–530. *In* Knipe DM, Howley PM, Cohen JI, Griffin DE, Lamb RA, Martin MA, Roizman B (ed), *Fields Virology*, 6th ed. Lippincott Williams & Wilkins, Philadelphia, PA.
4. **Khetsuriani N, Quiroz ES, Holman RC, Anderson LJ.** 2003. Viral meningitis-associated hospitalizations in the United States, 1988–1999. *Neuroepidemiology* **22:**345–352.
5. **Landsteiner K, Popper E.** 1908. Mikroskopische Praparate von einem Menschlichen und zwei Affenruckenmarken. *Wien Klin Wochenschr* **21:**1830.
6. **Enders JF, Weller TH, Robbins FC.** 1949. Cultivation of the Lansing strain of poliomyelitis virus in cultures of various human embryonic tissues. *Science* **109:**85–87.
7. **Dalldorf G, Sickles GM.** 1948. An unidentified, filtrable agent isolated from the feces of children with paralysis. *Science* **108:**61–62.
8. **Melnick JL, Shaw EW, Curnen EC.** 1949. A virus isolated from patients diagnosed as non-paralytic poliomyelitis or aseptic meningitis. *Proc Soc Exp Biol Med* **71:**344–349.
9. **Dalldorf G.** 1950. The Coxsackie viruses. *Bull N Y Acad Med* **26:**329–335.
10. **Gifford R, Dalldorf G.** 1951. The morbid anatomy of experimental Coxsackie virus infection. *Am J Pathol* **27:**1047–1063.
11. **Melnick JL, Ågren K.** 1952. Poliomyelitis and Coxsackie viruses isolated from normal infants in Egypt. *Proc Soc Exp Biol Med* **81:**621–624.
12. **Melnick JL.** 1954. Application of tissue culture methods to epidemiological studies of poliomyelitis. *Am J Public Health Nations Health* **44:**571–580.
13. **Robbins FC, Enders JF, Weller TH, Florentino GL.** 1951. Studies on the cultivation of poliomyelitis viruses in tissue culture. V. The direct isolation and serologic identification of virus strains in tissue culture from patients with nonparalytic and paralytic poliomyelitis. *Am J Hyg* **54:**286–293.
14. **Riordan JT, Ledinko N, Melnick JL.** 1952. Multiplication of poliomyelitis viruses in tissue cultures of monkey testes. II. Direct isolation and typing of strains from human stools and spinal cords in roller tubes. *Am J Hyg* **55:**339–346.
15. **Kibrick S, Meléndez L, Enders JF.** 1957. Clinical associations of enteric viruses with particular reference to agents exhibiting properties of the ECHO group. *Ann N Y Acad Sci* **67:**311–325.
16. **Weller TH.** 1953. The application of tissue-culture methods to the study of poliomyelitis. *N Engl J Med* **249:**186–195.
17. **Weller TH, Robbins FC, Stoddard MB.** 1952. Cultivation of coxsackie viruses in human tissue. *Fed Proc* **11:**486.
18. **Stobo J, Green I, Jackson L, Baron S.** 1974. Identification of a subpopulation of mouse lymphoid cells required for interferon production after stimulation with mitogens. *J Immunol* **112:**1589–1593.
19. **Committee on Enteroviruses.** 1962. Classification of human enteroviruses. *Virol* **16:**501–504.
20. **Committee on the ECHO viruses.** 1955. Enteric cytopathogenic human orphan (ECHO) viruses. *Science* **122:**1187–1188.
21. **Committee on the Enteroviruses.** 1957. The enteroviruses. *Am J Public Health Nations Health* **47:**1556–1566.
22. **Melnick JL, Chanock RM, Gelfand H, Hammon WM, Huebner RJ, Rosen L, Sabin AB, Wenner HA, Panel for Picornaviruses.** 1963. Picornaviruses: classification of nine new types. *Science* **141:**153–154.
23. **Melnick JL, Tagaya I, von Magnus H.** 1974. Enteroviruses 69, 70, and 71. *Intervirology* **4:**369–370.
24. **Schieble JH, Fox VL, Lennette EH.** 1967. A probable new human picornavirus associated with respiratory diseases. *Am J Epidemiol* **85:**297–310.
25. **Stanway G, Brown F, Christian P, Hovi T, Hyypiä T, King AMQ, Knowles NJ, Lemon SM, Minor PD, Pallansch MA, Palmenberg AC, Skern T.** 2005. Picornaviridae, p 757–778. *In* Fauquet CM, Mayo MA, Maniloff J, Desselberger U, Ball LA (ed), *Virus Taxonomy. Eighth Report of the International Committee on the Taxonomy of Viruses.* Elsevier Academic Press, Amsterdam.
26. **Knowles NJ** 2015, posting date. The picornavirus pages, http://www.picornaviridae.com/. Institute for Animal Health, Pirbright Laboratory. [Online.]
27. **Stanway G, Kalkkinen N, Roivainen M, Ghazi F, Khan M, Smyth M, Meurman O, Hyypiä T.** 1994. Molecular and biological characteristics of echovirus 22, a representative of a new picornavirus group. *J Virol* **68:**8232–8238.
28. **Hyypiä T, Horsnell C, Maaronen M, Khan M, Kalkkinen N, Auvinen P, Kinnunen L, Stanway G.** 1992. A distinct picornavirus group identified by sequence analysis. *Proc Natl Acad Sci USA* **89:**8847–8851.
29. **Coller BA, Chapman NM, Beck MA, Pallansch MA, Gauntt CJ, Tracy SM.** 1990. Echovirus 22 is an atypical enterovirus. *J Virol* **64:**2692–2701.
30. **Stanway G, Joki-Korpela P, Hyypiä T.** 2000. Human parechoviruses—biology and clinical significance. *Rev Med Virol* **10:**57–69.
31. **Ito M, Yamashita T, Tsuzuki H, Takeda N, Sakae K.** 2004. Isolation and identification of a novel human parechovirus. *J Gen Virol* **85:**391–398.

32. **Benschop KSM, Schinkel J, Luken ME, van den Broek PJM, Beersma MFC, Menelik N, van Eijk HWM, Zaaijer HL, VandenBroucke-Grauls CM, Beld MGHM, Wolthers KC.** 2006. Fourth human parechovirus serotype. *Emerg Infect Dis* **12:**1572–1575.
33. **Al-Sunaidi M, Williams CH, Hughes PJ, Schnurr DP, Stanway G.** 2007. Analysis of a new human parechovirus allows the definition of parechovirus types and the identification of RNA structural domains. *J Virol* **81:**1013–1021.
34. **Watanabe K, Oie M, Higuchi M, Nishikawa M, Fujii M.** 2007. Isolation and characterization of novel human parechovirus from clinical samples. *Emerg Infect Dis* **13:**889–895.
35. **Niklasson B, Kinnunen L, Hörnfeldt B, Hörling J, Benemar C, Hedlund KO, Matskova L, Hyypiä T, Winberg G.** 1999. A new picornavirus isolated from bank voles (Clethrionomys glareolus). *Virology* **255:**86–93.
36. **Yamashita T, Sakae K, Tsuzuki H, Suzuki Y, Ishikawa N, Takeda N, Miyamura T, Yamazaki S.** 1998. Complete nucleotide sequence and genetic organization of Aichi virus, a distinct member of the Picornaviridae associated with acute gastroenteritis in humans. *J Virol* **72:**8408–8412.
37. **Li L, Victoria J, Kapoor A, Blinkova O, Wang C, Babrzadeh F, Mason CJ, Pandey P, Triki H, Bahri O, Oderinde BS, Baba MM, Bukbuk DN, Besser JM, Bartkus JM, Delwart EL.** 2009. A novel picornavirus associated with gastroenteritis. *J Virol* **83:**12002–12006.
38. **Lentz KN, Smith AD, Geisler SC, Cox S, Buontempo P, Skelton A, DeMartino J, Rozhon E, Schwartz J, Girijavallabhan V, O'Connell J, Arnold E.** 1997. Structure of poliovirus type 2 Lansing complexed with antiviral agent SCH48973: comparison of the structural and biological properties of three poliovirus serotypes. *Structure* **5:**961–978.
39. **Stanway G, Hyypiä T.** 1999. Parechoviruses. *J Virol* **73:**5249–5254.
40. **Oberste MS, Maher K, Kilpatrick DR, Flemister MR, Brown BA, Pallansch MA.** 1999. Typing of human enteroviruses by partial sequencing of VP1. *J Clin Microbiol* **37:**1288–1293.
41. **Oberste MS, Maher K, Kilpatrick DR, Pallansch MA.** 1999. Molecular evolution of the human enteroviruses: correlation of serotype with VP1 sequence and application to picornavirus classification. *J Virol* **73:**1941–1948.
42. **Caro V, Guillot S, Delpeyroux F, Crainic R.** 2001. Molecular strategy for "serotyping" of human enteroviruses. *J Gen Virol* **82:**79–91.
43. **Norder H, Bjerregaard L, Magnius LO.** 2001. Homotypic echoviruses share aminoterminal VP1 sequence homology applicable for typing. *J Med Virol* **63:**35–44.
44. **Casas I, Palacios GF, Trallero G, Cisterna D, Freire MC, Tenorio A.** 2001. Molecular characterization of human enteroviruses in clinical samples: comparison between VP2, VP1, and RNA polymerase regions using RT nested PCR assays and direct sequencing of products. *J Med Virol* **65:**138–148.
45. **Mertens T, Pika U, Eggers HJ.** 1983. Cross antigenicity among enteroviruses as revealed by immunoblot technique. *Virology* **129:**431–442.
46. **Huovilainen A, Hovi T, Kinnunen L, Takkinen K, Ferguson M, Minor P.** 1987. Evolution of poliovirus during an outbreak: sequential type 3 poliovirus isolates from several persons show shifts of neutralization determinants. *J Gen Virol* **68:**1373–1378.
47. **Melnick JL, Wenner HA, Phillips CA.** 1979. Enteroviruses, p 471–534. *In* Lennette EH, Schmidt NJ (ed), *Diagnostic Procedures for Viral, Rickettsial, and Chlamydial Infections*, 5th ed. American Public Health Association, Washington, D.C.
48. **Kok TW, Pryor T, Payne L.** 1998. Comparison of rhabdomyosarcoma, buffalo green monkey kidney epithelial, A549 (human lung epithelial) cells and human embryonic lung fibroblasts for isolation of enteroviruses from clinical samples. *J Clin Virol* **11:**61–65.
49. **Patel JR, Daniel J, Mathan M, Mathan VI.** 1984. Isolation and identification of enteroviruses from faecal samples in a differentiated epithelial cell line (HRT-18) derived from human rectal carcinoma. *J Med Virol* **14:**255–261.
50. **Patel JR, Daniel J, Mathan VI.** 1985. A comparison of the susceptibility of three human gut tumour-derived differentiated epithelial cell lines, primary monkey kidney cells and human rhabdomyosarcoma cell line to 66-prototype strains of human enteroviruses. *J Virol Methods* **12:**209–216.
51. **Otero JR, Folgueira L, Trallero G, Prieto C, Maldonado S, Babiano MJ, Martinez-Alonso I.** 2001. A-549 is a suitable cell line for primary isolation of coxsackie B viruses. *J Med Virol* **65:**534–536.
52. **Saijets S, Ylipaasto P, Vaarala O, Hovi T, Roivainen M.** 2003. Enterovirus infection and activation of human umbilical vein endothelial cells. *J Med Virol* **70:**430–439.
53. **Pintó RM, Diez JM, Bosch A.** 1994. Use of the colonic carcinoma cell line CaCo-2 for in vivo amplification and detection of enteric viruses. *J Med Virol* **44:**310–315.
54. **Nsaibia S, Wagner S, Rondé P, Warter JM, Poindron P, Aouni M, Dorchies OM.** 2007. The difficult-to-cultivate coxsackieviruses A can productively multiply in primary culture of mouse skeletal muscle. *Virus Res* **123:**30–39.
55. **Schmidt NJ, Ho HH, Lennette EH.** 1975. Propagation and isolation of group A coxsackieviruses in RD cells. *J Clin Microbiol* **2:**183–185.
56. **Lee LH, Phillips CA, South MA, Melnick JL, Yow MD.** 1965. Enteric virus isolation in different cell cultures. *Bull World Health Organ* **32:**657–663.
57. **Romero JR, Selvarangan R.** 2011. The human Parechoviruses: an overview. *Adv Pediatr* **58:**65–85.
58. **Rossmann MG, He Y, Kuhn RJ.** 2002. Picornavirus-receptor interactions. *Trends Microbiol* **10:**324–331.
59. **Imamura T, Okamoto M, Nakakita S, Suzuki A, Saito M, Tamaki R, Lupisan S, Roy CN, Hiramatsu H, Sugawara KE, Mizuta K, Matsuzaki Y, Suzuki Y, Oshitani H, Perlman S.** 2014. Antigenic and receptor binding properties of enterovirus 68. *J Virol* **88:**2374–2384.
60. **Horstmann DM.** 1967. Enterovirus infections of the central nervous system. The present and future of poliomyelitis. *Med Clin North Am* **51:**681–692.
61. **Kono R.** 1975. Apollo 11 disease or acute hemorrhagic conjunctivitis: a pandemic of a new enterovirus infection of the eyes. *Am J Epidemiol* **101:**383–390.
62. **Welch JB, McGowan K, Searle B, Gillon J, Jarvis LM, Simmonds P.** 2001. Detection of enterovirus viraemia in blood donors. *Vox Sang* **80:**211–215.
63. **Welch J, Maclaran K, Jordan T, Simmonds P.** 2003. Frequency, viral loads, and serotype identification of enterovirus infections in Scottish blood donors. *Transfusion* **43:**1060–1066.
64. **Sawyer LA, Hershow RC, Pallansch MA, Fishbein DB, Pinsky PF, Broerman SF, Grimm BB, Anderson LJ, Hall DB, Schonberger LB.** 1989. An epidemic of acute hemorrhagic conjunctivitis in American Samoa caused by coxsackievirus A24 variant. *Am J Epidemiol* **130:**1187–1198.
65. **Hall CE, Cooney MK, Fox JP.** 1970. The Seattle virus watch program. I. Infection and illness experience of virus watch families during a communitywide epidemic of echovirus type 30 aseptic meningitis. *Am J Public Health Nations Health* **60:**1456–1465.
66. **Alexander JP Jr, Chapman LE, Pallansch MA, Stephenson WT, Török TJ, Anderson LJ.** 1993. Coxsackievirus B2 infection and aseptic meningitis: a focal outbreak among members of a high school football team. *J Infect Dis* **167:**1201–1205.
67. **Townsend TR, Bolyard EA, Yolken RH, Beschorner WE, Bishop CA, Burns WH, Santos GW, Saral R.** 1982. Outbreak of Coxsackie A1 gastroenteritis: a complication of bone-marrow transplantation. *Lancet* **1:**820–823.
68. **Keswick BH, Gerba CP, Goyal SM.** 1981. Occurrence of enteroviruses in community swimming pools. *Am J Public Health* **71:**1026–1030.
69. **Goyal SM, Gerba CP, Melnick JL.** 1979. Human enteroviruses in oysters and their overlying waters. *Appl Environ Microbiol* **37:**572–581.

70. **Clark CS, Bjornson AB, Schiff GM, Phair JP, Van Meer GL, Gartside PS.** 1977. Sewage worker's syndrome. *Lancet* **1:**1009. letter.
71. **Moore BE, Sagik BP, Sorber CA.** 1979. Procedure for the recovery of airborne human enteric viruses during spray irrigation of treated wastewater. *Appl Environ Microbiol* **38:**688–693.
72. **Feldman RA, Christopher S, George S, Kamath KR, John TJ.** 1970. Infection and disease in a group of South Indian families. 3. Virological methods and a report of the frequency of enteroviral infection in preschool children. *Am J Epidemiol* **92:**357–366.
73. **John TJ, Patoria NK, Christopher S, George S.** 1978. Epidemiology of enterovirus infections in children in Nagpur. *Indian J Med Res* **68:**549–554.
74. **Mukundan P, John TJ.** 1983. Prevalence and titres of neutralising antibodies to group B coxsackieviruses. *Indian J Med Res* **77:**577–589.
75. **Morens DM.** 1978. Enteroviral disease in early infancy. *J Pediatr* **92:**374–377.
76. **Ramsay ME, Begg NT, Gandhi J, Brown D.** 1994. Antibody response and viral excretion after live polio vaccine or a combined schedule of live and inactivated polio vaccines. *Pediatr Infect Dis J* **13:**1117–1121.
77. **Parks WP, Queiroga LT, Melnick JL.** 1967. Studies of infantile diarrhea in Karachi, Pakistan. II. Multiple virus isolations from rectal swabs. *Am J Epidemiol* **85:**469–478.
78. **Helfand RF, Khan AS, Pallansch MA, Alexander JP, Meyers HB, DeSantis RA, Schonberger LB, Anderson LJ.** 1994. Echovirus 30 infection and aseptic meningitis in parents of children attending a child care center. *J Infect Dis* **169:**1133–1137.
79. **Leonardi GP, Greenberg AJ, Costello P, Szabo K.** 1993. Echovirus type 30 infection associated with aseptic meningitis in Nassau County, New York, USA. *Intervirology* **36:**53–56.
80. **Mohle-Boetani JC, Matkin C, Pallansch M, Helfand R, Fenstersheib M, Blanding JA, Solomon SL.** 1999. Viral meningitis in child care center staff and parents: an outbreak of echovirus 30 infections. *Public Health Rep* **114:**249–256.
81. **Kew OM, Nottay BK, Lipskaya GY, da Silva EE, Pallansch MA.** 1984. Molecular epidemiology of polioviruses. *Rev Infect Dis* **6**(Suppl 2)**:**S499–S504.
82. **Kew O, Nathanson N.** 1995. Introduction: molecular epidemiology of viruses. *Semin Virol* **6:**357–358.
83. **Rico-Hesse R, Pallansch MA, Nottay BK, Kew OM.** 1987. Geographic distribution of wild poliovirus type 1 genotypes. *Virology* **160:**311–322.
84. **Centers for Disease Control and Prevention (CDC).** 2007. Progress toward interruption of wild poliovirus transmission—worldwide, January 2006-May 2007. *MMWR Morb Mortal Wkly Rep* **56:**682–685.
85. **Kopecka H, Brown B, Pallansch M.** 1995. Genotypic variation in coxsackievirus B5 isolates from three different outbreaks in the United States. *Virus Res* **38:**125–136.
86. **Oberste MS, Maher K, Kennett ML, Campbell JJ, Carpenter MS, Schnurr D, Pallansch MA.** 1999. Molecular epidemiology and genetic diversity of echovirus type 30 (E30): genotypes correlate with temporal dynamics of E30 isolation. *J Clin Microbiol* **37:**3928–3933.
87. **Brown BA, Oberste MS, Alexander JP Jr, Kennett ML, Pallansch MA.** 1999. Molecular epidemiology and evolution of enterovirus 71 strains isolated from 1970 to 1998. *J Virol* **73:**9969–9975.
88. **Hosoya M, Kawasaki Y, Sato M, Honzumi K, Kato A, Hiroshima T, Ishiko H, Suzuki H.** 2006. Genetic diversity of enterovirus 71 associated with hand, foot and mouth disease epidemics in Japan from 1983 to 2003. *Pediatr Infect Dis J* **25:**691–694.
89. **Lin KH, Hwang KP, Ke GM, Wang CF, Ke LY, Hsu YT, Tung YC, Chu PY, Chen BH, Chen HL, Kao CL, Wang JR, Eng HL, Wang SY, Hsu LC, Chen HY.** 2006. Evolution of EV71 genogroup in Taiwan from 1998 to 2005: an emerging of subgenogroup C4 of EV71. *J Med Virol* **78:**254–262.
90. **Ouzilou L, Caliot E, Pelletier I, Prévost MC, Pringault E, Colbère-Garapin F.** 2002. Poliovirus transcytosis through M-like cells. *J Gen Virol* **83:**2177–2182.
91. **Yin-Murphy M.** 1984. Acute hemorrhagic conjunctivitis. *Prog Med Virol* **29:**23–44.
92. **Wiechers DO.** 1987. Late effects of polio: historical perspectives. *Birth Defects Orig Artic Ser* **23:**1–11.
93. **Melchers W, de Visser M, Jongen P, van Loon A, Nibbeling R, Oostvogel P, Willemse D, Galama J.** 1992. The postpolio syndrome: no evidence for poliovirus persistence. *Ann Neurol* **32:**728–732. See comments.
94. **Arya SC.** 1997. Poliovirus genomic sequences in the central nervous systems of patients with postpolio syndrome. *J Clin Microbiol* **35:**334–335. letter; comment.
95. **Alexander JP Jr, Baden L, Pallansch MA, Anderson LJ.** 1994. Enterovirus 71 infections and neurologic disease—United States, 1977–1991. *J Infect Dis* **169:**905–908.
96. **Melnick JL.** 1984. Enterovirus type 71 infections: a varied clinical pattern sometimes mimicking paralytic poliomyelitis. *Rev Infect Dis* **6**(Suppl 2)**:**S387–S390.
97. **Chumakov M, Voroshilova M, Shindarov L, Lavrova I, Gracheva L, Koroleva G, Vasilenko S, Brodvarova I, Nikolova M, Gyurova S, Gacheva M, Mitov G, Ninov N, Tsylka E, Robinson I, Frolova M, Bashkirtsev V, Martiyanova L, Rodin V.** 1979. Enterovirus 71 isolated from cases of epidemic poliomyelitis-like disease in Bulgaria. *Arch Virol* **60:**329–340.
98. **Shindarov LM, Chumakov MP, Voroshilova MK, Bojinov S, Vasilenko SM, Iordanov I, Kirov ID, Kamenov E, Leshchinskaya EV, Mitov G, Robinson IA, Sivchev S, Staikov S.** 1979. Epidemiological, clinical, and pathomorphological characteristics of epidemic poliomyelitis-like disease caused by enterovirus 71. *J Hyg Epidemiol Microbiol Immunol* **23:**284–295.
99. **Figueroa JP, Ashley D, King D, Hull B.** 1989. An outbreak of acute flaccid paralysis in Jamaica associated with echovirus type 22. *J Med Virol* **29:**315–319.
100. **Gear JHS.** 1984. Nonpolio causes of polio-like paralytic syndromes. *Rev Infect Dis* **6**(Suppl 2)**:**S379–S384.
101. **Grist NR, Bell EJ.** 1970. Enteroviral etiology of the paralytic poliomyelitis syndrome. *Arch Environ Health* **21:**382–387.
102. **Hammon WM, Yohn DS, Ludwig EH, Pavia RA, Sather GE, McCloskey LW.** 1958. A study of certain nonpoliomyelitis and poliomyelitis enterovirus infections; clinical and serologic associations. *J Am Med Assoc* **167:**727–735.
103. **Mulford WS, Buller RS, Arens MQ, Storch GA.** 2004. Correlation of cerebrospinal fluid (CSF) cell counts and elevated CSF protein levels with enterovirus reverse transcription-PCR results in pediatric and adult patients. *J Clin Microbiol* **42:**4199–4203.
104. **Graham AK, Murdoch DR.** 2005. Association between cerebrospinal fluid pleocytosis and enteroviral meningitis. *J Clin Microbiol* **43:**1491.
105. **Landry ML.** 2005. Frequency of normal cerebrospinal fluid protein level and leukocyte count in enterovirus meningitis. *J Clin Virol* **32:**73–74.
106. **Cherry JD.** 2004. Enteroviruses and parechoviruses, p 1984–2041. *In* Feigin RD, Cherry JD, Demmler G, Kaplan SL (ed), *Textbook of Pediatric Infectious Diseases*, 5th ed, vol 2. Saunders, Philadelphia, PA.
107. **Rotbart HA, Brennan PJ, Fife KH, Romero JR, Griffin JA, McKinlay MA, Hayden FG.** 1998. Enterovirus meningitis in adults. *Clin Infect Dis* **27:**896–898.
108. **Romero JR.** 2002. Diagnosis and Management of Enteroviral Infections of the Central Nervous System. *Curr Infect Dis Rep* **4:**309–316.
109. **Baum SG, Koll B.** 2003. Acute viral meningitis and encephalitis, p 1288–1293. *In* Gorbach SL, Bartlett JG, Blacklow NR (ed), *Infectious Diseases*, 3rd ed. Lippincott Williams & Wilkins, Philadelphia, PA.

110. **Morens DM, Pallansch MA.** 1995. Epidemiology, p 3–23. *In* Rotbart HA (ed), *Human Enterovirus Infections*. ASM Press, Washington, D.C.
111. **Whitley RJ, Kimberlin DW.** 1999. Viral encephalitis. *Pediatr Rev* **20:**192–198.
112. **Steiner I, Budka H, Chaudhuri A, Koskiniemi M, Sainio K, Salonen O, Kennedy PG.** 2005. Viral encephalitis: a review of diagnostic methods and guidelines for management. *Eur J Neurol* **12:**331–343.
113. **Chan LG, Parashar UD, Lye MS, Ong FGL, Zaki SR, Alexander JP, Ho KK, Han LL, Pallansch MA, Suleiman AB, Jegathesan M, Anderson LJ, for the Outbreak Study Group.** 2000. Deaths of children during an outbreak of hand, foot, and mouth disease in sarawak, malaysia: clinical and pathological characteristics of the disease. *Clin Infect Dis* **31:**678–683.
114. **Abubakar S, Shafee N, Chee HY.** 1998. Outbreak of fatal childhood viral infection in Sarawak, Malaysia in 1997: inocula of patients' clinical specimens induce apoptosis in vitro. *Malays J Pathol* **20:**71–81.
115. **Ho M, Chen ER, Hsu KH, Twu SJ, Chen KT, Tsai SF, Wang JR, Shih SR, Taiwan Enterovirus Epidemic Working Group.** 1999. An epidemic of enterovirus 71 infection in Taiwan. *N Engl J Med* **341:**929–935.
116. **Rotbart HA.** 1995. Enteroviral infections of the central nervous system. *Clin Infect Dis* **20:**971–981.
117. **Lewis P, Glaser CA.** 2005. Encephalitis. *Pediatr Rev* **26:**353–363.
118. **Magnani JW, Dec GW.** 2006. Myocarditis: current trends in diagnosis and treatment. *Circulation* **113:**876–890.
119. **Kim K-S, Höfling K, Carson SD, Chapman NM, Tracy S.** 2003. The Primary Viruses of Myocarditis, p 23–53. *In* Cooper LTJ (ed), *Myocarditis from Bench to Bedside*. Humana Press, Totowa, NJ.
120. **Diamond C, Tilles J.** 2004. Pericarditis and Myocarditis, p 589–597. *In* Gorbach SL, Bartlett JG, Blacklow NR (ed), *Infectious Diseases*, 3rd ed. Lippincott Williams & Wilkins, Philadelphia, PA.
121. **Baboonian C, Treasure T.** 1997. Meta-analysis of the association of enteroviruses with human heart disease. *Heart* **78:** 539–543.
122. **Kono R, Sasagawa A, Ishii K, Sugiura S, Ochi M.** 1972. Pandemic of new type of conjunctivitis. *Lancet* **1:**1191–1194.
123. **Lim KH.** 1973. Epidemic conjunctivitis: discovery of a new aetiologic agent. *Singapore Med J* **14:**82–85.
124. **Yin-Murphy M.** 1972. An epidemic of picornavirus conjunctivitis in Singapore. *Southeast Asian J Trop Med Public Health* **3:**303–309.
125. **Yin-Murphy M, Lim KH.** 1972. Picornavirus epidemic conjunctivitis in Singapore. *Lancet* **2:**857–858.
126. **Yin-Murphy M.** 1973. Viruses of acute haemorrhagic conjunctivitis. *Lancet* **1:**545–546.
127. **Mirkovic RR, Kono R, Yin-Murphy M, Sohier R, Schmidt NJ, Melnick JL.** 1973. Enterovirus type 70: the etiologic agent of pandemic acute haemorrhagic conjunctivitis. *Bull World Health Organ* **49:**341–346.
128. **Mirkovic RR, Schmidt NJ, Yin-Murphy M, Melnick JL.** 1974. Enterovirus etiology of the 1970 Singapore epidemic of acute conjunctivitis. *Intervirology* **4:**119–127.
129. **Zaoutis T, Klein JD.** 1998. Enterovirus infections. *Pediatr Rev* **19:**183–191.
130. **Ikeda RM, Kondracki SF, Drabkin PD, Birkhead GS, Morse DL.** 1993. Pleurodynia among football players at a high school. An outbreak associated with coxsackievirus B1. *JAMA* **270:**2205–2206.
131. **Whiting DA, Smith MB.** 1969. The clinical appearance of hand, foot and mouth disease. *S Afr Med J* **43:**575–577.
132. **Itagaki A, Ishihara J, Mochida K, Ito Y, Saito K, Nishino Y, Koike S, Kurimura T.** 1983. A clustering outbreak of hand, foot, and mouth disease caused by Coxsackie virus A10. *Microbiol Immunol* **27:**929–935.
133. **Österback R, Vuorinen T, Linna M, Susi P, Hyypiä T, Waris M.** 2009. Coxsackievirus A6 and hand, foot, and mouth disease, Finland. *Emerg Infect Dis* **15:**1485–1488.
134. **Fujimoto T, Iizuka S, Enomoto M, Abe K, Yamashita K, Hanaoka N, Okabe N, Yoshida H, Yasui Y, Kobayashi M, Fujii Y, Tanaka H, Yamamoto M, Shimizu H.** 2012. Hand, foot, and mouth disease caused by coxsackievirus A6, Japan, 2011. *Emerg Infect Dis* **18:**337–339.
135. **Mirand A, Henquell C, Archimbaud C, Ughetto S, Antona D, Bailly JL, Peigue-Lafeuille H.** 2012. Outbreak of hand, foot and mouth disease/herpangina associated with coxsackievirus A6 and A10 infections in 2010, France: a large citywide, prospective observational study. *Clin Microbiol Infect* **18:**E110–E118.
136. **Centers for Disease Control and Prevention (CDC).** 2012. Notes from the field: severe hand, foot, and mouth disease associated with coxsackievirus A6 - Alabama, Connecticut, California, and Nevada, November 2011-February 2012. *MMWR Morb Mortal Wkly Rep* **61:**213–214.
137. **Mathes EF, Oza V, Frieden IJ, Cordoro KM, Yagi S, Howard R, Kristal L, Ginocchio CC, Schaffer J, Maguiness S, Bayliss S, Lara-Corrales I, Garcia-Romero MT, Kelly D, Salas M, Oberste MS, Nix WA, Glaser C, Antaya R.** 2013. "Eczema coxsackium" and unusual cutaneous findings in an enterovirus outbreak. *Pediatrics* **132:**e149–e157.
138. **Chung WH, Shih SR, Chang CF, Lin TY, Huang YC, Chang SC, Liu MT, Ko YS, Deng MC, Liau YL, Lin LH, Chen TH, Yang CH, Ho HC, Lin JW, Lu CW, Lu CF, Hung SI.** 2013. Clinicopathologic analysis of coxsackievirus a6 new variant induced widespread mucocutaneous bullous reactions mimicking severe cutaneous adverse reactions. *J Infect Dis* **208:** 1968–1978.
139. **Lott JP, Liu K, Landry ML, Nix WA, Oberste MS, Bolognia J, King B.** 2013. Atypical hand-foot-and-mouth disease associated with coxsackievirus A6 infection. *J Am Acad Dermatol* **69:**736–741.
140. **Miyamoto A, Hirata R, Ishimoto K, Hisatomi M, Wasada R, Akita Y, Ishihara T, Fujimoto T, Eshima N, Hatano Y, Katagiri K, Fujiwara S.** 2014. An outbreak of hand-foot-and-mouth disease mimicking chicken pox, with a frequent association of onychomadesis in Japan in 2009: a new phenotype caused by coxsackievirus A6. *Eur J Dermatol* **24:**103–104.
141. **Ben-Chetrit E, Wiener-Well Y, Shulman LM, Cohen MJ, Elinav H, Sofer D, Feldman I, Marva E, Wolf DG.** 2014. Coxsackievirus A6-related hand foot and mouth disease: skin manifestations in a cluster of adult patients. *J Clin Virol* **59:**201–203.
142. **Downing C, Ramirez-Fort MK, Doan HQ, Benoist F, Oberste MS, Khan F, Tyring SK.** 2014. Coxsackievirus A6 associated hand, foot and mouth disease in adults: clinical presentation and review of the literature. *J Clin Virol* **60:**381–386.
143. **Tunnessen WWJ.** 2004. Erythema infectiosum, roseola, and enteroviral exanthems, p 1220–1224. *In* Gorbach SL, Bartlett JG, Blacklow NR (ed), *Infectious Diseases*, 3rd ed. Lippipcott Williams & Wilkins, Philadelphia, PA.
144. **Neva FA, Enders JF.** 1954. Cytopathogenic agents isolated from patients during an unusual epidemic exanthem. *J Immunol* **72:**307–314.
145. **Modlin JF.** 1996. Update on enterovirus infections in infants and children. *Adv Pediatr Infect Dis* **12:**155–180.
146. **Verboon-Maciolek MA, Krediet TG, Gerards LJ, de Vries LS, Groenendaal F, van Loon AM.** 2008. Severe neonatal parechovirus infection and similarity with enterovirus infection. *Pediatr Infect Dis J* **27:**241–245.
147. **Portes SA, Da Silva EE, Siqueira MM, De Filippis AM, Krawczuk MM, Nascimento JP.** 1998. Enteroviruses isolated from patients with acute respiratory infections during seven years in Rio de Janeiro (1985–1991). *Rev Inst Med Trop Sao Paulo* **40:**337–342.
148. **Chung JY, Han TH, Kim SW, Hwang ES.** 2007. Respiratory picornavirus infections in Korean children with lower respiratory tract infections. *Scand J Infect Dis* **39:**250–254.

149. **Jacques J, Moret H, Minette D, Lévêque N, Jovenin N, Deslée G, Lebargy F, Motte J, Andréoletti L.** 2008. Epidemiological, molecular, and clinical features of enterovirus respiratory infections in French children between 1999 and 2005. *J Clin Microbiol* **46:**206–213.
150. **Oberste MS, Maher K, Schnurr D, Flemister MR, Lovchik JC, Peters H, Sessions W, Kirk C, Chatterjee N, Fuller S, Hanauer JM, Pallansch MA.** 2004. Enterovirus 68 is associated with respiratory illness and shares biological features with both the enteroviruses and the rhinoviruses. *J Gen Virol* **85:** 2577–2584.
151. **Imamura T, Oshitani H.** 2015. Global reemergence of enterovirus D68 as an important pathogen for acute respiratory infections. *Rev Med Virol* **25:**102–114.
152. **Midgley CM, Jackson MA, Selvarangan R, Turabelidze G, Obringer E, Johnson D, Giles BL, Patel A, Echols F, Oberste MS, Nix WA, Watson JT, Gerber SI.** 2014. Severe respiratory illness associated with enterovirus D68 - Missouri and Illinois, 2014. *MMWR Morb Mortal Wkly Rep* **63:**798–799.
153. **Centers for Disease Control and Prevention (CDC).** 2002. Certification of poliomyelitis eradication—European Region, June 2002. *MMWR Morb Mortal Wkly Rep* **51:**572–574.
154. **Centers for Disease Control and Prevention (CDC).** 2001. Apparent global interruption of wild poliovirus type 2 transmission. *MMWR Morb Mortal Wkly Rep* **50:**222–224.
155. **Deshpande J, Ram M, Durrani S, Wenger J.** 2005. Detecting polio through surveillance for acute flaccid paralysis (AFP). *J Indian Med Assoc* **103:**671–675.
156. **Centers for Disease Control and Prevention (CDC).** 2006. Resurgence of wild poliovirus type 1 transmission and consequences of importation—21 countries, 2002–2005. *MMWR Morb Mortal Wkly Rep* **55:**145–150.
157. **Burns CC, Diop OM, Sutter RW, Kew OM.** 2014. Vaccine-derived polioviruses. *J Infect Dis* **210**(Suppl 1)**:**S283–S293.
158. **Kew OM, Sutter RW, de Gourville EM, Dowdle WR, Pallansch MA.** 2005. Vaccine-derived polioviruses and the endgame strategy for global polio eradication. *Annu Rev Microbiol* **59:**587–635.
159. **World Health Organization.** 2013. *Polio Eradication & Endgame Strategic Plan 2013–2018.* WHO Press, Geneva.
160. **World Health Organization.** 2004. *Manual for the Virological Investigation of Polio WHO/IVB/04.10.* 4th ed. WHO Press, Geneva.
161. **Shulman LM, Manor Y, Azar R, Handsher R, Vonsover A, Mendelson E, Rothman S, Hassin D, Halmut T, Abramovitz B, Varsano N.** 1997. Identification of a new strain of fastidious enterovirus 70 as the causative agent of an outbreak of hemorrhagic conjunctivitis. *J Clin Microbiol* **35:**2145–2149.
162. **She RC, Crist G, Billetdeaux E, Langer J, Petti CA.** 2006. Comparison of multiple shell vial cell lines for isolation of enteroviruses: a national perspective. *J Clin Virol* **37:**151–155.
163. **World Health Organization.** 2001. *Manual for the Virological Investigation of Polio WHO/EPI/GEN/97.01.* WHO Press, Geneva.
164. **Pipkin PA, Wood DJ, Racaniello VR, Minor PD.** 1993. Characterisation of L cells expressing the human poliovirus receptor for the specific detection of polioviruses in vitro. *J Virol Methods* **41:**333–340.
165. **Wood DJ, Hull B.** 1999. L20B cells simplify culture of polioviruses from clinical samples. *J Med Virol* **58:**188–192.
166. **Rigonan AS, Mann L, Chonmaitree T.** 1998. Use of monoclonal antibodies to identify serotypes of enterovirus isolates. *J Clin Microbiol* **36:**1877–1881.
167. **Lin TL, Li YS, Huang CW, Hsu CC, Wu HS, Tseng TC, Yang CF.** 2008. Rapid and highly sensitive coxsackievirus a indirect immunofluorescence assay typing kit for enterovirus serotyping. *J Clin Microbiol* **46:**785–788.
168. **Zoll J, Rahamat-Langendoen J, Ahout I, de Jonge MI, Jans J, Huijnen MA, Ferwerda G, Warris A, Melchers WJ.** 2015. Direct multiplexed whole genome sequencing of respiratory tract samples reveals full viral genomic information. *J Clin Virol* **66:**6–11.
169. **Ullmann LS, de Camargo Tozato C, Malossi CD, da Cruz TF, Cavalcante RV, Kurissio JK, Cagnini DQ, Rodrigues MV, Biondo AW, Araujo JP Jr.** 2015. Comparative clinical sample preparation of DNA and RNA viral nucleic acids for a commercial deep sequencing system (Illumina MiSeq(®)). *J Virol Methods* **220:**60–63.
170. **Li L, Deng X, Mee ET, Collot-Teixeira S, Anderson R, Schepelmann S, Minor PD, Delwart E.** 2015. Comparing viral metagenomics methods using a highly multiplexed human viral pathogens reagent. *J Virol Methods* **213:**139–146.
171. **Alfson KJ, Beadles MW, Griffiths A.** 2014. A new approach to determining whole viral genomic sequences including termini using a single deep sequencing run. *J Virol Methods* **208:**1–5.
172. **Romero JR.** 1999. Reverse-transcription polymerase chain reaction detection of the enteroviruses. *Arch Pathol Lab Med* **123:**1161–1169.
173. **Sawyer MH.** 2001. Enterovirus infections: diagnosis and treatment. *Curr Opin Pediatr* **13:**65–69.
174. **Oberste MS, Pallansch MA.** 2005. Enterovirus molecular detection and typing. *Rev Med Microbiol* **16:**163–171.
175. **Popowitch EB, O'Neill SS, Miller MB.** 2013. Comparison of the Biofire FilmArray RP, Genmark eSensor RVP, Luminex xTAG RVPv1, and Luminex xTAG RVP fast multiplex assays for detection of respiratory viruses. *J Clin Microbiol* **51:**1528–1533.
176. **Nix WA, Oberste MS, Pallansch MA.** 2006. Sensitive, seminested PCR amplification of VP1 sequences for direct identification of all enterovirus serotypes from original clinical specimens. *J Clin Microbiol* **44:**2698–2704.
177. **Leitch EC, Harvala H, Robertson I, Ubillos I, Templeton K, Simmonds P.** 2009. Direct identification of human enterovirus serotypes in cerebrospinal fluid by amplification and sequencing of the VP1 region. *J Clin Virol* **44:**119–124.
178. **Oberste MS, Peñaranda S, Rogers SL, Henderson E, Nix WA.** 2010. Comparative evaluation of Taqman real-time PCR and semi-nested VP1 PCR for detection of enteroviruses in clinical specimens. *J Clin Virol* **49:**73–74.
179. **World Health Organization.** 2015. *Enterovirus Surveillance Guidelines.* World Health Organization, Copenhagen.
180. **Wylie TN, Wylie KM, Buller RS, Cannella M, Storch GA.** 2015. Development and evaluation of an enterovirus D68 real-time reverse transcriptase PCR assay. *J Clin Microbiol* **53:**2641–2647.
181. **Zhuge J, Vail E, Bush JL, Singelakis L, Huang W, Nolan SM, Haas JP, Engel H, Della Posta M, Yoon EC, Fallon JT, Wang G.** 2015. Evaluation of a real-time reverse transcription-PCR assay for detection of enterovirus D68 in clinical samples from an outbreak in New York State in 2014. *J Clin Microbiol* **53:**1915–1920.
182. **Centers for Disease Control and Prevention.** Enterovirus **D68:**2014. Real-Time RT-PCR Assay. http://www.fda.gov/downloads/MedicalDevices/Safety/EmergencySituations/UCM446784.pdf (accessed June 30, 2015)
183. **Weber B, Rabenau H, Cinatl J, Maass G, Doerr HW.** 1994. Quantitative detection of neutralizing antibodies against polioviruses and non-polio enteroviruses (NPEV) using an automated microneutralization assay: a seroepidemiologic survey. *Zentralbl Bakteriol* **280:**540–549.
184. **Bell EJ, McCartney RA, Basquill D, Chaudhuri AK.** 1986. Mu-antibody capture ELISA for the rapid diagnosis of enterovirus infections in patients with aseptic meningitis. *J Med Virol* **19:**213–217.
185. **Hodgson J, Bendig J, Keeling P, Booth JC.** 1995. Comparison of two immunoassay procedures for detecting enterovirus IgM. *J Med Virol* **47:**29–34.
186. **Terletskaia-Ladwig E, Metzger C, Schalasta G, Enders G.** 2000. Evaluation of enterovirus serological tests IgM-EIA and complement fixation in patients with meningitis, confirmed by detection of enteroviral RNA by RT-PCR in cerebrospinal fluid. *J Med Virol* **61:**221–227.
187. **Glimåker M, Samuelson A, Magnius L, Ehrnst A, Olcén P, Forsgren M.** 1992. Early diagnosis of enteroviral meningitis by

detection of specific IgM antibodies with a solid-phase reverse immunosorbent test (SPRIST) and mu-capture EIA. *J Med Virol* **36:**193–201.
188. **Goldwater PN.** 1995. Immunoglobulin M capture immunoassay in investigation of coxsackievirus B5 and B6 outbreaks in South Australia. *J Clin Microbiol* **33:**1628–1631.
189. **Day C, Cumming H, Walker J.** 1989. Enterovirus-specific IgM in the diagnosis of meningitis. *J Infect* **19:**219–228.
190. **Wang SY, Lin TL, Chen HY, Lin TS.** 2004. Early and rapid detection of enterovirus 71 infection by an IgM-capture ELISA. *J Virol Methods* **119:**37–43.
191. **Nibbeling R, Reimerink JH, Agboatwala M, Naquib T, Ras A, Poelstra P, van der Avoort HGAM, van Loon AM.** 1994. A poliovirus type-specific IgM antibody-capture enzyme-linked immunosorbent assay for the rapid diagnosis of poliomyelitis. *Clin Diagn Virol* **2:**113–126.
192. **Boman J, Nilsson B, Juto P.** 1992. Serum IgA, IgG, and IgM responses to different enteroviruses as measured by a coxsackie B5-based indirect ELISA. *J Med Virol* **38:**32–35.
193. **Swanink CM, Veenstra L, Poort YA, Kaan JA, Galama JM.** 1993. Coxsackievirus B1-based antibody-capture enzyme-linked immunosorbent assay for detection of immunoglobulin G (IgG), IgM, and IgA with broad specificity for enteroviruses. *J Clin Microbiol* **31:**3240–3246.
194. **Frisk G, Nilsson E, Ehrnst A, Diderholm H.** 1989. Enterovirus IgM detection: specificity of mu-antibody-capture radioimmunoassays using virions and procapsids of Coxsackie B virus. *J Virol Methods* **24:**191–202.
195. **Watanabe K, Oie M, Higuchi M, Nishikawa M, Fujii M.** 2007. Isolation and characterization of novel human parechovirus from clinical samples. *Emerg Infect Dis* **13:**889–895.
196. **Baumgarte S, de Souza Luna LK, Grywna K, Panning M, Drexler JF, Karsten C, Huppertz HI, Drosten C.** 2008. Prevalence, types, and RNA concentrations of human parechoviruses, including a sixth parechovirus type, in stool samples from patients with acute enteritis. *J Clin Microbiol* **46:**242–248.
197. **Li L, Victoria J, Kapoor A, Naeem A, Shaukat S, Sharif S, Alam MM, Angez M, Zaidi SZ, Delwart E.** 2009. Genomic characterization of novel human parechovirus type. *Emerg Infect Dis* **15:**288–291.
198. **Drexler JF, Grywna K, Stöcker A, Almeida PS, Medrado-Ribeiro TC, Eschbach-Bludau M, Petersen N, da Costa-Ribeiro-Jr H, Drosten C.** 2009. Novel human parechovirus from Brazil. *Emerg Infect Dis* **15:**310–313.
199. **Benschop K, Thomas X, Serpenti C, Molenkamp R, Wolthers K.** 2008. High prevalence of human Parechovirus (HPeV) genotypes in the Amsterdam region and identification of specific HPeV variants by direct genotyping of stool samples. *J Clin Microbiol* **46:**3965–3970.
200. **Lim KA, Benyesh-Melnick M.** 1960. Typing of viruses by combinations of antiserum pools. Application to typing of enteroviruses (Coxsackie and ECHO). *J Immunol* **84:**309–317.
201. **Benschop KSM, Schinkel J, Minnaar RP, Pajkrt D, Spanjerberg L, Kraakman HC, Berkhout B, Zaaijer HL, Beld MGHM, Wolthers KC.** 2006. Human parechovirus infections in Dutch children and the association between serotype and disease severity. *Clin Infect Dis* **42:**204–210.
202. **Benschop K, Molenkamp R, van der Ham A, Wolthers K, Beld M.** 2008. Rapid detection of human parechoviruses in clinical samples by real-time PCR. *J Clin Virol* **41:**69–74.
203. **Oberste MS, Maher K, Pallansch MA.** 1999. Specific detection of echoviruses 22 and 23 in cell culture supernatants by RT-PCR. *J Med Virol* **58:**178–181.
204. **Read SJ, Jeffery KJM, Bangham CRM.** 1997. Aseptic meningitis and encephalitis: the role of PCR in the diagnostic laboratory. *J Clin Microbiol* **35:**691–696.
205. **Shimizu C, Rambaud C, Cheron G, Rouzioux C, Lozinski GM, Rao A, Stanway G, Krous HF, Burns JC.** 1995. Molecular identification of viruses in sudden infant death associated with myocarditis and pericarditis. *Pediatr Infect Dis J* **14:**584–588.
206. **Legay V, Chomel JJ, Lina B.** 2002. Specific RT-PCR procedure for the detection of human parechovirus type 1 genome in clinical samples. *J Virol Methods* **102:**157–160.
207. **Corless CE, Guiver M, Borrow R, Edwards-Jones V, Fox AJ, Kaczmarski EB, Mutton KJ.** 2002. Development and evaluation of a "real time" RT-PCR for the detection of enterovirus and parechovirus RNA in CSF and throat swab samples. *J Med Virol* **67:**555–562.
208. **Joki-Korpela P, Hyypiä T.** 1998. Diagnosis and epidemiology of echovirus 22 infections. *Clin Infect Dis* **27:**129–136.
209. **Jokela P, Joki-Korpela P, Maaronen M, Glumoff V, Hyypiä T.** 2005. Detection of human picornaviruses by multiplex reverse transcription-PCR and liquid hybridization. *J Clin Microbiol* **43:**1239–1245.
210. **Noordhoek GT, Weel JFL, Poelstra E, Hooghiemstra M, Brandenburg AH.** 2008. Clinical validation of a new real-time PCR assay for detection of enteroviruses and parechoviruses, and implications for diagnostic procedures. *J Clin Virol* **41:**75–80.
211. **de Vries M, Pyrc K, Berkhout R, Vermeulen-Oost W, Dijkman R, Jebbink MF, Bruisten S, Berkhout B, van der Hoek L.** 2008. Human parechovirus type 1, 3, 4, 5, and 6 detection in picornavirus cultures. *J Clin Microbiol* **46:**759–762.
212. **Johansson S, Niklasson B, Maizel J, Gorbalenya AE, Lindberg AM.** 2002. Molecular analysis of three Ljungan virus isolates reveals a new, close-to-root lineage of the *Picornaviridae* with a cluster of two unrelated 2A proteins. *J Virol* **76:**8920–8930.
213. **Johansson ES, Niklasson B, Tesh RB, Shafren DR, Travassos da Rosa AP, Lindberg AM.** 2003. Molecular characterization of M1146, an American isolate of Ljungan virus (LV) reveals the presence of a new LV genotype. *J Gen Virol* **84:**837–844.
214. **Nix WA, Maher K, Pallansch MA, Oberste MS.** 2010. Parechovirus typing in clinical specimens by nested or semi-nested PCR coupled with sequencing. *J Clin Virol* **48:**202–207.
215. **Harvala H, Robertson I, McWilliam Leitch EC, Benschop K, Wolthers KC, Templeton K, Simmonds P.** 2008. Epidemiology and clinical associations of human parechovirus respiratory infections. *J Clin Microbiol* **46:**3446–3453.

Measles, Mumps, and Rubella Viruses

WILLIAM J. BELLINI, JOSEPH P. ICENOGLE, AND CAROLE J. HICKMAN

21

VIRAL CLASSIFICATION AND BIOLOGY

Measles

Measles virus is a single stranded, nonsegmented, negative sense RNA virus and the prototypic member of the *Morbillivirus* genus of the *Paramyxovirinae* subfamily of the *Paramyxoviridae*. The standard viral genome is 15,894 nucleotides in length and contains six genes and encodes eight proteins which include nucleoprotein (N), phosphoproteins (P) C and V, and matrix (M), fusion (F), hemagglutinin (H), and polymerase (L) proteins. The measles virion is spherical with a diameter ranging from 120 to 250 nm. The virus buds from the plasma membranes of infected cells and has an envelope composed of glycoproteins, the H and F proteins, and lipids. The H and F proteins appear as short surface projections and are responsible for receptor binding and virus entry into susceptible cells (1). Three cell surface receptors for wild-type measles virus have been identified and all interact with the H glycoprotein (2). The M protein is positioned under the virion envelope and anchors the nucleocapsids to the budding sites at the plasma membrane. Unlike the H and F surface proteins, the M is neither glycosylated nor transmembranous. The envelope encloses an elongated helical nucleocapsid in which protein units are spirally arranged around the nucleic acid. The nucleoprotein (N), phosphoprotein, and large polymerase protein, in conjunction with the virion negative strand RNA, comprise the ribonucleoprotein complex, the replicating, and transcriptional unit of measles virus (1, 3). Although measles virus has only a single serotype, it can be subdivided into 24 distinct genotypes based on the sequence variability of the last 450 nucleotides of the N gene. These sequences can vary by up to 12% among the genotypes and form the basis of the molecular epidemiology applied to tracking of transmission pathways, monitoring control measures, and distinguishing wild-type viruses from vaccine strains of measles (4, 5).

Mumps

Mumps virus is a single-stranded, negative sense, enveloped RNA virus in the *Paramyxoviridae* family, *Paramyxovirinae* subfamily, genus Rubulavirus. Mumps virions are pleomorphic but generally spherical structures, and range in size from 85 nm to 300 nm in diameter (1). Filamentous structures have also been observed. The viral genome is 15,384 nucleotides in length and encodes nine proteins from seven genes. The mumps genome is encapsidated by nucleoprotein (NP) and the phosphoprotein (P) and large (L) protein are associated with the encapsidated RNA to comprise the ribonucleoprotein complex. The envelope is a lipid bilayer membrane and contains the two surface glycoproteins, a haemagglutinin-neuraminidase (HN) and fusion (F) hemolysin protein, as well as a matrix (M) and a short hydrophobic (SH) membrane-associated protein. The SH protein may play a role in pathogenesis (6, 7). The gene encoding the SH protein is highly variable and has been used as the basis of genotyping mumps viruses for molecular epidemiological purposes (8). Genotypes show nucleotide variation of 2 to 4% within genotypes, and 6 to 19% between genotypes (9). There is one mumps virus serotype; however, distinct lineages of wild-type viruses cocirculate worldwide. The World Health Organization (WHO) currently recognizes 12 genotypes designated A to N (excluding E and M) which are based on the nucleotide sequences of SH and HN genes (10). The last two proteins, V and I, are nonstructural proteins. The V protein plays a role in interferon signaling and production, while the role of the I protein is not known.

Rubella

Rubella (German measles or 3-day measles) was first described in the 18th century and was accepted as a disease independent of measles and scarlet fever in 1881 (11). Although rubella had been largely ignored for 60 years, in 1941, N. McAlister Gregg first recognized that cataracts in children followed maternal rubella during gestation (12). The association between congenital rubella and a spectrum of significant birth defects including sensorineural hearing loss, cardiovascular abnormalities, cataracts, congenital glaucoma, and meningoencephalitis is now accepted. Rubella virus (RV) is recognized as the most potent infectious teratogenic agent yet identified (13).

RV is a 60- to 70-nm-diameter particle consisting of a core particle surrounded by a lipid envelope. The envelope contains two viral glycoproteins, E1 and E2, and the core contains the viral capsid protein (C) and an infectious single-stranded RNA genome of about 9,762 nucleotides (14). Rubella viral particles and some individual proteins have been studied by high resolution crystallographic and cryo-electron tomographic methods (15–17). The 5′ two-thirds of the genome codes for a polyprotein which is cleaved into nonstructural proteins necessary for virus replication.

doi:10.1128/9781555819156.ch21

Structural proteins (E1, E2, and C) are cleaved from a second polyprotein translated from a subgenomic mRNA produced during viral replication. The subgenomic RNA has the nucleotide sequence of the 3′ one-third of the genome. New virions are produced when genomic RNA, the E1 glycoprotein, the E2 glycoprotein, and the C protein assemble at cellular membranes.

A number of small, enveloped viruses having the same overall genetic organization and replication strategy as RV exist, and they are grouped into the family *Togaviridae*. The family consists of the *Rubivirus* genus, containing only RV, and the *Alphavirus* genus, containing about 25 other viruses, all of which are transmitted by arthropods (e.g., Chikungunya virus). RV has a restricted host range and humans appear to be the only species in which rubella circulates.

RVs currently circulating in the world contain RNAs that differ sufficiently that two clades of RVs, differing by about 10% in the nucleotide sequence, have so far been identified (14). Groups of related viruses within the clades have been classified as genotypes. At present, 12 genotypes and one provisional genotype of RVs have been recognized (18). Only minor immunologic differences exist among circulating viruses. Genotypes have been subdivided into different lineages, but there is no consensus on the criteria for lineages and a naming convention is lacking (19, 20).

EPIDEMIOLOGY

Measles

Man is the only known natural host for measles (rubeola) virus. Enders and Peebles first reported the successful isolation of measles virus in human and rhesus monkey kidney tissue cultures in 1954 (21). At that time there were more than 400,000 cases of measles reported each year in the United States. However, since virtually all children would acquire measles, the true number probably exceeded three million per year. In 1963, both an inactivated and a live attenuated vaccine (22) were licensed for use in the United States. The killed vaccine eventually proved less effective and children who received this material were at risk of developing an atypical severe form of the disease when subsequently exposed to live measles virus. In 1967, the inactivated vaccine was discontinued. After a further attenuated variant of the Edmonston B vaccine was introduced in 1968, the reported cases of measles took a dramatic downward turn. In 1960, the cumulative total number of cases was 399,852 from week 1 to 35. In 1970, the total was 39,365. The number of cases over the next decade decreased more than 10-fold and in 1983, there were 1,194 cases from week 1 to 35 (23). There was hope that 1983 would be the year in which measles would be eliminated from the United States but this goal was not accomplished. In fact, reported cases increased every year until 1986 when there were 6,282 cases. A major resurgence of measles (>50,000 reported cases and over 100 deaths) occurred between 1989 and 1991. This led to the current two-dose measles-mumps-rubella (MMR) schedule currently in place, i.e., the first dose given at 12 to 15 months of age and the second between 4 and 6 years (24). Epidemiologic and laboratory data suggested that the transmission of indigenous measles was interrupted for a 6-week period in the United States in 1993 (25, 26). That same year, the Childhood Immunization Initiative called for the elimination from the United States by 1996 of indigenous transmission of measles, rubella, congenital rubella syndrome, and three other childhood diseases. However, it was not until the year 2000 that measles was officially declared eliminated from the United States (27). During 2009 to 2014, 1,264 confirmed measles cases were reported in the United States and the annual median for this period was 130 case-patients (range, 55 to 667), compared with earlier postelimination years between 2001 and 2008, when the annual median number of cases reported was 56 cases-patients (range, 37 to 140). During 2009 to 2014, 865 (74%) of 1,173 U.S.-resident case-patients were unvaccinated and 188 (16%) had unknown vaccine status.

The maintenance of elimination of measles (and of rubella and congenital rubella syndrome [CRS]) through 2011 was certified by a panel of experts convened by the Centers for Disease Control and Prevention (CDC) at the request of the Pan American Health Organization (28). Imported cases of measles from endemic areas continue to be a cause for concern, particularly when such cases come in contact with unvaccinated groups (29, 30). Vaccination coverage must be maintained at a very high level (>95%) to sustain the elimination status of the United States. Measles remains endemic in most of the world, but the use of vaccination strategies that include a second dose of measles-containing vaccine have resulted in a drastic reduction in measles mortality (31). Even though measles is no longer endemic in the United States, the threat of reintroduction from endemic areas requires vaccination coverage at or near 95%, and disease surveillance capable of detecting imported measles cases that have the potential of initiating outbreaks in susceptible groups.

Mumps

Mumps occurs worldwide and humans are the only natural host of mumps virus infection, although nonhuman primates and rodents can be experimentally infected. Recently, a bat virus conspecific with the human mumps virus was identified as a potential reservoir (32). This virus was found to have a direct antigenic relatedness as well as a close genetic proximity to the human mumps virus. In addition, the surface glycoproteins of the bat mumps virus have recently been shown to be functionally related to those of human mumps viruses (33). It is not yet known if this virus is transmissible from bats to humans.

Although other viruses such as Epstein-Barr virus, human parainfluenza virus, adenovirus, enterovirus, human herpesvirus type 6, influenza A, and parvovirus can sometimes cause parotitis, mumps virus is the only known cause of epidemic parotitis. Mumps is transmitted by contact with infectious respiratory tract secretions or saliva either by direct person-to-person transfer, air-suspended droplets, or recently contaminated fomites. In comparison to measles and varicella, which can be transmitted by aerosol spread, mumps is less infectious. Mumps infection can be either clinically apparent or subclinical. The average incubation period is 16 to 20 days, but cases may occur from 12 to 25 days after exposure. The period of communicability can be from 7 days before the salivary gland involvement until 9 days thereafter. The virus is also excreted in the urine for as long as 14 days after onset of illness. Mumps infections in previously vaccinated persons likely result in decreased levels of virus shedding into the buccal cavity. This, and compliance issues with the current 9-day isolation period, has led to considerations by CDC and other public health agencies to reduce the number of days of patient isolation from 9 days to 5 days postsymptom onset (34). In healthcare settings, both standard and droplet precautions should be observed during the isolation period.

Mumps vaccine was licensed in the United States in 1967 and recommended for routine childhood immunization in 1977. Following implementation of a 2-dose MMR vaccine recommendation in 1989 for measles control, cases of mumps declined to extremely low levels, with an incidence of 0.1/100,000 by 1999. From 2000 to 2005, seasonality no longer was evident, and there were fewer than 300 reported cases per year representing a greater than 99% reduction in disease incidence since the prevaccine era. However, in 2006, a sizable mumps outbreak occurred in the Midwestern United States, with 6,584 reported cases (incidence of 2.2/100,000) (35). Most of the cases occurred among 18 to 24 year olds and most had previously received two doses of MMR vaccine. This outbreak was not merely a chance occurrence as it was followed by an outbreak from 2009 to 2010 in the northeastern United States and cases again were primarily seen among individuals that received two doses of MMR vaccine (3,502 cases). From 2011to 2015 several smaller scale outbreaks have occurred and case counts for these years range from 222 to 1,151 per year. Most cases occurred in close contact environments such as schools, camps, colleges, or were associated with sports teams. Mumps outbreaks among two-dose vaccine recipients are not unique to the United States or to populations vaccinated with the Jeryl Lynn vaccine strain (the only licensed mumps vaccine in the United States). In fact, mumps outbreaks among vaccinated populations have occurred in numerous countries, some of which utilize different strains of vaccine virus (36, 37).

In the prevaccine era, the peak incidence of mumps was between January and May and among children younger than 10 years of age. At present, mumps incidence peaks predominantly in late winter and spring, but the disease has been reported throughout the year. Vaccine coverage of 90% to 92% has been estimated to be necessary to achieve mumps elimination (38). Seroepidemiologic studies measuring mumps IgG in the United States using enzyme immunoassays indicate that 80% to 90% of adults have evidence of prior exposure to mumps, either through vaccination or previous mumps infection (39). Moreover, this value is likely an overestimate of those protected from disease, since mumps seroprevalence does not necessarily reflect mumps immunity and there is no antibody level that has been defined as a surrogate measure for immune protection from mumps (40). There is poor correlation between mumps neutralization titers and enzyme immunoassays which are frequently used to assess seroprevalence. Enzyme immunoassays predominantly detect nucleoprotein-specific antibody which does not correlate well with neutralization (41).

Rubella

RV was not isolated until 1962, largely because infected cells are difficult to identify in tissue culture (42, 43). After isolation of the virus, there was rapid progress in the development of a vaccine. Introduction of rubella vaccine in the United States (licensed in 1969) mostly through childhood immunization, immediately broke the 6- to 9-year epidemic cycle of rubella. Rubella incidence during epidemics is often 10-fold higher than in interepidemic periods (44). The last major U.S. rubella epidemic was in 1964 to 1965, when approximately 20,000 CRS cases occurred (45). A combined MMR vaccine was recommended for the United States in 1972. Rubella and CRS have been eliminated in the United States (28, 46). In the United States, most mothers of CRS children were born in countries without rubella immunization programs or with recently organized programs. Occasional outbreaks of rubella in some U.S. populations have not spread to under vaccinated populations, suggesting that herd immunity has been protective (47).

Postnatal transmission of RV is often by close contact with an infected individual, such as occurs in correctional institutions or day-care centers. Communicability is greatest between about day 14 and day 19 postexposure. Reinfection with RV can occur, but viremia is rare; reinfection of a pregnant woman poses low risk to the fetus (48). Indigenous rubella and CRS have also been eliminated from the Americas; rubella is targeted for elimination in Europe by 2015 and that region will likely continue to work on rubella control and an elimination goal (49). Rubella is poorly controlled in most of the world. The percentage of seronegative women of childbearing age varies depending on the vaccination program and on previous epidemics, and is sometimes above 25% (50). During rubella outbreaks in unvaccinated populations, lack of immunity in women of childbearing age typically results in one to two CRS cases per 1,000 births. About 100,000 to 200,000 CRS cases occur annually in the world. Nevertheless, based on successful control programs for rubella, rubella is targeted for elimination in five of six WHO regions by 2020 under the Global Vaccine Action Plan (49).

CLINICAL SIGNIFICANCE

Measles

Measles is spread through direct contact with infected droplets originating from a cough or sneeze. It is a highly contagious, acute biphasic disease with a prominent prodrome preceding the exanthemic phase. Susceptible persons intimately exposed to a measles patient have a 99% chance of acquiring the disease. Prior to the use of vaccines, more than 90% of the population had measles before 10 years of age. After an incubation period of 9 to 11 days there is an initial 3- to 4-day prodromal period characterized by fever, cough, coryza, and conjunctivitis. The incubation period in adults may last up to 3 weeks. A fever occurs 24 hours or less before other symptoms appear and these increase in severity, reaching a peak with the appearance of the rash on the fourth to fifth day.

Bluish-white lesions with a red halo, Koplik spots, appear on the buccal or labial mucosa in 50% to 90% of the cases, 2 to 3 days after the onset of the prodrome. These lesions are small, irregular red spots with bluish-white specks in the center and are located on the inner lip or opposite the lower molars and are pathognomonic for measles (51). They may be few in number early in the prodrome; however, they increase rapidly to spread over the entire surface of the mucous membranes. A lesion somewhat similar in appearance to Koplik spots has been reported with Coxsackie A9, A16, and A23 virus infections. The measles rash is first evident behind the ears or on the forehead. The lesions are red macules, 1 to 2 mm in diameter, which become maculopapules over the next 3 days. By the end of the second day, the trunk and upper extremities are covered with rash and by the third day the lower extremities are affected. The rash resolves in the same sequence, lasting approximately 6 days. The lesions turn brown and persist for 7 to 10 days and then are followed by a fine desquamation (52).

The most frequent complication of measles involves infections of the lower respiratory tract. Croup, bronchitis, bronchiolitis, and, rarely, giant cell interstitial pneumonia may occur. Otitis media is a common bacterial complication of measles. Prior to the advent of antibiotics, these complications contributed to a high number of fatalities and significant morbidity. Excluding pneumonia and otitis media, the most frequent serious complication of measles is postinfectious encephalitis (PIE). PIE occurs in 0.1% to 0.2% of measles patients during any stage of the illness, although it is most common 2 to 7 days after the onset of the exanthem. Death occurs in one to two out of every 1,000 reported cases in the United States. Other complications include thrombocytopenic purpura, appendicitis, myocarditis, and mesenteric lymphadenitis (53).

Subacute sclerosing panencephalitis (SSPE), also called Dawson's encephalitis, is a persistent measles infection of the central nervous system. SSPE is a progressive, invariably fatal, encephalopathy characterized by personality changes, mental deterioration, involuntary movements, muscular rigidity, and death. It usually begins 4 to 10 years after the patient has recovered from naturally acquired measles. Successful isolation of measles virus from brain and lymphoid tissues of SSPE patients (54, 55) clearly established the etiologic agent involved. In the prevaccine era, the incidence of SSPE was approximately 1:100,000 to 1,000,000 measles cases, although recent studies suggest that the incidence may have been about 19-fold greater (56). The introduction of live attenuated measles vaccine raised concerns that the vaccine virus might cause SSPE, but epidemiologic studies demonstrated a dramatic decrease in the frequency of this disease (57). Recent studies have clearly demonstrated that measles vaccine virus is not involved in the genesis of SSPE and that the use of measles vaccine not only is beneficial in preventing acute measles, but has all but eliminated SSPE from the United States (56, 58).

Transplacental infections have been associated with some fetal effects. There is an apparent increased frequency of premature labor, low birth weight, abortions, and stillbirths (59). Available evidence does not support an association between measles in pregnancy and congenital defects (60).

Atypical measles occurred in some children previously vaccinated with killed measles virus vaccines when they became infected with wild-type measles (61). Fever, a prodromal period, and subsequent rash characterized the disease. During the prodrome, some patients experienced malaise, myalgia, headache, nausea, and vomiting. Symptoms usually lasted for 2 to 3 days and frequently individuals had a sore throat, conjunctivitis, and photophobia along with nonproductive cough and pneumonia. Chest X-rays often showed patchy infiltrates. The rash produced was different from that of typical measles. It could be a mixture of macules, papules, vesicles, and pustules. Frequently, there was a petechial component, which began at the distal extremities and concentrated on the hands, wrists, ankles, and feet and then progressed centrally toward the trunk. Koplik spots were not reported and the face was rarely involved. Edema often occurred in the extremities. The appearance of atypical measles could be confused with Rocky Mountain spotted fever.

Among immunocompromised patients, measles can be severe and prolonged. This is particularly a risk for patients with certain leukemias, lymphomas, or human immunodeficiency virus infection with severely depressed CD4 counts (62).

Mumps

Mumps is an acute viral disease characterized by fever, swelling, and tenderness of one or more salivary glands, usually the parotid and sometimes the sublingual or submaxillary glands. Prodromal symptoms are nonspecific, consisting of myalgia, anorexia, malaise, headache, low-grade fever, and vomiting and may precede parotitis by several days. Parotitis reaches a maximum after 48 hours and typically lasts 7 to 10 days. Approximately 30% to 40% of mumps infections produce parotitis, 20% to 30% of infections are asymptomatic, and 50% are associated with nonspecific or respiratory symptoms. Parotitis may be initially unilateral, but becomes bilateral in about 65% of cases (63). The mumps virus replicates in the nasopharynx and regional lymph nodes, with a secondary viremia occurring late in the incubation period. The average incubation period is 16 to 20 days, but cases may occur from 12 to 25 days after exposure. During the viremic period, the virus spreads into the major target organs. Although the salivary glands are most commonly affected, the central nervous system, pancreas, liver, spleen, kidneys, and genital organs can also be involved. Complications of mumps may vary with sex and age, and can occur without parotitis. Severe complications, including deaths, are uncommon (64). In males, orchitis is the most common complication, occurring in approximately 20% to 30% of affected postpubertal males (65); orchitis is rare in prepubescent males. Orchitis usually occurs 1 to 2 weeks after parotitis and is most commonly unilateral with only 10% to 30% of cases presenting bilaterally (65, 66). Impairment of fertility is estimated to occur in about 13% of patients, although a greater percentage of patients with bilateral mumps orchitis may experience decreased fertility (80). Testicular atrophy is more common than impaired fertility with 30% to 50% of affected testicles showing some amount of testicular atrophy (67).

Previous vaccination appears to limit the severity of cases; complication rates among vaccinated persons are lower than rates reported during the prevaccine era (68–70). In a recent study, orchitis was reported in 11.3% of unvaccinated cases compared to only 4.4% of two-dose vaccinated cases during a U.S. outbreak in an Orthodox Jewish community (69). Similar results were seen during an outbreak in Israel, where the main complications included orchitis (3.8% of males of age >12 years) and meningoencephalitis (0.5%) (71). In postpubertal women, mastitis occurs in up to 30% and oophoritis occurs in approximately 5% of cases (63, 72). Mumps can cause sensorineural hearing loss (SNHL) in both children and adults. Hearing loss occurs suddenly 4 to 5 days after prodromal symptoms and is not related to the severity of the infection (73). Generally hearing loss is unilateral and reversible; however, serious and permanent hearing loss can occur (74). Aseptic meningitis and encephalitis can also occur, and their presence increases the risk of SNHL (75). Complications involving the CNS are common. Symptomatic aseptic meningitis occurs in up to 10% of mumps cases; patients usually recover without complications, though many require hospitalization. Mumps encephalitis is rare (1 to 2 per 10,000 cases), but can result in permanent sequelae such as paralysis, seizures, and hydrocephalus; the case-fatality rate for mumps encephalitis is about 1%. Mumps infection during the first trimester of pregnancy has been associated with spontaneous abortion, but there is no evidence that mumps during pregnancy causes congenital malformations. Pancreatitis, usually mild, occurs in 4% of cases. Less common complications include myocarditis, arthralgias, arthritis, thyroiditis, nephritis, endocardial fibroelastosis, thrombocytopenia, cerebellar ataxia, transverse myelitis, and ascending polyradiculitis. There have been reports of diabetes mellitus being associated with mumps (76), although this remains inconclusive. A recent retrospective cohort study among active component U.S.

military personnel did not find an increased risk of diagnosed type 1 diabetes and MMR (77).

Rubella

Postnatal rubella is usually a mild disease requiring little treatment. The disease is characterized by an acute onset and generalized maculopapular rash with mild fever (greater than 99°F) and may include arthritis or arthralgia (mostly in postpubertal females), lymphadenopathy (specifically postauricular and suboccipital nodes), and conjunctivitis. These symptoms can also be caused by other infections. The virus is found in throat, oral, and nasopharyngeal specimens and, with more difficulty, from blood and urine.

Symptoms of postnatal rubella usually begin at about day 14, but the disease may be asymptomatic. Joint involvement may be very painful and may, in rare cases, last a month. The virus has been reported to replicate in synovial membrane cell culture (78). Rare complications of postnatal rubella include postrubella encephalitis (about 1 in 6,000 cases). There are case reports of long-term RV replication in some diseases (79).

The clinical course of rubella in pregnant women is similar to that in nonpregnant women. RV infection, even subclinical infection, can be transmitted to the fetus. The likelihood of defects is highest during the first 11 weeks of gestation. Defects can be limited (e.g., sensorineural hearing loss) or profound (e.g., deafness-blindness). During maternal viremia, about days 7 through 16 postexposure, the placenta may be infected, particularly endothelial cells, and the resulting damage likely allows the virus to cross the placenta (80, 81). With infection in the first trimester, many fetal organs (heart, brain, and lens cells) can be infected. RV spread in the fetus likely occurs via the vascular system. Spontaneous abortion occurs in about 20% of infections acquired in the first 8 weeks of pregnancy. Neonates with CRS usually have RV-specific IgM but remain virus positive for months. Maternal rubella after the first trimester often leads to infection of fetal tissue, but fetal damage is often limited, presumably because organogenesis is complete, maternal IgG is transferred to the fetus, and a fetal immune response is present.

Late manifestations of CRS have been recognized. One disability is insulin-dependent diabetes mellitus. In a follow-up study, 40% of CRS patients from the 1964 rubella epidemic had developed evidence of overt or latent diabetes (82). Other endocrine disorders have been seen in small numbers of survivors of CRS (83).

Ocular consequences of CRS are observed during and after the neonatal period (84). CRS has recently been associated with Fuchs heterochromic iridocyclitis (85). A rare disability associated with CRS is progressive rubella panencephalitis (PRP) (86). Progressive deterioration of intellectual and motor function occurs with dementia close to the time of death. There is an intense immune response against rubella antigens, and high titers of rubella antibody are present in both serum and cerebrospinal fluid (CSF). No correlation has been made between the occurrence of PRP and the presence of rubella-associated defects or the severity of neonatal infection.

The pathology produced with congenital rubella in the fetus appears to result from a chronic viral infection with alterations of cell multiplication at critical points in organogenesis. This causes the hypoplastic organ development and other characteristic structural defects seen with this disease. The immune response may also contribute to permanent damage in the developing child either by an impaired immunity or by inflicting damage through inflammatory mechanisms (87).

Congenital rubella infection (CRI) results in both shedding of virus and IgM antibodies in the neonate. Diagnosis is not based on clinical symptoms and is based on detection of RV (RT-PCR or virus culture) or RV-specific IgM in such patients. If congenital defects characteristic of CRS are not present, the infant is diagnosed as having infection only (CRI) (88).

TREATMENT AND PREVENTION

Measles

Individuals having an illness compatible with measles should be isolated in order to limit contact with other potentially susceptible individuals or immunocompromised patients. The communicability of measles virus is extremely high. Therefore, any susceptible individuals who had direct face-to-face contact with the infectious individual should obtain prophylactic treatment. Measles vaccination may provide protection if given within 72 hours of exposure (89). The Advisory Committee on Immunization Practices (ACIP) supports readmission to school of all previously unimmunized children immediately following vaccination (90). Immune globulin, given within 6 days of exposure, can prevent or modify measles virus infection. It is indicated for susceptible, close contacts of measles patients at highest risk for severe complications, e.g., infants aged <12 months, pregnant women without evidence of measles immunity, and severely immunocompromised persons. If immune globulin is used, measles vaccine should be given 5 or 6 months later provided that the child is at least 12 months old. In the United States, children with severe measles, such as those who are hospitalized, should receive vitamin A. In other settings, treatment guidelines recommend vitamin A for all measles cases, particularly where malnourishment and know vitamin A deficiency is commonplace. The reader is referred to the most recent ACIP recommendations (91) regarding postexposure prophylaxis, since full discussion of such treatment is beyond the scope of this chapter.

The most effective and efficient means of preventing measles is immunization with MMR vaccine. Currently the ACIP recommends two doses of measles-containing vaccine be administered to children; the first dose given at 12 through 15 months and the second dose at 4 through 6 years of age before school entry. Two doses are recommended for persons at risk for exposure, such as college students or other posthigh school educational institutions, international travelers, and particularly healthcare personnel, and a single dose for other adults aged ≥ 18 years (91).

Mumps

Mumps has no specific treatment although symptoms can be relieved by using medicines such as analgesics to reduce pain and discomfort. Immune globulin (IG) is not effective postexposure prophylaxis for mumps and it is no longer available in the United States. Postexposure MMR vaccination does not prevent or alter the clinical severity of mumps and is not recommended (91).

Immunization with mumps-containing vaccine is the best way to prevent mumps infection. Mumps vaccine is included in the combination MMR vaccine. In the United States, the mumps vaccine was introduced in 1967 and recommended for routine use in 1977. Following implementation of the one-dose vaccine requirement mumps cases declined from an incidence of 50 to 251 per 100,000 to 2 per 100,000 in 1988. Despite high single-dose MMR coverage, sporadic outbreaks have occurred among highly vaccinated populations (92, 93). In 1991, a mumps outbreak was sustained in a population where 98% of individuals had been vaccinated with at least one-dose of MMR (92). Between December 1997 and May 1998, a mumps outbreak occurred in New York City. Among the 111 cases with known vaccination history, 92% had received at least one-dose of mumps-containing vaccine, and 62% had received two or more doses (94). In 1989 a second dose of MMR was recommended for purposes of measles control and mumps cases further decreased to 0.1 case per 100,000 in 2001 (95). Beginning in 2000 to 2005, seasonality was no longer apparent and there were less than 300 reported cases per year (incidence of 0.1/100,000) representing a 99% reduction in disease incidence since the prevaccine era.

The success of high two-dose MMR coverage led U.S. public health officials in 1998 to set a goal to eliminate endemic mumps transmission by 2010 in the United States. Unexpectedly, in 2006, the United States experienced the largest mumps outbreak in almost 20 years (6,500 cases), largely affecting Midwestern states. This outbreak was not a chance occurrence and was followed by a large outbreak (3,502 cases) among residents of Orthodox Jewish communities in 2009 to 2010 in New York City, New York, and New Jersey. A smaller outbreak occurred during this period in the U.S. territory of Guam (505 cases reported). In each of these outbreak settings, the majority of affected individuals had received two doses of MMR vaccine, reaffirming questions regarding the long-term effectiveness of the currently used mumps vaccines (40, 96). The U.S. *Healthy People 2020* goal for mumps has been changed to a disease reduction goal (i.e., to have fewer than 500 reported cases of mumps annually), rather than an elimination goal.

Mumps outbreaks among two-dose vaccine recipients are not limited to the United States or to use of the Jeryl Lynn vaccine strain (the only licensed mumps vaccine in the United States). In fact, mumps outbreaks among vaccinated populations have happened in numerous countries, some of which utilize different strains of vaccine virus (36, 37, 97). Mumps-specific antibody levels are known to decline with time and decreased vaccine effectiveness has been observed with increasing time postvaccination (98–101) suggesting that waning immunity is likely a contributing factor to these outbreaks. A study in Finland found that 21 years after vaccination, 24% of two-dose vaccine recipients had no measurable mumps antibodies (99). A cellular response, measured by lymphoproliferation or cytokine production, was detected but the role of cell-mediated immunity in protection is not clear. The immune response to mumps vaccination probably involves both the humoral and cellular immune response, but no definitive correlates of protection have been identified. An attempt to identify a correlate of immunity during an outbreak in the Midwest was not conclusive but did demonstrate that those who did not develop mumps had higher preexposure neutralization and enzyme immunoassay (EIA) titers than those who developed the disease (102).

A popular, yet controversial, explanation is that mumps vaccine elicits only a partially protective immune response against currently circulating wild-type mumps strains (103, 104). It has, however, been repeatedly shown that the most divergent strains can be neutralized *in vitro* with only slight variations in serum titers (104, 105). Interestingly, studies indicate that the frequency of mumps-specific memory B lymphocytes is very low compared to measles and rubella and that the predominant antibody response appears to be directed to the nucleoprotein, a nonneutralizing target (41). It is also significant that persons with documented mumps infections can be reinfected with mumps virus later in life (106, 107). This detail underscores the point that immunity to mumps is not well understood and breakthrough mumps infections following vaccination might reasonably be expected. Regardless of the reason(s) for such outbreaks, they indicate that more efficacious vaccines are likely required for control and elimination of mumps (68, 96, 108). Development of such vaccines will first require a better understanding of the immune response to mumps and the identification of a measurable correlate of immunity, if one can be defined.

Rubella

The means to control rubella and CRS is immunization. Safe, effective attenuated live-virus vaccines have been developed by serial passage in tissue culture. The virus used to produce the RA27/3 vaccine, which is used in the United States and many other countries, often as a component of the MMR vaccine, was passaged in cell culture between 25 and 33 times in WI-38 cells (109). Only minor immunologic differences exist among circulating viruses of all genotypes and immunity to RV of one genotype is sufficient to protect against clinical disease caused by other known RVs.

The RA27/3 vaccine produces a more wild-type immune response than a previously used vaccine, HPV77DE5, and induces 95% seroconversions of susceptible individuals. It produces the symptoms of mild rubella in 10% to 15% of vaccinees, occasionally even producing a rash. Although virus is shed from the upper respiratory tract, transmission of vaccine virus is very rare. Joint symptoms occur in about 14% of postpubertal women, and rarely, these develop into arthritis. Although pregnancy is a contraindication for vaccination, hundreds of seronegative women have been vaccinated during the first trimester and no fetal defects have been observed (110). Nevertheless, the ACIP recommends avoiding pregnancy for 28 days after receipt of rubella-containing vaccine (91).

Control of rubella through vaccination has resulted in elimination of rubella and CRS from the United States and the Americas. In the United States, the first dose of rubella vaccine should be given at 12 to 15 months of age, combined with MMR; the second dose of the MMR vaccine should be given at 4 to 6 years of age. Prenatal testing for immunity to rubella and counseling of seronegative pregnant women to avoid contact with rubella cases are an important part of CRS prevention. Pregnant women who do not have serologic evidence of rubella immunity or documented rubella vaccination should be vaccinated with rubella or MMR vaccine after completion or end of the pregnancy.

DETECTION/DIAGNOSIS

Measles

The clinical diagnosis of a case of measles can be made based on observation of rash, fever, and either cough, coryza, or conjunctivitis. Rapid and accurate laboratory confirmation is extremely important, however, because of the highly infectious nature of the virus and possible confusion with other rash-causing illnesses. The incidence of measles has decreased in many areas to the extent that medical personnel may be unfamiliar with the clinical presentation of the disease. The collection of specimens for both serological and molecular confirmation of clinically diagnosed measles is recommended at the time the patient is first seen in the healthcare facility. Rapid laboratory confirmation of clinically suspicious cases can be achieved using serum tested in a measles IgM assay, and/or respiratory or, less frequently, urine specimens in real time RT-PCR assays. Isolation of measles virus in cell culture from respiratory or urine specimens, seroconversion from measles IgG negative to positive, or a fourfold rise in measles IgG using a quantitative assay can also confirm a measles case. The latter methods are by no means rapid, and often exceed 7 days for the end result to be obtained (111).

Serum specimens are best collected 4 to 5 days postrash onset to assure optimal detection of measles IgM antibodies in enzyme immunoassays, but the timing varies with assay sensitivity (112, 113). Throat and nasopharyngeal (NP) swabs (preferably flocked Dacron or synthetic swabs) and nasopharyngeal aspirates should be collected as early as possible postrash onset to optimize RNA detection. In one study, measles RNA was successfully detected (93% to 100%) from nucleic acid extracts of throat swabs, NP swabs, and NP aspirates using RT-PCR if collected between 0 and 7 days after rash onset. Optimal detection in urine specimens was determined to be 4 to 7 days after rash onset, with detection of RNA in some specimens out to 16 days postrash onset (113).

Throat and NP swabs should be transferred to 1 to 3 ml of viral transport medium (VTM). Swabs should not be permitted to dry out before transfer to VTM. Urine specimens (10 to 50 ml) should be collected and centrifuged (800 × g) for 10 minutes at 4°C. The pelleted material containing sloughed cells from the urinary tract should be suspended in 2 to 3 ml of VTM or suitable fluid such as tissue culture medium or buffered saline with a protein source. All specimens should be kept at 4°C and shipped to the testing laboratory on wet ice within 24 hours for arrival within 48 hours. If longer time periods are required, all specimens should be frozen at −70°C or lower and shipped on dry ice to maintain specimen integrity.

The preferred laboratory confirmatory test is the detection of measles-specific IgM antibody in a single serum sample obtained during the acute phase of the disease (112, 114). For the detection of measles-specific IgM, commercial kits based on both indirect and IgM capture formats have been used (115–118). Though IgM capture is generally regarded as the more sensitive format, some of the commercial indirect IgM EIA kits have sensitivities and specificities that approach those of the capture format (119, 120). The availability of excellent immunoabsorbants capable of removing IgG from diluted serum specimens is largely responsible for the success of the indirect IgM methods. In elimination settings such as the United States, the performance characteristics of the available EIAs and other IgM tests are frequently not sufficient for measles confirmation. The false-positive rate is a compendium of factors including disease prevalence, assay specificity, cross-reacting agents and interfering substances (121–123).

In unvaccinated individuals, measles-specific IgG can appear a few days after IgM, but generally between 5 to 10 days postrash onset. Seroconversion from IgG negative to IgG positive or demonstration of a fourfold rise in IgG when tested in a quantitative IgG assay provides laboratory confirmation of measles, but requires the collection of paired serum specimens that are a few days or weeks apart, respectively (124). Traditional IgG antibody tests such as immunofluorescence antibody assay (IFA), hemagglutination inhibition (HI), plaque-reduction neutralization (PRN) (125), and EIA have been used extensively in the serologic confirmation of measles. However, because of the availability of sensitive and specific commercial kits, indirect EIAs have become the most widely used test format. Some of the available measles IgG kits are found to have sensitivities and specificities that compare favorably with the "gold standard" quantitative plaque-reduction neutralization (126, 127).

In highly vaccinated populations, it is advisable to simultaneously test the serum specimens collected from suspected cases for the presence of measles IgM and IgG. In this setting, the presence of both IgM and IgG in early specimens may have several interpretations depending upon when the serum was obtained postrash onset and the patient's vaccine history. Measles IgG avidity assays have been used to distinguish a primary IgG response (low avidity), that occurs days or weeks following vaccination or natural infection, from a mature IgG (high avidity) response that occurs months or years following vaccination or infection (128–131). Determinations of IgG avidity in some infections such as rubella (see below) are capable of influencing the interpretation of a positive rubella IgM test in individuals with a resident rubella IgG. Such suspect rubella cases have been demonstrated as IgM false positives, and can be ruled out as rubella cases on the basis of the high avidity rubella IgG. This is not the case with measles, since measles infections do occur in previously vaccinated persons. Low avidity measles IgG can rule in a measles case, particularly in the event that the IgM response is indeterminate or otherwise not interpretable.

Although the majority of measles cases in the United States are unvaccinated or have an unknown vaccination history, 8% to 12% of measles cases have been classified as reinfections, i.e., have a history of wild-type infection or vaccination with live-attenuated measles vaccine and subsequently were infected with wild-type measles. Measles infections years after receiving vaccine have been attributed to primary vaccine failure due to insufficient primary antigenic stimulation, or secondary vaccine failure due to a putative loss of protective antibody or waning (antibody) immunity, following a competent immune response to measles vaccine or wild-type infection. Measles IgG avidity measurements were found to distinguish between primary and secondary vaccine failure cases. Primary vaccine failure cases responded to infection, as would naïve individuals with an acute case of measles, i.e., making IgM and low avidity IgG. In contrast, persons known to have responded to vaccine in years past frequently made low levels of IgM and produced high avidity measles IgG in response to reinfection (132, 133). Recent studies have demonstrated that the high avidity measles IgG found associated with measles reinfections appears to occur a few days after rash onset. Measles-specific neutralizing titers can exceed 100,000 mIU, and strongly suggests that this response is a heightened B cell memory response triggered by measles infection (134, 135).

These suspect measles cases often presented with modified symptoms, such as reduced fever and modified rash illness in terms of appearance and duration and are reminiscent of cases referred to as "secondary vaccine failure" cases. While measles transmission from reinfection cases has been documented (136), confirmed measles reinfections in healthcare workers with numerous pediatric contacts resulted in no additional cases indicating that these modified infections pose a low transmission risk and little threat to the eradication effort (135). Nevertheless, good public health practice would dictate that these modified measles cases be monitored to assure that contacts remain uninfected, particularly unvaccinated contacts.

The plaque-reduction neutralization (PRN) assay which quantifies neutralizing antibodies that are directed against the surface glycoproteins of measles virus is more sensitive than HI or EIA (125). Since functional antibodies are detected, the PRN test provides the best serologic correlate for the assessment of immune protection. However, the PRN test is not suitable for routine serologic diagnosis because it is very labor intensive, requires paired serum samples, and takes 5 to 7 days to perform. A WHO working group has standardized the PRN assay for use in aerosol vaccination studies, so that the results of these assays can be compared between and among studies (137). As mentioned above, the PRN assay can be used as a means to quantify the neutralizing antibody response from reinfection cases, thus serving as a biomarker with high avidity measles IgG. Table 1 summarizes the possible interpretations of serologic testing of suspected measles cases referred to the laboratory.

The Vero/hSLAM cell line has been recommended for the isolation of measles virus in the WHO laboratory network. These cells are the result of the transfection of Vero cells with a plasmid encoding the gene for the human signaling lymphocyte activation molecule (hSLAM) (138). SLAM has been shown to be a receptor for both wild-type and laboratory-adapted strains of measles. Testing conducted to date indicates that the sensitivity of Vero/hSLAM cells for isolation of measles virus is equivalent to that of B95a cells, an Epstein-Barr virus-transformed B lymphoblastoid cell line derived from marmoset lymphocytes. The advantage to the Vero/hSLAM cells is that they are not persistently infected with Epstein-Barr virus, and therefore, are not considered as hazardous material like B95a cells. This provides a significant safety advantage for laboratorians and greatly facilitates international shipments. The disadvantage of the Vero/SLAM cells is that they must be cultured in medium containing geneticin (G418) to retain SLAM expression, thus increasing the cost of the cell culture medium. This cell line is available from commercial sources.

Intranuclear and intracytoplasmic inclusions and giant cells are characteristic CPE for cells infected with measles virus. Cytologic examination of various tissue specimens and secretions for these Warthin-Finkeldey giant cells can be used as a diagnostic procedure. Secretions are obtained by aspiration of mucus from the nose or by swabbing the nasal mucosa with a sterile, cotton-tipped applicator. Slides can be stained with either Wright stain or hematoxylin and eosin (H + E). Tissue samples may be fixed in 10% formalin, embedded in paraffin, sectioned, and then stained with H + E stain (53). Staining of tissue specimens with monoclonal antibodies to the measles nucleoprotein has been used to support the diagnosis of giant cell pneumonia, measles inclusion body encephalitis, and SSPE (56, 139).

Standard and real time RT-PCR (rRT-PCR) assays have been used for a number of years in research settings to detect measles virus RNA in clinical specimens and infected cells (140–143). The confirmation of clinically suspected measles cases in highly vaccinated or measles eliminated regions can be challenging. Therefore, both serologic testing and real time RT-PCR is now recommended for routine diagnosis of acute measles infections (111). Standard RT-PCR followed by genotyping analysis is critical for distinguishing rash illnesses caused by wild-type measles infections versus sequelae following vaccination where IgM testing is compromised by the concurrent or recent use of measles-containing vaccine as part of outbreak response or in settings of high vaccine coverage (144). Likewise, molecular detection methods can be used when cell culture isolation is not a practical alternative and/or when genetic characterization of the virus is required. RT-PCR has been particularly useful for the laboratory confirmation of measles inclusion body encephalitis, subacute sclerosing panencephalitis, and giant-cell pneumonia. Sequence determination obtained from the PCR products in conjunction with phylogenetic analysis has proven useful in suggesting the possible source of virus involved in outbreaks, tracking transmission pathways during outbreaks, and differentiating between vaccine and wild-type strains of measles (145).

Measles RNA is easily amplified from RNA extracted from infected cell culture and it has been possible to detect

TABLE 1 Interpretation of measles serological results

IgM result	IgG result	Previous infection history	Current infection	Comments
+	+ or −	Not vaccinated, no history of measles	Recent 1st MMR	Seroconvert*; low avidity IgG (if present)
+	+ or −	Not vaccinated, no history of measles	Wild-type measles	Seroconvert*, classic measles; low avidity IgG (if present)
+	+ or −	Previously vaccinated, primary vaccine failure	Recent 2nd MMR	Seroconvert*; low avidity IgG (if present)
−	+	Previously vaccinated, IgG+	Recent 2nd MMR	IgG level may stay same or boost
+ or −	+	Previously vaccinated, IgG+, secondary vaccine failure	Wild-type measles	May have few or no symptoms**; high avidity IgG, abnormally high IgG values if quantitative assay used e.g. plaque-reduction neutralization (PRN)
+	+	Recently vaccinated	Exposed to wild-type measles	Cannot distinguish if vaccine or wild-type, evaluate on epidemiologic grounds***
+ or −	+	Distant history of measles	Wild-type measles	May have few or no symptoms**; high avidity IgG, abnormally high IgG values if quantitative assay used (PRN)

*IgG response depends on timing of specimen collection (124).
**If so, do not consider contagious unless clinical presentation is consistent with measles.
***If IgM negative, helpful to rule out wild-type measles infection; nucleotide sequencing of viral RNA will distinguish wild type from vaccine.

measles RNA in nasal, urine, oral fluid, dried blood spots (146, 147), and sometimes serum samples by RT-PCR even when virus isolation has been unsuccessful (148, 149). Most recently, real time RT-PCR and semiquantitative molecular assays such as loop-mediated isothermal and quantitative RT-qPCR methods have been developed for measles using various target genes along the measles genome (5, 150, 151). Several of these assays claim sensitivities of 10 to 50 RNA copies per reaction.

Mumps

Clinical diagnosis of mumps infection can be made reliably when typical parotitis is evident at the time of patient examination. However, since parotitis may be caused by other viral and nonviral diseases or conditions, and as much as 20% to 30% of mumps infection can be asymptomatic, diagnosis by viral isolation, molecular detection, or serological techniques is preferable. Collection of both a buccal or oral swab specimen and a blood specimen is recommended from all patients with clinical features compatible with mumps.

Mumps virus can be detected from fluid collected from the buccal cavity, from urine, and from CSF. Detection from blood is not routinely performed or recommended. The early collection of buccal or oral swab specimens provides the best means of laboratory confirmation, particularly among suspected mumps patients with a history of vaccination. Urine samples are not preferred since they are less likely than oral specimens to contain sufficient virus for culture or detection by molecular methods. For virus detection, oral or buccal swabs should be collected as soon as mumps disease is suspected. Samples collected when the patient first presents with symptoms have the highest chance of having a positive result by RT-PCR. Buccal or oral swab specimens should be obtained by massaging the parotid gland area for 30 seconds prior to swabbing the area around Stensen's duct. A commercial product designed for the collection of throat specimens or a flocked polyester fiber swab can be used. Synthetic swabs are preferred over cotton swabs; the latter may contain substances that are inhibitory to enzymes used in RT-PCR. Flocked synthetic swabs appear to be more absorbent and permit more easy and efficient elution of samples. Swabs should be placed in 2 ml of standard VTM and remain in VTM for 10 minutes to 1 hour (4°C) with intermittent gentle vortexing before reaming the swab around the rim of the tube to express cells and fluid in the tube. The swab can be broken off and left in the tube or removed from the tube and discarded. Once collected, samples should be maintained at 4°C and shipped on cold packs (4°C) within 24 hours. If there is a delay in shipment, the sample is best preserved by freezing at −70°C. Frozen samples should be shipped on dry ice. Processing the swabs within 24 hours of collection will enhance the sensitivity of both the RT-PCR and virus isolation techniques.

For serologic assays, blood is best collected in a red-top or serum-separator tube. Serum specimens should be stored at 4°C and shipped on wet ice. Optimal timing for serum specimen collection varies depending upon the patient's vaccination history. If serum collected prior to day 3 of parotitis onset in an unvaccinated person is negative for IgM, testing a second sample collected 5 to 7 days after symptom onset is recommended since the IgM response may require more time to develop. Detection of mumps-specific IgM among previously vaccinated or previously infected persons can be challenging since patients may not have an IgM response or it may be delayed, transient, or not detected depending on the timing of specimen collection. The likelihood of detecting mumps is improved among cases with two doses of MMR, if serum is collected >3 to 10 days after parotitis onset (152, 153). The absence of a mumps IgM response in a vaccinated or previously infected individual presenting with clinically compatible mumps does not rule out mumps as a diagnosis. Failure to detect mumps IgM in previously vaccinated individuals has been well documented (154, 155).

Serologic assays currently in use include IgM capture assays, IFAs, indirect EIAs, and IgG assays. Most IgM assays perform well for diagnosis of primary mumps infection in unvaccinated persons (152, 156). Among previously vaccinated persons, acute-phase mumps specimens may contain significant levels of mumps IgG. The IFA format is particularly susceptible to interference by high levels of mumps IgG. Treatment of serum with an agent to remove human IgG antibody, such as GullSORB, or a similar IgG inactivation reagent, is necessary to avoid false-positive (rheumatoid factor) and false-negative IgM test results. A comparison of commercial IgM assays using virus culture-confirmed mumps cases indicated that IgM capture assays are more sensitive than indirect EIA or IFA for detection of mumps IgM. Fifty-one percent of serum samples were positive using the CDC IgM capture assay compared to 9% to 24% of serum samples by other assays. Among two-dose vaccine recipients, the ability to detect IgM improved among all assays when serum was collected at >3 days post parotitis onset (152). Table 2 summarizes the possible interpretations of serologic testing of suspected mumps cases.

A single serum sample tested for mumps-specific IgG is not helpful for diagnosing acute mumps infections since it does not differentiate between an active or past mumps infection. The presence of mumps-specific IgG, as detected using a serologic assay (EIA or IFA), does not necessarily predict the presence of neutralizing antibodies or indicate protection from mumps disease. A positive IgG result is expected among previously vaccinated persons as well as individuals with past mumps infections. Older persons or foreign-born individuals without a history of mumps illness or vaccination very frequently have detectable mumps IgG due to a previous subclinical infection. Parainfluenza viruses 1, 2, and 3, Epstein-Barr virus, influenza virus, adenovirus, and human herpesvirus 6 have all been noted to interfere with mumps serologic assays (157, 158).

Virus isolation is considered the gold standard for confirming mumps infection; however, sample quality must be maintained to ensure viability of the virus. Virus isolation from saliva, blood, urine, or CSF confirms the presence of recent mumps infection (159, 160). Virus may be detected when IgM antibodies or a rise in IgG titer are not detected. While primary monkey kidney cells are likely the most sensitive cells, availability, risk of infection with herpes B virus, and animal rights issues have resulted in the use of continuous cell lines. Growth of mumps virus has been noted to occur in primary human cell cultures; continuous cell lines such as HeLa and Vero are currently the cell lines of choice. Often it is necessary to grow the virus in culture to have adequate material for viral sequencing. Although mumps virus infection results in characteristic CPE consisting of large syncytia, some isolates vary in the intensity and frequency of the CPE and thus must be confirmed by IFA staining (161), by immunocytometric assay (162), or by molecular detection methods, such as RT-PCR (163, 164).

Rubella

Strong clinical identification of suspected rubella cases will dramatically improve the positive predictive value of

TABLE 2 Interpretation of mumps serological results

IgM result	IgG result	Previous infection history	Current infection	Comments
+ or −	+ or −	Unvaccinated, no history of mumps	Recent 1st MMR or wild-type mumps	IgM may be detected for weeks to months; retest if IgM negative ≤3 days post parotitis*; low avidity IgG (if present).
+ or −	+ or −	Previously vaccinated 1 dose	Recent 2nd MMR or wild-type mumps	High avidity IgG; approximately 50% of serum samples collected 1–10 days after symptom onset were IgM-positive; 50%–80% of serum samples collected >10 days after symptom onset were IgM-positive
+ or −	+ or −	Previously vaccinated 1 dose, primary vaccine failure	Recent 2nd MMR or wild-type mumps	Retest if IgM negative ≤3 days post parotitis**; Low avidity IgG (if present); second negative IgM result does not rule out mumps unless the IgG result is also negative
+ or −	+	Previously vaccinated two doses, IgG+	Wild-type mumps	Absence of a mumps IgM response in a vaccinated or previously infected individual presenting with clinically compatible mumps *does not rule out mumps*; detection of IgM improves somewhat when serum is collected ≥3 days or more after symptom onset**
Likely +	Likely +	Recently vaccinated first dose	Exposed to wild-type measles	Cannot distinguish if vaccine or wild-type, evaluate on epidemiologic grounds

*If serum sample collected ≤3 days post parotitis onset from an unvaccinated person is negative, testing a second sample collected 5–7 days after symptom onset is recommended, since the IgM response may require more time to develop.

**If an acute-phase serum sample collected ≤3 days after parotitis onset in a previously vaccinated person is negative for IgM, testing a second sample collected ≥10 days after symptom onset is recommended since the IgM response may require more time to develop.

laboratory confirmation. Postnatal rubella is often a mild disease and the defects occurring in CRS can occur for other reasons. Clinical diagnosis is further complicated because postnatal rubella is a rare illness in some countries, leaving many physicians with little experience with these diseases. Therefore, both suspected rubella and suspected CRS cases need to be confirmed with laboratory methods (165).

To identify (clinical) CRS cases and for surveillance purposes, the CDC has adopted the following procedure (88): CRS defects are divided into two groups. Group A consists of cataracts and congenital glaucoma, congenital heart disease (usually patent ductus arteriosus or peripheral pulmonary artery stenosis), loss of hearing, and pigmentary retinopathy. Group B consists of purpura, splenomegaly, jaundice, microcephaly, developmental delay, meningoencephalitis, and radiolucent bone disease. Two symptoms from group A or one from group A and one from group B result in a probable CRS case. The most common defects seen in CRS are hearing loss, cataracts, retinopathy, and congenital heart disease. Infants may present with a single defect, with hearing impairment being the most common. Therefore, in the United States, the presence of any defect consistent with CRS and laboratory data consistent with congenital rubella infection results in a CRS diagnosis. Detailed clinical descriptions of CRS are available and are useful in diagnosis.

Clinical specimens for culture of RV are usually throat swabs or nasopharyngeal secretions diluted into transport medium. The virus can also be isolated from a number of other specimens, including cataract tissue and urine (provided urine is removed immediately after the time allotted for virus attachment to cells). Urine is often a source of infectious virus from CRS patients. Specimens for virus detection should be stored at 4°C for short periods (∼2 days) or at −70°C for long periods. Virions lose infectivity at higher temperatures (e.g., 37°C) and are rapidly inactivated by mild heat (56°C), detergents, or lipid solvents. Specimens for serology or culture can be transported by standard methods (e.g., overnight carrier) at 4°C since both virions and rubella-specific antibodies are relatively stable at that temperature. RV-specific IgG can be detected in urine (166). Alternative specimens such as dried blood spots have been shown to be adequate for surveillance of rubella using IgM detection and virus detection (167).

The timing of specimen collection is especially important in postnatal rubella. Detection of RV-specific IgM is the laboratory diagnostic criterion typically used for rubella, but about 50% of rubella cases are IgM negative on the day of rash. Since postnatal rubella is a mild disease of short duration, special effort may be required to obtain a serum sample at least 5 days after rash, when most rubella patients are strongly IgM positive. Patients with CRS are IgM and virus positive for months; therefore, timing is less critical for these patients.

Laboratory diagnosis of both postnatal and congenital RV infections is by serologic and/or virus detection techniques. Antibodies specific for RV antigens can be detected by EIA, neutralization tests, latex agglutination, or HI methods.

A serum positive for RV-specific IgM is diagnostic for recent RV infection. An IgM capture assay is usually preferred because of fewer difficulties, such as false positives, but indirect EIAs that avoid such difficulties by proper removal of IgG by absorption techniques are acceptable (168). The major use of indirect EIAs for RV-specific IgG is prenatal screening for immunity to rubella. In addition, the IgG tests can be used to detect antibodies that have developed in response to a recent or past rubella infection or immunization or to identify those who have never been exposed to the virus.

Avidity tests have been used when IgM detection does not reliably indicate recent postnatal infection (e.g., inappropriate IgM testing during screening for immunity which can result in an incorrect positive result). Such tests are particularly useful in low prevalence settings, such as the

United States, in order to eliminate falsely positive IgM test results. Low avidity rubella IgG antibodies suggest recent postnatal infection (169). This test compares the ability of detergents or chaotropic agents to dissociate case IgG and control IgG from RV proteins. Both high and low avidity control sera should be used in each assay. Avidity tests are not widely available and vary in performance.

Since the clinical symptoms of postnatal rubella and CRS are dramatically different, it is not surprising that there are significant differences between the immune responses in patients with these diseases, beyond the timing of the IgM response and the persistence of rubella virus replication. These differences can be observed on Western blots, in which antibodies in sera from CRS patients often demonstrate different reactivity to rubella glycoproteins than those from postnatal rubella patients and this altered immune response continues (at least until 10 years of age) in CRS survivors (170).

Cell culture-adapted RV strains produce CPE in a variety of cells. Nonadapted viruses from clinical specimens typically do not produce CPE, and additional techniques are required to document the presence of RV in clinical specimens. Viruses in clinical specimens replicate in a variety of cell lines including Vero, RK13, BHK-21, and GMK-AH-1, and viral replication can be detected by IFA, immunocolorimetric assays (ICA), and RT-PCR (171).

The plaque reduction neutralization test is done when a quantitative assessment of the antiserum's capacity to neutralize is necessary, i.e., comparisons of vaccine potency as a result of various storage conditions, or the efficacy of several different rubella vaccines or lots of vaccine. The assay follows a format common to many tests. An immunocolorimetric neutralization assay for RV using a soluble substrate is a substantial improvement compared with plaque development. The detection portion of the assay can then be done with a microplate washer and dispenser, enhancing throughput by a factor of about three and reducing technician hands-on time by a factor of about six (172, 173).

Sensitive conventional RT-PCR assays and real time PCR assays have been developed which allow detection of RV RNA derived from tissue culture and directly in clinical specimens (20, 143). Not all RT-PCR protocols are sufficiently sensitive to be used directly with postnatal rubella/CRS clinical specimens. Assays that can reliably detect 3 to 10 copies of RV RNA per reaction are sufficiently sensitive. Real time and nested RT-PCR protocols usually meet this criterion. Many postnatal rubella cases are IgM negative before 4 to 5 days postrash and direct detection of viral RNA is the most sensitive test during this time period (174). No standard real time or RT-PCR protocol for detection of viral RNA has been established and there are currently a number of such tests being used. At present, real time and conventional RT-PCR are mostly used in national or global reference laboratories. The sensitivity of the RT-PCR system used to detect viral RNA from infected tissue culture cells is not critical, since the amount of rubella virus and available viral RNA has been substantially amplified by passage in tissue culture. Differentiating between wild-type virus and vaccine virus requires sequencing of small regions of the genome (about 700 nucleotides).

In postnatal infections, IgM is detectable in almost all cases 5 days after disease onset and persists for about 3 weeks. If a serum taken 4 or fewer days after disease onset is negative for RV-specific IgM, testing should be repeated with a serum taken about 5 days later. Alternatively, methods to detect RV RNA may be used. Since virus shedding in throat or nasopharyngeal specimens declines rapidly in the first week after onset of rash, many patients will present when virus shedding is low. In addition, the false-negative rates of such tests have typically not been determined. If acute and convalescent phase sera are available, a fourfold rise in RV-specific IgG is diagnostic for RV infection. For best results, such sera should be taken within 7 days of rash and 17 to 21 days after rash, respectively. Table 3 provides a summary of serologic interpretations for suspected postnatal rubella cases.

The utility of laboratory techniques is different for CRS than for postnatal rubella. When the risk of postnatal infection is low, the presence of IgM in infants less than 6 months of age is diagnostic of *in utero* infection. The percentage of cases of congenital rubella infections that are IgM

TABLE 3 Interpretation of suspected postnatal rubella serological results[Φ•]

IgM result	IgG result	Previous infection history	Current infection	Comments
+	+ or –	Not vaccinated, no history of rubella	Recent 1st MMR	Seroconvert*; low avidity IgG (if present)
+	+ or –	Not vaccinated, no history of rubella	Wild-type rubella	Seroconvert*, rubella; low avidity IgG (if present)
+	+	History of rubella or previously vaccinated	No known exposures or risk factors	May or may not have symptoms; high avidity IgG; probable false IgM positive
–	+	Previously vaccinated, IgG+	Recent 2nd MMR	IgG level may stay same or boost; high avidity IgG
+ or –	+	History of rubella or previously vaccinated	Rubella reinfection#	Usually few or no symptoms; high avidity IgG, fourfold or greater rise in IgG titers in acute and convalescent-phase sera run in parallel**
+	+	Recently vaccinated	Exposed to wild-type rubella	Cannot distinguish if vaccine or wild-type, evaluate on epidemiologic grounds***
–	–	No history of rubella or vaccination	Wild-type rubella	Specimen collected too early, second specimen collected at least 5 days postonset needed for confirmation

ΦInterpretations are for low incidence settings with some vaccinated individuals.
•Not all possible situations are listed. For example, IgM negative, IgG positive with no history of vaccination, and rubella are not listed.
*IgG response depends on timing of specimen collection.
**Timing of specimen collection is critical, useful only in cases with symptoms, sera must be titrated.
#Very rarely detected due to lack of symptoms in most cases.
***Nucleotide sequencing of viral RNA will distinguish wild type from vaccine.

positive declines between 6 and 12 months. Detection of IgM or RV RNA in a newborn with defects consistent with CRS confirms the diagnosis. Virus can be detected in CRS cases for up to 1 year and up to 3 years in some specimens, such as lens aspirates. Before rubella elimination, about half of the CRS cases in the United States were confirmed by RV-specific IgM and half by detection of virus. Infant IgG levels which are high or increasing in the first year of life, when maternal IgG declines, are consistent with congenital RV infection. Again, this is particularly useful when the risk of postnatal infection is low.

BEST PRACTICES

Measles

The following section summarizes laboratory "best practices" and the reader is referred to the source material for additional and more detailed information (90, 91, 111, 167). Measles was declared eliminated from the United States in 2000; elimination has been successfully maintained for well over a decade. Continued maintenance of elimination in an environment where frequent measles importations occur will require high levels of two-dose vaccination coverage, case-based surveillance, effective monitoring of vaccine exempt populations, and rapid and effective laboratory tests for case confirmation and classification.

The highly infectious nature of measles requires rapid investigation and response to persons showing the clinical signs of disease. While it is very important that laboratory confirmation be accurate and timely, epidemiologic investigations involving contact tracing and identification of an index case must not wait for laboratory confirmation, particularly in environments where there are known unvaccinated or a suspicion of possible unvaccinated individuals. It should be emphasized that in measles-eliminated populations, every suspect measles case should be investigated and, if possible, laboratory confirmed or excluded from consideration.

Regardless of the setting, the measles IgM assay has been the assay of choice for laboratory confirmation. Such assays must have excellent performance characteristics particularly when used in elimination settings. Measles is known to be most contagious 4 days before and 4 days following rash onset. Thus, specimens for viral culture or for viral RNA detection are optimally collected within this time frame. However, best practices strongly encourage the collection of specimens from suspected cases when first seen by a healthcare professional. It is recommended that both blood specimens for serology and respiratory specimens, i.e., throat swabs, nasopharyngeal swabs, or aspirates for molecular assays or virus isolation be collected to improve the probability of case confirmation. If possible, urine specimens should be collected, particularly from sporadic cases, to further enhance the possibility of virus and RNA detection.

Genotyping should be attempted from all measles cases, since identifying the circulating strain(s) of virus could aid in tracing pathways of transmission and contribute to surveillance data required for maintenance of elimination. Genotypic analysis is the only means to delineate measles vaccine reactions and adverse events from wild-type virus infections and their sequelae. In many instances, conventional RT-PCR and sequencing can be accomplished directly from the RNA extracts of specimens. Isolation of measles virus in cell culture is sometimes required to assure a sufficient source of viral RNA for eventual genotype analysis. Best practices in such cases would strongly recommend the use of the highly sensitive Vero/hSLAM cell line for propagation of measles.

The use of real time RT-PCR to detect measles RNA in extracts of appropriately collected respiratory specimens has been included among the laboratory tests capable of confirming a case of measles (111). It should be strongly emphasized, however, that failure to detect measles RNA with this method does NOT rule out a measles case. The vagaries of the techniques and timing of specimen collection, conditions of specimen storage and transport, as well as extraction procedures and performance characteristics of the real time RT-PCR assay, all affect the assay results. Many state public health laboratories have adopted the CDC real time RT-PCR protocol for detection of measles RNA (142).

In 2012, the CDC, in collaboration with the Association of Public Health Laboratories (APHL), established four Vaccine Preventable Disease Reference Centers (VPD RCs) at four State Public Health Laboratories. The VPD RCs were established to provide an efficient means of testing for measles (mumps, rubella, and other viral and bacterial VPDs) based on CDC developed and evaluated protocols that can be offered to state and local health departments that have limited budgets for testing for diseases of low incidence or low-volume testing. The RCs provide serological and molecular confirmatory testing and genetic characterization for other VPDs, surveillance support, and confirmation of vaccine adverse events in a shared-service model to 43 state (including the 4 RCs) and 10 local submitter public health laboratories. Results are contemporaneously transmitted to the submitting public health departments and CDC (175). Many public health departments have seen the merit in participating as submitter laboratories, rather than incurring the time and expense of establishing the molecular assays in their laboratories.

While the vast majority of U.S. measles cases occur in unvaccinated individuals, measles infections in previously vaccinated persons do occur. These infections can be difficult to confirm, since many do not have a classical presentation, and laboratory confirmation of measles infections in previously vaccinated persons may require specialized testing procedures consisting of PRN and measles IgG avidity. Although there are commercial measles avidity assays available, quantifiable IgG assays like the PRN are less available and would have to be evaluated and validated for use as an indicator of infection.

Mumps

Mumps is an acute viral disease characterized by fever, swelling, and tenderness of one or more salivary glands, usually the parotid gland. Treatment is symptomatic and vaccination with two doses of mumps-containing vaccine is the best method of prevention. Two doses of mumps vaccine are 88% effective at protecting against mumps. Clinical diagnosis of mumps can be made based on prodrome and parotitis; however, laboratory confirmation is desirable because prodromal symptoms are nonspecific and parotitis can result from other infections or conditions. Laboratory diagnosis is based on detection of viral nucleic acid or mumps-specific IgM antibodies in serum or saliva/oral fluid, or isolation of the virus from body fluids (excluding blood). Efforts should be made to obtain clinical specimens (buccal cavity/parotid duct fluids, throat swabs, urine, or CSF) for virus isolation and/or RNA detection by real time RT-PCR for all sporadic cases and at least some cases in each outbreak at the time of the initial investigation. Once a case has been identified, health departments should consider conducting case investigations and assessing the immune status of close contacts before laboratory results are known or before additional cases are identified, particularly in settings with high risk, such as households, schools, and summer camps. A mumps outbreak is defined as three or more cases linked by time and place. Negative laboratory results among vaccinated persons do not necessarily rule out the diagnosis of mumps, particularly if there is an outbreak of parotitis. Mumps outbreaks have occurred in highly vaccinated populations in high transmission settings such as schools, colleges, camps, and sports events. Mumps should not be ruled out on the assumption that individuals have evidence of mumps immunity because of vaccination.

Case contacts identified during the 2 days prior to symptom onset through 5 days after onset should be assessed for vaccination status, offered vaccine as appropriate, and instructed about signs and symptoms. Droplet precautions, in addition to standard precautions, in healthcare settings should be used. In the event of an outbreak, enhanced surveillance and public awareness should be instituted. If it is clear when and where transmission likely occurred, investigative efforts should be directed to these locations. Local or state health departments should contact area healthcare providers to apprise them of the outbreak and request reporting of additional suspected cases. During outbreaks, active surveillance for mumps should be conducted for every confirmed and probable mumps case and maintained for at least two incubation periods (50 days) following parotitis onset in the last case. To control the outbreak, the population at risk and the transmission setting should be defined and unvaccinated individuals should be identified and vaccinated if possible. If vaccination is contraindicated, individuals should be excluded from transmission settings to prevent exposure. Mumps vaccination of those already infected is not effective but vaccination may prevent infection in those not yet exposed or infected. During an outbreak, a second dose of mumps-containing vaccine may be considered for children aged 1 to 4 years and adults who have received one dose previously. If a nosocomial outbreak occurs, healthcare facilities should implement the two-dose recommendation for all healthcare personnel, including those who were born before 1957 and lack laboratory evidence of immunity or laboratory confirmation of disease. Healthcare personnel without evidence of immunity should be excluded from the 12th day after the first unprotected exposure to mumps through the 25th day after the last exposure.

Rubella

In the United States, best practices for investigation of cases of rubella and CRS are summarized in detail in frequently updated surveillance documents (88, 176). Important information to consider for best, nonlaboratory practices for postnatal rubella include the case definitions for suspected, probable, and confirmed rubella and the epidemiologic classification into internationally imported cases, U.S.-acquired cases, import-linked cases, imported virus cases, endemic cases, and unknown source cases. Guidance for case investigations, contact investigation, and outbreak control may also be important. Best (nonlaboratory) practices for CRS include the case definitions for suspected, probable, confirmed, and infection-only and congenital rubella infection-only cases and epidemiologic classification into internationally imported cases and U.S.-acquired cases.

Best practices for the laboratory are also important. Both IgM and IgG testing should be done with most sera for both suspected postnatal rubella and CRS cases, since the combination of test results often provides additional information useful for diagnosis. For example, with a rubella infection, the IgG is expected to become detectable by ELISA about 5 days after a rubella rash. Results from a serum sample taken at 8 days postrash that are positive for IgM but negative for IgG to RV would be inconsistent with the immune response to rubella; thus, the IgM result would be most suspect in this situation. A positive result for RV culture is obtained when growth is confirmed by real time or conventional RT-PCR using RNA extracted from the culture or when at least one cluster of infected cells is determined positive by IFA or ICA. When an IFA or ICA is used, the expected distribution of viral proteins within infected cells should be obtained (162). Direct detection of RV RNA by PCR-based protocols requires the laboratory to evaluate the significance of results from such tests. Multiple negative controls and amplified product in more than one specimen from a given patient will increase confidence of a positive diagnosis based on direct RT-PCR. When serum from a patient cannot be obtained, or when confirmation of serologic results is desired, direct detection of RV RNA by RT-PCR may be necessary, since it is more rapid than viral culture.

There is often a considerable burden on the laboratory in the diagnosis of rubella. For example, when primary RV infection is suspected for a pregnant woman, false-positive and false-negative results may lead to incorrect clinical decisions (177). Testing for recent infection with other viruses that cause clinically similar diseases (e.g., human parvovirus B19) is often prudent. Positive rubella results may be more believable if no other infection is found. Specimen retesting and testing of different specimen types with alternative methods (e.g., serology, RT-PCR, and viral culture) often yield consistent results and reduce the likelihood of false-positive results. False-positive results can occur even with IgM-capture ELISA. False-negative results can often be identified by testing more than one serum specimen from a patient obtained 1 week apart. If only a single specimen is available, it may be tested by multiple assays. For example, IgG avidity may resolve the diagnosis from a single serum sample that is IgM positive for both RV and human parvovirus B19.

Correct classification of suspected postnatal rubella and CRS is often primarily based on laboratory results rather than clinical presentation. Clinical, laboratory, and epidemiological information (e.g., international travel) all may enter into the final decision(s). A full description of classification criteria and recommendations should be consulted. One specific diagnostic situation should be noted. Standard TORCH (toxoplasmosis, other, rubella, cytomegalovirus, and herpes simplex virus) panels include testing for RV IgM, and clearly should not be used to determine immunity. Each state and territory has regulations or laws governing the reporting of rubella and CRS (and other conditions of public health importance).

Disclaimer: The statements and conclusions presented in this chapter are those of the authors and do not necessarily reflect the official position of the Centers for Disease Control and Prevention.

REFERENCES

1. **Lamb RA, Parks CL.** 2013. Paramyxoviridae, p 957–995. *In* Knipe DM, Howley PM (ed), *Fields Virology*, 6th ed, vol 1. Lippincott, Williams & Wilkins, Philadelphia.
2. **Yanagi Y, Takeda M, Ohno S, Hashiguchi T.** 2009. Measles virus receptors. *Curr Top Microbiol Immunol* **329:**13–30.

3. **Rima BK, Duprex WP.** 2009. The measles virus replication cycle. *Curr Top Microbiol Immunol* **329:**77–102.
4. **Rota PA, Brown K, Mankertz A, Santibanez S, Shulga S, Muller CP, Hübschen JM, Siqueira M, Beirnes J, Ahmed H, Triki H, Al-Busaidy S, Dosseh A, Byabamazima C, Smit S, Akoua-Koffi C, Bwogi J, Bukenya H, Wairagkar N, Ramamurty N, Incomserb P, Pattamadilok S, Jee Y, Lim W, Xu W, Komase K, Takeda M, Tran T, Castillo-Solorzano C, Chenoweth P, Brown D, Mulders MN, Bellini WJ, Featherstone D.** 2011. Global distribution of measles genotypes and measles molecular epidemiology. *J Infect Dis* **204**(Suppl 1)**:** S514–S523.
5. **Bankamp B, Byrd-Leotis LA, Lopareva EN, Woo GK, Liu C, Jee Y, Ahmed H, Lim WW, Ramamurty N, Mulders MN, Featherstone D, Bellini WJ, Rota PA.** 2013. Improving molecular tools for global surveillance of measles virus. *J Clin Virol* **58:**176–182.
6. **Xu P, Li Z, Sun D, Lin Y, Wu J, Rota PA, He B.** 2011. Rescue of wild-type mumps virus from a strain associated with recent outbreaks helps to define the role of the SH ORF in the pathogenesis of mumps virus. *Virology* **417:**126–136.
7. **Malik T, Shegogue CW, Werner K, Ngo L, Sauder C, Zhang C, Duprex WP, Rubin S.** 2011. Discrimination of mumps virus small hydrophobic gene deletion effects from gene translation effects on virus virulence. *J Virol* **85:**6082–6085.
8. **Jin L, Beard S, Brown DW.** 1999. Genetic heterogeneity of mumps virus in the United Kingdom: identification of two new genotypes. *J Infect Dis* **180:**829–833.
9. **Johansson B, Tecle T, Orvell C.** 2002. Proposed criteria for classification of new genotypes of mumps virus. *Scand J Infect Dis* **34:**355–357.
10. **Jin L, Örvell C, Myers R, Rota PA, Nakayama T, Forcic D, Hiebert J, Brown KE.** 2015. Genomic diversity of mumps virus and global distribution of the 12 genotypes. *Rev Med Virol* **25:**85–101.
11. **Cooper LZ.** 1985. The history and medical consequences of rubella. *Rev Infect Dis* **7**(Suppl 1)**:**S2–S10.
12. **Gregg NM.** 1941. Congenital cataract following German measles in the mother. *Trans Ophthalmol Soc Aust* **3:**35–46.
13. **Lee J-Y, Bowden DS.** 2000. Rubella virus replication and links to teratogenicity. *Clin Microbiol Rev* **13:**571–587.
14. **Abernathy E, Chen MH, Bera J, Shrivastava S, Kirkness E, Zheng Q, Bellini W, Icenogle J.** 2013. Analysis of whole genome sequences of 16 strains of rubella virus from the United States, 1961–2009. *Virol J* **10:**32.
15. **Prasad VM, Willows SD, Fokine A, Battisti AJ, Sun S, Plevka P, Hobman TC, Rossmann MG.** 2013. Rubella virus capsid protein structure and its role in virus assembly and infection. *Proc Natl Acad Sci USA* **110:**20105–20110.
16. **Battisti AJ, Yoder JD, Plevka P, Winkler DC, Prasad VM, Kuhn RJ, Frey TK, Steven AC, Rossmann MG.** 2012. Cryo-electron tomography of rubella virus. *J Virol* **86:**11078–11085.
17. **DuBois RM, Vaney MC, Tortorici MA, Kurdi RA, Barba-Spaeth G, Krey T, Rey FA.** 2013. Functional and evolutionary insight from the crystal structure of rubella virus protein E1. *Nature* **493:**552–556.
18. **World Health Organization.** 2013. Rubella virus nomenclature update: 2013. *Wkly Epidemiol Rec* **88:**337–343.
19. **Zhu Z, et al, Rubella Virology Surveillance Working Group.** 2015. Evolutionary analysis of rubella viruses in mainland China during 2010–2012: endemic circulation of genotype 1E and introductions of genotype 2B. *Sci Rep* **5:**7999.
20. **Namuwulya P, Abernathy E, Bukenya H, Bwogi J, Tushabe P, Birungi M, Seguya R, Kabaliisa T, Alibu VP, Kayondo JK, Rivailler P, Icenogle J, Bakamutumaho B.** 2014. Phylogenetic analysis of rubella viruses identified in Uganda, 2003–2012. *J Med Virol* **86:**2107–2113.
21. **Enders JF, Peebles TC.** 1954. Propagation in tissue cultures of cytopathogenic agents from patients with measles. *Proc Soc Exp Biol Med* **86:**277–286.
22. **Schwarz AJ.** 1962. Preliminary tests of a highly attenuated measles vaccine. *Am J Dis Child* **103:**386–389.
23. **Centers for Disease Control.** 1982. Countdown toward the elimination of measles in the U.S. *MMWR Morb Mortal Wkly Rep* **31:**447–478.
24. **Watson JC, Hadler SC, Dykewicz CA, Reef S, Phillips L, Centers for Disease Control.** 1998. Measles, mumps, and rubella—vaccine use and strategies for elimination of measles, rubella, and congenital rubella syndrome and control of mumps: recommendations of the Advisory Committee on Immunization Practices (ACIP). *MMWR Recomm Rep* **47** (RR-8)**:**1–57.
25. **Watson JC, Redd SC, Rhodes PH, Hadler SC.** 1998. The interruption of transmission of indigenous measles in the United States during 1993. *Pediatr Infect Dis J* **17:**363–366, discussion 366–367.
26. **Rota JS, Heath JL, Rota PA, King GE, Celma ML, Carabaña J, Fernandez-Muñoz R, Brown D, Jin L, Bellini WJ.** 1996. Molecular epidemiology of measles virus: identification of pathways of transmission and implications for measles elimination. *J Infect Dis* **173:**32–37.
27. **Katz SL, Hinman AR.** 2004. Summary and conclusions: measles elimination meeting, 16–17 March 2000. *J Infect Dis* **189**(Suppl 1)**:**S43–S47.
28. **Papania MJ, Wallace GS, Rota PA, Icenogle JP, Fiebelkorn AP, Armstrong GL, Reef SE, Redd SB, Abernathy ES, Barskey AE, Hao L, McLean HQ, Rota JS, Bellini WJ, Seward JF.** 2014. Elimination of endemic measles, rubella, and congenital rubella syndrome from the Western hemisphere: the US experience. *JAMA Pediatr* **168:**148–155.
29. **Gastañaduy PA, Redd SB, Fiebelkorn AP, Rota JS, Rota PA, Bellini WJ, Seward JF, Wallace GS, Division of Viral Disease, National Center for Immunization and Respiratory Diseases, CDC.** 2014. Measles—United States, January 1-May 23, 2014. *MMWR Morb Mortal Wkly Rep* **63:**496–499.
30. **Clemmons NS, Gastanaduy PA, Fiebelkorn AP, Redd SB, Wallace GS, Centers for Disease Control and Prevention (CDC).** 2015. Measles—United States, January 4-April 2, 2015. *MMWR Morb Mortal Wkly Rep* **64:**373–376.
31. **Perry RT, Gacic-Dobo M, Dabbagh A, Mulders MN, Strebel PM, Okwo-Bele JM, Rota PA, Goodson JL, Centers for Disease Control and Prevention (CDC).** 2014. Progress toward regional measles elimination—worldwide, 2000–2013. *MMWR Morb Mortal Wkly Rep* **63:**1034–1038.
32. **Drexler JF, Corman VM, Müller MA, Maganga GD, Vallo P, Binger T, Gloza-Rausch F, Cottontail VM, Rasche A, Yordanov S, Seebens A, Knörnschild M, Oppong S, Adu Sarkodie Y, Pongombo C, Lukashev AN, Schmidt-Chanasit J, Stöcker A, Carneiro AJ, Erbar S, Maisner A, Fronhoffs F, Buettner R, Kalko EK, Kruppa T, Franke CR, Kallies R, Yandoko ER, Herrler G, Reusken C, Hassanin A, Krüger DH, Matthee S, Ulrich RG, Leroy EM, Drosten C.** 2012. Bats host major mammalian paramyxoviruses. *Nat Commun* **3:** 796–806.
33. **Krüger N, Hoffmann Mm, Drexler JF, Müller MA, Corman VM, Sauder C, Rubin S, He B, Örvell C, Drosten C, Herrler G.** 2015. Functional properties and genetic relatedness of the fusion and hemagglutinin-neuraminidase proteins of a mumps virus-like bat virus. *J Virol* **89:**4539–4548.
34. **Bitsko RH, Cortese MM, Dayan GH, Rota PA, Lowe L, Iversen SC, Bellini WJ.** 2008. Detection of RNA of mumps virus during an outbreak in a population with a high level of measles, mumps, and rubella vaccine coverage. *J Clin Microbiol* **46:**1101–1103.
35. **Dayan GH, Quinlisk MP, Parker AA, Barskey AE, Harris ML, Schwartz JM, Hunt K, Finley CG, Leschinsky DP, O'Keefe AL, Clayton J, Kightlinger LK, Dietle EG, Berg J, Kenyon CL, Goldstein ST, Stokley SK, Redd SB, Rota PA, Rota J, Bi D, Roush SW, Bridges CB, Santibanez TA, Parashar U, Bellini WJ, Seward JF.** 2008. Recent resurgence of mumps in the United States. *N Engl J Med* **358:**1580–1589.
36. **Man W, Jin-Kou Z, Tao W, Li-Xin H, Chao M, Qi-Ru S, Hui-Ming L.** 2012. Mumps-containing vaccine effectiveness during outbreaks in two schools in Guangdong, China, 2012. *Western Pac Surveill Response J* **3:**29–32.

37. **Stein-Zamir C, Shoob H, Abramson N, Tallen-Gozani E, Sokolov I, Zentner G.** 2009. Mumps outbreak in Jerusalem affecting mainly male adolescents. *Euro Surveill* **14:**pii_19440.
38. **Anderson RM.** 1992. The concept of herd immunity and the design of community-based immunization programmes. *Vaccine* **10:**928–935.
39. **Kutty PK, Kruszon-Moran DM, Dayan GH, Alexander JP, Williams NJ, Garcia PE, Hickman CJ, McQuillan GM, Bellini WJ.** 2010. Seroprevalence of antibody to mumps virus in the US population, 1999–2004. *J Infect Dis* **202:**667–674.
40. **Date AA, Kyaw MH, Rue AM, Klahn J, Obrecht L, Krohn T, Rowland J, Rubin S, Safranek TJ, Bellini WJ, Dayan GH.** 2008. Long-term persistence of mumps antibody after receipt of 2 measles-mumps-rubella (MMR) vaccinations and antibody response after a third MMR vaccination among a university population. *J Infect Dis* **197:**1662–1668.
41. **Latner DR, McGrew M, Williams NJ, Sowers SB, Bellini WJ, Hickman CJ.** 2014. Estimates of mumps seroprevalence may be influenced by antibody specificity and serologic method. *Clin Vaccine Immunol* **21:**286–297.
42. **Parkman PD, Buescher EL, Artenstein MS.** 1962. Recovery of rubella virus from army recruits. *Proc Soc Exp Biol Med* **111:**225–230.
43. **Weller TH, Neva FA.** 1962. Propagation in tissue culture of cytopathic agents from patients with rubella-like illness. *Proc Soc Exp Biol Med* **111:**215–225.
44. **Horstmann DM.** 1975. Controlling rubella: problems and perspectives. *Ann Intern Med* **83:**412–417.
45. **Reef SE, Cochi SL.** 2006. The evidence for the elimination of rubella and congenital rubella syndrome in the United States: a public health achievement. *Clin Infect Dis* **43**(Suppl 3)**:** S123–S125.
46. **Centers for Disease Control and Prevention.** 2013. Three cases of congenital rubella syndrome in the postelimination era—Maryland, Alabama, and Illinois, 2012. *MMWR Morb Mortal Wkly Rep* **129:**226–229.
47. **Reef SE, Frey TK, Theall K, Abernathy E, Burnett CL, Icenogle J, McCauley MM, Wharton M.** 2002. The changing epidemiology of rubella in the 1990s: on the verge of elimination and new challenges for control and prevention. *JAMA* **287:**464–472.
48. **Best JM, Icenogle JP, Brown DWG.** 2009. Rubella, p 561–592. *In* Zuckerman AJ, Banatvala JE, Schoub BD, Griffiths PD, Mortimer P (ed), *Principles and Practice of Clinical Virology,* 6th ed. Wiley-Blackwell, Singapore.
49. **Lambert N, Strebel P, Orenstein W, Icenogle J, Poland GA.** 2015. Rubella. *Lancet* **385:**2297–2307.
50. **Mao B, Chheng K, Wannemuehler K, Vynnycky E, Buth S, Soeung SC, Reef S, Weldon W, Quick L, Gregory CJ.** 2015. Immunity to polio, measles and rubella in women of child-bearing age and estimated congenital rubella syndrome incidence, Cambodia, 2012. *Epidemiol Infect* **143:**1858–1867.
51. **Koplik H.** 1962. The diagnosis of the invasion of measles from a study of the exanthema as it appears on the buccal mucous membrane. *Arch Pediatr* **79:**162–165.
52. **Perry RT, Halsey NA.** 2004. The clinical significance of measles: a review. *J Infect Dis* **189**(Suppl 1)**:**S4–S16.
53. **Gershon A, Krugman S.** 1979. Measles virus, p 665–693. *In* Lennette EH, Schmidt NJ (ed), *Diagnostic Procedures for Viral, Rickettsial and Chlamydial Infections.* American Public Health Association, Washington, D.C.
54. **Horta-Barbosa L, Fuccillo DA, Sever JL, Zeman W.** 1969. Subacute sclerosing panencephalitis: isolation of measles virus from a brain biopsy. *Nature* **221:**974.
55. **Horta-Barbosa L, Hamilton R, Wittig B, Fuccillo DA, Sever JL, Vernon ML.** 1971. Subacute sclerosing panencephalitis: isolation of suppressed measles virus from lymph node biopsies. *Science* **173:**840–841.
56. **Bellini WJ, Rota JS, Lowe LE, Katz RS, Dyken PR, Zaki SR, Shieh W-J, Rota PA.** 2005. Subacute sclerosing panencephalitis: more cases of this fatal disease are prevented by measles immunization than was previously recognized. *J Infect Dis* **192:**1686–1693.
57. **Halsey NA, Modlin JF, Jabbour JT, Dubey L, Eddins DL, Ludwig DD.** 1980. Risk factors in subacute sclerosing panencephalitis: a case-control study. *Am J Epidemiol* **111:**415–424.
58. **Campbell H, Andrews N, Brown KE, Miller E.** 2007. Review of the effect of measles vaccination on the epidemiology of SSPE. *Int J Epidemiol* **36:**1334–1348.
59. **Eberhart-Phillips JE, Frederick PD, Baron RC, Mascola L.** 1993. Measles in pregnancy: a descriptive study of 58 cases. *Obstet Gynecol* **82:**797–801.
60. **Manikkavasagan G, Ramsay M.** 2009. The rationale for the use of measles post-exposure prophylaxis in pregnant women: a review. *J Obstet Gynaecol* **29:**572–575.
61. **Fulginiti VA, Eller JJ, Downie AW, Kempe CH.** 1967. Altered reactivity to measles virus. Atypical measles in children previously immunized with inactivated measles virus vaccines. *JAMA* **202:**1075–1080.
62. **Markowitz LE, Chandler FW, Roldan EO, Saldana MJ, Roach KC, Hutchins SS, Preblud SR, Mitchell CD, Scott GB.** 1988. Fatal measles pneumonia without rash in a child with AIDS. *J Infect Dis* **158:**480–483.
63. **Sullivan KM, Halpin TJ, Kim-Farley R, Marks JS.** 1985. Mumps disease and its health impact: an outbreak-based report. *Pediatrics* **76:**533–536.
64. **Azimi PH, Cramblett HG, Haynes RE.** 1969. Mumps meningoencephalitis in children. *JAMA* **207:**509–512.
65. **Beard CM, Benson RC Jr, Kelalis PP, Elveback LR, Kurland LT.** 1977. The incidence and outcome of mumps orchitis in Rochester, Minnesota, 1935 to 1974. *Mayo Clin Proc* **52:**3–7.
66. **Casella R, Leibundgut B, Lehmann K, Gasser TC.** 1997. Mumps orchitis: report of a mini-epidemic. *J Urol* **158:**2158–2161.
67. **Nickel WR, Plumb RT.** 1986. Orchitis, p 977. *In* Walsh PC, Gittes RF, Perlmutter AD, Stamey TA (ed), *Campbell's Urology,* 5th ed. WB Saunders Co, Philadelphia.
68. **Plotkin SA.** Commentary: mumps vaccines: do we need a new one? 2013. *Pediatr Infect Dis J* **32:**381–382.
69. **Barskey AE, Schulte C, Rosen JB, Handschur EF, Rausch-Phung E, Doll MK, Cummings KP, Alleyne EO, High P, Lawler J, Apostolou A, Blog D, Zimmerman CM, Montana B, Harpaz R, Hickman CJ, Rota PA, Rota JS, Bellini WJ, Gallagher KM.** 2012. Mumps outbreak in Orthodox Jewish communities in the United States. *N Engl J Med* **367:**1704–1713.
70. **Nelson GE, Aguon A, Valencia E, Oliva R, Guerrero ML, Reyes R, Lizama A, Diras D, Mathew A, Camacho EJ, Monforte MN, Chen TH, Mahamud A, Kutty PK, Hickman C, Bellini WJ, Seward JF, Gallagher K, Fiebelkorn AP.** 2013. Epidemiology of a mumps outbreak in a highly vaccinated island population and use of a third dose of measles-mumps-rubella vaccine for outbreak control—Guam 2009 to 2010. *Pediatr Infect Dis J* **32:**374–380.
71. **Zamir CS, Schroeder H, Shoob H, Abramson N, Zentner G.** 2015. Characteristics of a large mumps outbreak: clinical severity, complications and association with vaccination status of mumps outbreak cases. *Hum Vaccin Immunother* **11:**1413–1417.
72. **Philip RN, Reinhard KR, Lackman DB.** 1959. Observations on a mumps epidemic in a virgin population. *Am J Hyg* **69:**91–111.
73. **Hall R, Richards H.** 1987. Hearing loss due to mumps. *Arch Dis Child* **62:**189–191.
74. **Hashimoto H, Fujioka M, Kinumaki H, Kinki Ambulatory Pediatrics Study Group.** 2009. An office-based prospective study of deafness in mumps. *Pediatr Infect Dis J* **28:**173–175.
75. **Kanra G, Kara A, Cengiz AB, Isik P, Ceyhan M, Ataş A.** 2002. Mumps meningoencephalitis effect on hearing. *Pediatr Infect Dis J* **21:**1167–1169.
76. **Ratzmann KP, Strese J, Witt S, Berling H, Keilacker H, Michaelis D.** 1984. Mumps infection and insulin-dependent diabetes mellitus (IDDM). *Diabetes Care* **7:**170–173.
77. **Duderstadt SK. Rose CE Jr, Real TM, Sabatier JF, Stewart B, Ma G, Yerubandi UD, Eick AA, Tokars JI, McNeil MM.**

2012. Vaccination and risk of type 1 diabetes mellitus in active component U.S. Military, 2002–2008. *Vaccine* **30:**813–819.
78. **Lund KD, Chantler JK.** 2000. Mapping of genetic determinants of rubella virus associated with growth in joint tissue. *J Virol* **74:**796–804.
79. **Abernathy E, Peairs RR, Chen M-H, Icenogle J, Namdari H.** 2015. Genomic characterization of a persistent rubella virus from a case of Fuch' uveitis syndrome in a 73 year old man. *J Clin Virol* **69:**104–109.
80. **Webster WS.** 1998. Teratogen update: congenital rubella. *Teratology* **58:**13–23.
81. **Perelygina L, Zheng Q, Metcalfe M, Icenogle J.** 2013. Persistent infection of human fetal endothelial cells with rubella virus. *PLoS One* **8:**e73014.
82. **Menser MA, Forrest JM, Honeyman MC, Burgess JA.** 1974. Letter: Diabetes, HL-A antigens, and congenital rubella. *Lancet* **2:**1508–1509.
83. **Floret D, Rosenberg D, Hage GN, Monnet P.** 1980. Case report: hyperthyroidism, diabetes mellitus and the congenital rubella syndrome. *Acta Paediatr Scand* **69(2):**259–261.
84. **Boger WP III, Petersen RA, Robb RM.** 1981. Spontaneous absorption of the lens in the congenital rubella syndrome. *Arch Ophthalmol* **99:**433–434.
85. **Winchester SA, Varga Z, Parmar D, Brown KE.** 2013. Persistent intraocular rubella infection in a patient with Fuchs' uveitis and congenital rubella syndrome. *J Clin Microbiol* **51:** 1622–1624.
86. **Townsend JJ, Stroop WG, Baringer JR, Wolinsky JS, McKerrow JH, Berg BO.** 1982. Neuropathology of progressive rubella panencephalitis after childhood rubella. *Neurology* **32:**185–190.
87. **Rosenberg HS, Oppenheimer EH, Esterly JR.** 1981. Congenital rubella syndrome: the late effects and their relation to early lesions. *Perspect Pediatr Pathol* **6:**183–202.
88. **Centers for Disease Control and Prevention.** 2012. Congenital rubella syndrome, chapter 15. *In* Roush SW, Baldy LM (ed), *Manual for the Surveillance of Vaccine-Preventable Diseases*. U.S. Department of Health and Human Services, CDC, Atlanta, GA. http://www.cdc.gov/vaccines/pubs/surv-manual/chpt15-crs.html.
89. **American Academy of Pediatrics.** 1997. A summary of major changes in the Red Book. *In* Peter G (ed), *1997 Red Book: Report of the Committee on Infectious Diseases*, 24th ed. American Academy of Pediatrics, Elk Grove, IL.
90. **Centers for Disease Control.** 1989. Measles prevention: recommendations of the Immunization Practices Advisory Committee (ACIP). *MMWR Morb Mortal Wkly Rep* **38:**1–17.
91. **McLean HQ, Fiebelkorn AP, Temte JL, Wallace GS, Centers for Disease Control and Prevention.** 2013. Prevention of measles, rubella, congenital rubella syndrome, and mumps, 2013: summary recommendations of the Advisory Committee on Immunization Practices (ACIP). *MMWR Recomm Rep* **62** (RR-04)**:**1–34.
92. **Hersh BS, Fine PE, Kent WK, Cochi SL, Kahn LH, Zell ER, Hays PL, Wood CL.** 1991. Mumps outbreak in a highly vaccinated population. *J Pediatr* **119:**187–193.
93. **Briss PA, Fehrs LJ, Parker RA, Wright PF, Sannella EC, Hutcheson RH, Schaffner W.** 1994. Sustained transmission of mumps in a highly vaccinated population: assessment of primary vaccine failure and waning vaccine-induced immunity. *J Infect Dis* **169:**77–82.
94. **Whitman C.** 1999. Mumps outbreak in a highly vaccinated population. NY VAC SCENE 1[1]. The New York City Department of Health, New York, NY.
95. **McNabb SJ, Jajosky RA, Hall-Baker PA, Adams DA, Sharp P, Anderson WJ, Javier AJ, Jones GJ, Nitschke DA, Worshams CA, Richard RA Jr, Centers for Disease Control and Prevention (CDC).** 2007. Summary of notifiable diseases—United States, 2005. *MMWR Morb Mortal Wkly Rep* **54:**1–92.
96. **Kyaw MH, Bellini WJ, Dayan GH.** 2007. Mumps surveillance and prevention: putting mumps back on our radar screen. *Cleve Clin J Med* **74:**13–15.
97. **Atrasheuskaya AV, Kulak MV, Rubin S, Ignatyev GM.** 2007. Mumps vaccine failure investigation in Novosibirsk, Russia, 2002–2004. *Clin Microbiol Infect* **13:**670–676.
98. **LeBaron CW, Forghani B, Beck C, Brown C, Bi D, Cossen C, Sullivan BJ.** 2009. Persistence of mumps antibodies after 2 doses of measles-mumps-rubella vaccine. *J Infect Dis* **199:**552–560.
99. **Davidkin I, Jokinen S, Broman M, Leinikki P, Peltola H.** 2008. Persistence of measles, mumps, and rubella antibodies in an MMR-vaccinated cohort: a 20-year follow-up. *J Infect Dis* **197:**950–956.
100. **Cohen C, White JM, Savage EJ, Glynn JR, Choi Y, Andrews N, Brown D, Ramsay ME.** 2007. Vaccine effectiveness estimates, 2004–2005 mumps outbreak, England. *Emerg Infect Dis* **13:**12–17.
101. **Vandermeulen C, Roelants M, Vermoere M, Roseeuw K, Goubau P, Hoppenbrouwers K.** 2004. Outbreak of mumps in a vaccinated child population: a question of vaccine failure? *Vaccine* **22:**2713–2716.
102. **Cortese MM, Barskey AE, Tegtmeier GE, Zhang C, Ngo L, Kyaw MH, Baughman AL, Menitove JE, Hickman CJ, Bellini WJ, Dayan GH, Hansen GR, Rubin S.** 2011. Mumps antibody levels among students before a mumps outbreak: in search of a correlate of immunity. *J Infect Dis* **204:**1413–1422.
103. **Nöjd J, Tecle T, Samuelsson A, Orvell C.** 2001. Mumps virus neutralizing antibodies do not protect against reinfection with a heterologous mumps virus genotype. *Vaccine* **19:**1727–1731.
104. **Rubin S, Mauldin J, Chumakov K, Vanderzanden J, Iskow R, Carbone K.** 2006. Serological and phylogenetic evidence of monotypic immune responses to different mumps virus strains. *Vaccine* **24:**2662–2668.
105. **Rubin SA, Link MA, Sauder CJ, Zhang C, Ngo L, Rima BK, Duprex WP.** 2012. Recent mumps outbreaks in vaccinated populations: no evidence of immune escape. *J Virol* **86:**615–620.
106. **Gut JP, Lablache C, Behr S, Kirn A.** 1995. Symptomatic mumps virus reinfections. *J Med Virol* **45:**17–23.
107. **Yoshida N, Fujino M, Miyata A, Nagai T, Kamada M, Sakiyama H, Ihara T, Kumagai T, Okafuji T, Okafuji T, Nakayama T.** 2008. Mumps virus reinfection is not a rare event confirmed by reverse transcription loop-mediated isothermal amplification. *J Med Virol* **80:**517–523.
108. **Rubin S, Beeler J.** 2013. Mumps vaccines: do we need a new one? *Pediatr Infect Dis J* **32:**1156–1157.
109. **Reef S, Plotkin SA.** 2007. Rubella vaccine, p 79–94. *In* Banatvala J, Peckham C (ed), *Rubella Viruses, Perspectives in Medical Virology.* Elsevier, Amsterdam, The Netherlands.
110. **Castillo-Solórzano C, Reef SE, Morice A, Vascones N, Chevez AE, Castalia-Soares R, Torres C, Vizzotti C, Ruiz Matus C.** 2011. Rubella vaccination of unknowingly pregnant women during mass campaigns for rubella and congenital rubella syndrome elimination, the Americas 2001–2008. *J Infect Dis* **204**(Suppl 2)**:**S713–S717.
111. **Centers for Disease Control and Prevention.** 2013. Measles, p 1–21. *In* Roush SW, Baldy LM (ed), *Manual for the Surveillance of Vaccine-Preventable Diseases*, 6th ed. U.S. Department of Health and Human Services, CDC, Atlanta, GA.
112. **Helfand RF, Heath JL, Anderson LJ, Maes EF, Guris D, Bellini WJ.** 1997. Diagnosis of measles with an IgM capture EIA: the optimal timing of specimen collection after rash onset. *J Infect Dis* **175:**195–199.
113. **Woo GK, Wong AH, Lee WY, Lau CS, Cheng PK, Leung PC, Lim WW.** 2010. Comparison of laboratory diagnostic methods for measles infection and identification of measles virus genotypes in Hong Kong. *J Med Virol* **82:**1773–1781.
114. **Tuokko H.** 1984. Comparison of nonspecific reactivity in indirect and reverse immunoassays for measles and mumps immunoglobulin M antibodies. *J Clin Microbiol* **20:**972–976.
115. **Erdman DD, Anderson LJ, Adams DR, Stewart JA, Markowitz LE, Bellini WJ.** 1991. Evaluation of monoclonal antibody-based capture enzyme immunoassays for detection of specific antibodies to measles virus. *J Clin Microbiol* **29:**1466–1471.

116. **Mayo DR, Brennan T, Cormier DP, Hadler J, Lamb P.** 1991. Evaluation of a commercial measles virus immunoglobulin M enzyme immunoassay. *J Clin Microbiol* **29:**2865–2867.
117. **Hummel KB, Erdman DD, Heath J, Bellini WJ.** 1992. Baculovirus expression of the nucleoprotein gene of measles virus and utility of the recombinant protein in diagnostic enzyme immunoassays. *J Clin Microbiol* **30:**2874–2880.
118. **Samuel D, Sasnauskas K, Jin L, Gedvilaite A, Slibinskas R, Beard S, Zvirbliene A, Oliveira SA, Staniulis J, Cohen B, Brown D.** 2003. Development of a measles specific IgM ELISA for use with serum and oral fluid samples using recombinant measles nucleoprotein produced in Saccharomyces cerevisiae. *J Clin Virol* **28:**121–129.
119. **Arista S, Ferraro D, Cascio A, Vizzi E, di Stefano R.** 1995. Detection of IgM antibodies specific for measles virus by capture and indirect enzyme immunoassays. *Res Virol* **146:** 225–232.
120. **Ratnam S, Tipples G, Head C, Fauvel M, Fearon M, Ward BJ.** 2000. Performance of indirect immunoglobulin M (IgM) serology tests and IgM capture assays for laboratory diagnosis of measles. *J Clin Microbiol* **38:**99–104.
121. **Woods CR.** 2013. False-positive results for immunoglobulin M serologic results: explanations and examples, p 87–90. *In* Weinberg G, Woods RC (ed), *Pediatric ID Consultant, Oxford University Press on behalf of the Pediatric Infectious Diseases Society.* Oxford University Press, Oxford.
122. **Bellini WJ, Helfand RF.** 2003. The challenges and strategies for laboratory diagnosis of measles in an international setting. *J Infect Dis* **187**(Suppl 1)**:**S283–S290.
123. **Dietz V, Rota J, Izurieta H, Carrasco P, Bellini W.** 2004. The laboratory confirmation of suspected measles cases in settings of low measles transmission: conclusions from the experience in the Americas. *Bull World Health Organ* **82:**852–857.
124. **Helfand RF, Kebede S, Gary HE Jr, Beyene H, Bellini WJ.** 1999. Timing of development of measles-specific immunoglobulin IgM and IgG after primary measles vaccination. *Clin Diagn Lab Immunol* **6:**178–180.
125. **Albrecht P, Herrmann K, Burns GR.** 1981. Role of virus strain in conventional and enhanced measles plaque neutralization test. *J Virol Methods* **3:**251–260.
126. **Ratnam S, Gadag V, West R, Burris J, Oates E, Stead F, Bouilianne N.** 1995. Comparison of commercial enzyme immunoassay kits with plaque reduction neutralization test for detection of measles virus antibody. *J Clin Microbiol* **33:**811–815.
127. **Cohen BJ, Parry RP, Doblas D, Samuel D, Warrener L, Andrews N, Brown D.** 2006. Measles immunity testing: comparison of two measles IgG ELISAs with plaque reduction neutralisation assay. *J Virol Methods* **131:**209–212.
128. **Tuokko H.** 1995. Detection of acute measles infections by indirect and mu-capture enzyme immunoassays for immunoglobulin M antibodies and measles immunoglobulin G antibody avidity enzyme immunoassay. *J Med Virol* **45:**306–311.
129. **Narita M, Yamada S, Matsuzono Y, Itakura O, Togashi T, Kikuta H.** 1996. Immunoglobulin G avidity testing in serum and cerebrospinal fluid for analysis of measles virus infection. *Clin Diagn Lab Immunol* **3:**211–215.
130. **de Souza VA, Pannuti CS, Sumita LM, de Andrade Junior HF.** 1997. Enzyme linked immunosorbent assay IgG antibody avidity test for single sample serologic evaluation of measles vaccines. *J Med Virol* **52:**275–279.
131. **Mercader S, Garcia P, Bellini WJ.** 2012. Measles virus IgG avidity assay for use in classification of measles vaccine failure in measles elimination settings. *Clin Vaccine Immunol* **19:**1810–1817.
132. **Paunio M, Hedman K, Davidkin I, Valle M, Heinonen OP, Leinikki P, Salmi A, Peltola H.** 2000. Secondary measles vaccine failures identified by measurement of IgG avidity: high occurrence among teenagers vaccinated at a young age. *Epidemiol Infect* **124:**263–271.
133. **Pannuti CS, Morello RJ, Moraes JC, Curti SP, Afonso AM, Camargo MC, Souza VA.** 2004. Identification of primary and secondary measles vaccine failures by measurement of immunoglobulin G avidity in measles cases during the 1997 São Paulo epidemic. *Clin Diagn Lab Immunol* **11:**119–122.
134. **Hickman CJ, Hyde TB, Sowers SB, Mercader S, McGrew M, Williams NJ, Beeler JA, Audet S, Kiehl B, Nandy R, Tamin A, Bellini WJ.** 2011. Laboratory characterization of measles virus infection in previously vaccinated and unvaccinated individuals. *J Infect Dis* **204**(Suppl 1)**:**S549–S558.
135. **Rota JS, Hickman CJ, Sowers SB, Rota PA, Mercader S, Bellini WJ.** 2011. Two case studies of modified measles in vaccinated physicians exposed to primary measles cases: high risk of infection but low risk of transmission. *J Infect Dis* **204** (Suppl 1)**:**S559–S563.
136. **Rosen JB, Rota JS, Hickman CJ, Sowers SB, Mercader S, Rota PA, Bellini WJ, Huang AJ, Doll MK, Zucker JR, Zimmerman CM.** 2014. Outbreak of measles among persons with prior evidence of immunity, New York City, 2011. *Clin Infect Dis* **58:**1205–1210.
137. **Cohen BJ, Audet S, Andrews N, Beeler J, WHO working group on measles plaque reduction neutralization test.** 2007. Plaque reduction neutralization test for measles antibodies: description of a standardised laboratory method for use in immunogenicity studies of aerosol vaccination. *Vaccine* **26:** 59–66.
138. **Ono N, Tatsuo H, Hidaka Y, Aoki T, Minagawa H, Yanagi Y.** 2001. Measles viruses on throat swabs from measles patients use signaling lymphocytic activation molecule (CDw150) but not CD46 as a cellular receptor. *J Virol* **75:**4399–4401.
139. **Zaki S, Bellini WJ.** 1997. Measles virus, p 233–245. *In* Connor DH, Chandler FW, Schwartz DA, Manz DJ, Lack EE (ed), *Pathology of Infectious Diseases*, vol 1. Appleton and Lange, Stamford, CT.
140. **Nakayama T, Mori T, Yamaguchi S, Sonoda S, Asamura S, Yamashita R, Takeuchi Y, Urano T.** 1995. Detection of measles virus genome directly from clinical samples by reverse transcriptase-polymerase chain reaction and genetic variability. *Virus Res* **35:**1–16.
141. **Rota PA, Khan AS, Durigon E, Yuran T, Villamarzo YS, Bellini WJ.** 1995. Detection of measles virus RNA in urine specimens from vaccine recipients. *J Clin Microbiol* **33:**2485–2488.
142. **Hummel KB, Lowe L, Bellini WJ, Rota PA.** 2006. Development of quantitative gene-specific real-time RT-PCR assays for the detection of measles virus in clinical specimens. *J Virol Methods* **132:**166–173.
143. **Hübschen JM, Kremer JR, De Landtsheer S, Muller CP.** 2008. A multiplex TaqMan PCR assay for the detection of measles and rubella virus. *J Virol Methods* **149:**246–250.
144. **Hyde TB, Dayan GH, Langidrik JR, Nandy R, Edwards R, Briand K, Konelios M, Marin M, Nguyen HQ, Khalifah AP, O'leary MJ, Williams NJ, Bellini WJ, Bi D, Brown CJ, Seward JF, Papania MJ.** 2006. Measles outbreak in the Republic of the Marshall Islands, 2003. *Int J Epidemiol* **35:**299–306.
145. **Bellini WJ, Rota PA.** 1998. Genetic diversity of wild-type measles viruses: implications for global measles elimination programs. *Emerg Infect Dis* **4:**29–35.
146. **Katz RS, Premenko-Lanier M, McChesney MB, Rota PA, Bellini WJ.** 2002. Detection of measles virus RNA in whole blood stored on filter paper. *J Med Virol* **67:**596–602.
147. **Oliveira SA, Siqueira MM, Camacho LA, Castro-Silva R, Bruno BF, Cohen BJ.** 2003. Use of RT-PCR on oral fluid samples to assist the identification of measles cases during an outbreak. *Epidemiol Infect* **130:**101–106.
148. **Jin L, Brown DWG, Ramsay MEB, Rota PA, Bellini WJ.** 1997. The diversity of measles virus in the United Kingdom, 1992–1995. *J Gen Virol* **78:**1287–1294.
149. **Jayamaha J, Binns PL, Fennell M, Ferson MJ, Newton P, Tran T, Catton M, Robertson P, Rawlinson W.** 2012. Laboratory diagnosis, molecular characteristics, epidemiological and clinical features of an outbreak of measles in a low incidence population in Australia. *J Clin Virol* **54:**168–173.
150. **El Mubarak HS, De Swart RL, Osterhaus AD, Schutten M.** 2005. Development of a semi-quantitative real-time RT-PCR for the detection of measles virus. *J Clin Virol* **32:**313–317.

151. **Plumet S, Gerlier D.** 2005. Optimized SYBR green real-time PCR assay to quantify the absolute copy number of measles virus RNAs using gene specific primers. *J Virol Methods* **128:** 79–87.
152. **Rota JS, Rosen JB, Doll MK, McNall RJ, McGrew M, Williams N, Lopareva EN, Barskey AE, Punsalang A Jr, Rota PA, Oleszko WR, Hickman CJ, Zimmerman CM, Bellini W.J.** 2013. Comparison of the sensitivity of laboratory diagnostic methods from a well-characterized outbreak of mumps in New York City in 2009. *Clin Vaccine Immunol* **20:** 391–396.
153. **Krause CH, Molyneaux PJ, Ho-Yen DO, McIntyre P, Carman WF, Templeton KE.** 2007. Comparison of mumps-IgM ELISAs in acute infection. *J Clin Virol* **38:**153–156.
154. **Gut JP, Spiess C, Schmitt S, Kirn A.** 1985. Rapid diagnosis of acute mumps infection by a direct immunoglobulin M antibody capture enzyme immunoassay with labeled antigen. *J Clin Microbiol* **21:**346–352.
155. **Krause CH, Eastick K, Ogilvie MM.** 2006. Real-time PCR for mumps diagnosis on clinical specimens—comparison with results of conventional methods of virus detection and nested PCR. *J Clin Virol* **37:**184–189.
156. **Sakata H, Tsurudome M, Hishiyama M, Ito Y, Sugiura A.** 1985. Enzyme-linked immunosorbent assay for mumps IgM antibody: comparison of IgM capture and indirect IgM assay. *J Virol Methods* **12:**303–311.
157. **Davidkin I, Jokinen S, Paananen A, Leinikki P, Peltola H.** 2005. Etiology of mumps-like illnesses in children and adolescents vaccinated for measles, mumps, and rubella. *J Infect Dis* **191:**719–723.
158. **Barskey AE, Juieng P, Whitaker BL, Erdman DD, Oberste MS, Chern SW, Schmid DS, Radford KW, McNall RJ, Rota PA, Hickman CJ, Bellini WJ, Wallace GS.** 2013. Viruses detected among sporadic cases of parotitis, United States, 2009–2011. *J Infect Dis* **208:**1979–1986.
159. **Utz JP, Kasel JA, Cramblett HG, Szwed CF, Parrott RH.** 1957. Clinical and laboratory studies of mumps. I. Laboratory diagnosis by tissue-culture technics. *N Engl J Med* **257:**497–502.
160. **Utz JP, Szwed CF, Kasel JA.** 1958. Clinical and laboratory studies of mumps. II. Detection and duration of excretion of virus in urine. *Proc Soc Exp Biol Med* **99:**259–261.
161. **Reina J, Ballesteros F, Ruiz de Gopegui E, Munar M, Mari M.** 2003. Comparison between indirect immunofluorescence assay and shell vial culture for detection of mumps virus from clinical samples. *J Clin Microbiol* **41:**5186–5187.
162. **Chen M-H, Zhu Z, Zhang Y, Favors S, Xu WB, Featherstone DA, Icenogle JP.** 2007. An indirect immunocolorimetric assay to detect rubella virus infected cells. *J Virol Methods* **146:**414–418.
163. **Boriskin YuS, Booth JC, Yamada A.** 1993. Rapid detection of mumps virus by the polymerase chain reaction. *J Virol Methods* **42:**23–32.
164. **Palacios G, Jabado O, Cisterna D, de Ory F, Renwick N, Echevarria JE, Castellanos A, Mosquera M, Freire MC, Campos RH, Lipkin WI.** 2005. Molecular identification of mumps virus genotypes from clinical samples: standardized method of analysis. *J Clin Microbiol* **43:**1869–1878.
165. **Centers for Disease Control and Prevention.** 2001. Control and prevention of rubella: evaluation and management of suspected outbreaks, rubella in pregnant women, and surveillance for congenital rubella syndrome. *MMWR Recomm Rep* 50(RR-12)**:**1–23.
166. **Takahashi S, Machikawa F, Noda A, Oda T, Tachikawa T.** 1998. Detection of immunoglobulin G and A antibodies to rubella virus in urine and antibody responses to vaccine-induced infection. *Clin Diagn Lab Immunol* **5:**24–27.
167. **Centers for Disease Control and Prevention (CDC).** 2008. Recommendations from an ad hoc Meeting of the WHO Measles and Rubella Laboratory Network (LabNet) on use of alternative diagnostic samples for measles and rubella surveillance. *MMWR Morb Mortal Wkly Rep* **57:**657–660.
168. **Tipples GA, Hamkar R, Mohktari-Azad T, Gray M, Ball J, Head C, Ratnam S.** 2004. Evaluation of rubella IgM enzyme immunoassays. *J Clin Virol* **30:**233–238.
169. **Wandinger KP, Saschenbrecker S, Steinhagen K, Scheper T, Meyer W, Bartelt U, Enders G.** 2011. Diagnosis of recent primary rubella virus infections: significance of glycoprotein-based IgM serology, IgG avidity and immunoblot analysis. *J Virol Methods* **174:**85–93.
170. **Hyde TB, Sato HK, Hao L, Flannery B, Zheng Q, Wannemuehler K, Ciccone FH, de Sousa Marques H, Weckx LY, Sáfadi MA, de Oliveira Moraes E, Pinhata MM, Olbrich Neto J, Bevilacqua MC, Tabith Junior A, Monteiro TA, Figueiredo CA, Andrus JK, Reef SE, Toscano CM, Castillo-Solorzano C, Icenogle JP, CRS Biomarker Study Group.** 2015. Identification of serologic markers for school-aged children with congenital rubella syndrome. *J Infect Dis* **212:**57–66.
171. **Zhu Z, Xu W, Abernathy ES, Chen M-H, Zheng Q, Wang T, Zhang Z, Li C, Wang C, He W, Zhou S, Icenogle J.** 2007. Comparison of four methods using throat swabs to confirm rubella virus infection. *J Clin Microbiol* **45:**2847–2852.
172. **Lambert ND, Pankratz VS, Larrabee BR, Ogee-Nwankwo A, Chen MH, Icenogle JP, Poland GA.** 2014. High-throughput assay optimization and statistical interpolation of rubella-specific neutralizing antibody titers. *Clin Vaccine Immunol* **21:**340–346.
173. **Lambert ND, Haralambieva IH, Kennedy RB, Ovsyannikova IG, Pankratz VS, Poland GA.** 2015. Polymorphisms in HLA-DPB1 are associated with differences in rubella virus-specific humoral immunity after vaccination. *J Infect Dis* **211:** 898–905.
174. **Abernathy E, Cabezas C, Sun H, Zheng Q, Chen M-H, Castillo-Solorzano C, Ortiz AC, Osores F, Oliveira L, Whittembury A, Andrus JK, Helfand RF, Icenogle J.** 2009. Confirmation of rubella within 4 days of rash onset: comparison of rubella virus RNA detection in oral fluid with immunoglobulin M detection in serum or oral fluid. *J Clin Microbiol* **47:**182–188.
175. **Reisdorf E, Bellini WJ, Rota P, Icenogle J, Davis T, Wroblewski K, Hagan CN, Shult P.** 2013. *Public Health Reference Laboratories: A Model for Increasing Molecular Diagnostic Testing and Genotyping Capacity for Measles and Rubella, abstr, p 41*. Association of Public Health Laboratories Meeting. Association of Public Health Laboratories, Raleigh, NC.
176. **Centers for Disease Control and Prevention.** 2012. Rubella, chapter 14. *In* Roush SW, Baldy LM (ed), *Manual for the Surveillance of Vaccine-Preventable Diseases*. U.S. Department of Health and Human Services, CDC, Atlanta, GA. http://www.cdc.gov/vaccines/pubs/surv-manual/chpt14-rubella.html.
177. **Best JM, O'Shea S, Tipples G, Davies N, Al-Khusaiby SM, Krause A, Hesketh LM, Jin L, Enders G.** 2002. Interpretation of rubella serology in pregnancy—pitfalls and problems. *BMJ* **325:**147–148.

Gastrointestinal Viruses

MICHAEL D. BOWEN

22

Since the demonstration by Albert Kapikian and colleagues that a virus was the etiological agent of Norwalk gastroenteritis (1), multiple viruses have been recognized as significant causes of gastroenteritis in children and adults. Gastrointestinal viruses have been implicated in food and waterborne outbreaks of gastroenteritis, seasonal disease, and outbreaks in all age groups in both the developed and developing worlds. The major gastrointestinal viruses causing acute gastroenteritis are found in several virus families (Table 1) and will be discussed in greater detail in the following text.

VIRAL CLASSIFICATION AND BIOLOGY

Caliciviruses

The family *Caliciviridae* is divided into 5 genera, *Norovirus*, *Sapovirus*, *Lagovirus*, *Nebovirus*, and *Vesivirus*, with human pathogens found in the first 2 groups (noroviruses and sapoviruses). Caliciviruses infecting humans are nonenveloped round viruses 27 to 39 nm in diameter (Fig. 1). When viewed by electron microscopy (EM), noroviruses have a smooth round morphology, whereas sapoviruses have a "Star of David" appearance (1–3). The genome of noroviruses and sapoviruses is a single-stranded plus-sense RNA 7.3 to 7.5 kb in length (4, 5). The 5′ end of the genomic RNA is linked covalently to a small protein encoded by the viral genome VPg, and the 3′ end is polyadenylated (6). The genome of noroviruses and sapovirus genotypes GI, GIV, and GV encodes 3 open reading frames (ORFs), whereas the genome of sapovirus genotype GII encodes 2 ORFs (4, 5). Viral nonstructural proteins (NS1 through NS7) are encoded as a polyprotein in ORF1 located at the 5′ end of the genome in both viruses (6). The major capsid protein VP1 is also encoded in ORF1 in the sapoviruses but is expressed through ORF2 in the noroviruses (4, 5). Both viruses express the minor capsid protein VP2 in an ORF located nearest the poly-A tail at the 3′ end of the genome (ORF2 in sapoviruses, ORF3 in noroviruses) (6). Sapovirus genotypes GI, GIV, and GV also encode a small protein of unknown function in an ORF (ORF3) that overlaps the VP1 coding region of ORF1(4). VP1 comprises the viral capsid, which is characterized by structures resembling cups from which the virus family derives its name (calici originated from the Latin word for cup, *calyx*) (6). Standard systems for propagating human noroviruses and sapoviruses in cell culture have yet to be developed, so much of what is known about the biology and immunology of these viruses has been determined using surrogate animal strains or virus-like particles produced using recombinant expression systems (5, 7). Recent findings suggest that human B cells are the cellular target of noroviruses (8).

The *Norovirus* and *Sapovirus* genera derive their names from the prototype strains of each genus, Norwalk virus and Sapporo virus, respectively (1, 9). The genus *Norovirus* contains 1 species, *Norwalk virus* (Table 1), and 6 recognized genogroups, GI through GVI (6). Human pathogens are found in genogroups GI, GII, and GIV (Table 1) with 9, 19, and 1 individual genotypes causing human disease assigned to each genogroup, respectively (6). Genogroup GII, genotype 4 (GII.4), is the variant responsible for the majority of norovirus outbreaks (10) (Table 1). Genotype GI.1 is the prototype Norwalk virus strain (6). The genus *Sapovirus* contains 1 species, Sapporo virus, and 4 genogroups causing human disease, GI, GII, GIV, and GV (Table 1). The prototype Sapporo virus is designated genotype GI/1 (4).

Rotaviruses

Rotaviruses are members of the genus *Rotavirus* of the virus family *Reoviridae*. Within the genus there are 8 recognized groups: A, B, C, D, E, F, G, and H. Group A rotaviruses (RVA) are the primary human pathogen, although groups B, C, and H also have been infrequently reported to be associated with human disease (11, 12); see the section on other viruses. Groups D through G have been detected in animals only (12). RVA virions are approximately 100 nm in diameter, nonenveloped, and have a characteristic appearance when viewed by EM (Fig. 1.). Each virion is composed of 3 concentric shell layers with visible spikes protruding through the outmost shell (Fig. 1). The genome of rotaviruses contains 11 segments of double-stranded RNA (dsRNA) with a total genome size of approximately 17.5 kb. Each segment encodes one or two proteins, 6 structural proteins (VP1-4, VP6, VP7) and 5 or 6 nonstructural proteins (NSP1-5/6) (Table 2). The outer capsid is composed of VP7, a glycosylated protein (G-protein) shell from which spikes of VP4, a protease-sensitive protein (P protein), protrude and virions retaining the 3-layer configuration are referred to as triple-layered particles (13). VP6 comprises the inner capsid and

doi:10.1128/9781555819156.ch22

TABLE 1 Major gastrointestinal viruses causing acute gastroenteritis

Family	Genus	Species	Genogroup	Genotype/Serotype
Caliciviridae	*Norovirus*	*Norwalk virus*	GI	GI.1
			GII	GII.4
			GIV	
	Sapovirus	*Sapporo virus*	GI	
			GII	
			GIV	
			GV	
Reoviridae	*Rotavirus*	*Rotavirus A*	1	G1P[8]
				G3P[8]
				G4P[8]
				G9P[8]
				G12P[8]
			2	G2P[4]
Astroviridae	*Mamastrovirus*	*Mamastrovirus 1*	I	1
				2
				3
				4
				5
				6
				7
				8
		Mamastrovirus 6		MLB1
				MLB2
				MLB3
		Mamastrovirus 8	II	HMO-A
				VA-2
				VA-4
		Mamastrovirus 9		HMO-B
				HMO-C
				VA-1
				VA-3
Adenoviridae	*Mastadenovirus*	*Adenovirus F*		Ad40
				Ad41

is the dominant immunogenic protein. Particles that lack the outer shell but retain the inner capsid are termed double-layered particles. The core shell is composed of VP2, and, in the virion, the structure encloses equimolar amounts of the 11 genome segments as well as the VP1 and VP3 proteins (13).

After ingestion, RVA pass through the stomach to the small intestine where RVA virions bind to cellular receptors through VP4 and VP7. Some strains have been shown to bind sialic acid moieties of intestinal enterocytes and others bind nonsialated HBGA in a similar manner to caliciviruses (14, 15). The major human-infecting RVA strains bind either Lewis or H antigens (16). After binding, the VP4 is cleaved by intestinal proteases, such as trypsin, into VP5* and VP8*, and the protein undergoes conformational changes. The VP8* region at the head of the VP4 protein is important for cellular receptor binding, whereas the lipophilic VP5* region, which forms the stalk, is important for virion entry into cells (13). RVA replicate in viroplasms located in the cell cytoplasm and bud through the membranes of the endoplasmic reticulum. Virions are then released through cell lysis or released from the apical surface of polarized cells by vesicular transport (13, 17).

Serotyping and genotyping of the VP6 protein has led to the establishment of the RVA through RVH groups (12). Traditionally RVA strains have been classified using a binomial typing system based on the VP7 (G-type) and VP4 (P-type) proteins and genes. There are at least 27 G types (G1 through G27) and 35 P-types (P[1] through P[35]) described to date (11). The brackets used to enclose the P-type indicate that it is a genotype rather than a serotype. Human disease is caused primarily by genotypes G1P[8], G2P[4], G3P[8], G4P[8], and G9P[8], and G12P[8] appears to be an emerging genotype. In 2008 the Rotavirus Classification Working Group established a new classification system that assigns genotypes to all eleven genes (18). The new nomenclature is Gx-P[x]-Ix-Rx-Cx-Mx-Ax-Nx-Tx-Ex-Hx (where x = genotype number), which represents the genotypes for VP7-VP4-VP6-VP1-VP2-VP3-NSP1-NSP2-NSP3-NSP4-NSP5/6, respectively. The new nomenclature is particularly useful for classifying RVA for which whole genome sequence data are available. The segmented genome

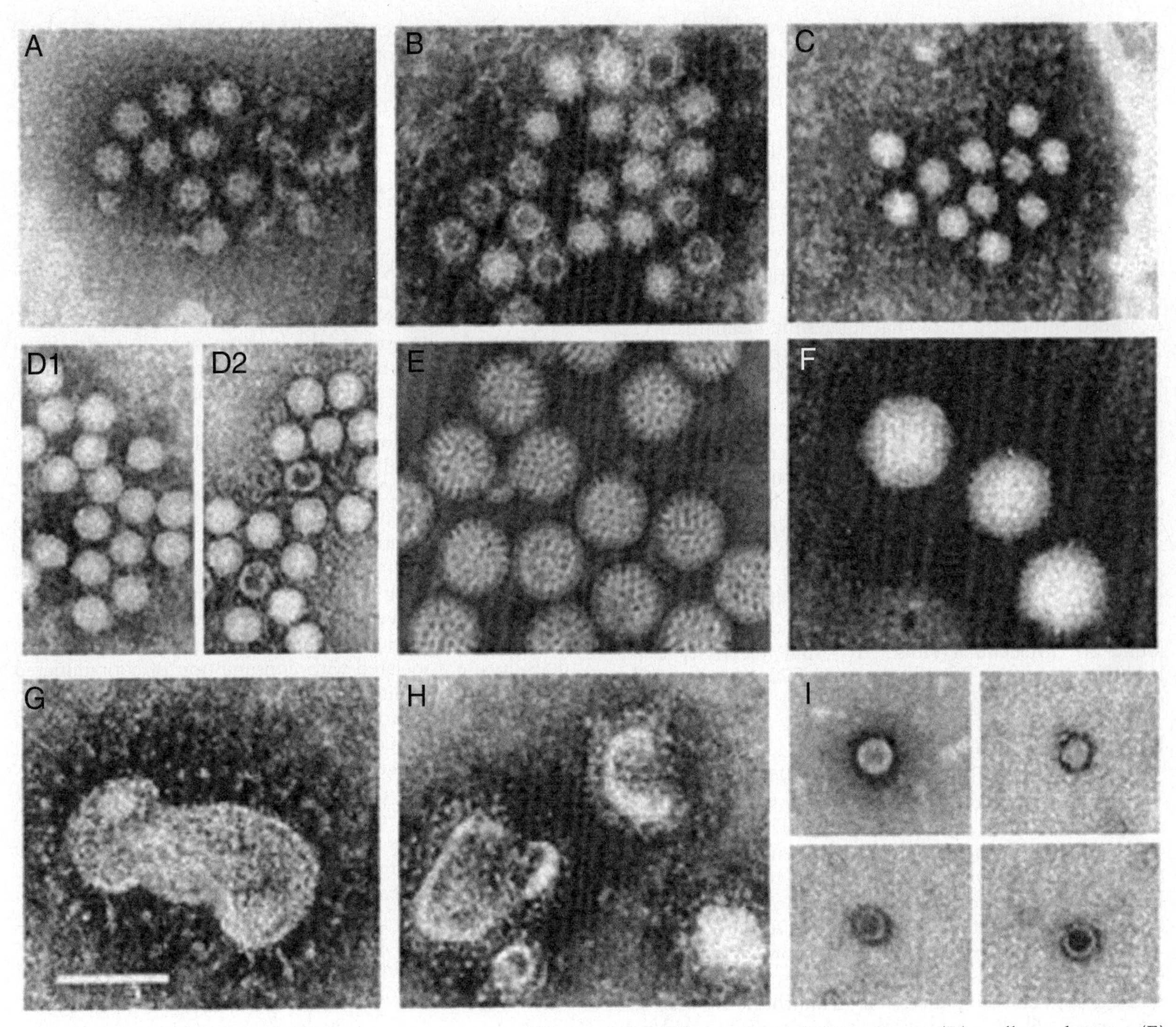

FIGURE 1 Electron micrographs of gastroenteritis viruses: (A) sapoviruses; (B) noroviruses; (C) astroviruses; (D) small round viruses; (E) rotaviruses; (F) adenoviruses; (G) coronaviruses; (H) torovirus-like particles; (I) picobirnaviruses. Bar = 100 nm.

TABLE 2 Rotavirus A genes and proteins

Segment no.	Gene	Protein
1	VP1	RNA polymerase
2	VP2	Core capsid shell protein
3	VP3	Methyltransferase
4	VP4	Protease-sensitive outer capsid spike protein
5	NSP1	Interferon antagonist
6	VP6	Inner capsid protein
7	NSP3	Translation enhancer
8	NSP2	NTPase
9	VP7	Glycosylated outer capsid shell protein
10	NSP4	Enterotoxin
11	NSP5	Phosphoprotein
11*	NSP6	

*not present in all strains.

of RVA permits different strains that infect the same cell to exchange segments through genetic reassortment, a major mechanism in RVA evolution.

Astroviruses

Astroviruses are members of the family *Astroviridae*. The family contains two genera: *Avastrovirus*, with 3 recognized species that infect birds, and *Mamastrovirus*, with 2 geno/serogroups and 19 species that have mammalian hosts (http://www.ictvonline.org/virusTaxonomy.asp). Of the strains infecting humans, *Mamastrovirus* species 1, which contains 8 genotypes (classical human astroviruses) and 3 genotypes (MLB1, 2, and 3), has been described for species 6 (Table 1). Species 8 and 9 encompass genotypes (HMO/VA strains) that infect humans but are closely related to strains from mink, sheep, and other mammals (19) (Table 1).

Astroviruses were first discovered by EM in 1975 in stool samples from children with gastoenteritis (20) and were subsequently given the name astroviruses due to their star-like appearance when viewed by EM following negative staining (21) (Fig. 1). Astroviruses observed in stool are approximately 28 to 30 nm in diameter and slightly larger when propagated in cell culture (22). The astrovirus genome consists of a single-stranded, positive-sense RNA approximately ranging in size from 6.2 kb to 7.7 kb plus a 3′ poly(A) tail approximately 30 adenines long (22, 23). The 5′ end of the genome is linked covalently to a virally-expressed VPg protein (24) and contains 3 ORFs designated ORF1a, ORF1b, and ORF2 (25). Full-length transcripts made from cloned astrovirus genomes are infectious (26). ORFs 1a and 1b express the nonstructural proteins, which are involved in gene transcription and genomic replication. These ORFs overlap by 10 to 148 bases, and the ORF1b protein is expressed through a ribosomal frameshift mechanism (22, 27). ORF2 encodes the viral structural proteins, and this gene is expressed as a polyprotein (VP90) via a positive-sense subgenomic RNA (23). Astroviruses have been shown to enter Caco-2 cells via a clathrin-dependent endocytic pathway (28). The process of astrovirus replication is not well understood at present but is known to involve production of a negative-sense intermediate RNA species that immediately precedes positive-strand RNA synthesis (29). VP90 has been shown to associate with membranes suggesting that these are the sites of initial particle assembly (30). The VP90 protein is cleaved by cellular capases into the capsid protein, VP70, which forms the capsid structures found on extracellular virus (31). Viral infectivity is enhanced by trypsin, which cleaves VP70 into 3 smaller structural proteins (32).

Enteric Adenoviruses

Adenoviruses were first isolated in the 1950s from explants of human tonsils and adenoids, the latter tissue serving as a source of the name "adenovirus" (33). The family *Adenoviridae* contains 5 genera with the adenoviruses of mammals assigned to the genus *Mastadenovirus* (http://www.ictvonline.org/virusTaxonomy.asp). Seven species—A, B, C, D, E, F, and G—comprise the human adenoviruses with at least 57 different serotypes. Adenoviruses are known to cause a variety of human diseases including respiratory illnesses, pharyngitis, keratoconjunctivitis, hemorrhagic cystitis, and meningoencephalitis, as well as gastroenteritis (34). Many adenovirus species can be detected in human stools, but their causal role in gastroenteritis is unclear (35). Two adenoviruses primarily associated with diarrhea are Species F types Ad40 and Ad41 (Table 1), but Species A types Ad12, Ad18, and Ad31 and Species C types Ad1, Ad2, Ad5, and Ad6 have also been implicated (36). There is a report of Species G type Ad52 causing diarrhea in adults (37).

Adenoviruses are nonenveloped viruses with an icosahedral capsid shell (Fig. 1) approximately 95 nm in diameter (38). The capsid is mostly composed of 240 trimers of hexon protein along with 12 pentamers of penton protein, which interact with trimeric fiber proteins (39). The adenovirus genome consists of a linear dsDNA molecule approximately 34 to 36 kb long (40). The genome encodes more than 40 proteins, only 12 of which have been shown to be structural components of the virus (38).

What is known about the replication of adenoviruses is derived from studies of nonenteric types. Various receptors have been identified for different adenovirus types, but the cellular receptor(s) for Ad40/41 has yet to be identified (41). The penton fiber serves as a cellular attachment protein. After entry into the cell via endocytosis, adenoviruses partially uncoat in the endosome and escape into the cytoplasm where further capsid disassembly occurs. Then the viral DNA and core proteins enter the cell nucleus where genomic replication takes place (39, 42). Newly synthesized viral proteins are transported from the cytosol to the cell nucleus for virus assembly (43). The final step in virus maturation involves processing of precursor structural proteins by a virally-encoded protease (44).

EPIDEMIOLOGY

Caliciviruses

Norovirus outbreaks occur year-round, though there is an increase in the winter months. Hence the name "winter vomiting disease." Noroviruses are spread primarily by fecal-oral transmission and are the leading cause of nonbacterial foodborne illness worldwide (45, 46). Noroviruses appear to be overtaking rotavirus as the most common cause of severe pediatric gastroenteritis in countries that have introduced rotavirus vaccination (47). In the United States, an estimated 19 to 21 million illnesses, 1.7 to 1.9 million outpatient visits, 400,000 emergency department visits, and 56,000 to 71,000 hospitalizations occur each year due to norovirus (48). Outbreaks occur frequently in closed populations such as restaurants, cruise ships, hospitals, daycare centers, and nursing homes, (45) and norovirus is commonly associated with traveler's diarrhea (49). Crowded conditions facilitate transmission. Waterborne transmission occurs as does disease associated with consumption of contaminated shellfish (50). Fomites can play a role in transmission due to the stability of noroviruses on environmental surfaces and their resistance to conventional cleaning agents (51). Norovirus is shed in feces at concentrations up to 10^{11} particles per gram of stool (52), and the 50% infectious dose for humans has been determined to be approximately 1320 genomic copy units (53). The virus is also shed in vomitus, and vomiting can transmit the disease as well as contaminate environmental surfaces (54).

The epidemiology of sapoviruses is similar to that of noroviruses except that sapovirus outbreaks tend to occur most often in children. Almost all children have serological evidence of infection by age 5 (55).

Rotaviruses

RVA are a significant cause of moderate to severe diarrhea in children under 5 years of age, and an estimated 453,000 deaths occur annually due to RVA acute gastroenteritis (AGE), primarily in Sub-Saharan Africa and the Indian subcontinent (56, 57). Most children are infected multiple times during the first few years of life. The primary infection tends to cause more severe disease than later infections, and the probability of reinfection decreases with each subsequent infection (58). These principles serve as the logical basis for RVA vaccination, in which the first infections are simulated by administration of live-attenuated vaccines (see the Treatment and Prevention section). RVA disease also occurs in adults with documented outbreaks in elderly populations, most frequently caused by genotype G2P[4] strains (59, 60).

In most parts of the world there is a seasonal pattern to RVA activity. In temperate regions most infections occur during the fall and winter months. In the tropics the period of peak activity is the dry season, though infections tend to occur throughout the year (61). RVA are transmitted most often by fecal-oral transmission and are shed by infected individuals at very high concentrations, up to 10^{11} particles per gram of stool. Combined with the fact that the infectious dose is very low (10^2 to 10^3 particles (62)), one can see how RVA are highly transmissible. RVA is environmentally stable and can also be transmitted via food and water, and there is some evidence of respiratory transmission (63). RVA are resistant to many disinfectants but can be inactivated by chlorine solutions and alcohol-based hand cleaners (64, 65).

Astroviruses

Astroviruses are transmitted primarily by the fecal-oral route. Most of the associated disease is in children under 2 years of age, though infections also occur in older children, adults, and the elderly (22). Astroviruses are found worldwide with typical prevalences of 2 to 9% in cases of acute nonbacterial gastroenteritis, though this rate of infection can be higher in rural areas and in developing countries (22, 66). Astroviruses are often detected along with other viral gastroenteritis viruses, such as noroviruses and RVA. Astroviruses are shed at very high concentrations by infected individuals, up to 10^{13} genome copies per gram of stool (67), and the infectious dose is presumed to be low. In temperate areas the highest incidence of disease is in the colder months, whereas in tropical areas the peak activity tends to be during the rainy season, but these seasonal patterns may be associated with classical human astroviruses only (22). Outbreaks of astrovirus diarrhea have been documented in closed communities such as schools, child care centers, hospitals, and elder care facilities (35). Foodborne outbreaks of astrovirus disease have been described, and astrovirus infection has been associated with shellfish consumption (68). Fomites and waterborne dissemination also are thought to play roles in transmission of astroviruses (69, 70). Astroviruses can be inactivated by application of free chlorine at concentrations of 0.5 to 2.5 milligrams per liter (68).

Enteric Adenoviruses

Enteric adenoviruses are shed at very high concentrations in feces, up to 10^{11} virus particles per gram of stool (71) and are presumably transmitted by the fecal-oral route. Ad40 tends to cause infections in children under 1 year of age, whereas Ad41 infections are more common in older children (72). There appears to be no seasonality to enteric adenovirus infections, though there may be a midsummer peak in some regions (73). Infection with enteric adenoviruses is thought to occur worldwide based upon serological studies (74). Multiple studies have detected enteric adenoviruses in the stools of children living in developed countries, with significantly greater detections of virus in cases of diarrheal disease compared with controls (71). In developing countries, enteric adenoviruses have been found to be significant causes of moderate to severe diarrhea in children under 5 years of age (56). Enteric adenoviruses have caused outbreaks of diarrhea in hospitals, orphanages, and child care centers (35).

CLINICAL SIGNIFICANCE

Caliciviruses

Noroviruses cause acute gastroenteritis known as "winter vomiting disease" (75), and the first outbreak associated with noroviruses occurred in Norwalk, Ohio, in 1968 (76). Sapoviruses were first associated with an outbreak of gastroenteritis among Japanese infants in 1979 (2). Both viruses produce a clinical syndrome characterized by nausea, abdominal cramps, vomiting, low-grade fever, and diarrhea (35). The diarrhea does not contain blood or mucus and can be loose or watery (77). Virus is shed in the stool and vomitus of infected individuals. The average incubation periods for noroviruses and sapoviruses are 1.2 and 1.7 days, respectively (78). Symptoms generally resolve within 12 to 72 hours (77). Shedding of noroviruses occurs an average of approximately two weeks after virus is first detected in the stool but can last as long as 60 days (79). Norovirus infections in infants and the elderly produce severe gastroenteritis with a longer clinical course that may necessitate hospitalization (80).

The mechanism by which caliciviruses produce gastroenteritis involves a reduction in the villus surface area of the duodenum along with increases in the permeability of tight junctions and epithelial cell apoptosis (81). There is transient malabsorption of lactose, D-xylose, and fat and a leaking of water and ions into the intestinal lumen (81, 82). Intestinal tissue shows little inflammation with the exception of an intraepithelial infiltrate of CD8+ lymphocytes (81). Whereas CD8+ lymphocytes may be involved in the pathogenesis of human norovirus disease, CD4+ lymphocytes and secretory/serum antibodies have been shown to play roles in protective immunity in the mouse norovirus model (83).

Noroviruses are known to bind to human histo-blood group antigens (HBGAs), which are found on the surface of mucosal epithelial cells and are secreted as free oligosaccharides into body fluids such as blood, milk, and saliva (16). HBGAs include A/B, H, and Lewis antigens, and norovirus genotypes differ in their antigen-binding affinity (84); e.g., GII.4 strains bind A/B and H antigens only, GII.5 strain MOH binds A/B but not H or Lewis antigens, and GII.9 strain VA207 binds Lewis antigens only (16). Bacteria expressing HBGAs are thought to serve as a stimulatory factor for norovirus infection of human B cells (8).

Rotaviruses

RVA are the primary cause of AGE in children worldwide. The disease spectrum ranges from asymptomatic infection to moderate to severe AGE. Death due to RVA is most often the result of dehydration and lack of prompt access to medical care, particularly in developing countries (85). After ingestion of RVA, there is an incubation period of 1 to 4 days (average 2 days) (78). Disease typically begins with fever and vomiting followed 1 to 2 days later by diarrhea (61), which is typically nonbloody. RVA infection produces viremia, which can last up to 8 days (86). Seizures and neurological involvement have been reported in rare cases (87). In severe cases, dehydration with shock, electrolyte imbalance, and metabolic acidosis can occur with fatal outcome if medical attention is not administered. In most cases, disease resolves in 3 to 7 days, though virus shedding can continue for weeks after symptoms have subsided.

RVA infect the mature enterocytes in the villi of the small intestine (61). Progeny virus is released from infected cells 10 to 12 hours after infection, and virus is shed at high concentrations into the gut and infects neighboring cells (88). In a murine model, RVA infection causes enterocyte death and proliferation of secretory crypt cells (89). The RVA encoded endotoxin, NSP4, is thought to produce extracellular leakage through changes in the tight junctions between enterocytes (90). NSP4 also causes increased chloride secretion and water loss with activation of calcium channels and induction of diarrhea (89). NSP4 is also thought to activate the enteric nervous system with induction of vagal afferent nerves that connect to regions of the brain associated with nausea and vomiting (91).

Immunity to rotavirus appears to be primarily antibody-mediated, though interferon types I and III are thought to play a role in limiting infection (92). RVA infection is a poor inducer of the T cell-mediated immune response, and neutralizing antibody titers correlate with protection (93). Neutralizing antibodies are both homotypic (i.e., recognize the immunizing strain genotype) as well as heterotypic (i.e., cross-react with other genotypes). IgM is a marker for acute RV infection (94). Secretory IgA is believed to be the primary immune effector, and serum IgA is used as a correlate for measuring immune protection (95, 96). RVA-specific B cells circulated in the blood of infected individuals express the gut-specific homing receptor α4b7, suggesting local protection (97).

Astroviruses

Astroviruses cause gastroenteritis primarily in young children. The incubation period ranges from 1 to 5 days with a median incubation period of 4.5 days (78). Disease typically lasts 2 to 3 days and is characterized by mild diarrhea, abdominal pain, vomiting, and low-grade fever (98). Diarrhea caused by astroviruses tends to be milder than that associated with RVA infection and is less apt to cause severe dehydration (23). Compared with calicivirus and RVA infection, vomiting occurs less frequently (22). Shedding of virus in the feces can occur from a few days to weeks post disease onset (99).

Astrovirus replicates in the tissues of the small intestine, and virus particles have been detected in the low villous epithelium, surface epithelium, and macrophages of the lamina propria (100). Infected intestinal tissues show minor histological changes, and inflammation does not appear to contribute to disease pathogenesis (22). The exact mechanism by which astrovirus induces diarrhea is not known, but studies using an avian animal model suggests that it occurs through induction of sodium malabsorption (101). Immunity to astrovirus is not well understood, though there is the suggestion that virus-specific T-cells are involved in the development of protective immunity and that natural killer cells are involved in the antiviral response (102). The capsid protein of astroviruses has been shown to block activation of the complement system through both the classical and antibody-mediated pathways (103).

Enteric Adenoviruses

After an incubation period of 8 to 10 days, cases of adenovirus gastroenteritis develop watery, nonbloody diarrhea with a duration of 7 to 8 days, which is typically accompanied by vomiting and sometimes by fever of short duration (104). Compared to RVA, enteric adenoviruses produce a milder infection with a longer course of diarrhea. Prolonged diarrhea, up to 2 weeks' duration, is common, especially with Ad41 (71). Little is known about the pathogenesis of enteric adenoviruses. Examination of a fatal case of Ad41 by electron microscopy revealed crystalline arrays of virus particles in nuclei of cells in the intestinal epithelium suggesting that this is the site of replication for the virus (105).

Infections in Immunocompromised Hosts

In the immunocompromised host, the course and severity of infection can differ markedly compared with disease in immunocompetent individuals. In congenital cases of severe combined immunodeficiency (SCID), RVA can cause chronic infections with prolonged courses of diarrhea months in duration (106). Live attenuated RVA vaccine strains have also been implicated in persistent infections of SCID cases with prolonged viral shedding (107). In immunosuppressed transplant patients, noroviruses can cause chronic diarrhea and prolonged virus shedding lasting months to years (77). In children with inherited immune deficiencies, norovirus is among the most common gastrointestinal pathogen detected, and shedding of virus has been shown to last longer than 9 months in over 50% of cases (108). The incidence of astrovirus diarrhea has been shown to be especially high in persons undergoing chemotherapy, tissue or organ transplantation or with HIV infection (102), and astrovirus VA1/HMO-C has been identified as a cause of encephalitis in immunocompromised hosts (109). Untyped adenoviruses are associated with diarrhea in HIV patients (110), and a nonenteric adenovirus type has caused an outbreak of diarrheal disease among hematopoietic stem cell transplant recipients (111). Other viruses known to cause severe gastroenteritis in immunosuppressed individuals include cytomegalovirus (CMV) and Epstein-Barr virus (EBV) (36).

TREATMENT AND PREVENTION

Currently there are no antiviral therapies for viral agents of gastroenteritis. Treatment is primarily supportive and designed to prevent and reverse dehydration caused by diarrhea and vomiting (112). Oral rehydration is used in cases of mild to moderate dehydration and intravenous therapy is utilized in cases of severe dehydration or when the patient is unable to drink fluids. The World Health Organization recommends the use of an oral rehydration solution (ORS) containing 75 millimoles per liter of sodium and 75 millimoles per liter of glucose (224 millimoles per liter of total osmolarity) for the treatment of noncholera diarrhea and the administration of additional liquid or ORS to maintain hydration following each episode of diarrhea (61). Other therapies shown to reduce the symptoms and shorten the course of gastroenteritis include zinc supplementation, probiotic regimens, bismuth subsalicylate, loperamide, and ondansetron (113).

With the exception of RVA, vaccines are not available for the gastrointestinal viruses. A live-attenuated human-rhesus reassortant RVA vaccine (Rotashield, Wyeth Laboratories, Inc., Marietta, PA) was licensed for use in the United States in 1998 but withdrawn from the market in 1999 after cases of vaccine-associated intussusception were reported (114). Two new live-attenuated RVA vaccines were subsequently licensed in the United States and are now in use worldwide. RotaTeq (Merck & Co., Inc. West Point, PA) is a pentavalent human-bovine reassortant vaccine in which human RVA G1-G4 and P[8] genes have been inserted into a bovine strain, WC3, through genetic reassortment (115). It was licensed for use in the United States in 2006. Rotarix is a monovalent G1P[8] vaccine derived by serial passage of a human RVA strain, 89–12, in cell culture (115). It was licensed by the U.S. Food and Drug Administration in 2008. Both vaccines have been shown to be safe and efficacious and are currently in use in more than 75 countries (http://sites.path.org/rotavirusvaccine/rotavirus-vaccines/#global-intro). Two additional vaccines, LLR (Lanzhou Institute of Biological Products Co. Ltd., Gansu, China) and ROTAVAC (Bharat Biotech International Ltd., Hyderabad, India) are live-attenuated RVA vaccines in use in China and India, respectively.

The disease and economic burden of noroviruses has made development of vaccines against these pathogens a priority area of research, which has been hampered by the lack of *in vitro* cultivation systems and animal models for human noroviruses. VLPs expressing the major capsid protein VP1 of norovirus have been expressed in a variety of systems and have been shown to function antigenically in a manner similar to authentic capsids (116). Phase I clinical trials assessing the safety and immunogenicity of GI.1 and GII.4 VLPs have been carried out (117, 118).

General infection control and prevention strategies used for the management of norovirus outbreaks can be applicable to other viral agents of gastroenteritis. These include identification of the etiologic agent, increased hygiene, especially thorough hand washing, cleaning and disinfection of contaminated surfaces, clothing, bedding and other items, case isolation, and removal of contaminated food and infected food handlers (35, 54).

DETECTION AND DIAGNOSIS

Caliciviruses

Serology

Antibody detection assays are not used primarily for diagnosis but rather for serosurveys of populations. Assays for detection of total Ig IgA, IgG, and IgM have been developed using norovirus virus-like particles (VLPs) expressed in baculoviruses in enzyme-linked immunosorbant assay (ELISA) and dissociation-enhanced lanthanide fluorescent immunoassay (DELPHIA) formats (49, 119, 120). Detection of a greater than 4-fold rise in specific IgG titers in paired sera or detection of IgM is considered diagnostic. Assays have also been developed for detection of specific IgA in stool specimens (121).

Electron Microscopy

Immunoelectron microscopy (IEM) was used to detect the prototype Norwalk virus in 1972 (1), but EM is rarely used today for diagnosis because it is too labor intensive and lacks the sensitivity of other diagnostic assays for caliciviruses. The limit of detection of EM is approximately 10^6 virus particles per gram of stool, and caliciviruses are typically shed at titers below this detection threshold during clinical illness.

Immunoassays

Several commercial antigen detection enzyme immuno assays (EIAs) using polyclonal, or pools of monoclonal, antibodies raised against VLPs or recombinant antigens are available today. The RIDASCREEN Norovirus 3rd Generation Antigen Enzyme Immunoassay (R-Biopharm AG) has been approved by the U.S. Food and Drug Administration for *in vitro* diagnostic detection of norovirus genogroups GI and GII. The sensitivity of antigen detection EIAs is low compared with PCR-based assays, but the analytical performance of EIAs for norovirus detection increases as the number of outbreak samples tested increases (51). While these assays are easy to run compared with RT-PCR-based assays, their lower sensitivity and the detection challenges presented by the large variety of norovirus genotypes render antigen detection EIAs a second choice for norovirus diagnosis if reverse transcription-PCR (RT-PCR) or real time RT-PCR (qRT-PCR) assays are available.

RT-PCR

RT-PCR assays have been used for diagnosis of caliciviruses since the 1990s and are, along with qRT-PCR, the primary method for diagnosis today. RT-PCR methods have targeted different gene regions in ORF1, ORF2, and the ORF1-ORF2 junction and the ORF3 of noroviruses (122). These assays, when run in RT-PCR or RT-PCR plus nested/heminested format, offer distinct improvements in sensitivity compared with EIA and permit sequencing of amplicons for strain determination. Capsid gene (norovirus ORF3) sequences have been shown to be most reliable for genotype assignment (123). The primary disadvantage of conventional RT-PCR is that a postamplification analytical step is required to detect amplicons (see the next section).

Real-Time RT-PCR

The qRT-PCR assays, typically using the TaqMan format, are the gold-standard method for norovirus detection due to their sensitivity, which is often in the range of 10 to 100 genomic copies, and ease of use. The most frequent target regions are ORF1 and the highly conserved ORF1-ORF2 junction (124–127). The advantages of qRT-PCR over conventional RT-PCR are that no post-run analytical step (i.e., agarose gel electrophoresis) is required and the assays can target short, highly-conserved regions of the calicivirus genome that can serve as targets for multiple genotypes and are less apt to change through genetic drift. Elimination of the postamplification analytical procedure reduces the chance of PCR amplicon contamination in the laboratory. In addition, internal process controls such as MS2 phage can be incorporated into multiplexed assays to monitor for inhibitory substances in specimens (125). The qRT-PCR assays can be used to quantitate viral load through the use of quantitative standards such as RNA transcripts. Norovirus qRT-PCRs assays have been used to test a variety of sample types including stool, vomitus, food, shellfish, water, sewage, and other environmental specimens (128–131).

Rotaviruses

Serology

Serology is used rarely today for diagnosis of RVA infection but has been used extensively to determine the efficacy of RVA vaccines and the immune response against specific serotypes/genotypes (132, 133). Neutralization assays have been used to measure antibodies directed against the VP4, VP6, and VP7 proteins (134). ELISA methods are used to measure titers of specific anti-RVA antibody classes in children and adults (135). Assays have been developed to measure secretory IgA titers in stool and breast milk samples (136, 137).

Cell Culture

Diagnosis of RVA infection by isolation of virus in cultured cells has been accomplished since the 1970s but is rarely practiced today due to the time, effort, and technical expertise required. RVA can be isolated and propagated in various cell lines derived from monkey kidneys (e.g., BSC-1, CV-1, MA-104) and human adenocarcinomas (Caco-2, HT-29) (138). Trypsinization of the virus is typically performed preinoculation and during culture to cleave the VP4 protein into VP8* and VP5* peptides in order to enhance and maintain viral infectivity (139, 140). The use of primary monkey kidney cells has been shown to result in greater efficiency of virus isolation compared with continuous cell lines such as MA-104 (141). RVA infection and growth can be observed in cultured cells through the use of plaque assays or fluorescent-focus assays (138). Growth of RVA in intestinal organoids derived from human stem cells and containing multiple cell types offers a novel method for isolation and cultivation of RVA (142).

Electron Microscopy

RVA are identifiable in stool specimens by electron microscopy (EM) due to their characteristic appearance (Fig. 1) and high concentration in feces (143). RVA in stool samples or grown in culture can be negatively-stained with uranyl acetate or phosphotungstic acid at pH 4.5 (144). The limit of detection for EM is approximately 10^5 to 10^6 particles per gram of feces (145). Methods for concentrating RVA in stool and IEM further increase the sensitivity of this method (146, 147). EM methods are still used today for visualization of RVA in stools but lack the sensitivity and high throughput of other available diagnostic methods.

Electropherotyping

The 11-segmented genome of RVA can be extracted directly from stool and visualized by gel electrophoresis (148, 149). Typically this is accomplished by phenol-chloroform extraction of a 10% stool suspension and subsequent electrophoresis of the extracted material on a 5 to 10% polyacrylamide gel. Individual RNA segments are then visualized by silver staining. Three primary mobility patterns (electropherotypes) have been described, "long", "short," and "super short," which are due to differences in the length of the 3′ nontranslated region of segment 11 (150). In strains with a short or super short electropherotype, the longer segment 11 migrates slower during electrophoresis and appears between segments 9 and 10. Most genotype G1P[8], G3P[8], G4P[8], G9P[8], and G12P[8] strains will exhibit the long electropherotype whereas G2P[4], G2P[4]-related strains, and other genotypes will display the short or supershort electropherotype. While labor intensive, electropherotyping is useful for screening at the genogroup level and can be used detect rotaviruses from other groups (RVB, RVC, RVD, etc.), which display unique mobility patterns that distinguish them from RVA. The sensitivity of this method is reported to equal that of EM and immunoassays (149).

Immunoassays

EIAs offer a simple, rapid, and sensitive method for routine laboratory detection of RVA antigen in stool specimens (151). Commercial EIA kits have been available since the 1980s. Most use a sandwich format with polyclonal or monoclonal antibodies in which RVA antigen is captured onto the bottom surface of a plastic well by one antibody and then is detected by a second enzyme-conjugated antibody. These assays can be run in 96-well formats and have limits of detection of approximately 10^5 to 10^6 RVA particles per gram of stool. The Premier Rotaclone (Meridian Bioscience, Inc., Cincinnati, OH), ProSpecT Rotavirus Microplate Assay (Oxoid, Ltd., Basingstoke, Hampshire, UK), and RIDASCREEN Rotavirus (R-Biopharm AG, Darmstadt, Germany) kits have been shown to be equivalent in performance (152). The Premier Rotaclone kit has been approved by the U.S. Food and Drug Administration for *in vitro* diagnostic use. Other kits have obtained approval for *in vitro* diagnostic use in Europe.

Lateral flow devices for RVA also offer a simple and rapid method for RVA antigen detection in stool. Most devices use an immunochromatographic format with capture antibodies mounted on capillary bed material and detector antibodies conjugated to gold, latex, or other nanoparticles. They are particularly convenient for testing single samples, whereas the EIA format is easier and faster for testing samples in batches or larger numbers. The ImmunoCard STAT! Rotavirus (Meridian Bioscience, Inc., Cincinnati, OH) has been approved by the U.S. Food and Drug Administration for *in vitro* diagnostic use. The Remel Xpect Rotavirus (ThermoFisher, Inc., Waltham, MA), the RIDAQUICK Rotavirus (R-Biopharm AG, Darmstadt, Germany), the Mascia Brunelli Rotavirus Card (Mascia Brunelli S.p.a., Milan, Italy), the VIKIA Rota-Adeno (bioMérieux, Marcy l'Etoile, France), Rota-Strip/Uni-Strip (Coris BioConcept, Gembloux, Belgium), and the Rapid Diagnostic Test-CE-Rotavirus-A Antigen (Beijing Macro-Union Pharmaceutical Co., Ltd, Beijing, China) have been approved for *in vitro* diagnostic use in Europe. The RIDAQUICK Rotavirus is available in both dipstick and cassette formats and in combination with a lateral flow assay for enteric adenoviruses.

RT-PCR

Conventional RT-PCR assays offer a high sensitivity testing format that is 10 to 100 times more sensitive than antigen detection assays. RT-PCR assays use either a one- or two-step reverse transcription followed by PCR amplification to achieve increased sensitivity for the detection of RVA. Gene targets of RT-PCR assays have included VP6 (153, 154) and VP7 (155, 156). Because RVA are typically shed in the feces at very high concentrations, antigen-detection EIAs usually are sufficient for diagnosis, and thus most clinical laboratories perform these assays instead of conventional RT-PCR for RVA. RT-PCR is done widely, however, in combination with heminested PCR to perform RVA genotyping of the VP4 (157–159) and VP7 (159–161) genes. In these methods, the genotype-specific products of the second round heminested PCR reaction are of different lengths and the G (VP7) and P (VP4) types of strains can be determined by analytical agarose gel electrophoresis. In addition, the products of the first round RT-PCR reactions can be sequenced to determine genotypes (162).

Real-Time RT-PCR

The qRT-PCR assays offer many advantages over traditional RT-PCR assays: increased sensitivity, higher throughput, and faster turnaround time as well as possible quantification of viral loads. The qRT-PCR assay does not require a post-run analytical step such as gel electrophoresis, and since the amplification products are not manipulated, the possibility of laboratory contamination with PCR products is greatly reduced. Multiple qRT-PCR assays have been developed for detection of RVA targeting VP2 (163),VP4 (164, 165), VP6 (166–170), VP7 (165, 171), NSP3 (172–177), and NSP4 (178) genes. Due to the genetic diversity of VP4 and VP6 gene segments, the NSP3 gene, encoded by genomic segment 7, has been shown to be the best target for detection of wide variety of RVA genotypes (172–177, 179), and multiple qRT-PCR assays have targeted a highly-conserved region near the 3′ end of the NSP3 gene (173–177). An NSP3-targeted qRT-PCR assay has been developed that does not require heat denaturation of the viral RNA prior to reaction setup (177). The qRT-PCR assays have been designed for detection of RVA vaccine strains in stool samples (180).

Astroviruses

Serology

Assays for detection of serological evidence of astrovirus infection have been developed in a variety of formats including ELISA, neutralization assay, and the luciferase immunoprecipitation system (LIPS) (181–184). Serology is useful for seroepidemiological studies of astroviruses but is not used generally for diagnostics today. Studies have shown high levels of seroprevalence against multiple astrovirus genotypes, which can reach 65% to greater than 90% in older children and adult populations (181–184).

Cell Culture

Astroviruses can be isolated and propagated in various cell lines such as Caco-2, HEK, T-PLC/PRF/5, and T84 if trypsin is included in the cell culture medium (185, 186) but, as for RVA, is rarely practiced today due to the time, effort, and technical expertise required. Integrated cell culture/RT-PCR (ICC/RT-PCR) and ICC/RT-PCR/nested PCR assays have been developed, which use cell culture to amplify astroviruses from clinical and environmental samples prior to detection by RT-PCR (187–189).

Electron Microscopy

EM has been used routinely for detection of astroviruses in stool specimens but is less sensitive than other detection methods (35). High concentrations of virus (10^6 to 10^7 particles per gram of stool) are needed for visualization of the characteristic starlike morphology because only around 10% of the virus particles exhibit this appearance (143, 190). Viral shedding at the necessary concentration for EM typically occurs only 12 to 48 hours after symptom onset (22) making diagnosis by EM difficult outside this time window. IEM enhances the sensitivity of EM and can permit serotyping of astrovirus strains but requires the availability of typing antibodies (191).

Immunoassays

EIAs using a sandwich enzyme capture format have been developed for both detection and serotyping of astrovirus antigens in stool specimens (192–194). Assays in both conventional EIA format as well lateral flow devices are available for astrovirus detection for diagnosis and epidemiologic studies. The ProSpecT Astrovirus (Oxoid, Ltd., Basingstoke, Hampshire, UK), DRG Astrovirus Ag ELISA (DRG International, Inc., Springfield, NJ), RIDASCREEN Astrovirus (R-Biopharm AG, Darmstadt, Germany), and the Serazym Astrovirus (Seramun Diagnostics, Heidesee, Germany) are EIA assays approved for *in vitro* diagnostic use in the European Union. Lateral flow devices include the InterMedical Astrovirus (feces) (InterMedical S.r.l., Villaricca, Italy) and CerTest Astrovirus (CerTest BioTec S.L., Zaragoza, Spain). The advantages of these assays are the commercial availability and ease of use, but they are not as sensitive as RT-PCR detection assays, with limits of detection around 10^5 to 10^6 particles per gram of stool (190, 195).

RT-PCR

RT-PCR assays offer increased sensitivity compared with EM and antigen detection EIAs. Conventional RT-PCR assays targeting ORF1, ORF2, and the 3′ untranslated region (3′-UTR) have been developed for classical human astroviruses, both pan-type as well as type specific (196). RT-PCR assays for detection of novel astroviruses (i.e., MLB, HMO) have also been developed in recent years (197–200). The limit of detection for astrovirus RT-PCR can be as low as 10 to 100 genome copies per gram of stool (190).

Real-Time RT-PCR

As is the case with caliciviruses and RVA, qRT-PCR has become the method of choice for rapid detection and quantitation of astroviruses due to its ease of use and sensitivity. The qRT-PCR assays have been developed for detection of astrovirus in clinical specimens (201–205). A trend in clinical diagnostics has been the development of qPCR/qRT-PCR assays for the simultaneous detection of astroviruses along with other agents of gastroenteritis (206–209) (See the Future Directions in Detection and Diagnosis section). The limits of detection for qRT-PCR assays can be less than 10 genome copies per gram of stool. In addition to qRT-PCR, isothermal loop mediated amplification (LAMP) assays have been develop for real-time detection of astrovirus (210–212).

Enteric Adenoviruses

Serology

Serology has been used to survey populations for evidence of enteric adenovirus infection but is rarely used for diagnosis. The prevalence of antibodies in children has been shown to rise with age. Serosurveys have shown that 33 to 50% of children exhibit serum antibodies to enteric adenoviruses after 6 months of age (71).

Cell Culture

Enteric adenoviruses were known as "fastidious" adenoviruses because they were more difficult to isolate than other adenoviruses and generally do not cause cytopathic effect in cell culture (213). Enteric adenoviruses can be isolated and cultivated in tertiary cynomolgus monkey kidney cells and human embryonic kidney (Graham) 293, Caco-2, HEp-2, and A549 cell lines (73, 213–215). A plaque assay for enteric adenoviruses has been developed using A549 cells (213).

Electron Microscopy

EM can be used to detect enteric adenoviruses in clinical samples but adenoviruses, typically types causing respiratory disease, are commonly detected in the stool of patients in the absence of gastrointestinal illness. Thus the detection of adenoviruses in the feces does not necessarily indicate enteric disease (40). The use of anti-Ad40/41 antisera in IEM permits specific identification of enteric adenoviruses.

Electropherotyping

Analysis of restriction endonuclease digests of viral DNA extracted from virus isolates by slab gel electrophoresis has been used to differentiate enteric adenoviruses from other adenovirus genome types and to characterize Ad40/41 strains (73).

Immunoassays

EIAs using monoclonal antibodies specific for Ad40/41 have been developed for detection of enteric adenovirus antigen in stool samples (216–218). Commercially available antigen detection EIAs include the DRG Adenovirus Ag ELISA (DRG International Inc., Springfield, NJ), the Adenovirus Antigen Capture ELISA (Virusys Corp., Taneytown, MD), the Adenovirus Antigen ELISA Kit (Abnova Corporation, Taipei City, Taiwan), the Adenovirus ELISA kit (Diagnostic Automation/Cortez Diagnostics Inc., Calabasas, CA), the Fecal Adenovirus Antigen ELISA Kit (Epitope Diagnostics, Inc., San Diego, CA), the Premier Adenoclone—Type 40/41 kit (Meridian Bioscience, Inc., Cincinnati, OH), the RIDASCREEN Adenovirus kit (R-Biopharm AG, Darmstadt, Germany), and the ProSpecT Adenovirus EZ Microplate Assay (Oxoid, Ltd., Basingstoke, Hampshire, UK). Commercially available lateral-flow antigen detection devices include the Adeno-Strip/Uni-Strip (Coris BioConcept,

Gembloux, Belgium) and the VIKIA Rota-Adeno (bio-Mérieux, Marcy l'Etoile, France).

PCR

PCR assays for detection of enteric adenoviruses in stool samples has been developed that target the EIB, hexon, and long-fiber genes (219–221). In addition, a panadenovirus typing assay that differentiates enteric adenoviruses from other types has been developed for use with virus isolates (222).

Real-Time PCR

Real-time PCR assays have been developed for detection of enteric adenoviruses in stool both in a single-target format (223) as well as a multipathogen assay format (166, 206, 208, 209, 224, 225). The multipathogen assay format for testing of stool for agents of gastroenteritis has been increasingly utilized (see the next section).

FUTURE DIRECTIONS IN DETECTION AND DIAGNOSIS

Multiple Pathogen Detection Assays

A trend in clinical diagnostics for infectious diseases has been a shift from specific-agent detection assays or culture methods to multiplexed PCR/RT-PCR assays for syndromic testing (226). Multiplexed PCR/RT-PCR assays offer a uniform testing format with excellent sensitivity and reproducibility and a short turnaround time when compared to conventional testing and individual PCR assays (227), especially those utilizing a qPCR/qRT-PCR format. These tests are potentially quite useful for screening of stool samples for agents of gastroenteritis because they can simultaneously screen for noroviruses, sapoviruses, RVA, astroviruses, and enteric adenoviruses, as well as bacteria and parasites. Commercial multiplexed PCR assays for detection of enteric pathogens have been developed for a variety of instrument platforms, and some have been approved for *in vitro* diagnostic use (226, 228). TaqMan array cards have been developed for simultaneous detection of 19 enteric pathogens including enteric viruses (229).

Next Generation Sequencing

Next generation sequencing (NGS) offers an unbiased approach to detection and characterization of enteric pathogens. With currently available diagnostics, enteric pathogens are identified in only 25 to 58% of cases of gastroenteritis (230), so samples from unexplained cases of gastroenteritis and outbreaks of diarrhea are candidates for NGS. Metagenomics studies using NGS identified novel astrovirus VA1 (231), bufavirus (232), picobirnaviruses, and circular DNA viruses (233) in cases of unexplained diarrhea. Multiple NGS platforms are available commercially and the technologies continue to evolve very rapidly (234). It is anticipated that the role of NGS will continue to expand in the study of gastroenteritis pathogen identification and will make its way into the clinical diagnostic laboratory as the technology becomes more automated and the costs decrease.

OTHER VIRUSES

The viruses listed in this section have been identified in stool specimens of gastroenteritis cases but their role as pathogens is less significant or has been less well established.

Rotaviruses B, C, and H

Rotaviruses B, C, and H (RVB, RVC, RVH) constitute other groups within the genus *Rotavirus*, family *Reoviridae*. Like RVA, all have genomes composed of 11 segments of dsRNA, which encode 6 structural genes and 5 nonstructural genes. RVB has been associated with gastroenteritis outbreaks in China, India, Bangladesh, and Nepal (235–239). RVC causes sporadic cases and outbreaks of gastroenteritis worldwide (240–247). RVH has been detected in adult gastroenteritis cases in China and Bangladesh (248, 249). RT-PCR assays have been designed for detection of RVB and RVC in stool samples (156).

Bufaviruses

Bufaviruses appear to be a new genus within the family *Parvoviridae*. Parvoviruses are small, nonenveloped viruses with single-stranded DNA (ssDNA) genomes approximately 5 kb in size (250). The prototype bufavirus was detected in pediatric cases of diarrhea from Burkina Faso (232) but have since been detected in the stools of cases of diarrheal illness from other parts of the world (233, 251–253). Studies have found low prevalences of bufavirus among subjects with diarrhea (i.e., 1 to 4%) (253), but the status of this virus as an etiologic agent of diarrhea remains to be determined.

Coronaviruses

Members of the family *Coronaviridae* are enveloped viruses with a single-stranded, plus-sense RNA genome 26 to 32 kb in length (254) (Fig. 1). Coronaviruses (Genus *Coronavirus*) are known gastrointestinal pathogens of animals and in humans cause primarily upper respiratory tract diseases. Two coronaviruses associated with severe respiratory disease, SARS-CoV and MERS-CoV, both cause vomiting and diarrhea in a significant percentage of cases (255). Reports in the literature document sporadic associations of other coronaviruses with gastrointestinal disease (256), but definitive proof of their etiologic role is lacking. Toroviruses (Genus *Torovirus*) are another group within the *Coronaviridae* (Fig. 1) that are suspect agents of human gastroenteritis (257, 258) but whose role as a definitive pathogen has yet to be established.

Picobirnaviruses

Picobirnaviruses (family *Picobirnaviridae*) are small, spherical viruses with a bisegmented dsRNA genome (Fig. 1). The larger RNA ranges from 1.75 to 2.6 kb in size and the smaller segment is 1.5 to 1.9 kb long (259). Picobirnaviruses have been detected in the stools of multiple mammalian species, reptiles and birds and have been found to cause persistent infections with longterm shedding usually without evidence of enteric disease (260). Picobirnaviruses have been detected in humans that are genetically similar to strains detected in cattle and pigs, suggesting that these viruses are potential zoonotic agents (259). In humans, case-control studies of picobirnavirus have either shown a slight increased incidence of infection in cases of diarrhea or have been inconclusive (261). Studies of picobirnaviruses have been hampered by the inability to culture these viruses *in vitro* and the lack of an animal model of infection (259).

Aichi Virus

Human Aichi virus is a member of the *Picornaviridae* family, genus *Kobuvirus*, species *Aichivirus* A (http://www.ictvonline.org/virusTaxonomy.asp). First associated with outbreaks of gastroenteritis associated with oyster consumption in Aichi

Prefecture, Japan, Aichi virus is a small, round nonenveloped virus approximately 30 nm in diameter (262) (Fig. 1). The genome is a plus-sense ssRNA approximately 8280 bases long, excluding a poly-A tail (263). Aichi virus gastroenteritis is characterized by diarrhea, abdominal pain, nausea, vomiting, and fever and may cause respiratory symptoms (263). The virus is thought to be transmitted by the fecal-oral route, either directly or indirectly through food or water (264). Serological studies from multiple countries have shown that 80% to 95% of persons aged 30 to 40 are seropositive for Aichi virus (263). An antigen detection ELISA and a qRT-PCR assay for detection of Aichi virus in stool have been developed (265, 266).

BEST PRACTICES

Gastrointestinal viruses will continue to be important human pathogens. As described in the detection and diagnosis section of this chapter, there are multiple methods for diagnosis of gastrointestinal viruses. Criteria for selection of assays to be used in the diagnostic laboratory include 1) the population being tested (children, adults, elderly, immunocompromised, etc.), 2) the likelihood of encountering specific viral pathogens, 3) equipment availability, 4) technical expertise for performing the testing and interpreting results, and 5) assay costs (267). Since gastrointestinal viruses are shed typically at very high concentrations in feces, the sample of choice for testing is stool. For caliciviruses, the current gold standard diagnostic method is qRT-PCR, and the populations to be tested include all age groups. For rotaviruses, astroviruses, and enteric adenoviruses, the availability of inexpensive commercial antigen detection assays makes them the current assays of choice when testing samples from pediatric populations, the elderly, and immunocompromised individuals. In the near future, multiplexed, multipathogen detection assays will come into greater use in clinical diagnostics and replace individual antigen-detection and PCR-based assays. These assays will be particularly useful when testing samples from populations in which infections with multiple pathogens are common (e.g., developing countries). NGS offers a powerful tool for pathogen detection and discovery, and metagenomics studies will help to determine the etiologic role of suspect agents of gastroenteritis (e.g., bufaviruses, picobirnaviruses) as well as identify novel viruses causing gastrointestinal disease. NGS will assume a greater role in clinical diagnostics in the near future as well.

REFERENCES

1. **Kapikian AZ, Wyatt RG, Dolin R, Thornhill TS, Kalica AR, Chanock RM.** 1972. Visualization by immune electron microscopy of a 27-nm particle associated with acute infectious nonbacterial gastroenteritis. *J Virol* **10:**1075–1081.
2. **Kogasaka R, Nakamura S, Chiba S, Sakuma Y, Terashima H, Yokoyama T, Nakao T.** 1981. The 33- to 39-nm virus-like particles, tentatively designed as Sapporo agent, associated with an outbreak of acute gastroenteritis. *J Med Virol* **8:**187–193.
3. **Farkas T, Zhong WM, Jing Y, Huang PW, Espinosa SM, Martinez N, Morrow AL, Ruiz-Palacios GM, Pickering LK, Jiang X.** 2004. Genetic diversity among sapoviruses. *Arch Virol* **149:**1309–1323.
4. **Hansman GS, Oka T, Katayama K, Takeda N.** 2007. Human sapoviruses: genetic diversity, recombination, and classification. *Rev Med Virol* **17:**133–141.
5. **Thorne LG, Goodfellow IG.** 2014. Norovirus gene expression and replication. *J Gen Virol* **95:**278–291.
6. **Green KY.** 2013. Caliciviridae: The Noroviruses, p. 582–608. *In* Knipe DM, Howley P (ed.), *Fields Virology* 6th ed., vol 1. Lippincott Williams & Wilkins, Philadelphia, PA.
7. **Debbink K, Costantini V, Swanstrom J, Agnihothram S, Vinje J, Baric R, Lindesmith L.** 2013. Human norovirus detection and production, quantification, and storage of virus-like particles. *Curr Protoc Microbiol* **31:**15K 15K.1.1-15K.1.45.
8. **Jones MK, Watanabe M, Zhu S, Graves CL, Keyes LR, Grau KR, Gonzalez-Hernandez MB, Iovine NM, Wobus CE, Vinjé J, Tibbetts SA, Wallet SM, Karst SM.** 2014. Enteric bacteria promote human and mouse norovirus infection of B cells. *Science* **346:**755–759.
9. **Chiba S, Sakuma Y, Kogasaka R, Akihara M, Horino K, Nakao T, Fukui S.** 1979. An outbreak of gastroenteritis associated with calicivirus in an infant home. *J Med Virol* **4:**249–254.
10. **Lindesmith LC, Donaldson EF, Lobue AD, Cannon JL, Zheng DP, Vinje J, Baric RS.** 2008. Mechanisms of GII.4 norovirus persistence in human populations. *PLoS Med* **5:**e31.
11. **Matthijnssens J, Ciarlet M, McDonald SM, Attoui H, Bányai K, Brister JR, Buesa J, Esona MD, Estes MK, Gentsch JR, Iturriza-Gómara M, Johne R, Kirkwood CD, Martella V, Mertens PP, Nakagomi O, Parreño V, Rahman M, Ruggeri FM, Saif LJ, Santos N, Steyer A, Taniguchi K, Patton JT, Desselberger U, Van Ranst M.** 2011. Uniformity of rotavirus strain nomenclature proposed by the Rotavirus Classification Working Group (RCWG). *Arch Virol* **156:**1397–1413.
12. **Matthijnssens J, Otto PH, Ciarlet M, Desselberger U, Van Ranst M, Johne R.** 2012. VP6-sequence-based cutoff values as a criterion for rotavirus species demarcation. *Arch Virol* **157:**1177–1182.
13. **Estes MK, Greenberg HB.** 2013. Rotaviruses, p. 1347–1401. *In* Knipe DM, Howley PM (ed.), *Fields Virology* 6th ed. Lippincott Williams & Wilkins, Philadelphia, PA.
14. **Ciarlet M, Ludert JE, Iturriza-Gómara M, Liprandi F, Gray JJ, Desselberger U, Estes MK.** 2002. Initial interaction of rotavirus strains with N-acetylneuraminic (sialic) acid residues on the cell surface correlates with VP4 genotype, not species of origin. *J Virol* **76:**4087–4095.
15. **Liu Y, Huang P, Tan M, Liu Y, Biesiada J, Meller J, Castello AA, Jiang B, Jiang X.** 2012. Rotavirus VP8*: phylogeny, host range, and interaction with histo-blood group antigens. *J Virol* **86:**9899–9910.
16. **Tan M, Jiang X.** 2014. Histo-blood group antigens: a common niche for norovirus and rotavirus. *Expert Rev Mol Med* **16:**e5.
17. **Jourdan N, Maurice M, Delautier D, Quero AM, Servin AL, Trugnan G.** 1997. Rotavirus is released from the apical surface of cultured human intestinal cells through nonconventional vesicular transport that bypasses the Golgi apparatus. *J Virol* **71:**8268–8278.
18. **Matthijnssens J, Ciarlet M, Rahman M, Attoui H, Bányai K, Estes MK, Gentsch JR, Iturriza-Gómara M, Kirkwood CD, Martella V, Mertens PP, Nakagomi O, Patton JT, Ruggeri FM, Saif LJ, Santos N, Steyer A, Taniguchi K, Desselberger U, Van Ranst M.** 2008. Recommendations for the classification of group A rotaviruses using all 11 genomic RNA segments. *Arch Virol* **153:**1621–1629.
19. **Jiang H, Holtz LR, Bauer I, Franz CJ, Zhao G, Bodhidatta L, Shrestha SK, Kang G, Wang D.** 2013. Comparison of novel MLB-clade, VA-clade and classic human astroviruses highlights constrained evolution of the classic human astrovirus nonstructural genes. *Virology* **436:**8–14.
20. **Appleton H, Higgins PG.** 1975. Letter: viruses and gastroenteritis in infants. *Lancet* **1:**1297.
21. **Madeley CR, Cosgrove BP.** 1975. Letter: 28 nm particles in faeces in infantile gastroenteritis. *Lancet* **2:**451–452.
22. **Bosch A, Pintó RM, Guix S.** 2014. Human astroviruses. *Clin Microbiol Rev* **27:**1048–1074.

23. **Mendez E, Arias CF.** 2013. Astroviruses, p. 609–628. *In* Knipe DM, Howley P (ed.), *Fields Virology* 6th ed., vol 1. Lippincott Williams & Wilkins, Philadelphia, PA.
24. **Fuentes C, Bosch A, Pintó RM, Guix S.** 2012. Identification of human astrovirus genome-linked protein (VPg) essential for virus infectivity. *J Virol* **86:**10070–10078.
25. **Jiang B, Monroe SS, Koonin EV, Stine SE, Glass RI.** 1993. RNA sequence of astrovirus: distinctive genomic organization and a putative retrovirus-like ribosomal frameshifting signal that directs the viral replicase synthesis. *Proc Natl Acad Sci USA* **90:**10539–10543.
26. **Geigenmüller U, Ginzton NH, Matsui SM.** 1997. Construction of a genome-length cDNA clone for human astrovirus serotype 1 and synthesis of infectious RNA transcripts. *J Virol* **71:**1713–1717.
27. **Marczinke B, Bloys AJ, Brown TD, Willcocks MM, Carter MJ, Brierley I.** 1994. The human astrovirus RNA-dependent RNA polymerase coding region is expressed by ribosomal frameshifting. *J Virol* **68:**5588–5595.
28. **Méndez E, Muñoz-Yañez C, Sánchez-San Martín C, Aguirre-Crespo G, Baños-Lara MR, Gutierrez M, Espinosa R, Acevedo Y, Arias CF, López S, Sandri-Goldin RM.** 2014. Characterization of human astrovirus cell entry. *J Virol* **88:**2452–2460.
29. **Jang SY, Jeong WH, Kim MS, Lee YM, Lee JI, Lee GC, Paik SY, Koh GP, Kim JM, Lee CH.** 2010. Detection of replicating negative-sense RNAs in CaCo-2 cells infected with human astrovirus. *Arch Virol* **155:**1383–1389.
30. **Méndez E, Aguirre-Crespo G, Zavala G, Arias CF.** 2007. Association of the astrovirus structural protein VP90 with membranes plays a role in virus morphogenesis. *J Virol* **81:**10649–10658.
31. **Méndez E, Salas-Ocampo E, Arias CF.** 2004. Caspases mediate processing of the capsid precursor and cell release of human astroviruses. *J Virol* **78:**8601–8608.
32. **Dryden KA, Tihova M, Nowotny N, Matsui SM, Mendez E, Yeager M.** 2012. Immature and mature human astrovirus: structure, conformational changes, and similarities to hepatitis E virus. *J Mol Biol* **422:**650–658.
33. **Rowe WP, Huebner RJ, Gilmore LK, Parrott RH, Ward TG.** 1953. Isolation of a cytopathogenic agent from human adenoids undergoing spontaneous degeneration in tissue culture. *Proc Soc Exp Biol Med.* Society for Experimental Biology and Medicine (New York, N.Y.) **84:**570–573.
34. **Wold WSM, Ison MG.** 2013. Adenoviruses, p. 1733–1767. *In* Knipe DM, Howley P (ed.), *Fields Virology* 6th ed., vol 1. Lippincott Williams & Wilkins, Philadelphia, PA.
35. **Farkas T, Jiang X.** 2009. Rotavirus, Caliciviruses, Astroviruses, Enteric Adenoviruses, and Other Viruses Causing Acute Gastroenteritis, p 283–310. *In* Specter S, Hodinka R, Young S, Wiedbrauk D (ed), *Clinical Virology Manual* 4th ed. ASM Press, Washington, DC.
36. **Wilhelmi I, Roman E, Sanchez-Fauquier A.** 2003. Viruses causing gastroenteritis. *Clin Microbiol Infect*: the official publication of the European Society of Clinical Microbiology and Infectious Diseases **9:**247–262.
37. **Jones MS II, Harrach B, Ganac RD, Gozum MM, Dela Cruz WP, Riedel B, Pan C, Delwart EL, Schnurr DP.** 2007. New adenovirus species found in a patient presenting with gastroenteritis. *J Virol* **81:**5978–5984.
38. **Smith JG, Wiethoff CM, Stewart PL, Nemerow GR.** 2010. Adenovirus. *Curr Top Microbiol Immunol* **343:**195–224.
39. **Gonçalves MA, de Vries AA.** 2006. Adenovirus: from foe to friend. *Rev Med Virol* **16:**167–186.
40. **Straus SE.** 1984. Adenovirus Infections in Humans, p 451–496. *In* Ginsberg HS (ed), *The Adenoviruses*. Plenum Press, New York.
41. **Chen RF, Lee CY.** 2014. Adenoviruses types, cell receptors and local innate cytokines in adenovirus infection. *Int Rev Immunol* **33:**45–53.
42. **Greber UF, Willetts M, Webster P, Helenius A.** 1993. Stepwise dismantling of adenovirus 2 during entry into cells. *Cell* **75:**477–486.
43. **Mangel WF, San Martín C.** 2014. Structure, function and dynamics in adenovirus maturation. *Viruses* **6:**4536–4570.
44. **D'Halluin JC.** 1995. Virus assembly. *Curr Top Microbiol Immunol* **199:**47–66.
45. **Koo HL, Ajami N, Atmar RL, DuPont HL.** 2010. Noroviruses: the leading cause of gastroenteritis worldwide. *Discov Med* **10:**61–70.
46. **Mead PS, Slutsker L, Dietz V, McCaig LF, Bresee JS, Shapiro C, Griffin PM, Tauxe RV.** 1999. Food-related illness and death in the United States. *Emerg Infect Dis* **5:**607–625.
47. **Payne DC, Vinjé J, Szilagyi PG, Edwards KM, Staat MA, Weinberg GA, Hall CB, Chappell J, Bernstein DI, Curns AT, Wikswo M, Shirley SH, Hall AJ, Lopman B, Parashar UD.** 2013. Norovirus and medically attended gastroenteritis in U.S. children. *N Engl J Med* **368:**1121–1130.
48. **Hall AJ, Lopman BA, Payne DC, Patel MM, Gastañaduy PA, Vinjé J, Parashar UD.** 2013. Norovirus disease in the United States. *Emerg Infect Dis* **19:**1198–1205.
49. **Ajami NJ, Kavanagh OV, Ramani S, Crawford SE, Atmar RL, Jiang ZD, Okhuysen PC, Estes MK, DuPont HL.** 2014. Seroepidemiology of norovirus-associated travelers' diarrhea. *J Travel Med* **21:**6–11.
50. **Bitler EJ, Matthews JE, Dickey BW, Eisenberg JN, Leon JS.** 2013. Norovirus outbreaks: a systematic review of commonly implicated transmission routes and vehicles. *Epidemiol Infect* **141:**1563–1571.
51. **Centers for Disease Control and Prevention.** 2011. Updated norovirus outbreak management and disease prevention guidelines. *MMWR Recomm Rep* **60:**1–18.
52. **Atmar RL, Opekun AR, Gilger MA, Estes MK, Crawford SE, Neill FH, Graham DY.** 2008. Norwalk virus shedding after experimental human infection. *Emerg Infect Dis* **14:**1553–1557.
53. **Atmar RL, Opekun AR, Gilger MA, Estes MK, Crawford SE, Neill FH, Ramani S, Hill H, Ferreira J, Graham DY.** 2014. Determination of the 50% human infectious dose for Norwalk virus. *J Infect Dis* **209:**1016–1022.
54. **Barclay L, Park GW, Vega E, Hall A, Parashar U, Vinje J, Lopman B.** 2014. Infection control for norovirus. *Clin Microbiol Infect* **20:**731–740.
55. **Nakata S, Kogawa K, Numata K, Ukae S, Adachi N, Matson DO, Estes MK, Chiba S.** 1996. The epidemiology of human calicivirus/Sapporo/82/Japan. *Arch Virol Suppl* **12:**263–270.
56. **Kotloff KL, Nataro JP, Blackwelder WC, Nasrin D, Farag TH, Panchalingam S, Wu Y, Sow SO, Sur D, Breiman RF, Faruque AS, Zaidi AK, Saha D, Alonso PL, Tamboura B, Sanogo D, Onwuchekwa U, Manna B, Ramamurthy T, Kanungo S, Ochieng JB, Omore R, Oundo JO, Hossain A, Das SK, Ahmed S, Qureshi S, Quadri F, Adegbola RA, Antonio M, Hossain MJ, Akinsola A, Mandomando I, Nhampossa T, Acácio S, Biswas K, O'Reilly CE, Mintz ED, Berkeley LY, Muhsen K, Sommerfelt H, Robins-Browne RM, Levine MM.** 2013. Burden and aetiology of diarrhoeal disease in infants and young children in developing countries (the Global Enteric Multicenter Study, GEMS): a prospective, case-control study. *Lancet* **382:**209–222.
57. **Tate JE, Burton AH, Boschi-Pinto C, Steele AD, Duque J, Parashar UD, Network WH-cGRS.** 2012. 2008 estimate of worldwide rotavirus-associated mortality in children younger than 5 years before the introduction of universal rotavirus vaccination programmes: a systematic review and meta-analysis. *Lancet Infect Dis* **12:**136–141.
58. **Velázquez FR, Matson DO, Calva JJ, Guerrero L, Morrow AL, Carter-Campbell S, Glass RI, Estes MK, Pickering LK, Ruiz-Palacios GM.** 1996. Rotavirus infections in infants as protection against subsequent infections. *N Engl J Med* **335:**1022–1028.
59. **Cardemil CV, Cortese MM, Medina-Marino A, Jasuja S, Desai R, Leung J, Rodriguez-Hart C, Villarruel G, Howland J, Quaye O, Tam KI, Bowen MD, Parashar UD, Gerber SI, Rotavirus Investigation Team.** 2012. Two rotavirus outbreaks caused by genotype G2P[4] at large retirement communities: cohort studies. *Ann Intern Med* **157:**621–631.

60. **Anderson EJ, Shippee DB, Tate JE, Larkin B, Bregger MD, Katz BZ, Noskin GA, Sederdahl BK, Shane AL, Parashar UD, Yogev R.** 2014. Clinical characteristics and genotypes of rotavirus in adults. *J Infect* 6:683–687.
61. **Parashar UD, Nelson EA, Kang G.** 2013. Diagnosis, management, and prevention of rotavirus gastroenteritis in children. *BMJ* **347:**f7204.
62. **Ward RL, Bernstein DI, Young EC, Sherwood JR, Knowlton DR, Schiff GM.** 1986. Human rotavirus studies in volunteers: determination of infectious dose and serological response to infection. *J Infect Dis* **154:**871–880.
63. **Zheng BJ, Chang RX, Ma GZ, Xie JM, Liu Q, Liang XR, Ng MH.** 1991. Rotavirus infection of the oropharynx and respiratory tract in young children. *J Med Virol* **34:**29–37.
64. **Lloyd-Evans N, Springthorpe VS, Sattar SA.** 1986. Chemical disinfection of human rotavirus-contaminated inanimate surfaces. *J Hyg (Lond)* **97:**163–173.
65. **Sattar SA, Abebe M, Bueti AJ, Jampani H, Newman J, Hua S.** 2000. Activity of an alcohol-based hand gel against human adeno-, rhino-, and rotaviruses using the fingerpad method. *Infect Control Hosp Epidemiol* **21:**516–519.
66. **De Benedictis P, Schultz-Cherry S, Burnham A, Cattoli G.** 2011. Astrovirus infections in humans and animals—molecular biology, genetic diversity, and interspecies transmissions. Infection, genetics and evolution: *J molec epidem and evol genet in infect dis* **11:**1529–1544.
67. **Caballero S, Guix S, El-Senousy WM, Calicó I, Pintó RM, Bosch A.** 2003. Persistent gastroenteritis in children infected with astrovirus: association with serotype-3 strains. *J Med Virol* **71:**245–250.
68. **Percival S, Chalmers R, Embrey M, Hunter P, Sellwood J, Wyn-Jones P.** 2004. 28 - Astrovirus, p 387–399. *In* Wyn-Jones SPCEHS (ed), *Microbiology of Waterborne Diseases*. Academic Press, London.
69. **Abad FX, Villena C, Guix S, Caballero S, Pintó RM, Bosch A.** 2001. Potential role of fomites in the vehicular transmission of human astroviruses. *Appl Environ Microbiol* **67:**3904–3907.
70. **Gofti-Laroche L, Gratacap-Cavallier B, Demanse D, Genoulaz O, Seigneurin JM, Zmirou D.** 2003. Are waterborne astrovirus implicated in acute digestive morbidity (E.MI.R.A. study)? *J Clin Virol*: the official publication of the Pan American Society for Clinical Virology **27:**74–82.
71. **Uhnoo I, Svensson L, Wadell G.** 1990. Enteric adenoviruses. *Baillieres Clin Gastroenterol* **4:**627–642.
72. **Uhnoo I, Wadell G, Svensson L, Johansson ME.** 1984. Importance of enteric adenoviruses 40 and 41 in acute gastroenteritis in infants and young children. *J Clin Microbiol* **20:**365–372.
73. **Wadell G, Allard A, Johansson M, Svensson L, Uhnoo I.** 1987. Enteric adenoviruses. *Ciba Found Symp* **128:**63–91.
74. **Albert MJ.** 1986. Enteric adenoviruses. Brief review. *Arch Virol* **88:**1–17.
75. **Zahorsky J.** 1929. Hyperemesis hiemis or the winter vomiting disease. *Arch Pediatr* **46:**391–395.
76. **Adler JL, Zickl R.** 1969. Winter vomiting disease. *J Infect Dis* **119:**668–673.
77. **Estes MK, Prasad BV, Atmar RL.** 2006. Noroviruses everywhere: has something changed? *Curr Opin Infect Dis* **19:**467–474.
78. **Lee RM, Lessler J, Lee RA, Rudolph KE, Reich NG, Perl TM, Cummings DA.** 2013. Incubation periods of viral gastroenteritis: a systematic review. *BMC Infect Dis* **13:**446.
79. **Teunis PF, Sukhrie FH, Vennema H, Bogerman J, Beersma MF, Koopmans MP.** 2015. Shedding of norovirus in symptomatic and asymptomatic infections. *Epidemiol Infect* **8:**1710–1717.
80. **Karst SM.** 2010. Pathogenesis of noroviruses, emerging RNA viruses. *Viruses* **2:**748–781.
81. **Troeger H, Loddenkemper C, Schneider T, Schreier E, Epple HJ, Zeitz M, Fromm M, Schulzke JD.** 2009. Structural and functional changes of the duodenum in human norovirus infection. *Gut* **58:**1070–1077.
82. **Karst SM, Zhu S, Goodfellow IG.** 2014. The molecular pathology of noroviruses. *J Pathol.* **2:**206–216.
83. **Zhu S, Regev D, Watanabe M, Hickman D, Moussatche N, Jesus DM, Kahan SM, Napthine S, Brierley I, Hunter RN III, Devabhaktuni D, Jones MK, Karst SM.** 2013. Identification of immune and viral correlates of norovirus protective immunity through comparative study of intra-cluster norovirus strains. *PLoS Pathog* **9:**e1003592.
84. **Tan M, Xia M, Chen Y, Bu W, Hegde RS, Meller J, Li X, Jiang X.** 2009. Conservation of carbohydrate binding interfaces: evidence of human HBGA selection in norovirus evolution. *PLoS ONE* **4:**e5058.
85. **Hagbom M, Sharma S, Lundgren O, Svensson L.** 2012. Towards a human rotavirus disease model. *Curr Opin Virol* **2:**408–418.
86. **Blutt SE, Matson DO, Crawford SE, Staat MA, Azimi P, Bennett BL, Piedra PA, Conner ME.** 2007. Rotavirus antigenemia in children is associated with viremia. *PLoS Med* **4:** e121.
87. **Lynch M, Lee B, Azimi P, Gentsch J, Glaser C, Gilliam S, Chang HG, Ward R, Glass RI.** 2001. Rotavirus and central nervous system symptoms: cause or contaminant? Case reports and review. *Clin Infect Dis* **33:**932–938.
88. **Greenberg HB, Estes MK.** 2009. Rotaviruses: from pathogenesis to vaccination. *Gastroenterology* **136:**1939–1951.
89. **Boshuizen JA, Reimerink JH, Korteland-van Male AM, van Ham VJ, Koopmans MP, Büller HA, Dekker J, Einerhand AW.** 2003. Changes in small intestinal homeostasis, morphology, and gene expression during rotavirus infection of infant mice. *J Virol* **77:**13005–13016.
90. **Tafazoli F, Zeng CQ, Estes MK, Magnusson KE, Svensson L.** 2001. NSP4 enterotoxin of rotavirus induces paracellular leakage in polarized epithelial cells. *J Virol* **75:**1540–1546.
91. **Hagbom M, Istrate C, Engblom D, Karlsson T, Rodriguez-Diaz J, Buesa J, Taylor JA, Loitto VM, Magnusson KE, Ahlman H, Lundgren O, Svensson L.** 2011. Rotavirus stimulates release of serotonin (5-HT) from human enterochromaffin cells and activates brain structures involved in nausea and vomiting. *PLoS Pathog* **7:**e1002115.
92. **Holloway G, Coulson BS.** 2013. Innate cellular responses to rotavirus infection. *J Gen Virol* **94:**1151–1160.
93. **Desselberger U, Huppertz HI.** 2011. Immune responses to rotavirus infection and vaccination and associated correlates of protection. *J Infect Dis* **203:**188–195.
94. **Xu J, Dennehy P, Keyserling H, Westerman LE, Wang Y, Holman RC, Gentsch JR, Glass RI, Jiang B.** 2005. Serum antibody responses in children with rotavirus diarrhea can serve as proxy for protection. *Clin Diagn Lab Immunol* **12:**273–279.
95. **Patel M, Glass RI, Jiang B, Santosham M, Lopman B, Parashar U.** 2013. A systematic review of anti-rotavirus serum IgA antibody titer as a potential correlate of rotavirus vaccine efficacy. *J Infect Dis* **208:**284–294.
96. **Velázquez FR, Matson DO, Guerrero ML, Shults J, Calva JJ, Morrow AL, Glass RI, Pickering LK, Ruiz-Palacios GM.** 2000. Serum antibody as a marker of protection against natural rotavirus infection and disease. *J Infect Dis* **182:**1602–1609.
97. **Gonzalez AM, Jaimes MC, Cajiao I, Rojas OL, Cohen J, Pothier P, Kohli E, Butcher EC, Greenberg HB, Angel J, Franco MA.** 2003. Rotavirus-specific B cells induced by recent infection in adults and children predominantly express the intestinal homing receptor alpha4beta7. *Virology* **305:**93–105.
98. **Walter JE, Mitchell DK.** 2003. Astrovirus infection in children. *Curr Opin Infect Dis* **16:**247–253.
99. **Mitchell DK, Monroe SS, Jiang X, Matson DO, Glass RI, Pickering LK.** 1995. Virologic features of an astrovirus diarrhea outbreak in a day care center revealed by reverse transcriptase-polymerase chain reaction. *J Infect Dis* **172:** 1437–1444.
100. **Moser LA, Schultz-Cherry S.** 2005. Pathogenesis of astrovirus infection. *Viral Immunol* **18:**4–10.
101. **Nighot PK, Moeser A, Ali RA, Blikslager AT, Koci MD.** 2010. Astrovirus infection induces sodium malabsorption and

redistributes sodium hydrogen exchanger expression. *Virology* **401:**146–154.
102. **Koci MD.** 2005. Immunity and resistance to astrovirus infection. *Viral Immunol* **18:**11–16.
103. **Sharp JA, Whitley PH, Cunnion KM, Krishna NK.** 2014. Peptide inhibitor of complement c1, a novel suppressor of classical pathway activation: mechanistic studies and clinical potential. *Front Immunol* **5:**406.
104. **Wood DJ.** 1988. Adenovirus gastroenteritis. *Br Med J (Clin Res Ed)* **296:**229–230.
105. **Whitelaw A, Davies H, Parry J.** 1977. Electron microscopy of fatal adenovirus gastroenteritis. *Lancet* **1:**361.
106. **Wood DJ, David TJ, Chrystie IL, Totterdell B.** 1988. Chronic enteric virus infection in two T-cell immunodeficient children. *J Med Virol* **24:**435–444.
107. **Bakare N, Menschik D, Tiernan R, Hua W, Martin D.** 2010. Severe combined immunodeficiency (SCID) and rotavirus vaccination: reports to the Vaccine Adverse Events Reporting System (VAERS). *Vaccine* **28:**6609–6612.
108. **Frange P, Touzot F, Debré M, Héritier S, Leruez-Ville M, Cros G, Rouzioux C, Blanche S, Fischer A, Avettand-Fenoël V.** 2012. Prevalence and clinical impact of norovirus fecal shedding in children with inherited immune deficiencies. *J Infect Dis* **206:**1269–1274.
109. **Brown JR, Morfopoulou S, Hubb J, Emmett WA, Ip W, Shah D, Brooks T, Paine SM, Anderson G, Virasami A, Tong CY, Clark DA, Plagnol V, Jacques TS, Qasim W, Hubank M, Breuer J.** 2015. Astrovirus VA1/HMO-C: an increasingly recognized neurotropic pathogen in immunocompromised patients. *Clin Infect Dis* **60:**881–888.
110. **Thomas PD, Pollok RC, Gazzard BG.** 1999. Enteric viral infections as a cause of diarrhoea in the acquired immunodeficiency syndrome. *HIV Med* **1:**19–24.
111. **Jalal H, Bibby DF, Tang JW, Bennett J, Kyriakou C, Peggs K, Cubitt D, Brink NS, Ward KN, Tedder RS.** 2005. First reported outbreak of diarrhea due to adenovirus infection in a hematology unit for adults. *J Clin Microbiol* **43:**2575–2580.
112. **Bok K, Green KY.** 2012. Norovirus gastroenteritis in immunocompromised patients. *N Engl J Med* **367:**2126–2132.
113. **Anderson EJ.** 2010. Prevention and treatment of viral diarrhea in pediatrics. *Expert Rev Anti Infect Ther* **8:**205–217.
114. **Dennehy PH.** 2008. Rotavirus vaccines: an overview. *Clin Microbiol Rev* **21:**198–208.
115. **Vesikari T.** 2008. Rotavirus vaccines. *Scand J Infect Dis* **40:**691–695.
116. **Tan M, Jiang X.** 2014. Vaccine against norovirus. *Hum Vaccin Immunother* **10:**1449–1456.
117. **Ramani S, Atmar RL, Estes MK.** 2014. Epidemiology of human noroviruses and updates on vaccine development. *Curr Opin Gastroenterol* **30:**25–33.
118. **Richardson C, Bargatze RF, Goodwin R, Mendelman PM.** 2013. Norovirus virus-like particle vaccines for the prevention of acute gastroenteritis. *Expert Rev Vaccines* **12:**155–167.
119. **Graham DY, Jiang X, Tanaka T, Opekun AR, Madore HP, Estes MK.** 1994. Norwalk virus infection of volunteers: new insights based on improved assays. *J Infect Dis* **170:**34–43.
120. **Gray JJ, Jiang X, Morgan-Capner P, Desselberger U, Estes MK.** 1993. Prevalence of antibodies to Norwalk virus in England: detection by enzyme-linked immunosorbent assay using baculovirus-expressed Norwalk virus capsid antigen. *J Clin Microbiol* **31:**1022–1025.
121. **Okhuysen PC, Jiang X, Ye L, Johnson PC, Estes MK.** 1995. Viral shedding and fecal IgA response after Norwalk virus infection. *J Infect Dis* **171:**566–569.
122. **Atmar RL, Estes MK.** 2001. Diagnosis of noncultivatable gastroenteritis viruses, the human caliciviruses. *Clin Microbiol Rev* **14:**15–37.
123. **Zheng DP, Ando T, Fankhauser RL, Beard RS, Glass RI, Monroe SS.** 2006. Norovirus classification and proposed strain nomenclature. *Virology* **346:**312–323.
124. **Kageyama T, Kojima S, Shinohara M, Uchida K, Fukushi S, Hoshino FB, Takeda N, Katayama K.** 2003. Broadly reactive and highly sensitive assay for Norwalk-like viruses based on real-time quantitative reverse transcription-PCR. *J Clin Microbiol* **41:**1548–1557.
125. **Rolfe KJ, Parmar S, Mururi D, Wreghitt TG, Jalal H, Zhang H, Curran MD.** 2007. An internally controlled, one-step, real-time RT-PCR assay for norovirus detection and genogrouping. *J Clin Virol*: the official publication of the Pan American Society for Clinical Virology **39:**318–321.
126. **Schultz AC, Vega E, Dalsgaard A, Christensen LS, Norrung B, Hoorfar J, Vinje J.** 2011. Development and evaluation of novel one-step TaqMan realtime RT-PCR assays for the detection and direct genotyping of genogroup I and II noroviruses. *J Clin Virol*: the official publication of the Pan American Society for Clinical Virology **50:**230–234.
127. **Miura T, Parnaudeau S, Grodzki M, Okabe S, Atmar RL, Le Guyader FS.** 2013. Environmental detection of genogroup I, II, and IV noroviruses by using a generic real-time reverse transcription-PCR assay. *Appl Environ Microbiol* **79:**6585–6592.
128. **Stals A, Mathijs E, Baert L, Botteldoorn N, Denayer S, Mauroy A, Scipioni A, Daube G, Dierick K, Herman L, Van Coillie E, Thiry E, Uyttendaele M.** 2012. Molecular detection and genotyping of noroviruses. *Food Environ Virol* **4:**153–167.
129. **Xue C, Fu Y, Zhu W, Fei Y, Zhu L, Zhang H, Pan L, Xu H, Wang Y, Wang W, Sun Q.** 2014. An outbreak of acute norovirus gastroenteritis in a boarding school in Shanghai: a retrospective cohort study. *BMC Public Health* **14:**1092.
130. **Zhou X, Li H, Sun L, Mo Y, Chen S, Wu X, Liang J, Zheng H, Ke C, Varma JK, Klena JD, Chen Q, Zou L, Yang X.** 2012. Epidemiological and molecular analysis of a waterborne outbreak of norovirus GII.4. *Epidemiol Infect* **140:**2282–2289.
131. **Rajko-Nenow P, Keaveney S, Flannery J, O'Flaherty V, Doré W.** 2012. Characterisation of norovirus contamination in an Irish shellfishery using real-time RT-qPCR and sequencing analysis. *Int J Food Microbiol* **160:**105–112.
132. **Ward RL, Kirkwood CD, Sander DS, Smith VE, Shao M, Bean JA, Sack DA, Bernstein DI.** 2006. Reductions in cross-neutralizing antibody responses in infants after attenuation of the human rotavirus vaccine candidate 89–12. *J Infect Dis* **194:**1729–1736.
133. **Vesikari T, Clark HF, Offit PA, Dallas MJ, DiStefano DJ, Goveia MG, Ward RL, Schödel F, Karvonen A, Drummond JE, DiNubile MJ, Heaton PM.** 2006. Effects of the potency and composition of the multivalent human-bovine (WC3) reassortant rotavirus vaccine on efficacy, safety and immunogenicity in healthy infants. *Vaccine* **24:**4821–4829.
134. **Ward R.** 2009. Mechanisms of protection against rotavirus infection and disease. *Pediatr Infect Dis J* **28**(Suppl)**:**S57–S59.
135. **Ray PG, Kelkar SD.** 2004. Measurement of antirotavirus IgM/IgA/IgG responses in the serum samples of Indian children following rotavirus diarrhoea and their mothers. *J Med Virol* **72:**416–423.
136. **Stals F, Walther FJ, Bruggeman CA.** 1984. Faecal and pharyngeal shedding of rotavirus and rotavirus IgA in children with diarrhoea. *J Med Virol* **14:**333–339.
137. **Bishop RF, Bugg HC, Masendycz PJ, Lund JS, Gorrell RJ, Barnes GL.** 1996. Serum, fecal, and breast milk rotavirus antibodies as indices of infection in mother-infant pairs. *J Infect Dis* **174**(Suppl 1)**:**S22–S29.
138. **Arnold M, Patton JT, McDonald SM.** 2009. Culturing, storage, and quantification of rotaviruses. *Curr Protoc Microbiol* **15:**15C.13.11–24.
139. **Babiuk LA, Mohammed K, Spence L, Fauvel M, Petro R.** 1977. Rotavirus isolation and cultivation in the presence of trypsin. *J Clin Microbiol* **6:**610–617.
140. **Estes MK, Graham DY, Mason BB.** 1981. Proteolytic enhancement of rotavirus infectivity: molecular mechanisms. *J Virol* **39:**879–888.
141. **Ward RL, Knowlton DR, Pierce MJ.** 1984. Efficiency of human rotavirus propagation in cell culture. *J Clin Microbiol* **19:**748–753.
142. **Finkbeiner SR, Zeng XL, Utama B, Atmar RL, Shroyer NF, Estes MK.** 2012. Stem cell-derived human intestinal

organoids as an infection model for rotaviruses. *MBio* **3:** e00159–e12.
143. **Madeley CR.** 1979. Viruses in the stools. *J Clin Pathol* **32:** 1–10.
144. **Nakata S, Petrie BL, Calomeni EP, Estes MK.** 1987. Electron microscopy procedure influences detection of rotaviruses. *J Clin Microbiol* **25:**1902–1906.
145. **Maes RK, Grooms DL, Wise AG, Han C, Ciesicki V, Hanson L, Vickers ML, Kanitz C, Holland R.** 2003. Evaluation of a human group a rotavirus assay for on-site detection of bovine rotavirus. *J Clin Microbiol* **41:**290–294.
146. **Gerna G, Passarani N, Battaglía M, Percivalle E.** 1984. Rapid serotyping of human rotavirus strains by solid-phase immune electron microscopy. *J Clin Microbiol* **19:**273–278.
147. **Gbewonyo AJ.** 1982. Rapid and reliable method for diagnostic electron microscopy of faeces. *J Microsc* **126:**191–195.
148. **Espejo RT, Calderon E, Gonzalez N.** 1977. Distinct reovirus-like agents associated with acute infantile gastroenteritis. *J Clin Microbiol* **6:**502–506.
149. **Herring AJ, Inglis NF, Ojeh CK, Snodgrass DR, Menzies JD.** 1982. Rapid diagnosis of rotavirus infection by direct detection of viral nucleic acid in silver-stained polyacrylamide gels. *J Clin Microbiol* **16:**473–477.
150. **Matsui SM, Mackow ER, Matsuno S, Paul PS, Greenberg HB.** 1990. Sequence analysis of gene 11 equivalents from "short" and "super short" strains of rotavirus. *J Virol* **64:**120–124.
151. **Dennehy PH, Gauntlett DR, Tente WE.** 1988. Comparison of nine commercial immunoassays for the detection of rotavirus in fecal specimens. *J Clin Microbiol* **26:**1630–1634.
152. **Gautam R, Lyde F, Esona MD, Quaye O, Bowen MD.** 2013. Comparison of Premier Rotaclone(R), ProSpecT, and RIDASCREEN(R) rotavirus enzyme immunoassay kits for detection of rotavirus antigen in stool specimens. *J Clin Virol*: the official publication of the Pan American Society for Clinical Virology **58:**292–294.
153. **Wilde J, Yolken R, Willoughby R, Eiden J.** 1991. Improved detection of rotavirus shedding by polymerase chain reaction. *Lancet* **337:**323–326.
154. **Iturriza Gómara M, Wong C, Blome S, Desselberger U, Gray J.** 2002. Molecular characterization of VP6 genes of human rotavirus isolates: correlation of genogroups with subgroups and evidence of independent segregation. *J Virol* **76:**6596–6601.
155. **Xu L, Harbour D, McCrae MA.** 1990. The application of polymerase chain reaction to the detection of rotaviruses in faeces. *J Virol Methods* **27:**29–37.
156. **Gouvea V, Allen JR, Glass RI, Fang ZY, Bremont M, Cohen J, McCrae MA, Saif LJ, Sinarachatanant P, Caul EO.** 1991. Detection of group B and C rotaviruses by polymerase chain reaction. *J Clin Microbiol* **29:**519–523.
157. **Gentsch JR, Glass RI, Woods P, Gouvea V, Gorziglia M, Flores J, Das BK, Bhan MK.** 1992. Identification of group A rotavirus gene 4 types by polymerase chain reaction. *J Clin Microbiol* **30:**1365–1373.
158. **Simmonds MK, Armah G, Asmah R, Banerjee I, Damanka S, Esona M, Gentsch JR, Gray JJ, Kirkwood C, Page N, Iturriza-Gomara M.** 2008. New oligonucleotide primers for P-typing of rotavirus strains: strategies for typing previously untypeable strains. *J Clin Virol*: the official publication of the Pan American Society for Clinical Virology **42:**368–373.
159. **Iturriza-Gomara M, Kang G, Gray J.** 2004. Rotavirus genotyping: keeping up with an evolving population of human rotaviruses. *J Clin Virol*: the official publication of the Pan American Society for Clinical Virology **31:**259–265.
160. **Gouvea V, Glass RI, Woods P, Taniguchi K, Clark HF, Forrester B, Fang ZY.** 1990. Polymerase chain reaction amplification and typing of rotavirus nucleic acid from stool specimens. *J Clin Microbiol* **28:**276–282.
161. **Das BK, Gentsch JR, Cicirello HG, Woods PA, Gupta A, Ramachandran M, Kumar R, Bhan MK, Glass RI.** 1994. Characterization of rotavirus strains from newborns in New Delhi, India. *J Clin Microbiol* **32:**1820–1822.
162. **Hull JJ, Teel EN, Kerin TK, Freeman MM, Esona MD, Gentsch JR, Cortese MM, Parashar UD, Glass RI, Bowen MD, National Rotavirus Strain Surveillance System.** 2011. United States rotavirus strain surveillance from 2005 to 2008: genotype prevalence before and after vaccine introduction. *Pediatr Infect Dis J* **30**(Suppl):S42–S47.
163. **Gutiérrez-Aguirre I, Steyer A, Boben J, Gruden K, Poljsak-Prijatelj M, Ravnikar M.** 2008. Sensitive detection of multiple rotavirus genotypes with a single reverse transcription-real-time quantitative PCR assay. *J Clin Microbiol* **46:**2547–2554.
164. **Min BS, Noh YJ, Shin JH, Baek SY, Min KI, Ryu SR, Kim BG, Park MK, Choi SE, Yang EH, Park SN, Hur SJ, Ahn BY.** 2006. Assessment of the quantitative real-time polymerase chain reaction using a cDNA standard for human group A rotavirus. *J Virol Methods* **137:**280–286.
165. **Kottaridi C, Spathis AT, Ntova CK, Papaevangelou V, Karakitsos P.** 2012. Evaluation of a multiplex real time reverse transcription PCR assay for the detection and quantitation of the most common human rotavirus genotypes. *J Virol Methods* **180:**49–53.
166. **Logan C, O'Leary JJ, O'Sullivan N.** 2006. Real-time reverse transcription-PCR for detection of rotavirus and adenovirus as causative agents of acute viral gastroenteritis in children. *J Clin Microbiol* **44:**3189–3195.
167. **Kang G, Iturriza-Gomara M, Wheeler JG, Crystal P, Monica B, Ramani S, Primrose B, Moses PD, Gallimore CI, Brown DW, Gray J.** 2004. Quantitation of group A rotavirus by real-time reverse-transcription-polymerase chain reaction: correlation with clinical severity in children in South India. *J Med Virol* **73:**118–122.
168. **Zhao W, Xia M, Bridges-Malveo T, Cantú M, McNeal MM, Choi AH, Ward RL, Sestak K.** 2005. Evaluation of rotavirus dsRNA load in specimens and body fluids from experimentally infected juvenile macaques by real-time PCR. *Virology* **341:** 248–256.
169. **Nordgren J, Bucardo F, Svensson L, Lindgren PE.** 2010. Novel light-upon-extension real-time PCR assay for simultaneous detection, quantification, and genogrouping of group A rotavirus. *J Clin Microbiol* **48:**1859–1865.
170. **Schwarz BA, Bange R, Vahlenkamp TW, Johne R, Müller H.** 2002. Detection and quantitation of group A rotaviruses by competitive and real-time reverse transcription-polymerase chain reaction. *J Virol Methods* **105:**277–285.
171. **Plante D, Bélanger G, Leblanc D, Ward P, Houde A, Trottier YL.** 2011. The use of bovine serum albumin to improve the RT-qPCR detection of foodborne viruses rinsed from vegetable surfaces. *Lett Appl Microbiol* **52:**239–244.
172. **Jothikumar N, Kang G, Hill VR.** 2009. Broadly reactive TaqMan assay for real-time RT-PCR detection of rotavirus in clinical and environmental samples. *J Virol Methods* **155:**126–131.
173. **Freeman MM, Kerin T, Hull J, McCaustland K, Gentsch J.** 2008. Enhancement of detection and quantification of rotavirus in stool using a modified real-time RT-PCR assay. *J Med Virol* **80:**1489–1496.
174. **Pang XL, Lee B, Boroumand N, Leblanc B, Preiksaitis JK, Yu Ip CC.** 2004. Increased detection of rotavirus using a real time reverse transcription-polymerase chain reaction (RT-PCR) assay in stool specimens from children with diarrhea. *J Med Virol* **72:**496–501.
175. **Zeng SQ, Halkosalo A, Salminen M, Szakal ED, Puustinen L, Vesikari T.** 2008. One-step quantitative RT-PCR for the detection of rotavirus in acute gastroenteritis. *J Virol Methods* **153:**238–240.
176. **Pang X, Cao M, Zhang M, Lee B.** 2011. Increased sensitivity for various rotavirus genotypes in stool specimens by amending three mismatched nucleotides in the forward primer of a real-time RT-PCR assay. *J Virol Methods* **172:**85–87.
177. **Mijatovic-Rustempasic S, Tam KI, Kerin TK, Lewis JM, Gautam R, Quaye O, Gentsch JR, Bowen MD.** 2013. Sensitive and specific quantitative detection of rotavirus A by one-step real-time reverse transcription-PCR assay without

antecedent double-stranded-RNA denaturation. *J Clin Microbiol* **51:**3047–3054.

178. **Adlhoch C, Kaiser M, Hoehne M, Mas Marques A, Stefas I, Veas F, Ellerbrok H.** 2011. Highly sensitive detection of the group A Rotavirus using Apolipoprotein H-coated ELISA plates compared to quantitative real-time PCR. *Virol J* **8:**63.
179. **Ward P, Poitras E, Leblanc D, Gagnon CA, Brassard J, Houde A.** 2013. Comparison of different RT-qPCR assays for the detection of human and bovine group A rotaviruses and characterization by sequences analysis of genes encoding VP4 and VP7 capsid proteins. *J Appl Microbiol* **114:**1435–1448.
180. **Gautam R, Esona MD, Mijatovic-Rustempasic S, Ian Tam K, Gentsch JR, Bowen MD.** 2014. Real-time RT-PCR assays to differentiate wild-type group A rotavirus strains from Rotarix (®) and RotaTeq(®) vaccine strains in stool samples. *Hum Vaccin Immunother* **10:**767–777.
181. **Mitchell DK, Matson DO, Cubitt WD, Jackson LJ, Willcocks MM, Pickering LK, Carter MJ.** 1999. Prevalence of antibodies to astrovirus types 1 and 3 in children and adolescents in Norfolk, Virginia. *Pediatr Infect Dis J* **18:**249–254.
182. **Koopmans MP, Bijen MH, Monroe SS, Vinjé J.** 1998. Age-stratified seroprevalence of neutralizing antibodies to astrovirus types 1 to 7 in humans in The Netherlands. *Clin Diagn Lab Immunol* **5:**33–37.
183. **Burbelo PD, Ching KH, Esper F, Iadarola MJ, Delwart E, Lipkin WI, Kapoor A.** 2011. Serological studies confirm the novel astrovirus HMOAstV-C as a highly prevalent human infectious agent. *PLoS One* **6:**e22576.
184. **Holtz LR, Bauer IK, Jiang H, Belshe R, Freiden P, Schultz-Cherry SL, Wang D.** 2014. Seroepidemiology of astrovirus MLB1. *Clin Vaccine Immunol* **21:**908–911.
185. **Lee TW, Kurtz JB.** 1981. Serial propagation of astrovirus in tissue culture with the aid of trypsin. *J Gen Virol* **57:**421–424.
186. **Brinker JP, Blacklow NR, Herrmann JE.** 2000. Human astrovirus isolation and propagation in multiple cell lines. *Arch Virol* **145:**1847–1856.
187. **Mustafa H, Palombo EA, Bishop RF.** 1998. Improved sensitivity of astrovirus-specific RT-PCR following culture of stool samples in CaCo-2 cells. *J Clin Virol*: the official publication of the Pan American Society for Clinical Virology **11:**103–107.
188. **Grimm AC, Cashdollar JL, Williams FP, Fout GS.** 2004. Development of an astrovirus RT-PCR detection assay for use with conventional, real-time, and integrated cell culture/RT-PCR. *Can J Microbiol* **50:**269–278.
189. **Chapron CD, Ballester NA, Margolin AB.** 2000. The detection of astrovirus in sludge biosolids using an integrated cell culture nested PCR technique. *J Appl Microbiol* **89:**11–15.
190. **Glass RI, Noel J, Mitchell D, Herrmann JE, Blacklow NR, Pickering LK, Dennehy P, Ruiz-Palacios G, de Guerrero ML, Monroe SS.** 1996. The changing epidemiology of astrovirus-associated gastroenteritis: a review, p 287–300. *In* Chiba S, Estes M, Nakata S, Calisher C (ed), *Arch Virol Suppl*, **12:**287–300.
191. **Noel J, Cubitt D.** 1994. Identification of astrovirus serotypes from children treated at the Hospitals for Sick Children, London 1981–93. *Epidemiol Infect* **113:**153–159.
192. **Herrmann JE, Hudson RW, Perron-Henry DM, Kurtz JB, Blacklow NR.** 1988. Antigenic characterization of cell-cultivated astrovirus serotypes and development of astrovirus-specific monoclonal antibodies. *J Infect Dis* **158:**182–185.
193. **Noel JS, Lee TW, Kurtz JB, Glass RI, Monroe SS.** 1995. Typing of human astroviruses from clinical isolates by enzyme immunoassay and nucleotide sequencing. *J Clin Microbiol* **33:**797–801.
194. **Moe CL, Allen JR, Monroe SS, Gary, Jr HE, Humphrey CD, Herrmann JE, Blacklow NR, Carcamo C, Koch M, Kim KH.** 1991. Detection of astrovirus in pediatric stool samples by immunoassay and RNA probe. *J Clin Microbiol* **29:**2390–2395.
195. **Dalton RM, Roman ER, Negredo AA, Wilhelmi ID, Glass RI, Sánchez-Fauquier A.** 2002. Astrovirus acute gastroenteritis among children in Madrid, Spain. *Pediatr Infect Dis J* **21:**1038–1041.
196. **Guix S, Bosch A, Pintó RM.** 2005. Human astrovirus diagnosis and typing: current and future prospects. *Lett Appl Microbiol* **41:**103–105.
197. **Finkbeiner SR, Le BM, Holtz LR, Storch GA, Wang D.** 2009. Detection of newly described astrovirus MLB1 in stool samples from children. *Emerg Infect Dis* **15:**441–444.
198. **Chu DK, Poon LL, Guan Y, Peiris JS.** 2008. Novel astroviruses in insectivorous bats. *J Virol* **82:**9107–9114.
199. **Medici MC, Tummolo F, Calderaro A, Elia G, Banyai K, De Conto F, Arcangeletti MC, Chezzi C, Buonavoglia C, Martella V.** 2014. MLB1 astrovirus in children with gastroenteritis, Italy. *Emerg Infect Dis* **20:**169–170.
200. **Kapoor A, Li L, Victoria J, Oderinde B, Mason C, Pandey P, Zaidi SZ, Delwart E.** 2009. Multiple novel astrovirus species in human stool. *J Gen Virol* **90:**2965–2972.
201. **Royuela E, Negredo A, Sánchez-Fauquier A.** 2006. Development of a one step real-time RT-PCR method for sensitive detection of human astrovirus. *J Virol Methods* **133:**14–19.
202. **Zhang Z, Mitchell DK, Afflerbach C, Jakab F, Walter J, Zhang YJ, Staat MA, Azimi P, Matson DO.** 2006. Quantitation of human astrovirus by real-time reverse-transcription-polymerase chain reaction to examine correlation with clinical illness. *J Virol Methods* **134:**190–196.
203. **Logan C, O'Leary JJ, O'Sullivan N.** 2007. Real-time reverse transcription PCR detection of norovirus, sapovirus and astrovirus as causative agents of acute viral gastroenteritis. *J Virol Methods* **146:**36–44.
204. **Dai YC, Xu QH, Wu XB, Hu GF, Tang YL, Li JD, Chen Q, Nie J.** 2010. Development of real-time and nested RT-PCR to detect astrovirus and one-year survey of astrovirus in Jiangmen City, China. *Arch Virol* **155:**977–982.
205. **Hata A, Kitajima M, Tajiri-Utagawa E, Katayama H.** 2014. Development of a high resolution melting analysis for detection and differentiation of human astroviruses. *J Virol Methods* **200:**29–34.
206. **van Maarseveen NM, Wessels E, de Brouwer CS, Vossen AC, Claas EC.** 2010. Diagnosis of viral gastroenteritis by simultaneous detection of Adenovirus group F, Astrovirus, Rotavirus group A, Norovirus genogroups I and II, and Sapovirus in two internally controlled multiplex real-time PCR assays. *J Clin Virol*: the official publication of the Pan American Society for Clinical Virology **49:**205–210.
207. **Liu J, Kibiki G, Maro V, Maro A, Kumburu H, Swai N, Taniuchi M, Gratz J, Toney D, Kang G, Houpt E.** 2011. Multiplex reverse transcription PCR Luminex assay for detection and quantitation of viral agents of gastroenteritis. *J Clin Virol*: the official publication of the Pan American Society for Clinical Virology **50:**308–313.
208. **Wolffs PF, Bruggeman CA, van Well GT, van Loo IH.** 2011. Replacing traditional diagnostics of fecal viral pathogens by a comprehensive panel of real-time PCRs. *J Clin Microbiol* **49:**1926–1931.
209. **Pang XL, Preiksaitis JK, Lee BE.** 2014. Enhanced enteric virus detection in sporadic gastroenteritis using a multi-target real-time PCR panel: a one-year study. *J Med Virol* **86:**1594–1601.
210. **Bai Z, Xie H, You Q, Pickerill S, Zhang Y, Li T, Geng J, Hu L, Shan H, Di B.** 2015. Isothermal cross-priming amplification implementation study. *Lett Appl Microbiol* **60:**205–209.
211. **Wei H, Zeng J, Deng C, Zheng C, Zhang X, Ma D, Yi Y.** 2013. A novel method of real-time reverse-transcription loop-mediated isothermal amplification developed for rapid and quantitative detection of human astrovirus. *J Virol Methods* **188:**126–131.
212. **Yang BY, Liu XL, Wei YM, Wang JQ, He XQ, Jin Y, Wang ZJ.** 2014. Rapid and sensitive detection of human astrovirus in water samples by loop-mediated isothermal amplification with hydroxynaphthol blue dye. *BMC Microbiol* **14:**38.
213. **Cromeans TL, Lu X, Erdman DD, Humphrey CD, Hill VR.** 2008. Development of plaque assays for adenoviruses 40 and 41. *J Virol Methods* **151:**140–145.

214. **Pintó RM, Diez JM, Bosch A.** 1994. Use of the colonic carcinoma cell line CaCo-2 for in vivo amplification and detection of enteric viruses. *J Med Virol* **44:**310–315.
215. **Perron-Henry DM, Herrmann JE, Blacklow NR.** 1988. Isolation and propagation of enteric adenoviruses in HEp-2 cells. *J Clin Microbiol* **26:**1445–1447.
216. **Wood DJ, Bijlsma K, de Jong JC, Tonkin C.** 1989. Evaluation of a commercial monoclonal antibody-based enzyme immunoassay for detection of adenovirus types 40 and 41 in stool specimens. *J Clin Microbiol* **27:**1155–1158.
217. **Herrmann JE, Perron-Henry DM, Blacklow NR.** 1987. Antigen detection with monoclonal antibodies for the diagnosis of adenovirus gastroenteritis. *J Infect Dis* **155:**1167–1171.
218. **Moore PL, Steele AD, Alexander JJ.** 2000. Relevance of commercial diagnostic tests to detection of enteric adenovirus infections in South Africa. *J Clin Microbiol* **38:**1661–1663.
219. **Rousell J, Zajdel ME, Howdle PD, Blair GE.** 1993. Rapid detection of enteric adenoviruses by means of the polymerase chain reaction. *J Infect* **27:**271–275.
220. **Allard A, Girones R, Juto P, Wadell G.** 1990. Polymerase chain reaction for detection of adenoviruses in stool samples. *J Clin Microbiol* **28:**2659–2667.
221. **Tiemessen CT, Nel MJ.** 1996. Detection and typing of subgroup F adenoviruses using the polymerase chain reaction. *J Virol Methods* **59:**73–82.
222. **Xu W, McDonough MC, Erdman DD.** 2000. Species-specific identification of human adenoviruses by a multiplex PCR assay. *J Clin Microbiol* **38:**4114–4120.
223. **Jothikumar N, Cromeans TL, Hill VR, Lu X, Sobsey MD, Erdman DD.** 2005. Quantitative real-time PCR assays for detection of human adenoviruses and identification of serotypes 40 and 41. *Appl Environ Microbiol* **71:**3131–3136.
224. **Jiang Y, Fang L, Shi X, Zhang H, Li Y, Lin Y, Qiu Y, Chen Q, Li H, Zhou L, Hu Q.** 2014. Simultaneous detection of five enteric viruses associated with gastroenteritis by use of a PCR assay: a single real-time multiplex reaction and its clinical application. *J Clin Microbiol* **52:**1266–1268.
225. **Feeney SA, Armstrong VJ, Mitchell SJ, Crawford L, McCaughey C, Coyle PV.** 2011. Development and clinical validation of multiplex TaqMan® assays for rapid diagnosis of viral gastroenteritis. *J Med Virol* **83:**1650–1656.
226. **Gray J, Coupland LJ.** 2014. The increasing application of multiplex nucleic acid detection tests to the diagnosis of syndromic infections. *Epidemiol Infect* **142:**1–11.
227. **Liu J, Kabir F, Manneh J, Lertsethtakarn P, Begum S, Gratz J, Becker SM, Operario DJ, Taniuchi M, Janaki L, Platts-Mills JA, Haverstick DM, Kabir M, Sobuz SU, Nakjarung K, Sakpaisal P, Silapong S, Bodhidatta L, Qureshi S, Kalam A, Saidi Q, Swai N, Mujaga B, Maro A, Kwambana B, Dione M, Antonio M, Kibiki G, Mason CJ, Haque R, Iqbal N, Zaidi AK, Houpt ER.** 2014. Development and assessment of molecular diagnostic tests for 15 enteropathogens causing childhood diarrhoea: a multicentre study. *Lancet Infect Dis* **14:**716–724.
228. **Reddington K, Tuite N, Minogue E, Barry T.** 2014. A current overview of commercially available nucleic acid diagnostics approaches to detect and identify human gastroenteritis pathogens. *Biomol Detect Quantification* **1:**3–7.
229. **Liu J, Gratz J, Amour C, Kibiki G, Becker S, Janaki L, Verweij JJ, Taniuchi M, Sobuz SU, Haque R, Haverstick DM, Houpt ER.** 2013. A laboratory-developed TaqMan Array Card for simultaneous detection of 19 enteropathogens. *J Clin Microbiol* **51:**472–480.
230. **Halligan E, Edgeworth J, Bisnauthsing K, Bible J, Cliff P, Aarons E, Klein J, Patel A, Goldenberg S.** 2014. Multiplex molecular testing for management of infectious gastroenteritis in a hospital setting: a comparative diagnostic and clinical utility study. *Clin Microbiol Infect* **20:** 460–467.
231. **Finkbeiner SR, Li Y, Ruone S, Conrardy C, Gregoricus N, Toney D, Virgin HW, Anderson LJ, Vinjé J, Wang D, Tong S.** 2009. Identification of a novel astrovirus (astrovirus VA1) associated with an outbreak of acute gastroenteritis. *J Virol* **83:**10836–10839.
232. **Phan TG, Vo NP, Bonkoungou IJ, Kapoor A, Barro N, O'Ryan M, Kapusinszky B, Wang C, Delwart E.** 2012. Acute diarrhea in West African children: diverse enteric viruses and a novel parvovirus genus. *J Virol* **86:**11024–11030.
233. **Smits SL, Schapendonk CM, van Beek J, Vennema H, Schürch AC, Schipper D, Bodewes R, Haagmans BL, Osterhaus AD, Koopmans MP.** 2014. New viruses in idiopathic human diarrhea cases, the Netherlands. *Emerg Infect Dis* **20:**1218–1222.
234. **Nguyen L, Burnett L.** 2014. Automation of molecular-based analyses: a primer on massively parallel sequencing. *Clin Biochem Rev* **35:**169–176.
235. **Hung T, Chen GM, Wang CG, Yao HL, Fang ZY, Chao TX, Chou ZY, Ye W, Chang XJ, Den SS, et al.** 1984. Waterborne outbreak of rotavirus diarrhoea in adults in China caused by a novel rotavirus. *Lancet* **1:**1139–1142.
236. **Fang ZY, Ye Q, Ho MS, Dong H, Qing S, Penaranda ME, Hung T, Wen L, Glass RI.** 1989. Investigation of an outbreak of adult diarrhea rotavirus in China. *J Infect Dis* **160:**948–953.
237. **Kelkar SD, Zade JK.** 2004. Group B rotaviruses similar to strain CAL-1, have been circulating in Western India since 1993. *Epidemiol Infect* **132:**745–749.
238. **Sanekata T, Ahmed MU, Kader A, Taniguchi K, Kobayashi N.** 2003. Human group B rotavirus infections cause severe diarrhea in children and adults in Bangladesh. *J Clin Microbiol* **41:**2187–2190.
239. **Alam MM, Pun SB, Gauchan P, Yokoo M, Doan YH, Tran TN, Nakagomi T, Nakagomi O, Pandey BD.** 2013. The first identification of rotavirus B from children and adults with acute diarrhoea in kathmandu, Nepal. *Trop Med Health* **41:** 129–134.
240. **Luchs A, do Carmo Sampaio Tavares Timenetsky M.** 2015. Phylogenetic analysis of human group C rotavirus circulating in Brazil reveals a potential unique NSP4 genetic variant and high similarity with Asian strains. *Mol Genet Genomics* **290:** 969–986.
241. **Castello AA, Argüelles MH, Rota RP, Humphrey CD, Olthoff A, Gentsch JR, Glass RI, Glikmann G, Jiang B.** 2009. Detection and characterization of group C rotavirus in Buenos Aires, Argentina, 1997–2003. *J Med Virol* **81:**1109–1116.
242. **Esona MD, Humphrey CD, Dennehy PH, Jiang B.** 2008. Prevalence of group C rotavirus among children in Rhode Island, United States. *J Clin Virol*: the official publication of the Pan American Society for Clinical Virology **42:**221–224.
243. **Jiang B, Dennehy PH, Spangenberger S, Gentsch JR, Glass RI.** 1995. First detection of group C rotavirus in fecal specimens of children with diarrhea in the United States. *J Infect Dis* **172:**45–50.
244. **Rodger SM, Bishop RF, Holmes IH.** 1982. Detection of a rotavirus-like agent associated with diarrhea in an infant. *J Clin Microbiol* **16:**724–726.
245. **Baek IH, Than VT, Kim H, Lim I, Kim W.** 2013. Full genomic characterization of a group C rotavirus isolated from a child in South Korea. *J Med Virol* **85:**1478–1484.
246. **Kumazaki M, Usuku S.** 2014. Epidemiological and genetic analysis of human group C rotaviruses isolated from outbreaks of acute gastroenteritis in Yokohama, Japan, between 2006 and 2012. *Arch Virol* **159:**761–771.
247. **Marton S, Deak J, Doro R, Csata T, Farkas SL, Martella V, Banyai K.** 2015. Reassortant human group C rotaviruses in Hungary. Infection, genetics and evolution: *J molec epidem and evol genet in infect dis* **34:**410–4. Pub Med
248. **Yang H, Makeyev EV, Kang Z, Ji S, Bamford DH, van Dijk AA.** 2004. Cloning and sequence analysis of dsRNA segments 5, 6 and 7 of a novel non-group A, B, C adult rotavirus that caused an outbreak of gastroenteritis in China. *Virus Res* **106:**15–26.
249. **Alam MM, Kobayashi N, Ishino M, Ahmed MS, Ahmed MU, Paul SK, Muzumdar BK, Hussain Z, Wang YH, Naik TN.** 2007. Genetic analysis of an ADRV-N-like novel rotavirus strain B219 detected in a sporadic case of adult diarrhea in Bangladesh. *Arch Virol* **152:**199–208.

250. **Berns K, Parrish CR.** 2013. Parvoviridae, p. 2437–2477. *In* Knipe DM, Howley P (ed.), *Fields Virology* 6th ed. Lippincott Williams & Wilkins, Philadelphia, PA.
251. **Yahiro T, Wangchuk S, Tshering K, Bandhari P, Zangmo S, Dorji T, Tshering K, Matsumoto T, Nishizono A, Söderlund-Venermo M, Ahmed K.** 2014. Letter: Novel human bufavirus genotype 3 in children with severe diarrhea, Bhutan. *Emerg Infect Dis* **20:**1037–1039.
252. **Väisänen E, Kuisma I, Phan TG, Delwart E, Lappalainen M, Tarkka E, Hedman K, Söderlund-Venermo M.** 2014. Bufavirus in feces of patients with gastroenteritis, Finland. *Emerg Infect Dis* **20:**1077–1079.
253. **Chieochansin T, Vutithanachot V, Theamboonlers A, Poovorawan Y.** 2015. Bufavirus in fecal specimens of patients with and without diarrhea in Thailand. *Arch Virol* **160:**1781–1784.
254. **Masters PS, Perlman S.** 2013. Coronaviridae, p. 825–858. *In* Knipe DM, Howley P (ed.), *Fields Virology* 6th ed. Wolters Kluwer Health/Lippincott Williams & Wilkins, Philadelphia.
255. **Hui DS, Memish ZA, Zumla A.** 2014. Severe acute respiratory syndrome vs. the Middle East respiratory syndrome. *Curr Opin Pulm Med* **20:**233–241.
256. **Esper F, Ou Z, Huang YT.** 2010. Human coronaviruses are uncommon in patients with gastrointestinal illness. *J Clin Virol*: the official publication of the Pan American Society for Clinical Virology **48:**131–133.
257. **Jamieson FB, Wang EE, Bain C, Good J, Duckmanton L, Petric M.** 1998. Human torovirus: a new nosocomial gastrointestinal pathogen. *J Infect Dis* **178:**1263–1269.
258. **Koopmans MP, Goosen ES, Lima AA, McAuliffe IT, Nataro JP, Barrett LJ, Glass RI, Guerrant RL.** 1997. Association of torovirus with acute and persistent diarrhea in children. *Pediatr Infect Dis J* **16:**504–507.
259. **Ganesh B, Masachessi G, Mladenova Z.** 2014. Animal picobirnavirus. *Virusdisease* **25:**223–238.
260. **Malik YS, Kumar N, Sharma K, Dhama K, Shabbir MZ, Ganesh B, Kobayashi N, Banyai K.** 2014. Epidemiology, phylogeny, and evolution of emerging enteric Picobirnaviruses of animal origin and their relationship to human strains. *BioMed Res Int* **2014:**780752.
261. **Ganesh B, Bányai K, Martella V, Jakab F, Masachessi G, Kobayashi N.** 2012. Picobirnavirus infections: viral persistence and zoonotic potential. *Rev Med Virol* **22:**245–256.
262. **Yamashita T, Kobayashi S, Sakae K, Nakata S, Chiba S, Ishihara Y, Isomura S.** 1991. Isolation of cytopathic small round viruses with BS-C-1 cells from patients with gastroenteritis. *J Infect Dis* **164:**954–957.
263. **Reuter G, Boros A, Pankovics P.** 2011. Kobuviruses - a comprehensive review. *Rev Med Virol* **21:**32–41.
264. **Khamrin P, Maneekarn N, Okitsu S, Ushijima H.** 2014. Epidemiology of human and animal kobuviruses. *Virusdisease* **25:**195–200.
265. **Drexler JF, Baumgarte S, de Souza Luna LK, Eschbach-Bludau M, Lukashev AN, Drosten C.** 2011. Aichi virus shedding in high concentrations in patients with acute diarrhea. *Emerg Infect Dis* **17:**1544–1548.
266. **Yamashita T, Sakae K, Ishihara Y, Isomura S, Utagawa E.** 1993. Prevalence of newly isolated, cytopathic small round virus (Aichi strain) in Japan. *J Clin Microbiol* **31:**2938–2943.
267. **Atmar RL, Estes MK.** 2010. Gastrointestinal Tract Infections, *Lennette's Laboratory Diagnosis of Viral Infections*, 4 ed. CRC Press, Boca Raton, FL.

Hepatitis A and E Viruses

GILBERTO VAUGHAN AND MICHAEL A. PURDY

23

Viral hepatitis refers to inflammation of the liver caused by several viral agents, including hepatitis A (HAV), B (HBV), C (HCV), D (HDV), and E (HEV) viruses. Generally, HAV and HEV are associated with acute, self-limited infection; however, severe and protracted disease can develop. These nonenveloped viruses have independently evolved interesting mechanisms aimed at evading the immune response and guaranteeing survival in the host. These viruses are phylogenetically unrelated but share several similarities. The genomes from both viruses are single-stranded, positive-polarity RNA. The viruses primarily infect the liver, cause acute infection, and are shed in feces. Moreover, the clinical manifestations from HAV and HEV infection are undistinguishable and require specific laboratory tests for each virus. Despite the similarities exhibited by these viruses, differences in the epidemiology of the respective diseases result in distinctive distribution patterns worldwide.

In this chapter, we discuss in detail the epidemiology, clinical features, and diagnostic assays, with emphasis in clinical settings, of both viruses.

VIRAL CLASSIFICATION AND BIOLOGY

HAV

Jaundice, as described in ancient Greek and Chinese literature and described by Hippocrates as "infectious icterus," was probably referring to viral hepatitis. The viral etiology of hepatitis A was originally established during the 1940s and 1950s by Krugman and colleagues (1). Subsequently, several studies showed that transmission of hepatitis A in humans occurred primarily by the fecal-oral route. In the 1970s, HAV was identified in the stools of acute cases, resulting in the characterization of the virus, development of serologic and molecular tests and eventually development of an extremely efficient vaccine [reviewed in (2)]. Importantly, the availability of methods capable of detecting HAV infections in a timely manner and the implementation of vaccination programs have led to important changes in the epidemiology of HAV-related disease by reducing significantly the number of acute cases.

HAV is an *Enterovirus* belonging to the *Picornaviridae* family (3), which shares several characteristics with other members of the family. However, HAV displays an array of unique features that have been sufficient to classify it as the only species in the genus *Hepatovirus*, although a newly discovery virus (*phopivirus*) infecting seals has been proposed as a second member of the genus (4). HAV is a nonenveloped virus that has evolved a unique strategy by which it hijacks cellular membranes, exiting the host cells fully cloaked in a lipid membrane (5). The HAV genome is a single-stranded, positive-polarity RNA approximately 7.5-kilobases in length (3, 6). It consists of three major regions, known as P1, P2, and P3, flanked by 5′- and 3′-untranslated regions (UTR) (Fig. 1). The P1 region encodes the structural proteins, while the nonstructural proteins are encoded by the P2 and P3 regions (7–9). The viral genome comprises a single open reading frame (ORF) encoding a large polyprotein of approximately 250 kilodaltons. Maturation of individual viral proteins and subsequent assembly of the viral particle requires the polyprotein to undergo co- and posttranslational enzymatic cleavage (3). P1 encodes for VP1, VP2, VP3, and VP4 proteins. VP1, VP2, and VP3 assemble into pentameric structures required for the assembly of the capsid. The P2 region encodes for the 2A, 2B, and 2C proteins. While the putative 2A gene of other picornaviruses shows proteinase activity, the HAV 2A protein is not enzymatically active. The role of HAV 2B is not known, but recent reports have shown that this protein alters membrane permeability (10). HAV 2C has been associated with helicase activity and contains the nucleotide triphosphate (NTP)-binding site (11). P3 encodes for 3A, 3B, 3C, and 3D proteins. HAV 3A, also known as preVpg, is responsible for anchoring the 3B protein to the membrane. In other picornaviruses, 3A serves as the membrane attachment molecule for the viral replicase allowing for RNA synthesis. Additionally, the 3A hydrophobic transmembrane domain acts as a signal to relocalize HAV 3C to the mitochondria (12). The 3B protein, also referred to as Vpg, is the adapter molecule ligated to the HAV RNA. HAV 3C is the trypsin-like serine HAV protease, and the 3D protein is the viral polymerase.

The virion of HAV is a small, spherical particle of 27 to 32 nanometers. The viral particle is an icosahedral capsid formed by 12 pentamers comprising multiple copies of the structural proteins (Fig. 1). In this viral particle, the major viral capsid protein, VP1, is stabilized by a network formed

doi:10.1128/9781555819156.ch23

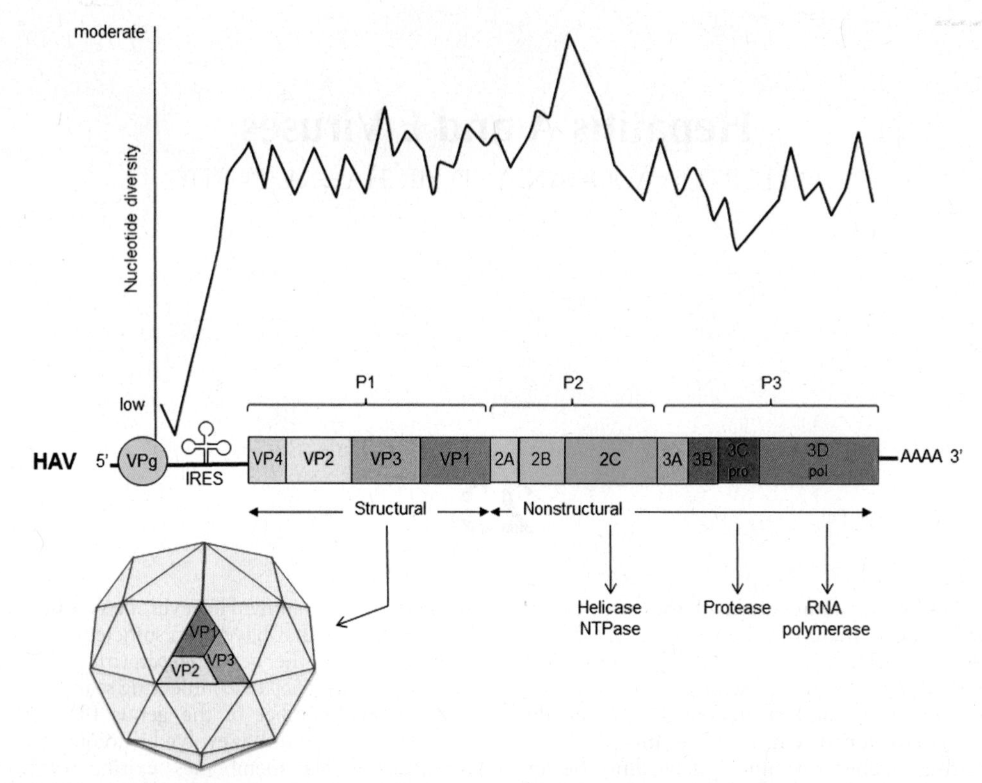

FIGURE 1 The composition of the HAV genome is depicted. All structural and nonstructural proteins are color coded and their known functions are also described. The nucleotide variability along the entire genome is also presented.

by the amino termini of proteins VP1 and VP3 (13). The result is a highly infectious particle, relatively stable to environmental elements; these characteristics facilitate the occurrence of outbreaks by allowing the virus to remain viable for extended periods of time in the environment. Viral inactivation can be attained by heating (above 185°F) and disinfection with sodium hypochlorite.

HAV is a relatively conserved virus with only one serotype. However, six genotypes (I through VI) and several subtypes have been recognized based on the nucleotide differences in the viral genome (14). Genotypes I, II, and III are further divided into subtypes A and B, and a novel subtype C has been recently proposed for genotype I (15). Humans are the main hosts for HAV; however, several nonhuman primates may also contract the infection. Genotypes I through III have been associated with infections in humans, while genotypes IV through VI are of simian origin.

HEV

Enterically transmitted non-A, non-B hepatitis (hepatitis E) was first described in 1957 (16). In the following years, a number of large outbreaks associated with contaminated water were described in developing countries (17, 18). The causal agent, HEV, was identified and sequenced in 1990 (19). Originally, hepatitis E was assumed to be confined to developing countries, with sporadic cases being identified in developed countries as travel-associated importation of the illness. As more information was obtained about the virus, it was determined that hepatitis E was also an autochthonous illness in developed countries. Additionally, while hepatitis E is usually a self-limiting disease in nonpregnant healthy individuals, it can be more severe (fulminant) in pregnant women (especially in the third trimester) and among individuals with pre-existing illness. Evidence from experimental infections of macaques indicates that hepatitis E is a dose-dependent illness, and while high doses of an infectious inoculum lead to the development of overt symptoms, a low dose can lead to subclinical infections (20). Further, it can cause asymptomatic, acute, and chronic hepatitis in immunocompromised individuals and can progress to liver fibrosis, cirrhosis, and end stage liver disease; severe neurological symptoms have also been seen (21, 22). Hepatitis E has been discovered among organ transplant patients, HIV-infected patients, and cancer patients undergoing chemotherapy (23). WHO estimates about 2 billion people, representing one-third of the world's population, live in endemic regions and are at risk of HEV infection.

HEV belongs to the family *Hepeviridae*. HEV is non-enveloped RNA virus with an icosahedral capsid about 27 to 34 nanometers in diameter, although recent evidence suggests that in the blood of infected individuals HEV may highjack host membranes, possibly as a means to escape immune detection (24). The viral genome has a 5′ 7-methylguanosine cap and a 3′ polyadenylated tail. The genome is a positive-sense single-stranded RNA of about 7.2 kilobases for variants infecting humans. The viral genome consists of three open reading frames (ORFs). The first ORF codes for genes necessary for viral replication, the second ORF codes for the viral capsid protein, and the third ORF codes for a phosphoprotein responsible for regulating cellular functions in, and viral release from, infected cells.

The virus has been isolated from humans, trout, chickens, camels, bats, bandicoots, rats, ferrets, foxes, rabbits, minks, moose, musk shrews, pigs, wild boars, deer, and mongooses. In spite of this diversity of hosts, not all of these variants can infect humans. Variants that infect humans belong to the genus *Orthohepevirus*, and the species *Orthohepevirus* A. *Orthohepevirus* A is organized into 7 genotypes (25).

Of the four recognized *Hepevirus (Orthohepevirus* A) genotypes, genotypes 1 and 2 are transmitted fecal-orally among humans, and genotypes 3 and 4 are transmitted to humans zoonotically from infected pigs, deer, wild boar, and possibly rabbits (25). It is not known if genotypes 5 and 6 can infect humans. Recently a transplant recipient from Somalia was found to be chronically infected with genotype 7, camel HEV. This individual farms camels and consumes camel milk and meat (26). It is not known if genotype 7 can be transmitted zoonotically to non-immunocompromised individuals.

EPIDEMIOLOGY

HAV

Worldwide, approximately 1.5 million infections occur annually (27). In endemic regions such as Latin America, the Indian Subcontinent, the Middle East, and Africa, HAV affects primarily young children (younger than 5 years of age). In nonendemic countries, large numbers of susceptible older individuals are infected owing to the lack of exposure during childhood. Improved sanitary conditions reduce the likelihood of fecal contamination and exposure in developed countries and regions transitioning from high to intermediate or low endemicity (28). The implementation of universal childhood vaccination is also responsible for the sharp decline of HAV infections in developed countries. As a consequence of successful vaccination programs, a shift in the age group and increased HAV-related disease is commonly observed in older individuals since they are more likely to develop manifestations. Moreover, the increased number of older susceptible individuals resulting from lower exposure rates facilitates the occurrence of HAV outbreaks in these regions. Other factors contributing to HAV transmission include crowding, poor hygiene, improper sanitation, and contamination of food and water.

The main risk factors associated with HAV infection are close contact with acute HAV cases, travel to endemic regions, men who have sex with men (MSM), and injection drug use. Other modes of transmission, such as blood transfusion and vertical transmission, are rare (2).

The most common mode of HAV transmission is by close personal contact between infected and susceptible persons. The prolonged shedding of HAV, before and after the onset of symptoms, facilitates person-to-person transmission. Epidemiological and genetic analyses of outbreaks have shown evidence of HAV person-to-person transmission in communities (29). Therefore, in closed institutions, such as schools, day care centers, and nurseries, the agglomeration of individuals, sharing of objects, inadequate hygienic conditions, and high proportion of susceptible individuals facilitate HAV transmission. Importantly, HAV transmission in day care centers or schools is commonly associated with household contacts and contact with infected Hispanic children (29).

Hepatitis A water- or foodborne outbreaks are characterized by a sudden increase in the number of jaundiced persons within a short period of time. Waterborne outbreaks of hepatitis A occur commonly among persons who drink contaminated water or swim in water contaminated by adjacent septic systems or sewage. On the other hand, foodborne transmission most frequently occurs when fecal matter from an HAV-infected food handler is transferred to food. HAV transmissions associated with contaminated food items, such as seafood, fruits, and vegetables, are well documented (30–35).Viral contamination of food items can occur at any point during harvesting, processing, preparation, or distribution. In the United States, more than 250 HAV outbreaks were identified from 1994 through 2004, highlighting the importance of HAV control in developed countries. Foodborne outbreaks, along with waterborne outbreaks, account for approximately 6.5% of total HAV cases in the United States. Moreover, foodborne outbreaks can affect large number of individuals as observed in the largest hepatitis A outbreak associated with consumption of raw clams in Shanghai where approximately 300,000 individuals were infected (36).

While HAV transmission occurs primarily in children in developing regions, in developed countries the virus spreads also via other modes of transmission, exclusive of adult populations. Thus, hepatitis A outbreaks among MSM have been reported in Europe and the United States (15, 37, 38). HAV transmission is not a direct result of sexual intercourse among MSM but instead is promoted by practices performed during intercourse leading to oral ingestion of viruses. Extended virus shedding facilitates virus spread via oro-anal and oro-genital contact. In the United States, approximately 4% of the reported risk factors for HAV infections are associated with MSM activity.

In addition to MSM, high rates of exposure to HAV and outbreaks are observed among injection drug users (IDUs) (39). Increased transmission of HAV among IDUs can be associated with poor sanitary and personal hygiene conditions, and factors related to lifestyle. Thus, frequent sharing of injection needles facilitates HAV transmission from viremic individuals in closed groups of IDUs.

Molecular epidemiologic approaches have helped us better understand the dynamics exploited by HAV to warrant persistence in different populations (28). Genotypic distribution and genetic relatedness studies have shown a characteristic distribution of viral lineages worldwide (2). Globally, genotype I is most prevalent, with subtype IA being more common than IB (2). HAV genotypes IA or IB have been shown to be the prevailing viral lineages in certain geographic regions; however, cocirculation is not uncommon (28). Subtype IA constitutes a major fraction of genotype I strains circulating in South and North America, Europe, Asia, and Africa, while subtype IB is predominant in the Middle East and South Africa, although it also circulates in South America (28). In the United States, strains belonging to subtype IA group in three major clusters that are associated with certain risk factors, suggesting that some viral lineages are tightly linked to certain modes of transmission (29). Genotype II isolates were originally identified in France and Sierra Leone; however, detection of this genotype is rare. Subtype IIA was suggested to have been originated in West Africa and to be endemic to Benin. Thus, it is not clear why genotype II has such a limited circulation around the world. Genotype III is more cosmopolitan and has been identified in Asia, Europe, and the United States (2). In India, all reported hepatitis A outbreaks to date have been caused by genotype IIIA infections, while in Japan, subtypes IIIA and IIIB cocirculate broadly with IA and IB strains. Interestingly, increased circulation of genotype IIIA strains has been recently reported in Korea, Russia, and Estonia, where genotype I was most prevalent (40–42). This might suggest that viral lineage replacement is taking place in these countries.

HEV

Genotypes 1 and 2 are transmitted to humans fecal-orally, usually through fecally comtaminated water and are responsible for sporadic cases and epidemics. Epidemics have been confined to developing countries. Some of these epidemics have infected tens of thousands of individuals. From 1988 to 1989 more than 110,000 individuals were infected during the Xinjiang, China, epidemic (18). Epidemics of fecal-orally transmitted hepatitis E occur in developing countries in Asia due to contamination of drinking water supplies. Outbreaks can also occur when water purification systems malfunction or when sewage is not properly isolated from drinking water supplies. In Africa outbreaks occur during humanitarian crises in refugee camps when access to clean water is limited. A study in India suggested that fecal viral excretions could be found in sewage year round, and HEV infections occurred even in nonepidemic periods (43). Cases of HEV genotype 1, identified in developed countries, have most often been attributed to importation either by immigrants from, or by travelers going to, endemic regions.

Genotypes 3 and 4 are transmitted zoonotically from infected hosts. Offal and meat from infected animals are the prime vehicles for transmission. Unlike genotypes 1 and 2, whose transmission is confined largely to epidemic regions, HEV genotype 3 infects humans worldwide, and genotype 4 has been found in China, Japan, Europe, and India. Genotypes 3 and 4 cause autochthonous hepatitis E in developed countries. Most of these infections appear to be subclinical. HEV RNA has been isolated from retail pig livers (44), uncooked pig sausage (45) and farmed rabbits (46). Although pigs are most often documented as the source of zoonotic transmission, there have been cases associated with consuming improperly cooked meat from infected deer and wild boar (47, 48), and there is a report of a woman in France infected with an HEV strain with characteristics very similar to HEV from rabbits (46). People who work with swine, such as veterinarians, farmers, other farm workers, and abattoir workers, exhibit higher HEV-IgG prevalence rates than populations without significant exposure to swine (48). The consumption of raw or undercooked mollusks is another potential source for zoonotic transmission (49).

The age-specific attack rate for hepatitis E is distinct from that for hepatitis A. In nonpregnant immune-competent individuals, hepatitis E is usually subclinical or a self-limiting acute hepatitis. Hepatitis E affects young- to middle-aged adults, although disease has occasionally been observed in children. The sex ratio for hepatitis E in children is 1:1, but among adults the disease occurs more often in males (50). Seroprevalence for anti-HEV IgG antibodies increases with age, with the highest rates being seen among the elderly. Case fatality rates in adults are 0.5 to 2%, except for pregnant women, where case fatality rates range from 10 to 30%. At present, the high mortality rates seen among pregnant women have only been seen with HEV genotype 1 and have not been confirmed with any other HEV genotype. Death is usually due to encephalopathy, hemorrhagic diathesis, or renal failure. Mothers can transmit the virus to the fetus. This can lead to premature birth, increased fetal loss, hypoglycemia, hypothermia, and acute hepatitis in the newborn. Transmission due to household contact is extremely low (17).

A few cases of transfusion-associated HEV have been described (51), and hemodialysis patients in endemic regions appear to be at higher risk of infection. HEV analytes have been detected in blood donors around the world (52–54), and blood donors who were viremic at the time of donation were detected in Europe (55); however, a study in the United States found that none of 1,939 blood donors tested was viremic (53). Donors who were viremic at the time of donation were asymptomatic. Attempts to screen the blood supply are complicated by the fact that neither ALT nor anti-HEV IgM testing correlate with the presence of HEV RNA. Thus nucleic acid testing is presently the only reliable method for the detection of HEV in blood (52). Additionally, the precise extent of this problem is not fully defined, but risk assessments are under way to determine whether donors should be screened. While chronic HEV infections have been identified among patients with HIV infection (21, 23), cohort studies have indicated that HEV is unlikely to be transmitted sexually, and HEV/HIV coinfection has not been seen as a common problem (56).

CLINICAL SIGNIFICANCE

HAV

Clinical manifestations of HAV infection vary greatly depending on age. Subclinical infection is commonly unrecognized unless biochemical or serologic testing is performed. Acute HAV infection is presented as an acute illness causing jaundice or elevated serum aminotransferases. Clinically, HAV infection may manifest as fatigue, malaise, vomiting, anorexia, fever, and abdominal pain. Jaundice may become evident, accompanied by pruritus, and, more infrequently, hepatomegaly, splenomegaly, lymphadenopathy, and rash. Symptoms and clinical laboratory abnormalities commonly resolve within two months after onset. Infection does not progress to chronic liver disease; however, relapsing, protracted disease, and prolonged excretion in feces of HAV have been observed. Clinical manifestations cannot be differentiated from other viral hepatitides, and, therefore, differential laboratory diagnosis is required. Atypical manifestations of HAV infection include relapsing hepatitis, prolonged cholestasis, and acute kidney damage. Disease can progress, for reasons not completely understood, from jaundice to acute liver failure. The term fulminant hepatitis (FH) has traditionally been used to describe patients without previous liver disease who rapidly undergo liver failure within four weeks of onset of symptoms. FH is characterized by rapid deterioration of hepatic function, resulting in encephalopathy, coagulopathy, and multiorgan failure. The clinical presentation of hepatic encephalopathy is variable and depends on the extent and rapidness of hepatic damage. Cerebral edema, which can reach up to 80% of FH patients, is a major cause of mortality. Orthotopic liver transplantation is often required; however, long-term survival post-transplantation is poor. While hepatitis B virus (HBV) infection is the most common cause for viral FH worldwide (approximately 1% of total cases), the incidence of HAV FH ranges from 0.015% to 0.5%. Little is known about the pathogenesis and underlying mechanisms involved in the development of FH due to HAV infection. Although spontaneous FH in chimpanzees has been reported (57), the absence of an animal model capable of faithfully reproducing the natural course of HAV FH disease has significantly hindered our ability to undercover the responsible mechanisms associated with disease severity.

HEV

The incubation period for hepatitis E ranges from 15 to 40 days with the average being longer than that observed for hepatitis A. The illness is usually divided into two phases. The preicteric phase lasts from 1 to 10 days (average, 3 to 4 days). Gastrointestinal symptoms include diarrhea, epigastric pain, nausea, hepatomegaly, splenomegaly, and vomiting. Additional symptoms include itching, headache, and fever and, less commonly, arthralgia and urticarial rash. The icteric phase begins with the appearance of jaundice, dark urine, and clay-colored stools. This phase can last from 12 to 15 days, and complete recovery usually takes place within one month. Elevations in serum alanine aminotransferase (ALT) and aspartate aminotransferase (AST) levels occur as a single peak preceding, or coinciding with, the onset of jaundice, and return to normal 6 to 42 days after onset of symptoms in uncomplicated cases (Fig. 2).

Liver biopsies show lobular disarray, enlarged portal ducts with an inflammatory infiltrate, ballooning of hepatocytes, and focal or bridging necrosis. There is marked canalicular and cytoplasmic cholestasis, and bile plugs are seen in the middle of pseudoductules (58).

In pigs experimentally infected with HEV isolated from humans, negative-strand HEV RNA, indicative of viral replication, has been detected in liver, lymph nodes, colon, small intestines, stomach, spleen, kidney, tonsils, and salivary glands (59). In humans, extrahepatic manifestations are associated with the brain, central nervous system, muscle tissue, kidney, pancreas, and placenta (21, 60, 61).

HEV is an underdiagnosed cause of chronic hepatitis in transplant patients and is not limited to liver transplantation. Patients with cleared hepatitis E infections pretransplantation can become reinfected posttransplantation. The incidence of chronic hepatitis E in transplant patients is 1 to 3% (60). Research indicates that hepatitis E infection is not due to transplantation but usually occurs because the patient is on immunosuppressive therapy and susceptible to infection posttransplantation. The immune suppression in these patients allows hepatitis E to become a chronic infection, and the risk of developing chronic hepatitis is higher than 60% in all types of transplant groups; however, the number of patients examined has been low, and further research is needed to confirm these results (23, 60). Additionally, hepatitis E can be misdiagnosed as drug-induced hepatitis.

TREATMENT AND PREVENTION

HAV

HAV treatment is established as supportive care. No specific antiviral medication is available for HAV treatment; however, amantadine and interferon-alpha have been reported to interfere with HAV replication. Amantadine inhibits HAV internal ribosome entry site (IRES)-mediated translation in hepatoma cells, but its usefulness in the treatment of HAV cases remains to be demonstrated (3, 62).

Appropriate patient management and early detection are critical to provide adequate support during the acute phase of the disease and also to prevent virus spread (63). Corticosteroid therapy does not significantly alleviate the symptomatology. Hospitalization is also recommended in case of coagulopathy, protracted vomiting, or encephalopathy. Liver function should be monitored to exclude progression to coagulopathy and liver failure. Management of HAV FH cases is determined by the complications that develop and the availability of liver transplantation. Hepatic failure, low serum aminotransferase, high serum bilirubin, and low albumin are associated with a poor outcome for HAV-related fulminant cases. Liver transplantation may be necessary if overall deterioration is observed.

HEV

In immune competent individuals, treatment remains supportive. At present there is no standard therapeutic protocol for the treatment of immunocompromised patients. The treatment of chronic hepatitis E among immunocompromised individuals has been most frequently studied in

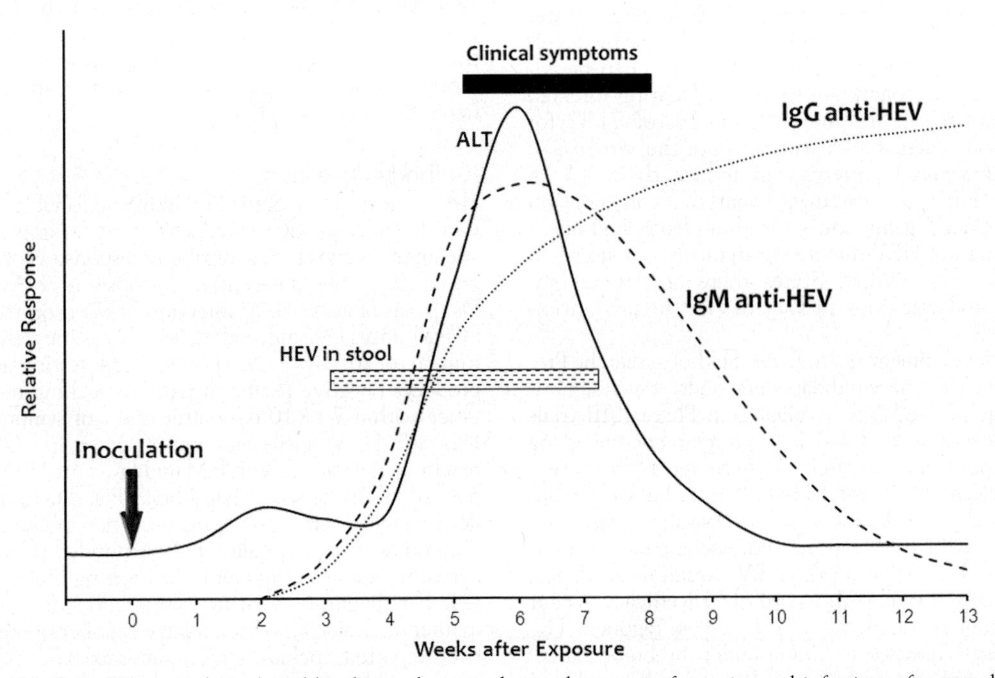

FIGURE 2 Virological, immunological, and biochemical events during the course of experimental infections of cynomolgus macaques with hepatitis E virus.

transplant patients. In transplant patients, reduction of immunosuppressants is suggested as the first intervention strategy to achieve viral clearance, but it should be used with caution because of the risk of rejection (23, 60).

Treatment of chronic hepatitis E with either ribavirin or pegylated interferon is effective, but the optimal dose and duration of therapy remain to be determined. Chronic hepatitis E has been successfully treated with ribavirin and pegylated interferon alone or in combination. In transplant patients, ribavirin monotherapy (600 to 800 milligrams per deciliter) for 3 to 5 months has been successful in achieving viral clearance at the end of therapy in 15 of 16 patients, with sustained virological response in 10 of 16 patients (3 patients were not available for follow-up testing). Pegylated interferon (135 micrograms per week for 3 to 12 months) was successful in achieving virological response in 4 of 5 patients at the end of therapy, with sustained virological response in 3 of these patients. In hematological patients, either ribavirin monotherapy or pegylated interferon alone achieved virological response in 3 of 3 patients after three months of therapy, with sustained virological response in 2 patients (1 patient was not available for follow-up testing) (60). One HIV-infected patient with chronic hepatitis E was treated with combination ribavirin and pegylated interferon for 9 months. This patient achieved a sustained virological response followed by normalization of liver function tests and reduction in inflammation and fibrosis (21). Another HIV-infected patient received 1200 milligrams per day of ribavirin for 3 weeks followed by rapid viral clearance with restoration of normal liver function, and a third received 800 milligrams per day of ribavirin for 3 months and achieved viral clearance in 2 months (64).

There is no treatment suggested for pregnant women with hepatitis E, although insufficient research has been conducted to examine this issue. In addition, ribavirin is a teratogenic agent.

Systematic guidelines for the prevention of hepatitis E have not yet been developed. In endemic regions, travelers should avoid drinking inadequately treated water or consuming uncooked or undercooked foods. In developed countries pork products and game meats can be adequately cooked to inactivate the virus. Heating infected pork products to an internal temperature of 71°C for 20 minutes has been shown to inactivate the virus (65). Immunocompromised patients can reduce their risk of exposure to HEV by abstaining from the consumption of pork products, game meats, and undercooked meat (60). Treatment of HEV-infected individuals was discussed previously. In the United States there are no recommended prophylactic drugs or vaccines for the prevention of hepatitis E.

Vaccine development is an active field of research. Presently, several vaccine candidates are under development with a couple of candidates evaluated in Phase II/III trials (66). The first is a 56 kilodalton protein, encompassing amino acid positions 112 and 607 from the HEV capsid, which underwent a successful Phase II trial, but no further progress on the development of this vaccine has been reported as of 2011 (66). The second, encompassing amino acid positions 368–606 from the HEV capsid, underwent a successful Phase III trial, with reports of 95% efficacy over at least 4.5 years after completion of a three-dose regimen. The drug is presently licensed in China under the brand name, Hecolin, (67, 68). Presently, no HEV vaccine has received FDA approval.

DETECTION/DIAGNOSIS

HAV

Biochemical Testing

Assessment of liver function by biochemical tests should be carried out to monitor cases of HAV infection. Cell blood count, prothrombin time, serum alanine aminotransferase (ALT), and aspartate aminotransferase (AST) concentrations are considered good markers of hepatocellular damage. High ALT:AST ratios are commonly observed in cases of HAV infection, with some exceptions including severe tissue damage, leading to an increase of AST blood levels. Lactic dehydrogenase (LDH) enzyme and bilirubin levels are also commonly elevated in hepatitis A cases with significant liver damage, peaking during the first weeks after onset of symptoms and resolving within 2 months. Other serum proteins, such as albumin, prothrombin, and fibrinogen, are also affected by HAV-related injury of the liver (3).

Acute cases of HAV infection exhibit normal or slightly reduced numbers of neutrophils and lymphocytosis; however, increased cell counts (12,000 cells per cubic millimeter) are usually considered a predictor of poor prognosis.

Occasional agranulocytosis, thrombocytopenia, red cell aplasia, aplastic, and hemolytic anemia can be observed in acute patients, the latter being associated with glucose-6-phosphate dehydrogenase deficiency (69).

Histological Characteristics

Changes in the parenchyma are usually observed during the acute phase of the disease and include accumulation of mononuclear inflammatory cells and hepatocyte degeneration (3). Apoptosis results in the migration of affected cells into the sinusoids, and phagocytosis by Kupffer cells. Activation of Kupffer cells leads to hypertrophy and hyperplasia. Inflammatory cells can be observed with some frequency in the adjacent parenchyma. Regeneration is promptly attained (8 to 12 weeks) after the convalescence phase, and it is characterized by significant reduction of the inflammatory response. Focal necrosis occurs in a small percentage of patients (5 to 10%), featuring neutrophil infiltration. Importantly, confluent hepatic necrosis can lead to fulminant hepatitis and death (3).

Antibody Detection

HAV infection is clinically indistinguishable from other viral hepatitides; therefore, serologic testing is required for adequate diagnosis. The antibody response is of critical importance in the differential diagnosis of HAV infection. Diagnosis of acute HAV infection is based on the detection of IgM anti-HAV antibodies, which can be detected in serum 5 to 20 days postexposure. IgM testing infrequently produces negative results in patients seeking medical assistance within 7 to 10 days after onset of symptoms. Serum IgG anti-HAV antibodies appear later in the infection, reaching high titers. Both IgM and IgG anti-HAV are usually present at the onset of symptoms (Fig. 3). IgM antibodies decay rapidly after the convalescence phase, while IgG antibodies stay in circulation for extended periods of time, persisting for years, apparently conferring lifelong immunity (3). For diagnosis in acute cases, IgM antibodies are the marker of choice for which a large number of methods have been reported, including radioimmunoassay, enzyme-linked immunosorbent assays, immunoblotting, etc. Several IgM anti-HAV enzymatic immunoassays are commercially avail-

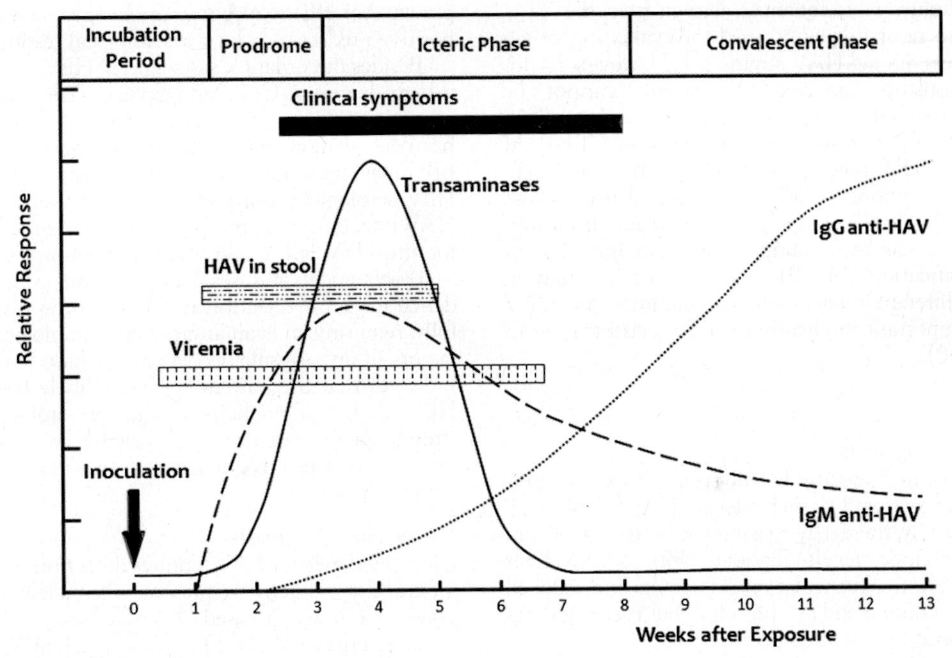

FIGURE 3 Virologic, immunological, and biochemical events during the course of experimental hepatitis A virus infection (adapted from reference (90) with permission.)

able [reviewed in (2)]. These diagnostic kits perform optimally during the first weeks postexposure when IgM antibodies are present at high titers. Importantly, some commercially-available IgM assays might detect antibodies as a result of vaccination in previously nonimmune individuals. Resolved HAV infection is commonly diagnosed by detection of IgG anti-HAV. Thus, the presence or absence of either antibody class can be used to differentiate between current and past infections.

Upon vaccination, antibodies to structural proteins are elicited. Interestingly, only a small percentage of vaccinees (8 to 20%) develop transient IgM anti-HAV response, while IgG anti-HAV is successfully elicited by nearly all immunized individuals. In-house assays based on nonstructural proteins are designed to differentiate between natural infections and vaccinees. Thus, different assay formats provide useful information indicative of the immune status of the patient.

Virus Isolation

HAV grows in several cell types of different origins; however, viral strains obtained from human samples require extensive adaptation. In cell culture, HAV produces relatively low viral titers, rendering this approach impractical for a clinical setting. HAV replicates in cell culture without obvious cell damage. Thus, due to the lack of a cytopathic effect, immunological or molecular assays are required to detect HAV proteins and RNA, respectively. Occasionally, cytopathic variants of HAV are observed, characterized by shorter replication cycles and higher viral yield (70).

Antigen Detection

HAV was originally identified by electron microscopy from fecal extracts using homologous antiserum. However, HAV antigen detection is not as straightforward. Virus capture in clinical samples is difficult because fibronectin can bind to HAV and mask antigenic determinants required for immunological detection (71). In addition, the HAV capsid has been detected circulating as immune complexes from experimentally infected chimpanzees, making virus capture even more difficult (72).

Molecular Detection

Nucleic acid detection methods are usually more sensitive than serologic assays for the detection of HAV in clinical, food, and environmental samples. A number of different molecular methodologies have been used to identify HAV, including restriction-fragment-length polymorphism, single-strand conformational polymorphism, southern blotting, reverse transcription-PCR (RT-PCR), etc. (2). While several of these methods have been used to identify HAV and viral antigen in liver tissue from infected primates, not all of them are suitable for a clinical setting.

Amplification of viral RNA by RT-PCR is currently the most sensitive and widely used method for detection of HAV RNA. Importantly, amplification and sequencing of HAV RNA are used to determine genetic relatedness between isolates and to identify sources of infection and transmission networks during occurrence of outbreaks (28, 73, 74). Different regions along the viral genome provide different degrees of resolution based on the sequence information contained within the region (28, 74). The 5′- and 3′- UTR are highly conserved regions and provide limited sequence information suitable for genotype classification (75, 76); therefore, both are preferred targets for HAV identification. Several small regions across the genome, including VP1, 2C, and 3D, display higher nucleotide variability than other regions along the HAV genome (Fig. 1). These regions have been used for specific identification of HAV strains in molecular surveillance and outbreak investigations (14, 29, 77).

HAV genotypes are identified by sequence analysis of different subgenomic regions, including the N-terminus of VP1, the VP1-P2A junction, and the full-length VP1 (28). Sequence variation in the VP1-P2A junction has been widely used to identify subtypes of HAV genotypes.

Phylogenetic analyses of sequences derived from the VP1-P2A region are most frequently used for evaluation of genetic relatedness among HAV strains (2). However, phylogenetic relationships among HAV strains cannot be accurately represented using small genomic regions (74). Comparative analysis of four genomic regions, VP1-P2A, VP1, VP2, and VP3, revealed significant variations in genetic relatedness among HAV strains assessed using these regions (14, 78). Moreover, analysis of sequences from subgenomic regions has been found to result in inconsistent genotype classification (14, 79). Thus, genetic information provided by different subgenomic regions from the HAV genome has important implications in the classification of viral strains (80).

HEV

Biochemical Diagnosis

Patients with acute hepatitis E will typically exhibit bilirubinuria, and, as stated earlier, elevated ALT and AST levels. Additionally, there may be a modest increase in serum alkaline phosphatase levels. Patients with severe liver damage have abnormal coagulation test results and reduced levels of serum albumin and prothrombin, but these analytes are not diagnostic for hepatitis E.

Serologic Diagnosis

IgM anti-HEV antibodies can usually be detected within 3 to 4 days after onset of jaundice and may persist for several months (average: 5 months). IgG anti-HEV antibodies appear simultaneously with the appearance of IgM or shortly after the appearance of IgM. IgG antibodies persist longer than IgM. IgG antibodies have been detected as long as 14 years after a hepatitis E outbreak but usually exhibit a rapid decline in titer 14 to 20 months after acute infection, although the antibodies are still detectable after this decline. There is some evidence that IgA anti-HEV antibodies may be useful as a marker of recent hepatitis E infection when used in conjunction with IgM assays (81).

Although serological assays for the detection of anti-HEV antibodies have been available for the past 20 years, head-to-head comparisons of these immunoassays have shown that many are discordant, requiring the clinician to be careful in interpreting the results of these assays. An analysis of six in-house and commercial anti-HEV IgM assays, using serum samples from patients with PCR-confirmed HEV infection, showed that sensitivity, specificity, and interassay agreement varied widely. Limits for analytic sensitivity were also found to be disparate among these assays (82). Another study compared two commercial IgG assays using WHO anti-HEV reference serum (83) and demonstrated substantial differences in the performance characteristics of these assays (84). A third comparison of eight commercial IgG and IgM assays used sera from immunocompetent and immunocompromised patients with PCR-confirmed HEV infections. Although these assays used antigens from HEV genotypes 1 and 2, several of these assays exhibited good sensitivity and specificity toward serum samples from patients infected with genotype 3, indicating no sensitivity problems in the detection of genotype 3. Clinical specificity varied from 84% to more than 99% and sensitivity ranged from 52% to 79% among the tested assays. One assay exhibited cross-reactivity toward samples from patients infected with Cytomegalovirus (CMV) or HAV (85). Additionally, neither ALT elevations nor IgM positivity correlate with the presence of HEV RNA (52, 86). The lack of a correlation between IgM positivity and the presence of HEV RNA in a serum specimen indicate a low positive predictive value for serological testing.

Besides discordant results among HEV serological assays, a high degree of CMV and Epstein-Barr virus (EBV) IgM cross-reactivity to HEV IgM (24% and 33%, respectively) has been demonstrated, and acute infection by CMV and EBV may cause false positivity for anti-HEV IgM in some HEV serological assays (85–87). Because CMV, EBV, and HAV infections may produce acute hepatitis, false positivity for anti-HEV IgM could result in diagnostic mistakes.

Biochemical and serological assays yield indirect evidence for HEV infection and need to be interpreted carefully, requiring an evaluation of epidemiological and clinical factors. If an assessment of a patient based on these factors indicates that the patient has most likely been exposed to HEV, then the predictive value of serological testing is strong, but if a patient is not likely to have been exposed to HEV, then a positive diagnostic test may have little evidentiary value (81).

Molecular Diagnosis

Direct evidence for HEV requires detection of HEV RNA in a clinical specimen. At present, all nucleic acid detection assays are research-based in-house assays. A 2011 study compared the results of 23 polymerase chain reaction (PCR) assays from 10 countries. No false positive results were seen. However, the study revealed differences in the sensitivity of the assays used. The analysis revealed a 100- to 1,000-fold difference in sensitivity between the assays evaluated. These results indicate the need for improved standardization of HEV RNA assays (88). PCR can be used to detect HEV RNA in serum and stool. HEV RNA can usually be detected in stool before serum, and RNA usually persists longer in stool than serum.

Loop-mediated isothermal amplification (LAMP) has also been used in the research laboratory to detect HEV RNA. Unlike PCR, which uses multiple cycles of temperature gradients, LAMP uses a single-step, single-temperature amplification process to amplify HEV RNA. One study using LAMP suggests that the LAMP assays may have 100-fold greater sensitivity than conventional PCR (89).

BEST PRACTICES

HAV

A number of methodologies have been developed for the detection of HAV. However, important advances in HAV detection have significantly improved our ability to adequately identify hepatitis A cases. Concomitantly, these advances have resulted in the obsolescence of several approaches, such as virus isolation, antigen detection, etc., in the clinical diagnosis of HAV. Availability of highly sensitive serological tests for identification of anti-HAV IgM provides excellent tools for the detection of acute cases in clinical settings and outbreak investigation. Conversely, detection of anti-HAV IgG antibodies remains important in epidemiological studies, but, considering the acute nature of HAV infection, its value in clinical settings is less prominent. Confirmation of serological findings using molecular approaches has been extremely important for understanding the global epidemiology of HAV and plays an important role in the confirmation of acute cases in the clinic. In conclusion, in clinical settings, IgM detection remains the tool of choice for the identification of acute HAV cases followed by molecular testing using real-time PCR-based assays targeting highly conserved regions.

HEV

There are no FDA approved assays for the detection of HEV. In a clinical setting, IgM detection can be used to detect acute cases, and IgG assays can detect post-acute hepatitis. However, because of the discordance between serological assays, direct detection of HEV RNA with PCR is the best method for detecting active HEV infection.

In the absence of PCR, biochemical and serological assays yield indirect evidence for HEV infection and need to be interpreted carefully, requiring an evaluation of epidemiological and clinical factors. If an assessment of a patient based on these factors indicates that the patient has most likely been exposed to HEV, then the predictive value of serological testing is strong, but if a patient is not likely to have been exposed to HEV, then a positive diagnostic test may have little evidentiary value (81).

REFERENCES

1. **Krugman S, Ward R, Giles JP, Bodansky O, Jacobs AM.** 1959. Infectious hepatitis: detection of virus during the incubation period and in clinically inapparent infection. *N Engl J Med* **261:**729–734.
2. **Nainan OV, Xia G, Vaughan G, Margolis HS.** 2006. Diagnosis of hepatitis a virus infection: a molecular approach. *Clin Microbiol Rev* **19:**63–79.
3. **Hollinger FB, Martin A.** 2013. Hepatitis A Virus, p 551–581. *In* Knipe DM, Howley PM (ed), *Fields Virology,* 6th ed. Lippincott Williams & Wilkins, New York.
4. **Anthony SJ, St Leger JA, Liang E, Hicks AL, Sanchez-Leon MD, Jain K, Lefkowitch JH, Navarrete-Macias I, Knowles N, Goldstein T, Pugliares K, Ip HS, Rowles T, Lipkin WI.** 2015. Discovery of a novel hepatovirus (*Phopivirus* of seals) related to human Hepatitis A Virus. *MBio* **6:**e01180–15.
5. **Feng Z, Hensley L, McKnight KL, Hu F, Madden V, Ping L, Jeong SH, Walker C, Lanford RE, Lemon SM.** 2013. A pathogenic picornavirus acquires an envelope by hijacking cellular membranes. *Nature* **496:**367–371.
6. **Cohen JI, Ticehurst JR, Purcell RH, Buckler-White A, Baroudy BM.** 1987. Complete nucleotide sequence of wild-type hepatitis A virus: comparison with different strains of hepatitis A virus and other picornaviruses. *J Virol* **61:**50–59.
7. **Probst C, Jecht M, Gauss-Müller V.** 1999. Intrinsic signals for the assembly of hepatitis A virus particles. Role of structural proteins VP4 and 2A. *J Biol Chem* **274:**4527–4531.
8. **Probst C, Jecht M, Gauss-Müller V.** 1998. Processing of proteinase precursors and their effect on hepatitis A virus particle formation. *J Virol* **72:**8013–8020.
9. **Schultheiss T, Kusov YY, Gauss-Müller V.** 1994. Proteinase 3C of hepatitis A virus (HAV) cleaves the HAV polyprotein P2-P3 at all sites including VP1/2A and 2A/2B. *Virology* **198:**275–281.
10. **Jecht M, Probst C, Gauss-Müller V.** 1998. Membrane permeability induced by hepatitis A virus proteins 2B and 2BC and proteolytic processing of HAV 2BC. *Virology* **252:**218–227.
11. **Gorbalenya AE, Koonin EV.** 1989. Viral proteins containing the purine NTP-binding sequence pattern. *Nucleic Acids Res* **17:**8413–8440.
12. **Yang Y, Liang Y, Qu L, Chen Z, Yi M, Li K, Lemon SM.** 2007. Disruption of innate immunity due to mitochondrial targeting of a picornaviral protease precursor. *Proc Natl Acad Sci USA* **104:**7253–7258.
13. **Rossmann MG, Arnold E, Erickson JW, Frankenberger EA, Griffith JP, Hecht H-J, Johnson JE, Kamer G, Luo M, Mosser AG, Rueckert RR, Sherry B, Vriend G.** 1985. Structure of a human common cold virus and functional relationship to other picornaviruses. *Nature* **317:**145–153.
14. **Costa-Mattioli M, Cristina J, Romero H, Perez-Bercof R, Casane D, Colina R, Garcia L, Vega I, Glikman G, Romanowsky V, Castello A, Nicand E, Gassin M, Billaudel S, Ferré V.** 2002. Molecular evolution of hepatitis A virus: a new classification based on the complete VP1 protein. *J Virol* **76:**9516–9525.
15. **Pérez-Sautu U, Costafreda MI, Lite J, Sala R, Barrabeig I, Bosch A, Pintó RM.** 2011. Molecular epidemiology of hepatitis A virus infections in Catalonia, Spain, 2005–2009: circulation of newly emerging strains. *J Clin Virol* **52:**98–102.
16. **Viswanathan R.** 1957. Infectious hepatitis in New Delhi (1955–56): a critical study; epidemiology. *Indian J Med Res* **45:**1–30.
17. **Bradley DW, Krawczynski K, Purdy MA.** 1993. Epidemiology, natural history and experimental models, p 379–383. *In* Zuckerman AJ, Thomas HC (ed), *Viral hepatitis: Scientific basis and clinical management.* The Bath Press, Avon.
18. **Teo C-G.** 2012. Fatal outbreaks of jaundice in pregnancy and the epidemic history of hepatitis E. *Epidemiol Infect* **140:**767–787.
19. **Reyes GR, Purdy MA, Kim JP, Luk KC, Young LM, Fry KE, Bradley DW.** 1990. Isolation of a cDNA from the virus responsible for enterically transmitted non-A, non-B hepatitis. *Science* **247:**1335–1339.
20. **Aggarwal R, Kamili S, Spelbring J, Krawczynski K.** 2001. Experimental studies on subclinical hepatitis E virus infection in cynomolgus macaques. *J Infect Dis* **184:**1380–1385.
21. **Dalton HR, Keane FE, Bendall R, Mathew J, Ijaz S.** 2011. Treatment of chronic hepatitis E in a patient with HIV infection. *Ann Intern Med* **155:**479–480.
22. **Kamar N, Bendall RP, Peron JM, Cintas P, Prudhomme L, Mansuy JM, Rostaing L, Keane F, Ijaz S, Izopet J, Dalton HR.** 2011. Hepatitis E virus and neurologic disorders. *Emerg Infect Dis* **17:**173–179.
23. **Zhou X, de Man RA, de Knegt RJ, Metselaar HJ, Peppelenbosch MP, Pan Q.** 2013. Epidemiology and management of chronic hepatitis E infection in solid organ transplantation: a comprehensive literature review. *Rev Med Virol* **23:**295–304.
24. **Nagashima S, Takahashi M, Jirintai S, Tanggis, Kobayashi T, Nishizawa T, Okamoto H.** 2014. The membrane on the surface of hepatitis E virus particles is derived from the intracellular membrane and contains trans-Golgi network protein 2. Arch Virol 159:979–991.
25. **Smith DB, Simmonds P, members of the International Committee on the Taxonomy of Viruses Hepeviridae Study Group, Jameel S, Emerson SU, Harrison TJ, Meng X-J, Okamoto H, Van der Poel WHM, Purdy MA.** 2015. Consensus proposals for classification of the family Hepeviridae. *J Gen Virol* **96:**1191–1192.
26. **Lee G-H, Tan B-H, Teo EC-Y, Lim S-G, Dan Y-Y, Wee A, Aw PPK, Zhu Y, Hibberd ML, Tan C-K, Purdy MA, Teo C-G.** 2016. Chronic Infection With Camelid Hepatitis E Virus in a Liver-transplant Recipient Who Regularly Consumes Camel Meat and Milk. *Gastroenterology* ahead of print.
27. **Wasley A, Fiore A, Bell BP.** 2006. Hepatitis A in the era of vaccination. *Epidemiol Rev* **28:**101–111.
28. **Vaughan G, Goncalves Rossi LM, Forbi JC, de Paula VS, Purdy MA, Xia G, Khudyakov YE.** 2013. Hepatitis A virus: Host interactions, molecular epidemiology and evolution. Infection, genetics and evolution. *Infect Genet Evol* **21:**227–243.
29. **Nainan OV, Armstrong GL, Han XH, Williams I, Bell BP, Margolis HS.** 2005. Hepatitis a molecular epidemiology in the United States, 1996–1997: sources of infection and implications of vaccination policy. *J Infect Dis* **191:**957–963.
30. **Amon JJ, Devasia R, Xia G, Nainan OV, Hall S, Lawson B, Wolthuis JS, Macdonald PD, Shepard CW, Williams IT, Armstrong GL, Gabel JA, Erwin P, Sheeler L, Kuhnert W, Patel P, Vaughan G, Weltman A, Craig AS, Bell BP, Fiore A.** 2005. Molecular epidemiology of foodborne hepatitis a outbreaks in the United States, 2003. *J Infect Dis* **192:**1323–1330.
31. **Dentinger CM, Bower WA, Nainan OV, Cotter SM, Myers G, Dubusky LM, Fowler S, Salehi ED, Bell BP.** 2001.

An outbreak of hepatitis A associated with green onions. *J Infect Dis* **183:**1273–1276.

32. **Wheeler C, Vogt TM, Armstrong GL, Vaughan G, Weltman A, Nainan OV, Dato V, Xia G, Waller K, Amon J, Lee TM, Highbaugh-Battle A, Hembree C, Evenson S, Ruta MA, Williams IT, Fiore AE, Bell BP.** 2005. An outbreak of hepatitis A associated with green onions. *N Engl J Med* **353:**890–897.
33. **Bialek SR, George PA, Xia GL, Glatzer MB, Motes ML, Veazey JE, Hammond RM, Jones T, Shieh YC, Wamnes J, Vaughan G, Khudyakov Y, Fiore AE.** 2007. Use of molecular epidemiology to confirm a multistate outbreak of hepatitis A caused by consumption of oysters. *Clin Infect Dis* **44:**838–840.
34. **Desenclos JC, Klontz KC, Wilder MH, Nainan OV, Margolis HS, Gunn RA.** 1991. A multistate outbreak of hepatitis A caused by the consumption of raw oysters. *Am J Public Health* **81:**1268–1272.
35. **Shieh YC, Khudyakov YE, Xia G, Ganova-Raeva LM, Khambaty FM, Woods JW, Veazey JE, Motes ML, Glatzer MB, Bialek SR, Fiore AE.** 2007. Molecular confirmation of oysters as the vector for hepatitis A in a 2005 multistate outbreak. *J Food Prot* **70:**145–150.
36. **Halliday ML, Kang LY, Zhou TK, Hu MD, Pan QC, Fu TY, Huang YS, Hu SL.** 1991. An epidemic of hepatitis A attributable to the ingestion of raw clams in Shanghai, China. *J Infect Dis* **164:**852–859.
37. **Bordi L, Rozera G, Scognamiglio P, Minosse C, Loffredo M, Antinori A, Narciso P, Ippolito G, Girardi E, Capobianchi MR, GEAS Group.** 2012. Monophyletic outbreak of Hepatitis A involving HIV-infected men who have sex with men, Rome, Italy 2008–2009. *J Clin Virol* **54:**26–29.
38. **Cotter SM, Sansom S, Long T, Koch E, Kellerman S, Smith F, Averhoff F, Bell BP.** 2003. Outbreak of hepatitis A among men who have sex with men: implications for hepatitis A vaccination strategies. *J Infect Dis* **187:**1235–1240.
39. **Spada E, Genovese D, Tosti ME, Mariano A, Cuccuini M, Proietti L, Giuli CD, Lavagna A, Crapa GE, Morace G, Taffon S, Mele A, Rezza G, Rapicetta M.** 2005. An outbreak of hepatitis A virus infection with a high case-fatality rate among injecting drug users. *J Hepatol* **43:**958–964.
40. **Mukomolov S, Kontio M, Zheleznova N, Jokinen S, Sinayskaya E, Stalevskaya A, Davidkin I.** 2012. Increased circulation of hepatitis A virus genotype IIIA over the last decade in St Petersburg, Russia. *J Med Virol* **84:**1528–1534.
41. **Tallo T, Norder H, Tefanova V, Ott K, Ustina V, Prukk T, Solomonova O, Schmidt J, Zilmer K, Priimägi L, Krispin T, Magnius LO.** 2003. Sequential changes in hepatitis A virus genotype distribution in Estonia during 1994 to 2001. *J Med Virol* **70:**187–193.
42. **Yun H, Lee HJ, Jang JH, Kim JS, Lee SH, Kim JW, Park SJ, Park YM, Hwang SG, Rim KS, Kang SK, Lee HS, Jeong SH.** 2011. Hepatitis A virus genotype and its correlation with the clinical outcome of acute hepatitis A in Korea: 2006–2008. *J Med Virol* **83:**2073–2081.
43. **Ippagunta SK, Naik S, Sharma B, Aggarwal R.** 2007. Presence of hepatitis E virus in sewage in Northern India: frequency and seasonal pattern. *J Med Virol* **79:**1827–1831.
44. **Feagins AR, Opriessnig T, Guenette DK, Halbur PG, Meng XJ.** 2007. Detection and characterization of infectious Hepatitis E virus from commercial pig livers sold in local grocery stores in the USA. *J Gen Virol* **88:**912–917.
45. **Kaba M, Davoust B, Marié JL, Colson P.** 2010. Detection of hepatitis E virus in wild boar (Sus scrofa) livers. *Vet J* **186:**259–261.
46. **Izopet J, Dubois M, Bertagnoli S, Lhomme S, Marchandeau S, Boucher S, Kamar N, Abravanel F, Guérin J-L.** 2012. Hepatitis E virus strains in rabbits and evidence of a closely related strain in humans, France. *Emerg Infect Dis* **18:**1274–1281.
47. **Tei S, Kitajima N, Ohara S, Inoue Y, Miki M, Yamatani T, Yamabe H, Mishiro S, Kinoshita Y.** 2004. Consumption of uncooked deer meat as a risk factor for hepatitis E virus infection: an age- and sex-matched case-control study. *J Med Virol* **74:**67–70.
48. **Meng X-J.** 2013. Zoonotic and foodborne transmission of hepatitis E virus. *Semin Liver Dis* **33:**41–49.
49. **Said B, Ijaz S, Kafatos G, Booth L, Thomas HL, Walsh A, Ramsay M, Morgan D, Hepatitis E Incident Investigation Team.** 2009. Hepatitis E outbreak on cruise ship. *Emerg Infect Dis* **15:**1738–1744.
50. **Balayan MS.** 1991. HEV infection: Historical perspectives, global epidemiology, and clinical features, p 498–501. *In* Hollinger FB, Lemon SM, Margolis HS (ed), *Viral Hepatitis and Liver Disease*. Williams & Wilkins, Baltimore, MD.
51. **Matsubayashi K, Kang JH, Sakata H, Takahashi K, Shindo M, Kato M, Sato S, Kato T, Nishimori H, Tsuji K, Maguchi H, Yoshida J, Maekubo H, Mishiro S, Ikeda H.** 2008. A case of transfusion-transmitted hepatitis E caused by blood from a donor infected with hepatitis E virus via zoonotic food-borne route. *Transfusion* **48:**1368–1375.
52. **Juhl D, Baylis SA, Blümel J, Görg S, Hennig H.** 2014. Seroprevalence and incidence of hepatitis E virus infection in German blood donors. *Transfusion* **54:**49–56.
53. **Xu C, Wang RY, Schechterly CA, Ge S, Shih JW, Xia N-S, Luban NLC, Alter HJ.** 2013. An assessment of hepatitis E virus (HEV) in US blood donors and recipients: no detectable HEV RNA in 1939 donors tested and no evidence for HEV transmission to 362 prospectively followed recipients. *Transfusion* **53:**2505–2511.
54. **Takeda H, Matsubayashi K, Sakata H, Sato S, Kato T, Hino S, Tadokoro K, Ikeda H.** 2010. A nationwide survey for prevalence of hepatitis E virus antibody in qualified blood donors in Japan. *Vox Sang* **99:**307–313.
55. **Vollmer T, Diekmann J, Johne R, Eberhardt M, Knabbe C, Dreier J.** 2012. Novel approach for detection of hepatitis E virus infection in German blood donors. *J Clin Microbiol* **50:**2708–2713.
56. **Keane F, Gompels M, Bendall R, Drayton R, Jennings L, Black J, Baragwanath G, Lin N, Henley W, Ngui S-L, Ijaz S, Dalton H.** 2012. Hepatitis E virus coinfection in patients with HIV infection. *HIV Med* **13:**83–88.
57. **Abe K, Shikata T.** 1982. Fulminant type A viral hepatitis in a chimpanzee. *Acta Pathol Jpn* **32:**143–148.
58. **Song D-Y, Zhuang H, Kang X-C, Liu X-M, Li Z, Hao W, Shi K-C, Hao F-M, Jia Q, Chen D-G, He Z-X, Ai Z-Z.** 1991. Hepatitis E in Hetian City: A report of 562 cases, p 528–529. *In* Hollinger FB, Lemon SM, Margolis HS (ed), *Viral Hepatitis and Liver Disease*. Williams & Wilkins, Baltimore, MD.
59. **Williams TPE, Kasorndorkbua C, Halbur PG, Haqshenas G, Guenette DK, Toth TE, Meng XJ.** 2001. Evidence of extrahepatic sites of replication of the hepatitis E virus in a swine model. *J Clin Microbiol* **39:**3040–3046.
60. **Kamar N, Rostaing L, Izopet J.** 2013. Hepatitis E virus infection in immunosuppressed patients: natural history and therapy. *Semin Liver Dis* **33:**62–70.
61. **Bose PD, Das BC, Hazam RK, Kumar A, Medhi S, Kar P.** 2014. Evidence of extrahepatic replication of hepatitis E virus in human placenta. *J Gen Virol* **95:**1266–1271.
62. **Yang L, Kiyohara T, Kanda T, Imazeki F, Fujiwara K, Gauss-Müller V, Ishii K, Wakita T, Yokosuka O.** 2010. Inhibitory effects on HAV IRES-mediated translation and replication by a combination of amantadine and interferon-alpha. *Virol J* **7:**212.
63. **Hoagland CL, Shank RE.** 1946. Infectious hepatitis; a review of 200 cases. *J Am Med Assoc* **130:**615–621.
64. **Robbins A, Lambert D, Ehrhard F, Brodard V, Hentzien M, Lebrun D, Nguyen Y, Tabary T, Peron JM, Izopet J, Bani-Sadr F.** 2014. Severe acute hepatitis E in an HIV infected patient: successful treatment with ribavirin. *J Clin Virol* **60:**422–423.
65. **Barnaud E, Rogée S, Garry P, Rose N, Pavio N.** 2012. Thermal inactivation of infectious hepatitis E virus in experimentally contaminated food. *Appl Environ Microbiol* **78:**5153–5159.

66. **Kamili S.** 2011. Toward the development of a hepatitis E vaccine. *Virus Res* **161:**93–100.
67. **Wei M, Zhang X, Yu H, Tang Z-M, Wang K, Li Z, Zheng Z, Li S, Zhang J, Xia N, Zhao Q.** 2014. Bacteria expressed hepatitis E virus capsid proteins maintain virion-like epitopes. *Vaccine* **32:**2859–2865.
68. **Zhu FC, Zhang J, Zhang XF, Zhou C, Wang ZZ, Huang SJ, Wang H, Yang CL, Jiang HM, Cai JP, Wang YJ, Ai X, Hu YM, Tang Q, Yao X, Yan Q, Xian YL, Wu T, Li YM, Miao J, Ng MH, Shih JW, Xia NS.** 2010. Efficacy and safety of a recombinant hepatitis E vaccine in healthy adults: a large-scale, randomised, double-blind placebo-controlled, phase 3 trial. *Lancet* **376:**895–902.
69. **Cuthbert JA.** 2001. Hepatitis A: old and new. *Clin Microbiol Rev* **14:**38–58.
70. **Cromeans T, Sobsey MD, Fields HA.** 1987. Development of a plaque assay for a cytopathic, rapidly replicating isolate of hepatitis A virus. *J Med Virol* **22:**45–56.
71. **Seelig R, Pott G, Seelig HP, Liehr H, Metzger P, Waldherr R.** 1984. Virus-binding activity of fibronectin: masking of hepatitis A virus. *J Virol Methods* **8:**335–347.
72. **Margolis HS, Nainan OV.** 1990. Identification of virus components in circulating immune complexes isolated during hepatitis A virus infection. *Hepatology* **11:**31–37.
73. **Vaughan G, Forbi JC, Xia GL, Fonseca-Ford M, Vazquez R, Khudyakov YE, Montiel S, Waterman S, Alpuche C, Gonçalves Rossi LM, Luna N.** 2014. Full-length genome characterization and genetic relatedness analysis of hepatitis A virus outbreak strains associated with acute liver failure among children. *J Med Virol* **86:**202–208.
74. **Vaughan G, Xia G, Forbi JC, Purdy MA, Rossi LM, Spradling PR, Khudyakov YE.** 2013. Genetic relatedness among hepatitis A virus strains associated with food-borne outbreaks. *PLoS One* **8:**e74546.
75. **Desbois D, Couturier E, Mackiewicz V, Graube A, Letort MJ, Dussaix E, Roque-Afonso AM.** 2010. Epidemiology and genetic characterization of hepatitis A virus genotype IIA. *J Clin Microbiol* **48:**3306–3315.
76. **Belalov IS, Isaeva OV, Lukashev AN.** 2011. Recombination in hepatitis A virus: evidence for reproductive isolation of genotypes. *J Gen Virol* **92:**860–872.
77. **Joshi MS, Walimbe AM, Chitambar SD.** 2008. Evaluation of genomic regions of hepatitis A virus for phylogenetic analysis: suitability of the 2C region for genotyping. *J Virol Methods* **153:**36–42.
78. **Costa-Mattioli M, Di Napoli A, Ferré V, Billaudel S, Perez-Bercoff R, Cristina J.** 2003. Genetic variability of hepatitis A virus. *J Gen Virol* **84:**3191–3201.
79. **Robertson BH, Jansen RW, Khanna B, Totsuka A, Nainan OV, Siegl G, Widell A, Margolis HS, Isomura S, Ito K, Ishizu T, Moritsugu Y, Lemon SM.** 1992. Genetic relatedness of hepatitis A virus strains recovered from different geographical regions. *J Gen Virol* **73:**1365–1377.
80. **Escobar-Gutiérrez A, Cruz-Rivera M, Ruiz-Tovar K, Vaughan G.** 2012. Genetic relatedness among Japanese HAV isolates, 2010. *J Clin Virol* **54:**287–288.
81. **Aggarwal R.** 2012. Diagnosis of hepatitis E. *Nat Rev Gastroenterol Hepatol* **10:**24–33.
82. **Drobeniuc J, Meng J, Reuter G, Greene-Montfort T, Khudyakova N, Dimitrova Z, Kamili S, Teo CG.** 2010. Serologic assays specific to immunoglobulin M antibodies against hepatitis E virus: pangenotypic evaluation of performances. *Clin Infect Dis* **51:**e24–e27.
83. **Ferguson M, Walker D, Mast E, Fields H.** 2002. Report of a collaborative study to assess the suitability of a reference reagent for antibodies to hepatitis E virus. *Biologicals* **30:**43–48.
84. **Bendall R, Ellis V, Ijaz S, Ali R, Dalton H.** 2010. A comparison of two commercially available anti-HEV IgG kits and a re-evaluation of anti-HEV IgG seroprevalence data in developed countries. *J Med Virol* **82:**799–805.
85. **Pas SD, Streefkerk RH, Pronk M, de Man RA, Beersma MF, Osterhaus AD, van der Eijk AA.** 2013. Diagnostic performance of selected commercial HEV IgM and IgG ELISAs for immunocompromised and immunocompetent patients. *J Clin Virol* **58:**629–634.
86. **Hyams C, Mabayoje DA, Copping R, Maranao D, Patel M, Labbett W, Haque T, Webster DP.** 2014. Serological cross reactivity to CMV and EBV causes problems in the diagnosis of acute hepatitis E virus infection. *J Med Virol* **86:**478–483.
87. **Fogeda M, de Ory F, Avellón A, Echevarría JM.** 2009. Differential diagnosis of hepatitis E virus, cytomegalovirus and Epstein-Barr virus infection in patients with suspected hepatitis E. *J Clin Virol* **45:**259–261.
88. **Baylis SA, Hanschmann K-M, Blümel J, Nübling CM,** on behalf of the **HEV Collaborative Study Group.** 2011. Standardization of hepatitis E virus (HEV) nucleic acid amplification technique-based assays: an initial study to evaluate a panel of HEV strains and investigate laboratory performance. *J Clin Microbiol* **49:**1234–1239.
89. **Lan X, Yang B, Li BY, Yin XP, Li XR, Liu JX.** 2009. Reverse transcription-loop-mediated isothermal amplification assay for rapid detection of hepatitis E virus. *J Clin Microbiol* **47:**2304–2306.
90. **Margolis HS, Nainan OV, Krawczynski K, Bradley DW, Ebert JW, Spelbring J, Fields HA, Maynard JE.** 1988. Appearance of immune complexes during experimental hepatitis A infection in chimpanzees. *J Med Virol* **26:**315–326.

Hepatitis B and D Viruses

REBECCA T. HORVAT

24

BIOLOGY AND CLASSIFICATION

Hepatitis was first described in the fifth century B.C. The earliest known outbreak of hepatitis occurred in Bremen, Germany, in 1883 among shipyard workers who received a smallpox vaccine stabilized with human serum. In 1950, viral hepatitis was referred to as either infectious hepatitis (hepatitis A) or serum hepatitis (hepatitis B) based on the epidemiologic characteristics of the diseases (1). This terminology was adapted by the World Health Organization (WHO) in 1973.

Hepatitis B (HBV) is an enveloped DNA virus in the *Hepadnaviridae* family and replicates by reverse transcription of an RNA intermediate (2). HBV is a 42 nm, partially double-stranded DNA virus that replicates in the nucleus of the host cell. HBV-infected cells have no cytopathic features because the virus causes little damage to the host cell (2, 3). The complete virus particle is known as the Dane particle (4). The viral nucleocapsid core is surrounded by a specific viral core protein (HBcAg) and encloses a single molecule of partially double-stranded circular DNA, hepatitis B e antigen (HBeAg), and a DNA-dependent polymerase (2).

The HBV genome is approximately 3,200 bases and contains overlapping genes. Replication occurs via an RNA intermediate. There are four open reading frames in the complete, minus strand. These genes encode the structural proteins (HBsAg and HBcAg), replicative proteins (polymerase and X protein), and regulatory proteins. The genome is compact, and most sequences are essential for productive infection (2).

The HBV polymerase lacks proofreading activity, which eventually leads to mutations which results in viral genetic heterogeneity. Because of these genetic changes there is a significant divergence in the genomes of HBV leading to a variety of genotypes. These genotypes are grouped based upon a divergence of 8 percent or more in the complete nucleotide sequence (5, 6). At this time HBV is classified into 10 genotypes (A to J) (2, 3).

HBV has a variety of genotypes which vary geographically (6). At this time there are 10 well-defined genotypes designated from A to J. HBV Genotype A is prevalent in Northern Europe, North America, India, and Africa. Genotypes B and C are prevalent in Asia, while Genotype D is detected in areas of Southern Europe, the Middle East, and India. HBV Genotype E is detected primarily in West Africa, while Genotypes F and H are found in Central and South America. Genotype G is more prevalent in France, Germany, and the United States. Genotype I has been detected in patients in Vietnam and Laos, while Genotype J was identified in patients from the Ryukyu Island in Japan (7–9).

HBV particles found in the sera of patients with active HBV infection have three morphologic entities in varying proportions. The most abundant forms are small, pleomorphic, spherical, noninfectious particles (17 to 25 nm in diameter). Less numerous are the tubular forms, which have diameters similar to those of the small particles. The third and least numerous particles is the complete HBV virion, with a diameter of 42 to 47 nm (4). There is no cell culture system that supports the growth of HBV.

The production of HBsAg exceeds what is needed for virion production, and this excess protein circulates in the blood of infected individuals as spherical and tubular particles and can be detected in clinical assays to identify active HBV infection (2). The antigen persists in the serum for variable periods after initial infection and in some patients can be as high as 10^{13} per ml.

Two additional HBV-specific proteins, HBcAg and HBeAg, play a key role in diagnostic testing. HBcAg and HBeAg have different antigenic specificities, and both can be distinguished from HBsAg. The HBcAg is a polypeptide encoded by the C gene of HBV and is translated from the pregenomic mRNA. The precore sequence within the C gene contains the start codon for HBeAg translation (2). Because of the different start codons the two proteins, HBcAg and HBeAg, are antigenically unrelated. HBeAg is a soluble protein or can be bound to albumin, a-1-antitrypsin, or immunoglobulin. It is a reliable marker for the presence of HBV virions which indicates high infectivity. In some HBV strains, a mutation at the end of the precore region results in a stop codon which prevents the translation of HBeAg. These precore mutants contribute to the pathogenesis of chronic HBV disease, leading to acute exacerbation of disease (10, 11). A variety of other HBV gene mutations have been observed in the core, core promoter, envelope, and polymerase regions. The envelope protein variants are applicable to clinical disease because some protein changes may not be recognized by the HBV vaccine. Thus some mutant viruses escape HBV immune responses. Consequently patients infected with these escape HBV may not have detectable

doi:10.1128/9781555819156.ch24

antibody to HBsAg and will not respond to HBV immunoglobulin (HBIG) therapy (12). The various antigens and antibodies associated with HBV infection in humans are listed in Table 1.

Hepatitis Delta Virus

In 1977, the delta antigen was recognized to be the component of a novel virus that was defective and required coinfection with HBV for its replication. The hepatitis delta virus (HDV) is a defective virus and requires the HBV surface antigen for transmission. HDV requires the HBV surface antigen (HBsAg) as its own viral coat (13, 14).

Host cells infected with HDV alone can replicate the viral genome, but viral particles are not released. HDV has an absolute requirement for the support of HBV for its own replication.

EPIDEMIOLOGY

HBV

HBV infections are prevalent around the world and represent a global public health problem. The WHO estimates that 2 billion people are infected with HBV, and approximately 600,000 people die each year as a result of acute or chronic HBV (15, 16). According to the WHO, HBV is 50 to 100 times more infectious than HIV-1 (http://www.who.int/csr/disease/hepatitis/whocdscsrlyo20022/en/index3.html#transmission).

HBV causes chronic hepatitis, cirrhosis, and hepatocellular carcinoma (HCC) worldwide. In most countries HBV infections are reportable to the public health authorities. Most of these HBV-infected individuals live in Asia or Africa (Figure 1). Approximately one-fourth of adults are infected with HBV as children and have serious complications from liver cirrhosis and/or HCC due to chronic HBV (17). The Centers for Disease Control and Prevention (CDC) estimates that as many as 1.4 million individuals in the United States are chronically infected with HBV (http://www.cdc.gov/hepatitis/statistics/2010).

In the United States, most infections are in adults, in whom the risk of chronic infection is lower. However, outside the United States, perinatal exposure is more common and leads to chronic HBV infections (18).

The most common mode of HBV transmission is vertical transmission (mother to child perinatally), early childhood infections from infected individuals, sexual activity (both heterosexual and male homosexual), injection drug use, or physical contact with infected body fluids (occupational exposure, contaminated blood products, etc.) (2). HBV is not transmitted from person to person by casual activities (2).

HDV

Approximately 15 to 20 million people worldwide are HDV carriers (13). Some areas of the world have a high prevalence of HDV-infected individuals, including countries around the Mediterranean, the Middle East, Central Asia, West Africa, the Amazon Basin, and the South Pacific Islands (13). In these areas of high endemicity, HDV appears to be transmitted by close person-to-person spread. How-

TABLE 1 Hepatitis B virus markers in different stages of infection and convalescence

	Molecular markers	Protein antigen markers		HBV-specific antibody markers			
				Anti- HBc			
Stage of infection	HBV DNA	HBsAg	HBe Ag	IgM	Total	Anti-HBe	Anti-HBs
Susceptible to HBV	–	–	–	–	–	–	–
Early incubation	+	–	–	–	–	–	–
Late incubation period	+	+	– / +	–	–	–	–
Acute infection	+	+	+	+	+	–	–
Recent infection[a]	– / +	–	–	++	+	+	+ ++
Remote infection[b]	– or very low	–	–	–	+	+/–	+
Remote infection or chronic infection with HBs Ag mutation[b]	– / +	–	–	–	+	+/–	–
HBsAg-negative acute infection	– $<10^3$ IU/ml	–	–	+	+	–	–
HbsAg variant infection	– / +	–	– / +	+/–	+	–	–
Immune active carrier	++ $>10^5$ IU/ml	+	–/+	– / +	+++	–	–
Inactive HBsAg carrier	– $<10^3$ IU/ml	+	–	–	+	+	–
Immune Tolerant carrier	+++	+	+	–	+	–	–
Vaccination response	–	–	–	–	–	–	+

HBsAg, protein found on the surface of HBV and on 20 nm diameter particles and tubular forms; HBcAg, antigen associated with 27 nm diameter core of HBV; HBeAg, protein that results from the proteolytic cleavage of the preCore/Core protein by cellular proteases and secreted as soluble protein in serum.

[a]This description can be applied to early convalescence and to individuals that remain HBV DNA positive for prolonged periods in the absence of HBsAg.

[b]Remote infection can be applied to individuals with anti HBc in the absence of other serological markers including DNA. These patients may or may not have anti-HBs. There is evidence that these patients may reactivate HBV during immunosuppression.

Geographic Distribution of Chronic HBV Infection — Worldwide, 2006*

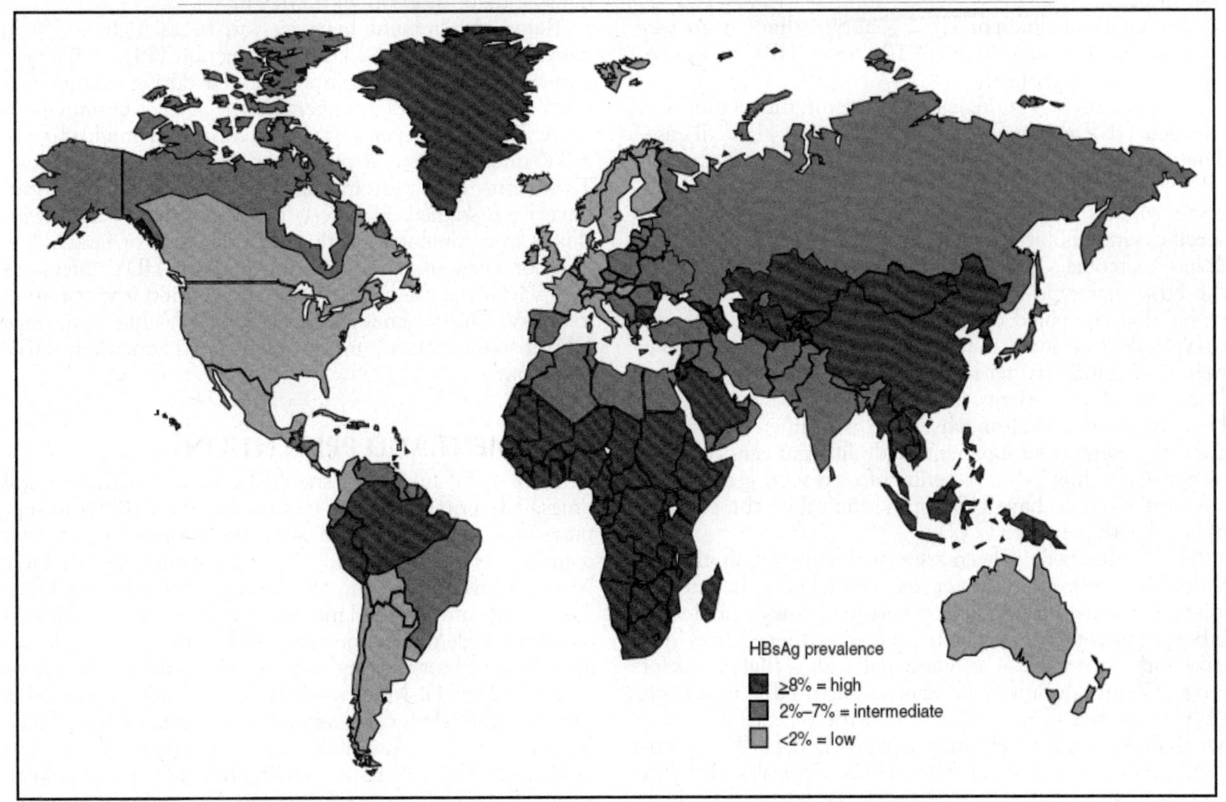

* For multiple countries, estimates of prevalence of hepatitis B surface antigen (HBsAg), a marker of chronic HBV infection, are based on limited data and might not reflect current prevalence in countries that have implemented childhood hepatitis B vaccination. In addition, HBsAg prevalence might vary within countries by subpopulation and locality.

Source: CDC. Travelers' health; yellow book. Atlanta, GA: US Department of Health and Human Services, CDC; 2008. Available at http://wwwn.cdc.gov/travel/yellowbookch4-HepB.aspx.

FIGURE 1 Worldwide distribution of chronic hepatitis B disease (CDC).

ever, many individuals acquire HDV through exposure to blood-contaminated needles and blood products (14).

CLINICAL SIGNIFICANCE

HBV

HBV infects hepatocytes, leading to an acute infection that resolves or a chronic infection lasting years. Some individuals with subclinical hepatitis have a mild disease without symptoms or jaundice. However, other patients have vague symptoms such as abdominal pain, nausea without jaundice (anicteric hepatitis), or nausea with jaundice (icteric hepatitis). HBV infections can result in the complete recovery of the patient, a chronic viral infection, or fulminant hepatitis with mortality. Newborns infected with HBV usually have chronic, asymptomatic infections, while older children and adults are typically symptomatic after a primary infection (2, 18).

The symptoms of acute HBV infection may be mild but sometimes include signs such as jaundice, dark urine, clay-colored stools, and hepatomegaly (2). Some patients experience weight loss, right upper quadrant pain, and a tender, enlarged liver (15). Acute HBV infections may be self-limited, and patients recover completely after specific antibodies (anti-HBs) clear the virus (18).

The disease outcome of acute HBV is age dependent, and most patients with acute disease are adults. Acute liver damage is caused by the host immune response to HBV infected hepatocytes (19). This results in massive necrosis, leading to permanent damage in the liver. Without a liver transplantation, fulminant hepatic failure is associated with high mortality. After transplantation, HBV reinfection of the "new" liver is common, resulting in injury to the new liver. HBIG and/or antiviral therapy can prevent this outcome (18).

Patients with perinatal acquired HBV are usually immune tolerant to HBV antigens. Thus there is an absence of severe liver disease despite high levels of virus (12). Patients who continue to have detectable HBsAg or detectable HBV DNA for 6 months after infection are considered to have chronic HBV (3, 16). Chronic carriers can remain positive for HBsAg indefinitely, although some HBsAg-positive patients spontaneously convert to HBsAg negative after the appearance of anti-HBs. Many of these patients continue to have detectable HBV DNA (2, 7).

Most HBV-infected individuals progress to the immune active phase, in which a liver biopsy shows inflammation with fibrosis. This pathology results from a persistent immune response to the HBV proteins on infected hepatocytes. The last phase of chronic HBV infection is the inactive carrier phase, characterized by less inflammation and normal liver enzyme levels. These patients have a low risk for HCC (7, 20).

A number of different HBV genotypes have been recognized. At this time there are 10 known HBV genotypes (designated A to J) (18).

Recently there is an increased interest in the relationship between HBV genotypes and progression of HBV disease. There is some variability in the clinical outcomes of chronic HBV infections. A number of factors can influence outcomes such as viral strain, host factors, HBV genotype, and specific viral mutations may be important in predicting disease outcome. There are reports recently that correlate the HBV genotype with clinical outcomes (21). A recent report that compared disease in patients infected with genotypes B or C noted that alanine aminotransferase levels were consistently higher in patients infected with genotype C compared to patients infected with genotype B (21). However, it is not clear why there is a difference in the clinical presentation of patients with different genotypes. A recent study shows that patients infected with genotype B are more likely to have HBeAg in serum than the patients infected with genotype C (21).

HBV variants have been associated with certain clinical features of HBV-related liver disease including hepatocellular carcinoma. For example genotype C is associated with HBeAg seroconversion which leads to advanced liver disease (21). Genotype A is associated with a higher risk of progression to chronicity in adult acquired HBV infections (21). Genotype D is associated with the precore mutation and HBeAg negative chronic hepatitis B (21). The genotypes prevalent in parts of West Africa, Central and South America—E, F, and H—respectively, are less well studied. Some HBV mutations especially in the basal core promotor have been associated with increased risk of fibrosis and cancer of the liver (21). The evaluation of genotype and viral variants should be used in predicting risk about liver related morbidity in patients with CHB.

Prevention of HBV

The most successful program to prevent HBV infection is the HBV vaccine. Safe and effective vaccines against HBV have been available since 1981. The complete HBV vaccination series is protective in >95% of infants, children, and young adults (18). The efficacy of HBV vaccination has been proven worldwide and is monitored by the WHO (22). Vaccine-induced protection lasts at least 20 years and may be lifelong (22). It is important that individuals with a known or suspected exposure to HBV should be given HBV vaccine as soon as possible after the exposure, in addition to HBIG. It is recommended that testing for anti-HBs be obtained 4 to 12 weeks following vaccination (23). Individual's that do not respond to the HBV vaccine have HBsAb levels that are < 10 IU/l. All nonresponders should be given the vaccine again with another full series of HBV vaccine (18).

HDV

HDV can be transmitted only to individuals who are infected with HBV already or when both HBV and HDV are transmitted together. A coinfection occurs when a naive individual is infected simultaneously with both viruses; coinfection occurs in only 2% of the cases (14). A superinfection occurs when an individual chronically infected with HBV is infected with HDV (14). Superinfection occurs in more than 90% of infected patients. Acute HDV superinfection has a greater risk of fulminant hepatitis and liver failure than HBV infection alone. Likewise, chronic HDV infection is associated with more rapidly progressing liver damage than infection with HBV alone.

Rates of fulminant hepatitis can be as high as 5% in patients with HBV and HDV coinfection (14). A biphasic clinical course is sometimes observed during coinfection. HDV infection does not increase the rate of chronicity of acute HBV. However HDV can convert a mild, chronic HBV infection into a rapidly progressive, severe disease. Treatments with antiviral agents that reduce HBV titers have been studied. However, it appears that therapy with IFN-α in combination with either ribavirin or lamivudine has not been useful in treating chronic HDV infections (14). However, the most effective prevention is vaccination for HBV. This vaccine also prevents HDV infection, since HDV cannot replicate in the absence of a concurrent HBV infection.

TREATMENT AND PREVENTION

Chronic HBV infections should be treated with antiviral drugs. The primary goal of treating chronic HBV is to suppress viral replication and slow the progression of liver damage (23). At present, seven therapeutic agents have been approved for use in the United States by the FDA. These drugs are classified into two categories, the interferons and nucleoside/nucleotide analogs. The first nucleoside analog approved for treating chronic HBV infection was lamivudine (23, 24). More recently, several new agents have become available for the treatment of chronic HBV. These include adefovir, entecavir, tenofovir, emtricitabine, and telbivudine (20, 24). Patients with HBV infection should be treated to prevent progression of the disease that can lead to cirrhosis, liver failure, and HCC (3, 7). Prevention of HDV is based on prevention of HBV, as HDV requires HBV for production and transmission. Effective medicines exist that can prevent the development of these conditions.

Therapy is currently recommended for patients with evidence of chronic or active HBV disease. The National Institutes of Health recommends nucleos(t)ide therapy for patients with acute liver failure, as well as cirrhotic patients who are HBV DNA positive. These recommendations also indicate therapy for patients with clinical complications, cirrhosis or advanced fibrosis with positive serum HBV DNA (24). Antiviral agents are available to treat HBV. Several agents are nucleoside analogue that inhibit the activity of the HBV polymerase such as adefovir. Another agent used for the treatment of HBV is entecavir (18, 15). This drug also inhibits the HBV polymerase and has a low rate of drug resistance. Telbivudine, a cytosine nucleoside analogue, is a second-line agent used in HBV therapy (15). This drug inhibits HBV DNA polymerase. A number of studies have shown that the clearance of HBV DNA after 12 months of therapy indicates successful antiviral therapy (18).

The recommended first-line antiviral agents for the treatment of HBV disease are pegylated interferon alfa (PEG-IFN-a), entecavir (ETV), and tenofovir disoproxil fumarate (TDF) (18). However a number of new drugs for treatment of HBV are under investigation. These new drugs have different mechanisms for eliminating HBV such as inhibiting encapsidation or entry of the virus to the host cell (20). Additionally, new therapies for HBV are under investigation including combinations such as tenofovir with emtricitabine (20).

Another new therapy is a nucleoside analogue (3TC) which inhibits the HBV polymerase. Treatment with this

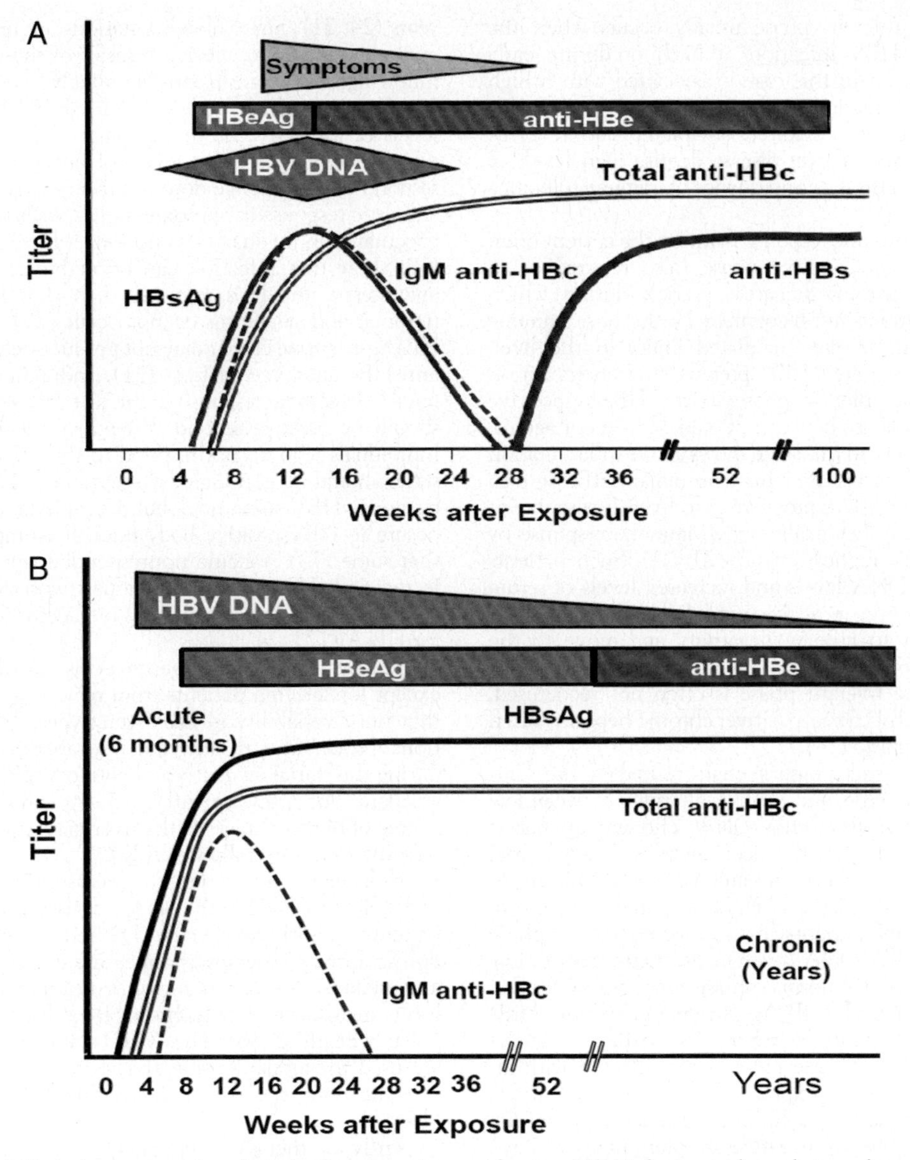

FIGURE 2 **(A)** Typical sequence of serologic markers in patients with acute HBV infection with resolution of symptoms; **(B)** typical sequence of serologic markers in patients with HBV infection that progresses to chronicity HBV. HBeAg is variably present in these patients.

drug shows an improvement in histologic damage in the liver. However, after 12 months of treatment with 3TC, nearly half of the patients have an HBV species with mutation (20).

The best method for preventing HBV infection is effective vaccination. The WHO has made eliminating HBV worldwide a goal by emphasizing global HBV vaccination (17, 18). At this time global immunization coverage is a goal still unmet and requires additional resources to eliminate the spread of HBV (18).

At this time HBV vaccination programs are ongoing in many areas of the globe. The HBV vaccine contains a recombinant HBV surface antigen (HBsAg) produced in yeast. A series of 3 injections are recommended. After effective vaccination, HBsAg antibody levels should be greater than 10 million IU/ml (22).

All newborns should be vaccinated against HBV including those born from mothers with HBV. Infants born to mothers with active HBV should receive HBV immunoglobulin (HBIG) (18, 22). All healthcare workers should also receive the HBV vaccination. Any healthcare worker that has a needle-stick accident from a patient with known HBV infection should be given one dose of the vaccine as soon as possible. These individuals should be monitored for the presence of active HBV infection.

DETECTION/DIAGNOSIS

HBV

The initial assessment for patients with a suspected viral hepatitis should include laboratory tests that measure serum transaminases, direct and total bilirubin, albumin total protein, a complete blood count, coagulation tests, and alpha fetoprotein. The specific laboratory tests to detect and monitor HBV infection are a mix of viral antigen detection, molecular measurements of HBV DNA, and serologic markers. The tests to diagnose new HBV infection are anti-HBc IgM, HBsAg, and HBV DNA (Table 1 and Figure 2).

The immune-tolerant phase usually occurs when the patient acquires HBV infection at birth or during early childhood. Infection in this case is associated with a high level of viral production and the presence of HBeAg. These markers indicate a high rate of viral replication (2, 3). There is an absence of liver disease despite high levels of HBV replication. This is a consequence of immune tolerance (3, 18).

As the host's immune response matures, the patient often moves to the immune-active phase (also referred to as HBeAg-positive chronic hepatitis phase), during which HBV-specific epitopes are recognized by the host immune system, leading to immune-mediated injury in the liver. Individuals who acquired HBV prenatally often transition from the immune tolerant phase to the HBeAg-positive chronic hepatitis phase between 20 and 30 years of age (7, 19). The liver biopsy in this stage shows active inflammation accompanied by fibrosis. Patients who remain HBeAg positive have a higher risk of progressing to liver disease due to the induction of a chronically active immune response by high rates of HBV replication (18, 21, 24). Such patients have high HBV DNA levels and increased levels of serum transaminases. However, as these individuals develop anti-HBe, they revert to HBeAg negativity and move to the inactive carrier phase (Table 1) (18, 24). The transition from the immune tolerant phase is often not recognized, since patients with HBeAg-positive, chronic hepatitis often remains asymptomatic (24).

The inactive carrier phase is characterized by the seroconversion to anti-HBe, and patients alternate between low and undetectable levels of HBV DNA. The seroconversion to anti-HBe is associated with a decrease in liver damage and the normalization of serum transaminase levels. Mild hepatitis may be noted on biopsy (24). Many patients remain in this phase for years. Patients in the inactive carrier phase have detectable HBV DNA in serum, at intermittent or low levels, and usually have normal serum transaminase levels.

A portion of inactive HBsAg carriers (about one-third) develop chronic hepatitis which recurs in the absence of HBeAg in their sera. These patients are infected with an HBV variant that cannot express HBeAg due to mutations in the HBV core gene (18, 20, 26). Patients with chronic hepatitis that are HBeAg negative are more likely to have more advanced liver disease in spite of lower serum HBV DNA levels (19, 20). HBsAg-positive patients can transmit HBV sexually, percutaneously, or prenatally. Individuals with detectable HBeAg pose the highest risk of transmitting HBV to others.

Patients who are negative for HBsAg, anti-HBc, and HBV DNA are not infected with HBV. In some individuals the presence of anti-HBc alone may be the only evidence of an active, occult HBV infection of remote origin. Patients infected with an HBsAg escape mutant test negative for HBsAg but are positive when tested for anti-HBc and HBV DNA (6, 9, 10).

Anti-HBs without anti-HBc develops in individuals who receive hepatitis B vaccine (which contains only HBsAg), and anti-HBs levels of ≥10 mIU/ml are considered protective (21). Due to the prevalence of vaccinated individuals, the detection of anti-HBc is used to evaluate past or current infection with HBV and identify individuals who should receive the HBV vaccination. Passive transfer of anti-HBs or anti-HBc may be observed in neonates of mothers with current or past HBV infections (21). However, passive antibody levels decline gradually over 3 to 6 months, while levels of antibody induced by infection are stable over many years (24, 21). Since blood donations are tested for HBsAg and total anti-HBc, passive transfer of these HBV markers following blood transfusions is unlikely.

Most individuals vaccinated for HBV have detectable levels of anti-HBs; however, some vaccinated people test negative due to waning levels of anti-HBs. They usually respond to a challenge dose of HBsAg vaccine with an anamnestic response in approximately 2 weeks (21). Studies of vaccinated individuals who no longer have detectable anti-HBs show that infection can occur but is blunted by the anamnestic anti-HBs response such that liver damage is minimal and symptoms do not occur (21). In contrast, an HBsAg-negative carrier may not produce detectable levels of anti-HBs after vaccination (21). Individuals who do not have a detectable response to the first series of HBV vaccine should be given a second three-dose vaccine series (18). Individuals who do not respond to the second vaccination series should be considered susceptible to HBV and should be given HBV immunoglobulin prophylaxis after any exposure to HBV-positive body fluid. It is important to note that some HBV vaccine nonresponders are chronically infected with HBV. Thus, individuals who do not have detectable anti-HBs after six doses of vaccine should be tested for HBsAg (23, 24).

Testing for the HBV genotype is usually not required except for selected patients from regions around the world that have variability in HBV genotypes. Testing for mutations associated with antiviral resistance is also not useful during the initial evaluation of patients (20, 33). The "gold standard" for assessing inflammatory activity (grade) and degree of fibrosis (stage) is the liver biopsy, which is a useful baseline for future follow-up (21).

Molecular assays are used to determine HBV DNA levels and help to establish the stage of disease and to monitor patients on antiviral therapy. The reduction of HBV DNA during antiviral therapy is a measure of treatment response and predicts histologic improvement. Increasing HBV DNA levels are associated with chronic liver disease, cirrhosis, and possibly death (2, 18). The WHO standards for HBV DNA are used to standardization the HBV DNA assays and results are now reported in international units per milliliter (14, 18).

Antiviral therapy is given to patients to prevent progression of liver disease (3, 14). During the course of therapy, treatment response is monitored using biochemical, virologic, serologic, and histologic results. Currently the most accurate monitor of virologic activity is the HBV DNA level using an assay with a wide dynamic range (18). The loss of HBsAg, seroconversion to anti-HBs, and long-lasting suppression of HBV DNA indicate a successful response to therapy (14, 18). Patients who appear to have suppressed HBV DNA levels are monitored periodically because relapse due to antiviral resistance is possible. The most reliable measure of a successful long-term treatment response is the sustained suppression of HBV DNA (18, 21).

When HBV DNA levels increase by 1 log10 in a patient taking antiviral treatment, it may indicate antiviral resistance (20, 24). It is not recommended that resistance testing be performed before starting therapy even if the patient has a very high viral level (14, 20). Resistance detection is useful only after a patient has been treated for several months and fails to show a reduction of at least 1 log10 in HBV DNA levels. They should be tested for the presence of resistant mutants to assist in selecting a new treatment. Antiviral resistance testing should also be performed to determine selection of resistant viral strains (2, 6, 14).

HBV infection is diagnosed by serologic and molecular markers detected in serum or plasma. All FDA-approved assays have specific specimen requirements defined in their package inserts. These requirements state the specimen types acceptable as well as describe the processing and storage of the specimen. In general, HBV antigens and antibodies are stable at room temperature for days, can be stored at 4°C for months, and can be frozen at −20 to −70°C for years. Although HBV markers are stable in serum stored at −70°C, repetitive freezing and thawing can lead to their degradation. The use of hemolyzed samples should be avoided due to the potential of these specimens to interfere with the detection signals used in immunoassays.

Useful information about many diagnostic tests can be found by consulting the FDA website and the specific year of test approval (http://www.fda.gov/MedicalDevices/ProductsandMedicalProcedures/DeviceApprovalsandClearances/Recently-ApprovedDevices/default.htm.

The HBV virion is very hardy and remains infectious for at least 7 days outside the host (18). Thus, spills or splashes should be cleaned using absorbent material and disinfected with an appropriate disinfectant. Decontamination should be carried out while wearing gloves. Laboratory personnel should regard all specimens as potentially dangerous. The Occupational Safety and Health Administration (OSHA) standards for occupational exposure to blood borne pathogens are designed to protect employees exposed to blood and other potentially infectious materials. OSHA mandates that all employees whose job requirements put them at risk for blood-borne pathogens be offered HBV vaccine at no cost. OSHA standards and additional safety recommendations can be found in the literature.

Microscopy

Microscopic detection of HBV does not play a role in the diagnosis of disease. However, a liver biopsy is often used to assess the extent of histologic involvement as well as the response to therapy. Histologic examination is useful for distinguishing among acute viral hepatitis, chronic hepatitis, and cirrhosis.

Antigen

Several HBV-specific proteins can be detected in patient specimens during infection. These are a marker of active viral replication. The presence of HBsAg and/or HBeAg in serum occurs during primary infection and during chronic HBV infection (Table 1). HBsAg is located on the outer surface of the HBV particles, while HBeAg is translated from the precore mRNA of HBcAg. The function of HBeAg and its role in disease have not been clearly identified; however, the detection of this protein in serum indicates high viral replication. Both HBsAg and HBeAg are made in large excess by infected host cells and can be detected in serum during active infection (3).

Diagnostic assays with high sensitivity and specificity are available to detect these HBV antigens. Table 1 lists the assays used to determine the stage of HBV disease. HBV antigens are detected using solid-phase assays based on capture with a monoclonal antibody and then detected with a second antibody attached to a signal. These assays use microparticles with different compositions and sizes and are performed on automated instruments. Antigen capture and detection reagents are specific for the major immunodominant region of HBsAg. Current detection methods use enzyme reactions, chemiluminescence, or fluorescence polarization to detect specific antigens.

HBsAg

The detection of HBsAg in serum plays an important role in establishing the diagnosis of HBV infection. Each HBsAg assay is approved by the FDA either for diagnostic use only, for testing donors of blood, organs, and tissue only, or for both applications. The "Name and Intended Use" section of the package insert should be consulted to determine what sample types have been approved for use with each HBsAg test.

The presence of HBsAg in the serum indicates that the patient is highly infectious. Patients who resolve an acute infection eventually produce anti-HBs. However, when HBsAg is present, anti-HBs can be negative in diagnostic tests because the antibody is bound to the HBsAg.

For all commercially available diagnostic assays, any specimens nonreactive for HBsAg are considered negative and do not require further testing. In contrast, specimens reactive for HBsAg are often repeated to verify positive results. These repeatedly HBsAg-positive results may be confirmed by a neutralization assay provided by the manufacturer consistent with FDA approval protocols. If the HBsAg-reactive serum is neutralized by the anti-HBs, then the specimen is considered positive for HBsAg. Conversely, if the anti-HBs does not neutralize the HBsAg, then the HBsAg test must be considered nonconfirmed and a new specimen should be requested and/or a recommendation that the patient be tested for other markers of HBV infection such as IgM anti-HBc or total anti-HBc should be made.

All HBsAg assays are capable of detecting subnanogram amounts of protein with no loss of specificity (5, 9). For diagnostic applications, this level of sensitivity is sufficient to detect the HBsAg in the sera of individuals with actively replicating HBV. However, a recent concern is that some assays cannot detect variants of HBsAg that have mutations within the major antigenic region of the protein. These mutant HBsAg can be missed by some diagnostic assays (10, 12), so initial testing should include the detection of antibodies to both HBsAg as well as HB core.

The major antigenic determinant on the HBsAg is designated the "a" determinant. This antigenic site is a conformational structure with a disulfide bond. The region between amino acids 124 and 147 is found within the major hydrophilic loop of the protein (2). There is a concern that diagnostic assays do not detect HBsAg with alterations within this major antigenic epitope, since some HBsAg assays use monoclonal antibodies that capture the HBsAg using this immunodominant epitope. These HBV strains are thus known as "escape mutants." The first escape mutant was described in a child born to an HBV positive mother who transmitted HBV to the child despite vaccination and HBIG (20). A single amino acid change altered the antigenic portion of the protein such that vaccine-induced antibody no longer recognized the antigen. This allowed the altered virus to persist in the infant. Subsequently, the patient remained positive for HBV DNA and HBsAg (with mutation) for longer than 12 years. Since that time, a number of other substitution mutants within the "a" determinant region of HBsAg have been recognized (26). Recent studies have evaluated HBsAg assays to determine their ability to detect well-defined HBsAg mutants and have found that some mutations in the HBsAg may be missed by diagnostic assays (9, 10, 19).

In response to the concern that blood donors with an HBsAg mutant are not to be identified by HBV antigen assays, most countries screen blood donors for anti-HBc in addition to screening for HBsAg. Blood donors in the United States are also tested for HBV DNA (14). Individuals with positive tests for HBV DNA and anti-HBc and/or patients with positive results for HBeAg and/or HBV DNA but negative for HBsAg could be infected with an escape mutant (6, 9, 10).

HBeAg

The detection of HBeAg in serum is a sign of rapid viral replication usually associated with high HBV DNA levels. HBeAg-positive patients are highly infectious. However, some HBV strains do not make the HBeAg due to a precore mutation. Patients infected with these mutant strains may have high HBV DNA levels in the absence of detectable HBeAg (26). Several commercial assays are available for the detection of HBeAg in serum and use principles similar to those of the HBsAg detection system; i.e., initially positive specimens should be retested and confirmed with neutralization.

Nucleic Acid Detection

Molecular assays to quantitate HBV DNA are used for the initial evaluation of HBV infections and the monitoring of patients with chronic infections during treatment. Additionally, blood donors are routinely screened for HBV DNA using qualitative tests to identify donors in the early stage of HBV infection.

A number of quantitative assays have been developed to detect and monitor HBV DNA levels in infected patients. Monitoring of HBV DNA levels provides information on the effectiveness of antiviral treatment. Several assays are available for these purposes and use a variety of different methods (3). Most of the available assays have a lower limit of detection, between 5 and 50 copies/ml, and can quantify levels up to1 million copies/ml. This wide range of quantitation allows for monitoring HBV DNA early after infection and identifying HBV infections resistant to antiviral therapy (2, 3).

Commercially available HBV DNA assays differ in their limits of detection, dynamic ranges, and methods used to measure DNA levels. An international HBV DNA standard was established in 2001 by the WHO in response to the need to standardize HBV DNA quantification (3). The WHO standard virus preparation is a high-titer genotype A preparation which has been assigned a potency of 106 IU/ml. The standard has established that 1 IU of HBV is equivalent to 5.4 genome equivalents. The WHO standard has allowed results for different HBV DNA assays to be reported in international units per milliliter. Most assays use serum or plasma as the specimen of choice. As with all FDA-approved diagnostic assays, the package insert should be followed for specimen type, processing, and storage. Repeated thawing and freezing (>2 times) should be avoided, as this may reduce assay sensitivity.

Most commercial assays quantitate HBV DNA and cover a 7-$\log^{10}$ to 8-$\log^{10}$ range, which permits the accurate evaluation of HBV levels above a million down to very low levels. Several studies have shown that a reduction of 2 log10 in HBV DNA levels in the first 6 months of antiviral therapy indicates treatment efficacy (15, 22, 24).

The viral load is determined by computerized analysis of individual results, which are compared to the standard curve.

Serologic Tests

Serologic tests for HBV-specific antibodies are used to determine the stage of HBV disease and to establish immunity after HBV vaccination. As the host mounts an immune response to the virus, the first antibodies to appear are IgM specific for HBc (IgM anti-HBc); this is followed by the appearance of total anti-HBc, anti-HBe, and finally anti-HBs (2). There are numerous commercial assays available for HBV serologic testing.

IgM Anti-HBc

IgM anti-HBc persists for several weeks to months after an initial infection. The detection of IgM anti-HBc indicates an infection of less than 6 months' duration (Table 1). During this stage of disease, the patient's liver enzymes may be elevated. A negative IgM anti-HBc excludes a recent, acute infection but does not rule out chronic infection (18, 26). The presence of IgM anti-HBc identifies patients who are acutely infected; such patients usually have high levels of HBV DNA (26).

Total Anti-HBc

A negative anti-HBc test indicates that a person does not have a history of infection with HBV. A positive result can indicate either an acute infection in which the patient is also HBsAg positive and IgM anti-HBc positive or a resolved (HBsAg-negative) or chronic (HBsAg-positive) HBV infection (Table 1) (3, 26). Total anti-HBc antibodies remain after IgM anti-HBc disappears and can be detected for many years. These antibodies persist longer than anti-HBs. Thus, total anti-HBc is the best marker for documenting prior exposure to HBV infection. Vaccines do not include HBc; thus, vaccination induces only anti-HBs. Therefore, anti-HBc should not be present in vaccinated individuals unless they were infected with HBV prior to vaccination (14, 21). Individuals positive for antibody to HBc without any other serologic evidence of HBV infections should be considered infected with HBV (18, 21). This serologic pattern is consistent with a remote, past HBV infection that has resolved and in the case of which the viral DNA levels are at times negative or very low (3, 24).

Commercial kits for detection of total anti-HBc are available and use a variety of different methods and instrumentation. All assays use recombinant HBc antigen for capture of antibody. Samples containing anti-HBc lead to a positive signal when anti-HBc is present. All anti-HBc assays have a test algorithm that involves testing initially reactive specimens in duplicate in an independent run. If one or both duplicates are reactive, the sample is reported as positive.

Anti-HBs

A negative result for anti-HBs in the absence of any other HBV specific antigen or antibody indicates that a person has not been infected with HBV, nor has the individual been vaccinated with HBV. A positive result is consistent with immunity to HBV due to an infection or from effective vaccination (Table 1).

Commercial assays for anti-HBs are solid-phase tests based on the sandwich principle. These assays provide a quantitative result to assist in determination of adequate immunity after vaccination. An initially reactive result for anti-HBs requires repeat testing in duplicate in an independent run. If one or both duplicates are reactive, the sample is positive. Anti-HBs quantitation panels are commercially available and should be used when validating anti-HBs assays.

Anti-HBs are key serologic markers for both vaccine-induced immunity and immune responses due to infection. As HBV vaccination has become more widespread, this serologic marker is used to monitor vaccine success. Both the WHO and CDC recognize a level of >10 mIU/ml of anti-HBs as an indication of protective immunity (2, 18).

Postvaccination testing for the presence of anti-HBs is not recommended. The CDC lists exceptions to this rule, which are infants born to mothers who are HBsAg positive, immunocompromised patients, dialysis patients, and to confirm successful vaccination in health care workers to prevent the transmission of HBV in the health care setting (18, 22, 26).

HBV escape mutants in which the HBsAg has mutated can result in HBV infections without detectable anti-HBsAg (26). Conformational changes in the major antigenic determinant of HBsAg result in reduced detection of the anti-HBsAg by diagnostic assays.

Anti-HBe

A positive test for anti-HBe indicates the resolution of acute infection and is associated with a decrease in viral replication (Table 1). During acute infection, these antibodies are bound to the HBe antigen; the antibody will not be detected until the HBeAg levels decrease. Patients who have recovered from acute HBV infection have detectable anti-HBe, anti-HBc, and anti-HBs. Interestingly, patients infected with HBeAg-negative strains of HBV still have anti-HBe (18).

Test algorithms for HBeAg and anti-HBe kits vary. Initially reactive samples in the anti-HBe assays are often retested. If the repeat test is reactive, the sample is reported as positive. The enzyme immunoassays have indeterminate or gray zones around the assay cutoffs. If the samples repeatedly yield values in this zone, additional testing should be performed with a new specimen.

Antiviral Testing

Therapy for HBV usually requires long-term treatment with nucleoside or nucleotide analogs which subsequently can lead to the development of antiviral resistance. HBV replicates through an RNA intermediate. The HBV RNA-dependent DNA polymerase is not precise during rapid replication cycles and does not proofread the final copies, leading to frequent errors. Some of these errors create resistant mutants, which are selected in the presence of antiviral agents (20). Over time the HBV strains with antiviral resistance become the major viral species. The current recommendation is that a patient who has a 1-log10 increase from the lowest HBV DNA level should be evaluated for the development of antiviral resistance (14, 22).

Recently the nomenclature of HBV antiviral resistance was standardized in order to track nucleotide changes associated with drug resistance and to recognize new mutations. Methods used to detect these mutations are available; however, the interpretation of results is not always straightforward. Some mutations predict resistance to multiple drugs. A single mutation at A181T is associated with resistance to lamivudine, adefovir, tenofovir, and telbivudine (10). In other situations genetic changes may not confer resistance when present alone but contribute to the resistance when additional mutations are present. For example, lamivudine resistance does not occur due to a single L180M mutation, but with the addition of a second mutation, such as A181T, resistance to lamivudine occurs. Thus, the detection of an L180M, A181T double mutation, which alters the position of a critical residue in the nucleotide binding pocket, is required for resistance to be apparent (6, 14). An additional concern is that patients treated with multiple drugs will have a combination of sequence changes that represent drug resistance mutations in addition to mutations that occur in order to balance the fitness of the HBV strains selected out after antiviral therapy (18, 22, 24). Many of the single nucleotide changes have no recognized phenotype and may represent the random genetic changes found in viruses that use a reverse transcriptase during replication.

There are some commercial assays that are available to detect mutations associated with antiviral resistance in HBV. These assays are generally performed in reference laboratories.

HDV

HDV infection is diagnosed by serologic and molecular markers and should be tested in patients who are infected with HBV from countries in which HDV is prevalent. Diagnosis of a coinfection with HBV and HDV is based on the detection of HBV DNA, and HDAg, anti-HDV IgM, and HDV RNA. A superinfection occurs when HDV infects a patient with chronic HBV. The diagnosis of HDV superinfection is made when anti-HDV is found simultaneously with HBsAg and anti-HBc in the absence of IgM anti-HBc. Anti-HDV is present indefinitely in patients with chronic HDV infection (15, 18).

DIAGNOSTIC BEST PRACTICES

During acute HBV disease, IgM anti-HBc is usually positive when serum transaminases are elevated. At this phase, HBsAg and/or HBeAg may not always be detectable. In a typical acute HBV infection, HBsAg can be detected 2 to 4 weeks before the liver enzyme levels become abnormal and before symptoms appear. Thus, when the patient presents with clinical signs, HBsAg may not be detected. However, HBV DNA can be detected 3 to 4 weeks before HBsAg, long before the onset of symptoms (21, 26). The levels of IgM anti-HBc eventually decline as the disease resolves or becomes chronic. Most patients with an active HBV infection which resolves have detectable anti-HBs shortly after the disappearance of HBsAg. Several published reports review the recommendations for the management of chronic HBV (21). These recommendations include the frequency of screening, diagnostic tests used to determine stage of disease and antiviral treatment. They also provide advice on laboratory tests needed to monitor a patient's treatment. The use of these guidelines has led to better treatment compliance and improved outcomes. Antiviral treatment should be started with one of the recommended first line therapies, such as interferon, entecavir, or tenofovir. Patients on treatment should be monitored regularly for serum HBV DNA, HBsAg, and liver function (21).

REFERENCES

1. **Bauer W and Wyman SM.** 1950. Infectious hepatitis, epidemic type. *N Engl J Med* **242:**261–264.
2. **Liang TJ.** 2009. Hepatitis B: the virus and disease. *Hepatology* **49**(Suppl)**:**S13–S21.
3. **Kuo A, Gish R.** 2012. Chronic hepatitis B infection. *Clin Liver Dis* **16:**347–369.
4. **Dane DS, Cameron CH, Briggs M.** 1970. Virus-like particles in serum of patients with Australia-antigen-associated hepatitis. *Lancet* **1:**695–698.
5. **Servant-Delmas A, Mercier-Darty M, Ly TD, Wind F, Alloui C, Sureau C, Laperche S.** 2012. Variable capacity of 13

hepatitis B virus surface antigen assays for the detection of HBsAg mutants in blood samples. *J Clin Virol* **53:**338–345.

6. **Lapiński TW, Pogorzelska J, Flisiak R.** 2012. HBV mutations and their clinical significance. *Adv Med Sci* **57:**18–22.
7. **Liu CJ, Kao JH.** 2013. Global perspective on the natural history of chronic hepatitis B: role of hepatitis B virus genotypes A to J. *Semin Liver Dis* **33:**97–102.
8. **Lin CL, Kao JH.** 2011. The clinical implications of hepatitis B virus genotype: recent advances. *J Gastroenterol Hepatol* **26** (Suppl 1)**:**123–130.
9. **Huang CH, Yuan Q, Chen PJ, Zhang YL, Chen CR, Zheng QB, Yeh SH, Yu H, Xue Y, Chen YX, Liu PG, Ge SX, Zhang J, Xia NS.** 2012. Influence of mutations in hepatitis B virus surface protein on viral antigenicity and phenotype in occult HBV strains from blood donors. *J Hepatol* **57:**720–729.
10. **Kajiwara E, Tanaka Y, Ohashi T, Uchimura K, Sadoshima S, Kinjo M, Mizokami M.** 2008. Hepatitis B caused by a hepatitis B surface antigen escape mutant. *J Gastroenterol* **43:**243–247.
11. **Zheng X, Ye X, Zhang L, Wang W, Shuai L, Wang A, Zeng J, Candotti D, Allain JP, Li C.** 2011. Characterization of occult hepatitis B virus infection from blood donors in China. *J Clin Microbiol* **49:**1730–1737.
12. **Peters MG.** 2009. Special populations with hepatitis B virus infection. *Hepatology* **49**(Suppl)**:**S146–S155.
13. **Rizzetto M, Ciancio A.** 2012. Epidemiology of hepatitis D. *Semin Liver Dis* **32:**211–219.
14. **Hadziyannis SJ.** 1997. Review: hepatitis delta. *J Gastroenterol Hepatol* **12:**289–298.
15. **McMahon BJ.** 2010. Recent advances in managing hepatitis B. *F1000 Med Rep* **2:**11.
16. **McMahon BJ.** 2010. Natural history of chronic hepatitis B. *Clin Liver Dis* **14:**381–396.
17. **Goldstein ST, Zhou F, Hadler SC, Bell BP, Mast EE, Margolis HS.** 2005. A mathematical model to estimate global hepatitis B disease burden and vaccination impact. *Int J Epidemiol* **34:**1329–1339.
18. **Sorrell MF, Belongia EA, Costa J, Gareen IF, Grem JL, Inadomi JM, Kern ER, McHugh JA, Petersen GM, Rein MF, Strader DB, Trotter HT.** 2009. National Institutes of Health consensus development conference statement: management of hepatitis B. *Hepatology* **49**(Suppl)**:**S4–S12.
19. **Shi YH, Shi CH.** 2009. Molecular characteristics and stages of chronic hepatitis B virus infection. *World J Gastroenterol* **15:**3099–3105.
20. **Liang TJ, Block TM, McMahon BJ, Ghany MG, Urban S, Guo JT, Locarnini S, Zoulim F, Chang KM, Lok AS.** 2015. Present and future therapies of hepatitis B: from discovery to cure. *Hepatology* **62:**1893–1908.
21. **Sunbul M.** 2014. Hepatitis B virus genotypes: global distribution and clinical importance. *World J Gastroenterol* **20:** 5427–5434.
22. **Kwon H, Lok AS.** 2011. Hepatitis B therapy. *Nat Rev Gastroenterol Hepatol* **8:**275–284.
23. **Lavanchy D.** 2012. Viral hepatitis: global goals for vaccination. *J Clin Virol* **55:**296–302.
24. **Uribe LA, O'Brien CG, Wong RJ, Gish RR, Tsai N, Nguyen MH.** 2014. Current treatment guidelines for chronic hepatitis B and their applications. *J Clin Gastroenterol* **48:** 773–783.
25. **Caspari G, Gerlich WH.** 2007. The serologic markers of hepatitis B virus infection—proper selection and standardized interpretation. *Clin Lab* **53:**335–343, 391–400.

Hepatitis C Virus

MELANIE MALLORY AND DAVID HILLYARD

25

Hepatitis C virus (HCV) is a single-stranded RNA virus that was not known as a causative agent of acute and chronic hepatitis until 1989. The development of blood transfusion technology in the mid-20th century led to its rapid spread into new populations with enormous disease burden due to cirrhosis and hepatocellular carcinoma (1, 2). Its discovery by reverse molecular genetics is one of the major achievements of modern medicine and a model for the discovery of many other pathogens during the past 25 years (3). Unlike hepatitis B virus (HBV) and human immunodeficiency virus (HIV), which establish permanent chronic infection, HCV is spontaneously cleared in some individuals and is also amenable to therapeutic cure. With advances in drug therapies, all HCV infections are potentially curable. The development of technologies for viral discovery and clinical detection and monitoring, as well as new classes of antiviral therapy, are important components of the HCV legacy.

VIRAL CLASSIFICATION AND BIOLOGY

HCV is a small 50-nM, enveloped virus with an uncapped positive-sense, single-stranded RNA genome that encodes a single 3010-amino-acid protein. It is a member of the *Flaviviridae* family and is grouped within the genus *Hepacivirus* along with two other hepatatrophic viruses, GB virus B (GB-B) and canine hepacivirus (CHV). The HCV genome resembles other flavivirus genomes with a single coding region flanked by 5′ and 3′ untranslated regions (UTR) (Fig. 1). The HCV 5′UTR is highly conserved among HCV genotypes but is distinct from other flaviviruses, making it the preferred target for all molecular detection tests. The region includes an internal ribosomal entry site (IRES) essential for initiating viral protein synthesis and two binding sites for the micro-RNA (miR)-122, which are required for efficient replication. The 3′UTR ends in a highly conserved, 998-nucleotide region that interacts with the 5′UTR for minus-strand priming during viral replication (4).

Three structural proteins (core, E1, E2) and seven nonstructural proteins (p7, NS2, NS3, NS4A, NS4B, NS5A, NS5B) are derived from a viral polyprotein in a process initiated by host proteases (Fig. 1). The core protein is highly conserved and binds to HCV RNA to form the nucleocapsid. Core protein may also play a role in transcriptional control and cellular transformation. The envelope (proteins E1 and E2) forms a noncovalent complex that helps scaffold the viral envelope. The Domain 1 of E2 interacts with the cellular receptor CD81. E2's hypervariable region 1 is a major target of neutralizing antibodies and is highly variable because of the pressure of immune selection (5). NS2 is a transmembrane protein that plays an essential role in polyprotein proteolysis and viral assembly. Autoprotease cleavage of the NS2/NS3 junction requires zinc stimulation and depends on activities of terminal domains of both NS2 and NS3 proteins. NS3 also has NTPase/helicase activity essential for replication. NS4A is a cofactor for NS3 and is also required for NS5A phosphorylation. NS4B binds to NS4A to indirectly effect proteolysis and replication and plays an essential role in the formation of HCV RNA replication complexes. NS5A has many proposed functions that affect cell signaling. It also contains a domain thought to confer interferon resistance called the interferon-alpha sensitivity-determining region (ISDR). NS5B is an RNA-dependent polymerase that first copies the HCV genome and subsequently creates a genomic plus strand from this intermediate. In addition to these HCV life cycle functions, both structural and nonstructural proteins have important immune-evasive roles (6).

HCV replication is highly error prone due to the viral polymerase's lack of proof-reading function. The high genome copy number present in most stages of infection also contributes to the selection of sequence variants. Alignment of HCV sequences reveals a complex pattern of conserved and variable regions across the genome. In 2005 a consensus proposal was adopted for classification of HCV variants into genotypes and subtypes that also designated formal rules for subsequence classification and naming. The most recent consensus update recognizes 7 distinct genotypes and 67 confirmed subtypes (7). In addition, analysis of HCV sequences from individual patients reveals the presence of closely related but distinct viral quasispecies with sequence variation up to 10% that evolve over the course of infection (8).

Both innate and adaptive immune responses play a key role in defense against HCV infection and contribute to manifestations of host disease. The innate immune response includes rapid release of interferon alpha and interferon beta from hepatocytes as an early viral defense. In addition, natural killer cells, which are abundant in the liver, produce interferon-gamma and other cytokines. The adaptive immune responses of antibody-producing B cells, cellular CD4

doi:10.1128/9781555819156.ch25

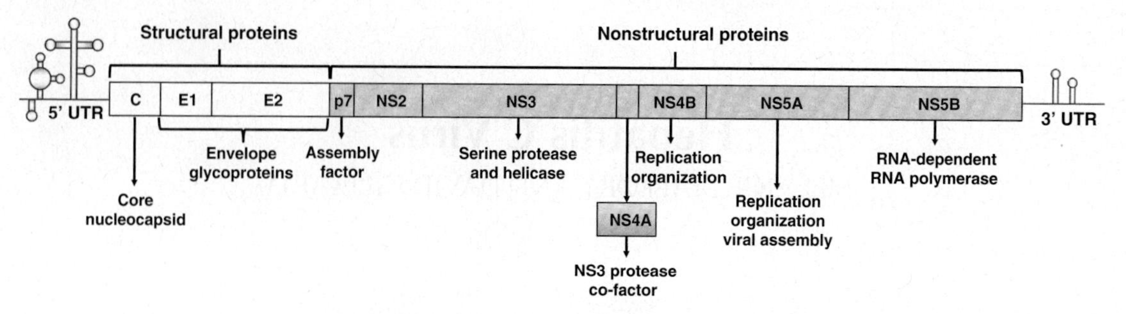

FIGURE 1 HCV Viral Genome Organization

T-helper cells, and CD8 cytotoxic T cells are also critical for controlling HCV infection. HCV antibodies appear approximately seven to eight weeks after infection although their role in viral neutralization is still unclear. HCV-specific T cells appear early in infection and are detectable for years during stages of chronic infection (9).

EPIDEMIOLOGY

Beginning in the 1940s, HCV spread widely throughout the world to populations in all social strata. The major contributors leading to this spread were the introduction of intravenous medical technologies, including transfusion and vaccination and an increase in intravenous drug use. Initiatives to control HBV also led to the first major reduction in new cases of HCV. New cases were further reduced with the introduction of targeted screening of blood products for HCV. Currently, new cases of HCV are skewed to specific at-risk populations, particularly injection drug users. Clusters of acute HCV have been seen in HIV positive populations, possibly as a result of sexual practices involving mucosal trauma and injection or noninjection drug use (10). HCV is transmitted by percutaneous inoculation, including intravenous injection, tattooing, acupuncture, and circumcision, by sexual intercourse, and from mother to child. Fluids including saliva, urine, and semen can contain HCV; however, transmission from these sources is rare. For all transmissions, the inoculum size and HCV concentration correlate with the likelihood of infection (11, 12).

It is estimated that more than 185 million people, or approximately 2.8% of the world's population, have been infected with HCV and that 130 to 170 million are chronically infected. Lack of surveillance resources has limited the accuracy of data for many global regions and for select populations in all countries. In general, the greatest burden is in developing countries including Africa, Latin America, and south-central regions of Asia where estimates of infection exceed 5% for many nations. In those areas, improper sterilization has often been associated with the spread of HCV (13). A campaign to eliminate schistosomiasis in Egypt from the 1950s to the 1980s employed IV injection of tartar emetic without needle sterilization and resulted in high rates of transmission of type 4 HCV. Unintended spread by public health campaigns has largely been corrected and most new cases are related to IV drug injection (14).

HCV prevalence in developed regions has been most systematically studied in the United States and Europe. Beginning in 1998, the National Health and Nutrition Examination Surveys (NHANES) program initiated testing of individuals from a cross-section of U.S. households. Data from 1988 to 1994 suggested that 1.8% of the U.S. population or approximately 3.9 million individuals were infected with HCV (15). The most recent studies by NHANES for the years 2003 to 2010 estimates an HCV prevalence of approximately 1%, or 2.7 million individuals (16). Based on these data, approximately 75% of HCV-infected individuals in the United States were born between 1945 and 1965. An analysis of veterans born within this time interval showed a prevalence of 10.3% compared to only 1.2% for those born after 1965 (17). With the advent of more effective treatment options, this "baby boomer" cohort has become the focus of intensive efforts to identify chronically infected individuals and bring them to therapy (18).

The traceability of HCV genotypes has been key to understanding its spread and formulating strategies for control. Accurate knowledge of HCV genotype distribution is also important for treatment and public health planning since many therapies are genotype specific. Genotypes 1 to 3 are the most common worldwide with genotype 1 virus dominant in the United States and Northern Europe. Phylogenetic analysis suggests that subtype 1b may have been spread earlier in blood products as compared to subtype 1a, which is associated with spread by drug use in these regions (13). Genotypes 4 and 5 are found mostly in Africa and genotype 6 is concentrated in Asia. Older studies of genotype and subtype distribution may not reflect recent population and HCV transmission dynamics. In particular the geographically clustered genotypes 4 to 7 have now spread to bordering countries and to North America and Europe (19).

HCV CLINICAL SIGNIFICANCE

HCV infection leads to chronic hepatitis in approximately 80% of cases, placing patients at risk for cirrhosis, decompensated cirrhosis, and hepatocellular carcinoma (20, 21). The time course for disease progression is variable and most patients are asymptomatic in early years of infection. In general, resolved infection is not associated with serious disease; however, chronic infection poses an increasing risk with time. The liver-related mortality of chronic HCV infection worldwide is estimated to be greater than 350,000 deaths per year (20). All-cause HCV mortality is significantly higher because of the frequent involvement of other organ-systems during the course of chronic infection. At this time, HCV-driven liver failure is the leading indicator for liver transplantation in the Western world (22).

HCV infection is usually mild and often asymptomatic. Symptoms may include fatigue, nausea, and abdominal pain, with accompanying jaundice and dark urine. In many cases the only indication of infection is the presence of elevated transaminases, which occurs 10 to 14 weeks after transmission. Fulminant infection occurs in only a small percentage of acute infections. The time course for HCV clearance is difficult to establish and depends on many factors, including severity of acute infection, host immunity, host genetics, and HIV coinfection. The likelihood of spontaneous resolution is approximately 25%, but estimates vary in different studies and populations. The time to viral clearance also varies, but clearance generally occurs within 6 months of infection and is not accompanied by significant disease. Likelihood of clearance is correlated with fewer overt symptoms of hepatitis, non-African descent, lack of HIV coinfection, and host capacity for interferon expression (21).

Chronic HCV infection is a slowly progressive disease characterized by high viremia, fluctuating inflammation, and steatosis. Although infections may remain indolent for many years, progression to cirrhosis occurs in approximately 5 to 10% of patients over a 20- to 30-year period. Progression rates are highly variable and difficult to predict. In the setting of established cirrhosis there is a 1 to 5% annual risk of hepatocellular carcinoma and a 3 to 6% annual risk of hepatic decompensation (23). Risk of progression to fibrosis with HCV is multifactorial and includes age at infection, male gender, alcohol consumption, obesity, insulin resistance, type 2 diabetes, and co-infection with HBV or HIV (23). Although most chronically infected individuals have persistent blood levels of HCV RNA in excess of 10^5 International Units (IU)/mL, there is no strong correlation of HCV viral load or genotype with fibrosis progression (24).

Extrahepatic complications frequently accompany HCV infection and manifest with a broad spectrum of disease affecting renal, cardiovascular, and central nervous systems. Between 15 to 30% of HCV patients have circulating cryoglobulins. Mixed cryoglobulinemia vasculitis causes inflammation in kidney, skin and other organs. Its progression is variable but may be life-threatening because of widespread vasculitis. The chronic stimulation of B cells during HCV infection has been linked to an increased risk of non-Hodgkins lymphoma. Likewise, treatment of low-grade, HCV-associated tumors has been associated with remission. Type 2 diabetes is associated with steatosis and inflammatory processes, and several studies have shown an association with chronic HCV infection (21, 25).

The composite effect of all conditions related to chronic HCV infection is profound and results in reduced health-related quality of life. Psychosocial and economic factors are also important consequences that can directly affect health. Patients commonly experience fear and anxiety regarding both health outcomes and stigmatization. Access to care and the cost of new highly effective but expensive therapies are factors that may profoundly affect total personal health (11).

TREATMENT AND PREVENTION

Treatment for HCV is based on either host immune modulation by interferon or disruption of viral replication by various classes of direct-acting antivirals (DAAs). Traditional interferon-based treatments employ pegylated α-interferon (peg-αINF) and ribavirin (RBV), a guanosine analog with broad antiviral activity (26). Current DAA targets include the NS3/4A protease complex, the NS5A packaging and assembly protein, and the NS5B polymerase. The goal of treatment is to achieve a sustained virologic response (SVR), which is defined as the absence of detectable HCV RNA 6 months after the end of therapy. An end therapeutic response (ETR), with no HCV RNA detected at the end of treatment, is not a reliable indicator of HCV cure because rebound is seen at varying frequencies for all therapies. Rates of SVR have increased dramatically, especially for new combination therapies, and now most patients can be cured, including those with more difficult-to-treat, advancing cirrhosis and late-stage decompensated liver disease. Selection and guidance of therapy is based on HCV viral genotype and monitoring the kinetics of response measured by quantitative HCV RNA viral load testing (27).

For many years the standard for HCV treatment was combination peg-αINF with RBV given for 6 months for genotype 2 or 3 infections and 12 months for other genotypes. With this therapy SVR rates of 70 to 80% are achieved for genotypes 2 or 3; however, other genotypes have SVRs of only 40 to 50%. These differences have been attributed to the association of genotype-specific sequence differences in the carboxyterminal part of the NS5A protein, including amino acids 2,209 to 2,248 (28, 29). Treatment is associated with significant side effects, and many patients discontinue treatment, resulting in even lower real frequencies of cure. Since 2011, recommended HCV therapies have incorporated DAAs, initially in combination with peg-αINF/RBV and more recently in interferon-free formulations. The first generation protease inhibitors, telaprevir and boceprevir, were used in combination with peg-αINF/RBV and were approved for treatment of type 1 virus only. They achieved SVR rates of ∼70% with much shorter treatment durations but with other significant side effects including life-threatening skin reactions with telaprevir (30). In 2013, the first NS5B inhibitor sofosbuvir was introduced for use with ribavirin in an all-oral formulation for treatment of genotypes 2 and 3 or with peg-αINF/RBV for treatment of genotypes 1 and 4, reducing treatment times again by about 50% (31).

Newly developed DAAs are used in combination and are capable of curing most HCV infection with regimens of 12 to 24 weeks. These include all-oral, interferon-free treatments with minimal side-effect profiles. Treatment algorithms are customized for different stages of disease progression, previous treatment history, and in many cases viral genotypes or subtypes. Due in part to the high cost of these drugs, patients are monitored for HCV viral load during therapy to assess compliance (27).

As HCV treatments have improved, the decision of whom to treat and when to treat has evolved. The unfavorable side-effect profiles and low SVRs of early therapies have led some patients to delay treatment in anticipation of better drugs. The arrival of dramatically improved HCV therapies has ignited debate over the need for timely therapy (32, 33). Early HCV therapy clearly mitigates future hepatic and extrahepatic disease; however treatment costs can exceed $150,000. Because HCV disease often progresses slowly over many years, rationing therapy to patients with most advanced hepatic fibrosis occurs in many settings and is linked to reimbursement issues. The case for routinely treating patients with earlier stages of disease recognizes the very high cost of treating late-stage HCV-driven disease and the better tolerability of early therapy, as well as ethical concerns (33).

The widespread elimination of HCV will be achieved only with the development of an effective and easily administered vaccine. HCV vaccine development is challenging because of HCV's intrinsically high genetic variability and ability to blunt cellular immune responses (34). Viral peptide, recombinant protein, and DNA vaccines have been developed but none have yet demonstrated suitability for either prophylactic or therapeutic use (35).

In the absence of an effective vaccine, HCV prevention is primarily based on screening blood products, eliminating unsafe medical practices, and reducing intravenous drug use–associated transmission. For populations with access to therapy, reducing the pool of infected individuals is also forecast to have a significant preventive impact (11). In developed nations, the current risk of acquiring HCV infection by blood transfusion is approximately 1 in a million (.0001%), and the window period for detecting HCV RNA has been reported to be as early as 3-5 days (36). Adherence to standard care practices is also critical for preventing transmission during routine phlebotomy and manipulation of intravenous solutions. Although transmission rates among injection drug users have declined, new infections still occur at significant frequencies (37). Needle exchange programs and other methods to reduce the numbers of new cases have been effective primarily in developed countries (38).

DIAGNOSIS, DETECTION, AND MONITORING

The diagnosis of acute or chronic HCV infection is based on serologic screening for anti-HCV antibodies and confirmation by detection of HCV RNA. Serologic tests for HCV have improved steadily due to the addition of recombinant antigens derived from HCV structural and nonstructural proteins. Third-generation enzyme immunoassays (EIA) use antigens from core, NS3, NS4, and NS5 and allow for the detection of anti-HCV antibodies approximately 4 to 6 weeks after infection (39). Chemiluminescence immunoassays have also been developed that offer improved specificity and precision in high-throughput, fully automated formats (40, 41). The sensitivity and specificity of current screening tests exceed 99%. Screening assays cannot distinguish active infection from resolved infection, nor can they detect HCV during early stages of infection. False-positives occur most frequently in low-risk populations, pregnant women, and in patients with immunologic disease. False-negatives occur at a rate of approximately 10% in patients with inadequate antibody response, such as immunocompromised patients and patients on long-term dialysis (42, 43).

Until recently, seropositive samples were confirmed by blot-based serologic tests. The recombinant immunoblot assay (RIBA) was used widely and incorporated antigens used in screening tests (44). This tool allowed for the identification of previously infected individuals who had a spontaneously cleared infection and therefore afforded valuable epidemiologic information. Algorithms for using signal-to-cutoff (S/CO) values from screening tests were developed based on the observation that high S/CO samples are strongly correlated with the presence of detectable HCV RNA (45). RIBA has high numbers of indeterminate results, and confirmation of high S/CO samples by HCV RNA testing is more efficient and cost effective than RIBA confirmation of all samples. Manufacture of the RIBA test was discontinued in 2013, and current guidelines call for HCV serologic confirmation by a sensitive HCV RNA detection method (46).

In attempts to better distinguish acute from chronic HCV infection, both anti-HCV IgM and avidity assays have been developed; however, these assays have not been adopted into clinical practice. Tests that detect core antigen alone have also been developed as a low-cost alternative for diagnosing acute infection. These tests have been shown to detect acute infection 1 to 2 days following the appearance of HCV RNA in serum and offer a low-cost, low-tech alternative in resource-limited settings. Due to its lower sensitivity of approximately 10^4 IU/ml HCV RNA, it is not recommended as a substitute for RNA testing (47, 48).

HCV RNA can be reliably detected 1 to 3 weeks after infection and approximately 4 weeks prior to appearance of anti-HCV antibodies. Its presence is the definitive marker for HCV disease. For confirmation of HCV seropositive samples, both qualitative and quantitative tests are routinely used. First-generation quantitative tests were less sensitive than qualitative tests and until recently until recently no quantitative tests were FDA-approved for diagnosis. Current (second- and third-generation) quantitative tests have equivalent or better sensitivity compared with qualitative tests and have the advantage of providing a baseline for monitoring patients during therapy, affording both convenience and cost savings (49).

The primary role of quantitative HCV RNA testing is to assess the efficacy of treatment. For peg-αINF/RBV therapy, the kinetics of HCV RNA decline is strongly predictive of SVR. Rapid viral response (RVR), defined as undetectable HCV RNA 4 weeks after initiation of therapy, is the strongest predictor of SVR for peg-αINF/RBV therapy. Slower responses to interferon-based therapies have also been indicators for extending treatment duration. Likewise, failure to achieve a greater than a 100-fold drop in viral load by week 12 predicts a <1% likelihood of SVR and is an indication to discontinue peg-αINF/RBV treatment. Although baseline viral load is also a predictor of therapeutic success, it is only one of many factors (49, 50). Treatment algorithms using the first-generation protease inhibitors boceprevir or telaprevir in combination with peg-αINF/RBV illustrated the subtleties of viral load use for response-guided therapy. These drugs were used in combination with peg-αINF/RBV. Decisions to extend treatment duration were based on quantifying HCV RNA at set time-points. If viral loads were above or below prescribed quantitative thresholds, therapy was continued or stopped (51). Current therapies with highly active DAAs call for treatment for defined intervals based on prior treatment history, stage of disease, and HCV genotype. Failure to achieve set endpoints of HCV viral load is also useful for assessing patient compliance (110).

HCV genotyping refers to both the taxonomic assignment of HCV genotypes and subtypes and to the analysis of the HCV genome to detect resistance-associated variants. Historically, the primary use of taxonomic genotyping has been for determining treatment duration for interferon-based therapies. Guidelines for treatment with peg-αINF/RBV call for 6-month treatment for genotypes 2 and 3 versus 12-month treatment for other types. In therapies that combine peg-αINF/RBV with DAAs, genotype-specific outcomes are also noted. In addition, differences in treatment outcome are seen between HCV subtypes 1a versus 1b. Genotype 1a/1b subtype differences are also noted for DAA-based therapies that do not include interferon (52). In general, subtype 1a virus is more difficult to treat than 1b in many therapies. Thus, current guidelines recommend using taxonomic genotyping and subtyping 1a/1b to select DAA therapy and its duration (110).

DAA-based therapies may be confounded by the presence of pre-existing resistance-associated variants and by resistance-associated variants selected on treatment. The likelihood of resistance emergence is reduced by selecting DAAs with high barriers to resistance, using DAAs with different resistance mechanisms in combination, and promoting patient compliance to prescribed therapy. Among DAA classes, the barrier to resistance is greatest for NS5B inhibitors, lower for NS3/4A inhibitors, and lower still for NS5A inhibitors (53).

Treatment with the first-generation protease inhibitors boceprevir or telaprevir along with peg-αINF gives rise to mutations in NS3/4A that reduce likelihood of SVR. Although second- and third-generation NS3/4A inhibitors have higher barriers to resistance, R155K and D168E/G/V variants can arise during treatment and affect outcomes. The mutation Q80K arises spontaneously and also following treatment with the second-generation protease inhibitor simeprevir during combination peg-αINF therapy. Resistance occurs at higher frequency in subtype 1a versus 1b infections and is associated with increased failure rates in some genotype 1a patients on simeprevir/peg-αINF therapy (54). Resistance testing to detect this mutation is done prior to initiation of therapy based on recommendations in the simeprevir package insert (55). The lower barrier to resistance for genotype 1a is due in part to the requirement for resistance of only two nucleotide transitions versus two transitions and two transversions for genotype 1b (56).

Mutations present in NS5A prior to treatment also have significance for clinical outcome (57). The number of drugs affected and the impact of these mutations appear to be greater for subtypes 1a compared to 1b (57, 58). NS5A mutations have a low fitness cost and persist for years. The type and number of mutations is also important in patients being retreated after failure with NS5A inhibitors (57). Resistance to NS5B inhibitors arises infrequently and no cross-resistance is seen among nucleotide and nonnucleotide classes.

Until recently, HCV resistance testing prior to initiating therapy has not been widely performed because of the evolving status of clinical data and lack of testing resource. Some reference laboratories now offer resistance testing for select indications. An example is the recommendation for testing for several potential mutations in NS5A prior to treatment of genotype 1a infection with the recently approved NS5A inhibitor Zepatier (elbasvir/grazoprevir) (59). Going forward, it is likely resistance testing will be done on a selective basis and consider factors such as prior treatment experience, stage of liver disease, and HCV genotype.

Several genome-wide association studies have revealed single nucleotide polymorphisms (SNPs) upstream of the IL28B gene on human chromosome 19 that are associated with spontaneous clearance of HCV and the host response to interferon-based therapies (60, 61). The IL28B gene encodes the type III α–interferon, IFN-λ3, which is important for antiviral immunity and targets receptors that are more localized to hepatocytes compared to type I IFN-α receptors. Patients on peg-αINF/RIB with the SNP rs12979860 CC genotype are twice as likely to achieve SVR compared with individuals with CT-heterozygous and TT-homozygous genotypes (62). The favorable C allele is prevalent in East Asian individuals who have high rates of response to interferon therapies and is less prevalent in individuals of African descent who have lower rates of response. Analysis of the codistribution of HCV and human IL28B genotypes has provided insights into the long-observed differences in SVR for interferon-based therapies. IL28 testing is not recommended for guidance of current interferon-based combination therapies.

HCV ASSAYS

A variety of assays are available for HCV screening, diagnosis, quantification, and genotyping. These tests employ many technologies including enzyme immunoassay or chemiluminescence immunoassay, transcription-mediated amplification, reverse-transcription PCR (RT-PCR), quantitative RT-PCR (qRT-PCR), hybridization-based identification, and sequencing.

Screening Assays

Screening assays for detecting HCV antibody generally employ chemiluminescent signaling and use recombinant antigens bound in solid phase to magnetic beads or microtiter wells (Table 1). Recombinant antigens are derived from conserved domains of the core, NS3, NS4, and NS5 regions of the HCV genome (43). These tests detect patients' HCV-specific antibodies using conjugate antibodies (antihuman IgG) with horseradish peroxidase or acridinium labels, which catalyze a luminol substrate. Light generated from luminol catalysis is compared with a signal threshold and the relative amount of light is reported as S/CO (63, 64). Assays have an indeterminate range for samples slightly below the cutoff. Samples with S/CO ratios in this range cannot be definitively identified as negative or positive and therefore require repeat and/or confirmation testing by another method. The use of the serology test samples for confirmatory RNA testing is controversial, with some labs supporting this practice and others requiring submission of separate serum and plasma samples. Prealiquoting from the serology test sample for subsequent molecular testing or requiring submission of separate samples are practices that reduce the possibility of false-positives due to contamination.

HCV RNA Detection and Quantification Assays

Several qualitative and quantitative assays are FDA-approved for HCV clinical testing (Table 2). These tests target the 5′UTR because of its high conservation, which allows for reliable amplification of most circulating HCV strains.

TABLE 1 Commercial anti-HCV screening assays[a]

Analyzer	Manufacturer	Method	Solid phase	HCV antigens
ADVIA Centaur	Siemens Corp., Washington, DC, USA	CLIA	Magnetic particles	c22-3 (Core), NS3, c200,NS5
Architect i2000SR	Abbott Laboratories, Abbott Park, IL, USA	CMIA	Paramagnetic particles	HCr43 (Core and NS3), c100-3 (NS4A)
AxSYM	Abbott Laboratories, Abbott Park, IL, USA	MEIA	Paramagnetic particles	HCr43 (Core), c200 (NS3), c100-3 (NS4A)
Elecsys	Roche Diagnostics, Indianapolis, IN, USA	ECLIA	Paramagnetic particles	Core, NS3, NS4
LiaisonXL	DiaSorin, Stillwater, MN, USA	CLIA	Paramagnetic particles	Core, NS3, NS4
VITROS Eci, VITROS 3600	Ortho-Clinical Diagnostics, Raritan, New Jersey, USA	CLIA	Microwell	c22-3 (Core), c200, (NS3 and NS4), NS5

CLIA= chemiluminescent immunoassay, CMIA= chemiluminescent microparticle immunoassay, MEIA= microparticle enzyme immunoassay, ECLIA= electroluminescence immunoassay

[a]This table includes examples of commonly used HCV screening tests commercially available in the United States and is not intended to be an all-inclusive list.

TABLE 2 Commercial HCV RNA tests[a]

Test	Manufacturer	Method	Lower limit of detection (IU/mL)	Quantitative dynamic range (IU/mL)
Qualitative				
Ampliscreen HCV Test v2.0	Roche Diagnostics, Indianapolis, IN	RT-PCR	<50	NA
APTIMA HCV RNA Qualitative Assay	Gen-Probe, Inc., San Diego, CA	TMA	5	NA
Quantitative				
Versant HCV RNA Assay v3.0	Siemens Corp., Washington, DC	bDNA	615	$615\text{-}7.7 \times 10^6$
COBAS TaqMan HCV Test v2.0 with High Pure System	Roche Diagnostics, Indianapolis, IN	Real-time RT-PCR	20	$25\text{-}3.0 \times 10^8$
COBAS AmpliPrep/COBAS TaqMan HCV Test	Roche Diagnostics, Indianapolis, IN	Real-time RT-PCR	18	$43\text{-}6.9 \times 10^7$
COBAS AmpliPrep/COBAS TaqMan HCV Test, v2.0	Roche Diagnostics, Indianapolis, IN	Real-time RT-PCR	15	$15\text{-}1.0 \times 10^8$
COBAS HCV (6800/8800 system)	Roche Diagnostics, Indianapolis, IN	Real-time RT-PCR	15	$15\text{-}1.0 \times 10^8$
RealTime HCV	Abbott Laboratories, Abbot Park, IL	Real-time RT-PCR	12	$12\text{-}1.0 \times 10^8$

NA = not applicable
[a]This table includes examples of commonly used HCV RNA tests commercially available in the United States and is not intended to be an all-inclusive list.

Internal controls are used to normalize for amplification inhibition or prevent reporting of results from significantly inhibited samples. Results are reported in International Units (IU) and are normalized to WHO standards. Qualitative assays previously had greater sensitivity than quantitative tests and were preferred for confirmation of anti-HCV antibody detection. The development of HCV quantitative tests with equivalent or better sensitivity has led to their widespread use for confirmatory testing and the discontinuation of some qualitative tests. The Aptima HCV RNA Qualitative Assay (Hologic, Inc., Bedford, MA, USA) has sensitivity equivalent to current quantitative tests and is approved for HCV diagnosis. Unlike most other qualitative assays that are PCR-based, the Aptima assay uses transcription-mediated amplification, an isothermal process that uses MMLV reverse transcriptase and T7 RNA polymerase to amplify RNA.

The FDA has approved several commercial tests for quantifying HCV RNA, including the Versant HCV RNA 3.0 assay (bDNA, Siemens Healthcare Diagnostics Inc. Tarrytown, NY), the COBAS AmpliPrep/COBAS TaqMan HCV Tests (Roche Molecular Systems, Inc., Branchburg, NJ) (58,65–67), and the RealTime HCV assay (Abbott Molecular Inc., Des Plaines, IL) (68) (Table 2). The APTIMA HCV RNA Qualitative Assay, COBAS AmpliPrep/COBAS TaqMan HCV Test v2.0, and COBAS HCV (6800/8800) tests also are approved as an aid in diagnosis. The Versant bDNA assay was one of the first to be developed and used hybridization technology in a microtiter well format, with signal amplifier probes forming a branched-chain structure and detection of linked alkaline phosphatase labeled probes by chemiluminescence. The final version 3.0 assay had a 615 IU/mL LLOD and a dynamic range of $615\text{-}7.7 \times 10^6$ IU/mL. The manufacturer has recently discontinued this assay because of the availability of more sensitive and rapid quantitative tests. Subsequent HCV RNA tests have all used target amplification in a variety of formats.

The Roche COBAS assays measure HCV RNA using qRT-PCR technology and TaqMan probes. This technology involves DNA polymerase using its 5′-3′ exonuclease activity to hydrolyze probe bound to the target during the elongation step of PCR. Prior to hydrolysis, the quencher is in close enough proximity to the fluorophore to prevent signal. Following hydrolysis, the fluorophore and quencher are spatially separated and fluorescence can be detected. The assay quantifies HCV RNA by comparing the measured amount of sample RNA to a known amount of internal calibrator included in each reaction. Although the assay was designed to universally detect all six HCV genotypes, several groups, mostly outside of the United States, reported that the first version of the assay under-quantified or failed to detect some HCV isolates. Some genotype 4 samples were particularly challenging for the assay, most likely because of polymorphisms at positions nt145 and nt165 in the 5′UTR (69–72). Roche released a second version of the assay with design improvements to address these issues (73–77). Other improvements to the version 2.0 assay include decreasing the required 1 mL sample input volume to 0.65 mL while decreasing the LLOQ from 43 IU/mL to 15 IU/mL. A version of the COBAS assay using manual extractions (the High Pure System) was used for many clinical trials and has a dynamic range of $25\text{-}3 \times 10^8$ IU/mL.

The Abbott RealTime HCV assay has a LLOD and LLOQ of 12 IU/mL and a dynamic range of $12\text{-}1\text{x}10^8$ IU/mL (68). Like the Roche COBAS assay, the Abbott RealTime HCV assay also uses qRT-PCR to measure HCV RNA, although it uses a different probe technology. Its probe has a 5′ fluorophore and 3′ quencher. In its unbound state, the probe is randomly coiled, which causes the fluorophore and quencher to be physically close together and thus prevents signaling. When the probe hybridizes to the target, the fluorophore and quencher are spatially separated and fluorescence can be observed. The Abbott RealTime HCV assay uses a two-point standard curve to quantify HCV RNA. The exogenous internal control to monitor sample processing and possible inhibition is a pumpkin gene.

The qRT-PCR assays are the method of choice for diagnosing active infection, assessing patient compliance, and monitoring therapeutic response due to their high sensitivity and broad dynamic range. It is important to note that different qRT-PCR assays may not detect some genotypes equally because of polymorphisms in probe and primer targets, although the assays have been designed to minimize genotype bias as much as possible (78). Their fully automated formats (available since 2007) also allow for relatively fast, high-throughput testing with minimal risks for carryover and sample-to-sample contamination. However, these tests require considerable technical expertise and expensive equipment, making them less suitable for resource-limited settings.

Genotyping Assays

Like the HCV RNA detection assays, genotyping assays have historically targeted the HCV 5′UTR. However, studies have shown the limitations of 5′UTR analysis for subtyping due to the region's high genetic conservation (79–81). Furthermore, some 1b subtypes and several subtypes of 6 have identical 5′UTR sequence and are indistinguishable by 5′UTR analysis (82) and other unusual subtypes may confound 5′UTR assays (83–85), leading to inaccurate genotype identification. In addition to the 5′UTR, most commercial assays now analyze more genetically variable regions of the HCV genome, including core and NS5B (86, 87). Sequencing outside of the 5′UTR (most commonly core and NS5B) is currently considered the gold standard for genotyping and provides much higher subtyping resolution. HCV nomenclature systems require core and NS5B sequencing to established new provisional subtypes (80, 88). However, genotyping by sequencing is not routinely performed except by select reference laboratories, in part because reliably amplifying and analyzing these highly variable regions across all genotypes is difficult. Several studies have highlighted the need for sequencing-based approaches to complement current commercial assays, although such sequencing is not performed routinely (89–93). Commonly used commercial assays include the Versant hepatitis C virus genotype 2.0 assay (LiPA 2.0) (Bayer Healthcare, Eragny, France), the Abbott HCV Genotype II IVD assay (Des Plaines, IL, USA), and the GenMark eSensor HCVg Direct assay (GenMark Dx, Carlsbad, CA, USA).

The Versant LiPA method involves hybridizing biotinylated PCR products to genotype-specific oligonucleotide probes immobilized in parallel lines on membrane strips. Hybridized products appear as purple/brown bands after the addition of alkaline phosphatase-labeled streptavidin followed by substrate (BCIP/NBT chromogen). The LiPA method requires less technical expertise and may be lower in cost than other methods. It may also detect subpopulations in mixed infections more sensitively than sequencing. The first version of the Versant LiPA assay was introduced in the 1990s and targeted solely the 5′UTR to identify genotypes 1 through 6 (94). The updated LiPA 2.0 assay primarily targets the 5′UTR but also interrogates the core region to more accurately identify subtypes 1a, 1b, and genotype 6. However, LiPA 2.0 still occasionally misclassifies common genotypes and can be confounded by rare subtypes (90, 95, 96). Lipa 2.0 may also occasionally fail to subtype genotype 1 specimens with reported frequencies of failure ranging from 2.2 to 7.4% (90,97–100).

The Abbott HCV GT II IVD assay (Des Plaines, IL, USA) is currently the only FDA-approved assay. It uses real-time PCR to amplify and detect portions of the 5′UTR and NS5B. The assay detects HCV genotypes 1 through 5 using probes targeting the 5′UTR amplicon and subtypes 1a and 1b using probes targeting the NS5B amplicon. Advantages of the GT II assay include its automated format, ease of use, and sensitivity. Like other commercial assays, previous studies have shown that the GT II assay fails to correctly genotype rare subtypes and fails to subtype in 4.3 to 9.1% of genotype 1 samples (86, 89, 93). Recently, Abbott introduced the HCV GT Plus RUO assay to serve as a resolution assay for samples with ambiguous GT II results by detecting subtypes 1a, 1b, and genotype 6 through interrogating the core region (101).

The GenMark eSensor method is also a hybridization-based assay but in an array format with numerous probes. The eSensor technology involves hybridizing PCR products to genotype-specific capture probes, which then hybridize to surface-bound, ferrocene-labeled signal probes. These signal probes are then detected electrochemically. Like the LiPA and Abbott assays, two versions of the GenMark HCV genotyping assay are available: the eSensor HCV Genotyping Test that targets the 5′UTR and the eSensor HCVg Direct Test that primarily targets the 5′UTR but also interrogates the core region. Like the other commercially available hybridization methods, the GenMark assay may be vulnerable to failure due to polymorphisms, but its array format may make it more sensitive than other methods (102).

A variety of hybridization-based commercial methods for genotyping are available. Current commercial assays satisfactorily genotype routine specimens. However, they may inaccurately genotype <5% of samples, including rare subtypes and intergenotype recombinants. Sequencing outside of the 5′UTR is a useful resolution method for determining the genotype of challenging samples.

Resistance Testing

Some reference laboratories have recently started offering Sanger and next-generation sequencing-based methods for detecting DAA resistance-associated variants. Currently, no commercial kits are available for resistance testing. In general, resistance tests are genotype-specific and thus require identifying taxonomic genotype prior to ordering resistance testing. Most resistance tests interrogate genotypes 1a and 1b only and more recently genotype 3. Furthermore, resistance testing is ordered on a per gene basis. For example, NS3/4A resistance testing is available for genotype 1a and 1b infections and detects Q80K along with other mutations associated with resistance to the protease inhibitors boceprevir, simeprevir, telaprevir, paritaprevir, and grazoprevir. NS5A resistance testing is available for genotypes 1a, 1b, and genotype 3 and detects resistance to DAAs such as ledipasvir and ombitasvir. NS5B resistance testing is available for genotypes 1a and 1b and detects mutations associated with polymerase-inhibitors such as dasabuvir and sofosbuvir. Resistance testing may become increasingly important for selecting treatment with newer and future drugs as well as for treatment-experienced patients who have failed previous therapies.

Point-of-Care Testing

Point-of-care (POC) tests for the detection of HCV antibodies have been developed using various technologies, including test strips, lateral flow (similar to ELISA), and immunochromatography. These tests play an important role in developing countries where access to equipment and facilities, a constant electricity source, and skilled technicians is often not available. With the recent availability of highly effective, interferon-free therapies, there has been an increased emphasis on potential applications for POC testing in developed countries as well, including its use as an inexpensive, rapid means for screening medium- and high-risk populations.

Sample types for POC tests include plasma, serum, whole blood, and oral fluid. Performance varies widely depending on factors including HCV genotype, co-infection with HIV, and storage and testing conditions (103–106). Studies have found little genotype bias for currently available POC tests

(103, 107, 108). Most POC tests generally have lower analytic sensitivity and detect HCV antibodies a few days to a few weeks later than standard immunoassays. One exception is the OraQuick HCV Rapid Antibody Test (OraSure Technologies Inc., Bethlehem, PA, USA), which performs similarly to third-generation EIA assays and is FDA-approved for rapid screening (103), making it potentially suitable for large-scale screening programs. Toyo anti-HCV test (Turklab, Turkey) may also perform sufficiently well for use in widespread screening (103, 109).

Developing countries and other resource-limited settings commonly use rapid, POC testing for HCV screening (103). Large-scale screening programs for these populations may lead to reduced transmission and improved access to care (109). As yet, POC molecular tests for HCV confirmation or monitoring have not been introduced; however, suitable technologies have been developed and are now in widespread use for other applications such as rapid respiratory testing. HCV genotyping requires varying degrees of sequence interrogation and this capability is not yet available on small portable devices. With the advent of pangenotypic interferon-free therapies, future testing may require only POC serologic screening and qualitative RNA confirmation. The goal of identifying and bringing to therapy larger proportions of at-risk populations might be well served by such testing.

REFERENCES

1. **Wise M, Bialek S, Finelli L, Bell BP, Sorvillo F.** 2008. Changing trends in hepatitis C-related mortality in the United States, 1995–2004. *Hepatology* **47:**1128–1135.
2. **Lavanchy D.** 2009. The global burden of hepatitis C. *Liver Int* **29**(Suppl 1)**:**74–81.
3. **Choo QL, Kuo G, Weiner AJ, Overby LR, Bradley DW, Houghton M.** 1989. Isolation of a cDNA clone derived from a blood-borne non-A, non-B viral hepatitis genome. *Science* **244:**359–362.
4. **Lindenbach B, Murray C, Thiel HJ, Rice C.** 2013. Hepatitis C virus, p 712–746. *In* Knipe D, Howley P (ed), *Field Virology, Vol.* Lippincott Williams and Wilkins, Philadelphia.
5. **Polyak SJ, McArdle S, Liu SL, Sullivan DG, Chung M, Hofgärtner WT, Carithers RL Jr, McMahon BJ, Mullins JI, Corey L, Gretch DR.** 1998. Evolution of hepatitis C virus quasispecies in hypervariable region 1 and the putative interferon sensitivity-determining region during interferon therapy and natural infection. *J Virol* **72:**4288–4296.
6. **Moradpour D, Penin F, Rice CM.** 2007. Replication of hepatitis C virus. *Nat Rev Microbiol* **5:**453–463.
7. **Smith DB, Bukh J, Kuiken C, Muerhoff AS, Rice CM, Stapleton JT, Simmonds P.** 2014. Expanded classification of hepatitis C virus into 7 genotypes and 67 subtypes: updated criteria and genotype assignment web resource. *Hepatology* **59:** 318–327.
8. **Gómez J, Martell M, Quer J, Cabot B, Esteban JI.** 1999. Hepatitis C viral quasispecies. *J Viral Hepat* **6:**3–16.
9. **Spengler U, Nischalke HD, Nattermann J, Strassburg CP.** 2013. Between Scylla and Charybdis: the role of the human immune system in the pathogenesis of hepatitis C. *World J Gastroenterol* **19:**7852–7866.
10. **van de Laar TJ, Paxton WA, Zorgdrager F, Cornelissen M, de Vries HJ.** 2011. Sexual transmission of hepatitis C virus in human immunodeficiency virus-negative men who have sex with men: a series of case reports. *Sex Transm Dis* **38:**102–104.
11. **Thomas DL.** 2013. Global control of hepatitis C: where challenge meets opportunity. *Nat Med* **19:**850–858.
12. **Seeff LB.** 1991. Hepatitis C from a needlestick injury. *Ann Intern Med* **115:**411.
13. **Lavanchy D.** 2011. Evolving epidemiology of hepatitis C virus. *Clin Microbiol Infect* **17:**107–115.
14. **Strickland GT.** 2006. Liver disease in Egypt: hepatitis C superseded schistosomiasis as a result of iatrogenic and biological factors. *Hepatology* **43:**915–922.
15. **Armstrong GL, Wasley A, Simard EP, McQuillan GM, Kuhnert WL, Alter MJ.** 2006. The prevalence of hepatitis C virus infection in the United States, 1999 through 2002. *Ann Intern Med* **144:**705–714.
16. **Denniston MM, Jiles RB, Drobeniuc J, Klevens RM, Ward JW, McQuillan GM, Holmberg SD.** 2014. Chronic hepatitis C virus infection in the United States, National Health and Nutrition Examination Survey 2003 to 2010. *Ann Intern Med* **160:**293–300.
17. **Backus LI, Belperio PS, Loomis TP, Mole LA.** 2014. Impact of race/ethnicity and gender on HCV screening and prevalence among U.S. veterans in Department of Veterans Affairs Care. *Am J Public Health* **104**(Suppl 4)**:**S555–S561.
18. **Chou R, Cottrell EB, Wasson N, Rahman B, Guise JM.** 2013. Screening for hepatitis C virus infection in adults: a systematic review for the U.S. Preventive Services Task Force. *Ann Intern Med* **158:**101–108.
19. **Messina JP, Humphreys I, Flaxman A, Brown A, Cooke GS, Pybus OG, Barnes E.** 2015. Global distribution and prevalence of hepatitis C virus genotypes. *Hepatology* **61:**77–87.
20. **Perz JF, Armstrong GL, Farrington LA, Hutin YJ, Bell BP.** 2006. The contributions of hepatitis B virus and hepatitis C virus infections to cirrhosis and primary liver cancer worldwide. *J Hepatol* **45:**529–538.
21. **Maasoumy B, Wedemeyer H.** 2012. Natural history of acute and chronic hepatitis C. *Best Pract Res Clin Gastroenterol* **26:**401–412.
22. **Tsoulfas G, Goulis I, Giakoustidis D, Akriviadis E, Agorastou P, Imvrios G, Papanikolaou V.** 2009. Hepatitis C and liver transplantation. *Hippokratia* **13:**211–215.
23. **Chen SL, Morgan TR.** 2006. The natural history of hepatitis C virus (HCV) infection. *Int J Med Sci* **3:**47–52.
24. **Pawlotsky JM.** 2010. More sensitive hepatitis C virus RNA detection: what for? *J Hepatol* **52:**783–785.
25. **Cacoub P, Comarmond C, Domont F, Savey L, Desbois AC, Saadoun D.** 2016. Extrahepatic manifestations of chronic hepatitis C virus infection. *Ther Adv Infect Dis* **3:**3–14.
26. **Feld JJ, Hoofnagle JH.** 2005. Mechanism of action of interferon and ribavirin in treatment of hepatitis C. *Nature* **436:**967–972.
27. **Pawlotsky JM, Feld JJ, Zeuzem S, Hoofnagle JH.** 2015. From non-A, non-B hepatitis to hepatitis C virus cure. *J Hepatol* **62** (Suppl)**:**S87–S99.
28. **Watanabe H, Enomoto N, Nagayama K, Izumi N, Marumo F, Sato C, Watanabe M.** 2001. Number and position of mutations in the interferon (IFN) sensitivity-determining region of the gene for nonstructural protein 5A correlate with IFN efficacy in hepatitis C virus genotype 1b infection. *J Infect Dis* **183:**1195–1203.
29. **Itakura J, Kurosaki M, Higuchi M, Takada H, Nakakuki N, Itakura Y, Tamaki N, Yasui Y, Suzuki S, Tsuchiya K, Nakanishi H, Takahashi Y, Maekawa S, Enomoto N, Izumi N.** 2015. Resistance-associated NS5A variants of hepatitis C virus are susceptible to interferon-based therapy. *PLoS One* **10:** e0138060.
30. **McHutchison JG, Everson GT, Gordon SC, Jacobson IM, Sulkowski M, Kauffman R, McNair L, Alam J, Muir AJ, PROVE1 Study Team.** 2009. Telaprevir with peginterferon and ribavirin for chronic HCV genotype 1 infection. *N Engl J Med* **360:**1827–1838.
31. **Keating GM.** 2014. Sofosbuvir: a review of its use in patients with chronic hepatitis C. *Drugs* **74:**1127–1146.
32. **Cardoso AC, Moucari R, Figueiredo-Mendes C, Ripault MP, Giuily N, Castelnau C, Boyer N, Asselah T, Martinot-Peignoux M, Maylin S, Carvalho-Filho RJ, Valla D, Bedossa P, Marcellin P.** 2010. Impact of peginterferon and ribavirin therapy on hepatocellular carcinoma: incidence and survival

in hepatitis C patients with advanced fibrosis. *J Hepatol* **52:**652–657.
33. **Attar BM, Van Thiel DH.** 2016. Hepatitis C virus: A time for decisions. Who should be treated and when? *World J Gastrointest Pharmacol Ther* **7:**33–40.
34. **Ansaldi F, Orsi A, Sticchi L, Bruzzone B, Icardi G.** 2014. Hepatitis C virus in the new era: perspectives in epidemiology, prevention, diagnostics and predictors of response to therapy. *World J Gastroenterol* **20:**9633–9652.
35. **Liang TJ.** 2013. Current progress in development of hepatitis C virus vaccines. *Nat Med* **19:**869–878.
36. **Selvarajah S, Busch MP.** 2012. Transfusion transmission of HCV, a long but successful road map to safety. *Antivir Ther* **17** (7 Pt B)**:**1423–1429.
37. **Bruggmann P, et al.** 2014. Historical epidemiology of hepatitis C virus (HCV) in selected countries. *J Viral Hepat* **21**(Suppl 1)**:** 5–33.
38. **Hagan H, Jarlais DC, Friedman SR, Purchase D, Alter MJ.** 1995. Reduced risk of hepatitis B and hepatitis C among injection drug users in the Tacoma syringe exchange program. *Am J Public Health* **85:**1531–1537.
39. **Glynn SA, Wright DJ, Kleinman SH, Hirschkorn D, Tu Y, Heldebrant C, Smith R, Giachetti C, Gallarda J, Busch MP.** 2005. Dynamics of viremia in early hepatitis C virus infection. *Transfusion* **45:**994–1002.
40. **Dufour DR, Talastas M, Fernandez MD, Harris B.** 2003. Chemiluminescence assay improves specificity of hepatitis C antibody detection. *Clin Chem* **49:**940–944.
41. **Kim S, Kim JH, Yoon S, Park YH, Kim HS.** 2008. Clinical performance evaluation of four automated chemiluminescence immunoassays for hepatitis C virus antibody detection. *J Clin Microbiol* **46:**3919–3923.
42. **Saludes V, González V, Planas R, Matas L, Ausina V, Martró E.** 2014. Tools for the diagnosis of hepatitis C virus infection and hepatic fibrosis staging. *World J Gastroenterol* **20:**3431–3442.
43. **Bendinelli M, Pistello M, Maggi F, Vatteroni M.** 2009. Bloodborne hepatitis viruses: Hepatitis viruses B, C, and D and candidate agents of cryptogenetic hepatitis, p 325–362. *In* Wiedbrauk DL (ed), *Specter S, Hodinka RL*, 4th ed. Clinical Virology Manual, ASM Press, Washington, DC.
44. **Martin P, Fabrizi F, Dixit V, Quan S, Brezina M, Kaufman E, Sra K, DiNello R, Polito A, Gitnick G.** 1998. Automated RIBA hepatitis C virus (HCV) strip immunoblot assay for reproducible HCV diagnosis. *J Clin Microbiol* **36:**387–390.
45. **Alter MJ, Kuhnert WL, Finelli L.** Guidelines for laboratory testing and result reporting of antibody to hepatitis C virus. Centers for Disease Control and Prevention. *MMWR Recomm Rep* 2003**52:**1–13,.
46. **Gong S, Schmotzer CL, Zhou L.** 2015. Evaluation of quantitative real-time PCR as a hepatitis C virus supplementary test after RIBA discontinuation. *J Clin Lab Anal, in press.*
47. **Veillon P, Payan C, Picchio G, Maniez-Montreuil M, Guntz P, Lunel F.** 2003. Comparative evaluation of the total hepatitis C virus core antigen, branched-DNA, and amplicor monitor assays in determining viremia for patients with chronic hepatitis C during interferon plus ribavirin combination therapy. *J Clin Microbiol* **41:**3212–3220.
48. **Hosseini-Moghaddam SM, Iran-Pour E, Rotstein C, Husain S, Lilly L, Renner E, Mazzulli T.** 2012. Hepatitis C core Ag and its clinical applicability: potential advantages and disadvantages for diagnosis and follow-up? *Rev Med Virol* **22:**156–165.
49. **Chevaliez S, Pawlotsky JM.** 2012. Virology of hepatitis C virus infection. *Best Pract Res Clin Gastroenterol* **26:**381–389.
50. **Ghany MG, Strader DB, Thomas DL, Seeff LB, American Association for the Study of Liver Diseases.** 2009. Diagnosis, management, and treatment of hepatitis C: an update. *Hepatology* **49:**1335–1374.
51. **Kohli A, Shaffer A, Sherman A, Kottilil S.** 2014. Treatment of hepatitis C: a systematic review. *JAMA* **312:**631–640.
52. **Pawlotsky JM.** 2014. New hepatitis C therapies: the toolbox, strategies, and challenges. *Gastroenterology* **146:**1176–1192.
53. **Cento V, Chevaliez S, Perno CF.** 2015. Resistance to direct-acting antiviral agents: clinical utility and significance. *Curr Opin HIV AIDS* **10:**381–389.
54. **Poveda E, Wyles DL, Mena A, Pedreira JD, Castro-Iglesias A, Cachay E.** 2014. Update on hepatitis C virus resistance to direct-acting antiviral agents. *Antiviral Res* **108:**181–191.
55. **OLYSIO Prescribing Information. Janssen Therapeutics.** Titusville, NJ, USA. Available at: http://www.olysio.com/shared/product/ olysio/prescribing-information.pdf Accessed 15 April 2016.
56. **Cento V, Mirabelli C, Salpini R, Dimonte S, Artese A, Costa G, Mercurio F, Svicher V, Parrotta L, Bertoli A, Ciotti M, Di Paolo D, Sarrecchia C, Andreoni M, Alcaro S, Angelico M, Perno CF, Ceccherini-Silberstein F.** 2012. HCV genotypes are differently prone to the development of resistance to linear and macrocyclic protease inhibitors. *PLoS One* **7:**e39652.
57. **Krishnan P, Beyer J, Mistry N, Koev G, Reisch T, DeGoey D, Kati W, Campbell A, Williams L, Xie W, Setze C, Molla A, Collins C, Pilot-Matias T.** 2015. In vitro and in vivo antiviral activity and resistance profile of ombitasvir, an inhibitor of hepatitis C virus NS5A. *Antimicrob Agents Chemother* **59:**979–987.
58. **Sarrazin C, Dragan A, Gärtner BC, Forman MS, Traver S, Zeuzem S, Valsamakis A.** 2008. Evaluation of an automated, highly sensitive, real-time PCR-based assay (COBAS Ampliprep/COBAS TaqMan) for quantification of HCV RNA. *J Clin Virol* **43:**162–168.
59. **Zepatier US Prescribing Information. Merck & Co Inc.**, Kenilworth, NJ, USA. Available at: https://www.merck.com/product/usa/pi_circulars/z/zepatier/zepatier_pi.pdf. Accessed 16 Apr 2016.
60. **Ge D, Fellay J, Thompson AJ, Simon JS, Shianna KV, Urban TJ, Heinzen EL, Qiu P, Bertelsen AH, Muir AJ, Sulkowski M, McHutchison JG, Goldstein DB.** 2009. Genetic variation in IL28B predicts hepatitis C treatment-induced viral clearance. *Nature* **461:**399–401.
61. **Tanaka Y, Nishida N, Sugiyama M, Kurosaki M, Matsuura K, Sakamoto N, Nakagawa M, Korenaga M, Hino K, Hige S, Ito Y, Mita E, Tanaka E, Mochida S, Murawaki Y, Honda M, Sakai A, Hiasa Y, Nishiguchi S, Koike A, Sakaida I, Imamura M, Ito K, Yano K, Masaki N, Sugauchi F, Izumi N, Tokunaga K, Mizokami M.** 2009. Genome-wide association of IL28B with response to pegylated interferon-alpha and ribavirin therapy for chronic hepatitis C. *Nat Genet* **41:**1105–1109.
62. **Thomas DL, Thio CL, Martin MP, Qi Y, Ge D, O'Huigin C, Kidd J, Kidd K, Khakoo SI, Alexander G, Goedert JJ, Kirk GD, Donfield SM, Rosen HR, Tobler LH, Busch MP, McHutchison JG, Goldstein DB, Carrington M.** 2009. Genetic variation in IL28B and spontaneous clearance of hepatitis C virus. *Nature* **461:**798–801.
63. **Ismail N, Fish GE, Smith MB.** 2004. Laboratory evaluation of a fully automated chemiluminescence immunoassay for rapid detection of HBsAg, antibodies to HBsAg, and antibodies to hepatitis C virus. *J Clin Microbiol* **42:**610–617.
64. **Jonas G, Pelzer C, Beckert C, Hausmann M, Kapprell HP.** 2005. Performance characteristics of the ARCHITECT anti-HCV assay. *J Clin Virol* **34:**97–103.
65. **Pittaluga F, Allice T, Abate ML, Ciancio A, Cerutti F, Varetto S, Colucci G, Smedile A, Ghisetti V.** 2008. Clinical evaluation of the COBAS Ampliprep/COBAS TaqMan for HCV RNA quantitation in comparison with the branched-DNA assay. *J Med Virol* **80:**254–260.
66. **Vermehren J, Kau A, Gärtner BC, Göbel R, Zeuzem S, Sarrazin C.** 2008. Differences between two real-time PCR-based hepatitis C virus (HCV) assays (RealTime HCV and Cobas AmpliPrep/Cobas TaqMan) and one signal amplification assay (Versant HCV RNA 3.0) for RNA detection and quantification. *J Clin Microbiol* **46:**3880–3891.
67. **Pyne MT, Hillyard DR.** 2013. Evaluation of the Roche COBAS AmpliPrep/COBAS TaqMan HCV Test. *Diagn Microbiol Infect Dis* **77:**25–30.
68. **Michelin BD, Muller Z, Stelzl E, Marth E, Kessler HH.** 2007. Evaluation of the Abbott RealTime HCV assay for

quantitative detection of hepatitis C virus RNA. *J Clin Virol* **38:**96–100.

69. **Akhavan S, Ronsin C, Laperche S, Thibault V.** 2011. Genotype 4 hepatitis C virus: beware of false-negative RNA detection. *Hepatology* **53:**1066–1067.
70. **Chevaliez S, Bouvier-Alias M, Castéra L, Pawlotsky JM.** 2009. The Cobas AmpliPrep-Cobas TaqMan real-time polymerase chain reaction assay fails to detect hepatitis C virus RNA in highly viremic genotype 4 clinical samples. *Hepatology* **49:**1397–1398.
71. **Germer JJ, Bommersbach CE, Schmidt DM, Bendel JL, Yao JD.** 2009. Quantification of genotype 4 hepatitis C virus RNA by the COBAS AmpliPrep/COBAS TaqMan hepatitis C virus test. *Hepatology* **50:**1679–1680.
72. **Halfon P, Martinot-Peignoux M, Khiri H, Marcellin P.** 2010. Quantification of genotype 4 serum samples: impact of hepatitis C virus genetic variability. *Hepatology* **52:**401.
73. **Chevaliez S, Bouvier-Alias M, Rodriguez C, Soulier A, Poveda JD, Pawlotsky JM.** 2013. The Cobas AmpliPrep/Cobas TaqMan HCV test, version 2.0, real-time PCR assay accurately quantifies hepatitis C virus genotype 4 RNA. *J Clin Microbiol* **51:**1078–1082.
74. **Pas S, Molenkamp R, Schinkel J, Rebers S, Copra C, Seven-Deniz S, Thamke D, de Knegt RJ, Haagmans BL, Schutten M.** 2013. Performance evaluation of the new Roche cobas AmpliPrep/cobas TaqMan HCV test, version 2.0, for detection and quantification of hepatitis C virus RNA. *J Clin Microbiol* **51:**238–242.
75. **Pyne MT, Mallory M, Hillyard DR.** 2015. HCV RNA measurement in samples with diverse genotypes using versions 1 and 2 of the Roche COBAS® AmpliPrep/COBAS® TaqMan® HCV test. *J Clin Virol* **65:**54–57.
76. **Vermehren J, Colucci G, Gohl P, Hamdi N, Abdelaziz AI, Karey U, Thamke D, Zitzer H, Zeuzem S, Sarrazin C.** 2011. Development of a second version of the Cobas AmpliPrep/Cobas TaqMan hepatitis C virus quantitative test with improved genotype inclusivity. *J Clin Microbiol* **49:**3309–3315.
77. **Zitzer H, Heilek G, Truchon K, Susser S, Vermehren J, Sizmann D, Cobb B, Sarrazin C.** 2013. Second-generation Cobas AmpliPrep/Cobas TaqMan HCV quantitative test for viral load monitoring: a novel dual-probe assay design. *J Clin Microbiol* **51:**571–577.
78. **Pawlotsky JM.** 2002. Use and interpretation of virological tests for hepatitis C. *Hepatology* **36**(Suppl 1)**:**S65–S73.
79. **Hraber PT, Fischer W, Bruno WJ, Leitner T, Kuiken C.** 2006. Comparative analysis of hepatitis C virus phylogenies from coding and non-coding regions: the 5′ untranslated region (UTR) fails to classify subtypes. *Virol J* **3:**103.
80. **Simmonds P, Bukh J, Combet C, Deléage G, Enomoto N, Feinstone S, Halfon P, Inchauspé G, Kuiken C, Maertens G, Mizokami M, Murphy DG, Okamoto H, Pawlotsky JM, Penin F, Sablon E, Shin-I T, Stuyver LJ, Thiel HJ, Viazov S, Weiner AJ, Widell A.** 2005. Consensus proposals for a unified system of nomenclature of hepatitis C virus genotypes. *Hepatology* **42:**962–973.
81. **Simmonds P, Smith DB, McOmish F, Yap PL, Kolberg J, Urdea MS, Holmes EC.** 1994. Identification of genotypes of hepatitis C virus by sequence comparisons in the core, E1 and NS-5 regions. *J Gen Virol* **75:**1053–1061.
82. **Murphy DG, Willems B, Deschênes M, Hilzenrat N, Mousseau R, Sabbah S.** 2007. Use of sequence analysis of the NS5B region for routine genotyping of hepatitis C virus with reference to C/E1 and 5′ untranslated region sequences. *J Clin Microbiol* **45:**1102–1112.
83. **Molenkamp R, Harbers G, Schinkel J, Melchers WJ.** 2009. Identification of two Hepatitis C Virus isolates that failed genotyping by Versant LiPA 2.0 assay. *J Clin Virol* **44:**250–253.
84. **Cento V, Landonio S, De Luca F, Di Maio VC, Micheli V, Mirabelli C, Niero F, Magni C, Rizzardini G, Perno CF, Ceccherini-Silberstein F.** 2013. A boceprevir failure in a patient infected with HCV genotype 1g: importance and limitations of virus genotyping prior to HCV protease-inhibitor-based therapy. *Antivir Ther* **18:**645–648.
85. **Martró E, González V, Buckton AJ, Saludes V, Fernández G, Matas L, Planas R, Ausina V.** 2008. Evaluation of a new assay in comparison with reverse hybridization and sequencing methods for hepatitis C virus genotyping targeting both 5′ noncoding and nonstructural 5b genomic regions. *J Clin Microbiol* **46:**192–197.
86. **Chevaliez S, Bouvier-Alias M, Brillet R, Pawlotsky JM.** 2009. Hepatitis C virus (HCV) genotype 1 subtype identification in new HCV drug development and future clinical practice. *PLoS One* **4:**e8209.
87. **Bouchardeau F, Cantaloube JF, Chevaliez S, Portal C, Razer A, Lefrère JJ, Pawlotsky JM, De Micco P, Laperche S.** 2007. Improvement of hepatitis C virus (HCV) genotype determination with the new version of the INNO-LiPA HCV assay. *J Clin Microbiol* **45:**1140–1145.
88. **Kuiken C, Simmonds P.** 2009. Nomenclature and numbering of the hepatitis C virus. *Methods Mol Biol* **510:**33–53.
89. **Benedet M, Adachi D, Wong A, Wong S, Pabbaraju K, Tellier R, Tang JW.** 2014. The need for a sequencing-based assay to supplement the Abbott m2000 RealTime HCV Genotype II assay: a 1 year analysis. *J Clin Virol* **60:**301–304.
90. **González V, Gomes-Fernandes M, Bascuñana E, Casanovas S, Saludes V, Jordana-Lluch E, Matas L, Ausina V, Martró E.** 2013. Accuracy of a commercially available assay for HCV genotyping and subtyping in the clinical practice. *J Clin Virol* **58:**249–253.
91. **Hong SK, Cho SI, Ra EK, Kim EC, Park JS, Park SS, Seong MW.** 2012. Evaluation of two hepatitis C virus genotyping assays based on the 5′ untranslated region (UTR): the limitations of 5′ UTR-based assays and the need for a supplementary sequencing-based approach. *J Clin Microbiol* **50:**3741–3743.
92. **Larrat S, Poveda JD, Coudret C, Fusillier K, Magnat N, Signori-Schmuck A, Thibault V, Morand P.** 2013. Sequencing assays for failed genotyping with the versant hepatitis C virus genotype assay (LiPA), version 2.0. *J Clin Microbiol* **51:**2815–2821.
93. **Mallory MA, Lucic DX, Sears MT, Cloherty GA, Hillyard DR.** 2014. Evaluation of the Abbott realtime HCV genotype II RUO (GT II) assay with reference to 5'UTR, core and NS5B sequencing. *J Clin Virol* **60:**22–26.
94. **Andonov A, Chaudhary RK.** 1995. Subtyping of hepatitis C virus isolates by a line probe assay using hybridization. *J Clin Microbiol* **33:**254–256.
95. **De Keukeleire S, Descheemaeker P, Reynders M.** 2015. Potential risk of misclassification HCV 2k/1b strains as HCV 2a/2c using VERSANT HCV Genotype 2.0 assay. *Diagn Microbiol Infect Dis* **82:**201–202.
96. **Cai Q, Zhao Z, Liu Y, Shao X, Gao Z.** 2013. Comparison of three different HCV genotyping methods: core, NS5B sequence analysis and line probe assay. *Int J Mol Med* **31:**347–352.
97. **Ciotti M, Marcuccilli F, Guenci T, Babakir-Mina M, Chiodo F, Favarato M, Perno CF.** 2010. A multicenter evaluation of the Abbott RealTime HCV Genotype II assay. *J Virol Methods* **167:**205–207.
98. **Noppornpanth S, Sablon E, De Nys K, Truong XL, Brouwer J, Van Brussel M, Smits SL, Poovorawan Y, Osterhaus AD, Haagmans BL.** 2006. Genotyping hepatitis C viruses from Southeast Asia by a novel line probe assay that simultaneously detects core and 5′ untranslated regions. *J Clin Microbiol* **44:**3969–3974.
99. **Shinol RC, Gale HB, Kan VL.** 2012. Performance of the Abbott RealTime HCV Genotype II RUO assay. *J Clin Microbiol* **50:**3099–3101.
100. **Verbeeck J, Stanley MJ, Shieh J, Celis L, Huyck E, Wollants E, Morimoto J, Farrior A, Sablon E, Jankowski-Hennig M, Schaper C, Johnson P, Van Ranst M, Van Brussel M.** 2008. Evaluation of Versant hepatitis C virus genotype assay (LiPA) 2.0. *J Clin Microbiol* **46:**1901–1906.
101. **Mokhtari C, Ebel A, Reinhardt B, Merlin S, Proust S, Roque-Afonso AM.** 2016. Characterization of Samples identified as hepatitis C virus Genotype 1 without subtype by

Abbott RealTime HCV Genotype II assay using the new Abbott HCV Genotype Plus RUO Test. *J Clin Microbiol* **54:** 296–299.

102. **Sam SS, Steinmetz HB, Tsongalis GJ, Tafe LJ, Lefferts JA.** 2013. Validation of a solid-phase electrochemical array for genotyping hepatitis C virus. *Exp Mol Pathol* **95:**18–22.
103. **Khuroo MS, Khuroo NS, Khuroo MS.** 2015. Diagnostic accuracy of point-of-care tests for hepatitis C virus infection: a systematic review and meta-analysis. *PLoS One* **10:**e0121450.
104. **Nyirenda M, Beadsworth MB, Stephany P, Hart CA, Hart IJ, Munthali C, Beeching NJ, Zijlstra EE.** 2008. Prevalence of infection with hepatitis B and C virus and coinfection with HIV in medical inpatients in Malawi. *J Infect* **57:**72–77.
105. **Smith BD, Jewett A, Drobeniuc J, Kamili S.** 2012. Rapid diagnostic HCV antibody assays. *Antivir Ther* **17**(7 Pt B)**:** 1409–1413.
106. **Smith BD, Teshale E, Jewett A, Weinbaum CM, Neaigus A, Hagan H, Jenness SM, Melville SK, Burt R, Thiede H, Al-Tayyib A, Pannala PR, Miles IW, Oster AM, Smith A, Finlayson T, Bowles KE, Dinenno EA.** 2011. Performance of premarket rapid hepatitis C virus antibody assays in 4 national human immunodeficiency virus behavioral surveillance system sites. *Clin Infect Dis* **53:**780–786.
107. **Daniel HD, Abraham P, Raghuraman S, Vivekanandan P, Subramaniam T, Sridharan G.** 2005. Evaluation of a rapid assay as an alternative to conventional enzyme immunoassays for detection of hepatitis C virus-specific antibodies. *J Clin Microbiol* **43:**1977–1978.
108. **Scheiblauer H, El-Nageh M, Nick S, Fields H, Prince A, Diaz S.** 2006. Evaluation of the performance of 44 assays used in countries with limited resources for the detection of antibodies to hepatitis C virus. *Transfusion* **46:**708–718.
109. **Chevaliez S, Poiteau L, Rosa I, Soulier A, Roudot-Thoraval F, Laperche S, Hézode C, Pawlotsky JM.** 2016. Prospective assessment of rapid diagnostic tests for the detection of antibodies to hepatitis C virus, a tool for improving access to care. *Clin Microbiol Infect* **22:**459.e1–459.e6.
110. American Association for the Study of Liver Diseases, 2016. HCV Guidance: Recommendations for Testing, Managing, and Treating Hepatitis C Available at: http://hcvguidelines.org/full-report-view. Accessed March, 15.

Herpes Simplex Viruses and Varicella Zoster Virus

SCOTT H. JAMES AND MARK N. PRICHARD

26

VIRAL CLASSIFICATION AND BIOLOGY

The herpesviruses are classified in the family *Herpesviridae* based on the characteristic large linear double-stranded DNA genome packaged within an enveloped icosahedral capsid. Members of this family have been identified in almost all animal species, and the specific herpesviruses are usually restricted to a single species. Nine human herpesviruses have been described: HSV-1, HSV-2, varicella zoster virus (VZV), Epstein-Barr virus (EBV), human cytomegalovirus (CMV), human herpesvirus 6A (HHV-6A), human herpesvirus 6B (HHV-6B), human herpesvirus 7 (HHV-7), and human herpesvirus 8 (HHV-8). All these viruses are structurally similar and encode a large number of biosynthetic enzymes involved in the synthesis of viral DNA as well as structure proteins that compose the capsid and tegument, as well as envelope glycoproteins that are required for infection.

These similarities, however, belie the biological diversity among the human herpesviruses, and they are further classified into three subfamilies. The *Alphaherpesvirinae* includes HSV-1, HSV-2, and VZV, all of which exhibit a relatively short replication cycle in primary human fibroblast cells and the ability to establish latent infections in sensory ganglia. *Betaherpesvirinae* replicate more slowly and establish latency in secretory glands and myeloid precursor cells and include CMV, HHV-6A, HHV-6B, and HHV-7. The *Gammaherpesvirinae*, EBV and HHV-8, generally replicate in lymphoblastoid cells and establish latent infections in lymphoid tissue. One common theme among all the herpesviruses is the establishment of lifelong infections that remain largely quiescent, yet become problematic in immunocompromised hosts. In this chapter, we will consider the *Alphaherpesvirinae* subfamily—HSV-1, HSV-2, and VZV.

Biological Characteristics of HSV

The replication of both HSV-1 and HSV-2 involves complex interactions with the host cell at the molecular level that drives an efficient replication cycle. The virus then targets sensory ganglia where it establishes and maintains an exquisitely controlled latent infection in which the viral genome remains largely silent and undetected by the host cell (1). Lytic infection initiates when viral glycoproteins in the viral envelope interact with receptors on the outer membrane of the host cell, resulting in fusion events that deliver the capsid into the cytosol. Translocation along microtubules conducts nucleocapsids to nuclear pores where they dock and release the viral genome into the nucleus. A temporally regulated cascade of transcription ensues, resulting in the expression of three kinetic classes of viral proteins that are sufficient to direct the replication of the viral genome in the nuclear replication compartment. Additional viral enzymes are required to cleave and package unit-length genomes from the newly synthesized DNA into newly assembled capsids. The nuclear lamina is then disrupted by viral-specific functions to release newly formed nucleocapsids into the cytoplasm where tegument proteins are acquired. Mature nucleocapsids are then enveloped in modified membranes of host origin in the golgi where viral glycoproteins are glycosylated to their mature form. Mature virions then egress from the host cell.

The replication of the viral DNA is of particular importance because this step is targeted by all the approved therapies for HSV infections. The genome of HSV is both large (152 kb) and complex, with unique long (UL) and unique short (US) segments bounded by a series of inverted repeats, such that several viral proteins are required not only to synthesize DNA but also to direct recombination, inversion, and concatamerization events that must occur prior to packaging of unit-length genomes (2). Both the UL and US segments invert during packaging resulting in four distinct isomers of the genome. DNA synthesis during lytic virus replication occurs in two distinct phases that require different viral gene products (3). The first phase of DNA synthesis begins with initiation events at three origins of replication and requires seven viral proteins: the DNA polymerase, its accessory processivity factor (UL42), the origin binding protein (UL9), the major single-stranded DNA binding protein (ICP8, UL29), and three subunits of the helicase-primase complex consisting of the UL5, UL52 subcomplex, and UL8. Each of these proteins performs basic functions that are conserved in other viruses as well as *Escherichia coli* and *Saccharomyces cerevisiae* (4).

Biological Characteristics of VZV

The genomic structure and replication cycle of VZV shares many characteristics with the other alphaherpesviruses, HSV-1 and HSV-2. Significant differences are observed, however, in the biological characteristics of the virus. One

doi:10.1128/9781555819156.ch26

important distinction involves altered replication characteristics in cell culture. First, VZV exhibits a restricted host cell range in cell culture that includes human and simian cells and a modestly extended replication cycle. More importantly, cell culture systems do not accurately reproduce the efficient assembly and release of progeny virus seen in the human host. One consequence of this characteristic is that the virus propagates rather slowly in cell culture by cell-to-cell spread, and the progeny virus is almost entirely cell associated. Virions that are released from the cell are poorly infectious and unstable, so the virus is generally maintained in infected cells. This aspect is critical to keep in mind when culturing and maintaining virus stocks in the diagnostic laboratory because the mishandling of specimens can result in precipitous drops in infectivity.

Replication of the viral genome has been well studied because it is targeted by all the approved therapies for VZV infections. The VZV genome is similar to that of HSV and is approximately 125 kb in length with UL and US segments bounded by repeat sequences. The inversion of the segments also results in four isomers, but two of the four isomers are underrepresented. The coding content is also comparable to HSV with 60 of the 66 genes having clear homologs in HSV. Of the six VZV unique genes, five are dispensable for replication. VZV encodes the core 40 genes that it shares with all the human herpesviruses, including clear homologs of the gene products required for DNA replication in HSV. This includes the ORF28 DNA polymerase and its accessory protein ORF16, the origin binding protein (ORF51), the ssDNA binding protein (ORF29), and three components of the helicase-primase complex encoded by ORF55, ORF6, and ORF52. Newly synthesized viral DNA is then cleaved and packaged into newly formed capsids in a concerted process that requires viral-specific proteins. The nucleocapsids egress the nucleus, tegument proteins are acquired, and the virus is enveloped in a process that is similar to HSV. The maturation and egress of VZV is orders of magnitude less efficient than HSV, but the precise differences that result in reduced infectivity are poorly understood.

EPIDEMIOLOGY

Herpes Simplex Viruses

Humans are the only known natural reservoir of HSV. Infections with HSV-1 and HSV-2 are common in both developed and undeveloped countries worldwide (5). Neither virus displays distinct seasonal variation in their incidence of infection. Acquisition of HSV results in lifelong infection with the establishment of latency in sensory neural ganglia followed by periodic clinical or subclinical reactivation of viral replication. Viral shedding occurs with both symptomatic and asymptomatic reactivation, a feature that contributes significantly to the spread of infection since transmission occurs by close contact with infected body fluids such as saliva, mucosal secretions, or vesicle fluid.

Surveillance of seroprevalence indicates that earlier acquisition of infection is seen with HSV-1 as compared to HSV-2, and in persons of lower socioeconomic status for both HSV-1 and HSV-2 (6, 7). The overall prevalence of HSV antibodies increases with age, indicating ongoing exposure to these viruses within the population. Over 90% of adults have acquired HSV-1 infection by their fifth decade of life, although only a minority develop clinically apparent disease at the time of acquisition (8). Both HSV-1 and HSV-2 can cause genital infection, but HSV-1 has typically been more associated with orolabial lesions, whereas HSV-2 has historically been associated with genital lesions. More recently, however, HSV-1 has become the more prevalent cause of genital herpes, responsible for 60% to 80% of genital herpetic infections in certain populations of young women (9, 10).

Previous studies indicated an increasing trend in HSV-2 seroprevalence rates in developed countries (11, 12) but more recent seroepidemiologic studies from the United States have demonstrated otherwise. Specifically, among persons 14 to 49 years old, the seroprevalences of HSV-1 and HSV-2 were approximately 58% and 17%, respectively, during the period spanning from 1999 to 2004, whereas a follow-up study from 2005 to 2010 showed that HSV-1 seroprevalence had declined to 54% and HSV-2 seroprevalence had not significantly changed (16%) (13, 14).

When considering HSV infections, the terminology of *first-episode primary*, *first-episode nonprimary*, and *recurrent* infections is commonly used. A *first-episode primary infection* indicates that an individual with no prior antibody to HSV-1 or -2 has newly acquired either virus. In contrast, a *first-episode nonprimary infection* occurs when a person with preexisting antibody to HSV-1 acquires an HSV-2 infection (or vice versa). Viral reactivation from latency and ensuing antegrade translocation of virus from sensory neural ganglia to skin and mucosal surfaces produces a *recurrent infection*.

Varicella Zoster Virus

Primary infection with VZV results in the cutaneous manifestation known as varicella or chickenpox. VZV is highly communicable by respiratory droplets as well as by direct contact with skin lesions, as demonstrated by attack rates as high as 30% in classroom settings and 90% in household contacts (15). After primary infection, VZV establishes lifelong latency in cranial nerves and dorsal root ganglia. Viral reactivation can occur decades later in the form of zoster, or shingles, a painful rash most commonly limited to a single dermatome. Zoster occurs in as many as 20% of individuals who have been previously infected with VZV. Its incidence increases with age and in individuals with impaired cell-mediated immunity (16).

Varicella can occur throughout the year but is more common in the late winter and spring, whereas zoster demonstrates no seasonal variability. Prior to the advent of vaccination against VZV infection, varicella was a near-universal childhood disease in the United States, as evidenced by antibodies being present in more than 90% of adults. The introduction of the varicella vaccine in 1995 has led to a significant reduction in the burden of VZV infection in the United States, with a greater than 90% reduction in the incidence of primary infection in childhood and no concomitant shift of disease burden to older populations (17).

CLINICAL SIGNIFICANCE

Herpes Simplex Viruses

Infections due to HSV can result in a wide spectrum of disease, ranging from asymptomatic or uncomplicated mucocutaneous involvement to severe disease involving multisystem dissemination or penetration into the central nervous system (CNS). The most common types of HSV infection are primary and recurrent infections of the orolabial or genital mucosa. Less frequent, but potentially-life threatening manifestations of HSV infection include herpes simplex encephalitis (HSE) and vertically transmitted neonatal infection. HSV can also cause serious ocular infections that have the potential to cause corneal scarring and vision loss.

Primary or first-episode nonprimary HSV mucocutaneous infections are often asymptomatic. Orolabial infections may also present with extensive oral involvement (gingivostomatitis), which is more often seen in young children, or with pharyngitis, which is more typical in adolescents and adults. Genital HSV infections can also present symptomatically with lesions most commonly found on the vulva, labia, vaginal introitus, or cervix in women and on the glans or shaft of the penis in men. Perianal involvement and lesions on the inner thigh are also possible with genital HSV infections. Cutaneous lesions typically present as painful erythematous papules that progress to vesicles of clear fluid. These fragile vesicles usually burst, but if they do not, pustules may develop due to an influx of inflammatory cells. After rupturing, lesions transition into shallow ulcers on an erythematous base. Lesions located on mucosal surfaces typically do not form vesicles but progress directly to ulcerations. Lesions occurring during recurrent infection are clinically indistinguishable from those that occur with primary or first-episode infections.

The total healing process for mucocutaneous lesions can last for as few as 7 to 10 days or as long as 21 days. A more severe illness with aseptic meningitis or systemic dissemination can also occur during primary or first-episode HSV infections, but these complications are rare in immunocompetent hosts. Immunocompromised hosts are subject to more frequent episodes of mucocutaneous reactivation, increased severity, prolonged duration of symptoms and viral shedding, and a greater risk of systemic dissemination.

Other cutaneous manifestations outside the orolabial and genital regions can also occur. HSV infection of the nail bed or tip of a finger is known as herpetic whitlow and is most common in health care professionals, but can also occur via autoinoculation in patients with orolabial or genital infections. Herpes gladiatorum indicates a cutaneous HSV infection transmitted during contact sports such as wrestling. Patients with skin breakdown due to eczema are at risk for an HSV infection known as eczema herpeticum.

Primary ocular HSV infections are more commonly caused by HSV-1 and are often asymptomatic or have self-limited, superficial involvement in the form of blepharitis, conjunctivitis, or keratitis. Reactivation from latency in the trigeminal ganglia can lead to disease recurrence in the region innervated by the infected dermatome and may involve tissues that were not affected by the primary infection, such as the cornea (herpetic keratitis).

HSE, usually due to HSV-1, is the most common cause of sporadic encephalitis in the United States. One-third of all HSE cases occur in patients less than 20 years old, and typical clinical presentations include fever, altered mental status, and focal neurologic symptoms (18). Hemorrhagic necrosis is characteristic of disease and is most typically localized to the temporal lobe. The mortality rate of untreated HSE exceeds 70%, with impaired neurologic outcomes in virtually all survivors.

Neonatal HSV infection occurs in an estimated 1 in 3,200 deliveries in the United States (19). Because of the increasing incidence of HSV-1 genital infections, the majority of neonatal HSV infections in many parts of the world now are caused by HSV-1 (20, 21). Mothers with recurrent genital HSV lesions pose less of a risk for transmission to an exposed neonate than do mothers with primary or first-episode nonprimary infections, likely due to the transplacental passage of protective antibodies. Mother-to-child transmission of HSV can occur during one of three time periods: *in utero* (5%), during delivery (85%), or postnatally via direct contact with an orolabial or other cutaneous lesion (10%) (18).

In utero (congenital) transmission of HSV is exceedingly rare and is characterized by a triad of cutaneous, neurologic, and ocular manifestations present at the time of birth. Neonatal HSV infection acquired perinatally or postnatally is categorized as skin, eye, and/or mouth disease, CNS disease, or disseminated disease. The natural history of neonatal HSV infections involves significant morbidity and mortality, particularly in CNS and disseminated disease.

Varicella Zoster Virus

Varicella (chickenpox) results from primary VZV infection and manifests with prodromal symptoms such as fever, headache, and malaise followed by the onset of a pruritic maculopapular rash that quickly progresses to vesicular lesions. Within 1 to 2 days of their appearance, most vesicles will have crusted over, leading to the development of scabbed lesions over the next few days. Lesions occur in successive crops on the face, trunk, and extremities such that patients often have concomitant lesions in different stages of development. New crops of lesions can continue to occur for up to 7 days. Secondary bacterial infections are the most common complication of varicella, but neurologic manifestations such as encephalitis and cerebellitis can also occur. Varicella in nonimmune adults is typically more severe than in children and may be associated with interstitial pneumonia up to 30% of the time. Primary varicella acquired during pregnancy can be associated with significant morbidity and mortality in both the mother and the fetus, including maternal pneumonia, spontaneous abortion, preterm labor, congenital varicella (infection of the fetus *in utero*), and neonatal varicella (transmission of VZV to a newborn whose mother develops varicella around the time of delivery).

Zoster (shingles) is a neurocutaneous reactivation of latent VZV and is characterized by a unilateral painful vesicular eruption in a dermatomal distribution. Eruptions are often heralded by pruritus and intense pain followed by the appearance of a maculopapular rash within 2 to 3 days. The maculopapular lesions quickly progress to vesicles, which may sometimes coalesce into small bullae. Crops of new lesions can develop within the same dermatome over the course of 3 to 5 days, ultimately crusting over and resolving over a period of about 2 weeks in most healthy patients. Ocular involvement is possible (herpes zoster ophthalmicus). Immunocompromised hosts may have a prolonged and more severe course, which may be complicated by cutaneous or visceral dissemination. The most common complication of zoster is postherpetic neuralgia, which is the development of chronic pain after the resolution of cutaneous lesions. The pain associated with postherpetic neuralgia can be constant or intermittent, can last for months to years, and in some cases can be debilitating.

TREATMENT AND PREVENTION

Current antiviral therapies for the treatment of both HSV and VZV infections all target the DNA polymerase. Specific pathways that yield active metabolites differ, yet they all ultimately inhibit the viral DNA synthesis by interfering with the viral DNA polymerases (Fig. 1). Acyclovir (ACV)

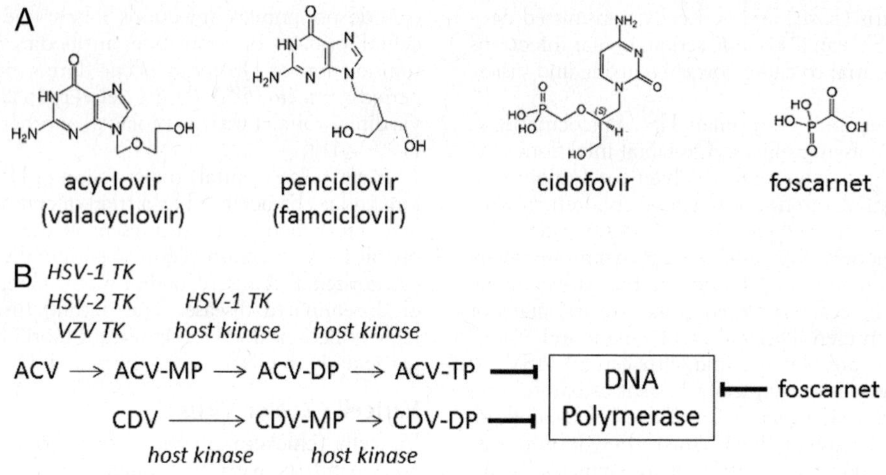

FIGURE 1 Antiviral therapies for HSV and VZV. (A) Therapies for the treatment of HSV and VZV are shown with prodrugs shown in parentheses. The pyrophosphate analog foscarnet is approved for resistant therapies. Cidofovir is also active against resistant isolates, but it has not been approved for that indication. (B) Acyclovir, the therapy of choice, is initially phosphorylated exclusively by the TK enzymes encoded by HSV-1, HSV-2, and VZV to the level of the monophosphate (ACV-MP). Cellular kinases further phosphorylate the drug to the active triphosphate metabolite (ACV-TP) that inhibits the viral DNA polymerase. Most drug-resistant viruses contain mutations in the TK that impair its ability to phosphorylate the drug and limits the concentration of the active metabolite. Neither foscarnet nor cidofovir require the TK such that both drugs are active against drug-resistant isolates of the virus.

was discovered 40 years ago and remains the therapy of choice for both viruses. The development of ACV and the orally bioavailable prodrug, valacyclovir (VACV), proved to be highly effective in the treatment of HSV and VZV infections (22, 23). A second therapy, famciclovir (FCV), is also highly effective against HSV infections, yet ACV remains the therapy of choice for serious infections. Since the advent of acyclic nucleoside analogs, the management of HSV infections has become both more successful and less toxic. Although second-line antiviral therapies for treatment-resistant HSV and VZV infections are available, they are of modest clinical utility because of associated toxicities (24). The pyrophosphate analog foscarnet (phosphonoformic acid, PFA) and the acyclic nucleoside phosphonate analog, cidofovir (CDV) generally remain active against ACV-resistant infections, but both must be administered parenterally and are associated with significant toxicities.

Mechanism of Action of Therapies for HSV Infections

It is important to understand the mechanism of action of therapies used to treat HSV infections because resistant infections occur in up to 13% of high-risk populations. The acyclic deoxyguanosine analog, ACV must be phosphorylated initially to the active triphosphate metabolite that acts as a suicide inhibitor of the viral DNA polymerase (Fig. 1). Once ACV enters an infected cell, it is phosphorylated initially by the HSV thymidine kinase (TK) encoded by the UL23 gene of HSV (25). Once this occurs, cellular kinases further phosphorylate the drug to yield the active ACV-triphosphate metabolite, which competes with deoxyguanosine triphosphate for incorporation into the elongating viral DNA strand. Since it lacks a 3′ hydroxyl analog, it acts as an obligate chain terminator. The orally bioavailable L-valyl ester prodrug of ACV, VACV, acts by a similar mechanism but has improved bioavailability and pharmacokinetic properties.

The drug FCV is a diacetylester prodrug form of penciclovir (PCV) and has a mechanism of action that is similar to ACV (26). Once FCV is converted to PCV, the nucleoside analog has a very high affinity for the viral TK and is rapidly phosphorylated to the monophosphate. It is then further phosphorylated by cellular enzymes to PCV-triphosphate metabolite that competitively inhibits the viral DNA polymerase. This does not necessarily result in chain termination, however, because it has the equivalent of a 3′ hydroxyl group on its acyclic side chain, which can allow for a limited amount of continued chain elongation.

Resistance to both drugs described above is primarily due to mutations in the viral TK enzymes, which result in greatly reduced phosphorylation of the drug to the active metabolite (24). Thus, most resistant infections are resistant to FCV, ACV, and ganciclovir, which is occasionally used off label for HSV infections. Mutations in the DNA polymerase can also impart resistance to the drugs but is much less common than mutations in the TK.

The pyrophosphate analog PFA also targets the viral DNA polymerase, but the mechanism of action is distinct from that of the nucleoside analogs (Fig. 1). This compound interacts directly with the pyrophosphate binding site of the DNA polymerase with a high affinity and inhibits the enzymatic activity of the enzyme (27). Importantly, it does not require phosphorylation by the viral TK, such that it remains active against most ACV-resistant isolates that exhibit reduced TK activity. It is approved for the therapy of resistant infections.

The acyclic nucleoside phosphonate analog CDV is a deoxycytidine monophosphate analog and is approved for the therapy of CMV retinitis, although it is often used off label to treat ACV-resistant HSV infections that do not respond to PFA (28). Since this compound is a monophosphate analog, it does not require an initial phosphorylation by the TK and undergoes two additional phosphorylation steps by cellular kinases to the active CDV-diphosphate metabolite, which is a deoxycytidine triphosphate analog (Fig. 1) (29). Since the drug is not phosphorylated by the viral TK, it typically retains full antiviral activity against ACV-resistant isolates. The incorporation of CDV-diphosphate into the nascent strand of DNA does not necessarily result in termination because of a second hydroxyl on the acyclic sugar moiety that allows further chain elongation. However, it alters the conformation of the DNA strand, particularly if two molecules are sequentially incorporated, making it a poor template for further DNA synthesis. Incorporation of CDV into viral DNA also renders the DNA refractory to amplification resulting in reduced sensitivity of nucleic acid amplification tests (NAATs) in specimens from patients treated with this drug.

Prevention of VZV Infection and Mechanism of Action of Therapies for VZV Infections

Vaccination is the best means to protect against disease associated with VZV infection. The Varivax live attenuated vaccine has been proven to be both safe and effective and has been recommended for routine use in children since 1995 (17). Since the recommendation of a second dose of the vaccine at age 12, the effectiveness against varicella has been estimated to be as high as 98% (30). Continued studies with the vaccine over an extended time frame will be required to gauge a potential impact on zoster. Protection against herpes zoster is also afforded by the same attenuated virus strain in the Zostavax vaccine and was approved for immunocompetent adults over 60 years of age in 2006. This vaccine is also safe and reduces both the burden of illness and the incidence of postherpetic neuralgia by more than 60% (31).

Prodrugs of ACV and PCV remain the therapy of choice for VZV infections (32). Systemic VZV infections in immunocompromised hosts have been shown to respond to ACV and it also reduces the incidence of dissemination and visceral disease in children with primary varicella (33). The mechanism of inhibition against VZV is similar to that described for HSV with minor differences. Both drugs are initially phosphorylated by the VZV TK homolog, although the active triphosphate metabolite of PCV is more stable than ACV-triphosphate and accumulates to higher intracellular levels. However, both drugs exhibit similar efficacy *in vitro* because the VZV DNA polymerase has a lower affinity for PCV-triphosphate than ACV-triphosphate. Both of these therapies limit the natural course of the disease, although their effect on attendant postherpetic pain is modest. Resistance to these drugs is rare and is typically detected only in immunocompromised hosts (34). Resistance typically maps to the viral TK that reduces the phosphorylation of the compound resulting in very low levels of the active metabolite. As with HSV, resistant infections can be treated successfully with either PFA or CDV, although their dose-limiting toxicities and lack of oral bioavailability limits their usefulness.

DETECTION/DIAGNOSIS

Herpes Simplex Viruses

Diagnosis of HSV infections is straightforward because the viruses are readily cultured and assays for the detection of antibodies, antigens, and viral DNA are highly sensitive. An algorithm for the evaluation of specimens sent for HSV or VZV testing is shown schematically in Figure 2. Clinical presentation consistent with HSV infection requires laboratory evaluation to confirm the etiology of the infection. Primary infections are typically evaluated by virus culture followed by detection of antigen by antibodies to type common antigens, or to type specific antigens to distinguish between HSV-1 and HSV-2 infections. NAAT-based assays are also capable of rapidly diagnosing infections and are increasingly used as costs are becoming more competitive with viral culture. Serology can also confirm primary infections, but is slow given the time required to mount an immune response to the infection. Recurrent infections complicate the serological diagnosis of infection, but they can be characterized by the presence of a robust IgG response with negligible levels of IgM antibody. Finally, neonatal infections and CNS infections are life threatening and require immediate therapy with ACV while diagnostic tests to detect viral DNA in cerebrospinal fluid (CSF) are performed. Over time, NAAT technologies will predominate over other assays as they become cheaper and easier to perform in the diagnostic laboratory.

Preanalytical

One important consideration in the diagnosis of HSV infections is specimens from patients on suppressive therapies for the infection, which can interfere with virus culture or NAAT in some cases. Swab specimens, mucosal secretions, CSF, and tear specimens among others may be acceptable specimens depending on how well methods have been characterized in the diagnostic laboratory. In neonates or when CNS infections are suspected, CSF is the preferred specimen type, and detection of viral DNA by NAAT remains the most sensitive method for the detection of encephalitis (35). Furthermore, HSV DNA can also be detected in whole blood in individuals with primary and recurrent infection although its prognostic value has not yet been established.

Methods used to collect specimens also require consideration. Acceptable swabs can be cotton, polyester, or rayon with a plastic shaft since both Dacron swabs and wooden shafts can inhibit some NAAT reactions. In most cases, swab specimens are collected from sites such as skin, lip, oral, tongue, throat, nasopharyngeal, genital, eye/conjunctiva, or anal/rectal/perineum. Ideally, fluid-filled lesions should be disrupted prior to collection and smaller, less-developed lesions are preferable because they contain more infectious virus. Swabs should be placed in universal transport medium and stored at temperatures 2 to 8°C if the test can be performed within 24 h, or frozen at −20°C if testing will be delayed. Specimens kept at ambient temperature for periods of <8 h may also be acceptable but should be avoided because the sensitivity of downstream assays may be reduced. CSF specimens should be collected in a sterile tube without the addition of transport media and should be stored as described above for swab specimens. Whole blood specimens should be collected in EDTA (purple top) tubes and should be maintained at 2 to 8°C. Freezing should be avoided prior to evaluation of these specimens. Serum can be collected in either gold-topped or plain red-topped tubes and stored under conditions described for swab specimens.

Analytical

Culture

Both HSV-1 and HSV-2 replicate well in many cell types, and virus culture remains an established and reliable means to diagnose both oral and genital lesions. A swab specimen in universal transport media or equivalent should be placed on ice until it can be cultured in the diagnostic laboratory. Shell vials containing monolayers of primary human fibroblasts or mixtures of African green monkey kidney cells are acceptable substrates for this assay. Since HSV replicates

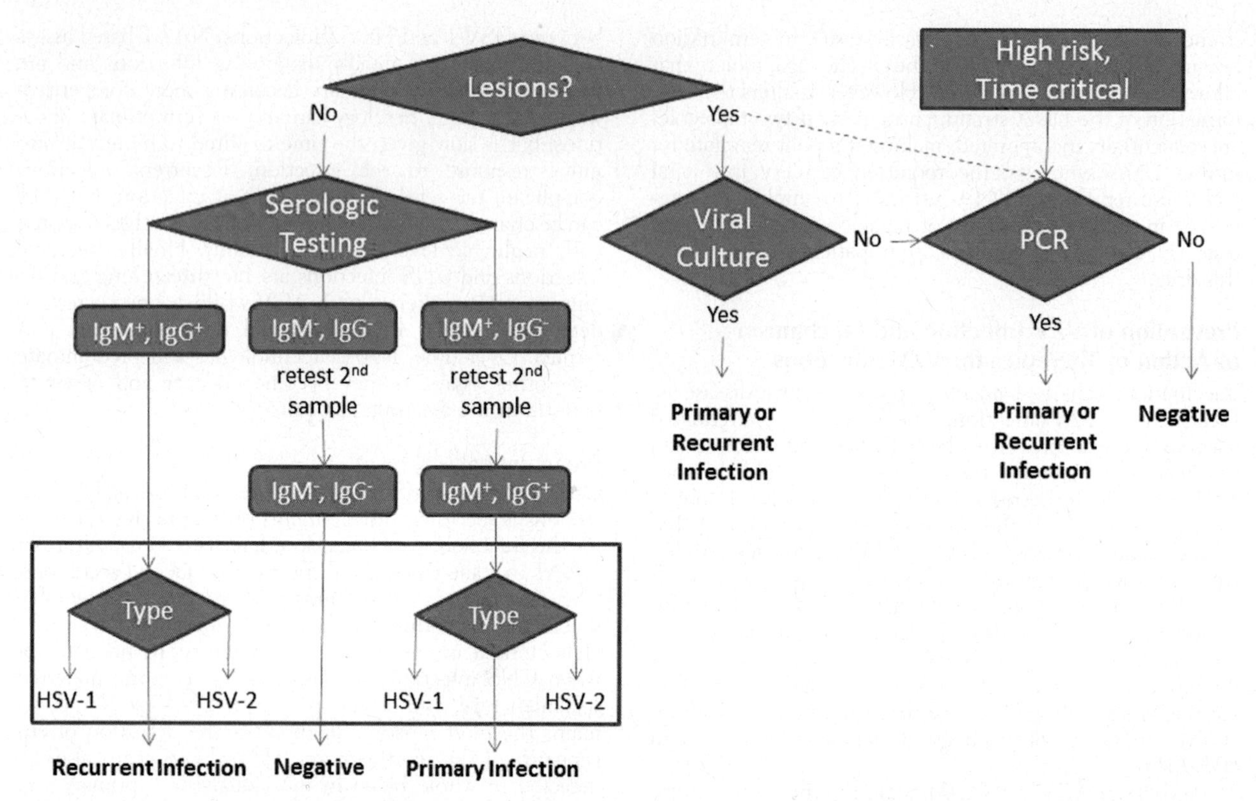

FIGURE 2 Algorithm for the diagnosis of HSV and VZV infections. If no lesions are present and time is not a critical factor, serological testing can identify patients infected with HSV or VZV. Further testing for HSV shown in the box can type the infection as being HSV-1 or HSV-2. If lesions are present, swab specimens can be cultured, and infected cultures can be further tested by fluorescent antibody to identify the infection as being HSV or VZV. For high-risk or time-critical specimens (or as an alternative to viral culture), PCR of specimens can be used to identify DNA from HSV or VZV. Many standard assays also distinguish between HSV-1 and HSV-2 at this stage.

with more rapid kinetics than VZV, visual cytopathology at 24 hours following inoculation is often observed with swabs from herpetic lesions. This still requires confirmation by other methods since swab specimens from lesions from patients with either varicella or zoster can also contain large quantities of virus that can also induce detectable cytopathology early in viral culture. When cytopathology is observed, or at day 7 following inoculation, viral antigens can be detected with HSV specific antibodies to confirm the identity of the infecting virus. Engineered cell lines and assays such as enzyme-linked virus-inducible system (ELVIS), are also effective in identifying HSV infection. This cell line uses a promoter with a specific response to HSV transcriptional factors produced as during HSV replication to drive the expression of β-galactosidase. The addition of the appropriate substrate for this enzyme results in the HSV-infected cells turning blue. (36).

Antibody

Serological testing is a proven technology to identify individuals infected with HSV-1 or HSV-2. Commercial assays, such as HerpeSelect-2 (Focus Technologies, Cypress, CA), are available and use purified antigens for each virus to diagnose infection and distinguish between the two viruses. Most assays utilize differences in the immune response to glycoprotein G (gG) of the two viruses with sensitivity and specificity reported to be in the range of 90% to 100% (37). Ideally, point-of-care assays could be used to identify those infected with the virus, yet many diagnostic laboratories do not perform the assay regularly and can report incorrect results from 46% to 83% of specimens.

Nucleic Acid Amplification Tests

With the advent of NAAT methods, detection of HSV DNA rapidly became the diagnostic method of choice for CSF specimens when HSE is suspected (38, 39). Acceptable specimens will vary somewhat between platforms. Generally, preferred specimens are swabs in universal transport media, CSF specimens, tear strips, or culture supernatants held at 2 to 6°C for <24 h. Well-characterized laboratory developed real-time PCR tests to detect viral DNA in specimens, such as mucosal secretions and CSF, have proven to be both more rapid and more sensitive to the "gold standard" consisting of culture of HSV followed by detection of specific viral antigen (40). Thus, NAAT diagnostic approaches should result in more rapid diagnosis and clinical decision-making compared to traditional culture-based approaches. For ocular infections, NAAT is also the standard of care and can distinguish between HSV, VZV, adenovirus, and *Chlamydia trachomatis* (41).

Many commercially available platforms and NAAT assays are available and have FDA-cleared assays for the detection of HSV-1 and HSV-2. The BD ProbeTec HSV Q^x assay (42), Focus Simplexa HSV 1&2 Direct assay (43), Luminex Aries HSV 1&2 assay, Lyra Direct HSV 1+2/VZV Assay (44, 45), Roche cobas 4800 HSV-1 HSV-2 Test, and the Seegene Anyplex HSV-1/2 Assay can all provide rapid and highly sensitive detection of HSV DNA (42).

While NAATs are not currently possible in most diagnostic laboratories, they offer significant advantages in terms of sensitivity and turnaround time to culture-based methods. The development of new "sample-to-answer" technologies have made monumental strides in recent years and future studies may provide support for NAAT performance in most diagnostic laboratories or even the point of care. Advances in this technology are welcome news to all who engage in diagnostic testing, yet it is important to be cognizant of the role of viral culture to monitor changes in the phenotypic characteristics of virus isolates over time and in infections that are refractory to antiviral therapy (46).

Postanalytical

As with all diagnostic assays, timely results from diagnostic assays are important to reduce turnaround time and facilitate rapid clinical decision making. Because of the potential for life-threatening infections with HSV, it is essential to notify the attending physician immediately with critical results such as the detection of HSV DNA in infants or any CSF specimen.

Varicella Zoster Virus

The laboratory diagnosis of VZV infection is important since it can cause rare life-threatening infections and disease manifestations that can, in some cases, be very similar to those of HSV. Generally, clinical presentation of lesions consistent with varicella or zoster is sufficient to diagnose infection, with confirmatory testing as appropriate. However, infections in immunocompromised hosts, suspected CNS infections, and infections in newborns following recent maternal infection need to be confirmed by laboratory diagnosis to confirm the identity of the virus. Diagnostic testing is also important in high-risk settings like nursing homes and day care centers.

Initial testing for varicella and zoster is generally viral culture followed by a direct fluorescent antibody (DFA) test. Serological tests for IgM antibody can be used to diagnose primary infection, but high levels of IgG can lead to false positives and a 4-fold rise in IgG titer is also required to confirm infections. Serum antibody titers prior to planned pregnancy also may be of value to identify seronegative women that would benefit from vaccination to prevent potential fetal infection late in pregnancy. In severe infections and neonatal infections that require immediate ACV therapy, evaluation of specimens by NAAT is required because of its high sensitivity, specificity, and rapid turnaround time.

Preanalytical

Acceptable specimen types and maintenance conditions for specimens are generally identical to those described for HSV above. However, one important distinction is for the handling of specimens for virus culture because the infectivity of VZV is markedly reduced when specimens are frozen. Specimens destined for culture should be maintained at 2 to 6°C and are typically stable for periods up to 72 hours. If frozen specimens are obtained they must be stored at −80°C and cultured immediately after the first thaw to maximize remaining viability of the virus.

Analytical

Culture

The "gold standard" for the diagnosis of VZV infections is viral culture followed by DFA. Because the virus is unstable after it has been released from infected cells, care must be taken to obtain acceptable specimens. A swab specimen in universal transport media or equivalent should preferably be refrigerated until it can be cultured in the diagnostic laboratory. It is critical that the specimen never be frozen because the sensitivity of subsequent culture will be greatly reduced. Shell vials containing monolayers of A549 cells, primary human fibroblasts (such as MRC-5 cells), or mixtures of African green monkey kidney cells and primary human fibroblasts are acceptable cell substrates. The virus replicates well in cell culture, but cytopathology is generally observed later after infection, with A549 cells being a superior cell substrate to MRC-5 or primary rhesus cells (47). Cell monolayers inoculated with swab specimens are generally stained with commercially available monoclonal antibodies in a direct fluorescent assay to confirm the identity of the virus to distinguish it from other viruses like HSV.

Antibody

Since most individuals born after 1995 have been vaccinated, serological testing is becoming less informative over time. Exceptions include women planning pregnancy so that they can be vaccinated to prevent transmission to the developing fetus late in pregnancy. Available assays include commercial kits such as the VZV-FAMa from Viran Clinical Diagnostics (Stevensville, MI) and VZV-IgG-ELISAPKS (medac GmbH, Hamburg, Germany), as well as the well-characterized laboratory-developed assay SERION ELISA classic VZV IgG assay (48).

Nucleic Acid Amplification Tests

As with HSV, rapid molecular assays for the detection of VZV DNA are increasing in popularity and may be of value in the management of transplant-associated infections (49). Laboratory-developed assays have been reported that provide rapid quantification of VZV DNA, although commercially quantitative assays are not yet available (50). However, a highly sensitive qualitative assay is commercially available on the LightCycler (Roche, Molecular Biochemicals) (51). The multiplexed Lyra Direct HSV 1+2/VZV Assay has also been cleared by the FDA (44, 45). Swab specimens evaluated by this assay are reported to have greatly improved sensitivity compared to culture on MRC-5 cells. As with NAATs for HSV, swab specimens in universal transport media are the preferred specimen type and should be held at 4 to 6°C for <72 hours. Other specimen types may be acceptable to laboratories that have validated against specimens such as ocular fluid to monitor retinal infections, CSF, and whole blood (52). Rapid NAATs are also likely to become available in the future and will likely offer increased sensitivity compared to viral culture.

Postanalytical

Reporting of critical results is required for certain infections particularly if viral DNA is detected in the CSF. Congenitally infected infants are also at risk, and the rapid reporting of results is essential to ensure that therapy can be provided in a timely manner. It is also important to consider that infections with VZV can be more severe in the very young and in adults.

BEST PRACTICES

Analytical methods used to diagnose both HSV and VZV infections are similar and can provide the essential information in a timely manner. Serology, virus culture, and NAAT-based methods are all well-established methods, and each have a potential role in the diagnostic laboratory (Fig. 2). While it is possible to detect viral antigens directly in patient specimens such as swabs and lesion scrapings, the best data from DFA-based systems is from cultured specimens and adds at least 24 hours to the turnaround time. In general, NAAT-based methods exhibit the highest sensitivity and have proven to be more sensitive than culture for the detection of both HSV (40) and VZV (51). As the cost and complexity of NAAT methods are reduced, their use will increase and will likely become the dominant technology for the detection of these infections. Nonetheless, culture, serology, and direct fluorescent methods are widely used and are preferred in some cases because of cost considerations.

Since the manifestations of HSV and VZV infection can range from asymptomatic to life-threatening infections, the most important consideration in the selection of a diagnostic method is the risk associated with the infections. In many cases rapid evaluation and reporting of data are of paramount importance and NAATs are essential. This is particularly important in disease states with increased risk of morbidity and mortality, including suspected viral infections in immunocompromised hosts and neonates and in patients with CNS involvement.

ACKNOWLEDGMENTS

We thank Stefanie Brown, Jennifer Potter, Richard Covington, and Sonya Nix for helpful discussions.

REFERENCES

1. **Whitley RJ, Roizman B, Knipe D.** 2007. Herpes simplex viruses, p 2501–2601. In Knipe DM, Howley PM, Griffin D, Lamb R, Martin M, Straus SE (eds), *Fields Virology*, 5th ed. Lippincott Williams & Wilkins, Philadelphia, PA.
2. **Ward S, Weller SK.** 2011. HSV-1 DNA replication, p 89–112. *In* Weller SK (ed), *Alphaherpesviruses: Molecular Virology*. Caister Academic Press, Norfolk, UK.
3. **Weller SK, Coen DM.** 2012. Herpes simplex viruses: mechanisms of DNA replication. *Cold Spring Harb Perspect Biol* **4:** a013011.
4. **Lehman IR, Boehmer PE.** 1999. Replication of herpes simplex virus DNA. *J Biol Chem* **274:**28059–28062.
5. **Smith JS, Robinson NJ.** 2002. Age-specific prevalence of infection with herpes simplex virus types 2 and 1: a global review. *J Infect Dis* **186**(Suppl 1):S3–S28.
6. **Nahmias AJ, Lee FK, Beckman-Nahmias S.** 1990. Sero-epidemiological and -sociological patterns of herpes simplex virus infection in the world. *Scand J Infect Dis Suppl* **69:**19–36.
7. **Tunbäck P, Bergström T, Andersson AS, Nordin P, Krantz I, Löwhagen GB.** 2003. Prevalence of herpes simplex virus antibodies in childhood and adolescence: a cross-sectional study. *Scand J Infect Dis* **35:**498–502.
8. **Corey L, Huang ML, Selke S, Wald A.** 2005. Differentiation of herpes simplex virus types 1 and 2 in clinical samples by a real-time taqman PCR assay. *J Med Virol* **76:**350–355.
9. **Bernstein DI, Bellamy AR, Hook EW III, Levin MJ, Wald A, Ewell MG, Wolff PA, Deal CD, Heineman TC, Dubin G, Belshe RB.** 2013. Epidemiology, clinical presentation, and antibody response to primary infection with herpes simplex virus type 1 and type 2 in young women. *Clin Infect Dis* **56:**344–351.
10. **Roberts CM, Pfister JR, Spear SJ.** 2003. Increasing proportion of herpes simplex virus type 1 as a cause of genital herpes infection in college students. *Sex Transm Dis* **30:**797–800.
11. **Corey L, Wald A, Celum CL, Quinn TC.** 2004. The effects of herpes simplex virus-2 on HIV-1 acquisition and transmission: a review of two overlapping epidemics. *J Acquir Immune Defic Syndr* **35:**435–445.
12. **Fleming DT, McQuillan GM, Johnson RE, Nahmias AJ, Aral SO, Lee FK, St Louis ME.** 1997. Herpes simplex virus type 2 in the United States, 1976 to 1994. *N Engl J Med* **337:**1105–1111.
13. **Xu F, Sternberg MR, Kottiri BJ, McQuillan GM, Lee FK, Nahmias AJ, Berman SM, Markowitz LE.** 2006. Trends in herpes simplex virus type 1 and type 2 seroprevalence in the United States. *JAMA* **296:**964–973.
14. **Bradley H, Markowitz LE, Gibson T, McQuillan GM.** 2013. Seroprevalence of herpes simplex virus types 1 and 2—United States, 1999–2010. *J Infect Dis* **209:**325–333.
15. **Kelley PW, Petruccelli BP, Stehr-Green P, Erickson RL, Mason CJ.** 1991. The susceptibility of young adult Americans to vaccine-preventable infections. A national serosurvey of US Army recruits. *JAMA* **266:**2724–2729.
16. **Weinberg JM.** 2007. Herpes zoster: epidemiology, natural history, and common complications. *J Am Acad Dermatol* **57** (Suppl)**:**S130–S135.
17. **Baxter R, Tran TN, Ray P, Lewis E, Fireman B, Black S, Shinefield HR, Coplan PM, Saddier P.** 2014. Impact of vaccination on the epidemiology of varicella: 1995–2009. *Pediatrics* **134:**24–30 .
18. **Kimberlin DW.** 2007. Management of HSV encephalitis in adults and neonates: diagnosis, prognosis and treatment. *Herpes* **14:**11–16.
19. **Brown ZA, Wald A, Morrow RA, Selke S, Zeh J, Corey L.** 2003. Effect of serologic status and cesarean delivery on transmission rates of herpes simplex virus from mother to infant. *JAMA* **289:**203–209.
20. **Kropp RY, Wong T, Cormier L, Ringrose A, Burton S, Embree JE, Steben M.** 2006. Neonatal herpes simplex virus infections in Canada: results of a 3-year national prospective study. *Pediatrics* **117:**1955–1962.
21. **Jones CA, Raynes-Greenow C, Issacs D.** 2014. Population-based surveillance of neonatal HSV infection in Australia (1997–2011). *Clin Infect Dis* **59:**525–531.
22. **Andrei G, Snoeck R.** 2013. Advances in the treatment of varicella-zoster virus infections. *Adv Pharmacol* **67:**107–168.
23. **Vere Hodge RA, Field HJ.** 2013. Antiviral agents for herpes simplex virus. *Adv Pharmacol* **67:**1–38.
24. **James SH, Prichard MN.** 2014. Current and future therapies for herpes simplex virus infections: mechanism of action and drug resistance. *Curr Opin Virol* **8:**54–61.
25. **Elion GB, Furman PA, Fyfe JA, de Miranda P, Beauchamp L, Schaeffer HJ.** 1977. Selectivity of action of an antiherpetic agent, 9-(2-hydroxyethoxymethyl) guanine. *Proc Natl Acad Sci USA* **74:**5716–5720.
26. **Vere Hodge RA, Sutton D, Boyd MR, Harnden MR, Jarvest RL.** 1989. Selection of an oral prodrug (BRL 42810; famciclovir) for the antiherpesvirus agent BRL 39123 [9-(4-hydroxy-3-hydroxymethylbut-l-yl)guanine; penciclovir]. *Antimicrob Agents Chemother* **33:**1765–1773.
27. **Öberg B.** 1989. Antiviral effects of phosphonoformate (PFA, foscarnet sodium). *Pharmacol Ther* **40:**213–285.
28. **De Clercq E, Holý A.** 2005. Acyclic nucleoside phosphonates: a key class of antiviral drugs. *Nat Rev Drug Discov* **4:**928–940.
29. **Cihlar T, Chen MS.** 1996. Identification of enzymes catalyzing two-step phosphorylation of cidofovir and the effect of cytomegalovirus infection on their activities in host cells. *Mol Pharmacol* **50:**1502–1510.
30. **Shapiro ED, Vazquez M, Esposito D, Holabird N, Steinberg SP, Dziura J, LaRussa PS, Gershon AA.** 2011. Effectiveness of 2 doses of varicella vaccine in children. *J Infect Dis* **203:**312–315.
31. **Gnann JW Jr.** 2008. Vaccination to prevent herpes zoster in older adults. *J Pain* **9**(1 Suppl 1)**:**S31–S36.
32. **Dworkin RH, Johnson RW, Breuer J, Gnann JW, Levin MJ, Backonja M, Betts RF, Gershon AA, Haanpaa ML, McKendrick MW, Nurmikko TJ, Oaklander AL, Oxman MN, Pavan-Langston D, Petersen KL, Rowbotham MC, Schmader KE, Stacey BR, Tyring SK, van Wijck AJ, Wallace MS, Wassilew SW, Whitley RJ.** 2007. Recommendations

for the management of herpes zoster. *Clin Infect Dis* **44**(Suppl 1)**:**S1–S26.

33. **Shepp DH, Dandliker PS, Meyers JD.** 1986. Treatment of varicella-zoster virus infection in severely immunocompromised patients. A randomized comparison of acyclovir and vidarabine. *N Engl J Med* **314:**208–212.
34. **Bryan CJ, Prichard MN, Daily S, Jefferson G, Hartline C, Cassady KA, Hilliard L, Shimamura M.** 2008. Acyclovir-resistant chronic verrucous vaccine strain varicella in a patient with neuroblastoma. *Pediatr Infect Dis J* **27:**946–948.
35. **Lakeman FD, Whitley RJ, National Institute of Allergy and Infectious Diseases Collaborative Antiviral Study Group.** 1995. Diagnosis of herpes simplex encephalitis: application of polymerase chain reaction to cerebrospinal fluid from brain-biopsied patients and correlation with disease. *J Infect Dis* **171:**857–863.
36. **Crist GA, Langer JM, Woods GL, Procter M, Hillyard DR.** 2004. Evaluation of the ELVIS plate method for the detection and typing of herpes simplex virus in clinical specimens. *Diagn Microbiol Infect Dis* **49:**173–177.
37. **Wald A, Ashley-Morrow R.** 2002. Serological testing for herpes simplex virus (HSV)-1 and HSV-2 infection. *Clin Infect Dis* **35**(Suppl 2)**:**S173–S182.
38. **Whitley RJ, Lakeman F.** 1995. Herpes simplex virus infections of the central nervous system: therapeutic and diagnostic considerations. *Clin Infect Dis* **20:**414–420.
39. **Kimberlin DW, Lakeman FD, Arvin AM, Prober CG, Corey L, Powell DA, Burchett SK, Jacobs RF, Starr SE, Whitley RJ, National Institute of Allergy and Infectious Diseases Collaborative Antiviral Study Group.** 1996. Application of the polymerase chain reaction to the diagnosis and management of neonatal herpes simplex virus disease. *J Infect Dis* **174:**1162–1167.
40. **Wald A, Huang ML, Carrell D, Selke S, Corey L.** 2003. Polymerase chain reaction for detection of herpes simplex virus (HSV) DNA on mucosal surfaces: comparison with HSV isolation in cell culture. *J Infect Dis* **188:**1345–1351.
41. **Thompson PP, Kowalski RP.** 2011. A 13-year retrospective review of polymerase chain reaction testing for infectious agents from ocular samples. *Ophthalmology* **118:**1449–1453.
42. **Van Der Pol B, Warren T, Taylor SN, Martens M, Jerome KR, Mena L, Lebed J, Ginde S, Fine P, Hook EW 3rd.** 2012. Type-specific identification of anogenital herpes simplex virus infections by use of a commercially available nucleic acid amplification test. *J Clin Microbiol* **50:**3466–3471.
43. **Binnicker MJ, Espy MJ, Irish CL.** 2014. Rapid and direct detection of herpes simplex virus in cerebrospinal fluid by use of a commercial real-time PCR assay. *J Clin Microbiol* **52:**4361–4362.
44. **Fan F, Stiles J, Mikhlina A, Lu X, Babady NE, Tang YW.** 2014. Clinical validation of the Lyra direct HSV 1+2/VZV assay for simultaneous detection and differentiation of three herpesviruses in cutaneous and mucocutaneous lesions. *J Clin Microbiol* **52:**3799–3801.
45. **Heaton PR, Espy MJ, Binnicker MJ.** 2015. Evaluation of 2 multiplex real-time PCR assays for the detection of HSV-1/2 and Varicella zoster virus directly from clinical samples. *Diagn Microbiol Infect Dis* **81:**169–170.
46. **Hodinka RL.** 2013. Point: is the era of viral culture over in the clinical microbiology laboratory? *J Clin Microbiol* **51:**2–4.
47. **Coffin SE, Hodinka RL.** 1995. Utility of direct immunofluorescence and virus culture for detection of varicella-zoster virus in skin lesions. *J Clin Microbiol* **33:**2792–2795.
48. **Sauerbrei A, Wutzler P.** 2006. Serological detection of varicella-zoster virus-specific immunoglobulin G by an enzyme-linked immunosorbent assay using glycoprotein antigen. *J Clin Microbiol* **44:**3094–3097.
49. **Kalpoe JS, Kroes AC, Verkerk S, Claas EC, Barge RM, Beersma MF.** 2006. Clinical relevance of quantitative varicella-zoster virus (VZV) DNA detection in plasma after stem cell transplantation. *Bone Marrow Transplant* **38:**41–46.
50. **Pevenstein SR, Williams RK, McChesney D, Mont EK, Smialek JE, Straus SE.** 1999. Quantitation of latent varicella-zoster virus and herpes simplex virus genomes in human trigeminal ganglia. *J Virol* **73:**10514–10518.
51. **Espy MJ, Teo R, Ross TK, Svien KA, Wold AD, Uhl JR, Smith TF.** 2000. Diagnosis of varicella-zoster virus infections in the clinical laboratory by LightCycler PCR. *J Clin Microbiol* **38:**3187–3189.
52. **Kalpoe JS, van Dehn CE, Bollemeijer JG, Vaessen N, Claas EC, Barge RM, Willemze R, Kroes AC, Beersma MF.** 2005. Varicella zoster virus (VZV)-related progressive outer retinal necrosis (PORN) after allogeneic stem cell transplantation. *Bone Marrow Transplant* **36:**467–469.

Cytomegalovirus

PREETI PANCHOLI AND STANLEY I. MARTIN

27

VIRAL CLASSIFICATION AND BIOLOGY

Cytomegalovirus (CMV) is one of 8 known herpesviruses to infect humans. Classified as a β-herpesvirus, along with human herpesvirus-6 and human herpesvirus-7, CMV is a large virus that was first isolated in humans from salivary glands in the 1950s (1). Since its discovery, CMV has been linked to a multitude of clinical syndromes in humans and is possibly the most important virus of clinical consequence among immunocompromised patients and pregnant women.

Like all herpesviruses, CMV has a linear, double-stranded DNA genome. Its size sets it apart, however, as it contains open reading frames to accommodate as many as 230 different proteins (2). The genome, similar to other β-herpesviruses, contains terminal repeat sequences that are complementary to each other. The DNA is found in the nucleoprotein core surrounded by matrix proteins (3). A lipid envelope surrounds the viral matrix and genomic nucleoprotein structure, containing numerous glycoproteins used for viral entry into the cell in ways that remain unclear. Many of the functions of the CMV proteins are still under active investigation, but some appear to interact with many cellular molecules associated with normal immune function, such as tumor necrosis factor-α receptors and other cytokines (4). These may be some of the reasons why CMV activity and immune modulation are so closely linked in numerous clinical scenarios (5).

CMV can infect a host of cell types from the myeloid lineage, as well as epithelial, neuronal, and smooth muscle cells (6). After endocytosis into the host cell, the genome undergoes uncoating and the DNA is transported to the nucleus, where synthesis of the polymerase and replication begins. CMV replication forms large nuclear inclusions, which can often be seen microscopically as a hallmark of active CMV infection. Like all herpesviruses, after CMV infects the cell, it becomes latent within a host of different organ system tissues. In the setting of immunosuppression such as that seen in patients with AIDS, patients who have undergone transplantation, or patients taking immunosuppressant therapies for autoimmune disease, reactivation of the virus can occur, leading to end-organ disease. Spread to subsequent hosts can be through intimate mucocutaneous contact, organ and bone marrow transplantation, blood transfusions and exchange of other biologic products, or from infected mother to unborn child *in utero*.

EPIDEMIOLOGY

Infection with CMV is fairly common. In a cooperative study by the World Health Organization (WHO) in the 1970s, the seroprevalence of CMV antibodies among healthy blood donors across the world ranged from 40 to 100%, with socioeconomic factors playing a role in risk for infection (7). Countries with higher rates of industrialization had a lower incidence of infection while developing countries had a much higher incidence. In the United States, age and ethnicity play a role in the risk of infection. Examination of tens of thousands of serum samples taken from the National Health and Nutrition Examination Survey (NHANES) show the lowest incidence at 36% among 6- to 11-year-olds and >90% among persons 80 years old or more (8). For NHANES 1999–2004, the overall age-adjusted CMV seroprevalence was 50.4%, with infection being higher among non-Hispanic black and Mexican-American children compared with non-Hispanic white children. CMV seropositivity overall is independently associated with older age, female sex, foreign birthplace, low household income, high household crowding, and low household education (9).

Neonates may be infected with CMV *in utero* when the pregnant mother exhibits viremia, or during exposure of secretions in the birth canal. This most likely occurs during primary infection of the mother while pregnant, though infection with novel strains of CMV may still be a concern in already-infected individuals. A prospective study of 205 postpartum seropositive women found that approximately one-third of the participants had CMV reinfection over a mean of 35 months of follow-up as determined by the appearance of new antibodies with strain-specific binding to unique polymorphic determinants on the envelope glycoprotein (10). A study specifically looking at mother-to-child transmission showed that reinfection with a different strain of CMV during pregnancy can lead to intrauterine transmission and subsequent symptomatic congenital infection (11). Despite these findings, the highest risk for congenital infection lies with primary infection of the mother during pregnancy. Meta-analyses looking at rates of transmission suggest a close to 30% rate in mothers with primary infection versus <2% in those already previously infected (12). Overall, CMV is the most common of congenital viral infections worldwide, though the overall prevalence is likely <1% of which the percentage of infected children with

doi:10.1128/9781555819156.ch27

CMV-specific symptoms at birth is just over 10% (13). The highest risk occurs when primary infection happens during the first half of the pregnancy (14).

The incidence of CMV disease, as opposed to just infection, has been most closely linked with the loss of protective cellular immunity and the reactivation of latent virus or transmission of virus to previously unexposed individuals undergoing transplantation. Prior to the development of highly active anti-retroviral therapy (HAART), CMV was considered the most common viral opportunistic infection in patients with AIDS. CMV retinitis, the most common manifestation of disease among patients with AIDS, occurred in up to 44% of cases, primarily in those with a CD4+ T lymphocyte count below 50 cells/mm^3 (15). In the modern era of HAART, the number of HIV-infected patients with at-risk CD4 counts is much lower, and the incidence of CMV retinitis is not entirely clear. One recent limited cohort evaluation of 1,600 patients found an overall incidence of CMV retinitis of 0.36/100 person-years (16).

Among transplant recipients, the incidence of disease varies considerably from organ to organ (17). Other factors, such as degree of immunosuppression and seropositive status of the donor and recipient, translate into the risk for disease, with the highest burden seen in the heavily immunosuppressed and in seronegative recipients receiving organs from seropositive donors (D+/R−). The best and largest studies on the overall incidence of disease have been performed in kidney transplant recipients, where rates can be as high as 60%, though with modern and extended prophylaxis and preventative measures, the rate can be as low as 5% (18). By comparison, patients undergoing allogeneic hematopoietic stem cell transplantation (HSCT) are most at risk when the recipient is already seropositive. Other factors subsequently increase the risk, such as receiving the transplant from a matched unrelated donor, older age, and the development of acute graft-versus-host disease (19). Prior to the use of prophylaxis or preemptive treatment regimens, HSCT recipients experienced a baseline risk of 70–80% reactivation of CMV (20). Historically, just after engraftment, pneumonia was the most common manifestation of disease in this population, but now is infrequent with the use of close monitoring, molecular diagnostics, and preemptive therapeutic approaches. As many as 30% may go on to develop CMV infection late after engraftment with the incidence of CMV disease being considerably less (21)

CLINICAL SIGNIFICANCE

Primary infection with CMV is usually asymptomatic. When symptoms do occur, they most commonly manifest as an infectious mononucleosis-like syndrome comparable to Epstein-Barr virus (EBV). Signs and symptoms include fever and a relative lymphocytosis that can last for weeks. Differences between primary CMV and EBV infection include a lesser degree of lymphadenopathy and splenomegaly, as well as less exudative tonsillitis with CMV (22). Other laboratory abnormalities beyond the atypical lymphocytosis may include anemia, thrombocytopenia, hemolysis, cold agglutinins, elevated rheumatoid factor, and positive antinuclear antibodies (ANA) (23). Organ-specific complications are rare in primary infection among immunocompetent individuals, but when they do occur, they can manifest in the form of gastrointestinal diseases such as diarrhea and abdominal pain, hepatitis, cardiac problems such as pericarditis and myocarditis, and neurological dysfunction such as encephalitis and Guillain-Barré syndrome (24). Other unusual associations with CMV have included inflammatory bowel disease and reactivation of CMV in patients who are critically ill from other causes in the intensive care unit. In both these scenarios, clinical outcomes may be worse with those that have evidence of underlying CMV activity (25, 26).

For newborn children who acquire CMV infection *in utero*, symptoms that do occur can be severe. Fulminant, overwhelming infection with hepatosplenomegaly, jaundice, rash, and multiple organ failure may occur. The neurological manifestations can be crippling, including microcephaly, motor limitations, and chorioretinitis. Classically, cerebral calcifications can be seen on routine radiography, though this is non-specific. For children who appear otherwise asymptomatic, concerns about long-term subtle neurologic sequelae abound and some studies suggest that those children may be more intellectually impaired in the long-term, have a higher dropout rate from school, and may be more likely to suffer hearing loss (27). Some estimates suggest that as many as one out of every 10 clinically asymptomatic patients may develop late sequelae including sensorineural hearing loss (28).

For the clinician taking care of adult patients, the clinical significance of CMV lies almost entirely in the reactivation of latent virus in the setting of immunosuppression. For patients suffering from AIDS, CMV disease primarily occurs as cellular immunity is lost. Retinitis is by far the most common manifestation of CMV disease in this population (29). Blindness can result after complete destruction of the retina, and patients may have characteristic white fluffy retinal infiltrates alongside areas of hemorrhage on fundoscopic examination. Other manifestations of CMV disease in AIDS patients may include gastrointestinal disease, encephalitis, and polyradiculopathy.

Among the transplant population, CMV disease may manifest as a non-specific flu-like illness, often with cytopenias in the setting of active CMV viremia, referred to as "CMV syndrome" (30). When CMV reactivation involves tissues, organ-specific manifestations can occur. The gastrointestinal tract is the most common location for end-organ disease involved with CMV after solid-organ transplant or HSCT, but almost any organ system can be included, with disease manifesting as pneumonitis, carditis, nephritis, cholecystitis, pancreatitis, and meningoencephalitis (30). Because of the complex interplay between CMV and the immune system, other indirect effects may manifest in this population as a result of CMV reactivation or acquisition. Phenomena such as acute and chronic allograft rejection, vascular disease, reactivation of other viruses, risk of other bacterial or fungal infections, mortality, and diabetes have all been linked in transplant patients with known CMV infection, viremia, or disease (31).

TREATMENT AND PREVENTION

In immunosuppressed patients at risk for CMV disease, there are two major approaches to CMV prevention. The first is universal antiviral prophylaxis and the second is targeted monitoring and preemptive therapy. With the first approach, an antiviral agent is given to the patient for an extended period of time after the transplant—usually 3–12 months, depending on the type of organ transplanted. The agent of choice is valganciclovir, a valine-esterized prodrug of ganciclovir with improved absorption from the gastrointestinal tract and improved efficacy compared to oral ganciclovir alone (32). With the preemptive approach, patients at risk of CMV disease after transplantation are monitored at fixed intervals with molecular-based assays looking for evidence of CMV activation before symptoms develop. Asymptomatic patients who go on to develop viremia are placed on antiviral therapy once a certain threshold is reached in order to prevent CMV disease from ever occurring. No preventive strategies have been studied in the HIV population, and antiviral prophylaxis beyond HAART is generally not employed.

The question of universal prophylaxis versus preemptive therapy remains an area of some controversy. Although preemptive therapy has the advantage of reducing drug-related costs and toxicity, and Cochrane database reviews have supported the strategy as beneficial in CMV prevention after solid-organ transplant (33), the indirect effects of CMV replication on the host may not be mitigated by the preemptive approach. In a large Swiss study, comparing the universal prophylaxis versus the preemptive strategy, patients who underwent the preemptive technique had a higher risk of subsequent allograft failure despite no significant difference in the rates of CMV disease (34). Other retrospective reviews have also supported a higher incidence of indirect effects via the preemptive approach, such as other infections and overall mortality (35). In HSCT populations, universal prophylaxis is often avoided in favor of the preemptive approach due to the limiting bone marrow suppressive effects of ganciclovir or valganciclovir.

Therapy for CMV disease is most often with ganciclovir—the antiviral bedrock of CMV therapy. Used primarily as an intravenous agent, ganciclovir is a synthetic analogue of guanosine. Upon entering the host cell, ganciclovir is phosphorylated to ganciclovir monophosphate by viral kinases produced by the aforementioned UL97 gene. Cellular kinases from the host convert the monophosphate molecule to a subsequent ganciclovir triphosphate, which acts as a competitive inhibitor of deoxyguanosine triphosphate and can be incorporated into the DNA of CMV by viral DNA polymerase, inhibiting further chain elongation. Studies in transplant recipients without life-threatening CMV disease, as well as in HIV-infected patients with CMV retinitis, show that valganciclovir can be equally efficacious for treatment of CMV disease compared to intravenous ganciclovir (36, 37). Ganciclovir can also be used as an intravitreal injection to treat retinitis (38).

Despite the efficacy of ganciclovir and its prodrug, it can be associated with an array of serious hematological adverse effects, including leukopenia and anemia. It is also considered a potential teratogen and mutagen. Alternatives to ganciclovir include foscarnet and cidofovir. Both of these drugs have been approved by the United States Food and Drug Administration (FDA) for treatment of CMV retinitis in AIDS patients, but other uses are off-label. These latter drugs unfortunately have no FDA-approved oral formulations available, though a lipid-conjugate oral prodrug of cidofovir called brincidofovir, or CMX001, is being tested (39). These drugs have the advantage of bypassing the need for phosphorylation by viral kinases and thus can be used in some ganciclovir-resistant strains of CMV. The downside is that they have even more significant toxicity problems than ganciclovir, predominantly through acute renal injury and electrolyte wasting. Other drugs are also currently in development for the treatment of CMV. Maribavir, leflunomide, and cyclopropavir all have activity against CMV, though *in vivo* studies remain limited. The novel compound, letermovir (AIC246) inhibits viral DNA processing by interacting with the viral UL56 subunit (40). The drug is undergoing Phase II trials for treatment and prophylaxis in both solid-organ and stem cell transplant recipients (41, 42).

The ultimate prevention of CMV disease would be through use of an effective vaccine. No such vaccine exists as of yet, but many studies are ongoing and of interest. The ASP0113 (formally TransVax) DNA vaccine is in Phase II trials and has been shown to limit the duration of viremia and the need for antiviral treatment in HSCT recipients (43). There is also a recombinant envelope glycoprotein B vaccine formulated with the adjuvant MF59 that seems to reduce by roughly 50% the incidence of primary CMV in young mothers (44). A CMV vaccine has yet to complete Phase III trial studies.

DETECTION AND DIAGNOSIS

Because of the ubiquity of CMV exposure, diagnosis of CMV infection and disease based on clinical grounds alone is often unreliable. Laboratory-based diagnosis is usually required to identify congenital and perinatal CMV disease, and to diagnose and monitor viral levels in immunosuppressed hosts. Diagnostic tests for CMV include serology, tests for active disease, including quantitative nucleic acid testing (QNAT), antigenemia, culture, and histopathology, as well as newer immunology assays reflecting the cellular immune response to CMV (Table 1).

Serology

Serologic tests that detect CMV antibodies (IgM and IgG) are widely available. Enzyme immunoassays are usually employed to detect anti-CMV antibodies. Various fluorescence assays and indirect hemagglutination and latex agglutination tests are also available for measuring antibody to CMV (44).

Detection of CMV IgG antibody is used for the diagnosis of CMV infection and not disease, since a majority of individuals are seropositive for CMV. Antibody tests of paired acute- and convalescent-phase serum samples showing a fourfold rise in IgG antibody and CMV IgM antibody can indicate active CMV disease.

The presence of CMV IgM is not solely indicative of primary infection. CMV IgM is detectable when a person (i) is newly infected; (ii) has been infected in the past but has been recently re-exposed to CMV; (iii) is undergoing reactivation of CMV infection that was acquired in the past; or (iv) has a false-positive test result. Thus, the presence of CMV IgM should not be used by itself to diagnose primary CMV infection (45). IgG avidity assays, which measure antibody maturity, have been shown to reliably detect recent primary CMV infection. Since the CMV IgG avidity increases with time, a low CMV IgG avidity suggests a primary CMV infection occurred within the past 2–4 months while a high CMV IgG avidity suggests that CMV infection occurred in the more distant past (46).

CMV serology is widely used for identification of primary infection in pregnant women and in classifying organ donors and recipients who have been infected with CMV. The likelihood of congenital infection and disability is highest for infants whose mothers were CMV-seronegative before conception and who acquire infection during pregnancy (47). CMV serology should be performed on all donors and recipients prior to transplant to detect CMV-IgM and IgG antibodies. Tests for anti-CMV IgG are recommended pre-transplant, as they have better specificity than IgM or combination IgG and IgM tests, neither of which should be used for pre-transplant screening, as false-positive tests for IgM may significantly decrease test specificity (48–51). The donor and recipient serostatus combination (D/R) is a key predictor of infection risk and management and it is

TABLE 1 Laboratory Tests for Cytomegalovirus Detection

Method	Principle	Sample type or processing details	Turnaround time	Results and clinical utility	Advantages	Disadvantages
Non-nucleic acid-based methods						
Serology	Detection of antibody against CMV (IgG, IgM)	Serum	6 h	CMV-IgG indicates past CMV infection (latent infection); CMV-IgM implies acute or recent infection; CMV IgG avidity more reliable than CMV-IgM.	Prognostication and risk assignment of patients and their donors prior to transplantation (CMV D+/−, D+/R+, D−/R+, or D−/R−)	Not useful for CMV disease diagnosis in transplant recipients due to attenuated and delayed antibody production; not useful for guiding duration of treatment
Histopathology	Demonstration of CMV-infected cells (enlarged cells with nuclear inclusions)	Tissue microscopy with H&E stain; may need *in situ* hybridization and immunohistochemical staining to increase sensitivity and specificity	24–48 h	Detection of CMV-infected cells indicates active tissue-invasive disease	Confirmatory test for tissue-invasive CMV disease when viral load in blood is negative (as it may be in gastrointestinal or neurological disease); highly specific	Need for invasive method to obtain tissue specimen; not generally used to monitor treatment response or risk of relapse
Virus culture						
Tube culture	Viral growth in human fibroblast cells is indicated by CPE	Cell culture facility; light microscopy	2–4 weeks	Detection of characteristic CPE indicates presence of virus	Highly specific for CMV infection; useful when viral load in blood is negative and for non-blood samples that may not be optimized by molecular methods; the viral isolate can be tested for phenotypic susceptibility	Prolonged processing time is not clinically useful in real-time clinical management; poor sensitivity; requires viable CMV; not generally used to monitor treatment response or risk of relapse due to poor sensitivity and long processing time
Shell vial assay	Viral growth with detection using monoclonal antibodies against viral antigens	Cell culture facility; immunofluorescence detection	16–48 h	Infectious foci detected by monoclonal antibody directed to immediate-early antigen of CMV (prior to the onset of CPE)	Highly specific for CMV infection; more sensitive and rapid than conventional tube cultures; useful when viral load in blood is negative and for non-blood samples that may not be optimized by molecular methods	Relatively low sensitivity compared to molecular methods; rapid decrease of CMV activity in clinical specimens; not generally used to monitor treatment response or risk of relapse

(*Continued on next page*)

TABLE 1 Laboratory Tests for Cytomegalovirus Detection (*Continued*)

Method	Principle	Sample type or processing details	Turnaround time	Results and clinical utility	Advantages	Disadvantages
Antigenemia assay	Detection of pp65 antigen	Polymorphonuclear cells; processing within 4–6 h; light microscopy or immunofluorescence	6 h	Number of CMV-infected cells per total number of cells (e.g., 5×10^4)	Rapid diagnosis of CMV infection; quantification (no. of positive cells) may indicate disease and infection severity, may be used as a guide for preemptive therapy, and may be used as a guide for treatment response and duration of treatment	Subjective interpretation of results; requires rapid processing; not useful in leukopenic patients; lack of standardization in no. of positive cells for various clinical indications
Nucleic acid-based molecular methods						
Nucleic acid amplification tests	PCR amplification and detection of CMV DNA or RNA	Various clinical samples (blood, other body fluids); various assays (commercial and laboratory-developed tests)	Few hours	Assays standardized to the WHO standard results reported as IU/ml; nonstandardized assay results reported as no. of CMV copies per volume of specimen or PCR	Highly sensitive and specific for rapid diagnosis of CMV; quantification (viral load) allows for individualized management of patients; used to indicate disease severity; used to guide preemptive therapy; used to assess the risk of CMV disease; used to guide duration of antiviral treatment; used as surrogate of disease relapse or infection with resistant virus	Currently without a widely accepted viral threshold for predicting CMV disease; lack of assay standardization limits portability of results—ongoing efforts at assay standardization may reduce this limitation; highly sensitive assay may detect latent CMV

CMV, cytomegalovirus; CPE, cytopathic effects; H&E, hematoxylin and eosin; WHO, World Health Organization.
(Adapted from reference 68 with permission.)

imperative that a test with high sensitivity and specificity be used close to the time of transplantation. In children younger than 12 months, passive transfer of antibodies can lead to false-positive serologic results (52). In these situations, cell-mediated immunity assays may be useful in establishing true serostatus in both transfused patients and patients younger than 12 months. In children younger than 12 months, culture or nucleic acid amplification tests of urine or throat swabs may be helpful to identify infected patients since children shed virus for long periods of time after primary infection.

Viral Culture

CMV is usually isolated in culture in human fibroblast cell lines. CMV culture can be slow, expensive, and insensitive. However, viral culture is highly specific for the diagnosis of CMV infection. Primary infection, reactivation, and reinfection during pregnancy can all lead to *in utero* transmission to the developing fetus. Seropositive humans may shed CMV in their secretions, especially during times of stress, rendering positive cultures that do not necessarily reflect active disease. Viral culture of stool for CMV has poor sensitivity, while urine and sputum are the specimens of choice for the diagnosis of congenital CMV when collected within 21 days of birth (53). CMV culture may take weeks before the virus can be detected. Shell vial centrifugation assay has a relatively more rapid turnaround time and has been used for the quantification of viremia and of CMV in bronchoalveolar lavage (BAL) fluid, but remains less sensitive compared to molecular assays (54, 55).

Culture of tissue specimens remains an important option for diagnosis of tissue invasive disease, particularly for gastrointestinal samples (i.e., colonic biopsies), where antigenemia or polymerase chain reaction (PCR) testing on blood may not always be positive even with invasive disease (56). Also, culture is still used in isolating CMV in non-blood samples because molecular methods are not yet optimized for these clinical specimens.

Histopathology

Immunohistochemistry for CMV should be routinely performed on all biopsy specimens where CMV is suspected to maximize diagnostic sensitivity. Identification of typical "owl's eye" inclusion bodies or viral antigens in various tissue (including liver, lung, kidney, gastrointestinal) biopsy material (57, 58) or in BAL specimens' cells is very specific for CMV disease. False negatives due to sampling errors are common, however. Tissue-invasive CMV disease, such as colitis or hepatitis, should be confirmed by immunohistochemistry or *in situ* DNA hybridization (59, 60). For quantitative nucleic acid testing, tissue samples should be normalized using a housekeeping gene.

Different antibodies have variable sensitivity, and results may vary between fresh and formalin-fixed paraffin-embedded tissue (61). Although histopathology confirms the presence of tissue-invasive disease, this entails a potentially risky procedure to obtain tissue for diagnosis. Its use has declined due to the availability of non- or less-invasive tests to document CMV infection in blood. However, histopathology is recommended in cases where another concomitant pathology (e.g., graft rejection) or co-pathogens are suspected, especially when patients do not respond to anti-CMV treatment. Histopathology may also be needed when CMV disease is suspected but CMV testing in the blood is negative.

Antigenemia Assay

Viremia is most commonly detected by either an antigenemia assay or a QNAT test (62). The CMV pp65 antigenemia test detects the CMV pp65 antigen in circulating leukocytes of fresh anti-coagulated blood or cerebrospinal fluid (CSF). The results are reported as the number of positive cells per total number of cells counted. This test has been shown to be helpful in the diagnosis of early CMV replication and disease, and in monitoring response to therapy (63–65).

Quantification of pp65 antigen by immunofluorescence has been used to initiate preemptive therapy in organ transplant recipients (66). It is relatively easy to perform although there are problems with a lack of assay standardization, including subjective result interpretation. Antigenemia has higher sensitivity than culture, but is less sensitive to NAT by PCR depending on the method used for antigenemia or PCR assay (56, 67). Limitations include the limited stability of leukocytes and the need to process the sample within a few hours (6–8 h) of collection, low throughput, and the approach not being amenable to automation. Also, in patients with neutropenia, the assay cannot be performed reliably with absolute neutrophil counts less than 1,000 cells/mm^3.

Nucleic Acid Testing (NAT)

Molecular tests that detect and quantify CMV DNA are the preferred methods for the diagnosis and monitoring of patients at risk for CMV disease. The clinical utility of CMV viral load is supported by a considerable body of literature on immunosuppressed patients and consensus guidelines for managing CMV in transplant patients (45). These assays can be used to determine when to initiate preemptive therapy, to monitor the response to therapy, to determine the duration of therapy, and to assess patients at risk of developing relapsing infections (53). Generally, the detection of CMV RNA is indicative of active CMV replication. However, the sensitivity may be lower since RNA molecules may be readily degraded *in vitro* and may yield false-negative results (56). Tests detecting CMV RNA are not in routine clinical use. In contrast, the detection of CMV DNA in blood components (DNAemia) may or may not reflect CMV replication. Though latent virus may be associated with low-level DNAemia and active replication may be associated with high levels of CMV DNA, distinguishing latent genome detection from active infection remains a significant diagnostic challenge.

The NAT testing requires specific equipment and specialized expertise. Real-time (named based on the ability of this method to evaluate amplification in real-time at each cycle as opposed to at the end of the reaction) QNAT assays are now considered the standard of care (53) because of their sensitivity, broad linear range in which to monitor patients, better precision, better accuracy, low risk of contamination, and high throughput. CMV QNAT is the main alternative option to antigenemia (68, 69). Several real-time platforms are available for testing CMV. The vast majority of the CMV NAT assays have been developed in-house (laboratory-developed tests or LDTs). Variation in test platforms, nucleic acid extraction, assay design, CMV primer, target and probe selection, specimen type (whole blood, serum, plasma, leukocytes, BAL, CSF), reaction and amplification protocols, quantification standards and controls (plasmids, whole virus) has led to considerable variability across different testing centers. This significant interassay quantification variability was demonstrated in an international multicenter study across 33 laboratories where inconsistency in viral load values for individual samples ranged from 2.0 to 4.3 $\log_{10}$ copies/ml. This lack of assay agreement has hampered the establishment of broadly applicable quantitative cut-off values for patient management decisions (68, 70). Because of this, the first WHO international standard for human CMV for nucleic acid amplification techniques, NIBSC code 09/162 (CMV WHO standard), was released for harmonization of QNAT tests in 2010 (71). This calibrator would allow clinical laboratories and manufacturers to standardize CMV QNAT tests and report the results in international units (IU) rather than copies per milliliter (30, 72). Using the WHO international standard for reporting, Razonable et al. (73) correlated viral load with CMV disease resolution in transplant patients. Patients with pretreatment CMV DNA of <18,200 (4.3 $\log_{10}$ IU/ml) were 1.5 times more likely to have CMV disease resolution. Also, CMV suppression (<137 [2.1 $\log_{10}$] IU/ml) was predictive of clinical response to antiviral treatment.

Widespread availability of commercial PCR tests that encompass all assay steps (nucleic acid preparation, reaction setup, calibration, amplification, and detection) and demonstrate reliable interlaboratory quantification as defined by agreement and precision may resolve these residual quantification disparities (73). Recently, a droplet digital QNAT (ddQNAT) platform that utilizes endpoint PCR and eliminates the need for calibration standards has been evaluated. While both digital and real-time QNAT PCR provide accurate CMV load data over a wide linear dynamic range, standard QNAT PCR showed less variability and greater sensitivity than the ddQNAT PCR in clinical samples (74). Further development of this technology may be warranted since the advantages of direct quantification by endpoint, limiting-partition PCR could obviate development and need of quantitative calibrators (74).

Two platforms have recently been approved by the FDA. The Cobas AmpliPrep/Cobas TaqMan (CAP/CTM) CMV test (Roche Diagnostics, Branchburg, NJ; approved July 2012) (75) and the Artus CMV Rotor-Gene Q (RGQ) MDX Kit (Qiagen, Valencia, CA; approved June 2014) (76) have been approved for monitoring CMV DNA levels in EDTA plasma of solid-organ transplant patients who are undergoing anti-CMV therapy and can be used to assess CMV viral load response to antiviral drug therapy utilizing the CMV WHO standards. These assays are currently not approved for monitoring hematopoietic stem cell transplant recipients for preemptive therapy and their use on this population would be considered off-label use. The CAP/CTM CMV test uses primers and probes targeting a conserved region of the CMV genome (UL54, virus encoded DNA polymerase) and has a linear quantification range from 137 and 9,100,000 IU/ml, with a limit of detection (LOD) of 91 IU/ml. The Artus CMV RGQ MDX Kit targets specific amplification of a 105 bp region of the CMV major immediate early (MIE) gene DNA and has a linear quantification range from 159 IU/ml to 7.94 × 107 IU/ml with an LOD of 77 IU/ml. These tests offer advantages of higher throughput, automation, and wide linear viral load range over the semiautomated CE/IVD labelled Cobas Amplicor CMV Monitor test (Roche Diagnostics, Branchburg, NJ) (77). The FDA approval of the above assays is an important step forward. A recent study showed good reproducibility in viral load values and precision of CAP/CTM CMV test results across 5 different laboratories over 4 orders of magnitude and could be valuable in prospective studies identifying clinical viral load thresholds for determining the risk of disease, for diagnosing disease, and for monitoring response to treatment across multiple labs when using the FDA-approved tests calibrated to the WHO standard (78). Additional studies will provide more information of the commutability of the first WHO CMV standard and to define the clinical thresholds (79). Regulatory authorities and professional organizations will likely require laboratories that continue to perform LDTs to recalibrate their test to the WHO international standard, which requires establishing collinearity with the WHO international standard, as well as reporting results as IU/ml. Based on the test precision and linearity, changes in viral loads of threefold (0.5 $\log_{10}$ copies/ml) may represent biologically relevant changes in viral replication. For lower viral loads where there is greater variability, a 5-fold change (0.7 $\log_{10}$ copies/ml) may be required to be considered significant.

Recent work in patients on treatment for CMV disease compared viral load testing of plasma versus whole blood real-time PCR that demonstrated good correlation but significant differences in absolute value and clearance kinetics (80). Since the viral load can vary significantly between plasma and whole blood specimens, one specimen type should be used in serially monitoring patients at high risk of CMV infection. CMV DNA is detected earlier and usually in greater quantitative amounts in whole blood (80, 81). CMV blood markers (culture, pp65 antigenemia, CMV plasma DNA, CMV RNA) and CD4 counts <75 cells/mm^3 remain risk factors for CMV disease in patients receiving HAART (83, 84).

Other body fluids and tissues, including biopsy, BAL, and CSF specimens, can be tested for CMV by QNAT, which may improve sensitivity, potentially with faster results compared to culture (85, 53). Several studies suggest QNAT on BAL specimens may be helpful in predicting pneumonitis (86) although some studies do not (87, 88). Due to its high sensitivity, the negative predictive value of PCR from tissue other than blood is high, so it can occasionally be used to rule out disease.

Three viral load measures have been demonstrated to correlate with the clinical outcome of CMV disease treatment. These include (i) pretreatment viral load; (ii) the rate of CMV decline during treatment; and (iii) attainment of viral suppression as a measure of successful therapy. The duration of treatment should be individualized, depending upon clinical and laboratory parameters such as the decline of CMV load in the blood as measured by rapid and sensitive molecular standardized testing. CMV load in the initial phase of active infection and the rate of increase in viral load both correlate with CMV disease in transplant recipients; in combination, they have the potential to identify patients at imminent risk of CMV disease. The rate of increase in CMV load between the last PCR-negative and the first PCR-positive sample was significantly faster in patients with CMV disease (0.33 $\log_{10}$ versus 0.19 $\log_{10}$ genomes/ml daily, *P was* <0.001) (89).

Measuring Host Immune Reactivity to CMV

A number of in-house and some commercially available immunological assays for the assessment of T cell immunity to CMV are being evaluated for their ability to predict the development of CMV disease (90–92). In recent years, an increasing number of reports have focused on gamma interferon-releasing assays (IGRAs) as the diagnostic standard for detecting cell-mediated immunity toward infectious agents in humans. In one study, cell-mediated immunity among donor CMV-seropositive, recipient-seronegative (D+/R−) patients determined using the Quantiferon-CMV assay (Cellestis Ltd, a Qiagen company) was shown to have clinical utility to predict if patients were at low, intermediate, or high risk for the development of subsequent CMV disease after prophylaxis (93). The CMV IGRA assays are not yet available for routine clinical use.

Diagnosis of Antiviral Resistance

Drug resistance generally occurs after prolonged drug exposure (from weeks to months) with incomplete viral suppression manifesting as increased viral loads or disease despite therapy. Laboratory testing should be used to confirm the occurrence of antiviral resistance and to provide guidance for alternative treatment options based on the type of resistance detected. Both phenotypic and genotypic assays have been developed to assess antiviral resistance (94). Routine phenotypic susceptibility testing of CMV culture isolates against antiviral drugs by the plaque reduction assay is impractical because of slow turnaround time and the lack of culture isolates in the era of molecular diagnostics. Culture is also laborious and subject to the selection bias introduced during growth of mixed viral populations and lacks the sensitivity to detect low-level resistance or minor resistant subpopulations.

Laboratory confirmation of drug resistance is now based on genotypic assays that rely on sequence analysis of PCR-amplified CMV DNA from clinical specimens for the presence of diagnostic mutations. For example, resistance mutations to ganciclovir have primarily been documented in 2 viral genes: UL97 (protein kinase) and UL54 (DNA polymerase). These genes can be sequenced in-house or commercially, and mutations known to convey resistance ("hot spots") can be identified. In a vast majority (>90%) of cases, ganciclovir resistance in clinical practice results in 1 to 7 mutations in the UL97 kinase gene (codons 400–670). Sequence variations in the CMV UL54 gene (codons 393–1000) are more complex and less common but may confer resistance to ganciclovir and cross-resistance to cidofovir or foscarnet. Variations in the UL54 gene typically occur after development of UL97 mutations and increase the overall level of ganciclovir resistance (95). There are few reports about multidrug-resistance with mutations in both the UL97 and UL54 genes in patients after allogeneic HSCT (96). Isolated UL54 mutation in the absence of UL97 mutation is rare. There is an increasing database of CMV sequence variants and special attention must be paid to the interpretation of new sequence variations. These must be carefully analyzed for reproducibility in serial specimens and treatment history by proximity to known gene mutations and by recombinant phenotyping involving targeted mutagenesis of a lab CMV strain for corroboration (96, 7). Standard dideoxy sequencing can detect an emerging resistance mutation when it exceeds ~20% of the sequence population and testing is more reliable if the CMV viral load in the specimen is at least 1,000 copies/ml (7). Evolving sequencing technology (pyrosequencing and next generation sequencing) may be able to detect mutant subpopulations at lower levels (96–98). The use of viral gene sequencing offers distinct advantages over other methods, including a rapid turnaround time, a broader range of antiviral resistance information, and the ability to provide information concerning new drugs as they become available.

BEST PRACTICES

When choosing CMV diagnostic testing, one must first take into account the patient's medical background, epidemiology, risk factors for disease, and presenting signs and symptoms. This, combined with the array of choices—from serology to culture to histopathology and the number of molecular-based assays available—makes for a daunting task. Because infection with CMV can be lifelong and the virus establishes latency, distinguishing between infection and disease adds to this challenge. Alternatively, for this reason, the clinician may find himself or herself in a situation where he or she was not concerned about CMV-related disease, but must now contend with an unexpected "positive" CMV diagnostic assay of unclear value. Knowing the clinical scenario for the patient becomes key.

Infectious Mononucleosis

When infectious mononucleosis is caused by the far more common EBV, patients are often diagnosed indirectly via the heterophile antibody assay. Immunocompetent patients with acute CMV infection will be heterophile-antibody-negative, and confirmation of the diagnosis of CMV can be challenging due to the sensitivity and specificity of CMV-specific serology. CMV IgM antibodies may persist in the host for months, leading to false-positive results in patients suspected of having acute infection, and CMV-specific IgG antibodies may not be detectable for several weeks after the start of the illness, though they persist for life once established (99). A single positive IgG antibody can only confirm past exposure to the virus. This is of value, however, in immunocompromised hosts or pregnant women in order to gauge risk of reactivation or acquisition of the virus in the future. Since therapy is otherwise not usually indicated in the immunocompetent host with acute CMV infection, and the disease is almost always self-limited in that setting, more advanced diagnostic testing is often not performed or even indicated.

Perinatal and Congenital CMV

Determining risk in a pregnant mother by serology also has its challenges. Seroconversion may be the gold standard, but this requires documented evidence of a prior negative serology and is often not done or available. Primary infection again cannot be relied upon fully by a positive CMV IgM alone. Avidity testing of CMV-specific IgG antibodies may be useful in these scenarios. The presence of an IgM and a low-avidity IgG is strongly suggestive of primary infection in the mother (100). The presence of a high-avidity CMV-specific IgG suggests past infection and thus a lower risk of transmission.

If infection is found in the mother, then prenatal evaluation of fetal infection can be done via testing of the amniotic fluid. Both culture and PCR-based testing can be done on amniotic fluid or from cordocentesis. The sensitivity of testing is likely higher using PCR testing compared to culture, but is <80% overall and limited by an inability to predict severity of disease and the overall invasive nature of amniotic fluid sampling (101, 102).

Serology for diagnosing CMV infection in the newborn also has limitations and is generally not relied upon. A positive CMV-specific IgG from a newborn may reflect passive transfer of maternal antibody to the child and not definitive evidence of infection, much less disease. CMV-specific IgM antibody testing may at times be positive in a newborn, strongly suggesting congenital infection, but false positive serological results may occur and the sensitivity of the assay even in symptomatic children with congenital CMV infection is low (103).

More direct evidence of the virus is often required and relied upon to confirm congenital CMV infection. Culture of the virus from the newborn can serve as confirmation of CMV infection, but many clinicians now rely on use of CMV DNA detection with PCR. Culture or PCR assays can be done with urine, saliva, tissue, and serum of newborns within the first 3 weeks (104), though PCR-based assays may be the most sensitive test overall in this setting (103).

CMV Infection and Disease in AIDS

The most common manifestation of CMV-related disease in HIV-infected patients with AIDS is retinitis. The diagnosis of retinitis is almost always based solely on physical examination by an ophthalmologist. Serology in the form of CMV-specific IgG is useful in these scenarios in order to identify past exposure, and thus, risk of reactivation in the setting of lower CD4+ T cell counts. CMV viremia as determined by DNA PCR assays can be common in this population, but does not predict progression to retinitis and cannot be used as a diagnostic tool in these situations (105). Cultures of CMV from the blood may not be helpful from a diagnostic standpoint for retinitis either, though AIDS patients with positive CMV blood or urine cultures may have a more rapid progression of retinitis and perhaps even a higher risk for mortality (106). Although not as well studied in HIV-infected patients compared to the transplant patient population, CMV viremia detected by DNA PCR assays alone also does not predict the development of CMV disease in this population, regardless of the type of organ involvement (107). Other manifestations of CMV disease in AIDS patients may include gastrointestinal disease, pneumonitis, and neurological disease such as encephalitis or polyradiculopathy.

The gold standard for diagnosing organ-specific tissue-invasive CMV disease in all other situations is the identification of CMV inclusions or CMV antigens by immunohistochemistry on biopsy specimens. Visual inspection through endoscopy of either the esophagus or colon may reveal plaque-like membranes, ulcerations, and erosions (108). Visual inspection alone, however, cannot confirm the diagnosis since patients with AIDS may have other pathogens causing similar findings. Infections such as *Mycobacterium avium* complex and *Cryptosporidium* may present with the same signs and symptoms. Tissue biopsy of affected areas revealing characteristic "owl's eye" intra-nuclear inclusions in the mucosal epithelium or mucosal crypts may be visible on routine tissue staining with hematoxylin and eosin. Using immunohistochemistry to stain for CMV-specific antigens in tissue may be of benefit in this population as well to confirm active CMV involvement.

For pneumonitis due to CMV, attempting to culture CMV from the respiratory tract is of little benefit, lacking both sensitivity and specificity. Asymptomatic HIV-infected individuals often shed CMV from the respiratory tract, and patients with severe respiratory disease may not have detectable CMV on culture (109, 110). Again, biopsy of tissue will be necessary, especially since other opportunistic infections in AIDS patients, such as *Pneumocystis* pneumonia or histoplasmosis, may present in similar fashion.

In cases of central nervous system involvement, relying on the isolation of CMV from CSF has become routine in diagnosing CMV disease. Studies are limited in this area, but small case series of patients with autopsy-proven disease suggest CMV DNA detection in the CSF correlates with active disease (111). The value of culture and antigen-based assays from the CSF is less clear.

CMV Infection and Disease in Transplant Recipients

Stratifying risk for disease is an important first step in diagnosing and managing CMV among transplant recipients. CMV-specific IgG serologies should be obtained on all organ donors and recipients prior to transplantation (112). Solid-organ transplant recipients who undergo seromismatch transplantation (D+/R−) are at highest risk for developing CMV disease post-transplant. For seromismatch transplant recipients, monitoring for seroconversion post-transplant may be of value in predicting protection from late-onset CMV disease (52). For pre-HSCT screening among patients with underlying hematologic malignancies and hypogammaglobulinemia, serologies can occasionally be falsely negative, however. Regardless of the transplant scenario, serologies by themselves should never be used alone for diagnosing active CMV disease (7).

For detecting active CMV in the transplant recipient, molecular-based assays have become the mainstay approach. Most centers rely on nucleic acid testing of serum, plasma, or whole blood to diagnose the presence of viremia. Results can vary from assay to assay across laboratories, and although harmonization of a universal quantifying unit has helped in this regard, it remains unclear what threshold of viremia should be used for clinical decision making. This means that the clinician who evaluates a patient with a positive serum, plasma, or blood PCR still needs to determine if the patient has CMV disease. Low-level viremia is commonly reported among immunosuppressed patients with no evidence of active disease, though higher viral loads do correlate with the increased risk for signs and symptoms of disease (89). Clinical parameters, symptoms, evidence of end-organ disease, transaminitis, and leukopenia can all be helpful in this regard, as well as serial monitoring to better gauge the trajectory of the viral load. Serial assays can also be useful from a monitoring standpoint in patients undergoing preemptive therapy or patients on active treatment for CMV disease. Because of interlaboratory variability, consistent use of a single assay is necessary to interpret serial measurements, though the international standard conversion may be helpful in these cases (70). Ultimately, trends in the quantity of viremia by PCR may be more helpful over time, particularly in the setting of understanding the patient's other ongoing risk factors for disease (immunosuppression, etc.). For patients on therapy, weekly monitoring for CMV by PCR is generally recommended, though the ideal time between intervals of testing remains unclear (53). Persistent viremia as determined by PCR testing can also be a predictor of risk for relapse (80).

For diagnosing tissue-invasive disease of an organ system, biopsy and histopathology remain the gold standard. Use of culture or nucleic acid testing from bodily fluids or tissue samples may be misleading since asymptomatic shedding of CMV is common in immunosuppressed patients with chronic latent infection. Even if higher viral loads in the serum correlate with risk of tissue-invasive disease, some patients with end-organ disease, particularly gastrointestinal CMV disease, may have low-to-undetectable levels in the serum (53). Measuring CMV nucleic acid or using culture methods on respiratory tract specimens can often reflect latent infection and asymptomatic shedding and should not be relied upon to confirm CMV pneumonitis.

For patients with repeated episodes of disease or extensive exposure to ganciclovir or valganciclovir, resistance may occasionally arise. Confirmation of resistance is through genotypic testing from the serum looking for known mutations that confer resistance. Guidelines for performing genotypic testing are still in their infancy and few data are available. Increases in viral loads during the first 2 weeks of treatment may not always be enough to indicate underlying resistance, though resistance testing is often performed and recommended in these circumstances (112). For patients with underlying risk factors for resistance who do not appear to be responding to treatment after at least 2 weeks, genotypic resistance testing may be more concretely warranted, though trying alternative therapies in the interim may be necessary in patients with severe disease (53).

Future of CMV Diagnostics and Testing

Appropriate use of CMV diagnostic assays is becoming an increasingly important focus of research. Despite advances in treating and suppressing HIV infection, more patients at risk for CMV disease are being increasingly recognized. Not only does CMV remain the most important virus among the ever-growing solid-organ and hematopoietic stem cell transplant recipient population, but CMV disease as a consequence of novel immunomodulating therapies in patients with malignancy and autoimmune disorders is becoming increasingly recognized. Beyond these special populations, the impact and sequelae of CMV as a congenital infection continues to be better understood. Future research will need to focus on all these populations to help better define optimal testing strategies, the role of CMV-specific immunity, and appropriate interventions of therapy. More research is needed to increase the development of novel antiviral therapies, improve our understanding of antiviral resistance among CMV, and increase our understanding of disease prevention through use of chemoprophylaxis, and hopefully, someday an effective vaccine.

REFERENCES

1. **Smith MG.** 1956. Propagation in tissue cultures of a cytopathogenic virus from human salivary gland virus (SGV) disease. *Proc Soc Exp Biol Med* **92:**424–430.
2. **Chee MS, Bankier AT, Beck S, Bohni R, Brown CM, Cerny R, Horsnell T, Hutchison CA, Kouzarides T, Martignetti JA, Preddie E, Satchwell SC, Tomlinson P, Weston KM, Barrell BG.** 1990. Analysis of the protein-coding content of the sequence of human cytomegalovirus strain AD169. *Curr Top Microbiol Immunol* **154:**125–169.
3. **Kalejta RF.** 2008. Tegument proteins of human cytomegalovirus. *Microbiol Mol Biol Rev* **72:**249–265.
4. **Lurain NS, Kapell KS, Huang DD, Short JA, Paintsil J, Winkfield E, Benedict CA, Ware CF, Bremer JW.** 1999. Human cytomegalovirus UL144 open reading frame: sequence hypervariability in low-passage clinical isolates. *J Virol* **73:**10040–10050.
5. **Miller-Kittrell M, Sparer TE.** 2009. Feeling manipulated: cytomegalovirus immune manipulation. *Virol J* **6:**4.
6. **Plachter B, Sinzger C, Jahn G.** 1996. Cell types involved in replication and distribution of human cytomegalovirus. *Adv Virus Res* **46:**195–261.
7. **Krech U.** 1973. Complement-fixing antibodies against cytomegalovirus in different parts of the world. *Bull World Health Organ* **49:**103–106.
8. **Staras SA, Dollard SC, Radford KW, Flanders WD, Pass RF, Cannon MJ.** 2006. Seroprevalence of cytomegalovirus infection in the United States, 1988-1994. *Clin Infect Dis* **43:**1143–1151.
9. **Bate SL, Dollard SC, Cannon MJ.** 2010. Cytomegalovirus seroprevalence in the United States: the national health and nutrition examination surveys, 1988-2004. *Clin Infect Dis* **50:**1439–1447.
10. **Ross SA, Arora N, Novak Z, Fowler KB, Britt WJ, Boppana SB.** 2010. Cytomegalovirus reinfections in healthy seroimmune women. *J Infect Dis* **201:**386–389.
11. **Boppana SB, Rivera LB, Fowler KB, Mach M, Britt WJ.** 2001. Intrauterine transmission of cytomegalovirus to infants of women with preconceptional immunity. *N Engl J Med* **344:**1366–1371.
12. **Kenneson A, Cannon MJ.** 2007. Review and meta-analysis of the epidemiology of congenital cytomegalovirus (CMV) infection. *Rev Med Virol* **17:**253–276.
13. **Dollard SC, Grosse SD, Ross DS.** 2007. New estimates of the prevalence of neurological and sensory sequelae and mortality associated with congenital cytomegalovirus infection. *Rev Med Virol* **17:**355–363.
14. **Stagno S, Whitley RJ.** 1985. Herpesvirus infections of pregnancy. Part I: cytomegalovirus and Epstein-Barr virus infections. *N Engl J Med* **313:**1270–1274.
15. **Hoover DR, Saah AJ, Bacellar H, Phair J, Detels R, Anderson R, Kaslow RA.** 1993. Clinical manifestations of AIDS in the era of pneumocystis prophylaxis. Multicenter AIDS Cohort Study. *N Engl J Med* **329:**1922–1926.
16. **Sugar EA, Jabs DA, Ahuja A, Thorne JE, Danis RP, Meinert CL, Studies of the Ocular Complications of AIDS Research Group.** 2012. Incidence of cytomegalovirus retinitis in the era of highly active antiretroviral therapy. *Am J Ophthalmol* **153:**1016–24.e5.
17. **Kusne S, Shapiro R, Fung J.** 1999. Prevention and treatment of cytomegalovirus infection in organ transplant recipients. *Transpl Infect Dis* **1:**187–203.
18. **Khoury JA, Storch GA, Bohl DL, Schuessler RM, Torrence SM, Lockwood M, Gaudreault-Keener M, Koch MJ, Miller BW, Hardinger KL, Schnitzler MA, Brennan DC.** 2006. Prophylactic versus preemptive oral valganciclovir for the management of cytomegalovirus infection in adult renal transplant recipients. *Am J Transplant* **6:**2134–2143.
19. **Marr KA.** 2012. Delayed opportunistic infections in hematopoietic stem cell transplantation patients: a surmountable challenge. *Hematology (Am Soc Hematol Educ Program)* **2012:**265–270.
20. **Gluckman E, Traineau R, Devergie A, Esperou-Bourdeau H, Hirsch I.** 1992. Prevention and treatment of CMV infection after allogeneic bone marrow transplant. *Ann Hematol* **64** (Suppl)**:**A158–A161.
21. **Ozdemir E, Saliba RM, Champlin RE, Couriel DR, Giralt SA, de Lima M, Khouri IF, Hosing C, Kornblau SM, Anderlini P, Shpall EJ, Qazilbash MH, Molldrem JJ, Chemaly RF, Komanduri KV.** 2007. Risk factors associated with late cytomegalovirus reactivation after allogeneic stem cell transplantation for hematological malignancies. *Bone Marrow Transplant* **40:**125–136.
22. **Klemola E, Von Essen R, Henle G, Henle W, Clinical Features in Relation to Epstein-Barr Virus and Cytomegalovirus Antibodies.** 1970. Infectious-mononucleosis-like disease with negative heterophil agglutination test. Clinical features in relation to Epstein-Barr virus and cytomegalovirus antibodies. *J Infect Dis* **121:**608–614.
23. **Horwitz CA, Henle W, Henle G, Snover D, Rudnick H, Balfour HH Jr, Mazur MH, Watson R, Schwartz B, Muller N.** 1986. Clinical and laboratory evaluation of cytomegalovirus-induced mononucleosis in previously healthy individuals. Report of 82 cases. *Medicine (Baltimore)* **65:**124–134.
24. **Eddleston M, Peacock S, Juniper M, Warrell DA.** 1997. Severe cytomegalovirus infection in immunocompetent patients. *Clin Infect Dis* **24:**52–56.
25. **Papadakis KA, Tung JK, Binder SW, Kam LY, Abreu MT, Targan SR, Vasiliauskas EA.** 2001. Outcome of cytomegalovirus infections in patients with inflammatory bowel disease. *Am J Gastroenterol* **96:**2137–2142.
26. **Limaye AP, Kirby KA, Rubenfeld GD, Leisenring WM, Bulger EM, Neff MJ, Gibran NS, Huang ML, Santo Hayes**

TK, Corey L, Boeckh M. 2008. Cytomegalovirus reactivation in critically ill immunocompetent patients. *JAMA* **300:**413–422.

27. **Hanshaw JB, Scheiner AP, Moxley AW, Gaev L, Abel V, Scheiner B.** 1976. School failure and deafness after "silent" congenital cytomegalovirus infection. *N Engl J Med* **295:**468–470.
28. **Goderis J, De Leenheer E, Smets K, Van Hoecke H, Keymeulen A, Dhooge I.** 2014. Hearing loss and congenital CMV infection: a systematic review. *Pediatrics* **134:**972–982.
29. **Gallant JE, Moore RD, Richman DD, Keruly J, Chaisson RE, The Zidovudine Epidemiology Study Group.** 1992. Incidence and natural history of cytomegalovirus disease in patients with advanced human immunodeficiency virus disease treated with zidovudine. *J Infect Dis* **166:**1223–1227.
30. **Kotton CN.** 2013. CMV: Prevention, diagnosis and therapy. *Am J Transplant* **13**(Suppl 3)**:**24–40, quiz 40.
31. **Fishman JA, Emery V, Freeman R, Pascual M, Rostaing L, Schlitt HJ, Sgarabotto D, Torre-Cisneros J, Uknis ME.** 2007. Cytomegalovirus in transplantation - challenging the status quo. *Clin Transplant* **21:**149–158.
32. **Pescovitz M.** 2006. Valganciclovir: what is the status in solid organ transplantation? *Future Virol* **1:**147–156.
33. **Owers DS, Webster AC, Strippoli GF, Kable K, Hodson EM.** 2013. Pre-emptive treatment for cytomegalovirus viraemia to prevent cytomegalovirus disease in solid organ transplant recipients. *Cochrane Database Syst Rev* **2:**CD005133.
34. **Manuel O, Kralidis G, Mueller NJ, Hirsch HH, Garzoni C, van Delden C, Berger C, Boggian K, Cusini A, Koller MT, Weisser M, Pascual M, Meylan PR, Swiss Transplant Cohort Study.** 2013. Impact of antiviral preventive strategies on the incidence and outcomes of cytomegalovirus disease in solid organ transplant recipients. *Am J Transplant* **13:**2402–2410.
35. **Kalil AC, Levitsky J, Lyden E, Stoner J, Freifeld AG.** 2005. Meta-analysis: the efficacy of strategies to prevent organ disease by cytomegalovirus in solid organ transplant recipients. *Ann Intern Med* **143:**870–880.
36. **Asberg A, Humar A, Rollag H, Jardine AG, Mouas H, Pescovitz MD, Sgarabotto D, Tuncer M, Noronha IL, Hartmann A, VICTOR Study Group.** 2007. Oral valganciclovir is noninferior to intravenous ganciclovir for the treatment of cytomegalovirus disease in solid organ transplant recipients. *Am J Transplant* **7:**2106–2113.
37. **Martin DF, Sierra-Madero J, Walmsley S, Wolitz RA, Macey K, Georgiou P, Robinson CA, Stempien MJ, Valganciclovir Study Group.** 2002. A controlled trial of valganciclovir as induction therapy for cytomegalovirus retinitis. *N Engl J Med* **346:**1119–1126.
38. **Smith CL.** 1998. Local therapy for cytomegalovirus retinitis. *Ann Pharmacother* **32:**248–255.
39. **Florescu DF, Keck MA.** 2014. Development of CMX001 (Brincidofovir) for the treatment of serious diseases or conditions caused by dsDNA viruses. *Expert Rev Anti Infect Ther* **12:**1171–1178.
40. **Härter G, Michel D.** 2012. Antiviral treatment of cytomegalovirus infection: an update. *Expert Opin Pharmacother* **13:**623–627.
41. **Stoelben S, Arns W, Renders L, Hummel J, Mühlfeld A, Stangl M, Fischereder M, Gwinner W, Suwelack B, Witzke O, Dürr M, Beelen DW, Michel D, Lischka P, Zimmermann H, Rübsamen-Schaeff H, Budde K.** 2014. Preemptive treatment of Cytomegalovirus infection in kidney transplant recipients with letermovir: results of a Phase 2a study. *Transpl Int* **27:**77–86.
42. **Chemaly RF, Ullmann AJ, Stoelben S, Richard MP, Bornhäuser M, Groth C, Einsele H, Silverman M, Mullane KM, Brown J, Nowak H, Kölling K, Stobernack HP, Lischka P, Zimmermann H, Rübsamen-Schaeff H, Champlin RE, Ehninger G, AIC246 Study Team.** 2014. Letermovir for cytomegalovirus prophylaxis in hematopoietic-cell transplantation. *N Engl J Med* **370:**1781–1789.
43. **Kharfan-Dabaja MA, Boeckh M, Wilck MB, Langston AA, Chu AH, Wloch MK, Guterwill DF, Smith LR, Rolland AP, Kenney RT.** 2012. A novel therapeutic cytomegalovirus DNA vaccine in allogeneic haemopoietic stem-cell transplantation: a randomised, double-blind, placebo-controlled, phase 2 trial. *Lancet Infect Dis* **12:**290–299.
44. **Pass RF, Zhang C, Evans A, Simpson T, Andrews W, Huang ML, Corey L, Hill J, Davis E, Flanigan C, Cloud G.** 2009. Vaccine prevention of maternal cytomegalovirus infection. *N Engl J Med* **360:**1191–1199.
45. **Schultz DA, Chandler S.** 1991. Cytomegalovirus testing: antibody determinations and virus cultures with recommendations for use. *J Clin Lab Anal* **5:**69–73.
46. Cytomegalovirus (CMV) and Congenital CMV Infection. Interpretation of Laboratory Tests. http://www.cdc.gov/cmv/clinical/lab-tests.html (accessed 8/19/14).
47. **Dollard SC, Staras SA, Amin MM, Schmid DS, Cannon MJ.** 2011. National prevalence estimates for cytomegalovirus IgM and IgG avidity and association between high IgM antibody titer and low IgG avidity. *Clin Vaccine Immunol* **18:**1895–1899.
48. **Bate SL, Dollard SC, Cannon MJ.** 2010. Cytomegalovirus seroprevalence in the United States: the national health and nutrition examination surveys, 1988-2004. *Clin Infect Dis* **50:**1439–1447.
49. **Weber B, Fall EM, Berger A, Doerr HW.** 1999. Screening of blood donors for human cytomegalovirus (HCMV) IgG antibody with an enzyme immunoassay using recombinant antigens. *J Clin Virol* **14:**173–181.
50. **Seed CR, Piscitelli LM, Maine GT, Lazzarotto T, Doherty K, Stricker R, Stricker R, Iriarte B, Patel C.** 2009. Validation of an automated immunoglobulin G-only cytomegalovirus (CMV) antibody screening assay and an assessment of the risk of transfusion transmitted CMV from seronegative blood. *Transfusion* **49:**134–145.
51. **Lazzarotto T, Brojanac S, Maine GT, Landini MP.** 1997. Search for cytomegalovirus-specific immunoglobulin M: comparison between a new western blot, conventional western blot, and nine commercially available assays. *Clin Diagn Lab Immunol* **4:**483–486.
52. **Humar A, Mazzulli T, Moussa G, Razonable RR, Paya CV, Pescovitz MD, Covington E, Alecock E, Valganciclovir Solid Organ Transplant Study Group.** 2005. Clinical utility of cytomegalovirus (CMV) serology testing in high-risk CMV D+/R− transplant recipients. *Am J Transplant* **5:**1065–1070.
53. **Kotton CN, Kumar D, Caliendo AM, Asberg A, Chou S, Danziger-Isakov L, Humar A.** 2013. Transplantation Society International CMV Consensus Group. Updated international consensus guidelines on the management of cytomegalovirus in solid-organ transplantation. *Transplantation.* **96**(4)**:**333–360.
54. **Ross SA, Novak Z, Pati S, Boppana SB.** 2011. Overview of the diagnosis of cytomegalovirus infection. *Infect Disord Drug Targets* **11:**466–474.
55. **Patel R, Klein DW, Espy MJ, Harmsen WS, Ilstrup DM, Paya CV, Smith TF.** 1995. Optimization of detection of cytomegalovirus viremia in transplantation recipients by shell vial assay. *J Clin Microbiol* **33:**2984–2986.
56. **Boeckh M, Huang M, Ferrenberg J, Stevens-Ayers T, Stensland L, Nichols WG, Corey L.** Optimization of quantitative detection of cytomegalovirus DNA in plasma by real-time PCR. 2004. *J Clin Microbiol.* 42(3):1142–1148.
57. **Razonable RR, Humar A, AST Infectious Diseases Community of Practice.** 2013. Cytomegalovirus in solid organ transplantation. *Am J Transplant* **13**(Suppl 4)**:**93–106.
58. **Chemaly RF, Yen-Lieberman B, Castilla EA, Reilly A, Arrigain S, Farver C, Avery RK, Gordon SM, Procop GW.** 2004. Correlation between viral loads of cytomegalovirus in blood and bronchoalveolar lavage specimens from lung transplant recipients determined by histology and immunohistochemistry. *J Clin Microbiol* **42:**2168–2172.
59. **Solans EP, Yong S, Husain AN, Eichorst M, Gattuso P.** 1997. Bronchioloalveolar lavage in the diagnosis of CMV pneumonitis in lung transplant recipients: an immunocytochemical study. *Diagn Cytopathol* **16:**350–352.

60. **Ljungman P, Griffiths P, Paya C.** 2002. Definitions of cytomegalovirus infection and disease in transplant recipients. *Clin Infect Dis* **34:**1094–1097.
61. **Halme L, Lempinen M, Arola J, Sarkio S, Höckerstedt K, Lautenschlager I.** 2008. High frequency of gastroduodenal cytomegalovirus infection in liver transplant patients. *APMIS* **116:**99–106.
62. **van der Bij W, Schirm J, Torensma R, van Son WJ, Tegzess AM, The TH.** 1988. Comparison between viremia and antigenemia for detection of cytomegalovirus in blood. *J Clin Microbiol* **26:**2531–2535.
63. **Schröeder R, Michelon T, Fagundes I, Bortolotto A, Lammerhirt E, Oliveira J, Santos A, Bittar A, Keitel E, Garcia V, Neumann J, Saitovitch D.** 2005. Antigenemia for cytomegalovirus in renal transplantation: choosing a cutoff for the diagnosis criteria in cytomegalovirus disease. *Transplant Proc* **37:**2781–2783.
64. **Baldanti F, Lilleri D, Gerna G.** 2008. Monitoring human cytomegalovirus infection in transplant recipients. *J Clin Virol* **41:**237–241.
65. **Gerna G, Baldanti F, Lilleri D, Parea M, Torsellini M, Castiglioni B, Vitulo P, Pellegrini C, Viganò M, Grossi P, Revello MG.** 2003. Human cytomegalovirus pp67 mRNAemia versus pp65 antigenemia for guiding preemptive therapy in heart and lung transplant recipients: a prospective, randomized, controlled, open-label trial. *Transplantation* **75:**1012–1019.
66. **Saracino A, Colucci R, Latorraca A, Muscaridola N, Procida C, Di Noia I, Santospirito VE, Santarsia G.** 2013. The effects of preemptive therapy using a very low threshold of pp65 antigenemia to prevent cytomegalovirus disease in kidney transplant recipients: a single-center experience. *Transplant Proc* **45:**182–184.
67. **Cardeñoso L, Pinsky BA, Lautenschlager I, Aslam S, Cobb B, Vilchez RA, Hirsch HH.** 2013. CMV antigenemia and quantitative viral load assessments in hematopoietic stem cell transplant recipients. *J Clin Virol* **56:**108–112.
68. **Razonable RR, Hayden RT.** 2013. Clinical utility of viral load in management of cytomegalovirus infection after solid organ transplantation. *Clin Microbiol Rev.* **26**(4):703–727.
69. **Boeckh M, Huang M, Ferrenberg J, Stevens-Ayers T, Stensland L, Nichols WG, Corey L.** 2004. Optimization of quantitative detection of cytomegalovirus DNA in plasma by real-time PCR. *J Clin Microbiol.* **42**(3):1142–1148.
70. **Pang XL, Fox JD, Fenton JM, Miller GG, Caliendo AM, Preiksaitis JK, American Society of Transplantation Infectious Diseases Community of Practice, Canadian Society of Transplantation.** 2009. Interlaboratory comparison of cytomegalovirus viral load assays. *Am J Transplant* **9:** 258–268.
71. **Fryer JF, Heath AB, Anderson R, Minor PD.** 2010. Collaborative study to evaluate the proposed 1st WHO International Standard for human cytomegalovirus (HCMV) for nucleic acid amplification (NAT)-based assays. WHOECBS (Expert Committee on Biological Standardization) Report. WHO/BS/10.2138.
72. **Kraft CS, Armstrong WS, Caliendo AM.** 2012. Interpreting quantitative cytomegalovirus DNA testing: understanding the laboratory perspective. *Clin Infect Dis* **54:**1793–1797.
73. **Razonable RR, Åsberg A, Rollag H, Duncan J, Boisvert D, Yao JD, Caliendo AM, Humar A, Do TD.** 2013. Virologic suppression measured by a cytomegalovirus (CMV) DNA test calibrated to the World Health Organization international standard is predictive of CMV disease resolution in transplant recipients. *Clin Infect Dis.* 56(11):1546–1553.
74. **Hayden RT, Gu Z, Ingersoll J, Abdul-Ali D, Shi L, Pounds S, Caliendo AM.** 2013. Comparison of droplet digital PCR to real-time PCR for quantitative detection of cytomegalovirus. *J Clin Microbiol* **51:**540–546.
75. COBAS® AmpliPrep/COBASO TaqMan® CMV Test (Roche Molecular Systems, Inc. Branchburg, NJ:2013). COBAS TaqMan CMV Test FDA approved package insert. http://www.accessdata.fda.gov/cdrh_docs/pdf11/P110037c.pdf
76. *artus*® CMV RGQ MDx Kit Test (Qiagen:2014). COBAS TaqMan CMV Test FDA approved package insert. http://www.accessdata.fda.gov/cdrh_docs/pdf13/P130027c.pdf
77. **Kerschner H, Bauer C, Schlag P, Lee S, Goedel S, Popow-Kraupp T.** 2011. Clinical evaluation of a fully automated CMV PCR assay. *J Clin Virol.* **50**(4):281–286.
78. **Hirsch HH, Lautenschlager I, Pinsky BA, Cardeñoso L, Aslam S, Cobb B, Vilchez RA, Valsamakis A.** 2013. An international multicenter performance analysis of cytomegalovirus load tests. *Clin Infect Dis* **56:**367–373.
79. **Caliendo AM.** 2013. The long road toward standardization of viral load testing for cytomegalovirus. *Clin Infect Dis* **56:**374–375.
80. **Lisboa LF, Asberg A, Kumar D, Pang X, Hartmann A, Preiksaitis JK, Pescovitz MD, Rollag H, Jardine AG, Humar A.** 2011. The clinical utility of whole blood versus plasma cytomegalovirus viral load assays for monitoring therapeutic response. *Transplantation* **91:**231–236.
81. **Razonable RR, Brown RA, Wilson J, Groettum C, Kremers W, Espy M, Smith TF, Paya CV.** 2002. The clinical use of various blood compartments for cytomegalovirus (CMV) DNA quantitation in transplant recipients with CMV disease. *Transplantation* **73:**968–973.
82. **Tang W, Elmore SH, Fan H, Thorne LB, Gulley ML.** 2008. Cytomegalovirus DNA measurement in blood and plasma using Roche LightCycler CMV quantification reagents. *Diagn Mol Pathol* **17:**166–173.
83. **Salmon-Céron D, Mazeron MC, Chaput S, Boukli N, Senechal B, Houhou N, Katlama C, Matheron S, Fillet AM, Gozlan J, Leport C, Jeantils V, Freymuth F, Costagliola D.** 2000. Plasma cytomegalovirus DNA, pp65 antigenaemia and a low CD4 cell count remain risk factors for cytomegalovirus disease in patients receiving highly active antiretroviral therapy. *AIDS* **14:**1041–1049.
84. **Erice A, Tierney C, Hirsch M, Caliendo AM, Weinberg A, Kendall MA, Polsky B, AIDS Clinical Trials Group Protocol 360 Study Team.** 2003. Cytomegalovirus (CMV) and human immunodeficiency virus (HIV) burden, CMV end-organ disease, and survival in subjects with advanced HIV infection (AIDS Clinical Trials Group Protocol 360). *Clin Infect Dis* **37:**567–578.
85. **Westall GP, Michaelides A, Williams TJ, Snell GI, Kotsimbos TC.** 2004. Human cytomegalovirus load in plasma and bronchoalveolar lavage fluid: a longitudinal study of lung transplant recipients. *J Infect Dis* **190:**1076–1083.
86. **Chemaly RF, Yen-Lieberman B, Chapman J, Reilly A, Bekele BN, Gordon SM, Procop GW, Shrestha N, Isada CM, Decamp M, Avery RK.** 2005. Clinical utility of cytomegalovirus viral load in bronchoalveolar lavage in lung transplant recipients. *Am J Transplant* **5:**544–548.
87. **Wiita AP, Roubinian N, Khan Y, Chin-Hong PV, Singer JP, Golden JA, Miller S.** 2012. Cytomegalovirus disease and infection in lung transplant recipients in the setting of planned indefinite valganciclovir prophylaxis. *Transpl Infect Dis* **14:**248–258.
88. **Riise GC, Andersson R, Bergström T, Lundmark A, Nilsson FN, Olofsson S.** 2000. Quantification of cytomegalovirus DNA in BAL fluid: a longitudinal study in lung transplant recipients. *Chest* **118:**1653–1660.
89. **Emery VC, Sabin CA, Cope AV, Gor D, Hassan-Walker AF, Griffiths PD.** 2000. Application of viral-load kinetics to identify patients who develop cytomegalovirus disease after transplantation. *Lancet* **355** (9220):2032–2036.
90. **Egli, A, Humar, A, Kumar, D.** 2012. State-of-the-art monitoring of cytomegalovirus-specific cell-mediated immunity after organ transplant: a primer for the clinician. *Clin Infect Dis*. **55**(12):1678–1689.
91. **Schmidt T, Schub D, Wolf M, Dirks J, Ritter M, Leyking S, Singh M, Zawada AM, Blaes-Eise AB, Samuel U, Sester U, Sester M.** 2014. Comparative analysis of assays for detection of cell-mediated immunity toward cytomegalovirus and M. tuberculosis in samples from deceased organ donors. *Am J Transplant*. Jul 10.

92. **Abate D, Saldan A, Mengoli C, Fiscon M, Silvestre C, Fallico L, Peracchi M, Furian L, Cusinato R, Bonfante L, Rossi B, Marchini F, Sgarabotto D, Rigotti P, Palù G.** 2013. Comparison of cytomegalovirus (CMV) enzyme-linked immunosorbent spot and CMV quantiferon gamma interferon-releasing assays in assessing risk of CMV infection in kidney transplant recipients. *J Clin Microbiol*. **51**(8):2501-2507.
93. **Manuel O, Husain S, Kumar D, Zayas C, Mawhorter S, Levi ME, Kalpoe J, Lisboa L, Ely L, Kaul DR, Schwartz BS, Morris MI, Ison MG, Yen-Lieberman B, Sebastian A, Assi M, Humar A.** 2013. Assessment of cytomegalovirus-specific cell-mediated immunity for the prediction of cytomegalovirus disease in high-risk solid-organ transplant recipients: a multicenter cohort study. *Clin Infect Dis* **56:**817–824.
94. **Gilbert C, Boivin G.** 2005. Human cytomegalovirus resistance to antiviral drugs. *Antimicrob Agents Chemother* **49:**873–883.
95. **Göhring K, Wolf D, Bethge W, Mikeler E, Faul C, Vogel W, Vöhringer MC, Jahn G, Hamprecht K.** 2013. Dynamics of coexisting HCMV-UL97 and UL54 drug-resistance associated mutations in patients after haematopoietic cell transplantation. *J Clin Virol*: **57**: 43-49.
96. **Hakki M, Chou S.** 2011. The biology of cytomegalovirus drug resistance. *Curr Opin Infect Dis* **24:**605–611.
97. **Sahoo MK, Lefterova MI, Yamamoto F, Waggoner JJ, Chou S, Holmes SP, Anderson MW, Pinsky BA.** 2013. Detection of cytomegalovirus drug resistance mutations by next-generation sequencing. *J Clin Microbiol* **51:**3700–3710.
98. **Chou S, Ercolani RJ, Sahoo MK, Lefterova MI, Strasfeld LM, Pinsky BA.** 2014. Improved detection of emerging drug-resistant mutant cytomegalovirus subpopulations by deep sequencing. *Antimicrob Agents Chemother* **58:**4697–4702.
99. **Chou S.** 1990. Newer methods for diagnosis of cytomegalovirus infection. *Rev Infect Dis* **12**(Suppl 7)**:**S727–S736.
100. **Lazzarotto T, Spezzacatena P, Pradelli P, Abate DA, Varani S, Landini MP.** 1997. Avidity of immunoglobulin G directed against human cytomegalovirus during primary and secondary infections in immunocompetent and immunocompromised subjects. *Clin Diagn Lab Immunol* **4:**469–473.
101. **Liesnard C, Donner C, Brancart F, Gosselin F, Delforge ML, Rodesch F.** 2000. Prenatal diagnosis of congenital cytomegalovirus infection: prospective study of 237 pregnancies at risk. *Obstet Gynecol* **95:**881–888.
102. **Azam AZ, Vial Y, Fawer CL, Zufferey J, Hohlfeld P.** 2001. Prenatal diagnosis of congenital cytomegalovirus infection. *Obstet Gynecol* **97:**443–448.
103. **Nelson CT, Istas AS, Wilkerson MK, Demmler GJ.** 1995. PCR detection of cytomegalovirus DNA in serum as a diagnostic test for congenital cytomegalovirus infection. *J Clin Microbiol* **33:**3317–3318.
104. **Demmler GJ, Buffone GJ, Schimbor CM, May RA.** 1988. Detection of cytomegalovirus in urine from newborns by using polymerase chain reaction DNA amplification. *J Infect Dis* **158:**1177–1184.
105. **Jabs DA, Gilpin AM, Min YI.Erice A, Kempen JH, Quinn TC.** 2002. HIV and cytomegalovirus viral load and clinical outcomes in AIDS and cytomegalovirus retinitis patients: Monoclonal Antibody Cytomegalovirus Retinitis Trial. *AIDS* **16:**877-887.
106. **Group SOC.** 1997. Cytomegalovirus (CMV) culture results, drug resistance, and clinical outcome in patients with AIDS and CMV retinitis treated with foscarnet or ganciclovir. Studies of Ocular Complications of AIDS (SOCA) in collaboration with the AIDS Clinical Trial Group. *J Infect Dis* **176:**50–58.
107. **Spector SA, Wong R, Hsia K, Pilcher M, Stempien MJ.** 1998. Plasma cytomegalovirus (CMV) DNA load predicts CMV disease and survival in AIDS patients. *J Clin Invest* **101:**497–502.
108. **Knapp AB, Horst DA, Eliopoulos G, Gramm HF, Gaber LW, Falchuk KR, Falchuk ZM, Trey C.** 1983. Widespread cytomegalovirus gastroenterocolitis in a patient with acquired immunodeficiency syndrome. *Gastroenterology* **85:**1399–1402.
109. **Wallace JM, Hannah J.** 1987. Cytomegalovirus pneumonitis in patients with AIDS. Findings in an autopsy series. *Chest* **92:**198–203.
110. **Mann M, Shelhamer JH, Masur H, Gill VJ, Travis W, Solomon D, Manischewitz J, Stock F, Lane HC, Ognibene FP.** 1997. Lack of clinical utility of bronchoalveolar lavage cultures for cytomegalovirus in HIV infection. *Am J Respir Crit Care Med* **155:**1723–1728.
111. **Arribas JR, Clifford DB, Fichtenbaum CJ, Commins DL, Powderly WG, Storch GA.** 1995. Level of cytomegalovirus (CMV) DNA in cerebrospinal fluid of subjects with AIDS and CMV infection of the central nervous system. *J Infect Dis* **172:**527–531.
112. **Boivin G, Goyette N, Rollag H, Jardine AG, Pescovitz MD, Asberg A, Ives J, Hartmann A, Humar A.** 2009. Cytomegalovirus resistance in solid organ transplant recipients treated with intravenous ganciclovir or oral valganciclovir. *Antivir Ther* **14:**697–704.

Epstein-Barr Virus

DERRICK CHEN AND BELINDA YEN-LIEBERMAN

28

VIRAL CLASSIFICATION AND BIOLOGY

In 1964, three researchers, Michael Anthony Epstein, Bert Achong, and Yvonne Barr, published in *The Lancet* their discovery of what would later be known as Epstein-Barr virus (EBV) (1). Before this breakthrough, Epstein had been studying chicken tumor viruses at the Middlesex Hospital in London. In 1961, Epstein attended a lecture by Denis P. Burkitt, a British surgeon who had been stationed in Uganda, in which he detailed the relationship between Burkitt lymphoma and the geographical patterns of temperature, rainfall, and altitude (2). Suspicious of a viral etiology, Epstein spent the next few years attempting to isolate viral material from lymphoma biopsy samples taken from tumors of Ugandan children sent weekly by Burkitt to London. Despite a switch to tissue culture and assistance from Yvonne Barr and Bert Achong, isolation of a virus was unsuccessful. On December 5, 1963, the sample sent by Burkitt was delayed due to inclement weather, which fortuitously resulted in viable, free-floating, lymphoma cells that astonishingly grew in culture and demonstrated viral particles by electron microscopy (1). Though this initial finding was met with skepticism, mounting evidence over the following decades eventually resulted in the acceptance of EBV as the cause of Burkitt lymphoma, and the World Health Organization and the International Agency for Research and Cancer declared EBV as a group 1 carcinogen in the 1990s (2). Approximately 90% of people worldwide are carriers of latent EBV, and it is estimated that the virus causes more than 200,000 cancers each year, primarily B-cell neoplasms, which account for 1.5% of all cancers (2). There are two types of EBV, EBV-1 (A type) and EBV-2 (B type), that are distinguished by differences found primarily in their latent genes. EBV in America and Europe are much more likely to be EBV-1, whereas EBV found in Africa may be EBV-1 or EBV-2.

EBV, also known as human herpesvirus 4, is a *Lymphocryptovirus* belonging to the subfamily *Gammaherpesvirinae* of the family *Herpesviridae*. It has a linear, 172-kb, double-stranded DNA genome with approximately 100 genes that is encased within an icosahedral capsid surrounded by tegument proteins underlying the outer lipid bilayer envelope. Embedded within the viral envelope are numerous proteins and glycoproteins including the following: gp350 (*BLLF1*), which binds to B cells; gp42 (*BZLF2*), which binds B cells and triggers fusion; gH/gp85 (*BDLF3*), which binds epithelial cells and triggers fusion; gL/gp25 (*BKRF2*), which serves as a chaperone for gH and activates/recruits gB; gB/gp110 (*BALF4*), which catalyzes membrane fusion; and BMRF-2 (*BMRF2*), which binds epithelial cells. Attachment to B cells is mediated by gp350 binding to B cell-specific complement receptor 2 (CR2 or CD21) or to complement receptor 1 (CR1 or CD35) (3). Certain epithelial cells, such as those found in the tonsils, have also been found to express low levels of CD21, which may allow for attachment. Alternatively, attachment to epithelial cells may occur via binding to specific epithelial cell surface integrins such as $\alpha v\beta 5$, $\alpha v\beta 6$, and $\alpha v\beta 8$ by the KGD binding motif of gH/gL. Fusion in B cells also utilizes gH/gL but requires it to be stably complexed with gp42, which interacts with HLA class II molecules on the surface of cells. For epithelial cells, however, gp42 can block the KGD binding motif, thus inhibiting viral attachment to epithelial integrins and subsequent fusion. Therefore, gp42 appears to have an important role in the switching between B-cell and epithelial cell tropism (3). gB is required for fusion in both B cells and epithelial cells. Within a few hours after attachment, the linear viral genome localizes to the host cell nucleus where it circularizes, mediated by terminal repeats, and persists in episomal form. A nonproductive infection of resting memory B cells is quickly established, and the viral lytic cycle is only seen in differentiated plasma cells or differentiated epithelial cells (3). Establishment and maintenance of an immortalized state is accomplished through various transcription programs involving at least 10 latent genes, including EBV nuclear antigens (EBNA-1, EBNA-2, EBNA-3A/EBNA-3, EBNA-3B/EBNA-4, EBNA-3C/EBNA-6, and EBNA-LP/EBNA-5), latent membrane proteins (LMP-1 and LMP-2A/B), and EBV-encoded RNAs (EBER-1 and EBER-2). EBV initially infects and activates naïve B cells of Waldeyer's tonsillar ring of nasopharyngeal and oropharyngeal lymphoid tissue to become proliferating blasts, which ultimately develop into memory B cells that primarily circulate between peripheral blood and Waldeyer's ring (4). EBV effectively evades immune detection by persisting in memory B cells and halting expression of virtually all protein-encoding genes. When an infected memory B cell eventually divides, expression of EBNA-1 alone is sufficient to permit viral episome replication by the host DNA polymerase. Several latency programs

doi:10.1128/9781555819156.ch28

have been described and linked to different diseases: type I seen in Burkitt lymphoma, in which only EBNA-1 is expressed; type II seen in Hodgkin lymphoma and nasopharyngeal carcinoma, in which EBNA-1 and LMP-1/2 are expressed; and type III seen in acute infectious mononucleosis and posttransplant lymphoproliferative disorder, in which all latent genes are expressed. Virtually all latently infected cells express EBERs in abundance, making EBER *in situ* hybridization a particularly useful clinical test for identifying and localizing EBV-infected cells. To allow for infection in new hosts, reactivation of latent virus occurs in sites that facilitate transmission, such as in Waldeyer's ring when memory B cells differentiate into plasma cells, thus permitting the spread of EBV through saliva. Switching from latent to lytic infection is dependent upon the transcription of BamHI Z fragment leftward open reading frame 1 (BZLF-1), which is followed by expression of early antigen (EA), synthesis of viral DNA and viral capsid antigen (VCA), and ultimately the lysis of the host cell and release of intact virions.

EPIDEMIOLOGY

Over 90% of people are infected by EBV at some point during their lifetime. Worldwide, primary EBV infection occurs within the first few years of life and does not usually manifest as infectious mononucleosis (IM); however, if EBV infection occurs later in life, often in adolescence and young adulthood, 25% to 50% of cases will result in an IM syndrome. Earlier infections tend to occur in resource-poor regions, while higher-resource regions tend to see relatively more EBV infections occurring at older ages. Younger age, health insurance coverage, higher household income, and education level have been associated with lower prevalence (5). Transmission occurs through exposure of the oropharynx to infectious saliva of an EBV-positive individual. EBV transverses the oropharyngeal mucosa to infect underlying B cells, such as those in the tonsils and adenoids, and disseminates via the lymphoreticular system during a 30- to 50-day incubation period (6). Other bodily fluids such as blood and semen have a questionable role in spreading the virus. EBV-seronegative solid organ or hematopoietic stem cell transplant recipients can be infected via transplantation from seropositive donors.

CLINICAL SIGNIFICANCE

Infectious Mononucleosis

IM, also known as "mono," the "kissing disease," or "glandular fever," presents with a triad of fever, lymphadenopathy, and pharyngitis in over 50% of cases (6). There may be a wide array of additional clinical features of IM which are systemic (malaise), oropharyngeal (airway obstruction, palatal petechiae), lymphoid (splenomegaly, splenic rupture), hepatic (hepatitis, jaundice), cutaneous (morbilliform rash, drug induced rash, Gianotti-Crosti syndrome, acute genital ulcers), cardiac (electrocardiogram abnormalities with T-wave changes, cardiac conduction defects, myocarditis, pericarditis), pulmonary (pneumonitis), renal (microscopic hematuria, interstitial nephritis, glomerulonephritis), neural (meningitis, encephalitis, mono- or polyneuritis, transverse myelitis), and hematological (thrombocytopenia, hemolytic anemia, neutropenia) (7, 8). Secondary infections, such as with β-hemolytic *Streptococcus*, may be a complication of IM.

The majority of IM syndromes are due to primary infection by EBV (9). However, other etiologies of this syndrome, such as infections by cytomegalovirus, human herpesvirus-6, human immunodeficiency virus, or toxoplasmosis, may be difficult to distinguish from EBV based on clinical presentation alone (9, 10). The sensitivities of various clinical features for identifying heterophile antibody-positive IM in patients with pharyngitis are as follows: fatigue (93%), any cervical adenopathy (87%), anterior cervical adenopathy (70%), headache (60%), inguinal adenopathy (53%), posterior cervical adenopathy (40%), palatal petechiae (27%), axillary adenopathy (27%), temperature ≥37.5°C (27%), and splenomegaly (7%); whereas the specificities of these clinical features for heterophile antibody-positive IM are splenomegaly (99%), palatal petechiae (95%), axillary adenopathy (91%), posterior cervical adenopathy (87%), temperature ≥37.5°C (84%), inguinal adenopathy (82%), any cervical adenopathy (58%), headache (55%), anterior cervical adenopathy (43%), and fatigue (24%) (10).

Lymphoproliferative Diseases Associated with EBV

Chronic Active EBV Infection

Chronic active EBV infection, as its name suggests, is a life-threatening condition characterized by persistent or recurrent infectious mononucleosis-like symptoms such as fever, lymphadenopathy, hepatosplenomegaly, and cytopenias. In the majority of EBV infections in immunocompetent individuals, EBV-infected cells are controlled, but not entirely eliminated, through an adaptive immune response, and most patients have an uncomplicated course (11). In some rare instances, however, EBV-infected T and NK lymphocytes proliferate uncontrollably despite an apparently functional immune system, and the pathophysiology behind this response remains unclear. Various terms and criteria have been used to describe this entity, and publications on this disease have reported varying clinical manifestations and outcomes. Proposed diagnostic criteria for chronic active EBV infection involve the following: persistent or recurrent infectious mononucleosis-like symptoms; unusual pattern of anti-EBV antibodies with raised anti-VCA and anti-EA and/or detection of increased EBV genomes in affected tissues, including the peripheral blood; and chronic illness that cannot be explained by other known disease processes at diagnosis (12). Infectious mononucleosis-like symptoms include prolonged or intermittent fever, fatigue, sore throat, headache, myalgia, arthralgia, lymph node tenderness and pain, swelling of lymph nodes, hepatosplenomegaly, and complications affecting the hematological, digestive, neurological, pulmonary, ocular, dermal, and/or cardiovascular systems.

Hemophagocytic Lymphohistiocytosis

In hemophagocytic lymphohistiocytosis, abnormal cytokine release due to dysfunctional T and NK cells results in histiocyte proliferation and activation. The histiocytes infiltrate into the reticuloendothelial system where they phagocytose blood cells, leading to cytopenias, fever, organomegaly, neurologic symptoms, and lymphadenopathy. Hemophagocytic lymphohistiocytosis can be categorized as being primary/genetic (e.g., familial form and forms associated with immune deficiency syndromes like Chediak-Higashi syndrome, severe combined immunodeficiency, and X-linked lymphoproliferative syndrome) or secondary/reactive (e.g., macrophage activation syndrome associated with autoimmune diseases, malignancy associated forms, and infection associated forms) (13). While bacterial, fungal, and parasitic infections may result in hemophagocytic lymphohistiocytosis, viral etiologies are the most common and EBV has been implicated in up to 74% of cases in which an infectious agent was identified (13). Diagnostic guidelines for hemophagocytic lymphohistiocytosis were proposed by the Histiocyte Society and are as follows: molecular identification of a gene mutation associated with hemophagocytic lymphohistiocytosis (e.g., *PRF1*, *UNC13D*, *STX11*, *STXBP2*, *RAB27A*, and *SH2D1A*) or the presence of five out of eight diagnostic criteria (fever, splenomegaly, cytopenia affecting two or more lineages, hypertriglyceridemia and/or hypofibrinogenemia, hyperferritinemia, hemophagocytosis in bone marrow/spleen/lymph nodes without evidence of malignancy, low or absent NK-cell activity, and elevated soluble CD25) (14). Upon diagnosis of hemophagocytic lymphohistiocytosis, potential underlying etiologies should be investigated.

X-Linked Lymphoproliferative Syndrome

X-linked lymphoproliferative syndrome, also named Duncan disease after the first identified family, most commonly (60% to 70%) is caused by a mutation in the *SH2D1A* gene, which codes for the signaling lymphocyte activation molecule (SLAM)-associated protein (SAP) (15). SAP protein is found on T cells, NK cells, and NKT cells, but is absent in most normal B cells. Its mutation leads to NK and T-cell dysfunction and failure to clear infected B lymphocytes. There are more than 70 *SH2D1A* mutations including missense, nonsense, and deletion mutations that result in deficient SAP protein expression (16). The type of mutation is not correlated with disease severity. Because *SH2D1A* is located on the long arm of the X chromosome (Xq25), X-linked lymphoproliferative syndrome predominantly affects males, but can also occur in female carriers due to skewed lyonization or other mechanisms (17). The syndrome typically occurs early in childhood, is triggered by EBV infection, and is characterized by fulminant infectious mononucleosis, dysgammaglobulinemia, and lymphoma. Patients are often in good health prior to EBV infection, but subsequently develop lymphoproliferation and hemophagocytic lymphohistiocytosis after infection (16). Other manifestations include lymphocytic vasculitis, aplastic anemia, lymphomatoid granulomatosis, and autoimmune conditions like colitis and psoriasis. Dysgammaglobulinemia and lymphoma can occur in EBV-positive or EBV-negative patients, and mortality for those with a history of prior EBV infection approaches 96% (16). Most patients die before 10 years of age. A diagnosis of X-linked lymphoproliferative syndrome should be considered in all male patients who present with fatal or near-fatal EBV infection, hemophagocytic lymphohistiocytosis at a young age, common variable immunodeficiency, or other hypogammaglobulinemia, or when there is a family history of X-linked lymphoproliferative syndrome (16).

Mutations in several other genes have been described to cause X-linked lymphoproliferative syndromes and related disorders, including X-linked inhibitor of apoptosis (*XIAP*), interleukin-2-inducible T-cell kinase (*ITK*), *CD27*, and *CORO1A* (15). In particular, XIAP deficiency, due to various deletions or point mutations in *XIAP* causing loss of protein function, is responsible for 20% to 30% of X-linked lymphoproliferative syndrome cases (it is also known as XLP type 2). The diseases caused by SAP, XIAP, ITK, CD27, and Coronin 1A deficiencies may have overlapping clinical characteristics, such as EBV susceptibility, but there are notable differences (15). SAP and XIAP deficiencies are both X-linked and are associated with hemophagocytic lymphohistiocytosis, while ITK, CD27, and Coronin 1A deficiencies are autosomal recessive and generally not associated with hemophagocytic lymphohistiocytosis. XIAP deficiency is associated with colitis and is not associated with B-cell lymphoproliferative disorders or lymphomas, in contrast to deficiencies in the other four proteins for which the opposite is true. Although *SH2D1A* and *XIAP* are both located at Xq25 and share certain clinical features, XIAP deficiency may be better classified as X-linked familial hemophagocytic lymphohistiocytosis (18, 19). Like SAP deficiency, the only cure for XIAP deficiency is hematopoietic stem cell transplantation. XIAP deficiency, however, tends to be less severe and has a better prognosis than SAP deficiency.

Posttransplant Lymphoproliferative Disorders

As the name indicates, posttransplant lymphoproliferative disorder (PTLD) is a complication of solid organ or hematopoietic stem cell transplantation involving lymphocytic or plasmacytic proliferations that occur in the setting of iatrogenic immunosuppression. EBV is positive in >90% of B-cell PTLD and >70% of T-cell PTLD (20). While EBV plays a significant role in the development of PTLD, EBV-negative PTLD does occur and is more often seen in adults and presents later (greater than 5 years) after transplantation compared to EBV-positive PTLD (21). Functioning T cells are important for controlling EBV infection, and levels of cytotoxic $CD8^+$ T lymphocytes correlate with the magnitude of circulating EBV and likelihood of developing PTLD (21). With less immunosuppression, patients have better cytotoxic $CD8^+$ T-lymphocyte function and are less likely to develop PTLD. In contrast, patients who receive anti-T lymphocyte antibodies like OKT3 and thymoglobulin are at increased risk for developing PTLD (21). For patients receiving allogeneic stem cell transplants, use of T cell-depleted stem cell transplants and administration of antithymocyte globulin are predictive for development of EBV lymphoproliferative disease (22). Primary EBV infection is a significant risk factor, and therefore pediatric patients are at greater risk than adults for the development of PTLD (23). In transplant recipients, EBV infection has a broad range of clinical manifestations from asymptomatic to infectious mononucleosis to life-threatening viral sepsis and PTLD.

EBV-Associated Lymphomas

EBV was first discovered in Burkitt lymphoma cells in 1964 (1). Since then, EBV has been associated with a wide variety of other lymphomas including Hodgkin and non-Hodgkin lymphomas. EBV-associated lymphomas are a diverse and heterogeneous group with variable frequencies of EBV-positivity. EBV is seen in 40% of classical Hodgkin lymphomas (>95% of HIV-related Hodgkin lymphomas are positive for EBV) and in 5% of all non-Hodgkin lymphomas (24). Further subclassification of non-Hodgkin lymphomas shows variable frequency of EBV (Table 1) (20, 24).

Epithelial Lesions Associated with EBV

Oral Hairy Leukoplakia

Oral hairy leukoplakia is a benign lesion found in patients with human immunodeficiency virus infection as well as those with immunosuppression. As its name suggests, oral hairy leukoplakia presents as unilateral or bilateral white to gray patches in the oral cavity that cannot be scraped off, forming prominent folds or projections that sometimes resemble hairs but more frequently have a corrugated or shaggy

TABLE 1 Frequency of EBV in non-Hodgkin lymphomas[a]

Lymphoma type	Frequency of EBV-related cases (%)
EBV-positive diffuse large B-cell lymphoma of the elderly	100
Diffuse large B-cell lymphoma associated with chronic inflammation	70–90
Primary effusion lymphoma	70–80
Plasmablastic lymphoma	60–70
HIV-associated oral type plasmablastic lymphoma	100
HIV-associated primary central nervous system lymphoma	95–100
Extranodal NK/T cell lymphoma, nasal type	95–100
Angioimmunoblastic T-cell lymphoma	80–90
Lymphomatoid granulomatosis	90–100
Richter syndrome	15
Endemic Burkitt lymphoma	95–100
Sporadic Burkitt lymphoma	20–30
HIV-associated Burkitt lymphoma	25–35
Peripheral T-cell lymphoma, not otherwise specified	40

[a]Modified and reprinted from the *Current Hematologic Malignancy Reports* and the *Journal of Molecular Diagnostics* with permission from the publishers (20, 24).

appearance (25). While this lesion occurs most frequently on the lateral border of the tongue (81% of cases), it can be found on the other areas of the tongue (8%), buccal mucosa (8%), and retromolar area (3%) (26). It is usually asymptomatic but can produce soreness, burning, mild pain, and alteration of taste. The differential diagnosis of oral hairy leukoplakia includes idiopathic leukoplakia, smoker's keratosis, frictional keratosis, acute pseudomembranous candidiasis, chronic hyperplasic candidiasis, "plaquelike" type of lichen planus, lichenoid reaction, white sponge nevus, and oral graft-versus-host disease (25).

Nasopharyngeal Carcinoma

Nasopharyngeal carcinoma is broadly classified by the World Health Organization as keratinizing squamous cell carcinoma (WHO type I) or nonkeratinizing carcinoma (WHO types II and III), with nonkeratinizing carcinomas further subdivided as either differentiated (WHO type II) or undifferentiated (WHO type III). Only 20% of nasopharyngeal carcinomas are keratinizing, while the remaining 80% are nonkeratinizing; it is this latter group that is strongly associated with EBV (27). The interplay between environmental and genetic factors is important in the etiology of nasopharyngeal carcinoma, which includes susceptibility in some individuals with particular human leukocyte antigen haplotypes, exposure to chemical carcinogens such as Cantonese salted fish at an early age, and latent EBV infection (27). EBV appears to promote the progression of nasopharyngeal carcinoma, rather than serve as an initiating factor as it does in EBV-associated B-cell tumors (27). Nasopharyngeal carcinoma is most common in southern China, accounting for about 20% of all adult cancers in the region, and is prevalent in Southeast Asia and other Asian countries, but it is much rarer in North America and Europe, having an incidence rate of <1 per 100,000 people (27). The incidence is two to three times higher in males than females, and peak incidence is in the sixth decade for high-risk populations. Patients commonly present with a neck mass from metastases to the cervical lymph nodes (28). Other symptoms include otitis media, deafness, tinnitus, epistaxis, and nasal obstruction.

Gastric Cancer

EBV accounts for approximately 10% of gastric carcinomas, and 70,000 to 80,000 patients are estimated to develop EBV-associated gastric carcinoma each year (29). There is a male predominance, preferential occurrence in the proximal stomach, and tendency toward multiple synchronous or metachronous tumors when compared to EBV-negative cases. Salty food intake and exposure to wood dust or iron filings has been linked to increased risk for EBV-associated gastric carcinoma (30). Grossly, these carcinomas are ulcerated or saucer-like with gastric wall thickening. Histologic examination reveals two patterns of carcinoma: a lymphoepithelioma-like type, which resembles nasopharyngeal carcinoma, and an ordinary type (29). EBV-associated gastric carcinomas, regardless of type, demonstrate increased tumor-infiltrating lymphocytes compared to EBV-negative cases. The lymphocytes are predominantly $CD8^+$ and $CD4^+$ T cells present in a 2:1 ratio, but only very few of these lymphocytes are infected with EBV. Rather, EBV latent infection is found in the vast majority of neoplastic cells, which defines EBV-associated gastric carcinoma (29).

Breast Cancer

Whether or not viruses are an etiologic agent for breast cancer has been a controversial topic. Research on EBV, as well as human papilloma virus and mouse mammary tumor virus-like sequences, as an etiologic agent of breast cancer has yielded varying results. In a meta-analysis of 24 studies, including 1,535 cases in which PCR was used for detecting EBV DNA in breast tissue, there was a significantly increased risk for breast carcinoma in individuals that tested positive for EBV (31). In 11 studies there was comparison to nonbreast cancer control groups, which when pooled demonstrated that EBV infection was associated with a 6.29-fold increased risk for breast carcinoma. EBV infection was especially prevalent in lobular carcinoma, as opposed to ductal or other types of breast carcinoma (31). While EBV may be implicated in lymphoepithelioma-like carcinomas in various other organs, such as the salivary gland, lung, stomach, and thymus, it has not been detected in lymphoepithelioma-like carcinomas of the breast (32). Caution should be taken when interpreting positive PCR results for EBV DNA because PCR can be positive due to the presence of EBV within the breast carcinoma cells or within infiltrating lymphocytes. However, some studies have shown that the prevalence of EBV DNA is higher in cancer cells than in infiltrating lymphocytes (31). In a separate analysis of 30 studies that utilized PCR and/or EBER *in situ* hybridization for detection of EBV, 25 studies demonstrated the absence of EBV in breast carcinoma, while only four studies showed EBV positivity (33). Two of the studies used both PCR and EBER *in situ* hybridization, which together showed that 15% of samples were positive by PCR, but only 8% showed EBV localization to the breast epithelium via EBER *in situ* hybridization. In a prospective study that included 81,807 women that were followed over nearly two decades, 2,349 developed invasive breast cancer (34). Questionnaires were administered to the women to determine whether they had ever had infectious mononucleosis and at what age. The results of this study did not identify a clear link between history of clinical infectious mononucleosis and risk of invasive breast cancer. Furthermore, a serologic study comparing 399 women with invasive breast cancer to 399 controls found that EBV seroconversion or changes in anti-EBV IgG were not significantly associated with breast cancer risk (35). Taken together, EBV does not appear to have a definitive role in the etiology of breast cancer in general, but may be associated with increased risk in specific subtypes.

Smooth Muscle Tumors Associated with EBV

EBV-associated smooth muscle tumors are essentially exclusive to immunosuppressed patients and can be seen in those with acquired immune deficiency syndrome or in the posttransplant setting. Similar tumors that do not occur in the setting of immunosuppression are consistently EBV negative (36). EBV-associated smooth muscle tumors are frequently multifocal due to separate transformation events giving rise to independent primary tumors (37). The biologic behavior is varied and ranges from benign to fatal, but prognosis is generally better than that of sporadic leiomyosarcomas.

Autoimmune Diseases Associated with EBV

EBV has been implicated in a number of autoimmune diseases, including multiple sclerosis, systemic lupus erythematosis, rheumatoid arthritis, autoimmune thyroiditis, inflammatory bowel diseases, insulin-dependent diabetes mellitus, Sjögren's syndrome, systemic sclerosis, myasthenia gravis, and autoimmune liver diseases; however, causation has yet to be proven (38). Multiple sclerosis serves as an example. Multiple sclerosis is a common and often debilitating inflammatory condition of the central nervous system with heterogeneous clinical presentation characterized by multifocal demyelinative lesions that are separated spatially and temporally. As a result of these lesions, patients may experience a variety of signs and symptoms including optic neuritis, internuclear ophthalmoplegia, trigeminal neuralgia, Lhermitte phenomenon, paresthesias, vertigo, nystagmus, paraparesis, ataxia, incontinence, sexual dysfunction, and fatigue. The cause of multiple sclerosis remains unknown, but it is widely believed to be an autoimmune disorder with various possible genetic and environmental risk factors. Abnormal findings in the cerebrospinal fluid including pleocytosis, increased protein, and oligoclonal bands, which are identified in multiple sclerosis, are typical of neurologic infections (39). Viral etiologies have been explored, including infection by EBV, but a definitive conclusion has not been reached. The association between EBV and multiple sclerosis has been suggested based on similar latitudinal distribution. However, not all studies demonstrated the same effect of latitude on EBV seropositivity, and in many of the studies, variations in latitude could not be easily separated from socioeconomic status (40). Reported history of past infectious mononucleosis, previous hospitalization for infectious mononucleosis, and EBV seropositivity were all associated with increased risk for multiple sclerosis (40). However, studies analyzing EBV viral load in peripheral blood mononuclear cells from patients with multiple sclerosis versus healthy carriers, EBER *in situ* hybridization performed on postmortem brains of patients with multiple sclerosis, and T-cell immune response to synthetic EBV peptides and to autologous B-cell lymphoblastoid cell lines have yielded discordant results (40). Despite much investigation, the role of EBV in multiple sclerosis is still unclear and requires further study. Proposed mechanisms for EBV involvement in autoimmune diseases include infection and immortalization of autoreactive naïve B cells, enhancement of innate immunity and cytokine production, transactivation of endogenous retroviruses, molecular mimicry and formation of cross-reactive autoantibodies, and bystander activation and epitope spreading (38). Alternatively, EBV latency could be perturbed by the proinflammatory environment in autoimmune diseases leading to its reactivation, thus resulting in the higher serum viral loads and detectable immune responses to EBV seen in these conditions, or increased serum or intrathecal antibody titers to EBV may be a byproduct of polyspecific B-cell activation (38).

TREATMENT AND PREVENTION

EBV infects over 90% of adults and is the cause of several clinical diseases and neoplasms. In 1976, a little over a decade after its discovery, Epstein proposed an EBV vaccine (41). Since then, progress has been slow, and while several patented EBV vaccines now exist, only a handful have been tested in clinical trials (42). Vaccinia virus constructs expressing gp220-350, vaccinia recombinant vector expressing EBNA-1 and LMP-2, adjuvant recombinant gp350, and HLA-B8-restricted EBNA-3A peptide epitopes have demonstrated safety and immunogenicity in phase 1 and/or phase 2 clinical trials (43). Although these findings are promising, bringing an EBV vaccine into phase 3 clinical trials poses a number of challenges, including obtaining funding, lack of an animal model except subhuman primates, optimization of antigen construct and adjuvant, selection of the ideal test population, and identification of what the vaccine hopes to achieve (43).

Infectious Mononucleosis

IM is primarily treated with supportive care, such as the use of acetaminophen or nonsteroidal anti-inflammatory agents to manage fever, throat discomfort, and malaise (44). Acetaminophen can be used as both an antipyretic and analgesic and is preferred by many clinicians over aspirin, because of the concern that aspirin may increase risk of splenic hemorrhage (45). Other analgesics that are recommended for IM include saltwater gargles, anesthetic throat lozenges, or viscous lidocaine hydrochloride. Codeine may also be used in patients that do not respond to other treatments. Adequate fluid and nutritional intake should be encouraged but may be difficult as many patients become anorexic during acute illness (45). Corticosteroids should only be administered to patients with severe complications because side effects such as immunosuppression and bacterial superinfection may result from routine use (46). Prednisone can be used to prevent airway obstruction from severe tonsillar hypertrophy, autoimmune hemolytic anemia, severe thrombocytopenia, or progressive neurologic disease (46). Antiviral therapy, such as acyclovir and valacyclovir, is not recommended (47). If patients are given antimicrobials, such as amoxicillin or other penicillin derivatives, a maculopapular rash may develop in up to 95% of cases (46). To minimize the risk of splenic rupture, patients should not participate in any sports for at least 3 weeks from illness onset and should avoid participation in strenuous contact sports for at least 4 weeks from illness onset (46). Most patients will recover from IM without any sequelae, and clinical and laboratory findings will resolve after approximately 1 month from diagnosis (47). Cervical adenopathy and fatigue may take longer to resolve, and fatigue lasting longer than 6 months with functional impairment has been reported (47).

Lymphoproliferative Diseases Associated with EBV

Chronic Active EBV Infection

Therapies such as antiviral (e.g., acyclovir and valacyclovir), immunomodulatory (e.g., interferon-alpha and interleukin-2), and immunosuppressive (e.g., corticosteroids and cyclosporine) treatments, rituximab, cytotoxic chemotherapy (e.g., EPOCH [etoposide, prednisone, vincristine, cyclophosphamide, and doxorubicin] with rituximab), and autologous EBV-specific cytotoxic T lymphocytes are often ineffective or only temporarily delay disease progression (48). Allogeneic hematopoietic stem cell transplantation offers definitive treatment for patients with chronic active EBV infection and is often curative. Nevertheless, death after transplantation is primarily due to progressive chronic active EBV disease (48).

Hemophagocytic Lymphohistiocytosis

There is no general consensus for treating secondary hemophagocytic lymphohistiocytosis as there is for the familial form (49). The use of rituximab, etoposide, and hematopoietic stem cell transplantation have shown success in improving outcomes. Treatment and prophylaxis for opportunistic infections is important for patients that develop neutropenia. Attempts to control the excess cytokines have included antithymocyte globulins, splenectomy, and plasma exchange or blood exchange transfusions (13). Notably, antiviral medications such as acyclovir, ganciclovir, and cidofovir are not particularly effective for treating EBV-associated hemophagocytic lymphohistiocytosis.

X-Linked Lymphoproliferative Syndrome

Treatment for X-linked lymphoproliferative syndrome depends on the clinical manifestations of the disease; that is, lymphomas should be treated with standard chemotherapy, hemophagocytic lymphohistiocytosis should be treated with disease-specific therapy, and dysgammaglobulinemia should be treated with immunoglobulin replacement therapy (16). Rituximab, which targets CD20-positive B cells, can be used to control the acute primary EBV infection. Myeloablative or reduced intensity allogeneic hematopoietic stem cell transplantation is the only curative treatment.

Posttransplant Lymphoproliferative Disorders

Treatment for PTLD generally first involves reduction of immunosuppression. Given the heterogeneity of PTLD and differing approaches to immunosuppression reduction, reported efficacy varies widely. Bulky disease, advanced stage, and older age were predictive for lack of response, while EBV seronegativity and B-cell histologic subtype did not affect outcome (23). The level of immunosuppression reduction needs to be weighed against the risk for acute rejection of the transplant. Surgical and radiation therapy can be used for localized disease or for managing symptoms and complications. Other therapies include antivirals (e.g., ganciclovir and acyclovir), rituximab, cytotoxic combination chemotherapy, and adoptive immunotherapy with cytotoxic T cells specific to EBV.

Epithelial Lesions Associated with EBV

Oral Hairy Leukoplakia

Treatment in some cases may be used to reestablish normal characteristics of the tongue, eliminate pathogenic microorganisms, improve patient comfort, and for cosmesis (50). Treatment strategies include surgery and systemic antiviral treatment, but topical agents have advantages of low cost, ease of use, few side effects, and efficacy over time (50).

Nasopharyngeal Carcinoma

Treatment commonly involves radiation therapy for early stage cancers and combined chemoradiation for more advanced stages. Neck irradiation can be given to all patients regardless of lymph node status, given the predilection of nasopharyngeal carcinoma for early regional lymph node spread.

Gastric Cancer

Current therapies are not specific for EBV-associated gastric carcinomas and include resection, radiation, and/or chemotherapy. EBV-associated gastric carcinomas tend to be lower stage and confer significantly better prognosis than EBV-negative cancer (51).

Smooth Muscle Tumors Associated with EBV

Optimal treatment is unknown but depends on the tumor localization and behavior (36). Potential therapies include surgical resection, antiviral therapy, reduction of immunosuppressive regimen, and chemotherapy.

DETECTION/DIAGNOSIS

Diagnosis of EBV infection in immunocompetent patients is generally made through serological testing. With acute primary infection causing infectious mononucleosis, there is a nonspecific increase in serum immunoglobulins, which includes heterophile antibodies, generally IgM antibodies, that are capable of binding to animal red blood cells. In the heterophile antibody test, the presence of heterophile antibodies in the patient plasma or serum is demonstrated by the agglutination of sheep, horse, goat, or bovine erythrocytes or agglutination of antigen-coated latex beads. Lateral flow immunoassays or enzyme-linked immunosorbent assays based on heterophile antibodies have been developed for rapid bedside and automated testing. Along with heterophile antibodies, there is a concomitant rise in EBV-specific antibodies in response to EBV infection. IgM antibodies to VCA and IgG antibodies EA and VCA are produced in response to acute infection, while IgG antibodies to the EBV nuclear antigen (EBNA) develop in the convalescent stage and persist with anti-VCA IgG years later as indicators of remote infection. Traditionally, indirect immunofluorescence assays were used, but less labor-intensive and automated techniques using enzyme-linked immunosorbent assays are now available. EBER-1 and EBER-2 are highly expressed in latently infected cells, at approximately one million copies per infected cell, making EBER *in situ* hybridization an especially efficacious test that is considered the gold standard for determining EBV status in biopsy specimens (24). Proper interpretation of EBER *in situ* hybridization requires performance of appropriate positive controls to demonstrate RNA preservation in the biopsy specimen and negative controls. Proteins synthesized by EBV can be detected and localized on biopsy specimens using immunohistochemical stains for EBNA-1, EBNA-2, LMP-1, LMP-2, BHRF-1, BZLF-1, and BMRF-1. With the exception of LMP-1 expression in Hodgkin lymphoma, immunohistochemistry is not as reliable as EBER *in situ* hybridization (24). Quantification of EBV DNA by real-time PCR on biopsy specimens can be informative, but results should always be interpreted in conjunction with histopathology findings, as benign EBV-infected lymphocytes incidentally present in the specimen can result in a positive test. Quantitative real-time PCR can also be performed on body fluids and blood samples and has been used for prognosis, as an adjunct in diagnosis, to monitor treatment, and to predict disease recurrence. Substantial differences exist when comparing results from various laboratories, though traceability to the first WHO International Standard for Epstein-Barr virus, approved in 2011, is expected to eventually reduce significant interlaboratory variability (52, 53).

Infectious Mononucleosis

Diagnostic testing for IM due to EBV is recommended for the following situations: febrile patients between 10 and 30 years of age who present with sore throat, fatigue, and splenomegaly, palatal petechiae, or posterior cervical, axillary, or inguinal lymphadenopathy; in younger or older patients, as an adjunct to the investigation of acutely raised transaminases or hemolytic anemia; or if IM is strongly suspected clinically or from contact history (9). Testing should include a complete blood count with differential count for lymphocytosis and atypical lymphocytes, heterophile antibodies in immunocompetent adults, and viral serology in patients less than 4 years of age and in immunocompromised patients. Repeat testing for heterophile antibodies, and consideration of EBV-specific viral serology to rapidly rule out IM, is recommended after 5 to 7 days if the initial test panel was not diagnostic of IM. If the second heterophile antibody test is also negative, identification of alternate etiologies, specifically cytomegalovirus and toxoplasmosis for pregnant or immunocompromised patients and human immunodeficiency virus in at-risk patients, is recommended when indicated (9). The Hoagland criteria are often cited for the diagnosis of IM and are as follows: presence of fever, pharyngitis, and adenopathy; $\geq$50% lymphocytosis with $\geq$10% atypical lymphocytes; and, confirmation by a positive serologic test (54). These criteria are specific but not sufficiently sensitive, and only approximately half of patients with symptoms suggestive of IM and positive heterophile antibody test would fulfill Hoagland's criteria (10). Instead, having a positive heterophile antibody test, or >20% atypical lymphocytes, or >10% atypical lymphocytes with an accompanying lymphocytosis of $\geq$50% strongly suggests IM, and no further testing is recommended (9).

The lymphocyte to white blood cell count ratio has been demonstrated to be a useful parameter in distinguishing patients with IM from those with bacterial tonsillitis, and having a ratio >0.35 has 90% sensitivity and 100% specificity for IM (55). It has been suggested that heterophile antibody testing should be ordered when the ratio is >0.35, but could be avoided when the ratio is <0.35. Alternatively, lymphocyte count alone of $\leq 4 \times 10^9$/L was found to have high sensitivity and specificity, 84% and 94% respectively, for negative monospot results, and could potentially be used to screen patients with IM symptoms, prior to heterophile or serological testing (56). The sensitivity and specificity increased to 100% and 97%, respectively, when the lymphocyte count was $\leq 4 \times 10^9$/L in patients 18 years of age and older (56). The monospot test was therefore recommended for adult patients with a lymphocyte count $>4 \times 10^9$/L, pediatric patients, and cases where there was a strong clinical suspicion for IM. However, these findings were not replicated in other studies in which using a lymphocyte to white blood cell count ratio >0.35 resulted in a sensitivity of 84% and specificity of 72% for IM, and using a lymphocyte count of $\leq 4 \times 10^9$/L to predict a negative monospot resulted in a sensitivity of 61.8% and specificity of 96.8% (57, 58). The exclusive use of either lymphocyte to white blood cell count ratio or lymphocyte count alone would therefore be insufficient for the diagnosis or exclusion of IM.

Aside from determining the white blood cell and lymphocyte counts, automated hematology analyzers are capable of flagging specimens for other quantitative or qualitative parameters. Depending on the analyzer, flags for blasts, variant lymphocytes, or both blasts and variant lymphocytes have sensitivities ranging from 41% to 43.4%, 15.8% to 72.4%, and 10.5% to 40%, respectively, and specificities ranging from 88.6% to 97.1%, 79.1% to 90.8%, and 96% to 98.1%, respectively, for identifying patients with a positive heterophile result (59). Manual review of Wright-Giemsa stained peripheral blood smears also demonstrates differences between heterophile-positive and heterophile-negative patients, with the former having a 40.9% mean of atypical lymphocytes and the latter having 9.15% (59). The percentage of Downey type I, Downey type II, Downey type III, and percentage of lymphocytes >15 μm were all significantly different between the two groups (p-value <0.0001 for all). The Downey type II lymphocyte morphology is significantly associated with EBV positivity, whereas the Downey type III and the Downey type I variants are associated to varying degrees with human herpesvirus-6 and cytomegalovirus positivity, respectively (60). Other features identified in peripheral blood smears included smudge cells and cloverleaf nuclei. Smudge cells and cloverleaf lymphocytes both had a high specificity for heterophile-positive patients, but low sensitivities of 30% and 15.5%, respectively (17). It should be noted that chronic lymphocytic leukemia, adult T-cell leukemia, and infection by human immunodeficiency virus, among other conditions, may demonstrate similar changes, so the overall clinicopathologic context must be considered. Apoptotic lymphocytes may demonstrate increased sensitivity for heterophile-positive patients, up to 88.9%, whereas only 3.75% of control peripheral smears contain apoptotic lymphocytes (61). Again, the overall clinicopathologic context in which apoptotic lymphocytes are found needs to be considered because upper respiratory viral infections and prolonged sample storage at room temperature can result in apoptosis.

For the detection of heterophile antibodies, newer assays have essentially replaced the classic Paul-Bunnell test due to its high sensitivity, specificity, and simplicity, in that there is no need for serum inactivation (62). While the Paul-Bunnell reaction is capable of quantitating the titer of heterophile antibody, there is little practical importance of a quantitative result because it does not correlate with disease severity, duration, or complications (62). There are currently numerous commercial tests available that detect heterophile antibodies. They overall have sensitivities ranging from 81% to 95% and specificities ranging from 98% to 100% (63). Approximately 10% of patients with EBV infection will be persistently negative for heterophile antibodies (64). Heterophile antibodies need time to increase to levels detectable by heterophile antibody-based assays, and within the first week of symptoms, up to 25% of patients may have a false-negative result (64). By the third week the false-negative rate drops to about 5%, so repeat testing is therefore suggested when there is still clinical suspicion despite an initial negative result (9). False-negative results occur more frequently in immunocompromised patients and in children, particularly children under 4 years of age in whom the heterophile antibody response is normally negative or undetectable, who should instead have EBV-specific antibodies measured (9). For patients with negative test results for heterophile antibodies with an absolute lymphocytosis or an instrument-generated atypical lymphocyte flag, 40% will still have positive EBV serologies; other etiologies include cytomegalovirus in 39%, human herpesvirus 6 in 25%, and *Toxoplasma* in 3% (60). In contrast, patients without absolute lymphocytosis or an instrument-generated atypical lymphocyte flag who were negative for heterophile antibodies frequently tested negative for anti-EBV IgM antibodies as well (60). False-positive heterophile antibody test results also occur and can be seen in patients with systemic lupus erythematosus, human immunodeficiency virus, *Toxoplasma*, babesiosis, malaria, rubella, leukemias, and lymphomas (8, 64).

Testing for EBV-specific antibodies may be useful in patients with negative heterophile antibody tests results who have symptoms suggestive of IM, in patients less than 4 years of age, and in immunocompromised patients (9). Because they can be detected earlier than heterophile antibodies, EBV-specific antibodies are also useful when a rapid rule-out of EBV infection is needed, as with someone wanting to quickly return to sports. EBV-specific antibodies have the added advantage of being able to identify acute and remote infections, even after heterophile antibodies have disappeared. IgM antibodies to EBV VCA are comparable to heterophile antibodies for ruling in EBV IM but are considered to be better than heterophile antibodies for ruling out EBV IM, to the extent that a negative anti-VCA IgM result is strong evidence against EBV IM as a diagnosis (9). Indirect immunofluorescence assays, used early in the research of EBV infection, are frequently considered the gold standard to which later developed enzyme-linked immunosorbent assays have been compared (65). Indirect immunofluorescence assays, however, are more labor intensive than microtiter systems that are automated or semiautomated and can have up to 9% of results that are uninterpretable (65). The sensitivities of various assays for EBV-specific antibodies range from 95% to 100%, while the specificities range from 86% to 100% (63). In immunocompetent patients, the status of three parameters, anti-VCA IgM, anti-VCA IgG, and anti-EBNA-1 IgG, can distinguish between several serological patterns associated with different clinical settings (Table 2) (66). Commonly in clinical practice, anti-VCA IgM, anti-VCA IgG, and heterophile antibodies are the only three analytes tested for diagnosing acute EBV IM or determining EBV status (67). However, when more analytes are used, such as testing anti-VCA IgM, anti-VCA IgG, anti-EA-D IgG, anti-EBNA-1 IgG, and heterophile antibodies, the sensitivity for diagnosing primary acute EBV infection is improved (67). Regardless of the number of analytes tested, the antibody combinations can generally be determined from a single sample, thus only a single blood draw is needed from the patient. This is advantageous over comparing changes in antibody titers from temporally separate samples, a strategy often employed for serologic assessment in other diseases, and decreases the burden to the patient, phlebotomist, and the laboratory. The results of serologic testing are not always clear-cut, however, and unusual patterns can be seen in up to 10% of patients (65). Furthermore, atypical antibody profiles may be present more often in immunocompromised patients, who generally have an increase in anti-VCA IgG and anti-EA IgG titers, and a decrease or loss of anti-EBNA-1 IgG titers (66). False-positive results for anti-VCA IgM have been reported for infections with cytomegalovirus, parvovirus B19, *Toxoplasma*, hepatitis A, and human immunodeficiency virus (66).

TABLE 2 EBV serological patterns and associated clinical settings[a]

Anti-VCA IgM	Anti-VCA IgG	Anti-EBNA-1 IgG	Interpretation
–	–	–	No immunity
+	–	–	Acute infection or nonspecific reactivity
+	+	–	Acute infection
–	+	+	Past infection
–	+	–	Acute or past infection
+	+	+	Late primary infection or reactivation
–	–	+	Past infection or nonspecific reactivity

[a]Modified and reprinted from the *World Journal of Virology* with permission from the publisher (66).

Although measuring EBV DNA in plasma has high sensitivity for early IM, serology remains the gold standard for diagnosis in immunocompetent patients (68). EBV viral load is commonly measured via real-time PCR technology, and there are currently several primers and probes available commercially (24). After primary EBV infection, levels of circulating EBV DNA within whole blood, plasma, serum, or memory B cells peak within 2 weeks and then become low or undetectable after weeks to months (24). A steady state of approximately 1 to 50 EBV DNA copies per million white blood cells in whole blood, and undetectable levels in plasma or serum samples, may take as long as a year or more to reach, during which time the levels of EBV DNA may rebound idiosyncratically (24). This decline in EBV DNA after onset of symptoms confers a very low negative predictive value for IM. Furthermore, asymptomatic patients or those experiencing reactivation may have positive EBV viral load measured in whole blood or peripheral blood mononuclear cells, thus limiting the specificity of EBV DNA for IM diagnosis (68). EBV DNA testing for IM is therefore best reserved for cases with inconclusive serological results and immunosuppressed patients. Detection of EBV DNA is also helpful for diagnosing IM in infants and young children, for whom serologic testing may be unreliable (27, 28). In younger patients, PCR can be more sensitive than serology because of the time required to mount an immunologic response (69, 70). PCR is 78.4% sensitive for diagnosing primary EBV infection, compared to the monospot test, which is only 54.5% sensitive in this population (70). For infants in the acute phase of IM, anti-VCA IgM was only positive in 25% of cases, but this increased to 80% for patients 4 years of age or older; anti-VCA IgG had higher rates of positivity in both groups, 79.2% for infants and 100% for patients 4 years of age or older (69). Nevertheless, some infants may not express any EBV-specific antibodies during the acute phase of IM, but are still positive for EBV DNA by PCR (69). A potential strategy for evaluating young patients with IM symptoms could therefore include initial serologic testing followed by EBV DNA testing when serologic results are negative or indeterminate.

Lymphoproliferative Diseases Associated with EBV

Chronic Active EBV Infection

Serological studies demonstrate elevated anti-EBV antibodies that generally consist of anti-VCA IgG ≥1:640 and anti-EA IgG ≥1:160, with anti-VCA IgA and/or anti-EA IgA commonly found. Recommended laboratory tests are as follows: detection of EBV DNA, RNA, related antigens, and clonality in affected tissue and/or peripheral blood via PCR, EBV *in situ* hybridization, immunofluorescence, and Southern blotting; histopathological and molecular studies via general histopathology, immunohistochemical staining, chromosomal analysis, and rearrangement studies; and immunological studies via general immunological tests, marker analysis of peripheral blood including HLA-DR, and cytokine analysis (12).

Hemophagocytic Lymphohistiocytosis

Serological testing can be used to determine if primary infection or reactivation of EBV is the cause of the disease. Most cases of EBV-associated hemophagocytic lymphohistiocytosis occur in pediatric patients with primary infection, and cases occurring from EBV reactivation are associated with worse outcome (13). Real-time PCR can also be performed to quantify EBV viral load, which has prognostic significance and can be used to monitor response to treatment.

X-Linked Lymphoproliferative Syndrome

Serologic testing for EBV in patients with X-linked lymphoproliferative syndrome can be unreliable as they may exhibit dysgammaglobulinemia and production of anti-EBV antibodies may be abnormal. Instead, EBV detection via PCR can be used. The most efficient laboratory method for identifying patients with X-linked lymphoproliferative syndrome is evaluation of SAP protein expression by flow cytometry. SAP deficiency is also suggested when there are decreased $CD27^+$ B memory cells, absence of NKT cells, and variably impaired NK cell lytic function (16). With the appropriate clinical setting and demonstration of SAP deficiency, confirmatory testing for *SH2D1A* mutation can be performed.

Posttransplant Lymphoproliferative Disorders

In patients with suspected PTLD, diagnostic workup includes complete blood count with differential, serum electrolytes/calcium/blood urea nitrogen/creatinine, liver function tests, uric acid, lactate dehydrogenase, quantitative immunoglobulins, EBV serologies (including anti-EBNA, -VCA, and -EA), peripheral blood EBV viral load, stool for occult bleeding, chest radiograph, computed tomography of neck/chest/abdomen/pelvis, and core needle/excisional biopsy of lesions with flow cytometric analysis of lymphocytes, EBER, and therapy-relevant markers (e.g., CD20) (21). EBV serologies are less helpful in immunocompromised patients who mount unreliable immune responses to EBV. Pathology is the gold standard for diagnosing PTLD, and the histologic spectrum of PTLD, as classified by the World Health Organization, can be divided into four categories: early lesions, polymorphic PTLD, monomorphic PTLD, and classical Hodgkin lymphoma-type PTLD (71). Early lesions include plasmacytic hyperplasia and infectious mononucleosis-like PTLD; they have preserved tissue architecture and are polyclonal proliferations or have a very small monoclonal B-cell population. Polymorphic PTLD demonstrates effacement of the tissue architecture, shows the full spectrum of lymphoid maturation (i.e., immunoblasts, variably sized lymphocytes, and plasma cells), and can consist of polyclonal or monoclonal cells that are often EBV-positive but does not fulfill diagnostic criteria for B or T/NK cell lymphomas. Monomorphic PTLD, in contrast, does meet diagnostic criteria for B-cell (e.g., diffuse large B-cell lymphoma, Burkitt lymphoma, plasma cell myeloma, plasmacytoma-like lesion, and others excluding indolent B-cell neoplasms) or T-cell neoplasms (e.g., peripheral T-cell lymphoma not otherwise specified, hepatosplenic T-cell lymphoma, and others), and is a monoclonal infiltrate that effaces normal tissue architecture. Finally, classical Hodgkin lymphoma-type PTLD fulfills the diagnostic criteria for classical Hodgkin lymphoma, containing a minority of malignant cells (i.e., Reed-Sternberg cells and variants) among inflammatory cells. Despite the utility of this classification system, multiple lesions simultaneously sampled from the same patient may exhibit differing histologic grades. It is important to remember that the spectrum of disease caused by EBV is continuous, and initially benign manifestations may progress to become more severe illnesses (21). EBER *in situ* hybridization performed on tissue samples is useful in identifying the presence of and localizing EBV to neoplastic cells. PCR for EBV DNA is not as informative for tissue samples because it does not distinguish between lesional and bystander lymphocyte positivity. Instead, EBV viral load measurement by real-time PCR performed on peripheral blood can be used to monitor patients that are at high risk for developing PTLD, such as EBV-seronegative recipients. High EBV viral loads may predict the development of PTLD, and almost all patients with PTLD demonstrate high levels of EBV DNA in whole blood and in plasma, although there is no consensus as to which is the preferred specimen type (72). Clinical trial guidelines often require frequent EBV testing in the first year after transplantation when PTLD risk is highest (24). EBV viral load has good sensitivity for EBV-positive PTLD but is not useful in EBV-negative PTLD. Because of this and its relatively poor specificity for PTLD, EBV viral loads are best suited for excluding a diagnosis of EBV-positive PTLD. There also has not been a consistent cutoff in the literature to define what constitutes "high" EBV viral load, so it is difficult to compare results produced by different laboratories. Within individual laboratories, results are more reproducible and trending EBV viral load is more meaningful. The World Health Organization approved the first WHO International Standard for Epstein-Barr virus in 2011, which perceivably should reduce significant interlaboratory variability (23). EBV viral load can be used to monitor therapy for PTLD because the level of circulating EBV serves as a gauge of tumor burden and can be followed during the course of treatment. Progressive disease and relapses have been documented, however, despite decreases in EBV viral load after therapy (23). Also, some patients may continue to demonstrate sustained EBV viral load elevation despite PTLD resolution.

EBV-Associated Lymphomas

Biopsy and microscopic examination is required for the diagnosis of these entities, and criteria for classification have been established by the World Health Organization (71). EBER *in situ* hybridization is the gold standard for identifying EBV infection and localizing the virus to specific cells. EBV viral load performed on patients with EBV-associated lymphomas demonstrates high levels of EBV DNA in the plasma or serum, in contrast to healthy carriers in whom EBV is restricted to the cellular fraction of blood (24). Improper storage or delayed testing of whole blood samples may lead to false-positive EBV viral loads in plasma or serum if EBV DNA is permitted to escape from the intracellular compartment. EBV DNA can often be detected before the diagnosis of lymphoma is made; levels of EBV DNA at diagnosis may be predictive of therapy efficacy and outcome, and EBV viral load in plasma can serve as a marker for tumor burden in EBV-associated lymphomas (24). For patients with primary central nervous system lymphoma, EBV DNA is present within the cerebrospinal fluid, which can be serially measured to monitor therapy.

Epithelial Lesions Associated with EBV

Oral Hairy Leukoplakia

A provisional diagnosis can be made when the appropriate clinical features of oral hairy leukoplakia are present, but when clinical features are unclear, sampling via cytological preparations or tissue biopsy is indicated. Hyperparakeratosis, acanthosis, keratinized "hairlike" projections, koilocytes with pyknotic nuclei and perinuclear halos in the prickle cell layer, intranuclear inclusions, few or no Langerhans cells, and limited to no inflammatory infiltrate in the lamina propria are histologic features seen in oral hairy leukoplakia, but they are not specific for this entity (25). When light microscopic and clinical findings correlate, a presumptive diagnosis can be made but a definitive diagnosis requires demonstrating the presence of EBV (25). A number of techniques such as PCR, *in situ* hybridization, immunohistochemistry, immunocytochemistry, and electron microscopy can detect EBV, but the gold standard for the diagnosis of oral hairy leukoplakia is EBER *in situ* hybridization (26). EBER *in situ* hybridization has been shown to be effective on punch biopsy as well as exfoliative liquid-based cytology (73). In the routine management of patients infected with human immunodeficiency virus, definitive diagnosis is seldom needed, and because oral hairy leukoplakia is generally asymptomatic, may spontaneously resolve, and has no premalignant potential, treatment is not often required (25).

Nasopharyngeal Carcinoma

Screening for nasopharyngeal carcinoma with IgA antibodies targeted against VCA and EA via immunofluorescence assay yields a sensitivity of 51% and specificity of 95% when both antibodies are identified (74). Subsequently developed enzyme-linked immunosorbent assays for nasopharyngeal carcinoma screening demonstrate improved test characteristics, with increased sensitivity of 95% and specificity of 94% for combined anti-VCA IgA and anti-EBNA IgA, and have the additional advantage of automation and large-scale testing (74). Given that the majority of people have been infected with EBV worldwide, however, the positive predictive values of serology-based tests are low and often less than 5% (75). Detection of plasma EBV DNA by real-time PCR has a sensitivity of 87%, specificity of 90%, and a positive predictive value of 30% for nasopharygeal carcinoma diagnosed within 1 year of follow-up in individuals who are positive for anti-VCA IgA and anti-EBNA-1 IgA (75). Plasma EBV DNA load in early stage nasopharyngeal carcinoma, however, has a lower sensitivity of 74% and the positive predictive value after 1 year of follow-up decreases to 8.3% (75). EBV DNA load may therefore be a useful adjunct to EBV-specific serology in screening for patients with advanced-stage nasopharyngeal carcinoma, but not for those with early disease or for predicting the development of nasopharyngeal carcinoma. Definitive diagnosis of nasopharyngeal carcinoma requires biopsy of the primary or metastatic masses. Premalignant lesions, such as dysplasia or carcinoma *in situ*, are not generally identified in the development of nasopharyngeal carcinomas, unlike carcinomas of the head and neck, uterine cervix, or other locations (76). EBV can be detected in biopsy specimens via EBER *in situ* hybridization, which demonstrates strong nuclear positivity in virtually all tumor cell nuclei in most cases of nonkeratinizing nasopharyngeal carcinoma. This method localizes EBV to the tumor cells, whereas PCR may be positive from bystander EBV-positive lymphocytes. Recently, EBV DNA load tested on nasopharyngeal brushing samples has been demonstrated to be useful in the diagnosis of nasopharyngeal carcinoma and may potentially obviate the need for more invasive and painful biopsy procedures (77). EBV DNA load in blood samples, but not in nasopharyngeal brushings, correlates with tumor stage, and high levels are predictive of poor prognosis. Both nasopharyngeal brushing and blood EBV DNA levels demonstrate significant reductions after therapy and hence may be suitable for follow-up monitoring after treatment (77).

Gastric Cancer

The gold standard for identifying this carcinoma is via EBER *in situ* hybridization, and gastric mucosal biopsies taken by upper gastrointestinal endoscopy can be used to make the diagnosis (30). EBV-specific antibodies, such as those targeting EBNA-1 and VCA, are elevated but not significantly different between those with EBV-associated gastric carcinoma and healthy controls (30).

Breast Cancer

The contribution of EBV to breast cancer is unclear. Symptomatology, serology, EBER *in situ* hybridization, and molecular methods have been used to suggest EBV involvement in breast cancer.

Smooth Muscle Tumors Associated with EBV

Histologically these tumors consist of monomorphic spindled smooth muscle cells with abundant eosinophilic cytoplasm arranged in short intersecting fascicles, with admixed or separate nodules of primitive round cells seen in approximately 50% of cases (34). Both cell types are positive for smooth muscle actin by immunohistochemistry. Infiltrating T lymphocytes are frequently seen, and occasionally necrosis and myxoid change are present. Mitoses may be present but generally do not exceed three mitotic figures per 10 high-power fields, and nuclear atypia tends to only be modest at most. The presence of EBV in tumor cell nuclei can be demonstrated via EBER *in situ* hybridization.

BEST PRACTICES

EBV is the most common cause of IM, which is generally self-limited. Lymphocytosis with atypical lymphocytes are classically seen in IM. Infection results in the formation heterophile antibodies that are not EBV-specific. However, in the correct clinical setting, a positive heterophile antibody result is sufficient to make the diagnosis of IM. In patients in whom IM is highly suspected but the heterophile antibody test is negative, repeat testing may provide the time necessary for heterophile antibodies to form. In immunocompromised patients or in young children, when heterophile antibodies are unreliable, EBV-specific antibodies targeting VCA, EBNA, and/or EA can be used. EBV-specific antibodies, along with EBV viral load, have also been used for diagnosis and prognostication of nasopharyngeal carcinoma. EBV viral load monitoring has been recommended for prevention of PTLD, particularly when a seronegative recipient is receiving an organ from a seropositive individual. Neoplasms associated with EBV infection are diagnosed based on cytological and histopathological criteria and utilization of appropriate ancillary studies including immunohistochemical staining, flow cytometric analysis, and chromosomal, rearrangement, and molecular studies. EBER *in situ* hybridization can be used to identify and localize EBV-infected cells.

REFERENCES

1. **Epstein MA, Achong BG, Barr YM.** 1964. Virus particles in cultured lymphoblasts from Burkitt's lymphoma. *Lancet* **1:** 702–703.
2. **Holmes D.** 2014. The cancer-virus cures. *Nat Med* **20:**571–574.
3. **Shannon-Lowe C, Rowe M.** 2014. Epstein Barr virus entry; kissing and conjugation. *Curr Opin Virol* **4:**78–84.
4. **Thorley-Lawson DA, Gross A.** 2004. Persistence of the Epstein-Barr virus and the origins of associated lymphomas. *N Engl J Med* **350:**1328–1337.
5. **Balfour HH Jr, Sifakis F, Sliman JA, Knight JA, Schmeling DO, Thomas W.** 2013. Age-specific prevalence of Epstein-Barr virus infection among individuals aged 6–19 years in the United States and factors affecting its acquisition. *J Infect Dis* **208:**1286–1293.
6. **Cohen JI.** 2000. Epstein-Barr virus infection. *N Engl J Med* **343:**481–492.
7. **Vetsika EK, Callan M.** 2004. Infectious mononucleosis and Epstein-Barr virus. *Expert Rev Mol Med* **6:**1–16.
8. **Di Lernia V, Mansouri Y.** 2013. Epstein-Barr virus and skin manifestations in childhood. *Int J Dermatol* **52:**1177–1184.
9. **Smellie WS, Forth J, Smart SR, Galloway MJ, Irving W, Bareford D, Collinson PO, Kerr KG, Summerfield G, Carey PJ, Minhas R.** 2007. Best practice in primary care pathology: review 7. *J Clin Pathol* **60:**458–465.
10. **Ebell MH.** 2004. Epstein-Barr virus infectious mononucleosis. *Am Fam Physician* **70:**1279–1287.
11. **Kimura H.** 2006. Pathogenesis of chronic active Epstein-Barr virus infection: is this an infectious disease, lymphoproliferative disorder, or immunodeficiency? *Rev Med Virol* **16:** 251–261.
12. **Okano M, Kawa K, Kimura H, Yachie A, Wakiguchi H, Maeda A, Imai S, Ohga S, Kanegane H, Tsuchiya S, Morio T, Mori M, Yokota S, Imashuku S.** 2005. Proposed guidelines for diagnosing chronic active Epstein-Barr virus infection. *Am J Hematol* **80:**64–69.
13. **Maakaroun NR, Moanna A, Jacob JT, Albrecht H.** 2010. Viral infections associated with haemophagocytic syndrome. *Rev Med Virol* **20:**93–105.
14. **Henter JI, Horne A, Aricó M, Egeler RM, Filipovich AH, Imashuku S, Ladisch S, McClain K, Webb D, Winiarski J, Janka G.** 2007. HLH-2004: diagnostic and therapeutic guidelines for hemophagocytic lymphohistiocytosis. *Pediatr Blood Cancer* **48:**124–131.
15. **Veillette A, Pérez-Quintero LA, Latour S.** 2013. X-linked lymphoproliferative syndromes and related autosomal recessive disorders. *Curr Opin Allergy Clin Immunol* **13:**614–622.
16. **Rezaei N, Mahmoudi E, Aghamohammadi A, Das R, Nichols KE.** 2011. X-linked lymphoproliferative syndrome: a genetic condition typified by the triad of infection, immunodeficiency and lymphoma. *Br J Haematol* **152:**13–30.
17. **Woon ST, Ameratunga R, Croxson M, Taylor G, Neas K, Edkins E, Browett P, Gane E, Munn S.** 2008. Follicular lymphoma in a X-linked lymphoproliferative syndrome carrier female. *Scand J Immunol* **68:**153–158.
18. **Marsh RA, Madden L, Kitchen BJ, Mody R, McClimon B, Jordan MB, Bleesing JJ, Zhang K, Filipovich AH.** 2010. XIAP deficiency: a unique primary immunodeficiency best classified as X-linked familial hemophagocytic lymphohistiocytosis and not as X-linked lymphoproliferative disease. *Blood* **116:**1079–1082.
19. **Filipovich AH, Zhang K, Snow AL, Marsh RA.** 2010. X-linked lymphoproliferative syndromes: brothers or distant cousins? *Blood* **116:**3398–3408.
20. **Dunleavy K, Roschewski M, Wilson WH.** 2012. Lymphomatoid granulomatosis and other Epstein-Barr virus associated lymphoproliferative processes. *Curr Hematol Malig Rep* **7:**208–215.
21. **Green M, Michaels MG.** 2013. Epstein-Barr virus infection and posttransplant lymphoproliferative disorder. *Am J Transplant* **13**(Suppl 3)**:**41–54, quiz 54.
22. **van Esser JW, van der Holt B, Meijer E, Niesters HG, Trenschel R, Thijsen SF, van Loon AM, Frassoni F, Bacigalupo A, Schaefer UW, Osterhaus AD, Gratama JW, Löwenberg B, Verdonck LF, Cornelissen JJ.** 2001. Epstein-Barr virus (EBV) reactivation is a frequent event after allogeneic stem cell transplantation (SCT) and quantitatively predicts EBV-lymphoproliferative disease following T-cell—depleted SCT. *Blood* **98:**972–978.
23. **Allen UD, Preiksaitis JK, AST Infectious Diseases Community of Practice.** 2013. Epstein-Barr virus and post-transplant lymphoproliferative disorder in solid organ transplantation. *Am J Transplant* **13**(Suppl 4)**:**107–120.
24. **Gulley ML, Tang W.** 2008. Laboratory assays for Epstein-Barr virus-related disease. *J Mol Diagn* **10:**279–292.
25. **Triantos D, Porter SR, Scully C, Teo CG.** 1997. Oral hairy leukoplakia: clinicopathologic features, pathogenesis, diagnosis, and clinical significance. *Clin Infect Dis* **25:**1392–1396.
26. **Braz-Silva PH, de Rezende NP, Ortega KL, de Macedo Santos RT, de Magalhães MH.** 2008. Detection of the Epstein-Barr virus (EBV) by in situ hybridization as definitive diagnosis of hairy leukoplakia. *Head Neck Pathol* **2:**19–24.
27. **Shah KM, Young LS.** 2009. Epstein-Barr virus and carcinogenesis: beyond Burkitt's lymphoma. *Clin Microbiol Infect* **15:**982–988.
28. **Colaco RJ, Betts G, Donne A, Swindell R, Yap BK, Sykes AJ, Slevin NJ, Homer JJ, Lee LW.** 2013. Nasopharyngeal carcinoma: a retrospective review of demographics, treatment and patient outcome in a single centre. *Clin Oncol (R Coll Radiol)* **25:**171–177.
29. **Fukayama M, Ushiku T.** 2011. Epstein-Barr virus-associated gastric carcinoma. *Pathol Res Pract* **207:**529–537.
30. **Iizasa H, Nanbo A, Nishikawa J, Jinushi M, Yoshiyama H.** 2012. Epstein-Barr Virus (EBV)-associated gastric carcinoma. *Viruses* **4:**3420–3439.
31. **Huo Q, Zhang N, Yang Q.** 2012. Epstein-Barr virus infection and sporadic breast cancer risk: a meta-analysis. *PLoS One* **7:** e31656.
32. **Jeong AK, Park SB, Kim YM, Ko BK, Yang MJ, Kwon WJ, Lee JH, Weon YC.** 2010. Lymphoepithelioma-like carcinoma of the breast. *J Ultrasound Med* **29:**485–488.
33. **Joshi D, Buehring GC.** 2012. Are viruses associated with human breast cancer? Scrutinizing the molecular evidence. *Breast Cancer Res Treat* **135:**1–15.
34. **Massa J, Hamdan A, Simon KC, Bertrand K, Wulf G, Tamimi RM, Ascherio A.** 2012. Infectious mononucleosis and risk of breast cancer in a prospective study of women. *Cancer Causes Control* **23:**1893–1898.
35. **Cox B, Richardson A, Graham P, Gislefoss RE, Jellum E, Rollag H.** 2010. Breast cancer, cytomegalovirus and Epstein-Barr virus: a nested case-control study. *Br J Cancer* **102:**1665–1669.
36. **Sprangers B, Smets S, Sagaert X, Wozniak A, Wollants E, Van Ranst M, Debiec-Rychter M, Sciot R, Vanrenterghem Y, Kuypers DR.** 2008. Posttransplant Epstein-Barr virus-associated myogenic tumors: case report and review of the literature. *Am J Transplant* **8:**253–258.
37. **Deyrup AT.** 2008. Epstein-Barr virus-associated epithelial and mesenchymal neoplasms. *Hum Pathol* **39:**473–483.
38. **Lossius A, Johansen JN, Torkildsen Ø, Vartdal F, Holmøy T.** 2012. Epstein-Barr virus in systemic lupus erythematosus, rheumatoid arthritis and multiple sclerosis—association and causation. *Viruses* **4:**3701–3730.
39. **Tselis A.** 2012. Epstein-Barr virus cause of multiple sclerosis. *Curr Opin Rheumatol* **24:**424–428.
40. **Lucas RM, Hughes AM, Lay ML, Ponsonby AL, Dwyer DE, Taylor BV, Pender MP.** 2011. Epstein-Barr virus and multiple sclerosis. *J Neurol Neurosurg Psychiatry* **82:**1142–1148.
41. **Epstein MA.** 1976. Epstein-Barr virus—is it time to develop a vaccine program? *J Natl Cancer Inst* **56:**697–700.
42. **Villegas E, Santiago O, Sorlózano A, Gutierrez J.** 2010. New strategies and patent therapeutics in EBV-associated diseases. *Mini Rev Med Chem* **10:**914–927.

43. **Balfour HH Jr.** 2014. Progress, prospects, and problems in Epstein-Barr virus vaccine development. *Curr Opin Virol* **6:** 1–5.
44. **Vouloumanou EK, Rafailidis PI, Falagas ME.** 2012. Current diagnosis and management of infectious mononucleosis. *Curr Opin Hematol* **19:**14–20.
45. **Odumade OA, Hogquist KA, Balfour HH Jr.** 2011. Progress and problems in understanding and managing primary Epstein-Barr virus infections. *Clin Microbiol Rev* **24:**193–209.
46. **Singer-Leshinsky S.** 2012. Pathogenesis, diagnostic testing, and management of mononucleosis. *JAAPA* **25:**58–62.
47. **Luzuriaga K, Sullivan JL.** 2010. Infectious mononucleosis. *N Engl J Med* **362:**1993–2000.
48. **Cohen JI, Jaffe ES, Dale JK, Pittaluga S, Heslop HE, Rooney CM, Gottschalk S, Bollard CM, Rao VK, Marques A, Burbelo PD, Turk SP, Fulton R, Wayne AS, Little RF, Cairo MS, El-Mallawany NK, Fowler D, Sportes C, Bishop MR, Wilson W, Straus SE.** 2011. Characterization and treatment of chronic active Epstein-Barr virus disease: a 28-year experience in the United States. *Blood* **117:**5835–5849.
49. **Chandrakasan S, Filipovich AH.** 2013. Hemophagocytic lymphohistiocytosis: advances in pathophysiology, diagnosis, and treatment. *J Pediatr* **163:**1253–1259.
50. **Brasileiro CB, Abreu MH, Mesquita RA.** 2014. Critical review of topical management of oral hairy leukoplakia. *World J Clin Cases* **2:**253–256.
51. **Camargo MC, Kim WH, Chiaravalli AM, Kim KM, Corvalan AH, Matsuo K, Yu J, Sung JJ, Herrera-Goepfert R, Meneses-Gonzalez F, Kijima Y, Natsugoe S, Liao LM, Lissowska J, Kim S, Hu N, Gonzalez CA, Yatabe Y, Koriyama C, Hewitt SM, Akiba S, Gulley ML, Taylor PR, Rabkin CS.** 2014. Improved survival of gastric cancer with tumour Epstein-Barr virus positivity: an international pooled analysis. *Gut* **63:**236–243.
52. **Rychert J, Danziger-Isakov L, Yen-Lieberman B, Storch G, Buller R, Sweet SC, Mehta AK, Cheeseman JA, Heeger P, Rosenberg ES, Fishman JA.** 2014. Multicenter comparison of laboratory performance in cytomegalovirus and Epstein-Barr virus viral load testing using international standards. *Clin Transplant* **28:**1416–1423.
53. **Abeynayake J, Johnson R, Libiran P, Sahoo MK, Cao H, Bowen R, Chan KC, Le QT, Pinsky BA.** 2014. Commutability of the Epstein-Barr virus WHO international standard across two quantitative PCR methods. *J Clin Microbiol* **52:** 3802–3804.
54. **Hoagland RJ.** 1975. Infectious mononucleosis. *Prim Care* **2:** 295–307.
55. **Wolf DM, Friedrichs I, Toma AG.** 2007. Lymphocyte-white blood cell count ratio: a quickly available screening tool to differentiate acute purulent tonsillitis from glandular fever. *Arch Otolaryngol Head Neck Surg* **133:**61–64.
56. **Biggs TC, Hayes SM, Bird JH, Harries PG, Salib RJ.** 2013. Use of the lymphocyte count as a diagnostic screen in adults with suspected Epstein-Barr virus infectious mononucleosis. *Laryngoscope* **123:**2401–2404.
57. **Lennon P, O'Neill JP, Fenton JE, O'Dwyer T.** 2010. Challenging the use of the lymphocyte to white cell count ratio in the diagnosis of infectious mononucleosis by analysis of a large cohort of Monospot test results. *Clin Otolaryngol* **35:** 397–401.
58. **Lennon P, O'Neill JP, O'Dwyer T, Fenton JE.** 2014. In reference to "use of the lymphocyte count as a diagnostic screen in adults with suspected Epstein-Barr virus infectious mononucleosis." *Laryngoscope* **124:**E447.
59. **Brigden ML, Au S, Thompson S, Brigden S, Doyle P, Tsaparas Y.** 1999. Infectious mononucleosis in an outpatient population: diagnostic utility of 2 automated hematology analyzers and the sensitivity and specificity of Hoagland's criteria in heterophile-positive patients. *Arch Pathol Lab Med* **123:** 875–881.
60. **Tsaparas YF, Brigden ML, Mathias R, Thomas E, Raboud J, Doyle PW.** 2000. Proportion positive for Epstein-Barr virus, cytomegalovirus, human herpesvirus 6, Toxoplasma, and human immunodeficiency virus types 1 and 2 in heterophile-negative patients with an absolute lymphocytosis or an instrument-generated atypical lymphocyte flag. *Arch Pathol Lab Med* **124:**1324–1330.
61. **Fisher MS Jr, Guerra CG, Hickman JR, Hensley RE, Doe RH, Dunn CD, Hall RB.** 1996. Peripheral blood lymphocyte apoptosis: a clue to the diagnosis of acute infectious mononucleosis. *Arch Pathol Lab Med* **120:**951–955.
62. **Basson V, Sharp AA.** 1969. Monospot: a differential slide test for infectious mononucleosis. *J Clin Pathol* **22:**324–325.
63. **Bruu AL, Hjetland R, Holter E, Mortensen L, Natås O, Petterson W, Skar AG, Skarpaas T, Tjade T, Asjø B.** 2000. Evaluation of 12 commercial tests for detection of Epstein-Barr virus-specific and heterophile antibodies. *Clin Diagn Lab Immunol* **7:**451–456.
64. **Hurt C, Tammaro D.** 2007. Diagnostic evaluation of mononucleosis-like illnesses. *Am J Med* **120:**911.e1–e8.
65. **Kreuzer C, Nabeck KU, Levy HR, Daghofer E.** 2013. Reliability of the Siemens Enzygnost and Novagnost Epstein-Barr virus assays for routine laboratory diagnosis: agreement with clinical diagnosis and comparison with the Merifluor Epstein-Barr virus immunofluorescence assay. *BMC Infect Dis* **13:**260.
66. **De Paschale M, Clerici P.** 2012. Serological diagnosis of Epstein-Barr virus infection: problems and solutions. *World J Virol* **1:**31–43.
67. **Klutts JS, Ford BA, Perez NR, Gronowski AM.** 2009. Evidence-based approach for interpretation of Epstein-Barr virus serological patterns. *J Clin Microbiol* **47:**3204–3210.
68. **Gärtner B, Preiksaitis JK.** 2010. EBV viral load detection in clinical virology. *J Clin Virol* **48:**82–90.
69. **Dohno S, Maeda A, Ishiura Y, Sato T, Fujieda M, Wakiguchi H.** 2010. Diagnosis of infectious mononucleosis caused by Epstein-Barr virus in infants. *Pediatr Int* **52:**536–540.
70. **Pitetti RD, Laus S, Wadowsky RM.** 2003. Clinical evaluation of a quantitative real time polymerase chain reaction assay for diagnosis of primary Epstein-Barr virus infection in children. *Pediatr Infect Dis J* **22:**736–739.
71. **Swerdlow SH, Campo E, Harris NL, Jaffe ES, Pileri SA, Stein H, Thiele J, Vardiman JW (ed).** 2008. *WHO Classification of Tumours of Haematopoietic and Lymphoid Tissues.* WHO Press, Geneva, Switzerland.
72. **Gulley ML, Tang W.** 2010. Using Epstein-Barr viral load assays to diagnose, monitor, and prevent posttransplant lymphoproliferative disorder. *Clin Microbiol Rev* **23:**350–366.
73. **Braz-Silva PH, Santos RT, Schussel JL, Gallottini M.** 2014. Oral hairy leukoplakia diagnosis by Epstein-Barr virus in situ hybridization in liquid-based cytology. *Cytopathology* **25:**21–26.
74. **Liu Y, Huang Q, Liu W, Liu Q, Jia W, Chang E, Chen F, Liu Z, Guo X, Mo H, Chen J, Rao D, Ye W, Cao S, Hong M.** 2012. Establishment of VCA and EBNA1 IgA-based combination by enzyme-linked immunosorbent assay as preferred screening method for nasopharyngeal carcinoma: a two-stage design with a preliminary performance study and a mass screening in southern China. *Int J Cancer* **131:**406–416.
75. **Ji MF, Huang QH, Yu X, Liu Z, Li X, Zhang LF, Wang P, Xie SH, Rao HL, Fang F, Guo X, Liu Q, Hong MH, Ye W, Zeng YX, Cao SM.** 2014. Evaluation of plasma Epstein-Barr virus DNA load to distinguish nasopharyngeal carcinoma patients from healthy high-risk populations in Southern China. *Cancer* **120:**1353–1360.
76. **Yoshizaki T, Ito M, Murono S, Wakisaka N, Kondo S, Endo K.** 2012. Current understanding and management of nasopharyngeal carcinoma. *Auris Nasus Larynx* **39:**137–144.
77. **Adham M, Greijer AE, Verkuijlen SA, Juwana H, Fleig S, Rachmadi L, Malik O, Kurniawan AN, Roezin A, Gondhowiardjo S, Atmakusumah D, Stevens SJ, Hermani B, Tan IB, Middeldorp JM.** 2013. Epstein-Barr virus DNA load in nasopharyngeal brushings and whole blood in nasopharyngeal carcinoma patients before and after treatment. *Clin Cancer Res* **19:**2175–2186.

Human Herpesviruses 6, 7, and 8

SHEILA C. DOLLARD AND TIMOTHY M. KARNAUCHOW

29

HHV-6, -7, AND -8

Human herpesvirus 6 (HHV-6A and -6B variants), human herpesvirus 7 (HHV-7), and human herpesvirus 8 (HHV-8) were the last three herpesviruses identified, from 1986 to 1994 (1–3). HHV-6 and HHV-7 are ubiquitous viruses showing >95% prevalence worldwide from early childhood, usually with asymptomatic primary infection. HHV-6 has been associated with many clinical conditions, but attempts to establish clear etiologic relationships with human disease have largely been confounded by the ubiquity of HHV-6 and the only relatively recent formal recognition of variants HHV-6A and HHV-6B. The most well-recognized clinical presentations of HHV-6 infections, roseola infantum and fever, occur as a result of primary infection mainly in otherwise healthy infants and young children. Primary infection in adults is uncommon. Severe syndromes typically result from HHV-6 reactivation in immunocompromised hosts. Most infections are caused by HHV-6B; disease attributed to HHV-6A is rare. Both of the HHV-6 variants and HHV-7 have been associated with disease in immunocompromised organ transplant recipients. HHV-8 stands out as being the only human herpesviruses with low prevalence in most of the world and is mainly known as the etiologic agent of Kaposi's sarcoma (KS). Like other herpesviruses, these viruses establish lifelong infections in their host and are maintained through a combination of latent (nonproductive) infections and intermittent or persistent lytic infections; consequently, they present diagnostic challenges.

DISCOVERY AND BIOLOGY

HHV-6

HHV-6 was discovered in 1986 by Salahuddin and colleagues, isolated from lymphocytes of immunocompromised patients (1). Based on *in vitro* cell tropism, antigenicity, nucleotide sequence, and epidemiology, two distinct variants (HHV-6A and HHV-6B) were identified and formally classified as distinct virus species in 2012 (4). HHV-6 is a beta-herpesvirus, and along with HHV-7, HHV-6A and HHV-6B are members of the *Roseola* virus genus. The historical lack of distinction between HHV-6A and HHV-6B in the literature and the dearth of laboratory assays able to distinguish between these viruses have hampered understanding of their respective biological behaviors and roles as agents of disease. Overall sequence divergence between HHV-6A and HHV-6B is <4%; while certain genome regions display elevated heterogeneity of 4% to 25%, with the major genetic differences lying in the terminal DR regions. HHV-6A within-strain sequence variation is <0.5%, while for HHV-6B strain variation is <1.0%.

In vivo tropism is broader than *in vitro* tropism and includes cells of lymphoid, epithelial, and neuronal origin; *in vivo* cell and tissue tropisms differ between HHV-6A and HHV-6B (4). *In vivo*, HHV-6B is more commonly present in peripheral blood mononuclear cells (PBMCs) of healthy adults in industrialized countries, and thus is most commonly the agent involved in post-transplant reactivation disease. HHV-6B is also detected in the GI tract of solid-organ transplant (SOT) recipients, has been documented in endodontic masses, and is the species found in adenoids and tonsils, especially in children with upper respiratory tract infection. HHV-6B is also found in salivary glands, lymph nodes, brain, and lung. Both HHV-6A and HHV-6B are found in a small percentage of vitreous fluids assayed, and both viruses are neurotropic, with evidence suggesting that HHV-6A is more neurovirulent (5, 6). HHV-6A is also found in skin, brain, and lung.

Typically, latent-form herpesvirus genomes are maintained as covalently closed circular episomes within the cell nucleus, and cells express a limited number of gene transcripts (latency-associated transcripts, LATs). HHV-6A and HHV-6B are unique among human herpesviruses in their ability to integrate their genomes into host cell chromosomes. Using fluorescent *in situ* hybridization, Luppi (7) first described this phenomenon in the study of PBMCs from three patients, a finding subsequently confirmed by many others. The mechanism of HHV-6 integration has not been fully elucidated but appears to involve a complex process of homologous recombination between viral genome terminal elements containing telomeric repeat sequences and homologous sequences found in the subtelomeric regions of host chromosomes. Integration occurs in lymphocytes, but also in the germline. The latter allows for vertical transmission of HHV-6, and residency of HHV-6 DNA in every somatic cell of the body. (8). Rather than representing a biological dead-end, chromosomal integration is an important—and some believe the sole—mechanism of HHV-6

doi:10.1128/9781555819156.ch29

latency. Chromosomally integrated HHV-6 (ciHHV-6) can reactivate and yield infectious virus *in vitro* (9) and *in vivo* (10). It is now clear that *in vivo* reactivation of ciHHV-6 results in production of viable virus and that even transplacentally acquired HHV-6 can originate from reactivation of maternal ciHHV-6 during pregnancy (11).

Immune system response and modulation: Interaction between virus and the host immune system is complex, and the immune response to HHV-6 infection remains incompletely understood. HHV-6 infection generates antibody against a wide variety of viral proteins including the major antigenic virion protein (U11); major glycoproteins gB (U39), gH (U48), and gQ (U100); gB, polymerase processivity factor (U27); late antigen REP protein U94; and membrane protein U24. Neutralizing antibodies have also been described, but these appear insufficient to control HHV6, as evidenced by the impact of T-cell depletion on virus reactivation. (12). The humoral response in cases of inherited ciHHV-6 differs from that seen in primary infection, and may leave ciHHV-6 individuals more susceptible than others to secondary infection by exogenous HHV-6 (8, 13).

HHV-7

HHV-7 was discovered by Frenkel and colleagues in circulating PBMC obtained from a healthy adult and was subsequently identified as a commensal inhabitant of saliva, with 75% of saliva specimens from adults carrying infectious virus (14). HHV-7 is closely related to the HHV-6 variants and shares many of their molecular and biological properties. HHV-7 genomes are more compact than HHV-6 genomes, with a length of approximately 145 kb. Like the HHV-6 variants, HHV-7 genomes are bounded by sequences that resemble mammalian telomeres. Sequence identity between HHV-7 and the HHV-6 variants is 40% to 60% for most genes. HHV-7 uses CD4 as its cellular receptor, as does HIV, and under appropriate conditions *in vitro*, HIV and HHV-7 can inhibit each other's growth by receptor competition (15).

HHV-8

HHV-8 was discovered as the etiologic agent of Kaposi's sarcoma (KS) as a result of several observations over 30 years. In the 1960s the disease patterns of KS in Africa indicated an infectious agent, which led investigators to culture herpesvirus-like particles from KS biopsy material (16). Observation of KS during the U.S. AIDS epidemic again indicated an infectious agent that was mainly spread through sexual contact (17). In 1994, investigators analyzed genetic material from a KS lesion to identify the eighth novel human herpesvirus, HHV-8, named Kaposi's sarcoma–associated herpesvirus (KSHV) (3), also named HHV-8 because of association with other diseases. HHV-8 joins Epstein-Barr virus, human papillomavirus, and hepatitis B virus in the small group of DNA viruses that cause cancer in humans.

Based on sequence and biological properties, HHV-8 is classified in the *Rhadinovirus* genus of the gammaherpesvirus subfamily. Its closest relative is EBV. The HHV-8 genome is approximately 160 kb long and encodes over 85 genes, including herpesvirus core genes and several genes of host derivation that encode proteins involved in cell-cycle regulation, signal transduction, and modulation of host immune responses (18, 19). As with all herpesviruses, HHV-8 infection lasts for a lifetime. Following resolution of primary infection, HHV-8 enters a latent state in which few viral genes are expressed and there is no production of infectious virus. Viral latency is an effective adaptation by herpesviruses to minimize exposure to the host immune system and thus escape the host antiviral response.

HHV-8 has a broad host range for virus entry and establishment of latency, but efficient replication and serial passage of clinical isolates in cell lines has not been demonstrated (20). Human B-cell lines naturally infected with HHV-8 have been generated from primary effusion lymphomas (PELs) and are in wide use as serodiagnostic reagents. In PEL cell lines, the virus is latent in most cells and can be induced with stimulants such as phorbol ester (tetradecanoyl phorbol acetate) and sodium butyrate. The high copy numbers of the kaposin and T1.1 transcripts during latent and lytic infection, respectively, make them useful targets for *in situ* hybridization.

EPIDEMIOLOGY AND TRANSMISSION

HHV-6

HHV-6 host range is restricted to humans and closely related nonhuman primates. HHV-6 is ubiquitous, with a global reach. The lack of type-specific serology assays means that the respective contributions of HHV-6A and HHV-6B are not defined, including whether or not there are ethnic and/or geographic differences in their circulation. In developed countries, HHV-6 seropositivity is >95% in adults. In the limited number of studies in which type-discriminating tests were used, HHV-6B appeared to be more prevalent (58%) than HHV-6A (25%) (21). Approximately 50% of children are seropositive by 1 year of age, increasing to 90% by age 2, with incidence of acquisition peaking between 9 and 21 months, and increasing to 95% by adulthood (22). HHV-6B almost always infects first, and HHV-6A infection is thought to be acquired later in life (23, 24). Although HHV-6A infection is largely asymptomatic, it has been associated with clinical illness (primary infection) in children from Africa and the United States and has been detected in the blood of hospitalized febrile children with HIV infection in Africa (25).

Transmission: Intrauterine, perinatal, and fecal-oral mechanisms of transmission have been suggested (24), and while these may contribute to HHV-6 spread, the preponderance of evidence indicates that saliva from asymptomatic adults and older children is the most important mechanism of virus transmission. (26, 27). An estimated 80% to 90% of the population intermittently sheds HHV-6 in saliva. HHV-6B DNA has been detected in salivary glands and in saliva (22). HHV-6 acquisition is associated with older siblings and parents but not with daycare attendance (28), indicating that close contact is needed for transmission. There is no evidence that HHV-6 can be transmitted through breast-feeding or blood transfusion (29, 30). Congenital infection occurs in approximately 1% of newborns; the majority, if not all, of infections are due to transmission of HHV-6 from ciHHV-6 (31).

HHV-7

HHV-7 is highly and widely prevalent in more than 90% of the human population and normally acquired early in life. The prevalence of HHV-7 reaches 50% to 70% by the age of 2, with primary infection generally following primary HHV-6 infection (32). Infectious HHV-7 is present in the saliva of most seropositive individuals; thus saliva is likely the primary route of transmission. Because HHV-7 seems to be much more readily available for transmission, it is a mystery why

HHV-6B is normally acquired earlier. In addition to saliva, HHV-7 has been detected in about 10% of breast milk specimens and 3% to 10% of cervical/vaginal fluids (33); thus additional paths of transmission are possible.

HHV-8

HHV-8 is unique among the other human herpesviruses in having a very low seroprevalence in much of the world; high HHV-8 infection rates of 40% to 60% are found only in sub-Saharan Africa where KS is a very common and aggressive cancer in men, women, and children. Moderate seroprevalence (10% to 30%) is observed in Eastern Mediterranean countries including Italy, Greece, and Israel. Low HHV-8 seroprevalence (<5%) is seen in North America and Northern and Western Europe, where the occurrence of KS is almost entirely limited to homosexual and bisexual men with AIDS and some organ transplant recipients (34).

There are four distinct clinical-epidemiological subtypes of KS: 1. Classical KS was first described over a century ago, occurring mostly in elderly men of Mediterranean, Eastern European, or Middle Eastern origin. It is a mild form of KS with lesions on lower extremities that run an indolent course and seldom cause death. 2. Endemic KS (or African KS) was first described in the 1960s among people living in sub-Saharan Africa. It is unrelated to HIV infection and affects adults and children. Endemic KS used to be the predominant form of KS in Africa but epidemic KS is now the most common form in Africa. 3. Epidemic KS (or AIDS-related KS) was first described in the 1980s among gay men and served as harbinger for the U.S. AIDS epidemic. Most cases have a rapid course with extensive skin and visceral involvement, although effective treatment for AIDS in the 1990s dramatically reduced KS incidence in developed countries. 4. Transplant-related (or iatrogenic) KS occurs in organ allograft recipients, discussed further in the following section on Transmission (34).

Transmission of HHV-8 appears to be via saliva for both casual and sexual transmission. In African countries where HHV-8 is endemic, transmission is clearly horizontal and casual, beginning early in childhood. In nonendemic countries, HHV-8 transmission is concentrated among men who have sex with men (MSM) with risk factors that are clearly sexual, and heterosexual transmission appears to be rare for reasons that are not understood (35).

Blood Transfusion and Organ Transplantation

The first evidence for bloodborne transmission of HHV-8 was a strong association between HHV-8 seropositivity and injection drug use (36, 37) followed by evidence for HHV-8 transmission by blood transfusions in Uganda (38). In the United States and many European countries where HHV-8 seroprevalence is low (<5%) and blood is universally leukoreduced, there is no evidence to date for HHV-8 transmission by blood transfusion and therefore no apparent need to promote universal screening of the blood supply.

Organ transplantation is an area of growing interest for HHV-8 because of increasing numbers of transplants overall and numbers of HIV+ patients receiving transplants. The prevalence of KS following organ transplantation increases with the HHV-8 prevalence in a region. For patients who are HHV-8 seropositive prior to surgery, the risk for developing KS is 14% to 30%. KS most often results from virus reactivation in the recipient but can also be donor-derived, which carries higher rates of severe illness and mortality. Regression of KS with less risk to the graft has been obtained by switching from calcineurin inhibitors (CNI) to sirolimus, also known as rapamycin, which is now considered the first-line treatment of KS for transplant recipients (39–41).

CLINICAL ASPECTS

HHV-6

HHV6 has been associated with many clinical conditions, but attempts to establish clear etiologic relationships with human disease have largely been confounded by the ubiquity of HHV-6, its ability to establish latency, its capacity for vertical transmission, and the relatively recent formal recognition of fundamental differences between HHV-6A and HHV-6B. The most well-recognized clinical presentations of HHV-6 infections, roseola infantum and fever, occur mainly in otherwise healthy infants and young children as a result of primary infection. Primary infection in adults is uncommon. Severe syndromes typically result from HHV-6 reactivation in immunocompromised hosts. Most infections are caused by HHV-6B; disease attributed to HHV-6A is rare.

Infections in Immunocompetent Hosts

Primary HHV-6 infection in immunologically normal hosts generally occurs early in life, usually after maternal antibody wanes at around 6 months of age. Most HHV-6 infections are self-limited, but severe and fatal cases have been reported. It is estimated that HHV-6 infections account for 20% to 40% of febrile presentations in pediatric emergency departments among children 6 to 24 months old (42, 43). Children are most commonly affected between 6 and 12 months of age. High fever (39.5–40°C) may be accompanied by irritability, malaise, and rhinorrhea. Other nonspecific symptoms and findings may also be present.

Roseola infantum (exanthem subitum; sixth disease): Roseola affects approximately one-third of children in the United States (43, 44). The typical presentation is of several days of high fever (39–40°C) and irritability, followed upon defervescence by a maculopapular rash beginning on the upper body and then spreading to the extremities, lasting 1 to 3 days. Other nonspecific symptoms may also be present. Most cases are relatively mild and self-limited. Viral DNA is detectable in blood by PCR for over a month after primary infection (42). Roughly 30% of febrile seizures in children < 2 years of age have been attributed to HHV-6. In a case series including 160 children with HHV-6 infection, febrile seizures occurred more commonly in HHV-6-infected children than those with fever due to other causes (43). Congenital infection is usually asymptomatic. One prospective study of 57 infants with congenital HHV-6 infection has suggested that infection may have a detrimental effect on neurologic development, although further study is warranted. (45)

Associations with Chronic Diseases: HHV-6 has been suggested as having a role as a causal agent or participant in an array of chronic conditions, including multiple sclerosis (MS) (46), temporal lobe epilepsy (47), and drug-induced hypersensitivity syndrome (DIHS) (47). The association between HHV-6 and many chronic diseases remains controversial and requires further investigation.

Infections in Immunocompromised Host

Viral activity in immunocompromised patients typically results from reactivation. Clinical findings may include pneumonitis, bone marrow aplasia, and persistent fever (24). HHV-6 can cause infection in cancer patients and those infected with HIV, but the most significant clinical presentations occur in hematopoietic stem cell transplant (HSCT) recipients and to a lesser extent the SOT population (48). During the post-transplant period, HHV-6 activity is detectable in up to 50% of HSCT and SOT recipients, most commonly 2 to 4 weeks after transplantation and is usually asymptomatic (49). Clinically apparent disease tends to be more severe in HSCT than in SOT recipients and includes fever and rash resembling graft-versus-host disease (GVHD), sinusitis, pneumonitis, graft suppression and rejection, encephalitis, and cytomegalovirus (CMV) reactivation (50). Although sometimes linked to GVHD, it is not clear whether HHV-6 activity is the cause of GVHD or is caused by GVHD.

Over 90% of HHV-6 detected following organ transplantation is HHV-6B (50). Whether ciHHV-6 causes disease in transplant patients is controversial and unresolved (51). In HSCT recipients, HHV-6 has been associated with delayed engraftment, delirium, cognitive dysfunction, and increased mortality (48, 52). The most common severe consequence of HHV-6 infection in HSCT recipients is encephalitis, occurring in about 1% of recipients with a mortality up to 40% (48, 53). Survivors often have persistent neurologic symptoms. Detection of viral DNA in CSF is diagnostic for HHV-6 encephalitis; in the post-HSCT setting CSF detection of HHV-6 DNA may also be clinically significant (54). Clinically apparent disease occurs in fewer than 1% of SOT recipients and most commonly presents as fever and bone marrow suppression, mimicking CMV disease (55).

HHV-7

It is likely that the full spectrum of HHV-7-associated disease has not been defined. To date, primary infection with HHV-7 has been associated with a subset (<10%) of roseola cases, with a subset of those developing neurologic complications such as febrile convulsions (32). The other main clinical aspect of HHV-7 is disease similar to that commonly associated with HCMV infections in solid organ transplant (SOT) recipients. Although the true incidence of HHV-7 related disease is not known, it is likely to occur in less than 1% of SOT patients. In most cases, HHV-7 has been detected in blood of patients with CMV disease, and the extent to which HHV-7 directly contributes to any of the clinical symptoms in patients with CMV disease is not yet clear (56, 57). HHV-7 has been implicated as having indirect effects, including a higher predisposition to developing CMV disease. HHV-7 may have immunomodulatory effects and may be a risk factor for other opportunistic infections in immunocompromised patients (57).

HHV-8

Similar to the betaherpesviruses (CMV, HHV-6A, HHV-6B, HHV-7) and the other gammaherpesvirus (EBV), primary infection with HHV-8 exhibits no symptoms or mild, nonspecific symptoms and thus is rarely documented. HHV-8 infection alone is not sufficient to cause KS, and cofactors that lead to development of KS have not been identified except that immunosuppression appears to play a role in all forms of KS.

Kaposi's Sarcoma

Epidemiologic, pathologic, and molecular data make it clear that HHV-8 is the etiologic agent of all forms of KS, characterized by reddish-brown lesions usually localized to the skin of the extremities and trunk (Fig. 1). Pathologically, KS lesions are characterized by cells with a spindle-like appearance, extensive networks of vascular slits, extravasated red blood cells, and purplish pigmentation (Fig. 2). Not a classic cancer, KS is a proliferative condition intermediate between the benign and malignant categories. The time from HHV-8 infection to onset of KS is 2 to 10 years for AIDS patients, compared to decades for classic KS (58).

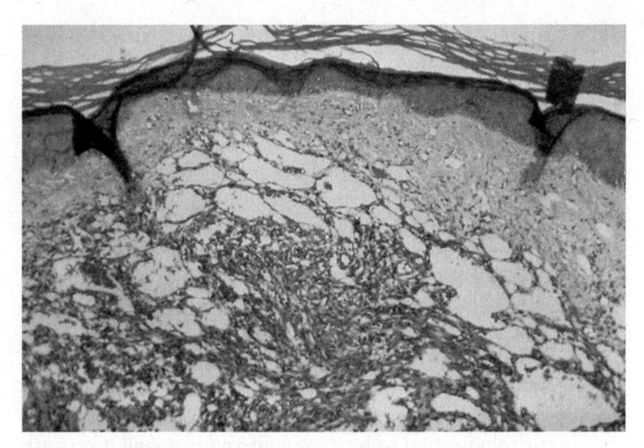

FIGURE 1 The photomicrograph depicts histopathologic changes seen in a human skin biopsy of Kaposi's sarcoma, most notably the appearance of the dermal layer with a cellular infiltrate and proliferation of vascular elements.

An important cofactor for development of AIDS- and transplant-associated KS appears to be severe immunosuppression. Cofactors for the development of classic and African KS are less clear, but immunosuppression secondary to age (classic) and malnourishment (African) is thought to play a role. KS remains the second most frequent tumor in HIV-infected individuals worldwide and is the most common cancer in sub-Saharan Africa. The prevalence of KS and the associated morbidity and mortality will remain high in Africa until treatment for HIV infection becomes widely available. The incidence of KS among AIDS patients in the United States and Europe has decreased dramatically with the availability of highly active antiretroviral therapy (HAART) that became available in the mid-1990s. Prior to HAART, approximately 20% of male patients experienced KS during the course of their AIDS. The incidence of KS in the United States and Europe could potentially rise as the HIV/ HHV-8 coinfected population ages. Primary Effusion Lymphoma and Multicentric Castleman's Disease: PEL

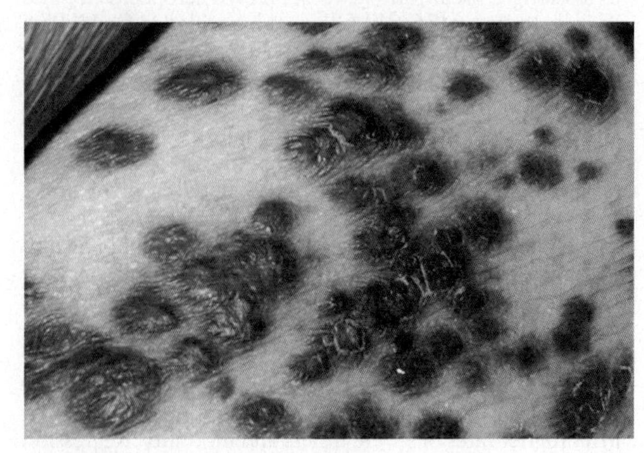

FIGURE 2 The skin of a patient displaying characteristic reddish-brown cutaneous plaques indicative of Kaposi's sarcoma.

belongs to a class of non-Hodgkin's B-cell lymphomas that arise without detectable tumor masses in body cavities. Almost all patients with PEL harbor HHV-8 (59). Half of PEL patients are coinfected with EBV, and EBV-positive PEL exhibits a different pattern of gene expression compared to EBV-negative PEL. At present there is no evidence that EBV-positive PEL has a different clinical outcome or responds differently to treatment. PEL is rare even in AIDS patients, constituting less than 1% of all AIDS-associated lymphomas in the United States (60). The relatively consistent growth of PEL cell lines in culture and the fact that many PEL cell lines can be induced to release infectious KSHV have made them invaluable for molecular studies in KSHV pathogenesis. All PEL cell lines depend on the presence of KSHV and the continued expression of KSHV genes for survival (59). Multicentric Castleman's disease (MCD) is a lymph node hyperplasia rather than a frank lymphoma and occurs primarily in AIDS patients (59). In HIV-negative individuals MCD is much less common and is less often associated with HHV-8. Numerous diseases have been studied for possible associations with HHV-8 including multiple myeloma, primary pulmonary hypertension, sarcoidosis, and angiosarcoma. However, at present only KS, PEL, and MCD have a clear etiologic link to HHV-8.

PREVENTION AND TREATMENT

HHV-6

The ubiquity of HHV-6 and nearly universal seropositivity in childhood make exposure-prevention strategies impractical and unlikely. Primary disease in children is self-limited, and treatment is supportive. Serious illness caused by HHV-6 acute infection or reactivation in immunocompromised hosts may successfully be managed through moderation/reversal of immune suppression if possible (61). No antiviral drugs are presently approved for the treatment of HHV-6 infection, and no guidelines currently exist for the treatment of acute HHV-6 infection. The International Herpes Management Forum and the American Society of Transplantation Infectious Diseases Community of Practice recommend antiviral therapy for cases of HHV-6 encephalitis (57). Ganciclovir (GCV), foscarnet (FOS), and cidofovir (CDV) each demonstrate *in vitro* activity against HHV-6 through inhibition of HHV-6 DNA polymerase; resistance to GCV has been described *in vitro* (57). Treatment of symptomatic illness in immunocompromised hosts with GCV and/or FOS may be beneficial, although the optimal course of therapy is unknown. Treatment of asymptomatic viremia is not indicated, as most episodes are transient and self-limited (57). Small clinical trials suggest that antiviral prophylaxis may be effective for preventing HHV-6 disease in HSCT patients, but current evidence is inadequate to support prophylaxis in HSCT or SOT recipients (57, 62). Artesunate has been used for treatment with reported effect (63). Immunotherapeutics directed at restoring or promoting T-cell responses to HHV-6B have shown encouraging results (64), but further study is warranted.

HHV-7

The ubiquity of HHV-7 and nearly universal seropositivity in childhood make prevention strategies unlikely and at this time unwarranted. Antiviral therapy might be of value for treatment of HHV-7-associated CMV disease in organ transplant recipients, but there have been no randomized and controlled trials assessing the efficacy of antiviral drugs for the prevention and treatment of HHV-7 infection after transplantation. Data on the susceptibility of HHV-7 to ganciclovir are conflicting, probably as a result of the lack of a widely accepted standard definition of HHV-7 disease, a specific antiviral agent for HHV-7, and a standard diagnostic assay to monitor response to treatment (57).

HHV-8

Safe-sex practices such as the use of latex condoms during sexual intercourse can reduce the risk of infection. Oral fluids are likely vehicles for transmission, but effective strategies have not been identified for reducing their exchange.

Dramatic declines in AIDS-associated KS began in 1996 with the introduction of highly active antiretroviral therapy (HAART) for the treatment of AIDS, which led to a 6- to 8-fold decrease in the incidence of new KS and regression of existing KS in a majority of patients (65, 66). The mechanism for the therapeutic effect of HAART on KS is largely indirect, by inhibition of HIV replication and subsequent recovery of the immune response to HHV-8. Despite the declines in the incidence of KS with HAART therapy, KS remains common among individuals with AIDS. Combined antiretroviral therapy (cART) remains key to KS treatment, and as sole therapy can achieve remission in 70% to 90% of patients with limited-stage disease (67). However, this usually follows the recovery of cell-mediated immunity, which can take several months. Therefore, in more advanced-stage or aggressive Kaposi's sarcoma, systemic chemotherapy in combination with cART is the established treatment strategy. The use of cytotoxic chemotherapy to treat AIDS-KS requires a careful balance of risk and benefit to achieve an anticancer effect without causing further immune compromise, especially in view of known drug interactions between cART and chemotherapeutic agents. Current strategies for chemo-, radio-, immuno-, systemic, and targeted therapies for KS are reviewed by La Ferla et al., and Robey and Bower (68, 69).

Organ transplantation: HHV-8 screening for the prevention of post-transplant KS should be considered because reduction of immunosuppressive therapy is risky for the allograft and, moreover, patients that develop visceral KS universally have a grim prognosis. HHV-8 serologic screening of donors and recipients in HHV-8 endemic regions and possibly of high-risk individuals (MSM, IDU) in non-endemic regions may be warranted, not to exclude the graft but to monitor HHV-8 viral load after transplantation to determine risk and avoid over-immunosuppression. (39, 70, 71). Therapeutic reduction of KS risk has been observed with the use of sirolimus, a mammalian target of rapamycin (mTOR) inhibitor. Accumulating experience among organ transplant studies points toward the potential of sirolimus to prevent the development of *de novo* malignancies of various types in the post-transplant period (39, 72).

HIV: The transplant community has seen the gradual acceptance of liver and kidney transplantation in carefully selected HIV-positive patients. Such patients have been showing 3- to 5-year survival rates for liver and kidney transplant similar to those of non-HIV-infected patients (73). Thus infection by HIV is no longer a contraindication for transplant in selected candidates with end-stage organ disease and has become more common. Organ transplantation is an appropriate therapeutic option for HIV-infected patients with end-stage organ disease who may be a group that warrants consideration for HHV-8 testing and monitoring for prevention of KS.

LABORATORY DIAGNOSIS

HHV-6

Classic roseola is diagnosed clinically. HHV-6 diagnostic testing by PCR is important in the management of SOT and HSCT recipients, especially in the diagnosis of encephalitis, as an adjunct diagnosis for DHIS, as part of diagnosing possible cases of measles, mumps, or rubella, and to prevent misdiagnosis of vaccine adverse events. Because HHV-6 is only one of several agents potentially causing a given clinical presentation (e.g., encephalitis), multiplex approaches may be appropriate to achieve operational efficiencies. As described in the following sections, discriminating primary from reactivated or active from latent infection remains a challenge in the laboratory. Health care providers should also be aware that because of a lack of standardized methods and definitions, caution is necessary in assessing the literature and in extrapolating results of extramural studies to local practice. The use of tests able to differentiate HHV-6A and HHV-6B is important and recommended in order to further understanding of the pathogenic roles and potential of each of these viruses (4) but is not clinically relevant.

Specimen Selection

Specimens include whole blood, plasma, serum, CSF, bronchoalveolar lavage (BAL), body fluid cell fraction (e.g., PBMCs), and tissue. (22, 27).

Virus Isolation

Culture of patient peripheral blood lymphocytes on PBMCs or on mitogen-stimulated human cord blood cells has proven successful, as has use of the immature T-cell lines HSB-2 (HHV-6A) and MT-4 and Mot-3 cell lines (HHV-6B) (74). HHV-6 is easily isolated from the blood of patients with roseola during the febrile first and second days of illness (75). Although recovery of HHV-6 in culture provides unequivocal evidence of active infection, because culture models are complex and labor intensive and no continuous cell line is recommended for isolation, culture is limited to research and drug evaluation settings.

Antigen Detection

Limited HHV-6, HHV-6A, and HHV-6B-reactive immune reagents have been developed by individual laboratories or are available from commercial vendors as research tools for laboratory-developed tests. Currently only lytic cycle-specific antibodies are available. Detection of viral antigen in PBMCs has been used to monitor HHV-6 in post-transplant patients. Immunohistochemical detection of HHV-6 antigen in biopsy material allows precise cellular localization of virus in tissue and can be important in the workup of SOT recipients with graft rejection (49).

Serology

The application of serology in the clinical setting is limited and interpretation of serologic results can be difficult. IgG seropositivity is an unhelpful diagnostic result in older children and adults. HHV-6 reactivation induces a rapid rise in IgG and an increase in IgM. Up to 5% of adults are IgM positive at any given time, through either reinfection or reactivation (76). Cross-reactivity of antibodies generated against CMV, HHV-6A, and HHV-6B due to shared epitopes or to heterotypic response are reported, affecting the performance of some assays (77).

Commercial serology assays in indirect immunofluorescence (IFA) and enzyme-linked immunosorbent (ELISA) formats are available, and studies comparing them differ in their conclusion as to which has the best specificity (78, 79). IgG avidity has proven useful in distinguishing primary from recurrent infection in SOT recipients but remains a research method (80). The high degree of nucleotide homology between HHV-6A and HHV-6B has stymied development of species-specific assays, and none are presently available outside of the research laboratory.

Nucleic Acid Detection and Monitoring

PCR is primarily useful for the diagnosis and evaluation of viremia and central nervous system (CNS) infection. HHV-6 DNA detection is commonly performed in CSF, whole blood, BAL specimens, bone marrow, and tissue biopsy relevant to end-organ disease. Caution is required in test selection and result interpretation, because neither detection nor quantitation values strictly predict disease. Much HHV-6 reactivation (thus DNA positivity) is asymptomatic and resolves spontaneously with clearance of virus without intervention. The possibility of ciHHV-6 further complicates result interpretation and its presence must also be considered—and excluded—to properly assess the significance of viral load results. Despite the many advances in this area, distinguishing between active, latent, and chromosomally integrated/inherited HHV-6 remains challenging, and clinically significant viral load threshold value definitions are lacking.

The gold standard for detecting and monitoring HHV-6 in patient specimens is real-time quantitative PCR (qRTPCR), for which primer and probe design can be readily exploited to develop species-differentiating tests (81, 82). A variety of well-validated PCR in-house tests are reported in the literature, and assay kits are commercially available. Genes targeted by these include U11, U22, U31, U38, U57, U67, U94, and U95.

Qualitative PCR

Qualitative HHV-6 DNA detection is of limited diagnostic value because it cannot distinguish active from latent infection. PCR on biopsy material may be helpful to quickly determine if viral DNA is present, but antigen detection is required to establish HHV-6 presence in tissue cells rather than in transiting inflammatory cells.

Quantitative RTPCR

Many laboratory-developed (LD) quantitative RTPCR (qRTPCR) assays are reported in the literature (83–89). A small number of qRTPCR assays is commercially available, and published evaluations are few. Two studies evaluating multicenter performance of molecular assays used to detect and quantify HHV-6 (88, 89) showed that results for qualitative detection were better than quantitative results but that agreement between assays was poor. A commutable international standard (such as available for CMV) (90) is needed to standardize qRTPCR results. Development and implementation of alternate detection and measurement technologies such as digital PCR (91) may also lead to progress in this area. Digital PCR has not yet been widely applied in the clinical virology arena but shows potential as a powerful diagnostic tool (92).

HHV-6 DNAemia

Previously, detection of HHV-6 DNA in plasma was thought to reflect active viral replication. However, plasma DNA has been shown to be derived from lysed PBMCs latently infected with HHV-6 and not to represent circulating virus (93). Thus, whole blood is an appropriate and convenient clinical specimen, and HHV-6 viral load measurement in peripheral blood is considered the most powerful method presently available for the diagnosis of active infection (94, 95). Awareness of a patient's white cell count is important, because the cellularity of blood directly influences viral load, and without correlation, results may be misleading (e.g., in leukopenia).

In primary infection, viral loads exceeding 5.5 $\log_{10}$ copies/mL of whole blood may be observed (96, 97). In healthy blood donors, HHV-6 viral loads were found to be low, in the range of 3.2 $\log_{10}$ to 3.5 $\log_{10}$ copies per mL and 1.09 $\log_{10}$ to 3.17 $\log_{10}$ copies per million cells (98). In HSCT recipients, viral loads are approximately 10 times higher than in the normal population, and symptoms attributed to HHV-6 have been associated with loads greater than 3 $\log_{10}$ copies per million PBMC or 4 $\log_{10}$ copies per mL (99, 100). It is reasonable and important to test transplant patients with end-organ disease (101). No benefit has been demonstrated by routine monitoring of HHV-6 DNAemia in SOT and in HSCT patients (62, 102).

HHV-6 DNA in CSF

CNS disease in the setting of HSCT (100) and in occasional instances, primary infection, justifies HHV-6 DNA testing of CSF. Provided that ciHHV-6 has been ruled out, the detection of HHV-6 in the absence of another etiologic agent is considered diagnostic in encephalitis. Normal CSF may contain up to 5 nucleated cells/μL, and RTPCR assays will detect ciHHV-6, potentially leading to a false diagnosis (103). Prevalence of HHV-6B in the CSF of immunocompetent patients with primary infection is low, and when detected, DNA is present in low concentration (mean viral load 2.4 $\log_{10}$ copies/mL) (6). In contrast, immunocompetent patients with suspected encephalitis and known to harbor ciHHV-6 have elevated concentrations of viral DNA (mean viral load was 4.0 $\log_{10}$ copies/mL (6). In HCT recipients, reported viral loads in encephalitis vary widely (600 to 288,975 copies/mL) peak median 16,600 copies/ml (104); >999,000 copies/mL (105).

HHV-6 DNA can also be detected in the CSF of asymptomatic HSCT recipients (median peak value 655 copies/mL, range 25 to 260,000 copies/mL) (53). Children under 3 years of age have detectable CSF HHV-6 in primary infection, even in the absence of encephalitis (106). Monitoring CSF viral load over the course of treatment has proven informative (107).

HHV-6 DNA in Saliva

Although saliva is not useful as a diagnostic specimen, some have evaluated the saliva as an HHV-6 monitoring tool. Zerr (22) detected and quantitated HHV-6 DNA in saliva of immunocompetent children following primary infection and found low viral burdens in the first week (median 1,700 copies/mL; in the first 4 weeks multiple sampling was required for detection) that increased by week 8 (100,000 c/mL) and remained detectable for at least 12 months (approximately 40,000 copies/mL). Nefzi (108) compared HHV-6 salivary viral load to conventional blood viral load in 50 leukemia patients. Detection frequency in saliva was significantly higher than in blood (P = 0.01 to P = 0.09), viral loads in saliva were higher than in blood, and increases in salivary vial load preceded onset of DNAemia. These findings appear to suggest that saliva may be a more sensitive specimen than peripheral blood for monitoring HHV-6 in these patients.

Viral Load Interpretations

Quantitative PCR assays can be highly variable, with performance influenced by sample selection, specimen processing, extraction method, amplification/detection reagents, amplification platform, and operator. This heterogeneity makes interlaboratory comparisons of viral loads difficult and precludes the establishment of universally meaningful clinical thresholds and treatment algorithms.

The presence of ciHHV-6 must also be considered to properly assess the significance of viral load results. CiHHV-6 produces persistently high levels of viral DNA in blood, tissues, and in specimens containing lysed cells (CSF, serum) and can lead to misdiagnosis of active infection. Ward showed that in immunocompetent ciHHV-6 subjects, viral loads were elevated in blood, serum, and hair follicles (7.0 $\log_{10}$ copies/mL, 5.3 $\log_{10}$ copies/mL, and 4.2 $\log_{10}$ copies per follicle cell, respectively) (81). When normalized, these data indicated that HHV-6 DNA was present in each sample type at ≥ 1 copy per cell. The accepted method for identifying ciHHV-6 is viral load testing of hair follicles or nails. As an alternative, detecting HHV-6 DNA at levels greater than 5.5 $\log_{10}$ copies/mL in whole blood is considered acceptable (103). However, viral loads greater than 5.5 $\log_{10}$ copies/mL in blood also occur in primary infection or reactivation (96, 109), and under these circumstances, persistence of elevated viral loads is used as the distinguishing characteristic of ciHHV-6 (81, 97). A number of hospital, university, and commercial laboratories offer quantitative HHV-6 DNA testing for the identification of ciHHV-6 (103). Information on commercial sources for ciHHV-6 testing is available through the HHV-6 Foundation (www.hhv6-foundation.org/clinicians/cihhv6.testing).

HHV-6 Transcript Detection

The detection and measurement of viral transcripts can be used to distinguish active replication from latency. Studies using reverse transcriptase PCR (RT-PCR) to detect HHV-6 mRNA in leukocytes have been reported (110). Assays target transcripts expressed at different stages of the replication cycle. Immediate early (U90), early (U12), and late (U100) mRNAs and others have been evaluated, and correlations have been made with active infection (109), with viremia (94), and between quantitated mRNA and viral load (95). Instability of mRNA and the generally low sensitivity of assays developed to date have prevented this technique from being useful in the routine diagnostic virology laboratory.

HHV-7

Antibody Tests

IFA, EIA, and immunoblot assays have been described for detecting HHV-7 antibodies. False-positive results can be obtained with the IFAs and EIAs due to cross-reactivity with HHV-6 (111). Immunoblot assays using antigens specific for each of the viruses do not cross-react (111), but reagents for such assays are not readily available.

The high prevalence of HHV-7 makes it difficult to interpret single positive serologic results and thus the detection of virus and viral DNA is more often used for clinical evaluations. The three circumstances where monitoring HHV-7 viral activity might have clinical value are in children experiencing febrile rash or neurologic illness following routine childhood vaccination, neurologic illness, and organ transplant recipients.

Methods available for detecting HHV-7 include culture, monoclonal antibodies, and PCR. Because viral culture is complex and labor-intensive, it has become less available in diagnostic laboratories. Several PCR systems for HHV-7 have been described, including quantitative and multiplexed methods (56, 112). Quantitative PCR from serum is likely to provide the most useful evidence for viral activity, but its use for this purpose is an area of research and specific guidelines for its application are not available. Because of the labor and other costs associated with screening for multiple agents, multiplexed PCR assays that allow screening for several agents simultaneously would provide greater efficiency for monitoring organ transplant recipients.

HHV-8

HHV-8 laboratory tests are not generally used to diagnose KS (or PEL or MCD) because KS is typically a clinical diagnosis with histological examination of a KS lesion biopsy. The major applications of HHV-8 diagnostic tests are for epidemiology, studies on transmission, prediction of KS development, monitoring efficacy of therapy, and investigation of KS risk and KS etiology in organ transplantation and new disease associations.

Serology

Serology is the method of choice to detect HHV-8 infection because most infected individuals will not have detectable levels of HHV-8 in their blood at any given time. In fact, HHV-8 DNA is found in the blood of only 30% to 50% of individuals with clinical KS and far less often among individuals with asymptomatic infection (113–115). Several serology tests have been developed for research and epidemiology in immunofluorescence assay (IFA) and enzyme-linked immunosorbent assay (ELISA/EIA) formats to detect antibodies against HHV-8 latent and lytic proteins. The majority of these tests show good concordance and sensitivity with sera from KS patients but show considerable discordance on sera from subjects without KS (116, 117). The difficulty with accurate detection of HHV-8 antibody is largely biological because humoral immune response to HHV-8 infection in immunocompetent people is often weak, with antibody levels hovering around the threshold of detection. Moreover, immune response involves several viral antigens, different subsets of which can develop in subjects during the course of HHV-8 infection (118, 119). Lastly, evaluation of HHV-8 test specificity is challenging since most infections are asymptomatic with virus at undetectable levels; thus it is very difficult to identify controls known to be HHV-8 uninfected.

Lytic and latent IFA

The first HHV-8 serology tests developed were immunofluorescence assays (IFA) based on HHV-8-infected PEL cell lines that are still used today for detection of antibodies to either lytic or latent viral proteins. PEL cells treated with a virus inducing agent such as tetradecanoyl phorbol acetate (TPA) or sodium butyrate induce HHV-8 into lytic replication and are used for the detection of antibodies against various lytic cycle proteins (120). The lytic IFA consistently shows the highest sensitivity for detection of antibodies in patients with KS, but the fact that early versions of the test reacted with 20% of U.S. blood donors was considered evidence of poor specificity. Later versions of the lytic IFA retained >95% sensitivity among patients with KS and dramatically reduced reactivity in U.S. blood donors to 3% (114). However, the lytic IFA is labor intensive with subjective interpretation and thus is not a practical choice for commercial or high-throughput testing. Untreated PEL cells are used for the detection of antibodies against the main latent cycle protein called latency-associated nuclear antigen (LANA) (121, 122). The LANA IFA by itself is relatively insensitive for identifying HHV-8 antibodies in KS patients and other groups but contributes to overall sensitivity and specificity in multiantigen tests.

ELISA

Several enzyme-linked immunosorbent assays (ELISA or EIA) have been developed based on immunogenic antigens expressed during the lytic cycle (K8.1, orf65) or latent cycle (orf73 and K12) (114, 117). The lytic cycle gene K8.1 encodes a glycoprotein located in the viral envelope with no homolog in other herpesviruses. Orf 65 is another lytic cycle gene that encodes a viral capsid protein that is the homolog to the EBV capsid protein BFRF3. Orf 73 encodes LANA required for replication and persistence of viral DNA. Several ELISA-based tests have been developed for HHV-8 based on the K8.1 and orf 65 viral proteins. The performance of most of these HHV-8 serology tests has been compared in several studies that test panels of sera from different demographic groups, with the following consistent results: lytic-antigen-based assays are more sensitive than latent-antigen-based assays, and multiple antigen tests are more sensitive than single antigen tests (114, 117, 123). Thus, most current testing involves two or more serology tests based on different HHV-8 antigens that show concordance or the use of multiple antigen tests. Commercial tests for HHV-8 are available and some have been included in assay comparison studies mentioned above and performed well in studies on high-risk populations, but none have yet emerged as broadly considered sensitivity and specific.

New tests in development are high-throughput, multiple-antigens formats and are providing encouraging results. A whole-virus ELISA developed by a Brazilian laboratory used a supernatant from sonicated, TPA-induced BCBL-1 cells and had a reported sensitivity of 97% and specificity of 86% by latent class analysis (124). ELISA testing has the advantage of being widely used and readily operational in any diagnostic laboratory. A bead-based multiplex assay is under development by the National Cancer Institute (NCI) that detects antibodies to six KSHV antigens; the three viral antigens most commonly used to date, K8.1, orf 65 and orf 73, together with three additional antigens determined to have immunogenic properties, orf 38, orf 59, and orf 61 (125). Bead-based assays are not yet as widely used, especially in lower resource settings, but have the advantage of flexibility for easily adding and removing different viral antigens as the test is further developed and for use in different populations.

Despite progress in the development of multiantigen-, high-throughput serology assays for HHV-8, the challenge remains of proving assay specificity in asymptomatic subjects with discordant test results. To address this, repeat measures from the same individuals over time could document increasing antibody titers or the presence of viral DNA and further establish accurate tests or testing algorithms.

Immunohistochemistry (IHC) and *In Situ* Hybridization (ISH)

IHC and ISH have proven useful to confirm the diagnosis of KS in lesions with nontypical pathologies or to rule out KS in the case of other spindle cell lesions. ISH has been useful to show viral gene expression in different cell types of infected tissue (126). IHC and ISH can more closely link viral expression to a given disease process to better understand viral pathogenesis and to allow for more targeted therapy (127, 128).

PCR

PCR is not the test of choice for diagnosis of HHV-8 infection because most infected individuals do not have detectable levels of viral DNA, even those with clinical KS. HHV-8 DNA has been found in 30% to 50% of peripheral blood mononuclear cells (PBMCs) from AIDS patients with KS, in 5% to 20% of PBMCs from HIV-infected subjects without KS, and is undetectable or rarely found in healthy HIV-negative HHV-8 seropositive individuals (113–115). However, HHV-8 PCR is invaluable to study transmission, correlates of viral load with disease, and response to antiviral therapy. Numerous PCR primer sets have been described against the following viral genes: orf 26 (3, 129), orf 25 (130), K6 (131), and orf 73 (132). HHV-8 DNA is present in high concentration in KS biopsy material, but collection of biopsy material is painful for the patient and is not necessary for routine diagnosis. Viral DNA when detectable is present in the saliva and whole blood, with a higher concentration in white blood cells, so these are the specimens of choice for many studies.

Quantitative PCR methods have revealed strong correlations between HHV-8 viral load and the presence and progression of KS (115,133–135).

DIAGNOSTIC BEST PRACTICES

HHV-6

A summary of the main methods and application for diagnosis of HHV-6 infection is shown in Table 1. As understanding of HHV-6A and HHV-6B pathogenesis and roles in human disease evolve and diagnostic test reagents and technologies improve, the type of and indications for testing for HHV-6 will also evolve. The reader is directed to the main text of this chapter as well as to the primary literature for detailed information.

Specimens submitted for analysis, diagnostic tests performed, and result interpretation must be appropriate for the given clinical context. As stated previously, whenever possible, the use of HHV-6A/HHV-6B differentiating assays is

TABLE 1 Laboratory diagnosis of HHV-6

Approach	Method	Application	Remarks
Virus isolation	Cell culture	Research, antiviral agent evaluation	Complex, labor-intensive methods; no continuous cell lines available. Not performed in diagnostic laboratories.
Antigen Detection			
PBMC	IHC	Monitoring infection in post-transplant patients	Qualitative and quantitative; nonstandardized methods, mostly LDTs; can be difficult to read.
Biopsy	IHC	In end-organ disease, cellular localization of HHV-6 in tissue	Nonstandardized methods, complex and labor-intensive; can be difficult to read.
Serology			
IgM	IFA, ELISA	Primary infection in children	Limited clinical utility due to methodological issues, antibody cross-reactivity, and virus prevalence. IgG avidity remains a research method.
IgG	IFA, ELISA	Seroconversion; seroprevalence	
Nucleic Acid Detection			
ISH	ISH	Identification of ciHHV-6; cellular localization of viral genome	Complex and labor-intensive methods; not performed in routine diagnostic laboratory.
Qualitative Detection	PCR, RTPCR	Detection of viral nucleic acid; identification of ciHHV-6 in hair follicles or nails	Limited clinical utility. Does not distinguish latent from active infection. Mainly LDTs; lack of method standardization.
Quantitative Detection and Monitoring	qRTPCR, ddPCR	Diagnosis and monitoring of infection	Gold standard method for HHV-6 detection. Nonstandardized methods, many LDTs, no viral thresholds defining clinical significance.
Transcript Detection	rtPCR	Detection of active replication	Non-standardized complex and labor-intensive methods; not performed in routine diagnostic laboratory.

ISH, *in situ* hybridization; IHC, immunohistochemistry; LDT, laboratory-developed test; IFA, immunofluorescence assay; ELISA, enzyme-linked immunosorbent assay; ciHHV-6, chromosomally integrated HHV-6; RTPCR, real-time polymerase chain reaction; qRTPCR, quantitative RTPCR; ddPCR, digital droplet PCR; rtPCR, reverse transcription PCR.

TABLE 2 Laboratory diagnosis of HHV-8

Approach	Method	Application	Remarks
Serology, IgG detection in blood	IFA, ELISA	Epidemiology of HHV-8-associated disease, not used for diagnosis of KS	Preferred method to detect HHV-8 infection. Nonstandardized methods, mainly LDTs, controversy regarding accuracy of various tests.
Antigen detection in KS biopsy	IHC	Cellular localization of HHV-8 in tissue	Nonstandardized methods, complex and labor intensive.
Viral DNA detection in blood or saliva, quantitative	qRTPCR	Research, monitoring response to KS therapy	Nonstandardized methods, many LDTs, no viral thresholds defining clinical significance.

IFA, immunofluorescence assay; ELISA, enzyme-linked immunosorbent assay; KS, Kaposi's sarcoma; LDT, laboratory-developed test; IHC, immunohistochemistry; qRTPCR, quantitative real-time polymerase chain reaction

strongly recommended (4). Important limitations remain in the interpretation of nucleic acid detection and quantitation test results and correlation of these with clinically significant pathogenicity. Standardization of molecular assays and establishment of broadly applicable viral load thresholds remain elusive at the time of writing. When molecular testing indicates the presence of HHV-6 DNA, viral load determinations must be used to exclude the possibility of ciHHV-6.

Primary Infection

Laboratory testing is rarely indicated in primary infection. Roseola is diagnosed clinically. Central nervous system infection resulting from primary infection is diagnosed by molecular detection of viral DNA in the CSF, following exclusion of more common etiologic agents.

Reactivation

Targeted testing is indicated based on clinical presentation and degree of suspicion of HHV-6 involvement in the transplant recipient.

HHV-7

The high prevalence of HHV-7 makes it difficult to interpret single positive serologic results and thus the use of PCR for detection of viral DNA is more often used for clinical evaluations. The three circumstances where monitoring HHV-7 viral activity might have clinical value are in children experiencing febrile rash or neurologic illness and in organ transplant recipients. Quantitative PCR from serum is likely to provide the most useful evidence for viral activity, but specific guidelines for interpretation of results is an area of research and not currently available.

HHV-8

A summary of the main methods and application for diagnosis of HHV-8 infection is shown in Table 2. Because KS is typically a clinical diagnosis, the major applications of HHV-8 diagnostic tests are not to diagnose KS but rather to study epidemiology, transmission, and disease progression and to monitor efficacy of therapy. Serology is the method of choice to detect HHV-8 infection because most infected individuals will not have detectable levels of HHV-8 in their blood at any given time. Most current serology testing involves two or more serology tests based on different HHV-8 antigens that show concordance or the use of multiple-antigen tests. Commercial tests for HHV-8 are available but none of the tests have yet emerged with broadly considered sensitivity and specificity. New tests in development although not yet commercially available have high-throughput, multiple-antigen formats and are providing encouraging results (see main text under Laboratory Diagnosis).

PCR is not the test of choice for diagnosis of HHV-8 infection because most infected individuals do not have detectable levels of viral DNA, even those with clinical KS. However, HHV-8 PCR is invaluable to study transmission, correlates of viral load with disease, and response to antiviral therapy.

ACKNOWLEDGMENTS

The authors would like to express their gratitude to Adam Wharton at the CDC for his invaluable assistance with literature searches for this chapter.

The findings and conclusions in this document are those of the authors and do not necessarily represent the official position of the Centers for Disease Control and Prevention or the Children's Hospital of Eastern Ontario.

REFERENCES

1. **Salahuddin SZ, Ablashi DV, Markham PD, Josephs SF, Sturzenegger S, Kaplan M, Halligan G, Biberfeld P, Wong-Staal F, Kramarsky B, Gallo RC.** 1986. Isolation of a new virus, HBLV, in patients with lymphoproliferative disorders. *Science* **234:**596–601.
2. **Frenkel N, Schirmer EC, Wyatt LS, Katsafanas G, Roffman E, Danovich RM, June CH.** 1990. Isolation of a new herpesvirus from human CD4+ T cells. *Proc Natl Acad Sci USA* **87:**748–752.
3. **Chang Y, Cesarman E, Pessin MS, Lee F, Culpepper J, Knowles DM, Moore PS.** 1994. Identification of herpesvirus-like DNA sequences in AIDS-associated Kaposi's sarcoma. *Science* **266:**1865–1869. see comments.
4. **Ablashi D, Agut H, Alvarez-Lafuente R, Clark DA, Dewhurst S, DiLuca D, Flamand L, Frenkel N, Gallo R, Gompels UA, Höllsberg P, Jacobson S, Luppi M, Lusso P, Malnati M, Medveczky P, Mori Y, Pellett PE, Pritchett JC, Yamanishi K, Yoshikawa T.** 2014. Classification of HHV-6A and HHV-6B as distinct viruses. *Arch Virol* **159:**863–870.
5. **De Bolle L, Van Loon J, De Clercq E, Naesens L.** 2005. Quantitative analysis of human herpesvirus 6 cell tropism. *J Med Virol* **75:**76–85.
6. **Ward KN, Leong HN, Thiruchelvam AD, Atkinson CE, Clark DA.** 2007. Human herpesvirus 6 DNA levels in cerebrospinal fluid due to primary infection differ from those due to chromosomal viral integration and have implications for diagnosis of encephalitis. *J Clin Microbiol* **45:**1298–1304.
7. **Luppi M, Marasca R, Barozzi P, Ferrari S, Ceccherini-Nelli L, Batoni G, Merelli E, Torelli G.** 1993. Three cases of human herpesvirus-6 latent infection: integration of viral genome in

peripheral blood mononuclear cell DNA. *J Med Virol* **40:** 44–52.
8. **Tanaka-Taya K, Sashihara J, Kurahashi H, Amo K, Miyagawa H, Kondo K, Okada S, Yamanishi K.** 2004. Human herpesvirus 6 (HHV-6) is transmitted from parent to child in an integrated form and characterization of cases with chromosomally integrated HHV-6 DNA. *J Med Virol* **73:**465–473.
9. **Arbuckle JH, Medveczky MM, Luka J, Hadley SH, Luegmayr A, Ablashi D, Lund TC, Tolar J, De Meirleir K, Montoya JG, Komaroff AL, Ambros PF, Medveczky PG.** 2010. The latent human herpesvirus-6A genome specifically integrates in telomeres of human chromosomes in vivo and in vitro. *Proc Natl Acad Sci USA* **107:**5563–5568.
10. **Goel P, Tailor P, Chande AG, Basu A, Mukhopadhyaya R.** 2013. An infectious HHV-6B isolate from a healthy adult with chromosomally integrated virus and a reporter based relative viral titer assay. *Virus Res* **173:**280–285.
11. **Gravel A, Hall CB, Flamand L.** 2013. Sequence analysis of transplacentally acquired human herpesvirus 6 DNA is consistent with transmission of a chromosomally integrated reactivated virus. *J Infect Dis* **207:**1585–1589.
12. **Gerdemann U, Keukens L, Keirnan JM, Katari UL, Nguyen CTQ, de Pagter AP, Ramos CA, Kennedy-Nasser A, Gottschalk SM, Heslop HE, Brenner MK, Rooney CM, Leen AM.** 2013. Immunotherapeutic strategies to prevent and treat human herpesvirus 6 reactivation after allogeneic stem cell transplantation. *Blood* **121:**207–218.
13. **Pantry SN, Medveczky MM, Arbuckle JH, Luka J, Montoya JG, Hu J, Renne R, Peterson D, Pritchett JC, Ablashi DV, Medveczky PG.** 2013. Persistent human herpesvirus-6 infection in patients with an inherited form of the virus. *J Med Virol* **85:**1940–1946.
14. **Wyatt LS, Frenkel N.** 1992. Human herpesvirus 7 is a constitutive inhabitant of adult human saliva. *J Virol* **66:**3206–3209.
15. **Lusso P, Secchiero P, Crowley RW, Garzino-Demo A, Berneman ZN, Gallo RC.** 1994. CD4 is a critical component of the receptor for human herpesvirus 7: interference with human immunodeficiency virus. *Proc Natl Acad Sci USA* **91:** 3872–3876.
16. **Giraldo G, Beth E, Haguenau F.** 1972. Herpes-type virus particles in tissue culture of Kaposi's sarcoma from different geographic regions. *J Natl Cancer Inst* **49:**1509–1526.
17. **Beral V, Peterman TA, Berkelman RL, Jaffe HW.** 1990. Kaposi's sarcoma among persons with AIDS: a sexually transmitted infection? *Lancet* **335:**123–128.
18. **Neipel F, Fleckenstein B.** 1999. The role of HHV-8 in Kaposi's sarcoma. *Semin Cancer Biol* **9:**151–164.
19. **Moore PS, Chang Y.** 2003. Kaposi's sarcoma-associated herpesvirus immunoevasion and tumorigenesis: two sides of the same coin? *Annu Rev Microbiol* **57:**609–639.
20. **Blackbourn DJ, Lennette E, Klencke B, Moses A, Chandran B, Weinstein M, Glogau RG, Witte MH, Way DL, Kutzkey T, Herndier B, Levy JA.** 2000. The restricted cellular host range of human herpesvirus 8. *AIDS* **14:**1123–1133.
21. **Wang FZ, Dahl H, Ljungman P, Linde A.** 1999. Lymphoproliferative responses to human herpesvirus-6 variant A and variant B in healthy adults. *J Med Virol* **57:**134–139.
22. **Zerr DM, Meier AS, Selke SS, Frenkel LM, Huang ML, Wald A, Rhoads MP, Nguy L, Bornemann R, Morrow RA, Corey L.** 2005. A population-based study of primary human herpesvirus 6 infection. *N Engl J Med* **352:**768–776.
23. **Dewhurst S, McIntyre K, Schnabel K, Hall CB.** 1993. Human herpesvirus 6 (HHV-6) variant B accounts for the majority of symptomatic primary HHV-6 infections in a population of U.S. infants. *J Clin Microbiol* **31:**416–418.
24. **De Bolle L, Naesens L, De Clercq E.** 2005. Update on human herpesvirus 6 biology, clinical features, and therapy. *Clin Microbiol Rev* **18:**217–245.
25. **Bates M, Monze M, Bima H, Kapambwe M, Clark D, Kasolo FC, Gompels UA.** 2009. Predominant human herpesvirus 6 variant A infant infections in an HIV-1 endemic region of Sub-Saharan Africa. *J Med Virol* **81:**779–789.
26. **Suga S, Yoshikawa T, Kajita Y, Ozaki T, Asano Y.** 1998. Prospective study of persistence and excretion of human herpesvirus-6 in patients with exanthem subitum and their parents. *Pediatrics* **102:**900–904.
27. **Fujiwara N, Namba H, Ohuchi R, Isomura H, Uno F, Yoshida M, Nii S, Yamada M.** 2000. Monitoring of human herpesvirus-6 and -7 genomes in saliva samples of healthy adults by competitive quantitative PCR. *J Med Virol* **61:**208–213.
28. **Rhoads MP, Magaret AS, Zerr DM.** 2007. Family saliva sharing behaviors and age of human herpesvirus-6B infection. *J Infect* **54:**623–626.
29. **Ward KN, Gray JJ, Efstathiou S.** 1989. Brief report: primary human herpesvirus 6 infection in a patient following liver transplantation from a seropositive donor. *J Med Virol* **28:**69–72.
30. **Kusuhara K, Takabayashi A, Ueda K, Hidaka Y, Minamishima I, Take H, Fujioka K, Imai S, Osato T.** 1997. Breast milk is not a significant source for early Epstein-Barr virus or human herpesvirus 6 infection in infants: a seroepidemiologic study in 2 endemic areas of human T-cell lymphotropic virus type I in Japan. *Microbiol Immunol* **41:**309–312.
31. **Hall CB, Caserta MT, Schnabel KC, Boettrich C, McDermott MP, Lofthus GK, Carnahan JA, Dewhurst S.** 2004. Congenital infections with human herpesvirus 6 (HHV6) and human herpesvirus 7 (HHV7). *J Pediatr* **145:**472–477.
32. **Hall CB, Caserta MT, Schnabel KC, McDermott MP, Lofthus GK, Carnahan JA, Gilbert LM, Dewhurst S.** 2006. Characteristics and acquisition of human herpesvirus (HHV) 7 infections in relation to infection with HHV-6. *J Infect Dis* **193:**1063–1069.
33. **Caserta MT, Hall CB, Schnabel K, Lofthus G, McDermott MP.** 2007. Human herpesvirus (HHV)-6 and HHV-7 infections in pregnant women. *J Infect Dis* **196:**1296–1303.
34. **Cohen A, Wolf DG, Guttman-Yassky E, Sarid R.** 2005. Kaposi's sarcoma-associated herpesvirus: clinical, diagnostic, and epidemiological aspects. *Crit Rev Clin Lab Sci* **42:**101–153.
35. **Engels EA, Atkinson JO, Graubard BI, McQuillan GM, Gamache C, Mbisa G, Cohn S, Whitby D, Goedert JJ.** 2007. Risk factors for human herpesvirus 8 infection among adults in the United States and evidence for sexual transmission. *J Infect Dis* **196:**199–207.
36. **Cannon MJ, Dollard SC, Smith DK, Klein RS, Schuman P, Rich JD, Vlahov D, Pellett PE, HIV Epidemiology Research Study Group.** 2001. Blood-borne and sexual transmission of human herpesvirus 8 in women with or at risk for human immunodeficiency virus infection. *N Engl J Med* **344:**637–643.
37. **Atkinson JO, Biggar RJ, Goedert JJ, Engels EA.** 2004. The incidence of Kaposi sarcoma among injection drug users with AIDS in the United States. *J Acquir Immune Defic Syndr* **37:** 1282–1287.
38. **Hladik W, Dollard SC, Mermin J, Fowlkes AL, Downing R, Amin MM, Banage F, Nzaro E, Kataaha P, Dondero TJ, Pellett PE, Lackritz EM.** 2006. Transmission of human herpesvirus 8 by blood transfusion. *N Engl J Med* **355:**1331–1338.
39. **Lebbé C, Legendre C, Francès C.** 2008. Kaposi sarcoma in transplantation. *Transplant Rev (Orlando)* **22:**252–261.
40. **Piselli P, Busnach G, Citterio F, Frigerio M, Arbustini E, Burra P, Pinna AD, Bresadola V, Ettorre GM, Baccarani U, Buda A, Lauro A, Zanus G, Cimaglia C, Spagnoletti G, Lenardon A, Agozzino M, Gambato M, Zanfi C, Miglioresi L, Di Gioia P, Mei L, Ippolito G, Serraino D, Immunosuppression and Cancer Study Group.** 2009. Risk of Kaposi sarcoma after solid-organ transplantation: multicenter study in 4,767 recipients in Italy, 1970–2006. *Transplant Proc* **41:** 1227–1230.
41. **Ariza-Heredia EJ, Razonable RR.** 2011. Human herpes virus 8 in solid organ transplantation. *Transplantation* **92:**837–844.
42. **Pruksananonda P, Hall CB, Insel RA, McIntyre K, Pellett PE, Long CE, Schnabel KC, Pincus PH, Stamey FR, Dambaugh TR, Stewart JA.** 1992. Primary human herpesvirus 6 infection in young children. *N Engl J Med* **326:**1445–1450.

43. **Hall CB, Long CE, Schnabel KC, Caserta MT, McIntyre KM, Costanzo MA, Knott A, Dewhurst S, Insel RA, Epstein LG.** 1994. Human herpesvirus-6 infection in children. A prospective study of complications and reactivation. *N Engl J Med* **331:**432–438.
44. **Tesini BL, Epstein LG, Caserta MT.** 2014. Clinical impact of primary infection with roseoloviruses. *Curr Opin Virol* **9:**91–96.
45. **Caserta MT, Hall CB, Canfield RL, Davidson P, Lofthus G, Schnabel K, Carnahan J, Shelley L, Wang H.** 2014. Early developmental outcomes of children with congenital HHV-6 infection. *Pediatrics* **134:**1111–1118.
46. **Voumvourakis KI, Kitsos DK, Tsiodras S, Petrikkos G, Stamboulis E.** 2010. Human herpesvirus 6 infection as a trigger of multiple sclerosis. *Mayo Clin Proc* **85:**1023–1030.
47. **Fotheringham J, Donati D, Akhyani N, Fogdell-Hahn A, Vortmeyer A, Heiss JD, Williams E, Weinstein S, Bruce DA, Gaillard WD, Sato S, Theodore WH, Jacobson S.** 2007. Association of human herpesvirus-6B with mesial temporal lobe epilepsy. *PLoS Medicine* **4:**0848–0857.
48. **Ljungman P, Singh N.** 2006. Human herpesvirus-6 infection in solid organ and stem cell transplant recipients. *Journal of Clinical Virology: The Official Publication of the Pan American Society for Clinical Virology* **37:**(Suppl 1) S87–91.
49. **Lautenschlager I, Razonable RR.** 2012. Human herpesvirus-6 infections in kidney, liver, lung, and heart transplantation: review. *Transpl Int* **25:**493–502.
50. **Zerr DM.** 2012. Human herpesvirus 6 (HHV-6) disease in the setting of transplantation. *Curr Opin Infect Dis* **25:**438–444.
51. **Lee SO, Brown RA, Razonable RR.** 2012. Chromosomally integrated human herpesvirus-6 in transplant recipients. *Transpl Infect Dis* **14:**346–354.
52. **Zerr DM, Fann JR, Breiger D, Boeckh M, Adler AL, Xie H, Delaney C, Huang ML, Corey L, Leisenring WM.** 2011. HHV-6 reactivation and its effect on delirium and cognitive functioning in hematopoietic cell transplantation recipients. *Blood* **117:**5243–5249.
53. **Zerr DM.** 2006. Human herpesvirus 6 and central nervous system disease in hematopoietic cell transplantation. *Journal of Clinical Virology: The Official Publication of the Pan American Society for Clinical Virology* **37:**(Suppl 1) S52–56.
54. **Hill JA, Boeckh MJ, Sedlak RH, Jerome KR, Zerr DM.** 2014. Human herpesvirus 6 can be detected in cerebrospinal fluid without associated symptoms after allogeneic hematopoietic cell transplantation. *J Clin Virol* **61:**289–292.
55. **Razonable RR.** 2010. Infections due to human herpesvirus 6 in solid organ transplant recipients. *Curr Opin Organ Transplant* **15:**671–675.
56. **Sassenscheidt J, Rohayem J, Illmer T, Bandt D.** 2006. Detection of β-herpesviruses in allogenic stem cell recipients by quantitative real-time PCR. *J Virol Methods* **138:**40–48.
57. **Razonable RR.** 2013. Human herpesviruses 6, 7 and 8 in solid organ transplant recipients. *Am J Transplant* **13**(Suppl 3)**:**67–77.
58. **Martin JN, Ganem DE, Osmond DH, Page-Shafer KA, Macrae D, Kedes DH.** 1998. Sexual transmission and the natural history of human herpesvirus 8 infection. *N Engl J Med* **338:**948–954.
59. **Cesarman E.** 2011. Gammaherpesvirus and lymphoproliferative disorders in immunocompromised patients. *Cancer Lett* **305:**163–174.
60. **Ablashi DV, Chatlynne LG, Whitman JE Jr, Cesarman E.** 2002. Spectrum of Kaposi's sarcoma-associated herpesvirus, or human herpesvirus 8, diseases. *Clin Microbiol Rev* **15:**439–464.
61. **Endo A, Watanabe K, Ohye T, Suzuki K, Matsubara T, Shimizu N, Kurahashi H, Yoshikawa T, Katano H, Inoue N, Imai K, Takagi M, Morio T, Mizutani S.** 2014. Molecular and virological evidence of viral activation from chromosomally integrated human herpesvirus 6A in a patient with X-linked severe combined immunodeficiency. *Clin Infect Dis* **59:**545–548.
62. **Fernández-Ruiz M, Kumar D, Husain S, Lilly L, Renner E, Mazzulli T, Moussa G, Humar A.** 2015. Utility of a monitoring strategy for human herpesviruses 6 and 7 viremia after liver transplantation: a randomized clinical trial. *Transplantation* **99:**106–113.
63. **Hakacova N, Klingel K, Kandolf R, Engdahl E, Fogdell-Hahn A, Higgins T.** 2013. First therapeutic use of Artesunate in treatment of human herpesvirus 6B myocarditis in a child. *J Clin Virol* **57:**157–160.
64. **Papadopoulou A, Gerdemann U, Katari UL, Tzannou I, Liu H, Martinez C, Leung K, Carrum G, Gee AP, Vera JF, Krance RA, Brenner MK, Rooney CM, Heslop HE, Leen AM.** 2014. Activity of broad-spectrum T cells as treatment for AdV, EBV, CMV, BKV, and HHV6 infections after HSCT. *Sci Transl Med* **6:**242ra83.
65. **Engels EA, Pfeiffer RM, Goedert JJ, Virgo P, McNeel TS, Scoppa SM, Biggar RJ, HIV/AIDS Cancer Match Study.** 2006. Trends in cancer risk among people with AIDS in the United States 1980–2002. *AIDS* **20:**1645–1654.
66. **Di Lorenzo G, Konstantinopoulos PA, Pantanowitz L, Di Trolio R, De Placido S, Dezube BJ.** 2007. Management of AIDS-related Kaposi's sarcoma. *Lancet Oncol* **8:**167–176.
67. **Carbone A, Vaccher E, Gloghini A, Pantanowitz L, Abayomi A, de Paoli P, Franceschi S.** 2014. Diagnosis and management of lymphomas and other cancers in HIV-infected patients. *Nat Rev Clin Oncol* **11:**223–238.
68. **La Ferla L, Pinzone MR, Nunnari G, Martellotta F, Lleshi A, Tirelli U, De Paoli P, Berretta M, Cacopardo B.** 2013. Kaposi's sarcoma in HIV-positive patients: the state of art in the HAART-era. *Eur Rev Med Pharmacol Sci* **17:**2354–2365.
69. **Robey RC, Bower M.** 2015. Facing up to the ongoing challenge of Kaposi's sarcoma. *Curr Opin Infect Dis* **28:**31–40.
70. **Riva G, Luppi M, Barozzi P, Forghieri F, Potenza L.** 2012. How I treat HHV8/KSHV-related diseases in posttransplant patients. *Blood* **120:**4150–4159.
71. **Le J, Gantt S, AST Infectious Diseases Community of Practice.** 2013. Human herpesvirus 6, 7 and 8 in solid organ transplantation. *Am J Transplant* **13**(Suppl 4)**:**128–137.
72. **Piselli P, Serraino D, Segoloni GP, Sandrini S, Piredda GB, Scolari MP, Rigotti P, Busnach G, Messa P, Donati D, Schena FP, Maresca MC, Tisone G, Veroux M, Sparacino V, Pisani F, Citterio F, Immunosuppression and Cancer Study Group.** 2013. Risk of de novo cancers after transplantation: results from a cohort of 7217 kidney transplant recipients, Italy 1997–2009. *Eur J Cancer* **49:**336–344.
73. **Stock PG, Barin B, Murphy B, Hanto D, Diego JM, Light J, Davis C, Blumberg E, Simon D, Subramanian A, Millis JM, Lyon GM, Brayman K, Slakey D, Shapiro R, Melancon J, Jacobson JM, Stosor V, Olson JL, Stablein DM, Roland ME.** 2010. Outcomes of kidney transplantation in HIV-infected recipients. *N Engl J Med* **363:**2004–2014.
74. **Mori Y, Yamanishi K.** 2007. HHV-6A, 6B, and 7: Pathogenesis, host response, and clinical disease, p 833–842. *In* Arvin A, Campadelli-Fiume G, Mocarski E, Moore PS, Roisman B, Whitley R, Yamanishi K (eds), *Human Herpesviruses: Biology, Therapy, and Immunoprophylaxis*, Cambridge University Press, Cambridge.
75. **Asano Y, Nakashima T, Yoshikawa T, Suga S, Yazaki T.** 1991. Severity of human herpesvirus-6 viremia and clinical findings in infants with exanthem subitum. *J Pediatr* **118:**891–895.
76. **Suga S, Yoshikawa T, Asano Y, Nakashima T, Yazaki T, Fukuda M, Kojima S, Matsuyama T, Ono Y, Oshima S.** 1992. IgM neutralizing antibody responses to human herpesvirus-6 in patients with exanthem subitum or organ transplantation. *Microbiol Immunol* **36:**495–506.
77. **Navalpotro D, Gimeno C, Navarro D.** 2006. Concurrent detection of human herpesvirus type 6 and measles-specific IgMs during acute exanthematic human parvovirus B19 infection. *J Med Virol* **78:**1449–1451.
78. **Dahl H, Linde A, Sundqvist VA, Wahren B.** 1990. An enzyme-linked immunosorbent assay for IgG antibodies to human herpes virus 6. *J Virol Methods* **29:**313–323.
79. **Sloots TP, Kapeleris JP, Mackay IM, Batham M, Devine PL.** 1996. Evaluation of a commercial enzyme-linked immuno-

sorbent assay for detection of serum immunoglobulin G response to human herpesvirus 6. *J Clin Microbiol* **34:**675–679.

80. **Ward KN, Gray JJ, Fotheringham MW, Sheldon MJ.** 1993. IgG antibodies to human herpesvirus-6 in young children: changes in avidity of antibody correlate with time after infection. *J Med Virol* **39:**131–138.
81. **Ward KN, Leong HN, Nacheva EP, Howard J, Atkinson CE.** 1571–1574. **Davies NWS, Griffiths PD, Clark DA.** 2006. Human herpesvirus 6 chromosomal integration in immunocompetent patients results in high levels of viral DNA in blood, sera, and hair follicles. *J Clin Microbiol* **44:**1571–1574.
82. **Cassina G, Russo D, De Battista D, Broccolo F, Lusso P, Malnati MS.** 2013. Calibrated real-time polymerase chain reaction for specific quantitation of HHV-6A and HHV-6B in clinical samples. *J Virol Methods* **189:**172–179.
83. **Nitsche A, Müller CW, Radonić A, Landt O, Ellerbrok H, Pauli G, Siegert W.** 2001. Human herpesvirus 6A DNA Is detected frequently in plasma but rarely in peripheral blood leukocytes of patients after bone marrow transplantation. *J Infect Dis* **183:**130–133.
84. **Razonable RR, Fanning C, Brown RA, Espy MJ, Rivero A, Wilson J, Kremers W, Smith TF, Paya CV.** 2002. Selective reactivation of human herpesvirus 6 variant a occurs in critically ill immunocompetent hosts. *J Infect Dis* **185:**110–113.
85. **Safronetz D, Humar A, Tipples GA.** 2003. Differentiation and quantitation of human herpesviruses 6A, 6B and 7 by real-time PCR. *J Virol Methods* **112:**99–105.
86. **Reddy S, Manna P.** 2005. Quantitative detection and differentiation of human herpesvirus 6 subtypes in bone marrow transplant patients by using a single real-time polymerase chain reaction assay. *Biol Blood Marrow Transplant* **11:**530–541.
87. **Boutolleau D, Duros C, Bonnafous P, Caïola D, Karras A, Castro ND, Ouachée M, Narcy P, Gueudin M, Agut H, Gautheret-Dejean A.** 2006. Identification of human herpesvirus 6 variants A and B by primer-specific real-time PCR may help to revisit their respective role in pathology. *J Clin Virol* **35:**257–263.
88. **Flamand L, Gravel A, Boutolleau D, Alvarez-Lafuente R, Jacobson S, Malnati MS, Kohn D, Tang YW, Yoshikawa T, Ablashi D.** 2008. Multicenter comparison of PCR assays for detection of human herpesvirus 6 DNA in serum. *J Clin Microbiol* **46:**2700–2706.
89. **de Pagter PJ, Schuurman R, de Vos NM, Mackay W, van Loon AM.** 2010. Multicenter external quality assessment of molecular methods for detection of human herpesvirus 6. *J Clin Microbiol* **48:**2536–2540.
90. **Mannonen L, Loginov R, Helanterä I, Dumoulin A, Vilchez RA, Cobb B, Hirsch HH, Lautenschlager I.** 2014. Comparison of two quantitative real-time CMV-PCR tests calibrated against the 1st WHO international standard for viral load monitoring of renal transplant patients. *J Med Virol* **86:** 576–584.
91. **Hall Sedlak R, Jerome KR.** 2014. The potential advantages of digital PCR for clinical virology diagnostics. *Expert Rev Mol Diagn* **14:**501–507.
92. **Leibovitch EC, Brunetto GS, Caruso B, Fenton K, Ohayon J, Reich DS, Jacobson S.** 2014. Coinfection of human herpesviruses 6A (HHV-6A) and HHV-6B as demonstrated by novel digital droplet PCR assay. *PLoS One* **9:**e92328.
93. **Achour A, Boutolleau D, Slim A, Agut H, Gautheret-Dejean A.** 2007. Human herpesvirus-6 (HHV-6) DNA in plasma reflects the presence of infected blood cells rather than circulating viral particles. *J Clin Virol* **38:**280–285.
94. **Ihira M, Enomoto Y, Kawamura Y, Nakai H, Sugata K, Asano Y, Tsuzuki M, Emi N, Goto T, Miyamura K, Matsumoto K, Kato K, Takahashi Y, Kojima S, Yoshikawa T.** 2012. Development of quantitative RT-PCR assays for detection of three classes of HHV-6B gene transcripts. *J Med Virol* **84:**1388–1395.
95. **Bressollette-Bodin C, Nguyen TVH, Illiaquer M, Besse B, Peltier C, Chevallier P, Imbert-Marcille BM.** 2014. Quantification of two viral transcripts by real time PCR to investigate human herpesvirus type 6 active infection. *J Clin Virol* **59:**94–99.
96. **Zerr DM, Boeckh M, Delaney C, Martin PJ, Xie H, Adler AL, Huang ML, Corey L, Leisenring WM.** 2012. HHV-6 reactivation and associated sequelae after hematopoietic cell transplantation. *Biol Blood Marrow Transplant* **18:**1700–1708.
97. **Hall CB, Caserta MT, Schnabel K, Shelley LM, Marino AS, Carnahan JA, Yoo C, Lofthus GK, McDermott MP.** 2008. Chromosomal integration of human herpesvirus 6 is the major mode of congenital human herpesvirus 6 infection. *Pediatrics* **122:**513–520.
98. **Géraudie B, Charrier M, Bonnafous P, Heurté D, Desmonet M, Bartoletti MA, Penasse C, Agut H, Gautheret-Dejean A.** 2012. Quantitation of human herpesvirus-6A, -6B and -7 DNAs in whole blood, mononuclear and polymorphonuclear cell fractions from healthy blood donors. *J Clin Virol* **53:**151–155.
99. **Boutolleau D, Fernandez C, André E, Imbert-Marcille BM, Milpied N, Agut H, Gautheret-Dejean A.** 2003. Human herpesvirus (HHV)-6 and HHV-7: two closely related viruses with different infection profiles in stem cell transplantation recipients. *J Infect Dis* **187:**179–186.
100. **Ogata M, Satou T, Kawano R, Goto K, Ikewaki J, Kohno K, Ando T, Miyazaki Y, Ohtsuka E, Saburi Y, Saikawa T, Kadota JI.** 2008. Plasma HHV-6 viral load-guided preemptive therapy against HHV-6 encephalopathy after allogeneic stem cell transplantation: a prospective evaluation. *Bone Marrow Transplant* **41:**279–285.
101. **Hill JA, Sedlak RH, Jerome KR.** 2014. Past, present, and future perspectives on the diagnosis of Roseolovirus infections. *Curr Opin Virol* **9:**84–90.
102. **Betts BC, Young JAH, Ustun C, Cao Q, Weisdorf DJ.** 2011. Human herpesvirus 6 infection after hematopoietic cell transplantation: is routine surveillance necessary? *Biol Blood Marrow Transplant* **17:**1562–1568.
103. **Pellett PE, Ablashi DV, Ambros PF, Agut H, Caserta MT, Descamps V, Flamand L, Gautheret-Dejean A, Hall CB, Kamble RT, Kuehl U, Lassner D, Lautenschlager I, Loomis KS, Luppi M, Lusso P, Medveczky PG, Montoya JG, Mori Y, Ogata M, Pritchett JC, Rogez S, Seto E, Ward KN, Yoshikawa T, Razonable RR.** 2012. Chromosomally integrated human herpesvirus 6: questions and answers. *Rev Med Virol* **22:**144–155.
104. **Zerr DM, Gupta D, Huang ML, Carter R, Corey L.** 2002. Effect of antivirals on human herpesvirus 6 replication in hematopoietic stem cell transplant recipients. *Clin Infect Dis* **34:**309–317.
105. **Seeley WW, Marty FM, Holmes TM, Upchurch K, Soiffer RJ, Antin JH, Baden LR, Bromfield EB.** 2007. Post-transplant acute limbic encephalitis: clinical features and relationship to HHV6. *Neurology* **69:**156–165.
106. **Hall CB, Caserta MT, Schnabel KC, Long C, Epstein LG, Insel RA, Dewhurst S.** 1998. Persistence of human herpesvirus 6 according to site and variant: possible greater neurotropism of variant A. *Clin Infect Dis* **26:**132–137.
107. **Hirabayashi K, Nakazawa Y, Katsuyama Y, Yanagisawa T, Saito S, Yoshikawa K, Shigemura T, Sakashita K, Ichikawa M, Koike K.** 2013. Successful ganciclovir therapy in a patient with human herpesvirus-6 encephalitis after unrelated cord blood transplantation: usefulness of longitudinal measurements of viral load in cerebrospinal fluid. *Infection* **41:**219–223.
108. **Nefzi F, Ben Salem NA, Khelif A, Feki S, Aouni M, Gautheret-Dejean A.** 2015. Quantitative analysis of human herpesvirus-6 and human cytomegalovirus in blood and saliva from patients with acute leukemia. *J Med Virol* **87:**451–460.
109. **Caserta MT, Hall CB, Schnabel K, Lofthus G, Marino A, Shelley L, Yoo C, Carnahan J, Anderson L, Wang H.** 2010. Diagnostic assays for active infection with human herpesvirus 6 (HHV-6). *J Clin Virol* **48:**55–57.

110. **Van den Bosch G, Locatelli G, Geerts L, Fagà G, Ieven M, Goossens H, Bottiger D, Öberg B, Lusso P, Berneman ZN.** 2001. Development of reverse transcriptase PCR assays for detection of active human herpesvirus 6 infection. *J Clin Microbiol* **39:**2308–2310.
111. **Black JB, Schwarz TF, Patton JL, Kite-Powell K, Pellett PE, Wiersbitzky S, Bruns R, Müller C, Jäger G, Stewart JA.** 1996. Evaluation of immunoassays for detection of antibodies to human herpesvirus 7. *Clin Diagn Lab Immunol* **3:**79–83.
112. **Fernández-Ruiz M, López-Medrano F, Allende LM, Andrés A, Paz-Artal E, Aguado JM.** 2014. Assessing the risk of de novo malignancy in kidney transplant recipients: role for monitoring of peripheral blood lymphocyte populations. *Transplantation* **98:**e36–e37.
113. **Boivin G, Côté S, Cloutier N, Abed Y, Maguigad M, Routy JP.** 2002. Quantification of human herpesvirus 8 by real-time PCR in blood fractions of AIDS patients with Kaposi's sarcoma and multicentric Castleman's disease. *J Med Virol* **68:**399–403.
114. **Pellett PE, Wright DJ, Engels EA, Ablashi DV, Dollard SC, Forghani B, Glynn SA, Goedert JJ, Jenkins FJ, Lee TH, Neipel F, Todd DS, Whitby D, Nemo GJ, Busch MP, Retrovirus Epidemiology Donor Study.** 2003. Multicenter comparison of serologic assays and estimation of human herpesvirus 8 seroprevalence among US blood donors. *Transfusion* **43:**1260–1268.
115. **Laney AS, Cannon MJ, Jaffe HW, Offermann MK, Ou CY, Radford KW, Patel MM, Spira TJ, Gunthel CJ, Pellett PE, Dollard SC.** 2007. Human herpesvirus 8 presence and viral load are associated with the progression of AIDS-associated Kaposi's sarcoma. *AIDS* **21:**1541–1545.
116. **Cannon MJ, Dollard SC, Black JB, Edlin BR, Hannah C, Hogan SE, Patel MM, Jaffe HW, Offermann MK, Spira TJ, Pellett PE, Gunthel CJ.** 2003. Risk factors for Kaposi's sarcoma in men seropositive for both human herpesvirus 8 and human immunodeficiency virus. *AIDS* **17:**215–222.
117. **Engels EA, Whitby D, Goebel PB, Stossel A, Waters D, Pintus A, Contu L, Biggar RJ, Goedert JJ.** 2000. Identifying human herpesvirus 8 infection: performance characteristics of serologic assays. *J Acquir Immune Defic Syndr* **23:**346–354.
118. **Chandran B, Smith MS, Koelle DM, Corey L, Horvat R, Goldstein E.** 1998. Reactivities of human sera with human herpesvirus-8-infected BCBL-1 cells and identification of HHV-8-specific proteins and glycoproteins and the encoding cDNAs. *Virology* **243:**208–217.
119. **Biggar RJ, Engels EA, Whitby D, Kedes DH, Goedert JJ.** 2003. Antibody reactivity to latent and lytic antigens to human herpesvirus-8 in longitudinally followed homosexual men. *J Infect Dis* **187:**12–18.
120. **Lennette ET, Blackbourn DJ, Levy JA.** 1996. Antibodies to human herpesvirus type 8 in the general population and in Kaposi's sarcoma patients. *Lancet* **348:**858–861.
121. **Kedes DH, Operskalski E, Busch M, Kohn R, Flood J, Ganem D.** 1996. The seroepidemiology of human herpesvirus 8 (Kaposi's sarcoma-associated herpesvirus): distribution of infection in KS risk groups and evidence for sexual transmission. *Nat Med* **2:**918–924.
122. **Simpson GR, Schulz TF, Whitby D, Cook PM, Boshoff C, Rainbow L, Howard MR, Gao SJ, Bohenzky RA, Simmonds P, Lee C, de Ruiter A, Hatzakis A, Tedder RS, Weller IV, Weiss RA, Moore PS.** 1996. Prevalence of Kaposi's sarcoma associated herpesvirus infection measured by antibodies to recombinant capsid protein and latent immunofluorescence antigen. *Lancet* **348:**1133–1138.
123. **Schatz O, Monini P, Bugarini R, Neipel F, Schulz TF, Andreoni M, Erb P, Eggers M, Haas J, Buttò S, Lukwiya M, Bogner JR, Yaguboglu S, Sheldon J, Sarmati L, Goebel FD, Hintermaier R, Enders G, Regamey N, Wernli M, Stürzl M, Rezza G, Ensoli B.** 2001. Kaposi's sarcoma-associated herpesvirus serology in Europe and Uganda: multicentre study with multiple and novel assays. *J Med Virol* **65:**123–132.
124. **Nascimento MC, de Souza VA, Sumita LM, Freire W, Weiss HA, Sabino EC, Franceschi S, Pannuti CS, Mayaud P.** 2008. Prevalence of, and risk factors for Kaposi's sarcoma-associated herpesvirus infection among blood donors in Brazil: a multicenter serosurvey. *J Med Virol* **80:**1202–1210.
125. **Labo N, Miley W, Marshall V, Gillette W, Esposito D, Bess M, Turano A, Uldrick T, Polizzotto MN, Wyvill KM, Bagni R, Yarchoan R, Whitby D.** 2014. Heterogeneity and breadth of host antibody response to KSHV infection demonstrated by systematic analysis of the KSHV proteome. *PLoS Pathog* **10:** e1004046.
126. **Katano H, Sato Y, Kurata T, Mori S, Sata T.** 2000. Expression and localization of human herpesvirus 8-encoded proteins in primary effusion lymphoma, Kaposi's sarcoma, and multicentric Castleman's disease. *Virology* **269:**335–344.
127. **Chadburn A, Wilson J, Wang YL.** 2013. Molecular and immunohistochemical detection of Kaposi sarcoma herpesvirus/human herpesvirus-8. *Methods Mol Biol* **999:**245–256.
128. **Benevenuto de Andrade BA, Ramírez-Amador V, Anaya-Saavedra G, Martínez-Mata G, Fonseca FP, Graner E, Paes de Almeida O.** 2014. Expression of PROX-1 in oral Kaposi's sarcoma spindle cells. *J Oral Pathol Med* **43:**132–136.
129. **White IE, Campbell TB.** 2000. Quantitation of cell-free and cell-associated Kaposi's sarcoma-associated herpesvirus DNA by real-time PCR. *J Clin Microbiol* **38:**1992–1995.
130. **Stamey FR, Patel MM, Holloway BP, Pellett PE.** 2001. Quantitative, fluorogenic probe PCR assay for detection of human herpesvirus 8 DNA in clinical specimens. *J Clin Microbiol* **39:**3537–3540.
131. **de Sanjosé S, Marshall V, Solà J, Palacio V, Almirall R, Goedert JJ, Bosch FX, Whitby D.** 2002. Prevalence of Kaposi's sarcoma-associated herpesvirus infection in sex workers and women from the general population in Spain. *Int J Cancer* **98:**155–158.
132. **Pauk J, Huang ML, Brodie SJ, Wald A, Koelle DM, Schacker T, Celum C, Selke S, Corey L.** 2000. Mucosal shedding of human herpesvirus 8 in men. *N Engl J Med* **343:** 1369–1377.
133. **Campbell TB, Borok M, Gwanzura L, MaWhinney S, White IE, Ndemera B, Gudza I, Fitzpatrick L, Schooley RT.** 2000. Relationship of human herpesvirus 8 peripheral blood virus load and Kaposi's sarcoma clinical stage. *AIDS* **14:**2109–2116.
134. **Engels EA, Biggar RJ, Marshall VA, Walters MA, Gamache CJ, Whitby D, Goedert JJ.** 2003. Detection and quantification of Kaposi's sarcoma-associated herpesvirus to predict AIDS-associated Kaposi's sarcoma. *AIDS* **17:**1847–1851.
135. **Engelmann I, Petzold DR, Kosinska A, Hepkema BG, Schulz TF, Heim A.** 2008. Rapid quantitative PCR assays for the simultaneous detection of herpes simplex virus, varicella zoster virus, cytomegalovirus, Epstein-Barr virus, and human herpes virus 6 DNA in blood and other clinical specimens. *J Med Virol* **80:**467–477.

Human Papillomaviruses

SUSAN NOVAK-WEEKLEY AND ROBERT PRETORIUS

30

Papillomaviruses are species specific; in humans they infect a number of sites such as the skin, mouth, anus, conjunctiva, and lower genital tracts of both males and females. The majority of infections, no matter the site, are typically asymptomatic and subclinical. Genital human papillomavirus (HPV) is the most common sexually transmitted disease. It has been established and accepted that oncogenic or high-risk HPV types are the main cause of cervical cancer in women and can cause other cancers such as vulvar, vaginal, penile, anal, and oropharyngeal cancer. Although persistent infection with a high-risk HPV type is necessary for the development of cervical cancer, many women will spontaneously clear the infection and are subsequently not at risk for developing cancer in the future. It is known that greater than 70% of cervical cancer cases are due to HPV types 16 and 18, and testing for HPV along with Pap smear testing is a widely accepted approach for cervical cancer screening. Several organizations have developed guidelines for cervical cancer screening in United States (US) including the United States Preventive Services Task Force (USPSTF), American Cancer Society (ACS), the American Society for Colposcopy and Cervical Pathology (ASCCP), and the American Society of Clinical Pathology (ASCP). Recent data suggest that HPV testing can be used as the primary screen for cervical cancer screening as algorithms continue to evolve around clinical patient management. Prevention of cervical cancer and other cancers is now feasible because of the availability of HPV vaccines.

VIRAL CLASSIFICATION AND BIOLOGY

Papillomaviruses are small, circular, nonenveloped double-stranded DNA viruses (∽55 nm in diameter) and taxonomically belong to the family *Papillomaviridae* (1). These viruses infect a number of hosts including various animals from birds to mammals and can cause benign wartlike lesions or produce malignant disease in the form of cervical or other cancers. The family *Papillomaviridae* contains approximately over 200 papillomavirus types, in five genera, infecting over 49 species, human, mammalian, reptilian and avian (2, 3). The five major genera include α-papillomavirus, β-papillomavirus, γ-papillomavirus, mu-papillomavirus, and nu-papillomavirus; classification is based on the nucleotide sequence of the open reading frame(ORF) coding for the L1 capsid protein (2). Classification of HPV into different genera is predicated on less than 60% nucleotide sequence similarity within the L1 portion of the genome for each HPV genus. Within a genus, different viral species typically share between 60 to 70% homology of the viral genome. For an HPV type to be novel, that virus must have less than 90% nucleotide similarity to any other HPV type known at the time of discovery (2). Nomenclature designation for papillomaviruses is determined by the papillomavirus study group, which is part of the International Committee on Taxonomy of Viruses (ICTV). It is estimated that almost 400 HPV types actually exist. HPV classification is managed by the International HPV Reference Center in Stockholm, Sweden. When the center receives new and novel HPV types, they are recloned and resequenced and if confirmed to be a new HPV type, assigned a type designation. All reference clones are listed on the following web site, www.hpvcenter.se, which is available for public access (2). HPV types continue to be discovered with γ-genus currently containing the most types [2]. L1 is the most conserved gene and has been used for many years to identify new HPV types (5). The current system of HPV type number designation, based only on L1 sequences, was agreed upon in 1995 at the Quebec Papillomavirus workshop (4).

Papillomaviruses contain a genome that is approximately 8 kilobases in size (ranging from 6953 to 8607 base pairs) (1, 3). Much is known about the molecular biology of HPV, and bovine papilloma virus type-1 has served as the prototype for many of these studies (6). Viral pathogenicity depends on a number of factors, as is true for other pathogens. These factors include the host immune system, the virus genotype, and the nature or tropism of the infected cell. What is important to note is that papillomaviruses replicate and assemble exclusively in the nucleus of the host cell. Cells infected in the host are the keratinocytes found in the basal layers of the stratified squamous epithelium (6). Zheng and Baker provide an excellent review regarding the molecular biology of HPV (6). Viral gene expression is tightly regulated, both at the transcriptional and post-transcriptional level. During viral gene expression there are six nonstructural gene ORFs expressed, which are E1, E2, E3, E4, E5, and E6 (Table 1). These ORFs are from the early region of the viral genome and are expressed in undifferentiated or intermediately differentiated keratinocytes. Those keratinocytes undergoing terminal differentiation express two structural viral capsid proteins, L1 and L2 (6), with L1 facilitating attachment to the host cell. Each gene described above has a

doi:10.1128/9781555819156.ch30

TABLE 1 The function of viral proteins

Core Proteins
E1—ATP dependent helicase. Role in papillomavirus genome replication.
E2—Coactivator of viral genome replication through the recruitment of E1 to the viral replication origin. Transcription facto or E6 & E7, also important for viral genome segregation.
L1—Major capsid protein. Assembles into pentomeric capsomeres, which are the primary components of the icosahedral virion shell.
L2—Minor capsid protein. Involved in encapsidation of viral DNA. Facilitates virus entry and trafficking to the nucleus.
E4—E4 gene is embedded within the E2 gene and is expressed abundantly as an E1^E4 fusion protein during the late stages of the viral life cycle. Binds to cytokeratin filaments and disrupts their structure. E4 is thought also to contribute to virus release and transmission.
Accessory Proteins
E5—Small transmembrane protein. In alpha-PVs, E5 interacts with the EGF receptor and activates mitogenic signaling pathways. Has a role in evading the immune response and apoptosis. Beta-, gamma-, and mu-PVs lack E5 gene.
E6—Drives cell cycle entry to allow genome amplification in the upper epithelial layers. E6 of high-risk alpha-PV types binds and degrades p53 and can also activate telomerase and contribute to transformation. Also involved in immune evasion. Some gamma-PVs lack the E6 gene.
E7—Drives cell cycle entry to allow genome amplification in the upper cell layers. E7 of high-risk alpha-PV types binds and degrades pRb and can induce chromosomal instability. E7 is necessary for cell transformation.
Both E6 and E7 have a number of cellular substrates, with the identity of these substrates differing between types (Holmgren et al. (116)).
The Function of Viral Proteins
All known papillomavirus encodes a group of "core" proteins that were present early on during papillomavirus evolution and that are conserved in sequence and in function among PV types. These include E1, E2, L2, and L1. The E4 protein may also be a core protein that has evolved to meet papillomaviruses' epithelial specialization. The accessory proteins have evolved in each papillomavirus type during adaptation to different epithelial niches. The sequence and function of these genes are divergent among types. In general, these proteins are involved in modifying the cellular environment to facilitate virus life cycle completion, contributing to virulence and pathogenicity. Knowledge of accessory protein function comes primarily from the study of alpha-papillomavirus types.

role, specifically E4, which continues to be expressed in the terminally differentiated keratinocytes while E1 and E2 are important in regulation of early transcription and viral replication (6). Oncogenes E5, E6, and E7 are responsible for transforming the cell into a cancerous cell (5,7–9). E1 along with E2, L1, and L2 ORFs are highly conserved within the papillomavirus family (5). E6 and E7 play an important role in inactivating two host cell tumor-suppressor proteins, p53 and pRb, respectively. In immunocompetent hosts alphapapillomaviruses are examples of a virus that is adept at host immune evasion. This is often reflected in the recalcitrant infections that can occur with cutaneous warts. Beta- and gamma-papillomaviruses typically cause visible lesions only when the host is immunocompromised. An excellent recent review by Egawa et al. summarizes viral tropism and the development of neoplasia in the host (3). Table 1 lists and describes core and accessory proteins that are important in the virus life cycle and pathogenicity (3).

EPIDEMIOLOGY

An Italian physician observed that during the period of 1760–1839 women who were sexually active had a higher frequency of cervical cancer than those who were not sexually active (26). It was deduced that the higher rate of cervical cancer was due to sexual contact. Electron microscopic visualization of viral particles associated with genital warts was achieved in 1949 (10). In 1967, Rownson and Mahy described warts caused by HPV (11). Among the approximately 200 or more HPV types, only certain types are oncogenic (3,12–14). It wasn't until 1970 that investigators suspected that cancerous and precancerous cervical changes were caused by viral-induced cellular transformation due to HPV. By the 1980s it was the prevailing opinion that virtually all cervical cancers were related to infection of the cervical epithelium by oncogenic types of HPV (15). Skin and genital warts were well known in ancient times (16). It is estimated that HPV is the cause of 72% of oropharyngeal cancers, 90% of all anal cancers, and 71% of vulvar, vaginal, or penile cancers (17).

Infection with HPV can occur at any age and has even been reported in children (18). Warts are the most common HPV infection in children and young adults. HPV risk in adults is highly correlated with the number of sexual partners a person has in a lifetime. It is estimated that at least 80% of adults who are sexually active will have at least one anogenital infection. It has been reported that a protective effect is seen in regard to HPV infection in women who have partners who have been circumcised (19). Condom use does appear to have a protective effect in college women in terms of acquiring new HPV infections and against cervical intraepithelial neoplasia (CIN) development (20, 21).

HPV is the most common sexually transmitted disease. Transmission of disease occurs primarily through abrasion of the skin and introduction of the virus and is associated with a variety of clinical conditions. There is conjecture though that infection can also occur via autoinoculation and via digital or fomite transmission as well (22). Men who have sex with men (MSM) are a reservoir for HPV infection and have a higher chance of developing anal cancer later in life than heterosexual men (23). In one study MSM had a higher anal HPV prevalence (84% versus 42%) mainly for multiple HPV types (≥ 3). In addition to having a higher prevalence, there was a higher incidence rate and lower clearance rate of anal HPV compared to rates in heterosexual males. Data on HPV in women appear to be lacking (24). According to the World Health Organization (WHO), cervical cancer is the second most common cancer in women living in less developed countries. In 2012 there were approximately 450,000 new cases of cervical cancer with an estimated 270,000 deaths worldwide (mostly in low- to medium- income countries).

The National Cancer Institute of the National Institutes of Health in the United States projected that in 2015 there will be approximately 12,900 cervical cancer cases in the United States and approximately 4100 deaths. This ranks cervical cancer below the 10 most common cancers in the United States. The percentage of women surviving at 5 years is estimated to be 68%. Typically there is an inverse relationship between age and the prevalence of HPV. In a study of 18,498 women between the ages of 15 and 74, the prevalence of HPV in women tested in 15 different areas on four continents showed considerable variation in the shape of the age-specific curves related to disease prevalence. The authors note that based on this study, the inverse relationship between disease prevalence and age does not always exist in all populations (25). Many studies continue to be performed assessing the HPV types that are most common in varying populations around the world. Recent population-based studies have shown that anal cancer rates are increasing, particularly in MSM. The populations identified at highest risk for anal cancer are HIV-positive homosexual and bisexual men.

CLINICAL SIGNIFICANCE

HPVs can infect both mucosal and cutaneous areas in humans, and many infections are asymptomatic. Table 2 lists the common genotypes associated with human diseases. The causative role of HPV in human cancers is nicely reviewed in a manuscript by Harald zur Hausen (26). HPV can cause a variety of cancers including cervical, anogential, head and neck, and cutaneous, as described above (26).

CUTANEOUS HPV

Cutaneous HPV infections are predominantly caused by HPV within the genus beta-papillomavirus and gamma-

TABLE 2 Diseases associated to human papillomaviruses in immunocompetent people[a]

Diseases in different anatomical location	HPV contribution	
	HPV genus	Common genotypes
Benign oral lesions		
Verruca vulgaris (common wart)	β-HPV	HPV types 1, 2, 4, and 7
Oral squamous cell papilloma	α-HPV	HPV types 6 and 11
Condyloma acuminata	α-HPV	HPV types 6 and 11
Focal epithelial hyperplasia (Heck's disease)	α-HPV	HPV types 13 and 32
Recurrent respiratory papillomatosis	α-HPV	HPV types 6 and 11
Potentially malignant oral disorders		
Oral lichen planus	α-HPV (23%)	HPV types 6, 11 and 16
Leukoplakia	α-HPV (63%)	HPV types 6, 11 and 16
Erythroplakia	α-HPV (50%)	HPV types 6, 11 and 16
Malignant oral disorders		
Oropharyngeal squamous cell carcinoma	α-HPV (47%)	HPV 16 (90%)
Oral cavity cancer	α-HPV (11%)	HPV 16 (96%)
Benign anogenital lesions		
Condyloma acuminata	α-HPV (90%)	HPV 6 (89%), HPV 11 (11%)
Atypical and low-grade squamous cell lesion of the cervix	α-HPV (55%)	HPV 16 (9%), HPV 6/11 (5%), HPV 31 (4%), HPV 33 (2%), HPV 18 (2%)
Premalignant anogenital lesions		
Cervical high-grade squamous cell lesion	α-HPV (84%)	HPV 16 (45%), HPV 31 (9%), HPV 33 (7%), HPV 18 (7%), HPV 58 (7%), HPV 52 (5%), HPV 35 (4%)
Malignant anogenital lesions		
Cervical squamous cell carcinoma and adenocarcinoma	α-HPV (>99%)	HPV 18 (10%), HPV 45 (6%), HPV 31 (4%), HPV 33 (4%), HPV 52 (3%), HPV 35 (2%), HPV 58 (2%)
Vaginal cancer	α-HPV (70%)	HPV 16 (54%), HPV 18 (8%)
Vulvar cancer	α-HPV (40%)	HPV 16 (32%), HPV 18 (4%)
Anal cancer	α-HPV (97%)	HPV 16 (75%), HPV 18 (3%)[16]
Penile cancer	α-HPV (45%)	HPV 16 (60%), HPV 18 (13%), HPV 6/11 (8%)
Benign skin lesions		
Common wart	β-HPV	HPV types 1 and 2
Flat warts	β-HPV	HPV types 3 and 10
Potentially malignant skin disorders		
Epidermodysplasia verruciformis	β-HPV	HPV types 5 and 8

HPV, human papillomaviruses
[a]Reprinted with permission from Grce M, Mravak-Stipetić M. 2014. Human papillomavirus-associated diseases. *Clin Dermatol* 32(2):253–258.

papillomavirus. These virus types are known to produce persistent subclinical infections typically infecting children at a very young age (27). Plantar, common, or flat warts are the cutaneous manifestation of HPV on the skin. These warts are typically a result of contact with infected skin or object harboring the virus. Warts such as this are known to occur in approximately 10% of children and young adults, with the highest incidence occurring between the ages of 12 and 16. Warts occur with greater frequency in girls versus boys and spontaneously clear in about 2 years in 40% of those infected. Cutaneous warts that are recalcitrant to therapy can be a reservoir for infection (28). Some beta-papillomaviruses cause nonmelanoma skin cancer in some individuals (29).

Epidermodysplasia verruciformis (EV) is a rare genetic disease that predisposes certain individuals to HPV infections of the skin. Onset of EV is typically seen in infancy or in young children and many patients seem to exhibit defects in their cell-mediated immune system. Mutations occur in the transmembrane channel gene TMC6 or TMC8 and result in the development of warts. EV is the first disease in which epidemiologists were able to correlate cancer and viral infection (30). Skin eruptions that are flat to papillomatous occur on the trunk, hands, and upper and lower extremities. They are typically reddish-brown pigmented plaque-like lesions. In areas that exposed to ultraviolet radiation there is a higher risk of developing malignancy, especially squamous cell carcinoma (SCC). The development of these cancers is typically between the ages of 20 and 40 and usually these patients are infected with multiple types of HPV.

MUCOSAL HPV

HPV infecting the mucosal areas in humans include the anogenital areas of both men and women and can include infection of the oral mucosa and oropharyngeal area. It is known that around 40 different HPV types can infect the genital area.

Genital Warts (Anogenital Warts, Condyloma and Genital Papilloma)

Anogenital warts are commonly caused by the non-oncogenic types of HPV and infections occur predominantly in young adults and populations that are sexually promiscuous. Most notably, HPV types 6 or 11 are the predominant types causing these infections, although other types such as 16, 18, 31, 33, and 35 can also produce anogenital warts and can be associated with types 6 or 11, resulting in coinfections. The aforementioned high-risk types can be associated with high-grade squamous intraepithelial lesions (HSIL), especially in those patients who are also infected with HIV. HPV types 6 and 11 can be associated with other body sites such as nares, conjunctiva, and larynx, as discussed below.

The incubation period for condylomas can be as short as 3 weeks or as long as 8 months, averaging 2.8 months. Depending on the size and general location of the anogenital wart or condyloma, the lesion can be painful and also pruritic. In terms of appearance they are often papillary (condyloma acuminatum), flat (condyloma planum), or pedunculated around the genital mucosal areas. The genital mucosal areas affected most often in men are the shaft of the penis, under the foreskin of the uncircumcised penis, and around the vaginal introitus in women. Warts are also seen within multiple other sites, such as within the anogenital tract or on the anogenital epithelium, including the perineum, perianal skin, vagina, cervix, urethra, anus, and scrotum (31). Intra-anal warts are observed predominantly in persons who have had anal-receptive intercourse. The Centers for Disease Control and Prevention (CDC) indicates that the incidence of genital warts has increased over the years in private clinics but stayed relatively flat in those patients seen in sexually transmitted disease (STD) clinics. Approximately 6 to 8% of patients seen in STD clinics had genital warts with a slightly higher percentage seen in MSM. Women have a much lower prevalence of genital warts, less than 3%, compared to hetero- or homosexual men (Centers for Disease Control and Prevention, www.cdc.gov/std/stats14/figures/52.htm). In women flat condylomas are more predominant in the cervix. These lesions can be visualized by the health care provider during a procedure called a colposcopy. Biopsies are taken and the confirmation of an HPV infection has classically been performed using histopathology.

Immunosuppression can have an impact on the increase in number and size of condylomas seen in populations mentioned previously. AIDS patients are known to have a high prevalence of condylomas. Condylomas in women can increase in number and size of lesions during pregnancy and regress after the birth of the child.

Respiratory Papilloma/Oropharyngeal Cancer

HPV is also reported to be associated with the malignant conversion of oropharyngeal papillomas into squamous cell carcinomas, and these types of infections have been observed in patients dating back to 1940 (32, 33). The initial reports of HPV-associated tongue and other oropharyngeal carcinomas was observed in 1985 (34). The rate of oropharyngeal cancer has increased and there are projections that the rate will surpass cervical cancer rates by the year 2020 (35). Typically oropharyngeal cancer is associated with a younger age at diagnosis, oral sex being the primary risk factor. The viruses recovered from both juvenile and adult onset disease are typically types 6 and 11, which are the same viruses that are responsible for genital warts (36). Early recognition is important relative to better outcomes and vaccination of young men and women will affect the numbers of cases in the future (35). In one study patients with HPV-positive oropharyngeal squamous cell carcinoma (OPSCC) have a 95% survival rate at 2 years versus a survival rate of 62% at 2 years for HPV-negative OPSCC (37). In one study that analyzed 116 cases of HPV-positive OPSCC, 87.9% were HPV type 16 (38). HPV can be transmitted from mother to child during childbirth, resulting in HPV colonization in the infant. Sanchez-Torices et al. indicated the rate of oropharyngeal colonization with HPV from a vaginal delivery was as high as 58% (39).

Cervical Cancer

Cervical cancer is one of the most common cancers in women. The International Agency of Research on Cancer (IARC) has been responsible for large studies that have ultimately helped established the link between HPV and cervical cancer in women. In addition to those studies, other data have shown the relative risk of cervical cancer is based on the genotype the woman is infected with. The PAP smear has been historically used to screen women for cervical cancer. LSIL or low-grade squamous intraepithelial lesions indicates mild dysplasia (CIN1) is usually caused by HPV and typically resolves spontaneously within a few years. HSIL or high-grade squamous intraepithelial lesions indicate moderate or severe cervical intraepithelial neoplasia or carcinoma *in situ*. These are the lesions that can lead to cervical cancer. HSIL corresponds to CIN2 or CIN3. By the 1980s it was clearly recognized that most cervical cancers were related to infection with specific oncogenic types of HPV (15).

The study by Bosch et al. in 1995 involved approximately 1000 women from 22 different countries with invasive cervical cancer. This cohort was very influential not only because of the fact that HPV was the causative agent of cervical cancer but also for the assignment of specific genotypes as being carcinogenic (40). This study was pivotal for several reasons and was used as the basis to assign risk to certain HPV genotypes for inclusion in diagnostic and research tests (40). There are approximately 15 to18 HPV types that are oncogenic, meaning that they have a high risk for the development of cervical cancer, but more recent studies have singled out types 16 and 18 as having an increased risk for causing high-grade disease. Based on studies, there is strong evidence for types 16, 18, 31, 33, 35, 45, 45, 51, 56, 58, and 59 as the cause of most cervical cancers in women. There is limited evidence for genotypes 26, 53, 66, 67, 70, 73, and 82 causing cervical cancer (41). Adding genotypes to cervical cancer screening tests must be done carefully. The detection of oncogenic types must be balanced with adding types that significantly affect the specificity of the assays as well (42).

It is becoming more widely accepted that women who harbor either genotype 16 or 18 would benefit from more aggressive management early on. Khan et al. looked at the 10-year cumulative incidence rate (CIR) for ≥CIN 3 in a screening population of women 30 years and older. In those women with negative cytology, the CIR for ≥CIN 3 for HPV 16 or 18 was 17% and 14%, respectively. The 10-year CIR was 3% for women infected with other than 16/18 genotypes and 1% for women that were high-risk HPV negative (43).

Other Cancers

Studies clearly demonstrate that some vulvar and penile cancers are caused by HPV, and there were found to be frequently positive for type 16 (44). In contrast to cervical cancer, only 50% of the vulvar squamous cell carcinomas are HPV positive (45). In one study of the vulvar cancers diagnosed, most were HPV type 16 (46, 47). The causation of the non-HPV cancers is speculative and remains a very important question. In contrast to vulvar cancers, 60 to 90% of cancers of the vagina are caused by HPV, similar to the percentage in anal and perianal cancers as well (45, 48). In 2000 the global incidence of anal and perianal cancers was relatively low, ranging from 0.1 to 2.8 cases per 100,000 (49). Sexual habits and men having sex with men (MSM) are increasing the incidence of these cancers. In the United States the CDC estimates that between 2004–2008 the incidence of anal cancer in women was 2.0 and in men was 1.1 per 100,000, respectively (50).

TREATMENT AND PREVENTION

HPV infections are not necessarily treated themselves; instead the treatment is directed at HPV- associated conditions that exist within the host. There are various treatment options that can be used for intraepithelial neoplasias and anogenital warts that vary based on the severity of disease.

Treatment of Anogenital Warts

The primary goal in treating anogenital warts is wart removal. There is no evidence that the presence of anogenital warts has an effect on the development of cervical cancer. Treatment is variable and consists of the following, which can be classified as patient-applied or physician-applied treatments. Patient-applied treatment includes the following: Imiquimod cream 3.75% or 5%, Sinecatechins ointment 15%, or Podofilox solution or gel 0.5%. Physician-administered therapy includes the following: podophyllin resin 10 to 25%, cryotherapy with liquid nitrogen or cryoprobe, trichloracetic acid (TCA) or bichloroacetic acid 80 to 90%, or surgical removal using tangential scissor excision, tangential shave excision, laser, curettage, or electrosurgery. It is important to note that many patients who have external anal warts might benefit by a digital examination of the anal canal using standard anoscopy or high resolution anoscopy, because internal genital warts are often found in patients that have external warts (31).

Treatment of Cervical Cancer and Pre-cancer

The persistence of HPV within cervical epithelial cells can result in precancerous lesions or cancer. Treatment of cervical cancer is directed at the macroscopic or pathological precancerous lesion. There is no antiviral therapy to eradicate these lesions. Precancerous lesions are detected through cervical cancer screening using Pap smear and/or HPV testing (31). Although the Pap smear is not diagnostic *per se*, it does direct the clinical management of the patient. For CIN 1 the recommended management would be follow-up with a physician to look for persistence or progression of a lesion. For precancers of a moderate or higher grade such as CIN 2, CIN 3, or adenocarcinoma *in situ*, there are several treatment options. These options can consist of removal of the abnormal area using laser, a loop electrosurgical excisional procedure (LEEP), or cold-knife conization. The abnormal area can also be destroyed using cryotherapy and laser vaporization as well. Cure rates are comparable among the procedures, and physicians can help guide the patient as to which procedure or approach is preferred (51).

Prevention

Currently at the time of this publication there are three vaccines for HPV for the prevention of certain types of HPV, all of which are approved by the U.S. Food and Drug Administration (FDA). HPV vaccines consist of recombinant HPV L1 capsid proteins from various HPV types as shown in Table 2. The prevention of up to 90% of genital warts and up to 70% of invasive cancer is now possible through the use of these vaccines in addition to other types of cancers that are caused by other HPV types now found in the 9-valent vaccine. The vaccines available today are the Quadrivalent and 9-Valent Gardasil vaccines (Merck & Co., Inc.) and the Cervarix vaccine (Glaxo SmithKline Biologicals). Gardasil is U.S. FDA approved and is a recombinant quadrivalent vaccine for types 6, 11, 16, and 18 and is approved for prevention of diseases associated with HPV infection (Table 3). Cervarix is also U.S. FDA approved and is a recombinant bivalent vaccine for types 16 and 18 (Table 3). The 9-valent Merck vaccine was only recently U.S. FDA approved in 2015. This vaccine includes HPV types 6, 11, 16, 18, 31, 33, 45, 52, and 58. The efficacy of the 9-valent vaccine was extensively studied in a randomized double-blind trial that included 14,215 previously unvaccinated females that were 16 to 26 years old. Each was given a dose of either the quadrivalent or the 9-valent vaccine on day 1 that was repeated at months 2 and 6. The 9-valent Gardasil vaccine was superior in reducing the risk of high-grade cervical, vulvar, or vaginal disease related to 31, 33, 45, 52, or 58 by 97% (0.1 vs. 1.6 cases/1000 person years) after a median of 40 months after final vaccination (52). Antibody responses were similar in the younger age group and in boys as compared to that seen in women 16 to 26 years of age.

TABLE 3 Comparison of human papillomavirus vaccines, United States (2015)

Vaccine	Cervarix[a] (Bivalent, HPV2)	Gardasil[b] (Quadrivalent, HPV4)	Gardasil 9[c] (9-valent, HPV9)
Manufacturer	GlaxoSmithKline	Merck and Co., Inc.	Merck and Co., Inc.
Year of Licensure (age range)	Females: 2009 (9–25 years) Males: not licensed for use	Females: 2006 (9–26 years) Males: 2009 (9–26 years)	Females: 2014 (9–26 years) Males: 2014 (9–15 years)
HPV Types (L1 protein composition)	HPV 16: 20 mcg HPV 18: 20 mcg	HPV 6: 20 mcg HPV 11: 40 mcg HPV 16: 40 mcg HPV 18: 20 mcg	HPV 6: 30 mcg HPV 11: 40 mcg HPV 16: 60 mcg HPV 18: 40 mcg HPV 31: 20 mcg HPV 33: 20 mcg HPV 45: 20 mcg HPV 52: 20 mcg HPV 58: 20 mcg
Adjuvant Type	AS04: 3-O-desacyl-4' monophosphoryl lipid A (50 mcg) adsorbed on to aluminum hydroxide (0.5 mg)	AAHS: amorphous aluminum hydroxyphosphate sulfate (225 mcg)	AAHS: amorphous aluminum hydroxyphosphate sulfate (500 mcg)
Dose, Route	0.5 mL, Intramuscular	0.5 mL, Intramuscular	0.5 mL, Intramuscular
Schedule	3 doses: 0, 1, and 6 months	3 doses: 0, 2, and 6 months	3 doses: 0, 2, and 6 months
Manufacturing	Insect cell line (*Trichoplusia ni*), infected with L1 encoding recombinant baculovirus	*Saccharomyces cerevisiae* (Baker's yeast), expressing L1	*Saccharomyces cerevisiae* (Baker's yeast), expressing L1

[a]Data from Cervarix [package insert]. Research Triangle Park, NC: GlaxoSmithKline Biologicals; 2015.
[b]Data from Gardasil [package insert]. Whitehouse Station, NJ: Merck & Co; 2015.
[c]Data from Gardasil 9 [package insert]. Whitehouse Station, NJ: Merck & Co; 2014.

The U.S. Advisory Committee on Immunization Practices (ACIP) has recently updated recommendations on HPV vaccination (53, 54). The recommendation for vaccination is a series of three doses. After the administration of the initial dose, the second dose should be given 1 to 2 months later. The third dose should be at least 6 months after the first dose. The safety and efficacy of the current vaccines is discussed in a recent publication by Kash et al. (55). Adverse reactions caused by the 9-valent vaccine were slightly more frequent reactions at the injection site than were seen with the quadrivalent vaccine (52). An additional adverse reaction that has been reported is a fall in blood pressure after the administration of the HPV vaccine in some patients. Vaccine acceptance for HPV has been less well received than other vaccines given to children and adolescents. In a survey in the United States from 2013 to 2014 in ages 13 to 17 only 33.6% percent of females and 41.7% of males were vaccinated with ≥1 dose of HPV vaccine. Cervical cancer and other HPV cancers have both a psychological impact on the patient and place a financial burden on the health care system. Several groups of investigators have tried to devise approaches to increase the vaccination rate among women. Although the rate of HPV vaccination varied by geographic regions in the United States, an improvement seemed to be noted in 2014, suggesting promise in increasing vaccination rates (56). Cohen et al. developed a video called "1-2-3 Pap" to address the vaccination problem in a rural part of one state and after implementation of the program saw improvement in the HPV vaccination rate (57). HPV vaccines are being studied for the prevention of skin cancer as well in some patients. In animal models HPV vaccines are shown not only to be immunogenic but also to prevent skin tumor formation (58).

DETECTION AND DIAGNOSIS

Because HPV is the predominant cause of cervical cancer, the Pap smear has been the test historically used to screen women for cervical cancer for over 50 years (59). Significant advances have been made with diagnostic tests that actually detect portions of the HPV genome. The literature on the topic of HPV is vast and includes newer studies on many of the assays that are available on the market today. It is beyond the scope of this text to cite and/or summarize all the relevant literature regarding diagnostic HPV assays now available. It is also beyond the scope of this chapter to summarize all relevant literature comparing the diagnostic tests to one another. The U.S. FDA-approved assays that are in use in many countries in the clinical laboratory setting today will be discussed below and are listed in Table 4. It is important that up-to-date literature be obtained prior to assessing the clinical or analytical performance of any of the assays described since the diagnostic field for HPV is constantly evolving. Laboratory-developed tests (LDTs) exist in some diagnostic and research laboratories but will not be covered in this chapter.

According to the American College of Obstetricians and Gynecologists (ACOG) Clinical Management Guidelines for Obstetricians and Gynecologists: Cervical Cytology Screening (Practice Bulletin #109, 2009) cervical cancer screening for women in the United States is recommended to begin at age 21. Algorithms for cervical cancer screening will be discussed in more detail in the section below on Best Practices. This section will focus on available diagnostic tests commonly used with the screening algorithms that include and promote the use of both Pap smear and HPV testing. Because the new guidelines recommend screening women every 3 to 5 years, the negative predictive value (NPV) of an

TABLE 4 U.S. FDA-approved assays for high-risk HPV testing

Assay	Specimen Type	HPV Type Detected	Genotyping
Qiagen, Hilden, Germany			(not FDA approved)
hc2 HPV DNA Test	PreservCyt, STM, Brush	16, 18, 31, 33, 35, 39, 45, 51, 52, 56, 58, 59, 68	16, 18, 45
Qiagen, Hilden Germany		Low-Risk types 6, 11, 42, 43, 44	(not FDA approved)
Cervista HPV	PreservCyt	16, 18, 31,33, 35, 39, 45, 51 52, 56, 58, 59, 66, 68	Reflex assay 16,18
Hologic, San Diego, CA			
Aptima HPV	PreservCyt	E6/E7 mRNA	Reflex assay 16, 18, 45
Hologic, San Diego, CA		16, 18, 31, 33, 35, 39, 45, 51, 52, 56, 58	
cobas 4800 HPV	PreservCyt	16, 18, 31, 33, 35, 39, 45, 51, 52, 56, 58, 59, 66, 68	Included, 16, 18
Roche Molecular, Pleasanton, CA			

HR = High Risk

HPV assay is extremely important. The better the NPV, the lower the risk women have for developing precancerous or cancerous lesions over that screening interval. Most new assays on the market today have the ability to genotype types 16 and 18. There is an excellent contemporary review on HPV assays by C. Ginocchio et al. within the *Manual of Clinical Microbiology* (Chapter 104) that addresses many non-FDA approved assays that are available or in development today (60).

Primary screening using HPV testing first as opposed to co-testing, PAP smear followed by HPV testing, has recently been U.S. FDA approved using the Roche cobas HPV Test. (Roche Molecular Diagnostics, Inc. Indianapolis, IN) (61). The adoption of HPV testing as a primary screen will have a significant impact on the volume of cytology testing and the current algorithms used by many institutions for cervical cancer screening.

Specimen Collection

Typically, depending on the assay used in the diagnostic laboratory sample collection, transport and storage requirements are located in the package insert of the respective commercial kit being employed. Using a collection device that is not approved for use by the manufacturer of the specific test would mean the test is being used off label from a diagnostic perspective and would essentially fall under the category of an LDT. In the United States the FDA is discussing the regulation of all LDTs used in the clinical laboratory environment even if the only variation in the assay is the use of an alternative nonapproved collection device. Each laboratory should rigorously validate the off-label use of any alternative collection device not tested by the manufacturer, in addition to any deviation from the protocol delineated in the package insert.

Most diagnostic assays used today for the detection of HPV from cervical specimens are approved with a liquid-based cytology (LBC) media. This allows for testing of both HPV and the Pap smear using the same vial. Exfoliated cells from the cervical epithelium are collected using a brush or spatula-brush device and placed into the LBC media for transport to the laboratory. Samples other than cervical, such as anal or oral lesions, are not FDA-approved for the commercial assays available today but some laboratories have validated off-label use with these sample types.

In most clinical settings HPV specimens are routinely collected by the provider. There are several reasons to encourage or promote the self-collection of HPV samples in women, but to reiterate, this method of collection would currently be considered "off label." This method would be used primarily in underserved populations where there is a need to increase HPV screening. There have been several studies that have focused on the self-collection of HPV specimens in women and found this approach to be promising (62, 63). The investigators in these studies reported that self-collection is comparable to physician-collected specimens and as sensitive in detecting CIN2+ when nucleic acid amplification tests (NAATS) are used for HPV detection (64, 65). In addition, alternative collection solutions for HPV screening in low-income countries has been performed on dried urine spots (DUS). The investigators used an in-house nested PCR assay and compared the results of DUS with those of freshly collected urine specimens (66). The sensitivity of the DUS was comparable to that of using fresh urine. Performance of urine testing for HPV has been compared to cervical cytology and has been reported in the literature (67).

Identification by DNA-Based Assays

Signal Amplification

The Hybrid Capture® 2 High Risk HPV DNA Test® (Qiagen, Hilden, Germany) (hc2) was the first commercial U.S. FDA-approved test that was available for HPV testing from clinical samples. hc2 is an *in vitro* nucleic acid hybridization assay with signal amplification using microplate chemiluminescence detection. The assay is a qualitative detection system for high-risk HPV types 16, 18, 31, 33, 35, 39, 45, 51, 52, 56, 58, 59, and 68 in cervical specimens. Specimens compatible with the assay that are validated by the manufacturer and U.S. FDA-approved for testing are the following: 1) Specimens collected with the Qiagen hc2 DNA Collection Device, 2) biopsy specimens collected in Specimen Transport Medium™ (STM) (Qiagen, Hilden, Germany), and 3) specimens collected using a broom-type collection device (placed in Thin Prep® Pap Test™ PreservCyt® Solution) (Hologic, Marlborough, MA). The early hc2 studies showed the value of HPV testing in women for the detection of cervical cancer, setting the path for development of additional assays and establishment of best practices for cervical cancer screening (68, 69). Manos et al. estimates that using an HPV based algorithm (including immediate colposcopy for ASCUS/HPV positive women) and repeat pap smears for all other women provides an overall sensitivity of 96.9% (95% Confidence Interval (CI),

88.3% to 99.5%) for detecting high-grade squamous intraepithelial lesions (HSILs) (70). The ALTS (ASCUS/LSIL Triage Study) Trial was a randomized multicenter clinical trial sponsored by the National Cancer Institute. The trial consisted of several study arms; the sensitivity and specificity of colposcopy, repeat cytology testing, and HPV testing using hc2 were compared for the detection of cervical intraepithelial neoplasia grade 3 (CIN3) in 3488 women. The sensitivity to detect CIN3 or above by testing for HPV DNA was 96.3% (Confidence Interval [CI] 91.6% to 98.9%). Sensitivity of a lower cytology triage (ASCUS) or above was 85.3% (CI 78.2% to 90.8%). hc2 had a greater sensitivity for detecting CIN 3 or above compared to a single Pap test (indicating ASCUS or greater) (69). Studies with hc2 paved the way for the initial guidelines incorporating HPV into the cervical cancer screening algorithms along with the Pap smear.

The hc2 assay can be automated using the Rapid Capture® System (RCS) Instrument Application. This instrumentation is useful for laboratories that have medium to higher HPV volumes in which the manual assay is too labor intensive for routine clinical use. For additional information regarding the procedure refer to the package insert at (https://www.qiagen.com).

The Cervista HR HPV assay (Hologic, Marlborough, MA), followed the hc2 test into the commercial market. The Cervista HR HPV assay detects similar HPV types compared to the Qiagen hc2 assay but also includes type 66 and is U.S. FDA approved for use with the Thin Prep® Pap Test™ PreservCyt® Solution (Hologic, Marlborough, MA). In addition to the high-risk screening assay, there is also a separate genotyping assay detecting genotypes 16/18, which is important based on the recent cervical cancer screening algorithms (71). This assay uses Invader® chemistry, which is a signal amplification method for the detection of HPV-specific nucleic acid sequences. The assay consists of two types of isothermal reactions. The primary reaction occurs on the DNA sequence targeted by the assay and the second reaction produces a fluorescent signal. Studies show that the Cervista assay is comparable to the hc2 assay from an analytical perspective (72, 73). Comparison in one study to the Aptima HPV Assay (Hologic, Marlborough, MA) showed less favorable agreement at 88%, which was mainly due to false-positive results with the Cervista assay (74). The false-positive results with the Cervista HPV assay had been previously documented (75). Bores et al. documented the value of adjusting the cutoff of the assay to improve on the specificity of the Cervista assay (76). For additional information regarding the procedure refer to the package insert at (http://www.hologic.com).

Amplification-Based Assays

The cobas HPV Test (Roche Molecular Systems, Inc., Pleasanton, CA) is a real-time PCR assay for the detection of high-risk HPV types. The test specifically identifies HPV 16, HPV 18, and the other high-risk (HR) types. The other HR types detected as a group include 31, 33, 35, 39, 45, 51, 52, 56, 58, 59, 66, and 68. The primers target a 200-nucleotide base pair sequence within the polymorphic L1 region of the HPV genome. As with the Cervista assay, this is an important feature since some cervical cancer screening algorithms suggest genotyping to help direct the physician with respect to patient management (71). There are several studies regarding the performance of the cobas HPV Test in the literature (61, 77–79). The ATHENA (**A**ddressing **THE** **N**eed for **A**dvanced HPV Diagnostics) HPV study compared the cobas 4800 HPV Test to the Qiagen hc2 Assay (79). The sensitivity of the cobas 4800 HPV assay was slightly higher than the hc2 assay for both CIN 2 and CIN 3 and specificities were very similar with no statistically significant difference. Sensitivity and specificity for ≥CIN2 was 90.0% and 70.5%, respectively, for the cobas 4800 HPV test and 87.2% and 71.1%, respectively, for the hc2 test. Sensitivity and specificity for ≥CIN3 was 93.5% and 69.3%, respectively, for the cobas 4800 HPV assay and 91.3% and 70.0% for the hc2 assay. Negative predictive value was similar for both assays, approaching 100% (79). The assay is automated using the cobas 4800 instrument. Acceptable specimens are cervical cells collected in cobas PCR Cell Collection Media (Roche Molecular Systems, Inc., Pleasanton, CA) or PreservCyt Solution (Hologic, Marlborough, MA). A FDA-approved clinical trial is currently under way for the BD SurePath™ Pap Test (Becton Dickinson, Sparks, Maryland) liquid-based cytology medium. As will be discussed below, the cobas HPV Roche Test is the only assay on the market FDA-approved for primary screening. If primary screening is adopted, it is recommended that HR HPV testing be performed. This is at least effective as 5-year cotesting (80). This will be discussed more in the section below on best practices. Refer to the manufacture package insert if additional information is needed for the assay.

The Aptima HPV (AHPV) (Hologic, Marlborough, MA) assay for high-risk papillomavirus testing (U.S. FDA approved) is a qualitative molecular test for E6/E7 RNA that detects the 13 types found in the hc2 assay in addition to HPV type 66 (81–83). Accompanying the screening assay is a genotype assay that detects HPV type 16 and HPV type 18 and/or HPV 45 but is a separate assay from the screening assay. During the clinical trial the AHPV assay was compared to colposcopy and histology for the detection of CIN 2 and CIN 3. For the CIN 2 assay, sensitivity, specificity, PPV, and NPV were 86.8%, 62.9%, 20.1%, and 97.8%, respectively. For CIN 3, sensitivity, specificity, PPV, and NPV were 90.2%, 60.2%, 9.4%, and 99.3%, respectively (84). Another study compared the cobas HPV test to the AHPV assay in a population of women who were found to be ASCUS on routine cytology screening (83). The assays were found to be similar in sensitivity but the specificity overall was slightly higher with the AHPV assay. Negative predicative value was equivalent (98%) for both assays (83). AHPV was also compared to the Qiagen hc2 assay in a routine screening population using split LBC samples (ages 30 to 60) (81). Sensitivity was slightly higher for hc2 but statistically equivalent when comparing the two assays. Sensitivity was overall better for CIN3 with the hc2 assay but among the three CIN3 that were missed, two of the specimens produced lesion-free cones and one was a non-high-risk type (81). The AHPV assay is FDA approved on the Tigris and Panther automation using the Thin Prep® Pap Test™ PreservCyt® Solution (Hologic, San Diego, CA). The Panther is a newer instrument that can automate HPV testing in small- to medium-size laboratories where the Tigris is often complementary to the higher volume laboratory where greater sample throughput is needed. For more detailed information on the assay refer to the manufacturer's package insert.

Genotyping

As shown in Table 4 there are several assays that can perform genotyping for HPV, some that are currently U.S. FDA approved. As described below, ASCCP recommends genotyping specifically for HPV 16 and HPV 18 in women with negative cytology but who are HPV positive on the screening test (85). Depending on the specific manufacturer, different approaches are available for genotyping, either a reflexive testing approach or a genotyping test integrated into the screening assay as described in Table 4. Other non-FDA-approved assays have been developed for broader-based genotyping applications detecting HR HPV but also some LR types (60).

Serology

There are no FDA-approved assays for the serological detection of human papillomaviruses, and serology is not used clinically to diagnose or monitor HPV infection in humans. Viral persistence may be due to failure in developing an appropriate antiviral response. Humoral immunity is not thought to clear HPV infection but may provide some protection against reinfection by certain HPV types (86). Some studies have shown that up to 40% of women with evidence of HPV DNA or lesions do not show detectable type-specific antibody responses (86–89).

Serological assays to measure a serum antibody response are very important in the development of HPV vaccines. Immunogenicity testing contributes to several aspects of each HPV vaccine not only in terms of development but also deployment of the vaccine. Immunogenicity is particularly useful to assess when increased utilization of the vaccine occurs in other patient populations who were not the intended target of the original vaccine (90). There are several assays that have been deployed in the evaluation of a serological response and used in recent vaccine trials. These assays include VLP (virus-like particle)-based ELISA assays, Luminex-based assays (Invitrogen), and neutralization assays (90, 91). The strengths and weaknessnesses of these assays are discussed in detail by Schiller and Lowy (92).

Antigen Detection/Alternative Methodologies

Although not the focus of this chapter, from a diagnostic perspective it is important to mention that there are several non-FDA-approved antigen-based methods available for the immunohistochemical or immunocytochemical detection of specific cellular proteins. These markers include MCM2, Ki-67, and $P16^{INK}$ tumor suppressor protein and have been evaluated in anal cancer diagnosis and as molecular markers for CIN2+ detection (93–95). The *Manual of Clinical Microbiology* provides an extensive list of *in situ* hybridization tests for the detection of HPV in clinical samples (60). To date none of these assays are FDA approved in the United States for routine use. Consensus from the College of American Pathologists (CAP) and the ASCCP in 2012 indicated that p16 can be used as an adjunct test to increase the diagnostic accuracy in some difficult cases (96). Other biomarkers such as miR-34a and miR-125b (miRNAs) have also been explored as biomarkers in cytological samples in cancer development. One study has shown that both these biomarkers are overexpressed in HPV-infected cells although further studies to demonstrate utility are warranted. Although miR-125b does seem to decrease as the cervical lesions progress, levels could still be used in invasive cervical cancer diagnosis (97). Another technology, called Raman spectroscopy, may be a promising biochemical tool in the field of cervical cancer diagnosis and potentially in the analysis of disease progression (98). Raman spectroscopy relies on the scattering of monochromatic light, usually from a laser beam source. The spectrum is unique to the molecular composition of what is being analyzed. Ramos et al. thoroughly review the advances of Raman spectroscopy in cervical cancer diagnosis (98).

BEST PRACTICES

The object of cervical cancer screening is the diagnosis of cervical cancer in asymptomatic women when the cancer is smaller, easier to treat, and less likely to have spread and prevention of cervical cancer by diagnosing its precursors, cervical intraepithelial neoplasia, CIN 2, or CIN3. The detection of the high-grade cervical intraepitheal neoplasia, CIN2 or CIN3, or carcinoma *in situ* positions the provider to select methods that will destroy the tissue with carcinoma *in situ* or cervical cells harboring the CIN 3 before they progress to invasive cancer. Cervical cancer screening has been effective in preventing morbidity and mortality from cervical cancer both because of the natural history of cervical neoplasia and the fact that cervical cells are relatively easy to obtain, either from self-collection of vaginal specimens or by a practitioner using a speculum-assisted observation and sample collection device. The natural history of cervical cancer is that of infection with one or more of the high-risk human papillomavirus (HR-HPV) genotypes, failure of the HR-HPV infection to be cleared, transformation of cervical cells to CIN 3, and finally development of cervical cancer (99). This natural history is suited to screening because the average durations between infection with HR-HPV and transformation to CIN 3 (estimated at 5.7 years) and between transformation to CIN 3 and development of invasive cervical cancer (3 to 17 years) are relatively long (100).

Cervical cancer screening was initially accomplished by cytologic interpretation of exfoliated cervical cells to detect changes suggestive of transformation (i.e., Pap smears). While cervical cancer screening with Pap smears is effective (101), interpretation of cervical cytology requires significant health care infrastructure, and the relatively low sensitivity of cervical cytology for CIN 2/3 or cancer (CIN 3+) requires a short screening interval. Because the exfoliated cells collected from the majority of women with CIN 3+ contain HR-HPV DNA (102), there has been enthusiasm for cervical cancer screening with HPV DNA tests. Because HPV type 16 and/or 18 DNA is found in exfoliated cells from 70.2% of women with invasive cervical cancer (102) and in 77.3% of exfoliated cells from women with CIN 3, there has also been enthusiasm for HPV DNA tests which allow genotyping.

The results of four randomized screening studies suggest that HR-HPV-based screening either with primary HR-HPV screening or cotesting with cervical cytology and HR-HPV tests provides 60 to 70% greater protection against invasive cervical carcinoma compared with cytology alone (103). This greater protection against invasive cervical cancer is realized because testing of cervical specimens for HR-HPV is 37% more sensitive for CIN 3+ than is cervical cytology with a cutpoint of low-grade squamous intraepithelial lesion (LSIL) or worse (104). Women with negative HR-HPV tests also have a lower cumulative risk of subsequent CIN 3+ than do women with negative cervical cytology (105, 106). In Dillner's series, the cumulative risk of CIN 3+ six years after a negative HR-HPV test was 0.27% [95% CI, 0.12% to 0.45%] (105) while the cumulative risk following a negative cytology was 0.97% [95% CI, 0.53% to 1.34%], while in Katki's series, the cumulative risk of CIN 3+ 5 years after a negative HR-HPV test was 0.17% and the cumulative risk after a negative cervical cytology was 0.36%, p=0.02 (106). The higher sensitivity and lower subsequent risk of CIN 3+ following a negative HR-HPV test allow an increased interval between screenings when HR-HPV tests are employed (107). As shown in Kulasingam's modeling study, the number of cancer cases associated with screening with cytology alone every 3 years starting at age 21 (N=8.5) is similar to that associated with screening with cytology every 3 years from ages 21 to 30 than with cytology and HR-HPV testing every 5 years after age 30 (N=7.44) (107).

The benefits of decreasing invasive cervical cancer by utilizing HR-HPV testing as a cotest with cytology are realized when women with negative cytology and positive HR-HPV tests are evaluated (108). Because the specificity of HR-HPV tests for CIN 3+ is 7% lower than that of cervical cytology with a cutpoint of LSIL (104), it is common to find women with negative cervical cytology and positive HR-HPV; between 3.4% and 11.3% of populations screened by cotesting with cervical cytology and HR-HPV tests have negative cytology and positive HR-HPV tests (109). If all women with negative cytology and positive HR-HPV tests were referred for colposcopy to exclude CIN 3+, the rate of colposcopy would increase by a factor of 2 to 3 times (109, 110). Determining how women with negative cervical cytology and positive HR-HPV tests should be evaluated is one of the current controversies within cervical cancer screening. In Kulasingam's modeling study mentioned above, women with negative cytology and positive HR-HPV tests were assumed to have repeat cotesting in 1 year and were referred for colposcopy if they remained HR-HPV positive (107). This strategy resulted in a higher number of colposcopies in women screened with cervical cytology every 3 years starting at age 21 (N=758) as compared with those screened with cervical cytology every 3 years between ages 21 and 29 and with cotesting with cervical cytology and HR-HPV testing every 5 years after age 30 (N=575) (107). The strategy of repeating cotesting in 1 year with referral if the HR-HPV test remains positive should be effective because CIN 3+ is associated with persistent rather than transient HR-HPV (99). The two problems with repeating the cotesting in 1 year are that many women with negative cytology and positive HR-HPV tests are lost to follow-up and that some women with invasive cervical cancer will have a delay in diagnosis. In the Kaiser Permanente study, 22.1% of the 1,726 women with negative cervical cytology and positive HR-HPV tests in 2007 did not have a subsequent HR-HPV test, and there may have been a delay in diagnosis of four cervical cancers (109). An alternative strategy for evaluating women with negative cytology and positive HR-HPV tests is to obtain genotyping on HR-HPV-positive specimens and refer those with HPV type 16 and/or 18 directly to colposcopy while repeating cotesting for those with HR-HPV other than type 16 or 18 (110). This strategy would likely follow fewer cervical cancers than would repeating cotesting at one year because 70% of the cervical cancers are positive for HPV type 16 and/or 18 (102). Unfortunately, about 20% of the women with negative cytology and positive HR-HPV tests have HPV type 16 and/or 18, and this group of women with HPV 16 and/or 18 includes only about two thirds of the CIN 3+ within women with negative cervical cytology and positive HR-HPV tests (111). The result is that the roughly 80% of women with negative cytology and positive HR-HPV tests that do not have HPV 16 or 18 will still require repeat cotesting in 1 year (111). The current consensus guidelines for management of women with negative cervical cytology and positive HR-HPV tests allow either repeat cotesting with cytology and HR-HPV in 1 year or immediate colposcopy for women with negative cytology and HR-HPV tests who test positive for HPV type 16 and/or 18, with repeat cotesting in 1 year for those without HPV type 16 or 18 (71).

Although HR-HPV tests are most useful in primary screening for cervical cancer (80, 103), they also are effective in triage of women with cervical cytology of ASCUS (112), in the evaluation of women with cytology of atypical glandular cells (AGC) (113), and as a test for cure following resection of CIN (114). Also worthy of mention is that the sensitivity of HR-HPV tests for CIN 3+ in self-collected vaginal specimens is similar to that in practitioner-collected endocervical specimens only if a PCR-based HR-HPV assay is employed (115).

REFERENCES

1. **Bernard HU, Burk RD, Chen Z, van Doorslaer K, zur Hausen H, de Villiers EM.** 2010. Classification of papillomaviruses (PVs) based on 189 PV types and proposal of taxonomic amendments. *Virology* **401:**70–79.
2. **Bzhalava D, Eklund C, Dillner J.** 2015. International standardization and classification of human papillomavirus types. *Virology* **476:**341–344.
3. **Egawa N, Egawa K, Griffin H, Doorbar J.** 2015. Human papillomaviruses; epithelial tropisms, and the development of neoplasia. *Viruses* **7:**3863–3890.
4. **Harari A, Chen Z, Burk RD.** 2014. Human papillomavirus genomics: past, present and future. *Curr Probl Dermatol* **45:** 1–18.
5. **de Villiers EM, Fauquet C, Broker TR, Bernard HU, zur Hausen H.** 2004. Classification of papillomaviruses. *Virology* **324:**17–27.
6. **Zheng ZM, Baker CC.** 2006. Papillomavirus genome structure, expression, and post-transcriptional regulation. *Frontiers in Bioscience.***11:**2286–2302.
7. **Münger K, Baldwin A, Edwards KM, Hayakawa H, Nguyen CL, Owens M, Grace M, Huh K.** 2004. Mechanisms of human papillomavirus-induced oncogenesis. *J Virol* **78:**11451–11460.
8. **Steenbergen RD, de Wilde J, Wilting SM, Brink AA, Snijders PJ, Meijer CJ.** 2005. HPV-mediated transformation of the anogenital tract. *J Clin Virol* **32:** Suppl 1:S25–33.
9. **zur Hausen H.** 2002. Papillomaviruses and cancer: from basic studies to clinical application. *Nat Rev Cancer* **2:**342–350.
10. **Strauss MJ, Shaw EW, Bunting H, Melnick JL.** 1949. Crystalline virus-like particles from skin papillomas characterized by intranuclear inclusion bodies. *Proc Soc Exp Biol Med* **72:**46–50.
11. **Rowson KE, Mahy BW.** 1967. Human papova (wart) virus. *Bacteriol Rev* **31:**110–131.
12. **Bosch FX, Lorincz A, Muñoz N, Meijer CJ, Shah KV.** 2002. The causal relation between human papillomavirus and cervical cancer. *J Clin Pathol* **55:**244–265.
13. **Muñoz N, Bosch FX, de Sanjosé S, Herrero R, Castellsagué X, Shah KV, Snijders PJ, Meijer CJ, International Agency for Research on Cancer Multicenter Cervical Cancer Study Group.** 2003. Epidemiologic classification of human papillomavirus types associated with cervical cancer. *N Engl J Med* **348:**518–527.
14. **Clifford GM, Smith JS, Plummer M, Muñoz N, Franceschi S.** 2003. Human papillomavirus types in invasive cervical cancer worldwide: a meta-analysis. *Br J Cancer* **88:**63–73.
15. **Bosch FX, de Sanjosé S.** 2003. Chapter 1: human papillomavirus and cervical cancer—burden and assessment of causality. *J Natl Cancer Inst Monogr* **2003(31):**3–13.
16. **Bäfverstedt B.** 1967. Condylomata acuminata—past and present. *Acta Derm Venereol* **47:**376–381.
17. **Centers for Disease CaP.** 2015. *Epidemiology and Prevention of Vaccine -Preventable Diseases*, 13th ed. Public Health Foundation, Washington, D.C..
18. **Antonsson A, Karanfilovska S, Lindqvist PG, Hansson BG.** 2003. General acquisition of human papillomavirus infections of skin occurs in early infancy. *J Clin Microbiol* **41:** 2509–2514.

19. **Castellsagué X, Bosch FX, Muñoz N, Meijer CJ, Shah KV, de Sanjose S, Eluf-Neto J, Ngelangel CA, Chichareon S, Smith JS, Herrero R, Moreno V, Franceschi S, International Agency for Research on Cancer Multicenter Cervical Cancer Study Group.** 2002. Male circumcision, penile human papillomavirus infection, and cervical cancer in female partners. *N Engl J Med* **346:**1105–1112.
20. **de Sanjose S, Almirall R, Lloveras B, Font R, Diaz M, Muñoz N, Català I, Meijer CJ, Snijders PJ, Herrero R, Bosch FX.** 2003. Cervical human papillomavirus infection in the female population in Barcelona, Spain. *Sex Transm Dis* **30:**788–793.
21. **Winer RL, Hughes JP, Feng Q, O'Reilly S, Kiviat NB, Holmes KK, Koutsky LA.** 2006. Condom use and the risk of genital human papillomavirus infection in young women. *N Engl J Med* **354:**2645–2654.
22. **Partridge JM, Hughes JP, Feng Q, Winer RL, Weaver BA, Xi LF, Stern ME, Lee SK, O'Reilly SF, Hawes SE, Kiviat NB, Koutsky LA.** 2007. Genital human papillomavirus infection in men: incidence and risk factors in a cohort of university students. *J Infect Dis* **196:**1128–1136.
23. **de Pokomandy A, Rouleau D, Ghattas G, Vézina S, Coté P, Macleod J, Allaire G, Franco EL, Coutlée F, Group HS, HIPVIRG Study Group.** 2009. Prevalence, clearance, and incidence of anal human papillomavirus infection in HIV-infected men: the HIPVIRG cohort study. *J Infect Dis* **199:** 965–973.
24. **Videla S, Darwich L, Cañadas MP, Coll J, Piñol M, García-Cuyás F, Molina-Lopez RA, Cobarsi P, Clotet B, Sirera G, HIV-HPV Study Group.** 2013. Natural history of human papillomavirus infections involving anal, penile, and oral sites among HIV-positive men. *Sex Transm Dis* **40:**3–10.
25. **Franceschi S, Herrero R, Clifford GM, Snijders PJ, Arslan A, Anh PT, Bosch FX, Ferreccio C, Hieu NT, Lazcano-Ponce E, Matos E, Molano M, Qiao YL, Rajkumar R, Ronco G, de Sanjosé S, Shin HR, Sukvirach S, Thomas JO, Meijer CJ, Muñoz N.** 2006. Variations in the age-specific curves of human papillomavirus prevalence in women worldwide. *Int J Cancer* **119:**2677–2684.
26. **zur Hausen H.** 2009. Papillomaviruses in the causation of human cancers—a brief historical account. *Virology* **384:** 260–265.
27. **de Koning MN, Quint KD, Bruggink SC, Gussekloo J, Bouwes Bavinck JN, Feltkamp MC, Quint WG, Eekhof JA.** 2015. High prevalence of cutaneous warts in elementary school children and the ubiquitous presence of wart-associated human papillomavirus on clinically normal skin. *Br J Dermatol* **172:**196–201.
28. **Bacelieri R, Johnson SM.** 2005. Cutaneous warts: an evidence-based approach to therapy. *Am Fam Physician* **72:**647–652.
29. **Quint KD, Genders RE, de Koning MN, Borgogna C, Gariglio M, Bouwes Bavinck JN, Doorbar J, Feltkamp MC.** 2015. Human beta-papillomavirus infection and keratinocyte carcinomas. *J Pathol* **235:**342–354.
30. **Burger B, Itin PH.** 2014. Epidermodysplasia verruciformis. *Curr Probl Dermatol* **45:**123–131.
31. **Workowski KA, Bolan GA, Centers for Disease Control and Prevention.** 2015. Sexually transmitted diseases treatment guidelines, 2015. *MMWR Recomm Rep* **64**(RR-03)**:**1–137.
32. **zur Hausen H.** 1977. Human papillomaviruses and their possible role in squamous cell carcinomas. *Curr Top Microbiol Immunol* **78:**1–30.
33. **Dalianis T.** 2014. Human papillomavirus (HPV) and oropharyngeal squamous cell carcinoma. *Presse Med* **43:**e429–e434.
34. **de Villiers EM, Weidauer H, Otto H, zur Hausen H.** Papillomavirus DNA in human tongue carcinomas. 1985 *Int J Cancer* **36(5):**575–578.
35. **Moore KA II, Mehta V.** 2015. The Growing Epidemic of HPV-Positive Oropharyngeal Carcinoma: A Clinical Review for Primary Care Providers. *J Am Board Fam Med* **28:**498–503.
36. **Mounts P, Shah KV, Kashima H.** 1982. Viral etiology of juvenile- and adult-onset squamous papilloma of the larynx. *Proc Natl Acad Sci USA* **79:**5425–5429.
37. **Fakhry C, Westra WH, Li S, Cmelak A, Ridge JA, Pinto H, Forastiere A, Gillison ML.** 2008. Improved survival of patients with human papillomavirus-positive head and neck squamous cell carcinoma in a prospective clinical trial. *J Natl Cancer Inst* **100:**261–269.
38. **Chaturvedi AK, Engels EA, Pfeiffer RM, Hernandez BY, Xiao W, Kim E, Jiang B, Goodman MT, Sibug-Saber M, Cozen W, Liu L, Lynch CF, Wentzensen N, Jordan RC, Altekruse S, Anderson WF, Rosenberg PS, Gillison ML.** 2011. Human papillomavirus and rising oropharyngeal cancer incidence in the United States. *J Clin Oncol* **29:**4294–4301.
39. **Sanchez-Torices MS, Corrales-Millan R, Hijona-Elosegui JJ.** 2015. Oropharyngeal perinatal colonization by human papillomavirus. *Acta Otorrinolaringol Esp* **67:**135–141.
40. **Bosch FX, Manos MM, Muñoz N, Sherman M, Jansen AM, Peto J, Schiffman MH, Moreno V, Kurman R, Shah KV.** 1995. Prevalence of human papillomavirus in cervical cancer: a worldwide perspective. International biological study on cervical cancer (IBSCC) Study Group. *J Natl Cancer Inst* **87:** 796–802.
41. **Bouvard V, Baan R, Straif K, Grosse Y, Secretan B, El Ghissassi F, Benbrahim-Tallaa L, Guha N, Freeman C, Galichet L, Cogliano V, WHO International Agency for Research on Cancer Monograph Working Group.** 2009. A review of human carcinogens—Part B: biological agents. *Lancet Oncol* **10:**321–322.
42. **Schiffman M, Khan MJ, Solomon D, Herrero R, Wacholder S, Hildesheim A, Rodriguez AC, Bratti MC, Wheeler CM, Burk RD, Group PEG, Group A, PEG Group, ALTS Group.** 2005. A study of the impact of adding HPV types to cervical cancer screening and triage tests. *J Natl Cancer Inst* **97:**147–150.
43. **Khan MJ, Castle PE, Lorincz AT, Wacholder S, Sherman M, Scott DR, Rush BB, Glass AG, Schiffman M.** 2005. The elevated 10-year risk of cervical precancer and cancer in women with human papillomavirus (HPV) type 16 or 18 and the possible utility of type-specific HPV testing in clinical practice. *J Natl Cancer Inst* **97:**1072–1079.
44. **Ikenberg H, Gissmann L, Gross G, Grussendorf-Conen EI, zur Hausen H.** 1983. Human papillomavirus type-16-related DNA in genital Bowen's disease and in Bowenoid papulosis. *Int J Cancer* **32:**563–565.
45. **Madsen BS, Jensen HL, van den Brule AJ, Wohlfahrt J, Frisch M.** 2008. Risk factors for invasive squamous cell carcinoma of the vulva and vagina—population-based case-control study in Denmark. *Int J Cancer* **122:**2827–2834.
46. **Alkatout I, Schubert M, Garbrecht N, Weigel MT, Jonat W, Mundhenke C, Günther V.** 2015. Vulvar cancer: epidemiology, clinical presentation, and management options. *Int J Womens Health* **7:**305–313.
47. **Ngan HY, Tsao SW, Liu SS, Stanley M.** 1997. Abnormal expression and mutation of p53 in cervical cancer—a study at protein, RNA and DNA levels. *Genitourin Med* **73:**54–58.
48. **Frisch M, Fenger C, van den Brule AJ, Sørensen P, Meijer CJ, Walboomers JM, Adami HO, Melbye M, Glimelius B.** 1999. Variants of squamous cell carcinoma of the anal canal and perianal skin and their relation to human papillomaviruses. *Cancer Res* **59:**753–757.
49. **Kleihues PS, Sobin LH.** World Health Organization classification of tumors. 2000. *Cancer* **88:**2887.
50. **Centers for Disease Control and Prevention (CDC).** 2012. Measles—United States, 2011. *MMWR Morb Mortal Wkly Rep* **61:**253–257.
51. **Wright TC Jr, Massad LS, Dunton CJ, Spitzer M, Wilkinson EJ, Solomon D, 2006 American Society for Colposcopy and Cervical Pathology-sponsored Consensus Conference.** 2007. 2006 consensus guidelines for the management of women with cervical intraepithelial neoplasia or adenocarcinoma in situ. *J Low Genit Tract Dis* **11:**223–239.
52. **Joura EA, Giuliano AR, Iversen OE, Bouchard C, Mao C, Mehlsen J, Moreira ED Jr, Ngan Y, Petersen LK, Lazcano-Ponce E, Pitisuttithum P, Restrepo JA, Stuart G, Woelber L, Yang YC, Cuzick J, Garland SM, Huh W, Kjaer SK, Bautista OM, Chan IS, Chen J, Gesser R, Moeller E, Ritter M,**

Vuocolo S, Luxembourg A, Broad Spectrum HPVVS, Broad Spectrum HPV Vaccine Study. 2015. A 9-valent HPV vaccine against infection and intraepithelial neoplasia in women. *N Engl J Med* **372:**711–723

53. **Petrosky E, Bocchini JA Jr, Hariri S, Chesson H, Curtis CR, Saraiya M, Unger ER, Markowitz LE, Centers for Disease Control and Prevention (CDC).** 2015. Use of 9-valent human papillomavirus (HPV) vaccine: updated HPV vaccination recommendations of the advisory committee on immunization practices. *MMWR Morb Mortal Wkly Rep* **64:**300–304.
54. **Markowitz LE, Dunne EF, Saraiya M, Chesson HW, Curtis CR, Gee J, Bocchini JA, Jr., Unger ER** Disease C, 2014. Human papillomavirus vaccination: recommendations of the Advisory Committee on Immunization Practices (ACIP). *MMWR Morb Mortal Wkly Rep* **63:**1–30.
55. **Kash N, Lee MA, Kollipara R, Downing C, Guidry J, Tyring SK.** 2015. Safety and Efficacy Data on Vaccines and Immunization to Human Papillomavirus. *J Clin Med* **4:**614–633.
56. **Reagan-Steiner S, Yankey D, Jeyarajah J, Elam-Evans LD, Singleton JA, Curtis CR, MacNeil J, Markowitz LE, Stokley S.** 2015. National, regional, state, and selected local area vaccination coverage among adolescents aged 13–17 years—United States, 2014. *MMWR Morb Mortal Wkly Rep* **64:** 784–792.
57. **Cohen EL, Head KJ, McGladrey MJ, Hoover AG, Vanderpool RC, Bridger C, Carman A, Crosby RA, Darling E, Tucker-McLaughlin M, Winterbauer N.** 2015. Designing for dissemination: lessons in message design from "1-2-3 pap." *Health Commun* **30:**196–207.
58. **Vinzón SE, Rösl F.** 2015. HPV vaccination for prevention of skin cancer. *Hum Vaccin Immunother* **11:**353–357.
59. **Mayrand MH, Duarte-Franco E, Rodrigues I, Walter SD, Hanley J, Ferenczy A, Ratnam S, Coutlée F, Franco EL, Canadian Cervical Cancer Screening Trial Study Group.** 2007. Human papillomavirus DNA versus Papanicolaou screening tests for cervical cancer. *N Engl J Med* **357:**1579–1588.
60. **Ginocchio CC, Smith JS.** 2015. Human papillomaviruses, p. 1783–1802 *In* Jorgensen JH, et al. (eds), *Manual of Clinical Microbiology*, 11th ed. Washington, D.C., ASM Press.
61. **Wright TC, Stoler MH, Behrens CM, Sharma A, Zhang G, Wright TL.** 2015. Primary cervical cancer screening with human papillomavirus: end of study results from the ATHENA study using HPV as the first-line screening test. *Gynecol Oncol* **136:**189–197.
62. **Dzuba IG, Díaz EY, Allen B, Leonard YF, Lazcano Ponce EC, Shah KV, Bishai D, Lorincz A, Ferris D, Turnbull B, Hernández Avila M, Salmerón J.** 2002. The acceptability of self-collected samples for HPV testing vs. the pap test as alternatives in cervical cancer screening. *J Womens Health Gend Based Med* **11:**265–275.
63. **Stewart DE, Gagliardi A, Johnston M, Howlett R, Barata P, Lewis N, Oliver T, Mai V, HPV Self-collection Guidelines Panel.** 2007. Self-collected samples for testing of oncogenic human papillomavirus: a systematic review. *J Obstet Gynaecol Can* **29:**817–828.
64. **Arbyn M, Verdoodt F, Snijders PJ, Verhoef VM, Suonio E, Dillner L, Minozzi S, Bellisario C, Banzi R, Zhao FH, Hillemanns P, Anttila A.** 2014. Accuracy of human papillomavirus testing on self-collected versus clinician-collected samples: a meta-analysis. *Lancet Oncol* **15:**172–183.
65. **Snijders PJ, Verhoef VM, Arbyn M, Ogilvie G, Minozzi S, Banzi R, van Kemenade FJ, Heideman DA, Meijer CJ.** 2013. High-risk HPV testing on self-sampled versus clinician-collected specimens: a review on the clinical accuracy and impact on population attendance in cervical cancer screening. *Int J Cancer* **132:**2223–2236.
66. **Frati ER, Martinelli M, Fasoli E, Colzani D, Bianchi S, Binda S, Olivani P, Tanzi E.** 2015. HPV Testing from dried urine spots as a tool for cervical cancer screening in low-income countries. *BioMed Res Intl* 1–5. Article ID 2833036.
67. **Tanzi E, Bianchi S, Fasolo MM, Frati ER, Mazza F, Martinelli M, Colzani D, Beretta R, Zappa A, Orlando G.** 2013. High performance of a new PCR-based urine assay for HPV-DNA detection and genotyping. *J Med Virol* **85:**91–98.
68. **Schiffman M, Solomon D.** 2003. Findings to date from the ASCUS-LSIL Triage Study (ALTS). *Arch Pathol Lab Med* **127:**946–949.
69. **Solomon D, Schiffman M, Tarone R, ALTS Study group.** 2001. Comparison of three management strategies for patients with atypical squamous cells of undetermined significance: baseline results from a randomized trial. *J Natl Cancer Inst* **93:** 293–299.
70. **Manos MM, Kinney WK, Hurley LB, Sherman ME, Shieh-Ngai J, Kurman RJ, Ransley JE, Fetterman BJ, Hartinger JS, McIntosh KM, Pawlick GF, Hiatt RA.** 1999. Identifying women with cervical neoplasia: using human papillomavirus DNA testing for equivocal Papanicolaou results. *JAMA* **281:** 1605–1610.
71. **Massad LS, Einstein MH, Huh WK, Katki HA, Kinney WK, Schiffman M, Solomon D, Wentzensen N, Lawson HW, 2012 ASCCP Consensus Guidelines Conference.** 2013. 2012 updated consensus guidelines for the management of abnormal cervical cancer screening tests and cancer precursors. *Obstet Gynecol* **121:**829–846.
72. **Boers A, Wang R, Slagter-Menkema L, van Hemel BM, Ghyssaert H, van der Zee AG, Wisman GB, Schuuring E.** 2014. Clinical validation of the Cervista HPV HR test according to the international guidelines for human papillomavirus test requirements for cervical cancer screening. *J Clin Microbiol* **52:**4391–4393.
73. **Wong AK, Chan RC, Nichols WS, Bose S.** 2008. Human papillomavirus (HPV) in atypical squamous cervical cytology: the Invader HPV test as a new screening assay. *J Clin Microbiol* **46:**869–875.
74. **Nolte FS, Ribeiro-Nesbitt DG.** 2014. Comparison of the Aptima and Cervista tests for detection of high-risk human papillomavirus in cervical cytology specimens. *Am J Clin Pathol* **142:**561–566.
75. **Kinney W, Stoler MH, Castle PE.** 2010. Special commentary: patient safety and the next generation of HPV DNA tests. *Am J Clin Pathol* **134:**193–199.
76. **Boers A, Slagter-Menkema L, van Hemel BM, Belinson JL, Ruitenbeek T, Buikema HJ, Klip H, Ghyssaert H, van der Zee AG, de Bock GH, Wisman GB, Schuuring E.** 2014. Comparing the Cervista HPV HR test and Hybrid Capture 2 assay in a Dutch screening population: improved specificity of the Cervista HPV HR test by changing the cut-off. *PLoS One* **9:**e101930.
77. **Castle PE, Stoler MH, Wright TC Jr, Sharma A, Wright TL, Behrens CM.** 2011. Performance of carcinogenic human papillomavirus (HPV) testing and HPV16 or HPV18 genotyping for cervical cancer screening of women aged 25 years and older: a subanalysis of the ATHENA study. *Lancet Oncol* **12:**880–890.
78. **Rao A, Young S, Erlich H, Boyle S, Krevolin M, Sun R, Apple R, Behrens C.** 2013. Development and characterization of the cobas human papillomavirus test. *J Clin Microbiol* **51:**1478–1484.
79. **Stoler MH, Wright TC Jr, Sharma A, Apple R, Gutekunst K, Wright TL, Group AHS, ATHENA (Addressing THE Need for Advanced HPV Diagnostics) HPV Study Group.** 2011. High-risk human papillomavirus testing in women with ASC-US cytology: results from the ATHENA HPV study. *Am J Clin Pathol* **135:**468–475.
80. **Huh WK, Ault KA, Chelmow D, Davey DD, Goulart RA, Garcia FA, Kinney WK, Massad LS, Mayeaux EJ, Saslow D, Schiffman M, Wentzensen N, Lawson HW, Einstein MH.** 2015. Use of primary high-risk human papillomavirus testing for cervical cancer screening: interim clinical guidance. *Gynecol Oncol* **136:**178–182.
81. **Iftner T, Becker S, Neis KJ, Castanon A, Iftner A, Holz B, Staebler A, Henes M, Rall K, Haedicke J, von Weyhern CH, Clad A, Brucker S, Sasieni P.** 2015. Head-to-head comparison of the RNA-based Aptima human papillomavirus (HPV) assay and the DNA-based hybrid capture 2 HPV test in a

routine screening population of women aged 30 to 60 years in Germany. *J Clin Microbiol* **53:**2509–2516.

82. **Castle PE, Eaton B, Reid J, Getman D, Dockter J.** 2015. Comparison of human papillomavirus detection by Aptima HPV and cobas HPV tests in a population of women referred for colposcopy following detection of atypical squamous cells of undetermined significance by Pap cytology. *J Clin Microbiol* **53:**1277–1281.
83. **Castle PE, Reid J, Dockter J, Getman D.** 2015. The reliability of high-risk human papillomavirus detection by Aptima HPV assay in women with ASC-US cytology. *J Clin Virol* **69:**52–55.
84. **Stoler MH, Wright TC, Jr., Cuzick J, Dockter J, Reid JL, Getman D, Giachetti C.** 2013. APTIMA HPV assay performance in women with atypical squamous cells of undetermined significance cytology results. *Am J Obstet Gynecol* **208:**144 e1-8.
85. **Saslow D, Solomon D, Lawson HW, Killackey M, Kulasingam SL, Cain J, Garcia FA, Moriarty AT, Waxman AG, Wilbur DC, Wentzensen N, Downs LS Jr, Spitzer M, Moscicki AB, Franco EL, Stoler MH, Schiffman M, Castle PE, Myers ER, American Cancer Society, American Society for Colposcopy and Cervical Pathology, American Society for Clinical Pathology.** 2012. American Cancer Society, American Society for Colposcopy and Cervical Pathology, and American Society for Clinical Pathology screening guidelines for the prevention and early detection of cervical cancer. *Am J Clin Pathol* **137:**516–542.
86. **Carter JJ, Koutsky LA, Hughes JP, Lee SK, Kuypers J, Kiviat N, Galloway DA.** 2000. Comparison of human papillomavirus types 16, 18, and 6 capsid antibody responses following incident infection. *J Infect Dis* **181:**1911–1919.
87. **Carter JJ, Koutsky LA, Wipf GC, Christensen ND, Lee SK, Kuypers J, Kiviat N, Galloway DA.** 1996. The natural history of human papillomavirus type 16 capsid antibodies among a cohort of university women. *J Infect Dis* **174:**927–936.
88. **Dillner J, Wiklund F, Lenner P, Eklund C, Frederiksson-Shanazarian V, Schiller JT, Hibma M, Hallmans G, Stendahl U.** 1995. Antibodies against linear and conformational epitopes of human papillomavirus type 16 that independently associate with incident cervical cancer. *Int J Cancer* **60:** 377–382.
89. **Kirnbauer R, Hubbert NL, Wheeler CM, Becker TM, Lowy DR, Schiller JT.** 1994. A virus-like particle enzyme-linked immunosorbent assay detects serum antibodies in a majority of women infected with human papillomavirus type 16. *J Natl Cancer Inst* **86:**494–499.
90. **Harper DM, Franco EL, Wheeler C, Ferris DG, Jenkins D, Schuind A, Zahaf T, Innis B, Naud P, De Carvalho NS, Roteli-Martins CM, Teixeira J, Blatter MM, Korn AP, Quint W, Dubin G, GlaxoSmithKline HPV Vaccine Study Group.** 2004. Efficacy of a bivalent L1 virus-like particle vaccine in prevention of infection with human papillomavirus types 16 and 18 in young women: a randomised controlled trial. *Lancet* **364:**1757–1765.
91. **Opalka D, Lachman CE, MacMullen SA, Jansen KU, Smith JF, Chirmule N, Esser MT.** 2003. Simultaneous quantitation of antibodies to neutralizing epitopes on virus-like particles for human papillomavirus types 6, 11, 16, and 18 by a multiplexed Luminex assay. *Clin Diagn Lab Immunol* **10:**108–115.
92. **Schiller JT, Lowy DR.** 2009. Immunogenicity testing in human papillomavirus virus-like-particle vaccine trials. *J Infect Dis* **200:**166–171.
93. **Wentzensen N, Follansbee S, Borgonovo S, Tokugawa D, Schwartz L, Lorey TS, Sahasrabuddhe VV, Lamere B, Gage JC, Fetterman B, Darragh TM, Castle PE.** 2012. Human papillomavirus genotyping, human papillomavirus mRNA expression, and p16/Ki-67 cytology to detect anal cancer precursors in HIV-infected MSM. *AIDS* **26:**2185–2192.
94. **Gustinucci D, Passamonti B, Cesarini E, Butera D, Palmieri EA, Bulletti S, Carlani A, Staiano M, D'Amico MR, D'Angelo V, Di Dato E, Martinelli N, Malaspina M, Spita N, Tintori B, Fulciniti F.** 2012. Role of p16(INK4a) cytology testing as an adjunct to enhance the diagnostic specificity and accuracy in human papillomavirus-positive women within an organized cervical cancer screening program. *Acta Cytol* **56:** 506–514.
95. **Waldstrøm M, Christensen RK, Ørnskov D.** 2013. Evaluation of p16(INK4a)/Ki-67 dual stain in comparison with an mRNA human papillomavirus test on liquid-based cytology samples with low-grade squamous intraepithelial lesion. *Cancer Cytopathol* **121:**136–145.
96. **Darragh TM, Colgan TJ, Cox JT, Heller DS, Henry MR, Luff RD, McCalmont T, Nayar R, Palefsky JM, Stoler MH, Wilkinson EJ, Zaino RJ, Wilbur DC, Members of LAST Project Work Groups.** 2012. The lower anogenital squamous terminology standardization project for HPV-associated lesions: background and consensus recommendations from the College of American Pathologists and the American Society for Colposcopy and Cervical Pathology. *J Low Genit Tract Dis* **16:**205–242.
97. **Ribeiro J, Marinho-Dias J, Monteiro P, Loureiro J, Baldaque I, Medeiros R, Sousa H.** 2015. mir-34a and mir-125b expression in HPV infection and cervical cancer development. *BioMed Res Intl* Article ID 304584.
98. **Ramos IR, Malkin A, Lyng FM.** 2015. Current advances in the application of raman spectroscopy for molecular diagnosis of cervical cancer. *BioMed Res Intl* 1–9. Article ID 561242.
99. **Schiffman M, Wentzensen N, Wacholder S, Kinney W, Gage JC, Castle PE.** 2011. Human papillomavirus testing in the prevention of cervical cancer. *J Natl Cancer Inst* **103:** 368–383.
100. **Mitchell MF, Tortolero-Luna G, Wright T, Sarkar A, Richards-Kortum R, Hong WK, Schottenfeld D.** 1996. Cervical human papillomavirus infection and intraepithelial neoplasia: a review. *J Natl Cancer Inst Monogr* (21):17–25.
101. **IARC Working Group on evaluation of cervical cancer screening programmes.** 1986. Screening for squamous cervical cancer: duration of low risk after negative results of cervical cytology and its implication for screening policies. *Br Med J (Clin Res Ed)* **293:**659–664.
102. **Coutlée F, Ratnam S, Ramanakumar AV, Insinga RR, Bentley J, Escott N, Ghatage P, Koushik A, Ferenczy A, Franco EL.** 2011. Distribution of human papillomavirus genotypes in cervical intraepithelial neoplasia and invasive cervical cancer in Canada. *J Med Virol* **83:**1034–1041.
103. **Ronco G, Dillner J, Elfström KM, Tunesi S, Snijders PJ, Arbyn M, Kitchener H, Segnan N, Gilham C, Giorgi-Rossi P, Berkhof J, Peto J, Meijer CJ, International HPV screening working group.** 2014. Efficacy of HPV-based screening for prevention of invasive cervical cancer: follow-up of four European randomised controlled trials. *Lancet* **383:** 524–532.
104. **Arbyn M, Sasieni P, Meijer CJ, Clavel C, Koliopoulos G, Dillner J.** 2006. Clinical applications of HPV testing: a summary of meta-analyses. *Vaccine* **24:** Suppl 3:S3/78–89.
105. **Dillner J, Rebolj M, Birembaut P, Petry KU, Szarewski A, Munk C, de Sanjose S, Naucler P, Lloveras B, Kjaer S, Cuzick J, van Ballegooijen M, Clavel C, Iftner T,**.2008. Long term predictive values of cytology and human papillomavirus testing in cervical cancer screening: Joint European cohort study. *BMJ* **337:**a1754.
106. **Katki HA, Kinney WK, Fetterman B, Lorey T, Poitras NE, Cheung L, Demuth F, Schiffman M, Wacholder S, Castle PE.** 2011. Cervical cancer risk for women undergoing concurrent testing for human papillomavirus and cervical cytology: a population-based study in routine clinical practice. *Lancet Oncol* **12:**663–672.
107. **Kulasingam SL, Havrilesky LJ, Ghebre R, Myers ER.** 2013. Screening for cervical cancer: a modeling study for the US Preventive Services Task Force. *J Low Genit Tract Dis* **17:**193–202.
108. **Sasieni P, Castle PE, Cuzick J.** 2009. Further analysis of the ARTISTIC trial. *Lancet Oncol* **10:**841–842, author reply 842–843.
109. **Pretorius RG, Belinson JL, Peterson P, Azizi F, Lo A.** 2013. Yield and mode of diagnosis of cervical intraepithelial neo-

plasia 3 or cancer among women with negative cervical cytology and positive high-risk human papillomavirus test results. *J Low Genit Tract Dis* **17**:430–439.

110. **Cox JT, Castle PE, Behrens CM, Sharma A, Wright TC, Jr., Cuzick J, Athena HPVSG.** 2013 . Comparison of cervical cancer screening strategies incorporating different combinations of cytology, HPV testing, and genotyping for HPV 16/18: results from the ATHENA HPV study. *Am J Obstet Gynecol* **208**:184 e1–e11.
111. **Goodrich SK, Pretorius RG, Du H, Wu R, Belinson JL.** 2014. Triage of women with negative cytology and positive high-risk HPV: an analysis of data from the SHENCCAST II/III studies. *J Low Genit Tract Dis* **18**:122–127.
112. **ASCUS-LSIL Triage Study (ALTS) Group.** 2003. Results of a randomized trial on the management of cytology interpretations of atypical squamous cells of undetermined significance. *Am J Obstet Gynecol* **188**:1383–1392.
113. **Pretorius RG, Peterson P.** 2015. High-grade squamous intraepithelial lesion cytology with negative high-risk human papillomavirus tests rarely diagnoses endometrial cancer. *J Low Genit Tract Dis* **19**:200–202
114. **Kong TW, Son JH, Chang SJ, Paek J, Lee Y, Ryu HS.** 2014. Value of endocervical margin and high-risk human papillomavirus status after conization for high-grade cervical intraepithelial neoplasia, adenocarcinoma in situ, and microinvasive carcinoma of the uterine cervix. *Gynecol Oncol* **135:** 468–473.
115. **Belinson JL, Du H, Yang B, Wu R, Belinson SE, Qu X, Pretorius RG, Yi X, Castle PE.** 2012. Improved sensitivity of vaginal self-collection and high-risk human papillomavirus testing. *Intl J Cancer* **130**:1855–1860.
116. **Doorbar J, Egawa N, Griiffin H, Kranjec C, Murakami I.** (2016) Human papillomavirus molecular biology and disease association. *Rev. Med. Virol* **25**:2–23.

Human Polyomaviruses

REBECCA J. ROCKETT, MICHAEL D. NISSEN, THEO P. SLOOTS,
AND SEWERYN BIALASIEWICZ

31

The expanding family *Polyomaviridae* infect a variety of different hosts (1). Generally, avian polyomaviruses are highly pathogenic and have a wide host range, while mammalian polyomaviruses have a limited host range and are asymptomatic in the immunocompetent host (2, 3). At present, 13 of these species are linked to human infection, JC polyomavirus (JCPyV) (4), BK polyomavirus (BKPyV) (5), WU polyomavirus (WUPyV) (6), KI polyomavirus (KIPyV) (7), Merkel cell polyomavirus (MCPyV) (8), human polyomavirus 6 (HPyV6) (9), human polyomavirus 7 (HPyV7) (9), Trichodysplasia spinulosa-associated polyomavirus (TSPyV) (10) and human polyomavirus 9 (HPyV9) (11), Malawi polyomavirus (MWPyV) (12), St. Louis polyomavirus (STLPyV) (13), human polyomavirus 12 (HPyV12) (14), and New Jersey polyomavirus (NJPyV-2013) (15). Zoonotic infections have not been reported, however they cannot be completely excluded.

VIRAL CLASSIFICATION AND BIOLOGY

In humans, polyomaviruses generally cause an initial asymptomatic infection in early childhood. They then become dormant, only reactivating and causing severe disease in the immunocompromised host. Human polyomaviruses have widespread seroprevalence in the general population and can be oncogenic in animal models, but, with the exception of MCPyV, evidence of human polyomaviruses involvement in the development of cancer is yet to be confirmed and is the matter of extensive scientific debate.

The International Committee on Taxonomy of Viruses' (ICTV) Polyomavirus Study Group has recently recommended the revision of the family *Polyomaviridae*, dividing the single genus into three genera—*Orthopolyomavirus*, *Wukipolyomavirus*, and *Avipolyomavirus*—essentially creating two mammalian genera and one avian genus. Division of the human genera is based on whole genome sequence homology of less than 81% to 84% (1). The recent discovery of highly divergent polyomaviruses in fish will mean that at least one other clade with these viruses as representatives will need to be created in the near future. Additionally, the early and late regions of many polyomaviruses exhibit evidence of ancient recombination events, which further obscure the true nature of the *Polyomaviridae* family tree (Fig. 1). These recent discoveries highlight how much more there is to learn about this growing virus family.

Structure and Genome Organization

The polyomavirus virion is a nonenveloped, icosahedral particle, which is 40 to 45 nm in diameter and contains 72 pentamers of the protein VP1. Each pentamer is associated with a single copy of VP2 or VP3 and forms a barrel-like structure. The virion contains a circular double-stranded DNA genome of approximately 5,000 base pairs (16). The genome is divided into three functional regions, the early and late coding region and the noncoding control region (NCCR) that contains transcription factor (TF) binding sites and the origin of replication. The early coding region produces nonstructural proteins including the small (STAg) and large tumor antigen (LTAg) and the late region encodes the three capsid proteins, VP1, VP2, and VP3. The origin of replication (ORI) separates these two coding regions, and replication commences bi-directionally from the ORI.(3, 16) Additional nonstructural proteins are encoded in some of the human polyomaviruses. For example JCPyV, BKPyV, and SV40 encode an agnoprotein within the 5'end of the late coding region.

The Polyomavirus Life Cycle

The mechanisms of polyomavirus cell entry, transportation through the cytoplasm, and nuclear entry are an ongoing area of research, however, it has become clear that multiple mechanisms are employed by the different polyomaviruses. The VP1 major capsid protein provides most of the surface residues that interact with the host cell, therefore forming one of the main determinants of tissue tropism, and through these interactions the virus binds and is imported into the cell (17). JCPyV uses a clathrin-dependent endocytosis pathway to gain entry into the host cell after binding to the 5HT2A serotonin receptor (18). BKPyV also utilizes caveolae mediated endocytosis, but binds to ganglioside receptors GD1b and GT1b (19). In general, once the virus is internalized, it is transported via vesicles to the endoplasmic reticulum during which time the VP1 capsid unravels, exposing the internal genome and VP2/3 proteins (20). The genome is then trafficked into the nucleus, where transcription and replication take place (17). Both transcription and replication are controlled by, and originate within, the

doi:10.1128/9781555819156.ch31

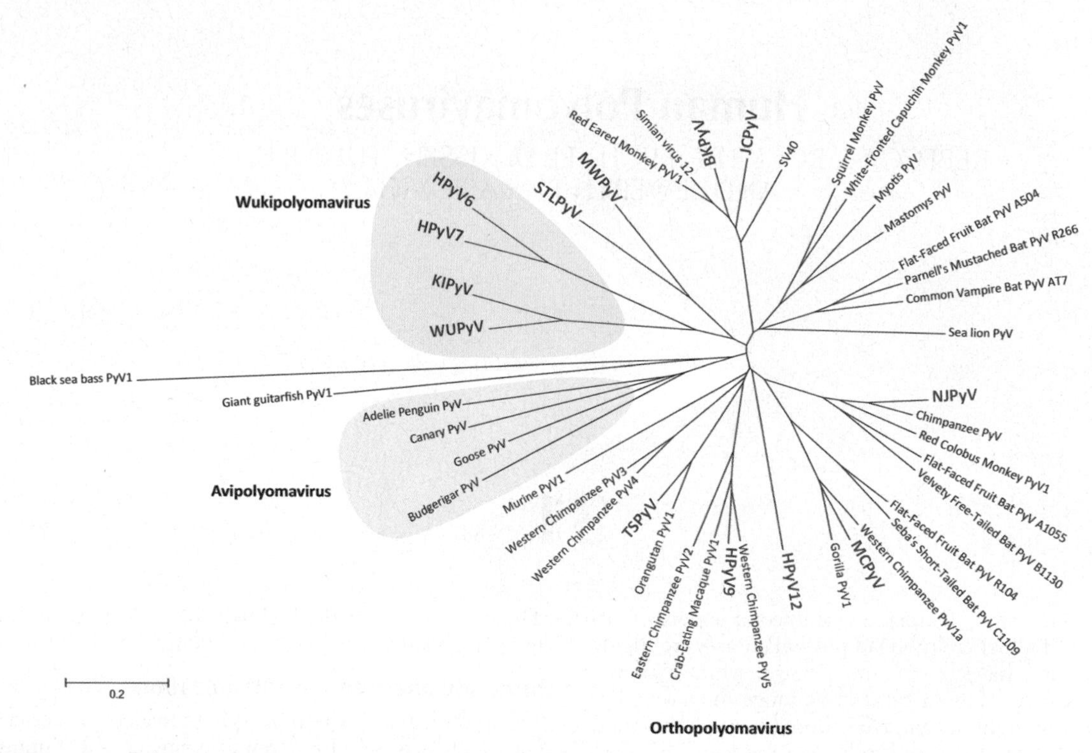

FIGURE 1 A radial phylogenetic tree generated using a Neighbour-Joining analyses of concatenated VP2, VP1, and LTAg gene sequences from a wide variety of mammalian, bird, and fish polyomavirus species. All known human polyomaviruses are shown in red, with recognized polyomavirus subclades highlighted.

NCCR, with initiation being dependent on the appropriate host TFs binding to the NCCR. Thus, the composition of the NCCR's TF binding sites also contribute to the virus' tissue tropism. Early transcription produces a single mRNA that is alternatively spliced into the small T antigen (STAg) and the large T antigen (LTAg) (16). Polyomaviruses do not possess the necessary DNA replication or translation machinery. Instead, they utilize the host cell replication machinery, and therefore require a method of driving the cell into S-phase (21). This is achieved through the interaction of the viral t-antigens with the host cell. The LTAg has an extensive array of effects on the host cell and its own replication. These include driving the host cell into S phase, a transcription activator and suppressor, helicase activity, and assistance in virion assembly (3). The STAg has an important role in promoting cell growth including viral replication, in part through its interaction with protein phosphatise 2A (22), which regulates numerous cell cycling processes through phosphorylation. Once viral replication begins, the infection enters late phase and the LTAg promotes the transcription of the late genes, while suppressing early gene expression (23). Transcription of the late genes produces a minimum of three proteins, VP1, VP2, and VP3, with the agnoprotein being produced only in JCPyV and BKPyV but not in WUPyV, KIPyV, or MCPyV (6, 7, 24). Similar to the early transcripts, the structural proteins are transcribed after alternative splicing of a single late region mRNA. After transcription, the structural proteins are translocated into the nucleus where the VP1 begins to self-assemble. Host-derived histones are associated with the viral genomes, which, along with the VP2/3 proteins, are then packaged into the assembling viral capsid. Exit of full assembled infectious polyomaviruses can occur through a number of processes including host cell lysis and increasing membrane permeability. Virus persistence and latency is not well understood, but may be influenced by production of self and host regulating viral miRNAs (25).

Polyomaviruses have demonstrated oncogenic potential in the laboratory setting, however to date, only MCPyV has sufficient proof of direct involvement in human cancer (8). The general pattern of oncogenesis involves the establishment of a nonproductive infection in which the t-antigens become constitutively expressed, thus driving the cell to persistent replication and chromosomal instability (3). Such nonproductive infections can arise from situations where the host cell's TF profile, or other replicative/transcriptional mechanisms, are partially incompatible with the virus, or if the viral genome becomes integrated into the host genome, thereby abrogating expression of the structural proteins and replicative function of the LTAg (3).

POLYOMAVIRUS EPIDEMIOLOGY

Seroepidemiology of Human Polyomaviruses

Seroprevalence of BK and JC Polyomaviruses

Many aspects of the epidemiology of the human polyomaviruses in the general population are still unknown, e.g., the source of infectious virus, the route of natural transmission, and the site of initial virus replication within the body. Since their discovery in 1971, BKPyV and JCPyV have been the most studied, and both viruses are endemic in almost all populations. Man is the sole host for both BKPyV and JCPyV, which circulate independently at both the individual and population levels. Individuals with antibody to BKPyV are often seronegative to JCPyV and vice versa (26). Neither BKPyV nor JCPyV produce a distinct, recognizable disease on primary infection, and disease is usually diagnosed by detection of viral DNA in various bodily compartments. However, an indication of a population's exposure to BKPyV and JCPyV may be more reliably made by assaying for antiviral antibodies.

BKPyV is a common infection in both developed and developing countries, showing a seroprevalence between 55% and 85% in studies covering all age groups of healthy individuals (27). However, in isolated populations, seroprevalence rates may vary from 0% to 100%, suggesting that this virus was absent or infrequently encountered in these populations (28). Generally, no difference was reported in seroprevalence between males and females.

Antibodies to BKPyV are typically acquired in early childhood (29). Elevated levels of maternal antibody, present at birth, are lost during the first few months of life, so that only 5% of infants between 4 and 11 months of age have anti-BKPyV antibody. Thereafter, seroprevalence rises rapidly from the second year of life to reach 83% by late childhood, consistent with antibody prevalence observed in adults. BKPyV-specific IgM antibody was detected in between 11% and 21% of healthy or unwell children under 5 years of age, but not in children younger than 12 months, which was consistent with BKPyV infection occurring in young children (29, 30).

In some studies of healthy adults or blood donors, BKPyV IgM antibody was not detected (31), whereas in others, a rate of between 3.6% and 19.5% was reported (28, 31, 32). However, the methods used to detect IgM antibody varied in sensitivity and specificity, and it is not known whether these results represent primary BKPyV infection or reinfection in adulthood, or are due to persistent IgM production in a few individuals. In the USA and England, the percentage of seropositivity for JCPyV in adults ranged from 44% to 77%, and between 85% and 92% in Brazil, Japan, and Germany (27). Some studies reported a significantly greater proportion of males with JCPyV antibodies compared to females (33, 34). More variation was apparent in the age at which JCPyV was acquired than BKPyV (27). Whilst in some populations most JCPyV infection was acquired in early childhood, in others the JCPyV seroprevalence rose more gradually, even continuing into old age. This may indicate that the transmission of JCPyV is more dependent on differences between cultures or socio-economic conditions than is BKPyV. In contrast to BKPyV, JCPyV antibody titres did not decrease with age, indicating that there may be continuing JCPyV stimulation throughout life, either by virus reactivation or reinfection (35). Low levels of JCPyV IgM antibody were reported in 15% of unselected blood donors in England (36), which may represent primary JCPyV infection in this population or sporadic virus reactivation.

Seroprevalence of the New Human Polyomaviruses

A number of studies have reported the seroprevalence of the novel human polyomaviruses in different populations. These results were largely based on ELISA assays employing recombinant viral VP1, VP1-GST capsomeres, or VLPs as the substrate antigen.

The observed seropositivity patterns of the new human polyomaviruses studied to date are shown in Table 1 (14, 28, 33, 37–47). These demonstrate that these viruses circulate widely in the general population and indicate that primary infections commonly occur in children and young adults. In a number of studies, age stratification of the seropositive specimens suggested an initial waning of immune response from loss of maternal antibodies, followed by rapid seroconversion during childhood. This increase of seropositivity with age for most of the viruses tested, except HPyV9, is compatible with the model of continuing primary infections throughout adult life. The differences in seroprevalence reported for the novel polyomaviruses between studies could represent true differences in different countries, but could also reflect differences in study populations, in techniques used for the detection of antibodies, and in cutoff definitions. In addition to seroprevalence data, Nicol et al. (39) also reported increasing antibody levels for MCPyV with age. This is consistent with reactivation of the virus at older ages, following waning immunity with age, and possible reactivation of infection. However, a decrease in antibody levels with age was observed for TSPyV suggesting that persistence and latency in immunocompetent individuals were less frequent with this polyomavirus (42). Such differences also suggest that the new human polyomaviruses may have different modes of transmission and capacities for persistence. Like JCPyV and BKPyV, routes of infection have not been determined for the new polyomaviruses, however limited evidence for MCPyV suggests intrafamilial saliva or skin contact as the source of transmission during childhood (48). Antibody studies of closely related viruses may be confounded with cross reactions of antibodies with related viral antigens. However, several studies have demonstrated the absence of cross reactivity against the different HPyVs, even between polyomaviruses with high VP1 homology (49), e.g., BKPyV and JCPyV (78%), KIPyV and WUPyV (66%), and MWPyV and STLPyV, which share 55% amino acid homology in the VP1 region (47).

Molecular Epidemiology

Sequence variation within each polyomavirus species is thought to be limited, primarily due to their reliance on the high fidelity host DNA polymerases. Nucleotide deletions and duplications have been observed most frequently in the regulatory region of the JCPyV and BKPyV genome and are typically associated with periods of immune dysregulation (27). It is thought that these changes are unique to the individual host and may not be transmitted. Limited nucleotide variation also occurs in the coding region of the polyomavirus genome, but these do often not result in amino acid changes. However, changes in the VP1 gene can result in antigenic variation in BKPyV, and generate multiple genotypes in JCPyV (50). Based on serological methods and molecular variations in the BC loop of the BKPyV- VP1 coding region, BKPyV isolates worldwide have been classified into four genotypes (I-IV) which can also be considered serotypes, with genotype I globally distributed, subtypes II and III found rarely throughout the world, and subtype IV is prevalent in Asia and parts of Europe (51). The main four BKPyV genotypes can be further categorized into 12 subtypes, of which subtypes Ib1 and Ib2 have recently been reported to behave as distinct serotypes (52). Prototype BKPyV (genotype I) is the most common detected type found in all categories of patients. Dual or multiple infections can also occur (27), with some evidence suggesting post-transplant nosocomial acquisition (52).

Only one serologically distinct JCPyV isolate (Mad-11) has been reported (53). However, genetically, very few amino acid changes have been observed between JCPyV strains in the BC loop of the VP1 coding region of the genome (54). Nucleotide changes do occur elsewhere in the

TABLE 1 Summary of seroprevalence rates for the new human polyomaviruses in different study populations

Virus	Age-group	Location	Seroprevalence	Reference
BKPyV	Adults	Global	55–85%	(33)
		Indian or aboriginal	0–100%	(28,33)
		Tribes in Brazil, Paraguay, and Malaysia		
JCPyV	Adults	USA, UK	44–77%	(33)
		Brazil, Japan, Germany	85–92%	(33)
WUPyV	Adults	USA	80%	(37)
		Germany	89%	(38)
KIPyV	Adults	USA	70%	(37)
		Germany	67%	(38)
MCPyV	Adults	Italy	66–87%	(39)
		Australia	68%	(40)
		USA	46–64%	(41)
	Adults (blood donors)	Italy	85%	(39)
HPyV6	Adults	Italy	69–83%	(39)
		Australia	76%	(40)
HPyV7	Adults	Australia	66%	(40)
TSPyV	Adults	Italy	70–90%	(39)
		Australia	81%	(40)
		The Netherlands	70%	(42)
	Adults (immunocompromised)	The Netherlands	89%	(42)
	Children (1–4 years)	Italy	31%	(39)
	Children (5–9 years)	Italy	70%	(39)
	Children (1–9 years)	The Netherlands	41%	(42)
	Children (1–4 years)	Finland	5%	(43)
	Children (6–10 years)	Finland	48%	(43)
HPyV9	Adults	Italy	27.5–42.4%	(44)
	Adults (>80 years)	Italy	70%	(44)
		Germany	20%	(45)
		Australia	24%	(40)
	Children	Germany	47%	(45)
MWPyV	Adults	Italy	41.8%	(46)
	Children (1–2 years)		26.9%	
	Children (2–4 years)		68.2%	
STLPyV	Adults	USA	91.2–95.2%	(47)
	Children (<0.5 years)		53.3%	
	Children (0.5–1 years)		37.9%	
	Children (1–2 years)		22.6%	
	Children (>2 years)		60–85.3%	
HPyV12	Adults (21–30 years)	Germany	27%	(14)
	Adults (>30 years)		15–33%	
	Children (2–5 years)		12%	
	Children (6–11 years)		26%	

JCPyV genome and genetic diversity in the intergenic region between VP1 and LT and within VP1 have so far identified eight major genotypes. These differ by 1% to 2.6%. Minor genetic variations are found consistently in different geographic areas, and on this basis, types 1, 2, 3, 7, and 8 can be further subdivided (55). The coevolution of JCPyV with humans dates back at least 50,000 years, with stable genotypes arising in geographically isolated populations. Exploration of the genetic diversity of the newly discovered human polyomaviruses has been severely limited, yet is important for investigating what role these potential genetic groups play in human disease.

To date, only one large-scale study has reported the characterization and genotyping of global WUPyV polyomavirus strains (56). This study examined 64 whole WUPyV genomes that were globally distributed, and reported the existence of three main WUV clades and five subtypes, provisionally termed Ia, Ib, Ic, II, IIIa, and IIIb. Overall nucleotide variation was low (0% to 1.2%).

CLINICAL SIGNIFICANCE

Clinically, most data are available for BKPyV and JCPyV. Although serologic studies show that infections in adults are ubiquitous with these viruses, little is still known about the clinical manifestations of primary infections. It is thought that the majority of primary infections present with few, or no, symptoms (57). Symptoms of upper respiratory tract infection were identified in one-third of children who seroconverted to BKPyV in one study (58) and in another study, serologic evidence for acute BKPyV infection was documented in 8% of children with upper respiratory tract illnesses (57). However, human polyomaviruses may cause significant disease in the immunocompromised host, with reactivation of latent to lytic infection resulting in viruria and viremia potentially leading to fatal disease.

BK Polyomavirus

Infections with BKPyV, like JCPyV, typically occur at a young age, and the viruses establish a life-long persistent infection in sanctuary sites such as the proximal renal tubule (49). BKPyV can cause a lytic infection that results in the destruction of tissue, similar to the well described cytopathic effect of the simian polyomavirus SV40. Uncontrolled BKPyV infection contributes to polyomavirus-associated nephropathy (PVAN) in patients who have received a renal transplant, and to haemorrhagic cystitis (HC) in patients who have received a haematopoietic stem cell transplant (59).

Urinary shedding has been reported to occur asymptomatically and intermittently in healthy individuals (60). In situations of altered immune function, including pregnancy, the rates and levels of urinary BKPyV shedding typically increase more dramatically than those for JCPyV, and urine cytology can appear as shedding of tubular epithelial or urothelial cells with intranuclear viral inclusions called "decoy cells" (61). Despite high urine viral loads, symptoms are absent in most solid organ transplant patients, suggesting that the cytopathic wear in the urothelial amplification compartment is well compensated.

HC is a BKPyV immune-reconstitution inflammatory syndrome (IRIS) with pathology characterized by a dominant inflammatory response to abundant PyV antigen, typically following a brisk recovery of the cellular response (62). PVAN, on the other hand, is a cytopathic inflammatory pathology characterized by high level viral replication and a significant inflammatory response to cytopathic lysis, necrosis with infiltrates of granulocytes, and lymphocytes. (63)

The role of BKPyV in human malignancies is controversial. The high seroprevalence and high detection rate of BKPyV and JCPyV by sensitive molecular tools in healthy individuals complicate studies and require a rigorous reevaluation of the published data. For BKPyV, the most convincing data of an oncogenic contribution have been presented in single case reports of urothelial malignancies and renal tubular malignancies, typically in the setting of kidney transplantation as illustrated by recently reported single cases (64).

There are also reports on the detection of BKPyV DNA in other tumors, e.g., colorectal tumors, lymphomas, pancreatic cancer, brain tumors, prostate cancer, and a range of sarcomas (65), and for non-UV light associated melanomas (66). The interpretation of the data in these reports have been various and controversial. However, based on "inadequate evidence" in humans and "sufficient evidence" in experimental animals, a WHO International Agency for Cancer Research Monograph Working Group recently classified BKV and JCV as "possibly carcinogenic to humans" (Group 2B) (67).

JC Polyomavirus

JCPyV is the causal agent of progressive multifocal leukoencephalopathy (PML), and although once rare, the incidence of PML has dramatically increased with the spread of AIDS. In the beginning of the HIV pandemic, it was estimated that 5% of HIV positive patients would go on to develop PML, although with the advent of combined antiretroviral therapy (cART), that incidence rate has dramatically dropped from 0.13% to 0.06% (68). Likewise, the introduction of cART reduced the one-year PML mortality rates from 82.3% to 37.6% (69). Despite the reduction in mortality rates however, PML remains the second leading cause of AIDS-related deaths, and has one of the worst prognoses of any AIDS-associated disease. Unfortunately, the surviving patients are often left with substantial neurological impairments. Occurrence of PML in hematological malignancy patients is uncommon, and its occurrence in transplant patients is rare (68). Since 2005, the use of monoclonal antibodies (MAb), such as natalizumab, to treat autoimmune disease has also led to an unexpected new source of PML. In multiple sclerosis (MS) patients, the incidence of PML is 0.0034%, with a reported higher frequency of seizures compared to AIDS-associated PML. Mortality in MS-associated PML has been reported to be up to 29%, depending on the time of diagnosis (70).

A further complication of PML seen in HIV positive patients and other immunosuppressed patients is the phenomenon of Immune Reconstitution Inflammation Syndrome (IRIS), where neurological symptoms paradoxically increase in response to restoration of immune system function. IRIS develops in 17% to 27% of HIV positive PML patients, typically after the start of cART, with risk factors including low starting CD4+ counts and rapid clearance of HIV viral load. IRIS also occurs in immunosuppressed patients after discontinuation of immunosuppression, and in MS patients, in which the IRIS can be triggered or even made worst by natalizumab therapy cessation, or attempts to remove the MAb from the bloodstream (68).

More recently, several additional neurological pathologies have been described associated with JCPyV, including JCPyV Granule Cell Neuropathy (JCV GCN), which infects the cerebellum and leads to dysmetria and ataxia (71), JCPyV encephalopathy, and JCPyV meningitis. The former two pathologies affect the same populations at risk of PML, but the latter is thought to possibly be a manifestation of primary JCPyV infection (68).

JCPyV's clinical impact is predominantly neuropathic, however, other body sites can be affected as well. JCPyV nephropathy in solid organ transplant recipients has been reported, but unlike BKPyV, these cases remain a rare occurrence (72). JCPyV has also been implicated in the development of human cancers including brain and gastrointestinal tumors, however to date, evidence of such involvement has been contradictory and contentious (68).

The Newly Discovered Human Polyomaviruses

On the basis of the sites where the polyomavirus DNA has been isolated, suspected sites of infection include the nasopharynx and lung for WUPyV and KIPyV, the skin for MCPyV, TSPyV, HPyV6, HPyV7, and HPyV9, the gastrointestinal tract for MWPyV and STLPyV, and the vascular endothelium for NJPyV.

These new human polyomaviruses are thought to be transmitted horizontally by direct contact or by aerosol or fecal-oral routes. However, the site of infection, the original site of polyomavirus identification, and the distant organs in which the virus can be detected might be distinct and difficult to relate to any pathophysiology (73). Therefore correlations between infection and respiratory symptoms or diarrheal illness remain circumstantial (74). Also, the current data remain confounded by the presence of numerous viral agents in the samples studied, such that no single entity has been identified as the causative agent (75).

Attempts to link the new polyomaviruses to clinical disease have focused on older individuals, in whom immunity is waning, or in the immunosuppressed. However, except for MCPyV and TSPyV, there has been no obvious association of the new human polyomaviruses with a clinical syndrome (76). Autoimmune diseases, in particular systemic lupus erythrematosis (SLE), have been proposed to result from a pathological response to "self," which may be triggered by human polyomavirus antigens (63).

Merkel Cell Polyomavirus

The key insight that prompted the hunt for a pathogen in MCC was the recognition that the incidence of MCC is higher in immunosuppressed recipients of organ transplants, and in patients with HIV/AIDS than in nonimmunocompromised individuals.

Patients with haematological malignancies, especially chronic lymphocytic leukaemia (CLL), are at an increased risk of developing MCC (77). There have been several reports of MCC developing in patients with autoimmune diseases such as psoriatic and rheumatoid arthritis, particularly when treated with immune-suppressive therapy. Additional risk factors for MCC include an age of greater than 60 years, and most MCCs occur in sun exposed areas of the skin. In addition, it has been reported that pharmacological use of statins and environmental exposure to arsenic increase the risk of developing MCC, although the mechanisms involved are unclear (78, 79). Clonal integration of the MCPyV genome has been identified in nearly all cases of MCC, and to date, is the only polyomavirus to have strong causal evidence for human tumorigenesis. MCPyV has been detected in other human cancers such as CLL and squamous cell skin carcinomas, although at low levels and not integrated into the genome (77). This is similar to JCPyV and BKPyV, which have been detected in a variety of gastrointestinal and genitourinary cancers (80). However, aside from one report of BKPyV integration into urothelial carcinoma (64), for the most part, there has been no evidence that these tumors contain integrated JCPyV or BKPyV genomes. SV40 has been considered as a zoonotic agent involved in human oncogenesis, but the low copy number (<1/1,000 cells), the inconsistent results between laboratories, and the near undetectable seroprevalence of the virus has not supported this hypothesis (81). Thus there is so far little evidence that MCPyV plays a role in the oncogenesis of tumors other than MCC. However, the role of MCPyV in MCC is further supported by the fact that MCC patients frequently have very high antibody titres to MCPyV as compared to the population in general, indicating that MCPyV infection indeed is associated with MCC (82). Other evidence of MCPyV involvement in MCC includes t-antigen protein expression within tumor cells, and MCC cell growth arrest and death upon knockdown of the viral t-antigen expression (83). High MCPyV antibody titres and viral loads in MCC patients have been suggested to be associated with better clinical outcomes (84), but the use of the virus as a marker for improved clinical outcomes remains controversial.

KI and WU Polyomaviruses

The first prevalence data for WUPyV and KIPyV were based on PCR detection of viral DNA in nasopharyngeal aspirate (NPA) samples, which showed that their prevalence was not significantly different for healthy individuals compared to those with respiratory symptoms, although some reports did show an increase in prevalence in immunosuppressed individuals (85). Recently, KIPyV DNA and capsid proteins were detected in the lung tissues of a young bone marrow transplant patient who died of respiratory failure (86). These results, in conjunction with the lack of any other microbial detection, would suggest that under some circumstances KIPyV may contribute to, or be capable itself of causing, respiratory disease. A study investigating viral infections in children under the age of two found that both WUPyV and KIPyV, when detected as the sole viral pathogen, were associated with upper respiratory tract disease (87). In parallel, other studies examined the prevalence of KIPyV and WUPyV at sites other than the respiratory site, e.g., in urine, blood, and stool, and showed KIPyV and WUPyV genomic DNA was not commonly detected in the urine of normal or immunosuppressed individuals, or in the blood, but WUPyV could be found at very low frequency (<1%) in stool samples (88). Also, KIPyV or WUPyV were not found in cerebrospinal fluid (CSF) of patients with or without PML (89). Several studies have investigated whether KIPyV or WUPyV are involved in tumor development and so far, the results have been negative for childhood brain tumors, and non-UV exposed melanomas (66). In one study, KIPyV VP1 DNA was detected in 9 out of 20 lung cancer cases of undefined origin, and in 2 out of 2 transplant patients (90). Only the C-terminal domain of the early region of KIPyV was successfully identified in two of the lung tissues, but possible integration was not analyzed. Thus, there is no compelling evidence to date for a role of KIPyV in oncogenesis, although the data may indicate the lung as a site of viral persistence.

Trichodysplasia Spinulosa Associated Polyomavirus

Trichodysplasia spinulosa (TS) is a rare skin disease in immunosuppressed patients characterized by papules and spines and alopecia in the face. In 2010, a new human polyomavirus (TSPyV) was detected in the plucked facial spines of a heart transplant patient with TS (10). Attempts were made to find TSPyV in other immunosuppressed patients not suffering from TS, and TSPyV was found in plucked eyebrows of 3 out of 69 (4%) examined patients, but at much lower viral loads than that seen in TS patients. TSPyV can also be found on the skin of healthy subjects at low viral loads, as well as occasionally in the blood and respiratory tract of immunocompromised patients, although with no apparent association to disease. However, TSPyV was not detected in 5 JCPyV positive cerebral spinal fluids, 20 BKPyV positive blood plasma, or in 20 KIPyV or WUPyV positive NPAs (91). Its presence and high seroprevalence in clinically healthy persons suggests that it causes a subclinical latent infection. Whether TSPyV is involved in other diseases or tumors remains to be investigated.

Human Polyomavirus 9

HPyV9 was detected in the serum of a renal transplant patient (92). More recently, it was shown that HPyV9 also resides in the skin (93), although no clear association with disease has been found. Nevertheless, additional studies may be necessary to identify possible other sites of HPyV9 replication and persistence, and its role, if any, in human disease. Interestingly, HPyV9 is related to African Green Monkey Polyomavirus, which can infect and immortalize human B cells (93). If HPyV9 has a similar tropism, it might have a causal relationship with B cell lymphomas.

Human Polyomavirus 6 and Human Polyomavirus 7

Both HPyV6 and 7 were originally found on the skin of healthy human volunteers (9), however no clear association with disease has been found. Recently, two cases of extensive velvety, pruritic rashes in lung transplant patients were reported, with both HPyV7 DNA and viral proteins being detected in the skin biopsies (94). Additionally, HPyV7 DNA and t-antigen proteins were found in the majority of thymic tumors, with the authors suggesting the virus may play a role in thymomagenesis (95).

Malawi, St. Louis, and New Jersey Polyomaviruses, and Human Polyomaviruses 10 and 12

MWPyV was identified by shotgun pyrosequencing of DNA purified from virus-like particles isolated from a stool sample collected from a healthy child from Malawi (12). A real-time PCR assay was then designed and used to screen 514 stool samples from children with diarrhea in St. Louis, USA. MWPyV was detected in 12 of these samples, but the index strain of these samples differed by 5.3% at the nucleotide level compared to the Malawi index case. In parallel, a virus termed human polyomavirus 10 (HPyV10) was found in chondyloma specimens from a patient with warts with hypoglobulinemia infections, and myeloathexis (WHIM) (96). Soon afterwards, a virus called MXPyV was described in the stools of Mexican children (97). Both HPyV10 and MXPyV are genetically nearly identical to MWPyV, and can be considered strains of the virus. MWPyV has been detected in respiratory, as well as stool samples, in immunocompromised and in immunocompetent children under the age of two years, with suggestions that it may play a role in respiratory disease (87). However, whether the virus elicits a chronic infection in the gastrointestinal system or is responsible for any disease in humans has yet to be fully elucidated.

HPyV12 has been first identified in resected human liver tissue (14). Investigation of organs, body fluids, and excretions of diseased individuals and healthy subjects with both HPyV12-specific nested PCR and quantitative real-time PCR revealed additional virus-positive samples of resected liver, cecum, and rectum tissues, and a positive fecal sample. Seroprevalences of 23% and 17%, respectively, were determined in sera from healthy adults and adolescents and a pediatric group of children. These data indicate that the virus naturally infects humans and that primary infection may already occur in childhood (14). No definitive association with disease is known to date.

Finally, NJPyV was detected in a 33-year-old pancreatic transplant recipient who developed weakness, retinal blindness, and necrotic plaques on her face, scalp, and hands (15). A muscle biopsy revealed micro-thrombosis and viral particles in swollen endothelial cell nuclei. *In situ* hybridization confirmed the presence of NJPyV in endothelial cells at sites of myositis and cutaneous necrosis, suggesting it may have tropism and cause disease in vascular endothelial cells. More research however is necessary to confirm such an association, or that with any other disease.

In conclusion, given the increasing use of immunosuppressive therapy after renal and haematopoietic stem cell transplantation, and specific monoclonal antibody therapies for multiple sclerosis (e.g., natalizumab), rheumatoid arthritis and other autoimmune diseases, and the ability of some polyomaviruses to cause disease in some individuals receiving this therapy, it is likely that the recognition of diseases resulting from polyomaviruses will increase.

TREATMENT AND PREVENTION

Lytic reactivations by JCPyV and BKPyV (e.g., PVAN & PML) can be serious, or even fatal, and therefore effective and safe antiviral therapies are desperately needed. Current antiviral therapies are limited by efficacy, tolerance, and safety. Adjunct therapies such as cidofovir, leflunomide, and intravenous immunoglobulins have been used (98), but the benefit is not documented in clinical trials. Routine monitoring for BKPyV in renal allografts, together with appropriate clinical management such as reduction of immunosuppressive therapy, has become a key recommendation (99). There is an increased risk of JCPyV-induced PML following natalizumab treatment for Crohn's disease and multiple sclerosis, therefore this treatment must also be carefully monitored.

Possible novel therapeutic strategies include small-molecule inhibitors of virus glycan binding or of T antigen functions (100), booster immunization in anticipation of immunosuppressive regimens to increase antibody levels, or enhancement of cell-mediated immunity in elderly patients. The increasing use of immunosuppressive therapies might lead to more cases of polyomavirus-induced disease, and could create a market for effective therapeutics that inhibit polyomavirus replication. In addition, new therapeutics that target the oncogenic activity of the MCPyV T antigens or boost the immune response would be extremely beneficial in MCC.

DETECTION AND DIAGNOSIS

Polyomavirus virions are considered to be hardy and can survive a wide range of harsh environments, including sewage (101), and temperatures exceeding 95°C (102). The PyV virion's robustness, as well as their stable double-stranded circular DNA genome, means that they may be able to persist in a detectable state within environmental and clinical samples for extended periods of time.

PCR and quantitative PCR (qPCR) are commonly relied upon for the detection of PyVs, both in the diagnostic and research setting. Caution should however be used when PCR assays are applied to certain types of samples in which collection quality or uniformity can be variable (e.g., respiratory swabs). In such instances, other markers of sample collection quality, such as human genome markers (e.g., RPL13A, ERV3) (103, 104) can be used to normalize the potentially variable samples.

Many of the newly discovered PyVs are poorly represented in public sequence databases, and as such, little is known about their genetic variability. Therefore, the use of two PCR assays targeting different parts of the PyV genome is recommended when feasible. Of note, the NCCR of many PyVs has been shown to be the primary area of genetic instability, particularly in the context of patient immunosuppression, and as such, the area should be avoided as a PCR target unless it has been sufficiently evaluated for conservation. Consensus PyV assays are not recommended for viruses that have potential to occur together in the same sample (e.g., JCPyV and BKPyV), as the presence of the more populous viral species DNA may partially or completely inhibit the amplification and detection of the secondary viral species' DNA (105). For use of PCR purely as a detection tool, it is important to note that like any other PCR, the smaller the PCR product is, the more likely it will successfully amplify at low template concentrations due to its inherent higher efficiency. Additionally, a smaller PCR product size will also minimize the impact of DNA fragmentation in formalin-fixed tissue samples.

Apart from specific PCR and qPCR assays, "Pan-Polyomavirus" PCR assays using degenerate primers have been utilized to detect both known and novel PyVs, including the discovery of human PyVs HPyV7 and HPyV9 (9). Rolling Circle Amplification (RCA) uses phi29 DNA polymerase to preferentially amplify circular DNA in an isothermal reaction and is ideally suited for use with PyV genomes. Because most applications of RCA to PyV genomes to date have used random hexamers (see details below), the specificity of the reaction is reduced, and as a consequence, the sensitivity suffers. In fact, a recent study showed that the use of random hexamers reduces the PyV amplicon yield in a mixed-template sample by at least 100-fold when compared with PyV-specific RCA primers (106).

BK Polyomavirus

Urine cytology has been commonly used to monitor transplant patients for BKPyV infection, with detection of BKPyV infected epithelial cells ("decoy cells") being a marker for virus reactivation and potential development of PVAN (107). Urines can be examined either neat or concentrated by centrifugation followed by either fixing and viewing by Papanicolaou staining (107), electron microscopy (EM) (107, 108), immunohistochemistry (IHC) (108), or phase-contrast microscopy (109). Decoy cells appear in both Papanicolaou-stained and phase contrast microscopy as cells with enlarged ground-glass like nuclei, and may also have altered chromatin, halos, or enlarged nucleoli (Fig. 2) (107, 109). Presence of large numbers of decoy cells in Papanicolaou-stained (>10/cytospin) (107) or phase-contrast microscopy (>1/×400 high-powered field) (109) were reported to be associated with PVAN development. Decoy cell monitoring has been suggested as a quick and cost-effective alternative to PCR for monitoring of transplant patients due to a near-100% PVAN negative predictive value (110). The urine cytology method however suffers from a low (21.2 to 27%) PVAN positive predictive value (108, 110), in part due to decoy cells also being produced as a result of other viral infections or malignancies.

In contrast, PCR of urine was found to have equivalent negative (100%) and higher positive (31%) predictive values than cytology, with viral loads greater than 2.5×10^7 copies/ml being associated with concurrent PVAN (111). Persistent viruria occurs in nearly all HC patients, but is also commonly shed in non-HC immunosuppressed patients, and therefore cannot be used independently as a diagnostic marker of HC (112). Processing of larger urine volumes (e.g., 1 ml), or of pelleting prior to nucleic acid extraction was reported to increase BKPyV PCR sensitivity by over 1.5 logs (113). Despite its cost-effectiveness and noninvasive nature, the utility of urine for monitoring does have drawbacks, including a wide fluctuation of viral loads, and a delayed clearance response to reduction of immunosuppression (114).

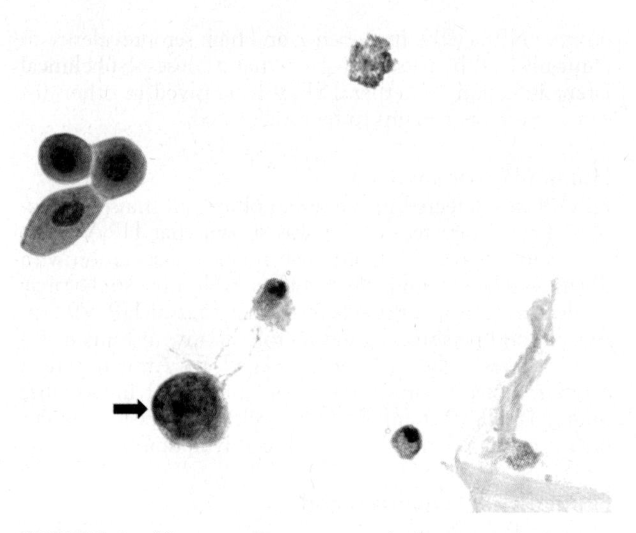

FIGURE 2 Urine cytology micrograph showing polyomavirus infected cells (Decoy cells). (Copyright 2010 Nephron. Permission is granted to copy, distribute, and/or modify this image under the terms of the GNU Free Documentation License Version 1.3.)

Monitoring of BKPyV viraemia by PCR has an improved PVAN positive predictive value (30% to 50%) compared to urine, with that figure rising to 90% when viral load is high (>1×10^6 copies/ml) (114). Viraemia by itself does not correlate to the development of HC, however decreasing BKPyV plasma loads have been correlated with increased HC recovery (115). For monitoring viraemia, and quantification of viral loads, standardized sample processing needs to be employed, as whole blood may give different BKPyV viral load results to that of serum.

A wide range of published and commercial conventional and quantitative real-time PCR assays exist, however all assay oligo sequences should be checked to ensure complete, or near-complete, homology with all BKV subtypes. Mismatches in the primer or probe sequences with a particular subtype will lead to inefficient binding, and possible underestimation of viral loads or outright false-negative results (116).

Rearrangements of the BKPyV NCCR can be evaluated through PCR and sequencing. Likewise, the amplification and sequencing of a fragment of the BKPyV VP1 gene is commonly used to type the 12 BKPyV subtypes, although neither the genotype nor specific NCCR changes have been associated with hemorrhagic cystitis, and thus, are primarily used for research purposes (117).

For immunohistological staining of biopsy material and urine cytology, a commercially available SV40 antibody (PAb 416, Merck Millipore, Germany), which cross-reacts with the BKPyV VP1 protein, is most commonly employed in diagnostic laboratories. Staining intensities can vary, but typically sufficient contrast remains to obtain a binary diagnosis (114).

Currently, no commercial serologic assays exist for BKPyV, however several research enzyme immunoassays (EIA) have been developed (29, 32).

JC Polyomavirus

JCPyV has been infrequently observed to be associated with PVAN and HC, and in these situations, diagnostic protocols similar to those for BKPyV can be followed. The predominant clinical complication of JCPyV infection, however, is PML and other less common neurological disorders. Originally, brain lesion biopsy was the diagnostic method of choice for suspected PML cases, and histopathological and IHC analysis of such biopsies is still considered the gold standard, with EM and *in situ* hybridization (ISH) providing further supporting evidence (118). JCPyV infection is typically seen in oligodendrocytes and astrocytes, with less frequent presence in neurons, and appears under EM as individual or crystalline arrays of 28 to 45 nm virions within the cell nucleus (68).

Typically IHC staining occurs within the cell nucleus, however some diffuse staining may also be present in the cytoplasm of cells with heavy nuclear staining (68). Commercial JCPyV antibodies exist and can be used for IHC, however it has been noted that the VP1 and LTAg proteins targeted by different antibodies may not be equally expressed, depending on the type of cell infected, and thus may affect the IHC sensitivity (68). Similar to BKPyV, commercially available JCPyV antibodies cross-react with SV40, with some also having been reported to lack sensitivity in formalin-fixed tissue (68). To mitigate the potential lowered antibody sensitivity and specificity, it has been proposed to use a dual-antibody approach targeting both the early and late proteins, which also has the additional benefit of informing on the stage of infection within the cell (68). ISH sensitivity can also be impacted by formalin-fixation (119), however its specificity may be higher than that of IHC due to its requirement for homology to large stretches of JCPyV genomic sequences (68). In cases of early or abortive infections, LTAg and STAg protein expression without the subsequent viral genome replication would occur. Consequently, in such situations, ISH protocols would have much less template to bind when compared to IHC and may explain the observed lowered ISH sensitivity when compared against IHC (120).

While brain biopsies may be the gold standard for PML diagnosis, they are also highly invasive and may be impossible to collect if the lesion is located in a difficult to reach portion of the brain. Because of these limitations, diagnosis which incorporates radiography, clinical presentation, and JCPyV PCR detection in the CSF has been favored as an acceptable alternative (68, 121). Using qPCR, higher viral loads in the CSF have been associated with poorer PML outcomes (122), however low JCPyV loads in CSF are more frequently found in MS patients with PML, as well as in HIV patients undergoing HAART (123). Both commercial and in-house published conventional, real-time, and quantitative PCR assays exist, but because of the potential for low copy numbers in CSF, it is critical that any assay chosen for testing at-risk patient groups has a high sensitivity. Using conventional PCR may only provide a sensitivity of 58% for detecting JCPyV in PML patient CSF (123), however, with highly sensitive real-time PCR methods, the figure can be improved to over 95% with detection limits of 10 viral copies/ml of CSF (124). Assay sensitivity can also be improved by increasing CSF NA extraction volume (e.g., 200 μl or greater), and eluting into smaller volumes (e.g., 50 μl). Particular attention should be paid to the collection quality of the CSF; a traumatic lumbar puncture or the presence of red blood cells in the CSF may contaminate an otherwise JCPyV negative CSF with viral particles if the patient in question happens to be viraemic. Collection quality is salient both for known PML patients, as the contamination can lead to overestimated viral loads, as well as in patients undergoing routine exclusion screening, which may lead to false-positive results due to the common presence of JCPyV even in healthy patients' blood (121). PCR can also be applied to brain biopsies, as it is more sensitive than ISH and IHC (120), however caution should be observed as JCPyV has been occasionally detected by PCR in the brain tissue of immunocompetent patients without PML as well (125).

Genetic recombination in the NCCR has been linked to increased virulence, and its presence in JCPyV genomes circulating in blood or urine may translate to a higher risk of disease. Screening for such NCCR changes has been typically accomplished through conventional PCR amplification, followed by Sanger sequencing (126), although recently a rapid method using duplex real-time PCR assay which can detect JCPyV at extremely low copy numbers, and discriminate between wild-type and rearranged (i.e., "virulent") NCCRs has been described (124). Other genetic changes in the agnoprotein and VP1 genes, linked to the development of neurological conditions distinct from PML, have been identified by standard PCR amplification and Sanger sequencing (89, 127).

The clinical utility of measuring the humoral response to JCPyV is still debated, with one emergent use being the prescreening and stratification of MS patients for risk of PML development prior to commencement of natalizumab therapy (128). Determination of cutoffs and the performance of the EIA assays have been highlighted as the primary issues when using such assays (118), with suggested solutions including a 2-step adsorption method to measure low-reactivity samples (128). Caution has to be observed with the serological results, as even with the improved 2-step protocol, some viruric patients would present as JCPyV seronegative (128). Complications which may contribute to misidentifying JCPyV positive patients as seronegative is the use of plasma-exchange samples which would reduce the antibody levels from blood (129), and the patient's use of mycophenolate mofetil, which will inhibit B-cell antibody production (130). To date, a number of research EIAs have been developed (33, 36), with only one commercial JCPyV EIA available (Stratify JCV; Focus Diagnostics, Cypress, CA).

For the remainder of the current human polyomavirus family, the clinical utility of diagnostic testing has not been established, in part due to the uncertainty surrounding many of their associations with disease. A number of detection methods have been described for these polyomaviruses, however most rely on PCR for specific virus identification.

WU and KI Polyomaviruses

These two viruses are commonly tested together as both are commonly detected in the respiratory tract. Detection methods primarily rely on PCR, either as conventional consensus assays (66), singleplex real-time PCRs (131), or multiplexed real-time PCR (132). Detections in the respiratory tract are not uncommon, but typically occur at high cycle threshold (Ct) values, which are indicative of low viral loads. Also, co-detection with other respiratory pathogens is very common, which confounds attempts to identify disease associations with these viruses. Other sample sites where

WUPyV and KIPyV have been detected include blood (88), lung tissue (86, 133), and feces (88). No commercial antibodies or serology assays were available at the time of writing.

MC Polyomavirus

PCR is the method of choice for the detection of MCPyV, with numerous qualitative and quantitative assays in existence (8, 88, 134). Particular caution needs to be taken with use of PCRs targeting MCPyV however, due to its propensity for environmental contamination (see below), and potential assay target disruption resulting from host genome integration events. MCPyV genomes are integrated within 80% of MCCs. These integration events result in seemingly random breakpoints across most of the virus genome, apart from the STAg and 5′ end of the LTAg genes, which are always conserved and appear critical for the ongoing oncogenic support of the tumor cell (135). Therefore, the use of PCR assays which target the conserved STAg region are recommended for use, particularly if MCC or other tumor tissues are being screened. Special attention needs to be paid to tumor tissue which has been formalin fixed, as the process will degrade and cross-link DNA to various degrees, depending on the method and length of fixation (136). Such degradation can lead to decreased viral load estimates or outright false-negative results in the case of samples containing low MCPyV copy numbers (137), and as such, DNA integrity and PCR inhibition should be monitored.

Detection of integration was originally achieved through traditional Southern blotting techniques using long radiolabeled MCPyV probes. This technique also allows the user to determine if the viral genome has been monoclonally or polyclonally integrated into the tumor genome (8). To identify the viral/host junction sequences, a modified Detection of Integrated Papillomavirus Sequences PCR (DIPS-PCR) can be used, which involves the restriction digestion of the virus and host genome, followed by ligation with adapters, adapter PCR, and subsequent sequencing (135). While this technique can also be used to detect integration events, it is arguably less objective than Southern blotting due to its semi-random nature and consequent difficulty in adequately controlling the reaction.

MCPyV can also been detected in blood (138), the respiratory tract (139), urine (140), healthy skin and eyebrow follicles (141), and many other internal organs (142). The significance of detection in most of these sites is unknown, apart from blood, in which monocytes have been suggested as reservoirs of MCPyV infection (138). Eyebrow follicles, along with skin, are presumed to be its primary tropism. MCPyV is shed from healthy skin and has been found to be present in 75% of the environmental swabs collected in one study (143). The prevalence of the MCPyV DNA on the person and in the environment must therefore be taken into account, and appropriate precautions taken against contamination during sample collection and processing, as well as assay setup if PCR will be used as the detection method.

Antibodies targeting the MCPyV LTAg and STAg have been used to detect viral protein expression in tumor tissue (144), which circumvents the risk of environmental contamination in DNA-based detection methods. The commercially available CM2B4 monoclonal antibody, originally created by Busam et al. (145), targets the non-truncated portion of the LTAg protein, and has been the most commonly used antibody for immunohistochemical (IHC) staining. In the majority of cases, CM2B4 staining correlates with MCPyV PCR positivity, with concordance ranging from 66.6% to 71.4% (145, 146). STAg has been found to be more commonly expressed than LTAg in MCCs however, and may therefore account for the observed CM2B4/PCR result discrepancy. Indeed, in some MCC cases, only the STAg was expressed (147), which suggests that a STAg-targeting antibody may be more sensitive for detecting viral presence in MCCs. The use of anti-MCPyV VP1 antibodies for IHC staining of tumor tissue is not recommended as the VP1 protein does not appear to be expressed in virus-positive MCCs (82).

TS Polyomavirus

TSPyV forms dense virion aggregates in the nuclei of inner hair root sheath cells of patients suffering from Trichodysplasia Spinulosa (TS) (148) when examined by EM. However the method is not sufficiently specific to conclusively identify TSPyV due to the similar morphological appearance of all polyomaviruses. Like many of the other newer human polyomaviruses, TSPyV is commonly detected using specific real-time PCR assays, although randomly-primed RCA followed by next generation sequencing has also been used (10). TSPyV can be detected in both lesions and nonlesion skin samples of TS patients, however the viral loads in lesions is multiple magnitudes higher (mean 1×10^6 copies/cell and >5 copies/cell, respectively). Detection from healthy subject skin is rare, and like the nonlesion skin, is of low viral load (149). Viral loads only marginally decrease after topical therapy and resolution of symptoms (10). For IHC staining, TSPyV is reportedly recognized not by the commonly used SV40 LTAg antibody (91) but rather by the polyomavirus Middle T antibody (150).

Human Polyomaviruses 6, 7, 9, 12, and MW, STL, NJ Polyomaviruses

For the remainder of the newly-discovered human polyomaviruses, real-time PCR has been the primary method of detection. Additionally, randomly-primed RCA has been used to amplify HPyV6 & 7, MWPyV, and STLPyV (9, 12, 13, 94) before detection by conventional PCR or full genome sequencing. ELISA assays have been developed for research (11, 13, 14, 40, 46), but HPyV9 has shown serological cross-reactivity with lymphotropic polyomavirus (LPyV) (11), a monkey-origin virus which has previously been detected by PCR and ELISA in human sera (151). HPyV9 and LPyV have close nucleotide sequence similarity, thus potentially leading to antigen cross-reactivity and suggesting that previous LPyV detection in the human population could be in fact off-target HPyV9 detection. Indeed, there is uncertainty over LPyV presence in the human population (152), which would suggest that HPyV9 ELISA cross-reactivity with LPyV should not be a major concern when examining human sera.

EM has been used to detect NJPyV as crystalloid virus assemblies localized within a muscle biopsy's atypical endothelial cell nuclei (15), while HPyV7 has been found in dense arrays within the keratinocytes of transplant patients suffering from unusual pruritic rashes (94).

BEST PRACTICE

As a general rule, irrespective of assay or clinical state being investigated, it is important to observe strict adherence to a consistent sample type (e.g., plasma or whole blood, but not both), particularly when using quantification methods to determine clinical management strategies.

BK Polyomavirus Associated Nephropathy and Hemorrhagic Cystitis

BK Polyomavirus Associated Nephropathy

PVAN is the major risk associated with BKPyV, as it progressively affects graft function and increases the risk of graft loss in <10% to 90% of patients (114). Because effective anti-polyomavirus drugs are not available, routine monitoring for BKPyV viraemia, together with appropriate clinical management has become a key recommendation (Fig. 3) (114). The detection of BKPyV helps to guide the reduction of immunosuppression, which allows for the restoration of BKPyV-specific cellular immune responses, curtailing of BKPyV replication in the graft, and clearance of viremia in 70% to 90% patients (153).

Although current screening algorithms focus on the prevention of BKPyV nephropathy via the detection of viremia and reduction in immunosuppression, the prevalence and consequences of long-term viruria have not been established. Recently, evidence emerged that prolonged BKPyV viruria may be an important contributor in the development of urothelial carcinoma in transplant recipients (64). This study highlighted the need for increased vigilance for bladder cancer detection in patients with persistent BKPyV in the presence of established risk factors, such as smoking and exposure to bladder toxins such as

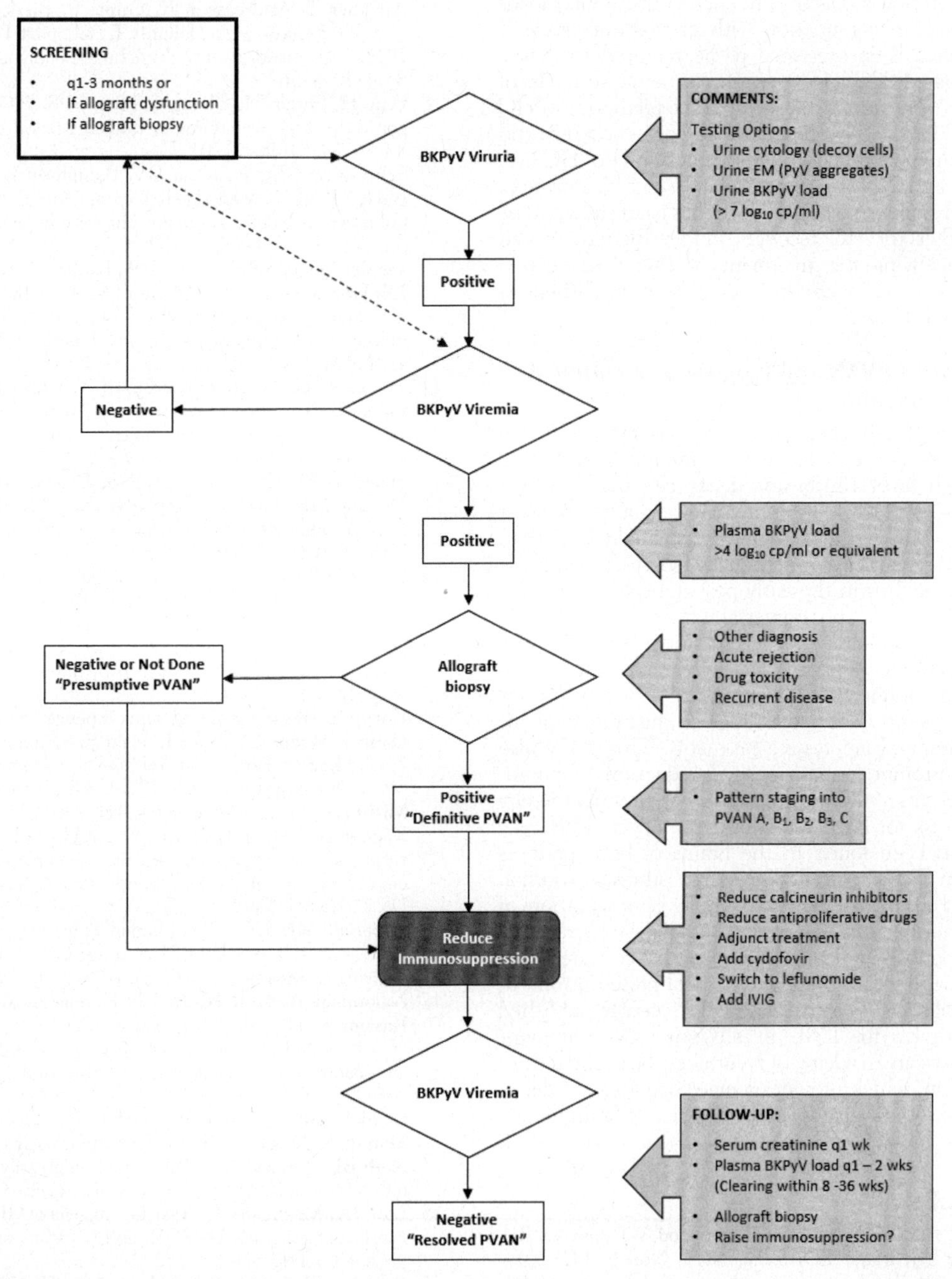

FIGURE 3 Proposed schematic for the screening and management of BKPyV replication and polyomavirus-associated nephropathy in kidney transplant patients (adapted from (114)).

cyclophosphamide or aromatic amines. However, the optimum management of BKPyV viruria is unknown and further study is required.

BK Polyomavirus Associated Hemorrhagic Cystitis

Examining several host- or HSCT-related characteristics, Cesaro et al. (154) showed that only BKPyV positivity in urine and/or blood was significantly associated with the development of HC. This supported other findings which showed that there was a quantitative correlation between BKPyV load and the development of post-engraftment HC, thus allowing this to be used as a guide for a pre-emptive treatment strategy (155).

Other causes of HC must be considered (155), however quantitative BKPyV markers can been used to identify the virus as the aetiological agent. Presence of urine viral loads >107 copies/ml, in conjunction with cystitis and grade II hematuria, have been suggested to be required to define BKPyV-associated HC (156), while urine viral loads of >9 × 106 copies/ml have been shown to be predictive of HC onset in children (154). Blood viral loads of >103 and >104 copies/ml have been found to predict onset of HC by a median of 10 and 17 days, respectively (157). Monitoring of plasma, rather than urine BKPyV load, has been shown to be a better marker for HC recovery (115). Evidence-based guidelines reviewing the treatment of BKPyV-associated hemorrhagic cystitis in children have been published by Harkensee et al. (157).

JC Polyomavirus PVAN and Progressive Multifocal Leukoencephalopathy

In rare cases, JCPyV has been identified as a cause of PVAN and may also be cleared after reduced immunosuppression. Although high level viruria and decoy cell shedding was common in these cases, JCPyV viremia was not a consistent feature (158). Also, JCPyV viruria was detected earlier than that observed for BKPyV. Although PCR screening for BKPyV polyomavirus in the early post-transplant period is now routine practice, no universal screening for JCPyV replication occurs. However, a recent study examining the role and clinical outcome of JCPyV viremia in renal transplant recipients found that screening for JCPyV in kidney transplant recipients was important, given the more frequent use of new immunosuppressive protocols. This study also showed that prompt recognition and reduction in immunosuppression can result in the clearance of viremia, thereby reducing the risk for PML and graft dysfunction (159).

JCPyV has been found in the brains of both patients suffering from PML, as well as those of otherwise normal individuals (160). Therefore, the simple demonstration of the virus, either in tissue or CSF, is insufficient to establish a definitive diagnosis of PML, and must be considered in conjunction with radiological and clinical presentation. To aid in the diagnosis, Berger et al. (121) recently published diagnostic criteria for PML in a Consensus Statement from the American Academy of Neurology, Neuroinfectious Disease Section, which provides a matrix to assist in determining diagnostic parameters in the absence of brain biopsy.

REFERENCES

1. **Johne R, Buck CB, Allander T, Atwood WJ, Garcea RL, Imperiale MJ, Major EO, Ramqvist T, Norkin LC.** 2011. Taxonomical developments in the family Polyomaviridae. *Arch Virol* **156:**1627–1634.
2. **Johne R, Müller H.** 2007. Polyomaviruses of birds: etiologic agents of inflammatory diseases in a tumor virus family. *J Virol* **81:**11554–11559.
3. **Imperiale MJ, Major EO.** 2007. Polyomaviruses, p. 2263–2298. *In* Knipe DM, Howley PM (ed.), *Fields Virology, 5th ed.* Lippincott Williams & Wilkins, Philadelphia, PA.
4. **Padgett BL, Walker DL, ZuRhein GM, Eckroade RJ, Dessel BH.** 1971. Cultivation of papova-like virus from human brain with progressive multifocal leucoencephalopathy. *Lancet* **1:** 1257–1260.
5. **Gardner SD, Field AM, Coleman DV, Hulme B.** 1971. New human papovavirus (B.K.) isolated from urine after renal transplantation. *Lancet* **1:**1253–1257.
6. **Gaynor AM, Nissen MD, Whiley DM, Mackay IM, Lambert SB, Wu G, Brennan DC, Storch GA, Sloots TP, Wang D.** 2007. Identification of a novel polyomavirus from patients with acute respiratory tract infections. *PloS Pathog* **3:**e64.
7. **Allander T, Andreasson K, Gupta S, Bjerkner A, Bogdanovic G, Persson MA, Dalianis T, Ramqvist T, Andersson B.** 2007. Identification of a third human polyomavirus. *J Virol* **81:**4130–4136.
8. **Feng H, Shuda M, Chang Y, Moore PS.** 2008. Clonal integration of a polyomavirus in human Merkel cell carcinoma. *Science* **319:**1096–1100.
9. **Schowalter RM, Pastrana DV, Pumphrey KA, Moyer AL, Buck CB.** 2010. Merkel cell polyomavirus and two previously unknown polyomaviruses are chronically shed from human skin. *Cell Host Microbe* **7:**509–515.
10. **van der Meijden E, Janssens RW, Lauber C, Bouwes Bavinck JN, Gorbalenya AE, Feltkamp MCW.** 2010. Discovery of a new human polyomavirus associated with trichodysplasia spinulosa in an immunocompromised patient. *PloS Pathog* **6:** e1001024.
11. **Trusch F, Klein M, Finsterbusch T, Kühn J, Hofmann J, Ehlers B.** 2012. Seroprevalence of human polyomavirus 9 and cross-reactivity to African green monkey-derived lymphotropic polyomavirus. *J Gen Virol* **93:**698–705.
12. **Siebrasse EA, Reyes A, Lim ESS, Zhao G, Mkakosya RSS, Manary MJJ, Gordon JII, Wang D.** 2012. Identification of MW polyomavirus, a novel polyomavirus in human stool. *J Virol* **86:**10321–10326.
13. **Lim ES, Reyes A, Antonio M, Saha D, Ikumapayi UN, Adeyemi M, Stine OC, Skelton R, Brennan DC, Mkakosya RS, Manary MJ, Gordon JI, Wang D.** 2013. Discovery of STL polyomavirus, a polyomavirus of ancestral recombinant origin that encodes a unique T antigen by alternative splicing. *Virology* **436:**295–303.
14. **Korup S, Rietscher J, Calvignac-Spencer S, Trusch F, Hofmann J, Moens U, Sauer I, Voigt S, Schmuck R, Ehlers B.** 2013. Identification of a novel human polyomavirus in organs of the gastrointestinal tract. *PloS ONE* **8:**e58021.
15. **Mishra N, Pereira M, Rhodes RH, An P, Pipas JM, Jain K, Kapoor A, Briese T, Faust PL, Lipkin WI.** 2014. Identification of a novel polyomavirus in a pancreatic transplant recipient with retinal blindness and vasculitic myopathy. *J Infect Dis* **210:**1595–1599.
16. **Imperiale MJ.** 2001. The Human Polyomaviruses: An Overview, p 53–71. *In* Khalilil K, Stoner GL (ed), *Human Polyomaviruses: Molecular and Clinical Perspective*. Wiley-Liss Inc.
17. **Nakanishi A, Itoh N, Li PP, Handa H, Liddington RC, Kasamatsu H.** 2007. Minor capsid proteins of simian virus 40 are dispensable for nucleocapsid assembly and cell entry but are required for nuclear entry of the viral genome. *J Virol* **81:**3778–3785.
18. **Elphick GF, Querbes W, Jordan JA, Gee GV, Eash S, Manley K, Dugan A, Stanifer M, Bhatnagar A, Kroeze WK, Roth BL, Atwood WJ.** 2004. The human polyomavirus, JCV, uses serotonin receptors to infect cells. *Science* **306:**1380–1383.
19. **Low JA, Magnuson B, Tsai B, Imperiale MJ.** 2006. Identification of gangliosides GD1b and GT1b as receptors for BK virus. *J Virol* **80:**1361–1366.
20. **Jiang M, Abend JR, Tsai B, Imperiale MJ.** 2009. Early events during BK virus entry and disassembly. *J Virol* **83:**1350–1358.

21. **Villarreal LP.** 1991. Relationship of eukaryotic DNA replication to committed gene expression: general theory for gene control. *Microbiol Rev* **55:**512–542.
22. **Khalili K, Sariyer IK, Safak M.** 2008. Small tumor antigen of polyomaviruses: role in viral life cycle and cell transformation. *J Cell Physiol* **215:**309–319.
23. **Farmerie WG, Folk WR.** 1984. Regulation of polyomavirus transcription by large tumor antigen. *Proc Natl Acad Sci USA* **81:**6919–6923.
24. **Khalili K, White MK, Sawa H, Nagashima K, Safak M.** 2005. The agnoprotein of polyomaviruses: a multifunctional auxiliary protein. *J Cell Physiol* **204:**1–7.
25. **Lagatie O, Tritsmans L, Stuyver LJ.** 2013. The miRNA world of polyomaviruses. *Virol J* **10:**268
26. **Taguchi F, Kajioka J, Miyamura T.** 1982. Prevalence rate and age of acquisition of antibodies against JC virus and BK virus in human sera. *Microbiol Immunol* **26:**1057–1064 .
27. **Knowles WA.** 2006. Discovery and epidemiology of the human polyomaviruses BK virus (BKV) and JC virus (JCV). *Adv Exp Med Biol* **577:**19–45.
28. **Brown P, Tsai T, Gajdusek DC.** 1975. Seroepidemiology of human papovaviruses. Discovery of virgin populations and some unusual patterns of antibody prevalence among remote peoples of the world. *Am J Epidemiol* **102:**331–340.
29. **Flaegstad T, Traavik T, Kristiansen BE.** 1986. Age-dependent prevalence of BK virus IgG and IgM antibodies measured by enzyme-linked immunosorbent assays (ELISA). *J Hyg (Lond)* **96:**523–528.
30. **Brown DW, Gardner SD, Gibson PE, Field AM.** 1984. BK virus specific IgM responses in cord sera, young children and healthy adults detected by RIA. *Arch Virol* **82:**149–160.
31. **Jung M, Krech U, Price PC, Pyndiah MN.** 1975. Evidence of chronic persistent infections with polyomaviruses (BK type) in renal transplant recipients. *Arch Virol* **47:**39–46.
32. **Gardner SD.** 1973. Prevalence in England of antibody to human polyomavirus (B.k.). *BMJ* **1:**77–78.
33. **Knowles WA, Pipkin P, Andrews N, Vyse A, Minor P, Brown DWG, Miller E.** 2003. Population-based study of antibody to the human polyomaviruses BKV and JCV and the simian polyomavirus SV40. *J Med Virol* **71:**115–123.
34. **Carter JJ, Madeleine MM, Wipf GC, Garcea RL, Pipkin PA, Minor PD, Galloway DA.** 2003. Lack of serologic evidence for prevalent simian virus 40 infection in humans. *J Natl Cancer Inst* **95:**1522–1530.
35. **Neel JV, Major EO, Awa AA, Glover T, Burgess A, Traub R, Curfman B, Satoh C.** 1996. Hypothesis: "Rogue cell"-type chromosomal damage in lymphocytes is associated with infection with the JC human polyoma virus and has implications for oncopenesis. *Proc Natl Acad Sci USA* **93:**2690–2695.
36. **Knowles WA, Gibson PE, Hand JF, Brown DW.** 1992. An M-antibody capture radioimmunoassay (MACRIA) for detection of JC virus-specific IgM. *J Virol Methods* **40:**95–105.
37. **Nguyen NL, Le BM, Wang D.** 2009. Serologic evidence of frequent human infection with WU and KI polyomaviruses. *Emerg Infect Dis* **15:**1199–1205.
38. **Neske F, Prifert C, Scheiner B, Ewald M, Schubert J, Opitz A, Weissbrich B.** 2010. High prevalence of antibodies against polyomavirus WU, polyomavirus KI, and human bocavirus in German blood donors. *BMC Infect Dis* **10:**215.
39. **Nicol JTJ, Robinot R, Carpentier A, Carandina G, Mazzoni E, Tognon M, Touzé A, Coursaget P.** 2013. Age-specific seroprevalences of merkel cell polyomavirus, human polyomaviruses 6, 7, and 9, and trichodysplasia spinulosa-associated polyomavirus. *Clin Vaccine Immunol* **20:**363–368.
40. **van der Meijden E, Bialasiewicz S, Rockett RJ, Tozer SJ, Sloots TP, Feltkamp MCW.** 2013. Different serologic behavior of MCPyV, TSPyV, HpyV6, HpyV7 and HpyV9 polyomaviruses found on the skin. *PloS ONE* **8:**e81078.
41. **Carter JJ, Paulson KG, Wipf GC, Miranda D, Madeleine MM, Johnson LG, Lemos BD, Lee S, Warcola AH, Iyer JG, Nghiem P, Galloway DA.** 2009. Association of Merkel cell polyomavirus-specific antibodies with Merkel cell carcinoma. *J Natl Cancer Inst* **101:**1510–1522.
42. **van der Meijden E, Kazem S, Burgers MM, Janssens R, Bouwes Bavinck JN, de Melker H, Feltkamp MCW.** 2011. Seroprevalence of trichodysplasia spinulosa-associated polyomavirus. *Emerg Infect Dis* **17:**1355–1363.
43. **Chen T, Mattila PS, Jartti T, Ruuskanen O, Söderlund-Venermo M, Hedman K.** 2011. Seroepidemiology of the newly found trichodysplasia spinulosa-associated polyomavirus. *J Infect Dis* **204:**1523–1526.
44. **Nicol JTJ, Touzé A, Robinot R, Arnold F, Mazzoni E, Tognon M, Coursaget P.** 2012. Seroprevalence and cross-reactivity of human polyomavirus 9. *Emerg Infect Dis* **18:** 1329–1332.
45. **Trusch F, Klein M, Finsterbusch T, Kühn J, Hofmann J, Ehlers B.** 2012. Seroprevalence of human polyomavirus 9 and cross-reactivity to African green monkey-derived lymphotropic polyomavirus. *J Gen Virol* **93:**698–705.
46. **Nicol JTJ, Leblond V, Arnold F, Guerra G, Mazzoni E, Tognon M, Coursaget P, Touzé A.** 2014. Seroprevalence of human Malawi polyomavirus. *J Clin Microbiol* **52:**321–323.
47. **Lim ES, Meinerz NM, Primi B, Wang D, Garcea RL.** 2014. Common exposure to STL polyomavirus during childhood. *Emerg Infect Dis* **20:**1559–1561.
48. **Martel-Jantin C, Pedergnana V, Nicol JTJ, Leblond V, Trégouët D-A, Tortevoye P, Plancoulaine S, Coursaget P, Touzé A, Abel L, Gessain A.** 2013. Merkel cell polyomavirus infection occurs during early childhood and is transmitted between siblings. *J Clin Virol* **58:**288–291.
49. **Kean JM, Rao S, Wang M, Garcea RL.** 2009. Seroepidemiology of human polyomaviruses. *PloS Pathog* **5:**e1000363.
50. **Kato K, Guo J, Taguchi F, Daimaru O, Tajima M, Haibara H, Matsuda J, Sumiya M, Yogo Y.** 1994. Phylogenetic comparison between archetypal and disease-associated JC virus isolates in Japan. *Jpn J Med Sci Biol* **47:**167–178.
51. **Zhong S, Randhawa PS, Ikegaya H, Chen Q, Zheng H-Y, Suzuki M, Takeuchi T, Shibuya A, Kitamura T, Yogo Y.** 2009. Distribution patterns of BK polyomavirus (BKV) subtypes and subgroups in American, European and Asian populations suggest co-migration of BKV and the human race. *J Gen Virol* **90:**144–152.
52. **Pastrana DV, Ray U, Magaldi TG, Schowalter RM, Çuburu N, Buck CB.** 2013. BK polyomavirus genotypes represent distinct serotypes with distinct entry tropism. *J Virol* **87:** 10105–10113.
53. **Walker DL, Padgett BL.** 1983. The epidemiology of human polyomaviruses. *Prog Clin Biol Res* **105:**99–106.
54. **Chang D, Liou ZM, Ou WC, Wang KZ, Wang M, Fung CY, Tsai RT.** 1996. Production of the antigen and the antibody of the JC virus major capsid protein VP1. *J Virol Methods* **59:**177–187.
55. **Agostini HT, Ryschkewitsch CF, Stoner GL.** 1998. JC virus Type 1 has multiple subtypes: three new complete genomes. *J Gen Virol* **79:**801–805.
56. **Bialasiewicz S, Rockett R, Whiley DW, Abed Y, Allander T, Binks M, Boivin G, Cheng AC, Chung J-Y, Ferguson PE, Gilroy NM, Leach AJ, Lindau C, Rossen JW, Sorrell TC, Nissen MD, Sloots TP.** 2010. Whole-genome characterization and genotyping of global WU polyomavirus strains. *J Virol* **84:**6229–6234.
57. **Sundsfjord A, Spein AR, Lucht E, Flaegstad T, Seternes OM, Traavik T.** 1994. Detection of BK virus DNA in nasopharyngeal aspirates from children with respiratory infections but not in saliva from immunodeficient and immunocompetent adult patients. *J Clin Microbiol* **32:**1390–1394.
58. **Goudsmit J, Wertheim-van Dillen P, van Strien A, van der Noordaa J.** 1982. The role of BK virus in acute respiratory tract disease and the presence of BKV DNA in tonsils. *J Med Virol* **10:**91–99.
59. **Kuypers DRJ.** 2012. Management of polyomavirus-associated nephropathy in renal transplant recipients. *Nat Rev Nephrol* **8:** 390–402.
60. **Polo C, Pérez JL, Mielnichuck A, Fedele CG, Niubò J, Tenorio A.** 2004. Prevalence and patterns of polyomavirus

urinary excretion in immunocompetent adults and children. *Clin Microbiol Infect* **10:**640–644.

61. **Gosert R, Rinaldo CH, Funk GA, Egli A, Ramos E, Drachenberg CB, Hirsch HH.** 2008. Polyomavirus BK with rearranged noncoding control region emerge in vivo in renal transplant patients and increase viral replication and cytopathology. *J Exp Med* **205:**841–852.
62. **Hirsch HH, Snydman DR.** 2005. BK virus: opportunity makes a pathogen. *Clin Infect Dis* **41:**354–360.
63. **Dalianis T, Hirsch HH.** 2013. Human polyomaviruses in disease and cancer. *Virology* **437:**63–72.
64. **Bialasiewicz S, Cho Y, Rockett R, Preston J, Wood S, Fleming S, Shepherd B, Barraclough K, Sloots TP, Isbel N.** 2013. Association of micropapillary urothelial carcinoma of the bladder and BK viruria in kidney transplant recipients. *Transpl Infect Dis* **15:**283–289.
65. **Abend JR, Jiang M, Imperiale MJ.** 2009. BK virus and human cancer: innocent until proven guilty. *Semin Cancer Biol* **19:**252–260.
66. **Giraud G, Ramqvist T, Ragnarsson-Olding B, Dalianis T.** 2008. DNA from BK virus and JC virus and from KI, WU, and MC polyomaviruses as well as from simian virus 40 is not detected in non-UV-light-associated primary malignant melanomas of mucous membranes. *J Clin Microbiol* **46:**3595–3598.
67. **Bouvard V, Baan RA, Grosse Y, Lauby-Secretan B, El Ghissassi F, Benbrahim-Tallaa L, Guha N, Straif K, WHO International Agency for Research on Cancer Monograph Working Group.** 2012. Carcinogenicity of malaria and of some polyomaviruses. *Lancet Oncol* **13:**339–340.
68. **Gheuens S, Wüthrich C, Koralnik IJ.** 2013. Progressive multifocal leukoencephalopathy: why gray and white matter. *Annu Rev Pathol* **8:**189–215.
69. **Khanna N, Elzi L, Mueller NJ, Garzoni C, Cavassini M, Fux CA, Vernazza P, Bernasconi E, Battegay M, Hirsch HH, Swiss HIV Cohort Study.** 2009. Incidence and outcome of progressive multifocal leukoencephalopathy over 20 years of the Swiss HIV Cohort Study. *Clin Infect Dis* **48:**1459–1466.
70. **Hoepner R, Faissner S, Salmen A, Gold R, Chan A.** 2014. Efficacy and side effects of natalizumab therapy in patients with multiple sclerosis. *J Cent Nerv Syst Dis* **6:**41–49.
71. **Dang X, Koralnik IJ.** 2006. A granule cell neuron-associated JC virus variant has a unique deletion in the VP1 gene. *J Gen Virol* **87:**2533–2537.
72. **Delbue S, Ferraresso M, Ghio L, Carloni C, Carluccio S, Belingheri M, Edefonti A, Ferrante P.** 2013. A review on JC virus infection in kidney transplant recipients. *Clin Dev Immunol* **2013:**1.
73. **Dalianis T, Garcea RL.** 2009. Welcome to the Polyomaviridae. *Semin Cancer Biol* **19:**209–210.
74. **Wattier RL, Vázquez M, Weibel C, Shapiro ED, Ferguson D, Landry ML, Kahn JS.** 2008. Role of human polyomaviruses in respiratory tract disease in young children. *Emerg Infect Dis* **14:**1766–1768.
75. **Mourez T, Bergeron A, Ribaud P, Scieux C, de Latour RP, Tazi A, Socié G, Simon F, LeGoff J.** 2009. Polyomaviruses KI and WU in immunocompromised patients with respiratory disease. *Emerg Infect Dis* **15:**107–109.
76. **Norja P, Ubillos I, Templeton K, Simmonds P.** 2007. No evidence for an association between infections with WU and KI polyomaviruses and respiratory disease. *J Clin Virol* **40:**307–311.
77. **Koljonen V, Kukko H, Pukkala E, Sankila R, Böhling T, Tukiainen E, Sihto H, Joensuu H.** 2009. Chronic lymphocytic leukaemia patients have a high risk of Merkel-cell polyomavirus DNA-positive Merkel-cell carcinoma. *Br J Cancer* **101:**1444–1447.
78. **Ohnishi Y, Murakami S, Ohtsuka H, Miyauchi S, Shinmori H, Hashimoto K.** 1997. Merkel cell carcinoma and multiple Bowen's disease: incidental association or possible relationship to inorganic arsenic exposure? *J Dermatol* **24:**310–316.
79. **Sahi H, Koljonen V, Böhling T, Neuvonen PJ, Vainio H, Lamminpää A, Kyyrönen P, Pukkala E.** 2012. Increased incidence of Merkel cell carcinoma among younger statin users. *Cancer Epidemiol* **36:**421–424.
80. **Vilkin A, Ronen Z, Levi Z, Morgenstern S, Halpern M, Niv Y.** 2012. Presence of JC virus DNA in the tumor tissue and normal mucosa of patients with sporadic colorectal cancer (CRC) or with positive family history and Bethesda criteria. *Dig Dis Sci* **57:**79–84.
81. **Poulin DL, DeCaprio JA.** 2006. Is there a role for SV40 in human cancer? *J Clin Oncol* **24:**4356–4365.
82. **Pastrana DV, Tolstov YL, Becker JC, Moore PS, Chang Y, Buck CB.** 2009. Quantitation of human seroresponsiveness to Merkel cell polyomavirus. *PloS Pathog* **5:**e1000578.
83. **Houben R, Shuda M, Weinkam R, Schrama D, Feng H, Chang Y, Moore PS, Becker JC.** 2010. Merkel cell polyomavirus-infected Merkel cell carcinoma cells require expression of viral T antigens. *J Virol* **84:**7064–7072.
84. **Touzé A, Le Bidre E, Laude H, Fleury MJJ, Cazal R, Arnold F, Carlotti A, Maubec E, Aubin F, Avril M-F, Rozenberg F, Tognon M, Maruani A, Guyetant S, Lorette G, Coursaget P.** 2011. High levels of antibodies against merkel cell polyomavirus identify a subset of patients with merkel cell carcinoma with better clinical outcome. *J Clin Oncol* **29:**1612–1619.
85. **van der Zalm MM, Rossen JW, van Ewijk BE, Wilbrink B, van Esch PC, Wolfs TF, van der Ent CK.** 2008. Prevalence and pathogenicity of WU and KI polyomaviruses in children, the Netherlands. *Emerg Infect Dis* **14:**1787–1789.
86. **Siebrasse E, Nguyen NL, Smith C, Simmonds P, Wang D.** 2014. Immunohistochemical detection of KI polyomavirus in lung and spleen. *Virology* **468-470C:**178–184.
87. **Rockett RJ, Bialasiewicz S, Mhango L, Gaydon J, Holding R, Whiley DM, Lambert SB, Ware RS, Nissen MD, Grimwood K, Sloots TP.** 2015. Acquisition of human polyomaviruses in the first 18 months of life. *Emerg Infect Dis* **21:**365–367.
88. **Bialasiewicz S, Whiley DM, Lambert SB, Nissen MD, Sloots TP.** 2009. Detection of BK, JC, WU, or KI polyomaviruses in faecal, urine, blood, cerebrospinal fluid and respiratory samples. *J Clin Virol* **45:**249–254.
89. **Dang X, Vidal JE, Oliveira AC, Simpson DM, Morgello S, Hecht JH, Ngo LH, Koralnik IJ.** 2012. JC virus granule cell neuronopathy is associated with VP1 C terminus mutants. *J Gen Virol* **93:**175–183.
90. **Babakir-Mina M, Ciccozzi M, Campitelli L, Aquaro S, Lo Coco A, Perno CF, Ciotti M.** 2009. Identification of the novel KI Polyomavirus in paranasal and lung tissues. *J Med Virol* **81:**558–561.
91. **Fischer MK, Kao GF, Nguyen HP, Drachenberg CB, Rady PL, Tyring SK, Gaspari AA.** 2012. Specific detection of trichodysplasia spinulosa–associated polyomavirus DNA in skin and renal allograft tissues in a patient with trichodysplasia spinulosa. *Arch Dermatol* **148:**726–733.
92. **Scuda N, Hofmann J, Calvignac-Spencer S, Ruprecht K, Liman P, Kühn J, Hengel H, Ehlers B.** 2011. A novel human polyomavirus closely related to the African green monkey-derived lymphotropic polyomavirus. *J Virol* **85:**4586–4590.
93. **Sauvage V, Foulongne V, Cheval J, Ar Gouilh M, Pariente K, Dereure O, Manuguerra JC, Richardson J, Lecuit M, Burguière A, Caro V, Eloit M.** 2011. Human polyomavirus related to African green monkey lymphotropic polyomavirus. *Emerg Infect Dis* **17:**1364–1370.
94. **Ho J, Jedrych JJ, Feng H, Natalie AA, Grandinetti L, Mirvish E, Crespo MM, Yadav D, Fasanella KE, Kuan S, Pastrana DV, Buck CB, Shuda Y, Moore PS, Chang Y.** 2014. Human polyomavirus 7-associated pruritic rash and viremia in transplant recipients. *J. Infect. Dis.* Sep 17. Pii: jiu524. Preprint:1–16.
95. **Rennspiess D, Pujari S, Keijzers M, Abdul-Hamid MA, Hochstenbag M, Dingemans A-M, Kurz AK, Speel E-J, Haugg A, Pastrana DV, Buck CB, De Baets MH, Zur Hausen A.** 2015. Detection of human polyomavirus 7 in human thymic epithelial tumors. *J Thorac Oncol* **10:**360–366.

96. **Buck CB, Phan GQ, Raiji MT, Murphy PM, McDermott DH, McBride AA.** 2012. Complete genome sequence of a tenth human polyomavirus. *J Virol* **86:**10887.
97. **Yu G, Greninger AL, Isa P, Phan TG, Martínez MA, de la Luz Sanchez M, Contreras JF, Santos-Preciado JI, Parsonnet J, Miller S, DeRisi JL, Delwart E, Arias CF, Chiu CY.** 2012. Discovery of a novel polyomavirus in acute diarrheal samples from children. *PloS ONE* **7:**e49449.
98. **Wu S-W, Chang H-R, Lian J-D.** 2009. The effect of low-dose cidofovir on the long-term outcome of polyomavirus-associated nephropathy in renal transplant recipients. *Nephrol Dial Transplant* **24:**1034–1038.
99. **Palazzo E, Yahia SA.** 2012. Progressive multifocal leukoencephalopathy in autoimmune diseases. *Joint Bone Spine* **79:**351–355.
100. **Yatawara A, Gaidos G, Rupasinghe CN, O'Hara BA, Pellegrini M, Atwood WJ, Mierke DF.** 2015. Small-molecule inhibitors of JC polyomavirus infection. *J Pept Sci* **21:**236–242.
101. **Bofill-Mas S, Rodriguez-Manzano J, Calgua B, Carratala A, Girones R.** 2010. Newly described human polyomaviruses Merkel cell, KI and WU are present in urban sewage and may represent potential environmental contaminants. *Virol J* **7:**141.
102. **Sauerbrei A, Wutzler P.** 2009. Testing thermal resistance of viruses. *Arch Virol* **154:**115–119.
103. **Li X, Yang Q, Bai J, Yang Y, Zhong L, Wang Y.** 2015. Identification of optimal reference genes for quantitative PCR studies on human mesenchymal stem cells. *Mol Med Rep*.**11:**1304-1311
104. **Alsaleh AN, Whiley DM, Bialasiewicz S, Lambert SB, Ware RS, Nissen MD, Sloots TP, Grimwood K.** 2014. Nasal swab samples and real-time polymerase chain reaction assays in community-based, longitudinal studies of respiratory viruses: the importance of sample integrity and quality control. *BMC Infect Dis* **14:**15.
105. **Bialasiewicz S, Whiley DM, Nissen MD, Sloots TP.** 2007. Impact of competitive inhibition and sequence variation upon the sensitivity of malaria PCR. *J Clin Microbiol* **45:**1621–1623.
106. **Rockett R, Barraclough KA, Isbel NM, Dudley KJ, Nissen MD, Sloots TP, Bialasiewicz S.** 2015. Specific rolling circle amplification of low-copy human polyomaviruses BKV, HpyV6, HpyV7, TSPyV, and STLPyV. *J. Virol. Methods* **215-216C:**17–21.
107. **Drachenberg CB, Beskow CO, Cangro CB, Bourquin PM, Simsir A, Fink J, Weir MR, Klassen DK, Bartlett ST, Papadimitriou JC.** 1999. Human polyoma virus in renal allograft biopsies: morphological findings and correlation with urine cytology. *Hum Pathol* **30:**970–977.
108. **Nickeleit V, Hirsch HH, Binet IF, Gudat F, Prince O, Dalquen P, Thiel G, Mihatsch MJ.** 1999. Polyomavirus infection of renal allograft recipients: from latent infection to manifest disease. *J Am Soc Nephrol* **10:**1080–1089.
109. **Fogazzi GB, Cantú M, Saglimbeni L.** 2001. 'Decoy cells' in the urine due to polyomavirus BK infection: easily seen by phase-contrast microscopy. *Nephrol Dial Transplant* **16:**1496–1498.
110. **Ranzi AD, Prolla JC, Keitel E, Brackmann R, Kist R, dos Santos G, Bica CG.** 2012. The role of urine cytology for 'decoy cells' as a screening tool in renal transplant recipients. *Acta Cytol* **56:**543–547.
111. **Viscount HB, Eid AJ, Espy MJ, Griffin MD, Thomsen KM, Harmsen WS, Razonable RR, Smith TF.** 2007. Polyomavirus polymerase chain reaction as a surrogate marker of polyomavirus-associated nephropathy. *Transplantation* **84:**340–345.
112. **Bedi A, Miller CB, Hanson JL, Goodman S, Ambinder RF, Charache P, Arthur RR, Jones RJ.** 1995. Association of BK virus with failure of prophylaxis against hemorrhagic cystitis following bone marrow transplantation. *J Clin Oncol* **13:**1103–1109.
113. **Pinto GG, Poloni JAT, Carneiro LC, Baethgen LF, Barth AL, Pasqualotto AC.** 2013. Evaluation of different urine protocols and DNA extraction methods for quantitative detection of BK viruria in kidney transplant patients. *J Virol Methods* **188:**94–96.
114. **Hirsch HH, Randhawa P, AST Infectious Diseases Community of Practice.** 2013. BK polyomavirus in solid organ transplantation. *Am J Transplant* **13**(Suppl 4)**:**179–188.
115. **Drew RJ, Walsh A, Ní Laoi B, Conneally E, Crowley B.** 2013. BK virus (BKV) plasma dynamics in patients with BKV-associated hemorrhagic cystitis following allogeneic stem cell transplantation. *Transpl Infect Dis* **15:**276–282.
116. **Randhawa P, Kant J, Shapiro R, Tan H, Basu A, Luo C.** 2011. The impact of genomic sequence variability on quantitative polymerase chain reaction assays for the diagnosis of polyomavirus BK infection. *J Clin Microbiol* **49:**4072–4076.
117. **Carr MJ, McCormack GP, Mutton KJ, Crowley B.** 2006. Unique BK virus non-coding control region (NCCR) variants in hematopoietic stem cell transplant recipients with and without hemorrhagic cystitis. *J Med Virol* **78:**485–493.
118. **Hirsch HH, Kardas P, Kranz D, Leboeuf C.** 2013. The human JC polyomavirus (JCPyV): virological background and clinical implications. *APMIS* **121:**685–727.
119. **Wüthrich C, Kesari S, Kim W-K, Williams K, Gelman R, Elmeric D, De Girolami U, Joseph JT, Hedley-Whyte T, Koralnik IJ.** 2006. Characterization of lymphocytic infiltrates in progressive multifocal leukoencephalopathy: co-localization of CD8(+) T cells with JCV-infected glial cells. *J Neurovirol* **12:**116–128.
120. **Muñoz-Mármol AM, Mola G, Fernández-Vasalo A, Vela E, Mate JL, Ariza A.** 2004. JC virus early protein detection by immunohistochemistry in progressive multifocal leukoencephalopathy: a comparative study with in situ hybridization and polymerase chain reaction. *J Neuropathol Exp Neurol* **63:**1124–1130.
121. **Berger JR, Aksamit AJ, Clifford DB, Davis L, Koralnik IJ, Sejvar JJ, Bartt R, Major EO, Nath A.** 2013. PML diagnostic criteria: consensus statement from the AAN Neuroinfectious Disease Section. *Neurology* **80:**1430–1438.
122. **Delbue S, Elia F, Carloni C, Tavazzi E, Marchioni E, Carluccio S, Signorini L, Novati S, Maserati R, Ferrante P.** 2012. JC virus load in cerebrospinal fluid and transcriptional control region rearrangements may predict the clinical course of progressive multifocal leukoencephalopathy. *J Cell Physiol* **227:**3511–3517.
123. **Marzocchetti A, Di Giambenedetto S, Cingolani A, Ammassari A, Cauda R, De Luca A.** 2005. Reduced rate of diagnostic positive detection of JC virus DNA in cerebrospinal fluid in cases of suspected progressive multifocal leukoencephalopathy in the era of potent antiretroviral therapy. *J Clin Microbiol* **43:**4175–4177.
124. **Ryschkewitsch CF, Jensen PN, Major EO.** 2013. Multiplex qPCR assay for ultra sensitive detection of JCV DNA with simultaneous identification of genotypes that discriminates non-virulent from virulent variants. *J Clin Virol* **57:**243–248.
125. **Ferrante P, Caldarelli-Stefano R, Omodeo-Zorini E, Vago L, Boldorini R, Costanzi G.** 1995. PCR detection of JC virus DNA in brain tissue from patients with and without progressive multifocal leukoencephalopathy. *J Med Virol* **47:**219–225.
126. **Delbue S, Sotgiu G, Fumagalli D, Valli M, Borghi E, Mancuso R, Marchioni E, Maserati R, Ferrante P.** 2005. A case of a progressive multifocal leukoencephalopathy patient with four different JC virus transcriptional control region rearrangements in cerebrospinal fluid, blood, serum, and urine. *J Neurovirol* **11:**51–57.
127. **Dang X, Wüthrich C, Gordon J, Sawa H, Koralnik IJ.** 2012. JC virus encephalopathy is associated with a novel agnoprotein-deletion JCV variant. *PloS ONE* **7:**e35793.
128. **Gorelik L, Lerner M, Bixler S, Crossman M, Schlain B, Simon K, Pace A, Cheung A, Chen LL, Berman M, Zein F, Wilson E, Yednock T, Sandrock A, Goelz SE, Subramanyam M.** 2010. Anti-JC virus antibodies: implications for PML risk stratification. *Ann Neurol* **68:**295–303.
129. **Goelz SE, Gorelik L, Subramanyam M.** 2011. Assay design and sample collection can affect anti-John Cunningham virus

antibody detection. *Ann Neurol* **69:**429–430, author reply 430–431.

130. **Smith KG, Isbel NM, Catton MG, Leydon JA, Becker GJ, Walker RG.** 1998. Suppression of the humoral immune response by mycophenolate mofetil. *Nephrol Dial Transplant* **13:**160–164.
131. **Bialasiewicz S, Whiley DM, Lambert SB, Gould A, Nissen MD, Sloots TP.** 2007. Development and evaluation of real-time PCR assays for the detection of the newly identified KI and WU polyomaviruses. *J Clin Virol* **40:**9–14.
132. **Antonsson A, Bialasiewicz S, Rockett RJ, Jacob K, Bennett IC, Sloots TP.** 2012. Exploring the prevalence of ten polyomaviruses and two herpes viruses in breast cancer. *PloS ONE* **7:**e39842.
133. **Teramoto S, Kaiho M, Takano Y, Endo R, Kikuta H, Sawa H, Ariga T, Ishiguro N.** 2011. Detection of KI polyomavirus and WU polyomavirus DNA by real-time polymerase chain reaction in nasopharyngeal swabs and in normal lung and lung adenocarcinoma tissues. *Microbiol Immunol* **55:**525–530.
134. **Foulongne V, Dereure O, Kluger N, Molès JP, Guillot B, Segondy M.** 2010. Merkel cell polyomavirus DNA detection in lesional and nonlesional skin from patients with Merkel cell carcinoma or other skin diseases. *Br J Dermatol* **162:** 59–63.
135. **Sastre-Garau X, Peter M, Avril M-F, Laude H, Couturier J, Rozenberg F, Almeida A, Boitier F, Carlotti A, Couturaud B, Dupin N.** 2009. Merkel cell carcinoma of the skin: pathological and molecular evidence for a causative role of MCV in oncogenesis. *J Pathol* **218:**48–56.
136. **Kassem A, Schöpflin A, Diaz C, Weyers W, Stickeler E, Werner M, Zur Hausen A.** 2008. Frequent detection of Merkel cell polyomavirus in human Merkel cell carcinomas and identification of a unique deletion in the VP1 gene. *Cancer Res* **68:**5009–5013.
137. **Pastrana DV, Wieland U, Silling S, Buck CB, Pfister H.** 2012. Positive correlation between Merkel cell polyomavirus viral load and capsid-specific antibody titer. *Med Microbiol Immunol (Berl)* **201:**17–23.
138. **Mertz KD, Junt T, Schmid M, Pfaltz M, Kempf W.** 2010. Inflammatory monocytes are a reservoir for Merkel cell polyomavirus. *J Invest Dermatol* **130:**1146–1151.
139. **Bialasiewicz S, Lambert SB, Whiley DM, Nissen MD, Sloots TP.** 2009. Merkel cell polyomavirus DNA in respiratory specimens from children and adults. *Emerg Infect Dis* **15:**492–494.
140. **Signorini L, Belingheri M, Ambrogi F, Pagani E, Binda S, Ticozzi R, Ferraresso M, Ghio L, Giacon B, Ferrante P, Delbue S.** 2014. High frequency of Merkel cell polyomavirus DNA in the urine of kidney transplant recipients and healthy controls. *J Clin Virol* **61:**565–570.
141. **Wieland U, Mauch C, Kreuter A, Krieg T, Pfister H.** 2009. Merkel cell polyomavirus DNA in persons without merkel cell carcinoma. *Emerg Infect Dis* **15:**1496–1498.
142. **Matsushita M, Kuwamoto S, Iwasaki T, Higaki-Mori H, Yashima S, Kato M, Murakami I, Horie Y, Kitamura Y, Hayashi K.** 2013. Detection of Merkel cell polyomavirus in the human tissues from 41 Japanese autopsy cases using polymerase chain reaction. *Intervirology* **56:**1–5.
143. **Foulongne V, Courgnaud V, Champeau W, Segondy M.** 2011. Detection of Merkel cell polyomavirus on environmental surfaces. *J Med Virol* **83:**1435–1439.
144. **Rodig SJ, Cheng J, Wardzala J, DoRosario A, Scanlon JJ, Laga AC, Martinez-Fernandez A, Barletta JA, Bellizzi AM, Sadasivam S, Holloway DT, Cooper DJ, Kupper TS, Wang LC, DeCaprio JA.** 2012. Improved detection suggests all Merkel cell carcinomas harbor Merkel polyomavirus. *J Clin Invest* **122:**4645–4653.
145. **Busam KJ, Jungbluth AA, Rekthman N, Coit D, Pulitzer M, Bini J, Arora R, Hanson NC, Tassello JA, Frosina D, Moore P, Chang Y.** 2009. Merkel cell polyomavirus expression in merkel cell carcinomas and its absence in combined tumors and pulmonary neuroendocrine carcinomas. *Am J Surg Pathol* **33:**1378–1385.
146. **Shuda M, Arora R, Kwun HJ, Feng H, Sarid R, Fernández-Figueras M-T, Tolstov Y, Gjoerup O, Mansukhani MM, Swerdlow SH, Chaudhary PM, Kirkwood JM, Nalesnik MA, Kant JA, Weiss LM, Moore PS, Chang Y.** 2009. Human Merkel cell polyomavirus infection I. MCV T antigen expression in Merkel cell carcinoma, lymphoid tissues and lymphoid tumors. *Int J Cancer* **125:**1243–1249.
147. **Shuda M, Kwun HJ, Feng H, Chang Y, Moore PS.** 2011. Human Merkel cell polyomavirus small T antigen is an oncoprotein targeting the 4E-BP1 translation regulator. *J Clin Invest* **121:**3623–3634.
148. **Elaba Z, Hughey L, Isayeva T, Weeks B, Solovan C, Solovastru L, Andea A.** 2012. Ultrastructural and molecular confirmation of the trichodysplasia spinulosa-associated polyomavirus in biopsies of patients with trichodysplasia spinulosa. *J Cutan Pathol* **39:**1004–1009.
149. **Kazem S, van der Meijden E, Kooijman S, Rosenberg AS, Hughey LC, Browning JC, Sadler G, Busam K, Pope E, Benoit T, Fleckman P, de Vries E, Eekhof JA, Feltkamp MC.** 2012. Trichodysplasia spinulosa is characterized by active polyomavirus infection. *J Clin Virol* **53:**225–230.
150. **Wanat KA, Holler PD, Dentchev T, Simbiri K, Robertson E, Seykora JT, Rosenbach M.** 2012. Viral-associated trichodysplasia: characterization of a novel polyomavirus infection with therapeutic insights. *Arch Dermatol* **148:**219–223.
151. **Delbue S, Tremolada S, Elia F, Carloni C, Amico S, Tavazzi E, Marchioni E, Novati S, Maserati R, Ferrante P.** 2010. Lymphotropic polyomavirus is detected in peripheral blood from immunocompromised and healthy subjects. *J Clin Virol* **47:**156–160.
152. **Costa C, Bergallo M, Terlizzi ME, Cavallo GP, Cavalla P, Cavallo R.** 2011. Lack of detection of lymphotropic polyomavirus DNA in different clinical specimens. *J Clin Virol* **51:**148–149.
153. **Gardner SD, MacKenzie EF, Smith C, Porter AA.** 1984. Prospective study of the human polyomaviruses BK and JC and cytomegalovirus in renal transplant recipients. *J Clin Pathol* **37:**578–586.
154. **Cesaro S, Facchin C, Tridello G, Messina C, Calore E, Biasolo MA, Pillon M, Varotto S, Brugiolo A, Mengoli C, Palù G.** 2008. A prospective study of BK-virus-associated haemorrhagic cystitis in paediatric patients undergoing allogeneic haematopoietic stem cell transplantation. *Bone Marrow Transplant* **41:**363–370.
155. **Leung AYH, Yuen K-Y, Kwong Y-L.** 2005. Polyoma BK virus and haemorrhagic cystitis in haematopoietic stem cell transplantation: a changing paradigm. *Bone Marrow Transplant* **36:**929–937.
156. **Rinaldo CH, Tylden GD, Sharma BN.** 2013. The human polyomavirus BK (BKPyV): virological background and clinical implications. *APMIS* **121:**728–745.
157. **Harkensee C, Vasdev N, Gennery AR, Willetts IE, Taylor C.** 2008. Prevention and management of BK-virus associated haemorrhagic cystitis in children following haematopoietic stem cell transplantation—a systematic review and evidence-based guidance for clinical management. *Br J Haematol* **142:** 717–731.
158. **Drachenberg CB, Hirsch HH, Papadimitriou JC, Gosert R, Wali RK, Munivenkatappa R, Nogueira J, Cangro CB, Haririan A, Mendley S, Ramos E.** 2007. Polyomavirus BK versus JC replication and nephropathy in renal transplant recipients: a prospective evaluation. *Transplantation* **84:**323–330.
159. **Rao N, Schepetiuk S, Choudhry M, Juneja R, Passaris G, Higgins G, Barbara J.** 2012. JC viraemia in kidney transplant recipients: to act or not to act? *Clin Kidney J* **5:**471–473.
160. **White MK, Khalili K.** 2011. Pathogenesis of progressive multifocal leukoencephalopathy—revisited. *J Infect Dis* **203:** 578–586.

Parvoviruses

RICHARD S. BULLER

32

VIRAL CLASSIFICATION AND BIOLOGY

The parvoviruses are a large group of DNA viruses capable of infecting a wide variety of both invertebrate and vertebrate hosts, including humans and their companion animals. Depending on the virus and the immune status of the host, human infection can range from overt disease to persistent asymptomatic infection. Current parvovirus taxonomy dates from 2004; however, in the intervening 10 years, as new viruses have been identified and more taxonomic data have become available, the current taxonomy is now considered outdated. A new taxonomic classification of the family *Parvoviridae* has therefore been proposed and is currently under review and likely to be instituted in the near future (1). The proposed taxonomy, based on the amino acid sequence of the NS1 and viral capsid proteins, maintains the family *Parvoviridae* and subfamily *Parvovirinae* for the vertebrate parvoviruses, but adds three new genera and a nomenclature change that impact the taxonomy of the human parvoviruses. A member of a proposed genus must now have >30% amino acid sequence identity to other members of the same genus but <30% identity to members of other genera (1). Criteria for inclusion in a viral species have also changed under the proposed taxonomy, with members of the same species having >85% amino acid sequence identity to other members of the same species and >15% amino acid diversity from members of other species (1). The proposed genera and the human viruses they contain are *Bocaparvovirus*: human bocaviruses 1–4 (HBoV1-4); *Dependoparvovirus*: human adeno-associated viruses (AAVs); *Erythroparvovirus*: human parvovirus B19-related viruses; *Protoparvovirus*: human bufaviruses; and *Tetraparvovirus*: human parvovirus 4–related viruses (Par4) (1).

The parvoviruses are small (~25 nm) non-enveloped, icosahedral viruses. The name is derived from the Latin word "parvus," meaning "small," as befitted their appearance on electron microscopy. The parvovirus genome consists of a single-stranded DNA (ssDNA) molecule of approximately 5,000 bases. As would be expected of such a small genome, parvoviruses have a relatively simple replication strategy compared to viruses with larger more complex genomes. Unlike other more complex DNA viruses, parvoviruses are unable to stimulate the cellular division needed for replication and therefore require actively dividing cells to infect. Exceptions to this requirement are the AAVs, which are incapable of independent replication and require the aid of a helper virus, typically an adenovirus, or less commonly a herpesvirus, to complete their replicative cycle (2). The replicative strategy of the autonomous parvoviruses varies among the many viruses but generally follows a pattern of transcription of two non-structural (NS) proteins involved in control of replication from the left side of the genome and structural VP proteins from the right side of the genome (3). One unusual aspect of parvovirus replication is the ability of some parvoviruses, including human parvovirus B19, to encapsidate both positive and negative strands of the genomic ssDNA. Another noteworthy feature of these viruses is that although their genomes are replicated by host cell polymerases, they appear to have relatively high mutation and recombination rates, a characteristic that has been attributed to the single-stranded configuration of their genomes (3). This mutability may play a role in the appearance of parvoviruses with expanded host ranges. For example, canine parvovirus type 2, the causative agent of a serious infection of dogs, is believed to have arisen in the 1970s as a result of the mutation of feline panleukopenia virus via an intermediate wild carnivore host (4). Readers interested in a more detailed description of the molecular biology of parvoviruses are referred to the excellent chapter by Berns and Parrish in *Fields Virology* (3).

Among the parvoviruses infecting humans, the following is a brief listing of important molecular characteristics of the different genera.

Erythroparvovirus: Parvovirus B19 was originally considered to have a relatively conserved genome but it is now known to exist as 3 distinct genotypes whose nucleotide sequences vary by 5–20% (5). The genotypes have geographic associations and implications for laboratory diagnosis.

Bocaparvovirus: There are 4 species of human bocaviruses, referred to as HBoV1-4, that are characterized by a nucleotide sequence divergence of 10–30%, with HBoV2 being further classified into strains A and B based on additional sequence divergence (6). Sequencing data of Kapoor et al. (6) further suggest that because HBoV1 sequences are relatively homogenous, this respiratory agent likely recently evolved from the more divergent GI tract HBoV2-4 viruses. Analysis of bocavirus genomes has also indicated that HBoV3 is likely a recombinant between HBoV1 and HBoV4 (7).

doi:10.1128/9781555819156.ch32

Dependoparvovirus: The human AAVs have been characterized as serotypes, with up to 11 serotypes identified as well as numerous genetic variants being identified in human samples (2). If the new proposed taxonomy is adopted, the AAVs will then be classified according to the previously mentioned criteria for genera and species (1). In the absence of a helper virus, the AAVs have the unusual ability to establish persistent, non-productive infections of cells that can involve chromosomal integration of the viral genome or its existence as an episome (3).

Tetraparvovirus: The human PARV4 viruses exist as 3 genotypes that each have minimal sequence diversity, suggesting the viruses may have entered the human population relatively recently via zoonotic transmission (8). When it was first identified, what is now called PARV4 genotype 2 was thought to be a different virus and was originally named PARV5 (9).

Protoparvovirus: Due to their recent identification, there is relatively little information available on the human bufaviruses. Initially two genotypes of human bufavirus were described (10) with a third being described in a subsequent report (11).

EPIDEMIOLOGY

Erythroparvovirus: Parvovirus B19, genotype 1 causes human infections with a global distribution. It has been studied the most thoroughly of the three genotypes and therefore more is known about the epidemiology of this virus than is about the other 2 genotypes. The primary mode of transmission of this virus appears to be via the respiratory route. In a study that would likely be difficult to perform now, human volunteers who were inoculated intranasally with diluted human plasma from a blood donor developed typical parvoviral B19 disease (12). In our laboratory in St. Louis, we found supporting evidence for this when we detected parvovirus B19 DNA by PCR in upper respiratory samples of individuals with B19 viremia (unpublished data). An interesting epidemiological quirk was identified when, after the discovery that erythrocyte P antigen was a cellular receptor for B19 virus (13), it was reported that those rare individuals lacking the erythrocyte P antigen were resistant to infection by B19 virus (14). Serosurveys indicate that by the age of 20 about 50% of individuals will have serological evidence of infection (15) with this rate rising to over 80% in some groups of older adults (16). Sporadic infections can occur throughout the year but in temperate regions increased activity is noted in the spring, with larger outbreaks occurring at intervals of several years (15). A study of 169 children admitted to a hospital in Israel with signs and symptoms compatible with B19 infection, such as fever, rash, and hematologic abnormalities, found that 13% had laboratory findings indicative of acute B19 infection (17). A study of 633 children with sickle cell disease, a population at high risk for severe complications from B19 infection, reported that over a 5-year period, 110 of the study participants who had been without serological evidence of prior B19 infection at the beginning of the study went on to develop evidence of infection during the study period (18). This same study also reported a 56% secondary attack rate for study participants with siblings who had acute B19 infection, highlighting the transmissibility of the virus (18).

Other less common modes of transmission include transfusion of blood and blood products, transplantation, and vertical transmission from mother to fetus. A review by Parsyan and Candotti (19) goes into greater depth on the issues surrounding B19 and blood transfusion. Here we state only that the extremely high titers of B19 viremia that can occur during infection and the ability of the virus to persist for prolonged periods of time after acute infection (20), are likely responsible for the detection of B19 in blood and blood product donations, including units with very high titers, that have been reported (21). These characteristics, combined with the fact that the virus is non-enveloped and resistant to inactivation procedures used to prepare blood factor concentrates (22), point to a need to develop new inactivation methods. The transmission of parvovirus by blood products has led to the development of screening protocols, where in Europe a level of $<10^4$ international units (IU)/ml is required for plasma pools (15), with a similar rule issued as a Guidance by the FDA in the United States in 2009 (23).

B19 infections after solid organ transplantation are uncommon but well described (24). The fact that B19 has been reported to persist for long periods of time in tissues (20, 25) suggests that this may be a route of infection for some organ transplant recipients.

Finally, B19 can be vertically transferred from an infected pregnant woman to her fetus via intrauterine infection. There is an estimated 25% risk of transmission to the fetus at one week post-maternal infection when viral titers are at their peak (26). In a review by de Jong et al. (26), the authors noted data showing that, among European women, the range of seronegative susceptible women of childbearing age was 26–44%, resulting in a risk of acquiring B19 infection during pregnancy of 0.61–1.24%, although it was noted that this risk could rise significantly during epidemics.

B19 genotypes 2 and 3 are much less common than the globally distributed genotype 1 and appear to have narrow current geographic associations. Although the original identification of genotype 3 viruses (initially referred to as V9) found they composed 11% of the circulating B19 viruses in France (5), they are now known to be primarily associated with some West African countries, where they have been found to comprise 100% of the strains in specimens from Ghana (27). A lower prevalence of genotype 3 has been reported in Brazil (28) as well as a rare instance of detection in a plasma donor in the United States (29).

The persistence of B19 genomes in the tissues of infected individuals has allowed for the development of a "bioportfolio" whereby it is possible to associate the age of a previously infected individual with a B19 genotype (30). To do this, the authors analyzed over 500 tissue samples from Europeans born during the period 1913–2000. They reported that genotype 1 viruses were found in all age groups, while genotype 2 was detected only in tissues from those born prior to 1973, suggesting that during the mid-20th century both genotype 1 and 2 circulated in Europe, a state that was then followed by the disappearance of genotype 2. None of the tissues were found to contain genotype 3 in this study (30). A subsequent study of autopsy specimens from Scotland reported similar results, with genotype 2 positive specimens being obtained from significantly older individuals than those harboring genotype 1 (31). Hübschen et al. (32) studied 166 sera from B19-infected individuals from 11 countries in Asia, Europe, and Africa. Confirming earlier studies, Genotype 1 was most prevalent, being detected in 92% of the samples, with only 2 African countries having genotype 3 positive samples and none found to contain genotype 2 genomes (32).

Bocaparvovirus: The human bocaviruses, so named for their relationship to bovine and canine viruses in the same genus, represent the first human parvoviruses recognized since the discovery of B19 virus in 1975. Since its first identification in respiratory specimens from Sweden (33), what is now known as human bocavirus 1 (HBoV1), has been found in respiratory specimens from around the globe. Being the first identified human bocavirus, more is known about the epidemiology of this virus than the more recently identified types 2, 3, and 4. As a caveat for interpreting human bocavirus epidemiology, it should be noted that all members of the group appear to be shed asymptomatically. Additionally, the fact that they are frequently found in the presence of other potential pathogens makes association with a specific disease problematic at times. HBoV1 appears to infect primarily young children in the 6–24 months of age range (34), with the virus less frequently detected in adults (35). The virus appears to cause year-round infections with some increase noted during winter months (34, 36). Reviews have noted studies showing ranges of 2–19% (34) or up to 33% (36) for the prevalence of HBoV1 DNA detection in respiratory specimens from cases of pediatric respiratory illness. Serosurveys performed with HBoV1 antigens confirm the ubiquitous nature of infection around the globe, with the majority of children being seropositive by the time they reach school age (15, 34, 36). Although HBoV1 is most commonly detected in respiratory specimens, HBoV1 DNA has also been detected in serum (37–39) and stool specimens (39–41), with one review of the literature noting that different studies have reported detecting the virus in 0.8–9.1% of stool specimens obtained from individuals suffering from gastroenteritis (42). With respect to detection in stool specimens, it is not clear whether HBoV1 is replicating and potentially causing disease in the gastrointestinal tract or whether virus present in respiratory secretions is simply being swallowed and excreted in the stool.

Following the identification of HBoV1 in respiratory specimens in 2005, 3 additional HBoV viruses were discovered in human stool specimens. The identification of HBoV2 was first reported in the stool of a Pakistani child in 2009 (43); HBoV3 in an Australian pediatric stool specimen (44); and HBoV4 reported in pediatric stool specimens collected from Africa and the United States (6). With the exception of rare reports of HBoV2 in respiratory specimens (45, 46), HBoV2-4 appear to be exclusively detected in stool specimens. In a review of the bocavirus literature, Jartti et al. (34) reported that the enteric bocaviruses are globally distributed, and that of the 3 types, HBoV2 was the most commonly detected bocavirus in stool specimens from cases of pediatric gastroenteritis at 26%, with HBoV3 at 3%, and HBoV4 least common at 2%. The viruses were also noted to be detected in adult specimens as well, but all at levels less than 5%. The association of the enteric HBoVs with disease is unclear, with some studies supporting a small role for HBoV2 as a cause of gastroenteritis and little or no data indicating HBoV3 and 4 as agents of disease (34, 36). The fact that there are serological cross reactions between the human parvoviruses has made seroepidemiological studies difficult to interpret in the absence of any methods able to increase specificity (47). It is unknown how the viruses are spread, but it is likely via a fecal oral route, a hypothesis supported by the detection of bocaviruses in surface waters and sewage (34).

Dependoparvovirus: Little is known about how AAVs are acquired. This fact, along with a lack of a clear disease association for these viruses, has resulted in little epidemiological data on these viruses beyond seroprevalence studies. Serological evidence suggests that human infection with various human AAVs is common (2, 48, 49). Calcedo et al. (50) reported on the prevalence of neutralizing antibodies to various AAV types in 10 countries on 4 continents and found neutralizing anti-AAV2 antibodies to be most common with a range of from 60% of specimens in Africa to 20% in the United States. Lower seroprevalences were found for AAV1, 7, and 8. It is worth noting that this and other recent studies are being performed not for epidemiological purposes, but rather for identifying AAVs with low seroprevalences for neutralizing antibodies in the human population in an attempt to find good candidate AAVs for use as gene therapy vectors.

Tetraparvovirus: Since its identification in the plasma of a patient with an acute viral syndrome in 2005 (51), PARV4 has been detected in blood and blood products from locations around the globe. The lack of a disease association with PARV4 infection has limited epidemiological studies to seroprevalance surveys and viral DNA detection studies in various populations. Matthews et al. (8) have recently published an excellent review of PARV4 that includes a summary of epidemiological data for the virus. Summarizing their review, there are three PARV4 genotypes; genotypes 1 and 2 (initially called PARV5) have been detected globally while genotype 3 appears restricted to sub-Saharan Africa (8). The review references a number of studies that indicate an association of PARV4 infection with hepatitis C virus infection, HIV infection, hepatitis B virus infection, and i.v. drug usage (8), suggesting a blood-borne or bodily fluid route of infection. However, other studies are also cited that suggest that not all acquisitions of the virus occur by a blood-borne route and that the seropositivity rates for PARV4 can vary by geographic area (8). The detection of PARV4 in nasal and fecal samples (52), plasma pools (53), and clotting factor concentrates (54) suggests other routes for infection. Another route was suggested by a study of 3 cases from Taiwan that reported PARV4 can be vertically transmitted from pregnant women to the fetus (55). As with parvovirus B19, PARV4 appears to be able to persist and can be detected in the tissues of some individuals (56, 57), suggesting that persistence in the tissues of infected individuals may be a characteristic of some members of this group of viruses.

Protoparvoviruses: The human bufaviruses are the most recently identified human parvoviruses, being first detected in rotavirus-negative stools of children <5 years of age with gastroenteritis from Burkina Faso, where they were found in 4% of the specimens (10). In this same study, the authors noted that there appeared to be 2 genotypes of the virus. To understand if the virus is more widely distributed, the authors also tested specimens from Chile and Tunisia where 1 of 63 (1.6%) of the Tunisian specimens contained bufavirus DNA, suggesting a wider distribution than just Africa (10). A study of 393 watery-diarrhea stool specimens collected from children <5 years of age in Bhutan identified 3 (0.8%) stools containing a bufavirus that appears to belong to a third genotype (11). Of additional interest, 2 of the 3 specimens did not contain other potential GI pathogens, indicating that this bufavirus strain may be capable of causing diarrhea (11). Väisänen et al. (58) tested 629 stools from patients of all ages in Finland with gastroenteritis and found 7 (1.1%) positive for bufavirus DNA. In contrast to previous studies that detected the virus in pediatric specimens, the authors reported that all positive specimens came from adults (median age 53). Of the positive specimens, 1 also contained norovirus RNA while the remaining 6 contained no other viral or bacterial pathogens, suggesting a role for the virus as a cause of gastroenteritis. However, the authors noted that the low level of viral DNA in the specimens ($<3.2 \times 10^4$ copies/ml of fecal supernatant) suggested that other factors may have been responsible for the gastroenteritis (58). Using viral metagenomics, Smits et al. (59) found 10 bufavirus genotype reads in 1 of 27 (3.7%) stool specimens from cases of acute diarrhea in the Netherlands. These initial studies indicate that human bufaviruses are widely distributed to at least African and European locations and that they are found at low frequencies in the stools of both adults and children with gastroenteritis. The route of infection of these viruses is unknown, although a fecal-oral route would seem likely. We will have to await further studies that elucidate the global distribution of these viruses and whether they are pathogenic or merely non-pathogenic members of the human virome.

CLINICAL SIGNIFICANCE

Erythroparvovirus: Of the human parvoviruses, the B19 viruses currently present the greatest clinical significance. First identified in 1975 (60), it can cause diseases ranging from acute, benign, febrile rash illnesses, to serious life-threatening conditions, with the outcome of infection determined by host factors as discussed below.

Infections in the immunologically and hematologically normal host: parvovirus B19 is the cause of erythema infectiosum (EI), sometimes also referred to as "fifth disease" after an outdated system of numbering febrile rash diseases of childhood. Although non-immune adults are also at risk, the disease is most common in children of school age and is characterized by fever and a "slapped cheek" appearing rash on the face. The disease is self-limited and considered benign with 25–50% of infected individuals being asymptomatic (61). Occasionally children will develop joint pain from B19 infection but polyarthropathy, sometimes lasting for months, is more common in those who become infected as adults, and is especially prevalent among women (62). Less commonly, B19 infection has been reported to produce illness with hemorrhagic purpuric or petechial rash (63). The role of B19 as the causative agent of EI was confirmed by inoculation and monitoring of 4 adult human volunteers (12). Because this study offers such a unique insight into B19 infection it will be briefly summarized here. Following intranasal inoculation of the virus, the following pertinent data were noted: (i) using dot-blot hybridization, viremia was first detected 6 days after inoculation and then persisted for about a week; (ii) anti-B19 IgM was detected during the second week followed by the appearance of specific IgG approximately 7 days later; (iii) a biphasic illness was noted in the adult volunteers with the first phase coinciding with viremia and characterized by fever, malaise, myalgia, headache, and itching; (iv) a second phase occurred in the female volunteers around 17 days after infection manifested by a pink maculopapular rash on the arms, legs, and trunk, which was then followed by joint pain. (v) Consistent with the fact that the virus infects red cell precursors in the bone marrow, reticulocytes were noted to disappear by days 8–10 post infection followed by their reappearance 2–4 weeks later. As would be expected, the volunteers were also noted to have significant drops in their hemoglobin levels (12).

Infection in the immunocompromised host: B19 infection in hosts that are unable to mount an effective immune response, primarily neutralizing antibody, to the virus frequently results in bone marrow suppression, chronic anemia, or pure red cell aplasia. Underlying conditions predisposing hosts to this include solid organ and bone marrow transplantation (24), HIV infection (64), chemotherapy (65), and congenital immunodeficiencies (66). Although cells of the erythroid lineage are most often affected, neutropenia and thrombocytopenia have also been reported (67). Prior to effective antiretroviral therapy, up to 25% of chronic anemia in HIV patients was due to B19 infection (67), with some cases still being reported in the face of effective HIV treatment (68). In addition to immunocompromised hosts, uncommon cases of persistent symptomatic B19 infection have also been reported in what appear to be immunocompetent individuals (61).

Infection in the host with hereditary anemia: in individuals with hereditary anemia, such as sickle cell disease, thalassemia, or hereditary spherocytosis, B19 infection can result in what is termed a transient aplastic crisis. Normal individuals infected with B19 become anemic but generally recover uneventfully because their bone marrow is able to rapidly compensate and produce more red blood cells. In the case of hereditary anemia, however, where red blood cells have abnormally shortened life spans, B19 infection can cause a life-threatening drop in hemoglobin levels that the patient is unable to compensate for (61, 69).

Infection during pregnancy: Parvovirus B19 is among the viruses that are capable of crossing the placenta and infecting the fetus. B19 infection of the fetus is the primary cause of a condition termed non-immune fetal hydrops, which is fetal edema that is not mediated by an immune mechanism like Rh disease. The underlying cause of the edema is B19-induced severe fetal anemia (26, 61). Outcome of fetal infection ranges from asymptomatic to fetal death and spontaneous abortion. Other less common outcomes include fetal central nervous system manifestations and thrombocytopenia (26).

Finally, there are some less common disease associations attributed to B19 infection such as meningoencephalitis (70, 71) and myocarditis (72–74), although the now known ability of B19 to persist in the tissues of infected individuals makes it harder to draw firm associations with some conditions.

Bocaparvovirus: Outside of the B19 viruses, because of their propensity to be shed asymptomatically and sometimes be detected in the presence of other pathogens, it becomes more difficult to ascribe clinical significance to the other human parvoviruses. Of the human bocaviruses, HBoV1 is most often detected in respiratory secretions. There is an extensive literature detailing studies attempting to link HBoV1 with respiratory tract illness that have been summarized in several reviews (15, 34, 36, 39, 42, 75). Although it appears that HBoV1 is responsible for some upper and lower respiratory tract illness, especially in young children, it has been difficult to interpret some of the data due to the ability to detect the virus at times in asymptomatic individuals and the fact that detection of another respiratory virus has been reported in up to 70% of symptomatic children with HBoV1 in respiratory tract specimens (36). The review by Jartti et al. noted an association of HBoV1 detection in respiratory specimens with wheezing, asthma, and bronchiolitis in children (34). The same review noted that studies rarely detect the virus in adult specimens and that immunosuppressed individuals do not appear to be at increased risk of illness (34). It appears that detection of HBoV1 DNA in serum may be a better indicator of acute infection and association with illness in children. In a study of 259 children with wheezing, it was reported that 48 (19%) seroconverted and that detection of HBoV1 DNA by PCR in serum correlated with seroconversion (76). Using seroconversion as a measure of acute illness, the authors noted HBoV1 associated–disease lasted longer than rhinovirus illness and was equal in severity to RSV infection (76). This same study noted that only 71% of PCR positive nasopharyngeal specimens had evidence of seroconversion leading to the conclusion that diagnosis of HBoV1 respiratory illness requires seroconversion or detection of DNA in serum, an important caveat for the interpretation of other studies (76). It is noteworthy that none of the current FDA-cleared commercial molecular multiplex respiratory infection panels contain HBoV1 as a target.

The association of clinical illness with infection by the gastrointestinal HBoV2-4 viruses is even less clear. The strongest association with gastroenteritis appears to be for HBoV2 infection, where 1 study reported 17% of children with acute gastroenteritis were found to have HBoV2 in their stools compared with 8% of healthy controls (44). Currently, there are no convincing data linking HBoV 3 or 4 to gastroenteritis or other human illnesses (6, 15, 36, 44).

Dependoparvovirus: Although infection with human AAVs is widespread, and the viruses have been detected in numerous human tissues, there is no evidence that these viruses cause human disease or worsening of the illnesses associated with the helper adenovirus infection (2). The primary area of interest with these viruses is in their use as vectors in human gene therapies.

Tetraparvovirus: At this time there is no evidence that the PARV4 viruses cause illness on their own or result in more severe pathologies associated with the blood-borne viruses such as HIV and hepatitis C and B that they are frequently found in association with (8).

Protoparvovirus: Beyond a weak association with the detection of bufavirus DNA in stools from cases of gastroenteritis, the pathogenicity of this group of human parvoviruses is currently unclear (10, 11, 58, 59).

TREATMENT AND PREVENTION

Since beyond basic infection control and public health hygiene measures, there are no specific treatment or prevention measures available for the non-B19 human parvoviruses, this discussion will be limited to the B19 viruses.

When erythema infectiousum is identified in an individual, usually a child, because the virus is transmitted via the respiratory route, it is recommended to exclude such children from daycare and school while febrile (77). Respiratory and contact isolation are also recommended for patients with hereditary hemolytic anemias that are admitted to hospitals with signs and symptoms consistent with B19 infection (77, 78).

Because of the ubiquitous nature of B19 infection, preparations of human immunoglobulin contain neutralizing anti-B19 antibodies, allowing for intravenous administration of these products to successfully treat persistent or other serious B19 infections (79–83). In an effort to avoid B19-related fetal loss, intravenous immunoglobulin has also been administered during pregnancy (84) and employed as an intrauterine therapy to treat the fetus directly (85).

Because of the serious nature of B19 infections in certain populations, and the risk to the fetus in non-immune pregnant women, efforts to develop a vaccine have been undertaken. A phase 1 trial of a recombinant B19 vaccine was shown to produce neutralizing antibodies in volunteers (86). However, a subsequent trial employing a virus-like particle vaccine was halted after 3 of 43 adult recipients developed cutaneous side-effects and the vaccine awaits further investigation (87).

As previously mentioned, in an effort to prevent transmission of B19 by blood products, requirements/recommendations have been developed that limit the concentration of B19 DNA to a level of $< 10^4$ IU/ml for plasma pools (15, 23).

DETECTION AND DIAGNOSIS

Excepting the B19 viruses, and perhaps HBoV1, making recommendations for the detection of human parvovirus infections and ascribing a diagnosis to their presence is problematic due to the lack of a clear disease association for these agents. At this time, detection of human parvovirus DNA, or the evidence of seroconversion for agents other than B19 or HBoV1, should be considered a research application and therefore not likely to be performed in the diagnostic laboratory. Readers interested in assays for the detection of DNA and/or seroconversion for these agents are referred to Table 1 for a list of appropriate references. Additionally, genesig (Southampton, UK), a company located in the UK with distributors around the globe, offers a commercial taqman-based "Quantification of Human bocavirus" product that they claim detects "all relevant strains and subtypes" of human bocavirus while the Luminex (Austin, TX) xTAG® Respiratory Virus Panel *FAST* Health Canada IVD and CE-IVD Europe-marked assay includes a bocavirus target.

HBoV1

As noted in the Clinical Significance section, HBoV1 has been linked to respiratory tract disease and rarely other conditions in young children. Application of a laboratory diagnostic test to detect HBoV1 may therefore make sense for individual cases or outbreaks of respiratory tract or other less common HBoV1-assicated infections in young children where it has not been possible to demonstrate the presence of other pathogens. The demonstration of HBoV1 infection relies on detection of viral DNA, seroconversion, or a combination of both tests. Other than the previously mentioned commercial genesig HBoV PCR assay, there are no commercial products for detection of HBoV nucleic acid or antibodies. Laboratories wishing to demonstrate infection will therefore have to develop their own assays or rely on published methods. Table 1 lists references for those interested in laboratory diagnostic methods for HBoV1.

Determining the significance of the qualitative detection of HBoV1 DNA in respiratory specimens can be difficult (75) due to the fact that the viral DNA has frequently been

TABLE 1 References for molecular detection and/or identification of antibodies for non-B19 human parvoviruses

Target	References
Adeno-associated viruses	(49, 50, 124, 125)
HBoV1	(37–39, 42, 47, 76, 90, 119, 126–129)
HBoV2	(6, 40, 43, 44, 47, 128, 130–133)
HBoV3	(6, 40, 44, 47, 128, 132, 134)
HBoV4	(6, 40, 47, 128)
PARV4	(9, 53, 57, 135–137)
Bufaviruses	(10, 58, 59)

detected in respiratory specimens from asymptomatic individuals or in the presence of another known respiratory pathogen (37, 76, 88–90). Data from several studies have indicated that acute HBoV1 infection, as determined by serology and/or high quantitative levels of viral DNA in specimens, is accompanied by the presence of viremia and that the presence of virus in the blood may more accurately indicate acute infection that correlates with symptoms, suggesting blood may be preferable to respiratory specimens for diagnosis of HBoV1 infection (37, 39, 76, 90, 91). However, it should be noted that the detection of HBoV1 viremia does not always correlate with illness, with about 5% of healthy blood donors and other asymptomatic individuals being reported to be viremic in some studies (38, 39). There are no standards for quantitation of HBoV1 DNA in clinical specimens; however, higher levels of HBoV1 DNA in clinical specimens have been reported to correlate with apparent acute symptomatic infection compared to specimens from asymptomatic individuals or specimens containing another pathogen (37, 39, 76, 90, 91), although 1 study reported no significant difference in viral loads between symptomatic and asymptomatic children attending daycare (89).

Serological diagnosis of HBoV1 infection, either by the detection of specific IgM or a rise in IgG, has also been reported using non-commercial laboratory-developed assays (Table 1). Due to cross reactions between different HBoV types, interpretation of HBoV1 serological assays may require removal of potential cross reactive antibodies (47).

B19 VIRUSES

The laboratory diagnosis of parvovirus B19 infection is effected primarily by serological and molecular methods. Methods employed less commonly in pathology laboratories include electron microscopy, *in situ* hybridization, and immunohistochemical techniques for the detection of B19 antigens in tissues (92, 93).

For immunocompetent individuals capable of mounting an immune response to the virus, serological methods can be used to determine acute infection or evidence of past infection. IgM antibodies arise 7–10 days after infection, persisting for up to 2–3 months, while IgG appears slightly later and is sometimes detectable for the life of the individual (94, 95). Detectable IgM is therefore generally considered to be a marker for acute infection while the presence of IgG alone signals past infection and immunity. The only commercial FDA-cleared serological assays for the detection of B19-specific IgG and IgM are the Biotrin (name changed to DiaSorin Ireland Ltd in October 2012) sandwich enzyme immunoassays performed in a 96-well plate format that won approval for sale in the United States in 1999. The assays have been reported to perform well with a sensitivity of 89% and specificity of 99% for detection of IgM (96, 97). The assays are also available outside the United States from DiaSorin in a chemiluminescence format performed on their Liaison analyzers. Additional B19 serological reagents in research-use-only formats are commercially available from a number of manufacturers and can be found by searching the http://www.biocompare.com/ website. To aid in the assessment of assays for the detection of anti-B19 IgG antibodies an international standard for parvovirus antibody has been formulated (98).

The diagnosis of B19 infection in immunocompromised patients relies on the demonstration of viremia, usually by PCR-based amplification methods (67). Nucleic acid detection can also be used to confirm a serological result or applied to other specimens such as amniotic fluid and fetal blood (99, 100), bone marrow (28, 101), cerebrospinal fluid (70, 71), joint fluid (102, 103), or tissue specimens (104, 105).

There are currently no FDA-cleared molecular assays for the detection of B19 DNA in clinical specimens. There is an extensive literature describing laboratory-developed assays as well as a number of commercial products currently available in formats that have not been cleared by the FDA (Table 2).

Although human parvovirus B19 is a well-established pathogen, the interpretation of tests for the detection of B19 DNA can be problematic at times. As was previously discussed, there are 3 genotypes of B19 virus with different geographic and temporal distributions. While the majority of B19 viruses in North America and Europe are genotype 1, genotype 2 and 3 viruses have been detected. It is therefore important to understand the ability of a particular assay, laboratory-developed or commercial, to detect non-genotype 1 viruses. Because it appears that genotype 2 viruses are currently circulating only infrequently, molecular assays capable of detecting genotype 1 and 3 viruses should be adequate for the diagnosis of most acute infections in immunocompetent individuals, keeping in mind that rare genotype 2 infections do occur and can be missed by genotype 1 and/or 3 assays (106). Because of the persistence of parvovirus in tissues following infection, it may be possible for genotype 2 viruses to be transmitted during organ transplantation or reactivated during periods of immunosuppression, therefore, negative results for assays that cannot detect genotype 2 viruses should be interpreted with caution in such patients (107). Likewise, because all three genotypes can be transmitted by blood products, screening of such products must employ methods that can detect all 3 genotypes (19, 108).

An unusual cause of a false negative B19 real-time PCR has been reported by Grabarczyk et al. (109) where 4 serum samples with extremely high levels of viremia (9.6×10^9 to $>10^{11}$ IU/ml) caused the real-time platform analysis software to produce false negative results. The problem was resolved by switching from an automated analysis algorithm to a manual method. The author has also observed this phenomenon in our laboratory on another platform, where specimens with Ct values less than 10 have been observed to cause analysis problems. In most cases, the problem has been resolved by diluting the nucleic acid extract and rerunning the assay (unpublished observations).

False positive B19 PCR results are another potential issue associated with the very high levels of viremia that are known to occur with the virus. Traditionally, concerns regarding amplicon contamination have been the driving force behind workflow design to reduce PCR contamination. However, the extraordinary high levels associated with B19 viremia raise concerns about the opportunity for pre-amplification contamination during specimen handling and nucleic acid extraction. Laboratories using automated instruments to extract specimens for B19 PCR should ask the manufacturer of their instrument for cross-contamination data for high titered specimens as well as perform regular B19 contamination surveys of specimen processing areas and instruments. Laboratories may also wish to consider implementing their own carry-over studies by alternating high- and low-positive specimens on their automated extractor on a regular schedule.

The somewhat surprising ability for human parvoviruses, including B19, to persist after acute infection, even in im-

TABLE 2 Laboratory-developed and commercial molecular assays for the detection of parvovirus B19 DNA

Type of Assay	Supplier or Reference
Commercial	**Focus Diagnostics** Parvovirus Primer Pair Analyte Specific Reagent real-time PCR assay for the detection of B19 VP1 gene utilizing a FAM labeled forward primer. For use on the Focus Diagnostics Integrated Cycler Instrument.
	altona Diagnostics RealStar Parvovirus B19 Kit 1.0 CE-marked real-time PCR assay for the detection and quantification of B19 DNA. The assay can be run on a variety of platforms.
	MBL SMITEST Parvovirus B19 DNA Detection Kit Ver.2 Detection of B19 by PCR followed enzyme immune assay hybridization detection of products.
	genesig Quantification of Human Parvovirus B19 Standard Kit Real-time PCR assay for the detection and quantification of B19 DNA. The assay can be run on a variety of platforms.
	Biomerieux Parvovirus B19 R-gene Real-time PCR assay for detection and quantification of B19 DNA. CE-marked assay that can be run on a variety of platforms.
	Sacace Biotechnologies Parvovirus B19 real-TM Quant Real-time PCR kit for detection and quantification of B19 DNA.
	Norgen Biotek Corp Norgen Parvovirus Kit Kit contains reagents for the isolation of DNA from specimens and primers for PCR amplification of B19 DNA with products detected by agarose gel electrophoresis.
	TIB Molbiol Parvovirus B19 Kit Real-time PCR assay for detection of B19 DNA using the Roche Diagnostics LightCycler instruments.
	DiaSorin IAM PARVO Detection of B19 DNA by Q-LAMP technology performed on the Liaison IAM platform.
	Abbott PLEX-ID Viral IC Spectrum Assay PCR-based assay combined with mass spectrometry to detect and identify 11 viral families, including parvovirus B19, in plasma specimens. CE-marked for use on the PLEX-ID instrument.
	Roche Molecular Diagnostics Cobas TaqScreen DPX Test Real-time PCR assay for detection and quantification of B19 DNA and the detection of hepatitis A RNA for use in human plasma screening. Test is run on the cobas platform and is CE-marked and available to FDA-authorized laboratories in the United States.
Laboratory-Developed Assays	(92, 94, 107, 138–144)

munocompetent individuals, is another factor to consider when interpreting the results of nucleic acid amplification tests for B19 DNA. Evidence has been found for the persistence of viral DNA after acute infection in immunocompetent individuals in the bone marrow (103, 110), synovial fluid (103), and tissues (25, 30, 31, 57), suggesting that caution be employed in interpreting positive results from these specimens. In addition to these sites, it is now apparent that at least in some immunocompetent people, acute infection can result in the persistence of detectable B19 DNA in the blood for periods of up to several years (20, 27, 111, 112). It appears that in immunocompetent individuals with persistent detection of B19 DNA in the blood, the level of B19 DNA is significantly lower than the high levels typically seen during acute infection, with levels in the $10–10^5$ IU or copies/ml reported (27, 111, 112). With the caveat that what constitutes a significant level of B19 DNA in blood has not been established, levels of B19 DNA seen below 10^5 IU or copies/ml, especially without evidence of recent compatible symptoms, may reflect persistent rather than acute infection. To aid in the interpretation of quantitative levels of B19 in blood, a second international standard for B19 DNA nucleic acid amplification assays has been prepared and is available (113). For laboratories that do not perform quantitative assays for B19 DNA, blood specimens that produce real-time PCR results with higher than normal crossing threshold values (reflecting lower levels of viremia) should be interpreted with caution and possibly considered for sending to a laboratory performing quantitative testing.

Qualitative and quantitative PCR have also been used to monitor B19 DNA in blood to determine the efficacy of immunoglobulin therapy in reducing the level of B19 viremia in certain patient populations. After 2 days of IVIg therapy, a 3 log drop was noted in B19 blood levels in an HIV patient with persistent B19 anemia (68) while another HIV patient with prolonged B19-induced anemia was noted to become B19 DNA negative in the blood after several rounds of IVIg therapy and the institution of highly active antiretroviral therapy (114). Similar monitoring of treatment by PCR has also been reported for a patient with common variable immunodeficiency (79), and cancer patients (115) where 1 patient was noted to have a 1–3 $\log_{10}$ reduction in B19 DNA serum levels following each dose of IVIg (116). In a case where gammaglobulin was administered into the peritoneal cavity of a B19-infected pregnant woman whose fetus was noted to have hydrops, it was reported that B19 DNA levels fell significantly in fetal ascites fluid and was accompanied by resolution of the fetal hydrops and anemia (85). Despite the fact that there are no standards for what constitutes a significant drop in B19 DNA following Ig treatment, the application of quantitative nucleic acid detection to monitor such treatment likely has benefit in helping physicians assess the efficacy of the treatment and to help determine an end point. When successive specimens have been taken from patients undergoing IVIg treatment, it makes sense to include all previous specimens and contemporary specimens in the same run in order to reduce the effect of inter-assay variation on quantification. For laboratories performing only qualitative B19 PCR assays, following real-time PCR crossing threshold values may allow for a useful approximation of increases or decreases in B19 DNA levels.

Finally, quantification of B19 DNA in plasma products to ensure that levels of B19 DNA do exceed 10^4 IU/ml in plasma products, whether mandated or recommended (113, 117), is effected through the use of quantitative nucleic acid amplification methods (118). However, these assays are usually employed in blood product facilities and are not usually employed in diagnostic laboratories.

BEST PRACTICES

For human parvoviruses other than B19 and HBoV1, there are no significant data that definitively link AAVs, HBoV2-4, PARV4, or bufavirus to human diseases, therefore, it is not possible to recommend best diagnostic laboratory practices at this time. Rather, laboratories interested in detecting evidence of infection by these viruses for research purposes will have to rely on laboratory-developed molecular or serological assays as discussed previously.

HBoV1

There is some moderate evidence suggesting that, in some instances, HBoV1 is responsible for some upper and lower respiratory disease in young children. Prolonged HBoV1 shedding in respiratory specimens, and its frequent detection in the presence of other known pathogens, suggests an approach for diagnosing HBoV1 respiratory disease. First, HBoV1 has rarely been identified in adults and older children and epidemiological studies suggest only children older than 6 months of age, when the maternal antibody wanes, and younger than 2 years, appear susceptible to HBoV1 disease (34, 36, 42, 75, 119) so it may make sense to confine testing to this age group. Second, HBoV1 DNA is frequently detected in the presence of other viral respiratory pathogens, especially rhinoviruses (36), and it is currently uncertain as to whether co-infection with HBoV1 causes more severe symptoms (75) so attaching clinical significance to the detection of HBoV1 as a co-infecting agent of respiratory disease cannot be recommended at this time. Third, as detailed earlier, there is evidence acute infection with HBoV1 is accompanied by viremia that is associated with the greater likelihood of the detection of HBoV1 DNA as the sole viral agent in respiratory specimens and the presence of respiratory symptoms (37, 76, 91), suggesting that testing serum may be reasonable to rule out HBoV1 as a pathogen. However, serum samples may not be available for testing and it may not be appropriate to collect serum from young children strictly to rule out an HBoV1 infection. Therefore, another option may be to employ quantitative PCR, where some studies have demonstrated that higher levels of HBoV1 DNA in respiratory specimens ($>2\times10^8$ copies/ml) were more likely to correlate with a lack of co-infecting pathogens and the presence of respiratory symptoms (34), suggesting quantitative HBoV1 PCR could be employed for respiratory specimens when serum is not available, although not all studies have confirmed a similar relationship between viral load and symptoms (34). Seroconversion to HBoV1 is also an option, either through showing an increase in IgG or the presence of IgM, to identify more potentially significant infections (76). However, the lack of a commercial HBoV1 serological assay and the need to account for the presence of cross reacting antibodies to other HBoVs (47) likely puts the use of serology to diagnose HBoV1 infections out of the reach of most diagnostic laboratories.

Finally, at this time, outside of research studies, the routine testing of stool specimens for the presence of HBoV1 as a cause of gastroenteritis cannot be recommended.

HUMAN PARVOVIRUS B19—ERYTHEMA INFECTIOSUM

In most cases, erythema infectiosum is a benign illness and does not require laboratory confirmation. In situations where laboratory diagnosis is desired, such as confirmation of illness in a child who may have exposed others at risk to more serious infections such as pregnant women or individuals with sickle cell disease, then the detection of IgM can be used to confirm acute infection (120, 121). IgM is usually detectable by the time the patient presents, but if there are concerns for an atypical presentation, or that the stage of the illness may be too early for the appearance of IgM, then it can be recommended to use PCR to try to detect the presence of B19 DNA in the blood (95), where the median level of B19 in the serum collected from 64 cases of acute erythema infectiosum has been reported to be 7.6×10^5 copies/ml, significantly lower than the range of 10^{10}–10^{13} copies/ml from 7 patients during the acute phase of aplastic crisis (122).

INFECTION IN IMMUNOCOMPROMISED PATIENTS

Because individuals with immunosuppression due to conditions such as hereditary immunodeficiencies, HIV infection, solid organ, or bone marrow transplant recipients may not produce detectable anti-B19 immunoglobulins, detection of B19 DNA, usually by nucleic acid amplification methods, is recommended for the diagnosis of B19 infection in immunocompromised patients. Because such patients may not present symptoms typical of acute B19 infection, such as rash and arthralgia, it is important to quickly investigate the possibility of B19 as the cause of chronic anemia or reticulocytopenia in such individuals. As B19 infection in some immunocompromised populations can also cause pancytopenia as well as be confused with recurrences of hematological malignancies (61), testing of patients manifesting these atypical manifestations may be warranted as well. The earlier that persistent B19 infection can be confirmed the earlier that consideration can be given to administer intravenous immunoglobulin treatment. Once immunoglobulin therapy is initiated, a quantitative B19 PCR assay can be used to monitor response to treatment (116). If a quantitative PCR is not available the laboratory can follow crossing points of real-time PCR assays to determine if viral DNA levels are decreasing in response to treatment. It is worth reiterating here that although genotype 2 viruses are now uncommon, there are reports of immunocompromised patients with persistent genotype 2 infections that were not detectable by PCR assays designed to detect genotype 1 viruses (107), underscoring the need to understand the reactivity of the PCR assay employed.

TRANSIENT APLASTIC CRISIS

Unlike hematologically normal individuals, patients with diseases of red blood cell production are unable to compensate for the bone marrow suppression caused by B19 infection, a condition that can lead to transient aplastic crisis and a dangerous life-threatening drop in hemoglobin levels that if identified can be treated by transfusion of red blood cells (69). Because high-titer viremia is usually present in patients presenting with transient aplastic crisis (95, 122), and the viremic state may precede the appearance of IgM, PCR detection of viral DNA in blood is preferred for diagnosis in these patients. Because these patients are usually immunologically normal, appearance of a neutralizing antibody resolves the viremia and induces life-long immunity (61, 95), thus, there is generally no need for following quantitative levels of B19 DNA in these individuals.

INFECTION DURING PREGNANCY AND FETAL HYDROPS

For cases where a pregnant woman may have been exposed to B19 infection, or has symptoms compatible with acute infection, a combination of maternal serology and PCR testing of maternal blood is recommended (26). This approach allows for determination of previous infection and immunity (IgM−, IgG+, PCR−) versus acute infection (IgM+, IgG+/−, PCR+). If evidence of acute infection is found, monitoring of the fetus for development of hydrops is warranted. Evidence of past infection and immunity implies that there is not a risk for development of *in utero* infection, while a non-immune woman who fails to seroconvert by 2 weeks post-exposure likely has not been infected (26). Low levels of B19 DNA in maternal blood should be interpreted with caution as it may represent distant infection that is not a danger to the fetus (26). If acute maternal infection is not recognized or suspected, consideration of B19 infection may not arise until fetal hydrops is detected. At the point hydrops is recognized, enough time may have elapsed for IgM levels to become undetectable, therefore PCR to detect viral DNA in maternal blood, fetal blood, or amniotic fluid is recommended (26, 123). Less commonly B19 infection has been linked to intrauterine fetal death as the result of infection later in pregnancy during the second or third trimester. Symptoms may not be present in the mother of the fetus but infection can be documented by detection of B19 DNA in placental or fetal tissues (61). While qualitative detection of B19 in fetal blood or amniotic fluid is diagnostic for infection, quantitation of viral DNA in maternal blood may be useful for assessing whether infection is acute or longstanding. Quantitative PCR may also have application to fetal specimens and has been employed to monitor fetal viral levels following injection of immunoglobulin into the fetal abdominal cavity (85).

LESS COMMON B19 ASSOCIATIONS

For less common or rare conditions such as possible B19-arthropathy or meningitis, PCR testing for viral DNA can be applied to joint or cerebrospinal fluid. Interpretation of positive results may be complicated by the ability of B19 to persist in joint fluid and the fact that it is not known if the high levels of viremia associated with acute infection can leak into the central nervous system (CNS) and then be detected in the cerebrospinal fluid of people without CNS disease.

Detection of B19 in tissue samples as a way to diagnose end-organ disease, such as myocarditis, is complicated by the ability of viral DNA to persist in the tissues of infected individuals. Therefore, detection of B19 DNA by PCR in tissue specimens should be considered supportive evidence for causality with the primary laboratory diagnosis being accomplished through histopathology and/or *in situ* hybridization (72, 105).

REFERENCES

1. **Cotmore SF, Agbandje-McKenna M, Chiorini JA, Mukha DV, Pintel DJ, Qiu J, Soderlund-Venermo M, Tattersall P, Tijssen P, Gatherer D, Davison AJ.** 2014. The family Parvoviridae. *Arch Virol* **159:**1239–1247.
2. **Flotte TR, Berns KI.** 2005. Adeno-associated virus: a ubiquitous commensal of mammals. *Hum Gene Ther* **16:**401–407.
3. **Berns KI, Parrish CR.** 2013. Parvoviridae, p 1768–1791. *In* Knipe DM, Howley PM (eds.), *Fields Virology*, 6th ed, **vol 2.** Lippincott Williams & Wilkins, Philadelphia, PA.
4. **Truyen U.** 1999. Emergence and recent evolution of canine parvovirus. *Vet Microbiol* **69:**47–50.
5. **Servant A, Laperche S, Lallemand F, Marinho V, De Saint Maur G, Meritet JF, Garbarg-Chenon A.** 2002. Genetic

diversity within human erythroviruses: identification of three genotypes. *J Virol* **76:**9124–9134.
6. **Kapoor A, Simmonds P, Slikas E, Li L, Bodhidatta L, Sethabutr O, Triki H, Bahri O, Oderinde BS, Baba MM, Bukbuk DN, Besser J, Bartkus J, Delwart E.** 2010. Human bocaviruses are highly diverse, dispersed, recombination prone, and prevalent in enteric infections. *J Infect Dis* **201:** 1633–1643.
7. **Cheng W, Chen J, Xu Z, Yu J, Huang C, Jin M, Li H, Zhang M, Jin Y, Duan ZJ.** 2011. Phylogenetic and recombination analysis of human bocavirus 2. *BMC Infect Dis* **11:**50.
8. **Matthews PC, Malik A, Simmons R, Sharp C, Simmonds P, Klenerman P.** 2014. PARV4: an emerging tetraparvovirus. *PLoS Pathog* **10:**e1004036.
9. **Fryer JF, Kapoor A, Minor PD, Delwart E, Baylis SA.** 2006. Novel parvovirus and related variant in human plasma. *Emerg Infect Dis* **12:**151–154.
10. **Phan TG, Vo NP, Bonkoungou IJ, Kapoor A, Barro N, O'Ryan M, Kapusinszky B, Wang C, Delwart E.** 2012. Acute diarrhea in West African children: diverse enteric viruses and a novel parvovirus genus. *J Virol* **86:**11024–11030.
11. **Yahiro T, Wangchuk S, Tshering K, Bandhari P, Zangmo S, Dorji T, Tshering K, Matsumoto T, Nishizono A, Söderlund-Venermo M, Ahmed K.** 2014. Novel human bufavirus genotype 3 in children with severe diarrhea, Bhutan. *Emerg Infect Dis* **20:**1037–1039.
12. **Anderson MJ, Higgins PG, Davis LR, Willman JS, Jones SE, Kidd IM, Pattison JR, Tyrrell DA.** 1985. Experimental parvoviral infection in humans. *J Infect Dis* **152:**257–265.
13. **Brown KE, Anderson SM, Young NS.** 1993. Erythrocyte P antigen: cellular receptor for B19 parvovirus. *Science* **262:**114–117.
14. **Brown KE, Hibbs JR, Gallinella G, Anderson SM, Lehman ED, McCarthy P, Young NS.** 1994. Resistance to parvovirus B19 infection due to lack of virus receptor (erythrocyte P antigen). *N Engl J Med* **330:**1192–1196.
15. **Brown KE.** 2010. The expanding range of parvoviruses which infect humans. *Rev Med Virol* **20:**231–244.
16. **Röhrer C, Gärtner B, Sauerbrei A, Böhm S, Hottenträger B, Raab U, Thierfelder W, Wutzler P, Modrow S.** 2008. Seroprevalence of parvovirus B19 in the German population. *Epidemiol Infect* **136:**1564–1575.
17. **Miron D, Luder A, Horovitz Y, Izkovitz A, Shizgreen I, Ben David E, Ohnona FS, Schlesinger Y.** 2006. Acute human parvovirus B-19 infection in hospitalized children: A serologic and molecular survey. *Pediatr Infect Dis J* **25:**898–901.
18. **Smith-Whitley K, Zhao H, Hodinka RL, Kwiatkowski J, Cecil R, Cecil T, Cnaan A, Ohene-Frempong K.** 2004. Epidemiology of human parvovirus B19 in children with sickle cell disease. *Blood* **103:**422–427.
19. **Parsyan A, Candotti D.** 2007. Human erythrovirus B19 and blood transfusion–an update. *Transfus Med* **17:**263–278.
20. **Musiani M, Zerbini M, Gentilomi G, Plazzi M, Gallinella G, Venturoli S.** 1995. Parvovirus B19 clearance from peripheral blood after acute infection. *J Infect Dis* **172:**1360–1363.
21. **Kooistra K, Mesman HJ, de Waal M, Koppelman MH, Zaaijer HL.** 2011. Epidemiology of high-level parvovirus B19 viraemia among Dutch blood donors, 2003–2009. *Vox Sang* **100:**261–266.
22. **Soucie JM, De Staercke C, Monahan PE, Recht M, Chitlur MB, Gruppo R, Hooper WC, Kessler C, Kulkarni R, Manco-Johnson MJ, Powell J, Pyle M, Riske B, Sabio H, Trimble S, Network USHTC, US Hemophilia Treatment Center Network.** 2013. Evidence for the transmission of parvovirus B19 in patients with bleeding disorders treated with plasma-derived factor concentrates in the era of nucleic acid test screening. *Transfusion* **53:**1217–1225.
23. 2009, Guidance for Industry: Nucleic Acid Testing (NAT) to Reduce the Possible Risk of Parvovirus B19 Transmission by Plasma-Derived Products. U.S. Department of Health and Human Services Food and Drug Administration Center for Biologics Evaluation and Research. [Online.]
24. **Eid AJ, Brown RA, Patel R, Razonable RR.** 2006. Parvovirus B19 infection after transplantation: a review of 98 cases. *Clin Infect Dis* **43:**40–48.
25. **Söderlund-Venermo M, Hokynar K, Nieminen J, Rautakorpi H, Hedman K.** 2002. Persistence of human parvovirus B19 in human tissues. *Pathol Biol (Paris)* **50:**307–316.
26. **de Jong EP, Walther FJ, Kroes AC, Oepkes D.** 2011. Parvovirus B19 infection in pregnancy: new insights and management. *Prenat Diagn* **31:**419–425.
27. **Candotti D, Etiz N, Parsyan A, Allain JP.** 2004. Identification and characterization of persistent human erythrovirus infection in blood donor samples. *J Virol* **78:**12169–12178.
28. **Sanabani S, Neto WK, Pereira J, Sabino EC.** 2006. Sequence variability of human erythroviruses present in bone marrow of Brazilian patients with various parvovirus B19-related hematological symptoms. *J Clin Microbiol* **44:**604–606.
29. **Rinckel LA, Buno BR, Gierman TM, Lee DC.** 2009. Discovery and analysis of a novel parvovirus B19 Genotype 3 isolate in the United States. *Transfusion* **49:**1488–1492.
30. **Norja P, Hokynar K, Aaltonen LM, Chen R, Ranki A, Partio EK, Kiviluoto O, Davidkin I, Leivo T, Eis-Hübinger AM, Schneider B, Fischer HP, Tolba R, Vapalahti O, Vaheri A, Söderlund-Venermo M, Hedman K.** 2006. Bioportfolio: lifelong persistence of variant and prototypic erythrovirus DNA genomes in human tissue. *Proc Natl Acad Sci USA* **103:**7450–7453.
31. **Manning A, Willey SJ, Bell JE, Simmonds P.** 2007. Comparison of tissue distribution, persistence, and molecular epidemiology of parvovirus B19 and novel human parvoviruses PARV4 and human bocavirus. *J Infect Dis* **195:**1345–1352.
32. **Hübschen JM, Mihneva Z, Mentis AF, Schneider F, Aboudy Y, Grossman Z, Rudich H, Kasymbekova K, Sarv I, Nedeljkovic J, Tahita MC, Tarnagda Z, Ouedraogo JB, Gerasimova AG, Moskaleva TN, Tikhonova NT, Chitadze N, Forbi JC, Faneye AO, Otegbayo JA, Charpentier E, Muller CP.** 2009. Phylogenetic analysis of human parvovirus b19 sequences from eleven different countries confirms the predominance of genotype 1 and suggests the spread of genotype 3b. *J Clin Microbiol* **47:**3735–3738.
33. **Allander T, Tammi MT, Eriksson M, Bjerkner A, Tiveljung-Lindell A, Andersson B.** 2005. Cloning of a human parvovirus by molecular screening of respiratory tract samples. *Proc Natl Acad Sci USA* **102:**12891–12896.
34. **Jartti T, Hedman K, Jartti L, Ruuskanen O, Allander T, Söderlund-Venermo M.** 2012. Human bocavirus–the first 5 years. *Rev Med Virol* **22:**46–64.
35. **Costa C, Bergallo M, Cavallo R.** 2009. Detection of Human Bocavirus in bronchoalveolar lavage from Italian adult patients. *J Clin Virol* **45:**81–82.
36. **Peltola V, Söderlund-Venermo M, Jartti T.** 2013. Human bocavirus infections. *Pediatr Infect Dis J* **32:**178–179.
37. **Allander T, Jartti T, Gupta S, Niesters HG, Lehtinen P, Osterback R, Vuorinen T, Waris M, Bjerkner A, Tiveljung-Lindell A, van den Hoogen BG, Hyypiä T, Ruuskanen O.** 2007. Human bocavirus and acute wheezing in children. *Clin Infect Dis* **44:**904–910.
38. **Bonvicini F, Manaresi E, Gentilomi GA, Di Furio F, Zerbini M, Musiani M, Gallinella G.** 2011. Evidence of human bocavirus viremia in healthy blood donors. *Diagn Microbiol Infect Dis* **71:**460–462.
39. **Karalar L, Lindner J, Schimanski S, Kertai M, Segerer H, Modrow S.** 2010. Prevalence and clinical aspects of human bocavirus infection in children. *Clin Microbiol Infect* **16:**633–639.
40. **Risku M, Kätkä M, Lappalainen S, Räsänen S, Vesikari T.** 2012. Human bocavirus types 1, 2 and 3 in acute gastroenteritis of childhood. *Acta Paediatr* **101:**e405–e410.
41. **Vicente D, Cilla G, Montes M, Pérez-Yarza EG, Pérez-Trallero E.** 2007. Human bocavirus, a respiratory and enteric virus. *Emerg Infect Dis* **13:**636–637.
42. **Chow BD, Esper FP.** 2009. The human bocaviruses: a review and discussion of their role in infection. *Clin Lab Med* **29:** 695–713.

43. **Kapoor A, Slikas E, Simmonds P, Chieochansin T, Naeem A, Shaukat S, Alam MM, Sharif S, Angez M, Zaidi S, Delwart E.** 2009. A newly identified bocavirus species in human stool. *J Infect Dis* **199:**196–200.
44. **Arthur JL, Higgins GD, Davidson GP, Givney RC, Ratcliff RM.** 2009. A novel bocavirus associated with acute gastroenteritis in Australian children. *PLoS Pathog* **5:**e1000391.
45. **Han TH, Chung JY, Hwang ES.** 2009. Human bocavirus 2 in children, South Korea. *Emerg Infect Dis* **15:**1698–1700.
46. **Song JR, Jin Y, Xie ZP, Gao HC, Xiao NG, Chen WX, Xu ZQ, Yan KL, Zhao Y, Hou YD, Duan ZJ.** 2010. Novel human bocavirus in children with acute respiratory tract infection. *Emerg Infect Dis* **16:**324–327.
47. **Kantola K, Hedman L, Arthur J, Alibeto A, Delwart E, Jartti T, Ruuskanen O, Hedman K, Söderlund-Venermo M.** 2011. Seroepidemiology of human bocaviruses 1–4. *J Infect Dis* **204:**1403–1412.
48. **Blacklow NR, Hoggan MD, Rowe WP.** 1968. Serologic evidence for human infection with adenovirus-associated viruses. *J Natl Cancer Inst* **40:**319–327.
49. **Chirmule N, Propert K, Magosin S, Qian Y, Qian R, Wilson J.** 1999. Immune responses to adenovirus and adeno-associated virus in humans. *Gene Ther* **6:**1574–1583.
50. **Calcedo R, Vandenberghe LH, Gao G, Lin J, Wilson JM.** 2009. Worldwide epidemiology of neutralizing antibodies to adeno-associated viruses. *J Infect Dis* **199:**381–390.
51. **Jones MS, Kapoor A, Lukashov VV, Simmonds P, Hecht F, Delwart E.** 2005. New DNA viruses identified in patients with acute viral infection syndrome. *J Virol* **79:**8230–8236.
52. **Drexler JF, Reber U, Muth D, Herzog P, Annan A, Ebach F, Sarpong N, Acquah S, Adlkofer J, Adu-Sarkodie Y, Panning M, Tannich E, May J, Drosten C, Eis-Hübinger AM.** 2012. Human parvovirus 4 in nasal and fecal specimens from children, Ghana. *Emerg Infect Dis* **18:**1650–1653.
53. **Fryer JF, Delwart E, Hecht FM, Bernardin F, Jones MS, Shah N, Baylis SA.** 2007. Frequent detection of the parvoviruses, PARV4 and PARV5, in plasma from blood donors and symptomatic individuals. *Transfusion* **47:**1054–1061.
54. **Fryer JF, Delwart E, Bernardin F, Tuke PW, Lukashov VV, Baylis SA.** 2007. Analysis of two human parvovirus PARV4 genotypes identified in human plasma for fractionation. *J Gen Virol* **88:**2162–2167.
55. **Chen MY, Yang SJ, Hung CC.** 2011. Placental transmission of human parvovirus 4 in newborns with hydrops, Taiwan. *Emerg Infect Dis* **17:**1954–1956.
56. **Corcioli F, Zakrzewska K, Fanci R, De Giorgi V, Innocenti M, Rotellini M, Di Lollo S, Azzi A.** 2010. Human parvovirus PARV4 DNA in tissues from adult individuals: a comparison with human parvovirus B19 (B19V). *Virol J* **7:**272.
57. **Schneider B, Fryer JF, Reber U, Fischer HP, Tolba RH, Baylis SA, Eis-Hübinger AM.** 2008. Persistence of novel human parvovirus PARV4 in liver tissue of adults. *J Med Virol* **80:**345–351.
58. **Väisänen E, Kuisma I, Phan TG, Delwart E, Lappalainen M, Tarkka E, Hedman K, Söderlund-Venermo M.** 2014. Bufavirus in feces of patients with gastroenteritis, Finland. *Emerg Infect Dis* **20:**1077–1079.
59. **Smits SL, Schapendonk CM, van Beek J, Vennema H, Schürch AC, Schipper D, Bodewes R, Haagmans BL, Osterhaus AD, Koopmans MP.** 2014. New viruses in idiopathic human diarrhea cases, the Netherlands. *Emerg Infect Dis* **20:**1218–1222.
60. **Cossart YE, Field AM, Cant B, Widdows D.** 1975. Parvovirus-like particles in human sera. *Lancet* **305:**72–73.
61. **Broliden K, Tolfvenstam T, Norbeck O.** 2006. Clinical aspects of parvovirus B19 infection. *J Intern Med* **260:**285–304.
62. **Woolf AD, Campion GV, Chishick A, Wise S, Cohen BJ, Klouda PT, Caul O, Dieppe PA.** 1989. Clinical manifestations of human parvovirus B19 in adults. *Arch Intern Med* **149:**1153–1156.
63. **Edmonson MB, Riedesel EL, Williams GP, Demuri GP.** 2010. Generalized petechial rashes in children during a parvovirus B19 outbreak. *Pediatrics* **125:**e787–e792.
64. **Fuller A, Moaven L, Spelman D, Spicer WJ, Wraight H, Curtis D, Leydon J, Doultree J, Locarnini S.** 1996. Parvovirus B19 in HIV infection: a treatable cause of anemia. *Pathology* **28:**277–280.
65. **Graeve JL, de Alarcon PA, Naides SJ.** 1989. Parvovirus B19 infection in patients receiving cancer chemotherapy: the expanding spectrum of disease. *Am J Pediatr Hematol Oncol* **11:**441–444.
66. **Blaeser F, Kelly M, Siegrist K, Storch GA, Buller RS, Whitlock J, Truong N, Chatila TA.** 2005. Critical function of the CD40 pathway in parvovirus B19 infection revealed by a hypomorphic CD40 ligand mutation. *Clin Immunol* **117:**231–237.
67. **Florea AV, Ionescu DN, Melhem MF.** 2007. Parvovirus B19 infection in the immunocompromised host. *Arch Pathol Lab Med* **131:**799–804.
68. **Morelli P, Bestetti G, Longhi E, Parravicini C, Corbellino M, Meroni L.** 2007. Persistent parvovirus B19-induced anemia in an HIV-infected patient under HAART. Case report and review of literature. *Eur J Clin Microbiol Infect Dis* **26:**833–837.
69. **Slavov SN, Kashima S, Pinto AC, Covas DT.** 2011. Human parvovirus B19: general considerations and impact on patients with sickle-cell disease and thalassemia and on blood transfusions. *FEMS Immunol Med Microbiol* **62:**247–262.
70. **Barah F, Vallely PJ, Chiswick ML, Cleator GM, Kerr JR.** 2001. Association of human parvovirus B19 infection with acute meningoencephalitis. *Lancet* **358:**729–730.
71. **Bonvicini F, Marinacci G, Pajno MC, Gallinella G, Musiani M, Zerbini M.** 2008. Meningoencephalitis with persistent parvovirus B19 infection in an apparently healthy woman. *Clin Infect Dis* **47:**385–387.
72. **Bültmann BD, Klingel K, Sotlar K, Bock CT, Baba HA, Sauter M, Kandolf R.** 2003. Fatal parvovirus B19-associated myocarditis clinically mimicking ischemic heart disease: an endothelial cell-mediated disease. *Hum Pathol* **34:**92–95.
73. **Enders G, Dötsch J, Bauer J, Nützenadel W, Hengel H, Haffner D, Schalasta G, Searle K, Brown KE.** 1998. Life-threatening parvovirus B19-associated myocarditis and cardiac transplantation as possible therapy: two case reports. *Clin Infect Dis* **26:**355–358.
74. **Pankuweit S, Moll R, Baandrup U, Portig I, Hufnagel G, Maisch B.** 2003. Prevalence of the parvovirus B19 genome in endomyocardial biopsy specimens. *Hum Pathol* **34:**497–503.
75. **Schildgen O, Müller A, Allander T, Mackay IM, Völz S, Kupfer B, Simon A.** 2008. Human bocavirus: passenger or pathogen in acute respiratory tract infections? *Clin Microbiol Rev* **21:**291–304. Table of contents.
76. **Söderlund-Venermo M, Lahtinen A, Jartti T, Hedman L, Kemppainen K, Lehtinen P, Allander T, Ruuskanen O, Hedman K.** 2009. Clinical assessment and improved diagnosis of bocavirus-induced wheezing in children, Finland. *Emerg Infect Dis* **15:**1423–1430.
77. 2008. *Control of Cummunicable Diseases Manual*, 19th ed. Amerian Public Health Association, Washington, DC, USA.
78. **Bell LM, Naides SJ, Stoffman P, Hodinka RL, Plotkin SA.** 1989. Human parvovirus B19 infection among hospital staff members after contact with infected patients. *N Engl J Med* **321:**485–491.
79. **Chuhjo T, Nakao S, Matsuda T.** 1999. Successful treatment of persistent erythroid aplasia caused by parvovirus B19 infection in a patient with common variable immunodeficiency with low-dose immunoglobulin. *Am J Hematol* **60:**222–224.
80. **Crabol Y, Terrier B, Rozenberg F, Pestre V, Legendre C, Hermine O, Montagnier-Petrissans C, Guillevin L, Mouthon L, Groupe d'experts de l'Assistance Publique-Hôpitaux de Paris.** 2013. Intravenous immunoglobulin therapy for pure red cell aplasia related to human parvovirus b19 infection: a retrospective study of 10 patients and review of the literature. *Clin Infect Dis* **56:**968–977.

81. **Frickhofen N, et al.** 1990. Persistent B19 parvovirus infection in patients infected with human immunodeficiency virus type 1 (HIV-1): a treatable cause of anemia in AIDS. *Ann Intern Med* **113:**926–933.
82. **Koduri PR, Kumapley R, Valladares J, Teter C.** 1999. Chronic pure red cell aplasia caused by parvovirus B19 in AIDS: use of intravenous immunoglobulin–a report of eight patients. *Am J Hematol* **61:**16–20.
83. **Kurtzman G, Frickhofen N, Kimball J, Jenkins DW, Nienhuis AW, Young NS.** 1989. Pure red-cell aplasia of 10 years' duration due to persistent parvovirus B19 infection and its cure with immunoglobulin therapy. *N Engl J Med* **321:** 519–523.
84. **Selbing A, Josefsson A, Dahle LO, Lindgren R.** 1995. Parvovirus B19 infection during pregnancy treated with high-dose intravenous gammaglobulin. *Lancet* **345:**660–661.
85. **Matsuda H, Sakaguchi K, Shibasaki T, Takahashi H, Kawakami Y, Furuya K.** 2005. Intrauterine therapy for parvovirus B19 infected symptomatic fetus using B19 IgG-rich high titer gammaglobulin. *J Perinat Med* **33:**561–563.
86. **Ballou WR, Reed JL, Noble W, Young NS, Koenig S.** 2003. Safety and immunogenicity of a recombinant parvovirus B19 vaccine formulated with MF59C.1. *J Infect Dis* **187:** 675–678.
87. **Bernstein DI, El Sahly HM, Keitel WA, Wolff M, Simone G, Segawa C, Wong S, Shelly D, Young NS, Dempsey W.** 2011. Safety and immunogenicity of a candidate parvovirus B19 vaccine. *Vaccine* **29:**7357–7363.
88. **Christensen A, Nordbø SA, Krokstad S, Rognlien AG, Døllner H.** 2008. Human bocavirus commonly involved in multiple viral airway infections. *J Clin Virol* **41:** 34–37.
89. **Martin ET, Fairchok MP, Kuypers J, Magaret A, Zerr DM, Wald A, Englund JA.** 2010. Frequent and prolonged shedding of bocavirus in young children attending daycare. *J Infect Dis* **201:**1625–1632.
90. **Wang K, Wang W, Yan H, Ren P, Zhang J, Shen J, Deubel V.** 2010. Correlation between bocavirus infection and humoral response, and co-infection with other respiratory viruses in children with acute respiratory infection. *J Clin Virol* **47:** 148–155.
91. **Christensen A, Nordbø SA, Krokstad S, Rognlien AG, Døllner H.** 2010. Human bocavirus in children: monodetection, high viral load and viraemia are associated with respiratory tract infection. *J Clin Virol* **49:**158–162.
92. **Doyle S.** 2011. The detection of parvoviruses. *Methods Mol Biol* **665:**213–231.
93. **Jordan JA.** 2011. Human Parvoviruses, p. 1636–1646. *In* Landry ML, Caliendo AM, Ginocchio CC, Tang YW, Valsamakis A (eds.), *Manual of Clinical Microbiology*, 10th ed., vol 2. ASM Press, Washington, DC.
94. **Peterlana D, Puccetti A, Corrocher R, Lunardi C.** 2006. Serologic and molecular detection of human Parvovirus B19 infection. *Clin Chim Acta* **372:**14–23.
95. **Young NS, Brown KE.** 2004. Parvovirus B19. *N Engl J Med* **350:**586–597.
96. **Butchko AR, Jordan JA.** 2004. Comparison of three commercially available serologic assays used to detect human parvovirus B19-specific immunoglobulin M (IgM) and IgG antibodies in sera of pregnant women. *J Clin Microbiol* **42:** 3191–3195.
97. **Doyle S, Kerr S, O'Keeffe G, O'Carroll D, Daly P, Kilty C.** 2000. Detection of parvovirus B19 IgM by antibody capture enzyme immunoassay: receiver operating characteristic analysis. *J Virol Methods* **90:**143–152.
98. **Ferguson M, Heath A.** 2004. Report of a collaborative study to calibrate the Second International Standard for parvovirus B19 antibody. *Biologicals* **32:**207–212.
99. **Bonvicini F, Manaresi E, Gallinella G, Gentilomi GA, Musiani M, Zerbini M.** 2009. Diagnosis of fetal parvovirus B19 infection: value of virological assays in fetal specimens. *BJOG* **116:**813–817.
100. **Lamont RF, Sobel JD, Vaisbuch E, Kusanovic JP, Mazaki-Tovi S, Kim SK, Uldbjerg N, Romero R.** 2011. Parvovirus B19 infection in human pregnancy. *BJOG* **118:**175–186.
101. **Prassouli A, Papadakis V, Tsakris A, Stefanaki K, Garoufi A, Haidas S, Dracou C.** 2005. Classic transient erythroblastopenia of childhood with human parvovirus B19 genome detection in the blood and bone marrow. *J Pediatr Hematol Oncol* **27:**333–336.
102. **Dijkmans BA, van Elsacker-Niele AM, Salimans MM, van Albada-Kuipers GA, de Vries E, Weiland HT.** 1988. Human parvovirus B19 DNA in synovial fluid. *Arthritis Rheum* **31:** 279–281.
103. **Nikkari S, Roivainen A, Hannonen P, Möttönen T, Luukkainen R, Yli-Jama T, Toivanen P.** 1995. Persistence of parvovirus B19 in synovial fluid and bone marrow. *Ann Rheum Dis* **54:**597–600.
104. **Dwivedi M, Manocha H, Tiwari S, Tripathi G, Dhole TN.** 2009. Coinfection of parvovirus b19 with other hepatitis viruses leading to fulminant hepatitis of unfavorable outcome in children. *Pediatr Infect Dis J* **28:**649–650.
105. **Molina KM, Garcia X, Denfield SW, Fan Y, Morrow WR, Towbin JA, Frazier EA, Nelson DP.** 2013. Parvovirus B19 myocarditis causes significant morbidity and mortality in children. *Pediatr Cardiol* **34:**390–397.
106. **Cohen BJ, Gandhi J, Clewley JP.** 2006. Genetic variants of parvovirus B19 identified in the United Kingdom: implications for diagnostic testing. *J Clin Virol* **36:**152–155.
107. **Liefeldt L, Plentz A, Klempa B, Kershaw O, Endres AS, Raab U, Neumayer HH, Meisel H, Modrow S.** 2005. Recurrent high level parvovirus B19/genotype 2 viremia in a renal transplant recipient analyzed by real-time PCR for simultaneous detection of genotypes 1 to 3. *J Med Virol* **75:** 161–169.
108. **Modrow S, Wenzel JJ, Schimanski S, Schwarzbeck J, Rothe U, Oldenburg J, Jilg W, Eis-Hübinger AM.** 2011. Prevalence of nucleic acid sequences specific for human parvoviruses, hepatitis A and hepatitis E viruses in coagulation factor concentrates. *Vox Sang* **100:**351–358.
109. **Grabarczyk P, Kalińska A, Sulkowska E, Brojer E.** 2010. False negative results in high viremia parvovirus B19-samples tested with real-time PCR. *Pol J Microbiol* **59:**129–132.
110. **Cassinotti P, Burtonboy G, Fopp M, Siegl G.** 1997. Evidence for persistence of human parvovirus B19 DNA in bone marrow. *J Med Virol* **53:**229–232.
111. **Cassinotti P, Siegl G.** 2000. Quantitative evidence for persistence of human parvovirus B19 DNA in an immunocompetent individual. *Eur J Clin Microbiol Infect Dis* **19:**886–887.
112. **Lefrère JJ, Servant-Delmas A, Candotti D, Mariotti M, Thomas I, Brossard Y, Lefrère F, Girot R, Allain JP, Laperche S.** 2005. Persistent B19 infection in immunocompetent individuals: implications for transfusion safety. *Blood* **106:**2890–2895.
113. **Baylis SA, Chudy M, Blümel J, Pisani G, Candotti D, José M, Heath AB.** 2010. Collaborative study to establish a replacement World Health Organization International Standard for parvovirus B19 DNA nucleic acid amplification technology (NAT)-based assays. *Vox Sang* **98**(3p2)**:**441–446.
114. **Mylonakis E, Dickinson BP, Mileno MD, Flanigan T, Schiffman FJ, Mega A, Rich JD.** 1999. Persistent parvovirus B19 related anemia of seven years' duration in an HIV-infected patient: complete remission associated with highly active antiretroviral therapy. *Am J Hematol* **60:**164–166.
115. **Katragadda L, Shahid Z, Restrepo A, Muzaffar J, Alapat D, Anaissie E.** 2013. Preemptive intravenous immunoglobulin allows safe and timely administration of antineoplastic therapies in patients with multiple myeloma and parvovirus B19 disease. *Transpl Infect Dis* **15:**354–360.
116. **Tang JW, Lau JS, Wong SY, Cheung JL, Chan CH, Wong KF, Wong A, Chan PK.** 2007. Dose-by-dose virological and hematological responses to intravenous immunoglobulin in an immunocompromised patient with persistent parvovirus B19 infection. *J Med Virol* **79:**1401–1405.

117. **Brown KE, Young NS, Alving BM, Barbosa LH.** 2001. Parvovirus B19: implications for transfusion medicine. Summary of a workshop. *Transfusion* **41:**130–135.
118. **Koppelman MH, Cuijpers HT, Wessberg S, Valkeajärvi A, Pichl L, Schottstedt V, Saldanha J.** 2012. Multicenter evaluation of a commercial multiplex polymerase chain reaction test for screening plasma donations for parvovirus B19 DNA and hepatitis A virus RNA. *Transfusion* **52:**1498–1508.
119. **Lindner J, Modrow S.** 2008. Human bocavirus–a novel parvovirus to infect humans. *Intervirology* **51:**116–122.
120. **Vafaie J, Schwartz RA.** 2004. Parvovirus B19 infections. *Int J Dermatol* **43:**747–749.
121. **Valentin MN, Cohen PJ.** 2013. Pediatric parvovirus B19: spectrum of clinical manifestations. *Cutis* **92:**179–184.
122. **Ishikawa A, Yoto Y, Tsugawa T, Tsutsumi H.** 2014. Quantitation of human parvovirus B19 DNA in erythema infectiosum and aplastic crisis. *J Med Virol* **86:**2102–2106.
123. **Enders M, Weidner A, Rosenthal T, Baisch C, Hedman L, Söderlund-Venermo M, Hedman K.** 2008. Improved diagnosis of gestational parvovirus B19 infection at the time of nonimmune fetal hydrops. *J Infect Dis* **197:**58–62.
124. **Friedman-Einat M, Grossman Z, Mileguir F, Smetana Z, Ashkenazi M, Barkai G, Varsano N, Glick E, Mendelson E.** 1997. Detection of adeno-associated virus type 2 sequences in the human genital tract. *J Clin Microbiol* **35:**71–78.
125. **Gao G, Vandenberghe LH, Alvira MR, Lu Y, Calcedo R, Zhou X, Wilson JM.** 2004. Clades of Adeno-associated viruses are widely disseminated in human tissues. *J Virol* **78:**6381–6388.
126. **Kahn JS, Kesebir D, Cotmore SF, D'Abramo A Jr, Cosby C, Weibel C, Tattersall P.** 2008. Seroepidemiology of human bocavirus defined using recombinant virus-like particles. *J Infect Dis* **198:**41–50.
127. **Kantola K, Hedman L, Allander T, Jartti T, Lehtinen P, Ruuskanen O, Hedman K, Söderlund-Venermo M.** 2008. Serodiagnosis of human bocavirus infection. *Clin Infect Dis* **46:**540–546.
128. **Kantola K, Sadeghi M, Antikainen J, Kirveskari J, Delwart E, Hedman K, Söderlund-Venermo M.** 2010. Real-time quantitative PCR detection of four human bocaviruses. *J Clin Microbiol* **48:**4044–4050.
129. **Lindner J, Karalar L, Zehentmeier S, Plentz A, Pfister H, Struff W, Kertai M, Segerer H, Modrow S.** 2008. Humoral immune response against human bocavirus VP2 virus-like particles. *Viral Immunol* **21:**443–450.
130. **Cheng WX, Jin Y, Duan ZJ, Xu ZQ, Qi HM, Zhang Q, Yu JM, Zhu L, Jin M, Liu N, Cui SX, Li HY, Fang ZY.** 2008. Human bocavirus in children hospitalized for acute gastroenteritis: a case-control study. *Clin Infect Dis* **47:**161–167.
131. **Norja P, Hedman L, Kantola K, Kemppainen K, Suvilehto J, Pitkäranta A, Aaltonen LM, Seppänen M, Hedman K, Söderlund-Venermo M.** 2012. Occurrence of human bocaviruses and parvovirus 4 in solid tissues. *J Med Virol* **84:**1267–1273.
132. **Santos N, Peret TC, Humphrey CD, Albuquerque MC, Silva RC, Benati FJ, Lu X, Erdman DD.** 2010. Human bocavirus species 2 and 3 in Brazil. *J Clin Virol* **48:**127–130.
133. **Xu ZQ, Cheng WX, Li BW, Li J, Lan B, Duan ZJ.** 2011. Development of a real-time PCR assay for detecting and quantifying human bocavirus 2. *J Clin Microbiol* **49:**1537–1541.
134. **Wang Y, Gonzalez R, Zhou H, Li J, Li Y, Paranhos-Baccalà G, Vernet G, Guo L, Wang J.** 2011. Detection of human bocavirus 3 in China. *Eur J Clin Microbiol Infect Dis* **30:**799–805.
135. **Lahtinen A, Kivelä P, Hedman L, Kumar A, Kantele A, Lappalainen M, Liitsola K, Ristola M, Delwart E, Sharp C, Simmonds P, Söderlund-Venermo M, Hedman K.** 2011. Serodiagnosis of primary infections with human parvovirus 4, Finland. *Emerg Infect Dis* **17:**79–82.
136. **Maple PA, Beard S, Parry RP, Brown KE.** 2013. Testing UK blood donors for exposure to human parvovirus 4 using a time-resolved fluorescence immunoassay to screen sera and Western blot to confirm reactive samples. *Transfusion* **53:**2575–2584.
137. **Väisänen E, Lahtinen A, Eis-Hübinger AM, Lappalainen M, Hedman K, Söderlund-Venermo M.** 2014. A two-step real-time PCR assay for quantitation and genotyping of human parvovirus 4. *J Virol Methods* **195:**106–111.
138. **Bergallo M, Costa C, Sidoti F, Novelli M, Ponti R, Castagnoli C, Merlino C, Bernengo MG, Cavallo R.** 2008. Variants of parvovirus B19: bioinformatical evaluation of nested PCR assays. *Intervirology* **51:**75–80.
139. **Bergallo M, Merlino C, Daniele R, Costa C, Ponzi AN, Cavallo R.** 2006. Quantitative competitive-PCR assay to measure human parvovirus B19-DNA load in serum samples. *Mol Biotechnol* **32:**23–29.
140. **Bonvicini F, Manaresi E, Bua G, Venturoli S, Gallinella G.** 2013. Keeping pace with parvovirus B19 genetic variability: a multiplex genotype-specific quantitative PCR assay. *J Clin Microbiol* **51:**3753–3759.
141. **Corcoran C, Hardie D, Yeats J, Smuts H.** 2010. Genetic variants of human parvovirus B19 in South Africa: cocirculation of three genotypes and identification of a novel subtype of genotype 1. *J Clin Microbiol* **48:**137–142.
142. **Daly P, Corcoran A, Mahon BP, Doyle S.** 2002. High-sensitivity PCR detection of parvovirus B19 in plasma. *J Clin Microbiol* **40:**1958–1962.
143. **Koppelman MH, van Swieten P, Cuijpers HT.** 2011. Real-time polymerase chain reaction detection of parvovirus B19 DNA in blood donations using a commercial and an in-house assay. *Transfusion* **51:**1346–1354.
144. **Manaresi E, Gallinella G, Zuffi E, Bonvicini F, Zerbini M, Musiani M.** 2002. Diagnosis and quantitative evaluation of parvovirus B19 infections by real-time PCR in the clinical laboratory. *J Med Virol* **67:**275–281.

Poxviruses

ASHLEY V. KONDAS AND VICTORIA A. OLSON

33

Variola virus (VARV), a member of the *Orthopoxvirus* genus, caused one of the most feared illnesses of mankind, smallpox. In 1798, Edward Jenner described that milkmaids with evidence of prior infection with cowpox (caused by *Orthopoxvirus Cowpox virus* [CPXV]) were immune to infection with smallpox (VARV). Smallpox vaccines, derived from *Orthopoxvirus Vaccinia virus* (VACV), were subsequently used extensively for routine vaccination against VARV. Through an intensive vaccination campaign, coordinated by the World Health Organization (WHO), naturally occurring VARV infections were declared eradicated in 1980. These modalities are also of interest in recognition and control of emerging zoonotic orthopoxviruses (*Monkeypox virus* [MPXV], CPXV, and VACV).

BIOLOGY OF HUMAN POXVIRUS PATHOGENS

Introduction

Classification

The first classification of poxviruses was based on a criterion which considered disease signs or symptoms as well as gross pathology. This resulted in the grouping of diseases which were characterized by pocks on the skin, including distinct diseases such as smallpox (VARV), chickenpox (*Varicella-zoster virus*), and syphilis (the spirochete *Treponema pallidum*). The classification of poxviruses was refined with the application of a more-stringent criterion based on morphological characterization of virions, cytoplasmic inclusion bodies, and light microscopy (Table 1). The infectious nature of the individual poxvirus virion was finally elucidated by Ledingham (1).

In 1953, Fenner and Burnet (2) wrote a review which summarized the characteristics of the poxvirus group. The poxvirus family (*Poxviridae*) is divided into two subfamilies: *Entomopoxvirinae* (poxviruses of insects) and *Chordopoxvirinae* (poxviruses of vertebrates). The vertebrate poxviruses were further subclassified into genera by comparing cross-protection in animal studies, cross-neutralization of virion infectivity in cell culture, and, through the analysis of genetic polymorphisms, in genomic viral DNA. The subfamily *Chordopoxvirinae* consists of 10 genera (and several unclassified viruses) which are based upon genomic comparisons: *Orthopoxvirus*, *Molluscipoxvirus*, *Parapoxvirus*, *Yatapoxvirus*, *Avipoxvirus*, *Capripoxvirus*, *Leporipoxvirus*, *Cervidpoxvirus*, *Crocodylidpoxvirus*, and *Suipoxvirus*. Four genera (*Orthopoxvirus*, *Molluscipoxvirus*, *Parapoxvirus*, and *Yatapoxvirus*) are known to contain species capable of infecting humans (Table 2).

Virion Morphology and Composition

Poxvirus virions are some of the largest animal viruses and can be visualized by light microscopy when tagged, with fluorescent dyes for example, although the details of the virion structure remain obscure. Based on electron microscopy, the virions appear to be oval or brick-shaped structures of about 200 to 400 nm in length, with axial ratios of 1.2 to 1.7 (Figure 1A). Each virion contains a noninfectious, linear, double-stranded DNA genome that varies from 130 to 300 kbp, depending on the poxvirus species (3). The virion has more than 100 polypeptides arranged in three structures (nucleosome core [N], lateral bodies [L], and membrane envelope [M]) (Figure 1B), as visualized by electron microscopy (EM) of virions subjected to thin sectioning, cryosectioning, and/or negative-staining procedures.

Poxvirus Life Cycle

A Single Cycle of Virus Replication

Poxviruses are unique among DNA viruses because the entire replication cycle occurs in the cytoplasm of the host cell. The majority of information concerning the replication of poxviruses has been obtained by examining VACV infections in immortalized cell lines and is depicted in Figure 2. There are two forms of infectious poxvirus: the mature virion (MV), which contains a single lipid membrane bilayer surrounding the dumbbell-shaped nucleosome core and lateral bodies, and the extracellular enveloped virion (EV), which consists of the MV particle wrapped in an additional lipid membrane bilayer (Figure 2). The duration of the replication cycle varies greatly between poxviruses. *Yaba monkey tumor virus* infection takes much longer (37 to 75 hours) to obtain maximum levels of progeny in cell culture compared to VACV infection (12 to 24 hours).

Poxvirus Replication and Spread in the Host

Poxviruses infect the host mainly through the cornified epithelium of the skin or the mucosa of the respiratory tract. Infection via the skin is probably by microscopic abrasions,

doi:10.1128/9781555819156.ch33

TABLE 1 Historic standards for diagnosis of poxviruses. (Adapted from reference 170.)

Genera species	*Orthopoxvirus*					*Molluscipoxvirus*	
	Cowpox virus	*Monkeypox virus*	*Variola virus*	*Vaccinia virus*	*Yatapoxviruses/Parapoxviruses*	*molluscum contagiosum virus*	*Herpesviruses*
Lesions on chorioallantoic membranes (CAMs)	Majority of pocks are flat, poorly defined, hemorrhagic; isolated white pocks	Small pocks with central hemorrhage; isolated larger white pocks	Mono-morphic white, sharply defined dome-like pocks	Large, white or grey flat pocks	Do not form	Do not form	Varicella Zoster does not form lesions; Herpes Simplex forms small white-ish lesions
Hemagglutination activity	Absent or weakly expressed	High	Absent or weakly expressed	Marked	Absent	Absent	Absent
Scarification reaction in rabbits	Papulo-pustular rash with hemorrhage and edema	Papulo-pustular rash, sometimes with a generalized process	Absent	Papulo-pustular rash	Absent	Absent	Absent
Intracellular inclusions	Cytoplasmic A and B type	Cytoplasmic B type	Cytoplasmic B type	Cytoplasmic B type	Cytoplasmic B type	Cytoplasmic, acidophilic granular masses (Molluscum bodies)	Nuclear

which allow access of the virus to the epidermal or dermal layer; all members of *Orthopoxvirus*, *Parapoxvirus*, *Yatapoxvirus*, and *Molluscipoxvirus* potentially infect their hosts in this manner. Epidemiological evidence strongly suggests that VARV was primarily transmitted in excretions from the mouth or nose (and less commonly from scab material). The exact area of the respiratory tract that is initially infected is not known. Human monkeypox (MPXV infection of humans) initial infection occurs via the oropharynx or nasopharynx, through abrasions of the skin, or possibly through the oral cavity (4).

Poxviruses replicate locally in epidermal cells, causing changes in the cellular structure, which can be detected with histochemical stains such as eosin and hematoxylin. Differential reactivity of these stains reveals the presence of at least three virus-specific staining patterns in infected cells. Areas of basophilic staining are referred to as B-type or Guarnieri inclusion bodies, and indicate sites of virus DNA synthesis (5). Additionally, at least three different orthopoxviruses (*Ectromelia virus*, Akhmeta virus, and certain strains of CPXV) form Marchal (6) and Downie (7) bodies which are acidophilic in character, and are now referred to as A-type inclusion bodies (8). *Molluscum contagiosum virus* (MCV) has unique intracytoplasmic, eosinophilic, granular inclusion called the molluscum body, which increases in size through the virus life cycle until the keratinocyte is devoid of any intracellular structure, except virions (Figure 3).

Virus replicated at the site of infection spreads to the draining lymph node via the lymphatics and possibly in infected cells. At this point, poxvirus infection follows one of two courses. Some viruses cause either a localized, self-limited infection with little spread from the original site of inoculation; such is the case with MCV or *Yaba monkey tumor virus*. Others cause a fulminant, systemic infection characterized by a generalized rash and high mortality rate, as with smallpox and human monkeypox. In certain cases, both of these disease patterns (localized and systemic) can be caused by the same virus but in different host species.

Transmission of virus between an index case and a susceptible host can originate from the primary site of infection, which is most likely the case in localized self-limited infections such as molluscum contagiosum, or from virus produced in the "end-organ" epithelia as a result of the secondary viremia. In the case of smallpox, infectious virus was released from lesions in the mouth, nose, and pharynx into the nasal and oropharyngeal secretions during the first week of the rash.

POXVIRUSES PATHOGENIC FOR HUMANS

Twelve poxviruses have been documented to infect humans (Table 2) (9). Except for VARV and MCV, the other poxvirus diseases are zoonoses. With the exception of VARV and MPXV, these zoonotic poxviruses fail to maintain human-to-human transmission. Most human poxvirus infections occur through minor abrasions in the skin. Orf and MCV cause the most frequent poxvirus infections worldwide with the incidence of molluscum contagiosum on the rise, especially as an opportunistic infection of late-stage AIDS patients (10).

Orthopoxviruses

Five orthopoxviruses have been shown to cause disease in humans. The most notorious is smallpox, which is exclusively a pathogen of humans. The time at which smallpox emerged as a human pathogen remains uncertain (11).

TABLE 2 Genera of vertebrate poxviruses which infect humans

Genus	Species	Geographic distribution	Reservoir host	Other known infected hosts
Orthopoxvirus				
	Akhmeta virus	Asia (Georgia)	Unknown	Cows, **humans**
	Camelpox virus[a]	Africa, Asia	Camels	None
	Cowpox virus[a]	Europe, western Asia	Bank voles, long tailed field mouse	Cats, cattle, **humans**, zoo animals
	Ectromelia virus[a]	Europe	Rodents	None
	Horsepox virus[a]	Central Asia	unknown	Horses
	Monkeypox virus[a]	Western and central Africa	Unknown—likely rodent	Monkeys, zoo animals, **humans**, prairie dogs, rodents
	Taterapox virus[a]	Western Africa	Gerbils	None
	Uasin Gishu virus	Eastern Africa	unknown	Horses
	Vaccinia virus[a]	unknown	unknown	**Humans**, rabbits, cows, river buffaloes
	Variola virus[a]	Eradicated (formerly worldwide)	**Humans**	None
	Raccoonpox virus	(Eastern) United States	Raccoons	None
	Skunkpox virus	(Western) United States	Skunks	None
	Volepox virus	Western United States	California vole	None
Molluscipoxvirus	*Molluscum contagiosum virus*[a]	Worldwide	**Humans**	None
Parapoxvirus	*Ausdyk virus*	Africa, Asia	Camels	None
	Bovine popular stomatitis virus[a]	Worldwide	Cattle (beef)	**Humans**
	Orf virus[a]	Worldwide	Sheep, goats	Ruminants, **humans**
	Pseudocowpox virus[a]	Worldwide	Cattle (dairy)	**Humans**
	Red deer poxvirus	New Zealand	Red deer	None
	Sealpox virus	Worldwide	Seals	**Humans**
Yatapoxvirus	*Tanapox virus*[a]	Eastern and central Africa	Unknown	**Humans**
	Yaba Monkey Tumor virus[a]	Western Africa	Unknown—likely primates	**Humans**

[a]Completely sequenced.
Adapted from Moss, B. 2013. Poxviridae, p. 2130–2159. *In* D. Knipe and P. Howley, P. (ed.), *Fields Virology*, Lippincott Williams, & Wilkins, Philadelphia, PA. Damon, I. 2013. Poxviruses, p. 2161—2184. *In* D. Knipe and P. Howley (ed.) *Fields Virology*, Lippincott Williams & Wilkins, Philadelphia, PA.

Retrospective genomic analysis of 45 VARV isolates has segregated strains into phylogenetic subgroups, which correlate with geographic origin (11, 12). Since the global eradication of smallpox in 1979, VARV has no longer circulated in nature (13). At this time, all known stocks of VARV are held in two WHO collaborating centers: the U.S. Centers for Disease Control and Prevention (CDC) in Atlanta, GA, and the State Center of Virology and Biotechnology (VECTOR) in Kotsovo, Russia. VARV is a select agent and subject to the U.S. Division of Select Agents and Toxins regulations. Although the WHO *ad hoc* Committee on Orthopoxvirus Infections has recommended the destruction of the remaining VARV stocks, this has been delayed pending the assessment of future scientific needs for live virus.

With the eradication of smallpox, the need for continued immunization (with VACV) was diminished. Currently, vaccination in the United States is limited to the controlled immunization of personnel that handle orthopoxviruses capable of infecting humans in the laboratory, the military, and select healthcare workers (14, 15). It is recommended that laboratory workers who directly handle cultures, animals, and/or materials contaminated or infected with non-highly attenuated VACV and their recombinants or other

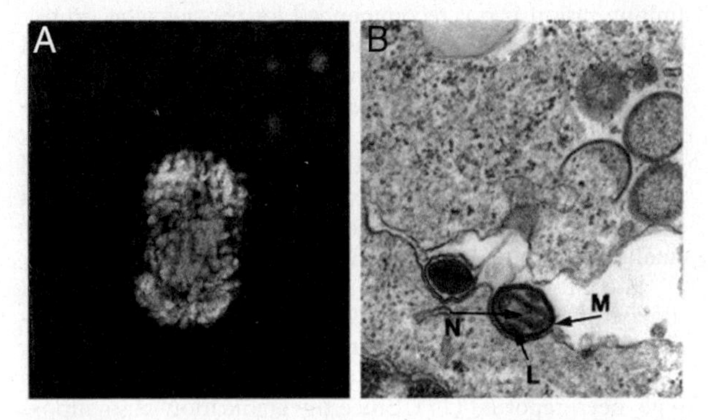

FIGURE 1 Morphology and structure of a poxvirus virion. (A) Electron micrograph of a negative-stained M form of a *Molluscum contagiosum virus* virion. Magnification, ×120,000. Note the textured surface. (B) Electron micrograph of a thin section of a *Cowpox virus* virion. N, nucleosome; L, lateral body; M, membrane. Note the immature forms of the virus in various stages of morphogenesis in the upper portion of the photograph. Magnification, ×120,000.

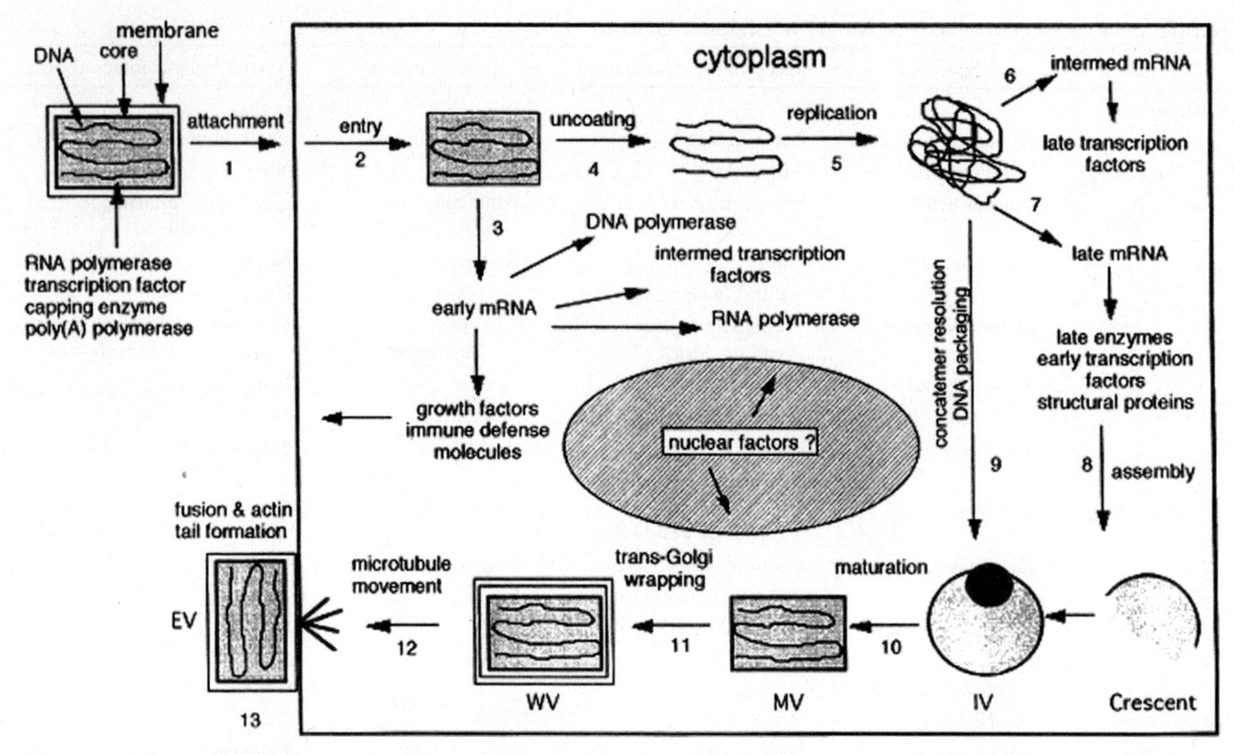

FIGURE 2 Replication cycle of *Vaccinia virus*. A virion, containing a double-stranded DNA genome, enzymes, and transcription factors, attaches to a cell (1) and fuses with the cell membrane, releasing a core into the cytoplasm (2). The core synthesizes early mRNA that is translated into a variety of proteins, including growth factors, immune defense molecules, enzymes, and factors for DNA replication and intermediate transcription (3). Uncoating occurs (4), and the DNA is replicated to form concatemeric molecules (5). Intermediate genes in the progeny DNA are transcribed, and the mRNA is translated to form late transcription factors (6). The late genes are transcribed, and the mRNA is translated to form virion structural proteins, enzymes, and early transcription factors (7). Assembly begins with the formation of discrete membrane structures (8). The concatemeric DNA intermediates are resolved into unit genomes and packaged in immature virions (IV) (9). Maturation proceeds to the formation of infectious intracellular MV (10). The MVs are wrapped by modified *trans* Golgi and endosomal cisternae (11), and the wrapped virions (WV) are transported to the periphery of the cell along microtubules (12). Fusion of the WVs with the plasma membrane results in release of extracellular EV (13). The actin tail polymerizes in the cytoplasm beneath the EV (13). Although replication occurs entirely in the cytoplasm, nuclear and cytoplasmic cell factors may be involved in transcription and assembly. (Reprinted from Moss, 2007 with permission from the publisher.)

orthopoxviruses capable of infecting humans (e.g., MPXV, CPXV, and others) be immunized with VACV vaccine. Immunization is not recommended for persons who do not directly handle virus cultures or who do not work with materials contaminated or animals infected with these viruses. Reimmunization should be carried out according to the Advisory Committee on Immunization Practices (ACIP) recommendations. Administering physicians should contact the CDC Drug Service and the CDC National Immunization Program for details on obtaining VACV smallpox vaccine and for advice on clinical questions (14).

The other three orthopoxviruses which cause human disease are also zoonotic. CPXV causes a disease presentation similar to that of the emerging vaccinia-like viruses. Disseminated human cowpox occurs rarely (16) but fatalities have been reported (17). Since the eradication of smallpox, monkeypox is typically the most severe *Orthopoxvirus* disease in humans. Clinical signs of human monkeypox were difficult to distinguish from smallpox; the primary distinctive feature was pronounced lymphadenopathy seen in monkeypox patients. Fortunately, MPXV is less efficiently spread from human to human (18) and has a lower case fatality rate than smallpox (4). Although MPXV is endemic in Africa,

FIGURE 3 Electron micrograph of a thin section of a *Molluscum contagiosum virus*-infected cell or molluscum body. All of the cellular organelles are beyond recognition, having been pushed to the periphery of the cell by the masses of virions. Magnification, ×3,000.

an outbreak of human monkeypox occurred in the United States in 2003. The epidemic was caused by infected African rodents imported into the United States (19). The discovery of the international dispersal of a zoonotic *Orthopoxvirus* supports the concept of orthopoxviruses as a potential emerging infectious diseases. Recently, a new *Orthopoxvirus* was identified, Akhmeta virus, to cause zoonotic infections within cattle herders in Georgia (country) (8).

Pathogenesis

Transmission

Prior to eradication, smallpox was believed to be transmitted person to person via the upper respiratory tract, with virus released in oropharyngeal secretions of patients who had a rash (4). Monkeypox person-to-person transmission is thought to occur via oropharygeal secretions but less efficiently ($R_{oSmallpox}$ = 3.5 to 6 versus $R_{oMonkeypox}$ = 0.8) (20, 21); however, an extended chain of six generations of confirmed human-to-human transmission has been documented (18). The exact mode of transmission of zoonotic orthopoxviruses from an animal source to humans is not known but may be via the oropharynx or nasopharynx or through abrasions of the skin (i.e., butchering of nonhuman species). CPXV infection is usually acquired by direct introduction of the virus from an animal source into minor abrasions in the skin; however, 30% of human infections have no known risk factor for infection or obvious route of inoculation (22). VACV and vaccinia-like viral infections are usually due to close contact with a recent smallpox (VACV) vaccinee or infected cattle, respectively (23–27). The two instances of Akhmeta virus infection in humans were reported in individuals who had contact with infected cattle (8). These observations represent the current knowledge on transmission of this newly identified *Orthopoxvirus*.

Lesion Histopathology

Orthopoxvirus lesions are characterized with epidermal hyperplasia; with infected cells becoming swollen and vacuolated and undergoing “ballooning degeneration.” The cells contain irregular, faint, B-type inclusion bodies. Skin lesions caused by CPXV in nonhuman animals contain A-type inclusion bodies in epidermal cells, sebaceous glands, and endothelial cells; however, similar inclusion bodies are not observed in the few human cowpox lesions examined (22). A-type inclusion bodies have also been observed within tissue culture propagated Akhmeta virus (8).

Epidemiology

Geographical Distribution

MPXV is found in the tropical rain forests of countries in western and central Africa. There are two clades of MPXV (28), with the less virulent strains occurring within West Africa. The Congo Basin clade has higher reported case fatality rates (up to 12%) and is found most notably in the Democratic Republic of Congo (formerly Zaire). Human-to-human transmission has only been documented with MPXV from the Congo Basin clade. The MPXV reservoir is unknown but is most likely one or more rodent species. Viable MPXV has been isolated from three African species imported into the United States: giant pouched rats (*Cricetomys* sp.), rope squirrels (*Funisciurus* sp.), and dormice (*Graphiuris* sp.) (29). In fact, viable MPXV has only been recovered twice in African wildlife; from a moribund squirrel discovered in Zaire and a deceased sooty mangabey (*Cercocebus atys*) in Cote d'Ivoire (30, 31).

CPXV is endemic to Europe and some western states of the former Soviet Union. Rodents (voles, wood mice, and rats) have been implicated as reservoirs of CPXV in Great Britain; with humans, cows, zoo animals, and cats as incidental hosts (22,32–36). Vaccinia-like viruses are known to be endemic to Brazil and India and are found in dairy cattle and buffalo herds (23–27); the possible wild animal reservoirs of these viruses are unknown. The novel Akhmeta virus has only been isolated twice from Georgia (country) in Asia (8). Further investigations will determine the geographic distribution of this virus.

Prevalence and Incidence

There are increasing reports of vaccinia-like viral infections associated with bovine contact (23–27) and human cowpox, usually transmitted by the domestic cat, cows, and rodents (22, 33, 36). However, MPXV infections are by far the most serious and prevalent human *Orthopoxvirus* disease. Reports document increasing cases of human monkeypox in the Democratic Republic of Congo: 51 PCR-confirmed monkeypox cases from 2001 to 2004 (37); 760 PCR-confirmed monkeypox cases from 2005 to 2007 (38). The reemergence/increase in monkeypox cases may be due to waning immunity following cessation of the immunization program, increased encroachment of larger human populations into the habitat of the animal reservoir, heightened surveillance, or a combination of these factors and possibly others. Furthermore, MPXV is an emerging zoonosis of potential worldwide concern. Human monkeypox has been reported in southern Sudan and there was an outbreak in the United States in 2003, which are outside of the known geographic range (39). These outbreaks reinforce the concept that poxviruses can be encountered outside their normal geographic range and may pose a threat of becoming established as agents of persistent zoonotic disease in novel ecologies.

Clinical Signs, Symptoms, and Severity

Approximately 12 days after infection with MPXV, fever and headache occur. This is followed 1 to 3 days later by a rash and generalized lymphadenopathy with illness lasting 2 to 4 weeks. Although the number of lesions is variable the rash typically appears first on the face and generally has a centrifugal distribution. Skin lesions begin as macules which progress over the course of days from papules to vesicles to pustules. At about 8 or 9 days after the onset of rash, the pustules become umbilicated and dry up, with a crust forming by 14 to 16 days after the onset of the rash. Most skin lesions are about 0.5 cm in diameter (4).

With human cowpox infection, usually a solitary lesion appears on the hands or face. The lesion appears as an inflamed macule and progresses through an increasingly hemorrhagic vesicle stage to a pustule which ulcerates and crusts over by the end of the second week, becoming a deep-seated, hard black eschar 1 to 3 cm in diameter (22). This lesion can be extremely painful and the patient can present with systemic symptoms, including pyrexia, malaise, lethargy, sore throat, and local lymphadenopathy. Complete recovery takes from 3 to 8 weeks. No human-to-human transmission has been reported. Complications can include ocular or generalized infections (35); the latter occur in patients with atopic dermatitis, allergic asthma, or atopic eczema and, in one case, was associated with death (22). Similarly, vaccinia-like viruses cause painful localized lesions usually on the hands or arms, usually without the hemorrhagic manifestation (23–27). Clinical presentation in the two documented Akhmeta virus cases was similar to that seen with human cowpox (8). Both patients experienced fever and lymphdenopathy with painful localized lesions present on the hands. One patient's lesions also developed into thick eschars.

Diagnosis and Differential Diagnoses

Previously, the diagnosis of human monkeypox typically required clinical (rash), epidemiologic (equatorial Africa), and laboratory (brick-shaped virion in scab material and/or orthopoxviral antigen, nucleic acid, or virus detected in lesions) findings. Although the rash with associated lymphadenopathy was usually pathognomonic, the sporadic nature of the disease and the similarity of the rash of human monkeypox to the rash of other infectious agents such as chickenpox (*Varicella-zoster virus*), tanapox, or syphilis limits diagnosis based on clinical presentation (4). In addition to the cocirculation of varicella viruses and MPXV in Africa, the spread of MPXV outside of its normal geographic distribution presents challenges to using clinical and epidemiology for diagnosis. Rapid identification of MPXV is essential due to its clinical similarity to smallpox and the increasing concern of potential bioterrorism. Laboratory diagnostic assays (real time PCR) are essential for rapid identification of human monkeypox cases.

Cowpox diagnosis is rarely based on clinical findings and usually requires laboratory results. VACV infections are usually identified based on epidemiologic and laboratory findings. Despite differences in rash presentation and/or evolution, generalized infection can be misdiagnosed as eczema herpeticum and localized infection is most frequently misdiagnosed as orf or milker's nodules, herpesvirus reactivation, or anthrax (22).

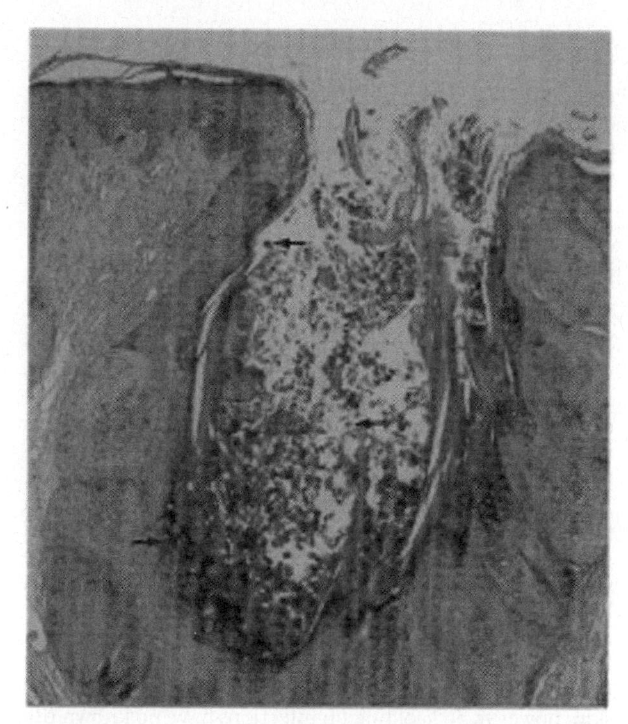

FIGURE 4 Histologic section of a molluscum contagiosum lesion. A hematoxylin- and eosin-stained wax section of a skin biopsy specimen showing hyperkeratosis and acanthosis of the epidermis. Note the hyperplasia associated with the lesion causes severe invagination of the epidermis without loss of integrity of the basal layer. Arrows indicate molluscum bodies. Magnification, ×100.

Molluscum Contagiosum Virus

Pathogenesis

Transmission

Molluscum contagiosum was first clinically recognized in 1817 (40). The incubation period of molluscum contagiosum is poorly understood, but epidemiologic studies suggest it ranges from weeks to months. Molluscum contagiosum is observed in children and adults. Nonsexual transmission is a consequence of infection by direct contact or through fomites (41). Transmission between persons in the absence of fomites requires intimate contact. Lesions can be commonly observed on opposing epithelial surfaces and the virus can be further spread by autoinoculation.

Lesion Histopathology

MCV has one of the most limited host cell tropisms of any virus, replicating only in the human keratinocyte of the epidermis (42). MCV replicates in infected keratinocytes and lesions become more prominent during the 9 to 15 days it takes the infected keratinocyte to reach the stratum granulosum. By light microscopy, these cells stain as hyaline acidophilic masses, are referred to as molluscum or Henderson-Patterson bodies, and are pathognomonic for disease (Figure 4). Higher magnification of the molluscum body reveals a cell almost entirely filled with virions (Figure 3). As a consequence of these hypertrophied, infected cells and hyperplasia of the basal cells, the molluscum lesion extends above the adjacent skin as a tumor and projects down into the dermis without breaking the basement membrane.

Epidemiology

Geographic Distribution

MCV has a worldwide distribution but is more prevalent in the tropics. Analysis of genomic DNA from MCV isolates has revealed the existence of at least four virus subtypes. Several studies suggest the distribution of subtypes can vary geographically (43–46).

Prevalence and Incidence

For non-sexually transmitted molluscum contagiosum, the disease is more prevalent in the tropics. For example, molluscum contagiosum was diagnosed in 1.2% of outpatients in Aberdeen, Scotland between 1956 and 1963, while 4.5% of an entire village in Fiji in 1966 had the disease (41, 47, 48). Genital molluscum contagiosum has been reported in higher numbers; an increase in cases of 400% in England between 1971 and 1985, and a 10-fold increase in the United States between 1966 and 1983 (48). Before effective retroviral therapies, molluscum contagiosum was a common and sometimes severely disfiguring opportunistic infection of human immunodeficiency virus (HIV)-infected patients (5 to 18%), especially those with severely depressed $CD4^+$-T-cell numbers (49).

Diagnosis

Clinical Signs, Symptoms, and Severity

Clinically, molluscum contagiosum presents as single or small clusters of lesions in immunocompetent individuals.

Lesions begin as pimples and become umbilicated, epidermal, flesh-colored, raised nodules of 2 to 5 mm in diameter. There are no other signs or symptoms associated with the disease. The lesions are generally painless, appearing on the trunk and limbs (except palms and soles) in the non-sexually transmitted disease. In children, disease can also be fairly common in the skin of the eyelids and can be complicated by chronic follicular conjunctivitis and later by a superficial punctate keratitis (50). As a sexually transmitted disease in teenagers and adults, the lesions are mostly on the lower abdominal wall, pubis, inner thighs, and genitalia. Lesions can persist for 2 weeks to 2 years, and reinfections are common. A semisolid caseous material can be expressed from the center of the lesion, is rich in molluscum bodies, and probably is responsible for disease transmission.

In immunocompromised individuals (especially in persons with HIV), molluscum contagiosum is typically not self-limiting and larger, more frequent lesions are observed, especially on the face, neck, scalp, and upper body, with multiple adjacent lesions sometimes becoming confluent. Molluscum contagiosum can be considered a cutaneous marker of severe immunodeficiency. Rarely, the disease will present as a large lesion (>5 mm in diameter) called "giant molluscum." Giant molluscum lesions (>5 mm) have been reported more frequently in individuals with HIV (49, 51), although the large lesions were observed in New Guinea prior to the recognition of HIV.

Diagnosis and Differential Diagnoses

The diagnosis of molluscum contagiosum is usually made clinically based on gross appearance of the lesions and their chronic nature. Laboratory confirmation is easily obtained by hematoxylin and eosin staining of a tissue section from a biopsy specimen or by a squash preparation of expressed material from the lesion (Figure 4). Several PCR RFLP and real time PCR assays have been described for *molluscum contagiosum virus* (46,52–54) which allow for subtyping when indicated.

Molluscum contagiosum (especially giant molluscum) can be confused with a number of other disorders such as keratoacanthoma, warty dyskeratoma, syringomas, hidrocystomas, basal cell epithelioma, trichoepithelioma, ectopic sebaceous glands, giant condylomata acuminata, chalazion, sebaceous cysts, verrucae, and milia or granuloma on eyelids (55–57). In immunodeficient patients, disseminated cutaneous cryptococcosis and histoplasmosis may resemble molluscum contagiosum (55). An inflamed molluscum lesion without the association of typical lesions may be mistaken for a bacterial infection.

Parapoxviruses

The parapoxviruses, including *Orf virus*, *Bovine papular stomatitis virus*, *Pseudocowpox virus* (milker's nodule), and *Sealpox virus*, cause occupational infections of humans, with orf infections being the most common. Wildlife activities (for example, skinning animals such as deer and reindeer) have also been sources of *Parapoxvirus* infection (58). It is believed that the majority of human *Parapoxvirus* infections go unreported, as many sheep farmers and rural physicians are aware of the diseases and make a diagnosis based solely on clinical findings and do not seek treatment. No human-to-human transmission of *Parapoxvirus* infections has been reported.

Pathogenesis

Transmission

Direct transmission of parapoxviruses has been observed as a consequence of contact with lesions on cattle, bottle-feeding lambs, animal bites to the hand, and contact with sheep and goat products during slaughter. Fomites found on items such as wooden splinters, barbed wire, or farmyard surfaces such as soil, feeding troughs, or barn beams have been implicated as sources for possible virus inoculation. No human-to-human transmission has been reported.

Persons in direct physical contact with eight different species of pinnipeds have reported an "orf-like" lesion after being bitten by infected animals (59–62). After the advent of reliable PCR tests for parapoxviruses (63), *Sealpox virus* was confirmed as the causative agent of the orf-like lesion (64).

Lesion Histopathology

Histopathological features of human orf and pseudocowpox lesions are indistinguishable and are similar to human lesions caused by bovine papular stomatitis and sealpox viral infections. Because orf infections are the most common human *Parapoxvirus* infections, only the histopathological features of human orf will be presented (65).

Epidermis

The most striking change in the epidermis during *Orf virus* infections is hyperplasia in which strands of epidermal keratinocytes penetrate into the dermis. Generally, a mild-to-moderate degree of acanthosis is detected, and parakeratosis is a common feature. Cytoplasmic vacuolation, nuclear vacuolation, and deeply eosinophilic, homogeneous cytoplasmic inclusion bodies, often surrounded by a pale halo, are also characteristic of the infection. An intense infiltration of lymphocytes, polymorphonuclear leukocytes, or eosinophils frequently involves the epidermis.

Dermis

A dense, predominantly lymphohistiocytic inflammatory cell infiltrate is present in all orf cases. Also there is marked edema both vertically and horizontally that may contribute to the overall papillomatous appearance. The most striking feature of the infected dermis is the massive capillary proliferation and dilation.

Epidemiology

Geographic Distribution

Orf in sheep and goat populations has been reported in Canada, the United States, Europe, Japan, New Zealand, and Africa. Pseudocowpox occurs in dairy herds of European-derived cattle found in all parts of the world. Bovine papular stomatitis is similarly distributed but is more often associated with beef rather than dairy animals. Seals and other pinniped populations worldwide have been found to harbor sealpox.

Prevalence and Incidence

In a 1-year New Zealand study, 500 meat workers from an at-risk population of 20,000 were infected with orf, with the highest risk (4%) of infection associated to those involved in the initial butchering of the sheep (66). Serologic surveys of orf-infected sheep and goat herds yielded orf antibody prevalences of up to 90%. The high seroprevalence of orf antibody in herds is believed to be associated with the highly stable nature of the orf virion, which contaminates the environment and causes reintroduction (67, 68).

Pseudocowpox has been found to be endemic in cattle herds in West Dorset, England (69). Pseudocowpox and bovine papular stomatitis are probably endemic in all European-derived dairy herds. *Sealpox virus* was identified as a unique species of *Parapoxvirus* in 2002 (63), but disease has been seen in pinnipeds found in Europe and North America since 1969 (70).

Diagnosis

Clinical Signs, Symptoms, and Severity

The clinical presentation of orf usually occurs 3 to 4 weeks postinfection. The human disease involves the appearance of single or multiple nodules (diameters of 6 to 27 mm) (65), which are sometimes painful, usually on the hands, and less frequently on the head or neck. Orf infection can also be associated with a low grade fever, swelling of the regional lymph node, and/or erythema multiforme bullosum. The orf lesion characteristically goes through a maculopapular stage in which a red center is surrounded by a white ring of cells which is surrounded by a red halo of inflammation; however, patients usually present later when the lesion is at the granulomatous or papillomatous stage 3 to 4 weeks following the initial infection. Resolution of the disease occurs over a period of 4 to 7 weeks (71), usually without complication; however, autoinoculation of the eye may lead to serious sequelae. Enlarged lesions can arise in humans suffering from immunosuppressive conditions, burns, or atopic dermatitis (71, 72). Also, lesion healing can be complicated by bullous pemphigoid (73). Reinfections have been documented (74).

Human pseudocowpox lesions usually appear on the hands and are relatively painless but may itch. The draining lymph node may be enlarged. Lesions are first observed as round, cherry-red papules, which develop into purple, smooth nodules of up to 2 cm in diameter and may be umbilicated. The lesions rarely ulcerate (74). The nodules are gradually adsorbed and disappear in 4 to 6 weeks (75).

Human bovine papular stomatitis lesions occur on hands, diminish after 14 days, and are no longer evident 3 to 4 weeks after onset (75). The lesions appear as circumscribed wart-like nodules which gradually enlarge until they are 3 to 8 mm in diameter (75).

Sealpox causes lesions clinically similar to orf and was initially identified in a marine mammal research technician (64).

Diagnosis and Differential Diagnoses

Diagnosis of *Parapoxvirus* infection is by clinical (lesion morphology), epidemiological evidence (recent contact with cattle, sheep, or pinnipeds), and EM of negative-stained lesion material (presence of ovoid particles with crisscross spindles) (65). Without knowing the animal source of the infection, orf cannot be differentiated from milker's nodule (pseudocowpox) based on clinical findings, histology, or EM (65). PCR-based diagnostic assays are important for identification of the *Parapoxvirus* (63,76–79).

Atypical giant orf lesions in patients who are immunocompromised or suffering from burns or atopic dermatitis may be confused with pyogenic granuloma (71, 72, 80).

Yatapoxviruses

The genus *Yatapoxvirus* has two members, *Tanapox virus* and *Yaba monkey tumor virus*, which are serologically related. Originally thought to be a third species of *Yatapoxvirus*, Yaba-like disease in monkeys is caused by the same virus that causes tanapox in humans (81–83) as evidenced by DNA restriction endonuclease analysis of genomic DNA (84) and by genomic sequencing (85, 86). *Yaba monkey tumor virus* has been isolated only from animal handlers, whereas *Tanapox virus* has been found to be acquired by humans in riverine or forested areas of Africa (81, 82, 87).

Pathogenesis

Transmission

Tanapox infection may occur via scratches or potentially by arthropod vectors. *Yaba monkey tumor virus* is a very rare infection of animal handlers at nonhuman primate facilities. There is no evidence for human-to-human transmission with either virus, and autoinoculation of virus to other areas of the body is not common.

Lesion Histopathology

Little is known concerning the pathology of yatapoxviruses except from the study of *Yaba monkey tumor virus* in nonhuman primate models (88).

Epidemiology

Geographic Distribution

Tanapox is endemic in equatorial Africa (89). The animal reservoir is not known. *Yaba monkey tumor virus* appeared in primate colonies, but has yet to be seen in nature.

Prevalence and Incidence

In the town of Lisa (population, 70,000) in northern Zaire, 264 laboratory-confirmed tanapox cases were observed between 1979 and 1983 (89). More recently, a case in a traveler returning from an extended stay in a forested area of the Republic of Congo was reported (87). There have been no reported human cases of *Yaba monkey tumor virus* in over 2 decades.

Diagnosis

Clinical Signs, Symptoms, and Severity

In most patients with tanapox, fever (38 to 39°C) commenced 1 to 2 days prior to skin eruptions and was frequently accompanied by severe headache, backache, and prostration. In most patients, only a single lesion was observed which developed on parts of the body not usually covered by clothes. When multiple lesions were observed, the lesion number ranged from 2 to 10. By the end of the first week after infection, the lesion is greater than 10 mm in diameter, with a large erythematous areola several centimeters wide surrounded by edematous skin. The lesions can develop into large nodules but more likely ulcerate without pus. The maximum diameter of the lesion is reached in the second week (89). Regional lymph nodes became enlarged with lesion development. Lesions, nodular in nature, usually disappeared spontaneously within 6 weeks, unless there was a secondary infection (89).

Diagnosis and Differential Diagnoses

Diagnosis of tanapox has historically been made by a combination of clinical (lesion character and number), epidemiologic (equatorial Africa), and laboratory (enveloped brick-shaped virions in lesion material) findings (89). Patients with multiple lesions can be misdiagnosed as having human monkeypox. PCR analysis of lesion material can be very useful in providing speciation of the poxvirus causing the infection (87, 90).

TREATMENT AND PREVENTION

Except for molluscum contagiosum, most current human poxvirus infections are zoonoses which fail to maintain human-to-human transmissions. Since MCV causes a benign, self-limiting disease in immunocompetent patients, there is a perceived lack of urgency for development of prevention strategies. Because of the chronic nature of molluscum contagiosum, curettage and cryotherapy of lesions are therapeutic options. Cryotherapy is relatively painless, cost-effective, and yields good cosmetic results; with patients infected with HIV, this treatment approach has the added advantage of mitigating the risk of disease transmission to medical personnel. Only removal of giant molluscum lesions usually results in scar formation. More-recent studies have suggested that other topical treatments (cidofovir and imiquimod) may be effective therapeutics for molluscum contagiosum and orf, even with immunocompromised patients (72,91–97).

Currently, the management of zoonotic *Orthopoxvirus*, *Parapoxvirus*, and *Yatapoxvirus* infections is largely supportive. There are no FDA approved and licensed systemic or topical chemotherapeutic agents commercially available to treat poxvirus infections. Vaccinia immune gamma globulin (VIG) has been useful in a number of human VACV infections (98–104) and is licensed for use in treatment of vaccine-related adverse events associated with direct viral replication. Prevention of secondary bacterial infections through the use of antibiotic ointments is also an option. In the case of cowpox, steroids are contraindicated and may exacerbate the illness (22). Current ongoing research has identified several potential antiviral therapies for treatment of serious orthopoxviral infection, some of which are in clinical trials (ST-246 and cidofovir) and undergoing review by the FDA (105, 106). Currently, the smallpox vaccine (Acambis 2000) licensed in the United States is non-attenuated VACV and has been stockpiled in the event mass vaccinations are required to contain an *Orthopoxvirus* pandemic. Research continues to identify an efficacious *Orthopoxvirus* vaccine that is safe for use by immune-impaired persons using highly attenuated strains of VACV. The replication-defective MVA strain was identified after serial passages (572 times) of VACV on chicken embryo fibroblasts (107, 108). MVA was first used as a vaccine at the end of the smallpox eradication campaign in Germany (109). Several studies have shown the safety of MVA in immunocompromised animal models and demonstrated that MVA vaccination provided protective immunity to MPXV challenge in nonhuman primate models (110–114). Although a good animal model for VARV infection does not exist, current research utilizing *in vitro* methods determined that MVA vaccination of humans induces an immune response capable of neutralizing VARV as efficiently as vaccination with replication-competent VACV (115). The United States is currently incorporating the MVA-derived vaccine (IMVAMUNE) to the national stockpile as it undergoes review for licensure by the FDA (116). Another possible vaccine candidate is the replication-competent attenuated VACV strain LC16m8, developed in Japan by passaging (45 times) VACV Lister through rabbit kidney cells (117). LC16m8 was selected for its decreased neurovirulence and was safely used to vaccinate over 50,000 children during the 1970s. In recent years, Japan has increased its national stockpile of LC16m8 and more detailed animal and clinical studies have been initiated (117–122). Other future candidate vaccines include NYVAC and dVV-L, attenuated VACV strains derived from specific deletions of genes (123–125).

LABORATORY PROCEDURES FOR DETECTING POXVIRUSES

Collection, Handling, and Storage of Specimens

In the United States, poxvirus diagnostic specimens should be evaluated by public health laboratories, many of which are members of the Laboratory Response Network (LRN). The LRN is a network of about 150 laboratories distributed throughout the United States and overseas that are equipped and trained to perform diagnostic assays for *Orthopoxvirus* infections (for more information on the LRN, go to http://http://www.bt.cdc.gov/lrn/factsheet.asp). An algorithm was created to assist clinicians in the identification of patients infected with smallpox and differentiation from other pustular rash illnesses (http://emergency.cdc.gov/agent/smallpox/diagnosis/pdf/spox-poster-full.pdf). The algorithm includes rule outs for common rash-like illnesses that may be confused with *Orthopoxvirus* infections and provides clinical clues that help differentiate them. In the absence of endemic smallpox, if an infection has a high suspicion of smallpox after careful clinical evaluation, the nearest government health department and the Federal Bureau of Investigation (FBI) must be immediately notified and samples sent simultaneously for VARV diagnostic testing at an LRN laboratory and the WHO Collaborating Center for Smallpox and other poxvirus infections at the CDC in Atlanta, GA. Final confirmation and characterization of VARV from initial smallpox cases would require additional examination (including DNA sequence analysis) at the CDC (for more details on smallpox definition and diagnostics, etc., go to http://emergency.cdc.gov/agent/smallpox/).

Appropriate biosafety level precautions must be taken for the handling, transport, and processing of suspect infected lesion material (84, 126). Guidelines have been published for sample collection from patients believed to be infected with an *Orthopoxvirus* (http://emergency.cdc.gov/agent/smallpox/vaccination/vaccinia-specimen-collection.asp and http://www.cdc.gov/ncidod/monkeypox/diagspecimens.htm). Skin lesions are typically the specimen of choice for diagnosis for any poxvirus infection. Virions are usually present in this material and can remain viable even after several weeks of storage with or without refrigeration. Because lesion material may be analyzed by several different methods, it is important that sufficient quantities of specimen are collected for submission. In brief, for suspected orthopoxiral, parapoxviral, and yatapoxviral infections, at least two lesion specimens should be obtained and may be acquired from lesions at the macular-papular, vesicular-pustular, and/or crusting stages. During the vesicular-pustular stage, the fluid (including cells from the base of the lesion where the virus is often found in high concentration) can be collected onto dry swabs, as a thick droplet or touch prep onto glass slides or on plastic-coated EM grids (126, 127). Carrier media should not be added to any of the specimens, as it may dilute specimens for EM. The specimens can be stored for a short time at 4°C; however, −20°C or −70°C storage is preferable for longer time frames. An in-depth guide for the collection and shipping of potential smallpox samples is available online (http://emergency.cdc.gov/agent/smallpox/response-plan/files/guide-d.pdf).

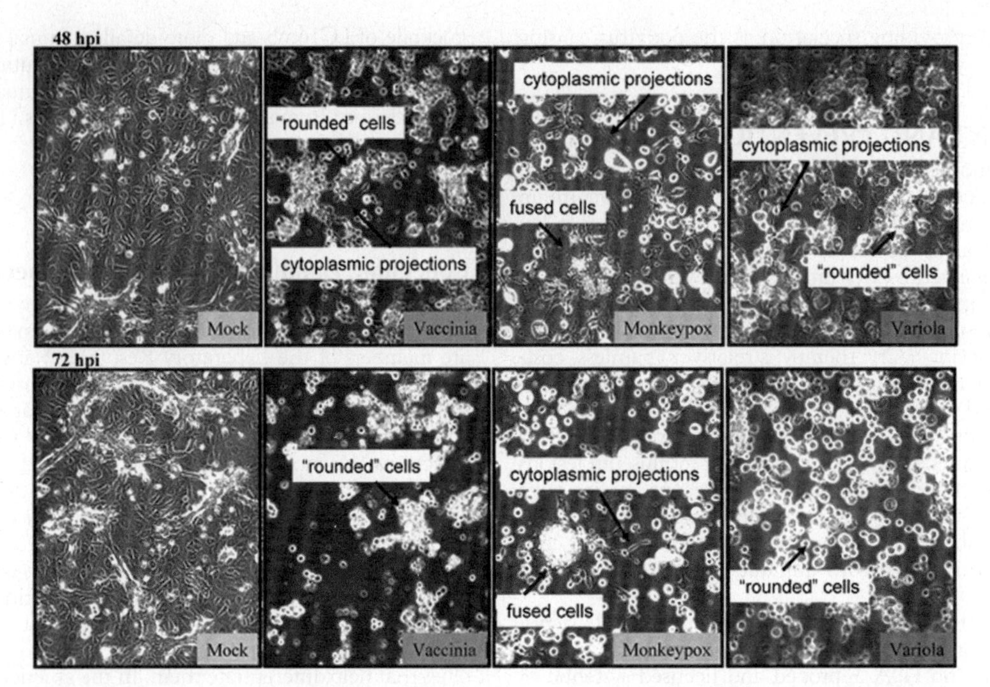

FIGURE 5 Cytopathic effect of *Orthopoxvirus* infection within tissue cell culture. African Green Monkey Kidney cells (BSC-40) were infected at a low multiplicity (multiplicity of infection = 0.01) to mimic what might be found within a clinical specimen. Cells were either mock infected or infected with one of the following orthopoxviruses: *Vaccinia virus*, *Monkeypox virus*, or *Variola virus*. Cells were observed daily, and photographs were taken. The number of hours postinfection (hpi) and characteristics of cytopathic effect are denoted. Magnification, ×10.

International and/or local shipping regulations must be followed when preparing suspect etiologic agents for shipment. Guidelines and links to these regulations can be found at http://www.cdc.gov/ncidod/monkeypox/pdf/specimenguide.pdf. The specimens should be initially packaged in a rigid receptacle matched to the size of the specimen with a leak-proof cap. It is critical to individually contain specimens to prevent cross-contamination. The primary receptacles should then be wrapped in sufficient absorbent material to absorb their entire contents should a leak occur and placed into a watertight, crush-resistant outer container. This secondary package should then be surrounded by dry or wet ice within an outer package that meets requirements of the U.S. Department of Transportation and International Air Transport Association (IATA) (http://www.iata.org/whatwedo/cargo/dgr/Documents/DGR52_PI650_EN.pdf) regulations.

Laboratory Diagnostic Methods

Identification and differentiation of poxvirus species have been carried out using a variety of analytical approaches, including virus growth, histology, EM, and serology; however, analysis of lesion material for poxvirus DNA by PCR is currently the most rapid and sensitive primary laboratory diagnostic technique. Viral culture analysis remains the sole method for determining whether the sample contains viable virus (Figure 5). In the absence of endemic smallpox, viral culture should never be attempted if a sample is suspected to contain VARV unless at one of the two WHO collaborating centers sanctioned for manipulation with live virus.

Histology

Histological examination of poxvirus lesions can be especially informative with both parapoxviruses and MCV. Parapoxviral lesions exhibit an epidermis which is hyperplastic with strands of epidermal keratinocytes penetrating into the dermis. The molluscum contagiosum lesion has pear-shaped lobules filled with intracytoplasmic Henderson-Patterson or molluscum bodies. In practice, histology is usually employed only for confirmation of MCV infections. This can be carried out via wax sections or a squash preparation. For the squash preparation, the keratotic dome-shaped molluscum lesion is placed on a regular slide and under a coverslip or second slide. The lesion is flattened and can be examined by light microscopy directly or after staining with Wright's or methylene blue stains. Round-to-ovoid molluscum bodies up to 37 by 27 μm are diagnostic of molluscum contagiosum. Alternatively, a biopsy specimen is fixed in 10% formal saline and submitted for wax embedding, sectioning at 5 μm, and staining with hematoxylin and eosin. Microscopic examination should provide a field of view similar to that shown in Figure 4.

EM

EM has been an important, dependable, and rapid method for diagnosis of a poxvirus infection when the results are considered along with clinical findings (Figure 1). The difference between genera (*Orthopoxvirus* and *Parapoxivrus*) can be easily ascertained by EM as well as other rash causing etiology but differences between species are not captured. If virus lesion material is not available, then serology is an alternative approach for establishing the etiology of the disease. Unfortunately, due to cross-reactivity, current serologic assays cannot differentiate between species.

DNA Analysis

Since the advent of PCR DNA amplification, primer pairs from the gene encoding the major A-type inclusion body protein (128) or hemagglutinin (127) have been used to distinguish among known *Orthopoxvirus* species. Importantly, this PCR analysis can be used directly on nucleic acid extracted from a scab or dried vesicle fluid (84, 127, 129). Additional PCR assays have also been validated for the identification of parapoxviruses such as *Sealpox virus* and *Orf virus* (63, 76, 77, 130, 131), the *Yatapoxvirus Tanapox virus* (87), and MCV (46, 52, 53).

Gel chromatography visualization of PCR product with or without subsequent RFLP analysis is a useful laboratory diagnostic for identification of poxviral infection to the species level despite being relatively time consuming (e.g., 8 to 12 hours). Other nucleic acid diagnostic approaches include random amplified polymorphic DNA fragment length polymorphism for *Orthopoxvirus* species and strain discrimination (132, 133). Microchips have also been utilized to identify poxvirus species by discernment of PCR amplified, fluorescence-labeled DNA fragments which hybridize to *Orthopoxvirus* species-specific immobilized DNA. In recent years, real time PCR of primary lesion material has been demonstrated to provide increased diagnostic sensitivity (often down to a few genome copies), increased specificity (can differentiate a single nucleotide polymorphism), and decreased time to result. Several different technologies exist for real time PCR, all utilizing primer pairs to amplify a short DNA sequence specific to the viral agent being studied. Most real time PCR assays also employ a probe that, upon binding to or amplification of the target DNA sequence, releases fluorescence (134), allowing measurement of DNA amplification. Many of the real time PCR assays for detection of orthopoxviral DNA (135–157) have been validated with a multitude of virus sample types and were instrumental in the diagnosis of monkeypox cases during the 2003 outbreak within the United States (149). Similarly, real time PCR assays for the detection of infections of MCV, parapoxviruses, and yatapoxviruses have also been developed (54, 78, 79, 90, 158). The efficacy of real time PCR assays has been demonstrated through analysis of a variety of specimens acquired from multiple sources and maintained under various conditions. However, vigilance is necessary to ensure the specificity of real time PCR assays is retained as new isolates within the highly similar *Orthopoxvirus* genus are identified (159).

Poxvirus Diagnostics in the Future

Real time PCR assays currently allow rapid and definitive diagnosis of the species of poxvirus causing an infection but still rely upon time-consuming processing of samples. Portable PCR machines have been designed for use in the field, requiring limited sample processing (160, 161). Robotic systems for nucleic acid extractions have become more prevalent, allowing more efficient extractions with less risk of sample contamination (162). Mass spectroscopy is being investigated as an adjunct to PCR for identification of amplified sequences (163–166). Future assays may rely upon pyrosequencing technologies, which can provide complete genomic sequencing, perhaps even from primary clinical material, in much less time than traditional Sanger sequencing (167, 168). Rapid sequencing not only allows for identification of the viral species but can also provide valuable information like subtyping or identifying drug resistance mutants (28, 169). Finally, possible antigen-capture methods may provide alternative protein-based approaches to poxvirus diagnostics in the future.

Disclaimer: The findings and conclusions in this report are those of the author(s) and do not necessarily represent the official position of the Centers for Disease Control and Prevention.

REFERENCES

1. **Ledingham JCG.** 1931. The aetiological importance of the elementary bodies in vaccinia and fowl-pox. *Lancet* **218:** 525–526.
2. **Fenner F, Burnet FM.** 1957. A short description of the poxvirus group (vaccinia and related viruses). *Virology* **4:**305–314.
3. **Moss B.** 2013. Poxviridae, p 2130–2159. *In* Knipe D, Howley P (ed), *Fields Virology*. Lippincott Williams and Wilkins, New York, NY.
4. **Jezek Z, Fenner F.** 1988. Human monkeypox, p 1–140. *In* Melnick J (ed), *Monographs in Virology*. S Karger AG, Karger, Basel, Switzerland.
5. **Guarnieri G.** 1892. Richerche sulla patogenesi ed etiologia dell'infezione vaccinia e variolosa. *Arch Sci Med (Torino)* **16:**243–247.
6. **Marchal J.** 1930. A hitherto undescribed virus disease of mice. *J Pathol Bacteriol* **33:**713–728.
7. **Downie AW.** 1939. A study of the lesions produced experimentally by cowpox virus. *J Pathol Bacteriol* **48:**361–379.
8. **Vora NM, Li Y, Geleishvili M, Emerson GL, Khmaladze E, Maghlakelidze G, Navdarashvili A, Zakhashvili K, Kokhreidze M, Endeladze M, Mokverashvili G, Satheshkumar PS, Gallardo-Romero N, Goldsmith CS, Metcalfe MG, Damon I, Maes EF, Reynolds MG, Morgan J, Carroll DS.** 2015. Three cases of human infection with a novel zoonotic Orthopoxvirus in the country of Georgia. *N Engl J Med* **372:**1223–1230.
9. **Damon I.** 2013. Poxviruses, p 2161–2184. *In* Knipe D, Howley P (ed), *Fields Virology*. Lippincott Williams and Wilkins, New York, NY.
10. **Schwartz JJ, Myskowski PL.** 1992. Molluscum contagiosum and human immunodeficiency virus. *Arch Dermatol* **128:** 1407–1408.
11. **Li Y, Carroll DS, Gardner SN, Walsh MC, Vitalis EA, Damon IK.** 2007. On the origin of smallpox: correlating variola phylogenics with historical smallpox records. *Proc Natl Acad Sci USA* **104:**15787–15792.
12. **Esposito JJ, Sammons SA, Frace AM, Osborne JD, Olsen-Rasmussen M, Zhang M, Govil D, Damon IK, Kline R, Laker M, Li Y, Smith GL, Meyer H, Leduc JW, Wohlhueter RM.** 2006. Genome sequence diversity and clues to the evolution of variola (smallpox) virus. *Science* **313:**807–812.
13. **Fenner F, Henderson DA, Aritz I, Jezek Z, Ladnyi ID.** 1988. *Smallpox and Its Eradication*. World Health Organization, Geneva, Switzerland.
14. **Rotz LD, Dotson DA, Damon IK, Becher JA.** 2001. Vaccinia (smallpox) vaccine: recommendations of the Advisory Committee on Immunization Practices (ACIP), 2001. *MMWR Recomm Rep* **50:**1–25; quiz CE21-27.
15. **Wharton M, Strikas RA, Harpaz R, Rotz LD, Schwartz B, Casey CG, Pearson ML, Anderson LJ, Advisory Committee on Immunization Practices, Healthcare Infection Control Practices Advisory Committee.** 2003. Recommendations for using smallpox vaccine in a pre-event vaccination program. Supplemental recommendations of the Advisory Committee on Immunization Practices (ACIP) and the Healthcare Infection Control Practices Advisory Committee (HICPAC). *MMWR Recomm Rep* **52**(RR-7)**:**1–16.
16. **Blackford S, Roberts DL, Thomas PD.** 1993. Cowpox infection causing a generalized eruption in a patient with atopic dermatitis. *Br J Dermatol* **129:**628–629.
17. **Czerny CP, Eis-Hübinger AM, Mayr A, Schneweis KE, Pfeiff B.** 1991. Animal poxviruses transmitted from cat to man: current event with lethal end. *Zentralbl Veterinarmed B* **38:**421–431.
18. **Learned LA, Reynolds MG, Wassa DW, Li Y, Olson VA, Karem K, Stempora LL, Braden ZH, Kline R, Likos A, Libama F, Moudzeo H, Bolanda JD, Tarangonia P, Boumandoki P, Formenty P, Harvey JM, Damon IK.** 2005. Extended interhuman transmission of monkeypox in a hospital community in the Republic of the Congo, 2003. *Am J Trop Med Hyg* **73:**428–434.

19. **Reed KD, Melski JW, Graham MB, Regnery RL, Sotir MJ, Wegner MV, Kazmierczak JJ, Stratman EJ, Li Y, Fairley JA, Swain GR, Olson VA, Sargent EK, Kehl SC, Frace MA, Kline R, Foldy SL, Davis JP, Damon IK.** 2004. The detection of monkeypox in humans in the Western Hemisphere. *N Engl J Med* **350:**342–350.
20. **Fine PE, Jezek Z, Grab B, Dixon H.** 1988. The transmission potential of monkeypox virus in human populations. *Int J Epidemiol* **17:**643–650.
21. **Gani R, Leach S.** 2001. Transmission potential of smallpox in contemporary populations. *Nature* **414:**748–751.
22. **Baxby D, Bennett M, Getty B.** 1994. Human cowpox 1969–93: a review based on 54 cases. *Br J Dermatol* **131:**598–607.
23. **de Souza Trindade G, da Fonseca FG, Marques JT, Nogueira ML, Mendes LC, Borges AS, Peiró JR, Pituco EM, Bonjardim CA, Ferreira PC, Kroon EG.** 2003. Araçatuba virus: a vaccinialike virus associated with infection in humans and cattle. *Emerg Infect Dis* **9:**155–160.
24. **Trindade GS, Lobato ZI, Drumond BP, Leite JA, Trigueiro RC, Guedes MI, da Fonseca FG, dos Santos JR, Bonjardim CA, Ferreira PC, Kroon EG.** 2006. Short report: isolation of two vaccinia virus strains from a single bovine vaccinia outbreak in rural area from Brazil: implications on the emergence of zoonotic orthopoxviruses. *Am J Trop Med Hyg* **75:**486–490.
25. **Trindade GS, Emerson GL, Carroll DS, Kroon EG, Damon IK.** 2007. Brazilian vaccinia viruses and their origins. *Emerg Infect Dis* **13:**965–972.
26. **Gurav YK, Raut CG, Yadav PD, Tandale BV, Sivaram A, Pore MD, Basu A, Mourya DT, Mishra AC.** 2011. Buffalopox outbreak in humans and animals in Western Maharashtra, India. *Prev Vet Med* **100:**242–247.
27. **Singh RK, Balamurugan V, Bhanuprakash V, Venkatesan G, Hosamani M.** 2012. Emergence and reemergence of vaccinia-like viruses: global scenario and perspectives. *Indian J Virol* **23:**1–11.
28. **Likos AM, Sammons SA, Olson VA, Frace AM, Li Y, Olsen-Rasmussen M, Davidson W, Galloway R, Khristova ML, Reynolds MG, Zhao H, Carroll DS, Curns A, Formenty P, Esposito JJ, Regnery RL, Damon IK.** 2005. A tale of two clades: monkeypox viruses. *J Gen Virol* **86:**2661–2672.
29. **Hutson CL, Lee KN, Abel J, Carroll DS, Montgomery JM, Olson VA, Li Y, Davidson W, Hughes C, Dillon M, Spurlock P, Kazmierczak JJ, Austin C, Miser L, Sorhage FE, Howell J, Davis JP, Reynolds MG, Braden Z, Karem KL, Damon IK, Regnery RL.** 2007. Monkeypox zoonotic associations: insights from laboratory evaluation of animals associated with the multi-state US outbreak. *Am J Trop Med Hyg* **76:**757–768.
30. **Khodakevich L, Jezek Z, Kinzanzka K.** 1986. Isolation of monkeypox virus from wild squirrel infected in nature. *Lancet* **1:**98–99.
31. **Radonić A, Metzger S, Dabrowski PW, Couacy-Hymann E, Schuenadel L, Kurth A, Mätz-Rensing K, Boesch C, Leendertz FH, Nitsche A.** 2014. Fatal monkeypox in wild-living sooty mangabey, Côte d'Ivoire, 2012. *Emerg Infect Dis* **20:**1009–1011.
32. **Hazel SM, Bennett M, Chantrey J, Bown K, Cavanagh R, Jones TR, Baxby D, Begon M.** 2000. A longitudinal study of an endemic disease in its wildlife reservoir: cowpox and wild rodents. *Epidemiol Infect* **124:**551–562.
33. **Pahlitzsch R, Hammarin AL, Widell A.** 2006. A case of facial cellulitis and necrotizing lymphadenitis due to cowpox virus infection. *Clin Infect Dis* **43:**737–742.
34. **Kurth A, Wibbelt G, Gerber HP, Petschaelis A, Pauli G, Nitsche A.** 2008. Rat-to-elephant-to-human transmission of cowpox virus. *Emerg Infect Dis* **14:**670–671.
35. **Becker C, Kurth A, Hessler F, Kramp H, Gokel M, Hoffmann R, Kuczka A, Nitsche A.** 2009. Cowpox virus infection in pet rat owners: not always immediately recognized. *Dtsch Arztebl Int* **106:**329–334.
36. **Ninove L, Domart Y, Vervel C, Voinot C, Salez N, Raoult D, Meyer H, Capek I, Zandotti C, Charrel RN.** 2009. Cowpox virus transmission from pet rats to humans, France. *Emerg Infect Dis* **15:**781–784.
37. **Rimoin AW, Kisalu N, Kebela-Ilunga B, Mukaba T, Wright LL, Formenty P, Wolfe ND, Shongo RL, Tshioko F, Okitolonda E, Muyembe JJ, Ryder R, Meyer H.** 2007. Endemic human monkeypox, Democratic Republic of Congo, 2001–2004. *Emerg Infect Dis* **13:**934–937.
38. **Rimoin AW, Mulembakani PM, Johnston SC, Lloyd Smith JO, Kisalu NK, Kinkela TL, Blumberg S, Thomassen HA, Pike BL, Fair JN, Wolfe ND, Shongo RL, Graham BS, Formenty P, Okitolonda E, Hensley LE, Meyer H, Wright LL, Muyembe JJ.** 2010. Major increase in human monkeypox incidence 30 years after smallpox vaccination campaigns cease in the Democratic Republic of Congo. *Proc Natl Acad Sci USA* **107:**16262–16267.
39. **Damon IK, Roth CE, Chowdhary V.** 2006. Discovery of monkeypox in Sudan. *N Engl J Med* **355:**962–963.
40. **Snell E, Fox JG.** 1961. Molluscum contagiosum venereum. *Can Med Assoc J* **85:**1152–1154.
41. **Postlethwaite R.** 1970. Molluscum contagiosum. *Arch Environ Health* **21:**432–452.
42. **Buller RM, Burnett J, Chen W, Kreider J.** 1995. Replication of molluscum contagiosum virus. *Virology* **213:**655–659.
43. **Porter CD, Archard LC.** 1992. Characterisation by restriction mapping of three subtypes of molluscum contagiosum virus. *J Med Virol* **38:**1–6.
44. **Nakamura J, Muraki Y, Yamada M, Hatano Y, Nii S.** 1995. Analysis of molluscum contagiosum virus genomes isolated in Japan. *J Med Virol* **46:**339–348.
45. **Agromayor M, Ortiz P, Lopez-Estebaranz JL, Gonzalez-Nicolas J, Esteban M, Martin-Gallardo A.** 2002. Molecular epidemiology of molluscum contagiosum virus and analysis of the host-serum antibody response in Spanish HIV-negative patients. *J Med Virol* **66:**151–158.
46. **Saral Y, Kalkan A, Ozdarendeli A, Bulut Y, Doymaz MZ.** 2006. Detection of Molluscum contagiosum virus (MCV) subtype I as a single dominant virus subtype in Molluscum lesions from a Turkish population. *Arch Med Res* **37:**388–391.
47. **Postlethwaite R, Watt JA, Hawley TG, Simpson I, Adam H.** 1967. Features of molluscum contagiosum in the north-east of Scotland and in Fijian village settlements. *J Hyg (Lond)* **65:**281–291.
48. **Porter CD, Blake NW, Cream JJ, Archard LC.** 1992. Molluscum contagiosum virus. *Mol Cell Biol Hum Dis Ser* **1:**233–257.
49. **Schwartz JJ, Myskowski PL.** 1992. Molluscum contagiosum in patients with human immunodeficiency virus infection. A review of twenty-seven patients. *J Am Acad Dermatol* **27:**583–588.
50. **al-Hazzaa SA, Hidayat AA.** 1993. Molluscum contagiosum of the eyelid and infraorbital margin—a clinicopathologic study with light and electron microscopic observations. *J Pediatr Ophthalmol Strabismus* **30:**58–59.
51. **Izu R, Manzano D, Gardeazabal J, Diaz-Perez JL.** 1994. Giant molluscum contagiosum presenting as a tumor in an HIV-infected patient. *Int J Dermatol* **33:**266–267.
52. **Nuñez A, Funes JM, Agromayor M, Moratilla M, Varas AJ, Lopez-Estebaranz JL, Esteban M, Martin-Gallardo A.** 1996. Detection and typing of molluscum contagiosum virus in skin lesions by using a simple lysis method and polymerase chain reaction. *J Med Virol* **50:**342–349.
53. **Thompson CH.** 1997. Identification and typing of molluscum contagiosum virus in clinical specimens by polymerase chain reaction. *J Med Virol* **53:**205–211.
54. **Trama JP, Adelson ME, Mordechai E.** 2007. Identification and genotyping of molluscum contagiosum virus from genital swab samples by real-time PCR and Pyrosequencing. *J Clin Virol* **40:**325–329.
55. **Janniger CK, Schwartz RA.** 1993. Molluscum contagiosum in children. *Cutis* **52:**194–196.
56. **Itin PH, Gilli L.** 1994. Molluscum contagiosum mimicking sebaceous nevus of Jadassohn, ecthyma and giant condylomata acuminata in HIV-infected patients. *Dermatology* **189:**396–398.

57. **O'Neil CA, Hansen RC.** 1995. Pearly penile papules on the shaft. *Arch Dermatol* **131:**491–492.
58. **Roess AA, McCollum AM, Gruszynski K, Zhao H, Davidson W, Lafon N, Engelmeyer T, Moyer B, Godfrey C, Kilpatrick H, Labonte A, Murphy J, Carroll DS, Li Y, Damon IK.** 2013. Surveillance of parapoxvirus among ruminants in Virginia and Connecticut. *Zoonoses Public Health* **60:**543–548.
59. **Wilson TM, Boothe AD, Cheville NF.** 1972. Sealpox field survey. *J Wildl Dis* **8:**158–160.
60. **Wilson TM, Cheville NF, Boothe AD.** 1972. Sealpox questionnaire survey. *J Wildl Dis* **8:**155–157.
61. **Hicks BD, Worthy GA.** 1987. Sealpox in captive grey seals (Halichoerus grypus) and their handlers. *J Wildl Dis* **23:**1–6.
62. **Roess AA, Levine RS, Barth L, Monroe BP, Carroll DS, Damon IK, Reynolds MG.** 2011. Sealpox virus in marine mammal rehabilitation facilities, North America, 2007–2009. *Emerg Infect Dis* **17:**2203–2208.
63. **Becher P, König M, Müller G, Siebert U, Thiel HJ.** 2002. Characterization of sealpox virus, a separate member of the parapoxviruses. *Arch Virol* **147:**1133–1140.
64. **Clark C, McIntyre PG, Evans A, McInnes CJ, Lewis-Jones S.** 2005. Human sealpox resulting from a seal bite: confirmation that sealpox virus is zoonotic. *Br J Dermatol* **152:**791–793.
65. **Groves RW, Wilson-Jones E, MacDonald DM.** 1991. Human orf and milkers' nodule: a clinicopathologic study. *J Am Acad Dermatol* **25:**706–711.
66. **Robinson AJ, Petersen GV.** 1983. Orf virus infection of workers in the meat industry. *N Z Med J* **96:**81–85.
67. **Mercer A, Fleming S, Robinson A, Nettleton P, Reid H.** 1997. Molecular genetic analyses of parapoxviruses pathogenic for humans. *Arch Virol Suppl* **13:**25–34.
68. **Lederman ER, Austin C, Trevino I, Reynolds MG, Swanson H, Cherry B, Ragsdale J, Dunn J, Meidl S, Zhao H, Li Y, Pue H, Damon IK.** 2007. ORF virus infection in children: clinical characteristics, transmission, diagnostic methods, and future therapeutics. *Pediatr Infect Dis J* **26:**740–744.
69. **Nagington J, Tee GH, Smith JS.** 1965. Milker's nodule virus infections in Dorset and their similarity to orf. *Nature* **208:**505–507.
70. **Wilson TM, Cheville NF, Karstad L.** 1969. Seal pox. Case history. *Wildl Dis* **5:**412–418.
71. **Robinson AJ, Petersen GV.** 1992. Parapoxviruses: their biology and potential as recombinant vaccines, p 285–327. *In* Poxviruses R (ed), *GL BMaS*, **vol 96**. CRC Press, Inc, Boca Raton, FL.
72. **Lederman ER, Green GM, DeGroot HE, Dahl P, Goldman E, Greer PW, Li Y, Zhao H, Paddock CD, Damon IK.** 2007. Progressive ORF virus infection in a patient with lymphoma: successful treatment using imiquimod. *Clin Infect Dis* **44:**e100–e103.
73. **Murphy JK, Ralfs IG.** 1996. Bullous pemphigoid complicating human orf. *Br J Dermatol* **134:**929–930.
74. **Becker F.** 1940. Milker's nodules. *JAMA* **115:**2140–2144.
75. **Carson CA, Kerr KM.** 1967. Bovine papular stomatitis with apparent transmission to man. *J Am Vet Med Assoc* **151:**183–187.
76. **Inoshima Y, Morooka A, Sentsui H.** 2000. Detection and diagnosis of parapoxvirus by the polymerase chain reaction. *J Virol Methods* **84:**201–208.
77. **Torfason EG, Gunadóttir S.** 2002. Polymerase chain reaction for laboratory diagnosis of orf virus infections. *J Clin Virol* **24:**79–84.
78. **Gallina L, Dal Pozzo F, Mc Innes CJ, Cardeti G, Guercio A, Battilani M, Ciulli S, Scagliarini A.** 2006. A real time PCR assay for the detection and quantification of orf virus. *J Virol Methods* **134:**140–145.
79. **Nitsche A, Büttner M, Wilhelm S, Pauli G, Meyer H.** 2006. Real-time PCR detection of parapoxvirus DNA. *Clin Chem* **52:**316–319.
80. **Tan ST, Blake GB, Chambers S.** 1991. Recurrent orf in an immunocompromised host. *Br J Plast Surg* **44:**465–467.
81. **Downie AW.** 1972. The epidemiology of tanapox and yaba virus infections. *J Med Microbiol* **5(4):**Pxiv.
82. **Downie AW, España C.** 1972. Comparison of Tanapox virus and Yaba-like viruses causing epidemic disease in monkeys. *J Hyg (Lond)* **70:**23–32.
83. **Downie AW, España C.** 1973. A comparative study of Tanapox and Yaba viruses. *J Gen Virol* **19:**37–49.
84. **Ropp S, Esposito JJ, Palumbo GJ.** 1999. Poxvirus infectiosn in humans, p 1131–1138. *In* Murray P (ed), *Manual of Clinical Microbiology*. American Society for Microbiology, Washington, DC.
85. **Lee HJ, Essani K, Smith GL.** 2001. The genome sequence of Yaba-like disease virus, a yatapoxvirus. *Virology* **281:**170–192.
86. **Nazarian SH, Barrett JW, Frace AM, Olsen-Rasmussen M, Khristova M, Shaban M, Neering S, Li Y, Damon IK, Esposito JJ, Essani K, McFadden G.** 2007. Comparative genetic analysis of genomic DNA sequences of two human isolates of Tanapox virus. *Virus Res* **129:**11–25.
87. **Dhar AD, Werchniak AE, Li Y, Brennick JB, Goldsmith CS, Kline R, Damon I, Klaus SN.** 2004. Tanapox infection in a college student. *N Engl J Med* **350:**361–366.
88. **Niven JS, Armstrong JA, Andrewes CH, Pereira HG, Valentine RC.** 1961. Subcutaneous "growths" in monkeys produced by a poxvirus. *J Pathol Bacteriol* **81:**1–14.
89. **Jezek Z, Arita I, Szczeniowski M, Paluku KM, Ruti K, Nakano JH.** 1985. Human tanapox in Zaire: clinical and epidemiological observations on cases confirmed by laboratory studies. *Bull World Health Organ* **63:**1027–1035.
90. **Zimmermann P, Thordsen I, Frangoulidis D, Meyer H.** 2005. Real-time PCR assay for the detection of tanapox virus and yaba-like disease virus. *J Virol Methods* **130:**149–153.
91. **Geerinck K, Lukito G, Snoeck R, De Vos R, De Clercq E, Vanrenterghem Y, Degreef H, Maes B.** 2001. A case of human orf in an immunocompromised patient treated successfully with cidofovir cream. *J Med Virol* **64:**543–549.
92. **Berman B, Poochareon VN, Villa AM.** 2002. Novel dermatologic uses of the immune response modifier imiquimod 5% cream. *Skin Therapy Lett* **7:**1–6.
93. **De Clercq E.** 2002. Cidofovir in the treatment of poxvirus infections. *Antiviral Res* **55:**1–13.
94. **Trizna Z.** 2002. Viral diseases of the skin: diagnosis and antiviral treatment. *Paediatr Drugs* **4:**9–19.
95. **Garland SM.** 2003. Imiquimod. *Curr Opin Infect Dis* **16:**85–89.
96. **Bikowski JB Jr.** 2004. Molluscum contagiosum: the need for physician intervention and new treatment options. *Cutis* **73:**202–206.
97. **Arican O.** 2006. Topical treatment of molluscum contagiosum with imiquimod 5% cream in Turkish children. *Pediatr Int* **48:**403–405.
98. **Goldstein JA, Neff JM, Lane JM, Koplan JP.** 1975. Smallpox vaccination reactions, prophylaxis, and therapy of complications. *Pediatrics* **55:**342–347.
99. **Bray M.** 2003. Pathogenesis and potential antiviral therapy of complications of smallpox vaccination. *Antiviral Res* **58:**101–114.
100. **Pepose JS, Margolis TP, LaRussa P, Pavan-Langston D.** 2003. Ocular complications of smallpox vaccination. *Am J Ophthalmol* **136:**343–352.
101. **Hopkins RJ, Lane JM.** 2004. Clinical efficacy of intramuscular vaccinia immune globulin: a literature review. *Clin Infect Dis* **39:**819–826.
102. **Wittek R.** 2006. Vaccinia immune globulin: current policies, preparedness, and product safety and efficacy. *Int J Infect Dis* **10:**193–201.
103. **Vora S, Damon I, Fulginiti V, Weber SG, Kahana M, Stein SL, Gerber SI, Garcia-Houchins S, Lederman E, Hruby D, Collins L, Scott D, Thompson K, Barson JV, Regnery R, Hughes C, Daum RS, Li Y, Zhao H, Smith S, Braden Z, Karem K, Olson V, Davidson W, Trindade G, Bolken T, Jordan R, Tien D, Marcinak J.** 2008. Severe eczema vaccinatum in a household contact of a smallpox vaccinee. *Clin Infect Dis* **46:**1555–1561.
104. **Lederman ER, Davidson W, Groff HL, Smith SK, Warkentien T, Li Y, Wilkins KA, Karem KL, Akondy RS, Ah-**

med R, Frace M, Shieh WJ, Zaki S, Hruby DE, Painter WP, Bergman KL, Cohen JI, Damon IK. 2012. Progressive vaccinia: case description and laboratory-guided therapy with vaccinia immune globulin, ST-246, and CMX001. *J Infect Dis* **206:**1372–1385.

105. **Toutous-Trellu L, Hirschel B, Piguet V, Schiffer V, Saurat JH, Pechère M.** 2004. Treatment of cutaneous human papilloma virus, poxvirus and herpes simplex virus infections with topical cidofovir in HIV positive patients. *Ann Dermatol Venereol* **131:**445–449.
106. **Bailey TR, Rippin SR, Opsitnick E, Burns CJ, Pevear DC, Collett MS, Rhodes G, Tohan S, Huggins JW, Baker RO, Kern ER, Keith KA, Dai D, Yang G, Hruby D, Jordan R.** 2007. N-(3,3a,4,4a,5,5a,6,6a-Octahydro-1,3-dioxo-4,6-ethenocycloprop[f]isoindol-2-(1H)-yl)carboxamides: identification of novel orthopoxvirus egress inhibitors. *J Med Chem* **50:**1442–1444.
107. **Mayr A, Hochstein-Mintzel V, Stickl H.** 1975. Abstammung, Eigenschaften und Verwendung des attenuierten Vaccinia-Stammes MVA. *Infection* **3:**6–14.
108. **Mayr A, Stickl H, Müller HK, Danner K, Singer H.** 1978. The smallpox vaccination strain MVA: marker, genetic structure, experience gained with the parenteral vaccination and behavior in organisms with a debilitated defence mechanism (author's transl). *Zentralbl Bakteriol [B]* **167:**375–390.
109. **Stickl H, Hochstein-Mintzel V, Mayr A, Huber HC, Schäfer H, Holzner A.** 1974. MVA vaccination against smallpox: clinical tests with an attenuated live vaccinia virus strain (MVA) (author's transl). *Dtsch Med Wochenschr* **99:**2386–2392.
110. **Stittelaar KJ, Kuiken T, de Swart RL, van Amerongen G, Vos HW, Niesters HG, van Schalkwijk P, van der Kwast T, Wyatt LS, Moss B, Osterhaus AD.** 2001. Safety of modified vaccinia virus Ankara (MVA) in immune-suppressed macaques. *Vaccine* **19:**3700–3709.
111. **Stittelaar KJ, van Amerongen G, Kondova I, Kuiken T, van Lavieren RF, Pistoor FH, Niesters HG, van Doornum G, van der Zeijst BA, Mateo L, Chaplin PJ, Osterhaus AD.** 2005. Modified vaccinia virus Ankara protects macaques against respiratory challenge with monkeypox virus. *J Virol* **79:**7845–7851.
112. **Edghill-Smith Y, Venzon D, Karpova T, McNally J, Nacsa J, Tsai WP, Tryniszewska E, Moniuszko M, Manischewitz J, King LR, Snodgrass SJ, Parrish J, Markham P, Sowers M, Martin D, Lewis MG, Berzofsky JA, Belyakov IM, Moss B, Tartaglia J, Bray M, Hirsch V, Golding H, Franchini G.** 2003. Modeling a safer smallpox vaccination regimen, for human immunodeficiency virus type 1-infected patients, in immunocompromised macaques. *J Infect Dis* **188:**1181–1191.
113. **Earl PL, Americo JL, Wyatt LS, Eller LA, Whitbeck JC, Cohen GH, Eisenberg RJ, Hartmann CJ, Jackson DL, Kulesh DA, Martinez MJ, Miller DM, Mucker EM, Shamblin JD, Zwiers SH, Huggins JW, Jahrling PB, Moss B.** 2004. Immunogenicity of a highly attenuated MVA smallpox vaccine and protection against monkeypox. *Nature* **428:**182–185.
114. **Earl PL, Americo JL, Wyatt LS, Eller LA, Montefiori DC, Byrum R, Piatak M, Lifson JD, Amara RR, Robinson HL, Huggins JW, Moss B.** 2007. Recombinant modified vaccinia virus Ankara provides durable protection against disease caused by an immunodeficiency virus as well as long-term immunity to an orthopoxvirus in a non-human primate. *Virology* **366:**84–97.
115. **Damon IK, Davidson WB, Hughes CM, Olson VA, Smith SK, Holman RC, Frey SE, Newman F, Belshe RB, Yan L, Karem K.** 2009. Evaluation of smallpox vaccines using variola neutralization. *J Gen Virol* **90:**1962–1966.
116. **Henderson DA, Arita I.** 2014. The smallpox threat: a time to reconsider global policy. *Biosecur Bioterror* **12:**117–121.
117. **Kenner J, Cameron F, Empig C, Jobes DV, Gurwith M.** 2006. LC16m8: an attenuated smallpox vaccine. *Vaccine* **24:**7009–7022.
118. **Morikawa S, Sakiyama T, Hasegawa H, Saijo M, Maeda A, Kurane I, Maeno G, Kimura J, Hirama C, Yoshida T, Asahi-Ozaki Y, Sata T, Kurata T, Kojima A.** 2005. An attenuated LC16m8 smallpox vaccine: analysis of full-genome sequence and induction of immune protection. *J Virol* **79:**11873–11891.
119. **Empig C, Kenner JR, Perret-Gentil M, Youree BE, Bell E, Chen A, Gurwith M, Higgins K, Lock M, Rice AD, Schriewer J, Sinangil F, White E, Buller RM, Dermody TS, Isaacs SN, Moyer RW.** 2006. Highly attenuated smallpox vaccine protects rabbits and mice against pathogenic orthopoxvirus challenge. *Vaccine* **24:**3686–3694.
120. **Saijo M, Ami Y, Suzaki Y, Nagata N, Iwata N, Hasegawa H, Ogata M, Fukushi S, Mizutani T, Sata T, Kurata T, Kurane I, Morikawa S.** 2006. LC16m8, a highly attenuated vaccinia virus vaccine lacking expression of the membrane protein B5R, protects monkeys from monkeypox. *J Virol* **80:**5179–5188.
121. **Meseda CA, Mayer AE, Kumar A, Garcia AD, Campbell J, Listrani P, Manischewitz J, King LR, Golding H, Merchlinsky M, Weir JP.** 2009. Comparative evaluation of the immune responses and protection engendered by LC16m8 and Dryvax smallpox vaccines in a mouse model. *Clin Vaccine Immunol* **16:**1261–1271.
122. **Kennedy JS, Gurwith M, Dekker CL, Frey SE, Edwards KM, Kenner J, Lock M, Empig C, Morikawa S, Saijo M, Yokote H, Karem K, Damon I, Perlroth M, Greenberg RN.** 2011. Safety and immunogenicity of LC16m8, an attenuated smallpox vaccine in vaccinia-naive adults. *J Infect Dis* **204:**1395–1402.
123. **Tartaglia J, Cox WI, Taylor J, Perkus M, Riviere M, Meignier B, Paoletti E.** 1992. Highly attenuated poxvirus vectors. *AIDS Res Hum Retroviruses* **8:**1445–1447.
124. **Paoletti E, Taylor J, Meignier B, Meric C, Tartaglia J.** 1995. Highly attenuated poxvirus vectors: NYVAC, ALVAC and TROVAC. *Dev Biol Stand* **84:**159–163.
125. **Coulibaly S, Brühl P, Mayrhofer J, Schmid K, Gerencer M, Falkner FG.** 2005. The nonreplicating smallpox candidate vaccines defective vaccinia Lister (dVV-L) and modified vaccinia Ankara (MVA) elicit robust long-term protection. *Virology* **341:**91–101.
126. **Nakano J.** 1979. Poxviruses, p 257–308. *In* NJ Schmidt (ed), *Diagnostic Procedures for Viral, Rickettsial, and Chlamydial Infections*, American Public Health Association, Washington. DC.
127. **Ropp SL, Jin Q, Knight JC, Massung RF, Esposito JJ.** 1995. PCR strategy for identification and differentiation of small pox and other orthopoxviruses. *J Clin Microbiol* **33:**2069–2076.
128. **Meyer H, Pfeffer M, Rziha HJ.** 1994. Sequence alterations within and downstream of the A-type inclusion protein genes allow differentiation of Orthopoxvirus species by polymerase chain reaction. *J Gen Virol* **75:**1975–1981.
129. **Meyer H, Damon IK, Esposito JJ.** 2004. Orthopoxvirus diagnostics. *Methods Mol Biol* **269:**119–134.
130. **Li Y, Meyer H, Zhao H, Damon IK.** 2010. GC content-based pan-pox universal PCR assays for poxvirus detection. *J Clin Microbiol* **48:**268–276.
131. **Bora DP, Venkatesan G, Bhanuprakash V, Balamurugan V, Prabhu M, Siva Sankar MS, Yogisharadhya R.** 2011. TaqMan real-time PCR assay based on DNA polymerase gene for rapid detection of Orf infection. *J Virol Methods* **178:**249–252.
132. **Stemmler M, Neubauer H, Meyer H.** 2001. Comparison of closely related orthopoxvirus isolates by random amplified polymorphic DNA and restriction fragment length polymorphism analysis. *J Vet Med B Infect Dis Vet Public Health* **48:**647–654.
133. **Shchelkunov SN, Gavrilova EV, Babkin IV.** 2005. Multiplex PCR detection and species differentiation of orthopoxviruses pathogenic to humans. *Mol Cell Probes* **19:**1–8.
134. **Klein D.** 2002. Quantification using real-time PCR technology: applications and limitations. *Trends Mol Med* **8:**257–260.
135. **Espy MJ, Cockerill FR III, Meyer RF, Bowen MD, Poland GA, Hadfield TL, Smith TF.** 2002. Detection of smallpox virus DNA by LightCycler PCR. *J Clin Microbiol* **40:**1985–1988.

136. **Sofi Ibrahim M, Kulesh DA, Saleh SS, Damon IK, Esposito JJ, Schmaljohn AL, Jahrling PB.** 2003. Real-time PCR assay to detect smallpox virus. *J Clin Microbiol* **41:**3835–3839.
137. **Kulesh DA, Loveless BM, Norwood D, Garrison J, Whitehouse CA, Hartmann C, Mucker E, Miller D, Wasieloski LP Jr, Huggins J, Huhn G, Miser LL, Imig C, Martinez M, Larsen T, Rossi CA, Ludwig GV.** 2004. Monkeypox virus detection in rodents using real-time 3′-minor groove binder TaqMan assays on the Roche LightCycler. *Lab Invest* **84:**1200–1208.
138. **Kulesh DA, Baker RO, Loveless BM, Norwood D, Zwiers SH, Mucker E, Hartmann C, Herrera R, Miller D, Christensen D, Wasieloski LP Jr, Huggins J, Jahrling PB.** 2004. Smallpox and pan-orthopox virus detection by real-time 3′-minor groove binder TaqMan assays on the roche LightCycler and the Cepheid smart Cycler platforms. *J Clin Microbiol* **42:**601–609.
139. **Nitsche A, Ellerbrok H, Pauli G.** 2004. Detection of orthopoxvirus DNA by real-time PCR and identification of variola virus DNA by melting analysis. *J Clin Microbiol* **42:**1207–1213.
140. **Nitsche A, Steger B, Ellerbrok H, Pauli G.** 2005. Detection of vaccinia virus DNA on the LightCycler by fluorescence melting curve analysis. *J Virol Methods* **126:**187–195.
141. **Nitsche A, Stern D, Ellerbrok H, Pauli G.** 2006. Detection of infectious poxvirus particles. *Emerg Infect Dis* **12:**1139–1141.
142. **Nitsche A, Kurth A, Pauli G.** 2007. Viremia in human Cowpox virus infection. *J Clin Virol* **40:**160–162.
143. **Olson VA, Laue T, Laker MT, Babkin IV, Drosten C, Shchelkunov SN, Niedrig M, Damon IK, Meyer H.** 2004. Real-time PCR system for detection of orthopoxviruses and simultaneous identification of smallpox virus. *J Clin Microbiol* **42:**1940–1946.
144. **Panning M, Asper M, Kramme S, Schmitz H, Drosten C.** 2004. Rapid detection and differentiation of human pathogenic orthopox viruses by a fluorescence resonance energy transfer real-time PCR assay. *Clin Chem* **50:**702–708.
145. **Aitichou M, Javorschi S, Ibrahim MS.** 2005. Two-color multiplex assay for the identification of orthopox viruses with real-time LUX- PCR. *Mol Cell Probes* **19:**323–328.
146. **Carletti F, Di Caro A, Calcaterra S, Grolla A, Czub M, Ippolito G, Capobianchi MR, Horejsh D.** 2005. Rapid, differential diagnosis of orthopox- and herpesviruses based upon real-time PCR product melting temperature and restriction enzyme analysis of amplicons. *J Virol Methods* **129:**97–100.
147. **Fedorko DP, Preuss JC, Fahle GA, Li L, Fischer SH, Hohman P, Cohen JI.** 2005. Comparison of methods for detection of vaccinia virus in patient specimens. *J Clin Microbiol* **43:**4602–4606.
148. **Fedele CG, Negredo A, Molero F, Sánchez-Seco MP, Tenorio A.** 2006. Use of internally controlled real-time genome amplification for detection of variola virus and other orthopoxviruses infecting humans. *J Clin Microbiol* **44:**4464–4470.
149. **Li Y, Olson VA, Laue T, Laker MT, Damon IK.** 2006. Detection of monkeypox virus with real-time PCR assays. *J Clin Virol* **36:**194–203.
150. **Scaramozzino N, Ferrier-Rembert A, Favier AL, Rothlisberger C, Richard S, Crance JM, Meyer H, Garin D.** 2007. Real-time PCR to identify variola virus or other human pathogenic orthopox viruses. *Clin Chem* **53:**606–613.
151. **Sias C, Carletti F, Capobianchi MR, Travaglini D, Chiappini R, Horejsh D, Di Caro A.** 2007. Rapid differential diagnosis of Orthopoxviruses and Herpesviruses based upon multiplex real-time PCR. *Infez Med* **15:**47–55.
152. **Aitichou M, Saleh S, Kyusung P, Huggins J, O'Guinn M, Jahrling P, Ibrahim S.** 2008. Dual-probe real-time PCR assay for detection of variola or other orthopoxviruses with dried reagents. *J Virol Methods* **153:**190–195.
153. **Kurth A, Achenbach J, Miller L, Mackay IM, Pauli G, Nitsche A.** 2008. Orthopoxvirus detection in environmental specimens during suspected bioterror attacks: inhibitory influences of common household products. *Appl Environ Microbiol* **74:**32–37.
154. **Putkuri N, Piiparinen H, Vaheri A, Vapalahti O.** 2009. Detection of human orthopoxvirus infections and differentiation of smallpox virus with real-time PCR. *J Med Virol* **81:**146–152.
155. **Li Y, Zhao H, Wilkins K, Hughes C, Damon IK.** 2010. Real-time PCR assays for the specific detection of monkeypox virus West African and Congo Basin strain DNA. *J Virol Methods* **169:**223–227.
156. **Schroeder K, Nitsche A.** 2010. Multicolour, multiplex real-time PCR assay for the detection of human-pathogenic poxviruses. *Mol Cell Probes* **24:**110–113.
157. **Shchelkunov SN, Shcherbakov DN, Maksyutov RA, Gavrilova EV.** 2011. Species-specific identification of variola, monkeypox, cowpox, and vaccinia viruses by multiplex real-time PCR assay. *J Virol Methods* **175:**163–169.
158. **Hošnjak L, Kocjan BJ, Kušar B, Seme K, Poljak M.** 2013. Rapid detection and typing of Molluscum contagiosum virus by FRET-based real-time PCR. *J Virol Methods* **187:**431–434.
159. **Kondas AV, Olson VA, Li Y, Abel J, Laker M, Rose L, Wilkins K, Turner J, Kline R, Damon IK.** 2015. Variola virus specific diagnostic assays: characterization, sensitivity, and specificity. *J Clin Microbiol* **53(4):**1406–1410.
160. **Raja S, Ching J, Xi L, Hughes SJ, Chang R, Wong W, McMillan W, Gooding WE, McCarty KS Jr, Chestney M, Luketich JD, Godfrey TE.** 2005. Technology for automated, rapid, and quantitative PCR or reverse transcription-PCR clinical testing. *Clin Chem* **51:**882–890.
161. **Ioannidis P, Papaventsis D, Karabela S, Nikolaou S, Panagi M, Raftopoulou E, Konstantinidou E, Marinou I, Kanavaki S.** 2011. Cepheid GeneXpert MTB/RIF assay for Mycobacterium tuberculosis detection and rifampin resistance identification in patients with substantial clinical indications of tuberculosis and smear-negative microscopy results. *J Clin Microbiol* **49:**3068–3070.
162. **Kalina WV, Douglas CE, Coyne SR, Minogue TD.** 2014. Comparative assessment of automated nucleic acid sample extraction equipment for biothreat agents. *J Clin Microbiol* **52:**1232–1234.
163. **Hartmer R, Storm N, Boecker S, Rodi CP, Hillenkamp F, Jurinke C, van den Boom D.** 2003. RNase T1 mediated base-specific cleavage and MALDI-TOF MS for high-throughput comparative sequence analysis. *Nucleic Acids Res* **31:**e47.
164. **Tost J, Gut IG.** 2005. Genotyping single nucleotide polymorphisms by MALDI mass spectrometry in clinical applications. *Clin Biochem* **38:**335–350.
165. **Emonet S, Shah HN, Cherkaoui A, Schrenzel J.** 2010. Application and use of various mass spectrometry methods in clinical microbiology. *Clin Microbiol Infect* **16:**1604–1613.
166. **Croxatto A, Prod'hom G, Greub G.** 2012. Applications of MALDI-TOF mass spectrometry in clinical diagnostic microbiology. *FEMS Microbiol Rev* **36:**380–407.
167. **Margulies M, et al.** 2005. Genome sequencing in microfabricated high-density picolitre reactors. *Nature* **437:**376–380.
168. **Moore MJ, Dhingra A, Soltis PS, Shaw R, Farmerie WG, Folta KM, Soltis DE.** 2006. Rapid and accurate pyrosequencing of angiosperm plastid genomes. *BMC Plant Biol* **6:**17.
169. **Halse TA, Edwards J, Cunningham PL, Wolfgang WJ, Dumas NB, Escuyer VE, Musser KA.** 2010. Combined real-time PCR and rpoB gene pyrosequencing for rapid identification of Mycobacterium tuberculosis and determination of rifampin resistance directly in clinical specimens. *J Clin Microbiol* **48:**1182–1188.
170. **Shchelkunov SN, Marennikova SS, Moyer RW.** 2005. Laboratory diagnostics of human orthopoxvirus infections, p 303–324. *In Orthopoxviruses Pathogenic for Humans.* Springer, New York.

Rabies Virus

ROBERT J. RUDD AND APRIL D. DAVIS

34

VIRAL CLASSIFICATION AND BIOLOGY

Rabies is the prototype virus of the *Lyssavirus* genus of the order *Mononegavirales*, family *Rhabdoviridae*. The *Mononegavirales* are characterized by a nonsegmented, negative-stranded RNA genome, encapsulated tightly into a ribonucleocapsid structure. The *Rhabdoviridae* are classified as a group based on a similar conical or bullet-shaped appearance by electron microscopy. The host range for *Rhabdoviridae* is highly diversified, including plants, arthropods, fish, and mammals (1). Previously known simply as the rabies and rabies-related virus group, the genus *Lyssavirus* is presently composed of genotype 1, classical rabies virus, and 14 other genotypes that are closely related antigenically and genetically and produce a clinical disease indistinguishable from rabies (2). However, lyssaviruses are serologically distinct from other rhabdoviruses.

Genotype 1 includes the majority of field viruses of global distribution in terrestrial mammals, insectivorous and hematophagous bats of the Western Hemisphere, and the laboratory and vaccine strains. With the exception of Australian bat lyssavirus, the distribution of the nonrabies lyssaviruses is restricted to the Old World. They include genotype 2, Lagos bat virus, isolated from African bats; genotype 3, Mokola virus, isolated from African rodents; genotype 4, Duvenhage virus, isolated from African bats; genotypes 5 and 6, European bat lyssaviruses (EBLV) 1 and 2, isolated from European bats; and genotype 7, Australian bat lyssavirus, isolated from Australian bats (3). Five additional lyssaviruses have been isolated from Eurasian bats and identified as new lyssavirus genotypes, which include Aravan, Khujand, Irkut, and Bokeloh bat lyssavirus and West Caucasian bat virus (4–6).

The rabies virus genome consists of a single-stranded, nonsegmented RNA molecule of negative sense polarity. It is approximately 12,000 nucleotides in length and has a molecular mass of approximately 4.6 by 10^6 kilodaltons (7). The viral RNA is transcribed into five polyadenylated, monocistronic mRNA species, corresponding to the five viral proteins. Unlike positive-sense genomes such as flaviviruses, which can be translated directly once inside the cell, the negative polarity of the rabies genome prevents direct translation into viral proteins, requiring an autonomous transcription step facilitated by the RNA polymerase (8), and the genome is therefore not infectious. This virion-associated RNA polymerase is "error prone," resulting in an absence of RNA proofreading, repair, and postreplication error-correction mechanisms in the cell. The lack of fidelity in the RNA polymerase of negative-strand RNA viruses is the main cause of nucleotide misincorporation in genome RNA, generating RNA sequence heterogeneity.

The structure of the rabies virion measures approximately 180 nanometers by 75 nanometers. The bullet-shaped virus is hemispherical at one end and usually planar at the other, where it buds from the surface membrane of an infected cell. The internal ribonucleoprotein complex (RNP) contains the viral RNA associated with three internal proteins: a large (190 kilodalton) RNA transcriptase, or L protein; a 55-kilodalton nucleoprotein, or N; and a 38-kilodalton noncatalytic polymerase-associated phosphorylated protein, P. Spike-like surface projections protrude 10 nanometers from the outer surface of the lipid bilayer. The viral envelope is composed of a 26-kilodalton matrix protein, M, and an envelope sheath consisting of lipids derived from the host cell plasma membrane and the surface spikes formed by a 67-kilodalton glycoprotein, G (9). The G protein is the viral antigen that induces the production of virus-neutralizing antibodies, conferring immunity against exposure (10). The induction of antibody and the conferred immunity is dependent upon the intact secondary and tertiary structure of the G protein (10). The G protein is responsible for the attachment of the virus to the cell and the properties determining its transport in a retrograde fashion into the central nervous system (CNS) (11).

Rabies virus is synthesized in the cytoplasm of infected cells and is released by budding through cell membranes (12). The virus is somewhat resistant to air drying and freeze-thaw cycles and is relatively stable at a pH of 5 to 10. However, it is labile at pasteurization temperatures and in UV light as well as in lipid solvents, ethanol, iodine disinfectants, and quaternary ammonium compounds (13).

The development of PCR gene amplification and direct nucleotide sequence analysis has substantially accelerated progress toward understanding the structure-function relationships of the various elements of the rabies virus (14). As noted above, the RNP functions in the transcription and replication of the virion. Accumulations of RNP result in intracytoplasmic inclusions within infected cells that can be detected by direct observation with histologic methods and

doi:10.1128/9781555819156.ch34

by antigen detection methods employing N protein–specific antibodies (15). The inclusion bodies are locations for viral transcription and replication (8).

Pathogenesis and Pathology

All mammals are susceptible to rabies infection. With very rare exceptions, rabies infection terminates in the death of each infected animal. Rabies virus has evolved a pathogenesis within the individual animal that facilitates the maintenance of the virus within the true host, the reservoir species population. Maintenance of the virus in the reservoir population by direct host-to-host transmission is dependent on simultaneous infection of the brain and salivary glands. It is the impact on behavior resulting from infection of the limbic system that can induce biting behavior, and concomitant infection of the salivary gland tissue allows infectious doses of virus in the saliva to serve as an infectious inoculum for bite transmission. This pathogenic pattern has resulted in the entrenchment of the virus in host populations and the continued risk to humans of exposure (16).

Rabies in vector animal species is similar in most aspects to the disease in humans, but the prodromal period is often followed by a stage of excitation with or without aggression either followed by, or intermixed with, periods of lethargy and depression. Progressive paralysis, which often begins at the site of exposure, may first be recognized in the posterior limbs or larynx. Paralysis of the throat may result in uncharacteristic vocalizations and the accumulation of copious and stringy saliva from the mouth. During the excitation period, animals may exhibit self-mutilation as well as heightened and inappropriate sexual behaviors. The clinical presentations of rabies in animals have often been characterized as either "furious" or "dumb," with the prior involving substantial agitation and aggression and the latter predominantly involving lethargy and paralysis. In some cases, the animal may manifest both forms at different times in the clinical course. Livestock may also demonstrate aggressive and heightened sexual behavior, but facial and pharyngeal paralysis; hypersalivation; bellowing; straining; and posterior paralysis, leading to a "sitting-dog" posture and then recumbency, are more common.

Empirical and laboratory evidence accumulated over centuries related to rabies cycles vectored by domestic and wild species supports the conclusion that rabies is transmitted most commonly via the bite of rabid animals. While aerosol transmission has been considered a possible mode of transmission of rabies to humans, very little data support such a conclusion. The apparent nonbite transmission to humans and animals by inhalation of infectious aerosols occurred in a cave with a bat population exceeding 10 million (17). Empirical data in this report verified aerosol transmission to caged animals in this cave environment. However, the conclusion that the two associated human cases in a cave setting were unquestionably the result of exposure to airborne virus may need to be reconsidered. There have been several cases of "cryptic rabies" in which exposure has not been established, and most are believed to be the result of a bite that goes unnoticed. This is in light of current observations that bat bites capable of rabies transmission may be associated with limited injury (18). Rabies transmission to humans has been reported to have occurred by aerosol in two laboratory accidents (19, 20). In both cases, the infected individuals worked with rabies virus in a setting that could have allowed alternative means of infection. In one study using recent technologies, the experimental aerosol transmission of rabies virus was documented in mice but failed to occur in two species of bats (21). Transmission by direct contamination of mucous membranes by saliva has been reported (22). Infection following the consumption of infected tissues was reported in dogs feeding on rabid fox carcasses in the Arctic (23) and in numerous laboratory studies (24). Human-to-human transmission has been reported following bites, mucous membrane exposure (25), and organ transplants from rabies victims (26). In 2004, four organ donation recipients developed fatal rabies infections following solid organ donations from a single donor (27). The donor was retrospectively determined to have been infected with rabies of bat origin. In the following year, three fatal rabies cases occurred in recipients of organs from a single donor in Germany (28). Additionally, a 2013 report published in the *Journal of the American Medical Association* describes the outcome of four organ transplant recipients who received organs from a single donor. One individual who received a kidney developed rabies following a long incubation period, yet the three other individuals remained healthy without a pretransplant rabies vaccination. The donor was retrospectively diagnosed with rabies identified as the raccoon rabies virus variant (29).

When an animal becomes infected following exposure from the bite of a rabid animal, the virus may invade peripheral nerves or nerve endings directly or may first be "amplified" by replication within striated muscle cells prior to infecting the nerve endings (30). It is unclear if the infection of myocytes at the site of exposure is an essential aspect of the pathogenesis of rabies or how this growth contributes to varying incubation periods. The early events of viral replication and muscle and nerve cell infection at the site of exposure occur without substantial stimulation of the immune system. The entry into a host cell is mediated by the viral G protein. The receptor(s) for rabies is complex and may vary with cell type. The putative receptor in muscle cells is the nicotinic acetylcholine receptor (31), but there is evidence that the neural cell adhesion molecule CD56 (32) and the low-affinity neurotrophin receptor may also serve as rabies virus receptors. It is likely that the virus is not limited to a single receptor for infection of mammalian cells (11). Cell entry may occur via fusion of the viral envelope with the cellular membrane or through coated pits and uncoated vesicles. Following entry into sensory or motor nerve endings the virus replicates and moves transneuronally, by retrograde axoplasmal flow, to the CNS and similarly within the CNS from first-order neurons to second-order neurons (33). The neurotropic nature of the virus and the lack of a viremia may limit exposure to the immune system, explaining the lack of an early antibody response (34).

Virus replication in the CNS occurs mainly in neurons, with extensive distribution in the brain and spinal cord. Rabies virus is described as nonlytic, as it does not generally cause host cell destruction. However, an inverse relationship between the induction of apoptosis (natural cell death) and the pathogenicity of rabies virus strains suggests that apoptosis may be a protective rather than pathogenic mechanism (34). Recognizable clinical signs of rabies generally do not appear until several replication cycles have occurred in the brain. Centrifugal spread occurs following infection of the brain via anterograde axoplasmic flow from the CNS to peripheral nerves and to some nonnervous tissues, including, most importantly, the salivary glands. This accounts for the appearance of rabies virus in some tissues and the saliva up to a few days before the recognized onset of rabies symptoms (24). Although infectious virus in the saliva is paramount for the maintenance of rabies in host populations, virus in the saliva may be sporadic during, and just prior to, the clinical period (35). Upon necropsy, rabies antigen may be demonstrated in the nerves innervating many tissues, including the buccal, nasal, and intestinal mucosa; the urinary bladder; epidermis; corneas; lungs; kidneys; heart; and adrenal medulla; and the brown fat (36).

Studies in animal models have demonstrated that multiple immune mechanisms are involved in the neuropathogenesis of rabies (37, 38). Early-death syndrome has been reported in animals and humans despite rabies vaccination prior to or immediately after exposure to rabies, with infected hosts often succumbing following an abbreviated incubation period when compared to naïve individuals.

In an *in vitro* model using a mouse macrophage cell line, the presence of rabies-neutralizing antibodies in concentrations below protective levels actually enhanced the ability of rabies virus to infect these cells (37). Immunosuppression has been demonstrated to have a sparing effect in some situations (38). Furthermore, virus replication may suppress the production of cellular neuropeptides and neurotransmitters, leading to the functional CNS failure and the fatal outcome of rabies infection (39).

The classical rabies pathogenesis described above results in a variable incubation period, typically 10 days to several months. However, few incubation periods have been greater than 1 year. The incubation period is followed by acute progression with clinical signs demonstrated by ataxia, aggression, and unusual vocalizations that progressively worsen and culminate in death.

Epizootiology of Animal Rabies

Rabies is maintained in bats, wild terrestrial carnivores, and domestic canine populations that serve as reservoirs and vectors of the disease. Specific variants of the virus are associated with each geographically and temporally defined wildlife cycle. The vectors are highly susceptible to the variant that has adapted to the population and are capable of transmission to conspecifics because of coincident aggressive behavior and infectious virus titers in their saliva. Despite the viral preference for conspecific transmission, spillover events from bats to skunks in the western United States have resulted in continued transmission among skunks (40). A sustained outbreak is also dependent upon an adequate vector population density and a host natural history that provides adequate opportunity for interspecific interactions within the characteristic clinical period. Rabies distribution in animals can be discussed in three general categories: domestic canine rabies, terrestrial wildlife rabies, and bat rabies.

Rabies virus maintained in domestic dog populations is responsible for 95% of all animal rabies cases reported globally (41) and still accounts for most of the zoonotic impact of the disease. Indeed, 90% of the human exposures to rabies and 99% of the human rabies deaths worldwide are attributed to canine rabies. Although the development of highly effective vaccines in combination with stray-dog-control programs have been proven effective in extinguishing dog rabies epizootics, dog rabies is still epizootic in most countries in Asia and Africa and some areas in South America. As a result of widespread rabies vaccination programs, the rabies incidence in Latin America and the Caribbean has decreased, and several countries (Uruguay, Chile, Costa Rica, Mexico, and Panama) and areas of Peru, Brazil, and Argentina are free of human rabies transmitted by dogs. However, rabies remains endemic in certain areas of Latin America and the Caribbean (42).

Previously, mass euthanasia of canine populations was performed to control human rabies infections. However, it is now well understood that large-scale euthanasia efforts do not significantly decrease human rabies infection. A recent report by the WHO details that the most effective way to combat human rabies virus infections attributed to canine rabies is through the mass vaccination of dog populations (2). Worldwide, there are an estimated 60,000 human deaths each year from dog-transmitted rabies and 10 million postexposure vaccination regimens administered as a consequence. Dog rabies remains a significant threat worldwide, with an estimated 3 billion humans at risk (2). Europe and North America controlled rabies in domestic dogs by stray-animal control and widespread vaccination programs during the 1950s and 1960s. To date, domestic dogs account for fewer than 5% of animal rabies cases in these regions; the majority are a result of spillover from bats or other wildlife or importation from a country with endemic canine rabies.

Molecular analyses have shed light on the origins and distribution of canine rabies cycles seen throughout the world. The most geographically widely distributed group of rabies viruses is referred to as the "cosmopolitan" lineage. Molecular epidemiologic studies reviewed by Nadin-Davis and Bingham (43) demonstrate that the genetic similarity of dog rabies variants (and some rabies viruses associated with wildlife rabies cycles) indicates that the virus variant was widely distributed during colonial times and that dogs of 14th-century Europe served as the source of these canine enzootics.

With the exception of some island nations and states, rabies in terrestrial wildlife is present worldwide. The disease in red foxes has been prevalent in subarctic and northern parts of North America, in subarctic Asia, and in central and eastern Europe. Raccoon, dog, and gray wolf rabies reservoirs occur in North Eurasia. In Africa, the dog rabies variant has established enzootic rabies in wolf, jackal, and wild dog populations. In South Africa, a distinct strain of rabies is maintained in the yellow mongoose. Surveillance and reporting in Asia is sporadic, but the disease is present in foxes, jackals, wolves, and most recently in ferret badgers. Rabies transmitted by dogs is reportedly a leading cause of death from infectious diseases in China. Furthermore, previously unrecognized wildlife vectors are important in rabies-related human mortality in China (44). In South America, distinct rabies variants exist in some terrestrial wildlife, but these variants are less understood due to the importance of dog rabies. Rabies is present in the Indian mongoose in Granada, Puerto Rico, Cuba, and the Dominican Republic of the Caribbean (45).

In North America, rabies in terrestrial species is maintained in geographically defined outbreaks with a single antigenically or genetically distinctive variant. The variant is maintained by intraspecific disease transmission within the population of a single predominant vector species, after which the outbreak is named. The infection of other species occurs, but these spillover cases have only rarely established sustained intraspecific transmission in another species. The sustained spillover event of bat variant into skunks in Flagstaff, Arizona, was subsequently identified as the result of multiple independent introductions of bat rabies virus into carnivores (46). The disease may persist in the vector population for decades, and the geographic area affected can grow, diminish, or shift gradually or rapidly. The establishment of a rabies epizootic in a new area in a susceptible population can also be a result of the human translocation of wildlife, as exemplified by the introduction of raccoon rabies into the mid-Atlantic states in 1977 (47).

Raccoon rabies was first recognized in Florida in the 1940s and became endemic in the Southeast. This intense outbreak presently affects areas of the 20 eastern states from Florida to Maine. The epizootic has been characterized by a rapid northeasterly spread, very large numbers of rabid raccoons in newly affected areas, and a subsequent cyclic nature, without ever completely dissipating in areas once affected. Spillover from this outbreak has occurred to an exceptional diversity of other mammalian domestic and wild species, most commonly to striped skunks. The high incidence of raccoon variant rabies virus in other species of animals found in the enzootic zone has presented challenges for both public health and wildlife officials (48). To date, two human rabies cases, in 2003 and 2013, can be directly attributed to the raccoon variant (29). The 2003 epidemiologic investigation identified no event during which the patient was exposed to raccoon virus. In February 2013, a patient died of rabies 18 months after receiving a deceased-donor kidney transplant. An investigation revealed that the donor, an outdoorsman and hunter, had been bitten by raccoons but failed to report the bites.

Skunk-vectored rabies enzootics exist across a broad region of North America, including most of the central United States, California, and some Canadian provinces. The viruses associated with this outbreak are actually composed of three genetically distinct lineages, maintained in the north central, south central, and Pacific coast areas (49). The distribution of skunk rabies is limited to those areas where the disease exists in the primary vector. Where rabies exists in other North American terrestrial vectors, there is significant spillover into skunks. The skunk cases in those areas result from infection with the outbreak-associated variant, such as the raccoon variant.

Rabies occurs in red foxes in Alaska and in several of the more remote regions of the Canadian provinces. The red fox variant is similar in these regions, as these outbreaks are vestiges of the southward spread of Arctic fox rabies that swept across the northern areas of the continent during the 1950s.

A relatively recent emergence of dog- and coyote-vectored rabies occurred in the southern regions of Texas, adjacent to the Texas-Mexico border. Beginning in 1988, the outbreak expanded northward to encompass most of South Texas. By 1999, oral rabies vaccination efforts had nearly extinguished this outbreak. The last case of canine rabies occurred along the Mexican border in March 2004. After 2 more years without an indigenously acquired case of canine variant rabies (50), the CDC reported that the United States was canine rabies free in September 2007.

The spillover of rabies into terrestrial mammal species other than the predominant vector species is present in all geographic areas that experience rabies in wildlife. The most problematic of these spillover events is rabies transmission to unvaccinated domestic animals, especially dogs and cats. The resulting rabid dogs and cats, usually in close proximity to their human caretakers, account for numerous human rabies deaths in regions where postexposure prophylaxis is unreliable or unavailable. In areas where biologics are available, exposure to rabid domestic animals results in the extensive administration of postexposure prophylaxis. Raccoon rabies has been endemic in New York State since 1990. During a 10-year period in New York State (2003 to 2012), there were 2,508 raccoons diagnosed as rabid. Other animal rabies cases during the same period that are classified as spillover events amounted to 1,464 cases, which includes 342 domestic animals, predominantly cats (51). Rabies in large rodents, particularly woodchucks and beavers, has been documented in areas where terrestrial rabies is present. Rabies is uncommon in small rodents and lagomorphs but has been occasionally confirmed in groundhogs; domestic rabbits; guinea pigs; and more rarely in prairie dogs, porcupines, chipmunks, and mice (52).

Rabies in bats and terrestrial mammals are largely independent cycles. The rabies variants isolated from bats are antigenically and genetically distinct from those associated with terrestrial rabies (53). In bats, classical rabies virus infections (genotype 1 lyssavirus) are limited in distribution to the Western Hemisphere, where these mammals are of major importance as reservoirs and vectors of rabies infection. Vampire bats are found throughout the more tropical areas of Latin America, and cattle losses from vampire bat–transmitted rabies range from 30,000 annually in Mexico alone to more than 1 million in all of Latin America (54). Human mortality from vampire bat–transmitted rabies continues to be reported. Sixty-two deaths were reported from exposure to hematophagous bats in Latin America during the 10-year period from 1993 to 2002, with local outbreaks in some rural areas causing significant human mortality (55).

Rabies is widespread in the insectivorous bats of North America. The disease has been confirmed in 49 states (Hawaii remains rabies free) and in the territory of Puerto Rico. Rabies virus has been identified in most of the 39 bat species indigenous to the United States. In 2012, the 1,680 laboratory-confirmed rabid bats from 51 states accounted for 27.3% of all United States animal rabies cases for the year. In the same year, 6 states reported rabies in bats but not in terrestrial species (56). Most laboratory-confirmed rabid bats in the nation are not identified to species, but among those that are, most cases occur in the big brown bat, the Mexican free-tailed bat, and the little brown bat, all widely distributed, common commensal species. Like rabies in terrestrial mammals, virus transmission in bats is mainly intraspecific, and distinct rabies variants can be identified antigenically or genetically for each specific bat species or to identify spillover into a heterologous bat species (57). A thorough analysis of bat rabies variants is hampered by a failure of many rabies diagnostic laboratories to identify the species of bats found to be rabid. Of the 27,572 bats sent to rabies diagnostic laboratories in the United States in 2012, only 13,451 (48.8%) were identified to species. Of the 1,683 rabid bats diagnosed in 2012, in 1,031 (61.2%) the bat species was undetermined (55). Associations of particular variants with the major colonial and migratory species have been described (56). Occasional spillover events resulting in the interspecific transmission of these variants are recognized in bats and, less frequently, terrestrial mammals. Although terrestrial mammals infected with bat variants may shed virus in their saliva during clinical disease (57), prior to 2001 there had been no direct observation that infection with a bat variant had established an outbreak in a terrestrial species. In 2001, an outbreak of rabies in skunks in Flagstaff, Arizona, emerged that was identified as the consequence of infection with a big brown bat variant of rabies virus (40).

While disease prevalence in samples from the random collection of asymptomatic bats at roost is generally estimated to be less than 1%, rabies positivity rates among bats submitted to public health laboratories for testing range from 5 to 15% (56). In New York State from 1987 to 2006, 3.2% (1,636 of 51,170) of bats submitted for public health testing were found to be rabid (R. J. Rudd, unpublished data).

However, there is a strong bias toward abnormal behavior and injured bats submitted to public health laboratories. Nevertheless, these rates are relevant for public health decisions because they accurately reflect the likelihood of rabies infection in individual bats encountered under common circumstances by people and pets.

The closely related EBLV 1 and 2 (genotype 5 and 6 lyssaviruses), the Australian bat lyssavirus (genotype 7 lyssavirus), and the less closely related Lagos bat virus (genotype 2 lyssavirus) and Duvenhage virus (genotype 4 lyssavirus) of Africa broaden the distribution of rabies-related encephalitis associated with bats. Lyssaviruses recently identified in Eurasian bats have increased the number of genotypes in the genus *Lyssavirus* to 14 (2). Current information on the prevalence, vectors, and distribution of the non–genotype 1 lyssaviruses is based on limited surveillance. Indeed, insufficient isolates from some of the novel lyssaviruses prevent further characterization via molecular or animal studies. Animal models have shown reduced protection with preexposure vaccination or conventional postexposure prophylaxis against four of these newly recognized lyssaviruses (58). These viruses are capable of producing rabies-like encephalitis in humans and other mammals and may eventually prove to be of greater epidemiologic importance as we learn more about their natural history. Molecular epidemiologic studies support the hypothesis that lyssaviruses existed in bats long before they were present in terrestrial carnivores. Previous studies report that early spillover events resulted in the current North American raccoon rabies variant and cycle, and a separate event gave rise to the existing cosmopolitan canine rabies lineage. A molecular clock model suggests that the time of the most recent common ancestor for current bat rabies virus variants in the Americas was the mid-1600s (59).

EPIDEMIOLOGY OF HUMAN RABIES

Rabies remains an important yet neglected disease in Africa and Asia. Disparities in the affordability, accessibility, and quality of postexposure treatment combined with the risks of exposure to rabid dogs result in a skewed distribution of the disease burden across society, with the major impact occurring on those living in poor rural communities, particularly children. Worldwide, data regarding the incidence of human rabies are often unreliable. Underestimates are common because human rabies in developing areas is largely a rural problem, and many rural cases go undiagnosed or unreported. Human rabies is still common in developing areas of Africa, Asia, and to a lesser extent certain areas of Latin America, with an estimated 35,000 to 60,000 annual deaths worldwide (2). Because most of these deaths are due to rabid dog bites and a lack of, or inadequate, rabies postexposure prophylaxis (PEP), the mortality from the disease is largely a problem of poverty and inaccessibility to health care.

In the United States, human rabies deaths decreased dramatically following the control of canine rabies in the early 1950s. There were 99 cases during the 1950s, 15 in the 1960s, 23 in the 1970s, 10 in the 1980s (60), 27 in the 1990s, 32 in the 2000s, and 11 between 2011 and September 2014. Of the 70 human rabies cases from 1990 through September 2014, 12 were acquired from dog bites occurring outside the United States. Of the 41 cases in which the infection was acquired in the United States and Puerto Rico, 2 were of a variant associated with indigenous domestic dogs, 2 were of a variant associated with raccoon rabies, 1 was of a variant associated with mongoose and dog rabies in Puerto Rico, and 37 were of a variant associated with insectivorous bats (55).

CLINICAL RABIES

Following the variable incubation period, the disease in humans is marked by a brief prodromal period of several days duration with complaints of nonspecific symptoms, including malaise, anorexia, fatigue, headache, and fever. Characteristically, during this period there is pain and paresthesia, or "tingling," at the site of exposure, which are usually the first rabies-specific symptoms (61). Behavioral manifestations may include apprehension, anxiety, irritability, and insomnia. Following the prodromal period, patients develop a rapidly progressive neurologic course, with a range of symptoms that may include disorientation, hallucinations, paralysis, nuchal rigidity, aerophobia, pharyngeal spasms, hydrophobia, hypersalivation, dysphagia, focal or generalized seizures, cardiac and respiratory arrhythmias, and hypertension, leading to coma and death (62). A review of 32 human rabies deaths in the United States from 1980 to 1996 (60) identified agitation and confusion, hypersalivation, hydrophobia or aerophobia, limb pain, and weakness as the most commonly observed signs of clinical rabies. The cases had a median clinical period of 19 days (range, 7 to 28 days). In 12 of the 32 cases, the disease was only diagnosed postmortem. In the absence of intensive care and secondary support therapies, death usually occurs in human rabies cases within 7 to 14 days of the onset of symptoms, generally from respiratory failure (63). In patients receiving intensive care, the disease will eventually and severely affect nearly every major organ system, and death occurs as a result of the cessation of cerebral and cardiovascular activity (64). Over the past 45 years, there have been seven well-documented human survivals from clinical rabies worldwide, with outcomes ranging from full recovery without sequelae, to some partial paralysis, to major residual neurologic impairment (65). Six of the seven patients had a history of some pre- or postexposure vaccination. Two of the seven cases, who survived with little or no sequelae, were infected with bat rabies virus. One of these cases occurred in 2004 in a Wisconsin teenager with a history of a bat bite on a finger 1 month prior to onset. This patient was maintained in a drug-induced coma and administered antiviral agents. Her treatment included the administration of ketamine, midazolam, phenobarbital, ribavirin, and amantadine. She improved and was released from the hospital with some neurologic sequelae but has experienced progressive neurologic improvement (66). The role of this strategy (Milwaukee protocol) and therapeutic regimen in this survival has not been clearly established (67).

The Milwaukee protocol has garnered considerable optimism; however, subsequent attempts to repeat this success have met with mixed results. According to 43 case reports logged on a central database, six patients treated by this protocol are registered as survivors (68). The application of a similar treatment has not been successful in four subsequent cases with fatal outcomes (69). A recent publication details the survival from clinical rabies of a 9-year-old child who subsequently died of complications due to dysautonomia (70). Although meeting the operational definitions of survival, the child died 76 days after presenting with rabies of vampire bat origin transmitted through a cat bite. This case describes the vast complexity challenging intensive care units when treating human rabies with the Milwaukee protocol. Efforts to utilize the developing technology of metabolomics show promise in advancing critical care and tactics designed to avert the inevitable complications that lead to dysautonomia in human rabies cases (71). Udow et al. postulated that human survival due to rabies of a bat variant origin versus a dog variant may account for a few documented cases of human rabies survival (72).

TREATMENT, PREVENTION, AND CONTROL

Human Rabies Prevention

The prompt management of potential human exposure to rabies is a critical component of human rabies prevention. All potential rabies exposures should be evaluated for rabies risk based upon the nature of the contact, the species of animals involved, the circumstances of the incident (e.g., behavior of offending animal, provoked or unprovoked encounter), the current local status of animal rabies, and the adequacy of surveillance. The rabies vaccination status of the animal should not be used to determine the potential for transmission. Not all biting animals need to be killed and tested. A dog, cat, or ferret that has bitten a person does not always need to be euthanized; nonlethal options can be considered by the owner and state health professionals. If the owner does not want to euthanize the animal and it is not demonstrating signs of a rabies infection, the animal can be confined and observed daily for 10 days. Numerous studies have been published describing viral-shedding patterns in the days preceding the onset of rabies-specific signs. The survival of the animal demonstrating no clinical signs for 10 days after the bite rules out the need for PEP. If, however, the animal dies or signs of rabies develop during the observation period, it must be immediately euthanized and examined. Similarly, when a 10-day confinement of the offending animal is not possible (if it is symptomatic or has died) or is inappropriate (the offending animal is a wild or exotic species or hybrid), the animal must be humanely euthanized and tested.

The Advisory Committee on Immunization Practices (ACIP) of the U.S. Department of Health and Human Services states that exposure to rabies occurs when infectious virus is introduced into bite wounds, open cuts in the skin, or onto mucous membranes. Any penetration of the skin by the teeth, regardless of location, must be considered a bite exposure. Nonbite exposures have occasionally led to rabies infection (27) putatively following exposure to infectious aerosols or solid organ or corneal transplants from rabies victims, which carries the greatest risk. Direct contamination of an open wound (abrasion or scratch) or mucous membrane with saliva or other potentially infectious material (e.g., neural tissue) from a rabid animal is also considered rabies exposure. Contact with blood, urine, or feces (including bat guano) or merely petting a rabid animal is generally not an indication of exposure. Rabies virus in saliva on environmental surfaces is quite labile; therefore, if the material on a surface is dry, it generally can be considered noninfectious (26). Generally, bites or other contact with a domestic rabid animal occurring more than 10 days prior to the recognized onset of signs of rabies in the animal are not considered potential rabies exposures.

Despite the existence of numerous and widespread terrestrial rabies outbreaks in the country during the period, a surprising 90% of human rabies cases in the United States from 1958 to 2004 have been attributed to bats or bat rabies variants. From 1953 to 2002, surveillance in the United States documented at least 39 human cases associated with bat rabies based on the patient's history or virus characterization. In just 9 of these cases, there had been a definite bat bite reported. In 11, there was known or likely contact with bats associated with an indoor bat encounter under circumstances in which a bat bite may have gone unrecognized (73). As a result, the ACIP has developed specific language regarding bat encounters and rabies treatment. "Rabies postexposure prophylaxis is recommended for all persons with bite, scratch, or mucous membrane exposure to a bat, unless the bat is available for testing and is negative for evidence of rabies. Post-exposure prophylaxis might be appropriate even if a bite, scratch, or mucous membrane exposure is not apparent when there is reasonable probability that such exposure might have occurred" (26). Furthermore, bat bites may result in very limited injury, but there is evidence that some bat rabies virus variants may be more likely to be transmitted by superficial dermal exposures (74). One must be cautious not to assume that merely being in close proximity to, or in the same room with, a bat constitutes an exposure. Particular concern must be directed to those situations in which contact was possible, but there is a reasonable probability a bite may have gone unnoticed. Examples of scenarios justifying consideration of rabies exposure include a sleeping person awakening to find a bat in the room or an adult witnessing a bat in the room with a previously unattended young child or mentally disabled person.

Because it is often very difficult to accurately reconstruct details immediately following an indoor bat encounter and evaluate the likelihood of a "reasonable probability" of exposure, it is prudent to recommend the capture and retention for testing of bats involved in incidents with the potential for human exposure. As rabies positivity rates among insectivorous bats encountered by the general public are typically 3 to 6% (75), more than 90% of the bats encountered in potential exposure situations will test negative for the virus, eliminating the need for postexposure investigations and avoiding the use of PEP. This practice will also help ensure that PEP is provided to those who have actually had contact with the relatively small proportion of encountered bats that are rabid.

Human Rabies Prophylaxis

The relatively long incubation period of rabies permits efficacious PEP. Modern biologics have afforded the potential for 100% success with proper wound treatment, prompt and appropriately administered immune globulin, and a course of vaccine. Postexposure vaccines have been hypothesized to confer protection largely because they prime an immune response in organs peripheral to the CNS, and the activated lymphocytes, CD4, antibody-secreting plasmocytes, and perhaps antibodies can migrate into the nervous system (76). However, a comprehensive understanding of the nature and sequence of events that confer protection requires further investigation. An immediate and thorough washing of bite wounds with soap, water, and a virucidal agent are valuable measures for the prevention of rabies (26). With the exception of patients with previous immunization, rabies PEP should always include passive immunization with human rabies immune globulin (HRIG) to neutralize virus at the site of exposure and active immunization with vaccine to produce neutralizing antibodies that develop in 7 to 10 days after vaccination is initiated. Active immunization also triggers a cell-mediated response that is critical to the success of PEP. The importance of the administration of HRIG to the success of PEP has been documented by numerous investigations of PEP failures (77). The HRIG is administered only once, at the beginning of the prophylaxis, but if there is a delay in administering the HRIG, it can be given through the seventh day after the administration of the first dose of vaccine. The recommended HRIG dose is 20 international units per kilogram of body weight. If anatomically feasible, the full dose should be infiltrated into and around the wounds. The remaining volume or the entire dose for treatment for exposures with no recognizable wounds should be injected intramuscularly at one or more sites distant from the site of vaccine inoculation. Two immunologically comparable vaccines grown in two differently derived cell culture systems are currently licensed for use in the United States: a human diploid cell vaccine (HDCV) and a purified chicken embryo cell vaccine (PCECV). Both are packaged for and administered as a 1-milliliter intramuscular dose for pre- or postexposure administration. For PEP, either of the vaccines can be used in a four-dose course, with 1-milliliter intramuscular injections given in the deltoid area (or for small children, the anterolateral aspect of the thigh) on days 0, 3, 7, and 14, commencing as soon as possible after the exposure is known (78). PEP for previously vaccinated individuals is comprised solely of two doses of vaccine administered as a 1-milliliter intramuscular injection in the deltoid region on days 0 and 3, commencing as soon as possible after the exposure is known. Two HRIG products are licensed for use as well, and both are anti-RIG (IgG) preparations concentrated by the cold ethanol fractionation from plasma of hyperimmunized human donors. HRIG is not administered to previously vaccinated patients. For this purpose, a previously vaccinated status applies only to individuals who have received one of the recommended pre- or postexposure regimens of the currently licensed vaccines or who have received another vaccine or regimen and had a documented adequate neutralizing antibody titer of 0.5 international units or greater or complete neutralization at 1 to 5 serum dilution on the Rapid Fluorescent Foci Inhibition Test (RFFIT).

Although several other efficacious cell culture vaccines are widely available outside the United States, nervous tissue vaccines and immune serum of equine origin are still employed in some areas. A purified RIG of equine origin has been used effectively in developing countries. Several alternative PEP schedules are employed outside the United States, including multisite regimens that are accelerated by the administration of more than one dose on the first day of treatment, by either intramuscular or intradermal inoculation (79). These regimens may induce early antibody response, which is beneficial where RIG is not available, and can reduce the amount of vaccine required.

Preexposure rabies vaccination is available for persons at risk of rabies exposure, such as veterinarians, animal control officers, animal handlers, rabies laboratory workers, others whose activities bring them into contact with a rabies vector species, and certain travelers. Preexposure vaccination does not preclude the need for PEP following a known exposure but eliminates the need for HRIG and reduces the number of vaccine doses to two. Furthermore, a previously vaccinated individual might be protected from inapparent exposures or when PEP is unavoidably delayed (26). The primary preexposure regimen consists of three 1-milliliter intramuscular injections, given one each on days 0, 7, and 21 or 28. Depending on the risk category, booster vaccinations may be recommended when rabies neutralizing antibody titer is less than 0.5 international units or at less than complete neutralization at a 1 to 5 serum dilution by RFFIT.

The routine determination of adequate serum rabies neutralizing antibody titer following primary pre- or postexposure vaccination is not required unless the person is immunosuppressed. Persons at the highest or continuous risk of inapparent rabies exposure, such as those working in rabies research or vaccine production laboratories, should have a titer determination every 6 months (80). Others in the frequent-risk category, such as those working with mammals in areas where rabies in animals is enzootic, should be serologically tested at 2-year intervals. A single booster vaccination is recommended when the titer is at less than complete neutralization at 1 to 5 serum dilution on the RFFIT or less than 0.5 international units. The ACIP recommends no routine preexposure booster doses following primary vaccination for certain travelers and animal workers in areas with minimal prospects of exposure (26).

Reactions following vaccination with the currently licensed cell culture vaccines and the administration of HRIG are less serious and less common than with previously available biologics. Mild local reactions such as pain, erythema, and itching are reported in 30 to 74% of vaccinees (81). Mild systemic reactions including headache, nausea, abdominal pain, muscle aches, and dizziness are reported in 5 to 40% of recipients. Guillain-Barre syndrome–like neurologic illness and other central and peripheral nervous system disorders temporally associated with HDCV administration have not been linked in a causal relationship with the vaccination (82). A delayed (2 to 21 days postinoculation) immune complex–like reaction was reported among 6% of patients receiving booster doses of HDCV. The reaction included hives sometimes accompanied by arthralgia, arthritis, angioedema, nausea, vomiting, fever, and malaise, but the reaction was not life threatening. No fetal abnormalities have been associated with rabies vaccination, and vaccination is not contraindicated during pregnancy (83).

Rabies Control

If the control of rabies is defined as the elimination of human mortality from the disease, then it can be achieved by success with individual or combined strategies that involve the elimination of the virus in animal populations, the elimination of human exposure to infected animals, the prevention of human infection by pre- or postexposure vaccination, or the development of an efficacious cure for clinical disease. Because rabies virus is a zoonotic disease with an adaptability to a wide variety of host populations, eradication of the virus, as has been achieved for smallpox, may not be a realistic goal. Control in domestic dog populations and some wildlife vectors in geographically defined areas has been accomplished by stray-dog control coupled with vaccination programs and by oral rabies vaccination campaigns in wildlife (84). The elimination of the disease in dog populations offers not only large reductions in human mortality but vast economic benefits to developing countries (85). However, resource limitations, an increasing movement of domestic and wild animals, and a greater knowledge of the significant role and distribution of rabies and rabies-related viruses in bats threaten to make even regional elimination unlikely for most of the world. Modern cell culture vaccines and purified rabies immune globulins of human origin have made safe and highly efficacious pre- and postexposure immunization a reality, yet resource limitations prevent the application of these methods for much of the world's population.

In North America, Europe, and other developed areas, great reduction in human mortality has been achieved by the virtual elimination of canine rabies outbreaks by vaccination programs (often compulsory) in association with leash and stray-dog measures. Further reduction has occurred from the development and availability of modern biologics for preexposure treatment and PEP. Other components of the rabies control effort include education of the public and health-care professionals regarding exposure avoidance, the proper management of potential exposure to rabies (including animal confinement and observation), prompt and accurate rabies diagnosis, and accessibility to prompt and proper PEP or preexposure vaccination when warranted.

Wildlife rabies control by vector population reduction has only rarely proven to be effective, and in North America such measures have not been encouraging. However, the reduction or elimination of wildlife rabies epizootics has proven to be achievable in some situations by wildlife vaccination. Oral rabies vaccination has controlled fox rabies in large areas of Europe and Ontario, Canada. A modified live virus oral vaccine has been used in Canada, and efforts in Europe have used baits containing either modified live virus or a recombinant, vaccinia-vectored vaccine. The baits have been distributed by hand, helicopter, or fixed-wing aircraft in one or two distribution campaigns per year in programs that have continued for many years. A recombinant vaccinia-rabies glycoprotein vaccine has proven safe and efficacious in laboratory and field trials (86) and was licensed by the U.S. Department of Agriculture in 1997 for use in oral rabies vaccination programs conducted by state and federal agencies to control rabies in raccoons. It has also been used with success to control coyote-vectored rabies in South Texas. With millions of vaccine-laden baits distributed annually in North America and with only a single human infection with the vaccinia recombinant vaccine reported, (87) the safety of this approach is well established. Improvements in baits and vaccines could increase the efficiency and efficacy of wildlife vaccination efforts, while at the same time making them more cost effective. Canine adenovirus-vectored and human adenovirus-vectored recombinant rabies glycoprotein vaccines are in development and evaluation, and other novel vaccine approaches are proposed (88). In Canada, the application of a human adenovirus-vectored recombinant rabies vaccine has shown great potential in reducing wildlife rabies (89). Advances in wildlife vaccination are now being extended to community dogs in less-developed nations. Integrated programs to eliminate dog rabies and human deaths in areas of Asia have recently shown great promise (90). Repeated mass dog vaccination campaigns appear to be the best avenue to control dog rabies in developing nations (91).

LABORATORY DIAGNOSIS

The rabies diagnostic laboratory has long played a prominent role in rabies control. Because the modern rabies laboratory can provide reliable results on the day of receipt of the specimen, the physician's decision to provide or withhold rabies treatment following a bite from a suspect animal is commonly based upon the laboratory tests performed on the animal's brain. Uniquely high standards of sensitivity and specificity are required of these tests, as a false-negative result could have the consequence of human mortality. Alternatively, if a delay in rabies testing is unavoidable, it may be appropriate to initiate rabies treatment. Subsequent negative results from a reliable laboratory would justify terminating PEP.

The rabies laboratory may support surveillance for the disease in wildlife to aid in the proper allocation and targeting of rabies control programs, such as the efforts to vaccinate wildlife with oral baits. Not all biting animals need to be euthanized and tested. Because of laboratory data and empirical knowledge of virus-shedding periods prior to the onset of signs of the disease, certain dogs, cats, and ferrets may be confined and observed for signs of rabies for 10 days following a bite to rule out rabies transmission (see Human Rabies Prevention).

Wild animals, especially bats, foxes, skunks, and raccoons, that have bitten or otherwise potentially exposed a human, should be euthanized and tested immediately. Because rabies is a disease affecting mammals, it is never necessary to test arthropods, amphibians, reptiles, or birds. In the United States, small rodents, including mice, rats, and squirrels, are essentially free of rabies, and therefore routine examination is not required; exceptions include rodents involved in unprovoked attacks in areas where rabies is endemic and larger rodents such as woodchucks, otter, muskrats, and beaver (92).

Diagnosis of Rabies in Animals

Since the development of the first human rabies vaccines, the accurate and timely diagnosis of rabies infections in animals has been essential to the prompt and successful postexposure prophylaxis of humans. Accurate diagnosis is essential to the maintenance of a disease surveillance system, which influences public health decisions determining animal quarantine, vaccination, control, and the related allocation of government resources. Thus, the serious nature of this zoonotic disease mandates the appropriate use of diagnostic techniques that will ensure the highest obtainable sensitivity and specificity. Adherence to the accepted standardized and validated protocol for the fluorescent antibody test in the postmortem diagnosis of rabies in animals is one essential step toward accomplishing that goal.

Rabies diagnosis can be achieved with 100% sensitivity only by the postmortem examination of brain tissue. Throughout most of the prolonged incubation period of rabies there is no reliable means to rule out infection; there is no rise in circulating antibody titer, and neither rabies virus, its antigens, nor rabies RNA can be reliably identified. This is due to the limited and unpredictable distribution of the virus and its proteins during the retrograde axoplasmal movement from the site of the exposure to the CNS. Rabies virus, its antigens, and rabies RNA do not move centrifugally away from the CNS during the incubation period (24). The testing of saliva collected antemortem from animals cannot be used with certainty to exclude the possibility of rabies infection and transmission by bites from the animal because virus may be present intermittently in saliva in infected animals (93). However, modern methods can always identify the presence of rabies virus in the brain of a rabid animal that dies or is euthanized during, or up to several days before, the onset of the clinical signs of the disease. Therefore, after a human exposure, when a decision is made to sacrifice and test the animal for rabies infection, it is never necessary to delay testing for further development of the disease to achieve a reliable diagnosis. Most importantly, the centrifugal spread of the virus from the CNS to the salivary glands (and therefore a potentially infectious bite) does not precede the appearance of demonstrable rabies antigen in the brain (24). Thus, a negative result of brain examination by acceptable methods ensures that the bite of the animal could not have caused an exposure to rabies.

Collecting, Preparing, and Submitting Rabies Specimens

The animal species, nature of the exposure, variant of rabies virus, and time and cause of death may affect the terminal distribution of rabies virus and its antigens in the brain. Generally, the brain stem and the cerebellum constitute the best diagnostic samples, and areas of each tissue are examined to provide a reliable negative report. Therefore, the intact head is the preferred diagnostic sample for the postmortem diagnosis of rabies in animals. Following the euthanasia guidelines of the American Veterinary Medical Association (94), an animal can be euthanized for rabies examination with barbiturate, a nonbarbiturate injectable, inhalants, or by other humane means that do not damage the brain. The specimen should be immediately preserved by cooling to 4°C and maintained at refrigeration temperature until it arrives at the laboratory. Should refrigeration not be possible, freezing is an acceptable but less desirable alternative. A single freeze-thaw cycle will not prevent reliable diagnosis, but freezing will make the dissection more difficult and may delay the test. However, repeated freeze-thaw cycles can damage the specimen. The head should be removed from the neck before the first vertebrae, using caution to avoid personal injury or the creation of infectious aerosols. Individuals capturing suspected rabid animals or handling and decapitating the carcass should receive appropriate safety training and rabies preexposure immunization (26). Bats should be submitted intact to avoid damage to the animal's CNS during decapitation. For large livestock species, including cattle and equines, a section of the brain stem and cerebellum can be removed through the foramen magnum by a veterinarian or specifically trained individuals. It is critical that a full cross section of the brain stem and all of the cerebellum be submitted for testing.

The specimen should be immediately transported to the rabies laboratory. Specimens may be shipped by a prompt parcel delivery service if properly packaged and labeled. Some state laboratories provide standard rabies specimen containers for shipping heads to the laboratory. Current guidelines applicable to the packaging of potentially infectious samples, to which shippers must strictly adhere, can be found at: https://www.iata.org/whatwedo/cargo/dgr/Documents/DGR52_PI650_EN.pdf. Included with the head should be a completed rabies specimen history form, if available from the diagnostic laboratory. If no form is available, provide all the significant information including the names, addresses, and telephone numbers of the owner, complainant, and all humans and animals in contact as well as information on the clinical observations, date of death or means of euthanasia, exact location of capture, and information on the person or agency to receive the report. Generally, reports of rabies-positive specimens are made immediately by telephone. Reporting practices vary widely, however, and the submitter should ascertain local practice by contacting the local, regional, or state health department.

Dissection and Sample Preparation

When the entire head is received at the laboratory for examination, the flesh is removed from the cranium, and an anterior and two lateral cuts are made in the cranium by chisel or saw to permit the calvarium to be reflected posteriorly. After removal of the meninges, the cerebellum and the brain stem are removed. An alternative method, useful for surveillance-only examinations, employs the removal of a core of brain tissue by the insertion of a soda straw or similar hollow tube into the foramen magnum and advancing forward to capture samples of the brain stem, cerebellum, and hippocampus that can then be forced out of the straw and used for slide and suspension preparation (95). Touch impressions or slip smears of each tissue are made on microscope slides. A 10% suspension of a mixture of the diagnostic tissues is prepared in suitable diluents for animal inoculation, cell culture virus isolation, or molecular detection. A sample of each brain tissue should be saved at −40°C to −80°C for further testing.

Direct Immunofluorescence for Rabies Antigen

With an achievable sensitivity and specificity approaching 100% and a routine turnaround time of less than a day, the microscopic examination of brain tissue stained with the direct immunofluorescence assay (DFA) is the gold standard for the diagnosis of rabies. Among the findings of a National Working Group on Rabies Prevention and Control was the need for a minimum national standard for the laboratory diagnosis of rabies (96). In response to this recommendation, a committee was formed of representatives from national and state public health laboratories to evaluate the procedures employed by rabies diagnostic laboratories in the United States. Both the National Working Group and this committee have as their goal the uniformity and the improvement of the overall quality of rabies testing through the formulation of guidelines and standards for equipment, reagents, training, laboratory protocols, quality assurance, and laboratory policy for rabies diagnosis. As a first step to attaining this outcome, the committee prepared a standardized protocol for the analytical phase of rabies testing using the DFA and evaluated the protocol by comparison testing 435 samples submitted to public health laboratories for rabies diagnosis. The standardized protocol was developed from published procedures and the collective laboratory experience of the committee members. The group recognized that a range of possible methods may achieve the desired outcome for some of the less critical steps in the diagnosis of rabies and that laboratory policy may be defined regionally in some cases. However, the goal of the group was to establish a single protocol by which all other methods could be validated by comparison. Furthermore, uniformity of procedures in the national laboratories performing these examinations permits the elucidation of diagnostic problems and solutions without the uncertainty provided by multiple variables. The recommendations included in this document should be closely followed to ensure a test of the highest sensitivity and specificity. Modifications or shortcuts in procedures could lead to false-positive or false-negative results and nonspecific reactions (97). The new standard protocol requires the routine use of two diagnostic antibody conjugates on each specimen, confirmatory protocols for the detection of minimal antigen distribution, and adequate sampling of brain tissues. The protocol has been placed on the websites maintained by the CDC (http://www.cdc.gov/rabies/pdf/rabiesdfaspv2.pdf).

Glass microscope slides containing slip smears or brain tissue impressions are fixed for 1 to 4 hours (or overnight) in acetone at −20°C to ensure permeability of the tissue to the diagnostic antibodies and to aid in tissue adherence to the slide. After removal from the acetone and air drying, the tissue is flooded with the diagnostic immunofluorescence reagent. These reagents are made by the conjugation of fluorescein isothiocyanate (FITC) to immunoglobulin G (IgG) that is specific for rabies nucleocapsid protein. The antibodies can be extracted from the antiserum of rabies-hyperimmune hamsters, goats, rabbits, or equines. The development of monoclonal antibody technology has permitted the production of highly specific and uniform diagnostic reagents. Employing a cocktail of these antibodies specific for different epitopes on the rabies antigen, a new generation of diagnostic conjugates has been developed for rabies diagnosis. After a 30-minute incubation at 37°C in a moist chamber incubator, excess diagnostic reagent is washed off in multiple saline baths. The FITC-conjugated antibodies remain attached to the tissue only where rabies nucleocapsid protein is present. Microscope slide preparations are examined using an incident light fluorescence microscope fitted with a high-intensity mercury or xenon lamp and appropriate excitation and barrier filters that present FITC-labeled structures as yellow-green fluorescent objects against a dark background. To obtain optimum results, rabies diagnostic laboratories should use the latest generation of microscopes and optics. Plan-apochromat objective lenses with magnifications of 20x and 40x are recommended for optimal results. Both dry and immersion lenses may be used successfully for rabies diagnosis. Immersion oil should have the same refractive index (1.515) as glass and should be formulated specifically for fluorescence applications and evaluated for autofluorescence. Objectives with a numerical aperture of 0.75 or greater should be employed for optimal light gathering, which is essential for rabies diagnosis. Rabies-specific staining in brain tissue appears as characteristic round or oval intracytoplasmic inclusions most prominent in the large neurons of the cerebellum, hippocampus, and brain stem (Fig. 1). The sensitivity of rabies immunofluorescence testing on fresh brain tissue can be comparable or superior to isolation procedures (98). Important factors in the avoidance of false-negative results in rabies diagnosis using DFA include strict adherence to a uniform methodology, the retention and training of qualified microscopists, the use of proper scientific controls for all procedures, the use of properly diluted high-quality diagnostic conjugate, and the optimization of lamp and microscope performance. The specificity of the procedure can approach 100% agreement with virus isolation when procedures and practices ensure the avoidance of cross contamination and are optimized to recognize, control, and ameliorate nonspecific fluorescence. When equivocal fluorescence staining patterns are suspected, select immunohistochemical procedures are employed to confirm the specificity of the staining. The use of an FITC-labeled IgG conjugate specific to an antigen other than rabies virus is one such procedure.

Proficiency testing for the diagnosis of rabies in animals using DFA is not mandated by the Clinical Laboratory Improvement Amendment of 1988. Voluntary rabies proficiency testing programs have been conducted periodically from 1973 to 1992 and each year since 1994. The performance of rabies diagnostic laboratories enrolled in proficiency testing in the United States has been excellent when panels of rabies-positive and rabies-negative test slides were evaluated by the DFA. An excellent consensus has been observed among participants for strong positive and negative test samples. Discrepancies have mainly occurred with very weakly positive slides. One of the most important factors for efficacy of the rabies DFA is the laboratory's recruitment and retention of properly trained and experienced microscopists (96).

A critical requirement for sensitive rabies diagnosis by immunofluorescence is a good-quality brain sample. Brain tissue exposed to chemical fixative, repeated freeze-thaw cycles, or elevated temperatures may result in denatured or masked rabies antigens, hampering recognition by the diagnostic reagents. Decomposition will affect the sensitivity of all rabies diagnostic procedures. Immunofluorescence tests may remain positive after the isolation of virus is no longer possible (99). Evidence of rabies infection by the DFA on decomposed or mutilated tissue fragments may support a valid rabies-positive report confirmable by isolation, immunohistochemistry, or molecular methods. A recent study has identified the ability of viral nucleic acid in decomposed specimens to remain detectable after DFA and virus isolation fail (100). One of the most difficult diagnostic decisions confronting the rabies laboratorian is to determine at what stage of decomposition it is no longer possible to issue a reliable negative rabies report. Certainly, once the CNS has become foul smelling and green in color, with some liquefaction, negative results are not reliable. Occasionally, even specimens with the appearance of only early decomposition on gross inspection may result in slides with each

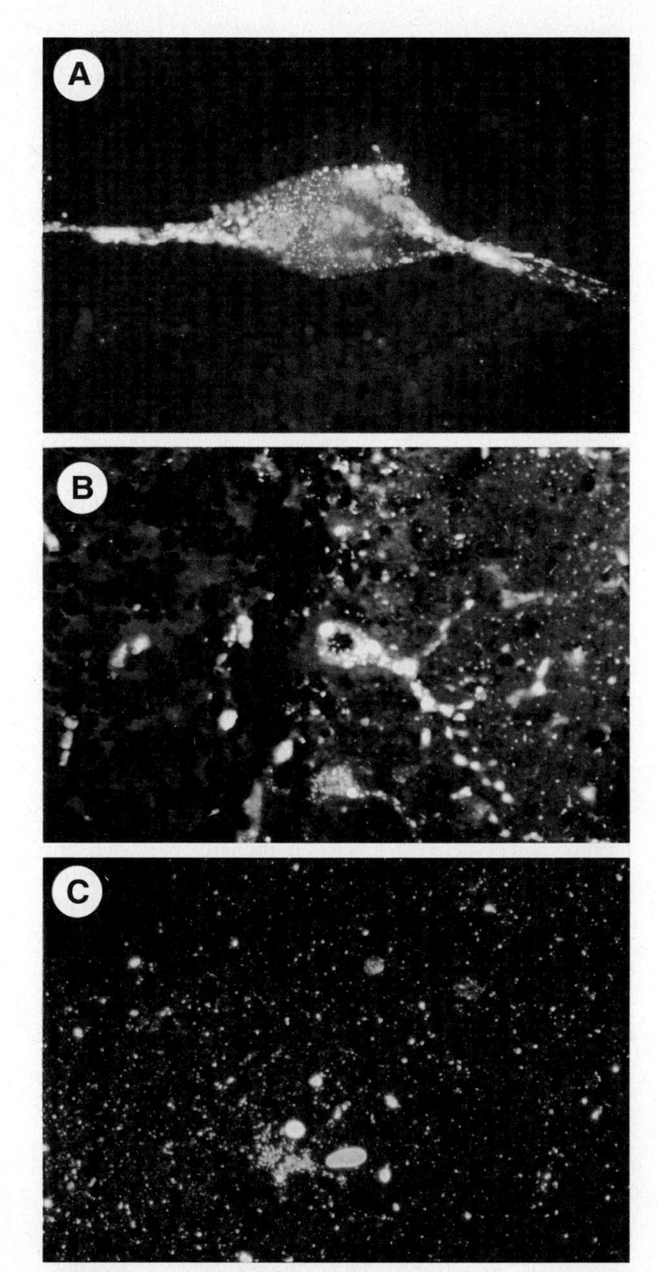

FIGURE 1 Immunofluorescence staining of nervous tissue from rabid animals. (A) White-tail deer spinal cord neuron. Magnification, x400. (B) Purkinje cell from a bovine cerebellum. Magnification, x200. (C) Slip smear of cerebellum from a rabid calf. Magnification, x200.

microscopic field so overgrown with autofluorescing decomposing bacteria that reliable examination is not possible. Mutilation of the tissue by trauma to the extent that the necessary sample regions of the cerebellum and brain stem are not available or unrecognizable also precludes a reliable negative result (101). This conundrum is especially pertinent with bats that are killed by trauma. Generally, it is not possible for lab personnel, based solely on a verbal description, to determine before the dissection whether or not a decomposed or mutilated specimen may be testable. Consequently, unless it is clear that the carcass is in the last stages of decomposition or is mutilated to the extent that no recognizable CNS exists, it is wise to submit compromised specimens to the laboratory for evaluation as to suitability.

Other Methods of Antigen Detection

Antigenic sites recognized by antinucleocapsid diagnostic reagents in the standard DFA may be masked by bonds created when tissues are fixed by exposure to formalin and other fixatives. As a result, the sensitivity of the procedure can be greatly reduced on fixed tissues. Advances have increased the sensitivity of direct and indirect immunofluorescence and immunohistochemical procedures applied to fixed tissue. Digestion of the tissue with enzymes, such as proteinase, prior to staining may expose antigenic sites (102). The use of avidin-biotin complex amplification and high-affinity monoclonal antibodies for the first label has improved the performance of these procedures so that they may be approaching the reliability of DFA on fresh tissues. A direct rapid immunohistochemical test employing a short formalin fixation of fresh or glycerol-preserved brain impressions and requiring no specialized microscopic equipment has been demonstrated to be of utility for testing under field conditions or for countries with limited diagnostic resources (103). Further evaluation is necessary before negative results on chemically fixed tissues can be used for public health decisions.

Histologic Examination

Negri bodies (15) are intracytoplasmic, acidophilic inclusion bodies that can be demonstrated best in the Purkinje cells of the cerebellum and the pyramidal cells of the hippocampus of many rabies-infected animals by microscopic examination of tissue stained with basic fuchsin and methylene blue or hematoxylin and eosin (104). The presence, distribution, and size of Negri bodies are related to the species of animal, variant of rabies virus, and duration of the clinical period before death or euthanasia. The demonstration of these pathognomonic inclusions is very specific and, coupled with evidence of an encephalitic inflammatory response, may provide a reliable positive rabies diagnosis when reported by an experienced pathologist. However, the sensitivity of the method is poor, with numerous histologic surveys indicating that 25% or more of rabid animals have no demonstrable Negri bodies, severely limiting the value of this diagnostic method for medical decisions. In most rabies diagnostic laboratories, Negri body detection for routine rabies diagnosis was replaced with the DFA during the early 1960s.

Virus Isolation

The consequences of false-negative results in cases of human exposure support a continued practice of utilizing rabies virus isolation as a backup confirmatory procedure, despite the proven sensitivity and reliability of the DFA. Virus isolation may be employed as a general quality-control procedure or just applied to instances of bites to humans from highly suspect animals. In either case, it serves to sustain confidence in the reliability of the DFA results and to exonerate the microscopist of the full burden of responsibility. The propagation of virus in the laboratory is also a critical component for identification of virus variants and the production of diagnostic reagents. *In vitro* and *in vivo* methods are both widely employed.

The mouse inoculation test (MIT) was introduced for diagnostic purposes in 1935 (105). It is a sensitive and reproducible procedure. When used as a confirmatory procedure for DFA results, portions of the same tissues that are used for the microscopic examination are ground into a suspension, employing a diluent of physiologic salt solution containing serum and antibiotic supplements. The suspension is inoculated intracerebrally into weanling Swiss albino mice. The mice are observed daily for 30 days for evidence of rabies infection. Mice that develop illness during the observation period are immediately euthanized and the brains examined by DFA. A valuable attribute of the MIT is its ability to detect small quantities of rabies virus even in very weakly positive specimens, and it can be successfully applied to mutilated and decomposed samples (99). Its weakness, beyond the inherent environmental and ethical issues with the use of live animals in the laboratory, lies in the typical 7- to 20-day period between inoculation and recognized illness in the mice. The limitation of the procedure results from the possibility that if treatment were withheld following a bite due to a false-negative DFA, detection by this backup method would occur after a period that would cause great concern about vaccine failure. The period can be shortened by the use of neonate mice, with daily sacrifice and DFA examination of numerous individual animals, but this greatly increases the labor-intensive nature of the procedure and can be prohibitive in a laboratory performing routine diagnosis on large numbers of specimens.

The delay associated with the demonstration of virus by the MIT can be avoided by the isolation and identification of rabies virus on a continuous cell culture (106). Tissue from the diagnostic regions of the brain of the suspect animal is ground into a suspension in a cell culture medium as a diluent. The suspension is incubated after the addition to the cells of a continuous cell line selected for its susceptibility to infection with rabies virus, generally a mouse neuroblastoma cell line. The test can be performed in tissue culture slides, 96-well plates, or Teflon-coated slides. As rabies virus does not produce a cytopatic effect in cell culture, it is necessary to stain for rabies antigen in the cells. After an appropriate incubation, the cell monolayers are rinsed with saline, fixed with acetone or a formalin-methanol mix, and examined by DFA. If infectious rabies virus is present in the brain tissue, characteristic intracytoplasmic inclusions of rabies antigen will be observed in fluorescent foci in the cells (Fig. 1). The sensitivity of the procedure is comparable to that of the DFA and the MIT (98, 99). Because results are available within a few days of receipt of the specimen, it serves as a much better means of confirming negative DFA results, as a false-negative DFA would be recognized in a period of time permitting the timely initiation of PEP.

Molecular Methods in Rabies Diagnosis

Molecular assays have become routine in many virology laboratories either as the primary or confirmatory test, motivating discussion as to the feasibility of replacing the rabies DFA with a real-time reverse transcriptase assay (RT-PCR) (107). Molecular assays are generally less labor intensive, require fewer employees, have increased sensitivity, and eliminate the need for highly skilled microscopists. These advantages have motivated some public health laboratories to consider replacing the DFA with RT-PCR. Currently in the United States, the use of real-time RT-PCR as a primary diagnostic assay has yet to be established, but the method may be valuable as a confirmatory test (108, 109). In addition, molecular methods, such as conventional and real-time RT-PCR, have been used in rabies virus research, yielding significant improvements in typing rabies virus variants and understanding rabies virus maintenance, transmission, and pathogenesis (110).

The early days of using PCR as a rabies diagnostic assay were problematic as false-positives results were a concern when using nested or heminested conventional assays, and false-negative results could occur as a result of primers that were not effective at detecting all the variants of rabies virus (107). The development of real-time RT-PCR has improved specificity and sensitivity over conventional PCR by eliminating the need for nested and heminested assays as primer and probe combinations, thereby decreasing the probability of false-positive results (111). The use of real-time RT-PCR in antemortem human rabies diagnosis is routinely employed to identify and quantify the viral load in diagnostic specimens, including cerebral spinal fluid, saliva, and nuchal biopsies (112).

The application of newer real-time RT-PCR assays in routine rabies diagnosis has been evaluated by several public health diagnostic laboratories (108, 109, 111, 112). To be considered for routine rabies diagnosis, real-time RT-PCR must match or surpass the sensitivity and specificity of the DFA, provide rapid results, and be cost effective. To compare the sensitivity of DFA and RT-PCR, Szanto (109) employed the two techniques on more than 700 reservoir species. All samples were negative via DFA, yet 10 of these DFA-negative samples were low positives (threshold cycle values approximately 40) via RT-PCR. Dupuis et al. collected 1001 rabies specimens submitted to the New York State Department of Health Rabies Laboratory for routine testing and evaluated the DFA, virus isolation, and RT-PCR (108). The RT-PCR assay employed an automated RNA extraction and template addition robot, decreasing the possibility for human error. In the Dupuis et al. study, specimens previously identified as DFA positive were positioned next to DFA-negative specimens for the purpose of identifying any potential for cross-contamination events. Additionally, the primer probe combination used in the study (RABVD1) demonstrated the sensitivity and specificity to identify all rabies variants found in the northeastern United States (112). Of the 1001 samples, 141 were positive by DFA; the same 141 samples were positive via RT-PCR, yet 1 negative DFA sample was positive using the molecular assay. This sample yielded a low threshold cycle value, and cross contamination was ruled out by returning to original tissue and repeating both the DFA and RT-PCR. Hoffman et al. evaluated the use of two RT-PCR assays employing different primers and probes for the diagnosis of rabies in several animal species found throughout the world (113). False-negative results occurred when only a single assay was performed; the risk of false-negative results was greatly reduced when both assays were used in the testing algorithm.

Relying on one primer/probe combination to identify the presence of rabies virus RNA in a diagnostic sample may be unwise. The diversity of rabies virus variants makes the design of a universal primer/probe combination challenging, and thus many successful assays include degenerate primers or more than one primer/probe pair. The use of degenerate primers often presents challenges in developing successful cycling conditions. Attempts to design a pan-lyssavirus RT-PCR assay illustrates the need for multiple primer/probe sets. Black et al. reported a successful assay capable of differentiating genotypes 1 through 6 with two primer sets and multiple probes (114). A study by Fischer et al. describes a "cascade approach" to detect and differentiate lyssavirus genotypes 1, 5, and 6 and Bokeloh bat lyssavirus (111). The first test in the cascade employs an intercalating dye to determine the presence or absence of rabies virus RNA. Following a positive result, the sample is subject to a primer/probe-based assay capable of determining the lyssavirus genotype.

Although a false-negative result is of significant concern in rabies diagnostic laboratories, false positives also pose considerable challenges and may result in the euthanasia of "exposed" animals, unwarranted postexposure prophylaxis, undue stress, and expense. As described previously, the potential for false positives has been diminished with newer molecular techniques. However, with the increased sensitivity comes the concern that false positives could occur in animals infected with other rhabdoviruses or infectious neurological pathogens. To ensure the sensitivity and specificity of the test, a specificity panel including several encephalitic viruses and rhabdoviruses, such as vesicular stomatitis virus, should be included in the development phase of the assay.

A common problem among rabies diagnostic laboratories is the submission of degraded tissue, which is compounded by the receipt of most samples in a rabies diagnostic laboratory at the warmest time of the year. Delays in shipping may quickly lead to degraded samples, resulting in tissues unsuitable for testing. Decomposition results in the breakdown of viral proteins, modifying the antigenic structure and thus decreasing the sensitivity of the DFA diagnostic conjugate. The viral proteins may be rendered too decomposed to be detected by the antibodies used in the DFA. Additionally, microbial contamination may be present in decomposed samples and affect the sensitivity of the DFA. The design of real-time RT-PCR and heminested PCR assays may be valuable in evaluating decomposed samples as these assays target small regions of the genome, usually 150 to 300 base pairs. Previous studies evaluating the use of real-time RT-PCR and heminested PCR in degraded tissue support the use of molecular assays as confirmatory tests of diagnostic specimens (115, 116). Beltran et al. (115) targeted a 159-base pair region of the N gene and was successful in detecting rabies RNA in tissues exposed to room-temperature conditions (20°C) for 120 days. Despite the potential value of molecular assays when confronted with unsatisfactory samples, the viral RNA may also be too degraded to result in a positive test. Because housekeeping genes can be affected by degradation at different rates than the virus, evaluating degradation based on host cell factors may not provide a direct correlation.

Although the literature describes the successful use of heminested RT-PCR and real-time RT-PCR in rabies virus diagnosis, there is a need for the standardization of methods used by rabies laboratories, otherwise interlaboratory results may not be comparable. To evaluate assay performance between laboratories, 16 European laboratories tested extracted RNA from multiple lyssavirus (rabies, EBLV 1 and 2) samples with the molecular assays available in their laboratories. Although several laboratories correctly detected the presence or absence of rabies in the samples by one or more molecular methods, the results were not always consistent among participating laboratories (111). The technologies, technical skill level, and reagents available among laboratories may not be equal and require considerable validation to confirm positive and negative results on rabies diagnostic samples. Additionally, not all laboratories have adequate space for the proper workflow required to prevent cross contamination. The separation of an upstream "clean" area and a post-amplification "dirty" laboratory space is essential due to the large amount of viral RNA typically found in positive rabies samples (108, 109). These data reveal the need to develop an assay that can be employed in rabies diagnostic laboratories and the need to implement a proficiency-testing program. The Wisconsin Division of Public Health rabies laboratory currently oversees the rabies diagnostic proficiency program in which participating laboratories receive identical samples to process via their DFA. Enrollment in this program allows laboratories to identify strengths and weaknesses in their programs and work toward corrective action. A similar program is needed for rabies diagnostic laboratories utilizing real-time RT-PCR as a confirmatory diagnostic test.

There are additional concerns that rabies diagnostic laboratories need to consider prior to adopting real-time RT-PCR as a primary or confirmatory test. Although the sensitivity of real-time RT-PCR may be greater than the DFA, the time required for the DFA is less than for real-time RT-PCR. Once the tissue is collected, which is accomplished by the same necropsy technique regardless of which diagnostic test will be used, processing a sample via DFA takes approximately 2 hours. The amount of time needed to complete the real-time RT-PCR procedure is approximately 4 hours but is dependent on the methods employed by individual laboratories.

The use of appropriate controls is essential to evaluate RNA extraction, contamination, and template integrity. A negative extraction control is necessary to discern contamination that could occur via aerosol, contaminated reagents, workspace, or equipment (107). Additionally, the use of an internal control during the RNA extraction step, such as green fluorescent protein, will allow the user to assess extraction efficiency and identify inhibition (108). Positive and negative samples should include a rabies-positive sample that is dissimilar to samples that are frequently tested in the diagnostic laboratory but has similar efficiency as local variants. This step is valuable for the identification of crossover contamination from a positive control. The inclusion of positive and negative controls in each assay is essential and will provide confirmation of a successful test. Evaluating template integrity is more difficult given that the expression of internal housekeeping genes (such as beta-actin) can be complicated by the numerous species of animals from which tissue is submitted for diagnostic testing (112). For use in a rabies diagnostic laboratory, an internal housekeeping gene would require a primer/probe set that could identify the gene of interest in a broad range of species.

Molecular assays have several applications in rabies laboratories and have greatly expanded our understanding of rabies epidemiology, pathogenesis, and transmission. However, there are important considerations that need to be addressed prior to implementing molecular assays as the primary diagnostic tool in rabies laboratories. The potential consequence of a false-negative or false-positive result cannot be overstated. The lack of treatment following a bite from a rabid animal that was misdiagnosed could have deleterious and potentially fatal results. Variability in the rabies genome is greater than at the protein level, and thus small changes in the viral RNA may affect the sensitivity of the assay. The DFA detects protein epitopes of the nucleocapsid, which is much more likely to be conserved than the small viral RNA molecules targeted by the molecular assays. Taken together, the establishment of a primary real-time RT-PCR diagnostic assay requires extensive research, collaboration, and validation.

Antemortem Diagnosis of Animal Rabies

It is possible to apply the methods described for the antemortem diagnosis of human rabies to suspect animals as well. Skin biopsy has been demonstrated to be particularly sensitive when applied to biopsy specimens taken from the snouts of terrestrial carnivores, which permits examination of the innervation of the tactile hairs (117). However, the same limitations apply to antemortem diagnosis in animals since all tests can remain negative well into or throughout the clinical period, and negative results do not rule out rabies infection. Rabies cannot be reliably diagnosed by current methods during most of the incubation period. Claims that molecular testing of the saliva of a biting animal can be used for decisions of bite management are false, as rabies virus may be shed in saliva sporadically prior to and during the clinical period (24). Therefore, these tests are of little value for public health decisions and should never be a substitute when circumstances require a 10-day observation or euthanasia and the examination of brain tissue.

Rabies Virus Variant Typing

Methods that characterize the antigenic and genetic attributes of rabies virus isolates and the rabies-related lyssaviruses now enable the laboratory to identify the virus variants responsible for epizootics as well as individual cases of rabies in animals and humans. Distinctive differences in the variants responsible for the major terrestrial outbreaks worldwide, the numerous bat rabies variants, the laboratory strains of rabies virus, and the rabies-related lyssaviruses are distinguishable by either antigenic or molecular typing. Rabies variant identification yields a greater knowledge of the epizootiologic relationships between the virus and the vectors, allowing the development of more effective animal-contact guidelines and rabies-control strategies. Reaction patterns in indirect DFA, employing panels of monoclonal antibodies specific for unique viral nucleocapsid epitopes, permit such discrimination (118). The immunofluorescence assays can be performed on brain tissue of the original rabies-infected tissue or mouse- or cell culture–passage virus. Genetic analysis permits more precise detail of the evolutionary relatedness of isolates, investigation of the spatial and temporal changes that may occur, and particularly, the measure of similarity among virus isolates. This is accomplished by the extraction, transcription, and amplification of the RNA of an isolate by RT-PCR and the subsequent sequence analysis of the cDNA nucleotide or amino acid sequence for the entire or partial nucleocapsid or glycoprotein genes. Sequence deposition on GenBank (119) allows for the sharing, comparison, and identification of variant identity. Using computer algorithms to perform pairwise comparisons, estimates of genetic identity can be calculated and expressed as percent homology among isolates. Molecular epidemiologic studies by many investigators are reviewed thoroughly by Trimarchi and Nadin-Davis (120), with a description of the complex relationships among genotype 1 rabies viruses and the other lyssavirus genotypes.

Rabies Antibody Assay

Assays to identify and quantitate rabies antibody in serum from humans and animals serve numerous functions in rabies diagnosis and control. Serologic testing for rabies-specific antibodies is performed on human serum to determine the response to pre- and postexposure vaccination and to determine the timing of booster vaccinations to maintain a rabies-immune status. Evidence of rabies antibody in serum and cerebrospinal fluid (CSF) is used in the antemortem diagnosis of rabies in humans. Antibody detection in human and animal vaccinees is used in vaccine efficacy trials and in the evaluation of field wildlife vaccination campaigns. Long quarantines for cats and dogs entering rabies-free countries can be reduced if an immunologic response can be confirmed by this means.

Because these antibody tests are generally employed to measure immune status, the most widely used assays measure neutralizing antibodies. Neutralization assays using varying serum dilutions and constant virus challenge are most commonly employed. Dilutions of heat-inactivated serum are combined with a standardized amount of rabies virus and then incubated for 1 hour at 37°C. The inoculation of mice or cell cultures is performed after incubation to demonstrate residual live virus. The mouse inoculation test developed in 1935 (105) is a very reproducible method, still employed in some laboratories, and is used as the standard to evaluate other procedures.

The commonly used cell culture virus neutralization techniques are those in which residual virus is demonstrated by immunofluorescence in the inoculated cell monolayers. The most widely employed test for human postvaccinal titer determination is the rabies fluorescent focus inhibition test (121). This technique determines the serum neutralization end point titer by a mathematical calculation of the number of viral-induced fluorescent foci remaining after incubation of the challenge virus at serum dilutions of 1 to 5 and 1 to 50. While results may be given as reciprocals of the calculated end point dilution, they are often expressed in terms of international units (IU) of neutralizing activity determined on each test by the titration of a standard reference serum. Alternative *in vitro* methods that determine the last dilution of the patient's serum that neutralizes the virus challenge, similar to the standard mouse test, are also utilized and generally report titers in international units (122).

Other serologic tests are employed to identify the presence of rabies antibodies, particularly for research and vaccine evaluation procedures. A high degree of sensitivity and specificity are reported with an enzyme-linked immunosorbent assay (ELISA) directed against whole virus for the measurement of antibodies (123). The rabies G protein should compose the immunosorbent for the ELISA. A competitive ELISA, using a neutralizing monoclonal antibody, reportedly achieves a high degree of sensitivity, freedom from the need for species-specific intermediate antibodies, and a measurement of neutralizing antibody (124). Because agreement between ELISA and neutralization tests is not always consistent, ELISAs should not be used to make human therapeutic decisions.

Diagnosis of Human Rabies

Despite the dire prognosis in rabies infection, testing should be done in all cases of acute, progressive human encephalitis of unknown etiology, even in the absence of a history of bite exposure. As a result of the efficacy of modern PEP regimens, human rabies cases in the United States and other developed nations are no longer commonly associated with vaccine failure. Also in these regions, rabies prophylaxis is provided in all cases of known exposures and even in most cases of suspected exposure to rabies. Therefore, human cases are most often identified in the absence of a clear history of a suspicious animal bite or other exposure. In numerous recent human rabies cases in the United States, the disease was not suspected or diagnosed during the clinical illness and was only recognized postmortem and sometimes after a lengthy delay (60).

The possibility of a successful outcome from a novel therapeutic approach to a human rabies infection was demonstrated in the recovery, following treatment with induced coma and antiviral drugs, of a Wisconsin teenager in 2004 (see the discussion in Clinical Rabies above). It is likely that a delayed diagnosis would reduce the efficacy of such treatments (66). Antemortem diagnosis is also a valuable tool to permit early identification and PEP of family and health-care staff potentially exposed by contact with the patient's saliva. Antemortem diagnosis also aids in patient management and allows the family to prepare for the high likelihood of a fatal outcome from the disease. The postmortem diagnosis of rabies in cases of fatal encephalitis of unknown etiology is critical to gain greater knowledge of the prevalence of rabies encephalitis in humans, the frequency of failure of pre- and postexposure vaccination, and the probable vectors and variants that pose the greatest risk to human health.

Early laboratory confirmation of a clinical diagnosis is essential for prompt initiation of PEP therapy. A retrospective study investigating 142 human rabies cases in North America, South America, Europe, Africa, and Asia reported that molecular assays were the most sensitive in identifying antemortem cases in humans. The best samples for antemortem diagnosis were saliva and skin biopsies (125). This report also noted that bat-acquired rabies is more often misdiagnosed than dog-acquired rabies. The most likely factor contributing to this variation is the often unrecognized exposure to rabies from a bat contact (126), whereas a dog bite is considerably more traumatic and memorable. Although brain biopsy would be the most sensitive antemortem diagnostic method, the risks associated with the procedure make its use uncommon. There are numerous less invasive intravitam tests that can assist in the diagnosis of a rabies infection. Magnetic resonance imaging may be helpful in the antemortem diagnosis of rabies (127). However, rabies virus, its antigens, and rabies RNA do not move centrifugally away from the CNS during the incubation period and only slowly during the clinical period (24). Similarly, humoral antibody responses may not occur during the incubation period and are generally not demonstrable until the second week of clinically recognizable illness. Antemortem diagnosis is therefore attempted by the analysis of numerous tissues by several methods searching for rabies-specific antibody in serum or CSF or viral antigen, live virus, or viral RNA in body fluids (saliva), peripheral nerves (skin biopsy), or epithelial cells (corneal impression). Antigen detection can best be accomplished by DFA performed on a full-thickness skin biopsy specimen taken from the nape of the neck and including several hair follicles (128). DFA can also be used to demonstrate rabies antigen in corneal impression slides (129). It is recommended that corneal impressions be taken by an ophthalmologist because of the risk of corneal abrasions. Virus isolation by MIT or cell culture inoculation can be applied to saliva and CSF. An antibody assay can be performed by a neutralization test, ELISA, or indirect fluorescent antibody test on serum and CSF. Care should be taken in the interpretation of the indirect fluorescent antibody test for rabies antibody determination in CSF (130) and the ELISA procedure (131) as false-positive results have been reported. Molecular methods can be included when investigating the presence of viral RNA in saliva, skin biopsy specimens, corneal impressions, or CSF (see Molecular Methods in Rabies Diagnosis, above). The demonstration of rabies antigen by DFA or rabies virus RNA in any solid tissue or saliva, the presence of rabies antibody in serum and CSF, or the isolation of rabies virus from any tissue is confirmatory of rabies infection. Antibody in serum alone may not be indicative of clinical rabies in a patient with a known or an unclear history of rabies vaccination or very recently treated with RIG.

The samples necessary to identify the presence or absence of antibodies are 1 milliliter or more of CSF and serum, submitted in a plastic tube or vial. The skin biopsy specimen can be submitted on a gauze sponge moistened with sterile physiologic saline and sealed in a small plastic container. Corneal impression slides should be submitted in a plastic slide container with the surface of the slide containing the impression clearly marked. A 1-milliliter sample of frank saliva should be collected in a plastic sputum jar. Alternatively, a buccal swab can be taken and submitted immersed in a tube containing 1 milliliter of sterile saline. All of these samples can be stored at −80°C and shipped on dry ice. Postmortem testing methods for human rabies are similar to those described for animals. If the patient dies and an autopsy is performed, the ideal samples for postmortem diagnosis are 1 cubic centimeter each of unfixed cerebellum and brain stem preserved by refrigeration or freezing. The storage of specimens in any fixative (formalin, methanol, etc.) is not advised and can impede testing and/or render the diagnosis unreliable. Samples must be shipped on dry ice packaged in compliance with applicable federal shipping guidelines for diagnostic specimens. International Air Transport Association guidelines that characterize rabies virus as a "biological substance, category B" in clinical specimens and as a "biological substance, category A" are followed only when the virus has been cultured.

The laboratory should be contacted immediately when human rabies is suspected. If original samples collected during the first week of symptoms do not disclose evidence of rabies infection, it must be understood that this does not conclusively rule out rabies and that repeat samples may be necessary because antemortem tests may remain negative well into the clinical period. A review of 32 human rabies deaths in the United States from 1980 to 1996 (60) showed that rabies was not suspected in 12 individuals and was only diagnosed after death. Of the remaining 20 cases, antemortem evidence of rabies was found by one or more tests in 18 cases. Antibody to rabies was detected in 10 of 18 patients tested, and virus was isolated from saliva in 9 of 15 cases. Rabies RNA was detected by RT-PCR in each of the 10 patients tested by this molecular method. However, nested PCR was required in almost all cases to detect the extremely small amounts of viral RNA in the samples collected antemortem. Antigen was demonstrated by DFA in nuchal skin from 10 of 15 patients tested and in corneal impressions from 2 of 7 patients tested. In another report (132), rabies RNA was detected in 4 of 9 saliva samples tested with a nonnested RT-PCR assay.

Diagnostic Best Practices

Rabies is often included in the differential diagnosis of an encephalitic patient, even in the absence of a bite from a rabies vector species. The application of new therapies, palliative patient care, and the treatment of individuals potentially exposed to the patient are reasons driving the need for a timely diagnosis of rabies in humans. Algorithms have been established by the International Encephalitis Consortium for guidance as specific etiologies are identified in fewer than 50% of cases, in part due to a lack of consensus on case definitions and standardized diagnostic approaches (133). Laboratory testing should be coordinated through local or regional public health departments as rabies diagnosis, especially antemortem human diagnosis, is performed in specialized laboratories at the state or federal level. Specific recommendations, including patient information forms for a possible human rabies case, may be found at the CDC website (134). Antemortem diagnosis is attempted by the analysis of numerous tissues by several methods searching for rabies-specific antibody in serum or CSF or viral antigen, live virus, or viral RNA in body fluids (e.g., saliva), peripheral nerves (e.g., skin biopsy), or epithelial cells (e.g., corneal impression). It is essential that all four diagnostic specimens (e.g., serum, CSF, saliva, and nuchal skin biopsy) be submitted on dry ice to the rabies diagnostic laboratory. Corneal impressions are optional but can be diagnostic if positive on the DFA procedure. The specimens should never be preserved in any fixative as this can destroy or reduce the sensitivity of diagnostic procedures. The submitter should contact the rabies laboratory by phone with the intent to submit rabies suspect specimens, as these conversations are essential in determining a timeline for laboratory results and confirmation that appropriate guidelines are followed. Negative results on specimens will not necessarily rule out rabies in the patient, as specimens can remain negative well into the clinical period, and tests may need to be repeated.

REFERENCES

1. **Wunner WH, Calisher CH, Dietzgen RG, Jackson RG, Kitajima AO, Lafon MF, Leong JC, Nichol ST, Peters D, Smith JS, Walker PJ.** 1995. Rhabdoviridae, p 275–280. In *Classification and Nomenclature of Viruses, Sixth Report of the International Committee on Taxonomy of Viruses*. Springer-Verlag, New York, NY.
2. **World Health Organization**, 2013. *No 2*. Insel Riems, Germany, Greifswald. WHO Expert Consultation on Rabies: First Report. *WHO Technical Report Series, no. 982. WHO, Geneva, Switzerland. World Health Organization. 2013. Rabies—Bulletin—Europe, Rabies Information System of the WHO Collaboration Centre for Rabies Surveillance and Research*, 37.
3. **Hooper PT, Lunt RA, Gould AR, Samaratunga H, Hyatt AD, Gleeson LJ, Rodwell BJ, Rupprecht CE, Smith JS, Murray PK.** A new lyssavirus: the first endemic rabies-related virus recognized in Australia. *Bull Inst Pasteur* 1997;**95:**209–218.
4. **Kuzmin IV, Hughes GJ, Botvinkin AD, Orciari LA, Rupprecht CE.** 2005. Phylogenetic relationships of Irkut and West Caucasian bat viruses within the *Lyssavirus* genus and suggested quantitative criteria based on the N gene sequence for lyssavirus genotype definition. *Virus Res* **111:**28–43.
5. **Hanlon CA, Kuzmin IV, Blanton JD, Weldon WC, Manangan JS, Rupprecht CE.** 2005. Efficacy of rabies biologics against new lyssaviruses from Eurasia. *Virus Res* **111:**44–54.
6. **Picard-Meyer E, Servat A, Robardet E, Moinet M, Borel C, Cliquet F.** 2013. Isolation of Bokeloh bat lyssavirus in *Myotis nattereri* in France. *Arch Virol* **158(11):**2333–2340.
7. **Tordo N.** 1996. Characteristics and molecular biology of rabies virus, p 28–52. *In* Meslin FX, Kaplan MM, Koprowski H (ed), *Laboratory Techniques in Rabies*, 4th ed. World Health Organization, Geneva, Switzerland.
8. **Wunner WH.** 2007. Rabies virus, p 23–68. *In* Jackson AC, Wunner WH (ed), *Rabies*, 2nd ed. Academic Press, New York, NY.

9. **Wiktor TJ, Macfarlan RI, Reagan KJ, Dietzschold B, Curtis PJ, Wunner WH, Kieny MP, Lathe R, Lecocq JP, Mackett M, Moss B, Koprowski H.** 1984. Protection from rabies by a vaccinia virus recombinant containing the rabies virus glycoprotein gene. *Proc Natl Acad Sci USA* **81:**7194–7198.
10. **Koprowski H.** 1991. The virus: overview, p 27–29. *In* Baer GM (ed), *The Natural History of Rabies*, 2nd ed. CRC Press, Boca Raton, FL.
11. **Lafon M.** 2005. Rabies virus receptors. *J Neurovirol* **11:**82–87.
12. **Murphy FA.** 1986. The rabies virus and pathogenesis of the disease, p 11–16. *In* Fishbein DB, Sawyer LA, Winkler WG (ed), *Rabies Concepts for Professionals*, 2nd ed. Merieux Institute, Miami, FL.
13. **Kaplan MM.** 1996. Safety precautions in handling rabies virus, p 3–8. *In* Meslin FX, Kaplan MM, Koprowski H (ed), *Laboratory Techniques in Rabies*, 4th ed. World Health Organization, Geneva, Switzerland.
14. **Tordo N.** 1996. Characteristics and molecular biology of rabies virus, p 28–52. *In* Meslin FX, Kaplan MM, Koprowski H (ed), *Laboratory Techniques in Rabies*, 4th ed. World Health Organization, Geneva, Switzerland.
15. **Lahaye X, Vidy A, Pomier C, Obiang L, Harper F, Gaudin Y, Blondel D.** 2009. Functional characterization of Negri bodies (NBs) in rabies virus-infected cells: evidence that NBs are sites of viral transcription and replication. *J Virol* **83:**7948–7958.
16. **Murphy FA.** 1986. The rabies virus and pathogenesis of the disease, p 11–16. *In* Fishbein DB, Sawyer LA, Winkler WG (ed), *Rabies Concepts for Professionals*, 2nd ed. Merieux Institute, Miami, FL.
17. **Constantine DG.** 1962. Rabies transmission by nonbite route. *Public Health Rep* **77:**287–289.
18. **Gibbons RV, Holman RC, Mosberg SR, Rupprecht CE.** 2002. Knowledge of bat rabies and human exposure among United States cavers. *Emerg Infect Dis* **08:**532–534.
19. **Winkler WG, Fashinell TR, Leffingwell L, Howard P, Conomy P.** 1973. Airborne rabies transmission in a laboratory worker. *JAMA* **226:**1219–1221.
20. **Centers for Disease Control and Prevention.** 1977. Rabies in a laboratory worker—New York. *MMWR Morb Mortal Wkly Rep* **26:**183–184.
21. **Davis AD, Rudd RJ, Bowen RA.** 2007. Effects of aerosolized rabies virus exposure on bats and mice. *J Infect Dis* **195:**1144–1150.
22. **Afshar A.** 1979. A review of non-bite transmission of rabies virus infection. *Br Vet J* **135:**142–148.
23. **Mansel GA.** 1951. *Royal Canadian Mounted Police Report*. G. Division, Port Harrison, Quebec.
24. **Charlton KM.** 1988. The pathogenesis of rabies, p 101–150. *In* Campbell JB, Charlton KM (ed), *Rabies*. Academic Press, Norwell, MA.
25. **Fekadu M, Endeshaw T, Alemu W, Bogale Y, Teshager T, Olson JG.** 1996. Possible human-to-human transmission of rabies in Ethiopia. *Ethiop Med J* **34:**123–127.
26. **Centers for Disease Control and Prevention**. 2008. Human rabies prevention—United States, 2008 recommendations of the Advisory Committee on Immunization Practices (ACIP). *MMWR Recommend. Rep* **57:**1–26, 28.
27. **Srinivasan A, Burton EC, Kuehnert MJ, Rupprecht C, Sutker WL, Ksiazek TG, Paddock CD, Guarner J, Shieh WJ, Goldsmith C, Hanlon CA, Zoretic J, Fischbach B, Niezgoda M, El-Feky WH, Orciari L, Sanchez EQ, Likos A, Klintmalm GB, Cardo D, LeDuc J, Chamberland ME, Jernigan DB, Zaki SR, Rabies in Transplant Recipients Investigation Team.** 2005. Transmission of rabies virus from an organ donor to four transplant recipients. *N Engl J Med* **352:**1103–1111.
28. **Johnson N, Brookes SM, Fooks AR, Ross RS.** 2005. Review of human rabies cases in the UK and in Germany. *Vet Rec* **157:**715–717.
29. **Vora NM, Basavaraju SV, Feldman KA, Paddock CD, Orciari L, Gitterman S, Griese S, Wallace RM, Said M, Blau DM, Selvaggi G, Velasco-Villa A, Ritter J, Yager P, Kresch A, Niezgoda M, Blanton J, Stosor V, Falta EM, Lyon GM III, Zembower T, Kuzmina N, Rohatgi PK, Recuenco S, Zaki S, Damon I, Franka R, Kuehnert MJ, Transplant-Associated Rabies Virus Transmission Investigation Team.** 2013. Raccoon rabies virus variant transmission through solid organ transplantation. *JAMA* **310:**398–407.
30. **Lewis P, Fu Y, Lentz TL.** 2000. Rabies virus entry at the neuromuscular junction in nerve-muscle cocultures. *Muscle Nerve* **23:**720–730.
31. **Thoulouze MI, Lafage M, Schachner M, Hartmann U, Cremer H, Lafon M.** 1998. The neural cell adhesion molecule is a receptor for rabies virus. *J Virol* **72:**7181–7190.
32. **Langevin C, Jaaro H, Bressanelli S, Fainzilber M, Tuffereau C.** 2002. Rabies virus glycoprotein (RVG) is a trimeric ligand for the N-terminal cysteine-rich domain of the mammalian p75 neurotrophin receptor. *J Biol Chem* **277:**37655–37662.
33. **Kelly RM, Strick PL.** 2000. Rabies as a transneuronal tracer of circuits in the central nervous system. *J Neurosci Meth* **103:**63–71.
34. **Jackson AC, Rasalingam P, Weli SC.** 2006. Comparative pathogenesis of recombinant rabies vaccine strain SAD-L16 and SAD-D29 with replacement of Arg333 in the glycoprotein after peripheral inoculation of neonatal mice: less neurovirulent strain is a stronger inducer of neuronal apoptosis. *Acta Neuropathol* **111:**372–378.
35. **Fekadu M, Shaddock JH, Baer GM.** 1982. Excretion of rabies virus in the saliva of dogs. *J Infect Dis* **145:**715–719.
36. **Debbie JG, Trimarchi CV.** 1970. Pantropism of rabies virus in free-ranging rabid red fox *Vulpes fulva*. *J Wildl Dis* **6:**500–506.
37. **King AA, Sands JJ, Porterfield JS.** 1984. Antibody-mediated enhancement of rabies virus infection in a mouse macrophage cell line (P388D1). *J Gen Virol* **65:**1091–1093.
38. **Smith JS, McCelland CL, Reid FL, Baer GM.** 1982. Dual role of the immune response in street rabiesvirus infection of mice. *Infect Immun* **35:**213–221.
39. **Fu ZF.** 1997. Rabies and rabies research: past, present and future. *Vaccine* **15**(Suppl):S20–S24.
40. **Leslie MJ, Messenger S, Rohde RE, Smith J, Cheshier R, Hanlon C, Rupprecht CE.** 2006. Bat-associated rabies virus in skunks. *Emerg Infect Dis* **12:**1274–1277.
41. **Fekadu M.** 1991. Canine rabies, p 367–378. *In* Baer GM (ed), *The Natural History of Rabies*, 2nd ed. CRC Press, Boca Raton, FL.
42. **Vigilato MA, Cosivi O, Knöbl T, Clavijo A, Silva HM.** 2013. Rabies update for Latin America and the Caribbean. *Emerg Infect Dis* **19:**678–679.
43. **Nadin-Davis SA, Bingham J.** 2004. Europe as a source of rabies for the rest of the world, p 259–280. *In* King AA, Fooks AR, Aubert M, Wandeler AI (ed), *Historical Perspective of Rabies in Europe and the Mediterranean Basin*. OIE, Paris, France.
44. **Zhenyu G, Zhen W, Enfu C, Fan H, Junfen L, Yixin L, Gangqiang D, Fontaine RE.** 2007. Human rabies cluster following badger bites, People's Republic of China. *Emerg Infect Dis* **13:**1955–1957.
45. **Everard CO, Everard JD.** 1985. Mongoose rabies in Grenada, p 43–70. *In* Bacon PJ (ed), *Population Dynamics of Rabies in Wildlife*. Academic Press, New York, NY.
46. **Kuzmin IV, Shi M, Orciari LA, Yager PA, Velasco-Villa A, Kuzmina NA, Streicker DG, Bergman DL, Rupprecht CE.** 2012. Molecular inferences suggest multiple host shifts of rabies viruses from bats to mesocarnivores in Arizona during 2001–2009. *PLoS Pathog* **8:**e1002786.
47. **Jenkins SR, Perry BD, Winkler WG.** 1988. Ecology and epidemiology of raccoon rabies. *Rev Infect Dis* **10**(Suppl 4): S620–S625.
48. **Wallace RM, Gilbert A, Slate D, Chipman R, Singh A, Cassie Wedd, Blanton JD. 2014.** Right place, wrong species: a 20-year review of rabies virus cross species transmission among terrestrial mammals in the United States. *PLoS One* **9:** e107539.
49. **Nadin-Davis SA, Muldoon F, Wandeler AI.** 2006. Persistence of genetic variants of the arctic fox strain of rabies virus in southern Ontario. *Can J Vet Res* **70:**11–19.

50. **Blanton JD, Hanlon CA, Rupprecht CE.** 2007. Rabies surveillance in the United States during 2006. *J Am Vet Med Assoc* **231:**540–556.
51. **New York State Rabies Diagnostic Laboratory.** 2012 Annual Report, http://www.wadsworth.org/rabies/AnnualSummaries/2012/2012_Rabies_report_for_web.pdf accessed December 21, 2015.
52. **Eidson M, Matthews SD, Willsey AL, Cherry B, Rudd RJ, Trimarchi CV.** 2005. Rabies virus infection in a pet guinea pig and seven pet rabbits. *J Am Vet Med Assoc* **227:**932–935, 918.
53. **Streicker DG, Turmelle AS, Vonhof MJ, Kuzmin IV, McCracken GF, Rupprecht CE.** 2010. Host phylogeny constrains cross-species emergence and establishment of rabies virus in bats. *Science*. **329:** 676–679.
54. **Stoner-Duncan B, Streicker DG, Tedeschi CM.** 2014. Vampire bats and rabies: toward an ecological solution to a public health problem. *PLoS Negl Trop Dis* **8:**e2867.
55. **Dyer JL, Wallace R, Orciari L, Hightower D, Yager P, Blanton JD.** 2013. Rabies surveillance in the United States during 2012. *J Am Vet Med Assoc* **243:**805–815.
56. **Nadin-Davis SA.** 2007. Molecular epidemiology, p 69–122. *In* Jackson AC, Wunner WH (ed), *Rabies*, 2nd ed. Academic Press, New York, NY.
57. **Trimarchi CV, Rudd RJ, Abelseth MK.** 1986. Experimentally induced rabies in four cats inoculated with a rabies virus isolated from a bat. *Am J Vet Res* **47:**777–780.
58. **Malerczyk C, Freuling C, Gniel D, Giesen A, Selhorst T, Müller T.** 2014. Cross-neutralization of antibodies induced by vaccination with purified chick embryo cell vaccine (PCECV) against different Lyssavirus species. *Hum. Vaccin. & Immunother.* Epub 2014 Oct 1.
59. **Badrane H, Tordo N.** 2001. Host switching in *Lyssavirus* history from the Chiroptera to the Carnivora orders. *J Virol* **75:**8096–8104.
60. **Noah DL, Drenzek CL, Smith JS, Krebs JW, Orciari L, Shaddock J, Sanderlin D, Whitfield S, Fekadu M, Olson JG, Rupprecht CE, Childs JE.** 1998. Epidemiology of human rabies in the United States, 1980 to 1996. *Ann Intern Med* **128:**922–930.
61. **Bernard KW.** 1986. Clinical rabies in humans, p 43–48. *In* Fishbein DB, Sawyer LA, Winkler WG (ed), *Rabies Concepts for Professionals*, 2nd ed. Merieux Institute, Miami, FL.
62. **Jackson AC.** 2013. Human Rabies, p 269–298. In Jackson AC, (ed), *Rabies*, 3rd ed. Academic Press, New York, NY.
63. **Jackson AC.** 2007. Pathogenesis, p 341–381. *In* Jackson AC, Wunner WH (ed), *Rabies*, 2nd ed. Academic Press, New York, NY.
64. **Fishbein DB.** 1991. Rabies in humans, p 519–549. *In* Baer GM (ed), *The Natural History of Rabies*, 2nd ed. CRC Press, Boca Raton, FL.
65. **Jackson AC.** 2013. Current and future approaches to the therapy of human rabies. *Antiviral Res* **99:**61–67.
66. **Willoughby RE Jr, Tieves KS, Hoffman GM, Ghanayem NS, Amlie-Lefond CM, Schwabe MJ, Chusid MJ, Rupprecht CE.** 2005. Survival after treatment of rabies with induction of coma. *N Engl J Med* **352:**2508–2514.
67. **Jackson AC.** 2005. Recovery from rabies. *N Engl J Med* **352:**2549–2550. Editorial.
68. Children's Hospital of Wisconsin: Rabies Registry Home Page: [http://www.mcw.edu/Pediatrics/InfectiousDiseases/PatientCare/Rabies.htm]
69. **Hemachudha T, Sunsaneewitayakul B, Desudchit T, Suankratay C, Sittipunt C, Wacharapluesadee S, Khawplod P, Wilde H, Jackson AC.** 2006. Failure of therapeutic coma and ketamine for therapy of human rabies. *J Neurovirol* **12:**407–409.
70. **Caicedo Y, Paez A, Kuzmin I, Niezgoda M, Orciari LA, Yager PA, Recuenco S, Franka R, Velasco-Villa A, Willoughby RE Jr.** 2015. Virology, immunology and pathology of human rabies during treatment. *Pediatr Infect Dis J* **34:**520–528.
71. **O'Sullivan A, Willoughby RE, Mishchuk D, Alcarraz B, Cabezas-Sanchez C, Condori RE, David D, Encarnacion R, Fatteh N, Fernandez J, Franka R, Hedderwick S, McCaughey C, Ondrush J, Paez-Martinez A, Rupprecht C, Velasco-Villa A, Slupsky CM.** 2013. Metabolomics of cerebrospinal fluid from humans treated for rabies. *J Proteome Res* **12:**481–490.
72. **Udow SJ, Marrie RA, Jackson AC.** 2013. Clinical features of dog- and bat-acquired rabies in humans. *Clin Infect Dis* **57:**689–696.
73. **Dimitrov DT, Hallam TG, Rupprecht CE, Turmelle AS, McCracken GF.** 2007. Integrative models of bat rabies immunology, epizootiology and disease demography. *J Theor Biol* **245:**498–509.
74. **Feder HM Jr, Nelson R, Reiher HW.** 1997. Bat bite? *Lancet* **350:**1300.
75. **Mondul AM, Krebs JW, Childs JE.** 2003. Trends in national surveillance for rabies among bats in the United States (1993–2000). *J Am Vet Med Assoc* **222:**633–639.
76. **Lafon M.** 2007. Immunology, p 489–504. *In* Jackson AC, Wunner WH (ed), *Rabies*, 2nd ed. Academic Press, New York, NY.
77. **Parviz S, Chotani R, McCormick J, Fisher-Hoch S, Luby S.** 2004. Rabies deaths in Pakistan: results of ineffective post-exposure treatment. *Int J Infect Dis* **8:**346–352.
78. **Rupprecht CE, Briggs D, Brown CM, Franka R, Katz SL, Kerr HD, Lett S, Levis R, Meltzer MI, Schaffner W, Cieslak PR.** 2009. Evidence for a 4-dose vaccine schedule for human rabies post-exposure prophylaxis in previously non-vaccinated individuals. *Vaccine* **27:**7141–7148.
79. **Wilde H, Shantavasinkul P, Hemachudha T, Tepsumethanon V, Lumlertacha B, Wacharapluesadee S, Tantawichien T, Sitprija V, Chutivongse S, Phanuphak P.** 2009. New knowledge and new controversies in rabies. *J Infect Dis Antimicrob Agents A Journal from Thailand* **26:**63–74.
80. **Centers for Disease Control and Prevention and the National Institutes of Health.** 2007. *Biosafety in Microbiological and Biomedical Laboratories*, 5th ed. http://www.cdc.gov/biosafety/publications/bmbl5/bmbl.pdf accessed 12/21/2015.
81. **Briggs DJ, Dreesen DW, Nicolay U, Chin JE, Davis R, Gordon C, Banzhoff A.** 2000. Purified chick embryo cell culture rabies vaccine: interchangeability with human diploid cell culture rabies vaccine and comparison of one versus two-dose post-exposure booster regimen for previously immunized persons. *Vaccine* **19:**1055–1060.
82. **Dreesen DW, Bernard KW, Parker RA, Deutsch AJ, Brown J.** 1986. Immune complex-like disease in 23 persons following a booster dose of rabies human diploid cell vaccine. *Vaccine* **4:**45–49.
83. **Chutivongse S, Wilde H, Benjavongkulchai M, Chomchey P, Punthawong S.** 1995. Postexposure rabies vaccination during pregnancy: effect on 202 women and their infants. *Clin Infect Dis* **20:**818–820.
84. **Slate D, Rupprecht CE, Rooney JA, Donovan D, Lein DH, Chipman RB.** 2005. Status of oral rabies vaccination in wild carnivores in the United States. *Virus Res* **111:**68–76.
85. **Shwiff S, Hampson K, Anderson A.** 2013. Potential economic benefits of eliminating canine rabies. *Antiviral Res* **98:**352–356.
86. **Rupprecht CE, Hanlon CA, Niezgoda M, Buchanan JR, Diehl D, Koprowski H.** 1993. Recombinant rabies vaccines: efficacy assessment in free-ranging animals. *Onderstepoort J Vet Res* **60:**463–468.
87. **Rupprecht CE, Blass L, Smith K, Orciari LA, Niezgoda M, Whitfield SG, Gibbons RV, Guerra M, Hanlon CA.** 2001. Human infection due to recombinant vaccinia-rabies glycoprotein virus. *N Engl J Med* **345:**582–586.
88. **Dietzschold B, Faber M, Schnell MJ.** 2003. New approaches to the prevention and eradication of rabies. *Expert Rev Vaccines* **2:**399–406.
89. **Rosatte RC, Donovan D, Davies JC, Allan M, Bachmann P, Stevenson B, Sobey K, Brown L, Silver A, Bennett K, Buchanan T, Bruce L, Gibson M, Beresford A, Beath A, Fehlner-Gardiner C, Lawson K.** 2009. Aerial distribution of

ONRAB baits as a tactic to control rabies in raccoons and striped skunks in Ontario, Canada. *J Wildl Dis* **45:**363–374.
90. **Lapiz SMD, Miranda MEG, Garcia RG, Daguro LI, Paman MD, Madrinan FP, Rances PA, Briggs DJ.** 2012. Implementation of an intersectoral program to eliminate human and canine rabies: The Bohol Rabies Prevention and Elimination Project. *PLoS Negl Trop Dis* **6:**e1891. Epub 2012 Dec 6.
91. **Townsend SE, Sumantra IP, Pudjiatmoko, Bagus GN, Brum E, Cleaveland S, Crafter S, Dewi AP, Dharma DM, Dushoff J, Girardi J, Gunata IK, Hiby EF, Kalalo C, Knobel DI, Mardiana IW, Putra AA, Schoonman L, Scott-Orr H, Shand M, Sukanadi IW, Suseno PP, Haydon DT, Hampson K.** 2013. Designing programs for eliminating canine rabies from islands: Bali, Indonesia as a case study. *PLoS Negl Trop Dis* **7:** e2372.
92. **Morgan SMD, Pouliott CE, Rudd RJ, Davis AD.** 2015. Antigen detection, rabies virus isolation, and Q-PCR in the quantification of viral load in a natural infection of the North American beaver (*Castor canadensis*). *J Wildl Dis* **51:** 287–289.
93. **Rupprecht CE, Hanlon CA, Hemachudha T.** 2002. Rabies re-examined. *Lancet Infect Dis* **2:**327–343.
94. **American Veterinary Medical Association.** 2013. *AVMA guidelines for the euthanasia of animals:* https://www.avma.org/kb/policies/documents/euthanasia.pdf accessed 12/21/2015.
95. **Barrat J.** 1996. Simple technique for the collection and shipment of brain specimens for rabies diagnosis, p 425–432. *In* Meslin FX, Kaplan MM, Koprowski H (ed), *Laboratory Techniques in Rabies*, 4th ed. World Health Organization, Geneva, Switzerland.
96. **Hanlon CA, Smith JS, Anderson GR, The National Working Group on Rabies Prevention and Control.** 1999. Recommendations of a national working group on prevention and control of rabies in the United States. Article II: laboratory diagnosis of rabies. *J Am Vet Med Assoc* **215:**1444–1446.
97. **Rudd RJ, Smith JS, Yager PA, Orciari LA, Trimarchi CV.** 2005. A need for standardized rabies-virus diagnostic procedures: effect of cover-glass mountant on the reliability of antigen detection by the fluorescent antibody test. *Virus Res* **111:**83–88.
98. **Webster WA, Casey GA.** 1996. Virus isolation in neuroblastoma cell culture, p 96–104. *In* Meslin FX, Kaplan MM, Koprowski H (ed), *Laboratory Techniques in Rabies*, 4th ed. World Health Organization, Geneva, Switzerland.
99. **Rudd RJ, Trimarchi CV.** 1989. Development and evaluation of an in vitro virus isolation procedure as a replacement for the mouse inoculation test in rabies diagnosis. *J Clin Microbiol* **27:**2522–2528.
100. **McElhinney LM, Marston DA, Brookes SM, Fooks AR.** 2014. Effects of carcase decomposition on rabies virus infectivity and detection. *J Virol Methods* **207:**110–113.
101. **Bingham J, van der Merwe M.** 2002. Distribution of rabies antigen in infected brain material: determining the reliability of different regions of the brain for the rabies fluorescent antibody test. *J Virol Methods* **101:**85–94.
102. **Warner CK, Whitfield SG, Fekadu M, Ho H.** 1997. Procedures for reproducible detection of rabies virus antigen mRNA and genome in situ in formalin-fixed tissues. *J Virol Methods* **67:**5–12.
103. **Lembo T, Niezgoda M, Velasco-Villa A, Cleaveland S, Ernest E, Rupprecht CE.** 2012. Evaluation of a direct, rapid immunohistochemical test for rabies diagnosis. *Emerg Infect Dis* **12:**310–313.
104. **Lepine P, Atanasiu P.** 1996. Histopathological diagnosis, p 66–79. *In* Meslin FX, Kaplan MM, Koprowski H (ed), *Laboratory Techniques in Rabies*, 4th ed. World Health Organization, Geneva, Switzerland.
105. **Webster LT, Dawson JR.** 1935. Early diagnosis of rabies by mouse inoculation. Measurement of humoral immunity to rabies by mouse protection test. *Proc Soc Exp Biol Med* **32:**570–573.
106. **Rudd RJ, Trimarchi CV.** 1989. Development and evaluation of an in vitro virus isolation procedure as a replacement for the mouse inoculation test in rabies diagnosis. *J Clin Microbiol* **27:**2522–2528.
107. **Hanlon C, Nadin-Davis SA.** 2013. Laboratory Diagnosis of Rabies, p 409–459. *In* Jackson, AC (ed), *Rabies: Scientific Basis of the Disease and Its Management*, 3rd ed. Elsevier Inc., San Diego, CA.
108. **Dupuis M, Brunt S, Appler K, Davis A, Rudd R.** 2015. Comparison of automated qRT-PCR and direct fluorescent antibody detection for routine rabies diagnosis in the United States. *J. Clin. Microbiol* **53:**2983–2989.
109. **Szanto AG, Nadin-Davis SA, Rosatte RC, White BN.** 2011. Re-assessment of direct fluorescent antibody negative brain tissues with a real-time PCR assay to detect the presence of raccoon rabies virus RNA. *J Virol Methods* **174:**110–116.
110. **Freuling C, Vos A, Johnson N, Kaipf I, Denzinger A, Neubert L, Mansfield K, Hicks D, Nuñez A, Tordo N, Rupprecht CE, Fooks AR, Müller T.** 2009. Experimental infection of serotine bats (*Eptesicus serotinus*) with European bat lyssavirus type 1a. *J Gen Virol* **90:**2493–2502.
111. **Fischer M, Wernike K, Freuling CM, Müller T, Aylan O, Brochier B, et al.** 2013. A step forward in molecular diagnostics of lyssaviruses—results of a ring trial among European laboratories. *PLoS One* **8(3):** e58372. doi:10.1371/journal.pone.0058372.
112. **Nadin-Davis SA, Sheen M, Wandeler AI.** 2009. Development of real-time reverse transcriptase polymerase chain reaction methods for human rabies diagnosis. *J Med Virol* **81:**1484–1497.
113. **Hoffmann B, Freuling CM, Wakeley PR, Rasmussen TB, Leech S, Fooks AR, Beer M, Müller T.** 2010. Improved safety for molecular diagnosis of classical rabies viruses by use of a TaqMan real-time reverse transcription-PCR "double check" strategy. *J Clin Microbiol* **48:**3970–3978.
114. **Black EM, Lowings JP, Smith J, Heaton PR, McElhinney LM.** 2002. A rapid RT-PCR method to differentiate six established genotypes of rabies and rabies-related viruses using TaqMan technology. *J Virol Methods* **105:**25–35.
115. **Beltran FJ, Dohmen FG, Del Pietro H, Cisterna DM.** 2014. Diagnosis and molecular typing of rabies virus in samples stored in inadequate conditions. *J. Infect. Dev. Ctries.* **8:**1016–1021.
116. **David D, Yakobson B, Rotenberg D, Dveres N, Davidson I, Stram Y.** 2002. Rabies virus detection by RT-PCR in decomposed naturally infected brains. *Vet Microbiol* **87:**111–118.
117. **Blenden DC, Bell JF, Tsao AT, Umoh JU.** 1983. Immunofluorescent examination of the skin of rabies-infected animals as a means of early detection of rabies virus antigen. *J Clin Microbiol* **18:**631–636.
118. **Smith JS, Reid-Sanden FL, Roumillat LF, Trimarchi C, Clark K, Baer GM, Winkler WG.** 1986. Demonstration of antigenic variation among rabies virus isolates by using monoclonal antibodies to nucleocapsid proteins. *J Clin Microbiol* **24:**573–580.
119. **Benson DA, Cavanaugh M, Clark K, Karsch-Mizrachi I, Lipman DJ, Ostell J, Sayers EW.** 2013. GenBank. *Nucleic Acids Res* **41**(D1)**:**D36–D42.
120. **Trimarchi CV, Nadin-Davis S.** 2007. Diagnostic evaluation, p 411–469. *In* Jackson AC, Wunner WH (ed), *Rabies*, 2nd ed. Academic Press, New York, NY.
121. **Smith JS, Yager PA, Baer GM.** 1996. A rapid fluorescent focus inhibition test (RFFIT) for determining rabies virus-neutralizing antibody, p 181–192. *In* Meslin FX, Kaplan MM, Koprowski H (ed), *Laboratory Techniques in Rabies*, 4th ed. World Health Organization, Geneva, Switzerland.
122. **Trimarchi CV, Rudd RJ, Safford M.** 1996. An in vitro virus neutralization test for rabies antibody, p 193–199. *In* Meslin FX, Kaplan MM, Koprowski H (ed), *Laboratory Techniques in Rabies*, 4th ed. World Health Organization, Geneva, Switzerland.
123. **Savy V, Atanasiu P.** 1978. Rapid immunoenzymatic technique for titration of rabies antibodies IgG and IgM. *Dev Biol Stand* **40:**247–253.

124. **Elmgren LD, Wandeler AI.** 1996. Competitive ELISA for the detection of rabies virus-neutralizing antibodies, p 200–208. *In* Meslin FX, Kaplan MM, Koprowski H (ed), *Laboratory Techniques in Rabies*, 4th ed. World Health Organization, Geneva, Switzerland.
125. **Udow SJ, Marrie RA, Jackson AC.** 2013. Clinical features of dog- and bat-acquired rabies in humans. *Clin Infect Dis* **57:**689–696.
126. **Messenger SL, Smith JS, Rupprecht CE.** 2002. Emerging epidemiology of bat-associated cryptic cases of rabies in humans in the United States. *Clin Infect Dis* **35:**738–747.
127. **Hemachudha T, Laothamatas J, Rupprecht CE.** 2002. Human rabies: a disease of complex neuropathogenetic mechanisms and diagnostic challenges. *Lancet Neurol* **1:**101–109.
128. **Rudd RJ, Rupprecht CE.** 2015. Use of a rapid skin biopsy technique for human rabies antemortem diagnosis, p 93–107. *In* Rupprecht C, Nagarajan T (ed), *Current Laboratory Techniques in Rabies Diagnosis, Research, and Prevention*, **vol 2**. Academic Press, San Diego, CA.
129. **Zaidman GW, Billingsley A.** 1998. Corneal impression test for the diagnosis of acute rabies encephalitis. *Ophthalmology* **105:**249–251.
130. **Rudd RJ, Appler KA, Wong SJ**. 2013. Presence of cross-reactions with other viral encephalitides in the indirect fluorescent-antibody test for diagnosis of rabies. *J. Clin. Microbiol.* **51:**4079–4082.
131. **Hanlon CA, Moore S, Rudd RJ, Wong SJ.** 2010. Emerging and zoonotic disease risk mitigation: rabies prevention as a template for best practices. *Int J Inf Dis* **14**.
132. **Crepin P, Audry L, Rotivel Y, Gacoin A, Caroff C, Bourhy H.** 1998. Intravitam diagnosis of human rabies by PCR using saliva and cerebrospinal fluid. *J Clin Microbiol* **36:**1117–1121.
133. **Venkatesan A, Tunkel AR, Bloch KC, Lauring AS, Sejvar J, Bitnun A, Stahl J-P, Mailles A, Drebot M, Rupprecht CE, Yoder J, Cope JR, Wilson MR, Whitley RJ, Sullivan J, Granerod J, Jones C, Eastwood K, Ward KN, Durrheim DN, Solbrig MV, Guo-Dong L, Glaser CA, Sheriff H, Brown D, Farnon E, Messenger S, Paterson B, Soldatos A, Roy S, Visvesvara G, Beach M, Nasci R, Pertowski C, Schmid S, Rascoe L, Montgomery J, Tong S, Breiman R, Franka R, Keuhnert M, Angulo F, Cherry J, International Encephalitis Consortium.** 2013. Case definitions, diagnostic algorithms, and priorities in encephalitis: consensus statement of the international encephalitis consortium. *Clin Infect Dis* **57:**1114–1128.
134. http://www.cdc.gov/rabies/specific_groups/doctors/ante_mortem.html (accessed 12/17/2015)

Arboviruses

LAURA D. KRAMER, ELIZABETH B. KAUFFMAN, NORMA P. TAVAKOLI

35

VIRAL CLASSIFICATION AND BIOLOGY

Arboviruses (**ar**thropod-**bo**rne viruses) are a biologically defined category of viruses that almost exclusively have RNA genomes. Most of the 100 medically important arboviruses belong to five families: *Togaviridae*, *Flaviviridae*, *Bunyaviridae*, *Rhabdoviridae*, and *Reoviridae*. A single genus in the family *Orthomyxoviridae* (*Thogotovirus*) and a single DNA virus in the family *Asfarviridae* (*African swine fever virus*) are also members of the group. The arboviruses include approximately 40 serological groups based on antigenic cross-reactivity. These viruses are unique in that they generally require cycling between disparate hosts (i.e., vertebrates and hematophagous arthropod vectors). Mosquitoes and ticks are the most common invertebrate vectors, while less common vectors include biting midges and sandflies. Some viruses, such as vertebrate-only flaviviruses (*Rio Bravo virus*, for one) and insect-only alpha- and flaviviruses (*Kamiti River virus*, for one) are not transmitted between disparate hosts.

Togaviridae and *Flaviviridae* are spherical, enveloped particles 60 to 70 nm and 40 to 50 nm, respectively, in diameter, with icosahedral symmetry, containing single-stranded positive sense RNA. The genera *Alphavirus* and *Flavivirus* contain three structural proteins, which are organized in the genome with distinctly different replication strategies, discussed later. *Bunyaviridae* are pleomorphic, enveloped particles, 80 to 120 nm in diameter, with helical symmetry, containing three-segmented single-stranded negative sense RNA. *Reoviridae* are naked, double-capsid, spherical particles, 60 to 80 nm in diameter, with icosahedral symmetry, containing 10 to 12 segmented double-stranded RNA, with seven structural proteins. *Rhabdoviridae* are enveloped, helical (bullet-shape), single-stranded negative sense RNA, 75 by 180 nm with four structural proteins. *Arenaviridae* and *Filoviridae* may also be considered arboviruses. The former are enveloped spherical particles containing two segments of negative-sense single-stranded RNA, the latter pleomorphic particles predominantly filamentous containing negative-sense single-stranded RNA. In this chapter, we will focus on the families *Togaviridae*, *Flaviviridae*, and *Bunyaviridae*.

Togaviridae

The family *Togaviridae* comprises four genera, *Alphavirus* (26 species), *Rubivirus* (one species), *Pestivirus* (three species), and *Arterivirus* (one species) (1). The alphaviruses (summarized in Table 1) include at least seven antigenic complexes: Western equine encephalitis (WEE), Eastern equine encephalitis (EEE), Venezuelan equine encephalitis (VEE), Middelburg, Ndumu, Semliki Forest, and Barmah Forest (2).

Alphavirus RNA is infectious and can initiate replication in a susceptible host. When alphavirus RNA is released into the cytoplasm of the host cell, the 49S RNA serves as mRNA for translation of three of the four nonstructural proteins and as template for transcription of replicative intermediate RNA. Replication takes place in the cytoplasm of the infected cell. The genome contains two open reading frames. The nonstructural proteins (nsPl, nsP2, nsP3, and nsP4) are translated directly from the 5′ two-thirds of the genomic RNA, and the structural proteins (capsid, E1, E2, and E3) at the 3′ end are translated from a subgenomic mRNA, 26S RNA. A junction region is situated between the structural and nonstructural regions. E2 contains the major neutralization epitope, while El contains the major fusion activity (3).

Alphavirus Classification

Old World and New World viruses appear to have diverged between 2,000 and 3,000 years ago (4). There are 31 different virus species in the genus *Alphavirus* (5) (http://www.ictvonline.org/virustaxonomy.asp), comprising the seven virus complexes listed above. *Eastern equine encephalitis virus* (EEEV) is the sole species in the EEE antigenic complex composed of four lineages; North and South American antigenic varieties can be distinguished serologically (6), geographically, epidemiologically, ecologically, genetically, and by level of pathogenicity. Because of clear biologic and genetic differences, South American EEEV (lineages II, II, and IV) was recently reclassified as a separate species named *Madariaga virus* (MADV) (7). North American EEEV (lineage I) is found mainly in the eastern United States, Canada, and the Caribbean Islands (8), and the isolates form a highly conserved lineage. In addition to MADV, there are old reports of EEEV isolations made in the Philippines, Thailand, the Asiatic region of Russia, and the former Czechoslovakia (9).

Members of the WEE complex represent a monophyletic group. The complex includes the prototype virus, *Western equine encephalitis virus* (WEEV), as well as *Buggy Creek virus*,

doi:10.1128/9781555819156.ch35

TABLE 1 Family *Togaviridae* (pathogenic to humans)

Genus	Subclassification	Representative viruses	Transmission cycle: Primary vector(s)	Transmission cycle: Primary amplifying hosts	Geographic distribution	Clinical syndrome
Alphavirus	WEE antigenic complex	WEEV	*Culex tarsalis*; *Aedes albofasciatus* (*S America*)	Primary: birds Secondary: mammals	Americas	Fever, headache, stiff neck, vomiting, lethargy. May progress to disorientation, seizures and coma
		SINV	*Culex* spp. and birds	Passerine birds	Africa, Asia, Europe, Australia	Rash, mild fever, arthritis
		WHAV	*Culiseta tonnoiri*; *Culex pervigilans*	Birds	New Zealand	Possibly influenza-like symptoms
	EEE antigenic complex	EEEV	*Culiseta melanura* (*Coquillett*)	Passerine birds	Americas	Encephalitis
	VEE antigenic complex	VEEV	Enzootic: *Culex* spp.; Epizootic: *Aedes*, *Psorophora*, *Mansonia* spp., others	Enzootic: rodents Epizootic: horses	Americas	Encephalitis
	Semliki Forest group	CHIKV	Sylvatic cycle: *Ae. africanus* Epidemic cycle: *Ae. aegypti* and *Ae. albopictus*	Sylvatic cycle: forest primates Epidemic cycle: humans	Africa, Latin America, India, SE Asia	Rash, arthritis
		RRV	*Ae. vigilax* and *Culex annulirostris*	Macropods; humans	Australia, South Pacific	Rash, arthritis
		MAYV	*Haemogogus* spp. mosquitoes	Primates, rodents, birds	South America	Rash, arthritis
		ONNV	*Anopheles funestus* and *Anopheles gambiae*	Humans, Sylvatic hosts unknown	Africa	Rash, arthritis
		SFV	*Ae. africanus*	Primates	Africa	Rash, arthritis
	Barmah Forest virus group	BFV	*Culex annulirostris/ Coquillettidia linealis*	Wallabies/ kangaroos	Australia	Fever, malaise, rash, joint pain, muscle tenderness

WEEV, *Western equine encephalitis virus*; SINV, *Sindbis virus*; WHAV, *Whataroa virus*; EEEV, *Eastern equine encephalitis virus*; VEEV, *Venezuelan equine encephalitis virus*; CHIKV, *Chikungunya virus*; RRV, *Ross River virus*; MAYV, *Mayaro virus*; ONNV, *O'Nyong Nyong virus*; SFV, *Semliki Forest virus*; BFV, *Barmah Forest virus*.

Fort Morgan virus, and *Highlands J virus* (HJV) in North America; *Aura virus* (AURAV) in South America; and *Sindbis virus* (SINV) and *Whataroa virus* (WHAV) in the Old World. AURAV is most distant by neutralization test (5). HJV is the sole representative of this complex on the east coast of the United States. WEE complex viruses have not been reported in tropical regions of Central America. Genetic variation among selected WEEV strains isolated in California since 1938 were analyzed by sequencing the E2 protein, and four major lineages were identified (10). WEEV arose by a single-recombination event within the E3 gene between a member of the EEEV lineage and the *Sindbis virus* (SINV) lineage, approximately 1,300 to 1,900 years ago. The envelope glycoproteins and a portion of the 3′ non-translated region are derived from a SINV-like ancestor, while the capsid and nonstructural proteins are derived from the EEEV ancestor. AURAV, a New World member in the WEE complex, SINV and WHAV, two Old World members, are not recombinants.

The VEE complex, the sister group of EEE, includes seven different species, of which *Venezuelan equine encephalitis* (VEEV) is the type species (11). Within the complex, six antigenic subtypes (I to VI) have been defined, and subtype I is composed of five varieties: AB, C, D, E, and F (5, 12). Each subtype maintains either an enzootic or an epizootic/epidemic life cycle (12–14). Most enzootic strains cycle between rodents as reservoir hosts and *Culex* sp. mosquito vectors, and although equines cannot be infected (15), human infections are increasingly seen (16). Epizootic species, subtypes IAB through IE, cause epidemics in horses and other animals of agricultural importance. Most human cases of VEEV result from spillover into the human population from an epizootic outbreak. Humans are only incidental and usually dead-end hosts. Other species in the complex include such viruses as *Everglades virus*, *Mucambo virus*, and *Pixuna virus* in subtypes II, III, and IV, respectively. Subtypes V and VI also exist as subtypes with single virus members.

Chikungunya virus (CHIKV) is a member of the Semliki Forest virus group, as is *Ross River virus* (Australia) and *Semliki Forest virus* (Africa). In 2013, the Asian genotype of CHIKV expanded its range to the New World after having been limited in distribution to the Old World before then (17).

Flaviviridae

The family *Flaviviridae* contains four genera: *Flavivirus*, which contains 67 human and animal viruses, *Hepacivirus*, *Pegivirus*, and *Pestivirus*. Only members of the *Flavivirus* genus are vector-borne, but not all flaviviruses are transmitted by arthropods. Flavivirus virions are small, generally ~50 nm in diameter, enveloped, and spherical, with a capsid surrounded by a host-derived lipid membrane containing two glycoproteins, the envelope (E) and the major antigenic determinant membrane protein (M). The viral RNA consists of single-stranded positive sense RNA ~11 kb in length, containing a single open reading frame that is translated into a single polyprotein. The polyprotein is co- or posttranslationally processed by cellular and viral proteases into three structural proteins, capsid (C), premembrane (prM), and envelope (E), at the 5′ end of the genome and seven nonstructural proteins (NS1, NS2A, NS2B, NS3, NS4A, NS4B, and NS5) at the 3′ end that are involved in RNA replication and virus assembly (18, 19). Flaviviruses bind to the surface of host cells via the large surface glycoprotein E and enter the cell through receptor-mediated endocytosis. Acidification within the endosome triggers fusion of the viral and cell membranes and release of the nucleocapsid into the cytoplasm. Viral RNA is uncoated and used for translation and processing of the polyprotein and replication of the genome. Immature noninfectious viral particles are assembled by budding of nascent nucleocapsids together with proteins prM and E into the lumen of the endoplasmic reticulum. Virus maturation occurs during transport through the host cell secretory system, by cleavage of protein prM by the host cell protease furin in the trans-Golgi network. Mature infectious particles are released from the cell by exocytosis (20, 21).

Flaviviruses

Members of the genus *Flavivirus* (summarized in Table 2) fall into several distinct ecologic groups: mosquito-borne (*Culex* or *Aedes*), tick-borne, agents that infect arthropods only, or those that infect vertebrates only (*Rio Bravo virus*) (22). The genus was originally divided serologically into eight antigenic complexes (23, 24). Molecular genetic analyses have upheld the original antigenic and ecologic groupings; that is, mosquito-borne viruses dividing into viruses maintained in *Culex*–bird transmission cycles and causing neurologic disease, *Aedes*–primate transmission cycles causing hemorrhagic disease, and the tick-borne flaviviruses forming a separate group. *West Nile virus* (WNV), an example of the *Culex* group viruses, is made up of approximately five lineages based on nucleotide sequence and phylogenetic analysis (25–27); however, lineage 3, represented by *Rabensburg virus*, from mosquitoes collected at the border of Czech Republic and Austria (28) appears to infect only mosquitoes naturally. *Dengue virus* (DENV), an example of the *Aedes* transmitted group, consists of four antigenically distinct serotypes, DENV-1, DENV-2, DENV-3, and DENV-4.

Tick-borne flaviviruses can be divided into mammalian, seabird, and Kadam tick-borne flavivirus groups (29). The mammalian tick-borne virus group was originally referred to as the tick-borne encephalitis serocomplex. Tick-borne viruses share a common ancestor with the other flaviviruses (30), and while closely related overall to each other, they are genetically divergent, widespread, and epidemiologically important (31, 32). *Powassan virus* (POWV) is the most genetically divergent member of the Tick-borne encephalitis (TBE) antigenic complex (24). The virus exists as two coexisting lineages, POWV and *Deer tick virus* (DTV) (33, 34). The genetic distance between these two viruses (16% nucleotide difference) suggests they diverged and evolved independently into two distinct ecological niches from a single origin (35). Viruses from both lineages have been responsible for human illness (34, 36). POWV and DTV appear to be genetically stable over time (34, 37) and distance (38), suggesting evolutionary constraint consistent with other arboviruses.

Bunyaviridae

The family *Bunyaviridae* comprises more than 300 viruses, with membership usually based on antigenic interrelatedness or morphological similarity. The family is divided into five genera: *Orthobunyavirus*, *Phlebovirus*, *Nairovirus*, *Hantavirus*, and *Tospovirus*, with the last of these infecting only plants (Table 3). Bunyaviruses are single-stranded RNA viruses. The RNA is negative sense and 13 kb in length. The virion is small, generally 90 to 100 nm in diameter, enveloped, and spherical with a helical capsid. Some bunyaviruses use ambisense coding (Nairovirus and Phlebovirus). Glycoprotein projections on the surface, 5 to 10 nm, contain the hemagglutinin and virus neutralization epitopes. The genome of bunyaviruses consists of three segments of single-stranded, negative-sense RNA, designated large (L), medium (M), and small (S). The L segment encodes a large polypeptide, the L protein, which has been shown to have replicase and transcriptase activities (39, 40). The M segment encodes a polyprotein that undergoes posttranslational proteolytic cleavage to give rise to virion surface glycoproteins G1 and G2, and a nonstructural protein called NSm (41–43). The S segment encodes two proteins, the nucleocapsid (N) protein and a nonstructural protein (NSs). These proteins are encoded in overlapping reading frames from the same mRNA (44). The virion surface glycoproteins have been implicated in many of the important biological properties of bunyaviruses, including virulence, attachment, cell fusion, and hemagglutination (45). Neutralizing antibodies are directed against epitopes on the G1 glycoprotein. The N protein encapsidates one copy of each RNA segment, forming nucleocapsids that each contain a few molecules of L protein. It is believed that the N protein induces the formation of complement-fixing antibodies upon infection of an appropriate mammalian host.

Bunyaviruses (summarized in Table 3) can cause severe disease: hemorrhagic fevers, as with *Crimean-Congo hemorrhagic fever virus* (CCHFV); encephalitides, as with *La Crosse virus* (LACV); or severe respiratory distress syndrome, as with hantaviruses.

EPIDEMIOLOGY

Arboviruses, which are found worldwide, have complex life cycles, with different families of arboviruses having exploited distinct niches with specific arthropod vectors and vertebrate hosts. Many arboviruses are zoonotic; that is, they

TABLE 2 Family *Flaviviridae*

Genus	Subclassification	Representative viruses	Transmission cycle: Primary vector(s)	Transmission cycle: Amplifying host(s)	Geographic distribution	Clinical syndrome
Flavivirus	Japanese encephalitis virus group	WNV	*Culex* spp. mosquitoes	Birds	Africa, Asia, Europe, Americas	Mild (fever, headache, arthralgia) to severe (meningitis, encephalitis)
		JEV	*Culex tritaeniorhynchus*	Birds, swine	Asia, Pacific	Mild (headache, cough, nausea) to severe (encephalitis)
		MVEV	*Culex annulirostris*	Water birds	Australia	Fever, headache, encephalitis
		SLE	*Culex* spp. mosquitoes	Birds	Americas	Fever, headache, meningitis, encephalitis
	Yellow fever virus group	YFV	Sylvatic cycle: *Haemogogus* spp. Epidemic cycle: *Aedes aegypti*	Sylvatic cycle: nonhuman primates Epidemic cycle: humans	Africa, South America	Fever, chills, body ache, jaundice, bleeding, multiple organ failure
	Dengue virus group	DENV-1 DENV-2 DENV-3 DENV-4	*Aedes aegypti* and *Ae. albopictus*	Humans	Worldwide in tropics	Mild (fever, headache, joint and muscle pain, rash) to severe (vomiting, bleeding, hemorrhage, organ failure)
	Spondweni virus clade	ZIKV	*Aedes* spp.	Nonhuman primates	Africa, SE Asia, Micronesia, South America	Asymptomatic or rash, fever, arthralgia, conjunctivitis
	Mammalian tick-borne virus group	TBEV	*Ixodes ricinus*, *Ix. persulcatus*	Rodents	Europe, Asia	Benign meningitis to severe meningo-encephalomyelitis
		DTV	*Ixodes scapularis*	White-footed mice	North America	Fever, headache, encephalitis
		POWV	*Ixodes cookie*, *Ix. marxi*	Woodchucks, squirrels	North America, Far East Asia	Fever, headache, encephalitis
		OHFV	*Dermacentor* spp.; *Ixodes persulcatus*	Rodents, especially muskrats and voles	Siberia	Chills, fever, bleeding, hemorrhage, encephalitis
		KFVD	*Haemaphysalis spinigera*	Porcupines, rats, squirrels, mice, shrews	South Asia	High fever, hemorrhagic symptoms, gastrointestinal bleeding

WNV, *West Nile virus*; JEV, *Japanese encephalitis virus*; MVEV, *Murray Valley encephalitis virus*; SLEV, *St. Louis encephalitis virus*; YFV, *Yellow fever virus*; DENV, *Dengue virus*; ZIKV, *Zika virus*; TBEV, *Tick-borne encephalitis virus*; DTV, *Deer Tick virus*; POWV, *Powassan virus*; OHFV, *Omsk hemorrhagic fever virus*; KFDV, *Kyasanur forest disease virus*.

are maintained in an amplification (or enzootic) cycle between mosquitoes and vertebrates (e.g., WNV and EEEV). Other arboviruses can be transmitted between mosquitoes and humans without an enzootic cycle (e.g., DENV and CHIKV). This latter group may also have a sylvatic enzootic cycle. Transmission is predominantly seasonal with the greatest intensity of transmission taking place in the summer and early fall as a consequence of the biology of their vectors. Mosquitoes are the major vectors, but ticks, phlebotamines, and *Culicoides* spp. may also be important. Transmission may be peroral through arthropod saliva containing virus, vertical (transovarial or transstadial), mechanical, or in some cases through co-feeding of uninfected arthropods in close proximity to infected ones.

Risk factors for infection with arthropod-borne viruses include exposure during outdoor activities or poor living conditions, travel to endemic areas of the world, and blood transfusion/organ transplantation, especially under immunosuppression. Accurate incidence of disease is difficult to assess since in general most cases are subclinical. Outbreaks and epizootics occur not only with seasonal changes in vector abundance, but also due to climate change, travel,

TABLE 3 Family *Bunyaviridae*

Genus	Subclassification	Representative viruses	Transmission cycle: Primary vector(s)	Transmission cycle: Amplifying host(s)	Geographic distribution	Clinical syndrome
Orthobunyavirus	California serogroup	LACV	*Aedes triseriatus*	Chipmunks and squirrels	North America	Mild (nausea, headache, vomiting) to severe (seizures, coma, paralysis, brain damage).
Nairovirus	Crimean-Congo hemorrhagic fever serogroup	CCHFV	Ixodid ticks, especially *Hyalomma*	Wild and domestic vertebrates, especially cattle, goats, sheep, and hares	Europe, Asia, Africa	Mild febrile syndrome to vascular leak, multiple organ failure, shock, and hemorrhage
Phlebovirus	Sandfly fever Sicilian serocomplex	RVFV	*Aedes* and *Culex* spp.	Livestock	Sub-Saharan Africa, Yemen	Mild (fever, muscle pains, headache) to severe (meningitis, encephalitis, hemorrhagic fever, eye disease)
	Sandfly fever Naples serocomplex	SFNV	Phlebotomine sandflies	Humans and small rodents	Mediterranean basin	Myalgia, headache, fever
		TOSV	*Phlebotomus* spp.	Humans and small rodents	Mediterranean basin, esp. Italy	Mild (fever, myalgia) to severe (meningitis, encephalitis)
	Bhanja virus serocomplex	HRTV	*Amblyomma americanum*	Unknown	Missouri, Tennessee, Oklahoma	Fever, lethargy, headache, myalgia, diarrhea, arthralgia
		SFTSV	*Haemaphysalis longicornis* ticks	Unknown	Northeast and Central China	Fever, vomiting, diarrhea, multiple organ failure, thrombocytopenia, leukopenia, and elevated liver enzyme levels

LACV, *La Crosse virus*; CCHFV, *Crimean-Congo hemorrhagic fever virus*; RVFV, *Rift Valley fever virus*; SFNV, *Sandfly fever Naples virus*; TOSV, *Toscana virus*; HRTV, *Heartland virus*; SFTSV, *Severe fever with thrombocytopenia syndrome virus*.

and increases in human population and urbanization. Climate change can induce an increase of vector populations, for example, expansion of the range of the mosquito *Aedes albopictus*, an important vector of DENV and CHIKV. Introductions of arboviral diseases to distant regions have been attributed to an increase in frequency and speed of travel by which infected mosquitoes or infected animals or humans may be transported via air travel or in cargo ships. Population expansion in many regions of the world is also an important factor in the epidemiology of arboviral disease. Expansion of urban populations into sparsely inhabited areas can result in a dramatic effect on the ecology of the region, allowing expansion or invasion of arbovirus vectors. Population gains result also in urban crowding, leading, for example, to increases in sewage and poor water drainage, which promote excellent breeding sites for mosquito vectors.

Arboviruses survive dry season in tropical environments or winter cold in temperate environments where vectors live throughout the entire year. Alternatively, virus can be perpetuated by vertical transmission from parent arthropod to progeny or survive by chronic infection of a vertebrate host.

Containment of the spread of arboviral diseases can be addressed best by surveillance (costly, labor intensive), vector control (insecticide resistance and cost of programs may be limiting), education and training, and immunization programs. Integrated approaches are important to consider because arboviral disease causes significant morbidity and mortality worldwide and high cost due to loss of human and animal life.

Alphaviruses

Eastern Equine Encephalitis Virus

North American EEEV was first isolated in 1933 by C. Ten Broek from the brain of an infected horse in Delaware during an equine outbreak (46). The virus is maintained in an enzootic cycle between passerine birds and ornithophilic mosquitoes, principally *Culiseta melanura*, in freshwater hardwood swamps in the Atlantic and Gulf Coast states and the Great Lakes region of the United States. Human cases are rare, but severe; infection is more common in equines, even though a formalin-inactivated equine vaccine (Ft. Dodge) is available.

Venezuelan Equine Encephalitis

The VEEV antigenic complex is comprised of six distinct subtypes, I to VI, which occur in the western hemisphere (47). Large outbreaks of epizootic subtypes IAB and IC occurred in Central and South America, emerging from subtype ID via a mutation in E2 glycoprotein (48). Horses are most severely affected, with approximately 50% developing encephalitis, but numerous and diverse animal species may become infected. Virus is maintained in an enzootic cycle through which it is transmitted predominantly between rodents and other smaller mammals and *Culex melanoconion* mosquitoes. Epizootic VEEV is transmitted predominantly by *Aedes* and *Psorophora* mosquitoes to horses and humans and other vertebrates. A licensed live attenuated vaccine, TC-83, exists for horses.

Chikungunya Virus

CHIKV is widespread in Africa, Asia, and since 2013, the Caribbean and the Americas; thus, it is now both a New World and Old World virus. Epidemics historically have been cyclic, re-occurring every 7 to 20 years. A large outbreak of the Central East South African (CESA) strain commenced in 2004 in Kenya, then spread to islands of the Indian Ocean and to Asia. In 2007, the first autochthonous European outbreak occurred in Italy (49, 50), facilitated by a mutation in the E1glycoprotein gene, which resulted in the A226V amino acid change that allowed the virus to be more efficiently transmitted by the invasive mosquito species, *Aedes albopictus*. The virus that invaded the western hemisphere is the Asian genotype as opposed to the CESA strain, although a CESA strain without the A226V mutation has also been detected in Brazil.

Sindbis Virus

SINV was first isolated in 1952 near Cairo, Egypt, and has since been found throughout the Old World. The virus is maintained in an avian–*Culex* spp. cycle. It consists of five genotypes, with isolates from Europe and South Africa (SINV-I), Australia (SINV-II), East Asia (SINVIII), Azerbaijan and China (SINV-IV), and New Zealand (SINV-V) (51). All human infections are from the SINV-I genotype, with cases reported from Sweden (Ockelbo disease), Finland (Pogosta disease), Western Russia (Karelian fever), and South Africa (Sindbis fever) (52).

Other Alphaviruses

Mayaro virus causes a dengue-like illness in South America; its enzootic cycle involves canopy mosquitoes and nonhuman primates. *Ross River virus* and *Barmah Forest virus* both occur in Australia and cause severe arthralgia. The natural host appears to be large marsupials in rural areas. WEEV is rarely detected anymore; similar to *St Louis encephalitis virus* and WNV in the western United States, its primary vector is *Culex tarsalis* and its host is wild birds.

O'Nyong Nyong virus, whose primary vector is anopheline mosquitoes, first appeared in 1959 in Uganda, causing an epidemic that spread into Kenya, Tanzania, and Zaire and affected over 2 million people. Its primary symptoms are polyarthritis, rash, and fever, and it is not fatal.

Flaviviruses

Dengue Virus

Dengue is the most prevalent arboviral disease of humans, present in pandemic proportions throughout most tropical and subtropical regions. Approximately 50% of the world population resides in areas at risk for infection. Infection by any one of four antigenically distinct serotypes, DENV-1, DENV-2, DENV-3, or DENV-4, results in lifelong protection against the infecting serotype, but protection is not cross-reactive among serotypes, and subsequent infection by another serotype can cause more severe forms of disease: dengue hemorrhagic fever or dengue shock syndrome. All four DENV serotypes are present in most endemic countries (hyperindemicity). The virus is maintained in an enzootic cycle between humans and predominantly *Aedes aegypti* mosquitoes. *Aedes aegypti* is an urban mosquito residing in household dwellings, and its population is expanding along with expansion of the human population. *Aedes albopictus*, a highly invasive species, can also transmit the virus.

DENV epidemics were reported first in the 1700s on three different continents, indicating that the virus was already widespread. Further expansion occurred during the 18th and 19th centuries along with escalation of the shipping industry (53). After World War II another expansion of the disease occurred as air travel became more common. In the last 50 years the incidence of DEN disease has increased 30-fold, and today more than 2 million cases occur annually (54). One study that used case records paired with risk maps estimated 390 million infections per year worldwide, of which 96 million were clinical or subclinical (55). The severity of infection varies, based on serotype, genotype, and human factors. Many infections being nonapparent and the mortality rate being comparatively low (5%) contribute to the global impact of the disease being underestimated. The overwhelming spread of DENV has been attributed to three principal drivers, namely urbanization, globalization, and lack of effective mosquito control (56). Without effective control, spread will continue, and it is imperative that efforts be made to mitigate the economic, social, and public health burden of the disease. To address these issues the WHO proposed a Global Strategy for Dengue Prevention and Control, 2012–2020 (54). Its objectives focus on reducing morbidity and mortality of the disease and developing a true estimate of disease burden. Mortality can be reduced to zero by early and efficient diagnosis, effective treatment (including intravenous rehydration), staff training, and expansion of capacity. Public health surveillance would play an important role in risk assessment and outbreak preparedness. Vector control is a critical component of prevention for all mosquito-borne arboviruses. It is interesting to note that effective vector control to eradicate *Ae. aegypti* initiated during construction of the Panama Canal to contain *Yellow fever virus* (YFV) was successful until the United States, followed by other countries, lost interest in the effort, after which a resurgence in DEN took place (56). Promising approaches under evaluation include Wolbachia-infected *Ae. aegypti* (57), release of males carrying a lethal gene (Oxitech), and vaccines in trial (58).

West Nile Virus

WNV is currently the most widely distributed arbovirus in the world, occurring on all continents except Antarctica. The virus was first isolated from the blood of a febrile woman in the West Nile district of Uganda in 1937 (59). Prior to 1996, occasional epidemics occurred in Africa, Eurasia, Australia, and the Middle East, but few cases of neurological disease were observed (60). Virus activity increased noticeably from 1996 to 1999 in the Mediterranean basin, southern Romania, and the Volga delta in southern Russia; then in 1999, WNV was introduced into the Western Hemisphere, where it spread rapidly in the United States, Canada, and Central and South America over the next 4 years. WNV is now the leading cause of encephalitis in the United States and Canada.

Worldwide, WNV is maintained in nature in an enzootic cycle between ornithophilic mosquitoes and competent susceptible avian hosts. Epizootics/epidemics occur in horses and humans, but these incidental hosts are considered dead-ends because the level of viremia is below the threshold required to infect mosquitoes. Genus *Culex* mosquitoes are the major vector globally; in the United States it has been estimated that *Culex* species are responsible for 80% of human WNV infections (61). Birds belonging to the order *Passeriformes* are the major amplification hosts. In 2000 alone, WNV infection in 55 species of birds was reported to ArboNet (62).

Lineage 1 WNV is the most geographically widespread. Lineage 2 contains strains isolated predominantly in sub-Saharan Africa and Madagascar, including the prototype 1937 Ugandan isolate. Until 2008, severe West Nile neurological disease in humans had been almost exclusively associated with lineage 1 strains, but since 2004, lineage 2 strains have been found circulating in Central, Southern, and Eastern Europe and have been responsible for major outbreaks in Hungary, Austria, Greece, Romania, Italy, and the Volgograd region of Russia (63–65). Using BEAST (Bayesian Evolutionary Analysis Sampling Trees) analysis May et al. (66) determined that Africa is the source of WNV introduced in all areas, usually by independent events. This analysis implies that both NY99 strain of WNV and the 1998 Israeli isolate widely believed to be its progenitor were introduced separately most likely from Africa. Migratory birds play a major role in the spread of WNV throughout the world. Migratory routes from Africa to Europe and Asia are well understood, and WNV and antibodies have been isolated from migrating species (67, 68). Successful spread of the virus by birds is dependent on avian viremia that is sufficiently high and long lasting to survive the journey and be infectious to mosquitoes.

Japanese Encephalitis Virus (JEV)

JEV is transmitted to humans through the bite of infected *Culex* species mosquitoes, primarily *Culex tritaeniorhynchus*. The virus is maintained in a cycle between mosquitoes and vertebrate hosts, especially pigs and wading birds. Humans are incidental or dead-end hosts because the concentration of JEV in their bloodstream is too low to infect mosquitoes. Virus transmission occurs primarily in rural agricultural areas, often associated with rice production and flooding irrigation. In some areas of Asia, these conditions can occur near urban centers. In temperate areas of Asia, virus activity is seasonal. Human disease usually peaks in the summer and fall. In the subtropics and tropics, transmission can occur year-round, often with a peak during the rainy season.

Yellow Fever Virus

YFV originated in Africa and spread to North and South American ports via the importation of infected *Ae. aegypti* vectors on sailing ships. YFV is now predominantly found in Africa and South America. The virus is maintained in an enzootic cycle between canopy dwelling mosquitoes and nonhuman primates and an urban cycle between *Ae. aegypti* and humans. An additional intermediate cycle, the savannah cycle, has also been identified in Africa between *Aedes* species mosquitoes and humans.

Other Important Flaviviruses Include Zika Virus, Murray Valley Encephalitis Virus (MVEV), and Tick-Borne Encephalitis Virus (TBEV)

TBEV in the United States (i.e., *Powassan virus*) is transmitted by several *Ixodes* species (e.g., *Ix. scapularis*, *Ix. cookei*, and *Ix. marxi*) and in Europe (i.e., Central European TBEV) by *Ixodes ricinus* (69). There have been rare reports of transmission through unpasteurized milk of Central European TBEV. *Zika virus* is an emerging flavivirus transmitted between *Aedes* species and nonhuman primates and is endemic to Africa and Asia (70). It was first detected outside this region in 2007 and occasional cases have been noted since then, as in French Polynesia, and in tourists returning from areas where the virus is active. In May 2015, a substantial outbreak occurred in Brazil, the first in the Americas (www.cdc.gov/zika). MVEV can be found in Australia and Papua New Guinea, where it is maintained in a bird-*Culex* cycle.

Bunyaviridae

Most bunyaviruses, with the exception of hantaviruses, are transmitted in nature by arthropods, most frequently mosquitoes but occasionally phlebotomine sandflies, midges, and ticks. Vertebrate reservoirs have been demonstrated for some viruses. In others, transovarial transmission is thought to play a dominant role in virus maintenance. Humans are not known to be a natural reservoir for any of these viruses, with the probable exception of *Sandfly fever virus*.

Orthobunyavirus

California serogroup viruses have been isolated in North and South America, Africa, Europe, and Asia. They are found in tropical, temperate, and Arctic regions. Based on serological relationships, viruses in the genus *Orthobunyavirus* have been divided into 16 serogroups and include more than 150 viruses. Although antigenically closely related, these viruses are maintained in nature in distinct cycles involving preferred arthropod vectors and vertebrate hosts. Transovarial transmission also is thought to be critical to the maintenance of the viruses, particularly in the Arctic. LACV is maintained in an amplification cycle between *Aedes triseriatus* mosquitoes and chipmunks (*Tamias striatus*) and squirrels (*Sciurus carolinensis*) as the principal vertebrate hosts.

Phlebovirus

Rift Valley fever virus (RVFV) was first isolated in 1930 as part of a large outbreak among sheep in East Africa (71). This virus is transmitted most commonly by mosquitoes of the *Aedes* species considered "reservoir/maintenance" vectors, and *Culex* species, considered "epidemic/amplifying" vectors, but it is also capable of infecting ticks and a variety of flies. RVFV is endemic in nearly all the countries in sub-Saharan Africa. Isolated human cases and large epidemics (domestic livestock and human) have been described in several countries, and the virus has most recently been detected in the Arabian Peninsula (72).

Nairovirus

CCHFV was first recognized in the Crimean peninsula during an outbreak of hemorrhagic fever among agricultural workers (73). The same virus was isolated in 1956 from a single patient in present day Democratic Republic of Congo, leading to the name (74, 75). Ixodid ticks of the genus *Hyalomma* are the principal vector. Although animals and humans can be infected, only the latter develop the disease.

The virus is transmitted by tick bite but human-to-human transmission of the virus is very characteristic.

Other Medically Important Bunyaviruses

Severe fever with thrombocytopenia syndrome virus (SFTSV) is an emerging tick-borne *Phlebovirus* isolated in China, Japan, and Korea that causes disease with a high fatality rate. The virus appears to be undergoing a recent population expansion (76).

CLINICAL SIGNIFICANCE

Arboviruses are important human and veterinary pathogens and are the cause of emerging and re-emerging diseases worldwide. Members of the genera *Alphavirus*, *Flavivirus*, and *Bunyaviridae* in particular cause infections with symptoms ranging from mild fever to respiratory illness, febrile illness, arthritis, encephalitis, hemorrhagic disease, shock, and even death.

Togaviridae; Alphaviruses

Alphavirus infections occur worldwide with New World alphaviruses in general causing rare cases of encephalitis in the Americas and large outbreaks of polyarthralgic illness in other parts of the world especially Asia and Africa. Alphaviruses are important pathogens of humans and horses and have the potential to be used as agents of bioterrorism. In this genus, EEEV and VEEV are the main causes of encephalitis (77), whereas CHIKV and SINV cause arthralgia, rash, and malaise.

Eastern Equine Encephalitis Virus

Among the alphaviruses, EEEV has the lowest incidence in humans but has the highest fatality rate (50% to 70%). Following an incubation period of 4 to 10 days, symptoms begin with a sudden onset of fever, headache, and muscle pain. In severe cases encephalitis occurs, which may be accompanied by fever, headache, vomiting, respiratory symptoms, seizures, hematuria, and coma.

Venezuelan Equine Encephalitis

Infections occur mainly in Central and South America. VEEV has the lowest fatality rate in humans (<1%), mainly in children. Epidemic strains of VEEV cause febrile illness, although most infections lead to mild clinical signs. The symptoms of neurological disease in patients infected with VEEV, especially children, include disorientation, ataxia, depression, and convulsions.

Chikungunya Virus

Outbreaks involving hundreds of thousands of individuals have occurred in the Indian subcontinent. In general, after an incubation period of 2 to 6 days, 95% of CHIKV infections are symptomatic and patients suffer from polyarthralgia and myalgia. The acute phase of infection includes a maculopapular rash, fever, and intense muscle and joint pain (78). Polyarthralgias and neurologic manifestations can last for weeks to years after the infection (79). In newborns and the elderly the virus has caused encephalitis, with symptoms including headache, altered mental status, seizures, sensory abnormalities, and motor dysfunction. Central nervous system (CNS) complications such as encephalitis and febrile seizures are common in children infected with CHIKV. The fatality rate is up to 10%, although in individuals over 65 it can be as high as 33% (78). Because CHIKV circulates in DEN-endemic regions and the symptoms mimic those of DEN, CHIKV was long underdiagnosed and its importance not appreciated (80). However, the 2006 Indian epidemic has shown that CHIK is a dangerous and emerging arboviral disease that can cause significant morbidity and mortality and huge economic losses (81). In addition, it has become established in the Caribbean and has the potential to become established in the Americas where it can be transmitted by *Ae. aegypti* and *Ae. albopictus*.

Sindbis Virus

Infections by SINV have mainly been reported from Northern Europe with large outbreaks involving thousands of cases in Finland (82). The infection is known as Pogosta disease in Finland (83), Ockelbo disease in Sweden (84), and Karelian fever in Russia (85). Sindbis fever is usually mild and resolves spontaneously. The symptoms include low-grade fever, rash, headache, arthritis, malaise, muscle pain, and fatigue (86, 87). Arthritis can persist for months to years (88).

Other Alphaviruses

Mayaro virus (MAYV) circulates strictly in South America and is closely related to CHIKV. It produces a similar debilitating disease characterized by rash and arthralgia. In addition, because it produces a DEN-like illness with symptoms including fever, chills, headache, eye pain, myalgia, and gastrointestinal signs, it is often misdiagnosed as DEN. However, unlike DENV, it can cause severe arthralgia that can persist for months (89) and it does not cause fatal disease.

Ross River virus (RRV) and *Barmah Forest virus* (BFV) are endemic throughout Australia and also cause regular outbreaks. RRV is responsible for the most significant numbers of disease cases in Australia. RRV infection is characterized by polyarthritis, arthralgia, myalgia, fatigue, fever, and rash (90). In addition, approximately half of those infected suffer from lethargy, joint pain, stiffness, and swelling for 6 to 12 months after infection. The symptoms of BFV infection are similar to RRV infection except that arthritis is more common with RRV and rash is heightened and seen more frequently in cases of BFV infection (91).

Flaviviridae; *Flavivirus*

In the genus *Flavivirus*, YFV, DENV, JEV, and WNV are of greatest interest because they cause the majority of human disease. Tick-borne flaviviruses, in particular TBEV, are among the medically most important arboviruses in Europe and Asia where 10,000 to 15,000 cases of tick-borne encephalitis are reported annually.

Dengue Virus

The number of dengue cases in the Americas increased fivefold between 2003 and 2013, according to the Pan American Health Organization/WHO. Between 2009 and 2012, over 1 million cases were reported annually, on average, with more than 33,900 severe cases and 835 deaths. One of the worst years for dengue in the hemisphere's history was 2013, with 2.3 million cases, including 37,705 severe cases and 1,289 deaths. By comparison, the number of cases reported region-wide in 2003 was 517,617.

Thousands of cases are annually reported in Africa, the Middle East, and Asia.

Dengue presents as a spectrum of symptoms and has been classified as dengue with and without warning signs and severe dengue. In young children dengue infection causes a nonspecific febrile illness. In older children and adults it causes a classic fever-arthralgia-rash syndrome. The onset of fever is abrupt and is accompanied by muscle and joint pain, retro-orbital pain, photophobia, and lymphadenopathy. Recovery is fairly rapid. The symptoms of dengue with warning signs (also referred to as dengue hemorrhagic fever) include abdominal pain, vomiting, mucosal bleeding, lethargy, fluid accumulation, and liver enlargement. Severe dengue (also referred as dengue shock syndrome) also includes severe plasma leakage leading to shock, thrombocytopenia, bleeding, and organ failure. Patients with warning signs and the more severe forms of disease require hospitalization, and therefore, it is important to recognize at-risk individuals during the triage process.

Severe dengue is strongly associated with secondary infections (92). The incidence of severe disease has increased due to the sympatric circulation of multiple DENV serotypes (hyperendemicity) and the antibody-dependent enhancement of macrophage infection by cross-reacting antibodies leading to severe disease (93, 94).

West Nile Virus

WNV is the leading cause of arboviral encephalitis in the United States (95), but the virus is distributed circumglobally (96). Most human WNV infections are asymptomatic although 20% to 30% of cases result in West Nile fever. West Nile fever develops after an incubation period of 3 to 14 days and is characterized by fever, headache, back pain, fatigue, arthralgia, and myalgia (97). In some cases anorexia, vomiting, diarrhea, retro-orbital pain, and pharyngitis have also been observed (98). In approximately half of West Nile fever cases, especially in children, a maculopapular rash is also observed. Serious cases can result in hepatitis, myocarditis, and pancreatitis (99). Neuroinvasive disease with a mortality rate of 10% is the outcome in less than 1% of West Nile infections. Older individuals are more susceptible to develop neuroinvasive disease and to die of the infection. In 60% of cases in which the CNS is infected, patients develop encephalitis; in the remaining cases, they develop meningitis (100). The symptoms of West Nile encephalitis may include muscle weakness, acute flaccid paralysis syndrome, seizures, or cerebellar ataxia (98, 99). Survivors of West Nile encephalitis frequently suffer long-term neurologic and cognitive impairment, including muscle weakness, insomnia, depression, confusion, headache, and myalgia.

Japanese Encephalitis Virus

In Asia an estimated 35,000 to 50,000 cases of viral encephalitis and 15,000 deaths are attributed to JEV each year (101). Most infections are asymptomatic, and the clinical disease can vary from minor to severe. The minor febrile illness can include headache, cough, nausea, vomiting, rigors, and diarrhea, whereas severe disease can include meningoencephalitis, meningitis, acute flaccid paralysis, and severe encephalitis. Seizures are common in children. In approximately 30% of severe cases long-term sequelae such as persistent motor deficits, behavioral problems, and learning difficulties are observed.

Yellow Fever Virus

Tens of thousands of human cases of YFV infection occur each year in Africa. Yellow fever presents with high fever, chills, headache, back and muscle aches, dizziness, anorexia, nausea, and vomiting. The disease may resolve or develop further to high fever, epigastric pain, jaundice, renal failure, bleeding, and death following prolonged shock.

Other Important Flaviviruses

Zika virus is closely related to YFV and is occasionally associated with human disease. The virus circulates in Africa, Asia, most recently in Brazil, and produces a flu-like illness associated with high fever, malaise, dizziness, anorexia, retro-orbital pain, edema, rash, conjunctivitis, lymphadenopathy, and gastrointestinal manifestations (102, 103). Human infections of MVEV are generally subclinical, although infections are more common and severe in children. Clinically the symptoms can include sudden onset of fever, nausea, headache, vomiting, seizures, and CNS dysfunction. The mortality rate is 15% to 31%. The disease caused by Kunjin virus (WNV lineage 1b) is generally similar to that caused by MVEV. However, it is typically milder and not as deadly (104). Both viruses are endemic in northwest Australia (105).

Medically the most important tick-borne flavivirus is TBEV, which has three subtypes distributed throughout Asia and Europe. The severity of disease caused by TBEV is dependent on the strain, infectious dose, and the age of the patient. Older patients are more prone to developing severe meningo-encephalomyelitis with neurological difficulties and possibly death. Children are more likely to suffer from benign meningitis. Disease caused by several other tick-borne flaviviruses have been reported, but these are in general restricted to local areas. Examples include Louping ill virus in the British Isles, Omsk hemorrhagic fever virus in Western Siberia, Powassan virus in North America and Far Eastern Siberia, and Langat virus in Southeast Asia.

Bunyaviridae

Many viruses in the *Bunyaviridae* family are significant pathogens of humans and cause illness ranging from mild infection to severe disease including hemorrhagic fevers, pulmonary disease, and encephalitis.

Crimean-Congo Hemorrhagic Fever Virus

After DENV, CCHFV is the second most widespread of all medically important arboviruses. Onset of symptoms is sudden with fever, headache, myalgia, photophobia, vomiting, diarrhea, and agitation followed by depression, lassitude, and hepatomegaly and occasionally hepatitis or liver or pulmonary failure. Petechiae are common, and in severe cases hemorrhagic manifestations are observed. CCHFV has a mortality rate of up to 30% in humans (106). In the United States historically there have been 42 to 174 cases of encephalitis reported each year, although this number may be an underestimation of actual cases.

La Crosse Virus

In general, LACV infection is a nonspecific febrile illness, but in a minority of children the disease progresses to acute encephalitis characterized by headache, fever, vomiting, stiff neck, and infrequently coma. Seizures occur in about half the cases, and epilepsy develops in 10% of cases. The case fatality rate for LACV encephalitis is approximately 0.3%, and approximately 2% of cases develop persistent paresis, cognitive deficits, and neurobehavioral sequelae such as hyperactivity.

Rift Valley Fever Virus

Outbreaks of RVFV with human involvement have occurred in sub-Saharan eastern Africa, South Africa, Sudan, Egypt, and West Africa including Mauritania (107, 108). Infection with RVFV can be asymptomatic or it can cause a range of diseases including retinitis, hepatitis, renal failure, encephalitis, hemorrhagic fever, and death (109, 110). Less than 1% of infections result in encephalitis. Symptoms of RVFV encephalitis include disorientation, drowsiness, severe headache, stiff neck, convulsions, paraparesis or hemiparesis, and coma (109, 111). The case fatality rate is approximately 10% to 20% (109).

TREATMENT AND PREVENTION

Apart from some exceptions, which are discussed below, there are generally no antivirals or vaccines available to protect or treat humans against arboviruses (112). The main method of disease prevention is therefore avoidance of infection by reducing exposure of humans to mosquito and tick vectors.

Treatment

Carbodine, which is a carbocylic analogue of cytidine, has been shown to have potential as an antiviral agent against VEEV (113), although cytotoxicity has been demonstrated. In addition, interfering RNAs have been shown to inhibit VEEV in cell culture (114), and their use is being investigated for developing antivirals for many other RNA viruses including CHIKV (115). There are currently no recognized antiviral therapies for CHIKV infection, although anti-inflammatory drugs are prescribed to treat the debilitating arthralgia.

Treatment for JEV infection involves supportive care because there are no specific antivirals available. Potential therapeutics including minocycline and curcumin are under investigation for their antiviral and neuroprotective results (116–118). Similarly, there are no antivirals against WNV or DENV. Some candidate therapeutic agents including ribavirin in combination with pegylated interferon drugs, humanized monoclonal antibodies and interferon alpha 2b, and immunoglobulin with high titer against WNV (Omr-IgG-aM) are under investigation (100, 119, 120).

There are no antivirals against bunyaviruses. However, ribavirin has been used off label to limit the severity of LACV infection because it is believed to inhibit infection by inhibiting RNA-dependent RNA polymerase.

Vector Control

Vector elimination and control is very difficult even in resource-rich nations. Vector control must rely on sustainable, community-based, integrated mosquito control, with limited reliance on insecticides. The best preventive measures for residents living in areas infested with *Ae. aegypti* is to eliminate the places where the mosquito lays her eggs, primarily artificial containers that hold water. Insecticide spraying has not proven very effective, and emergence of resistant mosquitoes is a major concern. Large-scale vector control measures can often be expensive, impractical, harmful to the environment, and controversial. There are some common methods of vector control, and these include controlling vector habitat, limiting contact with vectors, and using biologics or chemicals to control vector populations. To prevent an increase in mosquito populations, some simple steps to take can be performed by individuals and communities, such as preventing standing water, removing rubbish, and cutting down weeds and vegetation. To limit contact with vectors, it is recommended that individuals wear light-colored clothes that cover the body, use window screens and bed nets, and limit time spent outdoors especially at dusk when mosquitoes are most active. Large-scale aerial spraying using approved insecticides is performed in certain areas against WNV and EEEV with permission from the relevant authorities. However, this practice can be problematic because of the harmful effects on the environment and the development of resistance to insecticides.

Vaccination

Alphaviruses

Several vaccines are under development for the alphaviruses CHIKV, VEEV, EEEV, and WEEV, although only experimental vaccines are currently available for humans. Equine vaccines are in use for VEEV, EEEV, and WEEV in some countries and help reduce infections in horses. The vaccine for EEEV and WEEV consists of formalin-inactivated virus that is available as a double vaccine. The development of several CHIKV vaccines including safe nonreplicating vaccines; live attenuated vaccines (121), which can induce rapid and long-lived immunity after one dose; and chimeric vaccines (122) is underway. In the nonreplicating vaccine category are inactivated whole virus (123), DNA (124), virus-like particles (123), and adenovirus-based vectors (125).

Flaviviruses

JEV is the only encephalitis-causing virus for which vaccination of large human populations is promoted, particularly in Asia, where it has improved the control of the virus. Most vaccines are derived from infected mouse brain, although a cheaper inactivated cell culture–derived vaccine is also available (126–128). There are no human vaccines for WNV, although an equine vaccine is in use (129).

An effective human vaccine (17D) was developed for YFV in 1937 (130). Despite the availability of this vaccine, a significant number of infections occur in Africa on an annual basis. There are currently no licensed vaccines for DENV, although a phase 3 efficacy trial of a tetravalent recombinant, live, attenuated vaccine (CYD-TCV) looks promising (131). To ensure immunity to all serotypes of DENV, a vaccine is required that produces solid and long-lived tetravalent immunity. The problem with producing a vaccine against only one serotype is that there is a risk of infection with a different DENV serotype, which may increase the risk of antibody-dependent enhancement of immunized individuals. Chimeric vaccines in which immunogenic genes that encode envelope and membrane proteins from DENV are exchanged for equivalent genes in YFV have been developed and are being investigated (132, 133). The ChimeriVax-Dengue vaccine is a tetravalent vaccine composed of four recombinant, live, attenuated vaccines each using the 17D YFV as a replicative backbone to carry genes encoding structural proteins from one of the four DENV serotypes (134). This vaccine had an efficacy of 56% in a phase 3 trial that included 10,275 children and is expected to be licensed shortly based on the encouraging results (134).

A tetravalent vaccine prepared from monovalent DENV1-4 was attenuated following passage in primary dog kidney cells (135) and subsequently in fetal rhesus lung cells (136). The initial vaccine was tested in phase 1 and 2 clinical trials. Various formulations of the DENVax vaccine, which is also a live attenuated vaccine attenuated by serial passage, have been developed and are currently in clinical trials (137) . Directed mutagenesis is used in attenuation for the Tetra-Vax dengue vaccine (138). The mutations that are introduced are chosen to reduce reactogenicity and toxicity while maintaining immunogenicity. In addition to the live attenuated vaccines described thus far, inactivated whole virus vaccine that does not pose a risk of reversion to virulence has also been under investigation (139). Multiple expression systems have been used to produce recombinant subunit vaccine candidates mainly based on the E protein of DENV (140). Tetravalent formulations of the recombinant proteins have been evaluated as potential vaccine candidates (141). Viral-like particles (142) and virus vectors such as adenoviral vectors (143) have been studied as carriers of DENV vaccine antigens and have had some success in eliciting an antibody response in mice. Finally, DNA vaccines consisting primarily of the DENV E protein as the target antigen have shown some protective immunity in mice (144). As discussed, there have been many approaches to DENV vaccine development. Currently the live attenuated vaccines are in the most advanced stages of clinical trials and in all likelihood a tetravalent DENV vaccine will be licensed in the next 5 to 10 years (145).

Several vaccines have been developed based on inactivated purified TBEV (32). The use of these vaccines has successfully reduced the rate of tick-borne encephalitis cases in Russia, Austria, and elsewhere in Europe.

Bunyaviruses

There are no antivirals or vaccines for bunyaviruses except for vaccines that have been developed for veterinary use to protect against RVFV.

DETECTION AND DIAGNOSIS

Arboviral disease is fairly rare in the United States but more widespread globally. Many different agents, including fungal, bacterial, and viral agents, cause similar symptoms. In mild cases, nonspecific flu-like symptoms occur, and in severe cases, meningitis, encephalitis, and hemorrhagic fevers are seen. Therefore, determining a diagnosis based on symptoms alone is very difficult. It is crucial, however, to make a specific diagnosis because it helps determine treatment strategies and allows the prediction of virus activity and institution of control steps to prevent potential outbreaks. The most important prevention strategies are vector management, vaccination programs, and public awareness campaigns. The first step to establishing a diagnosis for arboviral disease is to determine an in-depth case history. This includes time of year, location, travel history, occupation, activities, contact with animals, ticks or mosquitoes, vaccination history, and immune status of the patient. In temperate climates, arboviral diseases usually occur in warmer months when arthropods are present. In warmer climates and in the tropics, infections can occur year round. Therefore, travel history and occupations that require outdoor work should prompt the need for arboviral testing. The selection of arboviruses to be tested will also depend on the location of the patient. As discussed above, many arboviruses are geographically distributed in certain regions of the world. Therefore, it would be prudent to test for a subset of arboviruses that are relevant to the location of the patient at the time of infection (Fig. 1), keeping in mind travel history (146).

The choice of specimen, timing of collection, transport and storage of the specimen are important factors in ensuring that a successful diagnosis is made. In general a blood or serum specimen is collected, and if there is CNS involvement then cerebrospinal fluid (CSF) is also collected. It is important to understand the stages of viremia and when to expect the appearance of IgM and IgG (147). During the viremic phase of the disease, viral nucleic acid can be detected in CSF and serum for approximately 1 to 5 days post onset (Fig. 2). Specimens collected during this acute phase of the disease are most appropriate for molecular analysis. Viral nucleic acid, especially RNA, is subject to degradation. Therefore, specimens should be refrigerated or kept on cold packs and delivered to the diagnostic laboratory in a timely fashion. In a primary infection, approximately 5 days post onset, IgM antibodies appear and remain detectable for over 3 months (Fig. 2). IgG antibodies appear weeks post onset, rise, and remain high for years (Fig. 2). In a secondary infection, IgG antibodies rise rapidly and remain high, while IgM antibodies are significantly lower. These timelines must be considered in the timing of specimen collection and in what test is performed and how it is interpreted (148).

Diagnostic testing for arboviruses have been performed for many decades and fall within two main categories of serology and viral detection. The traditional tests included virus isolation and identification and serological tests such as hemagglutination inhibition, complement fixation, plaque reduction neutralization test (PRNT), and indirect fluorescence antibody (IFA) tests, which detect the host's antibody response to the viral agent. As technology improved, new serological tests were developed to include IgM antibody-capture ELISA (MAC-ELISA), IgG ELISA, antigencapture ELISA, and microsphere immunoassay (MIA). In addition, molecular assays that detect the viral genome were utilized. Initially standard reverse transcription PCR assays (RT-PCR) and subsequently real-time RT-PCR assays, nucleic acid amplification tests (NAAT), and nucleic acid sequence-based amplification tests were added to the arsenal of tests to diagnose arboviral infections. In more recent years, sequencing methods, especially next generation sequencing, microarrays, and liquid chromatography electrospray ionization-tandem mass spectrometry (LC-ESI/MS/MS), have been used as arbovirus detection systems in more advanced laboratories. Each method has advantages and disadvantages, and it is rare that one method would be sufficient under any particular circumstance. In general, various combinations of tests are used depending on costs, the availability of reagents, equipment, and expertise and whether the testing is done in the field or at a laboratory.

Routine Diagnostics for Acute Arbovirus Infections

The significant burden of DENV and CHIKV disease worldwide and the rapid spread of WNV in the United States since 1999 has led to the availability of routine tests, mainly serological tests, for these viral agents.

IgM and IgG Antibody Detection

The most widely used approach for the diagnosis of WNV infection in humans is detection of specific IgM and/or IgG antibodies. Routine diagnostic laboratories, generally perform enzyme immunoassays (EIA) and immunofluorescence (IF)-based assays because they are rapid and reproducible and technically simple to perform, they can be automated, and they are specific if monoclonal antibodies are used.

Detection of IgM antibody, which is produced early in an immune response (149), is a valuable method for detecting acute infections. The detection of IgM antibodies in the CSF implies a recent infection. Specific IgM antibodies to most arboviruses are usually detectable within 10 days after onset of illness and can persist up to at least 1 year. Many separate MAC-ELISAs have been developed for the

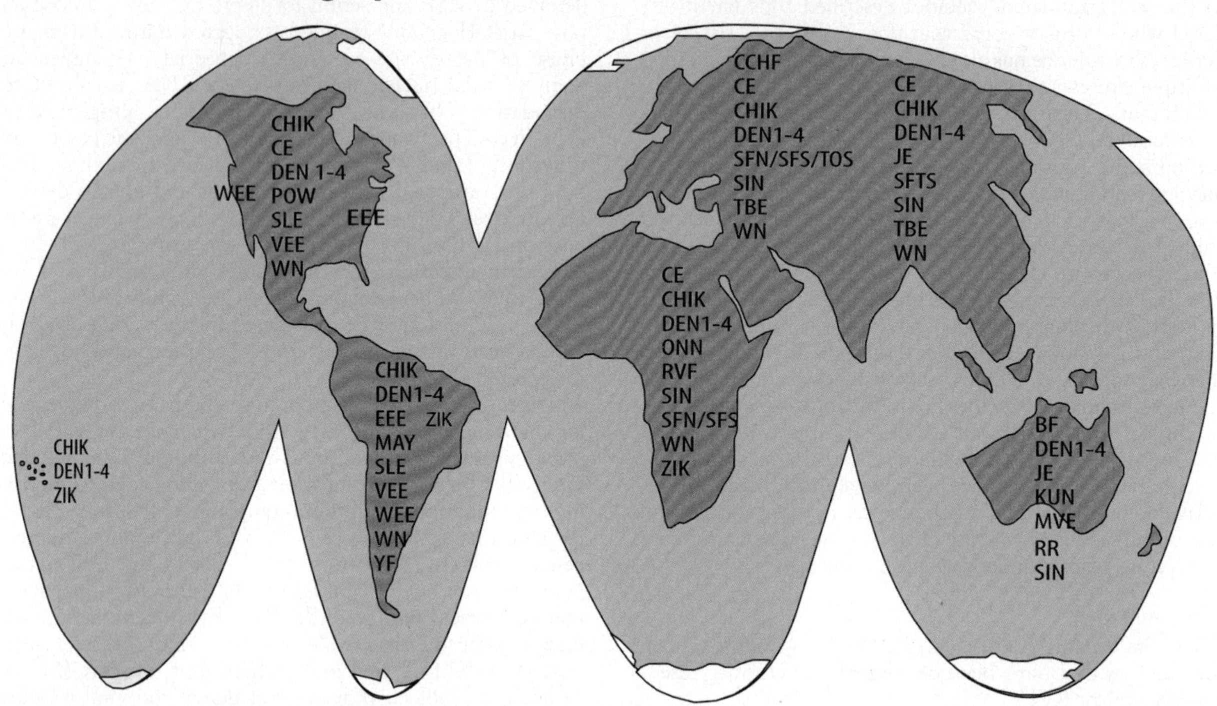

FIGURE 1 Geographic distribution of arboviral diseases. The choice of tests to be performed should be relevant to the location of the patient at the time of infection, with consideration for location of residence and travel history. Refer to Tables 1 to 3 for clinical syndromes. Virus abbreviations: BF, Barmah Forest; CCHF, Crimean-Congo hemorrhagic fever; CE, California encephalitis serogroup (in North America: La Crosse, Snowshoe hare, Jamestown Canyon; in Europe: Inkoo, Tahyna; in Middle East and Africa: Tahyna; in Asia: Snowshoe hare); CHIK, Chikungunya; DEN, Dengue; EEE, Eastern equine encephalitis; JE, Japanese encephalitis; KUN, Kunjin; MAY, Mayaro; MVE, Murray Valley encephalitis; ONN, O'Nyong Nyong; POW, Powassan; RR, Ross River; RVF, Rift Valley fever; SIN, Sindbis; SFN, Sandfly fever Naples; SFS, Sandfly fever Sicilian; SFTS, Severe fever with thrombocytopenia syndrome; SLE, St. Louis encephalitis; TBE, Tick-borne encephalitis; TOS, Toscana; VEE, Venezuelan equine encephalitis; WEE, Western equine encephalitis; WN, West Nile; YF, Yellow fever; ZIK, Zika. (Modified from a figure kindly provided by Robert Lanciotti, Center for Disease Control, Fort Collins, CO.)

detection of arboviruses (150–153), and the versatility of the assay lends itself to combining multiple assays (146). The latter uses broadly group-reactive monoclonal antibody conjugates together with virus-specific antigens to detect antibodies to different arboviruses within a genus in a single procedure.

Commercial ELISA assays for the detection of IgG and IGM antibodies against WNV have been available for a number of years. For example, the PANBIO WNV IgM assay and the Focus Technologies WNV IgM and IgG assays have U.S. Food and Drug Administration (FDA) approval for diagnostic purposes. These assays are comparable to IFA but require much less time and labor and are more suitable for testing large numbers of samples. An evaluation of the Focus MAC-ELISA showed that it had clinical sensitivity of 100%, clinical specificity of 97.1%, and an overall false-positive rate of 2.5% and therefore demonstrated acceptable performance (154). Microplate-based MAC-ELISAs have also been adapted to lateral flow tests in dipstick or cassette format. The modifications allow the assays to be used rapidly in the field (155, 156).

IgG antibodies are usually detectable 10 days after onset of symptoms and can last a lifetime. Arboviral IgM is more specific than IgG for arboviral antigens (157), and there is significant cross-reactivity between antisera and related heterologous antigens in the IgG ELISA. IgG ELISAs are to be used in tandem with MAC-ELISA to produce a clearer antibody profile for each specimen. For example, a positive IgG and a negative IgM in a sample collected after 10 days may indicate a previous infection at some point in time. A convalescent sample should be collected within 2 to 4 weeks after onset of illness for confirmatory testing.

The performance of the commercial assays should be continuously evaluated to detect changes in assay performance as the test population evolves (158). The European Network for the Diagnostics of Imported Viral Diseases collaborative Laboratory Response Network performed a quality assurance study for the serologic diagnosis of WNV infection in 2011 (159). The study found that the overall analytical sensitivity and specificity of in-house and commercial diagnostic tests for IgM detection were 50% and 95% and those for IgG were 86% and 69%, respectively (159). These results indicate that acute infections may be missed by serological tests and that flavivirus cross-reactivity remains a challenge.

IgM and IgG IF assays are slightly less sensitive than ELISA and MAC-ELISA. However, they remain a cost-effective and sensitive alternative for the serologic diagnosis of WNV (154). A specific and sensitive IFA test has been developed for YFV, making it a useful tool for rapid diagnosis of YFV during outbreaks and for epidemiological studies and serosurveillance after vaccination (160). In general

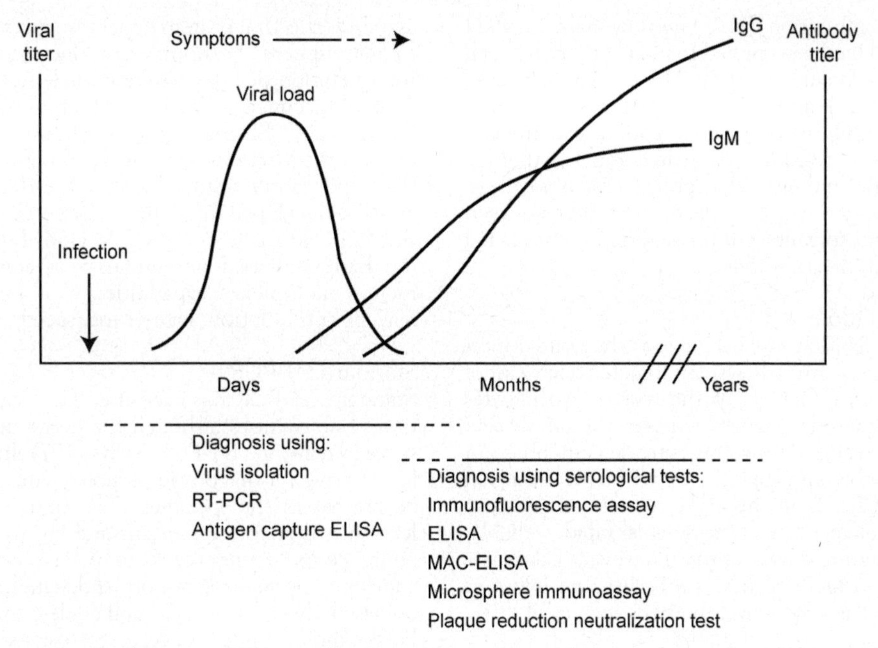

FIGURE 2 Viremia and antibody kinetics following arbovirus infection. As a guide for virologic and serologic tests, the solid lines below the graph represent the more common results, and broken lines represent reported ranges. The markers of DENV infection have been well studied and have proven useful for diagnostic methods (225). Viremia is generally short and spans the period of acute illness (0 to 6 days after onset). IgM appears as viremia declines, peaks approximately 14 days after onset of disease, and may persist up to 3 months. The structural protein, NS1 is expressed during the first 9 days of illness. IgG appears at the end of the first week of illness, slowly increases and may be detectable over the lifetime of the individual. In secondary infections, high levels of IgG are detectable even in the acute phase of illness, whereas IgM levels are significantly lower.

though, IFA tests are not definitive because of the serologic cross-reactivity between related arboviruses, especially flaviviruses.

For DENV detection the most commonly used method for detection is MAC-ELISA in a single specimen even though the result requires confirmation using paired sera. Multiple studies have investigated the diagnostic performance of commercial DENV IgM ELISAs including Panbio dengue virus IgM Capture ELISA (Alere, Australia), Standard Diagnostics dengue virus IgM ELISA (Standard Diagnostics, South Korea), and InBios DetectTM IgM Capture ELISA (InBios International, Seattle, WA) as well as the CDC MAC-ELISA (161, 162). Dengue IgM and IgG antibody-based rapid diagnostic tests have been used for rapid point-of-care diagnostics for the past two decades (163).

Commercial ELISA tests are also available for CHIKV. Abcam (Cambridge, MA) provides an Anti-Chikungunya virus IgM Human ELISA kit and Focus Diagnostics (Cypress, CA) performs IgM and IgG IFA assays.

Antigen Detection

In recent years ELISA-based tests targeting the NS-1 protein of DENV have become commercially available. The NS-1 antigen of DENV has been detected from day 1 to day 9 post onset of symptoms and is therefore a suitable target for detecting acute infection (164). The comparison of several commercially available assays showed the sensitivity of the assays to range from 51% to 81% and the specificity to range from 89% to 97% (165). The commercial test that detected NS1/IgM/IgG (SD BIOLINE^{TM} Dengue Duo, Standard Diagnostic Inc., Korea) simultaneously performed better than the other tests, although a negative result should not be used to rule out dengue for any of the NS-1 tests (165).

Epidemiologic Surveillance and Research Methods

The CDC and many of the state public health laboratories as well as research laboratories perform more specialized diagnostic testing that could be more expensive and require complex technical expertise or equipment. These laboratories often perform surveillance and/or act as reference laboratories and have different capabilities from commercial/hospital diagnostic laboratories. Reference laboratories often develop in-house assays during outbreaks and for surveillance and preparedness purposes.

Virus Isolation

Virus isolation is a direct detection method and requires the presence of replicating virus in patient specimens, such as serum, blood, CSF, or biopsy of the brain or brain tissue in a fatal case. Because of biosafety and biosecurity issues, many arboviruses such as EEEV have to be handled in biosafety level 3 (BSL3) laboratories and are considered federal select agents. Therefore, very few diagnostic laboratories have the facilities and capabilities to perform virus isolation. In addition, cells are expensive to purchase and maintain, and the process is time consuming. Virus can often be isolated from blood during the first 3 to 4 days of febrile illness when the viremia titers peak. Virus isolation is generally performed on Vero cells since the cell line is easy to culture, and most arboviruses can be amplified on these cells. As an example, alphaviruses show cytopathic effects on Vero cells within 24 hours of inoculation. Virus-positive samples are identified further by indirect IF assay and/or RT-PCR (166).

Hemagglutination Inhibition

Since the discovery of viral hemagglutination in 1941 by Hurst and the development of a hemagglutination and hemagglutination inhibition (HAI) test by Salk in 1944 (167), the method has been applied to many other virus/cell systems including arboviruses (168). The HAI assay is used to measure the level of antibodies in a patient's serum that will prevent the agglutination of susceptible erythrocytes (e.g., goose red blood cells) by inactivated antigens that are able to attach to the erythrocyte receptors on red blood cells. A fourfold or greater increase or decrease in titer between the acute phase and convalescent phase serum is considered to be diagnostic of infection (54).

Complement Fixation

The complement fixation test is based on the competition between two antigen–antibody systems for a fixed amount of complement leading ultimately to the lysis of erythrocytes (169). The antibodies detected by the test are mainly IgG antibodies that develop during the convalescent phase of illness. Therefore, this is an appropriate test for measuring an increase in antibodies from the acute to the convalescent phase. The complement fixation test is rapid, reliable, broadly reactive, and relatively cheap. However, it does not distinguish between IgG and IgM antibodies and is not as sensitive or specific as some newer methods such as ELISAs.

Plaque Reduction Neutralization Test

The PRNT is the most specific serological test and is therefore used for confirmation after a positive result is obtained in a less specific assay such as an ELISA (170). In the PRNT, serial dilutions of the patient's serum are mixed with viral antigen of known titer (171, 172). If the patient's serum contains neutralizing antibody, it will bind to the viral antigen and form an antigen–antibody complex. The mixture is used to inoculate a monolayer of cells. If the viral antigen is neutralized by antibody it will be unable to infect the cell monolayer, and plaques will not be observed. The turnaround time for plaque formation is usually 3 to 10 days for flaviviruses and 1 to 2 days for alphaviruses. A positive PRNT requires 50% to 90% reduction in titer, that is, the greatest dilution of serum that inhibits 50% to 90% of the plaques that would have formed had the serum not been added. A fourfold or greater rise in neutralizing antibody titer between acute and convalescent sera (collected 2 to 3 weeks apart) is sufficient to establish a confirmed case of arboviral infection. The disadvantage of performing PRNTs is that they are labor intensive, require training, have a long turnaround time, and require live virus—this is a problem for performing testing for viruses such as EEEV because the virus has to be handled in a BSL3 containment facility with select agent security. Some of the latter issues can be overcome by using attenuated chimeric viruses with equivalent antigenic makeup in a relatively benign backbone such as that of SINV (173) or the vaccine strain of YFV.

Microsphere Immunoassays

Microsphere immunoassays (MIAs) for detection of antibodies to arboviruses have been developed in the past decade and utilize the Luminex (Austin, TX) Multianalyte Profiling (LabMAP) technology. The method is based on color coding of microsphere beads to generate different bead sets. Each bead set can be coated with a specific probe that will recognize and detect a particular target. Laser technology is then used to detect the identity of each bead, and hence the probe linked to the bead and also the fluorescence of the reporter dye captured during the assay. It is theoretically feasible to perform a multiplex MIA in one tube that tests for any group of arboviruses found in a particular region, as long as the viral protein targets are specific and conjugated to microspheres with different fluorescent signals (174). Interpretation of MIA results is made in the context of the dates of specimen collection and illness onset, and the MIA and/or MAC ELISA results are from convalescent specimens. If the MIA result is positive, the specimen is submitted for confirmatory testing by PRNT. MIAs have been successfully developed for a number of viruses, including WNV, SLEV, POWV, DENV, and JEV (174–176). Advantages of the Luminex-based immunoassay as compared to ELISA include multiplexed capabilities, small sample volume, less reagent consumption, ease of use, speed, and sensitivity.

Standard RT-PCR

Standard RT-PCR has been used for decades for the detection of arboviruses although the assays may not be as sensitive as real-time RT-PCR assays (177). In order for PCR to be effective for diagnostic purposes, viral nucleic acid must be present in the specimen. The timing of specimen collection is therefore important and has to be during the viremic period (generally 2 to 6 days after infection). In addition, specimen transport and handling have to occur under conditions in which viral nucleic acid is not degraded. Furthermore, primers have to be selected in conserved regions of the genome because strains with mismatches in the primer binding sites will not be amplified efficiently, if at all.

Standard RT-PCR assays for the specific detection of individual arboviruses have previously been reported and should be performed as documented (177–180). Alternatively, group-specific RT-PCR assays for the detection of members of a specific genus followed by sequence analysis for identification of members of the group can be performed. For example, Sánchez-Seco et al. (181) and Pfeffer et al. (182) developed primers for the detection of alphaviruses and Scaramozzino et al. (183) developed flavivirus universal primer pairs in the conserved region of the NS5 gene that allow the detection of all flaviviruses.

A multiplex platform that combines PCR and mass spectrometry has been developed to differentiate a variety of agents that cause similar syndromes such as encephalitis/meningitis and hemorrhagic fevers (184). This MassTag-PCR method uses atmospheric pressure chemical ionization MS to read molecular weight reporter tags attached to PCR primers (184). The method can potentially be performed on small, portable instruments at a reasonable cost.

Real-Time RT-PCR

Real-time RT-PCR is a sensitive and specific method for detecting viral nucleic acid. The addition of a probe sequence in Taqman assays increases the specificity of real-time PCR compared to standard PCR. Real-time PCR is rapid, minimizes the possibility of contamination, and is amenable to high-throughput. In Taqman assays multiplexing is limited to the number of available fluorophores; therefore realistically an assay would only detect a maximum of three or four targets. The method is therefore efficient for detecting a small subset of viruses rather than discerning between a larger set of viruses. A multitude of assays have been developed to detect and differentiate among arboviruses and a selection of those will be discussed here. Del Amo et al. (185) developed a one-step real-time RT-PCR multiplex assay to detect WNV lineages 1 and 2 as well as Usutu virus. WNV also can be differentiated from JEV by using a one-step duplex Taqman RT-PCR assay targeting the nonstructural protein 2A gene using one primer pair and two probes for differential diagnosis (186). The use of intercalating dyes such as SYBR Green is an alternative to probe-based assays in real-time PCR. The use of intercalating dyes reduces specificity by allowing the detection of variants. For example, comparison of a Taqman and SYBR Green assay for the detection of WNV showed that the Taqman assay failed to detect single nucleotide variations in the probe region while the SYBR green assay detected all variants thus circumventing false-negative results (187). It should be noted that the disadvantage of using intercalating dyes is that they can generate false positive signals because they can also bind to nonspecific double-stranded DNA.

In 2012 the FDA approved a real-time PCR assay for the detection of DEN1 to DEN4 developed by the CDC (188). A clinically more sensitive real-time PCR assay for the detection and differentiation of DEN1-4 was reported in 2013 (189). In addition two multiplex assays were developed for the detection and differentiation of DEN1 to DEN4, WNV, and CHIKV (190). In this assay DEN1 to DEN4 were differentiated using high-resolution melting analysis performed on the same Lightcycler instrumentation following the thermal cycling for the multiplex (190). Assays have also been described that use degenerate primers in a real-time PCR assay to detect flaviviruses, followed by sequencing to identify the virus at the species level (191, 192). The disadvantage of using degenerate primers is that assay sensitivity is somewhat compromised.

Next-Generation Sequencing

Massively parallel DNA sequencing platforms, also termed next-generation sequencing technologies are being used more frequently in diagnostic laboratories for pathogen discovery and diagnostic applications (193, 194). Whole genome sequencing and phylogenetic analysis have been used to detect Highlands J virus in the brain of a Mississippi sandhill crane (195). The feasibility of using pyrosequencing to detect DENV in *Ae. aegypti* mosquito pools was investigated, and the method was found to be sufficiently sensitive to perform arbovirus surveillance (196). Similarly, massively parallel sequencing using the Personal Genome Machine (PGM, Life Technologies) has demonstrated the potential for this technology in assessing arbovirus ecology and evaluating novel control strategies (197).

In addition to whole genome sequencing, metagenomic sequencing, which involves high throughput sequencing of complex samples that includes nucleic acid from multiple organisms, has been used to characterize genomes of novel viruses (198–200). In this setting, the sequence of all the nucleic acid species of the sample are determined and compared with those in databases. Although these methods reduce time, labor, and cost associated with traditional sequencing, the costs remain prohibitive for routine diagnostics. Other challenges include the computationally intensive analysis, in some cases the lack of reference genome, sample preparation, as well as the requirement for adequate breadth and depth of coverage to detect low-level pathogens in specimens. Despite these challenges next-generation sequencing has been used in the characterization of new viruses, detection of unexpected viral pathogens in clinical specimen, monitoring of antiviral drug resistance, as well as in the investigation of diversity, evolution, and spread of viruses, and the technology is continuously being developed and improved.

Microarray Technology

Microarrays allow the simultaneous screening of hundreds and even thousands of targets at the same time. The initial development of a microarray is extremely labor intensive because a multitude of probes have to be designed and validated to ensure that they are suitable for detecting each target of interest. After each oligonucleotide is designed and synthesized, it is attached to the surface of a solid support (chip). The specimen containing the target is then hybridized to the probes. In the case of RNA detection, reverse transcription is performed to produce cDNA, and in some cases there is an initial amplification of the target to increase sensitivity. Probe–target hybridization is detected and quantified by detection of fluorophore-labeled target. Microarrays containing probes for various arboviruses among other agents have been developed in recent years (201–204). In general the use of this technology for diagnostic purposes is limited because the sensitivity requires improvement, the probes tiled on the arrays have to match the sequence of the pathogen, and the technology remains costly.

DNA microarrays have also been adapted to use for pathogen discovery. Wang et al. (205) designed a microarray containing the most highly conserved 70-mer sequences from every fully sequenced reference genome in GenBank. By physically recovering viral sequences hybridized to individual array elements and sequencing them, it is possible to detect new members of known virus families. More recently Berthet et al. (206) designed a resequencing DNA microarray (RMA) for the diagnosis of a variety of important pathogens including arboviruses.

Microarrays can be engineered to be used in the field. As such they need to be portable, compact, easy to use, able to withstand harsh conditions, and cost effective. Grubaugh et al. (207) reported the use of a portable microarray platform in Thailand that successfully identified flaviviruses, differentiated between mosquito genera, and detected mammalian blood meals. The relative lack of sensitivity of portable microarrays is an issue that needs to be addressed in order for these instruments to be more widely used.

Blood-Donor Screening for Arboviruses

Transfusion-associated transmission (TAT) of arboviruses has been documented, demonstrating the risk associated with infection via blood transfusions (208, 209). Nucleic acid testing (NAT) of donated blood for WNV was implemented in the United States in 2003. Initially minipools were tested, and individual specimens were subsequently tested in high prevalence regions (210). Despite that, some cases of TAT, including a recent fatality, have been reported (211). Two and a half billion people are potentially at risk of dengue infection. However, there is no FDA-approved test for blood supplies. In Puerto Rico an NS1-based test was initially used, and more recently NAT screening for DENV has been performed (212). NAT screening for CHIKV as well as photochemical pathogen inactivation has also been reported during the 2006–2007 epidemic in La Reunion (213). Other options include suspending whole blood donations during outbreaks, postdonation product quarantine, and deferral procedures for donors. Resource limitation, especially in less developed regions of the world, as well as test sensitivity are among the most important challenges for screening blood for arboviruses.

BEST PRACTICES

Many of the diseases caused by arboviruses present with similar signs and symptoms thus making a differential diagnosis challenging. As discussed earlier, it is important to obtain a pertinent travel history of the patient and determine which arboviruses may be relevant in the case. The appropriate diagnostic specimens to be collected if an arbovirus infection is suspected are serum and CSF. The application of PCR-based techniques in the early stages of disease is effective for the immediate identification of the etiologic agent because the viremia during this stage will facilitate viral genome detection. Initially, group-specific primers should be used to perform either conventional PCR or real-time PCR in order to detect a specific family of viruses and narrow down the diagnosis. Unfortunately group-specific primers are often degenerate and therefore less sensitive. Once a specific family has been identified, specific primers should be used to detect the etiologic agent at the species level. This can be done by conventional or real-time PCR or by sequencing the PCR fragment obtained following the group-specific PCR. Concurrently with molecular analysis, serology should be performed to detect and monitor the antibodies produced by the patient. MIA and MAC-ELISA are currently the most sensitive and valuable methods for diagnostic purposes. Even so, PRNT should be performed on acute and convalescent specimens for confirmation. The use of serologic tests is a problem in detecting infection in immunocompromised individuals because sensitivity of these tests may be compromised by immunosuppression. For example, in a fatal case of WNV viral RNA was detected in a chronically immunosuppressed transplant recipient by NAAT several weeks after the onset of symptoms, but IgM and IgG ELISA tests were negative in CSF and serum for 3 weeks post onset (214). Therefore, molecular testing in conjunction with serology is important in immunosuppressed patients with a clinical suspicion of arboviral encephalitis.

Rapid point-of-care diagnostic tests are helpful in the field and health care centers especially in low income countries where funding for more complex testing may not be available. For example, the use of the OnSite Chikungunya IgM Combo Rapid Test Kit (CTK Biotech, San Diego, CA) allowed prompt response at a tertiary facility that would otherwise not have diagnostic testing available during the CHIK outbreak in India in 2010 (215). In addition, rapid assays based on the identification of specific IgM are commercially available for the diagnosis of DEN infection in the early phase. These assays are rapid, easy to use, and do not require skilled personnel. However, their sensitivity and specificity are not ideal (216, 217).

Laboratories that have cutting edge technology and are not constrained by costs may use next-generation sequencing methods to diagnose a known virus or even for discovering new pathogens. Despite the many methodologies described in this chapter, it should be noted that many cases of arbovirus infection may remain undetected or be misdiagnosed. For example, despite performing extensive diagnostic testing, the etiology of diseases such as encephalitis and meningitis in many patients worldwide remains unknown (218–220). Furthermore, in regions where malaria and typhoid are endemic, many cases of arbovirus infections/coinfections are misdiagnosed because the clinical symptoms are very similar (221–223).

It is clear that many challenges remain for clinical laboratories and clinicians where arbovirus infections are concerned. Arboviruses are a significant public health concern and the disease burden caused by these viruses throughout the world is tremendous (224). Many viruses in the group are emerging and re-emerging and causing extensive epidemics. Therefore, more research is needed in the development of antiviral therapies and vaccine development. In addition, diagnostic tests need to be improved and surveillance systems need to be established in areas lacking surveillance systems and strengthened in other areas. Enhanced surveillance and monitoring will assist in improving outbreak response and in developing disease control strategies to decrease the burden of arboviral disease worldwide.

REFERENCES

1. **Westaway EG, Brinton MA, Gaidamovich SYa, Horzinek MC, Igarashi A, Kääriäinen L, Lvov DK, Porterfield JS, Russell PK, Trent DW.** 1985. Togaviridae. *Intervirology* **24:**125–139.
2. **Calisher CH, Karabatsos N, Lazuick JS, Monath TP, Wolff KL.** 1988. Reevaluation of the western equine encephalitis antigenic complex of alphaviruses (family Togaviridae) as determined by neutralization tests. *Am J Trop Med Hyg* **38:**447–452.
3. **Strauss JH, Strauss EG.** 1994. The alphaviruses: gene expression, replication, and evolution. *Microbiol Rev* **58:**491–562. (Erratum, 58:806, 1994.)
4. **Weaver SC, Hagenbaugh A, Bellew LA, Netesov SV, Volchkov VE, Chang G-JJ, Clarke DK, Gousset L, Scott TW, Trent DW, Holland JJ.** 1993. A comparison of the nucleotide sequences of eastern and western equine encephalomyelitis viruses with those of other alphaviruses and related RNA viruses. *Virology* **197:**375–390.
5. **Calisher CH, Karabatsos N.** 1988. Arbovirus serogroups definition and geographic distribution, p 19–57. *In* Monath TP (ed), *The Arboviruses: Epidemiology and Ecology*, vol 1. CRC Press, Boca Raton, FL.
6. **Casals J.** 1964. Antigenic variants of Eastern equine encephalitis virus. *J Exp Med* **119:**547–565.
7. **Arrigo NC, Adams AP, Weaver SC.** 2010. Evolutionary patterns of eastern equine encephalitis virus in North versus South America suggest ecological differences and taxonomic revision. *J Virol* **84:**1014–1025.
8. **Weaver SC, Hagenbaugh A, Bellew LA, Gousset L, Mallampalli V, Holland JJ, Scott TW.** 1994. Evolution of alphaviruses in the eastern equine encephalomyelitis complex. *J Virol* **68:**158–169.
9. **von Sprockhoff H, Ising E.** 1971. On the presence of viruses of the American equine encephalomyelitis in Central Europe. Review. *Arch Gesamte Virusforsch* **34:**371–380.
10. **Kramer LD, Fallah HM.** 1999. Genetic variation among isolates of western equine encephalomyelitis virus from California. *Am J Trop Med Hyg* **60:**708–713.
11. **Powers AM, Brault AC, Shirako Y, Strauss EG, Kang W, Strauss JH, Weaver SC.** 2001. Evolutionary relationships and systematics of the alphaviruses. *J Virol* **75:**10118–10131.
12. **Weaver SC, Ferro C, Barrera R, Boshell J, Navarro JC.** 2004. Venezuelan equine encephalitis. *Annu Rev Entomol* **49:**141–174.
13. **Weaver SC, Winegar R, Manger ID, Forrester NL.** 2012. Alphaviruses: population genetics and determinants of emergence. *Antiviral Res* **94:**242–257.
14. **Weaver SC, Bellew LA, Rico-Hesse R.** 1992. Phylogenetic analysis of alphaviruses in the Venezuelan equine encephalitis complex and identification of the source of epizootic viruses. *Virology* **191:**282–290.
15. **Walton TE, Alvarez O Jr, Buckwalter RM, Johnson KM.** 1973. Experimental infection of horses with enzootic and epizootic strains of Venezuelan equine encephalomyelitis virus. *J Infect Dis* **128:**271–282.
16. **Aguilar PV, Estrada-Franco JG, Navarro-Lopez R, Ferro C, Haddow AD, Weaver SC.** 2011. Endemic Venezuelan equine encephalitis in the Americas: hidden under the dengue umbrella. *Future Virol* **6:**721–740.
17. **Morrison TE.** 2014. Reemergence of chikungunya virus. *J Virol* **88:**11644–11647.

18. **Brinton MA.** 2002. The molecular biology of West Nile Virus: a new invader of the western hemisphere. *Annu Rev Microbiol* **56:**371–402.
19. **Lindenbach BD, Thiel HJ, Rice CM.** 2007. Flaviviridae: the viruses and their replication, p 1101–1152. *In* Knipe DM, Howley PM (ed), *Fields Virology,* 5th ed. Lippincott-Raven Publishers, Philadelphia.
20. **Mandl CW.** 2005. Steps of the tick-borne encephalitis virus replication cycle that affect neuropathogenesis. *Virus Res* **111:** 161–174.
21. **Mukhopadhyay S, Kuhn RJ, Rossmann MG.** 2005. A structural perspective of the flavivirus life cycle. *Nat Rev Microbiol* **3:**13–22.
22. **Gaunt MW, Sall AA, de Lamballerie X, Falconar AK, Dzhivanian TI, Gould EA.** 2001. Phylogenetic relationships of flaviviruses correlate with their epidemiology, disease association and biogeography. *J Gen Virol* **82:**1867–1876.
23. **Porterfield JS.** 1980. Antigenic characteristics and classification of Togaviridae, p 13–46. *In* Schlesinger RW (ed), *The Togaviruses: Biology, Structure, Replication.* Academic Press, New York.
24. **Calisher CH, Karabatsos N, Dalrymple JM, Shope RE, Porterfield JS, Westaway EG, Brandt WE.** 1989. Antigenic relationships between flaviviruses as determined by cross-neutralization tests with polyclonal antisera. *J Gen Virol* **70:** 37–43.
25. **Bondre VP, Jadi RS, Mishra AC, Yergolkar PN, Arankalle VA.** 2007. West Nile virus isolates from India: evidence for a distinct genetic lineage. *J Gen Virol* **88:**875–884.
26. **Go YY, Balasuriya UB, Lee CK.** 2014. Zoonotic encephalitides caused by arboviruses: transmission and epidemiology of alphaviruses and flaviviruses. *Clin Exp Vaccine Res* **3:**58–77.
27. **Suthar MS, Diamond MS, Gale M Jr.** 2013. West Nile virus infection and immunity. *Nat Rev Microbiol* **11:**115–128.
28. **Bakonyi T, Hubálek Z, Rudolf I, Nowotny N.** 2005. Novel flavivirus or new lineage of West Nile virus, central Europe. *Emerg Infect Dis* **11:**225–231.
29. **Grard G, Moureau G, Charrel RN, Lemasson JJ, Gonzalez JP, Gallian P, Gritsun TS, Holmes EC, Gould EA, de Lamballerie X.** 2007. Genetic characterization of tick-borne flaviviruses: new insights into evolution, pathogenetic determinants and taxonomy. *Virology* **361:**80–92.
30. **Thiel H-J, Collett MS, Gould EA, Heinz FX, Meyers G, Purcell RH, Rice CM, Houghton M.** 2005. Flaviviridae, p 981–998. *In* Fauquet CM, Mayo MA, Maniloff J, Desselberger U, Ball LA (ed), *Virus Taxonomy: Eighth Report of the International Committee on Taxonomy of Viruses.* Virology Division, International Union of Microbiological Societies, San Diego, CA.
31. **Demina TV, Dzhioev YP, Verkhozina MM, Kozlova IV, Tkachev SE, Plyusnin A, Doroshchenko EK, Lisak OV, Zlobin VI.** 2010. Genotyping and characterization of the geographical distribution of tick-borne encephalitis virus variants with a set of molecular probes. *J Med Virol* **82:**965–976.
32. **Gritsun TS, Nuttall PA, Gould EA.** 2003. Tick-borne flaviviruses. *Adv Virus Res* **61:**317–371.
33. **Ebel GD, Spielman A, Telford SR III.** 2001. Phylogeny of North American Powassan virus. *J Gen Virol* **82:**1657–1665.
34. **Kuno G, Artsob H, Karabatsos N, Tsuchiya KR, Chang GJ.** 2001. Genomic sequencing of deer tick virus and phylogeny of powassan-related viruses of North America. *Am J Trop Med Hyg* **65:**671–676.
35. **Telford SR III, Armstrong PM, Katavolos P, Foppa I, Garcia AS, Wilson ML, Spielman A.** 1997. A new tick-borne encephalitis-like virus infecting New England deer ticks, Ixodes dammini. *Emerg Infect Dis* **3:**165–170.
36. **Tavakoli NP, Wang H, Dupuis M, Hull R, Ebel GD, Gilmore EJ, Faust PL.** 2009. Fatal case of deer tick virus encephalitis. *N Engl J Med* **360:**2099–2107.
37. **Pesko KN, Torres-Perez F, Hjelle BL, Ebel GD.** 2010. Molecular epidemiology of Powassan virus in North America. *J Gen Virol* **91:**2698–2705.
38. **Leonova GN, Sorokina MN, Krugliak SP.** 1991. The clinico-epidemiological characteristics of Powassan encephalitis in the southern Soviet Far East. *Zh Mikrobiol Epidemiol Immunobiol* **(3):**35–39. (In Russian.)
39. **Jin H, Elliott RM.** 1991. Expression of functional Bunyamwera virus L protein by recombinant vaccinia viruses. *J Virol* **65:**4182–4189.
40. **Jin H, Elliott RM.** 1992. Mutagenesis of the L protein encoded by Bunyamwera virus and production of monospecific antibodies. *J Gen Virol* **73:**2235–2244.
41. **Gentsch JR, Bishop DL.** 1979. M viral RNA segment of bunyaviruses codes for two glycoproteins, G1 and G2. *J Virol* **30:**767–770.
42. **Fuller F, Bishop DH.** 1982. Identification of virus-coded nonstructural polypeptides in bunyavirus-infected cells. *J Virol* **41:**643–648.
43. **Elliott RM.** 1985. Identification of nonstructural proteins encoded by viruses of the Bunyamwera serogroup (family Bunyaviridae). *Virology* **143:**119–126.
44. **Elliott RM.** 1990. Molecular biology of the Bunyaviridae. *J Gen Virol* **71:**501–522.
45. **Schmaljohn AL, McClain D.** 1996. Alphaviruses (Togaviridae) and flaviviruses (Flaviviridae), *In* Baron S (ed), *Medical Microbiology,* 4th ed. The University of Texas Medical Branch at Galveston, Galveston, TX.
46. **Karabatsos N.** 1985. *International Catalogue of Arboviruses, 1985, Including Certain Other Viruses of Vertebrates,* 3rd ed. American Society of Tropical Medicine and Hygeine, San Antonio, TX.
47. **Young NA, Johnson KM.** 1969. Antigenic variants of Venezuelan equine encephalitis virus: their geographic distribution and epidemiologic significance. *Am J Epidemiol* **89:** 286–307.
48. **Weaver SC, Pfeffer M, Marriott K, Kang W, Kinney RM.** 1999. Genetic evidence for the origins of Venezuelan equine encephalitis virus subtype IAB outbreaks. *Am J Trop Med Hyg* **60:**441–448.
49. **Rezza G, Nicoletti L, Angelini R, Romi R, Finarelli AC, Panning M, Cordioli P, Fortuna C, Boros S, Magurano F, Silvi G, Angelini P, Dottori M, Ciufolini MG, Majori GC, Cassone A, CHIKV study group.** 2007. Infection with chikungunya virus in Italy: an outbreak in a temperate region. *Lancet* **370:**1840–1846.
50. **Charrel RN, de Lamballerie X.** 2008. Chikungunya virus in north-eastern Italy: a consequence of seasonal synchronicity. *Euro Surveill* **13:**13.
51. **Lundström JO, Pfeffer M.** 2010. Phylogeographic structure and evolutionary history of Sindbis virus. *Vector Borne Zoonotic Dis* **10:**889–907.
52. **Jupp PG.** 2001. The ecology of West Nile virus in South Africa and the occurrence of outbreaks in humans. *Ann N Y Acad Sci* **951:**143–152.
53. **Gubler DJ, Clark GG.** 1995. Dengue/dengue hemorrhagic fever: the emergence of a global health problem. *Emerg Infect Dis* **1:**55–57.
54. **(WHO) WHO.** 2012. *Global Strategy for Dengue Prevention and Control, 2012-2020.* WHO Press, Geneva.
55. **Bhatt S, Gething PW, Brady OJ, Messina JP, Farlow AW, Moyes CL, Drake JM, Brownstein JS, Hoen AG, Sankoh O, Myers MF, George DB, Jaenisch T, Wint GR, Simmons CP, Scott TW, Farrar JJ, Hay SI.** 2013. The global distribution and burden of dengue. *Nature* **496:**504–507.
56. **Gubler DJ.** 2011. Dengue, urbanization and globalization: the unholy trinity of the 21(st) century. *Trop Med Health* **39** (Suppl):S3–S11.
57. **Hoffmann AA, Montgomery BL, Popovici J, Iturbe-Ormaetxe I, Johnson PH, Muzzi F, Greenfield M, Durkan M, Leong YS, Dong Y, Cook H, Axford J, Callahan AG, Kenny N, Omodei C, McGraw EA, Ryan PA, Ritchie SA, Turelli M, O'Neill SL.** 2011. Successful establishment of Wolbachia in Aedes populations to suppress dengue transmission. *Nature* **476:**454–457.

58. **Thomas SJ.** 2014. Developing a dengue vaccine: progress and future challenges. *Ann N Y Acad Sci* **1323:**140–159.
59. **Smithburn KC, Hughes TP, Burke AW, Paul JH.** 1940. A neurotropic virus isolated from the blood of a native of Uganda. *Am J Trop Med Hyg* **20:**471–473.
60. **Hayes CG.** 2001. West Nile virus: Uganda, 1937, to New York City, 1999. *Ann N Y Acad Sci* **951:**25–37.
61. **Kilpatrick AM, Kramer LD, Campbell SR, Alleyne EO, Dobson AP, Daszak P.** 2005. West Nile virus risk assessment and the bridge vector paradigm. *Emerg Infect Dis* **11:** 425–429.
62. **Marfin AA, Petersen LR, Eidson M, Miller J, Hadler J, Farello C, Werner B, Campbell GL, Layton M, Smith P, Bresnitz E, Cartter M, Scaletta J, Obiri G, Bunning M, Craven RC, Roehrig JT, Julian KG, Hinten SR, Gubler DJ, ArboNET Cooperative Surveillance Group.** 2001. Widespread West Nile virus activity, eastern United States, 2000. *Emerg Infect Dis* **7:**730–735.
63. **Ciccozzi M, Peletto S, Cella E, Giovanetti M, Lai A, Gabanelli E, Acutis PL, Modesto P, Rezza G, Platonov AE, Lo Presti A, Zehender G.** 2013. Epidemiological history and phylogeography of West Nile virus lineage 2. *Infect Genet Evol* **17:**46–50.
64. **Bakonyi T, Ferenczi E, Erdélyi K, Kutasi O, Csörgő T, Seidel B, Weissenböck H, Brugger K, Bán E, Nowotny N.** 2013. Explosive spread of a neuroinvasive lineage 2 West Nile virus in Central Europe, 2008/2009. *Vet Microbiol* **165:**61–70.
65. **Kolodziejek J, Marinov M, Kiss BJ, Alexe V, Nowotny N.** 2014. The complete sequence of a West Nile virus lineage 2 strain detected in a *Hyalomma marginatum marginatum* tick collected from a song thrush (*Turdus philomelos*) in eastern Romania in 2013 revealed closest genetic relationship to strain Volgograd 2007. *PLoS One* **9:**e109905.
66. **May FJ, Davis CT, Tesh RB, Barrett AD.** 2011. Phylogeography of West Nile virus: from the cradle of evolution in Africa to Eurasia, Australia, and the Americas. *J Virol* **85:** 2964–2974.
67. **Rappole JH, Derrickson SR, Hubálek Z.** 2000. Migratory birds and spread of West Nile virus in the Western Hemisphere. *Emerg Infect Dis* **6:**319–328.
68. **Rappole JH, Hubálek Z.** 2003. Migratory birds and West Nile virus. *J Appl Microbiol* **94**(Suppl)**:**47S–58S.
69. **Dumpis U, Crook D, Oksi J.** 1999. Tick-borne encephalitis. *Clin Infect Dis* **28:**882–890.
70. **Kuno G.** 2011. Zika virus, p 313–319. *In* Liu D (ed), *Molecular detection of human viral pathogens*. CRC Press, Boca Raton, FL.
71. **Daubney R, Hudson JR, Gamham PC.** 1933. Enzootic hepatitis of Rift Valley Fever: an undescribed virus in sheep, cattle and man from East Africa. *East Afr Med J* **10:**2–19.
72. **Pepin M, Bouloy M, Bird BH, Kemp A, Paweska J.** 2010. Rift Valley fever virus (Bunyaviridae: Phlebovirus): an update on pathogenesis, molecular epidemiology, vectors, diagnostics and prevention. *Vet Res* **41:**61.
73. **Chumakov MP, Butenko AM, Shalunova NV, Mart'ianova LI, Smirnova SE, Bashkirtsev IN, Zavodova TI, Rubin SG, Tkachenko EA, Karmysheva VI, Reĭngol'd VN, Popov GV, Savinov AP.** 1968. New data on the viral agent of Crimean hemorrhagic fever. *Vopr Virusol* **13:**377. (In Russian.)
74. **Woodall JP, Williams MC, Simpson DI.** 1967. Congo virus: a hitherto undescribed virus occurring in Africa. II. Identification studies. *East Afr Med J* **44:**93–98.
75. **Simpson DI, Knight EM, Courtois G, Williams MC, Weinbren MP, Kibukamusoke JW.** 1967. Congo virus: a hitherto undescribed virus occurring in Africa. I. Human isolations–clinical notes. *East Afr Med J* **44:**86–92.
76. **Lam TT, Liu W, Bowden TA, Cui N, Zhuang L, Liu K, Zhang YY, Cao WC, Pybus OG.** 2013. Evolutionary and molecular analysis of the emergent severe fever with thrombocytopenia syndrome virus. *Epidemics* **5:**1–10.
77. **Steele KE, Reed DS, Glass PJ, Hart MK, Ludwig GV, Pratt WD, Parker MD.** 2007. Alphavirus encephalitides, p 241–270. *In* Dembek ZF (ed), *Medical Aspects of Biological Warfare*. Borden Institute (U.S. Army Walter Reed), Fort Detrick, MD.
78. **Queyriaux B, Simon F, Grandadam M, Michel R, Tolou H, Boutin JP.** 2008. Clinical burden of chikungunya virus infection. *Lancet Infect Dis* **8:**2–3.
79. **Gérardin P, Fianu A, Malvy D, Mussard C, Boussaïd K, Rollot O, Michault A, Gaüzere BA, Bréart G, Favier F.** 2012. Perceived morbidity and community burden of chikungunya in La Reunion. *Med Trop (Mars)* **72:**76–82. (In French.)
80. **Carey DE.** 1971. Chikungunya and dengue: a case of mistaken identity? *J Hist Med Allied Sci* **26:**243–262.
81. **Krishnamoorthy K, Harichandrakumar KT, Krishna Kumari A, Das LK.** 2009. Burden of chikungunya in India: estimates of disability adjusted life years (DALY) lost in 2006 epidemic. *J Vector Borne Dis* **46:**26–35.
82. **Sane J, Guedes S, Kurkela S, Lyytikäinen O, Vapalahti O.** 2010. Epidemiological analysis of mosquito-borne Pogosta disease in Finland, 2009. *Euro Surveill* **15:**15.
83. **Kurkela S, Manni T, Vaheri A, Vapalahti O.** 2004. Causative agent of Pogosta disease isolated from blood and skin lesions. *Emerg Infect Dis* **10:**889–894.
84. **Skogh M, Espmark A.** 1982. Ockelbo disease: epidemic arthritis-exanthema syndrome in Sweden caused by Sindbis-virus like agent. *Lancet* **1:**795–796.
85. **Lvov DK, Skvortsova TM, Berezina LK, Gromashevsky VL, Yakovlev BI, Gushchin BV, Aristova VA, Sidorova GA, Gushchina EL, Klimenko SM, et al.** 1984. Isolation of Karelian fever agent from *Aedes communis* mosquitoes. *Lancet* **2:**399–400.
86. **Kurkela S, Manni T, Myllynen J, Vaheri A, Vapalahti O.** 2005. Clinical and laboratory manifestations of Sindbis virus infection: prospective study, Finland, 2002-2003. *J Infect Dis* **191:**1820–1829.
87. **Turunen M, Kuusisto P, Uggeldahl PE, Toivanen A.** 1998. Pogosta disease: clinical observations during an outbreak in the province of North Karelia, Finland. *Br J Rheumatol* **37:** 1177–1180.
88. **Kurkela S, Helve T, Vaheri A, Vapalahti O.** 2008. Arthritis and arthralgia three years after Sindbis virus infection: clinical follow-up of a cohort of 49 patients. *Scand J Infect Dis* **40:**167–173.
89. **Coimbra TL, Santos CL, Suzuki A, Petrella SM, Bisordi I, Nagamori AH, Marti AT, Santos RN, Fialho DM, Lavigne S, Buzzar MR, Rocco IM.** 2007. Mayaro virus: imported cases of human infection in São Paulo State, Brazil. *Rev Inst Med Trop Sao Paulo* **49:**221–224.
90. **Condon RJ, Rouse IL.** 1995. Acute symptoms and sequelae of Ross River virus infection in South-Western Australia: a follow-up study. *Clin Diagn Virol* **3:**273–284.
91. **Flexman JP, Smith DW, Mackenzie JS, Fraser JRE, Bass SPH, Hueston L, Lindsay MD, Cunningham AL.** 1998. A comparison of the diseases caused by Ross River virus and Barmah Forest virus. *Med J Aust* **169:**159–163.
92. **Kalayanarooj S.** 2011. Clinical manifestations and management of dengue/DHF/DSS. *Trop Med Health* **39**(Suppl)**:**S83–S87.
93. **Halstead SB, O'Rourke EJ.** 1977. Antibody-enhanced dengue virus infection in primate leukocytes. *Nature* **265:**739–741.
94. **Halstead SB.** 2003. Neutralization and antibody-dependent enhancement of dengue viruses. *Adv Virus Res* **60:**421–467.
95. **Davis LE, DeBiasi R, Goade DE, Haaland KY, Harrington JA, Harnar JB, Pergam SA, King MK, DeMasters BK, Tyler KL.** 2006. West Nile virus neuroinvasive disease. *Ann Neurol* **60:**286–300.
96. **Ciota AT, Kramer LD.** 2013. Vector-virus interactions and transmission dynamics of West Nile virus. *Viruses* **5:**3021–3047.
97. **Hayes EB, Gubler DJ.** 2006. West Nile virus: epidemiology and clinical features of an emerging epidemic in the United States. *Annu Rev Med* **57:**181–194.
98. **Petersen LR, Marfin AA.** 2002. West Nile virus: a primer for the clinician. *Ann Intern Med* **137:**173–179.
99. **Solomon T, Vaughn D.** 2002. Clinical features and patho physiology of Japanese encephalitis and West Nile infections,

p 171–194. *In* Mackenzie J, Barrett A, Deubel V (ed), *Current topics in microbiology and immunology: Japanese encephalitis and West Nile infections.* Springer-Verlag, Berlin.

100. **Gyure KA.** 2009. West Nile virus infections. *J Neuropathol Exp Neurol* **68:**1053–1060.
101. **Tsai TF.** 2000. New initiatives for the control of Japanese encephalitis by vaccination: minutes of a WHO/CVI meeting, Bangkok, Thailand, 13-15 October 1998. *Vaccine* **18**(Suppl 2)**:**1–25.
102. **Olson JG, Ksiazek TG, Suhandiman, Triwibowo.** 1981. Zika virus, a cause of fever in Central Java, Indonesia. *Trans R Soc Trop Med Hyg* **75:**389–393.
103. **Duffy MR, Chen TH, Hancock WT, Powers AM, Kool JL, Lanciotti RS, Pretrick M, Marfel M, Holzbauer S, Dubray C, Guillaumot L, Griggs A, Bel M, Lambert AJ, Laven J, Kosoy O, Panella A, Biggerstaff BJ, Fischer M, Hayes EB.** 2009. Zika virus outbreak on Yap Island, Federated States of Micronesia. *N Engl J Med* **360:**2536–2543.
104. **Mackenzie JS, Smith DW, Broom AK, Bucens MR.** 1993. Australian encephalitis in Western Australia, 1978-1991. *Med J Aust* **158:**591–595.
105. **Russell RC, Dwyer DE.** 2000. Arboviruses associated with human disease in Australia. *Microbes Infect* **2:**1693–1704.
106. **Ergönül O.** 2006. Crimean-Congo haemorrhagic fever. *Lancet Infect Dis* **6:**203–214.
107. **Meegan J, Le Guenno B, Ksiazek T, Jouan A, Knauert F, Digoutte JP, Peters CJ.** 1989. Rapid diagnosis of Rift Valley fever: a comparison of methods for the direct detection of viral antigen in human sera. *Res Virol* **140:**59–65.
108. **Faye O, Diallo M, Diop D, Bezeid OE, Bâ H, Niang M, Dia I, Mohamed SA, Ndiaye K, Diallo D, Ly PO, Diallo B, Nabeth P, Simon F, Lô B, Diop OM.** 2007. Rift Valley fever outbreak with East-Central African virus lineage in Mauritania, 2003. *Emerg Infect Dis* **13:**1016–1023.
109. **Madani TA, Al-Mazrou YY, Al-Jeffri MH, Mishkhas AA, Al-Rabeah AM, Turkistani AM, Al-Sayed MO, Abodahish AA, Khan AS, Ksiazek TG, Shobokshi O.** 2003. Rift Valley fever epidemic in Saudi Arabia: epidemiological, clinical, and laboratory characteristics. *Clin Infect Dis* **37:**1084–1092.
110. **Bird BH, Ksiazek TG, Nichol ST, Maclachlan NJ.** 2009. Rift Valley fever virus. *J Am Vet Med Assoc* **234:**883–893.
111. **Soldan SS, González-Scarano F.** 2005. Emerging infectious diseases: the Bunyaviridae. *J Neurovirol* **11:**412–423.
112. **Steiner I, Budka H, Chaudhuri A, Koskiniemi M, Sainio K, Salonen O, Kennedy PG.** 2010. Viral meningoencephalitis: a review of diagnostic methods and guidelines for management. *Eur J Neurol* **17:**999–e57.
113. **Julander JG, Bowen RA, Rao JR, Day C, Shafer K, Smee DF, Morrey JD, Chu CK.** 2008. Treatment of Venezuelan equine encephalitis virus infection with (-)-carbodine. *Antiviral Res* **80:**309–315.
114. **O'Brien L.** 2007. Inhibition of multiple strains of Venezuelan equine encephalitis virus by a pool of four short interfering RNAs. *Antiviral Res* **75:**20–29.
115. **Parashar D, Paingankar MS, Kumar S, Gokhale MD, Sudeep AB, Shinde SB, Arankalle VA.** 2013. Administration of E2 and NS1 siRNAs inhibit chikungunya virus replication in vitro and protects mice infected with the virus. *PLoS Negl Trop Dis* **7:**e2405.
116. **Richardson-Burns SM, Tyler KL.** 2005. Minocycline delays disease onset and mortality in reovirus encephalitis. *Exp Neurol* **192:**331–339.
117. **Mishra MK, Basu A.** 2008. Minocycline neuroprotects, reduces microglial activation, inhibits caspase 3 induction, and viral replication following Japanese encephalitis. *J Neurochem* **105:**1582–1595.
118. **Dutta K, Ghosh D, Basu A.** 2009. Curcumin protects neuronal cells from Japanese encephalitis virus-mediated cell death and also inhibits infective viral particle formation by dysregulation of ubiquitin-proteasome system. *J Neuroimmune Pharmacol* **4:**328–337.
119. **Thompson BS, Moesker B, Smit JM, Wilschut J, Diamond MS, Fremont DH.** 2009. A therapeutic antibody against West Nile virus neutralizes infection by blocking fusion within endosomes. *PLoS Pathog* **5:**e1000453.
120. **Singh N, Levi ME, AST Infectious Diseases Community of Practice.** 2013. Arenavirus and West Nile virus in solid organ transplantation. *Am J Transplant* **13**(Suppl 4)**:**361–371.
121. **Edelman R, Tacket CO, Wasserman SS, Bodison SA, Perry JG, Mangiafico JA.** 2000. Phase II safety and immunogenicity study of live chikungunya virus vaccine TSI-GSD-218. *Am J Trop Med Hyg* **62:**681–685.
122. **Kim DY, Atasheva S, Foy NJ, Wang E, Frolova EI, Weaver S, Frolov I.** 2011. Design of chimeric alphaviruses with a programmed, attenuated, cell type-restricted phenotype. *J Virol* **85:**4363–4376.
123. **Tiwari M, Parida M, Santhosh SR, Khan M, Dash PK, Rao PV.** 2009. Assessment of immunogenic potential of Vero adapted formalin inactivated vaccine derived from novel ECSA genotype of Chikungunya virus. *Vaccine* **27:**2513–2522.
124. **Thiboutot MM, Kannan S, Kawalekar OU, Shedlock DJ, Khan AS, Sarangan G, Srikanth P, Weiner DB, Muthumani K.** 2010. Chikungunya: a potentially emerging epidemic? *PLoS Negl Trop Dis* **4:**e623.
125. **Wang D, Suhrbier A, Penn-Nicholson A, Woraratanadharm J, Gardner J, Luo M, Le TT, Anraku I, Sakalian M, Einfeld D, Dong JY.** 2011. A complex adenovirus vaccine against chikungunya virus provides complete protection against viraemia and arthritis. *Vaccine* **29:**2803–2809.
126. **Rojanasuphot S, Charoensuk O, Kitprayura D, Likityingvara C, Limpisthien S, Boonyindee S, Jivariyavej V, Ugchusak K.** 1989. A field trial of Japanese encephalitis vaccine produced in Thailand. *Southeast Asian J Trop Med Public Health* **20:**653–654.
127. **Fischer M, Lindsey N, Staples JE, Hills S, Centers for Disease Control and Prevention (CDC).** 2010. Japanese encephalitis vaccines: recommendations of the Advisory Committee on Immunization Practices (ACIP). *MMWR Recomm Rep* **59**(RR-1)**:**1–27.
128. **Duggan ST, Plosker GL.** 2009. Japanese encephalitis vaccine (inactivated, adsorbed) [IXIARO]. *Drugs* **69:**115–122.
129. **Kramer LD, Styer LM, Ebel GD.** 2008. A global perspective on the epidemiology of West Nile virus. *Annu Rev Entomol* **53:**61–81.
130. **Monath TP.** 1988. Yellow fever, p 139–231. *In* Monath TP (ed), *The Arboviruses: Epidemiology and Ecology.* CRC Press, Boca Raton, FL.
131. **Villar L, Dayan GH, Arredondo-Garcia JL, Rivera DM, Cunha R, Deseda C, Reynales H, Costa MS, Morales-Ramirez JO, Carrasquilla G, Rey LC, Dietze R, Luz K, Rivas E, Montoya MC, Supelano MC, Zambrano B, Langevin E, Boaz M, Tornieporth N, Saville M, Noriega F.** 2015. Efficacy of a tetravalent dengue vaccine in children in Latin America. *N Engl J Med* **372:**113–123.
132. **Guirakhoo F, Pugachev K, Arroyo J, Miller C, Zhang ZX, Weltzin R, Georgakopoulos K, Catalan J, Ocran S, Draper K, Monath TP.** 2002. Viremia and immunogenicity in nonhuman primates of a tetravalent yellow fever-dengue chimeric vaccine: genetic reconstructions, dose adjustment, and antibody responses against wild-type dengue virus isolates. *Virology* **298:**146–159.
133. **Guirakhoo F, Pugachev K, Zhang Z, Myers G, Levenbook I, Draper K, Lang J, Ocran S, Mitchell F, Parsons M, Brown N, Brandler S, Fournier C, Barrere B, Rizvi F, Travassos A, Nichols R, Trent D, Monath T.** 2004. Safety and efficacy of chimeric yellow Fever-dengue virus tetravalent vaccine formulations in nonhuman primates. *J Virol* **78:**4761–4775.
134. **Sinha G.** 2014. Sanofi's dengue vaccine first to complete phase 3. *Nat Biotechnol* **32:**605–606.
135. **Edelman R, Wasserman SS, Bodison SA, Putnak RJ, Eckels KH, Tang D, Kanesa-Thasan N, Vaughn DW, Innis BL, Sun W.** 2003. Phase I trial of 16 formulations of a tetravalent live-attenuated dengue vaccine. *Am J Trop Med Hyg* **69**(Suppl)**:** 48–60.
136. **Thomas SJ, Eckels KH, Carletti I, De La Barrera R, Dessy F, Fernandez S, Putnak R, Toussaint JF, Sun W, Bauer K,**

Gibbons RV, Innis BL. 2013. A phase II, randomized, safety and immunogenicity study of a re-derived, live-attenuated dengue virus vaccine in healthy adults. *Am J Trop Med Hyg* **88:**73–88.

137. **Osorio JE, Brewoo JN, Silengo SJ, Arguello J, Moldovan IR, Tary-Lehmann M, Powell TD, Livengood JA, Kinney RM, Huang CY, Stinchcomb DT.** 2011. Efficacy of a tetravalent chimeric dengue vaccine (DENVax) in Cynomolgus macaques. *Am J Trop Med Hyg* **84:**978–987.
138. **Durbin AP, Kirkpatrick BD, Pierce KK, Schmidt AC, Whitehead SS.** 2011. Development and clinical evaluation of multiple investigational monovalent DENV vaccines to identify components for inclusion in a live attenuated tetravalent DENV vaccine. *Vaccine* **29:**7242–7250.
139. **Putnak R, Barvir DA, Burrous JM, Dubois DR, D'Andrea VM, Hoke CH, Sadoff JC, Eckels KH.** 1996. Development of a purified, inactivated, dengue-2 virus vaccine prototype in Vero cells: immunogenicity and protection in mice and rhesus monkeys. *J Infect Dis* **174:**1176–1184.
140. **Coller BA, Clements DE, Bett AJ, Sagar SL, Ter Meulen JH.** 2011. The development of recombinant subunit envelope-based vaccines to protect against dengue virus induced disease. *Vaccine* **29:**7267–7275.
141. **Clements DE, Coller BA, Lieberman MM, Ogata S, Wang G, Harada KE, Putnak JR, Ivy JM, McDonell M, Bignami GS, Peters ID, Leung J, Weeks-Levy C, Nakano ET, Humphreys T.** 2010. Development of a recombinant tetravalent dengue virus vaccine: immunogenicity and efficacy studies in mice and monkeys. *Vaccine* **28:**2705–2715.
142. **Arora U, Tyagi P, Swaminathan S, Khanna N.** 2013. Virus-like particles displaying envelope domain III of dengue virus type 2 induce virus-specific antibody response in mice. *Vaccine* **31:**873–878.
143. **Khanam S, Pilankatta R, Khanna N, Swaminathan S.** 2009. An adenovirus type 5 (AdV5) vector encoding an envelope domain III-based tetravalent antigen elicits immune responses against all four dengue viruses in the presence of prior AdV5 immunity. *Vaccine* **27:**6011–6021.
144. **Lu H, Xu XF, Gao N, Fan DY, Wang J, An J.** 2013. Preliminary evaluation of DNA vaccine candidates encoding dengue-2 prM/E and NS1: their immunity and protective efficacy in mice. *Mol Immunol* **54:**109–114.
145. **McArthur MA, Sztein MB, Edelman R.** 2013. Dengue vaccines: recent developments, ongoing challenges and current candidates. *Expert Rev Vaccines* **12:**933–953.
146. **Martin DA, Muth DA, Brown T, Johnson AJ, Karabatsos N, Roehrig JT.** 2000. Standardization of immunoglobulin M capture enzyme-linked immunosorbent assays for routine diagnosis of arboviral infections. *J Clin Microbiol* **38:**1823–1826.
147. **Gea-Banacloche J, Johnson RT, Bagic A, Butman JA, Murray PR, Agrawal AG.** 2004. West Nile virus: pathogenesis and therapeutic options. *Ann Intern Med* **140:**545–553.
148. **Barrett ADT, Weaver SC.** 2012. Arboviruses: alphaviruses, flaviviruses and bunyaviruses, p 520–536. *In* Greenwood D, Barer M, Slack R, Irving W (ed), *Medical Microbiology*, 18th ed. Elsevier, London.
149. **Kuby J.** 1997. Antibodies, p 399. *In* Kuby J (ed), *Immunology*, 3rd ed. WH Freeman and Co, New York.
150. **Artsob H, Spence LP, Th'ng C.** 1984. Enzyme-linked immunosorbent assay typing of California serogroup viruses isolated in Canada. *J Clin Microbiol* **20:**276–280.
151. **Frazier CL, Shope RE.** 1979. Detection of antibodies to alphaviruses by enzyme-linked immunosorbent assay. *J Clin Microbiol* **10:**583–585.
152. **Niklasson B, Peters CJ, Grandien M, Wood O.** 1984. Detection of human immunoglobulins G and M antibodies to Rift Valley fever virus by enzyme-linked immunosorbent assay. *J Clin Microbiol* **19:**225–229.
153. **Karabatsos N, Lewis AL, Calisher CH, Hunt AR, Roehrig JT.** 1988. Identification of Highlands J virus from a Florida horse. *Am J Trop Med Hyg* **39:**603–606.
154. **Malan AK, Stipanovich PJ, Martins TB, Hill HR, Litwin CM.** 2003. Detection of IgG and IgM to West Nile virus. Development of an immunofluorescence assay. *Am J Clin Pathol* **119:**508–515.
155. **Prince HE, Tobler LH, Lapé-Nixon M, Foster GA, Stramer SL, Busch MP.** 2005. Development and persistence of West Nile virus-specific immunoglobulin M (IgM), IgA, and IgG in viremic blood donors. *J Clin Microbiol* **43:**4316–4320.
156. **Sambol AR, Hinrichs SH, Hogrefe WR, Schweitzer BK.** 2007. Performance of a commercial immunoglobulin M antibody capture assay using analyte-specific reagents to screen for interfering factors during a West Nile virus epidemic season in Nebraska. *Clin Vaccine Immunol* **14:**87–89.
157. **Westaway EG, Della-Porta AJ, Reedman BM.** 1974. Specificity of IgM and IgG antibodies after challenge with antigenically related togaviruses. *J Immunol* **112:**656–663.
158. **Malan AK, Martins TB, Hill HR, Litwin CM.** 2004. Evaluations of commercial West Nile virus immunoglobulin G (IgG) and IgM enzyme immunoassays show the value of continuous validation. *J Clin Microbiol* **42:**727–733.
159. **Sanchini A, Donoso-Mantke O, Papa A, Sambri V, Teichmann A, Niedrig M.** 2013. Second international diagnostic accuracy study for the serological detection of West Nile virus infection. *PLoS Negl Trop Dis* **7:**e2184.
160. **Niedrig M, Kürsteiner O, Herzog C, Sonnenberg K.** 2008. Evaluation of an indirect immunofluorescence assay for detection of immunoglobulin M (IgM) and IgG antibodies against yellow fever virus. *Clin Vaccine Immunol* **15:**177–181.
161. **Blacksell SD.** 2012. Commercial dengue rapid diagnostic tests for point-of-care application: recent evaluations and future needs? *J Biomed Biotechnol* **2012:**151967.
162. **Welch RJ, Chang GJ, Litwin CM.** 2014. Comparison of a commercial dengue IgM capture ELISA with dengue antigen focus reduction microneutralization test and the Centers for Disease Control dengue IgM capture-ELISA. *J Virol Methods* **195:**247–249.
163. **Blacksell SD, Jarman RG, Gibbons RV, Tanganuchitcharnchai A, Mammen MP Jr, Nisalak A, Kalayanarooj S, Bailey MS, Premaratna R, de Silva HJ, Day NP, Lalloo DG.** 2012. Comparison of seven commercial antigen and antibody enzyme-linked immunosorbent assays for detection of acute dengue infection. *Clin Vaccine Immunol* **19:**804–810.
164. **Alcon S, Talarmin A, Debruyne M, Falconar A, Deubel V, Flamand M.** 2002. Enzyme-linked immunosorbent assay specific to Dengue virus type 1 nonstructural protein NS1 reveals circulation of the antigen in the blood during the acute phase of disease in patients experiencing primary or secondary infections. *J Clin Microbiol* **40:**376–381.
165. **Osorio L, Ramirez M, Bonelo A, Villar LA, Parra B.** 2010. Comparison of the diagnostic accuracy of commercial NS1-based diagnostic tests for early dengue infection. *Virol J* **7:**361.
166. **Kauffman EB, Jones SA, Dupuis AP II, Ngo KA, Bernard KA, Kramer LD.** 2003. Virus detection protocols for West Nile virus in vertebrate and mosquito specimens. *J Clin Microbiol* **41:**3661–3667.
167. **Salk JE.** 1944. A simplified procedure for titrating hemagglutinating capacity of influenza virus and the corresponding antibody. *J Immunol* **49:**87–98.
168. **Clarke DH, Casals J.** 1958. Techniques for hemagglutination and hemagglutination-inhibition with arthropod-borne viruses. *Am J Trop Med Hyg* **7:**561–573.
169. **Atmar RL, Englund JA.** 1997. Laboratory methods for the diagnosis of viral diseases, p 59–88. *In* Evans AS, Kaslow RA (ed), *Viral Infections of Humans: Epidemiology and Control*, 4th ed. Plenum, New York.
170. **Beaty BJ, Calisher C, Shope RE.** 2014. Arboviruses, p 189–212. *In* Lennette EH, Lennette DA, Lennette ET (ed), *Diagnostic Procedures for viral, rickettsial and chlamydial infections*, 7th ed. American Public Health Association, Washington, DC.
171. **Beaty BJ, Calisher CH, Shope RE.** 1989. Arboviruses, p. 797–856. *In* Schmidt NJ, Emmons RW (ed), *Diagnostic procedures for viral, rickettsial and chlamydial infections*. American Public Health Association, Washington, DC.
172. **Calisher CH, Fremount HN, Vesely WL, el-Kafrawi AO, Mahmud MI.** 1986. Relevance of detection of immunoglob-

ulin M antibody response in birds used for arbovirus surveillance. *J Clin Microbiol* **24:**770–774.
173. **Johnson BW, Kosoy O, Wang E, Delorey M, Russell B, Bowen RA, Weaver SC.** 2011. Use of sindbis/eastern equine encephalitis chimeric viruses in plaque reduction neutralization tests for arboviral disease diagnostics. *Clin Vaccine Immunol* **18:**1486–1491.
174. **Basile AJ, Horiuchi K, Panella AJ, Laven J, Kosoy O, Lanciotti RS, Venkateswaran N, Biggerstaff BJ.** 2013. Multiplex microsphere immunoassays for the detection of IgM and IgG to arboviral diseases. *PLoS One* **8:**e75670.
175. **Wong SJ, Boyle RH, Demarest VL, Woodmansee AN, Kramer LD, Li H, Drebot M, Koski RA, Fikrig E, Martin DA, Shi PY.** 2003. Immunoassay targeting nonstructural protein 5 to differentiate West Nile virus infection from dengue and St. Louis encephalitis virus infections and from flavivirus vaccination. *J Clin Microbiol* **41:**4217–4223.
176. **Johnson AJ, Noga AJ, Kosoy O, Lanciotti RS, Johnson AA, Biggerstaff BJ.** 2005. Duplex microsphere-based immunoassay for detection of anti-West Nile virus and anti-St. Louis encephalitis virus immunoglobulin m antibodies. *Clin Diagn Lab Immunol* **12:**566–574.
177. **Lambert AJ, Martin DA, Lanciotti RS.** 2003. Detection of North American eastern and western equine encephalitis viruses by nucleic acid amplification assays. *J Clin Microbiol* **41:**379–385.
178. **Linssen B, Kinney RM, Aguilar P, Russell KL, Watts DM, Kaaden OR, Pfeffer M.** 2000. Development of reverse transcription-PCR assays specific for detection of equine encephalitis viruses. *J Clin Microbiol* **38:**1527–1535.
179. **Lee JH, Tennessen K, Lilley BG, Unnasch TR.** 2002. Simultaneous detection of three mosquito-borne encephalitis viruses (eastern equine, La Crosse, and St. Louis) with a single-tube multiplex reverse transcriptase polymerase chain reaction assay. *J Am Mosq Control Assoc* **18:**26–31.
180. **O'Guinn ML, Lee JS, Kondig JP, Fernandez R, Carbajal F.** 2004. Field detection of eastern equine encephalitis virus in the Amazon Basin region of Peru using reverse transcription-polymerase chain reaction adapted for field identification of arthropod-borne pathogens. *Am J Trop Med Hyg* **70:**164–171.
181. **Sánchez-Seco MP, Rosario D, Quiroz E, Guzmán G, Tenorio A.** 2001. A generic nested-RT-PCR followed by sequencing for detection and identification of members of the alphavirus genus. *J Virol Methods* **95:**153–161.
182. **Pfeffer M, Proebster B, Kinney RM, Kaaden OR.** 1997. Genus-specific detection of alphaviruses by a semi-nested reverse transcription-polymerase chain reaction. *Am J Trop Med Hyg* **57:**709–718.
183. **Scaramozzino N, Crance JM, Jouan A, DeBriel DA, Stoll F, Garin D.** 2001. Comparison of flavivirus universal primer pairs and development of a rapid, highly sensitive heminested reverse transcription-PCR assay for detection of flaviviruses targeted to a conserved region of the NS5 gene sequences. *J Clin Microbiol* **39:**1922–1927.
184. **Palacios G, Briese T, Kapoor V, Jabado O, Liu Z, Venter M, Zhai J, Renwick N, Grolla A, Geisbert TW, Drosten C, Towner J, Ju J, Paweska J, Nichol ST, Swanepoel R, Feldmann H, Jahrling PB, Lipkin WI.** 2006. MassTag polymerase chain reaction for differential diagnosis of viral hemorrhagic fever. *Emerg Infect Dis* **12:**692–695.
185. **Del Amo J, Sotelo E, Fernández-Pinero J, Gallardo C, Llorente F, Agüero M, Jiménez-Clavero MA.** 2013. A novel quantitative multiplex real-time RT-PCR for the simultaneous detection and differentiation of West Nile virus lineages 1 and 2, and of Usutu virus. *J Virol Methods* **189:**321–327.
186. **Barros SC, Ramos F, Zé-Zé L, Alves MJ, Fagulha T, Duarte M, Henriques M, Luís T, Fevereiro M.** 2013. Simultaneous detection of West Nile and Japanese encephalitis virus RNA by duplex TaqMan RT-PCR. *J Virol Methods* **193:**554–557.
187. **Papin JF, Vahrson W, Dittmer DP.** 2004. SYBR green-based real-time quantitative PCR assay for detection of West Nile Virus circumvents false-negative results due to strain variability. *J Clin Microbiol* **42:**1511–1518.
188. **U.S. Food and Drug Administration.** 2012. 510(k) substantial equivalence determination decision summary, K113336, p 22. U.S. Food and Drug Administration, Washington, DC.
189. **Waggoner JJ, Abeynayake J, Sahoo MK, Gresh L, Tellez Y, Gonzalez K, Ballesteros G, Guo FP, Balmaseda A, Karunaratne K, Harris E, Pinsky BA.** 2013. Comparison of the FDA-approved CDC DENV-1-4 real-time reverse transcription-PCR with a laboratory-developed assay for dengue virus detection and serotyping. *J Clin Microbiol* **51:**3418–3420.
190. **Naze F, Le Roux K, Schuffenecker I, Zeller H, Staikowsky F, Grivard P, Michault A, Laurent P.** 2009. Simultaneous detection and quantitation of Chikungunya, dengue and West Nile viruses by multiplex RT-PCR assays and dengue virus typing using high resolution melting. *J Virol Methods* **162:**1–7.
191. **Johnson N, Wakeley PR, Mansfield KL, McCracken F, Haxton B, Phipps LP, Fooks AR.** 2010. Assessment of a novel real-time pan-flavivirus RT-polymerase chain reaction. *Vector Borne Zoonotic Dis* **10:**665–671.
192. **Moureau G, Temmam S, Gonzalez JP, Charrel RN, Grard G, de Lamballerie X.** 2007. A real-time RT-PCR method for the universal detection and identification of flaviviruses. *Vector Borne Zoonotic Dis* **7:**467–477.
193. **Barzon L, Lavezzo E, Costanzi G, Franchin E, Toppo S, Palù G.** 2013. Next-generation sequencing technologies in diagnostic virology. *J Clin Virol* **58:**346–350.
194. **Capobianchi MR, Giombini E, Rozera G.** 2013. Next-generation sequencing technology in clinical virology. *Clin Microbiol Infect* **19:**15–22.
195. **Ip HS, Wiley MR, Long R, Palacios G, Shearn-Bochsler V, Whitehouse CA.** 2014. Identification and characterization of Highlands J virus from a Mississippi sandhill crane using unbiased next-generation sequencing. *J Virol Methods* **206:** 42–45.
196. **Bishop-Lilly KA, Turell MJ, Willner KM, Butani A, Nolan NM, Lentz SM, Akmal A, Mateczun A, Brahmbhatt TN, Sozhamannan S, Whitehouse CA, Read TD.** 2010. Arbovirus detection in insect vectors by rapid, high-throughput pyrosequencing. *PLoS Negl Trop Dis* **4:**e878.
197. **Hall-Mendelin S, Allcock R, Kresoje N, van den Hurk AF, Warrilow D.** 2013. Detection of arboviruses and other micro-organisms in experimentally infected mosquitoes using massively parallel sequencing. *PLoS One* **8:**e58026.
198. **Towner JS, Sealy TK, Khristova ML, Albariño CG, Conlan S, Reeder SA, Quan PL, Lipkin WI, Downing R, Tappero JW, Okware S, Lutwama J, Bakamutumaho B, Kayiwa J, Comer JA, Rollin PE, Ksiazek TG, Nichol ST.** 2008. Newly discovered Ebola virus associated with hemorrhagic fever outbreak in Uganda. *PLoS Pathog* **4:**e1000212.
199. **Masembe C, Michuki G, Onyango M, Rumberia C, Norling M, Bishop RP, Djikeng A, Kemp SJ, Orth A, Skilton RA, Ståhl K, Fischer A.** 2012. Viral metagenomics demonstrates that domestic pigs are a potential reservoir for Ndumu virus. *Virol J* **9:**218.
200. **Coffey LL, Page BL, Greninger AL, Herring BL, Russell RC, Doggett SL, Haniotis J, Wang C, Deng X, Delwart EL.** 2014. Enhanced arbovirus surveillance with deep sequencing: identification of novel rhabdoviruses and bunyaviruses in Australian mosquitoes. *Virology* **448:**146–158.
201. **Nordström H, Falk KI, Lindegren G, Mouzavi-Jazi M, Waldén A, Elgh F, Nilsson P, Lundkvist A.** 2005. DNA microarray technique for detection and identification of seven flaviviruses pathogenic for man. *J Med Virol* **77:**528–540.
202. **Chou CC, Lee TT, Chen CH, Hsiao HY, Lin YL, Ho MS, Yang PC, Peck K.** 2006. Design of microarray probes for virus identification and detection of emerging viruses at the genus level. *BMC Bioinformatics* **7:**232.
203. **Palacios G, Quan PL, Jabado OJ, Conlan S, Hirschberg DL, Liu Y, Zhai J, Renwick N, Hui J, Hegyi H, Grolla A, Strong JE, Towner JS, Geisbert TW, Jahrling PB, Büchen-Osmond C, Ellerbrok H, Sanchez-Seco MP, Lussier Y, Formenty P, Nichol MS, Feldmann H, Briese T, Lipkin WI.** 2007. Panmicrobial oligonucleotide array for diagnosis of infectious diseases. *Emerg Infect Dis* **13:**73–81.

204. **Xiao-Ping K, Yong-Qiang L, Qing-Ge S, Hong L, Qing-Yu Z, Yin-Hui Y.** 2009. Development of a consensus microarray method for identification of some highly pathogenic viruses. *J Med Virol* **81:**1945–1950.
205. **Wang D, Urisman A, Liu YT, Springer M, Ksiazek TG, Erdman DD, Mardis ER, Hickenbotham M, Magrini V, Eldred J, Latreille JP, Wilson RK, Ganem D, DeRisi JL.** 2003. Viral discovery and sequence recovery using DNA microarrays. *PLoS Biol* **1:**e2.
206. **Berthet N, Paulous S, Coffey LL, Frenkiel MP, Moltini I, Tran C, Matheus S, Ottone C, Ungeheuer MN, Renaudat C, Caro V, Dussart P, Gessain A, Desprès P.** 2013. Resequencing microarray method for molecular diagnosis of human arboviral diseases. *J Clin Virol* **56:**238–243.
207. **Grubaugh ND, Petz LN, Melanson VR, McMenamy SS, Turell MJ, Long LS, Pisarcik SE, Kengluecha A, Jaichapor B, O'Guinn ML, Lee JS.** 2013. Evaluation of a field-portable DNA microarray platform and nucleic acid amplification strategies for the detection of arboviruses, arthropods, and bloodmeals. *Am J Trop Med Hyg* **88:**245–253.
208. **Petersen LR, Busch MP.** 2010. Transfusion-transmitted arboviruses. *Vox Sang* **98:**495–503.
209. **Gan VC, Leo YS.** 2014. Current epidemiology and clinical practice in arboviral infections–implications on blood supply in South-East Asia. *ISBT Sci Ser* **9:**262–267.
210. **Busch MP, Caglioti S, Robertson EF, McAuley JD, Tobler LH, Kamel H, Linnen JM, Shyamala V, Tomasulo P, Kleinman SH.** 2005. Screening the blood supply for West Nile virus RNA by nucleic acid amplification testing. *N Engl J Med* **353:**460–467.
211. **Centers for Disease Control and Prevention (CDC).** 2013. Fatal West Nile virus infection after probable transfusion-associated transmission–Colorado, 2012. *MMWR Morb Mortal Wkly Rep* **62:**622–624.
212. **Lanteri MC, Busch MP.** 2012. Dengue in the context of "safe blood" and global epidemiology: to screen or not to screen? *Transfusion* **52:**1634–1639.
213. **Rasonglès P, Angelini-Tibert MF, Simon P, Currie C, Isola H, Kientz D, Slaedts M, Jacquet M, Sundin D, Lin L, Corash L, Cazenave JP.** 2009. Transfusion of platelet components prepared with photochemical pathogen inactivation treatment during a Chikungunya virus epidemic in Ile de La Réunion. *Transfusion* **49:**1083–1091.
214. **Koepsell SA, Freifeld AG, Sambol AR, McComb RD, Kazmi SA.** 2010. Seronegative naturally acquired West Nile virus encephalitis in a renal and pancreas transplant recipient. *Transpl Infect Dis* **12:**459–464.
215. **Arya SC, Agarwal N.** 2011. Rapid point-of-care diagnosis of chikungunya virus infection. *Asian Pac J Trop Dis* **1:**230–231.
216. **Blacksell SD, Bell D, Kelley J, Mammen MP Jr, Gibbons RV, Jarman RG, Vaughn DW, Jenjaroen K, Nisalak A, Thongpaseuth S, Vongsouvath M, Davong V, Phouminh P, Phetsouvanh R, Day NP, Newton PN.** 2007. Prospective study to determine accuracy of rapid serological assays for diagnosis of acute dengue virus infection in Laos. *Clin Vaccine Immunol* **14:**1458–1464.
217. **Dussart P, Petit L, Labeau B, Bremand L, Leduc A, Moua D, Matheus S, Baril L.** 2008. Evaluation of two new commercial tests for the diagnosis of acute dengue virus infection using NS1 antigen detection in human serum. *PLoS Negl Trop Dis* **2:**e280.
218. **Le VT, Phan TQ, Do QH, Nguyen BH, Lam QB, Bach V, Truong H, Tran TH, Nguyen V, Tran T, Vo M, Tran VT, Schultsz C, Farrar J, van Doorn HR, de Jong MD.** 2010. Virol etiology of encephalitis in children in Southern Vietnam: results of a one year prospective descriptive study. *PLoS Negl Trop Dis* **4:**e854.
219. **Casas I, Pozo F, Trallero G, Echevarría JM, Tenorio A.** 1999. Viral diagnosis of neurological infection by RT multiplex PCR: a search for entero- and herpesviruses in a prospective study. *J Med Virol* **57:**145–151.
220. **Dupuis M, Hull R, Wang H, Nattanmai S, Glasheen B, Fusco H, Dzigua L, Markey K, Tavakoli NP.** 2011. Molecular detection of viral causes of encephalitis and meningitis in New York State. *J Med Virol* **83:**2172–2181.
221. **Sang RC, Dunster LM.** 2001. The growing threat of arbovirus transmission and outbreaks in Kenya: a review. *East Afr Med J* **78:**655–661.
222. **Shears P.** 2000. Emerging and reemerging infections in Africa: the need for improved laboratory services and disease surveillance. *Microbes Infect* **2:**489–495.
223. **Baba M, Logue CH, Oderinde B, Abdulmaleek H, Williams J, Lewis J, Laws TR, Hewson R, Marcello A, D' Agaro P.** 2013. Evidence of arbovirus co-infection in suspected febrile malaria and typhoid patients in Nigeria. *J Infect Dev Ctries* **7:**51–59.
224. **LaBeaud AD, Bashir F, King CH.** 2011. Measuring the burden of arboviral diseases: the spectrum of morbidity and mortality from four prevalent infections. *Popul Health Metr* **9:**1.
225. **Peeling RW, Artsob H, Pelegrino JL, Buchy P, Cardosa MJ, Devi S, Enria DA, Farrar J, Gubler DJ, Guzman MG, Halstead SB, Hunsperger E, Kliks S, Margolis HS, Nathanson CM, Nguyen VC, Rizzo N, Vázquez S, Yoksan S.** 2010. Evaluation of diagnostic tests: dengue. *Nat Rev Microbiol* **8** (Suppl):S30–S38.

Animal-Borne Viruses

GREGORY J. BERRY, MICHAEL J. LOEFFELHOLZ, AND GUSTAVO PALACIOS

36

Zoonotic infections are infections of animals that can be transmitted to humans. There are more than 400 viruses with a zoonotic origin that can cause mild or severe clinical pathology in humans. This chapter will attempt to cover a handful of these viruses that have been especially relevant as human pathogens: arenaviruses, Ebola virus, Nipah virus, and Hantavirus. Arenaviruses consist of a number of species and collectively have a worldwide distribution. They cause mild to severe illness in humans. Filoviruses have a sylvatic epidemiology in Africa and can spill over into humans, where they usually cause severe illness with a high mortality rate. Filoviruses are efficiently spread person-to-person, as highlighted in the 2014 to 2015 urban epidemic in West Africa. Nipah virus has caused a relatively limited number of human cases in Southeast Asia. Nipah virus infection with central nervous system involvement is associated with high mortality. Hantaviruses have a worldwide distribution and cause renal and pulmonary syndromes in humans.

VIRAL CLASSIFICATION AND BIOLOGY

Arenaviruses

Arenaviruses are spherical or pleomorphic virions with a mean diameter of 110 to 130 nanometers. Their genome consists of two single-strand RNA molecules, with ambisense properties, named the L (large) and S (small) segments. The L segment encodes the Z matrix protein (which appears to be an important regulator of innate signaling pathways) and the RNA polymerase (which encodes the viral replicative capacity). The S segment encodes the glycoprotein precursor (GPC), which is encoded in the viral-sense strand and is crucial for cellular tropism since it contains the receptor binding site. The nucleocapsid (N) is encoded in the viral-complementary strand and is the main component of the ribonucleoprotein (RNP) complex.

The genus *Arenavirus* includes more than 25 viral species that are classified into two phylogenetically independent groups: Old World (lymphocytic choriomeningitis virus [LCMV]/Lassa complex) and New World (Tacaribe complex). The human pathogens among the Old World group consist of LCMV (discovered in 1933) (1), Lassa virus (1969) (2), and Lujo virus (2008) (3). Recent data indicate that different lineages of Lassa virus circulate in Africa (South Eastern lineage in Nigeria, North Western lineage in Sierra Leone, Guinea, and Ivory Coast). The New World group includes Junín (discovered in 1958) (4, 5), Machupo (1963) (5), Guanarito (1989) (6), Sabia (1993) (7), and Chapare (2004) (8) viruses.

Filoviruses

Filoviruses are filamentous, enveloped viruses with a non-segmented, negative-sense, single-strand RNA genome that is 19.1 kilobases long. The genome consists of 7 genes, which encode for the viral polymerase, structural proteins (virion envelope glycoprotein (GP), nucleoprotein (NP), and the VP24 and VP40 matrix proteins) and nonstructural proteins (VP30 and VP35). Whereas the GP Marburg virus produces one product, Ebola virus gives rise to a 60- to 70-kiloDalton soluble protein (sGP) and a full-length protein (GP) (9, 10). The viruses have a distinctive filamentous morphology under the electron microscope. Several features of their molecular organization and structure linked these viruses to members of the *Paramyxoviridae* and *Rhabdoviridae* families. The viruses have a central core formed by the RNP complex (RNA molecule bound by the NP and VP30, the VP35 and the L protein), which is covered by a lipid envelope derived from the cell plasma membrane where the three remaining structural proteins reside (GP on the outside and VP24 and VP40 located on the inner side of the membrane).

The family *Filoviridae* contains seven species, which are comprised of five Ebola virus species and one Marburg virus species, as well as one *Cuevavirus* species. A recent filovirus phylogenetic assessment by Petersen and Holder showed that Marburg virus appears to contain two distinct sympatric lineages. In contrast, Ebola viruses appear to show allopatric speciation (11).

Nipah Virus

Henipaviruses are pleomorphic, enveloped viruses with a negative-strand RNA genome. The genome contains six genes encoding glycoproteins F (fusion) and G (receptor-binding), matrix protein M, nucleoprotein N, RNA-dependent RNA polymerase L, and phosphoprotein P. Glycoproteins F and G are required for cell entry and additionally induce neutralizing antibodies. Serologic assays target the antibody response to these proteins (12). The genome is larger than that of other paramyxoviruses due to the extended open reading frame for the P gene (12). Nipah

doi:10.1128/9781555819156.ch36

virus is a member of the *Paramyxoviridae* family, *Henipavirus* genus. Two strains of Nipah virus have been described: Malaysian strain and Bangladesh strain. These differ at the molecular level, with about 8% amino acid variation between the P genes (13), and may be responsible for different epidemiology and host responses to infection.

Hantavirus

Hantaviruses are negative-strand RNA viruses with a genome consisting of the 6.5 to 6.6 kilobase large (L), 3.6 to 3.7 kilobase medium (M), and 1.7 to 2.1 kilobase small (S) RNA segments. These segments encode the L protein, the glycoprotein precursor (which makes viral glycoproteins Gn and Gc via cotranslational cleavage), and the nucleocapsid protein, respectively. The S segment can also encode a functional nonstructural protein (NSs) in some hantaviruses. Hantavirus particles are pleomorphic, lipid bilayer-enveloped and approximately 100 nanometers in diameter. Viral glycoproteins Gn and Gc form 6 nanometer spikes that protrude from the viral envelope (14). Hantaviruses are members of the family *Bunyaviridae*, genus *Hantavirus*. This genus includes a growing number of isolated viruses, including those that cause hemorrhagic fever with renal syndrome (HFRS) and hantavirus pulmonary syndrome (HFS). Several recently recognized species have been identified only by molecular genetic analysis.

EPIDEMIOLOGY

Arenaviruses

The epidemiology of arenaviral disease is intrinsically associated with the geographic distribution of the infected rodents and the nature of their contact with humans. Old and New World arenaviruses are endemic on the African and South American continents, respectively. The exception is LCMV, an Old World arenavirus, which has worldwide geographic distribution.

Natural host species of Lassa virus is *Mastomys* spp (multimammate rat) with a geographic distribution range limited to West Africa. Thus, Lassa fever is endemic in Nigeria, Guinea, Liberia, and Sierra Leone and is hyperendemic in some regions within Nigeria and Sierra Leone. Human case data and field studies of rodent populations provide evidence that the range of Lassa virus includes additional West Africa countries (15). Overall, Lassa fever has no seasonal peaks of incidence, although seasonality has been described in Nigeria (16). About 5 million people are at risk for Lassa fever infection. Several hundred thousand to up to 2 million infections and 5,000 to 10,000 deaths are reported annually. Nevertheless, underreporting is an issue, since over 50% of the population within the endemic area has serologic evidence of exposure (17). There exists a domestic and peridomestic cycle of transmission. The virus is transmitted through aerosol or direct contact with rodent excreta, oral secretions and, less commonly, by secondary person-to-person contact (urine of infected patients, sexual contact; i.e., direct contact), including transmission inside health care facilities.

The natural host species of LCMV is *Mus musculus* (house mouse) or *Mus domesticus* (common mouse), with worldwide distribution. Mice shed virus in large quantities in urine, feces, and respiratory secretions. A number of seroprevalence studies have been performed in the Americas and Europe, which report a prevalence of anti-LCMV antibodies generally around 5% in humans, ranging between 1% and 9%. Nevertheless, seroprevalence rates as high as 36% have been reported from Eastern Europe (18). LCMV-related disease is under-diagnosed/reported because laboratory testing is rarely performed. Outbreaks have been associated with exposure to pet hamsters (19). Human infections are acquired through aerosol or direct contact with rodent excreta and secretions. Person-to-person transmission has not been documented.

Lujo virus was isolated during an outbreak of hemorrhagic fever in South Africa that involved secondary nosocomial transmission. Including the index patient, there were five cases of Lujo virus infection, four of which were fatal (20). To date, no additional cases have been reported. The natural host species of Lujo virus is currently unknown.

The New World arenaviruses, Junín, Machupo, Guanarito, Sabia, and Chapare viruses, are endemic in Argentina, Bolivia, Venezuela, Brazil, and Bolivia, respectively. The natural host species of Junín, Machupo, and Guanarito viruses are *Calomys musculinus* (drylands vesper mouse), *C. callosus* (large vesper mouse), and *Zygodontomys brevicauda* (common cane mouse), respectively. The animal reservoirs of Sabia and Chapare viruses remain unknown. There are an estimated 300 to 1,000 cases of Junín virus infection annually (17). Machupo virus infections are both sporadic and outbreak related. While 19 cases were reported during the 1990s, over 200 cases were reported from 2007 to 2008 (17). To date, there have been 618 reported cases of Guanarito virus infection, as well as a single human case each of Sabia and Chapare virus infection; both were fatal (17). As with Old World arenaviruses, human infections are usually the result of accidental exposure to infected rodent excretions. Person-to-person transmission via direct contact, indirect contact, and aerosols has been described for New World arenaviruses, such as Machupo virus (21)

Filoviruses

Marburg virus was discovered in 1967 in Germany and Yugoslavia when workers became infected after handling kidneys or primary tissue cultures from monkeys that had been imported from Uganda. Ebola virus was first detected almost a decade later in 1976 during two simultaneous outbreaks in the countries of Sudan and Zaire (currently the Democratic Republic of Congo) with over 500 cases reported and fatalities of 53% and 88%, respectively. These outbreaks were subsequently found to be caused by two distinct species of the genus *Ebolavirus*. Another distinct species of Ebola virus, Reston ebolavirus, was first discovered in cynomolgus monkeys that had been shipped from the Philippines and quarantined in Reston, VA, and Perkasie, PA, in 1989 and 1990, respectively (22). Reston ebolavirus was also later isolated in Siena, Italy, (23) and Alice, TX, (22) from monkeys imported from the Philippines. A fourth distinct species of Ebola virus, Tai Forest virus, emerged in the Ivory Coast in 1992 and killed chimpanzees and caused disease in a human (24). The last species of the *Ebolavirus* genus, Bundibugyo ebolavirus was also isolated during an outbreak in Bundibugyo in western Uganda (25). Finally, more recently, and using only molecular methods, a new member of the family, *Lloviuvirus*, which was proposed to have its own genus (*Cuevavirus*), was discovered during a bat die-off in Cueva del Lloviu, Spain (26).

Since 1995, outbreaks of Ebola or Marburg have occurred in several African nations, including Uganda, South Sudan, the Democratic Republic of Congo, Angola, Gabon, Guinea, Liberia and Sierra Leone. The Ebola virus outbreak that started in late 2013 in Guinea, and that spread to Liberia and

Sierra Leone, has been the most extensive outbreak ever, with WHO reporting over 27,000 cases and over 11,000 deaths. Previous outbreaks occurred in secluded areas, limiting their ability to spread widely through the population. Unique epidemiological features of this outbreak involved individuals moving across borders and introducing the virus into densely populated urban centers, at which point exposures greatly increased.

There is strong evidence to suggest that fruit bats are the reservoir of filoviruses. Following a 2007 outbreak of Marburg virus among miners working in Kitaka Cave, Uganda, it was found that healthy Egyptian fruit bats living in the cave were also infected with a Marburg virus that closely matched the virus isolates from the infected miners (27). In the case of Ebola virus, three different species of fruit bats captured within close proximity to infected gorilla and chimpanzee carcasses were shown to be either serologically- or PCR-positive for the virus (28). Serological evidence of Ebola virus infections in bats has also been shown in other studies (29, 30). Additionally, direct exposure to fruit bats has been documented as a precursor to human Ebola virus outbreaks (31).

Ebola virus is spread through direct contact with blood, body fluids, or skin of infected individuals. The virus can also be harbored in breast milk (32), urine (33), aqueous humor (34), and semen after the virus is no longer detectable in the blood stream. In fact, Ebola virus RNA can be detected in the semen up to 101 days after symptom onset, as exhibited by a recent suspected sexual transmission to a Monrovian woman after unprotected sexual intercourse with an Ebola virus survivor (35).

Nipah Virus

Henipaviruses have an animal reservoir (fruit bats, *Pteropus*) and cause central nervous system and respiratory disease in humans. Nipah virus first emerged in Malaysia and Singapore in 1999. Ongoing cases have been reported in India and Bangladesh during the 2000s. Infections in Malaysia result in neurological symptoms and case fatality rates of approximately 40%; infections in Bangladesh show more respiratory symptoms and case fatality rates as high as 90%. As with other henipaviruses, fruit bats are also the natural reservoir of Nipah virus. Infection of nonbat hosts is believed to occur through contact with urine and feces from infected fruit bats. Human infections during Malaysia and Indonesia outbreaks are believed to have been acquired from infected pigs (36), whereas in Bangladesh and India, human infections were acquired following consumption of contaminated date palm sap (37). Of particular concern is the direct person-to-person transmission observed in Bangladesh. This has occurred through exposure to respiratory secretions as a result of caregiver contact with corpses and physician contact (with encephalitis patients) without gloves or mask (38).

Hantavirus

Hantaviruses cause two major types of illnesses, hemorrhagic fever with renal syndrome (HFRS) and hantavirus pulmonary syndrome (HPS) (39–41). These viruses are found in both Old World and New World rodents and are specific for certain rodent hosts; therefore viral infections are limited to the range of their rodent hosts (42). During the Korean War in the 1950s, more than 2000 UN troops were sickened by Korean hemorrhagic fever (43). Virus was subsequently isolated from the lungs of striped field mice (*Apodemus agrarius*) and identified as Hantaan virus (44), and the disease was renamed HFRS. Hantaan virus also causes disease in China and Eastern Russia. Dobrava-Belgrade virus, associated with the yellow-necked field mouse (*Apodemus flavicollus*), causes a severe form of HFRS in Greece, Russia, and the Balkans. A moderate form of HFRS is caused by Seoul virus, with a worldwide distribution given by its hosts, the Norway rat (*Rattus norwegicus*) and the black rat (*Rattus rattus*). The hantavirus infection causing the European form of HFRS, nephropathia epidemica, was identified as Puumala virus (45). In 1993, an outbreak of severe pulmonary illness (later named HPS) in the Four Corners region of the United States was identified as Sin Nombre virus (46). Several other hantaviruses have also been found to cause disease, including Amur virus in eastern Russia (47), Saaremaa virus in Europe (48), Andes virus in South America (49), and Sangassou virus in Africa (50). Overall, the distribution of HFRS is found throughout the world, while HPS has been found in the Americas. While there are fewer cases of HPS worldwide compared to HFRS, the mortality rate of HPS is very high at 20 to 60%. Humans typically become infected by exposure to rodent excrement, urine, or saliva, either by direct contact or contact with rodent nesting material. In the case of Andes virus, human to human transmission during the acute phase of HPS has also been documented (51). A recent outbreak of HPS in Yosemite National Park in the United States also highlighted the challenges of locating persons who disperse widely following a potential exposure (52).

CLINICAL SIGNIFICANCE

Arenaviruses

Lassa fever presents after a 1- to 3-week incubation period, showing signs and symptoms that include headache, fever, malaise, dry cough, back pain, sore throat, nausea and vomiting, mucosal bleeding, conjunctivitis, facial swelling, etc. Later in the disease, signs and symptoms include elevated liver enzymes, myalgia, and malaise. Infected persons often present with influenza-like or dengue-like disease, which confounds diagnosis. The Lassa fever case definition is fever, no signs of local inflammation, and no clinical response to antimalaria treatment or antibiotics. About 80% of infections are asymptomatic or mild, while 20% present with a severe multisystem involvement (shock, thrombocytopenia, leukopenia, mucosal bleeding), facial and neck edema, and pulmonary edema. The disease is more severe in pregnancy, where fetal loss is common. The overall case-fatality rate is 1% but can be higher in hospitalized persons and during epidemics. High Lassa virus titers are observed in blood, liver, spleen, lung, and adrenal gland. The viral load in blood is correlated with disease severity and prognosis (17). The liver is particularly affected, with findings of hepatocellular necrosis and cytoplasmic degeneration (17). Lassa virus infects the vascular endothelium, resulting in increased permeability. In severe cases this leads to edema, shock, and death. In contrast to hemorrhagic fever caused by filoviruses, disseminated intravascular coagulation does not occur in Lassa fever. In addition, a cytokine storm does not occur in Lassa virus infections (17).

The incubation period of lymphocytic choriomeningitis virus is 1 to 2 weeks. Influenza-like symptoms of headache and myalgia are common. Less common are central nervous system (CNS) involvement (meningitis or meningoencephalitis), arthritis, and rash. Up to one-third of infections are asymptomatic. Patients may have decreased leukocytes and platelets, mildly elevated liver enzymes, and lung infiltrates. Symptoms usually resolve in a few days. In some patients, the initial phase is followed by a second phase of CNS disease/aseptic meningitis (headache, fever, photophobia, vomiting, and nuchal rigidity). Marked cerebrospinal fluid (CSF) lymphocytosis is common. The entire disease course usually lasts 1 to 3 weeks. Recovery is usually complete, without sequelae. Rare neurological complications include transverse myelitis, Guillain-Barre-like syndrome, hydrocephalus, and hearing loss. Illness is rarely fatal (less than 1% mortality rate) in healthy adults.

Congenital infection with LCMV is usually severe and can result in spontaneous abortion and fetal death. The fetus usually acquires the virus transplacentally. Unlike other congenital infections, systemic manifestations are usually absent. LCMV is a neurotropic virus that spreads to the brain and causes pathologic effects such as microencephaly, hydrocephalus, periventricular calcifications, focal cerebral destruction, and gyral dysplasia (53). Impaired vision (bilateral chorioretinal scars) and brain dysfunction of varying severity occur in surviving fetuses (53).

Severe infection and a high case-fatality rate have also been reported in organ transplant patients (17). In a report of five clusters involving 17 organ recipients, 14 cases were fatal due to multisystem organ failure; hepatitis was a prominent feature (18). Encephalopathy, coagulopathy, thrombocytopenia, leukocytosis, fever, and graft dysfunction have been reported (54, 55). Inhalation results in lung infiltrates and edema, followed by hematogenous spread and viral replication in other organs, including meninges, choroid plexus, and ventricular ependymal linings. The host response results in tissue inflammation and symptoms, and immunosuppression results in sustained viremia.

Signs and symptoms of Lujo virus infection are very similar to those of Lassa fever. Infections are characterized by rapid onset of fever, headache, malaise, and myalgia followed by sore throat, rash, diarrhea, minor hemorrhage, neck and facial swelling, as well as other symptoms (20). Shock and multisystem organ failure lead to death.

New World arenaviruses can also cause hemorrhagic fevers in humans. The signs and symptoms have been summarized in a review article by McLay et al. (17). Mortality rates are similar, ranging from approximately 20 to 30%. High levels of interferon alpha are associated with fatal outcome in Argentine hemorrhagic fever (Junín virus infection). Junín virus-mediated pathology includes renal papillary necrosis and myocarditis, also observed in cases of Guanarito and Machupo virus infections (17).

Filoviruses

The clinical manifestations of Marburg and Ebola hemorrhagic fever are similar. They have an incubation period of 6 to 12 days with a range of 2 to 21 days (14, 56, 57). Onset is acute, and symptoms are nonspecific and include headache, fever, chills, myalgia, and loss of appetite. These initial symptoms are followed up by weakness, abdominal pain, sore throat, nausea, vomiting, and diarrhea. Conjunctival and pharyngeal injection are also common (14). During illness, a diffuse erythematous, nonpruritic maculopapular rash, that can desquamate and involve the face, neck, and trunk, may develop (58). Gastrointestinal symptoms are also common and include nausea, vomiting, abdominal pain, and watery diarrhea, leading to extensive dehydration, hypotension, and shock (56, 59, 60). It has also been shown that disseminated intravascular coagulation is commonly associated with fatal cases of disease (61). Disseminated infection and major organ necrosis of the lung, spleen, liver, kidneys, skin, and gonads are apparent in fatal infections (14). While hemorrhagic fever can be seen in Ebola virus disease (EVD) patients, data from the West Africa outbreak of 2014 to 2015 showed only approximately 20% of patients having unexplained hemorrhage. This commonly manifested as blood in the stool (about 6%), petechiae, ecchymoses, venipuncture site oozing, pregnancy-related hemorrhage, and/or mucosal hemorrhage (57, 62). EVD also has neurologic complications, including altered consciousness, stiff neck, and seizures suggestive of meningoencephalitis (33). Ocular involvement, including uveitis as long as 14 weeks after infection, has been described, with infectious virus being isolated from aqueous fluid at this late time point (34). There are also reports suggesting that asymptomatic infections may be more common than previously thought. One postoutbreak serosurvey showed that 71% of seropositive individuals were asymptomatic for EVD (63), while another study showed that 46% of close contacts of EVD patients were seropositive (64). Those who survive EVD typically begin to recover in the second week of the disease (57).

Nipah Virus

During the Malaysian outbreak, the incubation period of Nipah virus infections ranged from 2 to 30 days but usually lasted 1 to 2 weeks (65). Initial signs and symptoms are influenza-like (fever, headache, and myalgia), followed by encephalitis, with neurological signs including segmental myoclonus, areflexia, reduced levels of consciousness, and confusion. The incubation period during the Bangladesh outbreaks was between 6 and 11 days. Up to 70% of Bangladesh and Indian patients reported respiratory symptoms, compared with 14 to 27% of Malaysian and Singaporean patients (13). Most Bangladesh cases also had altered mental status. Nipah virus is detected in bronchiolar epithelial cells and is shed in upper respiratory tract secretions early in the illness. Lungs are characterized by alveolitis with hemorrhage, pulmonary edema, and aspiration pneumonia. The virus enters the bloodstream and then the CNS via the olfactory nerve, or hematogenous route. Pathologic changes in the CNS include vasculitis, thrombosis, parenchymal inflammation, and necrosis (66).

Hantavirus

Hantavirus infection is acquired from rodent hosts of the virus. Patients become infected by direct exposure to, and inhalation of, rodent urine, feces, or saliva, or through exposure to rodent bedding. Following exposure, hantaviruses have an incubation period of 2 to 4 weeks (67), with a reported incubation period as long as 7 weeks (68). HFRS has five phases: prodrome (3 to 7 days), hypotensive (7 to 10 days), oliguric (4 to 5 days), diuretic (7 to 11 days), and convalescent phase (weeks to months). HPS has four phases with a prodrome phase (3 to 6 days), cardiopulmonary phase (7 to 10 days), diuretic phase (1 to 3 days), and a convalescent phase (weeks to months) (14). In HFRS, the prodrome phase is characterized by fever and chills, malaise, headache, blurred vision, back and abdominal pain, anorexia, gastrointestinal involvement, facial flushing, conjunctival hemorrhage, and an erythematous rash. The hypotensive phase includes the onset of hypotension and acute shock, vascular leakage, and acute kidney failure. The oliguric phase is characterized by blood pressure returning to normal or high levels, urine output dropping, blood urea nitrogen and serum creatine increasing, and possible severe diffuse hemorrhage. The diuretic phase includes polyuria in excess of 3 liters a day (14, 69). In HPS, the prodrome phase of infection includes fever, chills, muscle aches, headaches, dizziness, nausea, abdominal pain, vomiting, anorexia, and diarrhea that vary in severity and is rapidly followed up by the cardiopulmonary phase in which capillary leakage and subsequent pulmonary edema can occur. The diuretic phase manifests as rapid resolution of pulmonary edema, fever, and shock. This is followed by the convalescent phase in which the patient continues to recover from illness (14, 69). Hantaviruses cause multisystem organ dysfunction syndrome by a process that is unclear. Thrombocytopenia, defects in platelet function, transient disseminated intravascular coagulation, and increased vascular fragility are all thought to play a key role in the early stages of the disease. The endothelium is targeted during infection, leading to microvascular damage, capillary engorgement, and subsequent leakage. Recent studies appear to indicate that the disease process is mostly immunopathologic and includes a strong humoral and cellular immune response (69, 70).

TREATMENT AND PREVENTION

Arenaviruses

Treatment of arenavirus infections is largely intensive supportive care with particular attention to fluid balance. Lassa fever has caused nosocomial infections in Africa and is the biosafety level (BSL)-4 agent most commonly imported into nonepidemic countries. Consequently, containment guidelines for the management of suspected Lassa fever cases to prevent secondary transmission have been established, which include the containment and monitoring of close contacts, stratified by risk. Ribavirin, a nucleoside analogue, is effective in vitro against arenavirus. It has been evaluated for the treatment of severe Lassa fever and reported to dramatically reduce the fatality rate, provided it is given early in the illness. The fatality rate without ribavirin is approximately 87% in Nigeria and 60% in Sierra Leone, while with early ribavirin therapy (within the first week of illness), the fatality rate is less than 10% (71). There are no data on ribavirin efficacy as a postexposure prophylaxis, and the drug is currently recommended only for definitive high-risk exposures (72). Ribavirin has also been utilized with apparent success against LCMV (in combination with reduced levels of immunosuppressive therapy) for organ transplant recipients and Lujo (a single non-fatal case received ribavirin treatment [20]). Candidate vaccines for Lassa fever have been developed. Lukashevich et al. describe the various approaches in detail (73).

Treatment of acute cases of Argentine hemorrhagic fever using immune plasma of human survivors reduced mortality from 15 to 30% to less than 2%, if given during the first 8 days of illness (74), and it is the current more widespread approach. An attenuated vaccine strain of Junín virus is available and has reduced the number of cases and mortality of Argentine hemorrhagic fever (17). There are limited data on the efficacy of ribavirin in the treatment of hemorrhagic fevers caused by New World arenaviruses. Ribavirin may have a beneficial effect in Argentine (75) and Bolivian (76) hemorrhagic fevers.

Filoviruses

While several drugs are currently in development for treatment of EVD or Marburg infections and have been used in specific circumstances, such as compassionate use, current treatment recommendations are supportive care, including providing intravenous fluids to maintain electrolyte balance, maintaining blood pressure and oxygen status, and treatment of additional infections. Several experimental drugs, vaccines, and other treatment options are currently under investigation and have been tested during the 2014 to 2015 outbreak in West Africa. Some antiviral drugs in clinical trials were repurposed for use in the outbreak. These include favipiravir, a nucleoside analog used to inhibit RNA viruses, and brincidofovir, an acyclic nucleotide analog used for DNA virus infections. Favipiravir has been shown to be effective in rodent models of EVD infection (77, 78) while brincidofovir showed in vitro activity against Ebola virus (79). Ebola-specific agents, ZMapp and TKM-Ebola, in combination with convalescent plasma, were also administered to select EVD patients as part of compassionate use administration (80, 81). Recombinant vesicular stomatitis virus-based vaccines for both Ebola and Marburg have also shown promise in nonhuman primate models (82). Additional Ebola-specific agents are also in development, including the nucleoside analog BCX4430, which inhibits viral RNA polymerase function (83), and phosphorodiamidate morpholino oligomers (PMOs), which are synthetic antisense oligonucleotide analogs that interfere with translational processes (84). Convalescent plasma and whole blood have also been used to treat patients during the West African outbreak, but the efficacy of this approach is not clear (85–87).

In order to prevent the spread of Ebola virus, strict infection control practices must be followed. Infection prevention and control guidelines have been published by both the United States Centers for Disease Control and Prevention (CDC) and WHO (88, 89). These include isolation of known or suspect EVD patients, hand hygiene, implementation of contact and droplet precautions, and the use of appropriate personal protective equipment (PPE). Proper donning and doffing of PPE are critical to prevent nosocomial transmission of infection (56, 90, 91).

Nipah Virus

Treatment of Nipah virus infection is primarily supportive. The efficacy of ribavirin is unclear (65). A recombinant replication-competent vesicular stomatitis virus-based vaccine that encodes glycoprotein has shown protection in African green monkeys (92). A human monoclonal antibody targeting viral G glycoprotein has also been effective for postexposure treatment of experimentally infected African green monkeys (93). Since there are neither vaccines nor antivirals for treatment and because henipaviruses are potentially able to transmit by the aerosol route, Nipah virus has been classified as a BSL-4 agent and is restricted to laboratories capable of handling such agents.

Hantavirus

There is no specific treatment for hantavirus infection, so care of patients is supportive therapy. Care includes hemodynamic and pulmonary support, with initiation of mechanical ventilation for respiratory failure, if necessary. Hypoxic patients should be given supplemental oxygen. Patients should be treated with broad-spectrum antibiotic therapy, until HPS is confirmed, to treat other pathogens that may be present, according to the differential diagnosis. Initial care recommendations also include antipyretics and analgesics. Due to the role of capillary leakage in hantavirus infections, vasopressors to manage hypotension and careful use of intravenous fluids are also recommended.

To prevent the spread of hantavirus infection, exposure to rodents and their excrement should be avoided. Homes should be rodent-proofed by sealing holes on the inside and outside with cement, or some other type of patching material, and steel wool or wire screen. Campers and hikers should avoid touching rodents and should limit their exposure to rodent droppings or places frequented by rodents. Areas with rodent activity should be thoroughly cleaned by soaking with disinfectant or chlorine solution. Gloves should be worn during cleaning, and careful attention should be paid to prevent generating aerosols (94).

DETECTION/DIAGNOSIS

Arenaviruses

With the exception of LCMV, laboratory testing for arenaviruses is generally limited to public health reference laboratories. Arenaviruses can be isolated early in the disease, which is normally followed by IgM and IgG antibody responses. Early diagnosis is important since treatment efficacy appears to correlate with early administration of therapy.

Lassa virus antigen (as detected by enzyme-linked immunosorbent assay [ELISA]) appears in the first week of illness and is replaced by IgM in the second week, with IgG appearing in the third week of illness. A combination of antigen ELISA and IgM ELISA was 88% sensitive and 90% specific for acute infection. Indirect fluorescent antibody (IFA) assay (detecting total antibody) sensitivity was 70% (95). IgM ELISA can be unreliable due to prolonged presence of IgM. RT-PCR for Lassa virus has also been described with limited clinical validation (96, 97). Cell culture requires BSL-4 practices. Virus isolation or detection of viral markers can be performed on blood, urine, or throat specimens. Immunohistochemistry can be performed on postmortem tissue. Currently there are no commercially available diagnostic tests for Lassa virus.

LCMV infections are usually diagnosed by serological methods (ELISA, IFA) for IgG and IgM. Commercial reference laboratories in the United States offer serologic testing. Indirect fluorescent antibodies were adopted in the 1970s as a more sensitive and specific alternative to complement fixation and neutralization assays (98) and are still used. LCMV can be isolated in cell culture. RT-PCR can be performed from blood or CSF (acute phase), but there is an absence of data demonstrating clinical diagnostic performance. National public health reference laboratories will perform serologic and virologic tests. In the United States, the Viral Special Pathogens Branch of the CDC will test serum and CSF for IgM and IgG.

The New World arenaviruses require BSL-4 practices (Junín can be handled at BSL-3 facilities, provided laboratory staff have been immunized and the laboratory exhaust is HEPA-filtered). Primary diagnostic methods are serology and molecular methods. During arenavirus hemorrhagic fevers, IgM is usually not present in the early stage of illness. An RT-PCR assay that detects Junín virus in whole blood specimens was 98% sensitive and 76% specific relative to seroconversion (99). It was speculated that most of the patients with false-positive PCR results had Junín virus infections that did not result in a detectable immune response. RT-PCR performed on serum was less sensitive and specific (62% and 71%, respectively) (99).

Filoviruses

For rapid and sensitive diagnosis of filovirus infections, RT-PCR is the gold standard (100). A commercially available multiplex PCR panel that includes Ebola virus was compared to a CDC RT-PCR assay both in specimens from Ebola patients. Overall agreement was 85% in whole blood and urine specimens (101). During a 2000 to 2001 Ebola outbreak in Uganda, RT-PCR was able to detect infection 24 to 48 hours before antigen-capture assays became positive (102). RT-PCR testing for Ebola and Marburg viruses is available in the United States in Laboratory Response Network (LRN) laboratories. The LRN is a network of laboratories, including state and local public health, veterinary, military, and international labs, that can respond to biological and chemical threats and other public health emergencies. LRN reference laboratories use standardized laboratory assay protocols. Recently developed rapid-antigen point-of-care tests for Ebola virus, tested on a limited number of patients, have also shown high sensitivity (100%) and specificity (92.2%) as compared to RT-PCR (103). Serologies can be performed for IgM for acute case diagnosis as well. Virus isolation is performed by inoculation of cell cultures, such as Vero, human diploid lung (MRC-5 or BHK-21), or MA-104 cells. This is followed by IFA, ELISA, or RT-PCR testing of inoculated cells to test for viral antigens or viral RNA. All isolation techniques must be performed in a BSL-4 laboratory. A consensus PCR for all filoviruses has also been described, allowing for surveillance for new virus strains and improved contact tracing in the case of an outbreak (104).

During the most recent 2014 to 2015 outbreak of Ebola virus in West Africa, several Ebola virus RT-PCR, and rapid antigen assays received Emergency Use Authorization from the Food and Drug Administration (FDA) for the presumptive identification of Ebola Zaire virus. This decision, which authorized the use of several different assays, was made because there were no FDA-approved/cleared tests to detect Ebola Zaire virus in clinical specimens (http://www.fda.gov/MedicalDevices/Safety/EmergencySituations/ucm161496.htm#ebola. Accessed 12 October 2015).

Nipah Virus

Serologic methods for the diagnosis of Nipah virus infections include enzyme immunoassay (EIA) and neutralization assay. IgM-capture and indirect IgG EIA can be performed on serum and CSF for simplicity and convenience. IgM antibody is reliably detected in serum 1 week after symptom onset and in CSF 10 to 15 days after symptom onset. IgG antibody is generally detected in serum 2.5 weeks after symptom onset (65). Neutralization assays with wild-type virus require BSL-4 conditions. The use of a vesicular stomatitis virus pseudotyped with Nipah virus proteins obviates the need for BSL-4 conditions when performing neutralization assays (12). Virus can be isolated from CSF, brain, spleen, lung, and kidney (but requires BSL-4 conditions). Nipah virus also grows well in Vero cells, and a cytopathic effect is usually observed within 3 days (105). Virus isolates can be identified by neutralization by specific antisera or RT-PCR. PCR targets include the M and N genes, coding for the matrix, and nucleoproteins, respectively. RT-PCR has been used for detection of viral RNA from serum, urine, and CSF (106). Members of the *Henipavirus* genus have a high mutation rate, challenging molecular-based tests (107). Consensus PCR assays have been developed for the detection of Nipah virus (107).

Hantavirus

The methods used for hantavirus diagnosis include serological assays on serum and plasma, viral antigen detection in tissue by immunohistochemistry, and nucleic acid amplification from sources such as blood or tissue. Serological assays

TABLE 1 Laboratory diagnosis of select animal-borne viruses

Virus	Diagnostic procedures	Optimum specimens	Remarks
Arenaviruses			
Old World			
Lassa virus	IgM-capture ELISA; antigen ELISA; IgG ELISA; IFA; RT-PCR	Serum (antibody); whole blood (antigen, RT-PCR)	Antigen is detected early in illness Antigen ELISA and IgM ELISA combination provides good sensitivity and specificity
Lymphocytic choriomeningitis virus	IgG ELISA; IgM-capture ELISA; IFA; RT-PCR	Serum (antibody); whole blood, CSF (RT-PCR)	Serologic testing available in most commercial reference laboratories
Lujo virus	RT-PCR	Not established	Can be performed at Viral Special Pathogens Unit at the CDC
New World (Chapare, Guanarito Junin, Machupo and Sabia)	IgG ELISA; IgM-capture ELISA; RT-PCR	Serum (antibody, RT-PCR); whole blood (RT-PCR)	RT-PCR most sensitive early in illness; IgM ELISA after one week of illness
Filoviruses (Ebola and Marburg)	RT-PCR; Rapid antigen test (lateral flow immunoassay); IgM-capture ELISA	Serum (antibody); whole blood (RT-PCR, antigen)	RT-PCR is the gold standard and is the most sensitive test early in illness.
Nipah virus	IgM-capture ELISA; IgG EIA; neutralization; RT-PCR	Serum, CSF (antibody and RT-PCR); urine (RT-PCR)	IgM detected in serum after 1 week of illness, in CSF after 2 weeks. IgG detected in serum after 2 and a half weeks illness
Hantaviruses	IgM-capture ELISA; IFA; WB; neutralization; RT-PCR	Serum, plasma	IgM capture ELISA is most widely used method in diagnosis. IgM is detectable at onset of symptoms. Virus not typically detectable in blood at symptom onset

Abbreviations: ELISA, enzyme-linked immunosorbent assay; IFA, indirect fluorescent antibody; RT-PCR, reverse transcription polymerase chain reaction; WB, Western blot

have been the most widely used assay since, at the time of disease manifestation, virus is usually not detectable in blood. These include IFA, immunoprecipitation, radioimmunoassay, Western blotting, high-density particle agglutination, hemagglutination inhibition, neutralization, and ELISA. Of the available methods, IgM-capture ELISA, which is performed by a variety of federal, state, and regional public health labs and by reference laboratories, is the most widely used method for HFRS and HPS diagnosis and has high sensitivity and specificity (108). All HFRS and HPS patients show elevated levels of anti-hantavirus IgM at the onset of clinical symptoms, or shortly thereafter (109, 110), that can persist longer than 2 months after the acute phase in HPS (111), and as long as 6 months in nephropathia epidemica (112). IgG is also frequently measurable during the acute phase of infection (110–114) and can remain elevated as long as 3 years following acute phase illness in HPS and as long as 10 years in HFRS. Some cross-reactivity between hantaviruses has been noted between antibodies directed against the nucleocapsid proteins and glycoproteins (113, 115, 116). IgG directed against hantavirus glycoprotein, Gn, is more hantavirus species-specific (110, 117). While nucleic acid amplification assays for the diagnosis of hantavirus infections have been developed (118), they are still currently considered research tools. Genetic methods, such as sequencing, have been useful for determining the taxonomy of different hantaviruses and in the development of diagnostic assays. Sequencing and subsequent analysis are typically performed on amplified product of the nucleocapsid protein gene or the Gc region of the glycoprotein precursor gene (14). Direct electron microscopic examination is limited in its diagnostic value but has been used postmortem to detect virions and hantavirus inclusion bodies (119, 120).

DIAGNOSTIC BEST PRACTICES

With the exception of LCMV, testing for viruses described in this chapter is generally restricted to public health and specific reference laboratories. Contact the local/state/regional public health laboratory for information on testing and specimen transport.

Arenaviruses

Whole blood and serum are common specimens for RT-PCR detection of arenaviruses (Table 1). RT-PCR testing of CSF is also warranted if neurological symptoms are present. RT-PCR sensitivity is highest with specimens collected early in illness.

Laboratory testing for suspected Lassa fever is important because the differential diagnosis is extensive. Rapid testing is justified because specific treatment is effective only when used early in the course of disease. A combination of antigen ELISA and IgM ELISA performed on blood specimens provides reasonably high sensitivity and specificity for acute infection (Table 1). However, the presence of IgM is not conclusive for recent infection. RT-PCR can also be used in conjunction with these methods.

LCMV is typically part of the differential diagnosis in congenital infections and must be distinguished from other TORCH (toxoplasmosis, other agents, rubella, cytomegalovirus, and herpes simplex) pathogens. Definitive diagnosis requires laboratory testing. The optimal test depends on the suspected mode of acquisition and symptoms. In acute acquired disease with CNS findings, virus isolation or PCR from CSF and IgM serology should be performed. In a congenital newborn infection, serologic testing (serum and CSF) for anti-LCMV antibodies should be performed; culture and PCR are likely to be negative because virus may be cleared.

The differential diagnosis of infections caused by New World arenaviruses includes malaria and leptospirosis. For optimal test sensitivity, consideration should be given to the time since symptom onset. RT-PCR is most sensitive early in illness, beginning at least from day 3 of fever onset (121), while serologic tests that detect IgM are more sensitive later in illness (Table 1).

Filoviruses

Early diagnosis of infection is critical in order to start isolation procedures to prevent further spread of infection and to initiate appropriate therapy as early as possible. Since the signs and symptoms of filovirus infections are nonspecific at the onset, infections, such as malaria, Lassa fever, influenza, and bacteremia, should be ruled out. RT-PCR is the test of choice for sensitive and rapid results (Table 1). Recently described rapid antigen tests for Ebola virus may have sensitivity approaching that of RT-PCR, but larger studies are needed to determine sensitivity and specificity throughout the course of illness (103).

Nipah Virus

Laboratory diagnosis of Nipah virus infection is usually achieved by the detection of IgG and IgM in serum and CSF (Table 1). Positive EIA results should be confirmed by neutralization assay (106). Biosafety classification generally precludes virus isolation. To date, performance data on RT-PCR assays are limited to analytical sensitivity (limit of detection) and analytical specificity (cross-reactivity among other viruses, including henipaviruses) (106). Clinical performance data are needed for RT-PCR.

Hantavirus

An early diagnosis is critical in the management of a patient with HFRS or HPS. In addition, in the case of Andes virus, it is crucial to institute proper isolation procedures to prevent nosocomial transmission. An early clinical diagnosis is difficult since the prodromal phase of infection is nonspecific (fever, muscle aches, nausea, and fatigue) and can be easily confused with other infections, such as influenza. Serological assays on serum and plasma are typically performed (Table 1). IgM-capture ELISA is the most widely used method for HFRS and HPS diagnosis and has high sensitivity and specificity (108). Importantly, virus is usually not detectable in blood at the time of symptom onset, limiting the usefulness of nucleic acid amplification assays for diagnosis at this stage.

REFERENCES

1. **Rivers TM, Scott TF.** 1936. Meningitis in man caused by a filterable virus: ii. Identification of the etiological agent. *J Exp Med* **63:**415–432.
2. **Buckley SM, Casals J.** 1970. Lassa fever, a new virus disease of man from West Africa. 3. Isolation and characterization of the virus. *Am J Trop Med Hyg* **19:**680–691.
3. **Briese T, Paweska JT, McMullan LK, Hutchison SK, Street C, Palacios G, Khristova ML, Weyer J, Swanepoel R, Egholm M, Nichol ST, Lipkin WI.** 2009. Genetic detection and characterization of Lujo virus, a new hemorrhagic fever-associated arenavirus from southern Africa. *PLoS Pathog* **5:** e1000455.
4. **Mettler N, Buckley SM, Casals J.** 1961. Propagation of Junin virus, the etiological agent of Argentinian hemorrhagic fever, in HeLa cell cultures. *Proc Soc Exp Biol Med* **107:**684–688.
5. **Webb PA.** 1965. Properties of Machupo virus. *Am J Trop Med Hyg* **14:**799–802.
6. **Tesh RB, Jahrling PB, Salas R, Shope RE.** 1994. Description of Guanarito virus (Arenaviridae: Arenavirus), the etiologic agent of Venezuelan hemorrhagic fever. *Am J Trop Med Hyg* **50:**452–459.
7. **Coimbra TLM, Nassar ES, de Souza LTM, Ferreira IB, Rocco IM, Burattini MN, Travassos da Rosa APA, Vasconcelos PFC, Pinheiro FP, LeDuc JW, Rico-Hesse R, Gonzalez J-P, Tesh RB, Jahrling PB. 1994.** New arenavirus isolated in Brazil. *Lancet* **343:**391–392.
8. **Delgado S, Erickson BR, Agudo R, Blair PJ, Vallejo E, Albariño CG, Vargas J, Comer JA, Rollin PE, Ksiazek TG, Olson JG, Nichol ST.** 2008. Chapare virus, a newly discovered arenavirus isolated from a fatal hemorrhagic fever case in Bolivia. *PLoS Pathog* **4:**e1000047.
9. **Volchkov VE, Becker S, Volchkova VA, Ternovoj VA, Kotov AN, Netesov SV, Klenk HD.** 1995. GP mRNA of Ebola virus is edited by the Ebola virus polymerase and by T7 and vaccinia virus polymerases. *Virology* **214:**421–430.
10. **Sanchez A, Trappier SG, Mahy BW, Peters CJ, Nichol ST.** 1996. The virion glycoproteins of Ebola viruses are encoded in two reading frames and are expressed through transcriptional editing. *Proc Natl Acad Sci USA* **93:**3602–3607.
11. **Peterson AT, Holder MT.** 2012. Phylogenetic assessment of filoviruses: how many lineages of Marburg virus? *Ecol Evol* **2:**1826–1833.
12. **Kaku Y, Noguchi A, Marsh GA, Barr JA, Okutani A, Hotta K, Bazartseren B, Fukushi S, Broder CC, Yamada A, Inoue S, Wang LF.** 2012. Second generation of pseudotype-based serum neutralization assay for Nipah virus antibodies: sensitive and high-throughput analysis utilizing secreted alkaline phosphatase. *J Virol Methods* **179:**226–232.
13. **Lo MK, Rota PA.** 2008. The emergence of Nipah virus, a highly pathogenic paramyxovirus. *J Clin Virol* **43:**396–400.
14. **Jorgensen JH, Pfaller MA, Carroll KC, American Society for Microbiology.** 2015. *In* Jorgensen J, Pfaller M, Carroll K, Funke G, Landry M, Richter S, Warnock D (ed.), American Society for Microbiology Press (ed.), *Manual of clinical microbiology*, 11th ed. Washington, DC.
15. **Sogoba N, Feldmann H, Safronetz D.** 2012. Lassa fever in West Africa: evidence for an expanded region of endemicity. *Zoonoses Public Health* **59**(Suppl 2):43–47.
16. **Khan SH, Goba A, Chu M, Roth C, Healing T, Marx A, Fair J, Guttieri MC, Ferro P, Imes T, Monagin C, Garry RF, Bausch DG, Network MRULF, Mano River Union Lassa Fever Network.** 2008. New opportunities for field research on the pathogenesis and treatment of Lassa fever. *Antiviral Res* **78:**103–115.
17. **McLay L, Liang Y, Ly H.** 2014. Comparative analysis of disease pathogenesis and molecular mechanisms of New World and Old World arenavirus infections. *J Gen Virol* **95:**1–15.
18. **Laposová K, Pastoreková S, Tomášková J.** 2013. Lymphocytic choriomeningitis virus: invisible but not innocent. *Acta Virol* **57:**160–170.
19. **Maetz HM, Sellers CA, Bailey WC, Hardy GE Jr.** 1976. Lymphocytic choriomeningitis from pet hamster exposure: a local public health experience. *Am J Public Health* **66:**1082–1085.
20. **Sewlall NH, Richards G, Duse A, Swanepoel R, Paweska J, Blumberg L, Dinh TH, Bausch D.** 2014. Clinical features and patient management of Lujo hemorrhagic fever. *PLoS Negl Trop Dis* **8:**e3233.
21. **Charrel RN, de Lamballerie X.** 2003. Arenaviruses other than Lassa virus. *Antiviral Res* **57:**89–100.
22. **Rollin PE, Williams RJ, Bressler DS, Pearson S, Cottingham M, Pucak G, Sanchez A, Trappier SG, Peters RL, Greer PW,**

Zaki S, Demarcus T, Hendricks K, Kelley M, Simpson D, Geisbert TW, Jahrling PB, Peters CJ, Ksiazek TG. 1999. Ebola (subtype Reston) virus among quarantined nonhuman primates recently imported from the Philippines to the United States. *J Infect Dis* **179**(Suppl 1)**:**S108–S114.

23. **Anonymous. 1992.** Viral haemorrhagic fever in imported monkeys. *Wkly Epidemiol Rec* **67:**142–143.
24. **Le Guenno B, Formenty P, Wyers M, Gounon P, Walker F, Boesch C, Boesch C.** 1995. Isolation and partial characterisation of a new strain of Ebola virus. *Lancet* **345:**1271–1274.
25. **Towner JS, Sealy TK, Khristova ML, Albariño CG, Conlan S, Reeder SA, Quan PL, Lipkin WI, Downing R, Tappero JW, Okware S, Lutwama J, Bakamutumaho B, Kayiwa J, Comer JA, Rollin PE, Ksiazek TG, Nichol ST.** 2008. Newly discovered ebola virus associated with hemorrhagic fever outbreak in Uganda. *PLoS Pathog* **4:**e1000212.
26. **Negredo A, Palacios G, Vázquez-Morón S, González F, Dopazo H, Molero F, Juste J, Quetglas J, Savji N, de la Cruz Martínez M, Herrera JE, Pizarro M, Hutchison SK, Echevarría JE, Lipkin WI, Tenorio A.** 2011. Discovery of an ebolavirus-like filovirus in europe. *PLoS Pathog* **7:**e1002304.
27. **Towner JS, Amman BR, Sealy TK, Carroll SA, Comer JA, Kemp A, Swanepoel R, Paddock CD, Balinandi S, Khristova ML, Formenty PB, Albarino CG, Miller DM, Reed ZD, Kayiwa JT, Mills JN, Cannon DL, Greer PW, Byaruhanga E, Farnon EC, Atimnedi P, Okware S, Katongole-Mbidde E, Downing R, Tappero JW, Zaki SR, Ksiazek TG, Nichol ST, Rollin PE.** 2009. Isolation of genetically diverse Marburg viruses from Egyptian fruit bats. *PLoS Pathog* **5:**e1000536.
28. **Leroy EM, Kumulungui B, Pourrut X, Rouquet P, Hassanin A, Yaba P, Délicat A, Paweska JT, Gonzalez JP, Swanepoel R.** 2005. Fruit bats as reservoirs of Ebola virus. *Nature* **438:** 575–576.
29. **Olival KJ, Islam A, Yu M, Anthony SJ, Epstein JH, Khan SA, Khan SU, Crameri G, Wang LF, Lipkin WI, Luby SP, Daszak P.** 2013. Ebola virus antibodies in fruit bats, bangladesh. *Emerg Infect Dis* **19:**270–273.
30. **Hayman DT, Yu M, Crameri G, Wang LF, Suu-Ire R, Wood JL, Cunningham AA.** 2012. Ebola virus antibodies in fruit bats, Ghana, West Africa. *Emerg Infect Dis* **18:**1207–1209.
31. **Leroy EM, Epelboin A, Mondonge V, Pourrut X, Gonzalez JP, Muyembe-Tamfum JJ, Formenty P.** 2009. Human Ebola outbreak resulting from direct exposure to fruit bats in Luebo, Democratic Republic of Congo, 2007. *Vector Borne Zoonotic Dis* **9:**723–728.
32. **Bausch DG, Towner JS, Dowell SF, Kaducu F, Lukwiya M, Sanchez A, Nichol ST, Ksiazek TG, Rollin PE.** 2007. Assessment of the risk of Ebola virus transmission from bodily fluids and fomites. *J Infect Dis* **196**(Suppl 2)**:**S142–S147.
33. **Kreuels B, Wichmann D, Emmerich P, Schmidt-Chanasit J, de Heer G, Kluge S, Sow A, Renné T, Günther S, Lohse AW, Addo MM, Schmiedel S.** 2014. A case of severe Ebola virus infection complicated by gram-negative septicemia. *N Engl J Med* **371:**2394–2401.
34. **Varkey JB, Shantha JG, Crozier I, Kraft CS, Lyon GM, Mehta AK, Kumar G, Smith JR, Kainulainen MH, Whitmer S, Ströher U, Uyeki TM, Ribner BS, Yeh S.** 2015. Persistence of Ebola virus in ocular fluid during convalescence. *N Engl J Med* **372:**2423–2427.
35. **Christie A, Davies-Wayne GJ, Cordier-Lasalle T, Blackley DJ, Laney AS, Williams DE, Shinde SA, Badio M, Lo T, Mate SE, Ladner JT, Wiley MR, Kugelman JR, Palacios G, Holbrook MR, Janosko KB, de Wit E, van Doremalen N, Munster VJ, Pettitt J, Schoepp RJ, Verhenne L, Evlampidou I, Kollie KK, Sieh SB, Gasasira A, Bolay F, Kateh FN, Nyenswah TG, De Cock KM, Centers for Disease Control and Prevention (CDC).** 2015. Possible sexual transmission of Ebola virus—Liberia, 2015. *MMWR Morb Mortal Wkly Rep* **64:**479–481.
36. **Parashar UD, Sunn LM, Ong F, Mounts AW, Arif MT, Ksiazek TG, Kamaluddin MA, Mustafa AN, Kaur H, Ding LM, Othman G, Radzi HM, Kitsutani PT, Stockton PC, Arokiasamy J, Gary HE Jr, Anderson LJ.** 2000. Case-control study of risk factors for human infection with a new zoonotic paramyxovirus, Nipah virus, during a 1998–1999 outbreak of severe encephalitis in Malaysia. *J Infect Dis* **181:**1755–1759.
37. **Rahman MA, Hossain MJ, Sultana S, Homaira N, Khan SU, Rahman M, Gurley ES, Rollin PE, Lo MK, Comer JA, Lowe L, Rota PA, Ksiazek TG, Kenah E, Sharker Y, Luby SP.** 2012. Date palm sap linked to Nipah virus outbreak in Bangladesh, 2008. *Vector Borne Zoonotic Dis* **12:**65–72.
38. **Sazzad HM, Hossain MJ, Gurley ES, Ameen KM, Parveen S, Islam MS, Faruque LI, Podder G, Banu SS, Lo MK, Rollin PE, Rota PA, Daszak P, Rahman M, Luby SP.** 2013. Nipah virus infection outbreak with nosocomial and corpse-to-human transmission, Bangladesh. *Emerg Infect Dis* **19:**210–217.
39. **Lee HW, Baek LJ, Johnson KM.** 1982. Isolation of Hantaan virus, the etiologic agent of Korean hemorrhagic fever, from wild urban rats. *J Infect Dis* **146:**638–644.
40. **Wong TW, Chan YC, Yap EH, Joo YG, Lee HW, Lee PW, Yanagihara R, Gibbs CJ Jr, Gajdusek DC.** 1988. Serological evidence of hantavirus infection in laboratory rats and personnel. *Int J Epidemiol* **17:**887–890.
41. **Peters CJ, Khan AS.** 2002. Hantavirus pulmonary syndrome: the new American hemorrhagic fever. *Clin Infect Dis* **34:**1224–1231.
42. **Plyusnin A, Morzunov SP.** 2001. Virus evolution and genetic diversity of hantaviruses and their rodent hosts. *Curr Top Microbiol Immunol* **256:**47–75.
43. **Sheedy JA, Froeb HF, Batson HA, Conley CC, Murphy JP, Hunter RB, Cugell DW, Giles RB, Bershadsky SC, Vester JW, Yoe RH.** 1954. The clinical course of epidemic hemorrhagic fever. *Am J Med* **16:**619–628.
44. **Lee HW, Lee PW, Johnson KM.** 1978. Isolation of the etiologic agent of Korean Hemorrhagic fever. *J Infect Dis* **137:** 298–308.
45. **Brummer-Korvenkontio M, Vaheri A, Hovi T, von Bonsdorff CH, Vuorimies J, Manni T, Penttinen K, Oker-Blom N, Lähdevirta J.** 1980. Nephropathia epidemica: detection of antigen in bank voles and serologic diagnosis of human infection. *J Infect Dis* **141:**131–134.
46. **Nichol ST, Spiropoulou CF, Morzunov S, Rollin PE, Ksiazek TG, Feldmann H, Sanchez A, Childs J, Zaki S, Peters CJ.** 1993. Genetic identification of a hantavirus associated with an outbreak of acute respiratory illness. *Science* **262:**914–917.
47. **Yashina LN, Patrushev NA, Ivanov LI, Slonova RA, Mishin VP, Kompanez GG, Zdanovskaya NI, Kuzina II, Safronov PF, Chizhikov VE, Schmaljohn C, Netesov SV.** 2000. Genetic diversity of hantaviruses associated with hemorrhagic fever with renal syndrome in the far east of Russia. *Virus Res* **70:**31–44.
48. **Vapalahti O, Mustonen J, Lundkvist A, Henttonen H, Plyusnin A, Vaheri A.** 2003. Hantavirus infections in Europe. *Lancet Infect Dis* **3:**653–661.
49. **Galeno H, Mora J, Villagra E, Fernandez J, Hernandez J, Mertz GJ, Ramirez E.** 2002. First human isolate of Hantavirus (Andes virus) in the Americas. *Emerg Infect Dis* **8:**657–661.
50. **Klempa B, Fichet-Calvet E, Lecompte E, Auste B, Aniskin V, Meisel H, Denys C, Koivogui L, ter Meulen J, Krüger DH.** 2006. Hantavirus in African wood mouse, Guinea. *Emerg Infect Dis* **12:**838–840.
51. **Padula PJ, Edelstein A, Miguel SD, López NM, Rossi CM, Rabinovich RD.** 1998. Hantavirus pulmonary syndrome outbreak in Argentina: molecular evidence for person-to-person transmission of Andes virus. *Virology* **241:**323–330.
52. **Roehr B.** 2012. US officials warn 39 countries about risk of hantavirus among travellers to Yosemite. *BMJ* **345**(sep10 1)**:** e6054.
53. **Bonthius DJ.** 2012. Lymphocytic choriomeningitis virus: an underrecognized cause of neurologic disease in the fetus, child, and adult. *Semin Pediatr Neurol* **19:**89–95.
54. **Fischer SA, Graham MB, Kuehnert MJ, Kotton CN, Srinivasan A, Marty FM, Comer JA, Guarner J, Paddock CD, DeMeo DL, Shieh WJ, Erickson BR, Bandy U, DeMaria A Jr, Davis JP, Delmonico FL, Pavlin B, Likos A, Vincent MJ,**

Sealy TK, Goldsmith CS, Jernigan DB, Rollin PE, Packard MM, Patel M, Rowland C, Helfand RF, Nichol ST, Fishman JA, Ksiazek T, Zaki SR, LCMV in Transplant Recipients Investigation Team. 2006. Transmission of lymphocytic choriomeningitis virus by organ transplantation. *N Engl J Med* **354:**2235–2249.

55. **Centers for Disease Control and Prevention (CDC).** 2008. Brief report: lymphocytic choriomeningitis virus transmitted through solid organ transplantation—Massachusetts, 2008. *MMWR Morb Mortal Wkly Rep* **57:**799–801.
56. **Schieffelin JS, Shaffer JG, Goba A, Gbakie M, Gire SK, Colubri A, Sealfon RS, Kanneh L, Moigboi A, Momoh M, Fullah M, Moses LM, Brown BL, Andersen KG, Winnicki S, Schaffner SF, Park DJ, Yozwiak NL, Jiang PP, Kargbo D, Jalloh S, Fonnie M, Sinnah V, French I, Kovoma A, Kamara FK, Tucker V, Konuwa E, Sellu J, Mustapha I, Foday M, Yillah M, Kanneh F, Saffa S, Massally JL, Boisen ML, Branco LM, Vandi MA, Grant DS, Happi C, Gevao SM, Fletcher TE, Fowler RA, Bausch DG, Sabeti PC, Khan SH, Garry RF, Program KLF, Consortium VHF, Team WCR, KGH Lassa Fever Program, Viral Hemorrhagic Fever Consortium, WHO Clinical Response Team.** 2014. Clinical illness and outcomes in patients with Ebola in Sierra Leone. *N Engl J Med* **371:**2092–2100.
57. **Centers for Disease Control and Prevention.**2015. Ebola Virus Disease (EVD) information for clinicians in U.S. healthcare settings. http://www.cdc.gov/vhf/ebola/healthcare-us/preparing/clinicians.html. Accessed 25 September 2015.
58. **Kortepeter MG, Bausch DG, Bray M.** 2011. Basic clinical and laboratory features of filoviral hemorrhagic fever. *J Infect Dis* **204**(Suppl 3)**:**S810–S816
59. **Bah EI, Lamah MC, Fletcher T, Jacob ST, Brett-Major DM, Sall AA, Shindo N, Fischer WA II, Lamontagne F, Saliou SM, Bausch DG, Moumié B, Jagatic T, Sprecher A, Lawler JV, Mayet T, Jacquerioz FA, Méndez Baggi MF, Vallenas C, Clement C, Mardel S, Faye O, Faye O, Soropogui B, Magassouba N, Koivogui L, Pinto R, Fowler RA.** 2015. Clinical presentation of patients with Ebola virus disease in Conakry, Guinea. *N Engl J Med* **372:**40–47.
60. **Chertow DS, Kleine C, Edwards JK, Scaini R, Giuliani R, Sprecher A.** 2014. Ebola virus disease in West Africa—clinical manifestations and management. *N Engl J Med* **371:** 2054–2057.
61. **Rollin PE, Bausch DG, Sanchez A.** 2007. Blood chemistry measurements and D-Dimer levels associated with fatal and nonfatal outcomes in humans infected with Sudan Ebola virus. *J Infect Dis* **196**(Suppl 2)**:**S364–S371.
62. **Jamieson DJ, Uyeki TM, Callaghan WM, Meaney-Delman D, Rasmussen SA.** 2014. What obstetrician-gynecologists should know about Ebola: a perspective from the Centers for Disease Control and Prevention. *Obstet Gynecol* **124:**1005–1010.
63. **Heffernan RT, Pambo B, Hatchett RJ, Leman PA, Swanepoel R, Ryder RW.** 2005. Low seroprevalence of IgG antibodies to Ebola virus in an epidemic zone: Ogooué-Ivindo region, Northeastern Gabon, 1997. *J Infect Dis* **191:**964–968.
64. **Leroy EM, Baize S, Volchkov VE, Fisher-Hoch SP, Georges-Courbot MC, Lansoud-Soukate J, Capron M, Debré P, McCormick JB, Georges AJ.** 2000. Human asymptomatic Ebola infection and strong inflammatory response. *Lancet* **355:** 2210–2215.
65. **Ksiazek TG, Rota PA, Rollin PE.** 2011. A review of Nipah and Hendra viruses with an historical aside. *Virus Res* **162:** 173–183.
66. **Escaffre O, Borisevich V, Rockx B.** 2013. Pathogenesis of Hendra and Nipah virus infection in humans. *J Infect Dev Ctries* **7:**308–311.
67. **Young JC, Hansen GR, Graves TK, Deasy MP, Humphreys JG, Fritz CL, Gorham KL, Khan AS, Ksiazek TG, Metzger KB, Peters CJ.** 2000. The incubation period of hantavirus pulmonary syndrome. *Am J Trop Med Hyg* **62:**714–717.
68. **Fritz CL, Young JC.** 2001. Estimated incubation period for hantavirus pulmonary syndrome. *Am J Trop Med Hyg* **65:**403.
69. **Peters CJ, Simpson GL, Levy H.** 1999. Spectrum of hantavirus infection: hemorrhagic fever with renal syndrome and hantavirus pulmonary syndrome. *Annu Rev Med* **50:**531–545.
70. **Muranyi W, Bahr U, Zeier M, van der Woude FJ.** 2005. Hantavirus infection. *J Am Soc Nephrol* **16:**3669–3679.
71. **McCormick JB, King IJ, Webb PA, Scribner CL, Craven RB, Johnson KM, Elliott LH, Belmont-Williams R.** 1986. Lassa fever. Effective therapy with ribavirin. *N Engl J Med* **314:** 20–26.
72. **Bausch DG, Hadi CM, Khan SH, Lertora JJ.** 2010. Review of the literature and proposed guidelines for the use of oral ribavirin as postexposure prophylaxis for Lassa fever. *Clin Infect Dis* **51:**1435–1441.
73. **Lukashevich IS.** 2012. Advanced vaccine candidates for Lassa fever. *Viruses* **4:**2514–2557.
74. **Maiztegui JI, Fernandez NJ, de Damilano AJ.** 1979. Efficacy of immune plasma in treatment of Argentine haemorrhagic fever and association between treatment and a late neurological syndrome. *Lancet* **2:**1216–1217.
75. **Enria DA, Maiztegui JI.** 1994. Antiviral treatment of Argentine hemorrhagic fever. *Antiviral Res* **23:**23–31.
76. **Kilgore PE, Ksiazek TG, Rollin PE, Mills JN, Villagra MR, Montenegro MJ, Costales MA, Paredes LC, Peters CJ.** 1997. Treatment of Bolivian hemorrhagic fever with intravenous ribavirin. *Clin Infect Dis* **24:**718–722.
77. **Oestereich L, Lüdtke A, Wurr S, Rieger T, Muñoz-Fontela C, Günther S.** 2014. Successful treatment of advanced Ebola virus infection with T-705 (favipiravir) in a small animal model. *Antiviral Res* **105:**17–21.
78. **Smither SJ, Eastaugh LS, Steward JA, Nelson M, Lenk RP, Lever MS.** 2014. Post-exposure efficacy of oral T-705 (Favipiravir) against inhalational Ebola virus infection in a mouse model. *Antiviral Res* **104:**153–155.
79. **Chimerix I.**2014. Chimerix announces emergency investigational new drug applications for brincidofovir authorized by FDA for patients with Ebola virus disease. http://ir.chimerix.com/releasedetail.cfm?releaseid=874647. Accessed 29 September 2015.
80. **Kraft CS, Hewlett AL, Koepsell S, Winkler AM, Kratochvil CJ, Larson L, Varkey JB, Mehta AK, Lyon GM III, Friedman-Moraco RJ, Marconi VC, Hill CE, Sullivan JN, Johnson DW, Lisco SJ, Mulligan MJ, Uyeki TM, McElroy AK, Sealy T, Campbell S, Spiropoulou C, Ströher U, Crozier I, Sacra R, Connor MJ Jr, Sueblinvong V, Franch HA, Smith PW, Ribner BS, Nebraska Biocontainment Unit and the Emory Serious Communicable Diseases Unit.** 2015. The use of TKM-100802 and convalescent plasma in 2 patients with Ebola virus disease in the United States. *Clin Infect Dis* **61:**496–502.
81. **Lyon GM, Mehta AK, Varkey JB, Brantly K, Plyler L, McElroy AK, Kraft CS, Towner JS, Spiropoulou C, Ströher U, Uyeki TM, Ribner BS, Unit ESCD, Emory Serious Communicable Diseases Unit.** 2014. Clinical care of two patients with Ebola virus disease in the United States. *N Engl J Med* **371:**2402–2409.
82. **Geisbert TW, Feldmann H.** 2011. Recombinant vesicular stomatitis virus-based vaccines against Ebola and Marburg virus infections. *J Infect Dis* **204**(Suppl 3)**:**S1075–S1081.
83. **BioCryst Pharmaceuticals I.**2014. BCX4430. http://www.biocryst.com/bcx_4430. Accessed 19 November 2014.
84. **Warren TK, Shurtleff AC, Bavari S.** 2012. Advanced morpholino oligomers: a novel approach to antiviral therapy. *Antiviral Res* **94:**80–88.
85. **Sadek RF, Khan AS, Stevens G, Peters CJ, Ksiazek TG.** 1999. Ebola hemorrhagic fever, Democratic Republic of the Congo, 1995: determinants of survival. *J Infect Dis* **179**(Suppl 1)**:**S24–S27.
86. **Jahrling PB, Geisbert TW, Geisbert JB, Swearengen JR, Bray M, Jaax NK, Huggins JW, LeDuc JW, Peters CJ.** 1999. Evaluation of immune globulin and recombinant interferon-alpha2b for treatment of experimental Ebola virus infections. *J Infect Dis* **179**(Suppl 1)**:**S224–S234.
87. **Mupapa K, Massamba M, Kibadi K, Kuvula K, Bwaka A, Kipasa M, Colebunders R, Muyembe-Tamfum JJ, Interna-**

tional Scientific and Technical Committee. 1999. Treatment of Ebola hemorrhagic fever with blood transfusions from convalescent patients. *J Infect Dis* **179**(Suppl 1)**:**S18–S23.
88. **World Health Organization.**2014. interim infection prevention and control guidance forcare of patients with suspected or confirmed Filovirus haemorrhagic fever in health-care settings. http://www.who.int/csr/resources/who-ipc-guidance-ebolafinal-09082014.pdf. Accessed 21 October 2014.
89. **Centers for Disease Control and Prevention.**2014. Infection prevention and control recommendations for hospitalized patients under investigation (puis) for Ebola virus disease (EVD) in U.S. hospitals. http://www.cdc.gov/vhf/ebola/healthcare-us/hospitals/infection-control.html. Accessed 28 September 2015.
90. **Centers for Disease Control and Prevention.**2014. Guidance on personal protective equipment to be used by healthcare workers during management of patients with Ebola Virus Disease in U.S. hospitals, including procedures for putting on (donning) and removing (doffing). http://www.cdc.gov/vhf/ebola/hcp/procedures-for-ppe.html. Accessed 21 October 2014.
91. **World Health Organization.**2014. Personal protective equipment in the context of Filovirus disease outbreak response. Technical specifications for PPE equipment to be used by health workers providing clinical care for patients. http://apps.who.int/iris/bitstream/10665/137410/1/WHO_EVD_Guidance_PPE_14.1_eng.pdf?ua=1. Accessed 03 November 2014.
92. **Prescott J, DeBuysscher BL, Feldmann F, Gardner DJ, Haddock E, Martellaro C, Scott D, Feldmann H.** 2015. Single-dose live-attenuated vesicular stomatitis virus-based vaccine protects African green monkeys from Nipah virus disease. *Vaccine* **33:**2823–2829.
93. **Geisbert TW, Mire CE, Geisbert JB, Chan YP, Agans KN, Feldmann F, Fenton KA, Zhu Z, Dimitrov DS, Scott DP, Bossart KN, Feldmann H, Broder CC.** 2014. Therapeutic treatment of Nipah virus infection in nonhuman primates with a neutralizing human monoclonal antibody. *Sci Transl Med* **6:**242ra82.
94. **Mills JN, Corneli A, Young JC, Garrison LE, Khan AS, Ksiazek TG, Centers for Disease Control and Prevention.** 2002. Hantavirus pulmonary syndrome—United States: updated recommendations for risk reduction. *MMWR Recomm Rep* **51**(RR-9)**:**1–12.
95. **Bausch DG, Rollin PE, Demby AH, Coulibaly M, Kanu J, Conteh AS, Wagoner KD, McMullan LK, Bowen MD, Peters CJ, Ksiazek TG.** 2000. Diagnosis and clinical virology of Lassa fever as evaluated by enzyme-linked immunosorbent assay, indirect fluorescent-antibody test, and virus isolation. *J Clin Microbiol* **38:**2670–2677.
96. **Vieth S, Drosten C, Lenz O, Vincent M, Omilabu S, Hass M, Becker-Ziaja B, ter Meulen J, Nichol ST, Schmitz H, Günther S.** 2007. RT-PCR assay for detection of Lassa virus and related Old World arenaviruses targeting the L gene. *Trans R Soc Trop Med Hyg* **101:**1253–1264.
97. **Drosten C, Göttig S, Schilling S, Asper M, Panning M, Schmitz H, Günther S.** 2002. Rapid detection and quantification of RNA of Ebola and Marburg viruses, Lassa virus, Crimean-Congo hemorrhagic fever virus, Rift Valley fever virus, dengue virus, and yellow fever virus by real-time reverse transcription-PCR. *J Clin Microbiol* **40:**2323–2330.
98. **Lewis VJ, Walter PD, Thacker WL, Winkler WG.** 1975. Comparison of three tests for the serological diagnosis of lymphocytic choriomeningitis virus infection. *J Clin Microbiol* **2:**193–197.
99. **Lozano ME, Enría D, Maiztegui JI, Grau O, Romanowski V.** 1995. Rapid diagnosis of Argentine hemorrhagic fever by reverse transcriptase PCR-based assay. *J Clin Microbiol* **33:**1327–1332.
100. **Leroy EM, Baize S, Lu CY, McCormick JB, Georges AJ, Georges-Courbot MC, Lansoud-Soukate J, Fisher-Hoch SP.** 2000. Diagnosis of Ebola haemorrhagic fever by RT-PCR in an epidemic setting. *J Med Virol* **60:**463–467.
101. **Southern TR, Racsa LD, Albariño CG, Fey PD, Hinrichs SH, Murphy CN, Herrera VL, Sambol AR, Hill CE, Ryan EL, Kraft CS, Campbell S, Sealy TK, Schuh A, Ritchie JC, Lyon GM III, Mehta AK, Varkey JB, Ribner BS, Brantly KP, Ströher U, Iwen PC, Burd EM.** 2015. Comparison of film array and quantitative real-time reverse transcriptase pcr for detection of zaire Ebola virus from contrived and clinical specimens. *J Clin Microbiol* **53:**2956–2960.
102. **Towner JS, Rollin PE, Bausch DG, Sanchez A, Crary SM, Vincent M, Lee WF, Spiropoulou CF, Ksiazek TG, Lukwiya M, Kaducu F, Downing R, Nichol ST.** 2004. Rapid diagnosis of Ebola hemorrhagic fever by reverse transcription-PCR in an outbreak setting and assessment of patient viral load as a predictor of outcome. *J Virol* **78:**4330–4341.
103. **Broadhurst MJ, Kelly JD, Miller A, Semper A, Bailey D, Groppelli E, Simpson A, Brooks T, Hula S, Nyoni W, Sankoh AB, Kanu S, Jalloh A, Ton Q, Sarchet N, George P, Perkins MD, Wonderly B, Murray M, Pollock NR.** 2015. ReEBOV Antigen Rapid Test kit for point-of-care and laboratory-based testing for Ebola virus disease: a field validation study. *Lancet* **386:**867–874.
104. **Zhai J, Palacios G, Towner JS, Jabado O, Kapoor V, Venter M, Grolla A, Briese T, Paweska J, Swanepoel R, Feldmann H, Nichol ST, Lipkin WI.** 2007. Rapid molecular strategy for filovirus detection and characterization. *J Clin Microbiol* **45:**224–226.
105. **Daniels P, Ksiazek T, Eaton BT.** 2001. Laboratory diagnosis of Nipah and Hendra virus infections. *Microbes Infect* **3:**289–295.
106. **Wang LF, Daniels P.** 2012. Diagnosis of henipavirus infection: current capabilities and future directions. *Curr Top Microbiol Immunol* **359:**179–196.
107. **Feldman KS, Foord A, Heine HG, Smith IL, Boyd V, Marsh GA, Wood JL, Cunningham AA, Wang LF.** 2009. Design and evaluation of consensus PCR assays for henipaviruses. *J Virol Methods* **161:**52–57.
108. **Feldmann H, Sanchez A, Morzunov S, Spiropoulou CF, Rollin PE, Ksiazek TG, Peters CJ, Nichol ST.** 1993. Utilization of autopsy RNA for the synthesis of the nucleocapsid antigen of a newly recognized virus associated with hantavirus pulmonary syndrome. *Virus Res* **30:**351–367.
109. **Clement J, McKenna P, Groen J, Osterhaus A, Colson P, Vervoort T, van der Groen G, Lee HW.** 1995. Epidemiology and laboratory diagnosis of hantavirus (HTV) infections. *Acta Clin Belg* **50:**9–19.
110. **Jenison S, Yamada T, Morris C, Anderson B, Torrez-Martinez N, Keller N, Hjelle B.** 1994. Characterization of human antibody responses to four corners hantavirus infections among patients with hantavirus pulmonary syndrome. *J Virol* **68:**3000–3006.
111. **Bostik P, Winter J, Ksiazek TG, Rollin PE, Villinger F, Zaki SR, Peters CJ, Ansari AA.** 2000. Sin nombre virus (SNV) Ig isotype antibody response during acute and convalescent phases of hantavirus pulmonary syndrome. *Emerg Infect Dis* **6:**184–187.
112. **Elgh F, Wadell G, Juto P.** 1995. Comparison of the kinetics of Puumala virus specific IgM and IgG antibody responses in nephropathia epidemica as measured by a recombinant antigen-based enzyme-linked immunosorbent assay and an immunofluorescence test. *J Med Virol* **45:**146–150.
113. **Elgh F, Linderholm M, Wadell G, Tärnvik A, Juto P.** 1998. Development of humoral cross-reactivity to the nucleocapsid protein of heterologous hantaviruses in nephropathia epidemica. *FEMS Immunol Med Microbiol* **22:**309–315.
114. **Padula PJ, Rossi CM, Della Valle MO, Martínez PV, Colavecchia SB, Edelstein A, Miguel SD, Rabinovich RD, Segura EL.** 2000. Development and evaluation of a solid-phase enzyme immunoassay based on Andes hantavirus recombinant nucleoprotein. *J Med Microbiol* **49:**149–155.
115. **Sjölander KB, Lundkvist A.** 1999. Dobrava virus infection: serological diagnosis and cross-reactions to other hantaviruses. *J Virol Methods* **80:**137–143.
116. **Hujakka H, Koistinen V, Kuronen I, Eerikäinen P, Parviainen M, Lundkvist A, Vaheri A, Vapalahti O, Närvänen A.** 2003. Diagnostic rapid tests for acute hantavirus infections:

specific tests for Hantaan, Dobrava and Puumala viruses versus a hantavirus combination test. *J Virol Methods* **108:**117–122.
117. **Chu YK, Rossi C, Leduc JW, Lee HW, Schmaljohn CS, Dalrymple JM.** 1994. Serological relationships among viruses in the Hantavirus genus, family Bunyaviridae. *Virology* **198:** 196–204.
118. **Mohamed N, Nilsson E, Johansson P, Klingström J, Evander M, Ahlm C, Bucht G.** 2013. Development and evaluation of a broad reacting SYBR-green based quantitative real-time PCR for the detection of different hantaviruses. *J Clin Virol* **56:**280–285.
119. **Zaki SR, et al.** 1995. Hantavirus pulmonary syndrome. Pathogenesis of an emerging infectious disease. *Am J Pathol* **146:**552–579.
120. **Hung T, Zhou JY, Tang YM, Zhao TX, Baek LJ, Lee HW.** 1992. Identification of Hantaan virus-related structures in kidneys of cadavers with haemorrhagic fever with renal syndrome. *Arch Virol* **122:**187–199.
121. **Drosten C, Kümmerer BM, Schmitz H, Günther S.** 2003. Molecular diagnostics of viral hemorrhagic fevers. *Antiviral Res* **57:**61–87.

Human Immunodeficiency Viruses and Human T-lymphotropic Viruses

JÖRG SCHÜPBACH

37

The human immunodeficiency viruses (HIV) and the human T-lymphotropic viruses (HTLV) originate from zoonotic transmission of ancestor retroviruses found in primates. HIV-1 and HIV-2 are the causative agents of AIDS; HTLV-1 causes adult T-cell leukemia/lymphoma and a broad spectrum of chronic inflammatory diseases, though only in few of those infected. The three principal questions in HIV diagnostics are (i) whether a person is HIV-infected and, if infected, (ii) with what exactly (viral properties) and (iii) how actively the virus is replicating (viral load). With regard to viral properties, identification of the virus type, i.e., HIV-1 or HIV-2, is essential for both the selection of a suitable viral load test for HIV RNA quantification in plasma and the decision with what to treat the patient. Answering these questions is important because current commercial viral load tests do not recognize HIV-2, and nonnucleoside reverse transcriptase inhibitors (NNRTI) are not effective against HIV-2 or group O viruses of HIV-1. Knowledge of preexisting viral mutations conferring resistance to antiretrovirals is another important point. As antiretroviral therapy is increasingly considered essential for all HIV infected persons, it is important to answer all relevant questions already at the timepoint of HIV diagnosis. Diagnosis of HTLV infection is performed accordingly, but with some important differences. HTLV never leads to viremia. Tests for viral RNA in plasma are thus useless, and all nucleic acid testing has to be performed on cells. The most important task is to differentiate between the pathogenic HTLV-1 and the virtually nonpathogenic HTLV-2.

VIRAL CLASSIFICATION AND BIOLOGY

Origin

The human immunodeficiency viruses (HIV) as well as the human T-lymphotropic viruses (HTLV) belong to the *Retroviridae* family.

The two types of HIV, HIV-1 and HIV-2, belong to the *Lentivirus* genus of the *Retroviridae*, more specifically to a subgroup of primate lentiviruses called simian immunodeficiency viruses (SIV). SIV naturally infect various species of Old World monkeys and the chimpanzee. They are categorized into five major lineages. Lineage 1 contains HIV-1, which meanwhile comprises four groups, M (main), O (outlier), N (non-M/non-O), and P (probable). The phylogenetic tree shows that certain group M isolates are more closely related to two isolates obtained from chimpanzees than to HIV-1 group O isolates. This proves that the HIV-1 epidemic is the result of zoonotic virus transmissions from chimpanzee to human (1). The origin of group M diversification, i.e., the beginning of the HIV-1 pandemic in man, is placed around 1930 (2, 3). Lineage 2 of the SIV contains the various isolates of HIV-2, which are related to viruses naturally prevalent in sooty mangabeys (SIVsm). Again, some isolates of HIV-2 differ less from SIVsm than from other human HIV-2 isolates. This demonstrates that the HIV-2 epidemic results from multiple simian-to-human cross-species transmissions. Transmission of the epidemic subtypes HIV-2 A and B may have occurred around 1940 (4).

The two types of HTLV that have a sizeable frequency in humans, HTLV-1 and HTLV-2 (also spelled HTLV I and HTLV II), belong to the group of primate T-lymphotropic viruses (PTLV) that, together with the bovine leukemia virus, form the genus *Deltaretrovirus* in the *Orthoretrovirinae* subfamily of the *Retroviridae*. Viruses related to HTLV-1 and HTLV-2 are found in many different species of Old World monkeys. Phylogenetic analysis separates the PTLV into four different branches, PTLV-1, -2, -3, and -4. Depending on whether PTLV are found in man or nonhuman primates, they are named either HTLV or STLV. HTLV-3 and -4 were identified in African hunters from Cameroon; for HTLV-4 a primate reservoir was recently identified in gorillas (5, 6). HTLV infection in humans thus has resulted from multiple cross-species transmissions of STLV in the past, and such zoonotic transmission may well be ongoing.

Retroviral Replication and Virus Diversity

Both HIV and HTLV are enveloped plus-strand RNA viruses with a diameter of about 110 nm. Infectious particles (virions) contain two identical copies of single-stranded RNA of about 9 to 10 kb. The genome is organized into three genes, *gag*, *pol*, and *env*, that are found in all retroviruses. Each genome further contains a set of regulatory and accessory genes important for virus replication. The *gag* gene codes for the structural proteins that form the nucleocapsid and the

doi:10.1128/9781555819156.ch37

matrix shell, *pol* for the viral enzymes (the primary targets for most of the current antiviral drugs), and *env* for the viral glycoproteins inserted as trimers into a lipid envelope derived from the host cell membrane.

For infection of a host cell, a virion binds via its envelope to virus receptor(s) located on the cell membrane. In the case of HIV, the viral envelope glycoprotein gp120 binds to the CD4 molecule. This is followed by several conformational changes in the gp120 trimer that enable further interaction of gp120 with a chemokine coreceptor, CCR5 or CXCR4, and insertion of the fusion domain of the viral transmembrane protein gp41 into the host cell membrane. This leads to fusion of the viral and cellular membranes and viral entry (reviewed by 7). In agreement with CD4 being the receptor, $CD4^+$ T lymphocytes are the main host cells for both HIV-1 and HIV-2.

In the case of HTLV-1, three molecules, Glucose Transporter 1, Neuropilin-1, and Heparan Sulfate Proteoglycans, appear to be involved successively in virus binding and entry by means of interaction with the viral glycoproteins gp46 and gp21. Despite the fact that there is no interaction with CD4, the main target cell for HTLV-1 in vivo is the $CD4^+$ T cell. In contrast, HTLV-2 predominantly infects $CD8^+$ T cells and appears to use a different receptor complex (8).

After infection of a host cell, the enzyme reverse transcriptase (RT) that is contained in all retroviruses synthesizes a cDNA of the viral RNA, degrades RNA from the cDNA-RNA heteroduplex and duplicates the cDNA strand. Regulatory sequences present at both ends of the viral genome are thereby complemented and partially duplicated in a manner that yields the long terminal repeats (LTR) that are located at both ends of the double-stranded viral DNA and act as promoters for the retroviral genes. The double-stranded DNA migrates into the nucleus where it is integrated into the host cell genome by the integrase (IN), another mandatory retroviral enzyme. The integrated retroviral DNA is called the provirus. The production of viral RNA, structural proteins, and enzymes involves the cellular enzymes associated with transcription and translation and also a number of viral regulatory and accessory proteins. New particles are assembled at the cell membrane and, while still immature and noninfectious, released by budding. For full maturation into infectious particles, the viral protease (PR) is required to cleave the *gag* and *gag-pol* precursor proteins into the different, functionally active subunit proteins.

In addition to replication by infectious cycles, retroviruses can also replicate by clonal expansion of the integrated provirus during successive rounds of mitosis of their host cells. This has the advantage that the virus has not to be OR to get OR to become expressed and thus is protected from the immune system. HIV appears to replicate predominantly by infectious cycles, but clonal expansion dominates for HTLV, although replication by de novo infection also occurs in the early period of infection. Depending on the dominating mode of replication, the degree of genetic diversity will differ among different retroviruses.

Characteristics of HIV Replication

In the case of HIV, at least 10^{10} HIV particles are produced newly every day, a single viral replication cycle lasts 1 to 2 days on average and the half-life of virus in plasma is in the order of one hour (reviewed by 9). Retroviral RTs do not possess a proofreading activity and thus have a high misincorporation rate. Additional errors may occur during transcription since RNA polymerase II does not proofread either. For the 9.5 kb HIV genome, the in vivo error rate amounts to 1 to 3 misincorporations per replication cycle (10). Given the high rate of HIV replication, every single mutation at every possible position of the genome could arise daily. Another mechanism contributing to sequence diversity is genomic recombination, which may occur after coinfection of a cell with two different viruses and encapsidation of both viral RNAs in the same particle (heterozygosity). Its frequency is estimated at 2 to 3 events per viral genome and replication cycle. Point mutation and recombination together lead to a rapid accumulation of virus variants in an infected person. Selective pressure factors, such as the local availability of host cell receptors or coreceptors, cellular or humoral antiviral immune responses, or antiretroviral drugs may then act on this pool of variant viruses, inhibiting the growth of some variants and favoring the replication of others that exhibit a better-suited phenotype. The outgrowth of such a group of viruses under selection pressure is called a quasispecies (11). The many quasispecies in each patient evolve both in time and space. It is estimated that the sequence variability in an infected, antiretroviral-therapy (ART)-naive person increases by about 1% per year. In a given patient, different quasispecies are present at different sites in the body, for example, in different parts of the skin, spleen, brain, or genital tract.

About 93 to 99% of the virus in the blood plasma of untreated HIV-1 infected patients originates from activated $CD4^+$ T lymphocytes that get infected, produce virus, and die with a half-life of only 0.7 ± 0.2 days (so-called productively infected $CD4^+$ T lymphocytes). An additional 1 to 7% of the virus in plasma originates from longer-lived cells (monocytes or macrophages, release of surface-bound virus from dendritic cells) that have a half-life of 14 ± 7.5 days. Less than 1% of the virus in plasma is produced by latently infected $CD4^+$ T-cells, which become activated and then start producing virus. This last compartment has a very slow decay rate with a half-life estimated at 6 to 44 months, or it may even not decay at all. Eradication of this compartment will not be possible without measures that activate the virus from its state of latency (12).

Characteristics of HTLV Replication

HTLV, in contrast to HIV, exhibits little genetic diversity. Although HTLV-1 comprises 7 different subtypes, a through g, representing as many zoonotic transmission events, the genetic distances within a given subtype amount to only a few percent. In a study of familial transmission of HTLV-1b, in which 7 infected family members in three generations were studied, all 7 HTLV isolates had identical *env* sequences, i.e., of the gene that is least conserved among retroviruses. A single mutation in the LTR accumulated within the 639 nt LTR-*env* sequence stretch studied over a period of 189 years, amounting to 1.1×10^{-6} nucleotide substitutions per site per year (13). The low degree of sequence variation indicates that reverse transcriptase plays only a minor role in HTLV replication. Indeed, HTLV spreads primarily by clonal expansion of infected cells. If de novo infection of cells or another individual occurs, it is by a highly targeted cell-to-cell transmission mechanism by means of virological synapses (14) and biofilm-like extracellular viral assemblies (15). Free virions have a poor infectivity and, much in contrast to HIV, HTLV virions are never found in the plasma. Instead, the blood contains very high levels of cells carrying the HTLV provirus, frequently exceeding 10% of all lymphocytes or 20% of the $CD4^+$ T-cells. Experiments in humanized mice infected with HTLV-1 and treated with RT inhibitors either simultaneously or with a one-week delay showed that the delayed treatment had no effect on the provirus load. This suggests that virus replication by infectious cycles occurs only during a short period after infection and that a pool of HTLV-1 infected cell clones is generated in this early period. Thereafter, clonal proliferation of infected cells appears to be predominant. In humans, many thousands of different clones can be found that are characterized by their unique site of provirus integration; they also differ considerably regarding their clone size. Oligoclonal expansion of HTLV-1 infected cells is frequently seen in patients with chronic infection or inflammatory disease of the central nervous system, and monoclonal expansion is a defining property of adult T-cell leukemia/lymphoma (ATLL). The processes that determine why most clones remain small, others grow in size to dominance, and some may eventually be transformed into highly malignant tumors are the target of intensive ongoing study (16, 17).

EPIDEMIOLOGY

Global Distribution of HIV

According to UNAIDS, between 33 and 37 million people were living with HIV in 2013. The global prevalence among individuals aged >15 years was estimated at 0.7 to 0.8%. New infections in 2013 amounted to 1.9 million in adults and to 240,000 in children. The region most affected by the epidemic is sub-Saharan Africa, where the overall prevalence among adults in 2013 was 4.7%. This accounts for almost two-thirds of the global total of HIV infection. Other world regions with a high prevalence include the Caribbean (1.1%) and Eastern Europe/Central Asia (0.6%), while the prevalence in other world regions is between 0.1 and 0.4%.

The extraordinary variability of HIV in conjunction with geographical compartmentalization has led to the development and geographical distribution of various distinctive clades, or subtypes, of viruses. HIV-1 is now composed of four phylogenetic groups: M, N, O, and P. Group M alone is responsible for the HIV pandemic. It is divided into subtypes A, B, C, D, F, G, H, J, and K. Genetic variation within a subtype may amount to 15 to 20%, whereas variation between subtypes is between 25 and 35% (18). Viral recombination, a possible consequence of infection of a person with more than one virus (coinfection or superinfection), has furthermore resulted in a great variety of so-called circulating recombinant forms (CRFs), which increasingly dominate the epidemic. To date, more than 65 CRFs have been defined for HIV-1 and one for HIV-2. According to a WHO study involving 23,874 HIV-1 samples from 70 countries, subtype C accounted for 50% of all infections worldwide in 2004. Subtypes A, B, D, and G accounted for 12%, 10%, 3%, and 6%, respectively. Subtypes F, H, J, and K together accounted for 1%. The circulating recombinant forms CRF01_AE and CRF02_AG each were responsible for 5% and CRF03_AB for 0.1%. Other recombinants accounted for the remaining 8% of infections. All recombinant forms together were responsible for 18% of infections (19). Isolates of HIV-1 group O, which are almost exclusively restricted to persons originating from Cameroon, Gabon, and Equatorial Guinea, differ as much from each other as do viruses from different subtypes of group M, but their limited number has so far precluded a definition of distinct subtypes. HIV-1 group N viruses were isolated from only a few individuals from Cameroon (20). A total of seven subtypes of HIV-2, two of which are epidemic (A and B) and five nonepidemic (C to G), have been defined, resulting from as many different simian-to-human transmissions (4).

HIV Transmission

HIV is transmitted predominantly by sexual intercourse, connatally from mother to child, postnatally by breast feeding or by parenteral inoculation, most importantly intravenous drug injection. Globally, the most frequent route of transmission is by sexual intercourse. The probability of HIV-1 transmission per 10,000 exposures amounts to 9,250 for blood transfusion, 63 for needle-sharing, 23 for percutaneous needle-stick, 138 for receptive anal intercourse, 11 for insertive anal intercourse, 8 for receptive penile-vaginal intercourse, and 4 for insertive penile-vaginal intercourse, while it is unquantifiably low for both receptive and insertive oral intercourse (21). In general, the risk is proportional to the viral load, as determined by RT-PCR for HIV-1 RNA (22, 23). Usually, a single virion is transmitted. The virus is not transmitted through casual contact in household settings, and there is no evidence for transmission by nonhuman vectors.

Sexual transmission is mediated by infectious HIV-1 particles and/or virus-infected cells in the semen or mucosal secretions. The risk of transmitting or acquiring infection varies greatly. Epidemiologic studies indicate that transmission is linked to viral shedding, i.e., the amount of infectious virus in genital fluids. This in turn is linked to the disease stage and is highest during acute infection and late-stage AIDS (23). Antiretroviral therapy (ART) can reduce HIV-1 shedding in semen and the female genital tract to undetectable levels, but virions can sometimes be found in semen even when they are undetectable in the blood plasma. Although some untreated infected individuals pose a low transmission risk—notably, no virus transmission was observed from individuals with less than 1,500 copies of HIV-1 RNA per milliliter of plasma or serum–others may be "supershedders" and highly infectious. Acutely infected individuals pose a particular risk (23). Moreover, other sexually transmitted diseases (STDs) markedly increase both viral shedding and the risk of acquiring HIV-1 infection (reviewed by 24 and 25).

Global Distribution of HTLV

The number of HTLV-1 infected persons worldwide is estimated to be at least 5 to 10 million, but due to incomplete epidemiological studies in the endemic regions the true number remains unknown. Prevalence rates of >1% are found in the Caribbean, Central Africa, and South Japan. In most other areas in the world, HTLV-1/2 infections are mainly found in high-risk groups (i.e., immigrants from endemic areas, their offspring or sexual contacts, and in patients and intravenou drug injection users attending sexually transmitted disease clinics). Also, a high rate of infection for both HTLV-1 and HTLV-2 was observed in the native Amerindian population in North America as well as in South America (26).

HTLV-1 currently comprises 7 different subtypes designated a through g. The "cosmopolitan" subtype HTLV-1a includes the prototype isolates from Japan and is found in many endemic areas worldwide. Its current worldwide distribution is thought to result from relatively recent human migration such as the European voyages of discovery of past centuries and the slave trade. Subtypes b, d, and f are still restricted to Central Africa. Subtype e is prevalent in South and Central Africa, subtype c is found in Melanesia, and subtype f was described in central African bushmeat hunters (5).

HTLV-2 comprises two main subtypes, a and b (27). Both are present in intravenous drug users in North America, Europe, and Asia and have been found sporadically in Africa. HTLV-2a is present in certain American Indian tribes of North, Central, and South America. Due to a high prevalence in isolated Amerindian populations, HTLV-2 was originally thought to be of New World origin. The discovery of endemic HTLV-2 infections in remote Pygmy populations and the identification of a simian virus closely related to HTLV-2 in bonobos suggest, however, that HTLV-2 rather has its origin in Africa. The molecular characterization of HTLV-2 isolates from Pygmies living in Cameroon and Congo also supports an ancient African origin of HTLV-2 (28).

HTLV Transmission

Like the HIVs, the HTLVs are transmitted by hetero- or homosexual intercourse, from mother to child or by parenteral inoculation. Live HTLV-infected cells are essential in all transmission modes. Mother-to-child HTLV transmission rate is 15 to 35%, similar to that of untreated HIV-1 infections and occurs predominately in the postnatal period through breast milk. Killing live cells from mother's milk by freezing and thawing abolishes viral transmission (29). Breast-milk transmission seems to be more efficient than for HIV-1 and occurs with a time- or dose-dependent frequency. In one study, overall transmission was 16%. It was 5% among infants breastfed for up to three months and 27% among those breastfed for longer than three months. Of 78 bottle-fed infants, 13% turned out to be infected, suggesting also connatal transmission (30). In other studies, however, connatal transmission was considerably less frequent (3%). Although HTLV-1 infected cells were detected by PCR in 2.5% of the cord bloods from HTLV-1–positive pregnancies, this was not associated with infection when the babies were formula-fed (31). Transmission by breast milk also depends on its provirus load (32).

HTLV transmission by blood products is, in contrast to HIV, strictly cell-associated; the virus is not transmitted by plasma or plasma-derived products (33). Recipients of contaminated blood seroconvert with a 40 to 60% probability and a median seroconversion time estimated at 51 days (34). Routine HTLV screening of blood donors is justified in countries with an elevated prevalence in the general population and has been implemented in Japan, North America, and several European countries.

CLINICAL SIGNIFICANCE

HIV-1 and HIV-2

HIV-1, first isolated in 1983 (35), is the virus responsible for the AIDS pandemic, as was conclusively shown in 1984 by a series of papers demonstrating its systematic serological and virological association with early and late stages of the disease (36–40). AIDS, first recognized as a new disease among homosexual young men in 1981 (41), is the result of a long-lasting continuous destruction of a patient's $CD4^+$ T lymphocytes induced by HIV. The underlying pathogenic mechanisms are not yet fully understood; for details on current knowledge consult recent reviews (42–44).

Important stations on the way to AIDS, as shown in Fig. 1, include (i) an initial hidden phase during which the virus spreads from the entry port to the various tissues and organs. All this occurs before the virus becomes detectable in the plasma and before clinical symptoms develop. (ii) This is followed by a period of unhindered replication mainly in activated $CD4^+$ T cells located in these sites, leading to a peak of viremia, which frequently exceeds 10^6 to 10^7 virions per milliliter plasma. The massive virus production induces an immune response both in the form of antibodies against all viral proteins and a $CD8^+$ T cell response directed against the viral antigens expressed on infected cells. The immune response also activates new $CD4^+$ T cells, which become readily infected and produce new virus, thus generating a vicious cycle. This phase of unhindered virus replication corresponds to the stage of acute or primary HIV infection (PHI) and is frequently associated with a flu-like disease. The acute phase comes to an end when the immune system has gained partial control of virus replication and the readily infectable activated $CD4^+$ T cells have been exhausted. This is manifested by rapidly falling virus concentrations in the

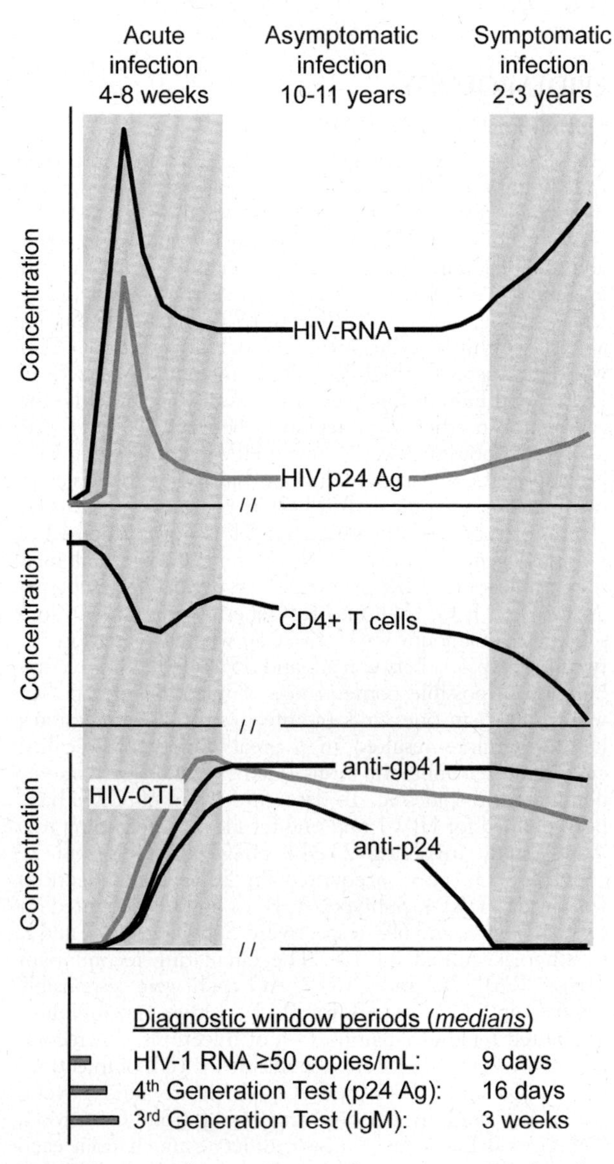

FIGURE 1 Virologic and immunologic parameters and median diagnostic window periods in the typical course of HIV-1 infection. For further description refer to sections III. Clinical Significance and V. Detection/Diagnosis.

plasma by two or three logs to a stable level called the "set point"; in the blood a transient decline of CD4$^+$ T cell numbers is seen. (iii) In the phase of chronic infection or "clinical latency", which may last for many years, the virus concentration in plasma remains either stable at the set point or increases slowly. The CD4$^+$ T cells decline slowly, by 70 cells/mm^3 per year on average. The patients in this clinically latent phase have no symptoms and usually do not know that they are infected. (iv) When the number of CD4$^+$ T cells has declined to a point below which immune control of adventitious infectious agents is no longer possible (below 200 cells/mm^3), opportunistic infections arise and tumors may develop; i.e., the final AIDS stage has been reached. This occurs about 10 years after infection on average. Any remaining immune control of the HIV infection itself is also lost, and plasma viremia increases again. If treatment with an effective combination of antiretrovirals is not instated at this point, AIDS will be fatal within a few years.

The process that is central to the pathogenesis of AIDS is the slow progressive destruction of the CD4$^+$ T cells. It has been known for a long time that cells productively infected with HIV-1 die of apoptosis, but this affects only a small proportion of the CD4$^+$ T cell total. Why apparently non-infected "bystander" cells are also destroyed has long remained unknown. Recent research now suggests that the infection of most CD4$^+$ T cells by HIV is abortive, allowing only the initial steps of reverse transcription and synthesis of a double-stranded proviral DNA. Migration of the proviral DNA into the nucleus and integration, however, appear to be rare events. The foreign viral DNA in these abortively infected cells now was found to activate a DNA sensor located in the cytoplasm. This leads to cell death due to an inflammatory process called pyroptosis, which is different from necrosis or apoptosis (45). Pyroptosis also causes immune activation, another hallmark of infection by HIV-1. Immune activation is thought to be responsible for the increased occurrence of atherosclerosis and metabolic syndrome in HIV-1 infected individuals.

HIV-2, discovered in 1986 in a patient with AIDS (46), is less pathogenic than HIV-1. Infection with HIV-2 occurs mainly in West Africa, but an increasing number of cases have been recognized in Europe, India, and the U.S. Rates of heterosexual and mother-to-child transmission of HIV-2 are low, and the virus rarely causes AIDS. Most HIV-2 infected individuals remain long-term nonprogressors, whereas most of those infected with HIV-1 progress. When clinical progression occurs, both diseases demonstrate very similar pathological processes, although progression to AIDS in HIV-2 occurs at higher CD4 counts and lower plasma viral loads. Immune activation is also absent. For more detailed information, consult (47) and (48).

HTLV-1 and HTLV-2

Most individuals infected with HTLV-1 remain disease free throughout their lifetime. In 2 to 6% of infected individuals, however, chronic disease may develop, usually after a long incubation time. Three characteristic disease entities have been etiologically linked with HTLV-1 infection in adults. They include adult T-cell leukemia/lymphoma (ATLL), a malignancy first described in 1977 (49, 50) and subsequently linked with HTLV-1 (51, 52); HTLV-1-associated myelopathy/tropical spastic paraparesis (HAM/TSP) (53, 54); and HTLV-1-associated uveitis (HAU) (55). Syndromes found associated with HTLV-1 in children include infective dermatitis (56) and HAM/TSP. Other manifestations less well linked with HTLV-1 infection include various inflammatory diseases like polymyositis, arthritis, infiltrative pneumonitis, Sjögren's syndrome, and, in children, persistent lymphadenopathy. A general susceptibility to infectious diseases is also frequent. For more detailed information consult 16, 17, 57, and 58.

HTLV-2 was originally isolated from a T-cell line derived from a patient with a T-cell variant of hairy cell leukemia (59). Subsequently, the virus was also isolated from a similar case, which upon closer examination demonstrated a coexistence of two different proliferative processes, namely, a CD8$^+$ T-cell leukemia with monoclonally integrated HTLV-2 and a B-cell hairy cell leukemia negative for integrated HTLV-2 (60). Epidemiological studies have excluded, however, that the typical B-cell form of hairy cell leukemia is associated with HTLV-2 (61). More recent studies have shown that HTLV-2, though persistently associated with elevated lymphocyte and platelet counts and with an increase in overall cancer mortality, does not cause hematologic disorders and is only sporadically associated with myelopathy (62).

TREATMENT AND PREVENTION

HIV

Antiretroviral Treatment

Eradication of HIV infection cannot be achieved with available antiretroviral regimens. This is due to the pool of latently infected CD4$^+$ T-cells that is established during early HIV infection and persists with a long half-life. Hence, the primary goals of therapy in HIV-infected patients are to reduce HIV-related morbidity and mortality, to improve quality of life, to restore and preserve immunologic function, and to maximally and durably suppress virus replication as measured by the viral load. Treatment with effective combinations of antiretroviral drugs has resulted in substantial reductions in HIV-related morbidity and mortality. Plasma viremia is a strong prognostic indicator of HIV disease progression. Reductions in plasma viremia achieved with such antiretroviral combination therapy account for substantial clinical benefits (63). Therefore, suppression of plasma viremia as much as possible for as long as possible is the goal of current antiretroviral therapy. The ambitious new goal of curing HIV infection will require "flushing out" of latently infected cells by agents capable of activating the virus from dormant cells; for more information see (64).

Due to the high variability of HIV and the generation of drug-resistant mutants, successful long-lasting suppression of virus replication can only be achieved with a suitable combination of antiretrovirals (ART). Such regimens are usually composed of at least three drugs, whereby one drug is selected from the groups of NNRTI, protease inhibitors (PI) or integrase inhibitors (INI), and the other two are chosen from the group of nucleoside (or nucleotide) analogue RT inhibitors (NRTI). NRTIs function as nucleoside-triphos-phates for RT-mediated cDNA synthesis and act as chain terminators; they are active against both HIV-1 (including group O) and HIV-2 and also against other retroviruses. NNRTI bind directly to the RT thereby blocking its active site either directly or indirectly. As NNRTI were developed against the RT of HIV-1 group M they are inactive against HIV-2 and group O of HIV-1, which appear to have a different active site configuration. PIs block the active site of the viral PR, thereby inhibiting the processing of the *gag-pol* and *gag* precursor proteins. INI block the active site of the HIV integrase. An entirely different class of drugs is directed against the cellular coreceptor, CCR5, which is required for virus entry. For information on FDA-approved or investigational drugs and all questions regarding ART consult http://aidsinfo.nih.gov/.

Prevention of HIV Transmission

Any measure that effectively reduces exposure to the viral inoculum present in genital secretions will contribute to reducing the extent of sexual transmission of HIV. Condom use is one of the central measures and reduces the incidence of new HIV infections by 80 to 95% when used consistently (65). Early identification, counseling, and treatment of infected individuals, including those in the highly infectious acute stage of infection, as well as contact tracing are also important. Prompt treatment of concomitant sexually transmitted diseases reduces inflammation and thus the number of infected, activated, and virus producing $CD4^+$ T lymphocytes and macrophages in genital secretions (25). Use of ART has been shown effective in reducing HIV-1 shedding in genital secretions, and a reduction in HIV-1 transmission from infected patients receiving ART has been observed in several studies. Expanded access to ART as a cost-efficient means to curb the growth of the HIV pandemic thus remains a promising strategy (44, 66).

Preventive measures that reduce the risk of a person for getting HIV infected sexually include barrier methods, but behavioral changes like abstinence, faithfulness, and partner reduction are of similar importance (67, 68). Male circumcision, which sizably reduces the infectable epithelial area of the penis, has been found associated with a 60% risk reduction (69, 70). Microbicide approaches, in particular those involving topical application of antiretroviral drugs like Tenofovir, have also shown promising results (71). Post-exposure prevention (PEP) of HIV transmission by systemic antiretroviral drug treatment has proven effective and is a standard procedure in developed countries (72); for actual recommendations consult http://aidsinfo.nih.gov/. Pre-exposure prevention (PrEP) with antiretroviral drugs like Tenofovir, a nucleotide reverse transcriptase inhibitor of high efficacy, is also advocated for high-risk individuals (73, 74). In contrast, vaccine development has not made significant progress and is many years away from routine application.

HLTV

Treatment of ATLL

Asymptomatic carriers of HTLV-1 or HTLV-2 are approached by watchful waiting; treatment is restricted to individuals with disease. Because in chronic HTLV infection the virus replicates predominantly by clonal expansion, the inhibition of viral replication by means of ART has no rational place in HTLV treatment. Nevertheless, the NRTI azidothymidine (AZT) has been used with success in combination with interferon alpha for treatment of ATLL. The effect of AZT, however, seems not to be due to its antiviral potency, but rather to induction of a telomerase, thereby reprogramming the cells to a p53-dependent senescence (75).

During the more than three decades since the recognition of ATLL a variety of treatment approaches has been evaluated. Clinical trials, mostly conducted in Japan, have demonstrated that combinations of chemotherapy can induce acceptable response rates, especially in the lymphoma subtype. However, the overall outcome remains poor due to a high rate of relapse. Similarly, the so-called indolent forms of ATLL, the smoldering and chronic subtypes, have a poor long-term prognosis. A worldwide meta-analysis showed that the combination of AZT and interferon alpha was highly effective in the leukemic subtypes of ATLL; it also improved the long-term survival of patients suffering from smoldering or chronic ATLL, as well as of a subset of patients with acute ATLL. A further increase of efficacy is expected from the addition of arsenic trioxide to this regimen. Allogeneic hematopoietic stem cell transplantation for patients that have achieved complete remission is a further option. Finally, a number of new agents including purine analogs, histone deacetylase inhibitors (which should improve virus expression in ATLL cells, thereby exposing them to the patient's antiviral cytotoxic T-cell responses), and monoclonal antibodies against CXCR4, are being evaluated in clinical trials. For more information consult (16, 58, 76).

Treatment of HAM/TSP

HAM/TSP is considered to be due to a CTL response directed against HTLV-expressing cells in the CNS. Therefore, immunosuppressive and immune modulating agents including corticosteroids, interferon alpha, and interferon beta 1a have been tried, albeit with limited success. The combination of interferon alpha and AZT was also ineffective.

Prevention of HTLV Infection

Overall, the diseases caused by HTLV, in particular ATLL, remain associated with a poor prognosis. Prevention of HTLV infection is thus of paramount importance. Public health intervention with the aim to provide education and counseling of high-risk individuals and populations is required. Avoidance of breast feeding and the introduction of HTLV screening of all blood donors have led to a significant decline of the carrier rate among the younger generation in Japan (77). Given the high kit costs for blood donor screening, a transfer of this strategy to resource-poor settings with HTLV-1 endemicity has so far not been possible. Blood transfusion still represents a risk of HTLV-1 infection for recipients in most African countries, as well as for other less developed areas. Prevention of mother-to-child transmission would likely have a significant impact on the incidence of HTLV-1-associated diseases, but the benefits of avoiding breastfeeding must be weighed against its risks, namely, malnutrition and increased infant mortality. Recommendations to prevent sexually transmitted infections are the same as for the prevention of HIV infection.

DETECTION/DIAGNOSIS

Diagnosis of HIV Infection

The three questions of principal interest in HIV diagnostics are (i) whether a person is HIV-infected and, if infected, (ii) with what exactly (viral properties) and (iii) how actively the virus is replicating (Table 1). With regard to viral properties, identification of the virus type, i.e., HIV-1 or HIV-2, is essential for both the selection of a suitable viral load test and the decision with what to treat the patient. Answering these questions is important because current commercial viral load tests do not recognize HIV-2, and NNRTI are not effective against HIV-2 or HIV-1 group O viruses. Knowledge of preexisting mutations conferring resistance to antiretroviral drugs is another clinically very important point. According to current opinion, antiretroviral combination therapy (ART) is increasingly considered essential for all HIV infected persons (see http://aidsinfo.nih.gov/), making it important to answer all medically relevant questions already at the timepoint of HIV diagnosis.

TABLE 1 The principal questions of HIV diagnosis and how best to answer them

1. Is an individual infected with HIV?	
HIV screening in individuals >18 months	Use approved HIV-1/2/O Ab + Ag combination test
Verification (confirmation) of HIV diagnosis	Two different tests of high diagnostic specificity are clearly reactive AND Two different samples have been tested, each with a clearly reactive or positive result
Neonatal/Pediatric HIV screening at <18 months	Initially perform an HIV screening test to see whether baby was exposed to maternal HIV infection. Approved HIV-1/2 differentiation immunoassay to determine the correct test for initial testing (HIV-1 or HIV-2); thereafter for HIV-1: HIV-1 RNA in plasma (or DNA-PCR in PBMC) and/or for HIV-2: HIV-2 DNA in PBMC by means of a high-input (MEGA) PCR
2. What properties does the virus have?	
Discrimination between HIV-1 and HIV-2, detection of double-infection HIV-1 plus HIV-2	Approved HIV-1/2 differentiation immunoassay
Confirmation of HIV-2 infection	High-input (MEGA) PCR for HIV-2 in PBMC
Resistance against antiretroviral medications	Genotypic resistance-testing of at least the PR and RT regions, optimally also the IN region
Identification of HIV-1 group O	Achieved most effectively via sequencing in the context of genotypic resistance testing Group O-specific DNA or RNA-PCR
3. How high is the viral load?	
HIV-1 quantification	Approved quantNAT for HIV-1 RNA
HIV-2 quantification	Approved quantNAT for HIV-2 RNA or sequence-independent quantification using the PERT assay
HIV-1 & HIV-2 double-infection	Approved quantNAT for HIV-1 PLUS quantNAT or qualNAT for HIV-2 RNA
Verification of a viral load of <1,000 copies/mL obtained using a commercial quantNAT	Use of a second, different quantNAT or sequence-independent testing using the PERT assay

Abbreviations: PR, protease; RT, reverse transcriptase; IN, integrase; PBMC, peripheral blood mononuclear cells; quantNAT, quantitative nucleic acid test; qualNAT, qualitative nucleic acid test.

Question 1: Is a Person Infected with HIV?

HIV infection can be detected by a variety of tests. Virus components that can be assayed include viral RNA or proviral DNA, viral proteins, in particular the p24 antigen and RT, the enzymatic activity of which can be detected by functional tests. Most frequently, however, HIV infection is diagnosed by tests that assess whether an individual has produced HIV-specific antibodies. Since retroviruses establish infections that persist for life, the demonstration of HIV-specific antibodies can be trusted to reflect ongoing infection, provided that it is consistent and directed against various viral antigens (and that the individual has not participated in an HIV vaccine trial). Thus, testing for HIV-specific antibodies is still an important tool for HIV diagnostics, at least in adults. In infants, solely the testing for virus components allows early diagnosis of the infection.

The diagnosis of HIV infection relies on commercially available test kits. There is a large number of well-standardized commercial diagnostic products of high sensitivity and specificity, which provide a continuously high standard of quality. They are usually better and yield more consistent results than research procedures developed by diagnostic laboratories. The use of good commercial tests is therefore strongly recommended. Using unregistered tests for screening or for certain types of supplemental testing is unlawful in many countries. In the U.S., refer to http://www.fda.gov/cber/products/testkits.htm for the actual list of U. S. Food and Drug Administration (FDA) approved commercial diagnostic tests. Commercial tests for diagnostic use in Europe need to be Communauté Européenne (CE)-marked.

Only very general descriptions of procedures are given in the following sections, since commercial test kits all contain detailed step-by-step instructions. For procedures that are not commercially available, the reader is directed to the referenced literature. The intent is to guide the reader through the multitude of available procedures and to discuss their strengths and weaknesses.

Window Periods in Early HIV Infection, HIV Screening in Adults

HIV-specific antibodies are produced within a few weeks after infection. They are induced by the various viral proteins produced during the phase of unrestricted HIV replication that is characteristic for the initial, acute phase of the infection (Fig. 1). The time to positivity in HIV antibody tests (i.e., to seroconversion) depends not only on the sensitivity of the test but also on the extent of virus replication, as shown by patients who, when diagnosed prior to seroconversion and immediately put on antiretroviral treatment, exhibit a delayed seroconversion (78).

In a study based on the first generation of HIV antibody screening assays developed three decades ago and using viral lysate for target antigen, HIV seroconversion was estimated to occur on average 45 days after infection; with 95% certainty the window period for 90% of individuals was less than 20 weeks (79). In comparison to this, the more recently developed tests have reduced the average window period as follows: third-generation anti-HIV-1/2 enzyme immunoassays for detection of antibodies (including IgM) by the double-antigen sandwich (DAGS) format, reduction by 20.3 d; use of p24 antigen or PCR for proviral DNA, reduction by 26.4 d; and PCR for viral RNA in plasma, reduction by 31.0 d (80). With third-generation antibody screening assays, the median window of seroconversion amounts to approximately 3 weeks, i.e., half of the infected individuals should become antibody positive within approximately 3 weeks after infection.

Compared to the 3-week median window of third-generation antibody assays, p24 antigen testing, or the use of fourth-generation combination assays that detect both HIV antibodies and p24 antigen, reduces the window by a further 5 days (i.e., to 16 days). RT-PCR and other sequence-specific approaches for amplification of HIV-RNA (but not HIV-DNA) are even more sensitive during the earliest stage of infection. In the most comprehensive study to date on early phase HIV-1 infection, HIV-1 RNA-based tests were already positive on an average of 5 days earlier than p24 antigen based tests (95% confidence interval of 3.1 to 8.1 days) (81). According to this initial modeling, the median time until detection of HIV-1 RNA amounted to 10 to 11 days after infection and for p24 antigen to 16 days after infection. A second model used in the same work came to the conclusion that, at a detection limit of 50 copies/ml, HIV-1 RNA became detectable 7 days earlier than p24 antigen, i.e., at day 9 post infection and, at a detection limit of 1 copy/ml, already at day 7 post infection (median values).

Note that median values are only indicative that at the given timepoints 50% of the HIV-infected individuals are positive, meaning that 50% of the individuals still have a negative result. Previous studies have shown that in rare cases seroconversion may be delayed (79, 82, 83). Given that the RNA-PCR, antigen, and antibody tests described in the above-mentioned study are always positive in rapid succession, as shown by the narrow 95% confidence intervals, even highly sensitive molecular-based HIV-tests will remain negative in these rare cases until shortly before the delayed seroconversion. Studies in animal models have shown that, prior to its entry into the bloodstream leading to its general distribution, virus can remain for varying lengths of time within lymphatic tissue at its point of entry into the body (83). In view of these results, it is clear that in humans the interval between infection and the start of a rapid and definitive increase in virus can also vary. This biological property has not changed despite our advances in viral diagnostics. In view of this, it remains impossible to waive a three-month waiting period prior to excluding HIV infection by a negative screening test result.

Due to the high costs of HIV RNA tests, it is not feasible to use these for HIV screening. In addition, these tests usually do not detect HIV-2, and they are also unable to detect HIV-1 infection in persons who have no detectable levels of HIV-1 RNA in their plasma at the time of diagnosis–which after all amounted to 0.7% in a recent study performed in Switzerland (84). As a consequence, if HIV-1 RNA tests were used for screening, their overall diagnostic sensitivity would be considerably lower than that of a good fourth-generation screening test. The latter therefore provide an overall optimal solution for detecting both early and late infections while still being cost-efficient.

Thus, whenever possible, FDA-approved or CE-marked combo screening tests capable of detecting not only HIV antibody against HIV-1 (including group O) and HIV-2, but also the p24 antigen, should be used for HIV screening (85–87). Antigen-testing within the framework of HIV screening is particularly important in order to facilitate identification of highly infectious individuals in the stage of acute or primary HIV infection (23, 88). In many European countries the use of combo tests for HIV screening has become mandatory more than a decade ago, and finally the U.S. also recommends these tests for screening.

Most combo tests have to be performed in the laboratory, but there is also one rapid point-of-care combination assay that is both CE-marked and FDA-approved, the Alere Determine HIV-1/2 Ag/Ab Combo. Unfortunately, this test has poor detection of the p24 antigen of most HIV-1 clades, including subtype B, compared to the laboratory-performed tests, as shown by a recent study performed with a panel of well-standardized recombinant p24-expressing virus-like particles that represent the entire width of genetic variation of HIV-1 (89). In agreement with these results, a study conducted in Swaziland showed that the Alere Determine HIV-1/2 Ag/Ab Combo had a diagnostic sensitivity of 0% for detection of acute HIV-1 infection, even when 10 million HIV-1 RNA copies per ml were present (90). Earlier studies came to the conclusion that HIV antibody detection by the Alere Determine HIV-1/2 Ag/Ab Combo was good, but that p24 antigen detection did not add sensitivity, but rather impaired the overall specificity of the test (91, 92). Clearly, there is also considerable variation in performance of instrument-performed combo tests, as shown by (89) and an earlier study that used culture-grown virus (93). Careful selection of these tests and thoughtful interpretation of negative results in cases of clear exposure to HIV are therefore of paramount importance.

When to Test after a Known or Suspected HIV Exposure

In persons with a history of recent HIV exposure and who already have symptoms of primary HIV infection (PHI), a combo screening test should be carried out immediately. A large proportion of the newly infected patients do not develop symptoms of PHI, however (94). Since patients are likely to benefit from early ART, as reviewed in (95), it is justified to perform a first combo test 16 to 20 days after the date of exposure, since this will identify 50% of the transmissions that occurred. If this test is negative, a second HIV screening test performed 3 months later is necessary to rule out infection, although a second test should be done immediately should symptoms suggestive of PHI present during this period or if risk of transmission is high. HIV-1 RNA testing should be used to confirm positive combo screening results in such cases; CDC guidelines also recommend HIV-1 RNA testing for resolving discrepant results of two combo screening tests. A negative result of a screening test performed three months after exposure definitively rules out that HIV transmission occurred. All patients should be made aware of the inherent high risk of virus transmission deriving from a PHI.

HIV Diagnosis in Infants and Children Who Are <18 Months

If the HIV status of the child's mother is unknown (e.g., in the case of an adopted child) an HIV screening test should be performed. In the case of a negative result, no further tests are required. However, in the case of a reactive result, it must be concluded that exposure to HIV has taken place and further testing is needed to determine the child's HIV status.

Since maternal IgG antibodies (including those directed against HIV) pass by means of active transplacental transport at high concentrations into the fetus, all babies born to HIV-infected mothers have HIV-specific antibodies and are thus diagnostically "HIV-positive." As a consequence, antibody testing for diagnosis of HIV infection cannot be used until the complete disappearance of these maternal antibodies, which should have taken place before 18 months of age. Evidence for HIV infection during this early period is based exclusively on tests that detect viral components. Because the quantification of HIV-1 RNA by commercial tests is routine in many laboratories, HIV RNA is typically tested for, although WHO recommends testing for HIV DNA (96). In general, if prevention of mother-to-child transmission (PMTCT) was applied, HIV-infected children may demonstrate PCR-negativity for both HIV-1 DNA and RNA in blood samples taken during their first and second months. However, test sensitivity during the child's third to sixth months is good (97).

In infants of any age, in the presence of symptoms of pediatric HIV infection, virus testing must be done immediately. In contrast, for asymptomatic children, a first virus test is generally recommended at age 6 weeks. The subsequent testing schedule depends largely on local conditions and financial constraints. Resource-rich countries with a low prevalence of HIV-exposed infants and no breastfeeding by HIV-positive mothers may repeat PCR testing at 4 and 6 months and perform a concluding antibody screening test at 18 months. For resource-poor settings with a high prevalence of maternal HIV infection and a predominantly breastfeeding population of infants, WHO recommends a rapid antibody screening test at 9 months for all HIV-exposed infants who were PCR-negative at 6 weeks. At 9 months, a majority of the uninfected infants will already have lost the maternal HIV-specific antibodies. In case of positivity in the rapid antibody test, a virus test has to be conducted for confirmation of the suspected transmission. A further PCR test has to be conducted 6 weeks after weaning (provided the child is still younger than 18 months); otherwise a concluding screening test is conducted at 18 months.

Several studies have shown that testing of heat-denatured plasma samples by signal amplification–boosted HIV-1 p24 antigen enzyme immunoassay (EIA) (also dubbed the "ultrasensitive p24 (Up24) assay") diagnoses pediatric HIV-1 infection with a sensitivity and specificity equivalent to that of tests for viral DNA or RNA, but at much lower expense (98, 99). WHO has reviewed the available literature in 2009 and has recommended the Up24 test for diagnosis of pediatric HIV-1 infection with the same strength of evidence as for PCR for HIV-1 RNA or DNA. Tests for HIV-1 DNA, RNA, or p24 antigen can also be conducted on dried blood spot specimens (96).

It should be noted that pediatric HIV-2 infection is not detectable using commercial kits for the quantification of HIV-1 RNA. Consequently, if the HIV type of the mother is unknown, the initial blood sample from a newborn must be tested first with an HIV-1/HIV-2 differentiation test in order that appropriate subsequent tests—for HIV-2 always for proviral DNA in peripheral blood mononuclear cells (PBMC)—can be implemented.

In the case of positive virus detection in newborns or infants, guidelines for test confirmation are the same as those to be followed for adults (see next section). However, when HIV-1 RNA in the first sample is detected with adequate certitude, as is the case when a sufficiently high virus concentration is detected and a follow-up sample has been successfully tested for genotypic-resistance, pediatric HIV infection is confirmed and all required virological information is in place for treatment. Borderline results ("positive, <20 copies/ml") must never be interpreted as positive; such cases are usually PCR-negative in follow-up.

Rapid Tests and Use of Alternative Specimens

Rapid tests can be performed with minimal or no laboratory equipment; they yield results within 30 minutes. Such tests may be useful in certain situations, e.g., in assessing the risk of HIV transmission in needlestick injuries and similar exposures to possibly HIV-contaminated materials, organ donations, or whenever a laboratory test result may not be available quickly. Rapid tests may be of different formats, including DAGS, indirect binding, Ig capture, agglutination, or chromatographic assay. The diagnostic sensitivity of some of these tests seems somewhat inferior to third-generation ELISA-based antibody tests, especially in seroconversion panels. Others, however, exhibit comparable diagnostic sensitivity and specificity and can therefore be recommended for diagnosing infections of 3 months' duration or longer.

Many persons infected with HIV are not tested until they develop symptoms of AIDS. Up to one-third of patients receive their HIV diagnosis within 2 months of progression to AIDS. The hope that such individuals could be motivated to be tested earlier has led to adapted testing strategies, particularly in the U.S. These recommend routine, "opt-out" testing in all healthcare settings (100). The shift in testing strategy also has led to the use of new test systems believed to be more attractive to the client. They include home-collection test systems, in which sample collection devices are ordered by phone and delivered by express courier. Blood is collected by finger pricking onto filter paper and sent to a designated laboratory for screening. Such testing systems have good sensitivity and specificity. Collecting a sufficiently large specimen may be the biggest problem, affecting 7 to 10% of the users. As an alternative, testing systems for other specimens, such as oral fluids or urine, also received FDA approval. Excellent sensitivity and specificity were reported in studies involving oral fluids collected from postseroconversion individuals. This also applies to FDA-approved test systems for urine samples. The FDA has also approved a rapid test system for oral fluids, whole blood, or serum, which is sufficiently easy to perform that testing at the point of care with a return of the result within 20 to 40 minutes has become possible. Extended studies of this device have reported a sensitivity and specificity comparable to that of other EIAs. The sensitivity of test systems utilizing specimens other than blood in early seroconversion remains untested, as standardized materials comparable to seroconversion panels are not available. The use of such alternative tests in recent exposure settings should therefore be avoided. True home tests that are sold over the counter have also been approved by the FDA. As with all other tests, a positive result of such tests is preliminary and follow-up confirmatory testing is needed.

Confirmation of the Diagnosis "HIV-Infected"

The central principle for reliability in the diagnosis "HIV-infected" is that (i) at least two different HIV-specific tests have been applied, both yielding positive results, and that (ii) two distinct, independently taken specimens have given unambiguous positive results, meaning that a first and a second sample must be to hand. Which combination of different tests is actually used is not that important, except that both tests need to be of high specificity. It is recommended that screening tests be combined with those tests that are required anyway in order to answer the diagnostic questions 2 and 3 (see below). Tests that determine viral properties or quantify the virus are thus also confirmatory assays. In virtually all cases, HIV infection can be unequivocally confirmed or ruled out using two samples and currently available tests.

Question 2. What Are the Properties of the Virus?

Differentiation of HIV-1 and HIV-2, Double Infections, and Detection of HIV-1 Group O Viruses

Infection with HIV-2 or Group O HIV-1 must be identified early on as these viruses are resistant to the entire class of NNRTI. In the case of HIV-2, a different method for determining the HIV RNA load is needed in addition.

Increased suspicion of HIV-2 infection exists when individuals present with an epidemiological link to West Africa (e.g., Ivory Coast, Ghana, Senegal, Guinea-Bissau, or Cameroon) or Portugal, which has historical colonial links with these regions. In Europe, HIV-2 infections amount to 0.1 to 1% of the number of HIV-1 infections (101), while in the U.S. only 0.01% of the 1.4 million HIV infections diagnosed during 1987 to 2009 were due to HIV-2 (102).

HIV type differentiation is carried out primarily by means of serological tests that are formatted as multiline or multispot assays, whereby the different lines or spots contain proteins and/or peptides of HIV-1 or HIV-2. In Europe, a CE-marked line immunoassay, the Inno-Lia HIV-I/II Score (Fujirebio) has long been used for both HIV confirmation and type differentiation (103). It is a form of "Western Blot of the second generation," whereby antibodies against five recombinant proteins or synthetic peptides of HIV-1 antibodies and the envelope glycoproteins gp105 and gp36 of HIV-2 are detected in a semiquantitative manner. This test is also being utilized in a national surveillance system for incident HIV-1 infection (104, 105). A more recently developed test, BioRad's Multispot HIV-1/HIV-2 Rapid Test, is now recommended for both HIV confirmation and type differentiation in a new testing algorithm recently introduced in the U.S. (106). Tests such as these are increasingly replacing the HIV Western blot (WB) introduced into diagnostics when HIV was discovered three decades ago. This is justified because over the years it has become clear that, in contrast to the continuously improved HIV screening tests, WB has remained a first-generation test with some well-known flaws, namely, an inferior sensitivity in seroconversion panels compared to third- and fourth-generation screening tests and a high proportion of indeterminate results due to the detection of cross-reactive antibodies. It has to be noted, however, that the new confirmatory/type differentiation tests, too, can only be used after seroconversion has occurred. They remain negative in all those cases that are screening-positive due to the detection of p24 antigen in fourth-generation combo tests. In the author's experience, this has amounted to about 5% of all new HIV-1 diagnoses in recent years.

If screening is performed by a fourth-generation combo screening assay, the reactants in a positive specimen are either HIV antibodies or the p24 antigen. In case of a negative confirmatory Inno-Lia or Multispot assay, it thus suffices to perform one of the CE-marked antigen-only assays that are available from different providers in Europe and have been carefully evaluated in recent studies (89, 93). If the p24 test is negative, the screening test was false positive, and no HIV-1 RNA is needed to confirm this. If antigen-only tests are unavailable, as seems to be the case now in the U.S., one has to perform a test for HIV-1 RNA instead. This carries no advantage, however, and is more expensive. Diagnostic utilization of PCR for HIV-1 RNA can safely be restricted to confirmation of a positive p24 antigen test prior to seroconversion.

The serological diagnosis of HIV-2 infection should be verified using a PCR approach. This approach should also be used in the case of any doubt, particularly when antibodies against the envelope region of both HIV-1 and HIV-2 are present at significant concentration. However, in the case of HIV-2 infection, the RNA in plasma is often undetectable, particularly in asymptomatic individuals. Therefore, for confirmation of HIV-2 infection it is necessary to resort to detection of proviral HIV-2 DNA in infected cells, requiring submission of EDTA blood for testing and not plasma alone. Furthermore, as the concentration of HIV-2 infected cells is often very low—less than 1 copy per µg of DNA—a very large DNA sample, meant for use in a high-input PCR ("MEGA-PCR"), must be used in order that a sufficiently high level of sensitivity can be achieved (107). As commercial nucleic acid tests for HIV-2 are not available, such testing has to be carried out in specialized reference laboratories. Note that the demonstration of HIV-1 RNA does not exclude simultaneous infection with HIV-2. In Switzerland, e.g., among a total of 3,851 HIV infections newly diagnosed from 2008 to 2013, we have found 21 HIV-2 infections (0.5%) and 4 dual infections HIV-1 plus HIV-2 (0.1%). Should a dual infection indeed be present, viral load testing must subsequently be undertaken separately for both HIV-1 and HIV-2.

Much harder to diagnose than HIV-2 are the viruses of HIV-1 group O, which may present in patients with epidemiological links to the West African countries Cameroon, Gabon, and Equatorial Guinea. Earlier versions of Roche's virus load test (i.e., the Amplicor HIV-1 Monitor Test, the COBAS AmpliScreen HIV-1 Test, vers. 1.5, and the COBAS TaqMan HIV-1 Test vers. 1) were unable to detect HIV-1 group O viruses. A search for such viruses could therefore be restricted to the few cases of ART-naïve HIV-1 infections that had an undetectable HIV-1 RNA in the Roche assay but were positive in the Abbott RealTime HIV-1. In view of the fact that the contemporary COBAS TaqMan HIV-1 Test vers 2 now also detects HIV-1 group O, this simple and efficient form of group O screening is no longer feasible. Nonetheless, diagnosis of these rare viruses is important because all NNRTI's are ineffective against group O. Today, group O viruses are most effectively diagnosed within the framework of genotypic resistance testing (GRT), provided that the GRT method has been validated for detecting group O, which is rarely the case. Therefore, if the HIV-1 RNA load is sufficiently high for conducting a GRT, but no amplification products can be generated in the GRT, this may among other possible causes indicate that a group O virus is present. This possibility could then be further investigated in a reference lab that has at its disposal group O specific RNA and DNA PCR tests for diagnosis of these rare viruses.

Detection of Drug Resistant Viruses and Determination of HIV Coreceptor Tropism

Knowledge of preexisting viral mutations conferring resistance is important for any antiretroviral therapy. In a review of 215 studies including 43,170 patients, transmission of drug-resistant HIV was found at a frequency of 12.9% in North America, 10.9% in Europe, 6.3% in Latin America, 4.7% in Africa, and 4.2% in Asia (108); for an interactive map of HIV-1 drug resistance in ART-naïve persons see http://hivdb.stanford.edu/surveillance/map/. As transmitted resistant mutants with time will reverse to the better replicating, drug-sensitive wildtype (while remaining preserved in the viral reservoir), the best time for drug resistance testing is at the time of HIV diagnosis—even if antiretroviral therapy is only to begin later. As a general rule, resistance testing has to cover those viral proteins, which are targeted by the antiretroviral drugs to be used. For a recent review see (109).

Genotypic resistance testing (GRT) meanwhile appears to prevail over phenotypic testing (PRT) due to lower price, faster turnaround, and superiority in detecting evolving resistance. PRT remains useful for determining the susceptibility of viruses with complex mutational patterns. This is particularly relevant for selecting a PI for salvage therapy because assessment of the clinical significance of many patterns of PI-resistance mutations can be difficult (109). One such assay, the Phenosense, is available for PRT of all classes of antiretrovirals at the U.S. company Monogram Biosciences.

GRT examines the population of viral genomes in a sample for the presence of mutations known to confer resistance. HIV resistance mutations databases and a wealth of further information are accessible online at http://hivdb.stanford.edu/ or http://www.hivfrenchresistance.org/. GRT involves RT-PCR for amplification of the population of relevant HIV-1 RNA sequences in plasma. The amplicons are sequenced, and the amino acid sequence is compared to a subtype B reference sequence. The differences are reported as a list of mutations and evaluated for mutations known to confer resistance (109). FDA-approved GRT kits include the Trugene HIV-1 Genotyping Kit and Open Gene DNA Sequencing System (from Siemens Healthcare Diagnostics) and the ViroSeq HIV-1 Genotyping System (Celera Diagnostics, Alameda, CA). These kits are expensive, however, and the manufacturers may lag behind in providing service for new antiretrovirals. Kits for HIV-2 or HIV-1 group O are unavailable. These shortcomings severely limit the usefulness of commercial kits. Many labs have therefore developed their own protocols and rely on publicly available interpretation algorithms and reporting systems.

The sensitivity of GRT ranges from 100 to 1,000 plasma HIV-1 RNA copies per ml. At low HIV-1 RNA, GRT is based on a very small number of assessed sequences, and mutations present as minority variants may be missed. As a general rule, standard GRT cannot detect variant sequences present in fewer than 20% of the total (109). Ultradeep sequencing should enable better detection of resistant minority variants.

More than 80% of the patients are initially infected solely with R5 tropic viruses that use the coreceptor CCR5. The small-molecule inhibitor maraviroc allosterically inhibits binding of HIV-1 gp120 to CCR5. IAS-USA and DHHS guidelines recommend maraviroc in combination with two NRTIs as an alternate regimen for first-line ART. Preexisting X4-tropic virus that binds to CXCR4 is the most common cause of maraviroc failure. Therefore, prior to using maraviroc, X4 virus should be excluded either phenotypically (Trofile assay, Monogram Sciences) or genotypically based on sequencing the V3 region of gp120. For more information refer to guidelines http://aidsinfo.nih.gov/guidelines/html/1/adult-and-adolescent-arv-guidelines/8/co-receptor-tropism-assays and (110).

Question 3. How High Is the Viral Load?

The concentration of HIV-1 RNA in plasma (viral load) is a predictor of CD4+ T-cell decline and disease progression and is utilized for monitoring the efficacy of ART (111). A test for the viral load is a must at the time of HIV diagnosis. Viral load determination, if clearly positive, contributes significantly to confirmation of the diagnosis, as sequence-based amplification tests are very specific–provided that carryover contamination has been eliminated. A high HIV-1 RNA load in plasma prior to full seroconversion is the hallmark of an acute HIV-1 infection. Caution regarding the diagnostic reliability should, however, be administered if a measured viral load is low (a few hundred copies/ml). For some viral load tests (bDNA), the lower limit of the quantification range was originally fixed in a way that 95% of uninfected controls were below it, i.e., 5% of the HIV-negative population were false positive. It is clear that tests with a specificity of only 95% must not be used for diagnostic purposes.

Tests for quantifying HIV RNA in plasma are available from various suppliers, such as Roche, Abbott, bioMérieux, and Siemens (consult http://www.fda.gov/cber/products/testkits.html). Those probably used most frequently today, Roche's COBAS TaqMan HIV-1 Test and Abbott's RealTime HIV-1 assay, are based on reverse-transcription (RT) TaqMan real-time PCR. Roche utilizes primers located in *gag* (and for the most recent version 2.0 additionally located in LTR) while Abbott amplifies a conserved sequence in *pol*. Both platforms now recognize group M and group O viruses of HIV-1; Abbott's was also shown to detect HIV-1 groups N and P. Earlier versions of the Roche test—remarkably still on the list of FDA-approved tests—do not recognize group O and may severely underdetect some strains of HIV-1 group M (84, 112, 113). The bioMérieux test (NucliSENS EasyQ) uses transcription-mediated amplification (TMA), an amplification process that mimics the retroviral replication cycle and is performed at a constant temperature (114). Siemens' Versant HIV-1 RNA 3.0 (bDNA) uses the bDNA probe amplification. This test is based on about 40 different probes that cover most of the *pol* gene and permit detection of the various subtypes of group M of HIV-1, but not of group O. For a recent comparative review of the performance of HIV viral load tests refer to (115). Due to differences between the technology platforms, plasma input volume, and ability to detect HIV subtypes, viral load monitoring of individual patients is best performed always using the same technology platform in order to ensure appropriate interpretation of changes.

For HIV-2, there are still no commercial tests for measuring the viral load. Therefore, one has to resort to in-house tests for HIV-2 RNA quantification that are performed in some reference laboratories or to the product-enhanced reverse transcriptase (PERT) test. The PERT assay measures the enzymatic activity of the retroviral reverse transcriptase (RT) enclosed in all retrovirus particles and has a sensitivity similar to RT-PCR, while being completely sequence independent and able to quantify all existing HIV-isolates (84, 116, 117).

Due to the highly variable nature of HIV, primer/probe binding to the target region occasionally can be suboptimal, resulting in underestimation of the viral load in tests based on sequence amplification such as PCR or TMA (112). At high virus concentrations this has negligible clinical consequences. Falsely low or even undetectable viral load due to underestimation may impact the interpretation of the success of ART, however, and increase the risk of inadvertent virus transmission as individuals with an undetectable viral load are considered practically noninfectious. Therefore, the author recommends that a viral load in an ART-naïve individual with fewer than 1,000 copies of HIV-1 RNA per ml should be verified by a second, alternative viral load test from a different manufacturer (84).

Diagnosis of HTLV Infection

The principles, tools, and problems of HTLV diagnosis are, with some modifications, the same as those for the diagnosis of HIV, but the task is simpler because HTLV infections never release virions into the plasma. Therefore, no tests for HTLV-RNA or HTLV-antigen need to be performed. Screening is based on tests for HTLV-specific antibodies by EIA or particle agglutination tests. Confirmatory tests are based on Western blot or line immunoassay. Since there are many indeterminate Western blot results, confirmation should usually be backed by PCR for proviral DNA. Since there is little sequence variation and because many cells carry the provirus, underdetection of HTLV as a result of sequence diversity is not a problem. As HTLV-1 is more pathogenic than HTLV-2, the diagnostic testing should clearly establish the type of HTLV.

Question 1: Is a Person Infected with HTLV?

Screening Tests for HTLV-1/2

As there is no HTLV viremia, there are no fourth-generation combo assays for HTLV, and antibody detection should be based on CE-marked third-generation tests of the double-antigen sandwich (DAGS) format that uses recombinant viral proteins for both antibody capture and tracing. DAGS is the preferred test format because such tests detect all antibody classes including IgM and thus provide the highest sensitivity in early infection (which cannot only exist in the case of HIV, but must also exist for HTLV, because sexual transmission or drug inoculation are well-known modes of transmission). Owing to the fact that the target antigens are recombinant and that two different contacts of the antibody molecule with the target antigen are required in order to generate a signal, DAGS-based screening tests are also more specific than earlier test generations. Particle agglutination assays, which also exhibit the DAGS format and are very popular in Japan, have a similar diagnostic performance. Outside the U.S., such tests are available from various manufacturers and also for automated test platforms (118); for FDA-approved first-generation tests that still use viral lysate for target antigen and do not detect IgM antibodies refer to http://www.fda.gov/cber/products/testkits.htm. Outside of the endemic areas, the specificity of HTLV screening tests is of high importance. In low-prevalence populations, a reactive result in an assay of low specificity is unreliable as it has a very low positive predictive value (PPV). For example, one study involving four different HTLV-1/2 EIAs found a specificity above 99% in three of the tests, but of only 93% in the fourth test. Even among a high-risk group with a prevalence of 311/100,000, the PPV varied from an unacceptably low 4% (for the test with 93% specificity) to a maximum of 44% for a test with 99.6% specificity. Provided that these assays exhibited the same specificity among blood donors (in whom the prevalence was 0.08/100,000), the PPVs would indicate that with the most specific assay only 2/10,000 reactive results would indicate a true infection. In contrast, with the least specific assay only 1/100,000 reactive results would indicate a true infection. These data illustrate the need for both a careful choice of screening assays and of confirmatory testing (119).

Confirmatory Tests for HTLV-1/2

Western blot and line immunoassay. Serologic confirmation of HTLV-1/2 infection requires the demonstration of antibodies to both *gag* (p24) and *env* (gp46 and/or gp68 proteins) by Western blot (120). WB kits are provided by various companies. Of particular value are strips that contain, in addition to the viral proteins derived from viral lysate, recombinant proteins representing TM (gp21) of HTLV-1 (which due to a high homology is also detected by antibodies from individuals infected with HTLV-2) and type-specific SU (gp46) of both HTLV-1 and HTLV-2. This increases the sensitivity since the concentration of the gp46 for HTLV-1 and gp68 for HTLV-2 *env* proteins on the strips is usually low in kits derived from lysate alone. Based on the pattern of the reactivity it is frequently, though not always, possible to decide whether infection by HTLV-1 or HTLV-2 is present. With such strips, intense reaction with the *gag* proteins p19 and p24 and the *env* recombinant proteins rgp21 and rgp46 of HTLV-1 satisfies positivity for HTLV-1, and intense reaction with p24, rgp21, and rgp46 of HTLV-2 satisfies positivity for HTLV-2 (121).

Furthermore, the use of line immunoassay (LIA) strips, which contain standardized concentrations of recombinant proteins and/or synthetic peptides of HTLV-1 and/or HTLV-2 at defined positions and thus can be considered as a form of second-generation WB, presents advantages with respect to both sensitivity and specificity (122).

Still, in many cases WB and LIA do not permit an unequivocal diagnosis, owing to a relatively high percentage of samples with indeterminate results. There is nonspecific reaction not only with lysate-derived natural *gag* proteins, but also with recombinant *env* proteins (Table 2). Some of these indeterminates may have very intense reactions with p19 or p24 and a variety of larger proteins such as p26, p28, p32, p36, p45, and p53 that are present on some WB. These proteins are also present in HTLV-1 infected cells and contain either a p19 or p24 moiety or both (123). Reaction with several of these *gag* proteins on a WB thus may look impressive, but signifies no more than reaction with a single epitope of p19 or, respectively, p24. Sometimes, intense *gag*

TABLE 2 Breakdown of procedures undertaken in a reference lab during repeat testing of HIV-positive samples with additional reactivity or high-negative results in HTLV-1/2 ELISA screening. (Reprinted from *Journal of Medical Virology* (126) with permission of the publisher.)

		ELISA		Western Blot[3]										PBMC	Culture Supernatants			Final Diagnosis
				Natural Proteins							Recombinant Proteins			PCR	PERT[4]	RT-PCR		
Sample	Risk Category[1]	Test[2]	OD/CO	p19	p24	p26	p28	p32	p36	p53	rp21E	rgp46E/HTLV-1	rgp46E/HTLV-2	HTLV		HTLV	HIV-1	
1	IDU	PL	25.77	+++	+++	-	++	-	+++	+++	+++	-	+++	+				HTLV-2
2	IDU	PL	8.09	++	+++	-	+	-	++	+	+++	-	+++	+				HTLV-2
3	HET	CR	7.76	+++	+++	+++	+++	-	-	-	+++	+++	-	+				HTLV-1
4	HET	PL	7.37	+++	+++	-	++	-	++	++	+++	-	+++	+				HTLV-2
5	IDU	CR	6.16	+++	-	-	-	-	-	-	+++	-	+++	+	+	+,+	-,-	HTLV-2
6	IDU	PL	3.34	++	+++	-	+	-	++	+	+++	-	+++	+				HTLV-2
7	IDU	AB	2.47	-	-	-	-	-	-	-	-	-	+	-	-,-	-,-	-,-	neg
8	IDU	AB	1.93	-	++	-	-	-	-	-	+++	-	+++	-	-,-	-,-	-,-	neg
9	IDU	CR	1.50	-	-	-	-	-	-	-	++	-	-	-	-	-	+	neg
10	HET	AB	1.35	-	-	-	-	-	-	-	-	+	-	-	-,-	-,-	-,-	neg
11	HET	AB	1.34	-	-	-	-	-	-	-	+	-	-	-	-,-	-,-	-,-	neg
12	HET	AB	1.30	-	-	-	-	-	-	-	-	-	+	-	-	-	+	neg
13	IDU	AB	1.26	-	-	-	-	-	-	-	-	++	-	-	-	-	-	neg
14	MSM	AB	1.25	-	-	-	-	-	-	-	-	+	+	-	-,-	-	-	neg
15	HET	AB	1.13	-	++	-	-	-	-	-	-	-	-	-	-	-	-	neg
16	MSM	AB	1.11	+++	-	++	-	++	-	-	-	-	+					neg
17	IDU	AB	1.07	-	+	-	-	-	-	-	-	-	-	-	-	-	+	neg
18	IDU	AB	1.06	-	+	-	-	-	-	-	-	-	-	-	-	-	-	neg
19	IDU	AB	1.01	-	-	-	-	-	-	-	-	+	-	-	-	-	-	neg
20	HET	AB	0.87	-	-	-	-	+++	-	-	-	-	-	-	-,-	-	-	neg
21	HET	AB	0.86	+++	-	++	++	-	++	-	-	-	-	-	-,-	-	-	neg
22	IDU	AB	0.83	-	-	-	-	-	-	-	++	+	+	-	-	nd	nd	neg

[1]MSM = homosexual contact; HET = heterosexual contact; IDU = intravenous drug use.

[2]PL = Platelia (Bio-Rad); CR = Cobas (Roche); AB = Abbott HTLV-I/II EIA.

[3]HTLV BLOT Version 2.4 (Genelabs Diagnostics, Inc). NB: Reaction intensities were subjectively rated as 'no reaction' (-); 'clearly visible, but weak' (+); 'intermediately strong' (++); 'strong' (+++). Reactions with natural proteins p21E and gp46E were all negative and are not listed.

[4]Symbols summarize results for supernatants sampled twice per week for at least 2 weeks. Symbols separated by a comma represent results of duplicate cultures.

patterns may also be combined with weak reaction to *env* rgp21 and/or rgp46^{I} or rgp46^{II}, but even this does not necessarily imply HTLV infection. Weak reactions with recombinant envelope proteins in all possible combinations may also be found in the absence of reaction with *gag* proteins. In many instances, confirmation by DNA-PCR is thus necessary. This is particularly true for areas or populations in which HTLVs are not endemic. Under such conditions, any suggestive serological result not strongly antibody-positive should be confirmed by DNA-PCR. Given the problems with nonspecific bands in WB, alternative diagnostic strategies based on testing with a combination of two sensitive and specific EIAs have been proposed (122). Thus, while maintaining a higher overall sensitivity than with the classical EIA-WB combination, an EIA-EIA strategy reduced the frequency of samples with indeterminate results to 2.5%.

DNA-PCR. PCR analysis for HTLV-1 and/or HTLV-2 DNA is necessary for all serologically indeterminates in which antibody reaction to *env* proteins (rgp21, rgp46^{I}, or rgp46^{II}) is present. Antibody reaction with *gag* proteins p19 and/or p24 alone, or rgp21 alone, has been found by PCR not to be associated with HTLV infection (124). PCR is performed on Ficoll-purified PBMC and frequently uses a sequence of the *tax* gene, which is conserved for both HTLV-1 and HTLV-2 and is amplified by primers designated SK43/SK44, while the product is detected by probe SK45 (125). Differentiation of HTLV-1 and HTLV-2 in samples positive in this initial "screening PCR" is then achieved by amplification of a type-specific region in *pol*. Primers SK110 and SK111 in combination with probe SK112 are used for detection of HTLV-1. The same primers in combination with probe SK188 are employed for HTLV-2. Alternatively, in-house PCR methods described by various authors can be used for both the screening step and the type differentiation (126). Real-time PCR methods also have been described by several groups.

Question 2: With What HTLV Is a Person Infected?

The only question to be answered here is whether the infection is caused by HTLV-1 or HTLV-2. This differentiation is important because only HTLV-1 has been firmly associated with disease causation, although luckily in only a low proportion of those infected. The differentiation can usually be made based on WB or LIA results, or if these are indeterminate, by type-specific DNA-PCR (127).

Question 3. How High Is the HTLV Proviral Load?

As a high HTLV-1 provirus concentration is associated with a higher probability for development of various HTLV-associated illnesses and the transmission risk by breast milk (32), measurement of the HTLV-1 provirus load is sometimes required, although it has not yet been established for disease monitoring as firmly as is the case with the HIV-1 virion load. Most ATLL cells contain only one provirus copy. As ATLL cells are derived from HTLV-1 infected cells, it is reasonable to conclude that most HTLV-1 infected cells contain also only one provirus. Provirus quantification by quantitative PCR thus can be used for enumeration of HTLV-1 infected cells in vivo. HTLV-1 provirus load in infected individuals differs more than 1,000-fold among asymptomatic carriers; particularly high concentrations are found in HAM/TSP patients (128). Quantitative real-time PCR methods for HTLV-1 have been described by various groups (127).

BEST PRACTICE

For diagnosis of HIV infection in a clinical setting the procedure described in Fig. 2 has stood the test of time. On a first specimen of the person undergoing testing, a fourth-generation HIV-1/2/O antibody plus p24 antigen combo screening test is performed, which, if reactive, yields the first positive result and the first HIV-positive material. On a second specimen drawn (EDTA-blood, 7 to 10 ml), all other required tests can be conducted: HIV-1/2 multispot or HIV-1/2 line immunoassay for confirmation of HIV infection and type differentiation. If again positive, this yields the second positive test and the second positive sample and in addition identifies the type of HIV (HIV-1 or HIV-2) to which the remaining tests should be directed. The next test in line will be for HIV load quantification, usually for HIV-1, by using one of the commercial virus load tests measuring HIV-1 RNA, rarely for HIV-2, which will require in-house tests for HIV-2 RNA quantification or PERT assay and should be performed in a specialized, reference-type laboratory. If the HIV-1 virus load is low (< 1,000 copies/ml), the author advises to confirm it by a second, alternative test (different brand). If the viral load is sufficiently high, a genotypic resistance test should be conducted, usually for HIV-1, exceptionally for HIV-2 (specialized, reference-type laboratory).

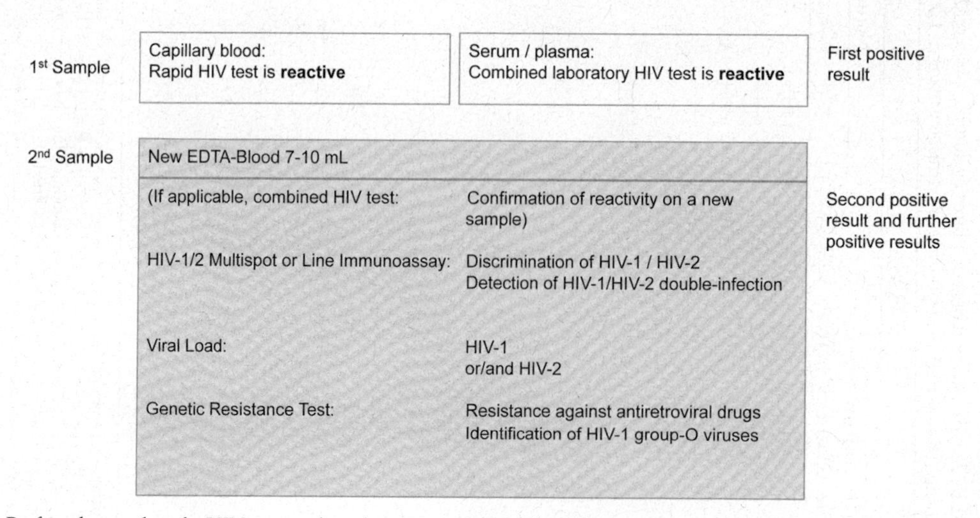

FIGURE 2 Preferred procedure for HIV testing describing tests conducted on the first and second sample. For a detailed description refer to section VI. Best Practice.

GRT should also lead to identification of rare HIV-1 group-O viruses. Thus, with just four tests all virological information needed for an optimal care of an HIV infected individual can be generated. This schedule applies to about 95% of all new HIV diagnoses. About 5% of the cases reactive in combo screening tests will not be confirmable or typable by multispot or line immunoassay because they are in the pre-seroconversion phase of infection and do not yet have HIV-specific antibodies. These cases can only be confirmed by tests for viral components (p24 antigen, HIV-1 RNA).

For diagnosis of HTLV infection one can follow an analogous approach. Screening should be done with a third-generation HTLV-1/2 antibody assay, as there is no viremia requiring simultaneous detection of HTLV antigen. For result confirmation and type differentiation by WB or LIA, a newly drawn EDTA-blood sample should be used, which also will permit qualitative or quantitative PCR for proviral DNA in PBMC if required. Note that HTLV RNA is not present in plasma.

REFERENCES

1. **Gao F, Bailes E, Robertson DL, Chen Y, Rodenburg CM, Michael SF, Cummins LB, Arthur LO, Peeters M, Shaw GM, Sharp PM, Hahn BH.** 1999. Origin of HIV-1 in the chimpanzee Pan troglodytes troglodytes. *Nature* **397:**436–441.
2. **Salemi M, Strimmer K, Hall WW, Duffy M, Delaporte E, Mboup S, Peeters M, Vandamme AM.** 2001. Dating the common ancestor of SIVcpz and HIV-1 group M and the origin of HIV-1 subtypes using a new method to uncover clock-like molecular evolution. *FASEB J* **15:**276–278.
3. **Korber B, Muldoon M, Theiler J, Gao F, Gupta R, Lapedes A, Hahn BH, Wolinsky S, Bhattacharya T.** 2000. Timing the ancestor of the HIV-1 pandemic strains. *Science* **288:**1789–1796.
4. **Lemey P, Pybus OG, Wang B, Saksena NK, Salemi M, Vandamme AM.** 2003. Tracing the origin and history of the HIV-2 epidemic. *Proc Natl Acad Sci USA* **100:**6588–6592.
5. **Gessain A, Rua R, Betsem E, Turpin J, Mahieux R.** 2013. HTLV-3/4 and simian foamy retroviruses in humans: discovery, epidemiology, cross-species transmission and molecular virology. *Virology* **435:**187–199.
6. **LeBreton M, Switzer WM, Djoko CF, Gillis A, Jia H, et al.** (2014) A gorilla reservoir for human T-lymphotropic virus type 4. *Emerging Microbes & Infections* **3**, e7 doi:10.1038/emi.2014.7. Published 22 January 2014.
7. **Wilen CB, Tilton JC, Doms RW.** 2012. HIV: cell binding and entry. *Cold Spring Harb Perspect Med* Aug **1:**2(8).
8. **Jones KS, Fugo K, Petrow-Sadowski C, Huang Y, Bertolette DC, Lisinski I, Cushman SW, Jacobson S, Ruscetti FW.** 2006. Human T-cell leukemia virus type 1 (HTLV-1) and HTLV-2 use different receptor complexes to enter T cells. *J Virol* **80:**8291–8302.
9. **Simon V, Ho DD.** 2003. HIV-1 dynamics in vivo: implications for therapy. *Nat Rev Microbiol* **1:**181–190.
10. **Coffin JM.** 1992. Genetic diversity and evolution of retroviruses. *Curr Top Microbiol Immunol* **176:**143–164.
11. **Wain-Hobson S.** 1992. Human immunodeficiency virus type 1 quasispecies in vivo and ex vivo. *Curr Top Microbiol Immunol* **176:**181–193.
12. **Siliciano JD, Kajdas J, Finzi D, Quinn TC, Chadwick K, Margolick JB, Kovacs C, Gange SJ, Siliciano RF.** 2003. Long-term follow-up studies confirm the stability of the latent reservoir for HIV-1 in resting CD4+ T cells. *Nat Med* **9:**727–728.
13. **Van Dooren S, Salemi M, Vandamme AM.** 2001. Dating the origin of the African human T-cell lymphotropic virus type-i (HTLV-I) subtypes. *Mol Biol Evol* **18:**661–671.
14. **Igakura T, Stinchcombe JC, Goon PK, Taylor GP, Weber JN, Griffiths GM, Tanaka Y, Osame M, Bangham CR.** 2003. Spread of HTLV-I between lymphocytes by virus-induced polarization of the cytoskeleton. *Science* **299:**1713–1716.
15. **Pais-Correia AM, Sachse M, Guadagnini S, Robbiati V, Lasserre R, Gessain A, Gout O, Alcover A, Thoulouze MI.** 2010. Biofilm-like extracellular viral assemblies mediate HTLV-1 cell-to-cell transmission at virological synapses. *Nat Med* **16:**83–89.
16. **Bangham CR, Cook LB, Melamed A.** 2014. HTLV-1 clonality in adult T-cell leukaemia and non-malignant HTLV-1 infection. *Semin Cancer Biol* **26:**89–98.
17. **Matsuoka M, Yasunaga J.** 2013. Human T-cell leukemia virus type 1: replication, proliferation and propagation by Tax and HTLV-1 bZIP factor. *Curr Opin Virol* **3:**684–691.
18. **Korber B, Gaschen B, Yusim K, Thakallapally R, Kesmir C, Detours V.** 2001. Evolutionary and immunological implications of contemporary HIV-1 variation. *Br Med Bull* **58:**19–42.
19. **Hemelaar J, Gouws E, Ghys PD, Osmanov S.** 2006. Global and regional distribution of HIV-1 genetic subtypes and recombinants in 2004. *AIDS* **20:**W13–23.
20. **Simon F, Mauclère P, Roques P, Loussert-Ajaka I, Müller-Trutwin MC, Saragosti S, Georges-Courbot MC, Barré-Sinoussi F, Brun-Vézinet F.** 1998. Identification of a new human immunodeficiency virus type 1 distinct from group M and group O. *Nat Med* **4:**1032–1037.
21. **Patel P, Borkowf CB, Brooks JT, Lasry A, Lansky A, Mermin J.** 2014. Estimating per-act HIV transmission risk: a systematic review. *AIDS* **28:**1509–1519.
22. **Quinn TC, Wawer MJ, Sewankambo N, Serwadda D, Li C, Wabwire-Mangen F, Meehan MO, Lutalo T, Gray RH, Rakai Project Study Group.** 2000. Viral load and heterosexual transmission of human immunodeficiency virus type 1. *N Engl J Med* **342:**921–929.
23. **Wawer MJ, Gray RH, Sewankambo NK, Serwadda D, Li X, Laeyendecker O, Kiwanuka N, Kigozi G, Kiddugavu M, Lutalo T, Nalugoda F, Wabwire-Mangen F, Meehan MP, Quinn TC.** 2005. Rates of HIV-1 transmission per coital act, by stage of HIV-1 infection, in Rakai, Uganda. *J Infect Dis* **191:**1403–1409.
24. **Kaul R, et al.** 2008. The genital tract immune milieu: an important determinant of HIV susceptibility and secondary transmission. *J Reprod Immunol.* 77(1):32–40.
25. **Galvin SR, Cohen MS.** 2004. The role of sexually transmitted diseases in HIV transmission. *Nat Rev Microbiol* **2:**33–42.
26. **Vrielink H, Reesink HW.** 2004. HTLV-I/II prevalence in different geographic locations. *Transfus Med Rev* **18:**46–57.
27. **Hall WW, Takahashi H, Liu C, Kaplan MH, Scheewind O, Ijichi S, Nagashima K, Gallo RC.** 1992. Multiple isolates and characteristics of human T-cell leukemia virus type II. *J Virol* **66:**2456–2463.
28. **Vandamme AM, Salemi M, Van Brussel M, Liu HF, Van Laethem K, Van Ranst M, Michels L, Desmyter J, Goubau P.** 1998. African origin of human T-lymphotropic virus type 2 (HTLV-2) supported by a potential new HTLV-2d subtype in Congolese Bambuti Efe Pygmies. *J Virol* **72:**4327–4340.
29. **Ando Y, Ekuni Y, Matsumoto Y, Nakano S, Saito K, Kakimoto K, Tanigawa T, Kawa M, Toyama T.** 2004. Long-term serological outcome of infants who received frozen-thawed milk from human T-lymphotropic virus type-I positive mothers. *J Obstet Gynaecol Res* **30:**436–438.
30. **Hirata M, Hayashi J, Noguchi A, Nakashima K, Kajiyama W, Kashiwagi S, Sawada T.** 1992. The effects of breastfeeding and presence of antibody to p40tax protein of human T cell lymphotropic virus type-I on mother to child transmission. *Int J Epidemiol* **21:**989–994.
31. **Hino S, Katamine S, Miyata H, Tsuji Y, Yamabe T, Miyamoto T.** 1996. Primary prevention of HTLV-I in Japan. *J Acquir Immune Defic Syndr Hum Retrovirol* **13**(Suppl 1)**:**S199–S203. Review. 34 refs.
32. **Li HC, Biggar RJ, Miley WJ, Maloney EM, Cranston B, Hanchard B, Hisada M.** 2004. Provirus load in breast milk and risk of mother-to-child transmission of human T lymphotropic virus type I. *J Infect Dis* **190:**1275–1278.

33. **Okochi K, Sato H, Hinuma Y.** 1984. A retrospective study on transmission of adult T cell leukemia virus by blood transfusion: seroconversion in recipients. *Vox Sang* **46:**245–253.
34. **Manns A, Hisada M, La Grenade L.** 1999. Human T-lymphotropic virus type I infection. *Lancet* **353:**1951–1958. Review. 82 refs.
35. **Barré-Sinoussi F, Chermann JC, Rey F, Nugeyre MT, Chamaret S, Gruest J, Dauguet C, Axler-Blin C, Vézinet-Brun F, Rouzioux C, Rozenbaum W, Montagnier L.** 1983. Isolation of a T-lymphotropic retrovirus from a patient at risk for acquired immune deficiency syndrome (AIDS). *Science* **220:**868–871.
36. **Popovic M, Sarngadharan MG, Read E, Gallo RC.** 1984. Detection, isolation, and continuous production of cytopathic retroviruses (HTLV-III) from patients with AIDS and pre-AIDS. *Science* **224:**497–500.
37. **Gallo RC, Salahuddin S, Popovic M, Shearer G, Kaplan M, Haynes B, Palker T, Redfield R, Oleske J, Safai B, et al.** 1984. Frequent detection and isolation of cytopathic retroviruses (HTLV-III) from patients with AIDS and at risk for AIDS. *Science* **224:**500–503.
38. **Schüpbach J, Popovic M, Gilden RV, Gonda MA, Sarngadharan MG, Gallo RC.** 1984. Serological analysis of a subgroup of human T-lymphotropic retroviruses (HTLV-III) associated with AIDS. *Science* **224:**503–505.
39. **Sarngadharan MG, Popovic M, Bruch L, Schüpbach J, Gallo RC.** 1984. Antibodies reactive with human T-lymphotropic retroviruses (HTLV-III) in the serum of patients with AIDS. *Science* **224:**506–508.
40. **Levy JA, Hoffman AD, Kramer SM, Landis JA, Shimabukuro JM, Oshiro LS.** 1984. Isolation of lymphocytopathic retroviruses from San Francisco patients with AIDS. *Science* **225:**840–842.
41. **Anonymous, Centers for Disease Control (CDC).** 1981. Pneumocystis pneumonia–Los Angeles. *MMWR Morb Mortal Wkly Rep* **30:**250–252.
42. **Coffin J, Swanstrom R.** 2013. HIV pathogenesis: dynamics and genetics of viral populations and infected cells. *Cold Spring Harb Perspect Med* **3:**a012526.
43. **Chereshnev VA, Bocharov G, Bazhan S, Bachmetyev B, Gainova I, Likhoshvai V, Argilaguet JM, Martinez JP, Rump JA, Mothe B, Brander C, Meyerhans A.** 2013. Pathogenesis and treatment of HIV infection: the cellular, the immune system and the neuroendocrine systems perspective. *Int Rev Immunol* 32:282–306.
44. **Maartens G, Celum C, Lewin SR.** 2014. HIV infection: epidemiology, pathogenesis, treatment, and prevention. *Lancet* **384:**258–271.
45. **Doitsh G, Galloway NL, Geng X, Yang Z, Monroe KM, Zepeda O, Hunt PW, Hatano H, Sowinski S, Muñoz-Arias I, Greene WC.** 2014. Cell death by pyroptosis drives CD4 T-cell depletion in HIV-1 infection. *Nature* **505:**509–514.
46. **Clavel F, Guyader M, Guétard D, Sallé M, Montagnier L, Alizon M.** 1986. Molecular cloning and polymorphism of the human immune deficiency virus type 2. *Nature* **324:**691–695.
47. **Nyamweya S, Hegedus A, Jaye A, Rowland-Jones S, Flanagan KL, Macallan DC.** 2013. Comparing HIV-1 and HIV-2 infection: lessons for viral immunopathogenesis. *Rev Med Virol* **23:**221–240.
48. **Campbell-Yesufu OT, Gandhi RT.** 2011. Update on human immunodeficiency virus (HIV)-2 infection. *Clin Infect Dis* **52:**780–787.
49. **Takatsuki K, Uchiyama J, Sagawa K, Yodi J.** 1996. Adult T-cell leukemia in Japan. *J Acquir Immune Defic Syndr Hum Retrovirol.* 13 Suppl 1:S15-9.
50. **Uchiyama T, Yodoi J, Sagawa K, Takatsuki K, Uchino H.** 1977. Adult T-cell leukemia: clinical and hematologic features of 16 cases. *Blood* **50:**481–492.
51. **Poiesz BJ, Ruscetti FW, Reitz MS, Kalyanaraman VS, Gallo RC.** 1981. Isolation of a new type C retrovirus (HTLV) in primary uncultured cells of a patient with Sézary T-cell leukaemia. *Nature* **294:**268–271.
52. **Yoshida M, Miyoshi I, Hinuma Y.** 1982. Isolation and characterization of retrovirus from cell lines of human adult T-cell leukemia and its implication in the disease. *Proc Natl Acad Sci USA* **79:**2031–2035.
53. **Gessain A, Barin F, Vernant JC, Gout O, Maurs L, Calender A, de Thé G.** 1985. Antibodies to human T-lymphotropic virus type-I in patients with tropical spastic paraparesis. *Lancet* **2:**407–410.
54. **Osame M, Usuku K, Izumo S, Ijichi N, Amitani H, Igata A, Matsumoto M, Tara M.** 1986. HTLV-I associated myelopathy, a new clinical entity. *Lancet* **1:**1031–1032. letter.
55. **Mochizuki M, Watanabe T, Yamaguchi K, Tajima K, Yoshimura K, Nakashima S, Shirao M, Araki S, Miyata N, Mori S, Takatsuki K.** 1992. Uveitis associated with human T lymphotropic virus type I: seroepidemiologic, clinical, and virologic studies. *J Infect Dis* **166:**943–944. letter.
56. **La Grenade L, Manns A, Fletcher V, Derm D, Carberry C, Hanchard B, Maloney EM, Cranston B, Williams NP, Wilks R, Kang EC, Blattner WA.** 1998. Clinical, pathologic, and immunologic features of human T-lymphotrophic virus type I-associated infective dermatitis in children. *Arch Dermatol* **134:**439–444.
57. **Satou Y, Matsuoka M.** 2013. Virological and immunological mechanisms in the pathogenesis of human T-cell leukemia virus type 1. *Rev Med Virol* **23:**269–280.
58. **Tsukasaki K, Tobinai K.** 2013. Biology and treatment of HTLV-1 associated T-cell lymphomas. *Best Pract Res Clin Haematol* **26:**3–14.
59. **Kalyanaraman VS, Sarngadharan MG, Robert-Guroff M, Miyoshi I, Golde D, Gallo RC.** 1982. A new subtype of human T-cell leukemia virus (HTLV-II) associated with a T-cell variant of hairy cell leukemia. *Science* **218:**571–573.
60. **Rosenblatt JD, Golde DW, Wachsman W, Giorgi JV, Jacobs A, Schmidt GM, Quan S, Gasson JC, Chen IS.** 1986. A second isolate of HTLV-II associated with atypical hairy-cell leukemia. *N Engl J Med* **315:**372–377.
61. **Hjelle B, Mills R, Swenson S, Mertz G, Key C, Allen S.** 1991. Incidence of hairy cell leukemia, mycosis fungoides, and chronic lymphocytic leukemia in first known HTLV-II-endemic population. *J Infect Dis* **163:**435–440.
62. **Ciminale V, Rende F, Bertazzoni U, Romanelli MG.** 2014. HTLV-1 and HTLV-2: highly similar viruses with distinct oncogenic properties. *Front Microbiol* **5:**398.
63. **O'Brien WA, Hartigan PM, Martin D, Esinhart J, Hill A, Benoit S, Rubin M, Simberkoff MS, Hamilton JD, Veterans Affairs Cooperative Study Group on AIDS.** 1996. Changes in plasma HIV-1 RNA and CD4+ lymphocyte counts and the risk of progression to AIDS. *N Engl J Med* **334:**426–431.
64. **Barouch DH, Deeks SG.** 2014. Immunologic strategies for HIV-1 remission and eradication. *Science* **345:**169–174.
65. **Weller S, Davis K.** 2002. Condom effectiveness in reducing heterosexual HIV transmission. *Cochrane Database Syst Rev* (1):CD003255.
66. **Montaner JS, Hogg R, Wood E, Kerr T, Tyndall M, Levy AR, Harrigan PR.** 2006. The case for expanding access to highly active antiretroviral therapy to curb the growth of the HIV epidemic. *Lancet* **368:**531–536.
67. **Green EC, Halperin DT, Nantulya V, Hogle JA.** 2006. Uganda's HIV prevention success: the role of sexual behavior change and the national response. *AIDS Behav* **10:**335–346, discussion 347–350.
68. **Shelton JD, Halperin DT, Nantulya V, Potts M, Gayle HD, Holmes KK.** 2004. Partner reduction is crucial for balanced "ABC" approach to HIV prevention. *BMJ* **328:**891–893.
69. **Gray RH, Kigozi G, Serwadda D, Makumbi F, Watya S, Nalugoda F, Kiwanuka N, Moulton LH, Chaudhary MA, Chen MZ, Sewankambo NK, Wabwire-Mangen F, Bacon MC, Williams CF, Opendi P, Reynolds SJ, Laeyendecker O, Quinn TC, Wawer MJ.** 2007. Male circumcision for HIV prevention in men in Rakai, Uganda: a randomised trial. *Lancet* **369:**657–666.
70. **Bailey RC, Moses S, Parker CB, Agot K, Maclean I, Krieger JN, Williams CF, Campbell RT, Ndinya-Achola JO.** 2007.

Male circumcision for HIV prevention in young men in Kisumu, Kenya: a randomised controlled trial. *Lancet* **369:** 643–656.

71. **Olsen JS, Easterhoff D, Dewhurst S.** 2011. Advances in HIV microbicide development. *Future Med Chem* **3:**2101–2116.
72. **Cardo DM, Culver DH, Ciesielski CA, Srivastava PU, Marcus R, Abiteboul D, Heptonstall J, Ippolito G, Lot F, McKibben PS, Bell DM, Centers for Disease Control and Prevention Needlestick Surveillance Group.** 1997. A case-control study of HIV seroconversion in health care workers after percutaneous exposure. *N Engl J Med* **337:**1485–1490.
73. **CDC.** 2007. CDC Trials of Pre-Exposure Prophylaxis for HIV Prevention.
74. **De Man J, Colebunders R, Florence E, Laga M, Kenyon C.** 2013. What is the place of pre-exposure prophylaxis in HIV prevention? *AIDS Rev* **15:**102–111.
75. **Datta A, Bellon M, Sinha-Datta U, Bazarbachi A, Lepelletier Y, Canioni D, Waldmann TA, Hermine O, Nicot C.** 2006. Persistent inhibition of telomerase reprograms adult T-cell leukemia to p53-dependent senescence. *Blood* **108:**1021–1029.
76. **Marçais A, Suarez F, Sibon D, Frenzel L, Hermine O, Bazarbachi A.** 2013. Therapeutic options for adult T-cell leukemia/lymphoma. *Curr Oncol Rep* **15:**457–464.
77. **Takatsuki K, Matsuoka M, Yamaguchi K.** 1996. Adult T-cell leukemia in Japan. *J Acquir Immune Defic Syndr Hum Retrovirol* **13**(Suppl 1)**:**S15–S19. Review. 38 refs.
78. **Kassutto S, Johnston MN, Rosenberg ES.** 2005. Incomplete HIV type 1 antibody evolution and seroreversion in acutely infected individuals treated with early antiretroviral therapy. *Clin Infect Dis* **40:**868–873.
79. **Petersen LR, Satten GA, Dodd R, Busch M, Kleinman S, Grindon A, Lenes B, The HIV Seroconversion Study Group.** 1994. Duration of time from onset of human immunodeficiency virus type 1 infectiousness to development of detectable antibody. *Transfusion* **34:**283–289.
80. **Busch MP, Lee LL, Satten GA, Henrard DR, Farzadegan H, Nelson KE, Read S, Dodd RY, Petersen LR.** 1995. Time course of detection of viral and serologic markers preceding human immunodeficiency virus type 1 seroconversion: implications for screening of blood and tissue donors. *Transfusion* **35:**91–97.
81. **Fiebig EW, Wright DJ, Rawal BD, Garrett PE, Schumacher RT, Peddada L, Heldebrant C, Smith R, Conrad A, Kleinman SH, Busch MP.** 2003. Dynamics of HIV viremia and antibody seroconversion in plasma donors: implications for diagnosis and staging of primary HIV infection. *AIDS* **17:** 1871–1879.
82. **Ciesielski CA, Metler RP.** 1997. Duration of time between exposure and seroconversion in healthcare workers with occupationally acquired infection with human immunodeficiency virus. *Am J Med* **102(5B):**115–116. Review. 11 refs.
83. **Busch MP, Satten GA** (1997) Time course of viremia and antibody seroconversion following human immunodeficiency virus exposure. Review. 43 refs. *Am J Med* 102:117-124; discussion 125-116.
84. **Vetter BN, Shah C, Huder JB, Böni J, Schüpbach J.** 2014. Use of reverse-transcriptase-based HIV-1 viral load assessment to confirm low viral loads in newly diagnosed patients in Switzerland. *BMC Infect Dis* **14:**84.
85. **Gürtler L, Mühlbacher A, Michl U, Hofmann H, Paggi GG, Bossi V, Thorstensson R, G-Villaescusa R, Eiras A, Hernandez JM, Melchior W, Donie F, Weber B.** 1998. Reduction of the diagnostic window with a new combined p24 antigen and human immunodeficiency virus antibody screening assay. *J Virol Methods* **75:**27–38.
86. **Weber B, Fall EHM, Berger A, Doerr HW.** 1998. Reduction of diagnostic window by new fourth-generation human immunodeficiency virus screening assays. *J Clin Microbiol* **36:** 2235–2239.
87. **Weber B, Meier T, Enders G.** 2002. Fourth generation human immunodeficiency virus (HIV) screening assays with an improved sensitivity for p24 antigen close the second diagnostic window in primary HIV infection. *J Clin Virol* **25:** 357–359.
88. **Yerly S, Vora S, Rizzardi P, Chave JP, Vernazza PL, Flepp M, Telenti A, Battegay M, Veuthey AL, Bru JP, Rickenbach M, Hirschel B, Perrin L, Swiss HIV Cohort Study.** 2001. Acute HIV infection: impact on the spread of HIV and transmission of drug resistance. *AIDS* **15:**2287–2292.
89. **Vetter B, et al.** 2014. Generation of a recombinant Gag virus-like-particle panel for the evaluation of p24 antigen detection by diagnostic HIV tests (accepted). PLoS One.
90. **Duong YT, Mavengere Y, Patel H, Moore C, Manjengwa J, Sibandze D, Rasberry C, Mlambo C, Li Z, Emel L, Bock N, Moore J, Nkambule R, Justman J, Reed J, Bicego G, Ellenberger DL, Nkengasong JN, Parekh BS.** 2014. Poor performance of the determine HIV-1/2 Ag/Ab combo fourth-generation rapid test for detection of acute infections in a National Household Survey in Swaziland. *J Clin Microbiol* **52:**3743–3748.
91. **Rosenberg NE, Kamanga G, Phiri S, Nsona D, Pettifor A, Rutstein SE, Kamwendo D, Hoffman IF, Keating M, Brown LB, Ndalama B, Fiscus SA, Congdon S, Cohen MS, Miller WC.** 2012. Detection of acute HIV infection: a field evaluation of the determine® HIV-1/2 Ag/Ab combo test. *J Infect Dis* **205:**528–534.
92. **Taegtmeyer M, MacPherson P, Jones K, Hopkins M, Moorcroft J, Lalloo DG, Chawla A.** 2011. Programmatic evaluation of a combined antigen and antibody test for rapid HIV diagnosis in a community and sexual health clinic screening programme. *PLoS One* **6:**e28019.
93. **Ly TD, Plantier JC, Leballais L, Gonzalo S, Lemée V, Laperche S.** 2012. The variable sensitivity of HIV Ag/Ab combination assays in the detection of p24Ag according to genotype could compromise the diagnosis of early HIV infection. *J Clin Virol* **55:**121–127.
94. **Henrard DR, Daar E, Farzadegan H, Clark SJ, Phillips J, Shaw GM, Busch MP.** 1995. Virologic and immunologic characterization of symptomatic and asymptomatic primary HIV-1 infection. *J Acquir Immune Defic Syndr Hum Retrovirol* **9:**305–310.
95. **Panel on Antiretroviral Guidelines for Adults and Adolescents.** 2014. Guidelines for the use of antiretroviral agents in HIV-1-infected adults and adolescents. *In:* Services DoHaH, (ed.), Department of Health and Human Services.
96. **WHO.** 2010. WHO recommendations on the diagnosis of HIV infection in infants and children. Recommendations and Annexes downloadable from http://www.who.int/hiv/pub/paediatric/diagnosis/en/.
97. **Burgard M, Blanche S, Jasseron C, Descamps P, Allemon MC, Ciraru-Vigneron N, Floch C, Heller-Roussin B, Lachassinne E, Mazy F, Warszawski J, Rouzioux C.** 2012. Performance of HIV-1 DNA or HIV-1 RNA tests for early diagnosis of perinatal HIV-1 infection during anti-retroviral prophylaxis. *J Pediatr* 160:60-66, e61.
98. **Nadal D, Böni J, Kind C, Varnier OE, Steiner F, Tomasik Z, Schüpbach J.** 1999. Prospective evaluation of amplification-boosted ELISA for heat-denatured p24 antigen for diagnosis and monitoring of pediatric HIV-1 infection. *J Infect Dis* **180:** 1089–1095.
99. **Fiscus SA, Cheng B, Crowe SM, Demeter L, Jennings C, Miller V, Respess R, Stevens W, Forum for Collaborative HIV Research Alternative Viral Load Assay Working Group.** 2006. HIV-1 viral load assays for resource-limited settings. *PLoS Med* **3:**e417.
100. **Branson BM, Handsfield HH, Lampe MA, Janssen RS, Taylor AW, et al.** 2006. Revised recommendations for HIV testing of adults, adolescents, and pregnant women in health-care settings. *MMWR Recomm Rep* 55:1-17; quiz CE11-14.
101. **Cazein F, Hamers F, Alix J, Brunet JB.** 1996. Prevalence of HIV-2 infection in Europe. *Euro Surveill* **1:**21–23.
102. **Centers for Disease Control and Prevention (CDC).** 2011. HIV-2 Infection Surveillance–United States, 1987–2009. *MMWR Morb Mortal Wkly Rep* **60:**985–988.

103. **Walther L, Putkonen P, Dias F, Biberfeld G, Thorstensson R.** 1995. Evaluation of HIV-1/HIV-2 immunoblots for detection of HIV-2 antibodies. *Clin Diagn Virol* **4:**67–79.
104. **Schüpbach J, Gebhardt MD, Tomasik Z, Niederhauser C, Yerly S, Bürgisser P, Matter L, Gorgievski M, Dubs R, Schultze D, Steffen I, Andreutti C, Martinetti G, Güntert B, Staub R, Daneel S, Vernazza P.** 2007. Assessment of recent HIV-1 infection by a line immunoassay for HIV-1/2 confirmation. *PLoS Med* **4:**e343.
105. **Schüpbach J, Niederhauser C, Yerly S, Regenass S, Gorgievski M.** Continuous Decline of Incident HIV-1 Infection by 50% in Switzerland 2008—2013 Shown by a HIV Line-Immunoassay-Based National Surveillance System. *PLoS Med* (Submitted).
106. **Branson BM, Ginocchio CC.** 2013. Introduction to 2013 Journal of Clinical Virology supplement on HIV testing algorithms. *J Clin Virol* **58**(Suppl 1)**:**e1–e134.
107. **Günthard HF, Huber M, Kuster H, Shah C, Schüpbach J, Trkola A, Böni J.** 2009. HIV-1 superinfection in an HIV-2-infected woman with subsequent control of HIV-1 plasma viremia. *Clin Infect Dis* **48:**e117–e120.
108. **Frentz D, Boucher CA, van de Vijver DA.** 2012. Temporal changes in the epidemiology of transmission of drug-resistant HIV-1 across the world. *AIDS Rev* **14:**17–27.
109. **Tang MW, Shafer RW.** 2012. HIV-1 antiretroviral resistance: scientific principles and clinical applications. *Drugs* **72:**e1–e25.
110. **Vandekerckhove LP, Wensing AM, Kaiser R, Brun-Vézinet F, Clotet B, De Luca A, Dressler S, Garcia F, Geretti AM, Klimkait T, Korn K, Masquelier B, Perno CF, Schapiro JM, Soriano V, Sönnerborg A, Vandamme AM, Verhofstede C, Walter H, Zazzi M, Boucher CA, European Consensus Group on clinical management of tropism testing.** 2011. European guidelines on the clinical management of HIV-1 tropism testing. *Lancet Infect Dis* **11:**394–407.
111. **Mellors JW, Rinaldo CR Jr, Gupta P, White RM, Todd JA, Kingsley LA.** 1996. Prognosis in HIV-1 infection predicted by the quantity of virus in plasma. *Science* **272:**1167–1170.
112. **Korn K, Weissbrich B, Henke-Gendo C, Heim A, Jauer CM, Taylor N, Eberle J.** 2009. Single-point mutations causing more than 100-fold underestimation of human immunodeficiency virus type 1 (HIV-1) load with the Cobas TaqMan HIV-1 real-time PCR assay. *J Clin Microbiol* **47:**1238–1240.
113. **Peeters M, Aghokeng AF, Delaporte E.** 2010. Genetic diversity among HIV-1 non-B subtypes in viral load and drug resistance assays. (accepted article); http://www3.interscience.wiley.com/journal/123591652/abstract. *Clinical Microbiology and Infection.*
114. **Muenchhoff M, Madurai S, Hempenstall AJ, Adland E, Carlqvist A, Moonsamy A, Jaggernath M, Mlotshwa B, Siboto E, Ndung'u T, Goulder PJ.** 2014. Evaluation of the NucliSens EasyQ v2.0 assay in comparison with the Roche Amplicor v1.5 and the Roche CAP/CTM HIV-1 Test v2.0 in quantification of C-clade HIV-1 in plasma. *PLoS One* **9:**e103983.
115. **Sollis KA, Smit PW, Fiscus S, Ford N, Vitoria M, Essajee S, Barnett D, Cheng B, Crowe SM, Denny T, Landay A, Stevens W, Habiyambere V, Perrins J, Peeling RW.** 2014. Systematic review of the performance of HIV viral load technologies on plasma samples. *PLoS One* **9:**e85869.
116. **Pyra H, Böni J, Schüpbach J.** 1994. Ultrasensitive retrovirus detection by a reverse transcriptase assay based on product enhancement. *Proc Natl Acad Sci USA* **91:**1544–1548.
117. **Bürgisser P, Vernazza P, Flepp M, Böni J, Tomasik Z, Hummel U, Pantaleo G, Schüpbach J.** 2000. Performance of five different assays for the quantification of viral load in persons infected with various subtypes of HIV-1. *J Acquir Immune Defic Syndr* **23:**138–144.
118. **Malm K, Kjerstadius T, Andersson S.** 2010. Evaluation of a new screening assay for HTLV-1 and -2 antibodies for large-scale use. *J Med Virol* **82:**1606–1611.
119. **Böni J, Bisset LR, Burckhardt JJ, Joller-Jemelka HI, Bürgisser P, Perrin L, Gorgievski M, Erb P, Fierz W, Piffaretti JC, Schüpbach J.** 2004. Prevalence of human T-cell leukemia virus types I and II in Switzerland. *J Med Virol* **72:**328–337.
120. Centers for Disease Control and Prevention and the USPHS Working Group: guidelines for counselling persons infected with human T-lymphotropic virus type I (HTLV-I) and type II (HTLV-II). 1993. *Ann Intern Med* **118:**448–454.
121. **Medrano FJ, Soriano V, Calderón EJ, Rey C, Gutiérrez M, Bravo R, Leal M, González-Lahoz J, Lissen E.** 1997. Significance of indeterminate reactivity to human T-cell lymphotropic virus in western blot analysis of individuals at risk. *Eur J Clin Microbiol Infect Dis* **16:**249–252.
122. **Thorstensson R, Albert J, Andersson S.** 2002. Strategies for diagnosis of HTLV-I and -II. *Transfusion* **42:**780–791.
123. **Schüpbach J, Kalyanaraman VS.** 1989. Detection of high concentrations of HTLV-1 p24 and a novel gag precursor, p45, in serum immune complexes of a healthy seropositive individual. *Int J Cancer* **44:**90–94.
124. **Defer C, Coste J, Descamps F, Voisin S, Lemaire JM, Maniez M, Couroucé AM, Retrovirus Study Group of the French Society of Blood Transfusion.** 1995. Contribution of polymerase chain reaction and radioimmunoprecipitation assay in the confirmation of human T-lymphotropic virus infection in French blood donors. *Transfusion* **35:**596–600.
125. **Kwok S, Lipka JJ, McKinney N, Kellogg DE, Poiesz B, Foung SK, Sninsky JJ.** 1990. Low incidence of HTLV infections in random blood donors with indeterminate western blot patterns. *Transfusion* **30:**491–494.
126. **Böni J, Bisset LR, Burckhardt JJ, Joller-Jemelka HI, Bürgisser P, Perrin L, Gorgievski M, Erb P, Fierz W, Piffaretti JC, Schüpbach J.** 2004. Prevalence of human T-cell leukemia virus types I and II in Switzerland. *J Med Virol* **72:**328–337.
127. **Waters A, Oliveira AL, Coughlan S, de Venecia C, Schor D, Leite AC, Araújo AQ, Hall WW.** 2011. Multiplex real-time PCR for the detection and quantitation of HTLV-1 and HTLV-2 proviral load: addressing the issue of indeterminate HTLV results. *J Clin Virol* **52:**38–44.
128. **Etoh K, Yamaguchi K, Tokudome S, Watanabe T, Okayama A, Stuver S, Mueller N, Takatsuki K, Matsuoka M.** 1999. Rapid quantification of HTLV-I provirus load: detection of monoclonal proliferation of HTLV-I-infected cells among blood donors. *Int J Cancer* **81:**859–864.

Chlamydiae

BARBARA VAN DER POL AND CHARLOTTE A. GAYDOS

38

CLASSIFICATION

The *Chlamydiaceae* are a family of small, metabolically dependent bacteria with a unique intracellular life cycle (1). As a result of the obligate intracellular growth of these organisms in eukaryotic cells, they are handled in ways more closely resembling the detection of viruses than bacteria. Thus, inclusion of these organisms in this manual is appropriate.

The taxonomy of the members of this family has evolved over the last 2 decades. Originally there was a single genus (*Chlamydia*) with 3 species: *C. trachomatis*, *C. psittaci*, and *C. pneumonia*. In 1999, revisions were proposed based on partial sequencing of the 16S and 23S rRNA subunits. At this time, a second genus (*Chlamydophila*) was created containing 6 species, which are predominately non-human pathogens: *C. abortus*, *C. caviae*, *C. felis*, *C. percorum*, *C. pneumoniae*, and *C. psittaci*. The original genus, *Chlamydia*, was reorganized to contain 3 species: *C. trachomatis*, *C. muridarum* (consisting of organisms formerly contained within *C. trachomatis*), and *C. suis* (2). However, this taxonomic reorganization was controversial and did not fully take into account the biological and pathogenic features of these organisms. Further, when full genome sequence data became available, the reorganization did not coincide well with the emerging picture of the evolution of these organisms (3, 4). Revised taxonomy has been accepted (5) to return to a single genus (*Chlamydia*) with 11 species: *C. abortus*, *C. avium*, *C. caviae*, *C. felis*, *C. gallinacea*, *C. muridarum*, *C. percorum*, *C. pneumoniae*, *C. psittaci*, *C. suis*, and *C. trachomatis*. For a recent review of the evolution of chlamydia, with a detailed discussion of taxonomy, see Bachmann et al. (6).

Of the 11 recognized species of chlamydia, *C. trachomatis*, a highly prevalent sexually transmitted organism, is the only species that is a strictly human pathogen. This organism will be the primary focus of this chapter. *C. pneumoniae* causes respiratory disease in humans as well as a variety of other hosts and will be briefly discussed. Psittacosis is a severe infection caused by *C. psittaci* that results primarily from interaction with (domesticated) birds. It is unclear whether the newer species (*C. avium* and *C. gallinacea*) play a role in this disease in humans. Cases of psittacosis are now extremely rare and have caused serious laboratory infections: samples from suspected cases should be handled by regional and federal public health agencies. Culture by local laboratories is specifically not recommended. Thus, *C. psittaci* and the remaining species, which may on rare occasions cause zoonotic infection in humans, will not be further described.

BIOLOGY

Both *C. trachomatis* and *C. pneumoniae* share the bi-phasic intracellular life-cycle that is unique to all *Chlamydiaceae*. Metabolically inert particles known as elementary bodies (EBs) are the infectious organisms. EBs are approximately 0.3 μm in diameter compared to other sexually transmitted pathogens such as the gonococcus (~1 μm) or trichomonads (~10 μm). Following adhesion to the eukaryotic cell surface, the EBs are ingested and contained within a phagosome, called an inclusion. The inclusion is successful at evading phagosome-lysosome fusion through mechanisms that are not wholly understood (7). Multiple inclusion bodies within a single cell may fuse or remain separate depending on the species and strain (8, 9). EBs transform into reticulate bodies (RBs), the metabolically active replicative form of the organism, in an asynchronous manner within the inclusion, beginning approximately 12 hours post-infection. Cellular energy sources and organelles are hijacked by the inclusion to meet the needs of chlamydia, which replicate via binary fission. Sometime after 24 hours post-infection, RBs begin to condense into the infectious EBs, which are then released into the intercellular milieu to infect new host cells. The entire life cycle takes place over a 48- to 72-hour period.

C. trachomatis exhibits at least partial tissue tropism associated with the different strains that comprise the species. Strains were initially categorized based on variations in antigens (serotyping) contained within the major outer membrane protein (MOMP). This method identified 15 serovars: A, B, Ba, C, D, E, F, G, H, I, J, K, L_1, L_2, and L_3. While most serovars can be found in a variety of epithelial cells, serovars A to C are most commonly found in isolates from the ocular conjunctiva, and are the causative agents of blinding trachoma. Serovars D to K are most commonly isolated from genital epithelial cells: the endocervical columnar epithelium or the cuboidal epithelium of the urethra. *Chlamydia* does not effectively infect or replicate in squamous epithelial cells. Lymphogranuloma venereum (LGV) (strains L_1, L_2, and L_3) are the most invasive strains of *C. trachomatis* and may cause genital ulcer disease, proctitis, or infection of the

doi:10.1128/9781555819156.ch38

inguinal lymph nodes. While the other strains of *C. trachomatis* require approximately 60 to 72 h for a single life-cycle, the LGV strains have a shortened reproductive rate of only 36 to 48 hours. Sequencing of the *ompA* gene, which encodes the MOMP protein, resulted in an expansion of the number strains, now referred to as genotypes (10). Variants within the D, I, J, and L_2 serovars have been described. Interestingly, the E serovar, which is the most commonly found serovar in epidemiologic studies conducted throughout the world, appears to have a highly conserved genome and does not have any identified *ompA* variants.

The genome of *C. trachomatis* is roughly 1 to 1.3 Mb in length, encoding approximately 900 proteins. The organism also contains a plasmid, often referred to as the cryptic plasmid given the limited understanding of its function, containing approximately 7.5 kb. Organisms contain 7 to 10 copies of this plasmid, and thus it has often served as a target for diagnostic nucleic acid amplification tests (NAATs). Interestingly, a phenomenon first observed in Sweden was where certain counties, using a single plasmid based assay, showed a decrease in positivity rates while other counties, using a different plasmid based assay, did not show any reduction in rates. Investigation revealed that *C. trachomatis* organisms in the counties with reduced rates had a mutation that had deleted a 377 bp area, thus effectively removing the amplification target binding site from the plasmid; therefore these strains were being missed by certain diagnostic assays. The strain of *C. trachomatis* with this mutation is often referred to as the "Swedish variant," or the new variant (nv *C. trachomatis*) (11–13). This is a rare example of the enrichment of a mutation in a population potentially in response to diagnostic, rather than antimicrobial, pressure. As strains without the mutation were diagnosed and treated, those with the mutation remained in the population and could be transmitted. Thus, vigilance in the epidemiology of *C. trachomatis* is warranted on a continuing basis as related to use of commercial NAAT assays.

C. pneumoniae has a similar intracellular life-cycle, but preferentially infects alveolar macrophages and the epithelial layer of airways. Multiple inclusions within a single host cell are occasionally seen. This species can be distinguished by culture and fluorescent staining using monoclonal antibodies as well as species-specific NAATs. They can also be identified based on morphology using electron microscopy due to the multiple inclusions that often have a "cluster of grapes" appearance, and the loose, pear-shaped EBs as opposed to the very dense, round EBs of *C. trachomatis* (14, 15).

The host immune response involves both CD4 and CD8 mediated cellular responses, as well as an antibody response that does not confer natural immunity in the short-term, as demonstrated by high re-infection rates (16). When exposed to certain antibiotics (e.g., penicillins), sub-lethal levels of cytokines (e.g., interferon gamma), or iron, chlamydia may enter into a persistent state *in vitro* with enlarged, non-replicating RBs that remain quiescent until the stressor mechanism is removed (17, 18). Following removal of the stressor, the RBs return to a normal replicative state and infectious EBs are produced. Chlamydia may induce low-level immune stimulation when undetected, or when exposed to non-lethal stressors, that results in scarring in the region of infected tissues. These responses are involved in the development of sequelae to untreated infection and thus caution will be required in the development of a potential vaccine.

EPIDEMIOLOGY

C. trachomatis

C. trachomatis is the etiologic agent of the most prevalent bacterial sexually transmitted infection (STI) in the U.S. and worldwide. The World Health Organization (WHO) estimates that approximately 100,000 million people have active infections globally at any given point in time (19). Infection with *C. trachomatis* is notifiable in the U.S., resulting in strong epidemiologic data regarding annual case rates, which are estimated to be 2 to 4 million annually (20). Nationally, representative surveys that include STI screening as a biomarker of sexual health in the population estimate the prevalence of this STI in the highest risk age groups to be 1.7% in women and men combined (21). Since its addition to the notifiable disease list, *C. trachomatis* has consistently been the most common of any reportable infectious disease organism with ≥25-fold more cases per year than the third most common disease (salmonellosis) according to the Centers for Disease Control and Prevention (CDC) (22).

Case rates are continuing to climb in the U.S. and in Western Europe. These increases are not attributable to changes in diagnostic methods or improved coverage of screening services, since similar increases are not being seen in *Neisseria gonorrhoeae* infection rates. Since gonococcal testing is paired with testing for chlamydial infection, at least in the U.S., epidemiologic artifacts related to testing methodology and screening coverage should be similar for the 2 organisms. Disparities by race and age are pronounced and are reproduced in representative national surveys, reducing the possibility that these are testing pattern artifacts. Prevalence is highest among women aged 15 to 25 (4.7%) and particularly among black women in this age group (13.5%) (21).

Trachoma, an ocular disease caused by *C. trachomatis*, is a significant cause of irreversible blindness due to scarring of the eyelid and cornea. Trachoma is endemic in 55 countries, resulting in approximately 3.8 million cases of blindness and 5.3 million cases of impaired vision throughout Africa and Southeast Asia (23, 24). According to the WHO, approximately 3% of cases of irreversible blindness worldwide (8 million people) is attributable to infection with *C. trachomatis*, and there are 84 to 100 million cases of infection that, if untreated, may lead to blindness in the coming years. Cases are found in the most resource constrained countries, predominately in tropical or subtropical climates. Children living in theses environments are often infected with this endemic disease by flies landing on or near the eyes. Scarring occurs as a result of the chronic, subclinical inflammation that is typical of untreated chlamydial infections. Blindness occurs in adults many years post-infection.

C. pneumoniae

C. pneumoniae is a respiratory pathogen causing infection that is predominately sub-clinical but may result in respiratory illness, including pneumonia. This infection is predominately spread through inhalation of airborne organisms or droplet secretions. Due to the difficulty associated with culture isolation of this organism, and until recently the lack of a commercial NAAT targeting *C. pneumoniae*, the epidemiology of this infection is based predominately on serological testing. Population based surveys conducted in various sites around the world suggest 40 to 70% of the general population has antibodies reactive to C. *pneumoniae*. With increasing age, the proportion of positive individuals continues to rise, making exposure nearly ubiquitous. Approximately 20% of people with community acquired pneumonia (CAP) were previously thought to have C. *pneumoniae*, but there are no strong data suggesting that this is the primary etiologic agent of CAP disease. The multifactorial nature of CAP, combined with the difficulty of diagnosing active C. *pneumoniae* infection, hinders our ability to fully understand the role C. *pneumoniae* plays in this disease.

CLINICAL SIGNIFICANCE

C. trachomatis

Infection with C. *trachomatis* may result in signs and symptoms consistent with other discharge-causing STIs such as urethral discharge, cervical discharge including mucopurulent discharge, pain or burning on urination, pelvic or scrotal pain/tenderness, or cervical friability. The urethral discharge in men, if present, tends to be clear rather than the cloudy discharge usually seen with gonococcal infection. The presence of symptoms is not clinically specific, but may be used for syndromic management of patients in resource constrained settings (25). While C. *trachomatis* accounts for 25 to 35% of cases of non-gonococcal urethritis in men, a large proportion of men (30 to 50%) and the majority of women ($\geq$60%) exhibit no signs or symptoms of infection (26). Thus the need for screening among asymptomatic populations is critical to control of this pathogen, in order to prevent serious sequelae.

Men may develop epididymitis or proctitis from untreated chlamydial infections. Among men who engage in receptive anal intercourse, rectal infection with genital strains or with LGV is common. Rates of infection with LGV have increased in Western Europe over the last decade (27); however, similar increases have not been observed in the U.S. While commercially available molecular assays will detect these infections, the assays do not distinguish between urogenital and LGV strains. Further diagnostic evaluation of patients suspected of having LGV infection may be necessary in order to determine appropriate treatment, since longer treatment is recommended for LGV strains. Untreated LGV may lead to serious complications including severe proctitis, rectal fistulas, or abscesses.

In women, who bear the largest burden from these silent infections, organisms may ascend from the lower genital tract to the endometrium or fallopian tubes. Pelvic inflammatory disease (PID) may result from untreated infection in as many as 15 to 25% of women (28). This results in a substantial burden on the healthcare system due to the high prevalence of undiagnosed disease in the population. While control efforts have not been largely successful in reducing the overall prevalence of disease, data from studies in North America and Western Europe suggest that rates of PID have been declining in countries with active control efforts (29–31). Scarring of the fallopian tubes may also result from long-term sub-clinical infection, increasing the likelihood of ectopic pregnancy or tubal factor infertility (32, 33). Due to difficulty determining and distinguishing the possible multiple etiologies of these reproductive complications, and differences in diagnostic classifications, the impact of chlamydia control efforts on these outcomes is difficult to determine with certainty.

Neonatal conjunctivitis (rapid onset within the first week of life) or chlamydial pneumonia (delayed onset at 2 to 4 months) (caused by C. *trachomatis* not C. *pneumoniae*) may occur when infants are delivered vaginally from women with undiagnosed/untreated infection. As a result of screening recommendations for pregnant women, and the implementation of neonatal eye drops at birth in developed countries, the occurrence of these complications is predominately limited to resource constrained settings.

In resource poor areas, primarily sub-Saharan Africa, repeated and untreated infection of the eyes with C. *trachomatis* (serotypes A to C) leads to scarring of the conjunctiva and inversion of the eyelid (trachoma). Inverted eyelashes abrade the cornea leading to scarring, which in turn results in blindness in adults. While this disease has been historically endemic in many areas of the world, trachoma elimination projects are under way across the globe. The WHO is actively engaged in a trachoma elimination project with the goal of eradication of this disease by 2020.

C. pneumoniae

Common outcomes of infection include upper respiratory symptoms, bronchitis, and clinically diagnosed pneumoniae (34). Untreated C. *pneumoniae* infection may remain asymptomatic for an extended, but unknown, duration. These sub-clinical infections have been epidemiologically linked with several significant outcomes including atherosclerosis (35), asthma (36), Alzheimer's disease (37), and reactive arthritis (38). Given the widespread exposure of this infection (based on serological epidemiologic studies) in the general population, such associations should be interpreted with appropriate caution. Studies evaluating treatment of C. *pneumoniae* in order to reduce long-term outcomes are highly confounded and have shown little impact. This is an area deserving of further research.

TREATMENT AND PREVENTION

C. trachomatis

Treatment

The CDC recommended regimen for treatment of chlamydial infections is (i) a single dose azithromycin, 1 gram orally or (ii) 100 mg of doxycycline taken orally twice per day for 7 days (34). In patients with low anticipated adherence, the use of azithromycin, preferably directly observed, is recommended over doxycycline. Alternate regimens for treatment, predominately intended to replace the doxycycline option, which is contraindicated during the third trimester of pregnancy, include erythromycin, ofloxacin, and levofloxacin. For additional treatment information, see the CDC STD Treatment Guidelines [www.cdc.gov/std/treatment].

Infections with LGV require prolonged treatment to ensure effective clearance of the organisms. Uncomplicated cases of LGV require a 21-day regimen of doxycycline, 100 mg taken orally twice daily. In men living with HIV, a test-of-cure is recommended following completion of the regimen to ensure eradication.

Prevention

Behavioral practices that are intended to reduce exposure to STI and HIV include delaying sexual debut, limiting the number of sexual partners, and use of barrier protection such as male and female condoms. There is no available vaccine effective against C. *trachomatis*. Annual screening, in the absence of signs or symptoms, is recommended for young women under the age of 25 as a prevention method intended to reduce the burden of disease at the population level (39, 40). Additional recommendations for screening apply to specific sub-populations such as pregnant women and men who have sex with men (MSM). Pregnant women should be screened during the first prenatal visit and again during the third trimester if risk of re-infection is high. The MSM population should be screened annually. Importantly, these men should have testing performed using both urine and ano-rectal samples. Studies have clearly shown that among men who engage in receptive anal intercourse, case finding rates are substantially increased (2- to 4-fold) by including ano-rectal testing (41). The majority of men with rectal infection do not have a concomitant urethral infection and thus cases are missed when testing only urine specimens from MSM. Finally, women under 36 and men <30 years of age should be screened on entry into correctional facilities (39).

The trachoma elimination projects have developed a prevention strategy through mass treatment in regions with endemic levels of disease (42). Positive results of single dose treatment using azithromycin in selected villages have supported the feasibility of this approach (43, 44). Continued global efforts are likely to reduce the incidence of disease and the sequelae of irreversible blindness in the coming years.

C. pneumoniae

Treatment

In general, patients with symptomatic infection caused by *C. pneumoniae* will be treated following recommendations designed for CAP. The regimens most likely to be effective specifically against *C. pneumoniae* include doxycycline, 100 mg orally twice daily for 14 to 21 days; tetracycline, 250 mg orally 4 times daily for 14 to 21 days; azithromycin, 1.5 g orally over 5 days; clarithromycin, 500 mg orally, twice a day for 10 days; levofloxacin, 500 mg, intravenously or orally, once a day for 7 to 14 days; or moxifloxacin, 400 mg orally, once a day for 10 days (45).

Detection/Diagnosis

Sample Collection

Sample collection is very similar for all of the assays for detection of *C. trachomatis* (see Table 1 for appropriate sample types for each class of assay). The CDC currently recommends only NAATs for the detection of chlamydia in urogenital samples, so sample collection is directed toward this class of assay. For women, vaginal swabs, self-obtained or clinician collected, and first-catch urine are minimally invasive samples. Endocervical sampling is more invasive and requires more clinician time. Self-collected vaginal specimens are obtained by instructing the patient to insert the swab as far as comfortable into her vagina and to rotate the swab. Care should be taken to emphasize the importance of not letting anything touch the swab prior to or following collection. This is critical to avoiding possible contaminants. Urine specimens should be first catch, preferably ≥1 hour since previous urination, with no prior cleaning of the vulvar area in order to maximize the sensitivity of the assays. Mid-stream samples are not appropriate for chlamydia-gonorrhea NAATs. Clinician collection of vaginal samples should be performed prior to insertion of a speculum if a speculum-assisted pelvic exam will be performed. Fluid and cells from the vaginal walls should be collected by rotating the swab around the entire circumference of the vagina. Endocervical samples can be collected using a Dacron swab on a plastic shaft, NAAT manufacturers' devices, or a Pap test collection device (cytobroom or cytobrush). For cervical swab collection, the swab should be inserted past the cervical os in order to collect the columnar epithelia cells of the endocervix. This sampling method was designed to collect the host cells in which chlamydia replicates for diagnosis by tissue culture. However, NAATs do not require viable cells and have a much lower analytical limit of detection and can therefore be used for testing mixed ecto- and endocervical cells obtained for Pap testing, and placed into liquid based cytology (LBC) medium. If using an LBC sample for chlamydia-gonorrhea NAAT diagnostics, collection should be consistent with the manufacturer's instructions for cytological sampling. NAATs performed from LBC

TABLE 1 Features of Various Classes of Assay for Detection of *C. trachomatis*

	NAATs	DFA	Culture	POC	Serology
Sample types	Vaginal swabs Endocervical samples Urethral swabs Rectal swabs* Oropharyngeal swabs* Conjunctival swabs* Urine	Endocervical swabs Urethral swabs Conjunctival swabs	Endocervical swabs Urethral swabs Rectal swabs Oropharyngeal swabs Conjunctival swabs Bronchoalveolar lavage**	Endocervical swabs Urine	Serum
Transport	Ambient	Ambient	Cold chain required	Ambient	Ambient
Storage	Stability in manufacturer's collection device ranges from 7–60 days at 2–30°C	Prior to staining, slides may be stored ambient or frozen indefinitely	2–8°C for up to 24 hours post-collection. Longer storage requires −80°C freezing	N/A	2–7 days at 2–8°C or frozen at ≤ −20°C indefinitely
Turnaround time	≥4 hours	30 minutes	2–4 days	20–100 minutes	4 hours

*No commercially available assay has claims for these sample types.
**For *C. pneumoniae* only.

should be performed following the NAAT package insert to determine if the sample for chlamydia diagnostics should be taken prior to, or following, preparation of the cytology slide. Laboratories should be aware of the potential for sample-to-sample contamination that may result from processing for cytology, since this is often done in a separate laboratory that is not operating under conditions designed to optimize samples for molecular testing.

Specimens from men for NAATs include first-catch urine, collected ≥1 hour after previous urination, and urethral swabs. Urethral swabs for NAAT may be collected according to the procedure recommended for culture, which involves insertion of a rayon swab on an aluminum shaft 2 to 3 cm into the distal urethra with rotation for at least 10 seconds. However, due to the sensitivity of NAATs, for symptomatic men the urethral sample may be a swab that collects exudate from near the urethral opening. Although the findings are mixed, several studies show that non-invasive glans, meatal, and penile shaft swab samples may provide acceptable sensitivity depending on the assay (46–49).

In addition to these genital samples, testing may be performed at non-genital sites using ano-rectal and oropharyngeal specimens (41, 50–59). These sample types are not cleared by the Food and Drug Administration (FDA) for use with any of the commercially available assays, but are necessary in order to support the CDC screening guidelines for MSM. Several reports of validation of these assays are available, suggesting that the performance of NAATs using these non-genital samples is excellent and better than culture (60, 61). Swabs used for endocervical or vaginal sampling are appropriate for use at both of these body sites. Ano-rectal samples may be clinician or patient collected (for NAATs only). Many men report being more comfortable with self-collection, so this strategy may enhance uptake of screening (62–64). The swab should be inserted approximately 2 to 3 cm and rotated in order to come into contact with the entire circumference of the ano-rectum. The swab does not need to be inserted beyond the first sphincter if used for NAATs. If used for culture, which is not recommended under routine circumstances, the swab should be clinician collected and taken from the rectum beyond the second sphincter. Stool is never an appropriate specimen for detection of *C. trachomatis*. Oropharyngeal samples should be collected by rotating the swab over and behind the tonsilar pillar. These samples should be collected by a clinician for culture (not recommended under routine circumstances) and can be self-obtained for NAATs (65).

Other specimen types include conjunctival, nasopharyngeal, and bronchoalveolar lavage (BAL). Conjunctival samples should be obtained using a rayon/dacron/nylon swab on an aluminum or plastic flexible shaft. The eyelid should be inverted, and the swab passed firmly over the conjunctiva in order to collect epithelial cells. While none of the NAATs has an FDA claim for ocular specimens, the assays appear to work quite well with this sample type (66, 67). Alternatively, this sample type may be used for culture or direct fluorescent antibody (DFA) staining. Nasopharyngeal swabs (for detection of *C. pneumoniae*) should be collected using the same swab type as the conjunctiva samples. The specimens should be obtained by insertion through the nasal passage, with firm contact with the nasopharyngeal epithelium. This is invasive and uncomfortable, but sensitive diagnosis will be compromised by inadequate specimen collection. Throat swabs have also been used by some researchers. BAL should be performed according to the procedures in place at clinical facilities. An aliquot of the lavage fluid should be placed into an appropriate transport medium depending on the assay requested: tissue culture or molecular testing.

Chlamydia Trachomatis

Until the 1970s, our understanding of chlamydial infections was based predominately on serologic evidence of past infections, which facilitated broad-scale epidemiologic studies that provided data suggesting that this organism is a significant human pathogen. Direct microscopy (e.g., using Gram's or other stains) was not practical due to the extremely small size of the EBs.

Organisms could only be cultured in a live egg-yolk system that was extremely labor intensive and logistically suited to research facilities rather than clinical laboratories. Tissue culture methods routinely used in virology labs were adapted to facilitate isolation of chlamydia by the end of the 1970s, and clinical laboratory diagnosis became available through reference and research laboratories. The strict requirements for adequate sample collection and maintenance of cold chain during transport hampered the ability of many providers to offer routine chlamydia testing. Non-culture based diagnostics were clearly needed, and several options were developed during the 1980s. The first test to be available was a fluorescent stain using monoclonal antibodies, that could be used directly on smears collected from the conjunctiva, the urethra, or the endocervix (68, 69). DFA staining reagents remain commercially available although the utility of this test is limited. Enzyme immunoassays (EIA) and enzyme linked immunosorbent assays (ELISA) were developed shortly after DFA (70). These antigen capture assays had the advantage of being able to detect non-viable organisms, and therefore the transport and storage conditions could be relaxed. Antigen capture chemistry was also applied to rapid test devices for use at point-of-care (POC). All of the antigen detection assays suffered from poor sensitivity (at best 85 to 90% of the sensitivity of culture), and in some instances poor specificity due to cross-reactive antigens found in patient samples.

The next significant advancement was the development of DNA-based assays that used hybridization probes to detect specific organisms (71–75). The major advantages to this class of diagnostic, compared to culture and EIA/ELISA, were (i) the ability to use a single swab (endocervical or urethral) specimen to test for both *C. trachomatis* and *N. gonorrhoeae* and (ii) the ability to use male urine as a specimen type. Since chlamydial and gonococcal infections are often identified as co-infections, and since collection of urethral swabs is painful, both features represented an improvement in clinical processes and thus the potential to offer increased screening. While there was an incremental, although not always statistically significant increase in the sensitivity of these assays compared to culture, specificity issues were reported (76, 77). Specificity is a particular concern when diagnosing STIs given the social implications of infection.

In the early 1990s, assays that used nucleic acid amplification of target sequences in order to identify both chlamydia and gonococcal infections were being developed and evaluated (78–81). Based on the performance of NAATs, previous evaluations of non-culture diagnostics were reviewed and the true sensitivity of these older assays was found to be lower than NAATs (77, 82). These assays (EIA/ELISA and DNA probe assays) should no longer be used and will not be discussed further (83). The earliest NAATs have now moved into second or third generation, with continuing improvement in sensitivity, specificity, sample type options, and automation.

Nucleic Acid Amplification Tests

Of the various classes of assay for *C. trachomatis*, NAATs provide the highest sensitivity and have excellent specificity as well. The estimates of sensitivity and specificity provided in package inserts, and in publications reporting clinical trial results, may underrepresent the true performance of these assays. This is particularly true for the older assays that were evaluated, compared to less sensitive methods such as culture, DFA, and EIA/ELISA. Based on analytical evaluations of the limits of detection, these assays provide positive results when as few as 10^{-1} to 10^{2} (depending on the assay) organisms are present in the specimen. The fact that assays are positive when testing dilutions that theoretically contain less than one organism is the result of using multiple targets (e.g., the plasmid, which is present in 7 to 10 copies, or rRNA, which is present in thousands of copies, per organism). The clinical sensitivity is likely >95% when organisms are captured with the sampling device. Further, although specificity was a concern with NAATs when they were first developed, studies using assays with essentially equivalent sensitivity have demonstrated that the specificity of these assays is excellent (usually >98%) (84–86).

Several NAATs are commercially available (with FDA clearance) and new products are in development in this rapidly evolving field of diagnostics. As a result, the description of assays here should not be considered to be exhaustive, but rather as illustrative of the technologies currently available. For specific assays, see the manufacturer's package insert for more detailed information and instructions.

Each commercially available assay provides a sample collection device and transport tube specific for that assay. The transport conditions and times in the package insert should be adhered to and closely followed in order to optimize the quality of the results. Urine samples generally require transfer of a portion of the urine into a transport tube that contains a preservative to stabilize the sample and reduce degradation. Once placed in transport tubes, samples are generally stable for transport at ambient temperature.

Once in the laboratory, samples have storage times ranging from 7 days to 2 years, depending on the assay. The assays share some common principles, which include nucleic acid extraction from the patient sample; target binding to primers; amplification; and detection of amplified product. All of the current generation assays utilize real-time outputs that allow detection to occur simultaneously with amplification. The majority of current NAATs are performed using semi- or fully automated systems that minimize technician hands-on time requirements, and maximize standardization of processes. Some of the assays include process controls that provide evidence of successful extraction and amplification, and some include sample adequacy controls that demonstrate the presence of human DNA in the sample. Both of these types of internal control are useful for verifying the interpretation of negative results. All of the currently available NAATs are based on a quantitative output, usually non-linear, that is interpreted by analyzer software to provide a qualitative result of "Positive" or "Negative." Several assays also provide a result option of "Equivocal," suggesting that the sample generated some low-level output, but did not cross the cycle threshold required to be called "Positive" or "Invalid," the latter of which indicates that internal controls did not satisfy assay requirements. Some manufacturers' package inserts suggest that patients should be re-sampled in these cases. Many laboratories have validated re-test procedures to minimize the need for return patient visits, particularly for "Equivocal" results.

All of the commercially available NAATs detect the LGV strains of infection; however, none of the assays provide results that distinguish between genital and LGV strains. Many laboratories offer laboratory developed tests (LDTs) that specifically target sequences in LGV strains using real-time or Light-Cycler technology. Positive specimens from patients for whom LGV is of concern should be sent to a laboratory offering a validated LGV LDT.

Specific assays vary based on the chemistry utilized for amplification and detection. The vast majority of laboratories in the U.S. are currently using one of the following assays (presented alphabetically): Amplicor CT/NG (Amplicor, Roche Molecular Diagnostics, Indianapolis, IN); Aptima Combo 2 CT/GC assay (AC2) and Aptima ACT assay (chlamydia only) (Gen-Probe/Hologic, San Diego, CA); cobas 4800 CT/NG assay (cobas, Roche Molecular Diagnostics, Indianapolis, IN); ProbeTec ET CT/GC assay (BDPT, BD Diagnostics, Sparks, MD); ProbeTec Qx CT/GC assay (Qx, BD Diagnostics, Sparks, MD); RealTime m2000 CT/GC assay (m2000, Abbott Molecular, Des Plaines, IL); and Xpert CT/NG (Cepheid, Sunnyvale, CA). The Amplicor and the BDPT are first generation tests and are quickly being replaced by the improved cobas and Qx assays, respectively, and thus will not be further described here.

The cobas (87–90), m2000 (91–93), and Xpert (94, 95) assays are all based on polymerase chain reaction using real-time detection via detection of fluorescent light emissions for multiplexed chlamydia and gonorrhea testing. The cobas and m2000 are similarly structured assays that utilize magnetic bead extraction of total DNA in the patient sample performed on a large liquid handling instrument. The extracted DNA is added to master mix, which contains primer sets specific for both *C. trachomatis* and *N. gonorrhoeae*. Both assays utilize 2 primer sets, targeting the plasmid for *C. trachomatis*, in order to detect both the original strains, as well as the nv *C. trachomatis*. Once 96-well plates are ready for amplification, they are transferred to a light cycler for simultaneous amplification and detection. Both assays provide internal control results to verify that extraction and amplification were successful. Due to automation, these assays are intended for batch testing and moderately high throughput. The entire process requires 4.5 to 6.5 hours depending on the assay.

The Xpert assay utilizes many of the same chemistries, but is performed in an individual cartridge on an instrument designed to manage all aspects of the assay automatically, once samples are loaded into the cartridge (96). The single amplification target is a highly conserved chromosomal target. Instruments are available to handle varying batch sizes from 1 to 80 cartridges, with random access so that samples can be individually tested as they arrive at the testing site. For each cartridge, the assay provides an internal control to assess extraction and amplification, a specimen adequacy control to verify the presence of human DNA in the sample, and a control to verify that reagents are reconstituted effectively by the instrument. The test is rated as moderately complex and not waived under the Clinical Laboratory Improvement Act (CLIA) regulations at this time, but may achieve that status in the near future. Runs require only 90 minutes, which will not qualify as a rapid

POC diagnostic, but has excellent "near patient" potential for use in specific settings.

The Qx assay uses an isothermal process called strand displacement to amplify and detect the target sequence (78, 97). This assay, which is run on a fully automated instrument, uses a single target located on the plasmid. DNA is extracted from patient samples using ferric oxide and magnetic separation technology. Samples are then exposed to primers and probes, and subsequently transferred to amplification wells where strand displacement amplification occurs at 52.5°C. As amplification occurs, fluorophores and quenchers attached to the probes are physically separated, allowing fluorescent emission that is detected by the analyzer. Results include an extraction control that verifies both the successful extraction of nucleic acids, and that amplification has not been inhibited. The assay requires approximately 4.5 hours and is run in 96-well batches, making this a moderately high throughput instrument. Chlamydia and gonorrhea tests are in separate wells, allowing the option of running chlamydia-only testing in those populations for which gonorrhea testing is not ordered. Thus the batch size will range from 46 patients (if performing both CT and GC) to 94, if testing only for chlamydia.

The AC2 assay differs from all other commercial assays in that the target for amplification is located on the 23S rRNA sub-unit (98–100). This target was chosen to enhance sensitivity given the 2 to 3 log increase in copy number/organism compared to the plasmid. The principle of the assay is based on magnetic target capture using sequence-specific oligonucleotides with linked magnetic particles. Following this target capture process, the unbound portions of the patient sample are removed from the reaction. Subsequent amplification of the target regions is performed through DNA-RNA hybridization, with measurement of the frequency of photon emissions over the course of the amplification process. This assay does not provide an extraction or amplification control based on the principle that all possible inhibitors are removed during the target capture process. The AC2 assay is a multiplexed chlamydia-gonorrhea test; however, the manufacturer also offers chlamydia or gonorrhea standalone tests, based on RNA targets that differ from those used in the AC2 assay, for labs that need to test for only one or the other organism. The assay requires 4 to 6 hours and can be performed on 1 of 3 automated systems, which provide varying levels of automation, and batch sizes that range from moderately high to high numbers of samples.

With all of the NAATs, the consideration of potential contamination is key to accurate test results in the clinical laboratory setting. Even though many of the next generation assays are partially or fully automated, and amplification is performed in a sealed environment, the potential for contamination remains. Environmental monitoring for potential amplicons (the product of the amplification process) is highly recommended. All surface areas and routinely used equipment (e.g., pipettors, etc.) should be tested for contaminants on a routine basis. Due to the high copy number of targets used for the RNA-based assay, it is very prudent for laboratories to monitor for sample-to-sample contamination in clinical settings where specimens are collected, to verify that no contamination is present on shared surfaces (e.g., exam tables, desks, etc.) (101). Laboratories should also monitor for laboratory environmental contamination by performing monthly "swipe tests" of laboratory surfaces and equipment areas. The cobas assay includes a process intended to minimize environmental contamination. This assay uses deoxyuridine triphosphate (UTP), rather than deoxythyamdine triphosphate (TTP), in the mix of nucleotide triphosphates that are the building blocks of the amplicons. The master mix preparation contains uracyl-n-glycosolase, which cleaves nucleic acid sequences whenever a uridine residue is encountered. This degrades any carryover contaminating amplicons that may be present from test contamination, and prevents them from serving as targets for primer binding. Once the amplification cycles begin, the enzyme is inactivated by the temperatures used for the actual amplification process.

As mentioned above, the specificity of these assays is quite high, and when performed in accordance with good laboratory practices, the accuracy of results will be optimized. However, as with any test, particularly those used to screen asymptomatic populations, laboratories should be aware that the positive predictive value (i.e., the probability that a positive result actually came from a patient with an infection), will vary based on the prevalence of infection in the population. For screening in low prevalence settings, for example prenatal screening, positive results should be interpreted by the clinician within the context of patient-specific information. Re-testing of original samples, or even re-sampling the patient for confirmation, is not recommended by CDC even in low prevalence populations, because the data do not provide a clear indication of how to interpret subsequent re-test results if they are inconsistent (84). Particularly for those assays that detect very low organism loads (ultra-sensitive assays), it is unclear whether these (true) positive results reflect exposure in the absence of active infection. Detectable dead organisms may result from many scenarios including unprotected intercourse with a recently treated partner; the individual's immune system may have inhibited successful colonization but organism DNA or RNA is still present; or the individual may have been treated (e.g., for an unrelated complaint) and the infection has been resolved but nucleic acids are still present. In other words, the strength of these assays, which is to say their excellent sensitivity, may sometimes present a weakness as well. All results should be interpreted by the clinician in the context of patient-specific information to which the laboratory has no access.

Direct Immunofluorescent Staining

DFA remains a useful diagnostic tool in very specific circumstances. The staining is simple to perform, but requires a highly trained microscopist and access to an epi-fluorescence microscope. Thus this is a high complexity test. Swabs collected from the infected tissue should be rolled onto a slide and allowed to air dry. The quality of sample collection and the smear are the most important aspects affecting the sensitivity of this test. After fixation, slides are stained with a fluorescently labeled monoclonal antibody. Products are available that are specific for *C. trachomatis* (anti-major outer-membrane specific) or are more broadly reactive (anti-LPS) across the genus, and will stain all chlamydia species. Slides are reviewed for the presence of punctate apple-green fluorescing organisms exhibiting appropriate morphology. Generally, results are reported as "Positive," "Negative," or "Insufficient Cellular Material," the latter indicating either poor sample collection or poor smear preparation. Positive slides should be reviewed by a second microscopist to verify that staining is not due to an interfering substance (e.g., powder residue from gloves during collection). This assay has very limited sensitivity (50 to 65%) compared to NAATs and suffers from poor specificity due to potentially autofluorescing substances in the sample. It is more accurate with eye swabs from babies with conjunctivitis.

Tissue Culture

Culture is quickly becoming a lost art for detection of *C. trachomatis*. Culture methods are still employed in some settings in order to maintain a repository of isolates, but this is predominately done in research or reference labs as was the case in the 1970s and 1980s. For details regarding cell line maintenance and media preparation, see the WHO laboratory manual (102). In general, culture was initially performed in shell vials containing a coverslip seeded with host cells, usually a mouse fibroblast line (McCoy cells) or a human endocervical line (HeLa cells). At the height of culture diagnostics, in order to accommodate the volume of testing required, many laboratories moved to 96-well microtiter plates. Confluent cell monolayers were inoculated with samples collected for culture (endocervical, urethral, rectal, oropharyngeal, or conjunctival swabs) with a media overlay that included cyclohexamide to inhibit eukaryotic protein synthesis. In this way, the monolayer would not overgrow during the growth cycle of the *C. trachomatis*. Cultures are typically allowed to incubate at 37°C for 48 to 72 hours post-inoculation, fixed with alcohol (ethanol or methanol depending on the staining method used), and stained. The most sensitive staining method is fluorescent antibody staining, and such reagents are available commercially. Giemsa and iodine staining are older options in settings that do not have access to fluorescent stains and a fluorescent microscope. Culture positive monolayers contain very large infected cells, with a chlamydial inclusion that fills the majority of the cell cytoplasmic space. By inoculating multiple monolayers, and staining only one of these, remaining monolayers may be harvested in order to freeze isolates for future research. The high degree of complexity (which leads to high cost), time to diagnosis, requirement for strict adherence to cold chain during transport, and the limited sensitivity of this assay (60 to 80% in the best controlled circumstances) all provide rationale for retiring this assay in most laboratories and restricting its use to research settings.

Point-of-Care (POC) Assays

POC are rapid (generally less than 30 minutes to 1 hour) assays that can be performed in the clinical setting while the patient waits. These assays have been available, in various formats, in the U.S. and globally since the late 1980s. Older assays were based on EIA/ELISA chemistries for antigen capture, and thus suffer from the same limitations of sensitivity. The intracellular nature of this organism requires lysing of the host cell to release the organisms for antigen capture. Low numbers of organisms captured on endocervical swabs and dilution of the number of organisms pose additional challenges to achieving adequate sensitivity. Many of these assays remain on the market today, but caution should be used in interpreting package inserts given that sensitivity estimates were calculated based on comparisons with culture in most cases (103–105). Thus a reported sensitivity of 85% means that, in reality, the assay can detect 85% of the 50 to 80% of samples that would have been culture positive and thus the actual sensitivity may be as low as 43 to 70%. While this sensitivity is unacceptable in most settings, there are circumstances under which the immediate treatment of infections may provide sufficient benefit to justify use of these assays (106, 107). Caution should be taken regarding the specificity of older POC assays. Routine assessment of the performance of these assays is not covered under any proficiency process in the U.S. At this time the CDC does not recommend that any of the older POC tests be used. The Health Technology Assessment Program performed an extensive cost-effectiveness review for the newer British CRT assay compared to the Clearview Chlamydia POC assay and standard-of-care PCR testing (108). That review reported from several studies that the pooled sensitivity of the CRT CT for vaginal swabs was 80%, and for first void urines was 77%. For Clearview, pooled estimates from 4 studies indicated the sensitivity for vaginal, cervical, and urethral swabs combined was 64%, and was 52% for cervical samples. This economic analysis reported that both POC tests were more costly and less effective than using NAAT assays.

Next generation POC tests are currently in development and evaluation from a number of researchers and manufacturers (105). The next generation assays utilize modern biotechnology, including molecular diagnostic strategies. Within the next 2 to 5 years, assays will be available that have the capacity to detect a single organism, or several related pathogens in a matter of 10 to 30 minutes. Microfluidics technology has played a significant role in miniaturizing and accelerating the target binding, amplification, and detection processes. Most of these newer POC assays will utilize small instrumentation (ranging from microwave to shoebox size, and potentially smaller in the future).

The Atlas Genetics platform in the United Kingdom is a promising chlamydia POC assay that may enter clinical trials soon (109). This PCR assay has a novel electrochemical detection method. The Velox technology contains an integrated fluidic card for sample processing and reagent handling. It incorporates a novel technique for detection of proprietary ferrocene electrochemical labels, and utilizes a low-cost reader instrument. One preclinical validation study using 306 archived clinical samples demonstrated a clinical sensitivity of 98.1%, and specificity of 98.0% (109).

The microwave accelerated metal enhanced fluorescence (MAMEF) assay is another new POC CT technology in development. The MAMEF assay can detect ~10 inclusion-forming units/ml of CT in less than 9 minutes, including DNA extraction and detection. Using a plasmid-based assay on archived clinical samples, sensitivity and specificity were 82.2% (37/45) and 92.9% (197/212), respectively (110). Target DNA sequences bind to an anchor probe covalently bound to the assay surface, and a fluorophore-labeled probe. If target CT DNA sequence is detected, metal enhanced fluorescence occurs through close proximity of the fluorophore to silver metallic nanoparticles.

These representative examples of advances demonstrate exciting technology on the horizon and such new POC technology will allow improved access to care and rapid, accurate treatment. They will also reduce the laboratory role in diagnosis of *C. trachomatis* infections if they can achieve CLIA waiver.

Chlamydia Pneumoniae

Acute infection with *C. pneumoniae* most commonly presents as CAP, which is often multifactorial. As a result, specific diagnostics are often not applied to these patients if they respond well to therapy. As mentioned above, the epidemiology of this pathogen is not well described and thus a need for routine diagnostic tools has not been established. The assays described in the sections below are predominately LDTs, or rely on reagents that are intended for investigational use only.

Serology

Three serological assays have routinely been used for assessment of chlamydial infection: micro-immunofluorescence (MIF) staining, complement fixation, and EIA (111). Complement fixation is appropriate for detection of exposure to *C. psittaci*, which should be restricted to use in regional or federal health agency laboratories and will not be discussed further here. MIF was the original assay that was used to perform epidemiologic studies to determine the overall disease distribution for both *C. trachomatis* and *C. pneumoniae* (112). The assay uses purified EB resuspended in yolk-sac and adhered to slides in micro-clusters. Each cluster contains 5 microspots of EBs, with representatives from the following sero-groups (based on sero-antigen relatedness): (i) group B contains strain types B, Ba, D, and E; (ii) group C contains A, C, H, I, J, and K; (iii) group F contains F and G; (iv) group L contains L_1, L_2, and L_3; and (v) contains *C. pneumoniae*. The micro-clusters containing each of these 5 microspots are incubated with patient serum in serial 2-fold dilutions. Patient antibodies, if present, bind to the microspots with varying degrees of specificity. If present, human antibodies are detected by a secondary anti-human fluorescently labeled reagent. Fluorescence may be type specific (e.g., binding to a single spot), species specific (e.g., binding to *C. trachomatis* spot(s) but not to *C. pneumoniae*, or the reverse), or generally reactive. Diagnosis of active infection is based on paired serum samples showing a ≥4-fold increase over time (usually with approximately 2 to 3 weeks between sample collection). While this provided the best evidence of active infection in the era preceding culture, and eventually non-culture based diagnostics, it has little utility for diagnosis of *C. trachomatis* infections today. Further, this assay is highly technical and requires organism stocks needed to make the microspots that few laboratories have access to today. Finally, the interpretation of the intensity of fluorescence needed to call a spot positive or negative was very subjective, and thus titration results varied widely between laboratories.

The MIF and assays utilizing EIA for antibody capture are still used in the assessment of lifetime exposure (not active infection specifically) to *C. pneumoniae*. Given the potential downstream consequences of infection, such as atherosclerotic disease, patient risk assessment may include *C. pneumoniae* antibody screening. The EIA methods utilize 96-well microtiter plates coated with whole chlamydial EBs or specific protein antigens to capture human antibodies. These assays provide a readout of "Reactive" or "Non-reactive," depending on the optical density of the reaction, or an approximate protein concentration. The assays may be used with or without titration. Interpretation of results is not clear since few studies have rigorously evaluated the clinical outcomes associated with the results generated by these assays.

NAATs

Similar to *C. trachomatis*, the most sensitive detection method for *C. pneumoniae* is NAAT. There is now one FDA cleared NAAT assay by Film Array (BioFire Diagnostics, Salt Lake City, UT), the use of which may help clarify the role of this organism as an etiologic agent in CAP and other respiratory infections (113, 114). Many laboratories have thoroughly described and validated LDTs that exhibit excellent performance characteristics (115, 116). Samples include nasopharyngeal and BAL specimens. Caution should be taken to recognize that neonatal chlamydial pneumonia is caused by *C. trachomatis* not *C. pneumoniae*, and requests for *C. pneumoniae*, NAAT should not be performed from this population. LDTs currently in use rely on real-time or light cycler technology to run PCR based assays. While the assays themselves have good sensitivity, specimen collection is problematic and thus the results of testing may often be falsely negative. Manufacturers are beginning to consider addition of *C. pneumoniae* to panels of respiratory pathogens that are detected in next generation molecular assays.

Culture

Tissue culture for isolation of *C. pneumoniae* is available in reference and research laboratories (117). Nasopharyngeal or BAL specimens provide adequate samples for culture if they are collected rigorously, if cold chain is maintained during transport, and if samples are inoculated onto tissue culture or frozen at ≥ – 70°C within 24 hours post collection. Since BAL samples are often spilt for a variety of tests performed in multiple sections of a pathology laboratory, this is often difficult to achieve.

HEp-2 cells, originally derived from human epithelial cells obtained from a laryngeal cancer, are the preferred cell line for isolation of *C. pneumoniae*. Caution should be taken to verify that the cell line has not been contaminated with HeLa cells as this is the case in some cell lines obtained from the American Type Culture Collection, the most common source of reference organisms and cell lines in the U.S. Monolayers are prepared as for *C. trachomatis* culture and are inoculated in similar fashion. Incubation is generally at 35°C rather than 37°C to optimize the sensitivity of the culture. Genus-specific immunofluorescent stains may be used after 48 to 72 hours post inoculation. Infected cells generally contain multiple inclusions throughout the cytoplasm unlike the single large inclusion formed by *C. trachomatis*. Staining with *C. trachomatis* species-specific stains (anti-major outer-membrane) should be negative, while genus-specific stains (anti-LPS) are positive. Further confirmation of the speciation requires research or LDT reagents. NAATs may also be used for speciation based on differing melting parameters. However, if this will be the final step in the culture process, NAATs should probably be used from the onset due to increased sensitivity and reduced time to results.

BEST PRACTICES

Chlamydia Trachomatis

The CDC Laboratory Guidelines for Detection of *C. trachomatis* and *N. gonorrhoeae* were developed based on extensive review of the available literature regarding assay performance (83). The recommendations developed from that review are presented here.

- NAATs are the recommended class of assay for both diagnostic and screening activities.
- There is no significant difference in the performance of these assays based on the presence or absence of symptoms in either men or women.
- For women, the optimum sample type is the patient obtained vaginal swab (118–123). This specimen provides sensitivity at least as high as any other specimen and may have other advantages. Most women prefer self-collection of vaginal swabs to endocervical sample collection, or to providing a urine specimen. Further, although the performance estimates for various sample types obtained from individual clinical trials are not

TABLE 2 Advantages and Disadvantages of Diagnostic Methodologies

	Advantages	Disadvantages	Optimal Application
NAATs	Highly sensitive and specific Do not require viable organisms As a result, expanded sample types (including non-genital specimens) and relaxed transport conditions are possible Results available in a single work shift	Some assays may detect species closely related to *N. gonorrhoeae* Organisms from exposures that may not lead to active infection may be detected Instrumentation and reagent costs remain high	Screening of asymptomatic patients in clinical settings as recommend by CDC, USPSTF, HEDIS, and other agencies Screening in non-clinical settings such as correctional facilities, schools, and work force entry programs Diagnostic testing for persons with signs or symptoms of disease Medical-legal testing for child or sexual abuse forensic evidence NAATs are ideal for detection of *C. pneumoniae*; however, no assays are commercially available
DFA	Rapid turnaround	Poor sensitivity and specificity Requires highly skilled microscopist	Ocular testing of neonates with suspected chlamydial conjunctivitis
Culture	Can verify active infection Highly specific Provides isolates for potential research applications such as typing and antimicrobial resistance testing	Requires specialized technical expertise available in few laboratories Need for viable organisms restricts potential sample collection sites to highly invasive endocervical or urethral specimens Poor sensitivity, which is heavily influenced by sample collection and transport Extended time to results	May be used for *C. pneumoniae* in order to obtain isolates For *C. trachomatis*, application should be restricted to reference or research laboratories in support of investigations such as recurrent infection despite treatment
Serology	Provides (limited) information regarding lifetime exposure	Assays are variable in quality Development of antibodies is dependent on duration of exposure and thus this may not be a useful biomarker in populations where routine screening and treatment are in place Duration of antibody response and specificity of response vary among individuals	Facilitates population based evaluations of chlamydial epidemiology

statistically different, a consistent trend suggests the rank order of vaginal swabs, endocervical swabs, urine, and endocervical samples in LBC. While urine samples may already be obtained in many clinical practices, these samples are often clean-catch midstream and thus inappropriate for use with NAATs. The LBC samples provide an opportunity to bundle several tests from a single specimen, but the sensitivity is somewhat less than that of the other sample types and the logistics of specimen handling through a cytology department prior to processing for chlamydia testing may result in delays and compromised specimen quality. Thus, from the perspectives of patient flow and testing quality (including sensitivity, specificity, and turnaround time) self-obtained vaginal swabs offer the best option. Finally, this sample type is also amenable to remote specimen collection in non-clinical settings, which may increase access to and utilization of services.

- The recommended sample type for both diagnosis and screening of men is urine. Male urine performs at least as well as urethral swabs and is much better tolerated by patients. Urethral swabs may actually be a disincentive for return visits for asymptomatic screening, which is a critical step in controlling the spread of this pathogen.
- MSM and women engaging in anal intercourse should be offered ano-rectal testing at least annually. The CDC recommends that testing be performed using NAATs, which require laboratories to perform in-house validation studies with this sample type.
- Culture capacity should be maintained in research and public health reference laboratories in the event that a need should arise for a repository of isolates.
- Serology, EIA/ELSIA, and DNA-probe technologies are no longer appropriate for detection of *C. trachomatis* in industrialized settings.

Table 2 provides a list of comparative advantages and disadvantages for each class of assay currently available and suggests appropriate populations and settings for use.

Chlamydia Pneumoniae

Diagnosis of C. *pneumoniae* continues to be hampered by the lack of commercially available diagnostic tools. Currently, for detection of acute infection, the one commercially available NAAT (BioFire, Film Array), or LDT NAATs are the best choice. This will change in the near future as this pathogen is added to more respiratory infection molecular assay panels. Serology remains a useful tool for assessing risk of future negative health outcomes in those cases where strong associations with C. *pneumoniae* infection have been shown (e.g., atherosclerosis). Serology is also a useful epidemiologic tool. Standardized assays that have regulatory approval would be a substantial benefit for this type of testing.

REFERENCES

1. **Hatch TP, Allan I, Pearce JH.** 1984. Structural and polypeptide differences between envelopes of infective and reproductive life cycle forms of Chlamydia spp. *J Bacteriol* **157:** 13–20.
2. **Everett KDE, Bush RM, Andersen AA.** 1999. Emended description of the order Chlamydiales, proposal of Parachlamydiaceae fam. nov. and Simkaniaceae fam. nov., each containing one monotypic genus, revised taxonomy of the family Chlamydiaceae, including a new genus and five new species, and standards for the identification of organisms. *Int J Syst Bacteriol* **49:**415–440.
3. **Stephens RS, Myers G, Eppinger M, Bavoil PM.** 2009. Divergence without difference: phylogenetics and taxonomy of Chlamydia resolved. *FEMS Immunol Med Microbiol* **55:** 115–119.
4. **Stephens RS, Kalman S, Lammel C, Fan J, Marathe R, Aravind L, Mitchell W, Olinger L, Tatusov RL, Zhao Q, Koonin EV, Davis RW.** 1998. Genome sequence of an obligate intracellular pathogen of humans: *chlamydia trachomatis*. *Science* **282:**754–759.
5. **Greub G.** 2013. International Committee on Systematics of Prokaryotes Subcommittee on the taxonomy of Chlamydiae: minutes of the closed meeting, 23 February 2011, Ascona, Switzerland. *Int J Syst Evol Microbiol* **63:**1934–1935.
6. **Bachmann NL, Polkinghorne A, Timms P.** 2014. Chlamydia genomics: providing novel insights into chlamydial biology. *Trends Microbiol* **22:**464–472.
7. **Fields KA, Hackstadt T.** 2002. The chlamydial inclusion: escape from the endocytic pathway. *Annu Rev Cell Dev Biol* **18:**221–245.
8. **Hackstadt T, Scidmore-Carlson MA, Shaw EI, Fischer ER.** 1999. The *Chlamydia trachomatis* IncA protein is required for homotypic vesicle fusion. *Cell Microbiol* **1:**119–130.
9. **Geisler WM, Suchland RJ, Rockey DD, Stamm WE.** 2001. Epidemiology and clinical manifestations of unique *Chlamydia trachomatis* isolates that occupy nonfusogenic inclusions. *J Infect Dis* **184:**879–884.
10. **Millman K, Black CM, Johnson RE, Stamm WE, Jones RB, Hook EW, Martin DH, Bolan G, Tavaré S, Dean D.** 2004. Population-based genetic and evolutionary analysis of *Chlamydia trachomatis* urogenital strain variation in the United States. *J Bacteriol* **186:**2457–2465.
11. **Won H, Ramachandran P, Steece R, Van Der Pol B, Moncada J, Schachter J, Gaydos C.** 2013. Is there evidence of the new variant *Chlamydia trachomatis* in the United States? *Sex Transm Dis* **40:**352–353.
12. **Unemo M, Clarke IN.** 2011. The Swedish new variant of *Chlamydia trachomatis*. *Curr Opin Infect Dis* **24:**62–69.
13. **Unemo M, Seth-Smith HM, Cutcliffe LT, Skilton RJ, Barlow D, Goulding D, Persson K, Harris SR, Kelly A, Bjartling C, Fredlund H, Olcén P, Thomson NR, Clarke IN.** 2010. The Swedish new variant of *Chlamydia trachomatis*: genome sequence, morphology, cell tropism and phenotypic characterization. *Microbiology* **156:**1394–1404.
14. **Grayston JT, Kuo C-C, Campbell LA, Wang S-P.** 1989. *Chlamydia pneumoniae* sp. nov. for Chlamydia sp. strain TWAR. *Int J Syst Bacteriol* **39:**88–90.
15. **Gaydos CA, Palmer L, Quinn TC, Falkow S, Eiden JJ.** 1993. Phylogenetic relationship of Chlamydia pneumoniae to Chlamydia psittaci and Chlamydia trachomatis as determined by analysis of 16S ribosomal DNA sequences. *Int J Syst Bacteriol* **43:**610–612.
16. **Brunham RC, Rey-Ladino J.** 2005. Immunology of Chlamydia infection: implications for a *Chlamydia trachomatis* vaccine. *Nat Rev Immunol* **5:**149–161.
17. **Beatty WL, Byrne GI, Morrison RP.** 1993. Morphologic and antigenic characterization of interferon gamma-mediated persistent *Chlamydia trachomatis* infection in vitro. *Proc Natl Acad Sci USA* **90:**3998–4002.
18. **Dean D, Suchland RJ, Stamm WE.** 2000. Evidence for long-term cervical persistence of *Chlamydia trachomatis* by omp1 genotyping. *J Infect Dis* **182:**909–916.
19. **World Health Organization.** 2012. *Global incidence and prevalence of selected curable sexually transmitted infections-2008.* WHO Press.
20. **Centers for Disease Control and Prevention.** 2012. Sexually Transmitted. *Dis Surveill*:2011. U.S. Department of Health and Human Services.
21. **Torrone E, Papp J, Weinstock H, Centers for Disease Control and Prevention (CDC).** 2014. Prevalence of *Chlamydia trachomatis* genital infection among persons aged 14–39 years–United States, 2007–2012. *MMWR Morb Mortal Wkly Rep* **63:**834–838.
22. **Adams DA, Jajosky RA, Ajani U, Kriseman J, Sharp P, Onwen DH, Schley AW, Anderson WJ, Grigoryan A, Aranas AE, Wodajo MS, Abellera JP, Centers for Disease Control and Prevention (CDC).** 2014. Summary of notifiable diseases–United States, 2012. *MMWR Morb Mortal Wkly Rep* **61:**1–121.
23. **Burton MJ, Mabey DCW.** 2009. The global burden of trachoma: a review. *PLoS Negl Trop Dis* **3:**e460.
24. **World Health Organization.** 2007. *Global Initiative for the Elimination of Avoidable Blindness: action plan 2006–2011.*
25. **World Health Organization.** 2003. *Guidelines for the management of Sexually Transmitted Infections.* WHO Press.
26. **Detels R, Green AM, Klausner JD, Katzenstein D, Gaydos C, Handsfield H, Pequegnat W, Mayer K, Hartwell TD, Quinn TC.** 2011. The incidence and correlates of symptomatic and asymptomatic *Chlamydia trachomatis* and *Neisseria gonorrhoeae* infections in selected populations in five countries. *Sex Transm Dis* **38:**503–509.
27. **Nieuwenhuis RF, Ossewaarde JM, Götz HM, Dees J, Thio HB, Thomeer MGJ, den Hollander JC, Neumann MHA, van der Meijden WI.** 2004. Resurgence of lymphogranuloma venereum in Western Europe: an outbreak of *Chlamydia trachomatis* serovar l2 proctitis in The Netherlands among men who have sex with men. *Clin Infect Dis* **39:**996–1003.
28. **Price MJ, Ades AE, De Angelis D, Welton NJ, Macleod J, Soldan K, Simms I, Turner K, Horner PJ.** 2013. Risk of pelvic inflammatory disease following *Chlamydia trachomatis* infection: analysis of prospective studies with a multistate model. *Am J Epidemiol* **178:**484–492.
29. **Bender N, Herrmann B, Andersen B, Hocking JS, van Bergen J, Morgan J, van den Broek IV, Zwahlen M, Low N.** 2011. Chlamydia infection, pelvic inflammatory disease, ectopic pregnancy and infertility: cross-national study. *Sex Transm Infect* **87:**601–608.
30. **French CE, Hughes G, Nicholson A, Yung M, Ross JD, Williams T, Soldan K.** 2011. Estimation of the rate of pelvic inflammatory disease diagnoses: Trends in England, 2000–2008. *Sex. Transm. Dis.* **38:**158–162.

31. **Rekart ML, Gilbert M, Meza R, Kim PH, Chang M, Money DM, Brunham RC.** 2013. Chlamydia public health programs and the epidemiology of pelvic inflammatory disease and ectopic pregnancy. *J Infect Dis* **207:**30–38.
32. **Tuffrey M, Alexander F, Inman C, Ward ME.** 1990. Correlation of infertility with altered tubal morphology and function in mice with salpingitis induced by a human genital-tract isolate of *Chlamydia trachomatis*. *J Reprod Fertil* **88:**295–305.
33. **Patton DL, Askienazy-Elbhar M, Henry-Suchet J, Campbell LA, Cappuccio A, Tannous W, Wang SP, Kuo CC.** 1994. Detection of *Chlamydia trachomatis* in fallopian tube tissue in women with postinfectious tubal infertility. *Am J Obstet Gynecol* **171:**95–101.
34. **Hammerschlag MR, Kohlhoff SA, Gaydos CA.** 2014. Chlamydia pneumoniae, p. 2174–2182. *In* Mandell GL, Bennett JE, Dolin R, Blaser MJ (eds.), *Principles and Practice of Infectious Diseases*, 8th ed. Elsevier, Sanders Inc., Philadelphia, PA.
35. **Cabbage S, Ieronimakis N, Preusch M, Lee A, Ricks J, Janebodin K, Hays A, Wijelath ES, Reyes M, Campbell LA, Rosenfeld ME.** 2014. *Chlamydia pneumoniae* infection of lungs and macrophages indirectly stimulates the phenotypic conversion of smooth muscle cells and mesenchymal stem cells: potential roles in vascular calcification and fibrosis. *Pathog Dis* **72:**61–69.
36. **Hahn DL, Schure A, Patel K, Childs T, Drizik E, Webley W.** 2012. *Chlamydia pneumoniae*-specific IgE is prevalent in asthma and is associated with disease severity. *PLoS One* **7:**e35945.
37. **Brandt J, Cader A, Semler L, Hammond C, Bell M, Devins M, Smith J, Joseph N, Velez M, Galluzzi K, Balin B, Appelt D.** 2014. Electron microscopy studies elucidate morphological forms of *Chlamydia pneumoniae* in blood samples from patients diagnosed with mild cognitive impairment (MCI) and Alzheimer's disease (AD) (LB81). *FASEB J* **28:** LB81.
38. **Carter JD, Espinoza LR, Inman RD, Sneed KB, Ricca LR, Vasey FB, Valeriano J, Stanich JA, Oszust C, Gerard HC, Hudson AP.** 2010. Combination antibiotics as a treatment for chronic Chlamydia-induced reactive arthritis: a double-blind, placebo-controlled, prospective trial. *Arthritis Rheum* **62:** 1298–1307.
39. **Workowski KA, Berman S, Centers for Disease Control and Prevention (CDC).** 2010. Sexually transmitted diseases treatment guidelines, 2010. *MMWR Recomm Rep* **59**(RR-12)**:**1–110.
40. **LeFevre ML.** 2014. Screening for Chlamydia and Gonorrhea: U.S. Preventive Services Task Force Recommendation Statement Screening for Chlamydia and Gonorrhea. *Ann. Intern. Med.* **161**(12): 1–30.
41. **Bachmann LH, Johnson RE, Cheng H, Markowitz L, Papp JR, Palella FJ Jr, Hook EW III.** 2010. Nucleic acid amplification tests for diagnosis of *Neisseria gonorrhoeae* and *Chlamydia trachomatis* rectal infections. *J Clin Microbiol* **48:** 1827–1832.
42. **Harding-Esch EM, Edwards T, Mkocha H, Munoz B, Holland MJ, Burr SE, Sillah A, Gaydos CA, Stare D, Mabey DCW, Bailey RL, West SK, PRET Partnership.** 2010. Trachoma prevalence and associated risk factors in the gambia and Tanzania: baseline results of a cluster randomised controlled trial. *PLoS Negl Trop Dis* **4:**e861.
43. **West SK, Munoz B, Mkocha H, Gaydos CA, Quinn TC.** 2011. Number of years of annual mass treatment with azithromycin needed to control trachoma in hyper-endemic communities in Tanzania. *J Infect Dis* **204:**268–273.
44. **Cajas-Monson LC, Mkocha H, Muñoz B, Quinn TC, Gaydos CA, West SK.** 2011. Risk factors for ocular infection with Chlamydia trachomatis in children 6 months following mass treatment in Tanzania. *PLoS Negl Trop Dis* **5:**e978.
45. **Hammerschlag MR, Kohlhoff SA.** 2012. Treatment of chlamydial infections. *Expert Opin Pharmacother* **13:**545–552.
46. **Moncada J, Schachter J, Liska S, Shayevich C, Klausner JD.** 2009. Evaluation of self-collected glans and rectal swabs from men who have sex with men for detection of *Chlamydia trachomatis* and *Neisseria gonorrhoeae* by use of nucleic acid amplification tests. *J Clin Microbiol* **47:**1657–1662.
47. **Leslie DE, Azzato F, Ryan N, Fyfe J.** 2003. An assessment of the Roche Amplicor *Chlamydia trachomatis/Neisseria gonorrhoeae* multiplex PCR assay in routine diagnostic use on a variety of specimen types. *Commun Dis Intell Q Rep* **27:**373–379.
48. **Pittaras TE, Papaparaskevas J, Houhoula DP, Legakis NJ, Frangouli E, Katsambas A, Tsakris A, Papadogeorgakis H.** 2008. Comparison of penile skin swab with intra-urethral swab and first void urine for polymerase chain reaction-based diagnosis of *Chlamydia trachomatis* urethritis in male patients. *Sex Transm Dis* **35:**999–1001.
49. **Dize L, Agreda P, Quinn N, Barnes MR, Hsieh Y-H, Gaydos CA.** 2013. Comparison of self-obtained penile-meatal swabs to urine for the detection of *C. trachomatis*, *N. gonorrhoeae* and *T. vaginalis*. *Sex Transm Infect* **89:**305–307.
50. **Kumamoto Y, Matsumoto T, Fujisawa M, Arakawa S.** 2012. Detection of *Chlamydia trachomatis* and *Neisseria gonorrhoeae* in urogenital and oral specimens using the cobas® 4800, APTIMA Combo 2® TMA, and ProbeTec™ ET SDA assays. *Eur J Microbiol Immunol (Bp)* **2:**121–127.
51. **Schachter J, Moncada J, Liska S, Shayevich C, Klausner JD.** 2008. Nucleic acid amplification tests in the diagnosis of chlamydial and gonococcal infections of the oropharynx and rectum in men who have sex with men. *Sex Transm Dis* **35:**637–642.
52. **Ota KV, Tamari IE, Smieja M, Jamieson F, Jones KE, Towns L, Juzkiw J, Richardson SE.** 2009. Detection of *Neisseria gonorrhoeae* and *Chlamydia trachomatis* in pharyngeal and rectal specimens using the BD Probetec ET system, the Gen-Probe Aptima Combo 2 assay and culture. *Sex Transm Infect* **85:**182–186.
53. **Peters RP, Verweij SP, Nijsten N, Ouburg S, Mutsaers J, Jansen CL, van Leeuwen AP, Morré SA.** 2011. Evaluation of sexual history-based screening of anatomic sites for *chlamydia trachomatis* and neisseria gonorrhoeae infection in men having sex with men in routine practice. *BMC Infect Dis* **11:**203.
54. **Bax CJ, Quint KD, Peters RP, Ouburg S, Oostvogel PM, Mutsaers JA, Dörr PJ, Schmidt S, Jansen C, van Leeuwen AP, Quint WG, Trimbos JB, Meijer CJ, Morré SA.** 2011. Analyses of multiple-site and concurrent *Chlamydia trachomatis* serovar infections, and serovar tissue tropism for urogenital versus rectal specimens in male and female patients. *Sex Transm Infect* **87:**503–507.
55. **Marcus JL, Bernstein KT, Stephens SC, Snell A, Kohn RP, Liska S, Klausner JD.** 2010. Sentinel surveillance of rectal chlamydia and gonorrhea among males–San Francisco, 2005–2008. *Sex Transm Dis* **37:**59–61.
56. **Hunte T, Alcaide M, Castro J.** 2010. Rectal infections with chlamydia and gonorrhoea in women attending a multiethnic sexually transmitted diseases urban clinic. *Int J STD AIDS* **21:**819–822.
57. **van der Helm JJ, Hoebe CJ, van Rooijen MS, Brouwers EE, Fennema HS, Thiesbrummel HF, Dukers-Muijrers NH.** 2009. High performance and acceptability of self-collected rectal swabs for diagnosis of *Chlamydia trachomatis* and *Neisseria gonorrhoeae* in men who have sex with men and women. *Sex Transm Dis* **36:**493–497.
58. **Baker J, Plankey M, Josayma Y, Elion R, Chiliade P, Shahkolahi A, Menna M, Miniter K, Slack R, Yang Y, Masterman B, Margolick JB.** 2009. The prevalence of rectal, urethral, and pharyngeal *Neisseria gonorrheae* and *Chlamydia trachomatis* among asymptomatic men who have sex with men in a prospective cohort in Washington, DC. *AIDS Patient Care STDS* **23:**585–588.
59. **Alexander S, Ison C, Parry J, Llewellyn C, Wayal S, Richardson D, Phillips A, Smith H, Fisher M, Brighton Home Sampling Kits Steering Group.** 2008. Self-taken pharyngeal and rectal swabs are appropriate for the detection of *Chlamydia trachomatis* and *Neisseria gonorrhoeae* in asymptomatic men who have sex with men. *Sex Transm Infect* **84:**488–492.

60. **Bachmann LH, Johnson RE, Cheng H, Markowitz L, Papp JR, Palella FJ Jr, Hook EW III.** 2010. Nucleic acid amplification tests for diagnosis of *Neisseria gonorrhoeae* and *Chlamydia trachomatis* rectal infections. *J Clin Microbiol* **48:**1827–1832.
61. **Goldenberg SD, Finn J, Sedudzi E, White JA, Tong CYW.** 2012. Performance of the GeneXpert CT/NG assay compared to that of the Aptima AC2 assay for detection of rectal *Chlamydia trachomatis* and *Neisseria gonorrhoeae* by use of residual Aptima Samples. *J Clin Microbiol* **50:**3867–3869.
62. **Dodge B, Van Der Pol B, Reece M, Malebranche D, Martinez O, Goncalves G, Schnarrs P, Nix R, Fortenberry JD.** 2012. Rectal self-sampling in non-clinical venues for detection of sexually transmissible infections among behaviourally bisexual men. *Sex Health* **9:**190–191.
63. **Rosenberger JG, Dodge B, Van Der Pol B, Reece M, Herbenick D, Fortenberry JD.** 2011. Reactions to self-sampling for ano-rectal sexually transmitted infections among men who have sex with men: a qualitative study. *Arch Sex Behav* **40:**281–288.
64. **Ladd J, Hsieh Y-H, Barnes M, Quinn N, Jett-Goheen M, Gaydos CA.** 2014. Female users of internet-based screening for rectal STIs: descriptive statistics and correlates of positivity. *Sex Transm Infect* **90:**485–490.
65. **Freeman AH, Bernstein KT, Kohn RP, Philip S, Rauch LM, Klausner JD.** 2011. Evaluation of self-collected versus clinician-collected swabs for the detection of *Chlamydia trachomatis* and *Neisseria gonorrhoeae* pharyngeal infection among men who have sex with men. *Sex Transm Dis* **38:**1036–1039.
66. **Dize L, West S, Williams JA, Van Der Pol B, Quinn TC, Gaydos CA.** 2013. Comparison of the Abbott m2000 RealTime CT assay and the Cepheid GeneXpert CT/NG assay to the Roche Amplicor CT assay for detection of *Chlamydia trachomatis* in ocular samples from Tanzania. *J Clin Microbiol* **51:**1611–1613.
67. **Yang JL, Hong KC, Schachter J, Moncada J, Lekew T, House JI, Zhou Z, Neuwelt MD, Rutar T, Halfpenny C, Shah N, Whitcher JP, Lietman TM.** 2009. Detection of *Chlamydia trachomatis* ocular infection in trachoma-endemic communities by rRNA amplification. *Invest Ophthalmol Vis Sci* **50:**90–94.
68. **Thomas BJ, Evans RT, Hawkins DA, Taylor-Robinson D.** 1984. Sensitivity of detecting *Chlamydia trachomatis* elementary bodies in smears by use of a fluorescein labelled monoclonal antibody: comparison with conventional chlamydial isolation. *J Clin Pathol* **37:**812–816.
69. **Stamm WE, Harrison HR, Alexander ER, Cles LD, Spence MR, Quinn TC.** 1984. Diagnosis of Chlamydia trachomatis infections by direct immunofluorescence staining of genital secretions. A multicenter trial. *Ann Intern Med* **101:** 638–641.
70. **Beebe JL, Masters H, Jungkind D, Heltzel DM, Weinberg A.** 1996. Confirmation of the Syva MicroTrak enzyme immunoassay for *chlamydia trachomatis* by Syva Direct Fluorescent Antibody Test. *Sex Transm Dis* **23:**465–470.
71. **Van Der Pol B, Williams J, Smith N, Batteiger B, Cullen A, Edens T, Davis K, Salim H, Chou V, Scearce L, Blutman J, Payne W.** 2001. Comparison of the semi-automated Digene Rapid Capture System TM with the manual system for the Hybrid Capture 2 Chlamydia Assay using endocervical samples., 14th meeting of the International Society for Sexually Transmitted Diseases Research Berlin, Germany.
72. **Beltrami JF, Farley TA, Hamrick JT, Cohen DA, Martin DH.** 1998. Evaluation of the Gen-Probe PACE 2 assay for the detection of asymptomatic *Chlamydia trachomatis* and *Neisseria gonorrhoeae* infections in male arrestees. *Sex Transm Dis* **25:** 501–504.
73. **Darwin LH, Cullen AP, Arthur PM, Long CD, Smith KR, Girdner JL, Hook EW III, Quinn TC, Lorincz AT.** 2002. Comparison of Digene hybrid capture 2 and conventional culture for detection of *Chlamydia trachomatis* and *Neisseria gonorrhoeae* in cervical specimens. *J Clin Microbiol* **40:**641–644.
74. **Schachter J, Hook EW III, McCormack WM, Quinn TC, Chernesky M, Chong S, Girdner JI, Dixon PB, DeMeo L, Williams E, Cullen A, Lorincz A.** 1999. Ability of the digene hybrid capture II test to identify *Chlamydia trachomatis* and *Neisseria gonorrhoeae* in cervical specimens. *J Clin Microbiol* **37:**3668–3671.
75. **Modarress KJ, Cullen AP, Jaffurs WJ Sr, Troutman GL, Mousavi N, Hubbard RA, Henderson S, Lörincz AT.** 1999. Detection of *Chlamydia trachomatis* and *Neisseria gonorrhoeae* in swab specimens by the Hybrid Capture II and PACE 2 nucleic acid probe tests. *Sex Transm Dis* **26:**303–308.
76. **Beebe JL, Sharpton TR, Zanto SN, Steece RS, Rogers C, Mottice SL.** 1997. Performance characteristics of the Gen-Probe Probe Competition Assay used as a supplementary test for the Gen-Probe PACE 2 and 2C assays for detection of *Chlamydia trachomatis*. *J Clin Microbiol* **35:**477–478.
77. **Newhall WJ, Johnson RE, DeLisle S, Fine D, Hadgu A, Matsuda B, Osmond D, Campbell J, Stamm WE.** 1999. Head-to-head evaluation of five chlamydia tests relative to a quality-assured culture standard. *J Clin Microbiol* **37:**681–685.
78. **Van Der Pol B, Ferrero DV, Buck-Barrington L, Hook E III, Lenderman C, Quinn T, Gaydos CA, Lovchik J, Schachter J, Moncada J, Hall G, Tuohy MJ, Jones RB.** 2001. Multicenter evaluation of the BDProbeTec ET System for detection of *Chlamydia trachomatis* and *Neisseria gonorrhoeae* in urine specimens, female endocervical swabs, and male urethral swabs. *J Clin Microbiol* **39:**1008–1016.
79. **Van Der Pol B, Quinn TC, Gaydos CA, Crotchfelt K, Schachter J, Moncada J, Jungkind D, Martin DH, Turner B, Peyton C, Jones RB.** 2000. Multicenter evaluation of the AMPLICOR and automated COBAS AMPLICOR CT/NG tests for detection of Chlamydia trachomatis. *J Clin Microbiol* **38:**1105–1112.
80. **Schachter J, Stamm WE, Quinn TC, Andrews WW, Burczak JD, Lee HH.** 1994. Ligase chain reaction to detect *Chlamydia trachomatis* infection of the cervix. *J Clin Microbiol* **32:**2540–2543.
81. **Chernesky MA, Lee H, Schachter J, Burczak JD, Stamm WE, McCormack WM, Quinn TC.** 1994. Diagnosis of *Chlamydia trachomatis* urethral infection in symptomatic and asymptomatic men by testing first-void urine in a ligase chain reaction assay. *J Infect Dis* **170:**1308–1311.
82. **Black CM, Marrazzo J, Johnson RE, Hook EW III, Jones RB, Green TA, Schachter J, Stamm WE, Bolan G, St Louis ME, Martin DH.** 2002. Head-to-head multicenter comparison of DNA probe and nucleic acid amplification tests for *Chlamydia trachomatis* infection in women performed with an improved reference standard. *J Clin Microbiol* **40:**3757–3763.
83. **Papp J, Schachter J, Gaydos C, Van Der Pol B.** 2014. Recommendations for the laboratory-based detection of *Chlamydia trachomatis* and *Neisseria gonorrhoeae*-2014. *MMWR Morb Mortal Wkly Rep* **63:**1–24.
84. **Schachter J, Chow JM, Howard H, Bolan G, Moncada J.** 2006. Detection of *Chlamydia trachomatis* by nucleic acid amplification testing: our evaluation suggests that CDC-recommended approaches for confirmatory testing are ill-advised. *J Clin Microbiol* **44:**2512–2517.
85. **Scragg S, Bingham A, Mallinson H.** 2006. Should *Chlamydia trachomatis* confirmation make you cross? Performance of collection kits tested across three nucleic acid amplification test platforms. *Sex Transm Infect* **82:**295–297.
86. **Chow JM, Bauer HM, Bolan G.** 2012. Repeating low-positive nucleic acid amplification test results for *Chlamydia trachomatis* and *Neisseria gonorrhoeae*: assessment of current practice in selected california public- and private-sector laboratories. *J Clin Microbiol* **50:**539.
87. **Van Der Pol B, Body B, Nye M, Eisenhut C, Taylor S, Liesenfeld O.** 2011. Vaginal swabs are the optimal sample for screening women for chlamydia and gonorrhea infection using the Roche cobas 4800 system. 21st Meeting of the European Congress on Clinical Microbiology and Infectious Diseases.

88. **Taylor S, Lillis R, Body B, Nye M, Williams J, Van Der Pol B.** 2011. Evaluation of the Roche cobas 4800 for detection of Chlamydia trachomatis and Neisseria gonorrhoeae in men. 19th Biennial meeting of the International Society for STD Research Quebec City, CAN.
89. **Rockett R, Goire N, Limnios A, Turra M, Higgens G, Lambert SB, Bletchly C, Nissen MD, Sloots TP, Whiley DM.** 2010. Evaluation of the cobas 4800 CT/NG test for detecting *Chlamydia trachomatis* and *Neisseria gonorrhoeae*. *Sex Transm Infect* **86:**470–473.
90. **Van Der Pol B, Liesenfeld O, Williams JA, Taylor SN, Lillis RA, Body BA, Nye M, Eisenhut C, Hook EW III.** 2012. Performance of the cobas CT/NG test compared to the Aptima AC2 and Viper CTQ/GCQ assays for detection of *Chlamydia trachomatis* and *Neisseria gonorrhoeae*. *J Clin Microbiol* **50:**2244–2249.
91. **Møller JK, Pedersen LN, Persson K.** 2010. Comparison of the Abbott RealTime CT new formulation assay with two other commercial assays for detection of wild-type and new variant strains of Chlamydia trachomatis. *J Clin Microbiol* **48:**440–443.
92. **Levett PN, Brandt K, Olenius K, Brown C, Montgomery K, Horsman GB.** 2008. Evaluation of three automated nucleic acid amplification systems for detection of *Chlamydia trachomatis* and *Neisseria gonorrhoeae* in first-void urine specimens. *J Clin Microbiol* **46:**2109–2111.
93. **Marshall R, Chernesky M, Jang D, Hook EW, Cartwright CP, Howell-Adams B, Ho S, Welk J, Lai-Zhang J, Brashear J, Diedrich B, Otis K, Webb E, Robinson J, Yu H.** 2007. Characteristics of the m2000 automated sample preparation and multiplex real-time PCR system for detection of *Chlamydia trachomatis* and *Neisseria gonorrhoeae*. *J Clin Microbiol* **45:**747–751.
94. **Gaydos CA, Van Der Pol B, Jett-Goheen M, Barnes M, Quinn N, Clark C, Daniel GE, Dixon PB, Hook EW III, CT/NG Study Group.** 2013. Performance of the Cepheid CT/NG Xpert Rapid PCR test for detection of *Chlamydia trachomatis* and *Neisseria gonorrhoeae*. *J Clin Microbiol* **51:**1666–1672.
95. **Tabrizi SN, Unemo M, Golparian D, Twin J, Limnios AE, Lahra M, Guy R, TTANGO Investigators.** 2013. Analytical evaluation of GeneXpert CT/NG, the first genetic point-of-care assay for simultaneous detection of *Neisseria gonorrhoeae* and *Chlamydia trachomatis*. *J Clin Microbiol* **51:**1945–1947.
96. **Gaydos CA.** 2014. Review of use of a new rapid real-time PCR, the Cepheid GeneXpert® (Xpert) CT/NG assay, for *Chlamydia trachomatis* and *Neisseria gonorrhoeae*: results for patients while in a clinical setting. *Expert Rev Mol Diagn* **14:**135–137.
97. **Taylor SN, Van Der Pol B, Lillis R, Hook EW III, Lebar W, Davis T, Fuller D, Mena L, Fine P, Gaydos CA, Martin DH.** 2011. Clinical evaluation of the BD ProbeTec™ *Chlamydia trachomatis* Qx amplified DNA assay on the BD Viper™ system with XTR™ technology. *Sex Transm Dis* **38:**603–609.
98. **Lowe P, O'Loughlin P, Evans K, White M, Bartley PB, Vohra R.** 2006. Comparison of the Gen-Probe APTIMA Combo 2 assay to the AMPLICOR CT/NG assay for detection of *Chlamydia trachomatis* and *Neisseria gonorrhoeae* in urine samples from Australian men and women. *J Clin Microbiol* **44:**2619–2621.
99. **Boyadzhyan B, Yashina T, Yatabe JH, Patnaik M, Hill CS.** 2004. Comparison of the APTIMA CT and GC assays with the APTIMA combo 2 assay, the Abbott LCx assay, and direct fluorescent-antibody and culture assays for detection of *Chlamydia trachomatis* and *Neisseria gonorrhoeae*. *J Clin Microbiol* **42:**3089–3093.
100. **Gaydos CA, Quinn TC, Willis D, Weissfeld A, Hook EW, Martin DH, Ferrero DV, Schachter J.** 2003. Performance of the APTIMA Combo 2 assay for detection of Chlamydia trachomatis and Neisseria gonorrhoeae in female urine and endocervical swab specimens. *J Clin Microbiol* **41:**304–309.
101. **Lewis N, Dube G, Carter C, Pitt R, Alexander S, Ison CA, Harding J, Brown L, Fryer J, Hodson J, Ross J.** 2012. Chlamydia and gonorrhoea contamination of clinic surfaces. *Sex Transm Infect* **88:**418–421.
102. **Van Der Pol B, Unemo M.** 2013. Chlamydial Infections, p. 55–72. *In* Unemo M (ed.), *Laboratory diagnosis of sexually transmitted infections, including human immunodeficiency virus*. World Health Organization, Geneva.
103. **Gaydos CA.** 2009. Can we climb out of the "pit" of poorly performing rapid diagnostic tests for chlamydia? *Sex Transm Infect* **85:**158.
104. **Rani R, Corbitt G, Killough R, Curless E.** 2002. Is there any role for rapid tests for *Chlamydia trachomatis*? *Int J STD AIDS* **13:**22–24.
105. **Gaydos C, Hardick J.** 2014. Point of care diagnostics for sexually transmitted infections: perspectives and advances. *Expert Rev Anti Infect Ther* **12:**657–672.
106. **Schachter J.** 1999. We must be realistic in evaluating rapid diagnostic tests. *Sex Transm Dis* **26:**241–242.
107. **Gift TL, Pate MS, Hook EW III, Kassler WJ.** 1999. The rapid test paradox: when fewer cases detected lead to more cases treated: a decision analysis of tests for *Chlamydia trachomatis*. *Sex Transm Dis* **26:**232–240.
108. **Hislop J, Quayyum Z, Flett G, Boachie C, Fraser C, Mowatt G.** 2010. Systematic review of the clinical effectiveness and cost-effectiveness of rapid point-of-care tests for the detection of genital chlamydia infection in women and men. *Health Technol Assess* **14:**1–97, iii–iv.
109. **Pearce DM, Shenton DP, Holden J, Gaydos CA.** 2011. Evaluation of a novel electrochemical detection method for *Chlamydia trachomatis*: application for point-of-care diagnostics. *IEEE Trans Biomed Eng* **58:**755–758.
110. **Melendez JH, Huppert JS, Jett-Goheen M, Hesse EA, Quinn N, Gaydos CA, Geddes CD.** 2013. Blind evaluation of the microwave-accelerated metal-enhanced fluorescence ultrarapid and sensitive *Chlamydia trachomatis* test by use of clinical samples. *J Clin Microbiol* **51:**2913–2920.
111. **Verkooyen RP, Willemse D, Hiep-van Casteren SCAM, Joulandan SA, Snijder RJ, van den Bosch JMM, van Helden HPT, Peeters MF, Verbrugh HA.** 1998. Evaluation of PCR, culture, and serology for diagnosis of *Chlamydia pneumoniae* respiratory infections. *J Clin Microbiol* **36:**2301–2307.
112. **Wang S.** 2000. The microimmunofluorescence test for *Chlamydia pneumoniae* infection: technique and interpretation. *J Infect Dis* **181**(Suppl 3)**:**S421–S425.
113. **Poritz MA, Blaschke AJ, Byington CL, Meyers L, Nilsson K, Jones DE, Thatcher SA, Robbins T, Lingenfelter B, Amiott E, Herbener A, Daly J, Dobrowolski SF, Teng DHF, Ririe KM.** 2011. FilmArray, an automated nested multiplex PCR system for multi-pathogen detection: development and application to respiratory tract infection. *PLoS One* **6:**e26047.
114. **Pierce VM, Elkan M, Leet M, McGowan KL, Hodinka RL.** 2011. Comparison of the Idaho Technology FilmArray System to real-time PCR for detection of respiratory pathogens in children. *J Clin Microbiol* **50:**364–371.
115. **Benitez AJ, Thurman KA, Diaz MH, Conklin L, Kendig NE, Winchell JM.** 2012. Comparison of real-time PCR and a microimmunofluorescence serological assay for detection of *chlamydophila pneumoniae* infection in an outbreak investigation. *J Clin Microbiol* **50:**151–153.
116. **Gaydos CA.** 2013. What is the role of newer molecular tests in the management of CAP? *Infect Dis Clin North Am* **27:** 49–69.
117. **Hammerschlag MR, Chirgwin K, Roblin PM, Gelling M, Dumornay W, Mandel L, Smith P, Schachter J.** 1992. Persistent infection with Chlamydia pneumoniae following acute respiratory illness. *Clin Infect Dis* **14:**178–182.
118. **Van Der Pol B, Taylor SN, Liesenfeld O, Williams JA, Hook EW III.** 2013. Vaginal swabs are the optimal specimen for detection of genital *Chlamydia trachomatis* or *Neisseria gonorrhoeae* using the Cobas 4800 CT/NG test. *Sex Transm Dis* **40:**247–250.
119. **Blake DR, Maldeis N, Barnes MR, Hardick A, Quinn TC, Gaydos CA.** 2008. Cost-effectiveness of screening strategies for *Chlamydia trachomatis* using cervical swabs, urine, and self-

obtained vaginal swabs in a sexually transmitted disease clinic setting. *Sex Transm Dis* **35:**649–655.

120. **Rose SB, Lawton BA, Bromhead C, Macdonald EJ, Lund KA.** 2007. Self-obtained vaginal swabs for PCR chlamydia testing: a practical alternative. *Aust N Z J Obstet Gynaecol* **47:**415–418.
121. **Hoebe CJ, Rademaker CW, Brouwers EE, ter Waarbeek HL, van Bergen JE.** 2006. Acceptability of self-taken vaginal swabs and first-catch urine samples for the diagnosis of urogenital *Chlamydia trachomatis* and *Neisseria gonorrhoeae* with an amplified DNA assay in young women attending a public health sexually transmitted disease clinic. *Sex Transm Dis* **33:**491–495.
122. **Schachter J, Chernesky MA, Willis DE, Fine PM, Martin DH, Fuller D, Jordan JA, Janda W, Hook EW III.** 2005. Vaginal swabs are the specimens of choice when screening for *Chlamydia trachomatis* and *Neisseria gonorrhoeae*: results from a multicenter evaluation of the APTIMA assays for both infections. *Sex Transm Dis* **32:**725–728.
123. **Chernesky MA, Hook EW III, Martin DH, Lane J, Johnson R, Jordan JA, Fuller D, Willis DE, Fine PM, Janda WM, Schachter.** 2005. Women find it easy and prefer to collect their own vaginal swabs to diagnose *Chlamydia trachomatis* or *Neisseria gonorrhoeae* infections. *Sex Transm Dis* **32:**729–733.

The Human Virome

MATTHEW C. ROSS, NADIM J. AJAMI, AND JOSEPH F. PETROSINO

39

The human microbiome is the collection of hundreds of trillions of microorganisms and their genomes that colonize the human body. Recent advances in sequencing technology have allowed these communities and their function to be characterized and defined in detail. While the bacterial members of the microbiome have received the bulk of the attention, the viral members, the human virome, promise to be no less important. This chapter describes these recent advances in the methods for metagenomic studies, highlights key early findings, and looks ahead to the remaining challenges to implementing viral metagenomics and human virome techniques in the clinical setting for diagnostic purposes.

INTRODUCTION

The human body plays host to hundreds of trillions of microorganisms; these have come to be known as the human microbiome. These microbes don't simply call our bodies home but comprise a symbiotic relationship that imparts numerous benefits for the human host (1–3). It is becoming apparent that the information encoded by the human genome is insufficient for maintaining health and that products produced by the microbiome are vital to protection from certain diseases (4). Humans can be thought of as a supraorganism, where our ultimate phenotype results from a combination of human encoded gene functions as well as the functions encoded by the genes from the bacteria, archaea, fungi, and viruses that comprise the human microbiome (5). Many recent studies have shown these colonizing populations in human adults are largely unique at the level of an individual (2, 6–8). It is therefore important to identify and understand the countless functions the human microbiome encodes in order to completely understand human physiology. Until relatively recently, the magnitude of the effect that the composition of the human microbiome has on our well-being has been generally underappreciated. It is now apparent that the human microbiome can influence numerous facets of our daily lives including immune system maturation, pathogen exclusion, digestion, nutrition, and many others (1–3, 9).

The microbiome of humans is not one homogenous mixture of microorganisms distributed throughout the body. The human body is divided into several very specific and unique niches that are inhabited by equally unique and individualized communities of microorganisms (e.g., oral cavity, gut, exposed skin, vagina, etc.). Paving the way for the development of the microbiome field was the introduction of next-generation sequencing (NGS). In 2005, 454 Life Sciences Corp. introduced pyrophosphate-based sequencing in picoliter-sized reaction volumes capable of generating ~250,000 100-base pair (bp) reads per run (10). While sacrificing some read length and base call accuracy, this technology represented a nearly 100-fold increase in quality bases sequenced per hour compared with Sanger's dye-terminator/capillary electrophoresis method. By 2008 the length of 454 sequence reads had increased to 500 bp, allowing for more confident read mapping and better sequence assemblies. By 2009, Illumina, Inc. released chemistry for its Genome Analyzer platform increasing output to 50 gigabases of sequence per machine run, a more than 60-fold increase in base yield per run over 454. Overall, these NGS methods drastically increased both the resolution with which scientists could examine microbial communities and the number of samples that could be analyzed simultaneously. Both pyrosequencing (454 Life Sciences) and sequencing by synthesis (Illumina) methods eliminated the requirement for cloning targeted DNA sequences prior to sequencing. Both of these methods allow for interrogation of DNA from mixed microbial communities without the need for cultivation of the organisms themselves, or any prior knowledge of the community members. The thousands of microbiome studies that followed NGS introduction greatly increased our understanding of the microbial communities that thrive on and in the human body and have revolutionized our understanding of the extensive effects the human microbiome has on human health.

The different microbiome compositions amongst individuals likely contribute to differential outcomes of treatment and disease progression across the human population. Members of the microbiome impart a significant portion of their physiological impact on the host via their metabolism of xenobiotic compounds including dietary compounds, antibiotics, and other drugs (11). Prominent examples include the inactivation of digoxin by *Eggerthella lenta* (12), and the interaction of *Helicobacter pylori* with L-DOPA, possibly reducing the amount of drug that actually arrives in the central nervous system (13). It is well established that

doi:10.1128/9781555819156.ch39

many viral infections alter an individual's risk for the subsequent contraction of certain infections. For example, infection with herpes simplex virus type 2 is associated with an increased risk of developing bacterial vaginosis (14), and infection with influenza A virus can increase the adhesiveness of several *E. coli* strains by altering the expressed glycosylation pattern of intestinal epithelial cells (15). Additionally, studies performed in animals demonstrate the effect of the gut bacterial composition on the success of certain viral infections. For example, Uchiyama et al. (16) demonstrated that ablation of gut bacteria through antibiotic treatment or the use of germ-free mice significantly reduced the infectivity of rotavirus, while simultaneously resulting in a more durable rotavirus mucosal/systemic humoral immune response. The significance of this latter observation is that the efficacy of certain vaccines may be boosted from pretreatment of individuals with antibiotics. Taken together, these and many other studies highlight the intimate interactions members of the microbiome have both with each other and with their host. The treatment of many infections, even if caused by a single etiologic agent, may be improved through evaluation of the microbiome composition of the infected individual.

TECHNOLOGIES FOR STUDYING THE HUMAN VIROME

Although the human microbiome consists of all bacteria, archaea, viruses, fungi, and other unicellular eukaryotes, the focus of this section is on the viral members of these communities, the human virome. Characterization of the human virome has lagged behind that of its bacterial counterpart largely because of technical limitations. For instance the bacterial and fungal components of the microbiome can be probed simply by amplifying and sequencing conserved ribosomal genes, a technique not applicable to viruses (17). While some putative marker genes have been identified for viruses, they rarely extend in scope beyond viral families, and those studies are hampered by the current paucity of viral genome sequence data available (18, 19).

Metagenomics is the study of a collection of genomes from a mixed community of organisms. Viral metagenomics is the study of the subset of nucleic acids from a sample that are of viral origin. This technique has revolutionized the field of viral ecology by providing researchers with a method to study the structure, function, and metabolic potential of viral communities in an unbiased, culture-independent, and high-throughput manner.

Viral metagenomics, in its simplest form, is applied by collecting an appropriate sample from an environment of interest, extracting nucleic acids, and sequencing with minimal laboratory manipulations (i.e., no culturing of the organisms to be studied). Prior to the introduction of nucleic acid sequencing, compositions of viral communities inhabiting the human body could be probed only via physical methods such as electron microscopy. The introduction of DNA sequencing provided scientists with the ability to probe communities of organisms genetically. However, preparatory processes including viral particle isolation, amplification in culture, and DNA fragment cloning prior to sequencing hampered throughput and limited the number of nucleic acid fragments that could be reasonably evaluated. Despite these limitations, viral metagenomics experiments on many disparate environmental habitats were carried out, revealing hints at the true complexity of the viral world including the enormous concentration of viruses in sea water, which contain an estimated $>10^{30}$ viral particles throughout the world's oceans (20).

Amplifying viruses in culture biases the detectable relative abundance and membership of these communities. To avoid this, early metagenomics experiments relied on the concentration of large volumes of primary sample to obtain sufficient material for cloning and sequencing. For example, Breitbart et al. (21), in one of the first papers to characterize the viral community of a human stool sample, concentrated viral particles present in 500 grams of stool down to a few milliliters by first resuspending the stool in 5 liters of phosophate-buffered saline (PBS), then concentrating the sample using tangential flow filtration. While this technique successfully bypassed the requirement of amplification in culture, it imposed processing steps incompatible with the throughput required for any study containing more than just a few samples.

The introduction of whole-genome shotgun (WGS) sequencing strategies removed many sample processing burdens imposed by Sanger sequencing, enabling the field of metagenomics to flourish. By allowing sequencing adapters to be added directly to free DNA fragments of interest, WGS and other NGS strategies eliminated the need for cloning prior to sequencing. This allowed massively parallel evaluation of unique DNA fragments, boosting the number of unique fragments evaluated from at best thousands with Sanger sequencing to billions with Illumina sequencing. These NGS technologies paved the way for applications of metagenomics techniques in viral discovery and community characterization by providing increased yield, decreased processing time, and decreased cost per sequenced base without the need for targeted amplification of marker genetic elements. This method has proven particularly adept at providing general overviews of viral community structure (21–24) as well as facilitating the discovery of many new mammalian viruses (25–27).

One of the biggest boons of WGS strategies is the lack of a requirement for any *a priori* knowledge about the organisms in the community being sequenced. One downside to this approach is that nucleic acids of viral origin typically comprise a very small proportion of the total extracted community. This manifests as an equally small portion of the finished sequence data being of viral origin. To overcome this drawback, samples are sequenced to greater depths, increasing the number of sequence reads from viral genomes. However, strategies utilizing targeted genetic sequences such as PCR or sequence-based capture assays have the potential to be cheaper and more sensitive. As viral genome databases become more complete, the development of targeted assays becomes viable, and the likelihood of missing viruses of interest diminishes. For example, Wylie et al. (28) created a capture-based viral nucleic acid enrichment system utilizing commonly occurring representative sequences derived from 337 viral reference genomes of species known to infect vertebrate cells. Their results showed that common sequences can successfully enrich the viral portion of total extracted nucleic acids while also enriching the portion of distantly related viruses that were not actually part of the reference genome database used to create their capture array. In the same vein, Briese et al. (29) constructed a probe library of nearly 2 million sequences derived from >200 genomes of viruses known to infect vertebrate cells. Through a method of probe hybridization to cDNA libraries followed by bead capture, the authors were able to achieve up to a 10^4-fold increase in viral sequences generated from blood and tissue homogenates, compared to established viral enrichment procedures. These WGS and viral genome enrichment methods will only continue to get better and more sensitive as characterization of the viral world continues.

COMPOSITION OF THE HUMAN VIROME

The human virome includes particles capable of infecting human, bacterial, archaeal, fungal cells, and integrated virus-derived genetic elements (21,30–35). These integrated elements have the potential to generate infectious particles, express proteins, and alter host-gene expression. Viruses and viral elements that comprise the human virome are in constant interaction with the human immune system, maintaining a dynamic equilibrium, and are the primary source of traditional viral pathogens (36, 37). Estimates of the size and membership of the human virome are rough at best; however, we can be certain we have explored little of it by sequence, and even less by function (38, 39). This is evidenced by the fact that large portions of viral metagenomics sequence datasets do not match any current database entries. To put this into perspective, the current number of viral reference genomes in NCBI RefSeq (~5000) is less than the number of virotypes expected in 100 liters of sea water (40). Adding to the complexity of studying the human virome are the ubiquity and rapid evolution of viruses (41, 42). Estimates place the error rate of viral replication at one per 10^3 to 10^8 bases copied, depending on genome type, resulting in the constant introduction of mutants during replication.

The majority of viruses found in the human virome (and likely on Earth) are bacteriophages, viruses that infect bacteria. These are of immense clinical importance because of their ability to transmit genes between their hosts (43, 44). This transmitted genetic material can increase pathogenesis through encoded toxins, spread antibiotic resistance, and possibly alter host functional abilities (45–49). This gives the resident bacteria access to a potentially massive pool of genetic material allowing quick adaptations, including alterations in virulence, to an ever-changing environment. The first evidence that phages could alter their host's virulence came with the discovery of platelet binding factors *pblA* and *pblB* present in tail fiber genes of oral phage (50). These proteins have been shown to contribute to *Streptococcus mitis* virulence in an animal endocarditis model, also demonstrating that the host effects of the oral microbiome are not restricted to the mouth. Additionally, bacteriophage particles that gain direct systemic access through a leaky gut stimulate production of pro-inflammatory cytokines, (e.g., IL-1 beta, TNF-alpha) (7, 51).

Phages can be major contributors to bacterial genomic DNA. For example more than half of the strain-specific DNA in *Lactobacillus johnsonii* was contributed by prophage sequences, phage genome sequences that have integrated into their host's chromosome (52). Prophages can also benefit bacterial host fitness in numerous ways. Integrated prophages can serve as anchor points for genome rearrangements (53, 54), silence nonessential genes (52, 55), and confer homoimmunity (56, 57). Additionally, induced prophages can cause lysis of closely related strains and introduce new fitness through conversion or transduction (58–60). Induction of prophages in the intestine may contribute to dysbiosis by changing the ratio of symbionts to pathobionts, a process termed by Mills et al. as community shuffling (38). This has been observed in the swine gut during antibiotic-associated prophage induction (61). Phages may also play a role in the dysregulated immune response of inflammatory bowel disease (IBD) patients to mucosal-associated bacterial populations (38). These observations suggest that the composition of the gut phage population affects human health both directly through interactions with the human immune system and indirectly through modulation of bacterial communities within the gut. The potential of the phage component of the human virome to affect human health and disease cannot be overstated, and the extent to which the virome helps define the host's genetic identity is likely underappreciated.

The intimate relationship between the gut microbiota and the immune system has been revealed through studies utilizing germ-free animals. The lack of microorganisms in the guts of these animals results in defects in the development of gut-associated lymphoid tissues and other morphological abnormalities, many of which are reversed following the introduction of gut microorganisms (62–64). While early studies focused on the effects of removing/reintroducing bacteria to the guts of germ-free animals, more recent studies have shown that certain viruses are capable of eliciting similar physiological responses. Cadwell (65) showed that monoassociation of germ-free mice with murine norovirus reversed many of the abnormalities observed in the guts of these animals and provided substantial protection against two models of intestinal damage. Other animal studies have also revealed that certain viral infections may protect the host against different diseases. For example, infection with lymphocytic choriomeningitis virus appears to prevent type 1 diabetes in nonobese diabetic mice (66). Interestingly, there is evidence that even plant viruses may impact human health. A study by Colson et al. (67) suggests that pepper mild mottle virus, a common plant pathogen of bell, hot, and ornamental peppers, may not simply be along for the ride through the human gut after ingestion but may interact with the human immune system and cause clinical symptoms.

The majority of the adult human population is chronically infected with members of the anellovirus, adenovirus, circovirus, herpesvirus, and polyomavirus families, among others (68). Members of these families are typically considered opportunistically pathogenic or even commensal, because chronic infection with many of these viruses appears clinically inconsequential. However, many chronic infections are so commonplace that their contribution to or effect on the human immune system and/or human health is obscured by the fact that they are normally present. Estimates place the number of permanent chronic systemic viral infections in the average healthy individual at around 10 (69). These often-present viruses range from traditional pathogens to those that are only pathogenic in small, susceptible portions of the population. However, the number of chronic viral infections is likely underestimated because methods for detecting the full diversity of the human virome have only recently become available, and many existing metagenomics studies have ignored RNA viruses altogether.

The idea of a commensal virus is relatively new. Recently developed molecular and sequencing techniques have revealed many new persistent viruses including polyomaviruses (70) and anelloviruses (71). Since they are not covered in detail elsewhere, we will briefly review the family Anelloviridae; Spandole et al. (72) have published an in-depth review. This family contains some of the most recently discovered viruses that chronically infect the majority of the human population. These include torque teno virus (TTV), torque teno mini virus (TTMV), and torque teno midi virus (TTMDV), discovered in 1997, 2000, and 2007, respectively. This family also contains many viruses that infect animals, and thanks to new sequencing technologies, novel anellovirus sequences are continually being described (73–75). Torque teno viruses are nonenveloped 30- to 50-nm particles with T = 1 icosahedral symmetry. Because of a lack of compatible cell systems, TTVs cannot currently be propagated *in vitro*, which has hampered structural and/or functional studies. The TTVs contain a circular single-stranded antisense DNA genome that varies in size from ∽2.8 kb for TTMV to ∽3.9 kb for TTV (76–78). The TTV genome contains a ∽1.2 kb well-conserved untranslated region (UTR) and a ∽2.6 kb coding section with three hypervariable regions. After the initial discovery of the TTVs, their detected prevalence in blood of populations throughout the world varied widely, from ∽5% in Brazil (79) to >90% in Pakistan, Japan, and Russia (71, 80, 81). More recent studies targeting the conserved UTR regions have raised the estimates of prevalence in blood, and TTV DNA has been detected at high rates in many other organs, tissues, and biological samples. This high prevalence led to numerous studies associating TTV with countless diseases and disorders including hepatitis, respiratory diseases, hematological disorders, autoimmune disorders, and even cancer (72). However, there are very few existing studies that attempt to discern possible mechanisms behind proposed TTV-associated pathologies. What is clear is that anelloviruses have infected many if not most mammals throughout their evolutionary history, and perhaps the absence of TTV infection could have greater consequences for human health than its presence.

Early viral metagenomics studies have revealed that humans, along with other mammals, house highly individualized and often novel communities of viruses including many currently unidentifiable sequences (82, 83). The extent of the impact the human virome has on the course of human disease progression and treatment outcomes remains to be determined. However, studies mentioned above, along with many others, demonstrate that this impact is likely much larger than is currently appreciated. One of the largest factors distinguishing the virome from other components of the human microbiome is systemic persistence. For example, herpesviruses can maintain life-long latent infections in numerous cell types after recovery from acute infection. The constant interaction between members of the virome and the immune system has the potential to alter host susceptibility and/or resistance to other infections or diseases. As new sequencing data is able to better define the human virome, identification of additional viral signatures within the human genome becomes possible. This information has the potential to significantly increase our understanding of our own genetics and the implications for human health and disease.

Our knowledge of the human virome is currently incomplete; in fact we have hardly generated a rough outline of its impact on human health. Outstanding barriers to a deeper understanding still include fundamental aspects of the human virome such as a clear quantitative and qualitative description of its membership. The effects on the immune system of not only the composition of the virome but also host age at exposure and other variables remain to be explored. The virome is an integral part of the human microbiome, and microbiome studies cannot provide a complete picture if its different components are studied in isolation.

LOOKING AHEAD

The public has been surprisingly receptive to the idea that the trillions of microscopic entities that inhabit each of us are not the enemy but are an integral part of what makes us what we are. The attitude for many has changed in the recent past away from efforts to constantly sterilize our external world and instead to implement rigid hygiene practices only where essential (e.g., hospital settings). Typing "antibacterial soap" into Google brings up the top suggestions "bad," "dangerous," "ban," etc., which suggests awareness extends beyond the scientific community. Efforts at deciphering the mysteries of the microbiome world have had a major impact on the way we view certain diseases, but the full implications of these microbial communities for our health remain to be discovered. We're beginning to understand the immense role that viruses can play in determining the composition of the human microbiome, as well as how individuals may respond differently to a particular disease treatment based on their microbiome composition. In light of the huge success intestinal probiotic products already enjoy, the future acceptance for microbiome manipulations as part of disease therapy seems like a foregone conclusion. Many groups are beginning to revisit bacteriophage as antimicrobial therapies for human diseases caused by bacterial strains harboring antibiotic resistance. However, more studies are needed to fully understand the interactions between hosts and their microbiomes before these relationships can be safely exploited.

REFERENCES

1. **Subramanian S, Blanton LV, Frese SA, Charbonneau M, Mills DA, Gordon JI.** 2015. Cultivating healthy growth and nutrition through the gut microbiota. *Cell* **161:**36–48.
2. **Weyrich LS, Dixit S, Farrer AG, Cooper AJ, Cooper AJ.** 2015. The skin microbiome: Associations between altered microbial communities and disease. *Australas J Dermatol*; Available from: http://www.ncbi.nlm.nih.gov/pubmed/25715969.
3. **Witkin SS.** 2015. The vaginal microbiome, vaginal anti-microbial defence mechanisms and the clinical challenge of reducing infection-related preterm birth. *BJOG* **122:**213–218.
4. **Madsen KL, Doyle JS, Jewell LD, Tavernini MM, Fedorak RN.** 1999. Lactobacillus species prevents colitis in interleukin 10 gene-deficient mice. *Gastroenterology* **116:**1107–1114.
5. **Turnbaugh PJ, Ley RE, Hamady M, Fraser-Liggett CM, Knight R, Gordon JI.** 2007. The human microbiome project. *Nature* **449:**804–810.
6. **Human Microbiome Project Consortium.** 2012. Structure, function and diversity of the healthy human microbiome. *Nature* **486:**207–214.
7. **Duerkop BA, Hooper LV.** 2013. Resident viruses and their interactions with the immune system. *Nat Immunol* **14:**654–659.
8. **Fouhy F, Ross RP, Fitzgerald GF, Stanton C, Cotter PD.** 2012. Composition of the early intestinal microbiota: knowledge, knowledge gaps and the use of high-throughput sequencing to address these gaps. *Gut Microbes* **3:**203–220.
9. **Corfe BM, Harden CJ, Bull M, Garaiova I.** 2015. The multifactorial interplay of diet, the microbiome and appetite control: current knowledge and future challenges. *Proc Nutr Soc* **74:**235–244.
10. **Margulies M, Egholm M, Altman WE, Attiya S, Bader JS, Bemben LA, Berka J, Braverman MS, Chen YJ, Chen Z,** et al. 2005. Genome sequencing in microfabricated high-density picolitre reactors. *Nature* **437:**376–380.
11. **Haiser HJ, Turnbaugh PJ.** 2013. Developing a metagenomic view of xenobiotic metabolism. *Pharmacol Res* **69:**21–31.
12. **Saha JR, Butler VP Jr, Neu HC, Lindenbaum J.** 1983. Digoxin-inactivating bacteria: identification in human gut flora. *Science* **220:**325–327.
13. **Niehues M, Hensel A.** 2009. In-vitro interaction of L-dopa with bacterial adhesins of Helicobacter pylori: an explanation

for clinicial differences in bioavailability? *J Pharm Pharmacol* **61:**1303–1307.

14. **Esber A, Vicetti Miguel RD, Cherpes TL, Klebanoff MA, Gallo MF, Turner AN.** 2015. Risk of bacterial vaginosis among women with herpes simplex virus type 2 infection: A systematic review and meta-analysis. *J Infect Dis* **212:**8–17.
15. **Aleandri M, Conte MP, Simonetti G, Panella S, Celestino I, Checconi P, Marazzato M, Longhi C, Goldoni P, Nicoletti M, Barnich N, Palamara AT, Schippa S, Nencioni L.** 2015. Influenza A virus infection of intestinal epithelial cells enhances the adhesion ability of Crohn's disease associated Escherichia coli strains. *PLoS One* **10:**e0117005.
16. **Uchiyama R, Chassaing B, Zhang B, Gewirtz AT.** 2014. Antibiotic treatment suppresses rotavirus infection and enhances specific humoral immunity. *J Infect Dis* **210:**171–182.
17. **Woo PCY, Lau SKP, Teng JLL, Tse H, Yuen KY.** 2008. Then and now: use of 16S rDNA gene sequencing for bacterial identification and discovery of novel bacteria in clinical microbiology laboratories. *Clin Microbiol Infect* **14:**908–934.
18. **Breitbart M, Miyake JH, Rohwer F.** 2004. Global distribution of nearly identical phage-encoded DNA sequences. *FEMS Microbiol Lett* **236:**249–256.
19. **Hambly E, Tétart F, Desplats C, Wilson WH, Krisch HM, Mann NH.** 2001. A conserved genetic module that encodes the major virion components in both the coliphage T4 and the marine cyanophage S-PM2. *Proc Natl Acad Sci USA* **98:** 11411–11416.
20. **Suttle CA.** 2005. Viruses in the sea. *Nature* **437:**356–361.
21. **Breitbart M, Hewson I, Felts B, Mahaffy JM, Nulton J, Salamon P, Rohwer F.** 2003. Metagenomic analyses of an uncultured viral community from human feces. *J Bacteriol* **185:** 6220–6223.
22. **Delwart E.** 2013. A roadmap to the human virome. *PLoS Pathog* **9:**e1003146.
23. **Lecuit M, Eloit M.** 2013. The human virome: new tools and concepts. *Trends Microbiol* **21:**510–515.
24. **Breitbart M, Salamon P, Andresen B, Mahaffy JM, Segall AM, Mead D, Azam F, Rohwer F.** 2002. Genomic analysis of uncultured marine viral communities. *Proc Natl Acad Sci USA* **99:**14250–14255.
25. **Grard G, Fair JN, Lee D, Slikas E, Steffen I, Muyembe JJ, Sittler T, Veeraraghavan N, Ruby JG, Wang C, Makuwa M, Mulembakani P, Tesh RB, Mazet J, Rimoin AW, Taylor T, Schneider BS, Simmons G, Delwart E, Wolfe ND, Chiu CY, Leroy EM.** 2012. A novel rhabdovirus associated with acute hemorrhagic fever in central Africa. *PLoS Pathog* **8:**e1002924.
26. **Lipkin WI, Firth C.** 2013. Viral surveillance and discovery. *Curr Opin Virol* **3:**199–204.
27. **Tan le V, van Doorn HR, Nghia HD, Chau TT, Tu le TP, de Vries M, et al.** 2013. Identification of a new cyclovirus in cerebrospinal fluid of patients with acute central nervous system infections. *MBio* **4:**e00231–13.
28. **Wylie TN, Wylie KM, Herter BN, Storch GA.** 2015. Enhanced virome sequencing through solution-based capture enrichment. *Genome Res* **25:**1910–1920.
29. **Briese T, Kapoor A, Mishra N, Jain K, Kumar A, Jabado OJ, Lipkin WI.** 2015. Virome capture sequencing enables sensitive viral diagnosis and comprehensive virome analysis. *MBio* **6:**e01491–e15.
30. **Breitbart M, Haynes M, Kelley S, Angly F, Edwards RA, Felts B, Mahaffy JM, Mueller J, Nulton J, Rayhawk S, Rodriguez-Brito B, Salamon P, Rohwer F.** 2008. Viral diversity and dynamics in an infant gut. *Res Microbiol* **159:**367–373.
31. **Kim MS, Park EJ, Roh SW, Bae JW.** 2011. Diversity and abundance of single-stranded DNA viruses in human feces. *Appl Environ Microbiol* **77:**8062–8070.
32. **Minot S, Grunberg S, Wu GD, Lewis JD, Bushman FD.** 2012. Hypervariable loci in the human gut virome. *Proc Natl Acad Sci USA* **109:**3962–3966.
33. **Minot S, Sinha R, Chen J, Li H, Keilbaugh SA, Wu GD, Lewis JD, Bushman FD.** 2011. The human gut virome: inter-individual variation and dynamic response to diet. *Genome Res* **21:**1616–1625.
34. **Reyes A, Haynes M, Hanson N, Angly FE, Heath AC, Rohwer F, Gordon JI.** 2010. Viruses in the faecal microbiota of monozygotic twins and their mothers. *Nature* **466:**334–338.
35. **Zhang T, Breitbart M, Lee WH, Run JQ, Wei CL, Soh SW, Hibberd ML, Liu ET, Rohwer F, Ruan Y.** 2006. RNA viral community in human feces: prevalence of plant pathogenic viruses. *PLoS Biol* **4:**e3.
36. **Foxman EF, Iwasaki A.** 2011. Genome-virome interactions: examining the role of common viral infections in complex disease. *Nat Rev Microbiol* **9:**254–264.
37. **Stelekati E, Wherry EJ.** 2012. Chronic bystander infections and immunity to unrelated antigens. *Cell Host Microbe* **12:** 458–469.
38. **Mills S, Shanahan F, Stanton C, Hill C, Coffey A, Ross RP.** 2013. Movers and shakers: influence of bacteriophages in shaping the mammalian gut microbiota. *Gut Microbes* **4:**4–16. t
39. **Reyes A, Semenkovich NP, Whiteson K, Rohwer F, Gordon JI.** 2012. Going viral: next-generation sequencing applied to phage populations in the human gut. *Nat Rev Microbiol* **10:** 607–617.
40. **Rohwer F, Thurber RV.** 2009. Viruses manipulate the marine environment. *Nature* **459:**207–212.
41. **Koonin EV, Dolja VV, Krupovic M.** 2015. Origins and evolution of viruses of eukaryotes: the ultimate modularity. *Virology* **479–480:**2–25.
42. **Sime-Ngando T.** 2014. Environmental bacteriophages: viruses of microbes in aquatic ecosystems. *Front Microbiol* **5:**355.
43. **Canchaya C, Fournous G, Chibani-Chennoufi S, Dillmann ML, Brüssow H.** 2003. Phage as agents of lateral gene transfer. *Curr Opin Microbiol* **6:**417–424.
44. **Kristensen DM, Mushegian AR, Dolja VV, Koonin EV.** 2010. New dimensions of the virus world discovered through metagenomics. *Trends Microbiol* **18:**11–19.
45. **Brüssow H, Canchaya C, Hardt W-D.** 2004. Phages and the evolution of bacterial pathogens: from genomic rearrangements to lysogenic conversion. *Microbiol Mol Biol Rev* **68:**560–602.
46. **Busby B, Kristensen DM, Koonin EV.** 2013. Contribution of phage-derived genomic islands to the virulence of facultative bacterial pathogens. *Environ Microbiol* **15:**307–312.
47. **Carrolo M, Frias MJ, Pinto FR, Melo-Cristino J, Ramirez M.** 2010. Prophage spontaneous activation promotes DNA release enhancing biofilm formation in Streptococcus pneumoniae. *PLoS One* **5:**e15678.
48. **Duerkop BA, Clements CV, Rollins D, Rodrigues JLM, Hooper LV.** 2012. A composite bacteriophage alters colonization by an intestinal commensal bacterium. *Proc Natl Acad Sci USA* **109:**17621–17626.
49. **Wang X, Kim Y, Wood TK.** 2009. Control and benefits of CP4-57 prophage excision in Escherichia coli biofilms. *ISME J* **3:**1164–1179.
50. **Willner D, Furlan M, Schmieder R, Grasis JA, Pride DT, Relman DA, et al.** 2011. Metagenomic detection of phage-encoded platelet-binding factors in the human oral cavity. *Proc Natl Acad Sci USA* **108** (Suppl 1):4547–53.
51. **Górski A, Wazna E, Dąbrowska BW, Dąbrowska K, Switała-Jeleń K, Międzybrodzki R.** 2006. Bacteriophage translocation. *FEMS Immunol Med Microbiol* **46:**313–319.
52. **Ventura M, Canchaya C, Pridmore D, Berger B, Brüssow H.** 2003. Integration and distribution of Lactobacillus johnsonii prophages. *J Bacteriol* **185:**4603–4608.
53. **Canchaya C, Desiere F, McShan WM, Ferretti JJ, Parkhill J, Brüssow H.** 2002. Genome analysis of an inducible prophage and prophage remnants integrated in the Streptococcus pyogenes strain SF370. *Virology* **302:**245–258.
54. **Van Sluys MA, et al.** 2003. Comparative analyses of the complete genome sequences of Pierce's disease and citrus variegated chlorosis strains of Xylella fastidiosa. *J Bacteriol* **185:**1018–1026.
55. **Coleman D, Knights J, Russell R, Shanley D, Birkbeck TH, Dougan G, Charles I.** 1991. Insertional inactivation of the Staphylococcus aureus beta-toxin by bacteriophage phi 13 occurs by site- and orientation-specific integration of the phi 13 genome. *Mol Microbiol* **5:**933–939.

56. **Bossi L, Fuentes JA, Mora G, Figueroa-Bossi N.** 2003. Prophage contribution to bacterial population dynamics. *J Bacteriol* **185:**6467–6471.
57. **Lin L, Bitner R, Edlin G.** 1977. Increased reproductive fitness of Escherichia coli lambda lysogens. *J Virol* **21:**554–559.
58. **Burke J, Schneider D, Westpheling J.** 2001. Generalized transduction in Streptomyces coelicolor. *Proc Natl Acad Sci USA* **98:**6289–6294.
59. **Hodgson DA.** 2000. Generalized transduction of serotype 1/2 and serotype 4b strains of Listeria monocytogenes. *Mol Microbiol* **35:**312–323.
60. **Schicklmaier P, Schmieger H.** 1995. Frequency of generalized transducing phages in natural isolates of the Salmonella typhimurium complex. *Appl Environ Microbiol* **61:**1637–1640.
61. **Allen HK, Looft T, Bayles DO, Humphrey S, Levine UY, Alt D, Stanton TB.** 2011. Antibiotics in feed induce prophages in swine fecal microbiomes. *MBio* **2:**e00260-11.
62. **Amit-Romach E, Uni Z, Reifen R.** 2008. Therapeutic potential of two probiotics in inflammatory bowel disease as observed in the trinitrobenzene sulfonic acid model of colitis. *Dis Colon Rectum* **51:**1828–1836.
63. **Ma D, Forsythe P, Bienenstock J.** 2004. Live Lactobacillus rhamnosus [corrected] is essential for the inhibitory effect on tumor necrosis factor alpha-induced interleukin-8 expression. *Infect Immun* **72:**5308–5314.
64. **Schultz M, Veltkamp C, Dieleman LA, Grenther WB, Wyrick PB, Tonkonogy SL, Sartor RB.** 2002. Lactobacillus plantarum 299V in the treatment and prevention of spontaneous colitis in interleukin-10-deficient mice. *Inflamm Bowel Dis* **8:**71–80.
65. **Cadwell K.** 2015. Expanding the role of the virome: commensalism in the gut. *J Virol* **89:**1951–1953.
66. **Oldstone MB.** 1988. Prevention of type I diabetes in nonobese diabetic mice by virus infection. *Science* **239:**500–502.
67. **Colson P, Richet H, Desnues C, Balique F, Moal V, Grob JJ, Berbis P, Lecoq H, Harlé JR, Berland Y, Raoult D.** 2010. Pepper mild mottle virus, a plant virus associated with specific immune responses, fever, abdominal pains, and pruritus in humans. *PLoS One* **5:**e10041. Internet
68. **Virgin HW.** 2014. The virome in mammalian physiology and disease. *Cell* **157:**142–150.
69. **Virgin HW, Wherry EJ, Ahmed R.** 2009. Redefining chronic viral infection. *Cell* **138:**30–50.
70. **zur Hausen H.** 2008. Novel human polyomaviruses—re-emergence of a well known virus family as possible human carcinogens. *Int J Cancer* **123:**247–250.
71. **Ninomiya M, Takahashi M, Nishizawa T, Shimosegawa T, Okamoto H.** 2008. Development of PCR assays with nested primers specific for differential detection of three human anelloviruses and early acquisition of dual or triple infection during infancy. *J Clin Microbiol* **46:**507–514.
72. **Spandole S, Cimponeriu D, Berca LM, Mihăescu G.** 2015. Human anelloviruses: an update of molecular, epidemiological and clinical aspects. *Arch Virol* **160:**893–908.
73. **Cibulski SP, Teixeira TF, de Sales Lima FE, do Santos HF, Franco AC, Roehe PM.** 2014. A novel anelloviridae species detected in Tadarida brasiliensis bats: First sequence of a chiropteran anellovirus. *Genome Announc* **2:**e01028-14.
74. **Mi Z, Yuan X, Pei G, Wang W, An X, Zhang Z, Huang Y, Peng F, Li S, Bai C, Tong Y.** 2014. High-throughput sequencing exclusively identified a novel Torque teno virus genotype in serum of a patient with fatal fever. *Virol Sin* **29:**112–118.
75. **Nishiyama S, Dutia BM, Stewart JP, Meredith AL, Shaw DJ, Simmonds P, Sharp CP.** 2014. Identification of novel anelloviruses with broad diversity in UK rodents. *J Gen Virol* **95:**1544–1553.
76. **Peng YH, Nishizawa T, Takahashi M, Ishikawa T, Yoshikawa A, Okamoto H.** 2002. Analysis of the entire genomes of thirteen TT virus variants classifiable into the fourth and fifth genetic groups, isolated from viremic infants. *Arch Virol* **147:**21–41.
77. **Ninomiya M, Nishizawa T, Takahashi M, Lorenzo FR, Shimosegawa T, Okamoto H.** 2007. Identification and genomic characterization of a novel human torque teno virus of 3.2 kb. *J Gen Virol* **88:**1939–1944.
78. **Okamoto H, Nishizawa T, Takahashi M, Tawara A, Peng Y, Kishimoto J, Wang Y.** 2001. Genomic and evolutionary characterization of TT virus (TTV) in tupaias and comparison with species-specific TTVs in humans and non-human primates. *J Gen Virol* **82:**2041–2050.
79. **de Oliveira JC, Nasser TF, Oda JM, Aoki MN, Carneiro JL, Barbosa DS, Reiche EM, Watanabe MA.** 2008. Detection of TTV in peripheral blood cells from patients with altered ALT and AST levels. *New Microbiol* **31:**195–201.
80. **Hussain T, Manzoor S, Waheed Y, Tariq H, Hanif K.** 2012. Phylogenetic analysis of Torque Teno Virus genome from Pakistani isolate and incidence of co-infection among HBV/HCV infected patients. *Virol J* **9:**320.
81. **Vasilyev EV, Trofimov DY, Tonevitsky AG, Ilinsky VV, Korostin DO, Rebrikov DV.** 2009. Torque Teno Virus (TTV) distribution in healthy Russian population. *Virol J* **6:**134.
82. **Delwart EL.** 2007. Viral metagenomics. *Rev Med Virol* **17:**115–131.
83. **Rosario K, Breitbart M.** 2011. Exploring the viral world through metagenomics. *Curr Opin Virol* **1:**289–297.

Human Susceptibility and Response to Viral Diseases

VILLE PELTOLA AND JORMA ILONEN

40

Host genetic variation in components of both specific and innate immune responses affects susceptibility to viral infections. Innate immunity provides the first line of defense, and the development of adaptive immunity is stimulated by innate responses. Pathogen recognition receptors (PRRs) initiate signaling pathways that result in the production of antiviral interferons and cytokines. Mutations or genetic variants (polymorphisms) have been recognized in several factors of innate immunity. Notably, human populations from distinct geographic areas have different frequencies of immune gene variants. The genetic susceptibility may vary from life-threatening manifestations of specific virus infections to a moderately increased frequency of nonsevere infections. Although the innate immunity is nonspecific by nature, the reactions are stereotypic for viral infections compared with bacterial infections. Even infections caused by specific viruses can be differentiated from each other based on the innate immune response. Host response pattern determination by expression analysis of a predefined set of genes is a novel strategy in the diagnosis of virus infections. Another strategy in differentiating viral and bacterial infections from each other could be the determination of a single marker, such as myxovirus resistance protein A (MxA), which is generally induced by viruses but not by bacteria. Host response analysis could also be used in monitoring infections and antiviral treatment, but applications for routine use are not yet available. Certain host gene variants correlate with the prognosis of infection. Currently, for instance, interleukin (IL) 28B genotyping is used to aid in hepatitis C treatment decisions.

INTRODUCTION

Primary immunodeficiencies are traditionally understood as rare syndromes with a markedly compromised defense against viruses, bacteria, or other pathogens. Patients with these conditions have severe or recurrent infections, which may be caused by atypical pathogens, and they do not respond to treatment as would be expected in otherwise healthy hosts. Depending on the type of immunodeficiency syndrome, the serum level of immunoglobulins and/or the number of certain immune cell types (e.g., lymphocyte subtypes or phagocytic cells) are usually decreased. The genetic basis of most of these syndromes is currently understood in detail.

More recently, with the use of modern molecular biology and sequencing technologies, a wide range of variations or mutations have been recognized in the genes coding for proteins that function in the immunologic response to viruses or other microbes (1). Many of these proteins are part of the innate immune system, whereas others regulate specific humoral or cellular responses or have more general functions, as in cellular metabolism. The clinical manifestations of such genetic variations may differ from classical primary immunodeficiency syndromes. In some cases, the variant types are also common in healthy populations, and disease associations are weak or unclear. Other genetic deficiencies are rare in healthy populations and are associated with severe infections, but the susceptibility pattern may be highly microbe specific (2). The phenotypic consequences of mutations are often highly variable.

The major histocompatibility complex (MHC), in human HLA genes, has a central role in ensuring that our species has during phylogenesis developed the ability to mount a specific immune response to the multitude of existing and emerging pathogens. The extreme variability observed at MHC loci is the result of selective force from infectious diseases (3, 4). The pattern of alleles common in various populations reflects their advantage in mounting efficient immunity to important local infections, as demonstrated by the high frequency of the HLA-B*53 allele associated with protection from severe malaria in western Africa (5). Subjects with certain class II haplotypes associated with autoimmune diseases, most importantly DR3/DQ2 and DR4/DQ8, may be especially efficient in clearing viral or other infections by producing specific cytokines. This has contributed to the survival of these haplotypes despite the associated tendency to present self-antigens.

Several mechanisms of innate immunity can be considered as redundant, or perhaps more precisely, as complementing each other. If one function is weak or inoperative, other signaling pathways can, to some extent, lead to similar responses. It has been suggested that accumulating dysfunctions may eventually lead to immunodeficiency (6). In the presence of increased infection pressure (i.e., living in an environment favorable for the transmission of viruses), even a minor weakness in innate immunity may result in recurrent infections. However, some aspects of innate immune functions are quite specific for certain

doi:10.1128/9781555819156.ch40

pathogens. Thus, a mutation in one immunologic gene may result in the patient being otherwise healthy and free of infections, but when the patient encounters a certain infectious agent, in particular for the first time in life, a severe manifestation may develop.

In addition to changes in the immune system, genetic polymorphisms can simply affect virus infection by preventing the virus from binding to its specific receptor and thus its cell entry. Examples are the rare persons with the p phenotype of erythrocyte P antigen who are resistant to parvovirus B19 infection because of a lack of this receptor on their erythrocytes and erythroid precursors (7). Furthermore, homozygosity for the nonsense mutation of the human secretor, the *FUT2* gene (G428A), correlates with complete resistance to norovirus gastroenteritis (8), and probably the best-known example is the resistance to HIV infection in subjects homozygous for deletion in co-receptor C-C chemokine receptor type 5 (CCR5) (9–11).

IMMUNE RESPONSE TO VIRAL INFECTIONS

Innate immune responses are critical in the first-line defense against viral infections. Type I interferons and other antiviral effectors limit the spread of infection and augment the generation of specific immune responses but simultaneously cause inflammation-related symptoms in the host. Viral proteins may suppress the immune functions by several mechanisms (9). The inflammatory responses need to be carefully balanced as a weak response would be ineffective for the control of infection, and an overreactive response would be directly harmful to the host. The recognition of cellular danger signals plays an important part in regulating immune responses and inflammation (10).

Most viruses initiate infection by binding to receptors on epithelial cells in the respiratory or gastrointestinal tracts, or in other locations. Epithelial cells are the first cells to respond to viral infection, and professional immune cells are also rapidly activated. Molecules that are present only in viruses, although not specific for any virus species, are recognized by PRRs. The pathogen-associated molecular patterns (PAMPs) in viruses include single-stranded and double-stranded DNA, 5' regions of RNA, and certain virus proteins (11, 12). PRRs include retinoic acid-inducible gene 1 (RIG-1)-like receptors (RLRs), Toll-like receptors (TLRs), C-type lectin receptors, and the receptors of the inflammasome (13). Of these, RLRs and TLRs are central in the initiation of antiviral responses. Although C-type lectins mostly operate in fungal immunity, this group of PRRs includes collectin subgroup surfactant proteins, which function in defense against virus infections in the respiratory tract (14). Inflammasome activation and the complement system have important roles both in the innate response and in the development of adaptive immunity (15).

RIG-1 and melanoma differentiation-associated protein 5 (MDA-5, also known as IFIH1, interferon induced with helicase C domain 1) recognize in the cytosol double-stranded RNAs (dsRNAs) or, in some cases, single-stranded RNAs (ssRNAs) that are specific for viruses. Many RNA viruses (for example, influenza viruses, paramyxoviruses, hepatitis C virus, and Japanese encephalitis virus) produce dsRNAs that are longer and preferentially recognized by RIG-1, whereas picornaviruses produce shorter dsRNAs that are recognized mainly by MDA-5 (16). When triphosphate or diphosphate at the 5' end of RNA is detected by RIG-1, a signaling cascade is initiated and proceeds through adaptor mitochondrial antiviral signaling (MAVS) and results in interferon production (12, 17).

In humans, TLRs comprise 10 numbered receptors, which reside on the cell surface or in the endosome. Different TLRs recognize specific microbial structures. All TLRs, except TLR3, initiate the signaling cascade via the myeloid differentiation primary response 88 (MyD88) protein. TLR3 and TLR4 activate the Toll-interleukin-1 receptor domain-containing adapter-inducing interferon beta (TRIF) protein. With important regulatory and cross-talking functions by several intermediary proteins, the signaling pathway from endosomal TLRs 3, 7, 8, and 9 via MyD88 or TRIF results in the production of interferons and inflammatory cytokines, whereas the pathway from cell surface TLRs 1, 2, 4, 5, and 6 via MyD88 ends in the production of inflammatory cytokines (18). Only in certain immune cells are interferons produced via this pathway. Endosomal TLRs principally recognize viral components: TLR3 recognizes dsRNA, TLR7 and 8 recognize ssRNA, and TLR9 recognizes unmethylated CpG-rich DNA. Cell surface TLRs 2 and 4 recognize Gram-positive and Gram-negative bacteria, respectively, but in addition to that, these TLRs also recognize viral proteins (19, 20).

Professional immune cells are activated directly by the virus or by cytokines and other mediators that are produced by virus-infected epithelial cells. Plasmacytoid dendritic cells can produce large amounts of interferons after the recognition of viral ssDNA by TLR7 in endosomes, independently of the RLR pathway (21). The inflammasome is activated in macrophages and dendritic cells and also in other cell types by signaling through TLR3 or other TLRs or directly by viral dsRNA. This results in caspase-1 activation and the production of IL-1β and IL-18 from their precursors (22). Inflammasome activation is critical in the induction of adaptive responses. T cells are important in clearance of virus and in recovery from the infection. Activated dendritic cells migrate to lymph nodes and present viral antigens to naïve T cells. They differentiate into $CD4^+$ (T helper) and CD8+ (cytotoxic T cells) effector cells. The cytotoxic T cells need help from cytokines produced by $CD4^+$ cells to develop to effector T cells, which possess cytotoxic antiviral activity.

Humoral responses develop rather late in the course of primary viral infections, but the production of neutralizing antibodies has an effect on virus clearance. The development of follicular and extrafollicular antigen-specific B cells with the help of $CD4^+$ T cells provides the host with the capability to respond rapidly to reinfections. The wide variety of antigenically different virus types within many virus species or families compromises the protection by neutralizing antibodies against reinfections. However, some degree of cross-protection usually exists. Both $CD4^+$ and $CD8^+$ T cells expanded during previous infections have been shown to be able to recognize relatively stable epitopes common to several enterovirus serotypes and various influenza A strains, thus differing from serotype and strain-specific epitopes recognized by neutralizing antibodies (23–25).

A delicate regulatory mechanism is needed to control the inflammatory response to virus infection. Regulatory T cells (Treg), and to some extent $CD4^+$ and $CD8^+$ T cells, produce IL-10 and other anti-inflammatory cytokines that are essential in preventing injury from excessive inflammatory reaction. Dendritic cells prevent T cells from uncontrolled activation.

HOST GENETIC FACTORS AND SUSCEPTIBILITY TO VIRAL DISEASES

Single nucleotide polymorphisms (SNPs), haplotypes, copy number variations, deletions, or larger defects in the genes of many proteins playing roles in the immune defense or in the pathogenesis of infection may make the host susceptible to viral infection or, in contrast, confer protection against infections. Examples of such genetic variants and associated viral infections are listed in Table 1. However, an exhaustive literature review of suggested genetic susceptibilities is not possible within this chapter.

Receptor Binding

Variations in cell surface receptors, which viruses use for attachment and internalization, can have an effect on the susceptibility of the host. The examples of genetic protection against infection due to lack of a parvovirus B19 receptor in subjects with a p phenotype of erythrocyte P antigen (7) and a norovirus receptor in homozygotes for *FUT2* (G428A) (8) were mentioned earlier. Also, poliovirus receptor and major group rhinovirus receptor (ICAM-1) have genetic variants, but the possible differences in susceptibility to infection between individuals with wild- or variant-type receptors are unclear (26). The effects of polymorphisms in HIV co-receptors CCR5 and CXCR4 on the transmission of the virus and on the development of disease have been characterized in detail. Homozygotes for a 32 base pair deletion (delta32) in the gene of CCR5 do not express this protein on the cell surface. These individuals are almost completely protected from HIV infection (27–29). Heterozygotes for this variant are not protected from being infected by HIV, but the progression of HIV disease is slow, and virus copy numbers remain low. Other variants in co-receptors have been detected, and they also have distinct effects on the pathogenesis of HIV infection. Interestingly, mutations in the HIV co-receptor CXCR4 cause an immunodeficiency syndrome named WHIM (warts, hypogammaglobulinemia, infections, myelokathexis) (30). This syndrome is characterized by susceptibility to human papillomavirus (HPV)-induced warts, bacterial infections, and neutropenia.

Retinoic Acid-Inducible Gene-Like Receptor and Toll-Like Receptor Variants

RLRs are important responders to virus infection. However, no genetic variants in RLRs with a clear effect on infection susceptibility have been reported. This may be partly due to the fact that the major function of the RLR pathway, which is the production of type I interferons, can also be achieved through other signaling pathways via TLRs or the inflammasome. Recent research has shown that despite a level of redundancy in the factors and the pathways of innate immunity, mutants in single genes may result in specific immune defects. Rare variants in the *IFIH1* gene encoding MDA-5 with a predicted loss-of-function effect are associated with a reduced risk of type 1 diabetes (31). Future research will show whether these or other variants in MDA-5, RIG-1, or related proteins make individuals susceptible to specific virus infections.

Genetic variants or mutants have been reported in several components of TLR-signaling pathways. Mutations that cause functional deficiencies of the central adaptor molecules MyD88 and interleukin-1 receptor-associated kinase 4 (IRAK-4) result in the impairment of TLR and IL-1 signaling. These immunodeficiencies manifest clinically as susceptibility to invasive bacterial infections (32). It has been suggested that in humans the redundancy in pathways of interferon and cytokine signaling provides normal immunity against viruses despite the deficient function of these TLR pathways. Clinical findings in humans are in contrast with studies in mice, which show that deficiencies in these signaling pathways lead to a susceptibility to viral infections (33). NF-kappa-B essential modulator (NEMO) is a moderating protein that is needed in signaling through TLR but also in other important pathways. Rare NEMO mutations result in severe immunodeficiency and other clinical features. These patients are susceptible to adenovirus, cytomegalovirus (CMV), and herpes simplex virus (HSV) infections, as well as to infections caused by pyogenic bacteria, mycobacteria, and fungi (34).

Children with TLR3 pathway dysfunction are at risk of neonatal HSV encephalitis. It is surprising that mutations either in TLR3, adaptor molecule UNC-93B, or in important proteins downstream in the signaling pathway (TRIF and tumor necrosis factor [TNF] receptor-associated factor 3 [TRAF3]), are all associated specifically with the manifestation of HSV infection as neonatal encephalitis (35–38). In addition to this disease, only coxsackie virus myocarditis has been reported in a patient with a TLR3 mutation (39). It needs to be noted that only a limited number of patients with these mutations have been reported, and other disease associations may be found in the future. Nevertheless, it seems that although TLR3 signaling functions in defense against several virus infections, other immune responses protect the host from most infections in case this pathway is impaired. A common minor TLR3 allele Leu412Phe confers strengthened proinflammatory cytokine production after stimulation and partial protection against HIV infection (40). Notably, no inactivating mutations in TLR7, 8, or 9 have been reported, suggesting that these endosomal TLRs are highly important in the protection against viruses. The TLR9 1635A/G variant correlates with HIV disease progression and possibly, with an increased risk of mother-to-child HIV transmission (41, 42).

TLR2 and 4 genes harbor several common SNPs. While TLR2 and 4 are predominantly involved in the response to bacterial infections, they also recognize viral proteins. Heterozygotes for a common variant in TLR4, Asp299Gly, have an impaired response to lipopolysaccharide. This variant was associated with severe respiratory syncytial virus (RSV) infection in one study (43) but not in another study (44). More recently, the risk of severe RSV in children with variant TLR4 was proposed to be dependent on the subgroup of the epidemic virus (45). Heterozygotic TLR2 variant Arg753Gln has a frequency of 5% in European populations (46). There is no data linking TLR2 polymorphisms with virus infections. However, the TLR2 variant has been associated with rhinovirus-induced susceptibility to *Haemophilus influenzae* infection in cell cultures and in mice (47).

C-Type Lectins

Dectin-1, which is probably the best characterized C-type lectin, is not much involved in viral infections. Instead, surfactant proteins, which belong to the group of collectins and are expressed in the respiratory epithelial and lung alveolar cells, are important in respiratory diseases and infections. While the major function of surfactant proteins is to decrease the surface tension on the lung epithelium, certain surfactant proteins have innate immune functions. Surfactant protein A (SP-A) binds to the carbohydrates of glycosylated RSV surface proteins and opsonizes and neutralizes the virus (48). Surfactant protein D (SP-D) also has antiviral activity in the lungs (49). The two genes encoding SP-A subunits and the gene for SP-D are polymorphic and have several SNPs that have functional effects. Variant SP-A and

TABLE 1 Examples of human genetic variants leading to susceptibility to or protection against viral infections

Affected protein or pathway	Gene or protein variant	Associated viruses	Consequences	References
Virus receptors				
Erythrocyte P antigen	p phenotype	Parvovirus B19	Protection	(7)
Human secretor α(1, 2)-fucosyltransferase	*FUT2* G428A	Norovirus	Protection	(8)
CCR5	32 base pair deletion	HIV	Protection	(27–29)
Toll-like receptor pathways				
NEMO	Various mutations leading to impaired NF-κB activation	Adenovirus, CMV, HSV	Severe bacterial, mycobacterial, fungal, and viral infections	(34)
TLR3 signaling	*TLR3*, *UNC93B*, *TRIF*, and *TRAF3* mutations	HSV	Neonatal encephalitis	(35–38)
TLR9	A1635G	HIV	Rapid disease progression	(41, 42)
TLR4	Asp299Gly	RSV	Severe infection depending on epidemic virus	(43–45)
C-type lectins				
Surfactant protein A	Certain alleles over- or underrepresented, differences between populations	RSV	Hospitalization for RSV bronchiolitis	(14, 50, 52)
Surfactant protein A	*SFTPA2* alleles rs1965708-C and rs1059046-A and haplotype 1A(0)	Influenza A/ H1N1/2009	Respiratory failure	(103)
Surfactant protein D	Met11Thr	RSV	Hospitalization for RSV bronchiolitis	(51)
Mannose-binding lectin	Heterozygotes, homozygotes, or combined variants at codons 52, 54, and 57	SARS coronavirus, HSV 2	Risk of (recurrent) infection	(57, 62)
Immune cell regulation				
X-linked inhibitor of apoptosis protein (XIAP), SLAM-associated protein (SAP)	Mutations in *SH2D1A* and *XIAP* genes causing X-linked lymphoproliferative syndrome	EBV	Life-threatening EBV infection, hemophagocytosis, lymphoma	(63)
Immunoglobulin production	Various genetic defects resulting in hypogammaglobulinemia	Enteroviruses, rhinovirus, and other respiratory viruses	Enterovirus meningitis, prolonged respiratory tract infections	(65, 66)
Other pathways				
Innate immune factors	*JUN*, *IFN5*, *NOS2A*, and other genes	RSV	Hospitalization for RSV bronchiolitis	(67)
Vitamin D receptor and vitamin D–binding protein	*VDR* Thr1Met, *VDBP* haplotype GC1s	RSV	Hospitalization for RSV bronchiolitis	(67, 68)
IL-10	rs1800896 GG	RSV	Severe infection	(99)
Interferon-induced transmembrane protein-3	*IFITM3* splice-site altering mutation	Influenza A virus	Severe infection in Chinese population	(72)
CXCR4	WHIM syndrome causing mutations	HPV	Warts, bacterial infections	(30)
IL28B	rs 12979860 CC	HCV genotype 1	Good prognosis with interferon and ribavirin treatment	(89)
HLA				
DR/DQ heterozygotes, DRB1*04, DRB1**13		HBV	Effective clearance of virus	(4, 88)
DRB1*07, DRB1*03		HBV	Virus persistence	(88)
DR/DQ heterozygotes, DRB1*11:01, DQB1*03:01		HCV	Effective clearance of virus	(4, 87)
DRB1*07		HCV	Virus persistence	(4)

Abbreviations: NEMO, NF-kappa-B essential modulator; TLR, Toll-like receptor; IL-10, interleukin 10; HLA, human leukocyte antigen; WHIM, warts, hypogammaglobulinemia, infections, myelokathexis; CMV, cytomegalovirus; HSV, herpes simplex virus; RSV, respiratory syncytial virus; SARS, severe acute respiratory syndrome; EBV, Epstein-Barr virus; HBV, hepatitis B virus; HCV, hepatitis C virus.

SP-D are overrepresented in children with severe RSV infection compared with the normal population, although findings from some studies are not entirely in line with other studies (14,50–52). Mannose-binding lectin (MBL), which belongs to the C-type lectins, is discussed under complement deficiencies.

Inflammasome Defects

Inflammasome activation and, as a result of caspase-1 activity, IL-1β and IL-18 release contribute markedly to the inflammatory response to viral infections. Genetically modified mice with deficiencies in the nucleotide-binding domain-like receptor family, pyrin domain-containing 3 (*NLRP3*) gene, or in other genes encoding inflammasome-related factors, are susceptible to severe influenza (53). In contrast, humans with genetic defects in the factors of inflammasome are not susceptible to any virus, but they develop inflammatory diseases. A portion of patients with Crohn's disease, which is a chronic inflammatory bowel disease, have a mutation in the nucleotide-binding oligomerization domain-containing 2 (*NOD2*) gene (54). Mutations in *NLRP* cause cryopyrin-associated periodic syndromes, which are characterized by uncontrolled inflammation and clinical manifestations such as fever episodes, arthritis, aseptic meningitis, sensorineural hearing loss, and urticaria (55).

Complement Deficiency

The complement system recognizes PAMPs, opsonizes and destroys pathogens, and modulates adaptive immunity. The classical, alternative, and lectin pathways of complement involve more than 30 proteins. Gene defects have been identified in many components of the complement. Most of these mutants are rare and result in autoimmune diseases (e.g., systemic lupus erythematosus), susceptibility to bacterial infections, or other clinical manifestations (e.g., hereditary angioedema or age-related macular degeneration). The gene *MBL2* encoding MBL has three SNPs in exon 1, which moderately reduces the serum MBL concentration in heterozygotic forms and strongly reduces the serum MBL concentration in homozygotic or in combined heterozygotic forms (56). Variant MBLs result in various degrees of lectin pathway dysfunction. MBL deficiency is very common, but the frequencies of the variant forms differ between populations. For example, 2 to 5% of the Finnish population are homozygotes, and one-third are heterozygotes (46). MBL deficiency causes an increased risk of bacterial infections, particularly in young children. Influenza viruses and coronaviruses bind to MBL, but an increased risk of respiratory viral infection in subjects with an MBL variant has been convincingly reported only for the severe acute respiratory syndrome (SARS) coronavirus (57). At the age of 2 months, rhinoviruses were detected as often in children with variant MBL as in those with wild type, but pneumococcal nasopharyngeal colonization occurred more often during rhinovirus infection in children with an MBL variant (58). An MBL variant was not associated with pneumococcal colonization without virus infection. These findings suggest a role for MBL in viral-bacterial coinfection. Supporting this suggestion, a variant MBL gene was not overrepresented among patients with fatal pandemic influenza, but a high proportion of those who died with methicillin-resistant *Staphylococcus aureus* coinfection as a complication of influenza had low-producing variants of MBL (59).

Ficolins are lectins that have similar functions in the complement lectin pathway as MBL. H- and L-ficolins can bind viruses, and H-ficolin inhibits the replication of influenza A viruses (60). However, the significance of ficolin deficiencies in relation to susceptibility to viral infections is currently unclear.

The activation of classical or lectin pathways of complement cleaves C2 and C4 components to form C3 convertase, which is needed for activation of the terminal pathway. There are usually two copies of both *C4A* and *C4B* genes. Having only one functional copy of either gene causes partial deficiency, and a lack of both copies causes total C4A or C4B deficiency. Total C4A or C4B deficiency has been reported in patients with recurrent oral mucosal HSV type 1 infection (61). MBL variant forms correlate with recurrent genital HSV type 2 infections (62).

Immune Cell Deficiencies and Hypogammaglobulinemia

Severe T-cell deficiencies predispose the patient to life-threatening CMV and other herpesvirus infections. These syndromes are caused by genetic defects that affect T-cell, B-cell, and/or natural killer (NK)-cell development and function. Respiratory viruses also cause severe infections in patients with combined T- and B-cell immunodeficiencies. Even the administration of live, attenuated virus vaccines (e.g., rotavirus or varicella vaccine) can result in an uncontrolled infection, and they should be avoided if a profound T-cell deficiency is present.

X-linked lymphoproliferative syndrome (XLP) is characterized by the defective regulation of T cells, B cells, and NK cells in male patients (63). XLP-1 is caused by defects in the Src homology 2 (SH2) domain-containing 1A (*SH2D1A*) gene and XLP-2 by defects in the X-linked inhibitor of apoptosis (*XIAP*) gene. *SH2D1A* codes for the signaling lymphocytic activation molecule (SLAM)-associated protein (SAP), which functions in intracellular signaling through SLAM-family receptors and has effects on the functions of many immune cells. The NK T-cell number is low, the memory B-cell number is reduced, the cytotoxic functions of NK cells and $CD8^{+}$ cells are impaired, and gamma globulin levels in serum are often low in XLP. Both forms of the disease typically present with fulminant Epstein-Barr virus (EBV) infection. EBV primarily infects B lymphocytes, which proliferate in an uncontrolled manner in these patients, leading to cytopenias, hemophagocytosis, or lymphoma. Lethality is high, and if the patient survives, combined immunodeficiency may develop. EBV infection is also known to be associated with nasopharyngeal carcinoma, with a high incidence in East Asia. Susceptibility to the disease is strongly associated with specific HLA class I alleles (64).

Hypogammaglobulinemia is a part of many clinically and genetically characterized immunodeficiency syndromes. The genetic basis of the common variable immunodeficiency is heterogenic and not fully understood. Bacterial infections and autoimmune diseases are predominant in the clinical picture. These patients have prolonged symptoms with respiratory virus and enterovirus infections, and the virus-shedding times can be much longer than in immunocompetent individuals (65, 66). The impaired clearance of viruses in patients with hypogammaglobulinemia is probably caused by defective humoral immunity, although other defects in immunologic functions may participate in the clinical manifestations.

Other Genetic Defects and Genome-Wide Studies

In addition to the genetic deficiencies depicted above, several other SNPs or genetic variations in humans have been studied in order to identify associations with susceptibility to virus infections. Results have often been inconclusive and inconsistent between different studies. The reasons for this may include a lack of controlling for environmental risk factors between case and control groups, genetic differences between study populations, reduced statistical power in multiple testing, and a lack of functional correlations of genetic findings.

In a microarray study investigating the effects of 384 SNPs in 220 candidate genes on the susceptibility of infants to RSV bronchiolitis, variants in innate immunity genes had the strongest disease associations, whereas chemotaxis, adaptive immunity, asthma-related, and cytokine gene variants had weaker associations at the allele or genotype level (67). Variants in the genes of vitamin D receptor and vitamin D–binding protein have an influence on the risk of RSV infection (67, 68).

In another study, genetic susceptibility to severe influenza infection was investigated using a microarray for 50,000 SNPs in 98 individuals with severe A/H1N1/2009 influenza and in 91 exposed but asymptomatic controls (69). SNPs in genes coding immunoglobulin Fc receptor (FCGR2A), a complement-binding protein (C1QBP), and a protein called replication protein A-interacting protein (RPAIN) were significantly overrepresented in subjects with severe influenza, but as Horby and colleagues conclude in their review, the study was underpowered, and these may be false-positive findings, leaving the question of genetic susceptibility to influenza unanswered (70). Later, recessive homozygosity for a possible splice-site altering mutation in interferon-inducible transmembrane protein 3 (*IFITM3*) was reported to correlate with mild influenza infection in outpatients compared with controls, but the association with severe influenza was not statistically significant (71). The C allele at this site is much more common in individuals of Chinese descent than in Caucasians. The CC genotype is strongly associated with severe influenza in the Chinese population, and it has been estimated to cause a 54% population-attributable risk (72).

OPPORTUNITIES FOR DIAGNOSIS, MONITORING, AND PREDICTING DISEASE OUTCOMES

Traditional serologic methods for detecting specific antibodies are in wide diagnostic use. In contrast, innate immune responses are based on the recognition of common patterns in pathogens, and the diagnosis of a specific virus infection by a single marker is hardly possible. However, innate responses are different in viral compared with bacterial or fungal infections. Furthermore, if a group of molecules is analyzed at the same time, pathogen-specific responses may be found. On this basis, diagnostic methods relying on the analysis of innate immune responses are under development, although not yet in routine clinical use.

Single-Molecule Markers for Viral Infection

Virologic diagnosis is used mostly in hospitalized patients with severe diseases. In the large majority of patients with milder illnesses, treatment decisions are based only on clinical examination, possibly with the aid of a few simple laboratory tests, such as white blood cell count, serum C-reactive protein (CRP), and urine analysis. A nonspecific surrogate marker of a viral infection might be useful in many settings where laboratory detection of the causative agent is not feasible. Currently, serum CRP and procalcitonin (PCT) levels and white blood cell or neutrophil counts are widely used in evaluating the risk of bacterial infection, despite their nonspecific nature. No such markers for viruses are in clinical use, although MxA has been suggested as a surrogate marker for virus infection.

MxA is induced by type I and type III interferons, which are expressed only in virus infections. It functions as a cytoplasmic GTPase and has antiviral activity (73). In contrast to interferons, its half-life is relatively long. Basal levels in healthy people are low. These features make it a suitable candidate marker for virus infections. MxA levels have been measured in isolated lymphocytes or directly in blood. Many respiratory and other viruses induce strong MxA responses during symptomatic infections, whereas in asymptomatic infections MxA levels are generally low (Fig. 1) (74). MxA is also induced by live virus vaccinations. MxA levels discriminate between viral and bacterial etiologies in febrile children, but large-scale studies of the clinical value of MxA measurement in identifying viral infections are lacking (75, 76). The effects of MxA measurements on the use of antibiotics, the need for other laboratory studies, the length of hospitalization, or other clinical end points should be evaluated. Notably, similar cost-benefit-related questions can be asked regarding the use of PCR or other routine methods for the detection of viruses.

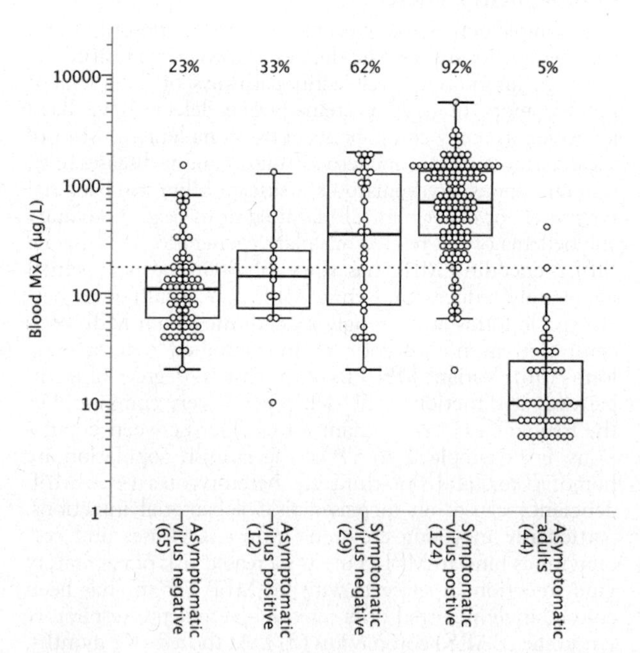

FIGURE 1 Myxovirus resistance protein A (MxA) blood levels in children 1 to 24 months of age according to virus detection and the presence of respiratory tract infection symptoms and in asymptomatic adults (*n* of subjects per group in parentheses). Respiratory viruses were diagnosed by reverse transcription real-time PCR for rhinovirus, enteroviruses, and respiratory syncytial virus (RSV) and by Seeplex RV12 multiplex PCR assay (Seegene, Seoul, Korea) for rhinovirus; RSV types A and B; adenovirus; influenza A and B viruses; parainfluenza virus types 1, 2, and 3; human metapneumovirus; and coronaviruses 229E/NL63 and OC43/HKU1. The boxes show median and interquartile range, and the whiskers show 10th and 90th percentiles. Percentages of measurements above the cut-off level of 175 micrograms per liter (dash line) are shown above the groups. (Reproduced from the *Journal of Clinical Virology* with permission of the original publisher [74].)

Host Response Patterns in Diagnostics

Ramilo and colleagues compared peripheral blood mononuclear cell gene-expression patterns in children with naturally acquired infections caused by influenza A virus, *Staphylococcus aureus*, *Streptococcus pneumoniae*, or *Escherichia coli* (77). Based on the expression of 35 genes, they could accurately differentiate children with influenza A virus infections from those with bacterial infections or from healthy children. Interferon-inducible genes were over-expressed in influenza and neutrophil-associated genes were over-expressed in bacterial infections. A few children had gene-expression patterns compatible with a viral-bacterial coinfection. Since this study, transcriptional profiling has been further developed as a diagnostic method for infectious diseases. Figure 2 shows the mapping of gene signatures specific for bacterial or viral infections. Zaas and colleagues studied gene-expression patterns in subjects experimentally infected with influenza A virus, RSV, or rhinovirus (78–80). They demonstrated virus-specific gene-expression patterns by using a set of 28 genes that differentiated individuals with symptomatic virus infections from subjects with asymptomatic infections. Hu and colleagues analyzed gene-expression profiles in febrile children with adenovirus, human herpesvirus type 6, or enterovirus infections; in children with bacterial infections; and in healthy controls (81). They could accurately differentiate febrile virus infections from bacterial infections. Furthermore, they showed that different expression profiles exist between symptomatic and asymptomatic infections and between specific virus infections. In other studies, host gene transcriptional profiling has been used in characterizing adults with influenza pneumonia, children with influenza A/H1N1/2009 infection, and dengue virus infections (82–84).

The use of host response transcriptional profiling for diagnostic purposes has the advantage of demonstrating pathogenic processes while at the same time identifying the causative microbe. In contrast, the direct detection of a microbe may leave its role as a causative pathogen unclear. The magnitude of the immunologic response may provide information regarding the severity of the disease. However, the use of transcriptional profiling for diagnostic purposes is currently technically demanding. Rapid and cheap methods that could be utilized widely in clinical work are not available at the moment.

Other diagnostic methods utilizing systems biology are under development. The mass spectrometry of serum proteins and the profiling of microRNA (miRNA) expression have been suggested for the detection of viruses based on stereotypic host responses (85). An area-under-the-curve value of 0.96 in receiver operator analysis was reported for combinations of miRNAs, including miR-17, miR-20a, miR-106a, and miR-376c, in the detection of avian influenza A/H7N9 virus infection in humans in China (86).

Monitoring Host Factors to Predict Virus Infections and Outcomes

Genetic defects that increase susceptibility to virus infections, as outlined above, often also increase the risk of severe disease manifestations or poor outcomes. Most studies in this field have deciphered the pathogenic mechanisms by identifying host genetic variants that influence the outcome, only rarely aiming to develop markers of virus infections that could be used in monitoring the disease process or in predicting the outcome. No host factor biomarker for monitoring the progress of virus infections is in clinical use. This is in contrast with the routine use of markers such as CRP and PCT in monitoring the treatment of bacterial infections. As a general marker of interferon response in virus infections, MxA could also be useful in the follow-up of the disease, but studies demonstrating such usefulness are lacking.

In some cases antiviral treatment can be modified according to host factor determinations that predict either the effectiveness of treatment or the risk of adverse effects.

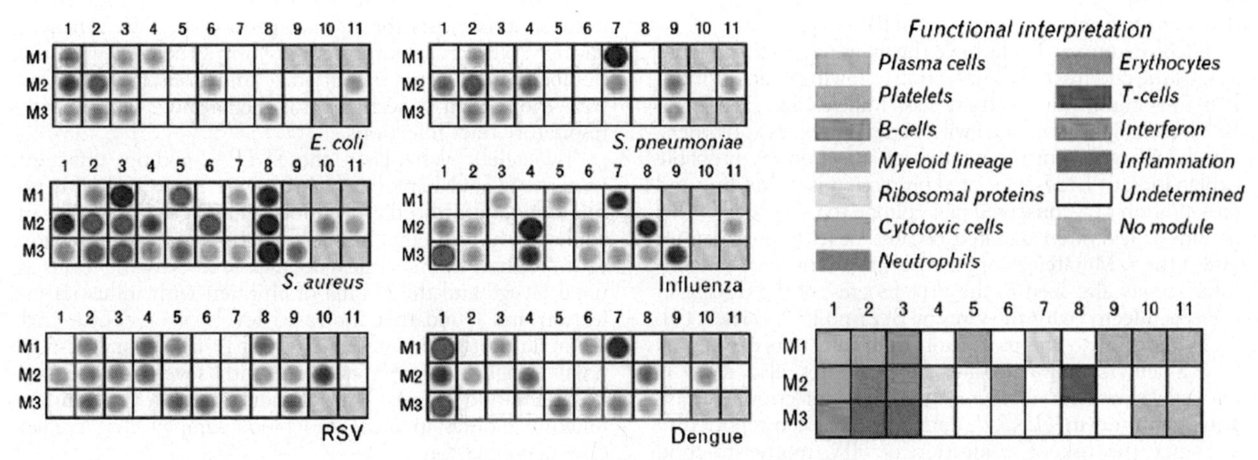

FIGURE 2 Transcriptional profiling of host response in the diagnosis of certain viral and bacterial infections. Patients had various types of acute bacterial infections, respiratory viral infections, or dengue. Expression levels of functionally related sets of genes are displayed on a grid, with the coordinates corresponding to 1 of 28 module IDs (e.g., Module M3.1 is at the intersection of the third row and first column). Mapping transcriptional changes at the module level identifies disease-specific biosignatures in patients with infectious diseases. Expression levels were compared between patients and appropriately matched healthy controls on a module-by-module basis. The spots represent the percentage of significantly overexpressed (red) or underexpressed (blue) transcripts within a module (i.e., set of coordinately expressed genes). Blank spots indicate that there are no differences in the genes included in that module between patients and healthy controls. Each pathogen induces a disease-specific biosignature that is easily identifiable. (Reproduced from *Current Opinion in Infectious Diseases* with permission of the original publisher [110].)

Hepatitis C Virus, Hepatitis B Virus, and HIV

The effect of HLA gene polymorphisms on the course of infections is obvious in chronic viral hepatitis and HIV. The clearance of both hepatitis B virus (HBV) and hepatitis C virus (HCV) infections has been shown to be more efficient in HLA-DR/DQ heterozygous than homozygous subjects, suggesting the importance of a broader presentation of viral peptides and a larger repertoire of specific T cells in heterozygotes (4). HLA-DRB1*07 is associated with the persistence of infection for both HBV and HCV, whereas HCV clearance is more efficient in subjects positive for DRB1*11:01 and DQB1*03:01 (87, 88). In the case of HBV, DRB1*03 was also associated with viral persistence, but DRB1*04 and DRB1**13 were found to favor viral clearance (88).

An example of a genetic marker that is useful in the prediction of outcome is the determination of IL28B alleles at position rs12979860 in consideration of HCV treatment with interferon and ribavirin. Caucasian patients with HCV genotype 1 infection with IL28B genotypes CC, CT, or TT at this position have 69%, 33%, and 27% probabilities for a sustained virologic response to treatment (89). Other SNPs near this region also influence outcomes. Interferon-induced gene expression in liver biopsy predicts treatment response independently of the IL28B genotype (90). The presence of IL28B variants, or another polymorphism in the same region activating the interferon lambda-4 (*IFNL4*) gene, correlates with the spontaneous resolution of HCV (91, 92). A genome-wide study identified gene variants that correlate with the progression of HCV infection to liver fibrosis (93). These genes function in control of apoptosis.

The risk of chronic carriage after HBV infection and the subsequent development of liver cirrhosis and hepatocellular carcinoma is strongly influenced by the genetics of the host. In addition to HLA genes, variants in genes coding MBL, interferon gamma, TLR4, TLR5, vitamin D receptor, and estrogen and androgen receptors have effects on the short-term or long-term prognosis of HBV infection (94, 95). There are inconsistencies between studies, however, and populations differ substantially in their genetic variations that protect from or predispose to HBV liver disease.

$CD8^+$ cytotoxic T cells have the main role in the defense against the progress of HIV infection following its acquisition. It is thus natural that certain class I alleles, like B*57 and B*27, seem to be associated with better outcomes and others, like B*35, with worse outcomes of infection. A probable explanation of B*27-associated protection is its binding and presentation of a conserved p24 epitope to $CD8^+$ cells. This peptide is less often mutated because of its importance for virus fitness. Mutated escape viruses are attenuated, which is interestingly also seen in the slow progress of the disease in subjects infected with the virus by B*27 positive carriers (4).

In addition to the protection from infection exerted by CCR5-delta32 homozygosity, heterozygotes also show a markedly slow natural progression of HIV infection. Certain polymorphisms in TLR2, 7, 8, and 9 have been reported to influence the risk of acquisition of HIV, mother-to-child transmission, viral loads, or disease progression (42, 96, 97). However, authors of a genome-wide $CD4^+$ T-cell mRNA expression and SNP study of HIV-infected individuals and elite controllers concluded that while host genetic variations modulate the immune responses, they do not have any strong influence on HIV control (98).

Genetic determinations are needed in some cases to guide the antiviral treatment of HIV. The risk of a serious hypersensitivity reaction to abacavir is highly increased in subjects with HLA-B*5701, and abacavir should only be used in the treatment of patients who are negative for this HLA type.

Respiratory Virus Infections

Host genetics and innate immune factor mRNA expression, or protein levels, have been studied in relation to respiratory virus infections in the context of identifying individuals with a risk of progressing to severe or complicated disease or acquiring later manifestations such as recurrent wheezing or asthma. In several studies reviewed above, genetic susceptibility was examined in groups of subjects hospitalized with a respiratory infection and compared with healthy controls. It should be noted that rather than showing susceptibility to acquiring infection, this setting identifies risk factors for severe disease because respiratory virus infections are highly frequent, and only rarely do infected subjects need hospitalization. For example, about 2 to 3% of infants need hospitalization because of RSV bronchiolitis, although almost all acquire RSV infection during early childhood.

Although hospitalization, as such, indicates severe disease, there are studies linking genetic variations to more severe disease within hospitalized patients. The IL-10 gene SNPs rs1800896 and rs1800890 are as common in children hospitalized with RSV infection as in healthy populations, but these polymorphisms correlate with the need for ventilator treatment for severe RSV (99). IL-10 is a regulatory and anti-inflammatory cytokine, which promotes Th2 effector $CD4^+$ T-helper cell-type responses and adaptive immunity and suppresses the production of pro-inflammatory cytokines. These functions appear to be important in recovery from RSV infection. In another study, while children with IL-10 genotype GG at rs1800896 are at risk for a severe manifestation of RSV bronchiolitis, the genotype AA was overrepresented in children hospitalized with bronchiolitis caused by rhinovirus or other viruses except RSV (100). After bronchiolitis, homozygotes for the major allele C at L412F of the *TLR3* gene had an increased risk of repeated wheezing episodes (101). Children with an IL6-174 polymorphism had 24% more upper respiratory tract infections than children with the normal genotype (102). An activating TNF-α polymorphism was not associated with the number of upper respiratory tract infections, but children with the variant gene had otitis media more often during respiratory tract infections.

Two allelic variants in the *SFTPA2*-coding surfactant protein A2 and haplotype 1A(0) were associated with respiratory failure and the need for mechanical ventilation in adults with pandemic influenza A/H1N1/2009 infection (103). Oshansky and colleagues measured cytokine levels in nasal lavage and the plasma of children with influenza infection and found that increased levels of monocyte chemotactic protein 3, interferon alpha 2, IL-10, and IL-6 in nasal lavage or plasma correlate with severe disease and hospitalization (104). They demonstrated distinct innate immune profiles in nasal and blood samples that predict disease progression.

Upper respiratory tract rhinovirus infections are highly frequent in children and adults. Some children develop wheezing illnesses during rhinovirus infections at an early age, and some develop asthma later in life, which may again exacerbate during virus infections. Bronchial epithelial cells from both asthmatic and healthy individuals produce interferon gamma–induced protein 10 (IP-10) and RANTES (chemokine ligand 5 or regulated on activation, normal

T cell expressed and secreted) after infection with rhinovirus 16 (105). In patients with acute asthma exacerbation, serum IP-10 levels can be differentiated between those with virus-induced acute asthma and those with asthma from another cause. This marker could be used to nonspecifically identify a viral trigger of acute asthma. Patients with asthma have been suggested to have an impaired interferon response to rhinovirus infection (106, 107). However, in a more recent study, the rhinovirus-induced production of interferon gamma and interferon beta was similar in bronchial epithelial cells obtained from subjects with or without asthma (108). More studies are needed before definite conclusions can be drawn, but taken together, these findings suggest that specific genotypic variants may correlate either with a respiratory tract infection caused by a certain virus, with a poor prognosis of the acute disease, or with long-term consequences such as the development of asthma.

CONCLUSIONS AND FUTURE CONSIDERATIONS

Genetic variations in innate immune factors have important effects on susceptibility to virus infections. The determination of genetic susceptibility could be used in predicting the risk of infections or in the selection of optimal treatment strategies. The clinical use of genetic tests for such purposes has thus far been limited, but clinical implementation of such testing will most likely increase in the near future. For example, current guidelines suggest determining the IL28B genotype in a subgroup of patients with HCV when planning treatment (109). Differences in the frequencies of genotypes between populations need to be considered when making generalizations about the findings of genetic studies.

The analysis of host innate immune responses in the diagnostics of viral infections is a novel, highly promising strategy. With or without the simultaneous detection of the causative agent by traditional methods, measurement of the specific host response can provide strong evidence of the causative role of the agent in the disease process. As new antiviral agents are developed, methods for monitoring treatment responses will be needed, and the follow-up of innate immune responses also holds promise for this purpose. Nevertheless, gene-expression pattern determinations are technically challenging, and routine methods are not available at the moment. There is a need also for clinically useful, simple biomarkers of viral infection.

REFERENCES

1. **Picard C, Fischer A.** 2014. Contribution of high-throughput DNA sequencing to the study of primary immunodeficiencies. *Eur J Immunol* **44:**2854–2861.
2. **Picard C, Casanova JL, Abel L.** 2006. Mendelian traits that confer predisposition or resistance to specific infections in humans. *Curr Opin Immunol* **18:**383–390.
3. **Doherty PC, Zinkernagel RM.** 1975. A biological role for the major histocompatibility antigens. *Lancet* **305:**1406–1409.
4. **Blackwell JM, Jamieson SE, Burgner D.** 2009. HLA and infectious diseases. *Clin Microbiol Rev* **22:**370–385.
5. **Hill AV, Allsopp CE, Kwiatkowski D, Anstey NM, Twumasi P, Rowe PA, Bennett S, Brewster D, McMichael AJ, Greenwood BM.** 1991. Common west African HLA antigens are associated with protection from severe malaria. *Nature* **352:**595–600.
6. **Bossuyt X, Moens L, Van Hoeyveld E, Jeurissen A, Bogaert G, Sauer K, Proesmans M, Raes M, De Boeck K.** 2006. Coexistence of (partial) immune defects and risk of recurrent respiratory infections. *Clin Chem* **53:**124–130.
7. **Brown KE, Hibbs JR, Gallinella G, Anderson SM, Lehman ED, McCarthy P, Young NS.** 1994. Resistance to parvovirus B19 infection due to lack of virus receptor (erythrocyte P antigen). *N Engl J Med* **330:**1192–1196.
8. **Thorven M, Grahn A, Hedlund KO, Johansson H, Wahlfrid C, Larson G, Svensson L.** 2005. A homozygous nonsense mutation (428G—>A) in the human secretor (FUT2) gene provides resistance to symptomatic norovirus (GGII) infections. *J Virol* **79:**15351–15355.
9. **Finlay BB, McFadden G.** 2006. Anti-immunology: evasion of the host immune system by bacterial and viral pathogens. *Cell* **124:**767–782.
10. **Collins SE, Mossman KL.** 2014. Danger, diversity and priming in innate antiviral immunity. *Cytokine Growth Factor Rev* **25:**525–531.
11. **Akira S, Uematsu S, Takeuchi O.** 2006. Pathogen recognition and innate immunity. *Cell* **124:**783–801.
12. **Goubau D, Schlee M, Deddouche S, Pruijssers AJ, Zillinger T, Goldeck M, Schuberth C, Van der Veen AG, Fujimura T, Rehwinkel J, Iskarpatyoti JA, Barchet W, Ludwig J, Dermody TS, Hartmann G, Reis e Sousa C.** 2014. Antiviral immunity via RIG-I-mediated recognition of RNA bearing 5′-diphosphates. *Nature* **514:**372–375.
13. **Wu J, Chen ZJ.** 2014. Innate immune sensing and signaling of cytosolic nucleic acids. *Annu Rev Immunol* **32:**461–488.
14. **Miyairi I, DeVincenzo JP.** 2008. Human genetic factors and respiratory syncytial virus disease severity. *Clin Microbiol Rev* **21:**686–703.
15. **Pang IK, Iwasaki A.** 2011. Inflammasomes as mediators of immunity against influenza virus. *Trends Immunol* **32:**34–41.
16. **Kato H, Takeuchi O, Sato S, Yoneyama M, Yamamoto M, Matsui K, Uematsu S, Jung A, Kawai T, Ishii KJ, Yamaguchi O, Otsu K, Tsujimura T, Koh CS, Reis e Sousa C, Matsuura Y, Fujita T, Akira S.** 2006. Differential roles of MDA5 and RIG-I helicases in the recognition of RNA viruses. *Nature* **441:**101–105.
17. **Tang ED, Wang CY.** 2009. MAVS self-association mediates antiviral innate immune signaling. *J Virol* **83:**3420–3428.
18. **Pandey S, Kawai T, Akira S.** 2015. Microbial sensing by Toll-like receptors and intracellular nucleic acid sensors. *Cold Spring Harb Perspect Biol* **7:**a016246.
19. **Bieback K, Lien E, Klagge IM, Avota E, Schneider-Schaulies J, Duprex WP, Wagner H, Kirschning CJ, Ter Meulen V, Schneider-Schaulies S.** 2002. Hemagglutinin protein of wild-type measles virus activates toll-like receptor 2 signaling. *J Virol* **76:**8729–8736.
20. **Kurt-Jones EA, Popova L, Kwinn L, Haynes LM, Jones LP, Tripp RA, Walsh EE, Freeman MW, Golenbock DT, Anderson LJ, Finberg RW.** 2000. Pattern recognition receptors TLR4 and CD14 mediate response to respiratory syncytial virus. *Nat Immunol* **1:**398–401.
21. **Diebold SS, Kaisho T, Hemmi H, Akira S, Reis e Sousa C.** 2004. Innate antiviral responses by means of TLR7-mediated recognition of single-stranded RNA. *Science* **303:**1529–1531.
22. **Pirhonen J, Sareneva T, Kurimoto M, Julkunen I, Matikainen S.** 1999. Virus infection activates IL-1 beta and IL-18 production in human macrophages by a caspase-1-dependent pathway. *J Immunol* **162:**7322–7329.
23. **Weinzierl AO, Rudolf D, Maurer D, Wernet D, Rammensee HG, Stevanović S, Klingel K.** 2008. Identification of HLA-A*01- and HLA-A*02-restricted CD8+ T-cell epitopes shared among group B enteroviruses. *J Gen Virol* **89:**2090–2097.
24. **Bengs S, Marttila J, Susi P, Ilonen J.** 2015. Elicitation of T-cell responses by structural and non-structural proteins of coxsackievirus B4. *J Gen Virol* **96:**322–330.
25. **Kreijtz JH, Fouchier RA, Rimmelzwaan GF.** 2011. Immune responses to influenza virus infection. *Virus Res* **162:**19–30.
26. **Karttunen A, Pöyry T, Vaarala O, Ilonen J, Hovi T, Roivainen M, Hyypiä T.** 2003. Variation in enterovirus receptor genes. *J Med Virol* **70:**99–108.

27. **Dean M, Carrington M, Winkler C, Huttley GA, Smith MW, Allikmets R, Goedert JJ, Buchbinder SP, Vittinghoff E, Gomperts E, Donfield S, Vlahov D, Kaslow R, Saah A, Rinaldo C, Detels R, O'Brien SJ, Multicenter AIDS Cohort Study, Multicenter Hemophilia Cohort Study, San Francisco City Cohort, ALIVE Study.** 1996. Genetic restriction of HIV-1 infection and progression to AIDS by a deletion allele of the CKR5 structural gene. *Science* **273:**1856–1862.
28. **Liu R, Paxton WA, Choe S, Ceradini D, Martin SR, Horuk R, MacDonald ME, Stuhlmann H, Koup RA, Landau NR.** 1996. Homozygous defect in HIV-1 coreceptor accounts for resistance of some multiply-exposed individuals to HIV-1 infection. *Cell* **86:**367–377.
29. **Samson M, Libert F, Doranz BJ, Rucker J, Liesnard C, Farber CM, Saragosti S, Lapoumeroulie C, Cognaux J, Forceille C, Muyldermans G, Verhofstede C, Burtonboy G, Georges M, Imai T, Rana S, Yi Y, Smyth RJ, Collman RG, Doms RW, Vassart G, Parmentier M.** 1996. Resistance to HIV-1 infection in caucasian individuals bearing mutant alleles of the CCR-5 chemokine receptor gene. *Nature* **382:** 722–725.
30. **Hernandez PA, Gorlin RJ, Lukens JN, Taniuchi S, Bohinjec J, Francois F, Klotman ME, Diaz GA.** 2003. Mutations in the chemokine receptor gene CXCR4 are associated with WHIM syndrome, a combined immunodeficiency disease. *Nat Genet* **34:**70–74.
31. **Nejentsev S, Walker N, Riches D, Egholm M, Todd JA.** 2009. Rare variants of IFIH1, a gene implicated in antiviral responses, protect against type 1 diabetes. *Science* **324:**387–389.
32. **Alsina L, Israelsson E, Altman MC, Dang KK, Ghandil P, Israel L, von Bernuth H, Baldwin N, Qin H, Jin Z, Banchereau R, Anguiano E, Ionan A, Abel L, Puel A, Picard C, Pascual V, Casanova JL, Chaussabel D.** 2014. A narrow repertoire of transcriptional modules responsive to pyogenic bacteria is impaired in patients carrying loss-of-function mutations in MYD88 or IRAK4. *Nat Immunol* **15:**1134–1142.
33. **von Bernuth H, Picard C, Puel A, Casanova JL.** 2012. Experimental and natural infections in MyD88- and IRAK-4-deficient mice and humans. *Eur J Immunol* **42:**3126–3135.
34. **Picard C, Casanova JL, Puel A.** 2011. Infectious diseases in patients with IRAK-4, MyD88, NEMO, or IκBα deficiency. *Clin Microbiol Rev* **24:**490–497.
35. **Casrouge A, Zhang SY, Eidenschenk C, Jouanguy E, Puel A, Yang K, Alcais A, Picard C, Mahfoufi N, Nicolas N, Lorenzo L, Plancoulaine S, Sénéchal B, Geissmann F, Tabeta K, Hoebe K, Du X, Miller RL, Héron B, Mignot C, de Villemeur TB, Lebon P, Dulac O, Rozenberg F, Beutler B, Tardieu M, Abel L, Casanova JL.** 2006. Herpes simplex virus encephalitis in human UNC-93B deficiency. *Science* **314:** 308–312.
36. **Pérez de Diego R, Sancho-Shimizu V, Lorenzo L, Puel A, Plancoulaine S, Picard C, Herman M, Cardon A, Durandy A, Bustamante J, Vallabhapurapu S, Bravo J, Warnatz K, Chaix Y, Cascarrigny F, Lebon P, Rozenberg F, Karin M, Tardieu M, Al-Muhsen S, Jouanguy E, Zhang SY, Abel L, Casanova JL.** 2010. Human TRAF3 adaptor molecule deficiency leads to impaired Toll-like receptor 3 response and susceptibility to herpes simplex encephalitis. *Immunity* **33:** 400–411.
37. **Sancho-Shimizu V, Pérez de Diego R, Lorenzo L, Halwani R, Alangari A, Israelsson E, Fabrega S, Cardon A, Maluenda J, Tatematsu M, Mahvelati F, Herman M, Ciancanelli M, Guo Y, AlSum Z, Alkhamis N, Al-Makadma AS, Ghadiri A, Boucherit S, Plancoulaine S, Picard C, Rozenberg F, Tardieu M, Lebon P, Jouanguy E, Rezaei N, Seya T, Matsumoto M, Chaussabel D, Puel A, Zhang SY, Abel L, Al-Muhsen S, Casanova JL.** 2011. Herpes simplex encephalitis in children with autosomal recessive and dominant TRIF deficiency. *J Clin Invest* **121:**4889–4902.
38. **Zhang SY, Jouanguy E, Ugolini S, Smahi A, Elain G, Romero P, Segal D, Sancho-Shimizu V, Lorenzo L, Puel A, Picard C, Chapgier A, Plancoulaine S, Titeux M, Cognet C, von Bernuth H, Ku CL, Casrouge A, Zhang XX, Barreiro L, Leonard J, Hamilton C, Lebon P, Héron B, Vallée L, Quintana-Murci L, Hovnanian A, Rozenberg F, Vivier E, Geissmann F, Tardieu M, Abel L, Casanova JL.** 2007. TLR3 deficiency in patients with herpes simplex encephalitis. *Science* **317:**1522–1527.
39. **Gorbea C, Makar KA, Pauschinger M, Pratt G, Bersola JL, Varela J, David RM, Banks L, Huang CH, Li H, Schultheiss HP, Towbin JA, Vallejo JG, Bowles NE.** 2010. A role for Toll-like receptor 3 variants in host susceptibility to enteroviral myocarditis and dilated cardiomyopathy. *J Biol Chem* **285:**23208–23223.
40. **Sironi M, Biasin M, Cagliani R, Forni D, De Luca M, Saulle I, Lo Caputo S, Mazzotta F, Macías J, Pineda JA, Caruz A, Clerici M.** 2012. A common polymorphism in TLR3 confers natural resistance to HIV-1 infection. *J Immunol* **188:**818–823.
41. **Bochud PY, Hersberger M, Taffé P, Bochud M, Stein CM, Rodrigues SD, Calandra T, Francioli P, Telenti A, Speck RF, Aderem A, Swiss HIV Cohort Study.** 2007. Polymorphisms in Toll-like receptor 9 influence the clinical course of HIV-1 infection. *AIDS* **21:**441–446.
42. **Ricci E, Malacrida S, Zanchetta M, Mosconi I, Montagna M, Giaquinto C, De Rossi A.** 2010. Toll-like receptor 9 polymorphisms influence mother-to-child transmission of human immunodeficiency virus type 1. *J Transl Med* **8:**1–5.
43. **Tal G, Mandelberg A, Dalal I, Cesar K, Somekh E, Tal A, Oron A, Itskovich S, Ballin A, Houri S, Beigelman A, Lider O, Rechavi G, Amariglio N.** 2004. Association between common Toll-like receptor 4 mutations and severe respiratory syncytial virus disease. *J Infect Dis* **189:**2057–2063.
44. **Paulus SC, Hirschfeld AF, Victor RE, Brunstein J, Thomas E, Turvey SE.** 2007. Common human Toll-like receptor 4 polymorphisms—role in susceptibility to respiratory syncytial virus infection and functional immunological relevance. *Clin Immunol* **123:**252–257.
45. **Löfgren J, Marttila R, Renko M, Rämet M, Hallman M.** 2010. Toll-like receptor 4 Asp299Gly polymorphism in respiratory syncytial virus epidemics. *Pediatr Pulmonol* **45:**687–692.
46. **Vuononvirta J, Toivonen L, Gröndahl-Yli-Hannuksela K, Barkoff AM, Lindholm L, Mertsola J, Peltola V, He Q.** 2011. Nasopharyngeal bacterial colonization and gene polymorphisms of mannose-binding lectin and toll-like receptors 2 and 4 in infants. *PLoS One* **6:**e26198.
47. **Unger BL, Faris AN, Ganesan S, Comstock AT, Hershenson MB, Sajjan US.** 2012. Rhinovirus attenuates non-typeable *Hemophilus influenzae*-stimulated IL-8 responses via TLR2-dependent degradation of IRAK-1. *PLoS Pathog* **8:** e1002969.
48. **Ghildyal R, Hartley C, Varrasso A, Meanger J, Voelker DR, Anders EM, Mills J.** 1999. Surfactant protein A binds to the fusion glycoprotein of respiratory syncytial virus and neutralizes virion infectivity. *J Infect Dis* **180:**2009–2013.
49. **Hillaire ML, Haagsman HP, Osterhaus AD, Rimmelzwaan GF, van Eijk M.** 2013. Pulmonary surfactant protein D in first-line innate defence against influenza A virus infections. *J Innate Immun* **5:**197–208.
50. **El Saleeby CM, Li R, Somes GW, Dahmer MK, Quasney MW, DeVincenzo JP.** 2010. Surfactant protein A2 polymorphisms and disease severity in a respiratory syncytial virus-infected population. *J Pediatr* **156:**409–414.e4.
51. **Lahti M, Löfgren J, Marttila R, Renko M, Klaavuniemi T, Haataja R, Rämet M, Hallman M.** 2002. Surfactant protein D gene polymorphism associated with severe respiratory syncytial virus infection. *Pediatr Res* **51:**696–699.
52. **Löfgren J, Rämet M, Renko M, Marttila R, Hallman M.** 2002. Association between surfactant protein A gene locus and severe respiratory syncytial virus infection in infants. *J Infect Dis* **185:**283–289.
53. **Thomas PG, Dash P, Aldridge JR Jr, Ellebedy AH, Reynolds C, Funk AJ, Martin WJ, Lamkanfi M, Webby RJ, Boyd KL, Doherty PC, Kanneganti TD.** 2009. The intracellular sensor NLRP3 mediates key innate and healing responses to influenza A virus via the regulation of caspase-1. *Immunity* **30:**566–575.

54. **Hugot JP, Chamaillard M, Zouali H, Lesage S, Cézard JP, Belaiche J, Almer S, Tysk C, O'Morain CA, Gassull M, Binder V, Finkel Y, Cortot A, Modigliani R, Laurent-Puig P, Gower-Rousseau C, Macry J, Colombel JF, Sahbatou M, Thomas G.** 2001. Association of NOD2 leucine-rich repeat variants with susceptibility to Crohn's disease. *Nature* **411:** 599–603.
55. **Levy R, Gérard L, Kuemmerle-Deschner J, Lachmann HJ, Koné-Paut I, Cantarini L, Woo P, Naselli A, Bader-Meunier B, Insalaco A, Al-Mayouf SM, Ozen S, Hofer M, Frenkel J, Modesto C, Nikishina I, Schwarz T, Martino S, Meini A, Quartier P, Martini A, Ruperto N, Neven B, Gattorno M, for PRINTO and Eurofever.** 2015. Phenotypic and genotypic characteristics of cryopyrin-associated periodic syndrome: a series of 136 patients from the Eurofever Registry. *Ann Rheum Dis* **74:**2043–2049.
56. **Garred P, Larsen F, Seyfarth J, Fujita R, Madsen HO.** 2006. Mannose-binding lectin and its genetic variants. *Genes Immun* **7:**85–94.
57. **Ip WK, Chan KH, Law HK, Tso GH, Kong EK, Wong WH, To YF, Yung RW, Chow EY, Au KL, Chan EY, Lim W, Jensenius JC, Turner MW, Peiris JS, Lau YL.** 2005. Mannose-binding lectin in severe acute respiratory syndrome coronavirus infection. *J Infect Dis* **191:**1697–1704.
58. **Karppinen S, Vuononvirta J, He Q, Waris M, Peltola V.** 2013. Effects of rhinovirus infection on nasopharyngeal bacterial colonization in infants with wild or variant types of mannose-binding lectin and Toll-like receptors 3 and 4. *J Pediatric Infect Dis Soc* **2:**240–247.
59. **Ferdinands JM, Denison AM, Dowling NF, Jost HA, Gwinn ML, Liu L, Zaki SR, Shay DK.** 2011. A pilot study of host genetic variants associated with influenza-associated deaths among children and young adults. *Emerg Infect Dis* **17:**2294–2302.
60. **Verma A, White M, Vathipadiekal V, Tripathi S, Mbianda J, Ieong M, Qi L, Taubenberger JK, Takahashi K, Jensenius JC, Thiel S, Hartshorn KL.** 2012. Human H-ficolin inhibits replication of seasonal and pandemic influenza A viruses. *J Immunol* **189:**2478–2487.
61. **Seppänen M, Lokki ML, Timonen T, Lappalainen M, Jarva H, Järvinen A, Sarna S, Valtonen V, Meri S.** 2001. Complement C4 deficiency and HLA homozygosity in patients with frequent intraoral herpes simplex virus type 1 infections. *Clin Infect Dis* **33:**1604–1607.
62. **Seppänen M, Lokki ML, Lappalainen M, Hiltunen-Back E, Rovio AT, Kares S, Hurme M, Aittoniemi J.** 2009. Mannose-binding lectin 2 gene polymorphism in recurrent herpes simplex virus 2 infection. *Hum Immunol* **70:**218–221.
63. **Filipovich AH, Zhang K, Snow AL, Marsh RA.** 2010. X-linked lymphoproliferative syndromes: brothers or distant cousins? *Blood* **116:**3398–3408.
64. **Tang M, Lautenberger JA, Gao X, Sezgin E, Hendrickson SL, Troyer JL, David VA, Guan L, McIntosh CE, Guo X, Zheng Y, Liao J, Deng H, Malasky M, Kessing B, Winkler CA, Carrington M, Dé The G, Zeng Y, O'Brien SJ.** 2012. The principal genetic determinants for nasopharyngeal carcinoma in China involve the HLA class I antigen recognition groove. *PLoS Genet* **8:**e1003103.
65. **Halliday E, Winkelstein J, Webster AD.** 2003. Enteroviral infections in primary immunodeficiency (PID): a survey of morbidity and mortality. *J Infect* **46:**1–8.
66. **Kainulainen L, Vuorinen T, Rantakokko-Jalava K, Österback R, Ruuskanen O.** 2010. Recurrent and persistent respiratory tract viral infections in patients with primary hypogammaglobulinemia. *J Allergy Clin Immunol* **126:**120–126.
67. **Janssen R, Bont L, Siezen CL, Hodemaekers HM, Ermers MJ, Doornbos G, van 't Slot R, Wijmenga C, Goeman JJ, Kimpen JL, van Houwelingen HC, Kimman TG, Hoebee B.** 2007. Genetic susceptibility to respiratory syncytial virus bronchiolitis is predominantly associated with innate immune genes. *J Infect Dis* **196:**826–834.
68. **Randolph AG, Yip WK, Falkenstein-Hagander K, Weiss ST, Janssen R, Keisling S, Bont L.** 2014. Vitamin D-binding protein haplotype is associated with hospitalization for RSV bronchiolitis. *Clin Exp Allergy* **44:**231–237.
69. **Zúñiga J, Buendía-Roldán I, Zhao Y, Jiménez L, Torres D, Romo J, Ramírez G, Cruz A, Vargas-Alarcon G, Sheu CC, Chen F, Su L, Tager AM, Pardo A, Selman M, Christiani DC.** 2012. Genetic variants associated with severe pneumonia in A/H1N1 influenza infection. *Eur Respir J* **39:**604–610.
70. **Horby P, Nguyen NY, Dunstan SJ, Baillie JK.** 2012. The role of host genetics in susceptibility to influenza: a systematic review. *PLoS One* **7:**e33180.
71. **Mills TC, Rautanen A, Elliott KS, Parks T, Naranbhai V, Ieven MM, Butler CC, Little P, Verheij T, Garrard CS, Hinds C, Goossens H, Chapman S, Hill AV.** 2014. IFITM3 and susceptibility to respiratory viral infections in the community. *J Infect Dis* **209:**1028–1031.
72. **Zhang YH, Zhao Y, Li N, Peng YC, Giannoulatou E, Jin RH, Yan HP, Wu H, Liu JH, Liu N, Wang DY, Shu YL, Ho LP, Kellam P, McMichael A, Dong T.** 2013. Interferon-induced transmembrane protein-3 genetic variant rs12252-C is associated with severe influenza in Chinese individuals. *Nat Commun* **4:**1–6.
73. **Haller O, Kochs G.** 2011. Human MxA protein: an interferon-induced dynamin-like GTPase with broad antiviral activity. *J Interferon Cytokine Res* **31:**79–87.
74. **Toivonen L, Schuez-Havupalo L, Rulli M, Ilonen J, Pelkonen J, Melén K, Julkunen I, Peltola V, Waris M.** 2015. Blood MxA protein as a marker for respiratory virus infections in young children. *J Clin Virol* **62:**8–13.
75. **Halminen M, Ilonen J, Julkunen I, Ruuskanen O, Simell O, Mäkelä MJ.** 1997. Expression of MxA protein in blood lymphocytes discriminates between viral and bacterial infections in febrile children. *Pediatr Res* **41:**647–650.
76. **Nakabayashi M, Adachi Y, Itazawa T, Okabe Y, Kanegane H, Kawamura M, Tomita A, Miyawaki T.** 2006. MxA-based recognition of viral illness in febrile children by a whole blood assay. *Pediatr Res* **60:**770–774.
77. **Ramilo O, Allman W, Chung W, Mejias A, Ardura M, Glaser C, Wittkowski KM, Piqueras B, Banchereau J, Palucka AK, Chaussabel D.** 2007. Gene expression patterns in blood leukocytes discriminate patients with acute infections. *Blood* **109:**2066–2077.
78. **Zaas AK, Burke T, Chen M, McClain M, Nicholson B, Veldman T, Tsalik EL, Fowler V, Rivers EP, Otero R, Kingsmore SF, Voora D, Lucas J, Hero AO, Carin L, Woods CW, Ginsburg GS.** 2013. A host-based RT-PCR gene expression signature to identify acute respiratory viral infection. *Sci Transl Med* **5:**203ra126.
79. **Zaas AK, Chen M, Varkey J, Veldman T, Hero AO III, Lucas J, Huang Y, Turner R, Gilbert A, Lambkin-Williams R, Øien NC, Nicholson B, Kingsmore S, Carin L, Woods CW, Ginsburg GS.** 2009. Gene expression signatures diagnose influenza and other symptomatic respiratory viral infections in humans. *Cell Host Microbe* **6:**207–217.
80. **Huang Y, Zaas AK, Rao A, Dobigeon N, Woolf PJ, Veldman T, Øien NC, McClain MT, Varkey JB, Nicholson B, Carin L, Kingsmore S, Woods CW, Ginsburg GS, Hero AO III.** 2011. Temporal dynamics of host molecular responses differentiate symptomatic and asymptomatic influenza a infection. *PLoS Genet* **7:**e1002234.
81. **Hu X, Yu J, Crosby SD, Storch GA.** 2013. Gene expression profiles in febrile children with defined viral and bacterial infection. *Proc Natl Acad Sci USA* **110:**12792–12797.
82. **Devignot S, Sapet C, Duong V, Bergon A, Rihet P, Ong S, Lorn PT, Chroeung N, Ngeav S, Tolou HJ, Buchy P, Couissinier-Paris P.** 2010. Genome-wide expression profiling deciphers host responses altered during dengue shock syndrome and reveals the role of innate immunity in severe dengue. *PLoS One* **5:**e11671.
83. **Herberg JA, Kaforou M, Gormley S, Sumner ER, Patel S, Jones KD, Paulus S, Fink C, Martinon-Torres F, Montana G, Wright VJ, Levin M.** 2013. Transcriptomic profiling in childhood H1N1/09 influenza reveals reduced expression of protein synthesis genes. *J Infect Dis* **208:**1664–1668.

84. **Parnell GP, McLean AS, Booth DR, Armstrong NJ, Nalos M, Huang SJ, Manak J, Tang W, Tam OY, Chan S, Tang BM.** 2012. A distinct influenza infection signature in the blood transcriptome of patients with severe community-acquired pneumonia. *Crit Care* **16:**1–12.
85. **Zaas AK, Garner BH, Tsalik EL, Burke T, Woods CW, Ginsburg GS.** 2014. The current epidemiology and clinical decisions surrounding acute respiratory infections. *Trends Mol Med* **20:**579–588.
86. **Zhu Z, Qi Y, Ge A, Zhu Y, Xu K, Ji H, Shi Z, Cui L, Zhou M.** 2014. Comprehensive characterization of serum microRNA profile in response to the emerging avian influenza A (H7N9) virus infection in humans. *Viruses* **6:**1525–1539.
87. **Hong X, Yu RB, Sun NX, Wang B, Xu YC, Wu GL.** 2005. Human leukocyte antigen class II DQB1*0301, DRB1*1101 alleles and spontaneous clearance of hepatitis C virus infection: a meta-analysis. *World J Gastroenterol* **11:**7302–7307.
88. **Yan ZH, Fan Y, Wang XH, Mao Q, Deng GH, Wang YM.** 2012. Relationship between HLA-DR gene polymorphisms and outcomes of hepatitis B viral infections: a meta-analysis. *World J Gastroenterol* **18:**3119–3128.
89. **Thompson AJ, Muir AJ, Sulkowski MS, Ge D, Fellay J, Shianna KV, Urban T, Afdhal NH, Jacobson IM, Esteban R, Poordad F, Lawitz EJ, McCone J, Shiffman ML, Galler GW, Lee WM, Reindollar R, King JW, Kwo PY, Ghalib RH, Freilich B, Nyberg LM, Zeuzem S, Poynard T, Vock DM, Pieper KS, Patel K, Tillmann HL, Noviello S, Koury K, Pedicone LD, Brass CA, Albrecht JK, Goldstein DB, McHutchison JG.** 2010. Interleukin-28B polymorphism improves viral kinetics and is the strongest pretreatment predictor of sustained virologic response in genotype 1 hepatitis C virus. *Gastroenterology* **139:**120–9.e18.
90. **Dill MT, Duong FH, Vogt JE, Bibert S, Bochud PY, Terracciano L, Papassotiropoulos A, Roth V, Heim MH.** 2011. Interferon-induced gene expression is a stronger predictor of treatment response than IL28B genotype in patients with hepatitis C. *Gastroenterology* **140:**1021–1031.e10.
91. **Duggal P, Thio CL, Wojcik GL, Goedert JJ, Mangia A, Latanich R, Kim AY, Lauer GM, Chung RT, Peters MG, Kirk GD, Mehta SH, Cox AL, Khakoo SI, Alric L, Cramp ME, Donfield SM, Edlin BR, Tobler LH, Busch MP, Alexander G, Rosen HR, Gao X, Abdel-Hamid M, Apps R, Carrington M, Thomas DL.** 2013. Genome-wide association study of spontaneous resolution of hepatitis C virus infection: data from multiple cohorts. *Ann Intern Med* **158:**235–245.
92. **Prokunina-Olsson L, Muchmore B, Tang W, Pfeiffer RM, Park H, Dickensheets H, Hergott D, Porter-Gill P, Mumy A, Kohaar I, Chen S, Brand N, Tarway M, Liu L, Sheikh F, Astemborski J, Bonkovsky HL, Edlin BR, Howell CD, Morgan TR, Thomas DL, Rehermann B, Donnelly RP, O'Brien TR.** 2013. A variant upstream of IFNL3 (IL28B) creating a new interferon gene IFNL4 is associated with impaired clearance of hepatitis C virus. *Nat Genet* **45:**164–171.
93. **Patin E, Kutalik Z, Guergnon J, Bibert S, Nalpas B, Jouanguy E, Munteanu M, Bousquet L, Argiro L, Halfon P, Boland A, Müllhaupt B, Semela D, Dufour JF, Heim MH, Moradpour D, Cerny A, Malinverni R, Hirsch H, Martinetti G, Suppiah V, Stewart G, Booth DR, George J, Casanova JL, Bréchot C, Rice CM, Talal AH, Jacobson IM, Bourlière M, Theodorou I, Poynard T, Negro F, Pol S, Bochud PY, Abel L, Swiss Hepatitis C Cohort Study Group, International Hepatitis C Genetics Consortium, French ANRS HC EP 26 Genoscan Study Group.** 2012. Genome-wide association study identifies variants associated with progression of liver fibrosis from HCV infection. *Gastroenterology* **143:**1244–1252.e12.
94. **Chatzidaki V, Kouroumalis E, Galanakis E.** 2011. Hepatitis B virus acquisition and pathogenesis in childhood: host genetic determinants. *J Pediatr Gastroenterol Nutr* **52:**3–8.
95. **Wu JF, Chen CH, Ni YH, Lin YT, Chen HL, Hsu HY, Chang MH.** 2012. Toll-like receptor and hepatitis B virus clearance in chronic infected patients: a long-term prospective cohort study in Taiwan. *J Infect Dis* **206:**662–668.
96. **Mackelprang RD, Bigham AW, Celum C, de Bruyn G, Beima-Sofie K, John-Stewart G, Ronald A, Mugo NR, Buckingham KJ, Bamshad MJ, Mullins JI, McElrath MJ, Lingappa JR.** 2014. Toll-like receptor polymorphism associations with HIV-1 outcomes among sub-Saharan Africans. *J Infect Dis* **209:**1623–1627.
97. **Oh DY, Taube S, Hamouda O, Kücherer C, Poggensee G, Jessen H, Eckert JK, Neumann K, Storek A, Pouliot M, Borgeat P, Oh N, Schreier E, Pruss A, Hattermann K, Schumann RR.** 2008. A functional toll-like receptor 8 variant is associated with HIV disease restriction. *J Infect Dis* **198:**701–709.
98. **Rotger M, Dang KK, Fellay J, Heinzen EL, Feng S, Descombes P, Shianna KV, Ge D, Günthard HF, Goldstein DB, Telenti A, Swiss HIV Cohort Study, Center for HIV/AIDS Vaccine Immunology.** 2010. Genome-wide mRNA expression correlates of viral control in CD4+ T-cells from HIV-1-infected individuals. *PLoS Pathog* **6:**e1000781.
99. **Wilson J, Rowlands K, Rockett K, Moore C, Lockhart E, Sharland M, Kwiatkowski D, Hull J.** 2005. Genetic variation at the IL10 gene locus is associated with severity of respiratory syncytial virus bronchiolitis. *J Infect Dis* **191:**1705–1709.
100. **Helminen M, Nuolivirta K, Virta M, Halkosalo A, Korppi M, Vesikari T, Hurme M.** 2008. IL-10 gene polymorphism at -1082 A/G is associated with severe rhinovirus bronchiolitis in infants. *Pediatr Pulmonol* **43:**391–395.
101. **Nuolivirta K, He Q, Vuononvirta J, Koponen P, Helminen M, Korppi M.** 2012. Toll-like receptor 3 L412F polymorphisms in infants with bronchiolitis and postbronchiolitis wheezing. *Pediatr Infect Dis J* **31:**920–923.
102. **Revai K, Patel JA, Grady JJ, Nair S, Matalon R, Chonmaitree T.** 2009. Association between cytokine gene polymorphisms and risk for upper respiratory tract infection and acute otitis media. *Clin Infect Dis* **49:**257–261.
103. **Herrera-Ramos E, López-Rodríguez M, Ruíz-Hernández JJ, Horcajada JP, Borderías L, Lerma E, Blanquer J, Pérez-González MC, García-Laorden MI, Florido Y, Mas-Bosch V, Montero M, Ferrer JM, Sorlí L, Vilaplana C, Rajas O, Briones M, Aspa J, López-Granados E, Solé-Violán J, de Castro FR, Rodríguez-Gallego C.** 2014. Surfactant protein A genetic variants associate with severe respiratory insufficiency in pandemic influenza A virus infection. *Crit Care* **18:**1–12.
104. **Oshansky CM, Gartland AJ, Wong SS, Jeevan T, Wang D, Roddam PL, Caniza MA, Hertz T, Devincenzo JP, Webby RJ, Thomas PG.** 2014. Mucosal immune responses predict clinical outcomes during influenza infection independently of age and viral load. *Am J Respir Crit Care Med* **189:**449–462.
105. **Wark PA, Bucchieri F, Johnston SL, Gibson PG, Hamilton L, Mimica J, Zummo G, Holgate ST, Attia J, Thakkinstian A, Davies DE.** 2007. IFN-gamma-induced protein 10 is a novel biomarker of rhinovirus-induced asthma exacerbations. *J Allergy Clin Immunol* **120:**586–593.
106. **Contoli M, Message SD, Laza-Stanca V, Edwards MR, Wark PA, Bartlett NW, Kebadze T, Mallia P, Stanciu LA, Parker HL, Slater L, Lewis-Antes A, Kon OM, Holgate ST, Davies DE, Kotenko SV, Papi A, Johnston SL.** 2006. Role of deficient type III interferon-lambda production in asthma exacerbations. *Nat Med* **12:**1023–1026.
107. **Wark PA, Johnston SL, Bucchieri F, Powell R, Puddicombe S, Laza-Stanca V, Holgate ST, Davies DE.** 2005. Asthmatic bronchial epithelial cells have a deficient innate immune response to infection with rhinovirus. *J Exp Med* **201:**937–947.
108. **Sykes A, Macintyre J, Edwards MR, Del Rosario A, Haas J, Gielen V, Kon OM, McHale M, Johnston SL.** 2014. Rhinovirus-induced interferon production is not deficient in well controlled asthma. *Thorax* **69:**240–246.
109. **Ghany MG, Nelson DR, Strader DB, Thomas DL, Seeff LB, American Association for Study of Liver Diseases.** 2011. An update on treatment of genotype 1 chronic hepatitis C virus infection: 2011 practice guideline by the American Association for the Study of Liver Diseases. *Hepatology* **54:**1433–1444.
110. **Mejias A, Suarez NM, Ramilo O.** 2014. Detecting specific infections in children through host responses: a paradigm shift. *Curr Opin Infect Dis* **27:**228–235.

APPENDIXES

Reference Virology Laboratories

Reference Virology Laboratory Testing Performed at the Centers for Disease Control

ROBERTA B. CAREY

APPENDIX 1

The virology laboratories at the Centers for Disease Control and Prevention (CDC) are situated organizationally across four infectious diseases centers. The laboratories are geographically located in Atlanta, Georgia; San Juan, Puerto Rico; Fort Collins, Colorado; and Anchorage, Alaska. A broad menu of tests are performed that are used for the diagnosis, confirmation, or typing of viruses that affect the population in the United States and globally.

The CDC Infectious Disease Laboratory Test Directory was created in 2013 and published on the CDC laboratory internet website to provide the most current information on laboratory services offered. The test directory lists all virology testing that can be requested by a clinical or public health laboratory. The directory provides the name of the subject matter experts, their phone number and email, whether prior approval is required before submitting the sample, and the sample requirements (sample type, volume, transport medium, labeling). The test description includes the shipping instructions, the methodology used to perform the test, expected turnaround time, and test limitations. CDC subject matter experts are available and encourage a consultation before laboratories submit their specimens. Table 1 details the current virology test orders, the CDC order number, and the method. For the latest list of test orders and contact information, refer to the current CDC Test Directory version online: http://www.cdc.gov/laboratory/specimen-submission/list.html.

TABLE 1 List of CDC Virology Test Orders

CDC test order name	CDC test code no.	Methodology[a]
Adenovirus Molecular Detection and Typing	10170	PCR, sequencing
Alkhurma Identification	10274	PCR, molecular typing
Alkhurma Serology	10285	ELISA
Arbovirus Isolation and Identification	10281	Cell culture
Arbovirus Molecular Detection	10280	RT-PCR
Arbovirus Neutralization Antibody	10283	Plaque reduction neutralization
Arbovirus Serology	10282	ELISA, MIA
Arenavirus (New World) Serology	10484	ELISA
Arenavirus (New World) Identification	10293	PCR, molecular typing
Arenavirus (Old World) Identification	10294	PCR, molecular typing
Congo-Crimean Hemorrhagic Fever Identification	10302	PCR, molecular typing
Congo-Crimean Hemorrhagic Fever Serology	10303	ELISA
Dengue Virus Diagnosis	10307	IgM ELISA, IgG seroconversion by ELISA, NS1 antigen, RT-PCR, viral isolation
Ebola Identification	10309	PCR, molecular typing
Ebola Serology	10310	ELISA
Enterovirus Detection and Identification	10312	Molecular methods
Epstein-Barr Virus Detection	10265	PCR
Hantavirus (North American) Identification	10319	PCR, molecular typing

(Continued)

doi:10.1128/9781555819156.app1

TABLE 1 List of CDC Virology Test Orders (*Continued*)

CDC test order name	CDC test code no.	Methodology[a]
Hantavirus (South American) Identification	10320	PCR, molecular typing
Hantavirus Serology	10321	ELISA
Hendra Serology	10324	ELISA
Hepatitis A Serology, NAT and Genotyping	10325	Antibody by chemiluminescence, HAV RNA, genotyping by NAT P2B sequencing
Hepatitis B Serology, NAT and Genotyping	10326	Antibody by chemiluminescence, HBV DNA, genotyping by NAT S gene sequencing
Hepatitis B Surface Antigen Confirmatory Test	10451	Neutralization
Hepatitis C Serology, NAT and Genotyping	10327	Antibody by chemiluminescence, HCV RNA, genotyping by NAT NS5B sequencing
Hepatitis D Serology, NAT and Genotyping	10328	Antibody by EIA, HDV RNA by real-time qRT-PCR, genotyping by sequencing
Hepatitis E Serology, NAT and Genotyping	10329	IgM and IgG by EIA, HEV RNA by real-time qRT-PCR, genotyping by sequencing
Herpes Simplex Virus 1 Detection	10258	PCR
Herpes Simplex Virus 1 Serology	10259	IgG antibody by EIA
Herpes Simplex Virus 2 Detection	10260	PCR
Herpes Simplex Virus 2 Serology	10261	IgG antibody by EIA
Herpesvirus Encephalitis Panel	10262	PCR
HIV Antigen/Antibody Combo	10485	EIA
HIV Monitoring (CD4)	10277	Fluorescence activated cell sorting
HIV-1/2 Laboratory Algorithm	10272	ELISA
HIV-1 Nucleic Acid Amplification (Qualitative)	10275	NAAT
HIV-1 Nucleic Acid Amplification (Viral Load)	10276	RT-PCR
HIV-2 Nucleic Acid Amplification (Qualitative)	10429	PCR
HIV-2 Serology	10273	HIV-1/2 differentiation assay, Western blot
Human Herpes Virus Detection and Subtyping	10266	PCR
Human Herpes Virus 7 Detection	10267	PCR
Human Herpes Virus 8 Detection	10268	PCR
Human Herpes Virus 8 Serology	10269	IgG antibody by IFA
Influenza Antiviral Resistance Diagnosis	10423	Pyrosequencing
Influenza Molecular Diagnosis	10421	Real-time PCR, sequence identification
Influenza Serology	10424	Hemagglutination inhibition, micro-neutralization
Influenza Surveillance	10422	Virus culture, hemagglutination inhibition
Junin Serology	10340	ELISA
Kyasanur Forest Disease Serology	10341	ELISA
Laguna Negra Serology	10342	ELISA
Lassa Fever Identification	10343	PCR, molecular typing
Lassa Fever Serology	10344	ELISA
Lymphocytic Choriomeningitis Identification	10345	PCR, molecular typing
Lymphocytic Choriomeningitis Serology	10346	ELISA
Machupo Identification	10347	PCR, molecular typing
Machupo Serology	10348	ELISA
Marburg Identification	10349	PCR, molecular typing
Marburg Serology	10350	ELISA
Measles and Rubella Detection and Genotyping	10243	Viral culture, real-time RT-PCR, sequencing
Measles and Rubella Serology	10247	Capture IgM, indirect IgG
Measles Avidity	10248	Avidity
Measles Detection and Genotyping	10240	Viral culture, real time RT-PCR, sequencing
Measles Neutralization Antibody (not for immune status)	10250	Neutralization-quantitative
Measles Serology	10244	CDC Capture IgM, indirect IgG

(*Continued*)

TABLE 1 List of CDC Virology Test Orders (*Continued*)

CDC test order name	CDC test code no.	Methodology[a]
MERS-CoV PCR	10488	PCR, sequencing
MERS-CoV Serology	10489	ELISA
Mumps Detection and Genotyping	10241	Viral culture, real-time RT-PCR, sequencing
Mumps Neutralization Antibody (not for immune status)	10351	Neutralization-quantitative
Mumps Serology	10245	CDC Capture IgM, indirect IgG
Nipah Virus Identification	10354	PCR, molecular typing
Nipah Virus Serology	10355	ELISA
Norovirus Genotyping	10356	PCR, sequencing
Norovirus Molecular Detection	10357	PCR
Norovirus Molecular Detection and Genotyping	10358	PCR, sequencing
Parechovirus Detection and Identification	10362	Molecular methods
Parvovirus B19 Molecular Detection	10363	PCR
Parvovirus B19 Serology	10364	IgG and IgM EIA
Picornavirus Detection and Identification (not hepatitis A or rhinovirus)	10374	Molecular methods
Polio Isolations, Intratypic Differentiation, Genotyping	10376	Cell culture, molecular methods
Polio Serology	10377	Neutralization
Poxvirus—Cowpox Specific Molecular Detection	10379	Real-time PCR
Poxvirus—Encephalitis Work-up (post-vaccinia, monkeypox)	10380	ELISA, real-time PCR
Poxvirus—Molluscum Contagiosum Specific Molecular Detection	10381	Real-time PCR
Poxvirus—Monkeypox Specific Molecular Detection	10382	Real-time PCR
Poxvirus—Orthopoxvirus Serology	10384	ELISA
Poxvirus—Pan-Poxvirus Molecular Detection (human infections)	10385	PCR
Poxvirus—Parapoxvirus Generic Molecular Detection	10383	Real-time PCR
Poxvirus—Parapoxvirus Molecular Detection	10386	Real-time PCR
Poxvirus—Sealpox Specific Molecular Detection	10387	Real-time PCR
Poxvirus—Smallpox (Variola Virus) Specific Molecular Detection	10388	Real-time PCR
Poxvirus—Tanapox Specific Molecular Detection	10389	Real-time PCR
Poxvirus—Vaccinia Specific Molecular Detection	10390	Real-time PCR
Puumala Serology	10391	ELISA
Rabies Antemortem Human Testing	10392	IgG and IgM by IFA, viral neutralizing antibodies by RFFIT, DFA for skin biopsy, RT-PCR, sequencing
Rabies Antibody—Pre/Post Exposure Prophylaxis	10393	Viral neutralizing antibodies by RFFIT
Rabies Confirmatory Testing (animal)	10394	DFA, direct rapid immunohistochemistry, RT-PCR, viral isolation antigenic typing, sequence analysis
Rabies Confirmatory Testing (human)	10395	IgG by IFA, DFA for skin biopsy, RT-PCR, sequence analysis, viral neutralizing antibodies by RFFIT, direct rapid immunohistochemistry, antigenic typing
Rabies Postmortem Testing (human)	10396	DFA, RT-PCR, direct rapid immunohistochemistry, viral isolation, sequence analysis, antigenic typing
Rabies Virus Genetic Typing	10397	RT-PCR, sequence analysis, viral isolation
Rabies Virus Typing—Central Nervous System Tissues	10398	DFA, IFA, isolation, sequencing analysis
Respiratory Virus Molecular Detection (not influenza)	10401	PCR
Rift Valley Fever Identification	10406	PCR, molecular typing

(*Continued*)

TABLE 1 List of CDC Virology Test Orders (*Continued*)

CDC test order name	CDC test code no.	Methodology[a]
Rift Valley Fever Serology	10407	ELISA
Rotavirus Antigen Detection	10408	EIA
Rotavirus Genotyping	10409	RT-PCR, sequencing
Rotavirus Molecular Detection and Genotyping	10410	Real time RT-PCR, RT-PCR, sequencing
Rubella Detection and Genotyping	10242	RT-PCR, real-time PCR, viral culture, sequencing
Rubella Serology	10246	Capture IgM and indirect IgG
Rubella Serology and Avidity	10249	CDC IgG avidity assay
SARS Molecular Detection	10412	PCR, sequencing
SARS Serology	10413	ELISA
Seoul Virus Serology	10414	ELISA
Tick Borne Encephalitis Identification	10415	PCR, molecular typing
Tick Borne Encephalitis Serology	10416	ELISA
Varicella Zoster Virus Avidity	10256	IgG avidity
Varicella Zoster Virus Detection	10254	PCR
Varicella Zoster Virus Genotyping	10257	PCR, sequencing
Varicella Zoster Virus Serology	10255	IgG and IgM by EIA

[a]DFA, direct immunofluorescent antibody; EIA, enzyme immunoassay; ELISA, enzyme immunosorbent assay; HAV, hepatitis A virus; HBV, hepatitis B virus; HCV, hepatitis C virus; HDV, hepatitis D virus; HEV, hepatitis E virus; IFA, indirect immunofluorescent antibody; MIA, microsphere immunoassay; NAAT, nucleic acid amplification test; NAT, nucleic acid testing; PCR, polymerase chain reaction; RT-PCR, reverse transcription polymerase chain reaction; RFFIT, rapid fluorescent focus inhibition test.

Public Health Laboratory Virology Services

JANE GETCHELL

APPENDIX 2

ROLE OF THE PUBLIC HEALTH LABORATORY

The public health laboratory (PHL) provides analytical data in support of clinicians, epidemiologists, and public health practitioners as an aid in the diagnosis and tracking of disease and in instituting and guiding control measures. PHLs are part of a system that includes those who order testing and those who use the test results, whether for the care of individual patients or for disease surveillance and control. This system includes health care and public health professionals at the local, state, and federal levels. Other members of the system are veterinary, agricultural, food safety, university, law enforcement, and military laboratories. With the 2014 Ebola outbreak in West Africa, Middle East respiratory syndrome coronavirus (MERS-CoV), ongoing transmission of influenza H5N1, and expanding ranges of Chikungunya (CHIK) and dengue (DEN) viruses, that system now extends worldwide.

The role of the state and local PHL is to ensure that essential state-of-the-art laboratory services are available. States and their local PHL partners may provide these services in house, or they may arrange for them to be provided by another laboratory. Much of the work performed in the PHL is done in support of various Centers for Disease Control and Prevention (CDC) programs in order to type, subtype, or characterize specific organisms of public health significance. As the state PHL often serves as a reference laboratory for clinical laboratories, CDC serves as the reference laboratory for PHLs. CDC, in collaboration with the Association of Public Health Laboratories (APHL), provides training, guidance on methods, and for some organisms, reagents and proficiency testing to states. PHLs in turn, provide training to clinical and other laboratories within their state or region.

To fulfill their role of ensuring that essential laboratory services are available, PHLs have worked together with CDC and APHL to establish "Public Health Reference Centers" for testing and characterization of certain organisms. For testing that is infrequently requested in a given state, the state or local PHL can refer specimens to these centers. Besides being more practical and economical, PHLs may choose to refer specimens to a reference center because they use the latest molecular technologies and equipment, which may not be available in every state.

TESTING PERFORMED AND SCOPE OF SERVICES

State and local PHLs continuously assess the organisms and infections for which they test, the methods they use, and the equipment they have to ensure that the services they provide are relevant, high quality, and cost effective. As a result, any comprehensive listing of the services provided by PHLs would quickly be out of date. Each state or local agency determines which tests to provide based on state and federal regulations and guidelines and on the needs of the population it serves. Consequently the test menus of PHLs vary widely and change over the course of a year. A listing of state PHLs and their addresses and websites is included at the end of this appendix (Table 1). The state PHL web site will have the latest information on tests performed and specimen submission instructions. In addition to state PHLs, local government agencies (counties and cities) may operate a health department and offer laboratory services. General information on PHL testing for specific viral infections is provided in the following sections.

Arboviruses

Arbovirus surveillance has been a mainstay of PHLs for many years, some testing human specimens only, but many testing mosquito pools, sentinel chickens, dead birds, and other animals. Arboviruses most commonly tested for are West Nile virus (WNV), St Louis encephalitis virus (SLEV), and Eastern equine encephalitis virus (EEEV), with some states also testing for LaCrosse virus and the California encephalitis virus group. Commercial assays are available for screening for WNV IgM antibody, and CDC has developed a WNV IgM/IgG antibody assay that is commonly used by PHLs. A 2014 Council of State and Territorial Epidemiologists (CSTE) assessment found that state and local health department capacity for WNV and other arbovirus surveillance and control have decreased substantially, and that some health departments had lost all mosquito monitoring capability and laboratory capacity to test for emerging arboviruses (1). All states belong to ArboNET, the national surveillance system developed by CDC and State Health Departments in 2000 in response to the emergence of WNV in 1999. Through this system all human arbovirus disease data are reported to CDC.

doi:10.1128/9781555819156.app2

TABLE 1

State and Territorial Public Health Laboratories (Reprinted with permission of the publisher from http://www.aphl.org/AboutAPHL/memberlabs/Pages/default.aspx)
Alaska Division of Public Health Laboratory 5455 Dr Martin Luther King Jr Avenue Anchorage, AK 99507 907.334.2100 907.334.2161 http://dhss.alaska.gov/Pages/default.aspx
Alabama Bureau of Clinical Laboratories 8140 AUM Drive Montgomery, AL 36117 334.260.3400 334.274.9800 http://www.adph.org/bcl
Arkansas Public Health Laboratory 201 South Monroe Street Little Rock, AR 72205 501.661.2220 501.661.2310 http://www.healthy.arkansas.gov/programsServices/healthlab/Pages/Services.aspx
Arizona Bureau of State Laboratory Services 250 North 17th Avenue Phoenix, AZ 85007 602.542.1188 602.542.0760 http://www.azdhs.gov/lab/index.htm
California Department of Public Health Laboratory 850 Marina Bay Parkway Richmond, CA 94804-6403 510.412.5846 510.412.5848 http://www.cdph.ca.gov/programs/lfs/Pages/default.aspx
Colorado Department of Public Health & Environment 8100 Lowry Boulevard Denver, CO 80230 303.692.3090 303.344.9989 http://www.colorado.gov/cs/Satellite/CDPHE-Lab/CBON/1251583470522
Dr. Katherine A. Kelly Public Health Laboratory, Connecticut Department of Public Health 395 West Street Rocky Hill, CT 06067 860.509.8500 860.920.6710 http://www.ct.gov/dph/site/default.asp

(Continued)

TABLE 1 *(Continued)*

State and Territorial Public Health Laboratories (Reprinted with permission of the publisher from http://www.aphl.org/AboutAPHL/memberlabs/Pages/default.aspx)
District of Columbia Public Health Laboratory 401 E Street SW Washington, DC 20024 202.727.8956 202.724.3927 http://dfs.dc.gov/page/public-health-laboratory-division-phl
Delaware Public Health Laboratory 30 Sunnyside Road Smyrna, DE 19977 302.653.2870 302.653.2877 http://www.dhss.delaware.gov/dph/lab/labs.html
Florida Bureau of Public Health Laboratories 1217 Pearl Street Jacksonville, FL 32202 904.791.1500 904.791.1567 http://www.floridahealth.gov/programs-and-services/public-health-laboratories/index.html
Georgia Public Health Laboratory 1749 Clairmont Road Decatur, GA 30032 404.327.7900 404.327.7919 http://dph.georgia.gov/lab
Guam Department of Public Health & Social Services 123 Chalan Kareta Mangilao, GU 96913-6304 671.735.7305 671.734.2103 http://dphss.guam.gov
Hawaii State Laboratories Division 2725 Waimano Home Road Pearl City, HI 96782 808.453.6652 808.453.6662 http://health.hawaii.gov/statelab
State Hygienic Laboratory at the University of Iowa 2490 Crosspark Road Coralville, IA 52241 319.335.4500 319.335.4555 http://www.shl.uiowa.edu

(Continued)

TABLE 1 *(Continued)*

State and Territorial Public Health Laboratories (Reprinted with permission of the publisher from http://www.aphl.org/AboutAPHL/memberlabs/Pages/default.aspx)
Idaho Bureau of Laboratories 2220 Old Penitentiary Rd Boise, ID 83712 208.334.2235 208.334.4067 http://www.healthandwelfare.idaho.gov/Health/Labs/tabid/99/Default.aspx
Illinois Department of Public Health Laboratory 535 West Jefferson Street Springfield, IL 62702 217.782.6562 217.524.7924 http://www.idph.state.il.us/about/laboratories/index.htm
Indiana Public Health Laboratory 550 West 16th Street Indianapolis, IN 46202 317.921.5500 317.927.7801 http://www.in.gov/isdh/22421.htm
Kansas Health & Environmental Laboratories 6700 SW Topeka Boulevard Topeka, KS 66620-0001 785.296.1500 785.296.1641 http://www.kdheks.gov/labs
Kentucky Division of Laboratory Service 100 Sower Blvd, Suite 204 Frankfort, KY 40601 502.564.4446 502.564.7019 http://chfs.ky.gov
Louisiana Public Health Laboratory 3101 West Napoleon Avenue Metairie, LA 70001 504.219.4664 504.219.4670 http://new.dhh.louisiana.gov/index.cfm/page/483
William A. Hinton State Laboratory Institute 305 South Street Jamaica Plain, MA 2130 617.983.6200 617.983.6210 http://www.mass.gov/eohhs/gov/departments/dph/programs/state-lab

(Continued)

TABLE 1 *(Continued)*

State and Territorial Public Health Laboratories (Reprinted with permission of the publisher from http://www.aphl.org/AboutAPHL/memberlabs/Pages/default.aspx)
Maryland Laboratories Administration Baltimore Science & Technology Park at Johns Hopkins 1770 Ashland Ave Baltimore, MD 21205 410.767.6500 410.333.5403 http://dhmh.maryland.gov/laboratories/SitePages/Home.aspx
Maine State Health & Environmental Testing Laboratory 221 State Street Augusta, ME 4333 207.287.2727 207.287.6832 http://www.maine.gov/dhhs/mecdc/public-health-systems/health-and-environmental-testing
Michigan Public Health Laboratory 3350 N Martin Luther King Jr. Blvd Lansing, MI 48909 517.335.8063 517.335.8051 http://www.michigan.gov/mdch/1,1607,7-132-2945_5103—,00.html
Minnesota Public Health Laboratory Division 601 North Robert Street Saint Paul, MN 55164 651.201.5200 651.201.5064 http://www.health.state.mn.us/divs/phl/index.html
Missouri State Public Health Laboratory 101 North Chestnut Street Jefferson City, MO 65101 573.751.3334 573.526.2754 http://health.mo.gov/lab
Mississippi Public Health Laboratory Training Office 570 East Woodrow Wilson Avenue Jackson, MS 39216 601.576.7582 601.576.7037 http://www.msdh.state.ms.us/msdhsite/_static/14,0,188.html
Montana Laboratory Services Bureau 1400 Broadway Helena, MT 59601 406.444.3444 406.444.1802 http://www.dphhs.mt.gov/publichealth/lab/index.shtml

(Continued)

TABLE 1 *(Continued)*

State and Territorial Public Health Laboratories (Reprinted with permission of the publisher from http://www.aphl.org/AboutAPHL/memberlabs/Pages/default.aspx)
North Carolina State Laboratory of Public Health 4312 District Drive Raleigh, NC 27601 919.733.7834 919.715.9243 http://slph.state.nc.us
North Dakota Division of Laboratory Services 2635 East Main Avenue Bismarck, ND 58502 701.328.6280 701.328.6280 http://www.ndhealth.gov/microlab/labservices.aspx
Nebraska Public Health Laboratory 984080 Nebraska Medical Center Omaha, NE 68198-5900 402.559.2440 402.559.7799 http://www.nphl.org
New Hampshire Public Health Laboratories 29 Hazen Drive Concord, NH 03301 603.271.4661 603.271.4783 http://www.dhhs.nh.gov/dphs/lab/index.htm
New Jersey Division of Public Health & Environmental Laboratories 3 Schwarzkopf Dr. Ewing, NJ 08628 609.984.2201 609.633.9601 http://www.state.nj.us/health/phel/index.shtml
New Mexico Department of Health 1101 Camino de Salud Albuquerque NM 87102 505.383.9000 505.383.9011 http://www.health.state.nm.us
Nevada State Public Health Laboratory–UNV 1660 North Virginia Street Reno, NV 89503-1738 775.688.1335 775.688.1460 http://dhhs.nv.gov

(Continued)

TABLE 1 *(Continued)*

State and Territorial Public Health Laboratories (Reprinted with permission of the publisher from http://www.aphl.org/AboutAPHL/memberlabs/Pages/default.aspx)
Wadsworth Center, New York State Department of Health 120 New Scotland Avenue Albany, NY 12201-0509 518.474.7592 518.474.3439 http://www.wadsworth.org
Ohio Department of Health Laboratories 8995 East Main Street, Building 22 Reynoldsburg, OH 43068 888.634.5227 614.644.4591 http://www.odh.ohio.gov/odhprograms/phl/lab/lab1.aspx
Oklahoma Public Health Laboratory 1000 NE 10th Oklahoma City, OK 73117-1207 405.271.5070 405.271.4850 http://www.ok.gov/health/Disease,_Prevention,_Preparedness/Public_Health_Laboratory
Oregon State Public Health Laboratory 3150 NW 229th Ave. Hillsboro, OR 97124-6536 503.693.4100 503.693.5602 https://public.health.oregon.gov/PHD/Directory/Pages/program.aspx?pid=6
Pennsylvania Bureau of Laboratories 110 Pickering Way Exton, PA 19341-1310 610.280.3464 610.594.9972 http://www.portal.state.pa.us/portal/server.pt/community/laboratories/14158
Puerto Rico Public Health Laboratory Department of Health Commonwealth of PR, Institute of Health Laboratories San Juan, PR 00936-8184 787.765.2929 787.274.5710 http://www.salud.gov.pr/Pages/default.aspx
Rhode Island State Health Laboratories 50 Orms Street Providence, RI 02904 401.222.5600 401.222.6985 http://www.health.ri.gov/programs/laboratory/index.php

(Continued)

TABLE 1 (*Continued*)

State and Territorial Public Health Laboratories (Reprinted with permission of the publisher from http://www.aphl.org/AboutAPHL/memberlabs/Pages/default.aspx)
South Carolina Bureau of Laboratories 8231 Parklane Road Columbia, SC 29223 803.896.0800 803.896.0983 http://www.scdhec.gov/health/lab
South Dakota Public Health Laboratory 615 East 4th Street Pierre, SD 57501 605.773.3368 605.773.6129 http://doh.sd.gov/Lab
Tennessee Department of Health: Laboratory Services 630 Hart Lane Nashville, TN 37243 615.262.6300 615.262.6393 http://health.state.tn.us/Lab/index.htm
Texas Department of State Health Services 1100 West 49th Street Austin, TX 78756 512.776.7318 512.458.7294 http://www.dshs.state.tx.us/lab/default.shtm
Unified Utah State Laboratories: Public Health 4431 South 2700 West Taylorsville, UT 84129-8600 801.965.2400 801.969.3704 http://health.utah.gov/els
Virginia Division of Consolidated Laboratory Services 600 North 5th Street Richmond, VA 23219 804.648.4480 804.371.7973 http://www.dgs.state.va.us/DivisionofConsolidatedLaboratoryServices/tabid/453/Default.aspx
Vermont Department of Health Laboratory 195 Colchester Avenue Burlington, VT 05402-1125 802.863.7335 802.863.7632 http://www.healthvermont.gov/enviro/ph_lab/PublicHealthLaboratory.aspx

(*Continued*)

TABLE 1 (*Continued*)

State and Territorial Public Health Laboratories (Reprinted with permission of the publisher from http://www.aphl.org/AboutAPHL/memberlabs/Pages/default.aspx)
Washington Public Health Laboratories 1610 NE 150th Street Shoreline, WA 98155 206.418.5400 206.418.5445 http://www.doh.wa.gov/PublicHealthandHealthcareProviders/PublicHealthLaboratories.aspx
Wisconsin State Laboratory of Hygiene 465 Henry Mall Madison, WI 53706 608.262.6386 608.262.3257 http://www.slh.wisc.edu
West Virginia Department of Health & Human Services 167 11th Avenue South Charleston, WV 25303 304.558.3530 304.558.6213 http://www.wvdhhr.org/labservices/index.cfm
Wyoming Public Health Laboratory 208 S. College Drive Cheyenne, WY 82002 307.777.7431 307.777.6422 http://health.wyo.gov/phsd/lab/index.html

Hepatitis B and C Viruses

The majority of states perform some type of testing for viral hepatitis. In the past, that has been primarily antibody testing. Since the publication of the recommended test algorithm for HCV in 2013 (2), as well as the CDC recommendation to test members of the cohort born between 1945 and 1965 (3) states are gradually adopting nucleic acid amplification testing (NAAT) to detect active HCV infection. A limited number of states also perform HCV genotyping.

HIV

Several rapid HIV antibody tests have been approved by the U.S. Food and Drug Administration (FDA) and are waived under the Clinical Laboratory Improvements of 1988 (CLIA). These sensitive and specific tests are often performed on finger stick blood at the point of care rather than in a laboratory setting. PHLs, as well as commercial reference laboratories, perform a second level of testing on serum or plasma to confirm the presence of antibody, to improve detection of HIV-2, and to detect early infection. Following guidance developed by CDC and APHL (4), most PHLs have implemented the recommended algorithm using a fourth-generation FDA-approved immunoassay that detects HIV-1 and HIV-2 antibody as well as HIV antigen. This is followed up using an FDA-approved supplemental test with the ability (as compared to the Western blot) to differentiate HIV-1 and HIV-2. In cases in which the antigen/antibody test and the supplemental test give discrepant results, a NAAT test is performed either in house or at one of the public health reference centers for NAAT testing that have been established for PHLs.

Human Papillomavirus (HPV)

Several FDA-approved tests are available to detect and type HPV nucleic acid, and a few PHLs have implemented these tests. The specific types detected vary by manufacturer. Methods are available to perform molecular testing on the same specimen that is submitted for chlamydia and gonorrhea testing, tests that are offered by the majority of PHLs. The U.S. Preventive Services Task Force published recommendations on the use of molecular tests in cervical cancer screening and prevention in 2012 (5). One state and one local PHL perform Pap smears.

Influenza Virus

Approximately 85 state and local PHLs perform influenza virus subtyping as part of the surveillance system established through the World Health Organization (WHO) in the 1940s. Each year, beginning in early fall, public health departments solicit specimens from patients presenting with influenza-like illnesses from sentinel health care providers, hospitals, and outpatient clinics. Specimens that are submitted through this program are tested using molecular methods and reagents provided by CDC to monitor the prevalence of circulating virus types and subtypes and provide other critical surveillance information for the state and the nation. Many states are also testing these specimens using FDA-cleared molecular respiratory virus panels for detection of additional respiratory viruses. These panels can also be used for influenza typing and subtyping as long as the manufacturer of the assay monitors and updates assay performance as circulating viruses change over time and the assay that is being utilized within the laboratory has been optimized to identify all currently circulating influenza viruses (6). As influenza virus prevalence and subtypes change throughout the year, PHLs may change their testing algorithm to optimize efficiency and throughput.

Additional testing methods used in PHLs include influenza virus culture, antiviral resistance testing by pyrosequencing, and influenza hemagglutination inhibition for subtyping of influenza A and B. It is important that at least some State PHLs as well as CDC maintain the ability to perform virus culture to provide isolates for validation and verification of new assays, antiviral resistance testing, antigenic characterization of the circulating viruses, and a backup method to PCR and to detect other respiratory viruses. The PHL is also responsible for referring representative specimens (and viruses if culture is performed) to CDC or a CDC-designated PHL for genetic and antigenic characterization throughout the year. In addition, PHLs should notify CDC immediately of specimens identified as unsubtypable using all available targets (influenza A, B, H1, H3, H1pdm 2009, etc.) and immediately refer these specimens to CDC (6). PHLs rarely perform rapid influenza diagnostic tests; however, they routinely monitor the performance of these assays and regularly report their findings.

Other Respiratory Viruses

PHLs have traditionally detected noninfluenza respiratory viruses using viral culture. However, these culture-based microbiological methods are increasingly being replaced by new FDA-cleared multiplex molecular methods available from multiple manufacturers. These newer test methodologies and platforms offer more rapid, and potentially more sensitive, diagnosis across a spectrum of respiratory pathogens and specimen types. They also provide more comprehensive respiratory pathogen surveillance and outbreak response. The advantages for patients and hospitals include more rapid triage, earlier appropriate treatment, and shorter periods of illness and hospitalization. However, molecular technologies can bypass culture and isolation of organisms, with no isolate available for characterization, susceptibility testing, or molecular epidemiology. In 2014 APHL published the results of a survey of PHLs that was conducted to determine what noninfluenza respiratory viruses were being detected and what methods were used for detection, including PCR, antigen testing, and virus culture (7). The results are shown in Figure 1.

Norovirus and Other Enteric Viruses

Molecular detection and typing of norovirus is conducted by many states in support of epidemiological investigations. This testing usually requires prior consultation with both epidemiology and the laboratory. RT-qPCR assays are the preferred laboratory method for detecting norovirus. These assays are very sensitive and can detect as few as 10 to 100 norovirus copies per reaction. They use different primers to differentiate genogroup I and genogroup II norovirus. RT-qPCR assays are also quantitative and can provide estimates of viral load. The assays may be used to detect norovirus in stool, vomitus, foods, water, and environmental specimens.

Conventional RT-PCR followed by sequence analysis of the RT-PCR products is used for norovirus genotyping. Typically, a partial region of the capsid gene, such as region D, is sequenced by laboratories participating in CaliciNet, a national laboratory surveillance network for norovirus outbreaks coordinated by CDC (http://www.cdc.gov/norovirus/reporting/caliciNet/index.html). CaliciNet, a network of federal, state, and local PHLs, was launched in 2009 to collect information on norovirus strains associated with gastroenteritis outbreaks in the United States. PHLs electronically submit laboratory data, including genetic sequences of norovirus strains, and epidemiology data from norovirus outbreaks to the CaliciNet database. The norovirus strains can be compared with other strains in the database, helping CDC link outbreaks to a common source, monitor norovirus strains that are circulating, and identify newly emerging norovirus strains. For outbreak samples that test negative for norovirus, PHLs can refer samples to one of several undiagnosed viral diarrhea reference centers that test samples for sapovirus, astrovirus, and rotavirus.

An increasing number of clinical laboratories and PHLs are using commercially available, FDA-cleared molecular gastrointestinal panels that test for a variety of enteric pathogens, including some viruses (e.g., Norovirus GI/II, adenovirus 40/41, and rotavirus). While these assays provide a sensitive method for rapid identification of the causative agent for patient management, they do not provide an isolate for public health surveillance or characterization of the organism for disease tracking and control. Public health is quickly trying to develop molecular methods for organism characterization using original specimen material or extracts, but until these methods are available and validated, clinical laboratories using molecular GI panels are urged to work with their public health partners to assure that vital surveillance data are not lost and that laborato-

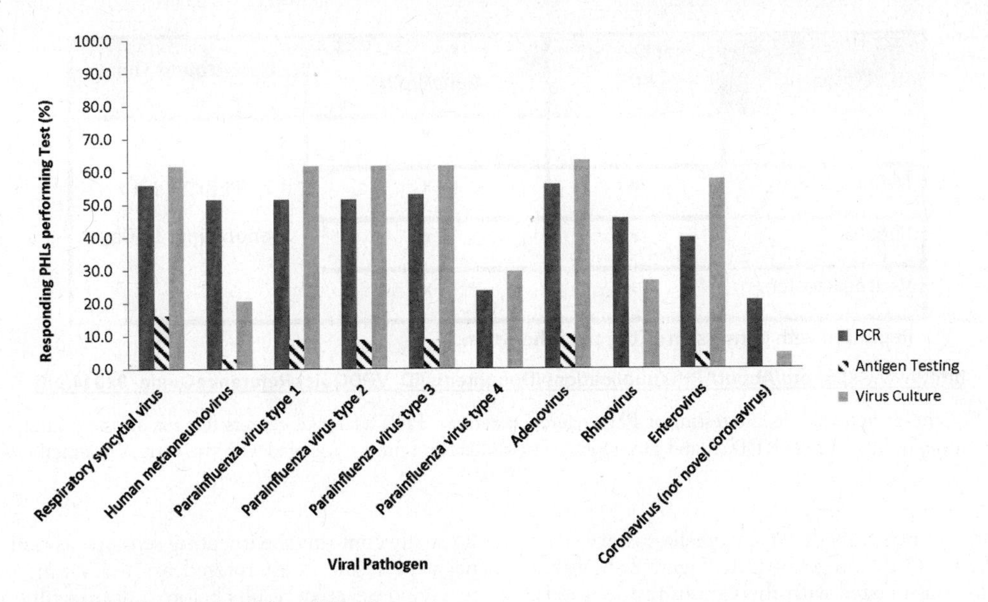

FIGURE 1 Noninfluenza respiratory virus testing at public health laboratories (PHLs). In 2014 the Association of Public Health Laboratories (APHL) published the results of a survey of PHLs that was conducted to determine what noninfluenza respiratory viruses were being detected and what methods were used for their detection, including PCR, antigen testing, and virus culture. The 10 viruses surveyed are shown on the *x* axis. PCR is indicated by the black bars, antigen testing by the checkered bars, and virus culture by the gray bars. The percentage of responding PHLs using the different test methods to detect each of the different viruses is shown on the y axis.

ries are in compliance with public health organism submission requirements.

Another major effect of culture independent diagnostic testing (CIDT) such as multiplex PCR is that such systems likely only test for what are thought to be the most common causes of acute gastroenteritis. CIDT could eliminate the ability to identify recognizable enteric pathogens because they are prevalent only in limited geographic locations. Furthermore, unlike conventional culture techniques, it offers no ability to detect new causes of diarrheal disease (8). However, next-generation CIDT methods such as deep or whole genome sequencing have been used to detect and identify new pathogens (9).

Poxviruses and Other Viruses Causing Rash Illnesses

The Laboratory Response Network (LRN), the nation's premier system for identifying, testing, and characterizing potential agents of biological and chemical terrorism, has developed protocols to be used in different situations involving patients with acute, generalized vesicular or pustular rash illnesses. Through the Public Health Emergency Preparedness grant, state and some large city PHLs are funded to perform molecular testing for nonvariola orthopox and orthopox viruses as well as other rash-causing viruses. However, not all states can test for variola. In these states, specimens from highly suspect cases of variola are sent to CDC. Specimens from all high-risk cases require consultation with CDC. Vaccinia can be readily grown in routine cell culture and definitively identified by referral to a specialty laboratory in the LRN and/or CDC.

Rabies Virus

All states test for rabies virus in animal brain material. Testing is usually conducted by the state PHL, but a few states have made alternate arrangements for testing to be done. Rabies is diagnosed using the direct fluorescent antibody (DFA) test, which looks for the presence of rabies virus antigens in brain tissue. While DFA remains the gold standard for rabies testing, molecular methods are becoming more common in aiding in the detection of rabies in animal brain material, either to confirm the results of the DFA or as the initial screening test.

State health departments should be the primary contact for physicians during consultation about possible human rabies cases. Rapid and accurate laboratory diagnosis of rabies in humans and other animals is essential for timely administration of postexposure prophylaxis. Within a few hours, a PHL can determine whether or not an animal is rabid and inform the responsible medical personnel. The laboratory results may save a patient from unnecessary physical and psychological trauma and financial burdens if the animal is not rabid.

Vaccine-Preventable Diseases (VPDs)

With the exception of rotavirus, testing for viral VPDs is performed at many PHLs. Those PHLs that have not established molecular capabilities for VPDs may use one of the PHL reference centers established for this purpose. State and local PHLs submit specimens from patients identified in conjunction with epidemiologists as a suspected, probable, or confirmed case to the designated reference laboratory. Eligible specimens may include those submitted for primary diagnostic purposes and/or for further case classification/characterization. Submitting PHLs provide specific information regarding specimen types, specimen volumes, storage/handling, test requisitioning, and shipping instructions. Other special instructions are also included (e.g., serology tests for measles may not be requested without concurrently requesting viral molecular testing). PHL reference centers

Viral Diseases	real time RT PCR	Genotyping	Turn Around Times
Measles	x	x	PCR: 2 days Genotyping: 10 Days
Mumps	x	x	
Rubella	x	x	
Varicella-zoster	x	x	

Reprinted with permission of the publisher from

http://www.aphl.org/AboutAPHL/publications/Documents/ID_VPDQuickReferenceGuide_92014.pdf

FIGURE 2 Vaccine-preventable disease testing at PHL reference centers. PHL reference centers test for measles, mumps, rubella, and varicella-zoster viruses using real-time RT PCR and genotyping. The turnaround times associated with the various test methods are shown in the figure.

test for measles, mumps, rubella, and varicella-zoster viruses using real-time RT-PCR and genotyping. Figure 2 shows the turnaround times associated with the various test methods.

Ebola Virus

For highly suspected Ebola virus cases, clinical laboratories should contact their state and/or local health department *before* contacting CDC. The PHL may provide guidance for safe handling and transport of specimens and facilitate specimen transport to CDC. The PHL may also be able to provide testing to rule out Ebola. In addition, CDC has provided training and reagents to many LRN reference PHLs so that states with increased numbers of travelers from West Africa have ready access to this testing.

Ebola virus is listed as a select agent. Biological agents that the Department of Health and Human Services (HHS) has determined to have the potential to pose a severe threat to public health and safety, such as Ebola virus, are regulated under the HHS Select Agent regulations (42 CFR Part 73). Ebola virus is also listed as a Tier 1 agent. (A subset of select agents and toxins have been designated as Tier 1 because these biological agents and toxins present the greatest risk of deliberate misuse with significant potential for mass casualties or devastating effect to the economy, critical infrastructure, or public confidence and pose a severe threat to public health and safety.) Entities that possess, use, or transfer Tier 1 select agents and toxins must adhere to the additional requirements detailed within the Select Agent Regulations (http://www.selectagents.gov/Regulations.html).

Chikungunya, Dengue Virus, and Middle East Respiratory Syndrome Coronavirus (MERS-CoV)

These newly emerging viruses are an increasing threat to the United States. In response to this threat, PHLs in collaboration with CDC have developed molecular methods to test for them. Several states, particularly those where imported cases are likely, have validated and implemented these methods.

Antibody assays for dengue virus are available commercially; however, PHLs more commonly use real time RT-PCR assays that are not yet commercially available to detect the dengue viral genome in serum. Because antibodies are detected later, RT-PCR has become a primary tool to detect virus early in the course of illness. Current tests are between 80% and 90% sensitive and more that 95% specific. A positive PCR result is definite proof of current infection, and it usually confirms the infecting serotype as well. However, a negative result is interpreted as "indeterminate." Patients receiving negative results before 5 days of illness are usually asked to submit a second serum sample for serological confirmation after the fifth day of illness.

Chikungunya virus testing is performed at CDC, a few state health departments, and one commercial laboratory. Laboratory diagnosis is generally accomplished by testing serum or plasma to detect virus, viral nucleic acid, or virus-specific immunoglobulin (Ig) M and neutralizing antibodies. When managing patients with acute onset of fever and polyarthralgia, especially travelers who have recently returned from areas with known virus transmission, contact your state health department for more information and to facilitate testing.

Most state, and a limited number of local PHLs in the United States are approved to test for MERS-CoV by using an rRT-PCR assay developed by CDC to detect active infection. This test is done under authority of an Emergency Use Authorization (http://www.cdc.gov/coronavirus/MERS/lab/lab/index.html) because there are no FDA-cleared/app roved tests available for this purpose in the United States. MERS-CoV serology tests (IFA, ELISA, microneutralization) are for surveillance or investigational purposes and not for diagnostic purposes—they are tools developed in response to the MERS-CoV outbreak and performed at CDC.

SUBMITTING SPECIMENS AND RECEIVING RESULTS

Most PHLs facilitate specimen submission by providing specimen collection kits that contain swabs for obtaining the specimen, containers in which to place the specimen, and specimen requisition forms to be completed by filling in the requested patient, provider, and specimen information and the specific test requested. These collection kits may be obtained by contacting the appropriate PHL. In many states collection kits may be requested electronically, test requisitions can be completed online, and only the information needed to link the specimen to its electronic record is submitted with the specimen. A variety of means are in place for transport of specimens to the PHL. Some states have courier systems that pick up specimens from locations across the state either daily or two to three times each week. Other states rely on commercial courier systems to deliver specimens overnight to the laboratory. Rabies specimens tend to be a special case because they are usually delivered directly to the laboratory by animal control officers, veterinarians, the general public, and others. The size of the animal that the PHL will accept also varies from state to state.

PHL test results are reported using a variety of means as well. Hard copy reports may be sent through the mail, reports may be faxed, or reports may be made available electronically to print out locally. PHLs may also call the submitter to report highly significant results. If the laboratory does not report its turnaround time, it will usually indicate the number of days each week that a particular test is performed. This information, together with specimen submission instructions, can be found on the state laboratory website. In addition to reporting results to the submitter, PHLs report significant results to the state health department for public health surveillance and disease control purposes. Some of these results (e.g., influenza) are also reported to CDC for purposes of disease tracking, vaccine development, and identification of new strains and to provide other information necessary for CDC to fulfill its public health mission.

Increasingly, PHLs are beginning to charge for the testing they perform to recover their costs. Although testing is still free at some states, those that do charge use a variety of means. Payers include the submitter, Medicare, private insurers, other government agencies, and health department programs.

CLINICAL LIBRARY AND PHL COLLABORATION

Clinical laboratories are a key partner with PHLs in ensuring that specimens are available for testing to provide critical information for public health surveillance, identification of new disease strains, emergence of drug resistance, trace back of disease agents, and initiation of control measures. Clinical laboratories also serve as the foundation of the LRN working with PHLs and federal agencies to detect and refer suspect select agents for definitive identification and characterization.

The LRN was formed in 1999 by CDC, APHL, and the Federal Bureau of Investigation. The LRN for Biological Threats Preparedness (LRN-B) is organized as a three-tier pyramid of laboratories. At the foundation are thousands of sentinel clinical laboratories, which perform initial screening of potential pathogens. Sentinel clinical laboratories quickly recognize, rule out, or refer potential biological threat agents to their designated local or state public health LRN reference laboratory. Clinical laboratories have been and continue to be an integral part of the LRN, and their engagement as active partners is the responsibility of the appropriate state or local public health LRN reference laboratory in partnership with the CDC. When sentinel clinical laboratories cannot rule out the presence of a biological threat agent, they refer specimens and isolates to an LRN reference laboratory. More than 160 state, local, and federal facilities compose the LRN-B reference laboratory tier, providing testing at varying levels (reference, standard reference, and advanced reference). These laboratories produce high-confidence test results that are the basis for threat analysis and intervention by both public health and law enforcement authorities. At the apex of the pyramid are national laboratories, such as those at the CDC and the Department of Defense with specialized testing capabilities (e.g., Biosafety Level 4 facilities, strain characterization) that can be leveraged when needed by the network. These laboratories test and characterize samples that pose challenges beyond the capabilities of reference laboratories and provide support for other LRN members during a serious outbreak or terrorist event.

PHLs provide training, education, and critical information to sentinel laboratories in preparation for and in response to health emergencies. To provide optimal health care and disease control for both individuals and the public, clinical laboratories and PHLs work together in an interdependent, collaborative laboratory system.

In 2010 APHL described 11 core functions of state PHLs. These core functions go beyond disease control and reference testing to laboratory improvement and regulation, policy development, public health–related research, training and education, environmental health and protection, and partnerships and communication. Ensuring these functions requires an alliance of laboratories and other partners that "operate in an interconnected and interdependent way to facilitate the exchange of information, optimize laboratory services, and help control and prevent disease and public health threats" (10).

REFERENCES

1. **Centers for Disease Control and Prevention.** 2014. National capacity for surveillance, prevention, and control of West Nile virus and other arbovirus infections—United States, 2004 and 2012. *MMWR Morb Mortal Wkly Rep* **63(13):**281–284.
2. **Centers for Disease Control and Prevention.** 2013. Testing for HCV infection: an update of guidance for clinicians and laboratorians. *MMWR Morb Mortal Wkly Rep* **62:**18.
3. **Centers for Disease Control and Prevention.** 2012. Recommendations for the identification of chronic hepatitis C infection among persons born during 1945–1965. *MMWR Morb Mortal Wkly Rep* **61(RR04):**1–18.
4. **Centers for Disease Control and Prevention.** 2014. Laboratory testing for the diagnosis of HIV infection: updated recommendations. http://www.cdc.gov/hiv/pdf/HIV testing algorithm recommendations-final.pdf.
5. **U.S. Preventive Services Task Force.** *Recommendation Summary.* March 2012. http://www.uspreventiveservicestaskforce.org/Page/Topic/recommendation-summary/cervical-cancer-screening.
6. **Association of Public Health Laboratories and the Centers for Disease Control and Prevention.** 2013. *Influenza Virologic Surveillance Right Size Roadmap.* Silver Spring, MD: Association of Public Health Laboratories,.
7. **Association of Public Health Laboratories.** 2014. *Non-influenza Respiratory Virus Survey.* Silver Spring, MD: Association of Public Health Laboratories.
8. **Janda JM, Abbott SA.** 2014. Culture-independent diagnostic testing: have we opened Pandora's box for good? *Diagn Microbiol Infect Dis* **80:**171–176.
9. **Yang J, Yang F, Ren L, Xiong Z, Wu Z, Dong J, Sun L, Zhang T, Hu Y, Du J, Wang J, Jin Q.** 2011. Unbiased parallel detection of viral pathogens in clinical samples by use of a metagenomic approach. *J Clin Microbiol* **49:**3463–3469.
10. **Association of Public Health Laboratories.** 2010. *The Core Functions of State Public Health Laboratories.* Silver Spring, MD: Association of Public Health Laboratories.

International Reference Laboratories Offering Virology Services

ARIEL I. SUAREZ AND CRISTINA VIDELA

APPENDIX 3

SUMMARY

National Reference Laboratories provide services and scientific expertise primarily to their respective nation, but also at the international level. Most countries have their own reference laboratory that operates inside the universities or research centers within the scope of a government public health agency. In the field of virology they provide services that contribute to the identification and control of viral diseases of public health importance. A comprehensive, although not necessarily complete list of reference laboratories outside of the United States on the continents of South and North America, Europe, Asia, Africa, and Oceania is provided in this appendix.

National Reference Laboratories (NRLs) and International Collaborating Centers contribute to improved health and prosperity of society by providing scientific expertise and laboratory services. Most nations have their own reference laboratory that operates under the scope of a government public health agency. While under national control, the benefits of NRLs are recognized beyond regional and national jurisdictions. Many NRLs provide services to the international community, collaborating with a number of organizations, universities, and institutions in the areas of epidemiology, public health, and research. Data generated by NRLs are reported to local and international surveillance networks. Within each NRL are laboratories that specialize in a particular content area or disease that is the focus of their scientific activities and reflects a particular health problem of the country. The scope of activity of the NRLs has adapted to the changing environment and the needs of society. Globalization, climate change, emerging infectious diseases, and trade, among other factors, require NRLs to offer services that meet the current and anticipated needs of the surrounding community. The World Health Organization (WHO), which plays a critical role in global health, has among its missions set standards, articulating policy options based on the evidence, providing technical support to countries, and monitoring and assessing health trends. Reference laboratories play an important role in pathogen surveillance and are selected by WHO as collaborating centers in different countries around the world. The WHO provides technology, training, and reagents to these laboratories for fulfilling the tasks of monitoring and diagnosis of diseases of global importance, including vaccine-preventable diseases such as polio, measles, and congenital rubella, and other diseases due to emerging and re-emerging pathogens, such as arbovirus, severe acute respiratory syndrome (SARS), pandemic influenza H1N1, H9N7 avian influenza, the new coronavirus Middle East respiratory syndrome (MERS), and the recent outbreak of the Ebola virus. WHO ensures that countries have rapid access to the most appropriate experts and resources for outbreak response through the Global Outbreak Alert and Response Network (GOARN). This network was created in 2000 to improve the coordination of international outbreak responses and to provide an operational framework to focus the delivery of support to countries. The viral diseases that are monitored by GOARN are dengue, yellow fever, Crimean-Congo hemorrhagic fever, Ebola hemorrhagic fever, Lassa fever, Rift Valley fever, avian influenza, hepatitis, virus Hendra and virus Nipah infection, SARS, and smallpox.

An example of how national reference centers collaborate at the international level is their contribution to the recent Ebola fever outbreak in Africa. The first cases were reported from the forested region of southeastern Guinea in Guéckédou prefecture near the border with Liberia and Sierra Leone. The Ebola viral etiology was confirmed on 22 March 2014 by the National Reference Centre for Viral Haemorrhagic Fevers (Institut Pasteur, INSERM BSL4 laboratory, Lyon, France).

NRLs should be a source of assistance and specialized information for local laboratories. The support must be based on scientific criteria and standards for the prevention, detection, and control of disease. In many cases, the NRLs themselves are responsible for the development of these standards and guidelines.

NRL functions should encourage and help sustain accreditation processes in the diagnostic virology laboratory, including organizing and conducting quality control programs. NRLs should also contribute to the development of technical regulations and diagnostic algorithms.

The primary mission of the NRLs is the prevention, control, and surveillance of diseases that pose a threat to public health and to advise the government health agency

doi:10.1128/9781555819156.app3

on appropriate measures to take. For these purposes, NRL services and responsibilities include the following:

- Implementing a system to receive samples and patient information from local and regional laboratories for pathogen confirmation and characterization, such as typing and subtyping of influenza viruses.
- Ensuring the quality of procedures provided by the NRL, and identifying the need for new and improved methodologies for the detection and prevention of both established and emerging pathogens.
- Evaluating and validating new technologies for dissemination throughout the national laboratory network.
- Providing technical and human resources training for all laboratories within the national laboratory network. Designing, implementing, monitoring, and ensuring training plans, providing continuing education to expand knowledge and skills and promoting the transfer of technology through the management of training programs.
- Developing policies and standards for the national laboratory network.
- Managing a laboratory quality assurance program for the national laboratory network.
- Implementing and maintaining a laboratory accreditation program for all levels of the national laboratory network based on a comprehensive quality management program.
- Supporting the national surveillance system for communicable, emerging, and re-emerging diseases.
- Providing laboratory-based surveillance data to public health authorities for the purpose of taking appropriate sanitary measures.
- Providing information to international health organizations to strengthen surveillance for events that may constitute a public health threat of international concern.

A list of NRLs offering diagnostic virology services is presented in this appendix ordered by continent.

AFRICA

Algeria

Viral Respiratory Unit
Institut Pasteur d'Algérie
Sidi-Fredj
Staoueli
Algiers
Algeria
Tel: (213) (0)21 37 68 50/51
Fax: (213) (0)21 39 02 57
Website: www.pasteur.dz

Specialties and services: Enterovirus, HIV, hepatitis, respiratory viruses, herpes virus, human papillomavirus, arboviruses

Cameroon

Laboratoire de Virologie
Centre Pasteur du Cameroun
Centre 1274
Yaoundé
Cameroon
Tel: (237) 99 65 47 67
Fax: (237) 22 23 15 64
Website: http://www.pasteur-yaounde.org

Specialties and services: Enterovirus, influenza, hepatitis, HIV, measles, hemorrhagic fever viruses

Central African Republic

Institute Pasteur de Bangui
BP 923
Bangui
Central African Republic
Tel: (236) 21 61 08 66
Fax: (236) 21 61 01 09
Website: http://pasteur-bangui.org/

Specialties and services: Arbovirus, hemorrhagic fever viruses, rabies, influenza

Côte d'Ivoire

Département des Virus Epidémiques
Institut Pasteur de Côte d'Ivoire
01 BP 490 Abidjan 01
Abidjan
Côte d'Ivoire
Tel: (225) 22 00 58 29
Fax: (225) 21 25 35 10
Website: http://www.pasteur.ci/

Specialties and services: Influenza, HIV, hepatitis

Ghana

Virology Department
Noguchi Memorial Institute for Medical Research (NMIMR)
University of Ghana, Legon
Accra
Ghana
Tel: (233) 302 501 178 9
Fax: (233) 21 502182
Website: http://www.noguchimedres.org/

Specialties and services: Influenza, HIV, polio, rotavirus

Madagascar

Institut Pasteur de Madagascar
Ambatofotsikely
101 Antananarivo
Madagascar
Tel: (261) 20 22 412 72 / 74
Fax: (261) 20 22 415 34
Website: http://www.pasteur.mg

Specialties and services: Influenza, rabies, HIV

Morocco

Laboratoire de Virologie
Institut National d'Hygiène
Ministère de la Santé
27 Avenue Ibn Batouta
Rabat
Morocco
Tel: (212) 537 76 11 21
Fax: (212) 37 772 067
Website: http://www.sante.gov.ma

Specialties and services: Polio, enterovirus, measles, rubella, influenza, rotavirus, HIV

Nigeria

College of Medicine
University College Hospital
Ibadan
Nigeria

Fax: (234) 02 241 1768
Website: http://com.ui.edu.ng
Specialties and services: Lassa fever, polio, African swine fever, measles, HIV, hepatitis B virus, hepatitis C virus, rubella, influenza, dengue

Senegal

Medical Virology Unit
Institut Pasteur de Dakar
36, Avenue Pasteur
Dakar
Senegal
Tel: (221) 33 839 92 00
Fax: (221) 33 839 92 10
Website: http://www.pasteur.sn/
Specialties and services: Influenza, rabies, polio, arboviruses, hemorrhagic fever viruses

South Africa

National Institute for Communicable Diseases/NHLS
1 Modderfontein Road
Sandringham, Johannesburg
South Africa
Tel: 27-11 386 6000
Fax: 27-11 882 0596
Website: http://nicd.ac.za
Specialties and services: Adenovirus, arboviruses, respiratory viruses, cytomegalovirus, herpes simplex virus, enterovirus, HIV, measles, mumps, rabies, varicella-zoster virus, hemorrhagic fever viruses

Tunisia

Institut Pasteur de Tunis
Laboratoire de Virologie Clinique
13, place Pasteur, B.P. 74
1002 Tunis, Belvédère
Tunisia
Tel: 216 71 843 755
Fax: 216 71 791 833
Website: http://www.pasteur.tn
Specialties and services: Polio, measles, HIV

Uganda

Uganda Virus Research Institute (UVRI)
Nakiwogo Road
Entebbe
Uganda
Tel: 256-414-320385/6
Fax: 256-414-320483
Website: http://www.uvri.go.ug/
Specialties and services: Influenza, arboviruses, HIV, hepatitis

NORTH AND SOUTH AMERICA

Argentina

Instituto Nacional de Enfermedades Infecciosas
ANLIS C.G. Malbran
Av. Velez Sarsfield 563
1281 Buenos Aires
Argentina
Tel: 54-11-4301-1035
Fax: 54-11-4301-1035
Website: http://www.anlis.gov.ar/inei/
Specialties and services: Enterovirus, Epstein-Barr virus, caliciviruses, rotavirus, influenza, hepatitis, HPV, hantavirus, mumps, measles, CMV, and parvovirus

Brazil

Instituto Oswaldo Cruz, Fundação Oswaldo Cruz
Departmento de Virologia
Avenida Brasil 4365, Marguinhos
Río de Janeiro
Brazil
Tel: 55-21 2598-4360
Fax: 55-21 25 73 95 91
Website: https://portal.fiocruz.br/
Specialties and services: HIV, influenza, measles, rubella, hepatitis, arboviruses

Canada

National Microbiology Laboratory, Health Canada
Canadian Science Centre for Human and Animal Health
1015 Arlington Street, R3E 3R2
Winnipeg, Manitoba
Canada
Tel: 1-204 789 6045
Fax: 1-204 789 2082
Website: https://www.nml-lnm.gc.ca/
Specialties and services: Hepatitis, respiratory viruses and influenza, chickenpox, measles, herpes simplex virus, human papillomavirus, and enteroviruses

Chile

Instituto de Salud Pública de Chile (ISPCH)
Vigilancia de Laboratorio de Virus Respiratorios
Subdepartamento Virología Clínica
Avenida Marathon 1000, Nunoa
Santiago de Chile
Chile
Tel: 56-2 350 7436
Fax: 56-2 350 7583
Website: http://www.ispch.cl/
Specialties and services: Arboviruses, poliovirus, norovirus, astrovirus, adenovirus, rabies, respiratory viruses, measles, cytomegalovirus, hepatitis, HIV

Colombia

Instituto Nacional de Salud (INS)
Avenida El Dorado, Carrera 50, Zona 6
Bogotá
Colombia
Tel: (+57-1) 220 7700
Website: http://www.ins.gov.co
Specialties and services: Dengue, HIV, hepatitis, respiratory viruses, rabies

Costa Rica

Centro Nacional de Referencia de Virología—INCIENSA
Apartado 4-2250
Tres Ríos, Cartago
Costa Rica
Tel: (+50-6) 2279-9911
Fax: (+50-6) 2279-5546
Website: http://www.inciensa.sa.cr
Specialties and services: Arboviruses, HIV, hepatitis, respiratory viruses, enterovirus, hantavirus, measles, rubella, rotavirus

Cuba

Instituto de Medicina Tropical "Pedro Kourí" (IPK)
Novia de Mediodia Km. 6
La Lisa, Ciudad de la Habana
Cuba
Tel: (+53-7) 202 0633
Fax: (+53-7) 204 6051
Website: http://instituciones.sld.cu/ipk/

Specialties and services: Dengue, HIV, influenza

Ecuador

Instituto Nacional de Investigación en Salud Pública—INSPI
Av. Julián Coronel 905, Guayaquil
Ecuador
Tel: (+59-3) 422 88097
Website: http://www.investigacionsalud.gob.ec/virologia/

Specialties and services: Arboviruses, HIV, hepatitis, influenza and respiratory viruses, polio, rubella, measles, cytomegalovirus, and Epstein-Barr virus

El Salvador

Laboratorio Central "Dr. Max Bloch"
Gabriela Mistral, Avenida del Prado, No. J-234
Colonia Buenos Aires I
San Salvador
El Salvador, C.A.
Tel: (+ 50-3) 2520-3000
Fax: (+50-3) 22 21 57 51
Website: http://ins.salud.gob.sv

Specialties and services: Influenza, arboviruses, HIV, rabies

Guatemala

Laboratorio Nacional de Salud de Guatemala
Laboratorio Central
Km. 22, Carretera al Pacífico, Barcenas, Villa Nueva
Guatemala, C.A.
Tel: (502) 6644-0569
Fax: (502) 6644-0569, ext. 241
Website: http://www.mspas.gob.gt/index.php/en/

Specialties and services: Arboviruses, HIV, hepatitis, influenza and respiratory viruses, polio, rubella, measles, rotavirus, rabies, and hantavirus

Honduras

Laboratorio Nacional de Vigilancia de la Salud—Sección de Virología
Secretaría de Salud
Colonia la Campaña
Tegucigalpa
Honduras
Tel: (+504) 2232 5840
Fax: (+504) 2239 7580
Website: http://www.salud.gob.hn

Specialties and services: Influenza, arboviruses

Mexico

Instituto Nacional de Diagnóstico y Referencia Epidemiológicos (INDRE)
Secretari de Salud
1er piso, Carpio 470, Colonia Santo Tomás, Delegación Miguel Hidalgo C.P. 11340
Ciudad de México
México
Tel: (+52-5) 341 1432
Fax: (+52-5) 341 0404
Website: http://www.spps.gob.mx/unidades-de-la-subsecretaria/indre.html

Specialties and services: Rubella, measles, mumps, parvovirus, Epstein-Barr virus, varicella-zoster virus, respiratory viruses

Nicaragua

Laboratorio de virología
Dirección de Microbiología
Centro Nacional de Diagnóstico y Referencia (CNDR)
Ministerio de Salud, Complejo Concepción Palacios
Managua
Nicaragua
Fax: (505) 2289-7723
Website: http://minsa.gob.ni

Specialties and services: Arboviruses, hepatitis, HIV, respiratory viruses, rabies

Panama

Instituto Conmemoratico Gorgas de Estudios de la Salud
Ave. Justo Arosemena y calles 35
Panamá
Republic of Panama
Fax: (507) 527-4889
Website: http://gorgas.gob.pa/

Specialties and services: HIV, arboviruses, respiratory viruses, rubella, measles, parvovirus, rotavirus

Paraguay

Laboratorio Central de Salud Pública,
Ministerio de Salud y Bienestar Social (MSPBS)
Avenida Venezuela y Tte. Escurra
Asunción
Paraguay
Tel: (021) 294-999 / 292-653
Fax: (+595-21) 294 999
Website: http://www.mspbs.gov.py/lcsp/

Specialties and services: Hepatitis C virus, human papillomavirus, measles

Peru

Centro Nacional de Salud Pública
Instituto Nacional de Salud
Av. Defensores del Morro No. 2268
Lima
Peru
Tel: (511) 748 0000
Website: http://www.ins.gob.pe

Specialties and services: Arboviruses, HIV, respiratory viruses, rubella, measles

Venezuela

Instituto Nacional de Higiene "Rafael Rangel" (INHRR)
Ciudad Universitaria,
Caracas
Venezuela
Tel: (58-212) 219-1654
Website: http://www.inhrr.gob.ve

Specialties and services: Respiratory viruses, herpes simplex virus, enterovirus, hemorrhagic fever viruses, hepatitis, HIV, Epstein-Barr virus, human papillomavirus, measles, rubella

Uruguay

Departamento de Laboratorio de Salud Publica
8 de Octubre 2720
Montevideo
Uruguay
Fax: 598 (2) 480 7014
Website: http://www.msp.gub.uy
Specialties and services: Dengue, HIV, influenza, hepatitis

ASIA

Afghanistan

Central Public Health Laboratory
Wazir Akbar Khan Rd
3rd floor of the main building of MoPH
Afghanistan
Website: http://moph.gov.af
Specialties and services: Influenza, polio, measles, rubella, respiratory viruses

Bangladesh

Institute of Epidemiology, Disease Control and Research (IEDCR)
Mohakhali, Dhaka 1212
Dhaka
Bangladesh
Tel: +880-2-9898796, 9898691
Fax: +880-2-9880440
Website: http://www.iedcr.org/
Specialties and services: Influenza, dengue, HIV, rotavirus, hepatitis B virus, hepatitis C virus

Cambodia—Phnom Penh

Institut Pasteur du Cambodge
Virology Unit
5, Blvd Monivong
Phnom Penh
Cambodia
Tel: (+855)-12-812-003
Website: www.pasteur-kh.org

China

Chinese Center for Disease Control and Prevention
155 Changbai Road
Changping District
Beijing
China
Tel: (+8610)-5890-0240
Fax: (+8610)-5890 0851
Website: http://www.chinacdc.cn/en
Specialties and services: Arboviruses, enterovirus, polio, measles, rubella, severe acute respiratory syndrome, influenza, human papillomavirus, HIV, hepatitis

Hong Kong

Centre for Health Protection
147C Argyle Street, Mongkok, Kowloon
China, Hong Kong Special Administrative Region
Tel: (+852)-319-8667
Fax: (+852)-2836-0071
Website: http://www.chp.gov.hk/en/index.html

India

National Institute of Virology
20-A Dr Ambedkar Road
P.O. Box 11
Pune 411001
India
Tel: 91-020-26127301/26006290
Fax: 91-020-26122669/26126399
Website: http://www.niv.co.in/
Specialties and services: Chandipura virus, Chikungunya virus, influenza, dengue, hepatitis, Japanese encephalitis virus, polio, rotavirus

Indonesia

Center for Biomedical and Basic Technology of Health
National Institute of Health Research and Development
Jl. Percetakan Negara 23
Jakarta Pusat 10560
Indonesia
Tel: (+62)- 0214261088
Fax: (021) 4243933
Website: http://www.litbang.kemkes.go.id/

Iran (Islamic Republic of)

Tehran University of Medical Sciences
School of Public Health
St. Enghelab Sq., Ghods St., Poursina St. (Northern Tehran University), School of Public Health Building
Tehran
Islamic Republic of Iran
Tel: (+98)-21-88989120
Fax: (+98)-21-88950595
Website: http://sph.tums.ac.ir/index.php?slc_lang=en&sid=8
Specialties and services: Influenza, polio, measles, rubella, respiratory viruses, hepatitis C virus, hepatitis b virus, HIV

Israel

Israel Center for Disease Control (ICDC)
Ministry of Health
2 Ben Tabai St
Jerusalem
Israel
Tel: 972-26551818
Fax: 972-2-6528079
Website: http://www.old.health.gov.il/english/
Specialties and services: Severe acute respiratory syndrome, HIV, influenza

Japan

National Institute of Infectious Diseases
Gakuen 4-7-1
Musashi-Murayama-shi
Tokyo
Japan
Fax: +81 42 565 2498
Website: http://www.nih.go.jp/niid/en/
Specialties and services: Japanese encephalitis virus, enterovirus, polio, measles, rubella, severe acute respiratory syndrome, influenza, human papillomavirus, HIV, hepatitis C virus

Malaysia

Virology Unit
Infectious Disease Research Centre
Institute of Medical Research
Jalan Pahang
Kuala Lumpur

Malaysia
Tel: 603-2616 2666
Fax: 603-2693 9335
Website: http://www.imr.gov.my/en/idrc.html
Specialties and services: Rabies, HIV, influenza, Japanese encephalitis virus and other arboviruses, acute flaccid paralysis, dengue, enterovirus, rubella, cytomegalovirus

Myanmar
National Health Laboratory
35, Mawkundaik Road
Dagon Township
Yangon
Myanmar
Tel: 95-1-371957
Fax: +95-1-371925

Nepal
National Public Health Laboratory
Teku Kathmandu
Nepal
Tel: 977-1-4252421
Fax: 977-42523755
Website: http://www.nphl.gov.np
Specialties and services: Hepatitis E virus, influenza, hepatitis B virus, hepatitis C virus, HIV

Pakistan
National Institute of Health
Chak Shehzad
Park Road
Islamabad
Pakistan
Tel: 92-051-9255110
Fax: 92-051-9255099
Website: http://www.nih.org.pk
Specialties and services: Middle East respiratory syndrome, coronavirus, influenza, polio, measles, rabies

Qatar
Department of Laboratory Medicine and Pathology
Hamad Medical Corporation
Doha
Qatar
Tel: 974-4439-2136
Fax: 974-4431-2751
Website: http://hgh.hamad.qa

Republic of Korea
National Institute of Health
Korea Center for Disease Control and Prevention
Osong Health Technology Administration Complex 643
Yeonje-ri, Gangoe-myeon, Cheongwon-gun
Chungcheongbuk-do 363-951
Seoul
Republic of Korea
Tel: 82-43-719-7700
Fax: +82 43 719 8190
Website: http://www.nih.go.kr
Specialties and services: Hepatitis C virus, hepatitis B virus, norovirus, enteroviruses, chickenpox, rotavirus, polio, influenza, severe acute respiratory syndrome, measles, mumps, rubella, adenovirus, parainfluenza virus, respiratory syncytial virus, tick-borne encephalitis virus

Singapore
National Public Health Laboratory
Communicable Disease Division
Ministry of Health
16 College Road
Singapore
Tel: 65-63259220
Fax: 65-63251168
Website: http://app.sgdi.gov.sg/

Sri Lanka
Medical Research Institute
Colombo 8
Western
Sri Lanka
Tel: 94 2 693532-34
Fax: 94 2 691495
Website: http://www.mri.gov.lk/
Specialties and services: Dengue, hepatitis, influenza, enterovirus, Japanese encephalitis virus

Syrian Arab Republic
Public Health Laboratories
Al Ghassani. Aleppo Street
Damascus
Syrian Arab Republic
Tel: 963 114451177
Fax: 963 114442153
Website: http://www.moh.gov.sy
Specialties and services: HIV, influenza, measles, mumps, rubella, polio

Thailand
National Institute of Health (NIH)
Department of Medical Sciences
Ministry of Public Health
88/7 Tiwanon Road
Nonthaburi
Thailand
Tel: 66-2589-9850
Fax: 66(2) 5915449
Website: http://nih.dmsc.moph.go.th
Specialties and services: Hepatitis, arboviruses, oncogenic viruses, respiratory viruses, intestinal viruses, nervous system and circulatory viruses

Turkey
Virology Reference and Research Laboratory
Ministry of Health
Public Health Institution of Turkey (PHIT)
Refik Saydam Campus,
Saglõk mah. Adnan Saygun str. No:55 F Block
06100 Sihhiye
Ankara
Turkey
Fax: +90 (312) 4582388

Vietnam
National Institute of Hygiene and Epidemiology
1 Yersin Street, Hai Ba Trung,
Ha Noi
Vietnam
Tel: 84-4 3971-6356
Fax: 84-4 3821-0853
Website: http://www.nihe.org.vn
Specialties and services: Influenza, HIV, dengue, severe acute respiratory syndrome, Japanese encephalitis virus

EUROPE

Albania

Institute of Public Health (Instituti i Shendetit Publik)
Aleksander Moisiu No. 80
Tirana
Albania
Fax: +355 (437) 0058
Website: http://www.ishp.gov.al/

Specialties and services: West Nile and dengue virus, HIV, influenza

Austria

Clinical Institute of Virology at Medical Institute of Vienna
Kinderspitalgasse 15
Building: Institutsgebäude, Zimmer 321
A-1090 Wien
Austria
Tel: 01/40490-79500, 01/40490-79551
Fax: 01/40490-9795
E-mail: Virologie@meduniwien.ac.at
Website: http://www.virologie.meduniwien.ac.at

Specialties and services: Influenza, hepatitis, measles, mumps, rubella, HIV, hemorrhagic fever, tick-borne encephalitis and other flaviviruses, hantaviruses

Belgium

Scientific Institute of Public Health
Service of Viral Diseases
Rue Juliette Wytsmanstraat 14
1180 Brussels
Belgium
Tel: +32 2 373 32 09
Fax: +32 2 373 32 86
Website: https://www.wiv-isp.be/Programs/communicable-infectious-diseases/Pages/EN-ViralDiseases.aspx

Specialties and services: Respiratory viruses, encephalitis viruses, emerging and re-emerging viruses, national reference center on influenza virus, measles, rubella, hepatitis viruses (B, C, D, and E), rabies, tick-borne encephalitis

Belarus

Republican Research & Practical Center for Epidemiology and Microbiology
23 Filimonova St.
220114 Minsk
Belarus
Tel: (+375-17) 267-32-67
Fax: (+375-17) 267-30-93
E-mail: belriem@gmail.com
Website: http://www.belriem.by/

Specialties and services: poliomyelitis, influenza, and other acute respiratory diseases, measles

Bosnia and Herzegovina

Clinical Center—University of Sarajevo
Institute for Clinical Microbiology
Bolnička 25
71000 Bosnia and Herzegovina
Fax: +387.33.22.69.60
Tel: +387.33.22.69.60 / +38.7.61.655.234
Website: http://www.kcus.ba/

Specialties and service: Measles, rubella, influenza, hantavirus

Bulgaria

National Centre of Infectious and Parasitic Diseases
26 Yanko Sakazov Blvd.
Sofia 1504
Bulgaria
Tel: +359 2 94 46 999
Fax: + 359 2 943 30 75
E-mail: ncipd@ncipd.org

Specialties and services: Enteroviruses and other diarrheal viruses, viral hepatitis, measles, mumps and rubella, herpes virus infections, Rikettsiae and Chlamidiae, influenza and acute respiratory diseases; National Confirmatory Laboratory of HIV/AIDS, arboviruses, retroviruses

Croatia

Croatian Institute of Public Health
Rockefellerova 12
Zagreb
Croatia
Fax: +385 (1) 468 3017

University Hospital for Infectious Diseases
"Dr. Fran Mihaljević"
Mirogojska 8, 10 000 Zagreb
Croatia
Tel: ++385 1 2826-222
Fax: ++385 1 4678-235
Website: http://www.bfm.hr

Specialties and services: Diagnostics and treatment of infectious diseases, viral hepatitis, HIV-infection, urinary tract infections, tropical and traveler diseases

Czech Republic

National Institute of Public Health
Institute of Epidemiology and Microbiology
Šrobárova 48
100 42, Praha 10
Czech Republic
Tel: 00420 26708 1111
Website: http://www.szu.cz/the-cem-national-reference-laboratories-nrls

Specialties and services: Enteroviruses, herpetic viruses, HIV/AIDS, influenza, measles, mumps, rubella, parvovirus B19, noninfluenza respiratory viruses, and viral hepatitis

Denmark

The Laboratory for Infectious Diseases
Statens Serum Institut
5 Artillerivej
DK-2300 Copenhagen S
Denmark
Tel: +45 3268 3268
Fax: +45 3268 3868
Website: http://www.ssi.dk/English/HealthdataandICT/National%20Reference%20Laboratories.aspx

Specialties and services: Influenza virus A and B and other respiratory viruses; HIV; tick-borne encephalitis virus; hepatitis A, B, C, D, and E viruses; herpes viruses; rubella virus; parvovirus; morbillivirus; enterovirus; parotitis virus; dengue virus; hantavirus; National WHO Reference Laboratory for Morbilli and Rubella; National WHO Reference Laboratory for Poliovirus

Estonia

Laboratory for Communicable Diseases
Health Board Kotka Str 2
Tallinn 11315
Estonia
Fax: +372 6943 651
Website: http://www.terviseamet.ee/en/laboratories.html
Specialties and services: Center for measles and rubella, polio, influenza

West-Tallinn Central Hospital
Centre for Infectious Diseases,
Paldiski mnt. 68
10617 Tallinn
Estonia
Tel: 650 7301
Fax: 659 8686
Website: http://www.ltkh.ee/
Specialties and services: HIV

France

Institute Pasteur
25-28 Rue du Docteur Roux
75015 Paris, Cedex 15
France
Specialties and services: Arbovirus, influenza virus, papillomavirus, rabies virus, enterovirus, hepacivirus, flavivirus, retrovirus

Unité de Biologie des infections virales émergentes (UBIVE)—IFR 128
21 Avenue Tony Garnier
69365 Lyon Cedex 07
France
Specialties and services: Hemorrhagic fever, hantavirus

Cayenne: Institut Pasteur
Institut Pasteur de la Guyane
23 Avenue Louis Pasteur—BP. 6010
97306 Cayenne Cedex
Guyane française
Website: http://www.pasteur.fr
Specialties and services: Hantavirus, arbovirus, influenza virus

Finland

National Institute for Health and Welfare
Department of Infectious Diseases Surveillance and Control
Mannerheimintie 166
P.O. Box 30
FI-00271 Helsinki
Finland
Tel: +358 29 524 6000
Virology Unit
Mannerheimintie 166
P.O. Box 30
FI-00271 Helsinki
Finland
Website: http://www.thl.fi/en/web/thlfi-en/about-us/organisation/departments-and-units/infectious-diseases/viral-infections
Specialties and services: Regional polio expert laboratory for the World Health Organization (WHO) as well as a national center for influenza

Greece

Hellenic Pasteur Institute
127 Vasilissis Sofias Avenue
11521, Athens
Greece
Tel: (+30) 210 6478800
Fax: (+30) 210 6425 038
Website: http://www.pasteur.gr/
Specialties and services: National Reference Laboratories for Infectious Diseases: National Influenza Reference Laboratory for Southern Greece; enteroviruses/polioviruses, measles, rubella

Georgia

Centre for Disease Control and Public Health
M. Asatiani str. 9
Tbilisi 0186
Georgia
Fax: +995 32 2 311485
Website: http://www.ncdc.ge
Specialties and services: HIV, polio, hepatitis, measles, rubella

Hungary

B. Johan National Center for Epidemiology
Virology Department
Gyáli út 2-6
H-1097 Budapest
Hungary
Tel: +36 1 476 1194
Fax: +36 1 476 1126
E-mail: oekfoigazgatosag@oek.antsz.hu
Website: http://www.oek.hu
Specialties and services: WHO National Poliovirus Reference Laboratory for influenza, respiratory viruses, hepatitis viruses, enteroviruses, exanthematous diseases (measles, rubella), viral zoonoses, herpes viruses, hanta, cytomegalovirus, dengue

Iceland

Landspitali—University Hospital
Armuli 1A
P.O. Box 8733
Reykjavik
Iceland
Fax: +354 (543) 5949
Website: http://www.landspitali.is/
Specialties and services: Influenza

Ireland

UCD National Virus Reference Laboratory
University College Dublin
Belfield, Dublin 4
Ireland
Tel: 01-716 4401
Fax: 01-269 7611
Website: http://nvrl.ucd.ie/
Specialties and services: HIV, rubella virus, parvovirus B19, enterovirus, mumps virus, herpes simplex virus, varicella zoster virus, measles, dengue virus, adenovirus, rotavirus, astrovirus, calicivirus, norovirus, coxsackie A16 virus, polyoma, hepatitis A virus, hepatitis B virus, hepatitis C virus, cytomegalovirus, HTLV, influenza, respiratory syncytial virus, parainfluenza, human metapneumovirus

Italy

Istituto Superiore di Sanità
Viale Regina Elena, n. 299
Roma 00161

Italy
Tel: 06 4990 1
Fax: 06 49 38 71 18
Website: http://www.iss.it
Specialties and services: Unit of viral diseases and live attenuated virus vaccines, arbovirus and hemorrhagic fevers, influenza

Germany

Robert Koch Institute
National Reference Centers and Consultant Laboratories in Infectious Diseases
Nordufer 20
D-13353 Berlin-Wedding
Germany
Tel: +49 (0)30 - 18754-0
Website: http://www.rki.de/EN/Home/homepage_node.html
Specialties and services: Electron microscopy, norovirus, rotavirus, polio virus, poxvirus, tick-borne encephalitis, enterovirus, influenza, measles, exanthemata virus

Friedrich-Loeffler-Institut
Federal Research Institute for Animal Health
Institute for Novel and Emerging Infectious Diseases
Headquarters Insel Riems
Südufer 10
17493 Greifswald-Insel Riems
Germany
Tel: +49 38351 7-0
Fax: +49 038351 7-1219, 7-1151, 7-1226
Website: http://www.fli.bund.de/en/startseite/friedrich-loeffler-institut.html
Specialties and services: Crimean-Congo hemorrhagic fever virus, equine encephalitis viruses (EEEV, VEEV, WEEV), filoviruses (Marburg/Ebola virus), hantavirus, Nipah/Hendra virus, Japanese encephalitis virus, Rift Valley fever virus, transmissible spongiform encephalopathies, West Nile virus

Kosovo

National Institute of Public Health of Kosovo
Department of Microbiology
Hospital area NN
10000 Prishtina
Kosovo
Fax: +381.38.55.08.58/+381.38.55.05.85
Website: http://www.niph-kosova.org
Specialties and services: Hemorrhagic fevers, respiratory diseases

Latvia

Riga Eastern Clinical University Hospital Laboratory
3 Linezera Str.
Riga LV1006
Latvia
Fax: +371 670145683
Website: https://www.aslimnica.lv
Specialties and services: WHO Influenza Center

Lithuania

National Public Health Surveillance Laboratory
Žolyno str. 36
LT—10210 Vilnius
Lithuania
Tel: (+370) 5270 9229
Fax: (+370) 5210 4848
E-mail: nvspl@nvspl.lt
Website: http://www.nvspl.lt
Specialties and services: WHO influenza, measles center

Luxembourg

Laboratoire National de Sante, Virologie
1, Rue Louis Rech
L-3555 Dudelange
Luxembourg
Tel: (+352) 28 100-514
Fax: (+352) 28 100-512
Website: http://www.lns.public.lu
Specialties and services: Influenza, hepatitis, HIV

Malta

The Infectious Disease Prevention and Control Unit
Health Promotion and Disease Prevention Directorate
5B, The Emporium,
Triq C. De Brocktorff,
Msida MSD 1421,
Malta
Tel: +35623266112/+35621332235
Fax: +35621319243
E-mail: disease.surveillance@gov.mt
Website: https://ehealth.gov.mt
Specialties and services: AIDS/HIV, acute viral encephalitis, cytomegalovirus, dengue, food-borne illness/infections, intestinal illness, hepatitis A, hepatitis B, hepatitis C, herpes simplex, influenza, new and emerging diseases, rotavirus; vaccine-preventable diseases: polio, measles, mumps, rubella, chickenpox, herpes zoster, influenza

Netherlands

Eijkman-Winkler Institute, Department of Virology, University Medical Center, Utrecht, The Netherlands
Erasmus MC
Viroscience
Virology Unit and WHO Center
Wytemaweg 80
3015 CN Rotterdam
Netherlands
Website: http://www.erasmusmc.nl/virologie/unit-research/?lang=en
Specialties and services: National Reference Lab for HIV; measles, mumps, influenza, exotic viral infections, HIV, respiratory syncytial virus, human metapneumovirus, hepatitis, herpes

Norway

Norwegian Institute of Public Health
Virology Department
PO Box 4404 Nydalen
N-0403 Oslo
Tel: +47 21077000
Fax: +47 22353605
Website: http://www.fhi.no/
Specialties and services: Influenza virus; polio/enterovirus; hepatitis A, B, C, D, and E; measles, mumps, and rubella; norovirus; severe acute respiratory syndrome; adenovirus; quantitation, identification, and characterization of virus strains; molecular epidemiology; quality control; basic virology

Portugal

Infectious Diseases Department
National Institute of Health Dr. Ricardo Jorge
Av. Padre Cruz

1649-016
Portugal
Tel: (+351) 217 519 200
Fax: +351 (21) 752 6400
Website: http://www.insa.pt
Specialties and services: National Influenza Reference Laboratory, polio virus, rubella, and measles virus

Institute of Hygiene and Tropical Medicine (IHMT)
Universidade Nova de Lisboa
Rua da Junqueira 100
Lisbon
Portugal
Tel: +351 21 3652600
Fax: 351 21 363 21 05
Website: http://www.ihmt.unl.pt
Specialties and services: Hepatitis C, rotavirus, HIV, flavivirus

Poland

National Institute of Public Health—National Institute of Hygiene
ul. Chocimska 24
00-791 Warsaw
Poland
Tel: +48 22 54 21 200
Fax: (48-22) 542-13-13
Website: http://www.pzh.gov.pl/
Specialties and services: National Polio Laboratory, measles/rubella, Herpertoviridae, mumps, adenovirus, hepatitis, unit of immunology of viral infections, unit of entero- and neuro-infections

Romania

Laboratory for Vector-borne Diseases and Medical Entomology
National Institute for Research & Development in Microbiology & Immunology
"Cantacuzino"
Splaiul Independetei 103
050096 Bucharest
Romania
Tel: 021.306.92.37
Fax: (lab): +40.21.3069.307 (LAB ITV)
Fax: 021.306.93.37
E-mail: laborator@cantacuzino.ro
Website: http://www.cantacuzino.ro/ro/index.php/contact/
Specialties and services: Laboratories for molecular epidemiology, hepatitis, arboviruses, enteroviruses, respiratory virus, zoonosis

Republic of Moldova

National Influenza Center
Str. Cosmescu 3, Chisinau
Republic of Moldova, MD 2029
Fax: +37 322 729725
National Center of Health Management
3, Cosmescu str., Chisinau,
Republic of Moldova, MD-2009
Tel.: +373 (22) 72–73–59
E-mail: ccm_secretariat@mednet.md
Website: http://www.ccm.md
Specialties and services: HIV and retrovirus, influenza

Russia Federation–Moscow

Laboratory of Environmental Virology
M.P. Chumakov Institute of Poliomyelitis and Viral Encephalitides
Russian Academy of Medical Sciences
Institute of Poliomyelitis
Kievskoye shosse, 27 km 142782
Moscow
Russian Federation
Tel: +7 (495) 841 90 07
Fax: +7 (495) 841 93 30
Website: http://www.poliomyelit.ru/

D.I. Ivanovsky Research Institute of Virology
Ministry of Health of the Russian Federation
16, Gamaleya str.
Moscow 123098
Russia
Tel: 7+499+1902842
Fax: 7+499+1902867
E-mail: lvovdk@virology.ru
Website: http://www.istc.ru
Specialties and services: Viruses with genome variability, national virus collection, genome structure of hemorrhagic fever virus, address liposomes in new drugs, influenza viruses in birds, conservation of prions and Chlamidiae, new generation of anti-HIV-1 compounds, vaccine against infections, bovine rhinotracheitis, genetic polymorphism of HIV-1

Russian Federation–St Petersburg

Research Institute of Influenza
Ministry of Health
Prof. Popov str. 15/17
St. Petersburg
Russian Federation
Fax: +7(812) 234 59 73, (812) 346 1270
Website: http://www.influenza.spb.ru

Serbia

Institute of Virology, Vaccines and Sera "Torlak"
Vojvode Stepe 458
11153 Belgrade
Serbia
Tel: + 381 11 397 66 74
Fax: + 381 11 247 18 38
Website: http://www.torlakinstitut.com
Specialties and services: National laboratory for influenza and other respiratory viruses, poliomyelitis and enteroviruses, rubella, morbilli, varicella and other rashes, viral hemorrhagic fevers, arboviruses

Slovakia

Public Health Authority of the Slovak Republic
Trnavská cesta 52
826 45 Bratislava 29
Phone: +421 2 492 84 111
E-mail: podatelna@uvzsr.sk
Website: http://www.uvzsr.sk/en/
Specialties and Services: Hepatitis, HIV, respiratory viruses, polio, measles.

Slovenia

Faculty of Medicine—University of Ljubljana
Institute of Microbiology & Immunology
Zaloska 4
1000 Ljubljana
Slovenia
Tel: 01 543 74 00

Fax: 01 543 74 01
Website: http://www.mf.uni-lj.si/en/
Specialties and services: Hepatitis, HIV, human papillomavirus, arboviruses and viral hemorrhagic fever, respiratory viruses

Spain

National Center of Microbiology, Carlos III Institute of Health
Carretera Majadahonda-Pozuelo Km 22
28220 Majadahonda, Madrid
Spain
Tel: 918 223 637, 918 223 000
Website: http://www.isciii.es/
Specialties and services: Reference center for arbovirus, enterovirus, gastroenteritis, influenza and respiratory viruses, hepatitis, human papillomavirus, retrovirus, viral serology

Sweden

Swedish National Institute of Public Health
Tel: 46 10-205 20 00
Fax: 46 8-32 83 30
171 82 Solna
Karolinska Institutet
SE-171 77
Stockholm
Sweden
Tel: +46-8-524 800 00
Fax: +46-8-31 11 01

Center for Infectious Disease Research
Nobels väg 18
171 82 Solna
Sweden
Tel: +46 8 457 23 00
Fax: +46 832 83 30
Website: http://www.ki.se/en
Specialties and services: Retrovirus, hepatitis, human papillomavirus, hantavirus, polyomaviruses, influenza

Global WHO HPV Reference Laboratory
Laboratory Medicine Skåne
Malmö University Hospital, Entrance 78
SE-20502 Malmö
Sweden
E-mail: joakim.dillner@med.lu.se

Switzerland

Centre National de la Grippe
Laboratorie de Virologie
Hôpital Cantonal Universitaire de Genève
1, av. de Beau-Séjour
1206 Genève
Switzerland
Fax: +41 (22) 372 40 88
Website: http://www.influenza.ch
Specialties and services: WHO reference lab Influenza

Swiss Retrovirus Reference Laboratory Institute of Immunology and Virology
University of Zurich
Gloriastrasse 30
CH-8028 Zürich
Switzerland
Regional WHO HPV Reference Laboratory,
Institute of Microbiology, CHUV
Bugnon 48
1011 Lausanne
Switzerland
E-mail: dnardell@hospvd.ch

Turkey

Refik Saydam National Public Health Agency (RSNPHA) Virology Laboratory—Department of Microbiology Reference Laboratories
G Block 1st Floor—Saglik Mahallesi Adnan Saygun Caddesi No:55
06100 Sihhiye
Ankara
Turkey
Fax (lab): +90.31.24.58.23.88
E-mail: gucank@gmail.com
Tel: +90.312.458.20.62
Website: http://www.toraks.org.tr/
Specialties and services: Hantavirus, arbovirus, hepatitis, influenza, mumps, enterovirus, viral gastroenteritis

Ukraine

L.V. Gromashevsky Institute of Epidemiology & Infectious Diseases
National Academy of Medical Science of Ukraine
5, M. Amosova Str.
Kyiv, 03680
Ukraine
Fax: +38 044 275 02 97
Website: http://www.uiph.kiev.ua/
http://duieih.kiev.ua/
Specialties and services: Influenza

United Kingdom

Public Health England
Wellington House
133-155 Waterloo Road
London
United Kingdom
Tel: 020 7759 2707/020 7759 2730

Virus Reference Department
PHE Colindale
61 Colindale Avenue
London NW9 5HT
United Kingdom
Tel: 020 8327 6017
Fax: 020 8205 8195
E-mail: vrdqueries@phe.gov.uk
Website: http://www.hpa.org.uk/
Specialties and services: WHO Collaborating Centre for Laboratory and Diagnostic: WHO Designated International Laboratory for SARS; WHO Global Specialized HIV Drug Resistance Laboratory; WHO Global Specialized Laboratory for Measles and Rubella; WHO National Laboratory for Polio and Influenza; WHO National Laboratory for Influenza; blood-borne viral pathogens, enteric and respiratory viruses, spongiform encephalopathy

OCEANIA

Australia

Virus Identification Laboratory
Victorian Infectious Diseases Reference Laboratory (VIDRL)
792 Elizabeth Street
Melbourne
Australia

Tel: 61 3 9342 9628
Fax: 61 3 9342 9666
Website: http://www.vidrl.org.au
Specialties and services: HIV, HBV, influenza, enterovirus, norovirus, measles

New Zealand

Institute of Environmental Science and Research
66 Ward Street
Wallaceville
Upper Hut 5018
Wellington
New Zealand
Tel: 64 4 529 0600
Fax: 64 4 529 0601
Website: http://www.esr.cri.nz
Specialties and services: Influenza, poliovirus, enterovirus, measles, respiratory syncytial virus, human metapneumovirus, human coronavirus, severe acute respiratory syndrome, coronavirus, enteric virus, cytomegalovirus, varicella-zoster virus, herpes simplex virus, mumps, rubella, dengue virus, Ross River virus, Barmah Forest virus, West Nile virus

Papua New Guinea

Institute of Medical Research
Eastern Highlands Province 441, Homate Street
Goroka
Papua New Guinea
Tel: 675-531 4200
Fax: 675-532 1998
Website: http://www.pngimr.org.pg/
Specialties and services: Influenza, rotavirus, HIV

OTHER USEFUL WEBSITES

International Association of National Public Health Institutes (IANPHI)

The IANPHI links and strengthens the government agencies responsible for public health. IANPHI improves the world's health by leveraging the experience and expertise of its member institutes to build robust public health systems.

Secretariat: Mexico
National Institute of Public Health (INSP) Cuernavaca Campus Av. Universidad No. 655 Col. Santa María Ahuacatitlán Cerrada los Pinos y Caminera CP. 62100 Cuernavaca Morelos, Mexico Tel: +52.777.329.3030; E-mail: secretariat@ianphi.org

Secretariat: USA
Emory University Global Health Institute 1599 Clifton Road Atlanta, Georgia, USA 30322 Tel: 1+404.727.1416; E-mail: cdusenb@emory.edu

Secretariat: France
Institut de veille Sanitaire (InVS) 12, Rue du Val d'Osée 94415 Saint-Maurice Cedex, France; Website: http://www.ianphi.org/

European Centre for Disease Prevention and Control (ECDC)

The ECDC was established in 2005. It is an EU agency with aim to strengthen Europe's defenses against infectious diseases. It is based in Stockholm, Sweden.
Postal address: ECDC 171 83 Stockholm, Sweden
Visiting address: Tomtebodavägen 11a, Solna, Sweden
Website: http://www.ecdc.europa.eu
Epidemic Intelligence duty e-mail: support@ecdc.europa.eu

EPISOUTH

The EpiSouth Network was established among countries of Southeast Europe, North Africa, and the Middle East to create a framework of collaboration on epidemiological issues for enhancing surveillance of communicable diseases and control of public health risks through communication, training, information exchange, and technical support to countries in the Mediterranean region.
EpiSouth Network Coordination Office c/o
Epidemiology of Communicable Diseases Unit
National Centre for Epidemiology, Surveillance and Health Promotion (CNESPS)
Istituto Superiore di Sanità
Viale Regina Elena, 299
00161 Rome
Italy
Fax: ++39 06 49904267
Website: http://www.episouthnetwork.org/

European Commission:
http://www.ec.europa.eu

World Organisation for Animal Health (OIE):
http://www.oie.int

World Health Organisation (WHO):
http://www.who.int

European Food Safety Authority (EFSA):
http://www.efsa.europa.eu

European Directorate for the Quality of Medicines and Healthcare (EDQM):
http://www.edqm.eu

Integrated Control of Neglected Zoonoses (ICONZ):
http://iconzafrica.com

Alliance for Rabies Control:
www.rabiesalliance.org

RABMED CONTROL:
http://www.rabmedcontrol.org

Predemics:
http://predemics.biomedtrain.eu

Discontool:
http://www.discontools.eu

Subject Index

I

J

K

L

Q

S